ENCYCLOPEDIA OF TROPICAL PLANTS

IDENTIFICATION AND CULTIVATION OF OVER 3,000 TROPICAL PLANTS

AHMED FAYAZ

FIREFLY BOOKS

To my son
Ibrahim Asham Ahmed
for his unconditional love and company.

A FIREFLY BOOK

Published by Firefly Books Ltd. 2011

First published in 2011 in New Zealand by David Bateman Ltd.,
30 Tarndale Grove, Albany, Auckland, New Zealand

First printing

Publisher Cataloging-in-Publication Data (U.S.)

Fayaz, Ahmed.
Encyclopedia of tropical plants : the identification and cultivation of over 3,000 tropical plants / Ahmed Fayaz.
[720] p. : col. photos.; cm.
Includes bibliographic references and index.
Summary: Survey of 3,000 tropical plant species arranged in phylogenic order using the latest nomenclature and systematics. Each species entry has a detailed botanical description, zone information, distribution and ecology. Additional information includes propagation and cultivation notes.
ISBN-13: 978-1-55407-489-1
1. Gardening – Encyclopedias. 2. Exotic plants – Encyclopedias. 3. Tropical plants – Encyclopedias. I. Title.
635.903 dc22 SB450.95.F39 2011

Library and Archives Canada Cataloguing in Publication

Fayaz, Ahmed, 1968-
Encyclopedia of tropical plants : the identification and cultivation of over 3,000 tropical plants / Ahmed Fayaz.
Includes bibliographical references and index.
ISBN 978-1-55407-489-1
1. Tropical plants–Encyclopedias. 2. Tropical plants–Identification. I. Title.
QK474.5.F39 2011 581.70913'03 C2010-900168-0

Published in the United States by
Firefly Books (U.S.) Inc.
P.O. Box 1338, Ellicott Station
Buffalo, New York 14205

Published in Canada by
Firefly Books Ltd.
66 Leek Crescent
Richmond Hill, Ontario L4B 1H1

Book design: Trevor Newman Design
Printed in China through Colorcraft Ltd., HK

CONTENTS

FOREWORDS

This overview book on tropical plants is one of the first publications of its sort to adopt the new Angiosperm Phylogeny Group (APG) classification of the orders and families of flowering plants. This book is not a complete treatment of the angiosperms, but it is a very handy reference for getting more information on commonly grown tropical plants from around the world. It has much more complete descriptive information than you typically find in such a publication, which should make it much more useful to a wider audience than any of its predecessors. A photograph of nearly all of over 3,000 plants is included. The combination of descriptive matter, photographs and APG classification should make it a useful reference book for general audiences as well as students of tropical botany.

PROFESSOR MARK W. CHASE FRS
Keeper of the Jodrell Laboratory
Royal Botanic Gardens, Kew
United Kingdom

We delight in the beauty and diversity of plants. Not long after I first met Ahmed Fayaz I invited him to look around our family garden. With camera constantly at the ready, he enthusiastically took photographs and shared his knowledge and excitement about the botany and beauty of plants. In this book Ahmed Fayaz shares his passion for plants by presenting the culmination of more than a decade of seeking out, photographing, researching and, in some cases, the discovery of thousands of tropical plants. The result of his endeavor is this beautifully photographed and researched encyclopedia that identifies through photographs and supporting descriptions thousands of tropical plants, some of which are not easily found and have not previously been recorded in photographs. There is a long and impressive history of plant collecting and identification, but as the biodiversity of the planet becomes ever more threatened it is even more essential that we know what plant species we have and, as Ahmed Fayaz has recognized, we also need to know how to grow and propagate them.

The plants in the book are arranged phylogenetically up to the generic level which makes finding plants and comparing closely related plants easy. Each plant has a botanical description and notes on either its propagation or a description of its natural environment. The book is comprehensively referenced and comes with an extensive glossary.

I marveled at the beauty of the photographs, the diversity of tropical plants and the energy and research that Ahmed Fayaz put into producing this book and recommend it as a timeless record of thousands of tropical plants.

PROFESSOR PETER D. KEMP
Head, Institute of Natural Resources
Massey University
Palmerston North
New Zealand

ACKNOWLEDGMENTS

It has been a long and arduous journey to make this book a reality. I would first like to thank the creator of all plants on this planet, for paving the path to create this work.

Since setting out literally 'into the sunset' some ten years ago, numerous friends and family members have been very supportive and have encouraged me to pursue further. I would especially like to thank my son Ibrahim Asham Ahmed for his company throughout the making of this book. I am also very thankful for his keen eye for new and interesting plants, while perusing the rainforests of tropical Asia. His help during the writing and proofreading of this book has also been huge. Growing up with my family, who are keen growers of plants, I was immediately drawn into the realm of plants at a very early age. I would like to thank my late grandmother Shameema Ali, who has taught me so much about tropical flora and its ethnobotany. There is a long list of the friends and family who have helped me to locate and, in some cases even taken me to, interesting plants deserve my earnest gratitude.

Along the way I have made an enormous number of acquaintances for whom I would like to extend my gratitude, especially those that work at the botanical gardens, natural reserves, etc. The guides at these are very helpful, always ready to direct you to new and interesting plants. I would specifically like to thank all the staff at Royal Botanical Gardens of Peradeniya (Sri Lanka), for allowing me to photograph their enormous collection of native, exotic and often rare tropical plants.

I would like to thank all those who took time out of their busy schedules to review this book. Scott Zona (Florida International University, Miami, Florida, USA) in addition to reviewing the order Arecales, has also provided me with numerous photos of palms and other plants. Thomas Couvreur (The New York Botanical Garden, New York, USA) reviewed the order Magnoliales, and also has provided me with numerous photos of African Annonaceae. I am indebted to Jennifer Tate (Institute of Molecular BioSciences, Massey University, Palmerston North, New Zealand) for reviewing the order Malvales, especially at such short notice. I would like to thank Mark W. Chase (Keeper of the Jodrell Laboratory, Royal Botanic Gardens, Kew, UK) for reviewing the huge chapter on orchids and for his helpful comments and suggestions. Brian Adam and Noel Adam (The Victoria Esplanade, Palmerston North, New Zealand) have been very helpful in locating, helping me identify uncommon plants and also reviewing the section on ferns.

I would like to thank all the photographers for providing me with such excellent photos, without which this book would not be as colorful and complete as it is.

Last but not the least, I would like to thank David Bateman Ltd for giving me the opportunity to bring this book to life. I would especially like to thank Tracey Borgfeldt for her patience and perseverance to get this book in its current form. I am also thankful to all the staff at David Bateman Ltd who are involved in this project.

AHMED FAYAZ, 2011

PHOTOGRAPHIC CREDITS

Photographs not taken by Ahmed Fayaz are acknowledged with initials at the end of the caption or beside the image.

AC	Ann Cameron, Sydney, Australia.
AH	Andreas Hörstemeier, Aachen, Germany.
CS	Caleb Slemmons, USA.
DM	Doug Moyer, Troutville, Virginia, USA.
EFdeV	E.F. de Vogel, Nationaal Herbarium Nederland, Leiden, The Netherlands.
F&KS	Forest & Kim Starr, Makawao, Hawai'i, USA.
GHIR	Grace Hirayama, Honolulu, Hawai'i, USA.
GL	Gustavo Luedemann, Brazil.
GS	Gunawan Saputra, Bandung, Jawa Barat, Indonesia.
HL	Hai Le, CanTho University, Vietnam.
HS	Himanshu Sarpotdar, Mumbai, India.
HSP	Shirley Sekarajasingham, Bainbridge Island, Washington, USA.
IAA	Ibrahim Asham Ahmed, Male', Maldives.
IK	Ido Kron, Jerusalem, Israel.
JB	Jeanie Barnett, USA.
JM	Jordan Metzgar, Alaska, USA.
JM/EdP	Jorge Moura, Ecos de Pedra, Portugal.
JML	John Michael Lynn, Homestead, Florida, USA.
JS	Jean-Jacques Segalen, Réunion.
JT	Jeff Tyler, Sacramento, California, USA.
KT	Kukiat Tanteeratarm, Patumthani, Thailand.
MD	Mike Davy, Vancouver, British Columbia, Canada.
MG	Martin Günther, Diedorf-Lettenbach, Germany.
MIW	Ming-I Weng, Taipei, Taiwan.
MM	Mauricio Montoya, Medellin, Colombia.
MS	Melanie Schori, Ohio University, Athens, Ohio, USA.
NDT	Nghiêm Đúc Trong, Ha Noi University of Pharmacy, Hà Nôi, Vietnam.
NE	Neil Ennis, Brisbane, Australia.
PL	Paul Latham, Scotland, UK.
PMER	Pierre Mercan, Switzerland.
RF	Ron Ferguson, Cairns, Australia.
RS	Ruta Schuld, The Netherlands.
RT	Ria Tan, Wild Singapore, Singapore.
SFA	Saffah Faroog, Male', Maldives.
SJ/SAJF	Serge Aubert, SAJF, Grenoble, France.
SP	Stefano Pagnoni, Italy.
SZ	Scott Zona, Florida International University, Miami, Florida, USA.
TAM	Tamara Schuster, Tel Aviv, Israel.
TC	Thomas Couvreur, The New York Botanical Garden, New York, USA.
TG	Tatiana Gerus, Brisbane, Australia.
TM	Thierry Meurgues, Châteauroux, France
TNR	Ton Rulkens, Mozambique.
TR	Tom Rau, Kaneohe, Hawai'i, USA.
TT	Thanh Tung, Vietnam.
YK	Yero R. Kuethe, Noordwijkerhout, The Netherlands.

Opposite: *Coelogyne cristata*.

INTRODUCTION

This work in part forms the culmination of over 10 years of research on tropical flora. The main aim of this book is to present a significant selection of tropical flora from around the world in an arrangement that makes the identification of plants much easier, and also helps the reader to understand better the relationships between various plants or groups of plants.

Over 3,000 tropical plants, representing all the major plant groups, are presented in this work. A selection of subtropical plants is also included, the majority often cultivated in the tropics. The plants described range from lower plants (mosses) to higher plants (flowering plants). The information is aimed at those studying tropical plants and also tropical plant growers. Many species described have never been published before in this form, making this book an invaluable resource for all tropical plant enthusiasts.

ARRANGEMENT OF THE PLANTS

The majority of works on plants are still presented in alphabetical order of genera. Only a handful of works are presented in the natural (or phylogenetic) order. For decades the major reference book on tropical plants has been *Tropica* by Alfred Byrd Graf (1978). With over 7,000 color photos of tropical plants, arranged in alphabetical order of families, this book has served as the standard identification guide to tropical plants. The plant descriptions are very short, however, and are placed at the back of the book. The excellent volume *Tropical Flowering Plants* by Llamas (2003) is also worth mentioning here, though, as with *Tropica*, it is also arranged in alphabetical order of plant families.

The plants in this book are arranged according to the latest Angiosperm Phylogeny Group classification (the APG III System), published in 2009 and based mostly on molecular plant systematics. Arranging plants in phylogenetic order brings closely related plants together, thus making plant identification much easier. The plants are initially divided into four parts, with each part then divided into chapters and arranged in phylogenetic order (Figure 1). The orders, families and subfamilies within each chapter are also arranged in phylogenetic order, as are the genera where phylogenetic studies are available. Otherwise, the genera and species are arranged in alphabetical order.

MOLECULAR PLANT SYSTEMATICS

Plant systematics is the biological classification of plants, and has been used since ancient times, for example the work done by Theophrastus (*Historia Plantarum*, written between the third and second century BC). Almost two millennia later, the work by de Jussieu (1789) marked a milestone in plant systematics by proposing a system for the classification of plants. The theory of evolution by Charles Darwin (1859) brought much needed order to plant systematics. Evolution takes place over long periods of time and cannot be observed directly. Biologists thus need to reconstruct phylogenetic trees (evolutionary trees that show relatedness among groups of organisms) by inferring the evolutionary relationships among extant organisms. Phylogenetic studies published until a few decades ago were mostly based on morphological (physical)

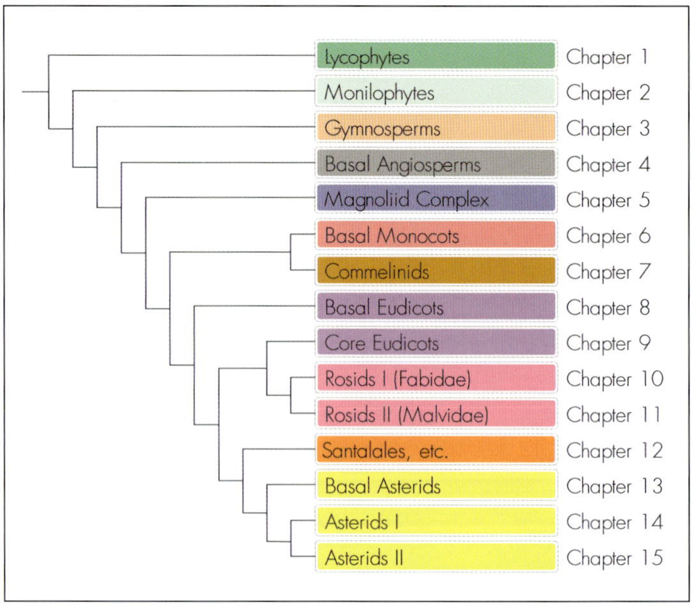

Figure 1 Arrangement of the plant groups according to APG III System (simplified).

characters, and the true evolutionary relationships of the plants were not revealed.

The application of molecular characters in plant systematics started about 25 years ago, leading to the birth of molecular systematics. Molecular systematics is defined as the use of mainly DNA and RNA sequences to infer evolutionary relationships among organisms (Judd et al. 2008). The molecular techniques are complex, and beyond the scope of this work. Comprehensive works on molecular systematics include Judd et al. (2008), Simpson (2006) and Stevens (2001 onwards). The latter is an online resource dedicated to plant systematics, particularly molecular and that of angiosperms, which is the main hub of the Angiosperm Phylogeny Group.

Molecular plant systematics is now becoming the norm for the classification of plants. Newly developed technologies enable faster analysis of plant material with significantly increased throughput. Sequencing of DNA is the most common method used in plant systematics, and has become both affordable and faster.

THE TROPICS

The tropics form a band around the Earth where the Sun shines directly overhead at least once during a solar year. Technically, this falls between ~23°S below the Equator to ~23°N above the Equator (Figure 2). The climate is usually warm, and the temperature rarely falls below 0°C (these usually in montane regions). The tropics also include most of the hottest places on Earth, such as the African savannas, where the temperature often rises above 50°C. Hardiness zones for the tropics range from 9–12 (minimum temperature -1 to 20°C).

The subtropical regions of the world lie between 23–40°N and 23–40°S (Figure 2), and contain a large number of plants that can thrive and are often cultivated in the tropics. Hardiness zones for this region range from 7–9 (minimum temperature: -18 to -1°C). Beyond the tropical and subtropical zones lies the

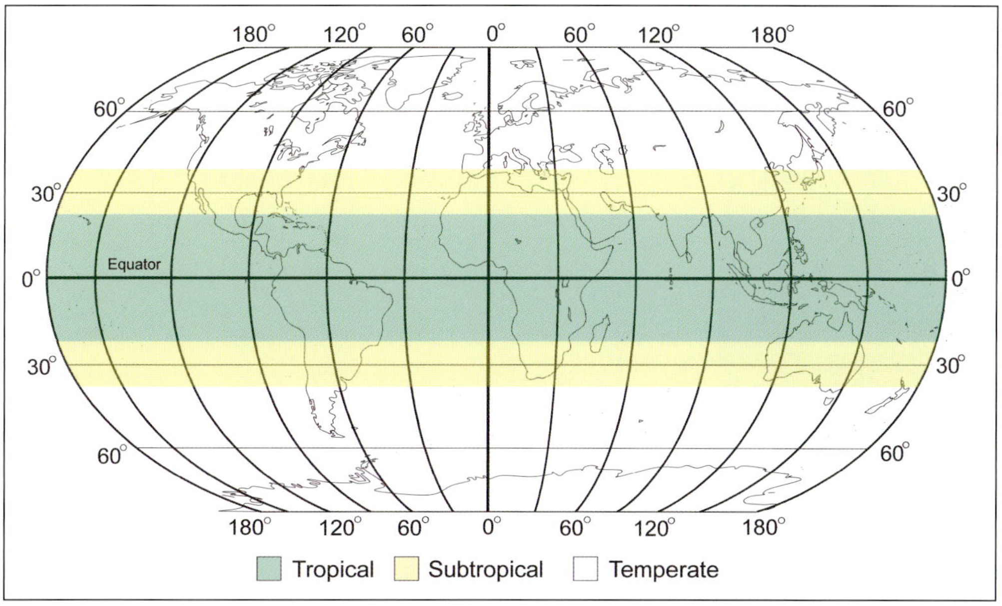

Figure 2 Climatic regions of the world (approximate).

temperate regions, with hardiness zones ranging from 1–6 (minimum temperature: -50 to -18°C).

The exact climate of these different regions varies considerably depending on the topography and other physical characteristics (such as altitude, distance from the coast, proximity to bodies of water, rainfall, wind, atmospheric pressure, etc.). For example, Mt Kilimanjaro in Tanzania lies on the Equator, but climatic conditions that are similar to temperate regions can be found as you climb towards the snow-covered peaks. The hardiness zones are thus a very imprecise measure, but they do give a generalized assessment of the climatic conditions.

Several vegetation types are found in the tropics and subtropics, the major types being rainforests, monsoon forests, humid forests, cloud forests, savannas and arid deserts. Rainforests (Figure 3) are characterized by exceptionally high annual rainfall of 1,750–2,000 mm spread throughout the year. Monsoon forests have a rainy season that lasts for several months and that is often associated with high winds. Annual rainfall of up to 10,000 mm has been recorded in these areas. Rainforests and monsoon forests consist mainly of broad-leaved plants, with leaves often designed to channel the water away from the plant.

Savannas and humid forests have an annual rainfall not exceeding 1,200 mm. Plants of savannas are usually of short stature, consisting mostly of grasses and others herbs, and shrubs. Cloud forests (Figure 4) are generally evergreen montane moist forests, characterized by a persistent, frequent or seasonal low-level cloud cover, usually at the canopy level. Plants in these areas have adaptations designed to collect as much moisture from the atmosphere as possible. Epiphytic species such as orchids and bromeliads make much use of this atmospheric moisture.

The tropical region also includes the arid deserts, where usually extremely high temperatures are experienced and annual rainfall is less than 250 mm. Daytime temperatures exceeding 45°C and night-time temperatures of 0°C have been recorded. Although harsh, these areas are home to a large number of diverse plant species (e.g. Apocynaceae of North Africa and the Arabian Peninsula). The desert plants have adaptations to store moisture, usually in abnormally enlarged plant parts such as trunks, branches and leaves.

Islands also have very unique climatic niches that range from tropical to subtropical. Larger islands such as Madagascar, New Guinea, New Zealand, Sri Lanka, etc., have a multitude of microclimates, ranging from lowland tropical rainforests to montane tropical forests (Figure 5). Mangroves are also a common vegetation type found on islands (Figure 6). It is interesting to note that in the high altitude alpine regions of New Guinea (in the tropical zone) climatic conditions are very close to those of temperate regions, while New Zealand (in the temperate zone) has climatic niches that are very similar to tropical and subtropical rainforests.

Figure 3 Tropical rainforest (Malaysia).

Figure 4 Cloud forest (Bolivia). HSP

PLANT DIVERSITY

The tropics are home to the majority of plant life on Earth, with extremely high diversity. A distinct feature of tropical regions is the dominance of evergreen angiosperms (flowering seed plants), in contrast to the dominance of deciduous gymnosperms (non-flowering seed plants, such as pines) in temperate regions. The diversity of plants is lowest near the poles, increasing exponentially towards the Equator, with the highest diversity in tropical rainforests.

Plants in tropical rainforests thrive with the availability of abundant moisture. Tropical montane forests and cloud forests also provide unique habitats for its constituent flora. These habitats are found across the globe from Malesia through Central Africa to South America.

Figure 5 A tropical montane forest (Horton Plains, Sri Lanka).

Certain plant groups also dominate the tropics, such as the orchids (family Orchidaceae), the largest family of plants, which have become a symbol of the tropics. The highly favorable conditions of the tropics have thus given rise to very high diversity in this family, as well as other major tropical flowering plant families, with very unique adaptations not seen elsewhere.

The abundance of suitable pollinators has also given rise to extraordinary flower forms and colors, as seen in many orchid genera (e.g., *Stanhopea*). The major habitats include tropical rainforests, cloud forests, montane forests, mangroves and tropical islands.

BIODIVERSITY HOTSPOTS

There are several plant biodiversity hotspots around the world, dominated mostly by tropical flora and home to nearly 50% of all known plants. The major hotspots include:

- The Amazon (South America)
- Cape Floristic Region (South Africa)
- Madagascar
- Tropical Andes (South America)
- Borneo (Malaysia/Indonesia)
- Western Ghats (South India)
- New Guinea (Indonesia/Papua New Guinea)

In addition to having high biodiversity, they also have a very large number of endemic species (species that are unique to that particular region).

Flora of tropical and subtropical islands also has a markedly high degree of endemism, attributed to their geographic isolation and the limited interchange with flora of the mainland or other neighboring islands. Madagascar, Seychelles, New Caledonia, New Zealand, Galapagos, and the islands of Malesia are noted for their high plant diversity and endemism.

PLANT CONSERVATION

Tropical plants are highly sought after, both in the tropics and other parts of the world, with an ever-growing interest in their cultivation and research. Unfortunately, many parts of the tropics are currently battling to keep their plant diversity and natural habitats intact, with growing demand for land and over-exploitation of numerous wild plants. The biggest challenge faced in the tropics is the competition between the growing human population and the environment, which is shared by humans, other animals and plants. People living in the tropics are highly dependent on the extremely high diversity of plants for edible fruits, medicine, and other domestic or commercial uses. Conservation of these tropical plants is thus very important for the sustainability of the people living in the tropics.

To understand the issues and challenges related to the conservation of tropical plants, one must understand their diversity, threats to them in the wild and their potential for cultivation. For any plant conservation effort to be successful, a sound knowledge of the plants being studied, including proper identification and an understanding of relationships between plant relatives is essential. A comprehensive identification guide for tropical plants is thus long overdue.

Figure 6 Mangrove forest (Maldives).

PLANT DESCRIPTIONS IN THIS BOOK

Detailed descriptions are given for all levels of hierarchy, including cultivation information under the genus description. The species are described in detail, and are accompanied by a color photograph.

The name of the species is followed by the author(s) and the place of publication. Where relevant and available, a selection of synonyms and common names is also given. The plant description is followed by the distribution (or geographical range) information. The ecology notes provide available information on the habitat, altitudinal range and phenology of the plant. Hardiness zones are also provided for all the plants in the book. Where available, further information regarding the plant is also provided, such as its toxicity, ethnobotany, and descriptions of subspecies or varieties.

Descriptions within parentheses denote uncommon or rare characteristics. For example, 'Trees, to 20(–25) m tall' indicates that the trees grow normally to a height of 20 m, but sometimes to a height of 25 m. (See Abbreviations below for more information on plant descriptions, and also the extensive Glossary, pages 687–698.)

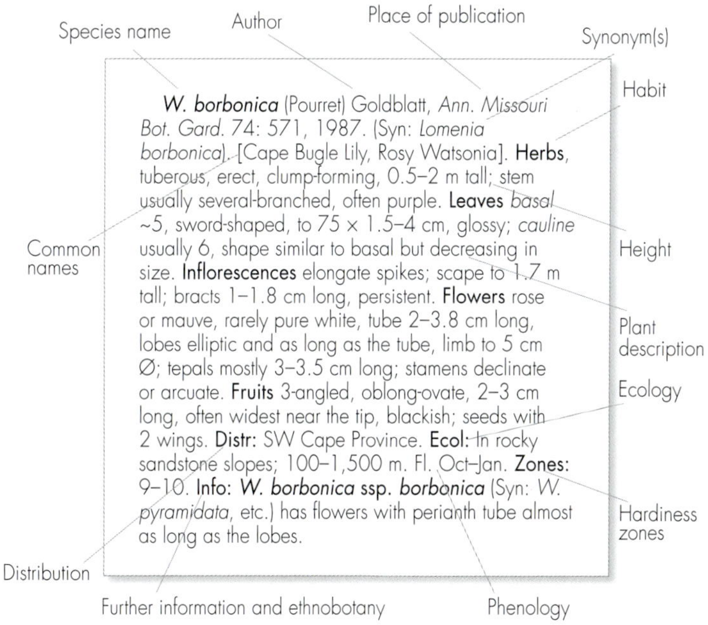

ABBREVIATIONS

Ø	diameter	Fl.	flowering
2-merous	in sets of 2s, e.g. 2, 4, 6, 8	Fr.	fruiting
3-merous	in sets of 3s, e.g. 3, 6, 9, 12, etc.	m	meter
cm	centimeter	mm	millimeter
dbh	diameter at breast height	sp.	species (pl. spp.)
Distr	Distribution	ssp.	subspecies
Ecol	Ecology	syn.	synonym(s)
f.	form, forma	var.	variety

PART 1
LYCOPHYTES, MONILOPHYTES, GYMNOSPERMS
[NON-FLOWERING PLANTS]

Opposite: *Cycas rumphii* ssp. *zeylanica* (male cone).

1 LYCOPHYTES [MOSSES]

LYCOPODIALES

FAMILIES 3, GENERA 5, SPECIES 1,280.

Roots have an endarch protoxylem (the first tracheary cells that develop within a patch of xylem, typically smaller and thinner than the later formed metaxylem). **Stems** have an exarch protoxylem (the protoxylem forms in a position exterior to the metaxylem, reverse that of the roots). **Lycophyll** (sporophytic leaf structure) present, and often heart-shaped (Simpson 2006, 2010; Stevens 2001 onwards). **Info:** Among the 3 families in this order, Lycopodiaceae is homosporous, Selaginellaceae and Isoetaceae are heterosporous. the Gametophytes of Lycopodiaceae are bisexual, mycorrhizal, and usually subterranean. They may be partly green and on or near the surface of the ground, or initially mycorrhizal and later photosynthetic. the Gametophytes of Selaginellaceae are unisexual (Judd et al. 2008).

LYCOPODIACEAE

[Club-moss].

Genera 3, **species** ~380. **Plants**, terrestrial, lithophytic or epiphytic; roots emerging near origin, or growing through cortex and emerging some distance from origin. **Horizontal shoots** present or absent, on substrate surface or subterranean, or forming stolons. **Upright shoots** simple or branched, usually conspicuously leafy at least at base, round or flat in cross section; abscising gemmae formed by reduced lateral shoots. **Lateral shoots** are present or absent, simple or branched, branching pattern dichotomous and sometimes pseudo-monopodial, round or flat in cross section; the leaves are uniform, dimorphic or trimorphic. **Leaves** on *subterranean* parts are flat, appressed, non-photosynthetic, and scalelike; on *aerial* parts appressed, needlelike to lanceolate to ovate, remote to dense, and imbricate. **Strobili** sessile or stalked, upright, nodding, or pendent. **Sporangia** are solitary, adaxial near leaf base or axillary; subtending leaves (sporophylls) unmodified and photosynthetic to much-modified and non-photosynthetic; sporangia are reniform to globose, with hundreds of spores; spores similar, trilete. **Gametophytes** are either subterranean and non-photosynthetic, or above ground and photosynthetic. **Distr:** Worldwide.

Lycopodiella

[Bog Club-moss].

Species 8–10. **Plants**, creeping on wet ground; roots emerging immediately on underside of stems. **Horizontal stems** on substrate surface, supine or arching. **Upright shoots** forming very leafy peduncles scattered along horizontal stems, 2–9 mm Ø, unbranched. **Gemmiferous branchlets** and **gemmae** absent. **Strobili** solitary, fully differentiated from peduncle or peduncle not differentiated, tip blunt to ± acute; peduncle leafy, leaves not in distinct ranks, not imbricate, usually monomorphic, linear-lanceolate, margins commonly with a few teeth; sporophylls generally longer than peduncle leaves. **Sporangia** nearly globose; spores rugulate, sides at equator convex, angles acute. **Gametophytes** photosynthetic, on substrate surface, pincushion-shaped; ring meristem absent. **Distr:** N temperate regions, tropical America. **Cult:** Propagate from spores; division and transplanting is difficult. Requires a fertile, humus-rich, moist soil in partial sun or shade; best grown on a moisture-retentive substrate.

L. alopecuroides (L.) Cranfill, *Amer. Fern J.* 71: 97, 1981. (Syn: *Lycopodium alopecuroides*). [Foxtail Bog Club-moss]. **Plants**, creeping on wet ground; roots emerging immediately on underside of stems. **Horizontal stems** strongly arching, 5–40 x 0.8–1.1 cm, rooting at tip; stems (excluding leaves) 2–4 mm Ø; leaves monomorphic, spreading to ascending, 5–7 x 0.5–0.7 mm, marginal teeth 1–7 per side; peduncles 1–3 per plant, 6–30 x 0.2–0.3 cm; strobili ⅓–½ total length; leaves spreading to ascending, 6–7 x 0.3–0.5 mm, marginal teeth 1–10 per side. **Strobili** 2–6 x 1.2–2 cm. **Sporophylls** wide-spreading, 6–7 x 0.5–0.9 mm, marginal teeth 1–5 per side in proximal half. **Distr:** N America, West Indies. **Ecol:** In bogs, marshes, etc.; 0–600 m. **Zones:** 10–12.

L. serpentina (Kunze) B. Ollg., *Op.Bot.* 92: 176, 1987. (Syn: *Lycopodium serpentinum*). [Bog Club-moss]. **Plants**, rhizomatous, perennial, 15–50 cm high; main stems prostrate, forking, rooting at intervals, to 15 cm long, giving rise to unbranched aerial stems to 5 cm long. **Sterile leaves** spirally arranged, ascending on prostrate stems, spreading on aerial stems; lamina linear, to 4 mm long, apex pointed, dull green tinged orange. **Strobili** on erect stipes, stalkless, solitary, terminal, to 2.5 cm long, yellow-green. **Distr:** Australia, New Caledonia, New Zealand. **Ecol:** In black peaty soil, granite, swamps, sometimes on moist forest margins. **Zones:** 8–12.

Lycopodium

Species 40. **Plants**, mainly trailing on ground; roots emerging from point of origin on underside of main stems. **Horizontal shoots** on substrate surface or subterranean, long-creeping. **Upright shoots** scattered along horizontal stem, 0.5–1.6 cm Ø, round or flat in cross-section, unbranched or with 1–4 lateral branchlets. **Leaves** not imbricate, linear to linear-lanceolate; on *horizontal* stems scattered,appressed,membranous; on *lateral* branchlets mostly 6-ranked or more, monomorphic with few exceptions, appressed, ascending to spreading, margins entire to dentate. **Gemmiferous branchlets** and **gemmae** absent. **Strobili** single and sessile or multiple and pedunculate, apex blunt to acute; peduncle, when present, conspicuously leafy; sporophylls extremely reduced, much shorter than peduncle or stem leaves. **Sporangia** reniform; spores reticulate, sides at equator convex, angles acute. **Gametophytes** non-photosynthetic, mycorrhizal, subterranean, flat and irregularly button-shaped, with ring meristem around circumference. **Distr:** Mainly temperate and subarctic, some in subtropical and tropical regions. **Cult:** Propagate by division or from spores. Requires a moist, gritty, acidic soil in shade and high humidity.

L. fastigiatum, R. Br., *Prodr.* 165, 1810. [Alpine Club-moss, Mountain Club-moss]. **Plants**; main stem or rhizome stout, creeping, subterranean, 15–60+ cm long. **Primary branches** rather distant, 7.5–36 cm long, stout, erect, rigid, often naked below, copiously fastigiately branched above; branchlets crowded. **Leaves** imbricated all around the branches, crowded, spreading and then incurved at the tips, linear or linear-lanceolate, 3–4 mm long, margins entire, apex acute or shortly mucronate, firm, midrib indistinct. **Strobili** spicate, peduncled, often 2 together, terminal on the upper branchlets or several in a central panicle, 2–5 cm long, rarely longer, ~4 mm Ø, erect, cylindric; bracts closely imbricate, broadly ovate at the base, narrowed upwards into a long spreading or recurved cuspidate point. **Distr:** Australia, Tasmania; New Zealand. **Ecol:** In hilly and montane regions; 0–1,500 m. **Zones:** 8–12.

L. scariosum Hook., *Icon. Pl.* 1: t. 89, 1836. [Creeping Club-moss]. **Plants**, terrestrial; main stems prostrate, branched, tough, bearing aerial stems; aerial stems much-branched, to 50 cm tall. **Sterile leaves** flattened in 1 plane, of 2 sizes; lateral leaves larger, slightly sickle-shaped, pointed, to 4 mm long, curling upwards; smaller appressed leaves on abaxial side of stem, linear, straight, pointed, to 2 mm long. **Strobili** terminal, erect, to 5 cm long, on sparsely leaved unbranched stalks, orange-brown, solitary, rarely paired. **Distr:** Australia, Borneo, New Zealand. **Ecol:** Terrestrial in thick, spreading patches on banks and in scrub in montane and subalpine regions. **Zones:** 8–12.

SELAGINELLACEAE

[Spike Moss].

Genus 1, **species** 750. *Familial description same as generic description.*

Selaginella

Species 750. **Plants**, terrestrial, lithophytic, or rarely hemi-epiphytic (initially terrestrial, becoming epiphytic) or epiphytic. **Stems** prostrate, creeping, decumbent, cespitose, climbing, or fully erect. **Rhizophores** usually present, stout or filiform; roots branching several times dichotomously from rhizophore tips. **Leaves** on *aerial* stems dimorphic or monomorphic, if monomorphic, then linear to narrowly lanceolate, highly overlapping, spirally arranged, if leaves on aerial stem dimorphic, then round or oblong to lanceolate, arranged in 4 ranks, 2 ranks of larger spreading lateral leaves and 2 ranks of smaller, appressed, and ascending median leaves, often with axillary leaf at base of each branching dichotomy. **Megasporangia** lobed to ovoid; megaspores tetrahedral, ovoid, or globose, variously sculptured, (127–)200–1360 µm Ø. **Microsporangia** reniform to ovoid; microspores tetrahedral, variously sculptured, 20–75 µm Ø. **Distr:** Worldwide, mainly tropical and subtropical regions. **Cult:** Propagate from spores or by dividing rhizomes. Requires moist, humus-rich soil, or a suitable substrate, in full or partial sun.

S. lepidophylla (Hooker & Greville) Spring, *Fl. Bras.* 1(2): 126, 1840. (Syn: *Lycopodium lepidophyllum*). [Dinosaur Plant, Resurrection Fern, Resurrection Plant, Rose of Jericho]. **Plants**, terrestrial or lithophytic, forming rosettes. **Main stem** spirally compact, branched, branches 2–3-forked, prostrate, flat when moist, curling inward when dry (ball-like), not articulate, weakly puberulent. **Rhizophores** borne on adaxial surface of stems, restricted to basal part of rosette, 0.3–0.5 mm Ø. **Leaves** thick and stiff; *lateral* leaves yellow to reddish on abaxial surface, green on adaxial surface, overlapping, ascending, ± deltate, 2–2.2 × (1–)1.7–1.8 mm, base cordate, pubescent, margins transparent, basally ciliate, apically dentate to ciliate, apex rounded; *median* leaves broadly ovate, 1.5–1.7 × 1.4–1.5 mm, base cordate to truncate, margins transparent, ciliate toward base, apex abruptly acuminate to obtuse. **Strobili** solitary, 0.3–1.2 cm; sporophylls monomorphic, deltate-ovate, slightly keeled. **Distr:** S USA (Texas) to Mexico. **Ecol:** In dry places; 900–2,000 m. **Zones:** 9–11.

ISOETACEAE

[Quillwort].

Genus 1, **species** ~150. *Familial description same as generic description.*

Isoetes

Species ~150. **Plants**, tufted, grasslike, heterosporous (megaspores and microspores different), perennial, evergreen aquatics to ephemeral terrestrials; rootstock brown, cormlike, lobed; roots arising along central groove separating each rootstock lobe, simple or dichotomously branched, containing eccentric vascular strand and surrounding lacuna. **Leaves** linear, simple, spirally or distichously arranged, dilated toward base, tapering to apex, containing 4 transversely septate longitudinal lacunae, a central collateral vascular strand; ligule inserted above sporangium. **Megasporophylls** and **microsporophylls** usually borne in alternating cycles; hardened scales and phyllopodia occasionally surrounding leaves. **Sporangia** solitary, adaxial; megasporangium with several to hundreds of megaspores; microsporangium with thousands of microspores. **Megagametophytes** white, endosporic, exposed when megaspore opens along proximal ridges; archegonia 1 to several. **Microgametophytes** 9-celled, endosporic, antheridium releasing 4 multitailed spermatozoids. **Distr:** Nearly worldwide. **Cult:** Propagate by division or planting mature, peeled quills in shallow water with sand at the bottom, which roots readily. Requires shallow water such as a tank in bright light.

I. piedmontana (Pfeiffer) Reed, *Phytologia* 12(6): 392. 1965. (Syn: *I. virginica* var. *piedmontana*). [Peidmont Quillwort]. **Plants**, short-lived; corm bilobed, transversely oblong to narrowly transversely oblong, 0.3–1 × 0.8–2.9 cm; roots dichotomous, arising from the circumbasal fossa. **Leaves** spirally arranged; *inner* leaves generally erect; *outer* leaves usually recurved or reflexed, 2.3–1.7 cm long; subulate portion green, 0.2–1.75 mm wide, triangular to semi-terete in cross-section, fibrous bundles present and variously distributed around perimeter; alae hyaline, 1–6.4 mm wide above sporangium, apices attenuate, sometimes oblique; leaf bases becoming brown to black indurated phyllopodia. **Sporangia** basal, ± elliptic to circular, 2.4–9.3 × 1.75–4.2 mm, becoming pigmented with age. **Ligule** triangular to deltate, generally auriculate, 1.5–1.7 × 1–1.4 mm, membranous, ephemeral. **Megaspores** trilete, white, wet or dry, 0.25–0.65 mm Ø, tuberculate to rugulate to cristate, girdle absent to narrow and papillate. **Microspores** monolete, brown to castaneous when mature, 27.7–47.3 µm long, generally spinulose. **Distr:** Endemic to SE USA. **Ecol:** In aquatic or terrestrial habitats. **Zones:** 7–11.

Right: *Lycopodiella alopecuroides*.
Below: *Lycopodium fastigiatum*.
Far right: *Lycopodium scariosum*.

MD

SJ/SAJF

2 MONILOPHYTES [FERNS]

OPHIOGLOSSALES

FAMILY 1, GENERA 4, SPECIES 80.

Root with 2–5 protoxylem poles; stem stele sympodial in construction; circular bordered pits present; leaf bases sheathing; vernation nodding; one or more sporophores associated with each tropophore (Stevens 2001 onwards). **Info:** This order is unique in that the leaf (or 'frond') consists of a *sterile* segment, which contains the photosynthetic lamina, and a *fertile* segment (Simpson 2006, 2010).

OPHIOGLOSSACEAE

[Adder's Tongue].

Genera 4, **species** 80. **Herbs**, perennial, terrestrial or epiphytic; roots without root hairs, unbranched or with a few narrow lateral branches, rarely dichotomously branched. **Stems** simple, unbranched, upright, with eustelic vascular tissue. **Leaves** 1(–2) per stem, with common stalk divided into *sterile* laminate, photosynthetic portion (trophophore) and *fertile* spore-bearing portion (sporophore); trophophore laminas compound to simple, rarely absent, veins anastomosing or free, pinnate, or arranged like ribs of a fan; leaf bases dilated, clasping, forming sheaths, open or fused, surrounding successive leaf primordia; primordia glabrous or with long, uniseriate hairs; indument absent or of widely scattered, long, uniseriate hairs, especially on petiole and rachises. **Sporophores** pinnately branched or simple. **Sporangia** exposed or embedded, 0.5–1.5 mm Ø, thick-walled, with 1,000s of spores; spores all one kind, trilete, thick-walled, surface rugate, tuberculate, baculate (with projecting rods usually higher than wide), sometimes joined in a delicate network, mostly with a ± warty surface. **Gametophytes** not green, usually fleshy, round or linear, subterranean, mycorrhizal. **Distr:** Nearly worldwide.

Botrychium

[Grape Fern, Moonwort].

Species 50–60. **Herbs**, terrestrial; roots occasionally branching laterally, yellowish to black, 0.5–2 mm Ø, smooth or with corky ridges, not proliferous. **Stems** upright, forming caudex to 5 mm thick; gemmae absent or minute, spherical. **Trophophores** ascending to perpendicular to stem, sessile or stalked; lamina linear, oblong, or deltate, simple to 5-pinnate, 4–25 x 1–35 cm; pinnae (reduced to segments in many species) spreading to ascending, fan-shaped to lanceolate to linear; margins entire to dentate to lacerate, apex rounded or acute; veins free, arranged like the ribs of a fan, or pinnate. **Sporophores** normally 1 per leaf, 1–3-pinnate, long-stalked, borne at ground level to high on common stalk; sporangial clusters with sporangia sessile to short-stalked, almost completely exposed, borne in 2 rows on pinnate sporophore branches. **Gametophytes** broadly ovate, unbranched, 1–3 mm x 0.1–1 cm. **Distr:** Nearly worldwide. **Cult:** Propagate from ripe spores, by surface sowing. Some species are reliant on specific fungi for germination. Requires a moist compost, in humid conditions and shade at first, later in partial or full sun.

B. lunaria (L.) Swartz, *J. Bot.* (Schrader). 1800(2): 110, 1801. (Syn: *B. onondagense*, *Osmunda lunaria*). [Common Moonwort, Moonwort]. **Plants**, terrestrial; rhizome erect, short and coarse; roots clustered, relatively thick and fleshy. **Trophophore** solitary; stalk 0–1 mm long; lamina dark green, oblong, 1-pinnate, to 7–15 x 4 cm, thick, fleshy; pinnae to 9 pairs, spreading, mostly overlapping except in shaded forest forms, basal pinna pair approximately equal in size and cutting to adjacent pair, broadly fan-shaped, undivided to tip, margins mainly entire or undulate, rarely dentate, apical lobe usually cuneate to spathulate, notched, approximate to adjacent lobes, apex rounded, venation like ribs of fan, midribs absent. **Sporophores** 1–2-pinnate, 0.8–2 times length of trophophore. **Distr:** All continents except Africa. **Ecol:** In open or sparsely wooded areas, and mesic woodlands. Sometimes in sparsely wooded sand dunes and moist well-drained places. **Zones:** 7–12.

Helminthostachys

Species 1. *Generic description same as specific.* **Cult:** Propagate from spores. Requires a well-drained soil in full sun.

H. zeylanica (L.) Hook., *Gen. Fil.* pl. 47, 1840. [Flowering Fern]. **Herbs**, terrestrial; rhizome very short, stout, fleshy, and erect in young plants, creeping underground in mature plants, with an ectophloic siphonostele, cortex without sclerenchyma; roots fleshy with endophytic fungi. **Fronds** tripartite; each division with 2–3 (or more) lobes; lobes lanceolate, 18–35 x 2–4 cm, herbaceous, margins entire or irregularly sinuate-toothed; lateral lobes often decurrent; stipe 20–35 cm long, with 2 small stipules at base of stipe; veins free, forking. **Fertile spike** rather short, with a long stalk, to 10 cm long; sporangia borne in clusters on short lateral branches, opening by a longitudinal slit. **Distr:** S China, India, Japan (Ryukus), Philippines, Sri Lanka, Taiwan; islands of the Pacific to New Caledonia and Australia. **Ecol:** Growing on edges of marshes, often in shallow water and sometimes in dry places. **Zones:** 10–12.

Ophioglossum

[Adder's Tongue].

Species 25–30. **Herbs**, terrestrial; roots unbranched, whitish-yellow to black, smooth. **Stems** upright, forming caudex, to 1.6 cm thick; gemmae absent. **Trophophores** erect to prostrate; lamina nearly sessile or stalked, lanceolate to cordate, simple and undivided, 0.04–10 x 0.03–4.5 cm, margins entire, apex rounded, acute, or apiculate, veins anastomosing, main areoles to 1.5 x 0.4 cm, but mostly less than 5 x 3 mm. **Sporophores** 1 per leaf, simple and unbranched, stalked, borne from ground level to well above ground at base of trophophore but usually absent, leaf made up of only trophophyll; sporangial clusters with sporangia in 2 rows, deeply sunken in simple, linear or oblong fleshy sporophore tip, tip usually ± apiculate. **Gametophytes** brown to white, narrowly linear, unbranched, 0.2–2 cm long, 1–3 mm Ø. **Distr:** Nearly worldwide, mainly tropical and subtropical regions. **Cult:** Propagate from spores. Requires a moist, well-drained soil in full or partial sun.

O. costatum R. Br., *Prodr.* 163, 1810. (Syn: *O. pedunculosum*, etc.). [Large Adder's Toungue]. **Herbs**, terrestrial; rhizome erect, short, cylindrical; roots numerous, long, fibrous. **Fronds** 1–2 (rarely more), erect, borne from the rhizome, 4–30 cm long, bases not persistent; common stipe above the ground 0.7–5 cm long. **Sterile fronds** sessile, ovate or ovate-lanceolate or ovate-rhomboid, 2–9 x 0.7–2.5 cm, base truncate or broadly cuneate, never cordate, apex acute or obtuse, rather fleshy, pale green, with a paler median band, venation double, except for elongated areoles in median zone; primary areoles with a finer network of secondary ones. **Sporophore** 1–6 cm long on a stipe 5–25 cm tall; sporangia 15–40 in each row. **Distr:** Africa and Madagascar through India and southern Asia to Malesia, Australia. **Ecol:** In moist grassy places, often near the margins of swamps, and often among rocks along streams; 0–600 m. **Zones:** 10–12.

O. petiolatum Hooker, *Exot. Fl.* 1: 56, 1823. [Stalked Adder's Tongue]. **Herbs**, terrestrial; roots dark brown, to 8 per plant, 0.8–1.3 mm Ø, producing proliferations; stem upright, 0.3–1 cm, 1.5–2.5mm Ø, 2–3 leaves per stem. **Trophophore** with stalk 0–3 mm long; lamina erect to spreading, usually plane or nearly so when alive, grey-green, dull, ovate to trowel-shaped, to 6 x 3 cm, fleshy, base cuneate to truncate to nearly cordate, apex acute, gradually contracted, apiculum mostly absent, venation coarse, reticulate, areoles large with few free or anastomosing included veinlets. **Sporophores** arising at ground level, 0.8–7 times length of trophophore; sporangial clusters to 4 x 0.35 cm, with up to 30 pairs of sporangia, apiculum 0.3–1.2 mm. **Distr:** North America, Mexico, northern South America, West Indies, Asia, Pacific Islands. **Ecol:** 0–500 m. **Zones:** 8–12.

PSILOTALES

FAMILY 1, GENERA 2, SPECIES 12.

Ordinal description same as familial.

Above: *Botrychium lunaria, Helminthostachys zeylanica, Ophioglossum costatum, Psilotum nudum.*

PSILOTACEAE

[Whisk Fern].
Genera 2, species 12. **Herbs**, perennial, terrestrial or more often epiphytic; roots absent, plant anchored by subterranean stems that bear gemmae. **Aerial shoots** erect or hanging, glabrous, simple or dichotomously branched. **Leaves** spirally arranged, sometimes 2-ranked; lamina scalelike, bractlike or awl-shaped to lanceolate, simple or once-forked, 1-veined or veins absent. **Sporangia** bi- or trilocular and lobed, large, sessile on or above the base of forked sporophylls; spores homosporous, bean-shaped, pale in color. **Gametophytes** subterranean, mycotrophic, fleshy, elongate, branched. **Distr:** Worldwide tropical regions.

Psilotum

[Whisk Fern].
Species 2 or 3. **Herbs**, terrestrial, sometimes epiphytic; aerial shoots often clumped, simple proximally, dichotomously branched distally, 3(–several)-ridged. **Appendages** minute, bractlike, borne distally on ridges of aerial shoots, sterile appendages subulate, those subtending synangia bilobed. **Synangia** ± globose, obscurely trilobed. **Distr:** Tropical and warm temperate regions. **Cult:** Propagate from spores. Requires a moist substrate in shade or partial sun.

P. nudum (L.) Palisot de Beauvois, *Prodr. Aethéogam.* 106, 112, 1805. (Syn: *Lycopodium nudum*). [Whisk Fern]. **Herbs**, terrestrial, epilithic or epiphytic; rhizome short-decumbent, dichotomously branched, rootless, to 1.5 mm Ø, with rhizoids; aerial shoots green, chlorophyllose, erect or pendent, to 50 cm long, 2–4 mm Ø, once or several times dichotomously branched, glabrous, angular or sulcate, with anomocytic stomata in the sulci. **Fronds** rudimentary, simple, lanceolate, adnate, spaced, estomate, to 1.5 mm long. **Sporangia** fused to form a sessile, bi- or trilocular, thick-walled synangium borne adaxially at base of forked sporophylls, to 2.5(–3) mm Ø, yellowish to greenish-yellow, each locule dehiscing with a longitudinal slit; spores numerous, achlorophyllaceous, elongate-ellipsoid, monolete, irregularly rugate, 60–75 µm long. **Distr:** Pantropical. **Ecol:** In low to mesic woods, thickets, swamps, hammocks and rocky slopes; 0–500(–1,250) m. **Zones:** 10–12.

EQUISETALES

FAMILY 1, GENUS 1, SPECIES 15.
Ordinal description same as familial.

EQUISETACEAE

[Horsetail Family].
Genus 1, species ~15. **Plants**, terrestrial to aquatic, perennial, rhizomatous; aerial stems (annual or perennial), to 8 m tall with most species <1 m tall, erect, hollow with a series of smaller carinal canals (under the ridges) and a central larger vallecular canal (under the valleys), ribbed, arising from creeping subterranean rhizomatous stems and producing whorls of branches (or without branches) at the swollen (jointlike) nodes; branches structurally similar to main stem; internodes with alternating vertical ridges and grooves externally. **Leaves** whorled; sheaths sometimes ± swollen; stem ridges traversing length of internode and continuing into sheaths, terminating in sheath teeth. **Sporangiophores** in cones, stalked, unbranched, peltate, hexagonal, whorled, arranged in compact strobili, terminal on the stems or branches; cone apices rounded or sharply pointed; sporangia numerous (5–10 per sporophyll) borne on the adaxial surface of the peltate heads of the sporangiophores, homosporous; sporangia dehiscing longitudinally; spores spherical, green, and with 4–6 straplike elaters wrapped around the spore that rapidly straighten and assist in spore dispersal. **Gametophytes** chlorophyllose, thalloid. **Distr:** Nearly cosmopolitan, absent in Australia, New Zealand and Antarctica. **Ecol:** In marshes, wetlands, etc.

Equisetum

[Horsetail, Scouring Rush].
Species 15. *Generic description same as familial.* **Cult:** Propagate from spores or by division of rhizomes. Requires a moist, well-drained compost in full or partial sun.

E. arvense L., *Sp. Pl.* 2: 1061, 1753. [Common Horsetail, Field Horsetail]. **Plants**, terrestrial or semi-aquatic; aerial stems dimorphic; vegetative stems green, branched, 0.2–0.6(–1) m tall, 1–5 mm Ø, center hollow; sheaths quadrate, 0.2–0.5(–1) × 0.2–0.5(–0.9) cm, teeth attenuate, dark, 4–14, narrow, 1–3.5 mm long, often cohering in pairs; branches in regular whorls, ascending, solid; ridges 3–4; valleys channelled; first internode of each branch longer than subtending stem sheath. **Fertile stems** brown, lacking stomates, unbranched, shorter than vegetative stems, with larger sheaths, fleshy, ephemeral. **Sporangia** cones, 0.5–4 cm long. **Distr:** Eurasia to Himalayas, C China, Korea, Japan, N America. **Ecol:** On riverbanks, marshes, weedy in some places; 0–3,200 m. **Zones:** 7–10.

E. ramosissimum Desfontaines, *Fl. Atlant.* 2: 398, 1799. (Syn: *E. thunbergii*, etc.). [Branched Horsetail]. **Plants**, terrestrial or semi-aquatic; rhizome subterranean, jointed, 3–6 mm Ø, roots borne in whorls at the nodes. **Stems** erect, to 1.5 m tall, to 5 mm Ø, branches borne in whorls at the nodes, up to 14 per whorl, to 13 cm long, to 1.5 mm Ø, main stem and branches ridged, with silicaceous tubercules or bands, with regular rows of paracytic sunken stomata in the sulci. **Fronds** membranous to thinly chartaceous, each with a single central vein, whorled, fused into a nodal sheath, to 1.1 cm long, ending in acuminate teeth, to 2 mm long, turning black with age, margins hyaline, estomate. **Sporangia** elongate, thin-walled, dehiscing longitudinally, borne on stalked, peltate sporangiophores forming a strobilus terminally on a main stem or branch, to 1.8 cm long, to 5 mm Ø, each sporangiophore bearing 6–9 sporangia; spores spheroidal, chlorophyllose, with 4 spathulate hygroscopic elaters. **Distr:** Africa, Asia, Europe. **Ecol:** In sandy and moist areas; ~150–800 m. **Zones:** 10–12.

MARATTIALES

FAMILY 1, GENERA 4, SPECIES 150.
Ordinal description same as familial.

MARATTIACEAE

Genera 4, species 150. **Ferns**, terrestrial or amphibious, eusporangiate. **Stem** mycorrhizal, generally unbranched, radial or dorsi-ventral, slender and creeping or ± stout and erect, often apparently massive since clothed with pairs of large, persistent photosynthetic stipules which protect the young croziers; roots fleshy, often produced aerially. **Fronds** clustered or ± distant; stipe sometimes grooved adaxially, with a pulvinus at base, another at base of the laminate portion of the frond and sometimes one or more pulvini in the middle region, sparsely to densely covered in non-clathrate, peltate (except in *Christensenia*) scales which may be borne on podia; lamina often very large and robust, simple to 4-pinnate (subpedate in *Christensenia*), with pulvini at the junctions of the different orders of pinnae and pinnules with rachises and also at the junctions of the different orders of rachises; rachises terete, sometimes winged; pinnae or pinnules thick, sparsely scaly. **Sporangia** arranged closely in elongate or circular sori, or synangia; spores bilateral or ellipsoid, monolete. **Gametophyte** green, lobed, the median region thickened; endophytic fungi present. **Distr:** Throughout the tropics and subtropics.

Angiopteris

Syn: *Archangiopteris*, etc.
Species ~150. **Ferns**; stem creeping and slender, or erect, sometimes massive. **Fronds** clustered, arising radially, 1–2-pinnate, to 7 m long; pinnae with pulvini at bases of rachillae, veins free, recurrent veins often present between true veins, running from margin towards costule. **Sporangia** dehiscing by a longitudinal slit; spores trilete, scabrate. **Young plants** much like *Archangiopteris*; lamina pinnate; stipe

Top row: *Angiopteris evecta, Osmunda claytoniana, Osmunda regalis, Hymenophyllum flabellatum.*
Bottom row: *Trichomanes reniforme, Lygodium japonicum, Anemia mexicana, Salvinia molesta.*

with pulvinus ~2.5 m above base (small bipinnate fronds lack this pulvinus); pinnae narrower than in *Archangiopteris* (~2 cm wide) and have recurrent veins. **Distr:** Australasia, Madagascar, and Oceania. **Cult:** Propagate from spores, offsets or ears at base of frond stalks. Requires a moist, well-drained, acidic soil in shade or partial sun.

A. evecta (G. Forst.) Hoffm., *Commentat. Soc. Regiae Sci. Gott.* 12: 29, t. 5, 1794. (Syn: *A. elongata, Polypodium evectum*). [Giant Fern, King's Fern, Mule's-foot Fern]. **Ferns**, robust; rhizome forming a massive, somewhat spherical trunk ~60–90 cm high, to 45 cm ∅. **Stipes** thick and fleshy, to ~2 m long, base swollen and bearing 2 flat, rounded, dark brown, leathery, stipulelike outgrowths ~5–10 cm long. **Fronds** 2–4-pinnate, glabrous, very large and spreading, to over 7 × ~2.5 m; pinnae ~30 cm wide; ultimate segments shortly stalked, linear, usually (8–) 10–13(–20) × (0.8–)1.5–2.5(–4) cm, base unequally wedge-shaped to ± rounded, especially on basal side, margins shallowly toothed, apex often long, narrow, acuminate, more deeply toothed; veins close, making a wide angle with the midrib, free, simple or once forked near the base, with pale, short translucent false-veins running between them. **Sori** without indusia, consisting of 3–7 pairs of sporangia clustered along the lateral veins near the margin. **Distr:** W Melanesia, Polynesia. **Ecol:** In moist forests and rainforests. **Zones:** 10–12.

OSMUNDALES

FAMILY 1, GENERA 3, SPECIES 16–36.
Ordinal description same as familial.

OSMUNDACEAE

[Royal Fern].
Genera 3, **species** 16–36. **Ferns**, terrestrial, herbaceous, frequently in clumps; stems creeping, beset with old petiole bases and black fibrous roots, scales absent; older stems seldom persisting. **Fronds** monomorphic or dimorphic; lamina 1–2-pinnate (rarely bipinnatifid); rachises grooved; pinnae monomorphic or dimorphic; indument of reddish to light brown hairs; veins dichotomous, running to margins. **Sori** absent; sporangia borne on slightly modified fertile segments of lamina also possessing fully expanded pinnae, or sporangia covering lamina lacking green expanded pinnae, clustered in marginal zones, indusia lacking; spores green, all similar. **Gametophytes** green, above-ground, obcordate to elongate. **Distr:** Nearly worldwide (temperate and tropical regions).

Osmunda

Species 10. **Ferns**, terrestrial; stems creeping, tips often somewhat erect. **Fronds** dimorphic; *fertile fronds* erect, often notably smaller than *sterile fronds* in length and width; lamina 1–2-pinnate; pinnae monomorphic to dimorphic, pinnatifid or pinnate. **Distr:** Nearly worldwide, tropical and temperate regions. **Cult:** Propagate from ripe spores or by division. Requires a rich compost near a water feature in full or partial sun.

O. claytoniana L., *Sp. Pl.* 2: 1066, 1753. [Interrupted Fern]. **Ferns**, terrestrial. **Fronds** pinnate-pinnatifid; petioles ~⅓ length of lamina, winged, with light brown hairs, becoming glabrate. **Sterile fronds** elliptic to oblong, ~0.5–1 m long; pinnae broadly oblong, lacking persistent tuft of hairs at base; ultimate segments with base truncate, margins entire, apex rounded. **Fertile fronds** with greatly reduced, sporangia-bearing medial pinnae that wither early, giving appearance of no middle pinnae (hence the vernacular name, interrupted fern). **Sporangia** greenish, turning dark brown. **Distr:** N America; Asia. **Ecol:** In rich, wooded slopes and moist ledges, rarely in forests. **Zones:** 3–10.

O. regalis L., *Sp. Pl.* 2: 1065, 1753. [Royal Fern]. **Ferns**, terrestrial, erect or semi-erect; stems 1–1.5 m tall, fibrous, brown-hairy, broken stipe bases persistent; crown usually single, rarely multiple. **Fronds** deciduous, ovate to triangular, bipinnate, 0.3–2 × 0.2–0.75 m, tough and leathery, yellow-green to dark green; secondary pinnae oblong, 2–7 × 0.8–1.8 cm, apices rounded, margins smooth or minutely toothed, base truncate, stalked; stipes and rachises yellow brown or pale brown, slightly hairy, with auriculate lobes at base. **Fertile fronds** terminal, completely replacing foliage fronds, much-reduced, bipinnate, erect, to 3 × 0.4 cm, densely covered in clusters of sporangia. **Sporangia** with spores released when green to yellow-brown. **Distr:** Africa, Asia, the Americas, Europe. **Ecol:** In swampy areas and streamsides, often forming large clumps and appearing shrubby; <2,000 m. **Zones:** 3–11.

HYMENOPHYLLALES

FAMILY 1, GENERA 9, SPECIES 600.
Ordinal description same as familial.

HYMENOPHYLLACEAE

[Filmy Fern].
Genera 9, **species** 600. **Ferns**, epiphytic, terrestrial,

or lithophytic; stems long-creeping, often threadlike and intertwining, or short-erect, protostelic, bearing brown hairs of 1–2 types; roots sparse or absent. **Fronds** small, 0.5–20 x 0.2–5 cm, often forming dense mats; lamina ovate or oblong to lanceolate, simple to decompound, entire or dentate, scales or simple and/or stellate hairs often borne on veins or leaf margins, veins free and divergent, occasionally present as unattached 'false' veins; petiole short, threadlike to wiry, often winged part or entire length. **Sori** marginal on vein ends, enclosed by 2-valved or conic involucres. **Sporangia** borne on moundlike receptacle or on elongate 'bristle,' sessile or short-stalked; annulus oblique; spores green, globose, trilete. **Gametophytes** filamentous or ribbonlike or a combination of both, much-branched, 0.2–1 cm, often bearing gemmae, persistent, clone-forming by vegetative reproduction. **Distr:** Worldwide in humid parts of the tropics and subtropics; a few in temperate regions.

Hymenophyllum
[Filmy Fern].
Species ~310. **Ferns**, epiphytic or lithophytic; stems long-creeping, inter-twining, threadlike; hairs brown, sparse; roots few, delicate. **Leaves** 1–3-pinnatifid, 2–6 x 0.5–1.5 cm; petiole short, threadlike, not winged; lamina with inconspicuous glandular hairs or prominent stellate hairs, margins entire to distantly dentate. **Soral involucres** 2-valved, the halves roundish to ovate. **Sporangial receptacle** a low mound of tissue included within involucre. **Gametophytes** persistent, ribbonlike, much-branched; gametophyte gemmae platelike or absent. **Distr:** Tropical regions worldwide. **Cult:** Propagate from spores. Requires a moisture-retentive substrate in shade.

H. flabellatum Labill., *Nov. Holl. Pl. Sp.* 2, 101, t. 250, 1806. (Syn: *H. nitens*, etc.). **Ferns**; rhizome slender, spreading, much-branched; clad, especially when young, in long yellow-brown hairs. **Fronds** with stipites rather distant to crowded; stipes slender, not winged, with a tuft of hairs at base, 2–10 cm long; rachis slender, winged only in upper part, ± clad in pale brown hairs; lamina ovate to narrow-oblong, yellowish to pale green, 3–15 cm long, 2–4-pinnatifid, final segments linear, obtuse. **Sori** usually on lateral segments, slightly immersed; indusium 1 mm Ø, suborbicular, 2-valved to more than halfway; valves entire or nearly so; receptacle included to slightly exserted. **Distr:** Australia, New Zealand, Tahiti, Tasmania. **Ecol:** In lowland forests, usually epiphytic. **Zones:** 8–11.

Trichomanes
[Bristle Fern].
Species ~320. **Ferns**, epiphytic or lithophytic; stems long-creeping or short and erect, clothed in masses of dark brown hairs of 2 or more types, including multicellular gland-tipped hairs and elongate, sometimes branched and often multicellular, rhizoidlike hairs; roots sparse or absent on creeping stems, numerous and wiry on erect stems. **Fronds** entire, lobed, or compound, 0.5–20 x 0.2–5 cm; lamina glabrous or with scattered, multicellular, gland-tipped hairs on veins; margins entire or minutely lobed, sometimes bearing dark stellate hairs; petiole short, wiry, often partially or wholly winged. **Soral involucres** conic. **Sporangia** sessile, formed at base of exserted bristle and carried outward by intercalary growth of bristle base. **Gametophytes** persistent, entirely filamentous or with proximal filamentous net producing aerial lamina with gemmiferous apices; gametophyte gemmae uniseriate. **Distr:** Nearly worldwide, mostly tropical, a few temperate. **Cult:** Propagate from spores. Requires a moisture-retentive substrate in shade.

T. reniforme G. Forst., *Flor. Ins. Austr. Prodr.* 84, 1786. (Syn: *Cardiomanes reniforme*). [Kidney Fern]. **Ferns**, forming large masses; rhizome rather stout, far-creeping, much-branched, with tufts of dark red-brown hairs at base of rather distant to approximate stipites. **Fronds** 3–10 x 4–13 cm, reniform, coriaceous when mature; sinus deep; juvenile fronds flabellately divided above, of adults sometimes shallowly sinuate; stipes stout, 5–20 cm long, not winged; fronds curl up in dry weather, recovering after rain. **Sori** often closely set along the margin, except in lower part and sinus; indusium ~2 mm long, cup-shaped to almost bell-shaped; receptacle stout, exserted. **Distr:** New Zealand. **Ecol:** Mainly terrestrial in lowland to montane forests. **Zones:** 8–11.

SCHIZAEALES

FAMILIES 3, GENERA 4, SPECIES 155.

Fronds with fertile and sterile lamina differentiated; lacking well-defined sori. **Sporangia** borne singly. **Spore** cases completely encircled near the tip by the annulus; annulus a group of thick-walled cells. **Distr:** Nearly worldwide. **Info:** This order is monophyletic (Hasebe et al. 1995; Pryer et al. 2001, 2004; Skog et al. 2002; Wikström et al. 2002).

LYGODIACEAE
[Climbing Ferns].
Genus 1, **Species** ~25. *Familial description same as generic.*

Lygodium
Species ~25. **Ferns**, terrestrial; stems subterranean, protostelic; indument of dark, dense hairs. **Fronds** vinelike, of indeterminate growth; pinnae reduced to short stalks, each bearing a pair of opposite pinnules, usually with an often dormant apical bud. **Sporangia** in 2 rows, 1 on each side of midvein of contracted, oblong, marginal lobes of ultimate segments, covered by hoodlike flap of tissue serving as indusium; spores tetrahedral-globose, trilete, rarely monolete. **Gametophytes** terrestrial, cordate, glabrous. **Distr:** Pantropical. **Cult:** Propagate by division of clumps or by planting fronds, which root readily.

L. japonicum (Thunberg ex Murray) Swartz, *J. Bot.* (Schrader) 1800(2): 106, 1801. (Syn: *Ophioglossum japonicum*). [Japanese Climbing Fern]. **Ferns**, stems creeping or climbing. **Leaves** to ~3(–30) m; petiole borne 2–7 mm apart, 10–35 cm. **Sterile pinnae** on 1.5–3.5 cm stalks, triangular to lanceolate, 2–3-pinnate, 6–15 x 5.5–15 cm; ultimate segments lanceolate, lobed or divided proximally; lobes usually acute at tip and directed toward apex; segment apices long-attenuate to acute; segments not articulate to petiolules, not leaving wiry stalks when detached; lamina tissue pubescent abaxially with short, curved hairs. **Fertile pinnae** on 1–2 cm stalks, lanceolate-triangular, 2–3-pinnate, 5–18 x 4–14 cm; ultimate segments ovate to lanceolate, fringed with fertile lobes, otherwise similar to sterile segments. **Distr:** E Asia. **Ecol:** In moist forests and waste areas; can become invasive in favorable conditions. **Zones:** 9–12.

ANEMIACEAE
Genera 2, **species** 119. **Ferns**, terrestrial or epilithic, small to large; stems compact or short-creeping, horizontal, solenostelic or dictyostelic, clothed with orange to reddish-brown hairs. **Fronds** erect (rarely forming a flat rosette), partially to entirely dimorphic. **Sporangia** in 2 rows on ultimate segments of fertile pinnae, sessile, oblong; annulus apical; spores tetrahedral-globose, with parallel or rarely anastomosing ridges. **Gametophytes** terrestrial, green, cordate with unequal lobes. **Distr:** Tropical and subtropical regions.

Anemia
Species 117. **Ferns**, terrestrial or epilithic, small to medium-sized; stems short-creeping, horizontal, clothed with dark hairs. **Fronds** partially dimorphic with sporangia restricted to erect, dissected, most proximal pair of pinnae arising from petiole just below sterile part of lamina or fronds fully dimorphic and lamina tissue lacking on fertile fronds; lamina 1–3-pinnate, papery to leathery, veins free (anastomosing), apex pinnatifid or less often with an odd terminal pinna; costae not or only shallowly grooved adaxially; stomata peri-, desmo- and polocytic. **Sporangia** on 2 skeletonised basal segments which are borne below or nearly at the level of the first sterile pinnae and often held erect, rarely on scarcely modified basal pinnae; spores trilete, globose to tetrahedral-globose, with prominent angles. **Distr:** Tropical and subtropical regions worldwide. **Cult:** Propagate from spores or by division. Requires a moist, well-drained soil in shade or partial sun.

A. mexicana Klotzsch, *Linnaea*. 18: 526, 1844. [Curly Grass Fern]. **Ferns**, terrestrial; stems ~2 mm Ø. **Fronds** partially dimorphic; lamina deltate-lanceolate, 1-pinnate, (15–)22–45 x 8–16 cm, somewhat leathery; petiole straw-colored, glabrate; pinnae 4–7 pairs, mostly alternate, lanceolate, base truncate, margins minutely serrulate, proximal margins often slightly excavate, apex acuminate, adaxially glabrous, abaxially hirsute with minute white hairs to glabrous. **Fertile pinnae** usually taller than sterile lamina; sporangia limited to proximal pair of pinnae; spores with ridges parallel and closely placed. **Distr:** USA (Texas), N Mexico. **Ecol:** On tropical forest floors and on lightly shaded limestone outcrops; 400–500 m. **Zones:** 9–11.

SALVINIALES
[Water Ferns].

FAMILIES 2, GENERA 5, SPECIES ~90.

Ferns with aerenchyma tissue often present in roots, shoots and petioles; plants heterosporous, spores with endosporous germination; monomegaspory. **Fronds** with fertile and sterile lamina differentiated; veins anastomosing; annulus absent. **Gametophytes** reduced. **Distr:** Subcosmopolitan. **Info:** This order is monophyletic (Hasebe et al. 1995; Pryer 1999; Pryer et al. 2001, 2004).

SALVINIACEAE
Genus 1, **species** 10. **Ferns**, small, floating aquatic; stems creeping, branched, bearing hairs but not true roots. **Fronds** in whorls of 3, with 2 fronds green, sessile or short-petioled, flat, entire, and floating, 1 frond finely dissected, petiolate, rootlike, and pendent (submerged). **Submerged fronds** bearing sori that are surrounded by basifixed membranous indusia (sporocarps). **Sporocarps** of 2 types, bearing either megasporangia that are few in number (~10), each with a single megaspore, or many microsporangia, each with 64 microspores; spores of 2 kinds and sizes, both globose, trilete; megagametophytes and microgametophytes protruding through sporangium wall; megagametophytes floating on water surface with archegonia directed downward; microgametophytes remaining fixed to sporangium

wall. **Distr:** Mostly tropical, N America, Mexico, West Indies, Central America, S America, Eurasia, Africa, Madagascar. **Cult:** Propagate from spores or by division. Requires a shallow, still or slow-moving water feature in full sun.

Salvinia

[Floating Fern, Water Spangles].
Species 10. *Generic description same as familial.*

S. molesta D.S. Mitchell., *Brit. Fern Gaz.* 10(5): 251–252, 1972. (Syn: *S. auriculata*). [African Payal, Aquarium Water-moss, Giant Salvinia, Kariba Weed, Salvinia, Water Fern]. **Ferns**, free-floating, fast-growing, mat-forming, branched, gregarious, annual or perennial; individual plants to 30 cm long with numerous fronds which usually form a mat to 2.5 cm thick; rhizome slender. **Fronds** produced in groups of 3, from the delicate stem, each group with 2 broadly ovate, entire, undivided, green, aerial fronds to 2.5 cm long situated on abaxial side of stem; the third, submerged, rootlike, situated ventrally, much divided, feathery, brown, to 25 cm long, resembles and functions as a true root, and bears the sporocarp (or spore-forming structures). **Sporocarps** globose, 2–3 mm Ø, on a short stalk 1 mm long; megaspores and microspores ~2 mm long, numerous, globular, covered with minute hairs. **Distr:** S America (S Brazil). **Ecol:** In still or slow-moving, fresh to slightly brackish waters. **Zones:** 10–12.

CYATHEALES

FAMILIES 8, GENERA 15, SPECIES >633.

Tree ferns, morphologically very diverse, without obvious defining morphological characters. **Stems** of some species are trunklike while the others have creeping rhizomes; some have only hairs on the stems and laminas, others have scales. sori are abaxial or marginal, either indusiate or exindusiate. **Sori** terminal on veins. **Sporangium** with stalk ~5 cells across; spores are globose or tetrahedral-globose, with a trilete scar; antheridium walls 5 cells across. **Gametophytes** green, cordate. **Distr:** Worldwide, especially diverse in the tropics and subtropics.

CYATHEACEAE

Genera ~5, **species** >600. **Tree ferns**, mostly arborescent; true scales present on the rhizomes, petiole bases, and sometimes on the laminas; stems with polycyclic dictyosteles, apices (and usually petiole bases) covered with large scales, sometimes also with trichomidia or hairs. **Fronds** usually large (to 5 m long); petioles with obvious, usually discontinuous pneumathodes in 2 lines; lamina 1–3-pinnate (rarely simple), veins simple to forked, free, rarely anastomosing (mostly in some *Cyathea*). **Sori** superficial (abaxial) or terminal on the veins and marginal or submarginal, round, exindusiate, or indusia saucerlike, cuplike, or globose and completely surrounding sporangia, or bivalvate. **Sporangia** maturing gradately, with oblique annuli; receptacles raised; paraphyses usually present; spores tetrahedral, trilete, variously ornamented. **Gametophytes** green, cordate. **Distr:** Pantropical.

Cyathea

Species >600. **Tree ferns**. **Fronds** with stipes smooth or spiny, sometimes with hairs; costa usually adaxially pubescent; veins free; indusium absent or scalelike to globose. **Distr:** Pantropical. **Cult:** Propagate from spores or basal suckers. Requires a fertile, moist, humus-rich soil in full or partial sun.

C. colensoi (Hook. f.) Domin, *Pterid.* 262, 1929. (Syn: *Alsophila colensoi*). [Creeping Tree Fern]. **Tree ferns**; caudex prostrate, hard, marked by scars of fallen stipites, ascending at tips, to 1+ m long and 8 cm Ø; rarely erect. **Fronds** with stipes 7–20(–30) cm long, rather stout, finely muricate, ± densely clad in pale shining subulate-attenuate paleae to 3 cm long, mingled in upper part with fulvous hairs; rachis deeply grooved, clad as are stipites; lamina broadly ovate, 0.3–0.6–1 × 0.15–0.6 cm, apex acute, membranous, brownish to yellowish-green; primary pinnae oblong-lanceolate, 15–30(–40) × 5–10 cm, apex acute to acuminate; secondary pinnae to 5 cm long, acute; pinnules oblong, 5–8 mm long, acute, lower distinct, upper confluent, acutely toothed or fertile pinnules lobulate; lobules 4–5 pairs, triangular, subacute, recurved. **Sori** <1 mm Ø pale brown; indusium absent. **Distr:** Endemic to New Zealand. **Ecol:** In montane to subalpine forests. **Zones:** 9–11.

C. medullaris (Forst. f.) Swartz, *Schrad. J. Bot.* 2: 94, 1801. (Syn: *Polypodium medullare*, etc.). [Black Tree Fern]. **Tree ferns**; caudex to 15+ m tall, to 30 cm Ø, except towards base much thickened by dense mass of rootlets, black, marked by hexagonal scars of fallen stipites. **Fronds** with stipes stout, black, to 1 m long, to ~8 cm Ø near base, tuberculate, copiously clad towards base with black linear paleae ~3 mm long; rachis (and secondary rachides and costae) stout, pale brown, muricate, clad in brownish hairs, later glabrate; lamina to ~6 × 2 m, 2–3-pinnate, coriaceous, adaxially dark green, abaxially pale green; primary pinnae to 1 m long, oblong-lanceolate, attenuate; secondary to 15 cm long, narrowly oblong, attenuate; pinnules ~1 cm long, lower distinct, upper confluent, narrowly oblong obtuse to acute, subfalcate, crenate-serrate to crenulately lobed, margins recurved, veins several times forked. **Sori** to ~16 per pinnule, 1+ mm Ø, usually copious. **Indusium** thin, at first covering sorus, splitting irregularly. **Distr:** New Zealand, Australia (Tasmania, Victoria), Pacific Islands. **Ecol:** In lowland forests. **Zones:** 9–11.

C. smithii Hook. f., *Fl. NZ.* 2: 8, t. 72, 1854. (Syn: *Hemitelia smithii*). [Soft Tree Fern]. **Tree ferns**; caudex to 8+ m tall, 30 cm Ø, sometimes forked, invested above with persistent pendulous withered fronds, widened towards base by a mass of rootlets. **Fronds** with stipes rather slender, 15–30 cm long, asperulous, dark, grooved, densely clad at base in chestnut to pale brown subulate-attenuate paleae, ~3 cm long; rachis stout, pale yellowish-green, finely muricate, ± paleate, at first densely clad in reddish hairs, later glabrate; lamina 1–2+ m long, membranous, soft, adaxially bright green, abaxially paler, 2–3-pinnate, oblong-lanceolate, apex acute to acuminate; primary pinnae lanceolate-oblong, 15–40 × 5–10 cm, apex acute to acuminate, with strigillose costae; secondary pinnae narrowly oblong, 3–6 cm long, margins crenate-serrate, apex acute, subfalcate. **Sori** ~1 mm Ø, copious; indusium hemispherical to nearly cup-shaped, on costal side of sorus. **Distr:** Endemic to New Zealand. **Ecol:** In lowland to montane forests. **Zones:** 9–11.

DICKSONIACEAE

Genera 3, **species** ~30. **Tree ferns**, terrestrial, mostly arborescent or with erect or ascending rhizomes; rhizomes with polycyclic dictyosteles, or solenostelic; stem apices and usually petiolar bases covered with uniseriate hairs. **Fronds** with lamina large, 2–3 pinnate, veins simple to forked, free. **Sori** abaxial and exindusiate or marginal and each with a bivalvate or cuplike indusium, adaxial (outer) valve formed by the reflexed segment margin and often differently colored. **Sporangia** with oblique annuli; receptacles raised; paraphyses often present, filiform; spores globose or tetrahedral, trilete. **Distr:** E Asia, Australasia, Neotropics, St Helena.

Dicksonia

Species ~25. **Tree ferns**, large, terrestrial; stem ascending or erect, rarely prostrate, trunklike. **Fronds** with lamina mostly large, firm, bipinnate and pinnatifid to quadripinnate, somewhat narrowed at base, veins pinnately forked. **Sorus** globose, on apices of veinlets; receptacle ± elevated. **Sporangia** subsessile or distinctly stalked; annulus oblique, indurated around the base, with lateral stomium; spores globose, sometimes with prolonged angles; exposure 2-layered, usually pitted; indusium coriaceous to submembranous. **Distr:** Tropical and southern temperate regions, with the highest diversity in New Guinea. **Cult:** Propagate from spores or offsets. Requires a moist, well-drained soil in shade or partial sun.

D. squarrosa (Forst. f.) Swartz, *Syn. Fil.* 136, 1806. (Syn: *Trichomanes squarrosum*, etc.). [New Zealand Tree Fern, Slender Tree Fern]. **Tree ferns**; rhizomes spreading from main stock to 1 m or more, giving rise to subsidiary erect caudices, often forming a grove; caudex slender, to ~10 cm Ø, to 6 m or more tall, invested by long-persistent bases of stipites;

Below: *Cyathea colensoi, Cyathea medullaris, Cyathea smithii, Dicksonia squarrosa.*

PL

Above: *Pteridium aquilinum, Adiantum capillus-veneris, Adiantum peruvianum, Adiantum raddianum.*

bearing dormant buds near base, or these developing into short or long branches. **Fronds** with stipes densely clad in deciduous dark filiform hairs to 4 cm long, roughened by enlarged persistent bases of fallen hairs; rachis clad in dark reddish-brown hairs when young, becoming rough with age; lamina ± oblong-lanceolate, 1–2.5 m long, 2–3-pinnate, coriaceous; primary pinnae deltoid-ovate, 25–50 cm long, acuminate; secondary rather closely set, 5–8 cm long, acute. **Fertile pinnules** closely set, confluent at base, 1–1.5 cm long; lobes strongly concavo-convex ~5 mm long, rounded, each bearing a sorus; lobes of barren pinnules shallowly concave, acute, usually sharply toothed; indusium rather delicate, concavo-convex. **Distr:** Endemic to New Zealand. **Ecol:** In lowland forests. **Zones:** 8–10.

POLYPODIALES

FAMILIES 15, GENERA ~210–257, SPECIES OVER 7,165.

Indusia laterally or centrally attached (indusia lost in many lineages); sporangial stalks 1–3 cells thick, often long; sporangial maturation mixed; sporangia each with a vertical annulus interrupted by the stalk and stomium; gametophytes green, usually cordate (sometimes ribbon-shaped in some epiphytes), surficial. **Distr:** Cosmopolitan. **Info:** This order is monophyletic (Hasebe et al. 1995; Pryer et al. 1995, 2001, 2004; Schneider et al. 2004).

DENNSTAEDTIACEAE

Genera ~20, **species** ~400. **Ferns**, large, terrestrial; rhizomes hairy, creeping or erect, with single or complex double solenosteles. **Fronds** large, tripinnate to 4-pinnate, with continuous apical growth in *Hypolepis*; veins free or anastomosing; stipes with undivided or dissected U-shaped vascular strands. **Sori** marginal, submarginal or superficial near the margin, small, subcircular to elongate; indusium absent, or ovate and shallowly cupped, or linear or with the leaf margin modified to form a pseudo-indusium, or both pseudo-indusia and true indusia present; soral paraphyses present or absent. **Spores** not green, tetrahedral or bilateral, monolete or trilete. **Gametophytes** green, cordate, with archegonia and antheridia borne on lower surface. **Distr:** Worldwide, mostly tropical.

Pteridium

Species 1. *Generic description same as specific.* **Cult:** Propagate from spores or by division. Requires a moist, well-drained soil in full or partial sun.

P. aquilinum (L.) Kuhn, *Reisen Ost-Afrika* 3, 3: 11, 1879. (Syn: *Asplenium aquilinum*, etc.). [Bracken, Brake Fern]. **Ferns**, terrestrial; rhizome subterranean, creeping, irregularly branched. **Fronds** widely spaced, erect, apical part often arching, 0.3–1.8(–3) x 0.6 m; coiled spirally when young, with dense chestnut hairs 1–1.5 mm long; stipe rigid, brown, adaxially sulcate, 15–40(–85) cm long; lamina triangular to oblong-ovate; pinnae ovate-triangular, to 40 x 15 cm, acute; pinnule segments linear to oblong. **Sori** linear; receptacle bare. **Sporangia** long-stalked, simple, triseriate below capsule, capsule elliptic in lateral view, annulus with (12–)15 indurated annulus cells, epistomium (3–)4(–6)-celled, hypostomium (3–)4(–5)-celled, stomium 2-celled; indusium marginal, outer linear, stramineous to hyaline, fimbriate, to 0.9 mm long, inner indusium partly developed; spores 64 per sporangium, brown, tetrahedral-globose, trilete, granulate, (34–)37.64(–42) µm Ø. **Distr:** Widespread in subSaharan Africa, western Indian Ocean region; temperate Europe. **Ecol:** In exposed open grasslands, etc. **Zones:** 4–11.

PTERIDACEAE

[Maidenhair Fern].

Genera ~50, **species** ~950. **Ferns**, perennial (rarely annual), on rock or terrestrial, of small (rarely large) stature; stems compact to creeping, branched or unbranched, dictyostelic, bearing hairs and/or scales. **Fronds** monomorphic to dimorphic, circinate or noncircinate in bud; petioles usually with persistent scales proximally, lacking spines; vascular bundles 1 to several, roundish or crescent-shaped in cross-section; lamina 1–6-pinnate; indument on petioles, rachises, costae, and laminas; veins pinnate or parallel in ultimate segments, simple or forked. **Sori** borne abaxially on veins, or sporangia densely covering abaxial surface (acrostichoid); receptacle not or only slightly elevated; indusia (when present) formed by reflexed, recurved, or revolute leaf margin (false indusium). **Sporangia** stalk of 2–3 rows of cells; annulus vertical, interrupted by stalk; spores 64 or 32 (rarely 16) per sporangium; spores all 1 kind, brown, black, or gray (rarely yellow), globose to globose-tetrahedral or trigonal. **Gametophytes** green, aboveground, obcordate to reniform, sometimes asymmetric, usually glabrous (rarely glandular-farinose); archegonia and antheridia borne on abaxial surface, antheridia 3-celled. **Distr:** Worldwide. **Cult:** Requires a fertile, moist but well-drained soil in a shady place. Propagation is by ripened spores or by division.

Adiantum

[Maidenhair Fern].

Species ~150–200. **Ferns**, terrestrial or lithophytic; stems short- to long-creeping or suberect, branched. **Fronds** monomorphic to somewhat dimorphic, usually densely clustered (rarely distant), 0.15–1.1 m; petiole chestnut brown to dark purple or blackish, adaxial groove single, glabrous, hispid, or strigose, vascular bundles1 or 2; lamina lanceolate, ovate, trowel-shaped, or fan-shaped, 1–4(–9)-pinnate proximally, membranous to papery, usually glabrous, adaxially dull or shiny, not striate; rachis straight or flexuous; ultimate segments subsessile to short-stalked, round, fan-shaped, rhombic, or oblong, 0.3–2.9 cm wide, base truncate to cuneate, free from costa, stalk dark, often lustrous. **Fertile segments** with marginal lobes recurved to form false indusia; veins of ultimate segments conspicuous, free, ± dichotomously forking near base, sometimes anastomosing, parallel distally; false indusia light grey-green or brown, narrow, 0.6–1 mm wide, marginal, concealing sporangia until sporangia dehiscence. **Sporangia** submarginal, borne along or sometimes also between veins on abaxial surface of false indusium, paraphyses and glands absent; spores yellow or yellowish-brown, tetrahedral-globose, trilete, rugulate to rugose or tuberculate, equatorial ridge absent. **Distr:** Nearly worldwide. **Cult:** Propagate from spores or by division. Requires a moist soil, rich in organic matter, in shade or partial sun.

A. capillus-veneris L., *Sp. Pl.* 2: 1096, 1753. (Syn: *A. capillus-veneris* var. *modestum*, etc.). [Common Maidenhair, Venus's Hair Fern, Southern Maidenhair]. **Plants**; stems short-creeping; scales golden brown to medium brown, concolored, iridescent, margins entire or occasionally with a single broad tooth near base. **Fronds** lax-arching or pendent, closely spaced, 15–75 cm; petiole 0.5–1.5 mm Ø, glabrous, occasionally glaucous; lamina lanceolate, pinnate, 10–45 x 4–15 cm, glabrous, gradually reduced distally; proximal pinnae 3(–4)-pinnate; rachis straight to flexuous, glabrous, not glaucous; segment stalks 0.5–3.5 mm, dark color extending into segment base; ultimate segments various, generally cuneate or fan-shaped to irregularly rhombic, about as long as broad, base broadly to narrowly cuneate, margins shallowly to deeply lobed, incisions 0.5–7 mm, occasionally ± laciniate, sharply denticulate in sterile segments, apex rounded to acute. **Indusia** transversely oblong or crescent-shaped, 1–3(–7) mm, glabrous. **Spores** mostly 40–50 µm Ø. **Distr:** Tropical America, West Indies; tropical to warm temperate regions of Eurasia and Africa. **Ecol:** On moist cliffs, banks, along waterways; 0–2,500 m. **Zones:** 8–11.

A. peruvianum Klotzsch, *Linnaea* 18: 555, 1845. [Peruvian Maidenhair Fern, Silver Dollar Fern]. **Ferns**; stem large and massive, very short-creeping. **Fronds** to 1 m long; petiole glabrous; rachis similar; lamina broadly ovate-triangular, to tripinnate, or small ones elongate-triangular or oblong, 1–2-pinnate; pinnae widely alternate, the basal largest, ultimate segments mostly ovate-trapeziform, bluntly acute or rounded (the terminal bluntly acute), glabrous, long-stalked, articulate, the dark color of the stalk not entering the segment, large. **Indusia** many, borne on the upper and

Above: *Hemionitis arifolia, Pellaea rotundifolia, Asplenium bulbiferum, Asplenium flaccidum.*

distal sides, few if any on the inner side, roundish to short-oblong or lunate. **Distr:** Tropical South America (Ecuador to Bolivia). **Ecol:** In rainforests, on rocky slopes or canyon sides in forests; 750–2,800 m. **Zones:** 10–12.

A. raddianum C. Presl, *Tentamen Pteridographiae* 158, 1836. (Syn: *A. amabile*, etc.). [Pacific Maidenhair Fern]. **Ferns**, terrestrial; rhizome short-creeping, provided with castaneous, linear or lanceolate scales 2–4 mm long, margins entire. **Fronds** crowded, often subcaespitose, petiolate, 16–60 × 6–18 cm; petiole 8–30 cm long, 0.8–2 mm thick, terete to flattened, or angular, lustrous, castaneous, atropurpureous or blackish, glabrous, or sparsely scaly near base; lamina delicate, membranous, glabrous, (2)3–4-pinnate, lanceolate to ovate or deltoid-ovate; rachis lustrous; pinnae stalked, 6–10 pairs, widely spaced, basal pair larger, with 3–7 pairs of pinnules, basal pinnule larger and pinnate; ultimate segments cuneate, subflabellate, distal margin shallowly lobed to deeply incised, sterile margins acutely or subacutely dentate, stalks not dilated or articulated apically, brownish to purplish, the dark color commonly entering the segment; veins free, subflabellate, 2–3-forked, tips each ending in a well-developed sinus. **Sori** few on each ultimate segment, often solitary on a segment lobe, usually orbicular-reniform, indusium sinuses narrow, often deep. **Distr:** Tropical America. **Ecol:** In humid, mixed forests; to 1,500 m. **Zones:** 11–12.

A. reniforme L., *Sp. Pl.*: 1094, 1753. **Ferns**, terrestrial; rhizome short with scales narrowly lanceolate, entire, formed of thin-walled-cells. **Fronds** tufted; stipe 15–30 cm long; lamina undivided, reniform or ± round, 1.5–8(–10) cm long and wide, usually with a deep basal sinus, lobes ± overlapping or in some young fronds ± truncate, margins entire or slightly lobed, subcoriaceous, glabrous except for woolly hairs sometimes present near the apex of stipe or in some variants lamina at first densely woolly all over; venation fan-shaped, the nerves usually bifurcate in lower third of limb. **Sori** numerous, closely placed or ± contiguous along the total circumference. **Distr:** Africa (DR Congo, Kenya, Malawi, Tanzania), Madagascar, Mauritius, Réunion; Portugal, Spain; Sri Lanka, China. **Ecol:** Usually in moist places; 500–2,100 m. **Zones:** 10–12.

Hemionitis

Species 8. **Ferns**, terrestrial; rhizome short, suberect, dictyostelic, clothed with narrow brown scales. **Fronds** crowded, small or medium-sized, herbaceous, somewhat dimorphic; sterile fronds on short stipes; fertile fronds on long stipes; lamina round or cordate, or palmately lobed, never compound or elongate, veins reticulate without free veinlets. **Sori** along all veins of fertile fronds; annulus of 14–20 thickened cells. **Spores** tetrahedral, reticulate-spinulose. **Distr:** Tropical America and Asia. **Cult:** Propagate from spores, plantlets (borne at frond bases) or by division. Require a moist compost in shade or partial sun.

H. arifolia (Burm. f.) T. Moore, *Ind. Fil.* 114, 1859. (Syn: *Asplenium arifolium*). [Heart Fern, Piggy-back Fern]. **Ferns**, terrestrial; rhizome short, suberect, clothed with narrow brown scales. **Fronds** with lamina dimorphous, simple; stipes maroon to black, scaly at the base. **Sterile fronds** narrowly ovate to oblong, 4–9 cm long, base deeply cordate, apex rounded, thickly coriaceous, abaxial surface with scattered scales and hairs like those on the stipe, with smaller and more numerous hairs on the margins; stipes 4–9 cm long. **Fertile fronds** usually about as long as the sterile or shorter, apical part narrower, oblong to triangular; stipes 15–30 cm long. **Sori** continuous along the veins, forming a network all over the surface. **Distr:** India, Indo-China, Malaysia, Myanmar, Philippines, Sri Lanka, Taiwan. **Ecol:** In hilly areas of tropical dry deciduous forests; common on forest slopes. **Zones:** 10–12.

Pellaea

Species ~35. **Ferns**, terrestrial or epilithic; rhizome short-decumbent, closely branched. **Fronds** up to 12 per plant, crowded, ceaspitose, erect; stipe rigid, brownish to black, nitid, terete, initially hairy and scaly, later glabrous, or distally short-pilose adaxially, hairs then ferrugineous; lamina thinly coriaceous to coriaceous, ovate, deltate, triangular, lanceolate or oblong-acute, to tripinnate, with up to 52 petiolated pinna pairs; rachis and lower order axes rigid, terete, brownish to black, nitid, usually glabrous; pinnae with petiole glabrous or hairy and scaly, opposite to alternate, proximally more widely spaced than distally, spaced or overlapping, simple to bipinnate, lower pinnae often directed towards the base, linear, ovate, rectangular-acute or broadly ovate, glabrous adaxially, abaxially usually glabrous; pinnules petiolate, alternate, spaced to overlapping, to 1-pinnate, ovate to oblong-cuneate, entire, adaxially glabrous, abaxially sparsely hairy; segments petiolate, alternate, cordate, hastate or pentagonal, entire, glabrous, articulated; venation obscure, repeatedly forked and free in sterile fronds. **Sori** forming a continuous line along near-marginal vascular commissure; sporangium short or long-stalked, simple, triseriate below capsule; indusium membranous to chartaceous, continuous, marginal, entire to shallowly crenulate. **Spores** brown, tetrahedral-globose, trilete, granulate, rugulose or elliptic and monolete, smooth granulose, 25–70 μm Ø. **Distr:** SE USA, Mexico; Africa, Madagascar; Indian Subcontinent. **Cult:** Propagate from spores or by division. Requires a rich, moist, well-drained soil in shade or partial sun.

P. rotundifolia (Forst.) Hook., *Sp. Fil.* 2: 136, 1858. [Button Fern, Round-leaved Fern]. **Ferns**, terrestrial; rhizomes creeping, scaly. **Fronds** pinnate, linear, 15–40 × 1.8–4 cm; stipes and rachises erect, dark brown, spreading-scaly; stipes 5–20 cm long; pinnae regularly spaced, shortly stalked, rounded or oblong, 0.8–2 × 0.5–1.3 cm, margins minutely toothed, scarcely inrolled, apices blunt and abruptly narrowed to a short sharp point, adaxially dark glossy green, abaxially paler. **Sori** continuous along upper and lower margins of pinnae, but absent from the apices; indusia lacking. **Distr:** Endemic to New Zealand. **Ecol:** In lowland montane regions, often in dry rocky places. **Zones:** 10–12.

ASPLENIACEAE

[Spleenwort].

Genera 1–10, **species** 700+. **Ferns**, terrestrial, lithophytic, or rarely epiphytic; stems erect or nearly erect, rarely long-creeping, scaly; steles radially symmetric or dorsiventral (with structurally distinct abaxial and adaxial aspects) dictyosteles. **Fronds** monomorphic, rarely almost dimorphic with fertile fronds taller and more erect than sterile ones; petioles with 1 vascular bundle X-shaped in cross-section or with 2 vascular bundles back to back and C-shaped; lamina extremely diverse, simple to 4-pinnate, commonly with tiny glandular hairs and a few linear scales, rarely with spreading hairs; veins free to anastomosing. **Sori** borne on veins, ± lunate to linear; indusia usually present, shape conforming to sorus and originating along 1 side of sorus. **Sporangia** with stalk of 1 row of cells, annulus vertical, interrupted by sporangial stalk; spores monolete; perispore typically winged, spiny, reticulate, or perforate. **Gametophytes** surficial, green, cordate. **Distr:** Worldwide.

Asplenium

Syn: *Ceterach, Phyllitis*. [Spleenwort].

Species ~700. **Ferns**, mostly terrestrial or growing on rocks, some epiphytic on tree trunks; rhizome creeping, or short and erect; roots fibrous, not proliferous or proliferous and producing tiny plantlets; stems erect, rarely long-creeping; scales clathrate, dark brown to black, basally attached; margins of bases of pinnae or pinnules continuous with rachis groove. **Fronds** simple, longitudinally cleft, trifoliate, or pinnate to decompound, sometimes proliferous at apex; venation free except in a few species where the tips of veins unite to form an intramarginal vein; petiole not articulate. **Sori** linear, borne on side of a vein; indusia linear, narrow, usually opening towards the midrib. **Distr:** Worldwide. **Cult:** Propagate from spores or by division of the plantlets. Needs moist, well-drained compost in partial shade.

A. bulbiferum Forst. f., *Prodr.*: 80, 1786. [Hen and Chickens Fern, Mother Spleenwort]. **Ferns**; rhizome short, stout, branching, often forming rounded masses, clad in translucent ovate-attenuate paleae ~1.5 × 0.3 cm; stipites tufted, basally with dense tuft of paleae. **Fronds** with stipes stout to slender, firm, (1–)10–30 × 0.5–1 cm, densely paleate, dark brown towards base, green apically, grooved, ± compressed; rachis narrowly winged, stout to slender, pale green, bearing numerous alternate pinnae; lamina submembranous, pale to dark green, ovate-oblong-lanceolate, acuminate, bipinnate to pinnatisect, (5–)30–80(–120) × (2–)10–20(–30) cm; pinnae ovate-lanceolate, shortly stalked, acuminate, spreading, (2–)10–15(–25) × (1–)3–5(–7) cm; costa winged; secondary pinnae decreasing in size to apex of primary pinna; lower ovate-oblong to lanceolate, obtuse, to 4 × 1.5 cm; upper pinnatisect, segments oblong, to 1.5 cm long, obtuse; margins crenate to crenate-serrate to subentire. **Sori** numerous, broad-oblong, submarginal, 2–5 mm long; indusium firm. **Bulbils** on adaxial veinlets, with fronds to 8 cm long before falling. **Distr:** Australia, New Zealand. **Ecol:** Mostly epiphytic (in Australia), on tree or tree fern trunks, in dense dark rainforests. **Zones:** 9–11.

A. falcatum Lam., *Encycl. Meth.* 2: 306, 1786. (Syn: *A. polyodon*, etc.). [Sickle Spleenwort]. **Ferns**, 20–60 cm tall; rhizomes short, erect, densely scaly; scales lanceolate, entire, subulate, blackish-brown, iridescent, polished. **Fronds** clustered; stipes 12–28 cm long, grayish-brown, bases densely scaly; lamina elliptic-rounded, 10–30 × 8–16 cm, imparipinnate; pinnae 2–5(–7) pairs, opposite or subopposite, long-stalked, *lower pinnae* 4–10 × 1.2–2.2 cm, rhomboid, broadly lanceolate, ± falcate, bases subsymmetrical cuneate, acroscopic sometime auriculate, margins irregularly serrate or with lacerate sharp serrations, apices long-acuminate, *upper pinnae* similar to lower ones, slightly smaller, terminal pinna larger, rhomboid-lanceolate to 3-forked, 5–11 × 2–5(–10) cm, veins prominent, abaxially main veins roundly raised. **Sori** linear, 1–3 cm long, dark brown, attached at veinlets, reaching margins; indusia linear, dark brown, thinly papyraceous, entire, opening to main veins, some opening to margin, persistent. **Distr:** China, India, Malaysia, Myanmar, Sri Lanka, Taiwan, Vietnam; S Africa; Australia and New Zealand. **Ecol:** On rocks by streamsides in dense forests; 320–750 m. **Zones:** 9–11.

A. flaccidum Forst. f., *Prodr.* 80, 1786. (Syn: *A. heterophyllum*, etc.). [Drooping Spleenwort, Hanging Spleenwort]. **Ferns**; rhizome short, stout, erect to shortly creeping, clad in brown linear-attenuate paleae ~1 cm long; stipites tufted to approximate, paleate at base, and with a few weaker paleae elsewhere when young. **Fronds** with rachis winged, bearing scattered paleae; laminae and pinnae diverse in size and shape, usually thick, long-pend. to short-erect; veins obscure, usually a single veinlet to each segment. **Sori** oblong, usually submarginal; indusium pale, firm. **Distr:** Australia, New Zealand, Pacific Islands, Tasmania; S Africa. **Ecol:** In lowland to montane forests. **Zones:** 9–11. **Info:** This species is highly variable across its native range.

A. lucidum Forst. f., *Prodr.* 80, 1786. (Syn: *A. obtusatum* var. *lucidum*). [Shining Spleenwort]. **Ferns**; rhizome stout, branching, often forming hard woody mass above ground, paleate, paleae brown, shiny, translucent, ovate-acuminate, to 3 cm long. **Fronds** with stipites tufted, basally paleate; stipes 10–40(–50) × 0.5–1 cm, stout, basally blackish, ± paleate, paleae slender, lanceolate-attenuate, to 2.5 cm long; rachis stout, narrowly winged, ± paleate, smooth, bearing distant subopposite to alternate pinnae; lamina ± lanceolate-oblong, 0.3–1 m × 15–35 cm, with 16–40+ pinnae, acuminate, thinly coriaceous, adaxially dark green and glossy; pinnae ± lanceolate-oblong, (5–)10–15(–20) × (0.2–)3(–5) cm, base obliquely cuneate, margins crenate-serrate, apex acuminate, abaxially ± paleate when young, long-stalked; veins evident, simple or forked. **Sori** oblique, linear, to 2 cm long; indusium firm. **Distr:** Australia (Norfolk and Lord Howe Islands), New Zealand. **Ecol:** In coastal, lowland and lower montane forests. **Zones:** 9–11.

A. nidus L., *Sp. Pl.* 2: 1079, 1753. [Bird's Nest Fern]. **Ferns**, large, epiphytic; rhizome short, stout, erect, covered with a mass of roots. **Stipes** very short, stout, bearing black, clathrate, easily abraded scales with long hairlike projection; fronds borne in a rosette. **Fronds** large, 1–1.5 m × ~10–15 cm, widest near middle, tapering to either end, veins nearly parallel, uniting at apices to form an intra-marginal vein. **Sori** borne along each veinlet on upper half of frond, reaching from costa midway to margin; indusia linear, flaplike, narrow, attached along lower edge, covering sori. **Distr:** Africa to Polynesia. **Ecol:** Usually epiphytic on tropical trees. **Zones:** 10–12.

BLECHNACEAE

[Chain Fern].

Genera ~10, **species** ~250. **Ferns**, perennial, mostly terrestrial, occasionally lithophytic or epiphytic; stems creeping to suberect or ascending, sometimes climbing (rarely arborescent), slender to stout, dictyostelic, scaly. **Fronds** monomorphic or dimorphic, large and coarse, usually 0.3–1+ m; petiole not articulate, usually >2 vascular bundles arranged in arc, usually scaly at least at base; lamina often anthocyanic (reddish) when young, pinnatifid (rarely simple) to pinnate-pinnatifid or bipinnate (rarely decompound), glabrous or occasionally bearing scales or capitate glands; rachis frequently grooved adaxially; veins of sterile leaves generally free, rarely anastomosing, veins of fertile leaves united to form sorus-bearing secondary vein parallel to costa or costule, sometimes anastomosing further. **Sori** elongate along secondary vein; indusia present (rarely absent), opening along costal side of fertile vein, frequently hidden by dehisced sporangia; sporangial stalk of 3 rows of cells; spores monolete, reniform; perine present, variously ornamented. **Gametophytes** green, cordate, sometimes bearing capitate hairs, antheridia and archegonia borne on lower surface. **Distr:** Mostly tropical and south temperate.

Blechnum

[Deer Ferns, Hard Ferns].

Species ~220. **Ferns**, terrestrial or rarely lithophytic; stems creeping to ascending or erect, slender to stout, sometimes climbing (rarely subarborescent); scales brown or black. **Fronds** monomorphic or variously dimorphic, cespitose to scattered; lamina pinnatifid to 1-pinnate, rarely simple or bipinnate; rachis and costae glabrous, scaly, or hairy abaxially; veins free, often forked. **Sori** borne on vascular commissures parallel to costae, 1 per side, normally uninterrupted, linear, continuous along length of costa; spores with perine smooth to variously winged or rugose. **Distr:** Mostly tropical, especially in the Southern Hemisphere. **Cult:** Propagate from spores, planted on moist sphagnum moss. Requires a fertile, moist, humus-rich soil in shade.

B. brasiliense Desv., *Ges. Naturf. Freunde Berlin Mag. Neuesten Entdeck.* 5: 330, 1811. (Syn: *B. corcovadense*). [Brazilian Blechnum]. **Ferns**, terrestrial; rhizome stout, erect, densely scaly; scales linear to lanceolate, light to dark brown or blackish. **Fronds** rigid, erect, monomorphic, 0.5–1.2 m × 20–40 cm; petiole 2–12 cm long, subterete abaxially, adaxially sulcate, light to dark brown, or blackish at base, densely scaly; lamina pinnate or very deeply pinnatisect, chartaceous to subcoriaceous, glabrous, narrowly elliptic, broadest at or near the middle, tapering gradually to a pinnatifid apex, reduced gradually and conspicuously at base; rachis stramineous to light brown, essentially glabrous, prominulous and terete or subquadrangular abaxially, sulcate adaxially; pinnae (or segments) numerous, most of them ascending, elongate-deltoid, base adnate to the rachis, margins undulate, apex acute, proximally shorter and more obtuse, lowermost 1–3 pairs often reduced, larger, middle ones 8–22 × 1–2.2 cm, veins distinct, usually 1-forked. **Sori** with indusium subentire, firm. **Distr:** S America. **Ecol:** 1,200–1,500 m. **Zones:** 9–11.

B. capense (L.) Schlecht., *Adumb. Fil. Bonae Spei.* 34, t. 18, 1825. (Syn: *B. procerum*, etc.). [Gully Fern]. **Ferns**; rhizome woody, ascending or creeping, to 40 cm long, densely paleate, paleae pale brown subulate-attenuate, ~2 cm long; stipites usually crowded. **Fronds** with stipes stout, pale brown, paleate basally or throughout, 10–20 (–50) cm long; rachis stout, grooved, adaxially densely paleate when young; pinnae alternate to subopposite, to 30 pairs. **Sterile lamina** adaxially dark green, abaxially paler, thinly coriaceous, harsh, broadly ovate-oblong to lanceolate-oblong, (0.1–)0.15–0.5(–1)+ m × (5–)10–30+ cm, apex acuminate; lower pinnae usually shorter, often obtuse; larger pinnae lanceolate to oblong, 7–15(–30) × 1–3 cm, base cuneately truncate to subcordate or auriculate, margins finely serrate, apex acuminate, adaxial costa grooved, abaxially prominent; terminal pinna to 20+ cm long. **Fertile lamina** lanceolate-oblong 15–50+ × 8–20 cm, pinnae linear, 5–25 × 0.3–0.5 cm, distant, base usually abruptly widened by

Below: *Asplenium lucidum, Asplenium nidus, Blechnum brasiliense, Blechnum chambersii.*

Top row: *Blechnum discolor, Blechnum gibbum, Blechnum novae-zelandiae* (sterile fronds).
Bottom row: *Blechnum novae-zelandiae* (fertile fronds), *Blechnum occidentale, Blechnum vulcanicum.*

sterile auricles; lower pinnae shorter, broader, ovate-oblong, ± sterile. **Sori** covering lamina abaxially, costa visible; indusium becoming erose. **Distr:** Australia, New Zealand, Tasmania, Pacific Islands; Malaysia; S America to W Indies; S Africa. **Ecol:** In lowland to subalpine forests. **Zones:** 9–12.

B. chambersii Tindale, *Fl. Sydney Reg.* 86, 1972. (Syn: *B. lanceolatum*). [Lance Fern]. **Ferns** terrestrial; rhizomes stout, erect. **Sterile fronds** with stipes 1–10 cm long, base scaly; lamina narrowly elliptic, pinnate, 13–50 × 1.5–12 cm, red-tinged when young, becoming dark green with age, glabrous; pinnae 15–40 pairs, longest ones near middle, oblong, 0.8–6 × 0.4–1.2 cm, reducing to short alternate flanges towards base, bases widened and adnate to rachis, margins slightly toothed, apices blunt or pointed. **Fertile fronds** slightly shorter than sterile fronds, but ± similar in width. **Sori** covering whole of undersurface. **Distr:** Australia, Fiji, New Zealand, Samoa, Tahiti. **Ecol:** In lowland to montane forests, usually near waterways. **Zones:** 9–12.

B. discolor (Forst.) Keyserl, *Pol. Cyath. Hb. Bung.* 66, 1873. [Crown Fern]. **Ferns**, terrestrial; rhizomes often forming short, woody trunks,producing stolons; stolons producing plantlets. **Sterile fronds** tufted, spread out in the form of a crown, abaxially paler than that of the fertile fronds; stipes 5–20 cm long, base scaly; lamina narrowly elliptic, pinnate, 0.2–1 m × 5–16 cm, leathery, adaxially bright green, abaxially paler, scaly; pinnae 35–60 pairs, longest ones at the middle, oblong, 2.5–8 × 0.6–1.4 cm, bases adnate to rachis, margins smooth, apices pointed. **Fertile fronds** growing stiffly upright in the center of the tufted sterile fronds, slightly longer than sterile fronds, pinnate; basal pinnae sterile; apical pinnae fertile; lower fertile pinnae with sterile bases and fertile apices. **Distr:** Endemic to New Zealand. **Ecol:** In open beech forests and scrubs. **Zones:** 8–11.

B. gibbum (Lab.) Mett., *Ann. Sci. Nat.*, Bot. sér. 4, 15: 68, 1861. [Dwarf Tree Fern, Miniature Tree Fern, Silver-lady Tree Fern]. **Ferns**, terrestrial, medium-sized, to 1.5 m high; rhizome erect, stout, often forming a small trunk with age; trunk to 30 cm high, blackish-scaly. **Fronds** with stipes and part of midrib black-scaly; midrib dull yellow or green. **Sterile fronds** evergreen, tufted in a rosettelike whorl, dimorphic; stipe swollen; lamina pinnate, ovate in outline, 0.6–1.3 m long, base narrowed, bright green; pinnae directed forwards. **Fertile fronds** erect, narrow. **Distr:** New Caledonia and South Pacific Islands. **Ecol:** Along rainforest streams, growing in clumps with 10 or more stems together; 100–400 m. **Zones:** 10–11. **Info:** *B. gibbum* resembles *B. brasiliense* but has black rhizome and stipe scales, dimorphic fronds, and usually narrower sterile pinnae.

B. novae-zelandiae T.C. Chambers & P.A. Farrant, *New Zealand J. Bot.* 36(1): 8, 1998. [Palm Leaf Fern]. **Ferns**, terrestrial; rhizomes short-creeping. **Sterile fronds** with stipes and rachises bearing brown scales with conspicuous dark centres; stipes 8–70 cm long; lamina elliptic or almost spathulate, 0.2–2.5 m × 8–60 cm, bright green, stiff, midribs scaly; pinnae crowded, 10–50 pairs, longest pinnae above middle, oblong to linear, 5–35 × 1–3 cm, markedly shorter, bases almost rounded, lobed and stalked, margins finely toothed, apices tapering. **Fertile fronds** equaling or shorter than sterile fronds, with a few sterile pinnae at base. **Distr:** Endemic to New Zealand. **Ecol:** In lowland to montane regions, on cliffs, scrubs and swamps. **Zones:** 8–11.

B. occidentale L., *Sp. Pl.* 1077, 1753. [Hammock Fern]. **Ferns**, terrestrial; rhizome erect or curved-ascending, to 12 cm long, 2 cm thick, ferruginous-paleaceous, stoloniferous. **Fronds** rigidly ascending, mostly 30–75 cm long; stipes of fertile fronds usually as long as the lamina, pale, laxly paleaceous abaxially, usually sulcate; lamina ovate to oblong-lanceolate, mostly 20–50 × 6–25 cm, basally pinnate, pinnatisect above, upper segments evenly reduced, narrowly joined, passing gradually into the acuminate or attenuate-subcaudate apex; pinnae numerous, alternate and falcate, linear, evenly attenuate from the middle or from the cordate or subcordate base (0.8–2 cm wide), basal ones usually short-stalked and deflexed, upper ones fully adnate, veins close, repeatedly bifurcate, margins scabrous, minutely cartilaginous-serrulate; leaf-tissue herbaceous or subcoriaceous, usually glabrous or nearly so abaxially. **Sori** costal, not quite reaching tip, continuous, indusia at first light vinaceous and meeting, at length reflexed and overlaid by sporangia. **Distr:** Mexico, Central America, West Indies. **Ecol:** In moist forest slopes. **Zones:** 9–12.

B. vulcanicum (Blume) Kuhn, *Ann. Lugd. Bat.* 4: 284, 1869, 284. (Syn: *Lomaria deflexa*, etc.). [Mountain Hard Fern]. **Ferns**; rhizome stout, erect, with bases of fallen stipites, densely paleate, paleae dark brown subulate-attenuate, ~1 cm long; stipites tufted. **Fronds** with stipes pale brown, stiff, basally paleate, ± compressed, 5–20(–35) cm long; rachis stout, grooved, bearing numerous pinnae. **Sterile lamina**

deltoid-oblong to -lanceolate, 10–35 × 5–12 cm, margins crenulate to subentire, thickened, apex acute to acuminate, coriaceous, dull green, ± softly pilose, pinnatifid apically; pinnae approximate, lowest ± similar in size, often deflexed, largest lanceolate-oblong to oblong, 3–6 × 0.5–1 cm, apex acute to obtuse, straight to falcate, attached by widened base, veinlets evident, usually forked. **Fertile lamina** to 40 cm long, with distant linear pinnae to 5 cm long, often arcuate, base widened. **Sori** covering all but costa; indusium becoming ± erose. **Distr:** Australia, New Zealand, Pacific Islands; Malaya. **Ecol:** In lowland to upper montane forests. **Zones:** 9–12.

Doodia

[Hacksaw Ferns, Rasp Ferns].

Species 12–15. **Ferns**, terrestrial, small to medium-sized; stem short, erect or ascending, sometimes stoloniferous, with dark scales. **Fronds** closely set; stipe well-developed; lamina very deeply pinnatifid to simply pinnate, with few exceptions reduced at base, with pinnatifid or caudiform apex; pinnae sessile or adnate, ovate to linar, herbaceous to coriaceous, mostly glabrous, margins sharply serrate-dentate, veins forked. **Fertile pinnae** sometimes slightly to moderately contracted. **Sori** short, on the cross-veinlets, sometimes somewhat embossed; indusium membranous, often dark. **Spores** ellipsoid, low-rugose or papillate, sometimes nearly plain. **Distr:** Australia, New Zealand, Pacific Islands east to Hawaii and Easter Island; few rare species in Sri Lanka, Java, Lesser Sunda Islands and New Guinea. **Cult:** Propagate from spores. Requires a moist, well-drained, acidic soil in shade.

D. australis (Parris) Parris, *Fl. Australia* 48: 710, 1998. (Syn: *D. media* ssp. *australis*). [Common Rasp Fern]. **Ferns**, terrestrial; rhizome erect or short- to long-creeping, producing stolons. **Fronds** sometimes slightly dimorphic; stipes and rachises brown-scaly; stipe 2–25 cm long, with tubercles; lamina pinnate, elliptic to narrowly elliptic, 8.5–60 × 1–15 cm, pink when young; pinnae 20–50 pairs, simple, lowest 1–3 pairs stalked, gradually becoming sessile towards middle of lamina, upper pinnae sessile to decurrent, longest pinnae 0.7–7.5 × 0.3–1 cm; rachis and pinna midvein without tubercles. **Sori** 0.7–2 (–3) × 0.5–1 (–2) mm, usually discrete, sometimes contiguous, rarely confluent laterally when mature, 1 or 2 rows per side, inner nearer midvein, outer nearer margin. **Distr:** Australia, New Zealand. **Ecol:** In rainforest margins, scrubs, often near creeks and among rocks. **Zones:** 9–11.

DRYOPTERIDACEAE

Genera 40–50, **species** 1,700. **Ferns**, perennial, terrestrial or on rock, occasionally hemi-epiphytic or epiphytic; stems creeping to erect, rarely arborescent, sometimes climbing, dictyostelic, bearing scales. **Fronds** circinate in bud, monomorphic or dimorphic; stipe usually not articulate, scales usually persistent at base; lamina simple to commonly 1–5-pinnate or more divided; veins pinnate or parallel in ultimate segments, simple or forked, areoles sometimes with included free veinlets; indument on lamina usually glands, hairs, and/or scales, especially on rachis and costae abaxially. **Sori** borne abaxially on veins or at vein tips, or sporangia acrostichoid and covering abaxially, if in discrete sori then variously shaped; receptacle not or only slightly elevated, indusium variously linear, falcate, or reniform, sometimes hoodlike, cuplike, or round. **Sporangia** with stalk of 2–3 rows of cells; annulus vertical, interrupted by stalk; spores all of 1 kind, usually not green, oblong or reniform, monolete, variously ornamented, 64 per sporangium. **Gametophytes** green, aboveground, cordate, glabrous or often bearing glands or hairs; archegonia and antheridia borne on lower surface, antheridia 3-celled. **Distr:** Worldwide.

Polystichum

Species ~160. **Ferns**, terrestrial or lithophytic, medium- to large-sized; rhizome short, erect, densely scaly; scales variable, becoming erose. **Fronds** tufted; stipes densely scaly at base; lamina 1–3-pinnate, rarely simple, triangular and truncate, or often narrowly oblong to lanceolate and ± tapering at base, or linear, often coriaceous; rachis sometimes bearing proliferous buds near apex; ultimate segments sharply pointed, sometimes with marginal teeth; veins free. **Sori** rounded, dorsal or subterminal on veins, indusiate, rarely exindusiate; indusium peltate, persistent or caducous. **Spores** monolete, ellipsoidal to somewhat globose, with prominent. **Distr:** Cosmopolitan. **Cult:** Propagate from spores or by division. Requires a moist, well-drained soil, rich in organic matter, in partial shade.

P. vestitum (Forst.) C. Presl, *Tent. Pterid.* 83, 1836. [Prickly Shield Fern]. **Ferns**, evergreen, 0.3–1 m high; rhizomes erect. **Fronds** with stipes a third or more the length of the lamina, mostly 10–50 cm long; stipes and rachises densely scaly; lamina lanceolate, narrowly elliptic or narrowly oblong, bipinnate, 0.2–1 m × 6–25 cm, adaxially dark glossy green, abaxially paler; pinnae 20–30 pairs, apices pointed; primary pinnae 3–15 × 1–2.5 cm; secondary pinnae with a basal acroscopic lobe, more deeply dissected on larger fronds, apices sharply pointed. **Sori** with indusia, peltate, medial, uniformly pale brown, almost flat. **Distr:** Endemic to New Zealand. **Ecol:** In lowland and montane regions, in tussock grasslands, scrublands and forests. **Zones:** 7–10.

Quercifilix

Species 1. *Generic description same as specific.* **Cult:** Propagate from spores. Requires a moist, well-drained soil in shade or partial sun.

Q. zeylanica (Houtt.) Copel., *Philip. Journ. Sci.* 37(4): 409–410, 1928. (Syn: *Ophioglossum zeylanicum*, etc.). [Oak Leaf Fern]. **Ferns**, terrestrial; rhizomes creeping or suberect, covered with narrow brown scales; stipes tufted, densely covered with long yellowish hairs and scattered lanceolate scales. **Sterile fronds** with terminal pinna, oblong, 5–10 × 2–4 cm, margins with large crenate lobes, with 1 pair of sessile lateral pinnae to 1.5 cm long, ± rounded but widened at base on the basiscopic side, adaxially glabrous, abaxial surface with ctenitis hairs along veins and margins; stipes to ~5 cm long. **Fertile fronds** with much constricted lamina, 1–3 mm wide, lobes linear with undulate or somewhat lobed margins; stipes to 15 cm long. **Sori** acrostichoid, exindusiate. **Sporangia** all over the veins and eventually spreading somewhat onto laminar surface forming acrostichoid sori; spores oblong, coarsely reticulate-tuberculate by shrinking of the epispore. **Distr:** S China, India, Malaysia, Polynesia, Sri Lanka, Taiwan, Vietnam. **Ecol:** In crevices of rocks and on clay banks. **Zones:** 10–12.

OLEANDRACEAE

Genus 1, **species** 40. **Ferns**, terrestrial, epilithic, or often scandent or epiphytic; rhizome long-creeping; stem elongate, with a perforated dictyostele and bearing peltate scales. **Fronds** simple, articulate to phyllopodia, or pinnate; pinnae articulate with the rachis, or the petiole also articulate; petioles usually remote, adaxially flattened or unisulcate, or virtually absent, vasculature consisting of 2 larger, ventral and a number of smaller, dorsal bundles, in transection arranged in about a semicircle; lamina simple or pinnate, apices variable. **Sori** terminal on the veins, less often dorsal, mostly round, exindusiate or with a reniform to suborbicular (but not pelate) indusium. **Spores** monolete, with strong sculpture derived from the perispore. **Distr:** Old World tropics to southern temperate.

Arthropteris

Species ~15. **Ferns**, epiphytic or lithophytic; rhizome slender, long-creeping; adaxially scaly, abaxially closely adhering to substrate with dense short hairs. **Fronds** simply pinnate, not markedly dimorphic; stipes scattered along rhizome, articulated to a phyllopodium; stipes and rachises grooved adaxially; pinnae articulated to rachis, generally lanceolate to falcate, anterior margin at base often auriculate, venation open, simple or forked; hydathodes prominent at vein endings on adaxial surface of pinnae. **Sori** superficial, usually in a regular row at vein endings and parallel with margins; indusium reniform to round, or absent. **Spores** spherical to ellipsoid, surface with irregular winglike folds.

Below: *Doodia australis, Polystichum vestitum, Quercifilix zeylanica, Arthropteris tenella.*

RT

Top row: *Acrostichum aureum, Microsorum diversifolium, Microsorum scandens.*
Bottom row: *Platycerium bifurcatum.*

Distr: Widespread in the tropics and temperate regions of the Southern Hemisphere. **Cult:** Propagate from spores. Requires a moist, well-drained soil, rich in organic matter, in partial shade.

A. tenella (Forst.) J. Sm., *Fl. Nov.-Zel.* 2: 43, t. 82, 1854. (Syn: *Polypodium tenellum*). [Jointed Fern]. **Ferns**, hemi-epiphytic, small to medium-sized; rhizomes long-creeping, adaxially densely scaly, glabrous with age. **Fronds** 12–30 cm long, tapering towards base and partly to apex; uppermost pinna pair and terminal pinna often enlarged; stipes 1–12 cm long, scaly; lamina elliptic or sometimes broadest at apex, pinnate, 10–40 × 5–15 cm, adaxially shiny dark green, abaxially paler, sparsely hairy and scaly; pinnae narrowly ovate, 3–8 × 0.5–1 cm, midribs prominent, apices tapering, margins scalloped, bases stalked and jointed to rachises. **Fertile pinnae** (1.8–)4–16 × (0.5–)1–2.3 cm; margins entire to crenate. **Sori** superficial, round, unprotected (indusium absent), close to the margin. **Distr:** Australia, New Zealand. **Ecol:** Epiphytic in rainforests, usually on tree trunks, occasionally reaching the canopy, or lithophytic in damp gullies. **Zones:** 9–12.

POLYPODIACEAE

Genera ~56, **species** ~1,200. **Ferns**, mostly epiphytic and epipetric, some terrestrial; rhizomes long-creeping to short-creeping, dictyostelic, scaly. **Fronds** with petioles cleanly abscising near their bases or not, leaving short phyllopodia; lamina monomorphic or dimorphic, mostly simple to pinnatifid or 1-pinnate (rarely more divided); veins often anastomosing or reticulate, sometimes with included veinlets, or veins free; indument lacking or of hairs and/or scales on lamina. **Sori** abaxial (rarely marginal), round to oblong or elliptic, occasionally elongate, or sporangia acrostichoid, sometimes deeply embedded; sori exindusiate, sometimes covered by caducous scales when young. **Sporangia** with 1–3 rows, usually long-stalked, frequently with paraphyses on sporangia or on receptacle; spores hyaline to yellowish, reniform, and monolete, or greenish and globose-tetrahedral, trilete; perine various, usually thin, not strongly winged or cristate. **Distr:** Pantropical; a few in temperate regions.

Acrostichum

[Leather Fern].

Species 3. **Ferns**, terrestrial; stems erect or creeping, branched; scales dark brown, concolored, linear-lanceolate, margins entire. **Fronds** slightly dimorphic, clustered, 1–5 m long; petiole brown, with a single adaxial groove, glabrous, smooth or with scale scars, with several abaxial vascular bundles and 2 adaxial vascular bundles; lamina lanceolate, pinnate, leathery, adaxially dull, not striate, abaxially glabrous or hispid, glabrous; rachis straight; pinnae stalked, free from rachis, narrowly oblong to lanceolate, 2–5 cm wide, base cuneate, margins plane; stalk green; fertile fronds bearing sporangia on most pinnae or on only more distal pinnae; veins of pinnae conspicuous, strongly anastomosing; false indusia absent. **Sporangia** spread abaxially, mixed with paraphyses (sori acrostichoid), with 64 spores; spores yellow, tetrahedral, minutely tuberculate or roughened, equatorial flange absent. **Distr:** Worldwide, warm and tropical regions. **Ecol:** In fresh- or saltwater habitats. **Cult:** Propagate from spores or by division. Requires shallow, fresh or brackish water in full sun.

A. aureum L., *Sp. Pl.* 2: 1069, 1753. [Coastal/Golden Leather Fern]. **Ferns**; stems creeping or ascending, frequently branched. **Fronds** ± arching, 1–3 m × 12–50 cm; rachis rounded abaxially, decidedly grooved adaxially; pinnae 24–30(–40), usually not overlapping, 10–34 × 1.3–7 cm; proximal pinnae always distant, frequently rounded at tip; costal areoles 3 or more times longer than wide; distal 6–12 pinnae of fertile leaves bearing sporangia. **Sporangia** spread abaxially on distal pinnae of fertile fronds; paraphyses stalked, ending with a single isodiametric, irregularly lobed cell. **Spores** (37–)57(–72) μm Ø, surface minutely tuberculate. **Distr:** Tropical and subtropical regions worldwide. **Ecol:** In coastal mangrove swamps, salt marshes, low hammocks, along lake and canal margins; 0–50 m. **Zones:** 10–12.

Microsorum

Species 50–60. **Ferns**, mostly epiphytic; rhizomes long, creeping or climbing, branches few, scaly when young, fronds generally well-spaced; stipes in 2 rows along adaxial surface of rhizome. **Fronds** with lamina entire or deeply lobed, veins prominent, smaller veins reticulate, much-divided and forming a network with free vein endings in the areoles. **Sori** round, irregularly scattered along veins on abaxial surface of lamina, or round to elliptic, in 1 or 2 rows along midvein of lamina lobes, deeply sunken in cavities that appear as raised bumps on upper surface lamina. **Distr:** Tropical Asia; Africa; Australia and Polynesia. **Cult:** Propagate

TG

RF

Above: *Platycerium superbum*, *Platycerium bifurcatum* ssp. *bifurcatum* var. *hillii*, *Pyrrosia rupestris*.

from spores or by division of rhizomes. Requires a moist, well-drained compost in full or partial sun.

M. diversifolium (Willd.) Copel, *Univ. Calif. Publ. Bot.* 16: 114, 1929. (Syn: *M. pustulatum*, etc.) [Hound's Tongue, Kangaroo Fern]. **Ferns**, epiphytic; rhizome often climbing, grey-green, fleshy (3–7 mm Ø), scaly when young; scales brown, appressed, shiny, peltate, clathrate, deciduous. **Fronds** widely separated, erect or weeping, variable in both size and shape, 5–50 cm long, firm; stipe often short, glabrous, scales clustered at swollen base; lamina narrowing abruptly at base, shiny, light green with clear, thickened margins, trimorphic: (1) simple, oblong to linear-lanceolate; (2) unevenly 3-pronged; (3) irregularly pinnatifid with long terminal segment and 1–8 pairs of lateral lobes, each 5–15 cm long; abaxial surface of central vein scattered with small, broad, inflated scales; veins prominent and branching repeatedly to give complex netted vein pattern with main veins dark, free veins ending in areoles; continuous wavy vein near and parallel to margin. **Sori** without indusia, globose, 2–4 mm Ø, slightly sunken (as raised circles adaxially), in 2 rows, each about halfway between midvein and margin. **Distr:** Australia, Tasmania; New Zealand, Kermadec Islands; New Caledonia. **Ecol:** In rainforests, occasionally as ground cover on forest floor. **Zones:** 10–12.

M. scandens (G. Forst.) Tindale, *Amer. Fern J.* 50: 241, t.20, 1960. [Fragrant Fern]. **Ferns**; rhizome creeping or climbing, densely scaly, dark green, wiry; scales brown erect, shiny, peltate, clathrate, persistent. **Fronds** separated, erect or often drooping, variable in both size and shape, 15–55 cm long, membranous to herbaceous; stipe short, glabrous, base scaly; lamina narrowing gradually at base, dull, dark green, variable in shape: (1) simple, linear-lanceolate, apex pointed; (2) irregularly pinnatifid with a long terminal segment and 4–20 pairs of lateral lobes of variable length, each 2–10 cm long; abaxial surface of central vein with a few narrow scales or scales absent; veins fine, branching to give a simple, netted vein pattern with free veins ending in areoles and towards margins. **Sori** without indusia, round to ovate, 1–2.5 mm Ø, slightly sunken (appearing as raised circles on upper surface), in 2 rows each nearer to margin. **Distr:** E Australia, Norfolk Island; New Zealand, Chatham Islands. **Ecol:** In moist, sheltered coastal ranges and foothills; 0–800 m. **Zones:** 10–12.

Platycerium

Species 15–18. **Ferns**, large, epiphytic, slow-growing; rhizome short-creeping, fleshy, densely scaly. **Fronds** dimorphic, both densely stellate-hairy when young; base (shield) leaves persistent, erect, broad, base appressed, thick and spongy, upper margin often lobed. **Foliage fronds** articulating from a phyllopodium, short stipitate to nearly sessile, entire to many times dichotomously divided into strap-shaped lobes, symmetrical or asymmetrical, midrib absent or partly developed, main veins dichotomous, often irregularly so, rarely with hydathodes, all frond surfaces hairy; sporangial area medial to apical on abaxial surface of the fertile frond. **Sporangia** exindusiate, borne on closely placed receptacles, sometimes acrostichoid, receptacles underlain by a specialized vascular system, capsule roundish to oblong, annulus vertical to very slightly oblique, pedicel short or long, (1)3 cells thick; paraphyses like the laminar hairs except the rays often shorter, glandlike in some species; spores without perispore, reniform, smooth, hyaline, rarely green, 64(8). **Gametophyte** cordate, early stages filamentous, margins with branched glandular hairs. **Distr:** Philippines, SE Asia, Indonesia, Australia, Madagascar, Africa, and South America. **Cult:** Propagate from spores or by division of suckers. Requires a fertile, moist, well-drained compost in full or partial sun. Best grown fastened to a substrate such as a tree trunk.

P. bifurcatum (Cavanilles) C. Christensen, *Index Filic.* fasc. 8: 496, 1906. (Syn: *Acrostichum bifurcatum*). [Common Elkhorn Fern, Common Staghorn Fern]. Ferns, epiphytic, highly variable; rhizome short with bud-forming roots, scales bearing long, tan, multicellular hairs. Basal fronds rounded, reniform, to 30 cm across, upper margin wavy-lobed. Foliage fronds erect or drooping, to 1 m long, base narrowly wedge-shaped, broadening upward, apical portion forked 2–3×, segments slightly unequal, strap-shaped, to 30 × 3.2 cm; petiole without scattered vascular bundles in central region. Sporangia covering tips or entire abaxial surface of the ultimate segments, usually extending around sinus. **Distr:** Australia (Lord Howe Island, NSW, Queensland), New Guinea. **Ecol:** In rainforests, swampy areas and open forests, mainly on tree trunks and sometimes forming large colonies on boulders and rock faces. **Zones:** 9–12. **Info:** ***P. bifurcatum* ssp. *bifurcatum* var. *hillii*** (T.Moore) Domin (Syn: *Alcicorne hillii*, *P. hillii*, etc.) [Northern Elkhorn] from NE Queensland (Australia) has the sori at the frond tips; sterile nest fronds branching at the tips, dark green below.

P. superbum, de Jonch. & Hennipman, *British Fern Gazette* 10(3): 114, 1970. [Giant Staghorn Fern, Large Elkhorn Fern, Staghorn Fern]. **Ferns**, epiphytic. **Sterile fronds** basal, rounded to kidney-shaped in the lower part, with upper margin deeply lobed, dividing twice at each division, spreading widely, fan-shaped, lobe tips rounded to pointed. **Fertile fronds** usually paired, symmetrical, to 2 m long, often divided 4–6 times and twisted, spreading, drooping or pendulous, lobes straplike, 1.5–4 cm wide at the ends, green. **Sori** in large patches where the fronds branch. **Sporangia** forming single patch below first fork. **Distr:** Australia (New South Wales, Queensland). **Ecol:** Epiphytic or occasionally lithophytic in tropical and subtropical regions, sometimes in coastal areas. **Zones:** 9–12.

Pyrrosia

Syn: *Cyclophorus*, *Neoniphopsis*, etc.
Species ~100. **Ferns**, epiphytic or epilithic, occasionally terrestrial, in small tufts or forming extensive colonies; rhizome creeping, paleate; scales peltate, dull and papery with broad, persistent bases; a thick coating of hairs protects the fern from excessive water loss and, together with special water-storage tissue, thus helping the plant survive drought periods. **Fronds** simple, small, fleshy, articulated to the rhizomes; lamina coriaceous, clad below in stellate paleae. **Sori** round to elongate, exindusiate, usually on included veinlets, with stellate paraphyses; veins obscured, irregularly anastomosing. **Sporangia** sessile to long-stalked; capsule 0.2–0.6 mm high, of the polypodiaceous type, glabrous; annulus with 15–22 indurated cells arranged in a regular 'bow', occasionally with 9–14 indurated cells situated on the apical side of the capsule; spores usually 64, very rarely 32 or 128 in each capsule. **Distr:** Mostly in Asia, Polynesia and Africa. **Cult:** Propagate from spores or by division. Requires a rich, moist soil in shade.

P. rupestris (R. Br.) Ching, *Bull. Chin. Bot. Soc.* 1: 49, 1935. (Syn: *Polypodium serpens*, etc.). [Rock Felt Fern]. **Ferns**; rhizome long-creeping, narrowly grooved ventrally, 0.5–1.33 mm thick, phyllopodia 0.5–2.5 cm apart; scales peltate, acumen light brown, entire to denticulate. **Fronds** distinctly dimorphic, ± stipitate. **Fertile fronds** with stipes 0.2–5(–11) cm long; lamina 1.3–11 × 0.3–1 cm, base gradually narrowed, apex acute to narrowly rounded. **Sterile fronds** with stipes to 2(–6) cm long; lamina 0.5–7 × 0.4–1.5 cm, base cuneate to attenuate, occasionally cordate-truncate, apex widely obtuse or rounded, tertiary veins forming ± irregular areole; hydathodes distinct, in a single marginal row, slightly sunken. **Sori** apical to all over the lamina, shortly spaced, slightly sunken; 2–3 in each soriferous areole, occasionally confluent, 1–2.5 mm wide. **Sporangia** stalked; capsule ± 0.3 mm high, with 16–21 indurate annulus cells; paraphyses mixed with sporangia, with short, straight rays; spores bisculpate. **Distr:** Australia, New Zealand. **Ecol:** Epilithic or epiphytic in rainforests and mountains. **Zones:** 10–12.

3 GYMNOSPERMS [NON-FLOWERING SEED PLANTS]

CYCADALES

[Cycads].

FAMILIES 2, GENERA 10, SPECIES 305.

Plants, long-lived, dioecious, perennial with subterranean or aerial woody stems, often palmlike, branched or unbranched, to 20 m tall; roots and stems with contractile tissue; mucilage copious, in canals. **Fronds** with petiole vascular bundles inverted omega shape; leaf vascular bundles amphicribral; epidermal cells with perforations; lamina large, pinnate; cataphylls present. **Pollen cones** (♂) with sporangia abaxial; microsporangia in synangia, many abaxial microsporangia/sporophyll, dehiscing by the action of the epidermis. **Seed cones** (♀) with megasporophylls, sterile portion terminal; pollen tube usually branched, growing away from ovule. **Seeds** with sarcotesta and inner fleshy layer, both vascularized. **Distr**: Throughout the tropics and subtropics. **Info:** The cycads is a surviving group of the most primitive cycads. The monophyly of the cycads is supported by molecular data (Treutlein and Wink 2002).

CYCADACEAE

Genus 1, **species** 100. **Cycads**, tree- or shrublike, evergreen, dioecious, mostly palmlike; trunk columnar, sometimes dichotomously branched at apex, rarely ovoid-bulbous and subterranean, clothed with bases of fallen fronds; bark often thickened and roughened. **Cataphylls** prominent, alternate with leaves, hairy, apex often rigid and pungent. **Fronds** apical, 1(–3)-pinnately compound, spirally arranged; new fronds erect (or somewhat inflexed and appearing coiled), with circinnate leaflets; petiole with spines (reduced basal leaflets), rarely unarmed, base swollen and hairy; leaflets numerous, alternate to subopposite, sometimes dichotomously branched, midvein present, margins usually entire. **Pollen cones** apical, cylindric or fusiform; microsporophylls numerous, scalelike, spirally and tightly arranged along axis of cone; microsporangia numerous, abaxially in groups; pollen tubes producing 2 motile sperm cells. **Megasporophylls** several to numerous, somewhat leaflike, alternating with flushes of leaves, arranged in a loose, 'conelike' crown surrounding apex of trunk, each with a linear fertile stalk and an apical, pinnatifid or subentire sterile lamina; ovules (1 or)2–5 on each side of stalk. **Seeds** drupelike, somewhat compressed; seed coat 3-layered, consisting of colored sarcotesta, woody sclerotesta, and membranous endotesta. **Distr:** E Africa and Madagascar, SE Asia to New Caledonia and Tonga.

Cycas

Syn: *Epicycas.*

Species ~60. *Generic description same as familial.* **Cult:** Propagate from seeds or rooted dormant buds (borne on the trunk). Requires a moist, well-drained soil in full or partial sun.

C. circinalis L., *Sp. Pl.* 2: 1188, 1753. (Syn: *C. nathorstii*). [Queen Sago Palm, False Sago, Fern Palm].**Cycads**, evergreen, to 8 m tall; trunk cylindrical, ~30–45 cm Ø, sometimes forked, covered with compact woody petiole-bases; crown with gracefully curved fronds; bark brown, smooth below, rough above, with diamond-shaped scars of fallen leaves. **Fronds** glossy, stiff, pinnate; 1.5–2.5 m long; leaflets 80–100 pairs, alternate, linear, 15–26 × 0.7–1.1 cm, margins flat, slightly revolute, apex blunt or acute, 1-nerved, prominent on both surfaces; petiole 40–60 cm long, with short distant spines; leaf-bases rusty tomentose. **Pollen cones** oblong-ovoid, ~33 × 7.6 cm, slightly stipitate; microsporophyll ovate-deltoid, 3.5–5 × 1.8 cm, prolonged into long rostrate acumen, clothed outside with brown tomentum, acumen glabrous. **Megasporophylls** 15–30 cm long, brown tomentose, sterile lamina rhomboid, 7.5 × 2.5 cm, margins strongly dentate or spinous-toothed, tapering into a long acumen; ovules 3–5 pairs above middle of stalk. **Seeds** subglobose or ovoid, 3.8 cm Ø, reddish-yellow. **Distr:** India (Malabar), Sri Lanka, Maldives; S China, Vietnam, Cambodia; Malay Archipelago to Pacific Islands; E tropical Africa, Madagascar. **Ecol:** In forests of intermediate regions; <450 m. **Zones:** 10–12.

C. media R. Br., *Prodr.* 1: 348, 1810. (Syn: *C. gracilis, C. kennedyana, C. normanbyana*, etc.). [King Sago Palm]. **Cycads**, stems arborescent, to 3(–6) m tall, ~10–18 cm Ø. **Fronds** bright green or deep green, highly glossy or semi-glossy, 0.9–2.3 m long, slightly keeled or flat; leaflets 160–300, with caducous orange tomentum, terminated by paired leaflets or a spine 0.3–5.5 cm long; petiole 30–70 cm long, sometimes spinescent; basal leaflets 7–15 cm long; median leaflets simple, strongly discolorous, 13–26 × 0.6–1 cm; median leaflets section flat; margins slightly recurved, apex aristate, not spinescent. Cataphylls linear, pungent, pilose, 8–12 cm long, persistent. **Pollen cones** ovoid, orange, 15–25 × 8–15 cm; microsporophyll lamina firm, 2.5–3.3 × 0.8–1.2 cm, fertile zone 1.6–2.2 cm long, sterile apex 0.8–1.1 cm long, deflexed, apical spine prominent, sharply upturned, 0.6–1 cm long. **Megasporophylls** 20–37 cm long, grey-tomentose or brown-tomentose; ovules 4–10, glabrous; lamina ovate to lanceolate, 3.5–10 × 1.7–3.2 cm, regularly dentate. **Seeds** flattened-ovoid, 3.1–3.8 × 2.6–3.2 cm; sarcotesta orange-brown, 3–4 mm thick. **Distr:** Australia (Queensland). **Ecol:** In open or closed forests, rainforests. **Zones:** 10–12.

C. revoluta Thunberg, *Verh. Holl. Maatsch. Weetensch. Haarlem.* 20(2): 424, 426–427, 1782. [Dwarf Sago Palm, Japanese Sago Palm]. **Cycads**; trunk to 3(–8) m tall, 45(–95) cm Ø, sometimes with numerous adventitious lateral branches or bulbils, apex tomentose; bark grey-black, scaly. **Fronds** 40–100+, 1-pinnate, 0.7–1.4(–1.8) × 0.2–0.25(–0.28) m; petiole subtetragonal, 10–20 cm long, with 6–18 spines; lamina oblong- or elliptic-lanceolate, strongly V-shaped, recurved, brown-tomentose when young; leaflets 60–150 pairs, straight to subfalcate, 10–20 × 0.4–0.7 cm, base decurrent, margin strongly recurved, apex acuminate, pungent, leathery, sparsely pubescent abaxially. **Cataphylls** triangular, 4–5 × 1.5–2.3 cm, densely brown-tomentose, apex acuminate. **Pollen cones** pale yellow, ovoid-cylindric, 30–60 × 8–15 cm; microsporophylls narrowly cuneate, 3.5–6 × 1.7–2.5 cm, apex rounded-truncate, cuspidate. **Megasporophylls** yellow to pale brown, 14–22 cm long, densely tomentose; stalk 7–12 cm long; sterile lamina ovate to narrowly ovate, 6–11 × 4–7 cm, deeply laciniate, with 21–35 lobes 1–3 cm long; ovules 2 or 3, densely pale-brown-tomentose. **Seeds** 2(–5), orange to red, obovoid or ellipsoid, somewhat compressed, (3–)4–5 × 2.5–3.5 cm, sparsely hairy; sclerotesta not grooved on sides. **Distr:** China, S Japan. **Ecol:** In thickets, sparse forests; <500 m. **Zones:** 9–12.

C. rumphii ssp. *zeylanica* J. Schust., *Pflanzenr.* (IV-I) Ht. 99: 75, 1932. [Ceylon Sago Palm, Bread Palm].**Cycads**, palmlike trees, evergreen, 2.5–8 m tall; trunk thick, rough, simple, sometimes forked when old. **Fronds** crowded at apex, 0.8–1.6 m long, pinnate, dark green, glossy; pinnae elongate-linear, 18–32 × 1.2–1.8 cm, apex acute-acuminate, coriaceous, 1-nerved, lower pinnules reduced to spines. **Pollen cones** shortly peduncled, ellipsoid-oblong, 32 × 11.5 cm, erect, reddish-tomentose; microsporophylls obovate-cuneate, 4–4.5 cm long, apex triangular, thickened, running into long upcurved acumen, rusty-tomentose, abaxially with stellately connected pollen-cells. **Seed cones** with megasporophylls very long-stalked, densely tawny villous, those of outer rows to 38 cm long, becoming shorter towards the center of lamina; lamina ovate to ovate-lanceolate, margins slightly toothed, terminating into a very long entire subulate acumen; ovules in the notches, 3–5 on each side of lamina. **Seeds** ellipsoid or subglobose, 5–7.5 × 2.5–3 cm, orange colored. **Distr:** India (Andaman and Nicobar Islands), S Myanmar, Thailand, Cambodia, Vietnam, Malaysia; New Guinea, N Australia; Mauritius, Madagascar and E Africa. **Ecol:** In moist regions. **Zones:** 9–12.

ZAMIACEAE

[Sago Palm].

Genera 9–10, **species** 200. **Plants**, superficially palmlike or fernlike, perennial, evergreen, dioecious; stems rarely polyxylic, subterranean with exposed

Top row: *Cycas circinalis, Cycas revoluta* (male cone), *Cycas media*.
Bottom row: *Cycas rumphii* ssp. *zeylanica* (male cone), *Cycas rumphii* ssp. *zeylanica* (seeds).

apex or aboveground, fleshy, stout, cylindric, simple or irregularly branched; roots with small secondary roots; hairs colored. **Fronds** pinnately compound, spirally clustered apically; leaflets flat (circinate), margins entire or dentate (spinose), leathery, venation dichotomous (netted), secondary veins regular, subparallel; petiole and rachis unarmed (rarely with stout spines). **Cones** axillary, appearing terminal, short-peduncled (sessile), disintegrating at maturity; sporophylls densely crowded, spirally arranged, often covered with indument. **Pollen cones** soon shed, generally smaller and more numerous than seed cones; sporophylls bearing many crowded, small microsporangia (pollen sacs) adaxially; pollen spheric. **Seed cones** persistent for a year or more, 1(–2) per plant, nearly globose to ovoid, tapering sharply or blunt at apex; sporophylls peltate, thickened and laterally expanded distally, bearing 2(–3) inverted ovules. **Seeds** angular, inner sarcotesta hardened, outer sarcotesta fleshy, often brightly colored. **Distr:** Throughout the tropics and subtropics.

Dioon

Syn: *Platyzamia*.
Species 11. **Plants**, dioecious, shrubby and palmlike; stems aerial or subterranean, pachycaulous, cylindrical, rarely with aerial branches; basal offsets (suckers) usually produced. **Fronds** many, leaf bases persistent or shedding to leave smooth bark in some species, pinnate, spirally arranged, interspersed with cataphylls, lower leaflets often reduced to spines; longitudinal and horizontal ptyxis erect; leaflets simple, inserted near margins of rachis, margins often spiny, pubescent when young, veins numerous bifurcating, parallel; petioles lacking prickles, usually swollen at base. **Pollen cones** with microsporophylls spirally aggregated into determinate, stalked ♂ cones, apex simple, sterile, flattened and upturned, microsporangia (pollen sacs) numerous, abaxial, opening by slits. **Seed cones** spirally aggregated into determinate, stalked ♀ cones; sporophylls with a simple dilated apex which is flattened and upturned; ovules 2(–3), inverted, stalked, orthotropous. **Seeds** subglobose to oblong or ellipsoidal, outer sarcotesta white or cream, fleshy. **Distr:** Mexico, Honduras and Nicaragua. **Cult:** Propagate from seeds or rooted offsets (borne at base of the trunk). Requires a humus-rich, moist, well-drained soil in full or partial sun.

D. edule Lindl., *Edward's Bot. Reg.* 29: misc. 59–60, 1843. (Syn: *Platyzamia rigida*, etc.). [Virgin Palm, Mexican Cycad]. **Plants**, arborescent; stem 4 m tall, 30–50 cm Ø; crown with 50–150 fronds. **Fronds** 1–2 m long, flat; leaflets 70–150, inserted at 90° to rachis, not overlapping, narrowly lanceolate, margins flat, entire, concolorous, bright green or bluish-green, semi-glossy to dull; basal leaflets reduced to spines; median leaflets 6–12 × 0.5–1 cm. **Cataphylls** 2–3 cm long, tomentose. **Pollen cones** ovoid to fusiform, 15–40 × 6–10 cm, pale brown; microsporophyll apex 3 × 2 cm. **Seed cones** ovoid, 20–35 × 12–20 cm, pale grey; megasporophyll apex 3.5 × 2.5 cm. **Seeds** ovoid, 2.5–4.5 × 2–3 cm, sarcotesta cream or white. **Distr:** Mexico. **Ecol:** In rocky habitats or on steep cliffs, of tropical deciduous forests or oak woodlands; 500–1,500 m. **Zones:** 10–12.

Encephalartos

Species 65. **Plants**, dioecious, shrubby and palmlike; stems often large, aerial or subterranean, pachycaulous, cylindrical, with persistent leaf bases; basal offsets produced in most species, aerial branching rare. **Fronds** usually many, spirally arranged, interspersed with cataphylls; lamina pinnate, longitudinal and horizontal ptyxis erect; leaflets simple, not articulated, inserted near margins of rachis towards the adaxial side, margins often spiny, dentate or lobed, venation bifurcating and parallel, midrib indistinct, pubescent, at least when young, lower leaflets often reduced to spines; petioles lacking prickles, usually swollen at base, sometimes with a distinctly differently colored collar. **Cones** with microsporophylls or megasporophylls spirally aggregated, determinate, each with a simple sterile apex often flattened or faceted but never produced into an upturned spine. **Pollen cones** with microsporophylls with numerous microsporangia abaxially; microsporangia opening by slits. **Seed cones** with megasporophylls with ovules 2(–3), sessile, orthotropous, inserted inside the thickened lamina and inverted. **Seeds** subglobose to oblong or ellipsoid, outer sarcotesta red, yellow, orange or brown, fleshy. **Distr:** Africa. **Cult:** Propagate from seeds or offsets. Requires a well-drained soil in full sun.

E. hildebrandtii A. Braun & C.D. Bouché, *Ind. Sem. Hort. Berol.* 8, 1874. (Syn: *Zamia hildebrandtii*, etc.). [Mombasa Cycad]. **Plants** arborescent; trunk to 6 m tall, 30 cm Ø. **Fronds** linear-oblanceolate, 2–3 × 0.6–0.65 cm, base tapered, dark green, glossy, ± keeled; rachis green, slightly curved, somewhat lax; petiole straight, with 1–6 prickles; leaflets lanceolate, weakly discolorous, not lobed, margins flat, heavily toothed; basal leaflets reduced to spines; median leaflets linear-lanceolate, 15–35 × 1.3–4.5 cm, apex acuminate, margins spiny, coriaceous, adaxially dark glossy green, obscurely striate with 26–40 parallel nerves abaxially. **Pollen cones** 3–8, cylindrical to subconical or fusiform, 20–50 × 5–9 cm, greenish-yellow to dull red; peduncle 5–25 cm long with scattered scale leaves. **Seed cones** cylindrical to ovoid, 28–60 × 15–25 cm, dull yellow, bulla of median scales deflexed, rhomboid, 3.5–5 × 2–3.5 cm. **Seeds** oblong, 3–3.8 × 1.5–2 cm, sarcotesta red, orange or yellow. **Distr:** Kenya, Tanzania. **Ecol:** In coastal savanna woodlands, riverbanks; <600(–1,200) m. **Zones:** 10–12.

E. lehmannii Lehm., *Pugill.* 6: 14, 1834. (Syn: *Zamia spinulosa*, etc.). [Karoo Cycad]. **Plants**, arborescent, often clustering; stem 2–3 m tall, 25–45 cm Ø, with persistent leaf bases. **Fronds** 1–1.5 m long, bluish or silvery, dull, strongly keeled; rachis bluish, straight with last third sharply recurved, strongly spirally twisted in most leaves; leaflets not overlapping, lanceolate, margins flat, upper and lower margins entire; median leaflets 12–20 × 1.5–2 cm; basal leaflets not reduced to spines; petiole straight, prickles lacking; leaf base collar prominent. **Pollen cones** 1, subsessile, slender, narrowly ovoid or fusiform, 25–35 × 8–10 cm, tapered towards base, yellow, green or brown, scales numerous, with a small suborbicular or quadrangular shortly strigillose-pubescent top. **Seed cones** 1, ovoid, 45–50 × 25 cm, green or brown. **Seeds** ovoid or oblong, 3–4.5 × 2.5–3 cm, sarcotesta red. **Distr:** E Cape Province, South Africa. **Ecol:** In arid shrublands on rocky ridges and steep hillsides. **Zones:** 9–12.

E. villosus Lem., *Ill. Hort.* 14, misc.: 79, 1867. (Syn: *Zamia villosa*, etc.). [Poor Man's Cycad, Woolly Cycad]. **Plants**, acaulescent; stem not well-developed above ground, 30 × 30 cm. **Fronds** 1.5–3 m long, dark green, highly glossy, recurved, densely woolly when young, glabrate; leaflets linear or lanceolate, strongly discolorous, margins flat, or incurved, margins with 1–3 teeth; median leaflets linear or narrowly oblong, 15–25 × 1.5–2 cm, ± straight, margins with usually 1–5 spines per side, sometimes entire; basal leaflets reduced to spines; rachis green, straight, stiff; petiole straight, with 6–12 prickles. **Pollen cones** 1–8, strongly aromatic, narrowly ovoid to cylindrical, 50–70 × 10–15 cm, yellow; peduncle to 20 cm long. **Seed cones** 1–4, oblong to ovoid, 30–50 × 20–25 cm, yellow; scales straight with the head much inflated; head elliptic with somewhat blunted margins, adaxial margin rounded and medially dentate, terminal facet deflexed. **Seeds** oblong to ovoid, 2.5–3 × 1.8–2 cm, sarcotesta red or crimson. **Distr:** SE coast of S Africa (E Cape Province through Natal to Swaziland). **Ecol:** In coastal woodlands; 0–100 m. **Zones:** 9–12.

Macrozamia

Species 40. **Plants**, dioecious, palmlike shrubs; stem aerial or subterranean, pachycaulous, cylindrical, with persistent leaf bases. **Fronds** few to many, pinnate, pubescent, at least when young, spirally arranged, interspersed with cataphylls, longitudinal and horizontal ptyxis erect, leaflets inserted near margins of rachis towards the adaxial side, often with a paler or differently colored basal gland; leaflets simple or dichotomously divided, with numerous parallel veins, midrib indistinct; lower leaflets often reduced to spines. **Cones** determinate, stalked, with apices, simple, often produced into upturned spines. **Pollen cones** with apex sterile; microsporophylls spirally aggregated, bearing numerous microsporangia (pollen sacs) abaxially; microsporangia opening by slits. **Seed cones** with megasporophylls spirally aggregated, appearing peltate with a simple dilated apex or lamina; ovules 2(–3), sessile, orthotropous, inserted on the inside of the thickened lamina and directed inverted. **Seeds** subglobose to oblong or ellipsoid, outer sarcotesta red, sometimes yellow, orange or brownish, fleshy. **Distr:** Endemic to Australia. **Cult:** Propagate from fresh seeds. Requires a well-drained soil in full or partial sun.

M. communis L.A.S. Johnson, *Proc. Linn. Soc. New South Wales* 84: 98, 1959. [Burrawang Palm, Australian Sago]. **Plants**, acaulescent, rarely caulescent; stem 0–1.5 m tall, 40–90 cm Ø; crown with 50–150 leaves. **Fronds** 1.4–2.5 m long, flat, pinnate; pinnae 70–150, dark green, semi-glossy; petiole 12–40 cm long, 1–1.6 cm wide at lowest pinna; basal pinnae reducing to spines; median pinnae simple, 16–36 × 0.8–1.3 cm, strongly discolorous, margins flat, apex entire, spinescent. **Pollen cones** narrowly ovoid to fusiform, 25–45 × 8–12 cm; microsporophyll lamina 2–4 × 1.5–2.5 cm, apical spine 0.5–5 cm long. **Seed cones** ovoid, 20–45 × 10–20 cm; megasporophyll with an expanded, peltate apex 1.5–4.3 × 0.3–0.85 cm, apical spine 0.8–8 cm long. **Seeds** flattened-ovoid, 3–4.5 × 2–3 cm, sarcotesta red. **Distr:** Australia (E NSW). **Ecol:** In coastal wet to dry sclerophyll forests. **Zones:** 9–11.

M. moorei F. Muell., *Australas. Chem. Druggist* 4: 84, 1881. (Syn: *Encephalartos moorei*). [Carnarvon Gorge Macrozamia]. **Plants**, caulescent; stem 2–7 m tall, 50–80 cm Ø; crown with 100–120 fronds. **Fronds** 0.1–2.5 m long, moderately keeled, with 120–220 pinnae, dark green to grey-green, dull

Top row: *Dioon edule, Encephalartos lehmannii, Encephalartos hildebrandtii* (seed cones), *Encephalartos villosus.*
Bottom row: *Macrozamia communis* (seed cone), *Macrozamia moorei, Zamia pumila.*

Above: *Ginkgo biloba* (male spikes), *Ginkgo biloba* (female spikes), *Ginkgo biloba* (seeds), *Pinus caribaea.*

to semi-glossy; basal pinnae reduced to spines; median pinnae simple, 20–35 × 0.5–1 cm, weakly discolorous, margins flat, apex entire, spinescent; petiole 2–10 cm long, 2.2–3 cm wide at lowest pinna. **Pollen cones** fusiform, 30–45 cm long, 10–19 cm Ø; microsporophylls 2.5–3.5 × 1.5–2 cm, apical spine 0.2–2.5 cm long. **Seed cones** ovoid, 40–80 cm long, 14–19 cm Ø; megasporophylls with an expanded, peltate apex 2–3 × 4–7 cm, apical spine 0.2–7 cm long. **Seeds** ovoid, 4–6 × 3–4 cm, sarcotesta red. **Distr:** Australia (C Queensland). **Ecol:** In open dry woodlands, in shallow, rocky, usually poor soils. **Zones:** 9–11.

Zamia

Species >50. **Plants**, dioecious, palm- or fernlike, shrubby; stems aerial or subterranean, pachycaulous, cylindrical or globose; dichotomous branching common in geophytic species, suckers produced in some species. **Fronds** few to many, bases mostly not persistent, new fronds emerging singly or in flushes; lamina pinnate, spirally arranged, interspersed with cataphylls, lower leaflets not reduced to spines; petioles often with prickles; leaflets simple, margins spiny in some species, with numerous bifurcating parallel veins and no distinct midrib, leaflets articulated, inserted near the edges of the rachis towards the adaxial side, lacking a differently colored basal gland, surfaces pubescent, at least when young, with branched and simple, transparent and colored hairs. **Pollen cones** with microsporophylls spirally orthostichous, aggregated, sterile apex simple, often flattened or faceted, sometimes weakly bilobed, never produced into an upturned spine; each microsporophyll bearing numerous microsporangia. **Seed cones** with megasporophylls spirally orthostichous, aggregated; sporophylls simple, appearing peltate with a simple dilated apex which is usually hexagonal, often flattened or faceted, sometimes weakly bilobed, never produced into an upturned spine; ovules 2(3), sessile, orthotropous, inserted on the inner surface of the thickened lamina and directed inwards. **Seeds** subglobular to oblong or ellipsoidal, sarcotesta red, orange, yellow or rarely white, fleshy. **Distr:** S, C and N America. **Cult:** Propagate from fresh seeds. Requires a well-drained soil in full or partial sun.

Z. pumila L., *Sp. Pl.*, ed. 2, 2: 1659, 1763. (Syn: *Encephalartos pumilus*, etc.). [Florida Arrowroot, Seminole Bread]. **Plants**, palmlike; stems subterranean and tuberous, 3–10(25) cm Ø, often dichotomously branched; cataphylls 1–2 cm long, sheathing at first, with a pair of inconspicuous stipules. **Fronds** 2–15; petioles with stipules, smooth; rachis smooth, bearing 5–30 pairs of leaflets; leaflets oblong, acute to somewhat rounded apically, often deeply cleft, with 10–15 distinct teeth in upper fourth, median ones 8–25 × 0.5–2 cm. **Pollen cones** 1–30, pedunculate, dark reddish-brown, cylindrical to ovoid-cylindrical, 3–15 × 0.8–2 cm, apex acute. **Seed cones** 1–5, pedunculate, dark reddish-brown, cylindrical to slightly ovoid, apex acuminate,6–15 × 4–6 cm. **Seeds** red to orange-red, ovoid, 1–2 cm long. **Distr:** C Cuba, S Puerto Rico, Hispaniola. **Ecol:** In dry, sandy pinelands and hammocks, rarely coastal dunes. **Zones:** 10–12.

GINKGOALES

FAMILY 1, GENUS 1, SPECIES 1.
Ordinal description same as specific.

GINKGOACEAE

Genus 1, **species** 1. *Familial description same as specific.*

Ginkgo

Syn: *Salisburia.*
Species 1. *Generic description same as specific.*
Cult: Propagate from seeds or half-hardened cuttings. Requires a moist, well-drained soil in full sun.

G. biloba L., *Mant. Pl.* 2: 313, 1771. (Syn: *Salisburia adiantifolia, S. biloba*). [Maidenhair Tree, Gingko]. **Trees**, to 40 m tall; trunk to 4 m Ø; bark light grey or grayish-brown, longitudinally fissured, especially on old trees; crown conical initially, finally broadly ovoid; long branchlets pale brownish-yellow initially, finally grey, internodes (1–)1.5–4 cm long; short branchlets blackish-grey, with dense, irregularly elliptic leaf scars; winter buds yellowish-brown, ovate. **Leaves** pale green, turning bright yellow in autumn, to 13 × 8(–15) cm on young trees but usually 5–8 cm wide, those on long branchlets divided by a deep, apical sinus into 2 lobes each further dissected, those on short branchlets with margins undulate, apex notched; petiole (3–)5–8(–10) cm long. **Pollen cones** ivory colored, 1.2–2.2 cm long; pollen sacs boat-shaped, with widely gaping slit. **Seeds** elliptic, narrowly obovoid, ovoid, or subglobose, 2.5–3.5 × 1.6–2.2 cm; sarcotesta yellow, or orange-yellow glaucous, malodorous when ripe; sclerotesta white, with 2 or 3 longitudinal ridges; endotesta pale reddish-brown. **Distr:** China. **Ecol:** In broad-leaved forests; 300–1,100 m. **Zones:** 3–11.

PINALES

[Conifers, Pines].

FAMILIES 7, GENERA 68, SPECIES 545.

Trees or **shrubs**, monoecious or dioecious, highly branched. **Leaves** opposite, sometimes clustered into short shoots (with adjacent nodes very short), rarely fascicled; lamina simple, linear, needlelike or awl-like (subulate), with a single vein. Reproductive organs with ♂ cones and ♀ cones, either on the same plant or less commonly on different plants; sporophylls are borne on the axis of the cone. Pollen cones (♂) with sporophylls producing ♂ sporangia (microsporangia) which in turn produces pollen grains. **Seed cones** compound, consisting of 'bracts' (modified leaves) borne on the axis; bracts subtending the seed-bearing structure 'ovuliferous scale' (modified lateral branch); **seeds** usually 2, on the adaxial surface of the ovuliferous scale; seed coat dry, not vascularized. **Distr:** Worldwide. **Info:** The monophyly of the conifers are strongly supported by molecular data (Stefanovic et al. 1998).

PINACEAE

Genera 10–11, **species** ~235. **Trees** (rarely shrubs), evergreen or deciduous, monoecious; branchlets often dimorphic; *long branchlets* with spirally arranged, sometimes scalelike leaves; *short branchlets* often reduced to slow growing lateral spurs bearing dense clusters of leaves at apex. **Leaves** (or needles) solitary or in clusters of (1 or)2–5(–8) basally subtended by a leaf sheath; lamina linear or needlelike, not decurrent. **Cones** unisexual. **Pollen cones** solitary or clustered, with numerous spirally arranged microsporophylls; microsporophyll with 2 microsporangia. **Seed cones** erect or pendulous, maturing in 1–3 years, dehiscent or occasionally indehiscent, with many spirally arranged ovulate scales and bracts; ovulate scales usually smaller than bracts at pollination, with 2 upright ovules adaxially; seed scales appressed, woody or leathery, variable, with 2 seeds adaxially, persistent or deciduous after cone maturity; bracts free or adnate basally with seed scales; seeds terminally winged or sometimes wingless. **Distr:** Northern Hemisphere.

Pinus

Species ~110. **Trees** or rarely shrubs, evergreen, with regularly whorled branches; branchlets strongly dimorphic; *long branchlets* bearing scalelike leaves and spreading leaf bundles; *short branchlets* with leaves in bundles of 2–5(–7). **Leaves** needlelike, slender or stout, straight or twisted, triangular, flabellate-triangular, or semi-orbicular in cross-section, stomatal lines several. **Pollen cones** usually borne in spikelike clusters at base of first year's branchlets, sessile, cylindric or ovoid. **Seed cones** pedunculate or subsessile, erect or pendulous, cylindric or ovoid, maturing in 2–3 years; seed scales spirally arranged, woody, exposed apex thickened and ridged, with a prominent protuberance, usually terminating in a spine or prickle, persistent; bracts minute; seeds variable, winged or not; wing adnate or articulated to seed. **Distr:** N Africa, Asia, Europe, N America. **Cult:** Propagate from seeds or by grafting. Requires a well-drained soil in full sun.

P. caribaea Morelet, *Rev. Hort. Cote d'Or.* 1:

107, 1851. [Caribbean Pine, Pitch Pine, Slash Pine]. **Trees**, to 40 m tall; trunk to 1 m Ø; bark grey or pale reddish-brown, fissured, shed in large, flat, wide plates; crown ovoid or irregularly shaped; branchlets initially green and glaucous, aging to orange-brown, producing a few short nodes each year; winter buds cylindric, scales white fringed at margin. **Needles** (2 or)3 per bundle, usually 4 or 5 per bundle on young trees, 15–30 cm × ~1.5 mm, dull green or pale yellow-green, stomatal lines present on all surfaces, resin canals (2 or)3 or 4(–8), internal, base with persistent sheath 1–1.5 cm, margin serrulate. **Seed cones** almost terminal, ovoid-cylindric, 5–10(–12) × 3–6 cm, often leaving a few basal scales; seed scales reflexed or spreading; apophyses lustrous, tan or reddish-brown, swollen, cross keeled; umbo slightly projecting, ending in a straight, minute prickle; seeds usually narrowly rhombic-ovoid, 6–7 mm long, mottled grey or light brown; wing dull grey, 2–2.5 cm long, usually remaining attached. **Distr:** Caribbean, Central America. **Ecol:** In lowland and hill pine forests, and mixed oak-pine forests, often growing among rocks, naturalizing abundantly in open and disturbed areas; (–650)1,000–1,500(–4,000) m. **Zones:** 9–12.

ARAUCARIACEAE

Genera 3, **species** 41. **Trees**, evergreen, dioecious or monoecious; trunk with pith relatively large, cortex containing resin. **Leaves** spirally arranged or decussate, sessile, decurrent. **Cones** unisexual. **Pollen cones** axillary or terminal on branchlets, solitary or clustered; microsporophylls numerous, spirally arranged, sessile; microsporangia 4–20, external, suspended, arranged in 2 rows, filiform. **Seed cones** solitary, terminal on branchlets, maturing in 2–3 years; bracts numerous, spirally arranged; ovulate scales degenerate or ligulate, connate with bracts on 1 side, each with 1 ovule at base abaxially; ovules connate with ovulate scales or with bracts; bracts of mature cones deciduous, flattened, woody or thickly leathery, bearing 1 seed in basal part; seeds connate with bracts or detached, flattened, winged or not. **Distr:** Malesia, SW Pacific Islands, New Zealand, South America. **Info:** This is one of the most primitive families of the living conifers and its monophyly is strongly supported by molecular data (Liu et al. 2009).

Agathis

Syn: *Dammara*.

Species 21. **Trees**, monoecious (dioecious when young), massive, evergreen, coniferous; branches usually whorled on young trees, becoming irregularly arranged on mature ones; branchlets deciduous, leaving a bare trunk with orbicular scars. **Leaves** spirally arranged on main branches, opposite or alternate on lateral branchlets, leaving cushion-shaped scars on falling, pinkish or reddish when young, finally dark green, highly variable in size and shape even on same branchlet, leathery, with numerous, indistinct, parallel, thin veins; petiole flattened, short. **Pollen cones** axillary, solitary, upright, cylindric, hard; microsporophylls densely arranged. **Seed cones** globose or broadly ovoid; bracts densely arranged, flabellate, apex thickened; seeds detached from bracts, with a lateral wing on 1 side and a small protrusion, the other side sometimes winglike. **Distr:** Indonesia, Malaysia, Papua New Guinea, Philippines; Australia, New Zealand, SW Pacific Islands. **Cult:** Propagate from seeds. Requires deep, moist soil in full to partial sun.

A. robusta (C. Moore ex F. Muell.) F.M.Bailey, *Syn. Queensl. Fl.* 498, 1883. [Kauri Pine, Queensland Kauri]. **Trees**, to 50 m tall; trunk usually very straight, slightly tapering, 1.8 m Ø, buttresses absent; bark smooth to slightly flaky. **Leaves** spirally arranged on orthotropic shoots but opposite to subopposite on plagiotropic shoots; lamina linear to elliptic, 5–13 × 1–4 cm, margins entire, stiff, veins fine, longitudinal, ± parallel; petioles ~0.3–1 cm long. **Cones** (strobili) ♂ and ♀ borne on the same plant. **Pollen cones** (♂) shortly pedunculate or almost sessile, usually axillary on slender leafy twigs, cylindric, 4–8.5(–10) × 0.7–0.9 cm; cone scales 600–1,300, each with 2–8 pollen sacs below. **Seed cones** (♀) globular to cylindric, 9–15 × 8–10.5 cm; scales 340–440, equatorial scales 3.4–4.1 × 3.9–4.6 cm; seeds narrowly cordate (excluding the wings). **Distr:** Australia (Queensland). **Ecol:** In dry marginal forests or rainforests; 0–900 m. **Zones:** 9–12.

Araucaria

Species 19. **Trees**, monoecious, occasionally dioecious; branches whorled or almost so; winter buds small. **Leaves** spirally arranged; lamina needlelike to broadly ovate or triangular, variable even on the same tree. **Pollen cones** tassel-like, axillary or terminal on branchlets, cylindric, ovoid, or ellipsoid; microsporophylls densely arranged. **Seed cones** borne near the top of the crown, erect, ellipsoid, or ovoid to subglobose; ovulate scales ligulate; bracts of mature cones large, woody, distal part thickened, distal margin sharply and transversely ridged, apex reflexed or upcurved, pointed; seed scales adnate with bracts in basal, seed-bearing part, detached in apical part, sometimes thickened and exposed; seeds connate with bracts, sometimes with 2 lateral wings formed from the bract. **Distr:** Papua New Guinea, Australia (including Norfolk Island), New Caledonia, S America. **Cult:** Propagate from fresh seeds or cuttings. Requires a moist, well-drained soil in full sun.

A. angustifolia (Bertol.) Kuntze, *Revis gen. pl.* 3(3): 375, 1898. (Syn: *A. brasiliana, Columbea brasiliana*). [Candelabra Tree, Prana Wood]. **Trees**, to 30+ m tall; trunk straight, terminated by a flat head of gaunt, horizontal branches with the branchlets in large terminal tufts; branches usually arranged in whorls of 4–8. **Leaves** of *sterile branches*, lanceolate, 3.2–5.7 cm long, green or glaucous, sometimes appearing as if in pairs; on *fertile branches* shorter, more densely and spirally arranged; in all types apex long-pointed, stiff, leathery, with stomata well-developed abaxially. **Pollen cones** borne on short shoots, axillary, catkinlike, dense, 7.5–10 × 1.2–2 cm. **Seed cones** broadly ovoid, 16.5 × 12.5 cm, narrowed from the middle upwards; scales terminated by stiff recurved appendages. **Seeds** bright brown, to 5 × 2 cm, 8 mm thick. **Distr:** Argentina, Brazil. **Ecol:** In montane regions. **Zones:** 8–11.

A. araucana K. Koch, *Dendrologie* 2(2): 206, 1873. (Syn: *A. imbricata, Pinus araucana*). [Chile Pine, Monkey Puzzle Tree]. **Trees**, evergreen, dioecious, to 50 m tall; crown parasol-like; trunk upright, cylindrical; bark very thick and wrinkled, breaking up into plates; branches in whorls of 5 to 7, long, cylindrical, bearing the leaves towards their end. **Leaves** spirally arranged, sessile, overlapping; lamina ovate-lanceolate, 2.5–3 × 1.5–2 cm, base brad, apex sharply pointed, leathery, rigid, dark-green. **Pollen cones** terminal, amentaceous, cylindrical, 8–12 cm long. **Seed cones** terminal, 10–15 cm across; ripened cones bearing 120–180 seeds, 4–5 × 1.5 cm. **Distr:** Argentina, Chile. **Ecol:** On sandy, rocky, usually volcanic soils; 600–1,400 m. **Zones:** 7–11.

A. bidwillii Hooker, *London J. Bot.* 2: 503. 1843. [Bunya-bunya, Bunya Pine]. **Trees** to 50 m tall; trunk to 1 m Ø; bark dark grey-brown, thick, flaking; crown pyramidal; branchlets dense, pendulous, green, glabrous. **Leaves** radially spreading, bright green, lanceolate, ovate-lanceolate, or triangular-ovate, flattened, slightly incurved, hard, thick, leathery, lacking a midvein but with numerous, parallel, thin veins, stomatal lines abaxial; leaves of same year dimorphic: those of *juvenile* trees and vegetative branchlets longer (0.7–2.8 cm) and more loosely arranged than those of *mature* trees and cone-bearing branchlets, those at middle of branchlets 2.5–6.5 cm, apex acuminate or acute. **Pollen cones** axillary, solitary, cylindric. **Seed cones** ovoid-subglobose, ~30 × 22 cm; bracts oblong-elliptic or oblong-ovate, margin relatively thick, wingless, apex triangular, reflexed; seed scales thickened, exposed at apex; seeds elongate-elliptic, wingless. **Distr:** NE Australia. **Ecol:** In moist coastal rainforests and steep forest margins; <1,000 m. **Zones:** 8–11.

A. columnaris Hook., *Bot. Mag.* 78, sub t. 4635, 1852. (Syn: *A. cookii, A. excelsa*). [Columnar Pine, Cook Pine, New Caledonian Pine]. **Trees**, columnar, to 30–60 m tall, often leaning; lateral branches short and crowded on the trunk. **Leaves** dimorphic; leaves of *juvenile* trees needlelike, spreading, upward-curving, to 7.6 mm long, gradually broadening as the tree grows; leaves on *adult* trees triangular, overlapping, 5–7.6 × 5–6.3 mm, apex incurved, acute (not needle-tipped). **Pollen cones** terminal, cylindrical, 5–10 × 1.5–2.3 cm, ± tapering toward apex, flexible; pollen scales widening to triangular apical portion, 0.8–1 × 0.5 cm. **Seed cones** globose or ovoid, 10–15 × 7.5–10 cm, spurs bent flat against cone; seed scales 3–3.6 cm long, lateral wings broad, membranous, apical spur 7.6 mm long. **Distr:** SE coasts of New Caledonia, Loyalty Islands to the east and the Isle of Pines. **Ecol:** On calcareous soils in open vegetation, often forming dense colonies on raised coral reefs. **Zones:** 10–12.

A. cunninghamii Aiton ex D. Don, *Descr. Pinus*, ed. 2. 3: t. 79, 1837. (Syn: *Eutassa cunninghamii*). [Colonial Pine, Dorrigo Pine, Hoop Pine, Moreton Bay Pine]. **Trees**, to 70 m tall; trunk to >1 m Ø; bark grey-brown or dark grey, rough, transversely split; crown tower-shaped when young, becoming flat-topped with age; lateral branchlets dense, drooping, almost pinnately arranged. **Leaves** dimorphic; leaves on *juvenile* trees and leaves of lateral branchlets loosely arranged, needlelike, falcate, subulate, or triangular, slightly curved, somewhat tetragonal or ridged, 0.7–1.7 cm × ~2.5 mm, stomatal lines weakly evident on abaxial surface, apex acute or acuminate; leaves on *mature* trees and cone-bearing branchlets densely arranged, overlapping, stretching upward, grey-green abaxially, glaucous, ovate to triangular, 0.6–1 × ~0.4 cm, ridged or not, midvein obvious or not, distal part of leaf tapering or slightly rounded, apex acute or obtuse. **Pollen cones** terminal, solitary, ovoid or ellipsoid. **Seed cones** ovoid or ellipsoid, 6–10 × 4.5–7.5 cm; bracts narrowly obovate, sharply ridged, with thin, lateral wing, apex caudate, thickened, acute, obviously reflexed; seed scales thin at apex; seeds ellipsoid, with a membranous, lateral wing. **Distr:** NE Australia, Papua New Guinea. **Ecol:** In rainforests, sometimes in swampy areas, often forming dense stands, rarely forming pure forests. **Zones:** 9–12.

A. heterophylla (Salisbury) Franco, *Anais Inst. Super. Agron.* 19: 11. 1952. (Syn: *Eutassa heterophylla*). [Norfolk Island Pine]. **Trees** to 50 m tall; trunk to 1.5 m Ø; trunk straight; bark dark grey, flaking; crown tower-shaped; branchlets spreading horizontally or drooping; lateral branchlets usually pinnately arranged, drooping. **Leaves** dimorphic; leaves of *juvenile* trees and lateral branchlets loosely arranged, spreading openly, 0.6–1.2 cm, 3- or 4-ridged, bright green, glaucous abaxially, subulate, usually laterally depressed, upcurved; leaves of *mature* trees and cone-bearing branchlets densely arranged, slightly openly spreading, broadly ovate or triangular-ovate, ± curved, 5–9 mm, widest at base, midvein obviously raised or not, apex obtuse. **Pollen cones** terminal, solitary. **Seed cones** subglobose, 8–12 × 7–11 cm, sometimes wider than long; apex of bracts triangular, flattened, acute, upcurved; seeds ellipsoid, slightly flattened, with

Top row: *Agathis robusta, Araucaria araucana, Araucaria bidwilli, Araucaria columnaris.*
Bottom row: *Araucaria angustifolia, Araucaria cunninghamii, Araucaria heterophylla.*

a lateral wing. **Distr:** Australia (Norfolk Island). **Ecol:** In lowland and coastal regions, subtropical broadleaf, palm and conifer forests. **Zones:** 8–12.

PODOCARPACEAE

Genera 17, **species** 250. **Trees** or **shrubs**, evergreen, dioecious or rarely monoecious. **Leaves** decussate, subopposite, or spirally arranged; lamina scalelike, subulate, or linear to elliptic, stomatal lines abaxial or present on all surfaces. **Pollen cones** terminal, solitary or clustered axillary, or borne in spikelike complexes; individual cones pedunculate or sessile; microsporophylls numerous, spirally arranged, with distinct adaxial and abaxial surfaces; microsporangia 2; pollen 2(or 3)-saccate (rarely non-saccate). **Seed cones** terminal or axillary, solitary, sometimes spikelike, with few to several spirally arranged bracts; all or only apical bracts fertile, smooth or warty; basal bracts sometimes fused and succulent forming a receptacle, or obsolete; ovule (inverted) or inclined. **Seeds** drupelike or nutlike, wholly or partly enveloped in a sometimes colored and succulent epimatium. **Distr:** Largely from the Southern Hemisphere, scattered, north to Japan, C America and the Caribbean.

Podocarpus

Syn: *Margbensonia.*
Species 100. **Trees** or **shrubs**, evergreen, dioecious. **Leaves** spirally arranged to subopposite, ± monomorphic, juvenile leaves similar to adult ones, often larger; lamina linear, lanceolate, or ovate-elliptic, >5 mm long, midvein single, often raised on 1 or both surfaces, stomatal lines present on abaxial surface. **Pollen cones** axillary, solitary or clustered, pedunculate or sessile; microsporophylls numerous, spirally arranged; microsporangia 2; pollen bisaccate. **Seed cones** usually axillary (rarely terminal), solitary (rarely >1); apical bracts fertile; basal bracts often fused to form a receptacle, sometimes obsolete; ovule 1 (rarely few), inverted. **Seeds** drupelike, dry, or leathery, with epimatium wholly enveloping seed, sometimes colored and succulent. **Distr:** Tropical and subtropical regions worldwide, also temperate regions in the Southern Hemisphere. **Cult:** Propagate from fresh seeds or cuttings. Requires a well-drained soil in full sun.

P. cupressina R. Br. ex Mirb., *Mém. Mus. Hist. Nat.* 13: 75, 1825. (Syn: *P. horsfieldii, P. imbricata*). [Javanese Podocarpus]. **Trees**, lofty; branchlets of 3 forms: (1) short, flat, featherlike, 2.5–5 cm long with a slender rachis and closely set distichous, horizontally spreading, linear falcate, acuminate, shiny leaves 0.8–1.2 cm long; (2) terete, filiform branchlets, clothed with minute, imbricate, ovate, acute leaves ~1.2 mm long; (3) stouter, terete, fruiting branchlets with subulate leaves 2.5–8 mm long. **Leaves** on older branches minute, quadrifarious, lanceolate, mucronate, keeled; on younger branches distichous, spreading, linear and falcate or compressed and tetragonous. **Seeds** globose ~6 mm Ø, on short decurved lateral branches. **Distr:** Malay Islands (Malacca, Penang), Myanmar. **Ecol:** In primary and secondary rainforests, often naturalized in rainforests; 200–3,000 m. **Zones:** 9–12.

P. elatus R. Br. ex Endl., *Syn. Conif.* 213, 1847. (Syn: *Nageia elata, P. ensifolius*). [Brown Pine]. **Trees**, to 30 m tall. **Leaves** *adult* shortly petiolate, linear-lanceolate, 5–15 × 0.75–1.6 cm, apex acute or obtuse, glossy dark green, midrib wide, 3 resin ducts present below vascular bundle. **Strobili** with ♂ and ♀ cones produced on separate trees, axillary. **Male cones** cylindrical, catkinlike, 3–5 cm long, 4–5 mm Ø, sessile, axillary in groups of 2–4; microsporophylls imbricate, broadly triangular, shortly apiculate. **Female cones** of 1 fertile scale with 1 (or 2) ovules, axillary on thick peduncles 0.3–1 cm long; receptacle 1.5–2.5 cm long, fleshy, black or purple at maturity, pruinose, with 2 deciduous, subtending bracts. **Seeds** 1.4–2.2 cm long, 1.2–1.5 cm Ø. **Distr:** Australia (Queensland, NSW). **Ecol:** In rainforests. **Zones:** 9–12.

CUPRESSACEAE

Genera 19, **species** ~125. **Trees** or **shrubs** evergreen, monoecious or dioecious. **Leaves** decussate or in whorls of 3, scalelike and often dimorphic with flattened facial leaves and keeled lateral leaves, or needlelike, especially in juvenile plants, often with an abaxial resin gland. **Pollen cones** terminal or axillary, solitary, maturing and shed annually; microsporophylls 6–16, decussate or whorled, each bearing (2 or)3–6(–9) pollen sacs. **Seed cones** usually terminal, solitary, globose, ovoid, or oblong, dehiscent or indehiscent when maturing in 1–3 years; bracts almost completely enveloped by cone scales, free only at apex; ovules 1-numerous per bract axil, erect; cone scales of mature cones 3–16, flat or peltate, woody, ± leathery, or succulent, 1–20-seeded; seeds winged or not. **Distr:** Worldwide.

Juniperus

Syn: *Sabina*.

Species 52. **Trees** or **shrubs**, evergreen, monoecious or dioecious; bark thin, exfoliating in long strips; branchlets not arranged in a plane, terete or 3-, 4-, or 6-angled. **Leaves** decussate or in whorls of 3, decurrent or non-decurrent; *juvenile leaves* always needlelike; *adult leaves* scalelike or needlelike, usually not dimorphic along branchlets but sometimes different on juvenile and adult branchlets, with 1 or 2 pale stomatal bands adaxially. **Pollen cones** yellow, ovoid or oblong; microsporophylls 6–16, each with 2–8 pollen sacs. **Seed cones** terminal or axillary, berrylike, globose or ovoid, indehiscent or slightly dehiscent when mature; cone scales connate or fused, succulent; each fertile scale bearing 1–3 seeds; free bract apex a small point. **Seeds** 1–6(–10) per cone, wingless, usually with resin pits. **Distr:** Northern Hemisphere. **Cult:** Propagate from fresh seeds, cuttings or by grafting. Requires a well-drained soil in full sun.

J. procera Endl., *Synops. Conif.*: 26, 1847. [African Juniper, African Pencil Cedar]. **Trees**, 20–25(–40) m tall; bark grey-brown, with shallow longitudinal fissures, exfoliating in thin papery strips; adult branchlets 0.5–1 mm Ø with opposite, decussate scale leaves, acute, hooded and keeled at the tip. **Leaves** *juvenile* in 3's, linear, spine tipped, decurrent on the branches, not jointed, to 1 cm long, margins rounded, gland linear; *intermediate* leaves paired, grading into the adult form. **Pollen cones** ellipsoid to subglobose, yellowish, 2–3 mm long, with 5–6 pairs of rounded obtuse to blunt apiculate scales with 2–3 pollen sacs. **Seed cones** of 3–4 pairs of scales, mature reddish-brown to purplish-black with blue bloom, subglobose to irregular, 4–8 mm Ø, with 1–4 flattened or triangular seeds. **Distr:** E Africa from E Sudan south to Zimbabwe, and the SW of the Arabian Peninsula. **Ecol:** In forests of montane regions and highlands, sometimes in savannas with stunted growth; 400–3,600 m. **Zones:** 10–12.

Thuja

Species 5. **Trees** or **shrubs**, evergreen, monoecious; branchlets arranged in a plane, flattened. **Leaves** 4-ranked, scalelike, base decurrent only on leading branchlets, dimorphic along branchlets: *facial pairs* obovate-rhombic; *lateral pairs* boat-shaped, ridged abaxially, <4 mm long, overlapping facial pairs, without conspicuous, white stomatal bands abaxially. **Pollen cones** with (4 or)6–10(–16) microsporophylls, each with (2 or)3 or 4 pollen sacs. **Seed cones** terminal, solitary, dehiscent when mature; cone scales 8 or 10, decussate, flat, only middle, larger 2 or 3 pairs fertile, each fertile scale bearing 1 or 2 seeds; free bract apex a very short mucro. **Seeds** flat, with 2 lateral, narrow wings. **Distr:** E Asia; E and W North America. **Cult:** Propagate from seeds or half-hardened cuttings. Requires a deep, moist, well-drained soil in full or partial sun.

T. orientalis L., *Sp. Pl.* 2: 1002, 1753. (Syn: *Platycladus orientalis*, etc.). [Eastern Arborvitae, Northern Whitecedar]. **Trees**, to 20+ m tall; trunk to 1 m Ø; bark reddish-brown to light grayish-brown, thin, flaking in long strips; crown ovoid-pyramidal when young, broadly rounded or irregular when old. **Leaves** 1–3 mm long, apex bluntly pointed; *facial leaves* rhomboid, with a conspicuous, linear, glandular groove at center abaxially; *lateral leaves* overlapping facial ones, boat-shaped, ridged, apex slightly incurved. **Pollen cones** yellowish-green, ovoid, 2–3 mm long. **Seed cones** when immature bluish-green, subglobose, ~3 mm Ø, when ripe reddish-brown, subovoid, 1.5–2(–2.5) × 1–1.8 cm; proximal 2 fertile cone scales 2-seeded, distal 2 fertile scales 1-seeded. **Seeds** grayish-brown or purplish-brown, ovoid or subellipsoid, 5–7 × 3–4 mm, slightly ridged. **Distr:** China, Korea, E Russia. **Ecol:** 300–3,300 m. Pollination Mar–Apr. Seed maturity Oct. **Zones:** 6–11.

TAXACEAE

Genera 6, **species** 30. **Trees** or **shrubs**, evergreen, dioecious or rarely monoecious. **Leaves** spirally arranged or decussate, linear or lanceolate, abaxial surface with 1 stomatal band on each side of prominent or inconspicuous midvein, resin canal present or absent. **Pollen cones** solitary in leaf or bract axils, or aggregated into spikelike complexes apically on branches; microsporophylls numerous. **Seed cones** solitary or paired in axils of leaves or bracts, pedunculate or sessile, with several overlapping or decussate bracts at base; ovule solitary, borne at apex of floral axis, erect. **Seeds** sessile or pedunculate, drupelike or nutlike, enclosed in a succulent, saccate or cupular aril; ♀ gametophyte tissue abundant. **Distr:** Mainly Northern Hemisphere.

Taxus

Species 8. **Trees** or **shrubs**, evergreen, dioecious; branchlets irregularly alternate, basal part with few or several persistent or early deciduous bud scales. **Leaves** 2-ranked, spirally arranged, sessile, or shortly petiolate; lamina linear, straight or falcate, base decurrent, adaxially midvein raised, abaxially with 2 pale grey, grayish-blue, or pale yellow stomatal bands. **Reproductive structures** axillary, solitary. **Pollen cones** pedicellate, globose, with overlapping bracts at base; microsporophylls 6–14, shield-shaped, each with 4–9 radially arranged pollen sacs. **Seed cones** subsessile, with several overlapping bracts, 2 or 3 distal pairs of which are decussate; aril red or orange when ripe, succulent. **Seeds** nutlike, enclosed within cupular aril but with apex exposed; hilum prominent. **Distr:** Mainly Northern Hemisphere. **Cult:** Propagate from seeds, cuttings or by grafting. Requires a well-drained soil in full sun.

T. sumatrana (Miquel) de Laubenfels, *Philipp. J. Biol.* 7: 151, 1978. (Syn: *Cephalotaxus sumatrana*, etc.). [Chinese Yew, Sumatra Yew]. **Shrubs** or **trees**, evergreen, to 14 m tall; trunk bark grayish-red, exfoliating in irregular flakes and leaving scars on the trunk, flakes ~1.5 mm thick; lenticels inconspicuous; outer bark ~0.4–1.6 mm, membranous or fibrous; inner bark 5–8 mm thick, pink, finely fibrous; freshly cut cambium and newly formed phloem colorless, transparent, becoming pale orange yellow on exposure. **Leaves** spirally arranged, spreading in 2 ranks; lamina linear-lanceolate, falcate, ~1.2–2.7 cm × 2–2.5 mm, base decurrent, apex abruptly pointed, adaxially yellowish-green, abaxially pale green. **Pollen cones** globular, with overlapping scales. **Seed cones** subcylindric, 2 × 1 mm; scales 5–6 seriate, conduplicate at base; seeds drupelike, angular, conical, 4 × 3 mm, tapered to apex, the fleshy arillate coat reddish at maturity. **Distr:** Afghanistan, Bhutan, China, India, Myanmar, Nepal, Taiwan, Tibet, Vietnam; Thailand, to Malesia. **Ecol:** In moist subtropical forests, moss forests; 400–3,100 m. **Zones:** 9–12.

GNETALES

FAMILIES 3, GENERA 3, SPECIES 96.

Plants dioecious; roots diarch (having 2 protoxylem points or groups). **Leaves** opposite, joined at base, with collateral buds. **Strobili** compound, micro- and megasporangium-bearing structures closely associated, bracts opposite; microsporangia in synangia, surrounded by a tubular 'bract', dehiscing apically by the action of the epidermis; ovules terminal, erect, integument with much-elongated beak, not vascularised, surrounded by vascularised connate structure. **Seeds** with outer fleshy and inner sclerenchymatous layer derived from the outer integument. **Distr:** Tropical regions worldwide, disjunct. **Info:** This order exhibits characters of both conifers (seeds not enclosed in an ovary) and angiosperms (vessels in wood, reproductive organs flowerlike, double fertilization) (Judd et al. 2008; Stevens 2001 onwards).

EPHEDRACEAE

Genus 1, **species** 65. *Familial description same as generic.*

Ephedra

Species 65. **Shrubs**, **subshrubs**, or **herbs**, dioecious, rarely monoecious; stems erect, procumbent, or occasionally climbing, jointed, much-branched, photosynthetic; branchlets opposite or whorled, green, terete, longitudinally grooved. **Leaves** opposite or in whorls of 3; lamina scalelike, basally ± connate, reduced to membranous sheaths, generally ephemeral, mostly not photosynthetic; resin canals absent. **Cones** terminal or axillary, ovoid or ellipsoid. **Pollen cones** solitary or clustered at nodes, composed of membranous bracts arranged in 2–8 decussate pairs or whorls of 3; proximal bracts empty; each distal bract subtending a ♂ flower composed of 2 basally fused, orbicular or obovate scales (false perianth), anthers sessile or stipitate on staminal column. **Seed cones** opposite or in whorls of 3 or 4 at nodes, composed of overlapping bracts; bracts arranged in 2–10 decussate pairs or whorls of 3, red and fleshy at maturity (rarely brown and membranous); proximal bracts empty; most distal bracts subtending an axillary ♀ flower composed of a pair of fused, leathery scales (false perianth) enclosing ovule with a single membranous integument prolonged into a slender, tubular micropyle. **Seeds** 1 or 2(or 3) per cone. **Distr:** E and N Africa, Asia, Europe, N America, S America. **Ecol:** In dry arid habitats. **Cult:** Propagate by division of clumps, separation of suckers, air-layering or from seeds. Requires a moist, well-drained soil in full or partial sun.

E. equisetina Bunge, *Mém. Acad. Imp. Sci. St.-Petersbourg*, Sér. 6, Sci. Math. 7: 501, 1851. (Syn: *E. shennungiana*). [Bluestem Joint Fir, Mongolian Ephedra, Desert Tea]. **Shrubs**, to 1(–1.5) m tall; woody stems well-developed, erect or partially procumbent, thick; herbaceous branchlets virgate, bluish-green or grayish green, pruinose, 1–1.5 mm Ø, rigid, internodes short, 1–3 cm × 1–1.5 mm, finely furrowed. **Leaves** opposite, brownish, 1.5–3 mm, connate for most of their length, free part bluntly triangular. **Pollen cones** solitary or in clusters of 3 or 4 at nodes, sessile or shortly pedunculate; bracts in 3 or 4 pairs, connate for third of their length; staminal column slightly exserted, anthers 6–8, sessile. **Seed cones** usually opposite at nodes, shortly pedunculate, elongate-ovoid or ovoid at maturity, 8–10 × 4–5 mm; bracts in 3 pairs, apical pair connate for most of their length, red and fleshy at maturity; integument tube to 2 mm long, straight or slightly curved, slightly exserted. **Seeds** usually 1, elongate-ovoid, 5–7 × 2.5–3 mm. **Distr:** Afghanistan, China, Kazakhstan, Kyrgyzstan, Mongolia, Russia, Tajikistan, Turkmenistan, Uzbekistan. **Ecol:** In dry rocky places; 800–3,000 m. **Zones:** 10–12.

WELWITSCHIACEAE

Syn: Tumboaceae.

Genus 1, **species** 1. *Familial description same as specific.*

MJW

Top row: *Podocarpus cupressina, Podocarpus elatus, Thuja orientalis, Taxus sumatrana.*
Bottom row: *Ephedra equisetina, Gnetum gnemon* (female spikes), *Gnetum gnemon* (fruits).

Welwitschia

Species 1. *Generic description same as specific.* **Cult:** Propagate from fresh seeds. Requires a deep, well-drained soil in full sun.

W. mirabilis Hook. f., Gard. Chron. 71, 1862. (Syn: *Tumboa bainesii, T. strobilifera, W. bainesii*). [Tree Tumbo]. **Plants**, dioecious, 0.5–1.8 m high, to 8.7 m wide, often forming colonies with the trunks sometimes fused; trunk subterranean, woody, short, thick, hollow, obconical (growing sideways after the growing tip dies), above ground part 30–90 cm Ø, apex with 2 leaf-bearing grooves and floriferous cushions, forming a raised rim (around the top of the hypocotyl); taproot greatly elongated. **Leaves** 3 pairs only, the second pair (from the seedling stage) persisting for the life, elongating from base, lying on the ground; lamina broadly strap-shaped, 2–4 m long, leathery, venation parallel. **Inflorescences** spikes (strobili), arranged in compound dichasial cymes (rarely solitary) arising annually from pits in the floriferous cushions; ♂ spikes (cones) salmon-colored, small, oblong conelike structures, bearing 40–70 axillary flowers in 4 rows, bracts connate, lowest pair or 2 pairs barren, flowers concealed by the bracts until the exsertion of the anthers; ♀ spikes (cones) blue-green, larger and more tapering, bearing 40–60 flowers in 4 rows, lowest 6–10 pairs of bracts increasing in size upwards, barren, lowest 2 or 3 pairs connate; nectar produced on both ♂ and ♀ strobili. **Flowers** unisexual or pseudo-bisexual; ♂ (pseudo-bisexual) with stamens 6, exserted, filaments connate into a short tube at base, anthers somewhat trilobed when mature, 3-celled, dehiscing by 3 slits from the summit, pollen ellipsoid, slightly coherent in irregular masses, ovule solitary, terminal, erect, orthotropous, imperfect, expanded at the tip into a stigmatiform disc; ♀ with ovule solitary, terminal, erect, orthotropous, perfect, integument single, produced into a straight micropylar tube, tube irregularly labiate or fimbriate. **Seeds** flattened, 3.6 x 2.5 cm, with a large papery wing (designed for wind dispersal), released when the ♀ cones disintegrate. **Distr:** SW Africa (Namib desert close to the ocean, in a narrow strip ~1,000 km long). **Ecol:** In coastal deserts. **Zones:** 9–11. **Info:** The plants can live for several hundred years (or even 2,000 years), with the leaves persistent but split and torn to ribbons due to wind and age. This plant is considered a living fossil, and the sole representative of this family. The plant uses the CAM pathway for carbon fixation in photosynthesis.

GNETACEAE

Genus 1, **species** 30. *Familial description same as generic.*

Gnetum

Species 30. **Vines**, evergreen, woody, less often erect **shrubs** or **trees**, dioecious or sometimes monoecious; stems with swollen nodes. **Stipules** absent. **Leaves** opposite, petiolate; lamina simple, margins entire, pinnately veined. **Reproductive organs** (flowers) unisexual, borne in whorled, spikelike cones ('spikes'), arranged in lax, dichasial cymes. **Cymes** terminal or lateral, sometimes arranged in dense, cauliflorous clusters on old stems. **Spikes** with many cupular to almost flat, annular, involucral collars, each formed by the fusion of a whorl of bracts. **Male spikes** with collars closely arranged and ± hiding axis (rarely laxly arranged), each collar with 20–80 flowers, often also with a whorl of sterile ♀ flowers, apical whorl with sterile ♀ flowers only; ♂ flowers with a cupular, succulent false perianth, usually ± obconical; stamens 2, filaments fused, exserted from false perianth; anthers opening by a common, apical slit. **Female spikes** solitary or several in a panicle, often cauliflorous; involucral collars widely separated, each with 4–12 flowers; ♀ flowers with a false perianth tightly enclosing ovule; ovule with 2 integuments, innermost integument elongated into a micropylar tube exserted from false perianth; outer integument with a fleshy, outer layer connate with false perianth and developing into a false seed coat, inner layer bony. **Seeds** drupelike, enclosed in a red, orange, or yellow, fleshy (rarely corky) false seed coat; ~ gametophyte tissue copious, succulent. **Distr:** Tropical regions. **Cult:** Propagate from seeds. Requires a rich, moist, well-drained soil in full or partial shade.

G. gnemon L., *Syst. Nat.*, ed. 12. 2: 637; *Mant. Pl.* 1: 125, 1767. [Gnemon Tree, Melinjo Nut]. **Shrubs** or **small trees**, dioecious, 10–15 m tall; bark grayish; crown narrow; trunk to 40 cm Ø; branches are noticeably swollen at base, becoming vinelike, green or yellowish-green. **Leaves** with petiole 0.5–1.8 cm; lamina drying yellowish-green, elliptic or oblong, 7.5–20 x 2.5–10 cm, base attenuate into petiole, apex acuminate or cuspidate, leathery or membranous, lateral veins inconspicuous, curved. **Male spikes** axillary, solitary, simple or once-branched, (1–)3–6 cm x 2.5–3 mm, involucral collars clearly separated, to 1 cm apart; each collar with 50–80 ♂ flowers and 5–15 globose sterile ♀ flowers, basal hairs inconspicuous. **Female spikes** similar to ♂, nodes each with 5–8 ♀ flowers. **Seeds** sessile or nearly so, yellow to orange-yellow or pink to red, ellipsoid, 1–2.5(–3.5) x 0.9(–1.5) cm, longitudinally ribbed, surface usually velvety, apex mucronate. **Distr:** China, India, Indonesia, Malaysia, Myanmar, Philippines, Thailand, Vietnam; Pacific Islands. **Ecol:** In tropical rainforests; 0–1,700 m. **Zones:** 10–12.

PART 2 ANGIOSPERMS

[FLOWERING SEED PLANTS]

Syn: ANGIOSPERMAE, ANTHOPHYTA, MAGNOLIOPHYTA.

Classes 2 (monocots and dicots), 45 orders, 457 families,species over 250,000. **Plants**, roots generally underground and serve to absorb water and nutrients; stems, of various types (e.g. bulbs, rhizomes, stolons, tubers). **Leaves**, simple or compound, alternate, opposite or whorled; lamina shapes from round to heart-shaped to oblong. **Flowers** (formed from 4 whorls of modified leaves – sepals, petals, stamens and pistils) with ♂ or ♀ reproductive organs; sepals (or calyx) are often small, usually green (but can be of any color), usually enclosing the flower before it opens; petals (or corolla) often brightly colored to attract pollinators, and may be simple to highly modified; stamens (♂ reproductive organs) produces microspores which turn into ♂ gametophytes, consisting of a stalk (filament) and a tip (anther) where the microspores are produced and turned into pollen; pistils (carpel or ♀ reproductive organs) consists of the ovary (where eggs are fertilized and developed into a seed), style (which carries the stigma) and stigma (receptacle where pollen is carried down the style) (Judd et al. 2008; Soltis et al. 2009; Stevens 2001 onwards). **Fruits** are the ripened (or mature) ovaries in which seeds develop and serve as protection, and as a means of dispersal for these seeds. **Distr:** Worldwide. **Ecol:** Aquatic or terrestrial. **Info:** The most distinctive feature being that they are flowering plants. They make up around 80% of all the extant plant species on Earth. Anthophyta is derived from 'anthe' (flower) and 'phyto' (plant), thus 'flowering plants'. Angiosperms are so named because the seeds are enclosed usually within a fruit.

Opposite: *Michelia doltsopa*.

4 BASAL ANGIOSPERMS

ANITA GRADE

ANITA grade refers to the evolutionary grade comprising the plant taxa Amborellales, Nymphaeales, Trimeniaceae and Austrobaileyales. These are thought to be the most basal angiosperm lineages, and so are of immense interest to systematic botanists.

AMBORELLALES

FAMILY 1, GENUS 1, SPECIES 1.

Shrubs or **small trees**; dioecious; vessel-less. **Leaves** with toothed leaf margins; vascular bundle sessile, branched towards apex. **Flowers** with pollen anaulcerate, 1 ovule per carpel, outer integument annular, embryo sac 9-nucleate (3 synergids), bipolar, antipodal cells die early, polar nuclei in chalazal region, stigma dry. Fruits drupelets; testa undistinguished; endosperm triploid.

AMBORELLACEAE

Genus 1. *Familial description same as generic.*

Amborella

Species 1. *Generic description same as specific.* **Cult:** Propagate from seeds. Requires a moist but well-drained soil in partial shade.

A. trichopoda Baill., *Hist. Pl.* (Baillon) 1: 328, 1869. **Shrubs** or **small trees**, 2–6 m tall, sprawling, evergreen, functionally unisexual and populations are dioecious, ♀-biased, and occur primarily in clumps; vessel-less. **Stipules** absent. **Leaves** alternate, 2-ranked; lamina simple, 8–10 cm long, margins serrated. **Inflorescences** terminal cymes or clusters, 2–30-flowered. **Flowers** small, greenish-yellow, 4–8 mm across, with both carpels and stamens, but only one sex develops fully and is fertile in the flowers of an individual plant (dioecious, functionally unisexual), the structures of the other sex remaining undeveloped; perianth of undifferentiated tepals arranged in a spiral (rather than in the whorls of more derived flowers); ♂ with 6–21 stamens and ♀ flowers 3–6 spirally arranged carpels and staminodes that mimic the fertile androecia in ♂ flowers. **Fruits** drupaceous red berries, obovoid, 5–8 mm long, with slightly compressed sides and apex, endocarp stony, 1-seeded. **Distr:** New Caledonia. **Ecol:** In moist, shaded understory of tropical montane rainforests; 200–1,000 m. Fl. Mar–May. Fr. Jun–Dec. **Zones:** 11–12. **Info:** This species represents a line of flowering plants that diverged about 130 million years ago from all the other extant species of flowering plants (Soltis et al. 2008), i.e. the oldest living angiosperm.

NYMPHAEALES

FAMILIES 3, GENERA 6, SPECIES 74.

Herbs, aquatic; lacks cambium and vessels; primary stem with ± scattered closed bundles. **Leaves** often large, floating; bracts absent. Flowers solitary, large, floral parts spiralate and indefinite; pollen boat-shaped. Fruits berrylike; **seeds** operculate, hilum outside operculum, endosperm scanty, perisperm copious, starchy. **Distr:** Worldwide; greatest diversity in N South America, C America, the Zambezian region of Africa and N Australia. **Ecol:** In quiet freshwater habitats throughout most of the world. **Info:** This order is considered to be a basal, or early diverging, group of angiosperms.

NYMPHAEACEAE

Genera 3, **species** 58. **Herbs**, perennial or rarely annual, aquatic, rhizomatous, tuberous; stem not floating; rhizomes erect or repent, branched or unbranched. **Leaves** arising from rhizome, simple, alternate, floating, immersed, or submersed, long petiolate but short petiolate on submersed vernal ones; lamina undivided, usually with a basal sinus, often peltate. **Inflorescence** axillary, long pedunculate, flowers solitary. **Flower** bisexual, hypogynous to epigynous, actinomorphic, entomophilous, mostly emergent; sepals 4–7, usually green, occasionally petaloid; petals numerous (rarely absent), distinct, usually showy, often transitional to stamens; stamens numerous; anthers introrse, dehiscing by longitudinal slits; connective sometimes appendaged; pistil 1, compound; carpels 5–many, partially or completely united, surrounding a sometimes projecting floral axis, ovary multilocular; placentation laminar; ovules numerous; styles absent or modified into abaxially projecting carpellary appendages, stigmas radiate on distal surface, often disclike. **Fruit** berrylike, many-seeded, irregularly dehiscent; seeds mostly arillate. **Distr:** Widespread in temperate and tropical regions.

Barclaya

Species 2–4. **Herbs**, aquatic, rhizomatous. **Leaves** with petiole inserted at notched end of lamina. **Flowers** perigynous to epigynous; distal portion of gynoecium cup-shaped, covered by the stigmatic tissue; sepals 4–5; stamens pendulous, attached to inner surface of corolla tube that extends above stigmatic cup. **Fruit** a berry; seeds spiny, without an aril. **Distr:** Indo-Malesia (La-ongsri et al. 2009). **Cult:** Propagate by division or from seeds. Requires a shallow small pond or similar.

B. longifolia Wall., *Trans. Linn. Soc. London* 15: 442, 1827. [Orchid Lily]. **Herbs**, aquatic, rhizomatous. **Leaves** linear to oblong, 12–28 × 2–6 cm, base cordate to subsagittate, margins crispate, apex obtuse, membranous, abaxially puberulous or glabrate; petioles 6–25 cm long, puberulous or glabrescent. **Inflorescence** with a flower solitary; peduncles 10–35 cm long; flower hardly expanding, emerging above water. **Flower** 1.5–2.5 cm long, purplish; sepals linear to elliptic, concave, 1.5–2.5 × 0.5–0.6 cm, obtuse, with midrib distinct distally and produced into a 3–5 cm long tail, adaxially purplish, abaxially greenish-purple; petals shorter than sepals, adaxially purplish, abaxially green, lobes broadly or narrowly oblong; inner stamens 2–2.5 mm long, curved inwardly; ovary superior, 6–12-loculed, style absent, stigmatic appendage curved over stigmas. **Fruit** a berry, 1–1.5 cm Ø; seeds ~1 mm Ø, brown, globose, spinulose, spines radiating in a tuft at micropylar region. **Distr:** E and S Thailand, Myanmar, E Malaysia, India. **Ecol:** In shallow waters along lakes, ponds and on stream margins. **Zones:** 10–12.

Nuphar

Syn: *Nymphozanthus.*

Species 10. **Herbs**, perennial; rhizomes repent, branched. **Leaves** dimorphic, either floating with thick leathery laminas and long petiole or submersed with thin, papery laminas and short petiole; lamina ovate to elliptic, venation primarily pinnate, base cordate, margins entire, not peltate. **Inflorescence** with a solitary flower. **Flower** hypogynous, emergent, perianth ± upright; sepals (4–)5(–7), yellow or orange, petaloid, oblong to obovate, persistent; petals numerous, yellow, small, stamenlike; stamens nearly as long as sepals, inserted at base of ovary, filament straplike, anthers yellow, connective unappendaged; carpels completely united, style absent, stigma sessile and radiate on flattened stigmatic disc, lacking marginal appendages. **Fruit** ovoid to urceolate, irregularly dehiscent; seeds smooth, not arillate. **Distr:** Widespread in northern subtropical to temperate regions. **Cult:** Propagate by division or from seeds. Best grown in a basket with rich soil mix placed in shallow still water, and in full sun.

N. lutea (L.) Smith, *Fl. Graec. Prodr.* 1: 361. 1809. (Syn: *Nymphaea lutea*). [Yellow Pond Lily]. **Herbs**, rhizomes stout, 3–8 cm Ø. **Leaves** with petiole ~50 cm long, glabrous; floating lamina elliptic, 15–30 × 10–22 cm, base cordate and basal lobes spreading, leathery, adaxially glabrous, abaxially glabrous or pubescent. **Inflorescence** with a solitary flower; peduncle to ~50 cm long, glabrous. **Flower** 4–5 cm Ø; sepals yellow, broadly ovate to orbicular, 2–3 cm; petals linear, 1–1.5 cm, apex truncate to rounded; anthers yellow, 4–7 mm long; stigmatic disc entire, 0.7–1.9 cm Ø, rays 5–25. **Fruit** ~2.5 cm Ø; seeds olive-green, ovoid, ~5 mm. **Distr:** Kazakhstan, Russia (Siberia); Africa, SW Asia, Europe. **Ecol:** Lakes and ponds. Fl. Jul–Aug. **Zones:** 4–11.

Above: *Amborella trichopoda, Nymphaea alba, Nymphaea nouchali, Nymphaea pubescens.*

Nymphaea

Species ~50. **Herbs**, perennial; rhizomes erect, ascending, or repent, branched or unbranched. **Leaves** mostly floating; lamina sometimes slightly peltate, base cordate to sagittate, margins entire to dentate, venation primarily palmate. **Inflorescence** with a solitary flower. **Flower** perigynous, floating or immersed, perianth spreading; sepals 4, greenish, not petaloid, inserted at base of ovary, sometimes persistent; petals 8 to many, large and showy, inserted on lateral surface of ovary, often grading into stamens; stamens shorter than sepals and petals, inserted on lateral surface of ovary, filament linear to ovate or obovate, anther connective appendaged or unappendaged; carpels partially or completely united, style absent or modified into abaxially projecting carpellary appendages, stigma sessile, radiate on cup-shaped stigmatic disc rimmed by carpellary appendages. **Fruit** irregularly dehiscent; seeds globose, ovoid, or ellipsoid, smooth or with pubescent longitudinal ridges, arillate. **Distr:** Widespread in temperate and tropical regions. **Cult:** Propagate by division. Best grown in a basket with rich soil mix placed in shallow still water, and in full sun. **Info:** Many species of this genus are cultivated for ornamental purposes worldwide.

N. alba L., *Sp. Pl.* 1: 510, 1753. [Water Lily]. **Herbs**, aquatic; rhizomes repent, sparsely branched, but not stoloniferous. **Leaves** suborbicular, 10–25 cm Ø, base deeply cordate and basal lobes subparallel or spreading, margins entire, papery, abaxially glabrous, scarcely peltate. **Flower** floating, (7–)10–20 cm Ø; calyx insertion on receptacle circular; sepals lanceolate, 3–5(–8) cm long, obscurely veined, caducous or decaying after anthesis; petals (12–)20–25(–33), white, ovate-oblong, 3–5.5(–8) cm, transition to stamens gradual; filament of inner stamens ± as wide as anthers, connective apically unappendaged; carpels completely united, walls between locules of ovary single, stigma rays (8–)14–20(–25), carpellary appendages triangular-tapered. **Fruits** semi-globose, 2.5–3 cm across; seeds ellipsoid, 2–3(–5) mm long, smooth. **Distr:** Kashmir, Russia (Caucasus); Africa, SW Asia, Europe. **Ecol:** Fl. Jun–Aug. **Zones:** 6–10.

N. nouchali N.L. Myanmarnn, *Fl. Indica*, 120, 1768. (Syn: *N. stellata*). [Blue Water Lily]. **Herbs**, aquatic; rhizomes erect, unbranched. **Leaves** elliptic-orbicular to orbicular, 7–15(–45) cm Ø, base cordate, basal lobes parallel to spreading, margin subentire to deeply crenate, papery, abaxially glabrous, peltate a few mm from base of sinus. **Flower** slightly emergent, 3–15 cm Ø; calyx insertion on receptacle circular; sepals lanceolate to oblong-lanceolate, 2.5–8 cm, slightly veined, persistent; petals 10–30, white tinged with purple, blue, or purple-red, linear-oblong to lanceolate, 4.5–5 cm, transition to stamens gradual; filament of inner stamens ± as wide as anther, connective apically appendaged; carpels only partially united, walls between locules of ovary double, stigma rays (8–)10–30; carpellary appendages triangular-tapered. **Fruits** globose, 1.5–4.5 cm Ø; seeds ellipsoid-globose, 0.5–1.3 mm, with longitudinal rows of hairs. **Distr:** Afghanistan, through India and Sri Lanka to Indo-China and Malesia, Taiwan; Australia. **Ecol:** In lowlands, particularly in dry regions, in ponds. Fl. Jul–Dec. **Zones:** 11–12. **Info:** The starchy rhizomes are often eaten roasted. Rhizomes and seeds used for medicinal purposes.

N. pubescens Willd., *Sp. Pl.* 2: 1154, 1799. (Syn: *N. lotus* var. *pubescens*). [White/Red Water Lily]. **Herbs**, perennial; rhizomes erect, producing slender stolons. **Leaves** ovate-elliptic to suborbicular, 15–26(–50) cm across, peltate more than 5 mm from base of sinus, base deeply ncordate, basal lobes subparallel, margins dentate and teeth acute to subspinose, papery, abaxially densely pubescent. **Flower** emergent, (2–)5–8(–15) cm Ø; calyx insertion on receptacle circular; sepals oblong, 5–8 cm long, conspicuously veined, caducous or decaying after anthesis; petals 12–14(–30), white, red, or pink, oblong, 5–9 cm long, transition to stamens abrupt; filament of inner stamens only slightly wider than anther, connective apically unappendaged; carpels completely united, walls between locules of ovary single, stigma rays 12–15(–30), carpellary appendages linear. **Fruit** ovoid to subglobose, 3.5–5 cm across; seeds ellipsoid to globose, 1–2 mm long, with longitudinal ridges. **Distr:** China, India and Sri Lanka to New Guinea. **Ecol:** Ponds in hills. Fl. Aug–Oct. **Zones:** 10–12.

Euryale

Species 1. **Herbs**, annual or short-lived perennials, aquatic; rhizomes erect, unbranched. **Leaves** submerged or floating; lamina broadly elliptic to orbicular, venation primarily radiate, juvenile ones with basal sinus, mature ones lacking sinus and only slightly indented basally, margins entire, centrally peltate. **Inflorescence** with a solitary flower. **Flower** epigynous, floating or often partially or entirely submerged, cleistogamous or perianth spreading; sepals 4, greenish, not petaloid, persistent; petals numerous, in ~5 series, showy, transition to stamens gradual; stamens shorter than sepals and petals, inserted at apex of ovary, filament linear, anther connective unappendaged; carpels 7–16, completely united, style absent, stigma sessile and radiate on cup-shaped stigmatic disc, lacking marginal appendages. **Fruit** irregularly dehiscent; seeds smooth, arillate. **Distr:** E Asia. **Cult:** Propagate from seeds. Best grown as an annual in a warm conservatory.

E. ferox Salisbury, *Ann. Bot.* 2: 74, 1805. [Prickly Water Lily, Fox Nut, Gorgon Plant]. **Herbs**, aquatic. **Leaves** *submerged* sagittate or elliptic, 4–10 cm long, base deeply cordate, not prickly; *floating* leaves to 1.3(–2.7) m Ø, base emarginate or sinuate, adaxially green, abaxially dark purple, subleathery, adaxially glabrous, abaxially sparsely pubescent, veins abaxially strongly ribbed, primary veins prickly on both surfaces; petiole prickly. **Inflorescence** with a solitary flower; peduncle stout, densely prickly. **Flower** to 5 cm Ø; sepals triangular-ovate, 1–1.5(–3) cm across, abaxially densely prickly, prickles retrorse; petals *outer* purple-violet fading to *inner* white, oblong-lanceolate, 1(–2.5) cm long; ovary 7–16-loculed, each locule with 6–8(+) ovules. **Fruit** dark purple, globose, 5–10 cm Ø, spongy, densely prickly; seeds black, 8–many, globose, 0.6–1 cm Ø; testa thick, rigid. **Distr:** Bangladesh, China, India, Japan, Kashmir, Korea, Russia (Far East). **Ecol:** In lakes, ponds. Fl. Jun–Aug. **Zones:** 8–11. **Info:** This species is cultivated for its starchy seeds.

Victoria

Species 2. **Herbs**, annual or short-lived perennial, aquatic; rhizome short, vertical. **Leaves** floating when mature, peltate, orbicular, usually 1.5–3 m across, margins upturned forming a continuous rim, abaxially with a network of prickly ribs; petioles to 8 m long, with stout prickles. **Flowers** epigynous, 30–50 cm across, nocturnal, white, changing to pinkish-red; sepals 4; petals 50–70; outer staminodia ~20; inner staminodia 25; stamens 150–200, anther dehiscence introrse; carpels 30–40, syncarpous, stigmatic cup surrounded by thermogenic carpellary appendages. **Fruits** large, prickly; seeds numerous, arillate. **Distr:** Tropical S America (Guyana to Argentina). **Cult:** Propagate from seeds. Best grown as annuals in large warm ponds, in full sun.

V. amazonica (Poepp.) J.C. Sowerby, *Ann. Mag. Nat. Hist.*, ser. 2 6: 310, 1850. (Syn: *Euryale amazonica, V. regia*). [Amazon Lily, Giant Water Lily, Water Platter]. **Herbs**, gigantic, aquatic, rooted. **Leaves** floating on the surface of the water; lamina orbicular, 1–3 m across, margins sharply upturned, forming a continuous rim 5–15 cm high, water collected on the leaf is drained through minute pores, red, adaxially red, prickled, corrugated along the veins; petiole prickly, attached at the center of the leaf, submerged, 7–8 m long, flexible. **Flowers** showy, cup-shaped, to 30–50 cm across, nocturnal, white, turning pinkish-red in the second day; petals 50–70; stamens numerous; carpels 30–40. **Fruits** large, prickly.
Distr: South America (Amazon Basin). **Ecol:** In shallow, still or slow-flowing water. **Zones:** 10–12. **Info:** This is the largest of all the water lilies. The flowers are pollinated by beetles, which are bribed with nectar to enter and feed within the flower. The leaves themselves are unique with the strong ribs enabling it to support weights up to 50 kg.

Top row: *Victoria amazonica, Illicium anisatum, Illicium floridanum, Illicium henryi, Kadsura japonica* (white form).
Bottom row: *Kadsura japonica* (red form), *Schisandra arisanensis, Chloranthus japonicus, Sarcandra glabra.*

AUSTROBAILEYALES

FAMILIES 3, GENERA 5, SPECIES 100.

Trees, **shrubs** or **lianas**; woody, vessels solitary. **Leaves** with petiole bundle(s) arcuate. Flowers bisexual, segments whorled or spiraled; perianth undifferentiated with an indeterminate number of parts, if differentiated then weakly so; filament laminar with 2 thecae widely separated; mature ♀ gametophytes contain 4 cells and 4 nuclei at maturity. **Fruits** berrylets; seeds with endosperm starchy. **Distr**: Worldwide. **Info**: This order is monophyletic and is supported by molecular and morphological characters (Qiu et al. 2000; Renner 1999; Soltis et al. 2000). Stamens and carpels similar to those of *Amborella*.

ILLICIACEAE

[Star Anise].
Genera 3, **species** 92. **Trees** or **shrubs**, evergreen, with unilacunar nodes, scattered spherical cells (ethereal) containing aromatic oils and branched sclereids, often with tannins; hairs simple. **Stipules** lacking. **Leaves** alternate and spiral (often appearing whorled), often clustered at tips of the shoots; lamina simple, entire, pinnate venation, punctate with pellucid dots. **Inflorescences** determinate, axillary or supra-axillary, 1–3-flowered. **Flowers** small, bisexual, radial, actinomorphic; receptacle convex to shortly conical; tepals usually numerous, distinct, imbricate, spirally arranged, outer usually sepal-like, inner ones petal-like, innermost ones sometimes minute and antherlike; stamens usually numerous (4–50), distinct, in one or more spiral series, apostemonous, filaments short and thick, poorly differentiated from anthers, connective tissue extended, anther dehiscence longitudinal; gynoecium apocarpous, carpels usually 5–21, distinct, unilocular carpels in a single whorl, ovaries superior with ± basal placentation, stigma extending down style on adaxial surface, style open, ovules anatropous, 1 per carpel; nectar produced at base of stamens. **Fruits** star-shaped aggregates of 1-seeded follicles (follicetum); follicles dehiscing elastically; seeds with smooth, hard coat. **Distr:** SE Asia, SE USA, Cuba, Hispaniola, Mexico. **Ecol:** Mainly in moist forests.

Illicium

Syn: *Badianifera*. [Star Anise].
Species 42. **Shrubs** or **small trees**. **Leaves** sometimes appearing whorled at ends of branches. **Inflorescences** with flowers erect to drooping; peduncle smooth to rugulose. **Flowers** with perianth white, yellow, yellow-green, pink, or red to maroon, often glandular; outer tepals often bractlike and reduced, sometimes ciliolate, inner tepals often much larger and ligulate, sometimes only slightly larger and broadly obtuse, innermost often reduced, sometimes transitional to stamens; pistils orbicular to deltoid, styles narrow and acute, often recurved. **Fruits** usually star-shaped follicles, dark brown, woody to leathery; seeds glossy, dark brown to tawny or golden. **Distr:** N America, West Indies, C America, NE South America. **Cult:** Propagate from half-hardened cuttings or by air-layering. Requires a moist but well-drained acidic soil in a shaded position.

I. anisatum L., *Syst. Nat.*, ed. 10. 2: 1050, 1759. (Syn: *Badianifera anisata*, etc.). [Japanese Sacred Anise, Japanese Star Anise]. **Shrubs** or **small trees**, to 8 m tall. **Leaves** irregularly alternate, or in lax clusters of 3–5; lamina narrowly obovate to elliptic, (4–)5–12 × (1.5–)2–4(–4.7) cm, base attenuate, margins undulate or narrowly recurved, apex obtuse to short-acuminate, coriaceous, secondary nerves usually 4–6 per side, ascending; petiole narrowly winged, 0.7–2 cm long. **Inflorescences** with flowers axillary, often crowded towards end of branchlets. **Flowers** with perianth-segments 17–24, membranous; outermost segments 2–6, suborbicular to oblong, 0.4–0.6(1.2) × 0.3–0.6 cm; largest segments (often 14–16) linear-oblong to oblong-obovate, 1.1–2.3 × 0.2–0.45 cm; innermost segments few, reduced, often lanceolate, 0.7–1 cm × 1.5–2 mm; stamens 17–25, biseriate, oblong, 2.5–3.5 mm long, filaments carnose, ligulate, 1.2–2 mm long; carpels usually 8(9–10), 3.5–5.5 mm long, ovary flattened-ellipsoid, 1.2–2.3 mm long, style subulate, 2.2–4 mm long. **Fruits** with follicles 6–8, 1.2–1.5 × 0.7–0.9 cm; seeds brown, 6.5 mm long. **Distr:** Japan, islands near S Korea. **Ecol:** 100–1,000 m. **Zones:** 7–11. **Info:** This species is highly toxic and the fruits are very similar to the non-toxic *I. verum*.

I. floridanum J. Ellis, *Philos. Trans.* 60: 524, 1770. (Syn: *I. mexicanum*, etc.). [Florida Anise, Poison Bay, Purple Anise, Stink Bush, Star Anise]. **Shrubs** or **small trees**, evergreen, to 3 m tall. **Leaves** elliptic to lanceolate, 5–21 × 1.5–6 cm, base cuneate, apex acute to acuminate, dark olive-green, strongly aromatic, leathery, smooth, shiny; petiole 0.6–2.6 cm long. **Inflorescences** with peduncles 1–11 cm long; bracteoles 3–6. **Flowers** 2.5–5 cm Ø, smelling of fish; tepals 20–33, red-maroon, rarely white or pink, inner tepals ligulate; stamens 25–50; pistils 11–21. **Fruits** aggregates of follicles (usually 10–15), collectively 2.5–4 cm Ø; seeds pale brown, dispersed by explosive dehiscence of the follicles. **Distr:** NE Mexico. **Ecol:** Along streams, in marshy areas, moist woods; 0–500 m. Fl. Apr–Jul. **Zones:** 8–11.

I. henryi Diels, *Bot. Jahrb. Syst.* 29: 323, 1900. (Syn: *I. silvestrii*, etc.). [Chinese Anise Tree]. **Shrubs** or **trees**, to 8(–12) m tall; perules small, inconspicuous.

Leaves in clusters of 2–5 at distal nodes; lamina oblanceolate, long-lanceolate or obovate-elliptic, 6–18 × 1.2–5(–6) cm, base cuneate, apex long-acuminate, leathery, midvein adaxially impressed, secondary veins 5–8 per side; petiole 0.7–2 cm long. **Inflorescences** axillary or subterminal; peduncle 1.5–5 cm long. **Flowers** with tepals 10–15, pink to dark red, oblong-elliptic to broadly elliptic, 0.7–1 × 0.4–0.85 cm; stamens 11–14, 2.2–3.5 mm long, filaments 1.2–2.3 mm long, anthers 0.9–1.2 mm long; carpels 7–9(–12), 3–5 mm long, ovary 1.5–2 mm long, style 2–3.3 mm long. **Fruits** with 7–9 follicles; follicles 1.2–2 × 0.5–0.8 cm; peduncle 1.5–5.5 cm long; seeds 6.5–7.5 mm long. **Distr:** China. **Ecol:** In forests, hills, thickets; 300–2,500 m. Fl. Apr–Jun. Fr. Aug–Oct. **Zones:** 8–11. **Info:** The fruits are poisonous.

I. verum J.D.Hooker, *Bot. Mag.* 114: t. 7005, 1888. (Syn: *I. san-ki*). [Star Anise]. **Trees**, to 15 m tall; perulae ovate, 2–3 mm. **Leaves** in clusters of 3–6 at distal nodes; lamina obovate-elliptic, oblanceolate or elliptic, 5–15 × 2–5 cm, base attenuate to cuneate, apex cuspidate to shortly acuminate, leathery to thickly leathery, secondary veins 5–8 per side; petiole 0.8–2 cm long. **Inflorescences** axillary or subterminal; peduncle 1.5–4 cm. **Flowers** with tepals 7–12, pink to dark red, broadly elliptic to broadly ovate, 0.9–1.2 × 0.8–1.2 cm; stamens (11–)13 or 14(–20), 1.8–3.5 mm long, filaments 0.5–1.6 mm long, connectives truncate, anthers 1–1.5 mm long; carpels usually 7–9(–11), 2.5–4.5 mm long, ovary 1.2–2 mm long, style longer than ovary. **Fruits** with ~8 follicles; follicles 1.4–2 × 0.7–1.2 cm; peduncle 2–5.5 cm long; seeds 0.7–1 cm long. **Distr:** China. **Ecol:** In forests; 200–1,600 m. Fl. Mar–May, Aug–Oct. Fr. Mar–Apr, Sep–Oct. **Zones:** 8–11. **Info:** This species is cultivated for perfumery, medicinal uses and also as a spice.

Kadsura

Species 16. **Vines**, woody, glabrous (except *K. induta*), dioecious or monoecious. **Leaves** elliptic, ovate, or obovate, base cuneate, broadly cuneate, truncate, or subcordate, margins denticulate to entire, apex acute to acuminate, papery to leathery. **Inflorescences** axillary or occasionally cauliflorous, generally solitary, sometimes in 2–4-flowered clusters. **Flowers** unisexual; tepals 7–24; ♂ with stamens 13–80, ± distinct, basally connate or sometimes tightly aggregated into a subglobose mass; ♀ with carpels 17–300, distinct; stigmatic crest forming subulate or laterally flattened 'pseudostyle' or modified as subpeltate or irregular 'pseudostigma', ovary with ovules 1–5(–11), pendulous or ventrally attached. **Fruits** aggregates of apocarps; receptacle ellipsoid or clavate; apocarps ripening red or yellow, subglobose, obovoid, or elongate-obovoid; seeds 1–5(–11) per apocarp, smooth. **Distr:** Mainly E and SE Asia.**Cult:** Propagate from seeds or semi-hardwood cuttings. Requires sheltered position and support for climbing, full or partial shade.

K. japonica (L.) Dunal, *Monogr. Anonac.* 57, 1817. (Syn: *K. matsudae*, *Uvaria japonica*). [Evergreen Magnolia Vine]. **Vines**, glabrous. **Leaves** mostly elliptic, 6–10.5(–12.5) × 2.5–5(–6.5) cm, base ± cuneate, margins subentire, denticulate, lobed-serrulate, serrulate, or serrate, apex ± acuminate, papery to subleathery, secondary veins 5–8 per side; petiole 0.6–2.3 cm long. **Inflorescences** ♂ with peduncles 1.1–2.6 cm long; ♀ with peduncles 1.3–3.9 cm long. **Flowers** with tepals 8–12, yellow, largest 0.6–0.9(–1.2) × 0.4–0.75 cm; ♂ with stamens 28–50, staminodes absent; ♀ with carpels 30–48. **Fruits** apocarps, red to purple, 0.55–1 × 0.55–0.75 cm; peduncles (2–)2.6–4(–6.1) cm long; seeds 1–3 per apocarp, reniform, 3–4 × 4.5–5.5 mm. **Distr:** Japan, Korea, Taiwan. **Ecol:** In forests, often near rivers; ~0–2,000 m. Fl. Apr–Oct. Fr. Oct–Dec. **Zones:** 7–10. **Info:** Often cultivated for ornamental and medicinal purposes, and the edible fruits.

Schisandra

Species 22. **Vines**, woody, glabrous or with trichomes, dioecious or monoecious. **Leaves** elliptic, ovate, or obovate, base cuneate to broadly cuneate, often decurrent on petiole, margins denticulate or entire, apex acute to acuminate, papery to leathery. **Inflorescences** axillary to bracts or leaves, usually solitary (sometimes paired or clustered in *S. propinqua* and *S. plena*). **Flowers** unisexual; tepals 5–20; ♂ with stamens 4–60, ± distinct or partially to fully connate to form a fleshy synandrium, pollen 3- or 6-colpate, distally syncolpate; ♀ with carpels 12–120, distinct, stigmatic crest forming a subulate 'pseudostyle', lacking 'pseudostigma', ovary with ovules 2(or 3), ventrally attached. **Fruits** aggregates of apocarps; apocarps ripening to red or rarely blackish, ellipsoid to obovoid; receptacle elongate; seeds (1 or)2(or 3), smooth to rugulose. **Distr:** E and SE Asia; N America (1 sp.). **Cult:** Propagate from seeds or half-hardened cuttings, suckers or by air-layering. Requires moist but well-drained soil in partial sun. **Info:** Most species in this genus are used for medicinal purposes.

S. arisanensis Hayata, *Icon. Pl. Formosan.* 5: 1, 1915. [Chinese Magnolia Vine]. **Vines**, glabrous throughout; young branches lacking wings; leaf-bearing branches elongated. **Leaves** elliptic to ovate, 5–10.5(–12.5) × 2.5–7 cm, base cuneate to broadly cuneate, often ± decurrent on petiole, margins denticulate to serrulate, apex acuminate to long-acuminate, papery to subleathery, not glaucous, secondary veins 4–6 per side; petiole 1–3.4 cm long. **Inflorescences** with flowers solitary, axillary to fugacious bracts at base of young shoots or axillary to leaves; ♂ with peduncle 1.8–4.5 cm long; ♀ with peduncle 3.5–4 cm long. **Flowers** with tepals 5–8(–13), pale yellow, yellow, orange, reddish-orange, or red, largest 0.4–1.2 × 0.35–0.85 cm; ♂ with stamens 8–30, ± distinct, pollen 6-colpate; ♀ with carpels 22–56, pseudostyle short, narrow. **Fruits** apocarps, red, 0.5–1.1 × 0.35–0.75 cm; peduncles 4–8 cm long; torus 3.5–15.5 cm; seed with testa usually ± smooth to rugulose. **Distr:** China, Taiwan. **Ecol:** In forests, shrublands; 200–2,300 m. Fl. Apr–Jun. Fr. Jun–Sep. **Zones:** 7–11. **Info:** Cultivated for medicinal purposes and the edible fruits.

CHLORANTHALES

FAMILY 1, GENERA 4, SPECIES 75. *Ordinal description same as familial.*

CHLORANTHACEAE

Genera 5, species ~70. **Shrubs**, **small trees**, or **perennial herbs**. **Stipules** small. **Leaves** simple, margins serrate or crenate, pinnately veined; petiole usually connate at base. **Inflorescences** spikes, heads or panicles; ♂ flowers many; ♀ flowers few. **Flowers** small, bisexual or unisexual; perianth present in ♀ flowers or absent; *bisexual flowers* with stamens 1 or 3, inserted on 1 side of ovary, filaments inconspicuous, connectives conspicuous, if stamens 3, connectives connate at base or free, anthers 1- or 2-loculed, longitudinally dehiscent, pistil 1, 1-carpelled, ovary inferior, 1-loculed, ovule 1, pendulous, orthotropous, style short or absent; *unisexual flowers*: ♂ with stamen 1; ♀ with perianth 3-toothed, calyxlike, connate to ovary. **Fruits** drupes, ovoid or globose, exocarp ± fleshy, endocarp hard. **Distr:** Tropical and subtropical regions of the world.

Chloranthus

Syn: *Nigrina*, *Tricercandra*.

Species 17. **Subshrubs** or **perennial herbs**. **Stipules** minute. **Leaves** opposite or whorled; lamina with margins serrated; petiole connected by a transverse ridge on stem. **Inflorescences** spikes or panicles, terminal or axillary. **Flowers** small, bisexual; perianth absent; stamens usually 3, rarely 1, on 1 side of apical part of ovary, basal part of connective confluent, or free and connected or overlapped at base, ovoid or lanceolate, sometimes elongated to linear, anthers 1- or 2-loculed, if stamens 3, central anther 2-loculed or occasionally absent, lateral anthers 1-loculed, if stamen 1, anther 2-loculed; ovary 1-loculed, ovule 1, pendulous, orthotropous, style usually absent, stigma truncate or parted. **Fruits** drupes, globose, obovoid, or pyriform. **Distr:** Temperate to tropical Asia. **Cult:** Propagate from seeds or rooted cuttings. Requires moist but well-drained fertile soil in full sun.

C. japonicus Siebold, *Nov. Actorum Acad. Caes. Leop.-Carol. Nat. Cur.* 14: 681, 1829. (Syn: *Tricercandra quadrifolia*, etc.). **Herbs**, perennial, 20–49 cm high; rhizomes many-noded, procumbent, aromatic; stems erect, fascicled, unbranched; nodes with scalelike leaves, triangular or ovate, 4–5 mm long, membranous. **Leaves** opposite, usually 4 clustered, apical, pseudo-whorled; lamina broadly elliptic or obovate, 8–14 × 5–8 cm, base broadly cuneate, margins sharply serrate, apex acute, papery, glandular-mucronate, adaxially shiny, glabrous, lateral veins 6–8 pairs; petiole 0.8–1.8 cm long. **Inflorescences** solitary spikes, terminal, 3–5 cm long; bracts triangular to semi-orbicular. **Flowers** white; stamens 3, connectives elongated, linear, ~5 mm, horizontally spreading or curved upward, connate at base, on outside of apical part of ovary, central connective without anther, lateral connectives with a 1-loculed anther each; ovary ovoid, style absent, stigma truncate. **Fruits** drupes, green, subglobose or obovoid, 2.5–3 mm across. **Distr:** Japan, Korea, Russia (Far East). **Ecol:** In forests, streamsides; 100–2,300 m. Fl. Apr–May. Fr. May–Jul. **Zones:** 4–10.

Sarcandra

Species 3. **Subshrubs**, glabrous; wood without vessels. **Stipules** small. **Leaves** opposite, usually in many pairs; lamina elliptic, ovate-elliptic, or elliptic-lanceolate, margins serrate except basally, glandular-mucronate on marginal teeth; petiole short, connate at base, forming a short sheath. **Inflorescences** spikes, terminal, usually branched, ± paniculate; bract 1, triangular, persistent; flowers sessile. **Flowers** bisexual; perianth absent; stamen 1, fleshy, baculate to depressed, anther 2(or 3)-loculed, thecae lateral to introrse; ovary globose or ovoid, style absent, stigma subcapitate or minutely spotted. **Fruits** drupes, globose or ovoid. **Distr:** SE Asia to India. **Cult:** Propagate from seeds or cuttings. Requires a moist but well-drained soil in full sun.

S. glabra (Thunberg) Nakai, *Fl. Sylv. Kor.* 18: 17, 1930. (Syn: *Ardisia glabra*, etc.). **Subshrubs**, evergreen, 0.5–1.5 l tall; stems cylindric, erect, glabrous, nodes swollen. **Stipules** subulate. **Leaves** elliptic or ovate to ovate-lanceolate, or broadly elliptic to oblong, 6–20 × 2–8 cm, base acute, cuneate, or broadly cuneate, margins serrate except basally, glandular-mucronate on marginal teeth, apex acute to acuminate, leathery or papery, glabrous, lateral veins 5–7; petiole 0.5–2 cm long. **Inflorescences** spikes, 1.5–4 cm long, usually branched, terminal; bracts triangular or ovate. **Flowers** yellowish-green; stamen 1, fleshy, baculate to terete or ovoid, anther 2-loculed; ovary globose or ovoid, style absent, stigma subcapitate or minutely spotted. **Fruits** drupes, shiny red or yellowish-red, globose or ovoid, ~4 mm Ø. **Distr:** Cambodia, China, India, Japan (including Ryukyus), Korea, Laos, Malaysia, Philippines, Sri Lanka, Taiwan, N Thailand, Vietnam. **Ecol:** In forests, thickets, grasslands, streamsides; ~0–2,000 m. **Zones:** 8–12.

5 MAGNOLIID COMPLEX

MAGNOLIALES

FAMILIES 6, GENERA 154, SPECIES ~2,922.

Trees, shrubs or climbers; vessels in multiples; secondary phloem stratified; pith septate. **Leaves** 2-ranked, conduplicate, midrib with adaxial plate of vascular tissue and P-type sieve elements that contain one to several protein crystals plus starch grains (or sometimes only starch grains). **Flowers** with perianth whorled, outer whorl usually 3-merous. **Seeds** usually have fleshy arils. **Info:** Monophyly of Magnoliales are supported by molecular data (Qiu, Chase, Les & Parks 1993; Mathews & Donoghue 2000; Qiu et al. 2000) and morphological data (Doyle & Endress 2000).

MYRISTICACEAE

Genera 20, species ~500. **Trees**, evergreen, monoecious or dioecious. **Leaves** alternate, arranged spirally or distichously, exstipulate; lamina simple, entire, veins pinnate, often pellucid-punctate. **Inflorescences** panicles, racemes, heads or cymes, axillary; bracts caducous; bracteoles basal. **Flowers** small, unisexual; perianth gamophyllous; lobes (2 or) 3–5, valvate; stamens 2–40, filaments connate into a staminal column or peltate staminal disc, anthers bilocular, extrorse, longitudinally dehiscent, adnate to column abaxially, or free; ovary superior, sessile, 1-locular, anatropous ovule 1, inserted near base, style short or lacking, stigma bilobed or lobes connate into a disc. **Fruits** with pericarp leathery-fleshy, or woody, dehiscent into 2 valves; seed 1, large, aril fleshy, entire or shallowly or deeply lacerate, testa of 3 or 4 layers, outer layer crustose, middle layer often woody and rather thick, inner layer membranous. **Distr:** China, tropical Asia to Pacific Islands; Africa; tropical America.

Horsfieldia

Syn: *Endocomia.*

Species ~100. **Trees**, evergreen, monoecious or dioecious. **Leaves** dispersed or distichous, papery or thinly leathery, often glabrous, abaxially not glaucous, tertiary veins reticulate, often obscure. **Inflorescences** ♂ often compound paniculate, flowers often aggregated into clusters, bracts caducous, bracteole absent; ♀ flowers pedicellate, sometimes shortly so, small. **Flowers** with perianth lobes (2 or)3–5; anthers 4–30, connate abaxially into globose or clavate sessile or stipitate synandrium sometimes concave at apex; ovary ovoid, glabrous or hairy, style absent. **Fruits** pericarps, often rather thick, glabrous or glabrescent; arils colorful, entire, rarely lacerate at apex; testa thinly woody. **Distr:** China, S Asia, from India to the Philippines, Papua New Guinea. **Cult:** Propagate from seeds or by air-layering. Requires any well-drained soil in full sun.

H. irya (Gaertn.) Warb., *Monog. Myrist.* 317, 1897. (Syn: *Myristica irya*). [Irya]. **Trees**, slender, 10–20(–30) m tall. **Leaves** elliptic-oblong to lanceolate, 8–26 x 2–7.5 cm, base rounded to attenuate, apex acute to acuminate, glabrous, veins 10–12 pairs; petiole 0.7–1.3 cm long; leaf-bud ~1(–2) cm long, rusty-brown-pubescent. **Inflorescences** ♂ 3–10 in clusters, densely tomentose, 5.5–10.5 x 2.5–7 cm, branched 3–4x; ♀ usually solitary or, more rarely, few together, densely tomentose, 2–6 cm long, branched 2–3x, pedicels 1–4 mm long. **Flowers** pale yellow, fragrant; ♂ with perianth subglobose or ellipsoid, 1–1.25 x 1.25–1.5 mm, anthers 6–9; ♂ with perianth ellipsoid or obovoid, 1.5–2 x 1.25–2 mm, split at anthesis into calyx lobes, ovary obovoid, glabrous. **Fruits** globose, 1.5–2 cm Ø, glabrous, sometimes finely granular, orange-brown to blackish; pericarp to 2 mm thick; pedicel 0.5–1 cm long. **Distr:** Andaman and Nicobar Islands, Indonesia, Myanmar, Sri Lanka, Thailand, Malay Peninsula to SE Asia, New Guinea, Solomon and Caroline Islands. **Ecol:** In primary and secondary forests of wet regions; <600 m. **Zones:** 11–12.

H. iryaghedhi (Gaertn.) Warb., *Monog. Myrist.* 332, 1897. (Syn: *Myristica iryaghedhi, M. horsfieldii*). [Ruk]. **Trees**, 5–25 m tall; branches long, drooping; twigs rusty-tomentose, later glabrescent. **Leaves** 10–28 x 4–12 cm, base rounded to attenuate, apex acute-acuminate, adaxially glabrous, abaxially glabrescent, lateral veins 9–15 pairs; petiole 1.5–2(–2.5) cm long; leaf bud 1–1.5 cm long, densely pubescent. **Inflorescences** ♂ clusters, ~70–85(+)-flowered, these being further grouped in 10–20 subglobose heads ~0.5–1 cm ♀ distributed along branches 6–15 x 4–10 cm, simple or more usually 2-branched, densely pubescent, flowers sessile or very rarely pedicelled;♀ much smaller, 1.5–4 x 1.5–2 cm, densely pubescent, pedicel very short or absent; common peduncle to 2 cm long. **Flowers** ♂ with perianth obovoid-oblong to obconic, ~2–2.5 x 1–1.5 mm, base tapering, glabrous, split at anthesis into 3(or 4) calyx lobes, anthers 3–5, erect, sessile; ♀ with perianth, glabrous, split at anthesis into (2 or) 3 calyx lobes, ovary broadly ellipsoid, short-pubescent. **Fruits** ellipsoid to oblong-ellipsoid,2.5–4 x 1.5–2.5 cm, rounded at both ends, rusty-yellow pubescent, dark brown. **Distr:** Sri Lanka. **Ecol:** In lowland rainforests; <500 m. **Zones:** 11–12.

Myristica

Species ~150. **Trees**, evergreen, sometimes with aerial roots at base. **Leaves** papery, abaxially white-papillose or rusty-pubescent, midvein often sunken adaxially, lateral veins connected at margin. **Inflorescences** densely racemose or subumbellate, axillary, often dichotomous or trichotomous; bracts caducous; bracteole basal, rarely deciduous; flowers pedicellate. **Flowers** with perianth (2 or)trilobed, urceolate or bell-shaped, rarely tubular; filaments connate into staminal column, anthers slender, 7–30, connate abaxially, synandrium often longer than basal stalk; ovary glabrous or hairy, style almost absent, stigma bifid. **Fruits** with pericarp thickly leathery, glabrous or pubescent; aril red, laciniate to base. **Distr:** China, S Asia from E India to the Philippines, New Guinea, Pacific Islands (W Polynesia). **Cult:** Propagate from seeds or by air-layering. Requires any well-drained soil in full sun.

M. ceylanica var. *cagayanensis* (Merr.) J. Sinclair, *Gard. Bull. Sing.* 23: 442, 1968. (Syn: *M. cagayanensis*). [Ceylon Nutmeg, Malaboda]. **Trees**, dioecious, evergreen, medium-sized, to 20 m tall; trunk 1 m Ø; branchlets dark reddish- or purplish-brown, much stouter than those of the typical variety; young leaves and inflorescences brown-pubescent. **Leaves** oblong, 15–25 x 6–8 cm, base obtuse, margins entire, apex short-acute, coriaceous, glabrous and concolorous on both surfaces, or slightly paler abaxially, lateral veins 13–18 pairs; petiole 2–3 cm long. **Inflorescences** axillary. **Flowers** campanulate, trilobed; stamens elongate, whorled. **Fruits** ellipsoid, 4–5 cm long, brown appressed-tomentose; seed 1, 3 cm long, covered with reddish, fleshy laciniate arils. **Distr:** Philippines, Taiwan. **Ecol:** 50–300 m. **Zones:** 10–12.

M. dactyloides Gaertn., *Fruct.* 1: 195, 1788. (Syn: *M. diospyrifolia*, etc.). [Perimavara]. **Trees**, to 10–27(–50) m tall; trunk buttressed to 2 m high, 0.5–1 m deep, stilt roots present with age. **Leaves** oblong- or elliptic-lanceolate or oblanceolate, (9–)13–22(–30) x (4–)6–11 cm, base acute or rounded to subcordate, apex acute, coriaceous, glabrous, veins 9–16 pairs; petiole 1.5–3 cm long. **Inflorescences** ♂ 1–1.5 cm long, woody, tomentose, simple or 3–4-branched; ♂ densely umbellate, light green inside, flowers pedicellate, numerous, bracteole orbicular-ovate, pedicel 3–4 mm long, rusty-tomentose; ♀ with flowers fewer per fascicle, sessile. **Flowers** ♂ with perianth divided, lobes broadly ovate, obtuse, staminal column ~4 mm long, anthers 10, ~2 mm long; ♀ with ovary oblong-globose, adpressed-pubescent. **Fruits** solitary or paired, oblong or ovoid, ~6 x 3 cm, densely orange- or rusty-tomentose, pericarp hard, ~3 mm thick; pedicel 1 cm long; seeds 2–3 x 1.25–1.8 cm, brown to black, very glossy, smooth, aril bright yellow, orange or red, fleshy. **Distr:** Sri Lanka, India. **Ecol:** In moist forests, streamsides; <1,500 m. **Zones:** 11–12.

M. fragrans Houttuyn, *Nat. Hist.* 2(3): 333, 1774. [Mace, Nutmeg]. **Trees**, small, to 10 m tall; branches slender, minutely pubescent, early glabrescent. **Leaves**

MIW

Top row: *Horsfieldia irya*, *Horsfieldia iryaghedhi* (fruits), *Myristica ceylanica* var. *cagayanensis*.
Bottom row: *Myristica dactyloides*, *Myristica fragrans* (flowers), *Myristica fragrans* (fruits).

elliptic or elliptic-lanceolate, 4–8 cm long, base broadly cuneate or nearly rounded, apex shortly acuminate, nearly leathery, glabrous, lateral veins 6–10 pairs; petiole 0.6–1.2 cm long. **Inflorescences** ♂ 2.5–5 cm, glabrous, shortly peduncled, simple or forked, flowers 4–8+, branches slender, pedicels 1–1.5 cm long, bracteole caducous; ♀ 1- or few-flowered, pedicels 0.8–1.2 cm long, bracteole inserted on base of perianth, leaving ringlike scar after abscission. **Flowers** ♂ with perianth trigonous-ovoid, 5–7 mm long, lobes 3(or 4), outside minutely tomentose, anthers 9–12, synandrium ~5 mm long, column ~2 mm long, apex sterile apex; ♀ with perianth ~ 6 × 4 mm, ovary ellipsoid, densely rusty-pubescent, style extremely short, stigmas 2, minute. **Fruits** 1 or 2, orange or yellow, pyriform or subglobose, 3.5–5 cm Ø; aril red, irregularly deeply lacerate; seeds ellipsoid, 2–3 × ~2 cm. **Distr:** Indonesia (Moluccas). **Ecol:** In lowland forests, usually on rich volcanic soils; to ~500 m. Fl. Feb. Fr. Apr. **Zones:** 11–12.

MAGNOLIACEAE

Subfamilies 2, **genera** 2, **subgenera** 3, **species** 227. **Trees** or **shrubs**, deciduous or evergreen, aromatic. **Stipules** early or tardily deciduous, often ochreate, leaving persistent annular scar around node. **Leaves** alternate, petiolate; lamina simple, unlobed or evenly 2–10-lobed, margins entire, pinnately veined. **Inflorescences** terminal, flowers usually solitary, sometimes paired, pedunculate; bracts spathaceous bracts 1 or 2. **Flowers** with perianth hypogynous, segments imbricate; tepals deciduous, 6–18, in 3 or more whorls of 3, ± similar or outer tepals sepaloid, inner tepals petaloid; stamens numerous, hypogynous, free, spirally arranged, filaments short, anthers introrse, latrorse, or extrorse, longitudinally dehiscent; pistils numerous, superior, spirally arranged on elongate receptacle (torus), stalked or sessile, free or ± concrescent, 1-locular, placentation marginal, placenta 1, ovules 1–2, style 1, short and recurved or large and winglike, stigma 1, terminal or terminal-decurrent or recurved. **Fruits** syncarps, conelike, consisting of aggregates of coalescent, woody follicles (or follicetums) or apocarps consisting of aggregates of indehiscent samaras (samaracetums); seeds 1–2 per pistil, arillate or without aril. **Distr:** Mostly in Asia, the Pacific Islands, and the Western Hemisphere. **Info:** Classification adapted below according to Figlar & Nooteboom (2004).

Subfam. MAGNOLIOIDEAE

Genus 1. *Subfamilial description same as generic.*

Magnolia

Syn: *Manglieta*, *Michelia*, *Talauma*, etc.
Species 218. **Trees or shrubs**, deciduous or evergreen; pith homogeneous or diaphragmed. **Stipules** adnate to or free from the petiole, early deciduous. **Leaves** distinctly alternate or sometimes crowded in terminal whorl-like clusters; lamina with base deeply cordate or auriculate or cuneate to abruptly narrowed or rounded, margins entire, apex obtuse or acute to acuminate, abaxially chalky-white or green to glaucous, pubescent or glabrous. **Flowers** bisexual, protogynous, appearing with or before leaves; tepals 9–21, 3–5-merous, petaloid, usually spreading, creamy white, rarely greenish, yellow, or orange-yellow, outermost tepals sepaloid, sometimes strongly reflexed, greenish; stamens on elongate torus, early deciduous, filaments white or purple, very short, anthers introrse or latrorse. **Fruits** follicles, persistent, coalescent, forming conelike aggregates, abaxially dehiscent; seeds with red, pink, or orange oily aril, extruded from follicles and suspended by funiculi. **Distr:** Temperate and tropical regions, Western Hemisphere, Asia (Himalayas, China, Japan, Taiwan, Malaysia, and Indonesia). **Cult:** Propagate from seeds or cuttings. Requires a humus-rich, well-drained soil in full to partial sun.

Subgen. Magnolia

Sections 8, **subsections** 7, **species** 146. **Branches** normally produced by syllepsis (except in subsect. *Oyama*). **Leaves** with vernation (prefoliation) conduplicate. **Flowers** terminal; stamens caducous during male phase (except in subsect. *Oyama*), anther dehiscence introrse, pollen large, usually >50 µm Ø; gynoecium sessile. **Fruits** ± ovoid or ellipsoid with fused carpels until dehiscence.

Sect. Magnolia

Species 15. **Branches** normally produced by syllepsis (except in subsect. *Oyama*). **Leaves** with vernation (prefoliation) conduplicate. **Flowers** terminal; stamens caducous during male phase (except in subsect. *Oyama*), anthers dehiscent introrsely, pollen large, usually >50 µm Ø; gynoecium sessile.

M. grandiflora L., *Syst. Nat.* ed. 10. 2: 1082, 1759. (Syn: *M. foetida*, etc.). [Southern Magnolia, Bull Bay]. **Trees**, evergreen, to 37 m tall; trunk ~2 m Ø; pith diaphragmed. **Stipules** 2, free, 4.5–13 × 1.5–3.5 cm, densely brown-silky, sometimes deeply notched. **Leaves** distinctly alternate, not in terminal whorl-like clusters; lamina narrowly to broadly elliptic or oblanceolate, (7.5–)13–20(–26) × (4.5–)6–10(–12.5) cm, base narrowly cuneate, apex abruptly tapered, acute to short-acuminate, rarely obtuse, thick-leathery, adaxially bright green, lustrous, glabrous, abaxially glabrous to densely red-brown felted. **Flowers** strongly lemony fragrant, 15–30(–45) cm across; spathaceous bracts 2, leathery, outer bract abaxially brown to grayish pilose, deeply notched, inner bract adaxially densely brown to grayish pilose, shallowly notched, larger; tepals creamy white; stamens (179–) 213–383(–405), 1.6–2.9 cm long, filaments purple; pistils (45–)55–81(–89). **Fruits** follicetums, cylindric to somewhat obovoid, 7–10 × 3.5–5 cm; follicles beaked; seeds lenticular to narrowly ellipsoid, (0.9–) 1.2–1.4 cm, aril red. **Distr:** S USA. **Ecol:** In wooded dunes; 0–120 m. **Zones:** 6–11.

Sect. Gwillimia

Subsections 2, **species** 17. **Stipules** adnate to petiole. **Leaves** evergreen. **Ovules** usually 2 per carpel.

M. coco (Lour.) DC., *Syst. Nat.* 1: 459, 1817. (Syn: *Liriodendron coco*, etc.). [Coconut Magnolia, Dwarf Magnolia]. **Shrubs to small trees**, 2–4 m tall, glabrous. **Stipules** adnate to petiole. **Leaves** elliptic, narrowly elliptic or ovate-elliptic, 7–17(–28) × 2.5–5(–9) cm, base cuneate, margins slightly undulate, apex long-acuminate, sometimes acute, coriaceous, adaxially dark glossy green, abaxially greenish, nerves 8–10 pairs; petiole 0.5–1.5 cm long with stipular scars. **Inflorescences** with flowers solitary, pendulous, nodding; peduncles 1.4–2 cm long, with 3 bract scars; pedicels 2–3 mm long. **Flowers** subglobose, 3–4 cm Ø, heavily scented, opening at night; tepals 9, unequal, obovate, dorsal faces convex, outer 3 greenish, thin-coriaceous, 2–2.6 × 1.2–1.7 cm, inner tepals 6, white, fleshy, oblong to obovate-oblong, 3–4 × 4 cm; stamens white, 5–7 mm long, filaments 1.5–2 mm long; gynoecium obovoid-ellipsoid, ~1 × 0.7 cm, carpels 8–12, 5–6 mm long, styles short; torus with the scars of perianth and stamens 4–5 × 3–5 mm. **Fruits** follicetums, ~3 cm long, ripe carpels subligneous. **Distr:** China, Hong Kong, Taiwan, N Vietnam. **Ecol:** In evergreen broad-leaved forests; 600–900 m. Fl. Apr–Oct. **Zones:** 8–11.

M. delavayi Franchet, *Pl. Delavay.* 1: 33, 1889. (Syn: *Lirianthe delavayi*, *M. carpunii*). [Chinese Evergreen Magnolia]. **Trees**, evergreen, to 8–12(–15) m tall; trunk to 80 cm Ø. **Leaves** ovate to ovate-oblong, 10–20(–32) × 5–10(–20) cm, base broadly rounded to slightly cordate, margins wavy, apex obtuse to rarely emarginate, leathery, adaxially covered with trichomes, later glabrescent, abaxially densely tomentose and white-powdery, secondary veins 11–16 per side; petiole 5–7(–10) cm long, densely villous. **Flowers** erect, fragrant, cupular, 15–20 cm Ø; tepals 9 or 10, outer whorl pale green, oblong, 6–8(–10) × 2–3(–4) cm, margin revolute, inner 2 whorls milky-white or red, obovate-spathulate, 8–11 × 2.5–3.5 cm, narrower; stamens ~210, 1.8–2.5 cm long, connective exserted, anthers divergent; gynoecium ovoid, 3–4 cm long, finely yellow-villous, acute, carpels ~100. **Fruits** ovoid-ellipsoid, 9–15(–20) cm; mature carpels narrowly ellipsoid, finely yellow-villous, dehiscing along dorsal suture, apical beak reflexed. **Distr:** China. **Ecol:** In forests; 1,500–2,800 m. Fl. Apr–Jun. Fr. Aug–Oct. **Zones:** 8–10.

M. pterocarpa Roxb., *Pl. Coast Corom.* 3: 62, 1820. (Syn: *M. sphenocarpa* [typographical error], *Talauma grandiflora*, etc.). [Himalayan Magnolia]. **Trees**, to 18 m tall; young branchlets tawny-pubescent, later glabrous. **Leaves** obovate to elliptic, 20–60 × 10–30 cm, base cuneate, margins revolute, apex rounded to obtuse, slightly pubescent, adaxially later becoming glabrous; petiole 2.5–5 cm long. **Flowers** fragrant, white (darkening to pale brown with age), 20 cm across; tepals 9–12, outer 3–6 tepals greenish, obovate, 5–10 cm long, inner 6 tepals 5–7.5 cm long, white, narrower than the outer tepals; stamens 1.5–2.5 cm long. **Fruits** follicetums, oblong, 12.5–25 cm long; seeds with testa reddish-orange. **Distr:** Assam, India, Myanmar, E Pakistan. **Ecol:** In forests; to ~1,000 m. Fl. Apr–May. Fr. Oct–Nov. **Zones:** 10–12.

Sect. Rhytidospermum

Subsections 2, **species** 10. **Branches** produced by syllepsis or prolepsis. **Leaves** deciduous. **Ovules** usually 2 per carpel.

Subsect. Rhytidospermum

Species 5. **Branches** produced mostly by syllepsis. **Leaves** of early season arranged in false whorls (flushing). **Stamens** caducous during ♂ phase.

M. officinalis Rehder & E.H. Wilson, *Pl. Wilson.* 1: 391, 1913. [Medicinal Magnolia]. **Trees**, deciduous, to 20 m tall. **Stipules** sparsely hairy, adnate to petiole. **Leaves** 7–8 crowded together; lamina large, obovate, 16.5–37.5(–46) × 9–20(–24) cm, base cuneate, apex acute, obtuse, emarginate to obcordate, thinly coriaceous, adaxially green, glabrous, abaxially pale green, glaucous, pubescent, later glabrescent, nerves 14–25(–40) pairs; petiole stout, densely hairy, 1.9–5 cm long. **Flowers** white, sometimes flushed with purplish-red, fragrant, 10–15 cm Ø; tepals 9–12(–17), thick-fleshy, outer 3 pale green outside, obovate, usually reflexed, inner tepals 6–9(–14), obovate to spathulate, 8–10 × 2.5–5 cm; stamens 3–4 cm long, filaments red, stout, 4–8 mm long; gynoecium ovoid, 2.5–3.5 × 1 cm. **Fruits** usually oblong-ovoid, 8–16 × 4.5–5 cm; ripe carpels 1–3 cm long; seeds ovoid to irregularly shaped, ± compressed, 0.8–1 × 0.6–0.9 cm. **Distr:** China. **Ecol:** In forests; 300–1,300 m. Fl. May–Jun. Fr. Aug–Oct. **Zones:** 6–10 **Info:** ***M. officinalis* ssp. *biloba*** has leaves with apices bilobed.

M. tripetala (L.) L., *Syst. Nat.* ed. 10. 2: 1082, 1759. (Syn: *M. frondosa*, etc.). [Umbrella Tree]. **Trees**, deciduous, often multitrunked, to 15 m tall; trunk to 87 cm Ø; pith homogeneous. **Stipules** (4–)6.6–9.4(–10) × 2.6–3.7 cm, abaxially red-glandular, sparsely pilose. **Leaves** crowded in terminal whorl-like clusters; lamina elliptic-oblong to narrowly obovate, or oblanceolate, (10–)26–57(–70) × (7.2–)10–30 cm, base cuneate to long-tapered, apex very short to long-acuminate or abruptly acute, rarely apiculate, thin, adaxially green, glabrous, abaxially densely pilose. **Flowers** malodorous, 5.5–25 cm across; spathaceous bracts 2, abaxially glandular; tepals spreading, creamy white, outermost whorl sepaloid, reflexed, greenish; stamens 81–103(–115), 0.8–1.7 cm long, filaments purple; pistils (45–)53–66(–73). **Fruits** follicetums, cylindric to ovoid-cylindric, 6–10 × 2–3.5 cm; follicles long-beaked, glabrous; seeds lenticular to nearly ovoid, 0.9–1.2 cm long, aril deep pink to red. **Distr:** SE USA (Appalachian Mountains). **Ecol:** In woods and ravines; 0–1,000 m. **Zones:** 5–9.

Subsect. Oyama

Species 5. **Branches** produced mostly by prolepsis. **Leaves** arranged 'normally' (no flushing). **Stamens** mostly persistent during ♂ phase.

M. sieboldii K. Koch, *Hort. Dendrol.* 4: 11, 1853. (Syn: *Oyama sieboldii*). [Oyama Magnolia]. **Trees**, to 10 m tall. **Leaves** obovate to broadly obovate, (6–)9–15(-25) × 4–9(–12) cm, base broadly cuneate, truncate, or subcordate, apex abruptly narrowly acute to shortly acuminate, membranous, abaxially glaucous, secondary veins 6–8 per side; petiole 1–4(–6.5) cm long, appressed-villous. **Flowers** appearing with leaves, cupular at first but plate-shaped at anthesis, 7–10 cm Ø, erect or slightly nodding, fragrant; tepals 9, white, subequal; outer 3 tepals oblong-obovate to obovate, 4–6 × 2.5–3.5 cm, base with white trichomes, apex rounded, inner 6 tepals smaller, base gradually narrowing to shortly clawed; stamens purplish-red, 0.9–1.1 cm long, connective not exserted, filaments 3–4 mm long, anthers ~6 mm long; gynoecium green, ellipsoid, ~1.5 cm long. **Fruits** red, obovoid to ellipsoid, 2–7 cm long; carpels narrowly ellipsoid, ~1 cm long, dehiscing into 2 valves, beaked; seeds cordate, 6–7 × 6–7 mm, testa red. **Distr:** China, Japan, Korea. **Ecol:** In forests; 1,600–2,000 m. Fl. Mar–Apr. Fr. Sep–Oct. **Zones:** 6–9.

M. wilsonii (Finet & Gagnep.) Rehder, *Pl. Wilson.* 1: 395, 1913. (Syn: *Oyama wilsonii*, etc.). [Wilson's Magnolia]. **Shrubs** or **small trees**, to 8 m tall. **Leaves** elliptic-ovate to oblong-ovate, 6.5–12(–20) × 3–5(–8) cm, base rounded to slightly cordate, apex acute to acuminate, papery, adaxially pubescent, abaxially appressed-villous; petiole (0.5–)1–3(–5) cm long, densely brown-villous. **Flowers** pendulous, appearing with leaves, cupular at first, plate-shaped at anthesis, 10–12 cm Ø, fragrant; tepals 9(–12), white; outer 3 tepals broadly spathulate to obovate, 4–6.5(–7.5) × 3–4.5(–5.5) cm, subequal to inner ones, base clawed, apex rounded; stamens purplish-red, 0.8–1.2 cm long, filaments red, 1.5–2 mm long, connective rounded or emarginate at apex; gynoecium green, ovoid-cylindric, 1.5–2 cm long, carpels ~1 cm long. **Fruits** red, becoming purple when mature, terete, 6–10 × 2–3 cm, pendulous; mature carpels beaked; seeds obovoid, ~6 mm long. **Distr:** SC China. **Ecol:** In forests; 1,900–3,000 m. Fl. May–Jun. Fr. Sep–Oct. **Zones:** 6–9.

Subgen. Yulania

Sections 2, **subsections** 6, **species** 94. **Branches** normally produced by prolepsis (less pronounced in tropical spp.). **Leaves** with vernation (prefoliation) conduplicate. **Flowers** with precocious flowering, generally before new leaves appear; terminal and/or on proleptic brachyblasts; stamens persistent during male phase, anther dehiscence introrse or latrorse, pollen small, usually <50 µm Ø; gynoecium sessile or stipitate. **Fruits** ± cylindrical in shape, often apocarps, longitudinally dehiscent, or occasionally with lignified carpels resulting in a more ellipsoidal shape which is dehiscent circumscissile.

Sect. Yulania

Subsections 2, **species** 14. **Leaves** deciduous. **Flowers** terminal or sometimes on proleptic brachyblasts; anther dehiscence latrorse; ovules usually 2 per carpel. **Fruits** cylindrical (not apocarpus).

Subsect. Yulania

Species 13. **Flowers** with pronounced precocious flowering; tepals white, pink, or purple (sometimes outer whorl sepaloid). **Fruits** cylindrical.

M. campbellii J.D. Hooker & Thomson, *Fl. Ind.* 1: 77, 1855. (Syn: *Yulania campbellii*, etc.). [Pink Tulip Tree]. **Trees**, to 30 m tall. **Leaves** elliptic, oblong-ovate, or broadly obovate, 10–23(–33) × 4.5–10(–14) cm, base rounded to broadly cuneate, usually unequal, apex acute to shortly acuminate, papery, adaxially deep green, glabrous, abaxially grayish green, white appressed-pilose, secondary veins 12–16 per side; petiole 1–5 cm long, pilose. **Flowers** appearing before leaves, slightly fragrant, 15–25(–35) cm Ø; tepals 12–16, dark red, pink, or sometimes white, obovate-spathulate to oblong-ovate, 6–14 × 4–6 cm, base gradually narrowed, clawed; outer 3 patent, reflexed, or pendulous; tepals of innermost whorl broadly ovate to suborbicular, 8–10 × 4–6 cm, erect, surrounding stamens and gynoecium; stamens 1–3 cm long, filaments purplish-red; gynoecium green, 2–3 cm long, stigmas red. **Fruits** follicetums, purplish-red turning brown, terete, 11–20 × 2.5–3 cm; mature carpels firmly connate; seeds cordate, 1–1.2 cm long. **Distr:** Bhutan, China, NE India (Assam, Sikkim), N Myanmar, Nepal. **Ecol:** In forests; 2,500–3,500 m. Fl. Mar–May. Fr. Jun–Jul. **Zones:** 7–9.

M. liliiflora Desrousseaux in Lamarck, *Encycl.*

PM

MD

MD

Top row: *Magnolia grandiflora, Magnolia coco, Magnolia delavayi.*
Middle row: *Magnolia pterocarpa, Magnolia officinalis, Magnolia officinalis* ssp. *biloba, Magnolia tripetala.*
Bottom row: *Magnolia sieboldii, Magnolia wilsonii, Magnolia campbellii.*

3: 675, 1792. (Syn: *Yulania liliiflora*, etc.). [Lily Magnolia]. **Shrubs**, to 3 m tall, usually caespitose. Leaves elliptic-obovate to obovate, 8–18 × 3–10 cm, base gradually narrowed, apex acute to acuminate, adaxially deep green, sparsely pubescent when young, abaxially grayish-green, secondary veins 8–10 per side; petiole 0.8–2 cm long. **Flowers** appearing with leaves, vase-shaped, erect, slightly fragrant; tepals 9–12, outer 3 tepals purplish-green, sepaloid, lanceolate, 2–3.5 cm long, caducous, tepals of inner 2 whorls purple to purplish-red outside and whitish inside, petaloid, elliptic-obovate, 8–10 × 3–4.5 cm, fleshy; stamens purplish-red, 0.8–1 cm long, connective exserted forming a mucro, anthers ~7 mm long, dehiscing laterally; gynoecium pale purple, ~1.5 cm long, glabrous. **Fruits** dark purplish-brown, cylindric, 7–10 cm long; mature carpels subglobose, apex shortly beaked. **Distr:** China. **Ecol:** In forest margins; 300–1,600 m. Fl. Mar–Apr. Fr. Aug–Sep. **Zones:** 6–11.

M. stellata (Siebold & Zucc.) Maximowicz, *Bull. Acad. Imp. Sci. Saint-Pétersbourg* Sér. 3, 17: 418, 1872. (Syn: *Yulania stellata*, etc.). [Star Magnolia]. **Shrubs**, 0.3–2.4 m tall. **Leaves** elliptic to obovate-elliptic, 7–12 × 2.5–4 cm, base cuneate, apex acuminate to caudate, glabrous, secondary veins 6–8 per side; petiole 0.3–1.2 cm long, glabrous. **Flowers** appearing before leaves, 5–7 cm Ø; tepals 12–15(–18), in 4 or 5(or 6) whorls, initially pale red becoming white with red on outside apically or only at middle,

Top row: *Magnolia liliiflora, Magnolia stellata, Magnolia acuminata, Michelia* x *alba.*
Bottom row: *Michelia champaca, Michelia compressa, Michelia doltsopa.*

oblanceolate to obovate-spathulate, 3.3–4.5 × 1.3–1.8 cm, apex rounded to nearly acute, fleshy; stamens 86–99, 0.7–0.9(–1) cm long, filaments 1–2(–3) mm long, anthers 6–7 mm long; gynoecium cylindric, 6–8 × 2–4 mm, carpels many, narrowly ovoid, ~2 mm long, distinct, glabrous, stigmas ~1.5 mm long. **Fruits** follicetums, reddish-green, knobby, 5 cm long, often falling immature; seeds orange-red. **Distr:** Japan. **Ecol:** In sparse forests, streamsides; 50–1,000 m. Fl. Feb. **Zones:** 5–9.

Subsect. Tulipastrum

Species 1. **Flowers** with tepals green to yellow (outer whorl sepaloid); precocious flowering less pronounced.

M. acuminata (L.) L., *Syst. Nat.* ed. 10. 2: 1082, 1759. (Syn: *Tulipastrum americanum*, etc.). [Cucumber Tree]. **Trees**, deciduous, single-trunked, to 30 m tall; trunk 1.26 m Ø. Stipules 3.2–4.3 × 1.4–1.6 cm, abaxially pilose. **Leaves** distinctly alternate, not in terminal whorls; lamina broadly ovate-elliptic, oblong to oblong-obovate, rarely somewhat rotund, (5–)10–25(–40) × 4–15(–26) cm, base cuneate to truncate or broadly rounded, often somewhat oblique, apex acuminate, adaxially green, usually glabrous, abaxially pale green to whitish, pilose to nearly glabrous. **Flowers** slightly aromatic, 6–9 cm across; spathaceous bracts 2, abaxially silky-pubescent; tepals erect, strongly glaucous to greenish or sometimes yellow to orange-yellow, outermost tepals reflexed, much shorter, green; stamens (50–)60–122(–139), 0.5–1.3 cm long, filaments white; pistils (35–)40–45(–60). **Fruits** follicetums, oblong-cylindric, often asymmetric, 2–7 × 0.8–2.7 cm; seeds heart-shaped, 0.9–1 cm long, aril reddish-orange. **Distr:** North Carolina, Georgia, Alabama. **Ecol:** In moist, rich, wooded slopes, stream banks and uplands; 30–1,500 m. Fl. May–Jun. **Zones:** 7–11.

Sect. Michelia

Subsections 4, **species** 80. **Leaves** evergreen. **Ovules** per carpel usually 2–6(–many).

Subsect. Michelia

Species ~63. **Flowers** mostly on proleptic brachyblasts, rarely terminal on long shoots; anthers latrorse dehiscence; gynoecium stipitate, ovules usually 2–6(–many) per carpel. **Fruits** cylindrical apocarps or occasionally ellipsoidal syncarps.

M.* × *alba (DC.) Figlar, *Proc. Int. Symp. Magnoliac.*: 21, 2000. (Syn: *Michelia* ×*alba*, etc.). [Hybrid Champaca, White Champaca, White Jade Orchid Tree]. **Trees**, to 17 m tall; trunk to 30 cm Ø. **Leaves** long-elliptic to narrowly ovate, 10–27 × 4–9.5 cm, thinly leathery, adaxially glabrous, abaxially sparsely puberulous, reticulate veins very conspicuous on both surfaces when dry, base cuneate, apex long-acuminate to caudate-acuminate; petiole 1.5–2 cm long, sparsely puberulous. **Flowers** very fragrant; tepals 10, white, lanceolate, 3–4 × 0.3–0.5 mm; staminal connective exserted, forming a long tip; gynophore ~4 mm long, gynoecium puberulous, carpels numerous, usually partly undeveloped, forming a sparsely follicular fruit as torus elongates when mature. **Fruits** usually does not mature. **Distr:** Indonesia. **Ecol:** Fl. Apr–Sep. **Info:** Probably a natural hybrid. **Zones:** 10–12.

M. champaca (L.) Baill. ex Pierre, *Fl. Forest. Cochinch.*: t. 3, 1880. (Syn: *Michelia champaca*, etc.). [Champaca, Champak]. **Trees**, medium to large, 9–50+ m tall; trunk 1.9 m Ø; twigs ascending; crown narrowly umbelliform. **Stipules** densely rusty-pubescent. **Leaves** elliptic, lanceolate to ovate-lanceolate, 10–20(–30) × 4.5–10 cm, base acute to broadly cuneate or rounded, apex long-acuminate to subcaudate, sparsely rusty-hairy becoming glabrous; petiole 2–4 cm long. **Flowers** fragrant, 5–7(–10) cm across; bracts coriaceous, rusty- or golden-hairy; tepals 12–15(–20), oblanceolate, 2–4 × 0.3–1.2 cm, subequal, pale yellow to yellow; stamens 5–7 mm long, filaments 1.5–2 mm long; carpels many, free, ovary 3–4 mm long, unilocular, 2–several-ovuled. **Fruits** follicles 1–2 cm long, brown; seeds 2–4 per carpel, 0.5–1.2 cm long, orange to brown, polished. **Distr:** Bangladesh, India, Indonesia, Malaysia, Myanmar, Nepal, Thailand, Vietnam. **Ecol:** In evergreen, broad-leaved forests; 200–1,600 m. Fl. Apr–Jul. Fr. Sep–Oct. **Zones:** 10–12. **Info:** Flowers used in the manufacture of perfumes. Two varieties described (var. *champaca* and var. *pubinervia*).

M. compressa Maxim., *Bull. Acad. Imp. Sci. Saint-Pétersbourg*, Sér. 3, 17: 417 (1872). (Syn: *Michelia compressa*, etc.). [Michelia]. **Trees**, to 17 m tall; trunk to 1 m Ø; stipular scars absent; brachyblasts 6–7 mm long, appressed-pubescent, with 1 or 2 bract scars. **Leaves** obovate-elliptic to narrowly elliptic, 5–7 × 2–3 cm, base cuneate, apex acute, acumen obtuse to sharp, thinly leathery, secondary veins 8–12 per side; petiole 0.8–1.2 cm long. **Flowers** with buds golden yellow, appressed-sericeous; tepals 12, pale yellowish-white to pale reddish-near base, narrowly obovate, 1.2–1.5 × 0.3–0.5 cm; stamens ~45, 5–6 mm long, connective exserted, 1–1.8 mm long, anthers 3.5–4 mm long; gynophore ~3 mm long, gynoecium ~4 mm long, with golden-yellow trichomes. **Fruits** 3–5 cm long; mature carpels ellipsoid to ovoid, 1.5–2 × 1.1–1.4 cm, dorsally dehiscent, apex mucronate; seeds 2–4 per carpel, pink. **Distr:** S Japan, Philippines, Taiwan. **Ecol:** In evergreen, broad-leaved forests; 200–2,600 m. Fl. Jan. Fr. Oct–Nov. **Zones:** 10–12.

M. doltsopa (Buch.-Ham. ex DC.) Figlar, *Proc. Int. Symp. Magnoliac.*: 21, 2000. (Syn: *Michelia doltsopa*, etc.). [Chinese Magnolia, Sweet Michelia]. **Trees**, to 30 m tall; young parts grayish-white appressed-pubescent; brachyblasts densely appressed-villous. **Leaves** elliptic, oblong-elliptic, or narrowly elliptic, 10–22 × 5–7 cm, base broadly cuneate or obtuse, margins slightly involute, apex acute to long-acute, abaxially grayish, densely pubescent, secondary veins 10–14 per side; petiole 1–2 cm long. **Flowers** with tepals white, narrowly obovate-spathulate, 5–7 × ~2.5 cm, base clawed, apex rounded; stamens 1.2–1.7 cm long, connective exserted, 1.5–2 mm long, anthers 0.8–1.2 cm long; gynophore 6–8 mm long, densely grayish, firmly puberulous; gynoecium narrowly ovoid, 1.5–2 cm long; carpels 0.9–1.2 cm long, densely grayish-pubescent, ovaries 4–5 mm long, styles 5–7 mm long. **Fruits** 4–7 cm long; mature carpels obovoid, ~1.5 cm long, apex acutely beaked. **Distr:** Bhutan, China, NE India, N Myanmar, Nepal. **Ecol:** In evergreen, broad-leaved forests; 1,500–2,400 m. Fl. Apr–May. Fr. Sep–Oct. **Zones:** 9–11.

M. figo (Loureiro) DC., Syst. Nat. 1: 460, 1817. (Syn: *Michelia figo*, *M. fuscata*, etc.). [Banana Shrub, Banana Magnolia, Port-wine Magnolia]. **Shrubs**, 2–3 m tall; young parts densely tomentose. **Leaves** narrowly elliptic to obovate-elliptic, 4–10 × 1.8–4.5 cm, base cuneate to broadly cuneate, apex obtusely acute, adaxially glossy and glabrous, abaxially glabrescent; petiole 2–4 mm long. **Flowers** erect, sweetly fragrant, 1.2–2 × 0.6–1.1 cm; tepals 6, pale yellow, margins sometimes red to purple, long-elliptic, 1.2–2 × 0.6–1.1 cm, fleshy and thick; stamens 7–8 mm long, connective exserted and forming a sharp tip; gynophore ~6 mm long, yellow-tomentose, gynoecium ~7 mm long, exceeding androecium, glabrous. **Fruits** 2–3.5 cm long; mature carpels ovoid to globose, apex with a mucronate beak. **Distr:** SE Asia, subtropical China. **Ecol:** In forests. Fl. Mar–May. Fr. Jul–Aug. **Zones:** 9–12.

M. nilagirica (Zenker) Figlar, *Proc. Int. Symp. Magnoliac.*: 23, 2000. (Syn: *Michelia nilagirica* var. *walkeri*, etc.). [Nilgiri Magnolia]. **Trees**, small, 8–12 m tall; twigs dark brown to black; brachyblast densely silky-pubescent. **Stipules** densely silky-hairy. **Leaves** (narrowly) elliptic to slightly obovate, 5–8 × 2–4.5 cm, base cuneate, narrowed into petiole, apex usually slightly acuminate, acumen 0–8 mm long, abaxially often glaucous, densely pubescent when young, nerves 7–10 pairs; petiole 0.5–1 cm long. **Flowers** cream-colored; tepals mostly in 3 whorls, often 9–12, mostly narrowly obovate, rarely broader; outer tepals 1.5–3 × 0.7–1.8 cm; stamens numerous, ~8 mm long, connective appendage ~1.5 mm long; carpels ~30, gynophore densely pubescent. **Fruits** with carpels at the end of the much enlarged gynophore, few to ~20, ovoid, 1–1.5 cm long, dehiscing into 2 valves. **Distr:** India, Sri Lanka. **Ecol:** In montane regions; 1,200–2,350 m. **Zones:** 10–12. **Info:** The good quality wood is used for timber (Nooteboom, 1987).

Subfam. LIRIODENDROIDEAE

Genus 1, species 2. *Subfamilial description same as generic.*

Liriodendron

Syn: *Tulipifera.*

Species 2. **Trees**, deciduous; pith diaphragmed. **Stipules** tardily deciduous on summer shoots, free. **Leaves** distinctly alternate, not in terminal pseudo-whorls; lamina evenly (2–)4–6(–10)-lobed, base rounded to shallowly cordate or truncate, apex broadly truncate or notched, adaxially lustrous, smooth, abaxially glaucous. **Flowers** protogynous, appearing with the leaves; tepals (7–)9, petaloid, tip recurved, greenish-yellow with feathered orange band or blotch near base, outermost tepals sepaloid, reflexed, green; stamens on a short torus, tardily deciduous, whorled, filaments to half the length of the extrorse anthers. **Fruits** syncarps, elongate, spindle-shaped, conelike, dry, indehiscent, with samaras deciduous; seeds adherent to dry endocarp. **Dist:** N America, E Asia. **Cult:** Propagate from seeds, by air-layering or grafting. Requires a fertile soil in full or partial shade.

L. tulipifera L., *Sp. Pl.* 1: 535, 1753. (Syn: *L. procera*, *Tulipifera liriodendron*). [Yellow Poplar, Tulip Poplar, Tulip Tree]. **Trees**, single-trunked, to 60 m tall; trunk to 3 m Ø; crown pyramidal becoming broader and more open with age. **Stipules** paired, elliptic to oblanceolate, 2–4.5 cm long. **Leaves** alternate; lamina simple, trapezelike, usually with 2 shallow upper lobes and 2 lateral lobes at broadest part, or sometimes squarrose and barely lobed, (6.5–)7.5–15(–23.5) × (8.5–)12.5–18.5(–25.5) cm, base rounded-cuneate, apex truncate to broadly emarginate, adaxially dark glossy-green, glaucous, abaxially pale-green; petiole 5–11.5 cm long. **Flowers** terminal, solitary, tuliplike, campanulate, 4–8 cm long; spathaceous bract 1, brownish, notched; tepals 9, erect, 3 outermost tepals green to glaucous, reflexed, 6 inner tepals yellowish-green with an orange blotch at base; stamens 20–50, 4–5 cm long, filaments white; pistils of 60–100 carpels arranged in a central column. **Fruits** woody samaras, elongated, 4.5–8.5 cm long, with numerous imbricate samaras 3–5.5 × 0.5–1 cm; samaras (1–)2-seeded, falling separately at maturity; receptacles with basal pistil persistent; seeds (1–)2. **Distr:** E North America. **Ecol:** In woodlands; 0–1,500 m. Fl. Nov–Dec. Fr. Jan–Mar. **Zones:** 4–10. **Info:** The whole plant is slightly poisonous. Cultivated widely in the subtropics.

HIMANTANDRACEAE

Genus 1, species 2. **Trees**, branchlets slender, subterete or faintly angled distally, densely lepidote. **Stipules** absent. **Leaves** alternate, aromatic, scaly when young and with translucent oil dots; lamina simple, margins entire and faintly recurved or plane, ± coriaceous, costa prominent abaxially, secondary nerves 7–6 per side, spreading, anastomosing toward margin, pinnate-nerved; petiole slender, rugulose. **Inflorescences** not well-defined, flowers borne on flowering shoots, axillary, flowers usually solitary, with 2(3) alternating bracts; bracts subcoriaceous, oblong, 1.3 cm long, occasionally foliaceous, with obscure axillary buds; pedicels similar to flowering shoots in texture. **Flowers** bisexual, white or red; calyx ovoid-conical, subcoriaceous, obtuse or umbonate, calyptrate, rupturing along an irregular line near base, glabrous inside; corolla similar to calyx slightly smaller, closely enveloped by calyx, calyptrate; outer staminodes ~7–23, 1- or 2-seriate, castaneous, carnose, sharply reflexed after anthesis; stamens numerous, several-seriate, closely appressed, similar to outer ones, longitudinally dehiscent; inner staminodes ~13–20, 1–3-seriate, similar to outer ones, linear-subulate, gradually narrowed to an acute apex, usually erect; carpels spirally arranged, free, laterally appressed, ovary ovoid or oblong-ellipsoid, styles plumelike, locule 1, ovule 1(2), anatropous. **Fruits** syncarps (berry- or drupelike), ellipsoid or subglobose, to 2.5 cm Ø, fleshy; seeds solitary (rarely 2). **Distr:** E Malaysia, Moluccas, New Guinea, N Australia, Solomon Islands, Sulawesi. **Ecol:** In rainforests.

Himantandra

Species 2. *Generic description same as familial.* **Cult:** Propagate from seeds. Requires a fertile soil in full or partial shade.

H. baccata (F.M. Bailey) Diels, *Bot. Jahrb.* 55: 128, 1918. (Syn: *Galbulimima baccata*). **Trees**, to 17 m tall. **Leaves** oblong-lanceolate, 7–11.5 × 2–4 cm, base acute to attenuate, apex subacute to shortly acuminate, thinly coriaceous, secondary nerves 7–10 per side; petiole 0.8–2 cm long, shallowly canaliculate. **Inflorescences** borne on flowering shoots, to 2 cm long at anthesis, vegetative portion angled, 0.5–1.3 cm long; pedicels shorter. **Flowers** with calyx to 1 cm long, slightly less in Ø at anthesis; outer staminodes ~7, 1-seriate, lanceolate, 4–6 mm long or probably longer after anthesis, acute; stamens ~40, several-seriate, 6–9 mm long, pollen sacs 1.5–2 mm long; inner staminodes ~13, 6–8 mm long, narrowed at base; carpels 7–10, ~2 mm long at anthesis. **Fruits** syncarps, berrylike; seeds with testa smooth, cartilaginous. **Distr:** Australia (Queensland). **Ecol:** In montane rainforests; 400–1,100 m. Fl. Sep–Oct. **Zones:** 9–11.

Below: *Michelia figo, Michelia nilagrica, Liriodendron tulipifera, Alphonsea hortensis.*

Top row: *Alphonsea zeylanica*, *Annona cherimola* (flowers), *Annona cherimola* (fruit), *Annona diversifolia* (fruits).
Bottom row: *Annona glabra* (flower), *Annona glabra* (fruit), *Annona montana* (flowers), *Annona montana* (fruit).

ANNONACEAE

Genera ~128, species >2,300. **Trees, shrubs,** rarely woody lianas, deciduous or evergreen; bark conspicuously fibrous; vessel elements with simple perforations; wood scattered with spherical cells containing ethereal oils (aromatic terpenoids). **Stipules** absent. **Leaves** alternate, 2-ranked, short-petiolate, aromatic, with pellucid dots; lamina simple, unlobed, margins entire, venation pinnate. **Inflorescences** determinate, terminal or axillary to leaf scars on old wood or to leaves on new shoots, flowers solitary or few-flowered fascicles, pedunculate; bracts or bracteoles present or absent. **Flowers** aromatic, radial, bisexual (rarely unisexual); receptacle becoming enlarged, elevated or flat; perianth hypogynous, segments valvate or imbricate; sepals persistent, (2–)3(–4), distinct or basally connate; petals either 6 in 2 unequal whorls of 3, with petals of outer whorl larger, petals of inner whorl fleshier, often with corrugate nectary zone, or petals 6–12(–15), nearly equal or unequal, veins impressed on inner face; stamens 10–20 or very numerous, hypogynous, spirally arranged, forming a ball or flat-topped mass, filament short, stout, anther linear to oblong-linear, extrorse, longitudinally dehiscent, poorly differentiated from anthers, connective apically elongate, connivent; pistils 1–many, superior, 1-carpellate, 1-loculed, distinct or connate to various degrees with at least stigmas distinct; carpels (1–)3–numerous, usually distinct, spirally arranged, placentation marginal, ovules 1–many per pistil, style short, thick, stigma terminal; nectarines or food tissues sometimes on inner petals. **Fruits** an aggregate of fruitlets (berrylike), distinct, 1–8(–12) per flower, or coalescent, forming fleshy syncarps, 1 per flower; seeds with raphe curving around top, often arillate, embryo minute, 1–many per pistil, endosperm ruminate, oily. **Distr**: Throughout the tropics. **Ecol**: In low altitude humid forests.

Alphonsea

Species ~30. **Trees**, leafy branches slender. **Leaves** symmetrical at base and decurrent along petiole for 1–2 mm, thinly coriaceous, adaxially glabrous, lateral veins 6–13 pairs; petiole 2–5(–7) mm long, broadly grooved with sharp edges. **Inflorescences** few-flowered, sessile or short-peduncled, internodal or leaf-opposed, rarely axillary, densely bracteate in the lower half; bracts triangular-ovate, to 1 mm long, pubescent outside, glabrous inside; pedicel to 5 mm long. **Flowers** pale yellow or orangish-yellow; sepals valvate in bud, broadly ovate, 1–2 × 1–2 mm, basally shortly fused or distinct, pubescent or tomentose outside, glabrous inside; petals 6, in 2 series, valvate in bud, ovate, 6–8 mm long, base saccate, apices recurved; outer petals spreading, densely pubescent or tomentellous outside, minutely puberulous inside except at base; inner petals slightly smaller, connivent but not coherent, puberulous outside, glabrous inside; stamens 15–50, loosely arranged, in several series, ovate or triangular, 0.6–1.5 mm long, obtuse; carpels (1–)3–6, distinct, with several ovules along the suture, densely pubescent, fulvous-hairy, stigma ± sessile. **Fruits** aggregates of fruitlets; fruitlets globose, 1.5–3 cm Ø, thick-walled, succulent or woody, several-seeded, yellowish or brownish; seeds adnate to fruit-wall. **Distr**: India and Sri Lanka through SE Asia to New Guinea. **Info**: Most species have a very limited area of distribution. **Cult**: Propagate from seeds. Requires a moist but well-drained fertile soil in shade.

A. hortensis H. Huber, *Flora of Ceylon* Vol. V: 28–29, 1985. (Syn: *A. lutea*). **Trees**, to 20 m tall; leafy branches yellowish-pubescent when young. **Leaves** ovate or ovate-elliptical, 7–11 × 3–4 cm, base obtuse, apex gradually becoming acuminate, adaxially shiny, glabrate at maturity except along midrib, abaxially often persistently pubescent, dry leaf not turning black, lateral veins 8–11 pairs, midrib often strongly verrucose adaxially. **Inflorescences** sessile or with a peduncle to 7 mm long; pedicels ferruginous-tomentellous. **Flowers** with sepals and petals obtuse, tomentose outside; stamens 42–50; carpels 4–6, stigma capitate, depressed, sessile. **Fruitlets** 1–5, broadly ovoid or almost globose, to 3 cm Ø, yellow or pale orange, brownish-tomentose, smooth, soft, juicy, acidic, stipitate; pedicel to 1 cm long; seeds 8–10 per carpel, compressed, brown, covered with the adherent pulp. **Distr**: Endemic to Sri Lanka. **Ecol**: In lowland forests. Fl. Mar. Fr. Jul. **Zones**: 10–12 **Info**: Known only from two wild collections in the 19th century, perhaps now extinct in the wild.

A. zeylanica Hook. f. & Thoms., *Fl. Ind.* 153, 1855. **Trees**, >8 m tall; leafy branches 0.8–1.2 cm thick, pubescent when young, soon glabrate. **Leaves** ovate-lanceolate or elliptic-lanceolate, 4–9(–12) × 1–2.5(–3.2) cm, both ends acute, long-acuminate, slightly shiny, midrib sparsely, rather persistently pubescent abaxially, otherwise glabrate at maturity; lateral veins 7–12 pairs. **Inflorescences** sessile; pedicels appressed-pubescent. **Flowers** with sepals and petals obtuse, outside densely appressed-pubescent; stamens 30–40; carpels 2–6, conical, stigma almost sessile. **Fruitlets** 1–6, globose, 1.5–2 cm Ø, densely fulvous-tomentose, hard and woody, faintly verrucose with flat warts, short-stipitate; pedicel to 6 mm long; seeds ~5 per carpel. **Distr**: Sri Lanka; rare. **Ecol:** In lower montane regions; 300–1,300 m. **Zones:** 10–12.

Annona

[Custard Apple].

Species >120. **Trees** or **shrubs**; trunks buttressed or not buttressed. **Leaves** persistent or deciduous to late deciduous; lamina leathery or membranous, glabrous to pubescent. **Inflorescences** axillary or supra-axillary, occasionally from axillary buds on main stem or older stems, flowers solitary or fascicled; peduncle bracteolate. **Flowers** with receptacle convex to ± globose or elongate, elevated; sepals deciduous, 3(–4), smaller than outer petals, valvate in bud; petals 6(–8) in 2 whorls, usually fleshy, those of outer whorl larger, those of inner whorl more ascending, distinctly smaller or reduced, rarely absent; nectaries present as darker-pigmented, usually corrugate zones adaxially near petal bases; stamens very numerous, packed into a ball, club-shaped, curved, connective dilated; pistils numerous, sessile, partially connate to various degrees with at least stigmas distinct, ovules 1(–2) per pistil, style and stigma club-shaped or narrowly conic. **Fruits** fleshy syncarps, usually ovoid to nearly globose, surface variable depending on orientation, structure, and relative connation of pistil apices; seeds usually 1 per pistil, ovoid to ellipsoid, beanlike, coat tough, often glossy, margins various, narrow. **Distr**: Mostly from the neotropics, N America, Africa and SE Asia. **Cult:** Propagate from seeds or by grafting. Requires a moist but well-drained humus-rich soil in full sun. **Info**: The seeds of all the species are poisonous, often used as an insecticide.

A. cherimola Miller, *Gard. Dict.* ed. 8: *Annona* No. 5, 1768. (Syn: *A. pubescens, A. tripetala*). [Cherimoya, Custard Apple]. **Trees** or **shrubs**, evergreen, briefly deciduous, 5–9 m tall; young parts rusty-hairy. **Leaves** alternate, 2-ranked, new leaves whitish abaxially; lamina mostly elliptic, 5–25 × 3–10 cm, base acute, apex rounded, leathery, hairless, dull, medium-green, veins paler adaxially; petiole stout 0.6–1 cm long, densely hairy. **Inflorescences** clusters, 2–3-flowered, or flowers solitary; pedicels densely rusty-hairy, 0.8–1.2 cm long. **Flowers** pale greenish-yellow, fleshy, 3 cm across, strongly fragrant; flower buds 1.5–1.8 cm long; outer petals 3, greenish, fleshy, oblong, downy; inner petals 3, smaller, pinkish with yellow or brown-hairy outside, whitish with purple spot and many stamens inside. **Fruits** syncarps, large, green, conical or heart-shaped, compound, 10–20 cm long, skin has the appearance of overlapping scales; ripens to brown with a fissured surface; pulp white, sweet; seeds brown or black, beanlike, glossy, 1–2 cm long. **Distr**: Andes region; Ecuador, Peru. **Ecol**: 700–2,300 m. **Zones:** 10–12. **Info**: Fruits often weigh over 2.5 kg. This species is among the best-tasting in the genus. *A. atemoya* [Atemoya] is a hybrid between *A. cherimoya* and *A. reticulata*.

A. diversifolia Safford, *Journ. Wash. Acad. Science* 2, 33: 471, 1911. (Syn: *A. macroprophyllata*). [Annona Blanca, Annona Illama, Ilama, White Annona]. **Trees**, erect, to 7.5 m tall; bark aromatic. **Leaves** aromatic when crushed, elliptic to obovate or oblanceolate, 5–15 cm long, adaxially glossy, abaxially dull, thin; new flush reddish-or coppery in color. **Inflorescences** with flowers usually solitary, borne on branchlets; bracts 1–2, leaflike, ± circular, 2.5–3.5 cm long, glabrous, clasping at base. **Flowers** maroon-colored, open to base; sepals small, rusty-hairy, narrow, blunt; outer petals minutely hairy, stamen-like, inner petals pollen-bearing. **Fruits** syncarps, cone-shaped, heart-shaped to round, to 15 cm long, surface smooth to rough and covered with ± prominent triangular spikes, skin color pale-green to deep-pink or purplish; seeds 25–80, hard, smooth, brown, cylindrical, 2 × 1 cm. **Distr**: SW Mexico, Pacific coast of Guatemala and El Salvador. **Ecol**: In coastal regions and foothills; 0–1,800 m. **Zones:** 10–12.

A. glabra L., *Sp. Pl.* 1: 537, 1753. (Syn: *A. palustris*). [Alligator Apple, Monkey Apple, Pond Apple]. **Shrubs** or **trees**, to ~15 m tall; trunks usually single-stemmed, sometimes multiple plants fuse together forming clumps. **Leaves** late deciduous; lamina ovate to elliptic, 5–15 × 6(–8) cm, base broadly cuneate to rounded, apex acute to short-acuminate, glabrous; petiole 1–2 cm long. **Inflorescences** axillary, flowers solitary; peduncle stout, to 2 cm long. **Flowers** with sepals reniform-cordate, 5–6 mm long, glabrous; outer petals cream-white, ovate-cordate, adaxially concave, 2.5–3 cm long, acute; inner petals cream-white, inside base deep purple, oblong-ovate, 2–2.5 cm long, base cupped, incurved-cuneate, corrugate; stamens linear, 3–4 mm long; pistils conical, connate. **Fruits** syncarps, pendulous, dull yellow blotched with brown, ± ovoid, 5–12 cm long, smooth with reticulate pattern formed by pistil boundaries, glossy; seed ellipsoid to obovoid, 1–1.5 cm long. **Distr**: West Indies, Mexico, C America, S America, W Africa, S Asia (Maldives, Sri Lanka, etc.). **Ecol**: Near waterways, brackish to fresh water. Fl. Throughout the year. **Zones:** 10–12. **Info**: Fruits edible but not eaten commonly, in the Maldives it is often made into a drink.

A. montana Macfad., *Flora Jamaica* 1: 7, 1837. (Syn: *A. marcgravii*, etc.). [Mountain Soursop, Wild Soursop]. **Trees**, to 5 m tall, glabrous. **Leaves** aromatic when crushed obovate or elliptic, 7–20 × 3–8 cm, base acute, apex shortly and obtusely cuspidate, charteaceous, lustrous, adaxially glabrous. **Inflorescences** clusters of 1–many flowers, cauliflorous on trunk and branches. **Flowers** with outer petals broadly ovate, gradually attenuate but not cuspidate, subacuminate, 3–3.5 × 2-3 cm; inner petals roundish-elliptic, nearly the same size as the outer, but thinner; claw 4–6 mm long. **Fruits** globose to ovoid-globose, to 15 cm Ø, with short, straight prickles; pulp yellow; seeds brown, ~ 2 × 1 cm. **Distr**: West Indies to S Brazil and Peru. **Ecol**: Fl. Apr–Aug. Fr. Jan–Apr. **Zones:** 10–12. **Info**: The fruit somewhat resembles that of *A. muricata*, but smaller and round. This species is not as tasty as the other more commonly cultivated species.

A. muricata L., *Sp. Pl.* 536, 1753. (Syn: *A. macrocarpa*, etc.). [Guanabana, Soursop]. **Trees**, to 4(–6) m tall; young branches hairy. **Leaves** oblong-obovate, 8–16 × 3.1–6.8 cm, apex acute to shortly acuminate, adaxially glossy dark green, glabrous, abaxially paler, minutely hairy to glabrous; petiole 0.4–1.3 cm long, glabrous. **Inflorescences** leaf-opposed or extra-axillary, 1–2(–3)-flowered; peduncle 2–5 mm long, woody; pedicels 1.5–2 cm long; bracteole basal, deltoid, densely hairy. **Flowers** with thick petals, yellowish; outer petals valvate, broadly ovate, 2.8–3.3 × 2.1–2.5 cm, base cordate, apex acute to acuminate; inner petals imbricate, ovate, 2.5–2.8 × 2 cm, base cuneate, apex obtuse, surfaces finely tomentose; stamens numerous, narrowly cuneate, 4.5 mm long; carpels linear, basally connate, ovary reddish-hairy, 1-ovuled, style short, stigma truncate. **Fruits** syncarps, ovoid, to 30 cm long, dark green, echinate, spines curved, fleshy; pulp white, juicy, aromatic; seeds glossy-black. **Distr**: Tropical America, W Indies; widely cultivated in Old World tropics. **Ecol**: Fl. & Fr. throughout the year. **Zones:** 10–12. **Info**: Fruits largest in this genus, sometimes weighing over 5 kg. The pulp is pleasantly sour and aromatic, made into drinks and desserts.

A. purpurea Moc. & Sessé ex Dunal, *Monogr. Fam. Anonacées* 64, 1817. (Syn: *A. involucrata*, etc.). [Annona, Negro-Head, Soncoya]. **Trees**, deciduous; trunk 45 cm Ø; branches rusty and woolly when young. **Leaves** alternate, almost sessile; lamina oblong-elliptic or oblong-lanceolate to oblong-obovate, 20–30 × 10–15 cm, margins undulate, apex acuminate, surfaces brown-tomentose. **Inflorescences** flowers usually solitary, emerging with new flush. **Flowers** fleshy, large, conical or globose, strongly scented, purple-brown, inside mottled with yellowish and purple colorations; calyx 3-parted, brownish-tomentose; outer petals 3, very fleshy, brownish-hairy outside; inner petals 3, creamy, thinner, white outside, purple inside. **Fruits** syncarps, ± ovoid, 15–20 cm across, surface covered with hard, thick, conical or pyramidal prickles each tipped with a curved hook, skin velvety hairy; pulp aromatic, sweet, mango-flavored with numerous yellowish-orange, soft fibers; seeds numerous, obovate, dark brown, 2.5 cm long. **Distr**: S Mexico through Colombia and Venezuela. **Ecol**: Fl. Mar–Jul. **Zones:** 10–12. **Info**: Fruits edible but of inferior quality.

A. reticulata L., *Sp. Pl.* 537, 1753. [Bullock's Heart, Custard Apple, Red Custard Apple]. **Trees**, to 8(–16) m tall. **Leaves** lanceolate to oblanceolate,

Below: *Annona muricata* (flower), *Annona muricata* (fruit), *Annona purpurea*, *Annona reticulata* (flowers).

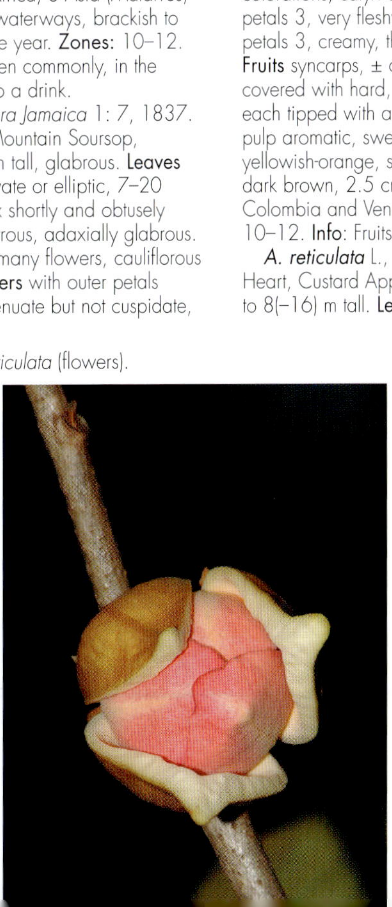

13.5–17 × 2.5–4.5 cm, base subrounded to cuneate, apex acute to acuminate, adaxially glabrous, abaxially sparsely hairy; petiole 1.2–1.5 cm long. **Inflorescences** leaf-opposed or extra-axillary, usually 2–3-flowered, occasionally branched and bearing more flowers; peduncle woody, stout; bract caducous; bracteole deltoid, densely hairy; pedicels 2.5–3 cm long. **Flowers** with sepals broadly deltoid, pilose, basally connate; outer petals ovate, 1.7–1.9 × 0.6–0.7 cm, basally concave inside, apex obtuse, thick, triquetrous, outside pilose, inside puberulent; inner petals minute, often absent; stamens linear, 1.5 mm long; carpels basally connate, ovary pilose dorsally, 1-ovuled, stigma elongated. **Fruits** syncarps, subglobose to roughly heart-shaped, 10–15 × 7.5–12.5 cm, ± smooth, areoles separated, surface reticulated, not tuberculate, reddish-brown, pulp white or cream, aromatic, granular; seeds smooth, black. **Distr**: Tropical America. **Ecol**: 0–1,500 m. Fl. Jun–Aug. Fr. Oct–Dec. **Zones**: 10–12. **Info**: Unripe fruits anthelmintic.

A. squamosa L., *Sp. Pl.* 1: 537, 1753. (Syn: *A. asiatica*). [Sweetsop, Sugar Apple]. **Shrubs** or **trees**, to ~8 m tall. **Leaves** with principal leaves late-deciduous; lamina narrowly elliptic to oblong or lanceolate, 5–17 × 2–5.5 cm, base broadly cuneate to rounded, apex acute to obtuse, dull green, surfaces glaucous, adaxially glabrate, abaxially pubescent; petiole 0.4–2.2 cm long. **Inflorescences** fascicles, or flowers solitary; peduncle slender, to 2 cm long. **Flowers** with deltate sepals, 1.5–2 mm long, apex acute, surfaces abaxially pubescent or glabrous; outer petals pale green above purplish-base, oblong, 1.5–3 cm long, base slightly concave, surfaces abaxially furrowed, pubescent, adaxially thickened, keeled; inner petals ovate, keeled, minute, nearly as long as stamens; stamens club-shaped, curved, 1–3 mm long; pistils conically massed, separable at anthesis, later connate. **Fruits** syncarps, pendulous, greenish-yellow, glaucous, mostly ± globose, 5–15 cm across, muricate; pulp whitish and granular; seeds ellipsoid to obovoid, 1–1.4 cm long, dark brown to black. **Distr**: West Indies; naturalized pantropically. **Ecol**: In dry sandy areas; 0–50 m. Fl. and Fr. throughout the year. **Zones**: 10–12. **Info**: The fruits are very sweet and have the best flavor in this genus.

Artabotrys

Syn: *Ropalopetalum*.

Species ~100. **Shrubs**, climbing or erect, rarely small trees, woody; plagiotropic shoots grow into vegetative shoots, recurved hooks/thorns or sympodial shoots with hooks and flowers (more hooks/thorns are produced in shade). **Leaves** alternate, glabrous; lamina simple, apex acuminate , thinly coriaceous, abaxially sparsely pubescent on midrib when young, both surfaces glabrous with age, lateral veins 7–16 pairs, venation prominent on both surfaces. **Inflorescences** extra-axillary cymes (rarely terminal), or flowers solitary; peduncle mostly thick and hooked; flowers pedicellate; bracts and bracteoles small, usually deciduous. **Flowers** bisexual; sepals 3, valvate, free or ± united at base, equal or smaller than petals; petals 6, in 2 whorls, valvate, subequal, free, usually concave at base and connivent over the stamens and carpels; stamens numerous, outer ones sometimes staminodes, linear-oblong or cuneiform, filament very short or absent; carpels numerous, free, ovoid or ellipsoid with 2 basal ovules. **Fruits** clusters of monocarps; fruitlets indehiscent, usually fleshy, cylindric or ellipsoid, mostly sessile, 1–2-seeded. **Distr**: Tropical Asia, Africa. **Cult**: Propagate from seeds, cuttings or by air-layering. Needs moist, fertile, well-drained soil in partial shade.

A. hexapetalus (L. f.) Bhandari, *Baileya.* 12(4): 147, 1964. (Syn: *A. odoratissimus, A. uncinatus*, etc.). [Climbing Ylang-ylang, Tail Grape]. **Shrubs**, large, evergreen, climbing by hooked peduncles, or semi-scandent; branches glabrous; young branches appressed-hairy. **Leaves** oblong-elliptic to oblanceolate, 10–18 × 2.8–4.8 cm, apex acuminate, subcoriaceous, glabrous, adaxially glossy, juvenile leaves brownish-hairy; petiole 2–7 mm long. **Inflorescences** with flowers usually solitary; peduncles hooked, flattened, leaf-opposed; pedicel 1.7–2 cm long. **Flowers** extremely fragrant; sepals 3, basally connate, ovate, 0.7–1 × 0.4–0.7 cm, apex acute, recurved, glabrate; petals 6, ± equal, ovate to oblong, 3.5–5.4 × 1–1.7 cm, base narrowed and concave, expanded above the concave part, apex subacute, flattened, yellowish-green to bright yellow, pubescent; outer petals with a minute ligulelike outgrowth; inner petals with a kneelike projection, concealing stamens and carpels; stamens spirally arranged, 2.5 mm long, cuneate, filaments short; ovary 5 mm long, glabrous, style short, stigma oblong, papillate. **Fruits** clusters of 1–10 baccate monocarps, sessile, obovoid, 3.7 × 2.2 cm, apiculate, yellow and fragrant when ripe, glabrous. **Distr**: S China, Java, S India, Sri Lanka. **Ecol**: Fl. Throughout the year. **Zones**: 10–12.

A. vanprukii Craib, *Bull. Misc. Inform. Kew* 1922: 226, 1922. **Shrubs**, climbing; branchlets at first densely adpressed ferruginous-pubescent, later glabrous; bark with lenticels inconspicuous. **Leaves** oblong-elliptic or obovate-elliptic, 12–15 × 4–5.5 cm, base rounded, cuneate-rounded, or rarely cuneate, sometimes slightly inequilaterate, apex obtuse, tip rounded, chartaceous-coriaceous or coriaceous, adaxially at first adpressed ferruginous-pubescent, later glossy, abaxially, nerves 9–12 pairs; petiole 0.5–1 cm long. **Inflorescences** with flowers usually paired; peduncles woody and hooked. **Flowers** with calyx 6 mm long, lobes 7 mm wide, densely puberulous; petals 2.7 cm long, limb 1.5–2 cm wide, both sides shortly adpressed pubescent; carpels glabrous, stigmas subequal. **Distr**: Thailand. **Ecol**: In mixed forests; ~400 m. **Zones**: 10–12.

A. zeylanicus Hook. f. & Thoms., *Fl. Ind.* 128, 1855. **Shrubs**, large, evergreen, climbing by the hooked peduncles, or semi-scandent; branches glabrous; young branches appressed-hairy; stems much compressed; leafy branches densely pubescent. **Leaves** elliptical, obovate, narrowly elliptical or narrowly obovate, 5–18.5 × 2.5–7 cm, usually 2–4 times as long as wide, base acute, obtuse, or rounded, apex short- to long-acuminate, adaxially shiny, lateral veins 7–12 pairs. **Inflorescences** with flowers solitary; pedicel ferruginous-tomentose. **Flowers** greenish; with sepals 3–4 mm long, shortly connate at base, tomentellose; outer petals 7–8 mm wide; carpels pubescent or tomentellous. **Fruits** clusters of monocarps; monocarps rounded-obovoid, 2–3 cm long, apices rounded and minutely apiculate, with some pubescence persisting at maturity; seeds ~1.3 cm long, dark brown, finely wrinkled. **Distr**: S Deccan Peninsula, Sri Lanka. **Ecol**: In moist primary and secondary forests; <1,000 m. **Zones**: 10–12.

Asteranthe

Species ~3. **Shrubs** or **small trees**, with indumentum of simple hairs. **Inflorescences** with flowers solitary or paired, extra-axillary, shortly petioled; bracteoles 2, eventually deciduous. **Flowers** hermaphrodite; sepals 3, valvate, free; petals 6, valvate, longitudinally folded in bud, quickly expanding and much larger than sepals, joined at the extreme base; stamens numerous, filaments very short, anther-cells linear, connective-prolongation truncate at apex, pubescent; carpels 10–22, free, with several biseriate ovules, style complicatedly laterally folded-in from both ends, stigmatose along upper margins with innermost parts of folds projecting. **Fruits** clusters of monocarps, indehiscent, subcylindrical, somewhat narrowed toward apex, subsessile, 3–10-seeded, slightly constricted between seeds, ± dry; seeds irregularly ellipsoid. **Distr**: Restricted to eastern parts of Kenya and Tanzania. **Cult**: Propagate from seeds. Requires moist but well-drained soil in full sun.

A. asterias Engl. & Diels, *Monogr. Afr. Pfl.* 6. 30, 1901. (Syn: *Uvaria asterias*). **Trees small** or **shrubs**, 1.5–8 m tall. **Leaves** obovate-elliptic, obovate-oblong, elliptic-lanceolate or oblong, (1.1–)4–16.5(–18) × (0.6–)2–6 cm, base subcordate, apex acuminate, rarely obtuse or acute, adaxially glabrous, shiny, abaxially densely hairy, soon pubescent or glabrescent; petiole 2–7 mm long. **Inflorescences** with flowers solitary; pedicels 3–8 mm long; bracteoles ovate-lanceolate, 2.5–4 × 1–1.5 mm, hairy. **Flowers** with an almost unpleasant scent; sepals triangular or ovate, 0.6–1.2 × 0.4–0.7 cm, acute or acuminate, densely hairy, persistent; petals green at first, then creamy-white apically, then deep crimson or purplish, and thickened at base, thickened area creamy-yellow, longitudinally striated and grooved, linear-oblong to linear-lanceolate or narrowly triangular, 2.5–6 × 0.7–1.5(–2) cm, base connate, apex acute, tapered, outside densely pilose basally; stamens cream, 1.5–2.5 mm long; carpels green, 10–22, oblong, 2–2.5 mm long, usually ~10-ovuled, densely hairy. **Fruits** clusters of monocarps, several, 2–4 × 0.8–1 cm, densely hairy; seeds yellow or brownish, 7–8 × 4–5 mm. **Distr**: Kenya, Tanzania, Zanzibar. **Ecol**: 0–60(–500) m. **Zones**: 11–12.

Cananga

Species 3. **Trees**. **Leaves** petiolate; lamina membranous. **Inflorescences** usually umbelliform clusters; peduncles borne from leaf axils or from defoliate nodes. **Flowers** large; sepals 3, valvate; petals 6, biseriate, valvate at first, subequal, elongate, plane; stamens numerous, linear, connective produced beyond anther and acute; torus convex, concave at middle; carpels numerous, attenuate to an oblong style bearing a capitate stigma, ovules numerous, biseriate. **Fruits** clusters of stipitate monocarps; seeds surrounded by pulp. **Distr**: Tropical Asia. **Cult**: Propagate from seeds (usually flowering in 3 years) or cuttings. Requires fertile, moist but well-drained soil in full sun.

C. odorata (Lam.) (Lam.) Hook.f. & Thomson, *Flora Indica* 1, 1820. (Syn: *Uvaria axillaris*, etc.). [Macassar Oil Tree, Ylang-ylang]. **Trees**, subcanopy, to 27 m tall; trunk 51 cm Ø, usually straight, bark smooth. **Stipules** absent. **Leaves** alternate (2-ranked); lamina simple, lamina oblong to ovate, 10–35 cm long, base asymmetric, margins entire, usually wavy, dark green; petiole, slender, 1–2 cm long. **Inflorescences** axillary clusters (usually 3-flowered) on older branches; pedicels 2.5–5 cm long. **Flowers** ~9 cm across, greenish maturing to bright yellow, strongly fragrant; calyx trilobed, lobes broad, pointed, yellow-green, spreading and slightly turned back; petals 6 (in 2 series), narrowly linear, 5–12.5 × 0.6–1.2 cm; stamens numerous, linear, 3 mm long, borne at base of ovary, connective produced into a lanceolate, acute process; ovaries numerous, style oblong. **Fruits** clusters of monocarps (~8–15 when ripe), 7.5–10 cm across; monocarps ovoid or obovoid, ~1.2 cm long, purple-black; seeds 6–12, light brown, pulp yellowish. **Distr**: From India to Fiji and Australia. **Ecol**: In secondary forests, riversides; <500 m. **Zones**: 10–12. **Info**: Flowers used in the manufacture of perfumes.

C. odorata var. fruticosa (Craib) J.Sinclair, *Sarawak Mus. J.* 5: 599, 1951. (Syn: *Canangium fruticosum*). [Dwarf Ylang-ylang]. **Shrubs**, 1–2.5 m tall. **Leaves** alternate; lamina simple, ovate or oblong, 10–18 × 5–9 cm, base acute-rounded, apex acute, light green, nerves 5–9 per side; petiole 1–2.5 cm long. **Inflorescences** terminal fascicles, or flowers solitary, opposite; pedicels 4–6 cm long. **Flowers** yellowish-

Top row: *Annona reticulata* (fruits), *Annona squamosa*, *Artabotrys hexapetalus*, *Artabotrys hexapetalus* (fruits).
Middle row: *Artabotrys vanprukii*, *Artabotrys zeylanicus* (flowers), *Artabotrys zeylanicus* (fruits), *Asteranthe asterias*.
Bottom row: *Cananga odorata*, *Cananga odorata* var. *fruticosa*, *Cleistopholis patens*, *Cyathocalyx zeylanica*.

green maturing to bright yellow, fragrant; sepals ovate, 1.2 × 0.6 cm; petals 15–20, in several whorls, each whorl with 3 segments, 5–9 × 0.5–1.8 cm. **Fruits** clusters of monocarps; monocarps 8–10, oblong, 1.5–1.8 × 1.2 cm, glossy green turning yellow when ripe. **Ecol:** Not known in the wild, perhaps of garden origin. Fl. Dec–Feb. Fr. Unknown. **Zones:** 10–12. **Info:** Cultivated widely in the tropics, mainly for the small stature. The flowers are larger and more attractive than *C. odorata*, but not as fragrant.

Cleistopholis

Species 3–4. **Trees** or **shrubs**, glabrous or with a few simple hairs. **Inflorescences** few-flowered fascicles, sessile or pedunculate and appearing umbellate, axillary; flowers pedicellate; bracts and bracteoles usually deciduous. **Flowers** bisexual; sepals 3, valvate, free; petals 6, in 2 series, free, those of outer series oblong-elliptic, valvate, mostly spreading, inner ones imbricate, much shorter, concave and recurved over androecium and gynoecium; stamens numerous, sessile, cuneiform, connective with a thick truncate prolongation; carpels 6–many, free, oblong or linear-oblong, 1–2-ovulate, styles mostly obsolete, stigma discoid or punctiform. **Fruits** clusters of monocarps; monocarps indehiscent, subglobose or ovoid or somewhat transverse, subsessile or distinctly stipitate, 1–2-seeded, if 2-seeded then slightly constricted between seeds; seeds globose or plano-convex. **Distr:** Tropical Africa. **Cult:** Propagate from seeds. Requires moist but well-drained soil in full sun.

MIW

Above: *Desmos zeylanica* (flower), *Desmos zeylanica* (fruits), *Enicosanthum acuminata*, *Fissistigma oldhamii*.

C. patens (Benth.) Engl. & Diels, E.M. 6: 35, t. 12/D, 1901. (Syn: *Oxymitra patens*). [Salt-and-oil Tree]. **Trees**, 7.5–20(–30) m tall; trunk to 10 m tall, 20–90 cm Ø. **Leaves** elliptic, oblong-lanceolate or obovate-oblong, slightly falcate, 2.5–25(–29) × 1.2–6.3 cm, base cuneate or rounded, ± decurrent, apex narrowly acuminate, subcoriaceous, discolorous, adaxially shiny, glabrous, lateral nerves 10–24; petiole channeled, 0.3–1 cm long. **Inflorescences** fascicles, 2–8-flowered; pedicels 0.8–2.5(–3.8) cm long; bracteoles ovate-lanceolate, ~2 × 1 mm, very deciduous. **Flowers** with sepals broadly ovate or round, 1.5–2 × 1.5–2 mm, ciliate; outer petals erect, green or greenish-yellow, obovate or obovate-oblong, 0.5–1.2 × 0.2–0.4 cm, mostly obtuse, subcoriaceous; inner petals broadly ovate or rounded, 2–2.5(–5) mm long, concave, ciliolate; stamens obpyramidal, 0.75–1 mm long, connective truncate. **Fruits** clusters of monocarps; monocarps 4–9, subglobose, ellipsoidal or transverse, 1.5–2.5 cm long, 1–2 seeded, glabrous, tuberculate; stipes ~7 mm long; fruiting pedicels 1.5–2.8 cm long, glabrous; seeds globose or ellipsoid, 1.2–1.5 cm Ø, tuberculate. **Distr:** W Africa to Angola, Central African Republic and Congo. **Ecol:** 750–1,110 m. **Zones:** 11–12.

Cyathocalyx

Syn: *Soala*.

Species ~36. **Trees**, monopodial. **Leaves** large, elliptical, ovate or ovate-lanceolate, apex acute or obtuse, thinly coriaceous, glabrous, lateral veins 8–12 pairs; petiole 0.8–1.2 cm long, narrowly grooved on ventral side, not swollen. **Inflorescences** 2-flowered or flowers solitary, leaf-opposed or borne slightly below nodes; bracts ovate, to 3 mm long, deciduous, densely rusty-tomentellous outside, glabrous inside; pedicels 1–1.2 cm long. **Flowers** pale apple-green, scented; sepals connate into an almost entire, cup-shaped calyx ~1 cm Ø, rusty-tomentellous outside, glabrous inside; petals 6, in 2 series, ovate-lanceolate to lanceolate, constricted above base, bases spathulate, closely appressed to stamens and carpel, spreading above, base golden-tomentose outside, glabrous inside; outer petals 4–5 cm long, distinct; inner petals slightly longer and narrower, fused basally concealing anthers; stamens numerous, closely arranged, 1.8–2.5 mm long, narrowly cuneate, connective broad. **Fruits** clusters of monocarps; monocarps ovoid, to 7 × 5.5 cm, sessile, rounded, glabrous, fleshy, yellow, turning blackish, indehiscent; seeds several, transversely compressed, pale-brown. **Distr:** India, Sri Lanka, SE Asia to Malesia. **Cult:** Propagate from seeds or by air-layering. Requires any well-drained soil in full sun.

C. zeylanica Champ. ex Hook. f. & Thoms., *Fl. Ind.* 127, 1855. **Trees**, very tall, almost columnar, with very sparse, laxy branches; bark smooth, pale brown. **Leaves** 10–24 × 5–9 cm, lateral veins with midrib at an angle of ~60°. **Flowers** with outer petals 0.9–1.8 cm long; stigma dark red. **Fruits** large, ovoid, not stipitate, rounded at both ends; seeds semi-circular in outline, 3.5–5 × 2–2.5 cm, 5–8 mm thick, smelling like nutmeg when dissected; pulp adherent to seeds. **Distr:** Sri Lanka, S Deccan Peninsula. **Ecol:** In disturbed rainforests and secondary vegetation, sometimes on foothills; <1,000 m. **Zones:** 11–12.

Desmos

Species ~25–30. **Treelets** or **trees**, erect or scandent, to 4 m tall; young branches appressed-pubescent. **Leaves** with lamina base not decurrent along petiole, membranous or coriaceous, abaxially glaucous, lateral veins 7–20+ pairs; petiole 0.3–1 cm long, narrowly grooved. **Inflorescences** axillary, terminal, internodal, leaf-opposed or from old wood, flowers paired or solitary; bracts ovate to narrowly lanceolate, 0.5–3 mm long. **Flowers** with sepals, 0.17–0.5(–1.2) cm long; petals 6, biseriate, 1–8 cm long, ovate-elliptical to lanceolate, distinct, flat; inner petals distinctly smaller; stamens numerous, closely arranged, 0.7–1 mm long, linear or narrowly wedge-shaped, connective broad, concealing anther-cells, truncate; staminodes absent; carpels ~20–60, appressed-pubescent, with several ovules along suture, stigma capitate. **Fruits** clusters of monocarps; monocarps mostly numerous, moniliform, apiculate, with (1–)2–4(–6) joints, glabrate, scarlet, succulent, indehiscent but disintegrating into ovoid joints, short-stipitate. **Distr:** India, Sri Lanka, SE Asia to New Caledonia and N Australia. **Cult:** Propagate from seeds or by air-layering. Requires any well-drained soil in full sun.

D. zeylanica (Hook. f. & Thoms.) Safford, *Bull. Torrey Bot. Club* 39: 506, 1912. (Syn: *Unona zeylanica*). **Treelets**, 0.5–4 m tall; branches few. **Leaves** narrowly elliptical to oblong, 12–21 × 2.5–5.5 cm, base mostly obtuse or rounded, apex obtuse, acute or faintly acuminate, thinly coriaceous, glabrous, lateral veins 10–15 pairs, at an angle of ~60° to midrib. with petiole 0.3–1 cm long, 1.3–2.5 mm thick; lamina. **Inflorescences** 2-flowered or flowers solitary, often borne from axils of fallen leaves; pedicels 0.7–1.7 cm long. **Flowers** reddish-brown or less commonly greenish-yellow; sepals triangular-ovate, ~2 × 2 mm; petals ovate-elliptical; outer petals 1.8–2.7 × ~1 cm; inner petals 1–1.2 × 0.7 cm; carpels ~20. Fruits clusters of monocarps; monocarps (1–)2–3-jointed, short-stipitate; joints (fruitlets) 0.8–1 × 0.6–0.7 cm. **Distr:** Endemic to SW Sri Lanka. **Ecol:** In dipterocarp forests of humid lower hill country; 200–800 m. **Zones:** 11–12.

Enicosanthum

Species ~16. **Trees**, to 35 m tall. **Leaves** obovate, obovate-elliptical or obovate-lanceolate, base rarely truncate, apex usually abruptly long-acuminate, rarely obtuse, thinly coriaceous, appressed-pubescent on midrib and lateral veins, lateral veins 11–14 pairs; petiole 3–8 mm long, shallowly grooved. **Inflorescences** cymes, 1–4-flowered, borne from axils of mostly fallen leaves or old wood; flowers ± sessile; bracts ovate, 3–8 mm long, obtuse, tomentose outside; pedicels 1.5–4 cm long, bracteate, tomentose. **Flowers** pale yellow, fragrant; sepals distinct, triangular, 0.8–1 × 0.8–1 cm, acute, tomentose outside; petals 6, biseriate, ovate or ovate-lanceolate, 2.5–3.5 × 1.5–2 cm, base excavate, gibbous above, flat and spreading distally, tomentellous outside; inner petals slightly shorter but wider; stamens very numerous, closely arranged, wedge-shaped, 1.5–2 mm long; staminodes absent; carpels numerous, distinct, strigose, with 1 basal ovule. **Fruits** clusters of monocarps; monocarps obovoid or almost globose, ~2 × 1.5–1.8 cm, yellowish-tomentose, stipitate, tardily dehiscent along suture, each half folding back and exposing the yellow-brown seed surrounded by a red, fleshy aril; stalk ~1 cm long. **Distr:** Sri Lanka and Myanmar through Malay Peninsula to Borneo and Philippines. **Cult:** Propagate from seeds or by air-layering. Requires any well-drained soil in full sun.

E. acuminata (Thw.) Airy-Shaw, *Kew Bull.* 1939: 277, 1939, 'acuminatum'. (Syn: *Polyalthia acuminata*). **Trees**, medium-sized; leafy branches 1.5–4 mm thick. **Leaves** 14–45 × 4.5–12.5 cm; petiole persistently tomentose. **Flowers** pale yellowish-green, fragrant, ± pendulous, 8–12 cm long; sepals distinct, triangular, 0.8–1 × 0.8–1 cm, acute, tomentose outside; petals 6, biseriate, ovate or ovate-lanceolate, 2.5–3.5 × 1.5–2 cm, base excavate, longitudinally furrowed, gibbous above, flat and spreading distally, tomentellous outside; stamens very numerous, closely arranged, 1.5–2 mm long; carpels numerous, distinct, strigose, with 1 basal ovule. **Fruits** clusters of monocarps; monocarps up to 30, exocarp bony. **Distr:** Endemic to Sri Lanka. **Ecol:** In lowland primary and secondary forests; <500 m. **Zones:** 11–12.

Fissistigma

Species ~75. **Shrubs**, climbing. **Leaves** alternate, simple, lateral veins distinct. **Inflorescences** terminal, flowers solitary, fascicled or panicled; pedicels bracteolate. **Flowers** with sepals 3, small, connate at base, pubescent; petals 6, in 2 series, valvate, outer ones somewhat larger than inner ones, trigonously ovate, concave; stamens numerous, spirally arranged, anthers 2-celled, longitudinally dehiscent, connectives broadly ovate; carpels numerous, free, pubescent, styles long, simple to bilobed; ovules 2–14. **Fruits** clusters of monocarps; monocarps berrylike, ovoid or globose. **Distr:** Tropical and subtropical regions. **Cult:** Propagate from seeds. Requires moist but well-drained soil in full sun.

F. oldhamii (Hemsl.) Merr., *Philip. J. Sci. Bot.* 15:

TC

TC

TC

Above: *Isolona cauliflora, Isolona heinsenii, Isolona hexaloba.*

134, 1919. (Syn: *Melodorum oldhamii*). **Shrubs**, climbing, evergreen, brown-tomentose. **Leaves** thick, oblanceolate or oblong, 9–12 × 3.5 cm, base cuneate, apex obtuse-acute, closely brown-pubescent beneath, entire, lateral veins 10–14 pairs; petiole 1 cm long. **Inflorescences** with flowers solitary or clustered, axillary or terminal; pedicels with 1 bracteole, brown-tomentose. **Flowers** pale yellow, ~1.2–1.8 cm long; sepals 3, deltoid, 5–7 × 5–6 mm, green, brown-tomentose, with prominent lines outside; petals 6 in 2 series, outer 3 ovate, 2.1–2.4 × 1.1–1.2 cm, valvate, pale yellow to golden, somewhat revolute, thicker, inner 3, 1.8–2.1 × 0.7 cm, involute and concave, yellowish-golden to brown above, pale reddish-violet below inside; stamens 240–288, linear, 1.5 cm long, anthers 2-celled, longitudinally dehiscent, connectives rounded, violet; carpels 24–36, ~4 mm long, ovary golden-hairy, styles pale white, free, ovules 10. **Fruits** clusters of monocarps; monocarps berrylike, globose, ~1.5 cm long, rusty-pubescent; seeds 4, hard, compressed. **Distr:** S China, Taiwan. **Ecol:** In broad-leaved forests; 300–1,100 m. **Zones:** 10–11.

Isolona

Species ~20. **Shrubs** or **trees**, glabrous or with an indumentum. **Inflorescences** with flowers solitary or paired, axillary, supra-axillary or borne on older leafless stems or trunk; flowers pedicellate or subsessile; bracteoles small, imbricate; buds subglobose. **Flowers** hermaphrodite; sepals 3, valvate, much smaller than petals; petals 6, in 1 whorl, valvate, equal, distinctly united at base; stamens numerous, cuneiform, filaments short, anther-cells lateral, extrorse, connective-prolongation broadened and truncate, those of inner stamens longer and more developed; carpels united in a 1-locular ovary, which is ovoid or ellipsoid and with numerous ovules, placentation parietal, styles obsolete, stigmas joined to form an irregular or peltate mass larger than ovary. **Fruits** syncarps, globose, ovoid or ellipsoid, pericarp often woody when dry, smooth or rugose, pulp resinous; seeds numerous, ellipsoid, testa often shiny. **Distr:** Tropical Africa. **Cult:** Propagate from seeds. Requires moist but well-drained soil in full sun.

I. cauliflora Verdcourt, *Kew Bull.* 25: 33, 1971. **Trees**, to 8 m tall; trunk 10 cm Ø. **Leaves** narrowly elliptic to obovate, 12–23 × 5–9 cm, base rounded, apex acuminate, acumen 1–2 cm long, coriaceous, glabrous, dark green, abaxially lighter, secondary veins 10–16(–18) pairs; petiole 3–5 mm long. **Inflorescences** cauliflorous, multiflowered; main flowering axis to 2 m long, borne from the trunk and growing on forest floor; pedicels 1.2–3.3 cm long, densely hairy, dark green to red. **Flowers** with sepals triangular, 3–6 × 2–4 mm, base truncate, margins hairy, apex acute to acuminate, densely hairy outside, green, pressed against tube, persistent; corolla light green when immature to light or dark red at anthesis, tube 4–6 mm long, 0.6–1 cm Ø, hairy, glabrous inside, lobes ± ovate, 1.3–3.2 × 0.4–1 cm, base narrowed, apex acute, hairy, coriaceous, 3 lobes spreading horizontally, the other 3 spreading vertically; stamens ~1.2 mm long; ovary ~2 × 2 mm, stigma 2.5–3 mm Ø. **Fruits** sometimes buried in forest soil, broadly ovoid, 2–4.5 × 2–4.3 cm, longitudinally 6(–7)-ribbed, coarsely rugulose, light to dark red or brown; fruiting pedicels 2.3–3 cm long, warty, dark red; seeds ellipsoid, 1–1.5 × 0.5–0.7 cm, pulp white, testa smooth, light-brown. **Distr:** SE Kenya, NE Tanzania. **Ecol:** In lowland rainforests, along waterways; 20–500 m. Fl. & Fr. May–Jun, Sep–Dec. **Zones:** 11–12.

I. heinsenii Engler & Diels, *Notizbl. Bot. Gart. Berlin-Dahlem* 2: 300, 1899. **Trees** or **shrubs**, to 6(–12) m tall; trunk 10–15 cm Ø. **Leaves** obovate to narrowly elliptic, (8.5–)15–22 × 3–8 cm, base broadly cuneate to rounded, apex acuminate, acumen ~1 cm long, papyraceous, glabrous, adaxially dark green, abaxially lighter, secondary veins 8–13 pairs; petiole 3–5 mm long. **Inflorescences** 1–2 rhipidia, on leafy branches; pedicels 0.5–1.2 cm long, light green. **Flowers** with sepals 3–5 × 2–3 mm, triangular, base truncate, apex acute, ± hairy, green, reflexed outward, deciduous; corolla green aging to light yellow, center of tube light red, tube 3–5 × 4–6 mm, cupulate, lobes lorate to narrowly oblong, 1.5–1.9 × 0.2–0.4 cm, base truncate, apex round to acute, spreading horizontally; stamens ~1 mm long, light yellow; ovary 1–2 × ~1 mm, stigma 3–4 mm long, light green. **Fruits** narrowly ellipsoid,4–10 × 1–3 cm, apiculate, apicule ~1 cm long, 6-ribbed, glabrous, heavily speckled with white and yellow; seeds irregularly ellipsoid, 1.5–2 × 1 cm, light-brown. **Distr:** E Tanzania. **Ecol:** In montane rainforests; (500–)1,000–1,440 m. Fl. Oct–May. Fr. Apr–May, Sep–Oct. **Zones:** 11–12.

I. hexaloba (Pierre) Engler, *Nat. Pflanzenfam. Nachtr.* zu 3(2): 161, 1897. (Syn: *Monodora hexaloba*, etc.). **Trees**, to 25–30 m tall; trunk to 50 cm Ø. **Leaves** ovate to narrowly elliptic, 10–28 × 3–11 cm, base acute, rarely obtuse, apex acuminate, acumen 1–2 cm long, coriaceous, shiny dark green, abaxially lighter, secondary veins 8–16 pairs; petiole 2–4 mm long. **Inflorescences** rhipidia 1–2, on leafy branches; pedicels (0.7–)1.3–2.5(–3) cm long. **Flowers** with sepals transversely elliptic to elliptic, 1–3 × 2–4 mm, base truncate, margins with hairy, apex shortly acuminate, glabrous, green, deciduous; corolla white to yellow or green becoming dark red, tube 0.4–1 × 0.4–0.7 cm, lobes elliptic or ovate, 0.6–2.5 × 0.4–1.2 cm, base narrowed, apex acute to rounded, glabrous, coriaceous, spreading; stamens 1.8–2 mm long; ovary ~ 1 × 0.5 mm, stigma ~2 mm Ø, glabrous, light red. **Fruits** broadly ovoid, 3–7 × 2.5–4 cm Ø, lumpy, ribbed, glabrous, light green becoming dark purple-red; seeds transversely ellipsoid, 0.8–1.5 × 0.4–0.6 cm. **Distr:** Tropical Africa. **Ecol:** In primary and secondary forests; 0–700 m. Fl. & Fr. throughout the year. **Zones:** 11–12.

I. zenkeri Engler, *Notizbl. Bot. Gart. Berlin-Dahlem* 2: 301, 1899. (Syn: *Diospyros oblongicarpa*). **Trees** or **shrubs**, to 7(–15) m tall; trunk to 15 cm Ø.

Below: *Isolona zenkeri, Mitrephora heyneana, Mitrephora maingayi.*

TC

KT

Leaves narrowly obovate to oblanceolate, 1.6–2.3 × 0.65–0.85 cm, base rounded to acute, apex abruptly acuminate, acumen 1–2 cm long, coriaceous, glabrous, secondary veins 11–13 pairs; petiole 2–5(–6) mm long. **Inflorescences** rhipidia 1(–2), on leafy branches, occasionally on old leafless branches; pedicels 3–7 mm long, brown-red. **Flowers** with sepals broadly ovate, 2–5 × 2–4 mm, base truncate, margins glabrous, apex acute, glabrous, brown-red or green, sometimes persistent; corolla light-green to light-yellow, becoming light red, tube 4–7 × 3–4 mm, glabrous, lobes narrowly oblong to lorate, 1.5–2.5 × 0.3–0.4 cm, apex acute, glabrous, coriaceous; stamens ~2 mm long; ovary ~3 × 1.5 mm, stigma ~2 mm Ø. **Fruits** globose to ellipsoid, 3–6.5 cm long, 1.5–3 cm Ø, smooth, yellow; seeds broadly ellipsoid, 1.5–2 × 0.8–1 cm, light-brown. **Distr:** Cameroon, Gabon. **Ecol:** In understory of rainforests; 0–800 m. Fl. & Fr. throughout most of the year. **Zones:** 11–12.

Mitrephora

Species 40. **Trees**, small. **Leaves** rather small, ovate or ovate-lanceolate, obtuse and base symmetrical, apex gradually tapered, tip obtuse, thinly coriaceous, adaxially glabrous, abaxially sparsely pubescent, lateral veins 5–7 pairs; petiole 3–6 mm long. **Inflorescences** internodal or leaf-opposed, flowers solitary or densely bracteate 2–3-flowered cymes; pedicel (1–)2–6 mm long, bracteate basally. **Flowers** dull yellow, sweetly scented; sepals ~ 1 × 1 mm, broadly triangular, shortly connate basally, appressed-pubescent outside, glabrous inside; petals 6, biseriate; outer petals elliptical-obovate, 0.8–1.3 cm long, apex obtuse, apiculate, spreading; inner petals 5–8 mm long, lozenge-shaped, base abruptly clawed, connivent, distal margins forming a vaulted cone, streaked or spotted with purple, densely puberulous outside, woolly inside; stamens numerous, ~5-whorled, 0.7 mm long, broadly wedge-shaped; carpels numerous, pubescent, several-ovuled, stigma subsessile. **Fruits** clusters of monocarps; monocarps ovoid, 1–1.6 × 0.8–1 cm, obtuse or rounded at both ends, yellow, pubescent, finely granular when dry, indehiscent; seeds 2–3, short-stipitate or almost sessile; stalk to 2 mm long. **Distr:** India and Sri Lanka, to SE Asia. **Cult:** Propagate from seeds or by air-layering. Requires any well-drained soil in full sun.

M. heyneana (Hook. f. & Thoms.) Thw., *Enum. Pl. Zeyl.* 8, 1858. (Syn: *Orophea heyneana*). **Treelets** or **trees**, to 12 m tall; bark blackish. **Leaves** ovate, 4–11 × 2–5 cm, both surfaces lustrous. **Inflorescences** with flowers usually solitary; peduncle to 6 mm long. **Flowers** with sepals usually persistent in fruit; outer petals 4–8 mm wide, inner petals 3.5–6 mm wide, thicker than the outer ones. **Fruits** clusters of monocarps; monocarps ovoid, 1–1.6 × 0.8–1 cm. **Distr:** S India, Sri Lanka. **Ecol:** On hills in dry and intermediate forests on rocks and along streams; 150–450 m. **Zones:** 11–12.

M. maingayi Hook. f. & Thomson, *Fl. Brit. India* 1: 77, 1872. **Trees**, 6–15 m tall; trunk 15–60 cm Ø; branchlets tomentose. **Leaves** ovate or oblong-acuminate, 7.5–14 × 3.8–5.6 cm, base rounded, coriaceous, glabrous, nerves 6–10 pairs; petiole 0.8–1 cm long. **Inflorescences** leaf-opposed, flowers several; peduncle red-velvety, with many bracts. **Flowers** large and showy, 2.5+ cm across, yellow with red spots; sepals connate ovate, acute, tomentose; petals with outer ones orbicular or obovate, edges undulate, pubescent inside, inner ones shorter cordate with a long linear claw; stamens numerous, appendages flat; ovaries narrowed into the styles, ovules 4. Fruits clusters of monocarps; monocarps ovoid, densely tomentose, pale green, 2.5 × 1.9 cm; stalks stout; seeds 4. **Distr:** Borneo, Malacca, S Vietnam. **Ecol:** In forest margins. **Zones:** 9–12.

Monodora

Species ~35. **Trees** and **shrubs**, sometimes climbing or scrambling, deciduous, glabrous with simple hairs. **Leaves** usually large and glabrous. **Inflorescences** solitary (rarely 2) with solitary (rarely in pairs or few-flowered cymes), pendulous; flowers are usually produced after the plant has shed all the leaves and the flowers last till the flush of new leaves are mature; **flowers** borne from the end of the branches, main branches and from axils of shed leaves. Flowers bisexual sepals 3, free, shorter than the petals; petals 6, usually in 2 whorls, inner whorl usually shorter, margins undulate; inner petals often united at apices; flower color often changes to darker shades before falling; carpels numerous, united to form a globose ovary with numerous ovules inside. **Fruits** syncarps, berrylike, globose or ovoid; seeds numerous, shiny. **Distr:** Tropical Africa. **Cult:** Propagate from seeds or by air-layering. Requires any well-drained soil in full sun.

M. angolensis Welw., *Ann. Conselho Ultram.* 1858: 587, 1859. (Syn: *M. gibsonii*). **Shrubs** or **trees**, 3–20 m tall, ± glabrous. **Leaves** oblong or elliptic to obovate or oblanceolate, (4.5–)6.5–18.5(–20) × (2–)2.8–7(–8.5) cm, base cuneate to ± rounded, apex acuminate or cuspidate, subcoriaceous; petiole 2–8 mm long. **Inflorescences** with flowers solitary or rarely paired, axillary or extra-axillary, pendulous; pedicels 1.5–6(–8) cm long. **Flowers** fragrant; sepals triangular-lanceolate, 0.5–1 cm long, subacute, recurved, margins ± undulate, sometimes ciliate, spotted with green or purple; outer petals widely spreading, greenish-yellow to orange with purplish-spots, whitish or pink-flushed toward base, ovate or oblong, 3–6 × 1.5–3 cm, margins undulate, apex obtuse to acute or attenuate; inner petals white or pink, often with 2 yellow spots at base, broadly ovate or sagittate, (0.4–)0.6–1.1 × (0.5–)0.9–1.5(–1.9) cm, obtuse or cuspidate, claw to 1 cm long, erect or ± connivent, margins glabrous or ± pubescent; stamens oblong or ± orbicular, ~0.5 mm long; ovary ± conic, ~1 mm long. **Fruits** globose or ovoid-conic, 5–9 × 4.5–7 cm; seeds ~1.3 cm long, brown. **Distr:** Tropical Africa. **Ecol:** In evergreen rainforests; to 1,000 m. **Zones:** 11–12.

M. carolinae Couvreur, *Adansonia* 28: 247, 2006. **Trees**, to 6 m tall; trunk to 15 cm Ø. **Leaves** narrowly elliptic to elliptic, 8–10 × 4–6 cm, base cuneate, apex acuminate, acumen 1–2 cm long, papyraceous, secondary veins 9–11 pairs; petiole ~4 mm long. **Inflorescences** with flowers solitary, leaf-opposed, pendulous; pedicels 1.5–3.5 cm long. **Flowers** with sepals triangular to ovate, 0.6–1.2 × 0.4–0.8 cm, base truncate, apex rounded, glabrous on both surfaces, green, reflexed, margins straight, densely hairy; outer petals oblong to elliptic, 1.5–2.5 × 0.6–1.2 cm, margins undulate, sometimes sparsely hairy, apex acuminate and reflexed upward at anthesis, both surfaces glabrous, creamy yellow with red spots; inner petals clawed ovate, 0.6–0.8 × 0.6–1.4 cm, base obtuse, apex acuminate, same color as outer petals, connivent by tip over receptacle; claw 3–5 mm long, glabrous; stamens in 6–7 rows, ~0.8 mm long, connective shield white-hairy; ovary 1–2 long, ~1 mm Ø; stigma ~2 mm Ø, glabrous, light yellow. **Distr:** SE Tanzania, NE Mozambique. **Ecol:** In moist semi-deciduous coastal forests; 50–800 m. Fl. Oct–Dec. **Zones:** 11–12.

M. globiflora Couvreur, *Adansonia* 28: 248, 2006. **Trees**, to 12 m tall; trunk to 30 cm Ø. **Leaves** obovate to narrowly elliptic, 12–21 × 4–7 cm, base cuneate to acute, apex acute to cuspidate, papyraceous, secondary veins 14–16 pairs; petiole 6–8 mm long. **Inflorescences** with flowers solitary, leaf-opposed, pendulous; pedicels 4–4.5 cm long. **Flowers** with sepals oblong, 0.8–1 × 0.6–0.8 cm, base truncate, margins entire, apex rounded, reflexed upward, light green, densely hairy, caducous; outer petals ovate, 2.8–3.2 × 1.8–2.3 cm, margins undulate, apex rounded, slightly reflexed, curved inward, creamy-yellow streaked with red slashes and spots; inner petals clawed, triangular, 0.7–0.8 × 1–1.2 cm, base truncate, apex acuminate to rounded, hairy, connivent over receptacle like a globe, bright yellow with red streaks; claw 3–5 × 2–4 mm, glabrous, bright yellow without red streaks; stamens in 5–6 rows, ~0.8 mm long, glabrous, white; ovary ~1.5 × 1 mm, stigma ~1 mm Ø, glabrous, white. **Fruits** irregularly globose, 4–5 × ~5 cm, green-reticulate; seeds 1.3–1.5 × 0.7–0.8 cm, light brown. **Distr:** Tanzania. **Ecol:** In montane rainforests; 1,700–2,000 m. Fl. Sep–Dec. Fr. Feb. **Zones:** 11–12.

M. grandidieri Baillon, *Adansonia* 8: 301, 1868. (Syn: *M. somalensis*, etc.). **Trees**, to 10–12 m tall; trunk to 15 cm Ø. **Leaves** obovate, 4–23(–25) × 3–10 cm, base cordate, apex acuminate, acumen ~1.5 cm long, coriaceous, secondary veins 9–14(–19) pairs; petiole 2–6 mm long. **Inflorescences** with flowers solitary, leaf-opposed, pendulous, borne from young shoots; pedicels (0.6–)2–5(–8) cm long. **Flowers** with sepals elliptic to narrowly elliptic,

Below: *Monodora angolensis*, *Monodora carolinae*, *Monodora globiflora*.

PL

TC

TC

TC

TC

TC

Top row: *Monodora grandidieri, Monodora hastipetala, Monodora myristica.*
Bottom row: *Monodora minor, Monodora tenuifolia.*

0.5–1.5(–2) × 0.5–0.7 cm, base truncate, margins undulate, apex acute, densely hairy outside, green to red brown with green veins; outer petals narrowly oblong to linear or spathulate, (1–)1.5–6(–7) × 0.3–1.2(–1.8) cm, base truncate, margins undulate, apex acute, green to yellow horizontally streaked with red-purple markings becoming bright white basally, spreading; inner petals clawed cordate, (0.4–)0.6–1(–1.2) × 0.5–1.5 cm, base cordate, margins undulate, apex acute, white with red-brown veins and yellowish to red margins, connivent by margins, claw incurved, 0.5–1.5 × 0.2–0.4 cm, white; receptacle 4–5 mm Ø, slightly convex; stamens in 5–7 rows, ~1.5 mm long; ovary ~ 2 × 1 mm, stigma 1.5–2 mm Ø, yellow. **Fruits** globose to obovoid,4.5–6.5 × 3.5–4.5 cm, minutely rugose; seeds transversely ellipsoid, 1–1.8 × 0.5–1.1 cm, pulp white, light brown. **Distr:** E Africa, S Somalia to S Mozambique. **Ecol:** In lowland rainforests, thickets; 0–900 m. Fl. & Fr. throughout the year. **Zones:** 11–12.

M. hastipetala Couvreur, *Adansonia* 28: 250, 2006. **Trees**, to 8 m tall; stem 5 cm Ø. Leaves narrowly obovate to narrowly elliptic, 10–12 × 3–4 cm, base acute to rounded, apex acuminate, acumen 1–1.5 cm long, glabrous, secondary veins 10–14 pairs, glabrous; petiole ~2 mm long. **Inflorescences** with flowers solitary, leaf-opposed, pendulous; pedicels 1.7–2 cm long. **Flowers** with sepals elliptic, 6–8 × 3–4 mm, margins entire, apex rounded to acute, glabrous, green, persistent in fruit; outer petals narrowly elliptic, 2–2.6 × 0.6–0.8 cm, base truncate, margins undulate, apex acute, glabrous, light-green, horizontal; inner petals clawed ovate, 1–1.7 × 0.4–0.7 cm, base truncate, margins straight, apex long-acuminate, densely hairy, white with purple base, connivent over receptacle; claw 4–5 mm long, glabrous, white to pale green, tinged purple apically; stamens 3-4-seriate, ~0.6 mm long, white; ovary ~ 1 × 0.5 mm, stigma ~0.8 mm Ø, glabrous, light-yellow. **Fruits** broadly ovoid, ~ 1.5–2.5 × 2–2.5 cm, glabrous, green with white dots; seeds transversely broadly ellipsoid, 1–1.2 × 0.9–1 cm, smooth, light-brown. **Distr:** E Tanzania. **Ecol:** In dry scrub and riverine coastal forests; 200–400 m. Fl. Oct–Nov. Fr. Nov. **Zones:** 11–12.

M. minor Engler & Diels, *Monogr. Afrik. Pflanzen-Fam.* 6: 88, 1901. **Trees**, to 6–7 m tall; to 5 cm Ø. **Leaves** usually obovate, 7–13 × 3.5–6 cm, base rounded to acute, apex acuminate, acumen ~1 cm long, coriaceous, glabrous, adaxially dark glossy-green glossy, abaxially paler, secondary veins 8–11 pairs; petiole 4–5(–7) mm long. **Inflorescences** erect above foliage, 2–3-flowered rhipidia, leaf-opposed or supra axillary, borne during leaf flush; pedicels 1–3 cm long. **Flowers** with sepals ovate, 0.6–1 × 0.3–0.5 cm, base truncate, margins undulate, apex rounded to acute, green-gray with red steaks toward base, strongly curved upward, deciduous, glabrous; outer petals ± ovate, 1.1–2.5 × 0.5–1.2 cm, base truncate, margins undulate, apex acute to rounded, glabrous, yellow streaked with red, glabrous; inner petals clawed, 0.5–0.8 × 0.6–1.3 cm, base cordate, apex acuminate to obtuse, inside densely hairy, light-yellow striped with purple and red, connivent at margins over receptacle; claw 3–5 × 2–3 mm, bright yellow; stamens 6–8-seriate, ~0.7 mm long; connective shield densely hairy, yellowish-white; ovary 0.8–1 × ~0.8 mm, stigma 1–1.5 mm Ø. **Fruits** ellipsoid to ovoid 2.5–3.5 × 1.5–2 cm Ø, apicule 3–4 mm long, smooth, glabrous, pale green with white dots; seeds 1–1.3 × 0.8–1 cm, dark brown. **Distr:** Coastal Tanzania, N Mozambique. **Ecol:** In coastal wet forests, thickets; 100–780 m. Fl. Oct–Feb. Fr. Feb. **Zones:** 11–12.

M. myristica (Gaertn.) Dunal, *Monogr. Anonac.* 80. 1817. (Syn: *Annona myristica*). [African Nutmeg, African Orchid Tree, etc.]. **Trees**, to 35 m tall, deciduous; trunk 2 m Ø. **Leaves** borne at ends of branches, distichous, alternate, short-petiolate; lamina oblong to obovate, 10–25 × 2–7.5 cm, base cordate, margins entire, adaxially shiny, bright green, abaxially paler. **Inflorescences** 1(2), opposite, flowers solitary; peduncles 10–18 cm long; bract green, variegated with yellowish-red. **Flowers** fragrant, 5–15 cm across, opening white with purplish-brown spots, turning yellow and the spots darkening to deep red; calyx with tube short, lobe deeply cleft into 3 unequal, colored sepals, margins crisped and wavy; corolla with tube short, reflexed, petals 6, biseriate; outer petals 3, oblong-ovate, margins crisped, wavy, same color as the calycine segments, bright-yellow, marked with rows of irregular spots of interrupted stripes of reddish-brown, the smaller calycine sepal sometimes variegated with greenish-yellow along middle, and deep red marginally; inner petals 3, shortly claw, cordate, convex, veined, yellowish-white outside, keeled, margins downy, shiny, pale-yellow, variegated with pale crimson spots; stamens closely set, ~11–12-seriate, anthers yellowish-white; gynoecium yellowish-white, ovules globular. **Fruits** subglobose berries, 20 × 15 cm; seeds numerous, aromatic, embedded in white, sweet smelling pulp. **Distr:** From Liberia to W Cameroon, to Uganda and Angola. **Ecol:** In evergreen and deciduous forests; 0–930 m. **Zones:** 11–12. **Info:** Seeds have the same aroma as the common nutmeg (*Myristica fragrans*) and often used for flavoring foods, and for medicinal purposes.

M. tenuifolia Benth., *J. Proc. Linn. Soc., Bot.* 5: 72. 1860. (Syn: *M. cabrae, Xylopia undulata*). [African Nutmeg]. **Trees**, deciduous, to 12–17 m tall. **Leaves** ovate, 7–15 × 3–6 cm, base acute, margins undulate, apex long-acuminate, thin. **Inflorescences** with flowers solitary, pendulous, borne at apices of the twigs and on old branches; pedicels short on branches, longer on twigs, 3–7 cm long. **Flowers** 5–12 cm across, opening white and maturing to golden yellow; sepals green; petals in 2 rows of 3; outer 3 petals oblanceolate, margins undulate, midvein prominent, base pure white, gradually becoming pale-yellow apically with dark reddish-variegation becoming darker; inner petals narrow, curved and joined at tips, lanceolate, 3 × 1 cm, margins incurved, white with reddish-markings, 2 white pubescences (almost globular appendages) halfway on the inner petals, margins ± pubescent toward apically; stamens closely set, filaments absent, anthers nearly spherical, bilouclar, creamish-white, translucent; gynoecium nearly spherical, creamish-white, stigma flattened, ovules globular. **Fruits** ovoid, 4–10 cm Ø; seeds many. **Distr:** Tropical Africa. **Ecol:** In primary, evergreen or fringe forests; 0–900 m. Fl. Dec–Feb, Jul–Aug. Fr. Apr. **Zones:** 11–12.

Mwasumbia

Species 1. *Generic description same as specific.* **Cult:** Propagate from seeds. Requires moist but well-drained soil in full sun.

M. alba Couvreur & D.M. Johnson, *Systematic Botany* 34(2): 270, 2009. **Trees**, 2–3 m tall; stem 3–5 cm Ø. **Leaves** obovate or rarely elliptic, 7–14 × 3–6 cm, base obliquely broadly cuneate to subcordate, apex acuminate to cuspidate, acumen 0.5–2.2 cm long, secondary veins 8–10 per side; petiole 1–2.5 mm long. **Inflorescences** 1-flowered rhipidia, axillary, sometimes cauliflorous; pedicels 1–2.2 cm long, densely hairy, green. **Flowers** bisexual, with 2 whorls of 3 free petals; sepals 3, free, basally slightly imbricate, broadly ovate, 4–3 × 4–5 mm, base subcordate, apex acute, densely hairy outside, glabrous inside, persistent; petals white to white-cream; outer petals elliptic, 1.2–1.8 × 1–0.75 cm, base truncate, apex obtuse, flat, spreading; inner petals narrowly oblong, 1.3–1.8 × 5–6 cm, base truncate, margins hairy, apex acute, glabrous on both sides, concave, erect; stamens 6–8-whorled, 2–2.7 mm long, filament <0.2 mm long, narrowed, glabrous, white-cream; torus ~1.8–2.2 mm long, 2–2.8 mm Ø at apex, short-cylindric; carpels 4(3?), 2–3 × ~1 mm, oblong, densely hairy, ovules 5–8, uniseriate, stigmas bilobed, bright yellow to white-cream. **Fruits** clusters of monocarps; monocarps ellipsoid, 1.8–2.5 cm long, 1–1.5 cm Ø, densely hairy; seeds ~2 per monocarp, spherical, flattened, ~ 1.1 × 1 cm, rugose, brown. **Distr:** Tanzania. **Ecol:** 200–500 m. Fl. Mar, Jun, Sep–Nov. Fr. Jun–Jul, Nov. **Zones:** 11–12.

Ophrypetalum

Species 1. *Generic description same as specific.* **Cult:** Propagate from seeds. Requires moist but well-drained soil in full sun.

O. odoratum Diels, *N.B.G.B.* 13: 270, 1936. **Shrubs** or **small trees**, deciduous and often flowering when leafless, 2.4–6(–9) m. tall. **Leaves** narrowly elliptic to obovate, 2–12 × 1.2–5.5 cm, base rounded, rarely cuneate, apex rounded, subacute or very obscurely and obtusely acuminate, thin, adaxially green, abaxially paler, often glaucous; petiole ± 5 mm long. **Inflorescences** with flowers solitary or in fascicles, on leafy shoots or mostly on older branches; pedicels 0.3–2.2 cm long, brown-hairy; bracteoles ~10, biseriate, to 2 × 2 mm. **Flowers** hermaphrodite, scented, petals irregularly crenulate; sepals 3, broadly triangular-reniform or ovate, 6–9 × 6–9 mm, thick, concave; petals and stamens cohering together and removable together; petals 6, imbricate above, open basally, 2 whorls of 3, subequal, clawed, pale-yellow or brownish; outer 3 semi-circular, ~ 0.35–0.55 × 0.8–1 cm, claw 3.5–5.5 × 3.5 mm, sericeous outside; inner 3 transversely oblong-subreniform, 4 × 8.5 mm, with a brush of fleshy brown setiform processes at junction of lamina and claw; claw 4–6 × 2.5–3 mm; stamens numerous (in hundreds), oblong, 0.6–0.8 mm long; carpels 10–15(–18), free, 4–6(–11)-ovulate, pilose, style cylindrical, stigma minute. **Fruits** clusters of monocarps; monocarps, 8–15, ± fusiform, 5–6.8 × 1.8–2.2 cm, 3-ribbed; seeds 5, oblong-ellipsoid, ± 2.5 × 1.4 cm, 8 mm. thick. **Distr:** Kenya, Tanzania. **Ecol:** In coastal woodlands and near lakes; 10–450 m. **Zones:** 11–12.

Polyalthia

Species ~120. **Trees** or **shrubs**, glabrous or with simple hairs. **Inflorescences** with flowers solitary, or 2 many-flowered cymes or fascicles, extra-axillary, axillary or leaf-opposed, occasionally on the trunk. **Flowers** mostly bisexual; sepals 3, valvate or slightly imbricate; petals 6, biseriate, valvate, subequal, often quite large and showy; stamens mostly numerous, filament very short, cuneiform, connective-tip expanded, truncate or slightly convex; carpels few to numerous, free, oblong, 1–5 basal or superposed ovules, stigma ± sessile, rectangular, globose or irregularly shaped. **Fruits** clusters of monocarps; monocarps few to many, indehiscent, ellipsoid, ovoid or subglobose, stipitate or subsessile; seeds 1–5. **Distr:** Mainly in S and SE Asia, and tropical Africa. **Cult:** Propagate from seeds, cuttings or by air-layering. Requires a moist, humus-rich soil in full sun.

P. cerasoides (Roxb.) Bedd., *Ic. Pl. Ind. Or.* 17, 1870. (Syn: *Guatteria cerasoides, Uvaria cerasoides*). **Trees**, small, 2–10 m tall. **Leaves** narrowly elliptical or narrowly ovate, 8–17 × 2.5–6 cm, base rounded, truncate or obtuse, larger leaves usually long-acuminate, smaller leaves often obtuse and not acuminate, membranous, lateral veins 7–10 pairs; petiole 2–4(–5) mm long. **Inflorescences** with flowers solitary, borne axillary or from axils of fallen leaves; peduncle 1–2.7 cm long, bracteate; bracts 1.5–5 mm long, occasionally foliaceous and 3 × 1.8 cm. **Flowers** green; sepals triangular-ovate, 3–4 × 2.5–3 mm, densely pubescent outside; petals ovate, 4–8 mm long, thinly puberulous outside, inner petals slightly larger; stamens 1–1.5 mm long; carpels numerous (~50), ovules 1, densely pubescent, stigma clavate, shortly stipitate. **Fruits** clusters of monocarps; monocarps numerous, ovoid-ellipsoid, 5–7.5 mm Ø, red, long-stipitate; stalk (0.7–)1.2–1.5 cm long, slender; seeds not separating from fruit-wall. **Distr:** India, Sri Lanka to Indo-China. **Ecol:** In dry forests. **Zones:** 11–12.

P. coffeoides (Thw. ex Hook. f. & Thoms.) Thw., *Enum. P1. Zeyl.* 399, 1864. (Syn: *Guatteria coffeoides*). **Trees**, 6–18 m tall. **Leaves** ovate to narrowly ovate, rarely elliptical, 8–16 × 2–6 cm, base obtuse, rounded or truncate, margins undulate, apex gradually tapering (rarely rounded or obtuse), membranous, glabrous, lateral veins 10–14 pairs; petiole (2–)4–6 mm long. **Inflorescences** flowers solitary or sometimes few-flowered clusters, borne from a common axillary peduncle <1 mm long, or several from tubercles on old wood; pedicels 0.8–1 cm long; bracts to 1 mm long, tomentellous. **Flowers** apple-green; sepals 2–2.5 mm long, broadly triangular to almost semicircular, tomentellous outside; petals oblong, tomentellous outside, outer ones to 2 × 0.3 cm, inner ones to 2.5 × 0.6 cm; stamens 1 mm long; carpels ~20, 1-ovuled, puberulous; stigma shortly stipitate. **Fruits** clusters of monocarps; monocarps ovoid, to 2 × 1.4 cm, bluntly pointed, dark purple, glabrous, shiny; stalk ~1 cm long; seeds 1, pale reddish-brown. **Distr:** S Deccan Peninsula, Sri Lanka. **Ecol:** In dry and intermediate forests; <500 m. **Zones:** 11–12.

P. korinti (Dunal) Thw., *Enum. Pl. Zeyl.* 398, 1864. (Syn: *Guatteria korinti*). **Treelets**, small, slender, 0.7–5 m tall. **Leaves** ovate, elliptic, ovate-lanceolate or elliptic-lanceolate, 4–17.5 × 1.5–5.5 cm, base acute or obtuse, apex gradually acuminate, thinly coriaceous, adaxially glabrous, abaxially sparsely pubescent, lateral veins 5–10 pairs; petiole 2–5 mm long. **Inflorescences** usually flowers solitary, rarely paired or several (from tubercles on old wood), axillary; pedicels (0.8–)1–2(–3) cm long; bracts to 1.2 mm long. **Flowers** green; sepals rounded-ovate to almost orbicular, 1.2–2.7 × 1.2–2.7 mm, outside appressed-pubescent; petals ovate or elliptic, (3–)4–7 mm long, outside puberulous, blackish when dry, inner petals larger; stamens ~1 mm long; carpels usually 7–14, 1-ovuled, densely pubescent; stigma capitate, almost sessile. **Fruits** clusters of monocarps; monocarps globose, 4–6 × 4–6 mm, bright crimson, very succulent, long-stipitate; seeds gray, opaque. **Distr:** S India, Sri Lanka. **Ecol:** In gallery forests; 0–300(–500) m. **Zones:** 11–12. **Info:** Fruits edible.

P. longifolia (Sonnerat) Thwait., *Enum. Pl. Zeyl.* 398, 1864. (Syn: *Guatteria longifolia, Uvaria longifolia*). [Mast Tree]. **Trees**, evergreen, to 13 m tall, columnar; trunk straight, branches drooping. **Leaves** lanceolate, 21.5–30 × 3.7–5.6 cm, margins undulate, tapering to a fine point, glabrous, adaxially glossy; petiole 4–8 mm long. **Inflorescences** 6–many-flowered, pendent, axillary, pubescent; peduncle to 1.1 cm long; pedicel 2–3 cm long, pubescent; bract minute, caducous; bracteole densely hairy. **Flowers** with sepals broadly deltoid, 2 × 2–3 mm,

Below: *Mwasumbia alba, Ophrypetalum odoratum, Polyalthia cerasoides.*

TC

TC

Above: *Polyalthia coffeoides, Polyalthia korinti, Polyalthia longifolia, Polyalthia suberosa.*

obtuse to subacute, ± free, pubescent; petals narrowly lanceolate, 1–1.9 × 0.25–0.35 cm, outer slightly smaller, tapering into a fine point, slightly undulate, pubescent, pale green to yellowish-green; stamens 1 mm long, anther subsessile, locules unequal; carpels free, ~1.5 mm long, ovary glabrous, stigma subsessile, papillate. **Fruits** clusters of monocarps; monocarps 10–15, baccate, ovoid, rounded at both ends, 3.7 cm long, purplish-black, 1-seeded; stipe ~1.2 cm long. **Distr:** India, Malaysia, Sri Lanka, Pakistan, Tropical East Africa. **Ecol:** Fl. Mar–May. **Zones:** 10–12. **Info:** The fruits are mainly eaten by fruit bats. The trunk is often used for building and for makings masts.

P. suberosa (Roxb.) Thw., *Enum. Pl. Zeyl.* 398, 1864. (Syn: *Uvaria suberosa, Guatteria suberosa*). [Corky Mast Tree]. **Treelets**, 1–4 m tall. **Leaves** narrowly obovate or oblong, 2–11 × 1–3.5 cm, base obtuse, rounded or truncate and strongly asymmetrical, apex obtuse or rounded, sometimes abruptly short-acuminate, membranous, lateral veins 9–12 pairs; petiole 1–4 mm long. **Inflorescences** with flowers solitary, internodal, halfway between 2 leaves, or rarely leaf-opposed; peduncle slender, 1.2–2.5 cm long; bracts 1–2.3 mm long, narrowly triangular. **Flowers** yellowish-green, mostly suffused with purple, pubescent outside; sepals 1–2 × 1–2 mm, broadly triangular; petals 3–6 mm long; outer petals elliptic, acute, flat, spreading; inner petals longer, oblong, obtuse, curved, erect; stamens 1 mm long; carpels numerous, 2-ovuled, pubescent, stigma triangular, short-stipitate. **Fruits** clusters of monocarps; monocarps to 18, almost spherical, 4–5 mm Ø, purple, long-stipitate; stalk 0.6–1 cm long, slender; seeds 1 or 2 per monocarp. **Distr:** India and Sri Lanka to Indo-China. **Ecol:** In coastal areas. **Zones:** 11–12.

Pseudartabotrys

Species 1. *Generic description same as specific.* **Cult:** Propagate from seeds. Requires moist but well-drained soil in full sun.

P. letestui Pellegrin, *Bull. Mus. Natl. Hist. Nat.* 26: 656, 1920. **Climbers**, to 20 m long; branchlets brown-pubescent, long-hairy. **Leaves** ovate-oblong, 6–15 × 3–6.5 cm, base rounded or subcordate, margins ± entire, ciliate, apex attenuate, acute-acuminate, membranous, hairy, adaxially bright green, nerves 9–10 per side; petiole subsessile, 2–3 mm long, curved, villous. **Inflorescences** opposite, flowers solitary; peduncles hooked, but not indurated; bract 1, foliaceous, suborbicular, 0.8–1 × 0.6–1.2 cm, base cordate, apex apiculate, surfaces villose. **Flowers** bisexual, velvety tomentose, 1–1.5 cm long; sepals 3, valvate, deltoid, 1.5 × 1.1 cm, acute, outside ferruginous, inside subglabrous; petals 6, valvate, 2-whorled, subequal, spathulate, 1 × 1 cm, base concave, deltoid, with long linear lobes, 4–5 × 0.3 cm, base villose, coherent around the reproductive organs; stamens numerous; staminodes absent; carpels numerous, 1-ovuled, basal, style filiform, 3–4 mm long, stigma villose. **Fruits** clusters of monocarps. **Distr:** Equatorial Africa (Gabon, Mayombe). **Ecol:** In humid forests. Fl. Sep–Nov. **Zones:** 11–12.

Rollinia

Syn: *Rolliniopsis.*
Species ~55–60. **Trees** or **shrubs**, very rarely lianas; leafy twigs pubescent. Leaves distichous, petiolate, estipulate; lamina simple, often elliptic to ovate, margins entire, abaxially densely to sparsely pubescent. **Inflorescences** with flowers solitary, or in few-flowered cymes, opposite leaves or extra-axillary from internodes; pedicels basally bracteate, articulate above. **Flowers** perfect, bisexual, radially symmetrical; sepals 3, small, white, valvate, free or connate at base; petals 6, biseriately valvate, connate at base forming a short globose tube; outer 3 petals with a spurlike process or with a vertical, laterally compressed wing; inner petals minute; stamens numerous, extrorse, connective dilated above anther; torus convex; carpels numerous, 1-ovuled, ovule basal, erect. **Fruits** aggregates of monocarps, syncarpous, often large, globose or ovoid. **Distr:** Tropical America, West Indies. **Cult:** Propagate from seeds or by air-layering. Requires any well-drained soil in full sun.

R. mucosa (Jacquin) Baillon, *Adansonia* 8: 268, 1868. (Syn: *R. orthopetala*, etc.). [Biribá, Biribarana, Cachiman, Sweetsop, Wild Cashina, Wild Sugar Apple, Wild Sweetsop]. **Trees**, to 20(–25) m tall; leafy twigs and petioles densely hairy, becoming glabrous. **Leaves** narrowly elliptic (rarely ovate or narrowly obovate), (6–)10–25(–35) × (2–)4–8.5(–13) cm, base acute to rounded, apex shortly acuminate, acumen acute to rounded, chartaceous, adaxially mostly dark green, abaxially paler, adaxially sparsely hairy, secondary veins 13–25 per side; petiole 0.5–2 cm long. **Inflorescences** supra-axillary to leaf-opposed, 1–3(–7)-flowered; peduncles and pedicels densely hairy; peduncles 1–4 mm long; pedicels 1–5.5(–6) cm long; bracts triangular, 1–4 mm long, acute, persistent. **Flowers** green to creamy, yellow when mature; sepals free, appressed, sometimes recurved, broadly to shallowly ovate-triangular, (1.5–)2–5(–7) × (1.5–)3–5(–6) mm, mostly acuminate; corolla tube (2–)3–4(–5) mm long, wings oblong-ovate to obovate, 0.7–1.5(–2.5) × 0.4–1(–1.6) cm. **Fruits** mostly globose to ovoid, composed of (30–)50–150 carpels, 2–12(–20) × 2.5–11(–15) cm, yellow, areoles pyramidal to conical from a broad base, straight to recurved; seeds 0.7–1.8 × 0.5–1(–1.3) cm. **Distr:** Neotropical. **Ecol:** In various forest types; 0–1,500 m. Fl. Apr–Aug. Fr. Apr–Oct. **Zones:** 10–12. **Info:** Fruits are edible with excellent flavor.

Sageraea

Species ~9. **Trees**, often large, 30+ m tall. **Leaves** large, narrowly elliptic to narrowly ovate, both ends obtuse, base not decurrent along petiole, apex mostly short-acuminate, coriaceous, surfaces glabrous, lateral veins 8–12 pairs; petiole 0.7–1.8 cm long. **Inflorescences** axillary, mostly few-flowered sessile

Below: *Pseudartabotrys letestui, Rollinia mucosa* (flowers), *Rollinia mucosa* (fruits).

cymes or sometimes flowers solitary, borne from the old wood; pedicels glabrous, 1–2 cm long. **Flowers** orange-yellow, sepals and petals imbricate in bud; sepals short, almost orbicular, margins ciliate, apices obtuse; petals 6, biseriate, distinct, almost orbicular, to 1.2 × 1.2 cm, very obtuse, concave, spreading, glabrous; inner petals slightly smaller; stamens ~20, broadly wedge-shaped; staminodes absent; carpels 3–6, several ovules arranged in 2 rows along suture, distinct, sparsely appressed-pubescent, tapering into a sessile, depressed, bilobed stigma; receptacle flat. **Fruits** clusters of monocarps; monocarps mostly 2–4, globose, 2–2.7 cm Ø, green, smooth, thin-walled, pulp yellow, almost sessile; stalk when present to 2 mm long; seeds several, compressed, 1.4 × 1.1 cm, pale reddish-brown. **Distr:** India and Sri Lanka to Philippines. **Cult:** Propagate from seeds or by air-layering. Requires any well-drained soil in full sun.

S. thwaitesii Hook. f. & Thoms., *Fl. Ind.* 93, 1855. (Syn: *Bocagea thwaitesii*). *Description same as for the genus.* **Trees**, medium-sized, leafy branches 1.8–4 mm thick. **Leaves** with lamina 14–35 × 6–12 cm. **Distr:** Endemic to Sri Lanka. **Ecol:** In lower montane forests; 300–800 m. **Zones:** 11–12. **Info:** The fruits are usually eaten by native monkeys.

Uvaria

Syn: *Pyragma, Uvariella*, etc.

Species ~110. **Lianas, woody climbers, scandent shrubs** or **treelets**, 2–6(–9) m tall; branches at first stellate-pubescent, later glabrous; branches often converted into leafy tendrils. **Leaves** oblong-elliptic, narrowly oblong or oblong-obovate, (5–)6–14(–16.5) × (1.6–)2.5–5.5 cm, base cuneate or rounded, apex obtuse, acute or shortly acuminate, subcoriaceous, margins and midrib ferruginous; petiole 0.2–1 cm long. **Inflorescences** terminal or extra-axillary, flowers solitary or rarely in short 2-flowered cymes; pedicels 0.7–2.5 cm long, ferruginous-tomentose; bracteoles deciduous. **Flowers** reddish, purple, yellow or creamish-white, scentless or slightly fragrant; calyx lobes 3, ovate or triangular-ovate (valvate in bud), 0.8–1.3 × 0.7–0.9 cm, obtuse or acute, surfaces tomentose; petals 6(–8), biseriate, yellow, yellow-green or yellow-brown, sometimes rose-colored at base, subequal, ovate to broadly ovate, 1.2–2.5 × 1.1–1.9 cm, tomentose outside, glabrous inside; stamens numerous, linear, ~1 mm long; carpels numerous, distinct, cylindrical or funnel-shaped, strigose, truncate at apex with stigma horseshoe-shaped (when viewed from above), ovules several along suture. **Fruits** clusters of monocarps; monocarps 20–80, globose, subglobose or ellipsoid, 1–1.5 × 0.8–1.2 cm; seeds 2, ovoid, flattened, 6–8 × 4 mm, brown-tomentose, finely rugulose; stipes 3–5.5 cm long. **Distr:** Old World tropics, Australia, New Caledonia. **Cult:** Propagate from seeds or by air-layering. Requires any well-drained soil in full sun.

U. grandiflora Roxb., *Fl. Ind.* 2: 665, 1824. (Syn: *U. purpurea*). **Climbers**, sturdy. **Leaves** obovate or obovate-oblong, 10–28 × 5–9 cm, rounded, base truncate or shallowly cordate, apex acuminate, firmly membranous, laxly pubescent or glabrate, tomentose, lateral veins 14–24 pairs; petiole 2–3 mm Ø. **Inflorescences** with flowers solitary, rarely in 2–3-flowered cymes, leaf-opposed; pedicels 2–4 cm long, tomentose; bracts foliaceous, obovate, to 3 cm long. **Flowers** large, vermilion-red turning purplish-with age, slightly fragrant; sepals 1.5–2.5 cm long, triangular, basally connate, tomentose outside; petals 6, spreading, obovate or oblong, 3–4 × 1–1.7 cm, tomentellous outside; stamens ~6 mm long, outer stamens flattened, 2.5 mm wide, sterile; inner stamens 1–1.5 mm wide, linear, constricted above anther-cells and produced into a flat appendage ~ 1 × 1 mm. **Fruits** clusters of monocarps; monocarps cylindrical, 4–5.5 cm long, obtuse, slightly constricted between seeds, stipitate, orange, fleshy, minutely tomentose; seeds pale-brown; pedicel 1.3–2.5 cm long. **Distr:** SE Asia. **Ecol:** In sparse forests; 400–1,000 m. Fl. Nov–Jan. **Zones:** 11–12.

U. narum (Dunal) Wall., *Num. List* nr. 6473. 1828. (Syn: *Unona narum*). **Climbers**, robust, 10+ m long. **Leaves** elliptic, narrowly elliptic or narrowly obovate, (4–)9–20 × (1.5–)2.5–7 cm, base obtuse or acute, rarely rounded, apex gradually long-acuminate, thinly coriaceous, surfaces glabrous, lateral veins 8–14 pairs; petiole 0.8–2.5 mm long. **Inflorescences** with flowers solitary, terminal or leaf-opposed; pedicels 1–3.5 cm long; bracts 1–3 mm long. **Flowers** with sepals, rounded-ovate or rather semi-circular, 5–8 × 8 mm, very obtuse, basally connate; petals 6–8, purple or red, ovate or ovate-oblong, rarely elliptic, 1–2(–3) × 0.5–0.8(–1.7) cm, basally connate, spreading, slightly reflexed, surfaces densely velvety; stamens all fertile, 1.7 × 0.7 mm, narrowly cuneate, apex truncate. **Fruits** clusters of monocarps; monocarps globose or ovoid, rarely shortly cylindrical, 0.8–3.5 × 0.8–2 cm, mostly 1–2(–6)-seeded, long-stipitate, bright orange or scarlet, smooth, glabrous, fleshy; stalk 2.5–5 cm long; seeds ovoid or compressed, ~1 × ~1 cm, smooth, orange-brown. **Distr:** Deccan Peninsula, Sri Lanka. **Ecol:** In humid lowlands; 0–1,400 m. **Zones:** 11–12.

Uvariodendron

Syn: *Uvaria* sect *Uvariodendron*.

Species ~16. **Shrubs** or **trees**. **Inflorescences** with flowers solitary or fascicled, axillary on leafy shoots and/or also on old wood, sessile or stalked; bracteoles 2–6, often forming an involucre in case of sessile flowers. **Flowers** hermaphrodite; sepals 3, valvate, free or basally united, in most species not enclosing petals in bud; petals 6, in 2 whorls of 3, outer valvate, and at length spreading, inner valvate or somewhat imbricate above, open below, often remaining connivent at apex long after flowers have opened, free, mostly very thick and fleshy, often brittle, subequal; stamens very numerous, linear, anther-cells lateral or extrorse, connective-prolongations capitate; carpels numerous, cylindric, with numerous mostly biseriate ovules or in one species practically in 1 row, style very short or more developed, consisting of a plicate or involute lamina, apical margin being the stigmatic surface. **Fruits** clusters of monocarps; monocarps indehiscent, ovoid, ellipsoid or broadly to narrowly cylindric, straight, subsessile or shortly stipitate, many-seeded, often constricted between seeds; seeds ± horizontal, ellipsoid. **Distr:** Tropical Africa. **Cult:** Propagate from seeds. Requires moist but well-drained soil in full sun.

U. kirkii Verdc., *Kew Bull.* 23: 518, 1969. **Shrubs** or **small trees**, 1.5–6(–10) m tall. **Leaves** elliptic, oblong-elliptic or narrowly elliptic, 7–17.5 × 2.2–7.5 cm, base rounded or broadly cuneate, apex mostly acuminate, coriaceous, adaxially glabrous, somewhat shiny, abaxially sparsely pubescent, lateral nerves 10–14, prominent; petiole 3–8 mm long. **Inflorescences** with flowers solitary or several together in fascicles, axillary or on old wood; bract 1, reniform, 3 × 4 mm; peduncles 1–2 cm long. **Flowers** with sepals reniform, 4–5 × 7–8 mm, somewhat acute, adpressed-silky outside; petals mostly cream tinged maroon inside or inner ones maroon with cream margins; outer petals elliptic or ovate-elliptic, 1.5–2.2 × 1–1.5 cm wide, apex subacute, ferruginous silky outside, densely puberulous inside; inner petals obovate-cuneate, strongly concavo-convex, 1.5 × 1–1.2 cm, base narrowed, apex subacute, pubescent outside; stamens 1.9 mm long; carpels 7–20, subcylindric, 7-angled, 2 × 1.2 mm, adpressed-pilose, ovules ~11, biseriate. **Fruits** clusters of monocarps; monocarps 7–20, ellipsoid or oblong, 1.6–4.5 × 1.3–2 cm, obtuse, ripening to dull orange, glabrous, 2–9(–12)-seeded, stipes very short; seeds yellow-brown, rounded-cuneiform, 1.2 × 0.85 cm. **Distr:** Kenya, Tanzania, Zanzibar. **Ecol:** 0–230(–340) m. **Zones:** 11–12.

Uvariopsis

Syn: *Tetrastemma*.

Species 12. **Trees** or **shrubs**. **Inflorescences** with flowers solitary or in fascicles, axillary, often on old wood, pedicellate, monoecious or dioecious; bracteoles 2, one basal the other median, ± persistent. **Flowers** unisexual (or at most a few stamens present in ♀ flowers); sepals 2, connate to form a bilobed calyx, much shorter than petals in bud; petals 4, valvate, in 1 whorl, eventually spreading, free or somewhat joined, rather thick, those of ♀ flowers mostly larger; stamens numerous, oblong to oblong-ovoid, filaments absent or very short, anther-cells extrorse, connective not or scarcely prolonged; carpels numerous, obovoid or subcylindric, with several to numerous biseriate ovules, stigma sessile, cylindric, depressed-obconic, lunate or horseshoe-shaped. Fruits clusters of monocarps, indehiscent, cylindric, ellipsoid or obovoid, stipitate, several-seeded, often somewhat constricted between seeds, usually succulent; seeds horizontal, without aril. **Distr:** W and C Africa, to Uganda and W Kenya. **Cult:** Propagate from seeds. Requires moist but well-drained soil in full sun.

U. lovettiana Couvreur & Q. Luke, *Blumea* 55(1): 70, 2010. **Trees**, monoecious, 3–7 m tall; trunk 7–30 cm Ø; young branches sparsely covered with short appressed hairs. **Leaves** narrowly elliptic, 15–28 × 4.5–7 cm, base acute, apex acute to acuminate, acumen 0.5–1 cm long, glabrous, coriaceous, adaxially dark green, abaxially lighter green, midrib sunken above, raised below, sparsely appressed-pubescent hairs when young and rapidly disappearing with age, secondary veins 11–14, looped toward margin, tertiary venation reticulate. **Inflorescences** with ♂ and ♀ flowers borne on leafy branches (no cauliflory observed), axillary, single; pedicels 3–5 cm long, sparsely puberulous, green; bract basal, 1 × 1 mm, hairy. **Flowers** with hairs conspicuously base dark red and apex translucent; sepals 2, depressed ovate, 2–3 × 4–5 mm, base fused, apex rounded, densely hairy outside, sparsely so inside, green; petals 4, free, in 1 whorl, ovate, 1.2–1.5 × 0.8–1 cm, 2–3 mm thick, base truncate, apex acute, coriaceous, sparsely hairy outside, glabrous inside, light green outside, white inside; receptacle globose, ~5 mm Ø; ♂ with numerous stamens, ~2 mm long, apex of connective not prolonged; ♀ with numerous tightly packed carpels, these 2.5–3 × ~1 mm, densely appressed-pubescent, ovules 4–5, uniseriate, stigma 0.2 mm Ø, sessile, glabrous, stamens 1–5, fertile, minute, ~1 mm long, present at base of receptacle. **Fruits** clusters of monocarps; monocarps 2–4, 2.5–3 × 1–1.5 cm Ø, oblong, glossy yellow turning orange red at maturity, very sparsely puberulous, stipe ~1 cm long, ~2 mm Ø, sparsely puberulous, apex rounded with the clear presence of stigma; fruiting pedicels 3–3.5 cm long, sparsely appressed-puberulous; seeds 2–4 per monocarp, uniseriate, ~ 1.5 × 1 cm, ~8 mm thick, testa smooth, very thin papery, peeling off, light cream. **Distr:** Endemic to Tanzania (E Arc Mountains). **Ecol:** On rocky soils in wet submontane forests; 1,300–1,700 m. **Zones:** 11–12.

Xylopia

Species ~100–160. **Trees, shrubs** or **subshrubs**, rarely climbing. **Inflorescences** with flowers solitary, fasciculate or in cymes, axillary; buds ovoid, subglobose or more usually narrowly conical and 3-angled; bracteoles 1–5, deciduous or persistent. **Flowers** hermaphrodite; sepals 3, valvate, ± united,

RF TC
TC TC

Top row: *Sageraea thwaitesii, Uvaria grandiflora, Uvariopsis lovettiana.*
Bottom row: *Uvaria narum, Uvariodendron kirkii, Xylopia aethiopica.*

much smaller than outer petals; petals 6, in 2 whorls of 3, free, thick, enlarged and ± concave at base; inner petals sometimes with a fleshy part covering stamens and gynoecium; stamens numerous, linear, outer and inner ones often flattened and staminodial, filaments articulated, sometimes partly united basally and enclosing gynoecium; carpels few to many or rarely solitary, free, ovoid-oblong or flattened, with 1–many 1–2-seriate ovules, styles linear or narrowly clavate, aggregated. **Fruits** clusters of monocarps; monocarps obovoid to cylindric, 1–8(–many)-seeded, ± fleshy; seeds collateral or biseriate. **Distr:** Pantropical. **Cult:** Propagate from seeds. Requires moist but well-drained soil in full sun.

X. aethiopica (Dunal) A.Rich., *Sagra, Hist. He Cub., Bot., Pl. Vase.* 1: 53, 1845. (Syn: *Unona aethiopica, X. eminii*). [African Grains of Selim, Ethiopian Pepper, Kani Pepper]. **Trees** or **shrubs**, tall, evergreen, aromatic, 5.4–30(–45) m tall; trunk sometimes buttressed. **Leaves** oblong, elliptic, ovate or oblanceolate, (6–)8–16.4 × 2.8–6.5 cm, base cuneate to rounded, decurrent, apex obtuse, acute or acuminate (acumen 2 cm long), coriaceous; petiole 3–6 mm long, blackish. **Inflorescences** with solitary flowers or 3–5-flowered fascicles; pedicels 0.4–1.1 cm long; bracteoles 2–3, cucullate, 1–4 × 2–2.5 mm. **Flowers** fragrant; sepals ovate-triangular, 3–5 × 3–5 mm, rounded to apiculate; petals cream, greenish-white or yellow; outer petals linear, 2.5–5.5 mm long, base concave, outside silky-pubescent; inner petals smaller, 2.7–4.5 cm × 1.5–2 mm; stamens linear, 1–1.5 mm long; carpels 24–32(–42), ovary cylindric, 1–1.5 mm long, 6–8-ovuled, styles aggregated into a narrow cone ~3–4.5 mm long. **Fruits** clusters of monocarps; monocarps (5–)16–24(–42), green to reddish, cylindrical, 1.5–6 × 0.5–0.7 cm, straight, 1–8-seeded, glabrous; seeds orange-red to black, cylindrical, 5–7 × 2–4 mm, aril papery, yellow. **Distr:** Senegal to Sudan, south to Angola, Congo, Zambia and Mozambique. **Ecol:** 800–1,200 m. **Zones:** 11–12.

LAURALES

FAMILIES 7, GENERA 91, SPECIES ~2,800.

Trees (predominantly) or **shrubs**, mostly woody. **Leaves** opposite; petiole bundle(s) arcuate. Inflorescences ± cymose. **Flowers** with a well-developed floral cup, or cupule, which initially encloses stamens and carpel(s), and later plays a role in seed-dispersal by supporting and displaying single-seeded fruits; basal ovules/carpel 1 (2 in Calycanthaceae). **Fruits** with fruitlets 1-seeded, indehiscent, hypanthium persistent. **Distr:** Tropical and subtropical regions, with a few genera in the temperate zone. **Info:** Laurales is a monophyletic order (Renner 1999).

CALYCANTHACEAE

[Strawberry Shrub Family].

Genera 3, **species** 6. **Shrubs**, deciduous, rhizomatous, aromatic. **Stipules** absent. **Leaves** opposite, petiolate; lamina with margins entire or, on fast-growing shoots, occasionally with a few teeth. **Inflorescences** terminal on short, leafy branches of current growth, flowers solitary. **Flowers** bisexual; hypanthium urn-shaped; tepals and stamens perigynous; inner tepals, stamens, and staminodes with white succulent food bodies at tips; tepals 15–30, fleshy, petaloid, spirally inserted on outside of hypanthium; stamens 5–30, spirally inserted at top of hypanthium; staminodes 10–25, on inner surface of hypanthium, locules opening by longitudinal slits; pistils 5–35, 1-carpellate, spirally inserted within hypanthium, ovules 2 per locule, distal ovule abortive, submarginal, style elongate, filiform, stigma dry, decurrent. **Fruits** achenes, 1–35, dark reddish-brown, cylindric, pubescent, enclosed within enlarged, fibrous persistent hypanthium (pseudocarp); seeds without endosperm. **Distr:** N America; China. **Info:** This family is predominantly temperate, but is included here as they are often cultivated in cooler regions of the tropics and subtropics.

Calycanthus

Syn: *Basteria, Sinocalycanthus*, etc.

Species 3. **Shrubs**, deciduous, aromatic; branchlets dichotomous; buds naked, hidden (or not hidden) by base of petiole. **Leaves** membranous. **Inflorescences** with flowers terminal on branchlets, solitary, not aromatic, distinctly pedicellate. **Flowers** with tepals numerous, distinctly dimorphic (or not); stamens numerous, filaments short, staminodes few; carpels numerous, with silky hairs. **Fruits** pseudocarps, campanulate, base with an attenuate long stalk, apex constricted; achenes oblong. **Distr:** China, N

America. **Cult:** Propagate from seeds, cuttings or by air-layering. Requires a moist soil in full or half-sun.

C. floridus L., *Syst. Nat.* ed. 10. 2: 1066, 1759. [Carolina Allspice, Sweetshrub]. **Shrubs**, to 3.5 m tall; lateral buds partially hidden by petiole base. **Leaves** elliptic, broadly elliptic, oblong or ovate, 5–15 x 2–6 cm, base acute to truncate, apex acute, acuminate, or blunt, surfaces abaxially green or glaucous, pubescent to glabrous; petiole 0.3–1 cm long, pubescent to glabrous. **Flowers** with hypanthium cylindric, ellipsoid, pyriform, or globose at maturity, 2–6 x 1–3 cm; tepals oblong-elliptic to obovate-lanceolate, 2–4 cm x 3–8 mm, acute; stamens 10–20, oblong. **Distr:** N America. **Ecol:** In exposed parts of upland hardwood and mixed pine forests, flooded plains, wooded slopes and stream banks. **Zones:** 5–9.

C. occidentalis Hooker & Arnott, *Bot. Beechey Voy.* 340, plate 84, 1841. (Syn: *Butneria occidentalis*). [California Spicebush, Sweetshrub]. **Shrubs**, to 4 m tall; lateral buds exposed. **Leaves** ovate-lanceolate to oblong-lanceolate or ovate-elliptic, 5–15 x 2–8 cm, base rounded to nearly cordate, apex acute to obtuse, abaxially green, pubescent to glabrous; petiole 0.5–1 cm long, pubescent to glabrous. **Flowers** with hypanthium campanulate or ovoid-campanulate at maturity, 2–4 x 1–2 cm; tepals linear to linear-spathulate or ovate-elliptic, 2–6 x 0.5–1 cm, rounded; stamens 10–15, linear to oblong-linear. **Distr:** USA (California, Washington). **Ecol:** Along streams and on moist canyon slopes; 200–1,600 m. **Zones:** 7–10.

Chimonanthus

Syn: *Butneria, Meratia.*

Species 6. **Shrubs** or **small trees**, erect, deciduous or evergreen; branchlets dichotomous, quadrangular to subterete. **Leaves** papery or subleathery, adaxially scabrous or ± smooth. **Inflorescences** axillary, subsessile to very shortly pedicellate, flowers solitary or paired. **Flowers** fragrant; tepals numerous, yellow, yellowish-white, or white and sometimes with purple markings, membranous, varying in size and shape from outer to inner but not distinctly dimorphic; stamens 5–8, arranged on cuplike receptacle, filaments filamentous but basally broad and connate, usually puberulent; staminodes few to numerous, puberulous, arranged inside stamens on receptacle; carpels 5–15, distinct, ovules 2 per carpel (usually 1 abortive). **Fruits** pseudocarps, urceolate, ovoid-ellipsoid, obovoid-ellipsoid, or campanulate, pubescent; achenes oblong, ellipsoid or reniform. **Distr:** China. **Cult:** Propagate from seeds or cuttings. Requires a fertile free-draining soil in full sun.

C. praecox (L.) Link, *Enum. Pl. Hort. Berol.* 2: 66, 1822. (Syn: *Butneria praecox*, etc.). [Wintersweet, Japanese Allspice]. **Shrubs** or **small trees**, 3–13 m tall. **Leaves** ovate, elliptic, oblong-elliptic, or oblong-lanceolate, 5–29 x 2–12 cm, base cuneate to rounded, apex acute, acuminate, or caudate, abaxially glabrous, secondary veins 4–6 per side; petiole 0.3–1.8 cm long. **Inflorescences** with flowers solitary or paired, appearing generally before leaves; pedicels 2–8 mm long. **Flowers** 1.5–4 cm Ø, sweetly fragrant; tepals 15–21, yellow, inner ones usually with purplish-red pigmentation, 0.5–2 x 0.5–1.5 cm; outer tepals orbicular to obovate, truncate or rounded; median tepals elliptic to oblong-elliptic, rounded to acute; inner tepals orbicular to oblong, base distinctly clawed, margins ciliate, apex rounded; stamens 5–8, 2.5–4 mm long, filaments broad; staminodes 2–15, subulate to linear-lanceolate, 2–3 mm long; carpels 5–15. **Fruits** pseudocarps, urceolate, ovoid-ellipsoid, or obovoid-ellipsoid, 2–6 x 1–2.5 cm, apical appendages 9 or 10; achenes 3–11, brown, ellipsoid to reniform, 1.5–1.65 x 0.5–0.5 cm. **Distr:** China. **Ecol:** In montane forests; 500–1,100 m. Fl. Oct–Mar. Fr. Apr–Nov. **Zones:** 6–10.

SIPARUNACEAE

Genera 2, **species** 75. **Shrubs** or **small trees**, sometimes lianas, evergreen; vessels with simple or scalariform and simple perforations, xylem without fiber tracheids. **Leaves** opposite or verticillate; lamina simple, margins entire, surfaces sometimes gland-dotted. **Inflorescences** cymes, spikes or panicles, monoecious or dioecious. **Flowers** actinomorphic to zygomorphic; hypanthium free; perianth with distinct calyx and corolla, segments 4–6(–8), or obscure, calyptrate; stamens (1)2–100, filaments free, anthers introrse, bisporangiate (in *Siparuna*) or tetrasporangiate (in *Glossocalyx*); gynoecium with carpels (3–)4–100, uniovulate, style free or apically connate. **Fruits** drupelets, enclosed in fleshy hypanthium; seeds erect, endosperm not oily. **Distr:** Tropical America; W Africa (1 sp.). **Info:** Plants accumulate aluminum (Takhtajan 2009).

Siparuna

Syn: *Angelina, Leonia,* etc.

Species 75. **Shrubs** or **small trees**, dioecious or monoecious, aromatic. **Leaves** usually opposite; lamina with margins entire to irregularly dentate, membranous to pergameneous, glabrous to densely hairy. **Inflorescences** with flowers pedicellate to subsessile, usually in axillary dichotomous or trichotomous cymes; ♂ flowers 3–many; ♀ flowers usually fewer. **Flowers** with hypanthium often surmounted by an annular fusion of tepal bases; ♂ with 4–8 subequal exappendiculate tepals united below to form an annulus interior to which is often a flat or hemispherical, apically perforate velum, stamens 1–many, scattered or in circular disposition, filaments often dilated or conduplicate, anthers oblong, valvate; ♀ similar to ♂ but larger, annulus and velum often present, carpels 4–many, styles elongate, usually exserted, often basally connate. **Fruits** drupaceous or coriaceous, permanently enclosed in accrescent hypanthium; seeds tuberculate, ultimately exposed by irregular bursting of aggregate fruit. **Distr:** Mexico, C America, S America. **Cult:** Propagate from seeds or rooted cuttings. Requires a moist but well drained compost in shade.

S. pauciflora (Beurl.) A. DC., *Prodr.* 162: 696, 1868. (Syn: *Citriosma pauciflora, S. cauliflora*). [Pasmo, Limoncillo]. **Shrubs** or **small trees**, dioecious, aromatic, to 10 m tall; trunk often branching from base; branchlets terete to quadrangular. **Leaves** opposite; lamina narrowly elliptic to obovate, (10–)15–40 x (6–)8–16 cm, basally cuneate to rounded, margins coarsely serrate to almost entire, apex acute to acuminate, usually membranous, lateral veins 10–16 pairs; petiole canaliculate, 1–6 cm long. **Inflorescences** ♂ of 5–15-flowered, subsessile axillary cymes, often cauliflorous, gray to yellow stellate-pubescent throughout, peduncles, 1–6 mm long; ♀ similar to ♂ but cymes usually sessile, 3–15-flowered; pedicels 0.2–1 cm long. **Flowers** ♂ with tepals connate, forming an undulate subentire annulus, velum conical, 4–5 mm Ø, apical pore 1–1.5 mm Ø, stamens (8–)12–30, outermost filaments dilated, ovate to orbicular, ~1 mm long, valvate anthers minute, introrse, usually slightly exserted; ♀ with tepals connate forming an undulate annulus, velum ultimately concave, many separate styles scarcely exserted through the orifice. **Fruits** aggregates, globose, capped by persistent annulus, yellowish-green, drying to black, 1–1.5 cm Ø, bursting irregularly exposing pinkish interior and several tuberculate seeds. **Distr:** Peru to Colombia and Costa Rica. **Ecol:** In forests of the Atlantic coast; <1,200 m. **Zones:** 9–12.

HERNANDIACEAE

Genera 5, **species** ~60. **Trees**, **shrubs** or **lianas**, with aromatic oils. **Leaves** alternate, exstipulate; lamina simple, digitately compound or 3–5-lobed. **Inflorescences** axillary, rarely terminal, corymbose or paniculate cymes, pedunculate, bracteate or ebracteate. **Flowers** regular, bisexual, or unisexual by abortion and plants monoecious or polygamous, rarely dioecious; perianth segments subequal, free or shortly united below, in two 3–5-merous, valvate whorls or in one 4–8-merous, imbricate whorl; stamens 3–5 in 1 whorl and opposite outer segments when perianth is double, or (2)4(7) when perianth is single, anthers bithecous; staminodes absent or ± glandlike, 1–2-whorled, usually outside the stamens, outer whorl paired at base of stamens, inner whorl alternating with stamens; ovary inferior, 1-locular, ovule solitary, pendulous, anatropous. **Fruits** dry, indehiscent, often winged, longitudinally ribbed or with lateral wings (in *Illigera*), or with 2 apical wings formed by accrescent perianth segments (in *Gyrocarpus*), or wingless and enclosed in an inflated capsule (in *Hernandia*); seeds solitary, without endosperm. **Distr:** Old and New World tropics.

Below: *Calycanthus floridus, Calycanthus occidentalis, Chimonanthus praecox, Siparuna pauciflora.*

SA/SAJF

MD

SZ

Hernandia

Syn: *Biasolettia.*

Species ~24. **Trees**, evergreen, monoecious. **Leaves** alternate, petiolate, peltately attached or not; lamina simple, broadly ovate or subcircular, 3–7-veined. **Inflorescences** lax panicles, apical, flowers pedicellate, ternately involucrate, central flower with a cupular involucel at base, lateral ones staminate; involucral bracts 4 or 5, subvalvate in bud. **Flowers** 3–5(or 6)-merous, unisexual; ♂ with tepals 6–8, in 2 series, subvalvate, stamens as many as and opposite outer tepals, filaments each with 2 free basal glands or glands connate; ♀ with tepals 8–10, in 2 series, subvalvate, staminodes absent, style short, surrounded by 4 or more free or connate glands, stigma dilated, irregularly toothed or lobed. **Fruits** enclosed in an inflated fleshy cupule; seeds globose or ovoid, testa thick, hard, ribbed, embryo thick, lobed, ruminate. **Distr:** E and W Africa; SE Asia, China; C America. **Cult:** Propagate from seeds, cuttings or by air-layering. Requires any soil in full sun.

H. nymphaeifolia (C. Presl) Kubitzki, *Bot. Jahrb. Syst.* 90: 272, 1970. (Syn: *H. peltata, H. sonora* [misapplied], etc.). [Jack in the Box]. **Trees**; bark smooth. **Leaves** peltate, ovate or orbicular-cordate, 20–40 × 15–30 cm, base rounded and cordate, apex acute, papery, glabrous, glossy, 3–7-veined; petiole nearly as long as lamina. **Inflorescences** axillary panicles, 3 together in a corymb, flowers arranged with lateral flowers ♂ and central one ♀; peduncle 6–20 cm long; involucral bracts 4, elliptic to obovate, 2–6 × 1–3.5 mm; ♂ flowers with pedicels tomentose; ♀ flowers sessile. **Flowers** ♂ with tepals 6, biseriate, stamens 3, filaments with 2 basal glands, ~1 mm long; ♀ with tepals 8, biseriate, style ~3 mm long, papillose, stigma dilated, irregularly toothed, surrounding glands 4, ~1 mm long. **Fruits** drupaceous, urn-shaped, costate, 3–4 cm Ø. **Distr:** Cambodia, Indonesia, Japan, Malaysia, Maldives, Philippines, Sri Lanka, Thailand, Vietnam; E Africa and east to the Pacific. **Ecol:** In sparse forests and near beaches (salt-tolerant); 0–500 m. Fl. and Fr. Jan–Dec. **Zones:** 11–12.

Illigera

Syn: *Gronovia, Henschelia.*

Species ~30. **Lianas**, climbing by leaf petioles, evergreen. **Leaves** alternate, petiolate; lamina trifoliolate (rarely 5-foliolate); leaflets petiolulate. **Inflorescences** panicles composed of axillary cymes, rarely terminal. **Flowers** 5-merous, bisexual; perianth segments (tepals) in 2 rows, valvate in bud; outer tepals (sepals) oblong or narrowly elliptic, rarely ovate-elliptic, 3–5-veined; inner tepals (petals) similar to outer ones, 1–3-veined; stamens 5, epigynous, filaments each at base with 2 appendages, inflated, membranous and shortly stipitate or appendages clavate and solid, anthers opening by valves, stamens alternating with mostly 5 small (interstaminal) glands; ovule pendulous from apex of locule, style filiform, stigma dilated into an undulate crest. **Fruits** 2–4-winged; wings brown when dry, broad, striate; seed solitary, with membranous testa. **Distr:** Tropical regions of Africa and Asia. **Cult:** Propagate from seeds. Requires moist but well-drained soil in full sun.

I. rhodantha Hance, *J. Bot.* 21: 321, 1883. **Lianas**; stems angulate; branchlets tomentose or villous. **Leaves** trifoliolate; leaflets ovate to obovate-elliptic or ovate-elliptic, 6–11(–16) × (2–)3–7(–9) cm, base subrounded or subcordate, apex obtuse or shortly acuminate, papery, lateral veins ~4-paired, conspicuous, veinlets scalariform; petiole 4–10 cm long, sparsely pubescent, tomentose, or villous; petiolules 0.3–1.5 cm long, densely brownish-tomentose. **Inflorescences** cymose panicles, axillary, narrow, densely brownish-tomentose. **Flowers** red; outer tepals purple-red, oblong, 0.8–1 cm long, abaxially sparsely pubescent; inner tepals similar to outer ones but slightly shorter, rose; stamens 6–9 mm long, pubescent, filaments straight in bud, 5–7 mm long, appendages on ~1 mm long stipe, overall ~2.5 mm long, membranous, abaxially with open mouth, apex dentate; interstaminal glands small; ovary densely brown pubescent, style 5–6 mm long, yellow-tomentose. **Fruits** 4-winged; large wings ligulate or suborbicular, 2.5–3.5 cm wide, smaller ones 0.5–1 cm wide. **Distr:** Cambodia, China, Laos, Thailand, Vietnam. **Ecol:** In valley forests, thickets, mountain slopes; 100–600(–2,100) m. Fl. (Jun–)Sep–Nov. Fr. Dec–May. **Zones:** 10–11.

LAURACEAE

Genera ~45, **species** 2,000–2,500. **Trees** or **shrubs** (*Cassytha* a twining parasitic perennial herb with leaves reduced or absent), mostly evergreen (deciduous in temperate regions), sometimes dioecious; bark and foliage usually aromatic. **Leaves** usually alternate, occasionally opposite or subopposite or verticillate, exstipulate; lamina simple, margins usually entire, rarely lobed (*Sassafras*), mostly pinninerved and subpalmately veined, or often triplinerved, usually punctate and leathery. **Inflorescences** usually axillary, occasionally subterminal, panicles, spikes, racemes, or pseudo-umbels. **Flowers** generally bisexual, sometimes unisexual, actinomorphic, mostly 3-merous, also 2-merous, small, greenish, yellowish, or white; perianth biseriate, of usually 4 or 6 basally connate usually undifferentiated sepal-like segments, deciduous or persistent; perianth tube usually persisting as a cupule at base of fruit; androecium typically of 4 whorls of 3 stamens each, adnate to perianth tube, innermost whorl rarely united, usually reduced to staminodes, filaments usually free, 3rd whorl usually with 2 sessile and distinct basal glandular protuberances, anthers basifixed, 2- or 4-celled at anthesis, those of 2 outer whorls mostly introrse, inner 3rd whorl extrorse, dehiscing by flaplike valves opening upward; pistil 1, ovary usually superior, 1-loculed, ovule solitary, anatropous, pendulous, placentation parietal, style 1, stigma 1, occasionally bi- or trilobed. **Fruits** drupes or berries, usually surrounded at base by enlarged and often persistent perianth tube seated on a large receptacle or pedicel; seed with large straight embryo. **Distr:** Tropical and subtropical regions, mostly in tropical SE Asia and tropical America.

Alseodaphne

Species >50. **Trees**, evergreen; terminal buds scaly. Leaves alternate, always clustered near apices of branchlets; lamina pinninerved, often turning black when dry. **Inflorescences** axillary, paniculate or racemose; bracts and bracteoles deciduous. **Flowers** bisexual, 3-merous; perianth tube short, lobes 6, subequal or outer 3 smaller, slightly dilated after anthesis but absent in fruit; fertile stamens 9, in 3 whorls, filaments of 1st and 2nd whorls glandless, those of 3rd whorl each with 2 glands at base, anthers 4-celled, cells of 1st and 2nd whorls introrse, those of 3rd whorl extrorse or upper 2 lateral and lower 2 extrorse; staminodes 3, of innermost whorl, very small, nearly sagittate; ovary partly immersed into shallow perianth tube, style often as long as ovary, stigma small, inconspicuous, discoid. **Fruits** black or purplish-black when mature, ovoid, oblong, or subglobose, sometimes nearly cylindric, fleshy, pulpy, always warty, apex truncate; pedicel red, green, or yellow. **Distr:** Cambodia, China, India, Indonesia, Laos, Malaysia, Myanmar, Philippines, Sri Lanka, Thailand, Vietnam. **Cult:** Propagate from seeds or by air-layering. Requires moist but well-drained soil in full sun.

A. semecarpifolia Nees, *Pl. As. Rar.* 2: 72, 1831. (Syn: *Persea semecarpifolia*). **Trees**, to 15 m tall; trunk 60 cm Ø, not buttressed; bark very rough, vertically deeply fissured; branchlets slender, smooth, glabrous, lenticellate. **Leaves** crowded near tips of branchlets; lamina cuneately obovate-elliptic to obovate or oblong, 6–20 × 2.5–7 cm, base cuneate, apex obtuse or rounded, rarely subacute, coriaceous, glabrous, surfaces minutely areolate-reticulate, adaxially glossy, abaxially paler or glaucous, midrib prominent, lateral veins 6–8 pairs; petiole 0.8–2.5 cm long, adaxially flat. **Inflorescences** panicles, very lax, longer than leaves, long-peduncIulate; pedicels filiform. **Flowers** with tube shallow; tepals glabrous or slightly puberulous, oval, obtuse, outer rather smaller. **Fruits** ellipsoid, 2–2.5 × 1 cm; pedicel long, swollen, dark red, warted. **Distr:** Peninsular India, Sri Lanka. **Ecol:** In forests of dry regions. **Zones:** 9–12.

Beilschmiedia

Syn: *Lauromerrillia.*

Species 300. **Trees** or **shrubs**, evergreen; terminal buds mostly conspicuous. **Leaves** opposite, subopposite, or alternate, leathery, thickly leathery, or papery, very rarely membranous, entire, pinninerved, veins usually conspicuous. **Inflorescences** short, mostly cymose-paniculate, sometimes as axillary

Below: *Hernandia nymphaeifolia* (flowers), *Hernandia nymphaeifolia* (fruits), *Illigera rhodantha* (flowers), *Illigera rhodantha* (fruits).

NDT

NDT

Above: *Alseodaphne semecarpifolia*, *Beilschmiedia roxburghiana*, *Cassytha filiformis* (flowers).

flower-bundles or subracemose, sometimes covered by imbricate caducous bracts when young; peduncles and pedicels enlarged or not enlarged after anthesis. **Flowers** small, bisexual; perianth tube short, perianth segments 6(or 8), equal or almost equal; fertile stamens (6 or 8 or)9, filaments of 1st and 2nd whorls glandless, those of 3rd whorl generally each with 2 stalked or stalkless glands at base, anthers 2-celled, anther cells of 1st and 2nd whorls introrse, those of 3rd whorl extrorse; staminodes of 4th whorl, ovoid, cordate, or deltoid, shortly stalked; ovary attenuate at apex into a style. **Fruits** baccate, ellipsoid, ovoid-ellipsoid, cylindric, obovoid, or subglobose; fruiting pedicels swollen or not swollen; perianth in fruit generally wholly deciduous. **Distr:** Mainly in tropical Africa, SE Asia, Australasia, America and China. **Cult:** Propagate from seeds or by air-layering. Requires moist but well-drained soil in full sun.

B. roxburghiana Nees, *Pl. Asiat. Rar.* 2: 69, 1831. (Syn: *B. fagifolia*). **Trees**, 10–15(–20) m tall; branchlets blackish-brown, compressed, conspicuously angled; terminal buds small, densely brownish-pubescent. **Leaves** opposite, sometimes alternate; lamina elliptic, narrowly elliptic, or elliptic-lanceolate, 9–14 × 3.5–5 cm, base broadly cuneate or subrotund, apex obtuse, obtusely acuminate, acute, or subrotund, papery or subleathery, surfaces minutely glandular-punctate, lateral veins 10–15 pairs, slender, interconnected on margins, abaxially elevated and reddish; petiole 1.5–2 cm long, slender. **Inflorescences** cymose panicles or racemes terminal or axillary, short, 5–15 cm, wholly densely yellowish-pubescent; pedicels ~1 mm long. **Flowers** small; perianth segments ovate, ~1.5 mm long; fertile stamens 9. Fruits ellipsoid, 4–5 × 2–3 cm, smooth, both ends subrounded, apex mucronate; pedicels robust, 0.5–2 cm long, always brown-maculate. **Distr:** Bhutan, India, Myanmar, Nepal, Thailand. **Ecol:** In evergreen broad-leaved forests on mountain slopes. Fl. & Fr. Aug. **Zones:** 10–12.

Cassytha

Syn: *Calodium*, *Volutella*.

Species 15–20. **Vines** twining, viscid, parasitic on trees and shrubs using haustoria; stem green or gray-brown, filiform, branched. **Leaves** reduced to minute scales. **Inflorescences** with flowers inserted in stalked or stalkless scalelike bracts, each with 2 bracteoles adnate to perianth base, all arranged into a spike, flowerhead, or racemes. **Flowers** small, bisexual (plants rarely dioecious or nearly dioecious due to abortion); perianth tube turbinate or ovoid, contracted on top after anthesis; perianth lobes 6, in 2 series, outer 3 very small; fertile stamens 9, rarely those of 2nd whorl reduced to narrow staminodes, filaments of 1st and 2nd whorls eglandular, those of 3rd whorl each with 2 subsessile glands; staminodes 3, of innermost whorl; ovary nearly not included in perianth tube when in flower, wholly included after anthesis due to dilated perianth tube contracted on top; style inconspicuous, stigma small or capitate, subsessile. **Fruits** included in dilated fleshy perianth tube, free; perianth tube with orifice and persistent lobes on top; seeds thinly membranous or leathery. **Distr:** Tropical regions worldwide, majority in Australia, few in Africa, one species pantropical. **Cult:** Propagate from seeds or cuttings. Requires a suitable host to climb on, and full sun.

C. filiformis L., *Sp. Pl.* 1: 35. 1753. [Love Vine]. **Vines**, herbaceous, perennial, stem green-brown, filiform, slightly woody, rusty pubescent when young but sparsely pubescent or glabrous when mature. **Leaves** reduced to minute scales. **Inflorescences** spikes ~2.5 cm long, densely rusty pubescent; bracts and bracteoles brown, minute, broadly ovate, ~1 mm, ciliate. **Flowers** white, small, less than 2 mm, sessile; perianth lobes 6, in 2 series, outer 3 small, orbicular, ciliate, inner 3 larger, ovate, all pubescent outside, subglabrous inside; fertile stamens 9, filaments of 1st whorl subpetaloid, those of 2nd and 3rd whorls filiform, of 3rd whorl each with 2 stalkless glands, others glandless, anthers 2-celled; staminodes 3, stalked; ovary ovoid, subglabrous, style short, slightly angled, stigma small, capitate. **Fruits** small, ovoid, included in dilated and fleshy perianth tube after anthesis, free, with persistent lobes on top. **Distr:** Tropical regions of Africa, Asia, and Australia. **Ecol:** In thickets or sparse forests on mountain slopes, becoming weedy in some places, salt-tolerant; 0–1,600 m. Fl. and Fr. May–Dec. **Zones:** 11–12. **Info:** Used for medicinal purposes.

Cinnamomum

Syn: *Camphora*, etc.

Species ~250. **Trees** or **shrubs**, evergreen; bark, branchlets and leaves aromatic; buds naked or perulate, if perulate, scales distinct or indistinct, imbricate. **Leaves** alternate, subopposite, or opposite, sometimes clustered at end of branchlets; lamina leathery, trinerved or triplinerved, also pinninerved. **Inflorescences** panicles, axillary, subterminal, or terminal, composed of (1–)3- to many-flowered cymes. **Flowers** yellow or white, small- to medium-sized, bisexual, rarely polygamous; perianth tube short, cupuliform or campanulate, lobes 6, subequal, entirely deciduous or upper halves deciduous but rarely entirely persistent after anthesis; fertile stamens 9, rarely less or more, in 3 whorls, filaments of 1st and 2nd whorls eglandular, those of 3rd whorl each with 2 stalked or stalkless glands near base, anthers 4-celled, rarely those of 3rd whorl 2-celled; staminodes 3, of innermost whorl, cordate or sagittate, stipitate; ovary always as long as style, style slender, stigma capitate or discoid, sometimes trilobed. **Fruits** fleshy, subtended by persistent perianth cup; perianth cup cupuliform, campanulate, or conical, truncate or undulate, irregularly denticulate or sometimes with 6 truncate lobe bases at apex. **Distr:** Tropical and subtropical Asia, Australia, Pacific Islands. **Cult:** Propagate from seeds or by air-layering. Requires moist but well-drained fertile soil in full sun.

C. camphora (L.) J. Presl, *Prir. Rostlin.* 2(2): 36, 1825. (Syn: *Laurus camphora*, etc.). [Camphor Laurel, Camphor Tree]. **Trees**, evergreen, large, to 30 m tall; trunk to 3 m Ø; whole plant strongly camphor-scented. **Leaves** alternate; lamina ovate-elliptic, 6–12 × 2.5–5.5 cm, base broadly cuneate or subrounded, margins cartilaginous, entire or sometimes undulate, apex acute, adaxially green or yellow-green and shiny, abaxially yellow-green or gray-green and glaucous, subleathery, triplinerved or sometimes inconspicuously 5-nerved, lateral veins 1–5(–7) pairs, conspicuously bullate adaxially; petiole slender, 2–3 cm long. **Inflorescences** panicles, 3.5–7 cm long, axillary; peduncle 2.5–4.5 cm long; pedicels 1–2 mm long. **Flowers** green-white or yellowish, ~3 mm long; perianth glabrous or puberulent outside, densely pubescent inside, tube obconical, ~1 mm long, lobes elliptic, ~2 mm long; fertile stamens 9, ~2 mm, filaments pubescent; ovary ovoid, ~1 mm long, style ~1 mm long. **Fruits** purple-black, ovoid or subglobose, 6–8 mm Ø. **Distr:** Japan, Korea, Vietnam. **Ecol:** Fl. Apr–May. Fr. Aug–Nov. **Zones:** 11–12. **Info:** Cultivated commercially for camphor.

C. verum J. Presl, *Prir. Rostlin.* 2(2): 36, 1825. (Syn: *Laurus cinnamomum*, *C. zeylanicum*). [Ceylon Cinnamon, Cinnamon]. **Trees**, small, evergreen, to 10 m tall; bark black-brown, inner bark with cinnamic aldehyde flavor; young branchlets gray, somewhat tetragonous, white-maculate; buds sericeous-puberulent. **Leaves** usually opposite; lamina ovate or ovate-lanceolate, 11–16 × 4.5–5.5 cm, base acute, margins entire, apex acuminate, leathery or subleathery, surfaces glabrous, adaxially green, shiny, abaxially greenish-white, triplinerved, midrib and lateral veins elevated on both surfaces, transverse veins and veinlets reticulate, conspicuously foveolate abaxially; petiole ~2 cm long, glabrous. **Inflorescences** axillary or terminal panicles, 10–12 cm long; peduncle and rachis sericeous-puberulent. **Flowers** yellow, ~6 mm long; perianth tube obconical, perianth lobes 6, oblong, subequal, gray-puberulent outside; fertile stamens 9, filaments hairy near base, those of 3rd whorl each with 2 glands, others glandless, anthers 4-celled; ovary ovoid, 1–1.5 cm, glabrous, style short, stigma discoid. **Fruits** ovoid, 1–1.5 cm long, black when mature; persistent perianth cupuliform, dilated, dentate, teeth truncate or acute. **Distr:** Endemic to Sri Lanka. **Ecol:** In moist low country; to ~700 m. **Zones:** 11–12. **Info:** Cultivated widely for the spice cinnamon (bark) and medicinal purposes.

SF

Above: *Cassytha filiformis* (fruits), *Cinnamomum camphora*, *Cinnamomum verum* (flowers), *Cinnamomum verum* (fruits).

Laurus

Species 2. **Trees**, evergreen, small. **Leaves** alternate; lamina leathery, pinninerved. **Inflorescences** pedunculate umbels enveloped by 4 decussate involucral bracts before anthesis, axillary, always in pairs, usually 1 or 3 in clusters on short racemes. **Flowers** unisexual; perianth tube short, lobes 4, subequal; ♂ with stamens 8–14, mostly 12, in 3 whorls, filaments of 1st whorl eglandular, but those of 2nd or 3rd whorls each with 2 reniform glands at the middle, anthers 2-celled, cells introrse, ovary infertile; ♀ with staminodes 4, alternate to perianth lobes, filaments each with 2 sessile glands and a lanceolate ligule between glands at apex, ovary 1-celled, style short, stigma slightly dilated, obtusely trigonous, 1-ovuled. **Fruits** ovoid, perianth tube not or slightly dilated, ruptured or not. **Distr:** Macaronesian islands, Mediterranean region. **Cult:** Propagate from seeds or half-hardened cuttings. Requires a fertile, well-drained soil in full sun.

L. nobilis L., *Sp. Pl.* 1: 369, 1753. [Bay Leaf, Laurel, Sweet Bay]. **Small trees** or **shrubs**, evergreen, to 12 m tall, dioecious. **Leaves** alternate; lamina oblong or oblong-lanceolate, 5.5–12 × 1.8–3.2 cm, base cuneate, margins slightly undulate, apex acute or acuminate, leathery, adaxially dark green, glabrous, abaxially greenish, lateral veins 10–12 pairs; petiole purple-red, 0.7–1 cm long, adaxially sulcate. **Inflorescences** umbels 1–3 in clusters of short racemes, globose; involucral bracts suborbicular, glabrous outside, sericeous inside; peduncle to 7 mm long, sparsely puberulent or subglabrous; ♂ flowers 5 per umbel, pedicels ~2 mm long, pilose. **Flowers** ♂ green, small, perianth tube short, densely pilose outside, lobes 4, broadly obovate or suborbicular, appressed villous, fertile stamens usually 12, in 3 whorls, filaments of 1st whorl eglandular, others with 2 reniform glands, ovary infertile; ♀ with staminodes 4, ovary 1-celled, style short, stigma slightly dilated, obtusely trigonous. **Fruits** dark purple when mature, ovoid. **Distr:** Mediterranean region. **Ecol:** Fl. Mar–May. Fr. Jun–Sep. **Zones:** 8–11. **Info:** The leaves are used as a spice worldwide, and has many medicinal uses.

Litsea

Syn: *Hexanthus*, *Malapoënna*, etc.
Species ~250. **Trees** or **shrubs**, evergreen or deciduous, dioecious. **Leaves** alternate, rarely opposite or verticillate; lamina pinninerved. **Inflorescences** umbels, or umbellate cymes or panicles, solitary or clustered in leaf axils; involucral bracts 4–6, decussate, persistent at flowering. **Flowers** unisexual; perianth tube long or short, segments usually 6, in 2 whorls of 3 each, equal or unequal, rarely lacking or 8; ♂ with fertile stamens 9 or 12, rarely more, in 3 or 4 whorls of 3 each, filaments of 1st and 2nd whorls usually eglandular, of 3rd and 4th biglandular at base, pistil rudimentary or absent; ♀ with staminodes as many as stamens of ♂ flowers, ovary superior, style conspicuous. **Fruits** berries, seated on persistent perianth tube; perianth tube ± enlarged, shallowly discoid or deeply cup-shaped or unaltered. **Distr:** Mainly in tropical and subtropical Asia, a few species in Australia and from N America to subtropical S America. **Cult:** Propagate from seeds or by air-layering. Requires moist but well-drained soil in full sun.

L. glutinosa (Loureiro) C.B. Robinson, *Philipp. J. Sci.* C 6: 321, 1911. [Indian Laurel, Pond Spice]. **Trees**, evergreen or deciduous, 3–15 m tall; young branchlets yellowish-tomentose. **Leaves** alternate; lamina obovate, obovate-oblong, or elliptic-lanceolate, 3.5–10(–26) × 1.5–11 cm, base cuneate, obtuse, or rotund, apex obtuse or rounded, tomentose when young, pinninerved, lateral veins 5–12 pairs; petiole 1–2.6 cm long, yellowish-tomentose. **Inflorescences** umbels solitary or several on short branchlets, few-flowered; short branchlets 2–4+ cm; peduncle 1–1.5 cm long; ♂ with pedicel tomentose. **Flowers** with perianth segments imperfect or lacking; fertile stamens often 15+, filaments pubescent, of 3rd whorl each with 2 long stipitate glands at base; rudimentary pistil glabrous. **Fruits** globose, 5–7 mm Ø; pedicels 3–6 mm long, slightly thickened apically. **Distr:** Bhutan, China, India, Myanmar, Nepal, Philippines, Thailand, Vietnam. **Ecol:** In forest margins, streamsides, thickets; 500–1,900 m. Fl. May–Jun. Fr. Sep–Oct. **Zones:** 9–12.

Persea

Species over 200. **Trees** or **shrubs**. **Leaves** alternate or rarely verticillate, petiolate; lamina entire and pinnately veined. **Inflorescences** axillary or pseudoterminal, solitary or in fascicles, paniculate to racemose (rarely capitate); bracts persistent or deciduous, puberulent and usually with many flowers; pedicels usually present. **Flowers** bisexual, perianth of 6 tepals in 2 whorls, campanulate to rotate; outer tepals usually shorter than the inner and glabrous on the inner surface or the outer tepals are equal to the inner tepals and puberulent on their inner surface, tepals often persisting in fruits, floral cup or tube absent or only slightly developed; androecium of 9 stamens and 3 staminodes, stamens of the outer whorls introrse and with long filaments, outer anthers (2)4-thecous, inner stamens with prominent filaments, staminodes usually large and conspicuous; pistil glabrous or puberulent, ovary globose to ellipsoid (occasionally stipitate), style slender, stigma small to discoid or triangular-peltate. **Fruits** berries, globose to ellipsoid or pyriform, 1–15 cm Ø. **Distr:** Mostly in Western Hemisphere (well represented in tropical and subtropical America); India, SE Asia, Australia, Mascarene Islands, the Azores, and the Canary Islands. **Cult:** Propagate from seeds, cuttings or by air-layering. Requires a well-drained humus-rich soil in full sun.

P. americana Miller, Card. Diet. ed. 8, 1768. (Syn: *Laurus persea*, *P. drymifolia*). [Avocado, Alligator Pear, Butter Fruit]. **Trees**, small to large, 5–40 m tall. **Leaves** alternate or subopposite; lamina narrowly to broadly ovate or ovate-elliptic, sometimes obovate or

Below: *Laurus nobilis*, *Litsea glutinosa*, *Persea americana*.

suborbicular, 6–22(–30) × 3–14(–19) cm, base acute to obtuse or rounded and often slightly asymmetric, margins entire, apex usually short-acuminate, secondary veins 5–9 per side; petiole 1–6 cm long. **Inflorescences** axillary to distal leaves, or in terminal clusters, 4–12 cm long, compact or loosely branched panicles. **Flowers** 5–8 mm long, campanulate; outer tepals ovate-elliptic to oblong, 4–6 × 1–3 mm, acute; inner whorl 4–7 mm long; outer stamens 3.5–4.5 mm long, inner stamens with slender filaments and stalked glands; staminodes 2–3 mm long; pistil 3–4 mm long, style long slender, stigma simple. **Fruits** berries, globose to pyriform or obovoid, 5–15 cm long, dark green; seed 2–5 cm Ø. **Distr:** Costa Rica, Cocos Island. **Ecol:** In evergreen and partly deciduous forests; 0–2,500 m. Fl. Jan–Mar. Fr. Feb–Nov. **Zones:** 9–12. **Info:** Widely cultivated for the edible fruits.

CANELLALES

FAMILIES 2, GENERA 9–13, SPECIES 75–105.

Trees or **shrubs**; indument absent; primary stem with continuous vascular cylinder; sieve tube plastids with starch and protein crystalloids. **Leaves** with petiole bundle(s) arcuate. **Flowers** with calyx and corolla distinct. **Fruits** berries. **Distr:** New Guinea, Australia, New Caledonia, and other islands of the SW Pacific; Africa, Madagascar; Mexico, S America. **Info:** The monophyly of the Canellales is supported by both morphological (Doyle & Endress 2000) and molecular (Hilu et al. 2003; Qiu et al. 1999, 2000, 2005; Soltis et al. 2000; Zanis et al. 2002, 2003) studies, and are sister to Piperales (Judd et al. 2008).

CANELLACEAE

[Canella, Wild Cinnamon].

Genera 5, **species** 13. **Trees** or rarely prostrate shrubs, evergreen, aromatic; pith homogeneous. **Stipules** absent. **Leaves** alternate; lamina simple, margins entire, unlobed pinnately veined, pellucid dots conspicuous or inconspicuous; petiole short. **Inflorescences** terminal and axillary, cymes, rarely racemes or flowers solitary, pedunculate. **Flowers** bisexual; perianth hypogynous, segments imbricate; sepals persistent, 3; petals 5(–12) in 1(–4) whorl(s); stamens (7–)10(–12), hypogynous, monadelphous, filaments connate, forming tube around pistil, anthers extrorse, longitudinally dehiscent; pistil 1, superior, 2–6-carpellate, ovary 1-locular, placentation parietal, placentas 2(–6), ovules 2–3 per placenta, style 1, generally short, stigma 1, usually 2–6-lobed. **Fruits** berries; seeds 2 or more, not arillate, endosperm oily (ruminate). **Distr:** Primarily tropical, Western Hemisphere, Africa, Madagascar.

Canella

[Canella, Wild Cinnamon].

Species 1. *Generic description same as specific.* **Cult:** Propagate from seeds or by air-layering. Requires a well-drained soil in full sun.

C. winterana (L.) Gaertner, *Fruct. Sem. Pl.* 1: 373, 1788. (Syn: *C. alba, Laurus winterana*). [Canella, Cinnamon Bark, Wild Cinnamon]. **Trees**, rarely large, sprawling shrubs, (3–)8–10(–15)m; bark whitish-gray; stems erect, rarely prostrate. **Leaves** obovate to oblanceolate, 5–15 × 2–5 cm, base acute, apex rounded, notched, or blunt, deep green, shiny, thick, leathery, abaxial surface pellucid-dotted, strongly aromatic when bruised, causes sharp burning sensation on tongue when bitten. **Inflorescences** 5–40-flowered, crowded toward end of stem; pedicels short. **Flowers** bisexual, protogynous, ~7 mm Ø; sepals green, imbricate, 2–3 mm long, fleshy; petals basally connate, 4.5–6 mm long, dark red to violet, base light red to yellow, thick and fleshy; stamens 10, filaments connate into tube surrounding pistil, tube protruding slightly beyond anthers, anthers light red, becoming yellow at anthesis, extrorsely dehiscent; pistil flask-shaped, ovary conic, style short, stigma bilobed, yellow. **Fruits** berries, green through red to dark purple with age, globose, fleshy; seeds 1–5, black, shiny, hard. **Distr:** Tropical N America (Florida), West Indies, and NE South America. **Ecol:** In coastal thickets, hammocks. Fl. May–Jul. **Zones:** 8–12.

PIPERALES

FAMILIES 4, GENERA 17, SPECIES 4,090.

Herbs or soft-woody plants; vessel elements with simple perforation plates; starch grains compound; primary stem with distinct vascular bundles; vessel elements in radial files, with simple perforation plates; wood with broad rays; nodes often swollen; stomata not paracytic. **Stipules** intrapetiolar. **Prophyll** single, adaxial. **Leaves** 2-ranked; lamina with leaf base sheathing, secondary veins palmate. **Flowers** ± 3-merous. **Distr:** Worldwide. **Info:** The monophyly is strongly supported by morphological (Doyle & Endress 2000) and molecular (Barkman et al. 2000; Graham & Olmstead 2000; Hilu et al. 2003; Mathews & Donoghue 2000; Nickrent et al. 2002; Qiu et al. 1999, 2000, 2005; Soltis et al. 2000; Zanis et al. 2003) studies.

PIPERACEAE

Genera 8 or 9, **species** 2,000–3,000. **Herbs**, **shrubs**, or **climbers**, rarely trees, usually aromatic; vascular bundles ± scattered in transverse section in a monocotyledonlike manner; tip of stem sometimes enclosed within a stipulelike sheath, prophyll, sometimes adnate to petiole, or absent. **Leaves** alternate, often opposite or whorled; lamina simple, base often asymmetric, palmately or pinnately veined. **Inflorescences** pedunculate spikes, rarely grouped into umbels, rarely racemes, leaf-opposed or axillary, rarely terminal, nearly always sessile; bracts small, usually peltate or cupular. **Flowers** small, bisexual, hermaphroditic, polygamous or dioecious, usually without perianth; stamens 1–10, filaments usually free, anthers bilocular, distinct or connate, longitudinally dehiscent; gynoecium 2–5-carpellate, connate; ovary superior, 1-locular, ovule 1, orthotropous, stigmas 1–5, sessile or with very short styles. **Fruits** small drupes or nutlets; pericarp fleshy, thin or dry, sometimes with sticky papillae or glochidiate spines. **Distr:** Tropical and subtropical regions, mostly in N and S America, rather fewer in Asia, a few in Africa.

Lepianthes

Species 2. **Herbs**, coarse, perennial or shrubby; stems erect, jointed and thickened at nodes. **Stipules** ovate-lanceolate, large, caducous. **Leaves** alternate, long-petioled; lamina ovate-cordate or orbicular-cordate, membranous, not fleshy, veins 4–8 pairs, palmate-arcuate, from base and one pair from midrib 1.5–3 cm above base; petiole distally cylindrical, proximally channeled and basally amplexicaul. **Inflorescences** spikes (2–)3–7, umbellately arranged on a common axillary peduncle; later mostly paired, sometimes 1 or 3, unequal in length and in number of spikes, erect; bracts densely fimbriate, peltately petioled, distinct, succulent, triangular with point directed downward. **Flowers** crowded, hermaphrodite; stamens 2; ovaries distinct, obovate-ovoid, triquetrous, truncate at apex, crowned by 3 cylindrical, recurved, deciduous stylules, style absent. **Fruits** not fleshy, sessile, brown or blackish when ripe. **Distr:** Primarily from the Americas, naturalized throughout the tropics. **Cult:** Propagate from seeds. Requires a very moist humus-rich soil in shade.

L. umbellatum (L.) Raf., *Sylva Tellur.* 85, 1838. (Syn: *Heckeria subpeltata*, etc.). [Shrubby Pepper, Monkey's Hand]. **Herbs**; stem 1–2(–3) m tall, slightly zigzag, coarsely warted and finely striate when dry, very minutely puberulous when young, soon glabrate. **Leaves** ~15–30 × ~15–30 cm, base deeply cordate, apex abruptly short-acuminate, densely dotted with minute, dark brown glands on both sides, cross veins 0.3–0.6(–1.5) cm distant, very faintly raised adaxially; petiole 14–32 cm long. **Inflorescences** spikes, 5–10(–12) cm long, 2.5–4 mm Ø; common peduncles 0.5–4 cm long, individual peduncles of 2–7 mm long; bracts 0.3–0.5 × 0.3–0.5 mm. **Fruits** top-shaped, 0.6–0.7 mm long. **Distr:** Tropical America, Africa, India, Sri Lanka, and SE Asia. **Ecol:** In shaded parts of disturbed and humid secondary forests. **Zones:** 10–12.

Macropiper

Species ~6. **Shrubs** or **small trees**, sometimes scrambling; branches often with swollen nodes. **Stipules** present. **Leaves** alternate, stipulate; lamina with margins entire, venation palmate; petiole sheathing the stem. **Inflorescences** unisexual, spikes or clusters, solitary or umbellate, axillary, flowers close-set, bracteate. **Flowers** unisexual, minute; ♂ with stamens (2–)3, anther-cells distinct; ♀ with carpels 3–4, ovary superior. **Fruits** small drupes, orange or red often connate to bracts and axis. **Distr:** New Zealand, Polynesia to New Guinea. **Cult:** Propagate from seeds or half-hardened cuttings. Requires a moderately fertile, well-drained soil in partial shade.

M. excelsum (Forst. f.) Miq., *Syst. Pip.* 221, 1843. (Syn: *Piper excelsum*). [Pepper Tree]. **Shrubs** or **trees**, aromatic, glabrous, to 6 m tall; branches ± zigzag, jointed and swollen at nodes, dark. **Leaves** opposite; lamina broadly ovate to suborbicular, 5–10 × 6–12 cm, base cordate, margins entire, apex rather abruptly narrowed to obtuse, subcoriaceous, dark to yellowish-green, sinus narrow to open, 5–7(–9) subpalmately nerved; petioles 1–4 cm long, adnate to stipules. **Inflorescences** unisexual, spikes, solitary or paired, 2–8 cm long, flowers very close-set; bract orbicular-peltate; peduncles ± 1 cm long; flowers sessile. **Flowers** minute; ♂ with stamens 2(–3); ♀ with 3(–4) stigmas. Fruits drupes, very close-set, 2–3 mm Ø, ± angled; exocarp succulent, yellow to orange, broadly obovoid. **Distr:** New Zealand and nearby islands; NSW (Lord Howe Island). **Ecol:** In lowland forests. Fl. & Fr. throughout the year. **Zones:** 7–11.

Peperomia

Species ~1,000. **Herbs**, annual or, usually, perennial, rooting from nodes toward base of stem and with erect or ascending flowering shoots; stems usually dwarf, fleshy; vascular bundles free, scattered; prophylls absent. **Leaves** alternate, opposite, or whorled; lamina with main lateral veins all basal. **Inflorescences** spikes, usually erect, terminal or axillary, rarely leaf-opposed, solitary, paired, or clustered; rachis as thick or slightly thicker than peduncle; bracts ± orbicular, peltate (sometimes oblong and/or not peltate). **Flowers** bisexual, very small, often within depressions in rachis, sessile; stamens 2, filaments short, thecae ± globose, ellipsoid, or cylindric; ovary 1-loculed, ovule 1, stigma 1, rarely 2-cleft, globose, apex obtuse or acute, beaked or brushlike, lateral or terminal. **Fruits** very small, sticky nutlets, often partly enclosed in pit in rachis, sometimes distinctly curved. **Distr:** Widely distributed in tropical and subtropical regions. **Ecol:** Mostly growing

Top row: *Lepianthes umbellatum, Macropiper excelsum, Peperomia argyreia, Peperomia caperata.*
Bottom row: *Peperomia griseoargentea, Peperomia obtusifolia, Peperomia pellucida.*

on trees or moss-covered rocks. **Cult:** Propagate by division, or from stem or leaf cuttings. Requires a moist but well-drained compost in partial shade.

P. argyreia E. Morr., *Belg. Hortic.* xvii., 2. t. 2, 1867. (Syn: *P. sandersii*, etc.). [Watermelon Plant]. **Herbs**, glabrous, rosette-forming, 15–25 cm tall; stems and petiole purplish. **Leaves** alternate; lamina peltate, broadly ovate, 8–15 cm long, base rounded, shortly and broadly acuminate, fleshy, adaxially light to dark green, longitudinally striped or blotched with silvery gray between nerves, abaxially paler, veins palmate, 8–11; petiole to 15 cm long. **Inflorescences** spikes, 3–5 in a panicle, very slender, to 10 cm long, white or pale green; flowering stalks red, sometimes bearing much smaller reduced leaves which are often not peltate; bracts circular, gland-dotted. **Flowers** minute, compact. **Fruits** subglobose-ovoid, smooth, stigma apical. **Distr:** Tropical America (Brazil). **Ecol:** Fl. throughout most of the year. **Zones:** 9–12.

P. caperata Yunck., *Kew Bull.* 12(3): 421, 1957. [Emerald Ripple]. **Herbs**, small, succulent, essentially glabrous; stems short, fleshy, ~3 cm high, 3–4 mm Ø, simple or branching from base. **Leaves** alternate or appearing clustered because of scarcely elongated internodes; lamina ovate or round-ovate, 2–3.5 × 1.5–2.8 cm, base rounded or truncate, peltate within 5 mm from the margin, or slightly to deeply cordate and narrowly to scarcely peltate, apex bluntly pointed, nerves 5–7-palmate, adaxially glossy vivid dark green and strongly plicate-bullate with bullae minutely hispidulous, abaxially paler and glabrous; petiole slender, to 7+ cm long, green flecked with minute red spots, dilated and clasping at base. **Inflorescences** spikes, from upper leaf axils, numerous, flowering portion white, mostly 4–6 cm long, loosely flowered; peduncles to 10+ cm long, rose-red flecked with deeper red; bracts rounded-peltate with somewhat irregular margin, glandular-dotted. **Flowers** with stamens 2, lateral; ovary obovoid-subturbinate with large, essentially apical, sometimes bifid, puberulent stigma. **Fruits** drupes, globose-obovoid. **Distr:** Origin unknown. **Ecol:** In tropical rainforests. **Zones:** 10–12.

P. griseoargentea Yuncker, *Kew Bull.* 1957: 421, 1958. (Syn: *P. hederifolia*). [Ivy Pepper, Silver-leaf Peperomia]. **Herbs**, erect, to 15 cm high. **Leaves** alternate; lamina round-ovate, 3.5–7 cm long, base cordate, apex rounded to acute, shiny, puckered to ± pleated and puckered, adaxially raised areas glabrous and silver-green, sunken areas purplish, abaxially paler and dull, secondary veins 7–9, palmate; petiole 10–15 cm long, pink striped with red, peltate. **Inflorescences** solitary, 5–12 cm long (excluding stalk), greenish-white; peduncle reddish, 10–14 cm long; flowers crowded, in discrete bands when young; bracts circular, gland-dotted. **Fruits** globose-ovoid, smooth, stigma apical, tufted, often 2-forked. **Distr:** Possibly Brazil. **Ecol:** Epiphytic in shaded parts of moist forests. **Zones:** 9–11.

P. obtusifolia (L.) A. Dietrich, *Sp. Pl.* 1: 154, 1831. (Syn: *Piper obtusifolium*, etc.). [American/Baby Rubber Plant, Florida Peperomia]. **Herbs**, perennial, rhizomatous, erect, decumbent, reclining, or somewhat scandent, simple or sparsely branched, 8–40 cm high, mostly glabrous. **Leaves** alternate; lamina highly variable in size and shape, prominently to obscurely pinnately veined, elliptic, broadly elliptic, spathulate to ovate, or very broadly obovate, 2.2–13 × 1.5–5.5 cm, base attenuate, mostly narrowly to broadly cuneate, rarely slightly acuminate, apex obtuse, rounded, retuse, or notched, lateral veins arching-ascending, originating from base to about middle of lamina, surfaces mostly glabrous. **Inflorescences** spikes mostly terminal, 1–2, densely flowered, 5.5–23 cm long. **Fruits** sessile, ellipsoid, both ends rounded, 0.8–1.1 × 0.4–0.5 mm, faintly longitudinally striate; mature fruiting spikes 3.5–5.5 mm Ø. **Distr:** USA (Florida), Mexico, West Indies, N South America. **Ecol:** In hammocks, epiphytic or terrestrial on decayed logs and humus; 0–20 m. **Zones:** 10–12.

P. pellucida (L.) Kunth, *Nov. Gen. Sp.* 1: 64, 1816. (Syn: *Piper pellucidum*). [Shiny Peperomia, Vietnamese Crab Claw]. **Herbs** annual, fleshy, 20–40 cm high, all parts glabrous; stems erect or ascending, branched, glabrous. **Leaves** broadly ovate or ovate-triangular, length ± equal to width, 1–3.5 cm long, base cordate, apex acute or obtuse, glabrous, membranous, translucent, veins 5–7; petiole 1–2 cm long. **Inflorescences** spikes, terminal or leaf-opposed, slender, 2–6 cm long, glabrous, flowers lax; bracts suborbicular, ~0.5 mm wide, stalk short; anthers subglobose; ovary ellipsoid, stigmas pubescent. **Fruits** nutlets, globose, ~0.5 mm Ø. **Distr:** Tropical America. **Ecol:** In wet places within forests, rock crevices, bases of cliffs, sometimes a weed of cultivation; near sea level to 200 m. Fl. Apr–Jul. **Zones:** 10–12.

Piper

Syn: *Chavica, Pothomorphe.*

Species 1,000–2,000. **Shrubs** or **climbers**, rarely herbs or small trees, aromatic; stems articulated, thickened at nodes; outer vascular bundles in a ring, inner bundles scattered in 1 or 2 series. **Prophylls** present, often ± connate to petiole, caducous, forming a prominent, ringlike stipular scar at each node. **Stipules** lanceolate, caducous. Leaves alternate, often aromatic; lamina often heart-shaped, margins entire, venation prominent; petiole not amplexicaul. **Inflorescences** leaf-opposed or rarely terminal spikes, rarely spikes grouped in apparently axillary umbels; bracts small, sometimes adnate to rachis, often peltate. **Flowers** mostly unisexual, dioecious, less often monoecious or bisexual, sessile; stamens 2–6, usually on rachis, rarely at base of ovary, anthers biloculed, 2–4-lobed; ovary distinct or sometimes immersed in rachis, ovule 1, stigmas (2 or)3–5. **Fruits** drupes, sessile or stalked, often red or yellow, ovoid, globose, or cuneate-obovoid and laterally compressed, rarely ellipsoid, usually glabrous; seed 1. **Distr:** Mainly in tropical regions. **Cult:** Propagate from cuttings or seeds. Requires moist, fertile soil in full or partial sun.

P. betle L., *Sp. Pl.* 1: 28, 1753. [Betel, Betle Pepper]. **Climbers**, dioecious, aromatic; stems rooted at nodes, slightly woody. **Prophylls** ~0.5–1.5 cm long. **Leaves** ovate to ovate-oblong, apical ones sometimes elliptic, 7–15 × 5–11 cm, base cordate, sometimes rounded in apical leaves, ± symmetric, apex acuminate, papery to ± leathery, adaxially glabrous, abaxially densely glandular, veins 7, apical pair arising 0.7–2 cm above base; petiole 2–5 cm long, powdery-pubescent. **Inflorescences** spikes, leaf-opposed; ♂ spikes nearly as long as lamina at anthesis; peduncle nearly as long as petiole; rachis pubescent; bracts orbicular, rarely obovate, 1–1.3 mm wide, peltate, ± sessile; ♀ spikes 3–5 × ~1 cm, longer in fruit; rachis fleshy, densely pubescent. **Flowers** ♂ with stamens 2; ♀ with ovary partly immersed in and connate to rachis, apex tomentose, stigmas usually 4 or 5. **Fruits** drupes fused to form terete, fleshy, reddish, compound fruit, apices tomentose, prominent. **Distr:** India, Indonesia, Malaysia, Philippines, Sri Lanka, Vietnam; Africa (Madagascar). **Ecol:** Fl. May–Jul. **Zones:** 8–12. **Info:** Widely used in tropical Asia for chewing with *Areca catechu* (areca nuts) and medicinal purposes. The flowers are edible with a spicy and pepperlike flavor.

P. kadsura (Choisy) Ohwi, *Acta Phytotax. Geobot.* 3: 81, 1934. (Syn: *Ipomoea kadsura*, etc.). [Kadsura Pepper]. **Climbers**, woody, dioecious; stems rooting at nodes, ridged. **Leaves** ovate or long ovate, 6–12 × 3.5–7 cm, base cordate to rounded, ± symmetric, apex acute or obtuse, ± leathery, adaxially glabrous, with uniformly scattered raised white glands, abaxially usually pubescent on veins, veins 5, apical pair arising up to 1.5 cm above base, others basal; petiole 1–1.5 cm long, sheathed at base only. **Inflorescences** spikes, leaf-opposed; ♂ spikes yellow, ascending, 3–5.5(–12) cm × ~2.5 mm, peduncles 0.6–1.5 cm long; rachis hispidulous, bracts yellow, orbicular, ~1 mm wide, peltate, margin irregular, abaxially roughly white-pubescent, ± sessile; ♀ spikes shorter than lamina, rachis and bracts as in ♂. **Flowers** ♂ with stamens 2 or 3, filaments short; ♀ with ovary globose, distinct, stigmas 3 or 4, linear, pubescent. **Fruits** drupes, brownish-yellow, globose, 3–4 mm ∅. **Distr:** Japan, S Korea, Taiwan. **Ecol:** In lowland forests, on trees or rocks; 200–1,500 m. Fl. May–Aug. **Zones:** 10–11.

P. longum L., *Sp. Pl.* 1: 29, 1753. (Syn: *Chavica roxburghii*). [Bengal Pepper, Indian Long Pepper, Jaborandi Pepper, Long Pepper]. **Herbs**, perennial, 30–60 cm high; stems jointed, slender, prostrate or ascending. **Leaves** of *sterile shoots* long-petioled, ovate, deeply cordate; leaves of *fertile branches* erect, narrowly ovate, elliptic, or oblong, 4–7 × 2–3 cm, base cordate, symmetrical or slightly asymmetrical, apex acute or shortly acuminate, thinly membranous, usually lacking silvery scales, arcuate veins (2–)3, all basal or upper pair fused with midrib for 0.5–1 cm, petioles 0.1–1.2 cm long. **Inflorescences** spikes, erect, with axis glabrous; peduncles 1.2–2.2 cm long; ♂ 3–6 cm long, 1.5–2 mm ∅; ♀ 1.4–3 cm long, at anthesis 3 mm ∅, in fruit to 8 mm ∅; bracts peltate, orbicular, narrowly stipitate, membranous; flowers crowded, dioecious. **Flowers** with stamens 2; stylules 3–4, 0.4–0.5 mm across. **Fruits** berries, concrescent, forming a fleshy spadix. **Distr:** China, India, Malaysia, Nepal, Sri Lanka, Vietnam; origin obscure, probably NE India. **Ecol:** In forests; ~600 m. **Zones:** 10–12. **Info:** Cultivated for medicinal purposes.

P. magnificum (Linden) Gentil, *Pl. Serr. Jard. Brux.* 152, 1907. (Syn: *Artanthe magnifica*, etc.). [Peruvian Pepper]. **Shrubs**, soft-wooded, mostly glabrous; stems with 6–8 frilled wings running longitudinally down the stem. **Leaves** elliptic to oblong and ovate, 15–22 × 7–11 cm, base shallow-cordate, apex subacuminate, pinnately nerved throughout, adaxially glossy dark green with midrib white, abaxially purple; petiole 1–5 cm long, broadly winged. **Inflorescences** spikes, bisexual, 1.5–5 cm long, 6 mm ∅, cuspidate, pendent; peduncles becoming equally long, winged, purplish; bracts rounded-subpeltate. **Flowers** with stigmas short, spreading. **Fruits** berries, round, with a slender style. **Distr:** Peru. **Ecol:** In lowland tropical forests. **Zones:** 10–12.

P. nigrum L., *Sp. Pl.* 1: 28, 1753. [Black Pepper, Common Pepper, Madagascar Pepper, Pepper Plant, White Pepper]. **Climbers**, woody; nodes enlarged and rooting, glabrous. **Leaves** ovate to ovate-oblong, rarely suborbicular, 10–15 × 5–9 cm, base rounded, usually slightly oblique, apex acute, thick, ± leathery, glabrous, veins 5–7(–9), apical pair arising 1.5–3.5 cm above base, alternate, others basal; petiole 1–2 cm long, glabrous. **Flowers** polygamous, usually monoecious, spikes leaf-opposed, to as long as leaves; peduncle nearly as long as petiole, glabrous; bracts spathulate-oblong, 3–3.5 × ~0.8 mm, adaxially adnate to rachis, only margin and broad, rounded apex free, shallowly cupular. **Flowers** with stamens 2, 1 on each side of ovary, filaments thick, short, anthers reniform; ovary globose, stigmas 3 or 4, rarely 5. **Fruits** drupes, red when ripe, drying to black when unripe, globose, 3–4 mm ∅, sessile. **Distr:** SE Asia. **Ecol:** Fl. Jun–Oct. **Zones:** 8–12. **Info:** Widely cultivated in the tropics for the commercial pepper.

P. ornatum N.E. Brown, *Gard. Chron.* II (22): 424, 1884. (Syn: *P. crocatum*). [Celebes Pepper, Ornamental Pepper, Ornate Pepper Vine]. **Climbers**, shrubby, sometimes scrambling, trailing, or as a ground cover, 3–5 m long; stems slender, glabrous, terete, rooting at the nodes; internodes 2.5–7.5 cm long. **Stipules** similar to those of *P. longum*, but are free from the petiole for the upper half of their length, ~1.2 cm long, free part terete-acuminate, caducous. **Leaves** directed to one side; lamina round, peltate, cordate or ovate-orbicular, 5–13 × 5–10 cm, apex abruptly contracted into an obtuse tip, acumen ~1.2 cm long, surface puckered, glabrous, lustrous green, mottled pink with pink veins, primary lateral nerves 7, curving upward, reaching apex; young leaves adaxially bright green, with pink spots, older leaves the spots dull becoming white; petiole 6.5–9 cm long, terete, slightly channeled, reddish-tinted when young. **Inflorescences** spikes, rarely formed in cultivation. **Flowers** minute, white. **Distr:** Indonesia (Sulawesi). **Ecol:** In areas with dappled light and high humidity. **Zones:** 10–12.

P. sintenense Hatusima, *Acta Phytotax. Geobot.* 4: 210, 1935. (Syn: *P. hispidum*). **Climbers**, to several m long, dioecious; stems rooting at nodes, densely roughly rusty pubescent when young. **Leaves** ovate or ovate-oblong, 3.5–5 × 2–3 cm, base cordate, slightly oblique, apex acute or obtuse, membranous, veins 5–7; apical leaves long-elliptic, oblong, or ovate-lanceolate, 7–11 × 3–4.5 cm, base oblique or semi-cordate, apex shortly acuminate; petiole 0.5–2.5 cm long, longest on stoloniferous branches, roughly pubescent, base sheathed. **Inflorescences** spikes, leaf-opposed; ♂ spikes 5.5–13 cm × 2–3 mm; bracts orbicular, 0.7–1 mm wide, peltate, abaxially glabrous, fascicled pubescent at insertion to rachis, stalk short; ♀ spikes 4–5.5 cm, bracts as in ♂ spikes. **Flowers** ♂ with stamens 2, filaments short, anthers subglobose; ♀ with ovary subglobose, distinct, stigmas 4, linear. **Fruits** drupes, obovoid, distinct, ~2 mm ∅. **Distr:** Taiwan. **Ecol:** In forests, usually on trees and rocks; 1,000–2,500 m. Fl. Mar–Jul. **Zones:** 10–11.

P. siriboa L., *Sp. Pl.* 29, 1753. (Syn: *Chavica siriboa*, *Piper betle* var. *siriboa*). [Betel, Amu Malaka]. **Climbers**, tall; nodes swollen; stems, petiole, and abaxial surface of leaves densely puberulous when young. **Leaves** of fertile branches with petiole 2–3.5(–5.5) cm long; lamina ovate, 12–16 × 8–12 cm, base cordate and not or slightly asymmetrical, apex abruptly short-acuminate, firmly membranous, scarcely shiny, dark green, without or irregularly with very minute silvery scales, veins usually 3 pairs, arcuate. **Inflorescences** spikes, to 8–10(–17 in fruit) cm long, pendulous; axes pubescent but concealed by the bracts; peduncles 3–7 cm long; bracts peltate, orbicular, broadly stipitate, margin membranous, free all round. **Flowers** crowded, dioecious; stylules 5–6, 0.8 mm across. **Fruits** berries, concrescent forming a fleshy spadix. **Distr:** Indonesia (Sunda Islands). **Ecol:** In tropical forests. **Zones:** 11–12.

P. zeylanicum Miq., *London J. Bot.* 4: 436, 1845. (Syn: *P. arcuatum*, *P. caninum*, *P. thwaitesii*). [Ceylon Pepper]. **Climbers**, terrestrial or epiphytic; stems glabrous, much swollen at nodes. **Leaves** of *fertile* branches with petiole 0.5–1.5(–2) cm long; lamina ovate, 2–6(–10) × 1.5–4(–5.5) cm, base obtuse, rounded, or very shallowly cordate and ± symmetrical, apex acute, usually coriaceous, surfaces glabrous, veins (1–)2(–3) pairs, arcuate, arising from base and one pair from midrib 3–6 mm above base. **Inflorescences** spikes, 1–4(–6) cm long; axes hairy, mostly flexuose and twisted in fruit; peduncles (0.2–)0.5–1.2(–2 in fruit) cm long, usually glabrous; bracts peltate, margins membranous, free all round. **Flowers** crowded, dioecious; stamens mostly 2; stylules 3–4, 0.7–0.9 mm across. **Fruits** berries, distinct, spherical or obovoid, 4–5 mm long, fleshy, orange. **Distr:** Endemic to Sri Lanka. **Ecol:** In moist primary and secondary forests; 1,300–2,400 m. **Zones:** 11–12.

ARISTOLOCHIACEAE

Genera ~5–8, **species** 450–600. **Herbs**, **shrubs**, rarely lianas, subshrubs, or trees; perennial, deciduous or evergreen; stems erect, twining or procumbent; most parts with aromatic oils. **Stipules** absent. **Leaves** alternate; lamina simple, usually pinnately veined, sometimes palmately 3–5-veined, margins usually entire, rarely 3–5-lobed; petiole usually present and well-defined. **Inflorescences** terminal or axillary, racemes, cymes, or corymbs, or flowers solitary; rarely fan-shaped cymes. **Flowers** bisexual, zygomorphic or actinomorphic; calyx enlarged, perianth usually with 1(2) petaloid whorl, mostly connate into a distinct tube, cylindric to campanulate or subglobose, limb rotate, urceolate, cylindric, or ligulate, 1–trilobed, lobes valvate; corolla usually reduced to scales or absent; stamens 6–12 in 1 or 2 series, filaments adnate to ovary or style column with anthers free, or filaments and anthers fully adnate to style column to form gynostemium, anthers biloculed; ovary inferior to superior, (4)6-loculed, carpels connate only at base or fully fused, ovules numerous, usually in 1 or 2 series, styles free or connate, column 3- or 6(5–20)-lobed. **Fruits** capsules, fleshy or dry, rarely siliquiform or follicular, regularly to irregularly loculicidal, rarely indehiscent; seeds many, often flattened. **Distr:** Primarily tropical and subtropical regions.

Aristolochia

Syn: *Isotrema*, *Siphisia*, etc.

Species ~400. **Shrubs** or **herbs**, rarely subshrubs or lianas, usually twining or climbing, rarely erect; perennial; roots often tuberous; stems woody or herbaceous. **Pseudostipules** sometimes present, true stipules absent. **Leaves** alternate, 2-ranked; lamina with margins entire or 3–7-lobed, membranous to

Top row: *Piper betle, Piper kadsura, Piper longum, Piper nigrum* (fruits).
Middle row: *Piper magnificum, Piper ornatum, Piper zeylanicum.*
Bottom row: *Piper sintenense, Piper siriboa.*

leathery, veins pinnate or palmately 3–7 from base; petiole sometimes very short. **Inflorescences** axillary fascicles or flowers solitary, sometimes produced from old woody stems. **Flowers** with uniseriate perianth, zygomorphic; calyx lobes fused, tube often enlarged at or near base to form a utricle, apically cylindric or funnel-shaped, rectilinear or curved, limb ligulate, discoid, or subpeltate, lobes 1–3; stamens 6 in 1 series, fully adnate to the style column to form gynostemium, filaments absent, anthers extrorse; ovary inferior, 6-loculed, 6-angled, carpels fully connate, gynostemium 3- or 6-lobed. **Fruits** capsules, dry, 6-valved, dehiscing acropetally or basipetally, or rarely indehiscent; seeds flat or plano-convex, with or without wing. **Distr:** Tropical, subtropical and temperate regions of the Old World, China, Australia; Americas. **Cult:** Propagate from cuttings or seeds. Requires moist but well-drained soil in full or partial sun. **Info:** Some species are cultivated for medicinal uses, even though poisonous.

A. bracteolata Lam., *Enc.* 1: 258, 1783. (Syn: *A. bracteata*). [Worm Killer]. **Herbs**; stems prostrate, slender, branched, glabrous. **Leaves** alternate; lamina reniform to ovate, ~2–5 cm, base deeply cordate, margins entire, minutely crisped, apex rounded to broadly acute, trinerved at base, adaxially glabrous, abaxially glaucous; petiole 0.5–2 cm long. **Inflorescences** with flowers solitary, pedicellate, axillary; pedicels to 1 cm long; bract basal, sessile, orbicular or subreniform. **Flowers** irregular, bisexual; perianth 1.2–1.5 cm long; base inflated, ovoid; tube cylindric with a trumpet-shaped mouth, light yellow within, with maroon stripes; lip oblong, as long as tube, maroon in color, margins revolute, pilose within, apex obtuse; stamens 6, anthers sessile, adnate to style; ovary inferior, 6-carpellary, 6-locular, syncarpous, style short, 6-lobed. **Fruits** capsules, oblong-ovoid, to 3.7 cm long, glabrous, apex blunt, 12-ribbed, dehiscence septicidal into 6 valves; seed deltoid with slightly cordate base, to 6 mm long. **Distr:** Sri Lanka, India, Arabia and tropical Africa. **Ecol:** Restricted to dry sandy regions and sand dunes. **Zones:** 10–12. **Info:** Used for medicinal purposes in parts of the native range.

A. californica Torrey, *Pacif. Railr. Rep.* 4(5): 128, 1857. [California Snakeroot, California Pipevine]. **Lianas**, twining, to 5 m long, woody; young stems ribbed, tomentose. **Leaves** ovate to reniform, 4–12 × 3–10 cm, base cordate, sinus depth 2–6 cm, apex obtuse to rounded, abaxially tomentose, venation palmate-pinnate; petiole 1–2.5 cm long. **Inflorescences** on new growth, axillary, flowers solitary; peduncle bracteolate, 1–4 cm long; bracteoles reniform, to 8 mm long. **Flowers** with calyx brown-purple, bent; utricle pendent, ellipsoid to narrowly ellipsoid, 0.5–5 × 1–1.5 cm; syrinx absent; tube bent, cylindric, 1–3 × 1–1.7 cm; annulus absent; limb purplish, trilobed, lobes 2 × 2 cm, glabrous; gynostemium trilobed, cylindric, 0.5–1 cm long; anthers 6; ovary 6-locular, 1–4 cm long. **Fruits** capsules, ovoid to cylindric, 5–6 × 1–3 cm, dehiscence basipetal; valves 6; septa entire, not attached to valves; seeds flat, triangular, 1–1.2 × 1 cm. **Distr:** California (USA). **Ecol:** In forest thickets, along streams; 50–700(–1,000) m. Fl. Apr–May. **Zones:** 8–10.

A. clematitis L., *Sp. Pl.* 2: 962, 1753. [Birthwort, Asarabacca]. **Herbs**, erect, to 1 m high; young stems ribbed, glabrous to puberulent. **Leaves** deltate to reniform, 4–6 × 5–6.5 cm, base cordate, sinus depth 1–2 cm, apex obtuse to rounded, surfaces abaxially glabrous, venation palmate; petiole 4–7 cm long. **Inflorescences** fascicles or flowers solitary, on new growth, axillary; peduncle not bracteolate, 1–4 cm long. **Flowers** with calyx yellow-brown, curved; utricle

SA/SAJF

MIW MIW

MIW

Top row: *Aristolochia clematitis, Aristolochia cucurbitifolia, Aristolochia foveolata.*
Bottom row: *Aristolochia heterophylla, Aristolochia indica* (flowers), *Aristolochia indica* (fruit).

angled upward, ovoid to globose, 5 × 2–3 mm, tube curved upward, cylindric, 1–3 × 0.2–0.3 cm, annulus absent, limb purple, 1-lobed, lobe funnel-shaped, 1–6 cm long, pilose; anthers 6; gynostemium trilobed, globose, 1–4 mm long; ovary 6-locular, 1–4 cm long. **Fruits** capsules, ± globose, 3 × 2–3 cm, dehiscence basipetal, valves 6, septa entire, not attached to valves; seeds flat, triangular, 1 × 1.2 cm. **Distr:** Europe. **Ecol:** In waste places; 0–100 m. Fl. Jun–Jul. **Zones:** 6–10.

A. cucurbitifolia Hayata, *Icon. Pl. Formosan.* 5: 137, 1915. **Shrubs**, climbing, slightly herbaceous; stems terete, striate, pubescent to subglabrous. **Leaves** ovate to ovate-cordate, 6–9 × 5–11 cm, base cordate, palmately 5–7-lobed, basal lobes ovate, 2–3 × ~1.5 cm, lateral lobes spathulate or oblanceolate, shorter and narrower than middle lobe, middle lobe spathulate-obovate, 2–9 × 1–3 cm, apex acute, subleathery, adaxially glabrous, abaxially sparsely pubescent, veins palmate, 3 pairs from base; petiole ~3 cm long, pubescent. **Inflorescences** axillary, flowers solitary; peduncles often pendulous, ~3 cm long, densely pubescent; bractlets oblong-ovate, ~ 1 × 0.5 cm, slightly amplexicaul, inserted at middle of pedicel. **Flowers** with calyx brownish, 5–6 cm long, tube horseshoe-shaped, abaxially pubescent, basal portion of tube ~ 2.5 × 0.7 mm, limb discoid, oblique, ~ 1.5 × 1 cm, shallowly bilobed, orifice with slightly raised ring; anthers oblong, ~1.5 mm; gynostemium trilobed. **Fruits** capsules, ovoid-fusiform, ~ 6 × 1.5 cm, dehiscing basipetally; seeds ovoid-triangular, ~ 4.5 × 5 mm. **Distr:** Taiwan. **Ecol:** In mixed forests; ~500 m. Fl. Apr. **Zones:** 10–11.

A. foveolata Merrill, *Philipp. J. Sci.*, C. 13: 280, 1918. (Syn: *A. kaoi*). **Herbs**, twining; stems terete, glabrous. **Leaves** ovate-lanceolate to lanceolate, 5–15 × 2–3 cm, base deeply cordate or auriculate, sinus 1–2.5 cm deep, lateral lobes overlapping or slightly incurved, apex acuminate, leathery, adaxially glabrous, abaxially densely puberulous along closely reticulate veinlets, veins palmate, 2–3 pairs from base; petiole 3–6 cm long, glabrous. **Inflorescences** racemes, axillary, 3- or 4-flowered or flowers almost fascicled, 2–3 cm long; pedicels ascending, 1–1.5 cm long, sparsely villous; bractlets ovate or lanceolate, 4–5 × 2–3 mm. **Flowers** with calyx purple-brown, tube rectilinear or slightly curved, abaxially glabrous; utricle globose, ~6 mm ∅, sessile, tube ~2 × 0.6 cm, limb unilateral, ligulate, ovate-oblong, ~2 × 1–1.6 cm; anthers elliptic, ~1 mm; gynostemium 6-lobed. **Fruits** capsules, cylindric to obovoid, 2–2.5 × 1.5 cm, dehiscing acropetally; seeds ovoid-deltoid, ~5 × 4 mm. **Distr:** Philippines, Taiwan. **Ecol:** In thickets on mountain slopes; 500–1,000 m. Fl. Jul–Aug. Fr. Aug–Oct. **Zones:** 10–12.

A. heterophylla Hemsl., *J. Linn. Soc. Bot.*, 26: 361, 1891. (Syn: *A. kaempferi* f. *heterophylla*, *A. schimadai*). **Herbs** or **shrubs**, scandent, twining. **Leaves** often polymorphic; lamina cordiform, oblong-cordate, 6–15 × 2–3 cm, base deeply cordate or auriculate, sometimes the 2 auricles widely separated and lamina appearing cordately trilobed, chartaceous, adaxially scabrous, abaxially pubescent or tomentose, pinnately nerved; petiole 2.5–7.5 cm long. **Inflorescences** with flowers solitary, axillary; pedicels 2–3 cm long; bracteole 1 clasping at midpoint. **Flowers** with perianth tube bent, U-shaped, ~2 cm long, limb trilobed, spreading, ~1.6 cm ∅; ovary 8 mm long, densely pubescent, style column trilobed at apex. **Fruits** oblong or ovoid-oblong, 4–5 cm long, 6-angular, dehiscing from apex toward base; seeds triangular or broadly ovoid, 5 × 4 mm, testa smooth. **Distr:** China, Taiwan. **Ecol:** In thickets and forests. **Zones:** 10–11.

A. indica L., *Sp. Pl.* 960, 1753. (Syn: *A. lanceolata*). [Indian Birthwort]. **Herbs**, perennial, stems twining, slender, base woody, glabrous. **Leaves** alternate; lamina variable, linear-lanceolate to obovate, 3–18.5 cm long, base shortly truncate, margins entire, flat to undulate, apex acute to acuminate, trinerved at base, surface glabrous; petiole 0.5–1.5 cm long. **Inflorescences** 4–10-flowered, axillary corymbs; pedicels 0.5–1.5 cm long, swollen toward apex; bract sessile, ovate, apex acuminate. **Flowers** irregular, bisexual; perianth to 3 cm long, base inflated, ovoid, ~0.5 cm ∅, tube cylindric with a funnel-shaped trunk, mouth oblique, trumpet-shaped, gradually passing into a short oblong, obtuse, glabrous, brownish lip; stamens 6, filaments absent, anthers placed around style; ovary 6-carpellary, 6-locular, syncarpous, style short, 6-lobed, lobes short and pointed. **Fruits** capsules, oblong-ovoid, glabrous, apex rounded-retuse, 12-ribbed, to 3 cm long, dehiscing septicidally into 6 valves; seed deltoid, 1 cm long. **Distr:** Bangladesh, India, Nepal, Sri Lanka. **Ecol:** To ~1,000 m. **Zones:** 10–12. **Info:** Used for medicinal purposes especially to treat centipede, scorpion and cobra bites.

A. kaempferi Willdenow, *Sp. Pl.* 4: 152, 1805. (Syn: *A. heterophylla*, *Isotrema chrysops*, etc.). **Shrubs**, climbing, somewhat herbaceous; stems terete, striate, villous, glabrescent. **Leaves** ovate to ovate-lanceolate, obovate-oblong, or linear, 3–18 × 0.3–8 cm, papery, abaxially sparsely white-villous, adaxially glabrous, veins palmate, 2–4 pairs from base, base shallowly cordate to auriculate, sinus 0.5–1 cm deep, margins lobed or entire, apex acute to acuminate; petiole 1.5–6 cm long, densely villous. **Inflorescences** with flowers solitary or paired, axillary; peduncles usually pendulous, 2–7 cm long, pubescent; bractlets brownish, ovate to lanceolate or subrotund, 0.5–1.5 × 0.3–1.5 cm. **Flowers** with calyx yellow-green and purple veins, throat yellow, 3–4 cm; tube horseshoe-shaped, abaxially villous, basal portion of tube 2–2.5 × 3–8 cm, limb discoid-subrotund, 2–3 cm Ø, trilobed, lobes broadly ovate; anthers oblong, ~1 mm long; gynostemium trilobed. **Fruits** capsules, cylindric or ovoid, 3–7 × 1.5–2 cm, dehiscing basipetally; seeds obovoid, 3–4 × 2–3 mm. **Distr:** China, Japan, Taiwan. **Ecol:** In forests, thickets. Fl. Mar–May. Fr. Jun–Aug. **Zones:** 10–11.

A. labiata Willd., *Mem. Soc. Nat. Mosc.* 2: 101, t. 6, 1809. (Syn: *Ambuya labiosa*, etc.). [Mottled Dutchman's Pipe, Rooster Flower]. **Lianas**, stout, evergreen, glaucous, to 10 m long. **Leaves** broadly cordate, 7–12 × 7–15 cm, apex obtuse, glabrous, adaxially green, abaxially gray. **Pseudostipules** large, sessile, amplexicaul, ruffled. **Inflorescences** with flowers solitary, axillary, ebracteolate. **Flowers** geniculate, mottled red, yellow, green and purple; utricle subglobose, 7 cm long, syrinx absent, tube straight, emerging from the side of the utricle at a sharp angle, 4 cm long, annulus absent, limb with 2 superposed lobes, upper lobe oblate-orbicular, narrowly clawed, ruffled, deflected and pendent, 13–15 × 14–18 cm, lower lobe stiffly erect, narrowly lanceolate, 10–15 cm long; gynostemium 6-lobed, lobes connivent-erect, 1 × 0.6 cm, anthers 6, equidistant. **Fruits** cylindric, 8 × 3 cm; seeds numerous, as in *A. ringens*. **Distr:** South America. **Ecol:** In forest margins. **Zones:** 10–12.

A. maxima Jacq., *Enum. Syst. Pl.* 30, 1762. (Syn: *A. oblongifolia*, *Howardia geminiflora*, etc.). [Florida Dutchman's Pipe]. **Lianas**, sprawling, subglabrous. **Leaves** oblong to spathulate, 6–12 × 3–7 cm, base truncate to shallowly subcordate, apex obtuse to apiculate, glabrous or glabrescent, abaxially with prominent, raised reticulate venation. **Pseudostipules** absent. **Inflorescences** racemose clusters, several-branched, axillary, bracteolate. **Flowers** geniculate, purple; utricle ovoid, 2.5 cm long, syrinx absent, tube bent, 2 cm long, limb 1-lobed, ovate, gradually expanding from the tube, 5–6 × 3–4 cm; gynostemium 6-lobed, 5 × 4 mm, anthers 6, equidistant. **Fruits** very large, pendent, 10–15 × 7–10 cm, dehiscence acropetal, septifragal, exposing latticed septa; hypanthium absent; seeds very numerous, flat, triangular, 1 × 1.5 cm, 1 mm thick. **Distr:** USA (S Florida), S Mexico to C America. **Ecol:** Along streams and in moist thickets. **Zones:** 7–12.

A. odoratissima L., *Sp. Pl.*, ed. 2. 2: 1362, 1763. [Fragrant Dutchman's Pipe]. **Vines**, glabrescent. **Pseudostipules** present, ovate, to 3.5 × 2.5 cm. **Leaves** often subhastate, sometimes narrowly ovate, 4–16 × 3–13 cm, base cordate, sinus 1–3.7 cm deep, apex acute to shortly acuminate, sometimes obtuse or emarginate, membranous, adaxially glabrescent, abaxially appressed-puberulous, basal primary veins (3)5; petiole 3–6.5 cm long, glabrescent. **Inflorescences** with flowers axillary, solitary; peduncle plus ovary 4.5–10 cm long, ebracteolate. **Flowers** with utricle obovoid, 1.5–2.5(–3) × 0.5–0.8 cm, tube narrowly funnelform, 1–2.5 cm long, 2–4.5 mm Ø, right angled with utricle, limb 1-lobed, ovate, 4–11 × 2.5–4.5 cm, right angled with tube, exappendiculate, apex acute, mucronate; gynostemium 3.5–5 mm long. **Fruits** capsules, narrowly cylindrical, slightly curved, 4.5–8 × 1–1.5 cm, rostrate; seeds broadly ovoid, 3–3.5 × 2–2.5 mm, unwinged. **Distr:** Mexico, C America, Antilles, Colombia, Ecuador, Venezuela, Brazil, Peru, Bolivia, Paraguay (Gonzalez, 1994). **Ecol:** In gallery and secondary forests; 30–400 m. **Zones:** 10–12.

A. passiflorifolia, A. Rich., *Fl. Cub. Fanerog.* 2: 195, 1850. (Syn: *A. trichostoma*, etc.). **Lianas** glabrescent. **Leaves** heteromorphic, subtriangular, or broadly to narrowly pandurate, medially constricted, 4–10 × 2–6 cm, base sagittate-auriculate to cordate to auriculate-hastate, apex acuminate to obtuse, smooth. **Pseudostipules** absent. **Inflorescences** with flowers solitary, axillary, ebracteolate. **Flowers** subgeniculate, purple marked with green; utricle ovoid, ~1 cm long, syrinx inequilateral, small, tube straight to arcuate, 1–1.5 cm long, limb 1-lobed, broadly spathulate, densely subulate-fimbriate, 2–3 × 1 cm, unappendaged; gynostemium 6-lobed, 2–3 × 2 mm, anthers 6, equidistant. **Fruits** cylindric, 3.5–4 × 2–2.5 cm, dehiscing acropetally, septifragally; hypanthium absent; seeds numerous, flat, 8 × 7 mm, 0.5 mm thick. **Distr:** Bahamas, Cuba. **Ecol:** In dry thickets. **Zones:** 10–12.

A. pistolochia L., *Sp. Pl.* 1: 962, 1753. (Syn: *A. fasciculata*). **Herbs**, suberect to erect, 20–80 cm high, branched mainly at base; stems cylindrical to angular, pubescent, becoming glabrous at base, simple or

Top row: *Aristolochia kaempferi*, *Aristolochia labiata*.
Bottom row: *Aristolochia maxima*, *Aristolochia odoratissima*, *Aristolochia passiflorifolia*, *Aristolochia pistolochia*.

MIW

SZ

SA/SAJF

SZ

branched. **Leaves** triangular-cordate, 1.7–2.1 × 1.5–1.9 cm, base ± cordate, 5-nerved at base, margins irregularly toothed, teeth cartilaginous, apex acute; auricles to ⅙ the length of midrib; petiole 1–3 mm long, pubescent, base completely stem-clasping. **Inflorescences** with flowers solitary, axillary, ebracteolate; pedicel pubescent, 0.7–1.4(–1.8) cm long. **Flowers** 2–5 cm across; perianth densely hairy outside, yellowish-green to yellowish-brown, limb dark purple; utricle sessile, globose, ~3 mm Ø, tube ± straight, 4–8 mm long, limb ovoid-lanceolate, 1.5–2 × 1–1.2 cm, margin slightly reflexed, apex obtuse to retuse; gynostemium with 6 anthers and 6 stigmatic lobes; ovary oblong-obovate, hexagonal, 3–4 × 1–2 mm. **Fruits** capsules, globose to pyriform, 2–3 cm Ø, opening from base; seeds triangular-ovoid, ~ 3 × 3 mm, dark brown, smooth, shiny. **Distr:** Morocco. **Ecol:** In dense bushes among rocks. **Zones:** 10–12.

A. ringens Vahl, *Symb. Bot.* 3: 99, 1794. [Gaping Dutchman's Pipe]. **Lianas**; stems woody. **Stipules** present, basal, heart-shaped, to 3 cm across. **Leaves** broadly heart-shaped, ~6–10 × ~8–15 cm, base broadly caudate, margins entire, apex rounded; **Inflorescences** with flowers solitary. **Flowers** with perianth to 25 cm long, base inflated, ovoid, to 3.5 cm Ø, tube funnel-shaped, to 3 cm long, gradually passing into an upper and lower lip, upper lip linear lanceolate, 8–15 cm long, lower lip spathulate, 3.5–5 cm long, white with a purple network of veins; anthers oblong, fused at base; stigma 6-lobed. **Fruits** globose-oblong, to 8 cm long; seeds flat, winged, ~1 cm long. **Distr:** West Indies, S America. **Ecol:** On roadsides, and in open areas of upland tropical forests. **Zones:** 11–12.

A. zollingeriana Miquel, *Fl. Ned. Ind.* 1(1): 1066, 1858. (Syn: *A. tagala* var. *kankauensis*, etc.). **Herbs**, twining; stems 6-angled, glabrous. **Leaves** triangular-ovate to rhombic-cordate, 5–7 × 5–7.5 cm, base shallowly cordate, apex acute, adaxially glabrous, abaxially densely puberulous, veins palmate, 2–3 pairs from base; petiole 1.5–3.5 cm long, glabrous. **Inflorescences** racemes, axillary, 3- or 4-flowered, 1–4 cm long; pedicels ascending, ~1 cm long, glabrous; bractlets ovate, ~ 1.5 × 1 mm. **Flowers** with calyx brownish-purple, 3–5 cm long, tube rectilinear or slightly curved; utricle globose, ~5 mm Ø, with stipe 2–3 mm long, tube 0.8–1 × 0.3–0.5 cm, limb ligulate, oblong, 1.5–2.5 × 0.3–0.5 cm, apex obtuse; anthers ovoid, ~1 mm long; gynostemium 6-lobed. **Fruits** capsules, cylindric, 5–6 × ~2 cm, dehiscing acropetally; seeds ovoid-deltoid, ~ 7 × 6 mm, winged. **Distr:** Indonesia, Japan, Malaysia, Taiwan. **Ecol:** In dense forests. Fl. Jul. Fr. Aug. **Zones:** 10–12.

Asarum

Syn: *Heterotropa, Hexastylis*, etc.
Species ~90. **Herbs**, perennial, deciduous, rhizomatous, without aerial stems. **Stipules** absent. **Leaves** alternate (sometimes appearing opposite because of crowding), 2-ranked; petiolate foliage leaves and sessile, triangular scale-leaves both present; lamina membranous or leathery, pubescent at least abaxially and on margins. **Inflorescence** terminal on rhizome, flowers solitary; bracts absent. **Flowers** with sepals distinct, usually mixture of white, green, tan, red, or purple, proximally touching valvately and forming well-defined false tube, externally usually villous, inside strigose, smooth or with weak longitudinal ribs, never with network of low ridges; vestigial petals present or absent; stamens 12, distinct, filaments longer than pollen sacs, terminal appendage of anther well developed; ovary inferior, 6-locular, styles connate in column. **Fruits** capsules, fleshy, dehiscence irregular; seeds ovoid, not winged, with fleshy appendage. **Distr:** N America, Eurasia. **Cult:** Propagate from seeds or by division. Requires a moist soil in partial to full shade.

A. canadense L., *Sp. Pl.* 1: 442, 1753. (Syn: *A. acuminatum*, etc.). [Canadian Wild Ginger]. **Herbs**; rhizomes horizontal, shallow, internodes 1–3.5 cm. **Leaves** with petiole 6–20 cm, crisped-hirsute; lamina not variegated, cordate-reniform to reniform, 4–8(–20) × 8–14(–21.5) cm, apex rounded or obtuse, surfaces usually appressed-hirsute. **Inflorescence** with flowers erect or ascending; peduncle 1.5–3 cm long. **Flower** with false calyx tube cylindric, externally tan or purplish, hirsute (often densely), internally white or pale green, occasionally mottled with purple, with white or purple hairs, distal portion of sepal spreading or reflexed at anthesis, 0.6–2.4 cm long, apex apiculate to acuminate or filiform-attenuate, abaxially green or purple, hirsute, adaxially purple, puberulent with crisped purple or pale hairs. **Distr:** USA. **Ecol:** In understory of deciduous forests; 0–1,300 m. Fl. Mar–Jul. **Zones:** 7–9.

A. epigynum Hayata, *Icon. Pl. Formosana* 5: 140, 1915. (Syn: *Geotaenium epigynum*). [Taiwanese Wild Ginger]. **Herbs**, rhizomes horizontal, 4–5 mm Ø, internodes 1–2 cm long. **Cataphylls** oblong, ~ 8 × 4 mm. **Leaves** paired; lamina uniformly colored, narrowly ovate or ovate-lanceolate, 4–9 × 2.5–6.5 cm, base cordate, lateral lobes 1–2.5 × ~2 cm, apex acuminate, adaxially sparsely pubescent on veins and margins, abaxially densely pubescent; petiole 4–6 cm long, villous. **Inflorescence** with peduncle declinate, 1–2 cm long. **Flower** with calyx slightly zygomorphic, greenish-purple, subrotate, 1.5–2 × 2–2.5 cm; sepals connate beyond attachment to ovary, adaxially dark red pubescent, abaxially densely pubescent, tube obconic, 0.5–1 × 0.7–0.8 cm, lobes spreading, triangular-reniform or ovate-rounded, ~ 0.8 × 1 cm; stamens 12, filaments shorter than anthers, anthers 0.8–1.2 mm long, connectives extended beyond anthers, ligulate, <0.3 mm long; ovary inferior, styles connate, short and thick, apex 6-cleft, stigmas terminal, radiating downward, decurrent. **Distr:** China, Taiwan. **Ecol:** In forests, valleys. Fl. Feb–Mar. **Zones:** 10–11.

A. hypogynum Hayata, *Icon. Pl. Formosana* 5: 144, 1915. (Syn: *A. grandiflorum*, etc.). **Herbs**, rhizomes vertical, 5–6 mm Ø, internodes <1 cm. **Cataphylls** ovate, ~ 4 × 5 mm. **Leaf** solitary; lamina oblong-triangular, 13–19.5 × 6.6–20 cm, base deeply cordate, lateral lobes 3–9 × 2.8–9.5 cm, apex acuminate or caudate, adaxially initially glabrescent, green with white spots, abaxially uniformly pale green, dotted with glands; petiole 10–18 cm long. **Inflorescence** with peduncle ascending, 3–6 cm long. **Flower** with calyx purple-green, campanulate-rotate, 3–3.5 × 3–4 cm; sepals connate beyond attachment to ovary, abaxially glabrous, tube subcylindric, ~ 0.9 × 1 cm, strongly constricted at throat, adaxially tessellate, orifice ring ~1.5 mm wide, lobes subrotund, ~ 1 × 2 cm, base with abundant glands and papillate-corrugate area; stamens 12, filaments much shorter than anthers; ovary superior or half-inferior, styles free, apex 2-cleft, stigmas subterminal. **Distr:** Endemic to Taiwan. **Ecol:** In forests, shady areas, mountain slopes; 1,000–2,000 m. Fl. Oct–Jul. **Zones:** 10–11.

A. macranthum J.D. Hooker, *Bot. Mag.* 114: t. 7022, 1888. (Syn: *A. albomaculatum*, etc.). **Herbs**, rhizomes vertical, 2–3 mm Ø, internodes <1 cm long. **Cataphylls** broadly ovate, 0.8–1 × 0.6–0.8 cm. **Leaf** solitary; lamina triangular-ovate, 10–13 cm, base deeply cordate, lateral lobes rounded, 1–1.5 × 2–2.5 cm, apex acute, adaxially greenish with yellow-green blotches, abaxially greenish, purple along principal veins; petiole streaked reddish-brown, 10–20 cm long, glabrous. **Inflorescence** with peduncle ascending to erect, ~9 mm long. **Flower** with calyx dark purple, campanulate, 3–3.5 × 5–6 cm; sepals connate beyond attachment to ovary, abaxially glabrous, tube obconic, ~ 1.7 × 2 cm, strongly constricted at throat, adaxially tessellate, orifice ring 4–5 mm wide, broadly laminate, lobes broadly ovate, ~ 2.5 × 2 cm, base with papillate-corrugate area, undulate along margin; stamens 12, oblong, filaments much shorter than anthers; staminodes 3, petaloid; ovary superior, styles free, stigmas lateral, linear-oblong. **Distr:** Taiwan. **Ecol:** In forests; 500–1,000 m. Fl. May. **Zones:** 10–11.

A. splendens (F. Maekawa) C.Y.Cheng & C.S.Yang, *Fl. Reipubl. Popularis Sin.* 24: 180, 1988. (Syn: *A. chingchengense, Heterotropa splendens*). [Showy Wild Ginger]. **Herbs**, rhizomes horizontal, 2–3 mm Ø, internodes ~1.5 cm long. **Cataphylls** oblong, ~ 2 × 1.5 cm. **Leaf** solitary; lamina ovate-cordate or nearly hastate, 6–10 × 5–9 cm, base auriculate to subcordate, lateral lobes 2–3.5 × 2–3.5 cm, apex acute, surfaces glabrous, adaxially often with white blotches; petiole 8–16 cm long. **Inflorescence** with peduncle recurved, ~1 cm long. **Flower** with calyx purplish-green, campanulate, 2–3 × 2.5–3 cm; sepals connate beyond attachment to ovary, abaxially glabrous, tube broadly cupular to subglobose, ~ 1.4 × 2 cm, slightly constricted at throat, adaxially tessellate, orifice ring narrow, lobes broadly ovate, ~ 2 × 2.5 cm, base with semi-circular corrugate-papillate area; stamens 12, filaments shorter; ovary superior, styles free, apex 2-forked or -notched, stigmas lateral. **Distr:** China. **Ecol:** In thickets, grasslands, moist mountain slopes; 800–1,300 m. Fl. Apr–May. **Zones:** 8–10.

Thottea

Syn: *Apama, Bragantia*.
Species ~25. **Subshrubs** or rarely shrubs, erect, woody at base. **Leaves** alternate; lamina with margins entire, pinnately veined or palmately 3–5-veined. **Inflorescences** short axillary racemes, cymes, corymbs or scorpioid cymes; bracts opposite to flowers. **Flowers** regular with uniseriate trilobed perianth, actinomorphic; calyx lobes fused, connate at base; tube rotate, spreading, campanulate, cupular, or urceolate, throat usually with thickened ring, lobes 3 or 4, valvate; stamens 8 or 9 in 1 series, or sometimes 6–36(–46) in 1 or 2 series, filaments short or absent, free or partly connate, adnate to style with anthers remaining free; ovary linear, inferior, 4-loculed, carpels fully connate, style short, 5–20-lobed, stigma 5–7-lobed, terminal or lateral. **Fruits** capsules, siliquiform, linear, elongate; seeds oblong, ellipsoid or broadly ovoid (usually 3-angled) with transversely rugose or glandular-punctate testa. **Distr:** Bangladesh, China, India, Malesia, Myanmar, Philippines, Sri Lanka, Thailand, Vietnam. **Cult:** Propagate from seeds or rooted cutting. Requires a fertile damp compost in filtered sun.

T. siliquosa (Lam.) Ding Hou, *Blumea* 27: 327, 1981. (Syn: *Apama siliquosa*, etc.). **Shrubs**, to 2 m tall, woody at base, branched; branchlets glabrous, often flexuous at nodes; nodes swollen. **Leaves** alternate, slightly aromatic when crushed; lamina broadly elliptic to lanceolate, 7–25 × 3–8 cm, base cuneate to oblique, margins entire, apex acuminate to acute, narrow, trinerved at base, main lateral veins arising halfway along the lamina, adaxially glabrous, abaxially minutely pubescent; petiole to 7 mm long. **Inflorescences** spikelike, arising at leaf axils, 1–2 per axil; pedicels short; bracts small, linear, pubescent. **Flowers** maroon; perianth campanulate, tripartite, fused at base, each segment heart-shaped; stamens 3, filaments fused at base, adnate to style, anthers basifixed, incurved; ovary inferior, hirsute, 4-carpellate, 4-locular, placentation axile, stigma 5–7-fid, cup-shaped. **Fruits** capsules, pendulous, quadrangular when mature, oblong, to 19 × 0.5 cm, loculicidal; seeds ~2 × 1 mm, trigonous, rugose. **Distr:** India, Sri Lanka. **Ecol:** In wet regions, growing under thickets, often along streams of rainforests. **Zones:** 11–12.

Top row: *Aristolochia ringens*, *Aristolochia zollingeriana*, *Asarum canadense*.
Middle row: *Asarum epigynum*, *Asarum hypogynum*, *Asarum macranthum*.
Bottom row: *Asarum splendens* (flower), *Thottea siliquosa* (flowers), *Thottea siliquosa* (fruits).

PART 3 ANGIOSPERMS: MONOCOTS

Orders 9, **families** 57, **species** ~60,000. **Plants** herbaceous, rhizomatous, sympodial or monopodial; root epidermis developed from outer layer of cortex; root vascular tissue oligo- to polyarch, medullated, lateral roots arise opposite phloem poles; vascular bundles in stem scattered, closed; vessel elements in root with scalariform and/or simple perforations; vessels in stems and leaves absent; sieve tube plastids with cuneate protein crystals alone; stomata paracytic (divisions of neighboring cells oblique); stems absent or present, when present, branched or unbranched. **Leaves** 1 to many; not differentiated into petiole and lamina, main venation parallel, developing both acropetally and basipetally from the base and converging toward the apex. **Inflorescences** 1–many, usually racemose. **Flowers** 3-merous (a character also shared by some basal angiosperms), polysymmetric, pentacyclic; tepals in 2 whorls, each member with 3 traces, median member of outer whorl abaxial, members of whorls alternating, similar, each providing a sector for the tepal tube when present; stamens equal in number and opposite each tepal member, anther and filament ± sharply distinguished, anthers subbasifixed; gynoecium opposite outer tepals, placentation axile, style hollow, short. **Fruits** loculicidal capsules; seeds testal, embryo long, cylindrical, cotyledon 1, terminal, plumule lateral, primary root unbranched, not very well developed, adventitious roots numerous (extremely rarely with a taproot), hypocotyl short, cotyledon with a closed sheath. **Distr:** Worldwide. **Info:** The most important morphological character (among many others) that separates the monocots from the rest of the angiosperms is the possession of a single cotyledon (vs. usually 2 in the rest of the angiosperms). The monophyly of the monocots is supported by morphological and molecular studies (Bharathan & Zimmer 1995; Chase 2004; Chase et al. 1993, 1995a, 1995b, 2000, 2006; Davis et al. 2004; Graham et al. 2006; Hilu et al. 2003; Savolainen et al. 2000a; Stevenson & Loconte 1995; Soltis et al. 1997, 2000, 2005).

Opposite: *Stanhopea tigrina*.

6 BASAL MONOCOTS

ACORALES

FAMILY 1, GENUS 1, SPECIES 2–4.

Ordinal description same as familial.
Distr: Eastern N America to E Asia. **Info:** The monophyly of Acorales is strongly supported by molecular studies. Acorales are sister to the rest of the monocots (Chase 2004).

ACORACEAE

Genus 1, **species** ~2–4. **Herbs** perennial, aquatic; rhizomes horizontal, creeping at or near surface, aromatic. **Leaves** green, narrow, straplike or sword-shaped, not differentiated into petiole and lamina, equitant, 2–ranked forming a fan (isobifacial), first leaf terete, unifacial, venation parallel along length of leaf, lateral basal nerves reticulate; sheath open. **Inflorescence** solitary, inconspicuous, densely flowered; spadix spicate, nearly cylindrical, tapering, apex obtuse, appearing lateral to the leaflike spathe (true spathe absent), distal sympodial leaf extending beyond spadix. **Flowers** small, bisexual, perfect (or weakly monosymmetric), pentacyclic; tepals 6, abaxial tepal large, bractlike, enclosing young flower like a hood; stamens 6, distinct, filaments linear, flat, anthers extrorse; ovaries (1–)3-locular, conical, ovules several per carpel (encased in mucilage), stigmas sessile, minute, style (essentially absent), broad, massive, stylar canal with exudates. **Fruits** berries, light brown to reddish with darker streaks; seeds 1–6(–14), oblong, embedded in mucilage. **Distr:** Temperate Northern Hemisphere, tropical Asia at higher elevations, and sporadically introduced into Southern Hemisphere.

Acorus

Species ~2–4. *Description same as for the family, given above.* **Distr:** Temperate Northern Hemisphere, tropical Asia; and sporadically introduced into Southern Hemisphere. **Cult:** Propagate by dividing the rhizomes. Requires wet soil in full sun. **Info:** *Acorus* was traditionally placed in the family Araceae, but recent studies indicate that it should be placed in its own family and order. Evidence based on DNA sequences fails to show any close relationships between *Acorus* and other genera of Araceae, and instead supports *Acorus* as the oldest extant lineage of monocotyledons (Duvall *et al.* 1993). The structure that has been called a 'spathe' in *Acorus* is not morphologically equivalent to the spathe of Araceae. Considerable confusion exists in the taxonomic literature over the status of *Acorus* in N America, whether it is native or introduced. *A. calamus* has been cultivated mainly for medicinal purposes since ancient times.

A. calamus L., *Sp. Pl.* 1: 324, 1753. [Cinnamon Sedge, Myrtle Sedge, Sweet-flag, Sweet Myrtle, etc.]. **Herbs** perennial, to 80 cm tall; rhizomes have a sweet, distinctive aroma and a bitter taste, stout, 1–2.5 cm broad. **Leaves** bright green, basally white with pink or red, midvein single (secondary midrib) prominently raised adaxially; *vegetative* leaves to 1.75 m long, sheathing base 22.1–66.5(–73.3) cm, distal part 0.32–0.96(–1.2) m × 0.5–2 cm, margins sometimes undulate or crisped; *sympodial* leaf (0.3–)0.35–1.6(–1.8) m long, sheathing base 0.16–0.76(–1) m long, distal part of leaf 0.13–0.86(–1) m × 0.4–1.9 cm. **Inflorescence** a spicate fleshy spadix, (3.8–)4.9–8.9 × 0.5–11 cm at anthesis, post-anthesis 5.5–8.7 × 0.6–1.26 cm. **Flowers** small, 3–4 mm across, yellowish-green; tepals oblong-obovate, ~2 mm long, surface with embedded raphides; filaments 2 mm long, flat, anthers ± orbicular, <1 mm long; ovary obconical, 3 mm long. **Fruits** berries, green, angular; seeds 1–3, oblong or obconical, 2 mm long. **Distr:** Asia, Africa, Europe, Indian Ocean Islands, Pacific Islands. **Ecol:** In wet open areas, marshes, along edges of quiet water; 0–2,200 m. Fl. Jun–Jul. **Zones:** 3–12. **Info:** Used widely for medicinal purposes.

ALISMATALES

FAMILIES 14, GENERA 166, SPECIES 4,490.

Herbs, rhizomatous; mycorrhizae absent; starch grains pteridophyte-type, amylophilic. **Flowers** with anthers extrorse, tapetum amoeboid, cells uninucleate; carpels with completely unfused canals, stigma dry. **Fruits** usually achenes or follicles; seeds with embryo large. **Distr:** Nearly worldwide. **Info:** The monophyly of Alismatales is supported by molecular data (Chase et al. 1993, 1995b, 2000, 2006; Duvall et al. 1993; Hilu et al. 2003; Käällersjö et al. 1998; Soltis et al. 2000).

ALISMATACEAE

[Arrowhead, Water Plantain].
Genera 12, **species** ~80. **Herbs**, annual or perennial, rhizomatous, stoloniferous, or cormous, caulescent, glabrous to stellate-pubescent; sap milky; roots septate or not septate. **Leaves** basal, submersed, floating, or emersed, sessile or petiolate, sheathing proximally; lamina with translucent markings of dots or lines present or absent, basal lobes present or absent; venation reticulate, primary veins parallel from base of lamina to apex, secondary veins reticulate. **Inflorescences** scapose racemes or panicles, rarely umbels, erect, rarely floating or decumbent, whorled (forming racemes) or whorls branching (forming panicles), bracteolate. **Flowers** bisexual or unisexual, if unisexual, ♂ and ♀ on same or different plants, hypogynous, subsessile to long-pedicellate; sepals persistent, 3; petals 3, deciduous, delicate; stamens absent, 6, 9, or to 30, distinct, anthers biloculed, dehiscing longitudinally; pistils 0 or 6–1,500 or more, distinct or coherent proximally, 1-loculed; placentation basal, ovules 1–2. **Fruits** achenes or follicles; seeds with embryo U-shaped, endosperm absent in mature seed. **Distr:** Nearly worldwide, primarily tropical and subtropical regions.

Hydrocleys

[Water Poppy].
Species 5. **Herbs**, aquatic; stolons often present, terete. **Leaves** *submersed* phyllodia (expanded petiole resembling and functioning like leaves), sessile; *floating* leaves long-petiolate, petiole terete, septate, lamina orbicular to oblong-lanceolate, base cordate (rounded), apex mucronate to obtuse. **Inflorescences** occasionally proliferating with stolons, leaves; scape septate; bracts subtending pedicel, distinct, elliptic to lanceolate, shorter than pedicel, delicate. **Flowers** with sepals persistent, erect, green, lanceolate, leathery, midvein absent (present), apex hoodlike; petals erect to spreading, yellow to white, orbicular (oblong-obovate), longer than (shorter than) sepals; stamens in (1–)4–5 series, outer often sterile; pistils terete, linear-lanceoloid, attenuate into style, style curved inward, apex papillose. **Fruits** ± terete, linear-lanceoloid, membranous, dehiscing along adaxial margins; seeds 50 or more, sparsely to densely glandular-pubescent. **Distr:** S America, the Lesser Antilles. **Cult:** Propagate by division. Requires deep, free-standing to fast-flowing water in partial shade.

H. nymphoides (Willdenow) Buchenau, *Abh. Naturwiss. Vereins Bremen.* 2: 2, 1869. (Syn: *Hydrocleys azurea*, *Limnocharis nymphoides*, *Sagittaria ranunculoides*, etc.). [Water Poppy]. **Herbs**, to 50 cm tall; stolons to 45 cm long. **Leaves** broadly ovate to orbicular, 1.4–11.9 × 0.9–10.6 cm, veins 5–9; petiole 1.5–40 cm long, sheathing base to 8.5 cm long. **Inflorescences** 1–6-flowered; peduncles to 30 cm long; bracts elliptic, 2–4.5 × 0.4–1 cm, obtuse; pedicels spreading, 3.5–17.5 cm long. **Flowers** ~6.5 cm across; sepals 1.3–2.8 × 0.7–1.3 cm; petals spreading, pale yellow to white with yellow base, 2.3–2.6 × 3.8–4.1 cm; stamens 20–25; staminodes 20+; pistils 5–8, 1 cm long. **Fruits** 1–1.45 cm × 2–3.5 mm, beak 3.5–5.5 mm long; seeds ~1 mm long, sparsely glandular-pubescent. **Distr:** C America, S America. **Ecol:** On margins of lakes and wet ditches; 0–100 m. Fl. Jun–Jul. **Zones:** 11–12.

Sagittaria

Syn: *Lophotocarpus*.

Species ~30. **Herbs**, perennial, rarely annual, aquatic, submersed, floating-leaved, or emersed; rhizomes often present, occasionally terminated by tubers; stolons often present; tubers white to brown, smooth. **Leaves** sessile or petiolate, petiole terete to triangular; lamina linear to obovate, base attenuate to hastate or sagittate, margins entire, apex round to acute, with translucent markings absent. **Inflorescences** racemes, panicles, rarely umbels, of 1–17 whorls, erect, emersed or floating, rarely submersed; bracts coarse or delicate, obtuse to acute; ♂ flowers pedicellate, distal to ♀ flowers; ♀ flowers mostly pedicellate, bracts lanceolate, obtuse to acute. **Flowers** unisexual, proximal rarely with ring of sterile stamens; receptacle convex; sepals recurved in ♂ flowers, recurved to erect in ♀ flowers, often sculptured, herbaceous to leathery; petals white, rarely with pink spot or tinge, entire; stamens 7–30, filaments linear to dilated; pistils to 1,500+, spirally arranged, not starlike, distinct, ovules 1, styles terminal. **Fruits** compressed, with abaxial wings and lateral wing. **Distr:** Mostly in Western Hemisphere; Europe, Asia. **Cult:** Propagate from seeds or by division. Requires deep, fertile, free-standing to fast-flowing water in full or partial sun.

S. lancifolia L., *Syst. Nat.* ed. 10. 2: 1270, 1759. [Bull Tongue Arrowhead]. **Herbs**, perennial, to 2 m tall; rhizomes coarse; stolons absent; corms absent. **Leaves** emersed; petiole terete, 44–58 cm; lamina linear to ovate or elliptic, 20.3–35 × 0.7–16 cm. **Inflorescences** racemes or panicles, of 6–13 whorls, emersed, 16–72 × 19–60 cm; peduncles 0.75–1.25 m tall; bracts connate ± to ¼ total length, lanceolate to ovate-lanceolate, 0.3–3.2 cm long, coarse; ♀ flowers pedicellate. **Flowers** to 3.3 ∅; sepals recurved to spreading, not enclosing flower or fruiting head; filaments cylindric, longer than anthers, pubescent; ♀ without ring of sterile stamens. **Fruits** achenes, in fruiting heads 0.5–1.2 cm ∅; achenes oblanceoloid, abaxially keeled, 1.6–2.5 × 0.8–1.1 mm, beaked, wings absent, glands 1, beak lateral, erect, 0.3–0.7 mm long; pedicels 0.6–3.8 cm long. **Distr:** N America, Mexico, West Indies, C America, S America. **Ecol:** In wetlands and coastal brackish marshes; 0–50 m. Fl. Jun–Aug. **Zones:** 8–12.

S. montevidensis Chamisso and Schlechtendal, *Linnaea*. 2: 156, 1827. (Syn: *Sagittaire spongieuse*). [Spotted Arrowhead]. **Herbs**, annual or perennial, to 1 m tall; rhizomes, stolons and corms present. **Leaves** submersed and emersed; *submersed* leaves absent or phyllodial, flattened, to 17 × 2 cm; *emersed* leaves with petiole triangular, 21–55 cm long, lamina hastate to sagittate, 2.5–17.5 × 0.6–22 cm. **Inflorescences** racemes or panicles, of 1–15 whorls, floating or emersed, 1.5–28 × 1.5–15 cm; peduncles 15–47 cm long; bracts distinct or basally connate, lanceolate to elliptic, 0.4–3.4 cm long, delicate; ♀ flowers pedicellate. **Flowers** 2–5 cm ∅; sepals spreading in ♂, erect in ♀, enclosing flower or fruiting head; filaments longer than anthers; ♀ with or without ring of sterile stamens. **Fruits** achenes, in fruiting heads 1.2–2.1 cm ∅; pedicels club-shaped, 0.5–4.2 cm long; achenes oblanceoloid, 2–4.3 × 0.7–1.5 mm, beaked, glands 1, beak lateral, horizontal, 0.4–0.8 mm long. **Distr:** N America, Mexico, S America. **Zones:** 8–12.

APONOGETONACEAE

[Cape Pondweed].

Genus 1, species 47. **Herbs**, aquatic, laticiferous, perennial, cormous, or rhizomatous; roots hydrophytic. **Leaf sheaths** with free margins, not persisting longer than lamina. **Leaves** alternate, floating and submerged leaves usually different, petiolate (sessile); lamina ovate to narrowly lanceolate (linear), parallel-veined, cross-venulate (tissue between veins breaks up as leaf matures, leaving a network of holes. **Inflorescences** terminal, flowers aggregated (commonly projected above the water in 1–3 spikes), subtended by spathe. **Flowers** minute, bisexual (unisexual), usually very irregular and zygomorphic; perianth present (absent); tepals 1(–6); stamens 6–18(–50) in 2–3(–4) series, not epitepalous, anthers distinct, dehiscing longitudinally; pistils 2–6(–9), distinct, ovules basal-marginal, anatropous, stigmas dry, papillate. **Fruits** follicles, non-fleshy, ± aggregated; seeds 4. **Distr:** N America, S Africa, tropical regions of Eastern Hemisphere.

Aponogeton

Species 47. *Generic description same as familial.*

A. crispus Thunb., *Nov. Gen.* 4: 73 ('*crispum*'), 1781. (Syn: *A. echinatum*). [Ruffled Sword, Wavy Swordplant]. **Herbs**, aquatic; tuber obovoid, to 5 cm ∅. **Leaves** *submerged* very variable, linear, narrowly lanceolate or narrowly ovate, to 50 × 4.5 cm, base (narrowly)cuneate, rounded or subcordate, apex rounded or cuneate, margins flat, undulate or crisped, parallel main nerves 5, 7 or 9; *floating* leaves lanceolate or (narrowly)ovate, very variable, to 20 × 5 cm, base rounded or (seldom) cordate, apex rounded, or cuneate, tip blunt, main nerves parallel,

Top row: *Hydrocleys nymphoides*, *Sagittaria lancifolia* (male flowers), *Sagittaria lancifolia* (female flowers).
Bottom row: *Sagittaria montevidensis* (male flowers), *Sagittaria montevidensis* (female flowers), *Aponogeton crispus*.

SZ

Above: *Aglaonema commutatum, Alocasia cucullata, Alocasia macrorrhizos, Alocasia odora.*

(7–)9–11. **Inflorescence** a spike, to 13 cm long, fragrant, densely or rather laxly flowered; peduncle to 75 cm long, widening toward the spike; spathe to 2.5 cm long, caducous. **Flowers** minute, facing all directions; tepals 2, to 2.25 × 1.75 mm, white, pink, or pale violet, 1-nerved; stamens 6, to 2.5 mm long, widened toward base, anthers violet, pollen yellow; ovaries 3, to 2 × 1 mm, ovules 2. **Infructescence** very dense. **Fruits** to 1.2(–1.8) × 0.5(–0.7) cm, terminally beaked. **Distr:** S India, Sri Lanka. **Cult:** Propagate from seeds or by division. Requires a still, shallow pond in full sun. **Info:** Commonly used in freshwater aquariums.

ARACEAE

[Arum].

Genera 105, **species** >3,300. **Herbs**, perennial, usually with milky or watery latex, rarely colored; rhizomes, corms, or stolons present; rhizomes vertical or horizontal, creeping at or near surface, sometimes branched; corms underground, starchy; stolons at or near surface; stems absent (sometimes above-ground or aerial). **Cataphylls** usually present. **Leaves** rarely solitary, alternate or clustered; lamina simple or compound (occasionally perforate), elliptic to obovate or spathulate, occasionally sagittate-cordate, >1.5 cm long, venation parallel, pinnate- or palmate-netted; petiole rarely absent, base sheathing. **Inflorescences** spadices, cylindric or ovoid, 3–900-flowered, various parts occasionally naked or with sterile flowers; flowers sessile; spathe rarely absent, persistent (sometimes only proximally) or deciduous, variously colored; ♂ and ♀ usually on same plants or functionally on different plants, ♂ flowers distal to ♀ when unisexual. **Flowers** bisexual or unisexual; perianth absent or present; stamens 2–12, distinct or connate in synandria; ovaries 1, 1–3(–many)-locular, sessile or embedded in spadix, styles 1, stigmas hemispheric, capitate, or discoid (sometimes strongly lobed). **Fruits** berries, distinct or connate at maturity; seeds 1–40(–many). **Distr:** Nearly worldwide, primarily tropical regions. **Ecol:** In wetlands or terrestrial, occasionally emergent or floating, often epiphytic or climbing.

Aglaonema

Species 24. **Herbs**, terrestrial, erect or ascending, evergreen; stems branched or unbranched, rooting, internodes distinct, green. **Leaves** spirally arranged; lamina undivided, usually oblong, oblong-lanceolate, midrib strong, primary lateral nerves >4–7 per side, secondary and tertiary lateral nerves parallel with primary ones, numerous; petiole sheathing. **Inflorescences** apical or nearly apical, usually several together; spathe initially tubular-involved, opening by a slit to nearly expanded, enveloping at least the ♀ portion of spadix; spadix cylindric, stipe short or absent; ♀ portion of spadix at base of rachis, persistent, few-flowered, much shorter than ♂ portion; ♂ portion above ♀, caducous, many-flowered. **Flowers** ♂ with stamens free, anthers sessile, staminodes between ♂ and ♀ portions of spadix, few, antherlike; ♀ with ovary 1(2)-loculed, each 1-ovuled, style thick, very short, stigma large. **Fruits** berries, 1-seeded; seed oblong, large, exalbuminous. **Distr:** NE India to Papua New Guinea. **Cult:** Propagate from stem cuttings, basal cuttings or by division. Needs fertile, moist and well-drained soil in partial shade. **Cult:** Propagate from stem cuttings, basal cuttings or by division. Needs fertile, moist and well-drained soil in partial shade.

A. commutatum Schott, *Syn. Aroid.*: 123, 1856. (Syn: *A. marantifolium* var. *commutatum*). [Chinese Evergreen]. **Herbs**, perennial, 0.2–1.8 **m** tall; stems erect, later decumbent, 0.5–6 cm Ø. **Leaves** alternate; lamina usually narrowly oblong-elliptic to lanceolate, usually 12–30 × 2.5–7.5 cm, base often unequal, margins wavy, apex shortly acuminate, dark green to yellow-green, variegated ashy-green around veins; petiole 1–1.5x as long as the lamina. **Inflorescences** 1–6, axillary; spathe oblong, pale green to white; spadix shortly stalked, slender, usually shorter than spathe. **Flowers** minute. **Fruits** berries, ellipsoid to ovoid, ripening from green, through yellow, orange to bright red, glossy. **Distr:** C Malesia (Philippines to NE Sulawesi). **Ecol:** In dense forests, valley forests and moist forest margins; to 1,500 m. Fl. Apr–Jul. Fr. Sep–Nov. **Zones:** 10–12. **Info:** Cultivated widely as an ornamental foliage plants.

Alocasia

Syn: *Colocasia, Ensolenanthe.*

Species ~70. **Herbs**, evergreen, perennial, often robust; corms or rhizomes forming the above-ground (pseudo)stem. **Leaves** ovate, peltate, base usually cordate, margins entire or undulate, apex acute. **Inflorescences** near apex, solitary or several together, flowers numerous; spathe constricted; sterile appendage of ♂ rachis well-developed, separated or not from the synandria by a constriction. **Flowers** minute, unisexual; ♂ with anthers linear, 3–8, sessile, united into a pyramidal synandria, dehiscent by a terminal pore; ♀ with ovary 1-celled, ovules several together, at or near base of locule, stigma discoid-globose, 3–4-lobed, terminating usually with a minute style or not. **Fruits** berries; seeds subglobose. **Distr:** tropical Asia and Malesia. **Cult:** Propagation is by division of offsets or from cut pieces of rhizome (preferably cross-sectional). Needs fertile and moist soil in partial shade.

A. cucullata (Lour.) G.Don, *Hort. Brit.*, ed. 3: 631, 1839. (Syn: *A. rugosa*, etc.). [Chinese Taro, Buddha's Palm, Hooded Dwarf Elephant Ear]. **Herbs**, somewhat robust, evergreen, clump-forming, to 1 m tall; stems erect, hypogeal, basally much-branched. **Leaves** broadly ovate-cordate, 10–40 × 7–28 cm, base shallowly cordate, apex acute, primary veins 4 per side, radiating from petiole, arching; petiole weakly D-shaped in cross-section, 25–30(–80) cm long, petiolar sheath reaching to about halfway, margins membranous. **Inflorescences** rarely produced, solitary or paired, axillary, subtended by membranous cataphylls; peduncle 20–30 cm long; spathe 9–15 cm long, green; lower spathe 4–8 × ~2.5 cm, limb narrowly cymbiform, 5–10 × 3–5 cm; spadix 8–14 cm long; ♀ flower zone cylindric, 1.5–2.5 × 0.7 cm; sterile interstice 2–3 × 0.3 cm; ♂ flower zone 3.4 × 0.8 cm, yellow; appendix narrowly conic 3.5 × 0.5 cm, yellowish. **Fruits** berries, rarely produced, subglobose, 6–8 mm Ø, ripening red. **Distr:** NE India (Sikkim) through to Taiwan, and south to Sri Lanka. **Ecol:** 0–900 m. **Zones:** 9–12.

A. macrorrhizos (L.) G. Don, *Hort. Brit.*, ed. 3: 631, 1839. (Syn: *Colocasia macrorrhiza*, etc.). [Elephant Ear, Giant Alocasia, Giant Taro, Wild Taro]. **Herbs**, erect, robust, rhizomatous, 1–4 m tall; pseudostems to 20 cm Ø, latex milky white. **Leaves** erect, very large; lamina ovato-sagittate, bluntly triangular, 0.6–1.2(–2.5) × 0.3–0.6(0.9) m, basally attached to petiole, base deeply cordate, lobes rounded, margins entire, wavy, leathery, adaxially glossy green, abaxially paler, secondary veins 7–12 pairs; petiole 1–1.5 m long, green. **Inflorescences** 2 or more per axil; peduncles 15–45 cm long; spathe oblong, 12.5–35 cm long, 3.8–6.5 cm Ø, constricted ~4 cm from base, white to yellow, tube oblong, lamina erect during flowering, glaucous, deciduous; spadix 10–32 cm long, fertile ♂ portion white, sterile appendix cream to tan, corrugated, apex tapered to ± rounded. **Flowers** minute; ♀ with ovaries pale green, ~3 mm Ø, stigma sessile, 3–5-lobed, lobes conic, yellow. **Fruits** ovoid, red. **Distr:** Indo-Malesia to Oceania. **Ecol:** 0–500 m. Fl. Aug–Sep. **Zones:** 11–12.

A. odora (Lindl.) K.Koch, *Index Seminum* (B) 1854(App.): 2, 1854. (Syn: *Colocasia odora*, etc.). [Asian Taro, Giant Upright Elephant Ear, Hardy Elephant Ear]. **Herbs**, massive, pachycaulous, evergreen, to 2.5 m tall; stem erect to decumbent, with short stolons terminating in tubercles arising from base; latex slightly milky. **Leaves** often clustered apically; lamina peltate, cordato-sagittate or cordato-ovate, to 1.3 × 1 m, base margins undulate, apex short acuminate, primary lateral veins 9–12 per side; petiole to 1.5 m long, petiolar sheath membranous. **Inflorescences** 2–3 together, subtended by membranous cataphylls; peduncles stout, ~35 cm long; spathe ~13–25 cm long, constricted near base, lower part ovoid, green, limb broadly oblong-lanceolate, 10–30 × 4–8 cm, greenish white; spadix shortly stipitate; ♀ flower zone 1–2 × ~1.5 cm; sterile interstice equaling ♂ zone, ivory, very slightly narrowed, synandrodia rhombo-hexagonal, ~2.5 mm Ø; ♂ flower zone cylindric, ~3–5 × ~2 cm, whitish, synandria rhombo-hexagonal, ~1.5 mm Ø; appendix elongate-conic, 3–5.5 × 1–2 cm, slowly tapering, white. **Flowers** ♀ with ovaries pale green, ~3 mm Ø, stigma sessile, weakly trilobed, lobes blunt, pale green. **Fruits** berries, globose, ~1 cm Ø, ripening scarlet; fruiting spathe ~6 cm long. **Distr:** Assam to Japan and Borneo. **Ecol:** Mostly in tropical rainforests; to 1,700 m. **Zones:** 8–12.

Amorphophallus

Syn: *Brachyspatha, Hydrosme*, etc.
Species 186. **Herbs**, perennial, small to massive terrestrial plants; rootstock tuberous or rhizomatous, ± globose. **Leaf** usually solitary, rarely paired or several, usually lasting one growing season; rachis naked, winged, leaflets numerous on proximal parts; sometimes bulbils develop on the leaves; petiole terete, uniformly colored or variously blotched. **Inflorescence** epigeal, rarely partly buried, solitary or simultaneous with or directly preceding leaf development, often sculptured and patterned as petiole; spathe elongate-triangular, triangular or ovate to broadly ovate, variously shaped, often cymbiform or campanulate, more rarely funnel-shaped, outside variously colored but often shades of brownish purple or whitish green, inside mostly paler; spadix sessile or shortly stipitate, flowers usually congested; ♀ zone cylindric, fusiform, conic or obconic, contiguous with ♂ zone or separated by a sterile zone; sterile zone usually with staminodes; ♂ zone cylindric, fusiform, conic or obconic; appendix rarely absent, erect, fusiform, triangular, ovate, subglobose or globose, often malodorous. **Flowers** ♂ with stamens (1–)3–6(–8); ♀ with pistil 1, ovary sessile or shortly stipitate, ± globose, 1–4-locular. **Fruits** berries, globose or elongate, orange-red, white, yellowish, blue; seeds ± globose, ovoid, or ellipsoid, usually with a distinct raphe. **Distr:** Tropical and S Africa to Pacific. **Info:** This genus produces some of the largest inflorescences in the plant kingdom. **Cult:** Propagation is usually by division of offsets or by seed. Needs moist compost in partial shade. Reduce watering after the leaves have wilted.

A. bulbifer (Roxb.) Blume, *Rumphia* 1: 148, 1837. (Syn: *Arum bulbiferum, Pythonium bulbiferum*, etc.). [Voodoo Lily]. **Herbs**, tuberous, bulbiferous; tuber globose. **Leaf** solitary, 30–45 cm across, much-divided; leaflets obovate or lanceolate, 7.5–20 cm long; petiole 1–1.3 m long. **Inflorescence** with peduncle 20–25 cm long, green and pink, streaked with green or black; spathe long-peduncled, 12.5–20 cm long, erect, pale pink or yellowish, clouded with pink, rose-pink inside, tube broad, limb rather long-ovate, cymbiform, apex obtuse; spadix sessile, very stout, shorter or slightly longer than the spathe, flower zone 7.5 cm long, ~1.5 cm Ø; appendage pale flesh-colored or white, as long as the inflorescence, broader than elongate, conelike or oblong, 7.5–10 cm long, apex rounded. **Flowers** with anthers short; stigma bilobed. **Distr:** S and NE India, Himalayas to Myanmar. **Ecol:** In tropical rainforests, forest margins, and in disturbed open areas; to 1,500 m. Fl. Aug–Oct. **Zone:** 12.

A. kiusianus (Makino) Makino, *Bot. Mag. Tokyo* 27: 244, 1913. (Syn: *A. konjac* var. *kiusianus*, etc.). [Voodoo Lily]. **Herbs**, tuberous; tuber depressed-globose, to ~20 cm Ø, to ~12 cm high. **Leaf** solitary; lamina ~60–90 cm Ø; leaflets elongate-elliptic to lanceolate, 6–20 cm long, 3–4.5 cm Ø, moderately to long acuminate, margin undulate, adaxially bright green with a narrow pale violet margin; petiole smooth, to ~65 cm long, ~4 cm Ø, glossy, dirty olive-green or grayish green, with narrowly elongate, oval or irregular whitish or greenish spots. **Inflorescence** solitary; peduncle as petiole, 0.4–1 m long, 1.5–4 cm Ø; spathe triangular, 9–25 cm long, ~4–13 cm Ø, base rounded, outside greenish pinkish or glossy dark purplish brown, with small, whitish spots; spadix subsessile, 9–22 cm long; ♀ zone cylindric or slightly conic, 2.5–4.5 × 0.5–2 cm; appendix fusiform-conic or elongate fusiform, 4–16 × 1–3 cm, acute or obtuse, entirely blackish or with scattered, tiny green spots, glossy, ± smooth, base grooved, naked or with scattered thin violet-brown hairlike staminodes; usually between ♂ and ♀ zone and appendix with hairs (staminodes). **Fruits** berries, blue. **Distr:** S China, Taiwan, Japan (S Kyushu). **Ecol:** In secondary forests, thickets; 150–900 m. Fl. Apr–Jun. Fr. May–Jul. **Zones:** 10–12. **Info:** The tuber is poisonous.

A. konjac C. Koch, *Woch. Gartn. Pflanz.* 1: 262, 1858. (Syn: *A. rivieri, Brachyspatha konjac*, etc.). [Devil's Tongue, Elephant Yam, Konjac]. **Herbs**; tuber depressed globose, to ~30 cm Ø. **Leaf** solitary; lamina highly dissected, to ~2 m Ø; leaflets elliptic, acuminate, 3–10 cm long, ~2–5 cm Ø, adaxially dull green; petiole to 1.5 m long, 3–5 cm Ø, smooth or with scattered punctiform warts at base, often entirely covered with large elongate, dark green confluent spots and smaller white dots. **Inflorescence** solitary, to ~1.1 m tall; peduncle as petiole; spathe elliptic-lanceolate to broadly ovate-triangular, ~10–55 cm long, apex acute, base and limb ± separated by a shallow constriction, outside base dirty pale brownish with blackish green spots, basal part blackish purple and densely verrucose, upper part whitish purplish, limb erect, undulate and/or longitudinally folded; spadix 2× as long as spathe; ♀ zone cylindric, 63 cm long, purple; ♂ zone to 82 cm long; appendix long-conic, 0.2–0.5(–1.1) m long, dark purple. **Fruits**

Below: (clockwise from top left) *Amorphophallus kiusianus, Amorphophallus bulbifer, Amorphophallus konja, Amorphophallus paeonifolius, Amorphophallus titanum.*

berries, red. **Distr:** E Himalayas to China, Philippines. **Ecol:** In forest margins, thickets; 830–1,200 m. Fl. Apr–Jun. Fr. Aug–Sep. **Zones:** 10–12.

A. paeoniifolius (Dennst.) Nicolson, *Taxon* 26(2/3): 337, 1977. (Syn: *A. campanulatus, Dracontium paeoniifolium*, etc.). [Elephant (Foot) Yam, Peony Leaf Voodoo Lily, Talingo Potato, Voodoo Lily]. **Herbs**, perennial; tuber depressed-globose, to 60+ cm Ø, 25+ cm high. **Leaf** solitary (rarely 2), erect, to 5+ m tall; lamina to ~3 m Ø, highly dissected, leaflets rounded to lanceolate, 3–35 × 2–12 cm, apex acuminate, green; petiole to ~2 m long, ~20 cm Ø, pale to dark green or blackish-green, blotched with dark dots, shallowly corrugate to strongly echinate-verrucose. **Inflorescence** malodorous; peduncle 3–20 cm long; spathe campanulate, 10–40 × 15–60 cm, base strongly convolute, outside pale green or brownish, spotted white, inside maroon; spadix 20–40 cm long, flowers congested; ♀ zone cylindric, 5–15 × 2.5–8 cm pale green; ♂ zone strongly obconic, 2–8 × 1.5–10 cm, base narrower, yellowish; appendix irregularly globose-conic, to ~ 20 × 20 cm Ø, pale or dark purplish brown, surface irregularly, shallowly or deeply folded. **Flowers** ♂ with stamens 2–6, elongate, 3–5 mm long, ivory white to yellow; ♀ with ovary ± globose, 4–5 mm Ø, base white, rest maroonish or pale green, style slender, 0.7–1.1 cm long, maroon, stigma large, capitate. **Infructescence** cylindric, 16–20 × 7 cm. **Fruits** berries, oblong-elliptic, orange to red, 2.5–3 × 1.7–2 cm; seed brown, smooth. **Distr:** Tropical and S Africa to Pacific. **Ecol:** In tropical forests; <750 m. Fl. Apr–May. Fr. Oct–Nov. **Zones:** 10–12. **Info:** Often cultivated for the edible tubers and the unique inflorescence.

A. titanum (Becc.) Becc., *Bull. Soc. Tosc. Ortic.* 1879: 46, 1879. (Syn: *A. selebicus, Conophallus titanum*). [Corpse Flower, Titan Arum]. **Herbs**, tuberous, massive, perennial; tuber massive, ± spherical, often weighing over 100 kg, >2 m Ø, seasonally dormant. **Leaf** solitary, umbrellalike; lamina trilobed, to 5 m across, deeply divided into numerous leaflets; petiole erect to 6 m tall, green, reticulated or mottled with white or creamish-white, glossy. **Inflorescence** with a fetid smell, lasting for only 3–4 days, 2+ m tall; spathe green or greenish purple, speckled with cream, deeply furrowed, enclosing the spadix until the maroon-purple collar unfurls at maturity, to 3 m Ø, rim golden, frilled, sides ribbed, inside rich crimson, base forming a chamber enclosing the flowers; spadix yellowish red to grayish yellow; ♀ zone cylindric, enclosed by the spathe; appendix conical, creamy, inflated. **Flowers** unisexual; ♀ malodorous when receptive. **Infructescence** massive, ± cylindrical, fruiting portion (apical) ~1–1.5 m long, densely covered with fruits; peduncle 1+ m tall, ~30 cm Ø. **Fruits** berries, ellipsoid, glossy red. **Distr:** Indonesia (W Sumatra). **Ecol:** On limestone hills of tropical rainforests; 200–1,200 m. Fl. May–Jul. **Zones:** 11–12.

Anthurium

Species 836. **Herbs**, terrestrial or epiphytic, erect or rarely climbing by means of adventitious roots; stems fleshy, elongate or very reduced. **Leaves** alternate; lamina simple or digitate; petiole elongate, covered by a cataphyll in the juvenile stages. **Inflorescences** with spathe usually chartaceous, reflexed, usually perennial, green, white, pink, or red; spadix sessile or short-stipitate, cylindrical or conical, with numerous sessile flowers from apex to base. **Flowers** bisexual, minute; tepals 4; stamens 4; ovary bilocular, with 1 or 2 pendulous ovules per locule, style short or absent, stigma discoid or lobate. **Fruits** berries, bilocular, fleshy, red, white, or violet; seeds oblong. **Distr:** Mexico to tropical America. **Cult:** Propagation is by division. Needs fertile well-drained but moist soil in partial shade.

A. andraeanum Linden ex André, *Ill. Hort.* 24: 43, 1877. (Syn: *A. andraeanum* var. *divergens, A. venustum*). [Flamingo Flower, Flamingo Lily]. **Herbs**, perennial; stem erect; cataphylls withered, persistent. **Leaves** alternate, pendent to spreading; lamina simple, ovate-oblong, deeply cordate, 15–35+ × 20 cm, coriaceous, primary lateral veins 7–9 per side; petiole 30–60 cm long. **Inflorescences** with spathe broadly ovate to subcircular, 5–15 × 5–12.5 cm, base cordate, apex obtuse to rounded, glossy, usually red, pink or white, coarsely bullate; spadix subcylindric to conic, 6–8 cm long, usually white, yellow or pink, decurved from vertically or obliquely held spathe; peduncle to 70 cm long. **Flowers** bisexual; tepals 4, decussate, cucullate. **Distr:** Colombia to Ecuador. **Fruits** berries. **Ecol:** Fl. throughout the year. **Zones:** 10–12. **Info:** This is one of the most widely cultivated house plants, and are commercially grown for cut flowers.

A. atropurpureum R.E. Schult. and Maguire, *Bot. Mus. Leafl.* 16: 60, 1953. **Herbs**, terrestrial (epiphytic); stem to 30 cm long. **Cataphylls** 1–7 cm long, narrowly acute, persistent. **Leaves** ± erect; lamina mostly elliptic, 11–75 × (3–)5–14 cm, base attenuate or acute, margins flat to undulate, apex acuminate, ± coriaceous, adaxially ± glossy, midgreen, abaxially matte to semiglossy, paler, adaxially midrib convexly raised at base; petiole 2–25 cm long; geniculum thicker and paler than petiole, 0.5–1.5 cm long. **Inflorescences** erect, shorter than or equaling leaves; peduncle (8–)15–48 cm long, 2–5 mm Ø; spathe spreading to reflexed, withering at anthesis, green, margins heavily purple-tinged to purple-red throughout, perpendicular to peduncle, linear-lanceolate, 2–15 × 0.5–2.2 cm, broadest near base, base acute, apex acute; spadix deep purple or maroon, erect, cylindroid to long-tapered, 2–14 cm long, 3–8 mm Ø near base, 2–4 mm at apex, spirals 4–8(–10)-flowered. **Flowers** rhombic or 4-lobed, 1.6–2.2 mm across. **Fruits** berries, dark purple to deep red, globose to ellipsoid, 5–6 × 4–6 mm; seeds 1–2. **Distr:** S tropical America. **Ecol:** In tropical rainforests, understory of white sand forests; ~150 m. Fl. Mar–Aug. **Zones:** 11–12.

A. coriaceum G. Don, *Hort. Brit.*, ed. 3: 633, 1839. (Syn: *Pothos coriaceous*, etc.). [Paddle-leafed Anthurium]. **Herbs**, terrestrial or epilithic, sometimes epiphytic; stem <20 cm long. **Cataphylls** broadly lanceolate, 12–14 cm long, acute to obtuse. **Leaves** erect-spreading, robust; lamina broadly elliptic to oblong, sometimes lanceolate, (0.3–)0.6–1.1 m × (6–)15–28(–39) cm, base acute to rounded, margins undulate, apex acute to rounded, stiff, leathery, midgreen; petioles (3–)19–60 cm long, to 1.5 cm Ø. **Inflorescences** erect; peduncle 11–44 cm long; spathe erect, broadly lanceolate-elliptic, (7.5–)10–24 × (1.7–)2.5–4.3 cm, green, coriaceous, acute to acuminate; spadix pale purplish to maroon, sessile, tapered, erect, ± straight, 10–31 × 0.8–1.2 cm, principal spiral with 13–23 flowers, 8–14 flowers in alternate spiral. **Flowers** rhombic, 2.5–3.1 × 1.8–2 mm; tepals smooth; lateral tepals 1.6–2 mm wide; stigma oblong-ellipsoid, 0.5 mm long; anthers 0.5–0.7 mm long. **Fruits** berries, white to pale lavender, apex greenish, obovoid-oblong, 7 mm long; seeds 2 per berry, 3–3.5 mm long. **Distr:** SE Brazil (Espírito Santo to Paraná). **Ecol:** Usually on coastal windward slopes; <500 m. **Zones:** 11–12.

A. scherzerianum Schott, *Oesterr. Bot. Wochenbl.* 7: 53, 1857. (Syn: *A. williamsii*, etc.). [Flamingo Flower]. **Herbs**, epiphytic or terrestrial; stems very short; cataphylls thick, to 1.5 cm long, apex caudate. **Leaves** spreading; lamina linear to elliptic or lanceolate, 5–26 × 1.5–6.5 cm, base obtuse to cuneate, apex narrowly acuminate, primary lateral veins 8–11 per side, collective vein arising from near base; petioles terete, 4–20 cm long, geniculum 4–7 mm long. **Inflorescence** erect; peduncle 14–52 cm long; spathe moderately thick, bright red-orange, elliptic to ovate, 3.7–12 × 2.4–6 cm, base cordate, apex abruptly short-acuminate; spadix pale orange to red, 2–8 cm long, ~4 mm Ø midway, tapered at apex and usually coiled, 2–3 flowers visible in the principal spiral, 5–6 flowers visible in the alternate spiral. **Flowers** square, 2 × 2 mm, sides straight to ± sigmoid; tepals glossy, sparsely papillate; lateral tepals ~1.3 mm long; stamens basal, filaments ~3 mm long, anthers white; pistil weakly emergent, somewhat translucent, stigma linear, 0.4–0.6 mm long. **Fruits** berries, orange to red. **Distr:** Costa Rica. **Ecol:** In montane rainforests; 1,300–2,100 m. **Zones:** 10–12.

Arisaema

Syn: *Muricauda*. [Jack-in-the-pulpit].
Species ~170. **Herbs**; corms (rhizomes) nearly globose. **Leaves** usually appearing with flowers, 1–2(–3), erect; lamina medium to dark green, sometimes glaucous adaxially, palmately or pedately (radiately) divided, not peltate; leaflets elliptic to broadly ovate or oblanceolate, base rounded to obtuse or attenuate, apex obtuse or acute to acuminate, primary lateral veins of each leaflet pinnate; petiole longer than lamina. **Inflorescences** with peduncle erect, nearly equal to leaves (to very short), apex not swollen; spathe variously colored or striped, distal part open at maturity, exposing tip to ½ or more of spadix appendage; spadix ± cylindric, surmounted by sterile appendage of variable shape; ♂ and ♀ on same or different spadix; ♀ flowers congested; ♂ flowers usually scattered, distal to ♀ flowers when both are present. **Flowers** unisexual; perianth absent. **Fruits** berries, not embedded in spadix, glossy orange to bright red; seeds 1–6, mucilage sometimes present. **Distr:** Mostly temperate Asia; N America, Mexico; Africa. **Ecol:** Terrestrial or in wetlands. **Cult:** Propagate from seeds or by division of tubers. Requires an organic compost containing loam and grit, in warm conditions and partial shade.

A. amurense Maximowicz, *Prim. Fl. Amur.* 264, 1859. (Syn: *A. robustum*, etc.). [Chinese Arisaema]. **Herbs**, tuberous; tuber depressed-globose, 6–7 cm Ø. **Leaves** 2(–3), trifoliolate or pedately 5-foliolate; leaflets obovate to narrowly oblong, 7–11 × 4–7 cm, base cuneate, margins entire, serrate, or denticulate, apex acute; lateral leaflets sessile, oblique, 9–15 cm long; petiole (5–)7–30 cm long, base sheathing, forming a pseudostem. **Inflorescence** with spathe greenish to dark-purple, usually with pale purplish stripes, tube funnelform to cylindric, 5 × 2 cm, mouth oblique-truncate; limb ovate-lanceolate to oblong, acuminate, 3–4 cm long; spadix unisexual, ♂ portion cylindric, 2 cm × 2–3 mm, synandria lax, anthers 2–3, locules globose; ♀ subconical, 2 × 1 cm, ovaries obovate, stigma orbicular, ovules 4; appendix stipitate, cylindric, robust, erect, 3–5.5 × 4–5(–7) mm, base truncate, apex obtuse. **Fruits** berries, red, long-conic; seeds 4, ovoid, red, smooth. **Distr:** S Russian Far East to N Korea. **Ecol:** In forests, by streams; 50–200 m. Fl. May. Fr. Sep. **Zones:** 5–10.

A. candidissimum W. W. Smith, *Notes Roy. Bot. Gard. Edinburgh* 10: 8, 1917. [Chinese Jack-in-the-pulpit]. **Herbs**; tuber depressed-globose, 3–5 cm Ø. Cataphylls 3 or 4, pinkish brown, sparsely spotted with whitish dots, 5–24 × 4–5 cm, membranous, obtuse, convolute, encircling petiole. **Leaf** solitary; lamina trifoliolate, papery; leaflets sessile, ovate or orbicular, apex rounded or acute; terminal leaflet 6–8 × 7–9 cm, base shortly cuneate; lateral leaflets slightly oblique, 5–6 × 4–8 cm; petiole green, 15–35 cm long, smooth. **Inflorescence** erect; spathe pale green or white with green or purple longitudinal lines, tube cylindric, 3–4 × ~2 cm, throat margins recurved,

Top row: *Anthurium andraeanum, Anthurium atropurpureum, Anthurium coriaceum, Anthurium scherzerianum.*
Bottom row: *Arisaema amurense, Arisaema candidissimum, Arisaema consanguineum, Arisaema fargesii.*

limb ovate to ovate-lanceolate, 5–6 × 3–4 cm, apex with a tail 2–3 cm long; ♂ spadix unisexual, ♂ zone cylindric, 1.6–2 cm long, appendix sessile, green to purple, cylindric, 3–4 cm × 2–3 mm, naked, from middle slightly recurved outward, or erect, apex acute; ♀ spadix with ♀ zone oblong, ~2.1 cm long, rounded, appendix suberect or erect, white or pale green, funnel-shaped, subcylindric, ~ 4.5 × 0.5 cm, naked, base attenuate, apex acute, obtuse. **Flowers** ♂ synandria of 2 or 3 staminodes, anthers yellow; ♀ with ovaries green, obovoid, stigma subsessile. **Distr:** China. **Ecol:** In forests, thickets; 2,200–3,300 m. **Zones:** 6–10.

A. consanguineum Schott, *Bonpl.* 7: 27, 1859. (Syn: *A. biradiatifoliatum*, etc.). [Chinese Cobra Lily]. **Herbs**; tuber depressed-globose, 2–7 cm Ø. **Cataphylls** 3, dark green with whitish markings, to 55 cm long, acute. **Leaves** 1(2); lamina radiate, leaflets 18–23, dull green, base cuneate, apex long-acuminate, caudate, tail filiform to 7–10 cm long; petiole to 1 m long, green, smooth, unmarked. **Inflorescence** with peduncle green, smooth, unmarked, to 75 cm long; spathe green, sometimes whitish-striped, tube cylindric, ~6–72 cm long, auriculate, mouth recurved, limb triangular-ovate, 8–12 × 4–8 cm, deep green, sometimes with purple margin beneath, pale green inside, acuminate, tail purple, filiform, to 35 cm long, pendulous; spadix unisexual, ♂ zone cylindric 4–4.5 cm long, rachis green, appendix erect, pale green, 3–5 cm long, ♀ zone conic, 3.5 cm long, appendix pale green, ~7 cm long. **Flowers** ♂ with synandria dark violet, anthers subsessile, 2–3; ♀ with ovaries green, obovate, stigma sessile, orbicular, pilose. **Fruits** berries, red. **Distr:** WC Himalayas to Taiwan. **Ecol:** In mixed forest, thickets; <3,200 m. Fl. Jun–Jul. **Zones:** 7–10.

A. fargesii Buchet, *Notul. Syst.* (Paris) 1: 371, 1911. **Herbs**, dioecious; tuber subglobose, 3–5 cm Ø, with a few tubercles. **Cataphylls** 3, brown, narrowed upward, obtuse or acute, inner one ~15 cm long. **Leaves** solitary; lamina trifoliate, glabrous, leaflets sessile, membranous; terminal leaflet rhombiform or ovate-oblong to ovate, 12–32 × 9–27 cm, base shortly cuneate or combined with lateral leaflets, apex acute; lateral leaflets obliquely elliptic, outer side semiovate, 9–23 × 6–16 cm; petiole 20–40 cm long. **Inflorescence** a spike; peduncle erect, fleshy, 18–26 cm tall; spathe purple, marked with whitish lines, tube nearly cylindrical, 4–8 cm long, 1.5–2 cm Ø, limb oblong-deltoid, long-acuminate, galeate, throat very distinctly auricled, 6–14 × 4–5 cm; spadix unisexual, >10 cm long, ♂ zone 2.5–3 cm long; appendix included in the spathe, 7.5 cm long. Flowers ♀ congested at base of spadix, ovary angular, tip somewhat rounded, ovules few, style very short, thick, stigma subsessile. **Distr:** China. **Ecol:** In forests, thickets; 900–1,600 m. Fl. May–Jun. **Zones:** 7–10.

A. lobatum Engler, *Bot. Jahrb.* 1: 487, 1811. **Herbs**, tuberous, dioecious; tuber globose, 2–4 cm Ø. **Cataphylls** 3, brown, spotted red, to 15–20 cm long. **Leaves** 1–2; lamina trifoliolate, adaxially green with pale green veins, abaxially pale greenish, with pink veins, apex acuminate; terminal leaflet oblong, elliptic, base narrowly attenuate or obtuse, petiolulate, 8–22 × 4–10 cm; lateral leaflets sessile, oblique, oblong, outer base auriculate, rounded, 3–23 × 2–8 cm; petiole yellow-green, purple-spotted, 17–35 cm long, basally sheathing. **Inflorescence** with spathe pale purplish or pale green with purple stripes, tube funnelform, 4–7 × 1–2.5 cm, throat obliquely truncate, limb elongate-lanceolate or ovate, 4–7 × 2.5–3 cm, base constricted, apex acuminate, tail 3–5 cm long, with whitish stripes inside, ± erect; spadix unisexual, ♂ zone cylindric, 1.5–2.5 cm long, ♀ zone cylindric or subglobose, 1–2 cm long; appendix pale green to whitish-green, erect, cylindric. **Flowers** ♂ with anther-cells violet; ♀ with ovaries green. **Distr:** C and S China. **Ecol:** In forests, thickets; 600–3,300 m. Fl. Apr–Jul. **Zones:** 8–10.

A. peninsulae Nakai, *Bot. Mag.* (Tokyo) 43: 537, 1929. (Syn: *A. angustatum* var. *peninsulae*). [Jack-in-the-pulpit]. **Herbs**; tuber depressed, 1.5–5.5 cm Ø; pseudostem 29–77 cm long. **Cataphylls** 3, brownish with purple spots, 3–10 cm long. **Leaves** 2; lamina pedate; leaflets 7–17(–23); leaflets narrowly elliptic, obovate-lanceolate, or ovate-oblong, base cuneate, margins entire or serrate, apex acuminate; terminal leaflet 9–18 × 3.5–9 cm; petiole 35–93 cm long. **Inflorescence** erect, 30–80 cm long; spathe green with white stripes or purplish to dark purple with whitish stripes, tube 5–8 cm long, mouth obliquely truncate, limb incurved, dark purple, oblong, 5–6 × 4–5 cm, acuminate; spadix unisexual, ♂ zone 1–1.5 cm long, ♀ zone 1.5–3 cm long; appendix erect, cylindric, 3.5–4 cm long, base truncate, apex obtuse. **Flowers** ♂ with synandria dense, sessile, anthers 2 or 3; ♀ with ovaries dense, narrowly ovoid, style short, stigma discoid. **Fruits** berries, reddish-yellow, ovoid-globose, ~8 mm long; seeds 2 or 3. **Distr:** Japan, Korea. **Ecol:** In forests; <500 m. Fl. May–Jun. Fr. Sep. **Zones:** 8–10.

A. ringens (Thunb.) Schott, *Melet.* 17, 1832. (Syn: *A. arisanense*, etc.). **Herbs**; tuber depressed-globose,

2–5 cm Ø; pseudostem 4–30 cm long, white to pale green, sometimes pale brownish-green. **Cataphylls** greenish to purplish. **Leaves** 2; lamina trifoliate; terminal leaflet ovate-lanceolate to rhomboid-elliptic; lateral leaflets obliquely ovate to lanceolate, apex filiform-caudate, usually with a threadlike tail, adaxially green, abaxially glaucous; petiole 10–35 cm long. **Inflorescence** with peduncle 3–6 cm long; spathe yellowish-green to green; tube cylindrical, gradually opening to an auriculate mouth, 3–5.5 × 1–2 cm Ø, yellowish-green to green outside, purplish-red with white to pale yellowish-green striations inside, lamina concave, revolute, saccate, apex caudate; spadix 3–8 cm long including appendage; appendage white, 2–6 cm long, attenuate, base truncate, apex obtuse, without modified sterile flowers. **Infructescence** erect. **Distr:** E China, S Korea, C Japan to Taiwan. **Ecol:** In lowland subtropical forests. Fl. Apr, Sep. **Zones:** 7–10.

A. speciosum (Wall.) Mart., *Melet. Bot.*: 17, 1832. (Syn: *Arum speciosum*). [Cobra-lily, Jack-in-the-pulpit]. **Herbs**, tuberous, dioecious; rhizome cylindric, 4–10 × 2–4 cm. **Cataphylls** 2–3, reddish-brown, to 15 cm long. **Leaf** 1; lamina trifoliate, green with narrow red margin; terminal leaflet ovate, oblong, 11–13 × 4–7 cm, base cuneate, apex acuminate or acute; lateral leaflets larger than terminal, oblong to lanceolate, 13–16 × 4.5–6.5 cm, base strongly asymmetric, apex abruptly shortly acuminate; petiole pale green, 15–50 cm long. **Inflorescence** with peduncle whitish, brown-dotted, 2.5–10 cm long; spathe dark purple with whitish longitudinal lines, tube cylindric, 4–10 cm long; spadix unisexual, ♂ portion cylindric, 2.5 cm long; ♀ portion subconical, 1.7 × 1.2 cm; appendix swollen at lower 4–5 cm, fusiform, pink to dark purple, 20–80 cm long. **Flowers** ♂ purple, synandria of 3–5 stamens; ♀ 1.7–1.2 cm long, ovaries dark purple, angulate-conic, stigma punctate. **Fruits** berries, red, obovate, obconic, angulate. **Distr:** Nepal to SC China. **Ecol:** In broad-leaf forests; 2,400–2,800 m. Fl. Apr–Jun. Fr. Oct. **Zones:** 8–10.

Caladium

Species 12. **Herbs**, terrestrial or lithophytic; stems short; rootstock tuberous, rarely rhizomatous. **Leaves** erect; lamina simple, ovate to arrow-shaped, variegated pink, white, red and green, secondary veins netlike; petiole often attached peltately. **Inflorescences** with spathe ovoid or globose, usually green, constricted between tube and lamina, spathe tube remaining closed, spathe lamina ovate or lanceolate, deciduous after flowering; spadix divided into ♂ and ♀ zones; appendix absent. **Flowers** unisexual, minute; tepals absent. **Fruits** berries, ovoid, white. **Distr:** C and S tropical America. **Cult:** Propagation is by division of the tubers. Needs fertile, moist, but well-drained soil in partial shade.

C. bicolor (Aiton) Vent., *Mag. Encycl.* 4: 464, 1800. (Syn: *C.* × *hortulatum*, etc.). [Angel Wings, Elephant's Ear, Fancy-leaved Caladium, Heart of Jesus]. **Herbs**, deciduous, stemless. **Leaves** tufted, alternate; lamina simple, ovate to triangular-ovate, appearing heart- or arrow-shaped, 10–60 × 7.5–40 cm, base usually cordate, adaxially dark green or spotted, blotched, or zoned with white, yellow, pink, red, or purplish red; petiole smooth, apex glaucous, attachment peltate, abaxially glaucous. **Inflorescence** 1–3, shorter than leaves; peduncle short, glaucous; spathe 6–15 cm long, white or greenish white, deciduous; spadix with ♂ zone spindle-shaped, to 10 cm long, obtuse, white, flowers tightly packed. **Flowers** minute. **Fruits** berries, ovoid or pear-shaped, white; seeds several, ovoid. **Distr:** Nicaragua to Argentina (Salta). **Ecol:** On limestone outcrops of tropical rainforests. Fl. Mar–May. **Zones:** 10–12.

Colocasia

Species 16. **Herbs**, terrestrial or sometimes semiaquatic; stolons with nodes produced at or near surface; corms underground (rarely above ground), tuberous. **Leaves** erect, several, clustered apically, usually appearing before flowers; lamina peltate, ovate- or sagittate-cordate, basal lobes rounded, apex mucronate, adaxially green to dark green or glaucous blue-green, primary lateral veins parallel, secondary lateral veins netted; petiole usually longer than lamina. **Inflorescences** bisexual; peduncle erect, shorter than leaves, apex not swollen; spathe tubular, green, orangish, opening basally and reflexing apically at anthesis, revealing the spadix; spadix slender, tapering, usually terminated by a sterile appendage, ♀ flowers at base of spadix, ♂ flowers at apex, with sterile flowers between ♀ and ♂ zones. **Flowers** minute, unisexual; perianth absent. **Fruits** greenish to whitish or red; seeds 0–5(–35), usually mucilaginous. **Distr:** SE Asia. **Cult:** Propagate from tubers. Requires fertile, moisture-retentive soil in full sun or partial shade. **Cult:** Propagate from tubers. Requires fertile, moisture-retentive soil in full sun or partial shade.

C. esculenta (L.) Schott, *Meletemata Botanica*. 18, 1832. (Syn: *Arum esculentum*, *C. antiquorum*). [Coco Yam, Elephant's Ear, Taro, Wild Taro]. **Herbs**, cormous; corms underground, starchy; stolons elongate, with nodes produced at or near surface, spreading horizontally. **Leaves** peltate for 2.5–7 cm, 17–70

Top row: *Arisaema lobatum*, *Arisaema peninsulae*, *Arisaema ringens*, *Arisaema speciosum*.
Bottom row: *Caladium bicolor*, *Colocasia esculenta*, *Cyrtosperma johnstonii*, *Dieffenbachia seguine*.

Above: *Dracunculus vulgaris*, *Epipremnum pinnatum* (juvenile form), *Epipremnum pinnatum* (adult form).

× 10–40 cm, apex mucronate, adaxially green to dark green or glaucous blue-green, usually with red or purple spot at point of petiole attachment; petiole green, often purple apically, 0.3–0.8(–1.8) m long, spongy and filled with air spaces. **Inflorescences** with spathe 20–35 cm long, tube green, lamina orange, reflexing apically at anthesis exposing spadix, >3× longer than tube; spadix 9–15 cm long. **Flowers** ♂ with sterile tip pale orange, stamens 3–6, connate; ♀ greenish, interspersed with white pistillodes, ovaries 1-locular, ovules 36–67; sterile flowers white to pale yellow. **Fruits** orange; seeds 1–1.5 mm long. **Distr:** tropical Asia to SW Pacific. **Ecol:** Usually forming large colonies near waterways and wetlands; 0–500 m. **Zones:** 9–12. **Info:** Cultivated widely for the edible corms.

Cyrtosperma

Syn: *Arisacontis*.

Species 12. **Herbs**, evergreen, creepers to treelike or climbing, slender to very large; rhizome creeping. **Leaves** simple, sagittate or hastate, sometimes divided; petioles and abaxial surface of major veins with prickles. **Inflorescence** with spathe convolute at very base, open above, ovate-lanceolate, persistent, usually dark in color, purplish to almost white, longer than spadix, apex sometimes narrowly elongated, usually hooded; spadix free, cylindrical, densely covered with bisexual flowers; appendix absent or with sterile organs. **Flowers** bisexual; perianth segments 4–6, free, imbricate, somewhat hooded at the truncate apex; stamens 4–6, filaments short, broad, anthers short, 2-celled; ovary ovoid, 1-loculed, stigma cushionlike, ovules 2–many, biseriate. **Fruits** berries, smooth; seeds 1–several, rounded-compressed. **Distr:** Malesia to Pacific. **Cult:** Propagate by division. Requires a very moist fertile soil or at edges of ponds, in full sun or partial shade.

C. johnstonii (N.E. Br.) N.E. Br., *Gard. Chron.*, n.s., 1882(2): 808, 1882. (Syn: *Alocasia johnstonii*). **Herbs**, terrestrial, climbing, evergreen. **Leaves** hastate, basal lobes pointed, margins reddish-brown, green with bright red veins, midvein more prominently colored, glossy, abaxially coppery (with minute carmine spots); petiole and abaxial surface of veins prickled; petiole ± 1.25–1.5 m long, mottled, spinose, spots rosy, brownish or black-red, prickles partly scattered, partly placed in obliquely and irregularly transverse lines, forming spiny crests. **Inflorescence** erect; spathe narrowly boat-shaped, ~20–40 cm long, subabruptly long-acuminate, inside greenish, outside black-purple, rather glossy, impresso-venose; spadix fleshy, cylindrical, ~12.5–25 cm long, brown or violet, flowers crowded; peduncle ~0.5–1.5 m long, rosy to black-red. **Flowers** bisexual; ovaries ovoid or subcylindrical, apex green, at length brown or violet, often 1(2)-ovulate, stigma roundish, broadly pulvinate, sessile, yellow or brown. **Distr:** Solomon Islands. **Ecol:** In moist rainforests, wetlands. Fl. throughout most of the year. **Zones:** 11–12.

Dieffenbachia

Species 57. **Herbs**, shrublike, monoecious; stem ± thick, base prostrate, upper parts erect, leafy apically; sap milky. **Leaves** large, entire, usually ± leathery, mottled, veins parallel to margins; petiole base sheathing stem. **Inflorescences** 1 to several, axillary; spathe oblong, edges overlapping basally, appearing tubular, persistent, somewhat constricted near middle, opening only above constriction, reclosing completely, usually green outside, paler inside; spadix cylindrical, divided into ♂ and ♀ zones, often separated by a sterile zone; ♂ zone club-shaped, white, free from spathe; ♀ zone ± fused to spathe. **Flowers** unisexual; tepals absent. **Infructescences** groups of berries; spathe usually colored, opening when fruit ripens. **Fruits** berries, globose or 2–trilobed; seeds 1–3. **Distr:** Mexico to tropical America. **Ecol:** In humid forests, forming large colonies. **Cult:** Propagate from stem cuttings, root cuttings or by division. **Info:** The milky exudates are poisonous, and can cause skin irritation on contact.

D. seguine (Jacq.) Schott, *Melet. Bot.*: 20, 1832. (Syn: *D. maculata*, etc.). [Dumb Cane, Dumb Plant]. **Herbs**, to 3 m tall; stem to 5 cm Ø. **Leaves** ovate-elliptic to oblong-ovate, 20–40 × 12–19 cm, base usually truncate to subcordate, dark green or white or yellow spotted and blotched; petiole to 30 cm long, green or white-spotted or white-streaked, sheathing about half its length, adaxially usually not grooved. **Inflorescences** erect to pendent; peduncle shorter than spathe, 3.8–15 cm long; spathe 25 cm long, green; spadix with ♂ zone partly projecting. **Infructescences** with spathe becoming orange; fruiting spadix pendent. **Fruits** orange-red. **Distr:** Caribbean, Tropical S America. **Ecol:** In the understory and on the margins of moist tropical forest. Fl. Aug–Oct. **Zones:** 10–12.

Dracunculus

Syn: *Anarmodium*.

Species 2. **Herbs**, tuberous, perennial, slender to robust; tuber compressed-globose to discoid; pseudostem formed from the long sheathing leaf bases. **Leaves** reniform, pedate, midlobe larger, lobes ± flat. **Inflorescence** solitary; peduncle robust to somewhat slender, terete, barely to exserted from apex of pseudostem; spathe usually large, tube oblong to elliptic-oblong, strongly convolute, limb ovate-lanceolate to oblong-triangular, acuminate, deciduous; spadix ± equaling spathe; appendix slender to very large, cylindric to flattened, stipitate; ♀ zone oblong basally; ♂ zone oblong to oblong-fusiform, separated from the ♀ zone by a narrow naked or pistillode-bearing interstice (in *D. vulgaris*). or contiguous with the ♀ zone (in *D. canariensis*). **Flowers** lacking perianth; ♀ with ovary oblong to oblong-fusiform, unilocular, few-ovulate, placentation apical and basal, style short; ♂ with stamens 3–4, filaments very short, conspicuous, fused. **Fruits** berries, ± obovoid; seeds few, globose-ovoid, laterally compressed, angled. **Distr:** Micronesia, S Europe to SW Turkey. **Cult:** Propagate from seeds or offsets. Requires a well-drained soil in full sun or partial shade. Water well during growth period only.

D. vulgaris Schott, *Melet. Bot.*: 17, 1832. (Syn: *Arum dracunculus*, etc.). [Black Arum, Black Dragon, Dragon Arum, Dragon Lily]. **Herbs**, terrestrial, tuberous, robust, usually solitary (clump-forming), perennial, to 2 m tall; pseudostem 0.5–1.5 m tall, pale to light gray-green with purple-black markings. **Leaves** 2–4, apical, deciduous; lamina reniform, pedate, leaflets 7–19+, elliptic, midlobes larger, to 45 cm long, bright green, often streaked, shiny, with nearly transparent windows. **Inflorescence** erect to (0.4–0.5)1.3 m tall; spathe to ~1.2 m tall, dark purplish-red inside, outside greenish, curling back with age; spadix to ~1.3 m tall, flowers hidden inside the basal tubular part of spathe, ♀ flowers basal, ♂ flowers separated by a very narrow, naked or pistillode-bearing zone; appendix dark purplish-red. **Flowers** lacking perianth; ♂ ~1 mm long, anthers 4; ♀ ~3 mm long, ovary with 4–5 ovules. **Fruits** berries, pear-shaped, ~1–2 cm long. **Distr:** SC Europe to W and SW Turkey. **Ecol:** In scerophyllous and mixed forests, wastelands, often weedy; 0–600 m. Fl. Nov–Jan. **Zones:** 9–11.

Epipremnum

Syn: *Anthelia*.

Species 15. **Lianas**, climbing with adventitious roots, slender to very robust, to 20 m long. **Leaves** homophyllous, lamina occasionally becoming increasingly pinnatifid and perforated in mature plants; lamina entire to regularly or irregularly pinnatifid, divisions pinnatifid to pinnatisect, membranous to stiffly chartaceous or coriaceous, occasionally midrib ± naked between segments, venation simple to pinnately compound; petiole canaliculate. **Inflorescences** solitary to several together; peduncle terete to laterally compressed; spathe canoe-shaped, beaked, ± flat at anthesis, soon deciduous, softly coriaceous, creamish-white, greenish or yellow; spadix ± cylindrical, sessile, rarely stipitate, bluntly tapering toward apex, flowers on uppermost and lowermost zones sterile or lacking stigma. **Flowers** bisexual, perianth lacking; filaments strap-shaped, anthers exserted; ovary ± cylindrical, often laterally compressed, irregularly angled, ovules 2–8, stigma punctate to linear. **Fruits** berries, pulp sticky; seeds curved or reniform. **Distr:** Tropical and subtropical Asia to Pacific. **Cult:** Propagate from tem cuttings, by air-layering or seeds. Requires support for climbing, high humidity and filtered light to full sun.

E. pinnatum (L.) Engler, *Pflanzenr.* IV (23B) 37: 60, 1908. (Syn: *Monstera pinnata*, etc.). [Dragon-Tail Plant, Money Plant, Taro Vine]. **Vines** or **hemi-**

epiphytes, non-woody, glabrous, climbing with aerial roots, 6–8 m long; stems with watery latex. **Leaves** alternate, chartaceous, dimorphic; *juvenile* leaves entire; *adult* leaves pendulous, pinnately compound, to 50 × 30 cm, base asymmetric, subcordiform, apex terminating in a leaflet or an acuminate segment, adaxially dull, glabrous, venation flat, abaxially dull, with prominent venation, with a line of domatia on both sides of the midvein, this sometimes with small spiniform outgrowths; petiole to 35 cm long, marginate along their length. **Inflorescences** axillary, solitary, erect; spathe almost cylindrical, to 18 cm long, opening along one side, fleshy, yellowish, acuminate at apex, deciduous; spadix sessile, ± cylindrical, 12–15 cm long. **Fruits** berries; seeds 1–2. **Distr:** Tropical and subtropical Asia to Pacific. **Ecol:** In humid to wet, evergreen or deciduous forests; 0–2,000 m. Fl. & Fr. Jan–Feb. **Zones:** 10–12.

Lasia

Species 2. **Herbs**, aquatic, perennial, evergreen; stem erect or ascending, with armed internodes. **Leaves** several, crowded at apex of stem; lamina sagittate-hastate, simple or divided; petiole long, aculeate, inside with longitudinal air-vessels; pulvinus well-developed, solid. **Inflorescences** apical, axillary, solitary; spathe constricted, ovoid-caudate, entirely enclosing the spadix, later caducous, lower part widened when mature opening by a slit, the caudate portion persistently involved; spadix sessile, cylindric, within widened part of spathe; peduncle long, aculeate. **Flowers** bisexual; tepals 4–6, free, with vaulted-truncate apex; stamens 4–6, anthers elliptic, longitudinally dehiscent, filaments much longer; ovary 1-celled, 1-ovuled, style very short, stigma suborbicular. **Fruits** drupaceous, exocarp thin, mesocarp finely fibrous, endocarp rather thick, soft; seed large, compressed. **Distr:** tropical Asia. **Cult:** Propagate by division or from seeds. Requires a very shallow, still water, such as margins of ponds, in full sun.

L. spinosa (L.) Thwaites, *En. Pl. Zeyl.*: 336, 1864. (Syn: *Dracontium spinosum*). **Herbs**, 1–2 m tall; stem with strong, nipple-shaped prickles. **Leaves** herbaceous, variable, to 35–65 × 20–60 cm, base constricted, lobes bifid, with 3–4 narrow lateral segments, margins entire or pedate, apex acuminate, primary nerves 2–4, venation smooth adaxially, abaxially with prickles; petiole aculeate, 0.3–1.25 m; pulvinus 1.5–3.5 cm long. **Inflorescence** with peduncle to 47 cm tall; spathe black-red outside, crimson inside, 18–35 cm long, caudate portion to 28 cm long, widened part much shorter, 3–10 cm wide; spadix cylindric, 3–5 cm long. **Flowers** with tepals oblong, 1.5–3 mm long, apex triangular-hooded, keeled; filaments ~1.5 mm long; ovary ovoid, ~1.5 mm high. **Fruits** obpyramidal, apex truncate, densely warty-aculeate, ~1 cm wide, sides unarmed; seed ovoid-cordate, 5–7 mm long. **Distr:** SE Asia, China to New Guinea. **Ecol:** In moist parts of tropical forests; <1,500 m. Fl. Jul–Sep. **Zones:** 9–12.

Monstera

Syn: *Serangium*, *Tornelia*.

Species 38. **Vines**, slender or robust, climbing using adventitious roots, evergreen; stems smooth or verrucose, elongate. **Leaves** alternate; lamina simple, margins entire or laciniate, usually perforated; petiole usually elongate, winged. **Inflorescence** with spathe boat-shaped, very thick, convolute, enclosing the spadix, deciduous; spadix cylindrical, thick, erect, fertile above, sterile below. **Flowers** bisexual, lacking perianth; stamens 4; pistil prismatic, truncate at apex, ovary bilocular, with 2 basal ovules per locule, stigma punctate or linear. **Fruits** berries, fleshy, green, white, yellow, or orange; seed usually 1, ovoid or cylindrical. **Distr:** Mexico to tropical America. **Cult:** Propagate from seeds or rooted cuttings. Requires a humus-rich, well-drained soil in full or partial sun.

M. deliciosa Liebm., *Vidensk. Meddel. Naturhist. Foren. Kjøbenhavn* 1: 19, 1849. (Syn: *M. borsigiana*, etc.). [Split-leaf Philodendron, Swiss Cheese Plant]. **Vines**, robust, climbing or creeping, hemi-epiphytic; stems to over 20 m long. **Leaves** circular or ovate, 25–90 × 25–75 cm, base cordate, margins regularly pinnately divided, leathery, perforations 1–2, oval near the midvein; petiole 0.3–1 m long. **Inflorescences** ± erect, solitary; peduncle 10–18 cm long; spathe hooded, broadly ovate, 15–30 × 7.5–18 cm when flattened, cream to pale yellow, leathery; spadix 10–25 × 2.5–4 cm. **Infructescence** aggregate of berries; fruiting spadix 15–25 × 4–6 cm; pulp yellow, spotted purple; seeds 0–2 cm long. **Distr:** Mexico to C America. **Ecol:** In tropical rainforests, forest margins and open areas of forests, often reaching the upper forest canopy; 300–2,000 m. Fl. Jun–Aug. **Zones:** 11–12. **Info:** The fruit (infructescence) has edible pulp, with a flavor similar to some tropical fruits.

Philodendron

Species 444. **Vines**, robust or slender, climbing by means of adventitious roots, or less frequently erect herbs; stems elongate, with watery, caustic sap. **Leaves** alternate; lamina simple, lobate, divided or pinnately compound; petiole elongate. **Inflorescences** with spathe convolute, enclosing the spadix, thickened, not adhering to spadix; spadix cylindrical, erect, almost sessile, with ♀ flowers on basal zone. **Flowers** unisexual, lacking perianth; stamens 2–6; ovary usually biloculed, locules with 1 or more ovules, stigma sessile, entire or lobate. **Fruits** fleshy; seeds >1, ovoid or ellipsoid. **Distr:** Mexico to tropical America. **Cult:** Propagate from seeds, cuttings or air-layering. Requires a moist but well-drained, humus-rich soil in full or partial sun. Water copiously during the growth phase.

P. bipinnatifidum Schott ex Endl., *Gen. Pl.* 1(3): 237, 1837. (Syn: *P. selloum*, etc.). [Cut-leaf Philodendron, Tree Philodendron]. **Herbs**, terrestrial; stems erect or decumbent, to 2.5 m long, 5–12.5 cm Ø. **Leaves** ovate, broadly ovate-cordate, or arrowhead-shaped, 0.7–1.3 × 0.6–1.2 m, usually bipinnately divided, with 10–12 segments per side, basal segment usually with 1–4 lobes; petiole about as long as lamina. **Inflorescences** usually 1(or 2), peduncle 5–10 cm long; spathe 15–33 cm long; ♀ zone basal, pale greenish, longer and thicker than fertile ♂ zone. **Fruits** pale yellowish-white; seeds ± cylindrical, ~3.5 mm long. **Distr:** SE and S Brazil to NE Argentina. **Ecol:** In rainforests, gallery forests, seasonally dry forests; 0–1,000 m. **Zones:** 10–12. **Info:** The ripened berries smell like papaya, pulp edible with a sweet taste.

Pinellia

Syn: *Hemicarpurus*.

Species 9. **Herbs**, perennial with cormlike tuber; bulbils produced at tuber, petiole or on base of leaf lamina. **Leaves** 2–5, tripartite, trilobate or pedate, leaflets oblong-elliptic, ovate-oblong, margins entire, intramarginal veins 3. **Inflorescence** solitary, appearing with leaves; spathe persistent, tube convolute, throat almost closed with a spacer inside, limb oblong, cymbiform; spadix monoecious, ♀ zone adnate to spathe tube from base to spacer on one side, bearing ovaries on the other side, ♂ zone free from spathe limb, placed above the spacer, shortly cylindric; appendix sessile, elongate-filiform, conic-cylindric, much longer than spathe. **Flowers** unisexual, perianth lacking; ♂ with stamens 2, fused into a synandria, filament short; ♀ with ovaries ovoid, 1-celled, 1-ovuled. **Fruits** berries, oblong-ovate, acute, 1-seeded. **Distr:** China to Temperate E Asia. **Cult:** Propagate from seeds, division of dormant tubers or where available axillary bulbils. Requires a moisture retentive soil in partial shade.

P. ternata (Thunberg) Breitenbach, *Bot. Zeitg.* 37: 687, 1879. (Syn: *Arum ternatum*, *Pinellia tuberifera*). [Crowdipper]. **Herbs**, tuberous; tuber globose, 1.2 cm Ø. **Leaves** 2–5, trifoliolate, leaflets adaxially green, abaxially greenish, oblong-elliptic or lanceolate, base cuneate, apex acuminate, terminal leaflet 3–10 × 1–3 cm, lateral leaflets 4–7.5 × 1.8–2.3 cm; petiole 15–20 cm long, base sheathing, bulbils present in sheath, on lower petiole and lamina base. **Inflorescence** with peduncle 25–35 cm long; spathe greenish, tube narrow-cylindric, 1.5–2 cm long, limb oblong, green, usually violet at margin, 4–51.5 cm long, obtuse or acute; spadix with ♀ zone 2 cm long, sterile zone 3 mm long, ♂ zone 5–7 mm long; appendix green to violet erect or sigmoid. **Flowers** minute. **Fruits** berries, ovoid, yellow-green, stigma persistent. **Distr:** China, Japan, Korea. **Ecol:** In grasslands, wastelands, secondary forests; <2,500 m. Fl. May–Jun. **Zones:** 7–10. **Info:** Tubers poisonous, but used for medicinal purposes.

Pistia

Syn: *Apiospermum*, *Limnonesis*, *Zala*, etc.

Species 1. *Generic description same as specific.* **Cult:**

Below: *Lasia spinosa*, *Monstera deliciosa*, *Philodendron bipinnatifidum* (inflorescence), *Philodendron bipinnatifidum* (habit).

Above: *Pinellia ternata, Pistia stratiotes, Pothos scandens.*

Propagate by division of plantlets. Requires a warm pond in partial to full sun.

P. stratiotes L., *Sp. Pl.* 2: 963, 1753. (Syn: *P. spathulata*). [Water Lettuce]. **Herbs**, aquatic, floating; stoloniferous, roots to 50 cm long, with short branches. **Leaves** 2–15(–20) cm, spongy, densely white-pubescent, surfaces light green to grayish-green, major veins 5–13(–15), nearly parallel, abaxially prominent. **Inflorescences** with spathe white to pale green, basally convolute, slightly constricted above middle, spreading apically, white to pale green, pubescent outside, glabrous inside; spadix adnate to spathe more than half its length, shorter than spathe; axis naked at base of ♂ part and sometimes extending beyond ♂ flowers; ♂ flowers (2–)6–8, in single whorl around central stalk; ♀ flowers solitary. **Flowers** ♂ with stamens 2, connate; ♀ with ovaries 1-locular, 4–5 mm long, ovules 4–15(–20), orthotropous, styles ~3 mm long, stigmas obtuse, hairy. **Fruits** with thin pericarp; seeds light brown, cylindric, 2 mm long. **Distr:** Tropical and subtropical regions worldwide. **Ecol:** In slow-moving waterways; 0–500 m. **Zones:** 9–12.

Pothos

Syn: *Gonirurus, Potha.*

Species ~70. **Climbers**, evergreen; climbing with roots. **Leaves** simple, entire, venation reticulate with a submarginal vein; petiole pulvinate, sometimes broadly winged. **Inflorescences** usually solitary, axillary, pedunculate with several persistent basal cataphylls; spathe persistent; spadix subglobose to elongated (and densely covered with flowers or spirally twisted (with scattered flowers or clusters of flowers), usually stipitate. **Flowers** small, bisexual, 3-merous; tepals 6 (3+3), subequal; stamens 6, equal; pistil trilocular with one subbasal ovule per locule, stigma subsessile, punctate. **Fruits** berries, 1–3-seeded, ellipsoid, green, ripening to bright red; seeds smooth. **Distr:** Indo-Malesian (outliers in Madagascar, China and Australia). **Ecol:** In lowland forests. **Cult:** Propagate from rooted cuttings or seeds. Requires a moist, humus-rich soil or compost, in partial shade. Needs support for climbing.

P. scandens L., *Sp. Pl.* 968, 1753. (Syn: *P. angustifolius*, etc.). **Lianas** or **climbers**, epiphytic, creeping; stems 4–6+ m long, branched. **Leaves** obovate, lanceolate to linear-lanceolate, 4–10 × 1–3 cm, base obtuse, apex acuminate, nerves 3 per side; petiole to 5 cm long, auriculate. **Inflorescences** on hanging branches, small, axillary, solitary; spathe very small, ovate and erect, or cymbiform, to 7 mm long, purple; peduncle 0.5–1.5 cm long, with cataphylls 6–8 at base; stipe deflexed, 4–6 mm long; spadix yellow-green or yellow, subglobose or ellipsoid, 4–5–6 × 5 mm, subglobose. **Fruits** berries, ripening to glossy yellow or scarlet to crimson red, 5–6 mm Ø. **Distr:** Madagascar to New Guinea. **Ecol:** In humid forests, rainforests; 0–1,500 m. Fl. throughout most of the year. **Zones:** 10–12.

Rhaphidophora

Species 97. **Lianas**, large, epiphytic, or creeping, often robust; stems thick, rooting at nodes. **Leaves** distichous; lamina entire, divided, or perforate, firm, pinnately nerved; petiole unilaterally grooved, shortly sheathing near base, with pulvinus short or long; raphides numerous in stem and leaf. **Inflorescences** apical, solitary or sometimes united into a rhipidium; spathe not constricted, initially inrolled and tubular, afterward spreading, firm, early caducous; peduncle short or absent; spadix cylindric, flowers numerous, bisexual or sometimes a few ♂. **Flowers** lacking perianth; stamens 4, anthers much shorter than filaments; style absent to shortly conical, stigma circular to linear, ovary 1- or incompletely biloculed, 2–many-ovuled, 1- or 2-seriate. **Fruits** baccate; seeds many, small, oblong to reniform. **Distr:** Tropical Africa to W Pacific. **Ecol:** Epiphytic or hemi-epiphytic on trees or creeping on rocks. **Cult:** Propagate from cuttings or seeds. Requires a moist, well-drained soil in shade or full sun.

R. decursiva (Roxb.) Schott, *Bonplandia* (Hannover) 5: 45, 1857. (Syn: *Pothos decursivus*, etc.). [Creeping Philodendron]. **Climbers**, robust; stem 3–5 cm Ø. **Leaves** *juvenile* orbicular, 16 × 13 cm, margins entire, apices abruptly acute; *mature* leaves oblong-ovate or ovate, 50–70 × 40–50 cm, base subcordate, margins irregularly and symmetrically pinnately divided to midrib, apex acute; pinnae (5–)9–15 per side, ~ 20 × 3–5 cm at midleaf, bases narrowed, apices truncate with ascending falcate tip, costa strong; petiole 20–40 cm long. **Inflorescences** solitary, axillary, erect; peduncle green, cylindric, robust, 11–20 × 1.5–3 cm; spathe initially involved, later spreading, soon deciduous, 13–20 × 10–12 cm, yellow; spadix sessile, gray-green, cylindric, 15–16 × 2–3 cm, base oblique. **Flowers** with filaments flat, ~5 mm long, anthers yellow; ovary hexagonal-conical, 5 × 3 mm, ovules many, stigmas yellow, slightly vertically oriented on a thin style ~1 mm long. **Infructescences** cylindrical, 15–20 × 5–5.5 cm. **Fruits** berries, green-white, obconical, ~1.8 cm long, congested, at maturity the upper 3–4 mm of the pistils fall away exposing the fragrant orange pulp with seeds and trichosclereids. **Distr:** Himalayas to Vietnam. **Ecol:** In submontane areas, monsoon rainforests, climbing on trees or rocks; 900–1,800 m. Fl. May–Aug. Fr. Jul–Sep. **Zones:** 8–11.

Sauromatum

Species 6. **Herbs**, protantherous or hysteranthereous; corms thick. **Leaves** solitary, pedate. **Inflorescence** with spathe long, cylindrical, limb long, acuminate, deciduous, appendage as long as the spathe; ♂ and ♀ flowers distant, separated by a small zone of neuter club-shaped organs. **Flowers** with anthers 4-locular, with a short stalk; ovary unilocular with 1–2 basal ovules. **Fruits** berries, 1-seeded. **Distr:** Tropical Africa, Asia. **Cult:** Propagate by division. Requires a moist but well-drained soil in full sun.

S. venosum (Ait.) Schott, *Prodr. Aroid.* 71, 1860.

Below: *Rhaphidophora decursiva* (inflorescence), *Rhaphidophora decursiva* (mature leaf), *Sauromatum venosum.*

(Syn: *Arum guttatum*, etc.). [Voodoo Lily]. **Herbs**, monoecious; corm to 13 cm across, globose-depressed. **Leaf** pedate, 7–11 lobed, lobes elliptic-lanceolate, 6–23 × 1.4–7 cm; petiole mottled black, 30–45 cm long. **Inflorescence** with spathe 32–44 cm long, tube ovoid, margins connate, much shorter than the limb, greenish, mottled purplish-black inside; peduncle 1.5–3.5 cm long; spadix shorter than spathe, appendage pale green, stout, 14–18(–21) cm long, obtuse; ♂ zone 1–13 cm long; *neuter* zone 1–1.2 cm long, neuter organs clavate, 4–5 mm long, yellow; ♀ zone 1.5–2 cm long. **Flowers** foul-smelling; ovary obovoid, stigma sessile. **Fruits** berries, red; seed ± ovoid, ~1.8 mm across. **Distr:** India, Pakistan. **Ecol:** In sunny places or as an undergrowth; 900–2,300 m. Fl. Apr–May. **Zones:** 9–12. **Info:** The leaves appear after flowering.

Scindapsus

Syn: *Cuscuaria*. [Ivy Arum].
Species 35. **Climbers**, herbaceous or somewhat woody; juveniles usually terrestrial. **Leaves** ovate to oblong-lanceolate, often large and sometimes variegated, margins entire or lobed, apex acuminate; petiole long, base somewhat sheathing, apically geniculate. **Inflorescences** with peduncles short; spathes boat-shaped, deciduous; spadix sessile or nearly so, cylindric, densely many-flowered, shorter than spathe. **Flowers** perfect, perianth none; stamens 4; ovary somewhat 4-angled, truncate, 1-celled, 1-ovuled, style absent. **Fruits** drupaceous, ± united, often large. **Distr:** Tropical and subtropical Asia, N Australia, W Pacific. **Ecol:** Climbing on trees. **Cult:** Propagate by cuttings or by air-layering. Requires a moist but well-drained soil in full or partial sun.

S. pictus Hassk., *Tijdschr. Natuurl. Gesch. Physiol.* 9: 164, 1842. (Syn: *Pothos argyraeus*, etc.). [Pothos Vine, Satin Pothos]. **Climbers**; stems 2–2.5 m long, when dry 4-angled and pale reddish-brown. **Leaves** falcately ovate-lanceolate to broadly ovate, 12.5–18 × 4–12.5 cm, base deeply cordate, apex acute or acuminate, thickly coriaceous, dull green, nerves slender; leaves on juvenile plants orbicular-cordate, green blotched with white; petiole 2.5–5 cm long, winged. **Inflorescences** with peduncles equaling petiole, ~2.5–5 cm long, very stout; spathe white, 4–6.5 cm long, cymbiform, beaked; spadix green. **Distr:** SE Bangladesh to W and S Malesia. **Ecol:** In humid forests; 250–850 m. **Zones:** 10–12.

Spathiphyllum

Syn: *Amomophyllum, Massowia*, etc.
Species 49. **Herbs**, terrestrial, hemi-epiphytic or lithophytic, rhizomatous. **Leaves** oblong to elliptic, or oblanceolate, secondary veins parallel, arising along midvein; petiole as long or longer than lamina, base clasping stem, apex swollen pulvinus-like. **Inflorescences** fragrant; peduncles longer than petioles; spathe slightly concave or flat, usually white to greenish; spadix uniform, shorter than spathe, stalk ± attached to spathe. **Flowers** bisexual; tepals 4–6. **Distr:** Mexico to tropical America, Malesia, W Pacific. **Cult:** Propagate by division of the rhizomes. Requires, fertile, moist, well-drained soil in full or partial sun.

S. cannifolium (Dryand. ex Sims) Schott, *Aroideae* 1: 1, 1853. (Syn: *Dracunculus cannifolius*, etc.). [Spathe Flower]. **Herbs**, terrestrial. **Leaves** 12.5–20 × 2.5–4 cm, with numerous pairs of secondary veins. **Inflorescences** with spathe large, ovate-oblong, 15 × 5 cm, shortly cuspidate, slightly exceeding the spadix, white inside with fine, often greenish veins and a submarginal collective vein, greenish outside, with darker green veins; spadix 7–12 cm long, white at anthesis, naked part of spadix short, 10 cm long, free from spathe, stipe 1.2–1.5 cm long. **Flowers** with perianth segments white at anthesis, connate into a cup; sepals truncate, as long as ovary; perigonium 4–6-angular; ovaries with locules 4-ovulate. **Distr:** Trinidad, S tropical America, Colombia. **Ecol:** In tropical wet forests, often along rivers; 200–1,000 m. Fl. throughout the year. **Zones:** 11–12.

S. wallisii Regel, *Gartenflora* 26: 323, 1877. [Peace Lily, Spathe Flower, White Anthurium]. **Herbs**, terrestrial, glabrous, to 50 cm tall; rhizome very short. **Leaves** in a loose rosette, ascending; lamina elliptic to oblong-elliptic, usually 12–24(–36) × 2.3–4(–10) cm, base attenuate, slightly asymmetric, margins entire, slightly wavy, glossy, chartaceous, larger secondary veins 8–10 pairs; petioles erect, ± cylindrical, 13–25 cm long, narrowly winged. **Inflorescences** ascending, taller than foliage; peduncle 20–45(–60) cm tall; spathe white, erect, ± hooded, ovate to oblong-elliptic, 7.5–11(–18) × 2.5–7.5 cm, base extending down 1.5–4 cm onto peduncle, apex acuminate; spadix erect, knobby, to 3(–5) cm long, stalk to 1.2 cm long. **Flowers** minute; tepals white; style conical, projecting above tepals. **Fruits** berries, green, conical ~4 mm long, produced throughout the spadix. **Distr:** Colombia to Venezuela. **Ecol:** Unknown in the wild. Fl. throughout most of the year. **Zones:** 11–12.

Syngonium

Species 35. **Herbs**, epiphytic or hemi-epiphytic, mostly climbing with aerial roots; stems with milky latex. **Leaves** simple or variously divided into 3–11 leaflets; petiole forming a leaf sheath at base. **Inflorescences** axillary, solitary or in groups of 1–11; peduncles erect in flower, pendulous when fruiting; spathe fleshy, convolute, notably narrow at middle, tube ellipsoid, lamina whitish to greenish, opening when mature; spadix shorter than spathe, erect, with ♀ flowers in a basal zone. **Flowers** unisexual; perianth absent; stamens 3–4, united into a synandrium; ovary with (1–)2(–3) locules, each 1(–2)-ovuled, stigma discoid or bilabiate. **Fruits** berries, united to form an ovoid syncarp; seed 1, obovoid or ovoid. **Distr:** Mexico to tropical America. **Cult:** Propagate from rooted cuttings or by air-layering. Requires a moist, humus-rich soil, in partial or full sun. Needs support for climbing.

S. podophyllum Schott, *Bot. Zeitung* (Berlin) 9: 85, 1851. (Syn: *S. podophyllum* var. *typicum*). [Arrowhead Vine]. **Vines**, evergreen, to 10 m long; stem cylindrical, rooting at nodes, glaucous, 1–2 cm ∅, with abundant milky sap. **Leaves** pedately divided; leaflets 3–11, united or free to base, base variously auriculate, margins sinuate, apex acuminate, coriaceous, outermost leaflets obovate, elliptic or lanceolate, 16–38 × 6–17 cm; petioles 15–60 cm long, nearly cylindrical, sheathed to entire length. **Inflorescences** 4–11 per axil, ascending; peduncles 8–9 cm long, slender; spathe ~10 cm long, convolute into an ellipsoid tube at base, lamina cream-colored, concave, ephemeral; spadix whitish, sessile, cylindrical with a constriction between the ♀ and ♂ zones. **Infructescences** syncarps, ovoid, red, reddish orange or yellow, 3–5.5 cm long. **Distr:** Mexico to tropical America. **Ecol:** In lowland rainforests and gallery forests; 100–500 m. Fl. Jan–Mar. **Zones:** 11–12.

Typhonodorum

Syn: *Arodendron*.
Species 1. *Generic description same as specific.* **Cult:** Propagate from seeds. Best grown at shallow ends of ponds, in full sun.

T. lindleyanum Schott, *Oest. Bot. Wochenbl.* 7: 70, 1857. (Syn: *Arodendron engleri, T. madagascariense*). [Water Banana]. **Herbs**, to 4 m tall; pseudostems 30+ cm ∅ at base. **Leaves** with lower part very broad, convolute, sheath forming pseudostem, whitish pink, striped and spotted purplish black; lamina held ± erect, triangular, to 1.4 × 0.85 m, base sagittate, green, basal lobes acute to rounded, separated by triangular sinus, apex acute, primary lateral veins 5–9 per side; petiole including sheath, to >3 m long, upper part green, eventually deciduous. **Inflorescences** with peduncle to 50 cm long, 4 cm ∅, green; spathe to 80 cm long, basal tube yellowish-green outside, becoming inflated and darker, inside white to pinkish, ~1/3 the length of expanded, cream-yellow, limp acuminate; spadix to 55 cm long, shorter than spathe, upper ♂ part and staminodial appendix together 3–4× longer than ♀ part, yellowish to white, deciduous with spathe-limb, ovary reddish to yellowish. **Fruits** berries, subglobose, ~4 cm ∅, yellow when ripe; seeds ~3 × 3 cm, brownish. **Distr:** Tanzania (Islands), W Indian Ocean. **Ecol:** In freshwater swamps near sea level. **Zones:** 10–12. **Info:** Cultivated occasionally in the tropics for ornamental purposes. The rhizomes and seeds are commonly eaten in Zanzibar and Pemba.

Zamioculcas

Species 1. *Generic description same as specific.* **Cult:** Propagate by division. Requires a moist but well-drained soil in partial or full sun.

Below: *Scindapsus pictus, Spathiphyllum cannifolium, Spathiphyllum wallisii, Syngonium podophyllum.*

Top row: *Typhonodorum lindleyanum* (inflorescence), *Typhonodorum lindleyanum* (habit), *Zamioculcas zamiifolia* (inflorescence).
Bottom row: *Zamioculcas zamiifolia* (habit), *Zantedeschia aethiopica*, *Limnocharis flava*.

Z. zamiifolia (Lodd.) Engl., *Pflanzenr.*, IV, 23B: 305, 1905. (Syn: *Caladium zamiifolium*, *Z. lanceolata*, *Z. loddigesii*). [Zanzibar Gem, Succulent Philodendron]. **Herbs**, terrestrial, perennial, evergreen; rhizome stout, creeping. **Leaves** radical, erect; lamina pinnate, 45–60 cm long; leaflets opposite or alternate, oblanceolate, 9–14 × 2.5–5 cm, cuneately narrowed to base, apex acute; petiole stout, base shortly sheathing. **Inflorescences** axillary, peduncle 3.8–6.5 cm long, surrounded by several bracts; spathe green, convolute, glabrous, tube subglobose, ~ 1.2 × 1.2 cm, limb horizontally spreading or somewhat reflexed, ovate-oblong, 3.8–5 × ~2.5 cm, acute; spadix free, sessile, monoecious, stout, 1.2 cm thick, constricted above ♀ zone (ovaries), very obtuse, whitish, sterile zone and appendix lacking, ♂ zone closely contiguous to ♀ zone, with the lower flowers abortive; ♀ zone short, cylindrical, 1.2 cm long; ♂ zone cylindric or subclavate, 2.3–2.8 cm long, obtuse. **Flowers** with perianth segments 4; ♂ with stamens 4, surrounding an abortive ovary, included, filaments free; ♀ with stamens lacking, ovary subglobose or ovoid, biloculed, locules 1-ovuled, style short, cylindric, stigma slightly exserted, discoid. **Distr:** E Tropical to S Africa (Kenya to KwaZulu-Natal). **Ecol:** In moist tropical forests, often in rocky areas and savannahs; to 850 m. Fl. Jul–Sep. **Zones:** 11–12.

Zantedeschia

Syn: *Richardia*.
Species 6–9. **Herbs**, terrestrial, acaulescent, perennial; stems tuberous or rhizomatous. **Leaves** several, radical, rosulate; lamina narrowly elliptic, cordate-sagittate or hastate, primary lateral veins excurrent into margins, finer venation striate; petiole with well-developed sheath, lacking apical pulvinus. **Inflorescences** appearing with leaves; peduncles erect, longer than petiole; spathe convolute, basally obconic, widely spreading, apically broadly ovate, cuspidate, white or brightly colored; spadix shorter than spathe, basally ♀, sometimes with staminodes scattered between pistils, apically ♂. **Flowers** unisexual, lacking perigone, densely congested; stamens free, anthers lateral, dehiscing by apical pores, connective slender; ovary 1–5-locular, each locule with 1–8 axile to subapical ovules. **Fruits** berries, fleshy, green or orange; seeds few–many, endosperm abundant. **Distr:** Predominantly in S Africa. **Cult:** Propagate by division, from basal offsets or by seed. Requires damp and boggy conditions in full or partial sun.

Z. aethiopica (Linn.) Spreng., *Syst. Veg.* 3: 765, 1826. (Syn: *Calla aethiopica*, etc.). [Calla, Calla Lily, Lily of the Nile, White Arum Lily]. **Herbs**, evergreen, tuberous, terrestrial, to 1.8 m tall, shorter (~90 cm) in cultivation; stem underground. **Leaves** sagittate, lanceolate, ovate-lanceolate to ovate, 37 ×18 cm, apex acute, bright green, posterior lobes to 12 × 8 cm; petiole to 60 cm long, sheath persistent, to 40 cm long, ligulate, free for 1 cm. **Inflorescence** with peduncle stout, as long as leaves or shorter, erect, inflorescence held well above the spreading leaves; spathe funnel-shaped, bright white, 12.5–25(–45) cm long, yellowish within below, margins revolute, tip narrowed; ♀ zone ~¼ the length of the spadix (~1–2 cm); ♂ zone to 6 cm long, yellow. **Flowers** with pistils green, intermixed with yellow staminodes. **Fruits** berries, yellow. **Distr:** S Africa. **Ecol:** Fl. Mar–Apr. **Info:** The leaves are toxic. **Zones:** 8–11.

LIMNOCHARITACEAE

Genera 3, **species** 7. **Herbs**, aquatic, with aerial leaves. **Leaves** petiolate; lamina ovate to orbicular. **Inflorescences** umbels, terminal; peduncle erect; bracts forming an involucre. **Flowers** with sepals persistent, green; petals white or yellowish, thin; stamens 8 or 9, or numerous, with sterile staminodes in outermost whorl, filaments flattened; carpels 6–9, or numerous, free, crowded into a head, laterally compressed, stigmas sessile. **Fruits** follicles; seeds numerous, strongly curved. **Distr:** Tropical to subtropical regions of N and S America; SE Asia to Malaya Islands, N Australia; N Africa.

Limnocharis

Species 2. **Herbs**, aquatic. **Leaves** aerial, petiolate; lamina ovate to orbicular. **Inflorescences** umbels, terminal; scapes erect; bracts forming an involucre; pedicels thick. **Flowers** with sepals persistent, green; petals yellowish, thin; stamens numerous, with sterile staminodes in outermost whorl, filaments flattened; carpels numerous, free, crowded into a head, laterally compressed, stigmas sessile. **Fruits** follicles; seeds numerous, strongly curved. **Distr:** Mexico to tropical America; naturalized in the Old World tropics. **Cult:** Propagate by seeds. Requires a shallow water pond, in full or partial sun.

L. flava (L.) Buchenau, *Abh. Natur. Verein Bremen* 2: 2, 1868. (Syn: *Alisma flavum*, *L. emarginata*). **Herbs**, aquatic. **Leaves** 6–28 × 4.5–20 cm, base

Above: *Apostasia wallichii, Cypripedium calceolus, Cypripedium fasciolatum, Cypripedium japonicum*. [All MG except *Apostasia wallichii* EFdeV.]

obtuse or subcordate, apex obtuse or retuse, veins 9–13, transverse veins numerous, parallel; petiole trigonous, 20–65 cm long, stout. **Inflorescences** umbels, 2–15-flowered; scapes trigonous, 20–90 cm long; bracts green, orbicular to broadly elliptic, 1.5–2 × 1–1.5 cm; pedicels 2–7 cm long. **Flowers** with petals yellowish but blackish toward base, broadly ovate to orbicular, 2–3 × 1–2 cm, apex rounded; stamens shorter than petals, filaments green; carpels yellowish green. **Fruits** clustered, follicles, suborbicular, 1.5–2 cm Ø; seeds brown. **Distr:** C America. **Ecol:** In marshes, shallow water; 600–700 m. Fl. Mar–Apr. **Zones:** 11–12.

ASPARAGALES

FAMILIES 24, GENERA 1,198–1,232, SPECIES 28,950–29,430+.

Herbs, **shrubs**, **trees**, **climbing vines** or **shrubs**, few saprophytic (lacking chlorophyll); rootstock often bulbous, rhizomatous, or cormous; stem often succulent. **Leaves** large and photosynthetic, absent or scalelike like. **Flowers** usually 3-merous; stamens 3 or 6, adnate to style or not; ovary inferior or superior. **Fruits** usually capsules (dry, hard, leathery, or triangular or nutlike), sometimes fleshy and berrylike. **Distr:** Worldwide, with the highest diversity in the tropics. **Info:** This order is monophyletic and is supported by cladistic analysis based on both morphological and molecular characters. A possible apomorphy uniting the order is the presence of seeds having a seed coat containing a black substance called phytomelan, although this character is lost in some taxa (including the largest family in the order – Orchidaceae), particularly those that have evolved fleshy fruits (Judd et al., 2008).

ORCHIDACEAE

Subfamilies 5, **Genera** 880, **species** 22,000–35,000. **Herbs** or rarely vines, perennial, rarely annual, strongly mycotrophic, epiphytic, terrestrial, lithophytic, or rarely aquatic or subterranean, usually green and photosynthetic, some without chlorophyll and saprophytic; roots subterranean or aerial, tuberoid or stolonoid, usually with spongy, multilayered velamen; stems erect or pendent or modified into creeping rhizomes, simple or sympodially or monopodially branched, delicate to stout, or thickened as corms or pseudobulbs, or greatly reduced, sometimes proliferous (especially diverse in sympodial orchids). **Stipules** absent. **Leaves** solitary, several, or reduced to scales, basal or cauline, alternate, distichous, or sometimes opposite or whorled, either convolute or duplicate, simple, sessile or petiolate; lamina articulate or not, plicate or conduplicate, cylindric, triangular, or laterally flattened, margins entire. **Inflorescences** terminal or lateral, racemes, spikes, panicles, or rarely cymose, erect or variously pendent, 1–many-flowered, lax or dense, flowering successively or simultaneously. **Flowers** bisexual rarely unisexual, epigynous, resupinate or not, pedicellate or sessile, 3-merous, usually bilaterally symmetric (rarely nearly radially symmetric), with abscission layer between pedicel and peduncle, rarely between ovary and perianth or ovary and pedicel; perianth of 6 tepals in 2 whorls, all petaloid or sepals sometimes greener and more foliaceous in texture; sepals alike or not, lateral sepals often connate (forming a synsepalum), or all 3 sepals variously connate and/or adnate or distinct and/or free; petals 3, median petal modified as lip, commonly larger or differing in form and color, lateral petals commonly but not always similar to sepals; nectaries of various sorts; extrafloral nectaries sometimes present on pedicels, bracts, or leaf sheaths; stamens usually 1–2(–3, if 3 the 3rd modified into a sterile staminode), all on side opposite lip, fully or partially adnate to style, forming column; pollen grains in monads or tetrads, usually in 2–8 pollinia, sometimes subdivided into small packets, rarely granular, sometimes pollinia with caudicles and/or stipes; gynoecium tricarpellate, connate, forming compound, inferior, 1- or 3-locular ovary, style variously adnate to filaments; stigmas usually trilobed, concave to convex, part of median stigma lobe modified into rostellum, often separating anther from fertile portions of stigma, commonly preventing or in some cases facilitating self-pollination; ovules numerous, anatropous, minute. **Fruits** capsules, opening (dehiscing) by longitudinal slits, rarely fleshy and indehiscent berries; seeds numerous (millions in some species), minute; endosperm absent. **Distr:** Worldwide, except Antarctica, with the highest diversity in tropical forests. **Info:** The majority of the species can be recognised by their strongly monosymmetric flowers, with often very distinct lips. This family forms approximately 10% of all extant flowering plant species, with newer additions on a daily basis. The classification adopted here is based on the classification outlined by Chase (2005).

Subfam. APOSTASIOIDEAE

Syn: Apostasieae.

Genera 2, **species** 16. **Herbs** terrestrial, base woody; roots elongate, fibrous, aerial then branched in soil, sometimes bearing root-nodules; rhizome scaly or absent; shoots branched or unbranched, erect, leafy. **Leaves** spirally arranged, leathery, plicate, spreading or suberect, green. **Inflorescences** terminal, racemose, few to many flowered; bracts persistent; pedicels obscure to short, twisted or not, with a torus. **Flowers** almost regular but usually slightly zygomorphic, concolorous; sepals spreading, entire, cuspidate, with thickened midribs; petals free, cuspidate; lip similar to petals or slightly broader; column short, straight to curved, style fused to base of filaments only; anthers 2 or 3, with 2 equal or unequal thecae each bilocular; pollen powdery, not forming pollinia; staminode present or absent; ovary trilocular, style free in upper part, cylindric, fleshy, stigma terminal, rounded to bi- or trilobed. **Fruits** capsular or berrylike, disintegrating into 3 valves; seeds ovoid to ellipsoidal with a terminal appendage. **Distr:** Sri Lanka, NE India to NE Australia and Japan.

Apostasia

Syn: *Adactylus, Mesodactylis, Niemeyera.*

Species ~7. **Herbs**, erect, glabrous; rhizome scaly, producing a few proplike roots; roots sometimes bearing tubers; stem simple or branched, rather slender, leafy. **Leaves** clustered or well-spaced, plicate, usually recurved along margins, forming a tubular awn at apex, base contracted and then dilating to form an enlarged amplexicaul stalk. **Inflorescences** terminal or subterminal, often arching or sometimes pendulous, usually branched; floral bracts relatively small. **Flowers** erect, non-resupinate, almost actinomorphic; sepals similar, cymbiform, slightly fleshy; petals similar, yellow to white, cymbiform; lip similar to petals though sometimes slightly larger; column erect or curved, with 2 fertile stamens, with or without a central staminode; anthers clasping style; staminode (if present) opposite to dorsal sepal, pollen not cohering to form pollinia; ovary triloculed, stigma terminal, capitate. **Fruits** capsules, narrowly cylindric, shallowly longitudinally 3-ribbed; seeds black, testa rigid. **Distr:** China, NE India, Nepal and Bhutan, north to S Japan, through SE Asia to New Guinea and N Australia. **Cult:** Propagate from seed. Requires a moist humus-rich soil in partial shade.

A. wallichii R. Brown, *Pl. Asiat. Rar.* 1: 75, 1830. (Syn: *A. gracilis*, etc.). [Yellow Grass Orchid]. **Herbs**, terrestrial, to ~60 cm tall; rhizome short; stem woody. **Leaves** many, often clustered basally; lamina linear to sublinear, 10–15(–18) × 0.7–1.2 cm, apex acuminate, with an awn 0.2–2 cm long; petiolelike base 1–4 cm long. **Inflorescence** arising from distal leaf axils, recurved, racemose or few-branched, 4–7 cm long, many-flowered; floral bracts lanceolate, 3–4 mm long; pedicel and ovary 1.3–1.8 cm long. **Flowers** 0.7–1 cm Ø; sepals yellow, linear to narrowly oblong-lanceolate, 4–6 × ~1.2 mm, 3–5-veined, mucronate; petals similar to sepals, yellow; lip similar to petals but usually slightly broader; column lacking wings; filaments ~0.5 mm long, anthers narrowly ovoid, 1.5–2.5 mm long, base sagittate; staminode subcylindric, adnate to style; style ~3.5 mm long, stigma apical, slightly dilated. **Fruits** cylindric, 1.2–2.5 cm long. **Distr:** Tropical and subtropical Asia

MG MG

Above: *Cypripedium formosanum*, *Paphiopedilum acmodontum*, *Paphiopedilum adductum*.

to N Australia. **Ecol:** In tropical wet forests, along streams; ~1,000 m. Fl. Aug. **Zones:** 7–12.

Subfam. CYPRIPEDIOIDEAE

Syn: Cypripedia, Cypripedieae, etc.
Genera 5, **species** ~180. **Herbs**, small to large terrestrial, lithophytic or rarely epiphytic. **Leaves** distichous, 1 to many, plicate or conduplicate, spreading or suberect, green, rarely checkered or tessellated with dark and lighter green or spotted with blackish maroon adaxially, thinly textured, fleshy or leathery, glabrous or variously pubescent, entire. **Inflorescence** terminal, 1–several-flowered; rachis hairy, glandular, or glabrous; bracts conduplicate; pedicel obscure to short. **Flowers** usually showy; dorsal sepal erect or hooded over lip, ovate, obovate, or elliptic, obtuse, acute, or acuminate, glabrous or pubescent outside, sometimes pubescent inside; lateral sepals usually fused to form a concave synsepalum ± similar to dorsal sepal, rarely free to base; petals free, spreading or pendent, sometimes spiralling, elliptic, ovate, linear, or oblanceolate, glabrous (pubescent) outside, margins ciliate, sometimes warty; lip deeply pouched, slipper-shaped or urn-shaped, with (without) ± pronounced incurved side lobes sometimes reduced to auricles, hairy inside, sometimes spotted inside, glabrous or hairy outside, front margin incurved or not, sometimes with short marginal horns, side lobes sometimes warty; column short, stalked, porrect; anthers 2, biloculed; staminode terminal, sessile or shortly stalked, often shieldlike, transversely reniform, oblong, ovate, obcordate, or linear, flat, convex or longitudinally conduplicate, glabrous to papillose or finely pubescent; ovary 1- or triloculed, glabrous or hairy. **Fruits** capsules, erect to pendent, 3-ribbed, cylindric to almost ellipsoidal, sometimes beaked apically; seeds ellipsoid, testa thin. **Distr:** Widespread in the temperate regions of Eurasia and N America, tropical Americas south to Brazil and Bolivia; China and tropical SE Asia to the Philippines, New Guinea, and Solomon Islands.

Cypripedium

[Lady's Slipper, Moccasin Flower].
Species ~50. **Herbs**, terrestrial, perennial. **Leaves** 2, radical or subopposite, or several and alternate, bases sheathing. **Inflorescence** terminal, solitary; flowers solitary or 2–several in a lax racemose spike; bracts large, foliaceous. **Flowers** resupinate, showy; sepals distinct or lateral sepals usually connate proximal to lip forming synsepalum; petals entire; lip sessile, inflated into a large ovoid or globose slipper or sac-shaped, with an adaxial orifice; pollinaria absent, loose granular pollen in 2 lateral anthers, dorsal anther a large subapical staminode; stigma free, 2–trilobed. **Fruits** capsules, ellipsoid to oblong-ellipsoid. **Distr:** Temperate and subalpine Northern Hemisphere to C America. **Cult:** Propagate from seeds. Requires a moist compost consisting of equal parts peat, loam, grit and forest bark. Plants need repotting annually.

C. calceolus L., *Sp. Pl.* 2: 951, 1753. [Yellow Lady's Slipper]. **Herbs**, terrestrial, 20–45 cm tall; rhizome stout; stem erect, glandular-hairy, with several basal sheaths, and 3 or 4 leaves above middle. **Leaves** elliptic or ovate-elliptic, rarely ovate-lanceolate, 7–16 × 4–7 cm, abaxially sparsely pubescent, apex acute or shortly acuminate. **Inflorescence** terminal, usually 1–2-flowered; floral bracts foliaceous, elliptic-lanceolate or ovate-lanceolate, 4–6(–10) × 1.5–4 cm; pedicel and ovary ~3 cm long, hairy. **Flowers** with sepals and petals maroon, lip yellow; dorsal sepal ovate or ovate-lanceolate, 2.5–5 × 0.8–1.5 cm, acuminate or cuspidate-acuminate; synsepalum similar to dorsal sepal, shallowly bilobed; petals twisted, linear or linear-lanceolate, 3–5 × 0.4–0.6 cm; lip deeply pouched, ellipsoid, 3–4 × 2–3 cm, outside glabrous, inner bottom hairy, incurved lateral lobes 3–4 mm wide; staminode suboblong-elliptic, 0.7–1 × 0.5–0.7 cm, abaxially carinate. **Distr:** China, Japan, Korea, Russia; Europe. **Ecol:** In forests, grasslands; 500–1,000 m. Fl. Jun–Jul. **Zones:** 3–10.

C. fasciolatum Franchet, *J. Bot.* (Morot). 8: 232, 1894. (Syn: *C. cheniae*, *C. langrhoa*, *C. wilsonii*). [Striped Lady's Slipper]. **Herbs**, terrestrial, 30–45 cm tall; rhizome stout, short; stem erect, with several basal sheaths and 3 or 4 leaves above. **Leaves** elliptic or broadly elliptic, 15–20 × 6–12 cm, glabrous, ciliate, apex shortly acuminate. **Inflorescence** terminal, 1(2)-flowered; peduncle pubescent apically; floral bracts foliaceous, elliptic or ovate, 7–10 × 3–6.5 cm, acuminate; pedicel and ovary 1.5–3 cm long, glandular-hairy. **Flowers** fragrant, large, to 12 cm Ø, pale to lemon-yellow, with maroon longitudinal stripes on sepals and petals; dorsal sepal ovate-elliptic or ovate, 5–6 × 2.8–3.5 cm, acuminate; synsepalum similar to dorsal sepal, 5–6 × 2–2.5 cm, shallowly bilobed; petals linear-lanceolate or broadly linear, 5.5–8 × 0.8–1.5 cm, acuminate; lip often ± ascending, deeply pouched, subglobose, 5–7 cm long, mouth often dentate, outside glabrous; staminode ovate-elliptic, 1.5–2 × ~1 cm. **Distr:** China. **Ecol:** In forests, grasslands; 1,600–2,900 m. Fl. Apr–Jun. **Zones:** 5–10.

C. formosanum Hayata, *Icon. Pl. Formosan.* 6: 66 (1916). (Syn: *C. japonicum* var. *formosanum*). **Herbs**, 30–40 cm tall; rhizome creeping, branched; stem erect, with several basal sheaths. **Leaves** 2, subopposite; lamina flabellate, 10–13 × 8–11 cm, base cuneate, margins undulate, minutely ciliate, apex mucronate, surfaces sparsely puberulent, adaxially with radiating veins. **Inflorescence** terminal, 1-flowered; peduncle glabrous; floral bracts foliaceous, ovate-lanceolate, 2.2–3 × 1–1.2 cm, acute; pedicel and ovary 1.8–2 cm long, densely pubescent. **Flowers** pendulous, white or pinkish; sepals and petals spotted with purplish-red at base; lip with purplish-red stripes and spots; dorsal sepal curving forward over lip, narrowly ovate, 4.5–5 × 1.6–2 cm, slightly pubescent, shortly acuminate; synsepalum elliptic-ovate, 4.5–5 × 2.5–3 cm, shallowly bilobed; petals oblong-lanceolate, 4.5–5 × 1.2–1.8 cm, acuminate or acute; lip pendulous, pouched, obovoid or ellipsoid, 4–6 × 3.5–4 cm, slightly grooved-margined around mouth; staminode ovate-triangular or ovate-sagittate, 1 × 0.6–0.7 cm. **Distr:** C. Taiwan. **Ecol:** In forests, thickets; 2,400–3,000 m. Fl. Apr–May. **Zones:** 5–11.

C. japonicum Thunberg, *Syst. Veg.*, ed. 14, 817, 1784. (Syn: *C. cathayanum*, *C. japonicum* var. *glabrum*). [Japanese Cypripedium]. **Herbs**, terrestrial, 35–55 cm tall; rhizome creeping; stem brown-tomentose, several basal sheaths several,. **Leaves** 2 (subopposite) or rarely 3 (nearly alternate); lamina flabellate, 10–16 × 10–21 cm, base subcuneate, margins undulate, basally hairy, minutely ciliate. **Inflorescence** terminal, single-flowered; peduncle brown-tomentose; floral bracts foliaceous, rhombic or ovate-lanceolate, 2.5–5 × 1–2(–3) cm; pedicel and ovary 2–3 cm long, hairy. **Flowers** pendulous; sepals and petals greenish-yellow, base spotted with purple; lip yellowish-pink to whitish, ± marked with purplish red spots and stripes; dorsal sepal narrowly elliptic, 4.5–5.5 × 1.5–2 cm, acuminate; synsepalum 4–5 × 1.5–2.5 cm, shallowly bilobed; petals obliquely lanceolate, 4–5 × 1–1.2 cm, acuminate; lip pendulous, pouched, subellipsoid or obovoid, 4–5 × 3–3.5 cm, margin of mouth strongly grooved; staminode elliptic, 1 × 0.6–0.7 cm, base shortly auriculate. **Fruits** subfusiform, 4.5–5 cm long. **Distr:** China, Japan. **Ecol:** In moist forests; 1,000–2,000 m. Fl. Apr–May. Fr. Jun–Oct. **Zones:** 5–10.

Paphiopedilum

Syn: *Cordula*, *Cypripedium*, etc.
Species ~85. **Herbs**, terrestrial, lithophytic, or epiphytic; rhizome short, rarely stoloniferous; stem short, enclosed in distichous leaf bases, rarely elongated. **Leaves** usually basal, 3–7, distichous, conduplicate toward base; lamina narrowly elliptic to suboblong, adaxially uniformly green or tessellated with dark and light green, abaxially pale green or sometimes spotted or flushed with purple at base or throughout. Inflorescence with a terminal a solitary flower, or several- to many-flowered; scape suberect to arching; peduncle usually hairy; floral bracts conduplicate. **Flowers** large, showy, variable in color; dorsal sepal often large, margin sometimes recurved; lateral sepals usually fused to form a synsepalum; petals variously shaped, suborbicular to spathulate; lip deeply pouched, inflated, globose, ellipsoid, or ovoid, basal portion narrowed and with incurved lateral lobes, hairy inside bottom; column short, with 2 lateral fertile stamens, with a terminal staminode above, and a stigma below; anthers biloculed, filament very short, pollen powdery

or glutinous; staminode variable in shape; ovary 1-loculed, stigma papillate, inconspicuously trilobed. **Fruits** capsules; seeds ellipsoid or spindle-shaped. **Distr:** S China to tropical Asia. **Ecol:** On shady forest floors of montane evergreen or seasonally deciduous forests; 0–2,300 m. **Cult:** Propagate from seeds, cuttings, offshoots or by division. Requires a terrestrial orchid compost with bark and limestone chips. The requirements for different species vary and is best to try and match the natural habitat it originated from. **Info:** The lip in this genus is effectively trilobed with the midlobe deeply saccate, other side-lobes infolded to give the lip an apparent claw. The lip acts as a trap for potential pollinators and is equipped with margins that prevent direct escape of insects trapped therein, but has a ladder of hairs leading from the pouch to the base of the lip for insects to escape through a gap on either side of the column passing beneath the stigma and an anther in succession (Cribb 1987).

P. acmodontum M.W. Wood, *Orchid Rev.* 84: 350, 1976. [Pointed Tooth Paphiopedilum]. **Herbs**, terrestrial; stem very short. **Leaves** distichous; lamina oblong-elliptic, apex tridenticulate, to 18 × 4 cm, basal margins sparsely ciliate, adaxially tessellated pale and dark green. **Inflorescence** erect, 1-flowered, to 25 cm long; peduncle sparsely pubescent, green with purple markings; bract ovate, 3 cm long, sparsely pubescent, ciliate. **Flower** with dorsal sepal ovate, 4 × 3.1 cm, acute, whitish or pinkish, lined with dark purple or purple-green veins; synsepalum ovate, 3.2 × 1.5 cm, acute, white, tinged purple with purple and green veins; petals spreading-reflexed, ligulate-oblanceolate, 4.3 × 1.4 cm, tridentate, sparsely ciliate and pubescent at base, purple abaxially, veined and spotted dark purple basally, abaxially green; lip deeply saccate, prominently auriculate, 4 × 2.3 cm, dentate or apiculate, minutely pubescent, bronze or olive-green with darker veins; staminode obovate, 1 × 0.9 cm, minutely pubescent, distal margin 3-toothed, pale green tinged with bronze or green. **Distr:** Philippines (Negros). **Ecol:** In montane areas near the coast; to 1,000 m. Fl. Jan–May, Aug. **Zones:** 10–12.

P. adductum Asher, *Orchid Digest* 47: 224, 1983. (Syn: *P. adductum* var. *anitum*, *P. anitum*). **Herbs**, terrestrial. **Leaves** ~6, oblong-ligulate to slightly oblanceolate, 23–26 × 3.9–4.2 cm, margins hyaline, apex obtuse, minutely tridentate, dark green. **Inflorescence** 2–3-flowered; peduncle arching, 20–29 cm long, purple, pubescent; bracts to 4 × 1.3 cm, green, maroon-veined; pedicel and ovary 3.5–5 cm long. **Flowers** ± glabrous; sepals pale greenish-yellow or white, strongly maroon-veined; dorsal sepal arching, ovate, 5–6.5 × 2.3–3.2 cm, acuminate, outside finely pubescent; synsepalum ovate, 5.2–6.9 × 2.3–3 cm, acuminate; petals pale yellow, basally maroon-spotted, apically maroon, arcuate-dependent, linear-tapering acute, 8–15 × 0.7–0.9 cm, apically finely pubescent; lip pale yellow, mottled and veined purple on front of pouch, porrect, 3.8–4.7 × 1.5–2.1 cm, grooved on back, with short auricles; staminode yellow with purple glands, rectangular, 8–9 × 4–7 mm, apex notched, sides hairy. **Distr:** Philippines (Mindanao). **Ecol:** On the forest floor, among decaying leaves; 1,250–1,350 m. Fl. Mar–Jun, Nov. **Zones:** 10–12.

P. appletonianum (Gower) Rolfe, *Orchid Rev.* 4: 364, 1896. (Syn: *Cordula appletoniana*, etc.). **Herbs**, terrestrial. **Leaves** 6–8, narrowly elliptic to oblong-elliptic, 7–25 × 2–4 cm, base sparsely ciliate, apex tridenticulate, subacute to obtuse, adaxially obscurely tessellated pale and darker green, abaxially purple-marked basally. **Inflorescence** erect, 1(–2)-flowered; peduncle 17–48 cm long, purple, shortly pubescent; bract lanceolate, 1.5–2.1 cm long, acute, green, ciliate; ovary and pedicel 3–6 cm long, pubescent. **Flower** 6–10 cm across; sepals greenish, green-veined; dorsal sepal basally purple-tinged; dorsal sepal ovate, 2.7–4.4 × 2–2.4 cm, base cordate, basal margins recurved, apiculate above; synsepalum elliptic-lanceolate, 2–3 × 1.1–1.5 cm, acuminate; petals spathulate, 4.4–5.8 × 1.2–1.8 cm, apically half-twisted, acute or tridenticulate, green with darker green and brown stripes, basally with maroon-black spots, purple above; lip ochre to pale purple with darker veins, 3.6–4.6 cm long, front margin bluntly dentate; staminode transversely elliptic, 7–9 × 8–9 mm, tridentate. **Distr:** Hainan to Indo-China. **Ecol:** In forests; 600–700 m. **Zones:** 10–12.

P. argus (Rchb. f.) Stein, *Orchid.-Buch*: 453, 1892. (Syn: *Cypripedium argus*, etc.). **Herbs**, terrestrial, erect, sympodial. **Leaves** 3–5, narrowly elliptic, 9–24 × 2–4.5 cm, basal margin ciliate, apex acute, minutely tridentate, adaxially mottled pale and dark green, abaxially base purplish. **Inflorescence** 1-flowered; peduncle 30–45 cm long, purple or green, purple-mottled, shortly pubescent; bract elliptic, 2–4.5 × 1.5–1.8 cm, subacute to acuminate, green, midvein purple, pubescent, ciliate; pedicel and ovary 5–8.5 cm long, green, pubescent. **Flower** ~6.5 cm Ø, dorsal sepal white green-veined and purple spotted toward base, synsepalum and petals white, green-veined, petals purple at apex, heavily maroon-spotted, lip green, pink-flushed, green-veined, lateral lobes maroon-spotted, staminode pale brownish-green, green-veined; dorsal sepal ovate, 3.4–4.7 × 3–3.9 cm, acute; synsepalum ovate-elliptic, 4–4.4 × 1.7–2.1 cm, acute; petals deflexed-recurved, ligulate, 4.5–6.5 × 1.1–1.8 cm, acute, basal margin bearing long purple-maroon hairs, hairs shorter apically; lip 4–4.5 × 2.1–2.5 cm, inside glandular-pubescent; staminode lunate, 0.7–0.9 × 1.1 cm, pubescent, apical side-lobes sickle-shaped. **Distr:** Philippines (Luzon, Negros and Visayas). **Ecol:** In montane forests, on forest floor; >600 m. **Zones:** 10–12.

P. armeniacum S.C. Chen and F.Y. Liu, *Acta Bot. Yunnan.* 4: 163, 1982. [Golden Slipper Orchid]. **Herbs**, lithophytic or terrestrial. **Leaves** 5–7, oblong, 6–12 × 1.8–2.3 cm, margins serrulate, apex obtuse or acute, tridenticulate, adaxially strongly marbled dark and light bluish-green, abaxially densely purple-spotted. **Inflorescence** 1-flowered; peduncle 24–26 cm long, green purple-spotted, brown-pubescent; bract conduplicate, lanceolate-ovate, ~1.5 cm long, subacute to acute, green, purple-spotted, pubescent; pedicel and ovary 3–3.5 cm long, green, pubescent, rostrate. **Flower** relatively large, bright golden-yellow, staminode red-veined, lip inside purple-spotted, pubescent; dorsal sepal ovate, 2.2–4.8 × 1.4–2.2 cm, subacute, basally pubescent; synsepalum ovate, 2–3.5 × 2 cm, outside bicarinate; petals ovate to subcircular, 2.8–5.3 × 2.5–3.3 cm, rounded to subacute, ciliate, basally pubescent; lip very inflated, thin-textured, large, 4–5 × 3.5–4 cm, margins incurved, white-pubescent and purple-dotted inside at base; staminode cordate, 1–2.2 × 1–2.2 cm. **Distr:** China (W Yunnan) to N Myanmar. **Ecol:** On limestone hills, near waterways. **Zones:** 9–10.

P. barbigerum Tang and F.T. Wang, *Bull. Fan Mem. Inst. Biol.* 10: 23, 1940. (Syn: *P. insigne* var. *barbigerum*). **Herbs**, lithophytic or terrestrial. **Leaves** 4–6, distichous; lamina linear, 12–19 × 0.7–1.3 cm, apex acute, minutely tridenticulate, adaxially green, sometimes very obscurely tessellated, abaxially paler, carinate. **Inflorescence** 1-flowered, suberect or arching; peduncle brownish-green, 10–14 cm long, densely purplish-hairy; floral bract suboblong, 1.4–3.2(–3.6) cm; pedicel and ovary 3–4.2 cm long, densely pubescent. **Flower** 6–7(–8) cm Ø, dorsal sepal white, central area green, red-brown, or blackish-brown, synsepalum whitish-green, petals brownish, margins paler, lip brownish, margins yellowish, staminode yellowish, umbo orange; dorsal sepal suborbicular, 2.8–3.4(–4.5) × 2.6–3(–4) cm, apical margin strongly undulate, obtuse or retuse; synsepalum elliptic, 2.5–3.3(–4) × 1.4–1.8 cm; petals spathulate or narrowly oblong, 3–4(–5) × 1–1.3 cm, margin undulate, obtuse; lip helmet-shaped, 3.5–4 cm long, pouch 2–2.5 × 1.5–2 cm, glabrous; staminode obovate-obcordate, 0.6–1 × 0.7–1 cm. **Distr:** S China to Vietnam. **Ecol:** On limestone cliffs or in open forests; 800–1,500 m. Fl. Sep–Oct. **Zones:** 7–10. **Info:** ***P. barbigerum* var. *lockianum*** (Syn: *P. coccineum*, etc.) endemic to Vietnam is paler in color. **Zones:** 7–10. ***P. barbigerum* var. *sulivongii*** from the limestone hills of Laos described in 2009 differs from the typical variety in the much larger flowers, similar in size to *P. barbigerum* var. *lockianum*. **Zones:** 7–10.

P. bellatulum (Reichb. f.) Stein, *Orchideenbuch*: 456, 1892. (Syn: *Cypripedium bellatulum*, etc.). [Egg-in-a-nest Orchid]. **Herbs**, terrestrial, usually clustered. **Leaves** 4–5, oblong-elliptic, 14 × 5 cm, apex rounded, minutely emarginate, dark green, adaxially mottled with green, adaxially purple-spotted. **Inflorescence** 1(–2)-flowered, very short, arcuate; peduncle 2–4.5 cm long, white-pubescent, purple or green, maroon-mottled; bract conduplicate, elliptic-ovate, 2.3–2.7 × 1.5–2.5 cm, acute to obtuse, pubescent, pale green, purple-spotted. **Flower** subcampanulate, 5.5–8 cm Ø, white or creamy-yellow, with large maroon spots, outside pubescent; dorsal sepal concave, broadly transversely elliptic to broadly ovate, 3–3.5 × 3.3–4.2 cm, obtuse to retuse; synsepalum deeply concave, cordate, 2–2.2 × 2.2–2.7 cm, obtuse to rounded; petals concave, broadly elliptic, 4.7–6.3 × 2.7–4.7 cm, rounded; lip narrowly ovoid to ellipsoidal, 2.5–4.2 × 1.5–2 cm, margins strongly incurved; staminode subquadrate-obcordate, apical tooth ~1 cm long; ovary ~3 cm long, green, maroon-spotted. **Distr:** W Burma, China, Thailand. **Ecol:** On mossy limestone outcrops; 1,000–1,500 m. **Zones:** 9–10.

P. callosum (Rchb. f.) Stein, *Orchideenbuch*: 457, 1892. (Syn: *Cordula callosa*, *Cypripedium callosum*). **Herbs**, terrestrial. **Leaves** 3–5, narrowly elliptic, oblong-elliptic or obovate, 10–20 × 3.2–4.8 cm, base ciliate, apex acute, tridenticulate, adaxially tessellated pale and dark. **Inflorescence** 1(–2)-flowered, 20–40 cm tall; peduncle purple, purple-pubescent, 12–25 cm long; bract ovate to elliptic, 1.5–2.8 × 1.5–2 cm, acute, sometimes purplish, ciliate; pedicel and ovary 3–6.5 cm long, green, purple-flushed, purple-pubescent. **Flower** large, 8–11 cm across, sepals white, purple-flushed below, purple and green veined; dorsal sepal broadly ovate to suborbicular, 4–5.5 × 4.2–6 cm, margins recurved, ciliate, apiculate; synsepalum concave, elliptic or

First row: *Paphiopedilum appletonianum, Paphiopedilum argus, Paphiopedilum armeniacum.*
Second row: *Paphiopedilum barbigerum, Paphiopedilum barbigerum* var. *lockianum, Paphiopedilum barbigerum* var. *sulivongii.*
Third row: *Paphiopedilum bellatulum, Paphiopedilum callosum, Paphiopedilum callosum* var. *warnerianum.*
Fourth row: *Paphiopedilum charlesworthii, Paphiopedilum ciliolare.*
Fifth row: *Paphiopedilum concolor, Paphiopedilum delenatii, Paphiopedilum dianthum.*
[All MG, except *P. appletonianum* and *P. delenatii* AF, *P. argus* MRF and *P. callosum* var. *warnerianum* PM.]

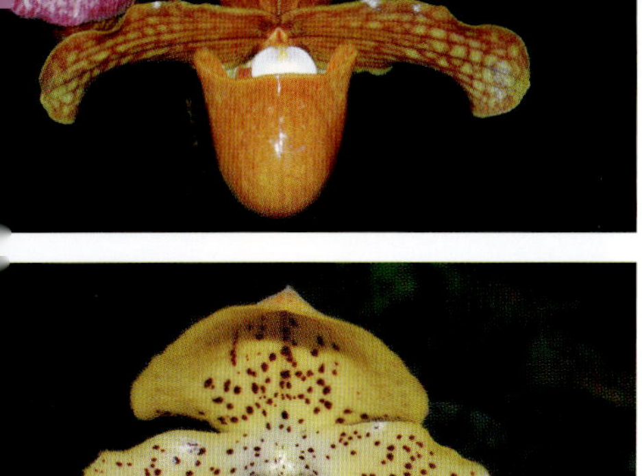

Top row: *Paphiopedilum emersonii, Paphiopedilum exul, Paphiopedilum fairrieanum.*
Middle row: *Paphiopedilum fowlei, Paphiopedilum glanduliferum, Paphiopedilum godefroyae.*
Bottom row: *Paphiopedilum godefroyae* (darker form), *Paphiopedilum hangianum, Paphiopedilum gratrixianum.*
[All MG, except *P. exul* PM, *P. glanduliferum* GS and *P. gratrixianum* AF.]

lanceolate, 2.7–3.2 × 1.6–2.5 cm, acute; petals white to yellow-green, apically purple, upper margin maroon-spotted, sometimes basally, sometimes reflexed, subsigmoid, ligulate, 4.6–6.8 × 1.2–1.8 cm, obtuse or rounded; lip green, maroon-flushed, 2.5–4.4 × 2–2.5 cm, side-lobes incurved, warty; staminode greenish, dark-veined, purple-flushed, lunate, 1.1 × 0.7 cm. **Distr:** Cambodia, Laos, Thailand, peninsular Malaysia. **Ecol:** In mossy forests; 300–1,300 m. **Zones:** 10–12.

P. charlesworthii (Rolfe) Pfitzer, *Bot. Jahrb. Syst.* 19: 40, 1894. (Syn: *Cypripedium charlesworthii*, etc.). **Herbs**, terrestrial. **Leaves** 4 or 5, distichous; lamina linear-oblong, 14–16 × 1.5–2.8 cm, apex acute, unequally bilobulate, adaxially green, sometimes tessellated, abaxially pale green, base black-spotted. **Inflorescence** 1-flowered, erect; peduncle 8–12 cm long, green, purplish-mottled, shortly pubescent; floral bract green-spotted or purple-mottled, subelliptic-obovate, 2.1–2.8 × 1.4–1.8 cm; pedicel and ovary 2.9–3.4 cm long, densely pubescent, 3-ribbed. **Flower** 6–8 cm Ø, dorsal sepal pinkish-white, dark-veined, synsepalum greenish, brownish-veined, petals pale green-yellow, densely brown-veined, lip pale yellow-brown, dark-veined, staminode white; dorsal sepal erect, suborbicular to transversely elliptic, 3.9–4.6 × 4.5–6.7 cm, subtruncate or mucronate, adaxially base purple-hairy; synsepalum elliptic-ovate, 3–3.5 × 1.6–2.3 cm; petals oblong-spathulate, 3.8–4.4 × 0.9–1.4 cm, adaxially base purple-villous; lip helmet-shaped, 3.4–4 cm long, pouch 2.2–2.5 × 2–2.4 cm, mouth very wide; staminode obovate, 0.9–1 × 0.8–0.9 cm, with a central umbo. **Distr:** Assam to China. **Ecol:** In evergreen broad-leaved forests; 1,200–1,600 m. Fl. Sep–Oct. **Zones:** 10–11.

P. ciliolare (Rchb. f.) Stein, *Orchid.-Buch*: 462, 1892. (Syn: *Cypripedium ciliolare*, etc.). **Herbs**, terrestrial, clustered. **Leaves** 4–6, oblong-elliptic to elliptic-oblanceolate, 10–16 × 2.6–5 cm, base sparsely ciliate, apex tridenticulate, acute to obtuse, adaxially mottled pale and darker green, abaxially base purple-tinged. **Inflorescence** 1-flowered, erect; peduncle 20–32 cm long, purple, shortly pubescent; bract ovate-lanceolate, 2–3 cm long, subacute; pedicel and ovary 6–8 cm long, purple-pubescent. **Flower** 7–9 cm across; sepals basally white, pale purple and green above, purple-veined; dorsal sepal ovate, 3.8–5.8 × 3–5.1 cm, acute or acuminate, ciliate; synsepalum elliptic-ovate, 2.5–3.2 × 1.8–2.2 cm, subacute; petals white, heavily spotted and veined dark-purple, slightly falcate, oblanceolate, 5–7.2 × 1.4–1.9 cm, subacute to obtuse, ciliate, upper margin warted; lip dark brownish-purple, 4.5–6 × 3–3.6 cm, side-lobes incurved, warted; staminode transversely reniform, 1.2–1.4 × 0.6–0.7 cm, base narrowly-incised, apex shallowly trilobed. **Distr:** Philippines (N Mindanao). **Ecol:** On forest slopes; 300–1,830 m. **Zones:** 10–12.

P. concolor (Lindl. ex Bateman) Pfitzer, *Nat. Pflanzenfam.* 2(6): 84, 1888. (Syn: *Cordula concolor, Cypripedium concolor*). **Herbs**, terrestrial or lithophytic, clustered. **Leaves** 4–6, oblong to elliptic-oblong, to 14 × 4 cm, apex rounded, minutely emarginate, adaxially tessellated dark and pale green, abaxially finely purple-spotted. **Inflorescence** 1–2(3)-flowered; peduncle to 8 cm long, white-pubescent, purple or green, purple-spotted; bracts ovate, 1–1.5 × 1–1.6 cm, pubescent, green, purple-spotted. **Flowers** large, pubescent, to 7 cm Ø, yellow, rarely ivory or white, finely purple-spotted; dorsal sepal broadly obovate, 2.5–3.5 × 3.3 cm, obtuse to emarginate; synsepalum concave, elliptic to ovate, 2.1–3 × 2.1–3 cm, obtuse to retuse; petals elliptic, 3.7–4.3 × 2.5–2.7 cm, rounded; lip ellipsoid, 3.4–3.8 × 1.4–1.5 cm, margins incurved, very fleshy; staminode shortly trullate, 1–1.3 × 1–1.2 cm; ovary 3–5 cm long, pubescent, dull purple or green, purple-spotted. **Distr:** China to Indo-China. **Ecol:** On limestone hills; 90–1,000 m. **Zones:** 10–12.

*P. **delenatii*** Guillaumin, *Bull. Soc. Bot. France* 71: 554, 558, 1924. (Syn: *Cypripedium delenatii*, etc.). **Herbs**, terrestrial, lithophytic, to 25 cm high. **Leaves** 5–7, distichous; lamina narrowly elliptic to oblong elliptic, 10–15 × 3–5 cm, apex minutely unequally trilobed, adaxially light green, veined, mottled dark green, abaxially spotted reddish-purple. **Inflorescence** with 1–2(3) flowers; peduncle erect, 15–25 cm long, reddish purple, white pubescent. **Flowers** fragrant, white-hairy, mostly pink, 8–10 cm across, dorsal sepal white to pale pink, staminode pink with a yellow center and base, darker pink around the yellow area; dorsal sepal ovate, 1.7–3.5 × 1.8–2.5 cm, margins with white hairs, apex acute; synsepalum ovate, 1.9–2.7 × 2 cm; petals ovate, 3–5 × 2.4–3.5 cm, obtuse; lip saccate, 2.5–3.5 × 1.8–3 cm, mouth narrow, base rounded, margins incurved, inside purple-spotted; staminode convex, ovate. **Distr:** S Vietnam. **Ecol:** In lowland and cloud forests; 750–1,500 m. Fl. Oct–Dec. **Zones:** 10–11.

*P. **dianthum*** Tang and F.T. Wang, *Bull. Fan Mem. Inst. Biol.* 10: 24, 1940. (Syn: *P. parishii* var. *dianthum*, etc.). **Herbs**, epiphytic, occasionally lithophytic. **Leaves** 5–8, ligulate, 45 × 4–7.5 cm, apex, asymmetric, rounded, green. **Inflorescence** arcuate to suberect, 30–50 cm long, (2–)3–6(–9)-flowered; peduncle 20–26 cm long, green, papillose; bracts conduplicate, broadly elliptic, 3–4 ×2–3.5 cm; pedicel and ovary 5 cm long, glabrous. **Flowers** to 13 cm long; sepals cream to green, dark-veined; dorsal sepal elliptic, 3.5–4.6 × 1.8–3 cm, basal margins recurved, apical margins incurved, apex obtuse; synsepalum ovate, 2.6–4 × 1.5–2.9 cm, margins recurved, apex acute, 2-keeled outside; petals green below, maroon-spotted, apical margins maroon, decurved-pendent, linear-tapering, 7–10.5 × 0.6–1.1 cm, basal margins undulate, lower basal margin maroon-spotted, ciliate, apical half spirally twisted, apex rounded; lip greenish or purplish-flushed, 3–4.5 × 1.5–2 cm, apex narrowed; staminode obcordate, 1.1–1.4 × 0.7–0.9 cm, cream, green-veined. **Distr:** SW China to N Vietnam. **Ecol:** 1,850–2,250 m. **Zones:** 9–11.

*P. **emersonii*** Koopowitz and P. J. Cribb, *Orchid Advocate* 12(3): 86, 1986. (Syn: *P. emersonii* f. *album*, etc.). **Herbs**, lithophytic. **Leaves** 5–7, distichous; lamina narrowly elliptic, 8–23 × 2.5–4(–5) cm, adaxially green or sometimes green-tessellated, abaxially pale green, basally purplish-spotted, leathery, basal margins ciliate. **Inflorescence** erect, 1-flowered; peduncle greenish, 8–15 cm long, pubescent; floral bract whitish, brown-flushed, elliptic, 2.8–3.8 cm; pedicel and ovary 2.5–3.5 cm long, greenish-white, white-pubescent. **Flower** slightly fragrant, 8–9(–10) cm Ø, white, puberulent, sometimes petal bases purplish-flushed, lip yellowish to whitish, inside purple-spotted, staminode yellow or whitish yellow, with red-brown markings; dorsal sepal elliptic-ovate, 3–5 × 2–3.5 cm, margin recurved, obtuse; synsepalum broadly elliptic or suborbicular, 3–4.8 × 3–4.8 cm, obtuse; petals obovate-elliptic to suborbicular, 3.5–6 × 2.5–5 cm; lip subglobose or obovoid-globose, 3.5–4.5 × 2–3 cm, apical margin involute; staminode 1.5–2 × 0.9–1 cm, with a central groove. **Distr:** China to N Vietnam. **Ecol:** In evergreen forests or thickets; 300–800 m. Fl. Apr–Jun. **Zone:** 11.

*P. **fairrieanum*** (Lindl.) Stein, *Orchid.-Buch*: 467, 1892. (Syn: *Cypripedium fairrieanum*, etc.). **Herbs**, terrestrial, clump-forming. **Leaves** arranged in a fan, 4–8; lamina linear-ligulate, 7.5–28 × 2–3 cm, margins serrulate toward apex, apex minutely unequally tridenticulate, obtuse or rounded, midgreen to dark green, adaxially faintly mottled, abaxially paler. **Inflorescence** 1(2)-flowered; peduncle 12–45 cm long, green, purplish-hairy; bract elliptic, 1–4 × 0.8 cm, obtuse, whitish, purple-pubescent. **Flower** with sepals white, green and purple veined, ± suffused on dorsal sepal, petals similar, lip yellow-green, dark-veined, staminode yellow, center green and purple veined; pedicel and ovary 3.5–6 cm long, purple, purple-hairy; dorsal sepal elliptic, 3.4–8 × 2.9–7 cm, ciliate, apical margins recurved, lateral margins undulate; synsepalum ovate, 2.4–3.5 × 1.7–2.5 cm, rounded; petals narrowly oblong, reflexed-decurved, S-shaped, 4–5 × 0.9–1.5 cm, acute, margins undulate, ciliate; lip deeply slipper-shaped, 3–4 × 1.6–2.25 cm, shortly apiculate, curved; staminode elliptic, 9 × 7 mm, tridentate. **Distr:** E Himalayas to Assam. **Ecol:** On crystalline limestone outcrops of sheltered grassy slopes; 1,400–2,200 m. **Zones:** 9–10.

*P. **fowliei*** Birk, *Orchid Digest* 45: 63, 1981. (Syn: *P. fowliei* f. *christianae*, etc.). **Herbs**, terrestrial, clustered. **Leaves** 4–6; lamina narrowly elliptic, to 18 × 4 cm, apex obtuse or subacute, minutely tridenticulate, adaxially bluish-green, obscurely tessellated with dark and light purple, basal margins ciliate. **Inflorescence** 1-flowered; peduncle to 32 cm long, pale purple, pubescent; bract ovate, 1.4–2.4 cm long, green, pubescent. **Flower** with sepals and petals white, green-veined, lip ochre to purple; dorsal sepal broadly ovate, 3–4.5 × 3.2–4.8 cm, acute to shortly acuminate; synsepalum concave, elliptic, 2.3–3.4 × 1.8–2.1 cm; petals recurved, ligulate, 4.6–5.5 × 1–1.4 cm, acute, ciliate, margins black-warted; lip 3.8–4.8 × 2.2–2.6 cm, apical margins ciliate, lateral lobe incurved, warted; staminode obovate, base auriculate, somewhat angled on sides, apically 3-toothed, side-teeth elongate-falcate, 0.8–1 × 0.7–1 cm; ovary 3.6–6.6 cm long, green, pubescent. **Distr:** Philippines (Palawan). **Ecol:** 600–950 m. **Zones:** 10–12.

*P. **glanduliferum*** (Blume) Stein, *Orchid.-Buch*: 468, 1892. (Syn: *Cypripedium glanduliferum*, etc.). **Herbs**, terrestrial (epiphytic), clustered. **Leaves** 4–6, linear-oblong, to 40 × 3–5.5 cm, basal margins sparsely ciliate, apex obtuse, tridenticulate, green, glabrous. **Inflorescence** 2–5-flowered; peduncle 30–50 cm long, purple to greenish-brown, shortly pubescent; bracts ovate, to 4.5 cm long, acute, glabrous, pale yellow-green purple-striped; pedicel and ovary 5–7 cm long, yellow-green, purple-striped. **Flowers** large, 12–16 cm across, sepals and petals yellow, maroon-veined; dorsal sepal ovate, 4.9–5.5 × 2.5–2.9 cm, acuminate; synsepalum similar, 5.3–5.5 × 2.8–2.9 cm; petals deflexed, linear-tapering, 9–10 × 1 cm, basal margins with maroon warts, ciliate, apically papillose, helically twisted; lip yellow, veined and ± flushed purple, subporrect, 5–5.5 × 1.8–1.9 cm, side-lobes narrow, acute, incurved; staminode convex, oblong to subquadrate, 1.7 × 1.2 cm, obtuse to retuse, sides hirsute. **Distr:** NW New Guinea and adjacent islands. **Ecol:** On limestone rocks; 0–200 m. **Zones:** 10–12.

*P. **godefroyae*** (Godefroy-Lebeuf) Stein, *Orchideenbuch*: 468, 1892. (Syn: *C. godefroyae*, etc.). **Herbs**, clustered, terrestrial or lithophytic. **Leaves** 4–6, oblong-elliptic, 14 × 2.9 cm, basal margins sparsely ciliate, apex obtuse, retuse, adaxially tessellated dark and pale green, abaxially usually purple-spotted. **Inflorescence** 1–2-flowered, erect; peduncle 4–8 cm long, purple, shortly pubescent; bract conduplicate, ovate, 1.5 cm long, subacute, purple, shortly pubescent. **Flowers** large, white or ivory-white, usually purple-spotted, ~4.6–7.6 cm across; dorsal sepal concave, broadly ovate, 2.5–3.3 × 2–3.7 cm, obtuse or rounded; synsepalum ovate, 2–3 × 1.7–3 cm, obtuse; petals oblong-elliptic, 3.5–5.5 × 2–3 cm, margins often undulate, somewhat recurved, apex rounded to retuse; lip ellipsoidal, 2.5–3.5 × 1.6 cm, margins strongly incurved; staminode transversely elliptic, 0.7 × 0.9–1.1 cm, apical margin 1–3-toothed, pubescent, ciliate; ovary 4 cm long, pubescent, purple. **Distr:** Peninsular Thailand and adjacent islands. **Ecol:** On limestone cliffs, often near the sea; 0–100 m. **Zone:** 12.

*P. **gratrixianum*** Rolfe, *Orchid Rev.* 13: 63, 1905. (Syn: *Cypripedium gratrixianum*, etc.). **Herbs**, terrestrial, often clumping. **Leaves** suberect, 4–7; lamina linear, to 30 × 2–2.3 cm, apex minutely trilobulate, acute, green, abaxially base purple-spotted. **Inflorescence** 1-flowered; peduncle erect, to 25 cm long, green, purple-hairy; bracts narrowly oblong-lanceolate to obovate, 3–4.5 × 1.2–1.5 cm, acute to obtuse, green, purple-spotted, glabrous; pedicel and ovary triangular in cross-section, 4–5.5 cm long, pale green, purple-hairy. **Flower** 7–8 cm across; dorsal sepal white above, pale green below, base heavily purple-spotted; synsepalum pale green, ovate to obovate, 4.8–5.2 × 4.4–4.6 cm, obtuse, outside purple-pubescent; petals glossy, yellow, flushed and veined purplish-brown, spathulate, 4.5–5.2 × 1.8–2.5 cm, obtuse, minutely ciliate, side margins reflexed; lip yellow, brown-flushed, apex tapered, 4–4.2 × 2.4–2.8 cm; staminode yellow, obcordate, 1.1 × 1.1 cm long, pustular, basally purple-hairy, with a central raised knob. **Distr:** Laos to Vietnam. **Ecol:** In primary wet, montane cloud forests; 900–1,200 m. Fl. throughout most of the year. **Zones:** 10–11.

*P. **hangianum*** Perner and O. Gruss, *Orchidee (Hamburg) Suppl.* 6: 3, 1999. (Syn: *P. singchii*, etc.). **Herbs**, lithophytic. **Leaves** 4–6, distichous, horizontally spreading; lamina narrowly oblong, 12–28 × 3.5–5.9 cm, basal margin purple-ciliate, apex obtuse, adaxially deep green, glossy, abaxially pale green, carinate, leathery. **Inflorescence** suberect or arching, 1-flowered; peduncle yellowish-green, purple-speckled, 8–20 cm long, white-pubescent; floral bract ovate-elliptic, 4.5–6 cm long, yellowish-green, purplish-speckled; pedicel and ovary 3–4.5 cm long, green, purplish-speckled, densely puberulent. **Flower** scented, puberulent, 11–14 cm Ø, yellowish, obscurely greenish reticulate-veined, petal bases purplish-flushed, inside bottom of lip densely purple-spotted, staminode purple-striped; dorsal sepal ± suberect over lip, broadly ovate-elliptic or elliptic, 4.8–6.2 × 3–4.6 cm, minutely ciliate, acute; synsepalum broadly elliptic, 4.8–6.4 × 3.6–5.7 cm, slightly bilobulate; petals broadly obovate-elliptic, slightly oblique, 5.5–7.3 × 3.5–6.4 cm, adaxially base white-villous, minutely ciliate; lip subglobose, 4–5 × ~3.4 cm, glabrous, apical margin involute; staminode broadly obovate-triangular, 1.6–2.1 × 1.8–2.2 cm, base clawed, apex obtuse-rounded. **Distr:** China (Yunnan) to Vietnam. **Ecol:** In rocky places or rock crevices, behind waterfalls; 600–800 m. Fl. Apr–May. **Zones:** 7–10.

*P. **haynaldianum*** (Rchb. f.) Stein, *Orchideenbuch*, 470, 1892. (Syn: *Cypripedium haynaldianum*, etc.). **Herbs**, terrestrial, lithophytic, rarely epiphytic, 1+ m tall. **Leaves** 6, straplike, ligulate or narrowly elliptic, to 45 × 5 cm, plain, dull deep-green, leathery. **Inflorescence** arching, 2–6-flowered, 50–75 cm long, flowers widely separated; flowers opening

Top row: *Paphiopedilum haynaldianum* f. *album, Paphiopedilum helenae, Paphiopedilum haynaldianum.*
Bottom row: *Paphiopedilum henryanum, Paphiopedilum hirsutissimum.*

successively; floral bracts oblong-lanceolate; bracts and ovary long-pubescent. **Flowers** 7.5–10(–15) cm across; dorsal sepal yellowish-green with large brown spots along veins, synsepalum whitish, greenish-veined, petals yellow-green, base with 8–12 large dark brownish spots, apically purple, lip greenish, purplish-tinged; dorsal sepal ovate 3.5–6.2 × 2.5–4 cm, basal margins reflexed, apex acute; synsepalum broadly ovate, 2.5–4.5 cm long; petals linear-spathulate, ± horizontal and then drooping at apex, 5.8–10 × 1–1.4 cm, slightly twisted; lip saccate, mouth 3.5–4.9 × 2–2.3 cm, lateral lobes infolded, of the same color, auricled; staminode obovate with a dorsal tooth; ovary 4 cm long. **Distr:** Philippines (in Luzon, Mindanao, Negros). **Ecol:** On limestone hills, serpentine cliffs; 0–1,500 m. Fl. Sep–Nov. **Zones:** 10–12. **Info:** ***P. haynaldianum* f. *album*** is a rare albino species with yellowish-green and white flowers. **Zones:** 10–12.

P. helenae Aver., *Orchids* 65: 1064, 1996. (Syn: *P. delicatum, P. helenae* f. *aureum*). **Herbs**, lithophytic. **Leaves** 2–4, distichous; lamina linear-oblong to linear-oblanceolate, 8–12.5 × 0.8–1.6 cm, apex acute or unequally bilobate, adaxially dark green, abaxially light green, base purple-speckled, leathery-fleshy. **Inflorescence** 1-flowered, arching or spreading horizontally; peduncle green, purple-spotted, 5–7 cm long, hairy; floral bract green, purple-spotted, broadly ovate, conduplicate, 1–1.3 × 0.7–1 cm, basally purple-hairy; pedicel and ovary 2.9–3.1 cm long, hairy. **Flower** 4.5–5 cm Ø, dorsal sepal yellowish or golden-yellow, whitish-margined, synsepalum yellowish-white, petals yellow-green or brown-yellow, midvein purple-red, petals and sepals basally blackish-hairy, lip yellow-green, purplish-tinged, staminode yellow-green, central umbo green; dorsal sepal broadly elliptic, 4.2–5 × 2–2.8 cm, obtuse-rounded; synsepalum suborbicular or broadly elliptic, 2.2–2.4 × 2–2.2 cm, obtuse; petals linear-spathulate or linear-oblong, 2.5–3.2 × 0.4–0.8 cm, obtuse or subtruncate; lip helmet-shaped, 2.5–2.8 cm, pouch 1.5–1.8 cm Ø, front convex-inflated; staminode obovate, 6–7 × 6–7 mm, with small bubble-shaped papillae. **Distr:** China to Vietnam. **Ecol:** On bushy cliffs of limestone mountains; 100–1,000 m. Fl. Sep–Nov. **Zones:** 7–10.

P. henryanum Braem, *Schlechteriana* 1(1): 4, 1987. (Syn: *P. chaoi, P. dollii*). **Herbs**, terrestrial or lithophytic. **Leaves** 3–6, distichous; lamina lanceolate or linear-ligulate, 12–23 × 1.1–1.8(–2.4) cm, apex obtuse, unequally bilobulate or minutely tridenticulate, adaxially dark glossy-green, abaxially paler, sharply keeled. **Inflorescence** 1-flowered, 12–17 cm long, green to purplish-pubescent; pedicel and ovary 3.5–5 cm long, purplish-pubescent. **Flower** (4)6–8 cm Ø, dorsal sepal yellowish, purplish-spotted, synsepalum similar in color with fewer spots, petals pink-violet to light purple-red, purple-spotted, lip pink-violet or light purple-red, rim yellowish, staminode yellowish, slightly purplish-flushed, with a green or brown-yellow umbo; dorsal sepal usually erect, broadly ovate to suborbicular-ovate, 3–3.5 × 3.2–3.8 cm, basal margin recurved, margin undulate, apex obtuse; synsepalum elliptic-ovate, 2.7–3.2 × 1.4–1.8 cm, slightly concave, minutely ciliate; petals horizontally spreading, spathulate, 3.2–3.9 × 1.4–1.8 cm, adaxially base purple-hirsute, minutely ciliate, apex obtuse or obscurely tridenticulate; lip helmet-shaped, 3.7–4.3 cm, pouch ellipsoid, 2.3–2.8 × 2.2–2.5 cm; staminode obcordate or broadly obovate, 6–8 × 7–8 mm, with bubblelike papillae, apex. **Distr:** China to N Vietnam. **Ecol:** On slopes, cliffs and forests; 600–1,400 m. Fl. Sep–Nov. **Zones:** 7–10.

P. hirsutissimum (Lindl. ex J D Hook. f.) Stein, *Orchideenbuch*: 470, 1892. (Syn: *Cypripedium hirsutissimum*, etc.). **Herbs**, terrestrial or lithophytic, stemless. **Leaves** 5–7, distichous; lamina linear-oblong to linear-ligulate, 23–44 × 1.4–2.2 cm, base channeled, equitant, costate, apex acute or obtuse, and obliquely bilobed, glabrous, carinate, leathery, green, abaxially purple-spotted. **Inflorescence** 1-flowered; peduncle 13–24(–30) cm long, terete, green, densely pubescent, purple-hairy; floral bract, villous, elliptic or ovate, sheathing, 0.8–1.5 cm long; pedicel and ovary 6–8.5 cm, pubescent. **Flower** 8–16 cm Ø, sepals ciliated; dorsal sepal broadly ovate or ovate-elliptic, 3.7–5 × 2.6–4 cm, margins at length reflexed, dark purplish-green, striated; synsepalum similar, slightly narrower; petals horizontally spreading, broadly spathulate, (4–)6–8 × 1.5–2.5 cm, slightly twisted apically, ciliated, purple, claw green, mottled and dotted with purple, margin undulate; lip helmet-shaped, 3–5 cm across, pouch 2.5–3(–3.5) × 1.5–2.5 cm; staminode subquadrate, 0.8–1 × 0.8–1 cm, convex, obtuse. **Distr:** Assam to S China. **Ecol:** On steep limestone cliffs; 1,200–1,800 m. Fl. Apr–May, Oct. **Zones:** 10–11. **Info:** ***P. hirsutissimum* var. *esquirolei*** from China to Indo-China, differs from the typical variety in its slightly larger flowers, with petals to 8 cm long, peduncle and ovary short-pubescent. **Zones:** 10–11.

P. insigne (Wall. ex Lindl.) Pfitzer, *Jahrb. Wiss. Bot.* 19: 159, 1888. (Syn: *Cypripedium chantinii*, etc.). **Herbs**, terrestrial. **Leaves** 5–6, ligulate, to 32 × 3 cm, apex minutely bilobulate, obtuse, green, abaxially base purple-spotted. **Inflorescence** 1-flowered; peduncle to 25 cm long, green, purple-pubescent; bract elliptic or oblong-elliptic, to 5 cm long, obtuse, glabrous, base purple-spotted. **Flower** 7–10 cm Ø; dorsal sepal green, margin white, in inner surface with raised maroon spots, ovate-elliptic to obovate-elliptic, 4–6.5 × 3–4 cm, apical margins incurved, apex obtuse; synsepalum pale green, brown-spotted, elliptic, 4–5 × 2–2.5 cm, apex subacute or obtuse; petals yellow-brown, veined red-brown, incurved slightly, spathulate, 5–6.3 × 1–1.8 cm, basally undulate on upper margin, apex obtuse; lip yellow marked with purple-brown, 4.5–5 × 3 cm; staminode yellow, obovate, 1 cm long, purple-hairy, with a central raised boss. **Distr:** NE India (Assam – Meghalaya). **Ecol:** On dolomitic limestone outcrops near waterfalls; 1,000–1,500 m. **Zones:** 10–12.

P. javanicum (Reinw. ex Lindl.) Pfitzer, *Jahrb. Wiss. Bot.* 19: 165, 1888. (Syn: *Cypripedium javanicum*, etc.). **Herbs**, terrestrial. **Leaves** spreading, 4–5, narrowly elliptic, 12–23 × 3.4–4 cm, apex minutely tridenticulate, obtuse, pale green, veined and lightly mottled darker green. **Inflorescence** 1-flowered; peduncle 16–36 cm long, purple, shortly white-pubescent; bract elliptic, 1.5–2.5 × 1.5–2.5 cm, margins and midvein ciliate, obtuse, pale green, very lightly spotted with purple; pedicel and ovary 4.3–5 cm long, rather sparsely pubescent on ridges. **Flower** 8–9.5 Ø; dorsal sepal ovate to elliptic, 3–3.8 × 2.5–2.9 cm, acute to retuse, shortly ciliate, outside pubescent, pale green, dark-veined, margin whitish-pink; synsepalum lanceolate, 2.5–2.6 × 1.1–1.3 cm, acute, outside pubescent, pale green; petals usually deflexed at ~45° to horizontal, narrowly oblong, 4.2–4.8 × 1.3–1.4 cm, obtuse, ciliate, pale green, basally finely maroon-spotted, apically pink-purple; lip 3.6–4 × 1.8–2 cm, outside shortly pubescent, side-lobes verrucose, bright green, dark-veined, often brown-flushed; staminode reniform, 0.8 × 1 cm, shortly pubescent. **Distr:** Sumatra to Lesser Sunda Is. **Ecol:** In primary montane forests; 750–2,100 m. **Zones:** 10–12. **Info:** ***P. javanicum* var. *virens*** (Syn: *P. virens*, etc.) from NE Borneo (Sabah to N Sarawak), can be distinguished by its bright green flower, more spathulate, almost horizontal and less heavily spotted petals, almost acuminate dorsal sepal, more pubescent ovary and convex staminode. **Zones:** 10–12.

P. liemianum (Fowlie) K. Karas. and K. Saito, *Bull. Hiroshima Bot. Gard.* 5: 57, 1982. (Syn: *P. victoria-regina* ssp. *liemianum*, etc.). **Herbs**, lithophytic. **Leaves** 4–7, narrowly oblong or oblong-elliptic, 16–24 × 3–5 cm, margins ciliate, apex obtuse or rounded, adaxially green, abaxially heavily purple-spotted in bands. **Inflorescence** many-flowered; peduncle 15–20 cm long, purple, finely pubescent; bracts elliptic, 1–1.7 cm long, obtuse, green, purple-spotted, ciliate. **Flowers** 8–9 cm Ø, sepals ciliate, outside purple-pubescent; dorsal sepal cream or white, center green, ovate, 2.8–3.4 × 2.4–3.7 cm, obtuse; synsepalum yellow-green, elliptic, truncate, 2.9–3 × 2.2 cm; petals ± horizontal, slightly reflexed, cream purple-spotted, linear, 4–4.6 × 0.7–1 cm, obtuse, apically half-twisted, ciliate; lip bulbously inflated toward apex, pink-purple, purple-spotted, 4–4.4 × 2.3–2.4 cm, margin pale yellow; staminode convex, ovate-subquadrate, 1–1.1 × 0.9–1 cm, base pubescent, apex subacute to shortly apiculate. **Distr:** N Sumatra; rare. **Ecol:** In limestone ravines; 600–1,000 m. **Zones:** 11–12.

P. lowii (Lindl.) Stein, *Orchideenbuch*: 476, 1892. (Syn: *Cypripedium lowii*, etc.). **Herbs**, epiphytic, terrestrial or lithophytic. **Leaves** 4–6, linear to narrowly elliptic, fleshy, dark green. **Inflorescence** erect to arching with 2–7-flowered, opening successively; peduncle 0.6–1 m long, green, purple-mottled, pubescent; floral bracts 1/3 the length of the ovary, yellow, purple-marked. **Flowers** nodding, 7.5–14 cm Ø; dorsal sepal without spots, 3.5 × 3 cm, greenish-yellow, base brownish-purple; lateral sepals 4 × 2.5 cm, yellowish-green; petals 8 × 2 cm, colored purple, narrowed toward base and colored yellow with purple blotches; lip brownish, 4 × 2.5 cm. **Distr:** Malaya, Sumatra, Java, Sulawesi and Borneo. **Ecol:** Epiphytic on large trees of lowland forests, and grasslands; 200–1,700 m. Fl. Oct. **Zones:** 10–12.

P. malipoense S.C. Chen and Z.H. Tsi, *Acta Phytotax. Sin.* 22(2): 119, 1984. (Syn: *P. angustatum*). [Jade Slipper Orchid]. **Herbs**, lithophytic, variable. **Leaves** 4–6, distichous; lamina coriaceous, oblong or narrowly elliptic, 10–20 × 3 cm, mottled dark and light green, heavily purple-spotted, abaxially keeled, apex acute to obtuse, asymmetric. **Inflorescence** 1(2)-flowered; peduncle green, spotted with purple, white hirsute, 30–60 cm long; floral bract narrowly ovate-lanceolate. **Flowers** 7.5–12 cm Ø, green with

Top row: *Paphiopedilum hirsutissimum* var. *esquirolei*, *Paphiopedilum insigne* (albino form), *Paphiopedilum insigne*.
Middle row: *Paphiopedilum javanicum*, *Paphiopedilum javanicum* var. *virens*.
Bottom row: *Paphiopedilum liemianum*, *Paphiopedilum lowii*, *Paphiopedilum malipoense*.

purple markings, scented (with a raspberry fragrance); dorsal sepals ovate, 4 × 3.5 cm, pale green, base with a few spots; petals pale yellowish-green with purple venation, elliptic, 7.5 × 3.5 cm; lip pale pink, 6.5 × 5 cm, outside white hirsute, inside purple-spotted; staminode glossy, white and dark purple. **Distr:** S China to N Vietnam; rare and endangered. **Ecol:** On limestone cliffs, in mixed forests, cloud forests; 450–1,600 m. **Zones:** 10–11.

P. mastersianum (Rchb. f.) Stein, *Orchid.-Buch*: 477, 1892. (Syn: *Cordula mastersiana, Cypripedium mastersianum*). **Herbs**, terrestrial. **Leaves** 4–6, oblong-elliptic, 18–30 × 3.5–4.3 cm, apex acute, tridenticulate, adaxially faintly tessellated dark and pale green. **Inflorescence** 1-flowered; peduncle 15–30 cm long, densely coarsely hairy, maroon; bract ovate-elliptic, 2–2.9 × 1.8–2 cm, acute, green, ciliate; pedicel and ovary 5.5–9.5 cm long, green with purple ridges, purple-hairy. **Flower** 8.5–9.5 cm Ø; dorsal sepal broadly ovate, 3–3.7 × 3.2–4.3 cm, obtuse, outside pubescent, shortly ciliate, cream, center bright green; synsepalum ovate, 2.3–2.9 × 1.7–2 cm, obtuse, greenish-yellow; petals slightly reflexed, oblong-spathulate, 4.5–5.5 × 1.3–2 cm, obtuse, broadly tridentate, shortly ciliate, glossy-green, spotted blackish-maroon toward base, apically brownish; lip 4.3–5.2 × 2.4–3 cm, pale rosy-brown, side-lobes brownish-spotted; staminode lunate, 0.8–1.1 × 0.8–1 cm, base divided, apical margin excised, shortly pubescent, side-lobes of apical margin incurved-falcate, pale green and brown. **Distr:** Moluccas (Lesser Sunda Islands to Maluku). **Ecol:** In montane forests, on steep slopes; 900–2,000 m. Fl. Apr–Jul. **Zones:** 10–11.

P. micranthum T. Tang and F.T. Wang, *Acta Phytotax. Sin.* 1: 56, 1951. (Syn: *P. globulosum*). [Silver Slipper Orchid]. **Herbs** terrestrial, lithophytic, rarely epiphytic, variable. **Leaves** 3–5, distichous; lamina oblong-elliptic, 15 × 3–5 cm, apex obtuse to subacute, adaxially pale green, mottled dark green, abaxially purple-spotted. **Inflorescence** 1(2)-flowered; peduncle 9–25 cm long, green, purple-spotted; floral bracts lanceolate to narrowly elliptic; pedicel and ovary 3.5–5 cm long, white-villous. **Flower** 3.5–8(–10) cm Ø, sepals and petals yellowish to whitish, sometimes pink-flushed, purplish-veined, lip pinkish, inside bottom purple-spotted, staminode white, purplish-speckles, apically yellow-flushed; dorsal sepal ovate, 1.5–2.5(–3.2) × 1.3–2.5 cm, acute, abaxially white-villous; synsepalum similar, abaxially obtusely bicarinate; petals broadly ovate to suborbicular, (2–)2.5–3.2 × (1.5–)2.6–3.5 cm, adaxially base white-villous, rounded; lip ovoid-globose or ellipsoid-globose, (2.2–)5–10 × (1–)4–5.5 cm, apical margin involute; staminode elliptic, conduplicate, 1–1.5 × 0.7–0.8 cm, acute, mucronate. **Distr:** China to NW Vietnam; endangered. **Ecol:** On steep limestone cliffs, rarely in montane cloud forests; 600–1,700 m. Fl. Mar–May. **Zones:** 10–11.

P. niveum (Rchb. f.) Stein, *Orchid.-Buch*: 478, 1892. (Syn: *Cypripedium niveum*, etc.). **Herbs**, terrestrial, often clustered. **Leaves** 4–5, strap-shaped to narrowly elliptic, 8–19 × 2.4–3.6 cm, apex rounded, minutely emarginate, adaxially mottled green, abaxially heavily purple-dotted, basal margins ciliate. **Inflorescence** 1- or 2-flowered, 6–25 cm long; peduncle to 20 cm long, purple, densely white-pubescent; bracts conduplicate, broadly ovate, 1.1–1.4 × 1–1.2 cm, obtuse, white to pale green, purple-spotted. **Flowers** 6 cm Ø, white, segment bases and front of lip often heavily purple-dotted, staminode center yellow; dorsal sepal broadly ovate, 2.7–3.1 × 3–5 cm, obtuse to emarginate; synsepalum concave, ovate, 2–2.8 × 1.5–2.2 cm, obtuse; petals elliptic, 3.3–3.9 × 2.2–2.6 cm, margins shortly ciliate, apex rounded; lip rather small, ovoid to ellipsoidal, 2.2–3 × 1.5–1.7 cm, margins incurved; staminode transversely subelliptic, 0.6–0.9 × 1–1.2 cm, apex 1–3-toothed; ovary 4.5 cm long, green purple-flushed, densely pubescent. **Distr:** Peninsular Thailand to N peninsular Malaysia. **Ecol:** In fissures on limestone outcrops. Fl. Jun–Jul. **Zone:** 12.

P. parishii (Rchb. f.) Stein, *Orchid.-Buch*: 479, 1892. (Syn: *Selenipedium parishii*, etc.). **Herbs**, epiphytic, occasionally lithophytic. **Leaves** 5–8, ligulate, 45 × 4–7.5 cm, apex, asymmetric, rounded, green. **Inflorescence** ± erect, 30–50 cm long, (2–)3–6(–9)-flowered; peduncle densely white-pubescent; bracts large, conduplicate, broadly elliptic, 3–4 × 2–3.5 cm, margins undulate, apex obtuse to subapiculate; pedicel and ovary 5 cm long, green, pubescent. **Flowers** to 13 cm long, sepals cream to green with darker veins, petals green below, spotted with blackish-maroon, margins apically dark maroon, lip yellow-green to green or purplish flushed; dorsal sepal elliptic, 3.5–4.6 × 1.8–3 cm, obtuse; synsepalum ovate, 2.6–4 × 1.5–2.9 cm, acute; petals decurved-pendent, linear-tapering, 7–10.5 × 0.6–1.1 cm, rounded, spirally twisted apically, basal margins undulate, with dark spots, ciliate; lip tapering, 3–4.5 × 1.5–2 cm; staminode obcordate, 1.1–1.4 × 0.7–0.9 cm, cream with dark green venation. **Distr:** Assam to China (W Yunnan). **Ecol:** Epiphytic on moss-

Top row: *Paphiopedilum mastersianum, Paphiopedilum micranthum, Paphiopedilum niveum.*
Bottom row: *Paphiopedilum parishii, Paphiopedilum philippinense, Paphiopedilum philippinense* var. *philippinense*. [Bottom row MG.]

GS

Top row: *Paphiopedilum philippinense* var. *roebbelinii, Paphiopedilum primulum, Paphiopedilum pupuratum.*
Bottom row: *Paphiopedilum rothschildianum, Paphiopedilum sanderianum, Paphiopedilum stonei.*

covered branches and boulders; 1,250–2,200 m. **Zones:** 7–11.

P. philippinense (Rchb.f.) Stein, *Orchid.-Buch:* 480, 1892. (Syn: *Cypripedium philippinense*, etc.). **Herbs**, terrestrial or lithophytic. **Leaves** to 9, ligulate, 20–50 × 2–5.5 cm, apex rounded, asymmetric, thick and coriaceous, V-shaped in cross-section. **Inflorescence** erect, 2–4-flowered, to 50 cm long; peduncle purple-pubescent; bracts elliptic, to 5 × 2 cm, acute; pedicel and ovary 4.5–6.5 cm long, purple, pubescent. **Flowers** variable in size, sepals white, dorsally maroon-striped, petals basally white or yellow, maroon above, basal half with marginal dark maroon warts, lip and staminode yellow; dorsal sepal ovate, 4–5 × 2–2.5 cm, acute; synsepalum similar, 4.5–5.3 × 2 cm; petals linear, 6–13 × 0.5–0.6 cm, apex tapered, ciliate; lip small, ovoid, 3.8 × 1.4 cm; staminode convex, cordate-subquadrate, emarginate, green-veined, sides purple-pubescent. **Distr:** Philippines to N Borneo. **Ecol:** On limestone cliffs; 0–500 m. Fl. Jan–Jun. **Zone:** 12. **Info:** ***P. philippinense* var. *philippinense*** is an albino variety. **Zone:** 12. ***P. philippinense* var. *roebbelinii*** endemic to Philippines (Luzon) has larger flowers, and petals pendent and longer (to 13 cm long). **Zone:** 12.

P. primulinum M.W. Wood and P. Taylor, *Orchid Rev.* 81: 220, 1973. (Syn: *P. liemianum* var. *primulinum*, etc.). **Herbs**, lithophytic. **Leaves** 4–7, narrowly oblong-elliptic, 15–17 × 2.5–3.8 cm, apex obtuse or rounded, green with pale longitudinal veins, lower margins and apex ciliate. **Inflorescence** many-flowered, flowers opening successively; peduncle 20–35 cm long, green, pubescent; bracts elliptic, 1.5–1.7 cm long, obtuse, green, ciliate; pedicel and ovary 4.7–5 cm long, green, finely pubescent. **Flowers** small, 6–7 cm Ø, primrose to yellow, sepals greenish; dorsal sepal ovate, ~ 2.5 × 2.5 cm, obtuse, back pubescent, ciliate; synsepalum ovate, 2.3–2.6 × 1.2–1.4 cm, truncate; petals spreading, linear-tapering, ~3.2 × 0.6–0.8 cm, obtuse, ciliate, apically twisted; lip bulbous toward apex, 3.2–3.5 × 1.6–1.9 cm; staminode oblong-ovate, 6–8 × 5–7 mm, truncate to shortly apiculate, base pubescent. **Distr:** N Sumatra. **Ecol:** In dwarf forests, on limestone hills; 400–500 m. **Zones:** 10–12.

P. purpuratum (Lindl.) Stein, *Orchideenbuch*: 481, 1892. (Syn: *Cypripedium purpuratum*, etc.). **Herbs**, terrestrial. **Leaves** 3–8, elliptic to oblong-elliptic, 7–17 × 2.3–4.2 cm, basal margins shortly ciliate, unequally bilobulate, apex acute, adaxially tessellated light and darker green, abaxially pale green. **Inflorescence** 1-flowered, erect; peduncle 12–20 cm long, slender, purple, purple-pubescent; bract narrowly ovate-elliptic, 1.5–1.7 × 0.7 cm, green purple- and white-pubescent; pedicel and ovary 2.5–4 cm long, green, purple-pubescent. **Flowers** 7–8 cm Ø; dorsal sepal ovate-cordate, 2.5–3.5 × 2.2–3.6 cm, apiculate or acuminate, white, heavily purple-maroon veined; synsepalum lanceolate, 2–2.7 × 1.2–1.5 cm, acute, green, dark-veined; petals ± horizontal, narrowly elliptic to oblong, 3.5–4.6 × 0.9–1.3 cm, margin ciliate, apex acute to obtuse, glossy-maroon, base greenish-white, heavily blackish-maroon spotted below; lip 3.3–4 × ~2 cm, side-lobes incurved, verrucose, brownish-maroon; staminode lunate, 0.8 × 0.8–1.1 cm, pubescent, pale ochre-purple, dark-veined. **Distr:** Hong Kong, China. **Ecol:** On wooded slopes near streams; 30–700 m. **Zones:** 6–10.

P. rothschildianum (Rchb.f.) Stein, *Orchid.-Buch:* 482, 1892. (Syn: *Cypripedium elliottianum*, etc.). **Herbs**, terrestrial or lithophytic, often growing in large clumps. **Leaves** several, linear to narrowly oblanceolate, 60 × 4–5 cm, acute, sparsely ciliate at base, green. **Inflorescence** 2–4-flowered, erect; peduncle to 45 cm long, purple, shortly pubescent; bracts ovate-elliptic, to 5.5 cm long, obtuse, ciliate and hairy on midvein, pale green or yellow, purple-striped. **Flowers** to 30 cm Ø; dorsal sepal ovate, 6.6 × 3.7–4.1 cm, acute to acuminate, ivory-white or yellow maroon-veined; synsepalum similar, ~5.7 × 3.3 cm; petals rounded, narrowly tapered, to 12.4 × 1.4 cm, ciliate, apex papillose, yellow or ivory-white, maroon-marked; lip subporrect, back grooved, ~5.7 × 2.2 cm, golden, heavily purple-suffused, side-lobes not auricular; staminode linear, 1.4–1.6 × 0.4–0.5 cm, apex bifid, geniculate, margins and base densely glandular-pubescent, pale yellow-green; ovary to 7.5 cm long, pale green, sparsely purple-spotted, glabrous. **Distr:** Borneo (Sabah: Mt Kinabalu); very rare. **Ecol:** On limestone cliffs; 600–1,200 m. **Zones:** 10–11.

P. sanderianum (Rchb.f.) Stein, *Orchideenbuch*: 482, 1892. (Syn: *Cordula sanderiana, Cypripedium sanderianum*). **Herbs**, lithophytic. **Leaves** 4–5, arcuate-pendent; lamina linear, to 45 × 5.3 cm, apex obliquely or obtusely bilobed, shiny, green, apical leaf reduced to a sterile bract at base of peduncle. **Inflorescence** horizontal or slightly ascending, 2–5-flowered; peduncle red-brown, purple-pubescent, 8–14 cm long; bracts elliptic-lanceolate, 4–5 cm long, acute, red-brown, margins and midvein ciliate. **Flowers** ~7 cm Ø, to 95 cm long, sepals yellow maroon-striped; dorsal sepal lanceolate, 4.8–6.5 × 1.3–2.5 cm, acuminate, slightly concave; synsepalum similar, 3.5–6 × 1–2 cm, 2-keeled; petals off-white to yellow, maroon-spotted, ribbonlike, 30–90 × 0.7–0.9 cm, pendent, tapering to apex, undulate and twisted, basal margins maroon-warted, basal half ciliate, apex minutely pubescent; lip yellow, subporrect, 4.8–5 × 1.5–2.5 cm, apex pointed, side-lobes incurved, very short, subacute; staminode yellow, convex, oblong, 1.3 × 1.1 cm, emarginate or obscurely tridenticulate, basal and side margins pubescent. **Distr:** Borneo (Sarawak); endangered. **Ecol:** On vertical limestone cliffs; 150–900 m. **Zones:** 11–12.

P. stonei (Hook.) Stein, *Orchideenbuch*: 487, 1892. (Syn: *Cypripedium platytaenium*, etc.). **Herbs**, lithophytic. **Leaves** ~5, ligulate, to 70 × 4.5 cm, apex rounded to obtuse, green. **Inflorescence** arcuate, to 70 cm long, usually 2–4-flowered; peduncle 17–35 cm long, glabrous, purplish; bracts lanceolate, 3.5–5.5 × 1.6–2.2 cm, acute or acuminate; pedicel and ovary 4–7 cm long, glabrous. **Flowers** to 12 cm Ø, sepals white lined with dark maroon; dorsal sepal ovate, 4.5–5.7 × 3–4.4 cm, acuminate; synsepalum elliptic-ovate, 3.7–5 × 2–3.4 cm, acuminate; petals yellow, lined and spotted maroon, ± maroon-flushed apically, arcuate-dependent, linear-tapering, 10–15 × 0.4–0.75 cm; lip pale yellow, flushed pink with darker veins, pointing forward, back grooved, 4.5–5.7 × 2.8 cm; staminode yellow, convex, suborbicular, 1.4 × 1.1 cm, margins coarsely hairy, apex truncate or incised. **Distr:** Borneo (S Sarawak). **Ecol:** At base of cliffs; 60–500 m. Fl. Apr, Jul–Dec. **Zones:** 11–12.

P. sukhakulii Schoser and Senghas, *Die Orchidee*, xvi. 110, 224, 1965. (Syn: *P. sukhakulii* f. *aureum*). **Herbs**, epiphytic or terrestrial, to 30 cm high. **Leaves** 6, narrowly elliptic, 13–20 × 4–5 cm, apex trifid, adaxially mottled pale and dark green. **Inflorescence** 1-flowered; peduncle to 25 cm long, maroon with white hairs; floral bract ovate, 1.5–2.5 cm long, green. **Flower** 14 cm Ø, dorsal sepal white or pale green, with 11–13 dark green stripes, synsepalum white, green-lined, petals pale yellowish-green, spotted with deep purple, lip pale green, veined and mottled with deep purple, lateral lobes purple-mottled, staminode pale yellow, center green-mottled; dorsal sepal cordate to ovate, 3.8–4 × 2.6–3 cm, acuminate; synsepalum smaller, lanceolate; petals narrowly oblong, slightly broader apically, ~6 × 1.5–2 cm, acute; lip deeply saccate, 4.5–5 × 2.3–3 cm, apex narrowed; staminode lunate, tridentate in sinus of ventral cusps. **Distr:** NE Thailand; endangered. **Ecol:** In shady forests; 240–1,000 m. Fl. Sep–Oct. **Zones:** 10–11. **Info:** ***P. sukhakulii* f. *aureum*** is an all green form.

P. superbiens (Rchb.f.) Stein, *Orchideenbuch*: 487, 1892. (Syn: *P. veitchianum*, etc.). **Herbs** terrestrial. **Leaves** 4–5, elliptic to oblanceolate-elliptic, 15–24 × 4–6.5 cm, base sparsely ciliate, apex minutely tridentate, adaxially tessellated dark and light green, abaxially base purple-flushed. **Inflorescence** erect, 1-flowered; peduncle 12–23 cm long, purple, purple-pubescent; bract elliptic-lanceolate, 2–3 cm long, acute, pubescent; pedicel and ovary 6–8 cm long, purple, shortly pubescent. **Flower** 6.5–8 cm Ø, sepals white, sometimes center green or purplish, green and purple veined; dorsal sepal ovate, 3.4–5.8 × 3–5.3 cm, acuminate, ciliate; petals white, sometimes purple-flushed, maroon-purple spotted, ± recurved-falcate, ligulate to ligulate-oblanceolate, 5.2–7.5 × 1.5–1.9 cm, acute to obtuse, apically half-twisted, ciliate, basal upper margins with raised spots; lip deep maroon, 4.6–6.5 × 2.6–3 cm, apex shortly ciliate, side-lobes warty; staminode transversely reniform, 0.9–1.2 × 1.4–2 cm, bluntly 3-toothed, pubescent. **Distr:** N and W Sumatra. **Ecol:** In coniferous forests, and on steep rock ledges; 800–1,300 m. **Zones:** 10–11.

P. thaianum Iamwir., *Orchid Rev.* 114: 278, 2006. **Herbs**, lithophytic or terrestrial. **Leaves** 3–5, distichous; lamina narrowly elliptic to strap-shaped, 3.5–9.5 × 1–2.4 cm, basal margins ciliate, apex rounded, minutely tridentate, adaxially tessellated with dark and pale green, abaxially heavily purple-spotted. **Inflorescences** 1(2)-flowered, 6–17(–23) cm long; peduncle 4–15 cm long, purple, densely white-pubescent; bract conduplicate, ovate, 6.9–8 × 3.5–3.9 mm, obtuse, pale green, purple-spotted. **Flowers** 3.3–4.2 cm Ø, petal bases pubescent, white, segment bases and front of lip sparsely purple-dotted, inside back of lip purple-spotted, staminode center greenish; dorsal sepal broadly ovate to round, 1.7–21 × 1.8–2.2 cm, obtuse; synsepalum concave, ovate, 1.3–1.8 × 0.8–1.6 cm, obtuse; petals elliptic, 2.2–2.6 × 1.4–1.5 cm, margins shortly ciliate, apex rounded; lip ovoid, 2–2.1 × 1–1.4 cm, margins incurved, lateral lobes auriculate, inflated, membranous; staminode very broadly or transversely obovate, 5.5–7.7 × 7.2–8.3 mm, apex 1- or 3-dentate; ovary 3.5–5.6 cm long, green, purple-flushed, densely pubescent. **Distr:** S Thailand. **Ecol:** In tropical rainforests, on limestone cliffs; 350–450 m. Fl. Apr–May. **Zones:** 10–12.

P. tonsum (Rchb.f.) Stein, *Orchideenbuch*: 488, 1892. (Syn: *P. braemii*, etc.). **Herbs**, terrestrial; stem short. **Leaves** ~6, distichous; lamina conduplicate, oblong-elliptic, to 20 × 4.5 cm, apex tridentate, adaxially green-veined, mottled darker green. **Inflorescence** erect, 1-flowered; scape to 35 cm long, shortly pubescent, green, densely purple-spotted; bract elliptic-ovate, 18–24 × 1–1.2 cm, acute, ciliate; pedicel and ovary ~6 cm long. **Flower** to 12(–14) cm Ø, sepals white, veined and tinged yellow-green, petals olive to yellow-green with darker venation and inner surface with a few black warts, lip olive-brown with pink-flushed, dark-veined; dorsal sepal obovate, ~4.5 × 4 cm, shortly apiculate; synsepalum ovate, ~3.5 × 1.5 cm, acute, shortly apiculate; petals slightly drooping, oblong-oblanceolate, ~6.5 × 2 cm, obtuse; lip saccate, ~5.5 × 3.3 cm; staminode subreniform, 1–1.1 × 1.3–1.4 cm, lower margin dentate, usually shallowly incised. **Distr:** N and W Sumatra. **Ecol:** In lower montane forests; 1,000–1,800 m. **Zone:** 10.

P. tranlienianum Gruss and Perner, *Caesiana* 11: 66, 1998. (Syn: *P. tranlienianum* var. *alboviride*, etc.). **Herbs**, terrestrial or lithophytic. **Leaves** 4–6, distichous; lamina narrowly oblong, 10–24 × 1.6–2.7 cm, apex unequally bilobulate or tridenticulate, adaxially deep green, margins paler, abaxially paler. **Inflorescence** 1-flowered, suberect or arching; peduncle green, 6–15 cm long, purplish-hairy; floral bract green purplish-spotted, ovate, 1.8–3 × 1.2–1.4 cm; pedicel and ovary 3–4 mm long, densely purple-pubescent. **Flower** 6–6.5 cm Ø, petals and lip pale green clearly flushed and veined with purple-brown, dorsal sepal white longitudinally striped with purple-brown basally, staminode pale yellow-green, with a green umbo; dorsal sepal suborbicular, 2.5–3.5 × 2.5–3.5 cm, minutely ciliate, basal margin recurved, apex acute; synsepalum pale green, ± veined with purple-brown, subovate, 2.2–3 × 1.7–2 cm; petals narrowly oblong, 2.7–3.9 × 0.8–1 cm, margin strongly undulate, white-ciliate, apex obtuse; lip helmet-shaped, pouch ellipsoid, 2.2–2.4 × 1.5–1.8 cm; staminode broadly obovate, ~ 1 × 1 cm. **Distr:** China (Malipo), N Vietnam. **Ecol:** In rocky thickets; ~1,000 m. Fl. Sep. **Zones:** 7–10.

P. venustum (Wall. ex Sims) Pfitzer, *Jahrb. Wiss. Bot.* 19: 165, 1888. **Herbs**, terrestrial. **Leaves** 4–5, elliptic to oblong-elliptic, 10–25 × 3–5.5 cm, apex obtuse, tricuspidate, rough, tessellated dark green and gray-green, abaxially densely purple-spotted, basally ciliate. **Inflorescence** 1(–2)-flowered; peduncle 10–23 cm long, purple, shortly hirsute; bract elliptic-lanceolate, 1.7–2.5 cm long, acute, sparsely pubescent; pedicel and ovary 4–5.9 cm long, shortly pubescent, green, ribs purple-spotted. **Flower** 8–9 cm Ø; sepals white, green-veined; dorsal sepal ovate, 3–3.8 × 2.3–2.8 cm, acute, outside hirsute; petals recurved, oblanceolate, 4.3–5.4 × 1.2–1.4 cm, subacute to obtuse, ciliate, white, green-veined, apically purple-flushed, with a few large maroon-black warts; lip 3.3–4.3 × 2.5–3.2 cm, side-lobes verrucose, yellow, tinged purple, green-veined; staminode reniform, 0.7–0.9 × 1–1.1 cm, minutely pubescent. **Distr:** NE India, E Nepal, Sikkim, Bhutan. **Ecol:** On cliff ledges above streams; 60–1,350 m. **Zones:** 10–11.

P. victoria-mariae (Sander ex Mast.) Rolfe, *Orchid Rev.* 4: 364, 1896. (Syn: *P. chamberlainianum* f. *victoria-mariae*, etc.). **Herbs**, lithophytic. **Leaves** oblong-ligulate, 30 × 6.5 cm, apex rounded, basally ciliate, green, adaxially mottled with dark green, abaxially base purple-flushed. **Inflorescence** many-flowered, to 1+ m tall; peduncle to 35 cm long, purple, pubescent; bracts elliptic-oblong, to 3 cm long, emarginate, green, ciliate; flowers appearing in succession, 1–2 at a time; pedicel and ovary to 6 cm long, green, brown-pubescent. **Flowers** 8–9 cm Ø, dorsal sepal pale yellow, center bright green,

Below: *Paphiopedilum sukhakulii, Paphiopedilum superbiens, Paphiopedilum thaianum.*

MG GS

Top row: *Paphiopedilum tonsum, Paphiopedilum tranlienianum, Paphiopedilum venustum.*
Middle row: *Paphiopedilum victoria-marie, Paphiopedilum vietnamense, Paphiopedilum villosum.*
Bottom row: *Paphiopedilum violascens, Paphiopedilum wardii, Paphiopedilum wilhelminae.*

petals green, flushed brown to reddish-purple, lip purple with margin yellow or green, staminode deep purple-brown, ends green; dorsal sepal broadly ovate to obovate, 2.5–2.9 × 2.3–3 cm, obtuse, ciliate, outside long-pubescent; synsepalum ovate, 2.1–2.8 × 1.5–1.7 cm, truncate, ciliate, outside long-pubescent; petals horizontal-reflexed, linear, 3.2–4 × 0.6–1.1 cm, obtuse or rounded, shortly ciliate, apically twisted; lip 3–4 × 2–2.5 cm, tapering to apex, side-lobes finely spotted; staminode trullate, apically curved, 1 × 0.7 cm. **Distr:** W Sumatra (Bukit Tinggi). **Ecol:** 1,500–2,000 m. **Zones:** 11–12.

P. vietnamense O. Gruss and Perner, *Orchidee* (Hamburg), Suppl. 5: 3, 1999. **Herbs**, lithophytic or terrestrial. **Leaves** to 20 × 7 cm, glossy, adaxially mottled dark green and gray-green, abaxially purple-flushed. **Inflorescence** 1- or 2-flowered, to 25 cm long; peduncle purple, white-pubescent. **Flowers** 10–12 cm Ø, white pubescent, sepals white or pink, flushed with purple-pink toward apex, petals white, pink or purple with a purple blotch on the front, spotted with purple inside, pouch intensely pigmented, staminode yellow with a white margin, center green; dorsal sepal ovate, to 5 × 3.2 cm, margins undulate, apex acute; synsepalum ovate, to 4.5 × 3.5 cm, notched; petals relatively large, to 6.5 × 4 cm, base narrowed, oblique, margins undulate, apex rounded; pouch relatively small, to 3.8 × 3.2 cm, ribbed, margins incurved; staminode ovate, to 2.5 × 1.8 cm, margins irregular. **Distr:** Vietnam; known from only a single locality. **Ecol:** In wet evergreen forests, on limestone; 350–550 m. **Zones:** 9–11. **Info:** Illegal trading of this species led to its decline in numbers within a few years of its discovery.

P. villosum (Lindl.) Stein, *Orchideenbuch*: 490, 1892. (Syn: *Cordula villosa, Cypripedium villosum*). **Herbs**, epiphytic (lithophytic), often clump-forming. **Leaves** 4–5, linear-ligulate, 14–42 × 2.5–4 cm, basal margins ciliate, apex unequally bilobulate, acuminate to acute, abaxially midgreen, base purple-spotted. **Inflorescence** 1-flowered, suberect to arcuate; peduncle 7–24 cm long, green, often purple-spotted, purple-villose bract elliptic, 3.7–6.5 × 3–3.8 cm, obtuse, green, maroon-spotted, glabrous; pedicel and ovary triangular in cross-section, 3–6 cm long, ochre, densely purple-villose. **Flower** 7.5–11.5 cm Ø, dorsal sepal green, margin white with central glossy deep maroon areas, synsepalum pale green, petals glossy reddish-brown, with a central maroon stripe, lip ochre, flushed with pink or reddish, staminode yellow, central knob yellow or green; dorsal sepal obovate, 4.5–6.5 × 3–3.5 cm, obtuse, basal margins reflexed; synsepalum adpressed to ovary, ovate, 3.8–5 × 1.8–2.6 cm, subacute; petals incurved, obovate-spathulate, 4.7–7 × 2.5–3 cm, retuse-emarginate, rounded; lip tapering to apex, 4–6 × 3–3.8 cm; staminode obcordate-obovate, ~1.6 × 1.4 cm, truncate, verrucose, hirsute. **Distr:** NE India, Myanmar, Thailand. **Ecol:** 1,100–2,000 m. **Zones:** 9–10.

P. violascens Schltr., *Repert. Spec. Nov. Regni Veg. Beih.* 1: 2, 1911. (Syn: *Cordula violascens*, etc.). **Herbs**, terrestrial (epiphytic). **Leaves** 4–6, elliptic to oblong-elliptic, 8–22 × 2.3–4 cm, apex minutely tridenticulate, obtuse, grayish-green, mottled above with darker green. **Inflorescence** 1(2)-flowered; peduncle 20–30 cm long, ciliate, pubescent; pedicel and ovary 4–6.8 cm long, green, purple-pubescent. **Flowers** 6–7.5 cm Ø, dorsal sepal white, green-veined, synsepalum green, dark-veined, petals white or greenish-white, apically heavily purple-flushed, lip green to ochre, staminode green, dark-veined; dorsal sepal broadly ovate, 2.7–3.2 × 2.9–3 cm, subacute,

MD

MIW

MIW

Above: *Phragmipedium longifolium, Galeola lindleyana, Vanilla albida.*

outside pubescent; synsepalum ovate, 2–2.6 × 1.5–1.6 cm, subacute, outside pubescent; petals deflexed at 45°, obliquely oblong, 3.5–4.4 × 1.5–1.8 cm, obtuse, minutely ciliate; lip deeply saccate, 3.8–5 × 2 cm, apical margin sparsely ciliate; staminode semicircular, 0.75 × 1.4 cm, truncate, shortly apiculate, pubescent. **Distr:** N and E New Guinea to Bismarck Archipelago (Manus Island). **Ecol:** In riverine forests; 200–1,200 m. **Zones:** 10–12.

P. wardii Summerh., *Gard. Chron.*, III, 92: 446, 1932. [Rainbow Orchid]. **Herbs**, terrestrial, clustered. **Leaves** 3–5, oblong-lanceolate, to 17 × 5.5 cm, apex acute, tridenticulate, adaxially irregularly mottled pale and dark bluish-green, abaxially purple-mottled. **Inflorescence** 1-flowered; peduncle 15–40 cm long, densely pubescent, purple; bract lanceolate, 2–3 cm long, acute, ciliate, green, purple-flushed; pedicel and ovary 4–6 cm long, green or reddish-purple, pubescent. **Flower** 7–10 cm Ø, dorsal sepal white, green-veined, petals greenish-white flushed brown-purple, dark maroon spotted, lip greenish or ochre, finely brown-spotted, staminode pale green, dark-veined; dorsal sepal ovate, 4–5 × 2–3 cm, acuminate, ciliate, outside pubescent; synsepalum lanceolate, 3.5–4.5 × 2–2.2 cm, acute, puberulent; petals spreading, somewhat pendulous, oblong-lanceolate to oblong, 5–6.5 × 1.5–1.7 cm, acute, setose-ciliate; lip 4–5 × 2 cm, side-lobes warty, apical margin obscurely ciliate; staminode lunate, 0.9–1 × 1.2–1.4 cm wide, finely pubescent. **Distr:** China (SW Yunnan) to Myanmar. **Ecol:** In shaded forest margins, or along streams; ~2,000 m. **Zones:** 9–10.

P. wilhelminae L.O. Williams, *Amer. Orchid Soc. Bull.* 10: 373, 1942. (Syn: *P. striatum*, etc.). **Herbs**, terrestrial, clustered. **Leaves** 4–6, linear-oblong, to 6–27 × 3–5.5 cm, basal margins sparsely ciliate, apex obtuse, tridenticulate, green, glabrous. **Inflorescence** 2–5-flowered; peduncle 30–50 cm long, purple to greenish-brown, shortly pubescent; bracts ovate, to 4.5 cm long, acute, glabrous, pale yellow-green, purple-striped; pedicel and ovary 5–7 cm long, glabrous or sparsely pubescent, pale yellow-green, purple-striped. **Flowers** small, sepals and petals yellow, maroon-veined, lip yellow, veined and flushed purple; dorsal sepal ovate, 4.9–5.5 × 2.5–2.9 cm, acuminate; synsepalum similar, 5.3–5.5 × 2.8–2.9; petals deflexed, linear-tapering, 5–7 × 1 cm, basal margins sparsely ciliate, often without warts, apically papillose, usually helically twisted; lip subporrect, 3.5–4.5 × 1.8–1.9 cm, side-lobes narrow, acute, incurved; staminode small, subquadrate, blunt. **Distr:** C New Guinea. **Ecol:** In highlands; 1,700–1,800 m. **Zones:** 10–12.

Phragmipedium

Species 23. **Herbs**, terrestrial, lithophytic or sometimes epiphytic; roots fibrous; stems short, leafy. **Leaves** distichous, 2-ranked (fanlike) but folded at base; lamina ligulate (strap-shaped), conduplicate, sulcate, coriaceous, mid to dark green. **Inflorescence** few- to many-flowered racemes or panicles, terminal, sometimes branched; peduncle and rachis hairy. **Flowers** large, showy, deciduous; sepals spreading; dorsal sepal free, erect or forming a hood; lateral sepals entirely united forming a synsepalum; petals free, broad and spreading, or narrow and pendent; lip sessile, inflated, calceolate, deep, margins inrolled; column short, stout, with 2 laterally placed fertile stamens, each bearing a 2-celled anther and a dorsally placed sterile petaloid staminode; pollen granular; ovary 3-celled. **Distr:** SW Mexico to S tropical America. **Cult:** Propagate from seeds. Requires moist but free-draining compost, in light to moderate shade.

P. longifolium (Warsz. and Rchb. f.) Rolfe, *Orchid Rev.* 4: 332, 1896. (Syn: *Cypripedium longifolium*, etc.). **Herbs**, terrestrial or lithophytic. **Leaves** 2-ranked, lower leaves shorter; lamina linear-lanceolate, to 80 × 1–4 cm, acuminate, adaxially dark green, abaxially paler. **Inflorescence** several-flowered, 35–75 cm tall; flowers opening in succession; peduncle purplish, covered by 3 sterile bracts; bracts adnate to the peduncle, distichous, lanceolate, 4–9 cm long, acute. **Flowers** 18–20 cm across, pale yellowish-green, marked with red or violet, dorsal sepal margins whitish, petals purple-edged, pouch yellow-green purple-flushed, staminode pale green; dorsal sepal curved forward, lanceolate, 3–5 × 1–1.6 cm, acute, side margins reflexed; synsepalum ovate, 2.5–6 × 1.7–2.5 cm; petals spreading or deflexed, linear, 8–12 × 0.4–0.8 cm, much-twisted, basal margins undulate, gradually tapered to apex, rounded; lip inflated, trilobed, 2.5–5 × 1.5–2 cm, side-lobes incurved, midlobe margins incurved; column 7 mm long; staminode obtriangular, 1 × 0.6 cm, upper margin purple-ciliate, eyebrow-shaped. **Distr:** Costa Rica to Ecuador. **Ecol:** In humid forests, especially on cliff slopes and landslides near rivers; sea-level to 1,600 m. **Zones:** 10–12.

Subfam. VANILLOIDEAE

Syn: Vanillinae.

Genera 15, **species** 180. **Herbs** small to large terrestrial autotrophic or holomycotrophic, monopodial or sympodial or scrambling or **climbing vines**; roots elongate, fleshy; rhizome short to elongate; stems erect or scrambling, unbranched or branched, very short to long, leafy. **Leaves** 1–many, sometimes reduced to scales, alternate, fleshy or leathery, conduplicate, spreading or suberect, ovate, elliptic, lanceolate, ligulate, or oblong, green, glabrous. **Inflorescences** terminal or lateral, 1–many-flowered, racemose or paniculate, rarely 1-flowered; rachis sometimes pubescent; bracts conduplicate; pedicels short; abscission layer between pedicel and ovary, sometimes with a distinct calyculus present. **Flowers** often showy; sepals free, subsimilar, glabrous or externally pubescent; petals free, glabrous; lip free or with side margins fused to column, trilobed or entire, usually with a callus of hairs, papillae, or backward facing scales, lacking a nectary; column elongate, slender, porrect, hooded at tip; anther 1, terminal, mobile; pollinia obscure, pollen loose, shed in monads or rarely tetrads; ovary locular or trilocular, glabrous or pubescent, stigma trilobed, concave, viscid. **Fruits** capsules, erect to pendent, 3-ribbed, cylindric to almost ellipsoidal, glabrous to pubescent; seeds spherical with a hard testa. **Distr:** Widespread in tropical regions of Africa, America, and Asia, with a few species reaching temperate areas. **Info:** The largest genus in this subfamily is *Vanilla* with about 100 species. Relationships within Vanilloideae are becoming fairly well resolved (Cameron 2004; Cameron and Molina 2006).

Galeola

Species ~10. **Herbs**, holomycotrophic, leafless; rhizomes subterranean, thick; stems scapose, erect, stout, usually yellowish brown, remotely nodose, with scale at node. **Inflorescences** terminal and also lateral in axils of upper scales, racemose or paniculate, with ± woolly tomentose rachis and flowers. **Flowers** yellowish or brownish, fleshy and spongy in texture, with abscission layer between perianth and ovary; sepals and petals similar, free, ± spreading; lip concave, sessile at base of column, base loosely embracing column; column short and stout, somewhat curved, dilated at apex; anther terminal, conical, with 2 protruding cells; pollinia 2, cleft, soft granular; stigma transversely broadened; rostellum short. **Fruits** capsules; seeds with thick coat and irregularly erose wing. **Distr:** Madagascar through tropical and subtropical Asia to Australia and Papua New Guinea. **Cult:** Propagate from seeds or by division. Requires a moist humus-rich compost in partial shade.

G. lindleyana (Hook. f. and Thomson) Rchb. f., *Xenia Orchid.* 2: 78, 1865. (Syn: *Cyrtosia lindleyana*, etc.). **Herbs**, holomycotrophic, terrestrial, rhizomatous; stem erect, reddish-brown, 1–3 m tall, nodes with broadly ovate scales. **Inflorescence** a panicle composed of terminal and lateral racemes; lateral raceme 2–5(–10) cm long, ~10-flowered, usually very shortly pedunculate; sterile bracts basal, ovate-lanceolate, 1.5–2.5 cm long; floral bracts ovate, 5–6 mm long, abaxially rusty-pubescent; pedicel and ovary often ± curved, 1.5–2 cm long, densely rusty-pubescent. **Flowers** yellow, ~3.5 cm Ø; sepals elliptic to ovate-elliptic, 1.6–2 × 0.9–1.1 cm, abaxially densely rusty-tomentose, carinate; lateral sepals slightly longer; petals broadly ovate to suborbicular, 1.2–1.4 cm wide, glabrous; lip concave, cup-shaped, nearly subglobose, ~1.3 cm Ø, adaxially papillate-hairy, with a basal callus, unlobed, margin shortly fimbriate; column clavate, ~7 mm long; anther cap papillate-spinescent. **Fruits** pale brown, suboblong,

8–12(–20) × 1.7–2.4 cm; seeds with broad wing, wings 1–1.3 mm wide. **Distr:** Himalayas to Taiwan, N Sumatra. **Ecol:** In sparse forests, thickets, moist rocky places; 700–3,000 m. Fl. May–Aug. Fr. Sep–Oct. **Zones:** 7–12.

Vanilla

[Vanilla].

Species ~100. **Herbs**, semiepiphytic; stems climbing to several meters, bearing leaf and root at each node. **Leaves** usually large and fleshy, base not sheathing. **Inflorescences** axillary, with few to many racemose flowers. **Flowers** usually large, with an abscission layer between ovary and perianth; sepals and petals subequal, free and spreading; lip with basal margins joined to sides of column, upper part widened and ± trilobed, often with hairy or brushlike appendages on disk; column slender, slightly curved near top; anther suberect, tilt downward on front of column; pollinia 2, granular, soft, often cleft, without caudicle or viscidium; stigma entire, slitlike under broad rostellum. **Fruits** elongate cylindrical, fleshy, often indehiscent; seeds rather large, with hard coats, wingless. **Distr:** Tropical and subtropical regions of N America, Mexico, West Indies, C America, S America, Africa, SE Asia, W Pacific Islands. **Cult:** Propagate from cuttings. Requires warm, very humid and well-lit conditions. Commercially, they are grown on poles or shrubs. Private collections are usually grown in baskets or against walls with suitable support for climbing.

V. albida Blume, *Cat. Hort. Buitenz.* 100, 1823. (Syn: *V. griffithii*, etc.). **Herbs**; stems to 3 m long. **Leaves** alternate; lamina elliptic-lanceolate, 15–20 × 4–8 cm, base obtuse-rounded, apex acuminate, thickly fleshy, coriaceous, glabrous, venation reticulate; petiole short, slightly embracing stem. **Inflorescences** shorter than leaves, usually 2–5-flowered; bracts triangular, 0.7–1 cm long; pedicel and ovary 3–4 cm long. **Flowers** whitish or tinged pale green, 5–6 cm Ø; sepals subequal, oblanceolate, 3.5–4 × 1 cm, spreading, recurved; petals similar to sepals, slightly oblique; lip pale green, outside pink-tinged, inside red-striped, trumpet-shaped, 3.5–4 × 1.5–2 cm, obscurely trilobed, free portion undulate on front margins, with many papillae arranged in 2 or 3 rows near apex; disk with white brushlike appendage; column subterete, ~2.5 cm long, with small wings; anther fleshy, 3 mm long; pollinia ovate, 1.2 mm long; stigma transversely slitlike. **Fruits** curved-cylindrical, ~3-ridged, ~10 cm long; seeds ovoid or globose, ~1.5 mm Ø. **Distr:** Malaysia, Thailand and Indo-China; Taiwan. **Ecol:** In evergreen forests; <1,200 m. **Zones:** 10–12.

Subfam. ORCHIDIOIDEAE

Syn: Neottioideae, Ophrydoideae, Spiranthoideae. **Genera** 208, species ~3,630 (divided between 8 tribes and 20 subtribes). **Herbs**, predominantly terrestrial, very rarely epiphytic; rootstock tuberous or fleshy; rhizomes short to long, rarely mycotrophic with leaves reduced to colorless sheaths. **Leaves** spirally arranged, 1–many, basal or arranged along stem, deciduous or rarely persistent, often sheathing at base, usually green, rarely variously colored and/patterned; pedicel often obscure. **Inflorescence** terminal, erect or arching, 1–many-flowered; peduncle usually terete; bracts linear, lanceolate, ovate, or elliptic, usually glabrous and green. **Flowers** usually resupinate; dorsal sepal free or often adnate to petals forming a hood; lateral sepals usually free, sometimes connate and oblique at base forming a spurlike mentum; petals entire or bilobed, often adnate to dorsal sepal; lip usually lowermost in flower, usually deflexed, entire, 3- or 5-lobed, or bipartite, occasionally callose or with 2 basal glands; nectary fusiform, clavate or cylindric; column basifixed or not; anther bilocular; pollinia 2 or 4; staminodes 2, lateral; ovary distinct, usually glabrous, lacking an abscission layer; stigma entire or bilobed. **Fruits** capsules, 3-ribbed, longitudinally dehiscent; seeds fusiform to almost cylindric, with thin testa. **Distr:** Mainly in temperate and tropical areas of both the Old and New Worlds.

TRIBE ORCHIDEAE

Disa

Syn: *Gamaria*, *Monadenia*, etc.

Species 179. **Herbs**, terrestrial, perennial, tuberous; tubers testicular, rarely absent; sterile shoots produced occasionally, to 15 cm long with several basal sheaths and several erect conduplicate acute leaves to 40 cm long; fertile shoots to 1 m tall. **Leaves** radical, basal or cauline; lamina linear to ovate, soft to rigid, lax to imbricate, monomorphic to dimorphic, green, grading into floral bracts or sharply distinct from floral bracts. **Inflorescence** racemose with 1–numerous flowers; bracts usually green, ovate, acute or acuminate. **Flowers** not resupinate, resupinate or doubly resupinate, 0.4–8 cm Ø, usually red or white, rarely blue or yellow or green; dorsal sepal erect, galeate, rarely cucullate or dish-shaped, usually ovate, with a spur obsolete–8 cm long, slender or rarely massive, rarely flattened or with a dorsal or ventral groove, horizontal, descending or ascending; lateral sepals lanceolate to broadly ovate, rounded to acuminate, often keeled, patent, sometimes ascending or descending; petals lanceolate to square, erect or reflexed, acute or truncate, apex curving over the anther or over the rostellum; lip usually lanceolate, rarely spathulate or broadly ovate, occasionally linear; rostellum trilobed, lateral lobes erect, square or hornlike, flat or canaliculate, free or fused to petals, central lobe usually small, fleshy; anther 2-celled. **Distr:** Tropical and S Africa, W Indian Ocean. **Cult:** Propagate by division. Requires intermediate to cool conditions with filtered sunlight and good ventilation. Best grown in a well-drained mixture of sand, peat or sphagnum moss (live moss is preferable). Water needs to be low pH (4) with few dissolved salts. **Zones:** 5–6.

D. uniflora P.J. Bergius, *Descr. Pl. Cap.*: 348, 1767. (Syn: *D. barellii*, etc.). [Pride of Table Mountain, Red Disa]. **Herbs**, terrestrial, perennial, tuberous, to 80 cm tall; stems stout, leafy. **Leaves** spreading, borne along the stem; lamina lanceolate to elongate-lanceolate, to 25 cm long, apex acute. **Inflorescence** a raceme, 1–3(–6+)-flowered, 20–30 cm tall; bracts acuminate. **Flowers** large and showy, 8–12 cm Ø, usually bright red, inside of hood orange, red-veined, sometimes yellow or pink; dorsal sepal erect, concave, 2–6 × 1.5–2 cm; spur 1–1.5 cm long; lateral sepals spreading forward, 6.5 cm long; spur to 1.5 cm long. **Distr:** SW Cape Province of South Africa (Table Mountain); endangered, strictly protected. **Ecol:** Terrestrial near waterways, rock crevices, etc.; 100–1,200 m. Fl. Jan–Mar. **Zones:** 9–11. **Info:** This species is solely pollinated by Cape Christmas Butterfly (*Aeropetes tulbaghia*).

Habenaria

[Rein orchid].

Species ~600. **Herbs**, perennial, terrestrial or semiaquatic, often rather succulent; roots scattered along basal portion of stem, both slender and tuberous, fleshy; tuberoids usually spheroid; stems leafy. **Leaves** basal, abruptly reduced to bracts or scattered, gradually reduced toward inflorescence, several, alternate, ascending to spreading; lamina conduplicate, bases sheathing stem. **Inflorescences** few- to many-flowered, terminal, lax to rather dense spicate racemes; bracts large and foliaceous to reduced; flowers pedicellate. **Flowers** resupinate, sometimes showy; petals bilobed, lateral lobe on abaxial margin; lip trilobed, spurred at base; pollinaria 2, pollinia 2; viscidia free; stigma with 2 fleshy processes flanking or encircling mouth of spur. **Fruits** capsules, ellipsoid. **Distr:** pantropical. **Cult:** Propagate from seeds or by division of dormant tubers. Requires a well-drained mixture of peat moss, fine bark and sand. Reduce watering during dormant period.

H. rhodocheila Hance, *Ann. Sci. Nat.*, Bot., sér. 5, 5: 243, 1866. (Syn: *Smithanthe rhodocheila*). [Red-lipped Habeneria]. **Herbs**, terrestrial, 8–35 cm tall; tubers oblong-cylindric, 2–4 × 1–2 cm, fleshy; stem erect, with 4–6 leaves below middle and 1–3 bractlike leaflets above leaves. **Leaves** linear-lanceolate to suboblong, 10–15 × 1.5–2 cm, base amplexicaul, apex acuminate. **Inflorescence** a raceme, loosely 2–10-flowered, 3–8 cm long; floral bracts ovate-lanceolate, 1.5–1.7 cm, shorter than ovary, apex acuminate; ovary twisted, terete, including pedicel 2–3 cm. **Flowers** with green sepals and petals and orange-yellow to red lip; dorsal sepal forming a hood with petals, erect, suborbicular, concave, ~9 × 8 mm, trinerved, apex obtuse; lateral sepals reflexed, oblong, 0.9–1 × ~5 mm, obtuse, 5-veined; petals spathulate-linear, ~8 × 2 mm, obtuse, 1-veined; lip spreading, ovate, 1.8–2 × ~1.5 cm, 4-lobed, base shortly clawed, spurred, lateral lobes spreading, oblong,

Below: *Disa uniflora*, *Habenaria rhodocheila*, *Satyrium coriifolium*.

PMER

Above: *Satyrium odorum, Anoectochilus chapaensis, Anoectochilus elwesii, Goodyera clausa.*

~7 x 5 mm, obtuse, midlobe bilobed, lobules nearly subovate, ~4 x 3 mm, apex obliquely truncate; spur pendent, dull yellow, cylindric, 2–3 cm x ~1 mm, apex often upcurved. **Fruits** fusiform, ~1.5 cm long, beaked. **Distr:** S China to peninsular Malaysia, Philippines. **Ecol:** In forests; 300–1,500 m. Fl. Jul–Aug. Fr. Oct–Nov. **Zones:** 10–12.

Satyrium

Syn: *Aviceps, Diplectrum, Hipporkis*, etc.
Species 83. **Herbs**, terrestrial, slender or robust, with sessile, undivided tuberoids; tuberoids globose to ovoid. **Leaves** *basal* or *cauline*, or on separate sterile shoots. **Inflorescence** few- to many-flowered; bracts ± herbaceous, often recurved. **Flowers** not resupinate, variously colored; tepals subsimilar, free or partly united at base; lip erect, hooded, 2-spurred or bisaccate at base; column slender, covered by the lip; anther basally firmly united with the column, reversed, so that base is upper most; pollinia 2, sectile, with caudicles and separate, naked viscidia; stigma, as a result of the rotation of the anther, on top of the latter, flat or hooded; rostellum between stigma and anther-cells, trilobed; ovary not twisted. **Distr:** Tropical and S Africa, W Indian Ocean, Indian Subcontinent to SC China. **Cult:** Propagate from seeds. Requires a well-drained terrestrial mixture consisting of coarse sand organic matter, which should be kept moist during growth period, with plenty of light.

S. coriifolium Sw., *Kongl. Vetensk. Acad. Nya Handl.* 21: 216, 1800. (Syn: *Diplectrum coriifolium, S. erectum*, etc.). **Herbs**, terrestrial, tuberous, robust, 20–75 cm tall; stem stout, erect, leafy, smooth, spotted throughout with brownish purple spots. **Leaves** several, base sheathing stem, margins rough and crenulate with numerous small cartilaginous teeth, apex with a small horny gland or blunt mucro, thick leathery; *basal leaves* ovate, to 15 cm long, base connate; *cauline leaves* narrower. **Inflorescence** a spike, densely many-flowered; bracts sessile, oblong, acute, concave, reflexed, erect initially. **Flowers** not resupinate, bright orange or bright yellow, segments joined at base, linear; sepals 1.3 cm long; lateral sepals spreading, 1.3 cm long; lip uppermost, bent forward, inflated, with a 2 mm long flap over the mouth, spurs 0.9–1.2 cm long, lying along ovary. **Distr:** South Africa (SW and S Cape Province). **Ecol:** In seasonally dry areas, on sandy soil or clay; 0–750 m. **Zones:** 8–11.

S. odorum Sond., *Linnaea* 19: 86, 1846. **Herbs**, terrestrial, tuberous, perennial, to 55 cm long. **Leaves** 2–7 (usually 4–5 basal and 1–2 cauline), spreading, gradual to abrupt transition from basal to sheathing (cauline) leaves; lamina oval to narrowly ovate, 10–15 x 6–8 cm, obtuse to acute, fleshy. **Inflorescence** a spike, densely 2–45-flowered, cylindrical, 50 cm tall; floral bracts relatively large, lanceolate, acute. **Flowers** not resupinate, yellowish-green, sometimes with a purple tinge, highly fragrant at night (clove to lemon scented); sepals obovate-oblong, obtuse; petals ligulate, obtuse; lip saclike with a narrow entrance and with slender spurs 1.3–1.8 cm long that are longer than the ovary. **Distr:** S Africa (SW Cape Province). **Ecol:** On rocky slopes in scrubs, on sandy soils and rock shelters; ~30 m. Fl. Aug–Nov. **Zones:** 8–11. **Info:** Pollinated by sunbirds in the wild.

TRIBE CRANICHIDEAE

Anoectochilus

Species ~25. **Herbs**; terrestrial; stem fleshy, creeping below, erect and leafy above. **Leaves** alternate; lamina ovate, petiolate, base oblique, dark green, with yellowish white venation, velvety above; petiole, conduplicate, forming sheaths at base. **Inflorescence** terminal, spicate, few-flowered. **Flowers** resupinate or not; sepals free, dorsal one smaller and forming a hood with petals, lateral ones spreading or appressed to base of lip; petals narrow, falcate, apex acuminate; lip adnate to base of column, tripartite, hypochile forming a spur, spur exserted beyond base of lateral sepals, mesochile clawed, claw canaliculate, bearing entire or toothed or pectinate flanges, epichile dilated and bilobed, lobes winglike; column short, erect or bent, with 2 big appendages on ventral side; anther 2-celled, sessile; pollinia 2, sectile, bipartite, attached to elliptic stipe; stigma 2, on lateral sides of column, near base of rostellum; rostellum erect or recurved, shallowly bifid at apex. **Distr:** tropical Asia, Australia and Polynesia. **Cult:** Propagate by division of well-established clumps. Requires humid conditions with moderate to high lighting. The substrate or free-draining soil mix needs to be kept moist after watering.

A. chapaensis Gagnepain, *Bull. Mus. Natl. Hist. Nat.*, sér. 2, 3: 679, 1931. [Jewel Orchid]. **Herbs**, 12–18 cm tall. **Leaves** 4 or 5; lamina ovate, oblique, 2–5 x 1.5–3.5 cm, apex acute, adaxially blackish-green, venation golden-red, abaxially pale green; pseudo-petiole 0.8–1.6 cm long. **Inflorescence** 2–7-flowered; peduncle 8–11 cm long; sterile bracts 2 or 3, sheathing, reddish; floral bracts reddish, ovate-lanceolate, 0.6–1 cm; pedicel and ovary narrowly cylindric, 1–1.3 cm long, pubescent. **Flowers** erect, not resupinate, white; sepals and petals ~6 x 3 mm, 1-veined; sepals pubescent outside; dorsal sepal ovate, cymbiform, acuminate; lateral sepals ovate-oblong, obtuse; petals ovate, obliquely falcate, caudate, incurved; lip white, Y-shaped, 1–1.2 cm; hypochile short; mesochile 5–6 x ~3 mm; flanges oblong, pectinate, divided into 3 or 4 narrow teeth; epichile longitudinally dilated, bilobed; lobes diverging widely, obovate, 8–9 x 6–7 mm, margin crenulate, apex truncate; spur conic, 4–5 mm long, obtuse, containing 2 calli; column short, with 2 broad lamellate wings. **Distr:** China (SE Yunnan), N Vietnam. **Ecol:** In humid parts of dense forests; 1,300–1,400 m. Fl. Jul–Aug. **Zones:** 6–11.

A. elwesii (C.B. Clarke ex Hook. f.) King and Pantl., *Ann. Roy. Bot. Gard.* (Calcutta) 8: 297, 1898. (Syn: *Cystopus elwesii*, etc.). [Jewel Orchid]. **Herbs**, terrestrial, rhizomatous, chlorophyllous, 15–25 cm tall; stem ascending or erect; leaf sheath tubular 0.5–2 cm long. **Leaves** 6–7, ovate to ovate-lanceolate, 1.5–5 x 1–3 cm, base petiolelike, apex acute adaxially deep green to dark greenish purple, sometimes with 3 reddish veins, abaxially reddish or greenish. **Inflorescence** laxly 2–4-flowered, pubescent; peduncle 8–11 cm long; sterile bracts 1–3; floral bracts ovate, ~5 mm long; ovary and pedicel twisted, cylindric-fusiform, ~1.3 cm long, pubescent. **Flowers** resupinate, sepals green or white, apex tinged purplish-red, pubescent; dorsal sepal forming a hood with petals, with 2 broad purplish red stripes, ovate, cymbiform, ~ 7 x 4 mm, apex acuminate; lateral sepals ovate, oblique, ~ 1 x 0.6 cm, base saccate; petals white, ovate, strongly oblique, falcate, ~ 7 x 4 mm, 1-veined, apex acuminate; lip white, Y-shaped, ~1.5 cm long, hypochile bisaccate, ~3.5 mm long, mesochile dark purple, 5–7 mm long, epichile dilated, ~ 1.2 x 1.4 cm; column ~4 mm long, with 2 lamellate wings. **Distr:** E Himalayas to Taiwan. **Ecol:** In evergreen and bamboo forests; 300–1500 m. Fl. Jul–Aug. **Zones:** 8–11.

Goodyera

[Lattice Leaf, Rattlesnake Plantain].
Species 105. **Herbs**, terrestrial, rhizomatous, scapose, glabrous except for rather sticky, multicellular hairs on peduncles, bracts, sepals, and ovaries; roots arising from nodes of rhizome, fibrous; stems erect, with rosette of leaves, not succulent. **Leaves** evergreen, >1, in basal rosette, petiolate; lamina commonly marked with white to pale green. **Inflorescences** terminal, 5–72-flowered spikes, erect; peduncles with sheathing bracts. **Flowers** resupinate, white, sometimes tinged green, ivory, or brown, sessile; sepals distinct, nearly equal; dorsal sepal and petals forming hood; lip free from column, fleshy, base concave to saccate, apex ligulate or pointed; anther 1, erect or inflexed; pollinia 2, sectile; rostellum notched or 2-pronged. **Fruits** capsules, erect, dehiscing along 3 ribs. **Distr:** Nearly worldwide, primarily SE Asia. **Cult:** Propagate by division of rhizomes. Requires a moist but well-drained mix of loam and leaf mould with leaf matter on the surface, in shaded and humid conditions.

G. clausa (A.A. Eaton ex Ames) Schltr., *Bot. Jahrb. Syst.* 45: 392, 1911. (Syn: *Epipactis clausa*). **Herbs**, terrestrial, rhizomatous, sympodial; stem upright, to 20 cm tall, succulent. **Leaves** 3, 2–4 cm apart; lamina ovate-lanceolate, to 3.5 x 1.3 cm, dark green with white or yellow veins. **Inflorescence** erect, ~4 cm long, hairy, to 20-flowered. **Flowers** white; dorsal sepal ovate to ovate-lanceolate broadly ovate to ovate-lanceolate, ~4 mm long, tip strongly reflexed; lateral sepals broadly ovate to ovate-lanceolate, 4 mm long; petals not symmetrical, dolabriform, 4 mm long, obtuse, widened into a round or axe-shaped plate above the claw; lip saccate at base, very broadly ovate, 2.5-3 mm long, somewhat cymbiform, obtuse, with unequal clavate hairs or emergences inside. **Distr:** Endemic to Philippines (Luzon, Negros). **Ecol:** Terrestrial in mossy forests, in damp, shady localities; ~2,200 m. **Zones:** 11–12.

Ludisia

Syn: *Dicrophyla, Haemaria, Myoda.*
Species 1. *Generic description same as specific.* **Cult:** Propagate by division of rhizomes (simply by breaking them). Requires a moist but well-drained mix of loam and leaf mould with litter on the surface, in shaded and humid conditions. **Info:** The single species of this genus is highly variable across its range.

*L. **discolor*** (Ker Gawl.) A.Rich., *Dict. Class. Hist. Nat.* 7: 437, 1825. (Syn: *Haemaria discolor*, etc.). [Jewel Orchid]. **Herbs**, terrestrial or lithophytic, 10–25 cm tall; stem erect or prostrate, with (2 or)3–5 leaves near base. **Leaves** ovate, oblong-lanceolate to elliptic, 3–9 × 1.7–3 cm, base narrowed into a pseudo-petiole, apex acute or mucronate, fleshy, adaxially blackish green with 5 silvery-red veins, abaxially pale red; pseudo-petiole 1.5–2.2 cm long. **Inflorescence** pubescent, with 2 or 3 sterile bracts, laxly few–10-flowered, 15–20 cm long; rachis 3–8 cm long; floral bracts reddish, ovate to ovate-lanceolate, ~1.5 cm long, membranous; pedicel and ovary 1.5–2 cm long, pubescent. **Flowers** asymmetrically twisted, white, red-tinged, ~7 mm Ø; dorsal sepal and petals forming a hood; dorsal sepal ovate-elliptic, cymbiform, 8–9 × 4.5–5 mm; lateral sepals obliquely ovate to subelliptic, 0.9–1 × 0.45–0.5 cm, abaxially with a very short keel toward apex, reflexed; petals obliquely ovate, 8–9 × 2–2.2 mm, obtuse; lip 0.9–1 cm long, base saccate, apex bilobed; hypochile containing 2 large fleshy calli; mesochile ~2 mm; epichile 5–6 mm wide; column ~5 mm long. **Distr:** S China to Sumatra. **Ecol:** In evergreen broad-leaved forests; 500–1,300 m. Fl. Feb–Apr. **Zones:** 11–12.

Macodes

Syn: *Argyrorchis, Pseudomacodes.* [Jewel Orchids].
Species 10. **Herbs**, terrestrial, evergreen; stems erect, fleshy; rhizome short, fleshy, creeping. **Leaves** several, rather fleshy, usually in loose rosettes, often decoratively marked with white or colored veins. **Inflorescence** erect, terminal, glandular, pubescent, racemose, few- to many-flowered; bracts small, often colored pink or brown. **Flowers** rather small, not resupinate, pubescent, not showy, asymmetric due to twisting of lip and column; dorsal sepal joined to petals; lateral sepals free, enclosing base of lip; petals much narrower than sepals; lip trilobed, with base saccate containing 2 small glands, lateral lobes short, rounded, midlobe narrow at base, spreading above; column short, twisted, with 2 thin parallel wings in front; pollinia 2. **Distr:** Vietnam and Malaysia to Pacific Islands. **Cult:** Propagate by division. Requires moist but well-drained mix of loam, sand and leaf litter, in shaded and humid conditions throughout the year.

*M. **petola*** (Blume) Lindl., *Gen. Sp. Orchid. Pl.*: 497, 1840. [Jewel Orchid]. **Herbs**, terrestrial, rhizomatous. **Leaves** few or several, mostly basal and close together; lamina ovate, to 6.5 × 4.2 cm, base narrowed into a petiole, apex shortly pointed, adaxially very dark green with 5 longitudinal, yellowish white veins and small similar colored cross veins; petiole 1–2.5 cm long. **Inflorescence** erect, to 25 cm long, to 15-flowered; peduncle pubescent, bearing 2 sterile bracts; fertile bracts shorter than ovary, to 8 mm long. **Flowers** reddish brown with a white lip; dorsal sepal ovate, ~5–6 cm, subacute; lateral sepals spreading, ovate-orbicular, 6 × 4 mm, obtuse; petals narrowly lanceolate, 3 mm long, acuminate; lip uppermost, asymmetric, trilobed, 5× 3 mm, base saccate with 2 glands, midlobe shortly narrowly clawed, transversely oblong above; column short, twisted, with 2 parallel wings in front, descending downward into the spur; rostellum cleft. **Distr:** Sumatra to Philippines. **Ecol:** In damp and lower montane forests, usually under shade; 100–1,500 m. **Zones:** 11–12. **Info:** The species is variable across its native range.

Pterostylis

[Greenhoods].
Species 179. **Herbs**, terrestrial with subterranean tubers, medium-sized, solitary or in colonies. **Leaves** usually in a basal rosette or borne along the stem, rosette either at base of flowering shoot or on a separate, sterile shoot. **Inflorescence** 1–few-flowered, erect. **Flowers** erect to nutant, somewhat tubular, usually hooded; dorsal sepal hooded, forming a galea with the adnate petals; lateral sepals fused basally, erect with short to long apical tails; lip ± hidden in flower, linear to ovate, ± entire, hinged to the column foot by an elastic strap, often with a penicellate appendage at base of the lip; column slender, with 2 wings at apex; pollinia 4, mealy. **Distr:** Predominantly in Australia and New Zealand, few in New Caledonia, Vanuatu and New Guinea. **Cult:** Propagate from seeds or by removal of extra tubers. Requires a terrestrial mix with regular watering in humid and cool conditions. Repot every 2 years to avoid overcrowding.

*P. **cardiostigma*** D. Cooper, *New Zealand J. Bot.* 21: 97, 1983. **Herbs**, terrestrial, 6–35(–40) cm tall; tuber heart-shaped. **Leaves** 5–7, cauline, sessile; lamina linear-lanceolate, 8–23 × 1–2 cm, apex acuminate, midrib red, adaxially slightly grooved, with 2(3) lateral yellow veins on either side, keel prominent, often red. **Inflorescence** erect, with a solitary flower. **Flower** green, erect, tall and narrow; dorsal sepal to 7 cm long, vertical in lower half, upper half steeply inclined or horizontal, tip acuminate, red; lateral sepals diverging, with long filiform red caudae overtopping galea by 2 cm; petals 0.5–1 cm shorter than dorsal sepal, long marginal strip exposed, tips red; galea width at petal tip 6 mm; lip long triangular, arched in upper third, tip acuminate, distal half red with retrorse pale hairs on upper surface and a darker red dorsal median ridge becoming paler toward base, lower surface with red median groove; column shorter than lip; ovary erect, 1–3 cm high. **Distr:** New Zealand (S North Island). **Ecol:** In mixed bush. Fl. Oct–Nov. **Zones:** 9–10.

*P. **curta*** R. Br., *Prodr.*: 326, 1810. (Syn: *P. neocaledonica*, etc.). [Blunt Greenhood Orchid]. **Herbs**, terrestrial, tuberous, forming colonies. **Leaves** 2–6, in a basal rosette; lamina ovate to oblong, 5–10 × 2–3 cm, margins crisped; petiole distinct. **Inflorescence** 1-flowered, erect, 10–30 cm tall. **Flowers** 2–3.5 cm long, whitish striped with green and flushed with green and brown; dorsal sepal concave, apex acute, curving forward; lateral sepals united in basal half, free parts 1.2–2.5 cm long, lying beside the hood but not overtopping it; petals as long as the dorsal sepal, obtuse; lip brown, shortly clawed, linear-ligulate, to 1.5–2 × 0.4–0.5 cm, apex obtuse and twisted, with a central fleshy ridge; column erect, wings with upper lobes short, erect, subulate. **Distr:** New Caledonia, E and SE Australia. **Ecol:** In open forests, especially near streams, forming dense colonies. **Zones:** 8–10.

*P. **montana*** Hatch, *Trans. and Proc. Roy. Soc. New Zealand* 77: 239, 1949. (Syn: *P. furcata* var. *linearis*). **Herbs**, terrestrial, tuberous, ~3–30(–35) cm tall (at flowering); stem erect, smooth, internodes much less than leaves. **Leaves** to 6, size variable from base upward, largest leaf on a stem to ~ 14(–18) × 1 cm, usually much smaller; lamina linear-lanceolate, apex acuminate, widest near sessile sheathing base and, except on smallest plants, ± overtopping flowers. **Inflorescence** 1-flowered. **Flower** with dorsal sepal ~2–3 cm long, apex ± horizontal, rather abruptly acute to shortly acuminate; lateral sepals diverging from a narrow angle, acuminate, separating from one another in young bud, not greatly exceeding galea in mature flowers; petals slightly shorter than dorsal sepal, broad almost to tip; lip oblong, slightly arched, broad almost to apex, apex subacute and often twisted to one side (usually to right), margins deflexed in about the prominent almost excurrent midrib; column about as tall as the lip; ovary usually erect, stigma short, broad, oval to cordate, inclined to be prominent. **Distr:** New Zealand. **Ecol:** In grasslands, scrubs, open forests and forest margins; to ~400 m. Fl. Sep–Dec. **Zones:** 8–10.

*P. **patens*** Colenso, *Trans. and Proc. New Zealand Inst.* 18: 270, 1886. (Syn: *P. banksii* var. *patens*). **Herbs**, terrestrial, 10–30(–50) cm tall. **Leaves** mostly

Below: *Ludisia discolor, Macodes petola, Pterostylis cardiostigma, Pterostylis curta.*

GS

4–6, becoming smaller upward; lamina linear-lanceolate, to ~ 5–25 × 1–1.5(–2) cm, ± keeled, apex acuminate to long-acuminate, usually overtopping flowers. **Inflorescence** with a solitary flower; pedicel and ovary usually erect. **Flower** with dorsal sepal 3–4(–5) cm tall, distal part usually ± horizontal, tip caudate and sometimes >2 cm long; lateral sepals diverging at a narrow angle, tips long-caudate, much overtopping galea, variously curved; petals usually much shorter than dorsal sepal, acuminate; lip elliptic-oblong, slightly arched, ± flat in transverse section, narrowing slightly toward apex, bent forward smoothly and symmetrically between lateral sepals, midrib only moderately prominent and becoming evanescent in the obtuse and faintly emarginate tip, margins narrowly rolled under; column about as tall as lip. **Distr:** New Zealand (North Island). **Ecol:** In forests and forest margins. Fl. Oct–Dec. **Zones:** 9–10.

P. pedunculata R. Br., *Prodr.*: 327, 1810. (Syn: *P. semirubra*). [Maroonhood Orchid]. **Herbs**, terrestrial, tuberous. **Leaves** rosetted, up to 6, prominently petioled; lamina ovate, to 6.5 × 1.8 cm. **Inflorescence** 1-flowered, slender, to 25 cm long. **Flower** translucent white, striped and suffused with pale to deep green, lateral sepals and galea apex brownish; galea 1.8 cm high, base strongly inflated, erect then abruptly curving forward, apex horizontal or pointing somewhat downward; dorsal sepal with the acute apex exceeding petals by 3 mm; lateral sepals held away from galea, sinus flat or weakly angled from side, V-shaped from front, free tips 1.5 cm long, loosely embracing galea, erect, filamentous, divergent; petals with apices acute; lip ovate, 5 × 2.5 mm, straight, tip just visible above sinus in set position. **Distr:** SE Australia, Tasmania. **Ecol:** On mountains and tablelands, in moist areas and scrubs. Fl. Jul–Oct. **Zones:** 8–10.

Sarcoglottis

Syn: *Narica*, *Synoplectris*, *Zhukowskia*.
Species 45. **Herbs**, terrestrial, glabrous or often pubescent; roots fleshy, tuberous. **Leaves** rosulate, often absent at flowering, ± petiolate; lamina often variegated. **Inflorescence** spicate, erect, laxly few- to several-flowered. **Flowers** small to medium-sized, rarely showy, often glandular pubescent on outer surface, mostly green, white or yellow, ± erect, not second; dorsal sepal and petals connivent, forming a hood over the column; lateral sepals deflexed, often long-decurrent onto the ovary at base; lip sessile or distinctly clawed, often narrow, concave, enclosing the column below, lacking calli or lamellate at base; column terete, ± decurrent at base; anther sessile or stipitate, ovate or oblong; pollinia powdery-granular, produced into a short caudicle; rostellum bidentate. **Distr:** Mexico to tropical America. **Cult:** Propagate from seeds or by division. Requires a moist terrestrial mix in shaded and humid conditions.

S. grandiflora (Hook.) Klotzsch, *Allg. Gartenzeitung* 10: 107, 1842. (Syn: *Dothilis grandiflora*, etc.). **Herbs**, terrestrial, tuberous, 54–78 cm tall. **Leaves** rosulate; lamina green, uniformly colored or green maculate, spathulate, channeled. **Inflorescence** 52–75 cm tall, terminal, simple, 14-flowered; peduncle sparsely hairy, 32–43 cm long; floral bracts green, lanceolate, 2.5–4 cm long, slightly hirsute; flowers shortly pedicellate. **Flowers** green, hirsute; dorsal sepal lanceolate, 2–2.3 × 0.3–0.5 cm, green, free, apex acute; lateral sepals green, free, lanceolate, acute apex, hirsute on the abaxial surface; petals hyaline, clawed, 1.5–2.2 × 0.2–0.3 cm, slightly greenish, with dark green lines; lip rhombic, 2.5–3.3 × 0.5–1 cm, cream-green with brown stripes; column cream-green 1.3–1.5 cm long. **Distr:** S tropical America to N Argentina. **Ecol:** Usually in shady places; ~300 m. Fl. & Fr. Jun. **Zones:** 8–11.

Spiranthes

Species 33. **Herbs**, perennial, terrestrial; roots fleshy, fasciculate, slender to tuberous; stems with foliaceous sheaths. **Leaves** basal or occasionally cauline. **Inflorescences** terminal spikes, flowers inserted in lax to dense, ± evident spiral, pubescent to glabrous, often glandular. **Flowers** resupinate, often somewhat apically recurved, white, cream, or yellow; perianth parts distinct or sepals basally connate, connivent; lip lanceolate to ovate or pandurate, rarely linear, fleshy to somewhat membranous, base with pair of calli, apex usually ± crisped; mentum or spur absent; column short, cylindric; anther cordate, apex acute or obtuse; pollinia clavate with slender viscidium; ovary sessile, cylindric. **Distr:** Eurasia to SW Pacific, N Africa, N and C America to Caribbean. **Cult:** Propagate by division. Requires a well-watered mix of loam and leaf mould in cool and well-lit conditions.

S. sinensis (Persoon) Ames, *Orchidaceae* 2: 53, 1908. (Syn: *Neottia sinensis*, etc.). [Ladies' Tresses, Pearl-twist]. **Herbs**, terrestrial, tuberous, 13–30 cm tall. **Leaves** 2–5, erect and spreading; lamina broadly linear to broadly linear-lanceolate, rarely narrowly oblong, 3–10 × 0.5–1 cm, base indistinct petiolelike, apex acute or acuminate. **Inflorescence** erect, 10–25 cm, glabrous, many-flowered, flowers spirally arranged; rachis 4–10 cm; floral bracts ovate-lanceolate, glabrous, apex long-acuminate; pedicel and ovary pale green, 4–5 mm. **Flowers** purplish red, pink, or white; dorsal sepal forming a hood with petals, narrowly oblong, cymbiform, ~ 4 × 1.5 mm, glabrous, apex subacute; lateral sepals lanceolate, slightly oblique, ~ 5 × 2 mm, glabrous, base slightly gibbose, subacute; petals rhombic-oblong, oblique, thin, obtuse; lip broadly oblong, 4–5.5 × ~2.5 mm, shortly clawed, base concave with 2 clavate glands, lateral margins erect, apical margin with strongly wrinkled teeth, apex obtuse and recurved; column erect, ~2 mm. **Distr:** E European Russia to SW Pacific. **Ecol:** In forests, thickets, wet grasslands; 200–3,400 m. Fl. Jul–Aug. **Zones:** 9–11.

TRIBE DIURIDEAE

Corybas

Syn: *Anzybas*, *Corysanthes*, *Singularybas*, etc.
Species 131. **Herbs**, terrestrial or rarely epiphytic, small; tuber solitary, globose to ovoid; stem slender, erect. **Leaf** solitary (rarely 2), sessile or base short and petiolelike, often appressed to ground; lamina cordate or broadly ovate, entire or rarely 3- or 5-lobed, with 1–3 main veins, venation white or red. **Inflorescence** with a terminal solitary flower (rarely flowers paired), nearly sessile, resupinate, held ± erect at leaf base; floral bracts 1 or 2. **Flower** with dorsal sepal hoodlike and curved over lip, basal margin incurved and forming a tube that embraces base of lip; lateral sepals and petals linear, narrow, free or connate at base; lip deeply channeled at base and forming a decurved tube with dorsal sepal, upper limb enlarged, spreading or reflexed, adaxially usually minutely papillose or hairy, with 2 spurs or auricles (rarely neither) at base; column short; anther terminal, erect, bilocular; pollinia 4 or 2 with cleft, granular-farinaceous, without caudicles, directly attached to sticky viscidium; rostellum dorsal; stigma entire, concave; ovary usually 6-ridged. **Fruit** strongly 6-ridged, borne on an elongated pedicel. **Distr:** Tropical and subtropical Asia to Pacific, subantarctic islands. **Cult:** Propagate by division of tubers. Requires a terrestrial mix with sandy loam and leaf compost (kept dry during active growth), in humid and cool conditions.

C. oblongus (Hook. f.) Rchb. f., *Beitr. Syst. Pflanzenk.*: 67, 1871. (Syn: *Corysanthes oblonga*, etc.). [Spider Orchid, Helmet Orchid]. **Herbs**, terrestrial, ~2–3 cm tall at flowering; stem ± erect. **Leaf** sessile; lamina ovate-oblong, ~1–4 cm long, base rounded or cordate, apex apiculate, green, membranous, ± veined with red or purplish red. **Inflorescence** with a solitary flower; floral bract about equaling ovary, foliaceous. **Flower** with dorsal sepal slightly longer than the lip, spathulate, apex broad, obtuse to apiculate; lateral sepals long, filiform, greatly exceeding lip; petals similar to lateral sepals but somewhat shorter; lip ~0.5–1 cm long, horizontal or sometimes tilted upward, auriculate at base, curved to form a cylinder with a wide circular mouth fringed with coarse teeth, short retrorse hairlike calli numerous, near midline on inner surface. **Distr:** New Zealand, Chatham Islands, Antipodes Islands. **Ecol:** On dry sunlit banks to damp shady places. Fl. Oct–Feb. **Zones:** 9–10.

Cryptostylis

Syn: *Chlorosa*, *Zosterostylis*.
Species 23. **Herbs**, terrestrial; rhizome horizontal or ascending, short, densely noded, producing aerial shoots from apical nodes. **Leaves** basal, elliptic to ovate-lanceolate, base long attenuate into erect petiolelike stalks, with or without loosely sheathing cataphylls at base. **Inflorescence** erect or ascending,

Below: *Pterostylis montana*, *Pterostylis patens*, *Pterostylis pedunculata*, *Sarcoglottis grandiflora*.

MG

Above: *Corybas oblongus, Cryptostylis taiwaniana, Microtis unifolia, Orthoceras strictum.*

terminal, racemose, several flowered, with sheathing cataphylls at base and scattered sterile bracts along peduncle; floral bracts sheathing; pedicel and ovary erect, arcuate. **Flowers** not resupinate; sepals and petals subsimilar, free, spreading, very narrow, margin often involute; petals often slightly shorter and smaller than sepals; lip attached to base of column, entire, tapering toward apex, basally expanded and embracing column, spurless; disk usually with longitudinal ridges or elongate calli; column short, with lateral wings; anther terminal, erect, 4-locular; pollinia 4, in 2 pairs, clavate, granular-farinaceous, attached to solitary viscidium; stigma entire, convex, fleshy; rostellum erect, broad and stout. **Distr:** Tropical and subtropical Asia to SW Pacific. **Cult:** Propagate by division. Requires a moist but well-drained, terrestrial mix, in humid and shaded conditions.

C. taiwaniana Masam., *Trans. Nat. Hist. Soc. Taiwan* 23: 208, 1933. (Syn: *C. arachnites* var. *philippinensis*, etc.). [Taiwanese Tongue Orchid]. **Herbs**, terrestrial, 15–25 cm tall. **Leaves** 1–3, ovate to ovate-oblong, 5–15 × 3–6 cm, base obtuse or rounded, apex acute, adaxially green with dark green blotches, abaxially pale green; petiolelike base with dark green or purplish-brown spots, 7–12 cm long. **Inflorescence** green with dark green or purplish-brown spots, to 25 cm, glabrous, with 2–5 sterile bracts; rachis 5–15 cm, 5–15-flowered; floral bracts green with purplish-brown spots, ovate-lanceolate, ~ 1.5 × 0.5 mm; ovary and pedicel 0.8–1 cm, glabrous. **Flowers** with sepals green, lanceolate, 0.8–1.5 cm × 2–3 mm, 3-veined, margin involute, acute; petals similar to sepals, green, 0.7–1.2 × ~0.2 cm, 1-veined, margin involute, apex acute; lip adaxially orange-purple, with scattered dark crimson spots, rhombic-oblong to obovate, 1.5–2 × 0.8–1.2 cm, 5-veined, base rounded, shallowly saccate, lateral margins slightly revolute, apex acute; column 2.3–2.5 mm long; wings rugose. **Distr:** Hong Kong, S Thailand to Philippines. **Ecol:** In broad-leaved forests; 100–500 m. Fl. Apr–Aug. **Zones:** 10–12.

Microtis

Species ~10. **Herbs**, small, terrestrial; tubers (tuberoids) globose to oval, often produced some distance from parent plant; stem erect, slender. **Leaf** usually solitary, elongated, terete-cylindrical, with closed sheath around stem in lower part, not articulate. **Inflorescence** terminal, spicate, with many small spirally arranged flowers; bracts minute. **Flowers** greenish to yellowish green; dorsal sepal broad, concave; lateral sepals spreading or recurved; petals usually narrower, curving or spreading; lip connate to base of column, sessile, ligulate, usually callous at base, spurless; column short, semiterete, with small auricle on both sides; anther dorsal, nearly erect, basally adnate to column, 2-celled, connective obscure, locules close, parallel or slightly divergent below; pollinia 2, sectile, with short stipe and basal viscidium; stigma entire, with rostellum shorter than anther. **Distr:** Mainly in Australia; a few species in Malaysia and E Asia. **Cult:** Propagate from seeds or by division. Requires a well-drianed terrestrial mix with leaf compost, in humid to dry, and well-lit conditions.

M. unifolia (Forst.) Rchb. f., *Beitr. Syst. Pflk.* 62, 1871. (Syn: *M. formosana*, etc.). [Onion Orchid]. **Herbs**, 10–30 cm tall; tuberoids underground, nearly globose, ~1 cm across; stems ~2 mm Ø. **Leaf** solitary, tubular-cylindrical, 10–20 cm long, 3 mm Ø. **Inflorescence** spike, terminal; peduncle 10–20 cm long; rachis 8–10 cm long, with more than 10 flowers ± loosely arranged; bracts lanceolate, 2 mm long, acuminate; pedicel and ovary 3–4 mm long. **Flowers** pale green, 2–3 mm Ø; dorsal sepal orbicular-ovate, 2.5 × 2 mm; lateral sepals elliptic or oblong, 2.5 × 1 mm, obtuse, recurved; petals elongate oblong, 2 × 0.6 mm, obtuse; lip ligulate, 2.5 × 1 mm, base truncate, with 2 globose calli above, margins ± wavy, apex bent downward; column 1 mm tall, auricles fingerlike. **Distr:** Japan and E Asia to Australia and New Zealand. **Ecol:** In moist sunny grasslands. **Zones:** 8–11.

Orthoceras

[Horned Orchids].

Species 2. **Herbs**, terrestrial, glabrous; tuber long-ovoid, usually produced close to the previous one. **Leaves** several, sheathing at base, narrow, followed by foliaceous cauline bracts. **Inflorescence** a raceme, erect, few- to many-flowered; floral bracts heavily textured and often exceeding ovary at flowering. **Flowers** with perianth glabrous; dorsal sepal uppermost, broad and arched over column; lateral sepals much longer, very narrow-linear, channeled; petals short, narrow and ± hidden under dorsal sepal; lip almost as large as dorsal sepal, sessile, trilobed, callus thick, median; column short, wings in form of 2 free lateral processes; anther dorsal, taller than stigma, pollinia ill-defined, pollen granular; stigma disk-shaped; rostellum prominent, acute. **Distr:** SE Australia to New Zealand and New Caledonia. **Cult:** Propagate from seeds or by division. Requires a well-drained terrestrial mix, in humid to dry, and well-lit conditions.

O. strictum R. Br., *Prodr.*: 317, 1810. [Bird's Mouth Orchid, Horned Orchid]. **Herbs**, terrestrial, to 70 cm tall (at flowering); tuber to 5 × 1 cm; stem stiffly erect. **Leaves** 2–5, tufted at base; lamina linear-lanceolate to linear, channeled, apex acute, shorter than stem. **Inflorescence** a raceme, erect, to ~20 cm tall, 2–12-flowered, well separated; axis often flexuous. **Flowers** with perianth green to dull reddish, often with varying amounts of the 2 colors; dorsal sepal 1–1.5 cm long and almost as wide, deeply concave, subacute; laterals to twice as long, ~0.5 mm Ø, semiterete, erect or diverging; petals narrowly oblong, flat, usually notched at tip; lip spreading or deflexed, of firm texture with smooth margins, trilobed, lateral lobes broad, oblique, midlobe larger, ovate, median callus near base, shortly conical, thick tip turned inward, often yellow, 2 smaller lateral basal calli usually present; processes of column-wings linear, ± papillose. **Distr:** New Caledonia, SE Australia to New Zealand (North Island). **Ecol:** In woodlands, open forests, scrubs and banks. Fl. Nov–Jan. **Zones:** 9–12.

Subfam. EPIDENDROIDEAE

Genera 650, **species** ~18,000 species. **Herbs**, terrestrial, epiphytic, lithophytic or scrambling lianas; stems thick. **Leaves** usually distichous, conduplicate (plicate). **Flowers** with anther incumbent, operculate; pollinia hard (sectile), waxy, entire, 2, 4 or 8, free or adhering at one end by caudicles or attached to 1 or 2 viscidia; apical part of median stigma lobe forms stipe. **Distr:** More or less worldwide, with the greatest diversity found in the tropics and represented very poorly in Australia. **Info:** The largest genera in this subfamily are *Bulbophyllum* with over 2,035 species, followed by *Epidendrum* with 1,500 species and *Dendrobium* with 1,400 species.

TRIBE SOBRALIEAE

Sobralia

Species ~35+. **Herbs**, reedy, terrestrial or epiphytic, often tall; stem simple. **Leaves** distichous, sessile, coriaceous or subcoriaceous; lamina cauline, strongly nerved, articulate with long sheaths, rigid and ± plicate, varying from oblong to ovate-lanceolate or oval. **Inflorescences** terminal or axillary racemes or heads, sometimes lateral (either a loose raceme or a branching panicle); bracts several, carinate, rigid, somewhat imbricate. **Flowers** usually large showy; sepals subequal, narrow, erect, connate at base; petals similar to but slightly broader than the sepals; lip adnate to the base of the column, entire or 2(–3)-lobed, with the basal half involute enclosing the column, the apex exserted, spreading, undulate or fimbriate; disk smooth or variously crested with calli; column elongate, footless, subincurved, semiterete, acutely angled or narrowly winged, trilobed at the apex; pollinia 8 (4 in each cell of the bilocular anther), granulose. **Fruits** capsules, oblong or elongate, rigid or fleshy. **Distr:** Tropical regions of Mexico, C and S America. **Cult:** Propagate by division. Requires a well-drained terrestrial mix containing bark, leaf compost and sand, in partial shade and humid conditions. Needs increased feeding and watering during the growth period.

S. decora Bateman, *Orch. Mex. and Guat.*, t. 26, 1841. [Bamboo Orchid]. **Herbs**, erect, reedlike, 35–75 cm tall; stem, lower surface of leaves and

MS

Above: *Sobralia decora, Sobralia macrantha, Tropidia curculigoides.*

leaf-sheaths ± black warty or hispidulous, leafy above. **Leaves** oblong-lanceolate to narrowly lanceolate, long-acuminate, 5.5–22 × 1.5–6.5 cm. **Inflorescence** terminal, 1–2 flowers, sessile or subsessile, subtended by short, scarious sheaths. **Flowers** pale lavender, and white; sepals strongly apiculate, apex recurved, dorsal sepal linear-oblong, 4–5 × 1.2–1.7 cm, lateral sepals linear-oblong, slightly falcate, 4–5 × 1–1.2 cm; petals oblong-elliptic, subacute to acute, 3.5–4 × 1–1.1 cm; lip tubular-involute in natural position, base cucullate, undulate-crisped and flared on the anterior margin, retuse or apiculate at the broadly rounded apex, rose-purple or lavender streaked with yellow and brown on the disk, when spread out obovate-rhomboid, 4–4.5 × 3–3.5 cm (near apex); column white, slender-clavate, 2–2.5 cm long, trilobed at the apex, lateral lobes slender-falcate, midlobe galeate. **Fruits** cylindrical, blackish, 6-ribbed, ~8 cm long. **Distr:** Mexico to Honduras. **Ecol:** In oak-pine forests; to 1,700 m. **Zones:** 9–12.

S. macrantha Lindl., *Sert. Orch.*, sub t. 29. 1836. [Candelaria]. **Herbs**, 0.45–2.1 m tall; glabrous or tuberculate on the leaf sheaths; stem leafy, often growing in large clusters from the same root mass. **Leaves** with lamina narrowly to broadly lanceolate, long-acuminate, spreading, rigid, 13–30 × 2–7.5 cm. **Inflorescence** 1-flowered; bract foliaceous, narrowly lanceolate, carinate, membranous, margins scarious, to 13 cm long. **Flower** very large, rose-purple, showy; sepals linear-oblong, 8–10.2 × 1.5–2.6 cm, subacute or acute, minutely apiculate, recurved; petals oblong-obovate, broadly rounded, 6.5–9 × 2.3–4 cm, apex inconspicuously mucronate, margins undulate-crisped above middle, near apex; lip very large, lower half forming a laterally compressed tube around column, round, apex deeply bilobed, expanding and undulate-crisped along margins above middle, white on the inside of the tubular basal portion, tinged with yellow at center, broadly oblong-obovate, 8–11 × 7 cm; column 3–3.5 cm long, slender-clavate; ovary stout, arcuate, pedicellate ~2.5 cm long. **Fruit** elongate-fusiform, recurved, 6-angled, ~11.5 × 2 cm. **Distr:** Costa Rica, Guatemala, Mexico. **Ecol:** On slopes and rock outcrops in open pine woodlands, and on rocks or sandy soils by streams; 0–3,400 m. Fl. Apr–Dec. **Zones:** 9–12.

TRIBE TROPIDIEAE

Tropidia

Syn: *Cnemidia, Govindooia*, etc.

Species ~20. **Herbs** terrestrial, autotrophic (rarely mycotrophic); rhizome ascending, stout; roots wiry, fibrous. Stem erect, rigid, branched or unbranched, noded, with sheaths at nodes. **Leaves** 2 to many, plicate; lamina ovate to narrowly lanceolate, base contracted into amplexicaul sheaths. **Inflorescence** a short raceme or panicle, terminal or axillary, several- or many-flowered. **Flowers** resupinate or not; dorsal sepal free; lateral sepals free or connate and embracing base of lip; petals free, similar to sepals or slightly smaller; lip cymbiform, slightly shorter than sepals, entire, with or without a short spur, sometimes constricted at middle, base loosely embracing column, distally expanded, acuminate, reflexed; column short; anther dorsal, erect; pollinia 2. **Distr:** Tropical and Subtropical Asia to SW Pacific, Florida to tropical America. **Ecol:** Fl. Jun–Aug. Fr. Oct. **Cult:** Propagate by division or from seeds. Requires a well-drained terrestrial mix containing leaf compost, in shaded and humid conditions.

T. curculigoides Lindley, *Gen. Sp. Orchid. Pl.* 497, 1840. (Syn: *T. assamica*, etc.). **Herbs**, terrestrial, 30–70+ cm; rhizome short, rigid; stem erect, internodes 2–4 cm long. **Leaves** many, often >10, distichous and well-spaced along stem; lamina narrowly elliptic-lanceolate to narrowly lanceolate, (10–)15–25 × (1–)2–4 cm, base contracted into amplexicaul sheath, apex long acuminate-caudate, papery. **Inflorescence** racemose, terminal or axillary, 10-flowered, crowded, 1–2.5 cm long; floral bracts imbricate, lanceolate, 0.4–0.7(–1) cm, conspicuously with longitudinal veins; pedicel and ovary 5–6 mm long. **Flowers** non-resupinate, greenish white; sepals lanceolate or oblong-lanceolate, 0.7–1 cm × 1.5–1.7 mm, apex long-acuminate; lateral sepals basally connate, apex recurved; petals oblong-lanceolate, 6–8 × ~1.3 mm; lip ovate-lanceolate or oblong-lanceolate, cymbiform, 6–8 mm long, concave-saccate at base, with 2 inconspicuous lamellae on disk, apex acuminate; column ~3 mm; anther ovoid, ~1.5 mm; rostellum erect, obovate, ~2 mm, apex bilobed-apiculate. **Fruits** suboblong, ~2 × 0.5 cm. **Distr:** E Himalayas to New Caledonia. **Ecol:** In forests; 200–1,000 m. Fl. Jun–Aug. Fr. Oct. **Zones:** 8–12.

TRIBE DENDROBIEAE

Bulbophyllum

Syn: *Cirrhopetalum, Megaclinum.*

Species ~2,000. **Herbs**, epiphytic, perennial; rhizomes creeping, long or short and obscure between pseudobulbs; pseudobulbs remote or closely aggregated, sometimes totally reduced and leaves arising from rhizome. **Leaves** solitary, ovate, oblong or linear, usually fleshy, coriaceous. **Inflorescence** with a solitary flower, or flowers umbellate or racemose; scapes from base of pseudobulb, rarely from rhizome; bracts minute. **Flowers** with sepals subequal, or dorsal one much shorter, lateral ones adnate to foot of column; petals usually much smaller than dorsal sepal; lip narrowed at base and hinged to foot of column, mobile, usually very small and strongly recurved; column very short, often slightly winged or with 2 aristate stylids at top, produced into a distinct foot at base; anther terminal, incumbent, 2-celled; pollinia 2 or 4, waxy, globose or ellipsoid, free or connected by viscid material but without caudicles; stigma entire, concave. **Fruits** capsules, fusiform. **Distr:** Tropical and subtropical, chiefly in Asia, a few in Africa and Australia. **Cult:** Propagate by division. Smaller epiphytic species are best grown on tree-fern slabs, and larger species are more suited for baskets and pots. Requires a moist but well-drained mix, in highly humid and well-lit conditions.

B. abbreviatum (Rchb. f.) Schltr., *Repert. Spec. Nov. Regni Veg. Beih.* 33: 198, 1924. (Syn: *Cirrhopetalum abbreviatum*, etc.). **Herbs**, epiphytic; pseudobulbs 1–2 cm apart, oblong and narrow, 1–1.2 × 0.3–0.4 cm, not angled. **Leaves** narrowly elliptic, 1.7–2.7 × 0.6–1.2 cm, base contracted into a short petiole. **Inflorescence** to 6 cm long; peduncle with 5 acute sheaths; bracts broadly ovate 4.1× 3 mm. **Flowers** glabrous, ~5 mm long; dorsal sepal obtuse; lateral sepals longer; petals oblong, 1.5 × 1 mm, obtuse; lip distinctly lobed, midlobe orbicular; column 1 mm high, foot 1.5 mm long, anther 0.7 mm long; ovary 1.5–2 mm long. **Distr:** Madagascar. **Ecol:** Epiphytic in forests; 700–1,000 m. **Zones:** 10–12.

B. ambrosia (Hance) Schlechter, *Repert. Spec. Nov. Regni Veg. Beih.* 4: 247, 1919. **Herbs**, epiphytic or lithophytic; pseudobulbs 3.5–8 cm apart, narrowly ellipsoid or obovoid to cylindric, 1.8–5 × 0.5–0.9 cm. **Leaf** elliptic, 3–9 × 0.8–2.1 cm, apex rounded to obtuse; petiole 3–9 mm long. **Inflorescence** erect to spreading, 3–11 cm, 1–flowered, rarely a subumbellate 2-flowered raceme; peduncle 1.7–8 cm long; bracts 3 or 4, longest 0.3–1 cm long; rachis to 5 mm long; floral bracts ovate, 2.5–5 × 1.5–4 mm, acute; flowers resupinate, moderately to widely, opening simultaneously; pedicel and ovary 0.8–1.8 cm long. **Flowers** with dorsal sepal recurved, (ovate-)triangular, 0.9–1.4 × 0.6–0.9 cm, thin, glabrous; lateral sepals free, oblique, elliptic to ovate, or triangular, 1–1.4 × 0.7–1.1 cm; petals recurved, triangular with lower margin decurrent along column foot, 5–8 × 2.2–4 mm, thin, glabrous, base broadly attached, margins entire, apex acute to acuminate; lip ± reflexed slightly, ± elliptic, 5.5–8 × 5–6 mm, very thick; column 1.8–3.2 mm long. **Distr:** China, Nepal, Vietnam. **Ecol:** In forests, thickets; 600–1,500 m. **Zones:** 10–11.

B. anguliferum Ames and C. Schweinf., *Orchidaceae* 6: 164, 1920. (Syn: *B. muluense*). **Herbs**, epiphytic; rhizome creeping, ~0.5 mm Ø; pseudobulbs 2–5 mm apart, ellipsoid, somewhat flattened, 2–5 × 3–4 mm. **Leaves** ovate to elliptic, 0.3–2.7 × 0.25–0.7 cm, rather thick; petiole to 6 mm long. **Inflorescence** 2.5–5 cm long, 1-flowered; peduncle 1.2–3 cm long; floral bracts tubular, 1.5–2 mm long; pedicel and ovary 0.8–1.1 cm long.

Above: *Bulbophyllum abbreviatum, Bulbophyllum ambrosia, Bulbophyllum anguliferum.* [All MG.]

Flowers with sepals translucent, very pale ochre or pale orange, orange-veined, often turning bright yellow toward the tip, petals of similar color, lip orange, or yellow with orange veins, or salmon-colored; dorsal sepal ovate, 0.65–1.2 × 0.18–0.6 cm, subobtuse, thin, glabrous; lateral sepals free, 0.8–1.3 × 0.4–0.75 cm; petals elliptic to obovate, 4–8 × 1.2–2.6 mm, obtuse to acute, glabrous; lip somewhat recurved, elliptic, 0.75–1.1 × 0.3–0.4 cm (0.4–0.9 cm when spread), margins erose, apex rounded, thin, glabrous; column 1.8–3 mm long; pollinia 2. **Distr:** Borneo (Sabah, Sarawak). **Ecol:** In wet or dry montane forests; 1,500–2,400 m. Fl. Mar–Aug, Nov. **Zones:** 11–12.

B. annandalei Ridl., *J. Straits Branch Roy. Asiat. Soc.* 82: 197, 1920. **Herbs**, epiphytic, miniature; pseudobulbs closely set, conical, 1.5 cm long, apex 1-leafed. **Leaf** thick, 10 × 1.8 cm, apex blunt. **Inflorescence** a loose umbel, 2–4(–7)-flowered; scape basal, 9–12 cm long. **Flowers** 2.5–5 cm long, deep crimson, faintly striped with pale brown (reddish stripes in Malaysian forms), lip yellow (in Thailand) or pinkish (in Malaysia); dorsal sepal ovate, 1.2 cm long, acute, sparsely fringed; lateral sepals 3 cm long, free above the broad dilated bases, greatest width above the base 7 mm; petals ovate, 7 mm long, edges fringed; lip curved, narrow, ~7 mm long. **Distr:** Peninsular Thailand to peninsular Malaysia. **Ecol:** 1,000 m. **Zones:** 10–12.

B. antenniferum (Lindl.) Rchb.f., *Ann. Bot. Syst.* 6: 248, 1861. (Syn: *Cirrhopetalum antenniferum*, etc.) **Herbs**, epiphytic, upright, sympodial; rhizome to 5 mm Ø; pseudobulbs 4-angled, 2–5 cm high, unifoliate. **Leaf** to 20 × 5 cm, leathery. **Inflorescence** almost as long as the leaf, upright, 1-flowered; flower borne singly along the rhizome and can appear from any position. **Flower** relatively large, to 12 cm high, ~2 cm wide, greenish brown with numerous small transparent spots (like opaque windows) on the dorsal sepal behind the column; dorsal sepal ovate to lanceolate for the first 4 cm, remaining 6.8 cm linear, concave, 4 mm wide; lateral sepals joined along the central edge most of their length, ~7.4 cm; petals very small, lanceolate, tipped with a small, hairlike appendage; lip recurved, hairy along edges, small, highly mobile, ~ 5 × 2.5 mm. **Distr:** Malesia to Solomon Islands. **Ecol:** Epiphytic on tree trunks and sometimes lithophytic on rocks; to 500 m. **Zones:** 10–12.

B. beccarii Rchb. f., *Gard. Chron.*, n.s., 1879: 41, 1879. (Syn: *Phyllorkis beccarii*). **Herbs**, epiphytic; rhizome to 20 cm Ø, encircling the trees on which it grows, much like a snake; pseudobulbs distant, egg-shaped, ~5 × ~5 cm. **Leaf** solitary, stiffly erect; lamina elliptic, to ~60 × 20 cm, base somewhat folded together, apex acute, very thick and rigidly leathery, usually yellowish-green. **Inflorescences** borne from the rhizome, near the pseudobulbs, short-pedunculate, pendulous, cylindrical, to 23 cm long, to 10 cm Ø, composed of several hundred small flowers. **Flowers** extremely and overpoweringly fetid, not opening widely, ~1.25 cm across; sepals oblong, yellowish, intricately netted inside with red; petals smaller, lanceolate, yellow, with a red median line; lip small, tongue-shaped, golden-yellow, with red stripes. **Distr:** Borneo. **Ecol:** Grows by wrapping itself around tree trunks in forests. **Zones:** 10–12. **Info:** This is one of the largest species in the genus.

B. biflorum Teijsm. and Binn., *Ned. Kruidk. Arch.* 3: 397, 1855. (Syn: *B. geminatum*, etc.). **Herbs**, epiphytic, upright, sympodial; rhizome ~2 mm Ø; pseudobulbs set close together, distinctly 4-angled, to ~4 cm high, 2.5 cm across, unifoliate. **Leaf** to 13 × 3 cm. **Inflorescences** 2-flowered, semipendulous, 9–10 cm long, borne from base of pseudobulb. **Flowers** 8–10 × 1.5 cm, white, cream or greenish-yellow with varying degrees of dark red or reddish-purple spotting, lip yellowish-brown with purple spots, with a peculiar odor; dorsal sepal ovate, 2.5 cm long, extended to a threadlike appendage with a thickened tip; lateral sepals linear, 7–8 × 0.7 cm, tapered, joined or sometimes separated, drawn out to slender filaments; petals small, 1 × 0.5 cm; lip highly mobile, oblong to ovate, to ~6 mm long, fleshy. **Distr:** Peninsular Thailand to W and C Malesia. **Ecol:** In brightly lit areas of woodlands; to 700–1,100 m. **Zones:** 11–12.

B. brevibrachiatum (Schltr.) J.J. Sm., *Bull. Jard. Bot. Buitenzorg*, II, 8: 23, 1912. (Syn: *Cirrhopetalum brevibrachiatum*). **Herbs**, epiphytic, upright, sympodial, often rambling; rhizome small, with pseudobulbs 3 cm apart; pseudobulbs round, ~1 cm Ø. **Leaves** to 12 × 2 cm. **Inflorescence** umbellate, to 12-flowered; peduncle erect, to 15 cm tall. **Flowers** 4 cm long, 6 mm wide, creamy yellow, spotted with pale red; dorsal sepal concave, 5 × 3.5 mm, margins with orange hairs; lateral sepals acuminate, 4 cm long, 3 mm Ø, joined for entire length; petals lanceolate, 4 × 3 mm, margins with small hairs; lip basally bright yellow, apically bright red, recurved, small, highly mobile, 3 × 1 mm. **Distr:** Philippines, Sulawesi. **Ecol:** Epiphytic on tree trunks; 500–1,400 m. **Zones:** 10–12.

B. brienianum (Rolfe) Merr., *J. Straits Branch Roy. Asiat. Soc., Spec. No.*: 184, 1921. (Syn: *Cirrhopetalum brienianum*, etc.). **Herbs**, epiphytic; pseudobulbs 2–3 cm apart, narrowly ovoid, to 2 cm long, strongly ridged, 1-leafed. **Leaf** elliptic-oblong, 12 × 2 cm, apex obtuse, coriaceous; petiole short. **Inflorescence** umbellate, 5–10-flowered, flowers arranged in a fan; scape suberect, elongate, 10–20 cm long, purple; bract subulate-lanceolate, acute. **Flowers** fragrant, dorsal sepal red-brown to dark brown-purple with fine yellow to brown-purple hairs on margins, lateral sepals pale yellow, base flushed with dark purple, lip dark purple, column pale yellow with a few purple spots; dorsal sepal ovate, elliptic-oblong, 5 mm long, concave, ciliate, tip slender, 2 mm long; lateral sepals connate to most their length, linear-elongate, narrow, 2.5 cm long, outer margins recurved; petals suborbicular, similar in size to dorsal sepal, margins ciliate; lip recurved, oblong-lanceolate, subobtuse; column short, arms small, rounded. **Distr:**

Below: *Bulbophyllum annandalei, Bulbophyllum antenniferum, Bulbophyllum beccari.* [All MG.]

Top row: *Bulbophyllum biflorum, Bulbophyllum brevibrachiatum, Bulbophyllum brienianum, Bulbophyllum cameronense.*
Bottom row: *Bulbophyllum carunculatum, Bulbophyllum catenarium, Bulbophyllum cootesii, Bulbophyllum cylindraceum.*

Arunachal Pradesh, W Malesia. **Ecol:** In lowland, warm and humid forests. Fl. Oct. **Zones:** 10–12.

B. cameronense Garay, Hamer and Siegerist, *Lindleyana* 11: 226, 1996. **Herbs**, epiphytic, miniature to small-sized; pseudobulbs and rhizomes encased in long, stiff bristles at base; pseudobulbs globose to subglobose, ~5.5 cm apart on the rhizomes. **Leaf** oblong-obovate, 31 × 12 cm, base narrowed into a petiole, margins entire, apex broadly acute, keeled in the middle. **Inflorescence** with peduncle short, 1-flowered; pedicel relatively long. **Flower** fragrant, pale yellowish-brown, sepals and petals 9-veined, veins purplish red; dorsal sepal lanceolate, 2.5 × 1 cm; lateral sepals ~2 × 1 cm, broader than dorsal, curved gradually with the apices pointing downward, apices rounded, base whitish; petals horizontal, linear-lanceolate, 2 × 0.5 cm; lip tongue-shaped, white with a bright orange spot in the middle; column basally white with purplish spots, apically orangish-yellow. **Distr:** Peninsular Malaysia. **Ecol:** On exposed tree trunks, with the flowers held away from the host trunk. **Zones:** 11–12.

B. carunculatum Garay, Hamer and Siegerist, *Lindleyana* 10: 174, 1995. **Herbs**, epiphytic, upright, sympodial, to 45 cm tall; rhizome short; pseudobulbs ovoid, unifoliate, to 6 cm tall. **Leaf** elliptic to obtuse, to 36 × 7 cm; petiole to 5 cm long. **Inflorescence** upright, to 12-flowered, to 45 cm tall; bracts tubular, to 5 cm long; flowers appearing in succession. **Flowers** golden yellow, 9 cm long, odor pungent; dorsal sepal ovate to oblong, to 4 × 1.5 cm, acute; lateral sepals falcate, to 5 × 1.5 cm; petals ovate, to 2.5 × 1.2 cm, acute; lip dark purple (almost black), 2.5 cm long, fleshy with wartlike texture, trilobed, lateral lobes short and toothed; column erect, ~7 mm long with 2 forward-pointing stelidia. **Distr:** Philippines; possibly also in Indonesia, New Guinea and Sulawesi. **Ecol:** In montane tropical rainforests; 500–1,500 m. Fl. Mar–Oct. **Zones:** 10–12.

B. catenarium Ridl., *Trans. Linn. Soc. London*, Bot. 4: 235, 1894. (Syn: *B. carunculilabrum*). **Herbs**, epiphytic; rhizome creeping; pseudobulbs 1–7 mm apart, ovoid, somewhat flattened, 1–7 × 1–3 mm. **Leaf** elliptic, 0.4–1.2 × 0.3–0.6 cm, thick; petiole 0.5 mm long. **Inflorescence** 1-flowered, 2.5–5.5 cm long; peduncle 2.5–3.2 cm long; floral bracts tubular, 1.5–2 mm long; pedicel and ovary 1–2.2 cm long. **Flower** scentless, sepals and petals bright yellow, occasionally with orange veins, lip very dark red or purple; dorsal sepal elliptic, 3–6.4 × 2–2.5 mm, margins erose, often ciliate in upper half, apex acuminate, very thin, glabrous; lateral sepals free, similar to dorsal sepal, 5–9 × 3.5–5.5 mm; petals ovate, 1–1.3 × 0.4–0.7 mm, margins glabrous or erose, apex obtuse, thin, surface glabrous; lip recurved, ovate, very thick, swollen apically, triangular, lateral lobes 1.5–2 × 0.8–1.6 mm, with 2 keels in basal half, margins ± glabrous, apex rounded, surface coarsely verrucose apically; column 1–2.5 mm long. **Distr:** Vietnam, peninsular Malaysia, Borneo. **Ecol:** In wet and mossy mixed forests; 1,700–2,500 m. Fl. Jun–Aug, Oct–Dec. **Zones:** 10–12.

B. cootesii M.A. Clem., *Austral. Orchid Review* 64: 6, 1999. **Herbs**, epiphytic, upright; rhizome short, 4 mm Ø; pseudobulbs ovate, 3 × 2.5 cm, wrinkled, unifoliate. **Leaf** linear 20 cm long including the petiole, 3.3 cm wide; petiole 3 cm long. **Inflorescence** at right angles to pseudobulb, several-flowered; peduncle 9 cm long with 3 short sheaths; sheaths 1.5 cm long; flowers opening sequentially; pedicel and ovary 2.5 cm long, 3–4 mm Ø. **Flowers** pale yellowish brown, veined with pale brown, 7 cm high, 5 cm wide (across the petals), tips of segments bright yellow, all segments glossy; dorsal sepal lanceolate, 4 × 1.5 cm, keeled along the middle; lateral sepals falcate, 4.2 × 0.8 cm, deeply ridged and somewhat folded (1.4 cm wide when flattened), joined at base to form a very short, mentum; petals lanceolate, 2.6 × 1.3 cm; lip basally white, front portion reddish brown, 1.5 × 0.5 cm, fleshy, midlobe oblanceolate, 2 short arms at the back of the lip. **Distr:** Endemic to Philippines (Mindanao). **Ecol:** Epiphytic on trees; 400–500 m. **Zones:** 11–12.

B. cylindraceum Lindley, *Gen. Sp. Orchid. Pl.* 53. 1830. (Syn: *B. cylindraceum* var. *khasianum*, *Phyllorkis cylindracea*). **Herbs**, epiphytic or lithophytic; rhizome creeping; pseudobulbs 3–5 × 3–4 mm. **Leaf** terminal, erect; lamina elliptic-oblong or obovate-lanceolate, 15–25 × 2–4 cm, apex obtuse, apiculate, fleshy or leathery; petiole 4–7 cm long. **Inflorescence** a raceme, pendulous, to 6.5 cm long, densely many-flowered; basal involucre spathelike, ~ 2.5 × 0.7 cm; scape basal, erect, 15–25 cm long; peduncle sheaths 2–3, tubular, 2–2.5 cm long; floral bracts ovate, ~4 mm long; pedicel and ovary ~3 mm long. **Flowers** pale purple, thick, not widely opening; dorsal sepal ovate-triangular, 5 × 2.2 mm, abaxially ± scaly, acute; lateral sepals obliquely ovate, ~ 5 × 3 mm, abaxially ± scaly, lower margins connate, obtuse; petals oblong-lanceolate, 2–4.4 × ~1.5 mm, thin, entire, acute or obtuse; lip recurved, ligulate, 3.5 × 2 mm, fleshy, grooved at base, adaxially with 3 keels, papillate, obtuse; column ~0.6 mm long. **Distr:** Nepal to Thailand. **Ecol:** On tree trunks, or rocks in forests; 1,400–2,400 m. Fl. Nov. **Zones:** 9–12.

B. dearei (Rchb. f.) Rchb. f., *Flora* 71: 156, 1888. (Syn: *B. godseffianum*, etc.). **Herbs**, epiphytic, upright, sympodial; rhizome to 5 mm Ø; pseudobulbs ~3 cm apart, ovoid, somewhat flattened, 1.8–5 × 1–2.8 cm, unifoliate. **Leaf** obovate to elliptic, 7.5–20 × 3–6.6 cm, leathery. **Inflorescence** upright, 1-flowered; pedicel and ovary, 0.65–2 cm long. **Flower** orange-yellow, marked with purple, lip reddish-purple; dorsal sepal ovate, 2.8–4.5 × 1.5–2.3 cm; lateral sepals free, triangular-ovate, 2.3–3.5 × 1.2–1.8 cm; petals triangular, to 3 × 1 cm, strongly reflexed backward; lip recurved, rather thick, ovate, 0.7–1.1 × 0.5–1 cm, trilobed, highly mobile, spurs 2, pointing backward, 4.5–7 mm long, midlobe ovate, to 1 × 0.8 cm; column 2.5–5 mm long. **Distr:** Borneo, peninsular Malaysia to W Philippines. **Ecol:** Epiphytic or lithophytic on hill forests and in lower montane forests; 700–1,200 m. Fl. May–Jul. **Zones:** 11–12.

B. echinolabium J.J. Sm., *Repert. Spec. Nov. Regni Veg.* 36, 7–11: 118, 1934. **Herbs**, epiphytic; rhizome to ~5 cm long; pseudobulbs ovoid to oblong-elliptic, 3–6 ×2–5 cm, olive-green, stained purple-brown. **Leaf** oblong to oblong-elliptic, 11–33 × 4–10 cm, base narrowed and cuneate, apex obtuse, coriaceous, mid green; petiole 1–5 cm long. **Inflorescence** erect; peduncle to 48+ cm long; bracts 3–4, ovate-elliptic, 3–6 cm long, acute, sheathing; rachis 6–10 cm long; floral bracts 2–2.5 × 1.5–1.6 cm, ovate, acute; pedicel with ovary 3.5–4 cm long, clavate. **Flowers** 17.5–25(–35) cm long, fleshy, odor unpleasant, sepals and petals pale yellowish-buff with cerise to lilac-pink nerves, lip reddish-apricot or sulfur-yellow with dark purple echinate portion, groove between keels whitish, whitish beneath, column pale yellow with dark purple stelidia, anther-cap pale pink; dorsal sepal narrowly lanceolate, 8.6–10.5 × 0.7–1.8 cm; lateral sepals obliquely lanceolate, 13.4–14 × 0.95–1.6 cm, base shallowly concave, apex caudate and linear-acuminate; petals obliquely subovate-lanceolate, falcate, ~ 6.2–7 × 0.8–1.4 cm, acuminate; lip versatile; column ~ 0.8–1.1 × 0.4–0.5 cm, foot 1–1.2 cm long; anther-cap conical-ovate, cucullate, ~ 5 × 4 mm, glabrous; pollinia 4; ovary obconical, 6-sulcate. **Distr:** Indonesia (Sulawesi). **Ecol:** In lower montane forests. Fl. Aug–Sep. **Zones:** 11–12.

B. elassoglossum Siegerist, *Amer. Orchid Soc. Bull.* 60: 865, 1991. **Herbs**, epiphytic, upright, sympodial, to 11 cm tall; rhizome slender, creeping, producing roots at the base of pseudobulbs, to 4.5 cm apart; pseudobulbs ovoid, ± tetragonal, to 3–3.5 cm high, 1.5 cm wide at base, unifoliate. **Leaf** elliptic to narrowly elliptic, to 8.5 × 2 cm. **Inflorescence** single, upright, hairlike, 1-flowered; dorsal sheath 1.1 cm long. **Flower** relatively large, showy, segments spreading, dorsal and lateral sepals white, heavily blotched with purple toward the base, petals and lip deep maroon; dorsal sepal elliptic, concave, to 2.4 × 1 cm, acute; lateral sepals oblong to elliptic, 2.3 × 0.7 cm, acute; petals elliptic, to 1 × 0.5 cm, rounded to obtuse, parallel with one another, lateral margins somewhat infolded and tips recurved; lip tongue-shaped, to 5 × 2 mm, fleshy; column short with erect, forward-pointing stelidia. **Distr:** Philippines. **Ecol:** Habitat unknown (described from a cultivated specimen in USA). Fl. May–Aug. **Zones:** 11–12.

B. elliae Reichb. f., *Ann.* 6: 263. 1861. (Syn: *Cirrhopetalum wightii*, etc.). **Herbs** epiphytic. **Leaf** 1, shortly petioled; lamina oval, linear-oblong or oblanceolate, 3.5–14 × 1–2.7 cm, apex emarginate, thickly coriaceous. **Inflorescence** umbels, 5–8-flowered, basal, red-speckled, with 2 or 3 scaly bracts; floral bracts 3–4.8 × 1.4–2 mm; pedicel and ovary 0.7–1.2 cm long. **Flowers** small, purplish, 5–7 mm across, dorsal sepal 5.5–7 × 3–4.2 mm, broadly ovate, obtuse, 5-veined, apically erose or dentate, bending over to form a hood over rostrum and petals; lateral sepals linear-ovate or linear-oblong, 1.3–1.5 × 0.38–0.5 cm, curved inward, red-speckled inside; petals ovate, 2.9–3.8 × 2–2.4 mm, acute, serrate; lip linear-oblong, tonguelike, 3.9–4.6 × 1.7–1.9 mm, thick, fleshy, strongly recurved, speckled red; column 1.8 × 1.3–1.7 mm; pollinia 4. **Fruits** oblong, cylindrical, 2 × 0.7 cm. **Distr:** Sri Lanka. **Ecol:** On trees of montane forests, wet evergreen forests; >854 m. Fl. Jan–May, Sep. **Zones:** 11–12.

B. facetum Garay, Hamer and Siegerist, *Lindleyana* 11: 227, 1996. **Herbs**, epiphytic, upright, sympodial; rhizome to 4 mm Ø, internodes 1 cm apart with the pseudobulbs spaced 5 cm apart on the rhizome; pseudobulbs ovoid, to 3.5 cm high, 2.5 cm Ø, unifoliate, with persistent bristles, to 3 cm long, at base of each pseudobulb. **Leaf** oblong to elliptic, 20 × 5 cm; petiole to 5 cm long. **Inflorescence** upright, ~6 cm long, borne from base of pseudobulbs, single-flowered; bract tubular at base of pedicel. **Flower** 7 cm across, creamy white, heavily spotted and tessellated with reddish brown; dorsal sepal ovate to elliptic, 4 × 1.5 cm, acute; lateral sepals obliquely ovate to oblong, to 3.3 × 2 cm; petals obovate to lanceolate, to 3.2 × 1.2 cm; lip heart-shaped, curved, ~1 × 0.8 cm. **Distr:** Endemic to Philippines (Luzon).

Top row: *Bulbophyllum dearei, Bulbophyllum echinolabium, Bulbophyllum elassoglossum.*
Bottom row: *Bulbophyllum elliae, Bulbophyllum facetum, Bulbophyllum falcatum.*

MG

JT MG

JT

JT

JT

JT MG

Above: *Bulbophyllum flavofimbriatum, Bulbophyllum foetidum, Bulbophyllum frostii.*

Ecol: Epiphytic on tree trunks, in shade; to 1,200 m. **Zones:** 11–12.

B. falcatum (Lindl.) Rchb. f., *Ann. Bot. Syst.* 6: 258, 1861. (Syn: *Megaclinium falcatum, Phyllorkis falcata*). **Herbs**, epiphytic; pseudobulbs bifoliate, ovoid or ellipsoid, 0.4–5 cm apart, 0.9–7 × 0.6–2 cm. **Leaves** broadly lanceolate to linear, 2–21 × 0.5–3 cm, apex obtuse, emarginate, coriaceous. **Inflorescence** 3–40 cm long, 8–60-flowered; peduncle 1–16 cm long, glabrous, with 4–8 scales; flowers distichous, placed on the flat rachis, 0.2–2(–3) cm apart; floral bracts ovate to oblong, 1.3–6.2 × 1–3.5 mm, acute; pedicel and ovary, 1.5–5 mm long, hairy. **Flowers** with sepals yellowish adaxially, abaxially whitish, yellowish or greenish, stained with dark red; dorsal sepal narrowly obovate to broadly obovate-lanceolate, ± spathulate, 2.3–9 × 0.8–4 mm, obtuse to acuminate; lateral sepals free, with dark red spots, broadly ovate to ovate oblong, 1.8–9 × 0.8–7 mm; petals ovate to broadly linear-lanceolate, 1.3–3.2 × 0.25–1 mm, rounded to acuminate; lip cream-colored, spotted with dark red or entirely dark red with a yellow base, deltoid to broadly ovate, 0.7–4 × 0.5–3 mm, obtuse; column 0.5–1.5 mm long; pollinia 4. **Fruits** ellipsoid, to 1.5 × 0.5 cm. **Distr:** W Tropical Africa to Uganda. **Ecol:** In lowland forests, lower montane forests; <1,800 m. Fl. Aug–May. **Zones:** 11–12.

B. flavofimbriatum J.J. Sm., *Bull. Jard. Bot. Buitenzorg*, III, 11: 143, 1931. (Syn: *Epicranthes flavofimbriata*). **Herbs**, epiphytic, pendulous, to 70 cm long, sympodial; rhizome ~1 mm Ø covered with persistent bracts; pseudobulbs 1.5 cm long, 5 mm Ø, deeply furrowed, unifoliate. **Leaf** ovate, to 5.5 × 2 cm, margins incurved, leathery. **Inflorescences** 1 or more, short, borne from the base of the pseudobulb. **Flowers** ~1 cm Ø, sepals yellowish green inside, spotted with brown outside; sepals ovate, to 8 × 3.5 mm; petals cream at base, with 10 black appendages attached to the base of petals by small filaments; lip red with greenish stripes, 3 × 1.5 mm, very fleshy, highly mobile. **Distr:** Borneo (Sabah and Kalimantan), Philippines (Luzon). **Ecol:** In montane and open forests; 500–1,500 m. **Zones:** 10–12.

B. foetidum Schltr., *Repert. Spec. Nov. Regni Veg. Beih.* 1: 748, 1913. **Herbs**, epiphytic; rhizome elongate, ~1.5–2.5 cm Ø; roots filiform, flexuose, glabrous; pseudobulbs 15–20 cm apart, oblong-cylindrical, 4–6 cm high, 2–2.5 cm Ø, unifoliate. **Leaf** erect; lamina elliptic, 35–45 × 12–15 cm, base narrowed, apex acuminate, glabrous. **Inflorescence** erect or arcuate, 30–45 cm tall, subdense, 8–15-flowered; bracts elliptic, acuminate, clasping the pedicels; pedicels long. **Flowers** malodorous, segments spreading, glossy; sepals oblong, 2.4 cm long, slightly apiculate; petals oblique, elliptic-lanceolate, 2.2 cm long, acuminate; lip ovoid-ligulate, obtuse, thick, fleshy, 2.2 cm long; column erect, thick; ovary pedicellate, cylindric, ~2.5 cm long, glabrous. **Distr:** New Guinea. **Ecol:** Epiphytic on tree trunks, occasionally on rocks; 400–1,000 m. **Zones:** 11–12.

B. frostii Summerh., *Bull. Misc. Inform. Kew* 1928, 2: 76, 1928. (Syn: *Cirrhopetalum frostii*, etc.). **Herbs**, epiphytic; pseudobulbs closely set on the creeping rhizome, compressed, ovoid, 1.5–2 cm long, 1-leafed. **Leaf** sessile; lamina elliptic, 3.5 × 2 cm, apex obtuse, carnose. **Inflorescence** few-flowered, flowers clustered; scape 1.5–2 cm long. **Flowers** unpleasantly scented during the warmer parts of the day, resembling Dutch clogs, 1–2 cm long, pale yellow-tan with dark reddish spots; dorsal sepal orbicular-ovate, 1.3 × 0.9 cm, strongly convex, base rounded, margins shortly ciliate, apex shortly acuminate, cupped, hirsute, projected over the column and lip; lateral sepals turned inward and joined, oblique, oblong-lanceolate, ~2 × 0.8–0.9 cm, subacuminate, 7-nerved; petals oblong, ~1 × 0.6 cm, slightly convex, margins erose-ciliate, apex obtuse, hirsute; lip strongly carnose, sagittate, curved, 7–8 mm long; column 5 mm long, obtuse, rostellum minute. **Distr:** Vietnam. **Ecol:** In evergreen lowland forests; ~1,450 m. Fl. Nov–Feb. **Zones:** 11–12.

B. gracillimum (Rolfe) Rolfe, *Bull. Misc. Inform. Kew* 1907: 412, 1907. (Syn: *B. psittacoides*, etc.). **Herbs**, epiphytic; rhizome creeping, terete, internodes 1–2 cm long; pseudobulbs ovoid, 2 cm long, unifoliate. **Leaf** subsessile, elliptic to elongate-elliptic, 12 × 2.3 cm, base ± contracted into a petiole 1–2 mm long, apex rounded, mucronate to emarginate, thick. **Inflorescence** a raceme, 8–16-flowered, axillary, flowers arranged in a ring with the segments facing outward; peduncle slender, purple, to 15–30 cm long; pedicel with 1 bract at base. **Flowers** usually bright crimson, sepals and petals with elongate tail; lateral sepals 4+ cm long, margins basally joined; lip violet-purple. **Distr:** Peninsular Thailand to SW Pacific. **Zone:** 12.

B. graveolens (F.M. Bailey) J.J. Sm., *Bull. Jard. Bot. Buitenzorg*, II, 8: 24, 1912. (Syn: *B. pachybulbum*, etc.). **Herbs**, epiphytic; pseudobulbs to 10 cm tall, set ~4 cm apart on the rhizome, apex unifoliate. **Leaf** to 60 × 10 cm. **Inflorescence** umbellate, ~20 cm across, with flowers arranged loosely, usually up to 12-flowered; pedicel and ovary ~5.5 cm long. **Flowers** 5–7.5 cm long, malodorous, long-lasting, yellow with a red center, glossy; sepals and petals pale greenish yellow to light orange, with or without fine dark purple spots, lip purplish red; dorsal sepal erect, oblong, ~3.7 × 1.3 cm; lateral sepals falcate, linear, 6.5 × 0.95 cm, apices rotated and adnate apically, projected forward; petals very small, ~7 mm long; lip mobile, strongly recurved, ~1.8 cm long; column erect, ~7 mm high. **Distr:** New Guinea to Australia (Queensland). **Ecol:** Epiphytic in rainforests; 100–600 m. **Zone:** 12.

B. guttulatum (J.D. Hooker) N.P. Balakrishnan, *J. Bombay Nat. Hist. Soc.* 67(1): 66, 1970. (Syn: *Cirrhopetalum guttulatum*). **Herbs**, epiphytic; rhizome creeping, ~3 mm Ø; pseudobulbs ovoid-conic or narrowly ovoid, 2–2.5 cm, 0.7–1 cm Ø, ribbed, with a terminal leaf. **Leaves** elliptic-oblong, 7–14 × ~3 cm, base contracted into petiole, apex obtuse-rounded, leathery; petiole 1–2 cm. **Inflorescence** umbel, often 2- or 3-flowered; scape erect, 10–15 cm, slender; peduncle with 2 or 3 tubular sheaths; floral bracts ovate-lanceolate, ~5 mm, apex acuminate; pedicels and ovary ~2.5 mm. **Flowers** yellow, spotted with red; dorsal sepal broadly ovate, concave, 0.8–1 × 0.6–0.7 cm, subtruncate and mucronulate; lateral sepals obliquely ovate-lanceolate, 1–1.1 × ~0.6 cm, ± narrowed and twisted near base; petals broadly ovate-triangular, ~ 4.5 × 4 mm, mucronate; lip recurved, subovate in outline, ~ 5 × 3.4 mm, adaxially with 3 longitudinal keels, base dilated and attached to end of column foot; column ~3 mm, with triangular, acute wings along lower margins. **Distr:** Bhutan, China, NE India, Nepal, Vietnam. **Ecol:** On tree trunks in broad-leaved forests; 800–1,800 m. Fl. Aug. **Zone:** 10.

B. hirundinis (Gagnepain) Seidenfaden, *Dansk Bot. Ark.* 29(1): 76, 1974. (Syn: *Cirrhopetalum hirundinis*, etc.). **Herbs**, epiphytic; rhizome creeping; pseudobulbs close to 2 cm apart, ovoid, 0.7–1.5 × 0.5–1 cm. **Leaf** subsessile, terminal; lamina narrowly elliptic, oblong, or ovate-ligulate, 1.3–5 × 0.6–2 cm, apex obtuse or acute, sometimes retuse, thickly leathery or fleshy. **Inflorescence** basal, erect, 3.5–13 cm long, umbel 3–5-flowered; peduncle slender, basal sheaths few; floral bracts lanceolate, ~3 mm long, acuminate; pedicel and ovary 6–7 mm long. **Flowers** yellow, tinged with purplish red; dorsal sepal ovate, concave, ~ 5 × 3 mm, margin fimbriate-ciliate, apex acute to subcaudate; lateral sepals linear, (1.6–)2–3 × 0.2–0.3 cm, twisted near base, lower edges ± connate forming an elliptic synsepal, apically free and divergent, entire or shortly ciliate; petals obliquely ovate-triangular, 3–4 × 1.5–2.5 mm, margins usually fimbriate-ciliate, acute; lip slightly recurved, ligulate, ~2.5 mm long, fleshy, glabrous, obtuse; column ~1.5 mm long, foot ~1.5 mm long. **Distr:** China, Taiwan, Vietnam. **Ecol:** Epiphytic on forest tree trunks; 500–3,000 m. **Zones:** 7–11.

B. insulsum (Gagnep.) Seidenf., *Dansk Bot. Ark.* 29: 141, 1973. (Syn: *Cirrhopetalum insulsum*, etc.). **Herbs**, epiphytic; pseudobulbs close, ovoid-conic, 2.5–3 cm, 0.8–1.5 cm Ø, unifoliate. **Leaf** apical, shortly petiolate; lamina narrowly oblong, 10–17 × 1.3–3 cm, base contracted into petiole, apex subacute, leathery. **Inflorescence** a raceme, elongate, sparsely 10-flowered; scape from base of pseudobulb, ~12 cm long, slender; peduncle with 2–4 sheathlike bracts; floral bracts lanceolate, 6–9 mm; pedicel and ovary shorter than floral bracts. **Flowers** yellowish-green with pale red spots; dorsal sepal ovate-lanceolate, 6–8 × ~2.5 mm, densely ciliate, apex acuminate and recurved; lateral sepals obliquely ovate-lanceolate, 0.7–1.1 × 0.2–0.3 cm, base adnate to column foot, apex long caudate; petals

JT GS

Top row: *Bulbophyllum gracillimum, Bulbophyllum graveolens.*
Bottom row: *Bulbophyllum guttulatum, Bulbophyllum hirundinis, Bulbophyllum insulsum.* [All MG.]

ovate, 2–4 × 1.5–2 mm, densely ciliate; lip ligulate, ~ 2.5 × 1.5 mm, adaxially with 3 ridges, central 1 yellow and lateral 2 yellowish-white, attached to end of column foot, immobile; column white, ~2 mm, with a foot ~2 mm. **Distr:** Taiwan, China to Vietnam. **Ecol:** In forests; 1,000–2,000 m. Fl. Jul–Sep. **Zones:** 8–11.

B. laxiflorum (Blume) Lindl., *Gen. Sp. Orchid. Pl.*: 57, 1830. (Syn: *Diphyes laxiflora*, etc.). **Herbs**, epiphytic; rhizome creeping, much-branched; pseudobulbs 0.5–1.5 cm apart, long-ovoid or cylindrical, 2.5+ cm tall, apex 1-leafed. **Leaf** shortly stalked; lamina elliptic, 12 × 2.5 cm, base petiolate, apex acute, leathery. **Inflorescence** capitulate, 12–20-flowered, flowers joined to the axis closely together; scape 7–12 cm long; pedicels slender, ~2 cm long. **Flowers** fragrant, white or pale yellowish, lip orange; sepals spreading equally, narrow, (0.9–)1(–1.6) cm long; petals narrow, 6 mm long, very short-hairy; lip very small, pointed, ~2 mm long; column arms long and slender. **Distr:** Indo-China to W and C Malesia. **Ecol:** In riverine, lowland hill and lower montane forests; 100–1,800 m. **Zones:** 10–12.

B. lemniscatoides Rolfe, *Gard. Chron.*, III, 1890: 672, 1890. (Syn: *B. comosum*, etc.). **Herbs**, epiphytic; pseudobulbs ovoid, 1.5–3 × 2 cm, brownish-green, shiny, bifoliate. **Leaves** linear-lanceolate, (5–)8–12 × (0.5–)1–1.5 cm, subacute. **Inflorescences** basal, densely up to ~30-flowered; peduncle ~13–30 cm long, terete; bracts 2–3, 5–8 mm long, remote, clasping, acute; rachis 3–5 cm long, pendulous; floral bracts narrowly elliptic, ~2–3 mm long, acuminate, pale brown; pedicel and ovary 2–2.5 mm long, hirsute. **Flowers** ~3 mm Ø, faintly scented, sepals very dark violet or purple with white hairs on outside, greenish near base, appendages whitish-banded pale mauve, sepals not opening widely, hirsute on outside; dorsal sepal ovate, 2 × 1.1–1.2 mm, obtuse, concave, versatile, appendage filiform, 8–9 mm long, borne below just apex; lateral sepals obliquely triangular-ovate, 2.4–2.5 × 2 mm, obtuse, concave, appendage similar to dorsal; petals linear to narrowly triangular, 1.9–2 × 0.5–0.6 mm, acuminate, finely toothed; lip crimson or purple, margins almost black, oblong to elliptic-oblong, 2 × 1 mm, obtuse, fleshy, concave; column white, 1 mm long, foot 1.1–1.2 mm long; anther cap yellow. **Distr:** Peninsular Malaysia, Thailand, Vietnam, Sumatra, Java and Borneo. **Ecol:** On mountainsides; ~1,500 m. Fl. Oct–Dec, May. **Zones:** 10–12.

B. lilacinum Ridl., *J. Linn. Soc., Bot.* 32: 276, 1897. (Syn: *B. lilacinum* var. *sorocianum*). **Herbs**, epiphytic; rootstock creeping, rhizomatus, 3–4 mm thick; pseudobulbs to 12 cm apart, narrowly ovoid, conic, 2.5–6 cm long, clothed below with membranous sheaths, unifoliate. **Leaf** shortly petiolate; lamina oblong-lanceolate, 12.5–20 × 2.5–4.5 cm, obtuse, fairly thick; petiole 6 mm long. **Inflorescence** raceme, 7.5 cm long, dense; scape arcuate, 10–20 cm long, thick with sheaths at base; bracts ovate-lanceolate, 3–5 mm long, acute or acuminate. **Flowers** rather small, lilac with purple blotches; dorsal sepal lanceolate to elliptic-ovate, 6 mm long, acute, recurved; lateral sepals longer, lilac with pink spots, connate, ovate, >8 mm long, acute, concave; petals much shorter than sepals, claw stout, lamina lanceolate to ovate, ~2 mm long, obtuse to acuminate, flat, straight; lip clawed, elliptic-oblong, ~4 cm long, obtuse, rather thick, with a finely 2-keeled disk, and with falcate-ovate rather blunt auricles; column rather long, foot as long as arms, lanceolate, acute, erect. **Distr:** Bangladesh to peninsular Malaysia. **Ecol:** In dipterocarp forests; ~1,200 m. Fl. Oct–Feb. **Zones:** 11–12.

B. lobbii Lindl., *Edwards's Bot. Reg.* 33: t. 29, 1847. (Syn: *B. claptonense*, etc.). **Herbs**, epiphytic; rhizome creeping, stout; pseudobulbs 4–8 cm apart, broadly ovoid, 3–5 × 2–3 cm, yellowish, unifoliate. **Leaf** elliptic to broadly elliptic, to 35 × 9 cm. **Inflorescence** 10–15 cm long, 1-flowered. **Flower** 7.5–10 cm across, color variable, usually pale brownish-yellow, flushed with red or brown, scented; dorsal sepal broadly lanceolate, to 5 × 1.7 cm, acuminate, veins crimson-dotted; lateral sepals shorter and broader than dorsal sepal, veins purplish, apical half brownish; petals horizontal, lanceolate, 3.5 × 1 cm, veins ~9, purplish; lip narrowly ovate, strongly recurved, ~ 1.3 × 1 cm, basal margins upcurved, purplish with a bright orange circular callus. **Distr:** NE India (Arunachal Pradesh), through SE Asia to Philippines. **Ecol:** In lowland and submontane forests; 100–2,000 m. Fl. throughout the year. **Zones:** 11–12.

B. longissimum (Ridl.) J.J. Sm., *Bull. Jard. Bot. Buitenzorg*, II, 8: 25, 1912. (Syn: *Cirrhopetalum longissimum*). **Herbs**, epiphytic; rhizomes creeping, stout; pseudobulbs well-spaced, ovoid or conical, to 4 cm long, basally clothed with ovate membranous sheaths, unifoliate; internodes 3.2–6.3 cm long. **Leaves** oblong, 9–15 × 2.5–4.5 cm, base narrowed, apex recurved, subacute, coriaceous, subsessile. **Inflorescence** umbellate, 4–7(–10)-flowered, arching or pendent, to 20 cm long; bracts 2–3, acuminate, sheathing; floral bracts oblong-lanceolate, 1–1.5 cm long; pedicels ~1–1.5 cm long. **Flowers** fragrant, short-lived, 20–30 cm long, white tinged or streaked with pink; dorsal sepal lanceolate or oblong-lanceolate, 1.7–1.9 cm long, incurved, concave, marked with 5 red longitudinal veins, ciliate; lateral sepals long connate, linear, 20–30 cm long, with distinct pink longitudinal streaks, ending in very long slender tails; petals falcate, oblong or oblong-lanceolate, somewhat acute, ciliate, 0.8–1 cm long; lip recurved, ovate-oblong, subacute, fleshy, channeled along the center, ~8 mm long; column oblong, ~8 mm long. **Distr:** Thailand, Burma, Malaysia, Borneo. **Ecol:** In lowland forests. Fl. Jul–Feb. **Zone:** 12.

B. macranthum Lindl., *Edwards's Bot. Reg.* 30: t. 13, 1844. (Syn: *Phyllorkis macrantha, Sarcopodium macranthum*, etc.). **Herbs**, epiphytic; rhizome very stout, 6 mm thick, woody, with large tufts of short roots at nodes; pseudobulbs ~1.3 cm apart, oblong wrinkled 2.5 cm long. **Leaf** elliptic 12–20 × 4.5–7.5 cm, obtuse, coriaceous; petiole stout, 2.5 cm long. **Inflorescence** single-flowered; peduncle very short; pedicel thick 7.5 cm long. **Flowers** 5 cm across, clove-scented; dorsal sepal lanceolate, acute, white with purple spots; lateral sepals larger, ± ovate, bases depressed, connivent, yellow with crimson spots, apices acuminate, dark brown; petals similar to dorsal sepal but slightly narrower; lip ~6 mm long, base quadrate with 2 truncate ears, apex acute curved, yellow tinted with red; column short and thick yellow, arms porrect, toothlike, acute; anther conic obtuse deep red. **Distr:** Assam to Indo-China and Solomon Is. **Ecol:** Epiphytic on tree trunks, in forests. **Zones:** 10–12.

B. macrobulbum J.J. Sm., *Bull. Dép. Agric. Indes Néerl.* 39: 4, 1910. (Syn: *B. balfourianum*). **Herbs**, epiphytic; pseudobulbs closely set, large, ovoid, 5.5–8 × 4–6 cm, cross-section subelliptic, obtuse, angulate, olive colored, rugose, 1-leafed. **Leaf** lanceolate, ~20(–60) × ~6.5 cm, base rounded, conduplicate, margins recurved, apex shortly acute and recurved, costa sulcate, glossy-green, margins and costa purplish. **Inflorescence** borne from the rhizome, ~2–4-flowered; peduncle ~7 mm long; bract ovate, base tubular, apex acuminate, concave; floral bracts ovate, ~1.5–1.6 × 1.2 cm, acute to acuminate, concave, ~13-nerved, membranous. **Flowers** large, carnose, malodorous, 10 cm across, yellow-green with dark brown spots, petals dark red at base, lip light purple; dorsal sepal oblique, erect, ovate, ~4 × 2.1 cm, canaliculate-subulate, acuminate, concave; lateral sepal decurved, falcate-oblong, ~2.4 × 1.65 cm, canaliculate-subulate, acuminate, concave, subcarinate; petals parallel, oblique-rhombic, base clawed, apex recurved, convex, lamina ovate-triangular, filiform, ~2.7 × 1.6 cm; lip strongly mobile, carnose, oblong-cordate, obtuse; pollinia 4; ovary obconic, 6-sulcate, ~7 mm long. **Distr:** New Guinea. **Ecol:** On moss-covered rocks or on trees at low elevations. **Zones:** 11–12.

B. makoyanum (Rchb. f.) Ridl., *Mat. Fl. Malay. Penins.* 1: 81, 1907. (Syn: *Cirrhopetalum makoyanum*). [Daisy Orchid]. **Herbs**, epiphytic, upright, sympodial; rhizome ~3 mm Ø; pseudobulbs 3 cm apart, ovoid, to 3 × 1.5 cm, unifoliate. **Leaf** elliptic to lanceolate, to 15 × 3 cm, base narrowed, apex obtuse, leathery, shortly petiolate. **Inflorescence** umbellate, 5–12-flowered; scape red, upright, slender, 21.5 cm tall, longer than the leaf; bracts lanceolate, 3 mm long, acuminate, purplish; ovary and pedicel 6 mm long. **Flowers** ~4 cm long, yellow with small red dots, fragrant; dorsal sepal ovate, concave, 4 × 2.5 mm, margins with small, dark red hairs; lateral sepals joined for entire 4 cm length, ~2 mm Ø, tapering gradually toward the tip; petals ovate, 4 × 2.5 mm, margins with dark red hairs; lip dull yellow, 4 × 1.6 mm, fleshy, highly mobile; column stout, foot sigmoid, arms broad acute. **Distr:** Peninsular Malaysia, Singapore, Borneo, Philippines. **Ecol:** In lowland forests; <300 m. **Zones:** 10–12.

B. maquilingense Ames and Quisumb., *Philipp. J. Sci.* 47: 208, 1932. **Herbs**, epiphytic, upright, sympodial; pseudobulbs conical, ~ 1.5 cm tall, 1 cm Ø, unifoliate. **Leaf** to 5 × 1.5 cm, leathery. **Inflorescence** erect, single-flowered, borne from the pseudobulbs. **Flower** 1.5 × 1 cm, pink, heavily mottled with maroon, only lateral sepals opening fully; dorsal sepal elliptic, 9 × 5 mm, hooded; lateral sepals obliquely oblong, 8 × 3 mm, pointing downward; petals elliptic, 6.5 × 3 mm, inside the hood formed by the dorsal sepal; lip maroon and white, highly mobile, narrow, 6 × 1.5 mm, fleshy. **Distr:** Philippines (Luzon). **Ecol:** Usually at base of small trees and shrubs; to 1,000 m. **Zones:** 11–12.

B. mearnsii Ames, *Philipp. J. Sci.*, C 8: 430, 1914. (Syn: *B. carinatum*). **Herbs**, epiphytic, upright, sympodial; pseudobulbs cylindrical, tapered, to 3 cm high, 2.3 cm Ø, unifoliate. **Leaf** to 25(–30) × 5.5 cm, leathery; petiole grooved, to 5 cm long. **Inflorescence** to 26 cm long, to 4-flowered; bracts several, funnel-shaped, 1 cm long; flowers opening in succession. **Flowers** golden-yellow mottled with small red spots, lip brownish-red; dorsal sepal lanceolate, to 3 × 1 cm; lateral sepals conspicuously ridged, triangular, to 4 × 1 cm; petals triangular, 1.5 × 0.7 cm, strongly reflexed toward the column; lip conduplicate, 1.4 cm long, fleshy, longitudinally grooved, covered with small warts, basal third smooth; column ~1 cm long, dividing about middle into 2 spreading lateral arms or wings, each wing deeply unequally cleft. **Distr:** Philippines. **Ecol:** In deep shade near forest floor; ~900(–2,000) m. **Zones:** 11–12.

B. membranifolium Hook. f., *Fl. Brit. India* 5: 756, 1890. (Syn: *B. insigne*, etc.). **Herbs**, epiphytic; rhizome ~3 mm thick; pseudobulbs conical to subcylindrical, apex narrow, rather distant, 2.5–4 cm long. **Leaf** elliptic-lanceolate, 15–25 cm long, base narrowed into a petiole, membranous. **Inflorescence** single-flowered; scape stout; sheath 1, lanceolate, acuminate. **Flower** ~4 cm across when expanded, yellowish green with red spots in lines; sepals ovate oblong or lanceolate, obtuse; lateral sepals deflexed; petals smaller, ovate lanceolate subacute; lip small, stipitate, tongue-shaped, decurved, apex verrucose, purple; column short, arms blunt, rather large, porrect. **Distr:** W Malesia to Philippines. **Ecol:** In swampy forests, lowland forests, lower mixed and open montane forests; 800–2,300(–2700) m. **Zones:** 10–12.

B. odoratum (Blume) Lindl., *Gen. Sp. Orchid. Pl.*: 54, 1830. (Syn: *Diphyes odorata, Phyllorkis odorata*). **Herbs**, epiphytic, upright, sympodial; pseudobulbs 2.5–8 cm apart, discoid, small, as slight thickenings on the rhizome, or non-existent, completely enveloped by several, imbricating, scarious sheaths, apex unifoliate. **Leaf** elliptic, to 25 × 6 cm, gradually narrowing into the elongate, conduplicate base, thick, leathery, medium to dark green. **Inflorescence** racemose, to 60–150-flowered, upright or pendent, 30–70 cm long; peduncle, terete, whitish, relatively thick, completely enveloped by several, tight, tubular, acuminate, imbricating sheaths. **Flowers** small, ~1.5 cm across, sweetly fragrant (sometimes only at night), creamish-white, yellowish or greenish; dorsal sepal triangular, 7 mm long, with the tips drawn out into tails; lateral sepals triangular, 1 cm long, with tips drawn out into tails; petals ovate, to 2 mm long; lip elliptic to ovate, 2 mm long, thick. **Distr:** Malesia. **Ecol:** In lowland and lower montane forests, rarely lithophytic; 0–1,000 m. **Zones:** 10–12.

B. ovalifolium (Blume) Lindley, *Gen. Sp. Orchid. Pl.* 49, 1830. (Syn: *Diphyes ovalifolia*, etc.). **Herbs**, small, epiphytic; rhizome creeping; pseudobulbs contiguous, nearly flattened-globose or ellipsoid, 0.4–1 × 0.5–0.8 cm, with a terminal leaf. **Leaf** subsessile or shortly petiolate; lamina elliptic, 1–3 × 0.45–1 cm, base contracted, apex obtuse, leathery. **Inflorescence** basal, erect, 3.2–6 cm, 1-flowered; peduncle filiform, 1–3 cm; floral bracts cupular, ~2 mm, acute; pedicel and ovary 1.5–2.3 cm long. **Flowers** cream-colored or yellow to red, usually with darker veins; dorsal sepal elliptic, 3.8–6.8 × 1.5–3 mm, apex acute-apiculate; lateral sepals elliptic, 0.45–1.2 × 0.2–0.8 cm, ± glabrous, acute to acuminate; petals obovate-oblong, 1–1.8 × 0.7–1 mm, obtuse; lip yellow, orange, or red, elliptic to ovate, 1.8–3.5 × 1.2–2.5 mm, fleshy, proximally concave, margins then folding downward, rounded, adaxial surface distally papillose to verrucose toward margins; column 1–1.8 mm long. **Distr:** China, Indonesia, Malaysia, Thailand. **Ecol:** In forests; ~2,400 m. Fl. May. **Zones:** 8–12.

B. piestoglossum J.J. Verm., *Blumea* 38: 152, 1993. **Herbs**, epiphytic, upright, sympodial; rhizome 4.5 mm Ø, covered with sheaths which disappear with age; roots borne from the internodes of the rhizome; pseudobulbs 5–6 cm apart, tapered, to 3.2 cm long, 1.5 cm Ø, unifoliate. **Leaves** oblong to elliptic, to 14 × 5.5 cm; petiole 2–2.3 cm long. **Inflorescence** upright, to 7 cm long, borne from the pseudobulbs with 3 basal sheaths, single-flowered. **Flower** to 5 cm Ø, orange-yellow with reddish brown stripes and spots; dorsal sepal lanceolate, 2.8 × 1.3 cm; lateral sepals greenish brown with some white spots, ovate, keeled, 3 × 0.8 cm; petals linear, 2.4 × 0.8 cm, apices pointed; lip deep yellow, abaxially brown, tongue-shaped, 1.5 × 0.7 cm, fleshy, highly mobile, curved. **Distr:** Endemic to Philippines (Luzon). **Ecol:** Epiphytic on tree trunks; 600–1,300 m. **Zones:** 11–12.

B. plumatum Ames, *Orchidaceae* 5: 184, 1915. (Syn: *B. jacobsonii, Rhytionanthos plumatum*). **Herbs**, epiphytic; rhizome long-creeping; pseudobulbs 2–3 cm apart, obliquely ovoid, 1.5–1.8 × 0.75–0.95 cm, quadrangular, unifoliate. **Leaves** narrowly elliptic, 7–9.5 × 1.5–2.9 cm, base shortly conduplicate, shortly petiolate, apex recurved, acute to shortly acuminate, thickly coriaceous, median nerve sulcate above, light green, waxy, tessellated pale purplish-red when young. **Inflorescence** subumbellate, 3–4-flowered, lateral, basal, ascending; peduncle to 10 cm long, greenish with purple flecks; sheathing bract subulate-apiculate, 7.5 mm long; floral bracts 8 mm long, greenish, spotted purple; pedicel and ovary ~3–4 mm long, greenish-white, purple-spotted. **Flowers** reddish-purple or yellow; dorsal sepal suborbicular or ovate-orbicular, 0.7–1 × 0.65–0.75 cm, shortly apiculate, acute, strongly concave; lateral sepals obliquely linear, 6.5–10 × 0.2–0.73 cm, acute, margin connivent for ~1.8 cm, free near the apex, pendulous; petals oblong or obliquely elliptic, 1.5–0.35 × 0.3 cm, rounded, obtuse, margin

Top row: *Bulbophyllum laxiflorum, Bulbophyllum lemniscatoides, Bulbophyllum lilacinum, Bulbophyllum lobbii.*
Second row: *Bulbophyllum longissimum, Bulbophyllum macranthum, Bulbophyllum macrobulbum.*
Third row: *Bulbophyllum makoyanum, Bulbophyllum maquilingense, Bulbophyllum mearnsii.*
Bottom row: *Bulbophyllum membranifolium, Bulbophyllum odoratum, Bulbophyllum ovalifolium.*

MG

GS

JT MG

JT

MG MG

MS MG

JT

JT

JT MG

Top row: *Bulbophyllum piestoglossum*, *Bulbophyllum plumatum* (dark form), *Bulbophyllum putidum*, *Bulbophyllum reptans*.
Bottom row: *Bulbophyllum retusiusculum*, *Bulbophyllum rothschildianum*, *Bulbophyllum saltatorium*, *Bulbophyllum sopoetanense*.

with numerous versatile, clavate, obtuse, filiform, vesiculate appendages; lip triangular-linguiform, 2–3 × 1.2 mm, acute, fleshy; column 2 mm long, fleshy. **Distr:** Peninsular Malaysia (Johor), Sumatra, Philippines (Mindanao). **Ecol:** In swampy lowland forests, hill forests; 1,000 m. Fl. Jul–Aug. **Zones:** 10–12.

B. putidum (Teijsm. and Binn.) J.J. Sm., *Bull. Jard. Bot. Buitenzorg*, II, 8: 27, 1912. (Syn: *Cirrhopetalum fascinator*, etc.). **Herbs**, epiphytic, 10 cm high; pseudobulbs clustered, elliptic or ovoid, to 2 cm high, glossy green or olive-green, apex 1-leafed. **Leaf** sessile; lamina oblong-elliptic, to 5 × 3 cm, glossy green. **Inflorescence** 1-flowered, to 10–14 cm tall; floral bract lanceolate, membranous. **Flower** to 20 cm long, yellow-green or pale green marked with purple or crimson, dorsal sepal fringed with fine purple hairs, lip purple; dorsal sepal ovate, 2.5–3cm long, fimbriate, apex pointed; lateral sepals joined for most of their length, tapering to become filiform, 10–18 cm long, caudate, tips free; petals linear-oblong, 2 cm long, curved downward, fringed with long paleolate or clavate ornaments; lip strongly curved, with 2 high keels from base to apex. **Distr:** India to Borneo and the Philippines. **Ecol:** In wet submontane and primary montane forests; 1,000–2,500 m. Fl. Sep–Nov. **Zones:** 10–12.

B. reptans (Lindley) Lindley, *Gen. Sp. Orchid. Pl.* 51, 1830. (Syn: *Phyllorkis reptans*, etc.). **Herbs**, epiphytic; rhizome creeping, branched, with tubular sheaths 0.7–1 cm long; pseudobulbs 5–9 cm apart, ovoid or ovoid-conic, 1–2 × 0.5–1 cm, apex 1-leafed. **Leaf** narrowly oblong, 7–8(–15) × 0.7–1(–1.4) cm, base contracted into petiole, apex obtuse and slightly emarginate, leathery; petiole 1–2 cm long. **Inflorescence** basal, erect, 12–14 cm long, slender; raceme 4–6 cm long, usually 3–6-flowered; peduncle with 3 or 4 tubular sheaths; floral bracts ovate-lanceolate, 3–6 mm long; pedicel and ovary 3–5 mm long. **Flowers** pale yellow with purplish-red stripes; sepals lanceolate, 0.7–1 × ~0.2 cm, subacute, thickly textured, abaxially carinate; lateral sepals lower edges basally connate, base adnate to column foot; petals ovate-elliptic or obovate, 3–4 × 1.5–2 mm, thinly textured, entire, obtuse-rounded; lip recurved, somewhat fleshy, entire, obtuse; column ~0.8 mm long, foot ~2 mm long. **Distr:** Bhutan, China, NE India, Myanmar, Nepal, Vietnam. **Ecol:** In evergreen broad-leaved forests; 1,000–2,800 m. Fl. Jan–Oct. **Zones:** 8–12.

B. retusiusculum Rchb. f., *Gard. Chron.* 1182, 1869. (Syn: *B. flavisepalum*, etc.). [Yellow Comb Orchid]. **Herbs**; pseudobulbs 1–3 cm apart, ovoid, 1–2 × 0.6–1 cm, greenish-yellow, obscurely rugose. **Leaves** sessile, linear-oblong, 3–7 × ~1–1.8 cm, base cuneate, apex obtuse or slightly retuse. **Inflorescence** an umbel, ~1.5 cm long, 5–8- flowered; scape ~4–7 cm long, yellowish green, with minute scales below; bracts ~2–3 mm long; pedicel and ovary 0.8–1 cm long. **Flowers** with dorsal sepal reddish, ovate, 3–4 × 1.5–2 mm, upper part thinner, with 3 dark veins, apex rounded to obtuse, lower part thickened and concave, with many small processes outside; lateral sepals orange to yellow, often connate along upper edge, linear-oblong, 1–1.5 cm × 2–3.5 mm, base oblique, with purplish veins; petals reddish, obovate to elliptic, 3 × 1.5 mm, trinerved, slightly oblique; lip yellowish-brown above, reddish below, hornlike, ~3 mm long, apically recurved; column purple, 1.5 mm long, with 2 short stylids. **Fruits** ellipsoid, ~1.5 mm long. **Distr:** S China, Himalaya, India, Myanmar, Taiwan, Thailand, Indo-China. **Ecol:** In forests; 500–1,500 m. **Zones:** 10–12.

B. rothschildianum (O'Brien) J.J. Smith, *Bull. Jard. Bot. Buitenzorg*, sér. 2, 8: 27, 1912. (Syn: *Cirrhopetalum rothschildianum*). **Herbs**, epiphytic; pseudobulbs well-spaced, ovoid, ~3 cm Ø, apex 1-leafed. **Leaf** subelliptic, usually 9–10 × 2–2.5 cm, apex obtuse and retuse; thickly leathery; petiole ~1 cm long. **Inflorescence** an umbel, 4–6-flowered; scape basal, erect, 20–24 cm long, stout; peduncle with 3 tubular sheaths, ~1 cm long; floral bracts lanceolate, ~1 cm long, acuminate; pedicel and ovary ~2 cm long. **Flowers** large, purplish red; dorsal sepal ovate, cymbiform, ~ 1.5 × 0.7 cm, abaxially sparsely papillate, margins fimbriate, apex caudate, fringes narrowly oblong, to 7 mm long; lateral sepals lanceolate, 15–19 cm long, abaxially densely papillate below middle, twisted near base, upper edges connate forming a synpetal ~2 cm wide, apex long-caudate; petals ovate-triangular, ~1 × 0.4 cm, abruptly contracted above middle and caudate, margins fimbriate; lip ligulate-elliptic, ~1 cm long, fleshy, ± conduplicate in basal half, basally attached to column foot, surface densely fimbriate-hairy; column ~5 mm long, with triangular, enlarged wings at middle. **Distr:** China, NE India. **Ecol:** In dense forests; 1,500–1,600 m. **Zones:** 10–11.

B. saltatorium Lindl., *Edwards's Bot. Reg.* 23: t. 1970, 1837. (Syn: *Phyllorkis saltatoria*). **Herbs**, epiphytic; pseudobulbs orbicular to narrowly ovoid, 0.7–5 × 0.6–3.3 cm, ± flattened, 1-leafed. **Leaf** narrowly elliptic to linear-lanceolate, 1.3–21 × 0.8–5.5 cm, apex rounded to subacute, slightly emarginate, occasionally oblique, thickly coriaceous. **Inflorescence** 3.5–84 cm long, 2–105-flowered, flowers distichous or scattered; peduncle erect to patent, 1.5–58 cm long, with 6–11 tubular scales to 1.3 cm long; floral bracts narrowly elliptic to oblong, 2–9.5 x 1–3.5 mm, obtuse to acute; pedicel and ovary 0.3–1 cm long. **Flowers** with dorsal sepal narrowly ovate-oblong to ovate-lanceolate, 0.5–1.5 x 0.18–0.38 cm, margins finely ciliate, apex acute; lateral sepals free, oblique, slightly falcate, 0.5–1.5 x 0.17–0.36 cm; petals not recurved, slightly falcate, linear, oblique, (2–)2.5–6(–7) x 0.15–0.6 mm, acute to acuminate; lip recurved, ovate-oblong to linear, 0.4–0.9(–1.2) cm x 0.7–1.8 mm, obtuse to acute; pollinia 4. **Fruits** obovoid, to 2.2 x 0.6 cm. **Distr:** Tropical Africa. **Ecol:** In primary forests; 0–750 m. **Zones:** 10–12.

B. sopoetanense Schltr., *Repert. Spec. Nov. Regni Veg.* 10: 181, 1911. (Syn: *B. conspectum*). **Herbs**, epiphytic, 15–22 cm high; rhizome elongate, flexuous, terete, rigid, ~3 mm Ø; pseudobulbs 3–4 cm apart, ~2.5 mm high, ~2.5 mm Ø, 1-leafed. **Leaf** erect, petiolate; lamina oblong-ligulate, 10–15 x $^2/_5$–3 cm, base cuneate, apex obtuse, glabrous, coriaceous; petiole sulcate, 3–5 cm long. **Inflorescence** a raceme, laxly 8–14-flowered, borne from the base of the pseudobulb, erect; peduncle terete, glabrous; bract oblong-lanceolate, acuminate, glabrous. **Flowers** glabrous; dorsal sepal ligulate, 1.3 cm long, acute; lateral sepals obliquely ovate, shortly acuminate; petals oblique, lanceolate-ligulate, ~9 mm long-obtuse, glabrous; lip curved, oblong-ligulate, ~1.1 x 0.45 cm; column subquadrate, angled, shortly apiculate, foot umbonate; ovary cylindrical, 5 mm long, glabrous. **Distr:** Borneo, Sulawesi. **Ecol:** Epiphytic near ground; 1,300–1,900 m. Fl. Dec. **Zones:** 11–12.

B. spathulatum (Rolfe ex Cooper) Seidenfaden, *Bot. Tidsskr.* 65: 347, 1970. (Syn: *Cirrhopetalum spathulatum*). **Herbs**, epiphytic; rhizome creeping, stout; pseudobulbs 10–11 cm apart from, narrowly ovoid, 3–4 x ~1 cm, with a terminal leaf. **Leaf** oblong, 10–18 x 2–2.4 cm, fleshy, apex obtuse; petiole ~1 cm, ± conduplicate. **Inflorescence** umbel >20-flowered; peduncle with 4 or 5 sheaths ~1 cm; scape from base of pseudobulb, 3–4 cm; floral bracts oblong-ovate, ~1 cm, apex subacute; pedicel and ovary 1.5–2 cm long. **Flowers** purplish-red; dorsal sepal subobovate, ~8 x 4 mm, apex subtruncate, slightly mucronate; lateral sepals ~1.8 cm, twisted near base, upper and lower edges connate to varying degrees forming a slipper-shaped synsepalum; synsepalum ~1.1 cm wide, outer surface finely papillate, apex rounded; petals narrowly oblong-lanceolate, ~ 5 x 1.3 mm, entire, obtuse; lip recurved, lanceolate, glabrous, ± conduplicate basally, apex obtuse; column ~2 mm, foot ~3 mm. **Distr:** China, India (Sikkim), Laos, Myanmar, Thailand, Vietnam. **Ecol:** In broad-leaved forests; 800–900 m. Fl. Oct. **Zones:** 9–12.

B. subumbellatum Ridl., *J. Linn. Soc., Bot.* 31: 274, 1896. **Herbs**, epiphytic; rhizome long and slender; pseudobulbs crowded or as much as 12.5 cm apart, cylindric, 5 cm long, 9 mm thick. **Leaf** lanceolate, 12.5–15 x 3.2 cm, acute, petiolate. **Inflorescence** racemes, short, 2–3-flowered, crowded; scape rather thick 10 cm long with 4 or 5 cuspidate ovate sheaths; bracts ovate, 6 mm long, cuspidate; pedicels 12.5 cm long. **Flowers** 5 cm long, greenish-yellow with red spots; dorsal sepal ovate, ~2.5 x 0.9 cm, acute; lateral sepals twisted; petals much smaller, broadly ovate, 9 x 6 mm; lip tongue-shaped, 1.25 cm long, stipitate, acute, curved; column short with a long slender foot, arms very short, hardly distinct. **Distr:** Peninsular Malaysia (Johor), Borneo (Sarawak). **Ecol:** In lowland swamp forests, on tree trunks and branches; 0–250 m. **Zones:** 10–12.

B. taiwanense (Fukuyama) K. Nakajima, *Biol. Mag. Okinawa* 10: 36, 1973. (Syn: *Cirrhopetalum taiwanense*). **Herbs**, epiphytic; rhizome creeping; pseudobulbs 0.5–1.5 cm apart, usually ovoid, 0.6–1.2 cm, ± longitudinally striped, with a terminal leaf. **Leaf** shortly petiolate; lamina narrowly oblong, 3–5 x 1–2 cm, fleshy, apex obtuse and retuse. **Inflorescence** a raceme, shortened, umbel-like, 0.6–1 cm long, densely 5–8-flowered; peduncle yellowish-green, longitudinally striped; scape basal, 6–10 cm long; floral bracts lanceolate, ~3 mm long, acuminate; pedicel and ovary ~1 cm long. **Flowers** not widely opening, yellowish-red; dorsal sepal ovate-lanceolate, concave, 6–8 x ~3 mm, ciliate, acuminate; lateral sepals free, divergent, obliquely lanceolate, 1.3–1.5 x ~0.2 cm, margins incurved in apical half, apex obtuse, sometimes ciliate; petals ovate, ~ 5 x 2 mm, membranous, apex acute, ciliate; lip slightly reflexed, hornlike, ~3 mm long, fleshy, adaxially with 1 longitudinal groove, base dilated and attached to end of column foot, apex obtuse; column semicylindric, ~1.5 mm long; foot curved, ~1.5 mm long. **Distr:** S Taiwan. **Ecol:** In forests; <1,000 m. Fl. Apr–May. **Zones:** 10–11.

B. uniflorum (Blume) Hassk., *Cat. Hort. Bot. Bogor.*: 39, 1844. (Syn: *Cirrhopetalum compressum*, etc.). **Herbs**, epiphytic; rhizome very short and stout; pseudobulbs cylindric, terete, slightly flattened, usually 2.5–5 cm long. **Leaf** 7.5–15 cm long, subsessile. **Inflorescence** single-flowered; scape very short, sheathed; pedicel 7.5–12.5 cm long. **Flower** 2 cm across, sepals orange-yellow, spotted with purple, lip tipped with red and petals with white; dorsal sepal ovate-oblong, obtuse, much large than lateral sepals; lateral sepals broadly ovate, pointing downward (uncinately incurved); petals linear, trinerved; lip with recurved and acute basal auricles. **Distr:** W and C Malesia. **Ecol:** Epiphytic on tree trunks of hill and lower montane forests or dense riverine forests; 600–1,800 m. **Zones:** 10–12.

B. vaginatum (Lindl.) Rchb. f., *Ann. Bot. Syst.* 6: 261, 1861. (Syn: *Cirrhopetalum vaginatum*, etc.). **Herbs**, epiphytic; rhizome long-creeping or pendulous, stout; pseudobulbs 3–5 cm apart, ovoid or conic, angular, 1.2–2.5 cm long, 1-leafed. **Leaf** oblong, 5–12 x 2.5 cm, base shortly petioled, apex blunt, emarginate or slightly cleft, thickly and densely coriaceous. **Inflorescence** (1–2)15-flowered, capitately racemose; scape 5–10 cm long, laxly covered with several sheaths; bracts lanceolate, 1.2 cm long, longer

Top row: *Bulbophyllum spathulatum, Bulbophyllum subumbellatum, Bulbophyllum taiwanense*.
Bottom row: *Bulbophyllum uniflorum, Bulbophyllum vaginatum, Bulbophyllum violaceolabellum*.

JT

JT

MG

JT

JT

MG

Top row: *Dendrobium aloifolium, Dendrobium amabile, Dendrobium anosmum.*
Bottom row: *Dendrobium amethystoglossum, Dendrobium antennatum, Dendrobium aphyllum.*

than the pedicels. **Flowers** pale yellow, 2.5+ cm across; sepals long-tailed, dorsal sepals and petals ciliate fimbriate or appendaged; dorsal sepal ovate, 9 mm long, margins shortly fringed or ciliate, apex acuminate, 5-nerved; lateral sepals lanceolate, 4–7 cm long, margins fringed toward the base, both margins joined near base, ends free, 5-nerved; petals oblong, 3 mm long, margins fringed, apex rounded, ciliate, faintly trinerved; lip oblong, 2 mm long, 2-ridged, columnar spurs long, slender, erect. **Distr:** Peninsular Thailand, W and C Malesia. **Ecol:** On trees in open trees, flowering gregariously, sometimes a single tree trunk may have thousands of flowers. **Zones:** 11–12.

B. violaceolabellum Seidenfaden, *Nordic J. Bot.* 1: 210, 1981. **Herbs**, epiphytic or lithophytic; rhizome stout, creeping; pseudobulbs 4–9 cm apart, ovoid, 1.5–3.4 cm, apex 1-leafed. **Leaf** oblong to obovate-oblong, 10–20 × 2.2–4.3 cm, apex obtuse, slightly fleshy or leathery; petiole conduplicate, 1.5–2 cm long. **Inflorescence** raceme shortened, umbel-like, often 3–5-flowered; peduncle with 3 or 4 tubular sheaths 0.8–1.3 cm long; scape from base of pseudobulb, erect, 19–26 cm long; floral bracts lanceolate, ~5 mm long, acuminate; pedicel and ovary ~2.2 mm long. **Flowers** widely opening, sepals and petals yellow and purplish-spotted, lip purple; dorsal sepal broadly ovate, 6–8 × ~6.2 mm, margin slightly undulate, apex obtuse; lateral sepals ovate-triangular, 8–9 × ~8 mm, base adnate to column foot, apex mucronate; petals ovate-lanceolate, 4–5 × ~3 mm, entire, aristate; lip strongly recurved, ligulate, ~5 × 3 mm, fleshy, base attached to end of column foot by a mobile joint, apex truncate; column yellow, ~2 mm long, with wings decurrent to foot, foot purple, ~5.5 mm long. **Distr:** China (S Yunnan), Laos. **Ecol:** In limestone areas of open forests; ~700 m. Fl. Apr. **Zones:** 7–11.

Dendrobium

Species over 1,400. **Herbs**, epiphytic, tufted; stems terete, slender, nodose, fleshy or not, erect or pendulous, often thickened and forming short pseudobulbs, sometimes partly dilated into several pseudobulbs, mostly surrounded by tight, persistent sheaths. **Leaves** sessile, often articulate with sheaths; lamina coriaceous, subcoriaceous or membranous, entire, flat or rarely subterete and laterally flattened. **Inflorescences** terminal or lateral, from nodes of stem, racemose or flowers solitary, rarely cymose; peduncle obscure or shorter than pedicel; bracts minute. **Flowers** resupinate; sepals subequal, dorsal one free, lateral ones often oblique at base and adnate to foot of column and forming spurlike or saccate mentum; petals more or less similar to dorsal sepal; lip various, attached to and usually incumbent on foot of column, contracted or clawed at base, often trilobed; lateral lobes erect, embracing column; column short, with prominent foot; anther terminal, incumbent, 2-celled; pollinia 4, ovoid or oblong, waxy, connate laterally in pairs in each cell, without caudicle or viscidium; rostellum usually hooked below. **Fruits** capsules, ovoid to oblong. **Distr:** Japan, Korea and China, through India, Malaysia and Indonesia to Papua New Guinea and Australia. **Cult:** Propagate from seeds, plantlets, stem cuttings (each with a dormant bud) or by division. Requires a permeable, coarse substrate (either potted, attached to bark or a tree) in full or partial sun.

D. aloifolium (Blume) Rchb. f., *Ann. Bot. Syst.* 6: 279, 1861. (Syn: *Aporum aloifolium*, etc.). **Herbs**, epiphytic; rhizome very short; stem erect to sprawling, 5.5–10.5 cm long, with 9–19 leaves. **Leaves** distichous, laterally compressed, held at ~45° to the stem; lamina ovate, to ~1.7 × 0.75 cm, apex acute, sharply pointed, succulent, thickly leathery; sheaths equitant, upper margin membranous. **Inflorescences** apical, borne from a compressed bundle of scales, 1-flowered; peduncle ~1 mm long, obscured by scales, cylindrical; pedicel and ovary descending, cylindrical, 3 mm long. **Flowers** very small, 4 × 3.5 mm, fleshy, stiff, sepals and petals off-white with longitudinal purplish stripes, lip white sometimes fringed faint purple, column greenish-white, anther green; sepals and petals strongly recurved; dorsal sepal oblong, 2.6 × 1.3 mm, abaxially broad-ridged, apex tridentate; lateral sepals 4 mm long, free part oblong, 2.4 × 1.6 mm, rounded, minutely retuse; petals linear, 2.2 × 0.7 mm, obtuse apiculate; lip porrect, 4 × 1.8 mm, obscurely trilobed, base subsaccate, midlobe subcircular ~1 mm long; column ~1 mm long; pollinia 4. **Distr:** Indo-China to New Guinea. **Ecol:** In lowland rainforests, hill forests; 0–550 m. Fl. throughout the year. **Zones:** 10–12.

D. amabile (Lour.) O'Brien, *Gard. Chron.*, III, 46: 393, 1909. (Syn: *Callista amabilis*, etc.). **Herbs**, epiphytic; pseudobulbs widening toward apex, somewhat longitudinally ribbed, to 70 × 1.5 cm, with (2–)4–6 leaves toward the apex. **Leaves** persistent; lamina ovate or oblong-lanceolate, 12 × 5 cm, apex acute, leathery, shiny, green. **Inflorescences** 1 or 2, laxly ~30-flowered, pendulous, from near the apex of older leafless pseudobulbs, 30 cm long; flowers lasting for ~10 days; floral bracts ~1.2 cm long; pedicel and ovary ~3.5 cm long, pink. **Flowers** to 4.5 cm Ø, sepals and petals pale pink or white, lip white with a bright orange patch toward the base; sepals united at base; dorsal sepal ovate, 1.4 × 1.2 cm; lateral sepals ovate, 2 × 1 cm; petals ovate, 2.2 × 1.6 cm, base clawed; lip broadly ovate, 2 × 2 cm, base deeply saccate enveloping the column, margin entire or minutely fringed; column ~3 mm long; mentum absent. **Distr:** Vietnam. **Ecol:** In lowland and submontane evergreen forests; 1,200–2,500 m. **Zones:** 10–11.

D. amethystoglossum Rchb. f., *Gard. Chron.* 1872: 109, 1872. (Syn: *Callista amethystoglossa*, etc.). **Herbs**, epiphytic, usually upright, sympodial; pseudobulbs to 1 m tall, 2 cm Ø. **Leaves** few, deciduous; lamina elliptic, 5–12 × 1.2–2 cm, acute, thin, light green, striped with purple when young. **Inflorescence** pendent or semipendent, usually arising on leafless canes, to 20-flowered, in a dense cluster, to 10 cm long. **Flowers** 3–4 cm Ø, white to cream or pale yellowish white, waxy, long-lasting, fragrant; dorsal sepal lanceolate, to 2 × 0.8 cm; lateral sepals lanceolate, to 3 cm long, joined to form a spur ~1.5 cm long; petals lanceolate, ~2 cm; lip lanceolate to linear-spathulate, ~2 × 0.4 cm, basal part white with a backward-pointing, fleshy, triangular ridge, apical half amethyst purple, tip usually white. **Distr:** Endemic to Philippines (N Luzon). **Ecol:** On trees or on limestone cliffs, of moist montane forests; 1,400 m. **Zones:** 9–12.

D. anosmum Lindl., *Edwards's Bot. Reg. Misc.* 32, 1845. **Herbs**; rhizome very short; pseudobulbs crowded, many-leaved, pendulous, cylindrical, 0.6–1.1 m long, unbranched. **Leaves** oblong to elliptic-ligulate, 10–15 × 2.2–3.5 cm, base cuneate, apex unequally bilobulate, soon deciduous. **Inflorescences** lateral, short, racemose; peduncles 0.5–0.7 cm long, covered with short, obtuse scales, 1–3-flowered; floral bracts similar to peduncle scales; pedicel and ovary slender, cylindrical, 3.5 cm long, glabrous. **Flowers** resupinate, ~6 cm Ø, outside whitish, inside violet-pink, lip with 2 darker blotches; dorsal sepal elliptic-ligulate, 3 cm long, apex obtuse, apiculate; lateral sepals obliquely elliptic-ligulate, slightly widened to base, 3 cm long, apex obtuse, apiculate, mentum 7 mm long, obtuse; petals obliquely elliptic, 3 × 1.4–1.6 cm, margins minutely subciliate-serrulate, apex apiculate; lip clawed, claw quadrangular, ~4 mm long, orbicular, 2.1 × 2.1 cm, base with a subquadrate callus; column 7 mm long, lateral lobules triangular, foot 7 mm long. **Distr:** Sri Lanka, Indo-China to New Guinea. **Ecol:** In lowland forests, especially along rivers; 0–650 m. Fl. throughout the year. **Zones:** 10–12.

D. antennatum Lindl., *Lond. J. Bot.* 2: 236, 1843. (Syn: *Callista antennata*). **Herbs** epiphytic; stems clustered on a short rhizome, 15–75 cm tall, somewhat fusiform. **Leaves** distichous, articulated at base to gray sheaths 2–2.9 cm long; lamina oblong-lanceolate to ovate-elliptic, 4–15 × 0.5–4 cm, apex acute, unequally bilobed, coriaceous to fleshy. **Inflorescences** 1 to several, 15–35 cm long, 3–15-flowered; bracts tubular, 3–4 mm long. **Flowers** fragrant, white with green or yellow-green petals and a purple- or violet-veined lip; pedicel and ovary 2.2–3.5 cm long; dorsal sepal oblong-lanceolate, 1.6–2.3 × 0.6–0.7 cm, acute or acuminate, recurved and often spirally twisted; lateral sepals oblong-lanceolate, 1.6–2.5 × 0.7 cm, acute or acuminate, recurved; mentum narrowly conical, 0.8–1 cm long; petals linear, 2.5–5 × 0.2–0.35 cm, acute, once- to twice-twisted; lip trilobed, 1.5–2.3 × 0.9–1.15 cm, side-lobes elliptic, rounded in front, midlobe circular to ovate, acute or apiculate, not recurved, callus of 5 longitudinal ridges; column 5–6 mm long. **Distr:** New Guinea to Solomon Islands and NE Australia. **Ecol:** To 1,200 m. **Zones:** 10–12.

D. aphyllum (Roxb.) C.E.C. Fisch., Fl. Madras: 1416, 1928. (Syn: *Cymbidium aphyllum*, etc.). **Herbs**, epiphytic or lithophytic; pseudobulbs long and slender, pendulous, 30–90 cm long, 0.6–1 cm Ø. **Leaves** soon deciduous, arranged in 2 rows along the whole stem, sessile; lamina linear-lanceolate or narrowly ovate, 5–13 × 2–3 cm, apex acuminate, thin. **Inflorescences** 1–3-flowered, abbreviated, borne on the nodes throughout the leafless old stems; bracts oblong, acute. **Flowers** 3–5 cm Ø, thinly textured and fragile, lasting 10–20 days, fragrant, sepals and petals white, pale pink or rosy-mauve, lip cream or pale yellow, base purple-veined or -marked; sepals oblong-lanceolate, 2.5–3 × 0.6–0.8 cm, acute; petals broadly oblong-elliptic, 2.5× 1.5 cm, obtuse; lip broad, suborbicular, 2.5 × 2.5 cm, tubular or convolute at base, shortly clawed, limb short, tip incurved, obscurely trilobed, pubescent above, margins erose, almost ciliate, callus at base of lip 3-ridged; column 9 mm long. **Distr:** Tropical and subtropical Asia to Queensland. **Ecol:** In tropical warm forests, valleys; 200–1,800 m. **Zones:** 9–12.

D. bensoniae Rchb. f., *Bot. Zeitung* (Berlin) 25: 230, 1867. (Syn: *Callista bensoniae*). **Herbs**, epiphytic; pseudobulbs stemlike, erect or pendulous, 0.3–1 m long, nodes swollen. **Leaves** deciduous; lamina linear, ~12 cm long, apex acute or emarginate. **Inflorescences** borne from the upper nodes of the leafless pseudobulbs, solitary or sometimes paired, 1–3-flowered, to 5 cm long; floral bract scalelike, 7 mm long; pedicel and ovary 2.5 cm long. **Flowers** 5–6.3 cm across, fragrant, white or creamish-white, lip orange-yellow with 2 large maroon or black-purple blotches at base; sepals oblong-lanceolate, acute; dorsal sepal 3 × 1 cm, apex obtuse, tip concave; lateral sepals 3.6 × 0.8 cm, apex pointed; petals round or oblong, 4.2 × 2 cm, base clawed, apex obtuse; lip concave, orbicular, forming a tube at base and arched below, 3.8 cm long, 6.5 cm at the widest point (when spread), tubular part 0.8–1 cm long, tomentose; column ~7 mm long. **Distr:** Assam to Thailand. **Ecol:** 450–1,550 m. **Zones:** 10–12.

D. bicaudatum Reinw. ex Lindl., *J. Linn. Soc.* 3: 20, 1859. (Syn: *D. antelope*, *D. minax*, etc.). **Herbs**, epiphytic; stems clustered, canelike, narrowly fusiform, dilated at or below middle, 8–43 × 1–2 cm, to 20-noded. **Leaves** 4–10, suberect, lanceolate or narrowly oblong-elliptic, to 9 × 2–2.4 cm, apex unequally and obscurely subacutely bilobed, thinly coriaceous, articulated to tubular sheaths 1.1–2.2 cm long. **Inflorescences** 8–15 cm long, few-flowered; bracts sheathing, tubular, 3 mm long; pedicel and ovary 1.5–2.4 cm long. **Flowers** yellow-green or green, sometimes suffused with brown or red especially on petals; dorsal sepal lanceolate, 1.5–2.1 × 0.4 cm, acute, recurved toward apex; lateral sepals falcate, obliquely triangular, 1.6–2.4 × 0.7–1 cm, acuminate; mentum obliquely conical, 7–9 mm long; petals linear, 2.3–5 × 0.25–0.3 cm, acute, slightly dilated above, not or only once-twisted; lip white or greenish, purple-veined, trilobed, 1.5–1.8 × 1–1.5 cm, lateral lobes obliquely triangular, rounded in front; midlobe transversely oblong, apiculate, 0.4 × 1 cm, sides incurved, callus of 5 fleshy ridges at base of midlobe, outer two often short and obscure; column 4–5 mm long. **Distr:** Sulawesi, Sulu Archipelago and Moluccas (Ambon, Ceram). **Ecol:** Epiphytic in rocky areas, sometimes lithophytic; 0–1,000 m. Fl. Feb–Jul. **Zones:** 10–12. **Info:** Flower morphology is highly variable across its native range.

D. bigibbum Lindl., *Paxton's Fl. Gard.* 3: 25, 1852. (Syn: *D. phalaenopsis*). [Butterfly Dendrobium]. **Herbs**; rhizome creeping; stem erect, slender, 13–18 cm long, few-leaved. **Leaves** narrowly ovate, to 8 × 1 cm, apex acute, pointed, light green, margins purple-suffused, thin, leathery. **Inflorescences** terminal or lateral from upper nodes, suberect to horizontal, to 15 cm long, slender, with ~4 well-spaced flowers; peduncles with 4 sheathing sterile bracts; floral bract small; pedicel and ovary 2 cm long. **Flowers** widely opening, to 5 cm Ø, sepals and petals slightly recurved, satiny pinkish-mauve with white margins, mentum green underneath, lip dark purple with deep purple veins on disk and midlobe, with 5 purple keels covered in short hairs on tips, column and anther purple; dorsal sepal erect, oblong, 2.1 cm long, obtuse, apiculate, slightly unequal; lateral sepals wide-spreading, narrowly elliptic, 2.3 cm long, acute, mucronate; petals spreading, broadly lanceolate, ~3 cm long, obtuse, apiculate; lip porrect, 3.3 × 3.3 cm, trilobed, lateral lobes large, erect and incurved to meet column, subquadrate, forward margins serrate, midlobe decurved, almost rectangular, 0.7 × 1 cm, truncate; disk with low crests, callus with 5 keels; column 4 mm long. **Distr:** New Guinea, Australia. **Ecol:** In seasonally swampy savanna; 0–500 m. Fl. May–Aug. **Zones:** 10–12.

D. bullenianum Rchb. f., *Bot. Zeitung* (Berlin) 20: 214, 1862. (Syn: *Callista intermedia*, etc.). **Herbs**, epiphytic, pendulous, sympodial; pseudobulbs to 1 m long, 1 cm thick, slender, slightly fractiflex, attenuate below, many-noded, sulcate stems covered with sheaths. **Leaves** deciduous, turning yellow before shedding; lamina ovate-oblong, to 10 × 1.5 cm,

Below: *Dendrobium bensoniae*, *Dendrobium bicaudatum*, *Dendrobium bigibbum*, *Dendrobium bullenianum*. [All MG except *D. bigibbum* AF.]

apex obtuse, unequally bilobed, submembranous. **Inflorescences** pendulous, short, borne at the leafless nodes of older dry stems, flowers densely clustered (almost globose), to 25-flowered. **Flowers** ~1.9 cm Ø, bright orange to yellow, petals and lip with red stripes; dorsal sepal lanceolate, to 8 × 3.5 mm; lateral sepals ovate, joined at base to form a short spur, ~1 cm long, overall 1.7 × 0.5 cm; petals lanceolate, to 8 × 3 mm; lip elliptic, ~1.6 × 0.5 cm. **Distr:** Endemic to Philippines. **Ecol:** In lowland evergreen forests; 300–1,000 m. **Zones:** 10–12.

D. calyculimentum R.S. Rogers, *Trans. and Proc. Roy. Soc. South Australia* 49: 261, 1925. **Herbs**, epiphytic; stems pendulous, branching, 0.9–1.2 m long, many-leaved. **Leaves** subpatent; lamina lanceolate, 1.1–3.7 × 0.25–0.4 cm, margins minutely serrate near apex, apex sharply acute, rigid. **Inflorescences** arising from apical part of stem, terminal and subterminal, racemose, 4–6-flowered; pedicel and ovary 1.4 cm long, slender. **Flowers** not resupinate, 1.4–1.5 cm long, dark rose at base, pale rose to whitish at the tips; dorsal sepal erect, ovate, slightly concave, narrowing at base, 5–6 × 2.8 mm, sharply acute; lateral sepals obliquely obcuneate, 1.4–1.5 cm long, connate at base for 4 mm, sharply acute; mentum curved, ~8 mm long, obtuse; petals erect, narrowly oblong, 5.5 mm long, margins minutely serrate near apex, apex acute; lip erect, elliptic-lanceolate, 1.3 cm long including the claw, 3.5 mm wide, base clawed, claw semicylindrical, 5 × 3 mm, 5-nerved, minutely serrate near apex, 7-veined, apex acute; column 3–4 mm long, foot ~0.9–1 cm long. **Distr:** Sulawesi, New Guinea, Solomon Islands. **Ecol:** In montane to subalpine forests; 1,000–3,150 m. Fl. Feb, Apr–Sep, Nov. **Zones:** 10–12.

D. canaliculatum R. Br., *Prodr.*: 333, 1810. [Antelope Orchid, Teatree Orchid]. **Herbs**, small, epiphytic; pseudobulbs fusiform to ovoid, few-noded, 3–12 × 1.5–3 cm Ø, clothed at first in sheathing bracts. **Leaves** 2–6, subterminal, semicylindrical, grooved on upper surface, linear, 5–20 × 1.2–1.6 cm, base articulated to a very short sheath, apex acute. **Inflorescences** 1–4, 10–40 cm long, densely many-flowered; bracts lanceolate, 3–4 mm long, acuminate; pedicel and ovary 1.2–2.5 cm long. **Flowers** small, 1.8–2.5 cm Ø, fragrant, sepals and petals white with yellow to brown apical halves, sometimes yellow or brown; dorsal sepal oblong, 0.9–1.5 × 0.3–0.4 cm, acute to obtuse, once-twisted; lateral sepals spreading, obliquely oblong, 0.7–1.5 × 0.3–0.4 cm, acute, twisted; petals suberect, linear to spathulate, 1.2–1.8 × 0.2–0.3 cm, obtuse to acute, not twisted to twice-twisted; lip white, lobes marked with purple, trilobed, 0.9–1.5 × 0.8–0.9 cm, lateral lobes obovate, rounded or truncate, midlobe sometimes recurved, ovate, acute or apiculate, callus of 3 to 5 ridges dissected toward apex, middle 3 raised at apex; column 3–4 mm long. **Distr:** NE tropical Australia, S New Guinea. **Ecol:** In savannas, often on exposed trees; <500 m. **Zones:** 10–12.

D. catenatum Lindley, *Gen. Sp. Orchid. Pl.* 84, 1830. (Syn: *D. tosaense*, etc.). **Herbs**, epiphytic; stems erect or pendulous, 3–35(–60) cm long. **Leaves** 2–5, ligulate-oblong or oblong-lanceolate, 3–21 × 0.5–1.5 cm, base slightly oblique, decurrent, apex obtuse, ± hooked, subacute, papery. **Inflorescences** 1–3, (1–)3–8-flowered; peduncle 0.2–1 cm; floral bracts pale white; pedicel and ovary yellowish-green, 2–4 cm long. **Flowers** spreading, sepals and petals yellowish-green, pale yellowish-white, or white, lip white or pale yellow, with a reddish to reddish-brown transverse mark, with a green or yellow basal callus, with purplish-red stripes on both sides below middle, disk with a purplish-red transverse spot above middle; column greenish-white or yellowish-green, with a purple spot on each side at apex, foot yellowish-green, purplish-striped, middle with a purple spot; dorsal sepal ovate-oblong or ovate-lanceolate, 1.2–1.7 × 0.4–0.7 cm, 3–5-veined; lateral sepals obliquely ovate-triangular or falcate-lanceolate, 1–1.3 cm long; petals oblong or ovate-oblong, 1.2–1.6 × 0.4–0.7+ cm wide; lip elliptic-rhombic, broadly ovate, or ovate-lanceolate, 1.3–1.7 × 0.8–1.1+ cm, base slightly cuneate, entire or obscurely trilobed. **Distr:** S Japan to EC Taiwan. **Ecol:** On mountains; ~1,600 m. **Zones:** 9–11.

D. ceraula Rchb. f., *Linnaea* 41: 40, 1876. (Syn: *D. gonzalesii*). **Herbs**, pendulous, sympodial; pseudobulbs to 15–40 cm long, 7 mm Ø; nodes tapered, increasing in size with age. **Leaves** deciduous, 6–10; lamina oblong-lanceolate, 7 × 2 cm, apex shallowly tridentate, papyraceous. **Inflorescences** with up to 4 flowers per node. **Flowers** to 3.5 cm Ø, color highly variable, ranging from pink to pale blue to creamy-white, lip maroon-striped in basal half; dorsal sepal narrowly oblong, 2.3 × 0.7 cm, apex tapered to a point; lateral sepals linear-lanceolate, 2.7 × 0.8 cm, joined at base forming a curved spur ~2.5 cm long; petals lanceolate, to 2.7 × 0.9 cm, acute; lip narrowly oblong or oblong-lanceolate, ~3.3 × 1 cm, base clawed, widest above middle; column short. **Distr:** Endemic to Philippines (Luzon). **Ecol:** In shaded parts of mountainous forests; >860 m. **Zones:** 11–12.

D. chrysanthum Wallich ex Lindley, *Edwards's Bot. Reg.* 15: t. 1299, 1829. (Syn: *Callista chrysantha*, etc.). **Herbs**, epiphytic or lithophytic; stems pendulous or curving, cylindric, 0.5–2 m, stout, 0.5–1.5 cm Ø. **Leaves** distichous along entire stem; lamina oblong-lanceolate, usually 13–19 × 1.5–4.5 cm, base sheathing, apex acuminate, papery. **Inflorescences** leaf-opposed, lateral, on upper part of mature leafy stems, fasciculate, almost sessile, 2–6-flowered; floral bracts ovate-triangular, ~3 mm, membranous; pedicel and ovary slightly compressed, 3.5–6 cm long. **Flowers** thickly textured, yellow, lip disk with a chestnut blotch on either side; dorsal sepal ± concave, oblong or elliptic, 1.5–2 × 0.9–1.1+ cm, obtuse, 7-veined; lateral sepals slightly concave, obliquely ovate-triangular, 1.5–2 × 1–1.2 cm; petals slightly concave, obovate, 1.6–2.2 × 1.1–1.4+ cm, rounded; lip concave, reniform or transversely oblong, ~ 1.8 × 2.2 cm, base abruptly contracted into short claw, apex rounded, adaxially densely barbellate, with an oblong callus; column ~4 mm, foot ~6 mm. **Fruits** cylindric, ~7 × 1.5 cm. **Distr:** Himalayas to S China and Indo-China. **Ecol:** In dense forests; 700–2,500 m. **Zones:** 9–11.

D. cruentum Rchb. f., *Gard. Chron.* 1884(1): 604, 1884. (Syn: *Callista cruenta*). **Herbs**, epiphytic; stems pseudobulbous, erect, terete to 50 cm long, base swollen, narrowed above; leaf sheaths with soft black hairs. **Leaves** deciduous, spaced along the upper part of the stem; lamina obliquely oblong or elliptic-oblong, 5–13 cm long, apex emarginate, leathery, abaxially hairy. **Inflorescence** 1–2-flowered, axillary, short; bracts lanceolate, convolute; pedicel and ovary 3.6 cm long, glaucous. **Flowers** lasting about a month, faintly fragrant, pale green or cream, with darker green venation; sepals triangular-ovate, acuminate, keeled; dorsal sepal 4.2 × 1.5 cm, held ± parallel to the lip; lateral sepals 3.9 × 1 cm, slightly recurved; petals linear, 3.6 × 0.6 cm, strongly recurved, acute; mentum rounded; lip trilobed, 3.2 × 1.5 cm, lateral lobes erect, oblong-falcate, 1.2 × 0.8 cm, crimson, midlobe ovate, 2.6 × 1.6 cm, apiculate, green with a tumid warted scarlet crest and 3 red ridges, basal margins scarlet, apical margins slightly ruffled; column green, margins crimson. **Distr:** Indo-China. **Ecol:** In warm, open forests; <1,000 m. **Zones:** 9–11.

D. crumenatum Swartz, *J. Bot.* (Schradder) 2: 237, 1799. (Syn: *D. kwashotense*, *Onychium crumenatum*). [Pigeon Orchid, White Dove Orchid]. **Herbs** epiphytic, tufted; stems slightly compressed, cylindric, 40–70 cm long, upper part slender, with 3 or 4 internodes dilated and fusiform above base, dilated portion of stem spindle-shaped, to 2 cm Ø, often longitudinally ridged. **Leaves** distichous on middle part of stem; lamina ovate-oblong, ~3–6 × 0.6–2.5 cm, apex obtuse and unequally bilobed, leathery, base with clasping sheaths. **Inflorescences** arising from distal leafless part of stem, flowers usually solitary; floral bracts elliptic, ~6 × 2.5 mm; pedicel and ovary pale red, ~1.5 cm long. **Flowers** lasting for 1 day, fragrant, spreading, sepals and petals white or sometimes pink-tipped, lip white, keels yellow, column and anther cap white; dorsal sepal ovate-lanceolate, 1.7–2.2 × ~0.5 cm, slightly obtuse; lateral sepals obliquely ovate-lanceolate, slightly larger than dorsal sepal; mentum narrowly conic, to 1.5 cm long; petals obovate-oblong, 1.7–2 × ~0.7 cm, subacute; lip obovate-oblong, 2.4–25 × 1.3–1.8+ cm, trilobed, lateral lobes erect, subobovate, subtruncate, midlobe obovate, ~1.2 × 1 cm, margin crenulate and crisped, apex mucronate; disk with 5 keeled ridges with denticulate margin; column ~3 mm long. **Fruits** ellipsoid, 2.5 cm long. **Distr:** Tropical and subtropical Asia. **Ecol:** In primary forests. Fl. Mar, Dec. **Zones:** 10–12.

D. cunninghamii Lindl., *Edwards's Bot. Reg.* 21: t. 1756, 1835. (Syn: *D. biflorum*, etc.). [Winika, Christmas Orchid, Bamboo Orchid]. **Herbs**, to 1.5 m long; stems canelike, thickened toward base, to 7 mm Ø; lateral branches numerous, ± drooping; bare parts mostly ~2 mm Ø, bright yellow, polished. **Leaves** with tubular leaf-sheaths, minutely papillose, overlapping, covering all younger stems; lamina narrowly linear, ~3–5 × 0.3 cm, dark at joint with sheath. **Inflorescences** 1–6-flowered, laterally from upper nodes; floral bracts tubular, short, inconspicuous; pedicels very slender, longer than ovary. **Flowers** ~2–2.5 cm Ø, glabrous, white, lip and column usually partly rose-pink to purplish or green; sepals elliptic, spreading, tips ± reflexed, lateral sepals connate under labellum and adnate to base of column; petals slightly broader; lip shorter, distinctly trilobed, lateral lobes small, often colored, inclined to stand parallel to one another, midlobe white, broad, subacute, minutely crenulate; disk with 4–5 pale-colored, low, longitudinal ridges terminating just above the short claw and near to a colored knob at end of column foot; column cylindric, very narrowly winged. **Distr:** New Zealand. **Ecol:** On well-lit tree trunks and branches, sometimes lithophytic. Fl. Dec–Feb(–Apr–Jun). **Zones:** 9–10.

D. cuthbertsonii F. Muell., *Trans. and Proc. Roy. Soc. Victoria* 24: 175, 1888. **Herbs**, epiphyte, or occasionally terrestrial, 2–8 cm high, usually tufted and clump-forming, sometimes to 12 cm Ø; pseudobulbs spheroid, ovoid or clavate, 0.5–8(–14) × 0.1–0.7 cm. **Leaves** 1–5, linear to broadly elliptic, sometimes ovate, to 0.5–4.2 × 1.5 cm, apex obtuse or acute, green to blackish-green, abaxially often purplish or purplish veined. **Inflorescence** 1-flowered, terminal (lateral), bracts ovate, tubular or cup-shaped, (1–)2.5–9 mm long, acuminate; flower held horizontal or pendulous; pedicel and ovary 1–3.2 cm long. **Flower** lasting up to 8 months, 2.2–4(–5) × ~3.3 cm, red, lip paler, with darker reddish-brown marks on margin, or flowers purple, pink, orange (rarely yellow or white), sepals and petals; dorsal sepal oblong to broadly elliptic, 1–2.1 × 0.3–1.2(–1.5) cm, obtuse; lateral sepals, oblong to broadly elliptic, 1.8–3(–3.4) × 0.4–1.2(–1.8) cm, apex obtuse; mentum 0.9–1.8 cm; petals obovate to rotund, 1–2 × (0.35–)0.6–1.6 cm, obtuse; lip obovate, 1.2–3 × 0.5–1.5 cm, adnate to column foot; column 5–7 × 5 mm. **Fruit** obliquely ellipsoid, to 2 × 1 cm. **Distr:** New Guinea, Bismarck Archipelago (New Ireland). **Ecol:** In mossy forest edges, exposed cliff faces, etc.; 750–3,450 m. Fl. throughout the year. **Zones:** 9–11.

Top row: *Dendrobium calyculimentum, Dendrobium canaliculatum, Dendrobium catenatum.*
Second row: *Dendrobium ceraula, Dendrobium chrysanthum, Dendrobium cruentum.*
Third row: *Dendrobium crumenatum, Dendrobium cunninghami, Dendrobium cuthbertsonii.*
Bottom row: *Dendrobium × delicatum, Dendrobium densiflorum, Dendrobium dichaeoides.*

D. × delicatum (F.M. Bailey) F.M.Bailey, *Queensl. Fl.* 5: 1527, 1902. **Herbs**, lithophytic, clump-forming; pseudobulbs cylindrical or slightly flattened, 10–50 × 1–4 cm, broadest at base, evenly tapered upward, covered with overlapping gray-brown bracts which are obvious when young. **Leaves** 2–6, at apex of pseudobulb, narrowly ovate to obovate, 5–18 × 1–3.5 cm, thick and leathery, dark green, occasional aerial growths produced from apical buds. **Inflorescences** racemes, 10–40 cm long, many-flowered, crowded; pedicels 1.5–2 cm long. **Flowers** 1.5–2.5 cm across, white, cream, pink or mauve, with pink or mauve blotches and suffusions, narrowly to widely open, waxy, fragrant; lip 1–1.5 cm long, curved, white with pink or purple markings. **Distr:** E Australia. **Ecol:** Fl. Aug–Oct. **Zones:** 9–11. **Info:** An uncommon natural hybrid between *D. speciosum* and *D. kingianum*, the parents themselves are highly variable.

D. densiflorum Lindl., *Pl. Asiat. Rar.* 1: 34, 1830. (Syn: *Callista densiflora*, etc.). [Pineapple Orchid]. **Herbs**, epiphytic; stems strongly clavate, swollen, ± pseudobulbous at base, 25–40 cm long, to 2 cm Ø, yellow, sheathless. **Leaves** 3 or 4, subterminal; lamina oblong-lanceolate, 8–17 × 2.6–6 cm, base not decurrent into clasping sheath, apex acute, leathery. **Inflorescence** subterminal, pendent, densely many-flowered; peduncle with 2–4 sheaths at base;

MS MG

MG

Above: *Dendrobium distichum, Dendrobium draconis, Dendrobium eriiflorum.*

rachis 6–16 cm; floral bracts broadly oblong or obovate, 1.2–1.5 × 0.6–1+ cm, membranous, strongly ~10-veined, revolute, obtuse; pedicel and ovary whitish-green, 2–2.5 cm. **Flowers** spreading, 3–4 cm Ø; sepals and petals pale yellow, lip golden-yellow, column and anther cap orange-yellow; dorsal sepal ovate, 1.7–2.1 × 0.8–1.2+ cm, obtuse, 5-veined; lateral sepals nearly as large as dorsal sepal, ovate-lanceolate, subacute, 5- or 6-veined; mentum semiglobose, ~5 mm wide; petals suborbicular, 1.5–2 × 1.1–1.5 cm, base clawed, main veins 3, margin erose above; lip orbicular-rhombic, to 1.7–2.2 × 2.2 cm, rounded, shortly clawed, densely lanate above, both sides embracing column below middle; column ~4 mm. **Distr:** C Himalayas to China (Hainan). **Ecol:** In evergreen broad-leaved forests; 400–1,000 m. **Zones:** 8–11.

D. dichaeoides Schltr., *Repert. Spec. Nov. Regni Veg. Beih.* 1: 507, 1912. **Herbs**; rhizome stem-like, creeping; pseudobulbs ± widely spaced, very densely leafy, ± clavate, creeping-pendulous; leaf sheaths entirely covering pseudobulb. **Leaves** patent, obliquely lanceolate-oblong, 1–1.5 × 0.3–0.5 cm, apex unequally bilobulate, apiculate, lobules aristate. **Inflorescences** lateral from the apical part of pseudobulb, racemose, short, headlike, densely 4–8-flowered; floral bracts ovate-lanceolate, apex acuminate; pedicel and ovary 7 mm long, clavate, glabrous. **Flowers** rose-red to purple red, not resupinate, ~1 cm long; dorsal sepal ovate-lanceolate, 6 mm long, apex subacute; lateral sepals obliquely ovate-lanceolate, widened at base, 6 mm long, apex subacuminate; mentum oblongoid, 5 mm long, obtuse; petals obliquely elliptic-lanceolate, 6 mm long, apex subacute; lip concave-ligulate, base slightly constricted, 8.5 × 3 mm, apex shortly acuminate; column short, foot ~5 mm long. **Distr:** New Guinea. **Ecol:** In montane forests; 1,000–2,300 m. Fl. throughout the year. **Zones:** 11–12.

D. distichum (C. Presl) Rchb. f., *Linnaea* 41: 39, 1876. (Syn: *Aporum distichum*, etc.). **Herbs**, epiphytic, semipendulous, sympodial, pseudobulbous; stems flattened, to 40 cm long, 3 cm wide across the leaves. **Leaves** triangular, 7–9 × 2.5–3 mm, base clasping the stem, giving a saw-tooth appearance on the stem, apex acute, thick. **Inflorescences** terminal; flowers borne among the dried bracts; pedicel and ovary 3–4 mm long, greenish-yellow, glabrous. **Flowers** ~5–6 mm across, glabrous, greenish-yellow with red stripes in the dorsal and lateral segments; dorsal sepal oblong-lanceolate, to 3–3.6 × 1.5 mm, 5-striped; lateral sepals ovate, to 3 × 2 mm, 5–6-striped; mentum ~1.8 mm long, saccate; petals oblong, ~3 × 1 mm, 3-striped; lip creamy-white, 1.2 × 2.4 mm, callus conspicuous; column ~1.3 mm. **Distr:** Endemic to Philippines. **Ecol:** In mangrove swamps and highlands; 0–900 m. **Zones:** 11–12. **Info:** *D. distichum* f. *album* lacks the colored stripes in the sepals and petals, and is rare (distribution unknown).

D. draconis Rchb. f., *Bot. Zeitung* (Berlin) 20: 214, 1862. (Syn: *D. eburneum*, etc.). **Herbs**, epiphytic; stems tufted, slightly fusiform or clavate, 8–9-noded, 30–45 cm long, 1–1.3 cm thick, sheaths covered with short black hairs. **Leaves** ligulate or lanceolate, 6–10 × 1–2 cm, apex acute or obliquely bilobulate, rather leathery, dark green. **Inflorescences** 2–5-flowered, borne from upper nodes; bracts minute, triangular lanceolate, acute. **Flowers** fragrant, waxy, long-lasting, 6–7 cm Ø, ivory-white, base of lip and inside of mentum lined with crimson or golden-yellow; dorsal sepal lanceolate, 4 × 1 cm, acuminate; lateral sepals subsimilar; mentum elongate, subulate, 2–3 cm long; petals oblong-lanceolate, 4 × 1.5 cm, acuminate; lip obscurely trilobed, rhombic above, 5.5 × 2 cm, base cuneate, lateral lobes small, rounded, subauriculate; midlobe oblong, margins undulate; disk with 3 elevated lines in basal half of lip. **Distr:** Assam to Indo-China. **Ecol:** 700–2,000 m. **Zones:** 9–11.

D. eriiflorum Griff., *Ic. Pl. Asiat.* 3: 316, 1851. (Syn: *Callista eriiflora*). **Herbs**, epiphytic, dwarf, to 25 cm high; stems clustered, base fleshy, attenuate above, 25 cm long, rectangular above, covered with loosely fitting sheaths, yellow. **Leaves** ~9 per stem, deciduous; lamina oblong or lanceolate, 7–12 × 1 cm, obtuse. **Inflorescences** 5–20-flolwered, 8 cm long, borne opposite the leaves; bracts minute. **Flowers** <1 cm long; sepals greenish-white to yellowish, petals white, lip yellow with purple veins, disk with 3 undulate green keels; dorsal sepal and petals lanceolate, 1–1.4 cm long, acuminate; lateral sepals triangular, falcate, 1.2–1.4 cm long, acuminate, incurved; mentum conical, incurved, obtuse; lip obscurely trilobed, obscurely rhombic, 6–8 mm long, lateral lobes with margins serrate-dentate, midlobe triangular, fleshy, margins crisped, callus 2-ridged at base of lip. **Distr:** Himalayas to Java. **Ecol:** Usually high up in trees where light levels are high; 1,000–2,000 m. **Zones:** 8–12.

D. erosum (Blume) Lindl., *Gen. Sp. Orchid. Pl.*: 86, 1830. (Syn: *D. aemulans*, etc.). **Herbs**, epiphytic, large, usually to 70 cm long; rhizomes abbreviated; stems pendent, clustered, many-noded, leafy. **Leaves** 2.5–3 cm apart; lamina lanceolate to linear-lanceolate, 4.5–7 × 0.5–0.8 cm, apex obtuse, unequal, glabrous. **Inflorescences** pendent, densely 5–15-flowered, globular, 2–3 cm long, borne from upper nodes of leafless canes, flowers erect-spreading; pedicel and ovary ~1.5 cm long, clavate, glabrous. **Flowers** not opening fully, rose-pink with segment apices paler; sepals subequal, oblong, 6 mm long, obtuse, glabrous; lateral sepals united and dilated forming a slightly flattened pouch, ~7.5 mm long, obtuse; petals oblique, elliptic, margin minutely ciliate, apex obtuse, glabrous; lip with base ovate-concave, dilated, forming a slipper, 1.1 × 0.3–0.4 cm, margins ciliate; column short, glabrous, foot elongated, lateral lobes triangular, midlobe lanceolate, acute. **Distr:** Indo-China to SW Pacific. **Ecol:** On large trees, near mangrove swamps and forests; 0–2,000 m. **Zones:** 8–12.

Below: *Dendrobium erosum, Dendrobium faciferum, Dendrobium falconeri, Dendrobium falcorostrum.* [All MG except *D. falcorostrum* AF.]

GS

MG

Above: *Dendrobium farmeri, Dendrobium finisterrae, Dendrobium fleckeri.*

D. faciferum J.J. Sm., *Bull. Dép. Agric. Indes Néerl.* 15: 10, 1908. (Syn: *Ceraia facifera*). **Herbs**, epiphytic; rhizome short; stem branched, slender, subterete, to ~1 m long, grayish-brown, striped greenish-yellow, base pseudobulbous, elongate-fusiform, ~5–7 cm long, ~4-noded, internodes ~14–21 cm long. **Leaves** many, narrowly lanceolate, ~4.8–9 × 0.7–1.2 cm, apex shortly and unequally bilobed, fleshy, adaxially shiny, abaxially darker; sheaths dirty violet. **Inflorescences** numerous, dense fascicles, 2–5-flowered, sessile, borne near apex of mature, leafless stems; pedicels pale yellowish-green, often tinged orangish-red. **Flowers** 1.3(–1.7) cm long, ~6.5 mm Ø, sepals basally connate; dorsal sepal oblong-ovate, 5.3 × 2.7 mm, slightly acuminate to obtuse, concave, trinerved; lateral sepals triangular, ~8.5 × 4 mm, obtuse, connate, forming a concave pouch, ~ 9 × 5 mm; lip clawed, trilobed, 6.3 mm long, lateral lobes erect, round, midlobe ovate-round, ~2.5 mm long, acuminate; pollinia 4; ovary ~3 mm long. **Distr:** Sulawesi to Maluku, Lesser Sunda Islands. **Ecol:** On trees, in tropical rainforests. Fl. Sep–Mar. **Zones:** 11–12.

D. falconeri Hook., *Bot. Mag.* 82: *pl. 4944*, 1856. (Syn: *D. erythroglossum*). **Herbs**; stems usually pendent, terete, much-branched, 30–40 cm long; pseudobulbs obliquely fusiform, 2–3 × 0.4–0.6 mm, with 1 or 2 nodes bearing many sheath-fibers. **Leaves** linear, 5 cm × 3 mm, base contracted, articulated with sheath, apex obtuse, shortly mucronate. **Inflorescences** racemes, lateral, arising from nodes of pseudobulb, 1(few)-flowered, with tubelike spathe at base; bracts ovate, hyaline; pedicel and ovary 1.5 cm long. **Flowers** showy, light purple, dark-purple toward apex of sepals and petals, ~3.5 cm Ø; dorsal sepal ovate-lanceolate, ~3 × 0.8 cm, base slightly contracted, apex acuminate, 7-nerved; lateral sepals linear-lanceolate, ~3.5 × 0.8 cm, base oblique, apex acuminate, adnate to column foot, forming saccate mentum ~5 mm long; petals ovate-rhomboid, 3.3–3.6 × 1.8–2.1 cm wide, base cuneate, apex acute, 5-nerved; lip ovate, 3.2–3.5 × 1.8 cm, base cuneate, apex acute, trilobed, lateral lobes yellow, semiorbicular, suberect, terminal lobe nearly white, apically purplish, triangular-ovate, 1.7–2 × 1.6 cm; disk shortly hirsute, with purplish patch between lateral lobes; column 4–5 mm long. **Distr:** E Sikkim to Taiwan. **Ecol:** In forests; ~1,000 m. **Zones:** 9–11.

D. falcorostrum Fitzg., *Sydney Morning Herald*, Nov. 18 1876, 1876. (Syn: *Thelychiton falcorostrus*, etc.). [Beech Orchid]. **Herbs**, epiphytic, forming large clumps; rhizomes thick short; pseudobulbs cylindrical or fusiform, erect to spreading, succulent, crowded, 10–50 × 1–1.5 cm, tapering gradually toward the ends, yellowish-green, ribbed, rooting only at base; roots smooth, creeping. **Leaves** 2–5, at apex of pseudobulb, spreading to erect; lamina narrow-elliptic to oblanceolate, 6–15 × 1.5–3 cm, conduplicate, thin, leathery, smooth, dark green. **Inflorescences** racemes, 4–20-flowered, 8–16 cm long; pedicels 1.5–2.5 cm long. **Flowers** 2.5–4 cm across, thick-textured, waxy, opening widely, strongly fragrant, sepals and lateral petals crystalline white to cream, lip white to cream with yellow and purple markings; dorsal sepal 1.6–2.5 × 0.4–0.9 cm; lip 1–1.5(–3.5) × 0.7–1.2 cm, curved, midlobe ending in a prominent point; column 3–4 mm long, foot 6–7 mm long. **Distr:** Australia (SE Queensland, NE New South Wales). **Ecol:** In cool-temperate rainforests, almost exclusively on *Nothofagus moorei*. Fl. Aug–Oct. **Zones:** 9–10..

D. farmeri Paxton, *Paxton's Mag. Bot.* 15: 241, 1849. (Syn: *Callista farmeri*). **Herbs**, epiphytic; stems erect, clavate or fusiform, 4–5-grooved or -angled, 25–45 cm, base slender. **Leaves** 2–4, elliptic or oblong, 10–20 cm long, apex acute or acuminate, coriaceous, striate. **Inflorescences** racemes, densely, many-flowered, lax, pendulous, cylindrical, to 20 cm long, lateral, borne from the nodes near apex of leafless and leafed stems; bracts small. **Flowers** 5 cm Ø, variable in color, sepals and petals white or delicate lilac-mauve, lip white with apex pink, disk yellow; sepals ovate-oblong, 2–2.5 × 1–1.2 cm, margins erose; petals much broader than sepals, shortly clawed, almost orbicular, 2–2.5 × 2 cm, rounded; mentum very short, rounded; lip shortly clawed, orbicular, 2.2 × 2.2 cm, lateral lobes inflexed, folds pubescent, margin erose; disk ochreous, minutely puberulous, callus small, fleshy, basal, with raised veins on each side. **Distr:** C Himalayas to Peninsular Malaysia. **Ecol:** In evergreen lowland forests and primary montane forests; 150–1,000 m. **Zones:** 10–12.

D. finisterrae Schltr., *Repert. Spec. Nov. Regni Veg. Beih.* 1: 495, 1912. **Herbs**, epiphytic, to 70 cm tall; pseudobulbs clustered, erect, clavate, 30–50 × 3 cm, ~5-noded below leaves, apex 2-leaved. **Leaves** suberect; lamina oblong-elliptic, to 23 × 9.5 cm, apex obtuse to subacute, coriaceous. **Inflorescences** erect, to 25 cm long, ~10-flowered; peduncles terete, bearing 3 short sheaths; bracts lanceolate, 2.5 × 0.6 cm, acute; pedicel and ovary 5 cm long. **Flowers** large, not opening widely, yellow, heavily spotted with red or maroon on sepals and petals, setose on pedicel, ovary and backs of sepals; dorsal sepal lanceolate, 3.2 × 0.8 cm, acuminate; lateral sepals obliquely lanceolate, 3.7 × 1.5 cm, acuminate; mentum incurved, 1.1 cm long; petals oblanceolate, acute to apiculate, 3 × 1.3 cm; lip yellow, heavily marked with maroon, trilobed, conduplicate, somewhat recurved, 2.7 × 3.2 cm, lateral lobes erect, subquadrate, truncate, midlobe broadly oblong-ovate, acute, callus 3-ridged; column 2.5 mm long. **Distr:** Papua New Guinea (Highland Provinces). **Ecol:** In montane forests; 900–2,100 m. Fl. Jan–Aug. **Zones:** 11–12.

D. fleckeri Rupp and C.T. White, *Queensland Naturalist* 10: 25, 1937. (Syn: *Thelychiton fleckeri, Tropilis fleckeri*). **Herbs**, epiphytic; stem pseudobulbous, very slender, 20–35 cm long, apex 2–4-leaved. **Leaves** small, lanceolate. **Inflorescences** racemes, short, borne between the leaves, 1–3-flowered. **Flowers** 2.5 cm Ø, with a

Below: *Dendrobium forbesii, Dendrobium friedericksianum, Dendrobium glomeratum, Dendrobium goldschmidtianum.*

GS

Above: *Dendrobium gracilicaule, Dendrobium hallieri, Dendrobium hercoglossum, Dendrobium heterocarpum.*

musky fragrance, color varies from greenish-yellow, through yellow and apricot, rarely almost brown, lip whitish at base, blending to pale yellow with brick-red to purple markings, midlobe margin with protruding dense hairlike white trichomes; dorsal sepal ovate, ~5 × 2.5 mm, acute; lateral sepals, falcate, broadly lanceolate, 7 × ~3.5 mm, acuminate; petals linear-lanceolate, ~5.7 × 1.2 mm, acute; lip 3.2 mm long, trilobed, lateral lobes erect, falcate, ~1.5 mm long, mid lobe channeled, ~2 × 2 mm; column ~2.5 mm long, arcuate. **Distr:** Australia (Cape York Peninsula, Queensland). **Ecol:** In montane forests; 800–1,550 m. **Zones:** 9–11.

D. forbesii Ridl., *J. Bot.* 24: 323, 1886. **Herbs**, large, epiphytic; pseudobulbs clustered, clavate, to 32 × 1.6–2.5 cm, 4–6-noded below leaves, apex 2-leaved. **Leaves** spreading; lamina elliptic or oblong-elliptic, 9–19 × 4.7–11 cm, base shortly and broadly petiolate, apex obtuse or rounded, coriaceous. **Inflorescence** erect, apical, 15–30 cm long, 7–20-flowered; peduncle almost covered by large inflated pale green sheaths; bracts large, lanceolate, to 3.8 × 1.3 cm, acuminate or acute; pedicel and ovary setose, 4 cm long. **Flowers** large, attractive, pure white or creamy white, sometimes with purple on side lobes of lip; dorsal sepal lanceolate, acuminate, 2.7–3.2 × 1 cm; lateral sepals obliquely lanceolate, 3–3.6 × 1.2–1.6 cm, acuminate, keel 2–3 mm wide, mentum incurved-conical, 1–1.3 cm long; petals obovate-spathulate, 3–3.5 × 1.6–2.3 cm, margins undulate, apiculate; lip 3-lobed, 2–2.6 × 2.7–3 cm, lateral lobes smaller than midlobe, erect, oblong, truncate, midlobe transversely oblong with a triangular apical lobule, callus 3-ridged; column short, 2.5 mm long, foot 1–1.2 cm long. **Distr:** Papua New Guinea. **Ecol:** In montane forests; 900–1,500 m. Fl. Oct–Dec. **Zones:** 9–11.

D. friedericksianum Rchb. f., *Gard. Chron.*, III, 2: 648, 1887. (Syn: *D. friedericksianum* var. *oculatum*). **Herbs**, epiphytic, medium-sized; stems pseudobulbous, light yellow, to 45 cm long, erect, then somewhat pendulous, ~1–1.5 cm Ø, basally slender, above cylindrical to clavate, internodes to 3.8 cm long, almost fully covered by a membranous sheath. **Leaves** suberect, borne in 2 ranks on the apical 2/3 of the pseudobulb, deciduous after a year; lamina 10–18 × 2 cm. **Inflorescences** racemose, 2–4-flowered, short, axillary, borne from the nodes near the apex of leafless canes; pedicel and ovary to 4 cm long, slender. **Flowers** long-lived, ~5–6 cm across, glossy, bright yellow, lip base with a pair of reddish blotches; sepals and petals, ± similar, linear-lanceolate, slightly twisted, margins undulate, apices slightly notched; dorsal sepal 3.2–3.8 × ~1 cm, apex slightly notched; lateral sepals to 3.2 × 1.3 cm; petals to 4 × 1 cm; lip broadly obovate, 2–2.5 × 3 cm, margin recurved and undulated, base with a short spur, ~4 mm long, rounded. **Distr:** SE Thailand to Cambodia. **Ecol:** In hot steamy lowlands, on tall trees at around 20 m from ground. **Zones:** 11–12.

D. glomeratum Rolfe, *Bull. Misc. Inform. Kew* 1894: 155, 1894. (Syn: *Chromatotriccum crepidiferum*, etc.). **Herbs**, epiphytic; stems slender, canelike, cylindrical, clustered, ~28–60 cm long. **Leaves** deciduous; lamina narrowly elliptic, 5–9 × 1.4–3 cm, apex minutely unequal, acute, adaxially shiny mid or dark green, abaxially matte, thinly coriaceous. **Inflorescences** racemes, loosely 4–6(or more)-flowered, axillary; floral bracts triangular-ovate to oblong-elliptic, 7–8 × 3–4 mm, pale green to brown, sometimes suffused pale purple; pedicel and ovary 2.5–3 cm long. **Flowers** wide-opening, 3.3–4.5(–6) cm Ø, bright rose-pink to magenta, lip and column orange to vermillion, anther-cap magenta; sepals and petals spreading; dorsal sepal elliptic, 2–3 × 0.9–1.5 cm, obtuse, ~5-nerved; lateral sepals obliquely ovate-triangular, anterior margin 3–4.5 cm long, posterior margin 2–3 cm long; mentum conical, 1.6–2 cm long, obtuse; petals elliptic or obovate-elliptic, 2–2.8 × 0.8–1.5 cm, obtuse, 5-nerved; lip 0.8–2.7 × 0.6–0.8 cm, attached with a broad claw to apex of column foot; column broadly oblong to quadrate, 2.5–0.5 × 0.4–0.5 cm; pollinia 4. **Distr:** Indonesia (Sulawesi to Maluku). **Ecol:** In lower montane forests; 1,500 m. Fl. Sep–Oct. **Zones:** 11–12.

D. goldschmidtianum Kraenzl., *Repert. Spec. Nov. Regni Veg.* 7: 40, 1909. (Syn: *P. miyake*, etc.). **Herbs**, epiphytic; stems erect or pendulous, cylindric, 40–60 cm long. **Leaves** lanceolate or ovate-lanceolate, 6–10 × 1.2–2 cm, apex acuminate, leathery; leaf sheath green, red-tinged, tightly embracing stem. **Inflorescences** multiple, lateral on usually bare penultimate and older canes, tufted, 5–25 mm, densely 6–10-flowered; floral bracts ovate-lanceolate, ~3 × 2.5 mm, acute, membranous; pedicel and ovary brownish-green, ~1.3 cm long. **Flowers** not opening widely, sepals, petals, and lip bright red, column yellow, foot yellowish-green, anther cap yellow, sepals and petals ± similar in size and shape; dorsal sepal elliptic, ~1 × 0.5 cm, obtuse, 5-veined; lateral sepals obliquely ovate, 5-veined, base oblique, apex acute; petals obliquely obovate-oblong, 3-veined, base contracted, apex acute; lip spathulate, 1.5–2.2 × 0.7–0.85+ cm, base narrowly clawed, margin entire, apex slightly obtuse; column ~2 mm long, foot ~1 cm long. **Distr:** Taiwan (Lan Yü, Lü Tao) to Philippines (Batan Is.). **Ecol:** In forests; 200–400 m. **Zones:** 10–12.

D. gracilicaule F. Muell., *Fragm.* 1: 179, 1859. (Syn: *Callista gracilicaulis, Thelychiton gracilicaulis*, etc.). [Leopard Orchid]. **Herbs**, epiphytic, forming clumps; rhizome much-branched; pseudobulbs 20–90 × 0.3–1 cm, base swollen, then constricted, then linear-cylindrical throughout, erect or deflexed, yellowish-green, becoming prominently ribbed, when young covered with brown, papery bracts. **Leaves** 3–6, at apex of pseudobulb; lamina ovate to lanceolate, 5–13 × 2–4 cm, apex often unequally bifid, dark green, thin-textured. **Inflorescences** racemes, 5–12 cm long, 5–30-flowered; pedicels ~1 cm long. **Flowers** 1–1.5 cm across, cup-shaped, not opening widely, fragrant, thick, waxy, often drooping, dull yellow inside, outside irregularly blotched with red-brown, rarely all yellow with no blotches, lip dull yellow; dorsal sepal 0.7–1 × 0.3–0.4 cm; lip 6–8 × 5–6 mm, curved with prominent lateral lobes; Column c. 3 mm long; column foot ~5 mm long. **Distr:** Australia (S Queensland to New South Wales), SW Pacific. **Ecol:** In rainforests and open forests, usually on trees but sometimes on rocks and boulders. Fl. Jul–Sep. **Zones:** 9–11.

D. hallieri J.J. Sm., *Bull. Jard. Bot. Buitenzorg*, II, 8: 40, 1912. **Herbs**, epiphytic; stems 30–45 cm long, 0.5–1.1 cm, leafy. **Leaves** oblong-elliptic, oblong-ovate or oblong, 4–9(–12.5) × 1.6–3.5 cm, base shortly contracted, sulcate, apex obliquely semiobtuse. **Inflorescences** erect, abbreviated, opposite, 2–7-flowered; peduncle and rachis 0.8–2.6 cm log, flexuose; floral bracts oblong-ovate, lowermost ovate, (1.5–)1.7–1.8 × 0.5–0.8 cm, acute to subulate-acuminate, concave, 9-nerved, nigrohirsute outside; pedicel and ovary yellowish, pale green below, 4–5.2 cm long, glabrous. **Flowers** to ~2.5 cm Ø, stiff and waxy, sweetly scented, pale primrose-yellow to greenish-white; sepals recurved; dorsal sepal oblong, 1.6–1.7 × 0.75–0.9 cm, apex narrowly obtuse, 5–7-nerved; lateral sepals obliquely triangular-ovate, (2.8–)3.1–3.2 × 1.3–1.8 cm, acute to acuminate; mentum narrowly conical, (2.3–)2.8–2.9 cm long, narrowly obtuse; petals obliquely ovate, 1.8–2 × 1.1–1.4 cm, ± obtuse; lip shallowly trilobed, creamy-white with rows of orange or scarlet warts on midlobe and red streaks between keels in throat, ~2.3–2.6 × 1.6 cm, midlobe ~ 0.6–0.7 × 1.1–1.2 cm, claw 5–9 mm long; column yellow 5 × 5 mm, foot orange-yellow, ~2.9 cm long; anther cap yellow and white. **Distr:** Borneo. **Ecol:** Near sea level. Fl. Jan, Jul–Aug. **Zones:** 11–12.

D. hercoglossum Rchb. f., *Gard. Chron.* 1886(2): 487, 1886. (Syn: *Callista hercoglossa*, etc.). **Herbs**, epiphytic; stems pendulous, 8–40 cm long. **Leaves** narrowly oblong or oblong-lanceolate, 4–10 × 0.4–0.8(–1.4) cm, base with tightly clasping sheaths, apex obtuse, unequally bilobed, leathery. **Inflorescences** often several, arising from old leafless stems, 2- or 3-flowered; peduncle green, 0.6–1 cm long; floral bracts ovate-lanceolate, small, 3–5 mm long, acute; pedicel and ovary pale pinkish-red, 1.2–1.5 cm long. **Flowers** spreading, sepals and petals pale pink, lip white, epichile pale pink, column white, anther cap purple; dorsal sepal ovate-oblong, 1.3–1.8 ×

GS MS GS JT

Above: *Dendrobium inflatum, Dendrobium ionopus, Dendrobium insigne, Dendrobium jenkinsii.*

0.5–0.8(+) cm, acute, 7-veined; lateral sepals slightly obliquely ovate-lanceolate, about as large as dorsal sepal, acuminate, 7-veined; mentum very short; petals obovate-oblong, 1.2–1.5 × 0.45–0.7(+) cm, acute, 3-veined; lip erect, ~1 cm long; hypochile subglobose, inner surface densely barbellate, apex densely fimbriate; epichile deltoid, relatively small, acute, glabrous; column ~4 mm long, lower part enlarged, foot ~2 mm long. **Distr:** S China to peninsular Malaysia, Philippines (Mindanao). **Ecol:** On tree trunks of dense mountain forests; 600–1,300 m. **Zones:** 8–12.

D. heterocarpum Wall. ex Lindl., *Gen. Sp. Orchid. Pl.*: 78, 1830. (Syn: *Callista aurea, D. atractodes,* etc.). **Herbs**, epiphytic, upright, but sometimes becoming semipendulous on older plants, sympodial; pseudobulbs to 50 cm tall, 1 cm Ø. **Leaves** deciduous; lamina to 10 × 2.5 cm. **Inflorescences** short, appearing from leafless nodes, 1–3-flowered; pedicel and ovary ~2 cm long. **Flowers** scented, spreading, ~5 cm Ø, sepals and petals white or creamy-yellow, lateral lobes of lip yellow with red striation, midlobe white or creamy-yellow, column white, purplish-red on sides, inner surface yellow, foot yellow; dorsal sepal oblong, to 2.5–3 × 0.8 cm, obtuse, 5-veined; lateral sepals oblong-lanceolate, to 2.5 × 1 cm; petals oblong-ovate, to 2.5 × 1.2 cm; lip ovate-lanceolate, nearly as long as sepals, inconspicuously trilobed, lateral lobes erect, middle part downcurved, adaxially with dense reddish-brown hairs, margin entire, midlobe acute; column ~3 mm long. **Distr:** China (S Yunnan) to tropical Asia. **Ecol:** In montane regions; 1,000–1,800 m. **Zones:** 9–11.

D. inflatum Rolfe, *Bull. Misc. Inform. Kew* 1895: 6, 1895. (Syn: *Anisopetala inflata*). **Herbs**, epiphytic, pendent, 60–80 cm long; stems densely foliate. **Leaves** 2-ranked, alternate; lamina oblong or ovate, 10–11 × 3–3.5 cm, base contracted, sheathing, apex long acuminate, membranous, bright green, midrib prominent. **Inflorescences** pendent, short, 2–4-flowered, usually borne on older leafless stems, flowers subopposite. **Flowers** white, short-lived, thinly textured, not opening fully; sepals ovate, 2.2 cm long, obtuse, glabrous; lateral sepals oblique, basally connate into a long, inflated spur ~1.5 cm long, obtuse, curved; petals oblique elliptic-spathulate, obtuse, glabrous; lip cuneate, trilobed, 1.1 × 1.3 cm, lateral lobes round, obtuse, margin serrulate-ciliate, midlobe semiorbicular, shortly acuminate; column short; ovary sessile, ~1.3 cm long, glabrous. **Distr:** Java to Bali. **Ecol:** On trees, in well-lit areas; 900–1,200 m. **Zones:** 11–12.

D. ionopus Rchb. f., *Gard. Chron.* 1882(2): 808, 1882. (Syn: *Callista ionopus, Eurycaulis ionopus*). **Herbs**, epiphytic, semipendulous, sympodial; pseudobulbs to ~60 cm long, 1 cm Ø. **Leaves** to 7 × 1.5 cm; young leaves purple-striped. **Inflorescences** several, 20-flowered, pendulous, borne from leafless pseudobulbs. **Flowers** 2 cm Ø, color variable, ranging from apple-green to bronze; dorsal sepal lanceolate, to 1.3 × 0.5 cm; lateral sepals ovate, joined to form a spur 2.5 cm long; petals lanceolate, 1.2 × 0.4 cm; lip quadrate, ~3 × 0.6 cm. **Distr:** Myanmar to peninsular Malaysia. **Ecol:** In montane forests; ~1,200 m. **Zones:** 10–12.

D. insigne (Blume) Rchb. f. ex Miq., *Fl. Ned. Indië* 3: 640, 1859. (Syn: *Grastidium insigne*, etc.). **Herbs**, epiphytic; rhizome short; stems crowded, erect, rigid, to 1 m long, many-leaved, internodes 2–2.5 cm long; leaf sheaths tubular. **Leaves** patent; lamina oblong-ovate, to 5.8 × 3.3 cm, apex unequally bidentate, carnose, rigid; apical leaves very small and deciduous. **Inflorescences** 2-flowered, lateral, very short; peduncle 9 mm long, compressed; spathe laterally compressed, broadly obtuse; floral bracts triangular, 2 mm long, concave; pedicels 7 mm long. **Flowers** ~3 cm Ø, fragrant, sepals and petals creamy-yellow with light brown tessellate blotches, lip whitish-yellow with pale brown markings; dorsal sepal lanceolate, 2 × 0.53 cm, concave, acute; lateral sepals oblong-triangular, angular-falcate, 1.3–1.8 × 0.95 cm, obtuse, concave, outside obtusely keeled; petals narrowly lanceolate, falcate, 1.65 × 0.4 cm, somewhat concave, acute; lip basally trilobed, 1.6 × 0.6 cm, inside hairy with 2 prominent keels, lateral lobes small, erect, midlobe curved, narrowly elliptic, 1.1 × 0.55 cm, base almost clawlike, narrowed; column 4 mm long; ovary 3 mm long, 6-grooved. **Fruits** pendulous, oblong-ovoid, triangular, lateral angles strongly winged, 4 × 1.7 cm. **Distr:** Maluku to Solomon Islands, Queensland (Torres Strait Islands). **Ecol:** In lowland forests, usually near the coast; 0–700 m. Fl. Several times a year. **Zones:** 9–12.

D. jenkinsii Wallich ex Lindley, *Edwards's Bot. Reg.* 25: t. 37, 1839. (Syn: *D. aggregatum* var. *jenkinsii*, etc.). **Herbs**, epiphytic; stems pseudobulbous, aggregated, appressed to substrate, ovoid, compressed, 1–2.5 cm long, 4-angled, ridged, with 2 or 3 nodes, 1-leaved. **Leaf** apical, 1–3 × 0.5–0.8 cm. **Inflorescence** lateral from pseudobulb, shorter than or nearly as long as stem, 1–3-flowered; peduncle ~1.4 cm long, glabrous; floral bracts lanceolate, 1.5–2.5 × 0.4–0.55 cm, acute, membranous; pedicel and ovary 4–5 cm long, slender. **Flowers** ~4 cm, uniformly yellow-orange; dorsal sepal oblong-elliptic, 1–1.2 × 0.5–0.6 cm, obtuse, 5-veined; lateral sepals narrowly ovate-elliptic, 1.3–1.4 × 0.4–0.5 cm, obtuse, 5-veined; mentum 4–5 mm wide; petals elliptic-ovate to orbicular, 1–1.6 × 0.5–0.9 cm, 5-veined; lip transversely obcordate, 1.5–2.2 × 1.7–2.8 cm (when spread out), adaxial surface pubescent centrally, entire, erose; column ~6 mm long, thick. **Distr:** Bhutan, China, NE India, Laos, Myanmar, Thailand. **Ecol:** In open forests; 700–1,300 m. **Zones:** 9–11.

D. johnsoniae F. Muell., *S. Sci. Rec.* 2: 95, 1882. **Herbs**, erect, epiphytic; pseudobulbs stout, subclavate, 4–6-noded below leaves, 20–30 × 0.8–1.1 cm, yellow, apex 4–5-leaved. **Leaves** elliptic or oblong-elliptic, 8–11 × 3.5–4.5 cm, obtuse, coriaceous. **Inflorescence** terminal or from below leaf bases, erect

Below: *Dendrobium johnsoniae, Dendrobium junceum, Dendrobium kauldorumii, Dendrobium kingianum.*

GS MG MG

or suberect, 17–27 cm long, few–many-flowered; bracts elliptic-lanceolate, 4–7 mm long, acute; pedicel and ovary 1.7–2.2 cm long. **Flowers** large, attractive, white with purple lines on lateral lobes of lip and base of column; dorsal sepal lanceolate, 2.6–2.8 × 0.7–0.8 cm, acuminate; lateral sepals obliquely lanceolate, 2.6–3.4 × 1–1.1 cm, acuminate; mentum obliquely conical, 8–9 mm long, somewhat noselike toward apex; petals narrowly clawed, rhombic above, 4.2–5 × 1.8–3 cm, acuminate; lip not recurved, trilobed, 3.5–4.6 × 2–3 cm, lateral lobes erect, subquadrate, margins erose, midlobe much longer, ovate-lanceolate, acuminate, base narrowed, callus fleshy, base obscurely 2-ridged; column 3 mm long. **Distr:** New Guinea, Solomon Islands. **Ecol:** In montane forests; 500–1,200 m. Fl. Apr–Nov. **Zones:** 10–12.

D. junceum Lindl., *Edwards's Bot. Reg.* 28(Misc.): 9, 1842. (Syn: *Callista juncea*, etc.) **Herbs**, large, epiphytic, upright, sympodial; pseudobulbs swollen at the base for 4 nodes (~10 cm), to 1.5 cm Ø, greenish-olive or reddish-brown, ridged; upper part of plant terete, pendent, branching, grows to 60 cm long, 2 mm Ø. **Leaves** terete, ~10–16 cm long, 2 mm Ø, apex tapering to a threadlike point, adaxially channeled, dark green. **Inflorescences** 1–2-flowered, to 4 cm long, born from the tufted bracts of the leaf axils or toward the end of pseudobulbs; flowers opening 1–2 at a time, with or without leaves. **Flowers** short-lived, fragrant, showy, to 2.5 cm Ø, green, lip maroon-veined; dorsal sepal lanceolate, to 1.5 × 0.5 cm; lateral sepals ovate, 2.5 × 0.6 cm, joined at base to form a short spur ~1.5 cm long; petals lanceolate, to 1.2 × 0.4 cm; lip flattened, trilobed, 2.7 × 1.4 cm, lateral lobes upright; column with ridges running toward the base. **Distr:** Borneo, Philippines (Luzon, Mindanao). **Ecol:** <500 m. **Zones:** 11–12.

D. kauldorumii T.M. Reeve, *Orchadian* 6: 223 ('*kauldorumi*'), 1980. **Herbs**, epiphytic, small to medium-sized; pseudobulbs clavate or fusiform, to 20 × 1.3 cm, apex 2–3-leaved. **Leaves** spreading; lamina elliptic, to 8 × 3 cm, apex obtuse, coriaceous, base purplish. **Inflorescence** erect, apical or from below lowest leaf, to 20 cm long, laxly 3–9-flowered; bracts ovate, 5 mm long, acute. **Flowers** spreading to pendent, sepals and petals white, lip pale green with darker venation, mentum, callus, anther cap, column foot base and ovary green; dorsal sepal oblong-lanceolate, 1.7 × 0.5 cm, subacute, lateral margins upturned; lateral sepals spreading, obliquely oblong-lanceolate, 1.8 × 0.8 cm, acute; mentum subconical, 7 mm long; petals oblanceolate, 1.7 × 0.5 cm, subacute; lip scarcely recurved, apically trilobed, 1.9–2.1 × 2 cm, lateral lobes erect-incurved, oblong, rounded in front, midlobe longer, ovate, acute, margins incurved, callus fleshy, apex bilobed, ~1 cm long; column short, 2.5 mm long, foot 6–7 mm long. **Distr:** Papua New Guinea. **Ecol:** On *Nothofagus* spp. in montane forests; 2,000 m. Fl. Sep. **Zones:** 11–12.

D. kingianum Bidwill ex Lindl., *Edwards's Bot. Reg.* 30 (Misc.): 11, 1844. (Syn: *Dendrocoryne kingianum*, etc.). [Pink Rock Orchid]. **Herbs**, lithophytic, forming large congested clumps; rhizome short, thick, much-branched; pseudobulbs 8–35 × 1–2 cm. **Leaves** 2–6, at apex of pseudobulb; lamina linear-ovate to obovate, 3–12 × 1–2 cm, thin-textured, adaxially dark green and somewhat shiny, abaxially paler; aerial growths produced from apical buds. **Inflorescences** racemes, 7–20 cm long, 2–15-flowered; pedicels 1–1.5 cm long. **Flowers** 1.5–2.5 cm Ø, variable in color from white with a lemon or purple lip through various shades of pink or red, mauve and purple, thick-textured, opening widely; dorsal sepal 0.9–1.6 × 0.4–0.7 cm; lip 0.8–1.5 cm long, spotted or blotched with red or mauve, curved, lateral lobes prominent and pointed; column ~3 mm long, foot 5–8 mm long. **Distr:** E Australia. **Ecol:** On rocks, boulders, cliff faces and crevices, rarely on trees, frequently forming extensive colonies. Fl. Aug–Nov. **Zones:** 8–10.

D. laevifolium Stapf, *Bot. Mag.* 149: t. 9011, 1924. (Syn: *Maccraithea laevifolia*, etc.). **Herbs**, epiphytic, tufted, 2.5–18 cm high; pseudobulbs cylindrical, fusiform, flask-shaped or ovoid, 0.7–4.5 × 0.2–1 cm, apex 2- or 3-leaved. **Leaves** linear, oblong-lanceolate or elliptic, (1.5–)5.5–14.5 × 0.5–2.5 cm, base subpetiolate, apex acute, mucronate, thick, green, sometimes purple-suffused abaxially; sheaths tubular to ~2 cm long, becoming fibrous with age. **Inflorescences** 1–2(–4)-flowered, mostly from leafless stems, terminal or from upper nodes, subsessile; pedicel and ovary (0.8–)1–2.8 cm long. **Flowers** 2–4.5 × 0.45 cm, color variable, cream, cream suffused pink, red, reddish-purple or purple-pink, lip similar in color or paler, sometimes orange, margin yellow, pink or orange, column pale green or pink; dorsal sepal elliptic-oblong to ovate-oblong, 1.2–2.5 × 0.5–1.1 cm, slightly keeled toward apiculate apex; lateral sepals oblong to ovate, 2–3 × 0.5–1 cm, apiculate, basal fused part 4–7 mm long, tubular to narrowly conic; mentum 0.8–2 cm long, tip obtuse; petals obovate, rotund or oblanceolate, 1.2–1.9 × 0.6–1.1 cm, apiculate; lip oblanceolate, boat-shaped, to 2.1 × 0.3 cm, papillose, margins upcurved, apex apiculate, with a crescent-shaped, shallowly transverse callus ~6 mm above base; column 4.5–5 mm long, foot 0.8–2 cm long; ovary 6-grooved. **Fruits** oblong-ovoid, 2.8 × 1 cm. **Distr:** New Guinea (Louisiade Archipelago) to Vanuatu. **Ecol:** In mossy rainforests; 360–2,100 m. Fl. Mar, Aug, Oct. **Zones:** 9–11.

D. lamellatum (Blume) Lindl., *Gen. Sp. Orchid. Pl.*: 89 (1830). (Syn: *Eurycaulis lamellatus*, etc.). **Herbs**, epiphytic; stems closely set, flattened, to ~12 × 3.5 cm, a few mm thick, racket-shaped, base narrowed, 2–3-leaved; sheaths large, membranous. **Leaves** to ~10.5 × 4 cm, base narrowed into a petiole, apex acute, adaxially dark purplish-green, abaxially purple, midrib adaxially channeled, abaxially elevated and paler; petiole 1.5–2 cm long. **Inflorescences** racemes, ~ 4-flowered, subapical, pendulous; pedicel and ovary 1.5 cm long, green, glabrous. **Flowers** 1.8 cm long, glabrous, colors variable; mentum 1.3 cm long, sepals and petals usually white with a pinkish tinge outside and greenish tinge centrally, becoming yellowish when old; dorsal sepal ovate, 6–6.5 × 7–9 mm, acute; lateral sepals broadly ovate, ~1 × 0.75–1 cm, acuminate; petals oblong, 8–9 × 5–5.5 mm, base narrowed, apex rounded; lip spathulate, ~0.9 × 1.2 cm, with a dull greenish patch and 3–5 keels inside, edges reflexed, ruffled. **Distr:** W Java. **Ecol:** 460–900 m. **Zones:** 10–12.

D. lawesii F. Muell., *Melbourne Chem. Druggist*, 1884. [Christmas Bell Orchid]. **Herbs**; rhizome short; stems crowded, pendulous, 7–65 cm long, leafy. **Leaves** (sub)patent, base twisted, spreading in 1 plane with stem; lamina ovate-lanceolate, 4.5–7 × 1–2 cm, margins near apex denticulate, apex acuminate, papyraceous. **Inflorescences** laterally from defoliated stem, racemose, very short, densely 1–8-flowered; floral bracts ovate, 2–5 mm long, acute; pedicel and ovary 2.5–3.5 cm long. **Flowers** 2.5–3.5 cm long, color variable, sepals and petals magenta, pink, bright red or whitish, lip yellowish at base, apically pink or whitish, all parts with darker veins; dorsal sepal broadly ovate-oblong, 7–8 × 3–4 mm, abruptly acute; lateral sepals broadly obliquely ovate-oblong, 7–8 mm wide, obtuse; mentum narrowly conical, 2.2–3 cm long, acute; petals lanceolate-oblong, 7–8 × 2.5–3.5 mm, acute; lip spathulate, 2.6–3 × 0.8–0.9 cm, base linear, adnate to column foot; column ~5 mm long, foot 2.2–3 cm long; pollinia 4. **Distr:** New Guinea, Louisiades, Solomon Islands. **Ecol:** In lower montane forests; 650–1,500 m. Fl. throughout the year. **Zones:** 9–11.

D. lichenastrum (F. Muell.) Kraenzl., *Pflanzenr.*, IV, 50: 289, 1910. (Syn: *Bulbophyllum lichenastrum*, etc.). **Herbs**, epiphytic or lithophytic, dwarf, creeping; rhizomes branching. **Leaves** prostrate; lamina ovate-oblong, obovate or circular, 0.4–1 × 0.3–0.8 cm, thick, fleshy. **Inflorescences** with flowers solitary. **Flowers** 4–7 mm Ø, sepals and petals hyaline, white, cream or pink, with variable number of red stripes, lip yellow or orange; dorsal sepal broadly ovate-orbicular, 3–5 mm long; lateral sepals obliquely triangular, 3–5 × 1.5–4 mm, acute to obtuse; petals linear-lanceolate, 2–5 mm long, obtuse or acute; lip oblong-subpandurate, 4–5 mm long, clawed, grooved at base, adaxially rugulose; column very short, 1 mm long. **Distr:** Australia (Queensland). **Ecol:** Epiphytic on trees in dense rainforests, or lithophytic on rocks in open forests; 0–1,950 m. Fl. throughout most of the year. **Zones:** 9–11.

D. linawianum H.G. Reichenbach, *Ann. Bot. Syst.* 6: 284, 1861. (Syn: *Callista linawiana*, *Ormostema purpurea*). **Herbs**, epiphytic; stems erect, cylindric, compressed, usually 25–30 cm long, longitudinally grooved. **Leaves** oblong, 4–7(–10) × 2–2.5 cm, base dilated into clasping sheaths, apex obtuse, unequally bilobed, leathery. **Inflorescence** 2–4-flowered, borne from old leafless stems; peduncle 7–8 mm; floral bracts ovate, 3–4 mm long, acute; pedicel and ovary to 5 cm long. **Flowers** spreading, large, white, sometimes upper part purplish-red, lip white, upper part purplish-red, disk with purplish-red stripes on each side at base, anther cap white; dorsal sepal oblong, 2.2–3.5 × 0.8–1 cm, 5-veined, slightly obtuse; lateral sepals obliquely oblong, base oblique, apex slightly obtuse, 5-veined; mentum narrowly conic, ~8 mm long; petals elliptic, 2.2–3.5 cm, broader than sepals, shortly clawed, obtuse; lip broadly oblong, sides embracing column, adaxially densely pubescent, shortly clawed, obtuse; column ~4 mm long, foot ~8 mm long. **Distr:** SE China, N Taiwan. **Ecol:** On tree trunks of montane forests; 400–1,500 m. **Zones:** 7–11.

D. lindleyi Steudel, *Nomencl. Bot.*, ed. 2, 1: 490, 1840. (Syn: *Epidendrum aggregatum*, etc.). **Herbs**, epiphytic; stems tufted, fusiform or ovoid-cylindric, laterally compressed, 1–5 × 0.5–1.5 cm, with 2–5 nodes, base contracted, internodes 4-ribbed. **Leaf** usually solitary, oblong, 3–8 × 0.6–3+ cm, base contracted, not decurrent into clasping sheaths, margins ± undulate, apex obtuse, emarginate, leathery. **Inflorescence** subterminal, to 27 cm, laxly 10-flowered; floral bracts narrowly ovate-triangular, ~2 mm long; pedicel and ovary yellowish-green, purple-tinged, 0.6–3 cm long. **Flowers** spreading, thinly papery, orange; dorsal sepal ovate-lanceolate, ~2 × 0.7–0.8+ cm, slightly obtuse; lateral sepals nearly equal to dorsal sepal; mentum semiglobose, ~5 mm across; petals broadly elliptic, ~2 × 1 cm, bluntly rounded; lip transversely oblong or subreniform, usually ~1.5 × 2 cm, each side embracing column below middle, entire, often emarginate; disk densely pubescent below middle; column ~4 mm long. **Distr:** China, Bhutan, India, Laos, Myanmar, Thailand, Vietnam. **Ecol:** On trees in sunny open forests; <1,000 m. **Zones:** 9–12.

D. lineale Rolfe, *Gard. Chron.*, III, 6: 381, 1889. (Syn: *D. cogniauxianum*, etc.). **Herbs**, epiphytic or lithophytic, erect; stems clustered, canelike, 0.5–3.2 cm long, base leafless. **Leaves** oblong-elliptic to elliptic, 13.5 × 8.5 cm, base articulated to tubular sheaths, apex unequally bilobed, coriaceous. **Inflorescences** 24–75 cm long, 17–20- or more-flowered; peduncles 10–15 cm long; pedicel and ovary 4.1–7 cm long. **Flowers** white or pale yellow with white, petals bluish or pink, lip purple- or violet-veined; dorsal sepal recurved, linear or linear-lanceolate, 1.8–2 × 0.5–0.6 cm, acuminate, often twisted; lateral sepals recurved, obliquely lanceolate-falcate, 1.8–2 × 1–1.3 cm, acute; mentum conical,

MG

GS MG

MG

MG

JT GS

GS

Top row: *Dendrobium laevifolium, Dendrobium lamellatum, Dendrobium lawesii.*
Second row: Dendrobium lichenastrum, Dendrobium lichenastrum (another form), *Dendrobium linawianum.*
Top row: *Dendrobium linguella, Dendrobium linguiforme, Dendrobium litorale.*
Bottom row: *Dendrobium lindleyi, Dendrobium lineale, Dendrobium loddigesii.*

0.8–1.1 cm long, curved; petals erect, spathulate, 2.2–3 × 0.4–0.6 cm, rounded or truncate, half- or not twisted; lip trilobed, 2–2.4 × 15–1.6 cm, lateral lobes obliquely elliptic, obtuse or rounded in front, margins erose, midlobe oblong, apiculate, margins erose, undulate, callus of three keels, all elevated on center of midlobe; column 6 mm long. **Distr:** New Guinea and adjacent islands. **Ecol:** In rainforests; 0–800 m. **Zones:** 10–12.

D. linguella Rchb. f., *Gard. Chron.* 1882(2): 552, 1882. **Herbs**, epiphytic or occasionally terrestrial; stems slender, erect (or pendulous), to ~0.6–1 m long, slender and wiry, ridges purplish. **Leaves** deciduous, borne in the upper parts of the stems; lamina lanceolate, ~10 × ~1.5 cm, base shortly petiolate, apex acute, adaxially green, abaxially paler, purplish-spotted. **Inflorescences** clustered, compact, 2–8-flowered, 3 cm long, borne in the apical 2/3 of mature leafless stems; pedicel and ovary mauve or pinkish, 2–2.8 cm long. **Flowers** 2–4 cm across, lasting ~2 weeks, waxy, entirely very pale mauve or mauve, anther deep purple; dorsal sepal triangular, 2.8 × 1.4 cm, acute; lateral sepals triangular, 3.5 × 2.4 cm, acute; mentum ~8 mm long, slightly projected forward; petals ovate, 2.4 × 1.4 cm, base narrowed, apex acute; lip hollow, 2.8 cm long, lateral lobes erect, hairy, with a prominent hairy callus at the base and a very short pointed tip. **Distr:** Indo-China to W Malesia. **Ecol:** On trees of riverine forests, lowland forests and foothill forests; 100–1,300 m. **Zones:** 9–12.

D. linguiforme Sw., *Kongl. Vetensk. Acad. Nya Handl.* 21: 247, 1800. (Syn: *Callista linguiformis*, *Dockrillia linguiforme*). [Thumbnail Orchid, Tongue Orchid]. **Herbs**, epiphytic or lithophytic, in spreading patches; rhizome ~0.4 cm thick, woody, branched. **Leaves** alternate, prostrate; lamina ovate, oblong or obovate, 2–4 × 0.8–1.5 cm, apex blunt, thick and fleshy, closely spaced, dark green, adaxially with a few longitudinal furrows. **Inflorescences** racemes, 5–15 cm long, 6–20-flowered. **Flowers** ~1 cm across, sepals and lateral petals white to cream, lip cream or white with purple markings; perianth segments slender, pointed, not spreading widely; dorsal sepal 1.5–2.2 × 0.3–0.4 cm; lip 5–6 × 3–4 mm, curved, surface with 3 long longitudinal keels; column ~2 mm long, foot ~3 mm long. **Distr:** E Australia, New Caledonia. **Ecol:** On trees or rocks in fairly open situations. Fl. Jun–Sep. **Zones:** 8–12.

D. litorale Schltr., *Repert. Spec. Nov. Regni Veg. Beih.* 1: 567, 1912. (Syn: *Aporopsis litoralis*, etc.). **Herbs**, epiphytic; rhizome very short; stems crowded, erect, slender, 2–4 internodes above base usually swollen, 4–6-angled, to 60 cm long, lower half many-leaved, slightly flattened, upper half bare. **Leaves** bilaterally flattened, obliquely linear, 4.5–6 × 0.3–0.4 cm, apex acute, coriaceous. **Inflorescences** 1-flowered, fasciculate, flowers appearing in succession, lateral, borne from bare upper part of stems; floral bracts ovate, minute, much shorter; pedicel and ovary 1.2 cm long, slender, glabrous. **Flowers** resupinate, 1.5 cm long, thin-textured, pale yellow, red-veined; dorsal sepal ovate, 5.5 mm long, obtuse; lateral sepals obliquely ovate, much widened at base, 5.5 mm long, obtuse; mentum oblongoid, 1 cm long, slightly curved; petals obliquely oblong-ligulate, 5.5 mm long, obtuse; lip trilobed apically, 1.3 ×1 cm, lateral lobes semirhombic, obtuse, midlobe semiquadrate, deeply bipartite; column short, foot linear-ligulate, 1 cm long. **Distr:** New Guinea, Australia (Queensland). **Ecol:** In lowland and beach forests; 10–15 m. Fl. Apr, Jul. **Zones:** 8–12.

D. loddigesii Rolfe, *Gard. Chron.*, ser. 3, 2: 155, 1887. (Syn: *Callista loddigesii*, etc.). **Herbs**, epiphytic; stems often pendulous, cylindric, 10–45 cm long. **Leaves** alternate along entire stem; lamina ligulate, oblong-lanceolate, or slightly obliquely oblong, usually 2–4 × 1–1.3 cm, base sheathing, apex acute and slightly hooked, leathery. **Inflorescence** lateral on upper part of old leafy stem, 1- or 2-flowered; peduncle 2–3 mm long; basal sheaths 1 or 2, cupular; floral bracts ovate, ~2 mm; pedicel and ovary 2–3 cm long. **Flowers** pale pink or purplish-red, lip adaxially golden-yellow at center, margins pale purplish-red; dorsal sepal ovate-oblong, 1.7–2 × ~0.7 cm, acute, 5-veined; lateral sepals lanceolate, 1.7–2 × 0.6–0.7+ cm, base oblique, apex acute, 5-veined; mentum semiglobose, ~5 mm across; petals elliptic, ~1.7–2 × 0.8–0.9+ cm, margin entire, apex slightly obtuse, 3–5-veined; lip suborbicular, 1.7–2 cm Ø, margin shortly fimbriate, apex emarginate, densely pubescent; column white, front lateral sides with red stripes, ~4 mm long; anther cap white, densely papillate. **Distr:** China, Hong Kong, Laos, Vietnam. **Ecol:** On tree trunks in montane forests, lithophytic in forests; 400–1,500 m. **Zones:** 9–11.

D. lowii Lindl., *Gard. Chron.* 1861: 1046, 1861. (Syn: *Callista lowii*). **Herbs**, epiphytic; stem erect, ~9-noded, 25–40 cm long, fairly thick, upper leaf sheaths covered with brown or black hairs. **Leaves** ~6, borne toward the apex of the stem; lamina narrowly ovate, 5–8 × 1.5–2.2 cm, apex obliquely bilobed, abaxially black-hairy. **Inflorescences** densely 2–7-flowered, short, subapical; bracts short, acute, black hairy. **Flowers** 5 cm Ø, fragrant, bright yellow, lip with keels orange-red; dorsal sepal broadly oblong, 2.5 × 1 cm, obtuse, apiculate; lateral sepals ovate-triangular, 2 × 1 cm, obtuse; petals broadly oblong, 2.8 × 1.5 cm, margins undulate, apex rounded; mentum attenuate, subulate, 3 cm long; lip deeply trilobed, 3 × 1.2 cm, lateral lobes basal, ligulate or oblong, 1.5 cm long, acute, midlobe linear-spathulate, margin crisped, decurved in front, rounded, bilobulate; disk 3–7-keeled, hairy; column with large, obtuse, stelidia. **Distr:** Borneo. **Ecol:** On trees in exposed areas; 900–1,000 m. **Zones:** 11–12.

D. maccarthiae Thw., *Bot. Mag.* 11: pl. 4886. 1855. [Vesak Orchid]. **Herbs**, epiphytic; stems slender, terete, pendulous, 16–60 cm long. **Leaves** clustered at apices of stems, 1 per node; lamina lanceolate, 4–8 × 0.9–1.3 cm, base sheathing internodes, apex acuminate, membranous, veins parallel, midrib prominent abaxially; leaf sheaths 2 cm long, green when young, appressed to internodes. **Inflorescences** drooping racemes, 9–11 cm long, pendulous, laxly 2–4-flowered; peduncle 2–3 cm long; floral bracts lanceolate, 0.6–0.5 × 0.3 cm, subulate, brown, papery; pedicel and ovary arched, 2.4–3.1 cm long. **Flowers** 5.5–6 × 7.5 cm, light violet-pink, lip paler but with a purple blotch in the center and on disk; dorsal sepal lanceolate, 3.5–4 × 1 cm, acuminate or acute, 7-veined; lateral sepals subfalcately lanceolate, 4–4.5 × 1.3 cm, acuminate or acute, bases adnate to column foot; petals oblong-lanceolate or ovate-oblong, 4–4.3 × 1.9 cm, acuminate or acute, 5-veined; lip rhomboid or subspathulate, 6 × 3 cm, lateral lobes rounded, convolute, midlobe broadly ovate, mucronate; column 4 mm high, apex 3-toothed; column and foot, 1.6 × 0.6 cm; pollinia 4. **Fruits** loculicidal, fusiform, 3.2 × 1.17 cm. **Distr:** Sri Lanka. **Ecol:** In tropical wet evergreen forests. Fl. May–Jul. **Zones:** 11–12.

D. macrophyllum A. Rich., *Voy. Astrol., Bot.*: 22, 1834. [Pastor's Orchid]. **Herbs**, epiphytic or lithophytic, medium-sized to large; pseudobulbs clustered, clavate or subclavate, rarely fusiform, to 50 cm long. **Leaves** (2–)3(+), suberect; lamina oblong or oblong-elliptic, 15–31 × 3.3–9 cm, apex obtuse, coriaceous. **Inflorescences** often densely many-flowered, erect, terminal, to 40 cm long; bracts lanceolate, 1.5–2.5 cm long, acute; pedicel and ovary setosely hairy, 2.5–4 cm long. **Flowers** variable in size, yellow or greenish, lateral lobes of lip purple-striped, midlobe with spots or stripes; sepals setose on outside; dorsal sepal oblong-ovate or oblong-lanceolate, 2.1–2.6 × 0.7–1 cm, acute; lateral sepals obliquely triangular, 2.3–2.6 × 1.2–1.4 cm, acute or acuminate; mentum obliquely conical, 1 cm long; petals oblanceolate or oblong-oblanceolate, 1.8–2.2 × 0.8–1.1 cm, margin somewhat undulate, apex acute; lip trilobed, strongly recurved, 1–2 × 1.6–2.7 cm, lateral lobes erect, subquadrate-flabellate, dilated above, truncate, midlobe transversely oblong, apiculate, conduplicate, callus white, obscurely 3-ridged, raised 2 mm at apex; column short, 3 mm long, apex with 2 hooklike appendages; foot 1 cm long. **Distr:** Malesia to W Pacific. **Ecol:** In forests; 0–1,700 m. Fl. throughout the year. **Zones:** 10–12.

D. macrostachyum Lindl., *Gen. et Sp. Orch.* 78, 1830. **Herbs**, epiphytic; stems slender, drooping, 30–70 cm long. **Leaves** distichous, spreading, sessile; lamina oblong or oblong-lanceolate, 4–6 × 0.9–2.3 cm, acute, many-veined, basal sheaths enveloping stem. **Inflorescences** clusters, 2–4-flowered, borne from leafless nodes; peduncles short; floral bracts oblong-ovate, 2.6 × 1.8 mm, acute, 3-veined; pedicel and ovary 1.3–1.7 cm long. **Flowers** greenish-yellow, 2–2.3 cm Ø; dorsal sepal oblong-lanceolate, 1.3

Below: *Dendrobium lowii*, *Dendrobium maccarthiae*, *Dendrobium macrophyllum*.

MG

GS

Top row: *Dendrobium macrostachyum*, *Dendrobium microbulbon*, *Dendrobium mirbelianum*, *Dendrobium mohlianum*.
Bottom row: *Dendrobium moniliforme*, *Dendrobium moschatum*, *Dendrobium nobile*, *Dendrobium otaguroanum*.

× 0.4 cm, acute, 5-veined; lateral sepals obliquely oblong-lanceolate, 1.4–1.5 × 0.45 cm, acute, 5–7-veined; petals lanceolate, 1.3–1.4 × 0.4 cm, acute, 3-veined; lip 1.7 × 1.2 cm, 5-veined, lateral lobes small, rounded, convolute, embracing column, midlobe quadrately ovate, apiculate, recurved, crenulate, ciliate; column 4 × 2.2 mm, pollinia 4. **Fruits** fusiform, 3 × 0.7 cm. **Distr:** India, Sri Lanka. **Ecol:** On trees, in wet and dry forests to subtropical montane forests; to 1,220 m. Fl. Mar, Jun.–Jul. **Zones:** 10–12.

D. microbulbon A. Rich., *Ann. Sci. Nat.*, Bot., II, 15: 19, 1841. (Syn: *Callista microbulbon*, etc.). **Herbs**, epiphytic; pseudobulbs small, crowded, ovoid, covered with a network of fibers formed by decaying leaves. **Leaves** soon deciduous, when present borne from the apex of the pseudobulb; lamina linear-oblong, 5–7.5 cm long, apex acute. **Inflorescence** a raceme, erect, 2.5–5 cm long, 4–8-flowered; bracts small, lanceolate, ~2.5 mm long, papery; pedicel and ovary ~1 cm long. **Flowers** ~1.2 cm across, sepals and petals white, lip pink with dark red veins; dorsal sepal narrowly ovate, ~6 × 2.5 mm, acute; lateral sepals narrowly ovate, 7 × 2.2 mm, acute; petals spathulate, base 5 × 2 mm, clawed, apex acute; lip oblong, 6 × 4 mm, lateral lobes almost entire, midlobe suborbicular, crenulate; column 2.5 mm long. **Distr:** S India (Western Ghats and Konkan). **Ecol:** Epiphytic on trees and shrubs. Fl. Jul–Sep, Dec–Mar. **Zones:** 10–11.

D. mirbelianum Gaudich., *Voy. Uranie, Bot.*: 423, 1829. (Syn: *Callista mirbelliana*, etc.). [Mangrove Orchid]. **Herbs**, epiphytic, erect; stems 0.5–3.2 m tall, canelike. **Leaves** oblong-elliptic to elliptic-ovate, 8–12.5 × 2.5–5.4 cm, base articulated to tubular sheaths, often purple-striped, apex obliquely bilobed, coriaceous. **Inflorescences** 4–30-flowered, suberect to spreading, 12–45 cm long; bracts deltoid to elliptic, 3–7 mm long, acute; pedicel and ovary 1–2.5 cm long. **Flowers** occasionally cleistogamous, greenish-yellow to olive-brown, lip veined with purple-brown; dorsal sepal erect, linear-lanceolate, 2.2–2.8 × 0.6 cm, acute or apiculate; lateral sepals spreading, obliquely lanceolate to linear-lanceolate, 2.4–2.8 × 0.6–0.7 cm, acute; mentum obliquely conical 7–8 mm long; petals spathulate, 2.2–5 × 0.4–0.7 cm, acute to subacute, not twisted or half-twisted; lip trilobed, 1.6–2.5(–3.8) × 1.2–1.5 cm, lateral lobes elliptic, rounded, margins erose, midlobe recurved, ovate, 0.9–1.5 cm long, acute, margins erose and undulate, callus white marked with purple or violet, with 5 erose ridges; column 4–5 mm long. **Distr:** Sulawesi to Solomon Islands. **Ecol:** In coastal forests, mangroves, rainforests, savannas; 0–650 m. Fl. throughout the year. **Zones:** 10–12.

D. mohlianum Rchb. f., *Bot. Zeitung (Berlin)* 20: 214, 1862. (Syn: *Chromatotriccum mohlianum*, etc.). **Herbs**, epiphytic; stems spreading or pendent, clustered, ribbed, to 50 cm long, base not swollen. **Leaves** distichous, sheathing stem at base; lamina lanceolate, 6–13 × 1–2.5 cm. **Inflorescences** racemose, in clusters of 4–6, borne laterally from leafless stem; peduncle 1–1.5 cm long; pedicels 1.5–2 cm long. **Flowers** orange to bright red; dorsal sepal and petals ovate, ~9 mm long; lateral sepals triangular, ~1.5 cm long, fused at base to form a mentum; lip obovate, 1.6 cm long, 1 cm wide at apex, apex cucullate, orange-red with or without purple lines; column ~3 cm long, white, with foot ~1.3 cm long. **Distr:** Solomon Islands to SW Pacific. **Ecol:** In montane and cloud forests; 430–3,100 m. **Zones:** 11–12.

D. moniliforme (L.) Swartz, *Nova Acta Regiae Soc. Sci. Upsal.*, ser. 2, 6: 85, 1799. (Syn: *Epidendrum moniliforme*, etc.). **Herbs**, epiphytic; stems erect, cylindric, usually 10–30+ cm long. **Leaves** several, alternate; lamina lanceolate, oblong, or narrowly oblong, 3–5(–7) × 0.5–1.2(–1.5) cm, base decurrent into clasping sheaths, apex obtuse, slightly unequally bilobed. **Inflorescences** 1 to several, 1–3-flowered; pedicel and ovary white or pink, 1–3 cm long, slender. **Flowers** sometimes fragrant, yellowish-green, creamy-white, or white tinged with pale purplish red, lip white, pale yellowish-green, or greenish-white, spots tinged with pale brown or purplish-red to pale yellow, often with a purplish-red, pale brown, or pale yellow, or yellowish-green central spot near base of midlobe, column white, foot often with purplish-red stripes at base, anther cap white or pale yellow; sepals and petals similar, ovate-oblong, oblong-lanceolate, or ovate-lanceolate, (1–)1.3–4 × (0.15–)0.3–1 cm, apex acute, acuminate, or obtuse; lateral sepals oblique, adnate to column foot; petals subelliptic, (2.3–)2.5–4 × 1–1.5 cm, 5- or 6-veined, acute; lip ovate-lanceolate, base cuneate, trilobed, lateral lobes embracing column, midlobe ovate to ovate-lanceolate. **Distr:** China to temperate E Asia. **Ecol:** On tree trunks in forests, lithophytic on rocks and cliffs; 600–3,000 m. **Zones:** 8–11.

D. moschatum (Buchanan-Hamilton) Swartz, *Neues*

MG

MG

Above: *Dendrobium pendulum, Dendrobium porphyrochilum, Dendrobium profusum, Dendrobium pugioniforme.*

J. Bot. 1: 94, 1805. (Syn: *Cymbidium moschatum*, etc.). **Herbs**, epiphytic; stems erect, cylindric, to 1 m tall. **Leaves** alternate along upper part of stem; lamina oblong to oval-lanceolate, 10–15 × 1.5–3 cm, base with clasping papery sheaths, apex acuminate or unequally bilobed, leathery. **Inflorescences** racemose, pendent, ~20 cm long, laxly to >10-flowered; peduncle ~5 cm long; basal sheaths 4, overlapping, cupular; floral bracts oblong, 1.2–2 × 0.3–0.5+ cm, obtuse, leathery; pedicel and ovary to 5 cm long. **Flowers** deep yellow (or pink or white with rose tips), disk with a pale purplish-brown blotch on either side, column yellow, anther cap purple; dorsal sepal oblong, 2.4–3.5 × 1.1–1.4 cm, obtuse, 6- or 7-veined; lateral sepals oblong, 2.4–3.5 × 0.9–1+ cm, base slightly oblique, apex slightly acute, 5-veined; petals obliquely broadly ovate, 2.6–3.5 × 1.7–2.3 cm, obtuse, 7-veined; lip rounded, slipperlike, ~2.4 × 2.2 cm, margin involute, cucullate, adaxially densely pubescent; column ~4 mm long, foot ~4 mm long. **Distr:** Himalayas to Indo-China. **Ecol:** On tree trunks in open forests; ~1,300 m. **Zones:** 9–12.

D. nobile Lindley, *Gen. Sp. Orchid. Pl.* 79, 1830. (Syn: *Callista nobilis*, etc.). **Herbs**, epiphytic or lithophytic; stems erect, cylindric, 10–60 cm long. **Leaves** oblong, 6–11 × 1–3 cm, base with clasping sheath, apex obtuse, unequally bilobed, leathery. **Inflorescences** arising from old stems before or after leaves have fallen, 2–4 cm long, 1–4-flowered; peduncle 0.5–1.5 cm long; basal sheaths several, tubular; floral bracts ovate-lanceolate, 0.6–1.3 cm long, membranous; pedicel and ovary pale purple, 3–6 mm long. **Flowers** large, all parts white, apex purplish-tinged, sometimes entirely pale purplish-red, with purplish-red striation on either side at base of lip, or lip disk with a central purplish-red spot, sometimes white; dorsal sepal oblong, 2.5–3.5 × 1–1.4 cm, obtuse; lateral sepals similar, base oblique, apex acute; petals ± obliquely broadly ovate, 2.5–3.5 × 1.8–2.5 cm, base shortly clawed, margin entire, apex obtuse; lip broadly ovate, 2.5–3.5 × 2.2–3.2 cm, base shortly clawed, embracing column, margin shortly ciliate, apex obtuse, densely pubescent; column green, ~5 mm long; anther cap purplish-red. **Distr:** Himalayas to S China and Indo-China. **Ecol:** On tree trunks in montane forests, mountain valleys; 500–1,700 m. **Zones:** 10–12.

D. otaguroanum A.D. Hawkes, *Lloydia* 20: 122, 1957. (Syn: *Sayeria otaguroan*). **Herbs**, epiphytic, small; pseudobulbs clustered, erect, 10–25 × 0.3–0.7 cm, 3–4-noded below leaves, yellow, apex 2–3-leaved. **Leaves** narrowly lanceolate to ovate, 5–7 × 1.2–2.1 cm, acute, coriaceous. **Inflorescence** erect, 2.5–6.7 cm long, 2–6(or more)-flowered; bracts elliptic, apiculate, 4 mm long; pedicel and ovary 1.7–2.5 cm long. **Flowers** subnutant, white with yellowish-green venation on lip; dorsal sepal ovate-lanceolate, 1.6–1.8 × 0.5 cm, acuminate; lateral sepals obliquely lanceolate, 1.6–2 × 0.8–1 cm broad, acuminate; mentum obliquely conical, 7–8 mm long; petals obovate, 1.6–1.5 × 0.8–1 cm, obtuse or apiculate; lip entire, obovate, 1.6–2.2 × 1.5–1.6 cm, apiculate, base narrowly clawed, callus fleshy, 3-ridged, over half the length of lip; column 2.5 mm long, foot 7–8 mm long; ovary somewhat rugulose. **Distr:** New Guinea. **Ecol:** In montane forests; 1,500–2,200 m. Fl. Jan, Mar, Oct. **Zones:** 11–12.

D. pendulum Roxb., *Fl. Ind.* ed. 1832, 3: 484, 1832. (Syn: *Callista crassinodis*, etc.). **Herbs**, epiphytic; stems forming tufted masses, ± arching or pendulous, 30–60 cm long, nodes 7–16, swollen, olive-green. **Leaves** deciduous; lamina lanceolate, semiamplexicaul, 10–12 × 1.5–2.5 cm, acute. **Inflorescences** borne from the nodes of the old stems, 1–3-flowered; bracts boat-shaped, papery; pedicel yellowish-green. **Flowers** fragrant, long-lasting, 4–7 cm across, sepals and petals white with purple apices, lip white with base yellow and margins purple; sepals oblong, 2–3.5 × 0.8–1.2 cm, obtuse, reflexed; petals broadly oblong, 1–3.2 × 1.2–1.6 cm, obtuse or rounded; lip shortly clawed, orbicular and shell-like, 2–2.5 × 2–2.6 cm, margins ciliolate, both surfaces pubescent; column 3–4 mm long; ovary pale purplish-red, 3–4 cm long. **Distr:** Assam to China (S Yunnan) and Indo-China. **Ecol:** On tree trunks in open forests; 760–1,600 m. **Zones:** 8–11.

D. porphyrochilum Lindley, *J. Proc. Linn. Soc., Bot.* 3: 18, 1858. (Syn: *Callista porphyrochila, D. caespitosum*). **Herbs**, epiphytic; stems erect, cylindric or narrowly fusiform, 1.5–4 × 0.2–0.4 cm, fleshy. **Leaves** 3 or 4, distichous, alternate; lamina narrowly oblong, to 4.5 × 0.6–1(+) cm, base contracted and enlarged into sheaths, apex acute, unequally bilobed, papery. **Inflorescences** solitary, proteranthous, apparently terminal, nodding, to 8 cm long, to 10(+)-flowered; floral bracts narrowly lanceolate, ~9 × 1 mm; pedicel and ovary hairlike, ~8 mm long. **Flowers** scented, small, not resupinate, spreading, thin, golden-yellow, or sepals and petals pale green, red-veined, lip deep purplish-brown, margin pale green, column white purple-tinged; dorsal sepal narrowly ovate-lanceolate, 8–9 × 1.8–2 mm, 3-veined, caudate-acuminate; lateral sepals narrowly lanceolate, about as long as dorsal sepal, slightly broader, 3-veined, base oblique, apex acuminate; mentum subglobose, small; petals narrowly elliptic, 6.5–7 × ~1.8 mm, acute; lip subrhombic or elliptic, concave, ~5 × 2 mm, entire, subacute; disk with 3 ± thickened central longitudinal ridges; column ~1 mm long, base dilated, foot ~1.4 mm long. **Distr:** Himalayas to China (Yunnan). **Ecol:** On tree trunks in mountain forests, lithophytic on rocks in forests; ~2,700 m. **Zones:** 8–10.

D. profusum Rchb. f., *Gard. Chron.* 1884(1): 510, 1884. (Syn: *Eurycaulis profusus*). **Herbs**, epiphytic, pendulous, sympodial; pseudobulbs to 50 cm long, slender, 6 mm Ø. **Leaves** deciduous; lamina lanceolate, to 7.5 × 2 cm. **Inflorescences** pendulous, to 7–15-flowered, borne on apical parts of mature leafless stems. **Flowers** 2.5 cm Ø, cream to yellow with pink stripes, outside mottled with reddish-pink; dorsal sepal ovate, 1.2 × 0.6 cm; lateral sepals triangular, 1.4 × 0.8 cm, forming a spur ~2.5 cm long; petals with a pinkish-red stripe running their length, ovate, 1.4 × 0.7 cm; lip trilobed, fiddle-shaped, 2 cm long, 9 mm wide (when flattened), midlobe inverted and heart-shaped, lateral lobes semicircular. **Distr:** Endemic to Philippines (Luzon). **Ecol:** Epiphytic in brightly lit habitats; 100–500 m. **Zones:** 11–12.

D. pugioniforme A. Cunn. ex Lindl., *Edwards's Bot. Reg.* 25(Misc.): 33, 1839. (Syn: *Callista pugioniformis*, etc.). [Dagger Orchid]. **Herbs**, epiphytic or lithophytic, hanging in large masses; stems sympodial, to 2 m long. **Leaves** mostly pendent; lamina ovate to narrowly ovate, 1–7 × 0.5–2 cm, base attenuate, apex acuminate, pungent, flattened, 1–3 mm thick, succulent. **Inflorescences** 1- or rarely 2- or 3-flowered, 4–7 mm long, borne toward apex of sympodial stems. **Flowers** non-resupinate, 2–2.5 cm Ø, white to light green with a pale lip bearing some bright purple or red markings on the lateral lobes and at the base of the midlobe; dorsal sepal and petals lanceolate, 0.8–1.2 × 0.2–0.3 cm, acuminate; lateral sepals lanceolate, 1.5 cm long; lip rhombic, trilobed, 1.3–1.8 × 0.6–0.9 cm, base ± cuneate, apex acute, lateral lobes rounded, obscure, midlobe broadly triangular, margins undulate, reflexed; disk 3-keeled, slightly undulate; column ~4 mm long, foot 4–5 mm long. **Distr:** E Australia. **Ecol:** In rainforests; to 1,150 m. **Zones:** 9–11.

D. purpureum Roxb., *Fl. Ind.* ed. 1832, 3: 484, 1832. (Syn: *Callista purpurea, Pedilonum purpureum*). **Herbs**, epiphytic; stems initially erect becoming pendulous, cylindrical to elongate-fusiform, (0.2–)0.5–1.1(–1.5) m long. **Leaves** erect-patent to patent, recurved; lamina lanceolate to elliptic, (5–)9–13.5 × 1.5–2.5(–4) cm, base rounded, apex somewhat obtuse, thinly leathery. **Inflorescence** dense, somewhat spherical, to 60-flowered, borne from leafless or occasionally leafy stems, erect-patent, to 2–5 cm long; bracts small; pedicel and ovary clavate, 1.1–1.8 cm long. **Flowers** 1.3–2 × 0.3–8 cm, not widely opening, bright to pale purple or white, sepal apices always green, lip apex sometimes pale green; dorsal sepal ovate-oblong, 4.4–7.2 × 2.2–7 mm, rounded to tapering, apiculate; lateral sepals similar, conical, 3.6–4.5 mm wide, obtuse, extended into mentum; petals obliquely elliptic-lanceolate, 4.4–5.5 × 1.5–2 mm, acute; lip constricted below middle, sides ± infolded, lamina ovate to oblong or elliptic, 6.9–9.4 × 2.4–4.6 mm; column including foot 1–1.6 cm long. **Distr:** Sulawesi to New Guinea (Kepulauan Kai, Aru Islands). **Zones:** 11–12.

D. ravanii Cootes, *Orchideen J.* 15: 70, 2008. **Herbs**, epiphytic, pendulous, sympodial; roots branching with a rugose surface; pseudobulbs stem-like, cylindrical, to 46 cm long, 3–6 mm Ø, leafy on the upper third, leaves 1 to 2 cm apart; internodes 2–3 cm long, lower ones covered with papery sheaths. **Leaves** deciduous after each season; lamina lanceolate, to 8.5 × 2.5 cm, thinly textured, shiny, distinct ridge down the center; leaf sheaths ~2 cm long, clasping the stem. **Inflorescence** pendulous, to 5 cm long, bearing numerous flowers 3.4 cm long (from the tip of the dorsal sepal to the tip of the mentum), ~3 cm across the widest point (lateral sepals); lip to 2.5 × 1 cm, midlobe ovate to quadrangular, margins undulate and slightly erose; column 3 × 3.5 mm. **Distr:** Endemic to Philippines (mountains of northern Mindoro). **Ecol:** Often close to the ground, in heavily shaded localities; 500–800 m. **Zones:** 11–12.

D. rhodostictum F. Muell. and Kraenzl., *Österr. Bot. Zeitschr.* 44: 300, 1894. (Syn: *D. madonnae, Sayeria rhodosticta*). **Herbs**, epiphytic or terrestrial, erect; pseudobulbs clavate, often markedly dilated above, 9–25 cm long, yellow, apex 2–4-leaved. **Leaves** suberect or spreading; lamina ovate-lanceolate to linear-lanceolate, 5–11 × 1–2.5 cm, apex acute to obtuse, minutely emarginate, coriaceous. **Inflorescences** apical or subapical, erect to arcuate, few-flowered; bracts elliptic, 3–4 mm long, acute; pedicel and ovary 2.5–3.5 cm long. **Flowers** subnutant, white with purple spots on lateral margins of lip; dorsal sepal ovate, acute or acuminate, 1.8–2.5 × 1 cm; lateral sepals falcate-lanceolate, 1.8–2.2 × 1–1.1 cm, acuminate; mentum obliquely conical, 0.8–1 cm long; petals obovate-spathulate, 3–3.5 × 1.5–2.4 cm, obtuse to apiculate; lip obscurely trilobed toward apex, broadly obovate, 1.9–2.6 × 1.8–2.8 cm, lateral lobes erect-incurved, semicircular, midlobe often obscure, callus 3-ridged above, raised and trilobed at base; column 2.5 mm long, foot 7–9 mm long. **Distr:** Papua New Guinea to Solomon Islands. **Ecol:** In montane forests; 800–1,200 m. Fl. Jan, May–Jun. **Zones:** 9–11.

D. sanderae Rolfe, *Gard. Chron.*, III, 45: 374, 1909. (Syn: *D. sanderae* var. *parviflorum*, etc.). **Herbs**, epiphytic, upright, sympodial; stems erect, somewhat thickened below middle, to 1+ m long, longitudinally striate, ~40-leaved; leaf sheaths blackish-hairy. **Leaves** ovate to narrowly ovate, 5–7 × 1.7–2 cm, apex bilobed, very stiff and leathery. **Inflorescences** short racemes, 3–4-flowered, to ~6 cm long, lateral. **Flowers** white with red or purple streaks at the base of the lip and on lateral lobes, 6–7 cm across; dorsal sepal lanceolate, 3.5–4 cm long, acuminate; lateral sepals oblong-lanceolate, 3.5 × 1.4 cm, keeled; petals broadly obovate to elliptic-obovate, 3.5–3.8 × 2–3 cm; lip trilobed, tubular at base, 4.2 × 3.2 cm, lateral lobes surrounding the column, midlobe truncate, ovate, 2.5 cm Ø, flattened; mentum spurlike, 2.2–2.5 cm long. **Distr:** Philippines. **Ecol:** In lowland and montane rainforests; 300–1,650 m. Fl. Jan–Oct. **Zones:** 9–11. **Info:** This species is variable in flower color and shape.

D. scabrilingue Lindl., *J. Proc. Linn. Soc.*, Bot. 3: 15, 1859. (Syn: *Callista scabrilinguis*, etc.). **Herbs**, epiphytic; stems pseudobulbous, tufted, erect, subclavate, fusiform or swollen toward apex, hirsute, 15–30 cm long, nodes ~10, with 4–6 leaf sheaths covered with black hairs; internodes 1.5 cm long, ~1.2 cm Ø. **Leaves** persistent; lamina ligulate or linear-oblong, ~6 cm long, apex broadly or obscurely, and unequally bilobed, coriaceous. **Inflorescences** with flowers paired, borne from leafing or leafless stems, 1–3-flowered; peduncle short; pedicels short; bracts very small, scarious. **Flowers** ~4 cm Ø, sweetly scented, long-lasting, ivory-white, with red or green veins on lateral lobes of lip which has a yellow midlobe, mentum green; sepals triangular-lanceolate,

Top row: *Dendrobium purpureum* (albino form), *Dendrobium ravanii, Dendrobium rhodostictum*.
Middle row: *Dendrobium sanderae, Dendrobium scabrilingue, Dendrobium schoeninum*.
Bottom row: *Dendrobium schrautii, Dendrobium secundum, Dendrobium serratilabium*.

GS

MG

MG

MG

MG

MG

MG

Top row: *Dendrobium shiraishii*, *Dendrobium smillieae*, *Dendrobium speciosum*, *Dendrobium spectabile*.
Bottom row: *Dendrobium stratiotes*, *Dendrobium striolatum*, *Dendrobium subclausum*.

1.9–2.2 × 0.7–0.8 cm, acute; mentum conical, obtuse; petals lanceolate to oblong-elliptic, 1.3–1.8 × 0.7–0.8 cm, acute, slightly reflexed; lip erect, trilobed, 1.8 × 0.8–0.9 cm, lateral lobes triangular, midlobe fleshy, elliptic, obtuse, verrucose; disk 5–7-keeled. **Distr:** Indo-China. **Ecol:** In lower montane forests; 600–1,200 m. **Zones:** 9–11.

D. schoeninum Lindl., *Gard. Chron.* 1846: 7, 1846. (Syn: *Callista beckleri*, etc.). [Pencil Orchid]. **Herbs**, epiphytic, forming sparse, erect or semierect clumps to 1 m long; rhizome thick, sparsely branched; stems ~0.3 cm thick, yellowish, tough. **Leaves** upright or spreading; lamina terete, 4-angled, 2–16 × 0.2–1.2 cm, thick and fleshy, erect, deeply grooved, dark green, often with black markings, decreasing successively in size. **Inflorescences** racemes, slender, 1–4-flowered. **Flowers** 2.5–3.5 cm across, white, pale green or mauve, with dark stripes on petals and sepals, and purplish markings on lip; sepals linear-lanceolate, 1.5–2.4 × 0.35 cm, acute; petals linear-lanceolate, 1.4–2.4 cm × 2–2.5 mm, acute; lip 2–3 × 0.3–1 cm, lateral lobes triangular, midlobe reflexed, white with purple markings, margins crisped, apex long an pointed. **Distr:** E Australia. **Ecol:** In lowlands swampy areas and along stream banks. Fl. Aug–Nov. **Zones:** 9–11.

D. schrautii Schildh., *J. Orchideenfr.* 13: 165, 2006. **Herbs**, epiphytic; stems pseudobulbous, to 30 cm long. **Leaves** lanceolate, 9–12.5 × 2–2.5 cm, base narrowed into a clasping sheath, apex acute. **Inflorescences** lateral, 1–2-flowered, borne in the upper nodes of the stems; pedicel and ovary 3–3.2 cm long, glaucous. **Flowers** 7 cm across, glabrous, waxy, sepals pure white inside, outside and especially toward apex bright yellow, petals pure white, lip center and callus orange, margins white to yellowish white, spur orange, ovary green; dorsal sepal ovate-caudate, concave, 2.5–3 cm long, ~1.5 cm across at widest point, abaxially keeled, apex appearing caudate, tail ~1.25 cm long, ~2 mm wide, reflexed; lateral sepals ovate-caudate, slightly concave, ~4 cm long, ~1.2 cm across at the widest point, abaxially keeled, apex appearing caudate, tail ~1.7 cm long, ~3 mm wide, not reflexed; petals broadly ovate, 3.2–3.5 × 1.7–1.9 cm, base slightly narrowed, apex acute, apical margins slightly inrolled; lip trilobed, 3.2–3.9 cm long, lateral lobes erect, enclosing the column, margins ruffled, midlobe quadrate, ~2–2.5 cm across, apiculate, margins ruffled, callus fimbriate; column ~8–9 mm long; spur ~2 cm long, forward-curved. **Distr:** Vietnam. **Ecol:** In evergreen forests; 15–1,600 m. Fl. Apr–Jul. **Zones:** 11–12.

D. secundum (Blume) Lindl., *Edwards's Bot. Reg.*: t. 1291, 1829. (Syn: *Pedilonum secundum*). [Toothbrush Orchid]. **Herbs**; stems crowded, stout, terete, grooved, to >1 m long. **Leaves** many, soon deciduous; lamina oblong to lanceolate, ~10 × 3.5–4.3 cm, apex obliquely obtuse. **Inflorescence** borne laterally from bare stems, densely many-flowered, with flowers all pointing in the same direction (secund), to ~11 cm long; bracts triangular, 3 mm long, acute; pedicel and ovary 1.4 cm long. **Flowers** lasting a few weeks, ~1.8 cm long, bright magenta, lip apically orange; sepals ovate-triangular, 7.5 × 3.7 mm, acuminate, 5-nerved; lateral sepals decurrent on column foot, front margins connate, forming a spurlike, somewhat inflated, 7.5 mm long obtuse mentum; petals lanceolate, 7.5 × 2.3 mm, subacute, trinerved; lip entire, linear-spathulate, channeled, 1.6 cm long, base adnate to column foot forming a spur, close to entrance of spur with a membranous, V-shaped appendage, margins in apical part incurved, apex subacute; column 3.5 mm long, foot 9 mm long. **Distr:** Indo-China to W and C Malesia. **Ecol:** In lowland rainforests. **Zones:** 10–12.

D. serratilabium L.O. Williams, *Bot. Mus. Leafl.* 5: 45, 1937. **Herbs**, epiphytic, pendulous, sympodial; pseudobulbs to over 40 cm long, thickened and wrinkled, new growths often appearing halfway along the pseudobulb. **Leaves** many, deciduous, lanceolate, to 8 × 2 cm; leaf sheaths yellow-streaked. **Inflorescences** pendulous, short, borne from leafless nodes, to 3-flowered. **Flowers** ~3 cm Ø, cream, outside of sepals striped with varying degrees of red; dorsal sepal lanceolate, to 2 × 0.5 cm; lateral sepals oblong-lanceolate, joined at base to form a spur ~2.5 cm long; petals elliptic-lanceolate, to 1.5 × 0.4 cm; lip spathulate, margin deeply toothed. **Distr:** Endemic to Philippines (Luzon). **Ecol:** Epiphytic in shaded places; 500–1,200 m. **Zones:** 11–12. **Info:** A variety with pure yellow or greenish yellow is also present in the locality of the type species.

D. shiraishii T. Yukawa and M. Nishida, *Lindleyana* 7: 127, 1992. **Herbs**, epiphytic, to 65 cm tall; pseudobulbs clustered, erect, clavate, strongly sulcate, dark greenish-brown, 17–43 cm long, apex 2–4-leaved. **Leaves** suberect; lamina oblong-elliptic, 15–29 × 4–12 cm, apex obtuse to subacute, dark green, coriaceous. **Inflorescences** erect, terminal, 28–37 cm long, ~10-flowered; pedicel and ovary muricate, 3.7–5.1 cm long. **Flowers** fleshy, widely opening, 5–7 cm Ø, sometimes weakly fragrant, sepals and petals greenish-yellow, outside of sepals maroon-spotted, petals with dark blackish-brown markings outside at base; sepals short-muricate on outer surface, keeled on midvein; dorsal sepal narrowly triangular, 3–3.5 × 1.1–1.2 cm, acute; lateral sepals obliquely lanceolate, 2.7–3 × 1–1.6 cm, acute; mentum incurved-conical, 1.2–1.3 cm long; petals obovate-spathulate, 2.9–3.3 × 1.4–1.6 cm, apiculate, deflexed; lip obscurely trilobed, somewhat recurved, 2.5–3 × 2.8–3.5 cm, dull red, apex greenish-yellow, callus white, 3-ridged, lateral lobes erect-incurved, triangular, midlobe ovate, apiculate; column yellowish-green, purple-spotted, 4–5 mm long, with 2 hooklike apical appendages; column foot 1–1.1 cm long; anther dull purple; ovary terete. **Distr:** W New Guinea. **Ecol:** In rainforests. **Zones:** 11–12.

D. smillieae F. Muell., *Fragm.* 6: 94, 1868. (Syn: *Callista smillieae*, etc.). [Bottlebrush Orchid]. **Herbs**, epiphytic; pseudobulbs 0.2–1 m long, distinctly fusiform, ribbed, somewhat spongy, when young clothed with white scarious sheaths. **Leaves** scattered in the upper half of stems; lamina oblong to lanceolate, 5–12 × 2–4 cm, thin-textured, often twisted. **Inflorescences** 4–15 cm long, erect, flowers numerous, tightly packed in a bottlebrushlike cluster; pedicels ~1 cm long. **Flowers** 1–2 cm across, not expanding widely, segments thick-textured, usually pink at base and paler toward the end, lip with a prominent green apex, rarely all white; sepals and petals oblong or spathulate, 1 cm long, shortly acute or obtuse; lip uppermost in flower, semicylindrical, 1–1.6 cm long, slightly trilobed; column dilated above, 3–6 mm long, foot 6–9 mm long. **Distr:** C Malesia to N Queensland. **Ecol:** In open forests, often forming large, robust clumps. Fl. Sporadic. Aug–Nov. **Zones:** 10–12.

D. speciosum Sm., *Exot. Bot.* 1: 17, 1804. (Syn: *Callista speciosa*, etc.). [King Orchid, Rock Orchid]. **Herbs**, epiphytic or lithophytic, forming medium to large, spreading clumps; rhizome thick, fibrous, much-branched; pseudobulbs, 0.1–1 m × 1–6 cm, variable in shape from linear to swollen at base, then tapered, straight or curved, yellowish, green or purplish, when young covered with bracts. **Leaves** 2–5, at apex of pseudobulb; lamina ovate to oblong or obovate, 4–25 × 2–8 cm, dark green, thick, leathery. **Inflorescences** racemes, 10–60 cm long, flowers numerous; pedicels 1–2 cm. **Flowers** 2.5–5 cm Ø, extremely variable, sepals and lateral petals white to deep dull yellow, lip white to deep dull yellow, finely dotted and striped with dark purple; dorsal sepal 2–4 × 0.3–0.6 cm; lip 0.9–1.3 × 0.9–1.3 cm; column 4–6 mm long; column foot 5–8 mm long. **Distr:** New Guinea to E and SE Australia. **Ecol:** On rocks or trees in rainforests, moist open forests. Fl. July–Oct. **Zones:** 9–11.

D. spectabile (Blume) Miq., *Fl. Ned. Indië* 3: 645, 1859. (Syn: *Latouria spectabilis*). **Herbs**, epiphytic; pseudobulbs clustered, canelike, to ~40 cm long, 4–6-leaved toward apex. **Leaves** suberect; lamina coriaceous, elliptic, to 23 × 4–8 cm, base shortly petiolate, apex obtuse. **Inflorescences** borne just below leaf bases, erect, 20–40 cm long, few–many-flowered; peduncle terete, with 4–6 small sheaths; bracts elliptic-lanceolate, 1–1.3 × 0.4 cm long, acute; pedicel and ovary 4–6 cm long. **Flowers** large, somewhat grotesque, yellow, heavily mottled with maroon on sepals and petals, lined with maroon on lip, callus white; dorsal sepal recurved, lanceolate, 3.6 × 0.5 cm, acuminate, margins undulate; lateral sepals recurved, lanceolate-falcate, 3.6 × 1.3 cm, acuminate, margins undulate; mentum obliquely conical, 1 cm long; petals linear-lanceolate, 4 × 0.6 cm, twisted, margins undulate, apex acuminate; lip trilobed, recurved, 4 × 2.2 cm, lateral lobes erect, subquadrate-semicircular, rounded in front, midlobe much longer than lateral lobes, lanceolate, acuminate, undulate-twisted; callus 3-ridged; column short, 3 mm long, foot 8.5 mm long. **Distr:** New Guinea, Solomon Islands. **Ecol:** In lowland swampy forests, mangroves, lower montane forests, sometimes also lithophytic; 0–1,200 m. Fl. May–Aug. **Zones:** 10–12.

D. stratiotes Rchb. f., *Gard. Chron.* 1886(1): 266, 1886. (Syn: *Callista stratiotes*, etc.). **Herbs**, epiphytic; stems canelike to somewhat fusiform, 0.6–1 m tall, basally slightly swollen. **Leaves** ovate to oblong-lanceolate, 8–13 × 2–5 cm, base articulated to tubular sheaths 1.8–3.2 cm long, apex obliquely bilobed, lobes subacute. **Inflorescences** suberect or erect, to 30 cm long, 4–8-flowered; bracts tubular, 4–5 mm long, subacute or obtuse; pedicels emerging 0.6–1 cm above axil of bract. **Flowers** large, overall length to 10 cm long, erect, sepals white, petals shiny, green or yellow-green apically; dorsal sepal recurved, linear-lanceolate, 3–3.4 × 0.6–0.9 cm, acuminate, twisted and slightly undulate; lateral sepals falcate, obliquely lanceolate, ~3.2 × 1.2 cm, margins undulate, apex acuminate; mentum pointing backward, slightly upcurved toward apex, narrowly conical, 1.5–1.6 cm long; petals erect or suberect, linear, 5–6.4 × 0.15–0.2 cm, acute, twisted 2–4 times; lip white, large, trilobed, 3.2–3.7 × 2.3–2.5 cm, lobes purple-veined, lateral lobes oblong, truncate to rounded, midlobe lacking or with a short claw, ovate, 1.8–2 × 1.5 cm, acuminate, callus of 3 or 5 ridges; column 6–7 mm long; ovary 1.8–2.6 cm long. **Distr:** Sulawesi, Maluku, Lesser Sunda Is. **Ecol:** In warm lowland forests; 100–300 m. Fl. throughout the year. **Zones:** 10–12.

D. striolatum Rchb. f., *Hamburger Garten-Blumenzeitung* 13: 313, 1857. (Syn: *Callista striolata*, *Dockrillia striolata*, etc.). [Rat-tail Orchid, Streaked Rock Orchid]. **Herbs**, epiphytic or lithophytic, sympodial; stems spreading to pendent, wiry, rooting mostly at base of sympodium but also occasionally from more apical stems, 1–3 cm long, ~1 mm Ø, apex 1-leafed; roots smooth, mostly creeping. **Leaves** spreading to pendent; lamina linear, terete, 4–11 cm long, 2–3 mm Ø, succulent, shallowly furrowed. **Inflorescences** 0.8–3.5 cm long, 1- or 2-flowered. **Flowers** with sepals and lateral petals cream to dull yellow with a few brown, red or purple striations at base; dorsal sepal 1.3–1.5 × ~0.2 cm; lip white, 0.7–1 × 0.3–0.5 cm; column 3–4 mm long, foot 4–6 mm long. **Distr:** SE Australia. **Ecol:** On rocks in rainforests; 0–1,000+ m. Fl. Sep–Nov. **Zones:** 9–11.

D. subclausum Rolfe, *Bull. Misc. Inform. Kew*: 361, 1894. **Herbs**; stems crowded, erect, slender, terete, flexuose in upper part, branching, 30–45 cm long; leaf sheaths verrucose-rough. **Leaves** linear-oblong, 2–4 × 0.25–0.4 cm, apex subobtuse. **Inflorescences** racemose, few-flowered, short; floral bracts oblong-lanceolate, keeled near apex, strongly concave, 4–7 mm long, apex acute; pedicel and ovary 2.2 cm long. **Flowers** long-lasting, ~2 cm long, orange to light red, dorsal sepal, petals and upper half of lateral sepals usually bright yellow; dorsal sepal ovate-lanceolate, 0.8 cm long, subobtuse; lateral sepals ovate-triangular, indistinctly keeled outside, mentum elongated-conical, 1.5 cm long; petals ovate-lanceolate, 9 mm long, subobtuse; lip clawed, narrow, strongly concave, 1.8 cm long, abruptly inflexed, with 3 folds, cucullate, margin fimbriate; column very short. **Distr:** Moluccas, New Guinea. **Ecol:** Epiphytic in montane rainforests and cloud forests, sometimes in deep shade of moss forests, terrestrial in montane grasslands, and lithophytic in open forests; 1,000–3,300 m. Fl. throughout the year, mostly in July. **Zones:** 11–12. **Info:** A form with yellow flowers, ***D. subclasum* var. *pandanicola***, is common in cultivation.

D. sulcatum Lindl., *Edwards's Bot. Reg.* 24: t. 65, 1838. (Syn: *Callista sulcata*). **Herbs**, epiphytic; stems intensely swollen, sides compressed, 15–25 cm long, deeply grooved and articulated; internodes ~6, compressed, green. **Leaves** 3–5, subterminal, often deciduous (especially in dry weather); lamina broadly elliptic to elliptic-ovate, 10–20 × ~4.5 cm, acuminate, coriaceous, trinerved. **Inflorescences** short racemes, 5–10(–20)-flowered, dense, compact, borne from leafless stems, dull orange or paler, ± hanging; bracts minute; pedicel and ovary ~2.5 cm long, slender. **Flowers** thin, uniformly golden-yellow, lip with red-purple markings or stripes at base; dorsal sepal oblong, ~2.5 × 0.9 cm, subacute, 5- or 6-veined; lateral sepals nearly equal to dorsal sepal; mentum conic, ~5 mm long, broad, obtuse; petals subobovate, ~2.4 × 1.1 cm, base contracted into a short claw, margin entire, apex acute, 5-veined; lip suborbicular to wedge-shaped, incurved, 1.8–2 × ~2 cm, adaxially densely pubescent near margin, base shortly clawed, margin minutely erose to ciliate-hairy, apex emarginate; column ~5 mm long. **Distr:** Sikkim to Indo-China. **Ecol:** On tree trunks in dense forests; 700–800 m. **Zones:** 8–11.

D. tetragonum A. Cunn. ex Lindl., *Edwards's Bot. Reg.* 25(Misc.): 33, 1839. (Syn: *Callista tetragona*, etc.). [Tree Spider Orchid]. **Herbs**, epiphytic, clump-forming; pseudobulbs 6–45 × 0.5–1.5 cm, semipendulous to pendulous, very slender, wiry at base, then thickened and markedly 4-angled,

Below: *Dendrobium subclausum* var. *pandanicola*, *Dendrobium sulcatum*, *Dendrobium tetragonum*, *Dendrobium thyrsiflorum*.

MG

MG

GS

JT

GS

Top row: *Dendrobium tobaense, Dendrobium toressae, Dendrobium trilamellatum.*
Bottom row: *Dendrobium unicum, Dendrobium uniflorum, Dendrobium vexillarius.*

usually brownish. **Leaves** 2–5, near the apex of the pseudobulb, oblong to ovate-lanceolate, 3–10 × 1.5–2.5 cm, dark green, often partially twisted. **Inflorescences** racemes, 1–3 cm long, 2–5-flowered, borne amongst the leaves or nodes lower down on the pseudobulb; pedicels 1.5–2.5 cm long. **Flowers** 4–28 cm Ø, extremely variable in size and color, very fragrant; perianth segments widely spreading, thin and spidery, green to greenish-yellow, or yellow with irregular reddish-purple blotches; lip 1–1.5 cm long, curved, with prominent lateral lobes. **Distr:** E Australia. **Ecol:** On trunks of rainforest trees, and along stream banks of open forests. **Zones:** 9–12.

D. thyrsiflorum H.G. Reichenbach ex André, *Ill. Hort.* 22: 88, 1875. (Syn: *Callista thyrsiflora*, etc.). **Herbs**, epiphytic; stems erect or ascending, cylindric, 12–46 cm, yellowish-brown, glossy. **Leaves** 3 or 4, alternate; lamina oblong or oblong-lanceolate, 9–16 × 2.4–5 cm, base narrowed into a petiole ~6 mm long, apex acute, leathery. **Inflorescence** arising lateral, on mature leafy stem, pendent, 10–16 cm long, densely many-flowered; pedicel and ovary with pale white stripes tinged with purple, 2.5–3 cm long. **Flowers** spreading, thin; sepals and petals white, lip golden-yellow, column white, foot pale yellow, anther cap white; dorsal sepal ovate, ~1.5 × 0.8 cm, entire, obtuse, 5-veined; lateral sepals slightly obliquely ovate-lanceolate, ~1.7 × 0.7 cm, entire, obtuse, 5-veined; petals suborbicular, ~1.4 × 1.2 cm, 7-veined, base with claw ~2 mm long, margin irregularly finely toothed, apex rounded; lip semiorbicular to triangular, ~1.5 × 1.9 cm, ± pubescent, bluntly rounded, base clawed, claw ~3 mm long with a ligulate structure; column foot ~4 mm long. **Distr:** Assam to Hainan and Indo-China. **Ecol:** On tree trunks in montane forests; 1,100–1,800 m. **Zones:** 9–11.

D. tobaense J.J. Wood and J.B. Comber, *Lindleyana* 8(3): 115, 1993. **Herbs**, epiphytic; pseudobulbs canelike, 20–45 cm long; leaf sheaths covered in black hairs. **Leaves** 8–9, oblong-elliptic, 3–8 cm long, apex bilobed, adaxially smooth, abaxially black-hairy. **Inflorescence** very short, several-flowered, borne from the base of the leaf sheath opposite the leaf, of leafy or leafless stems; flowers opening singly, successively. **Flowers** ~8 cm across, lasting for a month, opening a bright, vivid green with darker longitudinal lines, slowly turning to a rich golden-yellow color with age; dorsal sepal ~3 cm long, ~1.5 cm wide at base, abruptly narrowed to ~5 mm becoming 3 mm wide apically, apex acute; lateral sepals similar to dorsal sepal, base 2 cm wide; petals narrowly obovate, ~3.5 cm long, ~1 cm wide at base, gradually narrowing to ~2 mm wide, margins white, apex long acuminate, apically recurved; lip club-shaped, ~3.6 cm long, complex, lateral lobes erect pale orange, callus bright red, midlobe 2 cm long, basal claw ~6 mm long, yellowish green, apical part narrowly ovate, white; column white, erect, ~8 mm long. **Distr:** N Sumatra. **Ecol:** 1,000–1,500 m. **Zones:** 11–12.

D. toressae (F.M. Bailey) Dockrill, *Orchadian* 1: 64, 1964. (Syn: *Bulbophyllum toressae, Dockrillia toressae*, etc.). **Herbs**, epiphytic or lithophytic, miniature, forming dense mats; rhizomes producing pseudobulbs along its length; pseudobulbs vestigial, small, carrying many leaves. **Leaves** alternate, when young reddish, and covered in membranous sheaths; lamina canaliculate, adaxially channeled, 4–8 × 2–4 mm, basally rounded, succulent, adaxially channeled, sparkly due to the pits on the adaxial surface. **Inflorescences** very short, single-flowered, borne from the apex of the pseudobulb, at the base of the leaf; peduncle virtually absent. **Flowers** fragrant, long-lasting, fleshy, globular, ~6 mm across, tannish-pink with, lip yellow, column greenish, apex darker; dorsal sepal ~5 × 2 mm, obtuse; lateral sepals relatively large, falcate or oblique, 5 × 5 mm, obtuse; petals linear-lanceolate, 2.5 × 1 mm, obtuse; lip ~5 mm long, obtuse, recurved, fleshy; column 3 mm long, arching. **Distr:** Australia (NE Queensland). **Ecol:** On trees and shady rocks of rainforests and open forests; ~0–1,950 m. **Zones:** 10–12.

D. trilamellatum J.J. Sm., *Bull. Dép. Agric. Indes Néerl.* 19: 21, 1908. (Syn: *Cepobaculum semifuscum*, etc.). **Herbs**, epiphytic; stems narrowly fusiform, 30–50 cm long. **Leaves** clustered apically; lamina linear-lanceolate, to 14 × 1.65 cm, articulated to pale green sheathing bases striped with red, apex acutely bidentate or obliquely acute, fleshy-coriaceous. **Inflorescences** suberect to spreading, laxly few–many-flowered, to 50 cm long; peduncles to 16 cm long; bracts triangular, 2 mm long; pedicel and ovary 2–3 cm long. **Flowers** fragrant, sepals and petals yellow to yellow-brown, lip yellow lined with purple; dorsal sepal erect, linear-lanceolate, 2–3 × 0.4–0.6 cm, acute, spirally twisted; lateral sepals recurved, obliquely linear-lanceolate, 2.7–3.2 × 0.9–1 cm, acute, twisted; mentum obliquely conical, 0.8–1 cm long; petals suberect, linear, 2.2–3 × 0.3 cm, acute, 2–3-times twisted; lip porrect, trilobed, 1.6–2 × 0.7–1 cm, lateral lobes oblong-elliptic, obtuse, midlobe convex, ovate to circular, 0.7–1 × 0.7–1 cm, obtuse, callus of 3 ridges to apex of midlobe, dilated; column 5–7 mm long, apex denticulate. **Distr:** S New Guinea to N Australia (Queensland). **Ecol:** In savannas. Fl. Jun–Jul. **Zones:** 9–12.

D. unicum Seidenf., *Bot. Tidsskr.* 65: 332, 1970. **Herbs**, epiphytic; pseudobulb erect or becoming pendulous with age, slender, spindle-shaped, 25 × 1 cm, darkening to brown or black, 4–5-noded. **Leaves** 2–4 per pseudobulb; lamina linear-lanceolate, 4–6.5 × 0.5–1.5 cm, apex acute. **Inflorescence** borne near apex of old the pseudobulb, 2–4-flowered. **Flowers** 4–6 cm Ø, showy, sepals orange to flame-colored, lip striped or reticulate with purplish red, column white with apex orange; sepals and petals linear lanceolate, 2.22–3.5 × 0.3–0.7 cm, acute; lateral sepals forming with column foot a short, conical mentum, 5 × 7 mm; lip held on the upper side of the flower, pandurate-oblong, margin entire, apex subacute, as long or longer than sepals, 1.4 cm across, with 2 elevated keels at base; column very short, tridentate. **Distr:** Indo-China. **Ecol:** Epiphytic on low shrubs and trees, and lithophytic in open areas; 800–1,500 m. **Zones:** 10–11.

D. uniflorum Griff., *Not. Pl. Asiat.* 3: 305, 1851. (Syn: *D. quadrisulcatum*, etc.). **Herbs**, epiphytic, upright, sympodial; pseudobulbs erect, to 40 cm long, grooved, swollen, often marked with red stripes. **Leaves** distichous, lying on a single plane, twisted basally, deciduous; lamina elliptic to oblong-lanceolate, to 5 × 2 cm, apex bilobed, leathery, glossy. **Inflorescences** solitary, borne opposite the leaves; bracts ovate-lanceolate, acute. **Flowers** to 2 cm Ø, milky white, lip pale green with purplish-brown lines in the middle; dorsal sepal lanceolate, to 1.3 × 0.65 cm, reflexed; lateral sepals ovate to oblong, 2.2 × 0.8 cm, acuminate, reflexed; petals lanceolate, to 1.1 × 0.4 cm, reflexed; lip relatively large, trilobed, 2 cm long, 2.6 cm wide (when flattened), lateral lobes squarish, midlobe circular, recurved, with a notch at the center and 3 raised violet ridges; column broad, flat, in front with a raised line, arms very small, green. **Distr:** Indo-China to Malesia. **Ecol:** In evergreen lowland, and montane forests; 100–1,500 m. **Zones:** 9–12.

D. vexillarius J.J. Sm., *Bull. Dép. Agric. Indes Néerl.* 39: 12, 1910. (Syn: *Pedilonum vexillarius*). **Herbs**, tufted, epiphytic or terrestrial, erect to suberect, occasionally pendulous, 3–25(–40) cm high; pseudobulbs ovoid, obclavate to cylindrical, 0.5–30 cm long. **Leaves** erect to spreading (to pendulous); lamina linear-lanceolate to elliptic, 1–16(–24) × 0.2–1.8 cm, apex acute to obtuse, mucronate, green to purplish (red), sometimes finely purple-spotted. **Inflorescences** terminal or lateral, several, 1–7-flowered; pedicel and ovary 1.6–4 cm long. **Flowers** long-lasting (~6 months), 2.2–5 cm long, usually widely opening, to 4.25 cm Ø, variously colored, yellow to red, green, blue, violet, etc., lip green to almost black, apex usually red or orange or bi- to tri-colored with orange to yellow band below tip; dorsal sepal ovate to oblong-elliptic, 0.7–2 × 0.35–0.9 cm, obtuse to subacute, mucronate, keeled; lateral sepals obliquely triangular to ovate-triangular, 1.9–4.5 × 0.5–1.3 cm, obtuse to abruptly acuminate; petals obliquely oblong-obovate to subspathulate, sometimes subrhombic, 0.7–2 × 0.25–0.9 cm, obtuse to acute or apiculate; lip subtrilobate, sublinear, 1.8–3.9 × 0.3–0.7 cm; column 2.5–5.5 mm long, foot 1.4–3 cm long; ovary green to purplish or violet, 3-winged. **Fruits** ovoid, ~ 3 × 1.4 cm. **Distr:** Moluccas, New Guinea, Bismarck Archipelago. **Ecol:** In montane and alpine forests; (800–)1,200–3,500(–4,000) m. Fl. throughout the year. **Zones:** 9–11.

D. victoriae-reginae Loher, *Gard. Chron.* 1897(1): 399, 1897. (Syn: *D. coeleste*, etc.). [Blue Orchid]. **Herbs**, epiphytic; pseudobulbs long, slender, branching, forming a tangled mass, 0.3–0.4(–1) m long, covered with persistent sheaths, swollen at nodes, apically leafy. **Leaves** ~12, deciduous; lamina oblong or lanceolate, 3–8 × 1–2 cm, apex acute to acuminate, papery. **Inflorescences** 2–5(–12)-flowered, lateral, borne on old leafless stems; peduncle very short; bracts oblong, acute. **Flowers** 3 cm Ø, wide-opening, in various shades of blue-violet, usually with darker venation, column and bases of segments white; sepals oblong, 1.5–1.8 × 0.6 cm, obtusely acute; mentum short, straight, obtuse; petals obovate-oblong, 1.5 × 0.6 cm, margins minutely erose; lip obovate from a short linear base, slightly concave, shortly acute, denticulate and reflexed on basal margins with 3 elevated lines at base, 2–3.5 × 0.8–1.1 cm; disk sparsely pilose; column with recurved acute stelidia. **Distr:** Philippines. **Ecol:** In high mountains; >1,300 m. **Zones:** 9–11.

D. wardianum Warner, *Select Orchid. Pl.* 1: t. 19, 1862. (Syn: *Callista wardiana*, etc.). **Herbs**, epiphytic; stems pendulous, cylindric, (1.1–)2–2.6(–5) × 0.5–1+ cm, fleshy. **Leaves** oblong, 8–11 × 1.5–1.8+ cm, base with clasping sheath, apex slightly obtuse, hooked on one side, leathery. **Inflorescence** on old leafless stems, 1- or 2-flowered; peduncle 3–5 mm long; basal sheaths 2 or 3, broadly ovate, 3–5 mm long, papery, obtuse; floral bracts broadly ovate, 0.7–1 cm long, obtuse, papery; pedicel and ovary pale purple, ~2 cm long. **Flowers** fragrant, papery, sepals and petals white purple-tipped, lip purplish-striped on each side, disk with a pale yellow transverse semilunar central blotch, column white, front purple-striped, anther cap white; dorsal sepal ovate-lanceolate, 2.3–2.5 × 0.7–0.8+ cm, acute or slightly obtuse; lateral sepals suborbicular, equal in size, base oblique, apex acute; petals obliquely ovate, 2.3–2.5 × 1.3–1.4 cm, base contracted into a short claw, margin entire, apex obtuse; lip nearly broadly obovate, ~2.3 × 2 cm, base cuneate, margin ciliate, apex rounded, adaxially densely barbellate; column ~4 mm long. **Fruits** ovoid, ~ 3 × 1.3–1.6 cm. **Distr:** E Himalayas to China (Yunnan) and Indo-China. **Ecol:** On tree trunks in open forests; 1,300–1,900 m. **Zones:** 10–12.

Drymoanthus

Species ~2. **Herbs**, epiphytic, with roots attached widely over the bark of the host tree; stem mostly simple, often shorter than the leaves; internodes very short; old inflorescences and infructescences from previous years persistent. **Leaves** ± distichous; lamina elliptic to oblong. **Inflorescences** axillary racemes sometimes piercing the leaf sheath; floral bracts small and scalelike. **Flowers** minute; perianth with labellum uppermost; sepals and petals nearly equal, spreading, narrowly oblong; lateral sepals in part adnate to short column foot, in part to labellum but not meeting in front; lip fleshy, concave, continuous with short column-foot and not jointed to it, without a spur, midlobe small and not sharply defined from lateral lobes; column short, cylindric, foot at most barely as long as wide; anther narrowly joined to column; pollinia 4 in 2 pairs. **Fruits** capsules, with curled hairs; seeds numerous. **Distr:** E Australia and New Zealand. **Ecol:** On exposed trunks and branches of trees. **Cult:** Propagate from seeds. Best grown mounted on a tree-fern slab.

D. adversus (Hook. f.) Dockrill, *Australasian Sarcanthinae*, 32, t. 3, 1967. (Syn: *Sarcochilus adversus, S. breviscapa*). **Herbs**, forming a leafy tuft ~4–12(–15) cm Ø; roots usually the longest parts; stem to ~5 cm long, erect, lower part covered with imbricating old leaf bases. **Leaves** ~4–6(–9) at any one time; lamina elliptic-oblong, to ~ 6 × 1.5 cm, base meeting sheath along abscission line, apex acute, slightly twisted, minutely 2-toothed when young, green. **Inflorescences** racemes, borne among or below the leaves, sometimes both an upper and a lower one produced in a single season; flowers arranged opposite to irregularly alternate; pedicels with a small membranous bract. **Flowers** minute, perianth greenish-white red-flecked; sepals and petals subequal, ~3.5 × 1.5 mm; lip almost as long and broader, very fleshy, ± erect, smoothly convex externally, deeply channeled, midlobe distinct only as small cucullate tip, lateral lobes forming sides of clog-shaped structure and each with an internally projecting, partly transverse ridge; column much shorter, broadly cylindric. **Fruits** relatively large, ~1.5 × 0.35 cm. **Distr:** New Zealand. **Ecol:** On exposed trunks and branches of trees. **Zones:** 8–11.

Epigeneium

Syn: *Katherinea, Sarcopodium.*

Species 42. **Herbs**, epiphytic, small or medium-sized; rhizomes elongated; pseudobulbs ± ovoid, often enclosed by sheaths, 1- or 2-leaved at apex. **Leaves** oblong or obovate, coriaceous. **Inflorescence** racemose, 1- to several-flowered, at apex of pseudobulb, subtended by a spathaceous bract. **Flowers** medium-sized to large, often showy; dorsal sepal enclosing the column; lateral sepals larger, oblique, forming a mentum with column foot; petals triangular, long decurrent on mentum; lip pandurate-oblong to trilobed, callus lobulate or ridged at base; column short, foot long; pollinia 4, in 2 groups. **Distr:** Himalayas to S China, Indo-China to New Guinea. **Cult:** Grow in shallow baskets or mounted on slabs. Requires moderate shade, humid conditions and adequate watering throughout the year.

E. amplum (Lindley) Summerhayes, *Kew Bull.* 12: 260, 1957. (Syn: *D. amplum*, etc.). **Herbs**, lithophytic or epiphytic; rhizome usually branched; pseudobulbs ovoid or elliptic, 2–5 cm long, 2-leaved. **Leaves** elliptic or oblong-elliptic, 6–22.5 × ~5.5 cm, base contracted into petiole ~3 cm long, apex ± obtuse, retuse, leathery. **Inflorescence** 1-flowered; peduncle 1.5–2 cm long, enclosed in 2 sheaths; floral bracts 1–1.7 cm long; pedicel and ovary 4.5–5 cm long. **Flowers** large, spreading, yellowish-green, spotted with deep brown; dorsal sepal lanceolate, (3.5–)4.5–6.5 × ~0.8 cm, acute; lateral sepals falcate-lanceolate, ~4.5 × 1.1–1.5 cm, acute-acuminate; petals lanceolate, ~1.4 × 0.6 cm, acute-acuminate; lip pandurate, ~2.6 cm, trilobed, lateral lobes erect, small, midlobe subrhombic, ~6 × 6 mm, obtuse, mucronate; disk with 3 lamellae, median slightly longer; column ~1.5 cm long, stout, foot ~1.4 cm long. **Distr:** C Himalayas to S China. **Ecol:** In forests; 1,000–1,900 m. Fl. Nov. **Zones:** 8–11.

Below: *Dendrobium victoriae-reginae, Dendrobium wardianum, Drymoanthus adversus.*

MG

MG

GS MS

JT

MS

MS

Top row: *Epigeneium amplum, Epigeneium cymbidioides, Epigeneium stella-silvae.*
Bottom row: *Epigeneium treacherianum, Flickingeria interjecta, Malaxis densiflora* (habit), *Malaxis ramosii.*

E. cymbidioides (Blume) Summerh., *Kew Bull.* 12: 261, 1957. (Syn: *Callista cymbidiodes*, etc.). **Herbs**, epiphytic, small; rhizome fleshy, densely covered with rudimentary cataphylls; pseudobulbs well-spaced, ovoid, 3–4 cm high, fleshy, 2-leaved. **Leaves** oblong, 15 × 2+ cm, apex bilobed, obtuse, fleshy. **Inflorescence** erect or arcuate, to 10-flowered; bracts very small, hyaline, triangular, acute. **Flowers** pale yellow or straw-colored; lip yellow, suffused with rose or purple with 2 ± confluent golden spots at base of midlobe; sepals and petals oblong-ligulate or ligulate, 2.3–2.5 × 0.6–0.7 cm, acute, petals narrower; mentum short, rounded; lip trilobed, 1.5–1.8 × 1.3 cm, lateral lobes broadly semiobovate or obtriangular, rounded in front, midlobe shortly ovate-cordate, obtuse, broader than long, calli 3, basal; column short, lacking stelidia. **Distr:** Malay Peninsula, Sumatra to Java, Philippines (Luzon). **Ecol:** In moss forests; 2,200–2,800 m. **Zones:** 11–12.

E. stella-silvae (Loher and Kraenzl.) Summerh., *Kew Bull.* 12: 265, 1957. (Syn: *Dendrobium stella-silvae*, etc.). **Herbs**, epiphytic, miniature, 20 cm high; pseudobulbs short, ovoid, to ~3 × 2 cm, apex 2-leaved. **Leaves** 5–7 × 2 cm, thickly leathery. **Inflorescence** a raceme, terminal, to 25 cm long, erect, ~4-flowered; pedicel and ovary 2–2.5 cm long; bract ~1 cm long, brownish. **Flowers** 4–5 cm across, pure white, lip base to halfway yellow to orangish-yellow with dark orange or brownish venation; sepals linear-lanceolate, apex pointed; dorsal sepal 2.6 × 0.6 cm, base not narrowed; lateral sepals falcate, 2.3 × 0.6 cm, base slightly narrowed, widest at the middle; petals narrowly lanceolate, 2.4 × 0.6 cm, apex pointed; lip trilobed, 1.6 cm long, lateral lobes erect, enclosing the column, somewhat quadrate, 6 × 5 mm, rounded, midlobe spathulate, 1 × 0.8 cm, apex acuminate, narrowed into a fine tip; column ~6 mm long. **Distr:** Philippines (Luzon). **Ecol:** On montane forest slopes; 1,000–1,750 m. Fl. Jan–May. **Zones:** 11–12. **Info:** This species is very similar to *E. acuminatum* and often considered a synonym.

E. treacherianum (Rchb. f. ex Hook. f.) Summerh., *Kew Bull.* 12: 265, 1957. (Syn: *Dendrobium treacherianum*, etc.). **Herbs**, epiphytic, upright, sympodial; pseudobulbs crowded, flattened over the stout rhizome, ovoid, 4(–6)-angled, somewhat quadrate, 4 cm wide, 3 cm high, dull brownish green, with 2 sharp thornlike projections where the leaves have fallen. **Leaves** 2, oblong-elliptic to narrowly elliptic, 15 × 3 cm, apex retuse, very leathery. **Inflorescence** racemose, terminal, arching to pendent, 2–20-flowered, 10–15 cm long. **Flowers** fragrant, resupinate, 7.5 cm Ø, sepals and petals purplish red with the tips white, or purplish red gradually becoming paler to a white tip, lip purplish red; dorsal and lateral sepals linear, somewhat falcate, 5 × 1.2 cm, tapering to a narrow tip; petals linear-lanceolate, 4.5 × 1 cm, apex pointed; lip trilobed, 3 × 2.5 cm (when flattened), lateral lobes obtuse, midlobe ovate, narrowed into a point. **Distr:** Borneo, Philippines. **Info:** There are albino forms of this species. **Ecol:** In primary forests, in the canopy of large trees; 100–400 m. **Zones:** 11–12.

Flickingeria

Syn: *Desmotrichum, Ephemerantha.*

Species 65–70. **Herbs**, epiphytic; basal rhizome branching, creeping or prostrate; roots smooth; aerial shoots (sympodia) clambering, ascending or pendulous, terminating in a 1-leaved pseudobulb consisting of 1 internode; pseudobulbs usually ± laterally compressed, clavate or fusiform. **Leaves** apical, subsessile; lamina narrowly to oblong-elliptic, base slightly contracted, not sheathing, leathery. **Inflorescences** producing solitary flowers each on 1 or more short, subterminal, tufted peduncles, borne in front of or behind leaf base, or both. **Flowers** fragile, ephemeral, lasting less than a day, thin and membranous; sepals and petals acute; mentum conspicuous, deflexed, conic; petals narrower than sepals; lip trilobed, with a distinct hypochile, mesochile, and epichile, 2- or 3-keeled, lateral lobes erect, midlobe variable in shape, straight, curved or very undulate-pleated and transversely bilobulate; column short, with a long foot; pollinia 4 in 2 pairs, naked, subglobose. **Distr:** S and SE Asia, China; Australia, Pacific islands. **Cult:** Propagate from seeds, plantlets, stem cuttings (each with a dormant bud) or by division. Requires a permeable, coarse substrate (either potted, attached to bark or a tree) in full or partial sun.

F. interjecta (Ames) A.D. Hawkes, *Orquídea* (Rio de Janeiro) 27: 304, 1965. (Syn: *Dendrobium interjectum*, etc.). **Herbs**, epiphytic, upright, sympodial; rhizome to 3 mm Ø; stems appearing about every sixth node, carrying a pseudobulb; pseudobulb to 2.5 cm long, 7 mm Ø, unifoliate. **Leaves** terminal on the pseudobulb; lamina elliptic, to 4 × 0.9 cm, apex, obtuse leathery. **Inflorescences** short, flowers solitary, borne behind the leaves. **Flowers** 8 mm across, pale yellow, lip yellow in the apical third, and the rest purplish; dorsal sepal broadly ovate, to 4.5 × 3.5 mm; lateral sepals triangular, 9 × 5 mm; petals linear, 4.5 × 2 mm; lip trilobed, 1.25 × 0.6 cm, midlobe divided. **Distr:** Endemic to Philippines (Luzon, Leyte and Panay islands in the Visayas). **Ecol:** Epiphytic in bright areas; 100–700 m. **Zones:** 11–12.

TRIBE MALAXIDEAE

Malaxis

Syn: *Microstylis*. [Adder's Mouth].

Species ~600. **Herbs**, terrestrial, with fleshy globose or ellipsoid tuberoids; stems erect, with many cauline or basal leaves, with basal sheaths. **Leaves** sessile, base not jointed; lamina usually lanceolate or oblong. **Inflorescence** racemose or spicate, many-flowered, terminal. **Flowers** greenish or white; sepals free, subequal, dorsal one often connate with petals forming hood; petals narrow, entire or deeply divided; lip continuous to base of column, usually trilobed or tripartite, base produced into straight or curved spur; column stout; anther erect, basally connate with column, cells widely divergent below, with auricles or staminodes on outer side, base often produced into long tube; pollinia 2; stigma bilobed; rostellum broad, thick, rooflike. **Distr:** Widespread in tropical, subtropical and temperate regions. **Cult:** Propagate by division or from seeds. Requires a well-drained, moisture-retentive medium, in warm humid conditions.

M. densiflora (A. Rich.) Kuntze, *Rev. Gen. Bot.* 2: 673, 1891. (Syn: *Liparis densiflora*, etc.). **Herbs**, terrestrial; stems 8–15 cm long, base pseudobulbous; pseudobulbs at length globose, 1.5 cm Ø, annulate. **Leaves** 2 or 3, sessile or shortly petioled; lamina ovate to lanceolate, 5–7.5 cm long, apex acuminate or acute, margins wavy, 5-veined. **Inflorescences** racemes, densely flowered, 5–10 cm long; peduncle 2.5–7.5 cm long with a few deflexed, empty bracts; floral bract as long as the flower. **Flowers** yellow or greenish-yellow, very small; sepals linear-oblong, laterals deflexed, dorsal narrowest; petals linear, obtuse; lip subquadrate, short, cuneate or subreniform, pectinately toothed; column constricted in the middle, sides produced apically into short suberect horns; anther terminal, incumbent, biloculed; pollinia 4, waxy, in 2 pairs. **Fruits** small, pediceled, pyriform. **Distr:** South India (Nilgiri Hills), Sri Lanka. **Ecol:** In tropical humid evergreen forests; 213–366 m. **Zones:** 10–12. **Info:** Very closely related to *M. versicolor*, but differs with the 5-veined leaves, linear petals and the spreading arms of the column.

M. ramosii Ames, *Philipp. J. Sci.*, C 6: 45, 1911. (Syn: *Crepidium ramosii*, *Pseudoliparis ramosii*). **Herbs**, terrestrial; pseudobulbs inconspicuous. **Leaves** usually 2, subopposite; lamina pale green, obliquely elliptic-ovate, 5–8 × 3.5–4.5 cm, base with a sheathlike petiole, margins slightly crisped. **Inflorescence** erect; peduncle green, 10–12 cm long, slender; rachis 10- or more-flowered; floral bracts reflexed, very small; pedicel and ovary ~3.8 mm long. **Flowers** orange; dorsal sepal ovate, ~5 × 4 mm; lateral sepals obliquely obovate, ~4 × 2.8 mm; petals oblanceolate or obovate, slightly smaller than lateral sepals; lip superior, cordate, ~3 × 3 mm, base subcordate, with 2 small auricles on either side, unlobed, obtuse; disk with a pair of short dark ridges; column ~1.7 mm long, basal half connate to lip, apical part with short broad wings. **Distr:** Taiwan (Lan Yü) to Philippines. **Ecol:** In forests; 300–400 m. Fl. Aug–Nov. **Zones:** 10–12.

Oberonia

Syn: *Titania*.

Species 317. **Herbs**, epiphytic; stems often enclosed in leaf base, usually clustered, erect or pendent. **Leaves** iridiform and unifacial, rarely terete, often fleshy, base sometimes articulate, distichous, usually equitant, often imbricate at base, often dilated near base into flattened sheaths. **Inflorescences** terminal, racemose, unbranched, usually many-flowered, flowers often in whorls, pubescent or glabrous; peduncle often with sterile bracts below; floral bracts persistent. **Flowers** non-resupinate, flat, very small, usually only 1–2 mm Ø; sepals and petals free, spreading, often narrower than sepals, sometimes erose; lip usually spreading, sessile, entire or lobed, margin sometimes erose or fimbriate, occasionally concave at base, sometimes with calli, lateral lobes often embracing column; column erect, short, lacking a foot, wingless; anther cap terminal, bilocular; pollinia 4, in 2 unequal pairs, waxy; stigma transversely reniform to ovate; ovary glabrous or papillate. **Distr:** Tropical and S Africa to Pacific. **Cult:** Propagate from seeds or by division. Best grown on slabs of moisture-retentive tree fern or cork slabs with sphagnum moss covering the root ball, in warm, humid and shaded conditions.

O. mucronata (D. Don) Ormerod and Seidenfaden, *Contrib. Orchid Flora Thailand* 13: 20, 1997. (Syn: *Cymbidium iridifolium*, etc.). **Herbs**, epiphytic; stem short or inconspicuous. **Leaves** subbasal, 5 or 6, distichous-equitant; lamina laterally compressed, 6–16 × 0.6–1.5 cm, base articulate, apex shortly acuminate or obtuse, thick. **Inflorescence** densely many-flowered; rachis pendent, to 16 cm long; floral bracts nearly elliptic or oblong, 1.2–1.5 mm, margin erose; peduncle subterete, 20–25 cm, lower part narrowly winged on both sides, with a few very small sterile bracts below; pedicel and ovary ~1.5 mm long. **Flowers** reddish-brown; dorsal sepal ovate or broadly ovate, 1–1.3 × ~0.8 mm, obtuse; lateral sepals slightly narrower, acuminate; petals ovate-oblong, 0.9–1.1 × ~0.6 mm, margin irregularly erose; lip broadly ovate or nearly suborbicular, ~1.5 × 1.5 mm, base decurrent backward, inconspicuously trilobed, margin irregularly incised or fimbriate, apex bilobed; column short. **Fruits** ellipsoid, ~2.5 × 2 mm. **Distr:** China (Hainan) to tropical Asia. **Ecol:** On trees in forests; 1,300–1,400 m. Fl. & Fr. Aug–Dec. **Zones:** 8–12.

Stichorkis

Syn: *Cestichis*.

Species 70. **Herbs**, epiphytic, rarely terrestrial; rhizome creeping, covered by tubular scales; stems pseudobulbous, erect, 1-noded, clustered or not, covered when young by 2–5 sterile sheaths, apex 1-leafed. **Leaf** erect; lamina linear or oblanceolate, conduplicate. **Inflorescence** racemose, erect, flowers opening in succession; rachis laterally flattened, shorter than peduncle; floral bracts distichous, imbricate. **Flowers** non-resupinate, usually yellow, green, or orange, often translucent; sepals recurved; dorsal sepal free; lateral sepals sometimes fused for part or entire length; petals free, often linear, erect; lip often reflexed, entire or lobed, usually with a padlike callus, concave at base, lacking a spur; column incurved-arcuate, laterally winged near apex; anther attached by a slender filament, biloculate; pollinia waxy, ovoid, bilaterally flattened, 4, in 2 pairs, each with a small viscidium; stigma concave; rostellum thin-textured, blunt. **Distr:** W Indian Ocean, tropical and subtropical Asia to SW Pacific. **Cult:** Propagate by division or from seeds. Requires a freely draining bark-based medium. Creeping species can be mounted on tree fern slabs.

S. balansae (Gagnep.) Marg., Szlach. and Kulak, *Acta Soc. Bot. Poloniae* 77: 37, 2008. (Syn: *Liparis balansae*). **Herbs**, epiphytic; pseudobulbs congested, ovoid or ovoid, 1–1.5 × 0.4–0.7 cm, 1-leafed. **Leaf** oblanceolate or elliptic-oblanceolate, 10–17 × 1–2.8 cm, base contracted, apex acuminate, papery; petiole 2–4 cm long. **Inflorescence** laxly 3–5-flowered; 11–25 cm long; peduncle compressed-cylindric, narrowly winged; floral bracts narrowly lanceolate, 0.3–1.5(–2) cm long; pedicel and ovary 1.5–1.8 cm. **Flowers** green; dorsal sepal oblong-lanceolate, 1–1.1 × 0.2–0.29 cm, apex obtuse; lateral sepals obliquely ligulate-lanceolate to oblong-ligulate, 1–1.1 cm × 2.5–3 mm, subacute to obtuse; petals linear-lanceolate, 1–1.3 cm × 0.7–1.2 mm, 1-veined, subacute; lip flabellate-oblate or broadly obovate-orbicular, 0.8–1.2 × 1–1.3 cm, base contracted, with 2 calli, margin irregularly denticulate, apex rounded, mucronate; column arcuate, 4.3–5 mm long, apex winged; anther cap ~2 mm across. **Fruits** obovoid, 1–1.2 × 0.6–0.7 cm. **Distr:** China, Indo-China. **Ecol:** On trees or rocks in forests or along valleys; 500–2,200 m. Fl. Sep–Oct. **Zones:** 7–11.

S. condylobulbon (Rchb. f.) Marg., Szlach. and Kulak, *Acta Soc. Bot. Poloniae* 77: 37, 2008. (Syn: *Liparis condylobulbon*, etc.). **Herbs**, epiphytic; rhizomes distinct, rigid; pseudobulbs 2–4 cm apart; stems yellowish-green, fleshy; pseudobulbs cylindrical, 6–12 cm long, apex narrowed, base dilated, lustrous. **Leaves** 2, apical, articulated with pseudobulb; lamina oblanceolate-linear, 10–15 × 1–1.5 cm, base cuneate, apex acute. **Inflorescence** a raceme, terminal; peduncle and rachis ~10–20 cm long, nearly terete; bracts lanceolate, 3–5 mm long; pedicel and ovary 8 mm long. **Flowers** numerous, dense,

Below: *Oberonia mucronata. Stichorkis balansae, Stichorkis condylobulbon, Stichorkis crenulata.*

MS

MG

MG

MS

pale green or nearly white, with strongly reflexed sepals and petals; sepals oblong, 3.5 × 1 mm, obtuse; petals linear, 3.5 × 0.5 mm; lip ovate, 3–4 × 2–2.5 mm, base erect, margins thick, incurved, recurved downward from middle, apex bilobed, apical margin minutely ciliate; disk with 2 pointed calli at base; column green, 2 mm long, ± arcuate, apex inconspicuously winged, base dilated; anther green, margin white at base. **Distr:** Taiwan, Indo-China to SW Pacific. **Ecol:** 100–500 m. **Zones:** 9–12.

S. crenulata (Blume) Marg., Szlach. and Kulak, *Acta Soc. Bot. Poloniae* 77: 37, 2008. (Syn: *Liparis crenulata*, etc.). **Herbs**, epiphytic, medium-sized; pseudobulbs 6–12 cm apart, ovoid, 4–9 cm long, base enveloped by 4, imbricating sheaths, apex 2- or 3-leaved. **Leaves** lanceolate, 20–33 × 3.8–5.8 cm, base gradually narrowed, apex acute. **Inflorescence** longer than leaves, to 18–36 cm long, loosely many-flowered, semierect to horizontal; bracts lanceolate, foliaceous ~1.2 cm long; floral bracts triangular, acute. **Flowers** with sepals and petals brownish, reflexed; sepals lanceolate, obtuse; petals subsimilar, shorter; lip brick-red, recurved in the middle, ~1.2 × 1.2 cm, cordate-obovate, bilobed, lobes crenulate; column arcuate, obtuse. **Distr:** Sumatra to Java. **Ecol:** In wet submontane and montane forests; 650–2,200 m. **Zones:** 11–12.

TRIBE CALYPSOEAE

Chysis

Species ~7. **Herbs**, epiphytic; stems fusiform, thickened or subclavate (rarely ellipsoid), often loosely branched, entirely concealed by scarious sheaths. **Leaves** several, distichous, usually crowded near apex; lamina lanceolate-elliptic or elliptic-oblong, plicate, base and apex acuminate. **Inflorescences** racemose, lateral, loosely few- (to 12-)flowered. **Flowers** showy, white, yellow or rose-colored, concave; sepals free, spreading, about equally long; lateral sepals broader, oblique, forming a mentum with column foot; petals similar to dorsal sepal or slightly longer, oblique, usually oblong-spathulate; lip affixed to column foot, recurved, tubular-involute, trilobed near middle, lateral lobes erect, incurved, midlobe spreading or reflexed, entire or slightly bilobed in front; disk with callose veins or calli; column stout, arcuate, bialate, produced into a short foot; anther opercular, incumbent, 2-celled; pollinia 8, 4 per cell. **Distr:** Mexico to Peru. **Cult:** Propagate by division. Best grown in baskets in a coarse compost, or mounted, in intermediate conditions with partial shade. Water and fertilize freely during the growth period.

C. tricostata Schltr., *Notizbl. Bot. Gart. Berlin-Dahlem* 8: 123, 1922. [Baby Orchid]. **Herbs**, epiphytic; pseudobulbs 30–40 × ~3 cm, claviform. **Leaves** elliptic-lanceolate, ~ 30 × 6 cm. **Inflorescences** 15–22 cm long. **Flowers** with sepals and petals ± similar in shape and color, yellow or whitish outside, inside base white and apex yellow with brownish-red coloration; sepals elliptic-oblong or falcate-ovate, 2.6–3.5 × 0.7–1.2 cm; petals ~ cuneate or oblong-cuneate, 3.4 × 1 cm, apex obtuse or subtruncate; lip white, red-spotted, 2.1–2.5 cm, lateral lobes oblong, 8–10 × 6.5–7 mm, tapering into apex, midlobe obcordate, ~ 8 × 13 mm, margins undulate; disk with 3 keels, conspicuous, fleshy, joined to the apices, or just medial keel projected forward, ~1 mm. **Distr:** Costa Rica. **Ecol:** In humid rainforests; 1,000–1,200 m. **Zones:** 11–12.

Earina

Species ~10. **Herbs**, epiphytic or lithophytic; stems leafy, very long, canelike, enclosed within leaf sheaths, unbranched, arising from prostrate, sympodial axis bearing long roots (sometimes green). **Leaves** ± distichous, jointed to sheath which is ± flattened; lamina narrowly linear. **Inflorescence** terminal, a panicle of several racemes, produced in succession up one axis, each raceme with several to many crowded empty bracts near its base; floral bracts more widely spread, small. **Flowers** strongly fragrant; perianth glabrous, sepals and petals free, subequal, spreading; lip uppermost, ± lobed, broad above the narrow sessile base, lacking conspicuous calli, initially erect and embracing column; column short, ± cylindrical, very narrowly winged, foot very short or absent; anther terminal, operculate, incumbent, narrowly joined to column just below its apex, caducous; pollinia 4, ± waxy; stigma deeply concave. **Distr:** Fiji, New Caledonia, New Zealand, Samoa, Tahiti. **Ecol:** On forest trees, sometimes lithophytic. Fl. Sep–Nov(–May). **Cult:** Propagate from seeds. Best grown mounted on tree fern slabs, in humid conditions and in partial sun.

E. autumnalis (Forst. f) Hook. f., *Fl. New Zealand*, 1: 239, 1853. (Syn: *E. alba*, etc.). [Easter Orchid]. **Herbs**, epiphytic, to 1 m long; rhizome covered scale leaves; stems erect to drooping, clad in sheathing leaf bases, leaf sheaths overlapping, deeply split. **Leaves** narrowly linear, 4–12 × 0.5–0.8 cm, widest near base, broad almost to subacute apex, long-persistent. **Inflorescences** panicles, ~5–10 cm long, usually erect, racemes numerous, distichously arranged at short intervals up stiff central axis, each lateral stem to ~1 cm long with a few crowded empty bracts and 1–3 almost sessile flowers. **Flowers** with perianth ~1.3 cm Ø,

Top row: *Chysis tricostata, Earina autumnalis, Earina mucronata, Gastrodia sesamoides.*
Bottom row: *Gastrodia* sp., *Arundina graminifolia, Chelonistele sulfurea, Coelogyne asperata.*

MS MG

opening widely, waxy white, lip base yellow; sepals elliptic, slightly keeled, subacute; petals slightly broader and more obtuse; lip broad and obtuse, erect, half-enclosing column, transversely rhomboid, not lobed, distal margins slightly thickened, recurved, near base 2 short crescent-shaped ridges, leading to small, brightly colored pit; column much shorter, broadest at base, wings projecting in small lobes about level of stigma; pollinia clavate. **Distr:** New Zealand. **Ecol:** On forest trees, or growing on rocks. Fl. (Jan–)Feb–Apr(–Jun). **Zones:** 8–10.

E. mucronata Lindl., *Bot. Reg.* 20 (sub t. 1699), 1834. (Syn: *E. aestivalis*, *E. quadrilobata*). [Bamboo Orchid, Hanging Tree Orchid, Peka-a-waka]. **Herbs**, to 1 m long; rhizome slender; stem usually drooping, completely clad in sheathing leaf bases, leaf sheaths ~2–3 mm Ø, not split, overlapping. **Leaves** distichous, exposed surfaces pale with small dark spots; lamina narrowly linear, 6–15(–20) x 0.4–0.6 cm, widest near base, apex tapering gradually, tip acute, soon caducous. **Inflorescence** panicle, to ~10 cm long, drooping, racemes 2–6(–10), well-spaced on very slender axis, each 3–4 cm long; floral bracts ~4 mm long, barely overlapping, enclosing the short pedicels. **Flowers** with perianth ~1–1.2 cm across, opening widely, pale, slightly greenish-cream; sepals elliptic, subacute; petals slightly broader and more obtuse; lip much broader and more showy, yellowish to apricot, spreading widely, broadly oblong, broad proximal part connected by narrow isthmus with almost equally large distal lobe, near base 2 inconspicuous ridges leading to a small pit; column much shorter, narrowed toward base, wings obsolete; pollinia long-ovate. **Distr:** New Zealand. **Ecol:** On forest trees, sometimes lithophytic. Fl. Sep–Nov(–May). **Zones:** 8–10.

TRIBE GASTRODIEAE

Gastrodia

Species 52. **Herbs**, terrestrial, saprophytic without chlorophyll; rhizomes horizontal, tuberous, densely noded, often pubescent, with scales on node, rarely with coral-like roots; stem erect, scaly on nodes. **Leaves** absent. **Inflorescence** terminal, racemose, bearing few to many bracteate flowers; pedicels elongated in fruit or not. **Flowers** resupinate, fleshy; ovary pedicellate, sepals and petals united, forming a 5-lobed perianth tube, or forming a urceolate structure being split between lateral sepals, petal lobes located near sinus of sepal lobes, smaller than sepal lobes; lip adnate to apex of column-foot, embraced by perianth tube, or rarely adnate to perianth tube, not lobed or rarely trilobed, clawed at base, lamina often mobile, disk with or without lamellae or calli; column fairly long, erect on top of ovary, ± semicylindrical, straight, narrowly 2-winged near apex, often with foot at base, rarely with ventral appendage; anther terminal on column, incumbent, pollinia 2, soft, granular, without viscidium; stigma single, near base or rarely near top on ventral side of column; rostellum small or rarely absent. **Fruits** cylindrical or ellipsoid. **Distr:** Himalayas eastward to Japan, and southward through Malaysia to Australia and New Zealand. **Cult:** Propagate by division or tubers, or from seeds. Requires a well-drained organic compost in full or partial sun.

G. sesamoides R. Br., *Prodr.*: 33, 1810. [Cinnamon Bells, Potato Orchid]. **Herbs**, saprophytic, terrestrial, to 1(–1.2) m tall; rhizomes swollen, large, to 8 x 3 cm; stem erect, stout or slender; scale leaves widely spaced. **Inflorescence** erect, 1–20-flowered, ± drooping bringing dorsal sepal next to axis. **Flowers** tubular, ~1–1.6 cm long, 4–8 mm Ø, light brownish; perianth lobes slightly marginally thickened; lateral sepals connate almost to level of tip of lip; lip ovate-oblong, shortly adnate to perianth tube, long median calli much crested, margin well-developed, undulate, ± lobed, membranous above, thickened below, confined to free limb; column almost as tall as labellum, very narrowly winged throughout; anther short, operculate, filament smooth; stigma basal, far below anther. **Distr:** S and E Australia, New Zealand. **Ecol:** In open forests and scrubs; 30–1,100 m. Fl. Oct–Jan. **Zones:** 9–10.

G. sp. **Herbs**, saprophytic; rhizomes tuberous. **Inflorescence** a raceme, erect, to ~5-flowered; scape ~5(–10) cm tall, brown, somewhat succulent, ~2.5 mm Ø; rachis ~ 2.5 cm tall; bracts 2–3, imbricate, 5 mm apart, 2.8 mm long, membranous, pale brown; pedicels ~1 cm long; floral bracts triangular, ~3.6 mm long. **Flowers** 2 cm across the lateral sepals, dark chocolate-brown (appearing dark brown to black and camouflaged among the leaf litter), glossy, waxy, somewhat fleshy, lip green with midlobe bright orange, petals and sepals shortly basally connate, appearing campanulate; dorsal sepal narrowly ovate, ~ 1 x 0.72 cm, apex rounded, slightly recurved, held ± horizontally over the column; lateral sepals larger, spreading, broadly ovate, ~1.2 cm long, ~8 mm wide at middle, base concave, apex acute; petals linear, held beneath the dorsal sepal; lip 4 mm long; column ~2.5 mm long; ovary cylindrical, ~5.5 mm long, 2.5 mm Ø. **Fruits** cylindrical, ~ 1 cm long, ~5 mm Ø. **Distr:** Probably endemic to Sri Lanka (Galle District); very limited distribution. **Ecol:** In moist, low montane forests, forming small colonies. **Zones:** 11–12. **Info:** This species is under threat due to habitat loss to mainly deforestation (for conversion to tea plantations).

TRIBE ARETHUSEAE

Arundina

Species 2. **Herbs**, terrestrial, tall, pseudobulbous; stems tufted, reedlike, leafy throughout, thickened at base. **Leaves** ± distichous, grasslike, conduplicate, articulate, with overlapping sheaths. **Inflorescence** a raceme, terminal, simple or branched, gradually elongating; flowers produced in succession. **Flowers** resupinate; sepals narrow; dorsal sepal erect; lateral sepals usually close together behind lip; petals usually wider, spreading horizontally; lip sessile at base of column and embracing it; column elongate, slender, slightly dilated at apex; anther terminal, incumbent, mainly 2-celled; pollinia 8, relatively soft, laterally flattened, connected through caudicles in 2 pairs of 4, viscidium obscure; stigma entire. **Distr:** Tropical and subtropical Asia. **Cult:** Propagate by division or from seeds. Requires a well-drained compost rich in organic matter, in full or partial sun. Needs plenty of water during active growth.

A. graminifolia (D. Don) Hochreutiner, *Bull. New York Bot. Gard.* 6: 270, 1910. (Syn: *A. bambusifolia*, etc.). [Bamboo Orchid, Bird Orchid]. **Herbs**, tufted; stems to 1.5 m tall, grasslike, 1 cm Ø. **Leaves** linear or linear-lanceolate, grasslike, 10–28 x 1–2.3 cm, base narrow, apex acute or acuminate, attenuate, sheathing, 5–9-veined, thinly coriaceous, upper surface lustrous. **Inflorescences** racemes, terminal; peduncles <10 cm long, rachis green or yellowish-brown; flowers blooming successively; floral bracts scalelike, broadly ovate, 5–8 mm long, acute; pedicel and ovary ~2 cm long. **Flowers** rose-purple or pale pink, 5–7.5 cm across; sepals spreading, oblong to linear-lanceolate, ~3.5–4.5 x 1–1.3 cm, base contracted, apex acute, 9–11-veined, midrib and veins red; petals obovate to oblong-lanceolate, 3.5–4.5 x 1.5–2.1 cm, base strongly contracted, apex acute, 3-veined; lip cuneate-obovate, 3.5–5 x 2–3.7 cm (when spread out), shallowly trilobed, strongly undulate, lateral lobes rectangular, erect and embracing column, reflexed at apex, midlobe nearly square, 1.2 cm long, emarginate; disk with 3 yellow keels; column 2 cm long; anther semiovoid, slightly constricted near base; pollinia 8; stigma slightly convex. **Fruits** decurved, linear-oblong, 6-ribbed, 5–6 cm long. **Distr:** N India, Assam, China, Myanmar, Taiwan; Java, Malaya, Sumatra and Borneo. **Ecol:** In grasslands; <700 m. Fl. Oct–Dec. **Zone:** 12.

Chelonistele

Syn: *Sigmatochilus*.
Species 13. **Herbs**, epiphytic or lithophytic, small to medium-sized; rhizome creeping; pseudobulbs clustered or spaced, short and swollen to slender, terete or bilaterally flattened, apex 1- or 2-leaved. **Leaves** deciduous, petiolate, articulated at the apex of the pseudobulbs; lamina plicate, elliptic to linear. **Inflorescences** racemose, few- to many-flowered, erect to spreading, borne with or before the leaves develop, distichous or secund; bracts distichous, caducous, papery to membranous. **Flowers** small to medium-sized; sepals subsimilar, free; petals linear to ovate-lanceolate, spreading to recurved; lip deflexed, trilobed, sometimes obscurely so, concave or saccate at base, with a callus of 2 ± parallel keels on the upper surface; column elongate, spathulate, winged toward apex; pollinia 4, each attached to a small caudicle. **Distr:** W and C Malesia. **Cult:** Propagate by division or from seeds. Requires a well-drained compost in shade, with careful watering during the development of new growth.

C. sulphurea (Blume) Pfitzer, *Pflanzenr.*, IV, 50 II B 7: 137, 1907. (Syn: *Coelogyne sulfurea*, etc.). **Herbs**, epiphytic, upright, sympodial; pseudobulbs oval, 3–3.8 cm long, 1.5 cm Ø, apex 1–2-leaved. **Leaves** elliptic to obovate-lanceolate, 6–30 x 4 cm, apex acute, leathery; petiole 2.5–10 cm long. **Inflorescence** 4–12-flowered, flowers ± secund, arching, apical, appearing with new growth; peduncle 3–6 cm long; rachis 3.5–12 cm long; bracts ovate-oblong to ovate, 1.7–2.5 cm long; pedicel and ovary 1–1.6 cm long, cross-section triangular. **Flowers** 1.4 cm Ø, translucent, flesh-colored, greenish or pale pink, lip cream with a large central orange to yellow marking; dorsal sepal concave, ovate to lanceolate, 1–1.55 x 0.4–5.5 cm, acute; lateral sepals concave, ovate to lanceolate, 1–1.6 x 0.35–0.6 cm, base oblique, apex acute; petals linear, 1 x 0.12 cm, curved toward the pedicel; lip trilobed, to 1.5 x 1 cm, base saccate, 1.1–1.7 x 0.8–1.1 cm, lateral lobes triangular to oblong, midlobe broadly clawed, transversely oblong to reniform, callus of 2 keels; column spathulate, 0.7–1.2 cm long, apex winged and hooded. **Distr:** W peninsular Malaysia and Malesia to Philippines. **Ecol:** 600–2,700 m. **Zones:** 10–12.

Coelogyne

Species ~200. **Herbs**, epiphytic; rhizome creeping or ± pendulous, nodes dense; pseudobulbs ovoid to cylindric, usually covered with leathery sheaths, apex usually 2-leaved. **Leaves** often oblong to elliptic, usually thickly textured, usually base petiolate. **Inflorescence** erect or pendulous, usually several-flowered, rarely 1- or >20-flowered, flowers opening successively or simultaneously; floral bracts caducous, often large; with or without distichous-equitant, persistent sheaths at base or imbricate, sterile, convolute sheathing bracts at interface between peduncle and rachis. **Flowers** sometimes heteranthous; sepals similar, often concave; petals often linear, much narrower than sepals; lip often concave at base, trilobed or rarely unlobed, lateral lobes ± erect and embracing column, callus with longitudinal lamellae; column rather long, both sides winged, foot absent; pollinia 4, in 2 pairs; stigma concave. **Fruits** often ribbed or narrowly winged. **Distr:** Tropical and subtropical Asia to Oceania. **Cult:** Propagate by division. Requires a fairly coarse, soil-free compost,

MG

MS

MG

MG

Top row: *Coelogyne bilamellata, Coelogyne chloroptera, Coelogyne cristata, Coelogyne flaccida.*
Bottom row: *Coelogyne fuscescens, Coelogyne lawrenceana, Coelogyne lentiginosa, Coelogyne longipes.*

with plenty of water during active growth, in excellent light and intermediate conditions. Straggling species are best grown in hanging baskets.

C. asperata Lindl., *J. Hort. Soc. London* 4: 221, 1849. (Syn: *Cymbidium robustum*, etc.). **Herbs**, epiphytic, lithophytic or terrestrial, clump-forming; pseudobulbs erect, 1–6 cm apart, ovoid or conical to cylindrical, often slightly compressed, 5–25 cm long, green, sheathed when young. **Leaves** 2, oblong to lanceolate, <60 × 16 cm, margins entire, apex subacute to obtuse, green, thin, leathery, pleated; petiole furled, 2–25 cm long. **Inflorescence** single, 15–40-flowered, terminal, <32 cm long, arching; peduncle green, smooth, with 2 green bracts; floral bract obovate to lanceolate, 2–3 cm long; pedicel and ovary 3.5 cm long, cream to green-brown. **Flowers** 4–7 cm Ø, fragrant, waxy, creamy-white, lip with an orange-brown patch at base of midlobe, lateral lobes brown-tipped, with brown veins and markings; sepals not wide-spreading, ovate-elliptic, ~3.6–4.5 cm long, base concave, apex acute; petals wide-spreading, narrowly lanceolate to elliptic, ~3.4–4.1 cm long; lip creamy-white, trilobed, 2.9–4.2 cm long, lateral lobes erect; column ~2.5 cm long. **Distr:** Indonesia through Papua New Guinea, to the Solomon Islands. **Ecol:** In rainforests, grasslands; 0–1,600 m. Fl. Jun–Aug, Dec. **Zones:** 9–12.

C. bilamellata Lindl., *Fol. Orchid.* 5: 14, 1854. (Syn: *Panisea bilamellata, Pleione bilamellata*). **Herbs**, epiphytic, upright, sympodial; pseudobulbs pear-shaped, to 12 cm tall, 3 cm Ø. **Leaves** 2, ovate to oblong-lanceolate, to 20 × 5 cm, base gradually narrowed into a grooved petiole, apex acute, plicate, 5-nerved. **Inflorescence** pendulous, to 60+ cm long, somewhat zigzag, to 60-flowered, flowers opening in succession; floral bract chafflike and early deciduous (before anthesis. **Flowers** to 3 cm across, variable from pure white to creamish-brown; sepals oblong, to 1.5 × 0.5 cm, acute; petals linear, to 1.5 cm × 1.5 mm; lip trilobed, 1.6 cm long, 7.5 mm wide (when flattened), lateral lobes triangular, midlobe oblong to ovate with 2 ridges running lengthwise. **Distr:** Borneo, Philippines. **Ecol:** 60–1,000 m. Fl. throughout the year. **Zones:** 11–12.

C. chloroptera Rchb.f., *Gard. Chron.*, n.s., 1883(1): 466, 1883. **Herbs**, epiphytic, sometimes lithophytic, upright, sympodial; pseudobulbs pear-shaped, ~6 cm tall, shiny, yellowish-green, bifoliate. **Leaves** plicate, ~20 × 6 cm, base gradually narrowed, leathery, 7-nerved. **Inflorescences** 4–12-flowered, to 25 cm long, upright or sometimes arching, appearing with new growth. **Flowers** apple-green, waxy, fragrant; sepals lanceolate, to 2 × 0.7 cm; dorsal sepal hooded over the column; petals linear, to 2 × 0.2 cm, reflexed toward the pedicel; lip white and brown, trilobed, 1.9 cm long, 1.2 cm wide (when flattened), lateral lobes rounded, erect, tessellated with brown, midlobe semicircular, apically white, center brownish, 3 ridges run lengthwise along the whole length; column, curved, apex, brownish-orange. **Distr:** Philippines. **Ecol:** ~1,200 m. **Zones:** 9–11.

C. cristata Lindley, *Coll. Bot.* sub. t. 33, 1824. (Syn: *Pleione speciosissima*, etc.). **Herbs**, epiphytic; pseudobulbs oblong or ovoid, (1–)2.5–6 × (0.5–)1–1.7 cm, bifoliate, basal sheaths several, ovate or ovate-lanceolate, 2–6 cm, leathery. **Leaves** linear-lanceolate, (5–)10–17 × (0.4–)0.7–1.9 cm, base attenuate, apex long-acuminate, papery. **Inflorescence** heteranthous, 8–12 cm long, half-embraced in sheaths, raceme 5–7 cm long, 2–10-flowered; floral bracts not deciduous at anthesis, ovate-lanceolate, ± cymbiform, 2.5–3.5 cm, papery; pedicel and ovary 2–2.5 cm long. **Flowers** white, large; sepals lanceolate or oblong-lanceolate, 3–4 × 1–1.5 cm, abaxially ± carinate, acute; petals similar to sepals, 0.9–1.1 cm wide; lip ovate, concave, trilobed, lateral lobes erect, subovate, midlobe ± broadly obovate, 1.2–1.5 cm long, obtuse or emarginate, callus with 5 fimbriate lamellae extending from base of lip to base of midlobe; column 2–3 cm long, both sides winged. **Distr:** Himalayas to Bangladesh. **Ecol:** On large rocks along forest margins; 1,700–1,800 m. Fl. May. **Zones:** 9–10. **Info:** An all white variety is often seen in cultivation.

C. flaccida Lindley, *Gen. Sp. Orchid. Pl.* 39. 1830. (Syn: *Pleione flaccida*, etc.). **Herbs**, epiphytic; rhizome stout; pseudobulbs 2–3 cm apart, oblong or nearly cylindric, 6–12.5 × 1.5–3.2 cm, with several purplish basal sheaths 5–8 × 2.5–3 cm, apex attenuate, bifoliate. **Leaves** oblong-lanceolate to elliptic-lanceolate, 13–19 × 3–4.5 cm, leathery, apex subacuminate or shortly cuspidate; petiole 4.5–8 cm long. **Inflorescence** heteranthous, sparsely 8–10-flowered, 16–32 cm long, basal half enclosed in sheaths; rachis 11–21 cm long; floral bracts caducous, obovate-elliptic, ~ 2.5 cm long; pedicel and ovary 1.7–2 cm long. **Flowers** pale yellow to white, with yellow and pale reddish-brown spots on lip; dorsal sepal oblong or oblong-lanceolate, 2.1–2.4 × 0.6–0.7 cm; lateral sepals slightly narrower, abaxially ± carinate; petals linear-lanceolate, 2–2.2

MS

Above: *Coelogyne longirachis*, *Coelogyne miniata*, *Coelogyne mooreana*.

× 0.2–0.3 cm; lip subovate, 1.6–2 × ~1.5 cm, trilobed, lateral lobes erect, subovate, obtuse, midlobe suboblong, 6–7 × ~4 mm, margin ± crisped, callus with 3 crisped-incised lamellae; column slightly arcuate, ~1.3 mm long, winged. **Distr:** Nepal to China. **Ecol:** On trees in forests; 1,600–1,700 m. Fl. Mar. **Zones:** 9–11.

C. fuscescens Lindley, *Gen. Sp. Orchid. Pl.* 41, 1830. (Syn: *C. brunnea*, etc.). **Herbs**, epiphytic; pseudobulbs clustered, suboblong, slightly narrowed toward both ends, 2–3 × 0.5–0.7 cm, apex 2-leaved, base with brown sheaths 1.5–2 cm long. **Leaves** oblong-oblanceolate, 11.5–13.5 × 1.3–2 cm, base contracted into a petiole 1–2 cm long, apex obtuse or acute, leathery. **Inflorescence** proteranthous, rarely synanthous, 12–14 cm long; rachis 4–6 cm long, usually 2-flowered; floral bracts oblong-lanceolate, ± cymbiform, 2–2.5 cm long, membranous; pedicel and ovary 2–2.5 cm long. **Flowers** ~4 cm Ø, sepals and petals pale yellow or yellow-green, ± flushed with brown apically; sepals suboblong, 2.5–2.8 × 0.7–0.9 cm, acuminate or shortly cuspidate; petals linear, 2.2–2.4 × 0.25–0.3 cm; lip ovate, 2.5–2.8 cm long, base concave, trilobed, lateral lobes subovate, midlobe ovate, 1–1.3 × 0.6–0.7 cm, callus with 3 fleshy, undulate longitudinal lamellae; column ~2 cm long, both sides narrowly winged. **Distr:** Nepal to Indo-China. **Ecol:** ~1,300 m. **Zones:** 9–11.

C. lawrenceana Rolfe, *Gard. Chron.*, III, 1905(1): 227, 1905. (Syn: *C. fleuryi*). **Herbs**, epiphytic; pseudobulbs clustered, ovoid-oblong, 5–12 cm long, bifoliate. **Leaves** broadly lanceolate, 20–40 × 3.5 cm, plicate, 8-nerved, base gradually narrowed into petiole, apex acuminate; petiole elongate, grooved. **Inflorescence** initially erect then arching, 25 cm long, 1–6-flowered, flowers held above the foliage; scape terminal, borne from mature pseudobulbs; flowers opening successively. **Flowers** 10–12 cm across, waxy, somewhat translucent, sweetly fragrant, long-lasting, brownish, lip white with brownish coloration around middle; sepals elliptic, 5–6.5 cm long; petals linear, 5–6.5 × 10 cm, slightly reflexed; lip trilobed, 5–6.5 cm long, obscurely saccate, lateral lobes 3.5 cm long, erect, white, with a broad brown line below margin, emanated from base of midlobe, midlobe broadly ovate, margin wavy, base with lamella in 7 rows and 3 rows in the middle; column clavate, 3.5 cm long, winged. **Distr:** S Vietnam. **Ecol:** In primary montane forests; 800–1,000 m. **Zones:** 10–11.

C. lentiginosa Lindl., *Fol. Orchid.* 5: 3, 1854. (Syn: *Pleione lentiginosa*). **Herbs**, epiphytic; pseudobulbs linear-oblong, ellipsoid, cylindrical, or ovoid, 7.5–10 cm long, obtusely 4-angled, bifoliate. **Leaves** arching to suberect, elliptic-(ob)lanceolate, 15–20 cm long, plicate, base narrowed into petiole, apex acute; petiole short, grooved. **Inflorescence** a raceme, 4–8-flowered; scape with erect raceme 6–16 cm long, sheathed up to the flowers; bracts lanceolate, 1.2–1.6 cm long, persistent; flowers opening simultaneously. **Flowers** fragrant, ~4 cm across, pale greenish-yellow; sepals and petals subsimilar, narrowly lanceolate, 1.7–2.2 × 0.3–0.5 cm, acute; lip trilobed, slightly arcuate, 1.8 × 1.2 cm, lateral lobes oblong, obtuse, brown-speckled, margins brown, midlobe large, shortly broadly clawed, nearly semicircular, crenate, white with orange blotches at base, ridges not produced on to the midlobe; disk with 3 obscure short crenate ridges. **Distr:** Indo-China (Myanmar, Thailand and Vietnam). **Ecol:** In evergreen, deciduous forests; 500–1,300 m. **Zones:** 9–11.

C. longipes Lindley, *Fol. Orchid.* 5(Coelogyne): 10, 1854. (Syn: *Pleione longipes*). **Herbs**, epiphytic; rhizome stout; pseudobulbs nearly cylindric, 3–7.5 × 0.8–1.2(–1.5) cm, with 2 leaves at apex; leaf sheaths several, basal, bright yellow. **Leaves** oblong-oblanceolate, (10–)15–23 × (1–)1.8–3 cm, base cuneate, apex acuminate; petiole 2.5–6 cm long. **Inflorescence** 5–7-flowered, 7–15 cm, slender; pedicel and ovary 0.7–1 cm. **Flowers** white or pale yellow; sepals ovate-oblong, ~ 1 × 0.5–0.6 cm, 5-veined, apex shortly acuminate; petals narrowly linear or filiform, ~ 1 cm × 0.3 mm, attenuate toward apex; lip nearly broadly ovate, ~ 1 × 0.8–0.9 cm, trilobed, lateral lobes subovate, margins entire, midlobe broadly oblong or subelliptic, ~ 6 × 5 mm, margin ± crisped, apex truncate and emarginate, callus with 2 narrow longitudinal lamellae; column slightly arcuate, 5–6 mm long, winged. **Fruits** narrowly obovoid-oblong, ~ 2 × 0.6 cm. **Distr:** E Nepal to China (Yunnan) and Indo-China. **Ecol:** On trees in forests; 1,000–2,600 m. Fl. Jun. Fr. Mar. **Zones:** 8–11.

C. longirachis Ames, *Orchidaceae* 7: 88, 1922. **Herbs**, epiphytic; roots glabrous, 2 mm Ø; rhizome robust; pseudobulbs approximate, ~14 cm high, ~1.2 cm wide. **Leaves** 2, petiolate; elliptic-lanceolate, 24 × 4.5 cm, apex acute, attenuate; petiole 3 cm long. **Inflorescence** up to 60-flowered, 80 cm long, produced in succession, ~1 cm apart on the rachis, borne from young growth, arcuate; rachis fractiflex, usually ~56 cm long; bracts imbricate, 8 mm long; pedicel and ovary 1.7 cm long, in fruit 4.5 cm long. **Flowers** brownish-red; sepals usually narrowly elliptic, 7 × 6 mm, abruptly acute, subcarnose; petals linear, 1.1 × 0.1 cm, apices attenuate, acute; lip ~1.2 cm long, subcarnose, conspicuously trilobed, lateral lobes subquadrate, ~3.5 mm wide, margins unequally triangular-dentate, apices rounded, midlobe 6.5 ×3 mm, base slightly rounded, apex triangular-acute, callus tridentate which extends from the base of the column onto the lower third of the midlobe; column 8 mm long, carnose. **Distr:** Philippines (Mindanao). **Ecol:** Along streams in forests; 1,200 m. **Zones:** 11–12.

C. miniata (Blume) Lindl., *Gen. Sp. Orchid. Pl.*: 42, 1833. (Syn: *Pleione lauterbachiana*, etc.). **Herbs**, epiphytic; rhizome creeping; pseudobulbs 2.5 cm apart, cylindrical to ovoid, apex narrowed, sulcate, enveloped by a few basal sheaths when young. **Leaves** 1–2, elliptic to oblong-lanceolate, plicate, base abruptly narrowed into a petiole, acute, 7-nerved; petiole grooved. **Inflorescence** synanthous or proteranthous, base terete, stiff, erect 7.5–10 cm long, few-flowered; scape with base enveloped by several loose bracts; rachis fractiflex, short; floral bracts deciduous at anthesis; flowers opening about 4 at a time. **Flowers** 2.5 cm long, bright orange, glossy, fleshy and slightly translucent, not opening fully; sepals similar, ovate, 2.5 × 1 cm, acute; dorsal sepal, petals and lip not opening fully, touching each other; lateral sepals opening widely; petals linear, ~2.2 × 0.2 cm, acute; lip boat-shaped, 2.5 × 1.5 cm, acute; column short. **Distr:** Sumatra to Java, New Guinea. **Ecol:** On trees or lithophytic on large rocks, in lower and upper montane forests; 1,000–2,400 m. **Zones:** 11–12.

C. mooreana Rolfe, *Bull. Misc. Inform. Kew* 1907: 129, 1907. (Syn: *C. psectrantha*, etc.). **Herbs**, epiphytic; pseudobulbs clustered, ovoid-oblong, ~7 cm long, bluntly tetragonal, slightly furrowed, bifoliate. **Leaves** borne from apex of pseudobulb, arching; lamina narrowly linear-(ob)lanceolate, 24–50 × 3–4 cm, plicate, base gradually narrowed into a petiole, apex acute, 7-nerved. **Inflorescence** racemose, 3–8-flowered, erect, terminal, 30–40 cm tall, borne from newly emerging growth; bracts deciduous; pedicels 2.5–3.5 cm long. **Flowers** fragrant, white, lip white with orange patch at center; sepals elliptic-oblong, 4–5.5 cm long, acute; petals elliptic, 4–5.5 cm long, acute; lip trilobed, 3–3.5 cm long, base concave-saccate, lateral lobes oblong, obtuse, incurved and enclosing column, midlobe ovate, obtuse; column ~2.5 cm long. **Distr:** Vietnam. **Ecol:** In highland montane cloud forests; 1,300–2,000 m. **Zones:** 9–10.

C. nitida (Wall. ex D.Don) Lindl., *Coll. Bot.*: t. 33, 1824. (Syn: *C. conferta*, etc.). **Herbs**, epiphytic; pseudobulbs ovoid to oblong, 7–10 cm long, furrowed, glossy, bifoliate. **Leaves** petiolate; lamina narrowly elliptic-lanceolate, base narrowed into a petiole, apex acute, plicate, 8-nerved; petiole elongate, grooved. **Inflorescence** a raceme, erect to pendent, to 20 cm long, slightly fractiflex, terminal, 3–8-flowered, exserted, very long in fruit, borne before new growth; floral bracts deciduous; flowers opening simultaneously. **Flowers** 4 cm across, fragrant, white, lip white with confluent orange and yellow or brownish areas; sepals linear-oblong, 3.5 × 0.7 cm, acute to obtuse; petals narrowly oblong-lanceolate, 2.5 × 0.5 cm, subacute; lip white with confluent orange and yellow or brownish areas, lateral lobes serrulate, midlobe as long, broadly ovate, base serrulate, acute; disk 2-ridged; column 1.2 cm long. **Distr:** Nepal to Indo-China. **Ecol:** On trees, sometimes lithophytic on mossy rocks; 1,300–2,600 m. **Zones:** 8–10.

Top row: *Coelogyne nitida, Coelogyne ovalis, Coelogyne pandurata.*
Bottom row: *Coelogyne speciosa, Coelogyne tomentosa, Coelogyne trinervis.*

C. ovalis Lindley, *Edwards's Bot. Reg.* 24(Misc.): 91, 1838. (Syn: *C. decora*, etc.). **Herbs**, epiphytic; rhizome creeping; pseudobulbs 8–13 cm apart, nearly cylindric, 3–6 × 0.6–0.8 cm, apex 2-leaved, with 2 basal sheaths ~3 cm long. **Leaves** lanceolate to ovate-lanceolate, 6–12 × 2–3.7 cm, apex acuminate or obtuse, papery; petiole 5–10 cm long. **Inflorescence** few-flowered, often 1–2(–4) flowers opening successively; peduncle 5–12 cm long, base covered with several equitant tubular sheaths; bracts white at apex of rachis; floral bracts caducous; pedicel and ovary 0.6–1 cm long. **Flowers** greenish-yellow, lip with purplish-red stripes; sepals oblong-lanceolate, ~2 × 0.5–0.6 cm; petals filiform or narrowly linear, ~2 × 1 mm; lip suboblong-ovate, ~2 × 1.5 cm, trilobed, lateral lobes erect, subovate, midlobe elliptic, ~1 × 0.9 cm, margin fimbriate, callus with 3 longitudinal lamellae; column ~1.3 cm long, winged. **Fruits** subobovoid, ~2.5 × 1.2 cm. **Distr:** Himalayas to China and Indo-China. **Ecol:** On trees or rocks along rivers; 1,200–3,200 m. Fl. Aug–Nov. Fr. Sep. **Zones:** 8–11.

C. pandurata Lindl., *Gard. Chron.* 1853: 791, 1853. (Syn: *Pleione pandurata*, etc.). [Black Orchid]. **Herbs**, epiphytic; rhizome very stout; pseudobulbs very large, oblong or suborbicular, flattened, 7.5–12.5 × 3 cm, grooved. **Leaves** elliptic-lanceolate, 20–45 × 2–7 cm, rigid, 5–7-nerved; petiole short, stout. **Inflorescence** raceme, to 15-flowered; scape very stout, 15–30(–40) cm long; sheaths lanceolate, rigid, inner ± foliaceous; bracts 2.5 cm long, persistent. **Flowers** 7–13 cm across, waxy-green, highly fragrant; sepals 4–5 cm long; dorsal sepal largest, oblong-lanceolate, 11-nerved; lateral sepals linear-oblong, acute, 5-nerved, keeled; petals subspathulate, clawed; lip panduriform, to 4 cm long, mottled with red-purple, base cordate, lateral lobes recurved, almost enclosing the column, midlobe large, broad, subquadrate, retuse, contracted in the middle, margins crisped and wavy; column apically winged. **Distr:** W Malesia to Philippines. **Ecol:** In lowland rainforests, submontane forests, scrubs; 0–1,800(–2,700) m. **Zones:** 10–12.

C. speciosa (Blume) Lindl., *Gen. Sp. Orchid. Pl.*: 39, 1833. (Syn: *Pleione speciosa*, etc.). **Herbs**, epiphytic; pseudobulbs, crowded, ovoid 2.5–4 cm long, angled. **Leaves** apical, 1–2; lamina elliptic or lanceolate, 10–30 × 7.5 cm, plicate, acute; petiole 2.5–5 cm long. **Inflorescence** racemose, drooping or suberect, from axils of young leaves, 10 cm long, 1–3-flowered; bracts lanceolate acuminate, caducous. **Flowers** greenish, musky scented; sepals oblong, lanceolate, 5–6.5 cm long, acute, greenish or buff; petals narrowly linear, reflexed, 5 × 0.3 cm, acute; lip large, 2.4 × 3 cm, lateral lobes obtuse, toothed, midlobe broadly clawed, rounded, toothed, oblong, with 2 fringed crests, yellow blotched and marked red-brown, margin broad, white. **Fruits** ovoid, 6.5 cm long, with large wings. **Distr:** SW Malesia. **Ecol:** On trees in dense forests, sometimes terrestrial in rainforests; 0–700(–2,000) m. **Zones:** 9–11.

C. tomentosa Lindl., *Fol. Orchid.* 5: 3, 1854. (Syn: *C. massangeana*, etc.). [Necklace Orchid]. **Herbs**, epiphytic; pseudobulbs narrowly ovoid, 7.5–15 cm long, somewhat angled, wrinkled with age, unifoliate; sheaths lanceolate, very long, rigid. **Leaf** lanceolate or elliptic-obovate, 30–50 × 2.5–6.5 cm, plicate; petiole 5–7.5 cm long. **Inflorescence** raceme; bracts tomentose; rachis zigzagged; scape 30–45 cm long, pubescent; basal sheaths loose; bracts 2.5–3.5 cm long, subacute, persistent in flower. **Flowers** pale yellowish-brown, scented, short-lived, delicate; sepals lanceolate, 2.5–4 cm long; petals narrower; lip yellowish, lateral lobes narrow, streaked with brown, disk with 3 crisped ridges, the outer double and coalescing in a broad caruncled area on the small midlobe; ovary tomentose. **Distr:** Peninsular Thailand to W Malesia. **Ecol:** In river valleys; 1,150–2,100 m. **Zones:** 10–12.

C. trinervis Lindl., *Gen. Sp. Orchid. Pl.*: 41, 1830. (Syn: *C. cinnamomea*, etc.). **Herbs**, epiphytic; rhizome very stout; pseudobulbs closely set, large, ovoid, 5–7.5 cm long, curved or oblique, longitudinally 4-grooved, enveloped basally by a few sheaths, bifoliate. **Leaves** elongate-lanceolate, 25–45 cm long, trinerved; petiole 5–7.5 cm long. **Inflorescence** a raceme, short, 4–8-flowered; rachis fractiflex, slender; scape to 25 cm long, with short basal sheaths, arising from new growth; floral bracts lanceolate, caducous; flowers opening simultaneously. **Flowers** 3–4 cm across, pale creamish-white, fragrant but somewhat unpleasant; sepals lanceolate, 2.2 × 0.6 cm, acute; petals linear or lanceolate, 2.2× 0.3 cm, trinerved; lip brown, to 1.8 × 1.4 cm, trilobed, keels 7, 3 long, 4 short, lateral lobes short, broad, midlobe broadly clawed, subreniform, retuse, margin white; disk with 3 narrow ridges extending into the midlobe; column 1.3 cm long. **Fruits** narrowly pyriform, ~4 cm long, angles acute. **Distr:** Indo-China to W and C Malesia. **Ecol:** On trees or lithophytic on rocks; 100–1,600 m. **Zones:** 9–12.

Dendrochilum

Species ~265. **Herbs**, epiphytic; pseudobulbs tufted or well spaced on rhizome, elongate-ovoid, with single leaf at apex. **Leaves** petiolate, articulated; lamina narrow, coriaceous. **Inflorescences** many-flowered, flowers distichously arranged, terminal; peduncle arising from base of old pseudobulb, often producing new leaf before growth of new pseudobulb, slender, erect or flexuous. **Flowers**; sepals and petals subequal; lip subsessile, jointed with base of column, oblong or lyrate, often trilobed, lateral lobes smaller than midlobe; column thin, curved, hooded at apex, with narrow lateral arms and terminal wing around anther; anther incumbent, 2-celled; pollinia 4, ovoid, waxy, connected by short caudicles, viscidium small or obscure; stigma entire, often emergent. **Distr:** Mainland Asia east to the Philippines and New Guinea, particularly well represented in the montane areas of Sumatra and Borneo. **Cult:** Propagate by division. Requires a bark mix with sphagnum moss or similar free-draining medium, in intermediate conditions and partial shade. They do well when kept pot-bound for longer.

D. cobbianum Rchb. f., *Gard. Chron.* 1880(2): 748, 1880. (Syn: *Acoridium cobbianum*, etc.). **Herbs**, epiphytic, upright, sympodial; pseudobulbs clustered, narrowly conical, 4–8 cm long, tapered, yellowish, unifoliate. **Leaf** petiolate; lamina narrowly oblong, to 6–18 × 2.5–5.5 cm, midrib prominent. **Inflorescence** upright, flowering portion pendent, appearing with new growth, normally longer than leaf, to 80 cm tall, densely many-flowered, flowers borne 7 mm apart; floral bracts ovate, subacute, glumaceous, papery. **Flowers** 1.2 cm across, color extremely variable, some forms have all orange flowers, others have white to cream sepals and petals with bright orange labellum, sometimes flowers are icy-green, fragrant; dorsal sepal and petals lanceolate, to 5 × 2.5 mm; lateral sepals ovate, to 6 × 3 mm; lip triangular, to 7 × 3 mm. **Distr:** Endemic to Philippines. **Ecol:** On mossy tree trunks, sometimes lithophytic on rocks in open places; >1,200 m. **Zones:** 9–10.

D. convallariiforme Schauer, *Nov. Actorum Acad. Caes. Leop.-Carol. Nat. Cur.* 19(Suppl. 1): 429, 1843. (Syn: *Acoridium bistortum*, etc.). **Herbs**, epiphytic, upright, sympodial; pseudobulbs round or ovate, ~ 2.5 × 1.5 cm, unifoliate. **Leaf** lanceolate, to 10 × 2.5 cm, bright green, young leaf grayish. **Inflorescences** racemes, 1–3, arching, to 50-flowered, flowers spirally arranged, appearing with new growth. **Flowers** 0.8–1.2 cm across, pale to deep pink (dark salmon), mildly scented; sepals and petals similar in shape, lanceolate, to 5 × 2.5 mm; lip ovate to obovate, 3 × 2 mm. **Distr:** Endemic to Philippines. **Ecol:** In shaded areas of forests; 300–900 m. **Zones:** 11.

D. glumaceum Lindl., *Edwards's Bot. Reg.* 27(Misc.): 23, 1841. (Syn: *Acoridium glumaceum*, *Platyclinis glumacea*). [Hay-scented Orchid]. **Herbs**, epiphytic, upright, sympodial; pseudobulbs in dense, spreading masses, conical to ovoid, to 4 × 1.5 cm, clothed with large red sheathing scales, unifoliate. **Leaf** narrowly elliptic or broadly lanceolate, to 45 × 4 cm, plicate, base tapered into a petiole, apex obtuse, acute or acuminate, striated. **Inflorescence** arching, appearing with new growth, to 50 cm long; bracts lanceolate, convolute, white, apically tawny. **Flowers** 2 cm across, petals and sepals white, lip color variable, including yellow, orange and brown; sepals and petals spreading, apices acuminate; dorsal sepal linear-oblong, ~ 8 × 2 mm, apex tapered to a sharp point; lateral sepals linear-oblong, to 8 × 2 mm, tapered to a sharp point; petals lanceolate, to 6 × 2 mm; lip trilobed, lateral lobes small, midlobe oblong-ovate; column short, compressed, apex winged. **Distr:** Borneo to Philippines (Luzon, Mindanao, Visayas). **Ecol:** On trees, lithophytic on rocks; 500–2,300 m. **Zones:** 11–12.

D. javieriense Magrath, Bulmer and I.Shafer, *Lindleyana* 4: 135, 1989. **Herbs**, epiphytic, tufted; rhizome abbreviated; pseudobulbs clustered, terete-fusiform, 2–4.3 × 0.3–0.6 cm; cataphylls ~4, ± tubular, acute to acuminate, soon disintegrating into persistent fibers. **Leaves** linear, ~23 × 0.7–0.9 cm, base gradually attenuate and somewhat channeled, apex acute, rigid, nerves 5 distinct. **Inflorescence** densely ~15–many-flowered; peduncle 17.5–21.5 cm long; rachis 8.2–10 cm long; floral bracts broadly triangular-ovate, 5.4–6 × 5.9–6.1 mm, entire, acute; flowers opening from bottom of rachis; pedicel and ovary conical, 1.8–2.4 mm long. **Flowers** small, greenish-yellow to bright sulfur-yellow, glabrous, sepals and petals widely spreading; dorsal sepal oblong, 6.5–7.3 × 2.7–3 mm, entire, rounded-obtuse, trinerved; lateral sepals oblong, 6.5–8 × 3–3.3 mm, rounded-obtuse, 3–4-nerved; petals trulliform-elliptic to obovate, 4.8–5.5 × 2.5–3 mm, rounded-obtuse, trinerved. lip porrect, 1–1.2 × 2.2–2.4 mm, sessile, trilobed, obscurely trinerved, lateral lobes obliquely subspathulate, rounded-obtuse, midlobe very small, broadly triangular, acute; column subterete, 0.8–1 mm long; pollinia 4. **Distr:** Philippines (Luzon). **Ecol:** In

Top row: *Dendrochilum cobbianum, Dendrochilum convallariiforme, Dendrochilum glumaceum, Dendrochilum javieriense.*
Bottom row: *Dendrochilum smithianum, Dendrochilum tenellum, Dendrochilum uncatum, Dendrochilum wenzelii.*

MS MG

Top row: *Pholidota chinensis, Pholidota ventricosa, Pleione chunii.*

upper montane forests; 1,830–2,440 m. Fl. Oct–Nov. **Zone:** 11.

D. smithianum (Ames) L.O. Williams, *Philipp. J. Sci.* 80: 311, 1952. (Syn: *Acoridium smithianum*). **Herbs**, epiphytic, upright, sympodial; pseudobulbs closely set, elliptic to cylindrical, to 4 cm long, 1.2 cm Ø, tapered, unifoliate. **Leaf** linear to lanceolate, to 12.5 × 2.3 cm, leathery. **Inflorescence** arching, to 30-flowered; flowers start opening at the proximal end; bract ovate, glumaceous. **Flowers** ~4 mm across, sepals and petals white to cream, lip pink; dorsal sepal linear to lanceolate, to 4 × 1 mm; lateral sepals lanceolate, to 4 × 1 mm; petals lanceolate, to 3 × 1 mm, margins jagged, particularly near the apex; lip trilobed, lateral lobes pointed, longer than midlobe, rounded. **Distr:** Endemic to Philippines (Luzon). **Ecol:** ~500 m. **Zones:** 11–12.

D. tenellum (Nees and Meyen) Ames, *Philipp. J. Sci.*, C 2: 318, 1907. (Syn: *Acoridium sphacelatum, A. tenellum, D. junceum, D. sphacelatum*). [Grass Orchid]. **Herbs**, epiphytic, upright, sympodial; pseudobulbs 2 cm long, 2 mm Ø, tapered. **Leaves** grasslike, to 30 cm × 1 mm. **Inflorescence** semiarching, ~5 cm long, borne about ¾ of the way along the leaf, appearing with new growth, to 30-flowered. **Flowers** very small, ~3 mm Ø, cream-colored, lip orange to red; dorsal sepal oblong, to 2 × 1 mm; lateral sepals oblong-lanceolate, to 2 × 1 mm; lip trilobed, lateral lobes slightly longer than midlobe. **Distr:** Endemic to Philippines; widespread. **Ecol:** 1,000–1,500 m. **Zones:** 11–12.

D. uncatum H.G. Reichenbach, *Bonplandia* 3: 222, 1855. (Syn: *Platyclinis formosana*, etc.). [Golden Chain Orchid]. **Herbs**, epiphytic; pseudobulbs ± clustered, narrowly ellipsoid or narrowly ovoid, 2–4 × 1–1.5 cm, usually shiny. **Leaf** narrowly elliptic to linear-oblong, 8–15 × 1.5–3 cm, base cuneate, apex acuminate or acute; petiole 1–4 cm long. **Inflorescence** 20–30-flowered, 10–15 cm long, flowers distichously arranged, borne from rhizome near base of last pseudobulb, arching or nodding, slender; rachis often slightly flexuose; floral bracts ovate, 3–4 mm long; pedicel and ovary ~3 mm long. **Flowers** yellowish; sepals narrowly elliptic or oblong, 6–7 × 2–4 mm; petals oblong or elliptic, 5–6 × 2–3 mm; lip lyrate, ~5 × 3 mm, shortly clawed at base, trilobed; lateral lobes lunate, very small, midlobe obovate, ~3.3 × 2.5 mm; disk with 2 short longitudinal ridges near base; column ~2 mm long. **Fruits** globose, <1 cm Ø. **Distr:** Philippines, S Taiwan. **Ecol:** On trees in broad-leaved forests or thickets; 500–1,000 m. **Zones:** 11.

D. wenzelii Ames, *Orchidaceae* 5: 65, 1915. (Syn: *Acoridium wenzelii*). **Herbs**, epiphytic, upright, sympodial, pseudobulbs clustered; pseudobulb, narrow, ~2.5 cm long, 5 mm Ø, tapered, unifoliate. **Leaf** narrowly linear, to 30 × 0.5–0.7 cm, leathery. **Inflorescence** semiarching, flowers crowded, to 30-flowered, appearing with new growth. **Flowers** ~1 cm across, variable in color, greenish-yellow, orange, brown or dull red; dorsal sepal oblong, to 5 × 1.5 mm; lateral sepals broadly lanceolate, to 5 × 2 mm; petals oblong-lanceolate, to 3 × 1 mm; lip trilobed, midlobe forming a saclike structure. **Distr:** Endemic to Philippines (Luzon, Mindoro, Visayas); distribution limited. **Ecol:** In forests; 500–1,000+ m. **Zones:** 10–11.

Pholidota

Species 42. **Herbs**, epiphytic; rhizomes creeping; pseudobulbs close or well-spaced on rhizome, usually ovoid, apex with 1 or 2 leaves. **Leaves** shortly petiolate. **Inflorescence** racemes usually with flowers distichously arranged; peduncle arising from apex of rhizome, growing simultaneously with rudimentary pseudobulb below, slender; rachis often flexuous; bracts concave, enclosing pedicel and ovary; persistent or soon deciduous. **Flowers** small; sepals concave, lateral ones often keeled abaxially; petals similar to sepals; lip sessile on base of column, erect, concave, usually unlobed, saccate at base; column, stout, base with short foot, apex hooded over anther; anther terminal, incumbent, 2-celled; pollinia 4, 2 superposed pairs, waxy; stigma entire, often emergent. **Distr:** Tropical and subtropical Asia to SW Pacific. **Cult:** Propagate by division. Requires bark-based compost in intermediate conditions with partial shade, and plenty of water during growth phase.

P. chinensis Lindley, *J. Hort. Soc. London.* 2: 308, 1847. (Syn: *Coelogyne chinensis*, etc.). [Rattlesnake Orchid]. **Herbs**, epiphytic or lithophytic; pseudobulbs (3–)5–15 cm apart, narrowly ovoid-oblong, usually 1.6–8 × 0.5–2.3 cm, apex 2-leaved. **Leaves** obovate-elliptic, oblanceolate-elliptic, or suboblong, 5–22 × 2–6 cm, apex acuminate, acute, or shortly caudate; petiole 0.5–5 cm long. **Inflorescence** several to 20-flowered, basal, 12–38 cm long; rachis often arching; floral bracts persistent, oblong to broadly ovate, 1–1.7 × 0.6–0.8 cm, often ± conduplicate; pedicel and ovary 4–8 mm long. **Flowers** white or yellowish-tinged; dorsal sepal elliptic or ovate-elliptic, cymbiform, 0.7–1 × 0.45–0.6 cm; lateral sepals ovate-lanceolate, conspicuously carinate; petals lanceolate, 0.9–1 × 0.1–0.25 cm; lip broadly ovate, slightly trilobed, basal half saccate, lateral lobes suborbicular, midlobe ovate, 4–5 × 4–5 mm, mucronate; column 4–5 mm long, winged above middle. **Fruits** obovoid-ellipsoid, 1.5–3 × 1–1.6 cm, 6-ridged, 3 ridges narrowly winged. **Distr:** China, Myanmar, Vietnam. **Ecol:** In forests; 900–2,100 m. Fl. Apr–May. Fr. Sep–Jan. **Zones:** 9–11.

P. ventricosa (Blume) Rchb.f., *Bonplandia* (Hannover) 5: 43, 1857. (Syn: *Chelonanthera ventricosa, Coelogyne ventricosa*, etc.). **Herbs**, epiphytic, upright, sympodial, large-sized; pseudobulbs elliptic, to 15 × 3 cm, tapered. **Leaves** apical; lamina narrowly lanceolate, plicate, to 50 × 10 cm, base elongate, petiolate, apex acute, leathery, midvein distinct, 5-nerved. **Inflorescences** upright, to 50 cm long, to 60-flowered, slightly compressed, appearing with the new growth; floral bract large, at the back of each flower and does not prevent the flower from opening fully. **Flowers** pale yellow, sometimes with orange markings on the lip, 1.5 cm across; dorsal sepal lanceolate, to 8 × 5 mm; lateral sepals ovate, to 8 × 4 mm; petals linear, to 1 × 0.3 cm, strongly reflexed; lip to 1 × 0.6 cm, divided into 2 lobes. **Distr:** Vietnam, Malesia to New Guinea. **Ecol:** Epiphytic and terrestrial on mountains; 500–2,300 m. **Zones:** 11–12.

Pleione

Syn: *Gomphostylis*.

Species 21. **Herbs**, epiphytic or lithophytic herbs; pseudobulbs tufted, usually ovoid, apex unifoliate. **Leaves** produced from lower part of old pseudobulbs, growing with flower buds and matured after anthesis with new pseudobulb below, membranous or subcoriaceous, narrow, often plicate, articulate. **Inflorescences** with flowers terminal, 1 or 2 borne on a short peduncle; bracts rather large, deciduous. **Flowers** large and showy; sepals subequal, spreading, usually elongate-oblong; petals similar to sepals but slightly narrower; lip sessile at base of column, convolute near base forming tube, apex obscurely trilobed, disk with keels; column slender, erect, apex winged or hooded over anther; anther terminal, nearly erect on ventral side of column, 2-celled; pollinia 4, connected in 2 pairs; stigma entire. **Distr:** Himalayas to C China and Indo-China. **Cult:** Propagate by division of dormant pseudobulbs. Requires a rich, well-drained, terrestrial mix, in a cool climate.

P. chunii C.L. Tso, *Sunyatsenia* 1: 148, 1933. (Syn: *P. milanii*, etc.). **Herbs**, terrestrial or lithophytic; pseudobulbs usually clustered in large colonies, green or pale green, pyriform to conic, 2.5–4.5 × ~1.2 cm, apex 1-leafed. **Leaf** elliptic-lanceolate or narrowly elliptic, 6–20 × 2–3(–4.6) cm, apex acute. **Inflorescence** appearing with young leaf, erect, 1(2)-flowered; peduncle 5–7 cm, enclosed in tubular sheaths at base; floral bracts green, narrowly elliptic-oblanceolate, ~ 2.5 × 1–1.5 cm; pedicel and ovary 1.6–2 cm long. **Flowers** large, pale pink to rose-purple, often paler toward base, lip white with a longitudinal yellow or orange-yellow stripe at center and brown-red spotted toward apex; dorsal sepal narrowly elliptic or oblong-elliptic, 2.4–3.5 × 0.7–1 cm, subacute; lateral sepals obliquely elliptic, slightly shorter and wider, subacute; petals spreading, oblanceolate or spathulate, 2.4–3.3 × 0.5–0.9 cm, rounded or obtuse; lip broadly flabellate, 2–2.5 × 2.5–3 cm, obscurely trilobed, lateral margins erect, embracing column, apical margin dentate or irregularly erose; disk with 6 rows of long papillae; column 2–2.3 cm long, winged. **Distr:** S China. **Ecol:** In forests; 1,400–2,800 m. Fl. Apr–May. **Zones:** 9–10.

P. formosana Hayata, *J. Coll. Sci. Imp. Univ. Tokyo* 30(1): 326, 1911. (Syn: *P. pricei*, etc.). **Herbs**, lithophytic, epiphytic, or terrestrial; pseudobulb green or dark purple, compressed-ovoid or ovoid, 1.3–4 × 1.7–3.7 cm, unifoliate. **Leaf** elliptic or oblanceolate, 10–30 × 3–7 cm, base attenuate into a petiolelike stalk 3–4 cm, apex acute or obtuse, papery. **Inflorescence** erect, 1(2)-flowered; peduncle 7–16 cm, with 2 or 3 membranous sheaths at base; floral bracts linear-lanceolate to narrowly elliptic, 2.2–4 cm, acute; pedicel and ovary 1.5–2.7 cm long. **Flowers** sometimes slightly fragrant, rose-pink or occasionally white, lip often yellowish with reddish or brownish marks; dorsal sepal narrowly elliptic-oblanceolate or spathulate-oblanceolate, 4.2–5.7 × 0.9–1.5 cm, acute; lateral sepals narrowly elliptic-oblanceolate, slightly oblique, 4–5.5 × 1–1.5 cm, acute or subacute; petals linear-oblanceolate, 4.2–6 × 1–1.5 cm, subacute; lip broadly ovate-elliptic to orbicular when flattened, 4–5.5 × 3–4.6 cm, obscurely trilobed, apical margin lacerate, emarginate; disk with 2–5 entire or erose lamellae; column 2.8–4.2 cm long, obscurely toothed. **Fruit** fusiform, ~4 cm long. **Distr:** SE China, N and C Taiwan. **Ecol:** In montane forests; 600–2,500 m. Fl. Mar–Apr. **Zones:** 9–11.

P. grandiflora (Rolfe) Rolfe, *Orchid Rev.* 11: 291, 1903. (Syn: *Coelogyne grandiflora*, *P. pinkepankii*). **Herbs**, lithophytic; pseudobulb green, subconic, 3–4.5 × 1.2–1.5 cm, enclosed by fibrous remains of basal sheaths, unifoliate. **Leaf** lanceolate, papery, base attenuate into a petiolelike stalk, apex acute. **Inflorescence** erect, 1-flowered; peduncle 10–15 cm, covered by 3 or 4 membranous sheaths at base; floral bracts narrowly elliptic, 2.3–4 cm, exceeding ovary, apex obtuse. **Flower** large, white, sometimes with dark purplish-red or brown spots and streaks on lip; dorsal sepal oblanceolate, 5–5.5 × 1–1.5 cm, rounded; lateral sepals narrowly elliptic, slightly oblique, 4.8–5.5 × 1–1.5 cm, obtuse; petals oblanceolate, falcate, 5–5.5 × 0.8–1 cm, rounded; lip broadly ovate-elliptic to flabellate, 5–5.5 × 3–4 cm, obscurely trilobed, base narrowed into a claw ~1 cm, upper part of lateral margins and apical margin coarsely lacerate, apex emarginate; disk with 4 or 5(–7) irregularly lacerate lamellae; column arcuate, 3.8–4 cm, apex not toothed, column wings very narrow. **Distr:** China (S Yunnan) to NW Vietnam. **Ecol:** In forests; 2,600–2,900 m. Fl. May. **Zones:** 9–11.

P. hookeriana (Lindley) J. Moore, *Orch. Grow.-Man.* (ed. 6): 548, 1885. (Syn: *Coelogyne hookeriana*, etc.). **Herbs**, epiphytic or terrestrial; pseudobulb green or purple, ovoid to conic, 1–2 × 0.5–1 cm, unifoliate. **Leaf** elliptic-lanceolate or suboblong, 6–10 × 2–2.8 cm, base attenuate into petiolelike stalk 2–3 cm long, apex acute, papery. **Inflorescence** erect, 1-flowered; peduncle 6–10 cm long, with several membranous sheaths; floral bract suboblong, 1–1.7 × 0.4–0.5 cm, obtuse; pedicel and ovary 1–2 cm long. **Flower** small; sepals and petals pale purplish-red to nearly white, lip white with a yellow center and purple or yellowish-brown spots; dorsal sepal suboblong or oblanceolate, 2–3.5(–4.5) × 0.6–1 cm, acute; lateral sepals similar, lanceolate, falcate, slightly oblique; petals oblanceolate, 2–3.5 × 0.5–0.7 cm, acute; lip oblate or subcordate, 2.5–4 × 2.7–4.5 cm, obscurely trilobed, apical margin irregularly denticulate or almost entire, apex emarginate; disk with 7 rows of papillae extending along central veins; column arcuate, 1.5–2.6(–3) cm long, winged. **Fruits** suboblong. **Distr:** Bhutan, China, NE India, Laos, Myanmar, Nepal, Thailand. **Ecol:** On mossy rocks and cliffs at thicket margins; 1,600–3,100 m. Fl. Apr–Jun. Fr. Sep. **Zones:** 8–9.

P. praecox (Smith) D. Don, *Prodr. Fl. Nepal.* 37, 1825. (Syn: *Epidendrum praecox*, etc.). [Early Blooming Pleione]. **Herbs**, epiphytic or terrestrial; pseudobulb green, purplish- mottled, turbinate, 1.5–4 × 1–2.3 cm, (1)2-leaved. **Leaves** elliptic-oblanceolate to elliptic, 9–20 × 1.7–6.7 cm, base petiolelike, 2–6.5 cm long, apex acuminate, papery. **Inflorescence** 1(2)-flowered, appearing after leaves have withered, erect; peduncle 5–10 cm long, sheaths 3, papillate; floral bract oblong-oblanceolate, 2.5–3 × 1.5–2 cm, acute. **Flower** large, pink to purplish-red, rarely white, lip with a yellow callus, occasionally with purple spots; dorsal sepal suboblong-lanceolate, 5.5–6 × 0.7–0.8 cm, acute; lateral sepals slightly oblique, acute; petals linear-lanceolate, slightly falcate, 5–7 × 0.4–0.6 cm, acute; lip obovate-elliptic or elliptic, 4–5 × ~3 cm, obscurely trilobed, lateral lobes inconspicuous, midlobe lacerate on apical margin, apex emarginate; disk with 3–5 rows of papillae; column ± arcuate, 3.5–4.5 cm long, irregularly toothed. **Distr:** WC Himalayas to China (S Yunnan). **Ecol:** In forests, cliffs; 1,200–2,500(–3,400) m. Fl. Sep–Oct. **Zones:** 9–10.

Plocoglottis

Species 40. **Herbs**, terrestrial, medium-sized; rhizome short, creeping; pseudobulbs usually clustered, erect, cylindrical-tapering, apex unifoliate. **Leaf** erect, plicate, long-petiolate. **Inflorescence** a raceme, laxly few-flowered, basal, erect, unbranched; bracts small, persistent, flowers borne in succession. **Flowers** yellow, marked with red or purple; dorsal sepal and petals subsimilar but the petals narrower, free, spreading; lateral sepals adnate to base of column foot, otherwise free and spreading; lip convex, entire fleshy, short, joined to the sides and tip of column foot to form a short sac at base, attached by an elastic joint so that the lip springs up when triggered; column rather short, with a short foot; pollinia 4, waxy. **Distr:** Indo-China to Papuasia. **Cult:** Propagate from seeds or by division. Requires a slightly acidic, humus-rich soil in warm, humid conditions, in partial shade.

P. mindorensis Ames, *Philipp. J. Sci.*, C 2: 327, 1907. (Syn: *P. mcgregorii*, *P. wenzelii*). **Herbs**, terrestrial, rhizomatous, 60–80+ cm tall, very slender, not much thickened at base. long. **Leaves** narrowly oblong-lanceolate, ~ 30 × 2.5–5 cm, apex acuminate, very acute, tapering at both ends,

Top row: *Pleione formosana*, *Pleione formosana* (albino form), *Pleione hookeriana*.
Bottom row: *Pleione grandiflora*, *Pleione praecox*.

Above: *Plocoglottis mindorensis, Plocoglottis plicata, Thunia alba, Thunia alba* var. *bracteata.*

smooth; petioles ~20 cm long. **Inflorescence** a loose raceme, flowers numerous; scape ~80 cm long, pubescent, densely so near apex, clothed at intervals with tubular, lanceolate, pointed sheaths; bract rather rigid, triangular-lanceolate, ~6 mm long, acute, somewhat concave, pubescent; pedicels slender, pubescent, ~1 cm long. **Flowers** purplish, sepals externally pubescent; dorsal sepal oblong-lanceolate, 1.3 × ~0.5 cm long; laterals sepals falcate-lanceolate, ~1.2 cm long, 6 mm wide near middle; petals linear, falcate, 1.1 × 0.2–0.25 cm, obtuse; lip cuneate-quadrate, ~6 long, 9 mm wide at apex, in the middle of the anterior margin is a deflexed, triangular, 1 mm long tooth. **Distr:** Philippines. **Ecol:** In humid forests; 200–300 m. **Zones:** 11–12.

P. plicata (Roxb.) Ormerod, *Orchids*: 293, 2001. (Syn: *Dendrobium javanicum*, etc.). **Herbs**, terrestrial, medium-sized to large; rhizome stout, creeping; pseudobulbs cylindrical, 5–8 × 0.3–0.7 cm, several-noded, apex unifoliate. **Leaf** erect; lamina elliptic to ovate, 20–40 × 5.5–10 cm, base petiolate, apex acuminate; petiole slender, 5–14 cm long. **Inflorescence** laxly to densely few- to many-flowered, erect, lateral, borne from the lower nodes of the pseudobulb, to 50 cm long; peduncle and rachis terete, densely pubescent; bracts lanceolate, 0.6–1 cm long, acuminate; pedicel and ovary, 1.2–1.7 cm long, shortly pubescent. **Flowers** yellow with red spots on sepals and petals; sepals spreading, lanceolate, 1.8–2.2 × 0.3–0.4 cm, acuminate; lateral sepals slightly falcate and back-to-back; petals slightly oblique, lanceolate, 1.5–1.7 × 0.3 cm, acuminate; lip strongly deflexed, reflexed, recurved, convex, subquadrate, 0.7–1 cm long, shortly apiculate; column erect, 4–5 mm long. **Distr:** Malesia. **Ecol:** In rainforests; <500 m. **Zones:** 11–12.

Thunia

Species 5. **Herbs**, sympodial, terrestrial, sometimes lithophytic or epiphytic; stems erect, slightly nodose, clustered, erect or spreading, canelike, cylindrical, biennial, lower parts covered with persistent sheaths, upper part with sheathing leaf bases. **Leaves** distichous, sessile and sheathing below, deciduous; lamina thin-textured, glaucous. **Inflorescences** racemose, few-flowered, terminal on young leafy shoots, flowers produced in succession; floral bracts lanceolate, spathaceous, persistent. **Flowers** short-lived, large, showy and scented, not opening widely, glabrous; sepals and petals free, subsimilar; sepals lanceolate; lip trumpet-shaped, enclosing column, entire or obscurely trilobed, spurred at base, callus of several papillose ridges in apical half, apex frilled; column clavate, apex winged, foot absent; anther 2-celled, pollinia 4, clavate, deeply bilobed; stigma ventral, semicircular; rostellum a trilobed flap; ovary cylindrical. **Distr:** Indian Subcontinent to SC China and peninsular Malaysia. **Cult:** Canes should be potted in a terrestrial mix, watered and fertilized during the growth period. After flowering, plants should be kept dry in a cool, bright place.

T. alba (Lindl.) Rchb. f., *Bot. Zeitung* (Berlin) 10: 764, 1852. (Syn: *Phaius albus, T. marshalliana*). **Herbs**, 0.3–1.5 m tall. **Leaves** lanceolate or elliptic, 10–20 cm long, apex acuminate, thin-textured, pale green. **Inflorescence** to 30 cm long, nodding, 2–7-flowered; bracts boat-shaped, to 5 cm long, white. **Flowers** to 12.5 cm across, often not opening wide, fragrant; sepals and petals white, lanceolate to ovate, 6.5–7.5 × 1.5–3.3 cm, acuminate; lip orange or yellow with purple stripes, broadly elliptic to oblong, tubular at base, 5–6 × 4.5 cm, margin irregular and fringed, ridges branching, hairy; spur 1 cm long, shallowly bilobed, cylindrical, horizontal; column 2–2.5 cm long. **Distr:** Himalayas to SC China and peninsular Malaysia. **Ecol:** In forks of trees in submontane forests; 1,000–2,300 m. **Zones:** 9–10. **Info:** *T. alba* var. *bracteata* similar to typical species, except habit more robust, growing to 60 cm tall, flowers 4–10, each 7.5 cm across, lip <4 cm wide, oblong to ovate, ridges orange or purple, distributed in N India, Burma and Thailand.

TRIBE COLLABIEAE

Acanthephippium

Syn: *Acanthophippium*.

Species 13. **Herbs**, terrestrial; rhizome short; pseudobulbs crowded, long, clavate or conical, with 3–9 leaves near apex, nodes 1–3(–6). **Leaves** large, convolute, plicate, short-petioled, articulate. **Inflorescences** few-flowered, short and fleshy racemes arising from the base or the middle nodes of the pseudobulbs. **Flowers** perianth not widely spreading; pedicel thickened after anthesis; sepals cohering near base, tips recurved, free, lateral sepals much broader, adnate to the long curved foot of the column, connate at inner sides forming a large, urceolate, saccate mentum; petals narrow, erect, as long as the sepals, included in sepal tube, inserted on column foot, more or less spathulate; lip very small, enclosed in mentum, stipitate at apex of column foot, mobile, strongly trilobed, lateral lobes erect, midlobe recurved, disk lamellate with a few longitudinal ridges; column with a long recurved foot; anther terminal, incumbent, mainly 2-celled, 4-chambered; pollinia 8, waxy. **Distr:** Sri Lanka, the Himalaya, Indo-China to the Ryukyus and the Philippines, southward to Papua New Guinea and the Pacific Islands. **Cult:** Propagate from seeds or by division of pseudobulbs, and they do not like to be repotted often. Requires a moist, well drained terrestrial mix with medium bark and perlite. Water regularly during growing season and reduce watering during flowering season.

A. bicolor Lindl., *Bot. Reg.* 20: pl. 1730, 1835. **Herbs** large, terrestrial; pseudobulbs crowded, ovoid or fusiform, 5–14 cm long. **Leaves** 2 or 3, lanceolate to oval, 18–45 × 4–13.5 cm, base narrowed to sheathing petiole, apex acuminate, acute, abaxially strongly veined. **Inflorescences** racemes, 3–7-flowered,

Below: *Acanthephippium bicolor, Acanthephippium mantinianum, Acanthephippium pictum.*

MG

Above: *Ancistrochilus rothschildianus, Calanthe arcuata, Calanthe arisanensis, Calanthe aristulifera.*

apical; sheaths 6, imbricating, scaly; peduncle ~3 cm long, erect, with a sterile bract; floral bracts cymbiform, 3.5 × 3.5 cm; pedicel and ovary 4 cm long. **Flowers** large, yellow; petals and sepals 4.5 × 3.5 cm, apices purplish-red, fused to form a pitcherlike calyx-tube, gibbously inflated, base intruded, mouth oblique; dorsal sepal 3.2 × 1.1 cm, oblong, obtuse, fused with lateral sepals; lateral sepals subsimilar to dorsal sepal, fused along inner margins forming the base of the pitcher, apex obtuse and recurved; petals free, erect, fleshy, 3.1 × 1.5 cm, spathulate, obtuse; lip small, trilobed, 1.5 × 2 cm, jointed to a long, inflexed, recurved foot 4.5 cm long, lateral lobes erect, rounded, midlobe recurved, oblong-spathulate, obtuse; disk trilamellate; column 1.3 × 0.55 cm, semicylindrical; pollinia 8. **Distr:** Sri Lanka, South India (Pulney Hills). **Ecol:** In tropical shady jungles and wet evergreen forests; <610 m. Fl. Jan, Mar–May, Jul, Oct. **Zones:** 10–12.

A. mantinianum, L. Linden and Cogn., *J. Orchidées* 7: 138, 1896. (Syn: *Acanthophippium mantinianum*). **Herbs** terrestrial; pseudobulbs have a dark purplish and dark green colorations, translucent, ovoid, declining toward the apex with 3–5 fleshy leaves. **Leaves** deciduous; lamina plicate, oblong-elliptical to lanceolate, base tapering to a petiole. **Inflorescences** racemes, to 5-flowered, 25 cm long, borne from the new lateral growth. **Flowers** very fragrant, fleshy, 5 cm long, sepals yellow slightly shaded with green, spotted and blotched with purple, petals similar to sepals but with the lower portion whiter, lip waxy-white, thickened part of disk orangish-yellow with ridges spotted and lined with purple; dorsal sepal concave, linear, to 3.3 × 1.5 cm, unequally bilobed; lateral sepals obliquely oblong, basal 2/3 fused to form an obscurely bilobed mentum to 3.5 × 1.6 cm, apex slightly recurved; lip trilobed, lateral lobes erect, enclosing column, midlobe 2.9 × 1.1 cm, rounded, hinged halfway down its length. **Distr:** Philippines. **Ecol:** In shaded, mountainous regions; 500–1,500 m. Fl. Apr–May. **Zones:** 11–12. **Info:** This species is very closely allied to *A. bicolor*, but differing in its much more robust habit and flower color.

A. pictum Fukuy., *Bot. Mag. Tokyo* 40: 666, 1935. (Syn: *A. sylhetense* var. *pictum*). **Herbs**, terrestrial; pseudobulb fleshy, elongate-clavate, ~13 × 2.2–4 cm. **Leaves** usually 3; lamina elliptic, ~20–50 × 7–10 cm, lower leaf usually smaller, mainly 5–9-veined. petiole ~14 cm long. **Inflorescence** a raceme, 3–5-flowered, dense; bract ovate, ~3 × 2.5 cm; scape lateral from the new year's growth, basal, ~18 cm long; pedicel and ovary ~4 cm long. **Flowers** whitish-yellow, flushed with dark purple on upper part of inner side, with yellow or red veins apically outside, many-veined; sepals united into a swollen tube; dorsal sepals ~3.5 × 1.3 cm; lateral sepals ~4 × 2 cm; petals oblanceolate, ~3 × 1.5 cm; lip trilobed, with a very narrow claw ~2.2 × 0.6 cm, lateral lobes ~5 mm long, with 2 irregular ridges and 3 keels, midlobe tongue-shaped, ~1 cm long; column ~7 mm long, foot ~1.3 cm long, base adnate into a sac with lip and lateral sepals, pollinia 8. **Fruits** ellipsoid, ridged. **Distr:** Taiwan, Lanyu island and the Ryukyus, Japan. **Ecol:** Fl. Jun–Aug. **Zones:** 8–10.

Ancistrochilus

Species 2. **Herbs**, erect, semiterrestrial or epiphytic; pseudobulbs depressed, orbicular, onion-shaped. **Leaves** 2, deciduous; lamina lanceolate, acute, plicate. **Inflorescences** erect, 2–3-flowered, basal. **Flowers** with sepals subequal, spreading, lateral ones partially united at base to the very short column foot; petals similar, smaller and narrower than sepals; lip attached to short column foot, base subsaccate, strongly trilobed, lateral lobes erect, oblong, obtuse, midlobe basally narrowly triangular, long-attenuate above, recurved; disk with slightly raised lines, middle one more prominent; column clavate, winged, wings base extended into a short horizontal foot; pollinia 8. **Distr:** Nigeria and Cameroon to Uganda and Tanzania. **Ecol:** In tropical forests; 500–1,400 m. Fl. Aug–Oct. **Cult:** Propagate by division; repot when new growth appears. Requires high humidity; reduce watering after flowering. Best grown on slabs or in small baskets in a free-draining organic substrate. **Info:** Vegetatively the plants look similar to *Pleione*.

A. rothschildianus O'Brien, *Gard. Chron.* 41: 51, 1907. (Syn: *A. hirsutissimus*, etc.). **Herbs** erect, semiterrestrial or epiphytic; pseudobulbs conical or pear-shaped, to ~5 cm Ø. **Leaves** in pairs; elliptic-lanceolate, 10–40 × 2–7.5 cm, thin-textured, arising from the top of the pseudobulbs. **Inflorescences** 1(2), to 22 cm tall, 2–5-flowered. **Flowers** to 9 cm across; sepals and petals rose pink or lilac, lip deep magenta or purple; dorsal sepal oblanceolate, 2–3 × 1 cm, obtuse; lateral sepals oblanceolate, 2–3 × 0.9 cm, obtuse; petals oblanceolate,2–3 cm × 0.6 cm, obtuse; lip trilobed, 1.5 cm long, midlobe narrowly triangular, 1 × 0.4 cm, recurved, lateral lobes erect, rounded, greenish or brownish; column long, slender, slightly curved. **Distr:** Tropical West Africa, from Guinea and Sierra Leone to Uganda. **Ecol:** On large branches in evergreen forests; 500–1,100 m. Fl. Oct–Nov. **Zones:** 11–12.

Calanthe

Species over 150. **Herbs**, terrestrial, perennial; rhizomes distinct or obscure; pseudobulbs small, globose to clavate, usually covered by base of leaves, rarely large and conspicuous. **Leaves** arising from pseudobulbs, usually broad, rarely linear, thin and plicate, base narrowed into a long or short petiole or sheath. **Inflorescence** a raceme, arising from leaf axils or from base of pseudobulb, densely or loosely flowered; bracts small and persistent or large and deciduous. **Flowers** with wide spreading or connivent sepals and petals; sepals subequal, spreading or connivent; petals broad or narrow; lip short-clawed, base connate with column, forming a tube, rarely free and adnate to column base, spurred or spurless, limb usually trilobed, rarely unlobed; disk often ridged or with callose protuberances at base; column short or long, obliquely truncate, joined to base of lip for varying length or rarely free from lip; anther terminal, incumbent, conical or caplike, usually cordate, 2-celled; pollinia 8, clavate. **Fruits** oblong, often pendulous. **Distr:** Tropical and subtropical Old World to Pacific, Mexico to Colombia, Caribbean. **Cult:** Propagate by division. Requires a moisture-retentive but well-drained terrestrial compost, in partial shade.

C. arcuata Rolfe, *Bull. Misc. Inform. Kew* 1896: 196, 1896. (Syn: *C. arcuata* var. *brevifolia*, etc.). **Herbs**, terrestrial, 18–45 cm tall; pseudobulbs conic, ~1 cm Ø, sheaths 2 or 3. **Leaves** 3 or 4, basal; lamina narrowly elliptic-lanceolate or narrowly lanceolate, 15–28 × 0.7–3 cm, base petiolelike, sheathing, often forming a pseudostem 2–3 cm long, apex acute or acuminate. **Inflorescences** 1 or 2, axillary, erect, 30–50 cm, laxly ~10-flowered; rachis 10–25 cm long; floral bracts persistent, herbaceous, narrowly lanceolate, 1–1.8 cm long, glabrous; pedicel and ovary arcuate, 1.4–2 cm long, densely puberulent. **Flowers** white to yellowish-green, flushed reddish-brown on outer surfaces; sepals similar, narrowly lanceolate, 1.7–2.2 × ~0.4 cm, acuminate; petals linear, 1.7–2.2 × 0.2–0.3 cm, acuminate; lip adnate to column wings, white, apex purple-tinged, usually turning yellow later, 1.1–1.8 × 0.4–0.6 cm, trilobed (rarely unlobed), lateral lobes ovate-triangular or suboblong, ~7 mm long, apical margin sometimes toothed, midlobe elliptic-rhombic, 0.8–1 × 0.6–0.7 cm, base usually clawed, margin undulate, irregularly toothed; spur cylindric, ~5 mm long; column 4–5 mm long, thick, wings triangular; pollinia narrowly ovoid. **Fruits** subellipsoid, ~2 × 0.8 cm. **Distr:** C and S China, Taiwan. **Ecol:** In forests; 1,400–3,100 m. Fl. May–Sep. **Zones:** 7–11.

C. arisanensis Hayata, *J. Coll. Sci. Univ. Tokyo.* 30: 327, 1911. (Syn: *C. sasakii*). **Herbs**, 30–50 cm tall; pseudobulbs tufted, globose-ovoid, 1.5 cm long, with 2 or 3 nodes. **Leaves** 2 or 3, lanceolate, 30–40 × 4–7 cm, base cuneate, gradually narrowed into petiole, apex acuminate; petiole 5–15 cm long. **Inflorescences** 40–50 cm long, green or purple-tinged, rachis 5–10 cm long, laxly 5–10-flowered; bracts lanceolate, 1–2 cm long; pedicel with ovary 3–4 cm long, 6-ridged. **Flowers** white or purplish-tinged, widely opening; dorsal sepal lanceolate-ovate, 2–2.5 × 0.9–1.1 cm, acuminate; lateral sepals similar but longer and oblique, 2.5–2.7 × 1 cm, acute, base strongly contracted; petals linear-lanceolate, 1.8–2 ×

0.4 cm; lip adnate to column base, spurred, orbicular, 1.3–1.5 × 1.8–2 cm, trilobed, lateral lobes falcate, ovate, entire or slightly undulate, midlobe orbicular, 9 mm across; column 7–9 mm long. **Fruits** to 5 cm long, with sharp winglike ridges. **Distr:** Taiwan. **Ecol:** In forests; 1,000–2,000 m. **Zones:** 10–11.

C. aristulifera Rchb. f., *Bot. Zeitung* (Berlin) 36: 74, 1878. (Syn: *Alismorkis aristulifera*, etc.). **Herbs**, terrestrial, 28–55 cm tall; pseudobulbs subglobose, ~1 cm Ø, with 3 sheaths. **Leaves** 2–4, basal; lamina obovate-elliptic or elliptic, 15–30 × 4–8 cm, papery, abaxially densely puberulent, apex acute; petiolelike base (6–)27–30 cm, usually forming a pseudostem 13–20 cm long. **Inflorescences** 1–2, axillary, laxly to subdensely 8–20-flowered; scape 25–60 cm long, puberulent; rachis 6–25 cm long; floral bracts persistent, narrowly lanceolate, ~5 mm long, glabrous, acute; pedicel and ovary 1.5–2 cm long. **Flowers** white or pink, occasionally purple-flushed; dorsal sepal oblong-lanceolate, 1.2–1.7 × 0.5–0.8 cm, acuminate; lateral sepals oblong, oblique, 1.2–1.7 × 0.4–0.7 cm, acute; petals narrowly obovate or elliptic, 1.1–1.5 × 0.25–0.45 cm, subacute, glabrous, 3-veined; lip flabellate, 0.8–1.6 × 0.8–1.2 cm, trilobed, lateral lobes subsquare to suborbicular, falcate, 5–6 × 4–6 mm, obtuse-rounded, midlobe subsquare to oblate, 3–4 × 4–5 mm, emarginate; spur cylindric, 2–3.2 cm long; column ~6 mm long. **Distr:** S China, Japan, Taiwan. **Ecol:** In moist dense forests; 1,500–2,500 m. Fl. Feb–May. **Zones:** 9–11.

C. biloba Lindley, *Fol. Orchid.* 6: 3, 1855. (Syn: *Alismorkis biloba*, etc.). **Herbs**, terrestrial, to 60 cm tall; pseudobulbs cylindric, ~10 cm long. **Leaves** 2 or 3; lamina broadly elliptic, 12–17 × 7–8 cm, base petiolelike, sheathing, ~8 cm long, apex mucronate, papery. **Inflorescence** laxly >10-flowered, axillary, erect, ~50 cm long, densely puberulent, with 2 tubular sheaths below middle; floral bracts persistent, narrowly lanceolate, 1–1.7 cm long, acuminate, glabrous; pedicel and ovary 2–2.5 cm long, densely puberulent. **Flowers** large, purplish, lip yellowish; dorsal sepal broadly ovate, ~1.5 × 0.6–0.7 cm, aristate, abaxially puberulent, 5-veined; lateral sepals ovate, oblique, ~1.5 × 0.7–0.8 cm, aristate, abaxially densely puberulent, 5-veined; petals oblong, ~1.5 × 0.45 cm, aristate, glabrous, 3-veined; lip subreniform, ~1.5 × 2 cm, clawed, deeply bilobed, lobes subrhombic, margin erose, apex obtuse; disk without lamellae; spur narrowly conic, ~2 mm long, slightly obtuse; column ~5 mm long, thick, apex slightly dilated, glabrous. **Distr:** Nepal to Thailand. **Ecol:** In thickets on streamsides; ~1,800 m. Fl. Oct. **Zones:** 9–11.

C. ceciliae Rchb. f., *Gard. Chron.*, n.s., 1883: 432, 1883. (Syn: *Alismorkis wrayi*, etc.). **Herbs**, terrestrial, clustered. **Leaves** 4–5, lanceolate to elliptic, to 85 × 6–16 cm, apex acuminate, dark green; petiole to 12 cm long. **Inflorescence** 0.4–1 m tall, laxly to subdensely to 30-flowered; peduncle and rachis shortly pubescent; bracts lanceolate, to 1.5 cm long, acuminate. **Flowers** ~3 cm across, white to lilac, turning brownish-orange with age; sepals spreading, ovate lanceolate, 1.2–1.6 × 0.4–1 cm, acuminate; petals similar but slightly smaller; lip trilobed, 1.3–2 × 1.3–2.3 cm, lateral lobes spreading, ovate, midlobe deeply bilobulate, each lobule broader than lateral lobes, callus of 2 short verrucose ridges; spur straight, ascending, slender, cylindrical, 2.5–5 cm long; column 5 mm long. **Distr:** Myanmar to Java. **Ecol:** In lowland and montane forests; 400–2,000 m. **Zones:** 10–12.

C. davidii Franch., *Nouv. Arch. Mus. Hist. Nat. Paris*, s. 2. 10: 85, 1888. (Syn: *Alismorkis pachystalyx*, etc.). **Herbs**, 30–50 cm tall; pseudobulbs clavate, 2–4 cm long, ~1 cm thick, with 2–4 nodes. **Leaves** 6–10, nearly distichous, overlapping pseudobulbs at their bases; lamina linear, 20–60 × 1–4 cm, membranous. **Inflorescences** with numerous flowers, crowded; scapes erect, slender, 40–80 cm tall, terete; peduncle shortly pubescent; rachis ~30 cm long; bracts linear-lanceolate, 7–15 × 3–4 mm; ovary and pedicel 0.8–1.5 cm long, shortly pubescent. **Flowers** pale green or whitish, not resupinate; sepals and petals reflexed; sepals obovate or oblong, 7–9 × 3–5 mm, acute or obtuse; lateral sepals similar but often slightly wider; petals linear-spathulate, 7–9 × 2–3 mm, base cuneate, apex acute; lip white, spurred, broadly obtriangular, 0.5–0.9 × 1.1–1.5 cm, trilobed, lateral lobes falcate-oblong, margins entire or irregularly undulate, apex obtuse, midlobe obtriangular, deeply bilobed. **Distr:** India, the Himalaya, southern China, Japan, Taiwan. **Ecol:** In forests; 1,200–2,000 m. **Zones:** 8–11.

C. graciliflora Hayata, *J. Coll. Sci. Imp. Univ. Tokyo* 30(1): 329, 1911. (Syn: *C. graciliflora* var. *xuefengensis*, *C. hamata*). **Herbs**, terrestrial, to 70 cm tall; pseudobulbs subovoid, short, ~2 cm Ø, sheaths 3 or 4. **Leaves** not deciduous; lamina elliptic or elliptic-lanceolate, ~33 × 5.5–10 cm, base narrowed, apex acute; pseudopetiole ~10 cm long, forming a pseudostem 5–18 cm long. **Inflorescence** laxly many-flowered, axillary, to 70 cm tall, densely puberulent; rachis to 32 cm long; floral bracts persistent, lanceolate, 0.7–1 cm long, glabrous; pedicel and ovary 1.5–2 cm long, densely puberulent. **Flowers** yellowish-brown, lip white; sepals reflexed, similar, subelliptic, 1.2–1.3 × 0.5–0.6 cm, acute, glabrous, (3- or)4- or 5-veined; lateral sepals slightly narrower; petals obovate-lanceolate, 0.9–1.3 × 0.3–0.4 cm, glabrous, 3- or 4-veined, base shortly clawed, apex acute; lip adnate to column wings, ~1 × 1 cm, trilobed, lateral lobes obovate-cuneate, slightly oblique, ~4 mm long, obtuse-rounded or obliquely truncate, midlobe subquadrate, obovate, or transversely oblong, ~4 × 3 mm, dilated, emarginate and mucronate in sinus; disk with 4 brown spots and 3 parallel ridges; spur curved or straight, cylindric, 1–1.8 cm long, attenuate, inside densely puberulent; column ~4 mm long, glabrous. **Distr:** S China, N Taiwan. **Ecol:** In forests; 600–2,400 m. **Zones:** 8–11.

C. herbacea Lindley, *Fol. Orchid.* 6–7(Calanthe): 10, 1855. (Syn: *Alismorkis erytroglossa*, etc.). **Herbs**, terrestrial, 30–70 cm tall; pseudobulbs nearly narrowly ovoid or cylindric, 1–2.5 × 0.5–1.5 cm, with 2 sheaths. **Leaves** 3 or 4, lamina elliptic or elliptic-lanceolate, 15–30 × ~9 cm, base petiolelike 10–20 cm long, apex acute or acuminate. **Inflorescence** laxly ~10-flowered, axillary, erect, to 70 cm long; rachis 8–15 cm long; floral bracts persistent, lanceolate, ~1 cm long; pedicel and ovary ~3 cm long, puberulent. **Flowers** yellowish-green; dorsal sepal reflexed, obovate-lanceolate, ~1.5 × 0.7 cm, apex obtuse and mucronate, abaxially puberulent, 5-veined; lateral sepals reflexed, elliptic, oblique, ~1.5 × 0.6 cm, base clawed, acute; petals reflexed, subspathulate, ~1.2 × 0.2–0.25 cm, obtuse, glabrous; lip adnate to column wings, deeply trilobed, lateral lobes ovate, 6–7 × ~5 mm, obliquely truncate, midlobe bilobuled, lobules divergent; spur slightly arcuate, yellowish-green, 2–3 cm long; column white, ~7 mm long. **Distr:** Sikkim to N Taiwan. **Ecol:** In dense forests; 1,500–2,100 m. Fl. Jun–Aug. **Zones:** 8–11.

C. rosea (Lindl.) Benth., *J. Linn. Soc.*, Bot. 18: 309, 1880. (Syn: *Alismorkis rosea*, *Limatodis rosea*). **Herbs**, terrestrial, sometimes epiphytic, deciduous; pseudobulbs elliptic to fusiform, 10–20 cm long, acutely keeled and grooved. **Leaves** usually produced after flowering; lamina oblong- or elliptic-lanceolate, plicate, 20–45 cm long, apex acuminate, glabrous. **Inflorescence** a raceme, laxly 7–15-flowered, basal, erect at first, then arching, 30–45(–75) cm long, slender, pubescent; bracts ovate-lanceolate, 1.2–2.5 cm long, green, persistent; pedicel and ovary 3.8 cm long. **Flowers** large, ~4 cm across, rosy; sepals ovate-lanceolate; petals oblong, acute; lip ~4 cm long, base yellow edged with scarlet, lateral lobes narrowly convolute, midlobe spreading, large, obovate-oblong, entire; spur slender, shorter than the sepals and lip, villous; column very short, pubescent. **Distr:** Myanmar to Thailand. **Ecol:** ~100 m. **Zones:** 10–11.

C. sieboldii Decaisne ex Regel, *Ind. Sem. Hort. Petersb.* 80, 1868. (Syn: *C. discolor* var. *flava*, etc.). **Herbs**, 40–60 cm tall; rhizomes obscure; pseudobulbs closely aggregated, nearly globose, ~3 cm Ø, with 3 or 4 nodes. **Leaves** 2–4, long-petiolate, oblong, 20–30 × 7–15 cm, base cuneate, apex acute or obtuse; petiole ~20 cm long. **Inflorescences** 3–10-flowered, borne with young leaves and new pseudobulbs, ~30 cm long, flower bearing portion ~10 cm long; bracts lanceolate, 0.5–1 cm long; pedicel and ovary 1.5–2 cm long, shortly pubescent. **Flowers** golden-yellow, ± fragrant, 4–5 cm Ø; sepals subequal, elliptic-oblong, 2.5–3 × 1–1.2 cm, base slightly contracted, apex acute, glabrous inside, outside minutely hairy, 5-nerved; petals similar to sepals but smaller, 2.4–2.6 × 0.7–0.8 cm, base strongly contracted, apex acute, trinerved; lip orbicular, 2.5–3 cm across, spurred, trilobed, lateral lobes falcate, nearly entire, midlobe oblong-obovate or nearly rectangular, margins undulate, apex obtuse and shortly aristate, disk with 2 raised ridges at base and 3 slender ridges toward apex, spur ~7 mm long; column 6–7 mm long. **Distr:** S Japan, Ryukyu Islands; Taiwan. **Ecol:** In forests; ~1,500 m. **Zones:** 8–11.

C. tricarinata Lindley, *Gen. Sp. Orchid. Pl.* 252, 1833. (Syn: *Alismorkis tricarinata*, etc.). **Herbs**, terrestrial, 35–50 cm tall; pseudobulbs globose, ~2 cm Ø, sheaths 3. **Leaves** 3 or 4, basal; lamina elliptic or obovate-lanceolate, 20–30 × 5–11 cm, base petiole-like, sheathing, forming a stout pseudostem 4–15 cm long, 1–2.5 cm Ø, margins undulate, apex acute or acuminate, abaxially densely puberulent; sheaths large. **Inflorescence** axillary, erect, to 60 cm long; bracts sterile, several to many, ovate-lanceolate; rachis 3–20 cm, laxly few- to many-flowered; floral bracts persistent, ovate-lanceolate, 0.5–1 cm long; pedicel and ovary 1–2 cm long. **Flowers** pale yellow, lip reddish-brown; sepals similar, oblong-lanceolate, 1.6–1.8 × 0.5–0.8 cm, acuminate, base abaxially sparsely puberulent, 5–8-veined; petals obovate-lanceolate, 1.1–1.5 × 0.3–0.5 cm, acute or subobtuse, 3-veined; lip adnate to base of column wings, reddish-brown, 1–1.6 × 0.7–1.5 cm, spurless, trilobed; lateral lobes auriculate or nearly suborbicular, small, ~ 4 × 4–5 mm, midlobe reniform, 0.8–1 × 1–1.5 cm, margin strongly undulate; column ~6 mm long, thick. **Distr:** N Pakistan to temperate E Asia. **Ecol:** On grassy slopes, forests; 1,300–3,500 m. Fl. May–Jun. **Zones:** 6–10.

C. triplicata (Willem.) Ames, *Philipp. J. Sci. Bot.* 2: 326, 1907. (Syn: *C. brevicolumna*, etc.). [Black Orchid, Christmas Lily, Scrub Lily, White Crane]. **Herbs**, terrestrial, 50–60 cm tall; rhizomes indistinct; pseudobulbs clavate, 2–4 cm long, covered by leaf bases, nodes 3–5. **Leaves** 3–6, fasciculate, petioled; lamina oblong or oblong-lanceolate, 22–35 × 5–14 cm, base narrow, apex acute; petiole 8–14 cm long. **Inflorescences** ± corymbose; scapes 40–60 cm long, with 2 or 3 scales below; bracts large, ovate-lanceolate, 1.5–2 cm long; pedicel and ovary slender, 3–4 cm long. **Flowers** white; sepals and petals spreading; sepals obovate-oblong, 1.2–1.6 × 0.5–0.6 cm, acute or apiculate; lateral sepals wider and ± oblique; petals obovate-spathulate, 1.2 × 0.3–0.5 cm, base contracted, apex acute; lip 1–1.5 cm long, spurred, trilobed, lateral lobes linear-falcate or elongate-rectangular, 5–7 mm long, midlobe bilobuled, spur 1.2–2 cm long; column 5 mm long; pollinia yellowish. **Distr:** W Indian Ocean, tropical and subtropical Asia to Pacific. **Ecol:** In forests; <1,500 m. **Zones:** 9–12.

Top row: *Calanthe biloba, Calanthe ceciliae, Calanthe davidii.*
Middle row: *Calanthe graciliflora, Calanthe herbacea, Calanthe rosea.*
Bottom row: *Calanthe sieboldii, Calanthe tricarinata, Calanthe triplicat, Calanthe vestita.*

C. vestita Wall. ex Lindl., *Gen. Sp. Orchid. Pl.* 250, 1833. (Syn: *C. regnieri, C. vestita* var. *regnieri*). **Herbs**; pseudobulbs elongate, ovoid, bluntly angulate, constricted near middle, 5–12.5 cm long. **Leaves** appearing after the flowers; lamina broadly lanceolate, 30–45 cm long, tapering gradually to the acute apex, glabrous. **Inflorescences** racemes, loose, cernuous, villous, 30–45 cm long; peduncles erect, stout below, very hairy, basal; bracts large, ovate, 2–4 cm long; pedicels and ovary 4 cm long. **Flowers** rosy-white or greenish; sepals and petals subequal, elliptic; sepals ovate-lanceolate, 2.5–4 cm long, aristate, 5-nerverd; petals narrow to broadly oblong, apex, obtuse or retuse, apiculate, 5-nerved; lip variable, exceeding the sepals, large, flat, conspicuously 4-lobed, with a slender greenish spur at the base, lateral lobes cuneately obovate (sometimes absent), midlobe flabellately deeply obcordate, sinus acute; spur very long, slender, involute; column fleshy and usually deeply carmine-stained. **Distr:** Borneo. **Ecol:** Terrestrial on limestone in forests or hill forests; 600–1,000 m. Fl. Oct–Feb, May–Jul. **Zones:** 11–12.

Gastrorchis

Species 8. **Herbs**, terrestrial, rarely epiphytic; pseudobulbs small, 2- to 4-leaved. **Leaves** petiolate, thin-textured, plicate. **Inflorescence** arising from base of pseudobulb, unbranched. **Flowers** showy, turning blue-black if damaged; sepals and petals free, similar, usually ovate, often spreading; lip obscurely trilobed,

GS

Above: *Gastrorchis humblotii* var. *schlechteri, Nephelaphyllum pulchrum, Phaius tankervilleae, Spathoglottis kimballiana.*

base saccate, callus keeled; pollinia 8, in 2 groups of 4. **Distr:** Madagascar. **Cult:** Propagate from seeds. Requires a humus-rich, well-drained compost in shade. **Cult:** Propagate by division. Requires a humus-rich, free-draining compost, in intermediate to warm conditions under shade. Needs plenty of water during the growth period. **Info:** *Gastrorchis* differs from the closely related genus *Phaius* in having a saccate rather than a spurred lip and a raised, often hairy callus.

G. humblotii var. *schlechteri* (H. Perrier) Senghas ex Bosser and P.J.Cribb, *Orchids Madagascar*: 145, 1999. (Syn: *G. humblotii* var. *schlechteri*, etc.). **Herbs**, terrestrial, 60–80 cm tall; pseudobulbs ± globose. **Leaves** broadly ovate-lanceolate, 30–48 × 10–16 cm, base and apex acute, old leaves ± persistent; petiole 5–14 cm long. **Inflorescence** 18–20-flowered, basal, 60–90 cm tall; peduncle 25–40 cm long; sheaths persistent; floral bracts white, broadly ovate, apiculate. **Flowers** 6–8 cm across; sepals and petals broadly ovate, ~3 × 2.5 cm, apiculate, pure white, 7-veined; lip broad, 4 × 3.5 cm, distinctly trilobed, lobes dotted with bright red markings, margins fringed, midlobe transverse, 1.1 × 3.5 cm, much larger, edges corrugated or pleated, apex reddish becoming white in the middle, callus yellow forming a prominent 'V'; column 1.5 cm long; ovary ~3 cm long. **Distr:** Madagascar. **Ecol:** In humid evergreen forests and moss forests; 1,000–2,000 m. **Zones:** 10–12.

Nephelaphyllum

Syn: *Cytheris.*

Species 11. **Herbs**, terrestrial; rhizome creeping, fleshy, covered with thin sheaths; pseudobulbs erect, slender, cylindrical, unifoliate. **Leaf** erect or suberect, shortly petiolate; lamina ovate to cordate, fleshy, adaxially mottled dark and lighter green, abaxially often purple. **Inflorescence** lateral, erect, densely or laxly few-flowered, shorter than or slightly longer than the leaf. **Flowers** non-resupinate, medium-sized; sepals and petals often recurved or spreading, subsimilar, free, slender, acute; lip large uppermost in the flower, entire or trilobed, with a callus and with a short basal spur; column short, slightly winged; pollinia 8. **Distr:** E Himalayas to W and C Malesia. **Cult:** Propagate by division. Best grown in a shallow pan. Requires a free-draining compost (with minimal watering) containing bark and sphagnum moss, in shade.

N. pulchrum Blume, *Bijdr.*: 373, 1825. (Syn: *Tainia pulchra*). **Herbs**, terrestrial, rhizomatous, small; rhizome creeping, fleshy; pseudobulbs closely spaced, cylindric, unifoliate, 1–2 × 0.2–0.4 cm. **Leaf** erect; lamina ovate-triangular to cordate, 4–10 × 2.5–6 cm, apex acute or obtuse, adaxially mottled dark and light green, flushed with purple. **Inflorescence** densely 2–3-flowered, 3–5 cm long, elongating after flowering to 10 cm long; bracts lanceolate, 0.6–1 cm long, acuminate. **Flowers** erect, non-resupinate, sepals and petals pinkish, lip white or pale yellow, pinkish at base with a yellow callus; sepals and petals subsimilar, recurved, lanceolate or oblanceolate, 1.2–1.4 × 0.3–0.4 cm, acute; lip fleshy, oblong-elliptic, 1.5–1.7 × 1–1.2 cm, margin entire, apex emarginate, callus of 3 papillose ridges; spur clavate-orbicular, bilobed, 3–4 mm long; column 5 mm long; ovary 5–8 mm long. **Distr:** E Himalayas to Philippines. **Ecol:** In moist lowland and lower montane forests; 300–1,500 m. **Zones:** 9–12.

Phaius

Syn: *Gastrorchis, Phajus, Tankervillia*, etc.

Species 48. **Herbs**, terrestrial (rarely epiphytic); pseudobulbs or stems leafy, sometimes both pseudobulbs and stems present. **Leaves** arising from pseudobulbs or spaced along the stem with tubular sheaths enclosing the stems, petiolate; lamina plicate. **Inflorescence** a raceme, 1–few-flowered, arising from base of pseudobulb or lower part of stem. **Flowers** often large, showy; sepals and petals subequal, free, ± spreading; lip erect, spurred or saccate, upper surface with longitudinal ridges, joined to base of column or rarely fused with base of column, base narrow, apex enlarged, folded and ± embracing column, shallowly or indistinctly trilobed toward apex, lateral lobes erect or incurved, midlobe usually retuse, crispate; disk often keeled; column long, slender, dilated, apex ± winged, usually free from lip or rarely connate with base of lip; pollinia 8, waxy. **Distr:** Tropical Africa to Pacific. **Ecol:** In shady parts of moist forests. **Cult:** Propagate from seeds or by division. Requires a well-drained medium consisting of sand, bark and peat moss, in shaded light.

P. tankervilleae (Banks ex L'Hér.) Blume, *Ann. Mus. Bot. Lugduno-Batavum.* 2: 177, 1856. (Syn: *Limodorum tankervilleae, P. grandifolius*, etc.). [Nun Orchid, Nun's Hood Orchid, Swamp Orchid]. **Herbs**, terrestrial, highly variable; pseudobulbs ovoid or globose, ~2 cm across. **Leaves** 3–5, petiolate; lamina lanceolate, 50–80 × 10–15 cm, strongly plicate, acute. **Inflorescence** a raceme, 10–20-flowered, erect, stout, to 1 m tall, with many scalelike sheaths toward base; bracts large, elliptic, 4.5 cm long, deciduous. **Flowers** strongly fragrant (musky), whitish outside, pink inside, 8 cm Ø; sepals spreading, lanceolate or oblong, 4–5 × 1.5–2 cm, base contracted, apex acute; petals oblanceolate, as long as sepals but narrower; lip embracing column, 5 cm long, obtriangular, 4–5 cm wide, apex trilobed, undulate, villose inside, lateral lobes rounded, midlobe truncate, emarginate; spur 1 cm long; column ~2.5 cm long, apex dilated, narrowed downward, abaxially glabrous. **Distr:** Tropical and subtropical Asia to S Pacific. **Ecol:** In broad-leaved forests at low elevations. **Zones:** 10–12.

Spathoglottis

Species ~47. **Herbs**, terrestrial; pseudobulbs. globose to ovoid, sometimes depressed, each bearing a few leaves. **Leaves** long petiolate, elongate, not articulated; lamina narrow, plicate. **Inflorescences** racemes, borne from a basal leaf axil, tall and slender, many-flowered. **Flowers** with sepals and petals widely spreading, nearly equal; lip sessile, strongly trilobed, lateral lobes narrow, erect, midlobe clawed at base, dilated or widening and sometimes cleft at apex; disk with 2 callose protuberances near base; column slender, slightly curved, slightly enlarged toward apex, without foot at base; pollinia 8, waxy, in 2 groups of 4. **Distr:** Tropical and subtropical Asia to Pacific. **Cult:** Propagate by division or by seeds. Requires moist, fertile soil rich in humus or coco peat.

S. kimballiana Hook. f., *Bot. Mag.* 121: t. 7443, 1895. [Yellow Ground Orchid, Garden Orchid]. **Herbs**, terrestrial; pseudobulbs ovoid, 1–2 × 1–2 cm, enclosed in sheaths. **Cataphylls** 1–2, ~4–16 cm long, acuminate, imbricate. **Leaves** narrowly linear-lanceolate, ~27–70 × (1.5–)2–5 cm, apex long-acuminate, plicate, 5–7-nerved; petiole ~10–20 cm long, canaliculate. **Inflorescence** laxly 6–12-flowered, basal; peduncle ~30–80 cm long; non-floriferous bracts 8–12, ovate, 1.5–2.3 cm long, obtuse to acute, sheathing, imbricate; rachis ~5–9 cm long; floral bracts broadly ovate-elliptic, 1–1.5 cm long, obtuse, olive-green, red-stained; pedicel and ovary 2.5–4 cm long. **Flowers** 5–7 cm Ø, sepals bright yellow, mottled and splashed rose-red outside, petals bright yellow, lip similar, with dark red streaks on lower half of side lobes, calli and proximal portion of midlobe, column bright yellow, sepals and petals shortly hairy at base inside; dorsal sepal oblong-elliptic, 2.6–3.3 × 1.3–1.5 cm, obtuse; lateral sepals ovate-elliptic, 2.5–2.9 × 1.2–1.5 cm, obtuse; petals elliptic, 2.9–3 × 1.6–1.8 cm, obtuse; lip 2.4–2.5 cm long, lateral lobes 1.5 × 1.2 cm, midlobe 1.6–1.8 × 0.7–0.8 cm, calli 2, 4–5 mm long, fleshy, glabrous; column 2–2.3 × ~0.6 cm, curved. **Distr:** Borneo (Sabah), Philippines. **Ecol:** Among vegetation near riverbanks; to 1,500 m. Fl. Jan, Mar–Jun. **Zones:** 11–12.

S. plicata Blume, *Bijdr. Fl. Ned. Ind.* 6: pl. 76; 8: 401, 1825. [Boat Orchid, Ground Orchid, Philippine Orchid]. **Herbs**, to ~1 m tall; pseudobulbs ovoid or conic, ~3 cm long. **Leaves** 3–5, nearly distichous; lamina linear-lanceolate, 30–80 × 5–7 cm, base cuneate, apex acuminate or acute, plicate; petiole 10–20 cm long. **Inflorescence** ~10-flowered, opening successively; peduncle purplish, 50–80 cm long, scales many, tubular; bracts lanceolate, ~1.5 cm long; pedicel 1–2 cm long; ovary cylindrical, 1.5 cm long. **Flowers** pink to purple; sepals ovate, 1.5–1.8 × 0.8–1 cm, acute, 5–7-nerved, keeled outside; petals broadly obovate, 1.8 × 1.2–1.4 cm, base contracted, apex obtuse or rounded; lip 1.6–2.3 cm long, trilobed, lateral lobes falcate or elongate rectangular, 6–8 × 2 mm, erect below lower part of column, apex truncate, central lobe 1.3–1.5 cm long, lower part long-clawed, apical part dilated and flabellate, apex truncate, often

MS

Above: *Spathoglottis plicata, Spathoglottis tomentosa, Spathoglottis unguiculata.*

emarginate or bilobed, with 2 fleshy hirsute calli on disk near base; column 1.5 cm long, curved, apex slightly dilated. **Fruits** cylindric, to 10 cm long. **Distr:** Sri Lanka, India, Indo-China, the Philippines, Malay Peninsula, Maldives, Indonesia, Papua New Guinea and the Caroline Islands. **Ecol:** Sunny places at forests margins. **Zones:** 11–12.

S. tomentosa Lindl., *Edwards's Bot. Reg.* 31: t. 19, 1845. **Herbs**, terrestrial; stems tomentose. **Leaves** broadly lanceolate. **Inflorescence** a raceme, to 20-flowered, tomentose, multiflowered; scape tomentose; floral bracts 1.2 cm long, obtuse, concave, rigid, tomentose, yellowish-green; pedicel and ovary 3.2 cm long, tomentose, green. **Flowers** with sepals vey pale pink, tips green, petals pale pink (darker than sepals) with midline darker, lip pink, basal half yellow with orangish-brown blotches; sepals similar, 2 × 0.9 cm, obtuse or mucronate; petals oblong, 2 × 1.2 cm, base clawed, narrowed to ~2 mm wide, apex obtuse; lip 1.95–2.5 cm long, lateral lobes erect, falcate, to 1.2 × 0.6 cm projected forward to the same level as the column, midlobe ~1.95 cm long, mid part narrowed, 1.25 cm × 1.5 mm, apex widened to 1 cm across (somewhat reniform), into 2 lobes 4 mm long, apices rounded; column elongate, 1 cm long, curved, apex slightly dilated. **Distr:** Philippines (Mindanao). **Ecol:** 200–430 m. **Zones:** 11–12.

S. unguiculata (Labill.) Rchb. f., *Fl. Vit.*: 300, 1868. (Syn: *Limodorum unguiculatum*, etc.). **Herbs**, terrestrial; pseudobulbs ovoid, 4–6 cm tall, ~2.5–3 cm Ø at base. **Leaves** lanceolate, plicate, 40–60 × 2–2.5 cm, base attenuate into a petiole, apex acuminate. **Inflorescence** with 10- or more-flowered; scape with 4 sheaths at intervals; bracts broadly ovate, ~ 1.7 × 1.1 cm; pedicel and ovary, slender, ~3–4 cm long, ± hairy. **Flowers** rose to dark purple, hairy on outside of sepals, fragrant; dorsal sepal oblong-ovate, concave, ~2 × 0.9 cm; lateral sepals ovate, concave, slightly keeled, ~1.9 × 1 cm; petals broadly ovate, acute, margin slightly undulate; lip nearly round, 1–6 × 1–4 cm, trilobed, side-lobes very broad, oblong, 8 × 5 mm, broadest at apex; midlobe with a broad transversely elliptic apical lamina 8 mm across and a short claw ~ 8 × 1.5–2 mm, calli 2, broadly ovoid, hairy; column ~1 cm long. **Fruits** cylindric, 3 cm long, hairy on the ridges, stalk 2 cm long, hairy. **Distr:** New Caledonia, Vanuatu. **Ecol:** In woods and groves; ~200 m. **Zones:** 11–12.

TRIBE PODOCHILAEAE

Ceratostylis

Species ~100. **Herbs**, epiphytic; rhizome creeping or pendulous; stems simple or branched, tufted, sometimes terete and rushlike, 1-leaved, with thin, brown, often reticulate basal sheaths, sometimes entirely covered by membranous, reddish-brown sheaths, pseudobulbs absent. **Leaves** borne at apex of stem or branches, usually linear, narrowly lanceolate, or subcylindric, leathery, fleshy or subterete, rarely thinly textured, base articulate. **Inflorescence** terminal, usually with several clustered flowers, rarely reduced to a solitary flower. **Flowers** resupinate, small, rarely large and showy; sepals similar, erect, connivent; lateral sepals adnate at base to column foot forming a saccate or spurlike mentum, enclosing basal part of lip; petals narrower than sepals; lip adnate to apex of column foot by a long incumbent claw, slightly or partly thickened, spurless, entire or inconspicuously trilobed, usually with longitudinal calli; column short, dilated above, apex bilobed or with 2 spathulate erect arms, foot long; anther terminal, incumbent; pollinia 8, in 2 groups, waxy, sessile. **Distr:** tropical Asia to New Guinea and the Pacific Islands. **Cult:** Propagate by division. Best grown mounted or in shallow baskets. Requires moderately shady, humid conditions, and plenty of water during growth period.

C. retisquama Rchb. f., *Bonplandia* (Hannover) 5: 53, 1857. **Herbs**, epiphytic, semipendulous, sympodial; stems branched to 40 cm long, covered with overlapping, brown papery bracts. **Leaves** semiterete, channeled and curved, to 12 × 1 cm, leathery. **Inflorescences** with flowers solitary (rarely paired), borne from the base of the plant; peduncles short. **Flowers** to 4 cm Ø, surface enamelled, bright reddish-orange or brick-red, lip white; dorsal sepal lanceolate, 2 × 0.6–0.9 cm; lateral sepals oblong-lanceolate, to 2 × 0.5 cm, acute; petals oblanceolate, to 2 × 0.5 cm, acute; lip relatively small, 3 × 1.5 mm, base shortly clawed, apex ovate below, broadly cuneate-obtuse above, 7 mm long; disk bicarinate with a basal pubescent cushion; column 4 mm long, lateral arms erect. **Distr:** Endemic to Philippines. **Ecol:** Epiphytic at low elevations; to 500 m. Fl. Several times a year. **Zones:** 11–12.

Eria

Species 367. **Herbs**, epiphytic, lithophytic, or rarely terrestrial; rhizome creeping; stems pseudobulbous, of one distinctly enlarged internode, ovoid, 2–4 leaves toward apex; base covered by leaf sheaths. **Leaves** convolute in bud, articulate; lamina elliptic to narrowly elliptic, leathery, tapering at base. **Inflorescence** axillary, erect, many-flowered, pubescent; sterile bracts 2 or 3, imbricate; axis with brown stellate hairs; floral bracts brown, narrowly ovate to triangular. **Flowers** opening widely, usually cream-colored to pale yellow, sometimes with purple veins or purple markings on column; sepals narrowly triangular, abaxially hairy; lateral sepals slightly ventrally broadened at base, apex recurved; petals similar to sepals; lip simple or trilobed, callus absent or adorned with ridges; column short, foot incurved; pollinia 8; ovary angular in cross-section, sometimes winged. **Distr:** Tropical and subtropical Asia to Pacific. **Cult:** Propagate by division. Requires a bark-based compost, in high humidity and partial shade.

E. coronaria (Lindl.) Rchb. f., *Ann. Bot. Syst.* 6: 271, 1861. (Syn: *Coelogyne coronaria*, etc.). **Herbs**, epiphytic; rhizome creeping, often with funnel-shaped sheaths 6–7 mm long; pseudobulbs cylindric, 5–15 cm

Below: *Ceratostylis retisquama, Eria coronaria, Eria excavata.* [All MG.]

Above: *Eria paniculata*, *Eria pannea*, *Eria thao*. [All MG.]

long. **Leaves** 2, subterminal, sessile; lamina narrowly elliptic or obovate-elliptic, rarely ovate-lanceolate, 6–16 × 1–4 cm, apex acute or obtuse. **Inflorescence** 2–6-flowered, borne between leaves, 10–30 cm long; floral bracts lanceolate or linear, 5–8 mm long; pedicel and ovary ~1.5 cm long. **Flowers** white, lip purple-striped; dorsal sepal elliptic-lanceolate, ~1.7 × 0.5 cm, obtuse; lateral sepals falcate-lanceolate, ~1.5 × 0.5 cm, obtuse; petals oblong-lanceolate, ~1.7 × 0.45 cm, obtuse; lip oblong, 1.4–1.5 × 1.1–1.2 cm, trilobed, lateral lobes divaricate, suborbicular or suboblong, midlobe triangular or subsquare, ~5 × 4 mm, acute or subtruncate; disk with 3 entire or undulate lamellae running from base to midlobe, with 2–4 additional crenate or undulate lamellae on midlobe; column ~5 mm long. **Fruits** obovoid-cylindric, ~2 cm long. **Distr:** Himalayas to Indo-China. **Ecol:** Fl. May–Jun. **Zones:** 9–11.

E. excavata Lindley, *Gen. Sp. Orchid. Pl.* 67. 1830. (Syn: *E. flava* var. *rubida*, etc.). **Herbs**, epiphytic; pseudobulbs ~1 cm apart, cylindric, ~3 cm, sheathing at base, apex 4- or 5-leaved. **Leaves** elliptic-oblanceolate, 16–18 × 2–3.4 cm, base contracted into petiole, apex acute. **Inflorescence** axillary, erect, 15–18 cm long, reddish-brown pubescent, sparsely few-flowered; floral bracts lanceolate, 7–8 mm long, apex acuminate, abaxially brown-pubescent; pedicel and ovary 7–8 mm long, brown-pubescent. **Flowers** white, sepals and petals brown-pubescent abaxially; dorsal sepal subelliptic, ~9 × 3 mm, acute; lateral sepals falcate-lanceolate, ~9 × 4 mm, acute; petals elliptic, ~6 × 2 mm, acute; lip suborbicular, ~5 mm across, concave at base, trilobed from near base, lateral lobes ovate-triangular, small, obtuse, each with an erect callus, midlobe subreniform, ~4 × 5 mm, apex emarginate, warty-rugulose, with variable calli, usually bicallose basally, sometimes with 5 flabellate lamellate, central one forming an apiculum; column ~1 mm long, foot incurved. **Fruits** cylindric, 1–1.5 cm long. **Distr:** Nepal to Assam. **Ecol:** In broad-leaved forests along valleys; 1,700–2,100 m. Fl. Jun. **Zones:** 10–11.

E. paniculata Lindl., *Pl. Asiat. Rar.* 1: 32, 1830. (Syn: *Callista floribunda*, etc.). **Herbs**, epiphytic; 20–60 cm tall; stems nearly clustered, erect, terete, base slightly swollen, cylindric, 15–50 cm long. **Leaves** many, alternate along stem, sessile; lamina linear or narrowly lanceolate, 10–20 × 0.3–0.6 cm, apex acuminate. **Inflorescences** 1 or 2, many-flowered, subterminal, 10–15 cm long, with dense grayish-white cottony hairs; floral bracts ovate-lanceolate or subtriangular, 5–6 mm long; pedicel and ovary 2–3 mm long, densely grayish-white cottony. **Flowers** yellowish-green, sepals densely grayish-white cottony abaxially; dorsal sepal ovate-elliptic, ~3 × 1 mm, obtuse; lateral sepals ± obliquely triangular, ~3 × 2.5 mm, obtuse; petals oblong, ~3 × 1.5 mm, glabrous, broadly obtuse; lip subflabellate, base shortly clawed, distally trilobed, lateral lobes subovate-triangular, midlobe ± reniform, ~1.5 × 2.5 mm; disk with a white, dumbbell-shaped projection; column <1 mm long, foot ~2 mm long. **Fruits** cylindric, ~1 cm long, with grayish-white cottony hairs. **Distr:** C Himalayas to SC China and Borneo. **Ecol:** On trees in forests; ~800 m. Fl. Apr–Jun. **Zones:** 9–12.

E. pannea Lindl., *Edwards's Bot. Reg.* 28(Misc.): 64, 1842. (Syn: *E. odoratissima*, *E. teretifolia*, *Pinalia pannea*, etc.). [Flag Eria]. **Herbs**, terrestrial, small, densely white-tomentose when young; rhizome creeping; stems 2–5 cm apart, terete, 1–2 cm long, 3 or 4 leaves near apex. **Leaves** terete, slightly laterally compressed, 4–20 × ~0.3 cm, apex obtuse fleshy, adaxially grooved, often white-hairy along margins of grooves. **Inflorescence** 1–5-flowered, arising from between leaves, 3–5 cm long; floral bracts ovate-triangular, ~6 × 4 mm, obtuse; pedicel and ovary 0.7–1 cm long. **Flowers** vanilla-scented, pale yellow-green, lip dark purplish; sepals abaxially densely hairy; dorsal sepal oblong-elliptic, ~6 × 3 mm, obtuse; lateral sepals obliquely ovate-triangular, ~6 × 5 mm, obtuse; petals oblong, ~5 × 2 mm, surfaces sparsely white-hairy, obtuse; lip subobovate-elliptic, ~7 × 4 mm, base narrowed and with a linear callus, margin entire, apex obtuse-rounded, apex with 1 conspicuous elongate elliptic callus, apically slightly fleshy, surfaces white-tomentose; column ~1.5 mm long, abaxially sparsely white-hairy, foot ~4 mm long. **Distr:** E Himalayas to S China and W Malesia. **Ecol:** On trees or lithophytic on rocks in forests; 800–2,200 m. Fl. Apr–May. **Zones:** 9–12.

E. thao Gagnep., *Bull. Mus. Natl. Hist. Nat.*, II, 22: 503, 1950. **Herbs**, epiphytic, rhizomatous; pseudobulbs 1–3 cm apart, ovoid or globose, ~1.2 cm Ø, enclosed in 2 sheaths, apex unifoliate. **Leaf** elliptic or oblong-elliptic, (3–)5–10 × (1–)1.5–2 cm, base attenuate into a petiole 1.5–2 cm long, apex obtuse, leathery. **Inflorescence** 1-flowered, terminal, ~2 cm long, with dense reddish-brown cottony hairs; floral bracts broadly triangular, ~1 mm long; pedicel and ovary ~5 mm long, with dense reddish-brown cottony hairs. **Flowers** yellow, lip yellowish-red, purple-tinged; sepals with dense reddish-brown cottony hairs abaxially; dorsal sepal lanceolate-oblong, ~1.7 × 0.6–0.8 cm, obtuse; lateral sepals triangular-ovate, ~2 × 0.6–0.9 cm, obtuse; petals elliptic, ~1.5 × 0.5 cm, obtuse; lip obovate-rhombic, ~1.5 × 1 cm, trilobed, lateral lobes subtriangular-ovate, ~4 × 3 mm, midlobe suboblong, ~1 × 0.6 cm, margin thickened; disk with 3 lamellae; column ~6 mm long, apex dilated, shortly winged, foot ~8 mm long. **Fruits** ellipsoid-cylindric, ~3.1 × 1 cm, with sparse, reddish-brown cottony hairs. **Distr:** S Hainan to Vietnam. **Ecol:** On trees or lithophytic on rocks in forests; 600–1,200 m. Fl. Aug–Oct. Fr. Dec–Feb. **Zones:** 9–11.

Trichotosia

Species 78. **Herbs**, epiphytic, rarely terrestrial; stems long or short, leafy throughout except at base, usually throughout with reddish-brown, rarely white, hispid hairs, sometimes hairs restricted to leaf sheaths and inflorescences. **Leaves** 2–3, distichous on the elongate stem; lamina lanceolate to obovate, fleshy, bristly hairy. **Inflorescences** lateral, from any node, piercing through leaf sheath, short and few-flowered, or long, pendulous and many-flowered; floral bracts at right angles to rachis, large, concave, hairy. **Flowers** resupinate, not opening widely, small to medium-sized; sepals abaxially red-hairy; lateral sepals adnate to column foot forming a mentum; lip entire to obscurely trilobed; disk with or without keels, sometimes papillose; column with a foot; pollinia 8. **Distr:** S China to tropical Asia and SW Pacific. **Cult:** Propagate by division. Smaller species are best grown mounted, while larger species are more suited for pots with an epiphytic mix. Requires moderate humidity and shade.

T. vulpina (Rchb. f.) Kraenzl., *Pflanzenr.*, IV, 50(50): 141, 1911. (Syn: *Eria vulpina*, *Pinalia vulpina*). **Herbs**, epiphytic, semipendulous, sympodial, large; stems canelike, to 1 m long. **Leaves** 2-ranked; lamina oblong-lanceolate, to 10 × 2 cm, acute, adaxially silky, abaxially with reddish-brown hairs; leaf sheaths abaxially covered with reddish-brown hairs. **Inflorescences** pendulous, hairy, borne from near stem ends, to 10 cm long, to 8-flowered; peduncle hairy, 10–20 cm long, zigzagged; bracts hairy. **Flowers** 1 cm Ø, back of dorsal and lateral sepals covered with short brown hairs, petals cream, lip white, spur dark red; dorsal sepal elliptic, to 5 × 3 mm; lateral sepals triangular, to 8 × 5 mm, joined to form a short spur; lip oblong, tapered to an angled tip. **Distr:** Philippines to Vanuatu. **Ecol:** Epiphytic on trees; 50–1,600 m. **Zones:** 11–12.

TRIBE EPIDENDREAE

Subtribe Laeliinae

Arpophyllum

Species 4. **Herbs**, epiphytic; rhizomes creeping; stems erect, apex 1-leafed, covered below the leaf with many sheaths. **Leaf** leathery, very long, folded together at base, but not closed into a sheath. **Inflorescence** a spike, long, dense, cylindrical, with numerous flowers; flowers pediceled or sessile; peduncle terminal with 1 or 2 sheaths above the leaf; bracts very small. **Flowers** with sepals of nearly equal length, free, spreading, laterals slightly broader than the dorsal, concave at base, adnate to foot of column; lip attached to foot of column, concave and somewhat saccate at base, above the swelling contracted and incumbent, then erect, margin crenulate above; column erect, slightly

Above: *Trichotosia vulpina, Arpophyllum giganteum, Brassavola nodosa, Broughtonia negrilensis.*

bent, produced at base into a short foot; clinandrium slightly dilated, rostellum shorter; pollinia 8, 4 per cell. **Fruits** ellipsoidal, 3-keeled, 3-ribbed. **Distr:** Mexico to Venezuela, Jamaica. **Cult:** Propagate by division. Requires a coarse, free-draining compost, in moderate conditions with high humidity and good lighting.

A. giganteum Hartw. ex Lindl., *Ann. Nat. Hist.* 4: 384, 1840. [Giant Bottlebrush Orchid]. **Herbs**, over 60 cm tall; stem 14–25 cm long, compressed, joints 3 or 4, basal sheaths, increasing above to 11 cm long. **Leaves** ligulate, adaxially flat, abaxially keeled, 40–50 × ~2–3 cm. **Inflorescence** a spike, 14–15 cm long; peduncle 10–15 cm long; basal spathe 9–13 cm long; sterile bracts triangular, 1.3–1.5 cm long, lowest clasping; floral bracts 1.5 mm long; flowers sessile. **Flowers** pale purple; perianth 5 mm long; dorsal sepal oblong, 5.5 × 2 mm, obtuse; lateral sepals ovate-oblong, 6 × ~3 mm, obtuse, slightly oblique; lateral petals oblanceolate-linear, ~5.5 × 1.6 mm, margin erose above, apex rounded; lip ~5.5 mm long, not divided, hooded and erose above; column 3.5–4 mm long. **Fruits** 7–8 mm long. **Distr:** Mexico to NW Venezuela. **Ecol:** In rainforests; 700–1,850 m. **Zones:** 10–11. **Info:** 3 subspecies described.

Brassavola

Syn: *Epidendrum, Lysimnia, Tulexis.*

Species 20. **Herbs**, epiphytic or lithophytic; rhizome creeping, woody; pseudobulb narrow, cylindrical, unifoliate (rarely bifoliate). **Leaves** apical; lamina terete to flattened, coriaceous. **Inflorescence** 1- to several-flowered, terminal or lateral, racemose. **Flowers** large, fairly showy; sepals and petals subsimilar, spreading, linear to linear-lanceolate; lip clawed, tubular around the column below, spreading above, margins fimbriate or entire; column erect, short, mostly 2-winged; anther operculate; pollinia 8. **Distr:** Mexico to tropical America. **Cult:** Propagate by division. Can be grown in pots, baskets or mounted on slabs. Requires a well-drained, coarse, epiphytic mix, with good lighting.

B. nodosa (L.) Lindl., *Gen. Sp. Orchid. Pl.*: 114, 1831. (Syn: *Bletia nodosa*, etc.). [Lady of the Night]. **Herbs**, epiphytic or lithophytic, to 45 cm tall; pseudobulbous stem short, terete, 4–15 cm long, covered by scarious tubular sheaths, unifoliate. **Leaf** apical; lamina linear, 32 × 2–3 cm, acute, adaxially sulcate, very fleshy-coriaceous. **Inflorescence** terminal, to 20 cm long, 1- to several-flowered; peduncle terete; bracts triangular-lanceolate, 0.5–1 cm long, acute, scarious. **Flowers** showy, 7–8 cm across, fragrant at night; sepals and petals pale green; lip white, purple-spotted at base; sepals linear, attenuate, 5–9.5 × 0.3–0.5 cm; petals similar to sepals; lip with a tubular claw, spreading, ovate-orbicular above, apiculate-acuminate at apex, ± conspicuously venose, 2.5–5.5 × 2.2–4.5 cm; column small, 8 mm long. **Distr:** Mexico to tropical America. **Ecol:** In dry, deciduous forests, 50–400 m. **Zones:** 11–12.

Broughtonia

Species 6. **Herbs**, epiphytic; pseudobulbs clustered, flattened, subcircular, apex 1- or 2-leaved. **Leaves** conduplicate, coriaceous. **Inflorescence** many-flowered, terminal. **Flowers** showy, medium-sized; sepals and petals free, spreading, subsimilar; petals slightly wider than the sepals; sepaline spur long, adnate to the ovary; lip entire or somewhat trilobed, tubular, convolute below, enclosing the column, pilose on the adaxial surface; column short; pollinia 4, flattened, attached to caudicles. **Distr:** Greater Antilles (Caribbean). **Cult:** Propagate by division. Best grown mounted on a slab, raft or in small baskets (with a free-draining substrate). Water generously during growth period.

B. negrilensis Fowlie, *Orchid Digest* 25: 418, 1961. (Syn: *B. domingensis* [misapplied], *Laeliopsis domingensis* [misapplied]). **Herbs**, epiphytic; pseudobulbs, tightly clustered, ovoid, flat, 2.5–6 × 2.8 cm. **Leaves** 1–3, apical; lamina to 12–15(–20) × 1.2–2.5 cm. **Inflorescence** 0.3–1 m long, borne between the leaves, 6–15(–27)-flowered, flowers clustered, opening up to 3 at a time. **Flowers** 2.5–5(–7.5) cm long, mauve-pink, not opening widely, long-lasting, without fragrance; sepals and petals veined pink-lavender; dorsal sepal ovate; lateral sepals lanceolate; lip tubular, heavily veined with reddish-purple, veins originating from middle to undulate margins, throat yellow; column white, with thick wings, without appendages; nectary bulging from the side of ovary. **Distr:** Endemic to Jamaica (Negril Hills). **Ecol:** Epiphytic on trees (often on *Ceiba* spp.), occasionally terrestrial below trees harboring orchids, or lithophytic on limestone, usually in coastal areas; ~80–90 m. Fl. throughout most of the year. **Zones:** 11–12.

Cattleya

Species ~45. **Herbs**, epiphytic, sympodial; rhizome creeping ± elongate; stems thick, cylindric or clavate to strongly pseudobulbous, covered by several evanescent, scarious sheaths, bearing 1 or 2 leaves at apex. **Leaves** coriaceous or fleshy, oval to oblong. **Inflorescences** terminal, laxly racemose and few- to several-flowered above; peduncle ± concealed by a conspicuous, conduplicate spathe. **Flowers** usually large and showy, with spreading segments; sepals free, subequal; lateral sepals slightly oblique; petals usually much broader than the sepals, rarely subequal to them; lip free from the column but ± surrounding it below and forming a tube, simple or more commonly trilobed; column rather long, stout, arcuate; anther terminal, opercular, incumbent, 2-celled, with 4 waxy, compressed pollinia. **Distr:** tropical America (Mexico to Brazil, Bolivia and Argentina). **Cult:** Propagate by division. Best grown in pots. Requires a coarse bark-based medium with good drainage, and drying out between watering.

C. bicolor Lindl., *Edwards's Bot. Reg.* 22: t. 1919, 1836. (Syn: *Epidendrum bicolor*). [Bicolored Cattleya]. **Herbs**, epiphytic; pseudobulbs clustered, slender, to 30+ cm long, base swollen, jointed, grooved, apex 2-leaved. **Leaves** spreading; lamina oblong-lorate, 16 × 25 cm, apex obtuse, coriaceous, faintly streaked. **Inflorescence** 2–7-flowered, 20 cm long, subtended by a green- and purple-mottled spathe at base. **Flowers** slightly fragrant, to 9 cm across, sepals and peals coppery-brown or tawny-yellow, lip with lateral lobes pink, midlobe crimson, pink-edged, margin white; sepals oblong-lanceolate, 4.5 × 1 cm, acute; lateral sepals slightly deflexed, falcate; petals similar, obovate or broadly spathulate, 4.5 × 1.6 cm, apiculate; lip arcuate, reflexed, oblong-cuneate, 3.5 × 1.7 cm, margin recurved, apex bilobed; column exposed, very large, 3 cm long. **Distr:** WC and SE Brazil. **Ecol:** In coastal areas. **Zone:** 12.

C. forbesii Lindl., *Coll. Bot.*: t. 37, 1826. (Syn: *Epidendrum forbesii*, etc.). **Herbs**, epiphytic, medium-sized; pseudobulbs 15–25 cm long, slightly swollen, covered with membranous sheaths, apex 2-leaved. **Leaves** spreading; lamina oblong, 9–11 × 2.5–5 cm, apex rounded, notched, coriaceous. **Inflorescence** 1–5-flowered, 10–12.5 cm long, subtended by a basal sheath. **Flowers** 6–10 cm across, sepals and petals pale green or olive-yellow, ± tinged with purple-brown, lip with lateral lobes pink, center yellow, purple striped, midlobe with margin white, column white; dorsal sepal spreading, linear-lanceolate, 5 × 1 cm, subacute; lateral sepals similar but falcate-deflexed; petals similar to sepals but falcate, 4.5 × 0.9 cm, acute; lip trilobed, 4.5 × 3.6 cm, lateral lobes oblong, rounded, curved over column to form a tube, spreading, midlobe short, rotund, margin undulate-crisped; column clavate, semiterete, 2.5 cm long. **Distr:** Brazil (Rio de Janeiro to Santa Catarina). **Ecol:** In coastal areas. Fl. Jul–Oct. **Zone:** 12.

C. intermedia Graham ex Hook., *Bot. Mag.* 55: t. 2851, 1828. (Syn: *Epidendrum intermedium*, etc.). **Herbs**, epiphytic or lithophytic; stems numerous, jointed, 7.5–22 cm long, often deeply furrowed, covered with gray, withered, blunt, adpressed sheaths, green where exposed, terminated by 2 leaves. **Leaves** subopposite, spreading; lamina ovate-ligulate, 12.5 cm long, apex slightly notched, and mucronate, flat, fleshy, yellowish-green when young, aging to dark green. **Inflorescence** 1-flowered sometimes with an additional abortive bud (suggesting evolution from a 2-flowered inflorescence); spathe submembranous, green, 5 cm long. **Flower** pale lilac; dorsal sepal linear-elliptic, 6.5 cm long, margins reflexed, apex pointed and greenish; lateral sepals and petals nearly lanceolate, 5 cm long, falcate, margins undulate;

Top row: *Cattleya bicolor, Cattleya bicolor* (albino form), *Cattleya forbesii.*
Middle row: *Cattleya intermedia, Cattleya loddigesii, Cattleya loddigesii* (another color form).
Bottom row: *Cattleya luteola, Cattleya maxima, Cattleya velutina.*
[All MG except for *Cattleya intermedia* AF.]

lip as long as other segments, paler, with many erect papillae within column edges, apex trilobed, midlobe deep purple, mottled with the color of the lip; column boat-shaped, ½ the length of the lip; pollinia 4, in pairs, yellow, ovate. **Distr:** SE and S Brazil, Paraguay, Uruguay. **Ecol:** In coastal forests. **Zone:** 12.

*C. **loddigesii*** Lindl., *Coll. Bot.*: t. 37, 1826. (Syn: *E. loddigesii, E. violaceum*). **Herbs**, epiphytic or lithophytic; stems cylindrical, narrowing below, 20–30 cm long, apex 2-leaved. **Leaves** elliptic-oblong, 10–12.5 × 4–5.5 cm, apex obtuse, leathery. **Inflorescence** 2–9-flowered, to 30 cm long, subtended by a broad basal sheath. **Flowers** 8–11 cm across, waxy, fragrant, sepals and petals pale rose-lilac, lip as sepals and petals below, midlobe amethyst-purple, disk whitish passing to yellow below, column white; sepals elliptic-oblong, 4.5–6.5 × 2–2.5 cm, margins undulate, apex obtuse; lateral sepals subfalcate-deflexed; petals obliquely oblong-elliptic, 5.5–6 × 2–2.5 cm, margins undulate, apex obtuse; lip suborbicular, trilobed, lateral lobes erect, incurved, subrectangular, front margin dentate, midlobe spreading, margin crisped; column clavate, arcuate, each side winged, 3 cm long. **Distr:** SE Brazil to NE Argentina. **Ecol:** In coastal areas. **Zone:** 12.

*C. **luteola*** Lindl., *Gard. Chron.* 1853: 774, 1853. (Syn: *Epidendrum luteolum*, etc.). **Herbs**, epiphytic, dwarf, to 18 cm high; rhizome creeping; pseudobulbs club-shaped, ellipsoid or clavate-cylindrical, sulcate with age, 9–15 cm long, apex 1-leafed. **Leaf** oblong or oblong-elliptic, 6–17 × 3.8 cm, apex obtuse or notched. **Inflorescence** terminal, racemose, 2–6-flowered, shorter than leaf, base covered with a large scarious sheath. **Flowers** 4.5–5 cm across, yellow or yellow-green, lip spotted or streaked with red inside; sepals oblong or narrowly oblong, to 4.5 cm long, acute; lateral sepals obliquely oblong-elliptic; petals narrowly elliptic-oblong, slightly falcate, as broad as dorsal sepal; lip suborbicular, to 2.8 × 2.8 cm, obscurely trilobed in front, anterior margin crenulated; disk densely pubescent above middle; column short, arcuate, 2-winged, ~1.5 cm long. **Distr:** N Brazil, Ecuador to Bolivia. **Ecol:** In rainforests; 100–1,200(–2,000) m. **Zones:** 11–12.

*C. **maxima*** Lindl., *Gen. Sp. Orchid. Pl.*: 116, 1833. (Syn: *C. maxima* var. *aphlebia*, etc.). **Herbs**, epiphytic, erect, tufted, to 40 cm tall; pseudobulbs club-shaped, stout, 10–39 cm long, compressed, sulcate with age, almost covered with large membranous sheaths, apex 1-leafed. **Leaf** oblong to oblong-ligulate, to 25 × 7.5 cm, apex rounded or obtuse, notched. **Inflorescence** terminal, to 20 cm tall, laxly 5–15-flowered, base covered with a large sheath to 12 cm long. **Flowers** large, showy, scented, long-lasting, 12.5–16 cm across, petals and sepals rose to lilac, lip pale pink, purple-veined with a central yellow band, column white; sepals lanceolate, to 8.5 cm long, acuminate; lateral sepals shorter; petals oblong-ovate or oblong-elliptic, to 8.5 cm long, twice as wide as sepals; lip ovate-oblong, obscurely trilobed, 5–7 × 4 cm, lateral lobes forming a tube over column, midlobe spreading, margins crisped-undulate; column slender, obscurely angled, to 2.5 cm long. **Distr:** Colombia to Peru. **Ecol:** In seasonally dry coastal and inland forests; <1,800 m. **Zone:** 10.

*C. **velutina*** Rchb. f., *Gard. Chron.* 1870: 147, 1373, 1870. (Syn: *C. fragrans*, etc.). [Velvety Cattleya]. **Herbs**, epiphytic, medium-sized, often forming small clumps; pseudobulbs clustered, slender, reedlike, slightly compressed, to 45 cm tall, jointed, grooved, basally sheathed when young, apex 2-leaved; pedicel and ovary ~6 cm long, curved.

Leaves oblong to oblong-lorate, 20 × 6.5 cm, apex obtuse, coriaceous, rigid. **Inflorescence** 4–7(–12)-flowered, terminal, 12.5 cm long, usually erect. **Flowers** highly fragrant, long-lasting, 9–10 cm across, glossy, sepals and petals light orange or yellow-orange, ± densely spotted with purple or brownish-purple, lip whitish with midlobe purple-veined, disk golden-yellow; sepals oblong, to 6.5–7.5 × 1-8–2.7 cm, shortly acuminate, margins wavy, apices strongly recurved; petals spathulate, 7.3–7.8 × 3.5–4.5 cm, margins usually crisped, apices shortly acuminate, strongly recurved; lip trilobed, 6.9 cm long, lateral lobes short, rounded, enveloping the column, midlobe broadly heart-shaped, 4–4.5 × 4.8–5 cm, with numerous rough projections; column 3.2 cm long, arched. **Distr:** Brazil (Espírito Santo to São Paulo). **Ecol:** In exposed places; 300–800 m. Fl. Jun–Jul. **Zones:** 11–12. **Info:** This species is now thought to be extinct in the wild.

Encyclia

Species 152. **Herbs**, epiphytic; roots fleshy, glabrous; stems aggregate, erect, forming ovoid-pyriform pseudobulbs. **Leaves** apical, sessile, articulate; lamina linear-elliptic, leathery. **Inflorescences** apical, racemes to panicles, flowering only once; spathe absent. **Flowers** resupinate, pedicellate, simultaneous; sepals subequal; petals similar; lip appressed to column, adnate only at base, callus boat-shaped; pollinia 4, obovoid, laterally compressed, subequal; caudicles 4 in 2 pairs; column winged; rostellum semiorbiculate, entire, abaxially covered with viscous substance. **Fruits** ellipsoid, 3-ribbed, 1-locular. **Distr:** Florida, Mexico to tropical America. **Cult:** Propagate by division. Best grown on cork slabs, or in pots. Requires well-drained, bark-based medium in bright light.

E. cordigera (Kunth) Dressler, *Taxon* 13: 247, 1964. (Syn: *Cymbidium cordigerum*, etc.). [Large-lipped Encyclia]. **Herbs**, epiphytic, to 70 cm tall; pseudobulbs erect, clustered, conical-ovoid, 3–11 × 1.7–4.3 cm, apex obtuse or subacute. **Inflorescence** racemose or paniculate, 3–15-flowered, 15–75 cm long. **Flowers** large, fragrant, brown or purplish-brown, lip cream with 3 magenta streaks in center or pink; sepals and petals spathulate-oblanceolate, 2.6–3.3 cm long, margin undulate, apex obtuse or mucronate; lip adnate to base of column, 3.3–4 cm long, trilobed, lateral lobes oblong, 1.2–1.5 × 0.6 cm, clasping column, midlobe suborbicular to ovate, 2.3–3.2 × 2.1–3.4 cm, emarginate, callus elliptic, sulcate, passing into 3 fleshy veins on midlobe; column 1.5–1.7 cm long, dilated in the middle, midtooth oblong. **Distr:** Mexico to tropical America. **Ecol:** In dry forests, tropical rainforests and dense scrubs; 0–900 m. **Zones:** 11–12.

E. randii (Barb. Rodr.) Porto and Brade, *Rodriguésia* 1(2): 29, 1935. (Syn: *Epidendrum randii*). **Herbs**, epiphytic; pseudobulbs pyriform, base enveloped by several evanescent, imbricating sheaths, apex 1–2-leaved. **Leaves** narrowly cymbidiform, lanceolate-ligulate, smooth, coriaceous. **Inflorescence** paniculate, 10–20-flowered, terminal, 60 cm long, erect to arcuate; pedicel and ovary ~3 cm long. **Flowers** strongly scented, over 6.5 cm across, sepals and petals deep red-brown or sepia, sometimes with greenish overtones and a thin yellowish border, lip white or faintly pink-flushed, midlobe centrally veined with rose-purple to red; sepals ovate-lanceolate, 3.2 × 0.8 cm, acute; petals apically orbicular-spathulate, 2.8–3 × 1.3–1.4 cm, basal half widely clawed ~5 mm across, margins wavy; lip deeply lobed, 2.8–3 cm long, lateral lobes 1.3–1.6 × 2.3–2.4 cm, midlobe faintly reniform to broadly rounded, 1.6–1.8 cm wide, apex slightly notched; column ~1.8 cm long, winged, wings terminating in a rounded, auriclelike extension. **Distr:** W Brazil. **Ecol:** In lowland rainforests; 10–300 m. **Zones:** 11–12.

Epidendrum

Syn: *Amphiglottis*, *Tritelandra*, etc.

Species >1,000. **Herbs**, epiphytic, rarely lithophytic, cespitose; roots fleshy, glabrous; stems erect, repent, or hanging, canelike, simple or branching. **Leaves** alternate; lamina ovate to elliptic or lanceolate, fleshy-leathery; petiole tubular, sheathing, articulate. **Inflorescences** terminal (rarely lateral or basal), racemose to nearly corymbose or distichous, producing flowers only once, or during several years from same or new racemes from old axis, flowers pedicellate or not. **Flowers** resupinate or not, if not then distichous or spirally arranged, simultaneous or successive; sepals nearly equal; lip adnate to column throughout, forming nectary tube, tube penetrating ovary, occasionally producing nectar; column straight or slightly arched; clinandrium hood partly covering anther; anther 4-celled; pollinia 4, obovoid, laterally compressed, nearly equal; rostellum slit. **Fruits** ellipsoid to subglobose, sometimes with pedicel or beak, 3-ribbed, 1-locular. **Distr:** Tropical regions of Western Hemisphere; highest diversity in S America. **Cult:** Propagate by division. Best grown on cork slabs, or in pots. Requires well-drained, bark-based medium in warm conditions.

E. avicula Lindl., *J. Bot.* (Hooker) 3: 85, 1841. (Syn: *Lanium avicula*, etc.). **Herbs**, epiphytic, small; rhizome creeping, covered by loose, scarious, imbricating sheaths; stems ± remote, abbreviated, 1–3 cm long, pseudobulbous with age, apex obliquely bifoliate. **Leaves** sessile, spreading; lamina broadly elliptic to circular, 1.5–3.5× 1.5 cm, apex acute to rounded. **Inflorescences** racemose or paniculate, terminal, laxly several- to many-flowered; rachis, pedicellate-ovary and sepals densely white-tomentose. **Flowers** small, segments spreading, yellowish-brown, pale greenish-red or yellowish-green, ± red-spotted; dorsal sepal lanceolate or oblong-lanceolate, ~6–7.2 mm long, acute; lateral sepals broadly lanceolate or ovate-lanceolate, very oblique, complicate-acute, about as long as the dorsal sepal but much broader; petals linear, 5 mm long, acute or acuminate; lip adnate to column almost to apex, ovate-subrhombic, 4.5 × 4 mm, margin irregular, apex acuminate to apiculate; column stout, 5 mm long, dilated above, apex bidentate. **Distr:** W South America to Brazil. **Ecol:** Forming large clumps on dry rotten trunks, in sunny places of tall forests; 1,800 m. **Zone:** 12.

E. bracteolatum C. Presl, *Reliq. Haenk.* 1: 100, 1827. (Syn: *E. collare*, *E. leucopyramis*). [Small-bracted Epidendrum]. **Herbs**, epiphytic; stem fusiform-elongate, stout, deeply furrowed, 45 cm long, enveloped by dry, triangular, scarious sheaths, apex 2–3-leaved. **Leaves** spreading, canaliculate, narrowly elliptic, 2.5–7.5 cm long, base conduplicate, coriaceous. **Inflorescence** a raceme or few-branched panicle, laxly few-flowered, 40 cm long; rachis 5 cm long; floral bracts triangular-subulate, ~3 mm long, acute; pedicel and ovary ~3 cm long. **Flowers** fragrant at night, white, changing to yellow and brown with age; sepals elliptic-lanceolate, ~2 × 0.5 cm, margins undulate, apex obtuse to acute; petals linear-lanceolate, ~1.6 × 0.25 cm, margins undulate, apex obtuse to acute; lip adnate to column at its apex, ovate, 2.8 × 1.3 cm, margins entire, undulate, apex obtuse to truncate-retuse, decurved above middle, disk trilamellate at base, lateral lamellae bilobed, yellowish; column prominently cuniculate, ~1 cm long, clinandrium margin elevated, erect like a collar, denticulate along apical margin. **Distr:** C America to Ecuador. **Ecol:** In dry forests; 30–300 m. **Zones:** 11–12.

E. ciliare L., *Syst. Nat.* ed. 10, 2: 1246, 1759. (Syn: *Auliza ciliaris*, *Coilostylis ciliaris*, etc.). [Psider Orchid, Fringed Star Orchid]. **Herbs**, epiphytic, sometimes lithophytic, coarse, cespitose, to 50 cm tall; rhizome creeping; stems cylindrical to fusiform, 8–20 × 2 cm. **Leaves** 1(3), oblong-lanceolate, 10–28 × 2.5–8 cm, apex obtuse, fleshy, coriaceous, glossy. **Inflorescence** racemose, few- to many-flowered, with large imbricating basal sheaths, keeled, purple-spotted. **Flowers** fragrant, large, 10 cm long, yellowish-green, lip white; sepals linear-lanceolate, to 9 × 0.5 cm, acute or acuminate, membranous; petals similar to sepals, slightly smaller, obscurely falcate; lip adnate to basal half of column, deeply trilobed, lateral lobes narrowly semiovate, deeply and irregularly long-fringed on outer margins, to 4.5 × 0.4 cm, midlobe linear, rigid, to 6 cm long; disk 3-keeled, middle one extending the length of the lip; column clavate, arcuate, to 1.8 cm long. **Distr:** Mexico to tropical America. **Ecol:** On large trees, in montane forests and humid rainforests; 500–1,900 m. **Zones:** 10–12.

E. exasperatum Rchb. f., *Beitr. Orchid.-K. C. Amer.*: 87, 1866. (Syn: *E. chondranthum*, *Oerstedella exasperata*). **Herbs**, epiphytic, occasionally terrestrial, cespitose, erect, 0.2–2.15 m; stems covered with amplexicaul leaf-sheaths. **Leaves** elliptic to lanceolate to lanceolate-oblong, 4–13 × 0.5–4.5 cm, apex

Below: *Encyclia cordigera*, *Encyclia randii*, *Epidendrum avicula*, *Epidendrum bracteolatum*. [All MG except *Encyclia cordigera* AF.]

JT

MG

JT

MG

Top row: *Epidendrum ciliare, Epidendrum exasperatum, Epidendrum longipetalum, Epidendrum parkinsonianum.*
Bottom row: *Epidendrum peperomia, Epidendrum polyanthum, Epidendrum radicans, Epidendrum tomlinsonianum.*

bilobed, obtuse or acute. **Inflorescences** terminal, lateral or both, simple or paniculate. **Flowers** greenish or whitish, mottled with red or purple; dorsal sepal oblong-lanceolate, 0.8–1.4 × 0.3–0.6 cm, obtuse or acute, cucullate, fleshy, verrucose to mammillate dorsally, midnerve prolonged into a subterminal apicule; lateral sepals similar but elliptic-oblong and usually oblique; petals oblanceolate, 0.8–1.1 × 0.2–0.4 cm, obtuse; lip clawed, claw adnate to column, fleshy or callus-thickened, lamina 0.7–1.3 × ~0.7–1.3 cm, prominently trilobed, lateral lobes spreading, oblong, oblique, obtuse, 3–5 mm long, rarely lobulate or dentate, midlobe bilobulate, emarginate, lobules divergent to explanate, rarely dentate, obtuse; disk usually with a low callus; column provided with 2 apical stelidia. **Distr:** Costa Rica to Panama. **Ecol:** In very wet rainforests and cloud forests; 900–2,500 m. **Zones:** 11–12.

E. longipetalum A. Rich. and Galeotti, *Ann. Sci. Nat., Bot.*, III, 3: 22, 1845. [Long-petaled Epidendrum]. **Herbs**, epiphytic, sometimes lithophytic; stems short, ~5 cm long, gradually widening, enveloped by tubular, scarious sheaths. **Leaves** 2–3, borne near apex of stems; lamina elliptic to ovate, rather stiff, base clasping. **Inflorescence** terminal, racemose, unbranched, 50–75 cm long; peduncle long, erect to arcuate; flowers opening in succession over a long period; bracts many, elliptic. **Flowers** 4–5 cm long, green or bronze, lip white with yellow margins and rose-crimson stripes; sepals ovate, ~3.8 cm long; petals long and slender, ribbonlike, hanging down behind the lateral sepals becoming longer with age, 4–5 cm long, straight or twisted; lip almost round, obscurely trilobed, heart-shaped, highly glossy. **Distr:** Mexico. **Ecol:** Epiphytic, sometimes lithophytic, in cloud forests; 1,700–2,600 m. **Zones:** 9–11.

E. parkinsonianum Hook., *Bot. Mag.* 66: t. 3778, 1840. (Syn: *Brassavola pescatorii*, etc.). [Dagger-shaped Epidendrum]. **Herbs**, epiphytic, pendent, branching, to 2 m tall; pseudobulbs curved-ascending, terete, 6–10 cm long, 1–several-leaved; sheaths imbricate, to 15 × 3 cm. **Leaves** linear-lanceolate to narrowly lanceolate, 20–50 × 1–3.5 cm, acute to long-acuminate, fleshy-coriaceous, to 2.5 cm thick, flaccid, conduplicate, dark green, often purple-tinged. **Inflorescence** subfasciculate, 1–3-flowered; peduncle to 1.5 cm long; floral bracts short, ovate-lanceolate, 0.7–1.5 cm long, acuminate; pedicel and ovary curved, 10–14 cm long. **Flowers** large, showy, fragrant, sepals pale yellowish-green, often mauve or purplish-bronze outside, petals mostly white or yellowish, lip white or yellowish-orange; sepals spreading, linear-lanceolate to lanceolate, 5.6–8.5 × 1.1–1.7 cm, long-acuminate, margins revolute; lateral sepals subfalcate; petals spreading, linear-lanceolate, long-acuminate, falcate, 5.5–8 × 0.6–0.9 cm; lip adnate to column, deeply trilobed, 5.5–8 cm long, lateral lobes semicordate to semiovate, 2.5–4 × 1.5–1.8 cm, obtuse to subacute, midlobe linear, 3.5–5.5 × 0.2–0.4 cm, acuminate; disk with 2 erect-spreading flaplike obtuse keels; column arcuate, dilated upward, 2.5–3 cm long. **Fruits** fusiform-cylindrical, 3-ridged, 7–13 × 1–2 cm. **Distr:** Mexico to C America. **Ecol:** On trees and on rocks in oak-pine forests; to 2,000+ m. **Zones:** 10–11.

E. peperomia Rchb. f., *Bonplandia* (Hannover) 2: 20, 1854. (Syn: *E. gnomus, E. porpax, Nanodes porpax*, etc.). [Beetle Orchid]. **Herbs**, epiphytic, dwarf; stems clustered, leafy, to 10 cm long. **Leaves** distichous; lamina oblong, 2 × 0.9 cm, apex obtuse or emarginate, coriaceous-fleshy. **Inflorescence** 1-flowered, terminal, 3–4.5 long; bract sheathing, oblanceolate, 1–1.7 cm long, purple; pedicel and ovary 1.5–3.5 cm long, purplish. **Flower** fleshy, sepals and petals purplish-green, lip purple, margin yellowish; dorsal sepal narrowly elliptic, 1.2–1.5 × 0.4–0.5 cm, acute; lateral sepals obliquely oblong-lanceolate, 1.2–1.5 × 0.5 cm, acute; petals spreading, linear, 1–1.5 × ~0.1 cm, obtuse; lip deflexed-recurved, transversely oblong or transversely elliptic, 1-1.4 × 1.2–1.8 cm, slightly emarginate, shiny, callus of 3 small raised fleshy ridges at apex of column, spur united to ovary, cylindrical; column fleshy, dilated at apex, 7 mm long. **Distr:** Mexico to Peru and Venezuela. **Ecol:** In lower montane forests; 600–2,700 m. **Zones:** 11–12.

E. polyanthum Lindl., *Gen. Sp. Orchid. Pl.*: 106, 1831. (Syn: *E. hondurense*, etc.). [Many-flowered Epidendrum]. **Herbs**, epiphytic, erect, slender, 0.3–1.2 m tall; stem smooth, leafy, concealed by membranous leaf sheaths. **Leaves** articulate, erect-spreading; lamina linear to elliptic-lanceolate, rarely oblong-elliptic, to 22 × 5.5 cm, apex subobtuse to long-acuminate. **Inflorescence** a raceme, simple or compound, many-flowered, to 40 cm long; main rachis with whitish sheaths; branches 15 cm long; racemes 3–4 cm Ø; pedicel and ovary 0.5–1.5 cm long. **Flowers** variable in color, yellow-white, orange-yellow, greenish-yellow, brownish-yellow or reddish; sepals strongly ribbed along nerves, outside verruculose-papillose; dorsal sepal obovate-elliptic to elliptic-oblanceolate, 0.7–1.6 cm × 1.2–3.5 mm, rounded to obtuse; lateral

MG

Above: *Guarianthe bowringiana, Isabelia pulchella, Laelia anceps.*

sepals obliquely obovate to obliquely elliptic-obovate, 0.7–1.6 cm × 1.8–5 mm, obtuse to subacute; petals filiform to filiform-spathulate, 0.7–1.2 cm long, <0.5 mm wide; lip adnate to column, deeply trilobed, 6–8 × 7–9 mm, lateral lobes extremely variable, 1.5–3.5 × 3.5–5 mm, midlobe oblong-quadrate to oblong-cuneate, 3–6 × 2–3 mm; disk with 2 basal mammillate calli; column clavate, arcuate, 6–8 mm long. **Fruits** ovoid, 1–3 cm long, smooth or ± muricate. **Distr:** Mexico to Brazil. **Ecol:** In thickets, dense humid forests; to 2,700 m. **Zones:** 9–12.

E. radicans Pav. ex Lindl., *Gen. Sp. Orchid. Pl.*: 104, 1831. (Syn: *E. pratense*, etc.). [Crucifix Orchid, Rainbow Orchid]. **Herbs**, terrestrial or lithophytic, trailing, 0.3–1.5 m long; stems simple. **Leaves** >10, ovate-lanceolate, 2–9 × 1.2–2.5 cm, apex obtuse. **Inflorescence** apical, racemose, flowers 13–65 or more; peduncle 27–42 cm long, covered by 4–6 bracts, much shorter than the ovaries, 0.2–1.3 cm × 1–1.3 mm; flowers opening successively. **Flowers** resupinate, yellow to red (usually orange), lip usually more yellow toward the center and often with some red spots; sepals spreading elliptic, 1.4–1.6 × 0.5–0.6 cm, acute; lateral sepals oblique; petals spreading, elliptic, 1.4–1.6 × 0.5–0.6 cm, acute; lip 1.3–1.4 × 1.3–1.4 cm, trilobed, margins laciniate, lateral lobes semiovate, oblique, acute, midlobe rectangular, bilobed; column arched, 1–1.3 cm long. **Distr:** Caribbean, Mexico to Colombia. **Ecol:** Epiphytic or lithophytic in rainforests and cloud forests; 400–1,900 m. Fl. throughout most of the year. **Zones:** 9–12.

E. tomlinsonianum C.D. Adams, *Amer. Orchid Soc. Bull.* 35: 997, 1966. (Syn: *E. nutans* var. *obtusifolium*). **Herbs**, epiphytic; pseudobulbs absent; stem 0.4–1.3 m long. **Leaves** oblong-elliptic, 12–20 × 3–5.5 cm, base clasping and sheathing, apex obtuse, leathery. **Inflorescence** a panicle, terminal, longer than the leaves, laxly many-flowered, subnutant; spathaceous bracts much smaller than the leaves and passing above into the floral bracts; lower floral bracts linear, decreasing upward, uppermost triangular, 3–4 mm long, acuminate; pedicel and ovary 2–3 cm long. **Flowers** ~1.2 cm long, sepals and petals green, purple-tinged, lip and column dark purple; sepals oblanceolate, 1.3 × 0.5–0.6 cm, 5-nerved; dorsal sepal obtuse; lateral sepal broader and longer than dorsal, acute; petals subspathulate-linear, similar in length to sepals, barely 2 mm wide, obtuse; lip as long as the sepals, and as broad as long; column erect, club-shaped, 1.2–1.3 cm long, thick, clinandrium with a tooth on each side anteriorly. **Distr:** Jamaica. **Ecol:** On rocks in woodlands; ~600 m. Fl. Oct–Dec. **Zones:** 11–12.

Guarianthe

Species 4. **Herbs**, epiphytic, cespitose; pseudobulbs thickened, clavate, bifoliate. **Leaves** 2, conduplicate, coriaceous. **Inflorescence** terminal, racemose, from a prominent papery sheath. **Flowers** resupinate; sepals and petals similar; sepals free; petals broader than sepals, obovate; lip infundibuliform, surrounding the column, usually with broad, spreading limb; column clavate, 1–1.2 cm long, midtooth fleshy clasping anther; anther incumbent, terminal; pollinia 4, laterally flattened, with flattened caudicles; stigma concave; rostellum transverse; cuniculus type nectary quite small. **Distr:** Mexico to Venezuela, Trinidad. **Cult:** Propagate by division. Requires a moist, fertile, coarse free-draining compost in bright light, moderate humidity and intermediate to warm temperatures. Must be kept dry during the dormant period.

G. bowringiana (O'Brien) Dressler and W.E.Higgins, *Lankesteriana* 7: 38, 2003. (Syn: *Cattleya bowringiana*, etc.). **Herbs**, lithophytic, rarely epiphytic, to 60 cm high; pseudobulbs club-shaped, 20–60 cm long, base swollen, apex 2- or 3-leaved. **Leaves** narrowly oblong, 15–22 × 5–7.5 cm, apex obtuse, dark green. **Inflorescence** 5–15(–47)-flowered, to 25 cm long, subtended at base by 2 compressed sheaths. **Flowers** to 7 cm across, sepals and petals magenta with deeper-colored veins, lip pale rose-purple at base, deep purple around margin with white blotches edged with pale yellow in the throat, column white, base pink-tinged; sepals oblong or narrowly elliptic, 3.9–4.9 × 1–1.2 cm, acute; petals broader than the sepals, ovate-oblong, 4.1–4.9 × 2.4–2.8 cm, margins undulate, apex obtuse; lip ovate-oblong, 3.6–4 × 2.7–3.4 cm, basal half curved up around column to form a tube, apical margin undulate; column 1–1.2 cm long, narrowly winged on both sides. **Distr:** Mexico (Chiapas) to Honduras. **Ecol:** On exposed rocks and cliffs, ravines, along streams; 100–1,000 m. **Zones:** 10–12.

Isabelia

Species 3. **Herbs**, epiphytic, dwarf, creeping; rhizome short to long, slender; pseudobulbs clustered or well-spaced, small, ovoid, covered with fibrous basal sheaths (bracts), 1-leafed at apex. **Leaves** linear or needle-shaped. **Inflorescence** terminal, short, 1- or 2-flowered. **Flowers** small, white, pink, or magenta, sepals and petals free; sepals similar; petals either narrower or broader than the sepals; lip entire, saccate at base and appressed to the column, recurved above, somewhat thickened on the disk; column fleshy, terete; pollinia 8, pyriform, 4 per locule attached to a filiform caudicle. **Distr:** Brazil, NE Argentina. **Cult:** Propagate by division. Best grown on cork or tree-fern slabs, or in pots with sphagnum moss. Requires humid, shaded conditions with excellent ventilation.

I. pulchella (Kraenzl.) C. Van den Berg and M.W. Chase, *Lindleyana* 16: 109, 2001. (Syn: *Meiracyllium wettsteinii, Neolauchea pulchella*, etc.). **Herbs**, epiphytic, ~7.5 cm high; rhizome creeping, elongated, branched; pseudobulbs very spaced, narrowly ovoid, ~8 mm long, enveloped basally by fibrous sheaths, apex bearing a single leaf. **Leaf** slender, concave, appearing subterete, narrowly linear, fleshy, dark green, coriaceous. **Inflorescence** 1-flowered, about as long as the leaves, ~5 cm long, threadlike, erect to arching. **Flowers** red to magenta, 1.2–2 cm long; sepals and petals subconnivent; lip erect, with 2 sharply reflexed angles at base; ovary becoming visibly swallowed at the junction of the column foot where the base of the lip is partially fused forming a nectary; pollinia 4. **Distr:** S Brazil. **Ecol:** Epiphytic in Atlantic rainforests. **Zones:** 11–12.

Laelia

Syn: *Amalia, Amalias, Schomburgkia.*
Species 23. **Herbs**, terrestrial or epiphytic plants, with thickened, often hollow, orbicular or cylindric pseudobulbs; rhizome creeping; pseudobulbs compressed or terete. **Leaves** 1 or several; lamina broadly oval to oblong, not plicate, fleshy or coriaceous. **Inflorescence** terminal, a subsessile solitary flower or a simple or compound raceme (rarely a panicle); scape elongate. **Flowers** usually large and showy; sepals about equal, free, spreading, flat or undulate; petals similar to the sepals, usually larger or smaller; lip free or slightly adnate to the column, ± trilobed, lateral lobes usually convolute to enfold the column, mostly larger than the midlobe; disk smooth or adorned with lamellae; column mostly long, winged or wingless, usually toothed at apex; anther operculate, incumbent; pollinia 8, 4 in each cell of the anther, waxy, ovoid or laterally compressed. **Fruits** ellipsoid. **Distr:** Mexico to tropical America. **Cult:** Propagate by division. Requires a moist, coarse, bark-based medium with excellent drainage, in bright light.

L. anceps Lindl., *Edwards's Bot. Reg.* 21: t. 1751, 1835. (Syn: *Amalia[s] anceps, Bletia anceps, Cattleya anceps*). [Wand of Saint Diego, El Toro]. **Herbs**, epiphytic or lithophytic; pseudobulbs ovate-oblong, compressed, 7 × 2.5 cm, edges acute, ribs flattened, apex 1(2)-leaved. **Leaves** oblong-lanceolate, 15–22 × 3.4 cm, apex acute, leathery. **Inflorescence** 60–75 cm long, 2–5-flowered; scape with alternate, keeled sheaths along the length of the nodes. **Flowers** to 10 cm across, sepals pale rose-purple, petals darker, lateral lobes of lip pale rose, rich yellow striped and bordered with purple, midlobe rich crimson-purple, disk bright yellow, bordered white in front; sepals ligulate-lanceolate, 6 × 1.6 cm, apex acute, reflexed; petals oblong-elliptic, 6 × 2.6 cm, acute; lip trilobed, 4.5 × 3.8 cm, lateral lobes oblong, curved around the column, rounded, midlobe oblong, reflexed, undulate, blunt; disk with a thick longitudinal ridge; column semiterete, wingless. **Distr:** Mexico to Guatemala. **Ecol:** On tree trunks and branches, rocks. **Zones:** 9–11.

L. autumnalis (Lex.) Lindl., *Gen. Sp. Orchid. Pl.*: 115, 1831. (Syn: *Bletia autumnalis*, etc.). **Herbs**,

Above: *Laelia autumnalis, Laelia rubescens.* [Both MG.]

epiphytic, rarely lithophytic, to 1 m high; pseudobulbs ovoid, subconical or pyriform, tapering, curved, to 15 × 2.3 cm, ribbed, furrowed, apex 2- or 3-leaved. **Leaves** lanceolate, 12.5 × 18 × 3.5 cm, apex acute, leathery. **Inflorescence** laxly 5–15-flowered, 0.5–1 m long; scape jointed, stout, purplish, joints with scalelike bract. **Flowers** 7.5–10 cm across, sepals and petals bright rose-purple, lip with lateral lobes white, midlobe basally whitish, apically rose-purple, disk white with a yellow central line and margins purple-spotted, column purplish above, white below, anther yellow; sepals lanceolate, 5.5–6.5 × 1.3 cm, acuminate; petals ovate, 5–6.5 × 1.8 cm, margins undulate, apex acuminate; lip trilobed, 4 × 2.6 cm, lateral lobes erect, partly enclosing column, rounded, midlobe oblong, apex recurved, acuminate, disk 3-ridged, central ridge extended; column clavate, terete, 2.7 cm long. **Distr:** Mexico. **Ecol:** On trees and rocks, in highlands; 1,600–2,600 m. **Zones:** 9–11.

L. rubescens Lindl., *Edwards's Bot. Reg.* 26(Misc.): 20, 1840. (Syn: *Amalia rubescens*, etc.). [Guarita, Flor de Jesu]. **Herbs**, epiphytic or lithophytic, to 40 cm high; pseudobulbs ovoid or oblong, compressed, 3–7 × 2.5–4 cm, furrowed, dark green, clothed in papery sheaths when young, apex 1(2)-leafed. **Leaf** spreading or suberect; lamina oblong-ovate, 4.5–20 × 2–4.5 cm, leathery. **Inflorescence** 2–8-flowered, clustered, terminal, much longer than the leaf, to 75 cm tall; scape sheathed; bracts lanceolate, to 1.2 cm long, papery. **Flowers** fragrant, to 6 cm across, sepals and petals white, flushed with pink or pale purple, lip white, throat deep purple, base of midlobe yellow; sepals lanceolate, to 3.3 × 0.5 cm, acute; petals ovate-lanceolate, 2.7 × 0.9 cm, margins undulate, apex acute; lip trilobed, 2.5 × 1.3 cm, lateral lobes oblong, curved up around column, apex reflexed, midlobe spreading, oblong, margins undulate, apex acute, reflexed; disk 3-ridged, ridges apically pubescent; column semiterete, 0.8–1 cm long, apex with shallow notches between triangular teeth; pollinia 8. **Distr:** Mexico to C America. **Ecol:** In dense tropical forests; to 1,650 m. **Zones:** 10–11.

L. superbiens Lindl., *Bot. Reg.* 26: Misc. p. 46, 1840. (Syn: *Cattleya superbiens, Schomburgkia superbiens*). [Wand of Lord San Joseph]. **Herbs**, epiphytic, robust, stout, over 3.5 m tall; pseudobulb elongated, oblong-fusiform, 30+ cm long with large scarious sheaths. **Leaves** 1–2, apical; lamina oblong to oblong-lanceolate, to 30 × 6.5 cm, obtuse to acute, very coriaceous. **Inflorescence** a raceme, many-flowered; peduncle to 1.2+ m long, with tubular, closely appressed scarious sheaths; floral bracts large, lanceolate, to 12 cm long, acuminate; pedicel and ovary to 7.5 cm long. **Flowers** large, showy, purple, yellow-variegated; sepals spreading, linear-oblong, oblong-lanceolate or narrowly lanceolate, margins undulate, apex obtuse to acute, rose-mauve, basally paler, 5–7 × 1–1.5 cm; lateral sepals oblique; petals spreading, narrowly oblong to narrowly oblanceolate-oblong, 4.5–7 × 1.8 cm, margins undulate, apex broadly rounded to acute, oblique; lip free, distinctly trilobed above the middle, margins convolute, enfolding column, ovate-oblong to suborbicular-ovate, 4–5.5 × 2.8–3.5 cm, lateral lobes short, oblong, to 1.2 cm long, apex rounded, crisped, midlobe much larger, broadly obovate to obcordate, to 2 cm across, emarginate, edges wavy, crisped; disk yellowish, covered with 5–6 longitudinal prominent crisped and serrated lamellae; column 2.5–3 cm long. **Distr:** Mexico, Guatemala, Honduras. **Ecol:** In open or damp forests; <2,000 m. **Zones:** 10–12.

Leptotes

Species 9. **Herbs**, epiphytic, small, rarely more than 10 cm high; rhizomes creeping; stems very short, 1-leafed. **Leaf** erect; lamina subterete, apex acute, fleshy. **Inflorescences** racemose, laxly 1–few-flowered, terminal; peduncle short, slender, rigid; bracts small. **Flowers** medium-sized; sepals and petals subsimilar, free, ± spreading; lip free, fixed to the column at base, trilobed, lateral lobes with a short distinct claw, small and auriculate, enveloping the column, midlobe longer, margins entire, reflexed; disk naked or costate; column erect, short, fleshy, obscurely biauriculate above; pollinia 6, waxy, 4 large and 2 small. **Distr:** Brazil to NE Argentina. **Cult:** Propagate by division. Best grown on small cork or tree-fern slabs with excellent air-movement. Requires high humidity and bright light.

L. bicolor Lindl., *Edwards's Bot. Reg.* 19: t. 1625, 1833. (Syn: *L. glaucophylla*, etc.). **Herbs**, epiphytic, small; rhizome creeping, stout, elongate; stems short, fleshy, cylindric or subfusiform, erect or ascending, 1–3 × 0.4–0.8 cm, covered with several membranous sheaths. **Leaf** erect or spreading; lamina subcylindric, 5–10 × 0.5–0.8 cm, acute, dorsally deeply grooved, shiny, dark green. **Inflorescence** often 3-flowered, erect to arcuate, shorter than the leaves; flowers long-pedicellate. **Flowers** 3.5–5 cm across, fragrant, sepals and petals white, lip with lateral lobes white or greenish, midlobe deep purple and disk with a white apical margin, column green or purple; sepals linear-lanceolate, 1.6–2.2 × 0.4–0.5 cm, acute; petals linear-ligulate, 1.6–2.2 × 0.3 cm, very acute, membranous; lip erect, ± convex, 1.4–2 × 0.6–0.7 cm, trilobed, fleshy, lateral lobes small, round or subquadrate, midlobe sessile, narrowly ovate, abruptly acute; disk fleshy, 1-ridged; column obscurely 3-angled, 5 mm long, fleshy. **Distr:** Brazil to Paraguay. **Ecol:** In subtropical rainforests and coastal mountains; 500–900 m. **Zones:** 10–11.

Prosthechea

Syn: *Anacheilium*, etc. [Cockleshell Orchid]. **Species** >100. **Herbs**, epiphytic; roots fleshy, glabrous; stems aggregate or spaced, erect, forming flattened or fusiform pseudobulbs. **Leaves** 1–3, apical, sessile, articulate; lamina ovate to lanceolate, thinly leathery. **Inflorescences** paniculate racemes, terminal, flowers simultaneous or successive, produced only once, pedunculate or sessile. **Flowers** resupinate or not; sepals subequal; petals sometimes much narrower; lip appressed to column, adnate to proximal half, with callus; pollinia 4, obovoid, laterally compressed,

Below: *Laelia superbiens, Leptotes bicolor, Prosthechea brachychila.*

MG

Above: *Prosthechea chacaoensis*, *Prosthechea citrina*, *Prosthechea cochleata*.

subequal; caudicles 4 in 2 pairs; column not winged, apically 3- or 5-toothed; rostellum semiorbiculate, entire, covered abaxially with a viscous substance. **Fruits** ellipsoid to ovoid, prominently 3-winged, 1-locular. **Distr:** S Florida, Mexico to tropical America. **Cult:** Propagate by division. Best grown mounted on cork or tree-fern slabs, or in pots. Requires a well-drained, bark-based medium in bright light.

P. brachychila (Lindl.) W.E. Higgins, *Phytologia* 82: 376, 1997. (Syn: *Anacheilium brachychilum*, etc.). **Herbs**, epiphytic; rhizome tough, woody, without sheaths; pseudobulbs slightly compressed, fusiform, 10 cm long, lower half-clothed in sheaths. **Leaves** 2 or 3, apical; lamina ligulate, 25(–30) × 2.5 cm. **Inflorescence** to 10–flowered, to 25 cm long, borne on mature pseudobulbs, terminal; spathe acuminate; bracts obsolete. **Flowers** not resupinate, very fragrant; sepals and petals yellow, brown-spotted, ovate, acute; lip trilobed, fleshy, lateral lobes yellow-cream, rather variable in shape, midlobe with a splash of light maroon-red, callus yellowish-cream with fine hairs; anther fleshy, yellow. **Distr:** Venezuela, possibly also in Colombia and Ecuador. **Ecol:** In dense forests on the slopes of the Sierra Nevada, usually epiphytic on tree trunks along streams; ~2,000–2,800 m. Fl. Aug. **Zones:** 11–12. **Info:** This species is highly variable, especially in flower color and shape.

P. chacaoensis (Rchb. f.) W.E. Higgins, *Phytologia* 82: 377, 1997. (Syn: *Anacheilium chacaoense*, etc.). **Herbs**, epiphytic or lithophytic, 20–35 cm tall; pseudobulbs loosely clustered, ovoid or ellipsoid, 4–10 × 2–5 cm, somewhat compressed, smooth, pale grayish-green, apex 2–3-leaved. **Leaves** elliptic to narrowly elliptic, 6–30 × 1.5–5 cm, apex obtuse. **Inflorescence** 2–7-flowered, to 10 cm long, subtended by a basal sheath 3 cm long. **Flowers** very fragrant, not resupinate, creamy-white or pale green, lip marked with purple veins, to 4 cm across; sepals lanceolate or ovate-lanceolate, 1.4–2.2 × 0.45–1 cm, acute; petals broadly elliptic, 2 × 0.6–1 cm, clawed, apex obtuse, acute or acuminate; lip adnate to column at base, broadly ovate, concave, 1–2 × 1.2–2.5 cm, apex obtuse, acute or acuminate, callus pubescent; column stout, 0.65–1 cm long, midtooth at apex square or rounded, fleshy; ligule oblong, rounded. **Distr:** Mexico to Venezuela. **Ecol:** In humid forests; 0–1,100 m. Fl. Feb–Apr. **Zones:** 10–11.

P. citrina (Lex.) W.E. Higgins, *Phytologia* 82: 377, 1997. (Syn: *Cattleya citrina*, etc.). **Herbs**, epiphytic, to 32 cm long, often growing in pendent tufts; pseudobulbs in clusters, ovoid, conical or fusiform, 4–6 × 2–3 cm, 2–4-leaved. **Leaves** elliptic, 18–26 × 3–3.8 cm, apex obtuse or acute, grayish-green. **Inflorescence** pendent, 1- or 2-flowered, 6–20 cm long. **Flowers** sweetly fragrant, waxy, golden-yellow, lip with midlobe orange; sepals elliptic-oblong or oblong, 5–6.5 × 1.5–2 cm, obtuse or acute, fleshy; petals oblong to elliptic-obovate, 4.5–6.5 cm long, obtuse; lip adnate to column at base, oblong-obovate, 5.5–6.5 × 3.5–4.8 cm, apex obscurely trilobed, midlobe quadrate, obtuse or retuse, margins fleshy and crisped, callus fleshy, sulcate running onto base of midlobe; column 2.5–3.5 cm long. **Distr:** SW Mexico (to Durango). **Ecol:** In shaded areas of dry oak and pine forests; 1,300–2,600 m. **Zones:** 10–12.

P. cochleata (L.) W.E. Higgins, *Phytologia* 82: 377, 1997. (Syn: *Anacheilium cochleata*, etc.). [Octopus Orchid, Clamshell Orchid, Cockleshell Orchid]. **Herbs**, epiphytic, 20–60 cm tall; pseudobulbs stipitate, ellipsoid to pyriform, 5–20 × 2–4 cm, somewhat compressed. **Leaves** 2 or 3(4), elliptic or elliptic-lanceolate, 12–33 × 1.2–4 cm, apex acute. **Inflorescence** gradually elongated, 20–60 cm long; several flowers opened at once. **Flowers** not resupinate; sepals and petals similar, pale greenish-yellow, narrowly lanceolate, 1.5–4 × 0.2–0.5 cm, reflexed, twisted; lip white, base with blackish-purple stripes, the rest pale yellow-green to pale purple, deeply concave, broadly triangular-ovate, 0.9–1.7 × 1.1–1.7 cm, base deeply cordate, apex acute or apiculate; column pale greenish-yellow, 0.7–1 cm long. **Distr:** tropical America (S Florida, Caribbean, Mexico to N South America). **Ecol:** In humid forests; 450–1,000+ m. Fl. Jan–Mar, May–Jul. **Zones:** 9–11.

P. fragrans (Sw.) W.E. Higgins, *Phytologia* 82: 377, 1997. (Syn: *Epidendrum fragrans*, *Encyclia fragrans*, etc.). [Clamshell Orchid]. **Herbs**, epiphytic, erect, stout, to 40 cm high; rhizome creeping; pseudobulbs narrowly ellipsoid or fusiform, 5–11 × 3 cm. **Leaf** 1, apical, subtended by membranous sheaths; lamina ligulate-oblong to elliptic-lanceolate, 10–30 × 4.5 cm, obtuse, coriaceous. **Inflorescence** racemose, few- to several-flowered, to 17 cm long; sheaths keeled. **Flowers** non-resupinate, strongly scented, cream to pale greenish-white, lip white, with purple longitudinal stripes along veins; sepals spreading, elliptic-lanceolate, 1.5–3.5 ×0.5 cm, acuminate; petals clawed, ovate-lanceolate, 2–3 cm long, acute or acuminate; lip fleshy, erect, shortly adnate to column base, cupped, ovate, 1.5–2.4 × 0.9–1.7 cm, acuminate, disk nervosa with 2 small parallel calli at base; column clavate, 6–8 mm long, winged, dorsally keeled, apically tridentate. **Distr:** S Mexico to tropical America. **Ecol:** In very humid and pluvial forests; 0–1,500 m. Fl. Mar, May–Jul, Nov–Dec. **Zones:** 10–12.

P. vespa (Vell.) W.E. Higgins, *Phytologia* 82: 381, 1997. (Syn: *Anacheilium vespa*, *Encyclia vespa*, *Epidendrum vespa*). **Herbs**, epiphytic, erect, repent or cespitose, to 70 cm high; pseudobulbs 15–40 cm long, 0.5–2 cm Ø, cylindric to fusiform. **Leaves** 2–4, widely separated, frequently 1–3 cm apart, erect or spreading; lamina elliptic to oblong-lanceolate, 10–40 × 2–4.5 cm. **Inflorescence** a raceme, simple, 12–20 cm long. **Flowers** greenish or yellowish, lip cream or yellow; sepals similar, oblong to ovate to obovate, 0.9–1.35 × 0.3–0.65 cm, obtuse; petals oblanceolate to oblong-oblanceolate, 0.7–1.3 × 0.25–0.6 cm, obtuse, oblique; lip subrhombic, orbicular or ovate, 7–8 × 4–6 mm, fleshy, shortly clawed, claw short, adnate to column base, acute or obtuse, callus longitudinal at base; column thick, only slightly shorter than the lip. **Distr:** C America to Brazil,

Below: *Prosthechea fragrans*, *Prosthechea vespa*, *Sophronitis cernua*, *Sophronitis coccinea*.

MG

West Indies. **Ecol:** In very moist, cloud and rainforests; 550–1,650 m. **Zones:** 9–12.

Sophronitis

Syn: *Sophronia.*

Species 66. **Herbs**, epiphytic or lithophytic, dwarf; pseudobulbs clustered on the rhizome, each bearing a solitary leaf. **Leaf** suberect or erect; lamina coriaceous. **Inflorescence** terminal, 1–2 (or more)-flowered. **Flowers** showy, mostly scarlet, purple or orange-red with yellow on the lip, relatively large; sepals spreading, subsimilar, elliptic; petals spreading, broader than sepals; lip much smaller than the other segments, erect, sessile at base of column or shortly adnate to it, entire or trilobed, enveloping the column below; column thick, short, with a subpetaloid wing on each side of the stigmatic cavity; pollinia 8 in 2 series of 4 each. **Distr:** Brazil to N Argentina. **Cult:** Propagate by division. Requires a bark-based mix (for larger species) or sphagnum moss (for smaller species) in humid, partially shaded conditions.

S. cernua Lindl., *Bot. Reg.* 14: t. 1147, 1828. (Syn: *Cattleya cernua, Sophronia cernua*). [Nodding Sophronitis]. **Herbs**, epiphytic or lithophytic, small; rhizome robust, creeping; pseudobulbs densely clustered, subcylindric or ovoid, 2–3-noded at base, 1–2 × 0.5–1.2 cm, apex 1-leafed. **Leaf** small, ± spreading, broadly ovate or elliptic-ovate, apex obtuse or minutely apiculate, 2–2.4 × 1.2–1.8 cm, glossy, adaxially dark green, abaxially purple. **Inflorescence** erect, spreading, 2–5 cm long, 2–5-flowered. **Flowers** bright red, base of lip white or yellow, fleshy; sepals ovate-oblong, 1.2 × 0.5 cm, acute; petals narrowly ovate-rhomboid, 1–1.2 × 0.5–0.6, obtuse or minutely apiculate; lip connate at base to column, broadly triangular-ovate, 0.8–1 × 0.6–0.7 cm, apex abruptly acute; disk excavate at base. **Distr:** Brazil to NE Argentina. **Ecol:** Epiphytic on trees, sometimes lithophytic on rocks in coastal areas, high in the canopy of inland rainforest trees, and at sea level on rocky slopes and in mangrove forests. Fl. Nov–Jun. Fr. Sep. **Zones:** 11–12.

S. coccinea (Lindl.) Rchb. f., *Ann. Bot. Syst.* 6: 465, 1862. (Syn: *Cattleya coccinea*, etc.). [Scarlet Sophronitis]. **Herbs**, epiphytic or lithophytic, small; rhizome creeping, elongate, branching; pseudobulbs clustered, erect or ascending, terete, fusiform or oblong-ovoid, 1.5–4 cm long, 2–6 mm Ø, 1- or 2-noded below, apex 1-leafed. **Leaf** erect-spreading; lamina oblong-elliptic or narrowly ovate, 3–6 × 1–2.5 cm, apex obtuse or acute, fleshy, coriaceous. **Inflorescence** erect or somewhat spreading, 1-flowered, 3–6 cm long. **Flowers** very show, fleshy, vivid scarlet-red, base of lip yellow, column often pale purple; sepals spreading, broadly oblong, 1.7–2.2 × 0.7–1 cm; petals ovate-rounded, 2–3 cm long, obtuse; lip connate to column at base, broadly ovate-triangular, distinctly trilobed, 1.3–2 × 1.5–2.2 cm, lateral lobes erect, broadly triangular-rotund, obtuse, midlobe oblong-triangular, obtuse, concave; disk transversely cristate at base; column clavate, 6–9 mm long, apex slightly bilobed. **Distr:** SE Brazil to Argentina (Misiones). **Ecol:** In tropical coastal forests; 650–1,670 m. **Zones:** 9–11. **Info:** *S. coccinea* **f. *rossiteriana*** is yellow form.

S. crispata (Thunb.) C. Van den Berg and M.W. Chase, *Lindleyana* 15: 116, 2000. (Syn: *Laelia crispata*, etc.). **Herbs**, epiphytic or lithophytic, to 50 cm high; stems stout, cylindrical, 4–30 cm long, base pseudobulbous, ± tinged purple, 1-leafed. **Leaf** lanceolate or linear-lanceolate, 10–15 × 2.7 cm, apex obtuse, stiffly coriaceous, adaxially dark green, abaxially dark green or purplish. **Inflorescence** 5–9-flowered; scape 30–45 cm long, subtended by basal, compressed sheath. **Flowers** 3.2 cm across, bright yellow or pale orangish-yellow, column basally yellowish-green, apically white; sepals ligulate or oblong-ligulate, 2–3.5 × 0.6–0.9 cm; lateral sepals somewhat falcate; petals ligulate, 2.2–3.3 × 0.6–1.1 cm, subacute; lip narrowly trilobed, 2.4–3 × 1.8 cm, lateral lobes erect, semiovate, curved over column, midlobe oblong, crisped with 4 thickened nerves running to base of lip, blunt; column short, 9 mm long. **Distr:** Brazil (Minas Gerais). **Ecol:** On iron-ore outcrops; 800–1,250 m. **Zones:** 10–11.

S. kettieana (Pabst) C. Van den Berg and M.W. Chase, *Lindleyana* 15: 117, 2000. (Syn: *Cattleya kettieana*, etc.). **Herbs**, epiphytic or lithophytic, small, ~8 cm high; pseudobulbs globose to ellipsoid, 1–2.5 × ~1 cm, apex 1-leafed. **Leaf** obovate or somewhat boat-shaped, canaliculate, 1.5–4.5 × ~1 cm, base clasping, apex acute, succulent, to ~5 mm thick, deep to dull green, margins reddish, glossy. **Inflorescence** terminal, ~2–2.5 cm long, branched, to 5-flowered; bracts ~2 mm long, membranous. **Flowers** to ~3 cm across, pink, sepals and petals spreading; sepals elliptic, to 1.2 × 0.5 cm, apices recurved; petals elliptic-obovate, to 1.5 × 0.6 cm, apices recurved; lip trilobed, ~1 cm long, lateral lobes erect, enclosing the column, quadrate, to 4.8 × 3.5 mm, purplish-pink, midlobe with center white, yellow or pale pink, base narrowed, with margins crisped and reddish purple, sometimes almost absent; column ~3.5 mm long. **Distr:** Brazil (Minas Gerais). **Ecol:** In humid montane areas; 1,500–2,000 m. **Zones:** 9–10.

Top row: *Sophronitis coccinea* f. *rossiteriana*, *Sophronitis crispata*, *Sophronitis kettieana*.
Bottom row: *Sophronitis lundii*, *Sophronitis reginae*, *Sophronitis tenebrosa*.

MG

MG

MG

MG

Above: *Diodonopsis erinacea, Dracula chimaera, Dracula erythrochaete, Dracula sodiroi.*

S. lundii (Rchb. f. and Warm.) Van den Berg and m.W.Chase, *Lindleyana* 15: 117, 2000. (Syn: *Bletia lundii*, etc.). **Herbs**, epiphytic or lithophytic, small; pseudobulbs short, robust, oblong-fusiform, slightly compressed, 3.5–4 × 1.2–1.3 cm, apex 2-leaved. **Leaves** erect-spreading, slightly arcuate; linear, semicylindric, 3–9 × 0.4–0.5 cm, apex very acute, fleshy, adaxially deeply channeled. **Inflorescence** 1–3-flowered, terminal, ascending, ± arcuate, much shorter than the leaves; bracts minute. **Flowers** erect-spreading, white, lip veined with rose-purple; sepals subequal, ligulate-oblong, 2–2.2 × 0.4–0.5 cm, acute, fleshy; lateral sepals falcate; petals linear-lanceolate, 2–2.22 × 0.35–0.4 cm, margins scarcely undulate, apex acute; lip narrowly elliptic-ovate or subquadrate, 2–2.3 × 1.1–1.2 cm, deeply trilobed, fleshy; lateral lobes erect, semiovate, acute, midlobe semirotund, reflexed, margin crisped and lobulate; disk slightly puberulous and fleshy, 4-ridged; column short, broad, 7–8 mm long. **Distr:** Bolivia to Argentina (Misiones, Salta). **Ecol:** In coastal mountains, thickets and forests; 740–1,000 m. **Zones:** 8–10.

S. reginae (Pabst) C. Van den Berg and M.W. Chase, *Lindleyana* 15: 118, 2000. (Syn: *Laelia reginae*, etc.). **Herbs**, epiphytic, miniature, clustered; pseudobulbs conical, 1-leafed. **Leaf** boat-shaped, succulent, abaxially purple-tinged. **Inflorescences** 2–6-flowered, 5 cm long, held just above the leaves. **Flowers** 1.5–2.5 cm across, white or pale purplish, lip usually bright yellow; dorsal sepal lanceolate, 1.4 × 0.4 cm, acute; lateral sepals falcate-lanceolate, 1.2 × 0.5 cm; petals 1.2 × 0.4 cm, acute; lip 6 mm long, trilobed, lateral lobes curved up around the column, oblong, 2.5 × 5 mm, midlobe 2 × 2.5 mm, reddish, ruffled; column greenish, 5 mm long. **Distr:** Endemic to Brazil (Minas Gerais: Serra da Caraça). **Ecol:** On rock ledges; 1,200–2,000 m. **Zones:** 9–10.

S. tenebrosa (Rolfe) C. Van den Berg and M.W. Chase, *Lindleyana* 15: 118, 2000. (Syn: *Brasilaelia tenebrosa*, etc.). **Herbs**, epiphytic; stem clavate, to 18 × 3 cm, apex unifoliate. **Leaf** erect; lamina oblong-ovate to ligulate, to 28 × 6 cm, apex rounded, leathery. **Inflorescence** 2–4-flowered, 30 cm long. **Flowers** 14 cm across, sepals and petals bronze-green, lip rose-purple, base pale greenish with 2 very rich purple marks at center, column green; sepals lanceolate-ligulate, 8 × 1.6 cm, acute; lateral sepals slightly falcate-deflexed; petals obliquely elliptic-oblong, 7 × 2.5 cm, obtuse; lip trilobed, 6.5 × 4 cm, lateral lobes curved up around column, midlobe spreading, rounded, crisped-undulate; column 2–3 cm long. **Distr:** Brazil (SE Bahia to Espírito Santo). **Ecol:** Epiphytic on tree trunks and branches in shady areas of rainforests; 0–600 m. Rare in the wild and near extinction, with most of its habitat already lost. Fl. Apr–Jun, Dec–Feb. **Zones:** 10–11.

Subtribe Pleurothallidinae

Diodonopsis

Syn: *Pteroon.*

Species 5. **Herbs**, epiphytic, cespitose; ramicauls short. **Inflorescences** with flowers solitary or successively flowered, peduncles slender. **Flowers** small; sepals forming a short tube or cup, apices with tails; ovaries soft, burrlike. **Distr:** Costa Rica to W South America. **Ecol:** In cloud forests, humid forests, usually growing among mosses on twigs of small trees; (200–)650–2,000(–2,900) m. **Cult:** Propagate by division. Best grown either mounted or potted. Requires humid conditions with good air circulation, and never allowing it to dry out completely.

D. erinacea (Rchb. f.) Pridgeon and M.W. Chase, *Lindleyana* 16: 253, 2001. (Syn: *Masdevallia erinacea*, etc.). [Hedgehog Masdevallia]. **Herbs**, epiphytic; stems 0.2–1.3 cm long. **Leaves** narrowly linear, 2–5 × 0.2–0.5 cm. **Inflorescences** congested, flowers opening consecutively; peduncle 3–5 cm ling; pedicels 3–5 mm long. **Flowers** with sepals pale green or yellowish-green, speckled with red or magenta, apices yellowish-orange, petals white, lip yellow, magenta-spotted; sepals equal in length, externally carinate, hairy along the margins and veins, apices club-shaped; dorsal sepal connate to lateral sepals for 3 mm, broadly ovate, 4–6 × 5–7 mm, apex rounded, tail claviform, 5–7 mm long; lateral sepals very broadly ovate, 5–6 × 5–8 mm, connate for 4–5 mm, apices rounded, apices similar to that of the dorsal sepal; petals obovate, 2.5–3 × ~1.5 mm, apices rounded, minutely dentate, lip elliptic, 3.5–4.5 × 1.75–2 mm, apices rounded, minutely erose, apiculate; disk with a pair of verrucose calli, convergent. **Distr:** Costa Rica to Ecuador. **Ecol:** In very humid forests, pluvial forests; 700–1,800 m. **Zones:** 9–12.

Dracula

Species 123. **Herbs**, epiphytic, small to medium-sized; rhizome short, creeping, producing short, erect, 1-leafed shoots (ramicauls). **Leaf** erect; lamina obovate to oblanceolate, somewhat thin-textured, slightly plicate, with a prominent central midrib. **Inflorescences** basal, 1-flowered, erect or more commonly pendent; bract sheathing, shorter than the ovary. **Flowers** usually medium-sized, often pendent, flat to bell-shaped; sepals large, showy, subsimilar, connate for basal third or more, shortly hairy on inner surface, each with a long apical caudate tail; petals very much smaller than the sepals, entire, papillate at thickened apex, parallel to and as long as the column, bearing 2 short keels at apex; lip bipartite, hinged to column foot at base, basal part (hypochile) a fleshy oblong cleft claw, apical part (epichile) transversely elliptic or oblong, cochleate, adorned with radiating grooves from base; column short, semiterete, porrect, with a short fleshy foot; pollinia 2. **Distr:** SE Mexico to Peru. **Cult:** Propagate by division. Best grown in live sphagnum moss, preferably near a water feature. Requires cooler conditions with regular misting.

D. chimaera (Rchb.f.) Luer, *Selbyana* 2: 194, 1978. (Syn: *D. senilis*, etc.). **Herbs**, lithophytic or epiphytic; stems clustered, 3–6 cm long, subtended by 2–3 loose sheaths. **Leaves** erect; lamina narrowly oblanceolate, 15–30 × 2.5–5 cm. **Inflorescences** erect or spreading, 3–5-flowered, 15–60 cm long, flowers produced in succession; bracts tubular, 1–1.2 cm long; pedicels 1.2–1.4 cm long. **Flowers** pale yellow, heavily spotted cinnamon-red, maroon or blackish, lip white to pale yellow; sepals broadly ovate to triangular, 3–5.5 × 2.5–3 cm, connate for basal 1.5–1.8 cm, pubescent within, tails 6–12 cm long; petals spathulate or obovate, bilobed at apex, 3–5 × 1.5–2 mm, verrucose in apical part; lip saccate, ellipsoidal, 1.6–2 × 1–1.2 cm; hypochile large, 4–5 mm long; epichile 1.2–1.6 cm long, margins dentate; column 3–4 mm long, foot 1.5–2 mm long; ovary 0.9–1.1 cm long, subverrucose. **Distr:** W Colombia. **Ecol:** In cloud forests, 1,400–2,450 m. **Zones:** 10–11.

D. erythrochaete (Rchb. f.) Luer, *Selbyana* 2: 195, 1978. [Dracula Orchid, Red-tufted Orchid]. **Herbs**, epiphytic, variable; rhizomes short, creeping; stems short, erect. **Leaves** erect; lamina oblanceolate or linear-lanceolate, 15–20 × 1 cm, apex obtuse, tridenticulate, coriaceous, bright green. **Inflorescences** 2–3-flowered, apical, pendent to descending, flowers opening in succession; peduncle slender, terete, to 12.5 cm long; bracts ovate, sheathing below, apiculate, 1.2 cm long, pale green; pedicel and ovary 2–2.3 cm long. **Flowers** with widely spreading segments, sepals creamy-white yellow-tinged, spotted with crimson-purple, tails purple, petals white, lip white to flesh-pink, column pale yellow; perianth tube very broad and short; sepals pubescent within; dorsal sepal broadly ovate at base, 1.2–1.3 × 1.2–1.3 cm, abruptly narrowing into a slender erect tail 2.5–5.6 cm long; lateral sepals triangular-ovate, 1.3–1.4 × 1.3–1.4 cm, tails slender, 2.5–5.6 cm long; petals very small, oblong, 4 mm long, apex papillose, bilobed; lip saccate, clawed, spathulate, 1.3 × 0.5 cm, with 3 central lamellae and 5–6 smaller bifurcating ones, margins denticulate; column slender, 3–4 mm long, apex denticulate. **Distr:** Costa Rica to W Panama. **Ecol:** Epiphytic in cloud forests; 1,000–2,000 m. Fl. Jan–Jun, Oct. **Zones:** 11–12.

D. sodiroi (Schltr.) Luer, *Selbyana* 2: 197, 1978. (Syn: *Masdevallia sodiroi*). Red Bell Dracula]. **Herbs**, epiphytic or terrestrial; stems densely clustered, stout, 3–6 cm long, subtended by 2–3 sheaths. **Leaves** erect; lamina narrowly elliptic-oblanceolate, 15–25 × 1.2–1.6 cm, apex acute, thinly coriaceous. **Inflorescence** a raceme, 2–3-flowered, erect, to 25 cm long; peduncle and pedicel purple, subverrucose; pedicel 1–2 cm

Above: *Dryadella albicans, Masdevallia angulata, Masdevallia caloptera.*

long. **Flowers** campanulate, orange red with darker red central veins to the sepals, inside the cup with numerous white hairs; sepals obovate-obtriangular, 1.5–2 cm long, bases connate for 1.5–1.6 cm, apex truncate, tails 1.5–2 cm long; petals obovate-spathulate, 7 × 3 mm, margins erose, apex obtuse; lip oblong, 8 × 2 mm, obtuse, with 2 parallel lamellae on upper side; column semiterete, 7 mm long, yellow, foot 2 mm long; ovary 0.6–1 cm long, purple, subverrucose. **Distr:** Ecuador. **Ecol:** In cloud forests; 1,500–2,400 m. **Zone:** 10.

Dryadella

Species 53. **Herbs**, epiphytic or lithophytic, dwarf, clump-forming; rhizomes short, creeping; stems short, erect, unifoliate. **Leaf** suberect, petiolate; lamina narrowly elliptic to obovate, leathery, fleshy. **Inflorescences** 1–several-flowered, axillary, usually shorter than the leaves; floral bracts overlapping, small; flowers borne near the base of the plant. **Flowers** white, cream, greenish, or brownish, spotted and banded with maroon or purple, rather similar in most species, small, fleshy; sepals subsimilar, ovate or triangular, joined at the base and forming a cup, apex caudate, fleshy; lateral sepals with a transverse callus at the base; petals very much smaller than the sepals, short, broad, multiangled; lip hinged to the column foot, oblong, obtuse, with a slender claw and a shovel-shaped apical lamina, bicallose, auriculate at the base of the apical lamina; column semiterete, winged; pollinia 2, hard. **Distr:** Mexico south through C America to S tropical America. **Cult:** Propagate by division. Best grown in shallow pots with a fine organic substrate, or mounted on a slab with live moss. Requires moist and humid conditions.

D. albicans (Luer) Luer, *Selbyana* 2: 207, 1978. (Syn: *Masdevallia albicans*). **Herbs**, epiphytic, small, cespitose; secondary stems abbreviated, 0.6–1 cm long, with 2–3 close, tubular sheaths, unifoliate. **Leaf** erect; lamina narrowly linear, 3–6 × 0.2–0.3 cm, apex subacute, minutely tridenticulate, coriaceous. **Inflorescence** 1-flowered; peduncle ~5 mm long, on a secondary stem; floral bract thin, closely sheathing, 7 mm long; pedicel 1–1.5 cm long. **Flower** white with pale yellow tails; dorsal sepal connate to the lateral sepals for 1.5–2 mm, elliptic, concave, veins 3 prominent externally, apex acuminate into a thick tail 6 mm long, total length 1.3 × 0.55 cm; lateral sepals ovate, oblique, margins erose, apices acuminate into tails similar to dorsal sepal, basally connate above transverse carinae, each 1.2 × 0.3 cm; petals translucent pale green, oblong-subrhombic, apex slightly dilated, obscurely obtusely trilobed, 1.5 × 1.25 mm; lip white, elliptic, rounded; column with short, broad wings, apex denticulate, 2 mm long, foot nearly as long; ovary minutely glandular, 1.5–2 mm long. **Distr:** S Ecuador (Azuay). **Ecol:** On slopes of the Andes; 2,000 m. Fl. Jul–Aug. **Zones:** 9–12.

Masdevallia

Syn: *Alaticaulia, Byrsella, Spilotantha*, etc.
Species 571. **Herbs**, epiphytic, small, medium-sized or large, either cespitose or more rarely with a creeping rhizome; stem rather short or very short, not pseudobulbous, unifoliate above, concealed by 1 or several scarious, imbricating sheaths. **Leaf** usually coriaceous, generally long-narrowed toward the base. **Inflorescence** terminal, either 1–3-flowered near the apex or loosely several-flowered and racemose; peduncle terete or angled, usually 1–3-sheathed. **Flowers** with sepals generally ± connate at base into a cup or tube, usually terminating in an ± elongate tail or cauda; petals much smaller, usually narrow; lip very small, in size similar to petals, polymorphic, articulated to column foot, recurved or arcuate-spreading; column about as long as the lip, winged or margined above, usually with a short foot. **Distr:** Mexico to S tropical America. **Cult:** Propagate by division. Requires a moist mix with bark and perlite, in humid and shady conditions.

M. angulata Rchb. f., *Otia Bot. Hamburg.* 15, 1878. (Syn: *Byrsella angulata, M. burfordiensis*). [Sharp Angle Masdevallia]. **Herbs**, robust, epiphytic or lithophytic, clump-forming; stems to 6.5 cm long, unifoliate. **Leaf** apical, erect; lamina spathulate, 8–18 cm long, base cuneate into petiole, coriaceous; petiole stout, channeled. **Inflorescence** erect, horizontal or pendent, 2.5–6 cm long, 1-flowered; peduncle ~5 cm long, purple-spotted. **Flower** large, long-lasting, 5 cm across, fleshy, variable in color, with an unpleasant smell; sepals yellow, marked and flushed with purple-red, connate, forming a deep cup or tube ~1.9 × 1.9 cm, inside purple, outside whitish, free parts triangular, gradually narrowing into slender yellowish tails 1.5–3.7 cm long; petals very small, spathulate; lip ligulate, channeled in middle, densely spotted with rose-purple. **Distr:** S Colombia, NW Ecuador. **Ecol:** In wet montane forests, cloud forests; 600–2,600 m. Fl. Jan, Aug. **Zones:** 9–11. **Info:** The flowers often exude droplets of a sugary substance.

M. caloptera Rchb. f., *Gard. Chron.*, n.s., 2: 322, 1874. (Syn: *M. biflora* [misapplied], *Spilotantha caloptera*). [Beautiful-winged Masdevallia]. **Herbs**, epiphytic. **Leaves** oblong to lanceolate, 6–11 × 1.4–2 cm, apex obtuse to rounded, older leaves red-tinged. **Inflorescences** 2- or more-flowered; scape 3–18 cm long; flowers opening 2 or more at a time. **Flowers** small, white with purple streaks; sepals ~2 cm long, equally fused for ~6 mm, forming a tube sharply indented above mentum, margins finely toothed, tails slender and yellow; dorsal sepal almost circular, white with 2 purple lines; lateral sepals triangular to oblong, white with 2 purple lines; petals 4–5 mm long, toothed, white with a central purple stripe; lip oblong to shallowly trilobed, ~5 mm long, yellow and crimson; column ~4 mm long, green and purple. **Distr:** N Peru. **Ecol:** In wooded gullies; 2,750–3,200 m. **Zones:** 9–11.

M. caudata Lindl., *Gen. Sp. Orchid. Pl.*: 193, 1833. (Syn: *M. tricolor*, etc.). **Herbs**, epiphytic, tufted. **Leaves** suberect, petiolate; lamina oblanceolate, 12 × 3 cm, apex rounded, coriaceous. **Inflorescences** spreading, suberect or horizontal, 1-flowered; peduncle terete, 4–10 cm long. **Flowers** campanulate, perianth tube dilated with a pouch on lower side, variable in color, dorsal sepal white, yellow or yellow-green with darker nerves and purple spotting in apical half, lateral sepals rose-colored or white, tails light green, petals white, lip white purple-spotted, column white, margins purple-spotted; dorsal sepal concave, lamina ovate-elliptic, 2.3 × 2 cm, abruptly narrowed into a slender tail to 6.5 cm long; lateral sepals rounded-ovate with a basal saccate projection, 1.7 × 1.5 cm, apex abruptly contracted into slender deflexed tails 5–6.5 cm long; petals oblong, 5 mm long, with a flaplike extension basally, apex obliquely toothed; lip oblong, 6 × 4 mm, apical half recurved, with 2 fleshy ridges and an apical raised mound; column 7 mm long, foot arcuate, 7 mm long. **Distr:** Colombia to Venezuela. **Ecol:** In cloud forests; 1,800–2,800 m. **Zones:** 9–11.

M. coccinea Linden ex Lindl., *Orchid. Linden.*: 5, 1846. (Syn: *M. militaris*, etc.). [Banderitos, Little Flags]. **Herbs**, terrestrial or epiphytic, forming large tufts; ramicauls short, to 6 cm long, enveloped by 2 or 3 sheaths. **Leaves** erect; lamina oblong-lanceolate to spathulate, 15–23 × 2–3.4 cm, base petioled, apex rounded and minutely tridentate. **Inflorescence** 1-flowered, apical, slender, to 40 cm long, 3–4-noded; nodes with a bract, spotted. **Flowers** large, showy, variable in size and color, scarlet, magenta or magenta-purple to yellow or white; perianth tube cylindric-campanulate, 1–1.8 cm long, curved; dorsal sepal narrowly triangular-lanceolate or linear, 3–6 cm long, including the slender recurved tail; lateral sepals connate for basal third, free parts deflexed, incurved, obliquely oblong-ovate, 4.5–6 × 2.2–2.6 cm, apices with short tails; petals very small, linear-oblong, base auricled, keeled close to anterior margin, apex retuse or bilobed; lip oblong or linguiform, 0.6–1 cm long, bicarinate, base cordate, apex rounded. **Distr:** Colombia. **Ecol:** On rocky cliff sides; 2,400–2,800 m. **Zones:** 9–11.

M. ignea Rchb. f., *Gard. Chron.* 1871: 1482, 1871. (Syn: *M. ignea* var. *marschalliana*, etc.). **Herbs**, epiphytic or terrestrial; stems densely tufted, rigid, erect, base sheathed, 1-leafed. **Leaf** elliptic-lanceolate to oblong-lanceolate, 7.5 cm long, base cuneate and narrowed into a sulcate petiole 2.5–4 cm long, apex obtuse or notched, very coriaceous, adaxially bright deep green, abaxially paler. **Inflorescence** 1-flowered,

Above: *Masdevallia caudata, Masdevallia coccinea* (white form), *Masdevallia ignea, Masdevallia instar.*

slender, rigid, 30–38 cm long; lower sheath 2.5 cm long, upper ones spathaceous, compressed. **Flower** 3.8–5 cm long, horizontal, decurved; sepals united at base into a curved, gibbous tube, ~1.2–1.7 cm Ø, outside pale orange-red, inside bright dark orangish-red, lateral sepals with 3 darker or reddish veins; dorsal sepal abruptly contracted from a triangular ovate base to a long slender point, bent downward; lateral sepals elliptic-oblong, acute, obliquely incurved, convex above with recurved margins; petals very small, wholly included, appressed to sides of column, linear-oblong, obtuse, base auricled in front, white with a faint purple streak; lip equaling petals, included, linear-oblong, obtuse, recurved, cordate at base, white, tinged pale purple; ovary ~8 mm long. **Distr:** Colombia. **Ecol:** 3,350–3,650 m. **Zones:** 10–12.

M. instar Luer and Andreetta, *Phytologia* 39: 206, 1978. **Herbs**, epiphytic, medium-sized, cespitose; stems terete, 2–4 cm long, unifoliate, clothed by 2 brown sheaths. **Leaf** erect; lamina narrowly obovate to elliptic, 6–11 × 1.6–2.3 cm, base narrowly cuneate into a petiole 1.5–3 cm long, apex subacute, tridenticulate, coriaceous. **Inflorescence** 1-flowered; peduncle erect, slender, 6–11 cm long; floral bract close, tubular, 0.8–1.4 cm long; pedicel 1–2 cm long. **Flower** with sepals yellow, minutely flecked with red inside, tails slender, red-purple to orange-brown, petals white, lip cream, purple-spotted, column white, marked with purple; sepals connate basally for 4–5 mm into a shallow cup, tails 2–5 cm long; dorsal sepal ovate, 1.4–1.7 × 1.1–1.4 cm, subcarinate, obtuse to rounded, apex abruptly produced into erect tail; lateral sepals ovate, oblique, 1.2–1.5 × 1.1–1.3 cm, apices produced into deflexed tails; petals oblong, 6 × 2 mm, truncate, bilobed; lip erect, elliptic, 4.5–5 × 3 mm, constricted below, truncate, with a central callosity; column stout, semiterete, 4.5 mm long, foot 3 mm long, apex incurved; ovary 0.9–1 cm long. **Distr:** S Ecuador to NW Peru. **Ecol:** In cloud forests; 2,500–2,700 m. Fl. Jul–Aug. **Zones:** 9–11.

M. melanopus Rchb. f., *Gard. Chron.*, n.s., 2: 332, 1874. (Syn: *M. xanthodactyla*, etc.). **Herbs**, epiphytic, densely clump-forming, rather small; stem abbreviated, clothed with 1 or more sheaths. **Leaf** oblong-spathulate, ~4–12.5 × 1.5 cm, apex obtuse, minutely tridenticulate, gradually narrowed into a short, slender petiole. **Inflorescence** 3–8-flowered; peduncle much exceeding the leaves, to 25 cm long, slender, terete, loosely secund-flowered. **Flowers** small, white with minute purple dots, dark purple stain at base, and yellow tails; sepals connate below into an ovoid-campanulate cup which is gibbous below, to ~6 mm long, free portions triangular, very short, abruptly terminating in flattened tails ~1.3 cm long; petals very small, cuneate-spathulate, upper margins denticulate, apex broad, retuse-apiculate, with a longitudinal keel; lip about the same size as the petals, pandurate-oblong, 3–4 mm long, with the narrower anterior part bearing 2 longitudinal keels, apex rounded and ± apiculate. **Distr:** SW Ecuador to N Peru. **Ecol:** On trees in dense woods; 2,400–2,700 m. Fl. Oct–Nov. **Zones:** 10–11.

M. mendozae Luer, *Phytologia* 54: 382, 1983. **Herbs**, epiphytic, small, cespitose; secondary stems blackish, 1–1.7 cm long. **Leaf** 1, erect; narrowly elliptical, 2–4.5 × 1.2–1.6 cm, base cuneate into the ± blackish petiole 1–1.5 cm long, apex obtuse, coriaceous. **Inflorescence** 1-flowered, borne from a node low on the secondary stem; peduncle slender 3 cm long, with a bract near the base; floral bract close, tubular, 5 mm long; pedicel 5 mm long. **Flower** bright orange, tubular, sepals glabrous externally, shortly pubescent inside, carinate along midveins; dorsal sepal linear-oblong, curved, 2.7 × 0.8 cm, connate to lateral sepals for 2.2 cm to form a curved, cylindrical tube, free portion rounded, abruptly contracted into a 2 mm long, recurved tail; lateral sepals connate for 1.9 cm into an oblong, curved lamina 2.4 × 1.4 cm, free portions rounded, convex, abruptly contracted into tails 1.5 mm long; petals oblong, 6.5 × 1.75 mm, truncate, slightly trilobed, slightly carinate along lower margin becoming a large, thick, incurved tooth extending beyond the clawed base; lip oblong, 7.5 × 2.9 mm, truncate, slightly decurved, with a low callus; column semiterete, 5.5 mm long, foot 3 mm long; ovary 3–4 mm long. **Distr:** Ecuador (Zamora-Chinchipe). **Ecol:** In cloud forests; 1,800–2,300 m. **Zone:** 10.

M. schroederiana H.J. Veitch, *Gard. Chron.*, III, 8: 51, 1890. (Syn: *Reichantha schroederiana*). **Herbs**, epiphytic, medium-sized, densely tufted. **Leaves** petiolate; lamina oblanceolate, 15 × 2.5 cm, base gradually narrowed into the petiole, apex minutely tridentate; petiole ~5–7.5 cm long. **Inflorescence** 1-flowered, erect, slightly longer than the leaf; peduncle to 21 cm long. **Flower** relatively large, dorsal sepal rich orange-yellow, with 2 dark spots, lateral sepals violet on outside, inside white, tails of all sepals orange-yellow, petals and lip pale pink, spotted with rose-purple; perianth tube campanulate, 2 cm long, white lined with violet; dorsal sepal broadly ovate-triangular, tail erect, arching backward, 5–7.5 cm long; lateral sepals much longer than dorsal sepal, united almost to the apices of the laminas, 3 × 1.5 cm, tails deflexed, 5+ cm long; petals oblong, 8 mm long, margins wavy, fleshy; lip oblong, 9 mm long, base fleshy, bicarinate in center, apex recurved; column with apex denticulate. **Distr:** Costa Rica. **Ecol:** In cloud forests; 1,400–2,500 m. Fl. Nov–Feb, May–Jul. **Zones:** 11–12.

M. tovarensis Rchb. f., *Linnaea* 22: 818, 1850. (Syn: *Alaticaulia tovarensis, M. candida*). **Herbs**, epiphytic, cespitose. **Leaves** elliptic or lanceolate to obovate, to 15 × 2 cm, base narrowly cuneate into petiole, apex subacute, obtuse or rounded, thickly coriaceous, glossy green; petiole stout. **Inflorescence** racemose, to 4-flowered, all opening together; scape to 20 cm long, angled. **Flowers** 7–8 cm across, white, column purple, apex and base white, ovary green; sepals connate forming a tube ~6 mm, free parts

Below: *Masdevallia melanopus, Masdevallia mendozae, Masdevallia schroederiana, Masdevallia tovarensis.*

Above: *Pleurothallis truncata, Restrepia antennifera, Restrepia elegans, Restrepia trichoglossa.*

triangular to ovate and tapered into tails; dorsal sepal 4–5 cm long, finely tapered; lateral sepals triangular to ovate, 3.5–4.5 cm, tails slender, fused for 2–2.5 cm; petals linear, ~ 6 × 1.15 mm, acute, asymmetric; lip oblong, oblong-pandurate or spathulate, 5–6 × 2 mm, rounded or acute, recurved, papillose, acute, with 2 ± parallel ridges in the middle; column 4–4.5 mm long, finely toothed. **Distr:** N Venezuela. **Ecol:** In forests; 1,600–2,000 m. Fl. Dec–Mar. **Zones:** 11–12.

Pleurothallis

Species 625. **Herbs**, epiphytic or lithophytic, small- to medium-sized; rhizome creeping, giving rise to tufted stems; stems ramicaul, erect, short, jointed, usually 1-leafed. **Leaves** erect to spreading, sessile to petiolate; lamina fleshy to coriaceous. **Inflorescence** apical, short to relatively long, racemose, fasciculate or 1-flowered. **Flowers** small, occasionally secund; sepals ± subequal, spreading or ringent, ± connate; lateral sepals connate at base to completely united to apex, ± forming a short mentum with the column-foot; petals smaller than the sepals; lip small, simple to trilobed, mostly clawed; column ± winged, foot mostly distinct; pollinia 2, waxy, pyriform to spherical, stipe lacking, rostellum well-developed. **Distr:** Mexico to tropical America. **Cult:** Propagate by root division. Best grown in small pots with a well-drained mix (containing sphagnum moss or fine bark) or mounted on tree-fern slabs. Requires moist, humid conditions in shade.

P. truncata Lindl., *Companion Bot. Mag.* 2: 355, 1836. (Syn: *Humboldtia truncata, Orbis truncata*). **Herbs**, epiphytic, sometimes lithophytic, medium- to large-sized; ramicauls slender, erect, terete, elongate, enveloped basally by 2, close, tubular sheaths, and another 2, one at the middle and the other below middle, carrying at the apex a single leaf. **Leaf** elliptic or ovate-lanceolate, base attenuate to the short, conduplicate base, apex acute, glossy. **Inflorescences** 1 or more, racemose, 7.5–10 cm long, several- to many-flowered, apical, borne from a small spathe and holds the successively opening flowers tightly against the middle of the leaf; bracts membranous, convolute. **Flowers** very small, globular, ~3 mm across, ± closed, orange-colored; dorsal sepal ovate, large than other segments, hooded with the 2 lateral sepals; petals linear-oblong; lip short, involute; anther hemispheric; pollinia 2; ovary short. **Distr:** Ecuador. **Ecol:** Epiphytic on trees, in woods; 1,700–3,000 m. **Zones:** 9–10.

Restrepia

Species 52. **Herbs**, epiphytic or rarely lithophytic, small, tufted; stems short, unifoliate. **Leaf** suberect or erect; lamina conduplicate, coriaceous. **Inflorescences** 1 to several, terminal, 1-flowered; peduncle elongated; bract sheathing. **Flower** relatively large, narrow, pale green to yellow, ± suffused, spotted or striped crimson or purple; dorsal sepal linear-lanceolate, apex attenuate, slightly swollen; lateral sepals ± connate, often to the apex, cucullate below; petals similar to dorsal sepal but smaller; lip much smaller than the sepals, oblong to ± trilobed, bilobulate in front; column slender, arcuate, dilated toward apex; pollinia 4. **Distr:** S Mexico to W. South America and Venezuela. **Cult:** Propagate by division or from leaf cuttings. Best grown in sphagnum moss or mounted on cork or tree-fern slabs. Requires copious watering and humid conditions with with filtered light.

R. antennifera Kunth, *Nov. Gen. Sp.* 1: 367, 1816. (Syn: *Pleurothallis ospinae, R. hemsleyana*, etc.). **Herbs**, epiphytic, small, tufted; stems erect to 10 cm long, covered by 3 or 4 purple-spotted sheathing bracts. **Leaf** erect or erect-spreading; lamina ovate or elliptic, 5–5.5 × 2.5–3 cm, apex obtuse or rounded, coriaceous. **Inflorescence** axillary, 1-flowered; peduncle terete, slender. **Flower** relatively large, translucent, white to pale yellow-brown striped with dark purple, tails of sepals greenish, lip pale brown, maroon-spotted, column creamy-brown at apex, white below; dorsal sepal narrowly ovate below, long caudate above, 2–3 × 0.3 cm, tail swollen in apical half; lateral sepals united almost to the apex, narrowly elliptic, 2.3 × 0.6 cm, subacute; petals similar to dorsal sepal but smaller, porrect; lip trilobed at base, 1 × 0.3 cm, lateral lobes setose, suberect, midlobe oblong-pandurate, papillose; column clavate-arcuate, 5 mm long. **Distr:** W South America to NW Venezuela. **Ecol:** In forests; 1,500–3,500 m. Fl. Aug–Nov. **Zones:** 11–12.

R. elegans H. Karst., *Allg. Gartenzeitung* 15: 202, 1847. (Syn: *R. antennifera* ssp. *erythroxantha*, etc.). **Herbs**, epiphytic, cespitose, small, to 8 cm high; stems clustered, erect, ~4–6 cm long, covered by scarious, white sheaths, apex unifoliate. **Leaf** suberect; lamina elliptic or ovate-elliptic, 6 × 4 cm, margins slightly recurved, apex obtuse, minutely tridentate, coriaceous. **Inflorescence** 1-flowered, erect to spreading, axillary; peduncle long, terete, slender, 4–5 cm long; bracts tubular, white, amplexicaul. **Flower** narrow; dorsal sepal and petals translucent, white, lined with maroon, lateral sepals orange-brown, maroon-spotted, lip pale brown, maroon-spotted, column pale green, pink-flushed; dorsal sepal erect, lanceolate, long-attenuate, 2.5 × 0.3 cm, apex slender, clavate; lateral sepals united to just below apex, base saccate, 2.4 × 0.9 cm; petals similar to dorsal sepal but smaller and porrect; lip fleshy, flexuous, trilobed, 1.1 × 0.5 cm, truncate, lateral lobes erect, porrect, falcate-lanceolate, acuminate, midlobe oblong-oblanceolate, slightly constricted above base, verrucose; column slender, arcuate, clavate, 5 mm long. **Distr:** Colombia to NW Venezuela. **Ecol:** In wet, montane forests; 700–2,800 m. Fl. Apr–Jun. **Zones:** 11–12.

R. trichoglossa F. Lehm. ex Sander, *Sander's Orch. Guide*: 215, 1901. (Syn: *Pleurothallis filamentosa*, etc.). [Hairy-tongued Restrepia]. **Herbs**, epiphytic, 3–10 cm high; ramicaul concealed by 4–8, large, compressed, scarious, tubular, imbricate sheaths, apex 1-leafed. **Leaf** erect; lamina elliptic-ovate, 3–6 × 1.5–3 cm, base broadly cuneate to rounded, contracted into the twisted petiole base, apex minutely bidentate. **Inflorescence** 1-flowered, fasciculate; peduncle 3–6 cm long, borne near or beyond leaf apex; floral bract inflated. **Flower** with dorsal sepal midrib purple, synsepal yellow to yellow with finely lined with red-brown or purple, lip yellow or orange; dorsal sepal ovate at base, 1.6–2.2 cm × 1.5–2.5 mm, long-attenuate, long-claviform; lateral sepals connate forming a concave lamina, ovate, 1.5–2.2 × 0.5–0.6 cm, obtuse, sometimes slightly bifid; petals ± linear-triangular, 1.4–1.7 × 0.8–1 cm, filamentous appendage on one or both margins, distal half attenuated, claviform; lip ± oblong, 0.6–1.1 cm × 1–1.77 mm, tiny, base concave with a triangular appendage on each side, apex truncate to retuse, margins denticulate to fimbriate; column 3–4 mm long. **Distr:** Mexico (Chiapas) to Peru. **Ecol:** In very moist, cloud forests and rainforests; 300–3,200 m. **Zones:** 9–11.

Specklinia

Syn: *Empusella, Muscarella*, etc.
Species 125. **Herbs**, epiphytic, lithophytic or terrestrial, small, tufted or creeping; stems erect, enclosed in tubular sheaths, unifoliate. **Leaves** apical, usually petiolate; lamina almost round to ovate. **Inflorescences** unbranched; floral bracts tubular. **Flowers** with dorsal sepal free or joined to lateral sepals to form a cup; lateral sepals sometimes joined to form a synsepal; petals sometimes with margins fringed or toothed; column usually winged, with a foot; pollinia 2. **Distr:** Mexico to tropical America, including West Indies. **Cult:** Propagate by division. Best grown mounted with a little moss on the roots, or in pots with a fine bark-based compost. Requires cool to intermediate conditions with high humidity, moderate ventilation and good ventilation.

S. brighamii (S. Watson) Pridgeon and M.W. Chase, *Lindleyana* 16: 256, 2001. (Syn: *Pleurothallis brighamii, Sarcinula brighamii*). **Herbs**, epiphytic, forming dense clumps or tufts, to 12 cm high; secondary stems ~6 mm long. **Leaves** oblanceolate to linear-oblanceolate, 1.8–9 × 0.3–1.2 cm, base attenuated into a pseudopetiole, apex obtuse or acute, coriaceous. **Inflorescence** fascicles, 1–several-flowered, congested, flowers opening in succession; peduncle 3–12 cm long; pedicels 4–5 mm long. **Flowers** small, yellow, striped with red-brown or green, sepals glabrous; dorsal sepal oblong-oblanceolate, 0.6–1 cm × 1.2–3 mm, obtuse to acuminate; lateral sepals connate, oblong to oblong-ovate, 5–9 × 3–6 mm, acute; petals obliquely spathulate to obliquely oval, dilated, 2–3.5 × 0.7–1.7 mm, acute to acuminate;

JT

JT

Above: *Specklinia brighamii, Stelis emarginata, Stelis villosa, Bletilla striata.*

lip oblong-ligulate to oblong-lanceolate, 2.1–4 × 0.8–2 mm, obtuse, apex ciliate or denticulate, with an auriculate lateral lobe on each side near base of lamina, claw with a prominent callus. **Distr:** Mexico to Ecuador, Caribbean. **Ecol:** In wet forests; 0–950 m. **Zones:** 9–12.

Stelis

Species 887. **Herbs**, epiphytic or lithophytic, tufted; rhizomes horizontal, creeping; stems borne from the rhizomes, erect, clustered, simple, unifoliate, 2-jointed with a sheath to each joint and small basal sheaths. **Leaf** suberect, subsessile or petiolate; lamina base contracted into a short stalk, jointed near base, fleshy or coriaceous. **Inflorescences** 1 to many, racemes, axillary, slender, long, each enclosed by a spathaceous bract and smaller inner bracts, flowers numerous; floral bracts small, sheathing below, ± equaling the pedicel. **Flowers** minute to small; sepals ± subequal, laterally slightly oblique, ± connate, mostly spreading; petals and lip minute, fleshy. surrounding the column; column short, thickened above; rostellum ligulate; anther terminal, opercular, semiglobose; pollinia 2, waxy, pyriform, stipe lacking. **Fruits** small, ovoid or oblong, 3-cornered **Distr:** S Florida, Mexico to tropical America. **Cult:** Propagate by division. Best grown in sphagnum moss or in a fine bark mix. Creeping species can be mounted on tree-fern slabs. Requires moist, humid and shady conditions.

S. emarginata (Lindl.) Soto Arenas and Solano, *Icon. Orchid.* 5–6: t. 681, 2002. (Syn: *Epidendrum emarginatum*, etc.). **Herbs** epiphytic, small; stems erect or ascending, clustered, covered by tubular scarious sheaths, 1.5–12 cm long, apex unifoliate. **Leaf** suberect to erect, shortly petiolate; lamina elliptic to oblanceolate, 4–15 × 1.5–3 cm, apex obtuse, minutely retuse, fleshy. **Inflorescence** arcuate, 8–42 cm long, laxly many-flowered; bracts tubular-spathaceous, acute or acuminate, scarious, 3–7 mm long. **Flowers** cylindric-campanulate, greenish-yellow to brick-red; sepals united for over half their length into a cylindric sepaline tube; free parts spreading, elliptic-oblong to elliptic-lanceolate, 0.6–2.2 cm long including the tube, obtuse to acute; petals obovate, 1.5–2.5 × 1 mm, obtuse to rarely trilobed; lip trilobed toward apex, somewhat arcuate, canaliculate, dorsally carinate, 2–3.5 × 1–2 mm, lateral lobes suberect, rounded, midlobe ovate, margins crenulate, apex obtuse; column slender, semiterete, arcuate, 2–3 mm long, apex trilobed. **Distr:** Mexico to Peru. **Ecol:** In dry or wet forests; 1,800–3,500 m. Fl. Oct–Dec. **Zones:** 9–12.

S. villosa (Knowles and Westc.) Pridgeon and m.W.Chase, *Lindleyana* 16: 267, 2001. (Syn: *Effusiella villosa*, etc.). **Herbs**, epiphytic, growing in dense clumps or tufts; stem 2–7 cm long. **Leaves** erect, sessile; lamina elliptic, 3–5.5 × 1–2 cm, obtuse. **Inflorescence** subflexuose, 5–9 cm long, 1 or 2 flowers opening simultaneously; peduncle 2–3 cm long, spathe ~3 mm long; pedicels 4–7 mm. **Flowers** yellowish-green with purple markings, lip green marked with purple; sepals 5–7 mm long, margins with very distinct white pendent filaments, succulent, purple hairy inside; dorsal sepal concave, obovate-elliptic, ~ 7 × 2.5–3 mm, subacute; synsepal obovate, ~ 5 × 4 mm, apex bifid, obtuse; petals obovate-spathulate, 2.5–3 × ~1.25 mm, apex rounded; lip arched, oblong-clawed, ~1 mm long, thin. **Distr:** Mexico to C America. **Ecol:** In humid forests; 1,300–1,900 m. **Zones:** 9–11.

Subtribe Bletiinae

Bletilla

Species 5. **Herbs**, terrestrial; basal corms or pseudobulbous swellings on a short rhizome; stems dilated, slender, short, erect, leafy near the base or below the middle (sometimes apically), enclosed by tubular sheaths at base. **Leaves** 4–6, large, deciduous, convolute, mostly narrow, linear to lanceolate or oblong-elliptic, 3–40 cm long, plicate, sheathing at base. **Inflorescences** racemes 2–10-flowered, terminal, flowers often opening successively, simple or sometimes branched; peduncle slender, stiff, long; bracts caducous. **Flowers** with sepals and petals nearly equal, free, erect or spreading, often with white markings; lip adnate to base of column, trilobed, lateral lobes erect or incurved, slightly embracing column, midlobe slightly recurved, disk with many longitudinal ridges; column slender, slightly arching, winged on each side near top, without foot; anther terminal, nearly incumbent, 2-celled; pollinia 8. **Distr:** N Indo-China to temperate E Asia. **Cult:** Requires a rich, well-drained terrestrial mixture in bright light or full sun. Reduce watering when leaves are absent.

B. striata Rchb. f., *Bot. Zeitung (Berlin)* 36: 75, 1878. [Bamboo-leaf Orchid, Chinese Ground Orchid, Peacock Orchid, Urn Orchid, White Chicken]. **Herbs** small to large, 15–62 cm tall; pseudobulbs ovoid, compressed. **Leaves** broad, pleated, oblong-lanceolate, to 45 cm long. **Inflorescences** erect, 60 cm tall, up to 12 flowers, borne from new pseudobulbs, flowers opening in succession. **Flowers** 5 cm across, pink (rarely pale pink or white), fragrant, somewhat nodding; sepals and petals subsimilar, oblong-lanceolate or obovate-oblanceolate, to 3.5 × 0.8 cm, acute; lip trilobed to 3.5 × 2 cm, lateral lobes erect, semiovate, midlobe recurved, subquadrate, somewhat emarginate, disk with 5 parallel, flexuose lamellae. **Distr:** China, Himalayas (Tibet), Japan, Myanmar. **Ecol:** On cliffs or in mountain forests; 1,100–3,200 m. Fl. Mar–Jul. **Zones:** 7–11. **Info:** Often mistakenly placed in the genus *Bletia*, which is a New World genus.

TRIBE VANDEAE

Subtribe Polystachyinae

Polystachya

Species 150. **Herbs**, epiphytic, lithophytic, or

Below: *Polystachya bancoensis, Polystachya bella, Polystachya concreta, Polystachya fallax.*

JT

Above: *Abdominea minimiflora*, *Acampe rigida*, *Aerides leeana*, *Aerides multiflora*.

terrestrial, tufted; roots vermiform; stems leafy, slender or pseudobulbous, clustered or distant along rhizome. **Leaves** 1–several, conduplicate, thin (fleshy), distichous, coriaceous, contracted into sheaths at the base. **Inflorescences** terminal, simple or paniculate racemes with ± secund branches, few- to many-flowered; peduncles, enclosed by tubular, membranous sheaths. **Flowers** mostly not resupinate, generally small; dorsal sepal free; lateral sepals adnate to column foot forming mentum; petals smaller than sepals, free; lip trilobed (or simple), adnate to and articulate with column foot, forming part of mentum; disk usually callose, pubescent; column semiterete, short, foot distinct; stigmas confluent, transverse under rostellum, viscidium ovate to elliptic; anther terminal, incumbent, operculate, hemispheric, imperfectly 2-chambered; pollinia 4, globose or ellipsoid, hard, waxy, attached to single short stipe. **Fruits** fusiform, elongated. **Distr:** Africa, Asia, Australia, Pacific Islands; S USA, C America, West Indies. **Cult:** Propagate by division. Best grown in pots with a bark-based mix. Requires humid conditions in bright light.

P. bancoensis Burg, *Misc. Pap. Landbouwhoogeschool* 19: 23, 1980. (Syn: *P. affinis* var. *nana*). **Herbs**, epiphytic, dwarf, ~5 cm high; pseudobulbs broadly elliptic, compressed, 2–2.5× 1–1.5 cm, 1–2-foliate. **Leaves** elliptic to obovate-oblanceolate, 2–5 × 1–1.5 cm. **Inflorescences** simple, 1–4-flowered; rachis fractiflex; bracts suborbicular to ovate, apex abruptly acuminate to tridentate, 3–4 mm long. **Flowers** yellowish-brown, pubescent, sepals and petals with white straight hairs outside, glabrous inside, all segments with reddish-purple venation; sepals triangular, 5–6 × 4–5 mm, base wide, apex blunt; petals narrowly obovate, 5–6 × 2–3 mm, incurved and partially enclosing the column (slightly touching each other), apices recurved; lip 5–6 mm long, midlobes recurved (reaching the incurved petals), apex bilobed. **Distr:** W Tropical Africa (Ghana, Ivory Coast). **Ecol:** On fully exposed high branches of trees, in wet evergreen forests; 50–1,350 m. Fl. Nov. **Zones:** 11–12.

P. bella Summerh., *Kew Bull.* 14, No. 1 (African Orchids: XXVII): 137, 1960. (Syn: *P. obanensis*). **Herbs**, epiphytic, 15–20 cm tall; stems approximate, erect, cylindrical, clothed with old leaf sheaths, 3–4 leaves at the apex; pseudobulbs ellipsoid, 4 × 2 cm. **Leaves** elliptic- to oblong-ligulate, 5–16 × 2.7 cm, base narrowed to a petiole, apex rounded, unequal, slightly bilobed. **Inflorescences** many-flowered, terminal, simple to many-branched, usually 25 cm long, subdense; peduncles usually 10 cm long; floral bracts suberect, lanceolate, caudate-acuminate, 1.7 cm long, pubescent; pedicel and ovary 1 cm long, densely pubescent **Flowers** suberect, golden yellow; sepals slightly incurved, lanceolate to narrowly triangular-lanceolate, 1.2–1.4 × 0.3–0.35 cm, acute, slightly concave; lateral sepals oblique, lanceolate-triangular, 1.5–1.7 × 0.5–0.55 cm, acute, 3–4-nerved; petals linear-oblanceolate, 0.85–1.1 × 0.15 cm, acute; lip lanceolate, indistinctly trilobed, 9.5–1.15 × 0.48–0.55 cm, lateral lobes erect, slightly round-triangular, apex free, ~1 mm long, midlobe strongly recurved, lanceolate-triangular, 4.5–5 × ~2.5 mm, acuminate, fleshy; column slightly incurved, 2.5–3.5 mm long. **Distr:** Kenya, Uganda. **Ecol:** In forests; 1,800 m. **Zones:** 9–11.

P. concreta (Jacq.) Garay & Sweet, *Orquideologia* 9 (3): 206, 1974. (Syn: *Dendrobium polystachyum*, etc.). [Yellow Spike Orchid]. **Herbs**, epiphytic; stems stout, pseudobulbous, tufted, 15–25 cm long, clothed with old leaf sheaths. **Leaves** 4–6, alternate, distichous; lamina linear-oblong or oblanceolate, 10–20 × 1.2–3 cm, base sheathing, margins slightly wavy, apex obtuse coriaceous, shiny. **Inflorescences** panicles, racemes branching, terminal 6–7 cm long, many-flowered; peduncles 15–42 cm long; rachis faintly tomentose; floral bracts ovate, 5 × 3.6 mm, acuminate or acute, persistent; pedicel and ovary 6 mm long. **Flowers** pale yellow, 1 cm across, resupinate; dorsal sepal oblong-ovate, 3.9–4.1 × 2–2.1 mm, acuminate, 3-veined; lateral sepals triangular-ovate, 4.8–5 × 3.5–4 mm, acuminate, 3-veined; petals linear-oblong, 3.2–3.5 × 0.8 mm, subacute, 1-veined; lip cuneate-obovate, 4 × 4 mm, superior, joined to foot of column, 3-veined, lateral lobes falcate, subacute, midlobe broadly oblong, rounded, fimbriate; disk scurfy; column 1.9 × 1.3 mm; anther terminal, biloculed, 1 × 1.25 mm; pollinia 4. **Fruits** brown, small, fusiform, 1.2 × 0.35–0.6 cm. **Distr:** Pantropical. **Ecol:** In tropical wet evergreen forests; <1,700 m. Fl. throughout most of the year. **Zones:** 9–12.

P. fallax Kraenzl., *Bull. Misc. Inf. Kew* 1926: 292, 1926. **Herbs**, epiphytic, glabrous, 12–16 cm tall; pseudobulbs often in pairs, distantly placed, conical, 2.6–4.5 × 0.4–0.5 cm, shiny, clothed in 3–4 membranous basal cataphylls. **Leaves** articulated 4–5 mm above pseudobulb apex; lamina linear, 9.5–11.3 × 0.6–1 cm, base cuneate, apex acute to obtuse, coriaceous. **Inflorescences** racemes, simple, 4.4–7 cm long, usually 5-flowered, flowers opening successively; peduncles slender, 2.6–3.5 cm long, basal sheath membranous; bracts triangular, amplexicaul, 1.5–2 mm long. **Flowers** white, callus ochre-yellow, anther cap purple; dorsal sepal ovate-lanceolate, 0.75–1.15 × 0.25–0.32 cm, acuminate, trinerved; lateral sepals spreading, narrowly obliquely triangular, 0.8–1.4 × 0.5–0.7 cm, trinerved; mentum conical, to 5.5 mm long; petals spreading, linear-lanceolate, 6.5–8 × 1–2 mm, acute; lip strongly recurved, sessile, distinctly trilobed in middle, 1–1.25 × 0.4–0.6 cm, midlobe ovate-lanceolate, 5–7 × 1.4–3 mm, acuminate, lateral lobes porrect, ovate or elliptic, callus central or in basal half; column semiterete, 1–1.5 mm high, foot 4.5–5 mm long. **Distr:** Tropical Africa (Rwanda, Uganda, Zaire). **Ecol:** In rainforests; 1,350–1,600 m. **Zones:** 10–12.

Subtribe Aeridinae

Abdominea

Species 1. *Generic description same as specific.* **Cult:** Propagate by division. Best grown mounted on moist, mossy slabs. Requires humid conditions with partial shade.

A. minimiflora (Hook. f.) J.J. Sm., *Bull. Jard. Bot. Buitenzorg*, II, 25: 98, 1917. (Syn: *Saccolabium minimiflorum*, etc.). **Herbs**, epiphytic; stem erect to pendent, to 1.5 cm long. **Leaves** 3–7, spreading; lamina obliquely oblong-elliptic, 2–5.5 × 0.7–2 cm, base attenuate, apex unequally bidentate; sheath tubular, 2–3 mm long. **Inflorescences** usually simple, laxly many-flowered, ~7 mm wide; peduncle 1–3.5 cm long; sterile bracts 2 or 3; rachis 4–14 cm long; floral bracts narrowly elliptic to triangular, 1.5–1.6 mm long, subulate, acuminate; pedicel and ovary 1.5 mm long, dark green. **Flowers** ~3–3.2 mm long, quaquaversal, sepals and petals cinnamon-orange with blackish-purple warty spots, lip translucent greenish-yellow, with white tip and a reddish-brown subapical spot, column green, rostellum white, anther cap purple; sepals and petals 1-nerved; dorsal sepal erect, concave, oblong, 1.5–1.7 × 0.7–0.8 mm, obtuse, incurved; lateral sepals obliquely oblong, 1.4–1.6 × 0.8 mm, obtuse to acute; petals broadly subovate, obliquely rounded, concave, 1.4–1.7 × 1–1.3 mm, recurved and convex; lip immobile, 2.2–3 mm long, 1.5–1.7 mm deep, adnate to column, shallowly trilobed, laterally compressed, saccate portion irregular, lateral lobes erect, minute, triangular, midlobe triangular, obtuse; spur broadly conical, ~1 mm long; column ~1.4 mm long, foot absent; pollinia 4. **Distr:** Thailand, W Malesia to Philippines. **Ecol:** In lowland and hill forests on limestone; 300–900 m. Fl. May–Jul. **Zones:** 10–12.

Acampe

Species 6. **Herbs**, large, epiphytic or lithophytic; stems long, often branched near base; nodes many. **Leaves** distichous; lamina linear, apex often unequally bilobed, recurved, fleshy, coriaceous. **Inflorescences** racemes, short, lateral. **Flowers** fleshy, fragile, not resupinate; perianth not widely spreading; sepals and petals subequal, base ± narrowed; lip adnate to base of column, immobile, base shallow, saccate, unlobed or indistinctly trilobed, lateral lobes small, erect, midlobe large, fleshy, sac conic, without appendages inside, thick-walled, often short-hirsute inside or near mouth; column very short, nearly without foot; anther terminal, incumbent, opercular, with reduced partitions or nearly 1-celled; pollinia 4, unequal, in 2 groups, waxy, with a slender stipe and furcate viscidium; stigma entire; rostellum short, bifid. **Fruits** cylindrical. **Distr:** Tropical Africa to Philippines. **Cult:** Propagate by division. Best grown in a basket with a coarse bark-based mix. Requires excellent drainage and bright light conditions.

A. rigida (Buch.-Ham ex J. E. Sm.) Hunt, *Kew Bull.* 24: 98, 1970. (Syn: *A. longifolia*, etc.). **Herbs**, epiphytic; stems stout, 1 m or more long. **Leaves** linear, 25–30 × 4–5 cm, thickly coriaceous, yellowish-green, base decurrent into sheath, apex rounded, slightly oblique; sheaths tube like, embracing stem, 3 cm long. **Inflorescences** racemes, subcorymbose, lateral; bracts scalelike, 3–4 × 5–6 mm; pedicel and ovary 1.1–1.3 cm long. **Flowers** ~1.8 cm across, pale yellow with red transverse stripes; dorsal sepal oblanceolate, 1.3–1.5 × 0.7–0.8 cm, rounded; lateral sepals obovate, slightly oblique, 1.3–1.5 × 0.7–0.9 cm, obscurely keeled outside; petals oblanceolate falcate, 1.2–1.3 × 0.4–0.5 cm, base contracted, apex obtuse; lip fleshy, whitish above, yellowish below, 1 cm long, base saccate, lamina trilobed, lateral lobes erect, broadly triangular, midlobe ovate or oblong, recurved, base thickened; disk pubescent, sac shallowly conic, pubescent inside; column 2 mm high. **Fruits** nearly cylindrical, 5–6 cm long. **Distr:** India to Malaysia. **Ecol:** On exposed rocks and precipices along rivers; ~1,200 m. **Zones:** 9–12.

Aerides

Species ~20. **Herbs**, epiphytic, coarse, monopodial; roots many, thick; stems ascending, short to elongate, often stout, enclosed by leaf sheaths, nodes many, leafy. **Leaves** distichous; lamina linear-oblong to terete, base jointed and sheathing, apex bilobed, leathery, slightly fleshy. **Inflorescence** lateral, pendent, variable, racemose or paniculate, usually densely many-flowered. **Flowers** showy; sepals and petals similar, broad, spreading; lateral sepals decurrent on column foot; petals smaller than sepals; lip attached to end of column foot, usually immobile, spurred, trilobed, lateral lobes decurrent on column, erect, midlobe large or small, often erose; spur narrowly conic or horn-shaped, bent forward, usually with swellings or calli inside; column elongate, often broadened at apex, with a short or long foot; rostellum short or long, bifid; anther bilocular, ± beaked; pollinia waxy, 2, subglobular, grooved, attached by a long, slender stipe to a solitary, subcircular viscidium. **Distr:** India through Asia to New Guinea and the Philippines. **Cult:** Propagate from seeds or by division. Grow in baskets with a bark-based mix. Care should be taken when repotting to avoid damage to the roots.

A. leeana Rchb. f., *Gard. Chron.* 15: 656, 1881. (Syn: *A. jarckiana*, etc.). **Herbs**, epiphytic, to 40 cm tall; stems curved or recurved, 30–35 cm long. **Leaves** lorate, 12–20 × 2–3 cm, leathery, apex unequally bilobed, dorsally slightly carinate. **Inflorescences** subpendulous, densely many-flowered, 40–65 cm long; peduncles 30–50 cm long; pedicel and ovary white to pink. **Flowers** odorless, 1.2–1.3 × 1.8–1.9 cm, sepals and petals rose or white, spotted with light pink, midlobe of lip purple to magenta, lateral lobes pink spotted with magenta, spur magenta; dorsal sepal suborbicular-oblong, 7 × 4–5 mm; lateral sepals convex, suborbicular, 7 × 6 mm, subtruncate to rounded; petals oblong-suborbicular or ovate, 6.5 × 4 mm, rounded; lip fleshy, trilobed, lateral lobes incurved, overlapping, falcate-oblong, 3 × 2 mm, obtuse, denticulate, spur conical, incurved, 0.9–1 cm long, obtuse; column erect, fleshy, 8 mm long including. **Distr:** Philippines (Luzon). **Ecol:** ~800 m. **Zones:** 11–12.

A. multiflora Roxb., *Pl. Coromandel* 3: 68, 1820. (Syn: *A. affinis*, etc.). **Herbs**, epiphytic, large, erect to pendulous; stem stout, 10–25 cm long, many-leaved. **Leaves** distichous; lamina ligulate, deeply channeled and keeled, curved, 15–34 × 1.3–1.9 cm, apex bilobed. **Inflorescences** racemose or rarely branched, 15–34 cm long, densely many-flowered. **Flowers** 2–3.2 cm across, fragrant, white or rose-purple, sometimes spotted darker amethyst-purple at apex, lip light amethyst purple; sepals and petals subsimilar, oblong, elliptic-oblong or orbicular, 0.8–1.6 × 0.6–1 cm, rounded; lip clawed, geniculate, trilobed, convex, 1.5 × 1.1 cm, margins recurved, midlobe cordate or hastate-ovate, rounded, slightly convex above, callus fleshy, incurved, bilobed on the base of the midlobe, spur pointing forward, short, straight, compressed, 4–5 mm long; column beaked, with 2 rounded auricles on the foot. **Distr:** Himalayas to Indo-China. **Ecol:** In tropical and subtropical forests; 200–660 m. **Zones:** 8–11.

A. odorata Loureiro, *Fl. Cochinch.* 2: 525, 1790. (Syn: *Epidendrum odoratum*, etc.). **Herbs**, epiphytic; stems rather drooping, stout, branched. **Leaves** broadly lorate, 15–20 × 2.5–4.6 cm, thickly leathery, obtuse, unequally bilobed. **Inflorescences** numerous, nodding, racemose, 15–30 cm long, densely 20–30-flowered; floral bracts broadly ovate, 5–9 mm across, obtuse; pedicel and ovary 1–2.5 cm long, slender. **Flowers** fragrant, opening widely, 1.5–3 cm Ø, purple to almost white, often tipped and spotted purple, spur apex greenish-yellow; dorsal sepal elliptic, ~1 × 0.8 cm, obtuse-rounded; lateral sepals broadly ovate, ~1.2 × 0.9 cm, adnate at base to column foot, apex obtuse; petals subelliptic, 1.1–1.2 × 0.5–0.7 cm, base contracted, apex obtuse; lip immobile, spurred, trilobed, almost enclosing column, lateral lobes erect, obovate-cuneate, ~1.5 × 1 cm, distal margin irregularly toothed, apex obtuse-rounded, midlobe linear, ~1.2 × 0.3 cm, bilobed; spur bent forward, narrowly horn-shaped, ~1 cm long; column ~5 mm long, thick, foot ~9 mm long. **Distr:** India through SE Asia to Philippines. **Ecol:** In forests; 200–1,200 m. Fl. May. **Zones:** 10–12.

A. rosea Loddiges ex Lindley and Paxton, *Paxton's Fl. Gard.* 2: 109, t. 60, 1852. (Syn: *A. affinis* var. *rosea*, etc.). **Herbs**, epiphytic; stems 5–20 cm long, stout. **Leaves** narrowly oblong or lorate, 28–30 × 2–3.5 cm, fleshy, obtuse, unequally bilobed. **Inflorescences** often 1–3, densely 20–40-flowered, arising from peduncle sheath, pendulous, racemose or paniculate, 15–35 cm long; floral bracts ovate-lanceolate, 5–7 mm long; pedicel and ovary 1.2–1.4 cm long. **Flowers** white, spotted or tinged with purplish-red, particularly on lip, opening widely; dorsal sepal subobovate, ~1.2 × 0.7 cm, base contracted, apex obtuse; lateral sepals slightly oblique, ovate, ~8 × 7 mm, obtuse; petals obovate, ~1 × 0.7 cm, obtuse; lip deltoid to trullate, slightly laterally compressed, 1.4–1.6 × 0.8–1.1 cm, immobile, trilobed, acute, lateral lobes small, incurved over mouth of spur, midlobe hastate, triangular-acute; spur spreading horizontally, narrowly conic, ~5 mm long; column ~5 mm, foot ~1 mm long. **Fruits** subovoid, ~ 2 × 1.2 cm. **Distr:** E Himalayas to S China and Indo-China. **Ecol:** On forest tree trunks; 300–1,600 m. Fl. Jul. Fr. Aug–Oct. **Zones:** 8–11.

Amesiella

Species 3. **Herbs**, epiphytic, small; stems short; roots fleshy. **Leaves** several, distichous, fleshy, thick-textured. **Inflorescences** shortly pedunculate, racemose, few(usually 5)-flowered; bracts rigid. **Flowers** relatively large, white or cream; sepals and petals similar or subsimilar, spreading; lateral sepals adnate to column foot; lip inserted at apex of column foot, trilobed, spur long and narrow; column terete, erect, puberulent, with a distinct foot; rostellum trilobed; pollinia 2, entire; stipe oblong-linear; viscidium suborbicular. **Distr:** Philippines. **Cult:** Grow mounted or potted in a medium bark-based compost. Requires intermediate conditions with high humidity, good drainage and air movement.

A. monticola Cootes and D.P. Banks, *Orchids Austral.* 10(5): 26, 1998. **Herbs**, terrestrial, upright, monopodial, 2.5 cm tall by 16 cm across the leaves. **Leaves** linear to obovate, to 9.5 × 2.4 cm, apex unequally bilobed, leathery. **Inflorescences** to 3, bearing up to 5 flowers; pedicel pale green. **Flowers** ~6 cm Ø, pure white, sometimes with faint yellow to orange stripes in the lip; dorsal sepal elliptic, concave, with a strong ridge down center, to 2.5 × 1.8 cm; lateral sepals elliptic, with a strong ridge down center, 2.9 × 2.1 cm; petals elliptic, concave, to 2.5 × 2.1 cm; lip trilobed, lateral lobes semicircular, 7 × 5 mm, midlobe semicircular, appearing folded, 1 × 0.7 cm long; nectary tubular, to 11 cm long (when measured from tip of midlobe), 2.5 mm Ø. **Distr:** Philippine (Luzon). **Ecol:** On fringes of rainforests; 1,800–2,200 m. **Zones:** 11–12.

Below: *Aerides odorata*, *Aerides rosea*, *Amesiella monticola*.

GS

MS
MG

Above: *Arachnis flos-aeris, Armodorum sulingi, Ascocentrum ampullaceum.*

Arachnis

Syn: *Arachnanthe, Arrhynchium.* [Scorpion Orchids]. **Species** 11. **Herbs**, epiphytic, robust, monopodial; stems scrambling, elongate, often to several meters, branched, leafy. **Leaves** oblong, rigid, base sheathing, apex bilobed, somewhat fleshy or leathery. **Inflorescences** lateral, ascending or pendulous, rigid, often long and branched, few- to many-flowered; peduncle and rachis slender. **Flowers** fragrant, resupinate, often large and showy, usually widely opening, often scorpion- or spiderlike; sepals and petals similar, spreading, narrowly oblong to linear, usually widened toward apex; lateral sepals and petals often falcate-curved; lip short, trilobed, jointed to column foot by a short movable strap, base saccate or spurred, midlobe with a raised, central ridge or callus; column short, stout; stipe short, broad; viscidium broadly ovate; pollinia 4, appearing as 2 unequal masses, waxy. **Distr:** Indo-China to Solomon Islands. **Cult:** Propagate by division or from root stem cuttings. Best grown against a tree or mounted on long slabs. Requires a sunny position in humid to dry conditions.

A. flos-aeris (L.) Rchb. f., *Bot. Centralbl.* 28: 343, 1886. (Syn: *Arachnanthe flos-aeris*, etc.). [Scorpion Orchid]. **Herbs** terrestrial, lithophytic or epiphytic, large; stems terete, 1.2–2.5 m tall. **Leaves** ligulate or linear-oblong, 10–18 × 5 cm, apex notched. **Inflorescences** simple or paniculate, ascending or drooping, 0.6–1.6 m long; branches short, divaricate, 2–3-flowered. **Flowers** showy, 7.5–10 cm across, musk-scented, dark green or yellow, barred with maroon, lip with several orange lamellae, end of lateral lobes maroon; dorsal sepal erect, linear, to 7 × 1.6 cm; lateral sepals and petals narrowly linear-spathulate, falcate, shorter and broader than the dorsal sepal; lip trilobed, 1.5–2 × 1.4–2.2 cm, lateral lobes quadrate, recurved on upper surface, midlobe obovate, 1.3 cm long, acuminate, fleshy, shortly spurred below; column broad, truncate, 1.3 cm long. **Distr:** Thailand, W Malesia to Philippines. **Ecol:** In mangroves and along rivers; 0–1,000 m. **Zones:** 11–12.

Armodorum

Species 4. **Herbs**, epiphytic or terrestrial, scandent, monopodial, branching near base. **Leaves** distichous; lamina linear, coriaceous. **Inflorescences** few- to many-flowered, rigid, short or long, simple or branching. **Flowers** fleshy; sepals and petals free, subsimilar, spreading, narrowly spathulate; lip articulated to base of column by a strap of flexible tissue, fleshy, trilobed, lateral lobes erect, small, midlobe fleshy, dependent, ligulate or ovate, usually with 2 basal callus lobes almost obscuring the mouth of the spur; spur prominent; column short, fleshy; pollinia 4 in 2 pairs, flattened, subequal; stipe short, broad; viscidium broadly ovate to elliptic. **Distr:** Thailand to W Malesia. **Cult:** Propagate from cuttings. Best grown mounted or as hanging specimens. Requires plenty of water during growth period.

A. sulingi (Blume) Schltr., *Repert. Spec. Nov. Regni Veg.* 10: 197, 1911. (Syn: *Aerides sulingi*, etc.). **Herbs**, lithophytic or epiphytic; stems long, branching, usually rooting near base. **Leaves** linear, curved, 13–30 × 1.2–2.5 cm, apex bilobed. **Inflorescences** unbranched, 4–7-flowered; bracts scarious, broadly triangular, 5 mm long; pedicel and ovary 2.6–2.8 cm long. **Flowers** ~5 cm across, sepals and petals glossy brown, lip light yellow with brown stripes; sepals narrowly oblanceolate, 2.2–2.5 × 0.5–0.6 cm, obtuse; petals oblanceolate, 2 × 0.7 cm, obtuse; lip trilobed, 8–9 × 7 mm, lateral lobes erect, triangular 4 mm across, midlobe deflexed, convex, elliptic, obtuse, with a raised groove, white callus at base and 3 verrucose ridges in front; spur short, thick, parallel to lip midlobe, 3–4 mm long; column short, thick, 4–6 mm long. **Distr:** Sumatra to Bali. **Ecol:** In primary forests; 450–600 m. **Zones:** 11–12.

Ascocentrum

Species ~9. **Herbs**, epiphytic; stems usually short, rooting at nodes near base. **Leaves** usually distichous; lamina narrow, flat or nearly terete, sheathing at base. **Inflorescences** racemes, axillary, with many flowers facing in one direction. **Flowers** spreading or subspreading, sepals and petals similar, free; lip adnate to base of column, trilobed, base spurred, lateral lobes erect, rounded or triangular, midlobe larger, tonguelike, not flexible, spur narrow, pointing downward, without calli or appendages inside, with a small thickening at mouth near base of midlobe; column footless; anther terminal, opercular, with reduced partitions; pollinia 2, waxy, porate, attached by a slender stipe to a broad viscidium; stigma entire; rostellum bifid. **Distr:** S China, the Himalaya, Thailand, through Malaysia to Indonesia; Taiwan. **Cult:** Propagate by division of mature basal shoots. Best grown in pots or baskets, or mounted on slabs. Requires intermediate to warm conditions, plenty of water during growth period and bright light.

A. ampullaceum (Roxburgh) Schlechter, *Repert. Spec. Nov. Regni Veg. Beih.* 1: 975, 1913. (Syn: *Aerides ampullacea*, etc.). **Herbs**, epiphytic; stems erect, 1–4 cm, stout. **Leaves** 3 or 4, nearly basal; lamina narrowly oblong, 5–20 × 1–1.5 cm, flat, basally conduplicate, apex praemorse, thickly leathery, abaxially ± reddish-tinged, adaxially yellowish-green, purplish-spotted. **Inflorescences** often 2–4, erect, 5–7 cm, many-flowered; peduncle and rachis pale vermilion; floral bracts ovate-triangular, ~1 mm long, acute; pedicel and ovary pale vermilion, 1.3–1.5 cm long. **Flowers** deep pinkish-red, opening widely, 1.5–1.8 cm Ø, anther cap and pollinia purplish; sepals and petals broadly ovate, 7–9 × 4–6 mm, obtuse; lip trilobed, lateral lobes erect, subtriangular, very small, obtuse, midlobe narrowly oblong, or ligulate, 6–7 × 1–2 mm, obtuse to acute, base with 2 calli at spur entrance; spur pale, cylindric, 8–9 × ~2.5 mm, apical

Below: *Ascocentrum aurantiacum* ssp. *philippinense, Ascocentrum miniatum, Ascoglossum calopterum, Ceratocentron fesselii.*

half ± curved and slightly dilated; column ~2 mm long, thick. **Distr:** Nepal to Indo-China. **Ecol:** On tree trunks in broad-leaved evergreen forests; 1,100–1,500 m. Fl. Apr–May. **Zones:** 9–12.

A. aurantiacum* ssp. *philippinense Christenson, *Lindleyana* 7: 88, 1992. (Syn: *Saccolabium miniatum* var. *crinitum*). **Herbs**, epiphytic, upright, monopodial, to 25 cm tall, with vandaceous habit. **Leaves** to 16 × 1 cm, arching. **Inflorescences** upright, to 40-flowered. **Flowers** ~9 mm Ø, bright orange, side lobes and throat of spur darker orange, anther cap dark red; petals and sepals elliptic, to 7 × 3 mm; lip with midlobe tongue-shaped, lateral lobes to 2.5 mm high; spur noticeably curved forward. **Distr:** Endemic to Philippines. **Ecol:** Epiphytic in deeply shaded locations; 1,200 m. **Zones:** 11–12. **Info:** The Philippine plants differ from the original Sulawesi populations by being smaller in all measurements (Cootes 2001), and the flowers are lighter in color.

A. miniatum (Lindl.) Schltr. ex J.J. Sm., *Bull. Jard. Bot. Buitenzorg*, II, 14: 49, 1914. (Syn: *Gastrochilus miniatus, Saccolabium miniatum*). **Herbs**, epiphytic or terrestrial, 25–30 cm tall, often forming dense clumps; stem erect, stout, covered by persistent distichous leaf bases, to ~20 cm long. **Leaves** distichous; lamina ligulate, narrowly strap-shaped, conduplicate, 5–25 × 1–1.4 cm, apex dentate, thickly textured, often speckled with reddish-brown in strong lighting conditions. **Inflorescences** racemes, ± cylindrical to conical, erect, to 25 cm long, densely many-flowered, usually over 30-flowered. **Flowers** small, showy, ~1–1.5(–2) cm across, orange or scarlet; sepals elliptic-obovate, 8 × 3 mm, obtuse; petals similar to sepals but shorter; lip ligulate, 4 × 1.5 mm, subacute, deflexed; spur pendent, cylindrical, 0.7–1.2 cm long, straight, dependent, somewhat translucent at base, apically swollen; column short, terete; anther cap dark brownish-black. **Distr:** Assam to W Malesia and Philippines. **Ecol:** In deciduous forests; 0–1,200 m. **Zones:** 10–12.

Ascoglossum

Species 1. *Generic description same as specific.* **Cult:** Propagate by division or from seeds. Best grown mounted on a long slab. Requires a warm, humid and well-lit conditions.

A. calopterum (Rchb. f.) Schltr., *Repert. Spec. Nov. Regni Veg. Beih.* 1: 975, 1913. (Syn: *Cleisostoma cryptochilum*, etc.). **Herbs**, epiphytic, monopodial; stems ascending or arcuate, 40–50 cm long. **Leaves** arcuate-spreading; lamina straplike, 10–25 × 2–2.5 cm, bases sheathing to 2.5 cm long, apex unequally bilobed, obtuse. **Inflorescences** branching, 30–60 cm long; peduncle 20–30 cm long, purple; branches ~6-flowered; bracts triangular-ovate, 2–3 mm across; pedicel and ovary 1.3–1.8 cm long, purple. **Flowers** deep rose-purple, lip suffused with orange, bilaterally flattened; dorsal sepal reflexed, oblong-spathulate, 1.4 × 0.5 cm, obtuse; lateral sepals reflexed, narrowly spathulate, 1.5–1.6 × 0.6–0.9 cm, obtuse or acute; petals reflexed, narrowly elliptic, 1.1–1.2 × 0.4 cm, acute or obtuse; lip trilobed, 1.2 cm long, lateral lobes erect, attached to column, 4 × 2.5 mm, midlobe ligulate, 2 × 1 mm, acute; spur clavate, 5–9 mm long; column 5–6 mm long. **Distr:** Philippines to Solomon Islands. **Ecol:** On large trees in primary rainforests; 0–600 m. **Zones:** 11–12.

Ceratocentron

Species 1. *Generic description same as specific.* **Cult:** Propagate from seeds. Can be grown in pots, baskets, or mounted with the roots covered by sphagnum moss. Requires moist, humid and well-lit conditions.

C. fesselii Senghas, *Orchidee* (Hamburg) 40: 90, 1989. (Syn: *Tuberolabium calcaratum*). **Herbs**, epiphytic, upright, monopodial, miniature, 2.5 cm high, ~4 cm across. **Leaves** elliptic, 2.5 × 1 cm, basally conduplicate and clasping the stem, apex obtuse, very leathery, dark green. **Inflorescences** usually 2 per leaf axil, upright, 2–4-flowered. **Flowers** relatively large, to 1 cm Ø, very bright red or orange, lip white with pinkish-purple markings; dorsal sepal lanceolate, to 7 × 4 mm; lateral sepals ovate, to 5 × 4 mm; petals ovate, 5 × 4 mm; lip small, firmly attached to the column, with a little horn just below the entrance to the nectary, spur ~6 mm long. **Distr:** Philippines; limited distribution. **Ecol:** Epiphytic on tree trunks, in montane forests; ~1,000–1,200 m. **Zones:** 11–12.

Ceratochilus

Species 1. *Generic description same as specific.* **Cult:** Propagate from seeds. Best grown mounted. Requires intermediate conditions with high humidity and moderate shade.

C. biglandulosus Blume, *Bijdr.*: 358, 1825. **Herbs**, epiphytic, small, monopodial; stems usually short, 2–10 cm long. **Leaves** distichous; lamina V-shaped in cross-section, 1 cm long, 2.5 mm thick, fleshy. **Inflorescences** several, lateral, 1-flowered. **Flowers** relatively large, ~3 cm across, sepals, petals and spur white, turning pink to red as they age, lip green; sepals and petals free, spreading, petals narrower and shorter than the sepals; lip with a large saclike spur, hairy inside, and a short lamina lying in front of the spur; pollinia 4, in 2 pairs. **Distr:** Java. **Ecol:** Epiphytic on lichen-covered branches on isolated trees in montane grasslands; 1,000–2,000 m. **Zones:** 10–12.

Chiloschista

Species ~10. **Herbs**, epiphytic or lithophytic, usually stemless and leafless; roots green, photosynthetic, flat, scabrid. **Inflorescence** often pendulous, racemose, hairy or glabrous. **Flowers** ephemeral; sepals and petals subequal; lateral sepals and petals often adnate to column foot; lip ± clawed, hinged to apex of column foot, pouched or spurred, trilobed, lateral lobes erect, large, midlobe small; disk with a fingerlike callus, hirsute-pubescent; column semiterete, short, foot long; clinandrium shallow; stigma deep-set; rostellum deflexed, bifid; anther cap with 2 long filiform setae; pollinia 4, in 2 closely appressed, unequal, sessile pairs, stipe sublinear. **Distr:** Indian Subcontinent through SE Asia to Australia. **Cult:** Propagate from seeds. Best grown on a small slab or a small piece of bark. Requires warm, humid conditions with regular watering.

C. lunifera (Rchb. f.) J.J. Sm., *Orch. Java*: 553, 1905. (Syn: *C. indica*, etc.). [Thailand Moon Orchid]. **Herbs**, epiphytic, leafless, small; roots long, greenish; stem very short. **Inflorescence** erect or suberect to arching or pendulous, to ~20 cm long, 12(or more)-flowered; peduncle green, purple-spotted, pubescent, terete; bracts small, ovate, acute. **Flowers** 1.1–1.6 cm across, sepals and petals yellow or greenish-yellow, red-brown at center, lip white, yellowish on lateral lobes and below, lined with crimson on lateral lobes, column green; sepals and petals subsimilar, elliptic to ovate, 7–9 × 6 mm, rounded or obtuse; petals ciliate; lip trilobed, saccate at base, 8 mm long, 8 mm wide when spread, lateral lobes subquadrate, erect-incurved, midlobe smaller, blunt, emarginated; disk with an erect, fleshy, pubescent callus; column fleshy, 4 mm long, foot spotted with red, 4 mm long. **Distr:** Himalayas to Indo-China, Java. **Ecol:** In forests; 800 m. Fl. Feb–Jul, Oct–Dec. **Zones:** 9–12. **Info:** *C. lunifera* f. *alba* is yellow-colored form.

C. segawae (Masamune) Masamune and Fukuyama, *Bot. Mag.* (Tokyo) 52: 247 ('segawai'), 1938. (Syn: *C. hoi*, etc.). **Herbs**, epiphytic; stems inconspicuous, leafless; roots to 35 cm long. **Inflorescence** racemose, 6–15-flowered, often pendulous, 5–15 cm long; peduncle and rachis densely pubescent; floral bracts ovate-triangular, 3–5 mm long, membranous; pedicel and ovary 4–5 mm long, pubescent. **Flowers** slightly fleshy, whitish-green or yellow, sometimes with brownish-red patches near base of sepals and petals, lateral lobes of lip brown-striped; dorsal sepal nearly broadly elliptic, 3–5 × 2.5–4 mm, glabrous or sparsely pubescent abaxially, subobtuse; lateral sepals 4.5–6 × 3.5–4 mm; sepals slightly oblique, broadly ovate or elliptic, sparsely pubescent near abaxial base; petals elliptic, 4–6 × 3.5–4 mm, obtuse; lip 5–6 mm, base saccate, trilobed, lateral lobes erect, falcate-oblong, rounded, midlobe very small, thick, retuse, base with a longitudinal bilobed callus, pubescent, grooved; sac 2–3 mm long; column ~1 mm long, foot ~3 mm long; anther cap with 2 filiform appendages. **Distr:** S Taiwan. **Ecol:** On tree trunks in forests; 700–1,000 m. Fl. Apr–May. **Zones:** 9–10.

Chroniochilus

Species 4. **Herbs**, epiphytic, caulescent. **Leaves** thickly coriaceous. **Inflorescences** with flowers 2-ranked;

Below: *Ceratochilus biglandulosus, Chiloschista lunifera, Chiloschista lunifera* (albino form), *Chiloschista segawae*. [All MG except *Ceratochilus biglandulosus* GS.]

Above: *Chroniochilus virescens, Cleisocentron merrillianum, Cleisostoma racemiferum*. [All MG.]

peduncle short, glabrous. **Flowers** lasting 1 to several days; lip elastically attached to the column foot, sessile, saccate, conical, conical part solid, spur absent, callus of different substance from the rest of the lip, lateral lobes earlike, midlobe very fleshy; column foot distinct, long, with free end, jointed to the lip; pollinia 2, quite entire or with a very small notch, stipe linear-oblong. **Distr:** Peninsular Thailand to W Malesia. **Cult:** Propagate from seeds. Best grown mounted on a slab. Requires moist, humid conditions.

C. virescens (Ridl.) Holttum, *Kew Bull.* 14: 273, 1960. (Syn: *Sarcochilus virescens*). **Herbs**, epiphytic; stem very short, stout, erect. **Leaves** spreading; lamina lanceolate or narrowly strap-shaped, 3.8–6.6 × 1.2–1.8 cm, base narrowed, apex subacute or rounded, unequally and acutely bilobed, thickly fleshy. **Inflorescence** a raceme, 8-flowered, 2.5 cm long, flowers opening successively; peduncle flattened; bracts ovate; floral bracts broadly ovate, acuminate and apiculate, fleshy; pedicel and ovary ~1 cm long, stout. **Flowers** very fragrant, fleshy, 2.5 cm across, sepals and petals uniformly green, lip white; sepals ovate, acute, apiculate; dorsal sepal to 1.3 × 0.8 cm; petals narrower, lanceolate, 1–1.2 × 0.45–0.55 cm, acute; lip very short, trilobed, ~1.3 cm long, ~1 cm across lateral lobes, lateral lobes short, rounded, erect, reaching column, midlobe ~1.1 cm across, with an orange callus, spur short, broadly conic; column short and thick, ~4 mm long, foot as long, curved; anther orange, beak triangular. **Distr:** Peninsular Thailand to W Malesia. **Ecol:** In hill forests; 100–1,000 m. **Zones:** 11–12.

Cleisocentron

Species 5. **Herbs**, epiphytic. **Leaves** linear-ligulate. **Inflorescences** racemose, many-flowered, axillary. **Flowers** with lip firmly adnate to column, urceolate, lobulate above; column erect, cylindric, foot ± developed, decurrent on the dorsal wall of the lip as a thickened cushion, clinandrium shallow; stigma oval, vertical; rostellum distinct, porrect or arrect, bilobed with broad lobes; pollinia 2, globose, each deeply split into unequal sizes on a single, linear stipe, gland large, subquadrate. **Distr:** E Himalayas to N Borneo. **Cult:** Propagate from seeds. Best grown mounted on a tree-fern slab or cork slab with plenty of sphagnum moss. Requires humid conditions.

C. merrillianum (Ames) Christenson, *Amer. Orchid Soc. Bull.* 61: 246, 1992. (Syn: *Sarcanthus merrillianus, Robiquetia merrilliana*). **Herbs**, epiphytic, robust; stems 20–90 cm long, terete, enclosed in tubular leaf sheaths 1.3–4 cm long. **Leaves** *juvenile* linear-ligulate, 12–13 × 0.8–1.1 cm, base conduplicate, apex obtusely unequally bilobed or acute, thickly coriaceous; *mature* leaves terete, falcate or straight, 12–20 × 0.2–0.4 cm, base conduplicate, apex acute. **Inflorescences** 8–15-flowered, opposite leaves, flowers opening together; peduncle 0.5–1.5 cm long, with 2–3 obtuse bracts; rachis abbreviated, 6–7 mm long; floral bracts ovate, 1–2 mm long, acute; pedicel and ovary 1.5 cm long. **Flowers** ~1.5 cm long, long-lasting, lavender blue, lip with deep blue markings, column white or pale yellow; dorsal sepal ovate, strongly concave, 4–6 × 2.8 mm, broadly rounded, sometimes obliquely bilobed; lateral sepals obovate-oblong, oblique, 4.5–6 × 3 mm, apex slightly cucullate; petals oblong-spathulate, 3.5–5 × 2 mm, broadly rounded; lip trilobed, lateral lobes ~1 × 2–2.5 mm, midlobe triangular upcurved; spur cylindric, 1.1–1.2 mm long, curved, obtuse; column erect, cylindric, 3–4 mm long, fleshy. **Distr:** Borneo (Sabah: Mt Kinabalu). **Ecol:** In montane riverine forests; 1,100–3,000 m. Fl. Jul–Aug, Oct–Nov. **Zones:** 11–12.

Cleisostoma

Syn: *Sarcanthus*.

Species ~100. **Herbs**, epiphytic or rarely lithophytic, monopodial, small to medium-sized; stems erect or pendulous, rarely creeping, long or short, rigid, many-noded. **Leaves** few to many, distichous; lamina dorsiventrally flattened or terete, base usually jointed and sheathing, apex bilobed or acute. **Inflorescence** racemose or paniculate, lateral, axillary or opposite leaves, erect, horizontal, or pendent, few- to many-flowered; floral bracts small, much shorter than pedicel and ovary. **Flowers** small, ± fleshy; sepals free, usually spreading; petals free, often smaller than sepals; lip adnate to column by its base or back margins of lateral lobes, immobile, spurred at base, trilobed; disk often with longitudinal lamellae or ridges; spur conic, cylindric, or saccate, often longitudinally septate inside, rarely septum reduced to a narrow keel, usually with a callus on back wall and often with outgrowths on front wall closing entrance; column short, stout, foot short; rostellum small; pollinia 4. **Distr:** Tropical and subtropical Asia to W Pacific. **Cult:** Propagate by division or from seeds. Best grown in pots or mounted on slabs. Requires humid and well-lit conditions with copious watering.

C. racemiferum (Lindley) Garay, *Bot. Mus. Leafl.* 23: 173, 1972. (Syn: *Saccolabium racemiferum*, etc.). **Herbs**, epiphytic; stems erect, 5–20 cm long, branched. **Leaves** distichous; lamina flat, lorate, ~29 × 3–4 cm, apex obtuse, unequally bilobed. **Inflorescence** axillary, longer than leaves, many-branched, laxly many-flowered; floral bracts ovate-lanceolate, small, 1–2 mm long; pedicel and ovary ~4 mm long. **Flowers** with sepals and petals yellow with brownish-red spots, lip white; dorsal sepal suboblong, concave, ~3.5 × 2.5 mm, obtuse; lateral sepals slightly obliquely oblong, ~3.5 × 3 mm, base adnate to column foot, apex rounded; petals oblong, ~3 × 2 mm, obtuse; lip trilobed, lateral lobes erect, deltoid, midlobe spreading, deltoid, obtuse, with a keel running down to spur linking to rudimentary septum; spur conic, 2–2.5 mm long, callus inside spur ovate-triangular, base with papillate-hairy, slightly bilobed; column ~2.5 mm long, winged. **Distr:** E Nepal to SC China. **Ecol:** On tree trunks of open

Below: *Cleisostoma uraiensis, Dimorphorchis graciliscapa, Dimorphorchis lowii* (lower flowers), *Dimorphorchis lowii* (upper flowers).

MS

MG

GS GS

forests; 600–1,800 m. Fl. Jun. **Zones:** 8–10.

C. uraiense (Hayata) Garay and H.R. Sweet, *Orchids S. Ryukyu Is.* 156, 1974. (Syn: *C. viridescens*, etc.). **Herbs**, epiphytic; stems erect or pendulous, 20–40 cm long, 4–6 mm Ø, many leaved. **Leaves** distichous; lamina linear or narrowly oblong, 10–20 × 1.2–2 cm, apex obtuse and unequally bilobed. **Inflorescence** lateral, opposite to leaves, unbranched or with short branches, many-flowered; floral bracts deltoid, small, ~1 mm long; pedicel and ovary 3–5 mm long. **Flowers** yellowish-green, slightly fleshy; dorsal sepal subelliptic or obovate-oblong, concave, ~4 × 2.5 mm, obtuse; lateral sepals similar to dorsal sepal, concave; petals obliquely oblong, 2.5–3.5 × ~1.2 mm, obtuse; lip thickly fleshy, base spurred, trilobed, lateral lobes erect, falcate-triangular, small, midlobe suborbicular, ~1 mm long; spur ~2 mm long, front wall callus large, glabrous, slightly septate; column ~2 mm long; anther cap slightly narrowed toward apex; stipe slender; viscidium suborbicular. **Distr:** Japan (Ryukyu Islands), Philippines, S Taiwan. **Ecol:** On tree trunks in forests; ~200 m. Fl. Apr–Sep. **Zones:** 8–12.

Dimorphorchis

Species 5. **Herbs**, epiphytic, large; stems pendent, branching, rooting near base. **Leaves** distichous; lamina linear-ligulate, base sheathing, coriaceous. **Inflorescences** pendent, unbranched, laxly many-flowered, with a limp tomentose axis, flowers of 2 types on the same inflorescence. **Flowers** large, fleshy, fragrant; *basal* 1–3, very fragrant; *apical* (or the rest of the flowers) less fragrant, distinctly colored; sepals and petals subsimilar, spreading; lip fleshy, versatile, trilobed, saccate, midlobe with a raised median keel interrupted near base with a transverse groove, attenuated at apex; column short, fleshy; pollinia 4, in 2 pairs, subequal; stipe short, broad; viscidium broadly elliptic. **Distr:** Borneo. **Cult:** Propagate by division. Best grown in large containers with coarse materials such as charcoal. Requires frequent watering during the growth period.

D. graciliscapa (A.L. Lamb and Shim) P.J. Cribb, *Malesian Orchid J.* 2: 80, 2008. (Syn: *D. rossii* var. *graciliscapa*). **Herbs**, epiphytic, monopodial, robust, vandaceous, to 2 m long; stems simple, spreading to pendent. **Leaves** linear-ligulate, ~35–56(–70) × 3–6 cm, articulated to sheathing bases 4–7 cm long, apex unequally obliquely and obtusely bilobed, coriaceous, with some dark purplish-brown mottling and staining on the sheaths and leaves. **Inflorescences** 0.45–3 m long, pendulous, laxly usually 15–17-flowered, flowers spirally arranged, *basal* borne ~4–9cm apart, *distal* borne ~5–28 cm apart; peduncle and rachis tomentose; floral bracts ovate-elliptic, 1.2–2 × 1.2–1.4 cm, acuminate, base tomentose, brownish; pedicel with ovary 0.9–2 cm long, clavate, gently curving, olive-green, densely pubescent. **Flowers** dimorphic, resupinate, rigid and fleshy, sepals stellately pubescent on the reverse; *basal flowers* 2 or 3, 5.5 × 5.5–6 cm, strongly sweetly scented, orange-yellow to orange, with a few tiny red spots on the interior surface of the sepals and petals; *remaining flowers* 12–15, 4–5.5 cm high, 5(–6) cm across, unscented, creamy-white to pale yellow with small irregular red or purplish-brown spots or rings all over the interior surface of the sepals and petals; sepals and petals spreading; sepals free, oblong-ovate, obtuse; dorsal sepal 2.6–3 × 1.4–1.7 cm, apex reflexed; lateral sepals 2.8–3.2 × 1.3–1.6 cm, slightly undulate and with slightly incurved margins in distal flowers; petals elliptic, 2.1–2.8 × 1.4–1.8 cm, obtuse, margins of distal flowers undulate; lip 1.1–1.3 × ~1.5 cm, very shortly clawed, mobile, L-shaped in side view, fleshy, trilobed, lateral lobes erect, oblong, 5 × 5 mm; midlobe triangular, 4–5 × 3 mm, shortly apiculate, fleshy, callus keeled in front, ridge behind; column 0.6–0.8 × 0.8–1 cm; pollinia 4. **Distr:** N Borneo (Sabah). **Ecol:** In cool lower montane oak-chestnut forests; ~300 m. **Zones:** 11–12.

D. lowii (Lindl.) Rolfe, *Orchid Rev.* 27: 149, 1919. (Syn: *Arachnanthe lowii*, etc.). **Herbs**, epiphytic; stems long, pendent, stout, to 75 cm long, 1–2 cm Ø. **Leaves** ligulate, to 70 × 5 cm, base articulated and sheathing, apex unequally bilobed. **Inflorescences** pendent, laxly many-flowered, to 3 m long, flowers of 2 types; bracts lanceolate, to 2 × 0.7 cm; pedicel and ovary furfuraceous, ~1 cm long. **Flowers** *basal* yellow, purple-spotted, strongly fragrant, 4.5 cm high, 6 cm across, stellately hairy on outer surface, with sepals and petals overlapping; *apical* flowers yellow with large maroon blotches, 6 × 6.5 cm, with sepals and petals not overlapping and with margins undulate, stellately hairy on outer surface; sepals spreading-recurved, ovate-lanceolate, 3–3.5 × 1.5 cm acute; petals spreading-recurved, narrowly elliptic, 2.8–3.5 × 1.2–1.5 cm, margins undulate, apex acute; lip saccate, 1.2 cm long, trilobed, lateral lobes roundly triangular, 6 mm long, margins incurved, midlobe fleshy, at an obtuse angle to basal part of lip, laterally compressed, extending into an upcurved attenuated tip, callus oblong, strongly elevated; column terete, 7 mm long, pubescent. **Distr:** Borneo. **Ecol:** Epiphytic on tall trees by rivers; to 1,200 m. **Zones:** 11–12.

Diploprora

Species ~2. **Herbs**, epiphytic; stems, terete or slightly flattened, ascending or pendulous. **Leaves** distichous; lamina falcate-lanceolate, base articulated with sheath, ± fleshy. **Inflorescence** a raceme, lateral, few-flowered. **Flowers** with sepals subequal, widely spreading, keeled outside; petals narrower than sepals; lip adnate to base of column, sessile, not movable, consisting of a hypochile and an epichile, hypochile concave, broadly cymbiform, fleshy, margin with a thickened raising wall; disk thickened or with a large callus near center, epichile abruptly narrowed into a thin slender tip, ending in 2 narrow lobes; column short; anther terminal, opercular, 2-celled; pollinia 4, in 2 pairs of 1 large and 1 small, waxy, attached by a short stipe to a small viscidium; stigma entire; rostellum short, retuse. **Distr:** Himalaya, Myanmar, S China and Thailand; Taiwan. **Cult:** Propagate from seeds. Best grown mounted on a slab. Requires humid conditions.

D. championii (Lindl.) Hook. f., *Fl. Brit. India.* 6: 26, 1890. (Syn: *Cottonia championii*, etc.). [Cascading Orchid]. **Herbs**; stems 5–15 cm long; internodes 0.5–2 cm long, surrounded by persistent leaf sheaths. **Leaves** linear, falcate or linear-oblong, 8–12 × ~2 cm, apex acute, sometimes unequally bilobed, thick-fleshy, margins thin; sheath tubular, oblique at mouth. **Inflorescences** racemes, 5–8 cm long, 3–5-flowered; bracts scalelike, 2 mm long; pedicel and ovary 0.8–1 cm long. **Flowers** pale yellow, ~1.5 cm Ø; sepals oblong or oblong-oblanceolate, 7–9 × 3–4 mm, obtuse or rounded at apex; petals linear-oblong, 8 × 2–3 mm, rounded at apex; lip white or pale yellow, sometimes with yellowish or brownish stripes or hues, with an orbicular thick hypochile and a slender caudate tail (epichile), hypochile 4–5 mm Ø, thickened along margin, with a raised central callus, tail 3–5 mm long, with 2 narrowly acuminate apical lobes; column ~3 mm long. **Fruits** linear-fusiform, 4–5 cm long. **Distr:** S China, the Himalaya, India (Deccan Peninsula), Sri Lanka, Myanmar, Thailand, Vietnam, Taiwan. **Ecol:** In forests at low elevations. **Zones:** 8–12.

Esmeralda

Species 2. **Herbs**, epiphytic, erect or pendent, large; roots aerial, piercing through leaf sheaths along stem; stem scandent, terete, stout, with many nodes. **Leaves** distichous, twisted so that all the leaves lie in a single plane; lamina flat, usually oblong, base sheathing, jointed, apex unequally bilobed, thickly leathery. **Inflorescence** lateral, erect, racemose, few-flowered; floral bracts persistent, scarious, sheathing. **Flowers** large, showy, thickly textured; sepals and petals free, similar, spreading, overlapping, ovate to oblong; petals slightly smaller; lip free, pendent, movable, trilobed, lateral lobes small, midlobe reniform, clawed, margin crenate; disk with 2 calli; column short, stout, both sides winged, foot absent; anther cap terminal; pollinia 4, in 2 pairs, waxy, dorsally compressed, subglobular, unequal in size, attached by a triangular stipe to a large, saddle-shaped viscidium. **Distr:** C Himalayas to China (Hainan). **Cult:** Propagate from seeds. Requires a very coarse bark-based compost, with ample water and light.

E. clarkei Rchb. f., *Gard. Chron.*, n.s., 26: 552, 1886. (Syn: *Arachnanthe bella*, etc.). **Herbs**, epiphytic; stems to 1 m tall. **Leaves** distichous; lamina narrowly oblong, 13–25 × ~1.73 cm, apex obtuse, slightly unequally bilobed, leathery. **Inflorescence** to 32 cm long, unbranched, 3- or 4-flowered; peduncle 3–5 cm long; rachis 12–15 cm long; floral bracts broadly ovate, 1–1.7 cm across; pedicel and ovary

Below: *Diploprora championii*, *Esmeralda clarkei*. [Both MG.]

MG

Top row: *Euanthe sanderiana, Gastrochilus acaulis, Gastrochilus acutifolius.*
Bottom row: *Gastrochilus fuscopunctatus, Gastrochilus japonicus.* [Both MG.]

~3 cm long. **Flowers** slightly fragrant, opening widely, pale yellow with reddish-brown transverse stripes, 5.5–7.5 cm Ø, thickly textured; dorsal sepal oblanceolate-elliptic, ~ 3.5 × 1–1.2 cm, obtuse; lateral sepals obovate-elliptic, falcate, 3–3.2 × 0.9–1.2 cm; petals subspathulate, falcate, 2.8–3.1 × 0.9–1.1 cm, obtuse; lip pendent, ~1.7 cm long, trilobed; claw ~2 mm long, hypochile 6–8 × 5–6 mm, with 2 auriculate lateral lobes, mesochile narrow, 4–5 × 1–1.5 mm, epichile reniform, 8–9 × 5–6 mm, margin crenulate; disk with 7–9 radiating keels and 2 raised central calli; column clavate, 1.2–1.4 cm long. **Fruits** ellipsoid-cylindric, 3.5–5.5 × 1.5–2 cm. **Distr:** C Himalayas to China (Hainan). **Ecol:** Along valleys and on tree trunks of open forests; 500–2,100 m. Fl. Oct. Fr. Nov–Dec. **Zones:** 8–10.

Euanthe

Species 1. *Generic description same as specific.* **Cult:** Propagate from seeds. Best grown in pots with a coarse bark mixture. Requires intermediate to warm temperatures, bright light, plenty of water and fertilizer.

E. sanderiana (Rchb. f.) Schltr., *Orchideen*: 568, 1914. (Syn: *Esmeralda sanderiana*, etc.). **Herbs**, epiphytic, large, monopodial, to 1 m tall; stems elongate, enveloped in imbricating sheathing bases. **Leaves** distichous, articulated below to persistent leaf sheaths; lamina ligulate, 30–40 × 3 cm, apex unequally bilobed or truncate, coriaceous. **Inflorescence** erect to suberect, axillary, 6–10-flowered, mostly shorter than the leaves. **Flowers** 8–11.5 cm across, flat, very showy, pink and pale orange, lined and suffused with deep red, dorsal sepal pale rose, suffused with white, lateral sepals tawny-yellow with carmine-red anastomosing venation, petals similar in color to dorsal sepal, but with tawny blotch spotted with red on lower basal part, lip variable in color at base, mostly dull tawny-yellow, inside red-streaked, apex reddish-brown, column buff-yellow; sepals elliptic-suborbicular to broadly obovate, 5–6 × 4 cm, obtuse or rounded; petals ovate-rhombic, 4 × 3.5 cm, obtuse; lip trilobed, to 3 × 2 cm, base concave, lateral lobes rounded, erect, midlobe fleshy, shortly clawed, spreading, oblong-rounded, apex strongly recurved, 1.5 cm long; disk prominently 3-ridged; column very short, stout, 6 mm long. **Distr:** Philippines (Mindanao). **Ecol:** In coastal woodlands; 0–500 m. **Info:** *E. sanderiana* var. *albata* is an albino variety with dorsal sepal and petals almost white and lateral sepals almost greenish-yellow. **Zones:** 8–10.

Gastrochilus

Species 54. **Herbs**, small, epiphytic, without pseudobulbs; stems elongate or rather short, erect prostrate, rooting from whole length or near base; sheaths persistent, often closely clasping stem. **Leaves** many, distichous or nearly clustered; lamina fleshy, ± coriaceous, base often twisted with lower surfaces facing substrate. **Inflorescences** racemes, 2–6-flowered, lateral; peduncle short. **Flowers** with perianth widely spreading, fleshy; sepals and petals similar, oblong or ovate, free; lip firmly attached to base of column, consisting of a cuplike or conic sac or spur (hypochile) and a central limb (epichile or terminal lobe), sac adnate to base of column; limb spreading, perpendicular to sac, large or small, often with a thick central callus or cushion, margin entire or ciliate to fimbriate; column stout, truncate apically, without foot; anther terminal, opercular, 1-celled or with reduced partitions; pollinia 2, waxy. **Distr:** Japan, Taiwan and China, westward to the Himalayas and southward to Indonesia. **Cult:** Propagate from seeds. Best grown in small baskets, or mounted on a slab. Requires high humidity and shade.

G. acaulis (Lindl.) Kuntze, *Revis. Gen. Pl.* 2: 661, 1891. (Syn: *Cleisostoma acaule*, etc.). **Herbs**, epiphytic; stems very short, non-pseudobulbous, 1–2 cm long. **Leaves** 2–4, distichous; lamina linear-oblong, subfalcate, 4.5–7.5 × 0.75–1.1 cm, base petiolate, sheathing, apex unequally and obtusely 2-notched, coriaceous. **Inflorescence** racemes, 2- or 3-flowered, very short, decurved; peduncle stout, green with 2 brown, sterile bracts; floral bracts 2 × 3 mm, broadly ovate or semicircular, rounded; pedicel and ovary 8.4 mm long. **Flowers** 6 mm across, pale green, purple-spotted, lip white, sepals and petals 3-veined; dorsal sepal obovate-oblong or spathulate, 6 × 2.7 mm, concave or incurved, truncate; lateral sepals obovate-oblong or spathulate, to 5.6 × 2.8 mm, obtuse; petals 5.2 × 2.2 mm, similar to sepals, concave, rounded; lip broadly conic-saccate, 5 × 7 mm, lateral lobes short, midlobe triangular, acute, fimbriate; column bluntly conic-ovate, 1.7 × 2.4 mm, emarginate, bifid or truncate; rostellum bifid; pollinia 2. **Distr:** India, Sri Lanka. **Ecol:** In submontane, or tropical wet evergreen forests; ~900–1,500 m. Fl. Feb–May. **Zones:** 8–12.

G. acutifolius (Lindl.) Kuntze, *Revis. Gen. Pl.* 2: 661, 1891. (Syn: *Saccolabium acutifolium*, etc.). **Herbs**, epiphytic; stems slender, 10–30 cm long, base enveloped with leaf bases. **Leaves** distichous; lamina linear-oblong or oblong, to 12 × 2 cm, apex acute or acuminate, not notched, leathery. **Inflorescences** racemes, 10–20-flowered, densely congested and umbel-like or corymbose, 5 cm long; pedicel and ovary 1.2 cm long. **Flowers** 2 cm Ø, fragrant, greenish-yellow, sometimes with brown-purple spots, lip white; sepals and petals oblong, blunt; dorsal sepal 1 × 0.6 cm; lateral sepals 1 × 0.5 cm; petals 9 × 4 mm, sometimes very slightly falcate; lip 0.5 × 1 cm, trilobed, midlobe broadly triangular, 0.4 × 1 cm, adaxially papillose but not hairy, fringed; column ~4 mm high, curved forward. **Distr:** Nepal to Indo-China. **Ecol:** In open primary, mossy, semideciduous, mixed forests; 700–900 m. **Zones:** 9–10.

G. fuscopunctatus (Hayata) Hayata, *Icon. Pl. Formosan.*, Suppl. 6: 78, 1917. (Syn: *Saccolabium fuscopunctatum*). **Herbs**, epiphytic; stems often pendulous, 3–7 cm long. **Leaves** distichous; lamina oblong or falcate-oblong, 1.5–2.2 × 0.3–0.5 cm, apex acute and slightly trilobed, green, with dense purplish-red spots. **Inflorescence** subumbellate, often 1- or 2-flowered; peduncle 0.8–2 cm long, base with 1 or 2 sheaths; floral bracts ovate-triangular, 1–2 mm long, acute; pedicel and ovary ~7 mm long. **Flowers** with sepals and petals yellowish-green, brown-spotted,

Top row: *Grosourdya appendiculata, Holcoglossum amesianum, Holcoglossum flavescens.*
Bottom row: *Holcoglossum subulifolium, Holcoglossum wangii, Luisia magniflora.* [All MG.]

lip whitish, with a yellow and brown-spotted cushion on epichile; dorsal sepals elliptic, ~4 × 2.3 mm, obtuse; lateral sepals obliquely ovate-oblong, 3–4 × ~2 mm, subacute; petals subelliptic, ~3.2 × 2.2 mm, rounded; lip with an epichile and a saccate hypochile, epichile elliptic-orbicular, concave, ~2.5 × 2 mm, thickly fleshy, abaxially carinate, adaxially glabrous, with a central cushion, entire, apex obtuse, hypochile cupular, ± laterally compressed, ~5 × ~3.2 mm; column very short, both sides with purplish-red spots; anther cap narrowed into a beak. **Distr:** N and C Taiwan. **Ecol:** On tree trunks in dense forests; 1,000–2,500 m. Fl. Jan–Jul. **Zones:** 9–10.

G. japonicus (Makino) Schlechter, *Repert. Spec. Nov. Regni Veg. Beih.* 12: 315, 1913. (Syn: *Saccolabium japonicum*, etc.). [Yellow Pine Orchid]. **Herbs**, epiphytic; stems ascending or pendulous, 2–10 cm long. **Leaves** distichous, closely spaced; lamina ± falcate, linear-oblong, or linear-lanceolate, 5–14 × 1–2.5 cm, apex subacute, slightly hooked, fleshy. **Inflorescence** subumbellate, 4–7(–10)-flowered; peduncle 1.5–2 cm long; floral bracts ovate-triangular, 2–3 mm long, acute; pedicel and ovary 1–1.2 cm long. **Flowers** widely opening, sepals and petals yellowish-green, with purplish-red spots, lip whitish, yellow-tinged at apex of epichile, cushion yellow, purple-spotted; sepals similar, obovate-elliptic or subelliptic, 5–9 × 3–5 mm, obtuse; petals similar to sepals, slightly narrower, obtuse; lip with epichile subtriangular, 2–4 × 5–8 mm, hypochile cupular, laterally compressed, 5–7 × 4–6 mm; column pale purple, short. **Distr:** Hong Kong, Japan (Ryukyu Islands), Taiwan. **Ecol:** On tree trunks in forests; 200–1,500 m. Fl. Jun–Sep. **Zones:** 8–10.

Grosourdya

Species 9. **Herbs**, epiphytic, small, monopodial; stems very short, enclosed in basal leaf sheaths. **Leaves** few to several, basal, distichous; lamina flat, narrowly oblong to lanceolate, base jointed and shortly sheathing, apex often unequally bilobed. **Inflorescences** often many, lateral, usually shorter than leaves; peduncle longer than rachis, slender, both usually prickly hairy, with 1 or 2 flowers open at a time; rachis usually thickened, rather short; floral bracts persistent. **Flowers** ephemeral, opening widely, to 1.5 cm Ø; sepals and petals free, similar, spreading; petals narrower than sepals; lip mobile with end of column foot, spurred, ± trilobed, lateral lobes erect, narrow, midlobe often replicate on spur, sometimes with 2 basal lobules on both sides, with a small median tooth, giving a 4-lobed appearance; spur broad, usually spreading forward, rarely pendulous; column elongate, slender, bent forward; pollinia 2, waxy, subglobular. **Distr:** Indo-China to Malesia. **Cult:** Propagate from seeds. Can be grown in small pots or mounted on small slabs. Requires intermediate conditions with moderate shade and high humidity.

G. appendiculata (Blume) H.G. Reichenbach, *Xenia Orchid.* 2: 123, 1867. (Syn: *Dendrocolla appendiculata, Sarcochilus hirtulus*). **Herbs**, epiphytic; stem inconspicuous. **Leaves** usually 3–6, basal; lamina falcate-oblong, 7–10 × 1.4–1.9 cm, base attenuate, apex acute and unequally bilobed, somewhat fleshy. **Inflorescences** with rachis laxly 2- or 3-flowered; peduncle and rachis densely minutely black-hispid; peduncles 2–5, 3–4 cm long; floral bracts ovate-triangular, very small, fleshy, abaxially hairy; pedicel and ovary ~4 mm long. **Flowers** small, yellow, brown-spotted; sepals ovate-oblong, ~3 × 2 mm, acute; lateral sepals basally adnate to column foot; petals oblong, ~3 × 1 mm, obtuse; lip spurred, trilobed, lateral lobes erect, narrowly oblong, ~3 mm long, rounded and curved backward, midlobe with a very small apical lobule and 2 erect basal lobules, basal lobules parallel to lateral lobes; spur ~3 mm long; column ~2 mm long, winged, foot ~0.5 mm long. **Distr:** Indo-China to Malesia. **Ecol:** On tree trunks in evergreen broad-leaved forests. Fl. Aug. **Zones:** 8–12.

Holcoglossum

Species 14. **Herbs**, epiphytic, monopodial, small to moderate-sized; stems enclosed by persistent leaf sheaths, rooting mostly from base or lower part. **Leaves** many, distichous; lamina terete, semiterete, or broadly semiterete, fleshy, adaxially channeled, jointed and dilated into sheathing base. **Inflorescences** axillary, racemose, few- to many-flowered; rachis usually purple; floral bracts much shorter than pedicel and ovary. **Flowers** usually opening widely; sepals subsimilar, dorsally usually ± carinate; lateral sepals often slightly larger, oblique; petals similar to dorsal sepal; lip saccate or spurred, trilobed, lateral lobes erect beside entrance of spur, spotted, midlobe rather larger, often narrowed and with appendages at base; spur often cylindric and curved, attenuate toward tip, interior usually hispid papillose; column short, thick, winged, with a very short but distinct foot; rostellum distinct, sometimes reduced, bilobed; pollinia 2, waxy, globular, porate. **Distr:** Assam to Taiwan, with the highest diversity in China. **Cult:** Propagate from seeds. Best grown mounted. Requires intermediate conditions with moderate shade and high humidity.

H. amesianum (H.G. Reichenbach) Christenson, *Notes Roy. Bot. Gard. Edinburgh* 44: 255, 1987. (Syn: *Vanda amesiana*, etc.). **Herbs**, epiphytic, ascending; stems 2–5 cm long. **Leaves** 4–7, nearly basal; lamina broadly semiterete, 9–30 × 0.5–1 cm, acute, fleshy, adaxially channeled or conduplicate. **Inflorescence** ascending, 23–25 cm long, several-flowered; peduncle to 15 cm long, with 2–4 sheaths; pedicel and ovary whitish, 3–5 cm long. **Flowers** thinly textured, sepals and petals white, midvein pinkish; dorsal sepal elliptic, ~1.5 × 1 cm, base contracted, apex obtuse; lateral sepals obliquely ovate-elliptic, ~1.5 × 1.1 cm, obtuse; petals oblanceolate, ~1.4 × 0.9 cm, base contracted, apex obtuse; lip with 3 deep purple-red ridges above, spurred, trilobed, lateral lobes ascending, ovate-triangular, ~4 × 5 mm, ± adnate to base of column, obtuse, midlobe reniform-orbicular, ~1 × 1.4 cm, base with an erect square appendage at spur entrance, margin denticulate, apex retuse; spur narrowly conic, ~6 × 3–4 mm, obtuse, slightly curved backward at tip; column ~6 mm long, thick; anther cap white, narrowed, truncate. **Distr:** China, India, Laos, Myanmar, Thailand, Vietnam. **Ecol:**

In broad-leaved evergreen forests; 1,200–2,000 m. Fl. Dec–Mar. **Zones:** 8–11.

H. flavescens (Schlechter) Z.H. Tsi, *Acta Phytotax. Sin.* 20: 441, 1982. (Syn: *Aerides flavescens*, etc.). **Herbs**, epiphytic; stems 1–2 cm long, with several leaves. **Leaves** nearly basal, distichous; lamina semiterete, 4–8.5 × 0.3–0.4 cm, apex acute, fleshy or thickly leathery, adaxially broadly channeled or ± conduplicate. **Inflorescence** 1–3-flowered, suberect or arching, 3–5 cm long; floral bracts broadly ovate, 5–8 mm across; pedicel and ovary ~1.7 cm long. **Flowers** opening widely, white, yellow-tinged, anther cap white; dorsal sepal elliptic, 0.9–1.2 × 0.45–0.6 cm, base slightly contracted, obtuse; lateral sepals obliquely oblong, 1.3–1.4 × 0.5–0.6 cm, dorsally ± carinate, obtuse; petals elliptic, 0.9–1.2 × 0.42–0.55 cm, obtuse; lip spurred, trilobed, lateral lobes with brownish-purple stripes, erect, subovate or ovate-triangular, obtuse, midlobe broadly ovate-rhombic, 6–8 × ~7 mm, margin slightly undulate, apex obtuse-rounded or emarginate; callus basal, broadly ovate-triangular, yellow, strongly thickened, centrally channeled; spur bent forward, horn-shaped, ~7 mm long, narrowed toward tip; column striped purplish-red on either side, ~5 mm long, winged, foot ~2 mm long; rostellum large. **Fruits** ellipsoid, 2–3 × 0.6–0.7 cm. **Distr:** China. **Ecol:** In broad-leaved evergreen forests; 1,200–2,700 m. Fl. Mar–May. Fr. Aug–Sep. **Zones:** 8–10.

H. subulifolium (H.G. Reichenbach) Christenson, *Notes Roy. Bot. Gard. Edinburgh* 44: 255, 1987. (Syn: *H. auriculatum*, etc.). **Herbs**, epiphytic, ascending; stems 2–20 cm long. **Leaves** 3–5, nearly semiterete, 5–14 × ~0.3 cm, fleshy, adaxially channeled, apex acute. **Inflorescence** several-flowered, ascending, arching, ~14 cm long; floral bracts ovate, ~5 mm across; pedicel and ovary white, purple-tinged, 2–3 cm long. **Flowers** opening widely, white, thinly textured; dorsal sepal suboblong, ~1.7 × 0.6 cm, base contracted, apex acute; lateral sepals falcate-oblong, ~1.9 × 0.7 cm; petals oblong-lanceolate, slightly smaller than dorsal sepal; lip spurred, trilobed; lateral lobes brownish-yellow, tinged with pale purple, erect, triangular, acute; midlobe with yellowish-brown ridges, broadly triangular, ~1.2 × 1 cm, concave or shallowly cucullate, base with 3 ridges, margin incurved, undulate, irregularly denticulate; spur conic, 2–3 mm long, obtuse; column 4–5 mm long, foot very short; anther cap pale yellow slightly narrowed toward apex. **Distr:** China, Myanmar, Thailand, Vietnam. **Ecol:** In broad-leaved evergreen forests; 1,300–2,200 m. Fl. Mar–May.**Zones:** 8–11.

H. wangii Christenson, *Lindleyana* 13(2): 123. 1998. **Herbs**, epiphytic; stems ~2 cm. **Leaves** several, nearly basal; lamina semiterete, 30–58 cm × ~4 mm, fleshy. **Inflorescence** 3–5-flowered, ~5 cm long; rachis purple; floral bracts ovate, ~2 mm across; pedicel and ovary ~3.3 cm long. **Flowers** white, lateral lobes of lip yellow, purple-spotted; dorsal sepal oblong, ~2 × 0.7 cm, obtuse; lateral sepals oblique, elliptic, ~2.3 × 1 cm, base clawed and ± twisted, apex obtuse; petals elliptic-oblong, ~2.8 × 1 cm, base twisted and clawed, apex obtuse; lip spurred, trilobed, lateral lobes erect, unequally bilobulate, midlobe oblong-elliptic, ~1.8 × 1.5 cm, with 3–5 basal lamellae; spur cylindric, 0.8–1 cm long; column ~6 mm long. **Distr:** China, Vietnam. **Ecol:** On tree trunks in broad-leaved evergreen forests; 800–1,200 m. Fl. Oct–Dec. **Zones:** 8–12.

Luisia

Species ~55. **Herbs**, epiphytic; stems fairly long, rigid, terete, slender, surrounded by closely appressed tubular sheaths of leaves, tufted, not pseudobulbous; roots vermiform. **Leaves** well-spaced, articulated with sheath at base; lamina slender, terete. **Inflorescences** spikes or racemes, few-flowered, lateral, on short, axillary peduncle, drooping; rachis thickened. **Flowers** fleshy, perianth spreading; sepals and petals free, equal or petals longer and narrower; lateral sepals often keeled; petals similar to sepals; lip fleshy, sessile, fixed immovably to base of column, entire or obscurely trilobed, base concave or shallowly saccate, lateral lobes often erect, midlobe larger, often sulcate or longitudinally grooved, entire or with small lobules at apex; column short, thick, truncate, foot absent; anther terminal, operculate; pollinia 2, waxy, entire, porate or cleft, subglobose, globose or pyriform, connected by elastic thread to broad stipe or strap; viscidium perpendicular to stipe; stigma entire, broad, concave; rostellum broad. **Fruits** narrow, erect. **Distr:** Tropical and subtropical Asia to W Pacific. **Cult:** Propagate from seeds. Best grown on trees or mounted on slabs. Requires intermediate to warm conditions with good light and water.

L. magniflora Z.H. Tsi and S.C. Chen, *Acta Phytotax. Sin.* 32: 558, 1994. **Herbs**, epiphytic; stems ascending, to 30 cm long. **Leaves** 9–18 × 0.2–0.3 cm, apex obtuse. **Inflorescences** often 2- or 3-flowered, ~1 cm long; rachis stout; floral bracts broadly ovate, ~3 mm long, obtuse, fleshy; pedicel and ovary ~9 mm long. **Flowers** fleshy, yellowish to yellowish-green, tinged or spotted with purplish-red or brownish abaxially on sepals and petals, lip epichile brown or dark purple, column yellowish-green; dorsal sepal ovate-oblong, 0.9–1.2 × 0.5–0.6 cm, concave, obtuse; lateral sepals suboblong, 1–1.2 × 0.4–0.5 cm, margin incurved, ± embracing epichile, abaxially carinate-winged, becoming a long winged tail at apex; petals subelliptic, 1–1.1 × 0.5–0.55 cm, obtuse; lip 1–1.3 cm long, epichile and hypochile distinct; hypochile smaller, lateral lobes embracing column; epichile cordate, ~1 cm wide, adaxially papillate, margin recurved, apex emarginate; column ~5 mm long, thick. **Distr:** China (S Yunnan). **Ecol:** In open forests; 600–1,900 m. Fl. Apr–Jul. **Zones:** 8–10.

L. tenuifolia (L.) Blume, *Rumphia* 4: 50, 1848. (Syn: *Birchea teretifolia*, etc.). **Herbs**, epiphytic; stems tufted, terete, green, slender, drooping stems 17–75 cm long. **Leaves** terete, green, 8–28 cm long, apex constricted, beaked, bases sheathing; petiolar sheaths 2.8 cm long. **Inflorescences** close spiciform racemes, few-flowered, axillary; peduncles ~6 mm long; floral bracts broadly crescent-shaped, 2 × 1.8 mm, rounded, persistent; ovary with pedicels 1.2 cm long. **Flowers** 1.1 cm across, sharply reflexed, pale yellowish-purple, with a white-patched dark purple lip; dorsal sepal lanceolate-oblong, 6.2–7 × 3.1–3.2 mm, obtuse, 3- or 5-veined; lateral sepals lanceolate or oblong-lanceolate, 8.2–8.4 × 3.6 mm, subacute, concave, 4-veined; petals 9–9.8 × 1.8–1.9 mm, obtuse or rounded, 5-veined; lip 9.8 × 4.4–4.6 mm, panduriform, convex, base 2-auricled, margin crenate, apex bifid, 7-veined; column 3.9 × 2.4 mm, truncate, narrowed to a purple-blotched base; pollinia 2. **Fruits** slender, fusiform, 3 × 0.3 cm. **Distr:** India (Western Ghats), Sri Lanka. **Ecol:** In tropical evergreen forests; <915 m. Fl. Mar, May–Jun. **Zones:** 10–12.

L. teres (Thunberg) Blume, *Rumphia*. 4: 50, 1849. (Syn: *Epidendrum teres*, etc.). [Stick Orchid]. **Herbs**, epiphytic; stems ascending or pendulous, to 55 cm long, usually unbranched. **Leaves** quaquaversal; lamina terete, 7–13(–20) × 0.2–0.25 cm, apex obtuse. **Inflorescences** 1–7-flowered, borne from upper part of stem, ~1 cm long; floral bracts broadly ovate, ~2 mm long. **Flowers** fleshy, greenish-yellow or whitish, sometimes abaxially and apex of sepals and petals purple-tinged, lip marked with purple or purplish-brown, anther cap pale yellow; pedicel and ovary ~8 mm long; dorsal sepal ovate-oblong, 0.75–1.1 × 0.3–0.6 cm, obtuse; lateral sepals similar, but slightly longer, acute, dorsally carinate-winged, becoming apically winged; petals slightly falcate-elliptic, 0.7–1.3 × 0.2–0.4 cm, obtuse; lip 1–1.6 cm long, thickly fleshy, epichile and hypochile indistinct; hypochile slightly concave, base with subquadrate lateral lobes ~1 mm long; epichile spreading, subovate, large, adaxially with a fleshy longitudinal ridge near apex, furcately bilobed; lobules subtriangular, entire, with finely papillate hairs; column 3.5–4 mm long. **Distr:** China, Japan, Korea, Taiwan. **Ecol:** In forests; 1,200–1,600 m. Fl. Mar–May. **Zones:** 8–10.

L. tristis (G. Forst.) Hook. f., *Fl. Brit. India* 6: 25, 1890. (Syn: *L. teretifolia*, etc.). **Herbs**, epiphytic, rigid; stems branching, terete, 25–50 cm long; internodes 1–2 cm long. **Leaves** terete,8–25 cm long, base sheathing internodes, apex rounded, green. **Inflorescences** spikes, few-flowered, axillary, short-peduncled, emerging through petiolar sheaths; pedicels sharply recurved; floral bracts broadly triangular or ovate, 2 × 4 mm, rounded; pedicel and ovary 7 mm long. **Flowers** small, 7 mm across, purplish-green, lip dark purple; dorsal sepal oblong, 5.2–6 × 3.1 mm, rounded, concave, 5-veined; lateral sepals oblong-lanceolate, 6–6.8 × 3.2–3.5 mm, obtuse, concave with margins incurved, 5-veined; petals linear-oblong, 6.6–7.2 × 2.2–2.4 mm, obtuse, 3-veined, lateral veins branching from base, concave; lip 6.8 mm long, base saccate, hourglass-shaped, lateral lobes short, rounded, vertical, midlobe broadly ovate, or obscurely cordate, 5.2 mm across; column 3.8 × 2.4 mm, apex purple-spotted; stigma purple; pollinia 2. **Fruits** cylindrical, 3 × 0.6 cm. **Distr:** S China, tropical Asia to W Pacific. **Ecol:** In tropical wet evergreen forests; <750 m. Fl. Mar–Jun, Dec. **Zones:** 8–12.

Malleola

Species 32. **Herbs**, epiphytic, small, monopodial; stems enclosed in basal leaf sheaths, pendulous, usually elongate, sometimes slightly compressed. **Leaves** scattered along stem, several to many, distichous; usually narrowly elliptic to oblong, often purplish-flushed, oblique, base jointed, sheathing, apex unequally bilobed. **Inflorescences** 1–3, lateral, ascending or pendulous, mostly shorter than leaves, many-flowered; flowers facing in all directions. **Flowers** small, thinly textured; sepals and petals free, spreading; dorsal sepal often cymbiform; lateral sepals and petals flatter; lip firmly attached to column, immovable, with a variably shaped, spur cylindric, usually with interior ornaments, lateral lobes erect, broadly triangular, short, midlobe upcurved or down-curved, usually linear-lanceolate, sometimes conic, very small, base usually with a ridge or base thickened; column hammer-shaped, short, stout, foot absent, often bent backward at apex; pollinia 2, globose. **Distr:** Andaman Islands to Papuasia. **Cult:** Propagate from seeds. Best grown mounted on a small slab. Requires high humidity, moisture and shade.

M. baliensis J.J. Sm., *Bull. Jard. Bot. Buitenzorg*, III, 2: 118, 1920. **Herbs**, epiphytic; stem short, ~3.5 cm long, cross-section elliptic, internodes ~6 mm long. **Leaves** ~3–5, oblong to lanceolate, ~6.5–9.5 × 2.2–4 cm, apex briefly contracted, oblique, sometimes semiobtuse, denticulate, purple-suffused. **Inflorescence** densely flowered, 9 cm long; peduncle purple, pale dirty green, ~1–1.3 cm long; bracts reflexed, triangular, ~2–3.5 mm long, acute. **Flowers** 8.8 × ~6.4 mm, sepals, petals and spur golden, lateral sepals purple-suffused, lip white, sac golden with red stripes inside; dorsal sepal oblong-ovate, concave, ~5 × 2.3 mm, obtuse; lateral sepals deflexed, incurved, strongly oblique, spathulate-elliptic, concave, ~5 × 1–2 mm, obtuse; petals incurved, oblique, oblong-subelliptic, ~4.4 ×2 mm, obtuse; lip trilobed, lateral lobes erect, midlobe porrect, obtuse; column short; pollinia 2; ovary 6-sulcate, ~5 mm long. **Distr:** SE Java to Bali. **Ecol:** In deep tropical forests; 0–1,200 m. **Zones:** 11–12.

Above: *Luisia tenuifolia, Luisia teres, Luisia tristis, Malleola baliensis.*

M. dentifera J.J. Smith, *Bull. Jard. Bot. Buitenzorg*, III, 9: 191, 1927. (Syn: *Ascochilus vietnamensis*, etc.). **Herbs**, epiphytic; stems pendulous, terete, 3–10 cm long, 3–4 mm Ø, unbranched. **Leaves** many; lamina narrowly oblong, 4–7.5 × 0.5–1 cm, leathery, unequally bilobed or somewhat praemorse. **Inflorescences** opposite leaves, many flowered; floral bracts narrowly lanceolate, 1–2 mm long; pedicel and ovary ~8 mm long. **Flowers** ~8 mm Ø, sepals and petals pale brownish-yellow with 2 crimson stripes, lip purple and white, anther cap purple; dorsal sepal cymbiform, ovate, ~3 × 2 mm, acute; lateral sepals obliquely ovate, ~3 × 2 mm, ± concave; petals oblong, ~3 × 1.5 mm, acute; lip spurred, trilobed, lateral lobes erect, broadly ovate-triangular, ~ 1 × 1 mm, apex trilobuled, midlobe recurved, narrowly lanceolate, fleshy; spur subconic-oblong, ~5 × 2–3 mm. **Fruits** narrowly ovoid, 1–1.5 cm long. **Distr:** Indo-China to W Malesia. **Ecol:** On tree trunks in forests; 600–700 m. Fl. Jul. Fr. Aug–Sep. **Zones:** 10–12.

Micropera

Syn: *Camarotis*.

Species 20. **Herbs**, clambering, monopodial; stems long, with many long roots and leaves. **Leaves** many, flat, fleshy, oblong to linear, base sheathing, jointed. **Inflorescence** often borne opposite leaves, rather long, racemose, many-flowered. **Flowers** not resupinate; sepals and petals free, similar; lip conspicuously spurred or saccate, trilobed, lateral lobes broad, erect, midlobe smaller, fleshy; spur often ornamented at its entrance, usually with a longitudinal septum inside; column short, lacking a foot; pollinia 4, in 2 subequal pairs, stipe long. **Distr:** tropical Asia to W Pacific. **Cult:** Propagate from seeds. Requires intermediate to warm conditions with moderate light and watering throughout the year.

M. pallida (Roxb.) Lindl., *Edwards's Bot. Reg.* 18: t. 1522, 1832. (Syn: *Aerides pallida*, etc.). **Herbs**, epiphytic or lithophytic, monopodial; stem 0.6–1 m long, climbing, internodes ~2.5 cm long. **Leaves** many, spreading; lamina linear-ligulate, 15–20 × 1.5–1.8 cm, apex obtuse, coriaceous, abaxially keeled, unequally bilobed. **Inflorescence** racemose, laxly 20–25-flowered, axillary; rachis to 24 cm long; peduncle to 90 cm long, decurved; floral bracts persistent, elliptic-lanceolate, ~1 × 1 mm; pedicel and ovary ~6 mm long. **Flowers** ~8 mm Ø, white to creamish-white or light dull yellow; dorsal sepal linear, 7–8 × ~1.5 mm, distinctly 3-veined; lateral sepals linear, ~ 6 × 1.5 mm, joined together and adnate to claw of lip; petals linear, 7–8 × 1.2–1.5 mm; lip slipper-shaped, laterally compressed, lateral lobes large, rounded, sac with an exserted spine under the apex, spur conic-cylindric, 2–3 mm long; column slender, ~6 mm long. **Distr:** E India to W Malesia. **Ecol:** In swamp forests, mixed forests, evergreen broad-leaved dry forests; 0–1,000 m. Fl. Mar, Oct. **Zones:** 10–12.

Neofinetia

Syn: *Finetia*.

Species 3. **Herbs**, epiphytic, small, monopodial; stems erect, enclosed in leaf sheaths, many-leaved. **Leaves** distichous; lamina narrow, ± conduplicate toward base, dorsally carinate, base jointed and sheathing. **Inflorescence** axillary, racemose, laxly few-flowered. **Flowers** opening widely, medium-sized; sepals and petals free, similar; lip spurred at base, trilobed, lateral lobes erect, midlobe spreading, with appendages at base, spur slender, cylindric, long, sometimes slightly curved; column short, thick, winged, footless; anther cap apically narrowed; pollinia 2, cleft, waxy, globose. **Distr:** China to temperate E Asia. **Cult:** Propagate from seeds. Best grown in small pots with a fine bark mix or in sphagnum moss. Requires cool to intermediate conditions, with moderate shade and high humidity.

N. falcata (Thunberg) Hu, *Rhodora*. 27: 107, 1925. (Syn: *Orchis falcate*, etc.). [Samurai Orchid, Japanese Wind Orchid]. Herbs, epiphytic, 8–12 cm tall; stems 1–6 cm long, many-leaved. **Leaves** narrowly oblong-falcate, 5–12 × 0.7–1 cm, conduplicate, sheathing toward base, apex acute, thickly leathery. **Inflorescence** suberect, 2- or 3(-5)-flowered, 5–8 cm long; peduncle 1–3 cm long; floral bracts ovate-lanceolate, 7–9 mm long, acuminate; pedicel and ovary 2.8–5 cm long. **Flowers** fragrant, white, anther cap brown; sepals subobovate, 0.8–1 × 0.25–0.4 cm, obtuse; lateral sepals falcate-curved, dorsally carinate apically, recurved; petals oblanceolate or subspathulate, 0.8–1 × 0.22–0.3 cm, obtuse; lip fleshy, trilobed, lateral lobes oblong, 3.5–4 × 0.8–1 mm, obtuse, midlobe ligulate, 7–8 × 2–2.5 mm, retuse, with a triangular basal callus, adaxially with 3 slightly swollen ridges; spur arcuate, 3.5–5 cm long, slender, obtuse; column ~2 mm, wings triangular; anther cap triangular. **Distr:** China to temperate E Asia. **Ecol:** On tree trunks in forests; 1,500–1,600 m. Fl. Apr. **Zones:** 7–10.

Omoea

Species 2. **Herbs**, epiphytic, upright, monopodial, dwarf. **Leaves** thick, leathery. **Inflorescences** short, few-flowered, axillary. **Flowers** small; lip spurred, spur compressed, emarginate, lamina trilobed, lateral lobes erect, fleshy; anthers bilocular; pollinia 2, globose, bilobed, caudicles filiform. **Distr:** W and C Malesia. **Cult:** Propagate from seeds. Best grown mounted on a small slab with moss. Requires high humidity and moderate shade. **Info:** This genus is very similar to *Ceratochilus* and other related genera, with shared features.

O. philippinensis Ames, *Orchidaceae* 5: 232, 1915. **Herbs**, epiphytic, upright, monopodial; stems to 10 cm long, 3 mm Ø. **Leaves** to 3 × 0.3 cm, very leathery, thick, flattened, adaxially and abaxially grooved. **Inflorescences** short, to 6-flowered, axillary. **Flowers** with petals and sepals green with brownish tips, lip white, purplish-striped on lateral lobes; dorsal sepal lanceolate, concave, 4 × 1.5 mm, tip pointed; lateral sepals lanceolate, concave, to 4 × 1.5 mm, tip pointed; petals linear to 3 × 1 mm, tip pointed; lip

Below: *Malleola dentifera, Micropera pallida, Neofinetia falcata.* [All MG.]

Above: *Omoea philippinensis, Ornithochilus difformis*. [Both MG.]

trilobed with a spur, lateral lobes triangular, midlobe heart-shaped; column erect, pale yellow-white, anther cap pale yellow. **Distr:** Endemic to Philippines (Luzon). **Ecol:** Epiphytic in mossy forests; 800–2,000 m. **Zones:** 11–12.

Ornithochilus

Species 3. **Herbs**, epiphytic, monopodial, small; stems short, unbranched, rigid, enclosed in persistent leaf sheaths. **Leaves** distichous; lamina flat, base jointed and shortly sheathed, fleshy. **Inflorescences** paniculate, many flowered, axillary, equaling or exceeding leaves; floral bracts small, narrow. **Flowers** small to medium-sized; sepals and petals free; petals smaller and narrower than sepals; lip spurred from distal part of hypochile, epichile trilobed, margin ciliate or crenulate; disk with a pilose callus at spur entrance; column small, fleshy, lacking a foot, anther cap truncate; pollinia 4, in 2 pairs. **Distr:** Tropical and subtropical Asia. **Cult:** Propagate from seeds. Bes grown mounted on small slabs. Requires intermediate temperatures, moderate shade and high humidity throughout the year.

O. difformis (Wallich ex Lindley) Schlechter, *Repert. Spec. Nov. Regni Veg. Beih.* 4: 277, 1919. (Syn: *Aerides difformis*, etc.). **Herbs**, epiphytic; stems 2–4 cm long. **Leaves** 2 or 3, oblong-elliptic or obovate-elliptic, falcate, 7–19 × ~5.5 cm; basal sheaths ~1 cm long, embracing stem. **Inflorescences** laxly many-flowered; branches spreading, to ~21 cm long; sheaths tubular, overlapping, tubular, 3–6 mm long; rachis 9–25 cm long; floral bracts pale brown, triangular, ~2 mm long, acute; pedicel and ovary ~1 cm long. **Flowers** opening widely, 0.7–1 cm Ø, sepals and petals yellow to green, with red-brown striations, lip dark maroon with yellow markings at base and margin, spur maroon and green, column purplish-brown; dorsal sepal oblong, ~5 × 2 mm, obtuse; lateral sepals obliquely ovate-oblong, about as long as and wider than dorsal sepal, obtuse; petals narrowly oblong, ~4 × 1.2 mm, entire or scabrous, obtuse; lip shortly clawed, trilobed, lateral lobes suberect, subovate, midlobe slightly incurved, anchor-shaped, clawed, trilobulate; disk with 1 purplish-red, fleshy callus centrally; spur slightly curved, ~4 mm long; column ~2 mm long, hairy. **Distr:** Himalayas to S China and W Malesia. **Ecol:** On tree trunks in open forests; 500–2,100 m. Fl. May–Jul. **Zones:** 8–12.

Papilionanthe

Syn: *Vanda* sect. *Teretifoliae*.

Species 11. **Herbs**, epiphytic, scrambling, monopodial; stems erect to pendent, climbing or pendulous, elongate, terete, branching or unbranched, many noded, laxly many-leaved. **Leaves** terete, fleshy, adaxially channeled, jointed and sheathing at base; sheaths thickly leathery, tightly enclosing stem, persistent. **Inflorescence** axillary, 1- to several-flowered. **Flowers** flat, large; sepals and petals subsimilar, free, spreading; petals usually larger; lip continuous with column foot, immobile, spurred, trilobed, lateral lobes ± erect, either parallel to or embracing column, midlobe often dilated and 2- or trilobed at apex; spur funnel-shaped or conic to long horn-shaped; column subterete, short, fleshy, foot short; rostellum elongate, slender; pollinia waxy, 2, shortly cleft, attached by a broadly triangular or subquadrate stipe to a large viscidium. **Distr:** China, India, SE Asia and Malay Archipelago. **Cult:** Propagate by division of longer portions of stems. Grow in baskets or terrestrially in a coarse epiphyte mix, with support for climbing stems. Requires warm, humid conditions with full sunlight, plenty of water and fertilizer.

P. teres (Roxburgh) Schlechter, *Orchis* 9: 78. 1915. (Syn: *Vanda teres*, etc.). **Herbs**, epiphytic; stems scrambling, terete, usually to 1 m long, stout, branching. **Leaves** well-spaced; lamina terete, 8–18 × 0.4–0.5 cm, apex obtuse, fleshy. **Inflorescences** laxly 2–5-flowered, slightly longer than the leaf; peduncles stout, with 3 or 4 membranous sheaths; floral bracts broadly ovate, 4–6 mm long, obtuse, slightly fleshy; pedicel and ovary white, 2–3 cm long. **Flowers** opening widely, 4–10 cm Ø, thinly textured, sepals and petals white, tinged with pink or purplish, lip deep purple-red, spur yellowish-brown; dorsal sepal broadly elliptic, ~2.5 × 2 cm, obtuse-rounded; lateral sepals obliquely ovate-oblong, ~3 × 2 cm, base adnate to column foot, apex obtuse; petals suborbicular, large, ~3 × 2.7 cm, obtuse; lip spurred, trilobed, lateral lobes embracing column, subobovate, ~1.5 cm wide, rounded, midlobe spreading, obovate-triangular, ~2 cm, adaxially shortly hairy, deeply bilobed; spur conic, ~2 cm long; column ~5 mm long, foot ~5 mm long. **Distr:** Bangladesh, Bhutan, China (S Yunnan), India, Laos, Myanmar, Nepal, Thailand, Vietnam. **Ecol:** In open forests, forest margins; 500–900 m. Fl. May–Jun. **Zones:** 9–12.

P. vandarum (H.G. Reichenbach) Garay, *Bot. Mus. Leafl.* 23(10): 372, 1974. (Syn: *Aerides vandarum*). **Herbs**, epiphytic; stems pendent, branched, terete, to 60 cm long, flexuous, sheathed; sheaths 1.5–2.3 cm long. **Leaves** suberect, jointed; lamina terete, 9–25 × ~0.3 cm, apex acuminate, grooved. **Inflorescences** 1–4-flowered, leaf-opposed; peduncle ~2.2 cm long, sheathed; floral bracts triangular, 2–4 × 1.4–1.6 mm; pedicel and ovary 1.5–3.3 cm long, pedicel pink. **Flowers** 3.8–5 cm Ø, white with base of lip and spur purple-flushed; dorsal sepal obovate-oblong, 2–3.6 × 0.8–1.5 cm, margins undulate, apex obtuse; lateral sepals similar, ~3 × 1 cm; petals subrhombic, 2–2.5 × 1.1–1.9 cm, reticulately veined, base twisted, margins undulate, apex obtuse; lip trilobed, ~2.2 × 2.2 cm; lateral lobes erect, unequally bifid, flexuous, narrowly ovate, falcate, ~9 × 3 mm, apical margin dentate, midlobe clawed, obovate, ~9 × 9 mm, apex dilated and bilobulate, lobules orbicular, ~5 × 4 mm, serrate-denticulate; spur cylindric, base conic, straight, 1.8–2.5 cm long; column fleshy, ~6 mm long, foot long. **Distr:** Bhutan, S China, NE India, Myanmar. **Ecol:** Epiphytic in woodlands; 1,200–1,750 m. Fl. Feb–Mar, Jul–Sep. **Zones:** 9–11.

Below: *Papilionanthe teres, Papilionanthe vandarum, Parapteroceras quisumbingii*.

MG MG

Parapteroceras

Species 8. **Herbs**, epiphytic, monopodial; stems ascending or pendulous, enclosed in leaf sheaths. **Leaves** several to many, distichous; lamina flat, base, jointed, sheathing, apex unequally bilobed. **Inflorescences** racemose, lateral, arching to pendulous, several- to many-flowered; peduncle and rachis ± fleshy. **Flowers** small; sepals free; lateral sepals usually larger, base adnate to column foot; petals often smaller than sepals; lip immobile, attached to column foot, spurred, trilobed, lateral lobes on both sides of spur entrance, ascending, large, midlobe ± spreading forward, very small; spur often spreading horizontally, ± laterally compressed, narrowed toward tip, thick-walled; column short, thick, foot distinct, rostellum short, bilobed; pollinia 2, waxy, subglobose. **Distr:** Indo-China to S Pacific. **Cult:** Propagate from seeds. Best grown mounted on a slab. Requires intermediate conditions with high humidity and moderate shade.

P. quisumbingii (L.O. Williams) J.J. Wood, *Nordic J. Bot.* 10: 485, 1990. (Syn: *Saccolabium quisumbingii, T. quisumbingii, T. quisumbingii* var. *xanthum*). Herbs, epiphytic, monopodial, erect; stem abbreviated, 2–3 cm long. **Leaves** crowded, distichous; lamina ligulate, 8–15 × 1.5–3 cm, base constricted, articulated, apex unequally bilobed, leathery. **Inflorescences** about as long as the leaves, ~20-flowered, floriferous nearly to base; rachis somewhat thickened and winged; bracts hyaline, broader than long. **Flowers** to 1.5 cm Ø, petals and sepals pale green, lip white with pale purple blotches, lateral lobes edged with reddish-brown; dorsal sepal elliptic-oblong, 6 × 4 mm, obtuse, thickened, dorsally with a short apiculation toward apex; lateral sepals slightly oblique, ovate, 6 × 3 mm, base adnate to claw of lip, 4–5-nerved, dorsally with a short apiculation toward apex; petals narrowly obovate, 5–5.5 × 2–3 mm, obtuse, thickened, trinerved; lip fleshy, strongly saccate, trilobed, triangular in outline, ~6 mm long, lateral lobes squarish, slightly oblique, 2 × 2 mm, rounded, obtuse, with a short erect acute callus inside, midlobe, short, thickened and calluslike; sac or spur narrowed toward tip; column ~2 mm long, prolonged into distinct foot. **Distr:** Philippines (Luzon). **Ecol:** 300 m. **Zones:** 11–12.

Pelatantheria

Species 7. **Herbs**, epiphytic or lithophytic, monopodial; stems long climbing, rooting from nodes, usually slightly compressed trigonous, many noded, rigid, enclosed in persistent leaf sheaths, branched, leafy. **Leaves** many, usually densely distichous, flat, rarely subcylindric, leathery or slightly fleshy, base sheathing, jointed, apex bilobulate, rarely obtuse. **Inflorescence** racemose, axillary, usually very short, few-flowered. **Flowers** small or medium-sized, fleshy; sepals and petals subsimilar; sepals free; petals smaller; lip spurred at base, trilobed, lateral lobes erect, small, midlobe large, adaxially thickened and cushionlike centrally; spur narrowly conic, with a longitudinal septum or ridge on inner surface, and with a tough appendage on back wall; column short; pollinia 2. **Distr:** India to Japan and W Malesia. **Cult:** Propagate from seeds. Grow in pots with a medium bark-based compost, with a moss pole to support the stems. Requires intermediate conditions, moderate shade and high humidity.

P. insectifera (Rchb. f.) Ridl., *J. Linn. Soc.*, Bot. 32: 373, 1896. (Syn: *Sarcanthus insectifer*). **Herbs**, epiphytic, lithophytic or terrestrial, monopodial; stems robust, scandent, to 30+ cm long, leafy, producing many elongate fibers. **Leaves** subimbricate, distichously spreading; lamina oblong, to 5 × 1.6 cm, apex unequally obtusely bilobed, coriaceous. **Inflorescence** few-flowered, ~2 cm long, sessile; bracts small, deciduous. **Flowers** 1.3 cm across, thickly coriaceous, greenish-yellow, striped with red-brown, lip white and pink or purple, midlobe greenish; sepals ovate-oblong, 5–7 × 2.5–3 mm, obtuse; petals oblong, 6 × 2 mm, acute; lip trilobed, 1.2 × 0.6 cm, lateral lobes rounded, midlobe broadly triangular-ovate, acute or acuminate, fleshy; spur short, broadly conical, semiseptate, mouth with a pubescent ridge, dorsal callus large; column very short, stout, apex reclinate, with uncinate recurved arms; anther hemispheric; pollinia sulcate. **Distr:** Himalayas to W Indo-China. **Ecol:** In broad-leaved, evergreen, lowland forests; 200–1,000. Fl. Sep. **Zones:** 8–11.

P. rivesii (Guillaumin) Tang and F.T. Wang, *Acta Phytotax. Sin.* 1: 101, 1951. (Syn: *Sarcanthus rivesii*). **Herbs**, epiphytic or lithophytic; stems to 1 m long, often branched. **Leaves** elliptic, 3–4 × 1–1.5 cm, obtuse, unequally bilobed. **Inflorescence** 2–7-flowered, 0.5–1 cm long; peduncle very short, sheaths 2 or 3; pedicel and ovary ~8 mm long. **Flowers** thickly textured, sepals and petals pale yellow with 2 or 3 brown stripes, lip pink; dorsal sepal subelliptic, ~4 × 2.2 mm, obtuse; lateral sepals ovate-oblong, broader, obtuse; petals oblong, ~4 × 1.5 mm, obtuse; lip larger than sepals and petals, trilobed, lateral lobes erect, ovate-triangular, small, obtuse, midlobe broadly ovate-triangular, ~6 × 6 mm, base with a papillate callus on either side, apex slightly narrowed, obtuse; spur ~3 × 2.5 mm; column ~2 mm long. **Distr:** China, Indo-China. **Ecol:** In broad-leaved evergreen forests; 700–1,100 m. Fl. Oct. **Zones:** 8–11.

Pennilabium

Species 12. **Herbs**, epiphytic, small, monopodial; stems short, with a few close leaves. **Leaves** clustered; lamina flat, elliptic to oblong, ± fleshy, base often ± twisted, jointed and sheathing, apex unequally laterally bilobed. **Inflorescence** racemose, lateral, usually rather short, unbranched, with 1 or 2 flowers open at a time, placed in 2 rows; rachis somewhat thickened and complanate; floral bracts distichous; flowers opening successively, lasting 1 or 2 days. **Flowers** white, cream-colored, yellow, or orange, medium-sized; sepals and petals free, similar; petals slightly smaller, margin often dentate; lip adnate to column base, immovable, spurred, trilobed, internal callosities absent, lateral lobes either rather large, well-developed, truncate or reduced to small earlike lobes, often fimbriate or toothed, midlobe large, fleshy and solid or reduced to a small fleshy lobe; column short, slightly dorsiventrally compressed; stigma very large; pollinia 2. **Distr:** E Himalayas to W and C Malesia. **Cult:** Propagate from seeds. Best grown mounted on a small slab. Requires high humidity and partial shade.

P. struthio Carr, *Gard. Bull. Straits Settlem.* 5: 151, 1930. **Herbs**, epiphytic. **Leaves** falcate-lanceolate, to 6.5 × 1.3 cm, apex acute, abaxially channeled, glabrous, bright green. **Inflorescence** 1 cm long; rachis flat and widened, to 2.5 cm long, bearing 1 or 2 flowers at a time. **Flowers** with sepals spreading, translucent yellowish, spotted with dark red, petals colored similar to sepals; dorsal sepal elliptic, 1 × 0.5 cm, concave; lateral sepals similar to dorsal; petals 8.5 × 5 mm, apex slightly toothed; lip with lateral lobes white, 1 × 0.35 cm, base narrow, ends widened and diverging, margin fringed with hairs 1–1.5 mm long, the 2 lying side by side in one plane, midlobe fleshy, laterally flattened, 3.5 mm long; spur 1 cm long, nearly parallel to ovary. **Distr:** Peninsular Thailand to peninsular Malaysia. **Ecol:** In forests; 600–700 m. Fl. Jul. **Zones:** 11–12.

Phalaenopsis

Syn: *Doritis, Polychilos*, etc.

Species ~40. **Herbs**, epiphytic; roots elongate, flattened; stems short, enclosed by leaf sheaths. **Leaves** few, closely arranged, distichous, base articulated with sheath; lamina fleshy, coriaceous,. **Inflorescences** racemes, lateral, erect or ascending, branched or not; peduncles slender, much longer than leaves; bracts much smaller than pedicel and ovary. **Flowers** fairly large, long-lasting; sepals similar, spreading; petals usually wider than sepals, often clawed at base; lip clawed or sessile, continuous with short foot of column, trilobed, lateral lobes erect, midlobe spreading, sometimes with 2 slender tails at apex; disk with one or more appendages or callosities between lateral lobes or at base of midlobe; clinandrium shallow; anther terminal, operculate; pollinia 2, ovoid or globose, unequally cleft, attached by long stipes to viscidium, stipes narrow at base, widening upward; stigma entire, concave; rostellum well-developed, apex bifid. **Distr:** Tropical Himalaya, Malaysia, Philippines, Indonesia to Australia and Papua New Guinea. **Cult:** Propagate from seeds. Requires a bark or charcoal based media.

P. amabilis (L.) *Bl. Bijdr.* 7:294. 1825. (Syn: *Cymbidium amabile*, etc.). [Moon Orchid]. **Herbs**, robust epiphytic. **Leaves** variable, arching to pendent; lamina oblong-oblanceolate to elliptic-obovate, 50 × 10 cm, base folded, tapered, apex obtuse, obliquely minutely bilobed. **Inflorescences** scapose racemes or few-branched panicles, often producing secondary

Below: *Pelatantheria insectifera, Pelatantheria rivesii, Pennilabium struthio.* [All MG.]

branches; peduncle arching to pendent, terete, with distant appressed tubular sheaths; floral bracts inconspicuous, triangular, concave, scarious, to 5 mm long; pedicel and ovary to 5 cm long. **Flowers** showy, fragrant, membranous, white, lip and callus variously marked with yellow and red, alternately arranged in 2 ranks; dorsal sepal erect, elliptic-ovate, to 4 × 2.5 cm, obtuse-rounded, concave or convex; lateral sepals obliquely ovate-lanceolate to ovate-elliptic, to 4 × 2.5 cm, acute, concave; petals subrotund, to 4.5 × 5 cm, base cuneate-clawed, apex broadly rounded; lip trilobed, to 2.3 × 4.2 cm, lateral lobes obliquely elliptic, broadly clawed, obtuse-rounded, erect-incurved, midlobe forming a cross, basal lobules triangular, apex bluntly obtuse flanked by long, flexuous tendril-like appendages (cirrhi), callus uniseriate, shield-shaped; column straight, stout. **Distr:** Malesia to Australia. **Ecol:** In humid forests; 0–1,500 m. **Zones:** 10–12.

P. amboinensis J.J. Sm., *Bull. Dép. Agric. Indes Néerl.* 45: 23, 1911. (Syn: *P. hombronii*, etc.). [Amboin Island Phalaenopsis]. **Herbs**, epiphytic. **Leaves** usually 3 or 4, elliptic to obovate, to 24 × 10 cm, base tapered, apex obtuse-rounded. **Inflorescences** racemes or panicles, arching, to 45 cm long, each branch few-flowered; rachis fractiflex; bracts ovate, to 7 mm long, concave, acute, keeled; pedicel and ovary to 3 cm long. **Flowers** fleshy, creamy-white or yellow with transverse brown bars which frequently form ± concentric circles, lip brown-mottled; dorsal sepal elliptic-ovate, to 3 × 1.5 cm, acute-carinate; lateral sepals obliquely broadly ovate, to 3 × 1.5 cm, acute-carinate; petals elliptic, broadly cuneate, obtuse to subacute, subequal, slightly shorter than dorsal sepal; lip trilobed, to 2.2 × 1.8 cm, lateral lobes erect, triangular-oblong, midlobe oblong-elliptic, cuneate, obtuse, callus biseriate, lobes bluntly obtuse, anterior callus sulcate, bifid, divisions filiform; column white arching, to 8 mm long. **Distr:** Indonesia (Moluccas and Sulawesi). **Ecol:** In humid rainforests. **Zones:** 11–12.

P. aphrodite Rchb. f., *Hamb. Gartenz.* 18: 35, 1862. (Syn: *P. amabilis* var. *aphrodite*). **Herbs**; stems stout, <2 cm long. **Leaves** elliptic, oblong or oblong-falcate, 8–22 × 3–7 cm, base cuneate, sometimes oblique, apex acute, obtuse or rounded, thick, fleshy, sheaths broad. **Inflorescence** a raceme, many-flowered; peduncle 20–50+ cm long; rachis ± zigzag; bracts scalelike, 3–4 cm long; pedicel and ovary 3.5–4.5 cm long. **Flowers** white; dorsal sepal ovate-elliptic, 2.5–4 × 1.5–2.5 cm, base contracted, apex obtuse; lateral sepals obliquely ovate-lanceolate, 2.7–3.5 × 1.7–2.2 cm, base contracted, apex acute; petals rhombic-obovate, 2.8–3.5 × 3–3.8 cm, base narrowly clawed, apex rounded; lip trilobed, clawed, lateral lobes falcate-obovate, erect or ascending, 2 cm long, with red spots or lines near base, marked with yellow on distal edge, midlobe white, often yellow-tinged, rhombic-triangular, 2.8–3. 5 × 1.5–1.7 cm, base cuneate, apex acuminate, with 2 slender apical tails ~1.5 cm long; disk with a yellow, green-spotted forked callosity; column 1–1.2 cm long. **Fruits** fusiform, 3–4 cm long. **Distr:** Philippines, Taiwan. **Ecol:** In lowland forests. **Zones:** 11–12.

P. appendiculata C.E. Carr, *Gard. Bull. Straits Settlem.* 5: 16, 1929. (Syn: *Doritis appendiculata*, *Polychilos appendiculata*). **Herbs**, miniature, epiphytic; stems very short, concealed by leaf bases. **Leaves** 2–4, elliptic to oblong-elliptic, to 7 × 3.5 cm, base petiolate ~5 mm long, apex acute or obtuse. **Inflorescences** racemes, few-flowered, short, suberect; pedicel and ovary 5 mm long. **Flowers** white, sepals and petals violet-spotted; dorsal sepal obovate to suborbicular, 4 × 3.5 mm, deeply concave, margins erose-denticulate; lateral sepals obliquely broadly ovate to subrotund, 5 × 4.5 mm, shallowly concave, obtuse; petals sub orbicular, 4 × 3.5 mm, base broadly cuneate, margin erose-denticulate, apex obtuse-rounded; lip white, trilobed, lateral lobes fused with column foot, erect, oblong-linear, 5.5 × 7 mm, erose-denticulate, lateral lobes tipped with pale violet, anterior teeth pale yellow with violet spots, midlobe mobile, white with violet margins and streaks, posterior callus white with violet spots, anterior callus pale violet, the filiform extensions white, lateral rows of teeth violet with white apices; column short, white, with purple suffusion, apex of foot pale yellow bordered with pale violet. **Distr:** Peninsular Malaysia. **Ecol:** On mossy branches near waterways. **Zones:** 9–12.

P. bastianii O. Gruss and Röllke, *Die Orchidee* (Hamburg) 42(2): 76, 1991. (Syn: *P. bastianii* f. *flava*). **Herbs**, epiphytic. **Leaves** 2–10, obovate-elliptic, 15–23 × 5–7 cm, base cuneate, apex rounded and notched. **Inflorescences** racemes or panicles, 15–50 cm long, erect, branches 2–7-flowered; pedicel and ovary to 2 cm long. **Flowers** flat, glossy, 3.5–4 cm wide, greenish-white to yellow overlaid with large, transverse, reddish-brown bars; dorsal sepal elliptic to obovate, 1.6–1.8 × 0.65–0.75 cm, shallowly convex, acute; lateral sepals subsimilar, oblique, 1.5–1.7 × 0.7–0.8 cm, acuminate-subcarinate, divergent or subparallel; petals oblanceolate, 1.5–1.7 × 0.5–0.6 cm, obtuse; lip rose-purple trilobed, 9 × 9 mm, lateral lobes erect, subparallel, 3.5 mm long, apices falcate, pinkish-white with yellow or orange-ridged, midlobe elliptic-obovate, obtuse, 8 × 4 mm, central keel terminating in a raised subapical mound, callus biseriate; column nearly straight, 6–7 mm long, white or purple. **Distr:** Philippines (Sulu Archipelago). **Ecol:** In tropical forests; <200 m. Fl. Jan–Jun, Sep–Nov. v**Zones:** 11–12.

P. bellina (Rchb. f.) E.A. Christ., *Brittonia* 47: 58, 1995. (Syn: *P. violacea* var. *bellina*, etc.). **Herbs**, epiphytic. **Leaves** 3–7, elliptic-obovate, 20–25 × 7–12 cm, rounded, acute, waxy, shiny, pale green. **Inflorescences** racemes, rigid, arching, shorter than leaves, often resting upon them, 1 or 2 flowers open at a time; floral bracts ovate, to 7 mm long; pedicel and ovary 2.6 cm long. **Flowers** very fragrant, greenish-white to greenish-yellow, base of sepals and petals ± with purple suffusions and fine spotting, sepals keeled along back; dorsal sepal elliptic-lanceolate, to 3.6 × 1.5 cm, acuminate-carinate, concave; lateral sepals subfalcate, obliquely elliptic-ovate, to 3.5 × 1.7 cm, acuminate-carinate, basal inner edges purple; petals obliquely ovate, to 3 × 1.7 cm, acute; lip purple, trilobed, to 2.8 × 2.3 cm, lateral lobes yellow, oblong-ovate, anterior callus elongate, sulcate, terminally deeply bifid with narrowly linear divisions; column to 1.5 cm long. **Distr:** Malaysia (Malay Peninsula and Sarawak). **Ecol:** In shady, humid, lowland and swampy riverine forests; to 200 m. **Zones:** 11–12.

P. borneensis Garay, *Lindleyana* 10: 182, 1995. **Herbs**, epiphytic; stems erect, 3 cm long. **Leaves** distichous; lamina linear-oblanceolate, to 30 × 3–4 cm, base conduplicate, apex acute, unequal, coriaceous. **Inflorescences** axillary, arching, to 45 cm long, flowers produced sequentially; peduncle subterete, compressed, at middle with short, ovate-cucullate, acute bracts; rachis simple or branched, flattened, ± fractiflex, to 15 cm long; floral bracts in 2 rows, ovate-cucullate, 5 mm long. **Flowers** with segments spreading, yellow-green, with transverse brown spots; dorsal sepal elliptic, margins recurved, obtuse, dorsally mucronulate, 2.2 × 0.8 cm; lateral sepals similar, 2.2 × 1.1 cm; petals obovate-oblanceolate, 1.6 × 0.6 cm, acute; lip trilobed, 1.3 × 1.1 cm, lateral lobes broadly subquadrate, bearing a parallel callus, midlobe with base cuneate, transversely lunate, without a callus, margins erosulate, apex shortly mucronate; disk excavated with a central erect linear-triangular, acuminate structure above lamina with deeply bifurcated, filiform processes; column arching, 1 cm tall, with 2 prominent basal knees; ovary long-pedicellate. **Distr:** Borneo. **Ecol:** In lowland and riverine forests; 0–750 m. Fl. Dec–Sep. **Zones:** 11–12.

P. braceana (J.D. Hook.) E.A. Christ. *Selbyana* 9 : 169, 1986. (Syn: *Biermannia naviculare*, etc.). **Herbs**, miniature, epiphytic. **Leaves** 1 or 2, deciduous, rarely present; lamina oblong-elliptic, to 2.4 × 0.8 cm subacute. **Inflorescences** racemes, 4–6-flowered, ± erect; peduncle subequal to rachis; floral bracts lanceolate, 3–5 mm long, acuminate; pedicel and ovary to 1. 5 cm long. **Flowers** with segments spreading, variable in color, sepals and petals yellow, green, or greenish-bronze variously suffused with pink, sepals and petals subsimilar, subequal, elliptic, obtuse; dorsal sepal to 1.5 × 0.5 cm; lateral sepals to 1.5 × 0.6 cm; petals subspathulate, to 1.3 × 0.6 cm; lip dark rose-colored, trilobed, spurred, to 7 mm long, lateral lobes oblong, to 4 mm long, obtuse, midlobe oblong-oblanceolate, obtuse, convex, ± constricted at apex, spur cylindric, straight, to 3 mm long, callus biseriate; column cylindric, to 4 mm long, foot to 2 mm long. **Distr:** E Himalayas (Bhutan) to China. **Ecol:** At forest edges; 1,100–2,100 m. **Zones:** 8–10.

P. celebensis Sweet, *The Genus Phalaenopsis* 66, 1980. **Herbs**, miniature, epiphytic. **Leaves** oblong-obovate, to 17 × 6 cm, apex obtuse, dark green marbled with silvery-white. **Inflorescences** racemes, densely many-flowered; peduncle erect, to 40 cm long, rachis pendent, to 13 cm long; pedicel and ovary to 2.6 cm long. **Flowers** white, petals with a central brown suffusion, lip with base brown stripes composed of spots, base of lateral lobes yellow-suffused and brown-spotted, upper surface of column rose; dorsal sepal elliptic-suborbicular, to 1.3 × 1 cm, obtuse, variously concave, convex, or both; lateral sepals obliquely elliptic-ovate, to 1.4 × 1 cm, obtuse, divergent, basal margins revolute; petals obliquely transverse, obtuse-rounded, minutely emarginate, incurved, margins strongly revolute, to 0.9 × 1.2 cm; lip trilobed, lateral lobes obovate, to 5 × 3 mm, obtuse-rounded, each with a posterior ridge, midlobe elliptic-obovate, to 1 × 0.6 cm, obtuse, callus solitary, triangular, acuminate; column to 8 mm long, with a pair of fleshy basal knees. **Distr:** Indonesia (Sulawesi). **Ecol:** Fl. Jul–Aug. **Zones:** 11–12.

P. chibae T. Yukawa, *Ann. Tsukuba Bot. Gard.* 15: 19, 1996. (Syn: *Doritis chibae*, *Kingidium chibae*). **Herbs**, epiphytic. **Leaves** elliptic to obovate, to 11 × 4.5 cm, apex obliquely emarginate, lightly purple-suffused. **Inflorescences** racemes, to 13-flowered,

Top row: *Phalaenopsis amabilis, Phalaenopsis amboinensis, Phalaenopsis aphrodite.*
Second row: *Phalaenopsis appendiculata, Phalaenopsis bastianii.*
Third row: *Phalaenopsis bellina, Phalaenopsis borneensis.*
Fourth row: *Phalaenopsis braceana, Phalaenopsis celebensis, Phalaenopsis chibae.*
Bottom row: *Phalaenopsis cochlearis, Phalaenopsis corningiana.*
[All MG except *Phalaenopsis bastianii* AF.]

MG MG

Above: *Phalaenopsis cornu-cervi, Phalaenopsis deliciosa, Phalaenopsis doweryënsis.*

erect, dull purple, to 11 cm long, flowers opening simultaneously; floral bracts triangular, to 3 mm long, acute; pedicel and ovary terete, to 1.4 cm long. **Flowers** to 1.2 cm wide, pale yellow with chestnut-brown irregular markings, tips of sepals and petals unmarked, callus white with purple markings; dorsal sepal oblong, concave, 4.5 × 3.5 mm, obtuse; lateral sepals obliquely elliptic-ovate, 6 × 4.5 mm, obtuse; petals oblong, 5.5 × 2.5 mm, obtuse, concave toward apex; lip trilobed, saccate, densely pubescent behind callus, 6 × 8.5 mm, lateral lobes much smaller than midlobe, triangular, acute, 1 mm long, midlobe transverse, broadly reniform, 5 × 8.5 mm, obtuse-rounded, callus biseriate; column with a prominent foot, to 4 mm long; pollinia 4. **Distr:** Vietnam. **Ecol:** 400–600 m. **Zones:** 11–12.

P. cochlearis Holttum, *Orchid Rev.* 73: 409, 1964. (Syn: *Polychilos cochlearis*). **Herbs**, epiphytic; stem abbreviated, covered with persistent leaf sheaths. **Leaves** 2–4, oblong-obovate to oblong-elliptic, 13–22 × 5–8, apex acute to obtuse, thinly textured, nerves raised abaxially. **Inflorescences** suberect, branching; peduncle to 50 cm long, branches including fractiflex rachis 4–9 cm long; pedicel and ovary 1.5–2 cm long. **Flowers** white to pale greenish-yellow to yellow with 2 pale brownish or orange-brown bars at base of sepals and petals; dorsal sepal narrowly elliptic, 1.5–2 × 0.6–0.9 cm, obtuse, dorsally submucronate; lateral sepals obliquely ovate, 1.6–2.2 × 0.6–0.9 cm, margins revolute, acute, dorsally carinate; petals narrowly elliptic, 1.2–1.8 × 0.4–0.7 cm, acute, dorsally carinate; lip fleshy, 0.9–1.2 × 0.8–1 cm, lateral lobes white or cream with 3 broken cinnabar to orange-brown streaks on forward facing surfaces, midlobe primrose-yellow with several reddish-brown to orange-brown stripes on each side of median keel; column cylindric, 6–7 mm long. **Distr:** Borneo (Sarawak). **Ecol:** In forests on limestone; 400–600 m. Fl. Oct. **Zones:** 11–12.

P. corningiana Rchb. f., *Gard. Chron.*, n.s., 11: 620, 1879. (Syn: *P. cumingiana*, etc.). [Red Moon Orchid]. **Herbs**, epiphytic. **Leaves** elliptic-obovate, to 32 × 11 cm, base conduplicate, tapered, apex rounded, acute. **Inflorescences** racemes or panicles, arching to subpendent, to 30 cm long; pedicel and ovary to 4 cm long. **Flowers** fragrant, sepals and petals cream to greenish-cream overlaid with brown barring, bars transverse toward base of segments and typically longitudinally aligned apically, extreme individuals bear nearly solid red flowers; dorsal sepal oblong-elliptic, convex, to 4 × 1.3 cm, obtuse or minutely notched; lateral sepals obliquely oblong-obovate, to 3 × 1.5 cm, obtuse, shallowly convex; petals oblong-oblanceolate, to 3.5 × 1.2 cm, bluntly subacute; lip white with broad longitudinal purple stripes, trilobed, to 2 × 1.7 cm, lateral lobes erect, oblong-elliptic, truncate, midlobe oblong, obtuse, with a central raised keel, with a boss of dense trichomes at apex, callus apparently uniseriate; column white, arching, with erose-dentate hood. **Distr:** Borneo. **Ecol:** On huge mossy trees of limestone cliffs near waterfalls; 450–610 m. **Zones:** 11–12.

P. cornu-cervi (Breda) Blume and Rchb.f., *Hamburger Garten-Blumenzeitung* 16: 116, 1860. (Syn: *P. devriesiana*, etc.). [Deer Antler Orchid]. **Herbs**, epiphytic, robust; stems short, rhizomatous, enclosed in imbricating leaf bases. **Leaves** 5–6, lorate or oblong-ligulate to oblong-oblanceolate, 12–22 × 2–4 cm, apex obtuse, fleshy. **Inflorescences** racemes or panicles, 7.5 cm long, several-flowered, flowers several opening simultaneously; scapes to 40 cm long; bracts distichous, ovate-cucullate, to 5 cm long, dorsally keeled, subacute; pedicel and ovary 2.5 cm long. **Flowers** fleshy, waxy, yellow or greenish-yellow, spotted with brown blotches, lip white marked with parallel violet or reddish-brown stripes on lateral lobes and at base of column; dorsal sepal elliptic to oblanceolate, 1.6–2.3 × 0.5–0.8 cm, dorsally keeled, margins recurved, apex acute; lateral sepals similar, somewhat oblique, 1.9–2.3 × 0.7–0.9 cm, acute or subacuminate; petals lanceolate or elliptic-lanceolate, 0.7–1.8 × 0.5–0.6 cm, acute or obtuse; lip very fleshy, trilobed, 1.8 × 1 cm, lateral lobes erect, oblong, truncate, with a basal fleshy thickening, midlobe anchor-shaped, lateral lobules erose, callus a complex of 3 appendages in series, 2-pronged in front; column fleshy, arcuate, to 8 mm long, with a short foot. **Distr:** Indo-China to W Malesia and Philippines. **Ecol:** Epiphytic in lowland and riverine forests; 0–750 m. Fl. Apr–Aug. **Zones:** 11–12. **Info:** The very similar *P. lamelligera* (endemic to N Borneo) differs from *P. cornu-cervi* by having an 'm'-shaped lip, with larger intensive violet markings behind midlobe, and darker thicker blotches. Christenson (2001) reduced *P. lamelligera* to a synonym of *P. cornu-cervi*. Molecular studies suggest that there is significant genetic difference between *P. lamelligera* and *P. cornu-cervi* (Tsai, Huang and Chou, 2006). The study by Tsai, Huang and Chou (2006) also suggest that *P. cornu-cervi* is more closely related to *P. borneensis* (with very strong support), and *P. lamelligera* closely related to *P. pantherina* (with very strong support).

P. deliciosa H.G. Reichenbach, *Bonplandia* 2: 93, 1854. (Syn: *Doritis deliciosa*, etc.). **Herbs**, epiphytic; stems 1–1.5 cm long. **Leaves** 3 or 4, persistent at anthesis; lamina obovate-lanceolate or elliptic, 8–14.5 × 3–5.5 cm, margins undulate, apex acute, slightly curved, papery. **Inflorescence** often pendulous, 10–15 cm long, sometimes branched, densely several-flowered; floral bracts ovate-triangular, ~1 mm long, acute; pedicel and ovary yellowish-green, 4–5 mm long. **Flowers** whitish, purple-striped or -marked; dorsal sepal subelliptic, 6–7 × 3–3.5 mm, obtuse; lateral sepals obliquely ovate, 5.5–6 × 3.5–4 mm, base adnate to column foot, obtuse; petals subobovate, 5–5.5 × 2.7–3 mm, obtuse; lip trilobed, lateral lobes erect, obliquely elliptic-obovate, ~4 × 2.5 mm, rounded, with toothlike flaps, base with a conic spur, adaxially lamellate, midlobe spreading horizontally, obovate-cuneate, ~6 × 5 mm, apex deeply emarginate, basal appendage Y-shaped, furcately bilobed; column ~3 mm long. **Distr:** Indian Subcontinent to Malesia. **Ecol:** In forests; 300–1,600 m. Fl. Jul. **Zones:** 11–12.

P. doweryënsis Garay and Christenson in E.A.Christenson, *Phalaenopsis*: 115, 2001. **Herbs**,

Below: *Phalaenopsis fuscata, Phalaenopsis gibbosa, Phalaenopsis gigantea.* [All MG.]

epiphytic; stems short, compressed, 3–4 cm long. **Leaves** usually 2, ovate-elliptic to broadly elliptic, to 23 × 10 cm, apex obtuse, obliquely bilobulate. **Inflorescences** racemes, erect, shorter than leaves, to 20 cm long; peduncle somewhat compressed, sheaths remote; rachis to 3 cm long; floral bracts triangular-lanceolate, 3 mm long, acute; pedicel and ovary ~2 cm long. **Flowers** fleshy, sepals and petals greenish-yellow with brown spots and short transverse bars, column and lip white, lip marked with bright yellow, midlobe with reddish-brown parallel stripes; sepals similar, obovate, base cuneate, apex obtuse-rounded; dorsal sepal 1.8–2.3 × 0.9–1.3 cm; lateral sepals 1.5–2.2 × 0.9–1.3 cm; petals elliptic-obovate, 1.4–2.2 × 0.7–1.1, base cuneate, apex obtuse-rounded; lip trilobed, 1.3–1.5 × 1.4–1.5 cm, lateral lobes erect, obliquely triangular with a cushionlike callus, midlobe porrect, subquadrate, base rounded, apex obtuse-rounded, callus biseriate; column cylindric, base widened, 1 cm long. **Distr:** Borneo (Sabah). **Ecol:** In forests; ~150 m. **Zones:** 9–12.

P. fuscata Rchb. f., *Gard. Chron.*, n.s., 2: 6, 1874. (Syn: *P. denisiana, Polychilos fuscata*). **Herbs**, epiphytic. **Leaves** oblong-obovate to oblong-elliptic, to 30 × 10 cm, apex acute, fleshy, shiny. **Inflorescences** panicles, many-flowered, to 40 cm long; peduncle erect; pedicel and ovary 2–3 cm long. **Flowers** glossy, rigid, sepals and petals yellow with a coalesced field of brown spots at base, margins strongly revolute; dorsal sepal elliptic-ovate, to 1.7 × 1 cm, obtuse; lateral sepals obliquely elliptic, to 1.7 × 1 cm, obtuse; petals oblong-elliptic to elliptic-obovate, to 1.5 × 0.8 cm, obtuse; lip yellow with longitudinal brown stripes, trilobed, to 1.4 × 1.4 cm, lateral lobes subfalcate, oblong-elliptic, apex rounded, subtruncate, midlobe obovate-suborbicular, concave with a low median keel, obtuse-rounded, callus biseriate, posterior callus sulcate, bilobed, lobes divergent, anterior callus smaller, shallowly bifid; column yellow, erect, straight, to 9 mm long. **Distr:** W and C Malesia. **Ecol:** In shady lowland hill forests, epiphytic on trees near streams; 0–1,000 m. **Zones:** 10–12.

P. gibbosa H.R. Sweet, *Amer. Orchid Soc. Bull.* 39: 1095, 1970. (Syn: *Doritis gibbosa, Polychilos gibbosa*). [Pouched Phalaenopsis]. **Herbs**, epiphytic; stems concealed by imbricating leaf bases. **Leaves** elliptic to elliptic-obovate, to 12 × 4.5 cm, base cuneate, apex acute. **Inflorescences** racemes or panicles, usually 8-flowered, arching, to 15 cm long; peduncle terete; rachis slightly flattened, fractiflex; pedicel and ovary arching, to 4 mm long. **Flowers** white; dorsal sepal ovate-elliptic, concave, to 7 × 3.5 mm, obtuse-rounded; lateral sepals adnate to column foot, obliquely ovate, to 6 × 4.5 mm, subacute; petals narrowly ovate-elliptic, to 5 × 3 mm, obtuse-rounded; lip trilobed, linear, retrorse-falcate, to 2 mm long, acuminate, midlobe triangular-reniform, to 7 × 6 mm, with a pair of large, clear, bright yellow spots, obscurely trilobulate, lateral lobes erect, white with the leading edge brown-barred on yellow, lateral lobules rounded, central lobules subacute, callus biseriate, basal callus 4, threadlike; column to 4 mm long, bright yellow. **Distr:** Vietnam, Laos. **Ecol:** In broad-leafed, evergreen, lowland forests; 0–1,000. **Zones:** 10–11.

P. gigantea J.J. Sm., *Bull. Dept. Agric. Ind. Neerl.* 22: 45, 1909. (Syn: *P. gigantea* var. *aurea*, etc.). [Giant Phalaenopsis]. **Herbs**, epiphytic, robust. **Leaves** 5–6, pendent; lamina elliptic, to 68.5 × 25.5 cm, apex broadly rounded, leathery, pale olive-green, glossy. **Inflorescences** pendent, racemes or sparsely branched panicles, to 40 cm long, obscured by leaves, pale chalky-green; pedicel and ovary 4 cm long. **Flowers** showy, to 5 cm wide, sweetly fragrant, shallowly cupped, usually with overlapping segments, sepals and petals white or yellow, densely covered with transverse brown bars that align to form irregular concentric rings, sepals subsimilar, ovate-elliptic, obtuse; dorsal sepal to 3.5 × 2.5 cm; lateral sepals to 3.9 × 2.6 cm; petals elliptic, to 3 × 2.2 cm cuneate, obtuse; lip trilobed, white, 1.6 × 1.5 cm, lateral lobes erect, triangular-falcate, to 8 × 4 mm, teeth yellow-orange, midlobe with 6 purple longitudinal stripes, obovate, to 1.1 × 0.7 cm, rounded-obtuse, tip pointed, apical marginal irregularly dentate, callus biseriate; column white, to 1.1 cm long. **Distr:** Borneo (Sabah). **Ecol:** In lowland and hill forests, 0–900 m. **Zones:** 11–12.

P. hieroglyphica (Rchb. f.) H.R. Sweet, *Amer. Orchid Soc. Bull.* 38: 36, 1969. (Syn: *P. lueddemanniana* var. *hieroglyphica, Polychilos hieroglyphica*). **Herbs**, epiphytic, medium-sized, upright or pendent, monopodial; stem short, enveloped with imbricating leaf bases. **Leaves** few to many, oblong-ligulate, to 25 × 5 cm, plain green, glossy, coriaceous. **Inflorescences** racemose or paniculate, 3–4-flowered, upright, suberect or arching, simple or occasionally branched, to 30 cm long, flowers opening simultaneously; bracts small, ovate-cucullate. **Flowers** long-lasting, showy, stellate, to 4 cm across, sepals and petals cream or white, covered with brown or maroon hieroglyphiclike markings, lip white or pinkish; dorsal and lateral sepals elliptic-ovate, to 2 × 0.5 cm; petals similar in size and shape to other segments; lip narrow, trilobed, lateral lobes upright, midlobe oblong, obscurely toothed toward the tip. **Distr:** Philippines (Mindanao, Palawan, Polillo). **Ecol:** In shaded locations of warm and humid forests; to 500 m. **Zones:** 10–12. **Info:** Flowers are produced on old inflorescences in subsequent seasons if inflorescence is green and not dried out.

P. honghenensis F.Y. Liu, *Acta Bot. Yunnan.* 13: 373, 1991. (Syn: *Doritis honghenensis*). **Herbs**, miniature, epiphytic; roots minutely warty, forming flattened mats against the substrate; stems to 1 cm long. **Leaves** 1 or 2, obliquely elliptic, 5–7 × 1.5–2.5 cm, base sheathing, persistent, apex acute, spotted purple-red, ~5 mm long. **Inflorescences** racemes, 3–6-flowered, short-pedunculate, to 7.7 cm long; rachis 3.5–4.5 cm long; bracts broadly ovate; floral bracts ovate, 6 mm long; pedicel and ovary 1.6–2.2 cm long. **Flowers** to 2.9 cm across, sepals and petals rose-pink to pale green, lip purple; dorsal sepal oblong-elliptic, to 1.3 × 0.6 cm, obtuse; lateral sepals obliquely elliptic-ovate, to 1.3 × 0.6 cm, obtuse; petals spathulate, to 1.1 × 0.6 cm, obtuse; lip trilobed, minutely spurred, lateral lobes oblong, obliquely truncate, to 7 × 2 mm, midlobe oblong, to 8 × 2 mm, obtuse, ends tapered, spur to 1.5 mm long, callus biseriate; column to 9 mm long. **Fruits** clavate, ~6.2 × 4.5 cm. **Distr:** China (Yunnan) to Vietnam. **Ecol:** On tree trunks; 2,000 m. **Zones:** 8–10.

P. inscriptiosinensis Fowlie, *Orchid Digest* 47: 11, 1983. (Syn: *P. sinensis*). [Chinese Character Phalaenopsis]. **Herbs**, epiphytic; stems short. **Leaves** 3–5, elliptic-oblong, 8–16 × 4–8 cm. **Inflorescences** racemes, unbranched, decurved, 2–5-flowered, flowers 1 or 2 open at a time; peduncle terete, 0.7–1.1 cm long; pedicel and ovary 1.8–2 cm long. **Flowers** to 3.5 cm wide, white or pale yellow with ± greenish suffusion toward apices, sepals and petals convex with revolute margins and intricate transverse bands of cinnamon-brown; dorsal sepal oblong-elliptic, 1.7–1.9 × 0.7–0.9 cm, obtuse; lateral sepals subsimilar, divergent, 1.6–18 × 0.9–1.1 cm, apiculate; petals oblong-oblanceolate, to 1.3 × 0.6 cm, obtuse; lip trilobed, lateral lobes truncate, lateral corners continued as small bright yellow-orange teeth, to 4 mm long, midlobe white with dark red longitudinal stripes along lower 2/3, oblong, 8 × 4 mm, subacute, with a central carinate keel, callus biseriate; column hooded, white. **Distr:** Indonesia (Sumatra). **Ecol:** On forest trees. **Zones:** 11–12. **Info:** This species is confused with the morphologically different *P. sumatrana*, and this issue is yet to be resolved; see Christensen (2001) for a detailed discussion.

P. javanica J.J. Sm., *Bull. Jard. Bot . Buitenz.*, ser. 2, 26: 77, 1918. (Syn: *P. javanica* f. *alba, P. latisepala, Polychilos javanica*). [Javanese Phalaenopsis]. **Herbs**, epiphytic. **Leaves** numerous, elliptic-obovate, to 22 × 10 cm, base tapered, apex acute or obtuse, glossy. **Inflorescences** racemes, occasionally panicles, suberect, to 25 cm long; rachis fractiflex; pedicel and ovary to 2 cm long. **Flowers** fleshy, strongly cupped, to 3 cm wide, sepals and petals translucent, creamy-white with longitudinal brown stripes, stripes broken and composed of ± rectangular blocks, midlobe of lip rich purple, lateral lobes of lip and column white; dorsal sepal broadly elliptic, deeply concave, to 1.7 × 1 cm, obtuse, mucronulate; lateral sepals obliquely subrotund, obtuse, mucronulate, deeply concave, to 1.7 × 2.2 cm, channeled along midvein; petals broadly elliptic, to 1.4 × 1 cm, rounded, abruptly acute, convex; lip trilobed, to 1.8 × 1.4 cm, lateral lobes erect, falcate-lanceolate, bifid, midlobe oblong, acute, with a low central keel toward base, with a terminal knob bearing trichomes at apex,

Below: Phalaenopsis hieroglyphica, Phalaenopsis honghenensis, Phalaenopsis inscriptiosinensis. [All MG.]

callus biseriate; column arching, to 6 mm long. **Distr:** Indonesia (Java). **Ecol:** 700–1,000 m. Fl. Mar–May. **Zones:** 11–12.

P. kunstleri Hook. f., *Fl. Brit. India* 6: 30, 1890. (Syn: *Polychilos kunstleri*). **Herbs**, epiphytic. **Leaves** 2–4, oblanceolate-obovate, 24 × 8 cm, apex acute. **Inflorescences** racemes or panicles, arching to suberect, long-pedunculate, to 40 cm long; floral bracts ovate to 5 mm long, acute; flowers opening simultaneously; pedicel and ovary to 2 cm long. **Flowers** rigid, glossy, sepals and petals pale yellow with a coalesced field of large brown spots at base, margins strongly revolute, lip pale yellow with longitudinal brownish-red stripes, column very pale yellow; sepals elliptic, 2 × 1.1 cm, obtuse; petals subsimilar to the sepals, to 1.5 × 0.8 cm; lip trilobed, to 1.2 × 1 cm, lateral lobes oblong, retrorsely curved, with toothlike corners and a thickened calluslike center, midlobe elliptic-suborbicular, concave with a central longitudinal keel, callus biseriate; column stout, squat, bulbous, to 9 mm long. **Distr:** Myanmar to peninsular Malaysia. **Ecol:** In humid forests; at low to medium elevations. **Zones:** 10–12.

P. lindenii Loher, *J. Orchidées* 6: 103, 1895. **Herbs**, epiphytic, miniature. **Leaves** oblong-lanceolate to oblong-oblanceolate, to 25 × 4 cm, apex acute, minutely, obliquely bilobed, dark green, marbled with silvery-white. **Inflorescences** scapose racemes, densely many-flowered, occasionally branched, flowers opening simultaneously; peduncle suberect, rachis arching-pendent; floral bracts inconspicuous, to 2 mm long; pedicel and ovary to 3.5 cm long. **Flowers** with sepals and petals white with ± faint pink longitudinal stripes, lip white, basal half with 5 intense, bright pink longitudinal stripes, apical half suffused with pink in a solid patch, callus pale yellow with red-spotted, column pink, anther white; dorsal sepal oblong-elliptic, to 1.6 × 7 cm, obtuse-rounded; lateral sepals divergent, obliquely oblong-ovate, to 1.5 × 0.8 cm, acute; petals oblanceolate-subspathulate, to 1.4 × 0.8 cm, base cuneate, apex obtuse-rounded; lip trilobed, to 1.4 × 1.8 cm, lateral lobes oblong-oblanceolate, obtuse-rounded, incurved, midlobe broadly obovate-suborbicular, shallowly concave, apex broadly rounded, callus solitary; column arching, to 1 cm long. **Distr:** Philippines. **Ecol:** In moist montane forests; 1,000–1,500 m. Fl. Jul–Feb. **Zones:** 9–11.

P. lobbii (Rchb. f.) Sweet, *Genus Phalaenopsis* 53, 1980. (Syn: *P. parishii* var. *lobbii*, etc.). **Herbs**, epiphytic, miniature, forming clusters of stems by basal branching. **Leaves** broadly elliptic, to 13 × 5 cm, apex obliquely bilobed. **Inflorescences** erect, few- to 6-flowered racemes; floral bracts minute, elliptic, obtuse; pedicel and ovary to 1.5 cm long. **Flowers** white, swollen base of column with few irregularly distributed brown spots, lateral lobes of lip brown-spotted, midlobe white with 2 broad longitudinal brownish stripes; dorsal sepal oblong-elliptic, 1 × 0.5 cm, obtuse, concave; lateral sepals obliquely ovate to suborbicular, 8 × 7 mm, obtuse-rounded, slightly reflexed; petals obovate-subspathulate, 8 × 4 mm, obtuse-rounded; lip trilobed, lateral lobes erect, falcate, acute, midlobe kidney-shaped, rounded-obtuse, basal callus with 4 threadlike appendages; column short, arching, to 5 mm long. **Distr:** Bhutan, India, Myanmar, Vietnam. **Ecol:** Epiphytic on tree trunks of lowland evergreen forests and open forests, also on branches of gnarled trees in mixed mossy forests; 350–1,200 m. Fl. Dec–Jun. **Zones:** 9–11. **Info:** This species is quite variable in its native range.

P. lowii Rchb. f., *Bot. Zeit.* 20: 214, 1862. (Syn: *P. proboscidioides*, *Polychilos lowii*). **Herbs**, epiphytic or lithophytic, miniature; roots numerous, flattened, appressed to substrate. **Leaves** 1–5, articulated and deciduous; lamina ovate-lanceolate to oblong-lanceolate, to 9 × 3 cm, apex acute, adaxially dark green, abaxially purple-speckled. **Inflorescences** racemes to massive panicles, arching to pendent, few-flowered, to 30 cm long; floral bracts triangular, 3–4 mm long, acute; pedicel and ovary to 4 cm long. **Flowers** showy, membranous, white with variable degrees of pink suffusion, lip with lateral lobes white and teeth bright yellow-orange, midlobe purple with a narrow white border and 5 diffused longitudinal white stripes, column and rostellum white, pollinia purplish; dorsal sepal oblong elliptic to ovate-elliptic, 1.5–2 × 0.9–1.2 cm, obtuse; lateral sepals subequal, obliquely ovate to ovate-elliptic, margins revolute, apex acute or subobtuse; petals flabellate-reniform, 1.7–2.2 × 1.8–2.4 cm, base cuneate, apex broadly rounded; lip trilobed, 1.5 × 1 cm, lateral lobes erect, midlobe fleshy, obovate, acute, adaxially with a long keel, callus biseriate; column arching, ~8 mm long. **Distr:** Myanmar and adjacent W Thailand. **Ecol:** ~800 m. **Zones:** 10–11.

P. lueddemanniana Rchb. f., *Bot. Zeit.* 23 : 1 4, 1865. (Syn: *Polychilos lueddemanniana*, etc.). **Herbs**, erect, epiphytic. **Leaves** several, arching to drooping; lamina oblong-elliptic to elliptic-obovate, to 30 × 9 cm, base conduplicate and tapered, apex rounded, acute to obtuse. **Inflorescences** racemes, or panicles, each portion few-flowered, suberect, frequently producing apical plantlets; rachis 2-ranked; pedicel and ovary to 3.5 cm long. **Flowers** very fleshy, waxy, fragrant, very long-lasting, variable in size and color, to 6 cm wide, sepals dorsally keeled; dorsal sepal elliptic, to 3 × 1.4 cm, obtuse; lateral sepals obliquely elliptic-ovate, to 3 × 1.4 cm, subacute, channeled along midvein; petals broadly elliptic-ovate, cuneate to clawed, obtuse, subequal but smaller than sepals; lip trilobed, to 2.2 × 1.8 cm, lateral lobes erect, oblong, obliquely truncate, apices irregular, midlobe variable in shape, generally oblong to narrowly obovate, apex obtuse, abruptly constricted, convex, with a central raised longitudinal keel, with an apical linear raised pad densely covered with trichomes, callus biseriate; column to 1.2 cm long. **Distr:** Philippines. **Ecol:** In forests; <500 m. Fl. Mar–Apr. **Zones:** 11–12.

P. maculata Rchb. f., *Gard. Chron.*, n.s., 16: 134, 1881. (Syn: *Polychilos maculata*, etc.). [Spotted Phalaenopsis]. **Herbs**, epiphytic and lithophytic, miniature. **Leaves** oblong-ligulate to oblong-elliptic, to 21 × 4 cm, apex acute, adaxially waxy. **Inflorescences** arching racemes ± subequal to leaves to panicles several times longer than the leaves; pedicel and ovary 1.3 cm long. **Flowers** small, slightly cupped, usually with dorsal sepal arching over column, sepals and petals white to creamy-white with transverse reddish-brown bars, lip scarlet, column white; dorsal sepal oblong-elliptic, to 1.8 × 0.6 cm, acute-carinate; lateral sepals obliquely ovate-elliptic, to 1.5 × 0.6 cm, acute-carinate; petals narrowly oblong-oblanceolate, to 1.2 × 0.4 cm, obtuse; lip trilobed, 1 × 0.8 cm, lateral lobes erect, subquadrate, midlobe oblong-elliptic, obtuse, convex, with indistinct grooves, callus biseriate, posterior callus larger, sulcate, minutely bifid, anterior callus smaller, a shallowly notched structure; column slightly arched, 7 mm long. **Distr:** Malaysia (Pahang), E Malaysia (Sabah, Sarawak), Indonesia (Kalimantan Timur). **Ecol:** In deeply shaded areas; 0–1,000 m. **Zones:** 11–12. **Info:** *Phalaenopsis maculata* f. *flava* has deep yellow spots and lip.

P. malipoensis Z.J. Liu and S.C. Chen, *Acta Bot. Yunnan.* 27: 37, 2005. **Herbs**, epiphytic; aerial roots green, to 50 cm long; stem short, sheathed. **Leaves** subbasal, 3–5, deciduous, usually 2 at flowering; lamina oblong or elliptic, 4.5–7 × 3–3.6 cm, base broadly cuneate-rounded, sheathing, apex obliquely acute-obtuse, fleshy. **Inflorescences** often 3–4, racemose, laxly 5–10-flowered, basal, arching, 8–15 cm long; pedicel and ovary white, yellowish-tinged, 1–1.3 cm long. **Flowers** fully opening, 1.2–1.6 cm across, white or sometimes yellowish-tinged, lip white, disk orange, brown-tinged, column white with 1 or 2 crescent or semiorbicular brown blotches on ventral base; dorsal sepal oblong-elliptic, 7–9 × 3–4 mm, obtuse; lateral sepals obliquely ovate-elliptic, 6–7 × 4–5 mm, base adnate to column foot, sometimes dorsally carinate; petals spathulate or narrowly obovate, 6–8 × 2–3 mm, rounded; lip trilobed, spurless, lateral lobes erect, sublanceolate, 2–3 mm long, disk with 2 orangc and brownish-tinged conical calli, midlobe mobile, triangular, 4–5 × 6–7 mm; callus basally deeply forked, each callus arm divided into 2 filiform antennae to 3 mm long; column 4–5 mm long, foot 1–2 mm long. **Distr:** China (Yunnan). **Ecol:** Fl. Apr–May. **Zones:** 8–10.

P. mannii H.G. Reichenbach, *Gard. Chron.* 1871: 902, 1871. (Syn: *P. boxallii*, *Polychilos mannii*). **Herbs**, epiphytic; stems stout, 1.5–7 cm. **Leaves** 4 or 5, oblong-oblanceolate, 20–23 × 5–6 cm, base cuneate, apex acute. **Inflorescences** 1 or 2, ascending or pendulous, 5.5–30 cm long, sometimes branched, sparsely few- to many-flowered; floral bracts lanceolate, 0.6–1 cm long; pedicel and ovary 2–3 cm long. **Flowers** opening widely, long-lasting, thickly textured, waxy, glossy, sepals and petals yellow with dark brown spots and bars, lip midlobe white, column yellow; dorsal sepal obovate-lanceolate, 1.5–1.8 × 0.5–0.7 cm, acute; lateral sepals obliquely ovate-elliptic, 1.5–1.8 × 0.7–0.9 cm, acute; petals suboblong, 1.3–1.5 × 0.4–0.5 cm, acute; lip ~1 cm long, base with claw ~1 mm long, trilobed, lateral lobes erect, appressed, oblong-subquadrate, ~4 × 2 mm, obliquely truncate, midlobe transverse, anchor-shaped, margin fimbriate-erose, apex with trichomes, callus triseriate; column ~8 mm long, foot ~4 mm long. **Distr:** E Nepal to China (S Yunnan). **Ecol:** On tree trunks in broad-leaved evergreen forests; 900–1,400 m. Fl. Mar–Apr. **Zones:** 10–12.

P. mariae Burbridge ex R.Warner and H.Williams, *Orchid Album* 2: t. 80, 1883. **Herbs**, epiphytic, upright, monopodial. **Leaves** few, arching-pendent; lamina oblong to oblong-obovate, to 43 × 10 cm, apex acute. **Inflorescences** racemes or panicles, pendent, subequal to or longer than the leaves, to 61 cm long, branches pendent, many-flowered; pedicel and ovary to 2.5 cm long. **Flowers** showy, slightly fragrant, sepals and petals creamy-white overlain with large, bold, wide, transverse bars of rose, amethyst or reddish brown, lip rose to dark purple, column white; dorsal sepal oblong-elliptic to elliptic-obovate, to 2.2 × 1.1 cm, obtuse, with a short carinate-mucronulate extension, lateral margins ± revolute; lateral sepals obliquely oblong-elliptic, to 2.2 × 1.2 cm, obtuse, with a short carinate-mucronulate extension, with the lateral margins ± revolute; petals obliquely oblong-oblanceolate, to 1.7 × 0.9 cm, obtuse; lip trilobed, to 1.5 × 0.12 cm, lateral lobes erect, triangular-oblong, obliquely truncate, apex denticulate-serrulate, midlobe oblong, dilated below apex, obtuse, lateral margins irregularly denticulate, callus biseriate; column to 7 mm long. **Distr:** Borneo, Philippines. **Ecol:** In densely shaded areas; 100–3,300 m. **Zones:** 11–12.

P. minus (Seidenf.) E.A. Christ., *Phalaenopsis: A Monograph*, 54–55, 2001. (Syn: *Kingidium minus*). **Herbs**, epiphytic; stems short. **Leaves** elliptic, to 9 cm wide. **Inflorescences** racemes, laxly arching-pendent; flowers congested; pedicel and ovary ~1.3 cm long. **Flowers** with sepals and petals reflexed, white with pale brown longitudinal barring, inner halves of lateral sepals purple-spotted; dorsal sepal oblong, acute, concave, to 8 mm long; lateral sepals obliquely oblong-ovate, to 8 × 4 mm, obtuse; petals oblanceolate-subspathulate, 7 × 2 mm, obtuse; lip white with purple blotches at center, trilobed, to 1 × 1.1 cm, lateral lobes erect, suborbicular, broadly rounded, posterior corner

Top row: *Phalaenopsis javanica, Phalaenopsis kunstleri, Phalaenopsis lamelligera.*
Second row: *Phalaenopsis lindenii, Phalaenopsis lobbii, Phalaenopsis lowii.*
Third row: *Phalaenopsis lueddemanniana, Phalaenopsis malipoensis, Phalaenopsis maculata.*
Fourth row: *Phalaenopsis mannii, Phalaenopsis mariae, Phalaenopsis minus.*
Bottom row: *Phalaenopsis modesta.*
[All MG.]

Above: *Phalaenopsis pallens, Phalaenopsis pantherina, Phalaenopsis parishii.* [All MG.]

extended into falcate-triangular tooth, midlobe narrowly lanceolate, acute-acuminate, shallowly incurved, callus biseriate, posterior callus short, bifid, ~1 mm long, apex held over a shallow depression, anterior callus ligulate, apex bifid, ~2 mm long; column white with pale brown barring erect, hooded, to 5 mm long, with a pair of broadly rounded vertical wings lateral to stigma. **Distr:** Thailand. **Ecol:** Epiphytic in rainforests. Fl. Jul–Aug. **Zones:** 11–12.

P. modesta J.J. Sm., *Icones Bogor.* 3: 47, t. 218, 1906. (Syn: *P. modesta* ssp. *sabahensis*, *Polychilos modesta*). **Herbs**, epiphytic. **Leaves** 1–4, elliptic-obovate, to 23 × 6 cm, base subpetiolate and tapered, apex rounded to acute, glossy. **Inflorescences** usually racemes, sometimes sparsely branched, usually few-flowered, pendent, arching, usually shorter than leaves; rachis fractiflex; pedicel and ovary to 2 cm long. **Flowers** fragrant in early morning, sepals and petals white with variable purple transverse barring, sepals greenish toward apices, lip purple, lateral lobes with teeth yellowish-orange; dorsal sepal oblong-elliptic to obovate, to 1.6 × 0.7 cm, obtuse, convex; lateral sepals obliquely elliptic-ovate, to 1.6 × 0.9 cm, acute, apiculate-carinate; petals obliquely elliptic, to 1.5 × 0.5 cm, margins strongly revolute, apex acute; lip trilobed, to 1.4 × 1.4 cm, lateral lobes erect, oblong, subtruncate, posterior corner a retrorse, falcate hook, midlobe oblong-elliptic, rounded, with a central keel, callus biseriate; column white, to 9 mm long. **Distr:** Borneo (Sabah and Kalimantan). **Ecol:** Usually at base of host trees; 50–900 m. Fl. May–Jun, Sep–Nov. **Zones:** 11–12.

P. pallens (Lindl.) Rchb. f., *Ann. Bot. Syst.* 6: 932, 1864. (Syn: *Polychilos pallens*, etc.). **Herbs**, epiphytic. **Leaves** arching, elliptic to obovate, to 18 × 6 cm, base tapered, apex rounded-obtuse, often obliquely bilobed. **Inflorescences** racemes, 1–few-flowered, subequal to leaves; rachis fractiflex; pedicel and ovary to 2 cm long. **Flowers** rigid, waxy, glossy, pale yellow, sepals and petals with sparse, light brown transverse bars, lip white with variable pale pink suffusion on midlobe in longitudinal stripes, lateral lobes yellow to yellow-orange; dorsal sepal elliptic, shallowly convex, 2.3 × 1 cm, acute, tip recurved; lateral sepals obliquely elliptic-ovate, to 2.2 × 1 cm, to 2 × 0.9 cm, acute, channeled below apex, tip recurved; petals obliquely elliptic, acute, slightly constricted below apex; lip trilobed, to 1.7 × 1.4 cm wide, lateral lobes erect, oblong, truncate, midlobe narrowly obovate-oblong, obtuse-rounded, lateral margins irregularly fimbriate, callus biseriate, subequal, bifid; column white, to 8 mm long. **Distr:** Philippines. **Ecol:** In lowland forests; ~500 m. **Zones:** 11–12.

P. pantherina Rchb. f., *Bot. Zeit.* 22: 298, 1864. (Syn: *Polychilos pantherina*). [Panther Phalaenopsis]. **Herbs**, epiphytic; stems leafy, to 15 cm long. **Leaves** oblong to narrowly oblong-elliptic, obliquely obtuse, sublobulate, pale green, to 20 × 4 cm. **Inflorescences** pedunculate racemes or few-branched panicles, to 25 cm long; peduncle terete, to 20 cm long; floral bracts fleshy, distichous; pedicel and ovary to 1.5 cm long. **Flowers** with sepals and petals yellow with reddish brown spots and ± transverse bars; dorsal sepal elliptic-lanceolate, acute, with revolute lateral margins, to 3.3 × 0.9 cm; lateral sepals obliquely ovate, to 3 × 1.1 cm, acute, basal margins revolute; petals oblong-elliptic, to 1.9 × 0.7 cm, acute; lip white, trilobed, to 1.5 × 1 cm, lateral lobes erect, oblong, obliquely truncate, to 6 × 3 mm, midlobe lunate, clawed, to 1.2 cm wide, disk fleshy with trichomes, callus triseriate, variable; column with base dark red, apex yellow, to 8 mm long, with 2 basal knees. **Distr:** Borneo. **Ecol:** On tall trees of lowland to mixed montane forests; 0–800 m. **Zones:** 11–12. **Info:** This species is very similar to *P. cornu-cervi*. See Christensen (2001) for a detailed discussion.

P. parishii Rchb. f., *Bot. Zeitung* (Berlin) 23: 146, 1865. (Syn: *Biermannia decumbens*, etc.). **Herbs**, epiphytic, occasionally lithophytic. **Leaves** 5–8, ligulate, to 45 × 5–7.5 cm, apex asymmetric, rounded, green. **Inflorescences** (2–)3–6(–9)-flowered, 30–50 cm long; peduncle densely pubescent, 20–26 cm long, pale green, white-pubescent; bracts large, conduplicate, broadly elliptic, 3–4 × 2–3.5 cm, margins undulate, apex obtuse to subapiculate; pedicel and ovary 5 cm long, green, pubescent. **Flowers** to 1.3 cm long, sepals cream to green with darker veins, petals green below and spotted with blackish-maroon, dark maroon on margins and in apical half, lip yellow-green to green or purplish-flushed; dorsal sepal elliptic, 3.5–4.6 × 1.8–3 cm, basal margins recurved, apical margins incurved; synsepalum ovate, 2.6–4 × 1.5–2.9 cm, margins recurved, apex acute, outside 2-keeled; petals decurved-pendent, linear-tapering, 7–10.5 × 0.6–1.1 cm, basal margins undulate, maroon-spotted, ciliate, apically spirally twisted, rounded; lip tapered at apex, 3–4.5 × 1.5–2 cm; staminode obcordate, 1.1–1.4 × 0.7–0.9 cm, base umbonate, cream with dark green venation. **Distr:** E Himalayas to Indo-China. **Ecol:** On moss covered trees overhanging streams and on moss covered boulders; 500–2,200 m. Fl. Jun–Jul. **Zones:** 10–12.

P. philippinensis Golamco ex Fowlie and C.Z.Tang, *Orchid Digest* 51: 92, 1987. [Philippines Phalaenopsis]. **Herbs**, terrestrial or lithophytic. **Leaves** to 9, ligulate, 20–50 × 2–7 cm, V-shaped in cross-section, apex rounded, asymmetric, thickly coriaceous, dark green, mottled with silvery markings. **Inflorescence** 2–4-flowered, erect, to 50 cm long; peduncle purple-pubescent; bracts elliptic, 5 × 2 cm, acute, pubescent; pedicel and ovary 4.5–6.5 cm long, purple, pubescent. **Flowers** variable in size, sepals white, dorsal sepal striped with maroon, petals white or yellow at base, maroon above with marginal dark maroon warts in basal half, lip and staminode yellow; dorsal sepal ovate, 4–5 × 2–2.5 cm, acute; synsepalum similar to dorsal sepal, 4.5–5.3 × 2 cm; petals linear, tapering to apex, 6–13 × 0.5–0.6 cm, ciliate; lip small, rather ovoid, 3.8 × 1.4 cm; staminode convex, rather cordate-subquadrate, emarginate, yellow, veined with green, purple-pubescent on sides. **Distr:** Philippines (Luzon). **Ecol:** On limestone cliffs; 0–1,200 m. Fl. Jan–Apr. **Zones:** 11–12.

P. pulcherrima (Lindley) J.J. Smith, *Repert. Spec. Nov. Regni Veg.* 32: 366, 1933. (Syn: *Doritis pulcherrima*, etc.). **Herbs**, epiphytic, variable, branching at base, forming clumps; stem ~4–5 cm long. **Leaves** 3–6, oblanceolate to narrowly elliptic, 5–15 × 1.5–3 cm, apex obtuse to subacute. **Inflorescence** erect, unbranched, 20–60+ cm long, laxly many-flowered. **Flowers** showy, sepals and petals deep mauve-purple, lip with reddish lateral lobes, deep purple midlobe and disk with white lines; dorsal sepal narrowly elliptic, 1.2–2 × 0.7 cm, rounded; lateral sepals oblong, 1.2–1.7 × 0.8 cm, obtuse, connate at base to column foot forming a spurlike mentum 9 mm long; petals obovate, 1.2–1.7 × 0.7 cm, rounded; lip clawed, trilobed above, geniculate at middle, parallel basally with column foot, 1.4 × 1.7 cm, with a narrow erect-falcate lobe on either side of claw, lateral lobes erect, oblong, midlobe ligulate to oblong-apiculate, apiculate; disk with a callus between lobes on the claw; column 5–7 mm long, foot ~6 mm long. **Distr:** S China (Hainan) to W Malesia. **Ecol:** In rocky places, dense forests or thickets. Fl. Jul–Aug. **Zones:** 10–12.

P. pulchra (Rchb. f.) H.R. Sweet, *Amer. Orchid Soc. Bull.* 37: 1102, 1968. (Syn: *P. lueddemanniana* var. *pulchra*, *Polychilos pulchra*). **Herbs**, epiphytic, forming large clumps. **Leaves** arching; lamina oblong-elliptic to narrowly obovate, to 15 × 6 cm, apex acute to subobtuse. **Inflorescences** few-flowered, generally shorter than leaves; pedicel and ovary to 2.5 cm long. **Flowers** fleshy, glossy-lacquered, deep purple, lateral lobes of lip bright yellow; dorsal sepal elliptic to elliptic-oblanceolate, 2.5 × 1.3 cm, acute, slightly channeled apically, dorsally keeled; lateral sepals obliquely ovate, to 2.8 × 1.5 cm acute, slightly channeled apically, dorsally keeled, divergent; petals obliquely elliptic, to 2.3 × 1.1 cm, base tapered to almost clawed, subacute; lip trilobed, to 2.3 × 2 cm, lateral lobes erect, rectangular, truncate, corners extended in short teeth, midlobe cuneate-obovate, apex a fleshy, callus triseriate, posterior callus a transverse band of fleshy tubercles, middle callus short, bifid, anterior callus sulcate, bifid, subulate apices divergent; column white, to 1.3 cm long. **Distr:** Philippines (Luzon, Leyte). **Ecol:** On mountain slopes; 100–650 m. Fl. Aug–Nov. **Zones:** 11–12.

P. schilleriana Rchb. f., *Hamburger Garten-Blumenzeitung* 16: 115, 1860. (Syn: *P. schilleriana* var. *advena*, etc.). **Herbs**, epiphytic. **Leaves** oblong-elliptic to elliptic-obovate, to 45 × 11 cm, base tapered, apex obtuse, dark green, adaxially marbled with silvery-gray, abaxially purple-suffused. **Inflorescences** panicles, arching to pendent, flowers numerous (>250); pedicel and ovary to 4 cm long. **Flowers** membranous, fragrant, pink, callus and lateral lobes of the lip marked with red and yellow, midlobe of lip ± spotted at base; dorsal sepal elliptic to elliptic-ovate, to 3.5 × 1.6 cm, obtuse-rounded; lateral sepals, subequal to the dorsal sepal, obliquely elliptic-ovate, subacute, subparallel; petals rhomboid, cuneate-clawed, to 4 × 4 cm, margins slightly undulate, apex obtuse, broadly rounded; lip trilobed, to 2.2 cm × 3.4 cm, lateral lobes erect, spreading, obliquely elliptic, obtuse-rounded, midlobe elliptic-ovate, notched and flanked by anchorlike retrorse falcate lobules, callus uniseriate, peltate, channeled; column dark pink, stout, fleshy. **Distr:** Philippines. **Zones:** 10–12.

P. stobartiana H.G. Reichenbach, *Gard. Chron.*, n.s., 8: 392, 1877. (Syn: *Doritis stobartiana, Polychilos stobartiana*). **Herbs**, epiphytic; stems very short. **Leaves** 3 or 4, often deciduous, but present at anthesis; lamina ovate-lanceolate, oblong or elliptic, 7–11 × 3–3.4 cm. **Inflorescences** 1 or 2, often ascending, 7–37 cm long, unbranched, laxly 2–4-flowered; pedicel and ovary 4–4.3 cm long. **Flowers** opening widely, sepals and petals apple-green to dark olive-green, lip and column brilliant-rose; dorsal sepal oblong-elliptic, ~1.4 × 0.6 cm, base slightly contracted, apex obtuse; lateral sepals slightly oblique, ovate-elliptic, ~1.4 × 0.7 cm, subacute; petals elliptic-obovate, ~1.4 × 0.6 cm, obtuse; lip trilobed, base with a claw ~1 mm long, lateral lobes ~7 × 3.5 mm, narrow, midlobe obovate-elliptic, ~8 × 5 mm, narrowed toward base, apex mucronate, base with a deeply bilobed-tipped appendage, with a central longitudinal keel, callus concave; column ~7 mm long, base dilated, foot ~2 mm long. **Distr:** China (Yunnan) to Myanmar. **Ecol:** On tree trunks in forests; 1,300–1,400 m. Fl. May–Jun. **Zones:** 8–11.

P. stuartiana Rchb. f., *Gard. Chron.*, n.s., 15: 748, 1881. (Syn: *P. stuartiana* var. *bella*, etc.). **Herbs**, epiphytic. **Leaves** oblong-elliptic to oblanceolate-obovate, to 35 × 8 cm, base tapered, apex obtuse, dark green, adaxially silvery-gray marbled, abaxially purple-suffused. **Inflorescences** panicles, flowers numerous, arching-pendent, to 60 cm long; floral bracts ovate, to 1 cm long; pedicel and ovary to 4 cm long. **Flowers** membranous, faintly fragrant, white, inner halves of lateral sepals greenish-yellow with chestnut-brown spots, lateral lobes of lip with chestnut-brown spotting, midlobe with similar larger spotting; dorsal sepal elliptic, to 3.5 × 1.5 cm, base tapered-constricted, apex obtuse-rounded; lateral sepals obliquely elliptic-ovate, to 3.2 × 1.5 cm, subacute; petals subrhomboid, to 3.3 × 2.6 cm, base cuneate-clawed, apex obtuse, broadly rounded, occasionally shallowly notched; lip trilobed, to 2.5 × 3 cm, lateral lobes obliquely oblong-oblanceolate, obtuse-rounded, erect, spreading, midlobe elliptic-suborbicular, shallowly notched and flanked by a pair of anchor-like, retrorse, falcate lobules, callus yellow, uniseriate, peltate, channeled; column white, stout, straight. **Distr:** Philippines (Mindanao). **Ecol:** Epiphytic in lowland humid forests; 0–300 m. Fl. Jun–Jul, Nov–Feb. **Zones:** 11–12.

P. sumatrana Korth. and Rchb. f., *Hamburger Garten-Blumenzeitung* 16: 115, 1860. (Syn: *P. inscriptiosinensis*, etc.). [Sumatran Phalaenopsis]. **Herbs**, epiphytic. **Leaves** 5 or 6; lamina oblanceolate, 30 × 20 cm, apex obtuse, glossy. **Inflorescences** many; scapes 30 cm long; rachis flattened 4-angled, sometimes branched, ~15 cm long; bracts ovate cymbiform 6 mm long; pedicel and ovary to 2.5 cm long. **Flowers** 5 cm across, fragrant, extremely variable in color from white to green to wine-red, with few to many transverse brown bars or with barring absent, base of sepals and petals without markings, lip white with purple suffusion, column white; dorsal sepal oblong-elliptic, to 2.4 × 1.5 cm, base cuneate, apex acute-carinate, convex; lateral sepals obliquely elliptic, to 2.8 × 1.5 cm; petals oblong-oblanceolate, to 3.5 × 1.2 cm, acute; lip trilobed, to 2.5 × 1.5 cm, lateral lobes erect, oblong, oblique-falcate, midlobe oblong, obtuse, with a raised dorsal keel, apex a raised boss of dense trichomes, callus biseriate; column arching, 1.5 cm long, hooded. **Fruits** 15 cm long. **Distr:** Peninsular Thailand, Vietnam, S Sumatra. **Ecol:** In lowland to mixed montane forests, on trees and shady areas near streams; <700 m. **Zones:** 10–12.

P. taenialis (Lindl.) Christenson and Pradhan, *Indian Orchid J.* 1: 154, 1985. (Syn: *Aerides taenialis*, etc.). **Herbs**, miniature, epiphytic; roots numerous, flattened; stems very short, concealed bracts and leaf sheaths. **Leaves** 1 or 2, obliquely lanceolate, to 10 × 2 cm, apex acute, dark green. **Inflorescences** racemes, 6–8-flowered, pedunculate, to 10 cm long; pedicel and ovary to 1.6 cm long. **Flowers** to 2 cm across, sepals and petals pale pink, lip and anther rose-purple; dorsal sepal oblong, to 8 × 3 mm, obtuse; lateral sepals obliquely elliptic-ovate, to 8 × 4 mm, obtuse; petals elliptic-obovate, obtuse, to 8 × 3 mm; lip trilobed, spurred, to 8 mm long, lateral lobes erect, oblong, to 6 × 1.5 mm, obtuse, midlobe oblong-elliptic, to 6 mm long, obtuse, arching, callus biseriate; column to 4 mm long, strongly dilated at middle. **Distr:** Nepal, NE India, Bhutan, Myanmar, Thailand. **Ecol:** 1,000–2,500 m. Fl. Apr–May. **Zones:** 10–12.

P. tetraspis Rchb. f., *Xenia Orchid.* 2: 146, 1870. (Syn: *P. barrii, P. speciosa* var. *tetraspis*). **Herbs**, epiphytic. **Leaves** 4 or 5 (to 9); lamina elliptic-obovate, convex, to 20 × 8 cm, base tapered, apex, acute to obtuse. **Inflorescences** arching to subpendent, racemes or panicles, to 30 cm long; pedicel and ovary to 3 cm long. **Flowers** showy, fleshy, glossy, strongly fragrant; sepals and petals ± concave, variable in color, white with a few transverse purple bars at base; sepals subsimilar, oblong-elliptic to elliptic-obovate, to 2.8 × 1.2 cm, acute or obtuse; petals subsimilar to sepals, to 2.8 × 1 cm; lip white with faint purple suffusion, trilobed, to 1.8 × 1.6 cm, lateral lobes bright yellow, ovate-triangular, erect, apices irregularly denticulate, with a posterior falcate hook, midlobe oblong, obtuse-subacute, with a raised dorsal keel, apex produced into an elongate oblanceolate pad covered by dense trichomes, callus triseriate, posterior callus minutely 4-lobulate, median callus bifid; column usually white, straight, to 8 mm long, with a hood over anther bed.

Top row: *Phalaenopsis philippinensis, Phalaenopsis schilleriana, Phalaenopsis pulcherrima.*
Bottom row: *Phalaenopsis pulchra, Phalaenopsis stobartiana.*
[All MG except *Phalaenopsis pulchra* AF.]

Top row: *Phalaenopsis stuartiana, Phalaenopsis sumatrana, Phalaenopsis taenialis.*
Bottom row: *Phalaenopsis tetraspis, Phalaenopsis venosa, Phalaenopsis violacea.*
[All MG.]

Distr: Andaman and Nicobar Islands, Sumatra. **Ecol:** In warm forests; near sea level. **Zones:** 8–11.

P. venosa Shim and Fowlie, *Orchid Digest* 47: 125, 1983. (Syn: *Polychilos venos*). **Herbs**, epiphytic. **Leaves** 3–5, oblong-elliptic, 10–22 × 5–7.5 cm, margins shallowly undulate, apex acute, glossy. **Inflorescences** racemes or panicles, densely flowered, branches 2–5-flowered, erect, to 18 cm long; pedicel and ovary 2.8–3.2 cm long. **Flowers** shallowly cupped, 4–5 cm wide, unpleasantly fragrant, sepals and petals pale greenish-yellow, almost completely covered by broad transverse brown barring, bases of sepals and petals stark white forming a ring around column base, lip white with brown suffusion with callus and median keel on midlobe and with fine brown spotting on lateral lobes; dorsal sepal elliptic-ovate, 1.7–2.2 × 1.2–1.4 cm, subacute, bluntly apiculate; lateral sepals obliquely elliptic, 1.8–2.2 × 1–1.4 cm, acute-carinate; petals obliquely oblong-elliptic to ovate, 1.5 × 1.1 cm, abruptly subacute; lip trilobed, 1.8–2 × 1.3–1.5 cm, lateral lobes oblong-linear, obliquely truncate, midlobe elliptic-obovate, acute with a fleshy apicule, margins irregular toward apex, with a high central keel, 1 × 0.6 cm, callus biseriate; column to 9 mm long. **Distr:** Indonesia (Sulawesi). **Ecol:** ~900 m. **Zones:** 10–11.

P. violacea H. Witte, *Fl. Jard.* 4: 129, 1861. (Syn: *Polychilos violacea*, etc.). **Herbs**, pendulous, epiphytic. **Leaves** oblong-elliptic to obovate, 20–25 × 7–12 cm, base tapered, apex acute, rounded, succulent, pale to medium green. **Inflorescences** racemes, arching, sequentially 1- or 2-flowered, shorter than leaves but becoming subequal with age; rachis fractiflex; pedicel and ovary to 3 cm long. **Flowers** fleshy, fragrant, dark rose, apices of sepals and petals green, inner margins of lateral sepals and midlobe of lip darker, lateral lobes of lip and callus yellow; dorsal sepal oblong-elliptic, concave, 2–3.5 × 0.8–1.5 cm, acute-carinate; lateral sepals obliquely elliptic-ovate, strongly divergent, 2.2–3.5 × 1.1–1.7 cm, acute-carinate; petals oblique, upper half oblong, lower half elliptic-obovate, 2–3 × 0.7–1.7 cm, acute; lip trilobed, 2–2.8 × 1.6–2.3 cm, lateral lobes erect, oblong, truncate, midlobe elliptic-obovate, obtuse-rounded, with a dorsal keel ending in a raised subapical knob above and with a short, carinate keel beneath, callus biseriate; column to 1.5 cm long. **Distr:** Indonesia (Sumatra), Malaysia (Malay Peninsula). **Ecol:** Epiphytic in forests; ~150 m. **Zones:** 11–12.

P. wilsonii Rolfe, *Bull. Misc. Inform. Kew* 1909: 65, 1909. (Syn: *Doritis hainanensis*, etc.). **Herbs**, epiphytic and lithophytic, miniature, with massive root systems. **Leaves** deciduous; lamina oblong-elliptic, to 2.5 × 0.9 cm, acute. **Inflorescences** racemes, never branched, to 20 cm long, 10–15-flowered, elongate, arching, flowers opening simultaneously; floral bracts spreading, ovate, concave, to 8 mm long, acute; pedicel and ovary to 4 cm long. **Flowers** with sepals and petals pastel-pink grading to medium rose at their bases; dorsal sepal oblong-elliptic, cuneate, 2.2 × 0.8 cm, acute; lateral sepals obliquely elliptic, acute, subequal to dorsal sepal; petals oblanceolate-subspathulate, to 2.1 × 0.9 cm, obtuse-rounded; lip dark rose-purple, trilobed, to 2 × 1.4 cm, lateral lobes erect, oblong, falcate, obtuse, midlobe obovate, apex notched, convex, with a raised elliptic pad below apex, with a raised median at base, callus biseriate, subsimilar and subequal, sulcate, bilobed; column white, straight, stout, to 1 cm long. **Distr:** S China to Vietnam. **Ecol:** 800–2,200 m. **Zones:** 8–11.

Pomatocalpa

Species 26. **Herbs**, epiphytic; stems short or elongate. **Leaves** ± distichous, base articulated with short persistent sheath; lamina linear or falcate, apex obtuse or unequally bilobed. **Inflorescences** racemes, axillary, short or long, densely flowered. **Flowers** usually not resupinate; sepals and petals subequal, free, spreading; lateral sepals sometimes adnate to basal portion of column; lip adnate to base of column, with a saccate spur at base, trilobed at opening, lateral lobes triangular, midlobe fleshy, broadly triangular, often curved downward; spur ± contracted at opening, thickened on front wall, with a tonguelike appendage projecting from back wall; column stout, without foot; anther terminal, operculate; pollinia in 2 groups, each divided into unequal halves, connected by a slender stipe to a small viscidium; stigma entire, concave; rostellum bifid. **Distr:** Tropical and subtropical Asia to SW Pacific. **Cult:** Propagate from seeds. Grow larger species in pots or baskets with a free-draining, bark-based compost. Smaller species can be grown mounted. Requires warm temperatures, bright light and plenty of water throughout the year.

P. acuminatum (Rolfe) Schltr., *Repert. Spec. Nov. Regni Veg. Beih.* 4: 290, 1919. (Syn: *Cleisostoma acuminatum*, etc.). **Herbs**, epiphytic; stems 2–3 cm long. **Leaves** few, conduplicate toward base, articulated with a short sheath; lamina linear, 8–20 × 1.5–3 cm, apex unequally bilobed, yellowish-green. **Inflorescences** racemes, 6–10-flowered, densely aggregated or subcorymbose, 1.5 cm long; bracts ovate, 2.5–3 × 2–2.3 mm; pedicel and ovary 4–6 mm long. **Flowers** yellow, ~8 mm Ø, sepals with 2 dark purplish transverse bands, obovate-oblong, 4–5 × 2–3 mm, base slightly contracted, apex obtuse; lateral sepals slightly larger, oblique; petals with dark purplish patches near base, oblong-falcate, 4–5 × 1.5–2.5 mm; lip mainly with a yellow spur and a white midlobe, lateral lobes triangular, erect, 1 mm wide, midlobe broadly triangular or semiorbicular, 3 × 3.5 mm, base 2-keeled; spur dorsally compressed, 3 × 4 mm, with an appendage on back wall blocking the opening, appendage truncate, apex slightly serrate, base cuneate; column ~2 mm long. **Distr:** C and S Taiwan. **Ecol:** In lowland forests. **Zones:** 9–10.

P. diffusum Breda, *Gen. Sp. Orchid. Asclep.* 4: t, 1830. (Syn: *Cleisostoma cumingii*, etc.). **Herbs**, epiphytic, lithophytic or terrestrial, miniature to small-sized, monopodial; stem 15–25 cm long, ~2 cm thick. **Leaves** oblong to lorate, 6.5–23 × 3.8–5 cm, base narrowed, apex shortly, broadly and unevenly bilobed, flat, thickly coriaceous, glossy. **Inflorescences**

Top row: *Phalaenopsis wilsonii, Pomatocalpa diffusum, Pteroceras philippinense.*
Bottom row: *Pomatocalpa acuminatum.*
[All MG.]

racemose; peduncle erect, terete, 15–45 cm long, very stout, flexuous; branches 2.5–5 cm long; rachis very stout; sheaths annular; floral bracts minute, broad, triangular, acute, concave. **Flowers** ~4 mm Ø; sepals and petals obovate-oblong, greenish, margined or spotted with reddish-brown; lip with lateral lobes short, truncate, midlobe reniform-ovate, caruncled toward the base; spur inflated below, saccate, dorsal scale truncate, erose; column very short; rostellum inconspicuous; anther low, broad, strap of globose pollinia linear, gland small. **Fruits** fusiform, 2.5 cm long. **Distr:** Peninsular Thailand to Malesia. **Ecol:** On coral rocks, strand vegetation, grasslands, secondary forests and mixed dipterocarp forests; 0–800 m. **Zones:** 11–12.

Pteroceras

Species 25. **Herbs**, epiphytic, small, monopodial; stems mostly short. **Leaves** few to ~10, ± distichous; lamina flat, lorate, base jointed and amplexicaul-sheathing, apex usually unequally bilobed. **Inflorescences** solitary to several, lateral or axillary, penetrating leaf sheaths, erect or pendulous, racemose, with a few to many spirally or distichously alternating flowers. **Flowers** usually resupinate, small; sepals and petals free, spreading; lateral sepals often oblique and distinctly broader than petals, not adnate to column foot; petals narrower than sepals; lip hinged to end of column foot by a movable joint, trilobed, lateral lobes erect, rather long; midlobe very short and small, fleshy, base spurred; spur or sac pointing forward in line with column foot, without interior ornaments, although front wall sometimes quite fleshy; column short and stout, foot long; rostellum small, bilobed; pollinia 2. **Distr:** India to SC China and Malesia. **Cult:** Propagate from seeds. Best grown mounted on a small slab. Requires intermediate conditions with humidity. Avoid drying out completely.

P. philippinense (Ames) Garay, *Bot. Mus. Leafl.* 23: 194, 1972. (Syn: *Thrixspermum philippinense*, etc.). **Herbs**, epiphytic; stem short, usually to 2.5 cm long. **Leaves** articulate, clustered, distichous, ~7; lamina oblong or oblong-lanceolate, 6.5–15 × 1.5–2 cm, both ends attenuate, apex strongly and unequally bilobed, often falcate. **Inflorescence** a raceme, laxly few-flowered, ~2 cm long; scape lateral, hispid, ~7 cm long, erect; bracts tubular, ~2 mm long; pedicel and ovary ~1.5 cm long. **Flowers** whitish-yellow; dorsal sepal oblong-elliptic, 1 × ~0.3 cm; lateral sepals oblong-lanceolate, 1 × 0.35 cm, acute; petals oblong, 9 × ~2 mm, obtuse; lip adnate to column foot, lamina trilobed, lateral lobes erect, petaloid, oblong 4 × 1.5 mm, 2 mm wide at base, apex obtuse, interior margins with minute calli or papillae, midlobe denticulate; spur 8 mm long. **Fruits** ~8 cm long, ~2 mm Ø. **Distr:** Philippines. **Ecol:** In forests; 3–60 m. **Zones:** 11–12.

Renanthera

Species 20. **Herbs**, epiphytic or lithophytic, rarely terrestrial, robust, monopodial; stem long, often to several meters tall, climbing, sometimes branched, with many nodes and long internodes. **Leaves** many, distichous; lamina often oblong, flat, unequally bilobed, thickly leathery, base sheathing and lamina jointed to sheath. **Inflorescence** lateral, relatively long, paniculate, occasionally racemose, laxly many-flowered. **Flowers** resupinate, opening widely, predominantly red or orange, sometimes yellow, flat, medium-sized or large; sepals and petals free; dorsal sepal and petals similar, spreading; lateral sepals usually broader than dorsal sepal and petals, margins subparallel, undulate; lip attached to base of column base, immobile, much smaller than petals and sepals, saccate or spurred, trilobed, lateral lobes erect, adaxially with 1 callus at base, midlobe often ligulate and recurved, small, with lamellate basal calli; spur conic; column short and stout, without a foot; pollinia 4, in 2 pairs. **Distr:** SC China to tropical Asia. **Cult:** Propagate from cuttings with at least 3 roots. Best grown in wooden baskets or mounted on large cork slabs or tree trunks. Requires warm or intermediate temperatures, bright light, and plenty of water.

R. bella J.J. Wood, Orchid Rev. 89: 116, 1981. Herbs, epiphytic. **Leaves** semiterete, 10-12 × 1 cm. **Inflorescence** 10-15-flowered, unbranched, pendent; pedicel and ovary ~2.5 cm long. **Flowers** 5-6 cm across, red or deep pink with darker spots; dorsal sepal obovate, 2.5 × 1.25 cm, base narrowed, margins wavy, apex acute, slightly mucronate; lateral sepals obovate, convex, 4 × 1.3 cm, base clawed, claw ~6 mm long, margins wavy, apex truncate to slightly mucronate; petals narrowly ovate, 3.2 × 1 cm, base narrowed, margins slightly wavy, apex acute; lip ~8 mm long; column ~6 mm long. **Distr:** N Borneo. **Ecol:** In hill forests; 800–1,100 m. **Zones:** 8–12.

R. citrina Averyanov, *Orchids* 66: 1287, 1997. (Syn: *R. citrina* var. *sinica*, *R. sinica*). **Herbs**, epiphytic or lithophytic; stems suberect or ± scrambling, 20–40(–80) cm long, distichously many-leaved. **Leaves** narrowly oblong, 7–10 × 0.9–1.1 cm, thickly leathery, unequally obtusely bilobed. **Inflorescence** solitary, axillary, from upper part of stem, racemose, 5–10-flowered, 12–26 cm long; floral bracts ~1 mm long, membranous; pedicel and ovary 1.6–2.7 cm long. **Flowers** yellowish, sparsely spotted with purple-red; dorsal sepal narrowly oblong-spathulate, 1.8–2.2 × 0.3–0.4 cm, obtuse-acute; lateral sepals similar, twisted, 2.6–3.1 × 0.45–0.55 cm, base with a claw 5–6 mm long, margin strongly undulate, recurved; petals linear, 1.3–1.7 cm × 1.5–2.5 mm, obtuse; lip much smaller, trilobed, lateral lobes erect, ovate-lanceolate, 1.5–2 mm long, midlobe suborbicular, sub apically globose-saccate, ~2 mm long, base with 3 ridges, a shortly conic sac ~2 × 2 mm, and a pair of subquadrate lamellate calli; column 3.5–4 mm long. **Distr:** China (SE Yunnan), Vietnam. **Ecol:** On tree trunks or on rocks along valleys; 500–800 m. Fl. Apr–May. **Zones:** 8–11.

Rhynchostylis

Species ~5. **Herbs**, epiphytic; stems short and thick; pseudobulbs absent. **Leaves** sessile, distichous; lamina

Above: *Renanthera bella, Renanthera citrina, Rhynchostylis coelestis, Rhynchostylis gigantea.* [All MG except *R. gigantea* AF.]

long, channeled, unequally bilobed, tips toothed. **Inflorescences** erect or drooping, flowers densely crowded into long cylindric racemes. **Flowers** with sepals and petals spreading; petals smaller than sepals; lateral sepals adnate to column foot; lip not hinged, joined to column foot, with backward-pointing laterally flattened spur and forward-pointing unlobed or slightly trilobed lamina; column and column-foot both short, lip merging into them; rostellum and anther long-pointed; pollinia 2, cleft, on a long slender caudicle, gland small. **Distr:** Burma, India, Malaysia, Philippines, Thailand, Sri Lanka. **Cult:** Propagate from seeds. Best grown in wooden baskets or mounted. Requires warm temperatures, light to moderate shade and plenty of water.

R. coelestis (Rchb. f.) Rchb. f. ex H.J. Veitch, *Man. Orchid. Pl.* 7: 53, 1891. (Syn: *Saccolabium coeleste, Vanda pseudocaerulescens*). **Herbs**, epiphytic, monopodial; stem stout, to ~20 cm high, rarely more. **Leaves** few to numerous, alternate, closely arranged and imbricate at base; lamina straplike, 10–20 x 2 cm, adaxially channeled, abaxially keeled, apex unequally lobed, toothed, rigidly fleshy. **Inflorescences** racemes, densely ~50-flowered, on apical half, flowering portion ± cylindrical, axillary, erect to arcuate; pedicel and ovary furrowed, twisted, ~1.8–2 cm long, white or pale blue. **Flowers** waxy, fragrant, 1.5–2 cm across, dorsal sepal and petals white, lateral sepals broader and white with an indigo-blue apical blotch, lip basally white, apical half vivid indigo-blue; sepals and petals similar and subequal, oval-oblong, 8–9 x 5–6 mm, obtuse to truncate; petals with base narrowed; lip obovate-oblong, ~1.5 cm long, with a very compressed, slightly curved, saclike spur, 7–8 mm long; column very short, ~2 mm long, with a beaked dark blue anther. **Distr:** Indo-China. **Ecol:** In semideciduous and deciduous, dry, lowland forests and savannalike woodlands; 0–700 m. Fl. Apr–Aug. **Zones:** 9–11.

R. gigantea (Lindley) Ridley, *J. Linn. Soc., Bot.* 32: 356, 1896. (Syn: *Anota densiflora*, etc.). **Herbs**, epiphytic; stems erect, 4–13+ cm, stout, ~2 cm Ø. **Leaves** many, broadly lorate, 15–40 x 3.7–7.5 cm, base with amplexicaul sheaths, apex obtuse, unequally bilobed, fleshy. **Inflorescences** 2–4, usually arching, 14–30 cm long, densely many-flowered; rachis 10–20 cm long, thick; floral bracts often reflexed, broadly ovate, ~5 mm long. **Flowers** thickly textured, white, spotted with purplish-red; sepals similar, elliptic-oblong, 1.2–1.7 x 0.9–1 cm, obtuse; petals oblong, smaller than sepals, base contracted, apex obtuse; lip deep purplish-red, subobovate, ~1.7 x 1.2 cm, fleshy, base with a pair of ridges, margin trilobed, lateral lobes erect, subelliptic, midlobe very small, retuse; disk slightly papillate; spur narrowly conic, ~5 mm long, inside densely white-hairy; column purplish-red ~4 mm long, thick, foot very short or indistinct. **Fruits** obovoid, ~4 cm long. **Distr:** Cambodia, China (Hainan), Indonesia, Laos, Malaysia, Myanmar, Singapore, Thailand, Vietnam. **Ecol:** On tree trunks in open forests; ~1,000 m. Fl. Jan–Apr. Fr. Feb–May. **Zones:** 9–12.

R. retusa Blume, *Bijdr.* 286, pl. 49, 1825. (Syn: *Aerides guttatum*, etc.). [Foxtail Orchid]. **Herbs**, epiphytic stout; stems non-pseudobulbous, long, stout, 5–25 cm long. **Leaves** equitant; lamina linear, strap-shaped, 15–37 x 1.6–2 cm, recurved, apex unequally lobed, sheathing grooved above, coriaceous. **Inflorescences** racemes, axillary, stout, dense, compact, drooping, 20–30 cm long, 3.5–5 cm Ø; peduncles 9–10 cm long; ovary with pedicel 1.3 cm long. **Flowers** white, tinged violet-pink, 1.2–2 cm across; dorsal sepal ovate or ovate-oblong, 1.2 x 0.8 cm, obtuse or emarginate; lateral sepals obliquely and broadly ovate, 1.2 x 0.9 cm, acute or subacute; petals oblong-ovate, 1.2 x 0.5 cm; lip trilobed, clawed, claw deflexed, deeply saccate or spurred beyond, lateral lobes obscure, midlobe elongated, inflexed, cuneiform; disk bilobed, laterally compressed sac, or spur, tip rounded or emarginate; column 5 x 2 mm; pollinia 2. **Fruits** obovate or obconical, 2.8–4 x 1.5–1.8 cm, 6-ridged. **Distr:** China (Yunnan, SW Guizhou) to tropical Asia. **Ecol:** On trees, often by streams or forest margins. Fl. Jan, Apr, Jul, Nov. **Zones:** 9–12.

Robiquetia

Species 41. **Herbs**, epiphytic, monopodial, medium-sized; stem pendent, long, sometimes branching, leafy. **Leaves** oblong to elliptic, apex unequally bilobed. **Inflorescence** axillary, pendent, racemose or paniculate, densely many-flowered. **Flowers** small; sepals and petals similar, spreading; lip immovable, rather fleshy, trilobed, lateral lobes erect, small, sometimes fleshy, midlobe spreading, spur apical, inflated, with callosities on scales within; column short, stout, lacking a foot, rostellum large or small, stipe spathulate, often hooklike, long, viscidium small; pollinia 2, grooved. **Distr:** Tropical and subtropical Asia to W Pacific. **Cult:** Propagate from seeds. Grow smaller species on cork or tree-fern slabs, and larger species in small baskets. Requires humidity and partial shade.

R. cerina (Rchb.f.) Garay, *Bot. Mus. Leafl.* 23: 196, 1972. (Syn: *Malleola merrillii, R. merrillii, Saccolabium cerinum*). **Herbs**, epiphytic, semipendulous, monopodial, over 1 m long; stem flattened, ~2 cm wide. **Leaves** ± 2-ranked; lamina oblong, ~20 x 4–9 cm, base sheathing, apex unequally bilobed, lobes rounded, leathery, green-tinged. **Inflorescences** racemes, to 20 cm long, flowering part ~9–12 x 5 cm, pendulous, densely many-flowered, conical, flowers opening in succession; peduncle ~12 cm long; bracts 3, triangular, ~2 cm long, clasping. **Flowers** ~8 mm Ø, purplish; dorsal and lateral sepals broadly ovate, concave, to 5 x 4 mm; petals almost round, to 5 x 5 mm; lip trilobed, lateral lobes very small, midlobe with apex acute. **Distr:** Philippines. **Ecol:** On tree trunks; 350–3,200 m. **Zones:** 8–12.

Sarcochilus

Syn: *Parasarcochilus*, etc.

Species 25. **Herbs**, epiphytic or lithophytic; stems usually short, spreading or ascending. **Leaves** 2-ranked, often spreading and drooping; lamina with apex unequally bilobed. **Inflorescence** a raceme or spike, few- to many-flowered, flowers 2-ranked or facing all directions, opening 1 or few at a time. **Flowers** showy, fragrant, relatively short-lived, widely opening; sepals and petals free, subsimilar; lateral sepals ± adnate to column foot; lip hinged to apex of column foot, shallowly saccate, trilobed, lateral lobes large, usually narrow, erect, curved, midlobe small and fleshy, ± trilobed, basal sac almost filled by a callus attached along the midline of sac; column short, foot mostly well-developed, joined to lateral sepals; pollinia 4 in 2 pairs; stipe rather short; viscidium small, circular or elliptic. **Distr:** New Guinea to New Caledonia and E Australia. **Cult:** Propagate clumping species by division and epiphytic species from seeds. Grow epiphytic species mounted on tree-fern or cork slabs, and lithophytic species in pots with a coarse mixture containing gravel and perlite. Requires moisture, high humidity and good ventilation.

S. fitzgeraldii F. Muell., *Fragm.* 7: 115, 1870. (Syn: *S. fitzgeraldii* f. *aemulus*, etc.). [Ravine Orchid]. **Herbs**, epiphytic or lithophytic; stem to ~1 m long, branched, base covered with fibers of old leaf bases, apex with a tuft of drooping leaves. **Leaves** ~8, narrowly oblong-ligulate, 6–16 x 1 cm, apex acute, thin, rich dark green. **Inflorescence** to 15-flowered, to 22 cm long, horizontal to pendent, rarely erect; rachis shorter than peduncle, zigzagged; peduncle slender, terete. **Flowers** spreading, 3 cm across; sepals and petals white, pink or rarely pale crimson, barred with carmine spots in lower half, lip streaked with red or crimson inside, callus and midlobe a yellow or orange, spur yellow; sepals oblanceolate-spathulate, 1.2–1.7 x 0.6–0.7 cm, obtuse; petals oblanceolate, 1.2–1.7 x 0.5, obtuse; lip distinctly but unequally trilobed, 5 mm long, lateral lobes erect, falcate-elliptic, midlobe smaller, ± truncate; disk with a large callus, bilobed in front; spur conical; column 2–2.5 mm long, foot ~2–2.5 mm long. **Distr:** Australia (SE Queensland to NE NSW). **Ecol:** In tropical rainforests, often near waterways; 500–700 m. Fl. Oct–Nov. **Zones:** 9–12.

S. hartmannii F. Muell., *Fragm.* 8: 248, 1874. (Syn: *S. rubricentrus*, etc.). [Ravine Orchid, Cliff Orchid]. **Herbs**, lithophytic, rarely epiphytic, semierect; stems numerous, to 1 m long, usually much shorter, stout. **Leaves** 4–10, oblong-lanceolate, falcate, 10–20 x 1.5–2 cm, apex acute or obliquely emarginate.

Inflorescence erect, stout, longer than the leaves, to 16–25-flowered, 25 cm long; pedicels to 2 cm long, finely red-spotted. **Flowers** 1–3.5 cm across, white with crimson spots or blotches at center of perianth, rarely flowers wholly white or wholly crimson; sepals and petals ovate, obovate or oblong-elliptic, 0.8–1.6 × 0.3–0.8 cm, rounded; petals slightly narrower than sepals; lip small, trilobed, lateral lobes ovate to subquadrate, rounded, midlobe erect, very short, obtuse, disk with large bifid callus; spur subconical, 1–2 mm long, fleshy; column 1–2 mm long, foot almost at right angles, 1.5–3 mm long. **Distr:** Australia (SE Queensland to NE NSW). **Ecol:** On cliff faces on steep narrow ridges; 500–1,00 m. Fl. Oct–Nov. **Zones:** 9–11.

Sedirea

Species 2. **Herbs**, epiphytic, monopodial; stems short, covered by leaf bases and fleshy flexuous roots. **Leaves** nearly basal, distichous; lamina flat, slightly fleshy or thickly leathery, base jointed and shortly sheathing. **Inflorescence** racemose, axillary, laxly several-flowered; floral bracts broadly ovate, shorter than pedicel and ovary. **Flowers** opening widely, medium-sized; sepals and petals free, similar; lateral sepals adnate to column foot; lip very fleshy, attached at base to column foot by a movable joint, spurred, trilobed, lateral lobes erect, midlobe recurved, strongly clawed; spur long, parallel to midlobe, narrowed distally; column rather long, base with a short foot, or footless; stigma large, deeply concave, near middle of column; rostellum large, bilobed; pollinia 2, waxy, subglobular, cleft, attached by a common linear stipe to a large viscidium. **Distr:** China to temperate E Asia. **Cult:** Propagate from seeds. Grow in pots with epiphyte mix. Requires humidity and partial shade.

S. japonica (H.G. Reichenbach) Garay and H. R. Sweet, *Orchids S. Ryukyu Is.* 149, 1974. (Syn: *Aerides japonica*). [Nagoran, Nago Orchid]. **Herbs**, epiphytic; stems ~1 cm long. **Leaves** 4–6, nearly basal; lamina oblong or obovate-lanceolate, 6–13 × ~2.5 cm, apex slightly bilobed. **Inflorescence** pendulous, ~18 cm long, laxly 2–12-flowered; pedicel and ovary ~1.8 cm long. **Flowers** orange-scented, whitish-green, with 1–3 transverse bands of dull brown spots adaxially on lateral sepals, lip midlobe spotted purple-red, margins of lateral lobes purple-red, claw with purplish-red ridges adaxially; sepals oblong, ~1.7 × 0.75 cm, obtuse; lateral sepals slightly narrower than dorsal sepal, deflexed; petals oblong-ligulate, ~1.5 × 0.7 cm, obtuse; lip obscurely trilobed, lateral lobes subtriangular, very small, midlobe spathulate, ~1.5 cm long, adaxially concave, base clawed, margin irregularly crenate; spur spreading horizontally, nearly parallel to midlobe, ~1.3 cm long, with a suberect fleshy appendage at entrance; column elongate, ~1 cm long, apical half incurved, foot ~4 mm long. **Distr:** Korea, SC and S Japan to Ryukyu Islands. **Ecol:** On tree trunks in open forests; 600–1,400 m. Fl. Jun. **Zones:** 8–11.

S. subparishii (Z.H. Tsi) Christenson, *Taxon* 34: 518, 1985. (Syn: *Hygrochilus subparishii*). **Herbs**, epiphytic, miniature; stems 1–2 cm long. **Leaves** 3 or 4, nearly basal; lamina oblong or obovate-lanceolate, 5.5–19 × 1.5–3.4 cm, shallowly bilobed. **Inflorescence** to 10 cm long, laxly several-flowered; floral bracts ovate, 6–9 mm long; pedicel and ovary ~2.5 cm long. **Flowers** fragrant, slightly fleshy, yellowish-green, adaxially spotted with pale brown, abaxially whitish; sepals suboblong, 1.6–2 × 0.7–0.9 cm, acute, abaxially narrowly winged; lateral sepals slightly narrower; petals subelliptic, 1.5–1.8 × ~0.6 cm, acute; lip movable, trilobed, lateral lobes erect, suborbicular, margins slightly serrate, midlobe narrowly oblong, ~6 × 1.2 mm, fleshy, abaxially beaked near apex, adaxially with a laterally compressed conic callus at base and a high lamella extending from base to apex; spur spreading forward, nearly parallel to midlobe, horn-shaped, ~1 cm long; column ~1 cm long, base without a distinct foot, winged. **Distr:** S China. **Ecol:** On tree trunks in forests; 300–1,400 m. Fl. May–Jun. **Zones:** 8–11.

Smitinandia

Species 3. **Herbs**, epiphytic, monopodial, medium-sized; stems elongate, with many nodes, often rooting at nodes. **Leaves** distichous, usually well-spaced; lamina flat, rather narrow, slightly fleshy, base sheathing, jointed to sheath. **Inflorescences** racemose, many-flowered, lateral, piercing through sheath. **Flowers** small, somewhat fleshy; sepals conspicuously larger than petals; lip attached to base of column, immobile, with a distinct spur without inner appendages, its entrance ± closed by a fleshy transverse appendage at base of midlobe; column cylindric, short, base slightly dilated, without a foot; pollinia waxy, 4, each completely divided into 2 unequal pairs, flat, subglobular, viscidium ovate, stipe short. **Distr:** Himalayas to China (SW Yunnan) and Malesia. **Cult:** Propagate by division of mature basal shoots or from seeds. Requires intermediate to warm conditions with moderate shade and moisture throughout the year.

S. micrantha (Lindley) Holttum, *Gard. Bull. Singapore* 25: 106, 1969. (Syn: *Ceratochilus*

Top row: *Rhynchostylis retusa, Robiquetia cerina, Sarcochilus hartmannii, Sarcochilus fitzgeraldii.*
Bottom row: *Sedirea japonica, Sedirea subparishii, Smitinandia micrantha.* [All MG.]

Above: *Stereochilus dalatensis, Thrixspermum merguense, Thrixspermum pulchellum, Thrixspermum trichoglottis.* [All MG except *T. pulchellum* AF.]

micranthus, etc.). **Herbs**, epiphytic; stems suberect, terete, slightly flattened, 4–6 cm long. **Leaves** narrowly oblong, 9.5–11 × 1.4–2 cm, base subrounded, apex obtuse, unequally bilobed, slightly fleshy. **Inflorescences** racemes, 1–2, opposite, 9–11 cm long, densely many-flowered; pedicel and ovary ~4 mm long. **Flowers** opening widely, white, tinged with purple-red, 3–4 mm Ø, anther cap white with yellowish-brown midvein; dorsal sepal subobovate, ~3.5 × 1.5 mm, obtuse, 3-veined; lateral sepals slightly oblique, ovate-triangular, ~2.5 × 2 mm, acute, 3-veined; petals narrowly oblong, ~2 × 1 mm, margin ± scabrous, apex obtuse, 1-veined; lip trilobed, lateral lobes erect, subquadrate, ~1 mm long, midlobe obovate-spathulate, ~2.2 × 1.5 mm, base with a fleshy transverse appendage ± covering spur entrance, with 2 central ridges fused at apex, apex subacute; spur oblong-conic, slightly laterally compressed, ~2 × 1.5 mm, obtuse; column ~1 mm long. **Distr:** Himalayas to China (SW Yunnan) and peninsular Malaysia. **Ecol:** On tree trunks in forests; ~600 m. Fl. Apr. **Zones:** 8–12.

Stereochilus

Species 7. **Herbs**, epiphytic, monopodial; stems suberect, short or rather long, noded. **Leaves** several to many, distichous, jointed, sessile, base often sheathing; lamina flat, sometimes ± conduplicate, leathery. **Inflorescences** 1–3, axillary, racemose, lax, often pendulous, many-flowered; floral bracts very small, much shorter than pedicel and ovary. **Flowers** small, often opening widely; lateral sepals adnate to base of lip; petals smaller than sepals; lip adnate to column base, immovable, base saccately spurred, indistinctly trilobed, lateral lobes erect, small, midlobe larger; spur with longitudinal septum inside and often 1 or 2 calli on back wall; column rather stout, without a foot; rostellum subulate-lanceolate, rather long; pollinia 4, in 2 pairs, equal, ellipsoid-oblong to ovoid, waxy, with short but distinct caudicles, attached by a long, slender, somewhat clavate stipe to a small, ovate viscidium. **Distr:** E Himalayas to Indo-China. **Cult:** Propagate from seeds. Best grown mounted on a slab. Requires high humidity and moderate shade.

S. dalatensis (Guillaumin) Garay, *Bot. Mus. Leafl.* 23: 205, 1972. (Syn: *Sarcanthus dalatensis*). **Herbs**, epiphytic; stems to 10 cm long. **Leaves** to 12, distichous; oblong-elliptic, ~5 × 0.5 cm, apex rounded, distinctly V-shaped in cross-section, fleshy, dark green, often purple-spotted. **Inflorescence** laxly up to ~8-flowered, 3–4 cm long, axillary, borne from upper part of stem; floral bracts minute, ovate, rounded; pedicel and ovary ~1 cm long, glabrous. **Flowers** ~1 cm Ø; sepals and petals white, lip light purple, hypochile yellow inside, anther cap deep purple; sepals glabrous; dorsal sepal oblong-elliptic, ~5 × 3 mm, rounded; lateral sepals almost rhomboid, subspathulate, ~5 × 4 mm, 3-veined, rounded; petals spathulate, ~5 × 4 mm, 1-veined, rounded; lip ligulate, ~5 mm long, fleshy, lateral lobes obscure, rounded, midlobe ovate; spur pendent, cylindric, ~5 mm long; posterior callus papillose, sulcate in middle; column ~4 mm long; rostellum pendent, long; anther cap narrowly triangular. **Distr:** China (Yunnan), Thailand, Vietnam. **Ecol:** In evergreen forests; 15–1,600 m. Fl. Nov–Jul. **Zones:** 8–11.

Thrixspermum

Species 154. **Herbs**, epiphytic, erect or pendulous; stems either short with few leaves close together, or rather long with many well-spaced leaves. **Leaves** narrow, usually oblong, flat or conduplicate. **Inflorescences** racemes, axillary; bracts distichous or spiral; flowers opening successively, short-lived, rarely opening together and long-lasting. **Flowers** with sepals and petals subequal, ovate to elliptic; lip continuous or jointed to foot of column, concave or saccate at base, trilobed, lobes erect from foot of column forming an enclosed saclike structure, lateral lobes falcate, erect, embracing column, midlobe fleshy, with various trichomes and callosities on disk; column stout, with a distinct broad foot; anther terminal, operculate; pollinia 4, in 2 groups, each divided into unequal halves, or 2 and distinctly sulcate, waxy, attached to a large, broad stipe; stigma entire. **Fruits** linear-cylindrical. **Distr:** Tropical and subtropical Asia to W Pacific. **Cult:** Propagate from seeds. Best grown mounted on a tree-fern or cork slab. Requires humidity, moderate shade and watering throughout the year.

T. merguense (J.D. Hooker) Kuntze, *Revis. Gen. Pl.* 2: 682, 1891. (Syn: *Sarcochilus kusukusensis*, etc.). **Herbs**, epiphytic; stems 2–3 cm long. **Leaves** several, nearly basal; lamina narrowly oblong, 2–5 × 0.5–0.6 cm, base contracted, apex obtuse, slightly bilobed, fleshy. **Inflorescence** suberect or ascending, to 5 cm long, few-flowered, flowers opening successively, lasting nearly half a day. **Flowers** yellow, with red stripes on lateral lobes and orange-yellow on midlobe of lip; dorsal sepal oblong or ovate-oblong, ~5 × 2–3 mm, obtuse; lateral sepals slightly oblique, obtuse; petals obovate-oblong, 4–5 × 2–2.5 mm; lip triangular, ~3 × 5 mm, base saccate, margin trilobed, lateral lobes erect, large, obtuse, midlobe ~1 × 1.5 mm, truncate; disk with 2 conic calli and 3 tufts of white hairs; column ~2 mm long, foot short. **Distr:** Indo-China to W Malesia and Philippines. **Ecol:** On tree trunks in forests; to 700 m. Fl. Jun–Nov. **Zones:** 9–12.

T. pulchellum (Thwaites) Schlechter, *Orchis* 5: 57, 1911. (Syn: *Cylindrochilus pulchellus*, etc.). **Herbs**, epiphytic; stems non-pseudobulbous, flexuous, sharply and suddenly bent at terminal ends. **Leaves** sessile, distichous; lamina linear-oblong, straplike or falcate, 5–9 × 0.7–1.2 cm, apex rounded, obliquely bilobed, fleshy, coriaceous, adaxially grooved along midrib. **Inflorescences** spikes, short, terminal; peduncles 2.5–9 cm long, with 1 or 2 minute, sterile bracts; floral bracts 2.4–3.2 × 2.4–3 mm, acuminate, spiny, 1-veined. **Flowers** white, 1.2 cm across; dorsal sepal lanceolate-oblong, 7 × 3.6 mm, obtuse, 5-veined; lateral sepals broadly lanceolate-oblong, 7–8 × ~4 mm, subacute, 5-veined, adnate to column foot; petals lanceolate, 7 × 2.8 mm, 3-veined; lip orbicular-obovate, 5 × 8 mm, recurved, 7-veined, glandular-pubescent; lateral lobes of lip rounded and blunt, bent inward; midlobe short, blunt and sometimes bifid, with a thick, scurfy callus between mouth of sac and outer margin; column 3.2 mm tall, continuing into a foot; pollinia 2; ovary 6–9 mm long. **Fruits** long-cylindrical, 7.5 cm × 3.5 mm. **Distr:** Sri Lanka. **Ecol:** ~600 m. Fl. Apr–Jun. **Zones:** 11–12.

T. trichoglottis (J.D. Hooker) Kuntze, *Revis. Gen. Pl.* 2: 682, 1891. (Syn: *Sarcochilus trichoglottis*). **Herbs**, epiphytic; stems ascending, 3–7 cm long, nodes several. **Leaves** many, narrow, 3–5.5 × 0.5–1 cm, apex obtuse, slightly unequally bilobed, leathery. **Inflorescences** 2 or 3, densely several-flowered, usually 3–4 cm long, borne on lower part of stem, suberect, flowers opening successively; floral bracts persistent, densely arranged, ovate-lanceolate, 3–4 mm long, acuminate; pedicel and ovary ~5 mm long. **Flowers** yellowish-white, not opening widely, short-lived; dorsal sepal elliptic, ~4 × 2.5 mm, obtuse; lateral sepals obliquely ovate, ~4 × 2.5 mm, obtuse; petals elliptic, ~3.5 × 2 mm; lip trilobed, base with shallow sac, adaxially densely minutely papillate, lateral lobes erect, suborbicular, ~3 × 2 mm, apical margin with long clavate ciliae, midlobe ~1 mm long, thickened, ± conic-cucullate; column ~2 mm long, foot ~1 mm long. **Distr:** Assam to W Malesia. **Ecol:** On tree trunks in open forests; 700–800 m. Fl. Mar. **Zones:** 10–12.

Trichoglottis

Species 65. **Herbs**, epiphytic, monopodial; stems pendulous or climbing, short or long, slender, with few to many nodes, internodes elongated. **Leaves** many, distichous; lamina linear to elliptic, slightly fleshy or leathery, base sheathing and jointed, apex usually unequally bilobed. **Inflorescences** lateral, axillary, several to many or rarely solitary, often more than 1 per node, 1- to several-flowered; peduncle very short. **Flowers** small, lasting ~1 week, resupinate, opening widely, usually yellowish with light brown or purple markings; sepals and petals free; petals slightly smaller than sepals; lip immobile, spurred or saccate at base, trilobed, lateral lobes erect, midlobe sometimes further trilobed, often hairy or papillate; sac or spur often thickened, with a hairy ligulate appendage on back wall just below base of column; column short and stout, without a foot, often with small roughly hairy

Above: *Trichoglottis atropurpurea*, *Trichoglottis bipenicillata*, *Trichoglottis bipunctata*, *Trichoglottis latisepala*. [All MG except *T. latisepala* MS.]

stelidia; pollinia waxy, 4, appearing as 2 unequal masses. **Distr:** Tropical and subtropical Asia and NW Pacific. **Cult:** Propagate by division. Grow in pots with a free-draining coarse mixture, or mounted on a slab. Requires high humidity, regular watering and fertilizing.

T. atropurpurea Rchb. f., *Linnaea* 41: 30, 1876. (Syn: *T. brachiata*, etc.). **Herbs**, epiphytic, monopodial; stems erect or hanging, to ~60 cm long, sparsely branching near base, leafy. **Leaves** ascending; lamina oblong, to 8 × 4 cm, margins upcurved, apex obtuse, retuse, apiculate, rigidly leathery. **Inflorescences** 1(2)-flowered, short, borne opposite the leaves. **Flowers** faintly fragrant, heavily waxy, long-lived, to 6 cm across, very dark reddish-purple, almost black, lip bright pink, quite hairy along the central ridge, throat golden-yellow; dorsal sepal ovate, to 2.1 × 1.2 cm; lateral sepals ovate-lanceolate, to 2.2 × 1.5 cm; petals oblanceolate, to 2.2 × 0.8 cm; lip 5-lobed, 2.4 × 1.4 cm, central ridge hairy, distal 3 lobes forming a cross. **Distr:** Philippines. **Ecol:** In woodlands, mangrove swamps; 0–300 m. **Zones:** 11–12.

T. bipenicillata J.J. Sm., *Icon. Bogor.*: t. 125 A, 1903. **Herbs**, epiphytic, pendulous; stems to 70 cm long, internodes 1.5–3.5 cm long. **Leaves** oblong-ligulate to narrowly elliptic, 5–15.5 × 1.5–2.4 cm, base cuneate, apex obtusely unequally bilobed or acutely asymmetric. **Inflorescence** 1–3-flowered, abbreviated; peduncle 2–3 mm long; pedicel and ovary 1.6–2.3 cm long. **Flowers** 2 cm across, sweetly scented, sepals and petals pale to dark yellow with large irregular cinnamon or orange-brown blotches, or sometimes small spots, lip pale yellow with a cinnamon callus; dorsal sepal oblanceolate-spathulate, 1.5–1.7 × 0.4–0.5 cm, obtuse; lateral sepals oblanceolate, 1.3–1.5 × 0.4–0.5 cm, obtuse; petals narrowly oblanceolate, 1.5–1.6 × 0.3–0.35 cm, obtuse; lip 7–9 × 1.5–1.6 mm, lateral lobes falcate, 2.5–3 mm long, acute, midlobe oblong, 3 mm long, deflexed, narrowly triangular-ligulate, callus central, cuspidate, 2 mm long, porrect, base with a tuft of white hairs on each side, tongue ligulate, spathulate, apical margins ciliate; column 2.5–4 mm long; pollinia 4. **Distr:** Borneo (Kalimantan Barat, Sabah). **Ecol:** In mixed dipterocarp forests; 400–600 m. Fl. Apr–May, Aug. **Zones:** 11–12.

T. bipunctata (E.C. Parish and Rchb. f.) Tang and F.T. Wang, *Acta Phytotax. Sin.* 1: 101, 1951. (Syn: *Saccolabium bipunctatum*, etc.). **Herbs**, epiphytic, monopodial; stem slender, scandent or scrambling. **Leaves** linear or lorate, 10–15 × 1.2–1.6 cm, base narrow, clasping the stem, apex subacute, bidentate, costa stout abaxially. **Inflorescences** 1–2-flowered, borne from the nodes and held close to the stem and leaf base; peduncle very short; bracts obscure. **Flowers** fragrant, ~6 mm across, yellow and purplish, lip white; sepals oblong-lanceolate, ~3 × 1.5–2 mm, acute, trinerved; petals linear-oblong, falcate, 3 × 1.3 mm, 1-nerved; lip trilobed, 5.5 × 1.3mm, lateral lobes small, subfalcate, midlobe large, longer than the sepals, ovate-lanceolate, trinerved with a basal callus, dorsal scale membranous, toothed; spur very small, incurved, obtuse, saccate; column very short, ~1.5 mm long; anther short, very broad; pollinia subglobose, strap linear, gland small. **Distr:** Myanmar to peninsular Malaysia. **Ecol:** 50–300 m. Fl. any time of the year. **Zones:** 9–12.

T. latisepala Ames, *Philipp. J. Sci.*, C 4: 675, 1910. (Syn: *T. latisepala* var. *tricarinata*). **Herbs**, epiphytic, monopodial, large; stems elongate, slender, pendulous, 0.3–1 m long, 2.4 mm Ø, enveloped by imbricate leaf sheaths; leaf sheaths 1.5–1.7 cm long. **Leaves** linear-lanceolate, 7–12 × 0.7–1.5 cm, apex acute, coriaceous, distinctly purple in bright light. **Inflorescence** 2–6-flowered, short, borne opposite the leaves, flowers held close to the stem. **Flowers** very small, ~1 cm across, light pinkish-white; dorsal sepal oblong, 6 × 3 mm, obtuse, trinerved; lateral sepals oblong, shaped like a bird's wing in flight, 6 × 5 mm, subbipartite to bilobed, obtuse; petals oblong, subspathulate or oblanceolate, 5 × 2–2.5 mm, obtuse; lip trilobed, lateral small, lobes decurrent on the spur, obtuse, midlobe suborbicular to elliptic, rounded, callus pilose at base; column 3–3.5 mm long. **Distr:** Philippines. **Ecol:** In coastal forests; to 1,000 m. **Zones:** 11–12.

T. pusilla (Teijsm. and Binn.) Rchb. f., *Bonplandia* (Hannover) 4: 325, 1856. (Syn: *Vanda pusilla*, etc.). **Herbs**, epiphytic, miniature, monopodial; stems to 12.5 cm long. **Leaves** strap-shaped to oblong, to 12 × 1 cm, apex narrowed, oblique, acute, unevenly bidentate, canaliculate, V-shaped in cross-section. **Inflorescences** racemes, usually 3-flowered, ~4 cm long, flowers opening in succession; pedicel and ovary green, ~1.5 cm long. **Flowers** 2–2.5 cm across, fleshy, slightly fragrant or with a somewhat unpleasant scent, sepals and petals pale green to white, with horizontal purple stripes; sepals cuneate-ovate, acute; dorsal sepal 1 × 0.6–0.7 cm; lateral sepals falcate, 1.6 × 0.7–0.9 cm; petals ligulate, 1–1.25 × 0.4–0.45 cm, acute; lip trilobed, midlobe ovate, 1.5 × 1.2 cm, concave, lateral lobes erect, midlobe spathulate, ~1.2 cm across; column 4–5 mm long. **Distr:** Sumatra to Java. **Ecol:** On small branches in brightly lit areas of forests; 1,000–2,000 m. **Zones:** 11–12.

T. rosea (Lindley) Ames, *Enum. Philipp. Fl. Pl.* 1: 440, 1925. (Syn: *Cleisostoma breviracemum*, etc.). **Herbs**, epiphytic; stems clustered, pendulous, to 50 cm, internodes 1.5–2 cm. **Leaves** many, distichous; lamina narrowly lanceolate, 7–10 × 1.2–1.5 cm, base amplexicaul-sheathing, apex acute, slightly fleshy. **Inflorescences** several, borne opposite leaves, very short, 3–6-flowered; peduncles 3–6 mm long; floral bracts broadly triangular, ~2 mm long; pedicel and ovary ~5.5 mm long. **Flowers** white, sometimes tinged with pale yellow or purple; dorsal sepal oblong, 7–9 × 2.5–3.5 mm, acute; lateral sepals obliquely ovate, 6–8 × 4–5 mm, acute; petals falcate-oblanceolate, 7–9 × 2–3 mm, acute; lip somewhat fleshy, shortly spurred, lobes erect, deltoid, ~1 mm long, obtuse,

Below: *Trichoglottis pusilla*, *Trichoglottis rosea*. [Both MG.]

Top row: *Trichoglottis smithii, Trichoglottis triflora, Uncifera thailandica.*
Bottom row: *Tuberolabium kotoense, Tuberolabium rhopalorrhachis.*
[All MG.]

midlobe broadly ovate, 2–3 × 2.5–3.5 mm, base thickened, forming a fleshy appendage almost covering spur entrance, apex obtuse; spur obconic, ~2.5 mm long, with a ligulate appendage on upper back wall. **Distr:** S Taiwan. **Ecol:** On tree trunks in forests at low elevations. Fl. Apr–May. **Zones:** 9–10.

T. smithii Carr, *Gard. Bull. Straits Settlem.* 8: 125, 1935. (Syn: *T. appendiculifera, T. quadricornuta* [illegitimate name]). **Herbs**, epiphytic, scandent; stems 0.24–0.4(–2) m long. **Leaves** oblong-ligulate or oblong-elliptic, 4–5.5(–8) × (0.8–)1–1.6(–2) cm, apex obtusely unequally bilobed or retuse, midrib prominent, abaxially carinate, extended into an apical mucro. **Inflorescences** 1-flowered, abbreviated; peduncle ~3 mm long; pedicel and ovary 1.6–1.8 cm long. **Flowers** 2.5 cm across, slightly scented, sepals and petals creamy-white or pale greenish-yellow, with irregular transverse reddish-brown, reddish-orange or crimson markings, lip white with pale violet or deep magenta spots and streaks, base sometimes yellow-flushed, hairs white and violet, column white; sepals and petals spreading; dorsal sepal narrowly oblanceolate, 1.3 × 0.35–0.4 cm, obtuse to acute; lateral sepals obliquely narrowly elliptic, 1.4 × 0.45–0.55 cm, acute; petals linear-elliptic, 1.3–0.2 cm, acute; lip 0.9–1.1 × 0.4 cm, lateral lobes erect, oblong, 2.5 × 1.5 mm, midlobe oblong-cuneate, 6–7 × 3–3.5 mm, 3-toothed, fleshy, central fleshy keel 1.5–2 mm high, with dense pilose hairs; column 3 mm long, foot 2 mm long; pollinia 4. **Distr:** Borneo. **Ecol:** In open forests; to 1,300 m. Fl. Apr–Jun, Oct–Nov. **Zones:** 11–12.

T. triflora (Guillaumin) Garay and Seidenfaden, *Bot. Mus. Leafl.* 23: 209, 1972. (Syn: *Saccolabium triflorum*). **Herbs**, epiphytic, small; stem erect, 0.5–1.5 cm, enclosed in leaf sheaths. **Leaves** 6 or 7, distichous-equitant; lamina recurved, often V-shaped-conduplicate, 2–3.5 × 0.5–0.6 cm, apex obtuse, unequally bilobed. **Inflorescences** 1–3, axillary, 0.5–1 cm long, 1–3-flowered; peduncle with 1 or 2 sheaths at base; pedicel and ovary 2.5–3 mm long. **Flowers** not opening widely, yellowish-green, lip white, spotted with purple-red, slightly fleshy; sepals oblong-elliptic or oblong, ~3 × 1.2 mm, dorsally ± carinate, apex obtuse; lateral sepals slightly oblique; petals narrowly falcate-oblong, ~2.8 × 7 mm, dorsally carinate, apex obtuse; lip ~2.5 mm, trilobed, lateral lobes erect, suborbicular, midlobe ~1.5 × 2.5 mm, with 3 lobules nearly equal in size, lateral lobules subovate-triangular, large, obtuse, midlobule ovate-triangular, adaxially densely minutely papillate; spur parallel to ovary, tubular, ~2 × 1.2 mm, obtuse; column ~8 mm long. **Distr:** SW Thailand, Vietnam. **Ecol:** On tree trunks in forests; 1,100–1,200 m. Fl. Aug. **Zones:** 10–12.

Tuberolabium

Species 17. **Herbs**, epiphytic, small, monopodial; stems short, enclosed in leaf sheaths. **Leaves** few, linear-falcate or strap-shaped, flat, base jointed and sheathing. **Inflorescence** racemose, few- to many-flowered, lateral, often pendulous, elongate; peduncle short; rachis terete, sometimes clavate, fleshy, sulcate; floral bracts small. **Flowers** rather short-lived or lasting for about a 1 week, to 9 mm Ø, white, yellowish, or greenish with various purple, brownish-purple, or red markings, slightly fleshy; sepals and petals free, spreading; lateral sepals usually larger than dorsal sepal, base adnate to back wall of spur; petals smaller than sepals; lip base firmly connate to base of column, immovable, very fleshy, spurred, trilobed, lateral lobes on either side of spur entrance, very small, tooth-like, midlobe laterally compressed, very fleshy, with incurved margin; spur often ± laterally compressed, broadly conic, sometimes with protuberances at its entrance; column short, stout, foot absent; pollinia 2, waxy, subglobular to ovoid, entire. **Distr:** SE Taiwan to tropical Asia and Pacific. **Cult:** Propagate from seeds. Best grown mounted on a slab. Requires intermediate conditions with high humidity and moderate shade.

T. kotoense Yamamoto, *Bot. Mag. (Tokyo)* 38: 209, 1924. (Syn: *Saccolabium kotoense*). [Botel Tobago Orchid, Coconut-scented Orchid]. **Herbs**, epiphytic; stems short, indistinct. **Leaves** 2–7, ± basal; lamina oblong or oblong-elliptic, 6.5–16 × 2–4 cm, apex obtuse, slightly bilobed, fleshy or thickly leathery. **Inflorescence** densely many-flowered, pendulous, 10–15 cm long; floral bracts broadly ovate-triangular, 2–3 mm, sometimes denticulate toward apex; pedicel and ovary 5–8 mm long. **Flowers** small, white, often yellow-tinged, lip with lateral lobes purplish, midlobe white with a central purplish patch and 2 purple basal swellings, spur white or pale green, sometimes with purplish stripes or spots, column purplish, rostellum white; dorsal sepal obovate-elliptic, 4.5–5.5 × 2.2–3 mm, obtuse, concave; lateral sepals obliquely obovate, 5–6 × 2.5–4 mm, obtuse; petals oblong-spathulate, ~5 × 1.8 mm, obtuse; lip spurred, ~3 mm, trilobed, lateral lobes erect, subquadrate, 1–1.5 cm long, obtuse to rounded, midlobe thickly fleshy, adaxially slightly depressed at base and with 2 swellings; spur conic, laterally compressed, ~2 mm long; column 1–1.5 mm long. **Distr:** S Taiwan (Lan Yü). **Ecol:** On tree trunks in thickets. Fl. Dec–Feb. **Zones:** 9–10.

T. rhopalorrhachis (Rchb. f.) J.J. Wood, *Nordic J. Bot.* 10: 482, 1990. (Syn: *Dendrocolla rhopalorrhachis*, etc.). **Herbs**, epiphytic, upright, monopodial, miniature. **Leaves** linear, lanceolate or narrowly lanceolate, to 11 × 2 cm, apex unequally bilobed, each lobe acute, leathery. **Inflorescences** to 6-flowered, 2–4 cm long, axillary, thick, fleshy, upright, 2–4 flowers at a time; floral bracts close-set. **Flowers** ~8 cm across, fragrant, white or yellowish, base of sepals and petals striped with dark purple or red, lateral lobes of lip yellow and white, spur white and red; dorsal sepal elliptic to ovate, 6 × 3.5 mm; lateral sepals elliptic-ovate, 6 × 3.5 mm; petals oblanceolate, 5.5 × 2 mm; lip fleshy, trilobed, lateral lobes upright, midlobe triangular, porrect; spur laterally compressed. **Distr:** Indo-China to W and C Malesia; widely distributed. **Ecol:** Usually close to the ground in lower montane forests; 0–1,800 m. **Zones:** 9–12.

Uncifera

Species 6. **Herbs**, epiphytic, monopodial; roots not piercing leaf sheaths; stems usually pendulous, elongate, slender, leafy. **Leaves** overlapping, distichous; lamina flat, oblong to lanceolate, base amplexicaul, sheathing, jointed, apex unequally bilobed or praemorse, slightly fleshy. **Inflorescence** axillary, racemose, densely few- to many-flowered. **Flowers** thickly textured; sepals and petals similar, concave; lateral sepals slightly oblique; petals slightly smaller than sepals; lip funnel-shaped, spurred, trilobed, lateral lobes suberect, midlobe very small, fleshy, adaxially ± concave; spur long, hooked or curved, inside plain; column terete, bent over spur entrance, foot absent; anther cap conic; pollinia 2, waxy. **Distr:** Himalayas to SC China and Indo-China. **Cult:** Propagate from seeds. Best grown mounted on slabs. Requires humid conditions in bright light.

U. thailandica Seidenfaden and Smitinand, *Orchids Thailand.* 4(2): 828, 1965. **Herbs**, epiphytic; stems 5 cm long, enclosed by leaf sheath. **Leaves** 4–5, lamina ligulate, 7–12 × ~2–3.5 cm, unequally obtusely bilobed, fleshy. **Inflorescence** 10–15-flowered, gently curving to pendent, penetrating through leaf sheath; peduncle 4–6 cm long; rachis purple, 7–10 cm long; uneven bilobed; sterile bracts pale yellow; fertile bracts lanceolate, 5–6 cm long; pedicel and ovary ~1 cm long. **Flowers** with sepals light purple white-edged, petals light greenish with a purple central spot, lip white with a purple spot near apex of lamina; dorsal sepal concave, ~7 × 4.5 mm; lateral sepals ~7 × 5 mm; petals ~6.3 × 5 mm; lip trilobed, lateral lobes triangular with broad base, ~1 mm long, obtuse, midlobe small, with 2 small lobes turned backward; spur white, sigmoid, 7–8 mm long, apex turned backward toward ovary; column very short. **Distr:** China (Yunnan), Thailand. **Ecol:** On trunks or branches, in montane evergreen forests; ~1,400 m. **Zones:** 8–12.

Vanda

Species ~50. **Herbs**, epiphytic; stems fairly long, usually stout, internodes short. **Leaves** distichous; lamina flat, linear to lorate, becoming conduplicate toward base, apex often bilobed or variously toothed. **Inflorescences** racemes, few- to many-flowered, axillary, usually erect or ascending, simple, flowers well-spaced. **Flowers** large, fleshy, highly colored, long-lasting, often fragrant; sepals and petals subequal, widely spreading, base narrow, margins often twisted or wavy; lip attached to base of column, shortly spurred, trilobed, lateral lobes small, erect, midlobe large, usually with longitudinal keels, spur often laterally flattened, without internal appendages or callosities; column short, without a distinct foot; anther terminal, operculate, 2-celled; pollinia 2, waxy, porate or strongly cleft, attached to short broad stipe with large oval viscidium; stigma entire; rostellum short, broad, bifid. **Distr:** Tropical and subtropical Asia to the Pacific Islands and Australia. **Cult:** Propagate from cuttings with at least 3 roots. Best grown in pots with a coarse medium, or against tree trunks. Requires humid, intermediate to warm conditions, in bright light.

V. coerulea Griffith ex Lindley, *Edwards's Bot. Reg.* 33: sub t. 30, 1847. [Blue Vanda]. **Herbs**, epiphytic; stems 5–23 cm long. **Leaves** (7–)14–18 × 1.3–2(–3) cm, apex unequally bilobed, thickly leathery. **Inflorescences** 1–3, 20–42 cm long, sparsely (4–)6–14-flowered; peduncle 16–22 cm long; rachis 9–28 cm long; floral bracts broadly ovate, 0.8–1.1 × 0.5–0.9 cm, obtuse-acute; pedicel and ovary white, 4.5–6 cm long, blue-tinged. **Flowers** thinly textured, widely opening, 6–9 cm Ø, sepals and petals sky-blue, tessellated; sepals similar, broadly obovate, 3.5–5 × 1.7–3.5 cm, base clawed, apex rounded; petals elliptic-obovate, 3–4.5 × 1.8–2.5 cm, base clawed, apex rounded; lip sky-blue, shorter than sepals, fleshy, spurred; lateral lobes erect, narrowly falcate, adaxially white, yellow-spotted, 3.5–5 × 1–2 mm, subacuminate, midlobe deep blue, ligulate, 1.7–2.5 × 0.5–0.8 cm, base with a pair of calli, apex subtruncate and emarginate; disk with 3 longitudinal ridges; spur tubular, slightly conic, 5–7 × 2–3 mm, obtuse; column 4–6 mm long; operculum white. **Distr:** China (S Yunnan), NE India, Myanmar, Thailand. **Ecol:** In open forests, along rivers; 1,000–1,600 m. Fl. Oct–Nov. **Zones:** 8–12.

V. concolor Blume, *Rumphia* 4: 49, 1849. (Syn: *V. cruenta*, etc.). **Herbs**, epiphytic; stems 4–13 cm long. **Leaves** 15–30 × 1–3 cm, apex unequally bilobed, leathery. **Inflorescences** 1 or 2, 11–17 cm long, sparsely 4–8-flowered; peduncle 3–6 cm long; rachis 5–8 cm long; pedicel and ovary white, 2–4.5 cm long. **Flowers** thickly textured, widely opening, 3–4 cm Ø, fragrant, sepals and petals abaxially white, adaxially yellowish-brown, yellow-striped, tessellated; sepals subsimilar, lateral sepals slightly wider than dorsal sepal, oblong-obovate, 1.6–2 × 0.6–1.2 cm, base shortly clawed, obtuse; petals subspathulate, 1.5–1.9 × 0.8–1 cm, base clawed, apex rounded; lip fleshy, spurred, lateral lobes erect, inner surface white, purple-spotted, subfalcate, triangular or lanceolate, 4–5 × 2–3.5 mm, obtuse, midlobe yellowish-brown apically, yellow basal half, pandurate, 1.2–1.6 × 0.5–0.7 cm, apex bilobed; disk with 5 or 6 yellow papillate ridges, base often shortly hairy; spur white, cylindric, 7–8 mm long, subacute; column white, 6–7 mm long. **Distr:** China. **Ecol:** In forest margins; 700–1,600 m. Fl. Apr–May. **Zones:** 8–10.

V. lamellata Lindl., *Bot. Reg. Misc.* 66, pl. 24, 1838. (Syn: *V. amiensis*). **Herbs**; stems robust, to 10 cm tall. **Leaves** closely distichous; lamina linear, 15–20 × 2 cm, apex acute, bilobed, rigid, coriaceous, recurved. **Inflorescences** racemes, 5–15-flowered, erect; peduncles slender, 10–15 cm long. **Flowers** waxy, fragrant, color variable, mainly yellowish or greenish, with ± brown spots and longitudinal irregular stripes, ~3 cm Ø; dorsal sepal obovate to oblanceolate, 2 × 1 cm, base clawed, apex obtuse or retuse; lateral sepals falcate-obovate, 2 × 1.3 cm, base clawed, strongly oblique, apex rounded, with brownish-red patches or stripes; petals spathulate, 2 × 0.8 cm, base clawed, apex obtuse; lip yellowish-white, trilobed, base spurred, lateral lobes erect, 4 mm long, rounded, midlobe much larger, rectangular-ovate, 8 × 6 mm base truncate or slightly auricled, apex obtuse or rounded; disk with few stripes and 2 large erect keels, spur conic, ~4 mm long. **Fruits** ellipsoid, 3–4 cm long. **Distr:** Ryukyus, Philippines, Taiwan, Lanyu Island. **Ecol:** Usually on exposed rocks and tree trunks. **Zones:** 8–12.

V. roeblingiana Rolfe, *Kew Bulletin* 365, 1894. **Herbs**, epiphytic, upright, monopodial, to 40 cm tall; stem elongated, leafy. **Leaves** several, alternate; lamina linear-oblong, to 20 × 2 cm, recurved, apex oblique, truncate, leathery. **Inflorescences** horizontal, ascending or upright, 8–15-flowered, axillary, 17–30 cm long. **Flowers** 4–5 cm across, fragrant, long-lasting; dorsal sepal oblanceolate, to 2 × 0.8 cm; lateral sepals obovate, to 2 × 1 cm; lip relatively large, trilobed, lateral lobes small and upright, midlobe split, fan-shaped, margins fringed. **Distr:** Endemic to Philippines (Luzon). **Ecol:** Epiphytic in heavily shaded areas of montane regions; 1,500–1,600 m. **Zones:** 11–12.

V. tessellata (Roxb.) Lodd. ex G. Don, *Hort. Brit.* 372, 1832. (Syn: *Aerides tessellatum*, etc.). [Tessellated Vanda]. **Herbs** epiphytic; stems 30–60 cm long. **Leaves** distichous, equitant; lamina strap-shaped, 15–20 × 1.7–2.4 cm, recurved, thickly coriaceous, obtusely keeled, praemorse. **Inflorescences** racemes, 4- or 10-flowered, 15–20 cm long; peduncle stout, green; sterile bracts 2; pedicel and ovary 3.7 mm long. **Flowers** large, 5 cm across, gray, grayish-blue, buff, red or yellow, segments wavy and many-veined; dorsal sepal 2.4 × 1.4 cm, obovate, faintly trifid, margin; lateral sepals obovate, 2.5 × 1.8 cm, obtuse; petals obovate, 2.4 × 1.5 cm, obtuse or subacute; lip funnel-shaped, trilobed, lateral lobes erect, obliquely oblong, 1 × 0.45 cm, acuminate, acute, midlobe panduriform, 1.7 × 1.25 cm, constricted below the bifid apex; disk swollen with fleshy curved ridges; spur conical, 6–7 mm long, blunt, laterally compressed; column cylindrical, 9 × 5 mm, fleshy. **Fruits** clavate, ribbed, 7.5–9 cm long. **Distr:** Indian Subcontinent to Indo-China. **Ecol:** In tropical dry mixed evergreen forests. Fl. Jan, Mar–Aug, Dec. **Zones:** 9–12.

V. testacea (Lindl.) Rchb. f., *Gard. Chron.*, n.s., 8: 166, 1877. (Syn: *V. parviflora*, etc.). **Herbs**, epiphytic; stem 10–15 cm long. **Leaves** distichous, equitant, recurved; lamina linear-lorate, 10–26 × 1.3–1.9 cm, apex unequally lobed or toothed, coriaceous,

Below: *Vanda coerulea, Vanda concolor, Vanda lamellata.*

MG

MS

Above: *Vanda roeblingiana, Vanda tessellata, Vanda testacea, Vanda tricolor.*

keeled. **Inflorescences** racemes, many-flowered, erect, 6–15 cm long; peduncle stout, 5–25 cm long, red-tinged; pedicel and ovary 2.5 cm long. **Flowers** 1.8 cm across, cream-yellow, lip white, purple or reddish-pink; dorsal sepal obovate-spathulate, 0.6–1.1 × 0.27–0.52 cm, obtusely rounded, incurved; lateral sepals obovate-spathulate, 5.6–9.2 × 2.8–5.4 mm, obtusely rounded, incurved, twisted upward laterally; petals spathulate, concave, 0.5–1 × 0.2–0.5 cm, base narrow, twisted downward; lip with spur 9.5 mm long, sessile, trilobed, lateral lobes oblong, obtuse, erect, midlobe subquadrate-oblong, 6.4 × 3.5 mm, recurved, dilated, emarginate, crenulate, disk with 2 fleshy ridges; spur narrow, funnel-shaped, 3.1 mm long; column 2.8 mm high. **Fruits** clavate, 2.5-3 cm long. **Distr:** Indian Subcontinent to SC China. **Ecol:** In tropical dry mixed evergreen forests. Fl. Feb–Mar, Sep. **Zones:** 8–12.

V. tricolor Lindl., *Edwards's Bot. Reg.* 33: t. 59, 1847. (Syn: *V. suaveolens*, etc.). **Herbs**, epiphytic, lithophytic or terrestrial, erect, large; stems 0.5–1 m long, 1.4 cm thick. **Leaves** imbricate; lamina curved, ligulate, 37–45 × 3–4 cm, apex unequally bilobed. **Inflorescences** 6–10(or more)-flowered, ascending or spreading, shorter than leaves. **Flowers** large, showy, fragrant, 5–7.5 cm across, color variable, sepals and petals commonly light yellow, densely spotted with bright red-brown, white outside, lip whitish below, streaked with red-brown, bright magenta-purple above, spur white; sepals and petals subsimilar, obovate-oblong to orbicular-obovate, 2–3 × 2.2 cm, base clawed, margins undulate, fleshy; petals twisted at base; lip trilobed, to 2.5 × 1.8 cm, lateral lobes subquadrate, incurved, midlobe subpanduriform, deeply emarginate, convex and 3-ridged above; spur short, compressed, 9 mm long; column short, much swollen on each side at base. **Distr:** Laos, Java to Bali. **Ecol:** Epiphytic on branches in open forests, often on sugar palm trunks; 700–1,600 m. Fl. throughout most of the year. **Zones:** 9–12.

Vandopsis

Syn: *Fieldia.*

Species 4. **Herbs**, epiphytic, lithophytic, or terrestrial, monopodial; stems decumbent or pendulous, elongate, stout, rigid, sometimes branched. **Leaves** many, distichous; lamina flat, base jointed, leathery. **Inflorescences** racemose, many-flowered, long or short, axillary, suberect or pendulous, rarely shortly branched. **Flowers** resupinate, flat, medium to large, fleshy; sepals and petals similar, spreading; lip smaller than petals, firmly adnate to column, geniculately bent, ± channeled or gibbous at base, trilobed, lateral lobes often small, midlobe laterally flattened, rather long, fleshy, adaxially usually longitudinally ridged or keeled; column short, stout, foot absent; rostellum inconspicuous; pollinia waxy, 2, in unequal pairs, deeply grooved, sessile; stipe broadly oblong; viscidium transverse, prominent. **Distr:** S China to tropical Asia. **Cult:** Propagate from cuttings with at least 3 roots. Best grown in pots with a coarse medium, and allowed to climb a moss pole or tree trunk. Requires humid, intermediate to warm conditions, in bright light.

V. undulata (Lindley) J.J. Smith, *Natuurk. Tijdschr. Ned.-Indië* 72: 77, 1912. (Syn: *Stauropsis undulatus*, etc.). **Herbs**, epiphytic or terrestrial; stems ascending or pendulous, cylindric, to 1 m long. **Leaves** oblong, 9–12 × 1.5–2.5 cm, apex obtuse, unequally bilobed, leathery; basal sheaths persistent, amplexicaul, verrucose. **Inflorescence** laxly few-flowered, to 50 cm long, with a few branches; pedicel and ovary ~2.7 cm long, white. **Flowers** fragrant, 3–3.6 cm Ø, sepals and petals white, lip yellow to white at apex, lemon-yellow with purple-red basal striations, anther cap yellow; dorsal sepal subobovate, 2.5–4 × 0.8–1.2 cm, base clawed, margin crisped, apex obtuse; lateral sepals slightly recurved, ovate-lanceolate, 2.4–4 × 1.2–1.4 cm, margin crisped, apex subacuminate; petals similar to sepals but smaller, margin undulate, apex obtuse; lip shorter than petals, trilobed, lateral lobes erect, ± embracing column, subovate-triangular, ~5 mm, acute, midlobe laterally compressed, spathulate, ~6 × 5 mm, concave, truncate; disk 2-ridged; column ~4 mm long. **Distr:** Nepal to China (Yunnan). **Ecol:** In forests or on shrubby slopes; 1,500–2,300 m. Fl. May–Jun. **Zones:** 7–10.

Subtribe Aerangidinae

Aerangis

Species 51. **Herbs**, epiphytic, rarely lithophytic; stems erect, very short or rarely elongate, sometimes stemless. **Leaves** several, arranged distichously or fanlike, articulate with the leaf sheaths; lamina mostly obovate, obcuneate or oblanceolate, often twisted above sheath with leaves in 1 plane, apex unequally bilobed, coriaceous or rarely fleshy. **Inflorescences** racemose, rarely 1-flowered, axillary; rachis often zigzag; bracts small to large. **Flowers** with sepals and petals free, spreading to reflexed, white, ± flushed pink or green especially on the spur, or yellow, alternate, column white or rarely red; sepals and petals subsimilar, often lanceolate, less commonly elliptic, ± acuminate; lip ± similar to petals, reflexed with the lamina much shorter than the slender spur; column short, somewhat dilated toward base; rostellum simple, elongate; pollinia 2, ovoid, on a single elongate stipe; viscidium 1, circular to elliptic. **Distr:** Tropical and S Africa to Sri Lanka. **Cult:** Propagate from cuttings. Best grown mounted. Requires warm, humid and shady conditions.

A. hologlottis (Schltr.) Schltr., *Beih. Bot. Centralbl.* 36(2): 117, 1911. (Syn: *Angraecum hologlottis*). **Herbs**, epiphytic, stemless. **Leaves** oblong-lanceolate, falcate, 4.5–9 × 1.1–1.7 cm, apex unequally bilobed, coriaceous, adaxially dark green, shiny, abaxially paler, midrib prominent below and lateral veins reticulated. **Inflorescences** racemes, many-flowered, 13.5–16.5 cm long, bracteate, semierect, opening irregularly, first flowers to open at base or

Below: *Vandopsis undulata, Aerangis hologlottis, Sphyrarhynchus schliebenii.*

toward the middle of the raceme; peduncle 4.5–6.8 cm long, dark green; pedicel and ovary 7 mm long. **Flowers** white, 1.1–1.4 cm across; dorsal sepal lanceolate, 7.1 × 2.6 mm, acute, 3-veined; lateral sepals oblanceolate, 8 × 2 mm, acute, 3-veined; petals oblong-lanceolate, 7.2 × 2.6 mm, acute, 3-veined; lip lanceolate-spathulate, 8.4 × 3.4 mm, apiculate, 5-veined, with a cylindrical spur 5.4 mm long at base parallel to ovary; column 1.2 mm high, foot absent. **Fruits** fusiform, 1.5–1.8 cm long. **Distr:** SE Kenya to NE Tanzania (including Zanzibar), Mozambique, Sri Lanka. **Ecol:** In submontane or tropical wet evergreen forests; ~500 m. Fl. Jan–Apr. **Zones:** 10–12.

Sphyrarhynchus

Species 1. *Generic description same as specific.* **Cult:** Propagate from seeds. Best grown mounted. Requires intermediate temperatures, high humidity and moderate shade. Plants should not be allowed to dry out for long.

S. schliebenii Mansf., *Notizbl. Bot. Gart. Berlin-Dahlem* 12: 706, 1935. (Syn: *Angraecopsis schliebenii*). **Herbs**, epiphytic, dwarf; stem very short; roots flattened onto substrate, greenish-gray. **Leaves** elliptic or oblong-elliptic, up to 3.5 × 0.5 cm, apex unequally bilobed. **Inflorescence** spreading or pendulous, 2–7(or more)-flowered, to 3 cm long; peduncle slender; bracts several, short, black, sheathing. **Flowers** relatively large, white with a bright green central mark on the lip; sepals narrowly oblong-elliptic, somewhat reflexed, 1–1.4 × 0.3–0.4 cm, acute; petals narrowly oblong, 0.9–1.4 × 0.3–0.4 cm, acute; lip reflexed, ligulate, 4–6 × 2–3.5 mm, margins entire, apex acute; spur clavate, 6 mm long; column with a hammerlike rostellum, 1.5 mm long. **Distr:** Central Tanzania. **Ecol:** In montane forests; 900–1,600 m. **Zones:** 10–12.

Subtribe Angraecinae

Angraecum

Species ~200. **Herbs**, large or small, monopodial, epiphytic, lithophytic, rarely terrestrial; stems long or short, covered with old leaf bases. **Leaves** variable in shape, usually distichous; lamina with base articulated to a persistent sheath, coriaceous or fleshy. **Inflorescences** axillary, 1- to many-flowered, rarely branched; bracts usually small. **Flowers** from very small to very large, usually white, green or yellowish; sepals and petals subsimilar or subequal, free, usually spreading; lip entire or obscurely trilobed, concave, base ± with a long or short spur; column short, cleft in front; rostellum bifid; anther terminal, operculate, incumbent, convex, often produced in front, scarcely 2-celled; pollinia 2, globose, sulcate, stipitate, viscidia 2. **Fruits** oblong or fusiform. **Distr:** Africa, Comoro Islands, Madagascar (with >130 species), Mascarene Islands, Seychelles, Sri Lanka. **Info:** White-flowered, long-spurred species are moth-pollinated and almost all are strongly scented after dark. **Cult:** Propagate from seeds. Requires a shady situation and intermediate to warm conditions. Small and trailing species can be mounted, while the majority of the species can be grown in pots with standard epiphyte mix. Larger species may need support.

A. compactum Schlechter, *Beih. Bot. Centralbl.* 34. II. 339, 1916. **Herbs**, epiphytic; stems short, thick, sometimes elongated, to 30 cm long or more. **Leaves** ligulate, 6–14, 5–10 × 1.8–3.8 cm, apex unequally bilobed, thick, somewhat succulent, dark green, surface rough or shriveled, midrib prominent. **Inflorescences** 1–3-flowered, piercing leaf sheath on lower part of stem; peduncle 0.8–1 cm long, covered with 3 wide sheaths; bracts wider than long, keeled along the dorsal vein; pedicels flattened. **Flowers** sweetly scented (in the evening), white, spreading, flat or with reflexed petals; sepals oblong, 2–3 × 1.2 cm, obtuse; lateral sepals slightly longer and narrower; petals similar to sepals but shorter, sometimes partly reflexed; lip boat-shaped, very concave with a wide, funnel-shaped entrance to the spur, oblong, rounded at apex with a short apiculus, 3 × 2.6 cm; spur to 1 cm wide at the entrance, gradually narrowing then bent forward below the lip finally becoming straight, 11–13 cm long; ovary 6 cm long, triquetrous. **Distr:** Madagascar. **Ecol:** In humid evergreen forests; 700–2,000 m. Fl. Aug–Dec. **Zones:** 10–12.

A. sesquipedale Thouars, *Hist. Orch. Afr.* pl. 66 and 67, 1822. (Syn: *Aeranthes sesquipedalis*, etc.). [Christmas Star Orchid, Comet Orchid, Rocket Orchid, Star of Bethlehem Orchid]. **Herbs**, large, robust, epiphytic; stem usually single, upright or curving, to 1 m tall, sometimes branched. **Leaves** in 2 rows; lamina ligulate, 20–40 × 6–7 cm, apex unequally bilobed, lobes rounded, leathery, bluish or grayish-green, dull. **Inflorescences** 1–3(6)-flowered, axillary, erect or horizontal; peduncle 10–12 cm long. **Flowers** large, opening green and turning creamy-white within a few days; sepals lanceolate-acuminate, 7–11 × 2–3 cm, fleshy; lateral sepals keeled on the back; petals similar but slightly falcate, 7–8 × 2.5–2.8 cm; lip concave, pandurate, acuminate, with a ridge along the surface in front of the spur opening, 6.5–9 × 3.5–4 cm; spur thick but gradually attenuate toward the tip, 30–35 cm long (with only the distal end filled with nectar); column large, fleshy, with conspicuous rostellum lobes below the anther cap. **Distr:** Endemic to Madagascar. **Ecol:** In woodlands and forests, anthropic; 0–100 m. Fl. May–Nov. **Zones:** 10–12. **Info:** Charles Darwin hypothesized that there must be a moth with a proboscis long enough to reach the nectar in this flower. The moth (*Xanthopan morgani praedicta*) was 'discovered' over 40 years later in 1903. *A. sesquipedale* var. *angustifolium* is another variety of this species.

Jumellea

Species 58. **Herbs**, epiphytic or lithophytic; shorts very short to elongated, rarely branched. **Leaves** often distichous; lamina loriform, apex ± bilobed. **Inflorescences** always 1-flowered; peduncle mostly shorter than ovary and pedicel. **Flowers** white; sepals and petals subsimilar; lateral sepals ± united at base around base of spur; lip entire, contracted at base with 1 or 2 dilations above, with a central linear or tapering callus; spur short to very long, filamentous or slender; column short, fleshy; rostellum bifid; pollinia 2, each on a separate stipe or directly attached to a viscidium. **Distr:** Kenya to S Africa, W Indian Ocean. **Cult:** Propagate by division of mature plants. Miniature and pendent species are best grown mounted on tree-fern or cork slabs. Larger species can be grown in pots with a bark-based medium. Requires intermediate to warm conditions and high humidity.

J. comorensis (Rchb. f.) Schltr., *Beih. Bot. Centralbl.* 33(2): 428, 1915. (Syn: *Aeranthes comorensis*, *Mystacidium comorense*). **Herbs**, epiphytic, upright or pendent, monopodial; stems becoming elongated and branched with age, slightly flattened, 3–4 mm wide. **Leaves** 2-ranked; lamina ligulate-linear, 5–9.5 × 0.6–1.2 cm, base slightly contracted, apex shortly bilobed, glossy. **Inflorescences** piercing the sheath, axillary, single-flowered; peduncle short, covered with 3 flattened sheaths, upper sheath 1 cm long and overlapping the bract, 1.6–1.7 cm long; pedicel with ovary 7.5 cm long. **Flowers** fragrant, white, opaque; sepals similar to each other, linear, 2–3 × 0.4 cm; petals similar to sepals but narrowed below; lip contracted below middle into a narrow claw with a central keel, widened in the upper half, overall 3 × 1 cm; spur filiform, 9–15 cm long. **Distr:** Comoro Islands (Grande Comore). **Ecol:** In humid forests, on rough-barked trees; 0–600 m. Fl. Jan, Jul–Aug, Nov. **Zones:** 10–12.

TRIBE CYMBIDIEAE

Subtribe Cymbidiinae

Ansellia

Species 1. *Generic description same as specific.* **Cult:** Propagate by division. Best grown in a shallow pot with a coarse medium, or mounted on a slab. Repot only ever few years. Requires intermediate to warm conditions with plenty of water and bright light. **Info:** 2 subspecies described.

A. africana Lindl., *Edwards's Bot. Reg.* 30: t. 12, 1844. [Leopard Orchid]. **Herbs**, epiphytic, robust, forming large clumps; pseudobulbs canelike, fusiform-cylindrical, many-noded, yellow, 25–50+ long, 2–3 cm wide, 6–7-leaved. **Leaves** narrowly ligulate-lanceolate, 20–30 × 1.5–4 cm, plicate, apex acute. **Inflorescence** paniculate, many-flowered, to 50 cm long, borne near apex of pseudobulbs, branches 5–6, spreading. **Flowers** variable in shape, 3–5 cm across, pure yellow, or yellow spotted with maroon or

Below: *Angraecum compactum, Angraecum sesquipedale, Jumellea comorensis, Ansellia africana.*

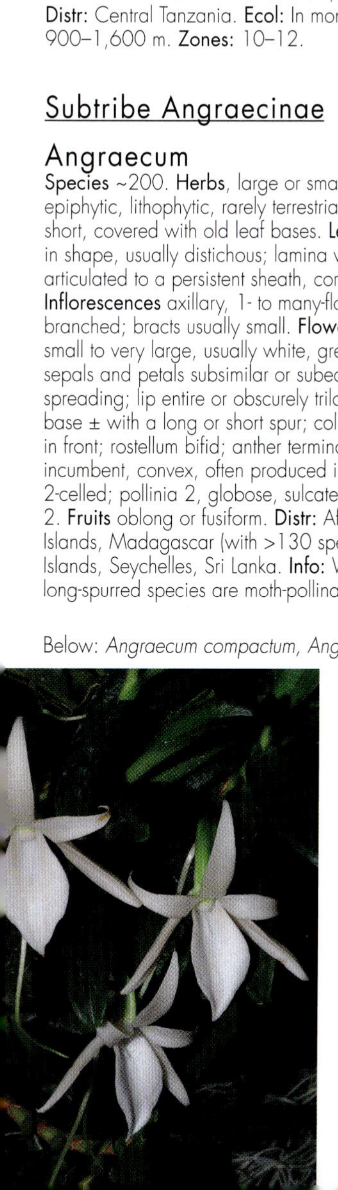

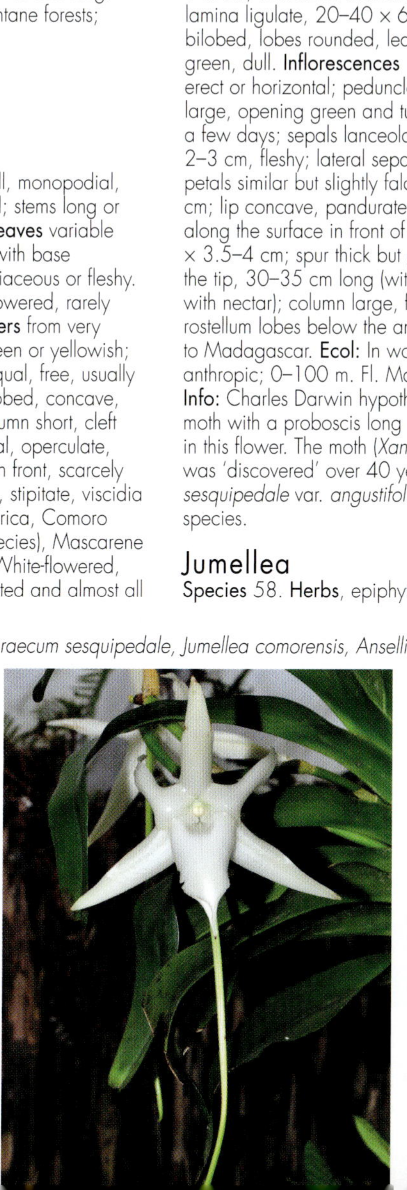

MG

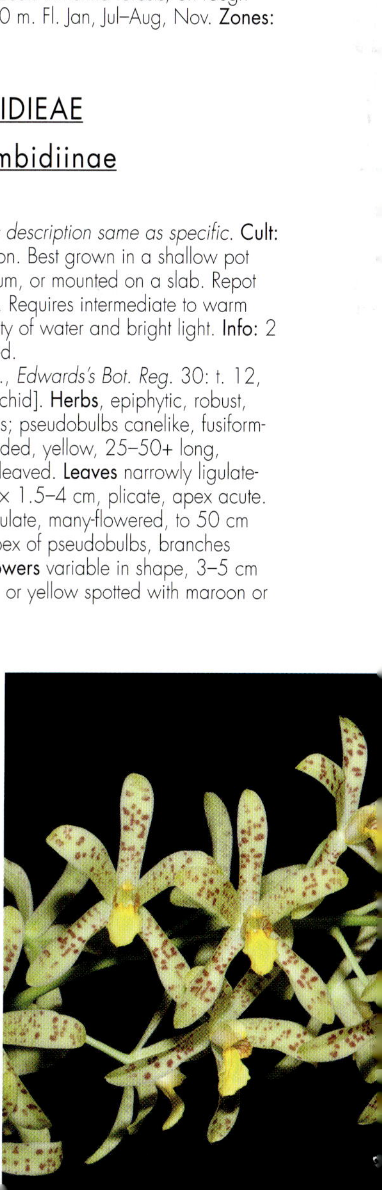

Above: *Grammatophyllum martae, Grammatophyllum measuresianum*. [Both MS.]

brown on sepals and petals, lip orange-yellow, lined with red-brown on lateral lobes, callus orange-yellow or yellow; sepals oblong-ligulate to narrowly elliptic, 1.5–3.5 × 0.5–1 cm, rounded or acute; petals similar but smaller; lip 2–2.5 cm long, trilobed, lateral lobes erect, narrowly oblong, midlobe round or oblong-oblanceolate, subacute, apiculate, disk with callus of 2 or 3 central keels; column arcuate, slender, 1–1.2 cm long. **Distr:** Tropical and S Africa. **Ecol:** In dry, open woodlands and forests, often in forks of trees; 0–700(–2,200) m. **Zones:** 10–12.

Grammatophyllum

Syn: *Gabertia, Pattonia.*

Species 11. **Herbs**, epiphytes, usually very large, sympodial, clump-forming; rhizomes short; stems stout; pseudobulbs canelike, with more than 1 node, few- to many-leaved. **Leaves** distichous, jointed on their sheaths; lamina linear or strap-shaped. **Inflorescences** racemes, basal, erect or pendent, many-flowered; scapes lateral, long, stout. **Flowers** usually large, showy; sepals and petals spreading, free, subsimilar or subequal; lip smaller, trilobed, concave, erect, adnate to base of column, side lobes embracing column, midlobe short recurved; column semiterete, foot absent, top incurved; anther broad, 2-celled; pollinia 2, subglobose, 2-cleft, attached one to each horn of a curved strap or gland. **Distr:** SE Asia, Malaysia, Indonesia, Philippines, Papua New Guinea, east to Fiji. **Cult:** Propagate by division or from seeds. Best planted in a pot or basket with excellent drainage using charcoal. Larger species can be grown on trees as well as terrestrially with good drainage. Requires ample, high humidity and regular watering. Low tolerance to disturbances.

G. martae Quisumb. ex Valmayor and D.Tiu, *Philipp. Orchid Rev.* A-V(3): 18, 1983. **Herbs**, erect, sympodial; pseudobulbs large, to 22 cm tall, 10 cm Ø. **Leaves** to 4; lamina oblanceolate, 60 × 10 cm, leathery. **Inflorescences** arching, to 1.5(–2) m long, with up to 80 flowers. **Flowers** 8.5 cm across, blotched with dark brown, sepal and petal margins green; dorsal sepal oblanceolate, to 4 × 2.5 cm; petals spathulate, to 4 × 2 cm; lateral sepals obovate, to 4 × 2 cm; lip trilobed, small. **Distr:** Endemic to Philippines (Negros Occidental on Negros Island in Visayas). **Ecol:** Grows epiphytic on trees; to 300 m. **Zones:** 11–12. **Info:** *G. martae* could be a tetraploid form of *G. elegans* (Cootes, Banks and Titmuss 2001).

G. measuresianum Sander, *Gard. and Forest* 2: 524, 1889. **Herbs**, erect, sympodial; pseudobulbs ovoid, to 25 cm tall, 6 cm Ø. **Leaves** 4–6 at apex of pseudobulbs; lamina broadly lanceolate, to 60 × 10 cm. **Inflorescences** upright to semiarching, to 1.5 m long, to 80-flowered. **Flowers** 5 cm across, base whitish to pale yellow, spotted varying degrees with dark chocolate-brown color, all floral segments reflexed in the apical third; dorsal sepal broadly lanceolate, to 3.8 × 1.8 cm; petals oblanceolate, to 3.5 × 1.4 cm; lateral sepals broadly lanceolate, to 3.8 × 1.2 cm; lip trilobed, 2 × 2 cm, lateral lobes flattened, encircling column, square at front, midlobe rectangular, 6 × 5 mm, apex rounded, rear of midlobe hairy. **Distr:** Philippines (Luzon, Mindoro, Palawan). **Ecol:** to 500 m. **Zones:** 11–12.

G. scriptum (L.) Blume, *Rumphia* 4: 48, 1849. (Syn: *Angraecum scriptum*, etc.). **Herbs**, epiphytic, large, 60–70 cm tall, clump-forming; pseudobulbs erect, ovoid, 8–25 × 4–8 cm, laterally compressed, slightly wrinkled, glossy green to pale green or yellow-green. **Leaves** 5–8, in 2 opposite ranks around pseudobulb, erect to suberect with largest leaves below apex, base V-shaped, <5 cm long; lamina elliptic to obovate, to 60 × 10 cm, midrib prominent, apex acute, rather thin, leathery. **Inflorescence** a raceme, 20–90-flowered, basal, erect or arching, 1–2 m long, flowers well-spaced, lowest ones deformed and often infertile, sometimes failing to develop; peduncle green, smooth, 0.6–1.5 cm Ø with 5 large green tightly sheathing bracts; floral bract triangular to lanceolate, ~1 cm long; pedicels to 4 cm long. **Flowers** *sterile* 1–6, often cruciform and lacking lip, with column tip eroded away, or else similar to normal flowers; *fertile* flowers not completely opening, 4–6 cm across, usually green to yellow with brown patches, markings and interior veins; sepals have external nectaries at their bases, not fully spreading, lanceolate, ~3.5 cm long, concave, acute; petals obovate-spathulate, ~3.4 cm long, obtuse; lip ~2.4 × 2.4 cm, trilobed, lateral lobes erect, rhombic, rounded, inner surface hairy at base, midlobe obovate, margins erose, apex truncate, small white downy-hairy on both surfaces; disk hairy, with 3 parallel pale yellow hairy keels; column ~2 cm long, white, decurved to sinuous, with small membranous wings around stigma; ovary to 1 cm long, twisted. **Distr:** Fiji, Solomon Islands through Papua New Guinea and Indonesia to the Philippines. **Ecol:** In coastal forests on exposed or well-lit tree trunks and branches. **Zones:** 11–12. **Info:** *G. scriptum* var. *scriptum* is a pure yellow form from Bougainville Island. *G. multiflorum* var. *tigrinum* is another common form with different flower segment colorations.

G. stapeliiflorum (Teijsm. and Binn.) J.J. Sm., *Orch. Java.* 487, 1905. (Syn: *Cymbidium huttoni*, etc.). **Herbs**, erect, sympodial, epiphytic; pseudobulbs to 12 × 5 cm. **Leaves** 2–3, ~30 × 5 cm. **Inflorescences** pendulous, ~10–flowered. **Flowers** 6 cm across; dorsal and lateral segment bases pale brown, heavily spotted with dark purple, petals almost solid dark purple, lip similarly marked to sepals; dorsal sepal ovate, to 3 × 2 cm; lateral sepals oblanceolate to obovate, to 3 × 1.5 cm, petals lanceolate, to 3 × 1.2 cm; lip trilobed, lateral lobes upright, midlobe small, almost round. **Distr:** Java, Malaysia, Papua New Guinea, Philippines, Sulawesi, Sumatra. **Ecol:** Epiphytic at low elevations; to 500 m. **Zones:** 11–12.

Galeandra

Species 38. **Herbs**, terrestrial (epiphytic), cespitose; stems reedlike or more often pseudobulbous. **Leaves**

Below: *Grammatophyllum scriptum, Grammatophyllum scriptum* var. *scriptum, Grammatophyllum stapeliiflorum*. [All GS.]

Above: *Galeandra baueri, Cymbidiella falcigera, Cymbidium atropurpureum.* [All MG.]

1–2 (several), rarely articulate, plicate, absent (present) at flowering. **Inflorescences** racemes, terminal, erect slightly arching to pendent; floral bracts lanceolate, scarious. **Flowers** resupinate, showy; sepals and petals membranous; lip sessile, simple (obscurely 2–3-lobate), base with conspicuous spur adnate to column foot, margins usually tomentose; spur shortly conic; disk basally 2(–4)-keeled; column slightly arcuate, shallowly winged, semiterete, produced into a conspicuous foot; anther terminal, incumbent, operculate, imperfectly biloculed; pollinia 2, yellow, spherical, cartilaginous. **Fruits** capsules. **Distr:** Indo-China, tropical and subtropical America. **Cult:** Propagate by division. Grow in pots with a free-draining bark mix. Requires intermediate conditions with moderate shade and plenty of water during the growth period.

G. baueri Lindl., *Ill. Orch. Pl.*: t. 8, 1830. (Syn: *G. cristata*, etc.). **Herbs**, epiphytic, often clump-forming, to 45 cm tall; stems ancipitous, fusiform, slightly zigzag, to 27 cm long, concealed by red-spotted, scarious leaf-sheaths. **Leaves** erect-spreading; lamina linear to linear-lanceolate, to 23 × 2 cm, apex acute or acuminate. **Inflorescence** racemose or branched, few-flowered; peduncle erect, bearing slender, spotted sheaths; bracts lanceolate, ~5 mm long, acuminate. **Flowers** showy, sepals and petals yellowish-brown, lip purple, base white or brownish; sepals narrowly oblanceolate, 2 × 0.5 cm, acute; petals similar to sepals, 1.8 × 0.5 cm; lip involute below, spreading above, margins crenulate, 5 × 3.5 cm, calls of 2 parallel ridges near base; spur tapering, curved, ~2 cm long; column arcuate, with a pair of ciliate side lobes at apex, rostellum dorsal, recurved, 1 cm long. **Distr:** French Guyana to N Brazil. **Ecol:** <800 m. **Zones:** 9–12.

Cymbidiella

Syn: *Caloglossum.*

Species 3. **Herbs**, epiphytic or terrestrial, large; rhizome short to elongate; pseudobulbs elongate. **Leaves** 8–40, distichous, arranged in a fan shape, persistent, articulated at apex of sheath; lamina plicate-venose, membranous. **Inflorescence** racemose or paniculate, branched or unbranched, many-flowered, borne from axils of lower leaves near base of pseudobulbs; bracts narrow. **Flowers** large, showy; sepals and petals free or joined to column foot; lip 3- or 4-lobed, lacking a spur, with a basal callus or lamellae; column with a short foot; pollinia 2, attached at base to a retractile stipe; viscidium rectangular or trapezoidal. **Distr:** Madagascar. **Cult:** Propagate by division. Best grown in baskets. Requires very warm conditions, with copious watering and good drainage. Avoid direct sunlight as it damages the foliage.

C. falcigera (Rchb. f.) Garay, *Orchid Digest* 40: 192, 1976. (Syn: *C. humblotii, Grammangis falcigera*, etc.). **Herbs**, epiphytic, robust, large; rhizomes creeping, stout; pseudobulbs 3–4, cylindric, to 30 cm long, covered by 7–40 leaves. **Leaves** distichous, arranged in a fan; lamina loriform, lanceolate, 25–60 × 1–3 cm, acuminate. **Inflorescence** paniculate, 0.8–1.5 cm long, many-flowered; bracts lanceolate, 2 cm long, acuminate. **Flowers** large, showy, 8 cm across, sepals, petals and lip pale green, lip marked especially on the lobes with black or purplish-black; sepals narrowly lanceolate, 4–5 × 1 cm, acute; lateral sepals somewhat falcate, recurved, apically keeled; petals forming a hood with the dorsal sepal, ovate-lanceolate, acute, 3.5–4 × 1.2–1.5 cm; lip trilobed, to 4 × 3 cm, lateral lobes erect, ovate, obtuse, midlobe reflexed, obovate to elliptic, margins undulate-crisped, apex acuminate, callus base bi-lamellate, prolonged into 3 verrucose keels in front; column 1 cm long, foot 4 mm long, channeled, anther cap purplish. **Distr:** N and NE Madagascar. **Ecol:** Along coastal areas, rivers in forests; 0–400(–1,450) m. **Zones:** 10–12.

Cymbidium

Species ~44. **Herbs**, terrestrial, epiphytic or rarely saprophytic; pseudobulbs ovoid or clavate, bearing tuft of leaves, covered with sheaths or overlapping leaf-bases. **Leaves** alternate or fascicled, erect or curved, petiolate or not; lamina linear or oblong-lanceolate, margins entire or minutely serrulate, coriaceous or membranous, never plicate. **Inflorescence** with a solitary flower or many racemose flowers; peduncle arising from lower part or base of pseudobulb, erect, arched or pendulous; bracts shorter or longer than pedicel and ovary. **Flowers** fairly large, often fragrant; sepals nearly equal, free and spreading, linear lanceolate, oblong or lanceolate; petals oblong or ovate oblong, not widely spreading, often covering column and lip; lip hinged to base of column or basally united with base and sides of column, ± trilobed, lateral lobes often erect and embracing column, midlobe usually recurved downward, disk with 2 longitudinal keels; column long, semiterete, apex enlarged; anther terminal; pollinia 2. **Fruits** usually fusiform. **Distr:** Tropical and subtropical Asia to Australia. **Cult:** Propagate by division. Best grown in pots with a coarse bark-based mixture. Requires cool to intermediate conditions, and must be kept moist throughout the year.

C. atropurpureum (Lindl.) Rolfe, *Orchid Rev.* 11: 190, 1903. (Syn: *C. pendulum* var. *atropurpureum*). **Herbs**, epiphytic, terrestrial or lithophytic, upright, sympodial; pseudobulbs not very prominent, clustered, enveloped completely by leaf-bearing sheaths. **Leaves** strap-shaped, to 60 × 2 cm, very leathery, deeply grooved. **Inflorescences** pendulous, to ~30-flowered, to 1 m long, flowers closely set. **Flowers** ~3 cm across, odor somewhat unpleasant, very dark purple, lip white with a few purple spots; dorsal and lateral sepals linear-oblong, 1.5 × 0.6 cm; petals similar to dorsal and lateral sepals; lip trilobed, to 2.5 × 1.5 cm, lateral lobes erect, shorter than the column, apices obtuse, midlobe broadly ovate. **Distr:** S Thailand, W and C Malesia. **Ecol:** In lowland and lower montane forests, lithophytic on vertical cliff faces; 0–2,200 m. **Zones:** 10–12.

Below: *Cymbidium bicolor, Cymbidium cyperifolium, Cymbidium ensifolium, Cymbidium erythrostylum.*

MG

MG GS

MG

Above: *Cymbidium finlaysonianum, Cymbidium lancifolium, Cymbidium lowianum.*

C. bicolor Lindl., *Gen. Sp. Orchid. Pl.*: 164, 1833. **Herbs**, epiphytic, robust, tufted; pseudobulbs narrowly ovoid, laterally compressed, enclosed by persistent leaf bases. **Leaves** 4–7, ligulate or linear-ensiform, 20–80 × 1–1.8 cm, apex oblique to irregularly and unequally bilobed or retuse, thickly coriaceous. **Inflorescence** racemose, 10–26-flowered, to 72 cm long; scape basal, arched or pendent, basal sheaths brown, scarious. **Flowers** showy, fragrant, lasting ~20 days, 3–7.5 cm across, petals with a central brown patch with a narrow yellow margin in the middle and margins broadly yellow, lip cream with maroon or red-brown spots, calluses yellow; sepals narrowly oblong, 1.4–2 × 0.3–0.5 cm, acute; petals narrowly elliptic, 1.2–1.7 × 0.3–0.4 cm, acute; lip 1.2–1.4 cm long, trilobed, lobes ovate, acute, base saccate, inside minutely papillose or puberulous, lateral lobes oblong, subundulate, obtuse; disk with 2 minutely papillose-puberulous lamellae; column 0.8–1.2 cm long. **Distr:** S China to tropical Asia. **Ecol:** In evergreen or semievergreen forests; 80–2,000 m. Fl. Mar–May. **Zones:** 8–12. **Info:** This species is closely related to *C. aloifolium*.

C. cyperifolium Wall. ex Lindl., *Gen. Sp. Orchid. Pl.*: 163, 1833. **Herbs**, epiphytic, terrestrial or lithophytic, autotrophic; pseudobulbs 1–3 × ~1 cm, enclosed in leaf bases. **Leaves** 9–13(–20), distichous-equitant, articulate 4–5 cm from base; lamina lorate, 0.3–1.2 m × (0.6–)1–1.3 cm, apex acute. **Inflorescence** 3–7-flowered, basal, erect, 20–50 cm long; pedicel and ovary 1.2–2.5 cm long. **Flowers** lemon-scented, sepals and petals usually yellowish-green or apple-green, with 5–7 longitudinal red-brown or purple lines; lip yellowish-green to pale yellow, with purple striations on lateral lobes and purple spots or patches on midlobe; sepals linear to linear-lanceolate, 2–3.7 × 0.4–0.8 cm, acuminate; petals narrowly ovate, 1.6–2.9 × 0.5–0.9 cm, acuminate; lip ovate or oblong, 1.4–2.2 cm long, slightly trilobed, lateral lobes small, midlobe strongly recurved, broadly ovate to ovate-elliptic, 0.9–1.3 × 0.8–1.2 cm, margin sometimes finely crenulate; disk minutely papillate on lateral lobes and toward apex of midlobe, with 2 longitudinal lamellae; column 1.1–1.6 cm long, narrowly winged. **Distr:** Himalayas to S China and Philippines. **Ecol:** In forests; 700–1,800 m. **Zones:** 8–12.

C. ensifolium (L.) Swartz, *Nov. Acta Soc. Sci. Upsal.* 6: 77, 1799. (Syn: *C. arrogans*, etc.). **Herbs**, terrestrial, autotrophic; pseudobulbs ovoid, 1.5–2.5 × 1–1.5 cm, enclosed in leaf bases. **Leaves** 2–4(–6), lorate, 30–60 × 1–1.5(–2.5) cm, articulate 2–4 cm from base, margin sometimes serrate toward apex, apex acute. **Inflorescence** 3–9(–13)-flowered, basal, erect, 20–35+ cm; pedicel and ovary 2–2.5(–3) cm long. **Flowers** fragrant, color variable, usually pale yellowish-green, purple-spotted, midlobe with a few red spots; sepals narrowly oblong or narrowly elliptic, 2.3–2.8 × 0.5–0.8 cm, acute; lateral sepals oblique; petals spreading horizontally, narrowly elliptic or narrowly ovate-elliptic, 1.5–2.4 × 0.5–0.8 cm, acute; lip subovate, 1.5–2.3 cm, lateral lobes erect, midlobe recurved, ovate, 0.8–1.2 × 0.7–1 cm, margin kinked; column slightly arcuate, 1–1.4 cm long, narrowly winged; pollinia 4. **Fruits** narrowly ellipsoid, 5–6 × ~2 cm. **Distr:** Tropical and temperate E Asia. **Ecol:** In open forests, thickets; 600–1,800 m. Fl. Jun–Oct. Fr. Dec–Feb. **Zones:** 8–12.

C. erythrostylum Rolfe, *Gard. Chron.* 1905(2): 427, 1905. (Syn: *Cyperorchis erythrostyla*). **Herbs**, epiphytic or terrestrial, ~45 cm high; pseudobulbs ovate-oblong, 2.5+ cm long, several-leaved. **Leaves** arching; lamina elongate-linear, 25–38 × 1.2 cm, base sheathed. **Inflorescences** 4–5-flowered, axillary, suberect, arching, 30–60 cm long; bracts lanceolate, 2.5–3.8 cm long, acute; pedicels 3.2–4.5 cm long. **Flowers** with sepals white, elliptic, concave, 4.5 × 2.5 cm, apiculate; lateral sepals keeled outside; petals white with ~3 rows of minute purple dots down the center of basal half, obovate-elliptic, shorter than sepals, apiculate; lip yellowish-white, deep toward the apex, lined throughout with red-purple, lines breaking up into dots at margins of lateral lobes, veins ± thickened, villous on disk and front lobe, broadly obovate, as long as petals, 3.8 cm wide, much narrowed toward base, apex trilobed, lateral lobes obtusely rounded, midlobe short, broad and obtuse; disk with 5 slender keels, thickened into a prominent trilobed crest; column clavate, arcuate, ~2.5 cm long, crimson; pollinia 4. **Distr:** Vietnam. **Ecol:** On trees, ground or rocks; ~1,500 m. **Zones:** 11–12.

C. finlaysonianum Lindl., *Gen. Sp. Orchid. Pl.*: 164, 1833. (Syn: *C. tricolor, C. wallichii*). **Herbs**, robust, epiphytic or lithophytic, to 70 cm tall, clump-forming; pseudobulbs shortly conical-ovoid, 6 × 3 cm, covered with leaf bases. **Leaves** 5, narrowly lanceolate, 0.3–1 m × 2–4 cm, base articulated to a pseudopetiole 14 cm long, apex rounded, unequally bilobed. **Inflorescence** 0.6–1.5 m long, slender, pendulous, laxly many-flowered; basal sheaths inflated. **Flowers** to 6 cm across; sepals and petals dull yellow, lip purplish or white, lateral lobes red-streaked, midlobe white, with a red-purple arcuate band near apex; disk with 2 red ridges running up to base of midlobe; column 1.8 cm long, foot very short. **Distr:** Indo-China to Malesia. **Ecol:** In exposed places of lowlands. **Zones:** 9–12.

C. lancifolium Hooker, *Exot. Bot.* 1: t. 51, 1823. (Syn: *C. aspidistrifolium*, etc.). [White Bamboo-leaf Orchid, Cut-grass Orchid]. **Herbs**, lithophytic or terrestrial, autotrophic; pseudobulbs, cylindric to narrowly fusiform, apex tapered, 2–7(–15) × 0.5–1(–1.5) cm, naked. **Leaves** 2–4, apical; lamina oblanceolate-oblong to narrowly elliptic, 6–17(–25) × 1.9–4(–6) cm, base contracted into a petiolelike articulate stalk 3–18 cm long, apical margin finely toothed, apex acuminate. **Inflorescence** (1 or)2–6-flowered, lateral, erect, 8–20+ cm long; pedicel and ovary 2–2.5 cm long. **Flowers** 2–2.5 cm ∅, sepals and petals white or pale green, midvein sometimes purplish-brown, lip white or pale green with purplish-brown markings; sepals oblanceolate-oblong, 2.2–2.7(–3) × 0.5–0.7 cm, acute; petals suboblong, 1.5–2.3 × 0.5–0.7 cm, acute; lip ovate-oblong, 1.5–2 cm, obscurely trilobed, lateral lobes erect, midlobe recurved, broadly ovate, 0.6–1.1 × 0.6–1.2 cm, lamellae 2, forming tube; column ~1.5 cm long; pollinia 4. **Fruits** narrowly ellipsoid, ~ 5 × 1.5 cm. **Distr:** In tropical and subtropical Asia. **Ecol:** In forests; 300–2,200 m. Fl. May–Aug. Fr. Oct–Dec. **Zones:** 8–12.

C. lowianum (H.G. Reichenbach) H.G. Reichenbach, *Gard. Chron.*, n.s., 11: 332, 1879. **Herbs**, epiphytic or lithophytic, autotrophic; pseudobulbs narrowly ellipsoid, 6–13 × 2–6 cm, enclosed in leaf bases. **Leaves** 5–7, lorate, 65–80 × 2–3.6 cm, articulate 6–9 cm from base, apex shortly acuminate or subacute. **Inflorescence** 10–20(or more)-flowered, basal, ± arching, 60–80 cm long; pedicel and ovary 3–4 cm long. **Flowers** 0.7–1.2 cm ∅, sepals and petals apple-green or yellowish-green with reddish-brown longitudinal venation, lip pale yellow to whitish, midlobe with a large V-shaped red to pale chestnut or pale brown patch at apex; sepals narrowly obovate-oblong, 4–5 × 1.4–1.6 cm, acute; petals narrowly obovate-oblong, 4–5 × 0.8–1 cm, acute; lip broadly ovate, 3.5–4 cm long, trilobed, lateral lobes erect, midlobe cordate, ~1.6 × 1.8 cm; disk with 2 short lamellae; column 2.7–3 cm long, winged; pollinia 2. **Distr:** China (Yunnan), Myanmar, Thailand. **Ecol:** In forests; 1,300–1,900 m. **Zones:** 8–11.

C. mastersii Griffith ex Lindley, *Edwards's Bot. Reg.* 31: t. 50, 1845. (Syn: *C. affine*, etc.). **Herbs**, epiphytic or lithophytic, autotrophic; pseudobulbs stemlike, usually 0.1–0.3(–1) m long, enclosed in distichous leaf bases. **Leaves** to 17+, lorate, 24–75 × 1.1–1.7(–2.5) cm, articulate 3–6(–10) cm from base, apex unequally bilobed, with a mucro in sinus, subleathery. **Inflorescences** 1 or 2, axillary, suberect, 25–45 cm long, 2–5(or more)-flowered; pedicel and ovary 4–5 cm long. **Flowers** almond-scented, not opening widely, 6–6.5 cm ∅; sepals and petals white, pink-tinged on outer surfaces, lip white with a central yellow patch at base of midlobe, lamellae yellow, sometimes purplish-red spotted; sepals narrowly elliptic or broadly lanceolate-oblong, concave, 4.5–5.5(–6) × 1–1.5(–2) cm, acute; petals broadly linear, 4.2–5 × 0.7–1 cm, acute; lip oblong-ovate, 4–4.5 cm, trilobed, lateral lobes erect, midlobe ovate, ~1.1 cm long, margin undulate; disk with 2 longitudinal lamellae; column ~3.5 cm long; pollinia 2. **Fruits** fusiform-ellipsoid, ~4 × 2.5 cm. **Distr:** China (Yunnan), India, Myanmar, Thailand. **Ecol:** In forests; 1,600–1,800 m. Fl. Oct–Dec. **Zones:** 8–12.

*C. **sinense*** (Jackson ex Andrews) Willdenow, *Sp. Pl.*, ed. 4, 111, 1805. (Syn: *C. chinense*, *C. fragrans*, etc.). [Chinese Cymbidium, Fours Seasons Orchid, New Year's Orchid]. **Herbs**, terrestrial, autotrophic; pseudobulbs ovoid, 2.5–6 × 1.5–2.5 cm, enclosed in leaf bases. **Leaves** 3–5, lorate, 0.45–1.1(–2) m × (1.5–)2–3 cm, articulate 3.5–7 cm from base, thinly leathery, deep green. **Inflorescence** 10–20(or more)-flowered, basal, robust, to 90 cm long; pedicel and ovary 2–2.5 cm long. **Flowers** usually strongly fragrant, variable in color, usually dark purple or purplish-brown, lip paler; sepals narrowly oblong or narrowly elliptic, 2.2–3.5 × 0.5–0.7 cm, acute; petals narrowly ovate, 2–2.7 × 0.6–1 cm, acute; lip ovate-oblong, 1.7–2.5(–3) cm, obscurely trilobed, lateral lobes erect, midlobe recurved, oblong-ovate, ~1.4 × 1 cm, margin slightly undulate; disk with 2 longitudinal lamellae; column 1.2–1.5 cm long, narrowly winged; pollinia 4. **Fruits** narrowly ellipsoid, 6–7 × 1.5–2 cm. **Distr:** Assam to Ryukyu Islands. **Ecol:** In moist forests, thickets along streamsides; 300–2,000 m. Fl. (Oct–)Nov–Mar. **Zones:** 9–11.

*C. **wilsonii*** (Rolfe ex De Cock) Rolfe, *Orchid Rev.* 12: 97, 1904. (Syn: *C. giganteum* var. *wilsonii*, *Cyperorchis wilsonii*). **Herbs**, epiphytic, autotrophic; pseudobulbs narrowly ovoid, bilaterally flattened, ~6 × 3 cm. **Leaves** 7, lorate, ~90 × 2.5 cm, articulate 6–11 cm from base, apex acute. **Inflorescence** ± arching, 25–70 cm, 5–15-flowered; pedicel and ovary 2.2–4.2 cm long. **Flowers** slightly fragrant, 9–10 cm Ø, sepals and petals green or yellowish-green with distinct reddish-brown speckles along veins, lip cream-yellow with dark reddish-brown striations on lateral lobes, with a submarginal V-shaped reddish-brown patch composed of dense spots and blotches along apical margin of midlobe; sepals narrowly obovate, 4.4–5.7 × 1.2–1.9 cm, acuminate; petals narrowly obovate, 4–5.3 × 0.7–1.3 cm, acuminate; lip broadly ovate-elliptic, trilobed, lateral lobes erect, ciliate, midlobe 1.5–1.8 cm long, margin undulate; disk papillate or puberulent; column 2.7–3.2 cm long, apically broadly winged; pollinia 2. **Distr:** China (S Yunnan) to Vietnam. **Ecol:** In forests; ~2,000 m. Fl. Feb–Apr. **Zones:** 8–11.

Porphyroglottis

Species 1. *Generic description same as specific.* **Cult:** Propagate from seeds or by division. Grow in pots or baskets. Requires humidity, moderate shade and plenty of water during growth period

*P. **maxwelliae*** Ridl., *J. Linn. Soc.*, Bot. 31: 290, 1896. **Herbs**, epiphytic, large; stems erect, cylindrical, leafy, to 2 m long. **Leaves** ascending, distichous, articulated to a sheathing base; lamina linear, 25–35 × 0.6–1 cm, apex acuminate. **Inflorescences** laxly few-flowered, lateral, erect to erect-arcuate, simple or branched, to 1.5 m long, flowers borne in succession with 2–3 open at a time; peduncle and rachis slender, dull pink; pedicel and ovary 2–2.5 cm long, dull pink. **Flowers** with sepals and petals dull salmon-pink, lip glossy maroon, apex yellowish; sepals and petals strongly reflexed, elliptic-oblong, 2–2.4 × 0.6–0.8 cm, obtuse; lip mobile, strongly convex, obovate, 1.5 × 0.6–0.9 cm, truncate, base and apex pubescent, attached by a straplike hinge to column foot, callus of obscure ridges at base of lip; column slender, arcuate, 1.5 cm long, with 2 lateral porrect wings in the middle, foot 4–5 mm long. **Distr:** W Malesia. **Ecol:** In the forks of trunks of large trees, in open areas and low-crowned heath forests; 0–500 m. **Zones:** 11–12.

Subtribe Catasetinae

Catasetum

Species 171. **Herbs**, mostly epiphytic, sometimes lithophytic or terrestrial, sympodial; stems short, developing into slender-ovoid or fusiform pseudobulbs, concealed when young by imbricating sheaths which are leaf-bearing above. **Leaves** several, distichous; lamina usually elliptic-lanceolate or oblanceolate, plicate. **Inflorescence** racemose, few- to many-flowered, basal, erect or pendent. **Flowers** monomorphic, dimorphic or trimorphic, monoecious, dioecious or rarely perfect; sepals and petals free, subequal or petals broader, subglobose-connivent or spreading; lip pendent or uppermost, very diverse, deeply saccate to almost flat, sessile, narrow or very broad, with entire, dentate, or deeply fimbriate margins; column erect, stout, footless, often elongate in ♂, with a pair of reflexed antennae or tendrils surrounding the stigma, often very short, in ♀ and without antennae; anther terminal, opercular, incumbent; pollinia 4, or 2 and deeply sulcate. **Distr:** Mexico to tropical America. **Cult:** Propagate by division. Best grown in baskets with a coarse bark-based compost. Requires humid, well-ventilated and bright conditions.

*C. **gnomus*** Linden and Rchb. f., *Xenia Orchid.* 2: 171, 1873. (Syn: *C. heteranthum*, etc.). **Herbs**, epiphytic; pseudobulbs robust, conical-fusiform, slightly compressed, 11–16 cm long, 3–4 cm Ø, 3-leaved. **Leaves** erect, spreading or recurved; lamina lanceolate, 20–25 × 3–4 cm, apex acute or shortly acuminate. **Inflorescence** erect or ascending, to 60 cm long, laxly many-flowered above, flowers nodding or pendulous; bracts short, fleshy, appressed, narrowly triangular, 4–6 cm long, acute. **Flowers** of 3 forms (♂, ♀ and hermaphrodite), slightly fragrant; ♂ green, purple-spotted, lip whitish, purple-spotted, column greenish-white, purple-spotted, sepals fleshy, broadly oblong, 5 × 1.5–1.8 cm, acute or shortly acuminate, petals lanceolate, 4.5 × 1.2 cm, acute, lip uppermost, fleshy, rigid, obtusely conical, deeply saccate, 2 cm long, margins undulate-denticulate or serrate, apex shortly acuminate, antennae 1.8–2 cm long, 1 straight, the other incurved toward apex; ♀ green, sepals and petals purple-spotted, lip pale green, margin pale yellow, sepals and petals as in ♂, lip uppermost, deeply saccate, margin dentate. **Distr:** S Venezuela to Brazil. **Ecol:** On trees near watercourses of lowland forests, and in wet montane forests; 200–1,200 m. Fl. Jul, Oct–Mar. **Zones:** 11–12.

Top row: *Cymbidium mastersii*, *Cymbidium sinense*, *Cymbidium wilsonii*.
Bottom row: *Porphyroglottis maxwelliae*, *Catasetum gnomus*, *Catasetum maculatum*.
[All MG except *Porphyroglottis maxwelliae* GS.]

C. maculatum Kunth, *Syn. Pl.* 1: 331, 1822. (Syn: *C. oerstedii*, etc.). [Spotted Catasetum]. **Herbs**, large, epiphytic; pseudobulbs conical-fusiform, clustered, many-noded, to 20 cm long, base covered with distichous sheaths when young. **Leaves** 3–5 per growth, suberect-spreading; lamina oblanceolate, to 50 × 9 cm, acute. **Inflorescence** arcuate, to 40 cm long, >9-flowered toward apex; bracts ovate, pale green, 1 cm long. **Flowers** non-resupinate, sepals and petals greenish-brown, marked and suffused with maroon, lip greenish-yellow, ± heavily suffused with maroon, column pale creamy-green, maroon-spotted; pedicel arcuate; ♂ with sepals elliptic to obovate, 4–4.5 × 1.7–1.8 cm, margins inrolled, apex acute, petals ovate-elliptic, 4 × 1.7 cm, margins ciliate, apex acute, lip deeply cucullate, somewhat conical, 2.2 cm deep, margins fringed, fleshy-coriaceous, margin with maroon hairs, column 2.2 cm long, 1 antenna porrect, the other strongly curved; ♀ with sepals and petals oblong-elliptic, 2 × 1 cm, subacute to obtuse, deflexed, lip deeply saccate, ~3 × 1.8 cm, margins shortly fimbriate, apex broad, column short, stout, fleshy. **Distr:** C America to N Brazil. **Ecol:** In coastal lowlands, usually near watercourses; 0–1,000 m. Fl. Jun–Jul, Oct–Dec. **Zones:** 9–12.

C. tabulare Lindl., *Edwards's Bot. Reg.* 30(Misc.): 40, 1840. (Syn: *C. tabulare* var. *brachyglossum*, etc.). **Herbs**, epiphytic, pseudobulbous. **Leaves** distichous; lamina pleated, ovate, 4–45 × 8 cm, base narrowed into a pseudo-petiole, apex acute, glabrous, adaxially deep green, abaxially paler. **Inflorescence** racemose, 2–6-flowered, suberect, 25–45 cm long, basal, borne from newly formed pseudobulbs; bracts triangular; pedicel and ovary 4.5–5 cm long, purplish. **Flowers** fleshy, extremely fragrant, opening wide, 7.5–10 cm across, fleshy, pale green with maroon blotches or solid maroon, lip with a conspicuous cream callus; sepals oblong-lanceolate, ~1.5 cm wide, base concave, apex acute; dorsal sepal 5.5 × 1.5 cm, acute; lateral sepals 5.8 × 1.5 cm, acute, margins folded; lip oblong, 4.8 × 2 cm, base concave, callus 3 × 1.2 cm; column ~1.4 cm long; antennae ~1.7 cm long. **Distr:** Colombia. **Ecol:** In tropical rainforests, riverine forests, swamps; to 1,000 m. Fl. May–Oct. **Zones:** 11–12.

C. tenebrosum Kraenzl., *Gard. Chron.*, III, 48: 229, 1910. (Syn: *C. tenebrosum* f. *smaragdinum*). **Herbs**, epiphytic; stems when young enveloped by distichously imbricating, scarious sheaths which are leaf-bearing above, at length thickened into an oblong-ovoid pseudobulb to 12 cm long. **Leaves** 6–8, congested; lamina oblanceolate to obovate-oblong, to 23.5 × 5.5 cm, gradually narrowed below, plicate, apex acute, often cuspidate. **Inflorescence** racemose, loosely 4–12-flowered, lateral, basal, erect-spreading or arcuate, to 28 cm long, raceme 6–15 cm long; peduncle to 12 cm long. **Flowers** small, segments spreading, dark brown to brownish-purple, lip honey-colored, greenish-tinged; dorsal sepal elliptic-oblong, to 3–3.5 × 1.4 cm, acute or shortly acuminate; lateral sepals similar, slightly oblique; petals very similar to dorsal sepal and connivent with it, obliquely oblong-elliptic; lip flat, broadly triangular-ovate, ~2.2 × 2.2 cm, base slightly cordate, entire, margins crenulate to dentate, apex obtuse, fleshy; disk slightly convex, with a central knoblike thickening, apex obtuse; column stout, to 2 cm long, apex abruptly incurved-rostrate, with a pair of small, decurved, parallel antennae below the stigmatic orifice. **Distr:** SE Ecuador to Peru, Brazil (Pará). **Ecol:** In riverine forests; 500–1,800 m. **Zones:** 10–11.

Cycnoches

Species 36. **Herbs**, epiphytic or terrestrial, large; pseudobulbs elongated, fusiform-cylindrical, leafy. **Leaves** plicate, membranous. **Inflorescence** racemose, few- to many-flowered, lateral, erect to arcuate-pendent. **Flowers** small to large, unisexual, dimorphic; sepals and petals subsimilar, free, spreading or reflexed; lip thinly textured to fleshy, subsessile to distinctly clawed, lanceolate to orbicular, entire to lobed, crested or fringed, mostly bearing calli; column elongated, slender to stout, fleshy, ± arcuate, clavate, lacking a foot; pollinia 2, waxy, ovoid, sulcate. **Distr:** Mexico to C and S tropical America. **Cult:** Propagate by division. Requires a well-drained compost in humid and moderately shady conditions.

C. warszewiczii Reichenbach f., *Bot. Zeitung* (Berlin) 10: 734, 1852. (Syn: *C. tonduzii*, *C. ventricosum* var. *warszewiczii*). **Herbs**, epiphytic, deciduous, large; pseudobulbs 15–70 × 2.5–4 cm. **Leaves** many, lanceolate, 15–45 × 2–10 cm, base narrowed, sheathing the stem, margins entire, apex acute. **Inflorescences** laxly flowered, pendent, 10–75 cm long. **Flowers** apple-green, very fragrant. ~7.5 cm across; sepals and petals spreading, ± similar, lip white, glossy; sepals and petals lanceolate or ovate, 4–7 × 1.2–3.5 cm; lip clawed, ovate, 3–4 × 2–3 cm, convex, apex subacute; column linear, 2.2–3.5 cm long, arcuate. **Distr:** S Nicaragua to Panama. **Ecol:** On trees and rotten logs, occasionally terrestrial near rotten logs, in cloud forests and very humid forests; 0–1,050 m. Fl. Sep–Oct. **Zones:** 10–12.

Subtribe Cyrtopodiinae

Cyrtopodium

Syn: *Tylochilus*.

Species 46. **Herbs**, terrestrial or epiphytic, cespitose; stems pseudobulbous, ovate-fusiform. **Leaves** deciduous, distichous, articulate; lamina plicate, leathery, apex acuminate. **Inflorescences** racemes to panicles, erect, borne from base of pseudobulbs; floral bracts large, showy, margins undulate. **Flowers** resupinate, showy; sepals and petals membranous, margins often undulate; lip adnate to apex of column foot, subsessile to clawed, trilobed; disk with tuberculate, cristate, or verrucose callus; column slightly arcuate, sometimes nearly club-shaped, semiterete, forming a foot; anther terminal, incumbent, operculate, imperfectly bilocular; pollinia 2, yellow, nearly triangular, sulcate, cartilaginous, attached to short, trowel-shaped viscidium. **Fruits** pendent, pear-shaped. **Distr:** S. Florida to tropical America. **Cult:** Propagate by division. Can be grown mounted or potted in a terrestrial mix.

C. andersonii (Lamb. ex Andrews) R.Br., *Hortus Kew.* 5: 216, 1813. (Syn: *Cymbidium andersonii*, etc.). [Cow Horn Orchid]. **Herbs**, epiphytic, large, 1.6+ m tall; pseudobulbs erect, spindle-shaped or fusiform-cylindric, 0.6–1 m long, 3–6 cm wide. **Leaves** narrowly linear-lanceolate, 30–60 × 2–5 cm, apex shortly acuminate, coriaceous. **Inflorescence** paniculate, many-flowered, branched, pyramidal in outline above, 0.9–2 m tall; bracts ovate, 3–5 cm long, acuminate, membranous, yellowish-green. **Flowers** showy, to 5 cm across, sepals and petals yellow to green-tinged, petals lemon-yellow, margins undulate, lip bright yellow, callus orange; sepals broadly ovate-elliptic, 1.8–2.2 × 1.3–1.7 cm, margins scarcely undulate, apex obtuse to shortly apiculate; petals broadly obovate, 1.8–2.3 × 1.2–1.5 cm; lip sessile, trilobed, 1.8–2 × 2.5–3 cm, lateral lobes obovate, rounded, midlobe shorter, broadly obovate-subquadrate; disk longitudinally grooved; column 5–6 mm long. **Distr:** Colombia to N South America and N Brazil, West Indies. **Ecol:** Terrestrial in sandy, rocky savanna, 300–1,500 m. **Zones:** 10–12.

Subtribe Oncidiinae

Ada

Species 16. **Herbs**, epiphytic and lithophytic, sympodial; pseudobulbs lacking or often with reduced development, pyriform, obovate, usually laterally flattened with leafy bracts, smooth, 1(–2) leaves at the apex. **Leaves** dark green, linear-lanceolate, elliptic-lanceolate or ovate-elliptic, distichous, folded at the base, apex asymmetrical, acuminate. **Inflorescences** 1 to many per plant, racemes, axillary, not branched, often longer than the subtending leaf; floral bracts, large, papery, not appressed to pedicel. **Flowers** spreading or opening at only the tips, red, bright orange, yellow, green or brown; sepals free, linear-lanceolate, pointing forward; petals similar but smaller; lip entire, acuminate, recurved, callus with 2 ridges, lacking a spur; pollinia 2, reniform to obovoid, ventral groove is well-developed, attached to a heart-shaped stipe, viscidium 1. **Fruits** cylindrical, not beaked. **Distr:** C and S America. **Cult:** Propagate from seeds or by division. Requires high humidity with cool to intermediate conditions and shaded sunlight.

A. aurantiaca Lindley, *Fol. Orchid.* 5 (Ada): 1, 1854. (Syn: *A. lehmannii*, etc.). **Herbs**, epiphytic, 30–45 cm tall; pseudobulbs to 10 cm long, subcylindrical, elliptic-pyriform, smooth, dark green, apex 1(–3)-leaved. **Leaves** broadly linear or linear-lanceolate, to 10–36 × 1–2.5 cm, dark green bases partially sheathed with reddish-brown mottled scales, apex acuminate. **Inflorescences** spikes, 7–18-flowered, terminal, oblong, to 20–35 cm long, arching or drooping, flowers distantly placed, subdistichous; pedicels 0.6–1.5 cm long. **Flowers** ± tubular, bright orange to red, ~3.5 cm long; perianth spreading above middle; sepals all similar, linear-lanceolate, rarely slightly falcate, 2–3.8 × 0.3–0.8 cm, acuminate; petals similar to sepals but smaller, 1.5–2.5 × 0.3–0.7 cm; lip about half the length of perianth, linear-lanceolate, 1–1.6 × 0.2–0.6 cm, ± parallel to column, callus of 2 membranous lamellae terminated by 2 erect, recurved toothlike masses; column short, thick, 4–5 mm high; pollinia 2; ovary elongated, clavate. **Distr:** Colombia, Ecuador, Venezuela. **Ecol:** In cloud forests; 1,500–2,500 m. Fl. Mar. **Zones:** 10–12. **Info:** Flowers adapted to pollination by hummingbirds.

A. keiliana (Rchb. f ex Lindl.) N.H.Williams, *Brittonia* 24 (1): 108, 1972. (Syn: *Brassia cinnamomea*, etc.). **Herbs**; pseudobulbs fairly large, pyriform, sheathing leaves numerous. **Leaves** linear-lanceolate, 18–60 × 2–4 cm. **Inflorescences** with floral bracts papery, broadly lanceolate, 1.7–3.5 × 0.7–1 cm; pedicels 2–3.5 cm long. **Flowers** with dorsal sepal lanceolate, 4–6 × 0.3–0.7 cm; lateral sepals falcate-lanceolate, 3.5–8 × 0.2–0.6 cm, acuminate; petals shorter than sepals, falcate, 2.2–5.5 × 0.3–0.7 cm, acuminate; lip ovate-oblong, 1.6–3.3 × 0.7–1.5 cm, apex long-acuminate, usually not reflexed, callus of 2 arching basal lamellae with 2 much smaller terminal toothlike recurved masses; column 0.4–0.6 cm high; stipe broadly obdeltoid. **Distr:** Colombia, Venezuela. **Ecol:** In damp forests; 1,000–1,800 m. Fl. Feb–Oct. **Zones:** 10–12.

Brassia

Species ~35–38. **Herbs**, epiphytic, ± cespitose; stems either reduced and covered with sheathing bracts or elongate with prominent, single-noded pseudobulb. **Leaves** 1(–2) per sympodium, apical; lamina conduplicate, oblong-lanceolate, dorsiventrally flattened, margins entire. **Inflorescences** racemes, lateral; bracts narrowly triangular. **Flowers** with sepals and petals distinct, free, long, attenuate-caudate; lip simple or obscurely lobed; callus prominent, bilamellate; column without appendages, footless;

MG

JT

MG

GS

Top row: *Catasetum tabulare, Catasetum tenebrosum.*
Second row: *Cyrtopodium andersonii, Cycnoches warszewiczii, Ada aurantiaca.*
Bottom row: *Ada keiliana, Brassia gireoudiana, Brassia verrucosa.*

stigmatic cavity ± round; rostellum not prominent, anther terminal, pollinia 2, waxy. **Distr:** Tropics and subtropics of Western Hemisphere. **Cult:** Propagate by division. Best grown in pots with a bark-based mix. Requires warm, moist and bright conditions.

B. gireoudiana Rchb. f. and Warsc., *Allg. Gartenzeitung* 22: 273–274, 1854. **Herbs**, epiphytic, scandent; pseudobulbs elliptic-ovoid, 6–14 × 2.5–4.5 cm, tapering, laterally compressed. **Leaves** (1)2, apical, petiolate; lamina elliptic-ovate, 20–35 × 2.2–5 cm, apex acute, apiculate, subcoriaceous, adaxially carinate. **Inflorescence** a raceme, 1–15-flowered, lateral, arcuate, neatly arranged on the axis, 0.75–1 m long, basal; pedicel and ovary 1.8–3 cm long. **Flowers** large and showy, 20–25 cm across, greenish-white, brown-dotted at base of sepals and lip, base of petals solid dark brown; sepals similar, free, linear-lanceolate, attenuate to apex; dorsal sepals erect, 10–13 × 0.4–0.5 cm; lateral sepals 11–20 × 0.4–0.5 cm; petals similar to sepals but shorter, linear-lanceolate, 5–10 × 0.4–0.5 cm, subfalcate; lip simple, rhombic, base narrowed forming a narrow claw united to base of column, 3.5–7 × 2–3 cm, margins slightly crisped, apex acuminate, callus at base with a fleshy pair of keels; column stout, 5–6 mm long. **Distr:** Costa Rica, Panama. **Ecol:** In semideciduous to evergreen premontane forests; 500–1,000 m. Fl. Feb–May. **Zones:** 10–12. **Info:** This species is often confused with *B. verrucosa* which differs by the lack of warts on the lip.

B. verrucosa Bateman ex Lindl., *Edwards's Bot. Reg.* 26: t. 36, 1840. (Syn: *B. brachiata*, etc.). **Herbs**, epiphytic, cespitose, clump-forming, to ~65 cm tall; pseudobulbs ovoid-conic, furrowed, 6–11 × 3.1–4 cm. **Leaves** 2, apical, shortly petiolate; lamina elliptic-lanceolate, 16–30 × 2.6–4.5 cm, base conduplicate, apex acute, coriaceous. **Inflorescence** a raceme, 6–10-flowered, lateral, erect or arcuate, 50–65 cm long, flowers distichous; pedicel and ovary 2.5–3 cm long. **Flowers** showy, pleasantly fragrant, to 25 cm high, yellowish-white with bars and spots of reddish-brown on the sepals and petals, lip with olive-green warts; sepals similar, free, linear-lanceolate, attenuate and somewhat canaliculate; dorsal sepal erect, 6.5–13.5 × 0.6–0.7 cm; lateral sepals 7.5–17.5 × 0.6–0.7 cm; petals linear-lanceolate, falcate, attenuate, 4–9 × 0.6–0.7 cm; lip obscurely trilobed, trullate-subpandurate, 3.5–5.2 × 2.4–3.2 cm, acuminate, verrucose below; callus a pair of basal keels; column erect, 7 mm long; pollinia 2. **Distr:** Mexico to N Brazil. **Ecol:** In evergreen to semideciduous cloud forests; 1,300–1,800 m. Fl. Dec–May. **Zones:** 9–12.

Caucaea

Species 9. **Herbs**, epiphytic; pseudobulbs clustered, narrowly ovoid, compressed, apex unifoliate, subtended by several rarely leaf-bearing, imbricate, acuminate sheaths. **Leaf** erect to spreading; lamina oblanceolate, coriaceous. **Inflorescences** erect, axillary, basal, unbranched, rarely 1-branched, laxly few–15-flowered; pedicel and ovary relatively long. **Flowers** small; dorsal sepal and petals free, subsimilar, spreading; lateral sepals united almost to apex, parallel to lip; lip obscurely trilobed at base, porrect, entire, with a basal callus, lacking a spur; column short, fleshy, basally winged but lacking a foot; pollinia 2, cleft; stipe small, oblong; viscidium small, circular. **Distr:** N and W South America. **Ecol:** In high-altitude cloud forests. **Cult:** Propagate by division. Best grown mounted on a slab. Requires humid, cool and shady conditions.

C. nubigena (Lindl.) N.H. Williams and M.W. Chase, *Lindleyana* 16: 283, 2001. (Syn: *Oncidium cucullatum* var. *nubigenum*, etc.). **Herbs**, epiphytic; pseudobulbs aggregated, narrowly ovoid to pyriform-cylindric, 3–10 cm high, compressed. **Leaves** (1–)2, linear-elliptic to narrowly oblong or oblanceolate, 5–26(–41) × 1.5(–4) cm, acute or acuminate. **Inflorescence** loosely racemose (rarely paniculate), 3–10-flowered; scape to 25 (–60+) cm long; peduncle to ~22.5(–53) cm long; pedicel and ovary 2–2.5 cm long. **Flowers** with segments wide-spreading, sepals and petals of variable color, either white and purple-spotted or brownish or olive with yellow margins; dorsal sepal oblong-oval or obovate-oblong, ~0.7–1.5 × ~0.8 cm, acute, cucullate; lateral sepals ± connate into a suborbicular-obovate lamina, apices free; petals obovate, acute or obtuse; lip much larger, obovate-suborbicular, pandurate-trilobed, 1.6–1.8 cm long, base sessile-cordate, apex retuse, yellowish-white rose-spotted, or crimson spotted with purple, basal portion transversely oblong, 0.9–1.6 cm wide, each side produced into a small, spreading, auriculiform lateral lobe, midlobe much larger, broadly flabellate to reniform, retuse, ~2.2(–2.8) cm wide; disk with a small, 3- or 5-lobulate basal callus; column, ~3 mm long, wingless above. **Distr:** W South America to Venezuela. **Ecol:** 2,500–4,000 m. **Zones:** 10–12.

Cischweinfia

Species 10. **Herbs**, epiphytic, small, cespitose; pseudobulbs clustered, ellipsoidal or ovoid-ellipsoidal,

Top row: *Caucaea nubigena, Cischweinfia dasyandra, Cochlioda rosea, Comparettia falcata.* [All MG except *Cochlioda rosea* AF.]
Bottom row: *Cuitlauzina pendula, Cuitlauzina pulchella, Cyrtochilum macranthum, Erycina pusilla.*

JT

MG MG
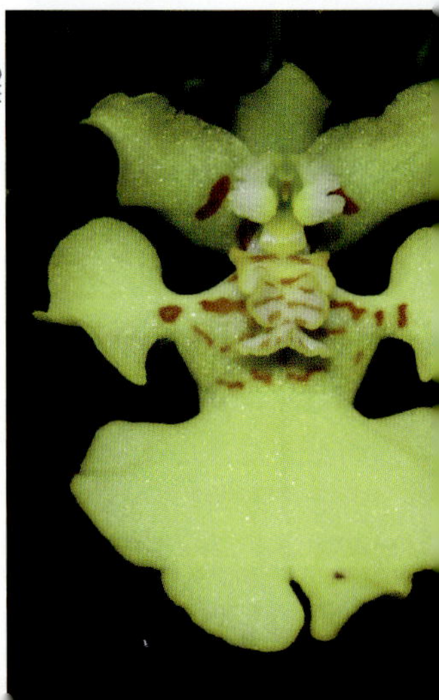

flattened, subtended by several leaf bearing sheaths, unifoliate. **Leaf** articulated at base, thinly textured. **Inflorescence** a raceme, lateral, shorter than the leaves, few-flowered, axillary from basal sheaths. **Flowers** small to medium-sized; sepals and petals free, subsimilar; lip basally adnate to column by the margins of the claw, pubescent inside; spur adnate to ovary for entire length; column with 2 auricles borne beneath the stigmatic cavity; clinandrium petaloid, hooded over the anther; pollinia 2, stipe slender, elongate, viscidium small. **Distr:** Costa Rica to W South America. **Cult:** Propagate by division. Best grown mounted or in well-drained pots. Requires moderate humidity and shade.

C. dasyandra (Rchb. f.) Dressler and N.H. Williams, *Amer. Orchid Soc. Bull.* 39: 991, 1970. (Syn: *C. glicensteinii*, etc.). **Herbs**, epiphytic, cespitose; pseudobulbs oblong, compressed, 3–8 × 0.6–1.1 cm, apex unifoliate, subtended by 2–3 foliaceous bracts. **Leaves** linear-lanceolate, 9–20 × 0.7–1.2 cm, base conduplicate, apex acute, abaxially keeled. **Inflorescence** a raceme, 1–4-flowered, 4–7 cm long, somewhat pendulous, basal; pedicel and ovary 1.3–2 cm long. **Flowers** 2.4–2.7 × 1.8–2.2 cm, sepals and petals green pink-suffused, lip white, pink-spotted, center yellow; sepals subequal, free, elliptic-lanceolate, 1.2–1.4 × 0.2–0.3 cm, acute; lateral sepals subfalcate and somewhat larger than the dorsal sepal; petals elliptic-oblong, acute, subfalcate, 1–1.2 × 0.2–0.3 cm; lip with basal margins folded around column, obovate to suborbicular, 1–1.2 × 0.9–1 cm, margins crenulate, apex emarginate, callus basal, glabrous, yellow, with 2 small keels; column 5–7 mm long, white; pollinia 2. **Distr:** Costa Rica to Ecuador. **Ecol:** In wet premontane forests; 700–1,100(–1,650) m. Fl. Jun–Sep. **Zones:** 10–12.

Cochlioda

Species 8. **Herbs**, epiphytic; stems very short, thickened into compressed pseudobulbs. **Leaves** 1 or 2, linear or oblong to lorate. **Inflorescences** 1 or 2, lateral, basal, erect, arcuate or pendent, loosely racemose, sometimes paniculate; floral bracts large or more commonly small and inconspicuous. **Flowers** scarlet to rose-red, with spreading segments; sepals subequal, free; petals broader than sepals, somewhat oblique; lip with an erect claw which is ± adnate to column, spreading, trilobed, lateral lobes rounded or oblong and often reflexed, midlobe porrect, entire or dilated above and retuse to bilobed in front; column erect and ± incurved, slender, often auricled above, with 2 stigmatic orifices; anther terminal, opercular, incumbent; pollinia 2. **Distr:** W South America. **Cult:** Propagate by division. Best grown in pots with free-draining organic matter or mounted on a moisture-retaining mossy substrate. Requires moderately humid, bright conditions with regular watering.

C. rosea (Lindl.) Benth., *Gen. Pl.* 3: 560, 1883. (Syn: *Mesospinidium roseum*, etc.). **Herbs**, epiphytic, small to medium-sized; pseudobulbs aggregated, ovoid to broadly ellipsoid, unifoliate or bifoliate, 3–5 cm high. **Leaves** variable, elliptic-oblong to linear, 5.8–20 × 1–2.5 cm, base cuneate, narrowed to a conduplicate, subpetiolate base, apex acute to obtuse. **Inflorescences** racemose to paniculate, loosely 5–20-flowered, lateral, basal, erect to flexuous or arcuate, 13–40 cm long; pedicel and ovary slender. **Flowers** rose-red, segments spreading, 2–3.5 cm Ø; dorsal sepal narrowly elliptic-lanceolate, to 1.1 × 0.4 cm, acute; lateral sepals similar, obliquely elliptic-lanceolate; petals lanceolate-elliptic, acute; lip adnate to lower third of column, ~0.95–1.6 cm long, sharply trilobed below or near middle, ovate, lateral lobes short, very obliquely ovate or rhombic, base cuneate, midlobe much larger, porrect, linear-oblong, linear-lanceolate or cuneate, apex obtuse or truncate; disk with 1 or 2 pairs of short fleshy keels near base; column stout, apex trilobed, ~6.5 mm high. **Distr:** Ecuador to Peru. **Ecol:** Epiphytic in wet montane forests; 1,500–2,200 m. Fl. Feb–Apr, Jun–Jul, Dec. **Zones:** 10–12.

Comparettia

Species 8. **Herbs**, epiphytic or lithophytic; rhizomes slender; stems small, pseudobulbous, unifoliate. **Leaf** coriaceous. **Inflorescence** basal, simple or branched, flowers distant. **Flowers** small, showy; sepals about equal in length, erect; dorsal sepal free; lateral sepals united, extended at base into a long spurlike mentum; petals as long as the dorsal sepal but broader; lip continuous with base of column, unguiculate, spreading, much larger and longer than the sepals and petals, produced at base into 2 long terete caudae that are enclosed in the mentum formed by the lateral sepals; column erect, wingless, footless; anther terminal, operculate, incumbent, 1-celled; pollinia 2, waxy. **Fruits** ellipsoid, 3-angled, long-beaked. **Distr:** Mexico to tropical America. **Cult:** Propagate by division. Best grown in small hanging baskets or mounted on small cork slabs. Requires high humidity and bright light.

C. falcata Poepp. and Endl., *Nov. Gen. Sp. Pl.* 1: 42, 1836. **Herbs**, epiphytic, medium-sized, variable; pseudobulbs clustered, oblong-cylindric, compressed, 1–4 cm long, apex unifoliate, covered by 1 to several pairs of distichous ± leaf-bearing sheaths. **Leaf** elliptic to oblong, 3–18 × 1–5 cm, apex acute to rounded. **Inflorescence** racemose, rarely paniculate, laxly 2- to many-flowered, basal, much longer than leaves, arcuate or nodding, to 90 cm long. **Flowers** deep rose-pink to pink, marked with white; dorsal sepal elliptic or oblong-ovate, to 1 × 0.5 cm, concave, acute; lateral sepals connate, oblong-ovate, concave, apex bidentate or entire, extended at base into a slender decurved spur 1.5 cm long; petals spreading, elliptic, 1.1 × 0.7 cm, subacute; lip obscurely trilobed, 1.1–1.7 × 1.1 cm, with 2 slender capillary spurs at base within the sepaline spur, lateral lobes auriculate, midlobe broadly oblong, spreading, emarginate; column small, clavate, 5 mm long. **Distr:** Mexico to tropical America. **Ecol:** In rainforests and cloud forests; 1,000–2,000 m. **Zones:** 9–12.

Cuitlauzina

Syn: *Osmoglossum, Palumbina.*

Species 5. **Herbs**, epiphytic; pseudobulbs ovoid, compressed, apex 2-leaved. **Leaves** subtended by a few imbricate sheaths; lamina ligulate, subcoriaceous. **Inflorescences** basal, pendent, simple, few- to many-flowered. **Flowers** showy, fragrant, ± flat; sepals free, spreading, subsimilar; petals free, spreading, similar to but broader than petals; lip deflexed, at an obtuse angle to column, narrowly clawed, apical lamina flat, transversely oblong-reniform, bilobed, with a prominent callus on claw; column short, erect, with 2 lateral wings and hooded above; pollinia 2, pyriform, attached by a stipe to a viscidium, anthers petaloid, unilocular. **Distr:** Mexico to C America. **Cult:** Propagate by division. Best grown mounted on cork slabs. Requires high humidity and bright light.

C. pendula Lex., *Nov. Veg. Descr.* 2: 33, 1825. (Syn: *Lichterveldia lindleyi*, etc.). **Herbs**, epiphytic; pseudobulbs clustered, ovoid, compressed, ancipitous, 10–15 × 5–7.5 cm, apex 2-leaved. **Leaves** broadly ligulate, 12–30 × 3 cm, apex acute, coriaceous. **Inflorescences** basal, pendent, simple, to 1 m long, laxly 2–22-flowered; pedicel and ovary 1–1.5 cm long. **Flowers** showy, very fragrant, flat, 6–7.5 cm across, sepals and petals white, sometimes base pink-tinged, lip white to rose-pink, callus yellow red-spotted; sepals obovate to ovate-elliptic, 2.3–2.6 × 1.5 cm, rounded or retuse; petals elliptic to subcircular, 2–2.8 × 1–2.5 cm, margins undulate, apex rounded to emarginate; lip narrowly clawed, apical part transversely reniform, 2.2–2.3 × 2.8–4 cm, emarginate; callus fleshy, base auriculate, concave, apex bilobed; column short, erect, 7–8 mm long, with 2 lateral wings near apex, obovate, erose. **Distr:** W and C Mexico. **Ecol:** In oak-pine forests; 1,400–2,200 m. **Zones:** 9–12.

C. pulchella (Bateman ex Lindl.) Dressler and N.H.Williams, *Selbyana* 24: 44, 2003. (Syn: *Odontoglossum pulchellum, Osmoglossum pulchellum*). **Herbs**, epiphytic, erect, often clump-forming; pseudobulbs clustered, ovoid to ovoid-ellipsoid, compressed, 6–10 × 3–4 cm, apex 2-leaved, subtended by several distichous, membranous, imbricate sheaths. **Leaves** erect, spreading; lamina linear, strap-shaped, 10–50 × 0.8–1.5 cm, apex acute, coriaceous. **Inflorescences** erect, 12–50 cm long, simple, laxly 3–10-flowered; bracts triangular-lanceolate, 0.8–1.3 cm long, acuminate; pedicel and ovary 1.2–1.5 cm long, bent in middle. **Flowers** not resupinate, fragrant, fleshy, white, pink-tinged below, callus yellow, red-spotted; dorsal sepal obovate or elliptic-ovate, 1–2 × 0.6–1.3 cm, shortly acuminate; lateral sepals basally united, ovate to lanceolate, 1–2 × 0.5–1 cm, acuminate; petals obovate to suborbicular-obovate, 1.3–2 × 0.7–1.5 cm, margins undulate, apiculate; lip strongly recurved, uppermost in flower, panduriform, 1–1.8 × 0.8–1.1 cm, margins undulate-crisped, obtuse, callus fleshy, basal half 3-ridged; column fleshy, stout, 4–5 mm long, wings trilobed. **Distr:** Mexico (Chiapas) to C America. **Ecol:** In montane cloud forests; 1,500–2,600 m. **Zones:** 9–12.

Cyrtochilum

Species 127. **Herbs**, epiphytic or lithophytic; rhizomes creeping; pseudobulbs often clustered, relatively large, round in cross-section, subtended by numerous long foliaceous bracts, apex leafy. **Inflorescences** basal, axillary, simple to branched, laxly many-flowered, elongate, erect to arcuate, sometimes several meters long and scrambling through surrounding vegetation. **Flowers** small to large, usually showy, often somewhat stellate, segments ± spreading, free, often clawed; sepals spreading, subsimilar; lateral sepals often slightly larger than the dorsal; petals free, spreading, usually slightly smaller than the sepals; lip rather fleshy, dependent, smaller than the sepals and petals, entire to obscurely trilobed, with a ± complex, basal ridged callus; column short, erect, with lateral wings near apex; pollinia 2, obovoid, sulcate. **Distr:** Caribbean, Costa Rica to Peru. **Cult:** Propagate by division. Best grown in pots with an epiphyte compost. Requires watering throughout the year.

C. macranthum (Lindl.) Kraenzl., *Notizbl. Bot. Gart. Berlin-Dahlem* 7: 95, 1917. (Syn: *Oncidium macranthum*, etc.). **Herbs**, epiphytic, large; pseudobulbs ovoid to conical, ± compressed, 7–15 cm long, subtended by 3–4 imbricate sheaths, apex 2-leaved. **Leaves** linear-oblong to oblanceolate, 25–55 × 2.5–5 cm, apex acute. **Inflorescences** paniculate, 0.6–3 m long, branches remote, to 30 cm long, each 3–5-flowered; pedicel and ovary 4.5 cm long. **Flowers** large, showy, 7–10 cm across, sepals honey-brown to dull yellow-brown, petals golden-yellow, lip white at base, yellow at apex, sides purple; sepals clawed, suborbicular-spathulate, 3.5–5 × 2.4–3.1 cm, rounded; petals shortly clawed, suborbicular-ovate, 3–4.5 × 2.5–3 cm, margins slightly crisped, apex rounded to obtuse; lip shortly clawed, flat triangular, 2.5–3 × 1.9–2.8 cm, obscurely trilobed, callus of 3 basally confluent lamellae and 3 smaller teeth in front; column stout, 1 cm long, with small, fleshy, flabellate wings, irregularly bilobed above. **Distr:** W South America. **Ecol:** In cloud forests; 2,800–3,000 m. **Zones:** 10–12.

SZ

JT

Above: *Ionopsis utricularioides, Mexicoa ghiesbreghtiana, Miltonia spectabilis.*

Erycina

Syn: *Psygmorchis, Stacyella.*

Species 7. **Herbs**, epiphytic, small to medium-sized; pseudobulbs with apex unifoliate, subtended by several distichous imbricate ± leaf-bearing sheaths. **Leaves** arranged in a fan shape; lamina laterally flattened. **Inflorescence** racemes or panicles, several-flowered, axillary, basal; bracts short, rigid. **Flowers** resupinate, showy, mainly yellow; dorsal sepal, convex below, acuminate and reflexed above; lateral sepals similar, slightly connate below, spreading; petals reflexed, similar to petals; lip much longer than other segments, trilobed, lobes subsimilar, calli 2 on disk, ligulate, parallel, erect; column with 2 ligulate arms on either side of stigma; anther, rostellum and stipe elongate; pollinia 2, ovoid. **Distr:** Mexico to S tropical America, Trinidad. **Cult:** Propagate by division. Best grown mounted on small slabs. Requires intermediate conditions with high humidity and shade.

E. pusilla (L.) N.H. Williams and M.W. Chase, *Lindleyana* 16: 136, 2001. (Syn: *Cymbidium pusillum*, etc.). **Herbs**, epiphytic, dwarf, 3–14 cm high; stems clustered, 3–5 mm long. **Leaves** arranged in a fan; lamina bilaterally compressed, conduplicate, oblong-lanceolate to oblong elliptic, 2.3–7 × 0.4–12. cm, apex acute. **Inflorescences** 1–6-flowered, axillary, erect or spreading, 3–6 cm long; peduncle bilaterally compressed; pedicel and ovary 5–8 mm long. **Flowers** large, 1.5 × 2.5–3 cm, sepals and petals yellow, base red-barred, lip yellow, marked with reddish-brown at base and on the callus; dorsal sepal concave, obovate-elliptic, 5 × 2 mm, obtuse; lateral sepals similar but connate at base, acute, keeled behind; petals spreading, oblong-ovate, 8 × 4 mm, margins slightly erose, crinkled, apex obtuse; lip spreading, trilobed, 1.3–2 × 1.3 × 2.3 cm, lateral lobes spreading, clawed, obovate, rounded, midlobe broadly clawed, 4-lobulate, callus of 3 raised plates; column erect, 2–3 mm long, winged. **Fruits** 1.5–2.5 × 0.8–1.2 cm. **Distr:** Mexico to S tropical America, Trinidad. **Ecol:** In very humid forests, cloud forests; 0–300(–750) m. Fl. Jan–Jul. **Zones:** 9–12.

Ionopsis

Syn: *Cybelion, Iantha, Konantzia.*

Species 6. **Herbs**, epiphytic; stems short, subcylindric, sheathing leaf bases with or without minute pseudobulbs. **Leaves** arranged in 2 ranks, approximate, bases imbricating; lamina narrow, coriaceous. **Inflorescences** 1–3, slender, short to elongate, erect or arching, simple or paniculate, subterminal or lateral, borne from leaf axils. **Flowers** small to medium-sized; sepals subequal, erect or spreading; dorsal sepal free; lateral sepals connate at base, produced into a short sac below the lip; petals subequal to dorsal sepal or broader; lip clawed, adnate to base of column, apical portion broadly expanded, entire or emarginate, conspicuously exceeding sepals; column short, stout, erect, lateral margins not winged, base without a foot; anther terminal, operculate, incumbent, 1-celled or imperfectly 2-celled; pollinia 2, waxy. **Distr:** S Florida, Mexico to tropical America. **Cult:** Propagate by division. Best grown mounted on branches. Requires regular misting and good ventilation.

I. utricularioides (Sw.) Lindl., *Coll. Bot.*: t. 39A, 1826. (Syn: *Cybelion pallidiflorum*, etc.). [Delicate Violet Orchid]. **Herbs**, epiphytic, 5–12 cm tall; sheathing bases sometimes enveloping a minute pseudobulb. **Leaves** approximate; lamina linear-lanceolate to oblong-lanceolate, 3.5–12 × 0.5–1.5 cm, apex acute, midvein forming a keel abaxially. **Inflorescences** 1–3, slender, erect or arching, simple or paniculate, 15–60 cm long, few- to many-flowered, borne from lateral or subterminal leaf axils. **Flowers** small to relatively large, lavender, rose-purple or white; sepals subequal, not spreading, oblong-lanceolate to elliptic-lanceolate, 3–6 × 1–2 mm, acute; dorsal sepal free, concave; lateral sepals connate at base and produced into a short, broad, obtuse sac; petals subequal to dorsal sepal, elliptic-ovate, 2.5–5 × 1.5–3 mm, acute, or obtuse, shortly apiculate; lip about twice as long as sepals, 0.5–1.2 × 0.5–1.2 cm, apex broadly spreading, deeply emarginate, bilobed, base contracted into a fleshy, biauriculate claw, apex of claw with 2 suborbicular, flattened calli; column ~2 mm long, foot absent. **Distr:** S Florida, Mexico to tropical America. **Ecol:** In tropical and subtropical lowlands. **Zones:** 9–12.

Mexicoa

Species 1. *Generic description same as specific.* **Cult:** Propagate by division. Grow in pots with coarse organic matter. Requires cool conditions.

M. ghiesbreghtiana (A. Rich. and Galeotti) Garay, *Bradea* 1: 424, 1974. (Syn: *Odontoglossum warneri*, etc.). **Herbs**, epiphytic or lithophytic, small, often clump-forming; pseudobulbs ovoid, conical or pyriform, 2–5 × 2 cm, grooved, apex bifoliate. **Leaves** linear or linear-lanceolate, 8–14 × 0.7–1.2 cm, apex acute. **Inflorescence** laxly 3–6-flowered, to 20 cm long, erect to arcuate; pedicel and ovary to 1.5 cm long. **Flowers** ~1.5 cm across, sepals and petals off-white or dull yellow, longitudinally striped with purple, lip yellow with an orange-yellow callus; sepals oblong to oblong-lanceolate, 1.5–1.6 × 0.4 cm, acute; petals porrect, ascending, oblong-obovate, 1.2–1.5 × 0.4 cm, subacute; lip deflexed, trilobed, 1.7–2 cm wide, lateral lobes spreading, obliquely elliptic, rounded in front, midlobe flabellate, bilobulate in front, lobules subquadrate, callus hairpin-shaped with a basal cup-shaped depression; column slender, arcuate, 5 mm long. **Distr:** C and SW Mexico. **Ecol:** In montane oak forests; 1,400–3,200 m. Fl. Feb–Jun. **Zones:** 9–12.

Miltonia

Syn: *Anneliesia, Macrochilus*, etc.

Species 11. **Herbs**, epiphytic; pseudobulbs, short, usually inconspicuous, compressed, base enveloped in few to many distichous, imbricate, foliaceous

Below: *Odontoglossum constrictum, Odontoglossum crispum, Odontoglossum cruentum.* [All MG.]

MG

Above: *Odontoglossum gloriosum, Oncidium brunleesianum, Oncidium cheirophorum.*

sheaths, apex 2-leaved. **Leaves** elliptic-lanceolate to narrowly linear, base contracted into short or elongate conduplicate petiole, coriaceous to subcoriaceous. **Inflorescences** 1–few-flowered, or sometimes many-flowered racemes, axillary from the base of the pseudobulb, often short, erect or arching; pedicels long, slender. **Flowers** medium-sized to large, showy, segments all in one plane; sepals spreading, free or shortly connate to lateral sepals, subequal; petals subequal to sepals or slightly wider; lip entire, broadly spreading, often bifid, base sessile or shortly to broadly clawed and affixed to base of column at a right angle, disk inconspicuous or sometimes lamellate; column short, rarely membranous and dilated, lacking a foot, apex biauriculate or bialate, truncate or bilobed; pollinia 2, waxy. **Distr:** Brazil to NE Argentina. **Cult:** Propagate by division. Grow mounted on cork or tree-fern slabs, or in shallow pots. Requires high humidity, warm temperatures and bright light. **Info:** Often hybridizes in the wild, with 6 hybrids described.

M. spectabilis Lindl., *Edwards's Bot. Reg.* 23: t. 1976, 1837. (Syn: *Oncidium spectabile*, etc.). **Herbs**, epiphytic; rhizome stout, creeping, scaly; pseudobulbs spaced, ovate-oblong, compressed, 7.5–10 × 2 cm, apex 2-leaved, ochreous-yellow. **Leaves** linear-ligulate, 10–15 × 1.3 cm, apex rounded, ochreous-yellow. **Inflorescence** 1-flowered, erect, to 25 cm long; peduncle covered by imbricating sheaths; bracts covering the ovary completely, to 7 cm long. **Flowers** 7.5 cm across, sepals and petals white or cream, rose-tinged toward base, lip purplish, with 6–8 darker radiating lines, margins white or pale rose, callus yellow, column rose-purple; sepals oblong-lanceolate, 3.5–4 × 1.6 cm, subacute or acute; petals similar to sepals but broader; lip large, spreading, obovate-orbicular, 4.5–5 × 4–4.6 cm, callus trilamellate, lamellae ending in small erect plates; column wings subtriangular. **Distr:** SE Brazil. **Ecol:** In cloud forests; ~800 m. Fl. May–Oct. **Zones:** 10–12.

Odontoglossum

Syn: *Symphyglossum*, etc.

Species 65. **Herbs**, epiphytic or lithophytic, medium-sized to large; rhizomes short; pseudobulbs ± compressed, covered by distichous, ± leaf-bearing sheaths, apex 1–3-leaved. **Leaves** rigid or flexible, coriaceous or fleshy. **Inflorescence** racemose or paniculate, 1- to many-flowered, basal, erect or arcuate. **Flowers** large and showy to small and inconspicuous; sepals and petals subsimilar, spreading; lateral sepals ± adnate; lip entire to trilobed, base parallel to column or somewhat adnate to it, lateral lobes erect or revolute, midlobe deflexed, spreading; disk lamellate, cristate or callose; column long, slender; pollinia 2, waxy, entire or sulcate, stipe straplike, viscidium small, ± elliptic. **Distr:** C and W South America to Guyana. **Cult:** Propagate by division. Grow in pots with a fine bark-based mix containing perlite and sphagnum mosss. Requires cool conditions and shade.

O. constrictum Lindl., *Edwards's Bot. Reg.* 29(Misc.): 17, 1843. (Syn: *Oncidium constrictum*, etc.). **Herbs**, epiphytic; pseudobulbs ovoid, ancipitous, 5–7.5 cm long, light green, apex 2-leaved. **Leaves** linear-lanceolate, 16–43 × 1.5–2 cm, apex subacute or acute. **Inflorescence** slender, arcuate, racemose or laxly paniculate, many-flowered, to 65 cm long; bracts ovate-triangular, 4 mm long, acute. **Flowers** 3.5–5 cm across, fragrant, sepals and petals yellow or greenish-yellow, blotched with red-brown, lip white with a rose blotch in front of the callus, column white; sepals spreading, linear-lanceolate, 1.8–2.4 × 0.45–0.6 cm, acute to acuminate; petals similar to sepals; lip obscurely trilobed in basal half, subpandurate, apiculate, 2.3–2.5 × 1 cm, midlobe deflexed, margins erose-undulate, callus bidentate, margins erose; column apex bicirrate. **Distr:** Colombia to N Venezuela. **Ecol:** On trees at edge of forests and in lower cloud forests; 1,700–2,400 m. Fl. Feb–Mar, Jun–Nov. **Zones:** 11–12.

O. crispum Lindl., *Ann. Mag. Nat. Hist.* 15: 256, 1845. (Syn: *O. alexandrae*, etc.). **Herbs**, epiphytic; pseudobulbs broadly ovoid, ancipitous, to 7.5 cm long, apex 2-leaved. **Leaves** suberect; lamina linear-lanceolate, 30–42 × 1.8–3.2 cm, apex acute. **Inflorescence** arching or pendulous, racemose or paniculate, many-flowered, to 50+ cm long; bracts small, lanceolate, cucullate, to 1.2 cm long. **Flowers** variable, 6–10 cm across, very showy, mostly white or white tinged rose, rarely pale yellow, ± spotted or blotched red or purple, disk of lip bright yellow; sepals ovate-lanceolate, 3.1–5 × 1.2–2.2 cm, margins undulate-crenulate, apex obtuse to acuminate; petals ovate-elliptic to oblong-elliptic, 3–4.5 × 1.7–3.8 cm, margins erose or dentate, undulate-crenulate, apex acute to acuminate; lip oblong or subpandurate, 2–3 × 1.2–1.5 cm, margins dentate and undulate, apex rounded, emarginate or apiculate, callus fleshy, with 2 diverging apical lobes; column arching slightly, 1.6–1.9 cm long, with 2 broad lacerate apical wings. **Distr:** Colombia. **Ecol:** On mountain slopes; 2,000–3,000 m. **Zones:** 11–12.

O. cruentum Rchb. f., *Xenia Orchid.* 2: 174, 1873. (Syn: *O. hrubyanum*, etc.). **Herbs**, epiphytic; pseudobulbs ovoid-pyriform, compressed, 2.5–7 cm long, apex 2-leaved, subtended by 2–3 leaf-bearing sheaths. **Leaves** suberect; lamina linear-ligulate, to 22 × 1.9–2.2 cm, apex acute. **Inflorescence** erect, slightly arching, racemose or slightly branching, to 45 cm long, loosely up to 28-flowered; bracts lanceolate, 1 cm long, acute. **Flowers** 5 cm across, sepals and petals greenish-yellow, inner surface with large maroon blotches, lip yellow with a central maroon blotch, callus white, column white, wings red; sepals linear-oblanceolate, 2.5–3.2 × 0.5–0.6 cm, acuminate; lateral sepals slightly twisted; petals narrowly elliptic-lanceolate, 1.8–2 × 0.6 cm, apiculate, apically sharply deflexed, margins reflexed, callus on claw fleshy, bifurcate above, with teeth on each side at base; column arcuate, clavate, 1–1.2 cm long, wings narrowly oblong. **Distr:** Ecuador to Peru. **Ecol:** In cloud forests; 2,000–2,500 m. Fl. Mar–Apr, Jul. **Zones:** 9–11.

O. gloriosum Linden and Rchb. f., *Bonplandia* (Hannover) 2: 278, 1854. (Syn: *O. gloriosum* f. *flaviflorum*, etc.). **Herbs**, epiphytic; pseudobulbs tufted, ovate or narrowly ovate, laterally complanate, 7.5–10 cm long, rugose, longitudinally grooved, partially enveloped by several pairs of imbricating (leafless below, leaf-bearing above), sheaths, tapering to a narrow apex, apex 2-leaved, becoming darker with age. **Leaves** broadly lanceolate, base conduplicate to form a petiolate base, apex acute, shorter than the inflorescence. **Inflorescence** a panicle, to 30-flowered, to 50 cm long, stout, much-branched. **Flowers** strongly fragrant, highly variable in shape, size and color, usually ~5–7 cm across, pale orangish-brown or greenish-yellow, ocellated or sometimes minutely spotted with dark chestnut-brown; sepals and petals nearly equal, lanceolate, margin wavy, apex acute; lip ligulate, base heart-shaped, abruptly carinate, margin smooth (not crisp), much attenuated in front; column clavate, hairy near apex with a solitary tooth in the middle. **Distr:** Colombia. **Ecol:** In cloud forests, in brightly lit areas; ~2,000–2,800 m. **Zones:** 11–12.

Oncidium

Species ~800. **Herbs**, epiphytic or terrestrial, ± cespitose; pseudobulbous, green, ovoid. **Leaves** persistent, 1–3, not recurved; lamina conduplicate, linear-lanceolate, dorsiventrally flattened, margins entire. **Inflorescences** paniculate racemes, sometimes over 2 m long; lateral branches to 10(–45) cm long; bracts narrowly triangular. **Flowers** resupinate; sepals and petals distinct and free, oblong-elliptic, obovate to nearly quadrangular or nearly circular, apex retuse to obtuse, sometimes apiculate; lip pandurate, trilobed; callus prominent warty, tuberculate-fringed; column with wings, arms, or fingerlike projections; stigmatic cavity ± round; rostellum not prominent. **Distr:** Tropical and subtropical regions of Western Hemisphere. **Cult:** Propagate by division. Best grown mounted on slabs or on trees. Requires intermediate to warm conditions with high humidity. **Info:** *Oncidium* is a large and probably paraphyletic assemblage of polymorphic species; further molecular studies will reveal its complex infrageneric relationships.

O. brunleesianum Rchb. f., *Otia Bot. Hamburg.*: 87, 1881. (Syn: *Baptistonia echinata*, etc.).

Top row: *Oncidium excavatum, Oncidium flexuosum, Oncidium leucochilum.*
Bottom row: *Oncidium longipes, Oncidium ornithorhynchum, Oncidium silvanum, Oncidium sphacelatum.*

[Hedgehog Oncidium]. **Herbs**, small, epiphytic; pseudobulbs narrowly ellipsoidal or subcylindric, 3–10 × 1–1.5 cm, compressed. **Leaves** 1–2, oblong-lanceolate, narrowly elliptic or obovate, 10–15 × 1.5–2.1 cm, apex acute. **Inflorescences** 1–2, subdensely few- to many-flowered, pendulous; branches 0–5, to 2 cm long, 2–6-flowered; pedicel and ovary 0.7–1.4 cm long. **Flowers** showy, nodding, sepals pale brownish-yellow to greenish-white with a pale brownish margin, petals whitish with transverse maroon bars, lip yellow with a purple midlobe and purple bars between lateral lobes; dorsal sepal hooded over column, obovate-elliptic, 1.2–1.3 × 0.5–0.6 cm, obtuse or subacute; lateral sepals porrect, united almost to apex, spathulate, 0.9–1.2 × 0.5–0.7 cm, bidentate; petals recurved-spreading, oblong-obovate, 0.9–1.2 × 0.5–0.6 cm, obtuse; lip trilobed, 1.3–1.6 × 1–1.5 cm, lateral lobes erect, elliptic, rounded in front, midlobe broadly ovate, mucronate, callus fleshy, tuberculate, center with 2 large white fleshy teeth; column 6–7 mm long. **Distr:** Brazil (Rio de Janeiro). **Ecol:** In montane rainforests; 50–1,200 m. **Zones:** 10–12.

O. cheirophorum Rchb. f., *Bot. Zeitung* (Berlin) 10: 695, 1852. (Syn: *O. cheirophorum* var. *exauriculatum*, etc.). **Herbs**, cespitose, epiphytic, to 15 cm tall; pseudobulbs smooth, green, marked with brown, pyriform to suborbicular, compressed, 1.5–3.5 × 1.5–2.5 cm; bracts 2–3, foliaceous. **Leaf** 1, shortly petiolate; lamina elliptic-lanceolate, 5–15 × 0.5–1.4 cm, base conduplicate, apex acute, with a prominent abaxial keel, thin. **Inflorescence** a panicle, lateral, 10–25 cm long; pedicel and ovary 6–9 mm long. **Flowers** greenish-yellow, 1–1.5 cm across; sepals dissimilar, unguiculate, 5–6 × 4–5 mm; dorsal sepal erect, obovate, strongly concave, obtuse, apiculate; lateral sepals shortly connate, reflexed, subquadrate, retuse; petals shortly unguiculate, reflexed, ovate, obtuse, apiculate, 5–6 × 4–5 mm; lip adnate to base of column, strongly trilobate, with a pair of fleshy basal glands, 1.1–1.2 × 1.2–1.4 cm, lateral lobes subquadrate, midlobe reniform, emarginate, callus complex below isthmus; column 2–3 mm long, with a pair of porrect wings; pollinia 2. **Fruits** ellipsoid, 1.5–1.6 cm long, with beak and persistent flower. **Distr:** Mexico to Colombia. **Ecol:** In montane forests; 600–1,700 m. Fl. Oct–Feb. **Zones:** 9–12.

O. excavatum Lindl., *Sert. Orchid.*: t. 25, 1838. (Syn: *O. aurosum*, etc.). **Herbs**, epiphytic, large; pseudobulbs numerous, aggregated, oblong-ovoid, 7–12(–18) cm long, 1- or 2-leaved, clothed with basal sheaths. **Leaves** linear-oblong or lorate, 30–50 × 2.5–4 cm, base long-attenuate, conduplicate, subpetiolate, apex acute. **Inflorescences** paniculate, with a few spreading-ascending, simple branches, flowers loose to subdense; scape erect or nodding, 0.6–1.5 m long; floral bracts small, narrowly triangular; pedicel and ovary 1.5–3 cm long. **Flowers** variable in size, segments spreading, yellow, barred and spotted with red-brown at lower half, sepals with margins ± undulate, apex abruptly acute; dorsal sepal narrowly obovate or oblanceolate-spathulate, 1.2–1.6 × 0.7–0.9(–1.2) cm, concave; lateral sepals free, oblong-oblanceolate, slightly longer than dorsal; petals oblong-obovate or triangular-obovate, retuse, apiculate; lip pandurate-trilobed, 1.2–2.2(–2.5) × 1 cm, lateral lobes small, midlobe much larger, disk at base with a convex callus of ~4 verrucose lines surrounded by minute tubercles; column ~5 mm high, with a pair of porrect wings. **Distr:** W South America. **Ecol:** In moist montane forests; 2,400–2,800 m. **Zones:** 9–11.

O. flexuosum Lodd., *Bot. Cab.* 5: t. 424, 1820. (Syn: *Ampliglossum flexuosum*, etc.). **Herbs**, epiphytic; pseudobulbs narrowly ovoid or ovate-elliptic, compressed, ancipital, 4–8 × 2–3 cm. **Leaves** (1)2, apical, spreading; lamina narrowly oblong-ligulate, 10–22 × 1.5–3 cm, apex obtuse or abruptly acute, slightly coriaceous. **Inflorescence** a panicle, many-flowered, erect or arcuate, to 1 m long; branches 4–10 cm long, each 10-flowered. **Flowers** small, variable in size, usually 2–3 cm across, sepals and petals yellow, brown-barred, lip bright yellow, marked with reddish-brown around callus, ± finely spotted with orange on lamina, callus yellow; dorsal sepal obovate-oblong, 3.5–4 × 2 mm, subrotund, concave; lateral sepals oblong, 4 × 1.5 mm, obtuse; petals obovate, 5 × 2.5 mm, subrotund; lip trilobed at base, 1.2–1.3 × 1.2–1.5 cm, lateral lobes auriculate, midlobe shortly clawed, broadly reniform above, deeply emarginate, callus fleshy, cristate, with 3 lacerate lobes in front; column short, fleshy, slightly furfuraceous, winged. **Distr:** E and S Brazil to NC Argentina. **Ecol:** In coastal mountains and along inland creeks and marshes; 500–800 m. Fl. Jun–Jul. **Zones:** 8–12.

O. leucochilum Bateman ex Lindl., *Edwards's Bot. Reg.* 23: t. 1920, 1837. (Syn: *Cyrtochilum leucochilum*, etc.). **Herbs**, epiphytic, to 3+ m tall; pseudobulbs ovoid to ovoid-ellipsoid, compressed, 5–13 × 3–6 cm. **Leaves** 1–2, apical or on rhizome; lamina ligulate, 10–60 × 1.5–4.5 cm, base conduplicate, apex obtuse to subacuminate, coriaceous. **Inflorescence** a panicle, many-flowered, rarely racemose, basal, to 3 m long; pedicel and ovary slender, 1.5–3 cm long. **Flowers** usually reddish-brown and white; sepals and petals similar, bright green or greenish-white, blotched with reddish-brown, oblong-elliptic to elliptic-oblanceolate, 1.32.3 × 0.5–1.2 cm, obtuse to shortly acuminate; lip white, sometimes tinged with pink or yellow, panduriform, deeply trilobed, 1.6–2.5 cm. long, lateral lobes semiovate to oblong or subquadrate, 4–8 × 2–3 mm, midlobe 1.4–2.3 cm wide, callus oblong, ~6 mm long, purple-tinged, with 5–9 teeth or filaments; column short, fleshy, 7–8 mm long, with pinkish or purplish apical wing on each side. **Distr:** SE Mexico to Honduras. **Ecol:** On trees in dry or humid forests; to 2,000 m. **Zones:** 10–12.

O. longipes Lindl., *Paxton's Fl. Gard.* 1: 46, 1850. (Syn: *Alatiglossum longipes*, etc.). **Herbs**, epiphytic; rhizome creeping; pseudobulbs clustered, oblong, tapering to apex, sheathed with brownish scales. **Leaves** 2, linear-oblong, base tapered, apex apiculate, fleshy, bright green. **Inflorescence** a raceme, several-flowered, 7.5–10 cm long, axillary; pedicels long. **Flowers** with sepals and petals spreading, dark reddish-brown inside, brownish-green outside; dorsal sepal spathulate, margins wavy, reflexed; lateral sepals narrower, united at base, deflexed; lip relatively large, bright, almost golden-yellow, with a broad reddish ring at base surrounding crest, trilobed, lateral lobes small, rounded, midlobe large, bilobed, margin fimbriated between lobes; crest slightly downy, oblong, fleshy disk, margin lobed, whitish, spotted; column short, with 2 winglike lobes beneath anther. **Distr:** SE and S Brazil to NE Argentina. **Ecol:** In primary forests. **Zones:** 8–12.

O. ornithorhynchum Kunth, *Nov. Gen. Sp.* 1: 345, 1816. [Bird's Beak Oncidium]. **Herbs**, epiphytic; pseudobulbs ellipsoid-conical, compressed, 6 × 3 cm, apex-2-leaved. **Leaves** somewhat arched, oblong-oblanceolate, to 24 × 3.5 cm, apex acute. **Inflorescences** pendulous or arcuate, paniculate, to 30 cm long. **Flowers** 2.2–2.5 cm across, sepals, petals and lip purple, pink or white, callus golden-yellow; dorsal sepal oblong-lanceolate, 8 × 2 mm, margin undulate, apex acute; lateral sepals joined at base, spreading at right angles to lip, falcate, 8 × 2.5 mm, subacute; petals oblong-lanceolate, 7 × 2.5mm, base shortly clawed, margin undulate, apex subacute; lip shortly clawed, obscurely trilobed to hastate, 1 × 0.8 cm, lateral lobes obscurely recurved, on upper and lower margins, midlobe bilobulate at apex, upper margins recurved; callus cristate, irregular; column erect, wings spreading, oblique. **Distr:** Mexico to Colombia. **Ecol:** In humid forests; <1,500 m. **Zones:** 9–12.

O. silvanum V.P. Castro and Campacci, *Orquidário* 5(3): 22, 1991. (Syn: *Baptistonia silvana*). **Herbs**, epiphytic, sympodial; pseudobulbs aggregated, cylindrical to club-shaped, 4.5–7 × 1–2 cm, tapered to apex. **Leaves** 1(–2), apical; lamina elliptic, 10–15 × 2–3 cm, base tapered to petiolelike stem, apex pointed, channeled along midvein. **Inflorescence** a spike, 6–20-flowered, 15–20 cm long, ascending at base, apically arching with the spike held pendent. **Flowers** showy; dorsal sepal golden-yellow at base, brownish-scarlet toward apex, egg-shaped, concave, forming a hood above column, 6 × 1 mm; lateral sepals golden-yellow, scarlet-striped, joined along inside margins, forming an elliptic, concave lamina; petals colored similar to sepals, club-shaped, 1.3 × 1 cm, bases narrow; lip trilobed, 1.3 × 1 cm, lateral lobes trapezoidal, 2.5 × 2.5 mm, blunt, midlobe reniform, 0.5 × 1 cm; column with 2 earlike appendages, ~6 mm long. **Distr:** Brazil (Bahia). **Ecol:** In montane rainforests; 200–500 m. Fl. Apr–May, Oct–Nov. **Zones:** 11–12.

O. sphacelatum Lindl., *Sert. Orchid.*: t. 48, 1841. (Syn: *O. sphacelatum* var. *majus*, etc.). [Popcorn Orchid]. **Herbs**, epiphytic, occasionally lithophytic, very large; pseudobulbs oblong, compressed, 10–15 × 2.5–3.5 cm, edges acute, apex 2–3-leaved. **Leaves** suberect; lamina linear-ligulate, 37–60 × 2–3.5 cm, apex acute, rigid. **Inflorescence** paniculate, many-flowered, 1–1.8 m long, basal, lateral, pendulous, much-branched. **Flowers** brightly colored, 2.5–3 cm across, mildly fragrant, sepals and petals with apices recurved in bud, dark chestnut-brown, yellow-barred, lip golden-yellow with a red-brown band before the callus, callus white, spotted yellow-orange, column white, margins crimson-spotted; sepals narrowly oblong, 1.3 × 0.5 cm, margin undulate, apex acute and reflexed; lateral sepals reflexed, similar to but larger than dorsal sepal; petals narrowly oblong-ovate, 1.5 × 0.6 cm, margin undulate, apex subacute or obtuse, reflexed; lip subpanduriform or obscurely trilobed, 1.6 × 1.7 cm, lateral lobes obscure, subquadrate, midlobe transversely oblong, margin undulate, apex emarginate or shortly apiculate, callus a fleshy plate, trilobed; column wings small, oblong, undulate. **Distr:** Mexico to C. America, SE. Venezuela. **Ecol:** <1,000 m. **Zones:** 9–12.

O. waluewa Rolfe, *Kew Hand-List Orch.*, ed. 2: 167, 1896. (Syn: *Baptistonia pulchella*, etc.). **Herbs**, epiphytic, sympodial, 10–12 cm high; pseudobulb narrowly oblong-cylindrical, laterally compressed, longitudinally grooved, unifoliate. **Leaf** erect or erect-spreading, apical; lamina lanceolate, 10 cm long, somewhat leathery, base folded longitudinally along midvein forming a petiolelike stem, apex sharply pointed. **Inflorescence** erect to nodding, flowers closely spaced; bracts greenish white. **Flowers** 1.5 cm across, strongly concave or cup-shaped, segments not spreading, white to greenish-yellow, petals transversely barred with pinkish-purple, lip lined and barred with pinkish-purple, callus purple; dorsal sepal concave, arching forward, hoodlike over column, oblong to spathulate, 0.8–1 × 0.35 cm, blunt; lateral sepals 7 mm long, joined along their inside margins forming a synsepal ~5 mm across, hidden behind lip, with 3 apical teeth; petals somewhat spreading, reflexed, 7 × 5 mm; lip 7 × 5 mm, joined to column at base, deeply trilobed, lateral lobes broadly triangular, tips rounded, midlobe reniform, with a linear callus containing minute conical projections; column short, wings narrowly linear. **Distr:** SE Brazil to Paraguay. **Ecol:** In montane rainforests; 1,000–2,000 m. **Zones:** 9–12.

Rhynchostele

Syn: *Cymbiglossum*, *Lemboglossum*.
Species 16. **Herbs**, epiphytic or terrestrial, cespitose to rhizomatous; pseudobulbs ovoid, usually strongly laterally compressed, subtended by 1–4 foliaceous bracts, apex 1–2-foliate. **Leaves** articulate; lamina conduplicate, elliptic to lanceolate, chartaceous to coriaceous. **Inflorescence** a scapose raceme, borne laterally from base of pseudobulb, 1–20-flowered. **Flowers** usually spreading, large, colorful; sepals lanceolate, usually attenuate; petals often shorter and broader than sepals, ovate to lanceolate, attenuate; lip free, claw with bilamellate callus; column slender, straight to slightly arcuate, broadened near stigma; pollinia 2, supported on a viscidium and stipe. **Distr:** Mexico to NW Venezuela. **Cult:** Propagate by division. Best grown in pots with a fine bark mix containing sphagnum moss. Requires humid, intermediate conditions with ample watering throughout the year.

R. cordata (Lindl.) Soto Arenas and Salazar, *Orquídea (Mexico City)*, n.s., 13: 148, 1993. (Syn: *Odontoglossum cordatum*, etc.). **Herbs** epiphytic; pseudobulbs borne on ascending rhizome, ovoid or subquadrate-oblong, 4–9 × 2–4 cm, with several imbricate sheaths, 2–4 are foliaceous, apex unifoliate. **Leaf** elliptic-lanceolate or elliptic-oblong, 8–30 × 1.6–5 cm, base conduplicate, apex apiculate. **Inflorescence** basal, erect, unbranched, 25–50 cm long, laxly 5–12-flowered; peduncle slightly flattened; bracts lanceolate, 2.5–3 cm long, acuminate; pedicel and ovary 3–4.5 cm long. **Flowers** showy, 4–7 cm across, sepals and petals yellow-green, with pale or reddish-brown spots, lip whitish with pale or reddish-brown spots; sepals elliptic-lanceolate to narrowly lanceolate, 2.4–5 × 0.4–1 cm, acuminate; petals ovate-lanceolate or elliptic-lanceolate, oblique, 2.5–4 × 0.7–1.1 cm, long-acuminate; lip shortly clawed, cordate-deltoid, 1.8–2.7 × 1.1–1.9 cm, margins entire, apex abruptly acuminate, callus united to claw, 7–8 mm long, fleshy, sides upcurved, apices bidentate; column 1.1–1.5 cm long, shortly hairy, yellow-green, winged. **Distr:** Mexico to NW Venezuela. **Ecol:** In humid forests; ~1,000–2,100 m. **Zones:** 9–12.

Below: *Oncidium waluewa, Rhynchostele cordata, Rhynchostele maculata.* [All MG.]

R. maculata (Lex.) Soto Arenas and Salazar, *Orquídea (Mexico City)*, n.s., 13: 150, 1993. (Syn: *Brassia oestlundiana*, etc.). [Spotted Rhynchostele]. **Herbs**, epiphytic; pseudobulbs oblong, compressed, 6–12 × 2.5–6 cm, with 1–2 foliaceous sheaths, apex unifoliate. **Leaf** lanceolate, 15–30 × 3–6 cm, apiculate. **Inflorescence** erect, basal, 2–7-flowered, 20–35 cm long; bracts lanceolate, 3–5 cm long, acuminate; pedicel and ovary 3.5–7.5 cm long. **Flowers** 4–7 cm across, sepals and petals yellow with blotches of pale to dark brown, lip yellow with brown or reddish-brown blotches, callus yellow, red-streaked; dorsal sepal linear-lanceolate to elliptic-lanceolate, 2.5–4 × 0.6–1.1 cm, acute; lateral sepals similar but 3–4.5 × 0.6–0.8 cm; petals oblong, elliptic or elliptic-obovate, 2.5–3 × 1.5–3 cm, acuminate; lip shortly clawed, triangular-cordate, 2.5–3 × 1.5–3 cm, acuminate; lip shortly clawed, triangular-cordate, 1.5–2.5 × 2–2.5 cm, margins, erose-denticulate, apex acuminate, callus on claw of lip, fleshy, boat-shaped, apex bilobed; column slender, clavate 1–1.5 cm long. **Distr:** C and SW Mexico to C America. **Ecol:** In humid and cloud forests; 2,000–2,700 m. **Zones:** 9–12.

Rodriguezia

Species 48. **Herbs**, epiphytic, small, sympodial; rhizomes elongated, scandent; roots long and fine; pseudobulbs set close together or well-spaced, small, short, ovoid, compressed, distant or approximate, usually covered (and hidden) by the leaf-bearing sheaths, apex 1–2-foliate. **Leaves** coriaceous. **Inflorescences** lateral, 1 to several, few- to many-flowered, racemose, unbranched, often pendent. **Flowers** medium-sized to large, ± showy; dorsal sepal and petals similar and forming a hood; lateral sepals spreading or variously connate, ± gibbose; lip clawed, ± shortly connate to the column base, produced into a short spur at base, lamina obovate; disk ± cristate; column slender, lacking a foot, apically dilated and winged; pollinia 2, waxy, attached by a narrow stipe to a small viscidium. **Distr:** S Mexico to S tropical America and Windward Islands. **Cult:** Propagate by division. Best grown mounted on cork or tree-fern slabs. Requires humidity and good ventilation.

R. decora (Lem.) Rchb. f., *Bot. Zeitung* (Berlin) 10: 771, 1852. (Syn: *Burlingtonia decora*, etc.). **Herbs**, epiphytic, small; rhizome slender, scandent, jointed; pseudobulbs ~12+ cm apart, ovoid, compressed, 2.5 cm long, 1-leafed at apex and 1 at base. **Leaves** linear-oblong, ~15 cm long, apex acute. **Inflorescences** racemose, 5–15-flowered, borne from the axils of basal leaves, erect, 30–40 cm long; bracts small, membranous, sheathing. **Flowers** 3.5 cm long, sepals and petals white spotted with purplish brown, lip white, column with purple apical horns; dorsal sepal elliptic-oblong, 1.2 × 0.4 cm; lateral sepals narrowly oblong, connate to above middle, 1.5 cm long, apiculate; petals elliptic-oblong, 1.4 × 0.6 cm, slightly concave, obliquely shortly acuminate; lip broadly clawed, suborbicular, 2.5–2.8 × 1.1–1.4 cm, bilobulate above, callus with 5 raised lines; spur 2 mm long, obtuse; column terete, prolonged at apex into 2 erect hairy horns, and 2 smaller paler ones. **Distr:** Brazil. **Ecol:** Epiphytic on trees in cool montane coastal rainforests, and savannahs; 600–800 m. Fl. throughout most of the year. **Zones:** 10–12.

R. venusta (Lindl.) Rchb. f., *Bot. Zeitung* (Berlin) 10: 771, 1852. (Syn: *Burlingtonia fragrans*, etc.). **Herbs**, epiphytic; pseudobulb, ovoid to ovoid-oblong, 2–3 × 1–1.5 cm, laterally compressed, longitudinally ridged, apex unifoliate. **Leaves** 2–3 (2 leaves usually basal and sheathing pseudobulb), distichous, twisted or recurved; lamina linear-lanceolate, channeled in middle, 7–15 × 1–2.5 cm, base conduplicate, long-attenuate, apex acute, coriaceous. **Inflorescence** loosely few-flowered, arching, to 7–18 cm long, axillary; pedicel and ovary linear-clavate, trigonal, 1–1.5 cm long. **Flowers** ~4 cm long, very fragrant, white with a rich golden-yellow disk on lip; sepals and petals narrowly elliptic, acute, membranous; dorsal sepal shortly obovate, 1.3–1.5 × 0.7–0.9 cm, acute, 7-nerved; lateral sepals connate, concave, 1.3–1.5 × 0.6–0.8 cm, acute; petals obovate-oblong, 1.5–1.7 × 0.6–0.7 cm, acute, 7-nerved; lip broadly obovate-obcordate or triangular-obcordate, 1.8–2.2 × 1.1–1.4 cm, base clawed, margins wavy, apex deeply emarginate, lobes ± rounded, membranous, glabrous, disk bicarinate, callus ~2 mm long; column erect, 8–9 mm long, apex bidentate; ovary short. **Distr:** Brazil to Ecuador. **Ecol:** In rainforests; 500–1,800 m. **Zones:** 9–12.

Tolumnia

Syn: *Antillanorchis*, *Jamaiciella*. [Equitant Oncidiums]. **Species** 23. **Herbs**, epiphytic or lithophytic, sometimes stoloniferous; pseudobulbs absent. **Leaves** distichous, forming a fan; lamina bilaterally flattened, triangular in cross-section (3-sided) or terete, fleshy. **Inflorescences** usually much longer than leaves, branched or unbranched. **Flowers** large, showy; lateral sepals partly joined, lying behind the lip; petals free, larger than sepals; lip much larger than other segments, trilobed, midlobe often bilobed, with a complicated fleshy callus at base; column with prominent wings; pollinia 2. **Distr:** Mexico to N South America, Florida to Caribbean. **Cult:** Propagate by division. Requires an open bark mixture, with high humidity and medium lighting.

T. gauntlettii (Withner and H.P. Jesup) Nir, *Lindleyana* 9: 149, 1994. (Syn: *Oncidium gauntlettii*, *Jamaiciella gauntlettii*, *Olgasis gauntlettii*). **Herbs**, epiphytic, sympodial, 4 cm high; rhizomes with several clusters of fanlike growths. **Leaves** 3–4 per growth, arranged in 2 rows; lamina linear, 4 × 0.5 cm, strongly keeled, adaxially shallowly grooved. **Inflorescence** 1–2-flowered, 8.5 cm long; peduncle basal, axillary; pedicel and ovary ~8 mm long. **Flowers** ~1.8 × 1.5 cm, dorsal sepal pale rose, lateral sepals cream, brown-streaked, lip yellow at center, lateral lobes pale lilac, midlobe apically rose-lilac, column greenish-yellow, wings lilac; dorsal sepal linear-lanceolate, 7 × 2, acute, curved forward over the column; lateral sepals linear, connate for most of their length; petals relatively wide spreading, somewhat recurved toward their tips, lanceolate, 8 × 3 mm, ± rounded; lip trilobed, ~1.2 × 1.1–1.2 cm,

Top row: *Rodriguezia decora*, *Rodriguezia venusta*, *Tolumnia gauntlettii*.
Bottom row: *Tolumnia guttata*, *Tolumnia pulchella*, *Tolumnia triquetra*.
[All MG except *Rodriguezia decora* AF.]

Top row: *Trichocentrum carthagenense, Trichocentrum lacerum, Trichocentrum lindenii.*
Bottom row: *Trichocentrum luridum, Trichocentrum pfavii, Trichocentrum pulchrum, Trichocentrum stacyi.*
[All MG except *Trichocentrum stacyi* AF.]

lateral lobe rounded, margins finely serrated, midlobe rounded, margins undulate, apical margin slightly notched; column ~3 mm long, winged. **Distr:** Jamaica. **Ecol:** In wet forests or near rivers in deep, shaded ravines; ~500 m. Fl. Sep, Dec. **Zones:** 11–12.

T. guttata (L.) Nir, *Lindleyana* 9: 149, 1994. (Syn: *Cymbidium guttatum*, etc.). **Herbs**, epiphytic, sympodial, tufted. **Leaves** 3–5, in 2 rows, bases overlapping and in 1 plane (fan-shaped); lamina linear, 7–15 × 0.6 cm, acutely 3-edged, with a channel on the inner side, edges sometimes minutely serrated. **Inflorescence** 30–60 cm long, axillary, many-flowered, flowers opening successively; peduncle erect; pedicel and ovary ~1.2 cm long. **Flowers** ~2.5 cm long, colors variable, sepals and petals brownish with purple or pink markings, yellow with bright brownish bars, or white with green and crimson markings, lip may be white tinted with pink or white with a red-brown blotch in front of the complex crest; lateral sepals and petals erect, widespread, oblong, margins wavy, keeled on the back; lateral sepals partly joined, synsepalum held behind the lip; lip widespread, trilobed, 1.2 cm across, lateral lobes oblong, midlobe reniform, margins scalloped or ruffled, with a yellow crest consisting of 7 projections; column white or greenish, wings white or tinged with pale pink, yellow-dotted. **Distr:** Mexico to Colombia and Caribbean. **Ecol:** In dry places; 0–300 m. Fl. throughout most of the year. **Zones:** 9–12.

T. pulchella (Hook.) Raf., *Fl. tellur.* 2: 101, 1837. (Syn: *Oncidium jamaicense*, etc.). **Herbs**, epiphytic, medium-sized; stems clustered, short, leafy. **Leaves** distichous, equitant; lamina lanceolate, falcate, 6–12 × 0.6–1.1 cm, apex acute, fleshy. **Inflorescences** densely many-flowered, to 50 cm long, branching; peduncle terete, to 40 cm long; pedicel and ovary 1.2–1.9 cm long. **Flowers** very showy, flat, sepals and petals off-white to purple, sometimes spotted, lip pale rose-pink to rich purple, callus yellow and white, finely spotted with red; dorsal sepal elliptic-oblanceolate, 6–8 × 2–2.5 mm, acute; lateral sepals fused almost to apex, 0.8–1.1 cm × 2.5–3 mm, tip bifid; petals spreading, obovate, 6–7.5 × 5–7 mm, obtuse; lip trilobed at base, dependent, 1.4–2 × 1.4–2 cm, callus of 3 fleshy ridges in the basal part of the lip; column 3–4 mm long, with obliquely ovate, falcate erect wings. **Distr:** Jamaica. **Ecol:** On slender tree trunks in forests; 600–1,000 m. **Zones:** 11–12.

T. triquetra (Sw.) Nir, *Lindleyana* 9: 149, 1994. (Syn: *Epidendrum triquetrum*, etc.). **Herbs**, epiphytic, sympodial; pseudobulbs very small, hidden by overlapping leaf bases. **Leaves** 3–4, in 2 rows; lamina linear, 7–12 × 1–1.2 cm, apex tapered, adaxially flat, green, reddish-brown in strong light. **Inflorescence** a spike, 5–14-flowered, to 27 cm long; peduncle jointed, basal, axillary, horizontal, producing accessory branches over time; pedicel and ovary ~1.5 cm long. **Flowers** long-lasting, slightly waxy, ~1.2(–2.3+) cm across, sepals cream with a reddish-brown or maroon blotch on inner surfaces, petals cream with maroon or reddish-brown streaks along the median line, lip white to cream, with variable maroon or reddish-brown streaks and spots and a small orange-yellow crest at center, column yellow; dorsal sepal ovate to lanceolate, 1–1.3 × 0.6 cm, acute, inclined over column; lateral sepals 0.8–1 cm long, joined for most of their length, synsepalum apically bilobed; petals widely spreading, 1–1.3 × 0.5–0.7 cm, ovate to lanceolate, bases slightly clawed, apices sharply pointed; lip heart- or shield-shaped, trilobed, lateral lobes rounded, earlike, midlobe with margin irregularly scalloped; column ~4 mm long, wings 2. **Distr:** Jamaica. **Ecol:** In pluvial areas; 150–300 m. Fl. throughout most of the year. **Zones:** 11–12.

Trichocentrum

Syn: *Lophiaris, Stilifolium*, etc.
Species 70. **Herbs**, epiphytic; stems very short, at length thickened into a small, fleshy pseudobulb; pseudobulbs 1-leaved, cespitose or separated. **Leaf** flat or equitant, oblong or obovate-oblong to elliptic (rarely linear), coriaceous. **Inflorescences** lateral, basal, 1–few-flowered. **Flowers** small to large, with spreading segments; sepals free, subequal; lateral sepals oblique; petals similar to sepals, especially to the dorsal sepal; lip adnate to the column, commonly more conspicuous than the rest of the flower, simple to trilobed or pandurate, acute to retuse or bilobed, produced at the base into a short or very long spur; column abbreviated, stout, footless, commonly with a pair of erect-spreading wings; anther terminal, opercular, incumbent; pollinia 2. **Distr:** S Florida, Mexico to tropical America. **Cult:** Propagate by division. Best grown mounted on small cork slabs. Requires intermediate to warm conditions with filtered light and regular watering.

T. carthagenense (Jacq.) M.W. Chase and N.H. Williams, *Lindleyana* 16: 137, 2001. (Syn: *Epidendrum carthagenense*, etc.). [Coot Bay

Above: *Trichopilia suavis, Bifrenaria harrisoniae, Bifreneria tyrianthina.*

Dancing-lady Orchid, Spreadeagle Lophiaris]. **Herbs**, epiphytic, usually large, to 2+ m long; pseudobulbs very small, <2.5 cm long, unifoliate, concealed by scarious sheaths. **Leaf** elliptic to oblong-elliptic or lanceolate, 9–40 × 3–7 cm, apex subobtuse to acute or subacuminate, thick. **Inflorescence** a panicle, many-flowered, spreading, much-branched (or rarely simple), basal, to 2+ m long; pedicel and ovary filiform, 1.5–2 cm long. **Flowers** small, showy, yellowish-white, blotched with lavender, magenta, or reddish-brown; sepals clawed, 0.8–1.3 cm long, margin undulate-crisped; dorsal sepal suborbicular, 5–8 mm. wide; lateral sepals broadly spathulate, 4–6 mm wide; petals broadly clawed, broadly obovate, 0.7–1.2 × 0.6–0.8 cm, margins undulate, apex broadly rounded; lip panduriform, trilobed, 0.9–1.6 × 0.7–1.4 cm, lateral lobes basal, semiorbicular, broadly rounded, margins strongly revolute, midlobe reniform to broadly flabellate, broadly rounded, ± retuse, callus near base of disk; column fleshy, 3–4 mm long, winged. **Fruits** ellipsoid, to 8 cm long. **Distr:** Caribbean, Mexico to N Brazil. **Ecol:** In low humid forests; 0–1,050 m. **Zones:** 10–12.

T. lacerum (Lindl.) ined. (Syn: *Cohniella stipitata*, etc.). **Herbs**, epiphytic, sympodial, clump-forming; pseudobulb ± obsolete, cylindrical or conical, 0.5–1 × 0.5–1 cm, apex truncate, enveloped in bracts, apex unifoliate. **Leaf** horizontal when young, pendent with age; lamina terete, 24–70 cm long, 0.6–1 cm Ø at base, longitudinally grooved, smooth, often spotted with reddish-brown. **Inflorescence** a panicle, with closely spaced branches at apex of scape, each branch 3–10-flowered; scape basal, 24–70 cm long, ascending. **Flowers** extremely variable in size, to 2.5 × 1.8 cm, sepals and petals yellow, richly marked with reddish-brown, lip bright yellow, basal half of isthmus reddish-brown; sepals free and spreading, base clawed; dorsal sepal ovate to spathulate, slightly concave, 4–6 × 2–4 mm; lateral sepals 5–7 × 3–4 mm; petals spreading, elliptic-oblong, 5–8 × 2–3 mm, margins undulate, apices blunt; lip fiddle-shaped, trilobed, 1–2 × 1–1.7 cm, lateral lobes narrow, linear-oblong, base contracted, apex blunt, midlobe semicircular, margins undulate, appearing bilobed; column 2.5 mm long, winged. **Distr:** C America to Colombia. **Ecol:** In deciduous forests; to 100(–600) m. Fl. Jan–Mar, Jul–Oct. **Zones:** 9–12.

T. lindenii (Brongn.)105 M.W. Chase and N.H. Williams, *Lindleyana* 16: 137, 2001. (Syn: *Oncidium lindenii*, etc.). **Herbs**, epiphytic, 20–25 cm high; pseudobulbs very small, enclosed by basal sheaths, apex 1(2)-foliate. **Leaf** linear, 20–25 × 5 cm, coarse, leathery, abaxially strongly keeled. **Inflorescence** many-flowered, to 2 m long, flowers opening in succession with inflorescence lengthening as new buds develop, few opening at a time; peduncle basal, branches very short. **Flowers** small, ~ 2 cm Ø, sepals and petals tan or pale yellow, with ± dense chocolate-brown spotting and mottling, lip dark red or purple-violet, waxy, callus bright yellow, terminating in several dark red knobs, column bright yellow, tip white, wings yellow, red-tipped; lip fiddle-shaped, ~7 mm across midlobe, decorated with a large, fleshy callus; column with wings projecting. **Distr:** Mexico to Guatemala. **Ecol:** In riverine and pluvial forests; ~300 m. Fl. Mar–Oct. **Zones:** 9–12.

T. luridum (Lindl.) M.W. Chase and N.H. Williams, *Lindleyana* 16: 137, 2001. (Syn: *Lophiaris cosymbephora*, etc.). [Dingy-flowered Trichocentrum]. **Herbs**, terrestrial, large, variable; rhizome very short; pseudobulbs very small, ovoid rarely to 3.5 cm high, unifoliate, concealed in sheaths. **Leaves** variable, oval to elliptic-oblong or oblong, 15–77 × 2.2–13 cm, base ± narrowed, apex acute, very thick, fleshy. **Inflorescence** erect or nodding, loosely paniculate, many-flowered above, 0.3–1.8 m long; peduncle stout, ~9 cm long, sheathed; rachis ± fractiflex; pedicel and ovary to 3.5 cm long. **Flowers** with segments spreading, margins undulate-crisped, usually dull yellowish or greenish, thickly brown-spotted; dorsal sepal spathulate, 1.2–1.9 × 0.75–1.2 cm, base unguiculately narrowed, apex rounded; lateral sepals free, ± similar to dorsal; petals oblong obovate, broadly clawed, shorter than dorsal sepal; lip slightly longer, much broader than other segments, suborbicular-obovate, pandurate-trilobed, base sessile, subcordate, apex retuse to bilobed, 1.4–2.1 cm long, basal portion ovate-subquadrate, midlobe cordate-reniform; column 4–5 mm high, winged. **Distr:** Mexico to N South America. **Ecol:** Epiphytic on trees in swampy woods and open forests, sometimes lithophytic on rocks; 0–1,400 m. Fl. Jul–Aug, Nov–Mar. **Zones:** 9–12.

T. pfavii Rchb. f., *Gard. Chron.*, n.s., 16: 70, 1881. **Herbs**, epiphytic, cespitose, pendent; pseudobulbs minute, rounded, ~5 mm long, covered by a brown sheath, unifoliate. **Leaf** sessile, oblong-elliptic to oblong-ovate, 7.2–12.5 × 1.4–3.8 cm, base conduplicate, apex obtuse to retuse, light green, fleshy. **Inflorescences** 2–4, pendent to suberect, shorter than leaves, frequently branched at base, mostly 2-flowered; peduncle terete, to 5.1 cm long, concealed by 2–3 spreading, ovate, concave bracts; pedicel and ovary linear-clavate, ~2 cm long. **Flowers** large, showy, tepals free, spreading, white with large median brown-rose blotch; sepals dissimilar, somewhat concave; dorsal elliptic-ovate, to 1.5 × 0.8 cm, rounded to emarginate; lateral sepals obliquely elliptic-lanceolate, 1.5 × 0.6 cm, obtuse, minutely apiculate; petals spathulate, to 1.4 × 0.6 cm, obtuse; lip white, marked basally by 1–2 rose-purple blotches, adnate to column, cuneate-flabellate, 2.2 × 1.6 cm, with cuneate claw 5 mm long, base with 2 lateral lobes, expanding to suborbicular lamina, producing with column base a saccate spur 3 mm long; column ~5 mm long, with rounded wings, margins brown-spotted; anther white; pollinia 2. **Distr:** C America. **Ecol:** Epiphytic in tropical rainforests; 600–1,300 m. Fl. Sep–Mar. **Zones:** 9–12.

T. pulchrum Poepp. and Endl., Poepp. and Endl., *Nov. Gen. Sp. Pl.* 2: 11, 1836. (Syn: *Oncidium undulatum, T. maculatum*, etc.). **Herbs**, small, epiphytic; rhizome very short, clothed by imbricating, sheaths; pseudobulbs minute, 1-leaved. **Leaf** oblong or oblong-elliptic, 9–10 × 2–2.5 cm, base ± long-cuneate, apex subacute, sometimes minutely mucronate, fleshy-coriaceous. **Inflorescences** lateral, basal, 1-flowered; peduncle 0.5–1 cm long, concealed by 2 or 3 imbricating sheaths; pedicel and ovary, linear-clavate, ~5 cm long. **Flower** large, white, variegated with blood-red; sepals spreading, membranous, concave; dorsal sepal ovate-elliptic or oval, ~ 2 × 1.2–1.3 cm, subacute, dorsally keeled; lateral sepals similar, slightly narrower, suboblique; petals similar to sepals, obtuse; lip distinctly longer than sepals, obovate, sessile, minutely emarginate in front, broadly cuneate below, anterior margins crenulate, sides reflexed, ~ 3 × 1.7–1.8 cm, produced at base into a linear-attenuate spur 4.5–5 cm long; disk thickly bicarinate; column stout, 3–4 mm long, with a pair of erect wings. **Distr:** S tropical America. **Ecol:** Epiphytic in wet montane forests; 1,000–1,600 m. Fl. Jun, Aug–Sep. **Zones:** 9–12.

T. stacyi (Garay) M.W. Chase and N.H. Williams, *Lindleyana* 16: 138, 2001. (Syn: *Cohniella stacyi*, etc.). **Herbs**, epiphytic, large, sympodial, pendent; pseudobulbs ovoid to cylindrical, 2.5 × 1.5 cm, concealed by several overlapping bracts, apex unifoliate. **Leaf** terete, 80 cm long, 7 mm Ø, stiffly leathery, slightly curved, apex sharply pointed, channeled along one side. **Inflorescence** approximately as long as leaves, spike slightly zigzag with a raceme at apex, 14–20-flowered, flowers crowded. **Flowers** widely spreading, 8 cm across, sepals and petals golden-yellow, heavily marked with irregular dark brown spots and blotches; dorsal sepal lanceolate-elliptic, 3.5 × 2 cm; lateral sepals obliquely spreading, lanceolate, 3.3 × 1.2 cm, base clawed; petals spoon- or club-shaped, 4 × 2 cm; lip intense yellow with reddish-brown spots and blotches, callus pale yellow, edges reddish, trilobed, 3 × 1.5 cm, lateral lobes bipartite; column intense yellow with reddish vertical markings, stout, with a pair of triangular wings at hooded apex. **Distr:** Bolivia, Peru. **Ecol:** In tropical rainforests; 450 m. **Zones:** 9–12.

Trichopilia

Syn: *Leucohyle, Pilumna.*

Species 38. **Herbs**, epiphytic, small to medium-sized; rhizome abbreviated, creeping; pseudobulbs usually conspicuous and cespitose, ± ancipitous, 1-leaved.

MG SZ

Above: *Cochleanthes discolor, Galeottia acuminata, Huntleya burtii.*

Leaf linear and subterete or more commonly rather broad and flat, fleshy or coriaceous. **Inflorescence** loosely 1(rarely to 7)-flowered above, lateral, basal, usually short, few-sheathed below. **Flowers** showy and often large, with spreading, often twisted, sepals and petals; sepals subequal, narrow, usually free, lateral ones rarely connate; petals very similar to dorsal sepal; lip varying from sessile to deeply adnate to column, simple to ± distinctly 3- or 4-lobed, concave, usually surrounding the column below; disk smooth, carinate or rarely callose; column ± elongate, usually dilated above, terminated by an ample, membranous, entire to trilobed and fimbriate clinandrium; anther opercular, incumbent, 1-celled; pollinia 2. **Distr:** Mexico to tropical America. **Cult:** Propagate by division. Grow in small hanging baskets or mounted. Requires intermediate conditions and moderate shade.

T. suavis Lindl. and Paxton, *Paxton's Fl. Gard.* 1: 44, 1850. (Syn: *T. kienastiana, T. suavis* var. *alba*). **Herbs**, epiphytic, erect; pseudobulbs approximate, subglobose or subquadrate, 4–8 × 2.5–6.5 cm, fleshy, compressed, apex 1-leafed, base covered with imbricate, papery bracts. **Leaf** oblong or elliptic, 10–40 × 4–8 cm, acute, coriaceous; petiole conduplicate. **Inflorescence** basal, short, arcuate or pendulous, 2–5-flowered, 3–4 cm long. **Flowers** large, attractive, fragrant; sepals and petals white or creamy-white, heavily spotted with rose-pink, with yellow or orange markings on the throat (rarely blotched red); sepals and petals subsimilar, free, spreading, lanceolate, 3–5.5 × 0.6–1.2 cm, margins undulate, apex acute; lip tubular, trilobed, obovate, 4.5–7 × 3–8 cm, lateral lobes rounded, convolute, anterior margins usually crisped and undulate, midlobe retuse or emarginate, margins crisped and reflexed; disk keeled; column elongate, apex fimbriate and 4-lobed. **Distr:** C America to Colombia. **Ecol:** Epiphytic on large branches and tree trunks of mossy woods; 1,000–1,700 m. Fl. Dec–May, Aug–Sep. **Zones:** 9–12.

Subtribe Zygopetalinae

Bifrenaria

Syn: *Cydoniorchis, Stenocoryne*, etc.
Species ~20. **Herbs**, sympodial epiphytes or terrestrials; pseudobulbs conical to ovoid, often 4-angled, 1–2-leaved at apex. **Leaves** thin-textured, pleated, rolled when young. **Inflorescences** arising at base of plant, short, erect or arching, unbranched, 1- to several-flowered. **Flowers** showy, scented, waxy; sepals and petals usually spreading, subsimilar; lateral sepals jointed to column-foot forming a spurlike mentum; lip trilobed, clawed, with a fleshy, ridged callus; column winged above; pollinia 4 in 2 pairs, on a short, separate stipe attached to a broad, oblong, folded viscidium. **Distr:** C and S tropical America. **Cult:** Propagate by division. Requires good light and high humidity. When growth completed, occasional misting necessary to avoid pseudobulb from wrinkling.

B. harrisoniae (Hook.) Rchb. f., *Bonplandia* (Hannover) 3: 217, 1855. (Syn: *Colax grandiflorus*, etc.). **Herbs**, epiphytic or lithophytic, 10–40 cm tall; pseudobulbs aggregate, conic, 3–9 × 1–3 cm, 4-angled, dark yellow to green, apex with dark ring, 1-leafed. **Leaf** lanceolate to ovate, to 8–38.5 × 3–12 cm, dark green, coriaceous; pseudo-petiole 0.5–10 cm long. **Inflorescence** 1–3-flowered, 2–20 cm long, erect; scape and floral bracts 1–3 × 0.5–3 cm; ovary and pedicel 2.5–6 cm long. **Flowers** 6–8 cm across, waxy, strongly scented, creamy-white, yellowish, pink to purple, lip pink to purple, rarely yellowish-white, red-striped, sepals and petals whitish or pale yellowish, often red-tinged; sepals ovate, truncate; dorsal sepal 2.3–4.5 × 0.9–2.7 cm; lateral sepals divergent to parallel to each other, 1.8–4.5 × 1.1–2.9 cm, spur with sepal segments superposed, 0.3–2.5 cm long; petals obovate, oblique to column, 1.3–3.9 × 1–2 cm; lip 2.2–5 × 0.7–4 cm, trilobed, purple to dark red, usually with darker veins, central lobe with a wavy margin, apex notched, hairy, callus yellow, hairy. **Distr:** SE and S Brazil. **Ecol:** In open forests, wet montane forests or woodlands; 200–1,800 m. Fl. Jul–Dec. **Zones:** 10–12.

B. tyrianthina (Lodd. ex Loudon) Rchb. f., *Xenia Orchid.* 1: 61, 223, 1854. (Syn: *B. tyrianthina* var. *magnicalcarata*, etc.). **Herbs**, epiphytic; 10–30 cm tall; pseudobulbs aggregate, conical, yellowish, 4–8 × 1.5–4 cm. **Leaves** lanceolate to ovate, 9.5–30.5 × 4.5–6.5 cm, coriaceous; pseudo-petiole to 3.2 cm long. **Inflorescence** 1–3 flowered, erect, 5–25.5 cm long; scape and floral bracts 1.1–2.7 × 0.7–2 cm; pedicel and ovary 3.5–6.5 cm long. **Flowers** strongly fragrant, pink to purple, lip white to pink-purple, red-striped, 5.5–7.5 × 6–7 cm; sepals ovate, truncate; dorsal sepal 3–4 × 1–3 cm; lateral sepals parallel to each other, 3.5–4.5 × 1.8–2.2 cm, spur with sepal segments superposed, 1.5–3 cm long; petals obovate, attenuate, oblique to column, 3–4.5 × 1.7–2.8 cm; lip trilobed, obovate, 3–4.5 × 2–3 cm, ± pubescent, margins undulate to crenate, lateral lobes involute, oblong, 1–2.8 × 0.5–1.8 cm, midlobe obcordate, recurved, claw 2–6 mm long, callus 1.5–2.5 cm long, anterior region bilobed; column 1.4–1.7 cm long, foot 1.8–3.2 cm long; anther 5 × 5 mm. **Distr:** E Brazil. **Ecol:** 1,000–1,300 m. Fl. Oct–Dec. **Zones:** 10–12.

Cochleanthes

Species 13. **Herbs**, epiphytic; stems short, leafy; pseudobulbs absent. **Leaves** numerous, distichous, arrange in a fan-shape; lamina with articulated to conduplicate sheaths near base, thinly membranous. **Inflorescence** axillary, 1-flowered, much shorter than the leaves. **Flowers** large, rather fleshy, showy, fragrant, whitish with a bluish or violet-marked lip; sepals subequal, free, spreading; lateral sepals attached to column foot; petals similar to sepals; lips articulate with column foot, trilobed or entire, shortly clawed, with a basal transverse fleshy ridged callus, sides of lip embracing column; column fleshy, incurved, with a short foot, wingless or with 2 short wings; pollinia 4, pyriform, on an ovate to elliptic, broad viscidium. **Distr:** tropical America. **Cult:** Propagate by division. Best grown in small pots with sphagnum moss or a fine bark mix. Requires high humidity and excellent ventilation.

C. discolor (Lindl.) R.E. Schultes and Garay, *Bot. Mus. Leafl.* 18: 324, 1959. (Syn: *Chondrorhyncha discolor*, etc.). **Herbs**, epiphytic, clustered; stems leafy, very short. **Leaves** distichous, ~5, suberect, spreading; lamina ligulate-oblanceolate, to 20 × 2–5 cm, acute. **Inflorescence** erect, basal, subtended by a basal sheath, 1-flowered; peduncle terete, to 10 cm long; bract pale green; thin-textured, 1.2 cm long. **Flower** showy, sepals pale green, petals creamy-green, tinged pale violet in the apical 2/3, lip dark violet-purple with darker venation, margins pale, callus shiny yellow, column white; sepals elliptic, 3.3–3.6 × 1.4–1.6 cm, obtuse or rounded; petals porrect, apex recurved, elliptic, 3.2 × 1.7 cm, rounded; lip involute below, shortly clawed, subquadrate-ovate, 3.6 × 4.2 cm, apical margins reflexed. apex emarginate, callus fleshy, porrect, of several ridges spreading fan-wise in front; column densely pubescent beneath, foot very short, bearing a small retrorse pubescent hook. **Distr:** C America to Venezuela, Cuba. **Ecol:** In lower montane forests; 700–1,850 m. **Zones:** 9–12.

Galeottia

Species 12. **Herbs**, epiphytic or terrestrial; rhizome with pseudobulbs set close together or far apart; pseudobulbs oblong, covered with papery sheaths when young. **Leaves** 2 or 3, plicate. **Inflorescences** few-flowered, basal, erect to arching, unbranched. **Flowers** resupinate, showy, sepals and petals spreading; lateral sepals saccate at base; lip fleshy, clawed, hypochile cup-shaped, epichile ovate, or trilobed, lateral lobes dentate, callus ridged; pollinia 4. **Distr:** S Mexico to S tropical America. **Cult:** Requires a well-drained compost. Need plenty of water until new growth appears.

G. acuminata (C. Schweinf.) Dressler and Christenson, *Lindleyana* 3: 211, 1988. (Syn: *Mendoncella acuminata, Zygopetalum acuminatum*). **Herbs**, large, terrestrial or epiphytic, variable. **Leaves** 2 or 3; lamina oblanceolate to cuneate-obovate, to 28 × 4.4–4.9 cm, plicate, base narrowed to a ± distinct petiole, apex acute, abaxially ~5-nerved; petiole conduplicate, to 4.5 cm long. **Inflorescence** a raceme, laxly 4–6-flowered, suberect, to 33 cm long; peduncle stout, with 2 remote, close sheaths; floral bracts ovate,

MG

JT

Above: *Scuticaria hadwenii, Zygopetalum maculatum, Zygostates lunata.*

deeply concave. **Flowers** large, segments spreading, pink and yellow or green and reddish-green; dorsal sepal lanceolate, 3.4–3.7 × 1.1–1.4 cm, acuminate; lateral sepals very obliquely lanceolate, as long as the dorsal sepal, but wider, long-acuminate; petals oblong-lanceolate, shorter and narrower than the sepals, acute or short-acuminate; lip shorter than the sepals, fleshy, clawed, trilobed with a reflexed anterior part, lateral lobes relatively small, erect, oblong, midlobe rhombic-ovate, ~2 × 1.7 cm, cuneate below, abruptly short-acuminate or sharply acute; disk with a deeply concave, trilobed callus; column stout, with a pair of porrect, oblong-ovate wings, ~1.1 cm long. **Distr:** Ecuador to Peru. **Ecol:** In forests; 2,200–2,400 m. **Zones:** 10–12.

Huntleya

Species 14. **Herbs**, epiphytic, tufted, sympodial, pseudobulbs absent. **Leaves** many, distichous, arranged in a fan-shape; lamina plicate, elongate-lanceolate, apex acute, membranous to thinly coriaceous, pale green. **Inflorescences** axillary, 1-flowered, shorter than leaves. **Flowers** large, fleshy, showy, often with a glossy sheen; sepals and petals subsimilar, spreading; lateral sepals and petals inserted on column foot; lip articulated apex of column foot, contracted basally, dilated above, clawed, basal callus transverse, long-fimbriate; column arcuate, apically dilated and subcrenulate, biauriculate above; pollinia 4, flattened, superimposed, attached by a short stipe to a flattened viscidium. **Distr:** C and S tropical America to Trinidad. **Cult:** Grow in well-drained pots or mounted on cork slabs or tree ferns. Requires shady and humid conditions.

H. burtii (Endres and Rchb. f.) Rolfe, *Orchid Rev.* 24: 236, 1916. (Syn: *Batemannia burtii*, etc.). **Herbs**, epiphytic. **Leaves** 8–10, arranged in a fan shape; lamina oblong, 30–40 × 3.2–4.2 cm, base conduplicate, articulate and sheathing, apex acute. **Inflorescence** 1-flowered, borne from lower leaf axils; peduncle 11–16 cm long. **Flowers** flat, to 12 cm across, waxy, long-lasting, sometimes fragrant, cream to pale yellow, whitish at center, petals with a red blotch near base, tips of petals and sepals glossy brown, lip brown, apex red-flushed, base white; sepals and petals 4–4.7 × 1.8–2.2 cm, heavily quilted; lip pubescent, trilobed, ~3 × 2.3 cm, lateral lobes with crest fimbriae purple, second line of fimbriae below plate white, midlobe ~2.6 × 2.3 cm, clawed, acute, callus crest semicircular, ~1 × 1 cm, adaxially fimbriate; column dorsally white, sometimes with fine purple striations ventrally, ~2 cm long, apical crest of column and border of column wings pale yellow-green or white, often with a purple spot lateral to stigma; pollinia 4, anther light yellow. **Distr:** C America. **Ecol:** In tropical wet and montane cloud forests; 300–1,200 m. **Zones:** 9–12.

Scuticaria

Species 10. **Herbs**, epiphytic, medium-sized, sympodial; rhizome short or creeping; stem pseudobulbous, small, 1-leafed. **Leaf** long, whiplike, fleshy, mostly pendulous, subterete or terete, sulcate, contiguous with stem. **Inflorescence** lateral, 1(–few)-flowered. **Flower** large or medium-sized, showy; sepals and petals free, subsimilar, erect-spreading; lateral sepals adnate to column foot forming a prominent mentum; lip sessile, attached to apex of column foot, articulated, broad, concave, deeply trilobed, lateral lobes large, erect, midlobe smaller; column erect, fleshy, semiterete, wingless, produced at base into a foot; pollinia 4, ovoid, waxy; stipites very short or obsolete, attached to transversely squamiform viscidia. **Distr:** S tropical America. **Cult:** Propagate by division or from seeds. Best grown on a vertical bark slab, to allow for the long pendent leaves, but can also be grown in a suitable hanging basket.

S. hadwenii (Lindl.) Planch., *Fl. Serres Jard. Eur.* 7: 239, 1852. (Syn: *Bifrenaria hadwenii*). [Wax Flower Orchid]. **Herbs**, epiphytic, medium-sized; stems clustered, pendulous, usually very short, knotty, swollen at base, ash-brown, cylindrical, ~6.5 cm long, furrowed, prominently jointed, 1-leafed. **Leaf** sharply pendulous, deflexed; lamina cylindrical, 22–45 cm long, 0.5–1 cm Ø, apex acute, adaxially narrowly sulcate, deep green. **Inflorescence** ascending or deflexed, 4–5 cm long, 1- or 2-flowered, sheathed at base by brownish, acute, scalelike bracts. **Flowers** nodding, fragrant, 6–8 cm long, sepals and petals fleshy, chestnut-brown or purplish, paler toward the apices, sometimes broken into blotches and bars on a green, yellow or greenish-yellow background, lip pale white or creamish-white, with a white margin and spotted with rose or red color, column pale creamish, purple-spotted, apex white; sepals spreading, oblong, 4–4.5 × 1.2–1.5 cm, acute; petals similar to sepals, slightly curving forward; lip broadly obovate or suborbicular, concave, 3–3.5 × 2.5–3 cm, downy within, callus an oblong plate, tridentate in front, base swollen; column 1.5 cm long. **Distr:** Guyana to Brazil. **Ecol:** In primary scrub forests; over 1,500 m. **Zones:** 10–12.

Zygopetalum

Syn: *Zygopetalon.*
Species 15. **Herbs**, epiphytic; pseudobulbs short, thick, leafy. **Leaves** 2 or more, apical, distichous; lamina membranous to rigid, elevated-venose to subplicate, elongate. **Inflorescences** 1 or more, lateral, from base of pseudobulb, 1- to several-flowered. **Flowers** large or small, usually showy; sepals subequal, spreading, free to shortly connate at base; lateral sepals inserted on column foot; petals similar to sepals; lip hinged to column foot, sessile, spreading, trilobed, lateral lobes small and spreading or rather large and erect, clasping the column, midlobe expanded, spreading, broadly ovate or suborbicular-obovate; disk transversely crested or prominently adorned with fleshy entire or lobed processes; column incurved, semiterete, wingless or shortly 2-winged at apex, produced into a short foot at base; anther terminal; pollinia 4. **Fruits** ovoid to oblong. **Distr:** Peru to Brazil and NE Argentina. **Cult:** Propagate by division. Grow in pots with a bark-based mix. Requires high humidity and excellent ventilation.

Z. maculatum (Kunth) Garay, *Orquideologia* 5: 189, 1970. (Syn: *Broughtonia maculata*, etc.). **Herbs**, terrestrial, large; pseudobulbs clustered, ovoid conical, 4–8 cm long, apex 3–5-leaved. **Leaves** oblong or elliptic-lanceolate, acute or acuminate, plicate, 17–50 × 2–6 cm. **Inflorescences** lateral, racemose, longer than leaves, laxly 3–5-flowered, to 40 cm long; bracts prominent, equaling pedicel and ovary. **Flowers** large, showy, variable, sepals and petals green, heavily blotched with crimson, lip white heavily lined with purple; sepals oblong-lanceolate, 3–3.5 × 1 cm, acute or subacute; petals similar to sepals; lip broadly obovate, trilobed at base, 3.5 × 3 cm; lateral lobes auriculate, midlobe large, cuneate-obovate to obcordate, margins undulate, apex retuse, callus basal, fleshy, bilobed, plurisulcate; column short, stout, dilated and winged, ~1.3 cm long. **Distr:** Peru to Brazil. **Ecol:** Terrestrial in shade on mountain ridges; 1,000–2,250 m. **Zones:** 9–12.

Zygostates

Syn: *Dactylostylis, Dipteranthus.*
Species 22. **Herbs**, epiphytic, dwarf; stems very short, erect to pendent, not pseudobulbous, covered with persistent leaf bases. **Leaves** articulated to persistent imbricate compresses sheathing bases; lamina linear-oblong to elliptic-oblong, apex blunt, coriaceous. **Inflorescences** racemose, laxly few- to many-flowered, erect to pendent; bracts small. **Flowers** membranous, white to greenish-yellow, translucent, delicate, opening widely; sepals free, spreading or recurved, similar; petals spreading, broader than sepals, with entire or toothed margins; lip entire, similar to petals, with margins entire or dentate, and a fleshy callus at base, spur absent; column slender, elongate, with 2 clavate basal staminodes; rostellum elongate, sigmoid-pendent; anther elongate; pollinia 4, waxy, attached by a slender sigmoid stipe to a small viscidium. **Distr:** S tropical America. **Cult:** Propagate by division. Best grown mounted on cork slabs. Requires high humidity with moderate to full sunlight.

Z. lunata Lindl., *Edwards's Bot. Reg.* 23: t. 1927, 1837. (Syn: *Dactylostylis fimbrata*, etc.). **Herbs**, epiphytic, dwarf; stem very short, to 1 cm long. **Leaves** several, linear to oblong-elliptic, obtuse, 4.5–11 × 0.7–2 cm, base articulated with persistent imbricate sheaths, apex obtuse. **Inflorescences** pendent,

JT

MG

Above: *Anguloa clowesii, Brasiliorchis gracilis, Brasiliorchis picta, Camaridum cucullatum.*

8–17 cm long, laxly to 30-flowered; peduncle and rachis very slender; bracts linear-lanceolate, 3–4.5 mm long, acuminate; pedicel and ovary 0.8–1.4 cm long. **Flowers** membranous, sepals white, petals and lip greenish-yellow; sepals elliptic, 3–4 × 2 mm, subacute; petals obovate-flabellate, shortly clawed, 4–5 × 5–6 mm, margins denticulate or erose; lip ovate, 5–6 × 3 mm, margins denticulate, apex acute, callus fleshy, basal, trilobed in front, outer lobe spreading; column slender, geniculate, 5 mm long including the rostellum. **Distr:** SE and S Brazil. **Ecol:** On mossy trees, in montane forests of the mist zone. **Zones:** 10–12.

Subtribe Maxillariinae

Anguloa

[Cradle Orchids, Tulip Orchids]

Species 9. **Herbs**, epiphytic or terrestrial, large; stems clustered, short, base sheathed, few-leaved above, developing into fleshy pseudobulbs; pseudobulbs ellipsoid, grooved. **Leaves** imbricate, deciduous at onset of new growth; lamina large, broadly lanceolate, plicate, base tapering to a grooved petiole. **Inflorescences** 1-flowered, occasionally 2-flowered, lateral, basal, erect, clothed with several loose sheaths. **Flowers** large, ± nodding, showy, scented, cup-shaped, fleshy; sepals and petals erect; sepals broad, subequal, subsimilar; lateral sepals oblique, base cucullate-rounded; petals similar to dorsal sepal but smaller; lip variably motile, trilobed, erect, parallel to column, lateral lobes large, often rounded in front, midlobe small, disk with a central longitudinal callus; column erect, stout, wingless, apex with a pair of triangular-lanceolate appendages in front, foot stout; pollinia 4. **Distr:** Venezuela, Ecuador, Colombia, Bolivia, Peru. **Cult:** Propagate by division. Requires cool to intermediate conditions, and best grown in containers. Water well to avoid the roots from getting dried out during growing season.

A. clowesii Lindley, *Edwards,s Bot. Reg.* 30(Misc.): 26, 1844. (Syn: *A. clowesii* var. *flava*). [Cradle Orchid, Tulip Orchid]. **Herbs**, terrestrial, large; pseudobulbs clustered, conical-pyriform, sulcate, slightly compressed, to 23 cm long, 7 cm Ø, 3–5-leaved. **Leaves** plicate, obovate, 45–80 × 20–26 cm, apex obtuse. **Inflorescences** 1-flowered, erect, to 30 cm long; peduncles stout, covered with large, slightly inflated sheaths. **Flower** large, 7–8 cm long, showy, fragrant, sepals and petals thick, rigid, lemon-yellow, lip fleshy rigid, pale creamy green or yellow, waxy, shiny, column white; sepals elliptic, concave, 6.5–7 × 5 cm, subacute or apiculate; petals elliptic, similar to sepals, slightly smaller; lip hinged to column foot, elliptic, boat-shaped, 4 × 2.5 cm, concave, apically trilobed, lateral lobes erect, obliquely triangular, acute in front, midlobe subrhombic, fleshy, shortly glandular-pubescent, basal callus porrect, shortly bifurcate, glandular-pubescent; column 2.5 cm long. **Distr:** Colombia, Venezuela. **Ecol:** 1,800–2,500 m. **Zones:** 10–11.

Brasiliorchis

Species 12. **Herbs**, epiphytic, rarely terrestrial, cespitose to climbing; pseudobulbs aggregated or distant, bifoliate, ridged to sulcate. **Leaves** with lamina conduplicate, straplike, mostly longer and stiffer in shady places, shorter and yellowish in sunnier conditions, margins entire. **Inflorescences** with solitary flowers borne laterally within a bract or leaf axil of rhizome of mature, immature, or apparently abortive shoots, several inflorescences are produced simultaneously from the base of the most recent pseudobulb; floral bracts always shorter than the pedicel and ovary, triangular, usually acute; pedicel normally well-developed, almost straight, lacking lateral appendages. **Flowers** food-deceptive, campanulate, showy, long-lasting (>10 days), fragrant; sepals (lack fibers, often with dark spots and darker externally) and petals nearly similar; dorsal sepal erect, concave; lateral sepals adnate to column foot; lip trilobed; column not winged, semiterete, with foot at base; pollinia 4, dorsally compressed, yolk-yellow; stigmatic cavity round; rostellum not prominent. **Fruits** ellipsoid to obovoid, apically dehiscent. **Distr:** E and S Brazil to NE Argentina. **Cult:** Propagate by division. Requires a coarse, bark-based mix in well-lit, humid conditions.

B. gracilis (Lodd.) R.B. Singer, S. Koehler and Carnevali, *Novon* 17: 96, 2007. (Syn: *Maxillaria gracilis*, etc.). **Herbs**, epiphytic, small, tufted; pseudobulbs clustered, ovoid-pyriform, compressed, 2–2.5 × 1–1.3 cm, apex 2-leaved. **Leaves** petiolate, conduplicate below; lamina linear-ligulate, 10–20 × 0.8–1.5 cm, apex acute, subcoriaceous. **Inflorescence** ascending, 1-flowered, to 10 cm long; peduncle slender, pale green or reddish; bract pale, 1.2–1.5 cm long. **Flowers** erect-spreading, fragrant, sepals and petals yellow, outer margins purple-flushed, lip yellow, lined or spotted with purple; sepals fleshy, erect-spreading, narrowly triangular, lanceolate, 1.7–2.2 × 0.5–0.7 cm, acute; petals fleshy, narrowly linear, lanceolate, 1.2–2 × 0.2–0.3 cm, very acute; lip rigid, erect, fleshy, deeply trilobed above middle, base furfuraceous, 1.2–1.5 × 0.7–1 cm, lateral lobes erect, semiovate, rounded, midlobe broadly oblong-linguiform, margins undulate-denticulate, apex shortly sulcate; disk with callus fleshy and ligulate. **Distr:** E and S Brazil. **Ecol:** In montane regions: >800 m. **Zones:** 11–12.

B. picta (Hook.) R.B. Singer, S. Koehler and Carnevali, *Novon* 17: 97, 2007. (Syn: *Maxillaria picta*, etc.). [Painted Maxillaria]. **Herbs**, epiphytic; pseudobulbs, clustered, dark green, ovate, obscurely furrowed, 4–7 × 2.5–4 cm, apex 1–2-leaved. **Leaves** linear-lanceolate, 20–35 × 2–5 cm, base petiolate. **Inflorescences** 4–8, produced independently from base of pseudobulbs, 8–20 cm tall, 1-flowered. **Flowers** ~5 cm across, erect-spreading, fragrant, sepals and petals pale yellow, purple-spotted, paler outside, lip yellowish or creamish-white, spotted and striped purple above, column and anther-cap black-purple; sepals oblong-ligulate, 3 × 0.6 cm, acute, fleshy; petals oblique, ligulate, 2.6–2.8 × 0.3 cm, acute, fleshy; lip erect-spreading, slightly recurved above, broadly oblong, deeply trilobed above middle, 2.4–2.5 × 1 cm, lateral lobes erect, semioblong, obtuse, midlobe oblong-linguiform, margins slightly undulate, subdenticulate, apex obtuse, disk with a furfuraceous, fleshy callus; column erect, 1–1.2 cm long. **Distr:** Tropical and subtropical America. **Ecol:** In forests; 1,100–1,400 m. **Zones:** 9–11.

Camaridium

Species ~80. **Herbs**, variable in growth habit, some species are cespitose, and others lack pseudobulbs completely and have monopodial shoots, a few species are dimorphic with juvenile sympodial growth and mature monopodial shoots (without pseudobulbs); pseudobulbs separated by rhizome segments of variable length. **Leaves** 1–many; lamina conduplicate, linear, lanceolate, or elliptic, margins entire. **Inflorescences** with solitary flowers borne laterally within a bract or leaf axil of rhizome of mature, immature, or apparently abortive shoots; floral bracts in almost all species are longer than the pedicel and ovary, and overlaps with the base of the dorsal sepal, lamina triangular, usually acute. **Flowers** mostly food-deceptive, some produce nectar; sepals and petals nearly similar, lacking fiber bundles, with a sparkling appearance; dorsal sepal erect, concave; lateral sepals adnate to column foot; lip trilobed or simple; column not winged, semiterete, column foot basal, short or long; pollinia 4, waxy; stigmatic cavity round; rostellum not prominent. **Fruits** pendent, ellipsoid to obovoid, apically dehiscent. **Distr:** Tropical and subtropical America. **Cult:** Propagate by division. Requires a coarse, bark-based mix in well-lit, humid conditions.

C. cucullatum (Lindl.) M.A. Blanco, *Lankesteriana* 7: 520, 2007. (Syn: *Maxillaria cucullata*, etc.). **Herbs**, clump-forming, variable; pseudobulbs ovoid or rhomboidal, compressed, 2.5–9.5 × 1–3 × ~1 cm, sheaths >10 cm long, unifoliate. **Leaf** linear to oblong-elliptic, 11–40 × 1–5 cm, base conduplicate, apex obtuse to rounded, obliquely retuse, coriaceous. **Inflorescences** several, from axils of basal bracts, 1-flowered; peduncles erect, 4.5–20+ cm long;

Above: Lycaste cruenta, Lycaste dowiana, Lycaste locusta.

pedicel and ovary 1.5–4 cm long. **Flowers** yellowish to almost black, often variously striped and spotted, usually a deep maroon, floral segments thick, rigid; sepals spreading, elliptic-lanceolate to linear-lanceolate, 2.3–4.5 × 0.4–1.1 cm, acute to acuminate, dorsally keeled; lateral sepals reflexed, ± oblique, adnate to column foot forming a mentum; petals broadly elliptic to narrowly lanceolate, 1.5–2.9 × 0.5–0.9 cm, acuminate, hooded over column; lip articulate with column foot, erect-arcuate, broadly elliptic to oblong-oblanceolate, 1.5–3 × 0.7–1 cm, trilobed, lateral lobes short, obcordate to bluntly triangular, ± clasping column, midlobe obovate to oblong-subquadrate, conduplicate, acute, disk with a spathulate basal callus; column 1.3–1.5 cm long. **Fruits** ellipsoid, 3–4.5 cm long. **Distr:** Mexico to C America. **Ecol:** In open or dense humid forests; <3,300 m. **Zones:** 10–11.

Lycaste

Species ~45. **Herbs**, epiphytic, terrestrial or lithophytic; stems short, base sheathed, pseudobulbous. **Leaves** many (often absent during flowering), apical; lamina large, plicate, many veined. **Inflorescences** up to 6, lateral, basal, erect or spreading-ascending, closely sheathed, 1(2–3)-flowered. **Flowers** large, waxy, often fragrant, nodding; sepals subequal, erect–spreading to ± revolute; lateral sepals broader, oblique, conjoined at base forming (with column foot) a saccate mentum; petals similar to sepals but smaller and thinner; lip loosely attached or articulate with column foot, sessile to unguiculate, shorter than sepals, trilobed, lateral lobes erect, broad to narrowly falcate, midlobe broad to narrowly elongate, spreading-decurved, entire to variously ciliate or fringed, occasionally pubescent, disk usually ± pubescent, callose in the middle; column long, arcuate, wingless to narrowly 2-winged above; anther terminal, operculate, incumbent, 1-celled; pollinia 4. **Fruits** oblong to fusiform or spindle-shaped, large, erect or pendent. **Distr:** West Indies and Mexico to Peru. **Cult:** Propagate by division. Requires well-drained bark mixed with peat moss, with much feeding and watering during the growth period.

L. cruenta (Lindl.) Lindl., *Edwards's Bot. Reg.* 29 (Misc.): 16, 1843. (Syn: *L. balsamea*, etc.). **Herbs**, to 50 cm tall; pseudobulb large, ovoid-oblong, compressed, to 10 × 5 cm, subtended by fibrous sheaths. **Leaves** several, apical; lamina elliptic-lanceolate to broadly elliptic, to 45 × 15 cm, apex acute to acuminate, plicate. **Inflorescences** several, lateral, basal; peduncles slender, 1-flowered, 7–17 cm long, with several sheaths ~2 cm long; pedicel and ovary 2–2.5 cm long. **Flowers** showy; sepals yellow-green, oblong-elliptic to ovate-triangular, 3–5 × 1.8–2.5 cm, acute, concave below, recurved above; petals bright yellow or orange-yellow, broadly elliptic, 3–4 × 2–2.3 cm, obtuse, sometimes retuse; lip hinged to column foot, yellow, maroon-flecked, base with a crimson blotch, trilobed, ~2.5 cm long, saccate portion usually whitish-hairy, lateral lobes broadly rounded, erect, midlobe suborbicular to broadly ovate, sometimes notched, decurved, margins crisped, adaxially pubescent, 0.8–1.3 × 0.7–1.2 cm, disk with a small central callus; column ~2 cm long. **Distr:** Mexico (Chiapas) to El Salvador. **Ecol:** In forests; to 2,200 m. **Zones:** 10–11.

L. dowiana Endres and Rchb. f., *Gard. Chron.* n.s., 2: 194, 1874. (Syn: *L. dowiana* var. *labello-alba*, etc.). **Herbs**, epiphytic; pseudobulbs ovoid, slightly compressed, 5–9 × 3–4.5 cm, rugose, spines ~0.5 mm long. **Leaves** 2 or 3 terminal, and 2 or 3 enveloping, persistent; lamina elliptic-lanceolate, 30–65 × 3–7 cm, plicate, base narrowed into a petiole, apex shortly acuminate. **Inflorescence** 1-flowered, with peduncles basal, thin, flexible, 6–12 cm long, enveloped by 4 inflated bracts, flowers opening sequentially. **Flowers** small, with sepals brown with green apices, or sepals green with brown coloration, petals and lip creamish-white or greenish-white, lip throat yellowish or orangish; sepals and petals oblong or lance-oblong, 2.8–4 × 1.5–1.8 cm; lip ~3 × 2.2 cm, trilobed, midlobe subcircular, ~1 × 1 cm, margins wavy, apex retuse, callus linguiform, raised. **Distr:** Costa Rica, Nicaragua, Panama. **Ecol:** In cloud forests and pluvial forests; 700–1,650 m. **Zones:** 10–11.

L. locusta Rchb. f., *Gard. Chron.* n.s. 11: 524, 1879. **Herbs**, terrestrial or epiphytic, variable; pseudobulbs aggregated, ovoid or oblong-ovoid to ellipsoid, 6–10 cm long, complanate, apex 1–3-leaved. **Leaves** spreading; lamina oblong-elliptic, to 55 × 6.6 cm, plicate, base cuneate, apex acuminate; petiole to 8 cm long. **Inflorescences** up to 7, lateral, basal, 1-flowered, to 29 cm long; floral bract surpassing ovary. **Flowers** fleshy, green, often column and margin of lip midlobe white; dorsal sepal ovate-elliptic or lanceolate-elliptic, concave, to 5 × 2.3 cm, acute; lateral sepals obliquely ovate or ovate-oblong, acute, equaling dorsal sepal, forming with column foot a narrowly conical mentum ~1.6 cm long; petals obliquely oblong-oblanceolate or elliptic-oblanceolate, shorter and narrower than sepals, acute or subacute; lip recurved, sharply trilobed, lower portion tubular-concave, cuneate below, elliptic-obovate, to 5 × 2.2 cm, lateral lobes small, obliquely ovate, acute or obtuse, midlobe larger, oblong-ovate, rounded to truncate, minutely retuse, fimbriate, disk callose; column ~2 cm long, apex auriculate. **Distr:** Peru. **Ecol:** On open dry rocks or rock walls; 2,000–3,000 m. **Zones:** 11–12.

Maxillaria

Syn: *Maxillariella*, etc.
Species >650 species. **Herbs**, epiphytic, rarely terrestrial, cespitose to climbing; stems reduced (elongate), usually terminated with a pseudobulb. **Leaves** 1–many; lamina conduplicate, linear, lanceolate, or elliptic, margins entire. **Inflorescences** with solitary flowers borne laterally within a bract or

Below: *Maxillaria cacaoensis, Maxillaria grandiflora, Maxillaria setigera.*

JT

MG

GS

Above: *Maxillariella houtteana, Maxillariella tenuifolia, Maxillariella variabilis, Mormolyca ringens.*

leaf axil of rhizome of mature, immature, or apparently abortive shoots; floral bracts triangular, usually acute. **Flowers**: sepals and petals nearly same; dorsal sepal erect, concave; lateral sepals adnate to column foot; lip trilobed or simple; column not winged, semiterete, with foot at base; pollinia 4, waxy; stigmatic cavity round; rostellum not prominent. **Fruits** ellipsoid to obovoid. **Distr:** Predominantly in tropical America. **Cult:** Propagate by division. Requires a coarse, bark-based mix in well-lit, humid conditions.

M. cacaoensis J.T. Atwood, *Selbyana* 19: 254, 1998. **Herbs**, epiphytic, cespitose, apparently densely so, to ~10 cm tall; pseudobulbs coarsely rugose when dry, semiorbicular, compressed, 2–3.5 × 2.5 cm, subtended by sheaths, apex 1-foliate. **Leaves** elliptic, 6–9 × 2–3 cm, apex obtuse, coriaceous, abaxially strongly keeled especially apically. **Inflorescences** short, ~3 cm long, concealed by inflated bracts. **Flowers** campanulate to somewhat spreading, yellowish-orange, bases of sepals and petals white, lips reddish; sepals ovate-lanceolate, 3.5 × 1.5 cm, acute; petals elliptic-oblanceolate, 2.8 × 1 cm, acute; lip hinged to column foot, sharply trilobate, lateral lobes rounded, clasping column, midlobe fleshy, ovate, ~1.8 × 1.4 cm, recurved to reflexed, bluntly acute to obtuse; calli 2, of 2 keels; column stout, ~1.4 cm long, foot 5–6 mm long; ovary ~1.5 cm long, concealed by subtending floral bract. **Distr:** Costa Rica. **Ecol:** In cloud forests; ~1,100 m. Fl. Jan–Feb. **Zones:** 10–11.

M. grandiflora (HBK.) Lindl., *Gen. and Sp. Orch.* 147, 1832. (Syn: *Dendrobium grandiflorum*, etc.). **Herbs**, large; pseudobulbs approximate, oblong-ovoid, strongly compressed, to 6 cm tall, apex unifoliate. **Leaves** elliptic-oblong to ligulate, 11–28 × to 5 cm, base cuneate, apex acute, coriaceous; petiole stout, strongly laterally compressed, 3–10 cm long. **Inflorescences** with 1 or 2 scapes, lateral, basal, 1-flowered, stout; peduncles 12–25 cm long, clothed with several tubular sheaths. **Flowers** large, showy, fleshy, milk-white; dorsal sepal ovate-oblong, concave, acute, 3.5–4.5 × 2 cm; lateral sepals widely spreading, triangular-ovate, acute, oblique, slightly longer and broader than dorsal sepal, forming a conical mentum; petals elliptic-lanceolate, subacute, slightly oblique, shorter and narrower than dorsal sepal; lip erect, parallel to column, sides erect, relatively short, ~2.5 × 1.5 cm, ovate, obtuse, lower margins entire, upper ones obscurely trilobed, disk densely furfuraceous, calli 3; column arcuate, 1.2–1.3 cm long. **Distr:** W South America to Venezuela. **Ecol:** 1,700–2,200 m. **Zones:** 10–12.

M. setigera Lindl., *Edwards's Bot. Reg.* 31(Misc.): 30, 1845. (Syn: *M. callichroma, M. leptosepala*, etc.). **Herbs**, epiphytic; pseudobulbs closely set, pyriform, smooth, 1-leafed at apex. **Leaf** erect; lamina narrowly linear, gradually narrowing below into petiolate base, coriaceous, rigid. **Inflorescences** several, erect, basal, 7.5 cm long, 1-flowered, borne on mature pseudobulbs, enveloped by a few bracts; bracts imbricate, tubular, acuminate. **Flowers** large, 10–15 cm across, sepals and petals yellow with white bases, slightly tinged with pinkish-red outside, lip white with brownish colorations at base and a few scattered hairs adaxially, center yellow, sides of the lower part are streaked with crimson veins, all segments with margins recurved; dorsal sepal ~6 × 2 cm; lateral sepals ~9 × 2 cm; petals linear, ~7.5 × 1 cm, terminating suddenly at the tip in a fine bristle. **Distr:** W South America to Guyana. **Ecol:** 40–670 m. **Zones:** 9–12.

Maxillariella

Species 45. **Herbs**, epiphytic, lithophytic or terrestrial, sympodial (monopodial), erect to pendent; pseudobulbs (when present) ovoid or oblong, laterally compressed. **Leaves** 1 or 2, sessile, articulate with sheaths or with apex of pseudobulb; lamina conduplicate, usually oblong or oblong-elliptic, broadly elliptic or obovoid (lanceolate), margins entire, apex acute to obtuse. **Inflorescences** with solitary flowers borne from axils of rhizome or stem sheaths or from base of pseudobulbs, erect or pendent; floral bracts usually much shorter than ovary. **Flowers** resupinate; campanulate (spreading); sepals and petals free, submembranous, usually without fibers; sepals subequal; lateral sepals often larger, oblique, and adnate to column foot forming a mentum, broadly triangular to obovate or acuminate; petals similar to sepals but smaller; lip free and hinged at apex of column foot, simple to weakly trilobed, glabrous without any obvious secretion or nectar; column cylindrical to hemi-cylindrical, straight to arcuate, column foot conspicuous; pollinia 4 in 2 superposed unequal pairs; stigma ventral, transverse to ovoid; rostellum transverse; ovary smooth, round in cross-section. **Fruits** ellipsoid to spherical, laterally dehiscent. **Distr:** Mexico to tropical America. **Cult:** Propagate by division. Requires a coarse, bark-based mix in well-lit, humid conditions.

M. houtteana (Rchb. f.) M.A. Blanco and Carnevali, Lankesteriana 7: 528, 2007. (Syn: *Maxillaria houtteana*). **Herbs**, epiphytic (plants smaller and pendent) or lithophytic (plants larger and erect); pseudobulbs linear-oblong, compressed, 5–6.5 cm long, smooth, pale green. **Leaf** solitary, sessile; lamina linear to linear-oblong, 15 × 1.2–2 cm, base conduplicate, apex acute to obtuse, emarginate, curved, adaxially dark green, deeply channeled, abaxially paler, keeled. **Inflorescence** few-flowered; peduncle 3.8–5 cm long; pedicel and ovary ~1 cm long. **Flowers** waxy, 2–2.5 cm across, long-lasting, color variable, sepals yellow outside, inside reddish, margins and tips usually yellow; sepals 1.2 × 0.45 cm, acute; dorsal sepal oblong; lateral sepals suberect, spreading and recurved, ovate-lanceolate; petals smaller than sepals, color similar to sepals, oblong, 8 × 3 mm, acute, incurved, tips connivent; lip oblong, 9 × 5 mm, rounded, emarginate, puberulous, margins entire to shallowly trilobed, disk stiffly callose, golden-yellow, distally spotted red-brown, basally purple-streaked; column yellow, spotted reddish-brown. **Distr:** Mexico to Venezuela. **Ecol:** In humid forests, montane rainforests; 1,700–2,300 m. Fl. throughout most of the year. **Zones:** 10–12.

M. tenuifolia Lindl., *Edwards's Bot. Reg.* 23: t. 1986, 1837. (Syn: *Maxillaria gracilifolia, M. tenuifolia*). [Coconut Orchid]. **Herbs**, epiphytic, densely clumping to ~75 cm across; stems numerous, straggling; pseudobulbs cylindric to fusiform, 2–5 × 1–2 cm, compressed, apex 1-leafed. **Leaf** linear, 12–50 × 0.2–0.8 cm, apex acuminate, thinly textured, subcoriaceous, light green. **Inflorescence** 1 per rhizome bract, 1–2 cm long; pedicel and ovary 3–4 cm long. **Flowers** fragrant, red or yellowish, red-spotted, lip yellow or white, red-spotted; sepals lanceolate-ovate, 2–3 × 0.6–0.9 cm; petals appressed to column, elliptic-lanceolate, 1.9–2.7 × 0.5–0.8 cm, falcate, apical margins erose; lip simple, arcuate-decurved, oblong-elliptic to oblong-pandurate, 1.5–2.2 × 0.5–1.1 cm, acute or retuse, lateral lobes erect, disk papillose, callus narrow; column arcuate, 1–1.5 cm long, foot 2 mm long. **Fruit** ellipsoid, 3–4 cm long. **Distr:** Mexico to C America. **Ecol:** On large trunks and branches in evergreen forests; 100–1,200 m. Fl. Mar–May. **Zones:** 10–12.

M. variabilis (Bateman ex Lindl.) M.A. Blanco & Carnevali, *Lankesteriana* 7:530 (2007). (Syn. *Maxillaria variabilis*). **Herbs**, epiphytic, straggly, to ~25 cm tall; stems elongate; pseudobulbs fusiform to cylindric, 1.5–4 × 1.25 cm, often glaucous, apex unifoliate. **Leaves** thin, usually shortly petiolate; lamina narrowly elliptic to lanceolate, 5–14 × 0.6–2 cm, apex obtuse to unequally bilobed. **Inflorescences** to ~2 cm long; pedicel and ovary 1.3–2 cm long. **Flowers** orange to yellow with red colorations, sometimes dark red or maroon; sepals, lanceolate to elliptic, 1–1.5 × ~0.5 cm, acute to obtuse, dorsal sepal concave and more elliptic, lateral sepal more lanceolate to falcate; petals oblanceolate, 1–1.4 × 0.4–0.5 cm, acute, strongly recurved; lip similar to sepals, yellow with dark, shiny red center, simple to slightly trilobate, ± straight, sides embracing column, elliptic-ovate, 1–1.3 × 0.5–0.65 cm, truncate to rounded or retuse, callus fleshy ligulate, shiny; column arcuate, 6–9 mm long, foot 1.5–2 mm long. **Distr:** Mexico to Panama. **Ecol:** In semideciduous forests; 600–1,800 m. Fl. Nov–Apr, Aug. **Zones:** 9–12.

Mormolyca

Species 7. **Herbs**, epiphytic, cespitose or rhizomatous and creeping; pseudobulbs, ovoid or elongate,

JT

Above: *Ornithidum sophrontis, Trigonidium egertonianum, Xylobium hyacinthinum.*

variously compressed, apex 1(–4)-leaved. **Leaves** ligulate, linear to elliptic-oblong, conduplicate, chartaceous to coriaceous. **Inflorescence** 1-flowered, lateral, basal; peduncle very slender, elongate, from axils of bracts subtending pseudobulbs. **Flower** usually with lurid colors; sepals free, subequal, similar or dissimilar, never connate, lateral sepals not forming a mentum; petals similar or dissimilar to but smaller than sepals; lip suberect, much shorter than sepals and petals, simple, trilobed, lateral lobes minute, erect (often resembling an insect), midlobe large, decurved, disk callose; column stout or slender, arcuate, wingless, ± footless; pollinia 4. **Fruit** ellipsoid. **Cult:** Same as for *Maxillaria*. **Info:** This genus is closely related to *Maxillaria*. *Mormolyca* can be easily distinguished from other genera by the apparent footless column.

M. ringens (Lindl.) Gentil, *Pl. Cult. Serres Jard. Bot. Brux.*: 124, 1907. (Syn: *M. lineolata, Trigonidium ringens*). **Herbs** epiphytic, subcespitose, to 40 cm tall, often in dense masses; pseudobulbs ovoid, 4 × 3 cm, sheaths reddish-brown. **Leaf** solitary, apical, shortly petiolate; lamina linear-ligulate to narrowly lanceolate, 9–35 × 1.5–3.5 cm, base conduplicate, apex obtuse to acute, adaxially strongly veined, coriaceous. **Inflorescences** several, lateral or from axils of basal sheaths, 1-flowered; peduncles 6–33 cm long; bracts brownish, 1–1.5 cm long; pedicel and ovary 3–4.5 cm long. **Flowers** fleshy; dorsal sepal concave, elliptic, 1.4–1.6 × 0.7–0.8 cm, acute to shortly acuminate; lateral sepals red-veined, convex, elliptic-falcate, 1.5–1.7 × 0.55–0.7 cm; petals convex, elliptic-falcate, twisted, 1.3–1.5 cm × 4–5 mm, obtuse, acuminate; lip porrect, parallel to column, velvety, elliptic-rhombic, 7–8 × 4–6 mm, trilobed below middle, basal callus maroon; pollinia 4. **Fruits** ellipsoid, 2–2.5 cm long. **Distr:** Mexico to C America. **Ecol:** In dense humid forests; 0–1,000 m. Fl. throughout the year. **Zones:** 10–11.

Ornithidium

Species 56. **Herbs**, mostly epiphytic, often lithophytic or terrestrial, monopodial, sympodial, or sympodial as juveniles with monopodial adult shoots; pseudobulbs often 1 or 2, basal, variable in shape. **Leaves** 1–2(–4), articulate, mostly sessile; lamina conduplicate, ligulate to lanceolate-elliptic. **Inflorescences** with solitary flowers, singly or in fascicles at base of pseudobulbs or from leaf axils, patent or erect; peduncle remotely sheathed; floral bracts inconspicuous, much shorter than pedicel and ovary. **Flowers** often showy, brightly colored, usually campanulate or subglobose, rarely with perianth patent; perianth fleshy (membranous), usually bright red, orange or yellow, sometimes greenish, cream, white or pinkish; sepals subequal, mostly elliptic or obovate-elliptic; petals similar and usually subequal to sepals but often smaller, narrower or both, oblong to obovate-elliptic; lip fleshier than other perianth parts, simple or trilobed, ecallose or with folds of lamina; column erect, often arcuate, usually semiterete, with a foot; pollinia 4, subequal; stigma ventral; rostellum transverse; ovary terete or trigonous, often much longer than peduncle. **Fruits** ellipsoid (trigonous), apically dehiscent. **Distr:** tropical America. **Cult:** Propagate by division. Requires a coarse, bark-based mix in well-lit, humid conditions.

O. sophronitis Rchb. f., *Bonplandia (Hannover)* 2: 18, 1854. (Syn: *Maxillaria sophronitis, M. ruberrima* var. *sophronitis*). **Herbs**, epiphytic, dwarf, creeping; rhizome slender, long, covered completely with brown sheaths; pseudobulbs ovoid, compressed, 2.5–3.5 cm apart on the rhizome, dull brown, 1.2 × 0.8 cm, 1-leafed at apex, subtended by 1 or 2 leaf-bearing sheaths. **Leaves** suberect-spreading; lamina oblong-elliptic, 2 × 0.9 cm, minutely apiculate, coriaceous. **Inflorescences** 1-flowered. **Flowers** relatively large, showy, sepals and petals bright orange-red, lip bright orange-yellow with creamy-yellow lateral lobes and margins, column pale green; sepals elliptic, 1.1–1.3 × 0.7 cm, acute or apiculate; mentum obscure; petals obovate-elliptic, 8 × 6 mm, apiculate; lip trilobed in basal half, 7 × 7 mm, clawed, lateral lobes erect, rounded, apical margins erose, inrolled, midlobe oblong-elliptic, margins erose, fleshy, papillose, callus raised behind, flat and smooth above; column 5 mm long. **Distr:** Venezuela. **Ecol:** In cloud forests; 750–1,700 m. **Zones:** 9–10.

Trigonidium

Species 14. **Herbs**, epiphytic or lithophytic, small to medium-sized; rhizome short or elongate; stems thickened into a cylindric-oblong, ovoid or oval, flattened, fleshy pseudobulb, bearing 1 or 2 leaves. **Leaves** linear to oblong or oblanceolate, coriaceous. **Inflorescences** lateral, basal, 1-flowered, several-sheathed. **Flower** with sepals subequal, lateral ones oblique, connivent into a tube through lower part, upper portion spreading or reflexed; petals much smaller than sepals, commonly fleshy thickened near apex; lip shorter than petals, trilobed, erect, lateral lobes erect and incurved, midlobe spreading or recurved, disk with a median, callose thickening at middle or through the lower part; column commonly shorter than the expanded lip, semiterete, wingless, footless; anther terminal, opercular, incumbent, 1-celled; pollinia 4, waxy. **Fruits** ellipsoid. **Distr:** Mexico to tropical America. **Cult:** Propagate by division. Requires a bark-based mix, high humidity and strong lighting.

T. egertonianum Bateman ex Lindl., *Bot. Reg.* 24: Misc. p. 73, 1838. [Dragon's Mouth]. **Herbs**, epiphytic, tufted, clump-forming; pseudobulbs ovoid-subglobose to ellipsoid, compressed, oblique, 4–9 × 2–3.5 cm, bifoliate. **Leaves** linear, 20–60 × 1–3 cm, base conduplicate, apex obtuse to acute, cuspidate, subcoriaceous, flexible. **Inflorescences** few to many, 1-flowered, axillary; peduncles 15–35 cm long; bracts to 5 cm long; pedicel and ovary 4–5.5 cm long. **Flowers** tubular-campanulate, sepals and petals yellowish-green to greenish with reddish-brown veins; dorsal sepal suborbicular-obovate to obovate-spathulate, 2.7–4.2 × 1.4–2 cm, base narrowly cuneate, apex broadly obtuse, dorsally carinate; lateral sepals obliquely oblong-elliptic, ± sigmoid, 2.8–4.2 × 1–1.4 cm, rounded to broadly acute; petals narrowly clawed, elliptic-lanceolate to narrowly elliptic, 1.4–2.1 × 0.35–0.7 cm, obtuse to acute, connivent, obliquely concave, callus purple-mammillate; lip green, marked with dark purple, trilobed, 0.8–1 cm long, lateral lobes semielliptic, obtuse, upturned, clasping column, venation blackish-purple, mid lobe short, very fleshy verrucose, sulcate, decurved-reflexed, disk with a linear-clavate callus; column ~7 mm long. **Fruits** erect, 4–4.5 cm long, glossy green. **Distr:** C and S America. **Ecol:** In wet forests; 0–1,000 m. **Zones:** 10–12.

Xylobium

Species >30. **Herbs**, epiphytic, rarely terrestrial, cespitose, rarely rhizomatous; pseudobulbs approximate, globose to cylindrical, 1–3-leaved. **Leaves** petiolate; lamina large, plicate. **Inflorescences** racemose, laxly to densely few- to many flowered, basal, lateral, simple, erect or arching, often short. **Flowers** with sepals and petals free, spreading, subsimilar, petals smaller; lip joined at bas to column foot, simple or trilobed, callus fleshy; column with a distinct foot; pollinia 4. **Distr:** Mexico, C and S America, West Indies. **Cult:** Propagate by division. Requires a free-draining substrate, in well-lit and high humidity conditions. Reduce watering after the growth period.

X. hyacinthinum (Rchb. f.) Gentil, *Pl. Cult. Serres Jard. Bot. Brux.*: 194, 1907. (Syn: *Maxillaria hyacinthina*). **Herbs**, epiphytic or terrestrial; pseudobulbs oblong or ovoid, 6 cm long, slightly compressed, smooth, becoming rugose with age, subtended by 2 scarious sheaths. **Leaves** 2, apical, petiolate; lamina oblong-lanceolate or ovate, 25–50 × 7.5–10 cm, apex acute or acuminate, plicate, coriaceous. **Inflorescence** a raceme, 10–20-flowered, basal, lateral, erect; peduncle 5–7.5 cm long, sheaths 3–4, membranous, acute; bracts linear-subulate; floral bracts inflated acuminate. **Flowers** fragrance similar to hyacinth, 3.5 cm across, pale yellow to creamish-white, faintly maroon-striped on petals and lips, midlobe of lip with deep maroon callus, column minutely maroon-spotted in front; dorsal sepal ~1.3 cm long; lateral sepals with a dorsal keel projecting at apex; petals suberect, broadly lanceolate, acuminate, base forming a mentum; lip cucullate, trilobed,

JT

MG

Above: *Peristeria elata*, *Peristeria pendula*, *Acineta densa*, *Gongora galeata*.

~1 cm long. **Distr:** Venezuela. **Ecol:** In dense forests; ~1,300 m. **Zones:** 11–12.

Subtribe Coeliopsidinae

Peristeria

Syn: *Eckardia*, *Peristera*.

Species 11. **Herbs**, epiphytic or terrestrial, robust; stems short, enveloped by imbricating sheaths, produced into fleshy pseudobulbs; pseudobulbs, stout, ovoid or subconic. **Leaves** deciduous, 1–5, apical; lamina broad, elliptic-lanceolate, often long, plicate, base contracted into a petiole. **Inflorescences** racemes, tall or short, erect or pendent, lateral, basal. **Flowers** fleshy, subglobose, relatively large; sepals fleshy, subequal, broadly concave, dorsal sepal usually free, laterals somewhat connate at base; petals subequal to the sepals, but smaller; lip very fleshy, with a broadly concave basal claw or hypochile which is adnate to or continuous with the base of the column, with or without a fleshy central callus, with or without elongate, erect, lateral wings, apical portion or epichile entire, articulated with the hypochile, inflexed or incumbent, lateral margins erect, spreading or retuse, disk with or without a fleshy ventricose or 2-keeled callus; column without a foot; anther terminal, operculate; pollinia 2, waxy. **Distr:** C and S America, Trinidad. **Cult:** Propagate from seeds or by division. Requires a well-drained medium in shade, with watering throughout the year.

P. elata Hook., *Bot. Mag.* t. 3116. 1831. [Dove Orchid, Holy Ghost Orchid, Panama Dove Orchid]. **Herbs**, terrestrial, erect; pseudobulbs stout, fleshy, broadly ovoid, subconic or subcylindric, 4–12 × 4–8 cm, bases enveloped in papery bracts. **Leaves** 3–5, apical, deciduous; lamina broadly lanceolate, 0.3–1 m x 6–12 cm, plicate, apex acuminate. **Inflorescences** racemes, tall, solitary, erect, unbranched 10–15(+)-flowered, 0.8–1.3 m long, opening 2–4 at a time; pedicels ~4 cm long. **Flowers** relatively large, fleshy, subglobose, waxy white, strongly fragrant; sepals subequal, fleshy, waxy white, broadly concave, dorsal sepal free, ovate, 2.5–3 × 2–2.5 cm, obtuse, laterals somewhat connate at base, ovate or suborbicular, 2.5–3 × 2.5–3 cm, shortly acute; petals elliptic-obovate, 2–2.5 × 1.5–1.8 cm, obtuse; lip very fleshy, claw or hypochile broad, continuous with base of column, lateral margins with ascending wings which are white, heavily spotted rose-red; column short, pure white, subconic, semiterete, 0.9–1.1 cm long; anther pure white, beaked. **Distr:** Colombia, Costa Rica, Panama, Venezuela. **Ecol:** In rainforests, often among rocks; 0–1,300 m. **Zones:** 10–12. **Info:** The lip and column combined looks like a white dove in flight. This is the national flower of Panama.

P. pendula Hook., *Bot. Mag.* 63: t. 3479, 1836. (Syn: *P. lentiginosa*, *P. maculata*). **Herbs**, epiphytic, large; pseudobulbs approximate, ovoid-conical to oblong-subconical, deeply sulcate, 8–15 cm long, apex 3–4-leaved. **Leaves** lanceolate-elliptic or oblong-lanceolate, to 80 × 12 cm, apex acute or acuminate; petiole slender, sulcate. **Inflorescences** racemose, densely 4–8-flowered, pendent, to 18 cm long. **Flowers** subglobose, large, fleshy, greenish-white outside, reddish-purple inside, purple-spotted, lip yellowish, purple-spotted; dorsal sepal ovate-subrotund, deeply concave, 2.8–3.5 cm long, obtuse; lateral sepals connate at base; petals similar to dorsal sepal but shorter and narrower; lip enclosed by lateral sepals, deeply trilobed, lateral lobes erect, basal, oblong-ovate, separated by a lunate or deeply sulcate V-shaped callus, midlobe ovate-rhombic or ovate-ligulate, obtuse, recurved; column stout, short, auriculate near summit. **Distr:** Trinidad to C and S tropical America. **Ecol:** Epiphytic in wet forests; 500–1,100 m. Fl. Mar, Jul–Aug, Dec. **Zones:** 10–12.

Subtribe Stanhopeinae

Acineta

Syn: *Lueddemannia*, *Neippergia*.

Species 17. **Herbs**, epiphytic, stout; pseudobulbs ovoid or subcylindric, often laterally compressed or furrowed. **Leaves** apical, 1–4; lamina broadly plicate. **Inflorescences** racemes, usually many flowered, elongate, pendulous or erect, from base of pseudobulbs. **Flowers** fleshy, relatively large; sepals subequal, broadly concave; dorsal sepal free; lateral sepals usually connate at base; petals subequal to sepals but usually smaller; lip trilobed, fleshy, concave, with a concave or subsaccate, basal claw or hypochile, lateral lobes large, erect, triangular or subreniform, apical lobe carinate, concave or spreading; disk fleshy, with variously shaped appendages; column erect, usually pubescent, somewhat arcuate, subclavate or narrowly winged, without a foot; anther terminal, operculate, incumbent, 1-celled or imperfectly 2-celled; pollinia 2, waxy. **Distr:** Mexico to W South America. **Cult:** Propagate by division. Best grown in hanging baskets. Requires a coarse bark-based compost and cool to intermediate conditions in moderate shade, with watering and fertilization during the growth period.

A. densa Lindl., *Paxton's Fl. Gard.* 1: 91, 1851. (Syn: *A. gymnostele*). **Herbs**, epiphytic, cespitose; pseudobulbs ovoid, ~ 8 × 4 cm, conspicuously grooved. **Leaves** ~4, terminal, subpetiolar; lamina plicate, ovate ~95 × 20 cm. **Inflorescence** a raceme, pendulous, 30–45 cm long. **Flowers** resupinate, fragrant, succulent, pale yellow or greenish-yellow, slightly crimson-spotted outside, lip yellow, spotted with broad blotches in lateral lobes, deep crimson in the space between the lobes; sepals ovate or obovate, concave, 3.2–3.7 × 2.2–2.8 cm; petals rhombic-ovate, 3.2–3.8 × 1.6–2.7 cm; lip 3.6–3.8 cm long, trilobed, hypochile 1.2–1.4 × 1.2–1.5 cm, deeply concave, lateral lobules reniform to obtriangular, 1–1.3 × 1.8–2.2 cm, midlobule rhombic-obovate, 0.9–1.2 × 0.7–1.1 cm; column arcuate, 2.5–2.7 cm long, puberulent; pollinia 2, viscidium ovate. **Distr:** C America. **Ecol:** In cloud forests; 1,600–2,000 m. Fl. Mar. **Zones:** 10–12.

Gongora

Syn: *Acropera*.

Species 73. **Herbs**, epiphytic; stems short, fleshy, pseudobulbous, bifoliate. **Leaves** large, plicate-venose, contracted at the base. **Inflorescence** a long raceme, basal, lateral, simple, laxly flexuose, reflexed; floral bracts small, narrow. **Flowers** rather large; dorsal sepal erect-spreading, adnate from base of column to near its apex; lateral sepals broader, adnate to foot of column, spreading and reflexed, strongly oblique; petals adnate to sides of column, spreading free from column near dorsal sepal, free part erect to spreading; lip continuous with foot of column, spreading to ascending, narrow, fleshy, lateral lobes thick, erect, variously horned to aristate, midlobe variously saccate to broadly compressed, apex bilobed or acute to long-acuminate; column erect to ascending, produced at base into a foot, semiterete above; anther terminal, operculate, incumbent, 1-celled to imperfectly 2-celled; pollinia 2, ovoid to narrowly oblong; ovary long-pedicellate. **Fruits** oblong to fusiform. **Distr:** Mexico, C and S America, Trinidad. **Cult:** Propagate by division. Best grown in a basket. Requires intermediate conditions with moderate shade, high humidity, and plenty of watering and fertilizer during growth period.

G. galeata (Lindl.) Rchb.f., *Xenia Orchid.* 1: 51, 1854. (Syn: *Acropera atropurpurea*, etc.). **Herbs**, epiphytic, rarely terrestrial or lithophytic, large; pseudobulbs clustered, ovoid, sulcate to pyriform, to 4.5 × 2.5 cm, covered with 2 or 3 membranous sheaths, apex 1- or 2-leaved. **Leaves** broadly lanceolate, plicate, 16–32 × 4–4.5+ cm, apex acute. **Inflorescence** mostly 10–15-flowered, basal, arcuate-pendulous, to 28 cm long, flowers arranged in 3 rows; bracts very small, purple; peduncle and rachis purple; pedicel and ovary curved and purple. **Flowers** buff to creamy green, faintly mottled orange below; sepals oblong-ovate, 2–2.5 × 1.2 cm, margins reflexed, apex obtuse; petals very short, oblong-falcate, 4–5 mm long, truncate and with 2 apical teeth; lip yellow-brown, articulate to column base, trilobed, 1.2 × 0.9 cm, lateral lobes large, inflexed, midlobe saccate; column yellow-green, purple-spotted, 1 cm long, margins winged. **Distr:** Mexico to Guatemala. **Ecol:** In rainforests and cloud forests; 600–1,800 m. **Zones:** 10–11.

JT

Top row: *Stanhopea hernandezii*, *Stanhopea jenischiana*, *Stanhopea oculata*.
Botto row: *Stanhopea tigrina*, *Stanhopea tigrina* var. *nigroviolacea*, *Stanhopea wardii*.

Stanhopea

Syn: *Ceratochilus*, *Stanhopeastrum*.
Species 58. **Herbs**, epiphytic, terrestrial or lithophytic; pseudobulbs ovoid or subcylindric, ± ridged, 1-leafed. **Leaf** usually large, broad, plicate, strongly veined, elliptic-lanceolate, base contracted into a slender sulcate petiole. **Inflorescence** 1-flowered or a 2–9-flowered raceme, short, pendent, lateral, basal; rachis enveloped in several broad, papery, imbricating bracts; pedicels long, each shielded by a broad, papery, spathaceous bract. **Flowers** large, showy, fleshy, often very fragrant; sepals free or with lateral ones slightly united at base, spreading-reflexed, subequal, lateral ones broadest; petals similar to sepals but smaller and narrower, usually undulate; lip affixed to or sometimes shortly connate with base of column, fleshy thickened, lateral lobes with erect margins and mostly thickened forming a variously shaped hypochile, hypochile usually globose or calciform, midlobe (when present) entire or variously divided, usually forming a simple or bicornuate mesochile and an epichile that is articulate with mesochile, epichile entire or trilobulate at apex and variously shaped; column long, erect, ± arcuate, wingless to broadly winged above, apparently without a foot; anther terminal, operculate, incumbent; pollinia 2, waxy. **Fruits** large, ellipsoidal. **Distr:** Mexico, C and S America, Trinidad. **Cult:** Propagate by division. Best grown in hanging wire baskets lined with moistened sphagnum moss and filled with a porous medium, to allow the pendent inflorescences to pierce through the basket. Requires ample watering and feeding during growth period.

S. hernandezii (Kunth) Schltr., *Beih. Bot. Centralbl.* 36(2): 490, 1918. (Syn: *Anguloa hernandezii*, etc.). **Herbs**, epiphytic or rarely lithophytic; pseudobulbs obliquely ovoid-conical, 6 × 3 cm, subtended by scarious sheaths, apex unifoliate. **Leaf** oblong to elliptic-lanceolate, 20–30 × 5–6.5 cm, apex acuminate; petiole sulcate, 10 cm long. **Inflorescence** lateral, 2–3-flowered; peduncle stout, covered by membranous, lepidote sheaths. **Flowers** large, fragrant, sepals and petals pale yellow or yellowish-orange, spotted with red-brown blotches, lip creamy white, marked with dark purple, column white, crimson-spotted; sepals ovate-oblong to ovate-elliptic, 5.5–6.5 × 2.5–3.8 cm, obtuse or rounded; petals oblong-lanceolate, recurved, 4–5 × 1.5–2 cm, acute; lip very fleshy, hypochile subrotund-globose, 2.5 × 3 cm, inner surface warty, mesochile with 2 incurved horns, arising from a broad fleshy tubercle, 2.5 cm long, epichile ovate, canaliculate, 2.5 × 1.5 cm, ± tridentate; column 4–5.5 cm long, ± narrowly winged. **Distr:** C and SW Mexico. **Ecol:** On trees and sloping rocks in forests; 300–1,900 m. Fl. Oct–Dec. **Zones:** 9–12.

S. jenischiana F. Kramer ex Rchb. f., *Bot. Zeitung* (Berlin) 10: 934, 1852. (Syn: *Anguloa grandiflora*, etc.). **Herbs**, epiphytic; pseudobulbs globose, longitudinally ribbed, apex 1-leafed. **Leaf** plicate, elliptic-lanceolate, 20–30 × 8–12 cm, base gradually narrowed into an elongate, channeled pseudo-petiole, margins undulate, apex acute. **Inflorescence** a raceme, 4–8-flowered, 30 cm long, pendent, basal, flowers borne at apex of inflorescence; bracts large, chartaceous. **Flowers** ~7 cm across, orangish or golden-yellow, maroon-spotted, petals and sepals reflexed, lip hypochile often with 2 purplish-red eye-spots; dorsal sepal 3.6 × 1 cm, margins recurved; lateral sepals ovate, 3.6 × 1.8 cm, margins incurved, apex acute; petals linear, 2 × 0.7 cm, strongly recurved; lip ~3.5 cm long, complex; column curved, ~3.6 cm long. **Distr:** Colombia, S Ecuador, Peru. **Ecol:** In montane forests with distinct dry seasons; 800–1,500 m. Fl. Jul–Oct. **Zones:** 10–11.

S. oculata (Lodd.) Lindl., *Gen. and Spec. Orch. Pl.* 158, 1832. (Syn: *Ceratochilus oculatus*, *S. bucephalus*). **Herbs**, epiphytic; pseudobulbs short, ovoid, subconic. **Leaves** single, apical, petiolate; lamina plicate, elliptic-lanceolate, 35–50 × 8–12 cm, apex acute, subcoriaceous. **Inflorescences** racemes, 2–4-flowered, short, pendulous. **Flowers** large, creamy white; sepals membranous, dorsal sepal free, erect, elliptic-lanceolate, concave, 6.5–7.5 × 1.8–3.5 cm, acute; lateral sepals connate at base, reflexed, obliquely elliptic-lanceolate, 6–7.5 × 2–3.5 cm, acute; petals membranous, reflexed, linear-lanceolate, 5–6.5 × 1.1–1.5 cm, acuminate; lip fleshy, elongate, 4.5–6.5 cm long; hypochile cymbiform, deeply concave, lateral margins with falcate thickenings, broadly elliptic-lunate, margins of canal to basal concavity closely appressed; mesochile short, inserted on apex of hypochile, with 2 lateral, slender, falcate, incurved, elongate horns; epichile articulated with apex of mesochile, obovate, abruptly apiculate; column elongate, arcuate, broadly winged above, 5–6.5 cm long. **Distr:** Mexico, C and S America. **Ecol:** In wet, humid forests; 400–1,500 m. Fl. Jun. **Zones:** 10–12.

S. tigrina Bateman ex Lindl., *Sert. Orch.* 1. 1838. (Syn: *S. cavendiski*, etc.). [Tiger Stanhopea]. **Herbs**, epiphytic; pseudobulbs ovate, deeply furrowed, somewhat quadrangular, 3.8 cm long. **Leaves** single, borne apical; broadly lanceolate, 30 × 10 cm, apex acute, shiny, coriaceous; petiole 10 cm long,

Above: *Astelia chathamica, Astelia fragrans, Astelia solandri, Curculigo orchioides.*

channeled. **Inflorescence** 2–4-flowered, short, pendulous, entirely clothed with scales. **Flowers** >15 cm across, strongly fragrant; sepals ovate-lanceolate, 10 cm long, straw-colored, marked with irregular longitudinal purple-brown blotches; petals narrower than sepals, much undulated, base marked with transverse bands of darker color; lip very large and fleshy, shiny, spotted, 7.5 cm long, inflated, underside slightly convex, margins turned slightly inward on upper side, widely separated from each other, interior of lip richly marked with various colors, front covered with many rows of tubercles, which terminate in 2 rough callosities at base of a pair of compressed falcate, sharp-pointed horns; column broadly semioblong, 7–8 cm long, winged. **Distr:** Mexico. **Ecol:** In forests. Fl. May–Jul. **Zones:** 10–11. **Info:** ***S. tigrina* var. *nigroviolacea*** (Syn: *S. nigroviolacea*) [Black Tiger Stanhopea] from Mexico (Veracruz, Hidalgo) has flowers which are heavily blotched with dark purple, and the channel between epichile and column is much narrower, suggesting a smaller pollinator.

S. wardii Lodd. ex Lindl., *Sert. Orch.* t. 20, 1838. (Syn: *S. aurea*, etc.). **Herbs** epiphytic or lithophytic, erect; pseudobulbs ovoid, short. **Leaves** single, apical; lamina broad, plicate, petiolate. **Inflorescences** racemes, short, elongate, pendent, basal. **Flowers** large, fragrant; sepals membranous, concave, pale yellow, brownish-purple dotted; dorsal sepal free, erect, elliptic-lanceolate, 4.5–6 × 1.5–2.5 cm, acute or apiculate; lateral sepals connate at base, reflexed, elliptic-lanceolate, 4.5–6 × 2–3 cm, acute; petals pale yellow, brownish-purple dotted, reflexed, margins undulate, lanceolate-acuminate, 4–5 × 0.8–1.5 cm; lip very fleshy, complexly 3-parted, 4.5–5 cm long; hypochile inflated, subsaccate, inserted on base of column, with 2 large, lateral, purple-brown spots, with a deeply emarginate, central constriction, lower surface with gibbose swelling, lateral margins with falcate thickenings, bases produced into 2 short, teeth; mesochile short, pale yellow, inserted on apex of hypochile, with 2 elongate, acuminate, incurved, falcate horns; epichile pale yellow, articulated to apex of mesochile, ovate, acute, concave, margins reflexed; column arcuate, broadly winged above, 4–4.5 cm long. **Distr:** C and S America. **Ecol:** In rainforests; 400–2,700 m. Fl. Aug–Sep. **Zones:** 10–12.

ASTELIACEAE

Genera 2–4, **species** 36. **Herbs**, terrestrial, epiphytic, perennial, ± rhizomatous, dioecious, dwarf to very large; stems contracted, rounded, solid; sap clear. **Leaves** spirally arranged, often rosette-forming, alternate, 2-ranked (distichous), base sheathing or not; lamina simple, margins entire, venation parallel, surface with peltate scales. **Inflorescence** a branched raceme or spike, terminal, lateral or axillary; bracts large; flowers pedicellate. **Flowers** unisexual or bisexual, relatively small, 3-merous, actinomorphic; tepals 6, basally connate (rarely free); stamens 6(3+3); pistil with carpels united, carpels 3, locules 3, placentation axile, ovary superior, with few to many ovules per carpel, style 1, divided or not, stigma dry, terminal, capitate or trilobed. **Fruits** berries, rarely capsular; seeds 2–10 per locule, black (phytomelan). **Distr:** New Zealand to New Guinea, Pacific Islands E to Hawai'i, Chile, the Mascarenes.

Astelia

Species ~25. **Herbs**, epiphytic or terrestrial, dioecious, tufted, with indumentum of scales; stems short, usually hidden. **Leaves** 3-ranked, sheath broad, closed; lamina linear to ensiform, usually keeled, nerves numerous, variously incrassate. **Inflorescences** terminal, panicles, sometimes much-reduced, flowers usually numerous; peduncle simple, leafless; spathes often large, foliaceous, each subtending a simple raceme or a branched subinflorescence of 2–5+ racemes; bracteoles membranous, usually adnate to short pedicels. **Flowers** small, ♂ usually larger than ♀; perianth membranous to fleshy, tube very short to well-developed, lobes triangular to strap-shaped; stamens in ♂ < tepals, anthers dorsifixed, mobile, in ♀ minute with flat empty anthers; ovary superior, 1- or trilocular, ovules few to many, smaller in ♂, style short, thick, stigmatic surfaces 3, well-separated or confluent; nectaries 3, slitlike, below and between stigmatic surfaces. **Fruits** berries; seeds black, ± glossy, sometimes with funicular hairs. **Distr:** Mascarenes, New Guinea, Oceania, Falkland Islands. **Cult:** Propagate by division. Requires a fertile, moisture-retentive soil in full or partial sun.

A. chathamica (Skottsb.) L.B. Moore, *NZ J. Bot.* 4: 228, 1966. (Syn: *A. nervosa* var. *chathamica*). [Silver Spear]. **Herbs**, robust, tufted. **Leaves** keeled near sheath, sheath base white, scaly; lamina 0.6–2 m × 4–10 cm, adaxially silvered-green with a metallic sheen, covered with clear pellicle, abaxial surface with a pale grayish indumentum. **Inflorescences** large, erect, most parts scaly, at least when young; racemes numerous, all spathes except the smallest subtending subinflorescences of 2–3+ racemes; lower spathes long; flowers pedicellate, usually well-spaced. **Flowers** ♂ pale, tepals to 8 × 2.5–3 mm, strongly reflexed, perianth tube very short, drooping around pedicel and exposing base of pistillode, outer tepals scaly externally; ♀ with tepals more scaly, longer and proportionately narrower than in related species, ovary trilocular, style well-developed. **Fruits** subglobose, ~1 × 1 cm, orange. **Distr:** New Zealand (Chatham Islands). **Ecol:** In wet peaty soil. Fl. Oct. Fr. Mar–Jul. **Zones:** 9–10.

A. fragrans Colenso, *Trans. and Proc. New Zealand Inst.* 15: 333, 1883. (Syn: *A. nervosa* var. *sylvestris, Funckia fragrans*). [Bush Astelia, Fragrant Astelia, Kahaka]. **Herbs**, robust, tufted, sometimes forming colonies. **Leaves** stiffly ascending in lower half, strongly keeled above sheath; lamina 0.5–2(–2.5) m × 2.5–7.5 cm, apex narrowly attenuate, adaxially green, scaly at first, costa prominent, sometimes reddish, abaxially finely scaly; sheath white-scaly. **Inflorescence** a panicle of racemes, erect, broad, mostly glabrous; peduncle stout, ± equal in length to panicle; lower spathes long, narrowly attenuate; racemes 15–30(–50); flowers pedicellate, usually well-spaced. **Flowers** with ♂ greenish-fawn, tepals ~5 × 2 mm; ♀ dark green, almost glabrous, tepals ~3 × 2.5 mm, tube closely appressed to about middle of ovary, tips dull reddish, recurved, ovary shiny, green, triloculed, style thick, ill-defined. **Fruits** subglobose, 4–9 × 4–9 mm, orange, ± red-flecked, fleshy. **Distr:** New Zealand. **Ecol:** Coastal to lower montane forests. Fl. Oct–Nov. Fr. Dec–May. **Zones:** 9–10.

A. solandri A. Cunn., *Compan. Bot. Mag.* 2: 374, 1837. (Syn: *A. cunninghamii*, etc.). [Kowharawhara, Perching Lily]. **Herbs**, epiphytic or terrestrial, robust, tufted. **Leaves** drooping; lamina strap-shaped, 1–2 m × 2–3.5 cm, base narrowed, keeled, tightly folded, covered in fine whitish felt, adaxially bright green, abaxially silvery, scaly when young. **Inflorescence** a panicle of numerous racemes, 15–40 cm long, drooping, scaly; racemes wide-spreading, more lax in ♂, all except the smallest spathes subtending subinflorescences of 3(–5) racemes; lower spathes long; peduncle 0.3–1 m long; flowers pedicellate, usually well-spaced. **Flowers** with perianth yellowish, pinkish or (especially ♂) maroon, divided almost to base; tepals 2–4(–5) mm long, spreading, tips lacking scales; ovary 1-loculed, green, style very short. **Fruits** globose, ~4–5 mm ∅, translucent green to yellowish or dull brown; seeds <2 mm long. **Distr:** New Zealand. **Ecol:** In wet lowland forests. Fl. Oct–Jun. Fr. Throughout most of the year. **Zones:** 9–10.

HYPOXIDACEAE

Genera 7–9, **species** 100–220. **Herbs**; rhizomes or corms covered with remnants of leaf bases. **Leaves** radical, mostly 3-ranked; lamina linear to lanceolate, plicate, venation parallel, conspicuous, often white-hairy. **Inflorescences** spikes, racemes, umbellate or headlike, sometimes reduced to a single flower. **Flowers** bisexual or ♂, regular, epigynous; perianth 3+3, free or united, sometimes forming a long narrow tube above ovary, ± persistent, mostly yellow; stamens mostly 3+3, filaments short, erect, inserted at base of perianth segments, anthers dorsifixed or basifixed, introrse, entire or basally bilobed, longitudinally dehiscent; ovary inferior, tricarpellate, triloculed, locules with few- to many ovuled, placentation axile, style 1, short, trifid. **Fruits** usually capsules, dehiscing via apical longitudinal slits, or fleshy and indehiscent, often crowned with persistent perianth; seeds small, globose, black. **Distr:** Warm temperate or seasonal tropical regions of the Southern Hemisphere.

MIW

Above: *Hypoxis aurea, Molineria capitulata, Doryanthes palmeri.*

Curculigo

Syn: *Molineria.*

Species 13. **Herbs**, perennial, often with tuberous rhizomes. **Leaves** several, basal, sessile or petiolate; lamina usually lanceolate, plicate, leathery or papery. **Inflorescences** racemose, spicate, or subcapitate; scapes axillary, erect or apically bent downward, long or short. **Flowers** bisexual or unisexual, perianth often yellow; segments spreading, subequal, sometimes basally connate into a tube; stamens inserted at base of perianth segments, filaments very short, sometimes subequaling anthers, anthers nearly basifixed or dorsifixed; ovary usually hairy, ovules 2–many, style columnar, slender, stigma trilobed. **Fruits** berries, indehiscent, apex sometimes beaked; seeds small, often striped. **Distr:** Tropical and subtropical regions worldwide. **Cult:** Propagate from seeds or by division. Requires moist, humus-rich compost in partial sun.

C. orchioides Gaertner, *Fruct. Sem. Pl.* 1: 63, 1788. (Syn: *C. orchioides* var. *minor*). [Golden-eye Grass]. **Herbs**, small; rhizomes erect, subcylindric, ~10 × 1 cm. **Leaves** sessile or shortly petiolate; lamina lanceolate to linear, 10–45(–90) × 0.5–2.5 cm, base tapering, apex narrowly acuminate, laxly pilose or glabrous. **Inflorescences** umbel-like racemes, 4–6-flowered; flowering stems nearly enclosed in base of sheathing petiole, 6–7 cm long, pilose; bracts lanceolate, 2.5–5 cm long, margins ciliate; pedicels ~2 mm long. **Flowers** with perianth yellow, segments oblong-lanceolate, 0.8–1.2 × 0.25–0.3 cm, outer ones sometimes abaxially laxly pilose; stamens about half as long as perianth segments, filament 1.5–2.5 mm long, anther 2–4 mm long; ovary narrowly oblong, to 7.5 mm long, pilose, stigma lobes longer than style. **Fruits** berries, subfusiform, 1.2–1.5 × ~0.6 cm, beak ~2.5 mm long. **Distr:** Tropical and subtropical Asia to W Pacific. **Ecol:** In shaded understory of forests, open grassy slopes; 0–1,600 m. Fl. & Fr. Apr–Sep. **Zones:** 8–12.

Hypoxis

Species 50–100 **Herbs**, perennial; rhizomes subglobose or tuberous. **Leaves** 3–20, basal, sessile. **Inflorescences** terminal, umbellate or racemose, few-flowered or flower solitary; flowering stems shorter than leaves, slender, usually pilose. **Flowers** with perianth segments free, persistent; stamens inserted at base of perianth segments, filament short; anther nearly basifixed; style short, stigma trilobed. **Fruits** capsules. **Distr:** Worldwide (except Europe). **Cult:** Propagate from offsets or fresh ripe seeds. Requires a well-drained soil in full or partial sun.

H. aurea Loureiro, *Fl. Cochinch.* 1: 200, 1790. [Golden Star Grass]. **Herbs**; rhizomes globose to cylindric, fleshy, covered with fibers. **Leaves** 4–12, linear, 5–30 × 0.2–0.6 cm, yellow-pilose, base membranous, apex narrowly acute, yellowish-brown pilose, strongly nerved. **Inflorescences** 1- or 2-flowered; flowering stems 2.5–10+ cm long, slender, pale brown pilose; bracts 2, setaceous, linear, small. **Flowers** with perianth yellow; segments 6, oblong, 6–8 mm long, brown-pilose, persistent; stamens 6, at base of perianth, shorter than perianth segments, filaments short; ovary 3–6 mm long, pilose, style short, columnar, stigma erect, trilobed. **Fruits** clavate, 0.6–1.2 cm, trivalved; seeds numerous, subglobose, tuberculate. **Distr:** Tropical and subtropical Asia. **Ecol:** In forest margins, moist grassy slopes; near sea level to 2,600 m. **Zones:** 8–12.

Molineria

Syn: *Aurota.*

Species 7. **Herbs**, perennial; stems tuberous, subterraneous, often stoloniferous. **Leaves** radical, long-petiolate, ± pilose. **Inflorescences** usually dense, with large pilose bracts. **Flowers** yellow; perianth segments subequal, densely villous outside; stamens 3+3, equal, filaments short, erect, subulate, anthers linear or lanceolate, apex entire, base sagittate, dorsifixed, adnate to filament on abaxial side; ovary beaked by very short perianth tube or not, densely pilose, each locule many ovuled, style subfiliform, glabrous, stigma capitate. **Fruits** berries, crowned by enlarged persistent perianth; seeds subglobose, testa shiny or densely verruculose. **Distr:** Tropical and subtropical Asia to Australia (NE Queensland). **Cult:** Propagate by division. Requires a humus-rich soil, in a humid, warm place, in full or partial sun.

M. capitulata (Lour.) Herb., *Amaryllidaceae*: 84, 1837. (Syn: *Curculigo capitulata*, etc.). [Palm Grass, Weevil Lily]. **Herbs**, to 1.8 m tall, stout; rhizomes tuberous, thick, with creeping, slender stolons. **Leaves** often 4–7; lamina oblong-lanceolate to suboblong, 40–90 × 5–14 cm, plicate, margins entire, apex acuminate, papery, sometimes pubescent; petiole 30–80 cm long. **Inflorescences** (10–)15–30 cm long, brown-villous; racemes nodding, capitate to subovoid, 2.5–5 cm across, densely many flowered; bracts ovate-lanceolate to lanceolate, 1.5–2.5 cm long, hairy; pedicels ~7 mm long. **Flowers** with perianth yellow, segments ovate-oblong, ~8 × 3.5–4 mm, obtuse, outer segments adaxially hairy, inner ones adaxially hairy on midvein or at base of midvein; stamens 5–6 mm long, filament <1 mm long, anther linear, ~5 mm long; ovary subglobose to oblong, hairy, style longer than stamens, slender, stigma subcapitate. **Fruits** berries, white, subglobose, 4–5 mm Ø, beakless; seeds black, irregularly striped. **Distr:** Tropical and subtropical Asia to NE Queensland. **Ecol:** In shady and moist places of forests; (300–)800–2,200 m. Fl. May–Jun. Fr. Aug–Sep. **Zones:** 9–12.

DORYANTHACEAE

Genus 1, **species** 2. *Familial description same as generic.*

Doryanthes

Species 2. **Herbs**, very large, rosette-forming; basal stem short; roots thick, fleshy. **Leaves** basal, spiraled;

Below: *Aristea africana, Aristea eckloni, Babiana angustifolia, Babiana tubulosa.* [All MD except *Aristea eckloni* AF.]

Above: *Chasmanthe floribunda*, *Chasmanthe floribunda* var. *duckittii*, *Crocosmia* × *crocosmiiflora*, *Crocosmia masoniorum*.

lamina dorsiventrally thickened, tip brown, terete, then convolute, withering and leaving a broad ragged apex. **Inflorescence** a compound raceme, elongate or globose, each flower subtended by a large bract; scape erect or arching, with thinly textured sheathing leaves. **Flowers** actinomorphic or slightly zygomorphic, 3-merous, bisexual; tepals 6, in 2 whorls, bases shortly united, spathulate, slightly expanded; stamens 6, filaments linear-subulate, basally enlarged, fused to tepals for half the length, then erect, anthers pseudo-basifixed (peltate but with apex of filament enclosed in a tube formed by the connective), bilocular, dehiscing by longitudinal slits; ovary inferior, trilocular, with numerous ovules, style 1, with 3 furrows, stigma 3-angled; septal nectaries present at base of tepals at top of ovary. **Fruits** capsules, loculicidal; seeds many per locule, brown, kidney shaped, compressed, with a papery wing. **Distr:** E Australia (NSW, Queensland). **Cult:** Propagate by division of clumps or suckers, or from seeds (which can take up to 10 years to flower). Requires a humus-rich, well-drained soil in full or partial sun.

D. palmeri W. Hill ex Benth., *Fl. Austral.* 6: 452, 1873. (Syn: *D. guilfoylei*, etc.). [Spear Lily]. **Herbs**, giant, rosette-forming; roots fibrous. **Leaves** basal, very numerous, spreading, recurved; lamina ensiform, sword-shaped, to 3 m × 10–20 cm, slightly ribbed, glabrous, tip brown, tubular, 10–15 cm long. **Inflorescence** thyrsoid, to 1.2 m long, compact, of many short few-flowered spikes surrounded by red-brown oblong acute bracts, inner bracts shorter than the perianth; scape 2.5–5 m long, clothed with lanceolate, short, erect bracts. **Flowers** scarlet or brownish red, 6–12 cm long; tepals 4–6 cm long, erect to slightly spreading; stamens shorter than perianth segments, filaments gradually narrowed upward, anthers 1.2–1.8 cm long, green-yellow later purple, filaments 2.4–3.6 cm long; ovary tubular, 3.8 cm long, style deeply grooved, base conical, stigmas very minute, radiating. **Fruits** capsules, ellipsoid to ovoid, 7–9 cm long; seeds 1.5–2.2 cm long. **Distr:** Australia (Queensland to NSW). **Ecol:** On exposed rocky outcrops in wet sclerophyll forests, on coastal mountain ranges. **Zones:** 9–12.

IRIDACEAE

Genera 66, species 2,025. **Herbs**, perennial, rarely shrubs or annuals, with rhizomes, bulbs, or corms. **Leaves** alternate, often 2-ranked, often oriented edgewise to aerial stem; lamina usually sword-shaped to linear, parallel-veined, base sheathing. **Inflorescence** sometimes a spike or reduced to a solitary flower, more often of monochasial, umbellate cymes, each enclosed in 2 opposed bracts (spathes) and termed a rhipidium, which may be solitary and terminal or numerous and variously arranged in racemes or panicles; bracts 1 to several. **Flowers** bisexual, showy, actinomorphic or zygomorphic; perianth segments 6, in 2 whorls, inner and outer ones equal or differing in size and/or color, tube filiform or trumpet-shaped; stamens (2 or)3, anthers extrorse; ovary inferior (or superior), triloculed, ovules few to many; placentation axile, style with filiform, slender, or petaloid branches. **Fruits** capsules, loculicidal; seeds with or without aril and wings. **Distr:** Nearly worldwide.

Aristea

Species ~50. **Herbs**; rhizome long or short; stems simple or branched. **Leaves** equitant, in a distichous basal rosette; lamina lanceolate, ensiform or linear. **Inflorescence** a little- to much-branched panicle; peduncle compressed or subterete; floral sheaths with 1, 2, or rarely clusters of many flowers; bracts membranous, often lacerate, green or rusty brown. **Flowers** usually bright blue, actinomorphic; perianth membranous, rotate, with short, narrow, cylindrical tube, segments spreading, subequal, obovate, cuneate or oblong, spirally twisting after anthesis; stamens inserted at throat of perianth tube, filaments short, free, filiform or basally dilated, anthers erect, ovoid or oblong; ovary oblong or cylindrical, locules many ovuled, style slender, erect or declinate toward one of the inner tepals, with 3 branches. **Fruits** capsules, ovoid or oblong, obtusely or acutely trigonous; seeds small, globose or angular. **Distr:** S and tropical Africa, Madagascar. **Cult:** Propagate by division or from seeds. Requires a humus-rich, moist but well-drained soil in full or partial sun.

A. africana (L.) Hoffmanns., *Verz. Pfl.-Kult.*: 40, 1824. (Syn: *A. cyanea*, *Moraea africana*, etc.). **Herbs**, rhizomatous, evergreen, small, usually 10–15 cm high, forming dense clumps over 30 cm across; stems erect, compressed, subdichotomously branched, sometimes simple, to 45 cm high. **Leaves** linear or straplike, 1.5–4 mm wide, midgreen. **Inflorescences** with spathes and bracts green with wide translucent margins, closely and deeply fringed and curled, sometimes rusty above, keels dark. **Flowers** blue; tepals free, spreading, mostly obovate, 0.8–1.2 cm long; style with 3 fringed lobes. **Fruits** capsules, short, 3-winged. **Distr:** SW Cape Province. **Ecol:** On sand flats and mountain slopes. Fl. Oct–Jan. **Zones:** 9–11.

A. ecklonii Baker, *J. Linn. Soc. Bot.* 16: 112, 1878. (Syn: *A. cyanea*, *A. dichotoma*, etc.). [Blue Flies, Blue Stars, Blue-eyed Iris, Blue Corn-lily]. **Herbs**, ~0.6–1 m tall, usually clump-forming; rhizome purplish-red, ~1 cm thick. **Leaves**, linear, conduplicate, 15–60 × 1–1.5 cm, bases purplish-red, veins prominent. **Inflorescence** a lax corymbose-panicle; scape terminal or laterally compressed, winged, with reduced leaves, flowers in clusters of 2–4; floral sheaths lanceolate, 1–1.4 cm long; pedicels 1–2 mm long; bracts lanceolate, 4–8 mm long. **Flowers** with perianth ~2 cm across, persistent, perianth tube 1–1.5 mm long, light green; outer segments lanceolate, 0.8–1 × 0.3–0.4 cm, 3-veined, greenish outside, bright purplish-blue inside; inner segments ovate or lanceolate, 1–1.2 × 0.5–0.7 cm, obtuse, deep purplish-blue, veins dark purple; filaments ~4 mm long, blue-purple, anthers ~3 mm long, yellow, dorsifixed; ovary trigonous, green, style ~1 cm long, blue-purple, suberect, apex shortly trilobed, stigmas capitate, blue-purple. **Fruits** capsules, oblong, 1.5–2.5 × 0.3–0.5 cm, brownish-black, shiny, with 3 deep furrows; seeds numerous, brownish-black, subcylindrical. **Distr:** Cameroon, Uganda to S Africa. **Ecol:** In montane grasslands. Flowers opening early morning and closing before noon. **Zones:** 9–11.

Babiana

[Baboon Flower].
Species 90. **Herbs** perennial; corms globose, aerial parts dying back annually; corms with fibrous tunics; stem simple or branched, sometimes subterranean, often puberulent. **Leaves** several, lower 2–3 *basal* and entirely sheathing (cataphylls); *foliage* leaves usually plicate. **Inflorescence** a spike; floral bracts green, or dry and brown, inner shorter than outer and forked, sometimes cleft to base. **Flowers** often in shades of blue or purple, sometimes yellow, red, pink or whitish, actinomorphic, or zygomorphic with stamens unilateral and arcuate; tepals united into a perianth tube, subequal or forming 2 lips, sometimes the upper tepal much enlarged; ovary oblong-globose, sometimes villous, style branches short, somewhat expanded above. **Fruits** capsules, oblong-ovoid, coriaceous; seeds blackish or dark-brown, several per locule, globose. **Distr:** S tropical and S Africa. **Cult:** Propagate by division or from seeds. Requires a rich, well-drained soil in full sun.

B. angustifolia Sweet, *Hort. Brit.*: 396, 1826. (Syn: *Acaste pulchra*). [Baboon Flower]. **Herbs**, tuberous, 10–20 cm high; stem inclined. **Leaves** alternate (basal, spreading fanlike), distichous; lamina simple, lanceolate, elliptic, ovate or obovate, (3.5–)11–22 × 0.4–2.2 cm, undissected, pleated, base tapering or oblique, narrowed, margins entire, apex acuminate or acute, hairy; pseudo-petiole (0.4–)7.5–30.5 cm long. **Inflorescences** spikes, inclined, many flowered; bracts herbaceous, brown-tipped, 1–2 cm long, inner bracts divided to base. **Flowers** zygomorphic, inverted, dark blue to violet, lower tepals with black or red markings; perianth tube curved, long, narrowly funnel-shaped, 1.1–1.7 cm long; stamens 3, filaments arched, 0.8–1.2 cm long, anthers 5–6 mm long; ovary syncarpous, hairy, triloculed, ovules 20–50 per locule. **Fruits** capsules, globose, dehiscent, 0.9–1 × 0.7 cm. **Distr:** S Africa (SW Cape Province). **Ecol:** In seasonally damp clay flats and lower slopes. Fl. Aug–Sep. **Zones:** 9–11.

B. tubulosa (Burm. f.) Ker Gawl., *Irid.Gen.* 154, 1827. (Syn: *Ixia tubulosa*). [Baboon Flower]. **Herbs**,

Above: *Cypella herbertii, Dietes bicolor, Dietes grandiflora, Dietes iridioides.*

cormous, annual, 7–45 cm high; stem mostly underground. **Leaves** alternate, distichous, sheathing, petiolate; lamina simple, linear to lanceolate, 21–26 × 0.3–0.6 cm, pleated, undissected, base tapering or oblique, margins entire, apex acuminate or acute, hairy; petiole 10–13 cm long. **Inflorescences** spikes, bracteate, 1–8-flowered; flowers sessile; bracts 2–4 cm long. **Flowers** predominantly white or cream, somewhat irregular; perianth 2-whorled; corolla 1.5–3 cm long, tube cylindrical, (4.5–)5.5–8.5 cm long, tepals 6, inner and outer tepals similar, all joined; stamens 3, adnate to the perianth, all alternating with the corolla parts (inner whorl), free of each other, filaments 2 cm long, anthers 5–7 mm long, basifixed, dehiscing via longitudinal slits; ovary syncarpous, 6.5–9 cm long, inferior, triloculed, ovules 20–50 per locule, style 1, with 3 branches. **Fruits** capsules, dehiscent, obovoid, 0.8–1 × 0.5 cm. **Distr:** S Africa (SW Cape Province). **Ecol:** On sandy flats and lower slopes. Fl. Sep–Oct. **Zones:** 9–11.

Chasmanthe

Species 3. **Herbs**, perennial, from corms; stems simple or branched. **Leaves** several; lamina plane, ensiform. **Inflorescence** a spike, axes usually slightly flexed, many flowered; bracts green or purplish, ± equal, coriaceous. **Flowers** unscented, zygomorphic; tepals connate into a tube, orange to scarlet, rarely yellow, unequal, dorsal tepal horizontal, other tepals spreading to reflexed; perianth tube slender, cylindrical proximally, abruptly expanded; stamens unilateral, extended horizontally below dorsal tepal, anthers parallel; style arching over filaments, branching opposite to or beyond anthers, branches 3, filiform, shallowly notched apically. **Fruits** capsules, irregularly globose, cartilaginous; seeds 2–4 per locule, globose, testa bright orange, sometimes fleshy. **Distr:** S Africa (Cape Province). **Cult:** Propagate from seeds or offsets. Requires a well-drained soil in full sun.

C. floribunda (Salisbury) N.E. Brown, *Trans. Roy. Soc. South Africa.* 20: 274, 1932. (Syn: *Antholyza floribunda*, etc.). [(South) African Cornflag]. **Herbs**, 0.6–1.2 m high; corm depressed-globose, 4–7 cm Ø; stems 1–2-branched. **Leaves** 8–10, mostly basal, these longer than cauline; lamina lanceolate, ± reaching base of spike, 1.8–3.5 cm wide. **Inflorescence** a spike, 25–40-flowered; outer spathe 1.3–1.5 cm long, apex obtuse to truncate; inner ± equaling outer, apex bifurcate; flowers distichous. **Flowers** with perianth tube 3.5–4.5 cm long, often twisted, base pouched; dorsal tepal horizontal, 2.8–3.3 × 0.7–0.9 cm, much exceeding others; lateral tepals patent or recurved, 1.2–1.5 × 0.4–0.7 cm; lower median tepal slightly smaller than laterals; filaments 5–5.5 cm, anthers 7–8 mm long; ovary 6–9 mm long, style branching shortly below and opposite to anthers, branches 0.7–1 cm long. **Fruits** capsules, 1–1.5 cm; seeds 5–7 mm Ø. **Distr:** S Africa (SW Cape Province). **Ecol:** In coastal and montane areas, grassy slopes; 0–100 m. Fl. Feb–May. **Zones:** 9–11. **Info:** *C. floribunda* var. *duckittii* [Yellow Adam's Rib, Yellow Cobra Lily] is often cultivated.

Crocosmia

Syn: *Curtonus.*

Species 8. **Herbs**, perennial, with large corms, aerial parts dying back annually; corms with fibrous tunics; stems terete, usually branched, branching often strongly divaricate. **Leaves** several, forming a fan, lower 2–3 sheathing stem base (cataphylls); *foliage* leaves plane or plicate, mostly basal; *cauline* leaves much-reduced. **Inflorescence** a spike, usually inclined or horizontal; floral bracts coriaceous, fairly short, the inner about as long as the outer. **Flowers** in shades of orange to reddish, actinomorphic to strongly zygomorphic with stamens arched under the upper tepal, this sometimes much larger than the others; perianth tube cylindric, funnel-shaped or curved with a narrowly cylindrical lower portion and broadly cylindrical upper portion; style exserted, branches filiform or apically broadened, apex entire or notched. **Fruits** capsules, globose to depressed, trilobed, ± coriaceous; seeds globose, hard, shiny, few per locule. **Distr:** Sudan to S Africa, Madagascar. **Cult:** Propagate by division of corms. Requires a moist, well-drained soil in full sun.

C. × crocosmiiflora (Lemoine) N.E. Brown, *Trans. Roy. Soc. South Africa.* 20: 264, 1932. (Syn: *Montbretia × crocosmiiflora*, etc.). [Montbretia]. **Herbs**, 0.5–1 m; corms globose, 1.5–2.5 cm Ø; stems usually 2–4-branched, often curving distally. **Leaves** 5–8, mostly basal, basal much larger than cauline; lamina lanceolate, to 90 × 0.8–2 cm. **Inflorescences** spikes, to 1.2 m long; spathes 0.6–1 cm long, outer apex acute, inner apex 2-keeled, bifurcate, flowers distichous. **Flowers** with tepals spreading; dorsal tepal weakly arcuate, center paler, sometimes with darker markings at base, subequal, outer whorl slightly larger than inner, 1.6–2.5 × 0.6–0.9 cm, perianth tube 1.2–1.5 cm long; filaments 1.5–2.2 cm long, anthers 6–8 mm long; ovary oblong, ~3 mm long, style arching over stamens, branching at or shortly beyond anther apices, branches ~4 mm long, apically bifid. **Fruits** capsules, ~8 mm long; seeds ~2.5 mm Ø. **Distr:** S Africa. **Ecol:** In forest margins and thickets. Fl. Jul–Aug. **Zones:** 6–11. **Info:** This hybrid is a cross between *C. aurea* and *C. pottsii*.

C. masoniorum (L. Bolus) N.E. Brown, *Trans. Roy. Soc. South Africa* 20: 264, 1932. (Syn: *Tritonia masoniorum*). [Swan Crocosmia, Giant Montbretia]. **Herbs**, tuberous, deciduous, 0.8–1.2 m high. **Leaves** arranged in a fan; lamina lanceolate, 0.45–0.6(–1) m × 2.5–5 cm, pleated, both ends tapered, glabrous. **Inflorescences** spikes, held horizontal, densely flowered, with the flowers borne in 2 rows, in a showy arc on the upper side of the spike, with all the flowers facing upward, axis not zigzag. **Flowers** bright orange to brilliant orange-red, 4–5 cm across, funnel-shaped; floral tube 1.8–2.5 cm long, slender, then abruptly expanded; tepals widely spreading, 1.8–3 × 0.5–1.2 cm; stamens suberect, borne at center of the flowers, gently arching, 3–3.5 cm long, projecting beyond the perianth lobes. **Distr:** E Cape Province to KwaZulu-Natal (Engcobo District). **Ecol:** On shaded, wet rock ledges near mountain streams. Fl. Dec–Jan. **Zones:** 7–10.

Cypella

Species 22. **Herbs**, perennial; rootstock a long bulb with pleated tunics. **Leaves** pleated. **Inflorescence** erect, several-flowered; scape usually branched, slender. **Flowers** short-lived; outer 3 perianth segments large, spreading; inner 3 perianth segments much smaller, usually erect and often hairy on the claw; segments free; filaments free; style longer than the stamens, branches sometimes petaloid. **Fruits** capsules; seeds angular. **Distr:** Peru, Brazil to N Argentina. **Cult:** Propagate by division or from seeds. Requires a rich, well-drained soil in full sun. Dormant bulbs should be stored in sand in a frost-free place.

C. herbertii (Lindl.) Herb., *Bot. Mag.* 53: t. 2637, 1826. (Syn: *Moraea herbertii*). [Goblet Flower]. **Herbs**, perennial, cormous; stem 30–50 cm long. **Leaves** linear, 15–30 × ~2 cm. **Inflorescence** usually much-branched, 25 cm long. **Flowers** actinomorphic, 4–6 cm across; outer 3 perianth segments larger, 3–4.5 cm long, dull orange-yellow, with a violet central stripe, claw concave; inner 3 perianth segments much smaller, 0.9–1.4 cm long, yellow, purple-spotted, rolled toward the apices; stigma purple. **Distr:** S Brazil to NE Argentina. **Ecol:** Fl. throughout most of the year. **Zones:** 8–10.

Dietes

Syn: *Naron.*

Species 6. **Herbs**, perennial, evergreen; rhizome creeping; stem usually erect, bearing leaves at lower nodes, with sheathing leaves at upper nodes; branching irregularly in upper half. **Leaves** several, ensiform, distichous; lamina lanceolate, coriaceous, midrib absent. **Inflorescences** terminal rhipidia (umbellate flower clusters enclosed in paired opposed spathes), flowers several per rhipidium, borne serially; spathes sheathing, coriaceous, outer spathe smaller than inner; pedicels pubescent above. **Flowers** actinomorphic; tepals free, all broadly unguiculate, limbs spreading; outer tepal claws pubescent; filaments usually free, occasionally united below, anthers linear, appressed to style branch; style dividing shortly above base into 3 broad petaloid appendages (crest), stigmas transverse, abaxial, below base of crests.

Above: *Ferraria crispa, Freesia alba, Iris bulleyana, Iris chrysographes.* [All MD except *Freesia alba* AF.]

Fruits capsules, cartilaginous, often irregularly grooved or smooth, tardily dehiscent or indehiscent, erect or pendent; seeds irregularly angled, relatively large. **Distr:** Ethiopia to S Africa, Lord Howe Islands. **Cult:** Propagate from seeds, or by division. Requires well-drained soil in full sun.

D. bicolor (Steud.) Sweet ex Klatt, *Linnaea* 34: 584, 1866. (Syn: *Moraea bicolor*). [Butterfly Iris, Peacock Flower, Yellow Wild Iris]. **Herbs**, rhizomatous, 0.8–1.2 m tall; stem erect, bearing short leaves on lower nodes and short paired opposite bractlike structures on upper nodes; stem bracts 0.9–2 cm long, acute. **Leaves** linear, 0.5–1 m × 0.6–1.2 cm, pale green, with a distinct, usually double central vein. **Inflorescences** terminal rhipidia; spathes paired, 3.4–4.5 cm long, obtuse to emarginate. **Flowers** yellow, usually with a dark brown nectar guide on outer tepals, guide sometimes lacking and tepal claw dotted; outer tepals 3.5 cm long, claw ~1.2 cm long, bearded and orange-speckled, limb horizontal, to 2.3 cm wide; inner tepals to 3.3 × ~1.8 cm, horizontal; filaments 6 mm long, anthers 4–8 mm long; ovary ~1 cm long, style ~2 mm long, branches 0.8–1 × ~0.9 cm; crests ~5 mm long. **Fruits** globose-truncate, to 2.5 cm long. **Distr:** S Cape Province to KwaZulu-Natal. **Ecol:** On banks of perennial streams and rivers. **Zones:** 9–11.

D. grandiflora N.E. Brown, *J. Linn. Soc.*, Bot. 48: 35, 1928. [Large Wild Iris, Fairy Iris]. **Herbs**, rhizomatous, 1–1.5 m tall; stem irregularly branched, lower nodes with short leaves, upper nodes with spathelike bracts 2.5–5 cm long. **Leaves** linear, 0.75–1 m × 1–1.5(–2) cm. **Inflorescences** terminal rhipidia; spathes 5–8 cm long, outer ones smaller. **Flowers** white with yellow nectar guides and a yellow beard on outer tepals; inner tepals marked with dark brown toward base, style branches pale mauve; outer tepals 4.5–6 cm long, claw 2–2.6 cm long, with a dense yellow beard down midline, limb horizontal to recurved, 2.5–3.5 cm wide; inner tepals (3.6–)4–4.5 × ~2.5 cm, limb horizontally spreading; filaments 1–1.3 cm long, tapered, anthers 0.7–1 cm long; ovary 1.3–1.6 cm long, style ~5 mm long, branches 1.2–2 × 0.6–0.8 cm, crests 1.2–1.5 cm long, erect. **Fruits** cylindrical, 2.8–4.5(–5) cm long, ridged and furrowed vertically. **Distr:** S Cape Province to KwaZulu-Natal. **Ecol:** In forest margins, along streams. **Zones:** 9–11.

D. iridioides (L.) Sweet, *Hort. Brit.*, ed. 2: 497, 1830. (Syn: *D. vegeta*, etc.). [African Iris]. **Herbs**, (15–)30–60 cm high, often tufted; stem bearing short leaves below, and sheathing bractlike leaves above, irregularly branched. **Leaves** narrowly lanceolate, 2.5–4(–6) × 6–15(–25) cm; sheathing leaves 2.5–3 cm long. **Inflorescences** rhipidia; spathes 3.5–5(–5.5) long; old inflorescences sometimes producing branches with a terminal cluster of leaves, and ultimately rooting. **Flowers** white with yellow nectar guides on outer tepals, claws of outer tepals orange-dotted, style branches blue or white flushed with blue above; tepals spreading to recurved; outer tepals 2.4–3.5 cm long, claw ± 1.6 cm long, limb 1.2–1.6 cm long; inner tepals 2.4–2.8 × 0.9–1.2 cm; filaments 5–9 mm long, anthers 3–6 mm long; style 2–3 mm long, branches 7–9 × 4–6 mm, crests ~5 mm long. **Fruits** ovoid-cylindric, erect, usually rostrate, 1.5–3 cm long. **Distr:** Ethiopia to S Africa. **Ecol:** In evergreen forests; 1,100–2,300 m. Fl. Jul–Jan. **Zones:** 9–11.

Ferraria

[Spider Iris].

Species 13. **Herbs**, cormous, deciduous, perennials; corm depressed-globose, persistent, new corms produced from base of flowering stem; stem erect, usually branched. **Leaves** unifacial, sword-shaped to linear or falcate, thick, leathery, succulent, usually glaucous. **Inflorescences** rhipidia, single, terminal on short branches, 2–6 flowers per rhipidium; spathes fleshy, leaflike, inner exceeding outer. **Flowers** fugacious or lasting as long as 3 days, often unpleasantly scented or sweetly scented (vanillalike), often dull-colored, in shades of brown to yellow, occasionally blue, usually mottled with darker color; tepal margins lighter or darker; tepals free, subequal or outer slightly larger, spreading, margins crisped; stamens united below into a column, diverging above, anthers appressed to style branches; ovary ovoid-truncate to narrowly oblong, tapering above, style filiform, bilobed. **Fruits** capsules, oblong to ellipsoidal, apex acute apex; seeds globose or angled. **Distr:** S Zaïre to S Africa. **Cult:** Propagate by division or from seeds. Requires a moderately fertile, well-drained soil in full sun.

F. crispa Burman, *Nova Acta Phys.-Med. Acad. Caes. Leop.-Carol. Nat. Cur.* 2: 199, 1761. [Sea Spider Iris]. **Herbs**, 0.4–1 m high; stems 2–3 per corm, to 45 cm long, usually much shorter, leafy and much-branched, covered by leaf sheaths. **Leaves** arranged in 2 ranks; *lower* leaves narrow, sword-shaped, 15–30 × 0.8–1.2 cm, with a thickened zone in the middle and a strong midrib; *uppermost* leaves bractlike, surrounding the flowers. **Inflorescences** rhipidia, numerous, mostly 2-flowered, borne from axils of leaflike bracts; bracts leaflike, boat-shaped, ~4 × 2 cm; pedicels sturdy. **Flowers** short-lived, cocoa-scented or somewhat fetid, dark brown with pale lines and spots, or pale yellow with brown lines and spots, 2.2–3 cm long, ~5 cm across, velvety textured; perianth lobes ~2 cm long, consisting of an oblong, erect claw and a slightly longer, spreading or slightly deflexed, strongly wavy lamina; anther lobes parallel; ovary spindle-shaped, without a beak, style branches fringed with numerous long, hairlike processes. **Distr:** Cape Province. **Ecol:** In coastal areas. Fl. Jul–Oct. **Zones:** 9–11.

Freesia

Syn: *Anomatheca.*

Species 16. **Herbs**, perennial, cormous; corms conic, ~1 cm Ø; tunics fibrous, fibers fine, pale, reticulate; stems simple or branched. **Leaves** several, sometimes prostrate; lamina plane, lanceolate (rarely ensiform or oblong). **Inflorescences** spikes, secund, usually strongly flexed basally, ± horizontal, several- to many flowered; bract green (pale straw-colored), coriaceous or membranous. **Flowers** often strongly fragrant, zygomorphic; tepals connate into a tube, mostly white or yellow, sometimes pink or red, subequal, outer whorl slightly larger than inner; perianth tube cylindric or funnel-shaped, short or long proximally, abruptly expanded into a short to long tube or flared distally; stamens asymmetric, unilateral, anthers usually parallel; style arching over filaments, dividing into 3 filiform branches. **Fruits** capsules, irregularly globose, usually rugose; seeds several per locule, globose, without wings; testa light to dark brown, shiny. **Distr:** Kenya to S Africa. **Cult:** Propagate by division of dormant corms. Requires a moderately fertile, well-drained soil in full sun.

F. alba (G.L. Meyer) Gumbleton, *Gard. Chron.*, ser. 3. 19: 392, 1896. (Syn: *F. refracta* var. *alba*). [Freesia]. **Herbs**, (5–)12–40 cm high; corm conical, tunics finely netted; stems erect, simple or branched, minutely pubescent. **Leaves** (4–)6–8, ascending or inclined; lamina lanceolate, slightly shorter than to ± equaling spike, apices acute. **Inflorescence** a spike, horizontal, (2–)4–6(–8)-flowered; spathes 5–8 mm long, outer usually longer than inner, firm-textured; bracts green and leathery. **Flowers** strongly scented, funnel-shaped; perianth tube gradually flared, distal portion (1.5–)2–4 cm across, basal portion slender, 0.6–1.2 cm across; tepals spreading, white, often with yellow markings on lower tepals, tepals and tube flushed with purple abaxially, oblong, 1.5–1.8 cm long; filaments included, 1.5–2.5 cm long, anthers 6–9 mm long, unilateral, sometimes parallel; ovary 2–3 mm long, style branching opposite anther apices. **Fruits** ~1 cm long, slightly papillose; seeds ~2 mm Ø. **Distr:** S Africa (SW Cape Province). **Ecol:** Mainly in coastal areas. Fl. Jul–Oct. **Zones:** 9–11.

Iris

Syn: *Pardanthopsis.*

Species 260. **Herbs**, perennial, usually with short or long rhizomes (or bulbs), sometimes with swollen storage roots; aerial stem developed, much shorter than or longer than leaves. **Leaves** mostly basal, 2-ranked, often oriented edgewise to aerial stem; lamina sword-shaped to linear. **Inflorescence** of rhipidia arranged in racemes, panicles, or solitary. **Flowers** relatively large, often fragrant, white, yellow, blue, violet, or purple; perianth tube usually trumpet-shaped; outer 3 perianth segments (falls) ornamented or not, with a

MD

MD

Above: *Iris domestica*, *Iris douglasiana*, *Iris fernaldii*, *Iris japonica*.

crested, bearded, or beardless claw (haft), and an expanded, usually reflexed limb (blade), the beard or crest (when present) extended along midvein of proximal part of limb adaxially; inner 3 perianth segments (standards) smaller, never ornamented, erect or spreading; stamens opposed to style branches and appressed to them, filaments free, flattened; style with 3 colored, expanded, petaloid branches, arching over stamens, apex of branches extended as 2 lobes (crests) projecting beyond flaplike, abaxial stigma. **Fruits** capsules, globose, ellipsoid, or cylindric, apex beaked or not; seeds pyriform, flattened, D-shaped, or irregularly angled, sometimes arillate. **Distr:** Temperate Northern Hemisphere to Philippines. **Cult:** Propagate by division. Requires permanently damp soil (bog irises) or moist but well-drained soil (woodland irises) in full or partial sun.

I. bulleyana Dykes, *Gard. Chron.* III, 47: 418, 1910. (Syn: *I. bulleyana* f. *alba*, *Limniris bulleyana*). [Siberian Iris]. **Herbs**; rhizomes creeping, thick. **Leaves** linear, 15–45 × 0.3–1 cm, grayish-green, midvein absent, base reddish. **Inflorescences** (1 or)2-flowered; scapes 10–40(–70) cm long, hollow, 2- or 3-leaved; spathes 2 or 3, green, margins reddish-brown, 5.5–12 × 0.8–1.2 cm; pedicel 2–6 cm long. **Flowers** pale lilac or bluish-violet to dark violet, rarely white, 6.5–7.5 cm Ø; perianth tube 1–1.2 cm long; outer segments obovate, with darker lines and markings on central, whitish or yellowish patch, 4.5–5 × ~2.5 cm; inner segments ± erect, light violet, lanceolate, ~4 × 1.5 cm; stamens ~2.5 cm, anthers milky white; ovary ~2 cm long, style branches ~3.5 cm long. **Fruits** cylindric, 4–5.5 × 1.5–1.8 cm, 3-angled, 6-veined, obtuse; seeds brown, semiorbicular, flat. **Distr:** Tibet to Myanmar. **Ecol:** In damp places such as streamsides; 2,300–4,300 m. Fl. Jun–Jul. **Zones:** 6–10.

I. chrysographes Dykes, *Gard. Chron.* III, 49: 362, 1911. (Syn: *Limniris chrysographes*, etc.). [Gold Net Iris]. **Herbs**; rhizomes shortly creeping, thick. **Leaves** linear, 25–70 × 0.3–1.2 cm, grayish-green, midvein obscure. **Inflorescences** 2-flowered; scape 25–50 cm, hollow; spathes 3, green, lightly tinted reddish-purple, lanceolate, 6.5–9 × 0.8–1.5 cm, acuminate. **Flowers** dark violet, velvety, 6–9 cm Ø; outer segments narrowly obovate, limb with golden-yellow basal stripes, 5.5–7 × 2.5–3.5 cm; inner segments obliquely spreading, narrowly lanceolate, ~6 × 1 cm; stamens 4–4.5 cm long, filaments purple, anthers violet; ovary 3–3.5 cm long, style branches dark purple, 4.5–5 × 0.6–0.8 cm. **Fruits** ellipsoid, 4–6 × 1.7–2 cm, 3-angled; seeds maroon-brown, semiorbicular. **Distr:** Tibet to NE Myanmar. **Ecol:** In forest margins, streamsides; <4,000 m. Fl. Jun–Jul. **Zones:** 8–10.

I. domestica (L.) Goldblatt and Mabberley, *Novon* 15: 129, 2005. (Syn: *Belamcanda chinensis*, etc.). [Leopard Flower]. **Herbs**, 0.6–1 m tall; rhizomes just below ground level, horizontal, flesh usually orange; stems usually 3–5-branched. **Leaves** 8–14 per stem; lamina narrowly lanceolate, 30–45 × 1–2 cm, glaucous. **Inflorescences** rhipidia, usually 3–6-flowered; spathes 1–2 cm. **Flowers** with tepals ascending proximally, spreading distally, light orange to reddish (rarely yellow), with large, irregularly scattered spots of darker pigment, lanceolate, 1.6–3.5 cm long; filaments (0.6–)1–1.2 cm long, anthers 6–8 mm long; ovary narrowly ovoid, 3-angled, (4–)6–8 mm long, style 3-angled proximally, thickened distally, dividing opposite anther apices, branches ~2 mm long. **Fruits** ovoid, 1.8–2.6 cm long, truncate; seeds persistent, ~4 mm Ø, shiny. **Distr:** Himalayas to Japan and Philippines. **Ecol:** On grasslands, rocky outcrops; ~0–2,200 m. Fl. Jun–Aug. **Zones:** 8–12. **Info:** See Goldblatt and Mabberley (2005) for a further discussion on the new name.

I. douglasiana Herbert G.A.W. Arnott, *Bot. Beechey Voy.* 395, 1840. (Syn: *Limniris douglasiana*, etc.). [Mountain Iris]. **Herbs**; rhizomes freely branching, forming large colonies, slender; stems 1–4-branched, solid, 15–70 cm long. **Leaves** *basal* linear, 4.5–10 dm × 2 cm, apex acute, yellow-green, occasionally darker green, sometimes flushed pink or red basally, prominently ribbed; *cauline* 1–3, reduced. **Inflorescences** (2–)3-flowered, branches 2–3-flowered; spathes opposite or separated, divergent, green, sometimes basally purple-flushed, lanceolate-acuminate, 6–12 × 0.7–1.2 cm; pedicel 2–5 cm long. **Flowers** with perianth deep red-purple, lavender, gray blue, cream, or white, with gold and blue or purple veins; floral tube 1.5–2.8 cm long, basally widened, bowl-shaped; sepals oblanceolate to obovate, 5–9 × 1.4–3 cm, base gradually attenuate, apex obtusely rounded; petals oblanceolate, 4.5–7 × 0.9–1.8 cm, base attenuate, narrowly clawed; ovary elliptic-oval, 3–4 cm long, style 1.7–3.5 cm long, crests overlapping, subquadrate, 1–2 cm long, margins coarsely toothed, stigmas triangular. **Fruits** 2.5–5 cm long. **Distr:** Oregon to California. **Ecol:** In open woods. Fl. Apr–Jun. **Zones:** 7–10.

I. fernaldii R.C. Foster, *Iridis Sp. Nov.* 1, 1938. (Syn: *Limniris fernaldii*). **Herbs**; rhizomes many branched, producing dense clumps; stems simple, solid, 20–40 cm long. **Leaves** *basal* to 40 × 0.7–0.8 cm, margins not thickened, often glaucous, gray green, usually brilliantly colored basally; *cauline* 2–several, spreading, sheathing stem, foliaceous, lamina not inflated. **Inflorescences** 2-flowered; spathes opposite, connivent, often basally red-flushed, broadly lanceolate, 5–9 × 0.6–1.1 cm, subequal; pedicel 0.9–2.2 cm long. **Flowers** with perianth pale creamy yellow, with gold or gray veins; floral tube funnelform, 3–6.2 cm long, spreading apically; sepals and petals with base gradually attenuate, broadly clawed; sepals horizontally spreading, claw with deep yellow median line, oblanceolate to spathulate, 4.5–7 × 1–2 cm; petals narrowly oblanceolate, 4.3–6 × 0.6–1.4 cm; ovary elliptical, 1.5–2.3 cm long, style 2.2–3 cm long, crests divergent, linear to narrowly oblong, 1–1.7 cm long, stigmas triangular, margins entire. **Fruits** oblong, roundly trigonal, beaked, 2.5–3.5 cm long; seeds dark brown, wrinkled. **Distr:** NW California. **Ecol:** In mixed evergreen forests; Fl. Apr–Jul. **Zones:** 8–10.

I. japonica Thunberg, *Trans. Linn. Soc. London*, Bot. 2: 327, 1794. (Syn: *Evansia japonica*, etc.). Herbs; rhizomes dimorphic, suberect, thick, creeping. **Leaves** basal; lamina sword-shaped, 25–60 × 1.5–3 cm, dark or yellowish-green, glossy or dull, basally reddish-purple, midvein absent. **Inflorescences** erect, with 5–12 apical branches; spathes 3–5, broadly lanceolate, 0.8–1.5 cm long, 2–4-flowered, apex obtuse; pedicel 1.5–2.5 cm long, stiff, persistent. **Flowers** pale bluish, 4.5–5.5 cm Ø; perianth tube 1.1–1.5 cm long; outer segments obovate or elliptic, 2.5–3 × 1.4–2 cm, limb spreading, with blue blotching around central, yellow patch around prominent, yellow crest, margin denticulate, undulate, apex retuse; inner segments obliquely spreading, elliptic or narrowly obovate, 2.8–3 × 1.5–2.1 cm, margin denticulate, undulate; stamens 0.8–1.2 cm long, anthers white; ovary 0.7–1 cm, style branches pale blue, terminal lobes fimbriate. **Fruits** ellipsoid-cylindric, 2.5–3 × 1.2–1.5 cm. **Distr:** Myanmar to China, Japan. **Ecol:** In open forest margins; 500–800 (–3,400) m. Fl. Mar–Apr. **Zones:** 8–10.

Ixia

Species 61. **Herbs**, perennial; corms with papery or fibrous tunics; stem simple or branched. **Leaves** mostly basal, distichous; lamina linear, lanceolate or threadlike, usually with a false midrib. **Inflorescence** a spike or panicle of spikes; spikes 2–many flowered, 2-ranked or spirally arranged; bracts longer than ovaries. **Flowers** with perianth gamophyllous, infundibuliform or salver-shaped, lobes equal; stamens 3, attached to the throat of the hypanthial tube or inside it; filaments free or connate at base; ovary biloculed, placentation axile, style threadlike, longer than stamens, stigma trifid. **Fruits** capsules, membranous, trivalved; seeds angular. **Distr:** S Africa (Cape Province). **Cult:** Propagate from offsets. Requires a well-drained soil in full sun. Corms need to be dried off after flowering.

I. monadelpha D. Delaroche, *Descr. Pl. Aliq. Nov.*: 22, 1766. [Wand Flower]. **Herbs**, 15–40 cm high; corm globose, 1.2–2 cm Ø, with fibrous tunics; stem mostly with 1 or 2 spreading branchlets, slender, terete, 15–40 cm long. **Leaves** 4–7, often spirally twisted; lamina linear to lanceolate, 8–30 × 0.3–1 cm, glabrous, moderately firm in texture, veins several. **Inflorescence** a spike, 4–12-flowered, flowers crowded, short; outer spathe-valve oblong, pale green,

Above: *Ixia monadelpha, Moraea aristata, Neomarica gracilis, Sisyrinchium striatum.*

tricuspidate, 6–8 mm long. **Flowers** blue, turquoise to purple, rarely whitish, usually with a greenish, blue or blackish center; perianth tube filiform or cylindrical, 1–1.8 cm long, limb campanulate, 1.2–2 cm long; tepals spreading, 1.3–2 × 0.6–1.1 cm long; stamens arising at throat of perianth tube, fully exserted, filaments ± united into a tube, dark-blue or black, anthers 7–8 mm long, erect, almost touching each other; style branches much shorter than the anthers. **Distr:** S Africa (Cape Province). **Ecol:** On wet sandy flats and lower slopes. Fl. Sep–Oct. **Zones:** 9–10.

Moraea

Syn: *Freuchenia*, etc.
Species ~200. **Herbs**, deciduous, geophytic, seasonal, perennial; stems usually aerial, occasionally entirely subterranean. **Leaves** 1–many, narrow, linear, falcate, coiled, terete or canaliculate, apex minute. **Inflorescence** a cymose-corymb, enclosed by 2(1) opposed spathes; flowers solitary or several per spathe, borne serially. **Flowers** long-lived or fugacious, mainly yellow to white; tepals 6 (occasionally 3), free or united, clawed, unequal to subequal, spreading to reflexed or inner erect; outer tepals unguiculate, larger than inner, marked with a conspicuous nectar guide; inner tepals petaloid and unguiculate, with a spreading limb, or entire and erect, or trilobed, cusplike, or absent; stamens opposite outer tepals and appressed to style branches, filaments occasionally free, anthers oblong to linear, occasionally exceeding stigma lobe; ovary ovoid, clavate, to linear, sometimes triangular. **Fruits** capsules, globose to obconic or oblong; seeds discoid or angular. **Distr:** Tropical and S Africa, Mediterranean to W Himalaya. **Cult:** Propagate by division or from seeds. Requires a fairly rich soil with excellent drainage, in full sun.

M. aristata (de la Roche) Ashcers.and Graebn., *Syn. Mitteleur. Fl.* 3: 518, 1906. (Syn: *Ferraria ocellaris*, etc.). [Peacock Moraea, Peacock Iris]. **Herbs**, 25–35 cm tall; corms ~2.5 cm Ø; tunics of pale fibers; stem glabrous, occasionally 1-branched. **Leaves** solitary, linear, glabrous, basal, exceeding the inflorescence. **Inflorescence** with flowers solitary; spathe herbaceous, dry above with dark brown, apex attenuate; inner spathe 5–7 cm long, outer spathe ~2.5–3.5 cm long. **Flowers** white with prominent nectar guides, variously marked with concentric crescents of black, blue violet, or green; outer tepals 3–3.5 cm long, claw heavily bearded, ~1.2 cm long, limb very broad, spreading, ~2 × 2.5 cm; inner tepals 1.5–2 cm long, tricuspidate, lateral lobes short, obtuse, central cusp long, spreading; filaments 3–4 mm long, free near apex only, anthers ~5 mm long; style branches ~7 mm long, crests 7 mm long, triangular. **Fruits** clavate-cylindrical, 1.5–2 cm long; seeds angled, testa spongy. **Distr:** S Africa (Cape Town); rare in the wild. **Ecol:** Fl. Sep. **Zones:** 9–11.

Neomarica

Syn: *Galathea*.
Species 21. **Herbs**, perennial; rhizomes short. **Leaves** distichous; lamina broadly linear, ensiform, flabellate, with a prominent costa. **Inflorescence** with peduncle compressed and leaflike, long-produced beyond the inflorescence, with 1 or few subsessile or pedunculate flower clusters. **Flowers** fugitive, blue, yellow, or white; perianth tube obsolete, segments of the 2 rows very unlike; outer segments obovate, spreading; inner segments much smaller, panduriform, convolute; stamens short, erect, filaments distinct; ovary clavate, 3-celled, ovules numerous, superposed, style subulate at base, cyathiform with 3 angles in the upper half to which the anthers adhere loosely; style crests lanceolate, with the transverse stigmas at their base. **Fruits** capsules, oblong, loculicidally trivalvate; seeds subglobose or angled by pressure, with a brown testa and corneous endosperm, surrounded by a fleshy aril. **Distr:** Mexico to Trinidad, Brazil to NE Argentina. **Cult:** Propagate by division or from seed.

N. gracilis (Herb.) Sprague, *Bull. Misc. Inform. Kew* 1928: 280, 1928. (Syn: *Moraea cultriformis*). [Walking Iris, White-and-Blue Iris]. **Herbs**, erect, or weak and reclining, 30–60 cm high. **Leaves** 30–80 × 1.5–2.5 cm, thin, deep green, with a conspicuous elevated costa. **Inflorescence** clusters, sessile or nearly so; peduncle about as long as the leaves, flat and leaf-like, bearing the flowers far below the apex, simple, tip of the leaf often taking root in the soil; spathes ~3.5 cm long, outer valves green, lanceolate, acuminate. **Flowers** to 5 cm across, delicate; outer segments obovate, pale yellow, sometimes white, with cross bars of yellow and brown near base; inner segments small, blue. **Fruits** ~2.5 cm long; seeds subtended by a red aril. **Distr:** WC and SE Brazil. **Ecol:** Mostly in dense wet forests; ~900 m. **Zones:** 10–12.

Sisyrinchium

Syn: *Bermudiana, Oreolirion*, etc.
Species 193. **Herbs**, annual or perennial; rhizomes none or very short; roots fibrous or often thickened and fleshy. **Leaves** linear or narrowly ensiform. **Inflorescences** with flowers usually several in a cluster, clusters solitary or fasciculate, rarely spicate or paniculate; peduncles flattened or terete, often winged, leafy or naked. **Flowers** blue or yellow, small; perianth tube very short or none, segments oblong, subequal, spreading from above the base; stamens inserted at base of perianth, filaments ± connate, anthers erect or versatile; ovary turbinate or globose, 3-celled, ovules numerous, super posed, style short, subulate, branches subulate, apex stigmatose. **Fruits** capsules, exserted from the spathe, subglobose or turbinate, loculicidally trivalvate; seeds numerous, minute. **Distr:** Hawai'ian Islands, temperate and subtropical America to Falkland Islands. **Ecol:** In montane regions. **Cult:** Propagate by division or from seeds. Requires a poor to moderately fertile soil in full sun.

S. striatum Sm., *Icon. Pict. Pl. Rar.* 2: t. 9, 1792. (Syn: *Moraea sertula*, etc.). [Satin Flower]. **Herbs**, perennial, rhizomatous, glabrous. **Leaves** *basal* equitant, sword-shaped, 25–30(–45) × 0.6–1.2 cm; *cauline* leaves below the inflorescence similar to basal ones, but gradually becoming shorter upward. **Inflorescence** 30–76 cm tall, 2-edged, narrowly winged, flowering portion ~30 cm long, flowers borne in terminal and lateral fascicles, each fascicle subtended by a 1-valved spathe, many flowered, each flower in axil of a transparent bract; scape 30–90 cm tall; pedicels ~1.2 cm long, exserted from spathe. **Flowers** with perianth campanulate, tube ~6 mm long, limb wide-spreading, ~2.5 cm Ø, fugacious; segments 6, in 2 series, somewhat obovate, abruptly tipped; outer segments 3, wider than inner ones, straw-yellow above, outside purple-striate, inside of tube red-spotted; stamens 3, united into a tube ~6 mm long, filaments free at apex, alternating with the 3 stigma lobes; ovary inferior, oblong. **Fruits** thin-walled, papery, trivalved, oblong, ~6 mm long; seeds numerous, nearly globose, surface irregularly veined. **Distr:** C Chile to W Argentina. **Ecol:** In coastal areas, coastal mountains and valleys; 500–2000 m. Fl. Apr–Aug. **Zones:** 8–10.

Sparaxis

[Blue Bonnets, Cape Buttercup, Harlequin Flower].
Species 15. **Herbs**, perennial, small, deciduous; corm globose to conical, tunics pale, fibers. **Leaves** several, unifacial, mostly basal; lamina lanceolate, apex usually obtuse. **Inflorescence** a spike; bracts dry, crinkled, translucent, irregularly streaked with brown markings; outer bracts usually tricuspidate, inner small; scape firm, erect, simple or branched. **Flowers** often large, variously colored, actinomorphic and rotate or zygomorphic and bilabiate; perianth tube either funnel-shaped, cylindrical lower part usually short; tepals subequal and spreading, or the uppermost largest and then erect or hooded, lower 3 smaller; stamens erect and symmetrically disposed, or unilateral and spreading, anthers straight or twisted; ovary globose, style branches short or long and filiform. **Fruits** capsules, barrel-shaped; seeds 8–15 per locule, globose, glossy. **Distr:** S Africa. **Cult:** Propagate by division or from seeds. Requires a fertile, moist, well-drained soil in full sun.

S. elegans (Sweet) Goldblatt, *J. S. African Bot.* 35: 233, 1969. (Syn: *Streptanthera elegans*). [Cape Buttercup]. **Herbs**, 9–20 cm high; corm globose with tunics of fine fibers; stem forked at base. **Leaves** 6–8, arranged in a fan-shaped basal rosette; lamina lanceolate, 7.5–10 × 0.6–0.8 cm, apex obtuse, glabrous, moderately firm in texture. **Inflorescence** a spike, 2–4-flowered; peduncle shorter than the leaves;

bracts obscurely toothed. **Flowers** actinomorphic, 4–5 cm across, salmon-colored, rarely white, with purple and yellow centers, unscented; perianth tube ~8 mm long, blocked below the style; tepals 1.8–2.2 × ~1.5 cm; stamens and style central, stamens half as long as the perianth segments, anthers sigmoid, style short, spathulate. **Distr:** W Cape Province. **Ecol:** In clay flats. Fl. Aug–Sep. **Zones:** 9–10.

Tigridia

Syn: *Cardiostigma, Colima, Fosteria*, etc.
Species 48. **Herbs**, cormous; corm ovoid with brown membranous tunics; stems terete, not winged, simple or branched. **Leaves** linear or lance-linear, plicate. **Inflorescences** with flowers solitary or few-flowered. **Flowers** often relatively large; perianth tube absent, segments in 2 series, dissimilar between the series, connivent in a cup at base, then spreading; filaments united to apex in a long cylindric column, anthers linear, erect-patent; ovary clavate, 3-celled, ovules numerous, superposed, style long, filiform, its branches divided into 2 slightly flattened, falcate forks. **Fruits** capsules, clavate-oblong, shortly trivalvate at apex; seeds angulate by pressure. **Distr:** Mexico to El Salvador, Peru to N Chile. **Cult:** Propagate from offsets or from seed.

T. pavonia (L. f.) DC., *Liliac.* 1: t. 6, 1802. (Syn: *Ferraria pavonia*, etc.). [Mexican Shell Flower, Tiger Flower]. **Herbs**, cormous, erect; corm rather large, 3–5 × 1.5–4 cm; stems leafy, mostly 0.3–0.6(–1.5) cm high, leafy. **Leaves** *basal* linear-lanceolate, 20–50 × 1–6 cm, base long-sheathing, several-nerved; *cauline* leaves 1–several. **Inflorescences** erect, spathes 1–2, few-flowered, 6–10 cm long; flowers long-pedicellate. **Flowers** 10–15 cm Ø, pink, red, orange, yellow or white, often spotted or blotched with darker colors; outer perianth segments obovate, 7–10 cm long, lamina spreading or reflexed; inner segments much shorter, lamina spreading or reflexed; staminal tube 4–7 cm long, projecting conspicuously beyond the flower cup, anthers erect, incurving. **Fruits** 3–7 × 1 cm. **Distr:** Mexico to El Salvador. **Ecol:** In moist forests, sometimes in oak forests, swampy thickets, or open meadows; 1,600–3,000 m. **Zones:** 9–10.

Trimezia

Syn: *Poarchon, Remaclea*, etc.
Species 41. **Herbs**, perennial, cormous; corms short, thickened, tunic splitting and shredding into coarse fibers. **Leaves** equitant; lamina flat, ± broadly linear, not plicate, thin, midrib prominent. **Inflorescence** with peduncle bearing a few long-stalked spathes, flowers more than 1 produced from each spathe; flowers pedicellate. **Flowers** actinomorphic; perianth tube absent; segments in 2 dissimilar whorls of 3, outer ones obovate with a broad concave claw, inner ones much smaller, convolute and ± deflexed; stamens short, erect, filaments free; ovary clavate, 3-celled, ovules many, superposed, style basally subulate, 3-branched above, branches ending in small tubercles or cusps, these overtopping the small horizontal stigmas. **Fruits** capsules, oblong, loculicidally trivalvate, exserted beyond the bracts. **Distr:** Lesser Antilles to S tropical America. **Cult:** Propagate from seeds. Requires a well-drained compost, and protection from winter.

T. steyermarkii R.C. Foster, *Rhodora* 64: 310, 1962. **Herbs**, erect, to 90 cm tall, forming dense clumps; corms ovoid, 2–4 cm Ø, tunics brown, fibrous. **Leaves** basal; lamina sword-shaped, 30–90 × 2–4 cm, thin, midvein evident. **Inflorescence** terminal, several-flowered; peduncles cylindrical, 7.5–18 cm long; spathe bracts paired, unequal, keeled, back edge crimped in middle, outer 2 cm long, inner one to 3–12.5 × 1+ cm; scapes often branched; pedicels coiled, persistent, glabrous, to 15 cm long. **Flowers** odorless, yellow with mottled brownish-purple bands toward base, ~3–5 cm across; tepals clawed and whorls differentiated, usually 3.2 × 1.8 cm; outer whorl larger; inner whorl darker yellow; ovary clavate, <1 cm long, glabrous. **Fruits** ~2.5 cm long. **Distr:** S Mexico to C America. **Ecol:** In wooded montane areas. **Zones:** 9–12. **Info:** This species is somewhat similar to *T. martinicensis*, except for the reduced size of style crests.

Watsonia

Syn: *Callanthus*, etc.
Species 52. **Herbs**, deciduous, perennial; corms depressed-globose. **Leaves** several, 2-ranked, basal and cauline; lamina sword-shaped to linear, midvein usually prominent. **Inflorescence** a spike or panicle taller than the leaves, flowers 2-ranked. **Flowers** zygomorphic, rarely actinomorphic, orange, red, rarely cream to pale yellow or white; corolla trumpet-shaped, base tubular, curved, flaring from base upward, 6-lobed above, lobes all equal or inner 3 wider; stamens 3, radially arranged, unilateral and arcuate or declinate, rarely symmetrically arranged; style threadlike, 3-branched, branches each 2-forked. **Fruits** capsules, oblong, globose or spindle-shaped. **Distr:** S Africa. **Cult:** Propagate from seeds or offsets. Requires a well-drained soil in full sun.

W. aletroides (Burm. f.) Ker Gawl., *Bot. Mag.* 15: t. 533, 1801. (Syn: *Antholyza aletroides*). **Herbs**, to 45 cm high; corm globose, 2.5 cm Ø, tunics of reticulated fibers, ending in long bristles at the top; stems 30–60 cm long (including the inflorescence), simple or branched, with a few much-reduced sheathing leaves. **Leaves** basal 4–6, ensiform to lanceolate, 15–30 × 0.3–1.2 cm, rigid in texture. **Inflorescence** a lax spike, 6–12-flowered, mostly unbranched; flowers nodding; bracts 1.1–2.4 cm long, clasping the stem. **Flowers** red, rarely pinkish or mauve; perianth tube elongate and recurved, 3.5–4.2 cm long; tepals ~1 cm long, barely spreading; stamens declinate, inserted at the bottom of the dilated part of the tube, anthers 6 mm long. **Fruits** spindle-shaped, tapering to a narrow tip. **Distr:** S Africa (SW Cape Province). **Ecol:** On grassy slopes. Fl. Sep–Oct. **Zones:** 9–10.

W. borbonica (Pourret) Goldblatt, *Ann. Missouri Bot. Gard.* 74: 571, 1987. (Syn: *Lomenia borbonica*). [Cape Bugle Lily, Rosy Watsonia]. **Herbs**, tuberous, erect, clump-forming, 0.5–2 m tall; stem usually several-branched, often purple. **Leaves** *basal* ~5, sword-shaped, to 75 × 1.5–4 cm, glossy; *cauline* usually 6, shape similar to basal but decreasing in size. **Inflorescences** elongate spikes; scape to 1.7 m tall; bracts 1–1.8 cm long, persistent. **Flowers** rose or mauve, rarely pure white, tube 2–3.8 cm long, lobes elliptic and as long as the tube, limb to 5 cm Ø; tepals mostly 3–3.5 cm long; stamens declinate or arcuate. **Fruits** 3-angled, oblong-ovate, 2–3 cm long, often widest near the tip, blackish; seeds with 2 wings. **Distr:** SW Cape Province. **Ecol:** In rocky sandstone slopes; 100–1,500 m. Fl. Oct–Jan. **Zones:** 9–10. **Info:** ***W. borbonica* ssp. *borbonica*** (Syn: *W. pyramidata*, etc.) has flowers with perianth tube almost as long as the lobes.

W. fourcadei J.W. Mathews and L. Bolus, *Ann. Bolus Herb.* 4: 26, 1925. (Syn: *W. stanfordiae, W. ryderae*). [Table Mountain Watsonia]. **Herbs**, tuberous, 1–2 m high. **Leaves** *basal* 8, sword-shaped, 60 × 1.5–4 cm; *cauline* leaf solitary, to 15 cm long. **Inflorescences** elongate spikes; bracts 1–2 cm long, shorter than the internodes; inner bracts deeply forked. **Flowers** mostly orange to red, rarely pink or purple; perianth tube elongate, 4–5.5 cm long, tubular above; tepals oblong-linear to obovate, 2.4–3.3 cm long; stamens arcuate, filaments 3.8–4.5 cm long. **Fruits** tapering above. **Distr:** S Africa (Cape Province). **Ecol:** In rocky sandstone slopes. Fl. Nov–Jan. **Zones:** 9–11.

W. wilmaniae J.W. Mathews and L. Bolus, *Ann. Bolus Herb.* 4: 114, 1927. (Syn: *W. starkeae, W. desmidtii*). **Herbs**, tuberous; stem 0.8–1.5 m high. **Leaves** basal 6, sword-shaped, to 50 × 1.6–4 cm. **Inflorescence** a spike, fairly dense, 4-branched; bracts to 0.7–2 cm long, slightly longer than the internodes. **Flowers** orange, red, or purple, perianth with darker stripes; perianth tube, elongate, slender, tubular above, 3–4 cm long; perianth lobes obovate, 1.8–2.2 × 1 cm; stamens arcuate. **Fruits** obovoid. **Distr:** SW Cape Province. **Ecol:** In rocky sandstone soils, often wet places. Fl. Nov–Jan. **Zones:** 9–11.

XERONEMATACEAE

Genus 1, **species** 2. *Familial description similar to generic.*

Xeronema

Syn: *Scleronema*.
Species 2. **Herbs**, evergreen, robust, tufted, glabrous; rhizome short, erect, hidden in massive fibrous remains of old leaves; tillers numerous. **Leaves** distichous, equitant, rigid; sheathing base long, thick, leathery and fleshy, unifacial. **Inflorescence** a raceme, terminal, flowers congested, secund; peduncle long with several large sheathing bracts; pedicels not articulate, subtended by floral bract. **Flowers** elongate, orange, crimson or red; tepals 6, free, ± equal, 1-nerved, at first erect, becoming reflexed and twisted and so persisting with the entangled staminal filaments until the capsule opens; stamens 6, highly exserted, filaments persistent, > tepals, glabrous, anthers oblong, versatile, dorsifixed, foveolate at insertion of filament, introrse, dehiscing by a longitudinal slit; ovary superior, ± stipitate, constricted about the middle, trilocular, each locule with several ovules, style > tepals, cylindric, stigma narrow. **Fruits** capsules, ± trigonous, broader below than above the equatorial constriction, loculicidal but valves usually separating widely only in upper part; seeds ± spiny papillose. **Distr:** New Caledonia, NW New Zealand. **Cult:** Propagate by division or from seeds. Requires well-drained soil in full sun. Water well during the growth period.

X. callistemon W.R.B. Oliv., *Trans. and Proc. New Zealand Inst.* 56: 1, 1926. [Poor Knights Lily]. **Herbs**, compact, clump-forming, sometimes several meters across; rhizome to 5 cm Ø, with numerous cordlike roots. **Leaves** 60–100 × 3–5 cm, very firm with numerous subequal nerves, margins thickened, apex subacute; sheath to 30 cm long, thick and very fleshy. **Inflorescence** a raceme, 15–35 cm long, inclined to peduncle, almost horizontal, flowers crowded; peduncle longer than leaves, mostly covered by 3–4 sheathing green bracts of decreasing size; pedicels 1–1.5 cm long, bracteate, abaxial ones bending upward, with all flowers upright. **Flowers** red; tepals 1–1.5 cm long, outer ~3 mm wide, inner ~1.5 mm; staminal filaments scarcely flattened, to 2× tepals, anthers to 4 mm long, pollen tangerine-colored; ovary ~4 × 2 mm, oblong-trigonous, stipitate, base surrounded by a thickened ring, nectar accumulating in a large viscid white drop on the concavity over each septum, style reaching anther-level, hardened and erect in fruit, ovules in 2 short rows of ~8 per locule. **Fruits** 0.9–1.1 cm long; seeds ovate, ~1.6 × 1 mm. **Distr:** New Zealand (Poor Knights Islands and Hen Island). **Ecol:** On sunny rocks, usually facing the sea; to 300 m. **Zones:** 10–11.

HEMEROCALLIDACEAE

Genera 19, **species** 85. **Herbs**, terrestrial. perennial; rootstock rhizomatous or tuberous; sap clear; stems elongate, contracted, rounded, solid. **Leaves** cauline or basal (in a rosette), alternate, 2-ranked (distichous) or spirally arranged, sheathing at base with sheath open; lamina simple, linear or filiform (grasslike),

Top row: *Sparaxis elegans*, *Tigridia pavonia*, *Trimezia steyermarkii*, *Watsonia aletroides*.
Middle row: *Watsonia borbonica*, *Watsonia borbonica* syn. *borbonica*, *Watsonia fourcadei*, *Watsonia wilmaniae*.
Bottom row: *Xeronema callistemon*, *Dianella ensifolia*, *Dianella nigra*.

Top row: *Hemerocallis fulva, Hemerocallis fulva* var. *aurantiaca, Hemerocallis lilioasphodelus, Aloe barberae.*
Bottom row: *Aloe brevifolia, Aloe ciliaris, Aloe ferox, Aloe hexapetala.*

margins entire, venation parallel, soft. **Inflorescence** terminal, bracteate, scapose, flowers solitary or in cymes, racemes or panicles; flowers with pedicels articulated; floral bracteoles present. **Flowers** bisexual, large and showy, actinomorphic or zygomorphic; perianth present, 3-merous; tepals 6; sepals separate or united; petals 6, united into a tube, basally united or mostly separate, yellow, orange to red; stamens 6 (3+3); pistil with carpels 3, united, ovary superior or inferior, styles 1, stigma terminal, capitate or trilobed, locules 1 or 3, placentation axile. **Fruits** capsules or berries; seeds numerous, 2–10 per locule, lacking tuft of hair, keeled or winged, surface ± smooth, round or ellipsoid, black (phytomelan). **Distr:** Papuasia to New Zealand and the Pacific, Australia, Europe to Asia, Malesia, India, Madagascar, and Africa, S America.

Dianella

Species 41. **Herbs** perennial, sometimes subshrubby, evergreen; rhizome generally branched, stout; stem simple or branched. **Leaves** subbasal, distichous, basally often equitant; lamina grasslike, rigid, midvein raised abaxially. **Inflorescence** a panicle, laxly branched, usually large, with several to many racemes or smaller panicles; bracts rather small; scape usually relatively tall, to 2 m, with few linear-lanceolate cauline leaves and a terminal panicle; pedicel articulate apically. **Flowers** usually nodding, rather small; tepals 6, free, 3–7-veined; stamens 6, inserted at base of tepals, filaments thickened, anthers basifixed, dehiscing by terminal pores; ovary triloculed, ovules 4–8 per locule, style slender, stigma small. **Fruits** berries; seeds black, often flattened. **Distr:** S Tropical Africa to Pacific, Bolivia. **Cult:** Propagate from seeds or by division. Requires fertile, moist soil in full sun.

D. ensifolia (L.) DC., *Liliac.* 1: t. 1, 1802. (Syn: *Eustrephus javanicus*, etc.). [Cerulean Flax Lily, Harebell]. **Herbs**, perennial, to 1.5 m high; rhizome creeping, horizontal, woody. **Leaves** distichous, sessile; lamina sword-shaped, gradually narrowed at both ends, 30–80 × 1–2.5 cm, leathery, midvein abaxially and margin usually scabrous, apex obtuse. **Inflorescence** a panicle, laxly branched, 10–40 cm long, usually with flowers borne distally; scape 1–2 m long, with several bractlike stem leaves 3–8 cm long; pedicel 0.7–2 cm long, usually arcuate. **Flowers** with tepals spreading, white, greenish white, yellowish, or bluish purple, linear-lanceolate to narrowly oblong, 6–7 × 3–3.5 mm; stamens shorter than tepals, filaments geniculate near middle, dilated distally; style ~6 mm long. **Fruits** berries, deep blue, subglobose, ~6 mm Ø, 5- or 6-seeded. **Distr:** S Tropical Africa (Chimanimani Mts) to tropical and subtropical Asia. **Ecol:** In forests, grassy slopes; 0–1,700 m. Fl. & Fr. Mar–Aug. **Zones:** 8–12.

D. nigra Colenso, *Trans. and Proc. New Zealand Inst.* 16: 339, 1884. (Syn: *D. reflexa, D. intermedia* var. *norfolkensis*). [Blue Berry, Ink Berry]. **Herbs**; rhizomes lateral; stem ± woody, ~5 mm Ø. **Leaves** distichous, equitant, sheath often reddish; lamina to 0.6(–1) m × 1–1.5 cm, tough, but not rigid, adaxially glossy, abaxially dull with numerous fine veins. Inflorescence a panicle, usually overtopping leaves, open, branchlets diverging; peduncle ± flattened; bracts ± sheathlike; pedicels 1.5–4 cm long, 3–4 in a group. **Flowers** greenish-white, inconspicuous; outer tepals ~3.2–4.5 × 1.5–1.7 mm, usually 5-veined, tip cucullate; inner tepals wider, 3(5)-veined, tip emarginate; staminal filaments ~3 mm long, upper part swollen into a ± globose, yellow, papillose struma, anther dehiscing by a slit; ovary ~1.5 × 1 mm, sessile, ovules 2 per locule. **Fruits** ~0.65–1.7 × 0.65–1 cm, globose to oblong, gray white and dull to strong violet-blue and glossy, pericarp spongy; seeds usually <6. **Distr:** New Zealand. **Ecol:** On dry hillsides among undershrubs. Fl. Dec. **Zones:** 8–10.

Hemerocallis

[Daylily].
Species 18. **Herbs**, perennial, scapose, clump-forming, rhizomatous, from fibrous or fleshy contractile roots often enlarged at ends; rhizomes spreading. **Leaves** many, basal, sessile, 2-ranked, bases sheathing; lamina long-linear, keeled, apex acuminate. **Inflorescences** 2, in terminal helicoid cymes, or solitary, rarely capitate; cymes 1–6(or more)-flowered. **Flowers** mostly diurnal and ephemeral, slightly irregular, showy; tepals 6, connate basally into a short, funnelform to campanulate tube, distinct parts imbricate, spreading, inner broader than outer; stamens 6, adnate to throat of perianth tube, filaments curved upward, distinct, unequal, anthers dorsifixed, bilocular, linear-oblong, dehiscence introrse; ovary superior, green, trilocular, conic, septal nectaries present, style curved upward, stigma indistinctly trilobed or capitate. **Fruits** capsules, leathery, dehiscence loculicidal; seeds in 2 rows in each valve, black. **Distr:** SE Alps, Siberia to temperate E Asia. **Cult:** Propagate by division. Requires a fertile, moist, well-drained soil in full or partial sun.

H. fulva (L.) L., *Sp. Pl.*, ed. 2. 1: 462, 1762. [Orange Daylily]. **Herbs**, usually deciduous in winter, 0.4–1.5 m tall; rootstock tuberous, globose-ellipsoid. **Leaves** linear, 50–90 × 1–2.8 cm, apex acute. **Inflorescence** of 2 helicoidal cymes; cymes 2–5(–10)-flowered; scape erect, hollow; sterile bracts scalelike or lanceolate; pedicel ~5 mm long. **Flowers** unscented, opening in morning and closing in evening, lasting for a day; perianth single, occasionally double (stamens petaloid), orange to reddish orange, tube 2–4 cm long, tepals spreading, with a purple or reddish-orange patch, 5–12 × 1–3 cm, margins sometimes crinkly undulate, inner tepals wider; filaments 4–5 cm long, anthers purplish-black, 7–8 mm long. **Fruits** ellipsoid, 2–2.5 × 1.2–1.5 cm. **Distr:** China to temperate E Asia. **Ecol:** In forests, thickets, grasslands, streamsides; 300–2,500 m. Fl. Jun–Nov. **Info:** ***H. fulva* var. *aurantiaca*** (Syn: *H. aurantiaca*) is an evergreen variety from SE China and Taiwan, with perianth single, orange to reddish-orange. **Zones:** 4–10.

H. lilioasphodelus L., *Sp. Pl.* 1: 324. 1753. (Syn: *Cameraria lilioasphodelus*, etc.). [Yellow Daylily]. **Herbs**, 70–80 cm tall, deciduous; rootstock sometimes tuberous. **Leaves** linear, 20–70 × 0.3–1.2 cm, apex acuminate. **Inflorescence** composed of 2–4(or 5) helicoidal cymes, branched; cymes 2–4(or 5)-flowered; bracts lanceolate, 2–6(–8) × 0.5–0.7 cm; pedicel 1–2 cm long; scape generally slightly shorter than leaves, solid, with sterile bracts; flower buds apically blackish-purple or green. **Flowers** fragrant, opening in the afternoon and lasting for 1–3 days; perianth lemon-yellow, tube 1.5–2.5 cm long; tepals spreading, 5–7 × 1.3–1.6 cm, inner ones slightly wider than outer; filaments 5–5.5 cm long, anthers yellow, ~8 mm long. **Fruits** ellipsoid, ~ 2.4 × 1.2 cm. **Distr:** SE Alps, Siberia to Korea. **Ecol:** In forests, thickets, grasslands; 100–2,000 m. Fl. Jun–Aug. **Zones:** 4–10.

ASPHODELACEAE

Genera 15 and **species** ~780. **Herbs**, often succulent, small to medium sized or sometimes large trees with leaves arranged in terminal rosettes; roots terete, slightly succulent, sometimes fusiform. **Leaves** lanceolate, spiral or distichous, linear, acuminate, terete, often succulent. **Inflorescences** racemes or panicles, pedunculate. **Flowers** actinomorphic or rarely zygomorphic, hypogynous, tepals 3+3, red, orange, yellow or white, gamo- or polyphyllous; stamens 3+3, filaments free, anthers dorsifixed, rarely basifixed; pistils tricarpellate, syncarpous, trilocular, axile placentation, 2–many ovules per locule, septal nectaries present, style distinct, stigma minute. **Fruits** capsules, dehiscent, rarely berry like; seeds completely covered by aril, sometimes winged, endosperm present. **Distr:** In arid and mesic tropical, subtropical and temperate regions of the Old World, with the greatest diversity in Southern Africa.

Aloe

Species ~400. **Herbs**, **shrubs**, or **trees**, usually with dense rosettes of very fleshy leaves; stems often very reduced but sometimes well developed or even with secondary thickening. **Leaves** rosulate, amplexicaul; lamina thick, succulent, always glabrous, margin mostly hard-dentate or spiny, apex sharply pointed. **Inflorescences** racemes, subterminal, ascending to erect, often branched; peduncle usually well-developed; bracts persistent, scarious; pedicels usually much shorter than perianth. **Flowers** with perianth usually red, orange, or yellow, rarely greenish or whitish, cylindric to 3-sided, sometimes slightly curved or with swollen base, segments usually connate to form a tube, very rarely nearly free, usually glabrous, apex spreading; stamens 6, inserted at base of perianth tube, usually exserted, filaments subulate, anthers dorsifixed; ovary triloculed, ovules many per locule, style filiform, stigma small. **Fruits** capsules, loculicidal; seeds 3-angled or flattened, often winged. **Distr:** S and tropical Africa (including Madagascar), tropical parts of the Arabian Peninsula. **Cult:** Propagate from seeds, stem cuttings or offsets. Needs fertile, well-drained soil in full sun.

A. barberae Dyer, *Gard. Chron.*, n.s., 1874(1): 566, 1874. (Syn: *A. bainesii*, etc.). [Tree Aloe]. **Trees**, succulent, caulescent, copiously branching dichotomously; stem erect, to 18 m tall, 1–3 m Ø at base. **Leaves** ± 20, densely rosulate; lamina ensiform, 60–90 × 7–9 cm, dull green, marginal teeth 2–3 mm long, dull white, brownish-tipped, firm horny, 1–2.5 cm apart. **Inflorescences** 40–60 cm tall, usually with 2 bracts, racemes cylindrical, slightly acuminate, 20–30 × 8–10 cm, dense; bracts linear-acuminate, 1 cm × 1 mm; pedicels 1 cm long. **Flowers** rose, rose-pink to orange, 3.3–3.7 cm, base rounded; tepals free for 2.8–3.2 cm; stamens exserted 1.5 cm. **Distr:** Mozambique to S Africa. **Ecol:** In dense bush and low forests. Fl. Jun–Aug. **Zones:** 10–12.

A. brevifolia Mill., *Gard. Dict. Abr.* ed. 6: 8, 1771. (Syn: *A. postgenita*, etc.). [Dune Aloe]. **Herbs**, succulent, acaulescent, suckering to form dense clumps. **Leaves** 30–40, densely rosulate; lamina lanceolate-deltoid, 6 × 2 cm, tip firm spine, glaucous, abaxially with a few soft prickles in median line or irregular in upper 1/3, marginal teeth 2–3 mm, whitish, 1 cm apart. **Inflorescence** 40 cm tall, simple, raceme conical, 15 × 7 cm, subdense; bracts ovate-lanceolate, 1.5 cm long; pedicels 1.5 cm long. **Flowers** pale scarlet, 3.8 cm long, base truncate, very slightly narrowed above; tepals free to base; ovary slightly enlarging to mouth. **Distr:** SW Cape Province. **Ecol:** In open bushvelds. **Zones:** 9–11.

A. ciliaris Haw., *Philos. Mag. J.* 10: 281, 1825. (Syn: *A. ciliaris* f. *gigas*, etc.). [Climbing Aloe]. **Climbers**, caulescent, succulent or semiwoody, perennial, branching, forming large masses; stems trailing or clambering, to 6 m long; internodes wide. **Leaves** terminal; lamina dark green, well-spaced, linear-lanceolate, 10–15 × 1.5–2.5 cm, base clasping stem entirely, margins with white triangular, firm teeth (1 mm long, 3 mm apart), these larger on margins of leaf sheaths, apex long-acuminate, leaf sheath 0.5–1.5 cm long, obscurely green-lineate, upper margin opposite lamina with fringe of cilia 2–4 mm long. **Inflorescences** racemes, subdense, 24–30-flowered, 20–30 cm long, usually unbranched, cylindrical, 8–15 × 4–5 cm, sometimes with a single bract; bract ovate-acuminate; pedicels ~5 mm long. **Flowers** large bright red or scarlet; corolla with green and yellow at mouth, 2.8–3.5 cm long, slightly clavate, base very shortly attenuate, gradually enlarging above the ovary; tepals free for 6 mm; stamens yellow. **Distr:** E Cape Province. **Ecol:** Among bushes in thickets. **Zones:** 9–11.

A. ferox Mill., *Gard. Dict.* ed. 8: 22, 1768. (Syn: *Busipho ferox*, etc.). [Bitter Aloe, Red Aloe]. **Shrubs**, caulescent, simple; stem to 3(–5) m long, with persistent dead leaves. **Leaves** 50–60, densely rosulate; lamina lanceolate-ensiform, to 1 m × 15 cm, marginal teeth ~6 mm long, reddish to reddish-brown, 1–2 cm apart, dull green, sometimes reddish-tinged, glabrous or with few–many irregular prickles. **Inflorescences** racemes, cylindrical, 50–80 × 9–12 cm, slightly acuminate, narrowing to ~6 cm at apex, very dense; bracts 5–8, ovate-acute, 0.8–1 × 0.3–0.5 cm; pedicels 4–5 mm long; buds horizontal. **Flowers** scarlet, sometimes orange, 3.3 cm long, slightly ventricose-clavate, base rounded, enlarging above; tepals free for 2.2 cm long; stamens exserted 2–2.5 cm long; ovary slightly narrowed at mouth. **Distr:** S Africa. **Ecol:** In dense, arid, tall bush and low forests. **Zones:** 9–11.

A. hexapetala Salm-Dyck, *Verz. Art. Aloe*: 28, 1817. (Syn: *A. speciosa*, etc.). [Tilt-head Aloe]. **Shrubs**, succulent, caulescent, simple or branched; stem to 4(–6) m tall, with persistent dead leaves. **Leaves** densely rosulate; lamina lanceolate-attenuate, 60–80 × 7–9 cm, dull glaucous-green, tinged bluish to reddish, with very narrow deep pink to pale reddish margin; marginal teeth 1 mm, pale red, ± 1 cm apart. **Inflorescences** ± 50 cm long, arcuate-erect, simple, racemes very dense, cylindrical, ~ 30 × 12 cm, slightly acuminate; bracts lanceolate-obtuse, to 2 × 1 cm, brownish; pedicels 5–8 mm long. **Flowers** changing color from red to greenish-white, greenish-tinged, 3–3.5 cm long, ventricose, base rounded, enlarged above; tepals free almost to base; stamens exserted to 1.6 cm; ovary narrowing slightly at mouth. **Distr:** Cape Province. **Ecol:** On rocky slopes; 760 m. Fl. Jul–Sep. **Zones:** 9–11.

A. plicatilis (L.) Mill., *Gard. Dict.* ed. 8: 7, 1768. (Syn: *Rhipidodendrum plicatile*, etc.). [Fan Aloe]. **Shrubs** to **small trees**, 3–5 m tall; stem stout, dichotomously branched; branchlets terminating in a cluster 12–16 leaves. **Leaves** distichous; lamina broadly linear to lorate, to 30 × 4 cm, base sheathing, margins slightly cartilaginous, upper third minutely dentate, sometimes almost entire, apex

Below: *Aloe plicatilis, Aloe polyphylla, Aloe striata, Aloe vera.*

F&KS

MD

Above: *Aloe zebrina*, *Bulbine frutescens*, *Gasteria carinata* var. *verrucosa*, *Agapanthus praecox*.

rounded, surfaces dull green to glaucous-green, flat or slightly convex. **Inflorescences** racemes, cylindric, slightly acuminate, 1 per leaf cluster, simple, to 50 cm long, flowering portion 15–25 cm long, laxly 25–30-flowered; peduncle basally flattened, apically terete, with a few sterile bracts below the raceme; lowest sterile bracts to 1.4 x 1.2 cm; pedicels ~1 cm long. **Flowers** 1–1.5 cm long; perianth scarlet, cylindric, slightly trigonous, to 5.5 cm long; outer segments free for 1/3 their length, obscurely 3–5-nerved; filaments lemon-yellow, anthers exserted 1–2 mm; ovary pale green, 8 x 6 mm, finely 6-grooved, stigma exserted 3–5 mm. **Distr:** SW Cape Province. **Ecol:** On sheltered steep rocky or sandstone slopes in mountain valleys; 150–700 m. Fl. Aug–Oct. **Zones:** 9–11.

A. polyphylla Pillans, *S. African Gard.* 24: 267, 1934. [Spiral Aloe, Crown Aloe]. **Herbs**, forming short-stemmed tight rosettes, 50–60 cm high, 50–60 cm across; stems to 10 cm long. **Leaves** 75–100, spirally arranged, clockwise or anti-clockwise, 5-ranked, each 15–30-leaved; lamina triangular, to 30 cm long, margins and apex armed, glaucous, thickly fleshy. **Inflorescences** panicles, 50–60 cm tall, branched; peduncle often thickened basally; peduncle and branches glaucous; bracts ~1 cm long, acuminate, adnate to branches; flowers erect in bud, becoming pendent at anthesis; floral bracts green; pedicels green, ~1 cm long. **Flowers** green with purple tips or pale red to salmon-pink, tubular, 3–4 cm long; lobes with apices slightly recurved; stamens slightly exserted. **Fruits** oblong. **Distr:** Lesotho. **Ecol:** In grasslands, steep rocky mountain slopes; 2,000–2,720 m. Fl. Aug–Dec. **Zones:** 8–11.

A. striata Haw., *Trans. Linn. Soc. London* 7: 18, 1804. [Coral Aloe]. **Herbs**, usually stemless (very old plants develop a short creeping stem, to 1 m long), succulent, decumbent, with persistent old leaves; stem usually unbranched, rarely branched with 2–6 rosettes. **Leaves** 12–20, spreading-incurved; lamina broadly lanceolate-attenuate, to 50 x 20 cm, somewhat fleshy, ± concave, margins entire, with pink to almost red border 2–3 mm wide, adaxially usually distinctly striate, glaucous to somewhat reddish-tinged, sometimes very obscurely spotted. **Inflorescences** panicles of racemes, 1–3, branched, to 1 m high; racemes capitate to slightly conical, ~6 x 6 cm; flower buds usually green-tipped; bracts narrowly deltoid-acuminate, ~5 mm long, thin, 3–5-nerved; pedicels 1.5–2.5 cm long. **Flowers** peach-red to coral-red; perianth ~3 cm long, base swollen, subglobose, to 6 mm Ø, constricted above ovary, slightly decurved apically; outer segments free for 6–8 mm, inner segments basally adnate to outer; filaments filiform-flattened, lemon-yellow, anthers exserted to 1 mm; ovary to 8 x 3 mm, finely 6-grooved, pale green, stigma exserted 1–2 mm. **Distr:** Namibia to Cape Province. **Ecol:** Fl. Jul–Oct. **Zones:** 9–11.

A. vera (L.) N.L. Burman, *Fl. Indica*. 83, 1768. (Syn: *A. barbadensis* var. *chinensis*, etc.). [Aloe, Aloe Vera]. **Herbs**, succulent; stems short, suckering freely to form dense clumps. **Leaves** subbasal, slightly distichous in seedlings and new shoots, erect, pale green, sometimes with pale spots in very young plants, linear-lanceolate, 15–35(–50) x 4–5(–7) cm, margins sparsely spiny dentate, apex 2- or 3-dentate-pointed. **Inflorescence** erect, 60–90 cm tall; peduncle to 2 cm thick; raceme 30–40 x 5–6 cm, sometimes with 1 or 2 ascending branches, flowers numerous; bracts whitish, broadly lanceolate, ~1 x 0.5–0.6 cm, acute, 5–7-veined; pedicels ~½ as long as bract. **Flowers** reflexed; perianth pale yellow mottled with red, slightly ventricose, 2.5(–3) cm, outer lobes free for ~1.8 cm, slightly recurved at apex; stamens exserted by 4–5 mm; style conspicuously exserted. **Fruits** 1.5 cm long. **Distr:** SW Arabian Peninsula. **Ecol:** On coastal rocks, sand plains and dunes; to 2,400 m. Fl. Feb–Jun. **Zones:** 9–12.

A. zebrina Baker, *Trans. Linn. Soc. London*, Bot. 1: 264, 1878. (Syn: *A. constricta*, etc.). [Zebra-leaf Aloe]. **Herbs**, perennial, solitary or suckering to form groups, acaulescent. **Leaves** in a compact rosette, spreading; lamina lanceolate, 15–35 x 6–7 cm, margins with stout pungent red-brown teeth 4–7 mm long, 1–1.5 cm apart, apex usually dried and twisted, adaxially dull dark green with transverse bands of whitish spots, abaxially usually obscurely spotted; sap drying purple. **Inflorescence** erect, 0.75–2 m high; peduncle 4–12-branched, lowermost branches often re-branched; branches suberect, subtended by scarious or sometimes leafy bracts 1–4 cm long; racemes laxly flowered, cylindric, 30–40 x 6 cm; bracts linear-lanceolate, 0.6–1.5 cm long, scarious; pedicels 0.6–1.2 cm long. **Flowers** with perianth dull-red to pinkish-red with segment margins paler, 2.5–3.5 cm long, ~8 mm Ø, abruptly constricted just above, then widening and becoming slightly decurved, cylindric-trigonous; outer segments apically free; stamens and stigma slightly exserted. **Fruits** oblong-ovoid, 2.5–3 x 1.4 cm, yellowish-gray; seeds ~3.5 x 6.5 mm, blackish-brown, wings pale brown. **Distr:** Angola, N Namibia, S Africa (NW Province). **Ecol:** On grasslands and in woodlands; 200–1,600 m. **Zones:** 10–12.

Bulbine

Syn: *Blephanthera*, etc.

Species 60. **Herbs**, perennial or annual, sometimes with a bulbous rhizome. **Leaves** basal, succulent; lamina linear, base sheathing and membranous, flat or channeled, tapering. **Inflorescence** a raceme. **Flowers** with tepals almost equal, narrowly elliptic to ovate, free or almost so, yellow with red-brown to brown central stripe, outer whorl slightly darker than inner whorl; stamens all or inner 3 only with a beard of clavate hairs at apex of filament, anthers usually dorsifixed; ovules 2–8 per locule, style simple, stigma entire or trilobed. **Fruits** capsules, globose to broadly ovoid; seeds black, brown or silvery gray, smooth or with transverse ridges, with or without wings. **Distr:** NE Tropical to S Africa, Australia. **Cult:** Propagate from seeds or by division. Requires a rich, moist, well-drained soil in full sun.

B. frutescens (L.) Willd., *Enum. Pl.*: 372, 1809. (Syn: *Anthericum frutescens*, etc.). [Bulbinella, Burn Jelly Plant, Cat's Tail, Snake Flower]. **Herbs**, geophytic, evergreen, perennial or subshrubby, succulent, 45 cm high, clump-forming, spreading often to 1 m wide; rootstock rhizomatous, roots numerous, wiry. **Leaves** fleshy, grasslike, forming a rosette; lamina cylindrical or subterete, 10–16 x 0.4–0.8 cm, green, thin, glabrous. **Inflorescences** racemes or spikes, dense, cone-shaped, 30–45 cm tall; pedicels slender. **Flowers** 1–2.5 cm across, star-shaped, orange, yellow or white; tepals 6; stamens bearded. **Fruits** small, round; seeds black. **Distr:** S Africa. **Ecol:** Sandy flats and slopes. Fl. Aug–Apr. **Zones:** 9–11.

Gasteria

Species 20. **Herbs**, slow-growing, perennial, glabrous, succulent, proliferating from base forming small dense groups. **Leaves** dimorphic, distichous when juvenile, usually becoming rosulate with age; lamina linear to lorate-lanceolate to triangular, margins wavy or rugulose, sometimes tuberculate-crenulate, apex mucronate, adaxially flat, abaxially convex, ± brittle, mottled, often with densely white-spotted; juvenile leaves lorate. **Inflorescences** 1–5, racemose or paniculate, terminal but appearing lateral, flowers ± secund, laxly arranged. **Flowers** 6-merous, pendent, pink to reddish, rarely white, apically white with green striae; tepals fused for the greater part of their length forming a curved tube, corolla gasteriform; stamens 6, usually included; ovary oblong-ovoid, triloculed, 6-grooved outside, style becoming shortly exserted. **Fruits** capsules, erect, oblong, 3-angled; seeds black, compressed, 2–8 mm long. **Distr:** S Africa. **Cult:** Propagate by division or from leaf cuttings. Requires a sandy, well-drained soil in full or partial sun.

G. carinata (Mill.) Duval, *Pl. Succ. Horto Alencon.*: 6, 1809. (Syn: *Aloe carinata*). **Herbs**, succulent, acaulescent, decumbent to erect, 3–18 cm tall, forming dense groups 15–80 cm across. **Leaves** distichous at first, becoming rosulate with age, erectly spreading; lamina triangular to triangular-lanceolate, 3–12 x 1–5 cm, margins cartilaginous, tuberculate-crenulate, rarely denticulate, tuberculate or smooth, spotted with raised or immersed white domed tubercles in obscure transverse bands; juvenile leaves distichous, lorate, tuberculate or smooth. **Inflorescence** racemose, 15–90 cm long, occasionally with a pair

Above: *Allium cepa*, *Allium sativum*, *Tulbaghia violacea*, *Amaryllis belladona* (white form).

of lateral branches. **Flowers** pink, perianth tips light pink to white with a central green stripe; perianth variable, narrowly ellipsoid, rarely globose-ellipsoid, 1.6–2.7 cm long, constricted into a tube 3–5 mm Ø; stamens oblong, included; ovary 6–7 × 2.5 mm, style 1.4 cm long. **Fruits** 1.9–2.3 cm long; seeds oblong, 3–4 × 2 mm. **Distr:** S Africa (SW Cape Province). **Ecol:** In thickets. Fl. Aug–Apr. **Zones:** 9–11. **Info:** *G. carinata* var. *verrucosa*, with leaves distinctly tuberculate and inflorescence 12–30 cm long.

AGAPANTHACEAE

Genus 1, **species** 9. *Familial description same as generic.*

Agapanthus

[Blue Lily].

Species 7. **Herbs**, perennial, deciduous or evergreen; rootstock rhizomatous, often exposed; roots fleshy. **Leaves** basal, 2-ranked; lamina straplike, slightly fleshy or leathery, slightly channeled, bases somewhat swollen and bulblike in deciduous species. **Inflorescence** an umbel, subtended by 2 large membranous bracts; bracts united along the margins in bud; peduncle naked. **Flowers** funnel-shaped or tubular; tepals united below into a short tube, blue or white, unscented; stamens inserted on the tepals at the top of the tube, declinate, unequal; ovary ovoid or oblong, style slender, declinate. **Fruits** capsules, 3-angled, loculicidal; seeds several per chamber, flat, winged, black. **Distr:** South Africa. **Cult:** Propagate from seeds or by division. Requires any well-drained soil in full or partial sun.

A. praecox Willd., *Enum. Pl.*: 353, 1809. (Syn: *Tulbaghia praecox*). [African Lily, Agapanthus, Lily of the Nile]. **Herbs**, evergreen, perennial, 0.5–1 m tall, usually forming dense clumps. **Leaves** up to 20, arching; lamina strap-shaped, 20–90 × ~5 cm, channeled. **Inflorescence** an umbel, dense, ± spherical, many flowered, flowers spreading; scape 0.9–1.2 m tall; pedicels 4–12 cm long. **Flowers** broadly funnel-shaped, 3.8–5 cm long; tepals thin in texture, white to medium blue, 3–7 cm long, united below for 0.7–2.6 cm, shortly tubular, lobes spreading, margins ± ruffled; stamens and style included; stamens mostly as long as the tepals. **Fruits** 3 angled, ellipsoid, papery; seeds numerous, flat, glossy, black. **Distr:** Cape Province to KwaZulu-Natal. **Ecol:** On stony slopes. Fl. Dec–Feb. **Zones:** 9–11.

ALLIACEAE

Genera 13, **species** 795. **Herbs**, perennial, bulbous, with or without a rhizome; scapigerous. **Leaves** mostly radical, spirally arranged; lamina cylindrical, semicylindrical, filiform to lanceolate or oval, fistular or not fistular, bases sheathing. **Inflorescence** a cymose umbel, enclosed in a spathe when young. **Flowers** with tepals 6, in 2 whorls, 1–7-nerved, free or connate; stamens 6, free or connate; ovary superior, trilocular. **Fruits** capsules, dehiscing loculicidally; seeds, compressed to subglobose. **Distr:** N and S America, N temperate Eurasia, Southern Africa.

Allium

Species 690. **Herbs** perennial, bulbiferous, sometimes with well-developed, usually aromatic; bulb covered with a tunic. **Leaves** sessile, very rarely narrowed into a petiole, with a basal sheath; lamina linear, linear-lanceolate, or lorate to orbicular-ovate, cross-section flat, angled, or semiterete to terete, fistulose or solid. **Inflorescence** an umbel, terminal or lateral, sometimes with bulbils, rarely flowerless and with bulbils only, enclosed in a spathelike bract before anthesis; pedicels sometimes bracteolate. **Flowers** bisexual, very rarely degenerating into unisexual (when plants dioecious); perianth segments free or united into a tube at base; filaments usually connate at base and adnate to perianth segments, entire or toothed; ovary with locules 1–several-ovuled, septa often containing nectaries; style simple, stigma entire or 3-cleft. **Fruits** loculicidal; seeds black, rhomboidal or spheroidal. **Distr:** Temperate and subtropical Northern Hemisphere, S Africa. **Cult:** Propagate from offsets, bulbils or seeds. Requires a fairly light, well-drained soil in full sun. Ample water is needed during growth period.

A. cepa L., *Sp. Pl.* 1: 30, 1753. (Syn: *Cepa vulgaris*, *Kepa esculenta*, *Porrum cepa*, etc.). [Onion, Scallion, Spring Onion]. **Herbs**, annual, bulbous, solitary or clustered; bulb globose to cylindric-ovoid, tunic purple-red, brown-red, pale brown-red, or yellow to pale yellow, papery to thinly leathery, entire. **Leaves** shorter than scape; lamina terete, 0.5–2 cm wide, fistulose. **Inflorescence** an umbel, globose, densely many flowered or with bulbils and a few flowers; spathe 2- or 3-valved, persistent; pedicels bracteolate; scape if developed, to 1 m long, terete, inflated below middle, fistulose, covered with basal leaf sheaths. **Flowers** with perianth off-white or white; segments with green or pale red midvein, oblong-ovate, 4–5 × ~2 mm; filaments equal, basally connate, partially adnate to perianth segments; outer ones subulate; inner ones broadened at base, sides 1-toothed; ovary subglobose, with concave nectaries covered by hoodlike projections at base, style slightly exserted. **Distr:** C Asia; a cultigen. **Ecol:** Fl. & Fr. May–Jul. **Zones:** 5–10. **Info:** Cultivated and consumed worldwide as a vegetable.

A. sativum L., *Sp. Pl.* 1: 296, 1753. (Syn: *A. controversum*, etc.). [Garlic]. **Herbs**, annual; bulb solitary, globose, usually consisting of several bulbils covered with a common tunic; tunic white to purple, membranous, entire. **Leaves** broadly linear to linear-lanceolate, shorter than scape, to 2.5 cm wide, apex acuminate. **Inflorescence** an umbel, with many bulblets and few flowers, 7–20 cm long; spathe deciduous; pedicels slender, longer than perianth; bracteoles ovate, rather large, membranous, apex acute; scape 25–50 cm, terete, covered with leaf sheaths for ~½ its length. **Flowers** with perianth usually pale red; outer segments ovate-lanceolate, ~4 × 1.4 mm; inner ones ovate, ~3 × 1.4 mm; filaments shorter than perianth segments, connate at base and adnate to perianth segments; ovary globose, style not exserted. **Distr:** C Asia; a cultigen. **Ecol:** Fl. Jul. **Zones:** 8–10. **Info:** Cultivated and consumed worldwide as a vegetable.

Tulbaghia

Syn: *Omentaria*. [Wild Garlic].

Species 26. **Herbs**, perennial, deciduous or evergreen, usually with an onion or garlic smell when bruised; rootstock a bulb or short rhizome. **Leaves** several, straplike, usually present at flowering. **Inflorescence** an umbel, 3–40-flowered, subtended by 2 large bracts, flowers ± nodding; bracts united along margins in bud, tapering ovate-lanceolate; peduncle erect, naked, often mottled gray or purple. **Flowers** radially symmetric, tubular or trumpet-shaped, often scented at night; tepals 6, united for half their length into a cylindrical tube surmounted by a corona of 3 or more free scales, sometimes united into a fleshy ring, white to purplish or green, corona sometimes orange or brown; stamens biseriate in upper part of tube or rarely upper series on corona, anthers sessile; ovary superior, ovoid, style short with a knoblike stigma. **Fruits** capsules, enclosed in withered corolla, erect, obovoid or turbinate, trilobed; seeds few per locule, triangular, black. **Distr:** Cameroon, SW Tanzania to S Africa. **Cult:** Propagate from offsets and seeds. Requires a moist, well-drained soil in full sun.

T. violacea Harv., *Bot. Mag.* 64: t. 3555, 1837. (Syn: *Omentaria violacea*, etc.). [Society Garlic]. **Herbs**, evergreen, strongly aromatic, 20–35 cm high; stems bulbous, borne from a rhizome, base covered in papery sheaths. **Leaves** suberect; linear, 15–30 cm × 3–6 mm, glaucous. **Inflorescence** an umbel, 6–30-flowered; scape 30–38 cm tall; pedicels 1–1.8 cm long. **Flowers** lavender; perianth tube cylindrical, 0.6–1.3 cm long; tepals purple, midrib darker, 0.6–1.2 cm long; corona of 3 fleshy scales, purple, 1.5–2.5 mm long, upper anthers inserted below the corona; ovary globose, style short. **Fruits** ovoid, 0.8–1.2 cm long. **Distr:** Cape Province to KwaZulu-Natal. **Ecol:** In forest margins and stream banks. Fl. Nov–Apr. **Zones:** 8–10.

AMARYLLIDACEAE

Genera 59, **species** 800+. **Herbs**, perennial, rarely shrubby or treelike, often with bulbs, corms, rhizomes, or tubers. **Leaves** basal or cauline; lamina often narrow, margins entire or spiny. **Inflorescence** a terminal spike, umbel, raceme, panicle, or

Above: *Clivia caulescens, Clivia gardenii, Clivia miniata, Clivia mirabilis.*

flowers solitary. **Flowers** bisexual, actinomorphic or zygomorphic, usually subtended by 1 to several spathaceous involucres; perianth segments 6, in 2 whorls, free or connate to form a short tube, with or without a corona; stamens 6, inserted at perianth throat or at base of segments, filaments sometimes basally connate, anther dorsifixed or basifixed, mostly introrse; ovary inferior, triloculed, ovules few to many per locule, placentation axile, style slender, stigma capitate or trilobed. **Fruits** capsules, usually loculicidal, sometimes dehiscing irregularly, rarely berries; seeds with endosperm. **Distr:** Tropical (temperate), especially in South America and Africa, also Mediterranean.

Amaryllis

Species 2. **Herbs**, perennial, bulbous. **Leaves** appearing after the flowers; lamina oblong, somewhat narrowed to base. **Inflorescence** an umbel, 5–many flowered, subtended by 2 large, equal, free spathes, which enclose the whole umbel in bud; flowers ascending to erect. **Flowers** with perianth bilaterally symmetric; corolla tube short, lobes 6, oblanceolate, somewhat clawed, acute, the 3 outer lobes each with a small, hairy, inwardly pointing appendage just below the apex; stamens 6, deflexed and then curving upward toward the apex; ovary triloculed, locules few-ovuled, style deflexed and then curving upward, stigma capitate. **Fruits** capsules, dehiscing irregularly. **Distr:** S Africa (Cape Province). **Cult:** Propagate by division. Requires a moist or dry soil in full or partial sun.

A. belladonna L., *Sp. Pl.*: 293, 1753. (Syn: *Coburgia belladonna*, etc.). [Naked Ladies, Belladona Lily]. **Herbs**, bulbous, to 90 cm high; bulb ~10 cm Ø. **Leaves** 5–11, spreading to erect, dry at flowering; lamina lanceolate, 50 × 1.5–7.5 cm, channeled, smooth, midrib distinct. **Inflorescence** an umbel, usually 6–12-flowered, occasionally with more flowers, flowers turned to one side or evenly spreading; scape to 60+ cm tall; spathes to 8 cm long. **Flowers** trumpet-shaped, 6–10 cm long, tube ~1.2 cm long, fragrant, white to pink, usually pale lemon at base, veins often dark pink; tepals 8–10 cm long, slightly longer than stamens and styles. **Distr:** SW Cape Province. **Ecol:** In seasonally moist places. Fl. Feb–Apr. **Zones:** 8–10.

Clivia

Species 6. **Herbs**, rhizomatous. **Leaves** 5–20 in a basal or cauline cluster, suberect or recurved; lamina strap-shaped, margins smooth or sometimes minutely serrated, apex tapering or broad, sometimes notched, longer than the inflorescence, midrib sometimes pale. **Inflorescence** an umbel, spreading or lax; scape fleshy, erect, compressed; spathe bracts 4 or more, soon withering; pedicels shorter than or as long as perianth. **Flowers** actinomorphic or slightly zygomorphic, narrowly to widely funnel-shaped, nodding to spreading, red, orange, and yellow; perianth tube short, tepals overlapping throughout or spreading in the upper half and reflexed toward the tip; stamens inserted in the throat, free, evenly spreading, as long as or slightly longer than tepals, anthers dorsifixed, yellow; ovary with locules 5–9-ovuled, style straight or slightly curved, as long as perianth or exserted, stigma minutely trifid. **Fruits** berries, indehiscent, subglobose to irregularly spindle-shaped, red with yellow flesh; seeds turgid, ± globose, ivory colored, embryo green. **Distr:** NW Cape and SE Africa. **Cult:** Propagate by division. Requires a moist compost in partial sun or shade.

C. caulescens R.A. Dyer, *Fl. Pl. South Africa* 23: t. 891, 1943. [Escarp Bush Lily, Forest Lily]. **Herbs**, evergreen, perennial, to 1 m high; rhizomes creeping; stem to over 1 m long, distinctly thickened, ~5 cm wide, bamboolike, internode length variable depending on the plant vigor. **Leaves** strap-shaped, to 0.3–0.4(–1+) m × 5 cm, margins entire, apex acute, dark green, leathery, glabrous. **Inflorescence** an umbel, often held above the foliage; peduncles erect, green with a purple coloration, thick, gradually tapered from halfway to apex; pedicels erect, very thin but bent due to the weight of the flower. **Flowers** drooping, tubular, deep orange to red, tepal tips yellowish green. **Fruits** berries, ripening to red, glossy. **Distr:** Limpopo to Swaziland. **Ecol:** In montane forests, terrestrial in moist, humus-rich places in the shade of forests or large boulders, sometimes epiphytic on trees or lithophytic on rocks (with stems usually ± pendent). Fl. any time of the year, mostly Sep–Nov. **Zones:** 10–11.

C. gardenii Hook., *Bot. Mag.* 81: t. 4895, 1855. (Syn: *C. gardenii* var. *citrina*). [Bush Lily]. **Herbs**, stemless, 40–60 cm high. **Leaves** 45–75 × 2.5–4 cm, margins entire or slightly toothed, gradually tapered to apex. **Inflorescence** an umbel, usually 10–20-flowered; scape 30–60 cm tall; flowers pendulous. **Flowers** tubular to narrowly funnel-shaped, often strongly curved, red, green-tipped, sometimes tinted with orange or yellow; perianth lobes 3.5–5 cm long; stamens equaling or exceeding the perianth, anthers 2.5–3 mm long. **Distr:** E Cape Province to KwaZulu-Natal. **Ecol:** In forests; ~650 m. Fl. Apr–Jun. **Zones:** 10–11.

C. miniata (Lindl.) Regel, *Gartenflora* 1864: 131, 1864. (Syn: *Vallota miniata*, etc.). [Bush Lily, Clivia, Fire Lily, Kaffir Lily]. **Herbs**, rhizomatous, to 80 cm high, stemless. **Leaves** in 2 opposite ranks; lamina strap-shaped, 0.2–0.9(–1.84) m× 3–7 cm, midrib absent. **Inflorescence** an umbel, to 30 cm across, 12–20-flowered, semipendent; peduncle short; pedicels 2.5–5 cm long. **Flowers** scented; corolla of fused tepals, tubular to funnel-shaped, 5–8 cm long, to 8 cm Ø, orange-red with a yellow throat, with 6 rounded lobes. **Fruits** berries, ovoid, ripening to red; seeds 3. **Distr:** S Africa. **Ecol:** In warm forests. Fl. throughout most of the year. **Zones:** 9–11. **Info:** A rare, wild yellow variety is often cultivated.

C. mirabilis Rourke, *Bothalia* 32: 1, 2000. [Bush Lily]. **Herbs**, to 80 cm tall; rhizome compact with many thick roots. **Leaves** 5–12 in a suberect basal cluster; lamina strap-shaped, 3–4 cm wide, smooth, dark green, often with a central whitish stripe down the upper surface. **Inflorescence** 20–35-flowered; pedicels reddish, lax, ~2.5 cm long. **Flowers** scarlet with green to yellow tips; tepals united into a slightly flared tube ~1 cm long, free for ~3 cm; stamens as long as tepals, style slightly exserted. **Fruits** scarlet, irregularly spindle-shaped. **Distr:** NW Cape Province (Bokkevel Mountains). **Ecol:** On loamy soils among rocks. Fl. Oct–Nov. **Zones:** 9–11.

C. nobilis Lindl., *Bot. Reg.* 14: t. 1182, 1828. (Syn: *Haemanthus cernuiflorus*, etc.). [Drooping Clivia, Red Bush Lily, Scarlet Clivia]. **Herbs**, rhizomatous, to 80 cm tall; roots fleshy, fascicled. **Leaves** fleshy, distichous, sheathing at base; lamina strap-shaped, 1 m × 25 cm, margins rough, apex distinctly notched, dark green, coriaceous. **Inflorescence** umbel, often with over 50 flowers; scape erect, plano-convex, border, furrowed toward the summit, 50–75 cm tall; pedicels long. **Flowers** cylindrical, pendent, overlapping, giving an umbrellalike appearance; perianth tubular, clavate, deciduous; tepals reddish orange to yellowish scarlet, fading toward tips, tips green, obtuse, darkest at the margin, imbricated in 2 rows, connate at base, outer ones shorter than inner; stamens 6, inserted near the mouth, equal, filaments smooth, anthers small, oval, greenish yellow, versatile; ovary inferior, greenish yellow, 3-celled, round, ventricose, ovules numerous, style filiform, stigma somewhat trilobed. **Fruits** berries, indehiscent, red; seed 1, smooth, transparent, oval. **Distr:** E Cape Province. **Ecol:** In forests and on rock faces. Fl. Jun–Oct. **Zones:** 9–11.

Crinum

[String Lily, Swamp Lily].
Species ~100. **Herbs** perennial, bulbiferous. **Leaves** basal, ligulate or ensiform, often large, with extensible fibers visible when torn. **Inflorescences** umbellate, few- to many flowered, rarely flower solitary; involucres large and broad; flowering stem solid; flowers pedicellate or sessile. **Flowers** with perianth actinomorphic or slightly zygomorphic, salverform or subsalverform, lobes oblong, lanceolate, or linear, straight or curved upward; stamens inserted at throat of perianth tube, filaments suberect to declinate, filiform; anther versatile, linear; ovary triloculed, inferior, ovules 2 to many per locule, style slightly decumbent, slender, stigma capitate, small. **Fruits** capsules, globose or subglobose, usually prominently beaked, dehiscing irregularly; seeds orbicular or angular, large, fleshy, testa often corky. **Distr:** Tropical and subtropical regions of both hemispheres, mainly in Africa. **Cult:** Propagate

SZ

Above: *Clivia nobilis, Crinum asiaticum, Crinum asiaticum* var. *asiaticum.*

from seeds or offsets. Requires well-drained soil in full or partial sun.

C. asiaticum L., *Sp. Pl.* 1: 292, 1753. (Syn: *Bulbine asiatica*). [Poison Bulb, Seashore Lily, Tree Crinum]. **Herbs**, to 2 m tall; bulb 5–15 cm Ø, narrowed into a cylindrical pseudostem to 50 cm long, clothed with old leaf-sheaths. **Leaves** 20–30, lower horizontal, upper semierect; lamina narrowly lanceolate, 0.9–2 m × 10–22 cm, base narrowed into a sheath, margins smooth, sometimes undulate, apex shortly acuminate, adaxially channeled along middle, coriaceous, bright green. **Inflorescences** umbels 1 or more, ~10–50-flowered; peduncles from axils of older leaves, 0.45–1 m long, green; bracts 2–13 × 3–4 cm, spathelike, acute; bracteoles linear, 6–12 cm long; pedicels 0.5–2.5 cm. **Flowers** fragrant; perianth salver-shaped, tube, cylindric, 7–13 × 0.4–0.5 cm, pale green, segments 5–12 × 0.5–1.2 cm, linear, revolute and recurved, apex acute or obtuse, white; filaments 3–8 cm long, white, purple distally, spreading, later deflexed, anthers 1.2–2 cm long, yellow, turning purple; ovary 1(–2) ovules per locule, style 8–18 cm long, white, purple distally, stigma with white papillae. **Fruits** ± subglobose, 2.5–5.5 cm Ø, pericarp fleshy, spongy; seeds 1–4, 3–4.5 cm long. **Distr:** tropical Asia. **Ecol:** On sandy seashores. **Zones:** 8–12. **Info:** The bulb is poisonous. The variety ***C. asiaticum* var. *asiaticum*** (Syn: *C. amabile, C. augustum*, etc.) [Giant Lily, Queen Emma Lily] is a very large herb reaching over 2 m in height, flowers are larger (to nearly 30 cm long) and the perianth segments are purplish on the outside.

C. jagus (J. Thomps.) Dandy, *J. Bot.* 77: 64, 1939. (Syn: *Amaryllis jagus*, etc.). [Swamp Lily, Spider Lily]. **Herbs**, 60–90 cm high; bulb stout, with a very short neck or neck absent. **Leaves** ± petiolate; lamina ± lanceolate, channeled, 20–75 × 3.8–11.5 cm, margins wavy, bright green, midrib thickened, parallel veins obvious. **Inflorescence** an umbel, 2–12-flowered; scape 25–75 cm tall, green; pedicels short or flowers sessile. **Flowers** fragrant, bilaterally symmetrical, funnel-shaped; corolla tube 10–20 cm long, limb much expanded and flattened, lobes spreading and ± overlapping, white sometimes tinged green; stamens and style downcurved; stamens white, anthers yellow soon turning brown (when the pollen is shed); style bright green, slender (relative to the stamens). **Fruits** ellipsoid, without a beak. **Distr:** Tropical Africa. **Ecol:** In swampy ground in forests. **Zones:** 10–12.

C. latifolium L., *Sp. Pl.* 291, 1753. (Syn: *C. esquirolii*, etc.). [Knotted Lily, Pink-striped Trumpet Lily, Milk and Wine Lily]. **Herbs**; bulb ovoid or subspherical, 8–15 cm Ø, narrowed into a stout pseudostem to 25 cm long. **Leaves** 12–20; lamina lorate, 0.6–1.25 m × 4–7 cm, margins rough, with distantly placed serrations, apex acuminate, bright green, glossy, fleshy, adaxially narrowly channeled along middle, abaxially keeled. **Inflorescence** an umbel, 10–20-flowered, axillary; peduncle 0.6–1.25 m long, green or purplish green. **Flowers** with perianth funnel-shaped, tube 9–13 cm long, purplish-green proximally, pale green distally, curved, limb 10–15 cm across, segments lanceolate or narrowly elliptic, 10–13 × 2.5–3.5 cm, acute or obtuse, retuse, often apiculate or shortly caudate, white, with broad purplish median stain from base to apex, darker outside; stamens declinate, filaments 5.5–9 cm long, white, purple distally, anthers 2.5–3 cm long, linear, yellow, purplish-gray after dehiscence; ovary oblong, with 5 or 6 ovules per locule, style 17–23 cm long, white, purple distally, stigma shallowly trifid, purple. **Distr:** Old World tropics. **Ecol:** On dry riverbeds and sandy soils. Fl. Jun–Aug. **Zones:** 9–12.

C. moorei Hook. f., *Bot. Mag.* 100: t. 6113, 1874. (Syn: *Amaryllis moorei*, etc.). [Cape Coast Lily]. **Herbs**, bulbous, to 1.3 m high; bulb narrowly ovoid, contracted into a long neck, neck 0.3–1.1 m long, proliferating offsets around base. **Leaves** numerous, erecto-patent; lamina ensiform, 0.6–1.4 m × 6.5–10 cm, widest near middle, flat, margins wavy, both ends tapering, apex obtuse, midrib thickened, parallel veins obvious. **Inflorescence** an umbel, usually 5–10-flowered, flowers nodding; spathes oblong-lanceolate, 15 cm long, concave, subacute, herbaceous, reflexed; scape 45–75 cm tall; pedicels 0.6–5.7 cm long. **Flowers** bilaterally symmetric, funnel-shaped or broadly campanulate, odorless; perianth tube curved, 7.5–10 cm long, lobes wide-spreading, 7.5–10 × 1.9–3.8 cm, white or white suffused with pink; stamens and style downcurved, style much shorter; stamens ~4 cm long, anthers ~1.2 cm long; ovary ~2.5 cm long. **Fruits** without a beak. **Distr:** E Cape Province to KwaZulu-Natal. **Zones:** 8–11.

C. zeylanicum (L.) L., *Syst. Nat.* ed. 12, 2: 236, 1767. (Syn: *Amaryllis zeylanica*, etc.). [Ceylon Crinum, Milk and Wine Lily]. **Herbs**; bulb subspherical, 6–15 cm Ø, narrowed into a cylindrical pseudostem to ~15 cm long. **Leaves** 9–12; lamina lorate or linear, 0.2–1 m × 2.5–7.5 cm, margins strongly undulate, apex acuminate, adaxially narrowly channeled along the middle, abaxially keeled. **Inflorescence** umbel, 6–12-flowered; scape to 75 cm or more; flowers sessile; bracts 6–9 cm long, purplish. **Flowers** fragrant, perianth white, broad, purplish-red keels (most pronounced adaxially) funnelform, tube 7.5–15 cm; perianth lobes oblong-elliptic, 7.5–18 × 1–2 cm; stamens declinate, filaments 6–8 cm long, white, purple distally, anthers 2–2.8 cm long, dull purplish-brown, turning black; ovary green, bluntly trigonous, locules 2–6-ovuled, style 16–20 cm long, white, purple distally, stigma bluntly trifid, papillose. **Fruits** subglobose or ovoid, 2–5 cm long, beak 8–10 cm long; seeds 2–6, warty. **Distr:** tropical Asia. **Ecol:** In fields, waste places, wet areas, sandy seashores; 0–600 m. Fl. Aug–Feb. **Zones:** 10–12.

Eucharis

Syn: *Microdontocharis.*

Species 17. **Herbs**, perennial; bulbs subterranean, tunicate, base producing offsets around the base.

Below: *Crinum jagus, Crinum latifolium, Crinum moorei, Crinum zeylanicum.*

Leaves radical, few or several, base abruptly narrowed into a slender petiole; lamina ovate or lanceolate. **Inflorescence** an umbel, few-flowered, flowers nodding; peduncle solid, mostly terete. **Flowers** large and showy, white, radially symmetric; perianth tube cylindric, short, with a dilated throat, ± strongly curved; segments oval or oblong, equal, spreading; stamens inserted in throat of perianth tube, shorter than perianth segments, filaments broadened at base and connate to form a corona; ovary inferior, triloculed, with 6–8 ovules in each loculus. **Fruits** capsules, loculicidal, deeply trilobed, leathery or thin-walled; seeds few, often single, large, rounded, black or blue. **Distr:** C America to Peru and N Brazil. **Cult:** Propagate by division or from seeds. Requires a humus-rich, moist, well-drained soil in full or partial sun.

E. amazonica Linden ex Planch., *J. Gén. Hort.* 12: 1216, 1857. [Amazon Lily, Eucharis Lily]. **Herbs**, evergreen, 30–60 cm tall, proliferating offsets at base; bulbs globose, to 7 cm Ø. **Leaves** 2–4, oblong to elliptic or oval, 25–35 × 10–15 cm, base cuneate or rounded, margins ± wavy, apex abruptly short-acuminate, adaxially dark glossy green, adaxially paler; petiole 15–30 cm long. **Inflorescence** an umbel, usually 4–6-flowered, taller than the leaves, flowers pendent; pedicels 1–2.5 cm long; scape to 50 cm tall. **Flowers** white, fragrant; perianth tube curved, 4.4–5.7 cm long, limb flaring, 5.7–8.9 cm across, segments oblong-ovate, 3–5 cm long, white, apex apiculate; filament flanges connate, free portion lanceolate; corona cylindrical, streaked green inside, ~1.2 cm long, 2.5 cm Ø; ovary oblong-ellipsoid, stigma deeply 3-forked. **Fruits** trilobed, green, 2.5 cm Ø; seed black. **Distr:** N Peru; cultivated widely elsewhere. **Ecol:** In forest understory; 1,300–2,000 m. Fl. Jan–Feb, May, Oct. **Zones:** 11–12.

Haemanthus

[Paintbrush Lily, Powder-puff Lily].

Species 22. **Herbs**, deciduous or evergreen, perennial; rootstock bulbous, with thick fleshy tunics or 2-ranked scales. **Leaves** 2, sometimes 1 or as many as 6, dry or green at flowering, erect to prostrate; lamina strap-shaped to elliptic, often barred with dark green or red, fleshy, sometimes hairy. **Inflorescence** umbel, compact to spreading; scape stout, somewhat fleshy; spathe bracts 4 or more, often fleshy, erect or spreading, white, pink, or red. **Flowers** actinomorphic, narrowly or broadly funnel-shaped, erect or spreading; perianth tube short, cylindrical or bell-shaped; tepals narrowly oblong to lanceolate; stamens inserted in throat, usually longer than tepals, anthers dorsifixed; ovary with 1 or 2 ovules per locule, style straight, as long as stamens, stigma minutely tricuspidate. **Fruits** usually fleshy, ovoid to globose, white, orange, or pink, often aromatic when ripe; seeds ovoid, fleshy, red, green, or ivory, embryo green. **Distr:** Southern Africa. **Cult:** Propagate from offsets or ripe seeds. Requires a fertile, compost-rich, well-drained soil in full sun.

H. albiflos Jacq., *Pl. Hort. Schoenbr.* 1: 31, 1797. (Syn: *Diacles ciliaris*, etc.). [April Fool]. **Herbs**, to 35 cm tall, evergreen; bulbs ovoid, to 8 cm across, often with green tunics. **Leaves** 2, 4, or 6, green at flowering, erect to prostrate; lamina strap-shaped, oblong or elliptic, 2.5–11.5 cm wide, smooth, hairy or fringed, unmarked. **Inflorescence** umbel, somewhat compact, up to 50-flowered; scape green, 5–35 cm tall; spathe bracts 4–8, slightly spreading, blunt, shorter or slightly longer than flowers, white with green veins; pedicels to 1 cm long. **Flowers** perianth funnel-shaped, 1.6–2.3 cm long; corolla tube 4–7 mm long; perianth segments spreading, oblong, 1–1.8 cm × 1–2.5 mm; stamens and style slightly exserted; ovary globose, to 3 mm Ø. **Fruits** ovoid, to 1 cm Ø, white to red. **Distr:** South Africa (Cape Province to KwaZulu-Natal). **Ecol:** In subtropical thickets, dry savannas, coastal and riverine scrubs; 100–800 m. Fl. Apr–Aug. **Zones:** 9–11.

H. coccineus L., *Sp. Pl.*: 325, 1753. [April Fool, March Lily, Paintbrush Lily, Blood Lily]. **Herbs**, variable in size, to 40 cm tall; bulbs with thick, cream tunics. **Leaves** usually 2, dry at flowering, suberect to prostrate; strap- to tongue-shaped, (2.5–)8.5–15(–21) cm wide, margins often rolled back, fleshy, smooth or hairy, barred with maroon or dark green beneath, sometimes on a whitish background. **Inflorescence** umbel, ± compact; scape stout, spotted or rarely plain red; spathe bracts (4–)6–9(–13), erect, overlapping, shorter to longer than flowers, bright red, stiff and fleshy. **Flowers** 1.4–3 cm long, scarlet; corolla tube 2–5 mm long; stamens and style exserted. **Distr:** Namibia to Cape Province. **Ecol:** In coastal scrubs and on rocky slopes. Fl. Feb–Apr. **Zones:** 9–11.

Hippeastrum

Syn: *Aschamia, Omphalissa*, etc.

Species 91. **Herbs**, perennial, scapose, glabrous; bulbs globose. **Leaves** few, basal; lamina liguliform, fleshy, parallel-veined, margins entire, apex tapering. **Inflorescence** umbellate, bracteate; bracts 2, scarious; scape hollow. **Flowers** fragrant, spreading to slightly

Top row: *Eucharis amazonica, Haemanthus albiflos, Haemanthus coccineus.*
Bottom row: *Hippeastrum papilio, Hippeastrum puniceum, Hippeastrum reticulatum.*

Above: *Hymenocallis caribaea, Hymenocallis littoralis, Pancratium zeylanicum, Proiphys amboinensis.*

drooping, syntepalous; perianth connate proximally, funnelform to campanulate, with minute corona reduced to small crown inserted on throat of tube; tepals 6 in 2 whorls of 3, outer slightly shorter than inner; stamens inserted on perianth tube, declinate, subequal, filaments slender; ovary inferior, ellipsoid, style slender, almost equaling tepals, stigma capitate or slightly trilobed. **Fruits** capsules, dehiscence loculicidal; seeds few to many. **Distr:** C and S tropical America. **Cult:** Propagate from offsets or fresh ripe seeds. Requires a well-drained soil rich in organic matter, in full or partial sun.

H. papilio (Ravenna) Van Scheepen, *Taxon* 46: 18, 1997. (Syn: *Amaryllis papilio*). [Butterfly Amaryllis]. **Herbs**, bulbous; bulbs ovoid, 3–8+ cm Ø; stems 30–60 cm tall. **Leaves** 5, evergreen, distichous; lamina lorate, to 50 × 4.8 cm, margins entire, apex acute, glabrous. **Inflorescence** an umbel, 2–3-flowered; scape 45–50(–68) cm tall. **Flowers** very large, to 19 cm tall, to 12.5 cm Ø, tube ~10 cm long, cream to pale green with conspicuous reddish-brown veins and streaks, upper 3 tepals heavily marked at center with veins toward the margin; tepals 10–12 × 3.3–5.8 cm; stamens 6, anthers elliptic, ~3–4 mm long, white, pollen white; pistil length ~12.5 cm long, stigma trifid. **Distr:** S Brazil. **Ecol:** Terrestrial or epiphytic in tropical rainforests. Fl. Oct, Jan. **Zones:** 8–12. **Info:** All parts of the plant are poisonous if ingested.

H. puniceum (Lamarck) Voss, *Vilm. Blumengärtn.* ed. 3. 1: 1033, 1895. (Syn: *Amaryllis punicea*). [Barbados Lily]. **Herbs**; bulbs globose or ovoid, stoloniferous, 4–10 cm Ø. **Leaves** 6–8, appearing after flowering; lamina strap-shaped, 30–60 × 2.5–5 cm, gradually narrowed to apex. **Inflorescences** 2–4-flowered; bracts (spathes) lanceolate, 5 cm long; scape terete, pale, 0.4–1 m tall; pedicels 3–7 cm long. **Flowers** slightly zygomorphic, ~10 cm long; perianth reddish to salmon, with whitish midstripe on adaxial surface of each outer tepal, tube 3 cm long; outer tepals lanceolate to subrhombic, 12 cm or more, apex acuminate. **Fruits** ellipsoid to ovoid, 2 cm across; seeds black, compressed, globose or subglobose. **Distr:** S tropical America; widely cultivated and naturalized throughout the tropics. **Ecol:** In lowlands to highlands; 0–700+ m. **Zones:** 10–12.

H. reticulatum (L'Hér.) Herb., *Bot. Mag.* 51: t. 2475, 1824. (Syn: *Amaryllis reticulata*, etc.). **Herbs**, bulbous, 25–35 cm high; bulb ± spherical, with a short neck. **Leaves** oblanceolate, to 30 × 2.5–5 cm, tapered at base into an almost terete stalk, thin, deep green with pale central marking. **Inflorescence** an umbel, 3–5-flowered; spathes to 5 cm long; flower stalks 2.5–4.5 cm long. **Flowers** with perianth funnel-shaped, 10–12.5 cm long, drooping, red-mauve, with a superimposed network of darker veins; perianth tube 1.3–2.5 cm long; stamens shorter than the perianth; style about as long as the perianth, stigma capitate, slightly trilobed, locules 1–3-ovuled. **Distr:** E Brazil to Argentina (Misiones). **Ecol:** In tropical rainforests. Fl. Apr–May. **Zones:** 9–12.

Hymenocallis

[Spider Lily].

Species 63. **Herbs**, perennial, scapose, from bulbs; bulb ovoid or globose, tunicate, often extending into neck of clasping, distichous leaf bases. **Leaves** 2–16, deciduous or evergreen, sessile, rarely petiolate; lamina narrowly to widely liguliform or oblanceolate, rarely ovate to elliptic. **Inflorescence** umbellate, 1–16-flowered, bracteate; flowers usually sessile, erect or slightly diverging; bracts 2–3, triangular, ovate, or lanceolate; floral bract narrowly lanceolate. **Flowers** large and starlike, fragrant; perianth connate basally into a short or long tube, surmounted by conspicuous staminal corona; tepals extending from base of corona, free portions reflexed or ascending, often distally recurved, linear; stamens adnate basally into showy funnelform or rotate corona, margins between free portions of filaments often dentate or lacerate, portions of filaments inserted on margin of corona, erect to incurved, filiform, anthers versatile, introrse, pollen yellow, often golden, or orange; ovary inferior, globose, ovoid, oblong, or pyriform, ovules 2–10 per locule, style exserted beyond stamens; stigma capitate. **Fruits** capsules, green, subglobose to elongate, trilocular, large, leathery; seeds large, green, fleshy. **Distr:** Tropical and subtropical America. **Cult:** Propagate by division of clumps, from offsets or fresh ripe seeds. Requires a damp, fertile, soil, in full or partial sun.

H. caribaea (L.) Herbert, *Appendix*: 44, 1821. (Syn: *Pancratium caribaeum*, etc.). [Spider Lily]. **Herbs**, bulbous. **Leaves** bifarious; lamina very narrowly oblanceolate or narrowly obovate, 0.3–0.6(–1) m × 5–7.5(–10) cm, base slightly narrowed, tapered to an acute or acuminate apex, glossy green, somewhat coriaceous. **Inflorescence** an umbel, 8–15-flowered; flowers nearly sessile; peduncle 30–95 cm long, compressed, with 2 lateral ridges, glaucous; bracts ovate to lanceolate, 3–6 cm long. **Flowers** fragrant; perianth tube 3.5–8.5 cm long, bluntly trigonous, slightly curved, glaucous green; segments 9–14 cm long, spreading, linear, white; corona funnel-shaped, 2–3.2 cm long, sometimes with a minute tooth between filaments, free part of filaments 3–6 cm long, white, dark green distally; anthers 2–2.5 cm long; ovules mostly 2 per locule, style ~13 cm long. **Fruits** globose, 1.5–2 cm Ø, lobed. **Distr:** Caribbean; widely cultivated and naturalized in the tropics. **Ecol:** In sandy coastal areas. **Zones:** 10–12.

H. littoralis (Jacq.) Salisb., *Trans. Hort. Soc. London* 1: 338, 1812. (Syn: *Pancratium littorale*, etc.). [Spider Lily]. **Herbs**; bulb subglobose, 5.5–10 cm across, narrowing to a false stem 4–10 cm long. **Leaves** 7–12, bifarious; lamina lorate to falcate, 0.4–1.3 m × 4–8 cm, apex acute, glossy green and channeled adaxially, with smooth edges. **Inflorescence** an umbel, 10–15-flowered; flowers nearly sessile; peduncle 0.5–1 m long, compressed, dark green, glaucous. **Flowers** with perianth tube 8–10 cm long, bluntly trigonous, glaucous green; segments 10–12.5 cm long, spreading, falcate, white, light green at apices; corona white, 2–2.5 cm long, with a single minute tooth between filaments, free parts of filaments 5.5–7.5 cm long, white proximally, dark green distally, declinate, anthers 1.5–2.5 cm long; ovary dark green, ovules mostly 4 per locule, style 13–22 cm long, stigma punctiform, dark green. **Distr:** Mexico to N Peru, Brazil; cultivated and naturalized in the tropics. **Ecol:** In lowland wet, marshy savannas, swampy areas, naturalized elsewhere on sandy sea shores and in thickets; near sea level. Fl. Jul–Aug. **Zones:** 10–12.

Pancratium

Syn: *Almyra, Chapmanolirion, Mizonia*

Species 21. **Herbs**, perennial, bulbiferous; bulbs subterranean. **Leaves** basal, sessile, bifarious; lamina linear or ligulate. **Inflorescences** umbellate, 1- to many flowered; involucres 1 or 2; bracteoles few, linear; pedicel often short; scape solid. **Flowers** with perianth subfunnelform; lobes 6, spreading, lanceolate or linear; stamens inserted at throat of perianth tube, filaments connate into a cup for most of their length but apically free, anthers versatile, linear; ovary trilocular, trigonous, with many ovules (<10), style filiform, stigma capitate, sometimes somewhat trilobed to -branched, small. **Fruits** capsules, trivalved, dehiscence loculicidal; seeds many, black, angular. **Distr:** Africa, Mediterranean to Malesia. **Cult:** Propagate from offsets or seeds. Requires a well-drained soil in full sun; salt-tolerant.

P. zeylanicum L., *Sp. Pl.* 290, 1753. [Rain Flower, Rain Lily, Sea Daffodil, Star Lily]. **Herbs**; bulb subglobose to ovoid, 3.5–7 cm across; pseudostem absent or to ~1.5 cm high. **Leaves** narrowly lanceolate or -oblanceolate, 3–12, 15–30 × 1.5–2 cm, margins smooth, fleshy, adaxially glossy green, abaxially slightly paler, glaucous. **Inflorescences** 1 or 2, flowers solitary; peduncles subterete or compressed, arising lateral to the leafy axis, 8.5–25 × ~1 cm; bract single, 3–6 cm long. **Flower** 6–12 cm Ø; perianth tube cylindrical, 2.5–8 cm long, very pale green, segments white, narrowly lanceolate, 5–9 × 0.8–1.2 cm, margins revolute, apex mucronate, adnate below to corona; corona membranous, obconical, white, 2–4 cm long, with 2 deltoid teeth ~1 cm long between filaments, free part of filaments 2.5–4 cm long, white, basally curved, anthers ~1.5 cm long, dull yellow; ovary oblong, dark green, locules many ovuled, style 8–12 cm long, white, stigma

Above: *Scadoxus multiflorus, Scadoxus multiflorus* ssp. *katharinae, Scadoxus puniceus, Sprekelia formosissima.*

very shallowly trifid. **Fruits** trigonous, ~2 × 1.5 cm; seeds several. **Distr:** Throughout tropical Asia. **Ecol:** In lowlands. Fl. throughout most of the year. **Zones:** 10–12. **Info:** Bulbs used for medicinal purposes.

Proiphys

Syn: *Eurycles.*

Species 4. **Herbs**, bulbs subterranean. **Leaves** radical, petiolate, reniform-cordate or elliptic, with the main veins connected by numerous parallel, very thin, pellucid cross-veins, thinly membranous when dry. **Inflorescence** an umbel, many flowered; peduncle solid. **Flowers** with perianth funnel-shaped, tube narrow, segments oblong to obovate, spreading or suberect, subequal; stamens inserted in throat of perianth tube, shorter than perianth segments, filaments broadened at base and connate to form a corona with 12 lobes, 2 between each pair of filaments, anthers medifixed; ovary with 2 ovules in each loculus, ascending from base. **Fruits** capsules, globose, somewhat fleshy, often 1-seeded. **Distr:** Malesia to N Australia. **Cult:** Propagate from offsets or seeds. Requires a moist, well-drained soil in partial shade.

P. amboinensis (L.) Herbert, *App. (to Bot. Reg. 7).* 42, 1821. (Syn: *Eurycles amboinensis, Pancratium amboinense*). [Cardwell Lily, Northern Christmas Lily]. **Herbs**; bulb 6–10 cm across. **Leaves** 2 or 3; lamina broadly cordate-reniform, 15–30 × 15–35 cm, base broadly and shallowly cordate, apex shortly acuminate; petiole 15–30 cm, channeled above. **Inflorescence** an umbel, 6–20-flowered; peduncle 30–50 cm, terete. **Flowers** with perianth tube 2.5–3.5 cm, pale green, segments 2.5–4 cm long, somewhat spreading, white; style white. **Fruits** with seed often replaced with a bulbil. **Distr:** Malesia. **Ecol:** On seashore and rocky places; 0–500 m. Fl. Apr–May, Nov–Dec. **Zones:** 10–12.

Scadoxus

Syn: *Choananthus, Demeusea, Nerissa.* [Blood Lily]. **Species** 9. **Herbs**, perennial; bulb with neck. **Leaves** 4–10(–12), usually in a basal rosette; lamina elliptic or ovate, midrib prominent. **Inflorescence** an umbel, compact, spherical, hemispherical or conical, borne from the side of the bulb, ~10–100-flowered, flowers usually erect; scape cylindrical, solid; bracts 4 to numerous. **Flowers** radially symmetric; perianth tube cylindrical, lobes 5; corona absent; stamens projecting, filaments adnate to mouth of perianth tube; ovary inferior, 3-celled. **Fruits** berries, fleshy, orange to red; seeds 1–3, pale. **Distr:** Cape Verde, tropical and S Africa, Arabian Peninsula. **Cult:** Propagate from bulbs and offsets. Requires well-drained soil in full sun.

S. multiflorus (Martyn) Raf., *Fl. tellur.* 4: 19, 1838. (Syn: *Haemanthus multiflorus*, etc.). [Blood Flower, Fireball Lily, Common Fireball, Powderpuff Lily, Small Fireball Lily, Snake Lily]. **Herbs**, deciduous; bulb to 6.35 cm Ø; pseudostem to 60 cm tall. **Leaves** 3–5, appearing after inflorescence, leaf bases rolled; lamina lanceolate to ovate, to 25 × 8 cm, base tapered to subcordate, apex ± pointed. **Inflorescence** an umbel, globose or hemispherical, 10–20 cm Ø; scape to 76 cm tall; bracts linear to lanceolate, drooping when flowers open. **Flowers** with corolla tube 0.5–1.25 cm long, lobes threadlike, 1.25–3.2 cm long, spreading, scarlet, fading to pinkish; anthers yellow or red. **Fruits** orange, 5–8 mm Ø. **Distr:** Cape Verde, tropical and S Africa to SW Arabian Peninsula. **Ecol:** In savanna woodlands, forest margins and secondary forests. **Zones:** 9–12. **Info:** This species has 3 subspecies described.

S. multiflorus* ssp. *katherinae (Baker) Friis and Nordal, *Norweg. J. Bot.* 23: 64, 1976. (Syn: *Haemanthus katharinae, H. multiflorus* ssp. *katharinae*). [Catherine Wheel]. Similar to *S. multiflorus* except: **Herbs**, to 1.2 m tall, flowerheads larger with more flowers, flowers deep pink. **Distr:** South Africa (SE to Zimbabwe). **Ecol:** In damp, warm forests. **Zones:** 9–12.

S. puniceus (L.) Friis and Nordal, *Norweg. J. Bot.* 23: 64, 1976. (Syn: *H. natalensis, H. puniceus*, etc.). [Red Paintbrush, Royal Paintbrush, Snake Lily]. **Herbs**, to 75 cm tall. **Leaves** 2–7, dry or green at flowering; bases petiolelike, sheathing to form a long false stem surrounded by spotted basal sheaths; lamina spreading, lanceolate to ovate, 5–15 cm wide. **Inflorescence** an umbel, dense, obconical to hemispheric; scape basally spotted; spathe bracts 5–8(–11), suberect, of variable shape, as long as or longer than flowers, green to red. **Flowers** 2–3.5 cm long, yellow-green to red; corolla tube 0.3–1.2 cm long; stamens and style slightly exserted. **Distr:** Ethiopia to S Africa. **Ecol:** In coastal and river valleys of subtropical thickets. Fl. Sep–Jan. **Zones:** 10–12.

Sprekelia

Species 2. **Herbs. Leaves** annual, bright green. **Inflorescence** with scape hollow, single-flowered; spathe bracts fused below. **Flower** extremely zygomorphic, red; tube very short, tepals narrow, 1 outer tepal dorsally erect and unguiculate, 2 inner tepals falcate and spreading laterally, remaining tepals declinate and convolute proximally, forming a pseudo-labellum that encloses the stamens and style; stamens fasciculate, declinate ascending; stigma trifid. **Fruit** a capsule; seeds black, D-shaped. **Distr:** Guatemala, Mexico. **Cult:** Propagate from offsets or fresh ripe seeds. Requires a moderately fertile, well-drained soil in full sun.

S. formosissima (L.) Herb., *Appendix*: 35, 1821. (Syn: *Amaryllis formosissima*, etc.). [Jacobian Lily]. **Herbs**, deciduous; bulb very large, with a tunic; stems sturdy, 30–45 cm tall, usually 1 stem per bulb, more stems in large bulbs. **Leaves** produced with or after flowers; lamina narrow, linear, straplike, apex pointed. **Inflorescence** with a solitary flower; scape hollow, 20 cm tall, arising from the basal rosettes of leaves; bract spathulate. **Flower** deep crimson, horizontal, bilaterally symmetrical, 10–15 cm across; perianth with 6 segments (tepals), uppermost segment wide, lateral segments narrow and strap-shaped, lowest segment

Below: *Tristagma uniflorum, Zephyranthes atamasco, Zephyranthes candida, Zephyranthes citrina.*

2-lipped, other 2 lower segments curled toward the base forming a tube, the 3 lower segments enclosing the stamens and pistil, bases of segments sometimes striped or blotched with yellow; stamens with filaments deep crimson, anthers linear; ovary inferior, style longer than the stamens. **Fruits** many seeded. **Distr:** Guatemala, Mexico. **Ecol:** On rocky hillsides and rocky outcrops; 1,000–2,400 m. Fl. May–Oct. **Zones:** 7–11.

Tristagma

[Star Flowers].

Species 30. **Herbs**, bulbous; some species with alliaceous odor. **Leaves** usually linear, flat, rarely canaliculate or keeled, terete or filiform. **Inflorescence** 1–8-flowered; scape slender, erect after anthesis; bracts united below at margins. **Flowers** sweetly scented; tepals 6, united below; anthers 6, arranged in 2 series in the tube, versatile or basifixed, filaments long to absent; ovary superior, locules several-ovuled. **Distr:** Brazil to S South America. **Cult:** Propagate from offsets or seeds. Requires a well-drained soil in full or partial sun.

T. uniflorum (Lindl.) Traub, *Pl. Life* 19: 61, 1963. (Syn: *Triteleia uniflora*, *Ipheion uniflorum*, etc.). [Spring Star Flower]. **Herbs**, bulbous; bulb ovoid, 1.5–2.2 cm across, tunic membranous; stem to 35 cm long, terete. **Leaves** 6–9, grasslike and mat-forming; lamina linear, 15–30 x 0.4–0.8 cm, glaucous, gray green, somewhat fleshy, smelling of onion when bruised. **Inflorescence** 1-flowered, scape as long as the leaves or slightly shorter; bracts brownish, scarious, forming a sheathing tube with 2 lobes, lobes lanceolate, 2.2–3.8 cm long, acute; pedicel 2.5–5 cm long. **Flowers** pure white or violet-blue, 4 cm across; perianth 2.5–3.5 cm long, tube 1.2–1.6 cm long, narrow; limb 3–4.5 cm across; lobes patent, rather longer than the tube, lanceolate-spathulate, acute; anthers and stigma just visible at the mouth of the tube; stamens biseriate in tube, filaments filiform; ovary sessile, style filiform, 0.8–1 cm long. **Distr:** NC Argentina to Uruguay. **Ecol:** Along banks of rivers and streams, often literally covering the area. Fl. Feb–Apr. **Zones:** 6–10.

Zephyranthes

Syn: *Atamosco*. [Rain Lily, Fairy Lily, Zephyr Lily].

Species ~70. **Herbs** perennial, scapose, bulbiferous; bulbs black or brown, ovoid or globose, covered with a tunic, sometimes with long neck. **Leaves** several, sessile, erect or recumbent, fascicled, with overlapping sheathing bases; lamina linear, smooth, rarely exceeding 1 cm in width. **Inflorescence** with scape, slender, hollow, spathaceous; involucre 1, basally tubular, apically 2-notched; flowers solitary, terminal, erect to declinate. **Flowers** actinomorphic; perianth subrotate to funnelform to salverform, connate basally into tube; tube 2–16 cm long; tepals 6, subequal; stamens 6, 3 long alternating with 3 short, inserted at throat or in tube of perianth, appearing equal to unequal, filaments inserted just above perianth tube, erect to slightly declinate, diverging except when shorter than tube, long-filiform to short-subulate, anthers submedially dorsifixed, usually parallel with floral axis, linear-oblong; ovary inferior, with many ovules, style filiform, stigma capitate, trilobed or 3-notched. **Fruits** capsules, subglobose or ± oblate, thin-walled, trivalved, loculicidal; seeds numerous, black. **Distr:** Warmer regions of the W Hemisphere (from S USA to S America). **Cult:** Propagate from seeds or offsets. Requires a fertile, moist, well-drained soil in full or partial sun.

Z. atamasco (L.) Herbert, *Appendix*. 36, 1821. [Atamasco Lily, Carolina Lily, Easter Lily, Naked Lady, Occidental Swamp Lily, Virginia Lily]. **Herbs** perennial, bulbiferous. **Leaves** shiny green, to 8 mm wide. **Inflorescence** with spathe (2–)2.2–3.2(–3.6) cm long; pedicel (0–)0.4–1.6(–3.4) cm, usually shorter than spathe. **Flowers** erect to slightly inclined; perianth mostly white, sometimes tinged or veined pink, more so with age, funnelform, (5.5–)6.6–9(–10) cm; perianth tube green, (0.8–)1–2(–2.1) cm, increasing Ø; tepals usually reflexed; stamens diverging, appearing equal, filaments filiform, (2.9–)3.2–4.4(–4.7) cm; anthers 3–6(–8) mm; style longer than perianth tube; stigma trifid, exserted more than 2 mm beyond anthers. **Distr:** N America. **Ecol:** In mixed forests, coastal plains; 0–700 m. Fl. Jan–May. **Zones:** 8–11.

Z. candida (Lindley) Herbert, *Bot. Mag.* 53: plate 2607, 1826. (Syn: *Amaryllis candida*, etc.). [Autumn Rain Lily, Peruvian Swamp Lily]. **Herbs** perennial, bulbiferous; bulbs ovoid, ~2.5 cm Ø, neck 2.5–5 cm. **Leaves** glossy, bright green; terete-linear, 20–30 cm x 2–4 mm, fleshy. **Inflorescences** with spathe (1.8–)2–4 cm; involucres red-brown; flowers solitary, erect, terminal; pedicel ~1 cm long. **Flowers** with white perianth, sometimes pinkish abaxially, subrotate, lobes ± free, 3–5 x ~1 cm usually with tiny scales near throat, apex obtuse to shortly acute; perianth tube green, 0.1–0.4 cm, increasing Ø; tepals not reflexed; stamens diverging, subequal, filaments filiform, 1–1.4 cm; anthers 5–8 mm; style slender, longer than perianth tube, stigma capitate, 3-notched, usually among or exserted less than 2 mm beyond anthers; pedicel (0.4–)1–2.5 cm, usually shorter than spathe. **Fruits** subglobose, ~1.2 cm Ø. **Distr:** S America (Argentina, Peru, Uruguay). **Ecol:** On coastal plains; 0–200 m. Fl. Jun–Nov. **Zones:** 9–11.

Z. citrina Baker, *Bot. Mag.* 108: plate 6605, 1882. [Citron Rain Lily]. **Herbs** perennial, bulbiferous. **Leaves** dull green; lamina to 4 mm wide. **Inflorescence** with spathe 1.6–2.6 cm. **Flowers** erect; perianth lemon yellow, funnelform, 3.1–5 cm; perianth tube green, 0.7–1 cm, increasing in Ø; tepals rarely reflexed; stamens diverging, in 2 distinctly subequal sets, filaments filiform, subequal, 1.2–2 cm; anthers 5–7 mm; style longer than perianth tube, stigma capitate, usually among or below anthers, not exserted more than 1 mm beyond anthers; pedicel 2.3–4.4 cm, usually longer than spathe. **Distr:** West Indies, C America, S America. **Ecol:** Moist, sandy loam; 0–100 m. Fl. Jul–Sep. **Zones:** 10–12.

Z. grandiflora Lindley, *Bot. Reg.* 11: plate 902, 1825. [Pink Rain Lily, Pink Storm Lily, Rose-pink Zephyr Lily]. **Herbs** perennial, bulbiferous. **Leaves** shiny green; lamina to 7 mm wide. **Inflorescence** with spathe 3–4 cm; flowers usually inclined. **Flowers** with pink perianth, funnelform, 5.6–9 cm; perianth tube green, (1.2–)1.8–2.3 cm, increasing in Ø; tepals reflexed; stamens diverging, appearing equal; filaments filiform, 2–3 cm; anthers 1.3–2.2 cm; style longer than perianth tube; stigma trifid, exserted more than 2 mm beyond anthers; pedicel 1.3–3.8 cm, usually shorter than spathe. **Distr:** Mexico, C America, West Indies. **Ecol:** Open areas in or near moist woods; 0–100 m. Fl. Apr–Jul. **Zones:** 10–12.

Z. rosea Lindley, *Bot. Reg.* 10: plate 821, 1824. [Pink Rain Lily, Fairy Lily, Cuban Zephyr Lily]. **Herbs** perennial, bulbiferous. **Leaves** dull green; lamina to 5 mm wide. **Inflorescence** with spathe 2–2.8 cm; pedicel 1.5–3.2 cm, usually exceeding spathe. **Flowers** erect to slightly inclined; perianth rose pink, spreading, funnelform, 2.7–2.8 cm; perianth tube green, 0.2–0.3 cm, increasing in Ø; tepals not reflexed; stamens diverging, in 2 slightly subequal sets with anthers overlapping ~90 percent, filaments filiform, 1–1.7 cm, anthers 3–6 mm long; style 1.8–2.7 cm, longer than perianth tube, stigma trifid, exserted ~1–5 mm beyond anthers. **Distr:** Mexico, West Indies, Central America. **Ecol:** Sandy loam; 0–50 m. Fl. Mar–Jul. **Zones:** 10–12. **Info:** This is the most widely cultivated species, now naturalized in many parts of the tropics and subtropics.

HYACINTHACEAE

Genera 41–70, species 770–1,000. **Herbs**, mostly bulbous; bulbs formed by foliage leaf bases or thickened cataphylls. **Leaves** 1–many, filiform, elliptic or circular. **Inflorescence** a raceme or rarely a spike or compound raceme, 1–many flowered, generally bracteate, scapose, simple. **Flowers** not articulated with the pedicel, actinomorphic, rarely zygomorphic, hermaphrodite, hypogynous, trimerous; tepals 3+3, polyphyllous or ± gamophyllous, white, blue, red, yellow, brownish or greenish; inner and outer tepals ± similar; stamens 3+3, sometimes 1 whorl sterile, sometimes epiphyllous, anthers introrse, dehiscing longitudinally; pistil, tricarpellate, syncarpous, trilocular, inner septal nectaries present, style with simple or sometimes thickened stigma. **Fruits** capsules with loculicidal dehiscence; seeds variable in shape and color. **Distr:** Predominantly Old World with the highest diversity in S Africa and Mediterranean; widely distributed in NW Europe, Central Asia and E Asia; 1 small subfamily extending W South America.

Albuca

Syn: *Albugoides*, *Igidia*, etc. [Slime Lily].

Below: *Zephyranthes grandiflora*, *Zephyranthes rosea*, *Albuca batteniana*.

MD

Above: *Eucomis bicolor, Eucomis comosa, Lachenalia aloides, Ornithogalum dubium.*

Species 106. **Herbs**, deciduous or rarely evergreen, perennial, bulbous; bulb globose, often large, subterranean or partially exposed, flesh sometimes pink or reddish. **Leaves** few to several, rarely 1, rarely dry at flowering; lamina linear to oblong or terete, hairless or rarely hairy, glandular-hairy or margins minutely hairy. **Inflorescence** a several- to many flowered raceme, sometimes subcorymbose with a shortened axis, flowers suberect, erect or nodding; bracts usually membranous or leafy. **Flowers** closing at night, often scented, cup-shaped but rarely weakly zygomorphic; tepals dimorphic, ± free, white to yellowish-green with broad green keels; outer tepals suberect or spreading; inner tepals erect and ± connivent around stamens and style; stamens with filaments joined to base of tepals; ovary ovoid to ellipsoidal, style subcylindrical, stigma 3-angled. **Fruits** capsules, ovoid, ± 3-angled; seeds many per chamber. **Distr:** Tropical and S Africa, Madagascar, Arabian Peninsula. **Cult:** Propagate from offsets or seeds. Requires a moist, well-drained soil, in full or partial sun.

A. batteniana Hilliard and B.L. Burtt, *Notes Roy. Bot. Gard. Edinburgh* 42: 247, 1985. (Syn: *Ornithogalum battenianum*). [Wild Coast Albuca]. **Herbs**, evergreen, perennial, to 60 cm high; bulb borne above ground (with only the roots embedded in the soil), green, scales firm, truncate and fibrous above. **Leaves** 6–12, arranged in a rosette; lamina lanceolate, to 12–30 × 2–3 cm, firm, succulent, bright green; petioles green with a white margin. **Inflorescence** a raceme, to 80 cm long, subsecund with an inclined peduncle, flowers numerous, erect on long pedicels. **Flowers** white with pale greenish keels, 2.5–4 cm long; outer tepals spreading-recurved, 3–4.2 × 0.7 cm; inner tepals cowled above, 2.5–3 × 0.7 cm; filaments 1.5–2 cm long, white, flattened at base, anthers oblong, versatile, outer anthers 4 × 1 mm, inner 7 × 2.5 mm; ovary obtusely trigonous, 0.8–1.2 cm long, style 1–1.3 cm long, stigma trilobate, white. **Fruits** with seeds flat and glossy. **Distr:** S Cape Province; rare. **Ecol:** On moist, coastal cliffs facing the sea. Fl. Jul–Oct. **Zones:** 9–11.

Eucomis

Syn: *Basilaea, Eucomea.*

Species 10. **Herbs**, perennial; bulbs usually large, ovoid, with shiny tunics. **Leaves** many, basal, strap-shaped to obovate, glossy, appearing with the flowers. **Inflorescence** a raceme, dense, erect, simple, with a tuft of green bracts at the top; peduncle terete. **Flowers** with perianth fused at base, persistent, lobes 0.6–1.7 cm long, spreading; stamens 6, attached near base of lobes, filaments broadened and fused below into a shallow cup, anthers versatile; ovary rounded to obovoid, 3-angled, with many ovules. **Fruits** triangular, inflated and membranous, or dry, hard and compact, opening by 3 slits; seeds rounded or ovoid, black or brown. **Distr:** Tropical and S Africa. **Cult:** Propagate from offsets or seeds. Requires a rich, well-drained soil in full or partial sun.

E. bicolor Baker, *Gard. Chron.*, n.s., 10: 492, 1878. [Pineapple Lily]. **Herbs**, perennial, to 45 cm high; bulb globose, ~5 cm across, seasonally dormant, tunic brown, membranous. **Leaves** 5–6, arranged in a basal rosette, sessile, suberect; lamina oblong or sword-shaped, 30–50 × 8–10 cm, margins wavy. **Inflorescence** raceme, dense, many flowered, erect, 8–10 cm long, flowers somewhat nodding; peduncle cylindrical, to 60 cm tall, with an apical tuft of sterile bracts; bracts 30–40, ovate or oblong, 5–8 cm long, acute, margins crisped, sometimes purple. **Flowers** with perianth pale green, lobes oblong, united at base, 1.2–1.5 cm long, with a distinct purple margin; stamens much shorter than the perianth segments, purple tipped, filaments lanceolate-deltoid, purplish, anthers small, oblong; ovary green, narrowed to a neck at base, style filiform as long as the ovary, stigma capitate. **Distr:** S Africa. **Ecol:** In damp meadows, on grassy stream banks, boulders, and in forests; to 2,800 m. Fl. Jun–Aug, Dec–Mar. **Zones:** 7–10.

E. comosa (Houttuyn) H. R. Wehrhahn, *Gartenstauden*, i. 154, 1929. (Syn: *E. punctata*). [Pineapple Lily]. **Herbs**, bulbous, 17 cm to 1 m tall; bulb ~7 cm across. **Leaves** suberect; lamina oblong-lanceolate, 30–77 × 3–8 cm, margins slightly undulate or crisped, abaxial surface speckled with numerous small dark purple spots near base. **Inflorescence** a raceme, densely many flowered, to 30 cm long; scape cylindrical, 45–70(–90+) cm long, spotted with dark purple; terminal bracts 8–20, suberect, much shorter than the inflorescence axis, lanceolate, 2.5–8 cm long, sometimes with margins purplish; pedicels 1.5–3 cm long. **Flowers** with perianth pale green, lobes ~1.2 cm long; ovary purple. **Fruits** firm, purple. **Distr:** S Africa (E Cape Province, Natal). **Ecol:** In grasslands or marshes. Fl. Dec–Feb. **Zones:** 8–10.

Lachenalia

[Cape Cowslip].

Species 115. **Herbs**, bulbous, scapigerous. **Leaves** 1–2(rarely more)-nerved, somewhat fleshy. **Inflorescences** racemes, lax or dense; flowers pedicellate or subsessile, cernuous, pendulous or suberect. **Flowers** yellow, red, greenish or whitish; perianth campanulate or tubular, persistent, tube short or very short, limb deeply 6-parted in 2 rows, segments lying close, the outer gibbous dorsally under the apex, the inner longer, narrowed below, recurved at apex; stamens attached to base of the perianth segments, often exserted; ovary sessile, 3-celled, ovules many or few, style filiform, erect, equaling or exceeding the stamens, stigma obtuse. **Fruits** capsules, ovoid, membranous, 3-cornered, 3-celled; seeds several per locule, subglobose to pear-shaped, black, glossy. **Distr:** Namibia to Cape Province and Free State. **Cult:** Propagate from offsets or by division of clumps. Requires a humus-rich, well-drained soil in full or partial sun.

L. aloides (L. f.) Engl., *Notizbl. Königl. Bot. Gart. Berlin* 2: 321, 1899. (Syn: *Phormium aloides*). [Cape Cowslip, Soldier Boys]. **Herbs**, bulbaceous, perennial, 5–30 cm high; bulb spindle-shaped, ~4–6 cm Ø. **Leaves** 1 or 2, lanceolate or strap-shaped, to 17.5–25 × 2.5 cm, plain or densely spotted or -mottled with green or purple. **Inflorescence** a spike, simple, erect, 15–30 cm tall; peduncle fleshy, green or reddish, glabrous, relatively thick, ~5–8 mm Ø; pedicels 5 mm long and usually horizontal, with flowers nodding or pendent. **Flowers** cylindrical, in combinations of orange, red, yellow, or greenish-blue, with greenish markings; perianth 2–3.5 cm long, inner lobes yellow at base shading to green with purple tips, all overlapping, outer lobes bright red at base, shading to yellow at tips and with green apical swellings; stamens as long as perianth, anthers included; style not projecting. **Distr:** SW Cape Province. **Ecol:** On granite and sandstone outcrops. Fl. May–Oct. **Zones:** 9–10.

Ornithogalum

Syn: *Avonsera*, etc.

Species 215. **Herbs**, slender to robust, all parts glabrous. **Leaves** erect; lamina lanceolate to filiform. **Inflorescences** racemes, 2–300-flowered; scapes 1–3 per plant; pedicels erect or spreading at anthesis, erect in fruit; bracts lanceolate to ovate-acuminate, with filiform apex, never spurred; flowering with the leaves. **Flowers** with perianth segments free, spreading, white or white with a dorsal green band, apex cucullate with glands; filaments free, white and flattened; ovary sessile or with a short carpophore, triloculed, ovules many per locule, style slender and as long as the ovary. **Fruits** capsules, subglobose; seeds black, sub to semicircular, irregularly compressed. **Distr:** Africa, Europe to Afghanistan. **Cult:** Propagate by division or from seeds. Requires a moist, well-drained soil in full sun.

O. dubium Houtt., *Handl. Pl.-Kruidk.* 12: 309, 1780. (Syn: *O. aureum*, etc.). [Snake Plant, Sun Star, Yellow Chincherinche]. **Herbs**, perennial, 10–50 cm high; bulb subterranean, outer tunics brown or dark black. **Leaves** 3–8, usually dry at flowering, ± erect or spreading; lamina linear to ovate, 5–20 cm long, yellowish green, margins minutely hairy. **Inflorescence** a subcorymbose, conical or cylindrical raceme, to 20-flowered. **Flowers** yellow to orange, or rarely white, often with a green or brown centers; tepals 1–2 cm long, margins minutely fringed; filaments fleshy, the inner with auricled or winged expansions folded around the ovary; ovary yellowish-green, style very short, much shorter than ovary. **Fruits** oblong, ~1.5 cm long. **Distr:** S Africa (S Cape Province). **Ecol:** On mountains, stony slopes and clay flats. Fl. Aug–Dec. **Zones:** 8–10.

Above: *Scilla peruviana, Veltheimia bracteata, Veltheimia capensis.*

Scilla

[Squill].

Species 92. **Herbs**, scapigerous, bulbous, perennial, deciduous; bulb globose or ovoid, tunics imbricate, sometimes fibrous above, rarely loose, brown. **Leaves** 2 to several, suberect or spreading; linear to ovate, sometimes hairy, sometimes with reddish markings toward base, never spotted. **Inflorescences** racemes, 1 or more, few- to many flowered, rarely branched, sometimes congested and subcorymbose; pedicels mostly long; bracts usually small, membranous, often auriculate, sometimes with an inner bracteole. **Flowers** cup- or bell-shaped, tepals 6, spreading or more erect and recurved above, united at base, blue to purple, rarely white to pink, usually unscented; stamens 6, filaments joined to base of tepals, free or united at base, linear-lanceolate; ovary obovoid to subglobose with 2–several ovules per locule, style slender, usually shorter than perianth. **Fruits** capsules, ovoid to obovoid; seeds few per locule, subglobose, black. **Distr:** France and N Spain to Asia Minor and Caucasus to Iran, Afghanistan, Pakistan, Kashmir, India and Sri Lanka, S Africa. **Cult:** Propagate by division of clumps when they become too dense. Requires moist, humus-rich, well-drained soil in full sun.

S. peruviana L., *Sp. Pl.*: 309, 1753. (Syn: *Hyacinthus peruvianus*, etc.). [Cuban Lily, Evening Lily, Hyacinth of Peru, Peruvian Lily]. **Herbs**, bulbous; bulb tunicate; stems to 30 cm high, elongating during flowering period. **Leaves** many, in a basal rosette, nearly evergreen; lamina broadly linear, 30 × 4 cm, base channeled, apex pointed, shiny. **Inflorescence** a dense raceme, to ~100-flowered, 15 cm Ø; peduncle relatively long, terminated with the inflorescence; pedicels 4 cm long, upper pedicels very short, ~8 cm long in fruit; bract 1, membranous. **Flowers** deep violet blue, star-shaped; perianth persistent, rotate-patent, segments lanceolate, acute; stamens shorter than perianth, broadly subulate, same color as perianth, anthers short, horizontally incumbent. **Distr:** W and C Mediterranean. **Ecol:** Fl. May–June. **Zones:** 8–11. **Info:** The specific epithet *'peruviana'* was mistakenly given by Linnaeus, as the plant was wrongly noted as being of Peruvian origin.

Veltheimia

[Sand Lily].

Species 2. **Herbs**, bulbous; bulb pear-shaped or globose. **Leaves** few to several; lamina lanceolate-oblong, margins undulate, many nerved. **Inflorescence** a raceme, densely many flowered; flowers shortly pediceled, pendulous; pedicels bibracteate at base. **Flowers** purplish-red; perianth tubular-subclavate, deciduous, limb 6-parted, regular, lobes short, 1-nerved, suberect; stamens attached to the middle of the perianth tube, subequal, the longer scarcely exserted; ovary sessile, oblong, 3-celled, tapering into a style, ovules 2–3 in each cell, style declinate-ascending, filiform, exserted, stigma shortly trilobed. **Fruits** capsules, dry membranous, obovate, sharply 3-angle-winged, loculicidal; seeds 2 per locule, black, rugose. **Distr:** South Africa (Cape Province). **Cult:** Propagate from offsets or fresh ripe seeds. Requires humus-rich, moist, well-drained soil or compost in full or partial sun. **Info:** The 2 species in this genus are very similar. The color of the leaves can be used to distinguish the species (*V. capensis* has grayish or glaucous leaves).

V. bracteata Harv. ex Baker, *J. Linn. Soc.*, Bot. 11: 411, 1870. [Forest Lily, Sand Onion]. **Herbs**, bulbous, ± evergreen or briefly deciduous, 20–40 cm tall; bulb tunics fleshy. **Leaves** lanceolate to oblanceolate, glossy green. **Inflorescence** a raceme, dense, ovoid or subcylindrical, many flowered; flowers spreading to nodding; floral bracts 1–3 cm long. **Flowers** pink or pale yellow, finely speckled with red, tube 2–3.5 cm long, tepals ovate, ~3 mm long. **Fruits** inflated, ellipsoidal, 3-winged. **Distr:** S Cape Prov. **Ecol:** In coastal scrubs. Fl. Aug–Sep. **Zones:** 9–11.

V. capensis (L.) DC., *Liliac.* 4: t. 193, 1807. (Syn: *V. viridifolia*, etc.). [Sand Lily]. **Herbs**, bulbous, deciduous, 20–40 cm tall; outer bulb tunics papery. **Leaves** lanceolate to oblong, sometimes narrowly oblong, margins strongly crisped, grayish or glaucous. **Inflorescence** a raceme, dense, ovoid or subcylindrical, many flowered, flowers spreading to nodding; floral bracts 1–3 cm long. **Flowers** pink or pale yellow, finely speckled with red, tube 2–3.5 cm long, tepals ovate, ~3 mm long. **Fruits** inflated, ellipsoidal, 3-winged. **Distr:** Cape Province. **Ecol:** On rocky slopes. Fl. Apr–Jul. **Zones:** 9–11.

AGAVACEAE

[Century Plant].

Genera 23, **species** 637. **Plants**, often robust, usually perennial, occasionally epiphytic, sometimes monocarpic or polycarpic, monoecious, dioecious, or polygamo-dioecious, small to gigantic, sometimes arborescent, usually scapose; stems subterranean or above-ground, sometimes branched. **Leaves** simple, annual or long-lived, in terminal rosettes or occasionally cauline, sessile or occasionally pseudo-petiolate; lamina linear, lanceolate, oblanceolate, ovate, or elliptic, fibrous, thin and flexible, thick and rigid or succulent, or fibrous, often glaucous, margins entire, serrulate, dentate, denticulate, corneous, or filiferous, apex rigid or flexible, sometimes pungent, often with short or long spine. **Inflorescences** terminal or axillary spikes, racemose or paniculate, sometimes umbellate, bracteate, often huge; bracts ascending or erect, occasionally reflexed, leaflike proximally, scalelike distally; pedicel usually distinct, articulate or not, rarely flowers sessile. **Flowers** 6-merous, bisexual or functionally unisexual; perianth of 2 similar petal-like whorls, semisucculent; tepals distinct or connate into tube, apex glandular or glandular-pubescent; stamens included or exserted, filaments often broadened and succulent, glabrous, pubescent, or papillose, anthers versatile, dehiscence longitudinal; ovary superior or inferior, trilocular or occasionally 1-locular, 3-angled,

Below: *Agave americana, Agave atrovirens, Agave attenuata.*

Above: *Agave sisalana, Beschorneria septentrionalis, Beschorneria yuccoides, Furcraea foetida.*

ovoid, or cylindrical, with axillary or rarely parietal placentation, style included or exserted, stigmas 1 or 3, trilobed or capitate. **Fruits** usually capsular; seeds 1–3(–many) per locule, 3-angled, hemispheric, ovoid or globose. **Distr:** Almost worldwide except very cold temperate regions.

Agave

Species 188. **Herbs**, perennial or monocarpic; stems very short or indistinct. **Leaves** in a basal rosette; lamina large, stout, leathery fleshy or somewhat woody, containing many fibers, margin usually spiny, rarely entire, apex tipped with a spine. **Flowering stems** branched or simple, tall, stout. **Inflorescence** terminal, a spike or panicle, mostly very large. **Flowers** with perianth tube short, lobes narrow, subequal; stamens inserted at throat or in tube of perianth, filament filiform, usually longer than perianth, anther versatile; ovary with many ovules, style slender, stigma trilobed. **Fruits** capsules, oblong, trivalved, loculicidal; seeds numerous, black, thin, flattened. **Distr:** S USA to Venezuela, Caribbean. **Ecol:** Mostly in arid and semiarid regions, rarely in moist forests. **Cult:** Propagate by seeds, bulbils or offsets when firms roots start to appear. Requires fertile well-drained soil in full sun.

A. americana L., *Sp. Pl.* 1: 323, 1753. (Syn: *Aloe americana*). [Century Plant]. **Herbs**, acaulescent or short-stemmed, commonly suckering, trunks <2 m tall; rosettes not cespitose, 1–2 x 2–3.7 m. **Leaves** erect, spreading to ascending, occasionally reflexed; lamina narrowly to broadly lanceolate, 0.8–2 m x 15–25 cm, margins nearly straight or undulate to crenate, armed, teeth single, 0.5–1 cm long, 1–4 cm apart, apical spine dark brown to grayish, conical or subulate, 2–6 cm long, smooth, rigid, surfaces light green to green or glaucous-gray, sometimes variegated or cross-zoned. **Inflorescences** paniculate, not bulbiferous; scape 5–9 m tall; bracts persistent, triangular, 5–15 cm long; lateral branches 15–35 cm long. **Flowers** erect, 7–10.5 cm long; perianth yellow, tube funnelform to cylindric, 0.8–2 x 1.2–2 cm, lobes erect, subequal, 2–3.5 cm long; stamens long-exserted, filaments inserted above mid perianth tube, erect, yellow, 6–9 cm long, anthers yellow, 2.5–3.5 cm long; ovary 3–4.5 cm long. **Fruits** oblong, 3.5–8 cm long, beaked. **Distr:** SW USA to Mexico. **Ecol:** In sandy places of desert scrubs; 200 m. Fl. Apr–Jul. **Zones:** 8–11.

A. atrovirens Karw. ex Salm-Dyck, *Hort. Dyck.*: 302, 1834. (Syn: *Littaea gracilis*, etc.) [Maguey]. **Herbs**, monocarpic, stemless, very large, succulent, to 8 m across and weighing over 2,000 kg; offsets produced. **Leaves** rosetted, very large, often weighing over 50 kg; lamina oblong spathulate, 2–3(–4) x 0.3–0.4 m, base contracted, margins with distant teeth, between which the margin is nearly straight, apex ascending, very thick, adaxially concave, abaxially convex and keeled, dark green or dull green; marginal teeth triangular or deltoid-cuspidate, ~1 cm long from low widened bases; apical spine decurrent, conical, 4–6(–10) cm long, gray. **Inflorescence** 10–12 m tall; peduncle stout; bracts leaflike, crowded, imbricating, ascending, lanceolate, 30–45 cm long; branches with short panicles; panicles somewhat rounded with the flowers held erect. **Flowers** red or purple at first, then later yellow, ~10 cm long; calyx green; corolla yellow; stamens exserted. **Distr:** Mexico (Oaxaca to Veracruz). **Ecol:** 1,800–3,400 m. **Zones:** 8–11. **Info:** Plant sap used to make a Mexican alcoholic drink called 'pulque'. The variety ***A. atrovirens* var. *mirabilis*** is among the largest in the genus with leaf rosettes reaching over 2.5 m in height, and is found in the central Mexican montane cloud forests, ~2,500 m altitude.

A. attenuata Salm-Dyck, *Hort. Dyck.*, 303, 1834 303. [Fox-tail Agave, Swan's-neck Agave]. **Herbs**, perennial, rosetted, to 1 m tall and 0.6–1.5 m across; stem 0–1.5 m tall, curving, smooth and pale grayish; base producing numerous offsets. **Leaves** 15–35, elliptic, 24–90 x 11–20 cm, widest at middle, grayish-green to pale yellowish-green, soft-succulent, both ends tapering, margins unarmed, occasionally very finely serrated, apex decurrent into a needlelike, soft tip, adaxially slightly concave to flat, thin. **Inflorescence** a dense racemose spike, to 2–3.7 m tall, unbranched, apex drooping and sometimes rising back again in a distinctive curve. **Flowers** ~5 cm long, pale cream to greenish yellow; tepals equal, recurved; stamens with filaments attached at tube mouth; ovary oblong, ~2.5 cm long. **Fruits** 2.5–4 cm long, shortly beaked. **Distr:** W and C Mexico. **Ecol:** On high, rocky outcrops; 1,800–1,950 m. Fl. Dec–Jan. **Zones:** 9–11.

A. sisalana Perrine, *Trop. Pl.* 87, 1838. [Sisal]. **Herbs**, short-stemmed, commonly suckering with elongate rhizomes; trunks 0.4–1 m; rosettes not cespitose, 1.5–2 x 1.5–2.5 m. **Leaves** spreading; lamina lanceolate, 0.9–1.3 m x 9–12 cm, margins straight, finely fibrous, typically unarmed, teeth single prickles when present, 1–2 mm long, apical spine dark brown, subulate, 2–2.5 cm long, surfaces green, slightly cross-zoned when immature, firm, adaxially plane, abaxially slightly convex. **Inflorescences** paniculate, open, often bulbiferous; scape 5–6 m tall; bracts persistent, triangular, 0.5–2 cm long; lateral branches 10–15(–25), ascending, >10 cm long. **Flowers** erect, 5.5–6.5 cm long; perianth greenish-yellow, tube urceolate, 1.5–1.8 x 1–1.2 cm, lobes erect, equal, 1.7–1.8 cm long; stamens long-exserted, filaments inserted in perianth tube, erect, yellow, 5–6 cm long, anthers yellow, 2–2.5 cm long; ovary 2–2.5 cm long, neck slightly constricted, 2–4 mm long. **Distr:** S Mexico. **Ecol:** In sandy places of desert scrubs; 200 m. Fl. Apr–Jul. **Zones:** 9–11.

Beschorneria

Species 8. **Plants** succulent, robust, perennial or polycarpic, acaulescent or rarely caulescent-arborescent, spreading freely via rhizomes. **Leaves** forming large rosettes; lamina usually lanceolate, fleshy, margins entire, marginal teeth or terminal spine absent. **Inflorescences** very large, often ± erect, upper half usually paniculate, flowers pendulous and in groups along the inflorescence; floral bracts brightly colored. **Flowers** tubular; sepals linear-spathulate, acute; stamens 6, filaments subulate, anthers bilocular, linear-lanceolate, base sagittately bilobed, apex obtuse emarginate, longitudinally dehiscent, mobile; ovary inferior, subclavate, obtuse, trilocular, style erect, trisulcate. **Fruits** capsules. **Distr:** Mexico. **Ecol:** In the high mountains and forests. **Cult:** Propagate from suckers or seeds. Requires a fertile, porous soil in a sunny location.

B. septentrionalis García-Mendoza, *Cact. Suc. Mex.* 33(1): 3–5, 1988. [False Red Agave]. **Plants**, acaulescent, rhizomatous; rosette cespitose. **Leaves** 10–20, recurved; lamina oblanceolate, 0.7–0.9(–1) m x (5–)6–9(–13) cm, base dilated, 1.8–2.5(–3.3) cm wide at basal constriction, margins finely denticulate, apex shortly acuminate, both surfaces glabrous, brilliant green. **Inflorescences** cymose-paniculate, 1.5–2.5 m tall, 90–130-flowered, flowers in groups of 2–4; partial inflorescences 4–7, 9–25(–50) cm long; scape carmine-red; scape bracts 4–5, oblanceolate, to 30 cm long, carmine-red; floral bracts 12–30, lanceolate to deltoid, reddish to translucent; pedicels (1–)3.5–4.5(–6) cm long. **Flowers** (5–)5.5–6(–6.5) cm long; tepals linear-oblong to oblong-spathulate, 2.5–3 x 0.2–0.8 cm, inside papillose, outside glabrous, carmine-red, margins and apices yellowish to greenish; filaments subulate, 2–4 mm shorter than tepals, papillose, anthers oblong-elliptic, 5–7 mm long; ovary slightly 6-sulcate, 2.5–3(–3.3) x 0.2–0.8 cm, carmine-red. **Fruits** ovoid, 3.5–5(–6.5) x 2.5–3.5 cm, green; seeds shiny, black. **Distr:** Mexico (Tamaulipas, Nuevo León). **Ecol:** In cloud forests; >1,400 m. **Zones:** 9–11.

B. yuccoides K. Koch, *Wochenschr. Gärtnerei Pflanzenk.* 2: 337, 1859. [Mexican Lily]. **Plants**, robust, perennial, polycarpic, to 2 m high; pseudostem of present in older specimens. **Leaves** lanceolate, 30–70 x 3–10 cm, narrowed below middle, dilated at the very base, apex narrowly acuminate, coriaceous, glaucous, gray green to blue-green. **Inflorescence** large, suberect, upper half forming a panicle of drooping flowers; scape 1–2 m long; branches rich coral red; bracts lanceolate, acuminate, membranous, deep rose colored. **Flowers** pendent, 4–8.5 cm long including the ovary; sepals linear-oblong, dark green with a yellow tinge, straight, approximating to form a tube; stamens scarcely exserted, filaments subulate, anthers linear; ovary inferior, cylindrical, narrower than the perianth, style filiform, dilated, trisulcate at base. **Distr:** Mexico (Hidalgo, Puebla, Veracruz). **Ecol:** In mountainous areas; 1,000–3,500 m. **Zones:** 9–11.

Furcraea

Syn: *Fourcroya, Funium, Roezlia.* [Mauritius Hemp].

Top row: *Phormium colensoi, Phormium tenax, Yucca gloriosa.*
Bottom row: *Polianthes tuberosa, Yucca rostrata.*

Species 23. **Plants**, massive, multiannuals, scapose; trunks usually absent, to 6 m when present; stems above-ground. **Leaves** up to 50, broad or narrowly lanceolate, to 3.3 m long, rigid or flexible, very fibrous, margins with small or large corneous teeth, mature apex a firm, blunt point. **Inflorescences** paniculate, to 13 m tall, flowers in clusters of 2–5, drooping frequently producing bulblets. **Flowers** with tepals spreading, distinct except at base, greenish or white, ovate to oblong, longer than stamens and styles; filaments abruptly expanded below middle; ovary inferior, style dilated and trilobed proximal to middle, stigma trilobed. **Fruits** capsules, globose to cylindrical, to 8 cm, infrequently produced; seeds many, black, flat, 2 rows per locule. **Distr:** Mexico to tropical America. **Cult:** Propagate from seeds, offsets or aerial plantlets. Requires any moderately moist and very well-drained, gritty soil in full sun.

F. foetida (L.) Haw., *Syn. Pl. Succ.*: 78, 1812. (Syn: *Agave foetida, F. gigantea*, etc.). [Furcraea, Giant Cabuya, Green Aloe, Maguey, Mauritius Hemp, Sisal]. **Plants**, very large, robust; stem short, ± 1 m tall. **Leaves** large, crowded; lamina broadly oblanceolate to lanceolate, to 1.6–2.4 m × 7–20 cm, thick, firm, smooth or striate, almost flat, margins hard, smooth and entire (at least in the distal half), sometimes undulate, abaxially somewhat roughened. **Inflorescence** large, paniculate; scape ~6–12 m tall. **Flowers** pale yellow to greenish-white, fragrant, tubular, 4–5 cm across; perianth segments 6, spreading, greenish white, 2.3–3 cm long; stamens included, filaments short, thick below; ovary 1.2–1.5 cm long. **Distr:** Costa Rica to N South America and S Caribbean. **Ecol:** In dry or moist areas, often on rocks, cliffs, and rarely in crotches of trees; 0–1,000 m. Fl. Sep–Dec. **Zones:** 8–12.

Phormium

Syn: *Chlamydia*.

Species 2. **Herbs**, perennial, tall, tufted; rhizome stout, expanding by budding. **Leaves** crowded in distichous fans; lamina linear-ensiform, equitant, strongly keeled, with numerous fine, longitudinal striations, with strong fibers. **Inflorescence** terminal; peduncle long, sheathed with bracts; bracts alternate, deciduous, progressively smaller, lower ones sterile, upper ones each subtending and entirely enclosing a relatively short, alternately branched, flowering lateral branch; smaller bracts scarious; pedicels articulate below flowers. **Flowers** bisexual, ± zygomorphic; tepals subequal, erect, forming a tube but connate only near base, ~12-nerved, marcescent; stamens > tepals, filaments glabrous, slightly flattened, inserted along the back of anther, inner filaments slightly longer, anthers linear-oblong; ovary superior, sessile, elongate, ovules numerous, style terete, stigma small, narrow. **Fruits** capsules, long, loculicidal; seeds numerous, flattened, almost winged, glossy, black. **Distr:** New Zealand, Norfolk Island. **Cult:** Propagate from seeds or by division. Requires a moisture-retentive soil in full sun.

P. colensoi Hook. f., *Rev. Hort.*, III, 2: 6, 1848. (Syn: *P. cookianum*, etc.). [Mountain Flax]. **Herbs**, perennial, tall, rhizomatous, tufted. **Leaves** mostly <2 m long, coriaceous, ± drooping, usually pale. **Inflorescences** to ~2 m tall; peduncle ~2–3 cm Ø, often arching, dark, terete, glabrous. **Flowers** 2.5–4 cm long, usually predominantly greenish, often with tones of orange or yellow; inner tepals with apices usually markedly recurved, 1 usually more than the other 2; ovary erect, but carpels sometimes twisted in a half turn from base to apex. **Fruits** 10(–20) cm long, pendulous, gradually narrowed to apex, twisted and becoming pale, fibrous, and more spirally curled with age; seeds ~0.8–1 cm long. **Distr:** New Zealand. **Ecol:** On coastal cliffs to mountain slopes, on shady faces in high country. Fl. Nov–Jan. Fr. Feb–Apr. **Zones:** 8–11.

P. tenax J.R. Forst. and G. Forst., *Char. Gen. Pl.*: 24, 1775. (Syn: *Chlamydia tenacissima*, etc.). [New Zealand Flax]. **Herbs**, perennial, tall, tufted. **Leaves** ± erect, at least in lower part; lamina 1–3 m × 5–12 cm, base thick and usually brightly colored, coriaceous. **Inflorescence** to 5–(6) m tall; peduncle ~2–3 cm Ø, usually erect, dark, terete, glabrous. **Flowers** 2.5–5 cm long, predominantly dull red; inner tepals with apices slightly recurved; ovary erect, carpels straight. **Fruits** usually <10 cm long, often much shorter, erect, trigonous, abruptly contracted to apex, not twisted, remaining firm and dark with age; seeds ± elliptic, platelike but ± twisted, ~ 0.9–1 × 0.4–0.5 cm. **Distr:** New Zealand, Norfolk Island. **Ecol:** In lowland swamps and intermittently flooded areas. Fl. Jan–Mar. Fr. Nov–Dec. **Zones:** 8–11.

Polianthes

[Tuberose].

Species 17. **Herbs**, perennial; rootstock tuberous. **Leaves** mostly basal, few; *basal* leaves lanceolate or linear, elongate, succulent but thin; *cauline* leaves mostly reduced and bractlike. **Inflorescence** with flowers mostly in pairs and racemose; pedicels articulate at apex. **Flowers** white or reddish, perianth with a long narrow funnelform tube, ± curved, segments short, subequal; stamens inserted about the middle of the tube, included, filaments filiform, short, anthers erect, linear, dorsifixed; ovary 3-celled, free at the apex, ovules numerous, superposed, style filiform, stigmas 3, ovate, falcate. **Fruits** capsules, crowned by the persistent perianth; seeds flat, testa lax. **Distr:** Mexico. **Cult:** Propagate from seeds or by division. Requires a very fertile, humus-rich, moist but sell-drained compost in full or partial sun.

P. tuberosa L., *Sp. Pl.* 453, 1753. [Tuberose]. **Herbs**, bulbous, usually 60–90 cm tall; produces offsets forming clumps. **Leaves** grasslike, thin, linear, slightly succulent, borne in sets of 6–9 arising from the tuber; *basal* leaves linear, pale green, 30–50 cm long, deeply channeled in lower half, spotted with dark brown, purple or red on the back; *cauline* leaves usually 8–12, smaller and shorter than basal ones. **Inflorescences** erect, to 1 m tall, forming a long lax spike. **Flowers** tubular, single or double, extremely fragrant (more at night), waxy white, often faintly flushed with pink; perianth funnel-shaped, 3.5–7.5 cm long, tube narrowed at base, segments 6, oblong-lanceolate, ~2.5 cm long, spreading, much shorter than tube; stamens 6 inserted within the floral tube; stigma 3-parted. **Distr:** Mexico; unknown in the wild. **Ecol:** Not known from the wild. Fl. Jun–Aug. **Zones:** 9–12. **Info:** Widely cultivated throughout the tropics and subtropics, for centuries.

Yucca

Syn: *Clistoyucca*, *Samuela*. [Spanish Bayonet]. **Species** 48. **Plants**, perennial, acaulescent or caulescent, sometimes arborescent, usually branching extensively from woody, subterranean or above-ground caudices, or single stems. **Leaves** sessile, in rosettes on caudices or at branch ends; lamina linear-lanceolate, expanded basally, usually rigid, occasionally fleshy, margins entire or denticulate, often filiferous and separating into elongated fibers, apex mostly sharp-pointed. **Inflorescences** erect or rarely pendent, paniculate or racemose, sometimes paniculate proximally and racemose distally, bracteate, occasionally pubescent; bracts ascending, erect, or rarely reflexed. **Flowers** bisexual; perianth campanulate or globose; tepals 6, similar, fleshy, distinct or connate at base, whitish to cream or tinged slightly with green or purple, occasionally pubescent; stamens 6, filaments flattened, as wide as anthers, smooth, papillose, or granular, fleshy; pistil obovoid or oblong-cylindrical, ovary superior, usually green, 3- or 6-locular with false septa, 6-lobed, style white to dark green, often thick, stigmas usually 3, sometimes 1 and subcapitate, white to pale green, 1–2 mm. **Fruits** erect or pendent, capsular or baccate; seeds many per locule, usually black, occasionally gray, flattened, round, rarely obovate or ovate. **Distr:** N and C America, Caribbean. **Ecol:** On coastal plains. **Cult:** Propagate from seeds, root cuttings or suckers. Requires well-drained loamy or sandy soil in full sun.

Y. gloriosa L., *Sp. Pl.* 1: 319, 1753. [Mound-lily Yucca, Palm Lily, Spanish Dagger]. **Plants**, forming colonies of rosettes, caulescent, arborescent, often branching; stems erect, to 5 m tall. **Leaves** erect or recurving; lamina lanceolate or sword-shaped, 0.4–1 m × 3.5–6 cm, flattened, concave distally, thin, rigid or flexible, glaucous at least when young, margins entire or roughly and minutely denticulate, green or blue-green. **Inflorescences** paniculate, arising partly within to well beyond rosettes, ovoid to ellipsoid, 0.5–1.2 × 0.45 m, glabrous or pubescent; peduncle scapelike, 0.9–1.5 m tall; pedicels to 2 cm long, often arching. **Flowers** pendent; perianth globose to campanulate; tepals distinct, white to creamy white or greenish-white, sometimes purple-tinged, elliptic to narrowly ovate, 4–5 × 2–2.5 cm; filaments ~2.6 cm long, hispid or slightly papillose, anthers ~3.5 mm long; pistil light green, ~3.6 cm, ovary sessile, ~2.8 cm, style ~9 mm long, stigmas separate. **Fruits** erect or pendent, baccate, indehiscent, 6-winged or 6-ribbed, elongate, 2.5–8 cm long, leathery; seeds ovate, 5–8 mm Ø. **Distr:** SE USA. **Ecol:** On coastal dunes; sea level. Fl. Mar–May. **Zones:** 7–12.

Y. rostrata Engelmann ex Trelease, *Rep. (Annual) Missouri Bot. Gard.* 13: 68, 1902. (Syn: *Y. linearis*, etc.). [Beaked Yucca, Big Bend Yucca]. **Plants**, solitary or forming colonies, caulescent, arborescent, 2.5–3.6 m tall (excluding the inflorescence), 18–32 cm Ø; rosettes each with >100 leaves; stems 1–several, erect, mostly simple, occasionally 1–3-branched. **Leaves** linear, often twisted, flat to concavo-convex, 25–60 × 1.2–1.7 cm, glaucous, smooth, margins minutely denticulate, lemon-yellow, hyaline, apex spinose, spine-tipped. **Inflorescences** paniculate, arising just within or beyond rosettes, ovoid, 0.3–1 m long; branches to 38 cm long, flowers pendent; bracts erect; peduncle sometimes scapelike, 0.3–1 m long, <2.5 cm Ø, glabrous or glabrescent. **Flowers** with perianth globose to campanulate; tepals distinct, white, narrowly ovate, 4.2–5.2 × 1.1–2 cm, apex sharply acuminate; filaments 1.7–2 cm long; pistil 2.5–3.5 cm long, style white, 0.6–1.4 cm long, stigmas lobed. **Fruits** erect, capsular, ovoid to ellipsoid, rarely constricted, 4–7 × 1.8–2.5 cm, dehiscence septicidal. **Distr:** SW Texas to Mexico (Chihuahua, Coahuila). **Ecol:** On rocky mountain slopes; 700 m. **Zones:** 7–12.

LAXMANNIACEAE

[Cabbage Tree].
Genera 14–15, **species** 178. **Trees** or **shrubs**, perennial, woody, terrestrial; rootstock rhizomatous; sap clear; stems elongate, rounded, solid. **Leaves** cauline or basal (rosette), alternate, 2-ranked (distichous) or spiral, petiolate, sheathing at base or not; lamina simple, linear, filiform (grasslike) or broad, margins entire, venation parallel, soft. **Inflorescences** racemes, spikes or panicles, terminal, bracteate; flowers pedicellate; flowers bracteolate. **Flowers** bisexual, perianth present, small, 3-merous, actinomorphic; tepals 6; sepals if differentiated, green or colored, separate; petals if differentiated, basally united or separate, white (to pinkish), greenish or brownish; stamens 6 (3+3); pistil with carpels 3, united, ovary superior, style 1, stigma terminal, capitate or trilobed, locules 3, placentation axile. **Fruits** capsules; seeds few (2–10) per locule; seeds ± smooth, round or ellipsoid, black (phytomelan). **Distr:** Predominantly Australian, also Madagascar, India, SE Asia to the Pacific, and South America.

Arthropodium

Species 9. **Herbs**, tufted, glabrous, spreading by sympodial rhizomes, or sometimes by cormlike tubers. **Leaves** linear to narrowly lanceolate. **Inflorescence** a terminal, lax, sometimes racemelike panicle; peduncle simple; lower bracts leaflike, sometimes 1–3 empty, those subtending flowers scarious, often in groups of 2–3; pedicel articulated about the middle and ± petaloid above that level. **Flowers** white or purplish; perianth spreading to slightly reflexed at flower, later standing erect; tepals free, subequal to unequal, trinerved; stamens 6, filaments hidden over part of their length by short hairs, anthers basifixed, introrse; ovary superior, subglobose, trilocular, style cylindric, stigma narrowly capitate. **Fruits** capsules, loculicidal; seeds black, not elongated, subangular; embryo curved. **Distr:** Madagascar, New Caledonia, Australia, New Zealand. **Cult:** Propagate from seed or by division. Requires a moist, well-drained soil in full or partial-sun.

A. cirratum (Forst. f.) R. Br., *Bot. Mag.* 49: t. 2350, 1822. (Syn: *Anthericum cirratum*). [Rock Lily]. **Herbs**, robust, evergreen, usually forming extensive colonies; rhizome half-woody, 1–2 cm Ø, internodes very short. **Leaves** 30–60 × 3–10 cm, narrowly lanceolate, subfleshy; veins numerous, fine, subequal; sheath short to long. **Inflorescences** exceeding leaves; peduncle stiff; panicle broad with strong laterals, occasionally branched again; main bracts green and leaflike; pedicels ~2 cm. long, white above the joint. **Flowers** mostly in uneven-aged groups of 2–3; perianth ~2–4 cm Ø, white; outer tepals narrow, inner of thinner texture, broader, margins slightly erose, apex obtuse; stamens < tepals, anther-tails purplish at top, white at middle and golden-yellow at free curled ends. **Fruits** ~8 × 6 mm, shallowly sulcate, valves rounded, widely dehiscing; seeds ~12 per locule, ~2 mm long, black, surface patterned. **Distr:** New Zealand. **Ecol:** Near the sea, usually on rock, often in colonies. Fl. (Sep–)Nov–Dec. Fr. Dec–Feb. **Zones:** 9–11.

Cordyline

Species ~20. **Plants**, treelike or shrubby; stems ± woody, usually few-branched, with conspicuous leaf scars distally. **Leaves** crowded at apex of stems, petiolate (or sessile); lamina elliptic-lanceolate to sword-shaped, veins essentially parallel but with lateral veins

Below: *Arthropodium cirratum*, *Cordyline australis*, *Cordyline fruticosa*, *Cordyline indivisa* (infructescence).

Above: *Eustrephus latifolius, Lomandra longifolia, Asparagus densiflorus, Asparagus setaceus.*

branching from midvein in proximal ½; petiole 10–30 cm, base amplexicaul. **Inflorescences** arising from axils of distal leaves, usually paniculate, large, many branched; pedicels usually short, articulate at or near apex. **Flowers** bisexual, usually tubular-campanulate or subcylindric; perianth with short tube, lobes in 2 whorls of 3; stamens 6, inserted in tube or throat of perianth, anthers versatile; ovary triloculed; ovules 2 to many per locule, style slender, stigma capitate, small. **Fruits** capsules, leathery, 1- to several seeded; seeds black, coated with phytomelan. **Distr:** S and SE Asia, Australia, Pacific Islands, S America. **Cult:** Propagate from seed, cuttings or aerial roots (by planting them upside down). Requires moist, but well-drained soil in full or partial sun.

C. australis (G. Forst.) Endl., *Prodr. Fl. Norfolk.*: 29, 1833. (Syn: *Terminalis australis*, etc.). [Cabbage Tree]. **Plants**, to 12–(20) m tall; stem in young unbranched plants 5–10 cm Ø, trunk of large many branched old trees to 1.5 m dbh; bark thick, rough and fissured. **Leaves** 0.3–1 m × 3–6 cm, base thickened, apically flat and thin, light green, midrib indistinct, nerves fine, equal and ± parallel. **Inflorescences** panicles, to 0.6–1.5 × 0.3–0.6 cm, branched to third or fourth order, branches coming off at about right angles and well-spaced; lower bracts green and foliaceous; ultimate racemes ~10–20 cm long; axes almost hidden by flowers; peduncle stout, to 4 cm Ø. **Flowers** sweetly scented; perianth ~5–6 mm long, white; tepals slightly keeled, free almost to base, reflexed; stamens about the same number as tepals; stigma shortly trifid. **Fruits** globose, ~4 mm Ø, whitish; seeds ~2.5 mm long, glossy, deeply notched on one side. **Distr:** New Zealand. **Ecol:** In forest margins, open places, abundant near swamps. Fl. Jan–Apr. Fr. Oct–Dec. **Zones:** 8–11.

C. fruticosa (L.) A. Chevalier, *Cat. Pl. Jard. Bot. Saigon.* 66, 1919. (Syn: *Terminalis fruticosa*, etc.). [Good Luck Tree, Hawai'ian Red Ti, Ti Plant]. **Plants** erect, shrubby; stems simple or sometimes branched, 1–3 m × 1–3 cm. **Leaves** petiolate; petiole 10–30 cm, channeled adaxially, base dilated, clasping stem and other petiole bases; lamina green or variously colored, oblong-lanceolate, elliptic-lanceolate, or narrowly oblong, 25–50 × 5–10 cm, midvein distinct and raised abaxially, apex aristate. **Inflorescences** panicles, 30–60 cm, branches spreading, 6–13 cm, many flowered, flowers subsessile or shortly pedicellate; pedicel (if distinct) to 4 mm, subtended by 3 bracteoles; bracteoles ovate, 2–3 mm, margin broadly membranous, apex cuspidate. **Flowers** with perianth reddish, yellowish, or bluish-purple; tube 5–6 mm; lobes erect or recurved, nearly as long as tube; stamens inserted in throat of perianth, scarcely exserted. **Fruits** reddish, several-seeded. **Distr:** Papuasia to W Pacific. **Ecol:** Fl. Nov–Mar. **Zones:** 10–12.

C. indivisa (G. Forst.) Endl., *Ann. Wiener Mus. Naturgesch.* 1: 162, 1836. (Syn: *Dracaena indivisa*, etc.). [Broad-leaved Cabbage Tree]. **Plants** shrublike or treelike; to ~8 m tall; stems massive, rather sparingly branched. **Leaves** broadly ensiform, to 1–2 m × 10–15 cm, drooping only in age, narrowed above base to a short petiole 3–5 cm long, margins laxly revolute, abaxially glaucous, midrib broad and conspicuous, often reddish, nerves unequal, especially adaxially, stronger ones often reddish like the midrib. **Inflorescences** panicles to ~ 0.6–1.6 m × 30 cm, tightly compact, usually (possibly always) with only first order branching from stout central axis; racemes ~10–20 cm long, ~2 cm Ø; axes hidden by crowded flowers; lower bracts broad and heavy; pedicels to 2–3 mm long; peduncle stout, short, ± hidden. **Flowers** ± fleshy; perianth ~7–8 mm long, tepals shortly connate, strongly recurved; stamens about equal in number to tepals, filaments narrow; stigma narrowly capitate. **Fruits** ~6 mm Ø, globose, bluish; seeds ~2 mm long. **Distr:** New Zealand. **Ecol:** In well-lit areas of forests. Fl. Dec–Jan. Fr. Jan–May. **Zones:** 9–10.

Eustrephus

Species 1. *Generic description same as specific.* **Cult:** Propagate from seeds. Requires a moist, well-drained soil in partial shade.

E. latifolius R. Br., Bot. Mag. 31, t. 1245, 1809. (Syn: *E. angustifolius*). [Wombat Berry]. **Shrubs** small or twining **climbers**, 1–5 m tall; roots fusiform, sometimes tuberous. **Leaves** non-resupinate, sessile or nearly so; lamina broadly ovate to lanceolate or narrowly linear, 2–20 × 0.2–5 cm, firm, longitudinally striate-nerved, apex usually acute, with scarcely distinct costa. **Inflorescences** axillary cymes, 1–6-flowered; pedicels filiform but rigid, persistent, 0.5–1.8 cm long; bract basal, ovate, scarious and imbricate. **Flowers** with perianth segments oblong, nearly equal, ~6 mm long; sepals elliptic-oblong, acute, 7–9-nerved, convex, firm, shortly hooded at apex; petals elliptic, thinner than sepals, flat, obtuse, bearing yellow or pellucid markings, fimbriate; stamens with filaments short, flat, connate at base. **Fruits** capsules, fleshy, yellow, globular or rarely pyriform, 0.7–2 cm Ø; seeds 8–12, subspherical to obtusely angled, strophiolate. **Distr:** East coast of Australia, New Caledonia; Malesia (S New Guinea). **Ecol:** On riverbanks, thickets, etc.; 0–3,000 m. **Zones:** 9–11.

Lomandra

Species 51. **Herbs**, perennial, dioecious. **Leaves** with distinctive sheathing basal margins; lamina linear or terete, apex entire or toothed, glabrous, papillose, warty or hairy. **Inflorescences** with ♂ and ♀ similar or dissimilar, spicate, racemose or paniculate, flowers separate or in dense or open clusters. **Flowers** small; ♀ with perianth usually larger than ♂, hardened in fruit, tepals free or fused, all similar or the outer tepals thinner; outer stamens sometimes inserted on rim of perianth tube; inner stamens inserted on perianth, close to or distant from the outer, their filaments ± different in length; ovary ± sessile, ovule 1 per locule, style short. **Fruits** capsules, globose, cylindrical or trigonous; seeds globose to ellipsoidal, brown, orange or reddish. **Distr:** New Guinea, New Caledonia, Australia. **Cult:** Propagate from fresh seeds or by division. Requires any well-drained soil in full sun.

L. longifolia Labill., *Nov. Holl. Pl.* 1: 92, 1805. (Syn: *Xerotes arenaria, X. longifolia*, etc.). [Spiny headed Mat-rush]. Herbs, perennial, tufted. **Leaves** flat or slightly concavo-convex, occasionally rolled, 0.5–1 m × (~2)5–7.5 mm, apex 2- or 3-toothed, central tooth in pronounced sinus or longer than laterals, glabrous; sheath margins intact or coarsely split, brown, orange-brown or reddish-brown, occasionally white. **Inflorescences** with ♂ and ♀ similar, simple or branched; branched often 2 per node, distichous, ~8–15 cm long; scape distinct, much shorter to much longer than rachis, flattened; flower clusters whorled; cluster bracts conspicuous, longer to much longer than flowers, pungent; bract and inner bract shorter than flower, truncate, whitish. **Flowers** ellipsoidal or cylindrical, sessile; ♂ 3–3.5 mm long, ♀ ~4.5 mm long; outer tepals shiny, scarious, purplish or yellow; inner tepals dull, fleshy, yellow or cream. **Distr:** E and SE Australia. **Ecol:** In sandy soils, swamps and wet places to montane regions, on banks of creeks, rocky hillsides, cliffs and open forests; 10–900 m. Fl. Mar–Jul. **Zones:** 8–12.

ASPARAGACEAE

Genera 2, **species** 165–295. **Shrubs**, **lianas** or **herbs**, perennial, rhizomatous, self-supporting, or climbing, twining anticlockwise (in *Asparagus*); plants hermaphrodite, monoecious, dioecious or polygamo-monoecious; principal photosynthesizing function transferred to stems (represented by axillary clusters of cladodes) from conventional leaves. **Leaves** alternate, much-reduced (to small bractlike structures), membranous (scales), ± sheathing, exstipulate; lamina simple, entire, parallel-veined. **Inflorescences** with flowers solitary, cymose, racemose or umbellate. **Flowers** small, regular, 3-merous, cyclic, hypogynous disk absent; perianth with tepals 6, free or joined, 2-whorled, sepaloid, or petaloid; similar in the 2 whorls, green, white or yellow; stamens 6, anthers dorsifixed, dehiscing via longitudinal slits, introrse; gynoecium tricarpellate, carpels isomerous with the perianth, pistil 3-celled, superior, ovary trilocular, ovules 2–12 per locule, placentation axile, style 1; ♀ with staminodes; floral nectaries present. **Fruits** berries, fleshy, indehiscent; seeds endospermic, testa with or without phytomelan. **Distr:** Mainly Old World, but very few in Australasia; few also in Mexico.

MIW

TG

Above: *Aspidistra attenuata, Aspidistra elatior, Beaucarnea recurvata.*

Asparagus

Syn: *Elide, Myrsiphyllum, Protasparagus*, etc.
Species 210. **Herbs**, perennial or subshrubs, dioecious or hermaphroditic, usually with short rhizomes; main stems erect or climbing, generally branched, with cladodes (leaflike stems) in axils of main stems and branches. **Cladodes** borne in clusters, rarely solitary, green, flat, 3-angled, or subterete. **Leaves** appressed to stem, not green, scalelike, base spurred, spurs often extended into spines. **Inflorescences** clusters, axillary, rarely a flowers solitary, sometimes racemes or umbels; pedicels articulate, subtended by membranous bracteoles. **Flowers** with perianth campanulate or subglobose, segments free or occasionally connate at base; stamens 6, filaments usually adnate to perianth segments in varying degrees, anthers dorsifixed; ovary triloculed, ovules few per locule. **Fruits** berries; seeds 1 to few. **Distr:** Europe and Africa to N Australia. **Cult:** Propagate by seeds or by division. Requires fertile, moist soil in partial shade.

A. densiflorus (Kunth) Jessop, *Bothalia*. 9: 65, 1966. (Syn: *Asparagopsis densiflora, Asparagus sprengeri*). [Asparagus Fern, Coarse Asparagus Fern, Emerald Fern, Foxtail Shrub]. **Subshrubs**, hermaphroditic; stems ± climbing, branched, to 1 m long; branches distinctly striate-ridged. **Cladodes** in fascicles of 1–5, linear, 1–3 cm × 1.5–2.5 mm, flat. **Leaf spur** spinescent; spine slightly hooked, 3–5 mm long, and sharp on main stems, very short and blunt on branches, woody. **Inflorescences** developing after cladodes, solitary or paired, axillary, each many flowered racemes or panicles 2–2.5 cm long; bracts linear, 2–5 mm long; pedicels ~2 mm long, articulate at middle. **Flowers** with perianth white; segments oblong-ovate, ~2 mm; stamens shorter than perianth, anthers minute. **Fruits** berries, red, 0.8–1 cm Ø, 1- or 2-seeded. **Distr:** S Africa. Fl. throughout the year. **Ecol:** In various coastal habitats, from dunes to open rocky places or woods. Fl. Jun–Jul. **Zones:** 7–12.

A. setaceus (Kunth) Jessop, *Bothalia*. 9: 51, 1966. (Syn: *A. plumosus*, etc.). [Asparagus Fern, Plumosa Fern Lily]. **Herbs**, hermaphroditic; stems climbing, much branched, to several meters long, slightly woody near base; branches spreading horizontally, with branchlets and cladodes arranged in 1 plane, frondlike (similar to that of a fern). **Cladodes** in fascicles of 10–13, 4–5 mm long, very slender, slightly trigonous. **Leaf spur** short, occasionally spinescent on main stems. **Inflorescences** developing after cladodes, flowers solitary or in clusters of 2 or 3; pedicel short, articulate at middle. **Flowers** with perianth white; segments widely spreading, lanceolate-oblong, ~7 mm long. **Fruits** berries, purplish-black, 6–7 mm Ø, 1–3-seeded. **Distr:** S Africa. **Ecol:** Fl. Jun. **Zones:** 7–12. **Info:** Cultivated widely for the ornamental fronds.

RUSCACEAE

Genera 26, **species** 475. **Herbs** rhizomatous to trees; stems sometimes with anomalous secondary growth, sometimes with resin canals, occasionally flattened and photosynthetic (phylloclades); hairs simple. **Stipules** absent. **Leaves** usually alternate and spiral, along the stem or in a basal rosette, occasionally petiolate, sheathing at base; lamina simple, margins entire, venation parallel. **Inflorescences** determinate, sometimes reduced to a single flower, terminal or axillary. **Flowers** bisexual, radial, usually small; tepals (4)6, distinct or commonly connate with the perianth urn- to bell-shaped, or wheel-shaped, petaloid, imbricate; stamens (4)6, filaments distinct or occasionally connate, often adnate to tepals; carpels (2)3, connate, ovary usually superior, placentation axile, stigma 1, wet, capitate to trilobed, ovules 2–few per locule; nectaries present in septa of ovary. **Fruits** usually berries (rarely drupaceous), few-seeded; seeds ± globose (rarely angled), seed coat without phytomelan. **Distr:** Temperate to tropical regions of mainly the Northern Hemisphere.

Aspidistra

Syn: *Macrogyne, Plectogyne*, etc.
Species ~55. **Herbs**, perennial, rhizomatous; rhizome creeping, elongate; nodes dense. **Leaves** solitary or 2–4-tufted, basal, erect, long petiolate; lamina many veined. **Inflorescences** terminal; scape usually very short, with 2–8 scales, 1(or 2)-flowered; flowers generally embraced by 1 or 2 bracts at perianth base. **Flowers** bisexual; perianth campanulate, urceolate, or cupular, fleshy, apically (4–)6–8(–10)-lobed; stamens as many as and opposite perianth lobes, usually inserted in proximal part of perianth tube, filaments very short or absent; anthers dorsifixed; ovary 3- or 4-loculed, ovules several per locule, style short, sometimes articulate, stigma usually peltate or mushroom-shaped, large, entire or lobed at margin. **Fruits** berries, globose to ovoid-ellipsoid, usually 1-seeded. **Distr:** China, India, Japan, Laos, Thailand, Vietnam. **Cult:** Propagate by division of rhizomes. Requires loam-based compost in partial shade. The leaves are easily scorched by sunlight but plants can withstand long durations of dryness and neglect.

A. attenuata Hayata, *Icon. Pl. Formosana* 2: 145, 1912. **Herbs**; rhizome terete, thickened, ~1 cm thick, usually covered with scales. **Leaf** solitary; lamina oblanceolate, 45–50 × 7–10 cm; petiole 30–40 cm. **Inflorescence** with scape erect, 3–5 cm; bracts 3–5, flower solitary. **Flower** with perianth tubular-campanulate, 2–2.5 cm, 7- or 8-lobed apically; lobes reflexed, speckled with black, ovate-deltoid, 7–8 mm; stamens 7 or 8, inserted at middle of perianth tube, subsessile, anthers narrowly oblong, ~2.5 mm; pistil 4–5 mm, ovary broadly conical, ~2 mm, style ~2 mm, stigma peltate, large. **Fruit** a berry, green, globose, 2–2.5 cm Ø. **Distr:** Taiwan. **Ecol:** In montane regions; 1,000–2,000 m. Fl. Jun. Fr. Jun. **Zones:** 9–11.

A. elatior Blume, *Tijdschr. Natuuri. Gesch. Physiol.* 1: 76, 1834. (Syn: *A. punctata* var. *albomaculata, Plectogyne variegata*). [Cast-iron Plant]. **Herbs**; rhizome subterete, 0.5–1 cm thick. **Leaves** solitary, spaced; petiole 5–35 cm, stiff; lamina sometimes with yellowish-white markings, oblong-lanceolate, lanceolate, or subelliptic, 20–45 × 6–10 cm. **Inflorescence** with flower solitary; scape 0.5–2 cm; bracts 2–4. **Flower** with perianth purplish to dark-purple, rarely pinkish, campanulate, (6–)8-lobed apically, tube 1–1.2 × 1–1.5 cm, lobes greenish at margin, deltoid, 6–8 × 3.5–4 mm, adaxially 4-keeled, keels purple-red, very thick, fleshy, smooth, central 2 up to 1.5 mm wide; stamens (6–)8, inserted at base of perianth tube, subsessile; anthers elliptic, ~2 mm; pistil ~8 mm, stigma peltate, orbicular, large, 1–1.3 cm Ø, 4-lobed at margin, lobes emarginate at apex. **Distr:** Japan (Osumi Islands). **Ecol:** Fl. Jan–Apr. **Zones:** 7–11.

Beaucarnea

Syn: *Pincenectitia*.
Species 9. **Trees**; trunks bare, usually enlarged at base; branches apical, few, densely leafy at the ends. **Leaves** elongate-linear, base dilated, margins entire or minutely serrulate, coriaceous. **Inflorescences** panicles, large, much-branched; flowers short-pedicellate and fasciculate in the axils of bracts; pedicels articulate below the flowers. **Flowers** small, dioecious; perianth campanulate, segments distinct, acute, entire, subequal, spreading at anthesis; stamens 6, attached to the base of segments, filaments filiform, anthers ovate or oblong, dorsifixed; ovary sessile, 1-celled. **Fruits** dry, trigonous, 3-winged, neither lobate nor inflated, irregularly rupturing at maturity; seeds trisulcate or trilobate. **Distr:** Mexico to C America. **Cult:** Propagate from seeds or offsets. Requires any dry and well-drained soil in full sun.

B. recurvata Lemaire, *Ill. Hort.* 8, Misc.: 59, 1861. (Syn: *Nolina recurvata*, etc.). [Elephant's Foot Palm, Ponytail Palm, Ponytail Tree]. **Trees**, to 9 m tall, evergreen, slow-growing; trunk bottle-shaped, surface smooth or ± plated, base swollen, tapering upward to slender stem; branches ending in rosette of leaves. **Leaves** recurved, 1–2 m × 2 cm, base widened, margins smooth, apex long-tapering into a filamentous tip. **Inflorescences** large, erect or ascending, 1–1.3 m long; peduncle short; bracts lanceolate; pedicels 3.8–5 cm long. **Flowers** minute, white; perianth 3.8 cm long. **Fruits** orbicular, 1–1.2 cm Ø, base and apex emarginate, pinkish red, long-pedicellate. **Distr:** Mexico (Oaxaca, Puebla, San Luis Potosí, Tamaulipas, Veracruz). **Ecol:** In dry rocky areas, and in forests. Fl. Mar–Nov. **Zones:** 9–12.

MIW

MIW

Top row: *Campylandra chinensis, Chlorophytum comosum, Chlorophytum orchidastrum, Disporum cantoniense.*
Bottom row: *Disporum kawakamii, Disporum shimadai, Disporum uniflorum.*

Campylandra

Syn: *Tilcusta.*

Species 16. **Herbs** perennial, rhizomatous, monopodial; rhizome ascending or less often creeping, thick, stout, sometimes slightly woody; stem very short or sometimes slightly elongate. **Leaves** basal or on short stem, usually distichous equitant, sometimes spaced, basally distinctly petiolate or not; lamina lorate to ovate-elliptic. **Inflorescence** with scape axillary, spike terminal, several to many flowered, sometimes with several sterile bracts apically; fertile bracts lanceolate to ovate, often longer than flowers, sometimes shorter. **Flowers** with perianth segments 6, connate into a tube for halfway, fleshy, sometimes with a ringlike, fleshy appendage in throat, lobes often spreading, sometimes incurved, sometimes fimbriate at margin; stamens 6, filaments proximally adnate to perianth tube, free part short to long, anthers positioned as high as or higher than stigma, dorsifixed; ovary triloculed, ovules 2–4 per locule, style 1, to 1(–3.5) mm, stigma small, trilobed. **Fruits** berries, 1–3-seeded. **Distr:** Bhutan, China, India, Nepal, Sikkim. **Cult:** Propagate from seeds or offsets. Requires any dry and well-drained soil in full sun.

C. chinensis (Baker) M.N. Tamura, S.Yun Liang and Turland, *Novon.* 10: 159, 2000. (Syn: *Tupistra chinensis*, etc.). **Herbs**, perennial; rhizome slightly elongate, terete, 1–1.5 cm thick. **Leaves** 4–8(–12), basal, distichous equitant; lamina linear-lanceolate to oblanceolate, 15–65 × 1.5–9.5 cm; petiole 3–8 cm, usually conduplicate. **Inflorescence** spike 2.5–14 cm, with several sterile bracts apically; peduncle 1–6(–15) cm; fertile bracts green, 4–7 mm. **Flowers** with perianth campanulate, 5–7 mm, tube 2–2.5 mm, lobes ovate, 3–5 × 2–4 mm; filaments proximally widened and adnate to perianth tube, widened part sometimes connate into a thin ring in perianth tube or throat, apical part incurved, 1–2 mm; ovary subglobose, 2–2.5 mm Ø, style to 1 mm, stigma trilobed. **Fruits** berries, purple-red at maturity, globose, 0.8–1 cm Ø. **Distr:** China, Taiwan. **Ecol:** Moist places in forests, hillsides along streams; 600–3,000 m. Fl. Apr–Jun. Fr. Sep–Nov. **Zones:** 9–11.

Chlorophytum

Syn: *Asphodelopsis, Hartwegia, Verdickia*, etc.

Species 193. **Herbs** perennial, rhizomatous; rhizome often short, inconspicuous, sometimes thick, elongate; roots usually ± thick or slightly fleshy. **Leaves** basal, subdistichous or fasciculate, sessile or petiolate; lamina usually linear to elliptic-lanceolate, conduplicate, base sheathing. **Inflorescence** a terminal raceme or panicle; scape axillary, proximally with bractlike cauline leaves; bracts small; pedicel articulate. **Flowers** bisexual; perianth usually white; tepals 6, free, 3–7-veined, persistent or marcescent; stamens 6, inserted at base of tepals, filaments filiform, usually slightly widened near middle, anthers nearly basifixed, introrse; ovary triloculed, ovules 1 to several per locule, style slender, stigma small. **Fruits** capsules, acutely 3-angled, loculicidal; seeds black, flattened. **Distr:** Tropical and subtropical regions of the Old World. **Cult:** Propagate from new plantlets, by division or from seeds. Requires any well-drained soil. The trailing species are best suited for hanging baskets.

C. comosum (Thunb.) Jacques, *J. Soc. Imp. Centr. Hort.* 8: 345, 1862. (Syn: *Anthericum comosum*, etc.). [Spider Plant]. **Herbs**, evergreen, highly variable; rhizome short, vertical or sometimes horizontal with elongated internodes; roots ± spongy, often with spindle-shaped tubers. **Leaves** rosulate, petiolate or not, ± lax; lamina lanceolate to strap-shaped, 0.1–0.6(–1) m × 1–3(–4) cm, margins sometimes scabrid, usually glabrous. **Inflorescences** 1 or 2, 10–50(–75) cm long, usually simple, sometimes ± paniculate; internode elongation variable, giving dense to extremely lax inflorescences; floral bracts 0.5–2 cm long, acute to acuminate; peduncles 5–20(–50) cm long, erect to ± drooping, with a few sterile bracts; small plantlets often produced from bracts (pseudo-vivipary); pedicels articulated, 0.4–1 cm long. **Flowers** with perianth slightly urceolate, whitish, often keeled greenish outside and apically; tepals patent to reflexed at anthesis, 4–7 × 2–3 mm, 3-veined, margins scabrid; stamens exserted, filaments fusiform, dilated above middle, scabrid to papillose, 3–5 mm long, anthers versatile; style straight, ~3–5 mm long. **Fruits** triquetrous to deltate, emarginate, 0.4–0.9 × 0.5–0.9(–1.1) cm; seeds saucer-shaped, slightly folded, 2–3 mm Ø. **Distr:** Tropical and S Africa. **Ecol:** Usually in rainforests; 0–2,450 m. **Zones:** 9–12.

C. orchidastrum Lindl., *Edwards's Bot. Reg.* 10: t. 813, 1824. (Syn: *C. macropodum*, etc.). [Marmalade Plant]. **Herbs**, stemless, to 1 m high; roots woolly tomentose, root fibers tuberous. **Leaves** 6–9, elliptic-oblong or lanceolate, to 30–60 × 1.2–10 cm, base narrowed into a distinct petiole, apex abruptly acuminate, strongly nerved, both surfaces glabrous and shiny; petiole 15–22 cm long, erect, channeled, winged, orange. **Inflorescences** panicles, usually branched, 30–45 cm long, flowers usually 2–3 per node, erect; bracts ovate-lanceolate, 0.6–1.2 cm long. **Flowers** pale green or white; perianth segments oblong-lanceolate, 3–9 mm long, subacute, 5-nerved; stamens 8 mm long, anthers 5 mm long; style 6 mm long. **Fruits** depressed-globose, 6 x 9 mm, trilocular, apex deeply lobed, 2–3-seeded; seeds ovate, 1.75 x 1 mm, black. **Distr:** W and C tropical Africa. **Ecol:** In forests. **Zones:** 10–12.

Disporum

Syn: *Drapiezia*.

Species 20. **Herbs** perennial, often shortly rhizomatous, sometimes long stoloniferous, often glabrous, sometimes scabrous; roots fleshy; **stem** erect, simple or branched in distal part, with 1 to several sheaths proximally. **Leaves** concentrated in distal part of stem, alternate, often shortly petiolate, sometimes sessile, linear to suborbicular, 3–7-veined. **Inflorescences** terminal or pseudo-lateral (terminal on a short, lateral branchlet opposite a leaf), umbellate or with flowers paired or solitary; bract absent. **Flowers** bisexual, often nodding, sometimes horizontal, tubular-campanulate to opening flat; tepals 6, free, white, greenish, yellow, pink, dark red, or dark purple, often saccate or spurred at base; stamens 6, inserted at base of tepals; filaments usually slightly flat, anthers basifixed to innate, extrorse; ovary trioculed; ovules 2–6 per locule, style filiform, trilobed to trifid apically with ± recurved lobes. **Fruits** berries, dark blue to black, 2(–6)-seeded; seeds globose or ovoid. **Distr:** Bhutan, China, India, Japan, Korea, Laos, Malaysia, Myanmar, Nepal, Russia, Sikkim, Thailand, Vietnam. **Cult:** Propagate by division or from seeds.

D. cantoniense (Loureiro) Merrill, *Philipp. J. Sci.* 15: 229, 1919. (Syn: *D. chinense* var. *pullum*, *Fritillaria cantoniensis*). [Cantonese Fairy Bells]. **Herbs**; rhizome creeping, thick, without stolon; stem usually branched distally, 0.5–1(–1.5) m. **Leaves** lanceolate to narrowly oblong-lanceolate, 5–12 x 1–5 cm; petiole 2–4 mm. **Inflorescences** terminal and pseudo-lateral, (2 or)3–10-flowered; peduncle usually distinct; pedicels 1–4 cm, usually papillose-scabrous. **Flowers** opening slightly; tepals purplish, oblanceolate, 1.5–2.5(–2.8) cm x 4–5 mm, base gibbous-spurred, apex subacute and without dense papillae; spurs 2–3 mm long; stamens 0.8–1.5 cm, included, filaments 0.8–1.1 cm long, anthers 3–4 mm long; ovary ~3 mm long, style 0.7–1.5 cm long. **Fruits** berries, 0.8–1 cm Ø, 2- or 3(–5)-seeded. **Distr:** Himalayas to S China and Malesia. **Ecol:** In forests, thickets; 700–3,000 m. Fl. May–Jul. Fr. Aug–Oct. **Zones:** 9–12.

D. kawakamii Hayata, *J. Coll. Sci. Imp. Univ. Tokyo.* 30: 365, 1911. (Syn: *D. taiwanense*, etc.). [Taiwanese Fairy Bells]. **Herbs**, rhizomatous; stem branched distally, 0.5–1.2 cm long. **Leaves** lanceolate to ovate, 10–17 x 1–5 cm, base broadly cuneate to rounded, margins slightly scabrous, apex acuminate; petiole 3–4 mm long. **Inflorescences** terminal and pseudo-lateral, 2–5-flowered; peduncle occasionally distinct; pedicels 2–4 cm long, subsmooth. **Flowers** tubular-campanulate; tepals cream or greenish purple-tinged near both ends, oblanceolate, 1.8–2.2 x 0.4–0.6 cm, somewhat pubescent abaxially, densely papillose apically, base gibbous-spurred; spurs ~2 mm long; stamens 1–2 cm long, included, filaments 0.8–1.2 cm, anthers 3–4 mm long; ovary ~3 mm long, style 0.8–1.4 cm long. **Distr:** Taiwan. **Ecol:** In evergreen forests; 300–1,700 m. Fl. Mar–May. **Zones:** 9–11.

D. shimadae Hayata, *J. Coll. Sci. Imp. Univ. Tokyo.* 30: 367, 1911. (Syn: *D. sessile* var. *shimadae*). **Herbs**; rhizome short; stem simple or branched distally, 15–50 cm long. **Leaves** with petiole absent to 2 mm long; lamina linear-lanceolate to oblong-lanceolate, 4–7 x 1–1.5 cm, base cuneate to rounded, apex acute. **Inflorescences** terminal, (1 or)2–5-flowered; pedicels 0.7–1.8 cm long. **Flowers** open campanulate; tepals yellow, rarely green apically, ovate-oblong to ovate-spathulate, 1.5–2.3 x 0.4–0.6 cm, very minutely papillose at margin proximally and at base adaxially, base spurred; spurs ~2 mm long; stamens 1.2–1.7 cm long, included, filaments 0.9–1.4 cm long, glabrous or sparsely papillose, anthers 2–3 mm long; ovary 2.5–3 mm long, style 1.1–1.7 cm long. **Distr:** Taiwan. **Ecol:** In grasslands; 500–1,100 m. Fl. Mar–Apr. **Zones:** 9–11.

D. uniflorum Baker ex S. Moore, *J. Bot.* 13: 230, 1875. (Syn: *D. flavens*, *D. sessile* ssp. *flavens*, *D. sessile* var. *pachyrrhizum*). [Spring Star]. **Herbs**; rhizome shortly creeping, 4–7 mm thick, with stolon 1–5 cm x 3–6 mm; stem simple or branched distally, 20–80 cm long. **Leaves** with petiole 0.5–1 cm long; lamina broadly elliptic to oblong-ovate, 4–9 x 1–6.5 cm, glabrous, base subrounded to broadly cuneate, apex shortly acuminate to acute. **Inflorescences** terminal, 1–3-flowered; pedicels 3–5 mm long. **Flowers** cylindric-campanulate; tepals yellow, spathulate-oblanceolate to obovate, 2–3 cm x 0.5–1 cm, base gibbous-spurred; spurs 1–2 mm; stamens 1.8–2.8 cm long, nearly included, filaments 1.5–2 cm long, minutely papillose proximally, anthers 4–8 mm long; ovary 4–5 mm, style 1.5–2.3 cm long. **Fruits** berries, blue-black, subglobose, 0.8–1 cm Ø. **Distr:** China, Korea. **Ecol:** In forests; 100–2,500 m. Fl. Mar–Jun. Fr. Jul–Nov. **Zones:** 9–10.

Dracaena

Syn: *Pleomele*.

Species 111. **Plants** treelike, shrubby, or subshrubby; stems simple or branched, ± woody. **Leaves** usually crowded toward apex of stems or spaced along distal part of stems, sessile or petiolate; lamina usually sword-shaped to elliptic-lanceolate, veins parallel from base, lateral veins absent; petiole to 8 cm long, base sheathing. **Inflorescences** terminal, branched, rarely simple, flowers clustered, sometimes solitary; pedicel articulate. **Flowers** bisexual; perianth cylindric, campanulate, or funnelform, tube short, lobes 6, similar; stamens 6, inserted in tube or throat of perianth, anthers versatile; ovary triloculed, ovules 1 or 2 per locule, style slender, stigma capitate or trilobed. **Fruits** berries, globose, 1–3-seeded; seeds without phytomelan coating. **Distr:** Tropical and subtropical regions worldwide. **Cult:** Propagate from seeds, stem cuttings, tip cuttings or aerial root cuttings. Requires rich, moist but well-drained soil in full sun.

D. angustifolia Roxburgh, *Fl. Ind.* ed. 1832. 2: 155, 1832. (Syn: *Pleomele angustifolia*). **Plants** shrubby, rhizomatous, 1–3 m tall; stems simple or few branched; internodes often longer than wide; bark grayish, smooth. **Leaves** spaced along distal part of stems, subsessile or indistinctly petiolate; petiole to 1 cm, base not completely covering internode; lamina nearly sword-shaped to linear-oblanceolate, 20–45 x 1.5–5.5 cm. **Inflorescences** terminal, branched, 30–50 cm; rachis glabrous, flowers in clusters of 2 or 3; pedicels 7–8 mm, articulate distally or near apex. **Flowers** with perianth greenish white, 1.9–2.3 cm, tube 7–8 mm, lobes 1.1–1.6 cm; filaments filiform, anthers 2–3 mm; style 5–8x as long as ovary. **Fruits** berries, orange, globose, 0.8–1.2 cm Ø, 1–2-seeded. **Distr:** Bhutan, Cambodia, China, India (including Andaman Islands), Indonesia, Laos, Malaysia, Myanmar, Papua New Guinea, Philippines, Taiwan, Thailand, Vietnam; N Australia. **Ecol:** In forests, thickets. Fl. Mar–May. Fr. Jun–Aug. **Zones:** 10–12.

D. aubryana Brongn. ex E.Morren, *Belgique Hort.* 10: 348, 1860. (Syn: *D. thalioides* [misapplied], *Draco thaliodes*, *Pleomele thalioides*, etc.). [Lance Dracaena]. **Shrubs**, low, unbranched, slender, erect or scrambling, 1.2–2.5 m high; stem unbranched. **Leaves** usually 2-ranked with leaf bases imbricated; lamina oblong to narrowly ovate, 20–50 x 5–9 cm, base abruptly narrowed into a pseudopetiole 20–40 cm long, extreme base sheathing, bright green, apex tapered into a threadlike tip, surface glossy, longitudinally ribbed. **Inflorescence** a raceme, terminal, to 15–90 cm long, erect, rarely branched, flowers arranged in clusters of 1–3 on short knobs along the rachis, flowering portion usually ~40 cm long; bracts present; pedicels short, apex jointed. **Flowers** white, greenish-white or yellowish, with reddish-brown or reddish-purple streaks outside, strongly fragrant, ~5 cm long; corolla tube shorter than lobes; stamens slightly exserted. **Fruits** deeply 1-trilobed, bright orange; seeds 1–3, dirty white. **Distr:** W and WC Tropical Africa to Uganda and Zambia. **Ecol:** In riverine forests and in shaded areas of undisturbed forests; 0–760 m. **Zones:** 10–12.

D. fragrans (L.) Ker Gawl., *Bot. Mag.* 27: t. 1081, 1808. (Syn: *Cordyline fragrans*, *Pleomele fragrans*, etc.). [Corn Plant, Forest Dracaena, Fragrant Dracaena, Mass Cane]. **Shrubs**, to 6 m tall; stem 2.5–10 cm Ø, rarely branched; stems mostly herbaceous

Below: *Dracaena angustifolia*, *Dracaena aubryana*, *Dracaena aubryana* (flowers).

or soft-wooded. **Leaves** spread along the stem when young, later in terminal rosettes, petiole absent; lamina sword-shaped or ribbonlike, 0.3–1 m × 2.5–10 cm, gradually narrowed to base, margins undulate, glossy green or banded with yellow or white. **Inflorescences** panicles of dense globular flower clusters, drooping, ~60 cm long; peduncle 4–10 cm long. **Flowers** 2 cm long, yellowish-white, with pink lines or tinged purple, very fragrant, opening at night; corolla with fused tepals, tubular, divided about halfway into 6 lobes, lobes oblanceolate, 1.5–2 cm long, tube shorter than lobes. **Fruits** berries, globose, 1.2–2 cm Ø, orange to red; seeds 1–3, rectangular to bean-shaped, whitish turning brown. **Distr:** Tropical Africa. **Ecol:** In dry lowland forests and upland montane forests, usually growing as understory forming dense stands; 0–1,500 m. **Zones:** 9–12.

D. reflexa Lam., *Encycl.* 2: 324, 1786. (Syn: *Pleomele reflexa*, etc.). [Malaysian Dracaena, Pleomele]. **Shrubs**, clump-forming, evergreen, to 4–6 m tall; stems arching and pliable, much-branched. **Leaves** spread along the stem in closely spaced whorls, sessile; lamina reflexed, strap-shaped, linear-lanceolate to elliptic, 12–20 × 1–2 cm, base flaring and clasping stem, margins wavy, pliant-leathery, apex abruptly pointed, glossy, dark or pale green, with margins yellow, venation parallel. **Inflorescences** loose panicles, erect or pendent, flowers single (not clustered), axillary to 1 lanceolate bractlet. **Flowers** with corolla greenish-white, 1.2–2 cm long, tube shorter than lobes. **Fruits** berries, globose, ~1.2 cm Ø, ripening to brownish-orange. **Distr:** W Indian Ocean. **Ecol:** In moist valleys, dry areas, coastal dune forests, and in coastal forests; 0–50 m. Fl. Apr–May. Fr. Jan. **Zones:** 10–12.

D. surculosa Lindl., *Bot. Reg.* 14: t. 1169, 1828. (Syn: *Draco surculosa*, *Pleomele surculosa*). [Gold Dust Dracaena]. **Shrubs**, slender, with bamboolike habit, branched, subscandent; stems flexuose, pale brown, annulate. **Leaves** opposite or ternately whorled; lamina elliptic, elliptic-oblong or elliptic-lanceolate, 7.5–12.5 cm long, base acute, membranous, many nerved, midrib distinct, adaxially bright green with irregularly scattered, orbicular, small and large white or cream spots, abaxially paler; petioles short. **Inflorescences** racemes, 5–7.5 cm long, ascending from the ends of the drooping branches, flowers in lax clusters of 3; pedicels slender, 4–6 mm long; floral bract 1, lanceolate, membranous, white, with 2 minute ones at base; peduncle short; bracts erect; rachis green. **Flowers** ~2.5 cm long; perianth pale green, tube very slender at middle, gradually enlarged to clavate base and infundibular limb, latter with 6 linear obtuse lobes as long as tube; stamens nearly as long as perianth lobes, anthers oblong; ovary ovoid, style very slender, stigma minute. **Fruits** berries, globose, vermilion-red, 1.2–1.6 cm Ø. **Distr:** W Tropical Africa to Cameroon. **Ecol:** 750–900 m. **Zones:** 10–12.

Maianthemum

Species 38. **Herbs**, terrestrial, aquatic, or epiphytic, 10 cm to 2.5 m tall; rhizome fleshy, densely clumped; leafy stem pendent, leaning, or upright, with 2–17 leaves distichously arranged. **Leaves** simple, clasping or with short (rarely long) petiole; lamina usually ovate, base rounded, margins flat or undulate, denticulate or not, apex acute to caudate, surfaces glabrous or pubescent. **Inflorescence** a panicle or a raceme, with 15–400 flowers, terminal, lateral branches or fascicles distichous or helical. **Flowers** usually perfect (rarely dioecious), trimerous or dimerous; pedicel usually subtended by a bract. **Flowers** with perianth spreading, cupuliform, or campanulate; tepals usually uniform, ovate, obovate, or triangular, 0.5 mm to 1.1 cm long, usually white (sometimes lavender to red, or green); stamens inserted at tepal base, anther tetralocular, introrse; ovary with 2 or 3 carpels, septal walls with nectaries, style usually <1.5 mm long, stigma bi- or trilobed, usually <1 mm wide. **Fruits** usually lobed, 0.4–1.2 cm across; seeds 1–12, usually spherical, 3–6 mm Ø, testa thin, pale brown. **Distr:** Temperate Northern Hemisphere to C America. **Cult:** Propagate by division or from fresh seeds. Requires a humus-rich, moisture-retentive soil, in partial sun.

M. racemosum (L.) Link, *Enum. Hort. Berol. Alt.* 1: 343, 1821. (Syn: *Convallaria racemosa*, etc.). [Large False Solomon's Seal]. **Herbs**, terrestrial, 0.75–1.25 m tall; rhizome linear, horizontal, sympodial, 3–4 × 0.8–1.4 cm. **Leaves** clasping and sessile, or with petiole 2–6 mm long, *scale* leaves 8–10, *foliage* leaves 7–12; *upper leaves* elliptic to ovate, 9–17 × 4.5–7.5 cm, *lower leaves* ovate, 9–14 × 5–8 cm, base rounded or tapered, margins undulate, apex acute or caudate, surfaces usually pubescent. **Inflorescence** a pyramidal panicle, 70–250-flowered; main axis 7–20 cm long, ribbed, green or yellow-green; pedicel 0.5–1 mm long. **Flowers** 3-merous; perianth inconspicuous, tepals spreading or reflexed, ~1 × 0.5 mm, white; stamens inserted at tepal base, filament petaloid to subpetaloid, 0.5–1 mm long, anther 0.5–1 mm long; ovary ~1 × 1 mm, style 0.1–0.3 mm long, stigma obscure. **Fruits** spherical to trilobed, 4–6 mm across, green, copper-spots when young, maturing to translucent red; seeds to 4,

Top row: *Dracaena fragrans*, *Dracaena reflexa*, *Dracaena surculosa*, *Maianthemum racemosum*.
Bottom row: *Polygonatum biflorum*, *Polygonatum odoratum*, *Sansevieria trifasciata*, *Sansevieria zeylanica*.

SZ

Top row: *Paris polyphylla, Trillium cernuum, Trillium chloropetalum.*
Bottom row: *Trillium ovatum, Trillium erectum.*

spherical, 2.5–4 mm Ø. **Distr:** N America to Mexico (Chihuahua). **Ecol:** In forests. **Zones:** 5–10.

Polygonatum

[Solomon's Seal].
Species ~60. **Herbs** perennial, rhizomatous, sympodial, usually terrestrial, rarely epiphytic; stem erect, arching, or sometimes ± scandent, simple. **Leaves** cauline, lateral and pseudo-terminal, alternate, opposite, or whorled, sessile or shortly petiolate, sometimes cirrose at apex. **Inflorescences** axillary, umbellate, corymbose, or racemose, or flowers solitary; bracts present or absent, when present herbaceous, membranous, or scarious, persistent or caducous; pedicel articulate at apex. **Flowers** bisexual, usually pendulous, rarely erect; perianth segments 6, valvate, connate proximally and forming a tube usually for at least ½ their length; corona absent; stamens 6, filaments proximally adnate to perianth tube, free part filiform or flattened, short to long, smooth, papillose or hairy, anthers basifixed to dorsifixed and versatile, introrse, base bilobed; ovary triloculed; ovules 2–8 per locule, style slender, stigma trilobed, small. **Fruits** berries, globose; seeds few to over 10, yellowish tan to brownish olive, globose or subglobose, 3–4.5 mm across. **Distr:** Temperate regions of the N Hemisphere, mainly from the Himalayas to Japan. **Cult:** Propagate by division of rhizomes or from seeds. Requires a rich, peaty soil in partial shade.

P. biflorum (Walter) Elliott, *Sketch Bot. S. Carolina.* 1: 393, 1817. (Syn: *Convallaria biflora*, etc.). [Giant Solomon's Seal]. **Herbs**; rhizomes deeply set, 1–3 cm thick; stems erect to arching, 0.5–2 m; sheathing bract usually absent. **Leaves** sessile to clasping, (5–)9–20(–25) x 3–9(–13) cm; lamina narrowly lanceolate to broadly elliptic, glabrous, prominent veins 1–19. **Inflorescences** in most leaf axils except distal 2–3 and proximal 2–7; peduncles flattened, not strongly reflexed, axillary 2–10(–15)-flowered, 6–9 cm long in fruit. **Flowers** with perianth whitish to greenish yellow, tube (1.3–)1.7–2.2 cm long, distinct tips gently spreading, 4–6.5 mm; stamens inserted near middle of perianth tube, filaments glabrous, sometimes minutely warty. **Fruits** berries, 0.8–1.2 cm long; fruiting pedicels 1–4 cm long. **Distr:** S USA, NE Mexico. **Ecol:** In dry to moist deciduous woods; 0–3,000 m. **Zones:** 8–11.

P. odoratum (Miller) Druce, *Ann. Scott. Nat. Hist.* 60: 226, 1906. (Syn: *Convallaria odorata*, etc.). [Solomon's Seal, Coral Bells]. **Herbs**; rhizome terete, 0.5–1.4 cm thick; stem arching, 0.2–0.5(–1) m long, glabrous, angled. **Leaves** 7–12, alternate; petiole short; lamina elliptic to ovate-oblong, 5–12(–20) x 3–6(–8) cm, apex acuminate and obtuse tipped, abaxially glaucous, often smooth, sometimes papillose-scabrous on veins. **Inflorescences** 1–4(–8)-flowered; peduncle usually 1–1.5 cm long; bracts small or absent; pedicel 0.5–1(–2) cm long. **Flowers** pendulous; perianth yellowish-green to white, cylindric to campanulate-cylindric, 1.3–2(–2.5) cm, lobes ~3 mm long; filaments filiform, smooth or verruculose, anthers ~4 mm long; ovary 3–4 mm long, style 1–1.4 cm long, included. **Fruits** berries, blue-black, 0.7–1(–1.2) cm Ø, 7–9-seeded. **Distr:** China, Japan, Korea, Mongolia, Russia, Taiwan; Europe. **Ecol:** In forests and shaded slopes; 500–3,000 m. Fl. May–Jun. Fr. Jul–Sep. **Zones:** 4–10.

Sansevieria

Syn: *Acyntha, Salmia.* [Bowstring Hemp, Devil's Tongue].
Species 62. **Herbs**, shrubby, perennial, erect, from thick rhizomes; stems very short, often stoloniferous. **Leaves** basal, erect; lamina flat, semiterete or cylindrical, often variegated and mottled, elongate, thick, succulent, rigid, fibrous. **Inflorescences** paniculate, racemose, or spicate, terminal, flowers clustered in alternate axils of scarious bracts; pedicels articulate. **Flowers** bisexual; tepals basally connate into tube, linear, equal, recurved at anthesis, white, yellowish-white, pale pink, or greenish-white, tips narrow; stamens inserted in top of perianth tube, filaments filiform, anthers medifixed; style filiform, stigma simple, slightly thickened, entire; ovary superior. **Fruits** berries, baccate; seeds 1–3, ± globose. **Distr:** Tropical and S Africa, Madagascar, SW Arabian Peninsula, India to Thailand, Sri Lanka. **Cult:** Propagate from suckers or by division. Requires any soil in full sun.

S. trifasciata Prain, *Bengal Pl.* 2: 1054, 1903. (Syn: *S. jacquinii*, etc.). [Bowstring Hemp, Mother-in-law's Tongue]. **Herbs**, strongly stoloniferous; stem very short; rhizome sympodial, yellowish-red. **Leaves** 2–6, narrowly lanceolate, 0.4–1.75 m x 2.5–9 cm, margins narrow, light green, yellow or reddish, apex acute to acuminate, dark green with conspicuous light or grayish-green, irregularly defined transverse bands. **Inflorescence** a raceme, erect, 40–75 cm long including the scape; flower clusters scattered or in groups; pedicels 6–8 mm long. **Flowers** with perianth 2.5–3 cm long, greenish-white, scented, segments broadening to a greenish tip; stamens 7–8 mm long, filaments 6–7 mm long, anthers 2 mm long; style

1.5–1.8 cm long, stigma 0.5–1 mm long. **Fruits** berries, ± globose, 7–9 mm Ø, orange; seeds usually 2, 6–7 mm long, ~5 mm Ø, creamish-brown. **Distr:** S Nigeria to WC tropical Africa. **Ecol:** In shrublands and rocky areas. Fl. Mar–May, Sep. **Zones:** 10–12.

S. zeylanica (L.) Willd., *Sp. Pl.* 2: 159, 1799. (Syn: *Cordyline zeylanica*, etc.). [Bowstring Hemp, Ceylon Bowstring, Devil's Tongue]. **Herbs**, clump-forming; rhizome horizontal, 1–1.5 cm thick, yellowish-white to orange, sympodially branching. **Leaves** 4–12 in a tuft, ± flat, ensiform or nearly terete, channeled, to ~1.1 m × 6 cm, tapered, acute, apiculate, tip spine-like, ~2 cm long, shiny, smooth, with transverse diffuse bands of green and dark green; outer leaves scale-like, 4–10 cm long. **Inflorescences** fascicles of 3–6, 30–60 cm long; pedicel 5–8 mm long; peduncle to ~25 cm long. **Flowers** with perianth tube cylindric, 0.7–1.2 cm long, yellowish or greenish-purple, lobes linear or narrowly lanceolate, 1–1.5 cm × 2.5–3 mm, revolute, pale greenish-white or purplish, tips darker, bases yellowish-green, apex obtuse, bearded; stamens yellowish, erect, filaments 1–1.2 cm long, anthers 2–3 mm long; style ~2.5 mm long, yellowish, stigma slightly trilobed, exserted. **Fruits** berries, globose, 0.6–1 cm across, orange; seeds 1 or 2. **Distr:** S India, Sri Lanka. **Ecol:** In dry lowlands. Fl. Jan–Jun. **Zones:** 10–12.

LILIALES

FAMILIES 11, GENERA 67, SPECIES 1,558.

Plants terrestrial. **Leaves** elliptical, base not sheathing, fine venation reticulate. **Inflorescence** terminal; tepals large; anthers extrorse; tepal nectaries alone, many tenuinucellate ovules/carpel, style long, stigma capitate (Stevens 2001 onwards). **Distr:** Worldwide; predominantly northern hemisphere. **Info:** Monophyly of the Liliales is supported by molecular characters (Chase et al., 1995a; Chase, et al., 2006).

MELANTHIACEAE

Genera 16, species 170. **Herbs**, terrestrial, perennials (rarely annuals), evergreen, mainly rhizomatous, or sometime tuberous, or rarely cormous, plants dioecious or monoecious; stems elongate, contracted, rounded, xylem without vessels. **Leaves** sheathing at base, simple, cauline, basal (rosette), alternate, whorled, 2-ranked (distichous), or spiral; lamina linear or filiform (grasslike), margins entire, venation parallel or palmate. **Inflorescence** terminal, bracteate, flowers solitary, or racemes, spikes, panicles or umbels, pedicellate; floral bracteoles present. **Flowers**: unisexual or bisexual, minute to large and showy, 3-merous, actinomorphic; tepals 6, free or joined, 2-whorled (3+3), sepaloid or petaloid, similar in the 2 whorls, white, yellow, brown-purple; stamens 6, 9 or 12, free, 2-whorled, anthers dorsifixed or basifixed, dehiscing via longitudinal slits, usually extrorse, occasionally introrse; carpels 3, pistil syncarpous, 3-celled, gynoecium superior or partly inferior, ovary triloculed, placentation axile, ovules 2–100(or more) per locule, styles 3 or 3-branched, stigma terminal, capitate or trilobed. **Fruit** a capsule or a berry; seeds few–numerous per locule, round or ellipsoid, compressed or flattened. **Distr:** Temperate East Asia and E North America to Peru.

Paris

Syn: *Daiswa*, *Euthyra*, *Kinugasa*.

Species ~24. **Herbs** perennial; rhizome slender or thickened; stem erect, simple. **Leaves** 4 to many, very rarely 3, in a terminal whorl, petiolate, lanceolate to ovate, with 3 main veins and anastomosing veinlets. **Flowers** bisexual, solitary, terminal, pedunculate; tepals 3–8, in 2 whorls, free; outer ones green, rarely ± white, ovate to lanceolate; inner ones linear or occasionally absent; stamens 8–24 or more, 2–8 × as many as tepals, filaments narrow, flat, anthers basifixed, often with convex connective apically; ovary subglobose, 1-loculed with parietal placentation or 4–10-loculed with axile placentation, style short, stigma lobes 4–10. **Fruit** a berry or a berry like capsule, indehiscent or loculicidal, several to many seeded. **Distr:** Temperate Eurasia to Indo-China. **Cult:** Propagate by division as the seeds take over 2 years to germinate (Hogan, 2003). Requires moist and humus-rich soil in partial sun. **Info:** Cultivated mainly for ornamental purposes. The rhizomes of many species are used medicinally in China.

P. polyphylla Smith, *Cycl.* 26: Paris no. 2, 1813. (Syn: *Daiswa polyphylla*). [Paris]. **Herbs**, 0.1–1 m tall; rhizome 1–2.5 cm thick. **Leaves** 5–10(–22); lamina variable, usually oblong to lanceolate, 6–15(–30) × 0.5–5 cm, base rounded to cuneate; petiole (0.5–)1–6 cm long. **Peduncle** 5–24(–65) cm. **Flower** with outer tepals (3 or)4–6(or 7), green or yellow-green, narrowly ovate-lanceolate to lanceolate, (3–4.5–7(–11) × 1–4 cm; inner ones usually yellow-green, narrowly linear, shorter or longer than outer ones, 1–1.5(–5) mm wide; stamens 2 ± as many as outer tepals, (6–)8–12(–14) or sometimes more; filaments 0.4–1 cm long; anthers 0.5–1.2 cm long; free portion of connective usually 0.5–4 mm long; ovary subglobose, ribbed, 1-loculed, sometimes tuberculate; style short, base enlarged, purple to white; stigma lobes (4 or)5. **Fruit** a capsule, globose; seeds red-arillate. **Distr:** Himalayas to China. **Ecol:** In forests, bamboo forests; 100–3,500 m. Fl. & Fr. Mar–Nov. **Zones:** 7–10. **Info:** This is a highly variable species with 9 described varieties.

Trillium

Species ~47. **Herbs**, perennial; rhizome usually stout; stem erect, simple, basally with a few brown, scalelike sheaths. **Leaves** 3, in a terminal whorl, sessile or shortly petiolate, rhombic-orbicular to ovate, with 3 or 5 main veins and anastomosing veinlets. **Inflorescence** solitary, terminal, pedunculate. **Flowers** bisexual; tepals 6, in 2 whorls, free; outer ones usually green, persistent; inner ones petaloid, rose-purple to white, slightly narrower than outer, withering or deciduous after anthesis; stamens 6, filaments short, anthers basifixed, linear, with very short connective; ovary ovoid to globose, triloculed, ovules several to many per locule, style short, deeply trilobed often to base. **Fruit** a berry, globose to ovoid; seeds several to many, ovoid, small. **Distr:** Afghanistan to temperate E Asia; N America. **Cult:** Propagate by division or from seeds. Requires humus-rich, moist, well-drained soil in full or partial sun.

T. cernuum L., *Sp. Pl.* 1: 339, 1753. (Syn: *T. cernuum* var. *macranthum*). [Nodding Trillium]. **Herbs**; rhizomes short, thick. **Scapes** 1–2(–3), round, 15–40 cm tall, slender, glabrous. **Bracts** often overlapping, sessile, umbrellalike, base petiolelike; lamina bright green, broadly rhombic-ovate to suborbicular, 5–15 × 6–15 cm, base attenuate, apex acuminate. **Flower** usually hidden beneath bracts, nodding, odorless; sepals spreading, green, lanceolate-ovate, 0.9–3 cm long, margins slightly raised, apex acuminate; petals usually strongly recurved above, white, rarely pale pink, oblong-lanceolate, 1.5–2.5 × 0.9–1.5 cm, thin-textured, margins entire, apex acuminate; stamens ± straight, 0.6–1.5 cm long, slender, filaments white, slender, anthers straight, pale lavender-pink or -gray, 2–6.5 mm long; ovary prominent, white to pinkish, pyramidal, strongly 6-angled, 0.3–1.2 × 0.3–1 cm, stigmas erect, recurved, distinct, white, not lobed adaxially, 3–8 mm long, fleshy, basally thickened, gradually tapered; pedicel 1.5–3 cm long. **Fruits** dark red, with a fruity fragrance, ovoid, to 3 cm Ø, fleshy, juicy. **Distr:** USA. **Ecol:** In deciduous forests, swamps; 30–600 m. Fl. Apr–Jul. **Zones:** 6–10.

T. chloropetalum (Torrey) Howell, *Fl. N.W. Amer.* 661, 1902. (Syn: *T. sessile* var. *chloropetalum*, etc.). [Giant Trillium]. **Herbs**; rhizomes ± erect, brownish, somewhat compressed-thickened, praemorse. **Scapes** 1–3, green, round, 20–65 cm tall, robust. **Bracts** held well above ground, sessile; lamina densely to weakly mottled in dark brownish-green, becoming more obscure with age, broadly ovate, 7–17.6 × 7.4–17.7 cm, obtuse-rounded. **Flower** erect, odor roselike, spicy; sepals green, lanceolate, 3.5–6.5 × 0.7–1.2 cm, margins entire, flat, apex obtusely rounded; petals long-lasting, erect, connivent, ± concealing stamens and ovary, yellow, bronze, maroon, brown, deep-purple, reddish-brown, pink, dark purplish-red, rarely greenish-white, oblanceolate to obovate, 6.5–10 × 1.5–2.5 cm, thick, base cuneate, margins entire, apex acute to almost truncate, erose; stamens erect, purplish, 1.7–2.6 cm long, filaments purple, ~4 mm long, anthers erect, straight, ± purple-brown, 1.3–2.2 cm long, connectives purple; ovary purple, ovoid, 6-angled, 0.6–1.2 cm long; stigmas small, divergent or erect, distinct, purple, subulate, 4–8 mm long, not fleshy. **Fruits** red-purple, fragrance not reported, ovoid, obscurely 6-angled, 2.5–3 cm, pulpy, juicy. **Distr:** USA (California). **Ecol:** Along ravines in redwood forests, thickets, open wooded slopes, open grasslands; 0–2000 m Fl. Feb–Apr, Nov. **Zones:** 6–10.

T. erectum L., *Sp. Pl.* 1: 340, 1753. [Birthwort, Red Trillium, Stinking Benjamin, Stinking Willie, Wake Robin]. **Herbs**; rhizomes short, thick, praemorse. **Scapes** 1–2, often with numerous offsets, round, 15–60 cm tall, ± robust, glabrous. **Bracts** sessile; lamina bright green, major veins prominent, broadly rhombic to ovate-rhombic, 5–20 × 5–20 cm, base attenuate, apex acuminate. **Flower** erect, malodorous; perianth open, flat; sepals flat to sulcate apically, green, often streaked or overlain with maroon, occasionally entirely dark maroon, lanceolate-acuminate, 1–5 cm long, texture leafy, margins entire, apex acuminate; petals spreading, dark reddish-brown, maroon, purple, or white, sometimes pale yellow, lanceolate, mostly ovate-lanceolate, 1.5–5 × 1–3 cm, heavy textured, apex acuminate; stamens erect to slightly recurved, 0.5–1.5 cm long, filaments white, pinkish, or dark purple, thin, anthers usually erect, dark maroon or yellowish; ovary dark purple to maroon, ovoid, elliptic to globose, 6-angled, 0.5–1 cm long, stigmas recurved, distinct, dark purple, subulate, 3–7 mm long, fleshy; pedicel straight, ± erect, 1–10+ cm long. **Fruits** dark maroon, slightly fragrant, ± globose to ± pyramidal, 1–1.6 × 1–1.5 cm, juicy. **Distr:** E North America. **Ecol:** In rich, moist upland forests, coniferous swamp borders; 200–700 m. Fl. Apr–May. **Zones:** 5–10.

T. ovatum Pursh, *Fl. Amer. Sept.* 1: 245, 1814. [Western White Trillium, White Trillium]. **Herbs**; rhizomes semierect to horizontal, short, stout, praemorse. **Scapes** 1–2, round, 20–50 cm long, ± slender, glabrous. **Bracts** sessile, subsessile, or short-petiolate; lamina medium green, sometimes blotched and mottled, main veins prominent, ovate-rhombic, 7–12 × 5–20 cm, expanding during anthesis, base rounded, apex acuminate. **Flower** erect or nodding, odorless; sepals green, lanceolate to oblong-lanceolate, 1.5–5 × 0.6–2 cm, margins entire, apex acute; petals erect-ascending, usually wide-spreading from base, exposing entire pistil, white or with pink or blush markings, fading to rosy pink, purple, or dark red, ± linear to widely obovate, 1.5–7 ×1–4 cm, thin, margins flat to undulate, apex acuminate; stamens prominent, slightly recurved-spreading to straight,

Above: *Gloriosa modesta*, *Gloriosa superba*, *Sandersonia aurantiaca*, *Luzuriaga radicans*.

1–1.8 cm; filaments white, slender, anthers yellow, 0.4–1.6 cm long, slender; ovary green or white, ovoid, 6-angled, 0.5–1.2 cm long, stigmas recurved, greenish-white or white, linear, 0.6–1 cm long, thin; pedicel 2–6 cm long. **Fruits** baccate, green or white, ± odorless, broadly ovoid, obscurely winged, 1.2–2.8 × 0.7–1.9 cm, pulpy moist. **Distr:** W North America. **Ecol:** In coniferous and deciduous forests, thickets; 60–1,600 m. Fl. Feb–Apr, Oct. **Zones:** 6–10.

COLCHICACEAE

Genera 5, **species** 170. **Herbs**, cormous, without scale leaves; coat brownish. **Leaves** radical, narrowly to broadly linear. **Scape** subsessile to sessile. **Flowers** hypogynous, hermaphrodite; perianth of 6 tepals, united to form a tube or split to the base; stamens 6, inserted at the bases of the perianth segments, included, anthers extrorse; carpels 3, ovary sessile, trilocular, styles 3, free to the base or united within the perianth tube, ovules many. **Fruits** capsular, dehiscence septicidal. **Distr:** Temperate to tropical regions worldwide, excluding S America.

Gloriosa

Species 10. **Herbs**, perennial, rhizomatous; rhizome stout, tuberous; stem sometimes branched, usually elongate and scandent. **Leaves** cauline, alternate, opposite, or whorled, subsessile, apex bearing a tendril. **Inflorescences** a cluster of few flowers, sometimes in a corymb; long pedicellate. **Flowers** usually large and showy; tepals 6, free, spreading or reflexed, persistent; stamens 6, inserted at base of tepals, filaments filiform, anthers versatile; ovary triloculed; ovules many per locule, style long, filiform, apically trilobed, adaxially stigmatic. **Fruits** capsules; seeds subglobose, testa bright red, spongy. **Distr:** S and tropical Africa, tropical Asia. **Cult:** Propagate by division of the tubers or from seeds. Requires any well-drained soil in full or partial sun.

G. modesta (Hook.) J.C. Manning and Vinn., *Taxon* 56: 178, 2007. (Syn: *Littonia modesta*, *L. keitii*). [Climbing Bell Lily]. **Herbs**, climbing using tendrils, tuberous, 1–2 m long; tuber 3 cm across, with 2 projecting lobes or horns pointing downward (new tubers formed from these projections) and a few capillary fibers in the center beneath; stems slender, terete, herbaceous, unbranched. **Leaves** dimorphic, spreading or reflexed, glabrous, sessile, striated with parallel nerves; *lower* leaves ternately verticillate, ovate-lanceolate or broadly lanceolate, tapering from a broad base to the apical filiform tendril; *upper* leaves alternate, smaller and usually narrower. **Inflorescences** with flowers solitary, axillary; peduncles rather short with flowers drooping. **Flowers** with perianth orange-yellow, lobes 6, erecto-patent (so that the perianth is subcampanulate), lanceolate or oblong, shortly but sharply acuminate, moderately concave, keeled at the back, inside at base with a small oblong nectariferous cavity; stamens 6, free, erect, hypogynous, as long as the pistil, much shorter than the sepals, filament subulate, anther oblong, subsagittate, basifixed; ovary ovoid, with 3 furrows and 3 obtuse lobes, triloculed, locules many ovuled, style about as long as the ovary, divided above into 3 rather long, recurved segments or branches, each tipped with an obtuse, downy stigma. **Distr:** Zimbabwe to S Africa. **Ecol:** In moist areas; 1,600 m. Fl. Dec–Feb. **Zones:** 9–11.

G. superba L., *Sp. Pl.* 1: 305, 1753. [Flame Lily, Glory Lily, Vine Lily]. **Herbs** perennial; rhizomes usually forked, ~1 cm in Ø, fleshy; stem scandent, 2–3 m or more, rather slender. **Leaves** alternate or occasionally also opposite, sessile or shortly petiolate, lanceolate to ovate-lanceolate, 7–13 cm, apex long caudate with a tendril. **Inflorescences** sometimes a corymb. **Flowers** nodding; pedicel 10–15 cm; tepals reflexed, bright red, proximally yellow-tinged, linear-oblanceolate, 4.5–5 cm × 7–9 mm, base slightly clawed, margin much crisped; filaments 3–4 cm; anthers ~1 cm; style 2.5–3.5 cm; stigma lobes 6–7 mm. **Distr:** Cambodia, India, Indonesia, Laos, Malaysia, Maldives, Myanmar, Nepal, Sri Lanka, Thailand, Vietnam; S and tropical Africa. **Ecol:** In forests, thickets; 900–1,300 m. Fl. Jul–Aug. **Zones:** 9–12. **Info:** The rhizomes are poisonous, containing alkaloid colchicine. In the Maldives the tubers are treated and made into very sought after sweet wafers.

Sandersonia

Species 1. *Generic description same as specific.* **Cult:** Propagate from offsets or from ripe seed. Requires a humus-rich, well-drained soil in full sun.

S. aurantiaca Hook., *Bot. Mag.* 79: t. 4716, 1853. [Christmas Bells, Chinese-lantern Lily]. **Herbs**; rhizome tuberous, ± globose, bilobed, abaxially with a few fibers; stem erect, to 75 cm long, leafy, simple, terete. **Leaves** sessile, scarcely sheathing, rather amplexicaulous, erect, subdistichous; lamina lanceolate, to 10 cm long, apex much and narrowly acuminate, striated with parallel veins; *lower* leaves rather distant; *upper* leaves almost secund, approximate and almost crowded. **Inflorescences** with flowers solitary, borne in the axils of the upper leaf axils; peduncle 2–5 cm long, slender, almost capillary, curved downward, flowers drooping. **Flowers** lantern-like, relatively large and showy; perianth orange, spherical-urn-shaped or subglobose-campanulate, 2–2.5 cm long, with 6 shallow furrows, thus appearing obtusely 6-angled, lobes fused except at the apices which curl outward, mouth somewhat contracted, 6-toothed, base of the perianth with 6 short nectar-bearing spurs (pointing upward); stamens 6, about a third as long as the perianth, borne at the base of the lobes, filaments longer than the ovary, filiform-subulate, anthers broad, oblong, introrse, 2-celled, opening via longitudinal slits; ovary superior, oblong-ovoid, obtusely trigonous with 3 deep longitudinal furrows, style as long as stamens, stigmas 3. **Fruits** capsules, with persistent flower. **Distr:** S Africa (E Cape Province to Swaziland). **Ecol:** In grasslands and forest margins. **Zones:** 9–11.

LUZURIAGACEAE

Genera 2, **species** 5. **Climbers** or **subshrubs**, perennial; stems ± woody, slender. **Leaves** 2-ranked, alternate, distichous, petiolate or sessile; lamina resupinate (twisted at base, thus 'upside down'), linear, lanceolate, or ovate, margins entire. **Inflorescences** with flowers solitary or sometimes aggregated into cymes. **Flowers** poly symmetric, bisexual, usually white (sometimes violet or white with violet colorations), 3-merous; perianth with tepals 6, free or almost joined, 2-whorled, petaloid, similar in the 2 whorls or different; stamens 6, anthers dorsifixed or basifixed, dehiscing longitudinally; carpels isomerous with perianth, pistil 3-celled, ovary superior, trilocular, style 1, stigma 1, placentation axile, ovules 3–9 per locule. **Fruits** berries, fleshy; seeds rounded, yellow, brown or black. **Distr:** Peru to Tierra del Fuego, Falkland Islands, New Zealand and Australia (NSW to Tasmania).

Luzuriaga

Syn: *Calcoa*, *Callixene*, *Enargea*.

Species 4. **Climbers** or **subshrubs**, glabrous, branched; stems jointed, lower nodes with small, membranous scales, upper nodes with small leaves. **Leaves** alternate, ± sessile, resupinate, usually small; lamina oblong to elliptic. **Inflorescences** axillary, flowers solitary, or few-flowered clusters; pedicels arising from a group of small membranous scales, bracteate. **Flowers** with perianth white, spreading; segments free subequal; stamens free, filaments glabrous, erect; ovary sessile, ovoid, trilocular, style terete, stigma obscurely trilobed. **Fruits** subglobose; seeds few, pale. **Distr:** New Zealand; S Chile to Falkland Is, Peru, Argentina (Patagonia). **Cult:** Propagate from seeds. Requires a damp, humus-rich soil in partial shade.

L. radicans Ruiz and Pav., *Fl. Peruv.* 3: 66, 1802. (Syn: *Enargea radicans*, *Luzuriaga cordata*). **Climbers**, 0.6–1 m tall; stems branched, flexuous, angled, green, with small sheathing persistent brown scales at the nodes. **Leaves** distichous, sessile, base twisted; lamina elliptic-oblong or linear-oblong, apex acute or apiculate, adaxially bright green, abaxially glaucous, nerves 3–6 pairs. **Inflorescences** with flowers solitary or 2–3-flowered clusters, flowers drooping, borne from small imbricating brown persistent bud-scales; peduncles 0.6–2.5 cm long, slender, with 1 or 2 small scalelike bracts. **Flowers** white or yellowish white with red lines and dots, variable in size, 3.8–5 cm across; perianth spreading, segments elliptic or lanceolate,

Above: *Alstroemeria pulchella, Bomarea multiflora, Lapageria rosea, Philesia magellanica.*

apex acuminate, the 3 inner rather large and thickened along the middle; stamens conniving into a conical column, filaments inserted at base of perianth segments, anther sessile, subulate-lanceolate, subacute, laterally dehiscent; ovary globose, style strict, angular, stigmas 3, minute, ovules biseriate. **Fruits** berries, globose, maturing through orange to red; seeds few. **Distr:** S Chile to S Argentina. **Ecol:** In forests, rooting up on mossy trunks of trees. **Zones:** 9–11.

ALSTROEMERIACEAE

Genera 3, **species** 165. **Herbs**, perennial, rarely annual, erect or twining, mostly glabrous, rhizomatous; phyllotaxis spiral. **Leaves** on elongated stem or crowded at its upper end, or on short stems forming rosettes near ground; lamina simple, margins entire, venation parallel or arched, usually twisted at base making the lamina ± inverted, sometimes reduced to sessile bracts. **Inflorescences** terminal, 2–many flowered, lax or dense, bracteate, umbellate, simple or branched, or flowers solitary; bracts leaflike or reduced, often forming a basal pseudo-whorl. **Flowers** bisexual, actinomorphic or zygomorphic, epigynous; tepals 3+3, petaloid, free, the 2 whorls similar or different, inner ones often basally narrowed into a nectariferous claw; stamens 3+3, free, anthers pseudo-basifixed, filaments filiform; ovary inferior, turbinate, syncarpous, tricarpellate, unilocular or trilocular, ovules usually many, style filiform with 3 stigmatic branches. **Fruits** usually dehiscent, dry, leathery, or somewhat fleshy capsules, loculicidal, trivalved, rarely berry like and indehiscent; seeds + globose, with or without a red sarcotesta. **Distr:** Tropical and temperate C and S America.

Alstroemeria

[Lily of the Incas].
Species 123. **Herbs**, perennial, erect, rhizomatous with storage roots. **Leaves** sessile, usually resupinate (twisted near base or along the lamina), concentrated toward apex of the stem, distributed along the stem, or rosetted near ground; lamina simple, linear, lanceolate, elliptic, or oblong, margins entire, both surfaces glabrous or adaxially papillose. **Inflorescences** terminal, umbellate, simple or branched cymes, 2–many flowered, rarely flowers solitary; bracts leaflike. **Flowers** zygomorphic, bisexual; tepals 6, petaloid, free, 2-whorled, similar or different, inner whorl often shorter, patterned, and basally narrowed into canaliculate nectariferous ducts; stamens 6, free, in 2 whorls, anthers pseudo-basifixed; ovary inferior, syncarpous, tricarpellate, trilocular, ovules many, anatropous, style apical, stigmas 3. **Fruits** capsules, dehiscent, loculicidal; seeds globose, without a sarcotesta. **Distr:** S Venezuela to S South America. **Cult:** Propagate by division when dormant, or from seed. Requires a moderately fertile, moist, well-drained soil in full sun.

A. pulchella L. f., *Suppl. Pl.*: 206, 1782. (Syn: *A. psittacina*, etc.). [Inca Lily, New Zealand Christmas Bell, Peruvian Lily, Parrot Lily]. **Herbs**, perennial, seasonally dormant, tuberous, to 1 m long; tubers very small, clustered, arranged like the spokes of a wheel; stems 20–90 cm long. **Leaves** held horizontal, lamina inverted, twisted at base (resupinate); lamina lanceolate to linear, 3–10(–15) × 1–3.5 cm, apex obtuse to bluntly acute. **Inflorescences** umbels, 4–6-flowered. **Flowers** green, red-flushed, throat streaked with red or maroon-flecked; tepals spathulate, overlapping to form a narrow tube 4–7 cm long; outer tepals 4–7 × 1.5–2 cm; inner tepals 4–7 × 0.7–1 cm; tepal margins unevenly ragged, slightly curled inward; stamens 3.5–6 cm long, anthers 4–6 mm long; ovary 1–1.5 cm long, ribbed, style 3.5–6.5 mm long, stigmas 3–5 mm long. **Fruits** globose, 1–1.5 cm Ø, with prominent ribs; seeds 2.5–3 mm Ø, reddish-brown. **Distr:** Brazil to N Argentina. **Ecol:** In disturbed places, often invasive; 0–100 m. **Zones:** 7–11.

Bomarea

Syn: *Vandesia, Wichuraea*, etc.
Species 122. **Herbs**, scandent or climbing, tuberous or rhizomatous, glabrous or pubescent; tubers large; roots slender; stem elongate, leafy. **Leaves** narrow or broad, many nerved, apex acute, abruptly short-petiolate; petiole short, twisted through 180°, so that the abaxial surface of the lamina faces upward. **Inflorescences** simple or compound umbels, terminal. **Flowers** large and showy, usually red and yellow, often spotted; perianth funnelform, segments distinct, biseriate, segments dissimilar between series; stamens attached to bases of the segments, about equaling them, filaments filiform, anthers oblong, basifixed; ovary 3-celled, ovules numerous, style filiform, stigma trifid. **Fruits** capsules, 3-celled, turbinate or subglobose, apex truncate, 6-costate, tardily dehiscent; seeds subglobose, covered with a bright red, pulpy testa. **Distr:** Tropical and subtropical America (mainly from Mexico to the Andes). **Cult:** Propagate by division or from seeds. Requires a humus-rich, moist, well-drained soil in partial sun.

B. multiflora (L. f.) Mirb., *Hist. Nat. Pl.* 9: 72, 1804. (Syn: *Alstroemeria caldasii, A. multiflora, B. caldasii, B. floribunda*, etc.). **Climbers**, tuberous, 2–2.5 m long; stem downy. **Leaves** oblong, 7–10 cm long. **Inflorescences** umbels, simple, drooping, 20–40-flowered. **Flowers** with perianth segments ± equal, ~2.5 cm long; outer segments tinged red; inner segments reddish-yellow with brown spots. **Distr:** Colombia to Ecuador. **Ecol:** 2,600 m. **Zones:** 9–11.

PHILESIACEAE

Genera 2, **species** 2. **Climbers** or **woody shrubs**. **Leaves** 1-nerved or ribbed, coriaceous, deciduous, reticulated. **Inflorescences** with flowers solitary. **Flowers** large, showy, hermaphrodite, scaly at base, either tripetaloideous with the calyx colored, membranous and short, or hexapetaloideous, with the sepals and petals equal and similar; stamens 6, inserted into the base of perianth, anthers linear, opening longitudinally; ovary superior, 1-celled, free, with 3 parietal placentas, style long, club-shaped, stigmas 3, ovules numerous, orthotropal, enveloped in mucilage. **Fruits** berries; seeds obovate, horny, embedded in pulp. **Distr:** Southern Hemisphere (S America, S Africa and Australia).

Lapageria

Species 1. *Generic description same as specific.* **Cult:** Propagate from seed. Requires a fertile, well-drained soil in partial sun.

L. rosea Ruiz and Pav., *Fl. Peruv.* 3: 65 (1802). (Syn: *L. hookeri, Philesia rosea*, etc.). [Chilean Bell Flower, Copihue]. **Climbers**, to 8 m long; stem terete, branched, naked below, sparsely scaly. **Leaves** alternate, short-petiolate; lamina ovate-lanceolate, base heart-shaped or rounded, margins entire, apex acuminate, coriaceous, stiff and leathery, dark glossy green, 3–5-nerved and reticulated. **Inflorescences** axillary, flowers solitary or in few-flowered clusters, flowers pendent; peduncles longer than petiole. **Flowers** deep rose-red, spotted with white, especially internally, trumpet-shaped, to 8 cm long; outer sepals spathulate, with protuberance at base; stamen and style shorter than perianth; inner sepals resembling outer ones, but broader and more spotted. **Fruits** capsules, egg-shaped, 3-sided. **Distr:** Chile. **Ecol:** Fl. Jun–Oct. **Zones:** 8–10. **Info:** Fruits edible. Varieties with white, pink or dark rose red flowers are often cultivated.

Philesia

Species 1. *Generic description same as specific description below.* **Cult:** Propagate from seed. Requires a fertile, well-drained soil in partial sun.

P. magellanica J.F. Gmel., *Syst. Nat.*: 1012, 1791. (Syn: *P. buxifolia*). [Philesia]. **Shrubs** or **subshrubs**, erect, much-branched, 1–1.5 m tall (or sometimes smaller in cultivation); branches alternate; branchlets angular, green, sparsely scaly. **Leaves** alternate, evergreen, petioled; lamina linear-oblong, 2.5–4 cm long, margins reflexed, apex mucronate, glabrous, coriaceous, penninerved; petiole articulated at first (often deciduous, leaving the persistent, short petiole). **Inflorescences** terminal, flowers solitary, drooping or pendent, borne on branches; peduncles exceedingly short, bracteate at base of flower. **Flowers** bisexual, large; calyx ~2 cm long; sepals 3, oblong, rather obtuse, imbricating, appressed, concave, scariose; corolla campanulate, 5–6 cm long; petals obovate-oblong, mucronate, bright rose-red, somewhat waxy, equal, concave, base united, each petal with a hard oblong depressed gland or nectary; stamens borne from the base of petals, filaments united into a tube below middle, then free, erect, equal, slightly shorter than petals, anthers erect, subsagittate; ovary small, ovate-trigonal, 1-celled, with 3 short parietal placentas, ovules numerous, style slightly longer than stamens, rather thick, stigma depressed-capitate, margins recurved, obsoletely trilobed. **Fruits** berries, ovate-subglobose, mucronate, surface rough; seeds many. **Distr:** Chile to S Argentina. **Ecol:** Fl. Sep–Oct. **Zones:** 9–11. **Info:** Due to a genetic mutation, double flowers are sometimes produced with a petaloid stamen.

SMILACACEAE

Genus 1, **species** 315. **Shrubs, herbs**, or **vines**, perennial, rhizomatous; stems erect or climbing, usually prickly, sometimes unarmed. **Leaves** opposite or alternate, usually bearing tendrils; lamina prominently 3-veined, reticulate between veins, usually leathery. **Inflorescences** umbellate (rarely racemose or spicate). **Flowers** unisexual, ♂ and ♀ on different plants; tepals 6, distinct, rarely united into a perianth tube; stamens 2–3-whorled, anthers 1-locular; ♀ flowers with staminodes, pistil tricarpellate; ovary biloculed, locules 1–2-ovuled. **Fruits** baccate; seeds 1–3. **Distr:** Pantropical to temperate.

Smilax

Species ~300. **Vines climbing** or **shrubs**, woody, less often suberect or herbs, dioecious, usually with short, thick rhizomes; stems and branches usually prickly. **Leaves** often with tendrils; lamina usually ovate to lanceolate, main veins 3–7, connected by a network of cross veins and veinlets; petiole usually narrowly winged proximally, with an abscission zone between winged portion and apex. **Inflorescences** borne in axil of leaf or scalelike bract, of 1(–3) umbels, or panicles, racemes, or spikes of umbels; peduncle sometimes with a scalelike prophyll at base. **Flowers** small; tepals 6, usually free (rarely connate); ♂ with stamens 6, rarely 8 or more, inserted at base of tepals, anthers mostly 1-loculed; ♀ with ovary triloculed, ovules 1 or 2 per locule, style very short, stigmas 3, staminodes absent to 6. **Fruits** berries, red to black, usually 1- or 2-seeded; seeds dark brown. **Distr:** Tropical, subtropical, and temperate regions of both hemispheres. **Cult:** Propagate from seeds or cuttings. Requires moist but well-drained soil in full sun.

S. bracteata C. Presl, *Reliq. Haenk.* 1: 131, 1827. (Syn: *S. lyi, S. stenopetala*). [Sarsaparilla Vine]. **Vines**, climbing, to 10 m long; stem branched, subterete, woody, sometimes sparsely prickly. **Leaves** with petiole 1–2 cm, narrowly winged for 2/5–1/2 its length; abscission zone distal; tendrils commonly present; lamina broadly elliptic to ovate-elliptic, 5–17 × 3–11 cm. **Inflorescences** racemes, with 3–6(–10) umbels, 3–7 cm, basally prophyllate; umbels of both sexes 12–25-flowered, base slightly thickened, globose, ~ 2 × 2 mm. **Flowers** ♂ with tepals olive green to dark red, ~ 5 × 0.7–1.3 mm; ♀ with tepals smaller than ♂, staminodes 3. **Fruits** berries, globose, 5–7 mm Ø. **Distr:** Japan (S Kyushu) to Indo-China and Caroline Islands. **Ecol:** In forests, thickets; 0–1,800 m. Fl. Nov–Feb. Fr. Jun–Aug. **Zones:** 8–10.

S. perfoliata Loureiro, *Fl. Cochinch.* 2: 622, 1790. (Syn: *S. ovalifolia, S. prolifera*). **Vines** climbing; stem branched, to 8 m, woody; stem and branches smooth or sometimes sparsely prickly. **Leaves** usually with developed tendrils; usually ovate, 6–17 × 3–12 cm; petiole 2–4 cm, broadly winged for 1/2–2/3 its length, wings 0.7–1.2(–2.5) cm wide, basally deeply clasping node, tapering toward apex; abscission zone distal **Inflorescences** of 10–30 umbels, 5–17 cm, basally prophyllate; rachis often slightly zigzagged; umbels of both sexes ± grouped in whorls of 2–4, densely many flowered, base ellipsoid, ~ 4 × 3 mm. **Flowers** ♂ with outer tepals 5–6 × ~1 mm, inner ones ~ 5 × 0.3 mm, stamens ~5 mm; ♀ with tepals slightly smaller than ♂ ones, staminodes 3. **Fruits** berries, globose, 4–6 mm Ø. **Distr:** Indian Subcontinent to Taiwan. **Ecol:** In forests, thickets; 0–1,500 m. Fl. Apr. Fr. Oct. **Zones:** 9–11.

LILIACEAE

Genera 16, **species** 635. **Herbs**, perennial, rhizomatous, bulbous, or cormous, rarely shrubby or treelike. **Leaves** basal and/or cauline, alternate, opposite, or whorled; lamina with parallel or rarely reticulate veins. **Inflorescence** a raceme, panicle, spike, umbel, reduced panicle, or other, or flowers solitary; bracts present or absent; bracteoles present or absent. **Flowers** bisexual, rarely unisexual, actinomorphic, rarely zygomorphic; perianth usually corollalike, 6-merous, rarely 4- or 8-merous, in 2 whorls; segments free (tepals) or united; stamens 6, rarely 3, 4, or 8, inserted opposite perianth segments, filaments free or adnate to perianth, rarely connate into a corona, anthers usually biloculed, basifixed or dorsifixed and versatile, introrse, latrorse, or extrorse, dehiscing usually by vertical slits; carpels usually connate for most or all of their length, rarely only at base, ovary superior, rarely semiinferior, triloculed, rarely 2- or 4-loculed, placentation axile, or rarely 1-loculed with a parietal placenta, ovules usually anatropous; nectaries septal, perigonal, or absent. **Fruits** capsules or berries; seeds usually numerous. **Distr:** North Temperate, especially E Asia and N America.

Calochortus

Syn: *Cyclobothra, Mariposa*.

Species 72. **Herbs**, perennial, sometimes bulbous; bulb coat membranous or fibrous-reticulate; stems scapelike or leafy, simple or branched, glabrous, often glaucous; bulblets sometimes borne in leaf axils. **Leaves** sessile, basal persistent or withering by flowering, solitary; lamina with base sometimes attenuate and petiolelike; *cauline* leaves 0–several, sometimes lowest leaves appearing as basal, reduced. **Inflorescences** monochasiate or ± umbellate, 1–many flowered, bracteate. **Flowers** with perianth globose to broadly campanulate; sepals 3, distinct, ovate to lanceolate, usually petaloid and glabrous; petals 3, distinct, usually longer and broader than sepals, sometimes clawed, usually hairy adaxially, bearing adaxial gland near base, often spotted to ± patterned; filaments widened at base, anthers usually basifixed or pseudo-basifixed, linear to oblong; ovary superior, style absent, stigmas 3. **Fruits** capsular, trilocular, 3-angled or 3-winged, linear, oblong, or globular, dehiscence septicidal; seeds many, in 2 rows per locule, irregular or flat, seed coat usually hexagonally reticulate. **Distr:** British Columbia to Guatemala. **Cult:** Propagate by division or from seeds. Requires a gritty, well-drained soil in full or partial sun.

C. luteus Douglas ex Lindley, *Edwards's Bot. Reg.* 19: plate 1567, 1833. (Syn: *Mariposa lutea*). [Cat's Ear]. **Herbs**; stems slender, 20–50 cm long. **Leaves** basal, withering, 10–20 cm long; lamina linear. **Inflorescences** subumbellate, 1–4-flowered; bracts 1–8 cm long. **Flowers** erect; perianth open, campanulate; sepals lanceolate-oblong, 2–3 cm long, attenuate; petals deep yellow, usually streaked red-brown proximally, often with median red-brown blotch, cuneate to obovate, 2–4 cm long, with a few slender hairs near gland; glands ± lunate to oblong, not depressed, covered with short, matted hairs; filaments 7–9 mm long, anthers linear-oblong, 4–6 mm long, apex obtuse or acute. **Fruits** capsules erect, lanceoloid-linear, angled, 3–6 cm long; seeds light beige, flat. **Distr:** USA (California). **Ecol:** In grasslands, open woodlands, mixed evergreen forests; 0–700 m. Fl. May–Jul. **Zones:** 5–10.

Cardiocrinum

Species 3. **Herbs** perennial, bulbiferous; bulb formed by swollen base of usually deciduous basal leaves; bulbils several, ovoid, covered with tunics; stem very tall, stout, glabrous. **Leaves** basal and cauline, petiolate, usually ovate-cordate, reticulate veined. **Inflorescence** a terminal raceme, few- to many flowered; bracts persistent or caducous. **Flowers** bisexual, tubular-funnelform, large; pedicel rather short; tepals 6, free, ± connivent; stamens 6, inserted at base of tepals, filaments flat, anthers dorsifixed, versatile, narrowly ellipsoid; ovary cylindric, triloculed, ovules many per locule, style elongate, stigma slightly trilobed. **Fruits** capsules, loculicidal; seeds reddish brown, flat, narrowly winged all round. **Distr:** Russian Far East to Himalaya. **Cult:** Propagate from seeds. Requires a nutrient-rich, humus-rich, moist, well-drained soil in full sun.

C. giganteum (Wallich) Makino, *Bot. Mag. (Tokyo)*. 27:125, 1913. (Syn: *Lilium giganteum*). [Giant Himalayan Lily, Giant Lily]. **Herbs** large, stem green, 1.5–3 m tall; bulbils 3.5–4 × 1.2–2 cm; stem erect, green or dark green, 1–3 m × 3–5 cm, hollow. **Leaves** on proximal ½ of stem larger, those on distal ½ much smaller, sometimes bractlike; petiole 15–

Below: *Smilax bracteata, Smilax perfoliata, Calochortus luteus, Cardiocrinum gigantea.*

MIW

20 cm; lamina ovate-cordate, 15–20 × 12–15 cm. **Inflorescence** a raceme, erect, 10–16-flowered; bracts caducous. **Flowers** with tepals white or tinged with green (abaxially), streaked with purple or purple-red adaxially, linear-oblanceolate, 12–15 × 1.5–2 cm, apex obtuse; stamens 6.5–7.5 cm; filaments slightly widened toward base; anthers ~8 × 2 mm; ovary 2.5–3 cm × 4–5 mm; style 5–6 cm. **Fruits** subglobose, 3.5–4 cm Ø; apex beaked; seeds ovate-deltoid, 4–5 × 2–3 mm. **Distr:** Himalayas to N Myanmar and C China. **Ecol:** In forests, hillsides; 1,200–3,600 m. Fl. Jun–Jul. Fr. Sep–Oct. **Zones:** 7–10.

Heloniopsis

Species 6. **Herbs**, perennial; rhizome short, thickened. **Leaves** basal, rosulate; lamina narrowly oblong to oblanceolate or obovate, base gradually narrowed to a petiole, margin smooth, sometimes minutely undulate, evergreen. **Inflorescence** an umbel or umbellate-raceme, 1–10-flowered, terminal; bract usually absent; scape borne from center of leaf rosette, erect, simple, hollow, with 2–8 scale leaves. **Flowers** bisexual, slightly nodding, spreading, funnelform; tepals 6, free, spathulate or linear-oblanceolate to oblong, adaxially often with a deep pocket of a nectary gland at base, persistent; stamens 6, often adnate basally to tepals, sometimes free, anthers lanceolate, dorsifixed, extrorse to latrorse; ovary superior, triloculed, ovules 60–180 per locule, style 1, slender, elongate, stigma capitate. **Fruits** capsules, trilobed, loculicidal; seeds small, linear, ends caudate. **Cult:** Propagate from seeds. Requires a nutrient-rich, humus-rich, moist, well-drained soil in full sun.

H. umbellata Baker, *J. Bot.* 12: 278, 1874. (Syn: *Helonias umbellata*, etc.). **Herbs**; rhizome short. **Leaves** usually oblanceolate or subspathulate, 13 × 2.2 cm, base attenuate, apex obtuse to acute with apiculus, relatively thick-textured. **Inflorescence** umbellate or subumbellate, 2–20-flowered; peduncle ~3–15 cm tall at flowering (to ~30 cm long at fruiting); pedicels 0.3–1.8 cm long. **Flowers** fragrant, funnelform; tepals oblanceolate, apex obtuse to subacute, sometimes slightly retuse, usually canaliculate below, with shallowly concave nectary at base, 0.6–1.2 cm × 2.5–3.7 mm, 3–5-veined, white or pink, turning green after flowering; stamens usually slightly longer than tepals, persistent, anthers ovate-oblong, apex confluent between thecae, base deeply bilobed, virtually bilocular, 1–3 mm long, purplish, filaments 0.4–1.2 cm long, inserted on base of tepals; pistil usually exceeding tepals and stamens, ovary globose, trilobed or trigonous, apex concave, base usually short stipitate, style terete, erect, 0.5–1.5 cm long, stigma discoid, sometimes slightly trigonous or trilobed, 1–2(–2.5) mm across. **Fruits** with numerous seeds (55–115 per locule), linear-fusiform, 3.5–5 mm long. **Distr:** Taiwan. **Ecol:** In shady wet or moist rocks and banks, sometimes in mountain meadows; 500–2,500 m. Fl. Jan–Jun. **Zones:** 9–11.

Lilium

Syn: *Martagon*.
Species 107. **Herbs**, perennial, bulbiferous; bulb of many imbricate, fleshy scales, without tunic; stem erect, leafy. **Leaves** alternate, rarely whorled, sessile or subsessile; lamina usually linear to linear-lanceolate; bulbils sometimes present in leaf axils. **Inflorescence** a raceme or solitary flower, very rarely an umbel or corymb, terminal; bracts leaflike. **Flowers** often funnelform or campanulate, sometimes tubular or cupular; tepals 6, free, usually connivent, sometimes strongly recurved or revolute, white, yellow, greenish, or reddish to purplish, nectariferous near base adaxially; nectaries usually narrowly grooved, sometimes fringed with papillae or hairs, rarely flat on outer tepals; stamens 6, filaments subulate or filiform, sometimes pubescent, anthers dorsifixed, versatile; ovary triloculed, ovules many per locule, style elongate, slender, stigma swollen, usually trilobed. **Fruits** capsules, loculicidally dehiscent; seeds many, arranged stacked, flat, narrowly winged all round. **Distr:** Temperate Northern Hemisphere to N Philippines. **Cult:** Propagate from offsets, detached bulb scales, bulbils or seeds. Requires a fertile, humus-rich, moist, well-drained soil in full or partial sun.

L. formosanum Wallace, *Garden.* 40: 442, 1891. [Formosa Lily, Saint Joseph's Lily]. **Herbs**, bulbous; bulb subglobose or broadly ellipsoid, 2–4 cm Ø, scales white or tinged yellow, lanceolate-ovate to lanceolate; stem sometimes tinged purple-red, 20–55 cm long, smooth or papillose. **Leaves** scattered, linear or narrowly lanceolate, 2.5–15 × 0.4–1.3 cm. **Inflorescences** with flowers solitary or sometimes several in subumbels. **Flowers** fragrant, funnelform, with a slender tube gradually expanding toward apex; tepals white, tinged purple-red abaxially; outer ones oblanceolate, 11.5–14.5 × 2.1–2.3 cm; inner ones spathulate, to 3 cm wide; nectaries green, rarely indistinctly papillose on both surfaces; filaments ~10 cm long, with minute protuberances near base; style ~6.5 cm long. **Fruits** 7–9 × ~2 cm. **Distr:** Taiwan. **Ecol:** On grassy slopes, seashores; 0–3,500 m. Fl. & Fr. Jun–Dec. **Zones:** 6–11.

Peliosanthes

Syn: *Lourya*, etc.
Species 16. **Herbs** perennial, rhizomatous, with thick roots; stem usually short, rarely elongate and procumbent. **Leaves** usually basal, rarely cauline, petiolate; lamina linear to elliptic-ovate, subplicate veined, with conspicuous, transverse veins between main, longitudinal veins, glabrous. **Inflorescences** with scape terminating in a raceme or reduced panicle; flowers solitary or in clusters of 2–5, subtended by a bract; pedicels articulate near apex; bracteole absent or 1. **Flowers** with perianth segments 6, united into a tube; stamens 6, filaments dilated, connate in a fleshy ring (corona), rarely free; corona flat or elevated; anthers subsessile; ovary inferior to semiinferior, triloculed, ovules 2–4(or 5) per locule, style shortly conical to columnar; stigma capitate to trilobed, small. **Fruits** bursting at an early stage and exposing young seeds; seeds blue at maturity, berry like, ellipsoid to globose. **Distr:** Himalayas to Taiwan and W Malesia. **Cult:** Propagate by division or from seeds. Requires a humus-rich, fertile, moist soil in shade.

P. teta Andrews, *Bot. Repos.* 10: t. 605, 1810. (Syn: *P. minor*, *P. tonkinensis*, *P. torulosa*). **Herbs**; stem short. **Leaves** 4–8; petiole (5–)20–30 cm, slightly compressed; lamina lanceolate to elliptic, (12–)16–23 × 2–4 cm, with 5 main veins, apex acute to acuminate. **Inflorescence** reduced panicle, (5–)10–15 cm; scape 3–20 cm, flowers in clusters of 2–5; pedicels (3–)5–8 mm; bracts lanceolate, 3–7 mm, membranous; bracteole 1, ~2 mm. **Flowers** with perianth purple, 6–8 mm Ø; tube 2–3 mm, mostly adnate to ovary, lobes oblong to ovate, 2.5–4 × 1.5–2 mm; corona purple, ~0.5 mm wide; anthers ~0.5 mm; style short, stigma capitate. **Fruits** berry like; seeds subglobose, 5–7 mm. **Distr:** Sikkim to S China and W Malesia. Fl. Jan. **Ecol:** In forests; ~600 m. **Zones:** 8–12.

Tricyrtis

Species 22. **Herbs** perennial, with short or sometimes long and creeping rhizomes; stems usually erect or ascending, sometimes branched distally. **Leaves** cauline, alternate, subsessile, usually ± amplexicaul. **Inflorescences** thyrses or thyrsoid, rarely racemes, flowers solitary. **Flowers** bisexual, showy; perianth campanulate or trumpet-shaped; tepals 6, free, white or yellow with purplish-spotted, usually recurved or reflexed distally, usually caducous; outer ones saccate or shortly spurred; stamens 6, inserted at base of tepals, filaments slightly flattened, proximally connivent to form a short tube, anthers dorsifixed, versatile, extrorse; ovary triloculed, ovules many per locule, style columnar, stigmatic lobes 3, spreading, apically cleft. **Fruits** capsules, broadly cylindric, 3-angled, septicidal; seeds many, ovate to orbicular, flattened, small. **Distr:** Himalayas to E Asia. **Cult:** Propagate by division or from seeds. Requires a humus-rich, moist, well-drained soil in full or partial sun.

T. formosana Baker, *J. Linn. Soc., Bot.* 17: 465, 1879. [Toad Lily]. **Herbs**; stem sometimes branched, usually flexuous, 25–80 cm long, glabrous or slightly pubescent distally. **Leaves** oblanceolate or narrowly elliptic-oblanceolate to obovate, 8–13 × 2.5–4.5 cm, adaxially glabrescent, abaxially pubescent particularly along veins, base usually narrowed and subcuneate, margins ciliate, apex acuminate. **Inflorescences** cymes, terminal and also axillary in distal part of stem, laxly several flowered; pedicels 1–6 cm long, pubescent or glabrous. **Flowers**

Below: *Helonias umbellata*, *Lilium formosanum*, *Peliosanthes teta*, *Tricyrtis formosana*.

MIW

MIW

trumpet-shaped; tepals purple-white, with purple spots adaxially, lanceolate, oblanceolate, or oblong, 2–4 × 0.4–1.1 cm, abaxially usually laxly pubescent; outer ones basally saccate; stamens usually included; ovary glabrous, style subequaling stigmatic lobes. **Fruits** 2.5–3.5 × 5–6 mm. **Distr:** Taiwan. **Ecol:** In forests, thickets, shaded places, roadsides; 0–3,000 m. Fl. & Fr. Apr–Nov. **Zones:** 5–10.

DIOSCOREALES

FAMILIES 5, GENERA 21, SPECIES 1,037.

Plants almost exclusively terrestrial, herbaceous or woody perennials; rootstock rhizomatous or tuberous; aerial stem often long, twining or trailing, often annual. **Leaves** variably arranged, often opposite or verticillate; leaf base simple or sometimes sheathing, appendages often present, these often filamentous and sometimes the petiole producing 2 tendrils; lamina usually simple and entire, sometimes pinnatisect, pinnatifid or palmately compound. **Inflorescences** highly variable, mostly axillary, sometimes terminal. **Flowers** variable, usually 3-merous, bisexual or unisexual; tepals often inconspicuous, mostly pale yellow, cream or green, rarely white or purple; stamens arranged in 2 isomerous whorls, filaments adnate or free; gynoecium is syncarpous with free stylodia or a single style. **Fruits** usually capsules, rarely baccate and indehiscent. **Distr:** Predominantly tropical, also in subtropical and warm temperate regions.

BURMANNIACEAE

Genera 9, **species** 95. **Herbs**, annual or perennial, mycotrophic and white or purplish, or autotrophic and green; roots subterranean, mycorrhizal, filiform or stoutly terete; stems monopodial, unbranched or with few branches, erect, slender; rhizomes present or absent. **Stipules** absent. **Leaves** present or absent, cauline or basal, alternate, sessile; lamina simple, often scalelike, margins entire. **Inflorescences** terminal cymes, sometimes appearing racemose or capitate, or flowers solitary; each flower subtended by a scalelike floral bract. **Flowers** with tepals 6, partially or wholly connate, tube persistent or caducous in fruit, limb lobes 3 in 1 whorl or 6 in 2 whorls, greenish, purple, or white, sometimes blue-tinged, often ribbed or winged, throat with or without annulus; stamens 3 or 6, sessile or subsessile; pistil 1, ovary inferior, 1- or trilocular, placentation parietal or axile, ovules numerous, style 1, 3-branched apically, stigmas 3. **Fruits** capsules, ovoid, obconic, or cup-shaped, dehiscence transverse, longitudinal by 3 valves, or irregular; seeds numerous, minute. **Distr:** Mostly tropical and subtropical, worldwide.

Burmannia

Species 57. **Plants** annual, autotrophic, at least partly green; rhizomes absent; roots filiform; stems usually unbranched, green. **Leaves** cauline, sometimes basal; lamina green, purplish, or pale, scalelike to linear. **Inflorescences** 2–25-flowered cymes or flowers solitary; floral bracts sometimes appearing imbricate if internodes of cyme rachis are very short; pedicels 0–1 mm long. **Flowers** erect, 3-ribbed to 3-winged; perianth persistent in fruit; annulus absent; stamens 3, proximal to inner perianth lobes, sessile; ovary trilocular; placentation axile. **Fruits** capsules, obovoid to ellipsoid, transversely dehiscent. **Distr:** Mostly in tropical and subtropical regions. **Cult:** Propagate by division or from seeds. Requires a humus rich compost in shade.

B. cryptopetala Makino, *Bot. Mag. (Tokyo)* 27: 3, 1913. (Syn: *B. cryptopetala* var. *daxikangensis*). **Saprophytes**, without chlorophyll, holomycotrophic, small; stem white, slender, 5–9 cm tall, usually unbranched; cauline leaves scalelike, alternate, triangular- or ovate-lanceolate, 3–5 × 1.5–2.5 mm; basal rosette leaves absent. **Inflorescences** terminal, flowers solitary or 2–4-flowered dichotomous cymes; bracts ovate-lanceolate, ~3 mm; flowers sessile or shortly pedicellate. **Flowers** whitish with yellow tepals, 1–1.2 cm; perianth tube 3–8 mm; wings 3, conspicuous, 0.5–2.5 mm wide; outer tepals triangular, 3, yellow, 1.5–3 mm, margin single, apex acute; inner tepals absent; stamens 3, sessile or subsessile, inserted at throat of corolla, connectives with an apical crest and a pendent basal spur or spurless; ovary ovoid or subglobose, 3.5–5 mm; style 3–4 mm, apically 3-branched, stigma spheroid. **Fruits** capsules, obovoid or subglobose; seeds numerous, ellipsoid. **Distr:** China, Japan, Taiwan. **Ecol:** Fl. & Fr. Jun–Nov. **Ecol:** In wet areas of dense forests and warm temperate broad-leaved forests; at low altitudes. **Zones:** 8–10.

DIOSCOREACEAE

Genera 4, **species** 870. **Herbs** twining or woody **vines**, rarely erect, small herbs; rootstock rhizomatous or tuberous; stem twining to left or right, pubescent or glabrous, sometimes prickly. **Leaves** alternate or opposite, petiolate; lamina simple or palmately compound, basal veins 3–13, interstitial veins reticulate; leaflets of palmately compound leaves often ovate or lanceolate. **Inflorescences** with flowers solitary, clustered, or in cymules, these in spikes, racemes, or thyrses, these sometimes grouped into panicles. **Flowers** usually unisexual (when plants dioecious, rarely monoecious), sometimes bisexual; ♂ with perianth lobes 6, in 2 whorls, basally connate or free, stamens 6, sometimes 3 reduced to staminodes or absent, inserted on perianth or receptacle, ovary rudimentary or absent; ♀ similar to ♂, staminodes 3, 6, or absent, ovary inferior, triloculed, ovules usually 2(rarely more) per locule, placentation axile, styles 3, free. **Fruits** capsules, berries, or samaras; seeds with a membranous wing or not. **Distr:** Predominantly tropical.

Dioscorea

Syn: *Rajania, Tamus, Testudinaria.*

Species >600. **Herbs**, twining climbers; rootstock rhizomatous or tuberous, variable in color, shape; aerial bulbils axillary or absent. **Leaves** alternate or opposite, petiolate; lamina simple or palmately compound, basal veins 3–9. **Inflorescences** arranged spirally in axillary, usually elongate spikes or racemes, or in small cymules in ± spikelike thyrses, often several together, sometimes in terminal or axillary panicles by reduction of subtending leaves; ♀ spikes, 3.5–10 cm long, few-flowered. **Flowers** unisexual (plants dioecious, rarely monoecious); ♂ with stamens 6, 3 sometimes reduced to staminodes or absent; ♀ with staminodes 3, 6, or absent. **Fruits** capsules, 3-winged, dehiscent apically at maturity; seeds with a membranous wing. **Distr:** Tropical, subtropical and temperate regions worldwide. **Cult:** Propagate from tubers or bulblets. Requires fertile, moist but well-drained soil in full sun.

D. alata L., *Sp. Pl.* 2: 1033, 1753. (Syn: *D. alata* var. *purpurea*, *D. purpurea*). [Guyana Arrow Root, Water Yam, White Yam]. **Climbers**, tuberous; tubers variable; stem twining to right, glabrous, ridged, with 4 narrow, membranous wings, prickly at base; bulbils present, variable in shape. **Leaves** alternate basally on stem, opposite distally on stem, simple; lamina green or purplish red, ovate, 6–15(–20) × 4–13 cm, papery, glabrous, base sagittate to deeply cordate, apex shortly acuminate or caudate petiole green or purplish-red, 4–15 cm. **Inflorescences** ♂ spikes solitary or a few together, 1.5–4 cm, sometimes forming a panicle, rachis zigzagged; ♀ spikes, solitary or 2 or 3 together. **Flowers** ♂ with outer perianth lobes broadly ovate, 1.5–2 mm, stamens 6; ♀ with staminodes 6. **Fruits** not reflexed, oblate, sometimes obcordate, 1.5–2.5 cm, wings 1.2–2.2 cm wide; seeds winged all round. **Distr:** SE Asia. **Ecol:** Fl. Nov–Jan. Fr. Dec–Jan. **Zones:** 10–12. **Info:** Cultivated widely for the starchy, edible tubers.

D. bulbifera L., *Sp. Pl.* 2: 1033, 1753. (Syn: *Helmia bulbifera*). [Aerial Yam, Air Potato, Otaheite Potato, Otaheite Yam]. **Climbers**, tuberous; tubers usually solitary, renewed annually, ovoid or pear-shaped, 4–10 cm thick; cork black; roots fibrous; stem twining to left, glabrous, smooth; bulbils purplish-brown with orbicular spots, globose or ovoid, variable in size, weight to 1 kg. **Leaves** alternate, simple; lamina broadly cordate, 8–15(–26) × 2–14(–26) cm, margin entire or slightly undulate, apex caudate-acuminate, glabrous; petiole 2.5–5.5 cm long. **Inflorescences** ♂ spikes, usually clustered in leaf axils or along leafless, axillary shoots, drooping, sometimes branched, flowers solitary, ± contiguous along rachis, bract and bracteole ovate; ♀ spikes, often 2+, similar to ♂ ones,

Below: *Burmannia cryptopetala, Dioscorea alata, Dioscorea bulbifera, Dioscorea esculenta.*

MIW

MS

Clockwise from above: *Tacca chantrieri, Tacca integrifolia, Tacca leontopetaloides, Tacca palmata* (flowers).

20–30 cm long. **Flowers** ♂ perianth purple, lobes lanceolate, stamens 6, inserted at base of perianth, filaments nearly as long as anthers; ♀ with staminodes 6. **Fruits** reflexed or drooping, straw-colored, densely purplish-dotted, oblong-globose, 1.5–3 cm long, glabrous; seeds oblong, 1.2–1.6 × ~0.5 cm. **Distr:** Tropical and subtropical Old World. **Ecol:** In mixed forests, often in dry places; 0–2,300 m. Fl. Jul–Oct. Fr. Aug–Nov. **Zones:** 9–12. **Info:** The tubers and aerial bulbs are bitter and inedible, but used medicinally. Some cultivars edible.

D. esculenta (Loureiro) Burkill, *Gard. Bull. Straits Settlem.* 1: 396, 1917. [Fancy Yam, Potato Yam, Lesser Asiatic Yam]. **Climbers**, tuberous; tubers usually 4–10, produced from apical branches of rhizome; cork light yellow, smooth; thorny roots present or absent; stem twining to left, with soft hairs, proximally prickly, distally so only at nodes. **Leaves** alternate, simple; lamina broadly cordate, to 15 × 17 cm, hairs especially abaxially, basal veins 9–13, base cordate, apex acute; petiole 5–8 cm long. **Inflorescences** ♂ spikes, solitary, dense, ~15 cm long, flowers usually solitary, rarely in cymules of 2–4, sessile or subsessile, bract ovate; ♀ spikes solitary, pendent, to 40 cm long. **Flowers** ♂ perianth shallowly cupular, puberulent, outer lobes broadly lanceolate, ~1.8 mm, inner ones slightly shorter than outer, stamens 6, inserted in perianth tube, slightly shorter than perianth lobes. **Fruits** ~3 cm long, base truncate, apex slightly emarginate, wings ~1.2 cm wide; seeds winged all round. **Distr:** India, Malaysia, Papua New Guinea, and Thailand. **Zones:** 10–12. **Info:** The most widely cultivated species of the genus. The tubers and aerial bulbs are starchy and edible. The tubers without thorny roots belong to *D. esculenta* var. *esculenta*, and those with a crown of thorny roots belong to *D. esculenta* var. *spinosa*.

TACCACEAE

Genus 1, species 12. *Familial description same as generic.*

Tacca

Species 12. **Herbs**, perennial; rhizomes cylindric to subcylindric or tubers globose to broadly ellipsoid-globose. **Leaves** radical; petiole erect, base sheathing; lamina with margins entire to deeply lobed. **Flowering stem** often a long scape. **Inflorescences** umbels; involucral bracts (2–)4–6(–12), in 2 whorls, elongate or broad; bracteoles linear, narrower than bracts. **Flowers** bisexual, actinomorphic; perianth tube adnate to ovary, 6-lobed, lobes in 2 whorls, petaloid; stamens 6, inserted on perianth lobes, filaments short, apex hooded or spathulate, anthers biloculed, introrse; ovary inferior, ovules many, anatropous, style short, stigma trilobed, usually petaloid, curved over style. **Fruits** berries or rarely trivalved capsules; seeds many. **Distr:** Pantropical, with the highest diversity in the Malesian-Pacific region. **Cult:** Propagate from seeds, by division of tubers or cuttings of tubers. Requires moist but well-drained humus-rich compost in full sun or shade.

T. chantrieri André, *Rev. Hort.* 73: 541, 1901. (Syn: *Tacca macrantha*, etc.). [Bat Head Lily, Bat Plant, Black Bat Flower, Tiger Whisker Plant]. **Herbs**; rhizomes subcylindric, thick. **Leaves** with petiole 10–30 cm; lamina oblong to oblong-elliptic, 20–50(–60) × 7–14(–25) cm, base rounded-cuneate to cuneate, apex shortly caudate, glabrous or abaxially pubescent. **Inflorescence** with scape long; involucral bracts 4, dark purple, outer 2 ovate-lanceolate, inner 2 broadly ovate; umbels 5–7(–18)-flowered. **Flowers** with perianth lobes purplish brown, lanceolate; stigma trilobed. **Fruits** berries, purplish-brown, ellipsoid, ~3 cm, fleshy, 6-ridged, with persistent perianth lobes; seeds reniform. **Distr:** Assam to S China and peninsular Malaysia. **Ecol:** In forests, valleys, along rivers; 200–1,300 m. Fl. & Fr. Apr–Nov. **Zones:** 10–12.

T. integrifolia Ker Gawler, *Bot. Mag.* 36: t. 1488, 1812. (Syn: *Ataccia integrifolia, T. cristata, T. laevis*). [White Bat Flower]. **Herbs**; rhizomes subcylindric, thick. **Leaves** oblong-lanceolate to oblong-elliptic, 50–55 × 18.5–21 cm, base cuneate, apex acuminate, sometimes caudate. **Inflorescence** with scape ~55 cm long; involucral bracts 4, outer 2 sessile, narrowly deltoid-ovate, inner 2 long petiolate, spathulate, thin. **Flowers** with perianth purplish-black; tube 1–2 cm long; lobes 6, in 2 whorls, outer ones narrowly oblong, inner ones broadly obovate; filaments spathulate at apex; style extremely short, stigma deeply trilobed. **Fruits** berries, narrowly ellipsoid, 4–5 × ~2 cm, fleshy, 6-ridged, with persistent perianth lobes; seeds irregularly ellipsoid-ovoid. **Distr:** Bhutan to W Malesia. **Ecol:** In forests, mountain slopes; 800–900 m. Fl. Jul–Aug. **Zones:** 10–12.

T. leontopetaloides (L.) Kuntze, *Revis. Gen. Pl.* 2: 704, 1891. (Syn: *Leontice leontopetaloides*, etc.). [Bat Flower, Indian Arrowroot]. **Herbs**; tubers globose to broadly ellipsoid-globose; cork dark gray to brown; transverse section white. **Leaves** broadly obovate, ovate, or oblong-ovate, palmately trilobed; lobes pinnately lobed. **Inflorescence** with involucral bracts 4–12; umbel 20–40-flowered; bracteoles to 25 cm.

Flowers with perianth pale yellow, pale yellowish-green, or dark purplish-green; outer lobes elliptic to ovoid, inner ones broadly to narrowly ovate. **Fruits** berries, drooping, globose, ellipsoid-globose, or ovoid-globose; seeds many, flattened globose, testa spongy. **Distr:** Africa, S and SE Asia; N Australia, S Pacific Islands. **Ecol:** Usually in coastal regions, rarely in heavy shade and in primary forests; 0–200(–1,100) m. Fl. Jan–May. **Zones:** 11–12. **Info:** Cultivated for the edible starch from the tubers.

T. palmata Blume, *Enum. Pl. Javae* 1: 83, 1827. (Syn: *T. angustilobata*, etc.). [Palm-leaved Bat Flower]. **Herbs**; tuber, ovoid, small, 2–4 cm Ø. **Leaves** with petiole slender, 20–30 cm long; lamina palmate 4–7-lobed, 15–20 cm Ø, thin; lobes oblong, elliptic-oblong to lanceolate, 15 × 2.5–7.5 cm, margins entire, apex acute to acuminate, lower two on each side more united than others. **Inflorescence** with scape slender, to ~30 cm long; involucral bracts 4, 2 outer ones broadly ovate, acuminate, 2.5 × 1.9 cm, nearly sessile, 2 inner ones ovate, 3.8 cm long, petioled, constricted at base; floral bracts none; flowers umbellate, 10 or more; pedicels 6 mm long. **Flowers** purple and green, 8 mm across; perianth tube campanulate, 4 mm long, lobes subequal, orbicular, base narrowed, apex obtuse, inner ones spathulate; stamens with filaments short, anther elliptic; style thick, arms recurved. **Fruits** globose or subglobose, fleshy, red, ~1 cm long; pedicels ~6 mm long. **Distr:** Indo-China to Malesia (including Misool Islands). **Ecol:** Usually in secondary vegetation and forest margins, sometimes in teak forests and bamboo groves; to ~1,000 m. Fl. & Fr. Nov–Jul. **Zones:** 9–12.

PANDANALES

FAMILIES 5, GENERA 36, SPECIES 1,345.

Herbs, shrubs, trees or climbers, terrestrial or sometimes epiphytic. **Flowers** unisexual and apetalous; pollen mostly spheroidal to ovoid or ellipsoidal, rarely reniform. **Distr:** Pantropical.

VELLOZIACEAE

Genera 9, **species** 240. **Herbs** perennial and sparsely branched shrubs; stems covered with persistent imbricate sheaths of fallen leaves; raphides sometimes present. **Leaves** in a basal rosette or apical tufts; lamina simple, linear-acuminate, to 1 m long, midrib present, margins saw-toothed. **Inflorescences** with flowers solitary to 12–15 per scape; flowers pedicellate, terminal. **Flowers** bisexual, rarely functionally unisexual and the plants dioecious, actinomorphic, often large; perianth segments 6, in 2 whorls, free or basally connate, sometimes with a small corona behind the anthers; stamens 6–48(–76), in 6 bundles when >6, lacerated staminal appendages often present, filaments epigynous or adnate to the perianth, sometimes flattened, anthers long; ovary inferior (rarely semiinferior), stigma large, erect or spreading. **Fruits** capsules, loculicidal, or dehiscent by apical, basal, or longitudinal slits; seeds numerous, small. **Distr:** S America; Africa, Madagascar to Arabia, China.

Vellozia

Species 143. **Shrubs**, xerophytic, to 6 m high, pachycaulous, slow-growing; stem branched, terete or trigonous, including the old leaf bases; adventitious roots produced near base of stem. **Leaves** persistent or caducous along a regular transverse line, sheaths persistent; lamina reflexed, not plicate. **Flowers** perfect, funnel-shaped; perianth tube equaling the ovary to several times longer; tepals subequal; stamens 6+, naked or basally appendaged, filaments terete, anthers basifixed, latrorse; style elongate, stigmas horizontal or somewhat reflexed, suborbicular, confluent at apex. **Distr:** Panama to Brazil. **Cult:** Propagate from seeds. Requires a well-drained soil mix.

V. bahiana L.B. Sm. and Ayensu, *Smithsonian Contr. Bot.* 30: 99, 1976. [Tree Lily]. **Shrubs**, 0.3–1.5 m tall; stem apically 1.2 cm Ø (including leaf sheaths). **Leaves** few-ranked; sword-shaped or linear, 20–28 × 1 cm, margins and midnerve laxly serrulate, attenuate to a narrowly obtuse apex, densely white-puberulent on all exposed surfaces, non-glutinous; sheaths subdensely imbricate with 1.5 cm of the apex exposed, 4 cm long. **Inflorescences** terminal, with flowers whorled near ends of branches, to 12 cm long. **Flowers** lavender; corolla trumpet-shaped; tepals elliptic, 5–6 cm long; stamens ~18, >3 cm long; phalanges irregular, unappendaged; ovary ovoid to subglobose, 8 mm long covered with tubercles (soft projections), style 4 cm long, stigmas terminal, orbicular, spreading. **Distr:** Brazil (Bahia). **Ecol:** In open montane forests; ~1,000 m. Fl. Apr–May. **Zones:** 10–11.

PANDANACEAE

Genera 4, **species** 885. **Trees**, **shrubs** or **woody lianas**, dioecious, terrestrial, evergreen; stems simple or often bifurcately branched, ringed with persistent annular leaf scars, often producing adventitious prop (stilt) roots. **Leaves** numerous, spirally arranged sessile; lamina simple, linear to lanceolate, keeled (carinate) beneath, glabrous, leathery, often lustrous or glaucous, base open-sheathed, amplexicaul, margin and midrib adaxially often serrate or spinulose, apex often long-acuminate. **Inflorescences** ♂ axillary and terminal, compound, bracteate, comprising of several-branched racemes or panicles with crowded flowers on spicate ultimate branches, bracts (spathes) often enlarged at apex, white or colored, flowers sessile (rarely pedicellate); ♀ terminal, solitary or in spikes, racemes or capitula, short, bracteate, with crowded flowers, often pendulous in fruit. **Flowers** ♂ with perianth lacking, pistillode sometimes present, stamens numerous, fasciculate, arising on rachises or spadix-branches, filaments smooth or papillose, seemingly branched, anthers basifixed, dehiscing by longitudinal slits; ♀ with or without staminodes, pistils free or aggregated and fused to adjacent pistils forming 1–many carpeled phalanges or clusters, ovary superior, 1–several-loculed, ovules solitary to numerous, anatropous, placentation basal or parietal, style absent or very short, stigmas 1 or more, subsessile, glandular-papillose. **Fruits** berries or drupes, when drupaceous syncarps, composed of 2–20 woody, or basally fleshy or fibrous, cylindrical to globose, connate, free or often crowded and clustered carpels, each cluster of carpels termed a 'phalange', with mesocarp fibrous and often pithy or hollow above, fibrous and fleshy below; seeds 1–numerous, minute, often fusiform, testa membranous. **Distr:** W Africa to the Pacific.

Freycinetia

Syn: *Jezabel*, *Victoriperrea*.

Species 304. **Shrubs**, scrambling or climbing, rarely herbs; aerial roots present. **Leaves** ± densely crowded; lamina linear or lanceolate, basally with membranous, marcescent or caducous sheath-margins (auricles). **Inflorescences** terminal or lateral, consisting of peduncled simple spadices (usually 2–5), arranged in an umbel or short raceme and initially enclosed by several, crowded, triseriate, green or colored, caducous spathes. **Flowers** unisexual, rarely bisexual; perianth absent; stamens densely congested on rachis; ovaries densely congested on rachis, usually surrounded by minute staminodes, 1-celled, ovules numerous, congested on 2 or more parietal placentas, stigmas 2 or more, separate or confluent. **Fruits** berries (baccate), with or without a hardened apex; seeds numerous. **Distr:** Sri Lanka to Ryukyu Islands and Pacific. **Cult:** Propagate by division or from seeds. Requires a humus-rich, moist compost in partial sun. Need support for climbing.

F. arborea Gaudich., *Voy. Uranie* 431, 1829. (Syn: *F. arnottii*, *F. longispica*). ['Ie 'ie]. **Climbers**, sometimes sprawling, woody. **Leaves** linear-lanceolate, 40–80 × 1–3 cm, apex gradually attenuate, coriaceous, midrib near apex minutely serrulate with prickles; auricles usually pale green, narrow, tapered, to 11 cm long. **Inflorescences** terminal; outer bracts green with base salmon-orange; intermediate bracts dark salmon-pink, tinged orange, fleshy, ovate-navicular, apical margins and dorsal costa minutely spinulose; inner bracts pink, innermost ones thick, almost linear; ♂ spadices in clusters of 3(4), yellowish-white, peduncles to 5 cm long, spikes ~10 cm long, with flowers densely crowded; ♀ spadices usually in clusters of 3(4), smooth, trigonous, peduncles 3–4 cm long. **Flowers** ♂ with filaments pale orange or buff, to 1.5 cm long, pistillodia present, minute; ♀ with stigmas (4)5–10(–12)+, usually in 2 rows, black, staminodia sometimes present. **Infructescences** with fruiting spadices oblong-cylindrical, 7.5–9.5(–15) cm long, 1–3.2 cm Ø. **Fruits** berries, red, oblong, compressed, to 1 cm long; seeds ellipsoid, to 1.5 mm long. **Distr:** Polynesia (Austral, Cook, Hawai'i, Marquesas, Society islands). **Ecol:** Mostly in humid forests; 300–1,500 m. **Zones:** 11–12.

F. banksii A. Cunn., *Companion Bot. Mag.* 2: 377, 1836. (Syn: *F. baueriana* ssp. *banksii*, *F. inclinans*). [Grass Tree, Kiekie]. **Climbers**, woody, perennial, climbing with aerial roots. **Leaves** ± tufted, spirally arranged; lamina to ~1.5 m × 2–2.5 cm, base sheathing, apex long-attenuate, margins and midrib abaxially, minutely and distantly spinulose. **Inflorescences** with spadices simple, in axils of several ± foliaceous bracts; bracts white to purplish; peduncle 1–4 cm long, glabrous, whitish; spadix at flowering ~ 7 × 1.5 cm, cylindrical or slightly flattened. **Flowers** ♂ with stamens several, filament long, anther ovate, ovary rudimentary; ♀ with 6–12 purplish staminodes at base of flattened, vertically elongated ovary ~ 2–4 × 1 mm, 2 mm high, long sides grooved between staminodes, stigmas ~6–12, sessile, arranged around a long groove. **Infructescences** to ~15 × 3 cm; peduncle stiff, woody. **Fruits** phalanges, ~1 cm long, laterally compressed, proximally thin-walled, woody toward apex; seed ~1 mm long, narrow, funicle slender. **Distr:** New Zealand. **Ecol:** In rainforests, coastal scrubs, on trunks of trees or in large masses on the ground. Fl. Dec–Feb. Fr. Sep–Nov. **Zones:** 9–11.

F. formosana Hemsley, *Bull. Misc. Inform. Kew* 1896: 166, 1896. [Taiwanese Climbing Pandanus]. **Lianas**, scandent or scrambling; stems procumbent or prostrate, to 10+ m; branches with aerial roots, glabrous, 0.5–5 cm wide. **Leaves** elongate, linear-lanceolate, 0.1–0.6(–2) m × 0.8–4 cm, base narrowed, margins sharply antrorsely serrate toward base and apex, sometimes entire in upper part, with scattered spines along midvein abaxially, apex acuminate, leathery. **Inflorescences** with spadices 2–4, racemose, on stout peduncle ~5 cm long; spathes yellow; ♀ peduncles 10–15 × 1.4–1.8 cm; spathes (bracts) to 6 cm long. **Infructescences** composed of numerous dense drupes, globose, cylindrical or ovoid, 2.5–11 × 1.5–2 cm. **Fruits** drupes, ovoid or irregularly angled with asymmetrical edges, ~ 0.5–1 × 0.4–1 cm; seed ± oblong. **Distr:** Ryukyu Islands to Philippines (Batan Island). **Ecol:** In thickets and cliffs near seashores. **Zones:** 9–12.

F. walkeri Solms, *Linnaea* 42: 92, 1878. (Syn: *F. radicans*). [Ceylon Climbing Pandanus]. **Climbers**, woody stemmed, adhering by clasping adventitious

Top row: *Vellozia bahiana, Freycinetia banksii* (fruits), *Freycinetia walkeri*.
Bottom row: *Freycinetia arborea, Freycinetia formosana.*

roots. **Leaves** linear-attenuate, mostly 40–75 cm long, rarely more, mostly 1.5–1.8 cm wide, base sheathing, darkening, with membranous auricles, these to 5 × 0.6 cm, truncate or ascending lobelike at apex, distally spinulose-margined, purplish. **Inflorescences** large, terminal; ♂ bracteate, inner bracts deltoid acute, bright red, to 10 × 4 cm, somewhat fleshy, spadices pseudo-umbellate, usually 3, peduncles slender, glabrous ~ 4.5 cm × 2–2.5 mm, spadix oblong to 3 × 0.5 cm; ♀ pseudo-umbellate, bracts as in ♂, spadices usually 3, sometimes 2 or 4, peduncles minutely appressed-hispidulous toward apex, to 5 × 0.45 cm, spadix cylindric, 4.5–6 × 1.62 cm, ripening red. **Flowers** without perianth; ♂ with stamens numerous, crowded, filaments 1.25–1.45 mm long, anthers short oblong-rounded, 0.5 mm long; ♀ with stigmas usually 2, rarely 3, confluent, round, 0.2 mm Ø. **Infructescences** steeply pyramidal, 5–6-angled, apex truncate. **Fruits** berries, ~8 mm long; seeds ± ellipsoid, 1 mm long. **Distr:** Sri Lanka. **Ecol:** 900 m. **Zones:** 11–12.

Pandanus

Species ~600. **Shrubs** or **small trees** of terrestrial, semiaquatic or epiphytic habitat, dioecious, evergreen; stems simple or branched (sometimes very short), erect, procumbent or prostrate, often with stiltlike, verrucose prop roots and aerial roots. **Leaves** terminal, sessile, densely arranged in corkscrew spirals, 3- or 4-seriate, simple, sessile; lamina linear or lanceolate, margin and midrib often spinose-serrate, sheathing at base. **Inflorescences** large, bracteate heads or spadices,

Top row: *Pandanus amaryllifolius, Pandanus ceylanicus.*
Bottom row: *Pandanus furcatus* (infructescence).

axillary or terminal; ♂ paniculate with spiciform branches subtended by secondary, usually colored, bracts (spathes), branches covered with numerous stamens, flowers not individually distinguishable; ♀ globose to cylindrical clusters or several-carpellate aggregates, flowers not individually distinguishable. **Flowers** always unisexual, perianth absent; ♂ with filaments smooth, anther oblong-linear, apex apiculate; ♀ with staminode absent, free or in phalanges, stigma sessile or style partly developed, stigma variously shaped (round, deltoid, linear, etc.), forked, dentate or spiny, carpels 1-ovuled, placentation subbasal. **Fruits** syncarps (comprising of an aggregation of individual connate, angled, fibrous phalanges or 'drupes'); drupes fibrous, woody, with persistent stigma, 1-seeded; seed large, albuminous. **Distr:** Exclusively in the Old World tropics. **Ecol:** From coastal seashores to over 500 m. **Cult:** Propagate from seeds, offsets, rooted suckers and detached limbs. Requires any moist, well-drained soil in full sun. Many species are salt-tolerant.

P. amaryllifolius Roxburgh ex Roxburgh, *Fl. Ind.* 3: 743, 1832. (Syn: *P. latifolius, P. odorus*). [Fragrant Pandan, Fragrant Screw Pine, Screw Pine, Screw Tree, Umbrella Tree]. **Shrubs** becoming a small tree when mature, evergreen, to 8 m tall; stems branched; aerial roots present; sucker shoots of small growth phase with stems slender, to 1–1.6 m × ~2–5 cm, decumbent and ascending. **Leaves** aromatic, somewhat glaucous beneath, keeled beneath but unarmed, margins entire except at apex, with very few minute prickles <1 mm long, apex with distinct twin lateral pleats, blades 25–75 × 2–5 cm, in mature phase lamina to 1.5–2.2 m × 7–9 cm. **Inflorescences** very large; ♂ to 1 m long, semipendent, yellowish-pink, spathes ~90 cm, spikes oblong, to 35 cm or more, upper ones much shorter, ~ 9–10 × 2 cm, of numerous crowded, flat staminal phalanges 1.5–2.5 mm wide; ~ very rare. **Flowers** very fragrant; ♂ with stamens mostly 3–6 per phalange, filament 0.5–1.5 × 0.4–0.6 mm, anther oblong, ~ 2.5 × 0.5 mm, apex bluntly convex, without or with a barely discernible apiculum. **Fruits** syncarps, large, rarely produced in cultivation **Distr:** Indonesia, Malaysia, Philippines, New Guinea (West Irian). **Ecol:** In tropical rainforests, naturalized widely in various habitats; 0–500 m. Fl. July–Aug (♂ flowers).**Zones:** 11–12. **Info:** Inflorescences are rarely produced in cultivation. Male inflorescences last for 2–3 months, filling the tropical evening air with fragrance. Fruits rarely produced are edible. Cultivated widely for the aromatic leaves used in cooking.

P. ceylanicus Solms, *Linnaea* 42: 16, 1878. (Syn: *P. furcatus, P. porrittianus, P. zeylanicus*). [w]. **Shrubs**, erect or ascending or partly prostrate, sparsely branched; stems pale grayish-tan, annulate by leaf scars about 1–1.5 cm apart, main trunk 2–3 m long; prop roots slender, to 40 × 1.5 cm, smooth or with a few whorls of blunt protuberances. **Leaves** linear-ensiform, 0.6–1.2 m × 2–3.5 cm, margins sharply prickly along whole length, at base prickles 3–5 mm long, apex gradually attenuate, coriaceous, adaxially dark green, abaxially paler and duller. **Inflorescences** terminal; ♂ racemose-spicate, at first suberect, later drooping and pendulous, white, 24–28 × 0.9 cm, trigonous, bearing about 9–11 spathaceous bracts, outer ones with green leaflike apices, inner ones pale yellow, narrowly ovate, boat-shaped, ~ 18 × 3 cm, spikes 5–7, ~7 × 2 cm, composed of numerous ♂ phalanges; peduncles 20–30 cm long. **Flowers** ♂ pale yellow, filaments to 1 mm long, anthers almost white, 2.5 mm long; ♀ bracteate, solitary ovoid-ellipsoid head, 3–4 cm long, 3 cm Ø, surrounded by pale yellow spathes, carpels pale green, stigmas white, carpels at anthesis 6–9 mm long, spathes similar to those of ♂, but longer or narrower. **Fruits** syncarps, subglobose, to 15 cm Ø, dark green, composed of approximately 500–1000 drupes, nodding; drupes 1-celled, oblanceoloid, 2.3–3 cm long, to 8–9 mm

Above: *Pandanus tectorius, Pandanus thwaitesii, Carludovica palmata.*

wide, 5–6-angled; seed ellipsoid, ~7 × 3 mm. **Distr:** Sri Lanka. **Ecol:** On rocky banks of rivers; ~500 m. Fl. May. **Zones:** 11–12.

P. furcatus Roxb., *Fl. Ind.* ed. 1832, 3: 744, 1832. (Syn: *Barrotia diodon*, etc.). [Ceylon Screw Pine, Himalayan Screw Pine]. **Trees**, small, to 4 m tall or more; not or scarcely branched. **Leaves** comparatively large; lamina linear, apex gradually attenuate; coriaceous, 2+ m × 8.5 cm, margins and midrib (beneath) armed with sharp prickles, ventral pleats smooth unarmed, margins toward base with small antrorse prickles, about 1.5–1.7 mm long, 5–8 mm apart, toward apex, prickles more crowded (1–3 mm apart) but not much smaller except at extreme apex where they are 0.5 mm long, midrib prickles similar to those of adjacent margin, but much more remote, to 3–3.5 cm apart, except along apex, where their spacing is similar, reticulations not or little evident, longitudinal nerves conspicuous, ~128, prominent on both surfaces, especially abaxially. **Inflorescences** not observed. **Fruits** syncarps, large, oblong, to 25+ cm long, 8 cm Ø; drupes numerous, closely packed, 4.5 cm long, ~1 cm wide at apex, persistent stigma, hooklike, abruptly bent. **Distr:** Nepal to W Indo-China. **Ecol:** In moist places, near waterways, usually in the shade of larger trees. **Zones:** 9–12.

P. tectorius Parkinson ex Du Roi, *Naturforscher* (Halle) 4: 250, 1774. (Syn: *Corypha laevis*, etc.). [Beach Screw Pine, Hala Screw Pine, Screw Pine, Wild Screw Pine]. **Trees** or **shrubs**, 3–10 m tall; stem erect or ascending, many branched, non-suckering. **Leaves** linear-ensiform, green, often glaucous beneath, spinose-serrate on margin and midvein beneath, to 1.8 m × 10 cm, abruptly long-acuminate at apex. **Inflorescences** ♂ to 60 cm, pedunculate, paniculately compound with ultimate spiciform branches; bracts 13–18, narrowly lanceolate, white, 10–60 × 1.5–4 cm, serrate on margin and midvein; spikes pendulous, 0.8–2 cm; ♀ capitate, solitary, globose to ovoid-ellipsoid, 10–25 cm, with numerous white spathes 15–30 × 1.4–2.5 cm, margin serrate. **Flowers** ♂ stamens 10(–25), racemosely fasciculate, filaments connate below, 1–3 mm, anthers linear, 3 × 0.6 mm, connective mucronate; ♀ ovule 1 per locule. **Fruits** syncarps, pendulous, orbicular or oblongoid, to 17 × 15 cm; segments (drupes) 40–80 per aggregate-head, each phalange comprising ~5–12 fused obconical carpels (locules) connate below and free above, fragrant, yellow to orange to yellowish-brown from base to apex, 4–10 × 1–6.5 cm; carpel-apices distinct, persistent stigma subsessile. **Distr:** Asia, China, Hong Kong, India, Macao, Taiwan; tropical Australia, Polynesia. **Ecol:** On seashores, sandy beaches. Fl. Jan–Jun. Fr. Oct. **Zones:** 9–12. **Info:** The male flowers are extremely fragrant. The fruits and seed kernels are edible.

P. thwaitesii Martelli, *Webbia* 1: 369, 1905. (Syn: *P. fetidus* var. *racemosus*, *P. humilis*). **Shrubs**; stems erect or decumbent-ascending, dichotomous but sparsely branched, to 2–4 m long, to 7–8 cm Ø, pale gray to whitish, smooth; leaf scars distant; prop-roots few, basal, rather short. **Leaves** well-spaced; lamina linear, gradually attenuate, to 2.2–2.8 m × 3.5–4.2 cm, adaxially medium dark, glossy green, abaxially dull, grayish. **Inflorescences** terminal, spicate; ♂ with rachis 35–40 cm long, bracts yellow, glossy, distal 9 decrescent, lower ones to 40 × 5 cm, apex abruptly acuminate-caudate, apical margins spinulose-denticulate, midrib keeled; most distal bracts acute, all ± navicular; spikes ~9, sessile, axillary, to 12–13 cm long, 2–2.5 cm thick, rachilla fleshy, densely set with separate crowded stamens; ♀ with rachis zigzag, >20 cm long, bracts as in ♂, heads 5 to 7, well-spaced, in flower ~2–2.5 × 1.2–1.6 cm. **Flowers** ♂ with filaments obsolete, anthers linear, 1–1.1 cm × 0.5 mm, apiculate; ♀ with numerous simple crowded carpels, styles produced spiniform with dorsal linear groovelike stigmas. **Fruits** syncarps, narrowly pyramidal, on pendulous peduncles, sessile, globose-ovoid, 9–12 cm long, pale glaucous yellowish-green, fetid-smelling; drupes 5–6-angled, basally connate, 3.5–4 cm long. **Distr:** S India, Sri Lanka. **Ecol:** In semidry forests; 500–600 m. **Zones:** 9–12.

CYCLANTHACEAE

Genera 12, **species** 225. **Herbs**, large and coarse, acaulescent, perennial, palmlike in appearance, or scandent shrubs with annulate vaginate rooting stems. **Leaves** distichous or spirally arranged, petiolate and flabellate; margins entire, bifid, or 2-parted, parallel-nerved. **Inflorescences** with flowers monoecious, those of each sex densely crowded in whorls or spirals on a thick spadix; spadices axillary, solitary, pedunculate, simple; peduncle short or elongate, subtended at the base by a bicarinate cataphyll; spathes 2–6, inserted on the peduncle and enclosing the inflorescence before anthesis, caducous. **Flowers** ♂ without a perianth, or the perianth stipitate, oblique, with a multidentate orifice, stamens numerous, filaments confluent with the perianth tube and with one another, free above, anthers adnate to filament, ovoid or linear, dehiscent by longitudinal slits; ♀ with perianth lacking or of 4 short, distinct or connate segments, these often confluent with those of adjacent flowers and forming a persistent multidentate annulus, often accrescent after anthesis and in fruit indurate, staminodia usually 4, short or long and flexuous, ovary free or immersed below in the axis of spadix and adnate to it, depressed-quadrate, truncate or 4-lobate at the apex, 1-loculed, style absent or short and pyramidal, stigmas solitary, or 4, depressed, and stellately spreading, ovules numerous, placentation parietal or pendulous. **Fruits** fleshy syncarps, composed of distinct or confluent berries crowned by the persistent depressed stigmas; seeds numerous, small, sessile or pendulous, on a short funicle, testa fleshy. **Distr:** C and tropical S America.

Carludovica

Syn: *Salmia*.

Species 4. **Herbs** or **shrubs**, acaulescent, terrestrial or epiphytic, often with greatly elongate and scandent stems rooting at the nodes. **Leaves** alternate or fasciculate, rarely distichous; lamina bifid or palmately parted, very rarely entire, segments multicostate, entire or dentate; petiole short or elongate, vaginate-sheathing at base. **Inflorescences** with spadix oblong or cylindric, flowers arranged in dense spirals, 4 ♂ ones surrounding a single ♀ flower; ♀ flowers free or immersed in the rachis and adnate to it; spathes 2 or more, caducous, concave. **Flowers** ♂ with perianth stipitate, oblique, fleshy, multidentate, stamens numerous, anthers didymous, oblong, exserted, dorsifixed; ♀ with perianth 4-angulate, 4-fid, or 4-parted, segments or lobes obtusely quadrate, persistent and indurate, staminodia 4, very long, filiform, deciduous, stigmas 4, sessile, stellately spreading, ovules very numerous. **Distr:** Tropical S America. **Cult:** Propagate from seeds. Requires a fertile, well-drained soil in a moist and humid environment with filtered sun.

C. palmata Ruiz and Pavon, *Syst. Fl. Peruv.* 291, 1798. (Syn: *C. gigantea*, etc.). [Panama-Hat Palm]. **Plants**, terrestrial, palmlike, acaulescent, densely cespitose and stoloniferous, to 5 m tall, often forming large clumps or colonies. **Leaves** fan-shaped, pleated, 3–4-parted, segments broadly cuneate, ~1 m wide, apex broad, deeply laciniate, many nerved; petiole terete, 1–2 m tall, slender, stiff, green. **Inflorescences** much shorter than the petioles, pedunculate; spathes 3, concave, 12 cm long, acute; spadix oblong, 10–30 cm long, acute, many flowered, red at maturity; peduncle short, thick. **Flowers** conspicuous due to numerous cream-colored sterile stamens (staminodes); staminodia very conspicuous at anthesis, as much as 15 cm long, hairlike, white; stigma ± ovate to orbicular, 1–1.5 mm long. **Distr:** SE Mexico to Bolivia. **Ecol:** In moist or usually wet, mixed, lowland or mountain forests, sometimes in open places, and secondary growth; to 800 m. **Zones:** 10–12.

7 COMMELINIDS [COMMELINOID MONOCOTS]

ORDERS 4, FAMILIES 31, GENERA 1,175, SPECIES 23,609.

Leaves with silica bodies. **Inflorescences** bracteate. **Flowers** with perianth fully bicyclic; stamens adnate to perianth, pollen starchy (Stevens 2001 onwards). **Distr:** Worldwide. **Info:** The monophyly of the commelinids is supported by molecular data (Chase et al., 2000, 2006; Graham et al., 2006).

ARECALES

FAMILY 1, GENERA 189, SPECIES 2,361.
Ordinal description same as familial.

ARECACEAE

Syn: Palmae.
Subfamilies 5, **Genera** ~1,914, **species** ~2,500. **Trees** or **shrubs**, rarely lianas, perennial, branched or unbranched, solitary or clustered; roots adventitious; stems woody, subterranean or terrestrial, creeping or erect (rarely climbing or growing in water), slender or massive, sometimes conspicuously enlarged and storing starch and water, smooth or covered with fibrous or prickly remains of leaf bases. **Leaves** spirally arranged; sheaths tubular, often forming a crownshaft, sometimes with ligular appendages; petiole terete, channeled, or ridged, unarmed or bearing prickles or marginal teeth; hastula (flap of tissue from petiole apex at junction with surface of lamina) absent or present adaxially, rarely present abaxially; lamina palmate, costapalmate (intermediate between palmate and pinnate), pinnate, or bipinnate (rarely undivided); plication (folding lengthwise into pleats or furrows) ʌ- or tent-shaped (reduplicate, splitting along abaxial ridges) or v-shaped (induplicate, splitting along adaxial ridges); segments lanceolate, linear, or cuneate (rhombic), glabrous or variously scaly, unarmed or bearing prickles (proximal segments modified into spines in *Phoenix*). **Inflorescences** solitary (rarely clustered) in bud, borne within, below, or above crown of leaves, paniculate, rarely spicate, usually branched to 1–5 orders; prophyll (1st bract on main inflorescence axis) 2-keeled; peduncular bract(s) (empty bract(s) between 1st prophyll and 1st bract subtending branch) present (rarely absent); flowers bisexual, unisexual with ♂ and ♀ on the same plants or on different plants, or both bisexual and unisexual on the same plant. **Flowers** solitary or variously clustered along rachillae of inflorescence, radially symmetric; perianth 1–2-seriate; sepals (2–)3(–4), distinct or connate; petals (2–)3(–4), distinct or variously connate; androecium with stamens (3–)6–34(–1,000), filaments distinct or connate or basally adnate to petals, anthers basifixed or dorsifixed, dehiscing latrorsely or introrsely; staminodes in ♀ flowers distinct or variously connate or adnate to pistil or petals; pistils 1 or 3, distinct or partially connate, each bearing 1 ovule and 1 stigma, or 1 pistil bearing 1–3 ovules and 3 stigmas, styles distinct or connate, short, stigmas dry; pistillode in ♂ flower present or absent. **Fruits** drupaceous or berrylike, stigmatic remains basal or apical, exocarp smooth, warty, prickly, or hirsute (corky or scaly), mesocarp fleshy or dry and fibrous, endocarp papery, leathery, or bony, sometimes with 3 germination pores (e.g. *Cocos nucifera*); seeds 1(–2+), free or adhering to endocarp, seed coat thin. **Distr:** Worldwide, especially warmer parts of C America, S America, SE Asia. **Ecol:** Mostly terrestrial, rarely associated with water (e.g. *Nypa fruticans*). **Info:** Monophyly of Arecaceae is strongly supported by molecular data (Asmussen et al. 2006; Hahn 2002). Previous molecular studies have divided the family into 7 subfamilies (Baker et al. 1999), while the current classification divides the family into 5 subfamilies (Asmussen et al. 2006; Hahn 2002; Lewis and Doyle 2001, 2002). The classification followed here is according to Dransfield et al. (2005).

Subfam. CALAMOIDEAE

Syn: Calamaceae, Lepidocaryaceae.
Genera 21, **species** 626. **Palms**, mostly lianas, rarely erect; breeding system various. **Internodes** usually well-developed. **Inflorescences** with axes sometimes adnate to the internode above; flowers in diads. **Flowers** with corolla valvate; gynoecium covered by reflexed scales, ovule basal, epitropous, funicle sometimes twisted, style branched. **Fruits** with endocarp thin (rarely thick); seeds 1–3, sarcotesta usually thick (Stevens 2001 onwards). **Distr:** Tropical regions, especially Sri Lanka to West Samoa and Fiji.

TRIBE LEPIDOCARYEAE

Eremospatha

Species 11. **Palms**, clustering, high-climbing, spiny, pleonanthic, hermaphroditic; stems suckering sympodially. **Leaves** varying from flabellate juvenile form to regularly or irregularly pinnate adult form, in adult and some juvenile stems terminating in a cirrus armed with reflexed thorns and more massive paired reflexed acanthophylls; leaf sheath unarmed, usually glabrous; ocrea well-developed, armed; leaflets few–many. **Inflorescence** axillary, shorter than the leaves. emerging from the ocrea, branching to 1 order; peduncle short; rachillae arranged subdistichously. **Flowers** hermaphrodite; calyx tubular with 3 low triangular lobes; corolla ± twice the length of the calyx, with 3 short triangular lobes; stamens 6, epipetalous, borne near the mouth of the corolla-tube, united basally, filaments free, very short, each bearing a small pair of anthers; ovary covered with reflexed scales and tipped with 3 stigmas, united for most of their length. **Fruits** somewhat oblong, covered in reflexed scales, and tipped by the stigmatic remains, calyx and corolla persistent; seeds 1–3, variously shaped, often somewhat lobed and grooved; testa thin; endosperm homogeneous; embryo lateral. **Distr:** W tropical Africa to Zambia. **Cult:** Propagate from seeds. Requires moist, fertile soil in full sun.

E. haullevilleana De Wild., *Ann. Mus. Congo Belge*, Bot., V, 1: 96, 1903. [Muka, Mbamba]. **Palms**, clustering, climbing, to 25+ m; stem to ~1 cm Ø. **Leaves** to 1 m long; leaflets 9–14 per side, irregularly arranged; proximal leaflets ~6–9 × 0.4–0.7 cm, densely black spiny-margined; leaflets at the middle spathulate or ovate, 11–17 × 1.2–3.5 cm, base cuneate, tip moderately to strongly praemorse; petiole absent or very short; juvenile leaf frequently broadly flabellate, to 30 × 20 cm, deeply notched, many-ribbed. Inflorescence to 35 cm long; peduncle to 10 cm long; rachillae distichous, ± horizontal, subopposite, lowermost ~6 cm long; flowers borne in rather remote subdistichous pairs. **Flowers** with calyx ~3 × 3 mm, with 3 faint lobes; corolla ~7 × 3 mm; stamens 6, borne ~3 mm below apex of corolla; ovary ~5 mm long. **Fruits** ovoid to almost cylindrical, tipped with stigmatic remains, 2.5–3 × 1.5–2 cm, with ~20 vertical rows of dull brown reflexed scales; seed 2–2.5 × 1.1–1.5 cm. **Distr:** WC and E tropical Africa. **Ecol:** In riverine and swamp forests; to ~1,300 m. **Zones:** 10–12.

Raphia

Species 20. **Palms**, monocarpic, monoecious, large, unarmed or with only the sheaths armed; stems erect, simple or dichotomously branched, densely annulate. **Leaves** in a terminal crown, equally pinnatisect; leaflets linear-lanceolate, acuminate, margins recurved at the base and setose or aculeolate, rachis not produced at apex. **Inflorescences** with spadices monoecious, large, pendulous, cylindrical, much-branched; branches and branchlets thick, compressed, the latter pectinately arranged, densely covered with cup-shaped bracts; common spathe none; flowers solitary in each bract, ♂ at base of ultimate branches of spadix, ♀ at apex. **Flowers** ♂ with calyx tubular, entire or minutely toothed, corolla curved, petals 3, linear-lanceolate, valvate, stamens 6–16, inserted at base of petals, filaments subulate or thickened and basally connate, anthers erect, linear, inserted near their bases, rudiment of ovary absent; ♀ larger than ♂, calyx as in ♂, corolla ± campanulate, 3-toothed, staminodes in a 6- to many-toothed cup adnate to the corolla, ovary triloculed, retrorsely scaly, stigmas sessile, recurved, subulate, ovule inserted slightly above base of locules. **Fruits** large, oblong or ellipsoid, rostrate, 1-seeded,

PL

Above: *Eremospatha haullevilleana, Raphia farinifera.*

covered with retrorsely imbricate scales, pericarp thick, endocarp spongy; seed laterally fixed, oblong, sulcate. **Distr:** Tropical and S Africa, W Indian Ocean, C America to Brazil. **Cult:** Propagate from seeds. Requires any fertile, moist soil with ample organic matter in partial shade, and tolerates boggy situations.

R. farinifera (Gaertn.) Hyl., *Lustgården* 31–32: 88, 1952. (Syn: *Sagus farinifera*, etc.). [Raffia Palm]. **Palms**, monocarpic, usually solitary, occasionally clustering, 15–24 m tall; trunk 1.8–9 m tall, 30–60 cm dbh; crownshaft absent; crown 15–24 m wide. **Leaves** pinnate, plumose, 9–21 m long; leaflets numerous, borne at different angles, linear-lanceolate, 1–2.4 m long, adaxially deep green, abaxially waxy-bluish, midrib orange, massive; petioles robust, orange, 1.8–6 m long. **Inflorescences** 3 m long, ropelike, terminal, narrow, with large overlapping bracts covering the flowers; bracts triangular, large; ♂ and ♀ flowers in the same inflorescence, with ♂ at the top and ♀ at base. **Flowers** unisexual, yellowish-white to orange; ♂ 4.5–5 mm long, with calyx tubular, trilobed, corolla with 3 lobes, stamens 6, filament 2–3 mm long; ♀ with calyx and corolla similar to ♂, petals 5–6.5 mm long, staminodes present, ovary 5.5 × 2.7 mm, stigmas 3. **Fruits** turbinate, 5–10 cm long, with overlapping, polished, mahogany-colored scales; seeds highly decorative with cream to brown patterns. **Distr:** Tropical Africa, N and E Madagascar. **Ecol:** In rainforests, riverine forests and swamps; 0–1,800 m. **Zones:** 10–12.

R. hookeri G. Mann and H. Wendl., *Trans. Linn. Soc. London* 24: 438, 1864. (Syn: *R. angolensis*, etc.). [African Wine Palm, Bamboo Palm, Mimbo Palm, Wine Palm]. **Palms**, solitary, rarely clustering; trunk 2–12 m tall, to 30 cm dbh, upper part of trunk clad with blackish fibers derived from the breakdown of leaf-petioles; crown heavy with fronds. **Leaves** variable, pinnate, to 13 m long, large, dark green, shiny, pinnae 120 pairs; leaflets alternate to irregular in 4 rows, soft, pendent, borne from the rachis at different angles, forming a fully plumose leaf, laminas linear, 0.3–0.6(–1.5) m × 2.5–5 cm, midrib prickly, bases obscured by large dense masses of both spinelike and long, curling dark gray to black fibers, adaxially glossy deep green, abaxially silvery-green; petiole 2–2.5 m long, dark green, gray or black, deeply channeled. **Inflorescences** monoecious, pendent, to 2.1 m long, ropelike, robust, circular in cross-section, with many congested branchlets, subtended by large triangular bracts; flower-bearing branchlets arranged fanlike on the inflorescence, 10–15 cm long, to ~1.2 cm thick; flowers arranged spirally on the branchlets. **Flowers** ♂ with stamens 15–18; ♂ with staminodes 15–18. **Fruits** variable, usually oblong-ellipsoid, scaly, conelike, 5–12 × 3.5–5 cm, with a prominent beak ~1.2 cm long; scales almost square to rhomboid-triangular. **Distr:** W and WC tropical Africa to Angola; naturalized in Madagascar. **Ecol:** In rainforests, swamps and regularly flooded areas.

TRIBE CALAMEAE

Metroxylon

Syn: *Coelococcus, Sagus.*

Species 7. **Palms**, monocarpic (except *M. amicarum* which flowers a number of years), monoecious; crown erect, shaped like a shuttlecock. **Leaves** arching or erect; lamina pinnate, 0.33–7 m long, unarmed or armed with single to multiple spines 1–46 cm long, in clusters or combs; leaflets 80–170 pairs, 0.7–2 m × 4–18 cm. **Inflorescence** large, paniculate, mainly terminal; rachis 2.3–7.7 m long, armed or unarmed; flowers borne on crowded spikes, spirally in pairs of ♂, bisexual and functional ♀ flowers. **Flowers** very similar among the species. **Fruits** globose, ovoid or pyriform, 1.5–8.3 × 2.3–10.6 cm, epicarp covered with 7–50 overlapping, green, golden yellow to dark brown, glossy scales. **Distr:** Maluku to W Pacific. **Cult:** Propagate from seeds, or clumping species from suckers (produced after the fruiting trunk has died). Requires moist, rich soil, in very humid, warm conditions.

M. sagu Rottb., *Nye Saml. Kongel. Dansk. Vidensk. Selsk. Skr.* 2: 527, 1783. (Syn: *M. inerme*, etc.). [Sago Palm]. **Palms**, monocarpic, clustering, 12–18 m tall; trunk 6–9 m tall, 30–60 cm dbh; crown shaped like a shuttlecock, 9–12 m Ø; crownshaft absent. **Leaves** erect but arching; lamina pinnate, 6 m long, V-shaped in cross-section; leaflets 1.2–1.5 m long, stiff, green; petiole 1.8 m long, base enlarged, covered with wavy rows of golden spines 8 cm long (rarely unarmed). **Inflorescence** 7.5 m tall, apical, with ♂ and ♀ flowers; branches tiered. **Flowers** unisexual. **Fruits** ovoid, 5–10 cm across, covered with glossy brown, overlapping scales. **Distr:** Maluku to New Guinea. **Ecol:** In swampy wooded areas; 0–700 m. **Zones:** 11–12. **Info:** Cultivated commercially for the production of sago (a starchy substance used as a source of food and for other uses).

M. vitiense (H. Wendl.) Hook.f., *Rep. Progr. Condition Roy. Bot. Gard. Kew* 1882: 68, 1884. (Syn: *Coelococcus vitiensis, Sagus vitiensis*). [Fiji Sago Palm, Soga]. **Palms**, robust, solitary, monocarpic, 5–16 m tall; trunk 36–50 cm dbh; crownshaft absent. **Leaves** erect to arching; lamina pinnate, 4–6 m long, base spiny; petiole green 1.8–8 m long, base enlarged, clasping the stem, spines brown, arranged in curved bands. **Inflorescence** suprafoliar, twice branched, 4 m long; first-order branches erecto-patent, 2–2.5 m long, second-order branches pendulous, 20 cm long, each with 8–9 rachillae (third-order branches). **Fruits** rounded to elliptic or ovate, 4.9–9 × 4–5.8 cm, epicarp covered in 25–27 rows of green, golden-yellow, or dark brown, reflexed scales, equatorial scales 0.9–1.2 cm long. **Distr:** Fiji. **Ecol:** In dense stands on alluvial coastal swamps and adjacent to rivers; 0–100 m. **Zones:** 11–12. **Info:** Fruits are dispersed in the wild by flying foxes, large parrots and small foraging mammals.

Below: *Raphia farinifera* (fruits), *Raphia hookeri, Metroxylon sagu, Metroxylon vitiense.*

SZ

TRIBE CALAMINAE

Calamus

Syn: *Rotang, Rotanga.*

Species 381. **Palms**, usually scandent, occasionally erect, tufted, dioecious, solitary or in small clumps, usually with very long stems, very prickly. **Leaves** pinnate, 1(–9), divided into numerous segments; rachis often spinous, with sharp hooked spines; leaf sheath usually very prickly, sometimes produced into a ligule or ochrea. **Inflorescences** axillary, much branched, armed, often with a sterile whiplike prolongation, flowers usually arranged in 2 ranks, often scorpioid, ♂ flower solitary, ♀ flower with a sterile ♂ flower in the same bracteole; bracts many, tubular, persistent. **Flowers** minute; ♂ with calyx 3-toothed, cup-shaped, coriaceous, petals 3, coriaceous, valvate, stamens 6, filaments small, anthers dorsifixed; ♀ with calyx 3-toothed, cup-shaped, corolla trifid, tubular below, valvate, staminodes forming a cup, ovary 3-chambered, covered with retrorse scales, stigmas 3. **Fruits** globose or ellipsoid, covered with scales and persisting stigmas at apex; seeds usually solitary, rarely 2 or 3. **Distr:** Tropical Africa, tropical and subtropical Asia to SW Pacific. **Cult:** Propagate from seeds. Requires high humidity, humus-rich, well-drained soil in full sun.

C. metzianus Schltdl., *Linnaea* 26: 727, 1855. (Syn: *C. rivalis, Palmijuncus rivalis*). [Yellow Rattan]. **Palms**, clustering, in thickets 2–3 m high, or climbing to 10–15 m high; stems with leaf sheath 2–3 cm Ø. **Leaves** pinnate, 1–1.5 × 0.7 m, tapering, emerging new leaf densely reddish-scaly; sheath green, with dark indumentum, armed with yellowish spines 2–2.8 cm long; ocrea 3–4 mm long, brittle and deciduous; petiole nearly absent; rachis armed with stout, pale, grapnel spines 0.8–1.3 cm long; leaflets bright green, 26–30 pairs, lanceolate, regular, 26–36 × 1.5–2.5 cm, terminal ones 11–12 × 1 cm, basal ones sometimes 40–45 cm long, sparsely bristled along midrib abaxially, and a few at leaflet tip. **Inflorescences** 1–2 m long, flagelliform; peduncle slender, pendulous; prophyll and bracts sparingly spiny; first order branches 60–65 cm long; floral bracteoles ciliate; peduncle, rachis and bracts covered in dark brown indumentum; rachillae 10–20, 3–5(6) cm long, strongly arching, bearing 20–32 closely packed distichous flowers. **Flowers** ♂ 5 mm long, scantily tomentose, calyx divided halfway, petals coriaceous, raised on a stalk, reaching the same level as the calyx; ♀ 6 mm long, sepals 4 mm long, petals striate, 5 mm long, covered with indumentum. **Fruits** ovoid, 1 × 0.8–0.7 cm, perianth not pedicelled; scales 7(–18) in a vertical row, gray-green, turning pale yellow when mature with a red-brown margin, not channeled in the center; stigmatic remains forming a prominent beak to 3(–7) mm long. **Distr:** SW India, Sri Lanka. **Ecol:** In lowlands; <1,000 m. **Zones:** 10–12.

C. moti F.M. Bailey, *Bull. Dept. Agric. Queensland* 13: 13, 1899. [Yellow Lawyer Cane, Wait-a-while]. **Palms**, clustering, climbing; stems canelike, slender, green, armed with many yellow spines and thorns. **Leaves** compound, 1–4 m long; sheathing base armed with yellow spines to 2 cm long, spiraled; rachis underside clothed in recurved yellow hooks; leaflets 80–100, 30–60 × 2–3 cm, margins finely spined, adaxially frequently clothed in fine brown spines ~5 mm long, abaxially sometimes clothed in spines. **Inflorescences** panicles, sparsely branched, axillary, long, slender, pendulous; ♂ and ♀ flowers borne on separate plants. **Flowers** 3–5 mm Ø, cream to white; calyx green or cream, sepals 3; petals 3, green or cream. **Fruits** drupes, creamish-green, cream or yellow, 0.8–1.4 × 0.8–1.3 cm; seed brownish-yellow, 0.7–1.2 cm long. **Distr:** Australia (NE Queensland: Mt Finnegan to the Clarke Range). **Ecol:** In humid rainforests; 0–1,000 m. Fl. Jul–Aug. Fr. Apr–Aug. **Zones:** 9–12.

C. quinquenervius Roxb., *Fl. Ind.* ed. 1832, 3: 777, 1832. (Syn: *Palmijuncus quinquenervius*). **Palms**, clustering, climbing, to 35–70 m tall or more; stem with sheath 4 cm Ø, cane ~2.5 cm Ø. **Leaves** with pinniferous part 2 m long; apical cirrus 1.5 m long; leaflets regularly arranged (often irregularly in juvenile plants), lanceolate, 30–43 × 3–5 cm, with 3–5 nerves, only adaxially with bristles, margins serrulate; rachis adaxially and margins with spines, abaxially with claw-spines; petiole adaxially and margin with spines; knee with 12.5 cm long flattened spines, spinule sparsely between the large spines. **Inflorescences** 1–1.2 m long; ♂ with 8 partial inflorescences; ♀ branching to 2 orders; spathe tubular, with claw-spine. **Infructescences** with fruiting perianth pedicellate; fruits ellipsoid, ~1.6–2.3 × 1.2–1.5 cm, with a conspicuous beak, scales in 14–17 vertical rows, grooved along the middle; seed oblong, 1.1–1.3 × 0.7 cm, pitted and deep chalazal fovea at the surface, endosperm homogeneous, embryo basal. **Distr:** Bangladesh. **Ecol:** In forests, 300–1,000 m. **Zones:** 10–12.

C. scipionum Lour., *Fl. Cochinch.*: 210, 1790. (Syn: *Palmijuncus scipionum, Rotang scipionum*). [Stick Rattan Palm]. **Palms**, climbing, massive and robust; stems clustered, climbing to 50+ m long, to 7 cm Ø, without leaf sheaths 3.5–5 cm Ø, nodes prominent, internodes very long, sometimes exceeding 1 m with sheaths 5 cm Ø. **Leaves** to ~2 m long; leaf sheaths green with whitish hairs and armed with scattered spines, sometimes spines absent; spines black-tipped, flattened, triangular, to 5 × 1.5 cm; ocreas present; knees present, conspicuous; flagella present, to 7 m long; leaf rachis to 2.3 m long, leaflets 23–30 per side, regularly arranged; leaflet lamina lanceolate; cirri absent. **Inflorescences** to 7 m long, flagellate, ♂ and ♀ superficially similar; partial inflorescences ~7, short to elongated, sometimes to 1.5 m long, with slender reflexed rachillae in ♂ and finely branched rachillae in ♀; bracts tubular; ♂ with strongly armed tubular spathes with rather distant strong hooks, spathes 3.8 cm long, dilated upwards, base bare, upper part with short hooks, branches 7.5–10 cm long, with alternate distichous spikes, 1.2–2.5 cm long; ♀ spadix very long, spathes hardly armed, edges fringed, flowers distichous. **Flowers** ♂ green, ~3 mm long, calyx campanulate, ribbed, with a scarious margin and 3 short ovate lobes, petals twice as long; ♀ with calyx similar to ♂ but gibbous at base, petals longer. **Fruits** ovoid, to 1.4 cm long, 0.9 cm Ø, with scattered pits, greenish. **Distr:** Peninsular Thailand to W Malesia, Philippines (Palawan). **Ecol:** Mostly in secondary forests;<400 m. **Zones:** 10–12.

C. thwaitesii Becc., *Fl. Brit. India* 6: 441, 1892. (Syn: *C. thwaitesii* var. *canaranus*). [Giant Rattan Palm]. **Palms**, climbing and clustering; stem robust, 30–50 m long, with leaf sheath 2–5 cm Ø; internodes 30–40 cm long. **Leaves** pinnate, 2–3 × 1 m; sheath 60 cm long, pale yellow-green, armed with black spines 3–4 cm long; leaflets dark green, broadly lanceolate, *young* leaves with 6–7 pairs, irregularly grouped, *mature* leaves with 30 pairs, 0.6–1 m × 3.5–5.5 cm; flagellum 4.5–6 m long, proximally armed with rings of black spicules, gradually replaced by stout 7-fid grapnel spines; petiole 20–30 cm long, armed with rings of black spines. **Inflorescences** 5–7 m long, flagelliform; peduncle pendulous, rachis armed with black-tipped spines; prophyll long, tubular, armed with spinules; first order branches 50–70 cm, ~60–80 cm apart; bracts and floral bracteoles funnel-shaped; rachillae flattened; ♂ 10–15 cm long, 20–30-flowered, distichous, sessile; ♀ 10–30 cm long, 25–30-flowered. **Flowers** ♂ 1 × 0.3 cm, calyx divided halfway, lobes broadly triangular, petals narrowly ovate, filaments broad, narrowed apically; ♀ ovoid, 0.8–1 cm long, sepals 4 mm long, bases thickened cushionlike, petals 2x as long. **Fruits** ovoid, 2–2.5 × 1 cm; scales 12 in a vertical row, dull orange; seed slightly ovoid, 1.2–1.5 × 0.7–1 cm, compressed. **Distr:** SW India, Sri Lanka. **Ecol:** In wet lowlands and lower montane forests; to 1,500 m. **Zones:** 10–12.

C. zeylanicus Becc., *Fl. Brit. India* 5: 466, 1893. [Ceylon Rattan Palm]. **Palms**, clustering, climbing, 40–50 m high; stems robust, with leaf sheath 4–5 cm Ø. **Leaves** pinnate, 1.5–2 × 1 m; sheath copper-colored, armed, ~1 cm long; rachis armed distally with 3–6-fid grapnel spines; leaflets 35–50 pairs, regularly arranged, linear, 45–65 × 2.5–3.8 cm, terminal leaflets 20–21 × 1.5 cm, bright green, shiny; cirrus to 1.7 m long, armed; petiole ~85 cm long, armed. **Inflorescences** with ♂ and ♀ dissimilar, usually 2–5, not flagelliform; ♂ to 1.8 m long, ♀ to 50 cm long; peduncles stout, ♂ pendulous, ♀ erect; prophyll and bracts sparsely armed; ♂ rachillae 2–3 cm long, with 8–15 closely packed distichous flowers per side; ♀ rachillae stout, 10–20 cm long, 5–8 cm apart with 10–20 flowers. **Flowers** ♂ oblong, 4–7 mm long, slightly curved, calyx thin, membranous, divided

Below: *Calamus metzianus, Calamus moti* (stem detail), *Calamus quinquenervius* (fruits), *Calamus scipionum.*

RF

MIW

Top row: *Calamus scipionum* (stems), *Calamus thwaitesii*, *Daemonorops draco*.
Bottom row: *Calamus zeylanicus*, *Daemonorops jenkinsiana*.

halfway into 3 ovate lobes, corolla longer, segments lanceolate, finely striate, pistillode divided into 3; ♀ 0.5–1.2 cm long, sepals separate, ovate, concave, finely striate, corolla longer, ~8 mm, similar. **Fruits** globose, 1.3–2 cm Ø; scales 6 in a vertical row, large, convex, dull copper-colored; beak 5 mm long; seed globular, finely pitted. **Distr:** SW Sri Lanka. **Ecol:** In wet lowland forests; <1,000 m. **Zones:** 11–12.

Daemonorops

Species 102. **Palms**, erect or climbing, solitary or clustering, pleonanthic (rarely hapaxanthic). **Leaves** pinnate, rachis apex extended to form a claw-shaped spiny cirrus; leaf-sheath tubular, flagellum absent, with scattered or seriate spines. **Inflorescences** with ♂ and ♀ superficially similar, spindle or tubular before flowering; ♂ dense panicles, rarely diffuse, rachilla bearing alternate distichous flowers, floral spathe scale-shaped, ♂ flowers solitary inside the floral spathe; ♀ panicles, usually dense, rachilla bearing each ♀ flower usually with a sterile flower. **Flowers** ♂ with stamens 6; ♀ larger than ♂, ovate, staminodes 6, base adnate to form a ring, ovary with scales, incomplete trilocular, ovules 3, anatropous, style short or conic, stigma 3. **Fruits** globose, ovate or ellipsoid, apex with beak, epicarp thin, covered with reflexed scales; seed usually 1, globose or slightly flattened. **Distr:** Tropical and subtropical Asia. **Cult:** Propagate from seeds. Requires moist but well-drained soil in full or partial sun.

D. draco (Willd.) Blume, *Rumphia* 2: 8, 1838. (Syn: *Calamus draco*, *C. draconis*, *D. propinqua*). [Dragon's Blood Palm, Rattan Palm]. **Palms**, climbing, robust, growing to 18 m long, clustering or clump-forming, armed with very sharp numerous dark-colored, flattened, elastic spines, often arranged in oblique rows, with their bases united; stems ~10 cm Ø; juvenile plants usually erect, becoming scandent and climbing with age. **Leaves** pinnate, rachis and leaf sheaths spiny; leaflets alternate, ensiform or narrowly elliptic, 30–45 × 2 cm, margins and midrib remotely armed with stiff, slender bristles, apices tapering, venation parallel. **Inflorescences** enclosed in large spathes; several, one to each of the 4 or 5 primary ramifications of the spadix, lanceolate and leathery, armed outside. **Flowers** ♂ and ♀ yellowish-white; calyx turbinate, ribbed, limb 3-toothed, often split into 3 portions due to the swelling of the ovary, and persistent in fruit; corolla 3-cleft, segments ovate-lanceolate, 2× as long as the calyx, and persistent; stamens 6, filaments very broad, and inserted into base of corolla, anthers prominent, filiform, and apparently abortive, yellow; ovary ovoid, style short, stigmas 3-cleft, segments revolute, glandular inside. **Fruits** ovoid, ripening to red. **Distr:** Peninsular Thailand to Sumatra, Borneo (Sarawak). **Ecol:** In lowland primary dipterocarp forests; ~50 m. **Zones:** 10–12. **Info:** A bloodlike resin is extracted from the immature fruits, often used for medicinal purposes.

D. jenkinsiana (Griff.) Mart., *Hist. Nat. Palm.* 3: 327, 1853. (Syn: *Calamus jenkinsianus*, etc.). [Major Jenkins's Rattan Palm]. **Palms**, clustering, high-climbing; stems robust, with leaf-sheath 3–4 cm Ø, internodes 15–20 cm long. **Leaves** pinnate, 2–3 × 0.5 m; sheath yellowish-green, covered with brown indumentum, armed with flattened deep brown to black spicules 3–4 cm long; petiole 50 cm long, tomentose, margins armed; rachis similarly armed, gradually replaced by reflexed grapnel spines toward leaf apex; leaflets regular, 40–50(–70) × 2–4 cm, terminal leaflets small; cirrus 2–2.5 m long, armed with bifid–trifid grapnel spines. **Inflorescences** with peduncle 3–6 cm long; prophyll tapering into a long beak, reddish; primary branches densely tomentose at base; ♀ rachillae 8 cm long, sinuous, with 6–7 ♀ flowers on each side. **Flowers** ♂ oblong in bud, 5 mm long, sepals 3, connate, calyx cupular, tip hairy, petals 3, oblanceolate, stamens 6, anthers subulate, connate and thickened at base; ♀ 5–5.5 mm long, sepals connate, truncate, calyx cupular, petals lanceolate, deeply divided, distinctly veined. **Fruits** globose, 1.8 cm Ø, scales ~18 in a vertical series, yellowish-brown, with distinct darker margins; seed globose, 1 cm Ø, minutely pitted, pits filled with dark resinous substance. **Distr:** E Himalayas to Taiwan and Indo-China. **Ecol:** In lower montane forests; to 100 m. **Zones:** 9–12.

Subfam. NYPOIDEAE

Genus 1, **species** 1. **Palms**, aquatic; stem dichotomously branched. **Inflorescences** with axes adnate to the internode above, with ♂ spikes and ♀ heads. **Flowers** ♂ with perianth undifferentiated, stamens 3, opposite outer perianth, connate, extrorse, pollen zonasulcate, pistillode absent; ♀ with staminode absent, pistils 3(4), margins conduplicate, outer integument 10 cells across, placentation laminar to submarginal. **Distr:** Sri Lanka to Malesia.

Nypa

Species 1. *Generic description same as specific.* **Cult:** Propagate from seeds. Best grown in a shallow pond, in full or partial sun; salt-tolerant.

Top row: *Sabal minor, Sabal mauritiiformis.*
Bottom row: *Nypa fruticans* (habit), *Sabal palmetto.*

N. fruticans Wurmb, *Verh. Batav. Genootsch. Kunsten* 1: 349, 1779. [Mangrove Palm, Nipa, Water Coconut]. **Palms**, aquatic, clustering; stem mostly subterranean, 45 cm Ø. **Leaves** 4–7, pinnate, 4–10 m long, erect, arching at the tip; sheath splitting early; petiole 1.2–1.5 m long, stout; leaflets dull green, 40–60 pairs, linear-lanceolate, 0.6–1.8 m × 5–9 cm, tip subulate, midrib yellowish. **Inflorescences** 1–2.5 m long, receptacle 10–15 cm long, terminating in a globular head of ♀ flowers with several lateral branches bearing short dense spikes of ♂ flowers; peduncular and rachis bracts bright yellow when fresh with green-striped outside. **Flowers** ♂ 4–6 mm long, cream; sepals acute, petals similar, stamens united into a column 4–6 cm long, flowers and branches quickly shrivelling up after shedding pollen; ♀ flower 1 cm long, lemon yellow, sepals acute, 4 mm long, carpels 3, woody, angular, 1 cm long, obscuring the perianth at maturity and closely packed in a terminal globose head 4 cm wide. **Fruits** 15–20 cm Ø, comprised of many closely packed 2–4 angled segments, each 9–15 × 5–10 cm; seed single, kidney-shaped, 4.5 cm Ø, grooved. **Distr:** Throughout the Asian and Pacific region (except peninsular India). **Ecol:** Strictly a mangrove palm, mostly in estuarine situations; salt tolerant. Fl. and Fr. throughout the year. **Zones:** 10–12. **Info:** The young endosperm is edible. This palm is unusual in having an underground stem which branches dichotomously to form new above-ground plants.

Subfam. CORYPHOIDEAE

Genera 45, **species** 477. **Stem** rarely branched (e.g. *Caryota*). **Leaves** palmate or costapalmate (pinnate), induplicate. **Inflorescences** various (terminal; adnate to the internode above; plant monoecious in tribe Caryoteae; flowers in triads). **Flowers** with corolla often valvate; gynoecium free, or connate by style, or style present, with 3 (in tribe Corypheae) or 1 (in tribe Sabaleae) stylar canals, or style absent (in tribe Caryoteae) (Stevens 2001 onwards). **Distr:** Pantropical (to warm temperate), few in South America.

TRIBE SABALEAE

Sabal

Syn: *Inodes.*

Species 16. **Palms**, solitary, dwarf, moderate, or tall, usually robust; stems aerial or subterranean, covered with leaf bases or clean, becoming striate or smooth with age. **Leaves** few to many; petiole split at base, unarmed; adaxial hastula well-developed, obtuse to acuminate-triangular; costa present; lamina weakly to strongly costapalmate; plication induplicate; segments lanceolate, ± basally connate (or in groups of 2 or 3 segments connate for nearly entire length), often bearing threadlike fibers between segments; apices acute or 2-cleft, stiff or lax. **Inflorescences** axillary within crown of leaves, paniculate, erect or arching beyond leaves, branched to 2 or 3(–4) orders; peduncular bracts 2–5, inconspicuous; rachillae glabrous; flowers sessile, borne singly along rachillae. **Flowers** bisexual, creamy-white, fragrant; perianth biseriate; calyx cupulate; trilobed; petals 3, imbricate, elliptic, obovate or spathulate, alternate with outer whorl of stamens, basally adnate to filaments; stamens 6 in 2 whorls; pistils 1, 1-carpellate, nectaries 3, ovules 3, but usually only 1 develops. **Fruits** drupes, berrylike, spheroid or lobed when >1 seed develops; seeds 1–3, oblate, glossy. **Distr:** N America, Mexico, West Indies, C America, S America. **Cult:** Propagate from seeds. Requires a fertile, well-drained soil in full sun.

S. mauritiiformis (H. Karst.) Griseb. and H. Wendl., *Fl. Brit. W. I.*: 514, 1864. (Syn: *S. allenii*, etc.). [Botan, Bayleaf Palm]. **Palms**, solitary, 5–15(–24) m tall; trunk 12–24 m tall, 25–30(–45) cm dbh; crownshaft absent; crown 3–4.5 m Ø. **Leaves** palmate, circular, 1.8–3 m Ø, adaxially dark green, abaxially silvery, adaxial hastula ~ 3.5–11 × 2–3 cm, acuminate, costa ~0.9–1.25 m; *adult* plants with lamina divided into long, limp segments with pendent ends; *juvenile* plants larger and divided into 2 or 3 very pleated and wedge-shaped segments; petioles ~1.3–3 m long. **Inflorescences** spikes, 1–3.5 m long; branches 70 cm long, usually extending just beyond leaf crown; peduncle ~1–2 m long. **Flowers** bisexual, ~3 mm across, cream; sepals connate into a tube, limb trilobed; petals 3, basally connate into a tube, apices divergent to suberect; stamens 6, filaments basally connate, adnate to corolla; pistil 1, ovary trilocular, stigma capitate, trilobed. **Fruits** ovoid, 1.2 cm Ø, ripening to black. **Distr:** Mexico to Venezuela, Trinidad. **Ecol:** In humid forests; 0–200+ m. **Zones:** 9–12.

S. minor (Jacq.) Pers., *Syn. Pl.* 1: 399, 1805. (Syn: *Brahea minima*, etc.). [Bush Palmetto, Dwarf Palmetto]. **Palms**, dwarf, solitary, 1.8–6 m tall; trunk usually subterranean, rarely above ground; crownshaft absent; crown 2.4–3 m Ø. **Leaves** 4–10, palmate or weakly costapalmate, 1.2–1.8 m across, not bearing fibers between segments, hastula obtuse, 0.8–4.7 cm; segments not filiferous, usually in groups of 2, 34–84 × 1.4–3.7 cm, stiff, dark green, apices weakly bifid. **Inflorescences** sparsely branched with 2 orders of branching, erect, much longer than leaves. **Flowers** bisexual, white, fragrant, 3.5–5.2 mm across. **Fruits** brownish-black, spheroid, 6.2–8.5 mm long, 6.4–9.7 mm Ø, pericarp thin; seeds 3.5–5.1 × 4.4–6.9 mm. **Distr:** S Oklahoma to SE USA. **Ecol:** Usually an understory palm, sometimes in swamps, flood plains and riverbanks. **Zones:** 8–11.

S. palmetto (Walter) Loddiges ex Schultes and Schultes f., *Syst. Veg.* 7(2): 1487, 1830. (Syn: *Chamaerops palmetto, etc.*). [Cabbage Palm, Palmetto]. **Palms**; stems usually aerial, 20–35 cm dbh. **Leaves** 15–30, strongly costapalmate, bearing threadlike fibers between segments; hastula acute to acuminate, 5.3–18 cm; segments 55–120 × 2.5–4.2 cm; apices bifid, 2-cleft. **Inflorescences** with 3 orders of branching, arching, equaling or exceeding leaves in length. **Flowers** bisexual, 4.1–6.7 mm across, cream-colored. **Fruits** black, spheroid, length 0.8–1.4 cm, 0.8–1.4 cm Ø; seeds 0.5–1 cm Ø. **Distr:** SE USA and West Indies (Bahamas, Cuba). **Ecol:** Hammocks, pinelands, riverbanks, dunes, tidal flats; 0–40 m. Fl. Apr–Aug. **Zones:** 8–12.

TRIBE CRYOSOPHILEAE

Coccothrinax

Syn: *Haitiella, Thrincoma, Thringis*. [Broom Palms, Silver Palm].

Species 14–50. **Palms**, small to medium-sized; stems solitary or occasionally cespitose, erect, slender, at first covered with leaf sheaths. **Leaves** with sheath fibers soft, netlike, eventually sloughing off to reveal smooth to fissured stem; abaxial hastula a low ridge or absent; adaxial hastula a crescent-shaped ridge to semi-cylindric excrescence; lamina palmate, plication induplicate (V-shaped), segments lanceolate, basally connate, cross veins obscure (conspicuous); petiole not split at base, unarmed. **Inflorescences** interfoliar, downcurved (long and arching), not extending beyond leaves, with 2 orders of branching; prophyll short, peduncular bracts several, sterile, tubular, distally expanded, silky-pubescent; primary branches subtended by smaller peduncular bracts; flowers borne singly along rachillae, short-pedicellate. **Flowers** bisexual; perianth 1-seriate, shallowly cupulate, lobes 5–7, apiculate; stamens 7–12, filaments acute, anthers dorsifixed, twisted when dry; pistil 1, 1-carpellate, glabrous, style slender, stigma funnelform. **Fruits** globose, stigmatic scar apical, exocarp purplish, smooth, slightly warty when dry, mesocarp fleshy, endocarp membranous; seeds globose. **Distr:** Caribbean Basin. **Cult:** Propagate from seeds. Requires a very well-drained soil in full or partial sun.

C. argentata (Jacquin) L.H. Bailey, *Gentes Herb.* 4: 223, 1939. (Syn: *C. garberi, C. jucunda, Palma argentata, Thrinax argentea, T. garberi*). [Florida Silver Palm, Silver Palm, Silver Thatch Palm, Silvertop Palm]. **Palms**, usually solitary, to 9 m tall; stems, to 6 m tall, 15 cm dbh, occasionally with vegetative sprouts forming on trunk, upper part usually with matted fibers; crown 2.4 m wide; crownshaft absent. **Leaves** palmate, circular in outline, to 1 m across; segments abaxially silvery, stiff or lax; petiole 60 cm long. **Inflorescences** erect and rising above the crown; rachillae stiff; pedicels conspicuous. **Flowers** bisexual, yellowish. **Fruits** globose, dark purple-black, 0.65–1.2 cm Ø. **Distr:** S Florida, West Indies (Bahama Islands). **Ecol:** On rocky, calcareous soils of coastal hammocks and scrubs; 0–10 m. **Zones:** 10–12.

Thrinax

[Peaberry Palm, Thatch Palm].

Species 7. **Palms**, small to medium; stem solitary, erect, ± smooth; leaf sheath producing soft fibers. **Leaves** with petiole split at base, unarmed; abaxial hastula a small crescent-shaped ridge or absent; adaxial hastula irregularly semicircular to nearly cylindric; lamina palmate; plication induplicate; segments lanceolate, basally connate; cross veins conspicuous (obscure). **Inflorescences** interfoliar, emerging through split leaf bases, arching beyond leaves, with 2 orders of pendent branches; prophyll short; peduncular bracts many, tubular; flowers bisexual, borne singly along rachillae, short- to long-pedicellate. **Flowers** with perianth 1-seriate, shallowly cupulate, lobes 5–7, apiculate; stamens 6–12, erect (inflexed in bud), filaments acute, anthers dorsifixed; pistils 1, 1-carpellate, glabrous, stigma funnelform. **Fruits** globose, stigmatic scar apical, exocarp white, smooth, slightly warty or rugose when dry, mesocarp mealy, endocarp membranous; seeds oblate-globose. **Distr:** N America (Florida), Mexico, West Indies (Bahamas), C America. **Cult:** Propagate from seeds. Requires a well-drained soil in full sun.

T. parviflora Sw., *Prodr.*: 57, 1788. [Broom Palm, Mountain Thatch Palm, Thatch Pole, Iron Thatch]. **Palms**, slender, erect, 3–10 m tall; stem to 15 cm dbh, enlarged at base. **Leaves** light green, 2.5 m long; lamina 1.3 m across, cleft almost to middle into 35–60 segments, scales absent on abaxial surface; segments mostly narrow, 0.3–1.6 m long, sometimes 3 cm wide; petiole 0.6–1.2 m long, thin, 2-edged, bearing many marginal fibers below; ligule broad, with a low upturned margin, produced into a prominent triangular point. **Inflorescences** clusters, to 1.5 m long; spadix ~1 m long; ultimate branchlets 4–8 cm long; flowers rather closely placed, conspicuously pedicellate, variable in size. **Flowers** ivory-white to yellowish; stamens 5–15, anthers longer than filaments, stigma expanded, ± urceolate. **Fruits** white, globose, 0.7 cm Ø, stalks short, shell brittle when dry, finely granulate; seed 4–5 mm Ø. **Distr:** Jamaica. **Ecol:** On exposed and eroded limestone hills with deciduous, open woodlands; <900 m. **Zones:** 10–12.

T. radiata Loddiges ex Schultes and Schultes f., *Syst. Veg.* 7(2): 1301, 1830. (Syn: *Coccothrinax radiata*, etc.). [Caribbean/ Florida/Jamaica/Sea/ Silk-top Thatch Palm]. **Palms**, solitary, to 10 m tall, fast-growing; stem erect, covered with old leaf bases and matted fibers. **Leaves** palmate; leaflets broad; sheaths with V-shaped cleft opposite petiole; hastula adaxially glabrous; lamina green, with minute, whitish dots on abaxial surface; segment apices lax; petiole bases forked. **Inflorescences** whitish turning green, spadix large, paniculate, to 2 m long; pedicels 1.5–2.3 mm long. **Flowers** bisexual, small, to 1.25 cm long, fragrant; stamens ~8. **Fruits** smooth, white, rounded, 0.7–1.4 cm Ø; fruiting pedicels short, conspicuous. **Distr:** USA (Florida), Honduras, SE Mexico, West Indies. **Ecol:** Pinelands and littoral hammocks and scrubs on limestone soils; 0–10 m. Fl. Jun–Jul. **Zones:** 10–12.

Chelyocarpus

Species 4. **Palms**, solitary or clustering, unarmed, hermaphroditic, monoecious. **Leaves** palmate, or costapalmate, induplicate; lamina deeply divided, each half divided into paired or irregularly grouped, single-fold segments, or divided into many-fold segments, split distally into bifid, single-fold, 1-ribbed segments; petiole not split basally, margins smooth. **Inflorescences** with rachillae sometimes fasciculate, flowers solitary. **Flowers** bisexual; sepals 2, 3 or 4, slightly connate or imbricate basally; petals similar to sepals; stamens 5, 6, 7, 8 or 9, filaments distinct, fleshy; carpels 3(1–4), follicular, ovule hemianatropous with aril. **Fruits** small to medium-sized, globose, epicarp smooth or coarsely corky-warted. **Distr:** W South America to N Brazil. **Cult:** Propagate from seeds. Requires a moisture-retentive soil in full or partial sun.

C. ulei Dammer, *Notizbl. Bot. Gart. Berlin-Dahlem* 7: 395, 1920. (Syn: *Tessmanniophoenix longibracteata*). **Palms**, understory; stem solitary, erect, to 8 m tall, 4–8 cm Ø. **Leaves** 10–15, palmate 1–1.6 m across, abaxially silver-white, deeply split into 5–12 leaflets, which in turn are partially split for up to half their length into 1-ribbed teeth; petiole 1–1.5 m long. **Inflorescences** 0.6–1 m long, basal part covered with 2 peduncular bracts, with a grayish-white waxy tomentum; flowering branches numerous, 7–20 cm long. **Flowers** 3–5 mm long, opening cream-colored, turning purplish. **Fruits** globose, 2–3 cm Ø, light brown, with corky-warty surface, 1-seeded. **Distr:** W South America to Brazil (Acre). **Ecol:** In lowland rainforests; <500(–900) m. **Zones:** 10–11.

TRIBE PHOENICEAE

Phoenix

Syn: *Dachel, Elate, Palma*. [Date Palms].

Species 14. **Palms**; stems solitary or clustered, erect or ascending (subterranean), slender to massive, often

Below: *Coccothrinax argentata, Thrinax parviflora* (fruits), *Thrinax radiata*.

SZ

SZ

clothed in old leaf bases. **Leaves** with sheath fibers soft; petiole not split at base, armed, base not split, not forming crownshaft; lamina pinnate; plication induplicate; segments lanceolate, in 1 or more planes; apices acute; basal segments modified into stout spines. **Inflorescences** axillary within crown of leaves, paniculate, ascending, much shorter than leaves, with 1 order of branching, alike in staminate and pistillate plants; prophyll often caducous, conspicuous, becoming boat-shaped, short; peduncular bracts absent; rachillae glabrous. **Flowers** ♂ borne singly along rachillae; calyx cupulate, trilobed, petals 3, free, valvate, stamens 6, free, pistillode inconspicuous or absent; ♀ borne singly on rachillae, calyx cupulate, trilobed, petals 3, imbricate, free, staminodial ring cupulate or deeply 6-lobed, pistils 3 (only 1 developing), distinct, stigmas small. **Fruits** drupes, berrylike, fleshy, exocarp blackish-brown, smooth, mesocarp fleshy or fibrous, endocarp papery; seeds 1, elongate, endosperm homogeneous, embryo lateral (basal), eophyll undivided, lanceolate. **Distr:** Eastern Hemisphere (including Canary and Cape Verde islands), S Europe, Africa (including Madagascar), S Asia, Philippines. **Cult:** Propagate from seeds, or suckers from suckering species. Requires any soil with excellent drainage in full sun.

P. canariensis Hort. ex Chabaud, *Prov. Agric. Hort. Ill.* 19: 293, 1882. [Canary Island Date Palm]. **Palms**, solitary, stout, 12–16 m tall; stems, erect, columnar, 55–140 cm dbh, covered with large closely placed leaf scars. **Leaves** many, large, widely spreading 5–6 m long, forming a dense crown; leaflets numerous, green, shining, ± regularly arranged or paired along rachis in one plane. **Inflorescences** elongate. **Flowers** ♂ small, c. 4 mm long, petals not acuminate; ♀ flowers with calyx almost as long as the petals. **Fruits** oblong-ellipsoid, ripening from green through yellow, orange to reddish-purple, 2.4–2.7× 1–1.2 cm, with coarse, fibrous, tasteless, dry flesh. **Distr:** Canary Islands. **Ecol:** In waste places and disturbed areas; 0–1,000 m. Fl. throughout the year. **Zones:** 9–11.

P. dactylifera L., *Sp. Pl.* 1188, 1753. [Date Palm]. **Palms**, solitary, 30+ m, suckers producing offsets and roots at base; crown large, open. **Leaves** large, glabrous, 1–5.1 m long, leaflet in many planes, 20–40 × 2–2.5 cm; pinnae strongly keeled; apex hard, spiny; lower leaflets transformed into spines, 4-ranked or the upper ones 2-ranked, 10–20 cm long, apex yellowish. **Inflorescences** covered by a hard, boatlike bract; ♂ much smaller, 12–25 cm long, sometimes larger, main stalk 60–90 cm long, flat, glabrous, rachilla arises from and near apex of main stalk; ♀ 0.9–1.2 m long, main stalk flat, 45–75 cm long, glabrous, rachilla ~30 cm long, spikelets numerous. **Flowers** ♂ sessile, white, sweet-scented, much larger than ♀, sepals trilobed, much smaller than petals, united, forming a cup, petals trilobed, valvate, 7–8 mm long, stamens 6, ~4 mm long, filaments short, subulate, anthers erect, pistillodes minute, 3, scalelike; ♀ rounded, green, distant, calyx lobes 3, united at base, forming a cuplike structure, upper portion thin and translucent, lower one hard, greenish, sepals ~4 × 2 mm, petals over 2× larger than sepals, rounded. **Fruits** cylindric, 2.5–5 × 1–1.5 cm, edible, fleshy, yellowish-brown to reddish-brown; seeds stony, acute at the apex, longitudinally grooved from one side. **Distr:** N Africa to Arabian Peninsula. **Ecol:** In oases and warm desert regions. Fl. Mar–Apr. Fr. Oct–Feb. **Zones:** 9–12. **Info:** This palm is among the earliest cultivated plants in the world. It is an important crop in their native range. Cultivated mainly for the edible, durable fruits; all parts of the palm are used for various purposes, including domestic and medical.

P. loureiroi Kunth, *Enum. Pl.* 3: 257, 1841. (Syn: *Phoenix humilis* var. *loureiroi*). [Mountain Date Palm]. **Palms**, erect, small, low growing, 0.3–3.5 m tall; stems often clustered, 20–25 cm dbh, usually clothed with remains of petiole. **Leaves** subglaucous, leaflets interruptedly fascicled along rachis, not prominently scaly beneath but with deciduous pale scales and persistent linear brown scales on midrib beneath, 25–55 × 2 cm. **Inflorescences** with peduncles elongating in fruits to 90 cm. **Flowers** ♂ ~8 mm long, petals blunt; ♀ flowers with calyx half as long as corolla, lobes very small, petals with short central lobe. **Fruits** 1.2–1.75 × 0.5–1.2 cm, oblong, orange when ripe, edible; seeds woody. **Distr:** Indian subcontinent to S China and Philippines. **Ecol:** On hilly regions; to 900 m. **Zones:** 10–12.

P. loureiroi* var. *loureiroi (Syn: *P. hanceana*, etc.). **Palms**, clustering or solitary, 2–5 m tall; trunk over 30 cm dbh. **Leaves** to 2 m long; leaflets solitary or in groups of 2 or 3, 4-farious, linear, 15–35 × 1–1.5 cm. **Inflorescences** ♂ with spathe 15–20 cm, brown, not splinted as 2 boat-shape limbs, peduncles over 60 cm long; ♀ branching short and robust, 7–15 cm long. **Flowers** ♂ nearly white, calyx 1–1.5 mm; ♀ 4.5 mm, calyx ~1 mm long, without triangular tooth. **Fruits** oblong, 1.5–2 cm long, purplish-black when ripe. **Distr:** Indo-China to S. China and Philippines. **Ecol:** In broad-leaved forests or conifer and broad-leaved mixed forests, 500–1,500 m. Fl. Apr–May. Fr. Jun–Oct. **Zones:** 10–12. **Info:** Fruits edible.

P. reclinata Jacquin, *Fragmenta Botanica* 1: 27, t. 24, 1801. (Syn: *P. comorensis*, etc.). [Senegal Date Palm, Wild Date Palm]. **Palms**, dioecious; stems multiple (clustering, thicket-forming), to 3–8 m tall, usually oblique (rarely erect), 10–15 cm dbh, with persistent leaf sheaths at 1–2 m below crown, with oblique leaf scars; exudate clear yellow. **Leaves** arcuate, 2–3 m long; leaflets with proximal ones modified into sharp spiny acanthophylls, distal ones arranged in a single plane, median leaflets in groups of 2–5, 25–45 × 2–3.6 cm, lower leaflets often fanned, abaxial surface with white indumentum. **Inflorescences** with prophyll ~70 cm long, greenish-yellow, brown-tomentose when young, 2-keeled, splitting ventrally; ♂ erect in bud, rachillae 20 cm long; ♀ elongating beyond prophyll on fruit maturity, rachillae 9–26 cm long. **Flowers** ♂ creamy white, scented; calyx cupule, trilobed; petals 6–7 mm long, apex thickened; stamens 6, sessile; pistillodes 3, minute; ♀ globose, 3–4 mm; carpels 3, usually 1 reaching maturity. **Fruits** orange to reddish-brown, ovoid-ellipsoid or almost obovoid, 1.8–2 × 0.9–1.2 cm, sweet; seeds 1.2–1.4 × 0.6 cm, deeply grooved. **Distr:** Africa, Madagascar. **Ecol:** In waste places and disturbed areas; 0–10 m. Fl. throughout year. **Zones:** 10–12.

P. roebelenii O'Brien, *Gard. Chron.* ser. 3. 6: 475, 1889. [Pygmy Date Palm]. **Palms**, clustering, usually solitary when cultivated, 1–3 m tall, rarely taller; stems to 10 cm Ø, with persistent triangular leaf bases. **Leaves** 1–1.5(–2) m; leaflets linear, 20–30(–40) × 5–10(–15) cm, soft, deep green, with gray scale-furfuraceous along veins abaxially, in bifarious, lower ones reduced to spines. **Inflorescences** with spathe 30–50 cm long, bilobed apex only; ♂ longer than ♀, partial inflorescence to 20 cm long. **Flowers** ♂ with calyces ~1 mm long, petals lanceolate, ~9 mm; ♀ subovate, ~6 mm, with conspicuous mucronate calyx tips. **Fruits** oblong, 1.4–1.8 × 0.6–0.8 cm, reddish when ripe, pulp thin. **Distr:** China (Yunnan) to N Indo-China. **Ecol:** Along riverbanks, 500–900 m. Fl. Apr–May. Fr. Jun–Sep. **Zones:** 10–12.

P. rupicola T. Anderson, *J. Linn. Soc., Bot.* 11: 13, 1869. [Cliff Date]. **Palms**, solitary, 6–9 m tall; trunk slender, naked, 4.5–6 m tall, 20–25 cm dbh; crownshaft absent; crown full and round. **Leaves** pinnate, 2.5–3 m long, quite glabrous; leaflets 45 cm long, bifarious and not fascicled, decurved, all in 1 plane, flaccid bright green, lower leaflets modified into ± flexible spines, midrib twisted so that leaflets at outer half are vertically oriented; petiole compressed. **Inflorescences** 60–90 cm long; spathe 30 cm long; spadices elongated, much compressed; ♀ flowers in fascicled spikes, near apex, spadix 1–1.3 m long, peduncled. **Flowers** unisexual, yellow. **Fruits** oblong, 1.9 cm long, shiny, yellow-orange ripening to purplish-brown; seed 1.25–1.7 cm long, groove dilated inwards. **Distr:** Bhutan to Darjeeling. **Ecol:** On hills and cliffs of rainforests; 120–450 m. **Zones:** 10–12.

TRIBE LIVISTONEAE

Brahea

Species 11. **Palms**, moderate, solitary, rarely clustered; sheath becoming fibrous, eventually splitting basally. **Leaves** 5–30, palmate or costapalmate, circular in outline, divided nearly to halfway into single-fold, deeply bifid, stiff or flexible segments; petiole toothed or unarmed. **Inflorescences** branched to 4 orders; flowering branches sometimes crowded; peduncular bracts 0–several; flowers solitary or in cincinni of 2–3. **Flowers** small, whitish, bisexual; sepals, petals and carpels often tomentose; sepals 3, distinct, imbricate; petals 3, tubular basally, valvate apically; stamens 6, epipetalous, filaments connate in a 6-lobed ring; carpels 3. **Fruits** globose to ovoid, greenish-yellow, yellowish-brown to brown, purplish-dark blue or black, usually smooth, sometimes slightly hairy. **Distr:** Mexico to C America. **Cult:** Propagate from seeds. Requires a moderately fertile, moist, well-drained soil in full sun.

B. edulis H. Wendl. ex S. Watson, *Proc. Amer. Acad. Arts* 11: 120, 1876. (Syn: *Erythea edulis*). [Guadalupe Palm]. **Palms**, solitary, robust, 4.5–12(–15) m tall, fast-growing; trunk to 9 m tall, 40–45 cm dbh, deep brown to dark gray, swollen at base; crownshaft absent; crown 4 m wide. **Leaves** ~20, palmate, semicircular to diamond-shaped, 0.9–1.8 × 0.9 m, divided to half its length into many stiff leaflets; leaflets 70–80, midgreen to deep green; petiole 1.5–1.8 m long, smooth with short thorns. **Inflorescences** spikes, shorter than leaves, 1.2–1.5 m long, arching, shortly branched. **Flowers** small, yellow, bisexual. **Fruits** globose, 2.5–3.5 cm Ø, ripening to black; pulp edible. **Distr:** Mexico (Guadalupe). **Ecol:** On rocky slopes and cliffs in desertlike vegetation; 400–1,000 m. **Zones:** 10–12.

Chamaerops

Species 1. *Generic description same as specific.* **Distr:** Mediterranean region. **Cult:** Propagate from seeds or suckers in clumping forms. Requires very good drainage with abundant moisture.

C. humilis L., *Sp. Pl.*: 1187 (1753). (Syn: *Phoenix humilis* [misapplied]). [Dwarf Fan Palm, European Fan Palm, Hair Palm, Mediterranean Fan Palm]. **Palms**, clustering or solitary, 1.8–6 m tall; trunk(s) 1–6 m tall, 30 cm dbh, covered with a dense mat of dark fibers and old leaf bases; crown 1.8–9 m wide; crownshaft absent. **Leaves** palmate, semicircular to circular, 90 cm

Top row: *Chelyocarpus ulei, Chelyocarpus ulei* (inflorescence), *Phoenix canariensis.*
Second row: *Phoenix dactylifera.*
Third row: *Phoenix loureiroi, Phoenix loureiroi* var. *loureiroi, Phoenix reclinata, Phoenix reclinata* (fruits).
Bottom row: *Phoenix roebelenii, Phoenix rupicola, Brahea edulis, Chamaerops humilis.*

SZ

SZ

TM

MD

Above: *Johannesteijsmannia altifrons, Johannesteijsmannia magnifica.*

across, deeply segmented, green to grayish-green or silvery, petioled; petiole with strong, forward-pointing spines. **Inflorescences** clusters, short, much-branched, usually hidden behind the petiole. **Flowers** small, cream to yellow, unisexual (with both ♂ and ♀ flowers on the same plant) or sometimes bisexual. **Fruits** round, 1.2 cm across, dark yellow or orange to reddish-brown when ripe. **Distr:** W and C Mediterranean including coastal N Africa. **Ecol:** On dry and hot rocky hills and foothills; to 1,000(–2,000) m. Fl. Apr–Jun. **Zones:** 8–10. **Info:** This species has been wrongly assigned to *Phoenix* in old literature.

Johannesteijsmannia

Syn: *Teysmannia*.

Species 4. **Palms**, solitary, acaulescent or short-trunked; stem decumbent or erect. **Leaves** diamond-shaped; petiole toothed; lamina subpinnately ribbed, lower margins toothed. **Inflorescences** short, curved, flowers solitary or in cincinni of 2–4; peduncular bracts to 7, slightly inflated. **Flowers** with calyx cup-shaped, lobes 3; corolla divided almost to base into 3 valvate lobes; stamens epipetalous, filaments broad, fleshy, connate basally to form a ring; ovule anatropous. **Fruits** rounded, mesocarp cracking to produce thick, corky warts; seed globose. **Distr:** Thailand, Malay Peninsula, Sumatra and western part of Borneo. **Cult:** Propagate from seeds. Requires a moist soil, preferably with organic matter, in shade.

J. altifrons (Rchb. f. and Zoll.) H.E. Moore, *Principes* 5: 116, 1961. (Syn: *Teysmannia altifrons*). [Diamond Joey Palm, Joey Palm]. **Palms**, solitary, 4–6 m tall (but variable in size and robustness, across the native range); stem subterranean, creeping; crown 4.5–7.5 m wide; crownshaft absent. **Leaves** to 30, palmate, forming an erect tussock; lamina elongated, diamond-shaped (rhomboid), 1.8–6 × 1.8 m, unsegmented, deeply pleated, margins coarsely toothed, adaxially light to deep green, abaxially pale green; petiole 1.8–3 m long, with fine spines. **Inflorescences** 60–90 cm long. **Flowers** bisexual, whitish, with a distinct odor. **Fruits** round, 5 cm across, corky, light brown. **Distr:** Peninsular Thailand to W Malesia. **Ecol:** In lowland rainforests; 300–1,200 m. **Zones:** 11–12. **Info:** Rare in the wild. Dried leaves often used as roofing material (as thatch), in its native range.

J. magnifica J. Dransf., *Gard. Bull. Singapore* 26: 75, 1972. [Blue Joey Palm]. **Palms**, solitary, 4–6 m tall; stem subterranean; crown 4.5–7.5 cm wide; crownshaft absent. **Leaves** palmate; lamina elongated diamond-shaped (rhomboid), 1.8–4 × 2 m, unsegmented, deeply pleated, adaxially light to deep green, abaxially densely covered with white pubescence; petiole 1.8–3 m long, with fine spines and 2 conspicuous yellow lines. **Inflorescences** 60–90 cm long. **Flowers** bisexual, whitish. **Fruits** round, 5 cm across, corky, light brown. **Distr:** Peninsular Malaysia (Selangor). **Ecol:** In low mountainous rainforests. **Zones:** 10–12. **Info:** Critically endangered in the wild. This species has the widest leaves in the genus.

Licuala

Species 149. **Palms**, clustering or solitary; stem with rings of leaf scars. **Leaves** ± rounded or flabellate, palmately divided into single-folded to several-folded cuneate-truncate segments or undivided; segments with apices truncate or toothed; petiole margins with spines; leaf sheath fibrous. **Inflorescences** interfoliar, simple or branched; spathe tubular. **Flowers** bisexual or unisexual; calyx cupular or tubular, trilobed or irregularly splitting; corolla deeply trilobed; stamens 6, filaments distinct or basally connate to form a conspicuous tube, with equal 6-toothed or trilobed staminate ring at apex; ovary developing from distinct or subdistinct 3-carpels, apically connate to form a slender style with dotlike stigma, ovule basifixed, anatropous. **Fruits** drupes, small, globose to ellipsoid, rarely narrowly oblong, epicarp membranous, smooth, rarely with corky-warted, mesocarp fleshy, somewhat fibrous, endocarp thin, crustaceous, with stigmatic remains at apex; seed globose, smooth or with large convolute intrusion of seed coat. **Distr:** Bhutan to Vanuatu.

L. glabra Griff., *Calcutta J. Nat. Hist.* 5: 329, 1845. **Palms**, small; stems solitary or rarely with basal shoots, rarely short and subterranean, more often to 2 m tall, 7 cm Ø. **Leaves** to 1.1 m across, divided into 6–33 leaflets, central one as wide or wider than the others, sometimes split into 2 or 3 lobes, sometimes borne on a short stalk; petiole to 2.2 m long. **Inflorescences** monoecious, shorter than the leaves, arching to 2 m long, with 4–11 branched partial inflorescences; partial inflorescence bracts tubular; flowering branches straight, glabrous, rarely with some hairs; flowers in groups. **Flowers** not hairy, cigar-shaped when young, calyx later cylindrical, 2.5 mm long, falling to leave a minute protruding stalk; calyx not lobed, margin becoming shallowly torn; petals as long as calyx, narrow. **Fruits** ellipsoid, to 1.5 cm long, 1 cm Ø, orange, persistent petals remaining erect as the fruits develop. **Distr:** Peninsular Thailand to peninsular Malaysia. **Ecol:** In lowland or montane rainforests; to 1,300 m. **Zones:** 11–12.

L. grandis H. Wendl., *Ill. Hort.* 27: t. 412, 1880. (Syn. *Pritchardia grandis*). [Ruffled Fan Palm, Vanuatu Fan Palm]. **Palms**, solitary, 4–6 m tall; stem 2.4–4 m tall, 8 cm Ø, upper parts with dense fiber and narrow leaf bases, lower parts gray with closely set rings or grooves; crown 2.4 m across; crownshaft absent. **Leaves** palmate, semicircular to broadly wedge- or diamond-shaped, 1 × 0.9 m, unsegmented, corrugated, deep green; petiole 90 cm long, with small spines at base. **Inflorescences** longer than leaves, clusters, branching to 3 orders, flowers arranged spirally and distantly, borne on short distinct spurs, each subtended by a minute triangular bract. **Flowers** bisexual, 5 × 3 mm, petals twice as long as sepals; staminal ring trilobed, lobes emarginate. **Fruits**

Below: *Licuala glabra, Licuala grandis, Licuala orbicularis.*

SZ

SZ

Above: *Licuala peltata*, *Licuala peltata* (inflorescence), *Licuala spinosa*, *Livistona chinensis*.

round, 1.2 cm across, red, in pendent clusters. **Distr:** Solomon Islands, Vanuatu. **Ecol:** In rainforests. **Zones:** 11–12.

L. orbicularis Becc., *Malesia* 3: 83, 1889. (Syn: *L. veitchii*). **Palms**, solitary, 1.8–2.4 m tall, 4–4.5 m wide; stem very short or subterranean; crownshaft absent. **Leaves** palmate, orbicular or nearly orbicular, 0.9–1.5 m wide, unsegmented, corrugated, adaxially deep glossy green, abaxially lighter green; petiole thin, 1.8 m long. **Inflorescences** erect to arching, 90 cm long, sparsely branched; rachis slender; rachillae densely flowered. **Flowers** small, bisexual, whitish or cream. **Fruits** round, 1.2 cm across, bright red. **Distr:** Borneo (Sarawak). **Ecol:** In the undergrowth of lowland rainforests. **Zones:** 11–12. **Info:** This palm is now nearly extinct due to habitat loss.

L. peltata Roxb. ex Buch.-Ham., *Mem. Wern. Nat. Hist. Soc.* 5: 313, 1824. **Palms**, solitary, to 2 m tall; stem ~10 cm Ø, often covered by old, black spinelike leaf bases. **Leaves** 5–10 at the crown; leaf sheath 50–60 cm long; petiole to 2.7 m long, green, almost glabrous, armed with flattened, recurved elongate spines, apices black, evenly spaced, to 1.5 cm long; lamina thick chartaceous, maximum span ~1.5 m, transverse veinlets obscure, scattered. **Inflorescences** arching well above the leaves, usually 2 or 3 per stem, to 3.5 m long; rachis 1.5–2.3 m long; peduncular bract 1, tubular, 33–36 cm long, green; rachis bracts similar to peduncular bract, apical one being empty and 7–9 cm long; flowers to 200 on longest rachilla, solitary. **Flowers** with calyx campanulate, 6–7 mm long, sericeous; corolla trilobed, 1–1.2 cm long, basally fused, sericeous, basally white, apically green; androecium 8–9 mm long; ovary 2–3 mm long, style 7–8 mm long. **Fruits** ellipsoid, 1.2–1.9 × 0.9–1 cm; seeds ellipsoid, 1.2–1.3 × 0.7–0.9 cm. **Distr:** India to S Thailand, peninsular Malaysia; Andaman and Nicobar Islands. **Ecol:** In lowland rainforests or drier forests, often on limestone soils; to 500 m. Fl. Feb–Mar, Jun–Jul. **Zones:** 10–12.

L. spinosa Wurmb, *Verh. Batav. Genootsch. Kunsten* 2: 474, 1780. [Mangrove Fan Palm]. **Palms**, cespitose, to 4 m tall; stems forming large clumps, 4–5 cm Ø, ringed by leaf scars. **Leaves** 13–17; leaf sheath to 40 cm long; ligule to 10 cm long, caducous; petiole to 2.5 m long, greenish, spines to 1 cm long; lamina to 1–1.5 m long, divided into 15–19 segments, adaxially dark green, shiny, thickly chartaceous; segments ± equal, 44–72 cm long. **Inflorescences** 2 or 3 per stem, 0.9–2.5 m long, extended above crown, with 5–9 partial inflorescences; peduncles 20–50 cm long; rachis 0.7–2 m long; rachis bracts dark green to yellowish-beige, tubular, to 4–8 cm long; rachillae 2–7, rigid, curving downward, longest rachilla 20–40 cm long; flowers to 200 per rachilla, solitary or in 2(–3)-flowered cincinni, flowers. **Flowers** with calyx cylindrical to cyathiform, ~3 mm long, golden-hairy; receptacle and calyx shortly fused; corolla light beige to white inside, 3.5–4 mm long, densely hairy; androecium, 2.5–3 mm long, fused to corolla; ovary ~1 mm long. **Fruits** globose, ~8 mm Ø. **Distr:** Indo-China to Philippines. **Ecol:** Often associated with mangroves, forming tussocks and also in forests. **Zones:** 10–12.

Livistona

Syn: *Saribus*, *Wissmania*. [Cabbage Palm, Footstool Palm].

Species over 25. **Palms**, small to large; stems solitary, erect, slender (rarely) to robust (>20 cm Ø), initially covered by persistent leaf sheaths, later becoming bare or covered with persistent petiole bases, ringed conspicuously or obscurely with leaf scars. **Leaves** with petiole not split at base, strongly usually armed; abaxial hastula minute or absent; adaxial hastula conspicuous; lamina palmate or costapalmate, plication induplicate, segments basally connate, lanceolate, not producing fibers between segments. **Inflorescences** axillary within crown of leaves, paniculate with 3–5 orders of branching, ± as long as leaves; prophyll membranous; peduncular bracts many, obscuring rachis; rachillae pubescent. **Flowers** mostly bisexual, borne singly along rachillae sometimes in small groups; perianth biseriate; calyx cupulate, trilobed; corolla trilobed, valvate; stamens 6, connate in short tube; gynoecium with pistils 3, distinct basally, glabrous, styles connate, slender. **Fruits** drupes, exocarp blackish, smooth, mesocarp fleshy, endocarp bony; seeds globose or ellipsoid, endosperm homogeneous, embryo lateral. **Distr:** Africa, Asia, Australia, Pacific Islands. **Cult:** Propagate from seeds. Requires fertile, neutral to acidic, well-drained soil in full or partial sun.

L. chinensis (N.J. Jacquin) R. Brown ex Martius, *Hist. Nat. Pal.* ed. 1.3: 240, 1839. (Syn: *Latania chinensis*, *Livistona mauritiana*). [Chinese Fan Palm]. **Palms**, stems usually short and stout, may attain a height of 13 m, profusely ringed; lower portion smooth, upper portion is surrounded by remains of petiole and reticulate fibrous network. **Leaves** flabellate, 1–1.8 m across, plicate, central part large ovate, divided into 60–90 segments about the middle; segments linear-lanceolate, acuminate, smooth, pendulous, apex deeply forked for ~30 cm; petiole large, 0.6–1+ m long, slender, green, glabrous, plano-convex, margins spiny only on lower portion or rarely lacking spines, spines small, green, pointed downward. **Inflorescences** light yellow when young becoming green with an unpleasant odor, repeatedly branched, bracts many, tubular, sheathing peduncle and branches. **Flowers** small, sessile in groups of 2–4 flowers; sepals imbricate; petals ~3 × 2 mm, joined at base. **Fruits** oval, 1.9 × 1.1 cm, ripening from green to bluish-green, closely packed along fruiting spikes. **Distr:** S Japan, Ryukyu Islands, Bonin Island. **Ecol:** In hammocks and mesic woods; 0–10 m. Fl. Feb–Mar. **Zones:** 8–12.

L. decora (W. Bull) Dowe, *Austrobaileya* 6: 979, 2004. (Syn: *Corypha decora*, etc.). [Ribbon Fan Palm, Weeping Cabbage Palm]. **Palms**, solitary, 10–21 m tall; trunk 12–15 m tall, 25 cm dbh, petiole bases usually present in lower 2 m of the trunk; crown 4.5 m wide; crownshaft absent. **Leaves** simple, palmate, 1.3–1.8 × 2.7 m, folded and curved along the midrib, with very deep green to grayish-green segments that are completely pendent for most of their length, 14–16-veined, glabrous, fork at segment apex 10–16 cm long; petiole 1.6–2.8 m long, spines absent or dark brown, ~4 mm long. **Inflorescences** panicles, axillary, 2.4 m long, arching out of the crown. **Flowers** bisexual, yellow; petals 3, ~2 mm long. **Fruits** round, 1.3–1.5 × 1.2–1.5 cm, black. **Distr:** Australia (E Queensland). **Ecol:** In coastal forests and swamp forests, gallery forests, littoral rainforests; 0–200 m. Fl. Sep. **Zones:** 9–12.

L. rotundifolia (Lamarck) Martius, *Historia Naturalis Palmarum.* 3: 241, plate 102, 1838. (Syn: *Corypha rotundifolia*, etc.). [Footstool Palm]. **Palms**, 15–20 m tall; trunk straight, smooth, marked with close annular scars. **Leaves** crowded at apex; petiole long and armed on the sides with sharp hard teeth; lamina orbicular, ~1 m Ø, pleated, base cordate, cleft into numerous 2.5–4 cm wide segments, those in the middle ~20 cm long, toward the sides longer, all cleft at apex into 2, lanceolate, acuminate, 3–5 cm long lobes, apices stiff. **Inflorescences** axillary, pendulous, to 1.5 m long, composed of sheathing partial spathes, dividing into 3 main axes, each axis bearing up to 4 orders of branching, branches spreading. **Flowers** small, sessile, numerous, ~2 mm long. **Fruits** globose, somewhat fleshy, ripening from green through yellowish-orange to black, ~1.5 cm Ø. **Distr:** Malay Peninsula, Java. **Ecol:** In disturbed hammocks and mesic woods; 0–10 m. Fl. May–Jul. **Zones:** 11–12.

L. saribus (Lour.) Merr. ex A.Chev., *Bull. Écon. Indochine*, n.s., 21: 501, 1919. (Syn: *Corypha saribus*, *Livistona hoogendorpii*, etc.). [Taraw Palm]. **Palms**, solitary, 15–18(–40) m tall; trunk 12–15 m tall, 30(–65) cm dbh, gray to reddish-brown, rough with old leaf bases; crown 4.5 m wide; crownshaft absent. **Leaves** palmate, circular, 1.2 m across, semi-segmented into a few broad divisions on juvenile plants, divided into slightly pendent segments halfway to the stalk on older plants, pale to deep green; leaflets 80–90, apices deeply split and pendent, threadlike; petiole 1.5–1.8 m long, sometimes reddish, with strong, large, backward-pointing spines. **Inflorescences** 1.5(–2.3) m long and mostly erect. **Flowers** bisexual, yellow; calyx trifid; stamens 6, anther cordate-oblong; pistils 3, style coalescent, stigmas connate or distinct. **Fruits** globose to ellipsoid, 2.5 cm long, 2 cm Ø, shiny, blue or grayish-purple. **Distr:** Indo-China to W and C Malesia. **Ecol:** In coastal swamps and rainforests; to 600 m. **Zones:** 9–12.

L. woodfordii Ridl., *Gard. Chron.* ser. 3 (1), 23: 177 ('*woodfordi*'), 1898. (Syn: *L. beccariana*). [Nggela Fountain Palm]. **Palms**, hermaphroditic; trunk to 16 m tall, 12–20 cm dbh; crown globose to broadly conical. **Leaves** 30–60; lamina costapalmate, regularly segmented, subcircular to circular, 0.6–1.7 × 0.45–0.9 m, divided for half its length, with 60–70 segments, rigid, adaxially shiny, midgreen, glaucous, abaxially paler, with fine powdery wax, apical lobes acuminate, semipendulous; hastula raised ~1 cm; petiole 1.1 m long, with white powder, margins unarmed or proximally with single, small, curved, green spines. **Inflorescences** branched to 3 orders, 1.2–2.7 m long; partial inflorescences 5–10, longest to 45 cm long; prophyll to 14 cm long; rachillae 4–6 cm long, basally with brown-purple tomentum, distally glabrous; flowers in clusters of 2–6, sessile. **Flowers** subglobose in bud, 1–1.5 mm long; sepals obtuse, ~0.3 mm long, imbricate, red; petals deltoid, 0.8–1 × 0.7–0.8 mm, red; style sharply tapered. **Fruits** globose, 0.7–1.2 cm Ø, reddish-orange to reddish-brown; seed globose, 6–9.5 mm Ø. **Distr:** Papua New Guinea, Solomon Islands. **Ecol:** In coastal forests; 0–120 m. Fl. Jan–Apr. Fr. May–Jul. **Zones:** 11–12. **IUCN:** VU.

Raphis

Species 12. **Palms**, small dioecious, 1.5(–3) m tall, forms dense clumps; stem slender, enclosed in a fibrous network. **Leaves** palmate, spaced along stem, midrib absent, palmately segmented nearly to base, induplicate, leaflets blunt-tipped, several ribbed, margin toothed; fibrous leaf-sheath at base of petiole. **Inflorescences** axillary, small. **Flowers** unisexual, calyx trilobed; petals connate; stamens 6, epipetalous; carpels 3, free. **Fruits** 1–3 per flower, small, 1-seeded berries; seeds thin, stony. **Distr:** Japan, S China, Indo-China to peninsular Thailand. **Cult:** Propagate from seeds. Requires fertile, humus-rich, neutral or slightly alkaline, moist, well-drained soil in full or partial sun.

R. excelsa (Thunb.) Henry ex Rehder, *J. Arn. Arb.* 11: 153, 1930. (Syn: *Chamaerops excelsa*, *R. flabelliformis*). [Dwarf Ground Rattan]. **Palms**, small, 0.9–1.2 m tall; stems many, grouped, to 8 cm dbh, densely covered by the leaf-sheath and coarse fiber. **Leaves** ~30 cm across, green on both surfaces, palmately divided almost to base into 6–9 segments, segments uniformly curving or drooping, apex broad, hastula fibrous; petiole ~20 cm long; base of leaf stalk with a sheath, sheath fibrous, dark brown, covering the stem. **Inflorescences** with flowers sessile. **Flowers** yellowish; ♂ flowers with calyx 3-toothed, cupular, corolla trilobed, stamens 6 and rudiments of carpels; ♀ flowers similar but calyx stalked, carpel separate, short styled. **Fruits** small, with soft exterior; seeds globose with homogeneous endosperm, ~3 mm across. **Distr:** S China and Indo-China. **Ecol:** In lowland forests or dry forests; to 1,000 m. Fl. Jun–Jul. **Zones:** 9–11.

Acoelorrhaphe

Syn: *Paurotis*. [Paurotis Palm].
Species 1. *Generic description same as specific.* **Cult:** Propagate from seeds. Requires a humus-rich, moist soil in full to partial sun.

A. wrightii (Grisebach and H. Wendland) H. Wendland ex Beccari, *Webbia*. 2: 109, 1907. (Syn: *Copernicia wrightii*, etc.). [Everglades Palm, Paurotis Palm]. **Palms**, densely clustering, 6–12 m tall; stems cespitose, erect, slender, to 4.5–7 m tall, 10–15 cm dbh, brown, covered in tardily deciduous leaf bases; crown 4.5–6 m wide; crownshaft absent; sheath fibers soft, eventually sloughing off revealing smooth to fissured stem. **Leaves** palmate, semicircular, 90 cm across, deeply segmented; segments lanceolate, stiff, basally connate, adaxially light to deep green, abaxially silvery-green, costa minute to absent; adaxial hastula conspicuous, abaxial hastula absent; petiole armed with stout hooked teeth, occasionally minutely dentate. **Inflorescences** panicles, axillary, within the crown, arching, longer than leaves, branching for 3(–4) orders; peduncular bracts 2, tubular, 2-keeled, long; rachillae pendulous, pubescent; flowers solitary, or in clusters of 2 or 3 along rachillae, sessile. **Flowers** bisexual; perianth biseriate; sepals 3, imbricate, briefly connate, margins ciliate; petals 3, basally connate, valvate, triangular with adaxial crest; stamens 6; pistils 3(–4), styles connate. **Fruits** globose, stigmatic scar apical; seeds 1 per fruit, globose. **Distr:** N America, SE Mexico, West Indies, C America, and N South America. **Ecol:** In lowland areas or near the sea, inundated by fresh or brackish water, thickets in savannas, or occasionally in pine woods. Fl. Apr–May. **Zones:** 8–12.

Pritchardia

Species 27. **Palms**, solitary, medium-sized, unarmed, flowering repeatedly; trunk smooth, columnar with grooved rings. **Leaves** costapalmate, usually deciduous; sheaths tomentose or glabrous, becoming fibrous, split opposite the petiole; petiole not toothed, usually tomentose, upper side concave; lamina segments 1-ribbed, divided only about halfway to

Top row: *Livistona decora, Livistona rotundifolia, Livistona saribus, Livistona woodfordii.*
Bottom row: *Raphis excelsa, Acoelorraphe wrightii, Pritchardia pacifica, Pritchardia thurstonii.*

SZ

F&KS

SZ

SZ

Above: *Serenoa repens, Washingtonia robusta, Copernicia rigida.*

the midrib and have a neat pleated appearance. **Inflorescences** borne among the leaves, 1–4 per axil, spikes or panicles branched to 3 orders; inflorescence bract 1, with several bracts on the peduncle. **Flowers** solitary, bisexual, small, cream to orange, bell-shaped; petals detaching as a cap; stamens 6. **Fruits** black to dark brown (or reddish-purple), smooth, flesh thin, firm, inside fibrous; seed endosperm not ruminate. **Distr:** SW and C Pacific. **Cult:** Propagate from seeds. Requires a humus-rich, well-drained soil in warm, humid, sunny conditions.

P. pacifica Seem. and H. Wendl., *Bonplandia* (Hannover) 10: 197, 1862. (Syn: *Washingtonia pacifica*, etc.). [Fiji Fan Palm, Pacific Fan Palm]. **Palms**, solitary, 7–12 m tall; crown 4 m across; trunk 6–9 m tall, 25–30 cm dbh, light gray to pure white, smooth for the most part; crownshaft absent. **Leaves** ~20, palmate, wedge- to diamond-shaped, forming almost a 120° sector, 0.9–1.8 × 1 m; leaflets 73–90, deeply pleated, stiff, rigid, dark green, covered with whitish-brown tomentum when young, later glabrous; petiole ~1–2.1 m long, white-tomentose, soon glabrous, base brown-fibrous. **Inflorescences** 1–3 per axil, slightly shorter than the petioles, 0.9–1.5 m long; branches short, yellow. **Flowers** bisexual; corolla with petals 3, yellow-green, 7–8 mm long. **Fruits** round, 0.6–1.2 cm Ø, glossy, dark brown to blackish; seed 7 mm Ø. **Distr:** Tonga (Vavau). **Ecol:** In coastal areas, often in tropical dry forests. Fl. any time of the year. **Zones:** 11–12. **Info:** The fruits are edible.

P. thurstonii F. Muell. and Drude, *Gartenflora* 36: 486, 1887. (Syn: *Washingtonia thurstonii*, etc.). [Lau Fan Palm, Thurston's Fan Palm]. **Palms**, solitary, 7.5–13.5 m tall; crown 3–4.5 m wide; trunk 6–9 m tall, 20–30 cm dbh, closely ringed; crownshaft absent. **Leaves** palmate, semicircular to diamond-shaped, 1.2–1.8 m across; leaflets 50–70, pleated, medium to deep green, both surfaces waxy-glaucous, abaxially with small elliptic scales, glabrous when mature; petiole 60–90 cm long. **Inflorescences** 1 per leaf axil, much longer than the leaves, 3 m long, projected out and down from the crown; peduncle 2.7 m long; panicle ~30 cm long; rachillae short, yellow. **Flowers** bisexual. **Fruits** round, 0.6–1.2 cm across, deep red to black or reddish-purple; seed 4 mm Ø. **Distr:** E Fiji (Sovu Islets, Vulaga, Ogea), Tonga ('Eua). **Ecol:** In coastal regions. **Zones:** 11–12.

Serenoa

Species 1. *Generic description same as specific.* **Cult:** Propagate from seeds. Requires any soil in full sun; salt-tolerant.

S. repens (W. Bartram) Small, *J. New York Bot. Gard.* 27: 193, 1926. (Syn: *Brahea serrulata*, etc.). [(Silver) Saw Palmetto]. **Palms**, clustered, shrubby, 1.8–9 m tall; stems subterranean (rarely to 5 m tall), procumbent or ascending, covered with leaf bases or becoming striate or smooth with age. **Leaves** palmate, plication induplicate; segments lanceolate, basally connate, apices acute or 2-cleft, adaxial hastula well-developed, obtuse, costa absent; petiole armed with fine teeth, sometimes only basally, base not split. **Inflorescences** axillary within crown of leaves, paniculate, ascending, about as long as leaves, with 2 or 3 orders of branching; main axis bearing 2 peduncular bracts above prophyll; rachillae pubescent; flowers borne singly or in pairs along rachillae, sessile. **Flowers** bisexual, perianth biseriate; calyx cupulate, trilobed; petals 3, imbricate, elliptic, reflexed; stamens 6 in 2 whorls, filaments narrowly triangular, basally connate; pistils 3, distinct basally, glabrous; ovules 3, only 1 developing into fruit, styles connate, glabrous; nectaries 3, septal. **Fruits** drupes, ellipsoid; seeds brown, ellipsoid, with conspicuous longitudinal raphe. **Distr:** SE USA. **Ecol:** Usually in woodlands, in dense colonies; 0–50 m. **Zones:** 8–11.

Washingtonia

Syn: *Neowashingtonia*. [Cotton Palm, Fan Palm, Petticoat Palm, Weeping Palm].

Species 2. **Palms**, solitary; stems erect, tall, partly or completely covered with old leaf bases and marcescent dry leaves forming conspicuous 'skirt' around stem. **Leaves** with sheath fibers soft; petiole split at base, margins armed with teeth, sometimes unarmed in tall plants; abaxial hastula absent; adaxial hastula irregularly shaped, margin becoming tattered, fibrous; costa prominent; lamina costapalmate; plication induplicate; segments lanceolate, basally connate, bearing fibers between segments, apices 2-cleft or irregularly tattered into fibers. **Inflorescences** axillary within crown of leaves, paniculate, arching well beyond leaves, with 2 orders of branching; prophyll leathery; rachis bracts very conspicuous, tubular at base, distally flattened and leathery, long; rachillae glabrous. **Flowers** bisexual, borne singly along rachillae, short-pedicellate; perianth biseriate; calyx cupulate, trilobed, apices and margins irregular; petals 3, long, chaffy, basally connate into tube; stamens 6, briefly adnate to petals; pistils 3, basally distinct, styles connate. **Fruits** drupes, blackish, ellipsoid, exocarp smooth, mesocarp thin, fleshy, endocarp thin; seeds ellipsoid. **Distr:** W North America, NW Mexico. **Cult:** Propagate from seeds. Requires a well-drained soil in full sun; drought-tolerant.

W. robusta H. Wendland, *Garten-Zeitum (Berlin).* 2: 198, 1883. (Syn: *Neowashingtonia robusta*, etc.). [Mexican Fan Palm, Sky Duster, Thread Palm, Washington Palm]. **Palms**, solitary, 21–30 m tall; stem slender, 20–27 m tall, 30–45(–80) cm dbh, sometimes covered in a 'skirt' of dead leaf bases, older parts grayish to white; crown 2.4–4 m across; crownshaft absent. **Leaves** palmate, semicircular to almost circular, 0.9–1.2 m across, divided into segments that extend more than halfway into the deep green lamina, tips pendent; petiole 1.2 m long, armed with reddish-tinged, slightly recurved spines 2.5–5 cm long. **Inflorescences** 6–7, dense spikes, paniculate, declined, 3 m long, exceeding leaves; primary branches 4–5, each bearing 2–5 separated thyrses, which are exceeded by their ligulate, chartaceous, spathoid bracts. **Flowers** small, white, fleshy; calyx 4.5 mm long; petals narrowly lanceolate, 6–7 × 1.5–2.2 mm; ovary long-obovate. **Fruits** fleshy, round, 0.6–1.2 cm across, dark brown to black when ripe, slightly wrinkled; seeds excavated on the raphal face. **Distr:** Mexico (C and S Baja Califórnia, W Sonora). **Ecol:** In arroyos and canyons at natural springs; to 350 m. Fl. May–Jul. **Zones:** 9–11.

Copernicia

Species 22. **Palms**, solitary; trunk sometimes covered with old leaf bases, scarred, or occasionally bare, base often swollen. **Leaves** palmate to shortly costapalmate, petiole lacking, short, or elongate, margins toothed; adaxial hastula short to long, unarmed or spiny; lamina wedge-shaped to orbicular divided $\frac{1}{4}$ to $\frac{1}{3}$ into single-fold segments, outermost bifid; dead leaves remaining on the palm creating a 'skirt' below the crown. **Inflorescences** branched to 6 orders; peduncular bracts 0–1; subsequent bracts tubular, often densely tomentose; flowers solitary or in cincinni of 2–4. **Flowers** with sepals 3, connate; corolla tubular, trilobed, thick-tipped, valvate; filament bases united in cupule. **Fruits** with endosperm ruminate; germination remote-tubular; eophyll entire. **Distr:** Caribbean, S tropical America. **Cult:** Propagate from seeds. Requires a well-drained soil in full or half-sun.

C. rigida Britton and P. Wilson, *Bull. Torrey Bot. Club* 41: 17, 1914. (Syn: *C. rigida* f. *fissilingua*). [Jata Palm]. **Palms**, 6–15 m tall; trunk slender, cylindric, 25–36 cm dbh, usually with a 'skirt' of persistent dead leaves at the top. **Leaves** stiffly erect; lamina wedge-shaped, 1.3–1.5 m long, margins armed mostly below middle with numerous recurved, straight, ascending, or sometimes hooked teeth 3–7 mm long, deeply grooved below middle, adaxially bright green, abaxially paler, margins of the grooves sometimes armed with teeth 1–4 mm long; petiole 10–15 × 10–14 cm, unarmed; ligule rigid, rhombic-ovate, 25–35 × 17–20 cm, margin armed with ascending, recurved, straight or

Above: *Arenga engleri* (flowers), *Arenga obtusifolia*, *Arenga porphyrocarpa*, *Arenga wightii*.

sometimes hooked teeth 0.3–1.2 cm long, coalescent with and decurrent on the short petiole. **Inflorescences** lax, longer than the leaves; branches slender, the ultimate ones densely clothed with short hairs; flowering branches 4.5–9 cm long, 1–3 mm ∅; spathes abruptly tapering to a long, slender, acuminate tip. **Flowers** small, brownish. **Fruits** subglobose, 1.5–1.9 × 1.4–1.6 cm, brown, shiny; calyx persistent beneath the fruit, lobes triangular; seed subglobose, 0.9–1.1 cm long. **Distr:** C and E Cuba. **Ecol:** In savannas and coastal salt marshes. **Zones:** 10–12.

TRIBE CARYOTEAE

Arenga

Syn: *Didymosperma*. [Sugar Palm].
Species 24. **Palms**, dwarf to large, solitary or clustered, monoecious or dioecious, solitary species monocarpic and clustered species hapaxanthic or pleonanthic, 2–20 m tall; stems unarmed or lightly spiny. **Leaves** pinnate, leaflets usually discolorous. **Inflorescences** bisexual or unisexual, sometimes multiple, spicate or branched to 1–2 orders. **Flowers** ♂ with sepals 3, imbricate, distinct, or briefly connate, corolla valvate, stamens 6, 9-merous, anthers elongate, latrorse, pollen sulcate, spiny; ♀ globose, sepals 3, distinct, imbricate, petals 3, connate basally, valvate, staminodes 3 or absent. **Fruits** globose, 1–3-seeded. **Distr:** Tropical and subtropical Asia to N Australia. **Cult:** Propagate from seeds, or by division of suckers in clumping species. Requires very moist but well-drained soil (tolerant of poor soils) in full sun. **Info:** Fruits are toxic and should be handled with care.

A. engleri Becc., *Malesia* 3: 184, 1889. (Syn: *A. tremula* var. *engleri*, *Didymosperma engleri*). [Dwarf Sugar Palm, Formosa Palm, Taiwan Arenga Palm, Taiwan Sugar Palm]. **Palms**, clustered, to 4.5 m tall; stems 3 m tall, 15 cm ∅; crown 5 m across, cloaked with delicate black stringlike fibers. **Leaves** triangular fishtail-shaped (pinnate), to 2.7 m long, dark olive-green, often twisted gracefully, giving them a slight spiralling appearance; leaflets borne from midrib of each bare rachis, adaxially dark-green to olive, abaxially silvery, 12.5–60 cm long, V-shaped (induplicate) cross-section, in a single plane. **Inflorescences** spicate, borne among leaves, unisexual or bisexual. **Flowers** red, orange or green, sweetly fragrant. **Fruits** deep purple, globular, to 2.5 cm ∅; seeds 1–3. **Distr:** Taiwan, Ryukyu Islands. **Ecol:** On slopes in dense forests. **Zones:** 9–12.

A. obtusifolia Mart., *Hist. Nat. Palm.* 3: 191, 1838. (Syn: *Saguerus langbak*). [Langkap, Sumatra Sugar Palm]. **Palms**, clustered, to 8–16 m tall; stems fibrous, to 30 cm dbh, when young clothed to the base with large, spreading fronds, with age becomes bare with prominent leaf scars; crownshaft absent. **Leaves** pinnate, 3–6 × 3 m; leaflets crowded, leathery, all in one plane, linear-lanceolate, 1.6 m × 8 cm, base attenuate, biauriculate, auricles inferior, apex obtuse, slightly bifid, adaxially bright green, abaxially silvery or grayish; petiole to 90 cm long, thick spiny. **Inflorescences** 1–1.5 m long, with the flowering starting from the lower nodes, continuing upwards. **Flowers** brownish. **Fruits** large, yellow turning blackish, egg-shaped, 3–5(–10) cm across, with a prominent sunken depression at apex. **Distr:** Peninsular Thailand, peninsular Malaysia, Sumatra and Java. **Ecol:** In coastal regions, on steep slopes of hills and in coastal and low elevation rainforests. **Zones:** 11–12.

A. porphyrocarpa (Blume ex Mart.) H. E. Moore, *Principes* 4: 114, 1960. (Syn: *Orania porphyrocarpa*, etc.). [Calappa Palm]. **Palms**, small, clumping, unarmed, 1.5–2.5 m tall; stems thin, canelike, brown; crownshaft absent. **Leaves** pinnate, 0.45–1.2 m long; leaflets ~9–17, variable, 45 cm long, margins entire or with several lobes and finely toothed, adaxially bright green, adaxially grayish-white to glaucous. **Inflorescences** sparsely branched, usually hidden behind the foliage; spadices lateral, congested,, 30–45 cm long, 3–7-branched; spathes 5–6, along the peduncle. **Flowers** unisexual; ♀ purplish. **Fruits** drupes, ellipsoid-oblong, 1.5 cm long, apex with persistent stigma reddish-purple to orange and shiny; seeds 3. **Distr:** Sumatra, Java. **Ecol:** In rainforests. **Zones:** 11–12.

A. wightii Griff., *Calcutta J. Nat. Hist.* 5: 475, 1845. (Syn: *Saguerus wightii*). [Wight's Sago Palm]. **Palms**, forming small clumps; stems few, crowded, black, fibrous, relatively short, 1–2.5 m tall, stout, soboliferous. **Leaves** erect, 4–8.5 m long; leaflets alternate, spreading in almost a flat plane on either side of the rachis, narrow, linear, to 0.9–1.1 m × 3.8–5 cm, base with a pair of prominent auricles, auricles very large and overlapping petiole margins unequally bilobed, sparingly toothed from the middle upwards, apex narrowed, adaxially dark green, abaxially white; petiole 2–2.5 m long. **Inflorescences** with spadices 1.2 m long; peduncles 60 cm long; ♂ buds acute. **Fruits** nearly round, 2 cm across, blackish; seeds usually 3. **Distr:** SW India (Nilgiri Hills). **Ecol:** In sheltered forests; ~900 m. **Zones:** 10–12.

Caryota

Syn: *Schunda-Pana*. [Fishtail Palm].
Species 13. **Palms**, solitary or clustered, monocarpic; stem slender to massive, smooth, with conspicuous nodal rings. **Leaves** bipinnate (1-pinnate in juvenile plants), plication induplicate; segments cuneate, in 1 plane, apices jagged and irregular, basal segments not modified into spines. **Inflorescences** initiated basipetally, first one appearing terminal, successive one borne axillary among leaves, and later ones below leaves, lowest ones often close to the ground, pendulous, paniculate, with 1 order of branching (spicate); prophyll small; peduncular bracts numerous, tubular; flowers sessile, borne in triads of 1 ♀ flower flanked by 2 ♂ flowers. **Flowers** unisexual; ♂ with sepals 3, imbricate, free; petals 3, connate basally, valvate; stamens numerous [6], free; pistillode absent. ♀ with sepals 3, imbricate, free, petals 3, connate for nearly ½ length, valvate, staminodial lobes present or absent, pistil 1, triloculate, ovules 1 per locule, styles indistinct, stigmas trilobed. **Fruits** berries, globular; exocarp purple, smooth; seeds globular. **Distr:** Tropical and subtropical Asia to Vanuatu. **Cult:** Propagate from seeds. Requires fertile, moist but well-drained soil in partial shade.

C. mitis Loureiro, *Fl. Cochinch.* 2: 569, 1790. (Syn: *C. furfuracea*, etc.). [Burmese Fishtail Palm, Clustered Fishtail Palm, Lesser Fishtail Palm]. **Palms**, small, clustering, 5–8 m tall; stem 8–15 cm ∅, stem green, surface covered in slightly white felt-tomentum. **Leaves** 3–4 m; leaflets cuneiform, extramarginal straight, intramarginal ½ arc or irregularly toothed, extended to form a tail or short point; petiole covered in brown-black felt-tomentum; sheath with brown-black net-fibers at the margins. **Inflorescences** 25–40 cm, densely spicate partial inflorescences; spathes covered with furfuraceous tomentum. **Flowers** ♂ with sepals entire at apex, densely ciliate; ♀ with sepals obtuse-rounded at apex, staminodes 3. **Fruits** globose, 1.2–1.5 cm, fuchsia when ripe; seed 1. **Distr:** China, India, Indonesia (Java), Malaysia, Myanmar, Philippines, Vietnam. **Ecol:** In valley forests. Fl. Apr–Jun. Fr. Aug–Nov. **Zones:** 9–12. **Info:** Inflorescences tapped for making sugar and alcohol.

C. ophiopellis Dowe, *Austral. Syst. Bot.* 9: 20, 1996. [Snake-skin Palm]. **Palms**, solitary, 6–8 m tall; trunk to 45 cm dbh, white with wide, dark rings of leaf base scars spaced 30 cm apart; crown sparse. **Leaves** 5–7, bipinnate, 3 m long; leaflets regularly spaced along the secondary rachises, deep green, margins smooth on one side and are shaped like an irregular triangle; leaf sheath and petioles deep green to almost black and irregularly circled with white or pinkish bands of tomentum, which extend to the rachis and secondary rachises. **Inflorescences** produced basipetally, to 1 m long, branched to 2 orders, axes dark green with dense dark brown scales; peduncle elongate, thin; prophyll laterally compressed, peduncular bracts 5–7, tubular, spirally arranged; flowers in triads throughout length of rachilla. **Flowers** ♂ pointed in bud, to 9 × 6 mm, sepals rounded, to 1.5 mm high, petals thick, woody, valvate, apically pointed, to 8 × 4 mm, purple outside, cream to light green inside, not opening widely, stamens 20–25, to 5 mm long, anthers linear, pistillode absent; ♀ inconspicuous during ♂ anthesis. **Fruits** subglobose to 1.6 cm ∅, dark crimson at maturity, predominantly 1-seeded; seed tear-shaped, to 9 mm long, black, glossy. **Distr:** Vanuatu. **Ecol:** In undergrowth of lowland

Top row: *Caryota mitis*, *Caryota ophiopellis*, *Caryota umbraculifera*.
Bottom row: *Caryota urens* (fruits), *Corypha rumphiana*.

and coastal rainforests, often found very close to the ocean; 0–600 m. **Zones:** 10–12. **Info:** The generic boundaries have recently become less clear with the discovery of this species, since it has features of both *Caryota* and *Arenga* (Dransfield and Uhl 1998) and considered a missing link between these 2 genera. Rare in the wild.

C. rumphiana Mart., *Hist. Nat. Palm.* 3: 195, 1838. (Syn: *C. albertii*, etc.). [Australian Fishtail Palm, Giant Fishtail Palm]. **Palms**, monocarpic, monoecious, solitary, to 20 m tall; trunk light gray or dark tan, to 45 cm dbh, rings widely spaced; crown to 6 m wide; crownshaft absent. **Leaves** bipinnate, 3–7 m long, dark green; pinnae regular, ± in a single plane; leaflets 245–360, fish-tail shaped, 10–30 × 10–30 cm, glabrous; petiole to 90 cm long, unarmed; petiolules to 40 cm long, covered with white hairs. **Inflorescences** large, pendent, usually produces ~5 in its life. **Flowers** 1–3.5 cm Ø, cream-colored; calyx cream, yellow or mauve; petals 3, cream or yellow, to 1.5 cm long. **Fruits** ovoid, reddish to blackish, to 2 cm across; seed 1. **Distr:** Philippines to Solomon Islands. **Ecol:** In lowland rainforests, semideciduous forests, gallery forests, monsoon forests and brackish or freshwater mangroves; to 600 m. **Zones:** 11–12.

C. urens L., *Sp. Pl.* 1189, 1753. [Jaggery Palm, Sago Palm, Solitary Fish Tail Palm, Toddy Palm, Wine Palm]. **Palms** solitary, monocarpic; trunk annulate, erect, to 12 m tall, 60–90 cm dbh. **Leaves** to 3.5+ × 3 m, not grouped in a terminal crown, but arising successively along trunk below summit; leaflets obliquely truncate, 1.2–1.5 m long, alternate, swollen at point of insertion, outer margins produced into a narrow point, abaxially bright green; petiole stout; sheath smooth, margin fibrous. **Inflorescences** axillary, very large, 3+ m long, pendulous, once-branched, branches of equal length; peduncles curved, thick, enveloped with large bracts. **Flowers** ♂ much larger than ♀ flowers, longer than broad, slender, 1.2 cm or more in length; calyx trilobed, sepal margins ciliate, corolla trilobed, valvate, woody, stamens 40–45, as long as the petals, anthers acuminate, basifixed, filaments 1–1.5 mm long, pistillode absent; ♀ with sepals as in ♂, petals 3, lobes valvate, short, staminodes 3. **Fruits** small, 2 cm Ø, with irritant juice; seeds 1–2. **Distr:** Tropical Asia (India, Myanmar, Sri Lanka). **Ecol:** Fl. Apr–May. **Zones:** 10–12. **Info:** Cultivated mainly for toddy and jaggery.

TRIBE CORYPHEAE

Corypha

Species 6. **Palms**, massive, monocarpic. **Leaves** very large, ± rounded, palmately divided; segments with a robust costa-nerve, apices with 2 lobes or teeth; petiole margins spiny, adaxially deeply channeled, abaxially rounded. **Inflorescence** terminal, very large cymose panicles, pyramidal, branching to 3 orders; flowers sessile with small scalelike bracteoles; spathe tubular. **Flowers** small, bisexual; calyx cup-shaped, trilobed, sometimes base extends to form a solid stalk; corolla lobes 3; stamens 6, filament basally connate, others subulate, anthers dorsifixed; ovary tricarpellate, 1 ovule per carpel, style short, subulate, stigmas 3. **Fruits** in clusters of 1–3, globose, with remnants of stigma; seed ± globose. **Distr:** Tropical Asia to N Australia. **Cult:** Propagate from seeds. Requires fertile, moist but well-drained soil in full sun.

C. umbraculifera L., *Sp. Pl.*: 1178, 1753. (Syn: *Bessia sanguinolenta*, *C. guineensis*). [Talipot Palm]. **Palms**, monocarpic, massive, robust, 18–25 m tall; trunk 50–60(–90) cm dbh, with dense ring of leaf scars. **Leaves** large, 2.5–5 × 1.5–2 m, palmate, deeply divided to form segments; segments 70–120, ensiform, shortly bilobed, 0.6–1 m × 7–9 cm; petiole 2.5–3 m long, 7–10 cm wide, robust, adaxially channeled, margins with shortly toothed, abaxially apex extended to form recurved costa-shaped rachis. **Inflorescence** single, terminal, very large, erect, panicle, 4–8 m tall, 30–35 partial inflorescences, lower partial inflorescences ~3.5 m long, upper ones ~1 m long, branching to 4 orders; pedicels short, 4 mm thick. **Flowers** 4 × 8 mm, creamy-white when fresh; calyx with broadly triangular lobes, to 2 mm long; petals boat-shaped, obtuse, 2 × 1.5 mm, margins usually inrolled; stamens with 2–3 mm long tapering filaments, the 3 antesepalous 1 mm wide at base, 3 antepetalous 1.5 mm wide at base; ovary globose, 1.5 mm Ø, distinctly 3-grooved, abruptly narrowed into style, this 1.5–2 mm long, slightly 3-grooved; stigma minute, scarcely differentiated. **Fruits** globose, 3.2–3.5 cm across; seed subglobose, or ovoid, 1.8–2 cm across, embryo apical. **Distr:** India, Sri Lanka. **Ecol:** Fl. Feb–Apr. Fr. May–Jun the following year. **Zones:** 11–12. **Info:** This is one of the largest species in this family. The palm matures in 40–75 years. After the fruits have ripened, the palm dies. Gregarious flowering is seen in Sri Lanka.

C. utan Lam., *Encycl.* 2: 131, 1786. (Syn: *C.*

elata, etc.). [Gebang Palm, Kennedy Palm]. **Palms**, monocarpic, to 25 m tall; trunk straight, spirally ridged, 18–22 m tall, 60 cm dbh. **Leaves** spirally arranged, lunate, 2.5–3 m long, 80–100-fid to about the middle, lobes ensiform, obtuse or bifid; petiole 2–4 m long, auricled, margins black with curved spines, 1–2 cm long. **Inflorescences** panicles, rounded or ovoid, branches spreading; spathes many; spadix ~1/4 the height of the trunk (~4.5–5.5 m tall), narrower than the crown, flowers in scattered fascicles on rather stout spreading branchlets of the spadix. **Flowers** 5–8 mm Ø, cream or yellow; calyx 3-toothed; petals 3, reflexed; ovary suddenly contracted into the short style. **Fruits** drupes, 2–3 cm Ø, shortly stipitate, olive-colored, smooth; pericarp friable, endocarp adnate to the testa; seed 1, brown, ~1.6 cm long. **Distr:** NE India to N Australia. **Ecol:** In riparian and gallery forests; 0–500 m. Fl. Aug–Oct. **Zones:** 9–12.

TRIBE BORASSEAE

Latania

Syn: *Cleophora*.

Species 3. **Palms**, medium-sized, solitary, mostly unarmed, dioecious. **Leaves** costapalmate, deciduous in mature plants, base split below petiole; lamina divided to halfway to base, thickly coriaceous, abaxially often woolly on ribs; petiole woolly, margins smooth or few-toothed. **Inflorescences** borne among the leaves; ♂ 2× branched, each first order branch with 1–14 branchlets; ♀ with only 1–2 branchlets per first-order branch; ♂ flowers borne in pits; ♀ flowers not borne in pits. **Flowers** ♂ with stamens 15–30+. **Fruits** oblong or obovoid, endocarp distinctively sculptured; seeds (1–)3, flattened-ovoid, pointed. **Distr:** Mascarenes. **Cult:** Propagate from seeds. Species growing together often produces intermediate hybrids. Requires a well-drained soil in full sun.

L. lontaroides (Gaertn.) H.E. Moore, *Principes* 7: 85, 1963. (Syn: *Cleophora lontaroides*, *L. commersonii*, etc.). [Red Latan Palm]. **Palms**, solitary, dioecious, 10–50 m tall; trunk 6–11 m tall, to 25 cm dbh, deep gray, with closely set rings or ridges; crownshaft absent; crown round, to 4 m wide; young leaves reddish. **Leaves** palmate, 1.8–3 m across, folded, adaxially deep green, abaxially grayish-green; petiole 1–1.5 m long. **Inflorescences** 1.5 m long, borne between the bases of the leaves; ♀ with branches 0.9–1.2 cm long. **Fruits** ovoid or globose, ~8 cm long, brown. **Distr:** Réunion. **Ecol:** In savannas. **Zones:** 11–12. **Info:** This species can be distinguished from *L. loddigesii* in having distinctly red petioles, red-margined and red-veined stiff leaflets when young.

Lodoicea

Species 1. *Generic description same as specific.* **Cult:** Propagate from seeds. Requires a fertile, deep, well-drained soil in full sun.

L. maldivica (J.F. Gmel.) Pers., *Syn. Pl.* 2: 630, 1807. (Syn: *Borassus sonneratii*, *Cocos maldivica*, *C. maritima*, *L. callypige*, *L. sechellarum*, *L. sonneratii*). [Coco-de-mer, Double Coconut, Maldivian Coconut, Sea Coconut]. **Palms** robust, often tall, solitary, unarmed, dioecious; stem erect, slightly expanded at base, inconspicuously ringed with leaf scars. **Leaves** induplicate, costapalmate, deciduous with age and due to its immense weight; sheath soon splitting opposite the petiole; petiole robust, deeply channeled adaxially, rounded abaxially, abaxial surface with minute black dots, irregularly tomentose, margins fibrous basally; hastulae absent; costa long, tapering reaching nearly to end of lamina; lamina about as long as petiole, divided 1/4–1/3 its length into single-fold segments, these shortly bifid, free ends often drooping, adaxially shiny, smooth, abaxially dull with thick indumentums along ridges, midribs prominent abaxially. **Inflorescences** interfoliar, massive, much shorter than the leaves, ± pendulous, unisexual; ♂ elongate, peduncle usually unbranched, ridged, terminating in a single rachilla or 2 or 3 digitately arranged rachillae, prophyll short, 2-keeled throughout its length, split ventrally but with a long closed triangular tip, bracts 1(rarely more), tubular, split ventrally above the center, rachillae massive, with large sterile, imbricate, cuplike bracts basally, above spirally arranged leathery bracts, connate laterally and distally with 60–70 flowers, bracteole fibrous; ♀ unbranched with prophyll and 2(–3+) tubular peduncular bracts, split ventrally, rachilla as an extension of the peduncle, with infertile sheathing bracts, fertile bracts completely sheathing the large ♀ flowers, few in number and sessile. **Flowers** ♂ with sepals 3, connate, in an asymmetric tube, apices imbricate, irregular, corolla with a long stalk and 3 elongate lobes, stamens 17–22, filaments short, anthers elongate, tips reflexed, pistillode columnar, trifid; ♀ very large, ovoid, sepals 3, distinct, imbricate, leathery, petals 3 and similar to sepals, staminodes triangular, basally connate with several (–11) pointed tips, gynoecium ovoid, tricarpellate, stigmas 3. **Fruits** very large, ovoid and pointed, 1(–3)-seeded, epicarp smooth, mesocarp fibrous, endocarp with 1–3 bilobed, thick, hard pyrenes; seed very large, bilobed, endosperm thick, hard, hollow, homogenous. **Distr:** Seychelles (Curieuse, Praslin). **Ecol:** In lowland rainforests, often in valleys. Fl. throughout most of the year. **Zones:** 11–12. **Info:** The seeds are the largest in the plant kingdom, weighing over 30 kg.

Borassus

Species 7. **Palms**, very tall, dioecious. **Leaves** massive, yellowish with broad split base; lamina fanlike, 0.9–1.2 cm across; leaflets induplicate, a thread hanging from each sinus; petiole armed with irregular spines; a hastula on both surfaces. **Inflorescences** very long, sheathed with several bracts; bracts tubular at base; ♂ not much branched, flowers in depressions within imbricating bracts; ♀ less branched than the ♂, flowers solitary. **Flowers** ♂ very small, sepals 3, wedge-shaped, imbricate, petals 3, smaller then sepals, imbricate, obovate-spathulate, stamens 6, filaments connate with corolla, anthers large, oblong; ♀ much larger, globose, perianth fleshy, sepals 3, imbricate, petals 3, small convolute, staminodes 6–9, ovary globose, 3–4-celled, ovule orthotropous. **Fruits** 15–20 cm Ø, wall fibrous; seeds 3, with a hard stony wall, endosperm homogeneous. **Distr:** Tropical and S Africa, Madagascar, tropical and subtropical Asia. **Cult:** Propagate from seeds. Requires a deep, porous, well-drained soil in full sun; drought-tolerant.

B. flabellifer L., *Sp. Pl.* 1187, 1753. (Syn: *B. flabelliformis*). [Palmyra Palm]. **Palms**, solitary, 20–25(–30) m tall; stem rough, blackish. **Leaves** 0.6–1.2 m long; segments 60–80, linear-lanceolate, induplicate; petiole semiterete, 0.6–1.2 m long, edges with hard irregular spines. **Inflorescences** ♂ 0.9–1.5 m long, with ~7 primary branches, secondary branches ~30 cm long, ~2 cm Ø; ♀ with flowering portion to 30 cm long, 2.5 cm Ø, flowers 8–16, spirally arranged. **Flowers** ♂ with sepals narrowly cuneate with truncate inflexed tips, petals shorter, obovate-spathulate, anthers subsessile, large; ♀ ~2.5 cm Ø, sepals fleshy, reniform, petals smaller, ovary subtrigonous, stigmas sessile, recurved. **Fruits** large, broadly ovoid, 15–20 cm Ø, mesocarp fibrous, fleshy and fragrant; pyrenes usually 3, obcordate, 6–7 mm across, black. **Distr:** India, Sri Lanka (far North). **Ecol:** In lowland, flat, disturbed areas, occasionally on hills and near wetlands; to 800 m. **Zones:** 10–12. **Info:** Cultivated often for toddy, edible fruits or for brewing alcoholic drinks.

Subfam. CEROXYLOIDEAE

Genera 8, **species** 43. **Inflorescences** with flowers solitary along the rachis. **Flowers** rarely with elongate calyx and corolla. **Seeds** rarely more than 3 (Stevens 2001 onwards). **Distr:** Mostly in Central and W South America, also NE Australia, Madagascar, Florida and the Antilles.

Pseudophoenix

[Cherry Palms].

Species 4. **Palms**, solitary; stems erect, robust, unarmed, smooth; leaf bases not completely tubular, forming partial crownshaft, unarmed. **Leaves** pinnate, plication reduplicate; segments irregularly spaced along unarmed rachis, in multiple planes, linear-lanceolate; apices acute; petiole unarmed. **Inflorescences** axillary within leafy crown, ascending, with 4 orders of branching; prophyll small, tubular, leathery; peduncular bracts 2, tubular, enclosing most of peduncle, leathery; flowers borne singly along rachillae, subtended by a minute bract. **Flowers** bisexual, sepals 3, connate, forming a shallowly triangular calyx; petals 3, ovate, basally slightly connate, persistent; stamens 6, basally adnate to corolla, filaments linear, anthers dorsifixed; pistils 1,

Below row: *Corypha utan*, *Latania lontaroides*, *Lodoicea maldivica* (habit), *Lodoicea maldivica* (male inflorescences).

Top row: *Lodoicea maldivica* (female inflorescences and fruits), *Borassus flabellifer, Pseudophoenix lediniana*.
Bottom row: *Pseudophoenix sargentii, Phytelephas macrocarpa.*

triloculed, ovules 1 per locule, styles indistinct, stigmas minute. **Fruits** drupes, globose, or bi- or trilobed; exocarp red, smooth; seeds 1–3, globose. **Distr:** S Florida to Caribbean, SE Mexico to Belize. **Cult:** Propagate from seeds. Requires any well-drained soil in full sun.

P. lediniana Read, *Gentes Herb.* 10: 189, 1968. [Cherry Palm]. **Palms**, solitary, 9–15 m tall; trunk 9–15 m, 30 cm dbh, light greenish-gray to almost white with regularly and widely spaced deep green rings; crownshaft 90 cm long, bluish to grayish-green, smooth, waxy, tapering; crown sparse, 4.5–6 m across. **Leaves** ~10, 4 m long, pinnate, finely divided; leaflets many, linear, somewhat limp (pendent when old), dark green, arranged in a single plane (even though the leaflets are borne at slightly different angles); petiole 15 cm long; petiolules short, thick. **Inflorescences** 1.2 m long, much-branched, borne among the leaves; branches short. **Flowers** small, bisexual, mixed with a few unisexual flowers, yellowish-white. **Fruits** round, 2.5 cm across, ripening to deep red. **Distr:** SW Haiti. **Ecol:** On dry, open and exposed hillsides. **Zones:** 10–12.

P. sargentii H. Wendl. ex Sarg., *Bot. Gaz.* 11: 314, 1886. (Syn: *Chamaephoenix sargentii*, etc.). [Buccaneer Palm, Cherry Palm, Hog Palm]. **Palms**, solitary, small, 3–8 m tall; trunk to 20 cm dbh, irregularly swollen, leaf scar rings closely spaced; crownshaft often bulging near the base, waxy. **Leaves** pinnate, ascending, 2 m long; leaflets slender, arranged in several ranks, basal pinnae longest, to 35 cm long; axis and insertion of the pinnae covered with brown scales; petiole silvery, 20–40 cm long, 4–6 cm wide at base. **Inflorescences** much-branched. **Flowers** unisexual, greenish, 5 mm long (with the pseudo-pedicel); perianth in 2 whorls of 3 lobes; stamens with filaments widened at base forming a cup attached to the petals; pistil with 3 conical glands at base, ovary sessile, triloculed, unilocular, stigmas 3. **Fruits** in fused clusters of 2 or 3, yellow turning dark red, waxy, 1-seeded; epicarp thin; mesocarp fleshy, yellowish, hard. **Distr:** S Florida to Caribbean, SE Mexico to Belize. **Ecol:** Usually near the sea, on well-drained, sandy, or limestone soils. Fl. Oct–Dec, Fr. Dec–Jan. **Zones:** 10–11.

Phytelephas

Species 6. **Palms**, glabrous; trunks, erect or inclined and rooting, moderately to very stout. **Leaves** terminal, many, pinnate. **Inflorescences** dioecious; basal spadices and peduncles squamate; ♂ large, ± pendulous, simple, ± scurvy, elongate; ♀ capitate, compact, erect. **Flowers** ♂ with stamens many, clustered, anthers basifixed; ♀ with sepals 3, imbricate, petals 5–10, rather fleshy, narrowly acuminate, staminodia many, ovary 4–9-celled, each cell with a single erect ovule, style terminal, stigma long-filiform. **Fruits** ± pressed together forming a large or small syncarp, 1-celled, ± succulent or lignescent, shortly, broadly and acutely tubercled, the hard nutlets loosely fibrous. **Distr:** Panama to S tropical America. **Cult:** Propagate from seeds. Requires a very moist, humus-rich soil, full or partial sun.

P. macrocarpa Ruiz and Pav., Syst. Veg. Fl. Peruv. Chil.: 301, 224, 1798. (Syn: *Elephantusia macrocarpa*, etc.). [Ivory Nut Palm]. **Palms**; trunk none or short and inclined. **Leaves** equally pinnate, crowded; rachis basally incrassate, sessile or nearly so; segments ~100 pairs, mostly exactly opposite, basally reduplicate, linear-lanceolate, attenuate, *medial* ~1.2 m × 3.5 cm, *upper* <80 cm long. **Inflorescences** with spadices axillary; ♂ ascending, fleshy, spiciform, 1+ m long, peduncle compressed, 25 cm long, rachis ~80 cm long, flowers sessile or subsessile, a cluster of many (150–280) stamens; ♀ with peduncle and spadix longer than ♂, the 12–20 sessile clusters apically compressed in an apical oval receptacle, bracteolate ovaries many, coalescent, the few central ones fertile; spathes 2, glabrate, the outer ~20 cm long, the inner linear-fusiform, 10 × 5 cm. **Flowers** ♂ with filaments 7 mm long, anthers 4 mm long; ♀ with style filiform, 5 cm long, stigmas 5 or 6, 5 cm long. **Fruits** in a globose infructescence (12–20 fruits), 25+ cm Ø, when mature, by mutual pressure 4–6-angulate-obpyramidate, apically muricate-tubercled; nutlets ~4 × 2.5 cm, perianth fragile. **Distr:** Peru to Bolivia and NW Brazil. **Ecol:** In lowland rainforests on alluvial soils, at low elevations. Fr. Feb. **Zones:** 9–12.

Ravenea

Syn: *Louvelia*.

Species 18. **Palms**, solitary, dioecious, armed; stem sometimes swollen at base; crownshaft absent; sheaths densely tomentose. **Leaves** spirally arranged, reduplicate, pinnate, ± erect; leaflets numerous, regularly arranged, single-fold, usually stiff, elongate, apex acute or acuminate, proximal few slender; petiole adaxially flattened. **Inflorescences** interfoliar, axillary, 1 or several within a single prophyll in a leaf axil, erect or spreading; ♂ and ♀ similar. **Flowers** congenitally open, more crowded in the distal part of the rachilla; ♂ symmetric to slightly asymmetrical, sepals 3, triangular, basally connate, petals 3, broadly ovate to ligulate, often fleshy, stamens 6, shortly adnate to sepals and petals; ♀ with sepals and petals similar to ♂, staminodes 6(–10), gynoecium ovoid, carpels 3, connate, each 1-ovulate, stigmas 3. **Fruits** globose or ellipsoid; seeds 1–3, globose or hemispherical. **Distr:** Madagascar and Comoro Islands. **Cult:** Propagate from seeds. Requires any well-drained soil in full or partial sun.

R. rivularis Jum. and H. Perrier, *Ann. Inst. Bot.-Géol. Colon. Marseille*, III, 1: 54, 1913. [Majestic Palm]. **Palms**, solitary; trunk cylindrical, 5–22 m tall, inflated toward middle, 36–50 cm dbh, near crown 15–18 cm Ø. **Leaves** 16–25, porrect to pendulous, slightly arching; rachis green, 1.2–1.7 m long; leaflets in 1 plane, 70–73 per side, 16–64 x 0.5–3.2 cm; sheath 30–50 x15–16 cm, green; petiole 6–20 cm long. **Inflorescences** ♂ multiple, 5–7, branched to 2 orders, 86–90 cm long, peduncle 32–45 cm long, green; ♀ solitary, erect or spreading, branched to 1 order, 1.3–1.5 m long, peduncle 40–60 cm long. **Flowers** ♂ with calyx basally connate, 0.6–1 mm across, free lobes to 1.3 mm long; ♀ with calyx basally connate, free lobes 1–1.3 mm long, triangular, acute, petals ovate or triangular, 3–3.5 mm long, acute. **Fruits** globose to slightly ellipsoid, bright red, 7.5–9 x 7–8.5 mm; seed 1, 5.5–6 x 5.5 mm. **Distr:** SW Madagascar. **Ecol:** In shallow, standing water on riverbanks, in deciduous or gallery forests; 350–750 m. **Zones:** 10–12.

Subfam. ARECOIDEAE

Genera 112, **species** 1,214. **Inflorescences** with prophyll and bract(s); plant usually monoecious with flowers in triads with central flower ♀, lateral flowers ♂, or dioecious with flowers in monopodial inflorescences. **Flowers** with 1 fertile pistil, styles present. **Distr:** Pantropical.

TRIBE CHAMAEDOREEAE

Chamaedorea

[Bamboo Palm].

Species 107. **Palms**, small, unarmed; stems clustered (rarely solitary), erect (rarely creeping or lianoid), slender; leaf sheaths tubular, unarmed, forming crownshaft. **Leaves** pinnate (undivided), with leaf segments regularly spaced along rachis, in 1 plane (many planes), plication reduplicate; segments linear-lanceolate, apical pair of segments sometimes wider than others. **Inflorescences** axillary below crown of leaves, ascending, with 1 order of branching (spicate or 2 orders); prophyll small; peduncular bracts 5–6, tubular, papery; rachillae green at anthesis, turning orange in fruit; ♂ flowers borne singly, partially sunken into fleshy rachillae; ♀ flowers borne singly, slightly sunken into fleshy rachillae. **Flowers** unisexual, sessile, ♂ and ♀ flowers on different plants; ♂ with sepals 3, briefly connate at base (distinct), petals 3, ovate, basally briefly connate (connate by tips), stamens 6, distinct, anthers dorsifixed, pistillode minute; ♀ with sepals 3, free, petals 3, free, ovate, staminodes 6, minute, pistil 1, triloculate, ovules 1 per locule, style indistinct, stigmas minute. **Fruits** drupes, globose; stigmatic scar basal, exocarp black, smooth; seeds globose. **Distr:** Mexico to C and S tropical America. **Cult:** Propagate from seeds or by division. Requires a moderately fertile soil rich in organic matter, in partial sun.

C. metallica O.F. Cook ex H.E. Moore, *Principes* 10: 45, 1966. [Metallic Palm]. **Palms**, solitary, 1.8–2.7 m tall; stem slender, 1–1.3(–2.4) m tall, 1.2 cm Ø, dark green with closely set darker rings and white spots, bears aerial roots at intervals along its length; crown 1 m wide, 60 cm high; crownshaft greenish. **Leaves** obovate, pinnate or undivided, held ± erect, ~30 cm long, apex bifid, deeply cleft into 2 fishtail-shaped wedges; lamina usually unsegmented, bluish-green, with an almost metallic, highly polished sheen, grooves greenish, apices cleft or notched giving a heartlike appearance; some plants have 3–8 S-shaped leaflets of unequal widths; petiole 5–15 cm long. **Inflorescences** branching near the ends; ♂ pendent, thin, branched; ♀ completely erect spikes. **Flowers** unisexual, globose, deep yellow or purplish to orange. **Fruits** round, 1.2 cm across, ripening to black, rachis and stalks bright orange, and swollen. **Distr:** Mexico (Veracruz, Oaxaca). **Ecol:** In rainforests. **Zones:** 10–11.

C. seifrizii Burret, *Notizbl. Bot. Gart. Berlin-Dahlem.* 14: 268, 1938. (Syn: *C. erumpens*) [Bamboo Palm, Reed Palm]. **Palms**, densely clustering, 4 m tall; stems 3 m tall, to 2.5 cm Ø, dark green, indistinctly ringed, leaf scar rings widely spaced, partly enclosed by papery leaf bases; crown 2 m wide; crownshaft loose and indistinct. **Leaves** pinnately divided, ~ 60 cm long; leaflets 12–24 pairs, a few of the distal pairs fused, broad to very narrow, linear-lanceolate, ~20 cm, uniform in width or terminal segments wider than other segment, dark green. **Inflorescences** panicles, erect, sparsely branched; flowers densely packed on the branches. **Flowers** small, yellow or greenish-yellow, unisexual, fragrant. **Fruits** globose, 0.5–1.2 cm Ø, ripening from green through orange to black. **Distr:** SE Mexico to Honduras. **Ecol:** In mesic hammocks and disturbed wooded areas; 0–10 m. **Zones:** 10–12. **Info:** Endangered in the wild.

C. tepejilote Liebm., *Hist. Nat. Palm.* 3: 308, 1849. (Syn: *C. wendlandiana, Nunnezharia tepejilote*, etc.). [Pacaya Palm]. **Palms**, solitary, sometimes clustering, 0.5–7.5 m tall; trunk 3–6 m tall, 1.2–8 cm Ø, dark green, rings widely spaced, dark gray, base with conspicuous roots supporting the trunk; crown 3 m wide; crownshaft loose and indistinct. **Leaves** 3–7, pinnate, 1.5–1.8 m long, arching; leaflets 12–23+ per side, broadly lanceolate, 25–60 x 2.2–10.5 cm, tips long, medium to dark green; rachis 70–140 cm long; petiole 18–60 cm long, sheath tubular. **Inflorescences** pendent, 60 cm long, branched, borne below the leaves; peduncles 7.5–23(–38.5) cm long, erect-divergent; ♂ racemose, branched, rachis 5–20.5 cm long, ~11–50+ branches, 6–18.5 cm long, divergent or pendent; ♀ racemes, rarely subpaniculate, branched, rachis 2.3–27.5 cm long, branches (3–)5–40, 2.5–22 cm, erect or flexuous. **Flowers** unisexual; ♂ ~1.5–2 mm, white to pale yellow, petals essentially suberect, without nerves or indistinctly ribbed; ♀ ~1–2 mm, greenish-white, with unpleasant odor. **Fruits** narrowly ellipsoid to obovoid, 0.8–1.6 x 0.5–0.8 cm, smooth, ripening to black. **Distr:** Mexico (Veracruz, Oaxaca, Chiapas) to W Colombia. **Ecol:** In rainforests; 0–1,600 m. **Zones:** 10–12.

TRIBE ORANIEAE

Orania

Species 18. **Palms**, solitary. **Leaves** distichous or spirally arranged; leaflets single-fold, discolorous. **Inflorescences** branched to 1–3 orders, protandrous; triads subdistichous or spirally arranged. **Flowers** ♂ asymmetrical, sepals 3, distinct or united, petals 3, distinct, valvate, stamens 3–30, anthers linear-oblong, basifixed, latrorse; ♀ with sepals and petals similar to ♂, staminodes 3–6. **Fruits** with epicarp smooth, embryo subapical or lateral, the shape of the fruits depend on the number of seeds. **Distr:** Madagascar, peninsular Thailand to New Guinea. **Cult:** Propagate from seeds. Requires very moist soil in full sun. **Info:** Most parts of the palms are poisonous.

O. regalis Zipp. ex Blume, *Rumphia* 2: 116, 1843. (Syn: *Arausiaca excelsa, O. aruensis*). **Palms**, large; trunk robust, straight, 10–15 m tall, to 60 cm dbh, light gray to tan, heavily ringed with widely spaced darker leaf base scars; crown hemispherical. **Leaves** ~10, spirally arranged; lamina 5 m long, stiff, not arching, pinnae regularly arranged; leaflets pendent, 0.7–1.2(–2) m long, 5–6(–15) cm wide at base, apex irregularly praemorse; petioles stout, 2 m long. **Inflorescences** branched to 2 orders, 1.2+ m long, congested; rachis 24–35 cm long, axes subglabrous with remnants of brown scales; rachillae 10–12 cm long, strongly flexuous, each bearing 46–84 flower clusters, of which 5–17 contain ♀ flowers. **Flowers** cream-colored; ♂ ~4.5 x 1.5 mm, stamens 3(4); ♀ 5 x 4 mm, staminodes 3, subulate. **Infructescences** hanging below the crown. **Fruits** depressed-globose, 4.3–4.8 x 4.7–5.3 cm, orange-colored. **Distr:** SW New Guinea. **Ecol:** In lowland rainforests as a subcanopy palm. **Zones:** 11–12. **Info:** This palm is very rare and has not been recorded in the wild for almost 130 years (Baker et al. 2000).

Below: *Ravenea rivularis, Chamaedorea metallica, Chamaedorea seifrizii, Chamaedorea tepejilote.*

SZ

SZ

Top row: *Orania regalis, Orania regalis* (inflorescence), *Roystonea dunlapiana, Roystonea regia.*
Bottom row: *Roystonea oleracea, Roystonea regia* (inflorescence).

TRIBE ROYSTONEAE

Roystonea

[Royal Palm].
Species 11. **Palms**, solitary; stems erect, >20 cm Ø, smooth, unarmed; leaf bases unarmed, forming crownshaft; crownshaft prominent, green, smooth. **Leaves** with petiole unarmed; lamina pinnate; plication reduplicate; segments linear-lanceolate, in more than 1 plane. **Inflorescences** axillary below the crown of leaves, paniculate, with 2 or 3 orders of branching; flowers sessile, in triads of 1 ♀ flower flanked by 2 ♂ flowers; prophyll tubular; peduncular bract greatly exceeding prophyll, leathery, splitting longitudinally on abaxial side and circumscissilly at base; rachillae covered with copious caducous dendritic trichomes, becoming glabrous. **Flowers** unisexual, self-pollinating; ♂ with sepals 3, imbricate, membranous, petals 3, valvate, stamens 6(–10), anthers dorsifixed, often twisting upon drying, pistillode minute, obscurely 3-cleft; ♀ globose to conic, sepals 3, imbricate, petals 3, basally connate, distally valvate, staminodes 6, basally connate, basally adnate to corolla, pistil 1, ovules 1, style indistinct, stigmas 3. **Fruits** drupes, fibrous, stigmatic scar basal, exocarp ripening from green to red to purplish-black at maturity, thin, leathery, mesocarp fleshy, oily, endocarp hard; seeds 1, nearly globose (obovoid), dorsiventrally compressed, abaxially attached to endocarp, endosperm homogeneous. **Distr:** Florida, Mexico, West Indies, C America, S America. **Cult:** Propagate from seeds. Requires a moist, well-drained, fertile soil in full sun.

R. dunlapiana P.H. Allen, *Ceiba* 3: 15, 1952. **Palms**, erect; trunk to 20 m tall, to 40 cm dbh. **Leaves** 10–15, lowest leaves hanging; crownshaft ~2 m long; rachis ~4 m long, middle segments 78–88 × 2.5–6.5 cm. **Inflorescences** ~1 × 1 m; peduncular bract ~2 m long, narrowly acuminate, as long as the crownshaft, widest in the middle; rachillae ~37 cm long, 1–1.3 mm Ø, stiff. **Flowers** ♂ white, sepals triangular, ~1 × 0.8–1.8 mm, petals elliptical to ovate, 2.9–5.6 × 2.1–2.4 mm, stamens 6, ~3.5 mm long, filaments awl-shaped, 2.4–4.8 mm long; anthers 1.8–2.7 mm long, purplish, pistillode minute; ♀ 3–4 mm long, sepals 1.5 mm long, petals 2.5–3 mm long, staminodes 6. **Fruits** oblong or obovoid, gibbous, 1.2–1.5 × 0.7–1 cm; epicarp purplish-black, stigmatic scar plain; endocarp ellipsoid, 1.1–1.4 × 0.65–0.74 cm; seed ellipsoid, 0.7–1 × 0.5–0.6 cm. **Distr:** SE Mexico to C America. **Ecol:** In lowland, inundated areas, tidal estuaries or mangrove swamps. Fl. Jul–Aug. **Zones:** 10–12.

R. oleracea (Jacquin) O.F. Cook, *Bull. Torr. Bot. Cl.* 28: 554, 1901. (Syn: *Areca oleracea, Oreodoxa oleracea*, etc.). [Cabbage Palm, South American Royal Palm, Venezuelan Royal Palm]. **Palms**, erect, 25–40+ m tall; trunk swollen at base, upward columnar. **Leaves** 3–4 m; leaflets ~100 or more, in one plane nearly basal and apical, usually in 2 planes at middle portion of mature plants, linear-lanceolate, 0.5–1 m × 5 cm, apex long-acuminate, irregularly bilobed. **Inflorescences** 90 cm or more, rachilla wavy curved, semiproducing from the spathe (spathe persistent in fruiting). **Flowers** ♂ 6 mm across. **Fruits** oblong-ellipsoid, convex at one side, purplish-black when ripe, 1.5–2 × 0.9–1 cm. **Distr:** Lesser Antilles to NE Colombia. **Ecol:** In rainforest clearings and low mountainous tropical rainforests. **Zones:** 11–12.

R. regia (Kunth) O.F. Cook, *Science*. ser. 2, 12: 479, 1900. (Syn: *Oreodoxa regia, R. elata, R. floridana*). [Cuban Royal Palm, Florida Royal Palm, Royal Palm]. **Palms**, to 30 m tall; stems gray-white, 35–40 cm dbh, smooth. **Leaves** with segments inserted on rachis in several ranks at divergent angles, giving the leaf almost a plumose appearance. **Inflorescences** to 1 m; rachillae 10–30 cm long, stiff. **Flowers** white; anthers pinkish. **Fruits** ripening from green through red to purplish-black at maturity, dorsiventrally compressed obovoid, ~1 cm long, ~8 mm Ø. **Distr:** S Florida, NC and SE Mexico to C America, Caribbean. **Ecol:** In tropical hardwood hammocks and mixed swamp vegetation in peat soils over limestone; 0–10 m. Fl. Jan–Jul. **Zones:** 9–12.

TRIBE COCOSEAE

Aiphanes

Syn: *Curima, Marara, Martinezia, Tilmia.*
Species 22. **Palms**, monoecious, acaulescent or caulescent, solitary or cespitose; stems to 20 m tall, armed with rings of black spines inserted in bands or spirals, sometimes unarmed with age. **Leaves** spirally arranged or distichous; sheath densely armed with spines; petiole armed like sheath or unarmed; lamina reduplicate, entire, or paripinnate; pinnae to 70 per side, regularly inserted or grouped, lanceolate to broadly cuneate, apex praemorse; midrib abaxially with 1–several spines. **Inflorescences** interfoliar, protandrous, erect or arching, often pendulous in

fruit, spicate or branched to 1(–2) orders; prophyll short, lanceolate, winged; peduncular bract unarmed or spiny; peduncle longer than rachis, ± spiny; rachillae to 300; fertile part of proximal rachillae with triads of 1 ♀ and 2 ♂ flowers, sometimes also with a few tetrads of 2 ♀ and 2 ♂ flowers, distally with dyads of ♂ flowers, or near apex with a few single ones; distal rachillae ♂ or with a few triads at base. **Flowers** unisexual, trimerous; ♂ with sepals free or shortly connate, petals free or shortly connate, ovate-acuminate, valvate, stamens 6, 2-whorled, pistillode small, trifid; ♀ larger than ♂, sepals free, ovate to reniform, petals basally connate, lobes acute-acuminate, valvate (or imbricate), staminodes 6, fused into a cup, ovules 3. **Fruits** globose or ellipsoid, red, rarely white, orange, or purple, glabrous, black-spinulose, or golden-spiny; seed 1, irregularly globose, brown. **Distr:** Tropical America. **Cult:** Propagate from seeds. Requires a fertile, moist, well-drained soil in partial sun.

A. horrida (Jacq.) Burret, *Notizbl. Bot. Gart. Berlin-Dahlem* 11: 575, 1932. (Syn: *A. caryotifolia*, etc.). [Ruffle Palm, Spine Palm]. **Palms**, solitary; stem 3–10(–15) m tall, 6–10(–15) cm Ø, armed with gray to black, to 10 cm long, flattened spines. **Leaves** 10–15; sheath 60–90 cm long, spiny; rachis 1.7–2 m long, armed with numerous black spines, to 5 cm long; pinnae 25–40 pairs, inserted in groups of (2–)4–6(–10); basal pinnae variable, 9–28 × 2–8 cm; middle pinnae 21–52 × 8–25 cm; apical pinnae 4–7 ribbed, 10–23 × 9–25 cm; petiole 0–25 cm long, armed. **Inflorescences** erect or arching; prophyll 1(–2)-branched, 30–35 × 3–8 cm; peduncular bract 0.8–1.4 m long, 4–8 cm wide; peduncles 0.3–1.5 m long, densely armed with black spines (to 5 cm long); rachis 20–75 cm long; rachillae 25–170; basal rachillae 8–40 cm long, flowers in triads. **Flowers** ♂ white, greenish-yellow, or yellow, sometimes fragrant, superficial, narrowly triangular, 1–2 mm long, petals 2.5–5.5 mm long, filaments 1–2 mm long, pistillode 0.5–1 mm long, trifid; ♀ superficial, or sunken into shallow pits in rachillae, sepals 1.5–5 mm long, imbricate, dark, petals 3–7 mm long, pistil sometimes covered spinulose. **Fruits** bright red, rarely orange or white, (1–)1.6–2.3 cm Ø, mesocarp orange. **Distr:** W South America (from Trinidad to Bolivia). **Ecol:** 0–300 m. **Zones:** 11–12. **Info:** Fruits edible.

A. minima (Gaertn.) Burret, *Notizbl. Bot. Gart. Berlin-Dahlem* 11: 558, 1932. (Syn: *A. acanthophylla*, etc.). [Coyure Palm]. **Palms**, solitary, (2–)5–18 m tall; stem 6–20 cm Ø, armed on internodes with rings of black, flattened spines. **Leaves** 10–20, spreading; rachis 1.3–4 m long; pinnae 18–34 pairs; basal pinnae 24–26 × 1–2 cm; middle pinnae 31–80 × 4–11 cm; apical pinnae 2- to several-ribbed, 25–34 × 9–22 cm; petiole 0.15–1.1 m long, armed with black, to 8 cm long spines. **Inflorescences** interfoliar, arching, branched; peduncular bract 0.6–1.9 m long, 1.5–8 cm wide; peduncles 0.28–1.3 cm long, 0.3–2.2 cm Ø, armed; rachis 0.25–1.5 m long, unarmed; rachillae 12–300, with peltate indument, often becoming glabrous; basal rachillae 10–50 cm long, flowers in triads at base, distally ♂; apical rachillae 5–15 cm long, ♂; dyads superficial. **Flowers** ♂ creamy-white to yellow, 3–4 mm long, sepals triangular, carinate, 0.6–3.5 mm long, petals oblong-acuminate, elongate, 3.4–6.1 mm long, filaments 0.9–1.8 mm long, pistillode to 1 mm high; ♀ creamy-white to yellow, 3–4 mm long, sepals broadly ovate, 1.2–1.8 mm long, petals ovate-acute, to 3.8 mm long, pistil ~3 mm high, glabrous. **Fruits** red, 1.2–1.6 × 1.4–1.7 cm. **Distr:** Lesser Antilles, including Dominican Republic, Puerto Rico, Barbados. **Ecol:** In subcanopy or understory of seasonal to wet forests. **Zones:** 11–12. **Info:** Seeds edible.

Attalea

Syn: *Orbignya*, *Scheelea*, etc.
Species 66. **Palms**, monoecious. **Leaves** pinnate, leaflets clustered in groups or evenly distributed along rachis; leaflets with apices inequilateral. **Inflorescences** either androgynous or staminate; peduncular bract large, woody, sulcate, enclosing the inflorescence in bud; ♂ branched to 1 order, rachillae with many flowers, in 1 or 2 rows, or spirally arranged in dyads; *androgynous* with sessile or short-pedicelled ♀ flowers. **Flowers** ♂ with sepals 3, short, triangular, petals 3, lanceolate-apiculate, valvate, flat, stamens 6–10, anthers straight, dehiscing longitudinally; ♀ 2–4 cm long, sepals 3, petals 3, both imbricate, stamens reduced to a prominent staminodial ring, ovary ovate, stigmas 3, apical. **Fruits** with exocarp fibrous, mesocarp pulpy and fibrous; seeds 1–3, irregular. **Distr:** Mexico to tropical America. **Cult:** Propagate from seeds. Requires a fertile, moist, well-drained soil in partial sun, protected from strong winds.

A. butyracea (Mutis ex L. f.) Wess. Boer, *Pittieria* 17: 312, 1988. (Syn: *Cocos butyracea*, *Scheelea butyracea*, etc.). [American Oil Palm, Jaci]. **Palms**, solitary, 18–30 m tall; young stems covered with old leaf bases and 30 cm thick mass of black fibers, making them appear 60–90 cm thick; old palms have much thinner trunks, dark gray to dark tan, leaf scar rings undulating; crown 6–12 m wide, leaves erect when young, nearly full and round when older;

Top row: *Aiphanes horrida*, *Aiphanes minima*, *Attalea butyracea*, *Attalea cohune* (juvenile form).
Bottom row: *Aiphanes minima* (trunk), *Attalea cohune* (fruits).

SZ

Above: *Astrocaryum alatum*, *Bactris gasipaes* (fruits), *Bactris gasipaes* (habit), *Bactris major*.

crownshaft absent. **Leaves** pinnate, 9 m long, juvenile leaves erect; leaflets 90 cm long, midrib invariably twisted so that leaflets at the outer half are vertically oriented. **Inflorescences** long, narrow sprays emerging from very large woody bracts; flowers mostly ♀ with a few ♂ in each cluster. **Flowers** yellow. **Fruits** ovoid, 5–8 cm long, ripening to yellowish-brown. **Distr:** S Venezuela to W South America. **Ecol:** In seasonal and wet forests, along river margins and open savannas; <300(–1,000) m. Fl. Jan–Feb. Fr. Aug–Sep. **Zones:** 10–12.

A. cohune Mart., *Voy. Amér. Mér.* 7(3): 121, 1844. (Syn: *Orbignya cohune*, *O. dammeriana*). [American Oil Palm, Cohune Palm]. **Palms**, solitary, similar but larger than *A. butyracea*; young stems covered with old leaf bases and 30 cm thick mass of black fibers, making them appear 60–90 cm thick; old palms have much thinner trunks, dark gray to dark tan, leaf scar rings undulating; crown 6–12 m wide, leaves erect when young, nearly full and round when older; crownshaft absent. **Leaves** pinnate, 9 m long; leaflets 90 cm long, midrib invariably twisted so that leaflets at the outer half are vertically oriented, apices lax. **Inflorescences** long, narrow sprays emerging from very large woody bracts; flowers mostly ♀ with a few ♂ in each cluster. **Flowers** yellow. **Fruits** ovoid, 5–8 cm long, ripening to yellowish-brown. **Distr:** SE Mexico to Nicaragua, Colombia. **Ecol:** In lowland tropical forests; <500 m. Fl. Dec–Feb, Fr. Feb–Mar, Jun–Jul. **Zones:** 10–12.

Astrocaryum

Species 18. **Palms**, medium- to large-sized, sometimes short and subterranean, and solitary or clustered; stems stout, some species with stems covered with sharp, black, flattened spines to 30 cm long, naked or covered with persistent, spiny leaf bases; spines on all parts are black and flattened. **Leaves** pinnate; leaflets 3–30, stiff and often ascending, always spiny, lamina usually linear, regularly arranged in a single plane or irregularly arranged in clusters and spread in different planes, sometimes incompletely split and leaves appearing almost simple, abaxially whitish. **Inflorescences** borne among leaves, branched to 1 order; peduncles and peduncular bracts are usually densely spiny; flowering branches numerous, with lower part bearing unisexual flowers in triads (2 laterals ♂ flowers with a central ♀ flower), upper parts with only ♂ flowers. **Flowers** ♂ with 3–12 stamens, usually 6. **Fruits** variously shaped, usually beaked, yellow, green-orange, or brown, smooth and spiny, sometimes fruits densely arranged into club-shaped infructescences, in some species the epicarp splits open at maturity revealing the brightly colored single seed. **Distr:** Mexico to E Brazil. **Ecol:** In lowlands; rarely to 1,200 m. **Cult:** Propagate from seeds. Requires a moist, humus-rich soil in full to partial sun.

A. alatum Loomis, *J. Wash. Acad. Sci.* 29: 142, 1939. [Coquillo Palm]. **Palms**, solitary, 6–9 m tall; trunk 3–4 m tall, distinctly ringed and mostly covered in closely set rows of black spines; crown 4.5–6 m wide; crownshaft absent. **Leaves** pinnate, 2.5–3 m long; leaflets broad, 90 cm long, grooved, adaxially deep green, abaxially silvery; petiole 60–90 cm long, spiny. **Inflorescences** spikes, clublike, bisexual, flowers closely set; spathes large, paddlelike, woody, spiny. **Flowers** unisexual, white. **Fruits** round, 2.5 cm across, yellowish-brown. **Distr:** C America to Colombia. **Ecol:** In lowland rainforests; <500 m. Fl. Sep. Fr. Mar–May, Nov. **Zones:** 10–12.

Bactris

Syn: *Guilielma*, *Pyrenoglyphis*, etc.
Species 73. **Palms**, small to large, monoecious, spiny; stems cespitose or solitary, usually aerial but sometimes short and subterranean. **Leaves** pinnate or (if simple) pinnately veined, reduplicate, with various types of spines; sheath open or closed, but not forming a crownshaft; pinnae regularly or irregularly arranged, spreading or lamina simple. **Inflorescences** interfoliar, branched to one order or with 1 rachilla; peduncle bearing a prophyll and 1(2) peduncular bracts; rachis bearing 1–numerous rachillae; flowers borne in regularly arranged triads, or triads irregularly arranged among paired or solitary ♂ flowers. **Flowers** ♂ with 3 sepals, these basally connate into a trilobed calyx, petals 3, basally connate, free and valvate above, stamens (3–)6(–12), pistillode minute or absent; ♀ with calyx trilobed, annular, cupular or tubular, corolla trilobed, tubular, staminodes minute or absent, gynoecium syncarpous, trilocular, triovulate. **Fruits** globose, ellipsoid, obovoid or ovoid, usually either orange-yellow or purple-black. **Distr:** Throughout the Neotropics. **Cult:** Propagate from seeds, or by division in clump-forming species. Requires a moist, well-drained, humus-rich soil in full or partial sun.

B. gasipaes Kunth, *Nov. Gen. Sp.* 1: 302, 1816. (Syn: *Guilielma gasipaes*). [Peach Palm, Pewa Palm]. **Palms**; stems solitary or cespitose, 4–18 m tall, 8–25 cm dbh, usually spiny. **Leaves** 7–20; leaf spines brown or yellowish-brown, to 1(–3) cm long, often in 3 longitudinal rows on petiole; sheath and petiole 0.7–1.3 m long, abaxially whitish-tomentose, occasionally densely spinulose; rachis 1.8–3.5 m long, tomentose; pinnae 90–141 per side, irregularly arranged in clusters of 2–5, spreading, linear; middle pinnae 0.5–1 m × 2–3(–4) cm. **Inflorescences** with peduncles 15–32 cm long, recurved, spinulose or not spiny; prophyll 12–25 cm long; peduncular bract 47–70 cm long, ± covered with blackish or brownish spines to 1 cm long; rachis 15–30 cm long; rachillae 40–77, 17–32 cm long; triads irregularly arranged among paired or solitary ♂ flowers. **Flowers** ♂ 4–6 mm long, deciduous, sepal lobes 1.2–2 mm long, petals 3.5–6 mm long, stamens 6, pistillode absent; ♀ 3.5–8 mm long, calyx annular, 1–3 mm long, corolla tubular, 3–8 mm long, staminodes minute. **Fruits** subglobose, obovoid, or broadly ovoid, 1.2–6.5 × 1.1–4.5(–6) cm, yellow, orange or red. **Distr:** Tropical C and N South America to Andean slopes. **Ecol:** In lowlands or montane rainforests; to 1,100 m. **Zones:** 9–12.

B. major Jacq., *Select. Stirp. Amer. Hist.*, ed. 2: 134, 1781. (Syn: *B. ovata*, etc.). [Black Roseau, Prickly Palm]. **Palms**; stems cespitose, clump-forming, 1–10 m tall, 2–6 cm ∅, internodes spiny. **Leaves** 3–10; leaf spines brown or black, ± terete, to 11 cm long; sheath 22–55 cm long; ocrea to 20 cm long; petiole 0.1–1.5 m long; rachis 0.7–1.8 m long; pinnae 24–46 per side, usually regularly arranged, linear; middle pinnae 25–62 × 1–3.5 cm. **Inflorescences** interfoliar; peduncles 15–40 cm long, recurved, spinulose or densely spiny; prophyll 13–30 cm long; peduncular bract 28–60 cm long, with blackish or yellowish-brown spines to 1(–2) cm long; rachis 0.5–4 cm long; rachillae (1–)5–10(–17), 9–23 cm long, ~2 mm ∅; triads irregularly arranged among paired or solitary ♂ flowers. **Flowers** ♂ 3–8 mm long, sepal lobes 1.5–3 mm long, petals 3–7 mm long, stamens 6, pistillode absent; ♀ 4–9 mm long, calyx tubular, 4–8 mm long, minutely spinulose, corolla tubular, 3–5 mm long, densely spinulose. **Fruits** subglobose, irregularly ellipsoid, ellipsoid-oblong, or broadly obovoid, 1.5–4.5 × 1–3.5 cm, brown or purple-black. **Distr:** Trinidad, Mexico to S tropical America. **Ecol:** In forests, often in coastal areas; to 600 m. **Zones:** 11–12.

Cocos

Species 1. **Palms**, monoecious; stems solitary, erect or ascending, often leaning, robust, unarmed, trunks with conspicuous nodal rings. **Leaves** with sheath coarse, clothlike, not forming crownshaft; petiole unarmed, base entire; lamina pinnate, unarmed; plication reduplicate; segments regularly arranged, apices acute. **Inflorescences** axillary within crown of leaves, paniculate, with 1 or 2 orders of branching, stiffly ascending; prophyll short; peduncular bract woody, beaked, splitting abaxially, becoming boat-shaped; rachillae glabrous; flowers sessile; flowers unisexual, borne in triads of 1 ♀ flower flanked by 2 ♂ flowers; ♂ flowers borne singly along distal portions of rachillae; ♀ flowers borne basally on rachillae. **Flowers** ♂ with sepals 3, free, imbricate, petals 3, valvate, stamens 6, pistillode trilobed; ♀ large, tepals imbricate, free, sepals 3, ± reniform, petals 3, staminode a thin ring at base of the large pistil, ovules 3, usually only 1 fertile. **Fruits** drupes, 10–50+ cm long, strongly to obscurely 3-angled, smooth, usually green, yellow or orange; mesocarp very fibrous, dry, endocarp thick, bony, with 3 basal germination pores; seed 1, very large,

Above: *Cocos nucifera* (flowers), *Cocos nucifera* (fruits).

hollow, fluid-filled, endosperm lining cavity of seed, homogeneous, oily. **Distr:** Originated from Malesia, but distributed pantropically, mostly between 20°N and 20°S of the equator. **Ecol:** Mostly on or near salty, brackish water and dry places; 0–500 m. **Cult:** Propagate from mature seeds. If the palm becomes too tall, air-layering can be used to reduce the height. Requires any moist soil in full sun; extremely salt-tolerant. **Info:** The name 'cocos' is probably derived from a Portuguese word meaning monkey, because of its nut bearing 3 germinating pores, resembling a monkey's face.

C. nucifera L., *Sp. Pl.* 2: 1188, 1753. (Syn: *Calappa nucifera, Cocos nana, Palma cocos,* etc.). [Coconut Palm]. **Palms**, solitary, to 25 m tall; stems erect or leaning, smooth. **Leaves** with segments inserted on rachis in 2 ranks. **Inflorescences** with bract persistent, peduncular, to 1 m long, woody. **Flowers** unisexual; ♂ flowers creamy yellow, 1.1–1.3 cm long. **Fruits** green, yellow, or bronzy red when immature, brown when mature, mesocarp dry, fibrous, endocarp brown; seeds bearing 3 germination pores. **Distr:** Pantropical. **Ecol:** In lowland tropical areas; 0–500(–1,000) m. Fl. & Fr. throughout the year. **Zone** 12. **Info:** Cultivated and naturalized widely throughout the tropics. Due to a genetic mutation, some specimens produce a very large number of branches (>18), usually near the crown. Another genetic mutation causes all parts of the palm to become sugary.

Elaeis

Syn: *Alfonsia, Corozo*. [Oil Palm].

Species 2. **Palms**, solitary; stems erect, robust, covered with persistent leaf bases or bare, unarmed; crownshaft absent. **Leaves** pinnate, unarmed, plication reduplicate, segments regularly arranged in multiple planes (rarely in 1 plane), apices acute to 2-cleft; petiole margins armed with lignified, indurate bases of midveins persisting as spines after lamina erode. **Inflorescences** within crown of leaves, densely paniculate, with 1 order of branching, either ♂ or ♀, partially obscured by leaf bases; peduncles short; prophyll short; peduncular bract woody, splitting abaxially; rachillae thick, apices stiff, sharp; flowers sessile, borne singly along rachillae. **Flowers** unisexual; ♂ flowers in pits in rachillae, sepals 3, free, petals 3, free, valvate, leathery, stamens 6, filaments briefly connate, anthers rectangular, pistillode with 3 minute lobes; ♀ with sepals 3, imbricate, free, petals 3, imbricate, free, staminodial ring bearing 6 short points, pistil 1; ovules 3, styles indistinct, stigmas 3. **Fruits** drupes, ovoid, exocarp orange-yellow and black, thin, mesocarp fleshy, oily, endocarp thick, bony, with 3 apical germination pores; seeds spheroid. **Distr:** C America, S America; Africa. **Cult:** Propagate from seeds. Requires a rich, moist, loamy soil in full sun.

E. guineensis Jacquin, *Select. Stirp. Amer. Hist.*: 280, 1763. [African Oil Palm]. **Palms**, erect, monoecious, over 10 m tall; stem 50 cm dbh. **Leaves** 3–8 m; leaflets reduplicate, linear-lanceolate, 0.7–1.2 m × 2–4 cm, lower leaflets on rachis reduced as a spiny. **Inflorescences** ♂ composing of several finger-shaped spikes; spikes 7–12 × 1 cm, densely flowered; ♀ inflorescences subcapitate, condensed, 20–30 cm, bracts large, 2 cm, apical spines 7–30 cm. **Flowers** ♂ with sepals and petals oblong, 4 × 1 mm, acute apex; ♀ with sepals and petals ovate or ovate-oblong, 5 × 2.5 cm, ovary ~8 mm. **Fruits** ovoid or obovoid, 4–5 × 3 cm, ripening from green to yellow to red or black; seed subglobose or ovoid, endocarp black, 3-sided, bearing 3 germination pores. **Distr:** Tropical Africa. **Ecol:** Fl. Jun. Fr. Sep. **Zones:** 11–12. **Info:** This palm produces edible oil, used industrially.

Lytocaryum

Species 2. **Palms**, solitary; stem small to medium-sized, sometimes slightly leaning or curved. **Leaves** 7–22, pinnate, gracefully spreading; sheath, petiole and rachis often covered with elongate gray or brown hairs or scales; leaflets regularly arranged and spread in the same plane, narrow, linear, abaxially gray. **Inflorescences** borne among the leaves, branched to 1 order; rachillae numerous; flowers arranged in triads with a central ♀ flower and 2 lateral ♂ flowers. **Fruits** 1-seeded; epicarp and mesocarp fleshy, splitting regularly into 3 sections at maturity. **Distr:** SE Brazil. **Cult:** Propagate from seeds. Requires a moist, well-drained, humus-rich soil in partial sun.

L. weddellianum (H. Wendl.) Toledo, *Arq. Bot. Estado São Paulo*, n.s., f.m., 2: 8, 1944. (Syn: *Cocos weddelliana*, etc.). [Miniature Coconut Palm]. **Palms**, solitary, 3–4.5 m tall; stem 1.8–3 m tall, 5–8(–15) cm Ø, mostly covered in dense, dark brown fibers and narrow tan leaf base stubs; crownshaft absent. **Leaves** arching; lamina pinnate, 60–90 cm long, divided into evenly spaced leaflets; rachis 65–85 cm long; leaflets regularly arranged in 1 plane, 50–60 per side, 10–30 × 0.5–1.5 cm, adaxially glossy dark green, abaxially silvery-gray; petioles 5–40 cm long. **Inflorescences** bisexual, 0.9–1.2 m long, arching, branches pendent, to 14 cm long, borne from narrow, woody, capsulelike bracts; rachis 32–36 cm long; rachillae 40–62 cm long; spathe glabrous or covered with short brownish hairs; flowers unisexual. **Flowers** yellowish. **Fruits** ovoid to ellipsoid, 1.7–2.3 × 1–1.7 cm, glaucous, green to orange, maturing from turquoise to brown and splitting when ripe. **Distr:** Brazil (Espírito Santo, Rio de Janeiro). **Ecol:** In lowland rainforests on steep slopes; to 1,000 m. **Zones:** 10–12.

Syagrus

[Queen Palm].

Species 32. **Palms**, solitary; stems erect, robust, unarmed, bearing conspicuous nodal rings; crownshaft absent. **Leaves** with petiole unarmed; sheath not forming crownshaft; lamina pinnate, unarmed;

Below: *Elaeis guineensis, Lytocaryum weddellianum, Syagrus romanzoffiana, Syagrus schizophylla.*

MD

SZ

SZ

Above: *Euterpe edulis, Calyptrogyne ghiesbreghtiana, Geonoma congesta, Pelagodoxa henryana.*

plication reduplicate; segments regularly arranged in multiple planes, apices acute to 2-cleft. **Inflorescences** axillary within crown of leaves, paniculate, 1 order of branching, ascending, becoming pendulous in fruit; flowers sessile, borne in triads of 1 ♀ flower flanked by 2 ♂; rachillae with ♂ flowers borne distally and ♀ borne basally; prophyll short; peduncular bract woody, becoming boat-shaped. **Flowers** unisexual; ♂ with sepals 3, connate, petals 3, free, valvate, leathery, stamens 6, free, anthers linear, pistillode with 3 minute lobes; ♀ borne basally on rachillae, massive, sepals 3, imbricate, free, petals 3, imbricate, free, staminode a low annular ring at base of the large pistil, ovules 1, style indistinct, stigmas 3. **Fruits** drupes, ovoid, to 4 cm Ø, exocarp orange. **Distr:** N America, West Indies (Lesser Antilles), S America. **Cult:** Propagate from seeds. Requires a moderately fertile, moist, well-drained soil in full to partial sun.

S. romanzoffiana (Chamisso) Glassman, *Fieldiana: Botany*. 31: 382, 1968. (Syn: *Arecastrum romanzoffianum*, etc.). [Queen Palm]. **Palms**, solitary, 13.5–27 m tall; stems 12–22.5 m tall, upper parts often green with conspicuous widely spaced rings, lower parts light gray to white and almost smooth; crown 3–4.5 m across; crownshaft absent. **Leaves** pinnate, plumose, 2.4–5 m long; leaflets numerous, medium to dark green, usually limp, 30–60 cm long. **Inflorescences** bisexual, pendent, 0.6–1.5 m, with long branches, bracts very large, boat-shaped; branched part of spadix up to 1.25 m long; branches to over 80, to 62 cm long; flowers unisexual. **Flowers** white; ♂ 0.7–1 cm long on upper part, 1.1–1.6 cm long on lower part; ♀ ~4.5–6 x 4–6 mm. **Fruits** ovoid, 2.5–3.5 cm, yellowish-orange, endocarp ovoid, brown, with 3 germination pores. **Distr:** S America (S Brazil and adjacent parts of Argentina, Paraguay, Uruguay). **Ecol:** In forests and woodlands; 0–30 m. Fl. throughout the year. **Zones:** 9–12.

S. schizophylla (Mart.) Glassman, *Fieldiana, Bot.* 31: 386, 1968. (Syn: *Arikury schizophylla*, etc.). [Arikury Palm]. **Palms**, solitary, occasionally clustering, 4.5–7.5 m tall; stem acaulescent, 2.4–4.5 m tall, 25 cm dbh, with narrow, wickerlike old leaf bases, 15 cm thick; crown 2.4–4 m across, hemispherical when mature; crownshaft absent. **Leaves** pinnate, 1.8 m long; leaflets narrow, 30 cm long, dark green to olive, stiff when young, pendent when old, apices oblique; petiole 60 cm long with short curved spines along the margins for most of the length. **Inflorescences** clusters; with spadix spiny, branched, 60 cm long; branches yellow, 28 cm long. **Flowers** white, unisexual; ♂ 2–5 mm and 5–7 mm; ♀ ovoid, 4.5–8 x 3–6 mm, tips acute or rounded. **Fruits** ellipsoid, red, 1-chambered, to 3 x 2.5 cm; seeds 1.7 x 1.4 cm, endosperm ruminate. **Distr:** Brazil (Bahia). **Ecol:** Sandy coastal forests. **Info:** Seeds edible. **Zones:** 10–12.

TRIBE EUTERPEAE

Euterpe

Species 7. **Palms**, solitary, slender, often tall, monoecious; trunk smooth but annulate; petiolar sheaths of leaves forming a conspicuous crownshaft. **Leaves** pinnate. **Inflorescences** with spadices inter- or infra-foliar; outer spathe short, apically open; inner spathe elongate, often rostrate; branches slender, few to many, spreading or pendent; flowers in groups of 3, sometimes partly or deeply immersed. **Flowers** with sepals imbricate; petals valvate, that of ♀ imbricate; stamens 6; staminodes usually obvious. **Fruits** small, subglobose, drupelike, 1-celled, stigmatic scar lateral or subapical, early somewhat succulent, soon hard. **Distr:** Tropical America. **Cult:** Propagate from seeds. Requires a moist, humus-rich soil in full to partial sun.

E. edulis Martius, *Hist. Nat. Palm.* 2: 33, t. 32, 1824. (Syn: *E. edulis* var. *clausa, E. espiritosantensis*). [Assai Palm, Cabbage Palm]. **Palms**; stems solitary, rarely cespitose with few stems, erect, 5–12 m tall, 10–15 cm Ø, base with a dense cone of reddish-brown roots. **Leaves** 8–15, spreading; sheath 0.8–1.4 m long, olive-green, sometimes reddish; rachis 1.5–3 m long, ± scaly; pinnae 38–62(–70) per side; basal pinnae 29–50 x 0.5–0.8 cm; middle pinnae 0.5–0.8(–1) m x 1.5–4 cm; apical pinna 15–35 x 0.6–1 cm; petiole 13–54 cm long, with reddish-brown or black scales. **Inflorescences** infrafoliar and ± horizontal; peduncles 4–8.5 cm long; prophyll to 1 m long; peduncular bract 0.6–0.7(–1) m long; rachis 45–69 cm long; rachillae 49–110(–120), 26–58(–75) cm long; flowers in triads, paired or solitary ♂ distally, triads often distichously arranged. **Flowers** ♂ 5–6 mm long, purplish, sepals deltate, 2 mm long, gibbous, petals lanceolate, 5 mm long, filaments 0.15–2.2 cm long, pistillode ~1 mm long; ♀ 3–4.5 mm long, sepals broadly ovate, 3–4 mm long, petals broadly ovate, 5 mm long. **Fruits** globose, 1–1.4 cm Ø. **Distr:** Brazil to Argentina (Misiones). **Ecol:** In rainforests; 0–1,000 m. **Zones:** 11–12.

TRIBE GEONOMATEAE

Calyptrogyne

Species 17. **Palms**, monoecious; stems obsolete or very short, erect, slender, solitary, unarmed. **Leaves** simple to pinnately compound, if simple, apex bifid and if pinnate, with leaflets arranged in 1 plane and often unequal; petioles beyond sheath ± short, unarmed. **Inflorescences** interfoliar, bisexual; prophyll ± slender, grooved, beaked; peduncle elongate, erect, with 1 bractlike prophyll, but inserted distally, caducous, leaving a conspicuous scar ring; flowers unisexual, in bracteate triads in pits. **Flowers** ♂ with sepals 3, free, petals 3, connate in basal half, stamens 6, filaments usually connate and adnate to the receptacle; ♀ with sepals 3, free, petals connate in a shallow tube, trilobed, lobes rounded, staminodes connate into a fleshy tube minutely 6-lobed, pistil 1, ovary triloculed, style elongate, triangular in section, stigmas 3, linear, reflexed. **Fruits** ellipsoid to obovoid, smooth, black, with basal stigmatic remnants. **Distr:** SE Mexico to C America. **Cult:** Propagate from seeds. Requires a moist, humus-rich soil in partial sun.

C. ghiesbreghtiana (Linden and H. Wendl.) H.Wendl., *Bot. Zeitung* (Berlin) 17: 72, 1859. (Syn: *Geonoma ghiesbreghtiana*). [Coligallo Palm, Vampire Palm]. **Palms**; stem obsolete or to 1+ m tall, to 3 cm Ø. **Leaves** simple or more often pinnately compound; rachis 0.22–1.4 m long, if lamina simple, apex bifid, if lamina pinnate, leaflets 2–11 per side, with the ones at the middle (23–)29–73 cm long; petiole beyond sheath 0.27–6(–1.2) m long. **Inflorescences** with peduncles (0.57–)0.95–1.8+ m long; bract ~(26–)37–74 cm long; rachilla (11–)20–39 cm long, stipe 0.9–8.4(–12) cm long. **Flowers** ♂ ~5 mm across, sepals glabrous (or apparently so), anthers 1.2–2.5 mm long; ♀ ~3.5–4 mm across. **Fruits** ovoid or ellipsoid to obovoid, 0.8–2.1 x 0.5–1.2 cm, purple to black. **Distr:** SE Mexico to C America. **Ecol:** In humid forests, rainforests and cloud forests; 0–1,500 m. Fl. throughout the year. **Zones:** 10–12.

Geonoma

Species 68. **Palms**, small, solitary or clustered. **Leaves** few to many, almost always arranged in a single plane. **Inflorescences** spicate or branched to 1–3(–4) orders; peduncular bracts (0–)1(–2), usually inserted near prophyll; rachilla pits decussate, spiral or whorled, with or without a distal lip. **Flowers** ♂ with stamens 3(6+), filaments connate in a stalk, tubular above, distally free, inflexed, connective divided with divaricate thecae or short and united, introrse; ♀ with petals basally connate, staminodes connate, tubular, truncate, or 6-lobed; mesocarp thin with narrow longitudinal fibers. **Distr:** Mexico to tropical America. **Cult:** Propagate from seeds. Requires a moist, well-drained soil in partial sun and high humidity.

G. congesta H. Wendl. ex Spruce, *J. Linn. Soc., Bot.* 11: 112, 1869. (Syn: *G. calyptrogynoidea*). [Cortadera, Suita Palm]. **Palms**; stems clustered, often in large clumps of 10–20 stems, rarely solitary, to 7 m tall, 2–4 cm Ø, greenish to light brown. **Leaves** ~15, usually erect, sometimes spreading; lamina to 1.7 m long, irregularly divided into 3 or more leaflets per side, narrow and broad ones intermixed, rarely lamina simple; petiole with very sharp margins. **Inflorescences** borne among or below the leaves, branched to one order, erect; peduncle usually <10 cm long; rachis very short, to 10 cm long; flowering branches 3–13, stiff, spreading candelabralike, to 30 cm long, 6–7 mm thick, bright orange in fruit, pits spirally and densely arranged. **Flowers** ♀ with staminodes united into a tube. **Fruits** globose or obovoid, rough, 1–1.5 cm Ø, green or black. **Distr:** C America to NW Colombia. **Ecol:** In lowland or premontane rainforests; <1,000 m. **Zones:** 10–12.

TRIBE PELAGODOXEAE

Pelagodoxa

Species 1. *Generic description same as specific.* **Cult:** Propagate from seeds. Requires a very moist soil in partial sun and protected from strong wind.

P. henryana Becc., *Rev. Hort.*, n.s., 15: 302, 1917. (Syn: *Pelagodoxa mesocarpa*). [Marquesas Palm]. **Palms**, solitary, 6–9 m tall; trunk 4.5–7.5 m tall, 15 cm dbh; crown 4–5.4 m across; crownshaft absent. **Leaves** pinnate, 1.8–2.1 × 0.6–0.9 m, undivided, adaxially deep green, abaxially grayish-green; petiole 60 cm long. **Inflorescences** interfoliar, 60 cm long, sparsely branched to 2 orders, with both ♂ and ♀ flowers; peduncular bract longer than prophylls; rachillae with triads in pits, ♂ flowers distally. **Flowers** unisexual, yellowish; ♂ with sepals 3, distinct, petals 3, basally connate, stamens 6, filaments basally united to pistillode, anthers medifixed, sagittate, latrorse, connective prolonged, pistillode pyramidal, staminodes 3, triangular, small. **Fruits** ovoid, <15 cm across, light brown, covered with pointed, corky, conelike projections; embryo basal. **Distr:** Marquesas Islands (Nuku Hiva). **Ecol:** In low valley rainforests. **Zones:** 10–12.

TRIBE ARECEAE

Actinorhytis

Species 1. *Generic description same as specific.* **Cult:** Propagate from seeds. Requires a well-drained soil in full sun.

A. calapparia (Blume) H. Wendl. and Drude ex Scheff., *Ann. Jard. Bot. Buitenzorg* 1: 156, 1876. (Syn: *Areca calapparia*, etc.). [Calappa Palm]. **Palms**, monoecious, slender, robust, solitary, erect, ~12–20 m tall; trunk to 30 cm dbh, smooth, gray-annulate; crown spreading with ~14 leaves; crownshaft 1.6 m long. **Leaves** pinnate, single-fold, lanceolate, 4–5 × 1.4 m; leaflets 94 per side, linear, 70–90 × 3.2–4 cm, apices acuminate, tip notched or equally 2-parted, recurved, midrib of leaflet yellowish adaxially; leaf sheath light green with sparse lines of reddish spots. **Inflorescences** infra-foliar, 1.3 m long, losing all major bracts, branched to 2 orders; major branches pinkish; primary branches 25, largest one 90 cm long; peduncle 10 cm long; spadix ascending with spathes; prophyll and peduncular bract enclosing inflorescence in bud, deciduous; rachillae bearing spiral superficial triads proximally, ♂ flowers distally. **Flowers** cream-colored; ♂ asymmetric, sepals and petals distinct, stamens 24–33, filaments distinct, inflexed, anthers elongate, medifixed, versatile, pistillode columnar; ♀ with sepals and petals imbricate, staminodes 3. **Fruits** ovoid, 6 cm across, orange-yellow to red; fruiting spadix spreading; seed 1. **Distr:** Papuasia. **Ecol:** In lowland coastal, tropical rainforests; 0–1,000 m. **Zones:** 11–12.

Adonidia

Species 1. *Generic description same as specific.* **Cult:** Propagate from seeds. Requires a moderately moist and fertile, well-drained soil rich in organic matter, in full or partial sun.

A. merrillii (Becc.) Becc., *Philipp. J. Sci.* 14: 329, 1919. (Syn: *Normanbya merrillii*, *Veitchia merrillii*). [Christmas Palm, Manila Palm]. **Palms**, solitary, monoecious, 6–15 m tall, fast-growing; trunk 6 m tall, base spreading, 10–15 cm dbh, light to dark gray, with closely set (or 5 cm apart) indistinct rings; crown 3–4 m wide; crownshaft 30–90 cm long, light to dark green or grayish-green, smooth, slightly bulging at base. **Leaves** strongly recurved; lamina pinnate, 2–2.7 m long, V-shaped in cross-section; leaflets 40–50, held obliquely, 30 cm long, apices jagged, stiff, adaxially deep green, abaxially paler; basal leaflets arranged in several ranks, becoming 1-ranked toward apex. **Inflorescences** 60 cm long, bisexual; flowers borne in triads on spikelike branches of the compound inflorescence; peduncle white; bracts 2, 1 inside the other, short-lived, papery. **Flowers** unisexual, both sexes alike, whitish. **Fruits** ovoid, plum-shaped, 2.5–4 cm long, bright red. **Distr:** Philippines (Palawan). **Ecol:** In lowlands. **Zones:** 10–12.

Archontophoenix

Species 6. **Palms**, solitary, monoecious, pleonanthic; stem tall and slender, with conspicuous rings of leaf scars. **Leaves** pinnate; leaflets linear-lanceolate, apex acuminate or 2-toothed, adaxially green, abaxially gray due to silvery scales, midrib prominent; leaf sheath tubular, forming a conspicuous crownshaft, usually swollen at base. **Inflorescences** infrafoliar, branched to 3 orders, partial inflorescence and rachilla curved, drooping; flowers in triads on the lower portion of rachilla; ♂ flowers borne on the upper portion of the rachilla, solitary or paired. **Flowers** ♂ asymmetrical, sepals and petals 3, distinct, stamens 8–24, pistillode trifid or columnar; ♀ smaller than ♂, sepals and petals 3, distinct, staminodes 3, toothlike, borne on one side of the gynoecium, gynoecium irregularly ovate, unilocular, uniovulate, stigma 3, recurved. **Fruits** globose to ellipsoid, reddish, stigmatic remains apical; seed ellipsoid to globose. **Distr:** E Australia. **Cult:** Propagate from fresh seeds. Requires well mulched soil in full or partial sun.

A. cunninghamia (H. Wendl.) H. Wendl. and Drude, *Linnaea* 39: 214, 1875. (Syn: *Ptychosperma cunninghamianum*, etc.). [Bangalow Palm, King Palm, Piccabeen Palm]. **Palms**, solitary, 15–25 m tall; trunk 15 m tall, 30 cm dbh, light gray to light brown; crown 4.5 m wide; crownshaft 90 cm tall, medium to dark green, slightly swollen at base. **Leaves** pinnate, 2.4–4.5 m long; leaflets 80–200, mostly pendent, closely spaced, 30–90 × 2–8 cm, adaxially medium to dark green, abaxially pale green, both sides glabrous, midrib usually twisting at midpoint so that the leaflets at the distal half are vertically oriented. **Inflorescences** spikes, many, pendent, branched, borne around the trunk just below the crownshaft, flowers of both sexes on the same inflorescence. **Flowers** small, unisexual, fragrant; calyx pink or pale purple; petals 3, ~5 mm long, pink or pale purple; stamens 9–16. **Fruits** drupes, round, 1–1.5 × 1–1.5 cm, red; seed 1, pale brown, 0.8-1 cm long. **Distr:** E Australia (Queensland, NSW). **Ecol:** In coastal areas. Fl. Feb–Jun. Fr. Dec–May. **Zones:** 10–11.

Areca

Species ~48. **Palms**, solitary or small clump-forming; stem(s) annulate, slender, smooth, unarmed, 0.5–25 cm Ø; leaf bases usually forming prominent, shiny, swollen crownshaft. **Leaves** pinnate, pinnae mostly narrow, often with 2 or more principal ribs, long acuminate, rather short stiff; leaflets thin, often confluent. **Inflorescences** much-branched, borne at base of crownshaft; ♀ in triads at base of rachillae, ♂ flowers small, variously arranged on upper part of rachillae. **Flowers** ♂ with sepals small, petals much larger, obliquely lanceolate, valvate, stamens 3–6(–12–24) or more, anthers basifixed, erect; ♀ larger than ♂, sepals and petals imbricate, petals acute, ovary 1-chambered, stigmas 3, very short, ovule 1, basal, erect. **Fruits** ovoid or oblong, exocarp fleshy, fibrous, 1-seeded; seeds with truncate base, embryo basal. **Distr:** NW India, Sri Lanka, Indo-China; Malesia to Solomon Islands and Queensland. **Cult:** Propagate from seeds. Needs a well-drained soil in full or partial sun.

A. catechu L., *Sp. Pl.* 1189, 1753. [Betel Nut Palm, Betel Palm, Areca Nut, India Nut]. **Palms**, solitary, slender, to 22.5 m tall, older portion of stem grayish, younger greenish; nodes prominently ringed with old leaf scars; crown shaft 75–90 cm long, glabrous, smooth, green, base slightly swollen. **Leaves** 8–10, 0.9–1.35 m long, arising from the top of the stem; leaflets unevenly divided, 2–4-ribbed, narrow, 50–55 in number, 30–45 cm long, glabrous, green; petiole unarmed, base enlarged into sheath. **Inflorescences** monoecious, much-branched, branches threadlike, long, having zigzagged depressions, upper part with small, sessile flowers, 2-ranked; bracts glabrous; ♂ flowers numerous; ♀ flower in triads with 2 ♂ flowers, few at base and axils of branches. **Flowers** creamy-white, lemon-scented; ♂ flowers numerous, sepals 3, small, petals 0.3–0.4 cm long, stamens 6, anthers linear, pistillode trilobed; ♀ with sepals 3, 1–1.3 cm long, petals 3, as long as the sepals, stigmas 3, sessile, ovary one-celled. **Fruits** drupes, oblong or ovoid, 4–5 cm long, orange, base enclosed by the perianth, outer wall fleshy, fibrous, 1-seeded. **Distr:** SE Asia. **Ecol:** Fl. Oct. Fr. Per. Dec–Jan. **Zones:**

Below: *Actinorhytis calapparia, Adonidia merrillii, Archontophoenix cunninghamia, Areca catechu.*

Above: *Areca triandra, Areca vestiaria, Basselinia pancheri, Carpentaria acuminata.*

11–12. **Info:** Cultivated widely for the edible nuts (seeds). This species does not tolerate brackish water, and is not suitable for planting close to the sea.

A. triandra Roxb. ex Buch.-Ham., *Mem. Wern. Nat. Hist. Soc.* 5: 310, 1826. (Syn: *A. polystachya*, etc.). [Australian Areca Palm, Burning Utan, Wild Areca Palm]. **Palms**, stems clustered, (1.5–)6–9 m tall; stems slender, green, 3–5 cm Ø, leaf scars forming white rings, base with stilt roots. **Leaves** slightly drooping, 1–1.8 m long; sheaths forming a swollen, green crown shaft; petiole slender; leaflets 0.3–1 m long, 3.5–5 cm wide. **Inflorescences** 15–30 x 5–15 cm; prophyll 30 cm long. **Flowers** ♂ cream-colored, with a strong lemon fragrance, 2.5 mm long; ♀ globose, cream-colored, 7.5 mm long. **Fruits** 2.5 cm long, with a prominently beaked tip, ripening to orange, then to scarlet. **Distr:** Tropical and subtropical Asia (Indian subcontinent, Sri Lanka, Indo-China, China, Malesia). **Ecol:** In mixed dipterocarp forests; 600–1,300 m. **Zones:** 9–12.

A. vestiaria Giseke, *Prael. Ord. Nat. Pl.*: 78, 1792. (Syn: *Pinanga vestiaria, A. langloisiana*, etc.). [Orange Collar Palm]. **Palms**, usually clustering, sometimes solitary, 3–6 m tall; crown 4.5 m wide; crownshaft ~90 cm tall, orange, suffused with darker reddish-orange or wholly dark reddish-orange. **Leaves** pinnate, 1.8–2.4 m long, arching, slightly V-shaped in cross-section; leaflets many, to 60 cm long, deeply grooved, stiff, deep green, terminal ones larger and usually fused; midrib and petiole usually deep yellow to orange; young parts sometimes reddish. **Inflorescences** small, ♂ and ♀ on the same inflorescence. **Flowers** unisexual, fragrant, orange. **Fruits** drupes, in pendent clusters, ellipsoid, 2.5 cm across, brownish-orange to reddish-brown or darker. **Distr:** Sulawesi to Maluku. **Ecol:** In lower montane rainforests. **Zones:** 11–12.

Basselinia

Species 12. Palms, solitary or clustered, erect. **Leaves** entire, bifid or variously pinnate; leaflets single- to multi-fold, apices acute. **Inflorescences** infrafoliar, branched to 1 or 2 orders; prophyll incomplete; peduncular bract exceeding prophyll; spiral triads superficial proximally, ♂ flowers distally. **Flowers** ♂ symmetrical, sepals and petals distinct, stamens 6, filaments inflexed, anthers dorsifixed, latrorse, pistillode equaling stamens in bud; staminodes 3. **Fruits** with stigmatic remains lateral to subapical; epicarp smooth; mesocarp with sparse fibers; endocarp vitreous, operculum rounded, basal. **Distr:** New Caledonia. **Cult:** Propagate from seeds. Requires a well-drained soil in partial sun.

B. pancheri (Brongn. and Gris) Vieill., *Bull. Soc. Linn. Normandie*, II, 6: 232, 1873. (Syn: *Kentia pancheri*, etc.). [Basselinia Palm, New Caledonia Palm]. **Palms**, solitary or clustering, 3–8(–15) m tall; trunk 3–5 m tall, 1–10 cm dbh, mostly green with widely spaced brown rings; crownshaft brown to purplish-brown, 18–45 cm high, 2–12 cm Ø, base bulging; crown variable. **Leaves** 7–8, spirally arranged, spreading; lamina 0.3–1.3 m long, pinnately divided, bifid or entire, erect, stiff, usually with very little splitting; leaflets irregularly distributed along the rachis in one plane, basal leaflets 9 x 0.7 cm, middle leaflets 24 x 6 cm, apical leaflets 18 x 6 cm; rachis 0.3–1 m long; petiole ~2–5 cm long, adaxially channeled, abaxially rounded. **Inflorescences** androgynous, branched to 2 orders, borne below crownshaft; prophylls orange-colored, boat-shaped, ascending at first; peduncle thick, 3.5 cm long; rachillae wriggly, ~10 cm long, primary branches 4–6, 9.5 cm long at apex and 13 cm long at base, densely flowered. **Fruits** glossy green to reddish or purplish-black, 5–6 mm Ø **Distr:** W and S New Caledonia. **Ecol:** Mostly in rainforests; 0–900 m. **Zones:** 10–12.

Carpentaria

Species 1. **Palms** solitary, fast-growing; stem slightly enlarged at base. *Generic description same as specific.* **Distr:** Tropical N Australia. **Cult:** Propagate from fresh seeds. Requires moist but well-drained soil in full sun.

C. acuminata (H. Wendl. and Drude) Becc., *Ann. Jard. Bot. Buitenzorg* 2: 128, 1885. [Carpentaria Palm]. **Palms**, solitary, to 30 m tall; stem smooth, gray, with rings of leaf scars, 25–30 cm dbh; crown small, ~5 m wide. **Leaves** pinnate, ~3(–3.5) m long, base smooth, sheathing stem to 60 cm long, short-petiolate, midrib arching; leaflets narrow, straplike, folded, giving a ∧-shape cross-section, surfaces glossy dark green. **Inflorescences** compound, with numerous slender branches (bearing triads of 1 ♂ and 2 ♀ flowers), borne on stem below leaves, initially covered by bracts 40 cm long; flowers sessile. **Flowers** cream; sepals 3; petals 3; ♂ with stamens 30, central pistillode with a long style; ♀ lacking stamens, carpels 3, united into a single ovary, maturing later than ♂. **Fruits** drupes, globular to ellipsoid, ~2 x 1.8 cm, with 3 persistent stigmas as lobes ~1 mm long at end of fruit, bright red, 1-seeded,. **Distr:** Australia (Northern Territory). **Ecol:** Solitary or in small groups, in monsoon or gallery forests; 0–20 m. Fl. throughout the year. **Zones:** 11–12. **Info:** The leaves are shed after 3 seasons, leaving a clean stem. The pulp of the fruit causes skin irritation.

Cyrtostachys

Species 11. **Palms**, solitary or clustered, moderate to tall, unarmed, pleonanthic, monoecious; stems erect, bare, conspicuously ringed with leaf scars, often bearing a mass of adventitious roots at base; crownshaft well-defined, formed from the tubular leaf sheaths. **Leaves** pinnate; leaflets regularly arranged, single-fold, acute or acuminate, often stiff, adaxially ± glabrous, abaxially paler. **Inflorescences** apparently protandrous, infrafoliar, branched to 3 orders, spreading; peduncle usually very short; prophyll enclosing inflorescences until leaf fall, tubular, 2-keeled, ± lanceolate, margins winged, splitting, soon caducous; flowers in triads, borne throughout

Below: *Carpentaria acuminata* (flowers and fruits), *Cyrtostachys renda, Cyrtostachys renda* (inflorescence).

the length of rachillae. **Flowers** ♂ with 3, distinct, imbricate, broad, strongly keeled sepals, petals 2× as long as sepals, basally united, stamens 9–15, filaments awl-shaped; ♀ ± same size as ♂, sepals 3, distinct, rounded, imbricate, petals 3, slightly larger than sepals, imbricate proximally, gynoecium unilocular, ellipsoidal with 3 short, recurved stigmas. **Fruits** broadly to narrowly ellipsoid, usually black, epicarp smooth, 1-seeded; seed globose or ellipsoidal. **Distr:** Peninsular Thailand to Papuasia. **Ecol:** In lowland rainforests; to ~500 m. **Cult:** Propagate from fresh seeds. Requires a moist soil in partial sun.

C. renda Blume, *Bull. Sci. Phys. Nat. Néerl.* 1: 66, 1838. (Syn: *Areca erythrocarpa*, etc.). [Lipstick Palm, Maharaja Palm, Sealing Wax Palm]. **Palms**, clustering, unarmed, 6–9 m tall; stem(s) 4.5–6 m tall, 8 cm Ø, upper parts green, lower parts light brown to light gray, with widely spaced darker rings; crown sparse, erect, 3 m wide; crownshaft 90 cm tall, slightly thicker, smooth and glossy, deep orange to deep red. **Leaves** ~10, pinnate, 1.5 m long; leaflets widely spaced, bright green, narrow, 30 cm long; petiole orange to red, 30 cm long. **Inflorescences** panicles, borne below leaves, branches greenish, bisexual (with ♂ and ♀ flowers in the same inflorescence). **Flowers** unisexual, small; ♂ yellowish-green, sepals 3, broad, distinct, imbricate, united at the very base, petals larger, broadly ovate, shortly acuminate, stamens 9–15, basally connate, pistillode long, narrow, trifid; ♀ slightly larger than ♂, sepals 3, distinct, rounded, imbricate, petals 3, slightly larger than sepals, proximally imbricate, asymmetric. **Fruits** ovoid, 2 × 1.2 cm, abruptly acuminate, ripening to black; seed 1, globose, 1.2 cm long. **Distr:** Peninsular Thailand to W Malesia. **Ecol:** In rainforests and swamps. **Zones:** 11–12.

Cyphokentia

Species 2. **Palms**, solitary, monoecious, moderate-sized; trunk expanded basally. **Leaves** regularly pinnate; leaf sheath tubular, white-waxy, forming a prominent crownshaft; pinnae acute, single-fold; petiole short, concave adaxially. **Inflorescences** infrafoliar, branched to 2–3 orders; prophyll completely encircling peduncle and enclosing the similar peduncular bract, both densely waxy, rachillae slender to stout, spreading to pendulous, triad bract low, bracteoles sepal-like. **Flowers** ♂ with 3 sepals, 3 petals, external stamens 3, internal stamens 3–9, pistillode columnar, as long as stamens in bud; ♀ with 3 sepals, 3 petals, gynoecium unilocular, uniovulate, with staminodes 3–6. **Fruits** ellipsoid or depressed-globose, epicarp smooth or papillate, stigmatic residue lateral, endocarp operculate; seed with homogeneous endosperm and basal embryo. **Distr:** New Caledonia. **Cult:** Propagate from seeds. Requires a moist, well-drained soil in full or partial sun.

C. cerifera (H.E. Moore) Pintaud and W.J.Baker, *Kew Bull.* 63: 72, 2008. (Syn: *Moratia cerifera*). [Moratia]. **Palms**, solitary, slow-growing; trunk to 18 m tall, 15–18 cm dbh, light to dark brown with distinct, closely spaced ridged rings of leaf-base scars in its upper parts; crownshaft cylindrical, 60 cm long, slightly bulging at base, slightly tapered above, deep orange but covered with a thick layer of pure white wax; crown open and nearly spherical. **Leaves** few, arched, recurving; ligule prominent; lamina 2.1 m long; leaflets borne from the rachis at an angle making the leaf appear V-shaped, densely inserted, overlapping, linear-lanceolate, long-tapering, stiff, deep yellowish to emerald-green; petiole <30 cm long, orange to yellow. **Inflorescences** borne below the crownshaft, large, 90 cm across, spreading at anthesis, rachillae grayish-green, stout, 0.7–1 cm Ø. **Flowers** ♂ dark red, sepals large, half the length of the petals, stamens 3–7 in the inner whorl, staminodes 3–4. **Fruits** depressed-globose, 1.2 cm across, reddish. **Distr:** NE and C New Caledonia. **Ecol:** In montane rainforests; 0–760 m. **Zones:** 10–12.

Cyphophoenix

Species 4. **Palms**, solitary, monoecious moderate-sized; trunk prominently ringed, internodes green when young, aging to gray. **Leaves** regularly pinnate, straight or arching, spreading; leaf sheath forming a prominent, tomentose crownshaft; pinnae stiff, lanceolate, single-fold, with prominent ramenta on ribs abaxially. **Inflorescences** infrafoliar, short-pedunculate, protandrous, prophyll shorter than peduncular bract, incompletely encircling peduncle; rachillae bearing prominent, rounded bracts subtending triads, bracteoles brown, rounded, sepal-like. **Flowers** ♂ with 3 sepals, 3 petals and 6 free stamens, anthers exserted and versatile at anthesis, pistillode prominent, conical or capitate; ♀ with 3 sepals, 3 petals, gynoecium unilocular and uniovulate, staminodes 3 on one side of gynoecium. **Fruits** symmetrically or eccentrically ovoid-ellipsoid with apical stigmatic remain, endocarp smooth or sculptured, operculate; seed with homogeneous endosperm and basal or subbasal embryo. **Distr:** New Caledonia. **Cult:** Propagate from seeds. Requires a moist, well-drained soil in full or partial sun; salt and wind tolerant.

C. alba (H.E. Moore) Pintaud and W.J. Baker, *Kew Bull.* 63: 67, 2008. (Syn: *Veillonia alba*). [White Crownshaft Palm]. **Palms**, solitary; trunk to 15 m tall, 15 cm dbh, light tan to light gray or almost white in older parts, green and waxy in the youngest parts, leaf-base scars ringed, widely spaced, deeply indented and dark brown; crownshaft 1.2 m long, bulging at base, with very thick, reddish-brown feltlike tomentum, on upper part only, lower part waxy-glaucous; crown sparse, rounded. **Leaves** spreading; 2.5 m long, ramenta on ribs abaxially light brown, lateral veins 2, prominent; leaflets linear-lanceolate, 1.2 m × 7.5 cm, long-tipped, thick, bright to deep green; petiole 60 cm long. **Inflorescences** much-branched, ropelike, coral-pink, covered with white waxy scales when in flower. **Flowers** ♂ with petals pulvinate, pistillode capitate. **Fruits** ovoid, brown or greenish-brown, stigmatic remains apical. **Distr:** NE New Caledonia. **Ecol:** In rainforests; 0–600 m. **Zones:** 10–12.

C. elegans (Brongn. and Gris) H. Wendl. ex Salomon, *Palmen*: 86, 1887. (Syn: *Kentia elegans*). **Palms**, solitary; trunk to 15 m tall, 15 cm dbh, younger parts green, older parts light to deep gray, with whitish rings of leaf-base scars; crownshaft bulbous, 60–90 cm long, markedly swollen at base, olive-green to light silvery-green, with thin, gray or brown tomentum covering the entire leaf sheath, not waxy; leaf crown semicircular because of the ascending leaves. **Leaves** arching, recurved; 1.8 m long; leaflets stiffly ascending, borne from the rachis at an angle giving the leaf a V-shape, lamina linear-lanceolate, 90 cm long, both surfaces deep yellow-green, ramenta on ribs abaxially dark brown, lateral veins not prominent. **Inflorescences** infrafoliar, short-pedunculate, with both ♂ and ♀ flowers. **Flowers** ♂ with petals not pulvinate, pistillode conical. **Fruits** ovoid, 1.2 cm long, red. **Distr:** NE New Caledonia. **Ecol:** In lower montane rainforests. **Zones:** 10–12. **Info:** This palm is now endangered in the wild.

Dictyosperma

Species 1. *Generic description same as specific.* **Cult:** Propagate from seeds. Requires a rich, very moist soil, in high humidity and full sun.

D. album (Bory) Scheff., *Ann. Jard. Bot. Buitenzorg* 1: 157, 1876. (Syn: *Areca alba*, *Linoma alba*). [Hurricane Palm, Princess Palm, Red Palm]. **Palms**, solitary, 9–12 m tall; trunk 7.5–10.5 m tall, 15 cm dbh, base spreading, gray to brown, rings closely set, darker, cut by vertical fissures; crown 4.5 m wide; crownshaft 1.2 m tall, smooth, white to green, bluish-green or reddish, swollen at base. **Leaves** pinnate, 2.5–4 m long; leaflets narrow, 60–90 cm long, deep green; petiole reddish in younger plants, in mature plants twisted, turning the ends of leaves on edge. **Inflorescences** panicles, 60 cm long, ♂ and ♀ flowers on the same inflorescence, branched, borne below leaves; bracts 2, thin, broad, one inside the other; flowers unisexual, borne in triads. **Flowers** large, yellowish to reddish, fragrant; ♂ with 6 stamens. **Fruits** ovoid, 1.7 cm long, ripening through deep purple to black, waxy; seed with operculum. **Distr:** Mascarene Islands in the Indian Ocean. **Ecol:** In coastal forests; 0–550 m. **Zones:** 10–12. **Info:** Variable species. Nearly extinct in the wild.

Dypsis

Syn: *Chrysalidocarpus*, *Neodypsis*, *Vonitra*.
Species 144. **Palms**, clustered; trunks erect, <15 cm dbh, with conspicuous nodal rings, unarmed, occasionally branching near base. **Leaves** with bases unarmed, yellowish, forming distinct crownshaft; lamina pinnate (rarely undivided), unarmed, reduplicate; segments lanceolate, evenly spaced, ascending,

Below: *Cyphokentia cerifera*, *Cyphophoenix alba*, *Cyphophoenix elegans*, *Dictyosperma album*. [All SZ except for *Dictyosperma album* AF.]

SZ

Top row: *Dictyosperma album* (inflorescence and crownshaft), *Dypsis ambositrae, Dypsis crinita.*
Bottom row: *Dypsis decaryi, Dypsis lutescens, Dypsis lutescens* (inflorescence).

apices acuminate. **Inflorescences** axillary below the crown of leaves, paniculate, branching in 3 orders of branching, flowers sessile, in triads of 1 ♀ between 2 ♂; prophyll small; peduncular bracts caducous, tubular. **Flowers** unisexual; ♂ with sepals 3, imbricate, free, petals 3, briefly connate basally, valvate, stamens 6, free, anthers dorsifixed, pistillode present; ♀ with sepals 3, imbricate, free, petals 3, imbricate, free, staminodes 6, minute, pistil 1, ovules 1, stigmas 3. **Fruits** drupes, ellipsoid, exocarp yellow, smooth; seeds ovoid. **Distr:** Tanzania (Pemba), Comoros, Madagascar. **Cult:** Propagate from seeds, or by division in clumping species. Requires a moist, well-drained soil in shade with high humidity.

D. ambositrae Beentje, *Palms Madagascar*: 195, 1995. [Ambositrae Palm]. **Palms**, clustering, in tufts of 2–3, sometimes solitary; trunk 3–7 m tall, ~12 cm Ø, internodes 10–20 cm long, pale brown to gray, green and ringed when young; crownshaft pale waxy gray-green. **Leaves** 7–11, spiral, gracefully arching; sheath 0.6–1 m long, green with a white bloom; rachis 2.1–2.8 m long; leaflets stiff with only the apices pendulous; proximal leaflets 0.7–1.4 m × 0.3–1.8 cm; median leaflets 0.9–1.1 m × 2.3–3 cm; petiole 9–30 cm long. **Inflorescences** interfoliar, branched to 2(3) orders; peduncle 0.7–1.2 m long; prophyll ~91 × 11.5 cm, narrowly 2-winged; peduncular bract deciduous, ~80 cm long; rachis 0.8–1 m long; all axes green with white bloom. **Flowers** ♂ with sepals broadly ovate, 2.2–2.5 × 1.6–2.2 mm, keeled, base gibbous, margins membranous, apex obtuse, petals basally connate, free lobes ovate or elliptic, 2.8–3 × 2.8–3.2 mm, acute, stamens 6, pistillode columnar, ~2.2 mm long; ♀ with sepals broadly ovate, 2.4–3 × 3–4.1 mm, petals hardly connate at base, broadly ovate, 3.5–4.1 × 4–5 mm, staminodes 6, 0.3–1.6 mm long, ovary asymmetrical, 2.7–4.8 mm long. **Fruits** ~1.4 × 1 cm. **Distr:** C Madagascar; critically endangered in the wild. **Ecol:** Mostly in riverine forests; 1,300–1,500 m. **Zones:** 10–12.

D. crinita (Jum. and H. Perrier) Beentje and J.Dransf., *Palms Madagascar*: 361, 1995. (Syn: *Vonitra crinita*). [Vonitra Palm]. **Palms**, clustering, in tufts of 2–4; trunk 4–15 m tall, 12–20(–30) cm dbh, slightly inflated at base, usually branched 2–3 times, rarely unbranched. **Leaves** 12–15, arching; sheath 54–56 cm long; rachis 2–3 m long, bright green; leaflets 50–60 per side, to 78 × 3 cm; petiole ~1 m long. **Inflorescences** many, interfoliar, branched to 2 orders; peduncle 1.2–1.8 m long; prophyll 52–59 cm long; peduncular bract, ~2 m long, beaked; rachis ~40 cm long. **Flowers** ♂ with sepals unequal, outermost hooded, 2–2.3 × 2.5–3.5 mm, petals purplish, ovate, ~2.5 × 1.8 mm, acute, stamens biseriate, pistillode ~1.5 mm; ♀ with sepals unequal, hooded, keeled, 1.8–2.8 × 4–6 mm, imbricate, petals imbricate, orbicular, 3.5–3.8 × 4–6 mm, rounded, staminodes 6. **Fruits** slightly ovoid, green, maturing to purple-black, 2–2.4 × 1.7–1.8 cm; seed 1.5–1.7 × 1.1–1.3 cm. **Distr:** N and NE Madagascar. **Ecol:** Near streams; 200–250 m. **Zones:** 10–12.

D. decaryi (Jum.) Beentje and J.Dransf., *Palms Madagascar*: 187, 1995. (Syn: *Neodypsis decaryi*). [Madagascar Three-sided Palm, Triangle Palm]. **Palms** solitary; trunk (0.5–)3–10(–12) m tall, 30–40 cm dbh; crown 4.5 m across; crownshaft absent. **Leaves** 3-ranked, 18–24, erect and stiffly arching; lamina pinnate, 1.5–3.25 m long; leaflets narrow, 60 cm long, grayish-green; petiole thick, 30 cm long. **Inflorescences** interfoliar, 1–1.8 m long, branched to 3 orders, flowers in triads; peduncle 50–58 cm long; prophyll 25–63 cm long; peduncular bract 40–55 cm long. **Flowers** ♂ with sepals 1.7–2.1 × 1.6–2 mm, concave, petals elliptic, 3.2–3.5 × 1.8–2.3 mm, stamens 6, slightly biseriate; ♀ with sepals broadly ovate, 2.4–2.8 × 2.6–3.3 mm, petals ~ 3.3 × 4.3 mm. **Fruits** ovoid, later subglobose, 1.5–2.2 × 1.2–1.9 cm, ripening to yellow or whitish; seed subglobose, 1.7–1.9 × 1.5–1.7 cm. **Distr:** SE Madagascar. **Ecol:** In semi-arid or dry forests and low montane forests; 80–600 m. **Zones:** 10–12. **IUCN:** VU. **CITES:** A 2.

D. lutescens (H. Wendl.) Beentje and J.Dransf., *Palms Madagascar*: 212, 1995. (Syn: *Chrysalidocarpus lutescens*, etc.). [Yellow Areca Palm]. **Palms**, clustering; stem 2–5 m tall, 10–15 cm Ø, smooth, gray, cylindrical with a basal mass of stemless rosettes; crownshaft swollen, yellow-green covered with caducous whitish wax and scales. **Leaves** green, ~1.5 m long, segments almost opposite, equally spaced, rachis yellow; leaflets 40–60 pairs, in bifarious, lanceolate, 35–60 × 1.2–2.5 cm, yellow-green, adaxially waxy glaucous, long tail acuminate, apex unequal, short bilobed; terminal leaflets gradually short, ~10 cm; petiole and rachis smooth, yellowish-green. **Inflorescences** arching, ~30 cm long, highly branched, main stalk arising from a tubular bract; flowers crowded. **Flowers** ♂ with sepals 3, petals 3, stamens free, exserted, ~4 mm long, anthers 2-celled,

Above: *Loxococcus rupicola*, *Nephrosperma van-houtteanum*, *Nephrosperma van-houtteanum* (flowers).

dorsifixed, pistillode much shorter than stamens; ♀ small, ~3 mm Ø, sepals 3, petals 3, coriaceous. **Fruits** subturbinate or obovate, dark yellow turning purple-black, ~1.5–2.5 × 1 cm long; seed subobovate, endosperm homogeneous. **Distr:** E Madagascar. **Ecol:** Fl. May. Fr. Jun–Aug. **Zones:** 10–12.

Loxococcus

Species 1. *Generic description same as specific.* **Cult:** Propagate from seeds. Requires a moist, well-drained, acidic soil in full sun.

L. rupicola (Thwaites) H. Wendl. and Drude, *Linnaea* 39: 185, 1875. (Syn: *Ptychosperma rupicola*). [Crook-fruited Palm]. **Palms**, solitary, unarmed; stem slender, dull green, 5–12 m tall, 10–12 cm dbh, closely ringed with leaf scars, slightly swollen at base; crownshaft 15–20 cm long. **Leaves** ~10, neatly abscising; lamina reduplicately pinnate, 1.4–2.5 × 1–1.2 m; leaflets 15–35 pairs, lanceolate, lower and middle leaflets 60–90 × 6–7 cm, apical leaflets 20–36 × 3.5 cm, dull green, coriaceous, apices truncate or oblique, adaxially glabrous, abaxially paler; petiole 30–50 cm long, copper-colored, scaly, margins undulate. **Inflorescences** 30–35 cm long; peduncle with base bulbous and 2 clasping wings, outer wing 8–10 cm long, inner wing 5–6 cm long; prophyll 30–35 cm long, narrowly ovate, beaked, laterally 2-keeled, bearing abundant green or reddish scales; rachis stiffly projecting upward, rachillae divaricate, rachis and rachillae glabrous, deep crimson at anthesis with flowers borne in triads. **Flowers** ♂ 1.5 cm long, fleshy, sepals keeled, minutely toothed, petals elongate, stamens 12, filaments short, slender, pistillode domed with 3 short slender appendages; ♀ globular, 1.2 cm Ø, sepals 4 mm long, petals twice as long as sepals. **Fruits** spherical, 2.5 cm Ø, reddish-brown at maturity; seed globose, 2 cm Ø. **Distr:** Endemic to SW Sri Lanka. **Ecol:** In rainforests; 300–1,500 m. **Zones:** 11–12. **Info:** Nuts used similarly as *Areca catechu*. **IUCN:** CR.

Nephrosperma

Species 1. *Generic description same as specific.* **Cult:** Propagate from seeds. Requires a well-drained soil in full sun.

N. van-houtteanum (H. Wendl. ex Van Houtte) Balf.f., *Fl. Mauritius*: 386, 1877. (Syn: *Areca nobilis*, *Oncosperma van-houtteanum*). **Palms**, solitary, to 12 m tall, spiny when young; trunk slender, gray, erect, becoming bare, conspicuously ringed with leaf scars; leaf sheaths tubular, becoming open, not forming a well-defined crownshaft. **Leaves** pinnate, neatly abscising, arching, to 3 m long, young plants with spiny petioles and these becoming less obvious with age; leaflets dark green. **Inflorescences** spikes, solitary, long, hanging clear of the crown, interfoliar; spadix once-branched with 2 spathes. **Flowers** ♂ symmetrical, sepals 3, distinct, imbricate, rounded, keeled, petals about 3–4 times as long as the sepals, valvate, ± boat-shaped, stamens ~40–50, filaments elongate, pistillode ovoid; ♀ globular, sepals 3, distinct, imbricate, ± rounded, tips valvate, staminodes 6, toothlike, gynoecium ± obpyriform, unilocular, uniovulate. **Fruits** ovoid to reniform, 1 cm Ø, ripening from green through yellow to red; seed ± globose to somewhat reniform, 1-seeded. **Distr:** Endemic to Seychelles. **Ecol:** In secondary forests, rocky slopes in open forests, moist tropical forests and ravines; 400 m. **Zones:** 11–12. **IUCN:** NT.

Normanbya

Species 1. *Generic description same as specific.* **Cult:** Propagate from seeds. Requires a moist soil in humid conditions and full sun.

N. normanbyi (W. Hill) L.H. Bailey, *Gentes Herb.* 2: 188, 1930. (Syn: *Ptychosperma normanbyi*, etc.). [Australian Black Palm, Black Palm, Queensland Black Palm]. **Palms**, to 15 m tall, unarmed, pleonanthic, monoecious; trunk erect, ringed with distinct leaf scars, vertically striate, gray, base bulbous; leaf sheaths forming a prominent crownshaft, pale, ashy gray, brownish at the top. **Leaves** pinnate, plumose, loosely arching; leaflets single-fold, irregularly arranged, divided to the base into 7–9 linear segments, with 1–3 large veins, apices praemorse; petiole short or lacking. **Inflorescences** infrafoliar, divaricate, somewhat pendulous in fruit, branched to 2 (or more) orders; prophyll tubular, rather narrow, laterally 2-keeled; peduncular bract like the prophyll, both deciduous; rachilla bracts subtending triads basally, paired and solitary ♂ flowers distally; floral bracteoles low, rounded; ♂ flowers bullet-shaped in bud, borne lateral to the ♀ on short, laterally flattened pedicels. **Flowers** ♂ symmetrical, sepals 3, distinct, imbricate, rounded, minutely toothed, petals 3, distinct, valvate, ovate, stamens 20–40, awl-shaped, anthers elongate, pistillode flask-shaped; ♀ ovoid, sepals 3, distinct, broadly imbricate, rounded, slightly fringed, petals 3, similar to sepals but longer, staminodes 3, broadly toothlike, gynoecium narrowly ovoid, unilocular, uniovulate, narrowing shortly above into 3 large stigmas. **Fruits** ovoid to obpyriform, 3–3.5 cm long, pointed distally, dull salmon-pink to purplish-brown; seed 1. **Distr:** Australia (N Queensland). **Ecol:** In rainforests, close to rivers and streams, often in swampy areas. **Zones:** 10–12.

Oncosperma

Species 5. **Palms**, pleonanthic and monoecious, tall, usually clustered, spiny; stem erect, ringed with leaf scars, frequently armed with scattered or dense robust spines, eroding with age; leaf sheaths armed, tubular, forming a crownshaft. **Leaves** reduplicately pinnate; leaflets numerous, single-folded, regularly arranged, or distinctly grouped, held in different planes, apices acute or acuminate; petiole densely armed with spines and covered with indumentum. **Inflorescences** solitary, infrafoliar, branching to 2 orders near base, to 1 order distally, protandrous, erect in bud, completely enclosed in a prophyll until anthesis; first-order branches pendulous; rachillae long, pendulous, glabrous, bearing spirally arranged, low, triangular bracts each subtending a triad, or in distal portion, subtending solitary or paired ♂ flowers. **Flowers** ♂ asymmetrical, sepals 3, acute, keeled, imbricate, briefly basally connate or distinct, petals 3, acute or acuminate, valvate, ± basally connate, stamens 6–9, briefly epipetalous or free, filaments short, thick, sometimes basally connate in a very short ring, anthers medifixed, basally sagittate, dehiscence latrorse, pistillode deeply trifid; ♀ globular, sepals 3, distinct, rounded, imbricate, petals 3, distinct, rounded, staminodes usually 6, gynoecium unilocular, uniovulate, globose or ovoid. **Fruits** spherical, epicarp smooth or pebbled. **Distr:** Sri Lanka to Philippines. **Cult:** Propagate from seeds or suckers. Requires a moist, well-drained soil in full sun.

O. tigillarium (Jack) Ridl., *J. Straits Branch Roy. Asiat. Soc.* 33: 173, 1864. (Syn: *Areca tigillaria*, etc.). [Nibung Palm]. **Palms**, to 24–30 m tall, densely clustering; stems 15–24 m, 15 cm Ø, light brown to light gray with many 5–10 cm long, black spines and very widely spaced dark, thin rings; crownshaft 1.5 m tall, pale green, cylindrical, with short black spines; crown 7.5–12 m. **Leaves** pinnate, 3 m long, arching; leaflets pendent, 60 cm long, light to dark green; petiole 60 cm long; both stalk and midribs with short black spines. **Inflorescences** clusters, 60–90 cm long, pendent, with both ♂ and ♀ flowers; branches ~55, yellow. **Flowers** yellow, unisexual; ♂ sepals ovate-triangular, with a thickened keel, almost free to base, petals lanceolate, cuspidate, ~6 mm long, stamens 6, filaments fleshy, triangular, adnate to the petals; pistillode large, conic with 3 processes; ♀ ~6 mm long, sepals orbicular, petals ovate, about as large as the globose pistil, stigmas triangular. **Fruits** drupes, globose, 0.6–2.5 cm across, dark purple or black when ripe; seed 1, scantily ruminate. **Distr:** SE Asia, Thailand to W Malesia. **Ecol:** In coastal swamps. **Zones:** 10–12.

Phoenicophorium

Species 1. *Generic description same as specific.* **Cult:** Propagate from seeds. Requires a moist soil in warm and shaded conditions.

P. borsigianum (K. Koch) Stuntz, *U.S.D.A. Bur. Pl. Industr. Invent. Seeds* 31: 88, 1914. (Syn: *Stevensonia borsigiana*, *S. grandifolia*, etc.). [Borsig's Palm, Thief's Palm]. **Palms**, solitary, 9–12 m tall; trunk slender, 6–9 m tall; crown 4–4.5 m across; crownshaft absent. **Leaves** spreading and recurved; lamina pinnate, cuneate-obovate or oblong, 1.8–2.1 m long, mostly undivided, usually with a V-shaped cleft at the end, heavily grooved on both surfaces; leaves and stalks of young plants covered with black spines; petiole

Top row: *Normanbya normanbyi, Oncosperma tigillarium, Phoenicophorium borsigianum, Pinanga cleistantha* (habit).
Bottom row: *Pinanga coronata, Pinanga disticha, Pinanga limosa, Pinanga malaiana.*

yellow to orange, 60 cm long. **Inflorescences** panicles, 90 cm across, with both ♂ and ♀ flowers, in triads with 1 ♀ flower between 2 ♂ flowers; spathes 2, lower 30–45 cm long, persistent, bristly; spadix 1–2 m long, axillary, ovoid, laxly branched, golden yellow; branches 30–45 cm long, slender, ± erect, densely flowered; peduncle 45–90 cm long. **Flowers** unisexual, yellow; ♂ 2 cm across, sepals imbricate, minute, reniform, petals ovate-lanceolate, acute, valvate; ♀ minute, globose, sepals orbicular, concave, imbricate, petals larger, ovary ovoid, 2-celled, stigmas sessile. **Fruits** drupes, ovoid, 1.2 cm long, ripening to dull red; seeds ovoid. **Distr:** Seychelles. **Ecol:** On hills and in valleys of rainforests. **Zones:** 11–12.

Pinanga

Syn: *Cladosperma, Ophiria, Pseudopinanga.*
Species 137. **Palms**, very small to robust, solitary or clustered, acaulescent or erect, unarmed, pleonanthic, monoecious; stem erect, very slender, ringed with leaf scars, occasionally stilt-rooted; leaf sheaths tubular, forming a well-defined crownshaft. **Leaves** undivided and pinnately ribbed, with or without an apical notch, leaflets 1 to several-fold, regularly to irregularly arranged, acute, acuminate or lobes; petiole present or absent. **Inflorescences** mostly infrafoliar, rarely interfoliar; prophyll thin, membranous, 2-keeled, enclosing inflorescence in bud; rachis usually longer than peduncle; rachis bracts triangular; rachillae bearing spiral or distichous triads throughout, or triads in 4 or 6 vertical rows or more. **Flowers** ♂ symmetrical, calyx copular with 3 triangular, frequently unequal lobes, petals 3 triangular, lobes unequal, stamens rarely 6, usually 12–68, pistillode absent; ♀ usually globose, symmetrical, much smaller than ♂, sepals 3, membranous, striate, imbricate, distinct, or proximally connate with 3 broad lobes, staminodes absent, gynoecium unilocular, uniovulate, globose, stigma usually convolute. **Fruits** globose, or ellipsoidal to spindle-shaped, usually bright crimson, scarlet, or black; fruiting rachillae brightly colored (orange or reddish); seed conforming to the fruit shape. **Distr:** Tropical and subtropical Asia to NW Pacific. **Cult:** Propagate from seeds or by division in clumping species. Requires a neutral, moist, well-drained soil with high humidity and ample water, in shade.

P. cleistantha J. Dransf., *Principes* 26(3): 127, 1982. **Palms**, clustering, dwarf, to 2.5 m tall; stems slender, arching-ascending. **Leaves** simple to pinnate; lamina entire or with only a few splits, mottled; rachis yellowish; petiole short. **Inflorescences** entirely concealed in the persistent bract. **Fruits** ellipsoid, red, glabrous. **Distr:** Peninsular Malaysia (Terengganu). **Ecol:** In damp lowland forests. **Zones:** 10–12. **Info:** This species is unusual in producing an inflorescence hidden entirely in the persistent bract. This rare species is now considered endangered in the wild.

P. coronata (Blume ex Mart.) Blume, *Rumphia* 2: 83, 1839. (Syn: *Areca coronata, P. kuhlii, Seaforthia coronata*, etc.). [Ivory Cane Palm]. **Palms**, clustering, 3–7.5 m tall; trunk 1.8–6 m tall, 5 cm Ø, dark green with dark brown rings; crown 4.5 m across; crownshaft 30 cm tall, slightly thicker than the trunk, yellow to light green, cylindrical; new growth often suffused with pink or red. **Leaves** 4–6, pinnate, 1.2–1.5 m long; leaflets narrow and pointed to wide with squared and jagged ends, light to deep green; petiole 30 cm long. **Inflorescences** bisexual, panicles, 30 cm long, rachises pinkish, aging to reddish. Flowers unisexual, white. **Fruits** ovoid, 1.2 cm long, ripening to glossy black. **Distr:** S Andaman Islands to Lesser Sunda Islands. **Ecol:** In low montane rainforests. **Zones:** 11–12.

P. disticha (Roxb.) H. Wendl., *Palmiers*: 253 1878. (Syn: *Areca disticha*, etc.). [Legong Palm]. **Palms**, clustering; stems several, slender, 0.9–1.3 m tall, ~4 mm Ø. **Leaves** usually simple, obcuneate, bilobed or with a few broad acuminate leaflets; leaflets 15–30 cm long, dark mottled with lighter patches, nerves prominent, numerous, 6–9 in lateral leaflets; petiole 10 cm long, slender, scurfy; leaf sheaths to 7.5 cm long. **Inflorescences** with spathes oblong, beaked, 2.5 cm long; spadix solitary (rarely branched), 3.2–10 cm long, deflexed, green; rachis slender, tomentose, flowers distichous, in triads with 1 ♀ with 2 ♂. **Flowers** ♂ 3 mm Ø, sepals short, rounded, petals ovate, acuminate, stamens ~15, filaments very short; ♀ with sepals orbicular, petals shortly cuspidate, ovate, pistil conic cylindrical, much longer than petals, stigma capitate. **Fruits** red, cylindric-ovoid or ellipsoid, 1–1.2 cm long, acute, persistent stigma discoid; seed elliptic with a few streaks of rumination. **Distr:** Peninsular Thailand to Sumatra. **Ecol:** Palm of forest undergrowth, often forming large clumps. **Zones:** 10–12.

P. limosa Ridl., *J. Straits Branch Roy. Asiat. Soc.* 44: 201, 1905. **Palms**, dwarf, usually solitary, 0.6–2 m high; roots stout; stem ~6 mm Ø, woody, distinctly ringed, internodes 2.5 cm long. **Leaves** oblong, 30–40 x 15 cm, base obcuneate, margins entire or divided into irregular broad leaflets with sides curving, apex deeply bifid, or more often lobed with 5 lobes,

Above: *Pinanga sarmentosa, Ptychosperma elegans, Ptychosperma macarthurii, Ptychosperma waitianum.*

lower ones acuminate or truncate with numerous short teeth, nerves numerous, prominent, pale green; leaf sheaths broad; petiole 7.5 cm long. **Inflorescences** with spadix deflexed, simple or with 3 branches; branches slender, 5 cm long; flowers 28, arranged alternately and distichously. **Flowers** with calyx lobes short, broad, orbicular; petals similar, slightly longer. **Fruits** small, oblong, ~6 mm long, drying to subacute, white, ribbed when dry; seed with testa fibrous, endosperm ruminate. **Distr:** Peninsular Malaysia. **Ecol:** In thickly wooded swamps and muddy areas in forests. **Zones:** 11–12.

P. malaiana (Mart.) Scheff., *Natuurk. Tijdschr. Ned.-Indië* 32: 175, 1871. (Syn: *Areca haematocarpon, Seaforthia malaiana*, etc.). [Malay Pinanga Palm, Malaya Sealing Wax Palm]. **Palms**, forming small clumps; stems 0.3–3.6 m tall, stout, 2.5 cm Ø, internodes 7.5 cm long. **Leaves** spreading; lamina pinnate, large, broad, 12–38 cm long; leaflets linear, 38 x 1.2–2.5 cm, with 2 strong ribs, apex acuminate, adaxially dark green, abaxially glaucous; petiole 45 cm long. **Inflorescences** with spathe oblong, 25 cm long; spadix compound, deflexed, stout, 30 cm long; branches 2–5; rachis broad, flattened, bright red in fruit, flowers arranged distichously. **Flowers** ♂ with sepals lanceolate, petals much longer, stamens 40, filaments very short; ♀ with sepals and petals very short, stigma discoid. **Fruits** drupes, 2 cm long, initially cream, ripening to black; seed ovoid, 2.5 cm long, base truncate, apex blunt, endosperm ruminate. **Distr:** Peninsular Thailand to Sumatra. **Ecol:** In forests. **Zones:** 10–12.

P. sarmentosa Saw, *Palms* 47: 141, 2003. **Palms**, clustering, pleonanthic; stem short, stoloniferous, stilt rooted, to ~20 cm high; internodes 0.5–1.2 cm long, 1–2.3 cm Ø; mature stolons 0.8–1 m long, creeping above the forest litter, terminating with plantlets; crownshaft to 22 cm long, 1–1.5 cm Ø. **Leaves** 4–5, pinnate, 70–90 cm long; sheaths tubular, 11–15 cm long, with dark reddish-brown indumentum; lamina variously divided into 2–8-fold leaflets, slightly mottled, apical margin pointed; leaflets 5–7 pairs, terminal leaflets 16–26 x 3–6 cm, proximal leaflets 25–35 x 1.5–5 cm, adaxially covered with scattered dark reddish indumentum, abaxially glabrescent; petiole 35–55 cm long; rachis slightly curving. **Infructescences** infrafoliar, spreading in various directions, 9–12 cm long; peduncle bright red, ± terete, 2–2.5 cm long; peduncular bracts scalelike, (1–)2; rachillae 2, rarely 3 bright red, rarely spicate, regularly zigzagging with fruits distichously arranged on the ridges of the folds, ~3 fruits/cm. **Fruits** green, maturing black, ellipsoid, ~1.3 x 0.8 cm; seed, 1.2 x 0.5 cm. **Distr:** Peninsular Malaysia. **Ecol:** In lowland dipterocarp forest. **Zones:** 11–12.

Ptychosperma

Species 28. **Palms**, solitary or clustering; stems to 15 cm Ø, unarmed. **Leaves** with sheaths green, unarmed, forming crownshaft; lamina pinnate, unarmed; plication reduplicate; segments in 1 plane, usually cuneate, apices jagged and irregular, cleft. **Inflorescences** axillary below crown of leaves, paniculate, with 2 or 3 orders of branching, pendulous; flowers sessile, in triads of 1 ♀ flower flanked by 2 ♂ flowers; prophyll short; peduncular bract tubular; secondary peduncular bracts often present, incomplete. **Flowers** unisexual; ♂ with sepals 3, imbricate, free, petals 3, valvate, free, stamens (9–)25–34(or more), in whorls of 3 or more, pistillode present, with slender style; ♀ with sepals 3, imbricate, free, petals 3, imbricate, free, staminode lobes or scales present, pistils 1, ovules 1, styles indistinct, stigmas 3. **Fruits** drupes, small, exocarp red (or black), smooth, mesocarp fleshy; endocarp bony, usually 5-lobed in cross section; seeds ovoid. **Distr:** Pacific Islands, Australia. **Cult:** Propagate from seeds. Requires a moist, well-drained soil, rich in organic matter, warm conditions, high humidity and shade.

P. elegans (R. Brown) Blume, *Rumphia.* [as '*eleganti*'] 2: 118, 1843. (Syn: *Archontophoenix elegans*, etc.). [Solitaire Palm]. **Palms**, solitary, 7.5–15 m tall; stems 6–10.5 tall, 12 cm Ø, light gray to white, with distinct, widely spaced dark green or brown rings; crown 2.4–4 m across; crownshaft 60 cm long, silvery-gray to silver-green, slightly bulging at base. **Leaves** pinnate, 1.8–2.4 m long, shallowly V-shaped in cross-section, with many leaflets; leaflets 60 cm long, apex jagged, adaxially dark green, abaxially pale green. **Inflorescences** with flowers in clusters, 60–90 cm long, 3–4x-branched, branches greenish yellow, axes orange-yellow in fruit. **Flowers** unisexual, white, fragrant; stamens usually 2—22(–30). **Fruits** globose to ellipsoid, 1.2 cm long, light to deep red; seed with endosperm deeply ruminate. **Distr:** Australia (E Queensland). **Ecol:** In moist organic soil over limestone in mesic hammocks and disturbed wooded areas; 0–10 m. **Zones:** 10–11.

P. macarthurii (H. Wendl.) H. Wendl. ex Hook. f., *Kew Report.* 55, 1882. (Syn: *Actinophloeus macarthurii, Kentia macarthurii*). [Cluster Palm, Hurricane Palm, Macarthur Feather Palm, Macarthur Palm]. **Palms**, stems clustering, with few to many stems, 2.5–12 m tall; stems 4.5–7.5 m tall, uniformly thickened, 5 cm Ø, older part of stem grayish-white, internodes 8.5–12.5 cm long; crown 4.5–7.5 m across; crownshaft 30–60 cm long, light green, slightly bulging at base. **Leaves** pinnate, 0.9–1.8 m long, divided into 20–28 leaflet pairs; leaflets smooth, glabrous, coriaceous, flat, green, unarmed, 25–60 x 5–7.5 cm, terminal leaflet absent; leaflets toothed at apex, like a fish tail; leaf sheaths forming a distinct crownshaft, ~45 cm long, unarmed, smooth, pale green; bracts 8.8–12.5 cm long, triangular, brownish-green; petiole 10–15 cm long, smooth, unarmed, concavo-convex. **Inflorescences** arising from node near base of crownshaft; branches yellowish-green. **Flowers** white, unisexual; ♂ with sepals 3, much smaller than petals, imbricate, coriaceous, smooth, glabrous, forming a cuplike structure, petals trilobed, coriaceous, green, ~7 mm long, stamens 22, filaments unequal, 3.5–6 mm long, anthers with nectar; ♀ with ovary 3-carpelled. **Fruits** densely packed along pendent, arching spikes, dark orange to deep red, ellipsoid with a small terminal beak, 1.2 cm long; seeds hard. **Distr:** Australia (NE Queensland), New Guinea (S PNG). **Ecol:** Fl. Feb. **Zones:** 10–12.

P. waitianum Essig, *Principes* 16: 39, 1972. [East African Wine Palm, Giant Palm, Ivory Coast Raffia Palm]. **Palms**, solitary or clustering, 3–5 m tall; trunk slender, to 2.5 cm Ø; crownshaft plump, 30 cm long, silvery-gray to nearly white, bulging at midpoint; crown hemispherical. **Leaves** 8, arching, spreading; lamina pinnate, caryotoid, 60–90 cm long, emergent leaves dark red; leaflets sessile, borne from the rachis in a single flat plane, widely spaced in the middle of the leaf, mostly ~10 cm apart, lamina triangular, broadly wedge- or fishtail-shaped, 15–25 x 14–18 cm, apices deeply cleft into 2 regularly notched leaflets with the distance between the rachis and the leaf sinus 10–15 cm long, inside margins jagged, both surfaces deep green; petioles 30 cm long. **Inflorescences** 30 x 30 cm, sparsely branched to 2 orders; rachillae ~15 cm long, sparsely flowered. **Flowers** unisexual. **Fruits** dark red to black. **Distr:** SE New Guinea. **Ecol:** In the understory of humid rainforests. **Zones:** 10–12.

Rhopalostylis

Species 3. **Palms**, solitary, erect; stem typically unbranched, not tapering, marked with leaf scars; crownshaft conspicuous, with large sheathing leaf bases. **Leaves** in adult rather sharply ascending, regularly pinnate, without spines, axis clad in peltate scales, pinnae subopposite and reduplicately inserted on axis; pinna long, narrowly lanceolate, acuminate, bearing peltate scales when young, midrib and usually one pair of lateral nerves prominent adaxially, margins not thickened; seedling leaves broad, bilobed. **Inflorescences** below leaves, shortly stalked, much and densely branched, enclosed within 2 caducous, boat-shaped spathes of nearly equal size, outer 2-winged, branches thick, divided several times, ultimate divisions long and closely and deeply notched to accommodate sessile glomerules each of 2 or 3 flowers subtended by a short bract. **Flowers** unisexual, one ♀ between two ♂, or distally the ♀ aborting leaving two ♂ together; ♂ asymmetric, soon falling, sepals much narrower than obliquely ovate petals, stamens 6, ovary narrow, ± columnar; ♀ smaller, globose-ovoid, sepals and petals broad; staminodes minute, ovary ovoid, 1-locular, stigmas terminal. **Fruits** symmetric, fibrous within; seeds hard. **Distr:** New Zealand (mainland, Kermadec and Norfolk Islands). **Cult:** Propagate from seeds. Requires

Top row: *Rhopalostylis baueri*, *Rhopalostylis sapida* (inflorescences), *Rhopalostylis sapida*.
Bottom row: *Verschaffeltia splendida* (infructescence), *Commelina bengalensis*.

a deep, moist soil in shade or partial sun. **Info:** *R. sapida* has the southernmost range of any palm genus.

R. baueri (Hook. f.) H. Wendl. and Drude, *Bot. Zeitung* (Berlin) 35: 638, 1877. (Syn: *R. baueri* var. *cheesemanii*, etc.). [Norfolk Island Palm, Norfolk Island Shaving Brush Palm]. **Palms**, solitary, 9–19 m tall; trunk 7.5–15 m tall, 20 cm dbh, light to dark green except the old parts, with very closely set tan to gray rings; crown shaped like a shuttlecock, 3–4.5 m across; crownshaft cylindrical, 90 cm tall, base bulging, light yellowish-green. **Leaves** erect and ascending, pinnate, 4 m long, V-shaped in cross-section, midrib stout, nearly white; leaflets many, 60–90 cm long, apices pointed, light to dark green. **Inflorescences** bisexual, 90 cm long, borne below the crownshaft; flowering branches numerous, tan to brown, to 35 cm long. **Flowers** unisexual, small; sepals relatively shorter, the longest one in ♂ rarely equal in size to the petals. **Fruits** ovoid or subglobose, ~1.3 × 0.2 cm, ripening to reddish white to brick-red; seed almost spherical. **Distr:** Kermadec Islands, Norfolk Island. **Ecol:** In low montane rainforests. Fl. and Fr. Throughout most of the year. **Zones:** 10–11.

R. sapida Wendl. et Drude, *Les Palmiers* 255, 1878. (Syn: *Areca sapida*, *Eora sapida*, *Kentia sapida*). [Feather Duster Palm, Nikau Palm, Shaving Brush Palm]. **Palms**, solitary, erect, to ~10 m tall; stem ~25 cm dbh, green between rather closely spaced leaf scars; crownshaft to 60 cm long, smooth and green, conspicuously bulging. **Leaves** to 3 m long; leaflets to 1 m long, closely set and ascending sharply; spathes ~30 × 15 cm, pink to yellow, smooth, caducous as first flowers open; ultimate branches to ~20(–30) cm long, ~1.5 cm Ø, with buds on, at first pale cream-colored; flower buds tightly packed, lilac. **Fruits** ~ 10 × 7 mm, elliptic-oblong, brick-red; seeds long-ovoid. **Distr:** N and C New Zealand, Chatham Islands. **Ecol:** In lowland rainforests. Fl. Nov–Apr. Fr. Feb–Nov. **Zones:** 9–11. **Info:** Found at latitude 44º S (Chatham Islands), making this the southernmost palm in the world. This palm very rarely produces multiple heads atop the trunk.

Verschaffeltia

Species 1. *Generic description same as specific.* **Cult:** Propagate from fresh seeds. Requires a moist, humus-rich soil in full or partial sun.

V. splendida H. Wendl., *Ill. Hort.* 12(Misc.): 6, 1865. (Syn: *Regelia magnifica*, *Stevensonia viridifolia*, etc.). [Seychelles Stilt Palm]. **Palms**, solitary, 15–21 m tall; trunk 12–16 m tall, 20 cm dbh, base thickened to 1.2–1.8 m, with a cone of stilt roots, apical parts covered with sharp, black, downward-pointing spines; crownshaft 60–90 cm tall, light green, slightly inflated at base; crown 4–5.4 m Ø. **Leaves** elliptic in outline, 1.5–2.4 m long; *juvenile* plants with leaf margins entire, apex notched; *mature* plants with leaves pinnately divided by wind into variously shaped segments (usually linear) and number. **Inflorescences** bisexual, 1.2–1.8 m long; branches several, thin, greenish-yellow. **Flowers** unisexual, small, yellow. **Fruits** ovoid, 2.5 cm Ø, brown, borne in pendent infructescences. **Distr:** Seychelles. **Ecol:** In mountainous rainforests. **Zones:** 11–12.

COMMELINALES

FAMILIES 5, GENERA 68, SPECIES 812.

Mycorrhizae absent. **Vessel elements** scalariform. **Seeds** with seed coat testal and tegmic; endosperm abundant (Stevens, 2001). **Info:** Monophyly of the Commelinales is supported by molecular data (Chase et al. 1995b, 2000, 2006; Davis et al. 2004; Soltis et al. 2000).

COMMELINACEAE

Genera 40, **species** 652. **Herbs**, annual or perennial, sometimes woody at base; stems with prominent nodes and internodes. **Leaves** alternate, distichous or spirally arranged, sessile or petiolate, leaf sheath prominent, open or closed; lamina simple, margins entire. **Inflorescences** usually of cincinni in panicles or solitary, sometimes shortened into heads, sometimes sessile with flowers fascicled, sometimes axillary and penetrating enveloping leaf sheath, rarely flowers solitary and terminal or axillary. **Flowers** bisexual, rarely unisexual, actinomorphic or zygomorphic; sepals 3, free or connate only at base, often boat-shaped or carinate, sometimes galeate at apex; petals (2 or)3, free, sometimes connate and tubular at middle and free at 2 ends, sometimes clawed; stamens 6, free, all or only 2 or 3 fertile, filaments glabrous or torulose villous, anthers parallel or slightly divergent, longitudinally

Above: *Commelina diffusa, Cyanotis arachnoidea, Cyanotis obtusa, Dichorisandra thyrsiflora.*

dehiscent, rarely dehiscent by apical pores; staminodes 1–3; antherodes 4-lobed and butterfly like, 3-sect, bilobed and dumbbell-shaped, or entire; ovary triloculed, or reduced to biloculed, ovules 1 to several per locule, orthotropous. **Fruits** capsules, loculicidal, 2- or 3-valved, rarely baccate and indehiscent; seeds few, large, endosperm copious. **Distr:** Tropical and temperate regions worldwide.

Commelina

Species ~170. **Herbs**, annual or perennial, often prostrate or suberect, glabrous to pubescent. **Leaves** usually narrowly elliptic, sometimes subfleshy, entire, sheathing at base, acute. **Inflorescence** usually 2-flowered, scorpioid cymes, subtended by short, hooded, funnel-shaped or falcate, folded spathes or involucral bracts; flowers of upper cymes small, often sterile, those of the lower cymes larger and fertile. **Flowers** with sepals 3, outer cucullate, free, inner 2 often connate below, obovate, sometimes persistent; petals 3, longer than sepals, clawed, ovate, inner 2 smaller than the outer one; perfect stamens 3, free, hypogynous, antisepalous; staminodes 2–3, filaments glabrous; ovary glabrous, 2–3-loculed, 2 of the locules 1–2-ovuled, 3rd locule, if present, either 1-ovuled or empty. **Fruits** capsules, loculicidal, posterior locule (cocci) indehiscent or absent, the 2 anterior cocci connate and indehiscent; seeds small, usually subtriangular, reticulately pitted. **Distr:** Mainly in tropical and subtropical regions. **Cult:** Propagate from cuttings or seeds. Requires any moist but well-drained soil in full sun.

C. benghalensis L., *Sp. Pl.* 1: 41, 1753. (Syn: *C. cavaleriei*). [Bengal Spiderwort, Dayflower, Tropical Spiderwort, Wandering-Jew]. **Herbs**, perennial; stems mostly creeping, ascending distally, diffuse, numerous branched, to 70 cm long, sparsely pubescent. **Leaf sheaths** sparsely hirsute-ciliate. **Leaves** with petiole distinct; lamina ovate, 3–7 × 1.5–3.5 cm, subglabrous. **Involucral bracts** borne opposite leaves, often several, aggregated at apex of branches, very shortly stalked, funnelform, 0.8–1.2 cm long, sparsely hairy, proximal margins connate, apex acute or obtuse. **Inflorescence** with proximal branch of cincinni with elongate peduncle and 1–3 exserted, infertile flowers, distal branch longer, with several included, fertile flowers. **Flowers** with sepals ~2 mm long, membranous; petals blue, 3–5 mm long. **Fruits** capsules, ellipsoid, 4–6 mm, 3-valved; posterior valve 1-seeded or seedless, indehiscent, other 2 valves each 2-seeded, dehiscent; seeds black, cylindric or semi-cylindric, ~2 mm long, rugose, irregularly reticulate, truncate at 1 end. **Distr:** Tropical and subtropical Africa and Asia. **Ecol:** In wet places; to 2,300 m. **Zones:** 9–12.

C. diffusa N.L. Burman, *Fl. Indica.* 18, plate 7, 1768. (Syn: *C. nudiflora*). [Birdbill/Climbing/Creeping/Spreading Dayflower]. **Herbs**, annual; stems creeping, branched, to >1 m long, glabrous or hispidulous throughout or in a line. **Leaf sheath** hispid or hispid-ciliate, with red lines. **Leaves** subsessile; lamina lanceolate or proximal ones oblong, 3–12 × 0.8–3 cm, glabrous or hispid. **Involucral bracts** borne opposite leaves, folded, ovate-lanceolate, 1–4 cm long, base cordate or rounded, apex usually acuminate. **Inflorescence** with cincinni dichotomously branched from base; 1 branch with 1.5–2 cm long peduncle and 1–4 long-exserted ♂ flowers; other branch with much shorter peduncle and 3–5 bisexual flowers included in involucral bracts; pedicels thick and curved, ~3 mm long. **Flowers** with sepals 3–4 mm, membranous; petals blue, 2 longer ones 4.2–6 mm long. **Fruits** capsules, oblong, trigonous, ~5 mm long, 3-valved, posterior valve 1-seeded, indehiscent, other 2 valves 2-seeded, dehiscent; seeds black, ovoid-globose, ~2 mm long, reticulate. **Distr:** Tropics and subtropics worldwide. **Ecol:** In forests, thickets; to 2,100 m. Fl. May–Nov. **Zones:** 9–12.

Cyanotis

Syn: *Amischophacelus.*

Species ~50. **Herbs**, annual or perennial, erect or creeping; rhizomes absent. **Leaves** alternate. **Inflorescence** with cincinni sessile, subtended by spathelike involucral bracts; bracts imbricate, biseriate, falcate-curved. **Flowers** actinomorphic; sepals free or connate only at base; petals connate and tubular in middle, free at both ends, purple, blue, or white, linear-lanceolate; stamens 6, all fertile, equal, filaments lanate, rarely glabrous, anther locules longitudinally dehiscent; ovary triloculed, ovules 2 per locule. **Fruits** capsules, 3-valved, obovoid- or broadly oblong, trigonous; seeds 1 or 2 per valve, cylindric or quadrangular, reticulate or pitted. **Distr:** Tropical and subtropical Africa and Asia to N Australia. **Cult:** Propagate from cuttings or seeds. Requires any moist but well-drained soil in full sun.

C. arachnoidea C.B. Clarke, *Monogr. Phan.* 3: 250, 1881. (Syn: *C. bodinieri, C. labordei*). [Grass of the Dew]. **Herbs**, perennial; roots fibrous; main stem undeveloped, short; fertile stems arising from beneath leaf rosette, diffuse, creeping, 20–80 cm long. **Leaves** in a basal rosette and cauline; rosulate lamina linear, 8–35 × 0.5–1.5 cm; cauline lamina on fertile stems much shorter, to 7 cm long, abaxially rather densely arachnoid. **Inflorescence** with cincinni often several, forming both terminal and axillary heads, sessile or on peduncle to 4 cm long; bracts 7–8 mm long. **Flowers** with sepals connate at base, linear-lanceolate, ~5 mm long, abaxially arachnoid; petals blue-purple, blue, or white, ~6 mm long; filaments blue arachnoid. **Fruits** capsules, broadly oblong, trigonous, ~2.5 mm long, densely hirsutulous at apex; seeds gray-brown, pitted. **Distr:** China, India, Laos, Myanmar, Sri Lanka, Taiwan, Thailand, Vietnam. **Ecol:** In humid ravines and rocks; to 2,700 m. Fl. Jun–Sep. Fr. Oct. **Zones:** 9–12.

C. obtusa (Trimen) Trimen, *Handb. Fl. Ceylon* 4: 312, 1898. (Syn: *C. arachnoidea* var. *obtusa*). **Herbs**, rosette-forming, perennial, densely hairy. **Leaves** *rosette* ± distichous; sheaths to ~3 cm long, usually split, oblong, ± conduplicate, 10–33 × 2–5 cm, margins sparsely to densely ciliate with woolly hairs, apex obtuse, surfaces maroon or occasionally green; *cauline* leaves similar, 2–7 × 1–1.5 cm. **Inflorescences** composed of numerous cincinni, terminal in clusters of 6; cincinni long- and short-stipitate; scapes usually much-branched, to 60 cm long; bracts slightly exceeding the cincinni; bracteoles lanceolate-oblong to lanceolate-ovate, 0.7–1 cm × 1.3–3 mm. **Flowers** bisexual; sepals lanceolate-oblong, 3.5–5 × 1–1.5 mm, densely lanate; corolla 0.6–1 cm wide, pink or lavender; stamens exceeding the corolla by 5–6 mm, filaments densely bearded with blue, pink or lavender hairs, anthers yellow or orange-yellow; style slightly shorter, bearded. **Fruits** capsules, 2.5–3 × 1.5 mm; seeds ovate to elliptic, 1–1.5(-1.9) × 0.7–1.1 mm, testa brownish or orangish to grayish. **Distr:** S India, Sri Lanka. **Ecol:** On rocky hills and outcrops; 180–600 m. Fl. Nov–Jan. **Zones:** 11–12.

Dichorisandra

Species ~30. **Herbs**; stems erect or climbing, rarely absent. **Leaves** alternate or ± 2-ranked; lamina thin-textured (not succulent). **Inflorescence** terminal panicle of cymes. **Flowers** ± bilaterally symmetrical; sepals free, unequal, upper one hooded, often colored like petal; petals white or violet-blue, often with white base; stamens usually 6, ± unequal, filaments hairless, anthers dehiscing via apical pores; ovary 3-celled, stigma pointlike. **Fruits** capsules; seeds embedded in fleshy, orange or red false aril. **Distr:** Tropical America. **Cult:** Propagate from cuttings or by division. Requires a moist soil in shade.

D. thyrsiflora J.C. Mikan, *Del. Fl. Faun. Bras.*: 3, 1820. [Blue Ginger, Queen Spiderwort]. **Herbs**, evergreen, clump-forming, spreading by underground rhizomes; stems erect, 1–2.5 m tall, usually unbranched. **Leaves** alternate, sessile; lamina elliptic-lanceolate, 20–30 × 5–8.3 cm, base tapering and sheathing stem, apex tapering, adaxially glossy deep green, abaxially sometimes purplish, glabrous. **Inflorescence** terminal, dense, cymes stalked. **Flowers** hooded, deep blue-violet, 1.25–1.9 cm Ø; sepals and petals ± similar; stamens 4, erect and touching, 2 spreading, anthers shorter. **Fruit** capsules, orange; seed with false red aril. **Distr:** E and S Brazil. **Ecol:** In very moist humid forests; ~1,500 m. Fl. throughout most of the year. **Zones:** 10–12.

Palisota

Syn: *Duchekia.*

Species 25. **Herbs**; stem simple, or nearly so, or hardly any. **Leaves** basal or in pseudowhorls, hairy when young; lamina with margins permanently and

Above: *Palisota mannii, Tradescantia pallida, Tradescantia spathacea, Tradescantia zebrina.*

densely clothed with ferruginous hairs. **Inflorescences** 1(rarely 2–3), quasi-terminal, panicles, elongate or dense, flowers in small cymes, many functionally ♂; lower pedicels of cymes often early caducous, so that upper part of cyme-peduncle appears knotted; bracts on axis of panicle short (except in *P. bracteosa*); bracteoles small or absent. **Flowers** with sepals 3; petals 3, small, obovate, white, sometimes tinged with rose or bluish; stamens 3, perfect, anther of interior whorl of stamens (apparently middle of 3) somewhat unlike that of the other 2, with curved or subdivergent cells; staminodes 2 or 3, with beaded hairs; ovary 3-celled, posticous cell often smaller, ovules 1–8 per cell. **Fruits** indehiscent, fleshy or succulent, often purple or red; seeds 2, superposed in each cell, or solitary, or more numerous and irregularly packed, pyramidal or trapezoid. **Distr:** Endemic to tropical Africa. **Cult:** Propagate from seeds. Requires a moist, well-drained soil in full or partial sun.

P. mannii C.B. Clarke in A.L.P.P. de Candolle and A.C.P.de Candolle, *Monogr. Phan.* 3: 132, 1881. **Herbs**; usually stemless, sometimes with stems 0.3–2 m tall, 4–6 mm Ø; young parts shaggy with fulvous or grayish hairs; upper leaves few. **Leaves** lanceolate or lanceolate-obovate, to 1.4 m × 6–12 cm, base acute, narrowed into petiole, margins densely hairy, acuminate, often nearly glabrate when mature; petiole to 40 cm long, base clasping the stem. **Inflorescence** oblong, very dense, 10–18 × 2.5–4 cm, with several hundred flowers; bracts small, concealed by flowers; pedicels 0–6 mm long, persistent. **Flowers** small, white, turning pale pink or lilac; ovary glabrous. **Fruits** berries, white, maturing to orange or reddish orange. **Distr:** S Nigeria to Tanzania. **Ecol:** In rainforests and forest swamps; 600–2,000 m. **Zones:** 10–12.

Tradescantia

Syn: *Rhoeo, Zebrina*, etc. [Spiderwort, Wandering-Jew, Spider Lily].
Species 74. **Herbs**, perennial; roots thin or tuberous. **Leaves** spirally arranged or 2-ranked; lamina sessile or rarely petiolate. **Inflorescences** terminal or terminal and axillary, pairs of cymes, cymes sessile, umbel-like, contracted, subtended by spathaceous bract; bract similar to leaves or differentiated, margins distinct; bracteoles persistent; pedicels very short or well-developed. **Flowers** bisexual, radially symmetric; sepals distinct (rarely basally connate), subequal; petals distinct (rarely basally connate), white to pink, blue, or violet, equal, rarely clawed; stamens 6, all fertile, equal, filaments bearded or glabrous; ovary trilocular, ovules (1–)2 per locule, 1-seriate. **Fruits** capsules, 3-valved, trilocular; seeds (1)2 per locule. **Distr:** Tropical and temperate America. **Cult:** Propagate by division, tip cuttings, rooted shoots, or from seeds. Requires a moist, well-drained soil in full sun; drought-tolerant.

T. pallida (Rose) D.R. Hunt, *Kew Bull.* 30: 452, 1975. (Syn: *Setcreasea pallida, S. purpurea*). [Purple Heart]. **Herbs**, perennial, succulent; stems suffused with purplish-violet. **Leaves** spirally arranged; lamina not variegated, suffused with purplish-violet, lanceolate-oblong to oblong-elliptic, (4–)7–15 × 1.5–3 cm (distal lamina wider or narrower than sheaths when sheaths opened, flattened), base symmetric, rounded to broadly cuneate, margins ciliate or ciliolate, apex acute, glabrous or glabrescent. **Inflorescences** terminal, often becoming leaf-opposed, pedunculate; peduncles (3.5–)4–13 cm long; bracts similar to leaves but usually greatly reduced; pedicels 4–9 mm long, densely white-pilose at summit. **Flowers** with sepals distinct, 0.7–1 cm long, pilose basally; petals ± basally connate, pink, clawed, 1.5–2 cm long; stamens epipetalous, filaments very sparsely bearded. **Fruits** capsules, 3.5 mm long, glabrous; seeds 2.5–3 mm long. **Distr:** Mexico. **Ecol:** Unknown in the wild. Fl. throughout most of the year. **Zones:** 9–12.

T. spathacea Swartz, *Prodr.* 57, 1788. (Syn: *Rhoeo discolor, R. spathacea*). [Moses-in-the-cradle, Oyster Plant, Boat Lily]. **Herbs**, erect or ascending, rarely rooting at nodes; stems unbranched, short. **Leaves** spirally arranged; lamina strap-shaped, to 35 × 5 cm (distal leaf blades wider or narrower than sheaths when sheaths opened, flattened), usually abaxially purple, adaxially green, leathery, succulent, glabrous. **Inflorescences** axillary, sessile, or pedunculate in axils well below shoot apex, cymes enclosed in pairs of boat-shaped spathes; flowers distinctly pedicellate; pedicels glabrous. **Flowers** with sepals distinct, white, 3–6 mm, glabrous; petals distinct, white, ovate, not clawed; stamens free, filaments bearded. **Fruits** capsules, 3- or (by abortion) bilocular, 3–4 mm; seeds 1 per locule, 3–4 mm. **Distr:** S Mexico, C America, West Indies. **Ecol:** Usually in hammocks. Fl. Jan. **Zones:** 10–12.

T. zebrina Hort ex Bosse, *Vollstandiges Handb. Blumengart.* 4: 655, 1849. (Syn: *T. pendula, Zebrina pendula*). [Inch Plant, Wandering Jew]. **Herbs**, decumbent. **Leaves** 2-ranked; lamina variegated, abaxially reddish-purple, adaxially striped green and white, lanceolate-elliptic to ovate-elliptic, 3–9 × 1.5–3 cm (distal lamina wider or narrower than sheaths when sheaths opened, flattened), base oblique, cuneate, apex acute to acuminate. **Inflorescences** terminal, consisting of pairs of sessile cymes enclosed in sheaths of spathaceous bracts, pedunculate; spathaceous bracts foliaceous, reduced; flowers subsessile. **Flowers** with sepals basally connate, 4–5 mm long; petals pink, clawed, claws basally connate forming tube; stamens epipetalous, filaments bearded. **Fruits** capsules trilocular, locules 2-seeded. **Distr:** Tropical America. **Ecol:** In disturbed places and hammocks. Fl. Sep–Feb. **Zones:** 10–12.

HAEMODORACEAE

Genera 14, **species** 116. **Herbs**, perennial or rarely annual, rhizomatous, often stoloniferous; sap often red or orange; stems simple. **Leaves** mostly basal, distichous; lamina linear, base sheathing, folded lengthwise distally, folded portions becoming fused. **Inflorescences** terminal, cymose or corymbose, many-branched, many-flowered, loose or condensed. **Flowers** with tepals persistent, 6 in 2 whorls of 3, distinct above level of insertion on ovary, ± equal, aestivation valvate; stamens 3 or 6; filaments elongate, anthers basifixed, bilocular, often basally sagittate, dehiscence introrse by longitudinal slits; pistil 1, compound, tricarpellate, ovary inferior or half-inferior to nearly superior, trilocular, style 1, undivided, elongate, stigma 1. **Fruits** capsular, loculicidally dehiscent; seeds discoid and peltate. **Distr:** Tropics and warm temperate regions.

Wachendorfia

Species 4. **Herbs**, rhizomatous, to 2.5 m tall; rhizomes red. **Leaves** linear, margins entire, plicate, glabrous to hirsute. **Inflorescence** a raceme, with numerous 3–4-flowered helicoid cymes; axes and bracts canescent. **Flowers** ~2 cm long, zygomorphic; tepals yellow to purplish, imbricate; stamens 3, free, anterior stamen curved to side opposite style, anthers subbasifixed; ovary superior, densely pubescent, ovule 1 per carpel, anatropous, hypotropous, style posteriorly curved, enantiomorphic. **Fruits** ovoid, trilobed, canescent; seeds ovoid, with a dark brown tomentum. **Distr:** S Africa (Cape Province). **Cult:** Propagate from fresh seeds or by division. Requires a moist soil in full sun.

W. thyrsiflora Burm., *Wachendorfia*: 2, 1757. (Syn: *W. elata*). [Blood Root]. **Herbs**, rhizomatous, 1–2(–2.5) m tall, robust, perennial, evergreen, clump-forming. **Leaves** basal, arranged in erect fans; lamina lanceolate, usually shorter than the stem, ~80 × 8 cm, conspicuously pleated along their length, hairless. **Inflorescence** a dense, cylindrical spikelike panicle, subtended by recurved papery bracts; peduncles tall, erect. **Flowers** golden-yellow; tepals 1.2–2.8 × 0.8–1.4 cm; stamens and style spreading, 0.9–2 cm long; flowers are usually left- or right-handed (promoting cross-pollination), left-handed flowers with style along with 1 of the 3 stamens deflected to the left, right-handed flowers with style along with 1 of the 3 stamens deflected to the right. **Distr:** S Africa (Cape Province). **Ecol:** In permanent swampy marshes, and streambanks. Fl. Sep–Dec. **Zones:** 10–11.

Xiphidium

Species 2. **Herbs**, stoloniferous, rather coarse, perennial; rhizomes stout, short or elongate, creeping and erect, subsucculent; stems leafy. **Leaves** equitant, distichous, sheathed at base. **Inflorescence** terminal, paniculate with the flowers secund in 2 rows on the branches; scape simple, flat, succulent, with numerous slender nerves, equitant; flowers pedicellate. **Flowers** small, white, glabrous; perianth 6-parted with spreading lobes; stamens 3, affixed to the bases of

Above: *Wachendorfia thyrsiflora*, *Xiphidium caeruleum*, *Eichhornia crassipes*, *Monochoria hastata*.

the inner segments, filaments short, anthers oblong; ovary free, superior, trilocular, with numerous ovules, styles filiform with apical stigmas. **Fruits** capsules, subglobose, fleshy, loculicidally 3-valvate; seeds numerous. **Distr:** Mexico to tropical America. **Cult:** Propagate by division, aerial plantlets or from seeds. Requires a moist, humus-rich soil in shade.

X. caeruleum Aubl., *Hist. Pl. Guiane.* 1: 33, 1775. (Syn: *Ixia xiphidium*, etc.). [Cola de Paloma, Dove Tail, Palmita, Soskia]. **Herbs**, terrestrial; rhizomes horizontally creeping, to 20 cm long and 1 cm thick, bearing slender stolons to 55 cm long, rooting at nodes; roots fibrous; stems erect mostly 20–70 cm tall, sometimes taller. **Leaves** ensiform, linear or linear-lanceolate, glabrous, 20–60 × 1.5–5 cm, base clasping, margins finely serrulate, apex acuminate. **Inflorescence** a thyrse to 30 cm long or more, branches (cincinni) 5–25-flowered, simple or sometimes once-branched; main axis and branches minutely pubescent; pedicels 2–3 mm long; small plantlets sometimes produced in axils of inflorescence bracts. **Flowers** glossy; perianth segments oblong, obovate or oblong-lanceolate, to 6–8 mm long, obtuse, glabrous, white to yellowish-white, sometimes pale bluish within. **Fruits** capsules, subglobose, 5–8 mm Ø, maturing from green to shiny orange, then turning purple-black; seeds many, black. **Distr:** Mexico to tropical America. **Ecol:** In moist, thickets, shaded banks and forest slopes; 0–500 m. **Zones:** 9–12. **Info:** Cultivated and naturalized widely in the tropics.

PONTEDERIACEAE

[Pickerel Weed].

Genera 9, **species** 33. **Herbs**, perennial, aquatic or of marshy habitats, often floating; vegetative stems indeterminate. **Leaves** vaginate; lamina broad or narrow. **Inflorescence** spikes, umbels or panicles, rarely flowers solitary, axillary from the rootstock or stem, spathaceous, usually decurved in fruit; bractlets minute or none; spathe with 2 valves, valves similar or, when different, lower one leaflike, upper one rarely with a small lamina. **Flowers** with perianth marcescent, salverform or funnelform, 6-parted, or with a 6-parted limb, nearly regular or zygomorphic, tube well developed; stamens 3 or 6, commonly unequal and dissimilar, anthers introrse, basifixed or versatile; ovary free, superior, 1- or triloculed, ovules numerous or by abortion 1, style slender or clavate, stigmas terminal, 3–6-lobate or 3–6-dentate. **Fruits** capsules, 1- or trilocular or 1-seeded and achenelike and enclosed in the fleshy, accrescent base of the perianth. **Distr:** Tropical regions, and also in some temperate regions, especially in the New World.

Eichhornia

[Water Hyacinth].

Species 7. **Herbs**, annual or perennial, rooting in mud or free-floating; vegetative stems submersed and growing to surface, or emersed and short; flowering stems emersed. **Sessile leaves** forming basal rosette or alternate on elongate vegetative stem; lamina with apex acuminate. **Petiolate leaves** floating or emersed; lamina cordate to ovate or round, apex obtuse, coriaceous. **Inflorescences** spicate, to 30-flowered, elongating over several days; spathes folded, with acuminate to caudate extension. **Flowers** open for 1 day only; perianth with tepals connate for more than ½ its length, blue, mauve, or rarely white, funnelform, limb lobes obovate to oblanceolate, longer than 2 cm, apex obtuse to acute, glandular-pubescent; stamens 6, distal 3 shorter than proximal 3, filaments curved upward near apex, purple, glandular-pubescent, anthers yellow, rounded to oblong; ovary incompletely trilocular, ovules many, style trilobed. **Fruits** capsular, elongate; seeds 10–200, ovoid, testa with longitudinal wings. **Distr:** Neotropics and Africa. **Cult:** Propagate by division or from seeds. Requires fertile, marginal to deep water in full sun.

E. crassipes (Martius) Solms, *Monogr. Phan.* 4: 527, 1883. (Syn: *Pontederia crassipes*, etc.). [Water Hyacinth]. **Herbs**, perennial, typically free-floating; vegetative stems condensed, except when branching; flowering stems erect, bending over after flowering, to 25 cm long, distal internode less than 4 cm. **Leaves** *sessile* in basal rosette; *petiolate* leaves floating or emersed, stipule 2.5–14 cm long, apex truncate; petiole at least slightly swollen, 3.5–33 cm long, lamina ovate to round, 2.5–11 × 3.5–9.5 cm. **Inflorescences** spikes, 4–15-flowered; spathes obovate, 4–11 cm; peduncle 5–12.5 cm, glabrous. **Flowers** opening individually within 2 hours after sunrise, wilting by dusk; perianth blue or mauve-blue, limb lobes obovate, 1.6–3.7 cm long, margins entire, central distal lobe with dark blotch in center and yellow spot within blotch; proximal stamens 2–3.5 cm, distal 1.4–1.9 cm long; style trilobed. **Fruits** capsules; seeds 11–14-winged, 1.1–2.1 × 0.6–0.9 mm. **Distr:** Tropical S America; naturalized elsewhere. **Ecol:** In ponds, ditches, canals, calm waters of rivers; 0–300 m. **Zones:** 9–12. **Info:** A very aggressive aquatic weed, which grows and multiplies rapidly, depriving those below water from oxygen.

Monochoria

Syn: *Gomphima*.

Species 7. **Herbs**, paludose or aquatic, perennial or under unfavourable conditions annual; stems erect or creeping. **Leaves** radical from vegetative stem, solitary from flowering stems, curvinerved. **Inflorescences** racemose or subumbelliform, enveloped in broad leaf sheaths when young; peduncle bracteate at base; bract large; flowers subsessile or shortly pedicellate. **Flowers** with perianth segments free nearly to base, spreading during anthesis and afterward spirally contorted, inner segments wider than outer; stamens 6, inserted at base of perianth, dimorphic, 1 with filament long, laterally with an erect, oblique tooth, anther basifixed, blue, large, the other 5 subequal, anthers basifixed, yellow, small; ovary triloculed, ovules numerous per locule, style filiform, stigma subentire or minutely trilobed. **Fruits** capsules, loculicidal, 3-valved; seeds small, longitudinally winged, transversely striate. **Distr:** Tropical and S Africa, India to Russian Far East and Australia. **Cult:** Propagate from seeds or by division. Requires a shallow pond in full sun.

M. hastata (L.) Solms, *Monogr. Phan.* 4: 523, 1883. (Syn: *Carigola hastata*, *Pontederia hastate*). [Arrow-leafed Monochoria, Hastate-leafed Pondweed]. **Herbs**, perennial, aquatic; vegetative stems often long and robust. **Leaves** *radical* with sheath broadened at base, lamina triangular or triangular-ovate, 5–15(–25) × 3–15 cm, base sagittate to hastate, apex acute to acuminate; petiole 30–90 cm long; *cauline* leaves with petiole 7–10 cm long. **Flowering stems** erect or obliquely so, 50–90 cm long. **Inflorescences** erect or suberect, remaining so after anthesis, subumbellate to shortly racemose, 10–40-flowered; peduncle distinctly shorter than associated leaf petiole; pedicels 1–3 cm long. **Flowers** with perianth segments bluish with green median vein and reddish blotch, ovate, 1–1.6 cm across; larger stamen with anther 5.3–6.5 mm long, smaller stamens with filaments filiform, anthers 3–4 mm long; style apex densely hairy. **Fruits** capsules, oblong, ~1 cm long; seeds brown, oblong, wings ~10. **Distr:** Tropical and subtropical Asia to N Australia. **Ecol:** Near waterways, ponds; 100–700 m. Fl. Aug. Fr. Mar. **Zones:** 9–12.

Pontederia

Syn: *Narukila*, *Umsema*, *Unisema*. [Pickerel Weed].

Species 6. **Herbs**, annual or perennial, rooted in mud; vegetative stems submersed and growing to surface, or rhizomatous; flowering stems submersed and growing to surface, or emersed and slightly constricted just below first node. **Sessile leaves** forming basal rosette, lamina thin. **Petiolate leaves** floating or emersed; lamina cordate to reniform, apex obtuse to acuminate. **Inflorescences** spicate, commonly 50 or more-flowered, elongating over many days; spathes folded, apex acute to acuminate; peduncle glandular-pubescent or pilose. **Flowers** open for 1 day only; perianth mauve, blue, or white, funnelform, limb lobes oblong to oblanceolate, glandular-pubescent or pilose, apex obtuse to acuminate; stamens 6, distal 3 shorter than proximal 3, filaments purple, glandular-pubescent, anthers yellow, ovoid to oblong; ovary trilocular, only 1 developing to maturity, ovule 1, style 1 or more. **Fruits** utriculate, ovoid, with spinulose, dentate or smooth longitudinal ridges; seed 1, ovoid, smooth. **Distr:** Western Hemisphere. **Cult:** Propagate from seeds or by division. Requires a shallow pond in full sun.

Above: *Pontederia cordata, Typha domingensis, Acanthostachys strobilacea, Aechmea chantinii.*

P. cordata L., *Sp. Pl.* 1: 208, 1753. (Syn: *Narukila cordata*, etc.). [Pickerel Weed]. **Herbs**, perennial, rooted in mud; vegetative stems contracted, rhizomatous; flowering stems erect, to 1.2 m long. **Leaves** *sessile* with lamina linear; *petiolate* leaves emersed, petiole distinctly constricted just below lamina, to 60 cm long, lamina lanceolate to cordate, 6–22 × 0.7–12 cm, stipule 7–29 cm long. **Inflorescences** spikes with up to several hundred flowers, 2–15 cm long; spathes 5–17 cm. **Flowers** with perianth mauve, tube 3–9 mm, limb lobes oblanceolate, 5–8 mm, distal central lobe with bilobed yellow spot; proximal stamens 0.7–1.3 cm long, distal 1.5–6.3 mm long; style trilobed. **Fruits** utricles, with dentate ridges, 4–6 × 2–3 mm. **Distr:** USA, Mexico, C America, S America. **Ecol:** In ponds and lake margins; 0–500 m. Fl. Mar–Nov. **Zones:** 9–12.

POALES

FAMILIES 17, GENERA 997, SPECIES 18,325.

Plants without mycorrhizae; vessels also in stem and leaves; epidermal cells with silica bodies; raphide crystals lacking. **Flowers** with perianth consisting of calyx and corolla; micropyle bistomal; style well-developed, separate or connate but strongly branched, stigmas small, dry. **Distr:** Worldwide. **Info:** Monophyly of the Poales is strongly supported by molecular data (Chase et al. 1995b).

TYPHACEAE

[Cat Tail].

Genus 1, **species** 29. **Herbs**, perennial, monoecious, of fresh to slightly brackish wetlands, often emergent, rhizomatous, caulescent in flower, smooth, glabrous; rhizome creeping, submerged, scaly; aerial stem erect, with or without nodes. **Leaves** radical or cauline, elongated. **Inflorescence** a terminal, cylindrical, superposed spike; ♂ above and ♀ below, zones separated or sometimes united; flowers numerous, densely crowded. **Flowers** ♂ consisting of 1–3 stamens and surrounded by 1 or more hairs, some representing the perianth, anthers basifixed, linear; ♀ usually intermixed with clavate pistillodes, covered with hairs and bracts or bracts absent, ovary superior, 1-locular, stipitate, stigma linear or lanceolate. **Fruits** follicles, dehiscing longitudinally by a slit; seed striate. **Distr:** In marshy and aquatic habitats worldwide.

Typha

Species 29. *Generic description same as familial.* **Cult:** Propagate from seeds or by division of rhizomes. Requires a wet soil, preferably in shallow ponds or damp mud banks, in full or partial sun.

T. domingensis Pers., *Syn. Pl.* 2: 532, 1807. (Syn: *T. angustata*, etc.). [Bullrush, Narrow-leaved Cumbungi, Southern Cat-tail]. **Herbs**, perennial, aquatic, monoecious, robust, 1.5–4 m or more tall, stem somewhat stout; rhizomes to ~2 cm ∅; stems to 2 cm ∅. **Leaf sheath** usually open at throat, margin free. **Leaves** 2 m × 0.5–2 cm; upper leaves with sheath not auriculate or only uppermost 1 or 2 leaves distinctly auriculate. **Inflorescence** with ♂ and ♀ spikes separated by 0.5–5 cm; ♂ spike 15–35 cm long, 0.5–1 cm ∅, covered with pointed or linear hairs, and bracts, pollen shed as single grains; ♀ spike usually 12–40 cm long, <2 cm ∅, cinnamon to brown with numerous spathulate (usually 4–8 cells across) floral bracts, peduncles ~1 mm long, stigmas linear, acuminate, markedly transversely concave, not fleshy. **Distr:** Pantropical. **Ecol:** In swamps, mangroves, brackish waters, margins of lakes and streams. Fl. throughout most of the year. **Zones:** 9–12.

BROMELIACEAE

Genera 57, **species** 1,700. **Herbs**, rarely shrubs, epiphytic, lithophytic, or terrestrial. **Leaves** spirally arranged, usually rosulate, sessile; lamina simple, veins parallel, base dilated, sheathing, margins often spinose serrate or sometimes entire. **Inflorescence** terminal or lateral, scapose or sessile, a panicle, raceme, spike, or capitulum, sometimes reduced to solitary, pseudo-lateral flowers; bracts usually brightly colored and conspicuous. **Flowers** bisexual or sometimes functionally unisexual, 3-merous; sepals and petals each 3, distinct, free or basally connate; petals often brightly colored, basal margin with a pair of scalelike appendages; stamens 6, in 2 whorls of 3, filaments free, connate, or collectively or individually adnate to petals, anthers 2-celled, dehiscing by longitudinal slits; gynoecium of 3 carpels united to form a compound, triloculed, superior or very often partly or wholly inferior ovary, ovules few to usually ± numerous in each locule, placentation axile, style terminal and often 3-parted, stigmas papillose. **Fruits** berries, sometimes septicidal capsules, rarely compound and fleshy; seeds usually winged or plumose, endosperm mealy. **Distr:** Predominantly in tropical and subtropical America; few in W Africa.

Acanthostachys

Species 2. **Herbs**, evergreen, perennial, epiphytic or lithophytic, stemless, rosette-forming, propagating by rhizomes. **Leaves** fasciculate, not contracted between sheath and lamina. **Inflorescence** simple, densely polystichous-flowered; scape well developed, arching; scape-bracts foliaceous; floral bracts well-developed. **Flowers** perfect, sessile; sepals free, nearly symmetric; petals free, bearing 2 scales; stamens of the second series highly adnate to petals, pollen ellipsoid, biporate; ovary in slight part superior, ovules caudate. **Fruit** a syncarp, resembling a small pineapple. **Distr:** Brazil to NE Argentina. **Cult:** Propagate from seeds, plantlets or by division. Requires fertile, humus-rich soil in full sun. Can also be grown as an epiphyte or in a rock garden.

A. strobilacea (Schult. and Schult.f.) Klotzsch, *Icon. Pl. Rar.* 1: 22, 1841. (Syn: *A. exilis*, *Hohenbergia strobilacea*). [Pinecone Bromeliad]. **Herbs**, flowering at 0.2–2 m high, erect at first but often whole plant becoming pendulous, propagating by short stout basal rhizomes. **Leaf sheaths** small, narrowly elliptic or suboblong, entire, dark castaneous. **Leaves** very few, fasciculate, erect at first, often secund-curved, covered beneath with a whitish membrane of fused scales; lamina linear, long-attenuate, to 1 m long, 1.2 cm wide at base, strongly channeled and involute, laxly serrate with spines 1.5 mm long, green. **Inflorescence** simple, very dense, ovoid or cylindric, 3–7 cm long, 2–3.5 cm ∅, bearing a small inconspicuous coma of reduced sterile bracts at apex; scape slender, cylindric; scape-bracts 2 at apex, foliaceous; floral bracts very broadly ovate with a spreading triangular-acuminate apex, densely serrulate to entire, broadly convex, thick, bright orange or red; flowers sessile. **Flowers** compressed, 2–2.5 cm long; sepals free, bright yellow, subtriangular, 0.8–1.1 cm long, acute, sparsely lepidote, lateral strongly carinate; petals yellow, erect, free, 1.6 cm long, bearing 2 scales above base; stamens included; ovary suborbicular, compressed, slightly superior with a broadly acute apex, ovules few, long-caudate. **Fruits** syncarp, small, pineapplelike. **Distr:** Brazil to Argentina (Misiones). **Ecol:** Epiphytic and lithophytic; 750–800 m. Zones 9–12. **Info:** Ripe fruits are edible and sweet-tasting.

Aechmea

Syn: *Gravisia*, *Macrochordion*, etc.

Species ~200. **Herbs**, perennial, epiphytic, lithophytic or terrestrial, evergreen, almost always acaulescent or medium-sized, frequently propagating by basal rhizomes. **Leaves** densely rosulate or fasciculate with the sheaths usually forming a tank; lamina with margins usually spinose-serrate. **Inflorescence** compound or simple, flowers distichous or polystichous, sessile or stipitate, sometimes showy; scape usually well developed. **Flowers** not showy, perfect; sepals free or connate, usually strongly asymmetric; petals regular, free, bearing 2 basal ± adnate appendages or these rarely ± aborted; stamens shorter than the petals, free or the second series ± adnate to the petals, unappendaged, anthers dorsifixed, pollen with 2, 3, 4, or numerous pores or sulcate; ovary wholly inferior, ovules mostly caudate. **Fruits** usually but slightly enlarged from the ovary; seeds small. **Distr:** Neotropical. **Ecol:** Mostly in rainforests. **Cult:** Propagate from offsets, when they are a couple of

Top row: Aechmea gamosepala, Aechmea mexicana, Ananas bracteatus.
Bottom row: Ananas bracteatus var. *tricolor, Ananas comosus, Ananas nanus.*

months old. Requires moist but well-drained, humus-rich soil or if planting in a tree, with coco peat or sphagnum moss, in a humid area with water filled into the center of the rosettes, in partial shade.

A. chantinii (Carrière) Baker, *Handb. Bromel.* 49, 1889. (Syn: *A. amazonica, Billbergia chantinii*). [Amazonian Zebra Plant]. **Herbs**, epiphytic, perennial, to 1 m high. **Leaves** 0.4–1 m long, densely punctulate-lepidote; sheaths large, broadly ovate, castaneous; lamina ligulate, apex acute or rounded and apiculate, 6–9 cm wide, abaxially usually white-banded. **Scape** erect, slender, white-flocculose; scape bracts lanceolate, acuminate, laxly serrulate, bright red, the lower lax and erect, the upper spreading and dense. **Inflorescence** laxly bipinnate, lower primary bracts similar to scape-bracts, 6–12 cm long, exceeding the spikes, serrulate, the upper abruptly reduced, resembling the floral bracts; spikes long-stipitate, narrowly lanceolate, densely 12-flowered; rachis geniculate, excavated next to the flowers; floral bracts distichous, broadly ovate, subtruncate, 1–1.3 cm long, slightly exceeding the ovary, strongly nerved, lepidote. **Flowers** to 3.2 cm long; sepals asymmetric, 1–1.2 cm long, equally shortly connate, muticous; petals obtuse, 2 cm long, orange; ovary glabrous. **Distr:** Colombia to Peru and Amazonian Brazil. **Ecol:** In forests; 100–1,160 m. **Zones:** 9–12.

A. gamosepala Wittm., *Bot. Jahrb. Syst.* 13(29): 3, 13, 1891. (Syn: *Ortgiesia gamosepala*). [Matchstick Plant]. **Herbs**, epiphytic, 40–75 cm high at flowering. **Leaves** 15–20, in a dense funnelform rosette, 25–55 cm long; sheaths narrow but distinct, sometimes tinged with blue; lamina ligulate, broadly rounded and apiculate, 3–5 cm wide, margins nearly entire or with a few minute spines near apex, covered with pale minute appressed scales especially abaxially. **Inflorescence** simple, slenderly cylindric, obtuse, sublax, 9–26 cm long; rachis slender, glabrous or nearly so; floral bracts narrowly triangular, attenuate, pungent, entire, nerved, the lower exceeding flowers, the upper shorter than ovary; flowers sessile, spreading or suberect; scape erect, slender, sparsely white-flocculose, soon glabrous; scape bracts erect, lanceolate or elliptic, acute, entire, membranous, deciduous. **Flowers** 1.5 cm long; sepals strongly asymmetric, rounded, 4–5 mm long without the 2.5–4 mm spine, half-connate, glabrous; petals erect, ligulate, obtuse, 9 mm long, purple or blue, bearing 2 fimbriate scales near base; stamens included; ovary slender, narrowed toward base. **Distr:** SE and S Brazil. **Ecol:** In moist lowland tropical forests. Fl. Jun–Nov. **Zones:** 8–12.

A. mexicana Baker, *J. Bot.* 17: 165, 1879. (Syn: *A. bernoulliana*, etc.). **Herbs**, epiphytic or terrestrial, from 0.7–1+ m high. **Leaves** many, densely rosulate; sheaths rather indistinct, ovate, brown, densely and finely brown-lepidote; lamina ligulate, 0.6–1.2 m × 6–12 cm, margins armed with straight spines to 2 mm long, apex acute to rounded-apiculate, pungent, finely pale-lepidote especially abaxially. **Inflorescence** paniculate, subcylindric to slenderly pyramidal, 30–70 cm long, furfuraceous; primary bracts linear to subfiliform, much shorter than axillary branches, membranous; branches spreading, basal ones much-divided, to 17 cm long; racemes laxly few-flowered; floral bracts filiform; flowers divergent to spreading; pedicels slender, 0.4–1.6 cm long; scape erect, stout, pale-furfuraceous; scape-bracts to 18 cm long, linear-lanceolate, entire, acuminate, membranous, stramineous, pale-lepidote, upper ones deflexed. **Flowers** with sepals broadly triangular-ovate, strongly asymmetric, mucronate, 6 mm long, free; petals ligulate, emarginate, 1–1.5 cm long, red or lilac, bearing 2 scales near base; epipetalous stamens strongly adnate to petals; ovary globose or ellipsoid, 6 mm long, often enlarged in fruit. **Distr:** Mexico to Ecuador. **Ecol:** 20–1,300 m. **Zones:** 9–12.

Ananas

Syn: *Ananassa*.

Species 9. **Herbs**, perennial, erect, simple, not stoloniferous. **Leaf sheath** scarcely enlarged. **Leaves** densely rosulate, scarcely enlarged at base, erect; lamina with margins spinose-serrate. **Inflorescence** densely strobiliform, erect, scapose, usually crowned with a tuft of sterile foliaceous bracts, often producing slips at base; flowers sessile. **Flowers** bisexual; sepals free, obtuse, slightly asymmetric; petals free, erect, violet or red, each typically bearing 2 slenderly funnelform scales; stamens included, pollen ellipsoid with 2 pores; ovaries inferior, coalescing with each other and with the bracts and axis to form a fleshy compound fruit, epigynous tube short, placentas apical, ovules caudate. **Fruit** a syncarp, fleshy. **Cult:** Propagate by division of offsets or by rooting the leafy rosette at the top of the fruit. Requires fertile, well-drained, humus-rich compost in full sun.

A. bracteatus (Lindley) Schultes filius, *Syst.* 7(2): 1286, 1830. (Syn: *A. bracteatus* var. *typicus*, etc.). [Red Pineapple, Wild Pineapple]. **Herbs**, rosette-forming, terrestrial. **Leaves** recurved; lamina concolorous, sword-shaped, 45 cm long, coarsely and laxly spinose-serrate, spines all ascending. **Inflorescence** with scape stout, usually short; floral bracts conspicuous, imbricate and covering ovaries, coarsely serrate, colored at maturity. **Flowers** small, yellowish-red with reddish-pink bracts; petals bearing scales. **Fruit** a syncarp, over 15 cm long at maturity, succulent. **Distr:** Ecuador and E Brazil to NE Argentina. **Ecol:** On coastal plains; 0–500 m. Fl. Dec–Jan. Fr. Mar–Apr. **Zones:** 9–12. **Info:** The variety ***A. bracteatus* var. *tricolor*** (Syn: *A. bracteatus* var. *rudis* subvar. *tricolor*, *A. bracteatus* var. *striatus*, *A. comosus* var. *tricolor*) with lamina variegated with longitudinal yellowish-white stripes, margins lined with pinkish-red spines; floral bracts reddish; flowers bluish.

A. comosus (L.) Merrill, *Interpr. Herb. Amboin.* 133, 1917. (Syn: *Bromelia ananas*, *B. comosa*). [Pineapple]. **Herbs**, highly variable with many forms and cultivars, to 1 m tall; not producing stolons, but axillary shoots from the stem and/or slips from the scape. **Leaves** densely rosulate; lamina concolorous, sword-shaped, not contracted toward base, margins coarsely, laxly and antrorsely spinose-serrate, coriaceous, abaxially densely lepidote. **Inflorescence** large, densely strobiliform, usually crowned with 1 or several rosettes of foliaceous bracts, many-flowered; scape to 1.8 m tall; floral bracts relatively inconspicuous, broadly elliptic to lanceolate, acuminate, soon exposing apices of ovaries, weakly serrulate or entire; flowers <50 to >150. **Flowers** numerous; sepals free, slightly asymmetric, apex obtuse; petals violet or reddish, free but connivent and tubelike, erect; stamens included. **Fruit** a syncarp, fleshy, subglobose, ovoid or elongate, formed from ovaries coalescing with each other, bracts and axis, over 15 cm long at maturity, fragrant; seeds small, absent or rarely produced. **Distr:** SE and S Brazil to Paraguay. **Ecol:** On coastal plains, white sand areas; 0–1,500 m. Fl. Apr–Jun. Fr. Jul–Sep. **Zones:** 9–12. **Info:** Cultivated widely in the tropics for the edible fruit.

A. nanus (L.B. Smith) L.B. Smith, *Bromel. Soc. Bull.* 12: 54, 1962. (Syn: *A. ananassoides* var. *nanus*, *A. microstachys* var. *nanus*). [Miniature Pineapple]. **Herbs**, terrestrial, perennial, 0.5–1 m tall. **Leaves** slender, recurved, arching; lamina sword-shaped, 20–70 cm long, margins antrorsely spinose-serrate, not fragile at apex, covered with white waxy scales. **Inflorescence** few-flowered, ~4 cm long; scape erect, 0.5–1m tall. **Flowers** small, lilac-purple to red; sepals 6 mm long; petals 1.5 cm long. **Fruit** a syncarp, 5–10 cm long, apex with a large coma of foliar bracts (often to 30 cm long); seeds few or none. **Distr:** Suriname, C and N Brazil. **Ecol:** In open forests; to 1,000 m. **Zones:** 9–12.

Billbergia

Syn: *Anacyclia*, *Cremobotrys*, etc.

Species 65. **Herbs**, epiphytic, lithophytic or terrestrial, ± stemless; offsets often present. **Leaves** rosulate or rarely fasciculate, sheaths large; lamina mostly ligulate, bases sheathing, margins spinose-serrate (rarely smooth), apex broad, often banded, firm-fleshy. **Inflorescence** simple or compound, apex usually pendent; scape erect or arching; scape bracts large, red, membranous; peduncle often mealy or scaly; flowers sessile or pedicellate; floral bracts small or absent. **Flowers** large, showy; sepals free, erect; petals free, often coiling, bearing 2 scales on the inner surface near base, claw long, lamina narrow; stamens exserted at anthesis, both series free or the second adnate to the petals up to the scales, pollen-grains with longitudinal folds when dry but no pores; ovary inferior, style exceeding the stamens, epigynous tube large, ovules many. **Fruits** berries; seeds without appendages. **Distr:** Mexico to tropical America. **Cult:** Propagate from offsets. Needs moist compost in partial sun in a humid situation.

B. nutans H. Wendl. ex Regel, *Gartenflora* 18: 162, 1869. [Bilbergia, Friendship Plant, Queen's Tears]. **Herbs**, epiphytic. **Leaves** usually fasciculate, 30–70 cm long (rarely rosulate and to 1 m long), the outer much reduced; sheaths narrowly ovate, densely white-scaly; lamina linear or very narrowly triangular, 0.6–1.7 cm wide, attenuate, margins laxly serrate with slender spines 1 mm long or sometimes entire, abaxially white-lepidote. **Inflorescence** pseudo-simple with very short branches, lax, usually few-flowered, glabrous; axis slender, geniculate; floral bracts minute, much shorter than ovaries, sometimes adnate to pedicels; flowers 2-ranked, sessile to shortly pedicellate, divergent or erect; scape decurved, very slender, glabrous; scape-bracts erect, densely imbricate, lanceolate, acute, rose. **Flowers** with sepals narrowly elliptic, obtuse, 1.5–2(–2.7) cm long, rose with dark blue margins; petals linear, obtuse, 3.3–4.6 cm long, equaling stamens, pale green with dark blue margins, bearing 2 large coarsely serrate scales at base; filaments of second series shortly connate with petals; ovary subellipsoid, 0.8–1.4 cm long. **Distr:** SE and S Brazil to N Argentina. **Ecol:** In forests; 700–1,000 m. **Zones:** 8–11.

B. pyramidalis (Sims) Lindl., *Bot. Reg.* 13: t. 1068, 1827. (Syn: *B. paxtonii*, etc.). [Summer Torch]. **Herbs**, epiphytic; rosette broadly tubular to funnel-shaped, to 45 cm high. **Leaves** few–13, erect, tips spreading or arching; lamina strap-shaped, 0.4–1 m x 4–6 cm, margins minutely spinulose-dentate, apex rounded or obtuse, apiculate. **Inflorescence** a dense capitula, corymbose-pyramidal, erect, whitish-mealy, to 15 cm long; scape erect, slightly longer than leaves; scape bracts overlapping, red to pink; flowers sessile or shortly pedicellate; floral bracts minute, ovate, acute. **Flowers** ~5 cm long; sepals pale red, shortly connate, slightly asymmetric, oblong, 1.3–1.8 cm long, obtuse or apiculate; petals red or red tipped with blue, tightly coiled after anthesis, ligulate, to 5.2 cm long, obtuse, bearing 2 fimbriate scales at base, slightly exceeding stamens; ovary whitish-mealy, subterete, 1.1–1.5 cm long. **Distr:** Windward Island to N and E Brazil. **Ecol:** In tropical rainforests; 0–1,700 m. Fl. Jul–Aug. **Zones:** 9–12.

Dyckia

Syn: *Garrelia*, *Prionophyllum*.

Species ~200. **Herbs**, small to large; rhizome thick, often repent. **Leaves** densely rosulate; sheaths usually ample, rather fleshy; lamina attenuate, not at all constricted at base, usually spinose-serrate, rigid. **Inflorescence** amply paniculate to simple; primary bracts inconspicuous; scape almost always conspicuous, usually lateral; floral bracts from shorter than pedicels to longer than flowers; flowers perfect and uniform or rarely dimorphic with some functionally unisexual, relatively small, yellow to red. **Flowers** with sepals imbricate, usually free, much shorter than petals; petals imbricate, centrally attached to filament tube, their lamina usually broad and distinct; stamens included to exserted, filaments basally connate, free or connate above the connection with petals; ovary wholly superior, usually glabrous, ovules many, each with a single wing. **Fruits** capsules, short and broad, septicidal and sometimes partially loculicidal as well; seeds numerous, broad, alate. **Distr:** S America; highest diversity in Brazil. **Cult:** Propagate from seeds or offsets. Requires fertile, well-drained soil in full or partial sun.

D. velascana Mez, *Mart. Fl. Bras.* 3(3): 476, 1894. **Herbs**, to 1 m high, densely colonial. **Leaves** over 50 cm long; lamina spreading, narrowly triangular, 4 cm wide, covered with pale appressed scales beneath, teeth antrorse, uncinate, 4 mm long; leaf sheaths suborbicular, 5 cm long, pale. **Inflorescence** paniculate, lax; primary bracts shorter than sterile bases of branches; branches short, very densely flowered, cinereous-lepidote-pubescent; scape erect, 6 mm Ø at apex, lepidote; scape-bracts erect, broadly ovate, lowest with narrowly triangular serrulate laminas, uppermost apiculate, much shorter than internodes; floral bracts reflexed, acuminate from a broadly ovate base, to 5 mm long, puberulous; pedicels very short. **Flowers** spreading, to 1.4 cm long, sulfur-yellow; sepals ovate, rounded and mucronulate, to 8 mm long, base ferruginous-puberulous, elsewhere glabrous and lustrous; petals erect, lamina subreniform, broadly rounded and minutely emarginate, not undulate; stamens exserted, free above common tube with petals, anthers sublinear, acute; styles 5–6 times shorter than ovary. **Distr:** NW Argentina. **Ecol:** On rocky slopes; 750–3,000 m. **Zones:** 8–11.

Neoregelia

Syn: *Aregelia*, *Karatas*, *Nidularium*, *Regelia*.

Species 97. **Herbs**, epiphytic. **Leaves** densely rosulate, usually spinose-serrate; sheaths usually large

Below: *Billbergia nutans*, *Billbergia pyramidalis*, *Dyckia velascana*, *Neoregelia concentrica*.

Top row: *Puya alpestris, Puya berteroniana, Puya chilensis, Puya mirabilis.*
Bottom row: *Tillandsia cyanea, Tillandsia ionantha, Tillandsia lindenii.*

and distinct; lamina from rounded and apiculate to long-attenuate. **Scape** short, completely enclosed. **Inflorescence** central, sunk in the center of rosette, simple in most species, very densely capituliform, umbellate or corymbose; flowers pedicellate. **Flowers** perfect; sepals strongly asymmetric, with the right side produced into a large wing; petals spreading, acuminate, violet, blue, white or rarely red; stamens and pistil included; ovules obtuse. **Distr:** Mainly from Brazil; 1 sp. from Guiana. **Ecol:** In rainforests. **Cult:** Propagate from offsets. Requires moist potting mix in partial sun.

N. concentrica (Vellozo) L.B. Smith, *Contr. Gray Herb.* 104: 78, 1934. (Syn: *Aregelia concentrica*, etc.). **Herbs**, usually large, propagating by stout basal rhizomes. **Leaves** 7–30, in a dense broadly funnelform rosette, 20–40 cm long; sheaths large, broadly elliptic, entire, both sides brown-scaly; lamina ligulate, broadly rounded and apiculate, 5–10 cm wide, margins usually densely serrate with stout black spines to 4 mm long, wholly green or with dark purple spots above or the inner leaves at times wholly purple, pale lepidote especially abaxially. **Inflorescence** sunk in the center of rosette, many-flowered; outer bracts broadly ovate, rounded and apiculate, entire, pale, membranous, apically lepidote; floral bracts linear, broadly acute or rounded and apiculate, ± incurved toward apex, entire, ecarinate, submembranous, pale green, sparsely lepidote; pedicels slender, 1–1.8 cm long. **Flowers** 4–5(–7) cm long; sepals free or shortly connate, slightly asymmetric, subelliptic, long-attenuate but never pungent, 2.2–3 cm long, green, glabrous; petals acuminate, white or blue, exceeding stamens; ovary slenderly ellipsoid, to 1.5 cm long. **Distr:** Rio de Janeiro and Guanabara, Brazil. **Ecol:** In rainforests; 0–800 m. **Zones:** 10–12.

Puya

Syn: *Pourretia.*

Species ~180. **Herbs**, perennial, stemless or more generally long-caulescent, simple or branched, stout, frequently flowering several meters high. **Leaves** densely rosulate, coriaceous; sheaths distinct, usually ample; lamina narrowly triangular, never constricted at base, almost always coarsely spinose-serrate. **Inflorescence** simple or paniculate. **Flowers** perfect, showy; sepals convolute, free, much shorter than petals, glabrous to tomentose or lanate; petals free, unappendaged or with a pair of vertical folds, the ample lamina generally spreading at anthesis, afterward twisted together in a spiral; stamens generally somewhat shorted than petals, filaments free, elongate; ovary superior or slightly inferior, glabrous, style long, slender. **Fruits** capsules, loculicidal and generally tardily septicidal as well; seeds large, bearing a dorso-apical wing. **Distr:** Costa Rica to tropical S America. **Cult:** Propagate from seeds; offsets are usually difficult to propagate. Requires a moist, well-drained soil in full sun.

P. alpestris (Poepp.) Gay, Fl. Chil. 6: 12, 1854. (Syn: *Pitcairnia alpestris*, etc.). [Sapphire Tower]. **Herbs**; caudex prostrate, branched; flowering shoot 1.2–1.5 m tall. **Leaves** relatively few in each rosette, arched-recurving; lamina 60 × 1.5–2.5 cm, marginal spines very lax, uncinate, 4 mm long, adaxially glabrous and bright green, abaxially densely white-scaly. **Inflorescence** sparsely and rather laxly bipinnate, 18–20 or much fewer branches, pyramidal; primary bracts like the upper scape-bracts, much shorter than axillary branches; racemes subsessile, to 20 cm long, laxly flowered on lower half or third, sterile on the remainder with numerous reduced bracts; scape stout; scape-bracts broadly oblong, reflexed, thin, ± deciduous; floral bracts elliptic, acute, much exceeded by the sepals at anthesis, membranous; pedicels slender, 1 cm long. **Flowers** with sepals narrowly oblong, acute or obtuse, ~2–2.5 cm long; petals elliptic, obtuse, ~5 cm long, blue-green, naked, glabrous; stamens and pistil included. **Distr:** SC Chile to Argentina (Neuquén). **Ecol:** On dry slopes. **Zones:** 8–10.

P. berteroniana Mez, *Monogr. Phan.* 9: 477, 1896. (Syn: *P. coarctata, P. coerulea*). [Chagual, Cardon, Magüey]. **Herbs**, robust; caudex prostrate, simple or branched, almost as large as that of *P. chilensis* (5 m long, 15–20 cm Ø), covered with bases of old leaves; flowering shoot to 4.5 m high. **Leaves** numerous in a dense rosette, arched-recurving; lamina to 1 m × 5 cm, marginal spines stout, uncinate, to 1 cm long, adaxially green and glabrous, abaxially white-scaly. **Inflorescence** amply bipinnate with 80–100 branches, slenderly pyramidal, to 1 m long, dense, persistently white-lanate; primary bracts like upper scape-bracts, much shorter than axillary branches; racemes subsessile, ~30 cm long, densely flowered on lower half, sterile on upper half with numerous reduced white or red bracts; scape 5.5 cm Ø, densely white-lanate when young; scape-bracts broadly lanceolate, reflexed, thin, ± deciduous; floral bracts elliptic, acute, membranous, white; flowers suberect; pedicels to 1.5 cm long. **Flowers** with sepals narrowly oblong, broadly acute or obtuse, 2.3 ×

0.65 cm, green, sometimes becoming glabrous with age; petals elliptic, obtuse, ~5 cm long, blue-green, naked, glabrous; stamens about equaling petals; pistil ~5 mm longer than stamens. **Fruits** ellipsoid, much shorter than sepals. **Distr:** C Chile. **Ecol:** On dry rocky slopes; 90–2,000 m. **Zones:** 7–10.

P. chilensis Molina, *Sag. Stor. Nat. Chili*: 160, 1782. (Syn: *Pitcairnia chilensis*, etc.). [Chagual]. **Herbs**, very large; caudex prostrate, simple or branched, to 5 m long, 15–20 cm Ø, woody, covered with bases of old leaves; flowering shoot to 4.5 m tall. **Leaves** numerous, densely rosulate, erect or stiffly spreading, slightly or not arching, sheaths ample; lamina 1 m × 5 cm, soon glabrous on both sides, strongly concave above, marginal spines stout, coriaceous, lowest often over 1 cm long. **Inflorescence** paniculate, bipinnate, 80–100-branched, 1+ m long, very dense, oblong-ovate in outline, sparsely tomentose, soon completely glabrous, flowers suberect at anthesis; primary bracts triangular-ovate, much shorter than axillary racemes, sparsely and finely serrate or entire, gray-brown; racemes subsessile, to 30 cm long, densely flowered on lower half, sterile on upper half with numerous reduced bracts; floral bracts narrowly elliptic, to 5 cm long, acute or acuminate, membranous, green; pedicels slender, ~1.5 cm long; scape erect, very stout, soon glabrous; scape bracts numerous, reflexed, lower foliaceous, upper broadly ovate, acute. **Flowers** with sepals elliptic-oblong, to 3.5 × 1.3 cm, broadly acute or obtuse, green; petals narrowly elliptic, ~5 cm long, varying from yellow to greenish-yellow, naked. **Fruits** ellipsoid, much shorter than sepals. **Distr:** C Chile. **Ecol:** On arid hillsides, mostly on north-facing slopes; 300–1,000 m. Fl. Jan–Feb, Apr–May. **Zones:** 7–10.

P. mirabilis (Mez) L.B. Sm., *Phytologia* 16: 461, 1968. (Syn: *Pitcairnia mirabilis*, *P. mirabilis* var. *tucumana*). **Herbs**, 0.3–1.5 m high. **Leaf sheaths** broadly ovate, very thick, forming an epigeous bulb, serrulate toward apex. **Leaves** persistent; lamina linear, 22–60 × 0.8–1.5 cm, base not contracted, margins coarsely spinose-serrate, apex acuminate, soon glabrous. **Inflorescence** laxly racemose, cylindric, 8–30 × 16 cm, few-flowered with a terminal coma of sterile bracts, wholly glabrous; scape stout, much exceeding the leaves, erect, glabrous; scape-bracts erect, lax, exceeding the internodes, broadly ovate, lower bracts with long spinose lamina, upper bracts with short ones; axis stout, straight; floral bracts suberect to spreading, broadly clavate-elliptic, acuminate, 3 cm long, exceeding the pedicels, nerved dorsally, laciniate-spinose; pedicels suberect at anthesis, becoming arching-decurved, 1.2 cm long. **Flowers** with sepals subtriangular, apex slightly asymmetric and obscurely mucronulate, ~6 cm long, ecarinate, thick, coriaceous; petals broadly linear, ~9.5 cm long, yellow-green, naked, contorted after anthesis; stamens included, anthers acuminate, 1.5 cm long. **Fruits** ~3 cm long, with persistent sepals. **Distr:** Bolivia to NW Argentina. **Ecol:** On rocky slopes; 750–2,590 m. **Zones:** 7–10.

Tillandsia

Syn: *Amalia*, *Renealmia*, *Viridantha*, etc.

Species 654. **Herbs**, usually epiphytic, stemless to long caulescent. **Leaves** mostly many-ranked, rosulate, or occasionally 2-ranked and/or laxly arranged; lamina linear to triangular or ligulate, margins entire, trichomes usually conspicuous. **Inflorescences** central, 1–many-flowered, 2-ranked; floral bracts mostly broad, conspicuous, rachis covered or exposed at anthesis. **Flowers** bisexual; sepals distinct or adaxial pair connate, usually symmetric; petals distinct; stamens included or exserted, filaments free; ovary superior. **Fruits** capsules, cylindric, dehiscent; seeds with basal, white, plumose appendage. **Distr:** Tropical and subtropical America. **Cult:** Propagate from offsets. Best grown mounted on a slab with regular misting, and green-leaved species can be grown in pots. Requires good ventilation and shade.

T. cyanea Linden ex K.Koch, *Wochenschr. Vereines Beförd. Gartenbaues Königl. Preuss. Staaten* 10: 140, 1867. (Syn: *T. lindeni* var. *tricolor*, etc.). [Kamehameha's Paddle]. **Herbs**, epiphytic; rosette funnel-shaped, to 40 cm high. **Leaves** recurving; lamina linear to narrowly triangular, to 36 × 1–1.9 cm, apex long-tapering, surfaces finely silvery-scaly or ± glabrous, sheath reddish. **Inflorescence** a spike, to 20-flowered, broadly elliptic or broadly oblong, flattened, to 16 × 7 cm, flowers borne singly, close to the inflorescence; peduncle usually hidden in rosette, sometimes projecting to 15 cm; bracts tightly overlapping, pinkish to rose. **Flowers** with petals clawed, lamina obovoid to ± diamond-shaped, deep blue-violet, with a white eye at the base; stamens included. **Distr:** Ecuador to NW Peru. **Ecol:** In forest trees; 200–1,000 m. **Zones:** 11–12.

T. ionantha Planch., *Fl. Serres Jard. Eur.* 10: 101, 1855. [Blushing Bride]. **Herbs**, epiphytic, stemless, usually forming dense masses. **Leaves** 30–40, stiffly erect; lamina narrowly triangular or subulate-acuminate, 5–6.5 × 0.5 cm, apex long-tapered, surfaces covered with coarse cinereous spreading scales, sheaths elliptic, about half as long as the lamina, outer leaves green, inner leaves red. **Inflorescence** sessile, terminal, all but the ends of the petals hidden, appearing simple with polystichous flowers but actually a reduced panicle, borne from the center of the rosette; scape absent or short; primary bracts lance-ovate, acute, equaling or exceeding the sepals, lepidote toward apex; floral bracts like the primary bracts but shorter than the sepals; flowers sessile, erect. **Flowers** with sepals free, lanceolate, acute, 1.6 cm long, thin; petals tubular-erect, over 4 cm long, violet; stamens exserted. **Fruits** subcylindric, 3 cm long. **Distr:** Mexico to C America. **Ecol:** In dry forests, 650–1,300 m. **Zones:** 11–12.

T. lindenii Regel, *Index Seminum* (LE) 1868: 92, 1869. (Syn: *T. lindeniana*, etc.) [Linden's Airplant]. **Herbs**, epiphytic, sometimes terrestrial or lithophytic, rosette-forming, perennial, evergreen, to 40 cm high. **Leaves** arching; lamina linear, grasslike, 30–60 cm long, recurved, deep green, sometimes colored with shades of red, coriaceous. **Inflorescence** elongate, flat, narrowly lanceolate, 20 × 5 cm, rising over 30 cm above the foliage; floral bracts showy, tightly overlapping, elliptic, 5+ cm long, pointed, greenish to orange-bronze, nerved; flowers borne from the center of the bracts; pedicels 3–4 cm long. **Flowers** relatively large, scented, blue to lilac, petals with a white basal patch; corolla funnel-shaped, segments spreading; sepals 3; petals 3, 2–3 × 2–3 cm. **Distr:** NW Peru. **Ecol:** In montane regions, often on exposed trees; 600–1,000 m. **Zones:** 9–11.

T. usneoides (L.) L., *Sp. Pl.*, ed. 2. 1: 411, 1762. (Syn: *Dendropogon usneoides*, *Renealmia usneoides*). [Black Moss, Long Moss, Spanish Moss]. **Herbs**, pendent in long festoons, to 3 m long; stems elongate. **Leaves** 4–8, 2-ranked, often twisted or contorted, gray to silver-gray, 1.5–3 × 0.1–0.2 cm, densely grayish-scaly; sheath pale, narrowly elliptic, 0.2–0.4 cm wide; lamina filiform, succulent, margins involute to nearly tubular, apex acute. **Inflorescences** with scape concealed within leaf sheath, appearing scapeless, pendent with shoot, ± 1 mm Ø, 1-flowered; floral bracts enveloping flower, erect, green, broad (covering all or most of rachis, rachis not visible at anthesis), ovate, not keeled, 4–5 mm long, thin-leathery, apex acute, surfaces densely grayish-scaly, venation even to slight; flowers apparently sessile. **Flowers** inconspicuous, fragrant; sepals free, ovate, 6–7 mm long, not keeled, thin, veined, apex acute, surfaces glabrous; corolla spreading, petals spreading, yellow-green, elliptic, to 1 cm long; stamens included; stigma included, simple-erect. **Distr:** S USA, Mexico, C America, West Indies, S America. **Fruits** to 2.5 cm long. **Ecol:** On trees, forming large clumps. **Zones:** 11–12.

JUNCACEAE

Genera 7, **species** 430. **Herbs**, very rarely shrublike, perennial or annual, tufted or with erect or creeping rhizome; stems erect, terete or laterally flattened. **Leaves** occasionally reduced to a basal sheath, without or almost without a lamina (cataphyll); leaf sheath often shortly prolonged on both sides into a pair of auricles at lamina juncture; lamina of basal and cauline leaves usually linear or filiform, grasslike and flat, or terete, glabrous except for pilose margin on some grasslike laminas. **Inflorescence** a panicle, corymb, or 1-flowered; flowers usually 1- or 2-bracteolate at base. **Flowers** bisexual or unisexual and plants dioecious, mostly wind-pollinated, regular, usually small; perianth segments (3 or)6, in (1 or)2 whorls, free, usually greenish to brownish or blackish, rarely white or yellowish, glumelike; stamens 3 or 6, if 6 then 3 opposite outer perianth segments, filaments thin,

Below: *Tillandsia usneoides*, *Juncus effusus*, *Carex trifida*.

Above: *Cyperus alternifolius, Cyperus involucratus, Cyperus papyrus, Cyperus rotundus.*

anthers basifixed, biloculed, dehiscing by longitudinal slits, pollen grains in tetrads; ovary superior, 1-loculed, or divided by 3 septa and triloculed, or incompletely septate, ovules 3 and inserted at base of ovary, or numerous and biseriate on 3 parietal placentas, stigmas 3, papillose. **Fruits** capsules, 1–3-valved, loculicidal; seeds globose, ovoid, or fusiform, small, sometimes appendaged; appendage caudate. **Distr:** Worldwide, especially Andes (3 endemic genera), S South America to New Zealand (2 genera).

Juncus

Syn: *Microschoenus, Tenageia.*

Species 300. **Herbs**, perennial or annual; rhizome shortly or long creeping; stems often tufted. **Leaves** either all basal or also cauline; leaf sheath open on 1 side, apex often auriculate; lamina grasslike, terete, or scalelike. **Inflorescences** terminal, or pseudolateral when basal involucral bract seemingly a continuation of stem, usually many-flowered; ultimate divisions dense heads, lax cymes, or 1-flowered; bracteoles present or absent. **Flowers** with perianth segments 6, in 2 whorls, papery to hyaline; stamens 3 or 6, attached to base of perianth segments, filaments filiform, anthers oblong or linear; ovary 1- or 3-loculed, placentas 3, parietal, sometimes inconspicuously prominent but more frequently protruding and sometimes connate at middle, ovules numerous per locule, style 1, stigmas twisted. **Fruits** capsules, 3-valved; valves usually bearing placentas at middle; seeds ovoid or oblong, minute, sometimes appendaged; appendage caudate. **Distr:** Cosmopolitan. **Cult:** Propagate from seeds or by division. Requires a wet soil or shallow water, in full or partial sun.

J. effusus L., *Sp. Pl.* 1: 326, 1753. (Syn: *J. conglomeratus* var. *effusus*). [Common Rush]. **Herbs**, perennial, densely tufted; rhizome shortly creeping, thick; stems terete, 25–90+ cm tall, 1–3(–4) mm Ø, striate; pith continuous. **Cataphylls** closely embracing stem, reddish-brown to chestnut-brown, sheathlike, 2–22 cm. **Inflorescences** pseudolateral, densely to laxly many-flowered; involucral bract erect, seemingly a continuation of stem, terete, 5–28 cm. **Flowers** with perianth segments usually pale brown, linear-lanceolate, 2–2.7 × ~0.8 mm, unequal with outer ones slightly longer than inner, apex acute; stamens 3(or 6), ~2/3 as long as perianth, anthers 0.5–0.7 mm, slightly shorter than filaments; ovary triloculed, style very short. **Fruits** capsules, ovoid to oblong, slightly depressed or not, subequaling or slightly longer than perianth, triseptate, apex obtuse; seeds ovoid-oblong, 0.5–0.6 mm, reticulate. **Distr:** Temperate N Hemisphere to W South America; Rwanda to S Africa, W Indian Ocean. **Ecol:** In forest margins, wet grasslands, lake margins, riverbanks, etc.; 200–3,400 m. Fl. Apr–Jul. Fr. May–Sep. **Zones:** 6–11.

CYPERACEAE

Genera ~70, **species** 4,000. **Herbs**, perennial, rarely annual plants, cespitose or rhizome creeping, often stoloniferous, sometimes tuberiferous or with bulbils; stem mostly scapose, solid, trigonous, without hairs, but scabrous along edges, often cylindrical, smooth. **Leaves** in 3 ranks, mostly basal or subbasal, with sheaths, with or without ligules; lamina usually glabrous, but apex and margins barbed, reduced in several groups. **Inflorescence** a spike, multiple spike or large compound, crateriform anthelodium; glumes spirally or distichously arranged. **Flowers** bisexual or unisexual, seldom dioecious; perfect flowers with or without bristlelike perianth, with 1–3 stamens, rarely more, and 2–3 stigmas; unisexual flowers composed of stamens in axil of glume, or gynoecium in axil of prophyll (utricle). **Fruits** nuts (achenes), trigonous or bi-convex, sometimes winged. **Distr:** Worldwide.

Carex

Species 1,792. **Herbs**, perennial, cespitose or not, rhizomatous, rarely stoloniferous; culms usually trigonous, sometimes round. **Leaves** basal and cauline, sometimes all basal; ligules present; lamina flat, V-shaped, or M-shaped in cross-section, rarely filiform, involute, or rounded, usually <2 cm wide, if flat then with distinct midvein. **Inflorescences** terminal, consisting of spikelets borne in spikes arranged in spikes, racemes, or panicles; bracts subtending spikes leaflike or scalelike; bracts subtending spikelets scalelike, very rarely leaflike; spikelets 1-flowered; scales 0–1. **Flowers** unisexual; ♂ without scales; ♀ with 1 scale with fused margins (perigynium) enclosing flower, open only at apex; perianth absent; stamens 1–3; styles deciduous or variously persistent, linear, 2–3(–4)-fid. **Fruits** achenes, biconvex, plano-convex, or trigonous, rarely 4-angled. **Distr:** Cosmopolitan. **Cult:** Propagate by division or from seeds. Requires a moisture-retentive soil in full or partial sun.

C. trifida Cav., *Icon.* 5: 41, 1799. (Syn: *C. incrassata, C. rekohu*). [Muttonbird Sedge, Tataka]. **Herbs**, very robust, forming dense, light green tussocks to 60 cm. Ø; culms 15–90 × 0.2–0.4 cm, trigonous, glabrous; basal sheaths cream, gray, or yellow-brown. **Leaves** overtopping inflorescence, 0.6–1.5 cm wide, double-folded, margins slightly scabrid with distant teeth; sheath only slightly broader than lamina, paler green. **Inflorescences** spikes, 6–9–(15), brown, very sturdy, 3.5–11 × 0.5–1.5 cm; upper 2–4 spikes ♂, occasionally with a few ♀ flowers intermingled, ± approximate, ± sessile; lower spikes ♀, lowest often compound, on stout erect peduncles; bracts leaflike; glumes linear-oblong, or lanceolate, emarginate, membranous, chestnut-brown, pale greenish-brown midrib produced into a very long hispid awn; utricles 4.5–6 × ~2 mm, subtrigonous, oblong-obovoid, turgid, distinctly nerved, straw-colored to light brown tinged, beak ~1 mm long; stigmas 3. **Fruits** nuts, ~2 mm long, trigonous, oblong-obovoid, dark brown. **Distr:** New Zealand, S Chile, subantarctic islands. **Ecol:** Usually on coastal cliffs or rock outcrops. **Zones:** 7–10.

Cyperus

Species ~600. **Herbs**, perennial or sometimes annual, to 1.5 m or more; rhizome often sturdy and long-creeping, producing stolons or tillers, sometimes tuberous; stem trigonous or terete, rarely septate. **Leaves** from shorter than to much longer than stem; lamina sometimes reduced, ligule mostly absent. **Inflorescence** variable, from a few spikes to a compact headlike structure or compound anthelodium; bracts of several lowest partial inflorescences foliose; spikes in ultimate distal partial inflorescences (cluster of spikes) spicately or digitately arranged, supported by glumelike bracts, and with binerved glumelike, often spongy prophylls; glumes from 3 to more than 50, distichous or, eventually, spirally arranged; sides of glumes often extending as rachis wings down spike rachis and enclosing flower and nut on opposite side of rachis; glumes or sometimes whole spikes caducous; whole spike axis sometimes swollen, articulated. **Flowers** bisexual; stamens 1–3; stigmas 2 or 3. **Fruits** achenes (nutlike), trigonous or bi-convex, flat side pressed against spike axis. **Distr:** Tropical and warm regions, worldwide. **Cult:** Propagate from seeds or by division. Requires moist or damp soil in full sun.

C. alternifolius L., *Mantissa Pl.* 28, 1767. [African Umbrella Plant, Umbrella Palm, Umbrella Sedge]. **Herbs**, tufted, perennial, 0.5–1.5 m; rhizome rather short, horizontal; stem 4–7 mm Ø, obtusely trigonous, smooth to finely scabrous. **Leaves** reduced; sheaths to more than 30 cm, coriaceous, upper yellowish or gray, lower reddish-brown or black-brown, mouth margin oblique; ligules absent; lamina absent. **Inflorescence** an anthelodium, compound, 3–13 cm across; bracts 18–22, to 30+ × 1.5 cm, foliose or upper filiform, green or grayish-green, acute, flat, margins narrowly recurved, margins and main nerves smooth or variously scabrous; primary branches to ~10 cm long, almost terete, tubular prophyll green or brownish; secondary anthelodia 1–3 cm long, with bracts to ~5 mm long; tertiary anthelodia occasional; cluster of spikes 0.5–1 cm long, globose, with 5–25 spirally arranged spikes; spikes 3.5–9 × ~2 mm, with 7–30 closely imbricating glumes, glumelike bract scarious, ~1.2 mm long, glumelike prophyll rounded, ~1.2 mm long; rachis 4-angled, ~0.5 mm wide; glumes ~1.5 mm long, cymbiform, lanceolate to elliptic; stamens 3. **Fruits** nuts, ~0.6–1 × 0.3–0.5 mm, ellipsoid or obovoid, trigonous, brownish. **Distr:** Madagascar, Mascarenes. **Ecol:** In wetlands and along rivers; <100(–300) m. Fl. Feb–Apr. **Zones:** 9–12.

RT

SZ

SZ

Above: *Flagellaria indica, Joinvillea ascendens, Bambusa chungii, Bambusa multiplex.*

C. involucratus Rottbøll, *Descr. Pl. Rar.* 22, 1772. [Umbrella Plant]. **Herbs**, perennial, cespitose, rhizomatous; culms trigonous, 0.3–1.5 cm × 1–5(–8) mm. **Leaves** without laminas. **Inflorescences** capitula, digitate, 1.5–3(–3.6) cm Ø; rays (14–)20–22, (2–)5–12(–20) cm long; secondary order rays 0.3–3(–4) cm long; tertiary order rays sometimes present, 0.3–2.5 cm long; bracts (4–)18–22, ± horizontal, flat, 15–27 × (0.15–)0.8–1.2 cm; spikelets 8–20, ovoid to linear-lanceoloid, compressed, 0.5–2.5 cm × 1.5–2 mm; floral scales 8–28, laterally whitish or light brown, ± hyaline, medially light brown, laterally without ribs, medially 3-ribbed, 2-keeled proximally, deltate-ovate, 1.6–2.4 × ~1.2–1.5 mm, apex acute. **Flowers** with stamens 3, anthers 0.7–1 mm long; styles 0.5–1 mm long, stigmas 0.6–1 mm long. **Fruits** achenes, brown, sessile or stipitate, broadly ellipsoid, 0.6–0.8 × 0.4–0.6 mm, obtuse, apiculate. **Distr:** E Africa. **Ecol:** In damp habitats; 0–100(–800) m. **Zones:** 9–12.

C. papyrus L., *Sp. Pl.* 1: 47, 1753. [Papyrus, Egyptian Paper Reed]. **Herbs**, perennial, rhizomatous, stout; culms roundly trigonous, 3–5 m × 1.5–4.5 cm, glabrous. **Leaves** without laminas. **Inflorescences** with spikes loosely cylindric, 1–2 × 0.6–1 cm; rays 40–100, drooping or arching, slender, 10–30 cm; secondary order rays 8–20 cm long; bracts 4–10, ± erect, V-shaped, 3–8 × 0.4–1.5 cm; secondary order bracts 2–5, (1.5–)4–16 cm × 0.5–2 mm; rachilla persistent, separating laterally, basally attached, wings 0.3–0.4 mm wide. **Spikelets** 6–30, slightly compressed, linear, ± quadrangular, 0.6–1 cm × 0.8–1 mm; floral scales 6–16, reddish beside 5-ribbed green medial part, white to hyaline near margins, ovate-elliptic, 1.8–2.2 × 1.2–1.5 mm, apex acute to obtuse. **Flowers** with anthers 0.8–1 mm long; styles 0.2–0.4 mm long, stigmas 0.8–1.2 mm long. **Fruits** achenes, pale brown, sessile, oblong, 0.8–1 × 0.4 mm. **Distr:** S Europe, SW Asia, Africa. **Ecol:** In streambanks, marshes; 0–30 m. **Zones:** 9–12. **Info:** The pith of the culms were used in ancient times to make paper.

C. rotundus L., *Sp. Pl.* 1: 45, 1753. [Coco Grass, Nut Grass, Purple Nut-sedge]. **Herbs**, perennial, stoloniferous; stolons (2–)5–12 × 0.1–0.2 cm, bearing tubers 0.3–0.8(–1.2) cm Ø, wiry; culms trigonous, 10–35(–40) cm × 0.7–3.4 mm, basally indurate, glabrous. **Leaves** V-shaped to flanged V-shaped, 5–30 cm × 2–6 mm. **Inflorescences** with spikes 1(–3), broadly ellipsoid, ~1.5–2.5 × ~2–3 cm; rays (3–)4–6(–7), 0.2–10 cm long; bracts (2–)3–5, horizontal to ascending, V-shaped to flanged V-shaped, 0.5–10 cm × 0.5–4 mm; rachilla persistent, wings 0.5–1 mm wide. **Spikelets** (2–)3–7(–12), compressed, linear, 0.4–4 cm × 1.3–1.8 mm; floral scales persistent, 6–36(–42), spreading or appressed, purple to reddish-brown, with narrow clear border and green midrib, 7–9-ribbed, ovate, (1.8–)2.6–3.4 × 2.2–3 mm, obtuse. **Flowers** with anthers 1–2.5 mm long; styles 1.3–3.5 mm long, stigmas (1.8–)2–3.3 mm long. **Fruits** achenes, ellipsoid, 1.4–1.7(–1.9) × 0.8–1 mm, obtuse, black. **Distr:** Americas, Africa, Asia, Australia. **Ecol:** In grasslands, wet or dry areas of mountain slopes, river margins; 0–2,100 m. Fl. & Fr. May–Nov. **Zones:** 9–12.

FLAGELLARIACEAE

Genus 1, **species** ~4. **Lianas**, high climbing, robust, glabrous; rhizome sympodial, diffuse; stems terete, solid, hard, apically usually equally branched; axillary buds absent. **Leaf sheath** tubular, closed, connected with lamina by a short pseudo-petiole. **Leaves** distichous, circinate; lamina grasslike, stomata paracytic, apex extended into tendril; tendril simple, involutely coiled, abaxially flattened, hard. **Inflorescences** terminal, paniculate. **Flowers** bisexual or rarely unisexual, sessile, actinomorphic, 3-merous, small; perianth segments 6, in 2 whorls, free, whitish, petaloid, membranous, persistent, 3 inner ones largest; stamens 6, in 2 whorls, exserted; filaments filiform; anthers basifixed, linear-oblong to linear, sagittate, biloculed, latrorse, dehiscing by longitudinal slits; pollen grains ulcerate and similar to those of grasses; ovary superior, obtusely 3-angled, triloculed, ovule 1 per locule, placentation axile, style very short, stigmas 3, linear-clavate. **Fruits** drupaceous with 1(or 2) seeds; seeds globose or ± flattened, endosperm copious, starchy. **Distr:** Tropical Africa, Asia, Australia, Pacific Islands.

Flagellaria

Species ~4. *Generic description same as familial.* **Cult:** Propagate by division of new shoots or from seeds. Requires a moderately fertile, moist soil in full or partial sun.

F. indica L., *Sp. Pl.* 1: 333, 1753. (Syn: *F. minor*, etc.). [Bush Cane, False Rattan, Supplejack, Whip Vine]. **Lianas**, perennial. **Leaf sheath** 2–7 cm long; pseudo-petiole 3–8 mm long, abaxially flattened. **Leaves** lanceolate to linear, 7–25 × 0.5–2 cm, base abruptly contracted, rounded, or slightly cordate. **Inflorescences** panicles, erect, irregularly branched, dense; bracts short, broad. **Flowers** small; tepals white, ovate to broadly so, 2–3 mm long, thinly membranous; stamens exserted; ovary narrow. **Fruits** drupes reddish when mature, globose, 4–6 mm Ø, smooth; seeds 1(or 2). **Distr:** S Tanzania to Mozambique and W Pacific. **Ecol:** In moist littoral forests, mangrove swamps, freshwater swamps; near sea level to 1,500 m. Fl. Apr–Jul. Fr. Sep–Nov. **Zones:** 9–12.

JOINVILLEACEAE

Genus 1, **species** 2. *Familial description same as specific.*

Joinvillea

Species 2. **Herbs**, erect, rhizomatous, perennial; stems terete or compressed, hollow except at nodes, unbranched, glabrous; leaf sheath open, covering internode, pubescent in upper part, glabrate with age. **Leaves** alternate, well-distributed along stem; lamina simple, grasslike, plicate in bud. **Inflorescences** panicles, pyramidal, 3–5-branched, 10–40 cm long; bracts of branches triangular, membranous, caducous; bracteoles triangular to subulate, much smaller than bracts. **Flowers** with tepals 6, in 2 imbricate series, scarious to chartaceous, spreading in fruit; outer tepals ovate to orbicular-ovate or deltate-lanceolate, acute to acuminate; inner tepals orbicular-ovate or deltate-lanceolate to ovate, retuse to acute or acuminate. **Fruits** green, becoming red, reddish-orange, or brownish-black at maturity; seeds 1–3, globose, rugulose. **Distr:** Malesia to Pacific. **Cult:** Propagate from seeds. Requires a moist, well-drained soil in full or partial sun.

J. ascendens Gaudich. ex Brongn. and Gris, *Bull. Soc. Bot. France* 8: 269, 1861. (Syn: *J. gaudichaudiana*). **Herbs**, erect, 1.5–5 m tall. **Leaves** narrowly elliptic, 45–80 × 4.5–16 cm, plicate with 7–13 pleats, surfaces with scattered bristles, abaxially also sparsely to moderately pubescent; auricles to ~1–4 cm long. **Flowers** with tepals as long as to twice as long as wide, usually splitting with age; outer tepals ovate to orbicular-ovate, 1.7–4 × 1.4–2.2 mm, with a central vein and rarely 1–2 inconspicuous lateral veins, apex mucronate and sometimes reflexed; inner tepals orbicular-ovate, 1.8–3.5 × 1.6–2.6 mm, truncate, sometimes becoming retuse with age. **Fruits** 3.3–5.8 mm Ø, styles not persistent. **Distr:** Hawai'ian Islands. **Ecol:** In wet forests; 30–1,250 m. **Zones:** 10–12.

POACEAE

Subfamilies 13, **genera** 668, **species** 10,035. **Herbs**, annual or perennial, or tall woody bamboos; flowering stems (culms) jointed, internodes hollow or solid; branches arising singly from nodes and subtended by a leaf sheath and 2-keeled prophyll, often fascicled in bamboos. **Leaves** alternately arranged, 2-ranked, differentiated into sheath, lamina, and an adaxial erect appendage at sheath/lamina junction (ligule); leaf sheath surrounding and supporting culm-internode, usually split to base, modified with reduced lamina in bamboos (culm sheaths); lamina divergent, usually long, narrow and flat, but varying from inrolled and filiform to ovate, veins parallel, sometimes with cross-connecting veinlets (especially in bamboos); ligule membranous or a line of hairs. **Inflorescence** terminal or axillary, an open, contracted, or spikelike panicle, or composed of lax to spikelike racemes arranged along an elongate central axis, or digitate, paired, or occasionally solitary; axillary inflorescences often many, subtended by spatheoles

Above: *Bambusa bambos*, *Bambusa vulgaris*, *Dendrocalamus asper*, *Dendrocalamus giganteus*.

(specialized bladeless leaf sheaths) and gathered into a leafy compound panicle; spikelets often aggregated into complex clusters in bamboos. **Spikelets** composed of distichous bracts arranged along a slender axis (rachilla); typically 2 lowest bracts (glumes) empty, subtending 1 to many florets; glumes often poorly differentiated from accompanying bracts in bamboos. Florets composed of 2 opposing bracts enclosing a single small flower, outer bract (lemma) clasping the more delicate, usually 2-keeled inner bract (palea); lemma often with apical or dorsal bristle (awn), glumes also sometimes awned. **Flowers** bisexual or unisexual; lodicules (small scales representing perianth) 2, rarely 3 or absent, 3 to many in bamboos, hyaline or fleshy; stamens 3 rarely 1, 2, 6, or more in some bamboos, hypogynous, filaments capillary, anthers versatile; ovary 1-celled, styles (1 or)2(rarely 3), free or united at base, topped by feathery stigmas, exserted from sides or apex of floret. **Fruits** normally dry indehiscent caryopses with thin pericarp firmly adherent to seed, pericarp rarely free, fleshy in some bamboos. **Distr:** Worldwide. **Info:** The classification followed here is that of Bouchenak-Khelladi et al. (2008).

Subfam. BAMBUSOIDEAE

[Bamboos].

Genera 84–101, **species** 940–1,320. **Bamboos**, usually arborescent, often fast-growing; culms perennial, woody, diffuse (rarely solitary); internodes usually hollow. **Leaves** pseudo-petiolate, often with inner and outer ligules; culm leaves often very different from the others. **Florets** with lodicules rarely 3, stamens (2–)6(–140), rarely basally connate, stigmas (1–)2–3. **Fruits** caryopses, rarely berries; embryonic leaf margins overlapping; first seedling leaf without lamina (Stevens 2001 onwards). **Info:** Most genera exhibit gregarious but unpredictable flowering, sometimes causing extraordinary events, such as in the case of *Melocanna baccifera* (Jeeva et al. 2009), where famine is often a result.

Bambusa

Syn: *Arundarbor*, *Bambus*.

Species 120. **Bamboos**, arborescent, occasionally shrubby or scrambling, 1–20 m tall; rhizomes short necked, pachymorph. **Culms** unicespitose, erect to pendulous, rarely subscandent; internodes terete; nodes not raised; branches several to many, often 1–3 dominant; branchlets of lower branches sometimes forming tough or weak thorns. **Culm sheaths** deciduous, rarely persistent; auricles usually conspicuous, always with marginal oral setae; lamina usually erect. **Leaves** variable in size, transverse veins inconspicuous. **Inflorescence** iterauctant, fully bracteate, subtended by a broad 2-keeled prophyll; pseudo-spikelets rarely solitary, usually several to many clustered to capitate on flowering branches. **Pseudo-spikelets** prophyllate; florets 2 to many, terminal floret sterile or imperfect, sessile; fertile glumes preceded by 1 or more gemmiferous, glumaceous, or spathaceous bracts and/or 1–3 empty glumes; rachilla internodes usually distinct and usually disarticulating with florets, falling separately; lemma broad, many veined; palea 2-keeled, apex acute or shortly bifid; lodicules 3 or 2, stamens 6, filaments free; ovary usually stalked, apex thickened and hairy, style solid, usually short, stigmas (1–)3, long, hairy, plumose. **Fruits** caryopses, terete, apex hairy; pericarp slightly thickened. **Distr:** Tropical and subtropical Asia. **Cult:** Propagate by division or by internode cuttings. Requires fertile, moist soils.

B. bambos (L.) Voss, *Vilmorin's Blumeng.*, ed. 3, i: 1189, 1895. (Syn: *B. spinosa*, etc.). [Giant Thorny Bamboo, Indian Thorny Bamboo, Spiny Bamboo]. **Bamboos**, perennial, cespitose; rhizomes short, pachymorph; culms erect, 20–30 m long, 10–15 cm Ø, woody, nodes with root thorns, internodes terete, glaucous; lateral branches dendroid. **Culm sheaths** deciduous, glabrous; ligule 1–2 mm high; lamina triangular, erect, hispid, acute, glabrous, outer margin hairy, oral hairs setose, 4–6 mm long; auricles erect. **Leaves** lanceolate, 7–18 × 1–1.8 cm, base broadly rounded, petiolelike, apex acuminate; petiole 3–5 mm long. **Synflorescences** bractiferous, clustered at nodes, with glumaceous subtending bracts; prophyllate below lateral spikelets. **Spikelets** comprising 5–7 fertile florets with diminished florets at apex, lanceolate, laterally compressed, 1.5–2 cm long; glumes several with 2–3 empty glumes, persistent, shorter than spikelet. **Florets** with fertile lemma ovate, 8–9 mm long; palea oblong, keels wingless; apical sterile florets resembling fertile though underdeveloped; lodicules 3, obovate, membranous, anthers 6, 4–5 mm long, stigmas 3, ovary umbonate. **Fruits** caryopses, with adherent pericarp, ellipsoid, 7–7.5 mm long. **Distr:** Africa, W Indian Ocean, tropical Asia, Indo-China, Malesia, S America. **Ecol:** In humid tropical regions, along riverbanks, river valleys, and in mixed moist deciduous forests; <1,000(–1,250) m. **Zones:** 9–12.

B. chungii McClure, *Lingnan Sc. Journ.* 1936, xv. 639, 1936. [Blue Bamboo, White Bamboo, White Powdery Bamboo]. **Bamboos**, perennial, cespitose; rhizomes short, pachymorph; culms erect, 12–18 m long, 3–5 cm Ø, woody, without nodal roots, internodes terete, thin-walled, 0.4–1 m long, glaucous; lateral branches dendroid. **Culm sheaths** yellow, pubescent; ligule fimbriate or ciliate; lamina narrowly ovate, reflexed, pubescent. **Leaves** cauline; sheaths glabrous, oral hairs setose, auricles falcate; lamina linear or lanceolate, 10–20 × 2–3.5 cm, base asymmetrical, petiolelike, apex acuminate, secondary veins 10–12, adaxially pubescent, abaxially hairy. **Synflorescences** bractiferous, clustered at nodes; with glumaceous subtending bracts; prophyllate below lateral spikelets. **Spikelets** comprising 4–5 fertile florets with diminished florets at apex, ovate, laterally compressed, 1–2 cm long; glumes several, comprising 1–2 gemmiferous bracts, 1–2 empty glumes. **Florets** with fertile lemma ovate, 9 mm long; palea with keels eciliate, apex truncate, or obtuse; apical sterile florets resembling fertile though underdeveloped; lodicules 3, membranous, anthers 6, stigmas 2–3, ovary umbonate. **Distr:** Temperate Asia. **Ecol:** In lowland hilly areas; 100–500 m. **Zones:** 8–10.

B. multiplex (Lour.) Raeusch. ex Schult., *Syst. Veg.* 7: 1350, 1830. (Syn: *B. aurea*, etc.). [Hedge Bamboo]. **Bamboos**, perennial, cespitose; rhizomes short, pachymorph; culms erect; 2–4 m long, 1–3 cm Ø, woody, without nodal roots, internodes terete, thin-walled, midgreen, lateral branches dendroid. **Culm sheaths** deciduous, auriculate, ciliate on shoulders; ligule 1–1.5 mm high; lamina triangular, 1–2 cm long, acute. **Leaves** 7–13 per branch; sheath oral hairs setose, 3–5 mm long, auricles falcate; lamina deciduous, lanceolate, 7–10 × 1–1.3 cm, base petiolelike, apex acuminate, surfaces dark green, glaucous, puberulous, abaxially hairy; petiole 1–2 mm long. **Synflorescences** bractiferous, clustered at nodes. **Spikelets** comprising 7–10 fertile florets with diminished florets at apex, lanceolate, laterally compressed, 3–4 cm long. **Florets** with fertile lemma lanceolate, 1.5 cm long; palea oblong; apical sterile florets resembling fertile though underdeveloped; lodicules 3, membranous, anthers 6, 6 mm long, stigmas 3, ovary umbonate. **Fruits** caryopses with adherent pericarp. **Distr:** Africa, NW Indian Ocean, tropical to temperate Asia, Malesia, Australasia, S America. **Ecol:** Mostly in lowland hills and riversides, sometimes on mountains; to 1,500 m. **Zones:** 9–12.

B. vulgaris Schrader ex J. C. Wendland, *Coll. Pl.* 2: 26, 1810. (Syn: *B. auriculata*, etc.). [Common Bamboo, Yellow-stemmed Bamboo]. **Bamboos**; clumps open; culms 8–15 m tall, 5–9 cm Ø, basally straight or flexuose, apically drooping; internodes deep green, 20–30 cm long, wall slightly thick; nodes with aerial roots and hairy rings, usually branching from lower nodes. **Culm sheaths** deciduous, densely brown-hairy, apex arched below lamina, concave below auricles; auricles conspicuous, ascending, oblong or reniform, 0.8–1 cm long; oral setae curved, fine; ligule 3–4 mm long, serrate; lamina deciduous, erect or deflexed, triangular, apex involute, sharply apiculate. **Leaves** narrowly lanceolate, 10–30 × 1.3–2.5 cm, glabrous. **Inflorescences** with pseudo-spikelets several, clustered at nodes, narrowly lanceolate to linear-lanceolate, slightly flattened, 2–3.5 × 0.4–0.5 mm.

Above: *Dendrocalamus longispathus, Melocanna baccifera, Pleioblastus fortunei, Pseudosasa hindsii.*

Florets 5–10; rachilla segments 1.5–3 mm long; glumes 1 or 2, apex apiculate; lemma 0.8–1 cm, apex apiculate; palea slightly shorter than lemma, keels ciliolate; lodicules 3, 2–2.5 mm long; anthers ~6 mm long, apex penicillate; style 3–7 mm long, slender; stigmas 3, short. **Distr:** SE Asia. **Ecol:** On riversides, open forests. **Zones:** 9–12.

Dendrocalamus

Species 40. **Bamboos**, arborescent, large-sized; clumps dense; rhizomes short-necked, pachymorph. **Culms** unicespitose, erect, or occasionally scrambling, apex usually pendulous; internodes terete; branches several to many, none to 3 dominant. **Culm sheaths** deciduous; ligule conspicuous; auricles often absent or small; lamina usually recurved, or erect. **Leaves** usually large; ligule conspicuous; auricles usually absent; venation not tessellate. **Inflorescence** iterauctant, fully bracteate, subtended by a narrow single-keeled prophyll, pseudo-spikelets clustered in soft or spiky globose mass at nodes of leafless flowering branches. **Spikelets** (pseudo) prophyllate, (1 or)2–8-flowered, with or without rachilla extension and rudiment, sessile, fertile glumes preceded by 1 or more gemmiferous bracts and 0–2 empty glumes; prophyll narrow, single-keeled; rachilla internodes usually abbreviated; florets falling together; lemma broad, nearly equal to palea, many-veined, sometimes long-mucronate; palea of lower florets 2-keeled, but rounded or imperfectly keeled in terminal floret if rachilla extension small or absent, apex acute or shortly bifid; lodicules 0–3; stamens 6, filaments usually free; ovary stalked, apex thickened and hairy; style very short, solid; stigmas 1(–3), long, hairy, plumose. **Fruits** caryopses, terete, apex hairy; pericarp slightly thickened. **Distr:** Tropical and subtropical Asia. **Cult:** Propagate from stem cuttings or by division. Requires a moist, humus-rich soil in full or partial sun.

D. asper (Schultes and J.H. Schultes) Backer ex K. Heyne, *Nutt. Pl. Ned.-Ind.*, ed. 2. 1: 301, 1927. (Syn: *Bambusa aspera*, etc.). [Rough Giant Bamboo]. **Bamboos**, perennial, cespitose; rhizomes short; pachymorph; culms 15–20 m tall, 6–10(–12) cm Ø; internodes 30–50 cm long; branches from ~9th node up, central branch dominant; culm sheaths deciduous, initially light green, leathery, apex rounded; ligule 0.7–1 cm long, margin with brown setae; auricles linear, ~2 × 0.7 cm, margin undulate; oral setae present; lamina reflexed, lanceolate. **Leaves** with leaf sheaths initially sparsely hairy, becoming glabrous; ligule truncate, ~2 mm long, entire or serrulate; lamina variable, lanceolate or oblong-lanceolate, (10–)20–30(–35) × (1–)3–5 cm, apex acuminate. **Spikelets** 6–9 mm long; florets 4 or 5, apical one sterile; glumes 1 or 2, ovate-lanceolate; lemma broadly ovate, margins ciliate; palea about as long as lemma, keels and margins ciliate, 1–3-veined between and 2-veined on either side of keels; anthers 6, 3–5 mm, apex apiculate; ovary and style hairy, stigmas 1. **Fruits** caryopses, with adherent pericarp. **Distr:** Temperate and tropical Asia (China, E Asia, India, Indo-China, Malesia). **Ecol:** In moist lowland areas; 400–500(–1,500) m. **Zones:** 9–12.

D. giganteus Munro, *Trans. Linn. Soc. London.* 26: 150, 1868. (Syn: *Sinocalamus giganteus*). [Giant Bamboo]. **Bamboos**; culms 20–30 m tall, 20–30 cm Ø; internodes 30–45 cm long; wall 1–3 cm thick; branches several. **Culm sheaths** deciduous, initially purple, thickly leathery, brown hairy; margins broadly rounded; auricles reflexed, wavy; oral setae absent; ligule 0.6–1.2 cm long, serrulate; lamina reflexed, ovate-lanceolate, 13–38 cm long. **Leaves** with ligule 1–3 mm long, serrulate; lamina usually oblong-lanceolate, to 45 × 10 cm, glabrous. **Inflorescence** branches pendulous, long; pseudo-spikelets 4–12(–25) cm long; spikelets 1–1.5 × 0.3–0.4 cm. **Florets** 5–8, apical one sterile; glumes 2, 3–4 mm long; lemma broadly ovate, margins ciliate, apex mucronate; palea about as long as lemma, 2-keeled, keels ciliate, 2-veined between and 1-veined on either side of keels; anthers ~6.5 mm long, apex apiculate; pistil ~1 cm long, shortly hairy, ovary ovoid, style long, stigma 1, purple. **Fruits** caryopses, oblong, apex obtuse, plumose. **Distr:** China, Myanmar. **Ecol:** In humid tropical highlands, sometimes naturalized in lowlands; to 1,200 m. **Zones:** 10–12.

D. longispathus (Kurz) Kurz, *Prelim. Rep. For. Veg. Pegu App. B* :94, 1875. (Syn: *Bambusa longispatha*). **Bamboos**, perennial, cespitose; rhizomes short; pachymorph; culms erect, 10–18 m long, 6–10 cm Ø, woody, without nodal roots, or with aerial roots from nodes; culm internodes terete, thick-walled, 25–60 cm long, glaucous, or gray, distally pubescent; lateral branches dendroid; buds or branches absent from lower quarter of culm. **Culm sheaths** persistent, 35–50 cm long, chartaceous, with dark brown hairs, apex concave; ligule dentate, or fimbriate; lamina reflexed, lanceolate, 25–40 × 2.5–3.5 cm, acuminate, pubescent. **Leaves** with leaf sheaths pubescent, hairs dark brown; ligule a ciliolate membrane; lamina lanceolate, 10–30 × 2.5–4.5 cm, base with a brief petiolelike connection to sheath, midrib conspicuous, venation with 16–20 secondary veins, glabrous, margins scabrous, apex acuminate. **Inflorescences** bractiferous, clustered at nodes, in dense, oblong clusters, 1.3–3.8 cm between clusters; fertile spikelets sessile, ovate, laterally compressed, 5–7.5 mm long, comprising 2–3 fertile florets; glumes several; florets with fertile lemma obovate, 4–6 mm long, flower lodicules absent, anthers 6, yellow, ovary umbonate. **Fruits** caryopses, with adherent pericarp, ovoid. **Distr:** Tropical Asia. **Ecol:** In mixed forests, evergreen primary forests and near streams. **Zones:** 9–12.

Melocanna

Species 2. **Bamboos**, arborescent, perennial, 2.5–20 m tall; rhizomes elongated, pachymorph; culms diffuse or in open clumps, erect; internodes terete, wall thin, nodal ridge not prominent; lateral branches many, subequal. **Culm sheaths** persistent, shorter than internodes, distally corrugate, with external ligule; auricles absent; lamina erect or basally erect, distally reflexed, sword-shaped, long. **Leaves** large, lanceolate, glabrous, transverse veins visible abaxially. **Inflorescences** fully bracteate, clustered at the nodes; clusters oblong subtended by spathaceous bracts with axillary buds at base of spikelet, leafless between clusters; fertile spikelets sessile, comprising of 5–6 fertile florets, with a barren rachilla extension. **Spikelets** lanceolate, laterally compressed, 1.2–1.8 cm long, breaking up at maturity, disarticulating below each fertile floret; rachilla internodes definite; glumes 2–4, persistent, shorter than spikelet; lower glume surface without pits; upper glume lanceolate. **Florets** with fertile lemma lanceolate, chartaceous, without keel, wingless, 5–9-veined, surface unwrinkled, without grooves, apex rostrate, mucronate; palea without keels; lodicules 2, ciliate, stamens 5–7, filaments free or irregularly connate, anthers obtuse, stigmas 2–4, ovary glabrous, with a steeplelike appendage. **Fruits** caryopses, pear-shaped, large, 5–13 cm long, apically long-beaked, pericarp thick, fleshy, without endosperm, usually viviparous. **Distr:** Bangladesh, India, Myanmar. **Cult:** Propagate by division or by internode cuttings. Requires fertile, moist soils.

M. baccifera (Roxb.) Kurz, *Prelim. Rep. Forest Pegu Appendix B*: 94, 1875. (Syn: *Bambusa baccifera, M. bambusoides*). [Berry Bamboo]. **Bamboos**, perennial; culms solitary; rhizomes elongated, pachymorph; culms erect, 10–20 m long, 5–9 cm Ø, woody, internodes terete, thin-walled, 30–60 cm long; lateral branches dendroid. **Culm sheaths** persistent, linear, reflexed coriaceous, pubescent, auricles absent. **Leaves** cauline, 5–15 per branch; leaf sheath oral hairs lacking; ligule an eciliate membrane; lamina lanceolate, 18–42 × 2–9 cm, base with a brief petiolelike connection to sheath, margins scabrous, venation with distinct cross veins. **Inflorescences** bracteate, clustered at nodes; secondary branches racemose; clusters oblong, 1–1.5 cm long, dense, with spathaceous subtending bracts; bracts 1–1.5 mm long, with axillary buds at base of spikelet; prophyllate below lateral spikelets, leafless between clusters; fertile spikelets sessile. **Spikelets** comprising of 5–6 fertile florets, lanceolate with a barren rachilla extension, laterally compressed, 1.5 cm long; glumes persistent; upper glume lanceolate. **Florets** with fertile lemma lanceolate, 5–7 mm long, rostrate, mucronate, chartaceous, 5–9-veined; palea without keels; lodicules 2, veined, ciliate, stamens 5–7, stigmas 2–4, ovary glabrous. **Fruits** caryopses, pear-shaped, viviparous, pericarp fleshy, 7.5–12.5 cm long. **Distr:** Tropical Asia (India and Indo-China). **Ecol:** On lower hills and plains, often near streams; to 500 m. Fl. gregarious (flowering cycle ~30 years); Mar–Jun. Fr. Jun–Jul. **Zones:** 9–12.

Pleioblastus

Species 40. **Bamboos**, small, shrubby or arborescent, spreading and loosely clumped; rhizomes leptomorph; culms pluri-cespitose, suberect to drooping, 0.5–8 m tall, 0.1–4 cm Ø; internodes slightly grooved above branches, glabrous, smooth or rough, pruinose below nodes; branches erect, initially 1–9 per node, long, subequal; sheaths and prophylls very persistent, papery, ± glabrous. **Culm sheaths** persistent to very persistent, nearly leathery; lamina deciduous, often reflexed. **Leaves** with sheaths persistent; lamina small to medium-sized, arrangement random or ± distichous, transverse veins distinct. **Inflorescences** open, racemose to paniculate, branching subtended by very small bracts or hairs, often with pulvini; spikelets exserted, 1–4 cm long, pedicels long, delicate, disarticulating below florets. **Florets** 4–8; glumes 1 or 2(–5), much shorter than first lemma, delicate, basally loose and usually subtending vestigial buds; lemma to 1 cm long; anthers 3; stigmas 3. **Fruits** caryopses. **Distr:** Temperate Asia to Indo-China. **Cult:** Propagate by division. Requires a rich, moist, well-drained soil in full (for variegated species) or partial sun.

P. fortunei (Van Houtte) Nakai, *J. Jap. Bot.* 9(4): 232, 1933. (Syn: *Bambusa fortunei*, *B. variegata*, etc.). [Dwarf Whitestripe Bamboo]. **Bamboos**, perennial; rhizomes elongated, leptomorph; culms zigzag, 0.6–1.2 cm long, 2–6 mm Ø, woody; internodes short, glabrous, nodal ridge flat or weakly prominent; branches absent or 1 per node. **Culm sheaths** persistent, glabrous. **Leaves** 4–7 per culm or branch; sheath glabrous; auricles absent; oral setae white, smooth; lamina longitudinally white to pale yellow striped, lanceolate, 6–15 × 0.8–1.4 cm, base broadly cuneate, apex acuminate, white pubescent, more densely so abaxially. **Inflorescences** comprising of a few spikelets; fertile spikelets with 6–7 fertile florets, with reduced florets at apex; spikelets linear, laterally compressed, 3 cm long; glumes 2, persistent, shorter than spikelet; lower glume lanceolate, 5 mm long; upper glume lanceolate, acute. **Florets** with fertile lemma ovate, 0.9–1 cm long, apex acuminate; apical sterile florets not developed, but resembling fertile; flower lodicules 3, anthers 3, stigmas 3. **Fruits** caryopses, with adherent pericarp. **Distr:** Temperate Asia, Australasia, New Zealand. **Ecol:** In woodland edges of forests. **Zones:** 5–10.

Pseudosasa

Species 19. **Grasses**, small and shrublike, or large and arborescent, spreading and densely clumped; rhizomes leptomorph. **Culms** pluri-cespitose, erect to drooping, 0.5–13 m tall, to 6 cm thick; internodes terete or moderately grooved; nodes not greatly swollen; supra-nodal ridge not evident; branch buds tall, prophylls 2-keeled, initially closed at front, without promontory; branches erect, initially 1–3 per node, short or long, central slightly dominant with basal nodes compressed, branches always fully sheathed, without replication of lateral branches, sheaths and prophylls ± glabrous. **Culm sheaths** deciduous to very persistent, tough; lamina erect or reflexed, narrowly triangular to strap-shaped; leaf sheaths persistent. **Leaves** relatively large, arrangement random, transverse veins distinct. **Inflorescence** an open raceme or panicle; branching subtended by large or small bracts. **Spikelets** 2–20 cm long; rachilla sinuous, disarticulation below florets. **Florets** 3–30; glumes 2, shorter than first lemma; lemma to 1 cm; palea 2-keeled; stamens 3; stigmas 3. **Distr:** China, Japan, Korea. **Cult:** Propagate by division. Requires a moist, fertile soil in full or partial sun.

P. hindsii (Munro) S.L. Chen and G.Y. Sheng ex T.G. Liang, *Fujian Bamboos*. 142, 1987. (Syn: *Arundinaria hindsii*, etc.). [Cane Bamboo]. **Grasses**; culms 3–5 m tall, ~1 cm Ø; internodes dark green, 20–36 cm long; branches 3–5 per node, erect. **Culm sheaths** persistent, sparsely white or light brown setose; auricles falcate; oral setae curved; ligule arcuate, ~3 mm long; lamina erect, broadly ovate-lanceolate. **Leaves** 4–9 per ultimate branch; sheath deciduous, glabrous or sparsely setulose, margins ciliolate; auricles absent; oral setae few; ligule truncate, rigid, 1–1.5 mm long; pseudo-petiole ~2 mm long; lamina linear-lanceolate or narrowly oblong, 7–22 × ~1.6 cm, base cuneate, one margin spinescently serrulate, other margin entire, apex acuminate, secondary veins 3–5-paired. **Inflorescence** composed of racemes; racemes bearing 2–5 fertile spikelets. **Spikelets** solitary; pedicels 0.2–1.5 cm long; fertile spikelets linear, laterally compressed; 2.3–5.5 cm long, comprising of 4–11 fertile florets. **Florets** with fertile lemma lanceolate or ovate, 1–1.4 cm long; palea 0.8–1 cm long; lodicules 3, 3 mm long, ciliate, anthers 3, 4.5–6 mm long, stigmas 3. **Fruits** caryopses. **Distr:** Temperate Asia, Australasia. **Ecol:** In coastal mountains. **Zones:** 6–12.

Thyrsostachys

Species 2. **Bamboos**, arborescent, moderately sized, densely clump-forming; rhizomes short-necked, pachymorph; culms unicespitose, erect, internodes terete; branches several to many, 1 dominant. **Culm sheaths** persistent, thickly papery; ligule short; auricles absent or small without bristles; lamina recurved or erect, triangular. **Leaves** deciduous; sheath small to moderate-sized, ligule short, auricles absent; lamina lanceolate. **Synflorescences** iterauctant, bracteate, comprising a cluster of 1–3 fertile pseudo-spikelets and 1 or 2 sterile ones subtended by a spathiform 2-keeled prophyll. **Spikelets** comprising 2 fertile florets, or 3 fertile florets; glumes 2–4; lemma papery; palea 2-cleft for third of its length. **Florets** with lodicules absent to 3, ciliate, stamens 6, long-exserted, anthers apiculate, ovary turbinate, stalked, style long, stigmas 1–3, feathery. **Fruits** caryopses, terete, smooth, glabrous, with persistent style base. **Distr:** Tropical and temperate Asia. **Cult:** Propagate by division or by internode cuttings. Requires fertile, moist soils.

T. siamensis Gamble, *Ann. Bot. Gard. Calc.* vii. 59, 1896. (Syn: *Bambusa regia*, etc.). [Monastery Bamboo, Umbrella-handle Bamboo]. **Bamboos**, perennial, cespitose; rhizomes short, pachymorph; culms erect, 7–13 mm long, 2–6 cm Ø, woody, internodes terete, thick-walled, 15–30 cm long; lateral branches dendroid. **Culm sheaths** persistent, chartaceous, pubescent; auricles without; lamina triangular. **Leaves** cauline, 4–12 per branch; sheath oral hairs lacking, or scanty; lamina deciduous, lanceolate, 7–15 × 0.6–1.2 cm, base briefly petiolate, margins scabrous, venation without cross veins, adaxially pubescent, abaxially hairy. **Synflorescences** bractiferous, fasciculate, with spathaceous subtending bracts 1–2 cm long; prophyllate below lateral spikelets; fertile spikelets sessile. **Spikelets** comprising 2 fertile florets with a barren rachilla extension, lanceolate, laterally compressed, 1.7 cm long; glumes persistent, similar, shorter than spikelet; lower glume oblong, membranous, without keels, pubescent; upper glume similar. **Florets** with fertile lemma oblong, 1.25 cm long, membranous, without keel, pubescent; palea chartaceous, keels ciliate; lodicules absent, anthers 6, 3.5 mm long, stigmas 1, ovary umbonate, glabrous. **Fruits** caryopses, with adherent pericarp, 6–8 mm long. **Distr:** Tropical and temperate Asia, Indo-China. **Ecol:** In mixed deciduous and teak forests, often found in pure stands, grows mostly on dry and poor soils; 300–400 m. **Zones:** 8–11.

Subfam. EHRHARTOIDEAE

Genera 17, **species** 120. **Grasses**. **Spikelets** with only one ♀ floret fertile and with basal ♀ or sterile florets; glumes very small. Florets with stamens (1–)6, styles separate almost from the very base (Stevens 2001 onwards). **Distr:** Mostly in the Southern Hemisphere.

Oryza

Species 24. **Grasses**, annual or perennial, tufted or shortly rhizomatous. **Culms** erect or ascending. **Leaves** mainly cauline; lamina broadly linear, flat; ligule membranous, sometimes long. **Inflorescence** a panicle, usually many-spiculate, often nodding, lower branches usually whorled, unbranched or sparsely branched, pedicels short. **Spikelets** with 3 florets, 2 lower florets reduced, sterile, upper floret fertile, strongly laterally compressed, disarticulating below sterile lemmas, persistent in cultivated species; glumes vestigial, remaining after disarticulation as a shallow lobed frill at pedicel apex; sterile florets reduced to 2 narrow lemmas at base of fertile floret; fertile lemma boat-shaped, keeled, leathery, closely papillose, sometimes spinulose, infrequently smooth, prominently 5-veined, apex awnless to long awned; awn straight; palea resembling lemma but narrower, 3-veined, apex beaked; stamens 6. **Fruits** caryopses, variable in shape. **Distr:** Africa, Asia, Australia; C and S America. **Cult:** Propagate from seeds. Requires moist soil in full or partial sun.

O. sativa L., *Sp. Pl.* 1: 333, 1753. (Syn: *O. formosana*, *O. sativa* var. *formosana*). [Rice]. Grasses, annual, aquatic, tufted; culms erect, rooting at lower submerged nodes, 0.5–1.5 m tall. Leaves with leaf sheaths slightly inflated below, upper sheaths tight, glabrous, auricles falcate, ciliate; lamina 25–60 × 0.5–2 cm, glabrous, smooth or scabrid on both sides, margins scabrid, apex acuminate; ligule 1–4 cm long. Inflorescences panicle, loosely contracted, to 30 cm long, nodding at maturity; branches 1–3 at lowest node, longest 2–12 cm long, axils bearded or glabrous; spikelets oblong to oblong-lanceolate, 0.7–1 cm long, persistent. Florets with sterile lemmas lanceolate, 1.5–4 mm long, acuminate; fertile lemma papillose, spinulose, acuminate; awn very variable, slender or stout, to 6+ cm, scaberulous, sometimes absent; anthers 1–3 mm long. Fruits caryopses, ovate or elliptic to cylindrical, 5–7 mm long, whitish-yellow to brown or black. Distr: SE Asia. **Ecol:** Cultivated widely throughout the tropics and warmer subtropical regions; 0–2,000 m. **Zones:** 9–12. **Info:** This is one of the most widely cultivated cereals in the world.

Subfam. POOIDEAE

Genera 189, **species** 3,300. **Herbs**, annuals, or tufted or rhizomatous perennials; culms hollow (rarely solid); aerial branching rare at most; micro-hairs absent. **Leaves** usually sessile, evergreen; lamina linear (rarely lanceolate), with a membranous ligule. **Inflorescence** with primary branches distichous. **Spikelets** with a definite or indefinite number of flowers, not branched; lemma usually 5-nerved; lodicules at most slightly vascularised; stamens usually 3, filaments free; styles separate almost from the very

Top row: *Thyrsostachys siamensis, Oryza sativa, Briza maxima, Briza minor.*
Bottom row: *Arundo donax, Coix lachryma-jobi, Cymbopogon citratus, Cymbopogon winterianus.*

base. **Fruits** caryopses, hilum often short (Dahlgren et al. 1985; Stevens 2001 onwards; Takhtajan 2009). **Distr:** Mostly northern temperate regions; few in tropical montane regions.

Briza

Species 21. **Grasses**, annual or perennial, slender. **Leaves** linear or broadly linear, flat. **Inflorescence** an open panicle; pedicels filiform. **Spikelets** ovate to rotund, laterally compressed or globose, florets several to many, glumes and florets horizontally spreading, disarticulating above glumes and between florets; glumes subequal, shorter than florets, narrowly ovate to orbicular, 3–5-veined, margins broad, scarious, apex hooded; lemmas tightly overlapping, orbicular to oblate, deeply concave, papery to leathery, 5–11-veined, back gibbous, usually slightly keeled, margins broad, membranous, appressed to floret above, base cordate, apex obtuse, cuspidate or bilobed; palea slightly or much shorter than lemma, lanceolate to orbicular, keels narrowly winged; stamens 1–3. **Fruits** caryopses, plumply elliptic; hilum round to elliptic, or linear. **Distr:** Europe, Africa, temperate and tropical Asia, Australasia, Pacific, N America, S America. **Cult:** Propagate from seeds. Requires a well-drained soil in full sun.

B. maxima L., *Sp. Pl.* 1: 70, 1753. [Blowfly Grass, Large Quaking Grass, Rattlesnake Grass]. **Grasses**, annual; culms solitary or loosely tufted, erect or geniculate at base, 20–60 cm tall. **Leaves** with sheaths smooth, glabrous; lamina thin, 4–20 × 0.3–0.8 cm, margins scabrid, otherwise smooth, apex acute; ligule 2–5 mm long, obtuse. **Inflorescence** a loose panicle, nodding, 7–10 cm long, sparingly branched with few pendent spikelets; branches inserted singly, scaberulous, sometimes with only 1 spikelet; pedicels hairlike, shorter or longer than spikelet, drooping. **Spikelets** ovate, ~1.2 × 1 cm, tinged reddish-brown, florets 7–20; lower glume 5–6 mm long, 5-veined, upper glume 6–7 mm long, 7–9-veined, margins purple or tawny, apex broadly rounded; lemmas very broadly ovate, wider than long, lowest 7–8 mm long, 7–9-veined, glabrous or appressed-pubescent; palea obovate, $^{1}/_{2}$– $^{2}/_{3}$ length of lemma, back glabrous, keels pubescent. **Distr:** N Africa, S Europe. **Ecol:** In dry or moderately moist disturbed areas; 0–500 m. Fl. & Fr. May–Aug. **Zones:** 7–10.

B. minor L., *Sp. Pl.* 1: 70, 1753. [Little Quaking Grass]. **Grasses**, annual; culms usually solitary, erect, slender, 20–60 cm tall. **Leaves** with sheaths thin, soft, smooth, loosely overlapping; lamina thin, 4–12 × 0.4–1 cm, adaxial surface and margins scabridulous, apex acute; ligule 3–6 mm long, subacute. **Inflorescence** a panicle open, erect, broadly ovate, 5–10 cm long, with many pendent spikelets; branches mostly paired, very slender, obliquely spreading, scabrid; pedicels hairlike, mostly longer than spikelets, flexuous. **Spikelets** broadly ovate, 3–5 × ~4 mm, pale green, florets 3–8; glumes broad, 2–2.5 mm long, 3–5-veined, apex rounded; lemmas very broadly ovate, wider than long, lowest 1.5–2 mm long, 7–9-veined, back smooth and glossy or pubescent, membranous margins very broad; palea ovate, slightly shorter than lemma, back appressed-pubescent or marginal hairs spreading and fringing keels, keels almost wingless, glabrous. **Distr:** N Africa, SW Asia, S Europe. **Ecol:** In dry or moderately moist disturbed areas; 0–500 m. Fl. & Fr. Jun–Sep. **Zones:** 7–11.

Subfam. ARUNDINOIDEAE

Genera 14, **species** 20–38. **Bamboos**, perennial, mostly tall reeds; culms hollow. **Fruits** caryopses, hilum short. **Distr:** Southern Hemisphere; widespread.

Arundo

Species 3. **Reeds**, perennial, usually tall, rhizomatous; culms usually robust, rarely slender, hollow. **Leaves** cauline; lamina broadly linear, tough, flat, base rounded or cordate; ligule membranous, margin minutely ciliolate. **Inflorescence** a large, copiously branched, plumose panicle of numerous spikelets. **Spikelets** wedge-shaped, laterally compressed. **Florets** 2–5, disarticulating above glumes and between florets; rachilla internodes glabrous; glumes subequal, narrow, as long as spikelet, membranous, 3–5-veined; floret callus laterally pilose; lemmas narrow, membranous, rounded on back, 3–7-veined, plumose below middle with spreading silky-white hairs, tapering to an entire or bidentate apex tipped by a straight awnlet; palea $^{1}/_{2}$– $^{2}/_{3}$ length of lemma body, keels glabrous or shortly ciliate. **Distr:** Mediterranean region, through Middle East and W Asia to China. **Cult:** Propagate by division. Requires any moist, well-drained, moderately fertile soil in full or partial sun.

A. donax L., *Sp. Pl.* 1: 81, 1753. (Syn: *Aira bengalensis*, etc.). [Giant Cane, Giant Reed, Spanish Cane, Wild Cane]. **Reeds**, robust, perennial; rhizomes thick, knotty, fibrous; culms very stout, erect, 2–6 m tall, 1–1.5 cm Ø, unbranched or with bamboolike clusters of slender branches from nodes. **Leaf sheaths** longer than internodes, usually glabrous except long pilose at mouth. **Leaves** cauline, 30–60 × 2–5 cm, margins scabrous, tapering to a slender filiform apex;

Above: *Pennisetum polystachion*, *Saccharum officinarum*, *Sorghum bicolor*, *Zea mays* (male inflorescence).

ligule a ciliolate membrane. **Inflorescence** a panicle, 30–60 cm long, dense, usually purplish; branches 10–25 cm long, ascending; spikelets 1–1.5 cm long. **Florets** 2–5; glumes narrowly lanceolate, 0.8–1.2 cm long, 3–5-veined, lower glume acute, upper glume sharply acuminate; lemmas linear-lanceolate, 0.8–1.1 cm, 3–7-veined, dorsal hairs 5–6 mm long, apex minutely bidentate with 1–2 mm long awnlet from sinus, lateral veins also shortly extended; palea ½ length of lemma body. **Distr:** Europe, tropical and temperate Asia, Africa, Australasia, Pacific, N America, S America. **Ecol:** On riverbanks and other damp places. Fl. and Fr. Oct–Dec. **Zones:** 7–11.

Subfam. PANICOIDEAE

Genera 206, **species** 3,245. **Grasses**, annual or perennial herbs, tufted or rhizomatous; culms usually solid, sometimes robust and woody; micro-hairs elongated, slender, thin-walled. **Leaves** broad or narrow; ligule is a ciliate membrane or a rim of hairs, sometimes absent. **Spikelets** dorsally compressed; 2-flowered, lower flower ♂ or sterile; rachilla absent. **Flowers** usually with 2 fleshy cuneate lodicules; stamens 1–3(–6); stigmas 2 (Stevens 2001 onwards; Takhtajan 2009). **Distr:** Tropical to temperate regions.

Coix

Species 4. **Grasses**, annual or perennial; culms robust, erect or decumbent, sometimes floating. **Leaves** cauline; lamina large, usually broad, flat; ligule membranous. **Inflorescences** many, fascicled in the upper leaf axils, each subtended by a globose or elongated, bony; each inflorescence comprising of 2 racemes, a ♀ sessile raceme, and a pedunculate ♂ raceme subtended by a prophyll and exserted; ♂ raceme deciduous at maturity, composed of imbricate spikelets borne in pairs or triads, 1(–2) sessile and 1 pedicelled, pedicelled spikelet often reduced in triads; ♀ raceme with 1 sessile fertile spikelet and 2 free stout pedicels, sometimes bearing vestigial spikelets. **Spikelet** ♂ with glumes subequal, herbaceous, lower glume flat on back, margins keeled, keels winged, upper glume boat-shaped, both florets ♂, lemma and palea hyaline; ♀ with lower glume broad, spikelet infolding, membranous with cartilaginous beak, upper glume narrower, keeled, lower floret reduced to a broad hyaline lemma, upper floret with hyaline lemma and palea, stigmas 2, elongate, exserted. **Fruits** caryopses, orbicular, ventrally furrowed, enclosed in a utricle. **Distr:** Tropical Asia. **Cult:** Propagate from seeds. Requires a moist to damp soil in full or partial sun.

C. lacryma-jobi L., *Sp. Pl.* 2: 972, 1753. [Jacob's Tears]. **Grasses**, annual; culms erect, robust, 1–3 m tall, nodes >10, branched. **Leaves** cauline; leaf sheaths shorter than internodes, glabrous; lamina linear-lanceolate, usually glabrous, 10–40 × 1.5–7 cm, midvein stout, base subrounded or cordate, margins scabrous, apex acute; ligule 0.6–1.2 mm long. **Inflorescences** with ♂ raceme 1.5–4 cm long. **Spikelets** in pairs with terminal triad, spikelets oblong-ovate, 6–9 mm long, glumes many-veined, lower glume winged on keels, wings 0.4–0.8 mm wide, wing margin ciliolate, anthers 4–5 mm long. **Fruits** utricles, ovoid to cylindrical, usually bony, shiny, 0.7–1.1 × 0.6–1 cm, white, bluish or gray-brown, sometimes with apical beak. **Distr:** Tropical and subtropical Asia to New Guinea. **Ecol:** In moist or marshy places. Fl. and Fr. Jun–Dec. **Zones:** 9–12. **Info:** Species highly variable across its native range.

Cymbopogon

Species ~70. **Grasses**, perennial, rarely annual; culms often tall, robust. **Leaves** aromatic, filiform to broadly linear; ligule scarious. **Inflorescence** a dense spathate compound panicle, each ultimate spatheole subtending a pair of short racemes on a short peduncle; spatheoles boat-shaped; each raceme borne on a short, flattened raceme base, often deflexed at maturity, lower raceme with one basal pair of homogamous spikelets below the fertile pairs; rachis internodes and pedicels linear, white-ciliate on margins, sometimes pedicel of homogamous pair swollen and fused to adjacent internode. **Sessile spikelet** dorsally compressed; callus obtuse, shortly bearded, inserted into internode apex; lower glume papery, flat or concave, sometimes grooved or wrinkled, 2-keeled, keels lateral, often winged above middle, with or without inter-carinal veins; upper glume boat-shaped, awnless; lower floret reduced to a hyaline lemma; upper lemma narrowly oblong, usually bilobed to near middle, lobes slender, ciliate, awned from sinus, occasionally subentire and awnless; awn geniculate, sometimes weakly, short, glabrous. **Pedicelled spikelet** ♂ or sterile, narrowly lanceolate, awnless. **Distr:** Tropics and subtropics of Africa, Asia and Australia. **Cult:** Propagate by division or from seeds. Requires fertile, moist but well-drained soil in full sun.

C. citratus (Candolle) Stapf, *Bull. Misc. Inform. Kew.* 1906: 357, 1906. (Syn: *Andropogon citratus*). [Citronella Grass, Lemon Grass, West Indies Lemon Grass]. **Grasses**, perennial, shortly rhizomatous, aromatic; culms tufted, robust, to 2 m tall, ~4 mm Ø, farinose below nodes. **Leaf sheaths** glabrous, greenish inside. **Leaves** glaucous, 30–90 × 0.5–2 cm, both surfaces scabrid, base gradually narrowed, apex long-acuminate; ligule ~1 mm long. **Inflorescence** spathate compound panicle, large, lax, to 50 cm long, drooping, branches slender; spatheoles reddish or yellowish-brown, 1.5–2 cm long; racemes 1.5–2 cm long; rachis internodes and pedicels 2.5–4 mm long, margins loosely villous; pedicel of homogamous pair not swollen. **Sessile spikelet** linear-lanceolate, 5–6 × ~0.7 mm; lower glume flat or slightly concave toward base, sharply 2-keeled, keels wingless, scabrid, veinless between keels; upper lemma narrow, entire and awnless, or slightly bilobed with ~0.2 mm long mucro. **Pedicelled spikelet** 4–5 mm long. **Distr:** Cultivated and naturalized in tropical Asia; origin unknown. **Ecol:** On dry or moderately moist hillsides; 0–1,200 m. Fl. & Fr. May–Jul. **Zones:** 9–12.

C. winterianus Jowitt ex Bor, *Oesterr. Bot. Z.* 112: 185, 1965. [Citronella Grass, Java Citronella]. **Grasses**, perennial from a shallowly rooted rhizome, aromatic; culms tufted, robust, to 2+ m tall. **Leaves** with sheaths glabrous, reddish inside; lamina relatively thin, drooping for ⅔ of their length, 0.4–0.8(–1) m × 1–1.5(–2.5) cm, abaxial surface glaucous, adaxial surface light green, margins scabrid, base narrow, apex long acuminate; ligule 2–3 mm long. **Inflorescence** a spathate compound panicle, large, lax, to 50 cm long, much-branched, main axis zigzag, finally nodding; spatheoles reddish-brown, 1.2–2.5 cm long; racemes 1.5–2.5 cm long; rachis internodes and pedicels softly ciliate on margins; pedicel of homogamous pair not swollen. **Sessile spikelet** elliptic-lanceolate, 4–5 × ~1 mm; lower glume flat or slightly concave, sharply 2-keeled, keels narrowly winged, 3-veined between keels; upper lemma very shortly bilobed, awn short or absent, not exserted from spikelet. **Pedicelled spikelet** 3.5–5 mm. **Distr:** Cultivated in tropical Asia; origin unknown. **Ecol:** On dry or moderately moist hillsides; 0–500 m. Fl. & Fr. Oct–Dec. **Zones:** 9–12.

Pennisetum

Syn: *Gymnotrix*.

Species ~80. **Grasses**, annuals or perennials; culms tufted or rhizomatous, prostrate to over 3 m tall. **Leaves** flat, folded or convolute; ligule a ciliate membrane. **Inflorescence** a spikelike panicle, cylindrical to subglobose; branches numerous, contracted into short clusters of one or more spikelets subtended by an involucre of bristles; involucres sessile or with a short basal stipe, deciduous with spikelets at maturity leaving peduncle stumps or scars on main axis; bristles slender, scabrous, sometimes plumose, simple or very rarely branched, very unequal, outer shorter, often innermost stouter and conspicuously exceeding the rest. **Spikelets** lanceolate to obovate, 2-flowered, dorsally compressed; disarticulation below the involucres, spikelets falling with involucre; glumes membranous, usually unequal, lower sometimes absent, upper to as long as spikelet; lower floret sterile or ♂, lemma membranous, palea membranous or absent; upper floret perfect, lemma chartaceous to indurate, palea of same texture; lodicules 0–2, minute when present; stamens 3, anthers penicillate at tip or glabrous; styles and stigmas completely connate to completely separate. **Fruits** caryopses, dorsally compressed, obovate, oblong or globose, hilum punctate, basal.

Above: *Zea mays* (female inflorescence), *Dactyloctenium aegyptium*, *Eleusine coracana*.

Distr: Throughout the tropics. **Cult:** Propagate from seeds or by division. Requires a moist, well-drained soil in full or partial sun.

P. polystachion (L.) Schultes, *Mant.* 2: 146, 1824. (Syn: *Cenchrus setosus*, etc.). [Blue Buffel Grass]. **Grasses**, annual or short-lived perennial, cespitose; culms 1–2 m tall, erect, much branched, branching mostly from upper and middle nodes. **Leaves** mostly cauline; sheaths with auricles 0.5–1.5 mm long, margins hyaline, glabrous, collar puberulent; ligule 1.3–2.9 mm long, membrane 0.4–1.5 mm long, cilia 0.6–2 mm long; lamina linear, flat, 10–50 × 0.3–1.5 cm, base narrowed, apex acuminate, variably pubescent. **Inflorescence** linear, 10–25 × 0.8–1 cm, yellow or purplish; axis angular with sharp decurrent wings below involucres, these densely packed; involucre with numerous bristles, densely ciliate in lower half, longest bristle 1–2 cm long. **Spikelet** narrowly lanceolate, 3–4.5 mm long; lower glume absent or reduced to a small triangular scale; upper glume as long as the spikelet, 5-veined, obtuse, ciliolate, apiculate; lower floret ♂ or neuter, lemma similar but slightly shorter, obtusely trilobed; upper floret $^{2}/_{3}$ spikelet length, cartilaginous, smooth, shiny, readily deciduous at maturity; anthers without hairs at tip; stigmas and styles free, stigmas purple. **Fruits** caryopses, elliptic, 1.5–1.6 × ~0.7 mm, obtuse. **Distr:** Throughout the tropics. **Ecol:** In tropical scrublands, lowland to montane swamp grasslands, roadsides and disturbed places, weedy in many regions; 0–2,400 m. Fl. anytime during the year. **Zones:** 10–12.

Saccharum

Syn: *Erianthus, Narenga, Ripidium.*

Species 35–40. **Grasses**, usually large, perennial, rhizomatous or tufted. **Culms** robust, to 7 m tall. **Leaves** cauline; lamina narrowly to broadly linear, midrib usually broad, white; ligule membranous, margin ciliolate. **Inflorescence** terminal, a large plumose panicle with elongate central axis, its branches bearing numerous hairy racemes; racemes fragile, sessile and pedicelled spikelet of a pair similar, both fertile; rachis internodes and pedicels filiform with cupular apex, pedicels resembling internodes but often shorter. **Spikelets** usually small, lanceolate, dorsally compressed or pedicelled spikelet more rounded on back; callus short, obtuse, bearded, often with long silky hairs surrounding the spikelet; lower glume membranous, thinly cartilaginous, or becoming leathery below, flat to broadly convex, veins indistinct, laterally 2-keeled; upper glume boat-shaped, resembling lower glume in texture and color; lower floret reduced to an empty hyaline lemma; upper floret bisexual, lemma entire, rarely 2-toothed, sometimes very narrow or small, with or without a short straight awn; stamens 2–3. **Distr:** Tropical and subtropical regions worldwide, mainly in Asia. **Cult:** Propagate from seeds or stem cuttings. Requires fertile, moist soil in full sun.

S. officinarum L., *Sp. Pl.* 1: 54, 1753. [Sugar Cane]. **Grasses**, perennial, forming tall clumps; culms 3–6 m tall, 2–5 cm Ø, 20–40-noded, solid, nodes glabrous, glabrous below inflorescence. **Leaf sheaths** glabrous, pilose at mouth. **Leaves** 0.7–1.5 m × 4–6 cm, usually glabrous, midrib large, white, margins sharply serrate, base rounded, apex acuminate; ligule 2–3 mm long, ciliate. **Inflorescence** a panicle, 0.5–1 m long, axis glabrous but pilose at nodes; racemes 10–25 cm long; rachis internodes 3–6 mm long, glabrous. **Spikelets** 3.5–4 mm long; callus hairs 2–3 times length of spikelet; lower glume oblong, uniformly firm throughout, buff-colored, back glabrous, margins membranous and ciliate above, apex acuminate; lower lemma oblong-lanceolate, subequal to glumes; upper lemma linear, awnless; lodicules glabrous; anthers 3. **Distr:** SE Asia, Pacific Islands. **Ecol:** In tropical rainforests; 500–3,000 m. Fl. anytime during the year. **Zones:** 9–12. **Info:** Widely cultivated for commercial sugar, and now for production of an alternative fuel.

Sorghum

Syn: *Andropogon* subg. *Sorghum.*

Species ~30. **Grasses**, perennial or annual, with or without rhizomes. **Culms** usually robust, erect. **Leaves** linear to linear-lanceolate; ligule a ciliate membrane. **Inflorescence** a large terminal panicle with elongate central axis; primary branches simple or branched, bearing short dense racemes of paired spikelets; racemes fragile; rachis internodes and pedicels slender, ciliate. **Sessile spikelet** dorsally compressed; callus obtuse, bearded, inserted into internode apex; lower glume usually leathery, shallowly convex, rounded on flanks, becoming 2-keeled and winged upward, usually hairy, apex membranous; upper glume boat-shaped, keeled upward; lower floret reduced to an empty hyaline lemma; upper lemma 2-toothed, awned from sinus or infrequently awnless; awn bigeniculate, glabrous; lodicules ciliate. **Pedicelled spikelet** well-developed or reduced to a glume, usually much narrower than sessile spikelet, awnless. **Distr:** Tropics and subtropics of the Old World; one species endemic to Mexico. **Cult:** Propagate from seeds. Requires any fertile, well-drained soil in full sun.

S. bicolor (L.) Moench, *Methodus.* 207, 1794. (Syn: *S. vulgare*, etc.). [Broom Corn, Millet, Sorghum]. **Grasses**, annual; culms erect, robust, 3–5 m tall, 2–5 cm Ø; nodes glabrous or pubescent; leaf sheaths glabrous or slightly farinose. **Leaves** linear or linear-lanceolate, 40–70 × 3–8 cm, glabrous; ligule subrounded, ciliate. **Inflorescence** a panicle, very variable, lax or dense, cylindrical or pyramidal to obovate in outline, to 60 cm long, main axis elongate to very short; primary branches ascending or spreading, lower branches sometimes almost as long as the panicle, stiff or pendulous; racemes tough at maturity, composed of 2–6 spikelet pairs. **Sessile spikelet** variable, broadly obovate to subglobose, 3.5–5.5 mm long; callus hispid; lower glume leathery to papery, glabrous to pilose, pale creamy-green to dark brown or blackish at maturity, upper lemma usually awned; awn 0.4–1.5 cm long. **Pedicelled spikelet** ♂ or barren, linear-lanceolate, persistent or deciduous. **Fruits** caryopses, large, often exposed between gaping glumes. **Distr:** Africa; widely cultivated in the tropics. **Ecol:** Fl. and Fr. Jun–Sep. **Zones:** 9–12.

Zea

Species 5. **Grasses**, annual. **Culms** robust, often tall with stilt roots, solid. **Leaves** large, broadly linear; ligule membranous. **Inflorescences** terminal and axillary; spikelets unisexual, separated into ♂ and ♀, not disarticulating at maturity, spikelets of a pair alike; ♀ axillary, enclosed in enveloping foliaceous sheaths; ♂ terminal, of many digitate or paniculate racemes. **Spikelets** ♀ all sessile in many longitudinal rows, partially sunk in the thickened, almost woody axis, glumes and lemmas chaffy, awnless, lower floret sterile, styles single, very long, silky, pendulous from inflorescence apex; ♂ paired, one spikelet subsessile, the other on a slender pedicel, papery, awnless, both florets ♂. **Fruits** caryopses, plump when mature, much larger than spikelet scales, very variable in shape and color. **Distr:** C America (4 spp.); warm parts worldwide (1 sp.). **Cult:** Propagate from seeds. Requires any soil in full sun.

Z. mays L., *Sp. Pl.* 2: 971, 1753. [Corn, Maize]. **Grasses**; culms erect, 1–4 m tall. **Leaves** 50–90 × 3–12 cm, glabrous or with tubercle-based hairs, margins scabrid, midvein stout; leaf sheaths with transverse veinlets; ligule ~2 mm long. **Inflorescence** ♂ (known as a 'tassel') composed of many digitate racemes, spikelets 0.9–1.4 cm long, unequally pedicellate, 1 pedicel 1–2 mm long, the other 2–4 mm long, glumes subequal, membranous, lower ~10-veined, margins ciliate, upper 7-veined, lower lemma and palea hyaline, subequal, upper lemma smaller than lower, anthers orange, ~5 mm long; ♀ cylindrical (known as a 'cob'), with 16–30 rows of spikelets, glumes equal, veinless, margins ciliate, florets hyaline. **Distr:** America; cultivated worldwide. **Ecol:** Fl. and Fr. May–Sep. **Zones:** 8–12. **Info:** This is one of the oldest cultivated plants and the third most cultivated crop in the world.

Subfam. CHLORIDOIDEAE

Genera 165, **species** 1,350. **Grasses**, annual or perennial, terrestrial, rarely aquatic, rhizomatous, mat-forming, sometimes decumbent herbs; culms

Top row: *Ensete ventricosum, Musa acuminata, Musa balbisiana.*
Bottom row: *Musa basjoo, Musa beccarii.*

solid or hollow. **Leaves** distichous or sometimes spirally arranged; adaxial ligule a fringe of hairs; abaxial ligule occasionally present as lone hairs. **Inflorescences** panicles, racemes or spikes, terminal or axillary. **Spikelets** bisexual, with 2 glumes; glumes tardily or rarely early deciduous; pales well-developed, lodicules 2 or sometimes absent, cuneate, free, fleshy; stamens 2–3; ovary usually glabrous, apical appendage absent; styles 2, free, stigmas 2. **Fruits** caryopses, hilum linear, endosperm hard, containing simple starch grains; embryo small. **Distr:** Tropical to warm temperate regions, especially in Africa and Australia. **Ecol:** Mostly in dry environments.

Dactyloctenium

Species 13. **Grasses**, annuals or perennials; culms tufted, sometimes stoloniferous, compressed. **Leaves** linear, flat or loosely folded; ligule membranous, often ciliolate. **Inflorescence** of paired or digitate spikelike racemes; racemes linear to narrowly oblong, spikelets sessile, biseriate, closely imbricate, uppermost abortive, rachis terminating in a bare pointed extension. **Spikelets** elliptic to ovate, laterally compressed, florets several, disarticulating above glumes but not usually between florets; glumes shorter than lemmas, keeled, 1-veined; lower glume smaller, sharply acute; upper glume with a stout oblique awn from just below the broadly rounded emarginate tip; lemmas membranous, 3-veined, glabrous, strongly keeled, acute to shortly awned and often recurved at apex; palea keels sometimes winged. **Fruits** grains, angular, ornamented, enclosed within a free hyaline pericarp which ruptures at maturity. **Distr:** Mainly from Africa to India; 1 sp. widespread. **Cult:** Propagate from seeds. Requires any soil in full sun; drought-resistant and salt-tolerant.

D. aegyptium (L.) Willdenow, *Enum. Pl.* 2: 1029 ['*aegyptiacus*'], 1809. (Syn: *Cynosurus aegyptius*, etc.). [Beach Wire Grass, Crowfoot Grass, Giant Button Grass]. **Grasses**, annual; culms slender to moderately robust, geniculately ascending to shortly stoloniferous and mat-forming, infrequently erect, 15–60 cm tall. **Leaves** with sheath margins ciliate; lamina flat, 5–20 × 0.2–0.6 cm, both surfaces tuberculate-pilose, apex acute or acuminate; ligule membranous, 1–2 mm long, margins ciliate. **Inflorescence** digitate, racemes 2–9, linear to narrowly oblong, often radiating horizontally. **Spikelets** broadly ovate, 3–4.5 mm long, florets 3–4; lower glume narrowly lanceolate, keel thick, hispidulous; upper glume elliptic to narrowly obovate, keel smooth, extended into a stout scabrid awn; lemmas ovate, 2.6–4 mm long, keel gibbous, hispidulous above middle, often extended into a stout cusp; palea equal to lemma, keels winged, wings ciliolate, tip 2-toothed. **Fruits** grains ~1 mm, broadly obtriangular, transversely rugose. **Distr:** Tropical and warm-temperate regions of the Old World. **Ecol:** In disturbed weedy places, especially on sandy soils. Fl. and Fr. May–Oct. **Zones:** 7–12.

Eleusine

[Goosegrass].

Species 9. **Grasses**, annuals or tussocky perennials; culms compressed. **Leaves** with sheaths strongly keeled; lamina linear, usually folded; ligule membranous, usually ciliate-fringed. **Inflorescence** of digitate or subdigitate spikelike racemes clustered at the top of the culm; racemes with closely imbricate, biseriate spikelets, terminating in a fertile spikelet. **Spikelets** laterally compressed, florets several, disarticulating between florets; glumes shorter than lemmas, persistent, 1–3(–7)-veined, keeled, awnless; lemmas membranous, 3-veined, glabrous, strongly keeled, keel thickened with 1–3 closely spaced additional veins, obtuse or acute. **Fruits** grains, ellipsoid to subglobose, trigonous in section, ornamented, pericarp free. **Distr:** Mostly in E and NE tropical Africa; 1 species a pantropical weed and one cultivated as a cereal. **Cult:** Propagate by division or from seeds. Requires any well-drained soil in full or partial sun.

E. coracana (L.) Gaertner, *Fruct. Sem. Pl.* 1: 8, 1788. (Syn: *Cynosurus coracanus*). [Finger Millet]. **Grasses**, annual; culms tufted, robust, erect or ascending, usually branched, 0.5–1.2 cm tall. **Leaves** flat, 30–60 × 0.6–1.2 cm, pilose or glabrous; leaf sheaths glabrous; ligule 1–2 mm long. **Inflorescence** subdigitate, racemes 5–20, stout, often incurved at maturity, 5–10 × 0.8–1.5 cm, hairy at base. **Spikelets** closely imbricate, ovate, 5–9 mm long, florets 6–9, not disarticulating at maturity; glumes lanceolate-oblong, scabrid along winged keel; lower glume 3-veined, 1.5–3 mm long; upper glume with additional veins in keel, 1.8–5 mm long; lemmas triangular-ovate, 2.2–4.7 mm long, keel 3-veined, scabrid and narrowly winged, subacute; palea narrowly ovate, keels scabrid, winged. **Fruits** grains, yellowish-brown, globose, finely striate-punctate. **Distr:** Old World tropics; introduced in the New World. **Ecol:** Fl. and Fr. May–Sep. **Zones:** 9–12. **Info:** Cultivated as a cereal crop.

Above: *Musa coccinea, Musa lasiocarpa, Musa lolodensis, Musa textilis.*

ZINGIBERALES

FAMILIES 8, GENERA 92, SPECIES 2,111.

Herbs, usually large and robust; no aerial stem except when flowering. **Leaves** distichous, with distinct petiole; lamina with midrib and fine cross venation. **Inflorescence** bracts large, persistent. **Flowers** large, monosymmetric, complex; perianth colored; perianth consisting of colored calyx surrounding the corolla, adaxial member of inner whorl not developed; anthers long, pollen inaperturate; gynoecium inferior, style long, stigma large, wet. **Fruits** capsular; seeds arillate, operculate, operculum testal. **Distr:** Predominantly tropical. **Info:** Monophyly of the Zingiberales is supported by molecular (Chase et al. 1995b, 2000; Soltis et al. 2000) and morphological (Kress 1995; Stevenson et al. 2000) data.

MUSACEAE

Genera 2, species 35. **Herbs**, treelike, perennial, growing from corms or rhizomes; true aerial stems absent. **Leaves** basal, in several ranks, differentiated into basal sheath, petiole, and lamina; sheaths overlapping, forming unbranched pseudostem, open, ligule absent; summit of petiole not differentiated; lamina with lateral veins parallel, diverging from prominent midrib. **Inflorescences** 1 per aerial shoot (extremely rarely 2, in *Musa*), projecting from tip of pseudostem, pedunculate racemes of 12–20-flowered monochasial cymes (cincinni); bracts of main axis enclosing cincinni. **Flowers** unisexual (proximal flowers ♀, distal flowers ♂), bilaterally symmetric; sepals and petals differentiated, sepals 3, petals 3, 3 sepals and 2 petals fused, remaining petal distinct; fertile stamens 5(–6), not petal-like; anthers bilocular; occasionally 1 rudimentary staminode; ovary inferior, tricarpellate, trilocular, all locules fertile; placentation axile; ovules many per locule; style terminal, filiform; stigma trilobed. **Fruits** berries; sepals not persistent in fruit; seeds without arils; endosperm copious; perisperm copious; embryo straight. **Distr:** Tropical Africa, Asia, Australia, and Oceania.

Ensete

Species 7. **Herbs**, monocarpic, usually large, single-stemmed; true stem remaining short until flowering; pseudostem composed of clasping leaf sheaths, tall, base swollen. **Leaves** large, rosulate; lamina oblong, base usually narrowed into a petiole; upper leaves passing into bracts. **Inflorescence** rosulate when young, becoming elongate and pendulous with age; bracts green, usually persistent; flowers many per bract, in 2 rows, those of the proximal bracts ♀ or hermaphrodite; flowers in the distal bracts ♂. **Flowers** with compound tepal linear, apex often trilobed, without 2 accessory teeth between lobes; free tepal usually wider than the compound one, often tricuspidate, sometimes entire; stamens 5. **Fruits** leathery berries, usually dry; seeds few, usually large, (0.5–)1+ cm Ø, (Cheeseman 1947). **Distr:** Tropical and S Africa to tropical and subtropical Asia. **Cult:** Propagate from seeds. Requires a moist, rich organic compost in full or partial sun.

E. ventricosum (Welw.) Cheesman, *Kew Bull.* 2: 101, 1947. (Syn: *E. edule*, etc.). [Abyssinian Banana, Ethiopian Banana, False Banana, Wild Banana]. **Herbs**, 4–9 m tall, pseudostem 1.5–5 m, often stained purple or purplish-brown; latex pale whitish, reddening on exposure. **Leaves** erect or spreading; lamina lanceolate to 5 × 1.5 m, midribs red or green. **Inflorescence** ♂ with rachis tapering; flowers 30–40 per bract, bud massive, to 35 × 13 cm, red, purple or green; ♀ rachis massive, bracts concealing fruits. **Flowers** usually hermaphrodite and ♂, occasionally ♀; hermaphrodite flowers with ovary 4–7 cm long, perianth variable, outer tepal either 1–trilobed, inner tepals 1–3, variable, 2-winged, apiculus to 1.5 cm long, stamens 0–5, 3.5 cm long, anthers violet to dark purple, style 2.5–4 cm long, stigma large, staminodes variable; ♂ biseriate, usually white with perianth tipped orange-yellow, outer tepal trilobed, 3.5–5.5 cm long, inner tepal serrate-apiculate, 1–1.5 × 1–1.7 cm, apiculus 0–1.3 cm long, stamens 5, 3–5 cm long, anthers violet to purple, staminode absent or minute, 0.1–1 cm long, style acicular 1–2 cm long. **Fruits** biseriate, 5–20+ in ~24 hands, 8–15 × 4.5 cm, orange; seeds (0–)15–25(–40), in edible but tasteless orange pulp. **Distr:** Tropical Africa. **Ecol:** In swamps, riverbanks. **Zones:** 9–12. **Info:** A variable species with a wide distribution (Baker and Simmonds 1953).

Musa

Species 63. **Herbs**, perennial, tufted, underground stems (corms) rhizomatous, stoloniferous (sympodial growth); stems remaining short until flowering; pseudostems to (0.5–)3–10 m tall, clustered and each pseudostem composed of closely packed leaf sheaths, to 60+ cm Ø, base slightly swollen. **Leaves** large, oblong or oblong-elliptic, (0.6–)2–3 × 0.3–0.6 m, unlobed, splitting with age; petiole long, base enlarged into a sheath. **Inflorescence** solitary (extremely rarely 2) terminal, erect, pendent, or subpendulous, not conical, lax; bracts green, brown, red, dull purple, or rarely yellow, flat or furrowed, convolute or imbricate in bud, usually deciduous; flowers unisexual in 1 or 2 rows per bract; ♀ flowers numerous, crowded in proximal bracts (with reduced stamens) or bisexual; ♂ flowers (with reduced gynoecium) in distal bracts (imbricate, forming budlike mass at apex of inflorescence); sometimes (in cultivated forms) all flowers functionally sterile. **Flowers** with compound tepal tubular, soon splitting to base on 1 side, apex 5-toothed, free tepal inserted within compound tepal and opposite to it, stamens 5, filaments short, anthers linear, biloculed. **Fruits** berries, fleshy, often elongate, cylindric, usually ± curved, weakly angled in cross-section, (10–)20–35(–48) cm long, soft, fleshy; seeds numerous (except in parthenocarpic forms), irregularly globose to lenticular. **Distr:** Tanzania (Pemba), tropical and subtropical Asia to W Pacific. **Cult:** Propagate from suckers, seeds or tissue culture. Requires humus-rich, moist soil in full sun. **Info:** A very important commercial food crop throughout the wet tropics. Sap of plant parts causes brown/black stains on clothing. The pseudostem of *M. mannii* is less than 1 m high while that of *M. ingens* is up to 15 m tall, making it the largest known herb.

M. acuminata Colla, *Mem. Reale Accad. Sci. Torino* (Mem. Gen. Musa) 25: 66, 1820. [Banana]. **Herbs**; pseudostems green, heavily blotched with brown or black, to ~4.8 m tall. **Leaves** oblong, 1.9–2.3 m × 50–70 cm, base cordate, asymmetric, adaxially green and pruinose, abaxially paler; petiole ~80 cm long. **Inflorescences** semi-pendent or vertically reflexed; bracts bright red to dark violet, sometimes yellow at apex, ovate, apex usually acute; pedicels short; bracts of ♂ flowers lanceolate or narrowly ovate, acute, abaxially yellow, red, or dull purple, adaxially yellow proximally, often yellow or dull purple distally; ♂ flowers ~20 per bract in 2 rows. **Flowers** unisexual; ♂ with compound tepal white or cream, 3.5–4 cm long, apex lemon yellow, apex of outer lobes with a hooklike, hairy appendage, free tepal shorter, apex emarginate, shortly apiculate; ♀ with deep yellow or orange stigmas. **Infructescence** pendent, ~1.2 m long. **Fruits** berries, incurved, green to yellow, ~10 cm long, white-setose, base curved and attenuate into a stalk; seeds numerous (in the wild), brown, depressed, 5–6 mm Ø, irregularly angled. **Distr:** Tanzania (Pemba), tropical and subtropical Asia. **Ecol:** In moist, shaded, disturbed areas, ravines and marshlands; to 1,200 m. Fl. throughout the year. **Zones:** 9–12. **Info:** Cultivated throughout the tropics for the edible fruits.

M. balbisiana Colla, *Mem. Reale Accad. Sci. Torino.* 25: 384, 1820. [Mealy Banana, Plantain]. **Herbs**; pseudostems clumped, yellow-green, often with large, black markings, ~6 m tall. **Leaves** ovate-oblong, ~2.9 × 0.9 m, base auriculate, asymmetric, adaxially green and sometimes slightly pruinose; petiole 60–75 cm long, margin open, ~2 cm wide, often closed when young. **Inflorescence** pendulous, ~2.5 m; peduncle and rachis glabrous; bracts of bisexual and ♂ flowers adaxially purple-red, abaxially brownish-purple to yellow-green and pruinose, ovate to lanceolate, persistent, apex obtuse, reflexed after flowering; bracts of ♀ flowers deciduous; ♂ up to 20 per bract, in 2 rows. **Flowers** with compound tepal adaxially pale purple, abaxially pale purple-white, 4–5 cm, striate, teeth yellow to orange; free tepal milky white, translucent, obovate, ~½ as long as the compound tepal, apex emarginate, shortly mucronate-apiculate. **Fruits** berries, arranged in a pendulous infructescence, with ~8 clusters (hands) each of 15 or 16 in 2 rows; berries gray-green, obovoid, ~13 × 4 cm, distinctly angled at maturity, base narrowed into a stalk ~2.5 cm, apex contracted or not into a short, angled column ~2 cm; seeds numerous, brown, oblate, 0.5–1 cm Ø, minutely warty. **Distr:** Sikkim to Papuasia. **Ecol:** In ravines of evergreen forests; ~1,100 m. **Info:** Cultivated for ornamental purposes and to feed pigs.

M. basjoo Siebold and Zucc. ex Iinuma, *Sômoku-Dzusetsu*, ed. 2: 3, t.1, 1874. [Japanese Banana, Snow Banana]. **Herbs**, perennial, suckering, 2.5–5 m tall; pseudostem cylindrical, green, glabrous, 15–20 cm Ø. **Leaves** oblong or oblong-lanceolate, 2–3 m × 30–70 cm, base rounded, asymmetric, thin, adaxially bright green, rarely glaucous; petiole robust, to 30 cm long. **Inflorescence** pendulous, 30–45 cm long, flowers arranged in 2 rows per bract; ♀ flowers in ~6 clusters, each 12–15-flowered; ♂ flowers in clusters of 8–20 per bract; peduncle ~30 cm long; lower (sterile) bracts leaflike; upper (floral) bracts broadly ovate or oblong-elliptic, soon deciduous, downy, yellow-green or brown-suffused outside, pale yellow and ridged inside. **Flowers** cream or white with yellow tips; ♀ with 5 fused segments forming a compound perianth segment, ~4–5 × 1 cm, apex 5-toothed, 1 free, posterior segment, ~4 cm long, staminodes 5, equaling perianth, style stout with stigma to 7 mm across; ♂ similar with 5 fertile stamens. **Fruits** berries, obovoid-trigonous (to 5-angled), ~7 × 2.5–3 cm, obtuse, ripening to yellowish-green, pulp white; fruiting pedicel 0–5 mm long; seeds black, warty, 6–8 mm long. **Distr:** Ryukyu Islands. **Ecol:** In tropical forests of volcanic islands. Fl. Jun–Nov. Fr. Aug–Jan. **Zones:** 9–12. **Info:** Cultivated for medicinal and ornamental purposes.

M. beccarii N.W. Simmonds, *Kew Bulletin* 14 (2): 198–212, 1960. [Ornamental Banana]. **Herbs**, small, freely suckering; pseudostems 1–1.5 m tall; sheaths bright green, edges brownish-purple. **Leaves** oblong-lanceolate, to 1 × 0.3 m, base slightly unequal, rounded or sharply cuneate, apex obtuse, bright green; petiole to 35 cm long, tightly clasping below. **Inflorescence** small; ♀ flowers in 2–5 'hands', basal, each with 1–3 uniseriate flowers, bracts deflexed, not rolled, deciduous, thinly leathery, light red, tips obtuse, green, glossy; ♂ bracts deciduous, lanceolate, 10 × 3–4 cm, obtuse, polished scarlet, tips greenish, 2–5 flowers per bract, flowers ♂ uniseriate. **Flowers** ♂ 4–5 cm long, abortive ovary 6 mm long, compound tepal 3.5–4.5 cm long, yellowish-white below, grading to green at tip, free tepal 2–3 × 1–1.2 cm, translucent near base, midrib green-tinged, stamens 5, 3.5–4 cm long, white; ♀ compound tepal 3–4 cm long, pale yellow at base grading to pale greenish-yellow at tip with 5 green minutely apiculate teeth, free tepal ovate-lanceolate, boat-shaped, 3 cm long, obtuse, crumpled-corrugate, translucent, yellow-tinge near base, ovary 5–6 cm long, style pale yellow, staminodes 5, nectar abundant. **Infructescence** small, erect, loosely packed, 2–5 hands of 1–3 fruits each. **Fruits** shortly pedicellate, erect, cylindrical, 5–15 × 2 cm; seeds subglobose, 4–5 mm Ø, light brown. **Distr:** Malaysia (Sabah). **Ecol:** In lowland rainforests. Fl. Jan–Jun. **Zones:** 10–12.

M. coccinea Andrews, *Bot. Repos.* 1: t. 47, 1799. (Syn: *M. uranoscopos*). [Red Dwarf Banana, Red Torch Banana, Scarlet Banana]. **Herbs** with pseudostems 1–2 m. **Leaves** with petiole 30–50 cm long, narrowly winged; lamina adaxially yellow-green, abaxially light yellow-green and not pruinose, oblong, 1.8–2.2 × 0.7–0.8 m, base rounded, noticeably asymmetric. **Inflorescence** erect; rachis glabrous; bracts adaxially pink, abaxially scarlet and conspicuously wrinkled; flowers 6 per bract, in 1 row. **Flowers** with tepals of ♀ flowers yellow; outer lobes of compound tepal cornered; free tepal equaling compound tepal, apex acute, finely toothed. **Fruits** berries obliquely pendulous on rachis, gray-white, straight, 10–12 × ~4 cm, not angled; stalk 3–3.5 cm long; seeds numerous. **Distr:** Vietnam. **Ecol:** In ravines and slopes; 0–600 m. Fl. Sep.–Nov. **Zones:** 9–12.

M. lasiocarpa Franch., *J. Bot.* (Morot) 3: 329, 1889. (Syn: *Ensete lasiocarpum*, *Musella lasiocarpa*, *M. splendida*). [Golden Lotus Banana]. **Herbs** perennial, tufted; rhizomes horizontal; pseudostems <60 cm tall, ~15 cm Ø at base, with persistent sheaths, base not swollen. **Leaves** narrowly elliptic, bilaterally symmetric, to 50 × 20 cm, base subrounded, expanded into a sheath, apex acute, glaucous. **Inflorescence** terminal, or axillary near base of pseudostem, erect, conical, dense, 20–25 cm long; bracts yellow to yellow-orange, scarious, persistent; flowers usually 8–10, in 2 rows per bract, persistent; flowers in proximal bracts ♀ (with reduced stamens) or bisexual; flowers in distal bracts ♂ (with reduced gynoecium). **Flowers** with compound tepal ovate-oblong, apex 5-toothed; free tepal emarginate and shortly cuspidate at apex; stamens 5. **Fruits** berries, trigonous ovoid, ~3 × 2.5 cm, densely hirsute; seeds numerous, brown to black-brown, oblate, 6–7 mm Ø, glabrous; hilum white, large. **Distr:** SC China (W and S Yunnan, S Guizhou) to N Vietnam. **Ecol:** Usually on slopes; 1,500–2,500 m. **Zones:** 9–11. **Info:** Cultivated for medicinal purposes and to feed pigs.

M. lolodensis E.E. Cheesman, *Kew Bulletin* 5(1): 27, 1950. **Herbs**; rhizomes short; pseudostem to 7 m tall, 60 cm Ø at base, green or sometimes blackish; sap variable, white, pink or violet. **Leaves** adaxially yellow-green, abaxially slightly paler; petiole green or often blackish on underside. **Inflorescence** with peduncle green; basal bracts long, curling backwards, soon deciduous; ♂ bud deep or pale pink to creamy-white; ♂ bracts, colored as in ♂ bud both inside and out, soon deciduous, ♂ flowers falling together from each bract, not separating; basal flowers ♀, with poorly developed, or entirely without staminodes; pedicels to 2 cm long. **Flowers** ♂ with free tepal almost as long as compound tepal; compound tepal cream with tips of lobes yellow; free tepal translucent white, somewhat concave, ovate with apex truncate. **Fruits** yellowish-green, glabrous, maturing to bright coppery-orange, pulp dull orange, dehiscent, skin splitting into irregular segments and curling backwards, ± ovoid about twice as long as wide; seeds ~6 mm Ø, irregularly and sharply angled. **Distr:** Indonesia, Papua New Guinea. **Ecol:** In lowland open areas. **Zones:** 11–12.

M. textilis Née, *Anales Ci. Nat.* 4: 123, 1801. [Manila Hemp]. **Herbs**; pseudostems clumped, 3–8 m × 12–30 cm. **Leaves** oblong, 1.2–2.4 × 0.2–0.4 m, margins tightly closed, submembranous, densely and finely wrinkled, adaxially bright green, adaxially glabrous, abaxially pruinose and often with large brown spots, midvein stout, secondary veins conspicuous, pinnate, parallel, base cordate, asymmetric; petiole 60–70 cm long. **Inflorescence** pendulous, long; bracts closely imbricate, reddish-brown to green, ovate, ~10 × 6 cm, deciduous; flowers 10–12 per bract, in 2 rows; flowers in basal 3–6 bracts ♀, those in distal bracts ♂. **Flowers** with compound tepal ~4 cm long, outer lobes cucullate, corniculate. **Fruits** berries, green, slightly trigonous, curved, 5–9 × 2–3 cm, inedible; seeds black, turbinate, ~7 mm Ø, multi-angled. **Distr:** Indonesia, Philippines. **Ecol:** In humid tropical rainforests; 300–500(–1,200) m. **Zones:** 11–12.

M. velutina H.(A.) Wendland and C.G.O. Drude, *Gartenflora*, 65: t. 823, 1875. [Fuzzy Banana, Hairy Banana, Pink Banana, Velvet Banana]. **Herbs**; pseudostems to 1.5(–3) m tall, to 7 cm Ø at base, yellowish-green. **Leaves** to 1 m × 35 cm, base unequal, apex truncate, adaxially shiny, dark green, abaxially paler, midrib abaxially red; petiole to 45 cm long, base winged. **Inflorescence** erect; peduncle red, 2 cm Ø, heavily white-pubescent; last leaf with a broadened petiole developing red color, following this a sterile persistent bract 15–20 cm long; basal flowers hermaphrodite, 3–5 per bract in a single row; fertile flowers 2–4; upper flowers ♂, ~5 per bract. **Flowers** *hermaphrodite* with compound tepal 3 cm long, bright orange, back pink-flushed, lateral lobes 4 mm long, dorsal appendage filiform, 3(–5) mm long, free tepal white, oval, emarginate, stamens usually <5 fertile, ≥1 being vestigial, ovary 3–3.5 × 1–1.5 cm, acutely trigonous or 4-sided; ♂ with compound tepal 3.5 cm long, orange-yellow, back pink-flushed, rudimentary ovary pink, lobes of tepal orange, ~2 mm long, free tepal oblong, 3.5 × 1 cm, stamens exserted. **Fruits** berries, schizocarpic, bright pink, ~7 × 3–4 cm, splitting at maturity, separating in irregular strips from apex to base, exposing white pulp and black seeds; seeds 4–6 × 2–3 mm. **Distr:** Assam, NE India. **Ecol:** In subtropical evergreen forests. Fl. Mar–Jul. Fr. Aug–Dec. **Zones:** 10–12.

HELICONIACEAE

[False Bird-of-paradise, Heliconia, Lobster-claw]. **Genus** 1, **species** 197. *Familial description same as generic.*

Heliconia

Species ~200. **Herbs**, perennial, from rhizomes; true aerial stems absent or weak and concealed in pseudostem. **Leaves** basal, or basal and cauline, 2-ranked, differentiated into basal sheath, petiole, and lamina; sheaths overlapping, forming unbranched pseudostem, open, ligule absent; summit of petiole not differentiated; lamina with lateral veins parallel, diverging from prominent midrib. **Inflorescences** 1 per aerial shoot, projecting from tip of pseudostem, pedunculate racemes of several- to many-flowered monochasial cymes (cincinni); bracts of main axis enclosing cincinni. **Flowers** bisexual, bilaterally symmetric; sepals and petals scarcely differentiated, sepals 3, petals 3, 2 sepals and 3 petals fused, remaining sepal distinct; fertile stamens 5, not petaloid, anthers bilocular; 1 rudimentary staminode opposite free sepal; ovary inferior, tricarpellate, trilocular, all locules fertile; ovules 1 per locule, style standing away from stamens and staminode, filiform, stigma capitate or trilobed. **Fruits** drupes, 1–3-seeded, sepals not persistent in fruit, blue (rarely red or orange) at maturity; seeds without aril. **Distr:** N America, Mexico, West Indies, C America, S America, and Oceania. **Cult:** Propagate by division, or from offsets or seeds. Requires a moist, well-drained, humus-rich soil in full or partial sun.

H. aurantiaca Ghiesbr. ex Lem., *Ill. Hort.* 9: t. 332, 1862. (Syn: *H. brevispatha*, etc.). [False Bird of Paradise]. **Herbs**, perennial, habit zingiberoid-type; stem underground, rhizomatous; pseudostem green, slender, 1–2 m tall, ~2 cm Ø. **Leaves** simple, 10–12 per stem; lamina linear to lanceolate, 6–35 × 3–10 cm, margins entire, apex acute to acuminate, midrib green; sheaths green to greenish-orange, petiolate; petioles ~3.5 cm long. **Inflorescence** 1 per shoot, erect, 10–20 cm long, each cincinni 5–7-flowered; cincinnal bracts remote, 3–8, subdistichous, first bract without flower, spiraled, ~4.5–9 × 3–3.5 cm, glabrous, orange with pale green tip, turning green with age, rachis orange; pedicels ~1 cm long, orange, bracteate. **Flowers** yellowish-white; perianth almost erect, 4.5–5.5 × 0.8–1 cm, glabrous, creamy-white or yellow; tepals 6, biseriate, ~5 cm long; stamens 5, filaments ~2.5 cm long, anthers

Top row: *Musa velutina, Heliconia aurantiaca, Heliconia bihai.*
Bottom row: *Heliconia indica, Heliconia latispatha, Heliconia psittacorum, Heliconia rostrata.*

~1 cm long; staminode ~6 mm long, acuminate, shallowly tridentate; carpels 3, ovary trioculed, ~4 mm Ø, off-white with yellow or green edge, ovule 1 per locule, style 1, 3.7 cm long, off-white, stigma yellow. **Fruits** schizocarps or drupes; seeds 1–3. **Distr:** SE Mexico to C America. **Ecol:** In forest understory; (0–)200–800 m. Fl. Dec–Jun. **Zones:** 10–12.

H. bihai (L.) L., *Mant.* 2: 211, 1771. (Syn: *Bihai bihai, Musa bihai*, etc.). [Macaw Flower, Red Lobster Claw, Red Palulu]. **Herbs**, perennial, musoid-type, perennial, glabrous; stem rhizomatous, underground; pseudostem 1.5–4 m tall, greenish-brown ~3.5 cm Ø. **Leaves** simple; lamina ovate-oblong, 0.5–0.6(–2) m × 20–22 cm, margins entire, apex shortly acuminate, adaxially midrib green and abaxially reddish-green; petioles ~32 cm long. **Inflorescence** erect, 1 per shoot, to 45 cm long, terminal, distichous; peduncle ~15 cm long, greenish-white; bracts 5–12, 14–17 cm long; cincinnal bracts bright rose-red with tip green, 1–1.5 cm apart; pedicels 1.3 cm long, off-white, glabrous. **Flowers** with tepals 6, in 2 series, 0.6–4 cm long; stamens 5, filaments 0.2–2.2 cm long, anthers 0.8–1 cm long, staminode ~5 mm long, acuminate, shallowly tridentate; carpels 3, ovary 3-celled, ~3 mm Ø, off-white, ovule solitary, style 1, 0.9–3.5 cm long, off-white, stigma off-white. **Fruits** schizocarps or drupes; seeds 1–3. **Distr:** Caribbean, Colombia to N South America. **Ecol:** In wet lowland forests or thickets; ~900 m. Fl. Dec–Jun. **Zones:** 10–12.

H. indica Lam., *Encycl.* 1: 426, 1785. (Syn: *Heliconiopsis indica*). [Blood Banana, Pink-striped Heleconia, Striped Blood Banana]. **Herbs**, robust, musoid-type, tufted, soboliferous and suckering, (3–)3.5–7.5 m tall; pseudostems 1.5–2(–4) m tall, purplish or brown mottled at base. **Leaves** bananalike, (1–)1.5–2.5 × (0.3–)0.5–0.6 m, abaxially glaucous at first, adaxially green to various shades of bronze, abaxially usually bronze; petiole (1–)1.5–2 m long. **Inflorescence** terminal, pendulous or erect, (0.2–) 0.6–1 m long; bracts (4–)7–15(–20), ± distichous or spirally arranged, spathaceous, 10–15(–20) cm long, greenish becoming orange; sometimes bracts empty, remainder 10–20-flowered; pedicels 1–2 cm long. **Flowers** hermaphrodite, greenish, yellowish or whitish, or green becoming red with age; perianth slightly upcurved, lobes 6, fleshy, acute, ~5–6 cm long; stamens 5, 3.2–5.5 cm long, filaments 1.8–3.8 cm long; staminode ovate to broadly ovate to ovate-elliptic, ± acuminate, 0.7–1 × 0.4–0.6 cm; ovary inferior, ± triangular obovoid, style filiform. **Fruits** fleshy, ± triangular prismatic or obovoid, 1.3–1.7(–2) × 1.1–1.5(–1.7) cm; seeds 3, to 1.7 × 1 cm. **Distr:** Maluku to SW Pacific. **Ecol:** In tropical forests. Fl. Sep–Nov. **Zones:** 11–12.

H. latispatha Bentham, *The Botany of the Voyage of H.M.S. Sulphur.* 6: 170, 1846. [Golden Lobster Claw]. **Herbs**, ± glabrous, 1.5–4 m tall; pseudostem well-developed, stout. **Leaves** long-petiolate, 3–5 per shoot; lamina 0.75–1.5 × 0.18–0.33 m, base rounded to subacute, usually oblique, apex shortly acuminate, adaxially green, abaxially slightly glaucous or green. **Inflorescences** erect, to 45 cm long, pedunculate; cincinnal bracts lanceolate, spreading, spirally arranged, 8–13 per inflorescence, middle bract with outer surface orange and/or red and glabrous, 14–19.5 × 5–7 cm, each cincinnus 10–15-flowered. **Flowers** with perianth green, yellow, or orange with dark green sepal margins, sometimes sparsely puberulent along fused sepal margins, 3.5–4.6 × 0.7–0.8 cm, base gibbous, nearly straight; sepal margins, fused, dark green, glabrous, sometimes sparsely puberulous. **Fruits** drupes, blue, 1–1.3 × 0.8–0.9 cm. **Distr:** Mexico, C America, S America. **Ecol:** In wet forests or thickets; 0–1,400 m. Fl. Sep. **Zones:** 10–12.

H. psittacorum L.f., *Suppl. Pl.* 158, 1782. (Syn: *H. humilis*, etc.). [False Bird of Paradise, Parakeet Flower, Parrot's Beak, Parrot's Flower]. **Herbs**, musoid-type, slender, glabrous, 1–2 m tall. **Leaves** often numerous, divaricate from the stem; lamina narrowly oblong, ~30 × 9 cm, base often oblique, rounded or subcordate, apex abruptly acuminate, thin, bright green. **Inflorescence** erect, sessile or pedunculate, 1–2 m tall; bracts 4–6, orange-yellow, lowest bract often ending in a small green lamina, others close together, 3–6 cm long, ~1 cm high at base, attenuate; flowers ~15 per bract; pedicels 1 cm long. **Flowers** with perianth straight, 3–5 cm long, yellow or orange with dark green spots distally, tip white, glabrous; ovary green to yellow-green, glabrous. **Fruits** 5 mm Ø or larger, blue. **Distr:** Lesser Antilles to S tropical America. **Ecol:** In wet lowland forests; ~500 m. **Zones:** 10–12.

H. rostrata Ruiz and Pavon, *Fl. Peruv.* 3: 71, 1802. (Syn: *Bihai poeppigiana*, etc.). [Hanging Lobster Claw, Painted Lobster Claw]. **Herbs**, rhizomatous, terrestrial, coarse, stout, usually 2–3 m tall. **Leaves** long-petiolate; lamina oblong, 1–1.5 × 0.25–0.35 m, apex abruptly shortly acuminate, abaxially glabrous, green or glaucescent, sometimes red-tinged. **Inflorescence** pendent, 30–60 ×15 cm; rachis terete, slender, very flexuous, densely brown-villous with short hairs; bracts cardinal-red or dull

JM/EdP

SZ

Top row: *Phenakospermum guyannense, Ravenala madagascariensis, Strelitzia nicolai.*
Bottom row: *Strelitzia reginae, Orchidantha fimbriata, Canna glauca.*

red, linear-lanceolate, widely spaced, divaricate or reflexed, long-attenuate, glabrous or nearly so, middle ones ~15 cm long and 2 cm high at base; pedicels stout, villous. **Flowers** with perianth pale yellow or greenish-yellow, composed of 6 fused tepals, mostly 4–5 cm long. **Fruits** drupes, 1–3-seeded. **Distr:** W South America. **Ecol:** In wet forests or thickets; 600–1,400 m. **Zones:** 10–12.

STRELITZIACEAE

Genera 3, **species** 6. **Trees** or **shrubs**, herbaceous or woody, perennial, terrestrial; pseudostem formed from leaf bases; roots rhizomatous or cormous; sap clear. **Leaves** petiolate, simple, cauline, alternate, 2-ranked (distichous), sheathing at base, sheath open; lamina broad, margins entire, venation pinnate-parallel. **Inflorescence(s)** terminal, lateral or axillary, bracteate, cymes, subtended by large spathe (leaflike bract), enclosed by distichous boat-shaped bracts; flowers pedicellate; bracteoles present. **Flowers** bisexual, 3-merous, zygomorphic, large and showy; perianth present; tepals 6 (3 petal-like sepals and 3 petals), 1 petal free, 2 united, sometimes resembling a barbed arrow, often colored; stamens 6, 1 modified into a petal-like staminode; pistil with carpels 3, united, ovary inferior, locules 3, placentation axile, style 1, stigma terminal, capitate or trilobed. **Fruits** capsules, woody; seeds numerous per locule with a tuft of hair, round or ellipsoid, arillate. **Distr:** Madagascar, southern Africa; S America.

Phenakospermum

Syn: *Musidendron.*
Species 1. *Generic description same as specific.* **Cult:** Propagate by division. Requires a fertile, moist, humus-rich soil in full or partial sun.

P. guyannense (A. Rich.) Endl. ex Miq., *Bot. Zeitung* (Berlin) 3: 345, 1845. (Syn: *Musidendron amazonicum*, etc.). [Red Traveller's Palm, South American Traveller's Palm]. **Herbs**, arborescent, large, rhizomatous, erect; leafy shoots 3.2–10.3 m tall; pseudostem palmlike, 3–7.4 m tall, 11–17 cm Ø, covered with overlapping sheathing leaf bases. **Leaves** distichous, in 1 plane, deciduous (12 persisting); lamina oblong 1–2.7 × 0.47–0.85 m, base usually cordate, apex obtuse or truncate, adaxially dark green, abaxially green and glaucous; petiole green, 0.37–1.92 m long, 2–3.2 cm Ø. **Inflorescence** terminal, robust, erect, to 3.65 m tall; peduncle green, glaucous, 1.2–2 m long, 7–18 cm Ø; rachis 2.5–9 cm Ø, internodes 6–24 cm long; bracts 3–10, distichous, greenish, waxy, 23–44 × 18–34 cm, each subtending a <25-flowered cincinnus; floral bract opaque, cream-colored, persistent, 17–34 × 5.8–8.5 cm; pedicels 1–1.2 cm long, cream-colored. **Flowers** hermaphroditic; perianth to 28 cm long, with 2 whorls united at base into a prolonged, solid tube above ovary, tube bicarinate, white, 9 × 1.8 cm; calyx with 1 free abaxial sepal, 17 × 3.5 cm, and 2 free, slightly imbricate, adaxial sepals, 18 × 3.5 cm, creamy-white; corolla with 2 connate abaxial petals, 14 × 2.5 cm, and 1 free adaxial petal, 12 × 3 cm, creamy white with thin dark green margins; stamens 5, filaments linear, 8.5 cm long. **Fruits** capsules, woody, loculicidal, 15 × 8 cm; seeds to 400, 0.8–1 × 0.6–0.8 cm, aril with bright reddish orange threads. **Distr:** Tropical N and C South America east of the Andes. **Ecol:** In forests bordering seasonally flooded savannas. Fl. Jun–Nov. **Zones:** 9–12. **Info:** The flowers are bat pollinated in the wild (Kress and Stone 1993).

Ravenala

Syn: *Urania.*
Species 1. *Generic description same as specific.* **Cult:** Propagate by division. Requires a fertile, moist, humus-rich soil in full or partial sun.

R. madagascariensis Sonn., *Voy. Indes Orient.* 2: 223, 1782. (Syn: *Heliconia ravenala*, etc.). [Traveller's Palm]. **Herbs**, palmlike, solitary or clump-forming; trunk woody, unbranched, to 20 m tall, 10 cm Ø. **Leaves** in 1-plane, 2-ranked, arranged flat and fanlike, erect, glaucous; lamina bananalike, oblong to narrowly elliptic, 3–4 m long, base truncate, apex rounded, leathery; petiole ± same length as lamina. **Inflorescences** cymose, lateral, compact, glaucous; spathes 6–12, 2-ranked; bracts lanceolate, 25–30 cm long, leathery. **Flowers** 20–25 cm long, greenish-white; tepals tapering linear-lanceolate; calyx segments 3; corolla segments 3, 2 of which are larger; fertile stamens 6, linear, ± recurved; style forked at apex. **Fruits** capsules, opening by 3 valves; seeds kidney-shaped, 0.6–1.2 cm long, blackish, aril bright blue, shredding. **Distr:** Endemic to N and E Madagascar. **Ecol:** In moist forests. **Zones:** 11–12. **Info:** Flowers open at night and are pollinated by bats.

Strelitzia

[Bird of Paradise, Crane Flower].
Species 5. **Herbs**, perennial, often treelike; trunks woody or stems absent, proliferating offshoots around base and clump-forming. **Leaves** 2-ranked; lamina simple, blue-green, glaucous, leathery. **Inflorescences**

with spathes 1–5; bracts boat-shaped, often colored; flowers opening 1 at a time. **Flowers** showy; tepals 6, outer 3 free, white or orange, inner 3 fused into arrow-shaped keel that encloses stamens and style; stamens linear, 5 fertile, 1 sterile; style threadlike, stigma trilobed. **Fruits** capsules, woody, 3-valved; seeds globose, loculicidal, hard, aril bright orange, hairlike. **Distr:** S Mozambique to S Africa. **Cult:** Propagate from seed or by division. Requires moist but well-drained soil in full sun.

S. nicolai Regel and K. Koch, *Gartenflora* 7: 265, 1858. (Syn: *S. alba* ssp. *nicolai*, *S. quensonii*). [Giant Bird of Paradise, Natal Wild Banana, White Bird of Paradise]. **Herbs**, very large, treelike, perennial; trunk woody, 3–8(–12) m tall, 15–20 cm dbh, smooth, producing offsets at base, forming large clumps. **Leaves** in 1 flat plane, fanlike (spreading to 4 m wide); lamina bananalike, oblong to elliptic, gray-green with a waxy coating, leathery. **Inflorescences** at base of leaves, 50 cm long, 1–few-flowered; spathes 1–5; bracts boat-shaped, grayish or purplish-blue, with deep red edges. **Flowers** ~20 cm across; outer tepals white or pale mauve; inner tepals (keel) bright blue. **Fruits** capsules; seeds black, with a tuft of bright orange woolly aril on the lobe. **Distr:** S Mozambique to KwaZulu-Natal, E Botswana. **Ecol:** Mostly in coastal dune vegetation and in evergreen forests near the coast. **Zones:** 10–12.

S. reginae Banks ex Aiton, *Hort. Kew.* 1: 285, 1789. [Bird of Paradise, Crane Flower]. **Herbs**, clump-forming, evergreen, perennial, stemless, erect, 1–2 m tall, clump-forming with numerous offshoots around base. **Leaves** ± 2-ranked, in 1 flat plane, erect to ascending; lamina simple, somewhat paddle-shaped, oblong or lanceolate, 22–75 × 5–15 cm, bluish-green or grayish-green, leathery, glaucous; petiole to 1.8 m long. **Inflorescences** erect, usually taller than the leaves; spathe bract 1, boat-shaped, 10–22 cm long, at right angle to stalk and held ± horizontally, 1–6-flowered, flowers opening 1 at a time over a period of weeks. **Flowers** 9–15 cm across; outer tepals to 15 cm long, yellow to orange; inner tepals (keel), consisting of 3 fused petals, to 15 cm long, deep blue. **Fruits** capsules; seeds several, with red aril at base. **Distr:** South Africa (Cape Province to KwaZulu-Natal). **Ecol:** In coastal areas. Fl. throughout most of the year. **Zones:** 10–12. **Info:** When the pollinating birds alight on the spathe and flower, the stamens protrude and spread pollen on their feathers.

LOWIACEAE

Genus 1, **species** 15. **Herbs**, rhizomatous; stem very short. **Leaves** basal, distichous; lamina lanceolate or oblong, with several pairs of longitudinal veins parallel to a distinct midvein; petiole well-developed, base sheathing. **Inflorescence** axillary or arising from rhizomes, cymose or reduced to a solitary flower, bracteate. **Flowers** bisexual, zygomorphic; sepals 3, lanceolate; petals 3, very unequal; middle petal enlarged to form a lip; lateral petals small, apex often aristate; stamens 5, filaments short, anthers 2-celled, dehiscing by longitudinal slits; ovary inferior, triloculed, apex elongate, ovules numerous per locule, placentation axile, style slender, stigmas 3, laciniate, zygomorphic and dorsiventral. **Fruits** capsules, 3-valved, loculicidal; seeds numerous, arillate, aril trilobed. **Distr:** Cambodia, China, Indonesia (Borneo), Laos, Malaysia, Vietnam.

Orchidantha

Syn: *Lowia*, *Protamomum*.
Species 15. *Generic description same as familial.* **Cult:** Propagate from seeds or by division. Requires a fertile, moist, humus-rich soil in shade.

O. fimbriata Holttum, *Gard. Bull. Sing.* 25: 243 (pl. 1), 1970. **Herbs. Leaves** to 1.3 m long; petiole with sheath to 30 cm long; lamina to 1 m × 15 cm, base narrowly decurrent, margins undulate, light green. **Inflorescence** few-branched; prophyll ~1.8 cm long; intermediate bracts to 7 cm long; floral bracts dull purple, to 15 cm long. **Flowers** with sepals reflexed, 11–14 × 2 cm, gradually narrowed to the mucronate apex, margin strongly recurved during anthesis, later flattening, dark dull purple throughout or green with purple base; lateral petals cream with dark purple tip and base, to 2.7 × 0.6 cm, narrowly triangular, apex filiform; lip dark purple-brown, 10–12 × 5 cm, narrow with inflexed sides at base, widening abruptly beyond the stigma to a creamy-white, irregularly grooved and folded lamina, usually trilobed towards apex; lateral lobes reflexed, midlobe smaller, straight, fleshy, smooth, mucilaginous; stamens 2.4 cm long, filaments 3 mm long, purple-flushed, anthers cream-colored; ovary purple, 1.2–2 cm long, distal part 4 mm Ø, style ~2 cm long, purplish, wiry, stigma ~2.4 cm long, base white, divided towards apex into 3 narrow, dark purple, lobes shiny, distally fimbriate for 5–6 mm. **Fruits** ~8.5 cm long; seeds hairy, ~1 cm long, aril lobes to 1.5 cm long. **Distr:** Malesia (West Malaysia). **Ecol:** In wet tropical forests. **Zones:** 11–12.

CANNACEAE

Genus 1, **species** 10. **Herbs**, perennial, from rhizomes; aerial stems present, unbranched. **Leaves** cauline, 2-ranked, differentiated into basal sheath, petiole, and lamina; sheaths overlapping, supporting stem, open, ligule absent; summit of petiole not differentiated; lamina with lateral veins parallel, diverging from prominent midrib. **Inflorescences** 1 per aerial shoot, terminal on leafy shoot, pedunculate racemes or panicles of flowers or of 2-flowered monochasial cymes (cincinni); bracts of main axis subtending flowers or cincinni. **Flowers** bisexual, asymmetric; sepals and petals differentiated, sepals 3, distinct, petals 3, basally connate; fertile stamens 1, petal-like, anther marginal, 1-locular; staminodes (1–)3–4, petaloid, showy, unequal, anterior staminode (lip) often broader than posterior staminodes; ovary inferior, tricarpellate, trilocular, all locules fertile, ovules few to numerous per locule, style standing away from stamens and staminode, petal-like; stigmatic area shaped as marginal callosity, style, stamen, and staminodes basally connate into tube. **Fruits** capsules, sepals persistent in fruit; seeds without aril. **Distr:** Tropical and subtropical Americas.

Canna

Species 10. *Generic description same as familial.* **Cult:** Propagation is by seeds or by division of rhizomes. Needs fertile, wet soil or compost in full sun.

C. glauca L., *Sp. Pl.* 1: 1, 1753. [Louisiana Canna, Water Canna]. **Herbs**; rhizomes far-creeping, 0.5–1.5 cm Ø. **Leaves** sheath and lamina glaucous; lamina narrowly ovate, 28–70 × 1.5–14 cm, base cuneate, apex very gradually narrowed to acute. **Inflorescences** racemes, simple or occasionally branched, bearing 2-flowered cincinni, >10 flowers per inflorescence; primary bracts 10–30 cm; secondary bracts 5–20 cm; floral bracts persistent, (broadly) ovate-triangular, 0.7–2.5 × 0.4–1 cm, apex entire or irregularly lobed, glaucous; bracteoles (broadly) ovate-triangular, 0.3–2 × 0.4–0.8 cm, apex entire or irregularly lobed; pedicels to 5 mm long. **Flowers** pure yellow, 7.5–10 cm long; sepals narrowly elliptic-triangular, 1.2–2 × 0.3–0.5 cm; petals erect, 5–9 cm long, tube 1–2 cm long, lobes narrowly ovate, 4–7 × 0.7–1.1 cm; staminodes 4, narrowly elliptic to narrowly ovate, 7.5–10 cm long, free part 0.5–2.3 cm wide, apex sometimes slightly notched; lip strongly reflexed, linear. **Fruits** capsules, globose to ellipsoid, 2–5 × 2–4 cm; seeds brown, ovoid, 0.7–1 × 0.6–0.8 cm. **Distr:** S USA, C America, S America. **Ecol:** In moist places; 0–100 m. Fl. Jun–Aug. **Zones:** 9–11.

C. indica L., *Sp. Pl.* 1: 1, 1753. (Syn: *C. edulis*). [Indian Shot, Queensland Arrowroot]. **Herbs**; rhizome developed, much-branched; stems stout, to 2.5 m long. **Leaves** ovate-oblong to oblong, 30–60 × 10–20 cm, adaxially green, abaxially and at margin green or purple; leaf sheath green or purple; petiole short. **Inflorescence** a raceme of cincinni, flowers 1 or 2 per cincinnus; bracts light purple, ovate, ~8 mm long. **Flowers** with sepals pale purplish-green, lanceolate, ~1.5 cm long; corolla tube apricot-yellow, ~1.5 cm long, lobes erect, red to apricot-yellow with purplish apex, lanceolate, ~4 cm long; staminodes 2(or 3), erect, red with apricot-yellow base, oblanceolate, 4–5.5 × 0.7–1 cm, one with apex emarginate; lip red, sometimes yellow-spotted, lanceolate, ~4.5 cm long, margin revolute, apex emarginate; fertile stamen with filament reddish apricot-yellow, lanceolate,

Below: *Canna indica, Calathea burle-marxii, Calathea inocephala, Calathea loeseneri.*

Above: *Calathea lutea, Calathea roseopicta, Calathea warszewiczii, Calathea zebrina.*

~4.2 cm long, anther ~9 mm long; ovary green, globose, ~6 mm Ø, warty, style red to apricot-yellow, narrow, ~6 cm long, base adnate to staminode column. **Fruits** capsules, broadly ovoid, 1.2–1.8 cm long. **Distr:** Tropical America. **Ecol:** Fr. Sep–Oct. **Zones:** 8–12.

MARANTACEAE

[Arrowroot, Maranta, Prayer Plant].
Genera 31, **species** ~530. **Herbs**, perennial, rhizomatous; aerial stems branched or unbranched. **Leaves** in a basal cluster, both basal and cauline, (or strictly cauline), alternate, differentiated into basal sheath, petiole (occasionally absent), and lamina; sheaths often overlapping, supporting stem, open, ligule absent; summit of petiole distinctly differentiated as pulvinus; lamina with lateral veins parallel, diverging from prominent midrib. **Inflorescences** 1–several per aerial shoot, terminal or terminal and lateral on aerial stem (occasionally arising directly from rhizome), scapose or pedunculate (subsessile), compound, headlike or spikelike to diffuse or much branched, bracteate; each bract enclosing pair of flowers or cincinnus of paired flowers (solitary flowers in *Monotagma* and *Monophrynium*). **Flowers** bisexual, asymmetric, paired flowers mirror images of each other; sepals and petals differentiated, sepals 3, distinct (rarely connate basally in *Megaphyrnium*), petals 3, corolla, staminodes, and style basally fused into tube; fertile stamens 1, occasionally petal-like, anthers 1-locular; staminodes (2–)3–4, outer staminodes (0–)1–2, petal-like, inner staminodes 2, one fleshy (callose staminode), 1 hooded (cucullate staminode), bearing 1(2 in *Thalia*) appendage(s), subterminal or medial, fingerlike or flaplike; ovary inferior, tricarpellate, trilocular, 1(–3) locule(s) fertile, placentation basal, ovules 1 per locule, terminal portion of style and stigma enclosed in cucullate staminode, explosively released, stout (appendaged in *Thalia*), stigma cup-shaped. **Fruits** fleshy (rarely dry) capsules, rarely indehiscent berries or achenes, sepals often persistent in fruit; seeds with aril present (rarely absent). **Distr:** Tropics, especially American; not in Australia.

Calathea

Species ~100. **Herbs**, perennial; stems leafy at base, simple or rarely branched. **Leaves** vaginate, on long or short petiole, the sheath not produced at the apex. **Inflorescences** spicate; spikes sessile or pedunculate; peduncle naked or bearing a single leaf or rarely 2 leaves; bracts usually spirally arranged and closely imbricate, rarely distichous; pairs of flowers usually more than 3 in each bract; bractlets membranous or indurate, rarely none. **Flowers** with sepals equal or nearly so, small or large; corolla tube equaling or much longer than sepals; outer staminode solitary, usually large; ovary 3-celled. **Fruits** capsules, small, cartilaginous-thickened at apex, sometimes tricorniculate, crustaceous or harder, 3-celled; seeds trigonous, convex dorsally, arillate. **Distr:** Tropical America. **Cult:** Propagate by dividing the clumps. Requires moist, but well-drained soil in partial shade.

C. burle-marxii H.A. Kenn., *Canad. J. Bot.* 60: 2365, 1982. [Ice Blue Calathea]. **Herbs**, spreading, 19–40 cm tall. **Leaves** 1–6 per shoot; sheath and petiole green, purple-tinged, pulvinus olive green; lamina oblong to obovate-oblong, 9–21 × 4.3–9.6 cm, base unequal, obtuse, apex truncate, glabrous, adaxially hairy, whitish-green with alternating large and small, dark green, sickle-shaped markings along veins, abaxially dark purple. **Inflorescences** 1–2 per shoot, rectangular, 2.5–6 cm long; bracts 12–14, herbaceous, white, translucent, appearing glabrous. **Flowers** pale violet to reddish-violet; corolla white; stamens white. **Distr:** Brazil. **Zones:** 11–12.

C. inocephala (Kuntze) T. Durand and B.D. Jacks., *Index Kew.*, Suppl. 1: 72, 1902. (Syn: *C. barbillana, Phyllodes inocephala*). **Herbs**, robust, without aerial stems, 1.5–3 m tall. **Leaves** 2–5, all basal (rarely 1 cauline leaf below the inflorescence; lamina (45–)70–120 × (22–)30–55 cm, base rounded to subtruncate, adaxially green, glabrous, abaxially dark gray-green, midrib tomentose; pulvinus 9–23 cm, olive-green, glabrous. **Inflorescences** 1 per stem, basal, ovoid, 6–12 × 5–9 cm; bracts >30, spirally arranged, soon dying to brown and papery. **Flowers** open; sepals 2–2.5 cm, brownish; corolla tube 2.4–3.2 cm, lobes cream to yellow, thick, hairy; staminodes yellowish to creamy. **Fruits** trilobate, succulent, orange, sepals absent; seeds blue. **Distr:** S Mexico to S tropical America. **Ecol:** In humid rainforests, in shade or near fallen trees; 0–200 m. Fl. Feb, Apr–Dec. **Zones:** 10–12.

C. loeseneri J.F. Macbr., *Publ. Field Mus. Nat. Hist.*, Bot. Ser. 11: 51, 1931. [Tulip Calathea]. **Herbs**, perennial, evergreen, 30–75 cm tall. **Leaves** forming an upright, rosette; lamina broadly ovate to elliptic, 15 × 5.5 cm, apex pointed, adaxially bright green with light green feathery markings along midvein, surface corrugated, glabrous; petiole ~13 cm long, stiff, densely hairy. **Inflorescence** a spike, subturbinate (inverted umbrellalike), at apex of scape, 7–8 × 2–2.5 cm; scape ~30–60 cm tall, minutely pilose, leafless; bracts subcoriaceous, puberulous, 2–2.5 cm long, apex rounded, white with pink at margins and apex; apical bracts sterile, showy, leaflike, erect, narrowly ovate-lanceolate, 4–5 cm long, acute to acuminate, persistent. **Flowers** small, white with a violet lip. **Fruits** capsules, ~7 mm long, glabrous. **Distr:** W South America to N Brazil. **Ecol:** Usually occurring at low elevations and tolerates seasonal flooding experienced in the region. **Zones:** 10–12.

C. lutea (Aubl.) E. Mey. ex Schult., *Mant.* 1: 8, 1822. (Syn: *C. cachibou*, etc.). [Havana Cigar]. **Herbs**, robust, 1.6–5 m tall. **Leaves** variable, mostly basal and 1(2) cauline; lamina elliptic or ovate, (0.18–)0.3–1.5(–2) × (0.12–)0.2–0.6 m, base rounded or attenuate, apex usually rounded and apiculate, abaxially whitish; petiole very long, callus often 13 cm

Below: *Donax canniformis, Maranta arundinacea, Maranta arundinacea* (flower), *Maranta leuconeura.*

Above: *Maranta leuconeura* (another form), *Maranta lietzei*, *Marantochloa purpurea*, *Pleiostachya pruinosa*.

long, glabrous. **Inflorescences** spikes, elongate, slightly compressed, bronze-colored, cylindrical, 9–30 × 2.5–6 cm; scapes often with 1 or 2 leaves; bracts 7–18, spiraled, distichous, broad, 4 cm long, coriaceous, bronze or brownish, apical margins erect, outside subglabrous, margin pilose, inside glabrous. **Flowers** with sepals linear, 0.6–1 cm long, obtuse, pinkish; corollas pale yellow, tube 2.5–3.4 cm long, curved, lobes elliptic, 1.5 cm long, sericeous, purplish-brown; outer staminode pale yellow, obovate, 2 cm long; ovary villous at base. **Fruits** capsules, obovoid, orange, base hairy, usually 1-seeded; seed 8 mm long, grayish, aril white. **Distr:** Tropical America. **Ecol:** In open swamps and along banks of streams, forming large colonies. **Zones:** 11–12.

C. roseopicta (Linden) Regel, *Index Seminum* (LE) 1869: 12, 1869. (Syn: *C. illustris*, etc.). [Rosy Prayer Plant]. **Herbs**, stemless, 25–50 cm tall. **Leaves** 1–3, basal; broadly elliptic to nearly circular, 15–25 × 13–21 cm, base rounded, adaxially dark green with pale green, white, or pink line along midvein, and narrow white or pink band along margin, abaxially dark red-purple, glabrous. **Inflorescence** spike, subcylindrical, 9 cm long; peduncle 5–15 cm long; sterile bracts forming an umbrella at apex; fertile bracts 15–20, spirally arranged, green, suborbicular, 3 cm long, pubescent with long hairs, sterile bracts narrower, glabrous. **Flowers** with sepals 1.5 cm long, nearly equaling corolla tube; corolla white; staminode slightly shorter, ± purple. **Distr:** W South America to N Brazil. **Ecol:** In rainforests. Fl. Aug–Oct. **Zones:** 10–12.

C. warszewiczii (L. Mathieu ex Planch.) Planch. and Linden, *Cat. Pl. Exot.* 10: 3, 1855. (Syn: *C. bullii*, etc.). [Jungle Velvet Plant]. **Herbs**, evergreen, slender, 0.5–1.2 m high; stems leafy; internodes 0.35–1 m long. **Leaves** 3–7 basal, and (1)2(3) cauline; lamina ovate, oblong-ovate or elliptic, 8–48 × 4–20 cm, base obtuse, broadly obtuse or rounded, apex abruptly or shortly acuminate, minutely puberulent to glabrous, adaxially dark green, variegated with white, light green or yellowish-green in the shape of fish tails along the midrib, with a row of hairs along the midrib, abaxially purple, velvety-tomentose; petiole 0.5–2 cm long, velvety-tomentose, purplish; pulvinus 0.9–6.3 cm long, purple (rarely green); sheaths 9–15 cm long, obtuse, minutely puberulent, sometimes purplish. **Inflorescences** spikes, 1 per stem, ellipsoid-turbinate to cylindrical, 3–5.5–3.5–7.5 cm long; peduncle 5–12 cm long, minutely pilosulose; bracts 4–16, loosely sheathing, obovate, 2–4 × 3–5 cm, margins recurved, apices shortly apiculate, membranous, white or yellowish and glabrous at base, apices tinged with red and purple with age, tomentose. **Flowers** closed, 2.6–4 cm long, widely exserted, white; corolla tube 2.6–3.5 cm long, white, lobes white; outer staminodes white; inner staminodes white to cream, callose. **Fruits** obovoid, with persistent sepals, becoming purplish with age; seeds gray. **Distr:** C America. **Ecol:** In moist lowland and foothill forests; 0–700 m. Fl. Jun–Feb. **Zones:** 10–12.

C. zebrina (Sims) Lindl., *Edwards's Bot. Reg.* 14: t. 1210, 1829. (Syn: *Maranta zebrina*, etc.). [Zebra Plant]. **Herbs**, robust, to 1.2 m high. **Leaves** with a hairless pulvinus, ~4 cm long, borne directly on the broadly winged sheath; lamina oblong-lanceolate, to 70 × 30 cm, adaxially velvety dark green with yellow stripes from the midrib to the margins giving a pinnate appearance, and shorter stripes bifurcating the ends of the green areas, abaxially red-purple. **Inflorescence** a spike, many-flowered, rather large, almost spherical or ovoid, ~10 cm long, dense; peduncle to 35 cm long; bracts broad, green, violet-tinted. **Flowers** ~4.5 cm long, dark purple and white. **Distr:** WC and E Brazil. **Ecol:** In coastal rainforests; 0–700 m. Fl. Feb–Jun. **Zones:** 11–12.

Donax

Syn: *Actoplanes*, *Ilythuria*.

Species 1. *Generic description same as specific.* **Cult:** Propagate by division. Requires a fertile, moist, humus-rich soil in shade or partial sun.

D. canniformis (G. Forster) K. Schumann, *Bot. Jahrb. Syst.* 15: 440, 1893. (Syn: *Thalia canniformis*, etc.). [Bamban]. **Herbs**, branched, 1.5–5+ m tall; stems with basal, canelike internodes 1–2.5 m long; leaf sheath to 8–20 cm long. **Leaves** broadly ovate to elliptic, 1–30 × 4–45 cm, base rounded to obtuse, apex acuminate, abaxially villous along midvein; petiole 8–20 cm, thickened into a cylindric pulvinus. **Inflorescences** often branched at base, to 20 cm long, slender; bracts 9–11, green, obovate to lanceolate, 2.5–3.5 cm, deciduous after anthesis; pedicel thickened in fruit. **Flowers** with sepals white, triangular-ovate, 3–3.5 mm long, glabrous; corolla tube 0.8–1 cm long, lobes linear, 1–1.4 × 0.2–0.3 cm; outer staminodes white, obovate, 1.2–1.4 × 0.5–0.6 cm; callose staminode yellowish, ~1.5 cm long, apex usually emarginate; cucullate staminode light yellow at apex; fertile stamen ~8 mm long, with a narrowly triangular appendage; ovary sericeous. **Fruits** white to pale cream, 1–1.5 cm Ø, dry, indehiscent; seeds 1 or 2, brown, globose to complanate. **Distr:** Taiwan (Lan Yü), Indo-China to W Pacific. **Ecol:** In wet places of secondary forests and dry bamboo scrubs. Fl. Jun. **Zones:** 9–12.

Maranta

Species 32. **Herbs**, terrestrial, prostrate, scandent, or upright, usually dying back to rhizome during dry season, 0.1–1.5(–1.8) m high; rhizomes occasionally swollen, storing starch; stems branched or unbranched with basal and cauline leaves to highly branched above elongate, canelike stem (internode) with few or no basal leaves. **Leaves** homotropic (rarely antitropic); sheath usually auriculate, not spongy; lamina (patterned) plain green, ovate to elliptic. **Inflorescences** usually 2–several per shoot, spikelike, unbranched; bracts persistent, subtending 2–6 pedicellate flower pairs, herbaceous; secondary bracts absent; bracteoles usually absent; prophylls keeled, membranous. **Flowers** self-fertilizing (or outcrossing), staminodes white (purple); sepals persistent in fruit, >5 mm long, herbaceous; corolla white, tube (0.4–)1.2–1.4 cm long, corolla lobes unequal; outer staminodes 2, petal-like; callose staminode apex usually petal-like; cucullate staminode with 1 appendage, medial (subterminal), flaplike (fingerlike); stylar movement in single plane; style unappendaged. **Fruits** capsules, 1-seeded, obliquely ellipsoid, pericarp relatively thin, dehiscent; seeds brown, ellipsoid, rugose; perisperm canal 1, distally branched; aril conspicuous, white. **Distr:** Tropical and subtropical regions of S Mexico, West Indies, C America, S America (to N Argentina). **Cult:** Propagate by division or from basal cuttings. Requires fertile, moist soil in partial sun.

M. arundinacea L., *Sp. Pl.* 1: 2, 1753. (Syn: *M. sylvatica*). [Arrowroot, Obedience Plant]. **Herbs**, erect, 0.3–1.3 m tall; rhizomes tuberous, thickened; stem often branched above. **Leaves** basal 4–8, cauline 1–8; sheath auriculate, 4.5–31 cm long, margins and apex densely pilose to nearly glabrous; petiole often absent in cauline leaves, 3.5–20 cm long; pulvinus 0.2–1.8 cm long, adaxially tomentose; lamina ovate, 3.5–35 × 3–11 cm, basal leaves largest, adaxially sparsely minutely pilose, abaxially glabrous to minutely pilose. **Inflorescences** with bracts 1–2(–3), 2.4–6 cm long; common pedicel of flower pair 2.3–5.5 cm long. **Flowers** with sepals 1–1.7 cm long; corolla white, tube curved, 1.2–1.4 cm long; staminodes white; ovary tan, densely pubescent, rarely glabrous or nearly glabrous. **Fruits** green or tinged red-brown, ellipsoid or obovoid, 8 × 4–5 mm, aril yellowish; seed pale red, rugulose. **Distr:** S Mexico, West Indies, C America, N South America (W Ecuador). **Ecol:** In moist or wet thickets, sometimes on brushy rocky slopes, 0–1,300 m. **Info:** Cultivated widely for the edible starch from the rhizomes. **Zones:** 11–12.

M. leuconeura E. Morren, *Ann. Hort. Belge Étrangère* 24: 323, 1874. (Syn: *Calathea leuconeura*, etc.). [Prayer Plant]. **Herbs**, prostrate or decumbent, much-branched at base. **Leaves** numerous, ± horizontal; lamina elliptic to oblong-elliptic, ~12 × 9 cm, base subcordate, apex truncate to rounded, tip shortly pointed, adaxially deep green, veins silvery-white or red, midvein often jaggedly striped yellow-green or whitish, or triangular-spotted at vein bases, abaxially green or purple; leaf sheath clasping, 3.3–9 cm long; petiole grooved, often absent; pulvinus to 6 mm long, hairy above. **Inflorescence** spike, solitary, very slender; bracts 2, ± distant, green or purple-tinged. **Flowers** with corolla white or violet;

MS

Top row: *Stromanthe thalia, Thalia dealbata, Thalia geniculata, Alpinia abundiflora.*
Bottom row: *Alpinia calcarata, Alpinia haenkei, Alpinia luteocarpa* (flowers), *Alpinia mutica.*

sterile stamens white, purplish-marked. **Distr:** WC and SE Brazil. **Ecol:** In rainforests. Fl. Oct–Dec. **Zones:** 10–11.

M. lietzei (E. Morren) C.H. Nelson, Sutherl. and Fern.Casas, *Fontqueria* 51: 3, 1998. (Syn: *Calathea lietzei, Phyllodes lietzei*). **Herbs**, clump-forming, 30–70 cm tall; stem long, erect, often branching, glabrous. **Leaves** 1–4 per shoot; lamina ovate to oblong-lanceolate, 9.5–18.5 × 3.5–7.5 cm, base rounded, margins wavy, apex acute, adaxially dark green with alternate, irregular gray-green fishtail bands (giving the appearance of a pinnate pattern), glabrous, abaxially dark purple; sheath and petiole dark green to purple; petiole short; pulvinus olive green to purple-brown, minutely downy, 2 cm long, often borne directly on the sheath. **Inflorescence** terminal, cylindrical, ~4.5–6 × 1.3 cm, subtended by a usually solitary leaf (at flowering); peduncle 40 cm long; bracts 3–6, 2-ranked, spirally arranged, dark green, ± glabrous, shiny; inflorescences later produce stalked plantlets. **Flowers** white; sterile stamens fleshy, petal-like apically. **Distr:** C America, NE Brazil. **Ecol:** In seasonally dry places and coastal forests. **Zones:** 10–12.

Marantochloa

Syn: *Clinogyne.*

Species 15. **Herbs**, erect or sprawling, glabrous or pubescent; stems erect or spreading, branched or rarely simple, bearing many leaves or rarely only one leaf, glabrous or pubescent. **Leaves** strongly asymmetric, usually homotropic, the rounded side with a more deeply colored marginal zone beneath. **Inflorescence** lax or congested, each branch with about 4 nodes, with 1 or 2 fascicled cymules at each node; each cymule backed by a prophyll; mesophylls present; cymule pedunculate or subsessile; flowers pedicellate arranged at different levels on the cymule. **Flowers** with outer staminodes 2, petaloid; staminode hooded with a spurlike appendage; ovary triloculed, often pilose. **Fruits** capsules, tardily dehiscent, smooth, ± pubescent, not fleshy; seed crescent-shaped, with a small basal aril. **Distr:** Tropical Africa, W Indian Ocean. **Cult:** Propagate by division or from basal cuttings. Requires fertile, moist soil in partial sun.

M. purpurea (Ridl.) Milne-Redh., *Bull. Soc. Roy. Bot. Belgique* 83: 21, 1950. (Syn: *Clinogyne arillata*, etc.). **Herbs**, erect or sprawling, 1.5–3 m high; stems branched. **Leaves** strongly asymmetric, ± ovate, to 43 × 18 cm, often much smaller, acumen normally to the right of the midrib, sometimes pruinose or purplish below; petiole with the sheathing part to 42 cm long, terete part below the calloused part to 38 cm long, calloused part to 4 cm long. **Inflorescence** lax, to 45 cm long, branched; rachis and bracts pink; primary bract 2.5–4 cm long, spreading, together with 2 cymules, each cymule 2-flowered, pedunculate which it envelopes. **Flowers** pale pink to deep purple, with the 2 inner lobes of staminode bright yellow, ~1.8 cm long; ovary shortly pilose. **Fruits** capsules, subglobose, ~8 mm Ø, pilose, bright red, perianth persistent; seed brown, smooth with whitish aril. **Distr:** Tropical Africa. **Ecol:** In swamp forests, or forests subject to flooding, gallery forests of the savanna; 1,110–1,500 m.

Pleiostachya

Species 3. **Herbs**, terrestrial, (0.9–)1.1–2.6 m tall. **Leaves** homotropic, variable, mostly basal, 1(or 2) on stem, 2-ranked; lamina ovate-oblong, apex slightly off-center; petiole often rigid at junction with sheath; pulvinus with basal rings. **Inflorescences** 1–12, narrowly ovate to oblong, laterally flattened; flowers paired, sessile, opening spontaneously; bracts 2-ranked, folded, strongly overlapping, stiff, persistent; bractlets 2 per flower. **Flowers** usually self-pollinating; sepals membranous, transparent and translucent; corolla tube elongate, 2.3–5 cm long; outer sterile stamen 1; ovary with 2 sterile cells and 1 fertile. **Fruits** capsules, oblong; seed 1, 3-angled, brown to dull black with shiny lines, aril basal, white. **Distr:** Mexico to Ecuador. **Cult:** Propagate from cuttings or by division. Requires a moist, humus-rich soil mix in shade.

P. pruinosa (Regel) K. Schum., *Pflanzenr.*, IV, 48: 165, 1902. (Syn: *Ischnosiphon pruinosus*, etc.). [Wheat Calathea]. **Herbs**, robust, clumping, (0.9–)1.1–2.5 m tall. **Leaves** variable, cauline 1 or 2; sheath green, tinged purple, hairy; pulvinus base with raised thickened ring, dark maroon, glabrous or hairy; lamina ovate-oblong, (23–)54–90 × (8.5–)20–35 cm, base rounded, apex asymmetric, adaxially dark green, abaxially dark red-purple, ± tomentose. **Inflorescences** 5–16(–23); 6–18 × 1.2–2.5 cm; bracts 2-ranked, flattened, 3.7–4.6 cm long, yellow-green, waxy. **Flowers** with sepals 2.6–3 cm long, white, translucent or transparent; corolla yellow, tube 2.5–2.9 cm long, lobe purplish; outer sterile stamen purple. **Fruits** tan; seed black, ~6 mm long. **Distr:** Mexico to Ecuador.

Ecol: In lowland evergreen rainforests in partial shade or exposed areas, sometimes in secondary and disturbed forests; 0–700(–900) m. **Zones:** 9–12.

Stromanthe

Species 15–20. **Herbs** perennial, tall and coarse, arising from thick rhizomes, often branched. **Leaves** large, thin but stiff and paperlike, cauline ones 2 or more. **Inflorescences** simply racemose or paniculate, open and lax or dense and congested, flowers in numerous pairs; bracts rather remote or sometimes crowded, complicate, finally deciduous; bractlets none. **Flowers** with sepals 3, free, lanceolate or broader; corolla tube short, lobes oblong; stamen tube very short; outer staminodia shorter or longer than inner ones; anther usually apiculate, filament appendaged; ovary 1-celled, smooth or tuberculate, glabrous or villous. **Fruits** capsules, 3-valvate; seed rugulose, arillate. **Distr:** Tropical America. **Cult:** Propagate by division or from cuttings.

S. thalia (Vell.) J.M.A. Braga, *Eugeniana* 21: 23, 1995. (Syn: *Heliconia thalia*, etc.). [Never-never Plant]. **Herbs**, robust, to 1.5 m tall; juvenile plants with basal leaves. **Leaves** antitropic, 3–10, 2-ranked, fanlike, clustered above the canelike stem; lamina ovate-lanceolate, to 53 × 12 cm, base obtuse to rounded, apex rounded and with a point, leathery, glabrous but margins with tiny hairs, adaxially dark glossy green, abaxially dark reddish-purple; leaf sheaths with apex lobed, light reddish and hairy along the margin; petiole absent in cauline leaves; pulvinus ~5 cm long, olive-green or tinged maroon. **Inflorescence** paniculate, with short branches; rachis, branches, bracts scarlet or bright red; scape long, rigid, with terminal inflorescences and stem leaves; sheaths broad, membranous, pinkish or deep red, margins often hairy. **Flowers** small; sepals scarlet; petals white; sterile stamens white. **Distr:** Brazil (Rio de Janeiro). **Ecol:** In understory of montane rainforests; ~1,000 m. Fl. Sep–May. **Zones:** 10–12.

Thalia

Species 6. **Herbs**, aquatic, coarse, upright, often dying back to rhizome in winter, 1–3.5 m tall, often waxy-pruinose; rhizome not evidently enlarged or specialized for starch storage; stems unbranched below inflorescence; leaves few, all basal or rarely with single cauline leaf borne above elongate (0.7–2.5 m long) internode. **Leaves** homotropic, petiolate; sheath not auriculate, spongy, containing prominent air spaces; lamina plain green, ovate to elliptic, thin but stiff. **Inflorescences** paniculate, branched, branches short and upright to elongate and usually arching; rachis internodes conspicuously zigzagged; bracts deciduous, both bracts and prophylls falling with flower if fruit not set, leaving proximal portion of rachis bare, each bract subtending 1 flower pair, herbaceous to leathery; prophylls not evidently keeled, membranous; secondary bracts absent; bracteoles absent. **Flowers** self-fertilizing, rarely outcrossing, pale to dark purple (corolla and staminodes); sepals persistent in fruit, 0.5–3 mm, membranous; corolla tube 1–6 mm, corolla lobes subequal to strongly unequal; outer staminode 1, petal-like, showy; callose staminode mainly fleshy, narrow apical rim petal-like; cucullate staminode with 2 appendages, subterminal, fingerlike; stylar movement helical when tripped; styles with 1 appendage, elongate, straplike. **Fruits** capsules, nutlike, 1-seeded, nearly globose to ellipsoid, exocarp membranous, pericarp thin, indehiscent; seeds dark brown, nearly globose or ellipsoid, smooth; perisperm canals 2, curved; aril reduced, at base. **Distr:** Tropical America extending northward into southern USA (5 spp.), C America (1 sp); Africa (1 sp). **Cult:** Propagate by division. Requires shallow water, or in moist to wet loamy soil in full sun.

T. dealbata Fraser, *Thaiszia*: t. 1 (1794). (Syn: *T. barbata*). [Alligator Flag, Hardy Canna, Powdery Thalia, Water Canna]. **Herbs**, 0.7–2.5 m. **Leaves** basal 2–5, cauline 0; sheath pruinose, glabrous; petiole pruinose, glabrous; pulvinus yellowish-brown to red or purple-brown, 0.6–2.5 cm, glabrous; lamina ovate, occasionally narrowly elliptic, 17–55 × 7–22 cm, firm, stiff-papery, base rounded, rarely obtuse, apex acuminate, abaxial surface pruinose, appearing whitish, glabrous, adaxial surface pilose to nearly villous in basal 0.5–1 cm of lamina, including midrib, at apex, and along margin of broader side. **Inflorescences** erect, tightly clustered, compact array, 9–31 × 7–18 cm; scapes 0.5–1.9 m; rachis pruinose; internodes 2–3 mm; bracts markedly pruinose, appearing whitish, red-brown to red-purple beneath waxy coating, orbiculate, strongly cupped, 0.8–1.5 cm, stiff, leathery, glabrous. **Flowers** with sepals 1.5–2.5 mm; outer staminode dark purple, 1.2–1.5 × ~0.6 cm; callose staminode base white to pale purple, margins and apex dark purple, apical rim reduced, petal-like. **Fruits** nearly globose to broadly obovoid, 0.9–1.2 × 0.8–1.1 cm; seeds dark brown to black, nearly globose to broadly ellipsoid, 0.8–1 × 0.7–0.9 cm. **Distr:** Southern USA. **Ecol:** In coastal plains, swamps, streamsides, ponds, etc. Fl. May–Sep. Fr. Jun–Oct. **Zones:** 9–12.

T. geniculata L., *Sp. Pl.* 2: 1193. 1753. (Syn: *T. divaricata*, *T. trichocalyx*, *T. welwitschii*). [Arrowroot, Fire Flag, Red-stemmed Thalia]. **Herbs**, 1–3.5 m tall. **Leaves** basal 2–6, cauline 0–1(–2); sheath green or occasionally red-purple, glabrous; petiole green or occasionally red-purple, glabrous; pulvinus caramel-colored, olive-green, or red-purple, 0.3–2.5 cm, glabrous; lamina ovate to narrowly ovate, 19–60 × 4–26 cm, firm, stiff-papery, base rounded to subtruncate, apex acute to acuminate, occasionally obtuse with acuminate tip, abaxial surface green, faintly pruinose, glabrous, adaxial surface glabrous. **Inflorescences** lax, broadly spreading to pendent, paniculate array, to ~0.6 × 1 m; scapes 0.8–2.5 m long; rachis not pruinose; internodes 0.5–2 cm long; bracts not pruinose, green or streaked or tinged with purple, narrowly ovate, 1.3–2.8 cm long, herbaceous, sparsely to densely villous. **Flowers** with sepals 0.5–2 mm long; outer staminode faint lavender to purple, 1.5–2 × 0.5–1 cm; callose staminode base yellow, apex purple, apical rim, reflexed, petal-like. **Fruits** ellipsoid, 0.9–1.2 × 0.6–0.7 cm; seeds smooth, dark brown to black, ellipsoid, 0.7–1 × 0.5–0.6 cm. **Distr:** Tropical Africa, SE USA to tropical America. **Ecol:** In lowlands, in ponds, swamps, etc. Fl. Jun–Dec. Fr. Aug–Jan. **Zones:** 9–12.

ZINGIBERACEAE

Genera 46–52, **species** 1,075–1,300. **Herbs**, perennial (annual), from tuberous rhizome (tuberous roots); aerial stems present or absent, unbranched. **Leaves** basal or basal and cauline, in 2 ranks, differentiated into basal sheath, petiole, and lamina, or petiole absent; sheaths often overlapping to form pseudostem, open, ligule present, summit of petiole not differentiated; lamina with lateral veins parallel, diverging from prominent midrib. **Inflorescences** 1 per aerial shoot, pedunculate racemes, or 2–7-flowered monochasial cymes (cincinni); bracts of main axis subtending flowers or cincinni. **Flowers** bisexual, bilaterally symmetric; sepals and petals differentiated, sepals 3, connate, petals 3, connate, sepals free from petals; fertile stamens 1, not petaloid, anthers 2-locular; staminodes 2, connate, forming petaloid structure lip, opposite fertile stamen, sometimes also with 2 petaloid or rudimentary distinct staminodes; ovary inferior, tricarpellate, triloculed, fertile, ovules many per locule, style filiform, stigma funnelform, sometimes slightly bilobed. **Fruits** capsules or berries, sepals often persistent; seeds arillate. **Distr:** Tropical and subtropical regions.

Alpinia

Species ~230. **Herbs**; rhizomes creeping, thick; pseudostems many, well-developed, rarely absent. **Leaves** many, rarely 1–4; lamina oblong or lanceolate. **Inflorescence**, terminal, panicle, raceme, or spike, dense or lax, covered by 1–3 spathulate involucral bracts when immature; bracts (when present) open to base, rarely hooded, each subtending 1 flower or a cincinnus of 2–many flowers; bracteoles open to base or tubular, rarely hooded, sometimes absent. **Flowers** with calyx usually tubular, sometimes split on 1 side; corolla central lobe ± hooded, usually wider than lateral lobes; lateral staminodes small or absent, subulate or toothlike, adnate to base of lip; lip often showy, usually larger than corolla lobes, sometimes inconspicuous, margin variously lobed or entire; filament present or absent, connective crested or not; ovary usually triloculed and placentation axile, stigma usually well expanded, sometimes clavate, rarely geniculate; stylodes often massive. **Fruits** capsules, usually globose, dry or fleshy, indehiscent or irregularly dehiscent; seeds numerous, often angled, arillate. **Distr:** Tropical and subtropical Asia; Australia, Pacific islands. **Cult:** Propagate from seeds or by division. Requires a fertile, moist soil in full or partial sun.

A. abundiflora Burtt and R.M. Sm., *Notes Roy. Bot. Gard. Edinburgh* 34: 179, 1975. (Syn: *Elettaria floribunda*, etc.). **Herbs**; stem leafy, 3+ m tall; ligule 1.5 cm long, obtuse, membranous. **Leaves** usually sessile, occasionally petioles to 2 cm long; lamina oblong or narrowly lanceolate, to 60 long, 12 cm wide (petioled leaves to 5.5 cm wide), base attenuate, apex acuminate. **Inflorescence** capitate, truncate or globose, to 10 × 10 cm, either separate from leafy stem on a leafless, sheath-covered, 20–40 cm peduncle or more rarely, terminal on a leafy shoot; sheaths narrowly lanceolate, 15 × 2 cm, often overtopping inflorescence, forming a showy red involucre; sterile bracts red, 1- or 2-seriate, ~3 × 4 cm; fertile bracts 3.5 × 2 cm, subtending 1–7 flowers; bracteoles open or tubular, ~2 × 1.5 cm. **Flowers** with calyx tubular, 1–1.5 cm long, saccate in upper half, unilaterally split; corolla tube slightly longer, dorsal lobe broadly ovate, 7 × 5 mm, laterals narrower; lateral staminodes absent; lip white, pink-striped, obovate, ~1 cm long, trilobed; anther to 5 mm long; ovary ~4 mm long, glabrous. **Fruits** capsules, spherical-oblong. **Distr:** S India, Sri Lanka. **Ecol:** In montane forest margins; to 2,000 m. Fl. Feb–Nov. **Zones:** 10–12.

A. calcarata Roscoe, *Trans. Linn. Soc. London, Bot.* 8: 347, 1807. (Syn: *Languas calcarata*). [Indian Ginger, Snap Ginger]. **Herbs**; pseudostems to 1.3 m tall. **Leaves** sessile; ligule 0.8–1.5 cm long, glabrous, apex obtuse; lamina linear-lanceolate, 20–32 × 2–3.5 cm, base attenuate, margins with well-spaced short bristles, apex acuminate, caudate-mucronate, glabrous. **Inflorescence** panicles <10 cm long; rachis slightly velvety; proximal branches 3- or 4-flowered; bracteoles oblong, to 1.7 cm long, membranous, apex obtuse; pedicel ~3 mm long. **Flowers** with calyx to 1.2 cm long, split on 1 side, pubescent, apex 3-toothed corolla tube white, ~9 mm long, lobes oblong, ~2.2 cm long; lateral staminodes red, subulate, ~3 mm long, adnate to base of labellum; lip white with rose red and purple streaks, obovate, 2.7–3.5 × 1.5–2 cm, emarginate, filament ~1.3 cm long, anther 5–7 mm long; ovary 3–4 mm Ø, sericeous. **Fruits** capsules, red, globose. **Distr:** China, India, Myanmar, Sri Lanka. **Ecol:** In forests. Fl. May. **Zones:** 9–12.

A. haenkei C. Presl, *Symb. Bot.* 1: 66, 1832. (Syn: *A. malaccensis*, etc.). **Herbs**, leafy shoots 1.5–

Top row: *Alpinia malaccensis, Alpinia purpurata.*
Bottom row: *Alpinia vittata, Alpinia zerumbet.*

2 m tall; stems red-mottled, hairy. **Leaves** ± lanceolate, 50–65 × 7.5–10 cm, margins undulate, hairy (appressed-hairy or silky) on both sides or sometimes glabrous. **Inflorescences** terminal, erect, 25–50 cm tall, ± densely hairy; bracts absent; bracteoles 3.8 cm long. **Flowers** with corolla white, silky-hairy, lobes 3–4.5 cm long, center red, yellow-mottled, margin yellow, ruffled. **Fruits** ± globose or subglobose, 1.9–2.5 cm Ø, pubescent. **Distr:** Java, Philippines. **Ecol:** In primary forests, at low to medium elevations. **Zones:** 11–12. **Info:** The fruits and seeds are used for medicinal purposes.

A. luteocarpa Elmer, *Leafl. Philipp. Bot.* 10: 3803, 1939. [Bamboo Ginger]. **Herbs**, forming small tussocks; stems erect or nearly so, ~1 m high; ligule short, truncate. **Leaves** sessile, many, in alternate rows, larger towards apex of stem, basal ones shorter but broader, sometimes bractlike; lamina narrowly oblong, 17 × 5 cm, base obtuse, margins entire, apex gradually tapering, acuminate, glabrous, adaxially dark green, abaxially green, purple-tinged or wholly purple. **Inflorescences** terminal on leafy stem, 1–3+ cincinni, each cincinni 1–2-flowered, subtended by a bract; flowers ± erect, but orientation irregular. **Flowers** glabrous, waxy; hypanthium green; calyx tubular, ~2–4 cm long, red, split on 1 side; corolla white, pink-tinged, outside the apices with a reddish midline; lip ± similar to other 2 petals, curved. **Fruits** capsules, dull yellow, elliptic to ovately elliptic or subglobose, 3 × 2 cm; seeds numerous. **Distr:** Philippines. **Ecol:** In woods along steep ravines; 0–600 m. **Zones:** 9–12. **Info:** This species does not have a Latin description and is poorly known, but cultivated or naturalized in many parts of SE Asia. *A. luteo-carpa* is the orthographic equivalent, and the original name under which it was first published.

A. malaccensis (N.L. Burman) Roscoe, *Trans. Linn. Soc. London*, Bot. 8: 345, 1807. (Syn: *Maranta malaccensis*, etc.). [Malacca Galanga, Ring Malacca]. **Herbs**, large, rhizomatous; pseudostems ~3 m or more tall; ligule 2-cleft, to 1 cm long, slightly tomentose. **Leaves** oblong-lanceolate or lanceolate, to 90 × 15 cm, base acute, apex acuminate, abaxially pubescent; petiole ~2 cm long, grooved. **Inflorescences** racemes, erect, to 35 cm long; rachis stout, densely yellow-pubescent; bracteoles white, broadly elliptic, 3.5–4 cm long; pedicel ~7 mm long, densely yellow-pubescent. **Flowers** with calyx campanulate, ~1.5 cm long, densely sericeous; corolla white, sericeous, tube ~1 cm long, lobes oblong-lanceolate, 2.5–3 cm long; lateral staminodes absent; labellum yellow with red stripes, ovate, ~3.5 cm long, apex emarginate; stamen ~2.4 cm long; ovary villous. **Fruits** capsules, yellow, globose, ~2 cm Ø, irregularly dehiscent. **Distr:** Assam to China (Yunnan) and peninsular Malaysia. **Ecol:** Fl. Apr. **Zones:** 10–12.

A. mutica Roxb., *Asiat. Res.* 11: 354, 1810. (Syn: *Renealmia mutica*, etc.). [Small Shell Ginger]. **Herbs**, rhizomatous; rootstock not tuberous; stem leafy, 1.3–2 m tall. **Leaves** alternate; lamina oblong-lanceolate to narrowly lanceolate, 30–45 × 5–7.5 cm, abaxially glabrous; petiole short. **Inflorescences** panicles, slightly compound, 15–23 cm long; rachis stout, very pubescent; lower branches 2–3, short; bracts oblong. **Flowers** large; calyx tube funnel-shaped, 1.25 cm long, white, bordered with pink; corolla segments oblong, 2.5 cm long, white; lip suborbicular, ~4 × 4 cm, trilobed, colorfully variegated red and yellow, base not spurred, margins much incurved. **Fruits** capsules, globose, 2.5 cm Ø, ripening to bright yellow or orange; seeds many per cell. **Distr:** S India, Indo-China to Malesia. **Ecol:** In forests. Fl. May–Jul. Fr. Oct–Feb. **Zones:** 10–12.

A. purpurata (Vieill.) K. Schum., *Pflanzenr.*, IV, 46: 323, 1904. (Syn: *A. grandis*, etc.). [Red Ginger]. **Herbs**, clump-forming, 1–5(–7) m tall; stems formed of overlapping leaf sheaths. **Leaves** alternate, ranked along stem; lamina oblong to narrowly elliptic, 30–75 × 10–20 cm, apex pointed; petiole short to absent. **Inflorescence** terminal, spike, conelike, 15–30 cm long, eventually becoming longer, showy; flowers often hidden by bright red (cultivars with white, pink and pale pink are commonly grown) bracts; bracts persistent, 3 cm long, becoming 4–6 cm long in fruit, spirally arranged. **Flowers** white, tubular, irregular, 2–2.5 cm long, ephemeral. **Fruits** dry, round, 2–3 cm across. **Distr:** Malesia to Pacific Islands. **Ecol:** In wet forests; 0–600 m. **Zones:** 11–12. **Info:** The old inflorescences usually produce plantlets, which can be used for propagation.

A. speciosa (Wendl.) Schum., *Fl. Kaiser-Wilhelmsl.* 29. 1887. (Syn: *A. nutans, Languas speciosa, Zerumbet speciosum*). **Herbs**, to 1 m tall, forming dense clumps. **Leaves** short petiolate; lamina base narrow, apex acuminate, to 70 × 10 cm, glabrous except on the puberulent margin; ligule 1 cm long or less. **Inflorescence** panicles to 30 cm long; bracts 2–3-flowered, broadly elliptic, whitish, about 3 cm long. **Flowers** with calyx almost 2 cm long, white; corolla white, tube 1 cm long, lobes broadly elliptic, obtuse, 3 cm long, apex red; lip broadly ovate, 4 cm long, subtrilobate, spotted and streaked with red and yellow. **Fruits** capsules globose, 2 cm Ø. **Distr:** China, Japan. **Ecol:** In lowlands and also at mid elevations. **Zones:** 9–11.

A. vittata W. Bull, *Cat.* 83: 4, 1873. (Syn: *A. sanderae, A. tricolor, Guillainia vittata*). [Variegated Ginger]. **Herbs**, terrestrial, rhizomatous, perennial, robust, erect; leafy shoots 2–3.6 m tall. **Leaves** sessile, simple, alternate, 2-ranked; lamina, elliptic-lanceolate, 38–64 × 10–15 cm, green variegated white. **Inflorescence** racemose panicle, pendent, 18–25 cm long; rachis reddish turning green; bracts 3.2 × 4.5 cm long, pink-variegated, persistent, enclosing 1–2(–5) flowers; bractlets 2–2.5 cm long, pink-variegated. **Flowers** with corolla white, lobes 3, fused to halfway, 1.5–2 cm long, often reflexed; lip white, oblong 1.5–2 cm long; stamen shorter than lip, white. **Fruits** capsules, many-seeded. **Distr:** Bismarck Archipelago to Solomon Islands. **Ecol:** Fl. throughout most of the year. **Zones:** 11–12.

A. zerumbet (Persoon) B.L. Burtt and R.M. Smith, *Notes Roy. Bot. Gard. Edinburgh.* 31: 204, 1972. (Syn: *Costus zerumbet, A. nutans*, etc.). [Shell Ginger, Pink Porcelain Lily]. **Herbs**, pseudostems 2–4.5 m long; ligule 0.5–1 cm long. **Leaves** lanceolate, 30–80 × 5–15 cm, base attenuate, apex acuminate; petiole 1–1.5 cm long. **Inflorescence** panicles, drooping, to 30 cm long, enclosed by 2 long sheaths when young; rachis purple-red, velvety; branches very short, 1- or 2(or 3)-flowered; bracteoles enclosing flower buds, white, apex pink, elliptic, 3–3.5 cm long, glabrous; pedicel 1–2 cm long. **Flowers** with calyx white, subcampanulate, ~2 cm long, split on 1 side; corolla tube shorter than calyx, lobes milky-white, oblong, ~3 cm long, apex pink; central lobe larger than lateral ones; lateral staminodes subulate, ~2 mm long; lip yellow, with purple-red stripes, broadly ovate-spathulate, 4–6 cm long, apex crisped; stamen ~2.5 cm long; ovary hirsute. **Fruits** capsules, vermilion, globose, ~2 cm Ø, ribbed, apex with persistent calyx; seeds angled. **Distr:** Tropical and subtropical Asia. **Ecol:** Fl. Apr–Jun. Fr. Jul–Oct. **Zones:** 10–12.

Boesenbergia

Syn: *Curcumorpha.*

Species 66. **Herbs**, rhizomatous; rhizomes tuberous or elongate. **Ligule** bilobed. **Leaves** basal or cauline, petiolate; ovate, oblong, or lanceolate. **Inflorescences** terminal on pseudostems or on separate shoots arising from rhizomes; bracts 2-ranked, apical one developing first, each subtending 1 bracteole and 1 flower. **Flowers** with calyx tubular; corolla tube exserted from calyx, slender, lobes subequal; lateral staminodes petaloid, usually wider than corolla lobes; lip obovate or broadly oblong, larger than corolla lobes and lateral staminodes, strongly concave, base narrow, grooved, margin crisped, apex entire or 2-cleft; filament erect, anther locules parallel, dehiscing by slits or apical pores, base without spur; ovary usually triloculed. **Fruits** capsules, oblong, 3-valved; seeds black, aril laciniate. **Distr:** Tropical and subtropical Asia. **Cult:** Propagate from seeds or by division. Requires a fertile, humus-rich, moist soil in full or partial sun.

B. rotunda (L.) Mansfield, *Kulturpflanze.* 6: 239, 1958. (Syn: *Curcuma rotunda*, etc.). [Chinese Key, Lesser Ginger, Tropical Crocus]. **Herbs**, to 50 cm tall; rhizomes bright yellow, ovoid-globose, strongly aromatic; roots robust; ligule 2-cleft, ~5 mm long. **Leaves** 3 or 4; leaf sheath red; petiole 7–16 cm long, channeled; lamina green on both surfaces, ovate-oblong or elliptic-lanceolate, 25–50 × 7–12 cm, glabrous except for sparsely hairy midvein abaxially, base rounded to cuneate, apex apiculate. **Inflorescences** terminal on pseudostems, appearing from within apical leaf sheaths, subsessile, 3–7 cm long; bracts lanceolate, 4–5 cm long. **Flowers** aromatic; calyx 1.5–2 cm long, apex 2-cleft; corolla pink, tube 4.5–5.5 cm long; lobes oblong, 1.5–2 cm long; lateral staminodes light pink, obovate, ~1.5 cm; lip white or pink, purple-striped, fiddle-shaped, 2.5–3.5 cm long, concave, margin slightly crisped, apex entire, filament short; connective appendage reflexed, 2-cleft, 1–3 mm long. **Distr:** India, Indonesia, Malaysia, Sri Lanka. **Ecol:** In dense forests; ~1,000 m. Fl. Jul–Aug. **Zones:** 11–12.

Curcuma

Species 91. **Herbs**, rhizomatous, tuberous; rhizomes branched, fleshy, aromatic. **Leaves** basal; lamina broadly lanceolate or oblong, rarely narrowly linear. **Inflorescence** a terminal spike on pseudostems or on separate shoots; peduncle erect; bracts connate halfway and forming pouches, apices free, each subtending a 2–7-flowered cincinnus, apical bracts often differently colored, large, sterile, forming an apical coma; bracteoles free to base. **Flowers** with calyx usually shortly tubular, split on 1 side, apex 2- or trilobed or -toothed; corolla funnelform, lobes ovate or oblong, subequal or central lobe longer, apex mucronate; lateral staminodes petaloid, basally adnate to filament and lip; lip with a thickened, central portion and thinner, lateral lobes overlapping with lateral staminodes; filament short, broad, anther versatile, base usually spurred; ovary triloculed. **Fruits** capsules, ellipsoid, 3-valved, dehiscent. **Distr:** Tropical and subtropical Asia to N Australia. **Cult:** Propagate by

Top row: *Boesenbergia rotunda, Curcuma aeruginosa, Curcuma alismatifolia, Curcuma aromatica.*
Bottom row: *Curcuma longa, Curcuma rhabdota, Curcuma zedoaria.*

SZ JB

SZ

Above: *Elettaria cardamomum* (seed pods), *Elettariopsis smithiae*, *Etlingera corneri*, *Etlingera elatior*.

division or from seeds. Requires humus-rich, moist but well-drained soil in partial shade.

C. aeruginosa Roxb., *Asiat. Res.* 11: 335, 1810. [Pink and Blue Ginger]. **Herbs**, terrestrial, 1.3–2 m high; rootstock large, tuberous; sessile tubers, ovoid, short, >2.5 cm thick, inside green-veined; stipitate tubers pure white inside. **Leaves** tufted; large, oblong, 60–90 × 15–23 cm, glabrous, green, faintly tinged with reddish-brown down the middle; petiole 60–90 cm long. **Inflorescence** erect, 15–20 cm long, 6.5–7.5 cm Ø; floral bracts green, ovate, 5 cm long, obtuse; bracts of coma (upper bracts) pink, oblong, longer. **Flowers** pale yellow, shorter than the bracts; corolla tube 2.5 cm long, lobes pale red; lip deflexed, 1.2 cm across. **Distr:** Indo-China to W Malesia. **Ecol:** In tropical rainforests (often on riverbanks or forest margins), teak forests and grasslands; 400–750 m. Fl. Apr–Jun. **Zones:** 10–12.

C. alismatifolia Gagnep., *Bull. Soc. Bot. France* 49: 259, 1903. [Siam Tulip]. **Herbs**, terrestrial, rhizomatous, 45–60 cm tall; rhizome white to yellowish, root tubers present. **Leaves** ~4, tufted; lamina narrowly elliptic, 25–36 × 5–7.5 cm, glabrous. **Inflorescence** borne from the leaf tuft, cylindrical 5–10 × (4–)12.5–15 cm; fertile bracts green or bronze-colored, 4–5 cm long, ± united, enclosing 4 flowers; coma bracts lavender-pink, free portion 5–6.4 cm long; bractlet white. **Flowers** with corolla ovate-elliptic, ~2.5 cm long, white; lip obovate, 1.9 × 1.5 cm, purple, base yellow-banded; lateral sterile stamen elliptic-lanceolate, 1.5 × 1 cm, white, anther 6.4 mm long. **Distr:** Indo-China. **Ecol:** In open grasslands. Fl. Jun–Aug. **Zones:** 10–11.

C. aromatica Salisbury, *Parad. Lond.* 1: t. 96, 1807. [Cochin Turmeric]. **Herbs**, rhizomatous, ~1 m tall; rhizomes yellow inside, elliptic or narrowly so, fleshy, aromatic; roots with fusiform tubers. **Leaves** oblong, 30–60 × 10–20 cm, base attenuate, apex narrowly caudate, adaxially glabrous, abaxially pubescent; petiole equaling lamina. **Inflorescence** a spike, cylindric, ~15 × 8 cm, borne on separate a shoot; fertile bracts pale green, ovate, 4–5 cm; coma bracts white, red-tinged, narrowly oblong, apex mucronate, pubescent. **Flowers** with calyx sparsely hairy, 0.8–1.5 cm long; corolla tube funnelform, 2.3–2.5 cm long, villous at throat, lobes pinkish-white, oblong, ~1.5 cm long; lateral staminodes yellowish, obovate-oblong, ~1.5 cm; lip yellow, obovate, ~2.5 cm long, apex emarginate; ovary villous. **Distr:** C and E Himalayas to Sri Lanka. **Ecol:** Fl. Apr–Jun. **Zones:** 9–12.

C. longa L., *Sp. Pl.* 1: 2, 1753. (Syn: *C. domestica*). [Turmeric]. **Herbs**, to 1 m tall; rhizomes many-branched, orange or bright yellow, cylindric, aromatic; roots tuberous at tip. **Leaves** oblong or elliptic, 30–45(–90) × 15–18 cm, base attenuate, apex shortly acuminate, glabrous, green; petiole 20–45 cm long. **Inflorescence** terminal on pseudostems; peduncle 12–20 cm long; spike cylindric, 12–18 × 4–9 cm; fertile bracts pale green, ovate or oblong, 3–5 cm long, apex obtuse; coma bracts spreading, white and green, sometimes tinged reddish-purple, apex acute. **Flowers** with calyx white, 0.8–1.2 cm long, puberulent, apex unequally 3-toothed; corolla pale yellow, tube to 3 cm long, lobes deltoid, 1–1.5 cm long, central one larger, apex mucronate; lateral staminodes shorter than lip; lip yellowish with a central yellow band, obovate, 1.2–2 cm long, anther spurred at base; ovary sparsely hairy. **Distr:** Of unknown origin. **Ecol:** Fl. Aug. **Zones:** 10–12. **Info:** Widely cultivated commercially for the spice turmeric. Have many medicinal uses.

C. rhabdota Sirirugsa and M.F. Newman, *New Plantsman* 7: 196, 2000. [Candy Cane, Striped Curcuma]. **Herbs**, terrestrial, rhizomatous, xerophytic, to 60 cm tall; shoots 1–2 per plant; leaf sheaths green, reddish-tinged, margins hyaline; ligule bilobed, 2–5 mm long. **Leaves** 3–7; lamina ovate, 5.5–16(–25) × 3.5–7.5(–10) cm, base cuneate to rounded, apex acuminate, adaxially mid to dark green, abaxially pale green; petiole 2–11.5 cm long, dark green, reddish-tinged. **Inflorescence** terminal on leafy shoot, held above the leaves on a scape 18–30 cm tall, reddish towards apex; bracts 8–12 of which 5–7 are fertile; fertile bracts 1.5–2 × 3–4 cm, reddish-brown with bluish main nerves and a blue spot on each side of the free, apical portion; coma bracts narrower, erect, to 4 × 2 cm, white with a broad, longitudinal, reddish-brown stripe along midrib abaxially; bracteoles keel-shaped. **Flowers** with calyx funnel-shaped, lobes 3, cucullate, 7.5–8 × 3 mm; corolla tube 1.4–1.7 cm long, white, lobes white, ventral lobes ~1 × ~0.5 cm, dorsal lobe 8.4–9 × 4.8–5 mm, cucullate; lateral staminodes obovate, ~1 × ~0.7 cm, violet with red ventral margin; lip bowl-shaped, or margins reflexed, narrow, ~1 × ~0.5 cm, distal margins pale violet, darkening towards ridges in throat, with bright yellow between these ridges and red margin adjacent to lateral staminodes; filament 3 × 3 mm, white with red spots on inner surface, apex narrowed; ovary trilocular, style slender, stigma funnel-shaped. **Distr:** Indo-China. **Ecol:** In open grasslands. Fl. Aug–Oct. **Zones:** 9–12. **Info:** This species is often confused with the similar species *C. gracillima*.

C. zedoaria (Christm.) Roscoe, *Trans. Linn. Soc. London* 8: 354, 1807. (Syn: *Amomum zedoaria*, etc.). [Rose Turmeric]. **Herbs**, terrestrial, rhizomatous, 0.6–1.2 m tall; rhizome colored inside, usually aromatic. **Leaves** tufted, usually 4–6, petiolate; lamina oblong-lanceolate or elliptic, 20–80 × 7.5–18(–38) cm, adaxially green or with purple coloration along midvein. **Inflorescence** on a separate shoot, 9–23 × 4–7.5 cm, peduncle 5–37 cm long; fertile bracts 2.5–7.5 cm long, joined halfway, enclosing 2–5 flowers, green or red-tinted, apex deflexed, coma bracts much longer, mostly free, usually violet, rose pink or purplish. **Flowers** with corolla 2.5–4.8 cm long, whitish or yellowish, lobes white, pink, or red; lip circular or obovate, 1.2–2 cm long, whitish or pale yellow, middle dark yellow; lateral sterile stamen 1.1–1.5 cm long, pale yellow; fertile stamen anther 5–8 mm long, base spurred. **Distr:** E Himalayas to Assam. **Ecol:** In open grasslands and bamboo thickets; <100 m. Fl. Apr. **Zones:** 9–12.

Elettaria

Syn: *Cardamomum*, *Matonia*.

Species 11. **Herbs**, rhizomatous; rhizome thick, horizontal, perennial; stem leafy, long. **Leaves** oblong-lanceolate. **Inflorescence** a panicle, borne directly from the rhizome, flexuous, decumbent; bracts bifid–trifid; flowers shortly pedicelled, bracteolate. **Flowers** with calyx cylindric, membranous, shortly lobed; corolla tube cylindric, trilobed, midlobe oblong, convex, lateral lobes narrower; lateral staminodes minutely dentate; lip obovate-cuneate; filament very short, anther cells contiguous, not crested; ovary 3-celled, ovules many, style filiform, stigma small, turbinate. **Fruits** capsules, globose or oblong, coriaceous, indehiscent; seeds small, angular. **Distr:** S India to W Malesia. **Cult:** Propagate from seeds or by division. Requires fertile, humus-rich soil in full sun.

E. cardamomum (L.) Maton, *Trans. Linn. Soc. London* 10: 254, 1811. (Syn: *Alpinia cardamomum*, etc.). [Cardamom, Lesser Cardamom]. Herbs, rhizomatous, evergreen, perennial, aromatic; rhizomes branching, producing upright stems; stem leafy 2–3 m tall. Leaves narrowly lanceolate, 30–60 × 7.5 cm, apex pointed, adaxially dark green, abaxially paler and pubescent. Inflorescences several per leafy stem, panicles, 30–60 cm long; bracts linear-oblong, 4–5 cm long, persistent. Flowers bisexual, white, 2 cm long; calyx 1.2 cm long; corolla tube shortly exserted, segments 1.25 cm long; lip longer than the corolla segments, white striped with violet and yellow margins. Fruits capsules, indehiscent, subglobose or oblong, marked with many fine vertical ribs; seeds 15–20, aromatic, brownish-black, hard, covered with a thin mucilaginous membrane. Distr: SW India (W Ghats). **Ecol:** In open areas of evergreen montane monsoon forests; 500–1,500 m. Fl. Nov–Jan. Fr. Jan–Apr. **Zones:** 10–12. **Info:** Commercially cultivated for the aromatic spice cardamom (dried unripe fruit), which is also used for medicinal purposes.

Elettariopsis

Species 17. **Herbs**, perennial, to 1 m tall; rhizomes creeping, slender, bearing pseudostems at intervals. **Leaves** mostly 1–8; lamina ovate, lanceolate, elliptic, or oblong; ligule entire or bilobed; petiole erect, long; **Inflorescences** arising from base of pseudostems, with flowers spaced along rachis or sometimes in an erect, dense heads; rachis prostrate or erect, simple or branched; bracts 1- or 2-flowered; bracteoles open, not tubular. **Flowers** with calyx white or pinkish, tubular, apex 2- or 3-toothed; corolla tube longer than calyx, slender, lobes 3, ovate-oblong or elliptic; lateral staminodes absent or very short; filament short and broad, connective appendage ± quadrate, lateral lobes not spreading; ovary triloculed, ovules numerous per locule, stigma obconical, ciliate; stylodes 2, slender. **Fruits** capsules, globose, glabrous. **Distr:** China (Hainan), Indo-China to New Guinea. **Cult:** Propagate from seeds or by division. Requires fertile, humus-rich soil in full sun.

E. smithiae Y.K. Kam, *Notes Roy. Bot. Gard. Edinburgh*, 40(1): 148, 1982. **Herbs**, perennial, rhizomatous, geophytic, aromatic; pseudostems 0.6–1 m tall. **Leaves** 3–6(–9), arranged in 2 ranks, with a lemonlike smell when crushed; lamina lanceolate, glabrous and glossy, grayish-green or emerald green; ligule entire or bilobed. **Inflorescences** subterranean, long, branched, few-flowered, flowers emerging among the leaf litter; partial inflorescences ~6–10 cm long; bracts 1–2 cm long; floral bracts lanceolate, pinkish; bracteole ~2 mm long. **Flowers** pale yellow, with orangish colorations towards the center; lip spathulate, ~4–5 cm long; stamens ~1 cm long, anthers 3–4 mm long; ovary ~5 mm long. **Fruits** capsules, spherical, 3 cm Ø, pale brown, slightly ridged. **Distr:** Peninsular Thailand to peninsular Malaysia. **Ecol:** In rainforests, among leaf litter, usually in well sheltered, shady areas. **Zones:** 11–12.

Etlingera

Syn: *Achasma*, *Nicolaia*, etc.

Species 98. **Herbs**; rhizomes creeping; pseudostems robust. **Leaves** petiolate; lamina lanceolate, large. **Inflorescence** arising directly from rhizomes, a spike or head with flowers arranged in 3 or 4 concentric circles on a flat receptacle, base surrounded by numerous sterile, involucral bracts; peduncle usually raised well above ground and long, or embedded in ground and short; bracts 1-flowered; bracteoles long tubular. **Flowers** with calyx tubular, membranous, split on 1 side, apex 3-toothed; corolla tube equaling or longer than calyx, lobes 3, much shorter than tube; lateral staminodes absent; lip tongue-shaped, ± trilobed, much longer than corolla lobes, base adnate to filament forming a distinct tube, central lobe colored, apex entire or bilobed, lateral lobes basally folded over stamen; stamen shorter than lip, free part of filament very short and wide, anther bent forward; ovary triloculed, ovules numerous per locule. **Fruits** capsules, fleshy, indehiscent, smooth, longitudinally ridged, or with obtuse warts in rows. **Distr:** Tropical and subtropical Asia to S Pacific. **Cult:** Propagate from seeds or by division. Requires moist, well-drained, fertile, humus-rich soil in full sun.

E. corneri J. Mood and H. Ibrahim, *Nordic J. Bot.* 20: 282, 2000. (Syn: *E. terengganuensis*). [Rose of Siam]. **Herbs**, perennial, erect, 3–4 m tall; rhizome short, creeping; pseudostem tomentose, 2.5–5 cm Ø. **Leaves** oblong-elliptic, 50–60 × 12–15 cm, base rounded, margins hyaline, apex shortly acuminate, both surfaces glabrous, adaxially dark green, abaxially green or purple, midrib puberulent below; petiole puberulent, ~5 mm long. **Inflorescences** heads, 7–8 cm long, crown 10–11 cm Ø; peduncle erect, ~40–75 cm long; bracts bright red, arranged like the petals of a rose, oblong to narrowly obovate, apex rounded to truncate, revolute, puberulent, glossy. **Flowers** with calyx strigose, white; petals oblong, ~1.8 × 0.3 cm, corolla tube strigose, white, 2.3–2.5 × 0.3 cm; lip erect, reddish with a broad white margin, slightly revolute, throat white, 1.6–2.5 × 1–1.4 cm, basally involute forming a tube; filament white; ovary glabrous, stigma triangular, pale violet, style filiform, white. **Fruits** capsules, ovoid, ~3.3–4 × 1.5 cm, triloculed, mostly upright, glabrous, glossy, white, ripening to orange-red. **Distr:** Peninsular Thailand to peninsular Malaysia. **Ecol:** In shady and swampy areas of evergreen forests; 200–500 m. **Zones:** 11–12.

E. elatior (Jack) R.M. Smith, *Notes Roy. Bot. Gard. Edinburgh*. 43: 244, 1986. (Syn: *Nicolaia elatior*, etc.). [Torch Ginger]. **Herbs**, robust, large; pseudostems closely grouped, to 5 m tall. **Ligule** bilobed, to 1.5 cm, glabrous. **Leaves** lanceolate, to 80 × 18 cm, glabrous; petiole 3–4 cm. **Inflorescence** a head, ~15 cm Ø, surrounded by involucral bracts, borne directly from the rhizome; peduncle erect, raised well above ground, 0.8–1(–1.5) m tall, robust; involucral bracts red, tapering toward apex, 2–3 × ~1 cm, glabrous; floral bracts similar to involucral bracts but pinkish, smaller; bracteoles tubular, ~2 cm, deeply split on 1 side. **Flowers** with calyx 3–4 cm, apex 3-toothed; corolla pink to red, sometimes white; labellum deep red with yellow margin; filament short, flat, whitish pubescent, anther red, longer than filament. **Fruits** on a fruiting head greenish or reddish, globose, ~2.5 cm Ø; seeds many, black. **Distr:** Indonesia, Malaysia, S Thailand. **Ecol:** In forests at low elevations. **Zones:** 11–12.

E. hemisphaerica (Blume) R.M. Sm., *Notes Roy. Bot. Gard. Edinburgh* 43: 245, 1986. (Syn: *Alpinia hemisphaerica*, etc.). [Black Tulip Torch]. **Herbs**, rhizomatous, very large, robust, clump-forming, 3–8 m tall. **Leaves** distichous, shortly petioled above the sheathing base; lamina oblong-lanceolate, 15–75 × 2.5–18 cm, base unequally rounded, margins fringed or sinuate, apex cuspidate, glabrous, adaxially green to bronze with a green tinge, abaxially purplish-red. **Inflorescences** heads, hemispherical or ± bowl-shaped, 5–12 cm Ø, 4–7.5 cm high, borne, flowers numerous, densely aggregated; peduncle robust, 45–60 cm tall, borne directly from the rhizome; sterile outer bracts erect at flowering, dark red, ovate, 5–10 cm long, appressed, margins paler; inner fertile bracts linear, nearly as long as the flowers. **Flowers** erect; calyx lobes lanceolate, green; corolla slightly curved, 4.5–5 cm long, dark red, lobes lanceolate; lip stiff, erect, deep red, margins bright yellow, apex emarginate; anther not distinctly crested; ovary 3-celled, ovules many per cell, style reaching the top of anther, stigma large, capitate. **Fruits** globose, obovoid, rarely spindle-shaped. **Distr:** Sumatra to Java. **Ecol:** In primary forests and rainforests, often in moist and shaded areas; 50–700 m. Fl. Oct–Dec. **Zones:** 11–12. **Info:** The flowers are used as a condiment.

Globba

Species ~100. **Herbs**; rhizomes creeping, slender; pseudostems erect, usually to 1.5 m tall, leafy. **Leaves** sessile or very shortly petiolate; ligule entire; lamina oblong, elliptic, or lanceolate. **Inflorescence** terminal, a thyrse or raceme, often lax; bracts each subtending a cincinnus of flowers, or flowers replaced by bulblets; bracteoles free to base. **Flowers** with calyx campanulate or turbinate, apex obtusely trilobed; corolla tube slender, lobes ovate or oblong, subequal, concave; lateral staminodes petaloid; lip reflexed, adnate to filament to form a slender tube above lateral staminodes and corolla lobes; filament long, curved, anther with or without appendages on each side; ovary 1-loculed. **Fruits** capsules, globose or ellipsoid, apex irregularly dehiscent; seeds small; aril white, lacerate. **Distr:** Tropical Asia; Australia (1 sp). **Cult:** Propagate from seeds or by division. Requires moist, well-drained, fertile, humus-rich soil in full sun.

G. campsophylla K. Schum., *Pflanzenr.*, IV, 46: 145, 1904. (Syn: *G. parviflora*). **Herbs**, slender, 50–60 cm tall. **Leaves** narrowly lanceolate, 9–15 × 1–2 cm, base abruptly narrowed to petiole, apex acuminate, acumen long, glabrous; petiole ~3 mm long; ligule very short, rounded, truncate, quite glabrous. **Inflorescence** panicle, 9 cm long, lax, with slender branches 3 cm long; bracts linear-oblong, rounded, glandular-dotted, 2 mm long; flowers borne on short articulations. **Flowers** white; calyx funnel-shaped, lobes 3, equal, ovate, keeled and shortly mucronate, glandular-dotted; corolla tube slender, 3× as long as calyx, lobes oblong, upper one hooded, all glandular-dotted; staminodes linear, longer, obtuse, glandular-dotted and pubescent; lip rather deeply bilobed, lobes linear, obtuse, glandular-dotted; anther strongly glandular-dotted on the back, spurs linear, acuminate, curved, much longer than anther. **Fruits** large, globose or oblong, ~1 cm long, glandular-dotted; seeds numerous, covered with a silky-white pubescence, and with a finely cut aril at base. **Distr:** Philippines. **Ecol:** In lowland damp thickets and forests, often on cliffs. Fl. Mar. **Zones:** 11–12.

G. globulifera Gagnep., *Bull. Soc. Bot. France* 48: 209, 1901. [Purple Globe Ginger]. **Herbs**, robust,

Below: *Etlingera elatior* (white form), *Etlingera hemispherica*, *Globba campsophylla*, *Globba globulifera*.

RF

MS

Top row: *Globba schomburgkii, Globba winitii, Hedychium coccineum.*
Bottom row: *Hedychium coronarium, Hedychium flavescens, Hedychium gardnerianum.*

45–60 cm tall; aerial bulbils formed in leaf axils and mature inflorescence, bulbils globose, glandular-dotted. **Leaves** 5, subsessile; lamina lanceolate, 20 × 3 cm, apex long-acuminate, adaxially glabrous, abaxially hairy. **Inflorescence** panicle, spikelike, dense, oblong to globular, 5–7.5 cm long, 2 cm Ø, pedunculate, flowers borne between the bracts; bracts tightly clustered, purple, 1 × 0.7–1 cm, subglabrous, ovate, obtuse. **Flowers** yellow, glandular-dotted; calyx tubular, teeth acute, unequal; staminodes lanceolate; lip divided to base; anthers ovate, 4-winged, acute; connective round; ovary. **Distr:** Indo-China. **Ecol:** In tropical rainforests. Fl. Jan–Feb. **Zones:** 10–12.

G. schomburgkii J.D. Hooker, *Bot. Mag.* 103: t. 6298, 1877. [Yellow Dancing Girl Ginger, Yellow Dancing Ladies]. **Herbs**; pseudostems 30–50 cm tall. **Leaves** 5 or 6; ligule short; petiole ~5 mm; lamina elliptic-lanceolate, 15–20 × 3–4.5 cm, glabrous, base obtuse, apex caudate. **Inflorescence** a pendulous raceme or thyrse, 3–11 cm long; bracts lanceolate, 0.6–1.2 cm long; cincinni (if present) 1–2.5 cm, laxly 2–many-flowered; bulblets present, globose or ovoid, 2–4 mm Ø, tuberculate. **Flowers** yellow or wholly reduced; calyx campanulate, 4–5 mm; corolla tube 0.8–1 cm long, pubescent, lobes ovate; lateral staminodes lanceolate, curved; lip yellow, base orange-spotted, narrowly cuneate, apex 2-cleft; filament curved, ~1 cm long, anther with 2 winglike appendages on each side; ovary tuberculate. **Distr:** Myanmar, Thailand, Vietnam. **Ecol:** In forests; ~1,300 m. Fl. Aug–Sep. **Zones:** 10–12. **Info:** 2 varieties described.

G. winitii C.H. Wright, *Bull. Misc. Inform. Kew* 1926: 173, 1926. [Dancing Girl Ginger, Dancing Lady Ginger]. **Herbs**; pseudostems 60 cm tall. **Leaves** oblong-lanceolate, 18(–80) × 5.5 cm, base cordate, apex acuminate, minutely puberulent; petiole 7.5 cm long, green. **Inflorescences** 1–3, panicles, axillary, pendent; bracts ovate, 3 × 1.8 cm, apex acute, initially bracts erect, but quickly becoming reflexed and appressed to rachis; pedicels 2.5 cm long. **Flowers** with calyx lobes oblong, obtuse, 3 mm long, magenta; corolla waxy yellow, thin, pubescent, 9 mm long, lobes 4 × 2 mm; lip shallowly bilobate, obtuse, 7 mm long; staminodes oblong, obtuse, 8 × 2.5 mm; filaments 2 cm long, anthers 2 mm long; ovary green. **Distr:** Myanmar to Thailand. **Ecol:** In moist shady banks of streams; to 400 m. Fl. Jul–Oct. **Zones:** 10–12.

Hedychium

Species 88. **Herbs**, terrestrial or epiphytic, with tuberous rhizomes, often forming dense colonies; pseudostems erect, leafy; ligule conspicuous. **Leaves** usually oblong or lanceolate. **Inflorescence** a terminal spike, densely numerous-flowered; bracts imbricate or lax, 1- or more-flowered; bracteoles tubular. **Flowers** with calyx tubular, usually split on 1 side, apex truncate or 3-toothed; corolla tube long, slender, lobes reflexed at anthesis, linear; lateral staminodes petaloid, larger than corolla lobes; lip suborbicular, large, apex usually 2-cleft; claw long or short; filament usually long, rarely absent, anther dorsifixed, base divaricate, connective appendage absent; ovary triloculed, placentation axile. **Fruits** capsules, globose, 3-valved; seeds numerous, aril lacerate. **Distr:** Madagascar, S China to tropical Asia. **Cult:** Propagate by division of the rhizomes or from seeds. Requires moist, well-drained, humus-rich soil in full or partial sun.

H. coccineum Smith, *Cycl.* 17: Hedychium no. 5, 1811. [Red Ginger Lily, Scarlet Ginger Lily]. **Herbs**, pseudostems 1.5–2 m. **Leaves** sessile; ligule 1.2–5 cm long; lamina narrowly linear, 25–50 × 3–5 cm, glabrous, base subrounded or attenuate, apex caudate-acuminate. **Inflorescences** spikes cylindric, usually dense, glabrous or sparsely villous; bracts oblong, 3–3.5 cm long, leathery, sparsely pubescent, rarely glabrous, 3-flowered, margin involute or rather flat, apex obtuse or acute. **Flowers** red; calyx ~2.5 cm long, sparsely pubescent especially at 3-toothed apex; corolla tube slightly longer than calyx, lobes reflexed, linear, ~3 cm long; lateral staminodes lanceolate, ~2.3 cm; lip orbicular, ~2 cm wide or smaller, apex deeply 2-cleft, filament ~5 cm long, anther 7–8 mm long; ovary sericeous, 2.5–3 mm long. **Fruits** capsules, globose, ~2 cm Ø; seeds red. **Distr:**, China, NE India, Myanmar, Nepal, Sikkim, Sri Lanka, Thailand. **Ecol:** In forests. Fl. Jun–Aug. Fr. Oct. **Zones:** 9–12.

H. coronarium J. König, *Observ. Bot.* 3: 73–74, 1783. (Syn: *H. coronarium* var. *baimao*). [Butterfly Lily, Garland Flower, White Butterfly Ginger, White Ginger]. **Herbs**, pseudostems 1–3 m tall. **Leaves** sessile; ligule 2–3 cm long, membranous; lamina oblong-lanceolate or lanceolate, 20–40 × 4.5–8 cm, adaxially glabrous, abaxially finely pubescent or thinly hairy, base acute, apex long acuminate. **Inflorescences** spikes ellipsoid, 10–20 × 4–8 cm; bracts imbricate, ovate, 4.5–5 × 2.5–4 cm, 2- or 3-flowered. **Flowers** white, fragrant; calyx ~4 cm long, split on 1 side; corolla tube ~8 cm long, slender, lobes lanceolate, ~5 cm long, central one spathulate, apex mucronate; lateral staminodes oblong-lanceolate,

~5 cm long; lip white, base pale yellow, obcordate, 4–6 × 4–6 cm, apex 2-cleft, filament ~3 cm long, anther ~1.5 cm long; ovary sericeous. **Distr:** Tropical Asia to Australia. **Ecol:** In forests. Fl. Aug–Dec. **Zones:** 11–12. **Info:** Cultivated widely for the extremely fragrant flowers and for medicinal purposes.

H. flavescens Carey ex Roscoe, *Monandr. Pl. Scitam.* t. 50, 1824. (Syn: *H. emeiense*, *H. panzhuum*). [Cream Garland Lily, Cream/Yellow Ginger]. **Herbs**, pseudostems 1–2 m. **Leaves** sessile; leaf sheath slightly pubescent; ligule 3–5 cm long, membranous; lamina elliptic-lanceolate or lanceolate, 20–50 × 4–10 cm, abaxially pubescent, base attenuate, margin membranous, apex caudate-acuminate. **Inflorescences** spikes oblong, 15–20 × 3–6 cm; bracts imbricate, oblong to ovate, 3–4.5 × 2–4 cm long, concave, 4- or 5-flowered; bracteoles tubular, membranous. **Flowers** yellow or yellow-white, fragrant; calyx 3.5–4 cm long, split on 1 side, apical margin entire; corolla tube 7–8.5 cm long, slender, lobes linear, 3–3.5 cm long; lateral staminodes wider than corolla lobes; lip erect, creamy-yellow with an orange patch at base, obcordate, longer than wide, apex bilobed; filament subequaling lip; ovary hairy, stigma funnelform, margin bearded. **Distr:** India, Nepal. **Ecol:** In forests; 500–800 m. Fl. Jul–Sep. **Zones:** 9–12.

H. gardnerianum Sheppard ex Ker Gawl., *Bot. Reg.* 9: t. 774, 1824. (Syn: *Gamochilus speciosus*, etc.). [Kahili Ginger, Salmon Butterfly Ginger, Salmon Ginger Lily]. **Herbs**, perennial, rhizomatous, 1–3 m tall. **Leaves** waxy, bluish-green; petiole 1–2 cm long; lamina ovate-elliptic, 20–45 × 10–15 cm, apex shortly tapering, glabrous or abaxially sparsely pubescent along midvein. **Inflorescences** cylindrical, 15–30(–45) cm tall; rachis exposed, glabrous; bracts ovate-elliptic, 2.5–5 cm long, spreading, margins inrolled, enclosing usually 2 flowers. **Flowers** fragrant; corolla lobes greenish-yellow to bright yellow, linear, 3.5–5 cm long; lip yellow, ~2.5 cm long, center tinged with orange; lateral stamens 2.8 cm long, yellow; fertile stamens reddish-orange, longer than labellum; ovary glabrous. **Fruits** capsules, ~1.5 cm long, fleshy, orange, arils ruby-red. **Distr:** Nepal to Indo-China. **Ecol:** In montane regions; to 2,400 m. **Zones:** 9–12. **Info:** Considered invasive in some parts of the world.

Kaempferia

Species ~50. **Herbs**, terrestrial; rhizomes fleshy, tuberous; roots often bearing small tubers; pseudostem short or obsolete. **Leaves** 1 to few; ligule usually small or absent; petiole short; lamina suborbicular to filiform, sometimes variegated or abaxially purple. **Inflorescences** terminal on pseudostems or on separate shoots arising from rhizomes, capitate, spirally few- to many-flowered; bracts 1-flowered; bracteoles small, apically bilobed or sometimes 2-cleft to base. **Flowers** with calyx tubular, split on 1 side, apex unequally 2- or 3-toothed; corolla tube equaling or much longer than calyx, lobes spreading or reflexed, lanceolate, subequal; lateral staminodes petaloid; lip usually white or lilac, sometimes marked with different color near base, showy, apically bilobed to 2-cleft to base; filament very short or absent, connective extended into crest exserted from throat of flower, entire or 2-cleft; ovary triloculed. **Fruits** capsules, globose or ellipsoid, pericarp thin; seeds subglobose to ellipsoid, aril lacerate. **Distr:** Tropical Asia. **Cult:** Propagate from seeds or by division. Requires moist, well-drained, fertile, humus-rich soil in partial sun.

K. elegans (Wallich) Baker, *Fl. Brit. India.* 6: 222, 1890. (Syn: *K. atrovirens*, *K. crawfurdia*, *K. pulchra*, *Monolophus elegans*). [Purple-flowered Resurrection Lily]. **Herbs**; rhizomes with fibrous roots bearing globose tubers. **Leaves** 2–4; petiole to 10 cm long; lamina oblong or elliptic, 13–15 × 5–8 cm, base rounded, apex acute, adaxially green, abaxially pale green. **Inflorescences** terminal on pseudostems, several-flowered; peduncles short; bracts green, oblong-lanceolate, ~5 cm long. **Flowers** pale purple; calyx ~2.5 cm long; corolla tube slender, ~5 cm long, lobes lanceolate, 1.7–2 cm long; lateral staminodes obovate-cuneate, ~1.2 cm long; lip apically 2-cleft to base, lobes obovate, 2–2.5 cm long; filament very short, anther ~3 mm long, connective appendage suborbicular, entire, ~6 mm Ø. **Distr:** Borneo, China, India, N peninsular Malaysia, Myanmar, Philippines, Thailand. **Ecol:** In mixed forests; ~330 m. Fl. Nov–Feb. **Zones:** 9–12.

K. rotunda L., *Sp. Pl.* 1: 3, 1753. (Syn: *K. longa*, etc.). [Hainan Resurrection Lily, Himalayan Crocus, Peacock Ginger, Resurrection Lily, Tropical Crocus]. **Herbs**; rhizomes with tuberous roots. **Leaves** 2–4, erect; ligule broadly triangular, 3–4 mm long; petiole 1–2 cm long, channeled; lamina lanceolate-oblong, 17–27 × 7.5–9.5 cm, base cuneate, adaxially usually variegated dark and pale green on both sides of midvein, abaxially purple-tinged. **Inflorescences** on separate shoots arising from rhizomes, appearing before pseudostems, sessile or shortly pedunculate, 4–6-flowered; bracts purple-brown; bracteoles ~2.3 cm long, apex 2-toothed. **Flowers** with calyx 4.5–7 cm long, split on 1 side, apex 3-toothed; corolla tube equaling calyx, lobes spreading, white, linear, ~5 cm long; lateral staminodes erect, white, lanceolate, ~5 × 1.7 cm, apex acute; lip lilac, suborbicular, apically 2-cleft to base, lobes downcurved, ~3.5 × 2 cm, apex acute, anther connective appendage erect, 2-cleft, fishtail-like; ovary 4–6 mm long, hairy. **Distr:** Indian subcontinent to S China. **Ecol:** In monsoon forests and open grasslands; 0–2,600 m. Fl. Apr. **Zones:** 9–12.

Roscoea

Species 22. **Herbs**, small, with annual pseudostems from erect, reduced rhizome; ligule a curved line at junction of tubular leaf sheath and leaf lamina. **Leaves** oblong or lanceolate; petiole absent. **Inflorescence** a spike, terminal on pseudostems; peduncle short and enclosed by leaf sheaths to long and exserted; bracts persistent, 1-flowered; bracteoles absent. **Flowers** with calyx tubular, split on 1 side, apex 2- or 3-toothed; corolla tube usually exserted from calyx, slender, wider at throat, central lobe erect, usually cucullate, lateral lobes spreading, narrower than central one, usually free; lateral staminodes erect, petaloid; lip reflexed or not, cuneate, large, apex emarginate to bilobed; filament short, anther linear, connective extended at base into a spur; ovary cylindric or oblong, triloculed, ovules numerous per locule, superposed, style filiform, stigma funnelform, margin ciliate. **Fruits** capsules, oblong, cylindric, or clavate, 3-valved; seeds small, arillate. **Distr:** N Pakistan to SC China. **Cult:** Propagate from seeds or by division. Requires moist, well-drained, fertile, humus-rich soil in partial sun.

R. auriculata K. Schumann, *Pflanzenr.* 20 (IV. 46): 118, 1904. (Syn: *R. purpurea* var. *auriculata*). [Hardy Nepalese Ginger]. **Herbs**, terrestrial, tuberous, erect, 20–40 cm tall. **Leaves** 3–7; leaf sheath purple, tubular, wholly closed; lamina lanceolate, 7.5–20 × 2–2.5 cm, base auriculate and clasping, apex acuminate, glabrous. **Inflorescence** several-flowered; peduncle enclosed by leaf sheaths; bracts linear-lanceolate, 2–2.5 cm long, membranous. **Flowers** usually bright purple or white; calyx ~3.5 cm long; corolla tube longer than calyx, lobes lanceolate, ~3.5 cm long; lateral staminodes erect, white, falcate, ~2 cm long, shortly clawed; lip reflexed, obovate, ~4.5 × 3 cm, apex 2-cleft; stamen ~2 cm, anther ~1 cm long, connective spur short. **Fruits** capsules, 2–3 cm long; seeds brown. **Distr:** E Nepal to Sikkim and SC Tibet. **Ecol:** In grasslands; 2,400–2,700 m. Fl. Jun–Aug.

R. purpurea Sm., *Exot. Bot.* 2: 97, 1806. (Syn: *R. procera*, etc.). [Purple Hardy Ginger]. **Herbs**; roots tuberous, in a fasciculated cluster; stems to 55 cm long, sometimes flushed purple, slender, leafy. **Leaves** 4–8, sessile; lamina lanceolate to oblong-ovate, 7–25 cm long, apex acuminated into a fine point. **Inflorescence** enclosed in the upper 2–3 leaf sheaths, usually 1 or 2 flowers open at a time. **Flowers** pale purple or white with dark purple markings; perianth tube exserted; upper perianth lobe erect, narrow, 3.2–6 × 1–2.8 cm; inferior lobes 2, linear-oblong, patent; lateral perianth lobes near base, apex deeply bilobed, lobes obtuse; lip 4.5–6.5 cm long, including the claw which is 0.8–1.2 cm long, entire or bilobed; staminodes 2.5–4 cm long, narrow; anther large, curved, with 2 conspicuous spurs at base; style terminating in a globose stigma, perforated and ciliated at the top. **Distr:** Himalayas. **Ecol:** In subtropical montane regions; to 3,000 m. Jan–Feb, Jun–Aug. **Zones:** 9–11.

Below: *Kaempferia elegans, Kaempferia rotunda, Roscoea auriculata, Roscoea purpurea.*

SZ

Above: *Zingiber officinale, Zingiber spectabile, Zingiber zerumbet, Chamaecostus cuspidatus.*

Zingiber

Species 142. **Herbs**, rhizomatous; rhizomes branched, tuberous, aromatic; pseudostems erect, leafy. **Leaves** distichous, held in a plane parallel to rhizome; lamina oblong, lanceolate, or linear; petiole swollen, cushionlike. **Inflorescences** conical, arising from rhizomes on peduncle clothed with scalelike sheaths, rarely breaking through leaf sheaths without peduncle; bracts closely imbricate, green or colored, 1-flowered, persistent; bracteoles not tubular. **Flowers** with calyx tubular, split on 1 side, apex 3-toothed; corolla tube slender, midlobe white or cream, usually wider than lateral lobes; lateral staminodes adnate to lip, forming a trilobed lip, midlobe retuse or cleft at apex; filament short, connective with elongate appendage wrapped around style; ovary triloculed, ovules numerous per locule, placentation axile, style slender, extending beyond anther locules, stigma not expanded. **Fruits** capsules, loculicidally or irregularly dehiscent; seeds black, covered by aril; aril white, margin irregularly lacerate. **Distr:** Tropical and subtropical Asia. **Cult:** Propagate from seeds or by division. Requires moist, well-drained, fertile, humus-rich soil in full sun.

Z. officinale Roscoe, *Trans. Linn. Soc. London, Bot.* 8: 348, 1807. (Syn: *Amomum zingiber, Z. sichuanense*). [Ginger]. **Herbs**; rhizomes branched, yellowish inside, thickened, fleshy, strongly aromatic; pseudostems 0.5–1 m tall. **Leaves** sessile; ligule slightly bilobed, 2–4 mm long, membranous; lamina lanceolate or linear-lanceolate, 15–30 × 2–2.5 cm, glabrescent. **Inflorescences** arising from rhizomes, ovoid, 4–5 × ~1.5 cm; peduncles to 25 cm tall; bracts pale green, sometimes margin yellowish, ovate, ~2.5 cm long, apex mucronate; bracteoles equaling bracts. **Flowers** with calyx ~1 cm long; corolla yellowish-green, tube 2–2.5 cm long, lobes lanceolate, ~1.8 cm long; lip with central lobe purple-striped and cream-blotched, oblong-obovate, shorter than corolla lobes, lateral lobes ovate, ~6 mm long, free nearly to base; stamen dark purple, anther ~9 mm long, connective appendage curved, ~7 mm long. **Distr:** Origin unknown; cultivated widely, especially in tropical Asia **Ecol:** Fl. Oct. **Zones:** 10–12. **Info:** Cultivated widely as a spice and for medicinal purposes.

Z. spectabile Griff., *Not. Pl. Asiat.* 3: 413, 1851. [Beehive Ginger, Golden Beehive Ginger, Nodding Gingerwort, Pinecone Ginger]. **Herbs**, to 3 m tall; rhizomes fleshy, creeping, seasonally dormant. **Leaves** numerous, simple, alternate, 2-ranked, subsessile, petiole to 1 cm long, swollen; lamina narrowly elliptic, oblong-lanceolate, 20–50 × 7.5–10 cm. **Inflorescences** spikelike, ovoid to cylindrical, 15–20(–30) cm tall, 7.5 cm Ø, many-flowered; scape leafless to 1 m tall; bracts dense, waxy, stiff, concave, ovate, greenish, turning yellow, deep red at maturity, margins rounded, revolute (tips inrolled). **Flowers** with corolla of fused petals, unequally trilobed, pale yellowish, usually 3–4 cm; lip obovate, bilobed, petaloid, 3 cm long, fused to lateral stamen for about half the length, purple-brown with yellow spots, enclosed by the whitish side lobes; stamen 1, anther crest dark maroon, curved ~1.25 cm long. **Fruits** capsules, 3-angled, enclosed within bracts. **Distr:** Thailand to peninsular Malaysia. **Ecol:** In moist tropical lowland forests and streamsides. Fl. throughout most of the year. **Zones:** 11–12.

Z. zerumbet (L.) Roscoe ex Smith, *Exot. Bot.* 2: 105, 1806. (Syn: *Amomum zerumbet*). [Bitter Ginger, Red Pinecone Ginger, Shampoo Ginger]. **Herbs**; rhizomes yellowish inside, tuberous; pseudostems 0.6–2 m tall. **Leaves** sessile or shortly petiolate; ligule 1.5–2 cm long, entire; lamina lanceolate or oblong-lanceolate, 15–40 × 3–8 cm, base narrowed, apex acuminate. **Inflorescences** arising from rhizomes, conical or ovoid-oblong, 6–15 × 3.5–5 cm, apex obtuse; peduncles 10–30 cm long, scalelike sheaths 5–7; bracts closely imbricate, green when young, red when old, slightly hairy, slimy adaxially; bracteoles ~1.5 cm long. **Flowers** with calyx 1.2–2 cm long, split on 1 side, apex 3-toothed; corolla tube 2–3 cm long, slender; lobes pale yellow, lanceolate, central one 1.5–2.5 cm long; lip pale yellow, ~1.5 × 2.5 cm, central lobe suborbicular or subobovate, 1.5–2 × ~1.5 cm, apex emarginate, lateral lobes obovate, ~1 cm long, free nearly to base; stamen ~1 cm long; connective appendage beaklike, ~8 mm long; ovary ~4 mm long, glabrous. **Fruits** capsules, ellipsoid, 0.8–1.2 cm long; seeds black. **Distr:** Tropical and subtropical Asia. **Ecol:** In moist places in forests. Fl. Jul–Sep. Fr. Oct. **Zones:** 9–12.

COSTACEAE

Genera 4, **species** ~120. **Herbs**, perennial, terrestrial, not aromatic; rhizomes ± fleshy; stems usually simple, often long, terete, sometimes spirally contorted, leafy; secondary branches (if present) breaking through leaf sheaths; leaf sheath tubular, closed; ligule present. **Leaves** spirally arranged, basal ones usually without a lamina; lamina simple, elliptic, longitudinally rolled in bud, glabrous or hairy, margins entire; petiole short. **Inflorescence** a conelike spike, terminal on leafy shoots or on separate, short, leafless shoots arising directly from rhizomes; bracts imbricate, with linear, nectariferous callus below apex, 1- or 2-flowered; bracteoles tubular or folded. **Flowers** bisexual, epigynous, zygomorphic; calyx tubular, 2- or trilobed or -toothed; corolla proximally tubular, distally trilobed, lobes unequal, imbricate in bud; staminodes 5, united into a lip equaling or much longer than corolla, often ± trilobed with crisped margin, never deeply lobed; fertile stamen 1, filament broad, anther borne below apex of filament or at apex with a connective crest, introrse, dehiscing via longitudinal slits; ovary inferior, 2- or triloculed, ovules many per locule, placentation axile, developed style 1, filiform, usually enclosed between anther locules, stigma with or without abaxial appendage; stylodes 2, reduced to septal, hollow nectaries at apex of ovary, or absent. **Fruits** capsules, 2- or 3-valved, loculicidally dehiscent; seeds numerous, black, arillate. **Distr:** Pantropical. **Cult:** Propagate from seeds or by division. Requires moist, well-drained, fertile, humus-rich soil in full sun.

Chamaecostus

Species 7. **Herbs**, low or diminutive, occasionally acaulescent rosettes, <1 m high; stem <1 cm Ø. **Leaves** elliptical, apex acuminate to long-acuminate. **Inflorescence** often capitate, not tightly compressed; bracts chartaceous to herbaceous, green or green-yellow often with deltate appendages; bracteoles membranous, tubular, adaxially bicarinate. **Flowers** with corolla and lip usually of the same color, yellow or orange to white, rarely vibrant red (in *C. lanceolatus*); calyx membranous, cylindrical, apex toothed; corolla tube exserted from calyx, lobes large, membranous, lanceolate; lip large, forming a long narrow tube that opens broadly to ovate at apex, greatly exceeding petals and bracts in length; stamen petaloid, apex reflexed, triangular, covering the opening to the nectary chamber; ovary trilocular, stigma cup-shaped. **Fruits** capsules, membranous, tardily dehiscent, persistent floral parts reddish-brown punctate. **Distr:** S Tropical America. **Cult:** Propagate from seeds, basal cuttings or by division of clumps. Requires a moist, well-drained, humus-rich soil in partial sun.

C. cuspidatus (Nees and Mart.) C.Specht and D.W.Stev., *Taxon* 55: 158, 2006. (Syn: *Costus cuspidatus*, etc.). [Fiery Costus]. **Herbs**, rhizomatous, to 50 cm tall; stems usually purplish-red, erect, stout, clothed with pale leaf sheaths. **Leaves** elliptic to elliptic-lanceolate, 10–20 × 4–8 cm, base produced into a petiole, apex elongate, narrowed, adaxially dark green, abaxially paler, red-tinged; sheath 4–5 cm ling, mouth red and ciliolate; petiole very short, broad, concave. **Inflorescence** terminal; bracts with apex leaflike, flowers 1 per bract; floral bracteole, tubular. **Flowers** orange, 6.5–7.5 cm across; calyx 2.5 cm long, tubular, unequally trifid; corolla narrowly funnel-shaped, twice as long as the calyx, lobes very variable, oblanceolate or oblong, ovate, obtuse or acute; lip ± circular, 5–7 cm across, margins undulate, expanded horizontally, irregularly crenate, bright orange; anther sessile, connective hooded, oblong, base cordate, apex abruptly contracted into an inflexed beak, golden-yellow, hairy outside; stigma small, urceolately funnel-shaped. **Fruits** capsules, membranous, tardily dehiscent, persistent floral parts reddish-brown punctate. **Distr:** E Brazil. **Ecol:** In shaded understory of tropical rainforests. Fl. Nov–Jan. **Zones:** 10–12.

Top row: *Cheilocostus speciosus, Costus erythrophyllus, Costus erythrophyllus* (flower), *Costus guanaiensis.*
Middle row: *Costus guanaiensis* var. *tarmicus, Costus ligularis, Costus productus.*
Bottom row: *Costus lucanusianus.*

Cheilocostus

Species 4. **Herbs**, >1.5 m tall, branching at higher nodes with secondary branching; branches breaking through leaf-sheaths. **Leaves** seasonally deciduous, attached to sheath by short petiole. **Inflorescence** an elongate spike, terminal either emerging directly from rhizome on a leafless stem or terminating a vegetative shoot; bracts chartaceous (to woody), red or brown, unappendaged; bracteole chartaceous, tubular, adaxially bicarinate. **Flowers** with calyx trilobed, apices pungent; corolla tube exserted from calyx, lobes large, membranous, lanceolate; lip large, obovate, thin, forming a long narrow tube, opening broadly at apex, greatly exceeding petals and bracts in length, never lobed, white, yellow to orange and sometimes red; stamen petaloid, tip reflexed, triangular, covering the opening of floral tube; ovary trilocular, ovules many, in 2 rows, stigma modified cup-shaped with small rounded dorsal appendage. **Fruits** capsules, 3-angled, lateral angles smaller, spreading, loculicidally dehiscent via 3 slits, not splitting to apex, dry; seeds angular, aril small, white, fleshy. **Distr:** Tropical and subtropical Asia and Queensland. **Cult:** Propagate from seeds, basal cuttings or by division of clumps. Requires a moist, well-drained, humus-rich soil in partial sun.

C. speciosus (J. König) C.Specht, *Taxon* 55: 159, 2006. (Syn: *Banksea speciosa, Costus speciosus,* etc.). [Can Reed, Crepe Ginger, Spiral Ginger, Wild Ginger]. **Herbs**; stems to 1–3 m tall, base slightly woody, apex branched and spirally twisted when old. **Leaves** oblong or lanceolate, 15–20 x 6–10 cm, base subrounded, apex acuminate or caudate-acuminate, abaxially densely sericeous; petiole

Top row: *Costus malortieanus, Costus spicatus.*
Bottom row: *Costus scaber, Dimerocostus strobilaceus.*

5–7 mm long. **Inflorescences** terminal, ellipsoid or ovoid, 5–15 cm long; bracts bright red, ovate, ~2 cm long, leathery, pubescent, apex sharply pointed; bracteoles pale red, 1.2–1.5 cm long. **Flowers** with calyx red, 1.8–2 cm long, leathery, apex trilobed; lobes reddish-black, rigid, apex densely sericeous; corolla tube ~1 cm long, lobes oblong-elliptic, ~5 cm long, apex white or red; lip white, trumpet-shaped, 6.5–9 cm long, apex toothed and crisped, edges overlapping; stamen petaloid, white with orange-yellow base, urceolate, ~4.5 × 1.3 cm, pubescent. **Fruits** capsules, red, globose, ~1.5 cm long, slightly woody; seeds black, glossy, ~3 mm long. **Distr:** Tropical and subtropical Asia, Queensland. **Ecol:** In forest margins, moist places; 0–1,700 m. Fl. Jul–Sep. Fr. Sep–Nov. **Zones:** 7–12.

Costus

Species ~90. **Herbs**; rhizomes horizontal, tuberous; stems sometimes branched, usually spirally twisted, leafy, rarely plants stemless. **Leaves** oblong to lanceolate. **Inflorescences** terminal or lateral on separate, short, leafless shoots arising from rhizomes, conical, densely many-flowered; bracts imbricate, 1- or 2-flowered. **Flowers** with calyx trilobed or 3-toothed at apex; corolla tube equaling or longer than calyx; lip obovate, large, margin incurved; stamen petaloid, anther locules linear; ovary triloculed, ovules many per locule, superposed, style filiform, stigma funnelform, stylodes absent. **Fruits** capsules, subglobose or ovoid, woody; seeds many, black; aril lacerate. **Distr:** Tropical and subtropical regions worldwide. **Cult:** Propagate from seeds, basal cuttings or by division of clumps. Requires a moist, well-drained, humus-rich soil in partial sun.

C. erythrophyllus Loes., *Notizbl. Bot. Gart. Berlin-Dahlem* 10: 707, 1929. [Blood-red Spiral Costus, Oxblood Ginger]. **Herbs**, rhizomatous, terrestrial, glabrous except the puberulent bracts, 0.5–1(–2) m high; ligules short, 1–3.8 cm long, entire, glabrous. **Leaves** mostly crowded toward the apex of the stem, shortly petiolate; sheath 0.8–2 cm across, green, glabrous; lamina obovate-oblong or narrowly elliptic, 15–25(–32) × 5.5–11 cm, distinctly 6–7-plicate, base cuneate, apex acute to acuminate, acumen 0–1 cm long, adaxially green, glabrous, abaxially purplish-red, glabrous to sparsely puberulous; petiole 0.5–1.5 cm long. **Inflorescences** basal spikes, subcapitate or ovoid, 4–8 × 3–6 cm, borne on a shoot 60 cm tall with 2–7 well-developed leaves; bracts obscurely lineate, 4 cm long; appendages foliaceous, triangular-ovate, 2.5–6 × 2–4 cm, green, puberulous. **Flowers** with calyx red, 1.1–1.6 cm long, puberulous, lobes triangular, 1.1–1.6 × 0.9–1.2 cm; corolla tubular, 7–8.5 cm long, white, glabrous; lip spreading, broadly ovate, 6–6.5 × 6–6.5 cm, white with a central yellow zone, reddish-striped, midlobe trilobed; stamen 4–5 × 1.2–1.6 cm, pinkish-white, apex irregularly tridentate, reflexed, anther 0.9–1 cm long. **Distr:** W South America to N Brazil. **Ecol:** In tropical rainforests, often near rivers. Fl. Jun–Jul. **Zones:** 9–12.

C. guanaiensis Rusby, *Bull. Torrey Bot. Club* 29: 694, 1902. (Syn: *C. sinningiiflorus*). [Cana de India, Canahuate]. **Herbs**, rhizomatous, stout, (0.5–)2–6 m tall; sheaths 1–4 cm Ø. **Ligule** truncate, 0.5–1(–1.5) cm long. **Leaves** narrowly ovate to narrowly obovate, (13–)20–65 × 5–15 cm, base usually cuneate, apex shortly acuminate; petiole 0.5–2 cm long. **Inflorescence** ovoid, 5–15(–30) × 4–6(–10) cm, obtuse, terminating a leafy stem, to 50 cm long; sheaths obliquely truncate, to 8.5 × 3.5 cm; bracts with exposed part green, covered part red (green), broadly ovate, 2–4(–6) × 1.5–5 cm; appendages triangular, 2–4(–10) × 1.5–2(–4) cm; bracteole red, often greenish-tipped, 2.5–4 cm long. **Flowers** with calyx red, (0.7–)1.4–1.8(–2.2) cm long, lobes triangular, (2–)3–6 mm long; corolla white, reddish-white, or yellowish-white, 7–10 cm long, tube ~2 cm long, lobes narrowly obovate, (5–)6–8 × 1.5–3.5 cm; lip white or yellowish-white, broadly obovate, 7.5–11 × 6–7 cm, lateral lobes usually red-striped, midlobe reflexed, blotched with yellow to orange at center; stamen narrowly elliptic, white to pinkish-white, 4–6 × 0.7–2 cm, obtuse or obscurely trilobed, anther 1–1.4 cm long; ovary 0.5–1 cm long. **Fruits** capsules, ellipsoid, 1.2–2.2 cm long; seeds black. **Distr:** Tropical America. **Ecol:** In rainforests and cloud forests; 0–1,600 m. **Zones:** 11–12. **Info:** ***C. guanaiensis* var. *tarmicus*** is a yellow-flowered variety.

C. ligularis Baker, *Fl. Trop. Afr.* 7: 298, 1898. (Syn: *C. fimbriatus*). **Herbs**, rhizomatous; leafy stem to 1.3 m long; sheath abaxially reddish. **Leaves** oblanceolate-oblong, 12–20 cm long, 7.5 cm wide above middle, base cuneate, apex acuminate, moderately firm, both surfaces glabrous; petiole scarcely any; ligule oblong, 2–3.8 cm long, obtuse, brown, membranous, pubescent. **Inflorescence** a globose cluster, many-flowered, sessile, borne terminal on the leafy stem; bracts orbicular, 6 mm across.

Flowers with calyx 6 mm long, apex toothed; corolla lobes over 2.5 cm long; lip 5(–8) × 5 cm, pink; fertile stamen with a 2.5 cm long process. **Distr:** Cameroon to Gabon. **Ecol:** In understory of tropical rainforests. Fl. Feb. **Zones:** 11–12.

C. lucanusianus J. Braun and K. Schum., *Ber. Deutsch. Schutzgeb.* 2: 151, 1889. (Syn: *C. dussii*, etc.). [African Spiral Flag, Sweet African Flag]. **Herbs**, rhizomatous, 1–2 m tall; sheaths glabrous, 0.8–1.3 cm Ø. **Leaves** narrowly elliptic, 12–26 × 4.5–7 cm, base rounded to slightly cordate, apex long-acuminate, adaxially glabrous, abaxially densely sericeous-puberulous; ligule truncate, 1–2 mm long, with a salient, ciliate rim at base; petiole ~5 mm long, glabrous. **Inflorescence** globose to cylindric, 4.5–7(–14) cm long, 5–6 cm Ø; bracts green, coriaceous, broadly ovate, 1.5–2.5 × 1.5–2.5 cm, obtuse, glabrous to sparsely puberulous towards the margins, callus 1–3 mm long; bracteole green, 1.5 cm long, glabrescent; callus 3 mm long. **Flowers** with calyx green, exceeding the bracts, 1.7–2 cm long, glabrescent, lobes deltate, reflexed, 5–7 mm long, puberulous towards the margins; corolla white, 4 cm long, glabrous, tube 1.5 cm long, lobes narrowly elliptic, 2.5 cm long; lip white, striped with dark purple, broadly obovate when spread out, ~4 cm long, margins undulate; stamen narrowly elliptic, 3.2 × 1–1.2 cm, anther 6 mm long; ovary 5–7 mm long, glabrous. **Fruits** capsules, globose, 1.4–1.5 cm Ø, glabrous; seeds black. **Distr:** W tropical Africa to Uganda. **Ecol:** In tropical forests. Fl. May–Jul. **Zones:** 11–12.

C. malortieanus H. Wendl., *Hamburger. Garten-Blumenzeitung* 19: 30, 1863. [Step-ladder Ginger]. **Herbs**, <1 m tall; sheaths reddish near insertion of petiole, to 2 cm Ø; **ligule truncate, <1 mm long. Leaves** obovate to elliptic, mostly 15–26 × 9–15 cm, base cuneate, apex shortly apiculate, light green marked with faint darker green bands, abaxially paler green or glaucous; petiole to 2 mm long, densely villose. **Inflorescence** globose to short-cylindrical, 4–9 cm long, ~4 cm Ø; bracts broadly ovate, obtuse, 3–4 × 3–4 cm, exposed part green, covered part reddish; bracteoles reddish, 1.4–2 cm long, glabrous. **Flowers** with calyx reddish to greenish, 5–9 mm long; corolla yellow to cream, 5–7 cm long, glabrous, tube 1–2 cm long, lobes narrowly obovate; lip yellow, broadly obovate when spread, 4–6 × 3.5–5 cm, margins slightly crenulate, lateral lobes usually red-striped, midlobe trilobulate; stamens white, purple-tinged, narrowly elliptic, 4.2–4.5 cm long, slightly trilobulate, anthers 6–7 mm long; ovary 5–8 mm long. **Fruits** capsules, ellipsoid, 1.2 cm long. **Distr:** C America. **Ecol:** In moist to wet lowland forests, rainforests; ~240 m. **Zones:** 10–12.

C. productus Gleason ex Maas, *Acta Bot. Neerl.* 24: 471, 1975. [Spiral Ginger]. **Herbs**, rhizomatous; sheath 0.8–1.5 cm across, glabrous or puberulous to strigose, membranous; ligule 2–4 cm long, unequally bilobed, lobes rounded to obtuse, puberulous to strigose, membranous. **Leaves** narrowly elliptic to narrowly obovate, 16–32 × 5–9 cm, base rounded, apex shortly acuminate, adaxially densely strigose to glabrous, costa strigose, abaxially densely strigose to glabrous; petiole 2–5mm long, densely villose to hirsute. **Inflorescence** ovoid, 6–11 × 3–4 cm, borne terminal on a leafy stem; bracts broadly ovate, to 2.5 × 3 cm, margins not breaking up into fibers, bright red to orange, lower ones greenish, chartaceous to coriaceous, glabrous to sparsely puberulous callus 0.6–1.2 cm long, all bracts with foliaceous reddish appendages; appendage broadly ovate to deltate, 0.5–2 × 0.5–2 cm, acute to obtuse; bracteole boat-shaped, to 1.7 cm long; callus to 4 mm long. **Flowers** with calyx glabrous to densely puberulous, lobes shallowly triangular to deltate, 1–1.5 cm long; corolla pale yellow to orange, glabrous, tube 1 cm long, lobes narrowly obovate, 3.5–4.2 cm long; lip small, tubular, lateral lobes involute, oblong-obovate when spread, 2.5–3.5 × 0.8–1.5 cm; stamen narrowly elliptic, 2.5–3 × 0.4–0.8 cm, apex rounded or obtuse, anther 6–8 mm long. **Distr:** Peru. **Ecol:** In humid forests; to 900 m. Fl. Jun–Jul. **Zones:** 11–12.

C. scaber Ruiz and Pav., *Fl. Peruv.* 1: 2, 1798. [Indian-head Ginger]. **Herbs**, slender, to 3 m tall; sheaths glabrous to puberulous, mostly 0.5–1.5 cm Ø; ligules obliquely truncate, 0.2–1.2 cm long. **Leaves** narrowly elliptic, 10–32 × 3–11 cm, base cuneate to rounded (rarely cordate), apex acuminate, adaxially with costa densely strigulose; petioles 0.2–1 cm long. **Inflorescence** ovoid to narrowly cylindrical, 4–10 cm long, to 22 cm long in fruit, 1.5–3.5 cm thick and to 4.5 cm in fruit; bracts orange-red to red, broadly ovate, 2–3.5 cm long, obtuse, coriaceous; bracteoles reddish, 0.9–1.2+ cm long. **Flowers** with calyx reddish, 3–7 mm long; corolla orange to yellow, 3.5–4 cm long, glabrous, tube ~1 cm long, lobes narrowly obovate, 3 cm long, lip yellow, oblong-obovate when spread out, 2–3 × 1.5–2 cm, lateral lobes inrolled, forming a tube; stamen narrowly elliptic, equaling or slightly exceeding the lip, red to orange-red, apex yellowish, 2–2.5 cm long, anther 5–7 mm long; ovary 3–8 mm long, glabrous to densely puberulous. **Fruits** capsules, ellipsoid to subglobose, 0.7–1.2 cm long; seeds black. **Distr:** Tropical America. **Ecol:** In humid to very humid forests; 0–1,250 m. Fl. Jun–Jul. **Zones:** 9–12.

C. spicatus (Jacq.) Sw., *Prodr.*: 11, 1788. (Syn: *Alpinia spicata*, etc.). **Herbs**, terrestrial, rhizomatous, to 2.5 m tall; sheaths 1–2 cm Ø, glabrescent; ligules truncate, 0.2–1 cm long. **Leaves** narrowly elliptic, 7–33 × 3.5–8.5+ cm, base rounded to cordate, apex shortly acuminate, glabrescent; petioles 0.2–1 cm long. **Inflorescence** ovoid to cylindrical, 5–27 × 3–4.5 cm; bracts greenish or reddish on exposed part, reddish on covered part, broadly ovate, 2–4 × 2–4 cm, margins of covered parts lacerating into fibers, apex obtuse, glabrous, coriaceous; bracteoles 1.7–3 cm long. **Flowers** with calyx 0.9–1.6 cm long; corolla yellow to pink, 4–5 cm long, glabrous, tube 1 cm long, lobes narrowly obovate, 3–5 cm long; lip yellow, broadly oblong-obovate when spread out, 2.5–5 × 2.5–5 cm, lateral lobes rolled inward, forming a slender tube, margins crenulate; stamen narrowly elliptic, 3–4 cm long, anther 7–8 mm long; ovary 4–9 mm long, sericeous (glabrous). **Fruits** capsules, ellipsoid, 1–1.5 cm long; seeds black. **Distr:** Caribbean (Hispaniola, Puerto Rico, and the Lesser Antilles). **Ecol:** In coastal wet forests, lowland thickets and open forests. Fl. Jun–Jul. **Zones:** 11–12.

Dimerocostus

Syn: *Mulfordia.*

Species 3. **Herbs**, very large, rhizomatous, unbranched. **Leaves** spirally arranged, with closed ligulate sheaths, congested near the tip of the elongate stem. **Inflorescence** spiciform, scarcely conelike as in *Costus;* bracts much shorter than the subtended clusters of flowers; bracteoles concrescent and tubular, closely investing the ovaries. **Flowers** with calyx tubular, 3-parted above; corolla trilobed; lip extremely showy; stamen 1, petaloid; ovary inferior, 2-celled. **Fruits** capsules, tardily dehiscent. **Distr:** C and S tropical America. **Cult:** Propagate from seeds, basal cuttings or by division of clumps. Requires a moist, well-drained, humus-rich soil in partial sun.

D. strobilaceus Kuntze, *Revis. Gen. Pl.* 2: 687, 1891. (Syn: *Costus uniflorus, D. uniflorus*). [Yellow Crepe Ginger]. **Herbs**, rhizomatous, 2–6 m tall; stem spiraled. **Leaves** congested toward stem apex; lamina ovate to obovate or oblong-oblanceolate, 12–50 × 3–12 cm, base narrowed into a pseudo-petiole, sheathing, apex acuminate, tip sharply pointed, glabrous, nerves densely puberulous-sericeous; sheaths glabrous to sparsely puberulent; ligule truncate, 1–2 mm long. **Inflorescence** strobilaceous, ovoid to cylindrical, spirally contorted, 10–40 × 4–7 cm, borne at apex of leafy stem; bracts numerous, closely sheathing, 2–3 cm long, green, each with a linear callus; bracteoles tubular. **Flowers** (2–)2.5–3.5(–4) cm long, often longer than bracts; calyx 2–3 cm long, coriaceous; corolla white, glabrous; lip showy, broadly bilobed, 9–11 × 8–13 cm, extended, yellow or white with an orange throat; stamen 3–4 cm long, white; ovary biloculed, subcylindrical, 2 cm long, both minutely sericeous. **Fruits** bilocular, crowned with persistent calyx; seeds bright brown, aril white, small. **Distr:** C and S tropical America. **Ecol:** In very humid forests and lowland thickets; 0–200 m. **Zones:** 10–12.

PART 4
ANGIOSPERMS: EUDICOTS
[TRICOLPATES]

Orders 38, **families** 328, **genera** 10,112, **species** 194,696. **Flowers** cyclic, with floral parts arranged in whorls, with the members of the individual whorls alternating; staminal filaments are usually slender, anthers well-differentiated, basifixed, microsporongenesis simultaneous, pollen tricolpate. **Info:** This is a very large group of plants, consisting of about 80% of all extant flowering plants (Judd et al. 2008; Soltis et al. 2009, Stevens 2001 onwards; Worberg 2009). This group was traditionally referred to as the 'Dicots'. Eudicots are further divided into Basal Eudicots and Core Eudicots. Core Eudicots comprises of Rosids (Rosids I and Rosids II) and Asterids (Asterids I and Asterids II). The monophyly of the Rosids is strongly supported by molecular data (Wang et al. 2009).

Opposite: *Leucospermum cordifolium.*

8 BASAL EUDICOTS [BASAL TRICOLPATES]

RANUNCULALES

FAMILIES 7, GENERA 199, SPECIES 4,445.

Young stems with separate bundles, vessels only in central part. **Leaves** with petiole bundles annular; leaf cuticle waxes as clustered tubules. **Micropyle** formed from both integuments. **Fruits** with perianth deciduous; (Stevens 2001 onwards; Wang et al. 2009). **Info:** The monophyly of Ranunculales is supported by molecular data (Chase et al. 1993), and also by combined molecular and morphological data (Wang et al. 2009).

BERBERIDACEAE

Genera 14, **species** 701. **Herbs**, perennial, or shrubs, rarely small trees, evergreen or deciduous, sometimes rhizomatous or tuberous; stems with or without spines. **Stipules** present or absent. **Leaves** alternate, opposite, or basal; lamina simple, or 1–3x pinnately or 2–3x ternately compound, venation pinnate or palmate. **Inflorescences** terminal or axillary, racemes, spikes, umbels, cymes, or panicles, or flowers fascicled or solitary; flowers pedicellate or sessile; bracteoles or bracts present or absent. **Flowers** bisexual, radially symmetric; perianth usually 2- or 3-merous, rarely absent; sepals 6–9, often petaloid, distinct, in 2 or 3 whorls; petals 6, distinct, flat, hooded, pouched, or spurred; nectary present or absent; stamens 6, opposite petals, anthers 2-celled, dehiscing by valves or longitudinal slits; ovary superior, 1-carpellate, ovules numerous, rarely solitary, placentation marginal or appearing basal, style present or absent, sometimes persistent in fruit as a beak. **Fruits** berries, capsules, follicles or utricles; seeds 1 to numerous, sometimes arillate, endosperm abundant. **Distr:** Mainly in the north temperate zone and on subtropical mountains.

Berberis

Species 600. **Shrubs**, evergreen or deciduous; branches gray, usually terete or becoming so with age, glabrous or tomentose, spinose or not; mature stems yellow, brown, red, purple, or black, young shoots usually green, initially sometimes partially or wholly reddish; spines simple or usually 3–5-fid. **Leaves** on short shoots, alternate, usually petiolate; lamina simple. **Inflorescences** flowers solitary or fascicled, racemes, umbels, or panicles; bractlets usually 3, caducous, scalelike. **Flowers** 3-merous, pale greenish, or shades of yellow, rarely red, purple or white; sepals 6(or 9), rarely 3 or 12; petals 6, base nectariferous; stamens opposite petals, anthers dehiscing by valves; ovary symmetrically club-shaped, ovules 1–12, rarely to 15, subbasal, styles short. **Fruits** berries, usually red, dark red, purple, or black, globose, ellipsoid, oblong, ovoid, or obovoid; seeds 1–10, light tan to red-brown or black, aril absent. **Distr:** Widespread in the N Hemisphere, especially diverse in the Himalayas and E Asia; Africa. **Cult:** Propagate from softwood cuttings, hardened cuttings or seeds. Requires a well-drained soil in full or partial sun.

B. wightiana Schneider, *Bull. Herb. Boissier* 5: 452, 1905. [Barberry]. **Shrubs**, to 1+ m tall; stems very furrowed, minutely pubescent. **Spines** 1.5–3 cm long, trifid. **Leaves** oblong-obovate, 2–4 × 1–1.4 cm, apex obtuse, margins entire, very rarely with 1–3 marginal spines, adaxially dull green, abaxially gray-pruinose, venation reticulate. **Inflorescences** racemose or subumbellate-racemose, 7–12-flowered, 2–4(–5) cm long; pedicels 0.8–1.8 cm long; outer sepals 3 × 1.25 mm, oblong-ovate; inner sepals obovate, 5 × 3.75 mm; petals obovate, 4 × 2.5 mm; stamens 3 mm long, shortly apiculate. **Fruits** berries, 9 × 4 mm, excluding 1 mm long style, narrowly obovoid. **Distr:** India and Sri Lanka. **Ecol:** In montane forest margins and open grasslands; 500–2,200 m. Fl. Apr–May. **Zones:** 10–12.

Nandina

Species 1. *Generic description same as specific.* **Cult:** Propagate from cuttings. Requires a rich, moist, well-drained soil in full sun.

N. domestica Thunberg, *Nov. Gen. Pl.* 1: 14, 1781. (Syn: *N. domestica* var. *linearifolia*). [Heavenly Bamboo, Sacred Bamboo, Southern Heaven Bamboo]. **Shrubs**, evergreen, erect, to 3 m tall, glabrous, few-branched, without rhizomes; branchlets reddish when young. **Leaves** alternate, 30–50 cm long, 2 or 3 times pinnately compound, estipulate, leaflets nearly sessile; lamina elliptic or elliptic-lanceolate, 2–10 × 0.5–2 cm, base cuneate, margins entire, apex acuminate, weakly leathery, glabrous, venation pinnate; petioles and petiolules swollen at base. **Inflorescences** panicles, terminal or axillary, erect, 20–35 cm long, flowers numerous; bracteoles present. **Flowers** bisexual, 3-merous, cream or white; sepals numerous, spirally arranged, ovate-triangular or ovate-oblong, 1–4 mm long; petals 6, larger than sepals, basally without nectaries, oblong, ~4.2 × 2.5 mm, obtuse; stamens ~3.5 mm long, anther connective prolonged; ovary ellipsoid, locules 1, ovules 1–3, style short, stigma entire or crisped; carpopodium 4–8 mm long. **Fruits** berries, red or purplish, globose, 6–8 mm Ø; **seeds** 1–3, grayish or brownish, oblate-spheroidal. **Distr:** China, India, Japan. **Ecol:** On streamsides in montane forests, thickets; to 1,000 m. Fl. Mar–Jun. Fr. May–Nov. **Zones:** 7–11.

LARDIZABALACEAE

Genera 9, **species** 36. **Shrubs**, mostly climbing, monoecious or dioecious. **Leaves** alternate, petiolate, exstipulate; lamina digitate or palmate, rarely pinnate compound; petiolules usually swollen at base. **Inflorescences** racemose, often axillary and fascicled. **Flowers** unisexual, sometimes polygamous regular, trimerous; sepals (or perianth) 3 to 6, in 1 or 2 whorls, imbricate or the outer valvate, inner often petaloid; petals 6, very small often reduced, nectariform or absent; stamens 6, free or basally connate, with anthers free, dithecous, extrorsely and longitudinally opening with connectives often protruding; staminodes 6 or absent in ♀ flowers; carpels 3(–15) in ♀ flowers, free, superior, diverging, each 1-locular, generally many-ovuled on parietal placenta, stigma oblique, sessile, very reduced or minute; pistillodes present or absent in ♂ flowers. **Fruits** fleshy follicles or berries, indehiscent or opening longitudinally by the adaxial suture at maturity, often flesh-colored; seeds ovoid or slightly reniform with copious endosperm and small embryo. **Distr:** E Asia, Himalayas and temperate S America (Chile).

Akebia

Species 5. **Climbers**, woody, deciduous or subevergreen, monoecious; stems twining. **Leaves** alternate or clustered on short branches, long-petiole; lamina palmately compound, 3–5(–7)-foliolate; leaflets with margins entire, sinuate, or undulate. **Inflorescences** axillary, racemose, sometimes umbellate; bracts subtending inflorescences, persistent, scaly; ♂ flowers borne apically on the inflorescence; ♀ flowers 1 or 2 basally on the inflorescence. **Flowers** with sepals usually 3(–6), purplish-red or greenish-white, subvalvate, reflexed at anthesis; petals absent; ♂ with stamens free, filaments very short, anthers incurved at anthesis, pistillodes small; ♀ larger than ♂ flowers, carpels 3–9(–12), free, cylindric, stigma capitate. **Fruits** fleshy follicles, dehiscent along abaxial suture; seeds numerous, in several rows within pulp, slightly compressed. **Distr:** China, Japan, Korea. **Cult:** Propagate from seeds (stratified for 4 weeks) or from softwood to half-hardened cuttings. Requires a moist, humus-rich soil in shade or partial sun.

A. quinata (Houttuyn) Decaisne, *Arch. Mus. Hist. Nat.* 1: 195, fig. 1(a-c), 1839. (Syn: *A. micrantha*, etc.). [Chocolate Vine, Five Leaf]. **Climbers**, deciduous to semi-evergreen, to 12 m long, woody, glabrous; stems grayish-brown, slender, cylindric; lenticels prominent, small, orbicular. **Leaves** with petiole 1.6–12.5 cm; leaflets mostly 5, petiolules 0.2–2.2 cm long; lamina oblong to ovate-elliptic, 0.7–8.2 × 0.4–4.2 cm, base rounded, margins entire, apex retuse. **Inflorescences** pendent, 4.5–12 cm long; pedicel with basal bracts; ♂ 4–15-flowered; ♀ (0–)1–5-flowered. **Flowers** fragrant; ♂ 1.2–1.6 cm Ø, sepals oblong to ovate or elliptic, 5–9 mm long, stamens 4–5 mm long; ♀ 2–3 cm Ø, sepals elliptic to ovate or nearly orbiculate, 1–1.6 cm long, pistils 3–7, 1 or more maturing. **Fruits** follicles, glaucous,

Above: *Berberis wightiana, Nandina domestica, Akebia quinata, Stauntonia hexaphylla.*

violet to dark purple, oblong, 5–15 cm long; seeds black, ovoid, pulp whitish. **Distr:** China, Japan, Korea. **Ecol:** In woodlands, forest margins; 0–1,500 m. Fl. Apr–May. Fr. Jun–Aug. **Zones:** 5–11.

Stauntonia

Syn: *Parvatia.*

Species ~25. **Climbers**, woody, evergreen, monoecious or dioecious. **Leaves** alternate, long petiolate; lamina palmately 3–9-foliolate or pinnately trifoliolate; petiolules unequal; leaflets with margins entire. **Inflorescences** axillary, to >10-flowered; bracts subtending inflorescences, scaly. **Flowers** with sepals 6, outer 3 valvate; petals absent or 6 and inconspicuous; ♂ with stamens 6, filaments usually connate into a tube, pistillodes 3, concealed within filament tube; ♀ with sepals usually slightly larger than those of ♂, staminodes minute, squamose, sessile, inserted at base of carpels opposite petals, carpels 3, straight, sessile, ovules in several rows, stigma terminal. **Fruits** single or in pairs, rarely trifascicled; seeds numerous, embedded in pulp. **Distr:** China, Hong Kong, N India, Japan, Myanmar, Taiwan, Vietnam. **Cult:** Propagate from seeds or cuttings. Requires moisture retentive but well-drained soil in a sheltered position, in semi-shade or in full sun.

S. hexaphylla Decne., *Ann. Sci. Nat., Bot.* sér. 2, 12: 105, 1839. (Syn: *Rajania hexaphylla, Rayana hexaphylla*). [Japanese Staunton Vine, Staunton's Climber, Stauntonia Vine]. **Climbers**, evergreen, to 10 m long, monoecious. **Leaves** palmately compound, to 30 cm long; leaflets (3–)5–7, elliptic to obovate, 6–15 × 2–5 cm, margins entire, apex acute, coriaceous, glabrous, adaxially light green, abaxially paler with prominent venation; petiole long. **Inflorescences** racemes, axillary. **Flowers** fragrant, ~2 cm long; sepals 6, fleshy, yellow-cream, tinged with purple-violet inside. **Fruits** ovoid, purplish, fragrant, fleshy, 5–10 × 2.5 cm, pulp white; seeds many, black. **Distr:** Myanmar, Hong Kong, Japan, Korea, Taiwan. **Ecol:** In thickets and woodlands, at low altitudes. Fl. Mar–May. **Zones:** 10–11. **Info:** The fruits are edible, watery with a sweet, honeylike flavor.

MENISPERMACEAE

Genera 70, **species** 420. **Shrubs** twining or rarely erect, or **small trees**, dioecious; wood in cross-section showing broad medullary rays. **Leaves** alternate, petiolate, exstipulate, sometimes peltate, without stipules; lamina usually simple, entire or lobed. **Inflorescences** various, many-flowered, flowers rarely solitary or geminate, axillary or borne on the leafless wood. **Flowers** small, actinomorphic, rarely slightly irregular; ♂ with sepals 3–12 or more, rarely 1, free or slightly connate, imbricate or valvate, petals 1–6 or absent, free or connate, usually imbricate, stamens 3–6 or indefinite, rarely 2, free or variously united; ♀ with sepals and petals generally as in ♂, sometimes not so numerous, staminodes present or absent, carpels 3–6 or more, rarely 1, free, ovules 2, soon reduced to 1 by abortion, attached to the ventral suture. **Fruits** drupaceous, with the scar of the style subterminal or near the base by excentric growth; seeds often curved and horseshoe-shaped, with uniform or ruminate endosperm, or without endosperm. **Distr:** Pantropical.

Anamirta

Species 1. **Lianas**, robust; almost entirely glabrous. **Leaves** palmately nerved at base, hairy domatia present in axils of nerves and main veins. **Inflorescences** cauliflorous; ♂ sometimes axillary, paniculate. **Flowers** ♂ usually with 2 minute outer sepals with 3+3 concave imbricate larger inner sepals, becoming reflexed, petals absent, synandrium consisting of a shortly stalked globular cluster of ~30–35 transversely dehiscing anthers; ♀ with sepals as in male, petals absent, staminodes 6, minute, carpels 3(–4) arising laterally from a central conical gynophore which greatly lengthens in fruit, stigma recurved. **Fruits** drupes, subreniform globose, pericarp very thin, endocarp woody with reticulate surface, with 2 small sublateral perforations on the ventral (concave) side; seeds with endosperm copious, embryo with divaricate, thin, foliaceous cotyledons. **Distr:** Indo-Malesia. **Cult:** Propagate from seeds or cuttings. Best grown against large trees. Requires a well-drained soil in shade or partial sun.

A. cocculus (L.) Wight and Arn., *Prod.* 1: 446, 1834. (Syn: *A. paniculata, A. toxifera, Menispermum cocculus*). [Crow Killer, Fish Berry, Levant Berry, Poison Berry]. **Lianas**, robust; young stems and petiole usually drying pale grayish straw-colored. **Leaves** with petiole 6–18(–26) cm, swollen at both ends, geniculate at base; lamina ovate to broadly ovate, base cordate to truncate (or broadly obtuse), apex usually acuminate, 16–28 × 10–24 cm, 3–5(–7) basal nerves with 4–5 pairs of lateral nerves which are linked together with scalariform veins. **Inflorescences** 16–40 cm long with lateral branches 2–5 cm long. **Flowers** ♂ with sepals white, yellow to pale green, outer 2 sepals scarcely 1 mm long, inner sepals broadly elliptic, 2.5–3 × 2 mm; synandrium 1.5–2 mm long; ♀ with sepals as in ♂, staminodes ~0.25 mm long, carpels 1.5–2 mm long. **Infructescences** with lateral branches to 15 cm long; carpophores (0.3–)0.6–1.6 cm long, continuous with pedicel 0.8–2 cm long. **Fruits** drupes, white, 0.9–1.1 cm long. **Distr:** SE Asia (excluding Borneo) to W New Guinea. **Ecol:** In lowland forests. Fl & Fr. Jan–Dec. **Zones:** 10–12. **Info:** The poisonous seeds have medicinal uses.

Cissampelos

Species 20–25. **Vines** or erect shrubs. **Leaves** ovate, cordate, or rotund, peltate or not, palmately veined. **Inflorescences** ♂ axillary, corymbose cymes, sometimes borne along a reduced shoot arising from axils of bracteal leaves, pedunculate; ♀ thyrsoid, elongate, composed of fascicles, bracts usually accrescent and foliaceous, overlapping. **Flowers** ♂ with sepals 4, obovate, often pubescent adaxially,

Below: *Anamirta cocculus* (habit), *Anamirta cocculus* (flowers), *Cissampelos pareira, Coscinium fenestratum.*

Top row: *Tinospora sinensis, Argemone mexicana, Eschscholzia californica.*
Bottom row: *Argemone ochroleuca, Bocconia frutescens.*

extended outward when blooming, petals connate into cup-shaped corolla, stamens 4(–10), connate into a peltate synandrium; ♀ with sepal 1, petals 1(–3), opposite to sepal, carpel 1, often pubescent. **Fruits** drupes, subglobose, slightly flattened, often pubescent, endocarp crustaceous or ± bony, horseshoe-shaped; seeds horseshoe-shaped, embryo elongate, terete, embedded in endosperm. **Distr:** Pantropical, mostly in Africa and America, few in Asia. **Cult:** Propagate from seeds or cuttings. Requires a well-drained soil in full or partial sun.

C. pareira L., *Sp. Pl.* 1031, 1753. (Syn: *C. hirsuta, C. subpeltata*). **Climbers**, slender, tomentose. [Velvet Leaf]. **Leaves** peltate, 2.5–12 × 2.5–11.5 cm, triangularly broad-ovate, or orbicular, obtuse, mucronate, base cordate or truncate, ± tomentose on both sides, ultimately becoming glabrous above and glaucous below; petiole pubescent. **Inflorescences** ♂ flowers in pedunculate branched cymes, clustered in the axil of a small leaf; ♀ flowers clustered in the axils of orbicular, whitish imbricate bracts, on 5–10 cm long racemes; pedicels filiform. **Flowers** minute; ♂ with sepals 4, obovate-oblong, hairy outside, petals 4, united to form a 4-toothed cup, hairy outside, stamens 4, filaments united, column short, anthers connate; ♀ with sepal 1, ovate-oblong, pubescent outside, petal 1, obtriangular subreniform, carpel 1, densely hairy, style shortly trifid. **Fruits** drupes, 4–6 × 3–4 mm, subglobose, compressed, hairy-pubescent, red when fresh, black when dry, endocarp transversely ribbed, tuberculate; seeds horseshoe-shaped. **Distr:** India and Pakistan. **Ecol:** Fl. Mar–Oct. **Zones:** 10–12.

Coscinium

Species 2. **Lianas**, large, with yellow wood and sap. **Leaves** often peltate, palmately nerved, adaxially often whitish-tomentellous. **Inflorescences** supra-axillary or ramiflorous, composed of a raceme of peduncled ± globose heads of flowers. **Flowers** ♂ with sepals 9, imbricate in 3 whorls, externally sericeous, petals 0, stamens 6, outer 3 free with 1-locular introrse anthers, inner 3 with connate filaments and with 2-locular latrorse anthers; ♀ with sepals as in ♂, petals 0, staminodes 6, carpels 3, densely pilose, style filiform, recurved. **Infructescences** with globose carpophore. **Fruits** drupes, subglobose, tomentellous, style-scar sublateral, endocarp covered with anastomosing fibrous ridges, condyle deeply intrusive, thickly clavate; seed subglobose, hollow, enveloping condyle. **Distr:** Sri Lanka, India to Indo-China and Malesia. **Cult:** Propagate from seeds or cuttings. Requires a moist, well-drained soil in full or partial sun.

C. fenestratum (Gaertn.) Colebr., *Trans. Linn. Soc. London* 13: 65, 1821. (Syn: *Menispermum fenestratum*). [False Calumba Root]. **Lianas**; branchlets brownish-tomentose at first, later glabrescent, becoming whitish. **Leaves** usually broadly ovate or ovate, sometimes ± oblong with basal lateral lobes, 11–33 × 8–23 cm, base broadly rounded, truncate or shallowly cordate, rarely broadly obtuse, apex acuminate; petioles 3–16 cm long, often much swollen at both ends, base geniculate, inserted up to 0.8(–2.7) cm from basal margin of lamina. **Inflorescences** arising singly or a few together, tomentose or tomentellous, globose heads, several-flowered, 6–7 mm Ø; peduncles 1–3 cm long, arranged in a raceme 5–11 cm long; ♂ flowers sessile, or pedicels to 1 mm long. **Flowers** ♂ with sepals densely sericeous-pilose outside, broadly elliptic to obovate, inner 3–6 spreading, yellow, 1.5–2 mm long, outermost smaller, 1–1.5 mm long; stamens 1 mm long; ♀ with staminodes claviform, 1 mm long, carpels 2 mm long. **Fruits** drupes, brown to orange or yellowish, 2.8–3 cm Ø; seed whitish. **Distr:** S India, Sri Lanka, Cambodia, W Malesia. **Ecol:** In lowland forests. **Zones:** 10–12.

Tinospora

Syn: *Campylus.*
Species >30. **Vines**, often with conspicuous aerial roots. **Leaves** not peltate, base cordate, sometimes sagittate or hastate, palmately veined. **Inflorescences** axillary or on old leafless stems, solitary or fascicled, racemose, cymose, or paniculate. **Flowers** ♂ with sepals usually 6 in 2 whorls, free, imbricate, outer 3 usually conspicuously smaller, membranous, petals (3 or)6, base clawed, often with lateral edges involute, stamens 6, filaments free (or connate), anthers subextrorse, dehiscing longitudinally and obliquely; ♀ with sepals and petals as in ♂ except petals often smaller, staminodes 6, carpels 3, curved-ellipsoidal, style short, fat and thick, stigma reflexed with short, pointed lobes. **Fruits** drupes, 1–3 borne on a short or columnar carpophore, style scar subterminal; seed half-moon-shaped. **Distr:** Tropical and subtropical Asia to Australia and the Pacific Islands; Africa, and Madagascar.

T. sinensis (Loureiro) Merrill, *Sunyatsenia.* 1:

193, 1934. (Syn: *Menispermum malabaricum*, etc.). [Malabar Gulbel, Chinese Tinospora]. **Vines**, deciduous, to 20+ m long, often producing very long aerial roots; stems slightly fleshy; lenticels raised, (2–)4(–6)-dehiscent. **Leaves** broadly ovate to subrotund, 7–14 × 5–13 cm, base ± cordate, margins entire, apex acutely acuminate, papery, palmately 5(–7)-veined at base; petiole (4–)6–13 cm long, puberulent. **Inflorescences** appear when plant is leafless; ♂ solitary or few fascicled, 1–4 cm or longer; ♀ solitary. **Flowers** ♂ with sepals 6 in 2 whorls, outer 3 oblong or subelliptic, 1–1.5 mm long, inner 3 broadly ovate, to 5 × ~3 mm, petals 6, rhomboidal, with claw ~1 mm long, lobes ~2 mm long, stamens 6, filaments ~4 mm long; ♀ with sepals and petals as in ♂, staminodes 6, to 1 mm long, carpels 3. **Fruits** drupes, red, subglobose, 7–9 × ~6 mm, surface irregularly tubercled or ridged. **Distr:** Cambodia, China, India, Nepal, Sri Lanka, Thailand, Vietnam. **Ecol:** In forest margins. Fl. Apr. Fr. May–Jun. **Zones:** 9–12.

PAPAVERACEAE

Genera 44, **species** 760. **Herbs**, annual, biennial, monocarpic perennial, perennial, or shrubby, non-mycorrhizal; laticifers (or elongated idioblasts) articulated or not, anastomosing or not. **Leaves** alternate or in a basal rosette, rarely opposite or whorled, usually without stipules; lamina entire to compound. **Inflorescences** racemes, panicles, dichasia, pseudo-umbels, or solitary flowers. **Flowers** actinomorphic, bisymmetric, or zygomorphic, always bisexual, usually 2-merous, rarely 3- or 4-merous; calyx caducous, green or petaloid; corolla choripetalous or quasi-sympetalous, very rarely absent; anthers opening by slits; ovary superior, syncarpous with 2 to several carpels, placentation parietal. **Distr:** Mainly in the N Hemisphere of both the Old and New Worlds, extending into C and S America, a few in Africa.

Argemone

Species 29. **Herbs**, annual, biennial, or perennial, usually stout, mostly branched, spiny, yellow-lactiferous; stems mostly erect. **Leaves** pinnate, lobes repand dentate, teeth apically spiny; 2 types, basal and cauline. **Inflorescences** with flowers solitary, terminal or arranged in cymes. **Flowers** 3-merous; receptacles narrowly conical; flower buds erect; sepals (2 or)3, caducous, apex with horned appendices; petals (4–)6, in 2 whorls, contorted or imbricate in bud, orange, yellow, yellow-white, or white, occasionally pink or mauve; stamens many, free, filaments usually filiform; ovary usually ovoid, carpels (3 or)4–6, united, ovules many, styles very short or absent, stigmas actinomorphic. **Fruits** capsules, usually spiny, apex divided for a short distance into 3–6 valves, rarely parting nearly to base; seeds many, spheroidal. **Distr:** Mainly in N America from USA to C Mexico and the West Indies, S America (9 spp.), Hawai'i (1 sp.); and the others scattered along the NW and W coasts of the Americas. **Cult:** Propagated easily from seeds. Requires any gravel type soil in full sun.

A. mexicana L., *Sp. Pl.* 1: 508, 1753. [Devil's Fig, Mexican Poppy, Prickly Poppy]. **Herbs**, annual or occasionally short-lived perennial, 0.3–1 m tall, usually stout; stems mostly short, branched, with sparse explanate fulvous spines. **Leaves** *basal* densely arranged, petiole 0.5–1 cm long, lamina broadly oblanceolate or obovate to elliptic, 5–20(–25) × 2.5–7.5(–8) cm, base cuneate, margins pinnatipartite, apex acute, glaucous with blue-green markings on veins, paler abaxially, glabrous, sparsely sharply spiny on veins, lobes undulate, dentate, teeth apically spiny; *cauline* leaves alternate, similar to basal leaves, but upper leaves smaller and sessile, often subamplexicaul. **Inflorescences** with flowers solitary, sometimes in few-flowered cymes; pedicels very short. **Flowers** with sepals 2(or 3), cymbiform, ~1 cm long, apex spurred, glabrous or sparsely spiny; petals 6, yellow or orange, broadly obovate, 1.7–3 cm long, base broadly cuneate, apex rounded; filaments ~7 mm long, anthers narrowly oblong, 1.5–2 mm long, coiling after dehiscence; ovary elliptic or oblong, 0.7–1 cm long, with explanate fulvous spines, styles very short, stigmas dark red, 4–6-lobed. **Fruits** capsules, oblong to broadly elliptic, 2.5–5 × 1.5–3 cm, sparsely fulvous-spiny, 4–6-valved dehiscing from apex; seeds spherical, 1.5–2 mm Ø, tessellate. **Distr:** C and tropical America. **Ecol:** Fl. and Fr. Mar–Oct. **Zones:** 9–12.

A. ochroleuca Sweet, *Brit. Fl. Gard.* 3: t.242, 1828. (Syn: *A. mexicana* var. *ochroleuca*). [Chicalote, Pale Mexican Poppy, Pale Prickly Poppy]. **Herbs**, prickly, glaucous, usually 35–40 cm tall; young stem whitish-purple or violaceous, fading with age; prickles 6–9 mm long, with broader bases. **Leaves** sessile, amplexicaul, *basal* leaves forming a rosette; *cauline* leaves alternate, variegated with white, 10–30 × 4–10 cm, sinuate-pinnatifid, segments incised dentate, with sharp yellowish spines on the margins, midrib and veins beneath, midrib and veins on lower side purplish-blue. **Inflorescences** with flowers solitary, axillary; sessile or on a short pedicel, subtended by 2–3 foliaceous bracts; bracts 2–4 × ~0.5 cm. **Flowers** white or creamish-white, becoming yellowish, 2.5–3 cm Ø; sepals 3, obovate oblong, 0.8–1 × 0.5–0.7 cm; petals 6, obovate, 2.8–3 × ~1.8 cm; stamens 0.7–1 cm long, anthers oblong, recurved; ovary ovate, 0.8–1 × 0.5 cm, covered with long pointed spines, stigma deeply dissected, 5-lobed, reddish-brown with an obscure style. **Fruits** capsules, oblong, 3–4 × 1.5–2 cm (with a prominent, 1–2 mm long style) covered with sharp, erect prickles; seeds black, rounded, 1.5–2 mm Ø, finely reticulate. **Distr:** Mexico. **Ecol:** Fl. Mar–May. **Zones:** 9–12.

Bocconia

Species 5. **Shrubs**, **small trees**, or **large herbs**, often glaucous, glabrous or pubescent, the sap yellow or orange. **Leaves** large, lobate, dentate, or entire. **Inflorescences** large terminal panicles, much-branched. **Flowers** small; sepals 2; petals none; stamens numerous or sometimes of definite number (8–24); ovary with 2 placentas, these sterile or bearing only a few ovules, only a basal ovule fertile, style short or somewhat elongate, stigma lobes oblong or linear, erect or recurved. **Fruits** capsules, ± stipitate, ellipsoid, dehiscent to the base, the valves recurved; seed usually 1, surrounded at base by a pulpy aril. **Distr:** Tropical America (from Mexico to Argentina, and West Indies). **Cult:** Propagate from seeds or suckers. Requires a humus-rich, moist, well-drained soil in full or partial sun.

B. frutescens L., *Sp. Pl.*: 505, 1753. [Bocconia, Parrot Weed, Plume Poppy, Sea Ox-eye Daisy, Tree Celandine, Tree Poppy]. **Shrubs**, usually 1.5–3 m high, simple or branched; young branches somewhat lanate-tomentose. **Leaves** oblong-elliptical, 15–35 × 10–20 cm, base usually truncate or rounded, margins pinnatifid, lobes subentire, short, very obtuse to rounded, repand-denticulate, adaxially glabrous or nearly, abaxially somewhat tomentose beneath or glabrate, often glaucous but sometimes green, upper leaves on the inflorescence with margins entire; petiole 1.5–3 cm long, tomentose, ± glabrescent. **Inflorescences** panicles, 25–40 cm long, laxly many-flowered; pedicels to 1 cm long. **Flowers** with sepals abruptly acuminate, pale, 0.6–1 cm long or often somewhat shorter; stamens ~16, nearly equaling sepals, anthers 4–5 mm long; style protruding, ~2 mm long, stigmatic branches longer, revolute. **Fruits** ellipsoid, 6–8 mm long, ends acute, long-stipitate; seed 6 mm long, muricate. **Distr:** Mexico to C America and the West Indies. **Ecol:** In moist thickets or forests; 0–2,800 m. **Info:** The roots and its juice used medicinally.

Eschscholzia

Species ~12. **Herbs**, perennial to annual, glabrous, slightly hyaline lactiferous. **Leaves** alternate, generally tripinnatipartite; lobules mostly linear, narrow. **Flowers** solitary on long pedicels, yellow or orange, occasionally white or cream, large; receptacles cupular-inflated, flowers perigynous; sepals 2, fused, usually herbaceous in one piece, oblong or conical, in bud margins joined to caplike, at flowering pushed off in one piece due to expanding corolla, apically acute or acuminate; petals 4, broad, from margin of cupular receptacle, imbricate or contorted in bud; stamens many, from margin of cupular receptacles, filaments short, anthers linear or oblong, usually longer than filaments, basifixed, 2-celled, introrse; ovary linear, 1-loculed, 2-carpellate, from bottom of cupular receptacles, styles very short; stigmas 2 or more, almost broadly linear. **Fruits** capsules narrowly terete, 10-ribbed, 2-valved, dehiscing from base to apex, later usually curved; seeds many. **Distr:** Pacific region of N America. **Ecol:** In deserts and grasslands. **Cult:** Propagate from seeds. Requires a gritty, well-drained soil in full sun.

E. californica Chamisso, *Horae Phys. Berol.* 73, 1820. [California Poppy]. **Herbs**, annual or short-lived perennial, glaucous, glabrous; stems erect to spreading, 30–60 cm high, longitudinally costate, with many branches spreading, dichotomous. **Leaves** *basal* several, 10–30 cm, petiolate, lamina tripinnatipartite, lobes various, acutely linear, acutely oblong, or obtusely spathulate-oblong, with 3 terminal lobes, middle mostly broad and short; *cauline* leaves similar to basal leaves, but progressively smaller and shortly petiolate. **Flowers** solitary, terminal; pedicel 5–15 cm long; receptacle concave, funnelform or nearly tubular, 3–4 mm Ø, cupular after flowering, margin undulately reflexed; calyx ovoid, ~1 cm long, apically shortly conical, caplike; petals 4, yellow, sometimes with orange flecks at base, deltoid-flabellate, 2.5–3 cm long; stamens >40, filaments filiform, basally inflated, ~3 mm long, anthers orange, linear, 5–6 mm long; ovary long, narrow, styles short, stigmas 4, subulate-linear, unequally long. **Fruits** capsules, narrowly terete, 5–8 cm long; seeds spherical, 1–1.5 mm Ø, tessellated. **Distr:** USA (California). **Ecol:** Fl. Apr–Aug. Fr. Jun–Sep. **Zones:** 6–11.

Papaver

Species 50–80. **Herbs**, annual, biennial, or perennial, rarely monocarpic; stems 1 or more, erect or ascending, terete, branched or not, very constricted or extended, usually setose, rarely glabrous, ivory malodorous lactiferous, with leaves or not. **Leaves** *basal* petiolate, usually glaucous adaxially, variously shaped, both surfaces setose, pinnatilobate, pinnatipartite, pinnatisect, or bipinnatifid, sometimes incised, serrate, or crenate, rarely entire; *cauline* leaves if present, similar to basal leaves, but usually sessile, sometimes amplexicaul. **Inflorescences** with flowers solitary, rarely in cymose racemes, pedunculate or sometimes scapose, erect; flower buds nutant, ovoid or spherical. **Flowers** with sepals 2(3), shedding as flowers open; petals 4(5 or 6), on short receptacles, often caducous, mostly red, rarely white, yellow, orange, or lavender, vivid and showy, usually obovate, in 2 whorls, outer whorl larger; stamens many, filaments white, yellow, green, purple, or red, occasionally blackish, mostly filamentous; ovary 1-loculed, superior, usually ovoid, carpels 4–18, united, setose or glabrous, ovules many, astylous, stigmas 4–18, actinomorphic, united into compressed or steepled disk covering ovary. **Fruits** capsules, narrowly terete, obovoid, or spherical, poricidal under actinomorphic

Above: *Papaver somniferum*, *Clematis smilacifolia*, *Naravelia zeylanica*.

stigmas; seeds many. **Distr:** Mainly in C and S Europe to temperate Asia, some in the Americas and Oceania, 1 in S Africa. **Cult:** Propagate annuals and biennials from seeds, and perennials from root cuttings or seeds. Requires moist but well-drained soil in full sun.

P. somniferum L., *Sp. Pl.* 1: 508. 1753. [Opium Poppy]. **Herbs**, annual, 0.3–0.6(–1.5) m tall. **Leaves** alternate; lamina ovate or oblong, 7–25 cm long, surfaces glabrous, glaucous, waxy, veins distinct, slightly raised, base cordate, margins irregularly undulate-serrate, apex acuminate to obtuse; *basal* leaves shortly petiolate but above sessile and amplexicaul. **Inflorescences** with flowers solitary; pedicel to 25 cm long, glabrous or rarely sparsely setose; flower buds nutant at first, erect at anthesis, oval-oblong or broadly ovoid, 1.5–3.5 × 1–3 cm, glabrous. **Flowers** deeply cup-shaped, 5–12 cm Ø; sepals 2, green, broadly ovate, margin membranous; petals 4, white, pink, red, purple, or various, often with a dark basal blotch, suborbicular or almost flabellate, 4–7 × 3–11 cm, margin undulate or variously lobed; stamens many, filaments white, linear, 1–1.5 cm long, anthers yellowish or cream, oblong, 3–6 mm long; ovary green, spherical, 1–2 cm Ø, glabrous, stigmas 5–12(–18), actinomorphic, united into a compressed disk, disk margin deeply divided, lobes crenulate. **Fruit** a capsule, brown when mature, spherical or oblong-elliptic, 4–9 × 4–5 cm, glabrous; seeds many, black or deep gray, adaxially alveolate. **Distr:** SE Europe, W Asia. **Ecol:** Fl. Mar–Aug. **Info:** All parts of the plant are poisonous. Cultivated for ornamental and medicinal purposes, especially for the pharmaceutical substances heroin, morphine, and other alkaloids. The seeds are used in cooking.

RANUNCULACEAE

Genera 62, **species** 2,525. **Herbs**, perennial or annual, sometimes subshrubs or herbaceous or woody vines. **Leaves** basal and cauline, alternate, rarely opposite or whorled, with or without stipules; lamina simple or variously compound, palmately nerved, rarely penninerved. **Inflorescence** a simple or compound monochasium, dichasium, simple or compound raceme, or flowers solitary. **Flowers** bisexual, sometimes unisexual, actinomorphic, rarely zygomorphic, hypogynous; sepals 3–6 or more, free, petaloid or sepaloid, imbricate or sometimes valvate in bud; petals present or absent, 2–8 or more, free, usually with nectaries; stamens numerous, rarely few, free, filaments linear or filiform, anthers latrorse, introrse, or extrorse, sometimes some sterile stamens becoming staminodes; carpels numerous or few, rarely 1, free, rarely connate to various degrees, ovary with 1 to many ovules. **Fruits** follicles or achenes, rarely capsules or berries; seeds small, with abundant endosperm and minute embryo. **Distr:** ± Worldwide, especially in northern temperate regions.

Clematis

Syn: *Archiclematis*, *Atragene*.

Species 300. **Vines**, woody or herbaceous and perennial, rarely erect shrubs, subshrubs, or perennial herbs. **Leaves** opposite, rarely fascicled or alternate, simple or compound; leaf or leaflet lamina palmately veined. **Inflorescences** cymose, sometimes solitary, usually pedunculate, 2-bracteate. **Flowers** bisexual or unisexual; sepals 4 or 5(–8), petaloid, spreading, ascending, or erect, usually valvate in bud; petals absent; stamens numerous, sometimes outer ones sterile and becoming linear or petaloid staminodes, anthers introrse; carpels numerous, usually pubescent or villous, ovaries 1-ovuled, ovule pendulous, style elongated after anthesis, rarely slightly or not elongated. **Fruits** achenes, usually ± bilaterally compressed, persistent style usually strongly elongated, plumose. **Distr:** Widespread throughout the world. **Cult:** Propagate from cuttings or layering. Requires fertile, humus-rich soil with the foliage preferably in the sun and the root system in a cool moist situation away from the sun.

C. smilacifolia Wallich, *Asiat. Res.* 13: 402, 1820. [Tropical Clematis]. **Vines**, woody; branches shallowly 8–20-grooved, glabrous. **Leaves** simple; petiole 3.5–6 cm long; lamina ovate to deltoid-ovate, 8–16.5 × 6.5–11 cm, leathery, surfaces glabrous, obscurely reticulate, base cordate to subcordate, sometimes ± peltate, margins entire, apex ± obtuse to acute; basal veins 5–7, abaxially ± prominent. **Inflorescences** cymes paniclelike, axillary, laxly flowered; peduncle 1–11 cm long, sparsely puberulous; bracts small, linear; pedicels 4–7 cm long, densely puberulous. **Flowers** 3–3.6 cm Ø; sepals 4 or 5, blue-purple, lanceolate-oblong, 1.6–1.8 × 0.5–0.7 cm, abaxially reddish-brown velutinous, adaxially glabrous, margin often revolute, apex ± obtuse; stamens 0.8–1.4 cm, glabrous, anthers narrowly oblong, 2.5–3 mm long, apex with a conspicuous 1–2 mm apicule; ovaries pubescent, style ~6 mm long, densely villous. **Fruits** achenes narrowly ovate, ~6 × 2 mm, puberulous; persistent style 5–8 cm long, plumose. **Distr:** Tropical and subtropical Asia to Malesia. **Ecol:** In forests, forest margins, scrubs; 900–2,300 m. Fl. Nov–Dec. Fr. Dec–Jan. **Zones:** 9–12.

Naravelia

Syn: *Atragene*.

Species ~9. **Climbers** or **woody vines**; stems terete. **Leaves** opposite; lamina trifoliolate or pinnately compound, exstipulate, terminal pinna generally transformed into a trifid tendril, basal 2 leaflets developed normally; lamina with margins entire, apex obtuse. **Inflorescences** many-flowered, axillary and terminal panicles; bracts present, bracteoles 2; pedicel 1–1.5 cm long, basally pilose. **Flowers** regular, bisexual, spreading; tepals 4–5, sepaloid, valvate, caducous; staminodes 8–14, petaloid, narrow, linear to club-shaped or clavate, elongating after fall of perianth; stamens numerous, filaments short, flattened; anthers linear, connectives prolonged beyond anthers; pistils hairy, carpels many, ovules unitegmic, 1 ovule per ovary, pendulous, styles persistent, plumose. **Fruits** achenes, with a short stalk, linear, slender, shortly stipitate, sometimes twisted at apex, terminated by the elongated feathery style. **Distr:** Indo-Malesian region, S Asia, SE Asia and S China. **Cult:** Propagate from seeds or cuttings. Requires a moderately rich, moist but well-drained soil in full sun.

N. zeylanica (L.) DC., *Syst. Nat.* 167, 1817. (Syn: *Atragene zeylanica*, *N. pilulifera* var. *yunnanensis*). [Narawel]. **Shrubs** climbing; young parts pubescent; stem striate, subglabrous. **Leaves** pinnately compound; lateral leaflet petiolules 1.5–2.5 cm; central leaflet petiolule 3–7 cm, shallowly striate; lamina ovate, 6–11 × 6–10 cm, base rounded to cordate, 5-veined from base. **Inflorescences** panicles, axillary or terminal; peduncle to 7 cm long, purplish below, silky-pubescent; bracts to ~1 cm long, narrowly oblanceolate, pubescent; pedicels, basally pilose, 1–3 cm long, with 2 opposite bracteoles ~1 mm long. **Flowers** fragrant, tepals 5–8 × 3–5 mm, ovate, concave, brownish-green, purplish below, velvety, caducous early; staminodes 12, in 4 groups of 3 each in the bud, 0.8–1.2 cm long, green, caducous; stamens many, erect, linear, 2.5–5 mm long, pale yellowish-green, anthers 2–3 mm long, yellow; pistils linear, carpels silky-hairy, style 1–2 mm long. **Fruits** achenes, fusiform, 15–20, linear, 2–4 mm long, with persistent style 4–5 cm long, spirally twisted, feathery. **Distr:** Bhutan, India to the Himalayas, Nepal, Sikkim, Sri Lanka. **Ecol:** In moist lowland forests. Fl. Sep–Jan. Fr. Nov. **Zones:** 9–12.

PROTEALES

FAMILIES 3, GENERA 82, SPECIES ~1,610.

Herbs, perennial, **shrubs** or **trees**. **Leaves** alternate or rarely opposite; lamina simple or compound. **Flowers** perfect, or occasionally polygamous or dioecious, 4-merous; perianth of 4 valvate distinct or partially united sepals; androecium of 4 stamens, one borne on each sepal (stamens opposite perianth parts); gynoecium a single carpel, carpels with 5 vascular bundles, ovary superior, often oblique; gynoecia with 1 or 2 pendent ovules per carpel may be synamorphic (Judd et al. 2008). **Fruits** indehiscent, or often follicular or capsular.

PMer

Above: *Nelumbo nucifera*, *Nelumbo nucifera* (white form), *Persoonia pinifolia*.

NELUMBONACEAE

Genera 1, **species** 2. *Familial description same as generic.*

Nelumbo

Syn: *Nelumbium*.

Species 2. **Herbs** perennial, aquatic; rhizomes branched, repent, forming swollen terminal storage tubers late in growing season; roots fibrous. **Leaves** arising from the rhizome on upper side, alternate, emersed or floating, usually standing above the water, long petiolate; lamina large, suborbicular, centrally peltate, veins radially extended; each leaf accompanied by 2 scale leaves and an ochrealike stipule forming a sheath around the terminal bud at its base. **Inflorescence** from axil of upper scale-leaf, flowers solitary, long-pedunculate. **Flower** large, bisexual, hypogynous, actinomorphic, entomophilous, borne above the water surface; perianth segments many, free and distinct, spirally arranged, outermost greenish, reduced and sepaloid, inner ones larger and progressively petaloid, caducous; stamens numerous, free, spirally arranged, filaments slender or dilated; pistils numerous, simple, ovary 1-loculed, consisting of many free carpels sunk individually in the large obconic flat-topped spongy receptacle, ovule solitary, styles very short, stigmas capitate, protruding. **Fruits** nutlike, indehiscent, embedded in the enlarged receptacle; seeds without endosperm and perisperm. **Distr:** E and S Asia, N Australia, C and N America. **Cult:** Propagate by division of rhizomes or from seeds. Requires shallow water, with the rhizomes placed in baskets or a rich soil mix. **Info:** Flowers pollinated by beetles.

N. nucifera Gaertn., *Fruct.* 1: 73. t. 19, f. 2. 1788. (Syn: *Nymphaea nelumbo*, etc.). [Lotus]. **Herbs** with creeping rhizomes, to ~5 cm across, white, brown-spotted, with internodes to >50 cm long; roots fibrous, white, arising all around the rhizome and confined to a zone just behind a node. **Leaves** orbicular or suborbicular, 25–90 cm Ø, margins entire, apex with a small mucro, adaxially dark blue-green, abaxially glaucous and light green, papery, flat, floating, water-repellent, ± concave or cuplike, veins prominent beneath; petiole 1–2 m long, terete, light-green, floating portion weak, raised portion rigid. **Inflorescence**, flowers solitary; lower scale leaves 5–6.5 cm long, upper 12–15 cm long; peduncles longer than petiole, glabrous or sparsely spinulate. **Flower** 10–23 cm Ø; tepals caducous, pink or white, oblong-elliptic to obovate, 5–10 × 3–5 cm; stamens slightly longer than receptacle; filament slender, anther linear, 1–2 mm long; connective appendage clavate, to 7 mm long, incurved; receptacle accrescent, turbinate, 5–10 cm Ø. **Fruits** oblong to ovoid, 1–2 × 0.7–1.5 cm, glabrous; pericarp thick, hardened. **Distr:** N Australia and warmer parts of SW Asia. **Ecol:** In ponds, lakes and reservoirs, sometimes in dry regions; 0–500 m. Fl. throughout the year. **Zones:** 9–11.

PROTEACEAE

Subfamilies 5, **Genera** ~80, **species** 1,600. **Trees** or **shrubs**. **Stipules** absent. **Leaves** alternate, rarely opposite or whorled, simple or variously divided. **Inflorescences** axillary, ramiflorous, cauliflorous, or terminal, simple or rarely compound, with flowers borne laterally either in pairs or sometimes singly, racemose, sometimes spicate, paniculate, or condensed into a head; bracts subtending flower pairs usually small, sometimes accrescent and woody; floral bracts usually minute or absent. **Flowers** bisexual or rarely unisexual and dioecious, actinomorphic or zygomorphic; perianth segments (3 or)4(or 5), valvate, usually tubular in bud; limb short, variously split at anthesis; stamens 4, opposite perianth segments, filaments usually adnate to perianth and not distinct, anthers basifixed, usually 2-loculed, longitudinally dehiscent, connective often prolonged; hypogynous glands 4 (or 1–3 or absent), free or variously connate; ovary superior, 1-loculed; ovules 1 or 2(or more), pendulous; style terminal, simple, often apically clavate; stigma terminal or lateral, mostly small. **Fruits** follicles, achenes, drupes or drupaceous; seeds 1 or 2(or few to many), sometimes winged; endosperm absent (or vestigial). **Distr:** Mostly in tropics and subtropics, predominantly in S Africa and Australia. **Info:** The classification followed here is that of Weston and Barker (2006).

Subfam. PERSOONIOIDEAE

Genera 5, **species** 110. **Cluster roots** absent. **Inflorescences** bracteate racemes or spikes or panicles of racemes. **Flowers** with staminal filaments largely or completely adnate to tepals, anthers free or basally (or completely) adnate to tepals; hypogynous glands present. **Fruits** not winged; cotyledons not auriculate (Weston and Barker 2006). **Distr:** Mostly Australia, also New Caledonia and New Zealand.

Tribe PERSOONIEAE

Persoonia

Syn: *Linkia*, *Pentadactylon*, *Pycnonia*.

Species 98. **Shrubs** or **small trees**. **Leaves** alternate or rarely opposite or whorled, sessile to shortly petiolate. **Inflorescences** terminal or lateral, abbreviated to elongate, anauxotelic or auxotelic; flowers subtended by scale leaves or leaves. **Flowers** with perianth actinomorphic or zygomorphic; tepals free but usually coherent towards base; stamens usually all fertile or rarely anterior stamen infertile, filaments adnate to tepals, sometimes free at tips, anthers straight or recurved, rarely flattened; hypogynous glands 2 or 4; anterior pair equal to posterior pair or much reduced or absent; gynoecium longer or shorter than stamens; ovary shortly stipitate or not contracted at base, style straight or slightly sinuous or prominently recurved; ovules 1 or 2. **Fruits** pyrenes, smooth; seeds 1 or 2, cotyledons 2–9. **Distr:** Australia. **Ecol:** <1,500 m. **Cult:** Propagate from seeds or young tip cuttings. Requires a slightly acidic, well-drained soil in full or partial sun.

P. pinifolia R. Br., *Trans. Linn. Soc. London* 10: 180, 1810. (Syn: *Linkia pinifolia*, etc.). [Mambara, Pine-leaved Geebung]. **Shrubs**, erect to spreading 2–4 m tall; bark smooth; hairs grayish, appressed to antrorsely spreading; young branchlets moderately hairy. **Leaves** linear, 3–7 cm × 0.5 mm, terete, often recurved, spreading, straight, innocuous, concolorous, not glaucous, sparsely to moderately hairy when immature, glabrescent. **Inflorescences** anauxotelic, 30–200-flowered; rachis 3.5–26 cm long. **Flowers** subtended by reduced leaves and scale leaves; pedicels 1–4 mm long, spreading, moderately hairy; tepals 8–9 mm long, obtuse to apiculate, moderately hairy on outside; ovary glabrous; ovule 1. **Fruits** pyrenes, ovoid, glossy. **Distr:** Australia (from Broken Bay to Cataract River in central-eastern NSW and west to the lower Blue Mountains). **Ecol:** In heath to dry sclerophyll forests on sandstone; 0–600 m. Fl. Dec–Mar. **Zones:** 9–11.

Subfam. PROTEOIDEAE

Genera 25, **species** 640. **Cluster roots** present. **Inflorescences** bracteate, basically racemose but often compound and often condensed to form spikes or capitula, sometimes with different orders of branching differing qualitatively in structure. **Flowers** with staminal filaments very slightly to completely adnate to tepals; ovule 1 (2 in a few flowers of some species of *Petrophile* and *Isopogon*). **Fruits** indehiscent; seeds with cotyledons not auriculate (Weston and Barker 2006). **Distr:** Africa south of the Sahara, especially the Cape region, Australia.

Tribe PROTEEAE

Protea

Species 115. **Small trees**, **shrubs** or acaulescent **perennials**; stems glabrous or hairy. **Leaves** alternate; lamina entire, coriaceous, hairy or glabrous. **Inflorescences** many-flowered, sessile or subsessile, terminal or lateral, usually solitary heads, enclosed in an involucre of numerous imbricate, coriaceous to scarious, glabrous or hairy, sometimes bearded, variously colored bracts. **Flowers** hermaphrodite, zygomorphic; perianth 4-merous, tubular in bud, slender, ± widened towards base, early divided into an anticous (abaxial) segment with a long very slender claw and a narrow oblong or linear limb and a posticous (adaxial) portion, consisting of fused posticous and lateral segments, claws forming a long anteriorly open sheath and their limbs a linear to oblong, concave or subtubular, equally or unequally

MD

Top row: *Protea aurea, Protea cynaroides, Protea repens.*
Bottom row: *Leucadendron argenteum, Serruria florida.*

trilobed lip, lobes either short, toothlike or produced into slender often filiform awnlike and very hairy processes, whole perianth variously hairy or glabrous; anthers sessile or subsessile, connective tip produced into a small fleshy gland; hypogynous scales 4, free, variously shaped, rarely absent; ovary long-hairy, style rigid, straight or curved, terete or laterally compressed, stigma slender, gradually passing into style or suddenly bent or kneed at the junction with the style, ovule 1. **Fruits** nuts, densely bearded, crowned by the persistent style. **Distr:** SW Cape, extending northwards into tropical Africa, few north of the equator. **Cult:** Propagate from seeds, cuttings or by grafting. Requires well-drained, gravelly, sandy or loamy soil in full or partial sun.

P. aurea (Burm. f.) Rourke, *J. S. African Bot.* 45(4): 471, 1979. (Syn: *P. longiflora*). [Common Shuttlecock Sugarbush, Long-bud Sugarbush]. **Shrubs** or **bushy trees**, glabrous, to 5 m tall, fast-growing, short-lived (~8 years); trunk single, prominent. **Leaves** elliptic or ovate to oblong, 4–10 × 2–4 cm, base cordate, slightly lobed and clasping the stem, leathery, shiny, hairy at first. **Inflorescences** heads, solitary, shuttlecock-shaped initially at anthesis, 9–12 × 6 cm; involucral bracts in 5–8 rows, silky, silvery, pink to creamish-green to dark pink or bright red, with brown margins; outer bracts broadly ovate, apex broadly cuspidate, surface and margins sericeous; inner series oblong or broadly ensiform, 3.5–9 × 1–1.4 cm. **Flowers** with receptacle convex; perianth white to occasionally pink, 8–10 cm long, straight, bilaterally symmetric with the 3 adaxial perianth segments fused from base, fourth perianth segment separates from perianth sheath as each individual flower opens; style 7–10.5 cm long; ovary ~3 mm long. **Fruits** nuts, densely hairy. **Distr:** S Africa (S Cape Region). **Ecol:** On mountain fynbos, in dense stands on cool, moist, southern slopes; 150–800 m. Fl. throughout the year. **Zones:** 8–11.

P. cynaroides (L.) L., *Mant. Pl. Altera* 190, 1771. [King Protea, King Sugarbush]. **Shrubs**, bushy, to 2 m high, sometimes acaulescent; underground rootstock produces new growth after veld fires; branches glabrous. **Leaves** subrotund and obtuse, or elliptic and acute, 6.5–13 × 5–9 cm, base cuneate, reticulately veined, punctate, glabrous; petiole to 11.5 cm long. **Inflorescences** heads, sessile, 12.5–20 × ~12.5–30 cm; involucral bracts 12–13-seriate, pale pink, rarely white; outer bracts ovate to ovate-lanceolate, acute, like the inner at first ± densely whitish- or grayish-tomentose; inner bracts lanceolate-oblong, acuminate or acute, mostly tomentose, exceeding the flowers. **Flowers** with perianth-sheath >5 cm long, outside pubescent above, pubescent to villous inside upwards; lip ~2.5 mm long, produced into 3 tomentose or villous awns, lateral awns ~5 mm long, median awn ~5 mm long; stamens all fertile, filaments ~6 mm long, flattened, anthers linear, ~1 cm long, apical glands oblong, ~2 mm long, obtuse; ovary oblong, ~5 mm long, whitish-hairy, style 6.5 cm long, laterally flattened, stigma filiform, ~1 cm long. **Distr:** S Africa (SW to SE coast of the Cape Region). **Ecol:** <1,000 m. **Zones:** 8–10.

P. repens L., *Mant. Pl. Altera* 189, 1771. [Real Sugarbush]. **Shrubs** acaulescent, to 5+ m high; stems erect, grayish-brown. **Leaves** sessile; lamina narrow, filiform to linear, 6.5–20 × 1.2–2.5 cm, apex acute, coriaceous, glabrous, smooth or ± tubercled and rough, yellowish to bluish-green. **Inflorescences** heads, shortly stipitate, 10 × ~7.5 cm; involucral bracts 10-seriate, white to deep pink, silky-pubescent outside, ciliolate; outer bracts ovate, obtuse; inner bracts much-elongated, oblong; innermost bracts narrowed downwards, equaling flowers. **Flowers** with perianth-sheath 5 cm long, dilated, pilose to villous outside, base glabrous; lip 2.3 cm long, densely pubescent to villous, 3-awned, lateral awns ~7 mm long, yellowish-hairy, median awn ~3.2 mm long; stamens all fertile, filaments swollen, anthers linear, 1.1 cm long; ovary obovoid, ~2.5 mm long, reddish-brown hairy, style 5 cm long, widened and flattened below, terete and tapered above, stigma 0.8–1 cm long; nectar copious. **Distr:** S Africa (SW to SW of the Cape Region). **Ecol:** Fl. Dec–Jun. **Zones:** 8–11.

Tribe LEUCADENDREAE

Leucadendron

Species 80. **Trees**, **shrubs**, or decumbent **undershrubs**. **Leaves** acicular, linear, lanceolate to obovate, margins entire, hardened at apex into a blunt (rarely acute) callus, coriaceous, glabrous, pilose or with silky-silvery indumentum. **Inflorescences** with ♂ flowers usually numerous, arranged in conical, globular or cylindric heads, each flower subtended by a bract, heads terminal, sessile or rarely pedunculate, sometimes surrounded by an involucre of imbricate bracts within upper leaves; ♀ flowers solitary, subtended by woody bracts aggregated in conelike heads; bracts erect or spreading, usually free. **Flowers** dioecious, regular; ♂ with perianth linear, limb thickened, segments at length usually separated to middle, differentiated into limb and claw, limb linear, oblong or elliptic, anthers sessile at base of limb, oblong or linear, hypogynous scales 4, pistil rudimentary; ♀ with perianth ± as in ♂, but usually more deeply divided, segments not so differentiated into limb and claw, staminodes usually linear, ovary ovoid, trigonous or compressed, ovule solitary. **Fruits** ovoid, transversely ellipsoid, trigonous or flattened and winged, sometimes emarginate. **Distr:** S Africa. **Ecol:** In shrublands and savannas. **Cult:** Propagate from seeds, cuttings, or by grafting. Requires a humus-rich, well-drained soil in full sun.

L. argenteum R. Br., *Trans. Linn. Soc. London* 10(1): 52, 1810. [Cape Silver Tree, Silver Tree, Silver-leaf Tree]. **Trees**, 5–9(–16) m tall; branchlets stout terete, densely pubescent. **Leaves** sessile; lamina lanceolate or elongate-lanceolate, 6.5–15 × 1.2–3.2 cm, margins entire, apex narrowed, subacute hardened, glabrous, rather thinly coriaceous, surfaces densely pubescent with silvery-silky hairs. **Inflorescences** ♂ terminal, solitary, ± globose, 3.8–5 cm Ø, floral bracts oblong, ~9 mm long, obtuse, coriaceous, ciliate, villous; ♀ similar to ♂ in position and shape, bracts very broadly ovate or semi-orbicular, rigidly coriaceous, velvety-tomentellous outside, glabrous and slightly shiny inside. **Flowers** ♂ with perianth tube ~8 mm long, cylindric, glabrous, segments ~8 mm long, limb oblong-linear, ~3.8 cm long, shortly and bluntly acuminate, villous outside, anthers linear, ~6 mm long, style ~1.3 cm long, pubescent below, stigma clavate-obtuse, ~2–3 mm long; ♀ with perianth tube inflated, segments 2 cm long, villous, limb oblong, 2.5 mm long, villous, staminodes linear, ~2 cm long, minutely glandular at apex, style ~1.7 cm long, filiform, glabrous. **Fruits** oblong-obovoid, 1.2 × 1 cm. **Distr:** S Africa (Cape Province). **Ecol:** On mountain slopes of the coastal region. **Zones:** 9–11.

Serruria

Species 51. **Shrubs**, leafy; branches erect or prostrate. **Leaves** usually crowded, often much dissected into cylindric acute segments, rarely entire. **Inflorescences** flowerheads in terminal, corymbs or panicles, on a common peduncle or solitary on simple axillary and terminal peduncles; heads often with a small involucre of barren bracts; floral bracts mostly villous, usually shorter than flowers; receptacle conical or subglobose, mostly hairy. **Flowers** hermaphrodite, actinomorphic; perianth segments usually free, usually villous or hirsute, differentiated into a slender claw and a short subacute ± oblong limb; stamens 4, anthers oblong; ovary 1-ovuled, style slender, straight. **Fruits** nuts, ovoid or subglobose, sometimes beaked. **Distr:** S Africa (SW Cape region). **Cult:** Propagate from seeds or cuttings. Requires a slightly acidic, gritty, well-drained soil in full or partial sun.

S. florida R. Br., *Trans. Linn. Soc. London* 10(1): 126, 1810. [Blushing Bride]. **Shrubs**, dense, bushy, 1–1.5 m high; branches erect or ascending, purplish, glabrous. **Leaves** pinnately or bipinnately divided, 4.5–6.5 cm long, rather broad at base, glabrous; ultimate segments very acute. **Inflorescences** heads, axillary and terminal, few together, pedunculate, 3.2–3.8 × 3.8 cm; peduncles 2.5–7.5 cm long, glabrous, bearing large lanceolate to ovate-lanceolate, white bracts; involucral bracts lanceolate to obovate, 2.5–3.8 × 1–1.8 cm long, acutely acuminate, membranous, white, becoming pinkish-yellow when dry; floral bracts linear or linear-lanceolate ~2 cm long, tapered to a fine subulate apex, very long-ciliate. **Flowers** with perianth tube ellipsoid, ~4 mm long, glabrous; segments 1.1 cm long, glabrous; limb oblong-linear, ~4 mm long, thinly villous; anthers ~4 mm long; ovary pubescent, surrounded by long hairs, style ~9 mm long, with a globose swelling at base, glabrous, stigma 2.5 mm long. **Fruits** oblong-ellipsoid, shortly beaked, rusty-pilose. **Distr:** S Africa (Cape Region). **Ecol:** On mountain slopes with granite soils, below the sandstone soils, and highly dependent on the fire ecosystem. Fl. Jul–Sep. **Zones:** 9–11.

Leucospermum

Species 48. **Shrubs** (erect or sometimes procumbent) or **small trees**, bisexual. **Leaves** usually very crowded; lamina with margins entire or with 1–16 subapical teeth, tip and teeth glandular, coriaceous, usually densely hairy. **Inflorescences** lateral, capitula, subtended by an involucre of imbricate bracts; heads solitary or in clusters of 2–3, rarely more, shortly peduncled or sessile, bracteate, all bracts of the head fertile or usually the outer barren, forming a ± distinct involucre; fertile bracts distributed over a cylindrical, conical or flat receptacle, like the barren ones persistent and often at length indurated, back densely tomentose; floral bracts scalelike. **Flowers** with perianth zygomorphic, cylindrical, straight or curved to posterior, ± hairy, usually yellow (rarely red) or reddish upwards, limb 4-partite; posterior and lateral tepals connate for most of their length, forming a sheath; anterior tepal distally free; stamens 4, anthers with connective always very marked on back of anther; hypogynous glands 4; styles yellow; pollen presenter usually swollen, often oblique, stigmas often red, terminal or vertical. **Fruits** nuts, smooth, whitish, often shiny, 1-seeded. **Distr:** S Africa (Cape Region), northeast to the Chimanimani Mountains, Zimbabwe. **Ecol:** In shrublands. **Cult:** Propagate from seeds or cuttings, or by grafting. Requires a well-drained soil in full sun.

L. bolusii Gand., *Bull. Soc. Bot. France* 48(Sess. Extraord.): 95, 1902. (Syn: *L. album*). [Gordon's Bay Pincushion]. **Shrubs**, rounded, evergreen, ~1.5 m high; stems erect, terete, pubescent. **Leaves** horizontal to erect, sessile; lamina small, ovate to elliptic, 2.5–4.5 cm long, bright green to dark green, margins reddish, thick and fleshy, hairy when young, later glabrous, with an apical reddish tooth. **Inflorescences** capitula, flat and rounded, ~2 cm Ø, 50–100-flowered; flower buds encased in pinkish-red bracts. **Flowers** small, tubular, creamish-white, fading to pale pink, strongly and sweetly fragrant; perianth tubular, segments spreading, recurved; style straight or slightly curved, 1.4–1.6 cm long, white, exserted, basally hairy, stigma globose, white to pale yellowish. **Fruits** capsules; seeds nutlike, falling off when ripe. **Distr:** S Africa (SW Cape Region). **Ecol:** On sandy slopes and flats. Fl. Sep–Dec. Fr. Nov–Mar. **Zones:** 8–11.

L. cordifolium Fourc., *Trans. Roy. Soc. South Africa* 21: 97, 1932. (Syn: *L. mixtum, L. nutans*). [Pincushion]. **Shrubs**, 1.2–1.5 m high; branches minutely woolly tomentose. **Leaves** ovate, oblong or subelliptic, 2.5–7 × 1.2–3.2 cm, base cordate, apex obtuse, rounded or truncate, entire or callously 2–4-toothed, dull green, almost glaucous, distinctly veined. **Inflorescences** heads, solitary, without a definite involucre of sterile bracts, 4–5 cm Ø; peduncle 1.5–2.3 cm long, tomentose; bracts numerous, imbricate, often recurved, ovate, apex acuminate, tomentose; receptacle cylindric, ~2.3 × ~0.8 cm Ø; floral bracts broadly ovate or obovate to obovate-cuneate, abruptly contracted into a subulate acumen, 1–1.5 cm long, densely tomentose below, acumen finely tomentose and ciliate; flower bud ~2.5 cm long. **Flowers** with perianth tube 0.7–1 cm long, adaxial and lateral claws permanently connate into a concave or flat sheath 1.5–1.8 cm long, abaxial claw free except towards apex, hairy; limbs oblong-ovate, ~4 cm long, subacute; stamens subsessile, anthers oblong-ovate, ~3 mm long; hypogynous scales linear, obtuse, ~2 mm long; ovary oblong, ~2 mm long, finely pubescent, surrounded by whitish hairs, 4–5 mm long; style ~5 mm long, acutely quadrangular above. **Fruits** subglobose, ~9 mm long, sparsely puberulous. **Distr:** S Africa (SW Cape Region). **Ecol:** In lowland to midland sandstone slopes. Fl. Jul–Nov. **Zones:** 8–11.

L. praecox Rourke, *Journal of South African Botany Suppl.* 8: 48, 1972. [Crimson-flame Pincushion, Large-tufted Pincushion, Mossel Bay Pincushion, Still Bay Pincushion, Witsand Pincushion]. **Shrubs**, rounded, robust, erect, (0.2–)1–3(–5) m tall, spreading to 2 m wide; stem single, stout; bark smooth. **Leaves** ovate or obovate to wedge-shaped, 3.5–7 × 1.5–3 cm, glabrous, apex with 5–11 teeth. **Inflorescences** globular, ~6 cm across, sessile, up to 4 per branch, terminal. **Flowers** ~3 cm long, yellow, becoming orange with age; perianth tube inflated; style 3.8–4.8 cm long **Distr:** S Africa (W Cape Region). **Ecol:** In coastal regions and up to 2 km inland, usually in shrublands with deep soils; 0–400m. Fl. May–Nov. Fr. Oct–Apr. **Zones:** 8–11. **Info:** Pollinated in the wild mainly by sugarbirds and sometimes by sunbirds.

L. reflexum Buek ex Meisn., *Prodr.* (DC.) 14(1): 254, 1856. [Rocket Pincusion]. **Shrubs** large or **small trees**, to 4 m tall; branches minutely grayish-woolly tomentose. **Leaves** rather crowded at base of the long flowering branches, more distant and exposing the stem above; lamina oblong or oblanceolate-oblong,

Below: *Leucospermum bolusii, Leucospermum cordifolium, Leucospermum praecox, Leucospermum reflexum.*

MD

Top row: *Knightia excelsa, Banksia ericifolia.*
Bottom row: *Lambertia formosa, Banksia integrifolia.*

margins (2)3-toothed or entire, apex obtuse, 2–4 × 6.5–11.5 cm, minutely but densely grayish-tomentose, obscurely veined. **Inflorescences** heads, usually solitary, peduncled, ~4.5–6.5 × 4.5–6.5 cm long; peduncle 2.5–5 cm long, stout, densely grayish-villous with ovate acuminate grayish-villous or tomentose bracts; receptacle oblong-cylindric, to 2.5 × 5 mm; floral bracts ovate, or (the inner) obovate-cuneate, gradually and long or (the inner) abruptly acuminate, to 1.3 cm long, tomentose above, densely villous towards the base or the inner very densely hirsute-tomentose up to the acumen; adult flower-bud to ~5 cm long. **Flowers** reflexed; perianth tube curved, gibbous, 1–1.3 cm long, finely villous above, glabrous at base, adaxial and lateral claws permanently united into a flattened-out sheath, villous, >2.5 cm long, abaxial claw very slender, limbs lanceolate-oblong, to ~8 mm long; hypogynous scales subulate, 2.5 mm long; ovary oblong, 2.5 mm long, finely grayish-pubescent, style to >7.5 mm long. **Distr:** S Africa (Cape Region). **Ecol:** In coastal areas. **Zones:** 8–11.

Subfam. GREVILLEOIDEAE

Genera 45, **species** 855. **Cluster roots** present. **Inflorescences** bracteate, racemes of flower pairs or panicles of such racemes, (occasionally simple racemes or panicles of racemes or highly reduced grevilleoid conflorescences). **Flowers** with staminal filaments basally to completely adnate to tepals. **Fruits** not winged; seeds with cotyledons auriculate or rarely peltate (auricles obscure in a few genera due to thickening or widening of the cotyledons) (Weston and Barker 2006). **Distr:** Australia and the SE Pacific to SE Asia, S India and Sri Lanka; S America; S Africa and Madagascar.

Tribe ROUPALEAE

Knightia

Species 1. **Trees** or **shrubs**. **Leaves** alternative. **Inflorescences** racemes, paired, pedicellate. **Flowers** perfect, regular, 4-merous; perianth cylindric, segments separating and curling spirally; stamens attached above middle, filaments very short, anthers long linear; hypogynous glands 4; ovary sessile, 1-celled, 4-ovuled, style long, persistent. **Fruits** follicles, woody; seeds winged at apex. **Distr:** New Caledonia, New Zealand. **Cult:** Propagate from fresh seeds. Requires a moist, well-drained soil in full or partial sun. **Info:** The New Caledonian species has foliaceous bracts.

K. excelsa R. Br., *Trans. Linn. Soc. Lond.* 10, 194, t. 2, 1810. [Bucket of Water Tree, New Zealand Bottlebrush, New Zealand Honeysuckle, Rewarewa]. **Trees**, to 30 m tall; fastigiately branched; trunk to >1 m dbh; final branchlets angled, clad in ferruginous tomentum. **Leaves** of *juveniles* thinly coriaceous, to ± 30 cm long, linear-lanceolate, acutely serrate, occasionally forked; leaves of *adults* very thick, 10–15(–20) × 2.5–4 cm, petiole stout, ~1 cm long, lamina narrowly oblong to narrowly obovate-oblong, rigid; bluntly, coarsely serrate, pubescent when young, occasionally forked. **Inflorescences** racemes, stout, to ± 10 cm long, densely flowered; pedicels densely clad in red-brown tomentum. **Flowers** with perianth densely clad in red-brown tomentum; ovary tomentose, style to 3 cm long, stout, clavate. **Fruits** follicles, 3–4 cm long, pubescent, tapering to persistent style; seeds ~1 cm long, terminated by wing ~1.5 cm long. **Distr:** New Zealand. **Ecol:** In lower montane forests. Fl. Oct–Dec. Fr. Dec–Feb. **Zones:** 9–11.

Lambertia

[Honeysuckle, Mountain Devil].
Species 10. **Shrubs** or **small trees**, pubescent, villous or glabrescent. **Leaves** in whorls of 3, rarely paired or 4 or 8; petiole short or absent. **Inflorescences** terminal or on short, lateral shoots and appearing axillary, sessile or subsessile, 1–7 (–19)-flowered; bracts numerous; outer very short; inner long and reaching halfway up perianth or exceeding it, red, yellow or green. **Flowers** actinomorphic or zygomorphic, bisexual; perianth tubular, often with 4 free segments,

Above: *Banksia serrata*, *Lomatia ferruginea*, *Lomatia hirsuta*.

pilose, hirsute or glabrous; limb 4-lobed, with 1 stamen in each lobe, becoming tightly revolute; stamens sessile or very shortly pedicellate; hypogynous glands (2–)4; ovary densely strigose or hirsute; style exceeding open perianth, narrow, glabrous or sparsely hirsute or pilose; stigma <1 mm long, with apical groove or pit, or bilobed, distal to slightly swollen pollen presenter. **Fruits** follicles, splitting when seeds ripe; seeds (1)2, narrowly winged, flat or slightly domed on 1 side, finely papillate, mottled black and brown. **Distr:** Australia. **Ecol:** In dry forests. **Cult:** Propagate from cuttings or seeds. Requires a well-drained sandy soil in full or partial shade.

L. formosa Sm., *Trans. Linn. Soc. London* 4: 214, 223, t. 20, 1798. (Syn: *Protea nectarina*, etc.). [Mountain Devil]. **Shrubs** to 2 m tall, often lignotuberous; branches ascending; young branches brown, villous. **Leaves** sessile or shortly petiolate; lamina narrowly obovate to linear or ovate to oblong, from 1 × 0.7 – 8 × 0.2 cm, strongly mucronate, glabrous above; hairs below dense, crisped, pale or red-brown; margins revolute. **Inflorescences** 7-flowered, usually solitary; bracts stiff, reaching to or above suture of perianth. **Flowers** zygomorphic, crowded, enclosed by bracts; perianth erect, ~4.5 cm long, red or pink, slightly dilated in middle, glabrous outside; sutures equally deep or the adaxial deepest; hypogynous glands fused; ovary hirsute; style sparsely pilose below. **Fruits** irregularly ovoid, 0.8–1 cm long; beak 3–5 mm long; 2 horns at base of suture to 1 cm long; surface variously spiny to almost smooth; seeds 2, with an irregular, annular wing. **Distr:** NSW, Australia. **Ecol:** In shrublands, woodlands, swampy or rocky areas. Fl. throughout the year, mainly Sep–Jan. **Zones:** 8–11. **Info:** A highly variable species.

Tribe BANKSIEAE

Banksia

Syn: *Isostylis*, *Sirmuellera*.

Species 76. **Trees** or **shrubs**, some with prostrate stems. **Leaves** alternate, sometimes whorled. **Inflorescences** many-flowered, dense cylindrical, ovoid or spherical; in 3 species reduced and headlike, with a basal involucre of narrow, usually hairy bracts that usually fall by anthesis. **Flowers**; flowers in pairs, each flower subtended by a bract and each pair by a larger bract; perianth of 4 ± equal tepals; pistil usually wiry; pollen presenter scarcely delimited to markedly modified, commonly ribbed; ovules 2. **Fruits** follicles, woody; seeds 1 or 2, with an intermediary 2-winged plate of similar outline and size, cotyledons with acute basal auricles. **Distr:** Endemic to Australia except *B. dentata* which also occurs on the islands to the north. **Cult:** Propagate from seeds. Follicles usually open spontaneously or with fire; drought- and fire-tolerant, sprouting by epicormic shoots. Follicles can be heated in an oven to obtain seeds. Requires a well-drained, sandy soil in full sun.

B. ericifolia L. f., *Suppl. Pl.* 127, 1782. (Syn: *B. ericaefolia*, etc.). [Heath Banksia, Heath-leaved Banksia, Lantern Banksia]. **Shrubs** or **small trees**, to 6 m tall, without lignotuber; bark smooth with lenticels, becoming corky; stems sparsely hairy. **Leaves** alternate, crowded; lamina narrowly linear, 0.9–2 cm × 0.7–1 mm, truncate but mucronate, margins revolute, with a tooth each side of apex, adaxially sparsely hirsute, glabrescent, abaxially tomentose; petiole 1–2 mm long. **Inflorescences** 7–22 cm long; involucral bracts narrow, pubescent, persistent. **Flowers** golden-brown to pale yellow, often mauve at base; styles golden, orange or orange-red; perianth 1.9–2.2 cm long including limb of 2.5–3 mm long, pubescent outside, hirsute inside along margins; pistil hooked, 3–3.5 cm long, glabrous; pollen presenter ovoid, 0.5 mm long; old perianths persistent but styles soon falling. **Fruits**, follicles numerous, oblong to rhombic, 1.3–2.2 × 0.4–1.3 cm, ridged along each side and convex on top, hirsute; seed obovate, 1.7–2 cm long; seed body cuneate, 0.9–1 × 0.4–0.5 cm, smooth outside, swollen and somewhat ridged inside. **Distr:** Australia (NSW). **Ecol:** In open forests, shrub-woodlands and heaths. Fl. Apr–Jun. **Zones:** 9–11.

B. integrifolia L. f., *Suppl. Pl.* 127, 1782. (Syn: *Sirmuellera integrifolia*, *Isostylis integrifolia*). [Coast Banksia, White Banksia, White Bottlebrush]. **Trees**, to 25 m tall, fire tolerant; bark roughly tessellated; stems pubescent and pilose, glabrescent, pale brown. **Leaves** in whorls of 3–5; lamina narrowly obovate to narrowly elliptic, 4–20 × 10–2.6 cm, margins not or slightly recurved, entire, apex obtuse or emarginate, adaxially dull green, pubescent, glabrescent, abaxially closely white-woolly; petiole 0.4–1 cm long. **Inflorescences** 5–12 cm long; involucral bracts 0.2–1 cm long, tomentose. **Flowers** pale yellow, including styles; perianth 2.2–2.5 cm long including limb of 3.5–4.5 mm long, closely pubescent outside, glabrous inside; pistil straight or slightly curved, 2.7–3.2 cm long, glabrous; pollen presenter 0.5–1 mm long, scarcely thickened; old flowers soon deciduous. **Fruits** follicles, to ~60, narrowly elliptic, 0.7–1.5 × 0.3–0.6 cm; valves semi-elliptic, smooth, tomentose, glabrescent; seed obovate, 1–2 cm long; seed body ± cuneate, 0.6–1 cm long, ± smooth. **Distr:** Australia (widespread along the east coast and inland to the nearby mountains, from Proserpine, Queensland, to Port Phillip Bay, Victoria). **Ecol:** In various habitats and climatic regions, from lowland dunes to montane areas. Fl. May–Jun. **Zones:** 8–11.

B. serrata L. f., *Suppl. Pl.* 126, 1782. (Syn: *B. conchifera*, etc.). [Saw Banksia]. **Trees**, to 16 m tall, sometimes a robust shrub of 1–3 m high, fire-tolerant; bark thick, verrucose, friable, gray-brown; stems tomentose and hirsute. **Leaves** with petiole 1–2 cm long; lamina broadly oblong to narrowly obovate, 7–22 × 2–4 cm, margins serrate, slightly recurved, apex truncate, surfaces tomentose, glabrescent except abaxial pits. **Inflorescences** terminal, 7–15 cm long; involucral bracts tomentose, falling early. **Flowers** creamy-gray, styles cream; perianth 3.8–4.2 cm long including limb of 7–8 mm long, pubescent outside, glabrous inside; pistil curved, 4.7–6 cm long, glabrous, or papillose towards base; pollen presenter fusiform with thickened base, 2.5–3 mm long, finely 8-ribbed; old flowers persistent. **Fruits** follicles, to 30, prominent, broadly elliptic, 2.5–3.5 × 1.5–2.2 cm, smooth, tomentose; seed obovate, 3–3.4 cm long; shallowly pitted outside, verrucose inside. **Distr:** E Australia. **Ecol:** In woodlands, also as a low shrub in tall shrublands. Fl. Jan–Jun. **Zones:** 9–11.

Tribe EMBOTHRIEAE

Lomatia

Syn: *Tricondylus*.

Species 12. **Trees** or **shrubs**; young growth rusty-tomentose. **Leaves** alternate, simple or compound, entire or divided, sometimes heteromorphic. **Inflorescences** axillary or terminal, pseudo-racemose, laterally bearing flower pairs, basitonic or acrotonic, solitary or aggregated. **Flowers** usually fragrant, bisexual; bracts small, caducous or absent; perianth zygomorphic, slit on lower side in bud; limb globular, recurved in bud; segments eventually separating; anthers sessile, pollen-presenter a flat, oblique disk; hypogynous glands 3; style long, bent at 90° to ovary, ovary on a long gynophore, glabrous, ovules many in 2 rows, imbricate upwards. **Fruits** follicles, ovoid with persistent style, coriaceous, opening flat; seeds ± triangular, flat, covered with greenish-yellow granular material, with a translucent wing, surrounded by marginal raphe. **Distr:** 3 species endemic to S America and 9 species endemic to E Australia (from N Queensland to Tasmania). **Cult:** Propagate from fresh seeds or cuttings. Requires a moist, well-drained, acidic soil in full or partial sun.

L. ferruginea R. Br., *Trans. Linn. Soc. London* 10(1): 200, 1810. [Fern Tree]. **Shrubs** or **small trees**, much-branched, 3–6 m tall; younger parts densely covered with a ferruginous velvetlike pubescence. **Leaves** shortly stalked, bipinnatisect, or sometimes tripinnatisect at base, bipinnatipartite toward apex, spreading, leathery, oblong-ovate, usually ~20–40 × 10–20 cm, almost glabrous; pinnae in 5–15 pairs, oblong-lanceolate, usually 4–8 cm long, shortly stalked; segments ovate-lanceolate, 1–2 cm long, entire, or with 1–6 ± spinescent lobes. **Inflorescences** axillary, erect–patent; pedicels 4–6 mm long. **Flowers** ~1–1.3 cm long, calyx golden-yellow and scarlet outside, bright scarlet with yellow at base and apex of

TG

Above: *Embothrium coccineum, Alloxylon flammeum, Telopea speciosissima.*

segments inside, covered with ferruginous adpressed hairs outside; calyx segments oblong, 4.5 mm wide with 3 nerves; gynoecium 1 cm long, style short. **Fruits** follicles, oblong, stalked, slightly curved, 2.5–3 cm long, apex acuminate; seeds with an oblong, obliquely truncate wing at apex. **Distr:** Argentina, Chile. **Ecol:** In moist shady places, such as in gorges; 0–500 m. Fl. Dec–Jan. **Zones:** 9–11.

L. hirsuta (Lam.) Diels ex Macbr., *Publ. Field Mus. Nat. Hist.*, Bot. Ser. 13(2): 369, 1937. (Syn: *Embothrium hirsutum*, etc.). [Radal]. **Shrubs** or **small trees**, 2–15 m tall; crown spherical; bark light-gray, thin with dark patches; young branches rust-colored, hairy. **Leaves** alternate; lamina simple, ovate, 5–16 × 2–12 cm, margins very unequally serrate or crenate, thickly coriaceous, adaxially lustrous and dark green, abaxially brighter with prominent veins; petiole 2–5 cm long, slightly channeled, glabrous, often colored. **Inflorescences** racemes, axillary, 8–16-flowered, reddish-villous or rusty, to 15 cm long; peduncle hairy. **Flowers** yellowish-white; perianth 1 cm long; tepals 4; stamens 4, placed over the tepals; gynoecium unicarpellate, pluriovulate. **Fruits** follicles, grayish-black, pedicellate, oblong, slightly curved, 2–5 cm long; seeds with a truncate wing. **Distr:** Argentina, Chile, Ecuador, Peru. **Ecol:** Mostly in submontane regions; 0–3,000 m. Fl. Oct–Dec. **Zones:** 9–12.

Embothrium

Syn: *Oreocallis*.

Species 1. *Generic description same as specific.* **Cult:** Propagate from seeds, cuttings or basal suckers. Requires a well-drained soil in full sun.

E. coccineum J.R. Forst. and G. Forst., *Char. Gen. Pl.* 8, 1775. [Chilean Fire Bush]. **Shrubs**, evergreen or occasionally deciduous (in cooler climates), to 6 m tall, or trees 10–20 m tall, sometimes suckering forming dense groves; trunk to 70 cm dbh; branches terete, brown, woody, willowlike; young branchlets scaly. **Leaves** shortly petiolate; lamina oblong-oval to elliptical, 5–12 × 2–4 cm, base tapered, margins entire, apex very obtuse and mucronate, firm, coriaceous, glabrous, adaxially dark green, abaxially paler, penniveined; petioles ~1.2 cm long. **Inflorescences** dense racemes, terminal, sessile, many-flowered; rachis green; pedicels red, 1.2–2.5 cm long, erecto-patent. **Flowers** bright scarlet, rarely yellow or white, 2.5–4.5 cm long; in bud ~5 cm long, tubular, curved upwards, the closed segments of the perianth forming a globose head, eventually separating from apex into 4, spathulate, reflexed, and often spirally twisted lobes, exposing the stamens and style, apices concave, with sessile, oblong anthers imbedded; ovary elongated, cylindrical, shortly stipitate, stipes with adaxial prominent glands, tapering gradually into the red exserted style, style oblong-fusiform or oblique at the tip, stigma oblong, yellow, tipped with green. **Distr:** Argentina, Chile. **Ecol:** On sandy, light and moist soils, at high altitudes up to the timber line; 400–1,500 m. Fl. Sep–Oct. **Zones:** 9–11.

Alloxylon

Species 4. **Trees**, small to tall. **Leaves** alternate; *adult* simple or pinnate, entire (some pinnatisect in *A. pinnatum*), tapered to petiole; *intermediate* leaves simple, pinnate or sinuate-lobed to pinnatifid. **Conflorescences** axillary or terminal, racemose to corymbose, on basitonic or acrotonic seasonal growth units, usually aggregated into distinct paniculate bunches (except in *A. pinnatum*), without involucral bracts. **Flowers** strongly zygomorphic, bisexual, weakly incurved in bud, orange-red, crimson or pink, floral orientation diagonal; receptacle oblique; tepals coherent in a split tube after anthesis; hypogynous glands fused, crescent-shaped to horseshoe-shaped; pollen presenter strongly oblique. **Fruits** follicles, woody, canoe-shaped after dehiscence; seed wing with raphe on inner side. **Distr:** E Australia, S New Guinea and Aru Islands. **Ecol:** Usually in rainforests. **Cult:** Propagate from seeds. Requires a moderately fertile, moist, well-drained soil in full or partial sun, sheltered from strong winds.

A. flammeum P.H. Weston and Crisp, *Telopea* 4: 503, 1991. [Queensland Waratah, Red Silky Oak, Tree Waratah]. **Trees**, to 33 m tall; trunk 60 cm dbh; ferruginous-tomentose on upper parts of branchlets, petiole bases, conflorescence axes, bracts, pedicels and perianths. **Leaves** *adult* simple, petiolate, narrowly elliptic or narrowly obovate, 8–25 × 2–4.5 cm, base tapered to petiole, apex obtuse or occasionally acute; *intermediate* leaves strongly pinnatifid, to 50 cm long. **Conflorescences** corymbose, 10–52-flowered, aggregated in bunches of 2–20 or more, to 7.5 cm Ø; **bracts subulate; pedicels 2.5–4.3 cm long.** **Flowers** bright orange-red; tepals 3.5–5 cm long. **Fruits** follicles, 7–10 cm long; seeds 5 cm long, 1-winged, wings ~4 cm long. **Distr:** Australia (Queensland, Atherton Tableland). **Ecol:** In rainforests; 700–820 m. Fl. Aug–Oct. **Zones:** 10–11.

Telopea

Syn: *Hylogyne*.

Species 5. **Shrubs**, rarely small trees, usually multistemmed from a lignotuber. **Leaves** when mature simple, entire or toothed, tapered or abruptly contracted to petiole; intermediate leaves simple or sinuate-lobed, pinnatifid or pinnatisect, occasionally present on flowering seasonal growth units. **Inflorescences** terminal, capitate on basitonic seasonal growth units, not aggregated, with red to pink involucral bracts. **Flowers** strongly incurved in bud, red, crimson, scarlet or pink, rarely white or yellow; floral orientation diagonal; receptacle oblique; tepals coherent in a split tube after anthesis (separating in *T. truncata*); hypogynous glands fused, crescentic to horseshoe-shaped; pollen presenter strongly oblique. **Fruits** follicles, woody, nearly flat after dehiscence; seeds winged with internal raphe. **Distr:** SE Australia. **Cult:** Propagate from seeds or cuttings. Requires a deep, well-drained, acidic soil in full or partial sun.

T. speciosissima (Sm.) R. Br., *Trans. Linn. Soc. London* 10: 198, 1810. (Syn: *Embothrium speciosissimum*, etc.). [Waratah]. **Shrubs**, erect to 3 m tall, with 1–few stems, slender, sparingly branched, glabrous except at bract tips. **Leaves** when *mature* narrowly obovate to narrowly spathulate, 8–28 × 2–6.5 cm, base tapered to very attenuate, margins serrate, with 1–3 teeth in basal halves of margins, rarely entire, apex acute to truncate, coriaceous, but usually not harsh to the touch, veins prominently raised, surfaces minutely granulate when dried (due to sclereids in mesophyll); *intermediate* leaves pinnatifid to pinnatisect. **Inflorescences** few, very broadly to depressed-ovoid, 90–240-flowered, 4–11 cm Ø (excluding bracts); anthesis acropetal; involucral bracts conspicuous, mostly 5–9 cm long, bright red, glabrous except tips. **Flowers** usually scarlet; pedicels 0.3–2.5 cm long; tepals 1–3.4 cm long, cohering along margins in a slit tube; style incurved. **Fruits** follicles, 8–15 cm long. **Distr:** Australia (NSW, along the coast from Ulladulla to Watagan Mountains, and in the Blue Mountains). **Ecol:** In eucalypt forests as understory. Fl. Aug–Nov. **Zones:** 9–11.

Stenocarpus

Syn: *Cybele, Agnostus*.

Species ~25. **Trees** or **shrubs**. **Leaves** alternate, scattered or spirally arranged; lamina simple, variously lobed or compound. **Inflorescences** terminal or in upper leaf axils, umbellate, sometimes together in umbels or racemes; bracts caducous or absent. **Flowers** zygomorphic, bisexual; tube opening on lower side, all 4 segments finally separating, limb globular, recurved; anthers broad, sessile, connective not extended, pollen presenter a flattened, oblique disk; hypogynous glands united into a disk or cup, sometimes very reduced; ovary stipitate, ovules numerous, in 2 rows. **Fruits** follicles, mostly narrowly oblong or cylindrical, or flattened and semicircular (in 1 species), splitting down 1 side; seeds numerous, in 2 rows, highly compressed, winged. **Distr:** Mostly from New Caledonia; 9 species from Australia with 2 species extending to New Guinea and Aru Is. **Cult:** Propagate from seeds or by grafting. Requires a moist, well-drained soil with organic matter, in full or partial sun. **Info:** *S. reticulatus* have exclusively simple leaves, all other species have variously lobed, divided or compound juvenile leaves. Most Australian species have simple (undivided) mature leaves.

S. sinuatus (Loudon) Endl., *Gen. Pl. Suppl.* 4(2): 88, 1848. (Syn: *Agnostus sinuata*, etc.). [Firewheel

Above: *Stenocarpus sinuatus, Buckinghamia celsissima.*

Tree]. **Trees**, to 30 m tall; branchlets terete, ferruginous-tomentose, glabrescent. **Leaves** variable from simple to deeply pinnately lobed, petiole to ~2.5 cm long; lamina with margins undulate, discolorous, dark glossy green above, paler beneath, main veins and reticulations well-defined particularly beneath; *lobed leaves* with lamina to 48 cm long; *simple leaves* with lamina obovate, oblanceolate to elliptic, to 23 cm long. **Inflorescences** umbels of 6–20-flowered, terminal or in upper leaf axils, minutely hairy; peduncles simple, 4–10 cm long, sometimes several arising from a common axis; pedicels 1.4 cm long. **Flowers** 2.5–3.8 cm long, bright red, with a prominent, globular, yellowish limb in bud; hypogynous gland prominent, encircling ~$^2/_3$ of base of gynophore. **Fruits** follicles, cylindrical, 5–10 cm long, grayish-brown, with short rusty hairs; pedicel to 4 cm long overall, forming a kneelike joint at junction of pedicel and gynophore; seeds ± oblong, to ~3.6 cm long including wing. **Distr:** Australia (NSW to NE Queensland); New Guinea. **Ecol:** In rainforests or open areas. Fl. Mar–Jun. **Zones:** 9–12.

Buckinghamia

Species 2. **Trees**, to 30 m tall; branchlets terete. **Leaves** simple, spiral; lamina with margins entire or sometimes lobed on coppice shoots or juvenile plants. **Conflorescences** terminal or axillary, racemelike or paniculate; involucral bracts caducous; floral bracts absent. **Flowers** zygomorphic, bisexual, strongly perfumed; receptacle straight; perianth recurved; tepals separating at anthesis; stamens 4, adnate to perianth, connective short, pollen presenter a broad, oblique disk; hypogynous glands fused into a horseshoe-shaped structure, anterior; ovary stipitate, glabrous, suture opposite anterior tepal, locule 1, style slender, recurved, stigma small, ± central, ovules 4. **Fruits** follicles, bilaterally symmetrical, broadly ovate, ± striate, dehiscing along upper suture, 2–3 × 1.5–2 cm; seeds flat, ± rhomboid, with a narrow marginal wing. **Distr:** Australia (NE Queensland). **Cult:** Propagate from seeds. Requires moist, well-drained, loamy soil in full or partial sun.

B. celsissima F. Muell., *Fragm.* 6: 248, 1868. [Ivory Curl, Spotted Silky Oak]. **Trees**, small or medium-sized, to 10 m tall; trunk cylindrical; new growth pink to red. **Leaves** alternate; lamina variable, elliptic, or ovate to broadly lanceolate, 8–16 × 3–7 cm, margins entire or deeply 2–5-lobed, apex acute to obtuse, adaxially dark green, abaxially ± glaucous; petiole 1–2 cm long, base swollen. **Conflorescences** dense, cylindrical, 10–27 × 4–5 cm, terminal or axillary, gray- to brown-puberulous, flowers paired; pedicels 6–8 mm long. **Flowers** irregular, bisexual, sweetly fragrant, cream or whitish, initially pinkish; perianth bent back, lobes 4, spreading; tepals 0.7–1 cm long, gray- to brown-puberulous on outer surface, glabrous inside; stamens 4, fused to the perianth; hypogynous gland with margins wavy; ovary stalked, 1-loculed, 4 ovules per locule, style 1.5–2 cm long. **Fruits** follicles, 2–3 cm long, 1–4-seeded, opening on 1 side; seeds brown, winged, with cream markings. **Distr:** Australia (NE Queensland). **Ecol:** In rainforests. Fl. Jan–Apr. **Zones:** 10–12.

Hakea

Syn: *Conchium, Mercklinia.*
Species 149. **Shrubs** or **small trees**, with 2-armed hairs; young growth glabrous or hairy. **Leaves** sessile, sometimes cupped about stem, or with attenuate base resembling petiole; lamina simple or compound, flat or terete, margins toothed or entire, surfaces similar. **Inflorescences** developing within a cone of involucral bracts or not, usually axillary, rarely from older wood or terminal, few-flowered umbelliform racemes on short rachis, sometimes many-flowered on elongated rachis; flowers pedicellate, usually paired, straight or curved in bud. **Flowers** with tepals splitting to base or remaining fused and splitting adaxially only; pollen presenter erect, oblique or lateral, discoid or conical with narrow or broad basal flange; hypogynous gland semicircular or a curved flap, rarely absent; pistil glabrous, ovary stipitate or subsessile, 2-ovulate. **Fruits** modified follicles, variously woody, often beaked, sometimes horned, usually tardily dehiscent; valves splitting fully or partly down one or both sides; seed occupying whole valve or marginal, collateral, flattened on one side, with a distal to encircling wing clasped between valves. **Distr:** Australia. **Cult:** Propagate from seeds. Requires a fertile, well-drained soil in full sun.

H. salicifolia (Vent.) B.L. Burtt, *Bull. Misc. Inform.* 33, 1941. (Syn: *Conchium salicifolium*, etc.). [Willow-leaved Hakea]. **Shrubs** upright tall, or **small trees**, 3–5 m high; branchlets with several prominent longitudinal ribs, ± deep red, lenticellate, glabrous; bark brown, attractive, peeling. **Leaves** narrowly elliptic, 8–15 × 0.4–1.7(–2.7) cm, base narrowly attenuate, apex usually acute or acuminate, more rarely obtuse, blackened apically but scarcely mucronate, moderately appressed-sericeous with white and ferruginous hairs when young, rapidly glabrescent, young leaves darker. **Inflorescences** single umbels, 16–28-fowered, in upper axils; rachis 1–1.5 mm long; pedicels 4.5–7 mm long; involucral buds 3 mm long, subglabrous externally. **Flowers** white; perianth 2–3.5 mm long, glabrous, glaucous; inner surface of tepals sometimes with dense glandular-verrucose covering above ovary; pistil 6–6.5 mm long. **Fruits** single or paired, obliquely ovate, 2.3–3.5 × 1.3–2.3(–3) cm in median view, basally attenuate, with raised black pusticules or 1–5 mm high blunt and black-topped warts, beak smooth or with blunt-topped warts, horns often eroded; seed 1.7–2 cm long. **Distr:** Australia (NSW, Queensland); widespread. **Ecol:** In tall forests, woodlands, rainforest margins and scrubs in moist places. **Zones:** 8–11.

Grevillea

Species ~200. **Shrubs** or **trees**, bisexual, evergreen. **Leaves** alternate, simple to pinnately compound or pinnately divided. **Inflorescences** usually terminal, racemes or umbels, flowers in pairs along the rachis, with hairs usually attached in the middle. **Flowers** bisexual, radially or bilaterally symmetrical; perianth tube straight and slender or curved, apical part of perianth usually expanded and bearing sessile stamens; anthers ovate, connective not produced beyond anthers; hypogynous disk usually a single semicircular gland on the upper side of the receptacle or rarely completely circular or absent; ovary usually stipitate, ovules 2, laterally attached near middle of locule wall, style filiform, often protruding from the slit in the underside of the perianth tube before perianth opens completely, stigma on the terminally expanded style. **Fruits** follicles, coriaceous or woody, often with the long style persisting, usually opening along the upper margin; seeds flat or round, usually bordered by a wing. **Distr:** Predominantly from Australia; 9 species in New Caledonia; 2 species in Malaysia. **Cult:** Propagate from half-hardened cuttings, seeds or by grafting. Requires a well-drained, loamy soil in full or partial sun.

G. robusta A. Cunn. ex R. Br., *Port. Nov.* 24, 1830. [Silky Oak, Silver Oak]. **Trees**, straight, evergreen, 10–25 m tall. **Leaves** 15–33 cm long; leaflets 7–19, 3–12 cm long, sessile, entire or pinnatifid, adaxially olive-green, abaxially silver-gray silky-hairy; margin recurved. **Inflorescences** racemes, 5–15 cm long, appearing on the old wood, solitary, 2 or a few forming a panicle; pedicel 1–1.5 cm long, glabrous, leaving a permanent white lenticular scar. **Flowers** with hooded sepals 1.5–2 cm long, initially all fused together except on one side, later fused in twos basally and apically, free in the middle, these pairs in turn free from each other or slightly fused above, orange-yellow to orange or golden-yellow to lemon-yellow with a dark red inner base; stamens sessile; connective not produced beyond anther cells; anthers ~1 mm long; disk semi-annular; gynophore ~2–3 mm long; ovary glabrous; style lemon-yellow, 1–2.5 cm long, dilated at apex and bearing a greenish-yellow 1 mm long stigmatic cone. **Fruits** follicles, 2-seeded, 1.5–2. cm long, ~1 cm broad, silver-gray to olive-green, dehiscent; seeds 1–1.5 × 0.5–1 cm broad, broadly winged, thin, ovate, non-endospermic. **Distr:** Australia. **Ecol:** Fl. Mar–Apr. **Zones:** 8–12. **Info:** The seeds are highly toxic.

G. victoriae F. Muell., *Trans. Philos. Soc. Victoria* 1: 107, 1855. [Royal Grevillea]. **Shrubs**, erect to spreading, 1–3 m tall; branchlets terete, subsericeous

Top row: *Hakea salicifolia, Macadamia integrifolia.*
Bottom row: *Grevillea robusta, Grevillea victoriae.*

or rarely tomentose. **Leaves** elliptic or narrowly elliptic to lanceolate, rarely oblanceolate, (2–) 6–12 (–20) × 1–4.5 cm, margins flat or shortly recurved, adaxially smooth, lateral veins conspicuous, abaxially densely sericeous to subsericeous or subtomentose. **Inflorescences** terminal or axillary, decurved to pendulous, pedunculate, simple or few-branched; unit conflorescence regular, conico-cylindrical or a loose ovoid cluster, many-flowered; floral rachis (1–)2–6(–9) cm long. **Flowers** with perianth densely subsericeous or shortly tomentose outside, bearded inside; pistil 1.7–2.6 cm long, minutely pubescent near the apex; pollen-presenter lateral or almost so, the base not decurrent to the style, the face flat to convex. **Fruits** follicles, 1.7–2 cm long. **Distr:** Australia (SE NSW and montane Victoria). **Ecol:** In high montane forests, woodlands and heaths. Fl. Aug–Jan (mainly Oct–Jan). **Zones:** 8–11.

Tribe MACADAMIEAE
Macadamia

Species 8. **Trees**, to 40 m tall, sometimes multi-stemmed, ± buttressed. **Leaves** simple; adult leaves mostly verticillate, ± petiolate; margins entire or spinose. **Inflorescences** (conflorescences) simple or in whorls on stalks; pseudo-racemes axillary and/or terminal. **Flowers** ± slightly zygomorphic, bisexual, each pair subtended by a ± caducous bract; pedicels free; tepals 4, straplike, dilated distally, coherent to free, becoming circinately reflexed, cream, cream-brown or pink; filaments 4, partially adnate to tepals; anthers 4, with connective slightly exceeding anthers; hypogynous gland cylindrical, glabrous; ovary glabrous to sericeous, sessile to shortly stipitate, ovules 2, orthotropous; style terete to slightly quadrate, stigma ovoid, clavate. **Fruits** follicles, globular, with an apical horn, tardily dehiscent along a ± distinct, longitudinal suture; seeds globular to broadly ovoid; testa brown, smooth or wrinkled; embryo with cotyledons, cream, globose, sweet, bland or bitter. **Distr:** E Australia (7 endemic spp.); Indonesia (Sulawesi; 1 endemic sp.). **Cult:** Propagate from seeds, cuttings or by grafting. Requires humus-rich well-drained soil in full or partial sun. **Info:** The fruits (nuts) of *M. integrifolia* and *M. tetraphylla* are edible, and are cultivated commercially.

M. integrifolia Maiden and Betche, *Proc. Linn. Soc. New South Wales* 21: 624, 1897. (Syn: *M. ternifolia* var. *integrifolia*). [Macadamia Nut, Poppel Nut, Queensland Nut, Smooth-shelled Macadamia Nut]. **Trees** 5–18 m tall. **Leaves** when mature in whorls of 3; lamina ovate to obovate or oblong to oblanceolate, 5–15 × 2–6.5 cm, base very shortly attenuate, margins weakly undulate, entire to spinose, apex acute to obtuse, with a rounded tip or occasionally retuse, ± mucronate, adaxially glossy, abaxially paler, leathery and stiffly coriaceous, glabrescent, midrib prominent abaxially, flattened adaxially; petiole 0.4–1.8 cm long. **Conflorescences** axillary, simple; rachis 8.5–25 cm long, sparingly puberulous; bracts of flower pairs ovate, concave, 0.5–5 mm long; pedicels 3–4 mm long. **Flowers** with tepals 0.5–1.1 cm long, white, cream or yellow, pilose; filaments 4.5–7 mm long, adnate to tepals for 3–5.5 mm; ovary 0.5–2 mm long, style 0.75–1.1 cm long, base brownish-villous. **Fruits** globose, 3–4 × 2–4.5 cm, pericarp 2–6 mm thick; seeds usually 1, 2.5–3.1 × 2.4–3 cm wide, testa 0.6–1 cm thick, bony, smooth. **Distr:** Australia (NSW and SE Queensland, Australia). **Ecol:** In remnant rainforests. Fl. Apr–May. Fr. Jul–Aug. **Zones:** 9–12.

M. ternifolia F. Muell., *Trans. and Proc. Philos. Inst. Vic.* 2: 72, 1858. (Syn: *Helicia ternifolia, M. lowii, M. minor*). [Queensland Nut]. **Trees**, small, multi-stemmed, to 8 m tall. **Leaves** in whorls of 3 (in adult trees); lamina narrowly ovate, 9–12.5 × 2–3.5 cm, base attenuate, margins somewhat undulate, antrorsely spinose, apex acute, mucronate, rigidly coriaceous, glabrous, adaxially dull, abaxially paler, midrib prominent abaxially, slightly sunken adaxially; petiole 0.4–1 cm long. **Conflorescences** simple; rachis 5–18 cm long; bract oblong, 1–5 mm long. **Flowers** with tepals 6–8.5 mm long, pink; filaments 4–7 mm long, adnate to tepals for 0.5–6 mm, anthers 1–2 mm long; ovary 0.4–1 mm long, style 0.5–1 cm long. **Fruits** 1.4–2.2 × 1.3–2.2 cm, pericarp ~1.5–3.5 mm thick; seeds globose to broadly ovoid, ~1.6 × 1.2 cm, testa ~1 mm thick, smooth. **Distr:** Australia (Queensland). **Ecol:** In forests. **Zones:** 9–11.

M. tetraphylla L.A.S. Johnson, *Proc. Linn. Soc. New South Wales* 79: 15, 1954. [Macadamia Nut, Poppel Nut, Prickly Macadamia, Queensland Nut, Rough-shelled Macadamia]. **Trees** or **shrubs** 3–18 m tall, often branched near base. **Leaves** when mature in whorls of (3–)4(–5); lamina oblong to oblanceolate, 7–30 × 1.4–6 cm, base truncate to ± cordate, attenuate to abruptly attenuate, margins

Top row: *Macadamia ternifolia*, *Macadamia tetraphylla*.
Bottom row: *Gevuina avellana*.

weakly undulate, antrorsely spinose, serrate, apex acute to shortly acuminate or subacute, adaxially semi-glossy, abaxially paler, rigidly coriaceous, sparsely pilose, glabrescent, sometimes mucronate, midrib prominent on both sides; petiole 0–4 mm long. **Conflorescences** axillary, simple; rachis 5.5–38 cm long, densely brownish-yellow tomentose; bracts of flower pairs subulate to linear, ovate, 0.2–1.4 mm long; pedicel 2–3 mm long. **Flowers** with tepals 0.55–1.5 cm × 0.35–0.6 mm, cream, yellow or pink, free or coherent; filaments 0.4–1 cm long, adnate to tepals for 3–8 mm; ovary shortly stipitate, 0.2–1.2 × 0.4–1 mm, style 0.7–1.2 cm long, ovary and base of style brownish-yellow pubescent. **Fruits** globose, 2.4–5 × 1.5–5 cm, pericarp 2–9 mm thick, apex apiculate; seeds usually 1, 2.6–3 × 1.6–2.4 cm. **Distr:** Australia (N NSW to SE Queensland); rare. **Ecol:** In rainforests. Fl. Mar–Apr. Fr. Nov–Dec. **Zones:** 9–11.

Gevuina

Species 1. *Generic description same as specific.* **Cult:** Propagate from seeds or by softwood cuttings. Requires a moderately fertile, well-drained soil in partial shade.

G. avellana Molina, *Sag. Stor. Nat. Chili* 184, 1782. [Chilean Hazelnut]. **Shrubs** or **trees**, 10–12 m tall; crown compact; branches stout, pubescent, young parts densely reddish-hairy. **Leaves** large, 45 cm long, adult leaves imparipinnate to tripinnate, terminal leaflet often trilobed; leaflets 30, margins serrate, glossy, dark green. **Conflorescences** racemes, lateral, flowers paired, each pair subtended by a scalelike bract; common peduncle present; flowers shortly pedicellate or sessile on common peduncle. **Flowers** tubular, 2.5 cm long, white to ivory-white; perianth zygomorphic; hypogynous glands 2, anterior, free; sepals fused; pollen presenter swollen, strongly oblique; carpel orientation antero-posterior, ovules 2, orthotropous, style strongly curved, stigma lateral. **Fruits** drupaceous, 1.2–2.3 cm Ø, initially green, maturing through red to brown or black; seed not winged, globular, enclosed in a tough shell. **Distr:** S Chile, Argentina. **Ecol:** In rainforests and disturbed places; 400 m. **Zones:** 9–10.

9 CORE EUDICOTS [CORE TRICOLPATES]

ORDERS 33, FAMILIES 308–313, GENERA 9,822, SPECIES 188,467.

Distr: Worldwide. **Info:** There are 9 families still unplaced (Stevens 2001 onwards).

GUNNERALES

FAMILIES 2, GENERA 2, SPECIES ~45.

Plants dioecious, perennial herbs or poikilohydric shrubs, secretes resin; sieve tube plastids with protein crystalloids and starch; pith with sclerenchymatous diagrams. **Leaves** alternate or opposite; lamina with toothed hydathodal margins, secondary veins palmate. **Inflorescences** with terminal flowers. **Flowers** small, epigynous or hypogynous; perianth absent; anthers latrorse; ovary 1- or 3–4-locular, stigma at most weakly secretory. **Fruits** drupes or septicidal capsules (Stevens 2001 onwards; Kubitzki 2007). **Distr:** Circum S Pacific, Africa and Madagascar.

GUNNERACEAE

Genera 1, species 40–50. **Herbs**, annual and diminutive or small to perennial and gigantic (South America), monoecious or sometimes dioecious, rhizomatous or stoloniferous, sometimes pachycaulous, sometimes mat-forming. **Leaves** mostly in a basal rosette or spirally arranged, exstipulate or ligulate; lamina simple, or sometimes palmately lobed and serrate. **Inflorescences** simple or compound, axillary or pseudo-terminal, erect, racemes, spikes or panicles; lower flowers often ♀, upper flowers ♂, those between them perfect, or ♂ and ♀ flowers in separate inflorescences (in Tasmanian spp.); flowers bracteate or not, crowded. **Flowers** small, unisexual, rarely bisexual; ♂ with sepals 0–2, petals 0–2, hooded, stamens 1–2; ♀ with sepals usually 2, often minute, petals 2, ± hooded, carpels 2, united, ovary inferior, ovoid to globular, unilocular, ovule 1, pendulous, styles 2, slender, completely stigmatic, shortly fimbriate, stigmas dry. **Fruits** drupaceous (nuts), 1-seeded. **Distr:** S America, Juan Fernandez Islands, Hawai'i, S Africa, Madagascar, Malesia to S Pacific, New Zealand, Tasmania. **Ecol:** In cool and humid areas of tropical to subantarctic regions; 0–3,000+ m.

Gunnera

Species ~60. *Generic description same as familial.* **Cult:** Propagate from seeds or by division. Requires a moisture-retentive soil in full sun.

G. manicata Linden ex André, *Ill. Hort.* 20: 156–157, 1873. (Syn: *G. brasiliensis*). [Dinosaur Food, Giant Rhubarb]. **Herbs**, perennial, forming broad crowns of pale fur, to 2 m high, spreading to 3+ m. **Leaves** held erect, umbrella-shaped (directing rainwater to its root), deciduous; lamina usually to rounded to kidney-shaped, pleated, 3.4 × 1.5–2 m, deep green, prickly, especially abaxially on the veins; petiole thick, succulent, to 2.5 m long, prickly. **Inflorescences** spikes, dense, stiff, conical, with flexible lateral branches 9.5–11(–15+) cm long, axis 3–4 mm ∅; rachis 3–3.3 cm ∅; pedicels thick, thorny. **Flowers** minute, greenish to pinkish, hermaphrodite or unisexual, to varying degrees; perianth rudimentary, with or without petals; sepals minute; stamens 1 or 2; ovary inferior, ovule 1, styles 2. **Fruits** drupes, minute, ~2.5 mm long, red-green. **Distr:** SE Brazil (Serra do Mar Mountains), Colombia. **Ecol:** In cloud forests and montane rainforests, mostly in moist to boggy areas; high elevations. Fl. Jul–Aug. **Zones:** 7–11.

DILLENIALES

FAMILY 1, GENERA 10, SPECIES 300.
Ordinal description same as familial.

DILLENIACEAE

Genera 10, species 300. **Trees**, **shrubs**, or **woody climbers**, rarely herbs; hairs sclerified and/or silicified, sometimes fasciculate. **Stipules** absent. **Leaves** usually spirally arranged, rarely opposite; lamina simple, rarely pinnatisect or pinnately compound, margin entire or serrate, leathery, herbaceous, or membranous; petiole sometimes winged. **Inflorescences** with flowers solitary or in racemes, panicles or cymes. **Flowers** bisexual, rarely unisexual, actinomorphic, occasionally zygomorphic; sepals (3 or)4–5(–18), imbricate, persistent, usually leathery or fleshy, sometimes accrescent in fruit; petals (2 or)3–5(–7), white, yellow, or red, imbricate, crumpled in bud; stamens numerous (or 1–10), free, or basally united fascicles, anthers basifixed; staminodes often present; gynoecium (1 or)2–7(–20)-carpellate, superior, carpels free or hemi-syncarpous, stylodia free, apically stigmatic, ovules 1 to numerous, anatropous, or basal when ovules 1 or 2. **Fruits** follicles, poly follicles, berries, or capsules, or indehiscent and enclosed by fleshy sepals; seeds 1 to numerous. **Distr:** Pantropical and extending into warm-temperate Australia.

Acrotrema

Species ~9. **Herbs**, perennial, stemless with woody rootstock or with small, horizontal, woody rhizome. **Leaves** all radical or on a short stem; lamina simple or pinnatisect or pinnatipartite; petiole with sheathing, membranous, caducous wings. **Inflorescence** a terminal raceme, sometimes reduced to a single flower, with membranous bracts. **Flowers** 5-merous, yellow; stamens 15 to many, usually in 3 bundles alternating with carpels; carpels usually 3, slightly coherent on the adaxial side or in the center, styles linear or subulate, recurved, ovules 2–6(–20). **Fruits** of 3 follicles, irregularly dehiscent; seeds 1–15, with a white, membranous aril, testa crustaceous, pitted. **Distr:** Sri Lanka, India (Deccan Peninsula), Myanmar, peninsular Thailand, Malay Peninsula. **Cult:** Propagate from seeds. Requires a moist, humus-rich soil in shade.

A. walkeri Wight ex Thw., *Enum. Pl. Zeyl.* 3, 1858. (Syn: *A. uniflorum* var. *walkeri*, *A. walkeri* var. *argyroneuron*). **Herbs**, perennial; rhizome very small, woody. **Leaves** narrowly obovate-oblong, 4.8–12.5 × 1.5–4.8 cm, base subauriculate, margins crenate-dentate to sharply toothed, ciliate, adaxially deeply plaited and laxly hairy, dull abaxially red and hairy on the veins, frequently variegated with white along the midrib and veins adaxially; petioles short, 6–8 mm long, winged. **Inflorescences** racemes, axillary, short, laxly few- to many-flowered; pedicels 3–5 cm long, laxly hairy or villous. **Flowers** yellow, 1.8–2 cm across; sepals 5, ovate, ~7 × 2 mm, patent, ciliate-hairy without; petals 5, obovate, 0.8–1 × 0.4–0.5 cm; stamens ~15, filaments in 3 bundles, dilated at the apex, anthers erect, ± linear; carpels 3, ± united with 2–many ovules. **Fruits** on maturity irregularly dehiscent; seeds with membranous arils. **Distr:** Sri Lanka. **Ecol:** In montane forests of the wet zone; 610–1,300 m. Fl. Jun–Oct. **Zones:** 11–12.

Dillenia

Syn: *Wormia*, *Lenidia*.
Species ~65. **Trees**, rarely shrubs, evergreen or rarely deciduous; bark red, gray, or brown. **Leaves** alternate; lamina simple, to 50 cm long, with or without persistent or deciduous amplexicaul wings. **Inflorescences** with solitary flowers or in few- to many-flowered racemes. **Flowers** large, showy; sepals (4 or)5(–18), fleshy, often accrescent; petals (4 or)5(–7) or absent; stamens numerous, inner often with longer anthers than outer, anthers dehiscing by 2 (sub)apical pores or less often by 2 longitudinal slits, connective narrow, linear; carpels partially joined, 4–20, adnate to conical receptacle, ovules 1 to numerous per carpel. **Fruits** enclosed by enlarged sepals, carpels dehiscing along ventral suture or indehiscent; seeds usually only 1 or few per carpel, arillate or not. **Distr:** Madagascar, through Indian Ocean Islands to Fiji and Australia, with the highest diversity in tropical Asia. **Cult:** Propagate from seeds or cuttings. Requires moist but well-drained soil or compost in full or partial sun.

D. indica L., *Sp. Pl.* 1: 535, 1753. (Syn: *D. elliptica*, *D. speciosa*). [Elephant Apple]. **Trees**, evergreen, to 30 m tall; trunk ~1.2 m dbh; bark reddish-brown, exfoliating; young branchlets brown-pubescent, glabrescent; leaf scars prominent. **Leaves** with petiole narrowly winged; lamina oblong or obovate-oblong, 15–40 × 7–14 cm, margins serrate, secondary veins (20–)30–40(–70) pairs, parallel. **Inflorescences** with flowers solitary, terminal. **Flowers** white, fragrant, 12–20 cm across, >5 cm Ø in bud; sepals 5, approximately rounded, 4–6 cm Ø, thickly fleshy; petals white, obovate, 7–9 cm; stamens in 2 distinct groups, outer very numerous, slightly curved in bud, inner ~25, apically reflexed outward in bud; anthers dehiscing with 2 pores; carpels 16–20, stylodia spreading, ovules many per carpel. **Fruits** aggregates, globose, 10–15 cm Ø, indehiscent, persistent sepals fleshy, slightly swollen; seeds 5 or more per carpel, exarillate, black, flat, margins hairy. **Distr:** Tropical Asia to Malesia. **Ecol:** In valleys, streamsides, forest margins. **Info:** The fruits are edible (with a lemony taste) and often used in making curries. **Zones:** 9–12.

D. ovata Wall. ex. (Hooker f. & Thomson), *Fl. Ind.* 1: 70, 1855. **Trees**, small, bushy, evergreen, 12–15 m tall; trunk 1.2 m dbh, knobbly; crown deep, with ~5 m of the trunk clear; bark reddish or brown, flaky; twigs slender, softly hairy. **Leaves** with petiole 2.5–5 cm, soft, hairy, margins not winged; lamina ovate or elliptic, (7–)10–20(–30) × (5–)7–12(–15) cm, base abruptly rounded and often asymmetric; margins finely toothed, apex blunt or slightly tipped, abaxial surface velvety-hairy, secondary nerves 18–25 pairs; sapling leaves 0.5–1 m × 25–30 cm, apex pointed, secondary nerves ~50 pairs. **Inflorescences** with flowers solitary, terminal on leafy twigs; pedicels slender, hairy. **Flowers** yellow, ~16 cm across; petals lemon-yellow; stamens in 2 distinct groups. **Fruits** round, ~6 cm across, dull yellow; seeds blackish-brown. **Distr:** N Malaya, Indo-China, Sumatra. **Ecol:** In open woodlands, open and secondary forests at low altitudes. Fl. Mar–Jun. Fr. Jun–Feb. **Zones:** 10–12.

D. reticulata King, *Journ. As. Soc. Beng.* lviii. II: 367, 1889. **Trees**, medium-sized or large, deciduous, to 36 m tall; trunk ~3 m dbh; buttresses if present, tall and upright, with conspicuous stilt roots; bark grayish-brown or reddish-brown; branchlets stout, grayish-brown, glabrous or hairy, with prominent leaf scars. **Leaves** elliptic or elliptic-oblong, (10–)25–45 × (10–)18–30 cm, base truncate-cordate, margins entire to slightly undulate or toothed, very coriaceous, adaxially glabrous except nerves, abaxially pubescent midribs and nerves very prominent, reticulations conspicuous, nerves 32–36 pairs; petiole 4.5–10 cm long; sapling leaves to 1.27 m long, base rounded, densely soft velvety-hairy. **Inflorescences** panicles, 10–12-flowered, terminal, sometimes lateral, leaf-opposed, forming a loose cluster at ends of leafless twigs or with new leaves. **Flowers** ~8 cm across; petals yellow, oblong, 2.5–12.5 cm long; carpels 9, puberulous. **Fruits** indehiscent, greenish-yellow, globular, slightly flattened; seeds glossy black with rudimentary aril. **Distr:** Malaya, Sumatra, Borneo. **Ecol:** In lowland swampy and dry areas; <180 m. Fl. Mar–May. Fr. Apr–Jul. **Zones:** 11–12.

D. retusa Thunb., *Trans. Linn. Soc. London* 1: 200. t. 19, 1791. (Syn: *D. integra*, etc.). **Trees**, evergreen trees, to 20 m tall; trunk 70 cm dbh; branches sympodial, glabrescent. **Leaves** obovate, 8–21 × 5–12 cm, base acute, decurrent, margins entire or slightly dentate, apex rounded to retuse, subcoriaceous, 10–14-nerved, adaxially glabrous, abaxially strigose on nerves and midrib; petiole 1.3–2.2 cm long. **Inflorescences** terminal, usually (1)2–3-flowered, simple racemes, axis 2–4 cm long; bracts triangular, 4–8 × 1–2 mm; pedicels 1–3.5 cm long. **Flowers** 6–8 cm across; sepals 5, fleshy, circular to oval, 2 outermost ones smaller, 1.6 × 1.4 cm, 3 innermost ones 2 × 2 cm, glabrous within, densely silky-hairy outside, all densely ciliate at margin; petals 5, white, obovate-spathulate, 3.5 × 2.5 cm; stamens ~180, linear, gradually longer towards center, outermost ones ~6.5 mm long, innermost ones ~1.1 cm long, apically strongly reflexed, anthers white; carpels 5–6, lanceolate, 8 × 2.5 mm, glabrous, each 22–34-ovuled, styles spreading, cylindric, 7–8 mm long. **Fruits** pseudocarps, indehiscent, orange, 2–3 cm Ø; seeds ovoid, 2.5 × 2 mm, exarillate. **Distr:** Sri Lanka, India. **Ecol:** In primary rainforests; <450 m. Fl. throughout the year. **Zones:** 10–12.

D. suffruticosa (Griff.) Martelli, *Malesia* 3: 163, 1886. (Syn: *Wormia burbidgei*, etc.). **Shrubs** or **small trees**, spreading, evergreen, to 8 m tall; trunk often rooting from lower branches forming thickets; branches sympodial, ± glabrescent. **Leaves** elliptic to ovate or obovate, 10–45 × 5–26 cm, base obtuse, margins entire to dentate, apex rounded to obtuse, adaxially deep green, abaxially pale green, 7–27-nerved; petiole 2–6.5 cm long with persistent, 0.4–1.5 cm wide wings merging into lamina. **Inflorescences** racemes, 4–18-flowered, terminal, to 30 cm long, simple or branched to 1(2) order(s); bracts triangular, 0.6–1.5 × 0.3–0.5 cm, caducous; pedicels 0.8–3 cm long. **Flowers** 8–12 cm Ø; sepals 5, obovate, 1.5–2.2 × 0.8–1.2 cm, fleshy, glabrous to densely villose outside; petals 5, obovate-spathulate, 4–5 × 2.5–3 cm, bright yellow; androecium with ~100 staminodes on outer side, staminodes linear,

Top row: *Gunnera manicata, Acrotrema walkerii, Dillenia indica.*
Bottom row: *Dillenia ovata, Dillenia reticulata, Dillenia retusa.*

Above: *Dillenia suffruticosa, Dillenia triquetra, Schumacheria castaneifolia.*

4–6 mm long, yellow, apex obtuse, stamens ~175, outermost ones slightly curved, ~8 mm long, innermost ones apically reflexed outward, 1.3 cm long, stamens of intermediate length in between; carpels 5–8, elliptic, 5 × 2 mm, glabrous, each 7–10-ovuled, styles spreading, filamentous, ~1 cm long. **Fruits** pseudocarps, dehiscent; seeds obovoid, 3 × 2 mm, aril scarlet. **Distr:** Malay Peninsula, Sumatra and Borneo. **Ecol:** In wet lowlands; <900 m. Fl. throughout the year. **Zones:** 11–12.

D. triquetra (Rottb.) Gilg, *Nat. Pflanzenfam.* [Engler & Prantl] 3(6): 123, 1893. (Syn: *D. aquatica*, etc.). **Trees**, 5–8(–20) m tall; trunk 30 cm dbh. **Leaves** subcoriaceous, oval to oblong or obovate, 6–30 × 4–19 cm, base obtuse to rounded, margins entire to distinctly undulate, apex rounded or often ± retuse to truncate, adaxially glabrous, 5–14-nerved; petiole 3–13 cm long, wings caducous, linear-lanceolate, 3–8 mm (near base), 0.5–1.5 mm (at apex). **Inflorescences** terminal, simple or 1(2)-branched racemes, 3–20-flowered, to 25 cm long, axis 2–4 mm Ø, glabrescent; bracts triangular to lanceolate, to 1.5 × 0.4 cm; pedicels 0.7–4 cm long. **Flowers** 4–7 cm Ø; sepals usually 5, thick, oval, 1.7 × 1.3 cm; petals 5, obovate-spathulate, 3 × 2 cm, white; stamens 125–140, 8 mm long; carpels 4–5(–7), elliptic-oblong, 3–3.5 × 1.5–2 mm, glabrous, each 7–12-ovuled, styles 6–8 mm long. **Fruits** pseudocarps, ovoid, ~2.5 × 20 cm; seeds obovoid, 5 × 3 mm, dark-glossy brown, aril membranous. **Distr:** Sri Lanka and Madagascar. **Ecol:** In rainforests; 0–1,200 m. Fl. throughout the year. **Zones:** 11–12.

Schumacheria

Syn: *Pleurodesmia.*

Species 3. **Shrubs**, erect or scrambling, or small trees, evergreen; branches mostly zigzag, glabrous, with completely amplexicaul leaf-scars. **Leaves** alternate, exstipulate; lamina simple, repando-serrate, serratures mucronate, coriaceous, glabrous; petiole channeled above, with completely amplexicaul wings enclosing terminal bud in young leaves, subsequently splitting in leaf-opposed center and persistent with leaves. **Inflorescences** simple or panicled spikes, axillary or terminal, sessile or nearly so, bibracteate. **Flowers** with sepals 5, 2 exterior, 3 interior, imbricate, persistent in fruit; petals 5, imbricate, yellow, little larger than sepals, early caducous; stamens numerous, monadelphous, connate filaments forming a semicylindrical column at abaxial side of flower, anthers linear, elongate, 2-celled, connective with apical mucro; carpels 3 (bicarpellate flowers are quite common), free, densely covered with sclerified hairs, with a single basal ovule, style terete, curved. **Fruits** with fruiting carpels dehiscent, spreading starlike; seeds globular at base, with small, membranous aril. **Distr:** Sri Lanka. **Cult:** Propagate from rooted cuttings or seeds. Requires a moist, well-drained, humus-rich soil in shade or partial sun.

S. castaneifolia Vahl, *Skr. Naturhist. Selsk.* 6: 122, 1810. (Syn: *S. castaneifolia* var. *vahlii*, etc.). **Shrubs**, large, often straggling, or small trees, to 8 m tall; branches cylindric, hoary, at length glabrate. **Leaves** variable, usually ovate or ovate-oblong or elliptic, 10–20 × 5–12 cm, base obtuse, rounded or retuse, margins sinuate-dentate, serratures mucronate, apex obtuse to acute, 18–25-nerved, glabrous, puberulous on abaxial veins; petiole 0.5–1.2 cm long, channeled above. **Inflorescences** usually terminal spreading panicles, sometimes axillary in upper few leaves, to 40 × 30 cm, many-flowered (to 500+); bracts sericeous. **Flowers** 1–1.4 cm Ø; sepals 5, coriaceous, sericeous, 6–8 × 4–5 mm, 3 outer longer than 2 inner; petals usually 4, crisped, oblong, caducous; stamens ~25–30, filament ~1 mm long, anthers ~4 mm long; carpels ~2 mm long, styles 4 mm long. **Distr:** Sri Lanka. **Ecol:** In primary rainforests; 100–760 m. Fl. throughout the year. **Zones:** 11–12.

SAXIFRAGALES

FAMILIES 16, GENERA 112, SPECIES 2,470.

Branching from the previous flush in woody members; cuticle waxes as clustered tubules. **Leaves** with glandular teeth, broadening distally and with apical foramen, higher order lateral veins joining it. **Flowers** 5-merous, segments distinct; hypanthium often present; floral apex concave early in development; anthers basifixed, with basal pit, sagittate; carpels free, at least apically, micropyle bistomal, (outer integument largely dermal in origin), styluli short, stigmas decurrent, at most slightly wet. **Fruits** dry; seeds ± exotestal (Judd et al. 2008; Stevens 2001 onwards). **Distr:** Worldwide. **Info:** The monophyly of Saxifragales is supported by molecular data (Hilu et al. 2003; Jian et al. 2008).

HAMAMELIDACEAE

Genera 27, **species** 82. **Shrubs** or **trees**, evergreen or deciduous, hermaphroditic, andromonoecious, or monoecious; indumentum usually of stellate hairs or stellate or peltate scales; buds perulate or naked. **Stipules** cauline, minute to large, usually paired (rarely solitary and enclosing bud, or apparently absent). **Leaves** 2-ranked (distichous) or spirally arranged, rarely subopposite or opposite; lamina simple or palmately lobed, pinnately veined or palmately 3–5-veined; petiole usually well-defined. **Inflorescences** usually spikes or heads, rarely racemes or (condensed) thyrses or panicles, axillary or terminal; flowers bracteate, often bracteolate. **Flowers** bisexual or unisexual, usually actinomorphic, (2–)4–5(–7)-merous; calyx free to connate; petals absent or 4 or 5, yellow, white, greenish or red, often ribbonlike, caducous; stamens 4, 5, or many; ovary biloculed, ovules mostly 1 per carpel, styles and stigmas 2. **Fruits** capsules, dehiscing septicidally, septifragally, or loculicidally and 4-valved; seeds 1 to many per carpel. **Distr:** Tropical to temperate, especially E Asia to Australia, excluding South America.

Exbucklandia

Syn: *Bucklandia*, etc.

Species ~4. **Trees**, evergreen; branchlets stout, nodes distinct. **Stipules** large, leathery, coherent, caducous, leaving annular scar at each node. **Leaves** alternate, long-petiolate; lamina simple or, in saplings and young shoots, palmately 3–5-lobed, margins entire, thickly leathery, venation palmate. **Inflorescences** capitate, sometimes in compound racemes, usually axillary, pedunculate, 7–16-flowered; flowers initially enclosed by a pair of stipules. **Flowers** bisexual; hypanthium adnate to ovary base; petals absent or 2–5, white, linear, straight in bud; stamens 10–15, filaments subulate, length variable, anthers oblong or ovoid; ovary semi-inferior, ovules 5 or 6 per locule, styles somewhat elongated, stigmas decurrent, caducous in fruit. **Infructescences** globose or subglobose. **Fruits** capsules, loculicidally dehiscent, locules 4; exocarp usually smooth, sometimes tuberculate; seeds 5–7 per locule, upper 4 or 5 sterile and wingless, lower 1 or 2 fertile and narrowly winged. **Distr:** Bhutan, China, India, Indonesia, Laos, Malaysia, Myanmar, Nepal, Sikkim, Thailand, Vietnam. **Cult:** Propagate from seeds. Requires a moist, well-drained soil in full or partial sun.

E. populnea (R. Brown ex Griffith) R.W. Brown, *J. Wash. Acad. Sci.* 36: 348, 1946. (Syn: *Bucklandia populnea*, etc.). [Malayan Aspen]. **Trees**, 16–20(–30) m tall; branchlets pubescent. **Stipules** elliptic or obovate, 2–3 × 1–2 cm. **Leaves** broadly ovate-rounded or sometimes palmately trilobed in young leaves, 10–17 × 9–13 cm, base broadly cuneate, margins entire, abaxially glabrous, palmately 5–7-veined; petiole 3–6 cm, longer in young leaves, glabrous. **Inflorescences** capitate, sometimes arranged in racemes; heads 8–12-flowered; peduncle 1–2 cm long, pubescent. **Flowers** with petals usually absent (rarely 2–3 mm long); stamens ~5 mm long, filaments slender, glabrous; ovary yellow-brown pubescent; styles 3–4 mm, erect. **Fruits** capsules, ellipsoid, 7–9 × 5–6 mm, smooth, dehiscing above middle; seeds narrowly winged. **Distr:** Bhutan, China, India, Indonesia, W Malaysia, Myanmar, Nepal, Sikkim, Thailand, Vietnam. **Ecol:** On slopes in evergreen forests; 820–1,900 m. Fl. May–Jul. Fr. Aug–Oct. **Zones:** 8–12.

SZ

MD

Top row: *Exbucklandia populnea*, *Exbucklandia populnea* (flowers), *Aeonium arboreum*.
Bottom row: Aeonium arboreum (dark purple form), *Aeonium lancerottense*.

CRASSULACEAE

[Stonecrop].

Genera 34, **species** 1,370. **Herbs**, **subshrubs**, or **shrubs**; stems mostly fleshy. **Stipules** absent. **Leaves** alternate, opposite, or verticillate; lamina usually simple, margins entire or slightly incised, rarely lobed or imparipinnate. **Inflorescences** terminal or axillary, cymose, corymbiform, spiculate, racemose, paniculate, or sometimes reduced to solitary flowers. **Flowers** usually bisexual, sometimes unisexual (when plants dioecious or rarely gynodioecious), actinomorphic, (3 or)4–6(–30)-merous; sepals almost free or basally connate, persistent; petals free or connate; stamens as many as petals in 1 series or twice as many in 2 series; nectar scales at or near base of carpels. **Fruits** follicles, sometimes fewer than sepals, free or basally connate, erect or spreading, membranous or leathery, 1- to many-seeded; seeds small, endosperm scanty or not developed. **Distr:** Africa, America, Asia, Europe.

Aeonium

Species 35. **Shrubs**, rarely perennial herbs, succulent, not viviparous, (0.1–)0.7–1(–2) m, glabrous or pubescent; stems erect, branched (rarely simple), woody or fleshy. **Leaves** persistent, crowded in rosettes at ends of branches, alternate, sessile, not basally connate; lamina oblong-lanceolate or obovate, laminar to terete, 3–15 cm long, base not spurred, margins ciliate, fleshy. **Inflorescences** cymes, terminal; flowers pedicellate, erect or spreading. **Flowers** (6–)7–12(–16)-merous; sepals basally connate, all similar; petals spreading or erect, distinct or nearly so, cream or bright yellow; nectaries mostly rectangular; stamens twice as many as sepals, filaments adnate to corolla base; pistils erect, distinct or nearly so, ovary base rounded, styles 2+ times shorter than ovary.

Fruits erect; calyx and corolla not circumscissile in fruit; seeds ellipsoidal, ribbed, finely cross-ribbed. **Distr:** Atlantic Islands (Canary Islands, Cape Verde Islands, Madeira), SW Morocco, E Africa, S Arabia. **Cult:** Propagate from basal suckers, cuttings or seeds. Requires a gritty soil with excellent drainage in full sun; drought-tolerant.

A. arboreum (L.) Webb & Berthelot, *Hist. Nat. Îles Canaries.* 3(2,1): 185, 1840. (Syn: *Sempervivum arboreum*, etc.). [Schwarzkopf, Tree Aeonium]. **Shrubs** or **subshrubs**, perennial, 1(–2) m high, without prop roots; stems ± erect or ascending, often in groups, few-branched, 1–4 cm Ø, fleshy, without reticulations; bark nearly smooth. **Leaves** 50–75, in dense rosettes 10–25 cm Ø; lamina concave or flattish, oblong-oblanceolate, 5–15 × 1–4.5 cm, 1.5–3 mm thick, shiny, bright green (rarely purple-black), firm, marginal cilia spreading to ascending, wavy, surfaces glabrous. **Inflorescences** cymes, dense, conic to ovoid, 10–25 × 10–15 cm, puberulent; peduncle 5–20 cm long, puberulent; pedicels 0.2–1.2 cm long, puberulent. **Flowers** rotate, 2 cm Ø; calyx puberulent; petals 9–11, spreading, bright yellow, narrowly oblong to lanceolate, 5–7 × 1.5–2 mm, acuminate; filaments glabrous. **Distr:** Canary Islands (Gran Canaria). **Ecol:** 200–1,200 m. **Zones:** 9–11.

A. lancerottense (Praeger) Praeger, *J. Bot.* 66: 222, 1928. (Syn: *Sempervivum lancerottense*). **Subshrubs**, perennial, densely branched, to 60 cm high; branches ascending, 0.7–1.5 cm Ø, glabrous, reticulate. **Leaves** in rosettes 10–18 cm Ø, flattish; lamina obovate- or oblanceolate-spathulate, 5–9 × 1.5–4 cm, 3–6 mm thick, base cuneate, margins with some wrinkled cilia in the upper part, apex acute, caudate, green to yellowish-green, margins and apex often reddish-variegated, glaucous, glabrate. **Inflorescences** dome-shaped, 8–30 × 8–25 cm; peduncle 6–20 cm long; pedicels 1–3.5 mm long, glabrous. **Flowers** 7–8-merous; sepals puberulent to glabrate; petals lanceolate, 6–9 × 1–1.5 mm, acuminate, whitish, median part pinkish-variegated; filaments glabrous. **Distr:** Canary Islands (Lanzarote). **Ecol:** 200–600 m. **Zones:** 9–11.

A. percarneum (R.P. Murray) Pitard & Proust, *Îles Canaries.*, 191, 1909. (Syn: *Sempervivum percarneum*, etc.). **Subshrubs**, perennial, few-branched, to 1 m high; branches ascending, 0.7–2 cm Ø, glabrous, reticulate. **Leaves** in rosettes 8–20 cm Ø, flattish; lamina obovate- to oblanceolate-spathulate, 4.5–10 × 2–4 cm, 3–6 mm thick, base cuneate, margins often weakly denticulate, often with curved cilia to 1 mm long, apex acute, caudate, glaucous, glabrate, dark green, margins often reddish-variegated. **Inflorescences** dome-shaped, 10–30 × 10–25 cm; peduncle 8–20 cm long; pedicels 1–3 mm long, puberulent. **Flowers** 8–10-merous; sepals puberulent; petals lanceolate, 7–8 × 1.2–1.8 mm, acuminate, whitish, median part pinkish-variegated; filaments sparsely puberulent. **Distr:** Canary Islands (Gran Canaria). **Ecol:** ~100–1,300 m. **Zones:** 9–11.

A. tabuliforme (Haworth) Webb & Berthelot, *Hist. Nat. Îles Canaries.* 3(2,1): 185, 1840. (Syn: *Sempervivum tabuliforme*, etc.). [Saucer Plant]. **Subshrubs**, biennial or perennial, low-spreading, solitary or very rarely cespitose, usually monocarpic; stems stout, smooth, short to almost absent. **Leaves** in a flat rosette 9–40 cm Ø, imbricate and tightly appressed to each other; lamina obovate to oblanceolate, 4–20 × 2–4 cm, 3–6 mm thick, base cuneate, margins fimbriate with hairs to 2 mm long, apex rounded, mucronate, puberulent when young, becoming glabrate, pale to bright green. **Inflorescence** 15–30 × 12–30 cm; peduncle 12–30 cm long; pedicels 0.2–2 cm long. **Flowers** 7–9-merous; sepals pubescent; petals elliptic, 6–7 × 1.5–2 mm, acute, pale yellow; filaments glabrous. **Distr:** Canary Islands (N Tenerife). **Ecol:** <850 m. **Zones:** 9–11.

A. urbicum (C. Smith ex Hornemann) Webb & Berthelot, *Phytogr. Canar.* 3(2,1): 194, 1841. (Syn: *Sempervivum urbicum*, etc.). [Dinner Plate, Saucer Plant]. **Subshrubs**, perennial, monocarpic, unbranched or rarely few-branched, to 2 m high; branches ascending, to 6 cm Ø, glabrous, rough when young, reticulate. **Leaves** in rosettes 15–32 cm Ø, flattish; lamina obovate-spathulate or oblanceolate, 8–22 × 3–5.5 cm, 4–7 mm thick, base cuneate, margins with ± straight cilia 0.5–1 mm long, apex acuminate, yellowish-green to dark green, variously reddish-variegated. **Inflorescences** dome-shaped, 15–75 × 10–45 cm; peduncle 3–15 cm long; pedicels 2–6 mm long, glabrous. **Flowers** 8–10-merous; sepals glabrous; petals lanceolate, 0.7–1 cm × 1.2–2 mm, acuminate, whitish, occasionally reddish-variegated; filaments glabrous. **Distr:** Canary Islands (Tenerife). **Ecol:** <1,900 m. **Zones:** 9–11.

Bryophyllum

Species 32. **Herbs**, rarely subshrubs or shrubs; roots fibrous; stems usually erect. **Leaves** opposite, rarely 3-verticillate, petiolate; lamina pinnately compound, rarely simple or pinnately lobed (or simple and bearing bulbils along margin). **Inflorescences** terminal, cymose, many-flowered, flowers usually pendulous. **Flowers** bisexual, 4-merous, brightly colored, large; calyx tubular or rarely campanulate, tube sometimes basally dilated; corolla usually purple-red, tubular to salverform, equaling or longer than calyx, lobes shorter than or scarcely longer than tube; stamens twice as many as petals, inserted below middle of corolla tube, usually near base, filaments equaling corolla tube; nectar scales entire or emarginate; carpels erect, free, styles long. **Fruits** follicles, many-seeded. **Distr:** Africa, Madagascar. **Cult:** Propagate from stem cuttings, leaf cuttings or seeds. Requires a moderately fertile, gritty soil in full sun.

B. pinnatum (L. f.) Oken, *Allg. Naturgesch.* 3(3): 1966, 1841. (Syn: *Kalanchoë pinnata*, etc.). [Air Plant, Cathedral Bells, Chandelier Plant, Life Plant]. **Herbs**, 0.4–1.5 m tall, glabrous; stems usually branched. **Leaves** pinnately compound with 3–5 leaflets, 10–30 cm long; leaflets oblong to elliptic, 6–8 × 3–5 cm, margins crenate, apex obtuse; petiolules 2–4 cm long. **Inflorescences** paniculate, terminal, 10–40 cm long, many-flowered. **Flowers** pendulous; calyx tubular, 2–4 cm long; corolla reddish to purple, to 5 cm long, base sparsely ciliate, lobes ovate-lanceolate; stamens inserted basally on corolla; nectar scales oblong. **Fruits** follicles, included in calyx and corolla tube; seeds striate. **Distr:** Africa. **Ecol:** Fl. Jan–Mar. **Zones:** 11–12. **Info:** Cultivated for ornamental and medicinal purposes.

Crassula

Species 195. **Shrubs** (usually small and soft-wooded), herbs (annual or aquatic), succulent, perennial. **Leaves** opposite, spread along stem or crowded in rosettes, leaf pairs often united at base, evergreen; lamina with margins entire, often thick, fleshy. **Inflorescences** terminal, umbellate or paniculate, branched, composed of cymose clusters; bracts leafy. **Flowers** 3–5-parted, star-shaped, usually white to red or pinkish, less often yellow or greenish; sepals 4–7 or 12, free or shortly united at base, fleshy, often of unequal size; petals 4–7 or 12, free or shortly united at base, spreading or tubular, apices recurved; stamens 4 or 12, in 1 whorl; carpels free, each 1–few-ovuled. **Fruits** clusters of 5 dehiscent follicles. **Distr:** S Africa to SW Arabia. **Cult:** Propagate from leaves, stem cuttings or seeds. Requires a well-drained, humus-rich compost in full or partial sun.

C. arborescens (Miller) Willdenow, Sp. Pl. 1(2): 1554, 1798. (Syn: *C. arborea*, *C. cotyledon*). [Jade Plant, Jade Tree, Money Tree, Pink Joy]. **Shrubs**, erect, branched, to 1.5(–3) m high, usually with a single main to 6 cm Ø; branches succulent, gray-green, leaf-bearing branches 0.7–1 cm Ø, older branches with yellowish-brown peeling bark, distinctly articulate at nodes. **Leaves** ascending, deciduous when old; lamina obovate to obovate-orbicular, base cuneate, margins reddish, apex obtuse, mucronate or not, glaucous with powdery bloom, often spotted with conspicuous hydathodes. **Inflorescences** thyrses, terminal, round-topped, 5–8 cm tall, with numerous dichasia; peduncle 1.5–3 cm long; pedicels 1–1.2 cm long. **Flowers** with sepals broadly triangular; corolla stellate, 1.8–2 cm Ø, light pink or white; petals lanceolate, 0.9–1 × 0.25–0.3 cm; filaments 5–6 mm long, anthers <1 mm long, purplish. **Fruits** small, 6 mm long, with 3–5 separate, ovate, sharply tipped follicles. **Distr:** S Africa (W & E Cape Region, S KwaZulu-Natal). **Ecol:** In succulent karoo, dry rock outcrops on warm slopes. Fr. Nov–Jan. **Zones:** 10–11.

C. multicava Lemaire, *Rev. Hort.* 97, 1862. (Syn: *C. quadrifida*, *Septimia multicava*). [Cape Province Pygmyweed, Fairy Crassula]. **Herbs**, variable, decumbent to erect, 10–20 cm high (excluding the inflorescence), branched from the base forming dense clumps; roots fibrous; branches 0.5–1 cm Ø, green. **Leaves** shortly petiolate, ascending-spreading; lamina ovate, flat, 3–6 × 2.8–4.8 cm, base cuneate to cordate, margins entire, apex acute with conspicuous hydathodes, green (rarely abaxially purplish); petiole to 1.5 cm long. **Inflorescences** ascending pyramidal thyrses, 18–22 cm tall, with numerous dichasia (up to 5 nodes); peduncle to 12 cm long; pedicels 3–8 mm long. **Flowers** with sepals linear, 1.5 mm long; corolla 4- or 5-merous, 1 cm long, stellate, petals 5 × 2 mm, white to light pink; filaments 2 mm long. **Distr:** S Africa (E Cape Region, KwaZulu-Natal). **Ecol:** In coastal forests, valleys and savannas. **Zones:** 9–11.

Echeveria

Syn: *Oliveranthus*, *Oliverella*, *Urbinia*.
Species 140. **Herbs** to **subshrubs**, glabrous to hirsute; stem none or tall, branching or not. **Leaves** rarely scattered along the stems, usually (ob)lanceolate and mucronate, often glaucous or highly colored. **Inflorescences** racemose, paniculate or rarely spicate thyrsoids, or cymose with 1 to several cincinni; pedicels usually with 1 to several minute bracteoles. **Flowers** with sepals reflexed to appressed but usually somewhat expanding, almost completely free, equal to strongly unequal; corolla cylindrical to pentagonal to urceolate; petals imbricate (sometimes valvate), white through yellow and orange to red, rarely green, inner surface usually with nectar cavity at base; stamens 10, 5 attached at top of nectar cavities, 5 at top of corolla tube between petals; carpels basally connate, erect at anthesis; stylodia slender. **Fruits** widely divergent follicles; seeds reticulate or smooth. **Distr:** S USA, Mexico, C and S America. **Cult:** Propagate from seeds, offsets or by division. Requires a gritty, well-drained soil in full sun.

E. agavoides Lem., *Ill. Hort.* 10: misc. 78, no. 2, 1863. (Syn: *Cotyledon agavoides*, etc.). [Lipstick Echeveria, Moulded Wax, Red Edge Echeveria].

Top row: *Aeonium percarneum, Aeonium tabuliforme, Aeonium urbicum, Bryophyllum pinnatum.*
Second row: *Crassula arborescens, Crassula multicava, Echeveria agavoides, Echeveria elegans.*
Third row: *Echeveria multicaulis, Hylotelephium spectabile, Kalanchoë delagoensis.*
Bottom row: *Kalanchoë beharensis, Kalanchoë blossfeldiana.*

MD

MD

MD

MD

Herbs, rosette-forming; stem usually 5–10 cm tall, 2.5–3 cm Ø, usually unbranched; rosette 7–15(–35) cm Ø. **Leaves** ~20, arranged in a dense rosette; lamina ovate or ovate-deltoid, 3–7 cm long, ~3 cm near base, apex acute or acuminate, ~5 mm thick, spine-tipped, margins rounded and hyaline, very rigid, surfaces green, sometimes reddened along margins or apex. **Inflorescences** mostly with 2 cincinni, few-flowered, 20–30 cm long; bracts secund-racemose; pedicels 0.8–2(–3) cm long, slender. **Flowers** with calyx small, spreading, sepals 5, small, ovate, <5 mm long, somewhat spreading; corolla conoid-urceolate, 1–1.4 cm long, 5–8 mm Ø near base, 0.5–1 cm Ø at mouth, basally terete, pinkish to orange-red, united for 1/4 their length, apices spreading, yellow; stamens all borne on the corolla, attached above it base. **Distr:** Mexico. **Ecol:** On rocky outcrops; at higher altitudes. Fl. Dec, May–Jul. **Zones:** 9–11.

E. elegans Rose, *N. Amer. Fl.* 22(1): 22, 1905. (Syn: *E. perelegans*). [Hen and Chicks, Mexican Gem, Mexican Snowball, White Mexican Rose]. **Herbs**, variable, 0.8–1 m high; stem nearly lacking, branching. **Leaves** numerous, densely rosulate; lamina oblong-obovate or obovate-cuneate, 3–7.5 × 1–3.5 cm, margins translucent or reddish, apex acute to truncate, mucronate, adaxially flat to concave, thick, turgid, nearly white or greenish-blue, glaucous, sometimes purple-tinged. **Inflorescence** a cincinnus or secund raceme, 5–7-flowered, 10–20 cm tall, pinkish, with 8–12 pinkish leaflike bracts; pedicels 0.6–1.4 cm long; flower buds broadly oblong, acute. **Flowers** with sepals brightly colored, very unequal, appressed to expanded, 0.5–1 cm long, basal margins often toothed; corolla conoid-urceolate, 1–1.8 × 0.8–1.2 cm, rose to pink, slightly glaucous, lobes, distinct nearly to base, often yellowish or greenish at apices, apices spreading but connivent with age; stamens all borne on the corolla, attached just above its base, 2/3 its length; scales broad; carpels distinct, tapering into slender styles. **Distr:** Mexico (Hidalgo, San Luis Potosí, Guanajuato, Querétaro). **Ecol:** In mountainous habitats. **Zones:** 9–11.

E. multicaulis Rose, *Contr. US Nat. Herb.* 8: 294, 1905. [Copper Roses, Copperleaf]. **Subshrubs**, caulescent; stem much-branched from the base, 0.25–1.2 m long, granular-roughened, naked below, apices leafy; flowering branches brightly colored. **Leaves** in subrosettes ~7–9 cm Ø; lamina obovate-cuneate, 3–4 × 1.2–3 cm, apex obtuse to truncate, mucronate, fleshy, dark green, shiny, margins and apex reddish; flowering branches with scattered oblanceolate leaves. **Inflorescences** racemes, spikes or thyrses, 6–15-flowered, compact, to 25 cm long, granular-roughened; flowers subtended by small brightly colored bracts; pedicels ~1 cm long, upper ones shorter. **Flowers** with sepals ascending, narrow, to 6 mm long, acute; corolla campanulate, pentagonal, ~1–1.2 cm long, 5 mm Ø near base, 1 cm Ø at the mouth, carmine to scarlet, yellowish inside. **Distr:** Mexico. **Ecol:** On steep rocky slopes; at high altitudes. Fl. Sep.–May. **Zones:** 9–11.

Hylotelephium

Species ~33. **Herbs**, perennial, glabrous or hairy; roots fibrous or tuberous, often carrot-shaped; rhizome short, fleshy or woody; young branches not covered with scales; flowering stem arising from lateral bud of previous year, annual and abscising from base or persisting and basally woody. **Leaves** cauline, alternate, opposite, or 3–5-verticillate; lamina flat, glabrous, spurless. **Inflorescences** terminal, sometimes also subterminal and thus paniculate, cymose, often corymbiform, sometimes umbel-like, many-flowered, bracteate. **Flowers** bisexual or unisexual, unequally (4 or)5-merous, sometimes reduced; sepals usually shorter than petals, base subconnate and spurless; petals usually free, purple, red, pink, or white, occasionally yellowish or greenish; stamens twice as many as petals, inserted basally on corolla; nectar scales oblong-cuneate to linear-oblong, apex rounded or subemarginate; carpels free, base narrow, substipitate. **Fruits** follicles, erect, many-seeded; seeds narrowly winged. **Distr:** Asia, Europe, N America. **Cult:** Propagate by division, from cuttings or seeds. Requires a moist, well-drained, gritty soil in full sun.

H. spectabile (Boreau) H. Ohba, *Bot. Mag.* (Tokyo). 90: 52, 1977. (Syn: *Sedum spectabile*). [Ice Plant]. **Herbs**, perennial; stem erect, 30–70 cm high, spreading 40–80 cm wide. **Leaves** opposite or triverticillate; lamina narrowly elliptic-oblong, oblong-ovate, ovate, or broadly so, 2–10 × 0.8–5 cm, base attenuate, margins entire, ± undulately dentate, or dentate, apex obtuse to acute. **Inflorescences** cymes, 3-branched, flat-topped, dense, 7–11 cm across. **Flowers** small, pink to red, ~1 cm Ø; sepals linear-lanceolate to broadly lanceolate, ~1 mm long, acuminate; petals purplish-red to purple, lanceolate to broadly so, 4–5 mm long; stamens longer than corolla, 6–8 mm long, anthers violet; nectar scales rectangular, 1–1.2 mm long, emarginate; carpels narrowly ellipsoid, ~3 mm long; styles ~1.2 mm long. **Fruits** follicles, erect. **Distr:** China, Korea. **Ecol:** In forest margins, rocky slopes on low mountains. Fl. Aug–Sep. Fr. Sep–Oct. **Zones:** 6–11.

Kalanchoë

[Widow's Thrill].

Species 145. **Herbs**, sometimes biennial, subshrubs, or shrubs; roots usually fibrous. **Leaves** opposite, petiolate or sessile, usually amplexicaul; lamina with margins entire, dentate, crenate, or leaves pinnate. **Inflorescences** terminal, cymose, sometimes also with subterminal cymes and thus paniculate, many-flowered; bracts small. **Flowers** bisexual, erect, 4-merous; sepals free or basally subconnate, triangular to lanceolate, usually shorter than corolla tube; corolla yellow (or white, red, pink, or orange), salverform, tube subquadrangular or basally inflated, urn-shaped, base slightly narrowed, lobes longer than tube; stamens twice as many as petals, inserted near middle of corolla tube, filaments unequal in length, usually very short; nectar scales linear to suborbicular; carpels erect, styles short or long. **Fruits** follicles; seeds many, ellipsoid. **Distr:** Old World (Africa, Asia). **Cult:** Propagate from stem cuttings, leaf cuttings or seeds. Requires a moderately fertile, gritty soil in full sun. **Info:** Most members of this genus reproduce asexually, often producing very large number of plantlets on the leaf margins (Garces et al. 2007).

K. beharensis Drake, *Bull. Mus. Hist. Nat.* (Paris) 41, 1903. [Donkey's Ear, Madagascar Felt Bush, Velvet Bush, Velvet Elephant's Ear, Velvet Leaf]. **Shrubs**, large, to 1.5 m high, sometimes small trees to 6 m tall in the wild; stem slender, knotted, glabrous below, woolly hairy towards apex. **Leaves** decussate, opposite; lamina large, triangular-lanceolate, 10–45 × 5–30 cm, fleshy, velvety-tomentose, gray, margins toothed, adaxially concave (especially in older leaves), abaxially convex, margins doubly crenate, tips of teeth with dark brown hairs, olive-green, surfaces with dense woolly brown hairs, hairs on stems, younger leaves and petioles white; petiole 3–4 cm long (longer with age). **Inflorescences** terminal or axillary, 50–60 cm high, forming lax and intricately branched corymbs, hairy; pedicels short. **Flowers** insignificant, yellowish, urn-shaped; calyx 7 mm long, 4-anged at base, lobes oblong, acuminate; corolla tube urn-shaped, 7 mm long, lobes oblong, rounded, 4 mm long. **Distr:** Madagascar. **Ecol:** In xerophytic forests on various soils. Fl. Jun–Jul, Dec–Mar. **Zones:** 10–12.

K. blossfeldiana Poelln., *Repert. Spec. Nov. Regni Veg.* 35: 159, 1934. [Flaming Katy]. **Herbs** or **subshrubs**, perennial, 10–40+ cm high, glabrous, smooth; shoots erect, few-branched. **Leaves** alternate, borne on upper stems; lamina simple, elliptic to ovate, oblanceolate, 2–12.5 × 2.5–7 cm, margins red, entire up to halfway, then sinuate, scalloped and crenate to apex, succulent, dark green, glossy; petiole grooved, 2.5 × 0.8 cm. **Inflorescences** panicles, terminal, spreading, many-flowered, often to 6-flowered. **Flowers** 1.2–1.3 cm long, 0.9–1 cm Ø, scarlet; sepals lanceolate, 4–5 mm long; corolla of fused petals, red or sometimes yellow, salverform, tube 0.6–1.1 × 1–2 cm, greenish outside, lobes 4, spreading, about half as long as the tube, oblong, variously colored (yellow, orange, pink, red or purplish); stamens and pistil just inside corolla throat. **Fruits** composed of 4 follicles (formed from separate ovaries). **Distr:** Madagascar. **Ecol:** In shade of forests; 1,600–2,400 m. **Zones:** 10–12.

K. delagoensis Ecklon & Zeyher, *Enum. Pl. Africa. Austal.*, 3: 305, 1837. (Syn: *Bryophyllum delagoense*, etc.). [Chandelier Plant]. **Herbs**, biennial or ± perennial, robust, glabrous, 0.2–2 m tall, often forming dense stands; stem simple, erect, terete. **Leaves** ternate to apparently opposite or alternate, sessile, usually straight, erect to spreading; lamina subcylindrical, adaxially ± grooved, 1–13 cm long, 2–6 mm Ø, base attenuate, apex with 2–9 small teeth with numerous bulbils, reddish-green to grayish-green with reddish-brown spots. **Inflorescences** thyrses, compact, rounded, many-flowered, 10–25 cm across; pedicels slender, 0.6–2 cm long, flowers pendent. **Flowers** with calyx campanulate, reddish to green, red-striped, tube 2.5–6 mm long, lobes triangular-lanceolate, 0.5–1 × 0.3–0.6 cm, very acute; corolla red, violet, pale orange to yellow, red-suffused, tube funnelform, 2.2–4 cm long, lobes spreading, oblong-obovate, 0.7–1.2 × 0.6–0.9 cm, obtuse or truncate; stamens inserted below middle of corolla tube, all included; carpels ovate-oblong, 5.5–6.5 mm long, style 1.9–2 cm long; nectar scales semicircular to square, 0.7– 2 mm long, rounded. **Fruits** follicles; seeds 0.6–2.5 mm long. **Distr:** Madagascar. **Ecol:** In grasslands, rocky slopes. **Zones:** 11–12.

Sedum

Species 420. **Herbs**, annual or perennial, mostly glabrous; rootstock usually not developed, occasionally plants rhizomatous; stems erect or decumbent, sometimes fasciculate or mosslike, fleshy, glabrous or hairy, base rarely woody. **Leaves** alternate, opposite, or verticillate; lamina with base often spurred, margins normally entire (rarely 4–8-dentate and remotely dentate). **Inflorescences** terminal or axillary, cymose, often corymbiform, 1–many-flowered. **Flowers** usually bisexual, rarely unisexual, mostly unequally (3–)5(–9)-merous; sepals and petals free or basally connate; petals free or almost so, mostly yellow, less often white or reddish; stamens usually twice as many as petals, (1)biseriate, antepetalous ones adnate to petal base or slightly above; nectar scales entire or apex emarginate; carpels usually as many as petals, occasionally fewer, free or basally widened and -connate, styles short or long. **Fruits** follicles; seeds few–many, smooth or papillate, less often striate. **Distr:** Mainly in the Northern Hemisphere, extending to the Southern Hemisphere in Africa and S America. **Cult:** Propagate by division, from cuttings or seeds. Requires a moist, well-drained, gritty soil in full sun.

S. lydium Boissier, *Diagn. Pl. Orient.* 1(3): 17, 1843. [Least Stonecrop, Stonecrop]. **Herbs**, perennial, glabrous, small, tufted to cushion-forming, 15–30 cm wide, ascending, 5–10 cm high; branches creeping and rooting, with perennating sterile rosettes. **Leaves** alternate, imbricate, sessile; lamina linear, terete, ~6 mm long, apex obtuse, fleshy, green, red-tipped. **Inflorescences** dense corymbs, (7–)10–20-flowered,

MD

Top row: *Sedum lydium, Sedum morganianum, Leea guineensis, Leea guineensis* (red form).
Bottom row: *Leea indica, Leea rubra, Cissus discolor, Cissus heyneana.*

1–2 cm Ø; flowers pedicellate. **Flowers** small, 5-merous; sepals ~2 mm long, acute; petals ~3 mm long, white with midvein red. **Fruits** capsules, erect, glabrous. **Distr:** W Turkey. **Ecol:** In damp, mossy places on mountains. **Zones:** 9–11.

S. morganianum E. Walther, *Cactus and Succulent Journal* 10: 35, 1938. [Burro's Tail, Donkey's Tail, Lamb's Tail]. **Herbs**, perennial; stems prostrate, trailing or hanging, 0.3–3 m long, covered with thick succulent leaves. **Leaves** alternate, densely arranged along stems; lamina ± cylindrical or spindle-shaped, curved, 1.2–2.5(–3.2) cm long, to 8 mm Ø, apex pointed, bluish-green, glaucous. **Inflorescences** clusters, 1–6-flowered, terminal or from axils of leaves near stem tips. **Flowers** rose-purple, rose-pink to brick-red, ~1.2 cm long, not opening fully; petals 5; stamens 10, 2-whorled; carpels 5, longer than stamens. **Distr:** Mexico. **Ecol:** Unknown in the wild. Fl. Apr–Nov. **Zones:** 9–12. **Info:** Cultivated widely for ornamental purposes, usually grown in hanging baskets.

VITALES

FAMILIES 1, SUBFAMILIES 2, GENERA 15, SPECIES 882.

Vessel elements with simple perforation plates; raphide bundles present. **Leaves** with glandular teeth. **Flowers** with stamens equal in number to and opposite petals, and from a common primordium; nectary gynoecial; 2 apotropous ovules/carpel, style short. **Fruits** berries, calyx deciduous; seeds perichalazal, vascular bundle ± surrounding seed, ± ruminate, testa multiplicative, exotesta fleshy; embryo minute. **Distr:** Pantropical and warm temperate. **Info:** Molecular data still do not link Vitaceae unambiguously with any other core Eudicot, but they are definitely not nested within any group (Stevens 2001 onwards). They were placed sister to core rosids, with only moderate support by Soltis et al. (2000), and with stronger support by Jansen et al (2006a), Ravi et al. (2007) and Zhu et al. (2007).

VITACEAE

Subfamilies 2, **genera** 15, **species** 882. **Climbers**, woody, sometimes vines, rarely small succulent trees, hermaphroditic or polygamo-monoecious to polygamo-dioecious; stems unarmed, sometimes with conspicuous lenticels, or bark sometimes shredding; branches often swollen at 3–7-lacunar nodes. **Stipules** (0)2, often caducous. **Tendrils** simple, bifurcate to trifurcate, or 4–12-branched, usually leaf-opposed, rarely tendrils absent. **Raphide sacs** present in parenchymatous tissues. **Leaves** simple, lobed or unlobed, or digitately or pedately compound to 1–3-pinnately compound, alternate, distichous, variously toothed, commonly with multicellular, stalked, caducous spherical structures known as 'pearl' glands. **Inflorescences** panicles, corymbs, or rarely spikes, flowers small, with prophylls, often leaf-opposed, pseudo-terminal, or axillary. **Flowers** actinomorphic, hypogynous, 4 or 5(–7)-merous; calyx with 4 or 5(–7) small teeth or lobes or a continuous ring; petals valvate, 4 or 5(–7), free or basally connate, or distally connate forming a calyptra; stamens 4 or 5(–7), antepetalous, anthers introrse, longitudinally dehiscent, tetrasporangiate, rarely bisporangiate; floral disk intrastaminal, ring-shaped, cupular, or gland-shaped; ovary superior, biloculed, locules 2-ovuled, placentation axile, apotropous or anatropous, bitegmic, style simple, connate, stigma discoid or capitate, rarely 4-lobed. **Fruits** berries; seeds 1–4. **Distr:** Worldwide, but mostly in tropical and subtropical regions.

Subfam. LEEOIDEAE

Genus 1, **species** 34. **Herbs** to **trees**; raphides barbed. **Stipules** borne along petiole margin, sheathing. **Leaves** spirally arranged; lamina with teeth with small glandular apex, one lateral vein continues its course above the tooth. **Inflorescences** terminal. **Flowers** with corolla basally connate; stamens adnate to corolla, connate, with a lobed tube; nectary disk absent; gynoecium 3(4), (semi-inferior), odd member abaxial, locules divided. **Fruits** rarely with raphides in the seed coat. **Distr:** Most Indo-Malesian, few Africa and Madagascar.

Leea

Syn: *Nalagu.*
Species ~34. **Shrubs**, **small trees**, **scramblers**, or rarely large perennial herbs; stems unarmed or with rows of prickles; tendrils absent. **Stipules** sheathing petiole margins with conspicuous, persistent or caducous stipular wings. **Leaves** 1–4-pinnate to trifoliolate or simple; leaflets glabrous to pubescent, margins crenate, serrate or dentate, teeth with small glandular apex, abaxially usually with specialized multicellular, stellate or globular caducous 'pearl' glands. **Inflorescences** paniculate, often corymbiform, terminal or axillary, erect or pendulous. **Flowers**

Above: *Cissus latifolia*, *Cissus quadrangularis*, *Cissus rotundifolia*.

hermaphroditic, (4 or)5-merous; calyx campanulate, lobes triangular, tips glandular; petals valvate, apically often cucullate, reflexed at anthesis, basally connate; floral disk tubular, intrastaminal; stamens 4 or 5, antepetalous, anthers bilocular; ovary superior but sometimes partly sunken in disk, 2–3(–5)-carpellate, 4–6(–10)-loculed, locules 1-ovuled, style elongate, stigma discoid, capitate. **Fruits** berries, rather dry, depressed subglobose, purple, black, or orange; seeds endotestal, endosperm ruminate with roughly 5 ingrowths. **Distr:** Widespread in tropical and subtropical Asia, extending into the Himalayan region and Australia, 2 spp. in Africa and Madagascar. **Cult:** Propagate from cuttings or seeds. Requires a moist, fertile soil in full or partial sun.

L. guineensis G. Don, *Gen. Hist.* 1: 712, 1831. (Syn: *L. manillensis*). [West Indian Holly]. **Shrubs** or **small trees**; branchlets terete, almost glabrous. **Leaves** 2- or 3-pinnate; leaflets oval-elliptic to long- and roundly lanceolate, 5–15 × 2.5–8 cm, base broadly cuneate, or rarely suborbicular, margins with acute teeth, apex acuminate, glabrous, lateral veins 6–11 pairs, abaxial veinlets conspicuous but not protruding; petiole 6–13 cm long, central petiolules 1.5–4 cm long, lateral petiolules 0.5–1.5 cm long, glabrous. **Inflorescences** corymboselike compound dichasia, ~50 cm Ø; pedicels very short or nearly none, sparsely with papillary hairs; buds ~3 mm long. **Flowers** with calyx tube cupulate; sepal triangular, acute, glabrous; petals 5, elliptic, red; stamens 5, filaments 1.2–1.6 mm long, anthers yellow; ovary ovate, stigma expanded slightly. **Fruits** berries, subglobose, ~8 mm Ø. **Distr:** Tropical and subtropical Asia to Malesia; Africa, Madagascar. **Ecol:** In forests, shrublands. **Zones:** 10–12. **Info:** The cultivar 'Burgundy' has dark purple stems and foliage.

L. indica (N.L. Burman) Merrill, *Philipp. J. Sci.* 14: 245, 1919. (Syn: *L. umbraculifera*, *Staphylea indica*). [Bandicoot Berry]. **Shrubs** to **small trees**, erect; branchlets terete, longitudinal ridges obtuse, glabrous. **Stipules** broadly obovate, 2.5–4.5 × 2–3.5 cm, rounded, glabrous. **Leaves** 2- or 3-pinnate, glabrous; leaf axis 14–30 cm long, glabrous; leaflets elliptic to elliptic-lanceolate, 6–32 × 2.5–8 cm, base usually rounded, margins usually irregularly dentate, apex acuminate or caudate, lateral veins 6–11 pairs; petiole 13–23 cm long, central petiolules 2–5 cm long, lateral petiolules 2–5 mm long. **Inflorescences** opposite, compound dichasial or umbelliform; peduncles 1–2 cm long, brown-hairy; involucre elliptic-lanceolate, ~8 × 3–8 mm, acuminate, glabrous; bracts deciduous; pedicels 1–2 mm long, pubescent. **Flowers** with calyx tube urceolate, sepals triangular, glabrous; petals elliptic, 1.8–2.5 mm long, glabrous, white or greenish-white; staminodial tube 0.5–1 mm long, lobes minute; stamens 5; ovary globose, style very short. **Fruits** berries, 0.8–1 cm Ø, 4–6-seeded. **Distr:** Tropical and subtropical Asia; N Australia, Pacific Islands. **Ecol:** In forests, shrublands; 200–1,200 m. Fl. Apr–Jul. Fr. Aug–Dec. **Zones:** 10–12.

L. rubra Blume ex Spreng., *Syst. Veg.* 1: 670, 1824. [Hawai'ian Holly, Red Leea, West Indian Holly]. **Shrubs**, ~0.3–2 m tall; young branches minutely scaly pubescent, ultimately glabrous. **Leaves** 2- to 3-pinnate; main rachis angled, sometimes slightly winged; leaflets 3–7, ovate or oblong, 5–9.5 × 2.5 cm, base rounded or cuneate, margins serrate, apex acuminate, membranous, dark green, main nerves 6–10 pairs, winged and crisped, young leaves with minute black hairs along their sides, otherwise glabrous; petiole <5 mm long. **Inflorescences** cymes, dense 5–7.5 cm across, dark red; peduncles 1.3–6.5 cm long, furfuraceous rusty-puberulous, condensed, much-branched; bracts and bracteoles absent. **Flowers** crimson; staminal tube with lobes notched; stamens white. **Fruits** berries, 6 mm Ø. **Distr:** E Bengal, Java, Malaysia, Myanmar, Singapore. **Ecol:** Near tidal rivers; 300 m. **Zones:** 10–12. **Info:** This species is often confused with *L. coccinea*.

Subfam. VITICOIDEAE

Genera 14, **species** 825. **Vines** or **lianas** (rarely stout trees, herbs; rootstock swollen) climbing by leaf-opposed tendrils; raphides smooth; (cuticle waxes as tubular rodlets). **Leaves** opposite, spirally arranged or 2-ranked, teeth with gland broadening distally and with foramen, veins from above and below. **Inflorescences** leaf-opposed (terminal). **Flowers** with corolla rarely connate by papillae and calyptrate, rarely tapetum plasmodial; disk annular or as 5 glands; gynoecium biloculed, transverse or vertical, (style long). **Fruits** with raphides in the seed coat (Stevens 2001 onwards). **Distr:** Pantropical and warm temperate.

Cissus

Species ~350. **Lianas**, woody or subwoody, hermaphroditic or polygamo-monoecious. **Tendrils** unbranched or bifurcate, sometimes racemosely branched. **Leaves** simple or palmately compound. **Inflorescences** compound dichasia or umbels, leaf-opposed. **Flowers** 4-merous; calyx cupular, glabrous; petals falling off separately; stamens 4; disk well-developed, margin undulate or slightly 4-lobed; style conspicuous, stigma undivided or 2-divided. **Fruits** berries, obovoid or ellipsoid, 1- or 2-seeded; seeds obovoid-elliptic or elliptic, base rostrate, apex rounded, chalazal knot at or near base, cross-section of endosperm M-shaped. **Distr:** Mainly in tropical regions of Africa, Asia, Australia, C & S America, and N America (Mexico). **Cult:** Propagate from stem cuttings or seeds. Requires a well-drained soil in full or partial sun.

C. discolor Blume, *Cat. Gewass. Buitenz.* 39, 1823; *Bijdr*, 181, 1825. (Syn: *Vitis discolor*). [Begonia Treebine, Mottled Cissus, Painted Cissus, Rex Begonia Vine]. **Climbers**, to 5 m long, glabrous or minutely velvety; stems red, slender, 5–6-angled. **Stipules** persistent. **Tendrils** forked. **Leaves** simple; lamina ovate-oblong to lanceolate, 6–25 × 2–8 cm, base deeply cordate, rounded or cuneate, margins serrate, apex acuminate, membranous, adaxially glabrous, dark green, variegated with silvery white, red and pale pink, abaxially purple or burgundy red, nerves 5–7 pairs; petiole 0.5–3.2 cm long. **Inflorescences** cymes, umbellate, lax, pubescent or compact, 1.25–1.5 cm across; pedicels red. **Flowers** minute, 2.5 mm long, red outside; petals pale yellow, green inside. **Fruits** pyriform, globose, 5 mm across, blackish-red. **Distr:** SE Asia through Indonesia to Australia. **Ecol:** On limestone rocks, on trees near forest margins; 500–1,500 m. **Zones:** 10–12.

C. heyneana Steud., *Nom., Bot.* ed. 2. 372, 1840. (Syn: *C. glauca*, etc.). **Lianas**, large, glabrous, fetid-smelling when bruised; bark usually white, terete, reddish when young, bright green to glaucous later. **Stipules** orbicular, ~1.5 mm long, evanescent, margins fimbriate. **Tendrils** simple, becoming robust, bearing minute bract near base when young. **Leaves** very variable, juvenile ones palmate, those of flowering stems palmate to simple, all thick, *palmate* to 30 × 25 cm, *simple* to 21 × 14 cm but usually smaller especially on flowering lateral shoots; petiole 5–15 cm on main shoots, terete, basally swollen, articulated with lamina; *palmate* laminas 5-lobed, lobes triangular, acute, mucronate; *simple* laminas ovate, base obtuse to truncate; all laminas with base subpeltate, margin spinulose-serrate, costae ~4 or 5 opposite to subopposite pairs. **Inflorescences** sweetly scented, terminal or terminating lateral shoots from nodes flanked by 2 sterile ones, ~25–30 cm long, with ~6–10 leafy nodes, leaves smaller nearer apex and opposite cymes; cymes ~2.5 × 3 cm, corymbose, pseudo-trichotomous, with 2 orders of branching; bracts minute, erose-fimbriate; peduncle 0.8–1.2 cm long. **Flowers** with hypanthium ~5–8 mm long; calyx cupular ~2 × 2 mm, red, margin entire to obscurely lobed; petals ~2.5 mm long, obtuse, valvate, red, apex thickened; anthers oval-rotund, yellow, filaments narrowly triangular; disk 4-lobed, yellow. **Fruits** ~2 cm Ø, globose, purple-black, foul-tasting; seed ovoid, ± angled. **Distr:** S India, Sri Lanka. **Ecol:** In lowland forests. Fl. Apr–Oct. **Zones:** 11–12.

C. latifolia Lam., *Enc.* 1: 39, 1783. (Syn: *C. acuminata*, etc.). **Lianas**, large; stems corky, dumbbell-shaped in cross-section; leafy stems terete, glaucous, pruinose; new growth finely silky-hairy. **Stipules** persistent, triangular-cordate to subrotund, ~2–2.5 mm long, margins scarious. **Tendrils** bifid to pinnately branched. **Leaves** cordate-ovate on main shoots, smaller and narrower on flowering ones, to 25 × 25 cm, usually broader than long; petiole 3–11 cm long on main shoots, terete to ± angled; base forming ± deep sinus, margins subrepand, spinulose-serrate, apex abruptly acuminate, costae in ~5 or 6 opposite

Above: *Cissus verticillata*, *Tetrastigma nilagiricum*, *Vitis vinifera*.

to subopposite pairs. **Inflorescences** borne on tendril-less flowering shoots, apparently terminal as well as lateral, ~20–30 cm long with ~8 leafy nodes, leaves increasingly ovate, smaller and bases truncate the more distal, opposite cymes, shoot tips apparently vegetative; cymes umbellate, ~11 × 8 cm, branched to 2 orders, terminal cymules, lax, axes silky-hairy; bracts triangular, ~0.5 mm long; peduncle 3–5 cm long. **Flowers** with hypanthium ~8 mm long, calyx cupular, ~2 × 2 mm, entire to obscurely lobed; petals ~2.5 mm long, obtuse, valvate, yellow-green; disk yellow. **Fruits** ~1 cm Ø, purplish, 1-seeded. **Distr:** S India, Sri Lanka. **Ecol:** In lowland forests. Fl. Oct–May. **Zones:** 11–12.

C. quadrangularis L., *Syst. Nat.* ed. 12, 2: 124, 1767. (Syn: *C. edulis*, *V. quadrangularis*). [Winged Treebine, Veldt Grape]. **Lianas**, succulent, glabrous, deciduous, to ~8 m long; apical stems ~1–1.5 cm Ø, quadrangular in cross-section, constricted at nodes, green, faces concave, angles almost winged. **Stipules** ovate, to 5 × 5 mm, obtuse, pale green, evanescent. **Tendrils** simple, becoming woody, bearing minute bract in proximal half. **Leaves** variable, evanescent, ovate or broadly ovate to rounded-deltoid, ~2–5 cm long, base truncate, margins distantly spinulose-crenate, apex obtuse, thick, abaxially scattered-punctate, veins ~3 pairs; petioles ~0.5–1 cm long. **Inflorescences** cymes, ~2–3 × 5–6 cm, asymmetric, subumbellate with usually 3 rays, 1 sometimes further branched, tips bearing congested cymules of sweetly scented flowers opening 1 or 2 at a time; peduncles ~0.7–1.5 cm long, erect. **Flowers** with hypanthium ~6–8 mm long; calyx subcupular, ~2 mm Ø, reddish; petals ovate, ~2.5 mm long, acute, recurved, pale green; anthers oval-rotund, yellow, evanescent; disk 4-lobed, bright yellow, rapidly fading to cream. **Fruits** berries, ~0.7–1 cm Ø, subglobose, maroon, with persistent calyx and style; seed 1, ~5 × 4 mm, weakly obovoid, smooth. **Distr:** Tropical Africa, Arabia, Madagascar, Mascarenes, India, Sri Lanka. **Ecol:** In dry regions. Fl. Aug–Jun. **Zones:** 10–12.

C. rotundifolia (Forsskål) Vahl, *Symb. Bot.* 3: 19, 1790. [Arabian Wax Cissus]. **Climbers** vigorous, to 10 m long, becoming ± woody; stems succulent, 4–5-ridged with age, wings often becoming corky, to 2.5 cm wide, pubescent or glabrescent. **Stipules** ± semicircular, to 4 mm long, glabrous, early caducous. **Tendrils** bifid or not. **Leaves** simple or occasionally lobed, nearly circular, 5–8(–10) cm Ø, base cordate, margins nearly entire to scalloped, apex obtuse, brittle-fleshy, glaucous bluish-green; petiole to 1 cm long, pubescent. **Inflorescences** cymes, leaf-opposed and terminal, lax, longer than leaves; peduncle 3 cm long; pedicels 4–5 mm long, lengthening to 1 cm in fruit, appressed-pubescent; flower bud 3.5 × 1.5 mm. **Flowers** very small, greenish to whitish; calyx 1 mm long, entire; petals green, yellowish or whitish, 3–3.5 mm long, glabrous; ovary glabrous, style 0.5 mm long. **Fruits** berries, ellipsoid, 1.5 × 1.3 cm, red or purplish; seeds 1–2, 9 mm long, smooth with a single crest. **Distr:** S Arabia through eastern Africa southwards to Mozambique and the Transvaal. **Ecol:** Mostly in coastal areas, sometimes extended inland in dry bush and woodlands. **Zones:** 10–12.

C. verticillata (L.) Nicolson & Jarvis, *Taxon* 33: 727, 1984. (Syn: *Viscum verticillatum*, *C. sicyoides*). [Possum Grape Vine, Princess Vine, Season Vine]. **Vines**, non-woody, climbing with tendrils, to >10 m long; stems fleshy, with abundant watery latex, cylindrical when young, flattened when mature, to 5 cm Ø, nodes swollen. **Stipules** 2.5–3.5 mm long, auriculate. **Tendrils** opposite leaves, simple or bifurcate, to 25 cm long, spiralling. **Leaves** alternate; lamina ovate, 5–12 × 3.8–6.5 cm, base cordiform, margins revolute, denticulate, apex acute or rounded, coriaceous, adaxially green, shiny, abaxially dull green; petiole 2–5 cm long, sulcate. **Inflorescences** compound cymes, opposite leaves; pedicels ~3 mm long, yellowish-green or reddish. **Flowers** with calyx yellowish-green or reddish, 0.7–1 mm long; petals 4, yellowish or pink, oblong-lanceolate, 2–2.5 mm long, deciduous; disk annular, yellow, 0.5–0.8 mm high. **Fruits** globose, fleshy, 0.7–1 cm Ø, shiny, dark violet or black; seeds 1 or 2. **Distr:** Southern N America to northern S America, including the Antilles. **Ecol:** In disturbed areas, weedy. Fl. & Fr. throughout the year. **Zones:** 9–12.

Tetrastigma

Species ~100. **Climbers**, woody, rarely herbaceous, usually polygamo-dioecious. **Tendrils** unbranched or bifurcate, sometimes palmately branched. **Leaves** usually palmately 3–5-foliolate or pedately 5–7-foliolate, rarely simple. **Inflorescences** polychasia, umbels, or a compound cymes. **Flowers** 4-merous; petals spreading, free, usually hooded or galeate, sometimes corniculate; stamens short, abortive in ♀ flowers; disk well-developed in ♂, inconspicuous in ♀; style conspicuous or inconspicuous, stigma usually 4-divided, rarely irregularly divided. **Fruits** berries, globose, ellipsoid, or obovoid, 1–4-seeded; seeds elliptic, obovoid-elliptic, or obtriangular, abaxial surface with a linear to orbicular chalazal knot, adaxial surface with a linear raphe, cross-section of endosperm T- or M-shaped. **Distr:** Asia to Oceania. **Cult:** Propagate from stem cuttings or seeds. Requires a well-drained soil in full or partial sun. **Info:** Some species of this genus serve as hosts for the gigantic-flowered, parasitic *Rafflesia* spp.

T. nilagiricum (Miq.) Shetty, *Kew Bull.* 44: 477, 1989. (Syn: *Cissus lanceolaria*, etc.). **Lianas**, to 25 m long, though flowering when much smaller; stems strongly verrucose, lenticellate. **Stipules** ovate to lanceolate, 1–2 mm long, acuminate, papery, evanescent. **Tendrils** simple, persistent, woody with age. **Leaves** to 25 × 25 cm, 3- or 5-foliate, persistent; leaflets broadly lanceolate to obovate, 3–13 × 1.5–7 cm, apical leaflet largest, base acute, margins coarsely serrate especially distally, teeth apices glandular, apex acute to subacuminate, coriaceous, glabrous, laterals asymmetric; petiole 3–9 cm long; petiolules 0.3–1.5 cm long. **Inflorescences** with ♂ longer than ♀; branches to 3 orders, first 2 (3–5-branched) subumbellate, 3rd a raceme or cymule of pedicellate flowers; bracts triangular, ~1 mm long; pedicels minute, articulated with elongated hypanthium (pseudo-pedicel). **Flowers** small; calyx salverform, margin ± irregularly 4-lobed, ± ciliolate; petals 0.7–1 mm long, acute, white. **Fruits** ~1.2 cm Ø, spherical, creamy-white; seeds 2–4, transversely rugose. **Distr:** S India, Sri Lanka. **Ecol:** In forest margins; <2,000 m. Fl. Feb–May. **Zones:** 11–12. **Info:** At lower altitudes, plants usually have simple leaves.

Vitis

Syn: *Ampelovitis*, *Spinovitis*.
Species ~60. **Lianas**, woody, usually polygamo-dioecious, rarely hermaphroditic; bark usually shedding. **Stipules** usually caducous. **Tendrils** leaf-opposed, usually bifurcate. **Leaves** simple, often lobed, sometimes palmately compound. **Inflorescences** thyrses. **Flowers** 5-merous; calyx saucer-shaped; sepals minute; petals united at apex and shed as a cap (calyptra) at anthesis; stamens opposite to petals, undeveloped and abortive in ♀ flowers; disk conspicuous, 5-lobed or ring-shaped; pistil 1, style slender, stigma slightly expanded. **Fruits** berries, globose, 2–4-seeded; seeds obovoid or obovoid-elliptic. **Distr:** Mostly in temperate regions, extending to the subtropics with the highest species diversity in China and E N America. **Cult:** Propagate from cuttings, seeds or by air-layering.

V. vinifera L., *Sp. Pl.* 1: 202, 1753. [Grape Vine]. **Lianas**; branchlets terete, with longitudinal ridges, glabrous or sparsely pilose. **Stipules** caducous. **Tendrils** bifurcate. **Leaves** simple, conspicuously 3–5-lobed or cleft; lamina oval, 7–18 × 6–16 cm, base deeply cordate, notch rounded, 2 sides usually overlapping to nearly so, margins 22–27-toothed on each side, teeth large, irregular, and sharp, basal veins 5, lateral veins 4 or 5 pairs, apex of midlobes acute; petiole 4–9 cm long. **Inflorescences** panicles, opposite, basal branches well-developed, 10–20 cm long; peduncles 2–4 cm long; pedicels 1.5–2.5 mm long, glabrous; buds obovoid, 2–3 mm long, apex rounded. **Flowers** with calyx glabrous, undulate; filaments filiform, 0.6–1 mm long, anthers yellow; pistil entirely abortive in ♂ flowers, ovary oval, style short, stigma expanded. **Fruits** berries, globose or elliptic, 1.5–2 cm Ø; seeds obovoid to obelliptic, apex subrounded. **Distr:** SW Asia and SE Europe. **Ecol:** Fl. Apr–May. Fr. Aug–Sep. **Zones:** 6–12. **Info:** Cultivated widely in both hemispheres.

10 ROSIDS I [FABIDAE]

ORDERS 8, FAMILIES 79–80, GENERA 2,096, SPECIES 51,362.

Endosperm scanty (Stevens 2001 onwards). **Info:** The monophyly is very strongly supported by molecular data (Wang et al. 2009).

ZYGOPHYLLALES

FAMILIES 2, GENERA 27, SPECIES 305.

Vessel elements with simple perforation plates; rays (predominantly) uniseriate. **Leaves** with stomatal orientation transverse. **Flowers** with pollen colpate; style present. **Fruits** with seeds ± exotestal; endosperm absent (Stevens 2001 onwards). **Distr:** Worldwide. **Info:** The position of Zygophyllales in Rosids is moderately supported by molecular data (Soltis et al. 2000).

ZYGOPHALLACEAE

Genera 22, **species** 285. **Shrubs**, **shrublets**, or perennial **herbs**, rarely annuals, frequently strong-scented and resinous; branches often with articulate nodes. **Stipules** distinct or connate, often persistent. Leaves opposite or sometimes alternate; lamina simple or pinnate, often fleshy. **Inflorescences** axillary or terminal, solitary, racemose, or cymose. **Flowers** bisexual, actinomorphic or zygomorphic; sepals 4 or 5, imbricate or valvate; petals 4 or 5, imbricate or valvate; stamens 4 or 5 or up to 3× as many as petals, with or without appendages, anthers versatile, introrse, longitudinally dehiscent; ovary usually (2 or)3–5(–12)-locular, ovules 1 or several in each cell, pendulous or ascending. **Fruits** drupes, capsules, or schizocarps, leathery or fragile. **Distr:** Worldwide.

Guaicum

Species 6. **Trees** or **shrubs** with hard resinous wood. **Stipules** small, triangular. **Leaves** opposite; leaflets 2–14 pairs, base oblique, margins entire, leathery. **Inflorescences** dichasial cymes in umbellate fascicles, or flowers solitary, borne between the stipules; flowers pedicellate, ebracteate. **Flowers** actinomorphic, hypogynous, blue, purple or white; sepals 4–5, free, unequal, deciduous, imbricate; petals 4–5, free, broadly obovate, imbricate; stamens 8–10, inserted on the disk, unappendaged, filaments filiform, anthers oblong; carpels 2–5, ovary stipitate, obovate or clavate, 2–5-loculed, each locule with 8–10 pendulous ovules, style simple. **Fruits** capsules, 2–5-loculed, leathery, 2–5-lobed or winged, rounded or compressed, septicidally dehiscent; seeds ovoid, 1 per locule, testa slimy, embryo straight. **Distr:** S Florida and Mexico to northern South America, West Indies. **Cult:** Propagate from seeds. Requires a fertile, moist soil in full or partial sun.

G. officinale L., *Sp. Pl.* 381, 1753. [Holy Wood, Lignum Vitae]. **Trees** small, 6–10 m tall; branches knotted at nodes, older rugose, with deeply furrowed bark, young ones minutely pubescent. **Leaves** abruptly paripinnate; leaflets sessile, 2–3 pairs, elliptic-rotund or obovate, 1.5–4.5 × 1–2.5 cm, apex obtuse, uppermost pair largest, lowest pair unequal; pedicel 1.5–2 cm long. **Flowers** showy, ~2.5 cm across, bluish-purple, fading to white; sepals 5, 2 outer very broadly ovate, 3 inner broadly oblong, hairy, purplish-green; petals 5, shortly clawed, 1.2–1.3 × 0.6–0.7 cm, obovate, mucronate, hairy, bluish-purple; stamens 10, free, equal, filaments 6–7 mm long, bluish-purple; disk annular, fleshy, with nectariferous glands; ovary bicarpellate, syncarpous, stipitate, biloculed, laterally compressed, clavate with 2 lateral ridges and 2 central furrows, style twisted. **Fruits** capsules, compressed, retuse, mucronate, winged, orange-yellow, 1-seeded by abortion; seed pendulous, perisperm red, testa black. **Distr:** S America and West Indies. **Ecol:** Fl. Mar–Oct. **Zones:** 10–12. **Info:** This species has the highest known density of wood.

Tribulus

Species 25. **Herbs**, annual, biennial or perennial, prostrate; stems branched. **Stipules** membranous, lanceolate or subulate. **Leaves** opposite, abruptly pinnate. **Inflorescences** with flowers solitary, axillary, long-pedunculate. **Flowers** with sepals 5, lanceolate, herbaceous, caducous; petals 5, spreading, larger than sepals, yellow or orange, imbricate, obovate; stamens 10, with glands, outer 5 longer and opposite petals, inner 5 shorter and alternate with petals, filaments filiform, anthers cordate or oblong; ovary sessile, 5-celled, surrounded at base by an urceolate (or ring shaped) 10-lobate disk, styles united into a short thick column, stigmas 5, ± connate, ovules 5–10 per cell, obliquely pendulous, anatropous. **Fruits** schizocarps, depressed, (4)5-angulate, usually covered with short thick spines, separating at maturity into 5 bony carpels, leaving no central axis, carpels divided by transverse septa into 3–5 one-seeded compartments; seeds obliquely pendulous, oblong-obovate. **Distr:** Worldwide, but mainly in temperate regions. **Cult:** Propagate from seeds. Requires moist, well-drained soil in shade or full sun.

T. terrestris L., *Sp. Pl.* 1: 387, 1753. [Caltrop, Devil's Thorn, Puncture Vine, etc.]. **Herbs**, annual, prostrate, glabrous, pubescent, or hispid; branches 20–60 cm. **Leaves** opposite, even-pinnate, 1.5–5 cm, with 6–16 leaflets; leaflets oblong to obliquely oblong, 0.5–1 × 0.2–0.5 cm, base slightly oblique, margins entire, apex acute to obtuse. **Inflorescences** with flowers solitary; pedicels shorter than leaves. **Flowers** yellow, ~1 cm Ø; petals spreading; sepals persistent; stamens inserted on base of disk, with scalelike appendages; ovary 5-angled, 5-locular, with 3 or 4 ovules per locule; stigma 5-parted. **Fruits** schizocarps, 4–6 mm long, hard, pubescent or glabrous, with 5 carpels with 2 hardened 4–6 mm long spines at mid margin, surface spiny or prickly. **Distr:** Worldwide. **Ecol:** In sandy and dry areas, wastelands, hillsides, somewhat salt-tolerant; 0–3,300 m. Fl. May–Aug. Fr. Jun–Sep. **Zones:** 10–12. **Info:** Cultivated mainly for medicinal purposes.

CELASTRALES

FAMILIES 3 OR 4, GENERA 94, SPECIES 1,355.

Vessel elements with simple perforation plates. **Stipules** present. **Inflorescences** cymose. **Flowers** small, micropyle bistomal. **Distr:** Tropical to temperate regions worldwide. **Info:** The position of Celastrales is very weakly supported by molecular data (Soltis et al. 2000).

CELASTRACEAE

Genera 92, **species** 1,250. **Trees**, erect or scandent **shrubs**, **lianas**, evergreen or deciduous; stems sometimes producing rootlets, sometimes thorny, rarely with buttressed trunks. **Stipules** small and caducous, or estipulate. **Leaves** alternate or opposite, petiolate; lamina simple, margins entire, crenate, serrate, or dentate, venation pinnate. **Inflorescences** few- to many-flowered, cymose, thyrsoid, racemose, fasciculate, or flowers solitary, rarely paniculate, axillary or terminal. **Flowers** actinomorphic, bisexual or unisexual; perianth (3 or)4- or 5-merous; sepals and petals free; disk intrastaminal, stamens on disk, or extra-staminal, annular, margins upturned, pulvinate, or cupular, conspicuous, rarely obscure, entire, lobed, or angular; stamens 3–5, alternate with petals, anther (1 or)2-celled, basifixed to dorsifixed; ovaries 3–5, completely or incompletely 2–5-locular, placentation axile, ovules erect, axile, or pendulous, ovules (1 or)2(or 3 or more) per locule, style terminal, simple, short to absent, stigma simple or lobed. **Fruits** loculicidally dehiscent capsules, schizocarps of 2–5 indehiscent mericarps, drupes, berries, or winged-samaras, rarely indehiscent capsules or nuts with lateral style, rarely prickly, laterally winged; seeds 1–12, smooth or occasionally furrowed, sometimes winged. **Distr:** Largely tropical, but also temperate.

Brexia

Syn: *Thomassetia, Venana.*

Species 1. **Trees** or **shrubs**, glabrous. **Stipules** minute, caducous. **Leaves** alternate, entire or

Top row: *Guaicum officinale, Tribulus terrestris, Brexia madagascariensis.*
Bottom row: *Cassine glauca, Cassine melanocarpa.*

dentate to spinose-dentate, petiolate. **Inflorescences** condensed pedunculate cymes or false umbels or fascicles, sometimes cauliflorous. **Flowers** with sepals (4)5, free, imbricate, deciduous; petals (4)5, free, imbricate, spreading, deciduous; stamens (4)5, slightly perigynous, inserted between and united with the disk lobes, with filaments dorsally compressed, slightly dilated near base, anthers oblong-sagittate, versatile; disk (4)5-lobed, each lobe bearing 3–4(–6) stiff filaments; ovary narrowly ovoid-pyramidal, 5–10-angled, completely or incompletely 5(6–7)-locular, with numerous horizontal ovules in 2 rows, style thick, simple, stigma shallowly 5(6–7)-lobed. **Fruits** drupaceous, woody, 5(6–7)-locular or eventually 1-locular, each loculus numerous-seeded; seeds obovoid-oblong, angular, with black testa. **Distr:** Madagascar, Seychelles and coastal E Africa. **Cult:** Propagate from seeds or by air-layering. Requires moist, well-drained soil in shade or partial sun. **Info:** This genus was previously placed in the family Escalloniaceae. The single species in the genus is highly variable across its range and often considered as up to 6 or 9 separate species.

B. madagascariensis (Lam.) Ker-Gawl., *Bot. Reg.* 9, t. 730, 1823. (Syn: *Venana madagascariensis*, etc.). **Shrubs** or **small trees**, much-branched, evergreen, 2–6(–10) m tall. **Leaves** very variable, with different types of leaves on the same plant, narrowly oblong or linear-oblong to broadly obovate, 3.5–35 × 2–7.6 cm, base rounded or cuneate, margins ± entire, obscurely crenate or spinously toothed, apex broadly rounded or even retuse; petiole 1–2 cm long. **Inflorescences** (1–)3–12(–17)-flowered; peduncles 1–9 cm long, mostly flattened or even ± winged, 1.5–5 mm wide, sometimes with a leaflike bract ~1 × 1 cm; pedicels 0.6–1.8 cm long; bracts scaly, small. **Flowers** with calyx mostly 0.8–1.2 cm wide, sepals ~2.5 × 3.5–4 mm, rounded; petals thick, greenish- or yellowish-white, elliptic-oblong, 1.2–1.7 × 0.9–1.2 cm, obtuse; ovary 5-angled. **Fruits** ovoid, oblong-fusiform or cylindrical, 4–10 × 1.9–3 cm, sometimes tapering, prominently 5-ribbed; seeds brown or blackish, irregularly compressed-ellipsoid, 4.5–7.5 × 3–3.5 mm, keeled, minutely rugulose in ridges. **Distr:** Comoros, Madagascar, Mozambique, Seychelles. **Ecol:** In sclerophyllous coastal scrubs, mangrove swamps and fringing forests near the sea; 0–40 m. Fl. Aug–Oct. **Zones:** 11–12.

Cassine

Syn: *Elaeodendron*.

Species 60. **Shrubs** or **trees**. **Leaves** alternate or opposite. **Stipules** small, caducous. **Inflorescences** cymes, axillary. **Flowers** small, bisexual, rarely unisexual, 4-5-merous; calyx 4–5-lobed; petals 4–5; stamens 4–5, on or under the edge of the disk, anthers subglobose; ovary basally slightly united with disk, 2–5-celled, style very short, stigma obscure or bilobed, ovules erect, 2 per cell. **Fruits** dry or fleshy indehiscent drupes, cells 1(–2)-seeded; seeds exarillate. **Distr:** Tropical and subtropical Asia, America, Australia and S Africa. **Cult:** Propagate from seeds or by air-layering. Requires moist, well-drained soil in full or partial sun.

C. glauca (Rottb.) Kuntze, *Revis. Gen. Pl.* 114, 1891. (Syn: *Elaeodendron glaucum, E. roxburghii, Mangifera glauca*). [Ceylon Tea]. **Trees**, large, 15–20 m tall; branchlets reddish; bark red inside, exudate watery. **Leaves** opposite or subopposite, 3–16 × 1.5–7.5 cm, elliptic, ovate-oblong or obovate, glabrous, dark green, adaxially shining, abaxially glaucous; petiole 0.5–1 cm long, channeled. **Inflorescences** cymes, axillary, dichotomous, 7.5–12 cm long; peduncle 2.5–6.5 cm long, often red. **Flowers** 5 mm ∅, whitish; calyx 4–5-lobed, lobes rounded, margin membranous; stamens included, filaments recurved, anthers with a bilobed connective; ovary adnate to the disk, style very short, persistent. **Fruits** drupes, dry, ovoid, 1–1.5 cm long, 1-celled, 1-seeded. **Distr:** Pakistan, India, Sri Lanka and Malaysia. **Ecol:** In open forests in the dry regions. Fl. Apr–Jul. Fr. Jul–Oct. **Zones:** 10–12. **Info:** Leaves used medicinally and the exudate (gum) used as an adhesive.

C. melanocarpa (F.Muell.) Kuntze, *Revisio Generum Plantarum* 1: 114, 1891. (Syn: *Elaeodendron melanocarpum, Barringtonia sphaerocarpa*, etc.). [Black Olive Plum, False Olive]. **Trees**, dioecious, to 3–13 m tall; bark gray-cream, smooth. **Stipules** minute. **Leaves** simple, 6–16.5 × 2–8.5 cm, margins entire or toothed, glabrous, dark green, midrib prominent; petioles 0.7–1.1 cm long. **Inflorescences** panicles or racemes, axillary or terminal. **Flowers** with calyx white, cream or green; petals 3 or 4, cream or green, ~2 mm long. **Fruits** drupes, 1.5–2.5 × 1.2–2.5 cm, black, shiny; seeds 1–3 within a stony endocarp, endocarp 1.3–2.3 cm long. **Distr:** E Australia. **Ecol:** In rainforests and monsoon forests, also in beach scrubs, riparian and vine thickets; 0–1,100 m. Fl. Dec–Sep. Fr. throughout the year. **Zones:** 10–12.

Gymnosporia

Syn: *Celastrus* sect. *Gymnosporia*.

Species 70. **Shrubs** or **small trees**, usually erect,

Above: *Gymnosporia emarginata* (flowers and fruits), *Gymnosporia emarginata* (fruits and seeds), *Pleurostylia opposita*, *Salacia reticulata*.

glabrous or pubescent; branches often thorny. **Leaves** alternate or fasciculate, estipulate; lamina with margins entire or serrate. **Inflorescences** axillary, fasciculate, cymose, or flowers solitary. **Flowers** bisexual, (4 or)5-merous; disk fleshy, annular, (4 or)5-angled or lobed, intrastaminal; anthers longitudinally dehiscent, introrse; ovary (2 or)trilocular, base confluent with disk, ovules erect, 2 per locule. **Fruits** capsules, obovoid or nearly globose, leathery, loculicidally dehiscent; seeds usually 3–6, ellipsoid, aril basal to enveloping seed. **Distr:** Tropical and subtropical regions. **Cult:** Propagate from seeds or by air-layering. Requires moist soil rich in organic matter, in shade or partial sun.

G. emarginata (Willdenow) Thwaites, *Enum. Pl. Zeyl.* 409, 1864. (Syn: *Maytenus emarginata*, etc.). [Vingar]. **Shrubs**, armed, 2–5 m tall; spines 2–3.5 cm long. **Leaves** obovate to oblanceolate, rarely elliptic-oblong, 2.5–8 x 1.5–4 cm, margins usually entire, apex rounded or obtuse, usually emarginate, glabrous, chartaceous to coriaceous; petiole 3.5–8 mm long. **Inflorescences** cymes, fascicled, axillary, or terminal on short lateral shoots, often on spines, much-condensed; bracts deltoid, shortly fimbriate; pedicels slender, 3.5–5 mm long. **Flowers** 4 mm Ø, white; calyx lobes deltoid, acute, sometimes obtuse, slightly erose or shortly laciniate, ~1 mm long; petals obovate-oblong or ovate, 2–2.5 x 1–1.5 mm, entire, obtuse; disk fleshy, sinuate-lobed; stamens usually 2–3 mm long, inserted slightly beneath disk margin, anthers broadly ovoid, ~1 mm long, sometimes very small and abortive, slightly apiculate or obtuse; ovary triloculed, narrowed into style, ovules 2 per cell, stigmas 3, distinct. **Fruits** capsules, broadly obovoid or subglobose, 1–1.2 x 0.8–0.9 cm, 3-celled; seeds ovoid or ellipsoid, 2.5–3.5 x 2–2.5 mm, red, aril fleshy. **Distr:** S India, Sri Lanka, SE Asia, Malesia and N Queensland. **Ecol:** In dry regions, open scrub forests. Fl. Oct–Jan. Fr. Dec–Apr. **Zones:** 10–12.

Pleurostylia

Species ~5. **Trees** or **shrubs**, evergreen, glabrous. **Leaves** opposite, decussate, stipulate; lamina with margins entire, subcoriaceous to coriaceous. **Inflorescences** cymose, few-flowered, axillary. **Flowers** bisexual, 5-merous; disk fleshy, cupular, 5-lobed, intrastaminal; anthers longitudinally dehiscent, introrse; ovary conical, bilocular or reduced to 1-locular, ovules erect, 2–8 per locule. **Fruits** drupes, indehiscent, ellipsoid or obovoid, bony, with a prominent persistent lateral style; seeds 1 or 2, ovoid. **Distr:** Old World tropics and subtropics. **Cult:** Propagate from seeds or by air-layering. Requires moist, well-drained soil in full or partial sun.

P. opposita (Wallich) Alston, *Handb. Fl. Ceylon.* 6(suppl.): 48, 1931. (Syn: *Celastrus oppositus*). **Trees** or sometimes **shrubs**, to 15 m tall; branchlets subterete or obscurely 4-angled. **Leaves** ovate to obovate-oblong, rarely suborbicular, 3–8 x 1.5–5.2 cm, base cuneate, margins entire, apex obtuse, acute, rarely acuminate, very rarely retuse, papery, sometimes slightly recurved, veins 5 or 6 pairs, thin and obvious; petiole short, 2–5 mm long. **Inflorescences** cymes, 1- or 2-branched, few-flowered; peduncles 2–3 mm long; pedicels 1–2 mm long. **Flowers** green; calyx lobes rounded or subreniform, ~0.5 x 0.75–1 mm; petals elliptic or broadly ovate, 1.25–1.5 x ~1 mm, reflexed at anthesis, obtuse or slightly acute; disk margin slightly crenate; stamens attached just abaxially on disk margin, filament subulate, fleshy, slightly flattened, anther slightly ovoid; ovary adnate to disk at base, style very short, stigma capitate. **Fruits** drupes, indehiscent, ellipsoid or sometimes slightly obovoid, 5–7 x 4–5 mm. **Distr:** Tropical Asia through Malesia to Australia (Queensland) and Pacific Islands (New Caledonia). **Ecol:** In lower primary and secondary rain forests; <700 m. Fl. May–Dec, Fr. Jul–Jan. **Zones:** 10–12.

Salacia

Syn: *Courondi*.

Species 150. **Shrubs** climbing or scrambling, or **small trees**, glabrous or glabrescent; branchlets subterete, internodes usually dilated or slightly flat. **Leaves** opposite to subopposite, rarely alternate, exstipulate; lamina with margins entire, crenate, or serrate. **Inflorescences** axillary or cauliflorous, thyrsoid, cymose, or fasciculate. **Flowers** bisexual, 5-merous; disk fleshy, pulvinate or cupular, entire or lobed, extra-staminal; stamens (2 or)3, anthers longitudinally , obliquely, or transversely dehiscent, extrorse or apical; ovary (2 or)trilocular, whole or mostly concealed in disk; ovules axile, 2–9 per locule. **Fruits** berries; spheroid, fleshy or subligneous; seeds 2–12, angular, exalbuminous, surrounded by mucilaginous arils. **Distr:** Tropical regions worldwide. **Cult:** Propagate from seeds, cuttings or by air-layering. Requires a moist, well-drained soil in shade or partial sun. Climbing species best grown on trees.

S. reticulata Wight, *Ill. Ind. Bot.* 1: 134, 1839. [Marking Nut Tree]. **Climbers**, woody, or scandent shrubs, dichotomously branched; stem relatively tough; young parts glabrous. **Leaves** opposite, elliptic or oval, 4.2–11.6 x 2.2–5.4 cm, base narrowed, margins entire or very shallowly crenate-serrate, apex obtuse or with a short point, glabrous, coriaceous, adaxially shiny, abaxially paler; petiole 0.7–1.2 cm long. **Inflorescences** 2–8-flowered, on axillary tubercles or peduncles; pedicels 4–5 mm long, glabrous. **Flowers** greenish-yellow, 6–7 mm Ø; calyx entire or scarcely lobed, glabrous; petals 5, ovate or oblong, 3–3.5 mm long, obtuse, spreading; disk intrastaminal, annular-pulvinate, thick, 1 mm wide, slightly lobed; stamens 3, inserted on rim of disk, 1.5 mm long, filaments broad at base, anthers transversely dehiscent; ovary ovoid, triloculed, 2–6 ovules per cell, style short. **Fruits** drupaceous, globose, 1.6–3.8 cm Ø, tuberculate, bright pink-orange; seeds 1–4, almondlike, yellow. **Distr:** India (Malabar), Sri Lanka. **Ecol:** In moist wet secondary forests; to 1,500 m. **Zones:** 11–12.

OXALIDALES

[Wood Sorrel].

FAMILIES 7, GENERA 60, SPECIES 1,815.

Herbs, **shrubs** or **trees**. **Leaves** often compound. **Flowers** with sepals and petals 5 or 6, twice as many as anthers; carpels 5(4), ovarial portions ± united, ovules bitegmic/tenuinucellate. **Fruits** capsular or baccate; seeds often with aril attached (often a feature associated with animal dispersal). **Distr:** Pantropical and temperate Southern Hemisphere. **Info:** Floral characteristics are further discussed in Matthews & Endress (2002). The monophyly of Oxalidales is strongly supported by molecular data (Soltis et al. 2000).

CONNARACEAE

Genera 12, **species** 180. **Trees** small, **shrubs**, or **vines**, erect or scandent, evergreen or deciduous. **Leaves** alternate, exstipulate, petiolate, odd-pinnate, trifoliolate, or 1-foliolate; leaflets subopposite or alternate, margins entire, rarely lobulate, usually leathery. **Inflorescences** terminal, pseudoterminal, or axillary, paniculate or racemose, bracteate. **Flowers** bisexual, rarely unisexual, actinomorphic, small; sepals (4 or)5, free or united at very base, imbricate or valvate, persistent and clasping base of fruit; petals (4 or)5, free, rarely slightly connate at middle, imbricate or valvate, rarely circinate; stamens 5–10, in 2 whorls, those opposite petals often shorter and abortive, filaments free or shortly basally connate, anthers longitudinally dehiscent; disk thin or absent, surrounded by base of stamens; carpels (1–)5(–8), free, 1-celled, hirsute, style subulate or filiform, stigma subcapitate, simple or bilobed, ovules 2 per carpel (1 sometimes sterile), collateral, erect. **Fruits** usually solitary follicles, sessile or stipitate, dehiscent or sometimes indehiscent; seed 1(or 2), erect, usually arillate, rarely exarillate, testa thick, aril colored, fleshy. **Distr:** Pantropical, especially Old World.

Connarus

Syn: *Anisostemon*, etc.

Species 80. **Lianas**, **climbing shrubs** or **small trees**; branchlets glabrous. **Leaves** alternate, exstipulate, petiolate, odd-pinnate or trifoliolate, rarely 1-foliolate; leaflets subopposite or alternate, margins entire, usually with pellucid glands. **Inflorescences** mostly terminal, paniculate, large, or rarely subcymose or racemose in distal leaf axils. **Flowers** bisexual, fragrant, sepals, petals, and stamens usually glandular; sepals (4 or)5, broadly imbricate or subvalvate in bud, slightly basally connate, thick and fleshy, persistent; petals 5, free, subequalling or slightly longer than sepals, often coherent at middle, imbricate in bud; stamens ~10, 5 long ones opposite sepals, 5 short ones opposite petals and often abortive, filaments basally connate, anthers oblong; carpel 1, ovary globose, 1-loculed, pilose, ovules 2, collateral, style slender, proximally pilose, distally glandular-pubescent, stigma uniform, oblique. **Fruits** follicles, slightly compressed, sometimes narrowed at base into a long, slender stipe, striate, longitudinally dehiscent, calyx persistent, apex obtuse and usually shortly beaked, pericarp woody or leathery; seed 1, reniform, testa mostly black-purple, glossy, aril cupular, crenulate or laciniate, fleshy, endosperm absent. **Distr:** Pantropical, especially in Africa, America, and SE Asia. **Cult:** Propagate from seeds or by air-layering. Requires a moist, well-drained soil in full sun.

C. monocarpus L., *Sp. Pl.* 675, 1753. (Syn: *Canarium zeylanicum*, etc.). **Shrubs**, scandent, to 5 m tall; bark reticulate, striate, minutely pustular, lenticellate; young parts ferruginous-tomentose. **Leaves** pinnate; leaflets 3 or 5, ovate oblong, 12.4 × 4.4 cm, base ± obtuse or rounded, apex, subacute to caudate, often twisted, glabrous, adaxially glossy dark green, abaxially paler, brown-scaly; young flush coppery-brown; rachis and petiole 4.5 cm long; petiolules to 5 mm long. **Inflorescences** paniculate cymes, terminal, erect, irregularly pyramidal, ferruginous-tomentose; bracts <1 mm long; pedicels 3–5 mm long. **Flowers** small; sepals narrowly triangular or oblong, to 3.5 mm long, margins ciliate; petals linear-oblanceolate, to 8 mm long, base cuneate to tapered, apex acute to slightly rounded, inner surface glandular, margins ciliate; stamens fertile, 2-whorled, outer filaments to 8 mm long, others 2 mm long; ovary subglobose, densely pubescent, brownish, style slender. **Fruits** follicles, irregularly obovoid, fusiform, compressed, 4.5 × 2 cm, apiculate, base tapered, with persistent calyx and petals, valves leathery, green through yellow to scarlet red; seed 1, ovoid, 2 × 1.2 cm, compressed, reniform, glossy, purplish-black, base with a large lobulate, bright yellow aril. **Distr:** S India, Sri Lanka. **Ecol:** In secondary forests. Fl. Jul–Sep. **Zones:** 11–12.

CUNONIACEAE

Syn: *Baueraceae, Davidsoniaceae, Eucryphiaceae*. **Genera** 27, **species** 280. **Trees**, **shrubs** or occasionally climbers. **Stipules** often large and prominent, fused across the stem at base of petioles, soon caducous, usually leaving distinctive scars. **Leaves** opposite or whorled, rarely spirally arranged; lamina palmate, pinnate or trifoliate, rarely simple. **Inflorescences** small capitula or extended clusters, rarely solitary and axillary. **Flowers** mostly bisexual, small, regular; sepals 4–6(–10), occasionally basally fused; petals same number as, and alternating with sepals, usually smaller or absent, occasionally united at base; stamens (4–)8–20, free; carpels 2–5, free or united, ovary mostly superior with (1)2–5 chambers, rarely half-inferior, containing numerous ovules, placentation axile or apical. **Fruits** generally capsules, sometimes nuts or follicles, occasionally drupelike. **Distr:** Mainly temperate and tropical Southern Hemisphere, but very few in Africa and Madagascar.

Davidsonia

Species 3. **Trees**. **Leaves** alternate, imparipinnate; rachis and jagged; lamina with prominent teeth, hairs rigid, caducous, mildly urticating. **Inflorescences** elongate panicles or racemes, axillary, terminal, and adventitious, borne along branches and trunk; floral maturation basipetalous. **Flowers** with a single perianth whorl, lobes fused half their length; nectary of small segments. **Fruits** drupaceous, with fleshy, edible pulp containing 4 seeds. **Distr:** Endemic to Australia. **Info:** Fruits edible, sour. **Cult:** Propagate from seeds. Requires a moist, humus-rich, well-drained soil in shade or full sun.

D. pruriens F. Muell., *Fragm. Phyt. Austr.* (Mueller) 6: 4, 1867. [Davidson's Plum]. **Trees**, to 18 m tall. **Stipules** 5–8 mm long. **Leaves** alternate or opposite, compound; rachis 6 cm to 1.5 m long, toothed with small leaflike growths; leaflets 7–19, 5–46 × 2.5–16 cm, adaxially densely to sparsely hairy, abaxially hairy, lateral petiolules 1–5 mm long. **Inflorescences** panicles or racemes, axillary or mostly cauliflorous or ramiflorous. **Flowers** bisexual, 3–5 mm Ø; sepals 4 or 5, green or pink; petals absent. **Fruits** drupes, purple, blue or blackish with a bloom, obovoid, 3–5.5 × 3.2–6 cm, usually sparsely hairy, fleshy pink-red; seeds 2 though usually only 1 is firm and fertile. **Distr:** NE Australia (coastal NSW, N Queensland). **Ecol:** In rainforests; 0–1,095 m. Fl. & Fr. throughout most of the year. **Zones:** 10–12. **Info:** Although very acidic when ripe, the plumlike fruits are edible and make excellent jams and jellies.

OXALIDACEAE

Genera 6, **species** 770. **Herbs**, annual or perennial, sometimes shrubs or trees. **Stipules** absent or small. **Leaves** alternate or whorled, basal or cauline, pinnate or palmate; leaflets often folded together at night, margins always entire. **Inflorescences** umbellate, cymose, or racemose, or flowers solitary. **Flowers** bisexual, regular, 5-merous, usually heteromorphic, heterostylous; sepals 5, distinct or basally connate, imbricate, rarely valvate; petals 5, sometimes slightly basally connate, convolute; stamens 10, in 2 whorls of 5, outer whorl usually shorter, opposite petals, filaments basally connate, anthers 2-celled, slits longitudinal; ovary superior, carpels 5 and fused, placentation axile, each locule with (1 or)2 to several ovules, styles 5, distinct, stigmas capitate or shortly 2-cleft. **Fruits** loculicidal capsules, or berries; seeds often with basal aril involved in explosive ejection of seed from capsule, endosperm fleshy. **Distr:** Tropical and subtropical regions of both hemispheres, extending into temperate regions.

Averrhoa

Species 2. **Trees** or **shrubs**. **Stipules** absent. **Leaves** alternate or subopposite, odd-pinnate; leaflets nearly opposite. **Inflorescences** axillary or rameal, cymes or panicles. **Flowers** small or medium-sized, regular, fragrant; sepals 5, red, semifleshy, basally connate or imbricate; petals 5, white, pink, or purple, contorted; stamens 10, all fertile or 5 without anthers, united at base; ovary 5-lobed, 5-celled, ovules numerous, several per locule; styles 5, distinct, stigmas capitate. **Fruits** berries, baccate, indehiscent, oblong, usually 5-lobed; seeds few or numerous, naked or arillate, endosperm minute. **Distr:** Old World tropics. **Cult:** Propagated easily from cuttings and by air-layering, or from seeds. Requires a fertile, very moist soil in partial or full sun.

A. bilimbi L., *Sp. Pl.* 1: 428, 1753. [Cucumber Tree, Pickle Tree]. **Trees**, 5–6(–15) m tall; young parts and petioles rusty pubescent. **Leaves** aggregated at apex of branches, odd-pinnate, 15–40 cm long; leaflets 10–20 pairs, oblong to lanceolate, 3–5 × ~2 cm, base obliquely rounded, apex acuminate, surfaces somewhat pubescent; petiole 2–4 mm long. **Inflorescences** ramiflorous, paniculate, many-flowered. **Flowers** with sepals ovate-lanceolate, ~4 mm long, pubescent; petals reddish-purple, 1.3–1.8 × ~0.3 cm; stamens all fertile. **Fruits** berries, greenish-yellow to yellow, oblong, 5–10 cm, terete to obscurely 5-angled, fleshy; seeds few, flat. **Distr:** Tropical SE Asia. **Ecol:** In secondary vegetations. Fl. Apr–Dec. Fr. Jul–Dec. **Zones:** 11–12. **Info:** Widely cultivated for the edible but sour fruits (often used in pickles and curries) and for ornamental purposes. The petals are also edible with a sweet-sour taste.

A. carambola L., *Sp. Pl.* 1: 428, 1753. [Carambola, Chinese Star Fruit, Five-angled Fruit, Star Apple, Star Fruit]. **Trees**, 3–12(–15) m tall, densely branched; young parts finely pubescent or glabrous. **Leaves** odd-pinnate, 7–25 cm long; petiole 2–8 cm; leaflets (3–)5–13, ovate to elliptic, 3–8 × 1.5–4.5 cm, base obliquely rounded, apex acute to acuminate, abaxially pubescent to nearly glabrous; petiolules 1–2.5 mm long. **Inflorescences** axillary or ramiflorous, panicles or cymes, flowers numerous; branches and flower buds crimson. **Flowers** small; sepals narrowly elliptic, 3–5 mm long, base sparingly pubescent; petals white with purple markings or pink to red with darker markings, 6–9 × 3–4 mm; shorter stamens sterile, occasionally 1 or 2 fertile; ovary pubescent. **Fruits** berries, yellow to yellow-brown, oblong, 7–13

Below: *Connarus monocarpus, Davidsonia pruriens* (fruits), *Averrhoa bilimbi, Averrhoa carambola* (flowers).

× 5–8 cm, deeply (3–)5(or 6)-ribbed, stellate in cross-section, very fleshy; seeds numerous, blackish-brown. **Distr:** SE Asia. **Ecol:** In secondary open forests; <1,000 m. Fl. Apr–Dec. Fr. Jul–Dec. **Zones:** 11–12. **Info:** Fruits edible with a sweetly sour taste.

Biophytum

Species 50. **Herbs**, annual or perennial; stem sometimes woody at base, simple or dichotomously branched. **Stipules** filiform. **Leaves** crowded or almost whorled apically on stems, even-pinnate; leaflets opposite, sessile, apices rounded and mucronate, apical leaflet reduced to an awn; petiole basally swollen. **Inflorescences** umbellate, pedunculate or sessile. **Flowers** with sepals lanceolate, persistent in fruit, apex acuminate; petals yellow, white, or rarely purple; ovary subglobose. **Fruits** capsules, ovoid to oblong, loculicidally dehiscent, sometimes splitting to base into 5 spreading valves, each valve with several seeds; seeds brown, often with small tubercles. **Distr:** Tropical and subtropical regions worldwide. **Cult:** Propagate from seeds. Requires a moist, well-drained soil in full or partial sun.

B. reinwardtii (Zucc.) Klotzsch, *Reise Mossamb. Bot.* 1: 85, 1862. (Syn: *B. sensitivum, Oxalis reinwardtii*). **Herbs**, erect, to ~16 cm tall; stems stiff, usually unbranched, sometimes very shortly branched apically; often red or reddish-green. **Leaves** crowded at shoot apices, 3–8 cm long; pinnae 10–20 pairs, oblong to broadly ovate, 0.2–1 × 0.2–0.5 cm, base cordate, adaxially dark green, abaxially paler, ciliate, overlapping; petiole 0.3–1 cm long, reddish-pubescent; rachis 2–7 cm long, pubescent. **Inflorescences** umbellate, to ~20-flowered; peduncles to ~10 cm long, reddish; bracts narrowly lanceolate, to ~2 mm long, purplish; bracteoles linear, 1–1.5 mm long; pedicels to ~7 mm long. **Flowers** with sepals green, narrowly lanceolate, 2–5 mm long, acute, 3–5-veined; petals yellow, spathulate, 0.6–1 cm long, rounded, shallowly emarginate, limb red-veined; stamens with *longer* filaments 3.5–4 mm long, glandular-hairy, *shorter* filaments 1–1.5 mm long, glabrous, anthers pale yellow; pistil 2–2.5 mm long, ovary 5-lobed, light green. **Fruits** capsules, broadly ovoid, reddish, 3–5 mm long, glandular-hairy; seeds ovoid, 1–4 per locule, ~1 mm long, brown. **Distr:** India, Malaysia, Sri Lanka. **Ecol:** In disturbed places, sometimes lithophytic on rock faces and cliffs. **Zones:** 11–12.

Oxalis

Species ~700. **Herbs**, annual or perennial, usually with tubers, bulbs, or rhizomes; stems erect, creeping, or absent (rosette). **Stipules** usually absent or very small. **Leaves** radical or alternate; lamina digitately compound. **Inflorescences** solitary, cymose, or umbellate; peduncle long, with 2 small bracts at apex, bracts appearing to be at the middle of the peduncle in solitary-flowered species. **Flowers** with sepals distinct, imbricate; petals yellow, red, pink, or white, convolute, sometimes slightly basally connate; stamens all fertile, filaments basally connate or distinct; ovary locules with 1 to several ovules. **Fruits** capsules, loculicidally dehiscent; seeds with an outer fleshy coat which bursts elastically. **Distr:** Mainly in tropical and subtropical regions of both hemispheres, extending into temperate regions. **Cult:** Propagate from cuttings or seeds, and by division in bulbous and perennial species.

O. barrelieri L., *Sp. Pl.* ed. 2, 624, 1762. [Lavender Sorrel]. **Herbs** or **subshrubs**, erect, to 1 m tall; stem with leafy internodes ~0.5–8 cm long. **Stipules** linear, ~3 mm long or absent. **Leaves** alternate to subopposite; leaflets usually ovate to elliptic-oblong, 1–5.5 × 0.8–2.5 cm, base obtuse to rounded, lateral leaflets asymmetric at base, apex bluntly obtuse to acute, sparsely puberulent; petioles 1–4 cm long. **Inflorescences** cymes, (1–)3–11(–30)-flowered, longer than leaves, usually axillary; peduncles to 6.5 cm long, glabrescent, terminated by a flower and with 2 unbranched secondary branches to 3 cm long, bracts and bracteoles 1–2 mm long; pedicels 1.5–3.5(–5) mm long, articulate above base. **Flowers** pink, sepals greenish, linear to narrowly oblong, 3–4 × 1–2 mm, acute, minutely mucronate; petals obovate to spathulate, 7–9 × ~3.5 mm, retuse, pink distally, white or yellow in lower half; *long* stamens ~3 mm long, ligulate at middle, *shorter* stamens 1.5–2 mm long; pistil 3.5–4 mm long, ovary ovoid, glabrous, locules 3- or 4-ovuled, styles puberulent. **Fruits** ovoid, 5–9 × 4 mm, calyx persistent; seeds ovoid, 1.5–2 × ~1 mm, 6–8-ridged. **Distr:** Tropical and subtropical America. **Ecol:** In evergreen lowland forests; 0–1,000(–2,000) m. Fl. throughout the year. **Zones:** 10–12.

O. debilis H.B.K., *Nov. Gen. Sp.* 5: 236, t. 466, 1822. [Lilac Oxalis, Pink Woodsorrel]. **Herbs**, bulbous; bulb globose, ovoid, or oblong, 1–2 cm long, covered with scales and leaf bases. **Leaves** trifoliolate; leaflets broadly obovate to suborbicular, obcordate, 1.2–5.5 × 1.5–6 cm, apex broadly rounded distally and emarginate to retuse, deeply notch, adaxially glabrous to sparsely puberulent, abaxially densely puberulent; petioles to 30 cm long, petiolules 0.5–1 mm long. **Inflorescences** cymes bifid (sometimes twice bifid), 6–15-flowered, to 30(–45) cm long; peduncles similar to petioles, bracteoles 1–3 mm

Top row: *Averrhoa carambola* (fruits), *Biophytum reinwardtii, Oxalis barrelieri, Oxalis debilis.*
Bottom row: *Oxalis massoniana, Aristotelia serrata, Crinodendron hookerianum.*

MD

Above: *Elaeocarpus angustifolius, Elaeocarpus grandis* (flowers), *Elaeocarpus grandis* (fruits), *Elaeocarpus serratus.*

long; pedicels to 3 cm long. **Flowers** pinkish to red-violet; sepals narrowly ovate-elliptic or oblong, 4–7 × 0.7–1.5 mm, glabrescent, usually with 2 brownish antherlike calli ~0.5 mm long outside at apex; petals 2–3 times longer, obovate, base narrowed; *long* stamens 4.5 mm long, *short* stamens ~3 mm long, glabrous, basally connate; pistil ~7.5 mm long, ovary glabrous, locules 2–12-ovuled. **Fruits** cylindric, ~1 8 mm long, persistent styles 2 mm long; seeds ovoid, with 12 longitudinal ribs and 12 transverse ridges. **Distr:** S America. **Ecol:** In evergreen areas, weedy in some places; to 1,500 m. Fl. throughout most of the year. **Zones:** 10–12.

O. massoniana T.M. Salter, *Journal of South African Botany* 2: 3, 1936. [Wood Sorrel, Yellow Oxalis]. **Herbs**, deciduous, winter-growing, dwarf geophyte, 4–6 cm high; bulbs oblong, 1–2 × 0.4–0.6 cm; stem simple, 2–4 cm long, dull red. **Leaves** *cauline*, glaucous or light green, *upper* leaves subsessile, *lower* leaves shortly petiolate; leaflets 3, sessile, linear, 4–6 × 1–1.5 mm, margins involute, sometimes conduplicate, ciliate, adaxially glabrous, abaxially pilose; petioles scalelike, striate, 0.5–1 mm long. **Inflorescences** 1-flowered, axillary, slender, 2–3 cm long, reddish-brown; bracts 2, alternate, subulate, 2–3 mm long, striate, dull red. **Flowers** with sepals broadly lanceolate, 4–5 mm long, ciliate, dull red, striate; corolla widely campanulate, 1.6–1.8 cm long, tube funnel-shaped, light to bright yellow, tepals cuneate-obovate, 1–1.2 × 0.7–0.9 cm, upper third brick-red, lower two-thirds light to bright yellow; stamens included, connate, *short filaments* 1.8–3 mm long, *long filaments* 3–6 mm long, apices sparsely pilose, slightly gibbose; ovary ovoid, 1–1.5 mm long, dotted with numerous calli, locules 1-ovuled, style included, sparsely pubescent. **Fruits** capsules, globose; seeds light green. **Distr:** S Africa (Cape Region). **Ecol:** In shallow pockets of sandstone rock sheets. Fl. May. **Zones:** 9–11.

ELAEOCARPACEAE

Genera 12, **species** 605. **Trees** or **shrubs**, evergreen or semievergreen. **Leaves** alternate or rarely opposite, stipulate or not, petiolate; lamina simple, pinnately veined. **Inflorescences** axillary or terminal, racemose, corymbose, paniculate, or sometimes fascicled or flowers solitary. **Flowers** bisexual or polygamous, 4- or 5-merous, actinomorphic, with or without bracts; sepals 4 or 5, basally connate or free, usually valvate; petals 4 or 5, sometimes absent, valvate or imbricate, margin laciniate or rarely entire; stamens 8 to numerous, filaments free and borne on disks, anthers 2-celled, dehiscent by apical or longitudinal slits, awned or tipped with hairs at apices; disk circular or glandularly lobed; ovary superior, 2- to several-loculed, placentation axile, ovules 2 to several per locule, style connate or free. **Fruits** drupes or capsules. **Distr:** Tropical, especially Papuasia–Australia, Madagascar.

Aristotelia

Species 5. **Trees** or **small shrubs**. Stipules present but small. **Leaves** mostly opposite; lamina simple, margins toothed. **Inflorescences** racemes or flowers solitary, axillary. **Flowers** bisexual, rarely unisexual (in New Zealand), 4–5-merous; sepals valvate, petals usually lobed or toothed; disk thickened; stamens numerous or 4–5, inserted on disk, anthers opening by apical slits; ovary 2–4-celled, ovules 2 per cell, styles subulate. **Fruits** berries; seeds with embryo straight. **Distr:** Australia, New Zealand, New Hebrides, S America. **Cult:** Propagate from seeds or cuttings. Requires fertile, moist but well-drained soil in full or partial sun.

A. serrata (J.R. et G. Forst.) W.R.B. Oliver, *Trans. NZ. I.* 53, 1921, 365. (Syn: *Dicera serrata, Friesia racemosa, A. racemosa*). [Wineberry]. **Trees**, dioecious, to ~10 m tall; trunk ~30 cm dbh; bark of branchlets rather light to dark red, pubescent. **Leaves** opposite to subopposite; lamina broad-ovate, 5–12 × 4–8 cm, apex acuminate, glabrate (pubescence long-persistent on veins below), membranous, margins deeply, doubly, irregularly, sharply serrate, adaxially light or dark green, abaxially paler green or purplish; petioles slender, to ~5 cm long, pubescent. **Inflorescences** paniculate, 6–10 cm long; flowers numerous; pedicels slender, pubescent ~0.5–1 cm long. **Flowers** 4–6 mm Ø; sepals 4, ovate, ~3 mm long, pubescent; petals 4, trilobed, often deeply, ~9 mm long; stamens numerous, on glandular disk, minutely pubescent; ovary 3–4-celled, styles 3–4. **Fruits** berries, ~5 × 4 mm or sometimes larger, brighter or dark red or almost black; seeds ~8, angled. **Distr:** New Zealand. **Ecol:** In lowland to montane forests, often forming thickets. Fl. Sep–Dec. Fr. Nov–Jan. **Zones:** 8–10.

Crinodendron

Species 4. **Trees** and **shrubs**, to 30 m tall; trunk to 1 m dbh. **Leaves** opposite, subopposite or alternate, occasionally whorled; lamina simple, oblanceolate, obovate or ovate, margins toothed, leathery; petiole short, often caniculate. **Inflorescences** 1–2-flowered, axillary; pedicels long, flowers hanging. **Flowers** bell-shaped, white or red; sepals fused into a ring that is soon shed; corollas emerging through the calyx, petals 5, fleshy, apices mostly with 3 teeth; stamens mostly 15–20, borne in 2 whorls; ovary superior, 4–6-loculed, ovules 4–15 per locule. **Fruits** capsules, tough, loculicidal, subovoid, 3–4-winged or wingless; seeds 1–5 per locule, ovoid or ellipsoid. **Distr:** S America (Argentina, Chile). **Cult:** Propagate from seeds or half-hardened cuttings in a sandy potting mix. Requires a moist, well-drained, fertile soil in full or partial shade.

C. hookerianum Gay, *Fl. Chil.* i. 341, 1845. (Syn: *Tricuspidaria lanceolata, C. eriocladum*). [Chilean Lantern Tree]. **Shrubs** or **small trees**, 3–8 m tall; young shoots gray-pubescent. **Leaves** opposite, occasionally alternate or whorled; lamina narrowly elliptic to obovate, 1.5–10 × 0.7–2.1 cm, margins entire at base and toothed above, apex pointed, leathery, dark green, abaxially veins hairy; petioles 1.5–9 mm long. **Inflorescences** with flowers solitary; pedicels 2–8 cm long. **Flowers** with calyx reddish, sericeous, 5-lobed, segments 5–7 × 4–8 mm; petals 5–6, red 1.4–2.4 × 0.3–0.8 cm, adaxially hirsute, abaxially glabrous, weakly trilobed, lobes 0.5–3 mm long; disk ~6 mm Ø, 10-lobed, lobes 2 × ~1 mm; stamens 15, filaments red, 0.7–1.3 cm long; ovaries pilose, 5-loculed, ~9 × 4 mm, ovules 12–15 per locule, styles 0.1–1.5 cm long. **Fruits** subglobose, 1.5–4.5 × 1.1–2.1 cm; seeds brown, elliptic, 2–14 per locule, 7–9 × ~5 mm. **Distr:** Chile. **Ecol:** In understory of wet forests. Fl. May–Jun. Fr. Aug–Feb. **Zones:** 8–10.

Elaeocarpus

Syn: *Ganitrus, Monocera.*

Species ~360. **Trees**, rarely shrubs. **Stipules** linear or rarely leaflike, caducous, rarely persistent. **Leaves** alternate or spirally arranged; petiole usually long and swollen at both ends; lamina with margins serrate or entire, pinnately veined. **Inflorescences** axillary, racemose. **Flowers** bisexual, 4- or 5-merous; sepals 4 or 5, valvate, abaxially usually pubescent; petals 4 or 5, white, free, margin laciniate, rarely entire or lobed; stamens 8 to numerous, filaments short, anthers 2-celled, dehiscing from apical slits, with awn or hairs at tip; disk usually glandularly 5–10-lobed, rarely circular; ovary superior, 2–5(–7)-loculed, ovules 2–12 per locule, style linear or subulate. **Fruits** drupes, 1(or 5)-loculed; endocarp hard, bony, surface usually lacunose; seeds usually 1 per locule. **Distr:** Old World tropics. **Cult:** Propagate from seeds or cuttings. Requires fertile, moist but well-drained soil in full or partial sun.

E. angustifolius Blume, *Bijdr.* 120, 1825. (Syn: *E. ganitrus, E. subglobosus*). [Bead Tree, Blue Olive Berry, Fairy Petticoats, Fringe Bells, Lily of the Valley Tree, Native Olive, Scrub Ash]. **Trees**, to 30 m tall; trunk to 1.7 m dbh. **Leaves** obovate-oblong to oblong-lanceolate, 9–14 × 3–4.5 cm, base broadly cuneate, margins serrate, apex acute or obtuse, papery or membranous, villous when young, glabrescent when mature, lateral veins 10–13 per side, impressed adaxially; petiole 1–1.5 cm long. **Inflorescences** racemes, in axils of current and fallen leaves, 2–4 cm long; peduncles villous; pedicels ~5 mm long. **Flowers** villous; sepals 5, lanceolate, ~5 × 1.5 mm; petals 5, about as long as sepals, laciniate to half their length, lower half villous; stamens 25; ovary 5-loculed, densely tomentose, style ~5 mm long. **Fruits** drupes, globose, 1.7–2 cm Ø, 5-loculed; seed 1 per locule, endocarp hard, rugulose. **Distr:** Tropical Asia to Malesia, Australia and Pacific Islands. **Ecol:** In broad-leaved montane and rainforests; 400–1,300 m. Fl. Aug–Sep. Fr. Sep–Nov. **Zones:** 9–12.

E. grandis F. Muell., *Frag. Phyt. Austr.* 2: 81, 1861. [Blue Fig, Blue Marble Tree, Blueberry Ash, Blue Quandong, Silver Quandong]. **Trees**, to 35 m tall; trunk buttressed at base. **Stipules** 2 mm long. **Leaves** simple, narrowly elliptic, 8–15 × 2–4.5 cm, base acute, margins finely serrated, apex acute, the very tip obtuse, glabrous, domatia as large foveoles; petioles 1–2 cm long, pulvini absent. **Inflorescences** racemes, axillary or ramiflorous. **Flowers** fragrant; calyx green, white or cream; petals 5, green, white or cream, ~5 mm long. **Fruits** drupes, purple or blue, often with a metallic sheen, spherical, 1.5–3.3 cm across, glabrous; pedicels 1.2–2 cm long; seeds 2–5 within a sculpted, woody, 5–7-sutured endocarp. **Distr:** E Australia. **Ecol:** In coastal and montane rainforests; 0–1,100 m. Fl. Oct–Mar. Fr. throughout most of the year. **Zones:** 9–12. **Info:** Often cultivated for the edible fruits, timber and for ornamental purposes.

E. serratus L., *Sp. Pl.* 515, 1753. (Syn: *E. oblongus, E. perim-kara*). [Ceylon Olive]. **Trees**, rather small, to ~10–12 m tall; trunk to 30 cm dbh; young parts pubescent. **Leaves** obovate-oval, 5–10 cm long, base acute, margins shallowly crenate-serrate with dents in the incisions, apex obtuse, rounded, pale olive green, lateral veins ~4–5, with glandular thickenings in axils, connected by marginal network of veins; petioles 1–1.5 cm long. **Inflorescences** racemes, flowers numerous, axillary, (4)7–8 cm long; pedicels as long as sepals. **Flowers** with sepals lanceolate, 4–6 mm long, acute; petals white, ~5–6 mm long, base broadly cuneate, cut more than halfway down into numerous filiform lacinae; stamens 25–35, filaments short, anthers linear, black, 1 valve tipped with a brush of hair half as long; disk hairy, orange; ovary hairy, 3-celled. **Fruits** drupes, ovoid, ~3 cm long, bluntly pointed, smooth, dull greenish-yellow, pulp copious, acid, edible; stone ovoid-oblong, thick-bony, strongly tubercled, 1-celled, 1-seeded. **Distr:** Indo-Malesia. **Ecol:** In low altitude, moist forests; to ~900 m. **Zones:** 11–12. **Info:** Often cultivated for the edible fruits.

Sloanea

Syn: *Echinocarpus.*

Species ~120. **Trees** or **shrubs**; buds usually pubescent. **Stipules** caducous. **Leaves** alternate or clustered at twig apices, petiolate; lamina simple, margins entire or serrate, pinnately veined. **Inflorescences** with flowers solitary or in racemes, axillary, long stalked. **Flowers** usually bisexual; sepals 4 or 5, valvate or imbricate, ovate, slightly connate at base; petals 4 or 5, rarely absent, obovate, imbricate, margin entire or dentate at apices; stamens numerous, inserted on disk, filaments short, anthers opening by vertical slit, connective usually elongated as beak; disk broad, thick; ovary 3–7-loculed, ovules several per locule, furrowed, pubescent, style free or connate, subulate. **Fruits** capsules, globose or ovoid, prickly, septicidally dehiscent into 3–7 valves, exocarp woody, thick, endocarp thin, leathery, usually detached from exocarp when dry; seeds 1 to several, pendulous, usually arillate in lower half, endosperm fleshy, copious. **Distr:** Primarily in the tropics and subtropics. **Cult:** Propagate from seeds or by air-layering. Requires a moist, well-drained soil in full sun.

S. javanica (Miq.) Szys. ex K. Schum., *Nat. Pflanzenfam.* III. 6: 5, 1890. (Syn: *Phoenicospermum javanicum*). **Trees**, (10–)20–30 m tall; trunk fluted, buttresses to 5 m high. **Stipules** ± linear-subulate, 2–4(–8) mm long. **Leaves** ovate to elliptic, rarely ± obovate, very variable in size, 8–23 × (3–)6–12(–13) cm (juvenile leaves to 50 × 21 cm), base broadly cuneate to rounded or truncate, margins usually entire, apex usually acute and often acuminate, coriaceous; petioles (1–)2–6(–7) cm long. **Inflorescences** cymes, axillary, pedunculate, usually 2- or 3-flowered, rarely 1-flowered, felted-pubescent; peduncle (2–)3–5 cm long; bracts 2–5, narrow or subulate, sometimes laciniate; pedicels 1–3 cm long. **Flowers** usually 4-merous; perianth reflexing and ± easily detaching; sepals incurved, fairly tough, ovate to ± elliptic, ~1.1 × (0.4–)0.5–0.7 cm, acute, densely hairy; petals similar or oblong, to 1.6 × 0.2 cm; stamens 70–140, filaments 2–4 mm long, densely pilose; ovary 4-cornered, 3–4 mm long, pilose to felty, style 3 mm long, usually 4-loculed, locules 12–16-ovuled. **Fruits** smooth, orange-brown to reddish, globose-ovoid, to 5.5 cm long, valves 4, purplish or pink inside; seeds 1(–2), 1.9–2.8 × 1.3–1.8 cm, aril thick, red or orange, waxy, testa brownish-black. **Distr:** Malay Peninsula, Malesia. **Ecol:** In mixed dipterocarp forests; <800 m. Fl. May–Jun. **Zones:** 11–12.

MALPIGHIALES

FAMILIES 39, GENERA 716, SPECIES 15,935.

Leaves with margins toothed, teeth with single vein running into a congested and often deciduous apex; most families predominantly tricarpellate. **Flowers** with stigma dry, placentation axile (Judd et al. 2008; Stevens 2001 onwards). **Info:** The position and monophyly of Malpighiales is strongly supported by molecular data (Chase et al. 1993; Davis et al. 2005; Davis & Chase 2004; Hilu et al. 2003; Soltis et al. 2000).

ERYTHROXYLACEAE

Genera 10, **species** ~300. **Shrubs** or **trees**. **Stipules** intrapetiolar. **Leaves** alternate or rarely opposite, simple; lamina margins entire. **Inflorescences** axillary, fascicles or cymes. **Flowers** usually bisexual, regular, 5-merous, often heterostylous; sepals 5, basally connate, with imbricate or valvate lobes, persistent; petals 5, distinct, imbricate, usually with a scale on inner face at base; stamens 5, 10, or 20, 1- or biverticillate, filament bases usually connate into a tube, anthers elliptic, 2-celled, with longitudinal slits; ovary superior, connected with 3–5 carpels, 3–5-locular, with 1(or 2) axile, ovules pendulous, anatropous to hemitropous, placentation axile, styles 1–3 or 5, distinct or somewhat connate, stigmas oblique. **Fruits** capsules, or a 1-seeded drupe; seeds with straight embryo. **Distr:** Tropical and subtropical regions, especially S America.

Erythroxylum

Species ~230. **Shrubs** or **small trees**, usually glabrous. **Stipules** intrapetiolar, often imbricating on short branches. **Leaves** alternate, often subdistichous, simple. **Inflorescences** axillary, flowers solitary or fascicled. **Flowers** small, often heterostylous; sepals 5(or 6), usually basally connate; petals 5, white or pink, clawed, usually with a ligule on inner face at base, imbricate; stamens 10, filament usually basally connate into a glandular or non-glandular tube; ovary 3(or 4)-locular, with 1(or 2) ovules in single fertile locule, styles distinct or connate, stigmas oblique. **Fruits** drupes, 1-seeded; seed with testa thin, endosperm variable in quantity or absent, embryo straight. **Distr:** Tropics and subtropics, especially S America. **Cult:** Propagate from seeds or cuttings. Requires a moist, well-drained soil in shade or full sun.

E. coca Lam., *Encycl.* 2: 393, 1786. (Syn: *E. bolivianum, E. peruvianum*). [Coca, Cocaine Plant]. **Trees, small,** or **shrubs**, branched, glabrous, to 2.5 m tall; branches slender; bark with warty lenticels. **Stipules** paired, triangular, persistent. **Leaves** alternate, simple; lamina broadly elliptic or narrowly obovate-elliptic, 2–8 × 2–4 cm, base acute, apex obtuse to acute, mucronulate, adaxially dark green, abaxially paler and glaucous, longitudinal nerves 2 or 4, slender, more distinct when young; petiole short. **Inflorescences** clusters, axillary, 6~~–12-flowered;~~ pedicels 3–4 mm long; bracteoles paired, triangular. **Flowers** pale yellow or yellowish-green, small; petals elliptic, ~4.5 mm long; flowers dimorphic, one with stamens of equal length and 3 free styles 2 mm long, the other with stamens of unequal length and 3 free styles 4 mm long. **Fruits** drupes, oblong or oblong-ovoid, 0.7–1 × 0.3–0.45 cm, pointed, succulent, red; seed 3-ridged, 6 × 1.6 mm. **Distr:** Tropical S America. **Ecol:** In lowlands; <500 m. Fl. throughout the year. **Zones:** 10–12. **Info:** The whole plant is toxic. Cultivated mainly for medicinal purposes.

E. moonii Hochr., *Bull. Inst. Bot. Buitenzorg* 22: 54, 1905. (Syn: *E. acuminatum, E. lucidum, Sethia acuminata*). **Shrubs** or **small trees**, to 5 m tall, much branched; branches glabrous; bark pale brown, lenticellate; twigs often with a few deciduous bracts. **Stipules** finely serrate, caducous. **Leaves** elliptic-lanceolate, 3.5–9 × 1.2–2.8 cm, base acute, margins entire, apex obtusely caudate-acuminate, glossy; petiole 2.5–3.5 mm long. **Inflorescences** axillary, flowers 1 or 2 together; pedicels 0.8–1 cm long; bracteoles 2, triangular. **Flowers** white ~2 mm long, dimorphic; calyx campanulate, 1.5 × 1.5 mm, persistent, lobes 5, triangular, acute; petals elliptic-oblong, ~2 mm long, obtuse, slightly exserted, ligulate; ligule trifid, near base on inner side, persistent; stamens 10, unequal, staminal tube as long as calyx, margin dentate; ovary ovoid, triloculed, styles 3, connate almost throughout, stigma capitate. **Fruits** drupes, ellipsoid-oblong, slightly falcate, 1.3 × 0.3 cm, furrowed, glabrous, green, shiny, 1-seeded. **Distr:** India, Sri Lanka. **Ecol:** In evergreen forests; <1,000 m. Fl. & Fr. Aug–Dec. **Zones:** 11–12.

RHIZOPHORACEAE

Genera 16, **species** 149. **Trees** or **shrubs**, evergreen, without spines, often with aerial roots; stem nodes swollen. **Stipules** interpetiolar, sheathing terminal bud, caducous. **Leaves** simple, opposite or distichous, petiolate; lamina margins entire, serrulate near apex, or completely serrulate, leathery, usually glabrous. **Inflorescences** axillary, dense cymes. **Flowers** bisexual, actinomorphic; hypanthium present (or absent); calyx lobes 4–16, inserted on rim of hypanthium, free or scarcely base-connate, valvate, persistent in fruit; petals usually as many as sepals, free, usually caducous, margin entire, lacerate, or 2-cleft; stamens twice as many as calyx lobes, anther locules 4 to many, longitudinally dehiscent or by an adaxial valve; ovary inferior or half-inferior, carpels 2–5(–20), 2-8-loculed, ovules usually 2 per locule, pendulous, style 1, stigma entire, capitate, or lobed. **Fruits** pulpy or leathery, indehiscent; seeds 1 to few, viviparous; seedling (propagule) 7–80 cm when shed. **Distr:** Pantropical.

Bruguiera

Species 6. **Trees**, producing kneelike pneumatophores; trunk basally buttressed, often with a cone of stilt roots at base; leaf scars showing evidence of 3 vascular bundles. **Leaves** entire. **Inflorescences** cymes, 1–5-flowered, pedunculate. **Flowers** generally nodding; calyx accrescent, tube extending beyond ovary, lobes 8–14(–16), subulate-lanceolate; petals as many as calyx lobes, usually bilobed, fringed with hairs and/or bristles; stamens 2× as many as petals, in unequal pairs opposite petals, each pair enclosed by inrolled petal margins, anthers introrse, dehiscing lengthwise; ovary inferior, 2–4-loculed, ovules 2

RT

Above: *Sloanea javanica, Erythroxylum coca, Erythroxylum moonii, Bruguiera cylindrica.*

per locule, style lobes 2–4. **Fruits** campanulate; seed usually solitary, viviparous, hypocotyl terete or obscurely ribbed. **Distr:** E Africa, China, SE Asia, N Australia, Indian Ocean islands, Pacific Islands. **Cult:** Propagate from seeds. Species require a well-aerated soil with fresh or brackish water, in shade or full sun.

B. cylindrica (L.) Blume, *Enum. Pl. Javae.* 1: 93, 1827. (Syn: *Rhizophora cylindrica*, etc.). [Mangrove]. **Trees**, 10–15 m tall; trunk dbh 15–25 cm; bark gray, smooth. **Stipules** 2.5–3.5 cm long. **Leaves** elliptic, 7–17 × 2–8 cm, base cuneate, apex acute, thin, secondary veins abaxially remote and thin, reticulate veins mostly obscure; petiole 1–3.5 cm long. **Inflorescences** cymes, pedunculate, 2- or 3-flowered; pedicel 1–4 mm long. **Flowers** greenish, <2 cm long; calyx tube 4–6 × ~2 mm, not ribbed, smooth, lobes 7 or 8, ± as long as tube; petals white but soon turning brown, 3–4 mm long, bilobed, outer margins usually basally fringed with white hairs; stamens 1.5–2.5 mm long; disk in open flowers not entirely lining calyx tube; style 3–4 mm long. **Fruits** with fruiting calyx tube ~1 cm long, slightly ribbed, lobes recurved; hypocotyl cylindric, often curved, 8–15 × ~0.5 cm. **Distr:** Tropical Asia to Malesia and Pacific Islands. **Ecol:** In mangrove forests; sea level. **Zones:** 9–12.

B. gymnorhiza (L.) Savigny, *Encycl.* 4: 696, 1798. (Syn: *Rhizophora gymnorhiza*, etc.). [Burma Mangrove, Large-leaved Orange Mangrove, Oriental Mangrove]. **Trees**, 6–20 m tall; trunk 30–60 cm dbh; bark gray, deeply fissured. **Stipules** often reddish, ~4 cm long, interpetiolar, caducous. **Leaves** elliptic-oblong, 8–21 × 4–7(–9) cm, base cuneate to rarely obtuse, apex acute, leathery, glabrous; petiole 2–4.5 cm long. **Inflorescences** with flowers solitary, axillary; pedicel 1–2.5 cm long. **Flowers**, red or pinkish-red, ~3 cm long; calyx lobes 10–14, linear, 1.5–2 cm long, glabrous; petals 12–14, 1.3–1.5 cm long, outer margin fringed with white silky hairs, sinus bristles 3 or 4 per petal, 2–3 mm long, exceeding petals; stamens twice as many as petals, 0.8–1.1 cm long, filaments filiform; disk cup-shaped; ovary inferior, trilocuded, style filiform, ~1.5 cm long, stigma lobes 3 or 4. **Fruits** adnate to calyx tube, ~2.5 mm long, calyx persistent on fruit or hypocotyls ribbed only apically; seed 1, viviparous; hypocotyl cigar-shaped, slightly angular, 15–25 × 1.5–2 cm. **Distr:** E Africa, Indian Ocean islands, Madagascar; tropical and subtropical Asia to Malesia, N Australia and Pacific Islands. **Ecol:** In mangrove forests; sea level. Fl. May–Jun. **Zones:** 10–12.

Carallia

Syn: *Barraldeia, Diatoma*, etc.
Species ~10. **Shrubs** or **trees** (a single species occasionally with small stilt-roots at base of stem). **Stipules** lanceolate. **Leaves** decussate, petioled; lamina with margins entire to dentate or serrate, papyraceous to thinly coriaceous, black-dotted abaxially. **Inflorescences** clusterlike condensed or lax, dichotomously or trichotomously branched, sometimes reduced to 2, or even a single flower, sessile or peduncled; bracteoles 2, free, caducous or at least basally connate and persistent. **Flowers** with calyx lobes 5–8, deltoid; petals 5–8, clawed; stamens twice as many as petals, free, persistent, of 2 lengths, shorter ones opposite petals; disk annular, fleshy; ovary inferior, 5–8-loculed, ovules 2 per locule, style filiform, stigma discoid, capitate, or obscurely lobed. **Fruits** pulpy, 1-loculed; seeds 1(–5), arillate, not viviparous. **Distr:** Madagascar; S and SE Asia, China; New Guinea, Pacific Islands, N Australia. **Cult:** Propagate from seeds or by air-layering. Requires a moist soil in full or partial sun.

C. brachiata (Loureiro) Merrill, *Philipp. J. Sci.* 15: 249, 1920. (Syn: *Petalotoma brachiata*, etc.). [Billabong Tree, Caralla Wood, Freshwater Mangrove]. **Trees**, to 10 m tall; branches brown, corky; stipules 1–2.5 cm. **Leaves** elliptic, obovate, oblanceolate, or rarely suborbicular, 5–15 × 2–10 cm, papery to thinly leathery, base cuneate, margins entire, serrate, or denticulate, apex acute to shortly acuminate; petiole ~1 cm. **Inflorescences** 1–6 cm, often resinous, usually shining from secreted resin; flowers shortly pedicellate or sessile; bracteoles 2 or 3, partially connate into a cup. **Flowers** with calyx lobes 6 or 7, deltoid, 3–4 mm; petals white, suborbiculate, ~1.5 mm Ø, apically emarginate and unevenly lacerate; stamens ~2 mm; ovary bulbous, ~2 mm, style ~2 mm, stigma discoid, apically 4–8-lobed. **Fruits** pink to red globose, ~5 mm Ø, glossy; seeds reniform. **Distr:** Tropical Asia through Indo-China to Malesia, N Australia and Pacific Islands; Madagascar. **Ecol:** In rainforests, evergreen forests; 1,800 m. Fl. Feb–Oct. **Zones:** 10–12.

Ceriops

Species 2. **Shrubs** or **small trees**; trunk often with stilt roots clustered around base. **Stipules** lanceolate. **Leaves** crowded near shoot apex; lamina with veins obscure on both surfaces. **Inflorescences** condensed cymes, pedunculate, 2- to many-flowered; bracteoles 2, partly connate. **Flowers** with calyx deeply 5- or 6-lobed, tube adnate; petals 5 or 6, white, each embracing 2 stamens, cohering at base, margin with uncinate trichomes; stamens twice as many as calyx lobes; disk cupular, shallowly lobed; ovary semi-inferior, triloculed, ovules 2 per locule, style short, stigma entire or shortly lobed. **Fruits** ovoid or conical; seed 1, albumen absent; hypocotyl clavate, ridged and grooved, apically tapering. **Distr:** E Africa, S and SE Asia, N Australia, China, Indian Ocean islands, Madagascar, New Guinea, Pacific Islands. **Cult:** Propagate from seeds or by air-layering. Requires a moist soil in full or partial sun.

C. tagal (Perrottet) C.B. Robinson, *Philipp. J. Sci.* 3: 306, 1908. (Syn: *Rhizophora tagal*, etc.). [Spurred Mangrove, Tagal Mangrove, Yellow Mangrove]. **Shrubs** or **trees**, 2–5 m tall; trunks with buttresses or small stilt roots; bark brownish. **Stipules** lanceolate,

Below: *Bruguiera gymnorhiza, Carallia brachiata, Ceriops tagal, Rhizophora apiculata.*

MS

RT

SZ

RT

RT

Above: *Rhizophora mangle, Rhizophora mucronata, Rhizophora stylosa, Reinwardtia indica.*

1–2 cm long. **Leaves** obovate to obovate-oblong, 4–9 × 2–5 cm, base cuneate, apex obtuse to sometimes slightly emarginate; petiole 1–3 cm long. **Inflorescences** 4–10-flowered; peduncle 1–2 cm long; pedicels to 2 mm long. **Flowers** with calyx 6–7 mm long, erect or slight spreading in flower, widely spreading or reflexed in fruit; petals oblong, ~4 mm long, margins fringed with trichomes and slightly cohering, apex with 3 club-shape terminal appendages; stamen filaments 3–5 mm long, shorter than calyx lobes, anthers <1 mm long. **Fruits** with hypocotyl 15–30 cm long, apex sharply pointed. **Distr:** Tropical and subtropical Asia to Malesia and N Australia; E Africa, Indian Ocean islands, Madagascar. **Ecol:** In mangrove forests; sea level. **Zones:** 10–12.

Rhizophora

Species 8–9. **Trees** or **shrubs**, with aerial roots. **Stipules** reddish, sessile, leaflike, lanceolate. **Leaves** opposite or distichous; lamina glabrous, midvein extended into a caducous point, margins entire or serrulate near apex. **Inflorescences** axillary, dense cymes; bracteoles forming a cup just below flower. **Flowers** with calyx tube adnate to ovary, persistent, lobes 5–8; petals 4, lanceolate; stamens 8–12, filaments much shorter than anthers or absent, anthers introrse, locules many, dehiscing by an adaxial valve; ovary inferior, biloculed, apically partly surrounded by a disk, free part elongating after anthesis, style 1, sometimes very short, stigmas 4. **Fruits** brown, ovoid, ovoid-conic, or pyriform; fertile seed 1 per fruit, germination viviparous, hypocotyl protruding to 78 cm before propagule falls. **Distr:** Tropical and subtropical regions worldwide. **Cult:** Propagate from seeds. Requires a well-aerated soil with fresh, brackish or salt water, in shade or full sun.

R. apiculata Blume, *Enum. Pl. Javae.* 1: 91, 1827. (Syn: *R. candelaria*). [Tall-stilt Mangrove]. **Trees** or **shrubs**, 3–6(–10) m tall; bark gray, usually with vertical fissures. **Stipules** 4–8 cm long. **Leaves** elliptic-oblong to sublanceolate, 7–16 × 3–6 cm, base broadly cuneate, apex acute to apiculate, abaxial midvein reddish; petiole 1.5–3 cm long, usually reddish-tinged; **Inflorescences** 2-flowered cymes; peduncle to 1 cm long; flowers sessile. **Flowers** with calyx lobes ovate, concave, 1–1.4 cm long, apex acute; petals lanceolate, flat, 6–8 mm long, membranous, glabrous, white; stamens mostly 12, 4 adnate to base of petals, 8 adnate to sepals, 6–7.5 mm long, anthers nearly sessile, apex apiculate; ovary largely enclosed by disk, free part 1.5–2.5 mm long, style ~1 mm long. **Fruits** ~2.5 × 1.5 cm, apical half narrower; hypocotyl cylindric-clavate, ~3.8 × 1.2 cm. **Distr:** Tropical Asia to Malesia, N Australia and Pacific Islands; E Africa. **Ecol:** In mangrove forests; sea level. Fl. & Fr. throughout the year. **Zones:** 10–12.

R. mangle L., *Sp. Pl.* 1: 443, 1753. [Red Mangrove]. **Shrubs** or **small trees**, glabrous, 3–7 m tall; young branches reddish; conspicuous prop or stilt roots produced by trunk and lower branches. **Leaves** opposite; lamina simple, ovate to obovate or broadly elliptic, 6.5–15 cm long, leathery and somewhat fleshy, margins entire and slightly rolled under. **Inflorescences** axillary, cymose, usually 2–5-flowered; peduncles and pedicels thick; flower buds yellow, 4-angled. **Flowers** bisexual, white and yellow, soon pendulous, wind pollinated; calyx greenish-yellow, campanulate with 4 lobes, lobes 0.8–1.2 cm long, fleshy; corolla white to pale yellow, petals 4, 7–8 mm long, hairy; stamens 8, anthers 4–6 mm long, sessile, honeycomblike; ovary partially inferior, style thick, bilobed, stigma receptive after stamens and petals fall. **Fruits** berries, greenish-brown, cone-shaped, 1.5–2.5 cm long; seed 1, viviparous; hypocotyl 15–25 cm long. **Distr:** Pantropical. **Ecol:** In mangroves, forming impenetrable thickets along the shorelines. **Zones:** 11–12.

R. mucronata Lamarck ex Poiret, *Encycl.* 6: 189, 1804. (Syn: *R. longissima*). [Asiatic Mangrove]. **Trees**, to 27(–30) m tall; trunk to 70 cm dbh. **Stipules** 5.5–8.5 cm long. **Leaves** broadly elliptic to oblong, 8.5–16 × 5–10 cm, base cuneate, apex ± blunt to ± acute; petiole 2.5–4 cm long. **Inflorescences** 2–4(–8)-flowered cymes; peduncle 2–5 cm long; flowers sessile. **Flowers** with calyx lobes ovate, 0.9–1.4 × 0.5–0.7 cm; petals lanceolate, 7–9 mm long, fleshy, partly embracing stamens, margins pilose; stamens 8, 4 borne on base of petals, 4 borne on sepals, 6–8 mm long, anthers sessile; ovary emerging far beyond disk, free part elongate-conic, 2–3 mm long, style 0.5–1.5 mm long. **Fruits** dirty brownish-green, elongate-ovoid, 5–7 × 2.5–3.5 cm, basally often tuberculate, apically slightly contracted; hypocotyl cylindric, 30–60 cm long. **Distr:** Old World tropics. **Ecol:** In mangrove forests; sea level. **Zones:** 11–12.

R. stylosa Griffith, *Not. Pl. Asiat.* 4: 665, 1854. (Syn: *R. mucronata* var. *stylosa*). [Small-stilted Mangrove]. **Trees**, often <8 m tall; bark reddish or pale gray, rough. **Leaves** obovate, 6.5–11 × 3–4(–5.5) cm, base broadly cuneate, apex mucronate; petiole 2–3 cm long. **Inflorescences** 2- to many-flowered; peduncle 1–5 cm long; pedicels 0.5–1 cm long, terete; bracteoles brown, connate. **Flowers** with calyx lobes lanceolate to oblong-lanceolate, 0.9–1.2 × 0.3–0.5 cm; petals to 1.2 cm long, involute, margin densely villous; stamens usually 8, filaments short but distinct, anthers 5–6 mm long; ovary emerging beyond disk, free part shallowly conic and <1.5 mm long; style 4–6 mm long, stigma lobes 2. **Fruits** green, conic, 2.5–3 × ~2 cm; hypocotyl cylindric, 30–40 cm long, apex acute. **Distr:** Asia to N Australia and Pacific Islands. **Ecol:** In mangrove forests; sea level. **Zones:** 9–12.

LINACEAE

Genera 10–12, **species** 300. **Herbs** or **shrubs**. **Stipules** small or absent. **Leaves** alternate or rarely opposite; lamina simple, margins entire. **Inflorescences** cymes or racemes. **Flowers** bisexual, regular; sepals (4 or)5, distinct or basally connate, imbricate, persistent; petals as many as and alternate with sepals, distinct or basally connate, convolute, often basally clawed, with 2–5 extra-staminal nectary glands or a disk; stamens (4 or)5 or 10(or 15), in 1 whorl, alternate or opposite sepals, often some reduced to staminodes, filament bases connate into a tube; ovary superior, with 2–5 carpels or seemingly 4–10-loculed by intrusion of a false septum, with 1 or 2 ovules per locule, placentation axile, styles as many as carpels, filiform, distinct or basally connate, stigmas subcapitate. **Fruits** usually septicidal capsules or drupes; seeds with straight oily embryo and thin endosperm. **Distr:** Worldwide, but mostly in temperate regions.

Reinwardtia

Species 1. **Shrubs**. **Stipules** minute, subulate, caducous. **Leaves** alternate. **Inflorescences** axillary and terminal cymose fascicles or flowers solitary and axillary. **Flowers** with sepals 5, lanceolate, persistent, margins entire, apex acuminate; petals 4 or 5, yellow, much longer than sepals, confluent, fugacious; stamens 5, hypogynous, basally connate; staminodes 5, alternating with and inserted between stamens; glands 2–5, adnate to staminal ring; ovary 3(or 4)-celled, cells biloculed by intrusion of a false septum, with 1 ovule per locule, styles 3(or 4), filiform, distinct or basally connate, stigmas subcapitate. **Fruits** capsules, splitting into 6(–8) 1-seeded mericarps; seeds with a membranous wing. **Distr:** E, S, and SE Asia. **Cult:** Propagate from soft-tip cuttings or seeds. Requires a moist, well-drained, humus-rich soil in full or partial sun.

R. indica Dumortier, *Commentat. Bot.* 19, 1822. (Syn: *R. trigyna*, etc.). [Golden Dollar Plant, Golden Girl, Yellow Flax]. **Shrubs**, to 1 m tall; branches gray, glabrous. **Leaves** elliptic to obovate-elliptic, 2–8.8 × 0.7–3.3 cm, base cuneate, margin entire or crenate, apex acute to subrounded and with an apiculate tip, papery; petiole 0.8–2.5 cm. **Flowers** 1.4–3 cm Ø; sepals 0.9–1.2 × ~0.3 cm, distinct; petals yellow, 1.7–3 × ~1.3 cm, distinct but basally confluent; stamens ~1.3 cm, anthers ~2 mm; staminodes subulate; styles 3 or 4, 0.7–1.8 cm. **Fruits** capsules, globose, splitting into 6 or 8 mericarps; seeds reniform. **Distr:** Bhutan, China, N India, Kashmir, Laos, Myanmar, Nepal, Pakistan, Thailand, Vietnam. **Ecol:** In forests, mountain slopes, thickets, along trails, ravines, often in calcareous soil; 500–2,300 m. Fl. & Fr. Apr–Jan. **Zones:** 9–11. **Info:** Cultivated for medicinal and ornamental purposes.

AH

MS

Above: *Rafflesia kerrii, Rafflesia manillana, Suregada lanceolata.*

RAFFLESIACEAE

Genera 9, **species** 27. **Herbs**, fleshy, parasitic on roots, stems, or branches of various hosts; vegetative body thalloid or reduced to myceliumlike tissue within host. **Leaves** absent. **Inflorescences** terminal, usually 1-flowered, rarely a spike; bracts scalelike, several at base of inflorescence. **Flowers** usually unisexual (plants dioecious or monoecious), bisexual in *Mitrastemon*, actinomorphic, minute to extremely large (*Rafflesia arnoldii* of Indonesia produces a flower weighing up to 10 kg and to 1 m Ø), 4–10-merous; perianth lobes well-developed or absent, basally connate, rarely free; stamens 5 to very numerous, in 1–3 series, fused with the rudimentary pistil into a gynostemium or in bisexual flowers connate into a cylinder; free filaments absent, anthers biloculed, dehiscent longitudinally or by apical pores; ovary superior, semisuperior, or inferior, absent in ♂ flowers, 1-loculed with many parietal placentas; ovules numerous, anatropous, integuments 1 or 2; style 1 or none, stigma discoid, capitate, or with many lobes. **Fruits** berries; seeds minute, very numerous, testa hard, endosperm present. **Distr:** Mostly in tropical and subtropical regions.

Rafflesia

[Corpse Flower].
Species 20. **Herbs**, parasitic, rootless, stemless, leafless; chlorophyll absent. **Bud** cabbagelike, bursting through the host's bark, maturing usually in ~9–12 months. **Flower** 15 cm to 1 m across, weighing up to 10 kg, usually with a strong fetid aroma, lasting only a few days, color variable but usually reddish with paler or whitish blotches, fleshy, unisexual (♀ usually rare), imperfect, usually producing large amounts of nectar; petals 5; central area with numerous tubercles; disk spiked, to which either stigma or stamens are attached (depending on the gender of the flower); perigon lobes 5; ramenta only inside the flower tube; perianth connate, uniseriate; stamens 12–40, extrorse, adnate to a central column in the ♂ flowers; ♀ flowers with ovary inferior. **Fruit** round, to 15 cm across, flesh smooth; seeds numerous, minute, testa hard. **Distr:** Thailand, south and westwards in peninsular Malaysia, Sumatra, Borneo, Java, Philippines. **Ecol:** In primary and secondary lowland tropical rainforests, parasitic on *Tetrastigma* spp. (Vitaceae); 500–700 m. **Cult:** Difficult to cultivate, due to their requirement of specialized hosts. Seeds can be embedded in the roots of a compatible host species. Requires a humid, tropical climate. **Info:** This genus produces some of the largest flowers in the plant kingdom. The only structure of the plant outside the flower consists of strands of funguslike tissue that grow inside the host vine, and inconspicuous when not in flower.

R. kerrii Meijer, *Blumea* 30(1): 212, 1984. [Bua Phut, Bua Tum]. **Herbs**, parasitic. **Bud** reddish-brown, glossy; maturing in 8–9 months. **Flower** 50–90 cm Ø, weight 6–8 kg, smell carrionlike at a distance outside flower (produced by perianth lobes), inside the chamber of perianth cup with a faintly fruity fragrance (the 2 scents separated by the diaphragm), unisexual, flowers lasting 5–7 days; orifice 12–20 cm across. **Cupule** adnate to base of flower, 15 cm Ø. **Perigone lobes** 5, dark red, ovate, ~30 × 30 cm, rounded; tube ~30 cm Ø. **Diaphragm** ~45 cm Ø, opening ~20 × 20 cm; ramenta inside on inside surface of diaphragm. **Disk** with processes on upper surface, ~3–5 cm long, spiny and branched; column thick demarcated by a sulcus, ~20 cm Ø; anthers on the underside of the collar of disk. **Distr:** S Thailand, peninsular Malaysia; rare. **Ecol:** In rainforests, also in primary or logged lowland forests, parasitic on roots of *Tetrastigma tuberculatum*; 500–1,000 m. Fl. Jan–Apr(–Jul). This is the largest species in the Malayan Peninsula (and Thailand). **Zones:** 11–12.

R. manillana Teschem., *Boston J. Nat. Hist.* 4: 65, 1844. (Syn: *Rafflesia cumingii*). [Stinking Corpse Lily]. **Herbs**, parasitic. **Bud** ~7 cm Ø; windows on the underside of the diaphragm at the lower row ~3 × 3 mm; middle row ~4 × 4–5 mm; upper row ~3 × 4–5 mm. **Flower** 6.5–8 cm high, 11–16(–24) cm wide (with cupule). **Cupule** 2–2.5 cm high, 5–6 cm wide. **Bracts** in 3 whorls, outer whorl 1–2 × 2.5–3 cm, middle whorl 2–2.5 × 2.5–3.5 cm, inner whorl ~4 × 4–5 cm. **Perigone lobes** 4–5 × 4–5.5 cm; warts 5–7 across widest part, closely spaced. **Diaphragm** incurved, 7–12 cm across, 2–3 cm broad, opening 4–5 cm across; upper surface whitish in bud, turning to the same color as perigone lobes as it matures. **Perigone tube** to 2.5–3.5 cm high, 5–9 cm wide; ramenta on lower part of tube stalked. **Disk** 3.5–4 cm Ø, 4–8 mm thick; rim to 2 mm high, slightly raised, hairy; processes 14–15, 3–5 mm high, wedge shape (in ♀) or conical (in ♂), hairy on top edge. **Distr:** Philippines. **Ecol:** In lowland primary forests and growing on the host vine *Tetrastigma* sp. **Zones:** 11–12.

EUPHORBIACEAE

Genera 218, **species** 5,735. **Trees**, **shrubs** or **herbs**, sometimes scandent and twining, mostly with milky sap. **Stipules** often present. **Leaves** chiefly alternate, sometimes opposite or verticillate, simple or rarely digitately compound, sometimes palmate-lobate, dentate or entire. **Inflorescences** highly variable in form. **Flowers** usually small but sometimes large, unisexual, monoecious or dioecious, generally regular; perianth sometimes none, usually small, often dissimilar in flowers of the 2 sexes, either a calyx or a calyx and a corolla, the segments free or united, imbricate or valvate in bud; ♂ with stamens sometimes indefinite, often as many as sepals or fewer, sometimes only 1, filaments free or united, rudimentary ovary present or none; ♀ with ovary usually 3(sometimes 1–4+)-celled, styles as many as carpels, free or connate, entire, cleft, or laciniate, ovules 1 per cell, or 2. **Fruits** generally capsular and separating into 2-valvate cocci, or the fruits drupaceous and indehiscent; seeds commonly as many as ovules. **Distr:** Pantropical and warm temperate.

Subfam. CROTONOIDEAE

Hairs often stellate or lepidote; laticifers articulated or not. **Seeds** carunculate/arillate or not (Stevens 2001 onwards). **Distr:** Pantropical.

TRIBE GELONIEAE

Suregada

Syn: *Erythrocarpus, Gelonium, Owataria.*
Species ~35. **Trees** or **shrubs**, dioecious (very rarely monoecious); latex not evident; indumentum absent or short and simple. **Stipules** small, connate, caducous, leaving annular scars at nodes. **Leaves** alternate, short-petiolate; lamina simple, entire or rarely sparsely serrulate, venation pinnate, usually pellucid-punctate. **Inflorescences** glomerules or cymes, leaf-opposed. **Flowers** ♂ with sepals 5 or 6, free, imbricate, sometimes glandular abaxially, petals absent, disk extra-staminal or intrastaminal, annular or dissected, stamens 6–60, free, anthers longitudinally dehiscent, pistillode absent; ♀ with sepals 4–8, imbricate, sometimes glandular abaxially, petals absent, disk annular, staminodes sometimes present, ovary (2 or)3(or 4)-locular, ovules 1 per locule, stigmas bifid (rarely multifid), spreading. **Fruits** capsules or sometimes indehiscent and drupelike; seeds not carunculate, covered by pale sarcotesta, endosperm present, cotyledon broad and flat. **Distr:** Tropical Africa, Asia, and Oceania. **Cult:** Propagate from seeds or by air-layering. Requires a moist, well-drained soil in full sun.

S. lanceolata (Willd.) Kuntze, *Rev. Gen.* 619, 1891. (Syn: *Gelonium lanceolatum*). **Trees**, small or shrubs, 2–9 m tall; young branches glabrous. **Leaves** alternate, petiolate; lamina 3.5–11 × 1.5–5 cm, obovate-oblong to oblanceolate or broadly elliptic, margins entire, apex rounded-obtuse, glabrous, shiny, veins prominent; petiole 3–5 mm long, glabrous. **Inflorescences** axillary, extra-axillary or leaf-opposed; ♂ flowers many, shortly pedicellate, in small, usually leaf-opposed, glomerulate clusters, clusters to ~0.6 cm Ø, pedicels 1.5–2 mm long; ♀ flowers solitary, or to 3 or 4, pedicellate on short, rather stout peduncles, pedicels 1.5–2 mm long. **Flowers** ♂ with calyx lobes ~1.5 mm long, shortly ciliate, broadly ovate, stamens 26–34, filaments 2 mm long, filiform, glabrous; ♀ with sepals 1.5–2 × 2 mm, rotund, shortly ciliate, sparsely, finely pubescent, ovary ~1 mm Ø, spherical, 3-celled, 1 ovule per cell. **Fruits** capsules, 0.6 × 1–1.1 cm, shallowly trilobed, gray-green; seeds 5 × 4 mm,

MD

Top row: *Manihot esculenta* (male flowers), *Manihot esculenta* (female flowers), *Manihot grahamii*.
Bottom row: *Manihot esculenta* (fruits), *Manihot glaziovii*.

subglobose, grayish. **Distr:** India, Sri Lanka. **Ecol:** In intermediate and secondary forests, scrubland and alluvial flats; to 800 m. Fl. Jan–Mar. **Zones:** 11–12.

TRIBE MANIHOTEAE

Manihot

Syn: *Janipha, Mandioca, Manihotoides*.
Species ~60. **Trees**, **shrubs**, or **herbs**, monoecious (rarely dioecious), sometimes with fleshy root-tubers; stems and branches with prominently large leaf scars; latex white; indumentum absent. **Stipules** small, usually caducous. **Leaves** alternate; lamina usually palmately lobed or dissected, sometimes uppermost subentire, apex glandular or stipellate; petiole long. **Inflorescences** terminal or pseudo-axillary, mostly racemose-paniculate, usually bisexual, with long peduncles. **Flowers** apetalous; ♂ flowers with calyx synsepalous, sepals 5, imbricate, petaloid, disk intrastaminal, usually 10-lobed, stamens 8–15, in 2 series, filaments free, delicate, anthers bilocular, longitudinally dehiscent, connectives hairy at apex, pistillode small or absent; ♀ flowers with sepals 5, free or connate, sometimes deciduous, disk annular, sometimes with staminodes, ovary 3(or 5)-locular, ovules 1 per locule, styles short, free, mostly multifid or lacerate (rarely bifid), stigmas broad. **Fruits** capsules, 3-valved, columella often persistent; seeds carunculate, testa dry, crustaceous, endosperm fleshy; cotyledon broad and flat. **Distr:** Tropical America, mostly in Brazil. **Cult:** Propagate from cuttings. Requires moist soil in full sun.

M. esculenta Crantz, *Inst. Rei Herb.* 1: 167, 1766. (Syn: *Jatropha manihot*, etc.). [Cassava, Manioc, Tapioca]. **Shrubs**, erect, 1.5–5 m tall; young parts green; root tubers terete, large, sometimes branched. **Stipules** triangular-lanceolate, 5–7 mm long, entire or with 1 or 2 bristly segments. **Leaves** palmately 3–9-lobed, 5–20 cm long, lobes oblanceolate to narrowly elliptic, 8–18 × 1.5–4 cm, margins entire, apex acuminate, lateral veins 5–15; petiole 6–35 cm long, slightly peltate, inserted <5 mm from margin. **Inflorescences** racemes, terminal or axillary, 5–8 cm long; bracts oblong-lanceolate; pedicels 4–6 mm long. **Flowers** ♂ with calyx ~7 mm long, purple-red, divided to ± middle, lobes long-ovate, 3–4 × ~2.5 mm, hairy inside, stamens 6–7 mm long, anthers white-pubescent at apex; ♀ flowers with calyx ~1 cm long, lobes oblong-lanceolate, ~8 × 3 mm, ovary ovoid, longitudinally 6-angled, stigmas recurved, plaited. **Fruits** capsules, ellipsoidal, 1.5–1.8 × 1–1.5 cm, longitudinally 6-winged, scabrous; seeds slightly triangular, ~1 cm long, testa crustaceous, smooth, with spot-stripes. **Distr:** Brazil. **Ecol:** Fl. Sep–Nov. **Zones:** 9–12. **Info:** Cultivated throughout the tropics for the edible tubers. All parts of the plant contain highly poisonous, cyanide-related compounds. Cooking removes the toxicity of these compounds. The tubers are the part mainly used. The leaves are often eaten in the Maldives after frying with fish and spices.

M. glaziovii Müll. Arg., *Fl. Bras.* 11(2): 446, 1874. (Syn: *M. carthaginensis* ssp. *glaziovii*). [Ceara Rubber Tree]. **Trees** or **shrubs**, to 14 m tall; branches herbaceous, glabrous; bark with abundant latex. **Stipules** lanceolate, 4–6 mm long, gray-green, laciniate. **Leaves** palmately 3–7-lobed, 10–30 × 15–25 cm, lobes obovate, elliptic, rhomboid, or spathulate, 7–10 × 3.5–7 cm, margins entire, apex acute to shortly acuminate, papery or membranous; petiole 4–25 cm long, with many longitudinal grooves, peltate, inserted ~5 mm from margin. **Inflorescences** panicles, 7–9 cm long; bracts lanceolate, ~2.5 mm long. **Flowers** ♂ with calyx 0.9–1.1 cm long, 5-lobed, lobes oblong, ~5 × 2–3 mm, obtuse, glabrous inside, disk shallowly cupular, 10-lobed, glabrous, anthers oblong, ~2.5 mm long; ♀ with calyx 1–1.1 cm long, 5-lobed, ovary ellipsoidal, glabrous. **Fruits** capsules, subglobose or globose, ~2 cm Ø, warty, wingless, rugose; seeds depressed-ovoid, ~1.5 cm long, with shallow brown spot-stripes. **Distr:** Brazil. **Ecol:** Fl. Sep–Oct. Fr. Nov–Dec. **Zones:** 10–12. **Info:** Cultivated mainly for commercial rubber.

M. grahamii Hook., *Hooker's Icon. Pl.* 11: t. 530, 1843. (Syn: *Janipha loeflingii* var. *multifida*, etc.). [Wild Cassava]. **Shrubs** or **small trees**, thicket-forming, deciduous, monoecious, 2–7 m high, much-branched, mostly glabrous. **Leaves** alternate; lamina deeply palmately incised (almost to the base) into 5–13 lobes; lobes ± oblong, 5–15 × 1–2 cm, apex acute to acuminate, central lobes flared toward the apices then abruptly narrowing to a sharp point, glabrous, abaxially ± glaucous; petioles 5–20 cm long. **Inflorescences** racemes, spreading, 3–10 cm long; pedicels 0.5–2.5 cm long. **Flowers** yellowish-green, purplish-striped. **Fruits** capsules, rounded, 3-sided, 1.4–2 cm Ø; seeds 3, ~1 cm long, gray-mottled. **Distr:** SE Brazil to N Argentina. **Ecol:** Habitat unknown. Fl. Sep–Dec. Fr. Feb. **Zones:** 9–11.

TRIBE MICRANDREAE

Hevea

Syn: *Caoutchoua, Siphonia*.
Species ~10. **Trees**, monoecious; exudate milky latex. **Stipules** small, caducous. **Leaves** alternate, spirally

Above: *Hevea brasiliensis* (fruit), *Croton aromaticus*, *Croton laccifer*, *Croton tiglium*.

arranged, or subopposite at apex of branches; lamina palmately compound, usually trifoliolate; leaflets entire, petiolulate; petiole long, apex glandular. **Inflorescences** many-flowered panicles of cymes, in axils of leaves or fallen leaves, central flower of each cyme ♀, others ♂. **Flowers** ♂ with buds subglobose or ovate-globose, sepals 5-dentate or 5-lobed, lobes valvate, disk small, 5-lobed or dissected, rarely indehiscent, stamens 5–10, filaments connate into column longer than anthers, anthers sessile, in 1 or 2 series; ♀ with sepals as in ♂, disk dissected or obsolete, ovary 3-celled, ovules 1 per cell, styles usually absent, stigmas stout. **Fruits** capsules large, usually 3-valved, epicarp subfleshy, endocarp woody; seeds oblong-elliptic, maculate-striate, caruncle absent, cotyledon broad and flattened. **Distr:** Amazonian S America. **Cult:** Propagate from seeds or by air-layering. Requires a well-drained soil in full sun.

H. brasiliensis (Willdenow ex A. Jussieu) Müll. Arg., *Linnaea.* 34: 204, 1865. (Syn: *Siphonia brasiliensis*). [Rubber Tree]. **Trees**, large, to 30 m tall; exudate milky. **Leaves** with leaflets 3, elliptic, 10–25 × 4–10 cm, base cuneate, margins entire, apex acute to acuminate, surfaces glabrous, lateral veins 10–16, reticulate veins prominent; petiole 5–15 cm, apex usually with 2(–4) glands. **Inflorescences** panicles, axillary, to 16 cm long, gray-white pubescent. **Flowers** ♂ with sepals ovate-lanceolate, ~2 mm long, stamens 10, in 2 series, anthers bilocular, longitudinally dehiscent; ♀ with sepals as in ♂, but larger, ovary (2 or)3(–6)-locular, stigmas 3, sessile. **Fruits** capsules, ellipsoidal, 5–6 cm Ø, with 3 longitudinal grooves, apex rostrate, base retuse, epicarp thinner, with reticulate stripes, endocarp thicker, woody; seeds ellipsoidal, gray-brownish, maculate-striate, very decorative. **Distr:** Brazil. **Ecol:** Fl. May–Jun. Fr. Aug–Sep. **Zones:** 10–12. **Info:** Cultivated widely in the tropics for the commercial rubber and wood.

TRIBE CROTONEAE

Croton

Syn: *Tridesmis*.

Species 1,199. **Trees** or **shrubs**, rarely subshrubs, monoecious or more rarely dioecious; latex clear or reddish, often scanty or apparently absent. **Stipules** present or absent. **Leaves** alternate, rarely opposite or nearly verticillate; lamina simple, margins entire or denticulate, often with glands, venation pinnate or palmate; petiole with or without apical paired glands. **Inflorescences** terminal or axillary, racemelike thyrses, often with proximal solitary ♀ flower and distal ♂ flowers, often in clusters. **Flowers** ♂ with sepals mostly 5, imbricate to valvate, free, petals 5, smaller or nearly equal to sepals, usually 5, and opposite sepals, stamens 10–20, filaments free, inflexed in bud, anthers muticous, pistillode absent; ♀ with sepals 5, often narrower than in ♂, imbricate to reduplicate-valvate, persistent, sometimes accrescent, petals 5, smaller than in ♂, mostly reduced or obsolete, ovary trilocular, ovules 1 per locule, styles 3, longer and slender, free or nearly so, once to several times bifid. **Fruits** capsules, dehiscing into three 2-valved cocci, more rarely ± indehiscent and fleshy, drupelike or berrylike; seeds usually smooth, carunculate, testa crustaceous. **Distr:** Tropical and subtropical regions worldwide, with the highest diversity in the neotropics. **Cult:** Propagate from scarified seeds or cuttings. Requires humus-rich, moist but well-drained soil in full or partial sun.

C. aromaticus L., *Sp. Pl.*: 1005, 1753. (Syn: *Cascarilla aromatica*, etc.). **Shrubs** or **small trees**, to 3 m tall; young branches scabrid or tomentose. **Leaves** alternate; lamina broadly ovate to ovate-lanceolate, shortly acuminate, 3.5–13 × 2–11 cm, base cordate or broadly rounded, apex usually acute, occasionally tip filamentous, appearing mucronate, margins irregularly shallowly dentate, coarsely pubescent, less so above, 3-veined from base, abaxially olive-green, 2 glands at base stipitate; petiole 1–3.5(–6) cm long, densely hairy. **Inflorescences** racemose, 7–18(–24)+ cm long, pubescent, flowers numerous, shortly pedicellate, laxly distributed; bracts ~1–2 mm long, linear-lanceolate; pedicels 2–7 mm long, densely pubescent; ♀ flowers few at base of inflorescence. **Flowers** white; ♂ with sepals narrowly ovate, 2–3 × 1–1.5 mm, densely pubescent, petals equaling sepals, white-fringed, stamens ~20+, filaments slender, 2–3 mm long; ♀ with ovary densely hairy, ~1–1.5 mm Ø, styles ~3–6 mm long, divided to base into ~5 filamentous branches. **Fruits** capsules, ~1 cm Ø, globose, obscurely trilobed, minutely scabrid. **Distr:** S India, Sri Lanka. **Ecol:** In dry scrublands. **Zones:** 11–12.

C. laccifer L., *Sp. Pl.*: 1005, 1753. (Syn: *Aleurites laccifera*, etc.). **Shrubs** or **small trees**, to 10 m tall; young branches densely yellowish-hairy. **Leaves** alternate; lamina broadly ovate to obovate, 4–11(–16.5) × 2.5–8.5(–13) cm, base rounded to narrowly subtruncate, usually irregularly serrate-dentate, apex usually rounded to obtuse, surfaces softly yellowish-hairy, 3-veined from base, 2 glands at base, orange-brown, shortly stipitate; petiole 0.8–2.4(–7) cm long, densely hairy. **Inflorescences** 5–10(–20) cm long, densely hairy, flowers laxly distributed on mature inflorescences, solitary or in small clusters otherwise; bracts 1.5–2 mm long, narrowly linear; ♂ flowers with pedicels 2–5 mm long, slender, densely hairy; ♀ flowers few at base of inflorescence, pedicels to ~4.5 mm long, somewhat stouter than ♂. **Flowers** white; ♂ sepals ovate, ~2.5 × 1.5 mm, stamens 15–24, filaments to 4 mm long; ♀ sepals ~3 × 1.5 mm, ovary ~2 mm Ø, subglobose, densely hairy, styles ~2.5 mm long, divided almost to base. **Fruits** capsules, 0.8–1 cm Ø, subglobose, densely hairy; seeds ellipsoid, 5 × 3.5 mm, dull brown. **Distr:** India, Sri Lanka. **Ecol:** In forests and forest margins; <800 m. **Zones:** 11–12.

C. tiglium L., *Sp. Pl.* 2: 1004, 1753. (Syn: *Alchornea vaniotii*, etc.). [Purging Croton]. **Treelets**, to 7 m tall; young branches green, sparsely hairy later glabrous. **Stipules** subulate, 1.5–4 mm long, caducous. **Leaves** ± ovate, 5–15 × 2–7 cm, base cuneate or broadly so, usually rounded, margins serrulate or subentire, apex acute or acuminate, sometimes long-acuminate or caudate-acuminate, papery, glabrous or glabrescent, with discoid glands, basal veins 3(–5); petiole 2.5–6 cm long, subglabrous. **Inflorescences** racemes, terminal, 8–20 cm long; bracts subulate, ~2 mm long. **Flowers** ♂ with buds subglobose, sparsely hairy or glabrescent; ♀ flowers with sepals oblong-lanceolate, ~2.5 mm long, glabrescent, ovary densely hairy, styles bipartite. **Fruits** capsules, ellipsoidal, oblong-ovoid, or subglobose, 1–2 × 1–2 cm, sparsely hairy or glabrescent; seeds elliptic or oblong-ovate, 0.8–1.2 × 0.6–0.7 cm, gray-brown. **Distr:** Tropical & subtropical Asia. **Ecol:** In montane sparse forests, shrublands; 300–700 m. Fl. Jan–Jul. **Zones:** 10–12. **Info:** Most parts of the plant are very poisonous, especially the seeds, but are used for medicinal purposes.

Paracroton

Species 3. **Trees**, stellate-hairy or lepidote; branchlets terete. **Stipules** triangular to subulate. **Leaves** alternate; lamina elliptic, oblong-elliptic, obovate to oblanceolate, base cuneate, margins glandular-dentate or serrate, apex acuminate or cuspidate, lateral nerves 6–20 pairs; petiole slightly pulvinate at apex and base, biglandular at apex. **Inflorescences** ♂ terminal and axillary, rarely leaf-opposed, narrowly thyrsoid, fasciculate or on short densely bracteate cymules, fragrant, bracts triangular, subulate or deltoid; ♀ terminal, thyrsiform, practically racemose. **Flowers** ♂ fragrant; calyx cupular, lobes 5, petals 5, free, oblong or spathulate, white, disk glands 5, 7 or 10, stamens 12–32, the outer free, the inner united, pistillode absent; ♀ solitary or 2–3 together at each node, calyx as in ♂, 5-lobed, lobes orbicular, obovate or triangular, dorsally appendiculate (all or 1–2) as in ♂, petals 5, coherent, oblong or triangular or orbicular, ovary subglobose, trilocular, locules 1-ovuled, styles 3. **Fruits** capsules globose or slightly depressed, trilocular; seeds suborbicular, large. **Distr:** Peninsular India and Sri Lanka; Malesia. **Cult:** Propagate from scarified seeds or cuttings. Requires humus-rich, moist but well-drained soil in full or partial sun.

P. pendulus* ssp. *zeylanicus (Thwaites) N.P. Balakr. & T. Chakrab., *Kew Bull.* 48: 719, 1993. (Syn: *Ostodes zeylanica*, etc.). **Trees**, 10–30 m tall. **Leaves** oblong to elliptic-oblong or obovate to oblong-oblanceolate, (11–)16–50 × (3–)5–15 cm, base usually cuneate or acute, margins glandular-dentate or -serrate, apex usually acuminate (acumen 1–3 cm long) to cuspidate. **Inflorescences** ♂ terminal and axillary, fascicled, 40–65 cm long, few-branched (lowermost

Above: *Paracroton pendulus* ssp. *zeylanicus*, *Jatropha curcas*, *Jatropha gossypiifolia*, *Jatropha integerrima*.

branch 25–45 cm long); ♀ terminal, thyrsiform, flowers 2–3 together or solitary, densely reddish-furfuraceous or tomentellous, 18–50 cm long. **Flowers** ♂ with calyx 4.5–7 × 7–8 mm, densely reddish-furfuraceous outside, lobes (4–)5, orbicular to spathulate, 3–5 × 3–6 mm, petals 5, oblong or spathulate-obovate, 6–9 × 2.5–4 mm, disk glands 5 or 10, 0.7–1 × 1–1.5 mm, stamens 14–32; ♀ densely tomentellous, calyx 4.5–8 × 5–8 mm, lobes 5, orbicular, ovate or triangular, 3.5–6 × 2–6 mm, dorsally appendiculate (all or 1–3), petals 5, oblong, triangular or orbicular-obovate, 6–8 × 2–4 mm, deciduous, ovary subglobose or trigonous-ovoid, 2.5–3.5 × 3–4 mm, styles 3, 1–3 mm long. **Fruits** capsules, globose, unlobed or scarcely trilobed, 2.5–3(–3.5) cm Ø; seeds suborbicular, ~1.9 × ~1.7 cm, dark brown. **Distr:** SW India, Sri Lanka. **Ecol:** Fl. & Fr. Jan–Dec. **Zones:** 10–12.

TRIBE JATROPHEAE

Jatropha

Species ~175. **Trees**, **shrubs**, subshrubs, or herbs; stems with laticifers, latex clear, white, or reddish; indumentum simple, sometimes glandular. **Stipules** persistent or deciduous, entire or setiform, or a series of stalked glands. **Leaves** alternate, unlobed or palmately lobed or parted, usually with glands at apex of petiole; venation palmate (or pinnate). **Inflorescences** with flowers monoecious or dioecious in paniculate dichasia. **Flowers** ♂ with sepals 5, imbricate, slightly basally connate, petals 5, imbricate, free, sometimes coherent or connate at base, disk glands 5, free or connate into ring, stamens 8–12, sometimes more, in 2–6 series, filaments at least partly connate, sometimes inner filaments connate into a column, pistillode filamentous or absent; ♀ flowers with sepals 5(or 6), free, imbricate, usually persistent in fruits, disk annular, lobed, or dissected, sometimes with staminodes, ovary 2- or 3(or 4 or 5)-locular, ovules 1 per locule, styles unlobed or bifid, sometimes dilated. **Fruits** capsules; seeds caruncułate, episperm crustaceous, endosperm fleshy, cotyledon broad and flat. **Distr:** Predominantly in tropical and subtropical America; Africa. **Cult:** Propagate from cuttings or seeds. Requires a fertile, well-drained soil in full or partial sun.

J. curcas L., *Sp. Pl.* 2: 1006, 1753. (Syn: *Manihot curcas*). [Barbados Nut, Physic Nut, Purging Nut]. **Shrubs** or **treelets**, 2–5 m tall; exudate watery latex; bark smooth; branches glaucous-gray, glabrous, sparsely lenticellate, pith large. **Stipules** small. **Leaves** rotund to ovate, 7–18 × 6–16 cm, papery, nitid green and glabrous adaxially, gray-green and along nerves puberulent to glabrous abaxially, base cordate, apex acute, veins palmate, 5–7; petiole 6–18 cm long. **Inflorescences** axillary, 6–10 cm long; bracts lanceolate, 4–8 mm long; ♀ flowers with pedicels elongate. **Flowers** ♂ with sepals 5, ~4 mm long, basally connate, petals oblong, green-yellow, ~6 mm long, connate to middle, hairy inside, disk glands 5, nearly terete, stamens 10, outer 5 filaments free, inner filaments connate in lower part; ♀ with sepals free, ~6 mm long, petals and disk glands as in ♂, ovary triloculed, glabrous, styles bifid at apex. **Fruits** capsules, ellipsoidal or globose, 2.5–3 cm across, yellow; seeds ellipsoidal, 1.5–2 cm long, black. **Distr:** Tropical America. **Ecol:** Fl. Sep–Oct. Fr. Oct–Dec. **Zones:** 10–12.

J. gossypiifolia L., *Sp. Pl.* 1006, 1753. [Bellyache Bush, Cotton Leaf, Red Leaf]. **Herbs**, annuals or perennials, usually herbaceous throughout but sometimes suffrutescent below, to 1 m tall; branched. **Stipules** dissected into numerous gland-tipped filiform divisions. **Leaves** long-petiolate, membranous, petiole bearing numerous branched hairs with gland-tipped divisions; lamina 7–15 × 7–15 cm, base cordate, margins deeply 3–5-lobate, ± pubescent or glabrous, lobes acute or acuminate, denticulate, glandular-ciliate; young leaves bronze-red. **Inflorescences** small pedunculate cymes; bracts linear-oblong, margins glanduliferous. **Flowers** small, greenish; sepals ovate, glandular-ciliate, pubescent outside, 5–7 mm long; petals 5, obovate, purple, somewhat longer than sepals; stamens usually 8; ovary pubescent. **Fruits** capsules 1 cm Ø, trisulcate, glabrate; seeds oblong, brown, caruncułate. **Distr:** S Mexico, Honduras and Salvador to Panama; West Indies; S America. **Ecol:** In moist or dry fields or thickets, frequently on sand or gravel along streams; 200–400 m. **Zones:** 10–12.

J. integerrima Jacq., *Sel. Stirp. Pl. Amer.* 256, t. 183, f. 47, 1763. (Syn: *J. hastata*). [Peregrina, Rose-flowered Jatropha]. **Shrubs** or **small trees**, 1–4 m tall; leafy stems becoming reddish-brown. **Stipules** 0.5–1 mm long, glandlike or broadly triangular. **Leaves** oblong or ovate-oblong to ovate, 6–15 × 3–8(–13) cm, sometimes with 1–2 short (~7 mm long) or prominent (1–2 cm long) lateral lobes along margin in proximal half of lamina, base rounded and truncate or slightly cordate, margins entire or basally sparsely glandular, apex acuminate; petiole 3–8 cm long. **Inflorescences** terminal or pseudo-axillary, often bisexual, 15–20 cm long; peduncles 10–14 cm long, cymose branches 2–7 cm long; bracts 0.3–1.2 cm long, margins dark-glandular; pedicels 5–9 mm long. **Flowers** ♂ with calyx 3–5 mm long, lobes 1–2 mm long, triangular with blunt apex, petals 1.2–2.2 × 0.8–1 cm, obovate, dark red or rose, stamens 10, staminal column ~4 mm long, filaments ~3 mm long; ♀ with perianth similar to ♂, ovary 2–2.5 mm long, glabrous, stylar column ~2.5 mm long, style branches ~3 mm long, bifid. **Fruits** 1.2–1.3 cm Ø, subglobose, apex truncate, longitudinally trisulcate; seeds ~1 × 0.5–0.58 cm, 2-parted. **Distr:** Cuba. **Ecol:** In dry, sandy and disturbed areas. Fl. throughout most of the year. **Zones:** 11–12.

J. multifida L., *Sp. Pl.* 2: 1006, 1753. [Coral Bush, Coral Plant, Guatemala Rhubarb, Physic Nut]. **Shrubs** or **treelets**, 2–3(–6) m tall; stems glabrous. **Stipules** divided into forked setiform, to 2 cm long. **Leaves** orbicular in outline, 10–30 cm across, margins palmately 9–11-lobed, lobes entire, adaxially green, abaxially gray-green, surfaces glabrous, venation pinnate; petiole 10–25 cm long. **Inflorescences** terminal; peduncles 13–20 cm long; pedicels short; flowers dense. **Flowers** ♂ with calyx 2–3 mm long, lobes 5, rotund, glabrous, sepals 5, spathulate, red, ~4 mm long; stamens 8, filaments basally connate, anthers elongate; ♀ flowers with calyx as in ♂, sepals 6–7 mm long, red, ovary glabrous, styles 3, connate in lower half. **Fruits** capsules, ellipsoidal to obovate, ~3 cm long, glabrous. **Distr:** Tropical and subtropical America. **Ecol:** Fl. Jul–Dec. Fr. Sep–Feb. **Zones:** 9–12.

J. podagrica Hooker, *Curtis's Bot. Mag.* 74: t. 4376, 1848. [Bottle Plant]. **Shrubs**, erect, to 2 m tall; stem woody, swollen at base or lower part; branches stout and short, fleshy, sparsely lenticellate. **Stipules** spiniform, persistent leaf scar large and prominent. **Leaves** with petiole 8–10 cm, glabrous; lamina peltate, rotund to elliptic, 8–18(–25) × 6–16 cm, nitid green adaxially, gray-green abaxially, glabrous on both surfaces, base truncate or obtuse, margins entire or shallowly 2–6-lobed, apex obtuse, palmate veins 6–8. **Inflorescences** terminal, long-pedunculate, branches short, red. **Flowers** with calyx ~2 mm long, sepals rotund, ~1 mm across; petals obovate-oblong, ~6 mm long, red; stamens 6–8, basally connate; ovary glabrous, styles 3, basally connate, apex bifid. **Fruits** capsules, ellipsoidal, 1.3–1.8 × ~1.5 cm, with 3 longitudinal grooves; seeds ~1.1 cm long, smooth. **Distr:** C America. **Ecol:** Fl. & Fr. throughout most of the year. **Zones:** 10–12.

Joannesia

Syn: *Anda*, *Andicus*.

Species 2. **Trees**, large, evergreen or briefly deciduous; young parts usually rusty-tomentose. **Leaves** palmate; leaflets 3–7. **Inflorescences** panicles, terminal. **Flowers** monoecious, yellowish to white; calyx campanulate, 5-toothed; petals 5; ♂ with stamens 8, filaments unequal, monadelphous at base; ♀ with style bifid, stigmas 2. **Fruits** large, fleshy, with a large 2-celled stone. **Distr:** S America. **Cult:** Propagate by air-layering or from seeds. Requires a well-drained soil in full sun.

J. princeps Vell., *Alogr. Alkalis*: 199, 1798. (Syn: *Anda brasiliensis*, etc.). [Arara Nut Tree, Cutieira]. **Trees**, large, evergreen, sometimes briefly deciduous, lactescent, 15–30 m tall; young parts reddish-pubescent. **Leaves** alternate, palmately compound, exstipulate, initially rusty-pubescent, later glabrous;

Above: *Jatropha multifida, Jatropha podagrica, Joannesia princeps, Aleurites moluccana.*

leaflets usually 5–6, ovate, 5–10(–20)+ cm long, margins entire, apex acuminate, adaxially shiny, abaxially dull; petiole to 7.5 cm long, shorter on lower branches, 2 sessile glands at petiolule junctions, petiolules to 2.5 cm long. **Inflorescences** bisexual, panicles, terminal, 10–15 cm long; ♂ flowers pedicellate, bracts 2, glands 2; ♀ flowers sessile. **Flowers** ~2.5 cm across, petaliferous, white; calyx campanulate, yellowish; petals 5, alternating with calyx teeth, lanceolate in ♂, rounded in ♀; ♂ with stamens 8; ♀ with styles 2, ovary ovoid. **Fruits** large, ovoid, resembling a coconut, ash-colored, slightly 4-angled with 4 prominent ridges, 12–15 × 8–10 cm, pericarp thick, dehiscing into 4 segments, stone large, tough; seeds 2–3, 2–3 cm Ø. **Distr:** E Brazil. **Ecol:** In rainforests of coastal mountains; 230–515 m. Fl. Mar–Apr. Fr. Nov–Dec. **Zones:** 10–12. **Info:** The oil from the seeds (poisonous) and the bark are used for medical purposes.

TRIBE ALEURITIDEAE

Aleurites

Syn: *Camirium.*

Species 2. **Trees**, evergreen, monoecious; indumentum of stellate hairs; sap watery. **Stipules** minute, soon lost. **Leaves** alternate; lamina simple, entire or 3–5-lobed, palmately 3–5-veined; petiole long, biglandular at apex. **Inflorescences** terminal, branched, conical, lower bracts subtending clusters of ♂ flowers, each major axis terminated by solitary ♀ flower; buds subglobose. **Flowers** ♂ with calyx splitting irregularly into 2 or 3(or 4) valvate lobes, petals 5, white or cream, disk glands 5, thick, stamens 15–32 in 3 or 4 series, outer ones free, inner ones united into column; anthers bilocular, introrse, pistillode absent; ♀ similar to ♂ but slightly larger, pedicellate, ovary 2(or 3)-locular; ovules 1 per locule, styles bifid. **Fruits** drupes, subglobose, exocarp thinly fleshy, indehiscent, endocarp woody, 1–2(–5)-seeded; seeds compressed globose, woody. **Distr:** Tropical and subtropical regions of Asia and Oceania. **Cult:** Propagate from seeds or mature cuttings. Requires any fertile, well-drained soil in full sun. Some species are drought-resistant.

A. moluccana (L.) Willdenow, *Sp. Pl.* 4: 590, 1805. (Syn: *A. triloba*, etc.). [Candle Berry, Candle Nut Tree, Indian Walnut, Varnish Tree]. **Trees**, evergreen, to 20 m tall; bark dark gray, shallowly dehiscent to nearly smooth. **Stipules** minute, subulate. **Leaves** usually ovate to elliptic-lanceolate (entire or up to 5-lobed), usually 14–20 × 7–17 cm, base usually broadly cuneate or obtuse, with 2 adaxial glands, margins entire or shallowly repand-dentate, apex acute to acuminate, papery; petiole 6–12(–22) cm long. **Inflorescences** panicles, terminal, (10–)15–20 × 10–14 cm. **Flowers** ♂ with calyx lobes ovate, 3–3.5 × 2–3 mm, acute, petals narrowly oblong-spathulate, 6–8 × ~2 mm, obtuse, yellowish-white, stamens 15–20, inner filaments connate into hairy column, ~4 mm long; ♀ with calyx and petals as in ♂ but petals ~1.5 mm wide, disk ~2 mm across, flat, ovary 2(or 3)-locular, styles 2. **Fruits** drupes, subglobose or slightly obliquely globose, 4–5 × 4–6 cm; seeds 1 or 2, broadly ovoid, ~2.5 × 2.8 cm. **Distr:** Tropical & subtropical Asia to NE Queensland. **Ecol:** In mixed evergreen forests; 100–1,000 m. Fl. Apr–Oct. Fr. Oct–Dec. **Zones:** 10–12. **Info:** The seeds are rich in volatile oil which burns like a candle, hence the common name. The oil also has medicinal and commercial uses.

A. rockinghamensis (Baill.) P.I. Forst., *Muelleria* 9: 8, 1996. (Syn: *A. moluccana* var. *rockinghamensis*). [Candlenut]. **Trees**, large, to 40 m tall; branches spreading-pendulous and ascending; bark smooth, gray. **Stipules** ~1 mm long. **Leaves** simple, unlobed or trilobed, 11–40 × 7–30 cm; petiole 3–21 cm long, glands at apex, 0.5–2 mm Ø. **Inflorescences** robust, 7.5–38 × 7.5–30 cm, axes densely fulvous- to ferruginous-pubescent; bracts 3–5 × 0.8–2 mm; ♂ flowers with pedicels 0.5–1.5 cm long, pubescent; ♀ flowers with pedicels 2–5.5 mm long, pubescent. **Flowers** ♂ 0.5–1.2 × 0.6–1.2 cm, calyx lobes 3–5.8 × 2.5–5 mm, petals 0.45–1 × 0.2–0.4 cm, stamens 20–32; ♀ 0.8–1.2 × 0.8–1.2 cm, calyx lobes 4–7 × 3–6 mm, petals 0.9–1.3 × 0.25–0.4 cm, ovary subglobose, trigonous (rarely tetragonous) in cross-section, 3(4)-loculed, 2–3 × 2.5–3.5 mm, densely tomentose, styles bifid, 2.6–3.2 mm long. **Fruits** broadly ovoid-subglobose with 6(8) longitudinal ridges, 4–6.5 × 5.5–8 cm; seeds subglobose to obovoid, 2–2.7 × 2–2.5 cm, white to cream, brown-mottled. **Distr:** Papua New Guinea to NE Queensland. **Ecol:** In rainforests; <1,200 m. **Zones:** 10–12.

Reutealis

Species 2. **Trees**, deciduous, monoecious or dioecious. **Leaves** alternate; lamina simple, ovate to cordate, membranous, glabrous, basal veins 5–7, biglandular; petiole long. **Inflorescences** terminal, thyrsoid, many-flowered, densely gray-tomentose; bracts cucullate, deciduous; pedicels very short. **Flowers** ♂ with calyx spatheate-cupular, bi- or trilobed; petals 5, uniformly colored, obovate-spathulate, basally sericeous-pubescent, stamens 7–13, biseriate, anthers large, elliptic, extrorse, connectives broader; ♀ with calyx and petals as in ♂, ovary ovoid, 3- or 4-locular, sericeous, styles 3 or 4, bifid. **Fruits** drupes, large, subglobose, 3- or 4-locular, 3- or 4-angled, 3- or 4-seeded. **Distr:** China, Philippines. **Cult:** Propagate from seeds or mature cuttings. Requires any fertile, well-drained soil in full sun. Some species are drought-resistant.

R. trisperma (Blanco) Airy Shaw, *Kew Bull.* 20: 395, 1967. (Syn: *Aleurites trisperma*, etc.). [Otaheite Walnut, Lumbang Tree]. **Trees**, dioecious, sometimes monoecious, to 15 m tall; trunk to 35 cm dbh. **Leaves** ovate or ovate-cordate, 12–14 × 12–13 cm, base usually deeply cordate with rounded lobes, sometimes rounded or obtuse on old trees, apex gland-tipped,

Below: *Aleurites rockinghamensis, Reutealis trisperma* (flowers), *Reutealis trisperma* (fruits).

TG

IAA

Above: *Vernicia montana, Garcia nutans, Baloghia inophylla.*

abaxially with hair-tufts in many vein axils; petiole 14–15 cm long. **Inflorescences** ♂ flat-topped, with many cymelets; ♀ flat-topped, racemose, or thyrsoid. **Flowers** ♂ with calyx cylindric, ~1.3 cm long, bi- or trilobed, glabrous, petals 5, yellowish-white, obovate, obovate-spathulate, or obovate-lanceolate, 0.8–1.6 × 0.3–0.5 cm, stamens 10–13, biseriate, outer filaments free, inner ones basally connate; ♀ with ovary 3-celled, ovoid-globose-trigonous, densely hairy, styles bifid. **Fruits** capsules, depressed-globose to obtusely trigonous, pericarp 4–5 × 5–6.5 cm, longitudinally 3-angular, rugose, densely velutinous-pubescent. **Distr:** Philippines. **Ecol:** Fl. Apr. Fr. Oct. **Zones:** 11–12. **Info:** Cultivated mainly for medicinal uses and oil from the seeds.

Vernicia

Syn: *Dryandra*.

Species 3. **Trees**, monoecious or dioecious; indumentum of simple or T-shaped hairs, often sparse. **Stipules** caducous. **Leaves** alternate; lamina simple, entire, palmately veined; petiole long, with 2 adaxial glands at apex. **Inflorescences** terminal, much-branched, thyrsoid, cymules of bisexual inflorescences each with several ♂ flowers and a terminal ♀ flower; bracts inconspicuous. **Flowers** ♂ with buds ovoid or subglobose; calyx ± spatheate, ± regularly bi- or trilobed, petals 5, white or reddish-white to purple, base clawed, disk segments 5, subulate; stamens 8–12, biseriate, outer filaments free, inner ones longer and basally connate; ♀ flowers with sepals and petals as in ♂, disk obscure or absent, ovary densely pubescent, 3(–8)-locular, ovules 1 per locule, styles 3–5, bifid. **Fruits** drupes, large, subglobose, beaked at apex, indehiscent or slightly dehiscent at base, 3-seeded; exocarp crustaceous; seeds without caruncle, seed coat woody. **Distr:** Myanmar to Indonesia, China, Japan. **Cult:** Propagate from seeds or mature cuttings. Requires any fertile, well-drained soil in full sun. Some species are drought-resistant.

V. montana Loureiro, *Fl. Cochinch.* 2: 586, 1790. (Syn: *Aleurites montana*, etc.). [Mu Tree, Mu Oil Tree]. **Trees**, evergreen, to 20 m tall, usually dioecious; bark brown; branches lenticellate. **Leaves** broadly ovate, 8–20 × 6–18 cm, base cordate to truncate, margins entire or 2–5-fid, apex acute to acuminate, sinus usually with cupular glands, pubescent when young; petiole 7–17 cm long, apex with 2 glands. **Inflorescences** produced with new leaves, usually unisexual. **Flowers** ♂ with calyx ~1 cm long, bifid or trifid, glabrous, petals obovate, 2–3 cm long, base clawed, white or purple-red at base and with purple-red stripes, stamens 8–10, filaments hairy, outer filaments free, inner filaments connate below middle; ♀ with calyx and petals as in ♂, ovary densely brown-pubescent, trilocular, styles 3, bipartite. **Fruits** drupes, ovoid, 3–5 cm ∅, longitudinally 3-angular, between angles with sparsely reticulate wrinkles, 3-seeded; seeds compressed-globose; testa thick, verrucose. **Distr:** China, Myanmar, Taiwan, Thailand, Vietnam. **Ecol:** In open forests; <1,600 m. Fl. Apr–Jun. Fr. Jul–Oct. **Zones:** 9–12. **Info:** The oil obtained from the seeds are used in the manufacture of paints and varnishes.

Garcia

Species 2. **Trees** or **large shrubs**. **Leaves** alternate, exstipulate, petiolate; lamina with margins entire, firm-membranous, penninerved. **Inflorescences** terminal, 1–3-flowered. **Flowers** unisexual, rather large, petaliferous; ♂ calyx membranous, globose in bud, valvately ruptured at anthesis into 2–3 segments, petals 8–12, narrow, sericeous, longer than calyx, glands of disk free or basally connate, glabrous, stamens numerous, inserted on a convex pilose receptacle, filaments free; ♀ with calyx like the ♂, caducous, petals fewer than ♂, hypogynous disk deeply lobate, ovary 3-celled, cells 1-ovuled, style very short, stigmas thick, reflexed, emarginate-bifid. **Fruits** capsules, large, separating into bivalvate cocci, endocarp somewhat ligneous; seeds globose, not caruncluate. **Distr:** C America to S America, West Indies. **Cult:** Propagate from seeds or mature cuttings. Requires any fertile, well-drained soil in full sun.

G. nutans Rohr, *Skrivt. Naturh. Selsk. Kjoebenhavn* 2: 217. pl. 9, 1792. [Drooping Garcia]. **Trees**, to ~9 m tall; branchlets, petiole, and lower leaf surface finely puberulent at first, soon glabrate. **Leaves** oblong or oblong-obovate, mostly 10–15 × 4–6 cm, base acute or obtuse, apex abruptly acuminate or shortly acuminate; petiole slender, 2–5 cm long. **Inflorescences** cymes, 1–3-flowered; peduncles very short; pedicels almost twice as long as the calyx; flower buds globose. **Flowers** nutant; calyx 5 mm long, pubescent; petals lanceolate, acute, 0.8–1.2 cm long, whitish-sericeous, purplish inside; filaments villous to middle; ovary pubescent. **Fruits** capsules, glabrous, tricoccous, 2.5 cm across; seeds 1.2 cm ∅. **Distr:** W and S Mexico, Salvador, Costa Rica, West Indies, Colombia, Venezuela. **Ecol:** In forests; 300–500 m. **Zones:** 10–12.

TRIBE CODIAEAE

Baloghia

Species 15. **Trees** or **shrubs**, dioecious or monoecious, latex present; buds scaly. **Stipules** obsolete. **Leaves** opposite or alternate, glabrous, mostly with 2 embedded marginal glands 3–5 mm above base, pinnately veined, translucent dots visible in lamina with a lens. **Inflorescences** racemose, terminal, ♂ and ♀ flowers mixed or on separate plants; bracts deciduous. **Flowers** with calyx lobes 4 or 5, imbricate, shortly fused, persistent; petals 4 or 5, free, longer than sepals; ♂ flowers with stamens ~40–50, filaments shortly united at base; ♀ flowers with ovary trilocular, ovule 1 per locule, styles 3, free or shortly united, deeply bifid, stigma obscure. **Fruits** capsules, ± globose, trilobed, somewhat fleshy at first, becoming hard and separating from the persistent axis into three 1-seeded units; seeds oblong, with caruncle small or absent. **Distr:** Australia to New Caledonia. **Cult:** Propagate from seeds or by rooted cuttings. Requires a moist, well-drained soil in partial sun.

B. inophylla (G. Forst.) P.S. Green, *Kew Bull.* 41: 1026, 1986. (Syn: *Croton inophyllus*, *B. lucida*, etc.). [Brush Bloodwood, Ivory Birch, Scrub Bloodwood]. **Shrubs** or **trees**, to 7+ m tall, with red sap. **Leaves** opposite; lamina elliptic to broadly elliptic, sometimes slightly oblanceolate, (4–)5–11(–15) × (2–)3.5–6(–7) cm, base acute or rounded, with a marginal embedded gland on each side 1–5 mm above petiole, margins thickened, apex rounded, abruptly acuminate or acute, coriaceous, glabrous, shiny; petiole 0.4–1 cm long. **Inflorescences** terminal. **Flowers** with calyx lobes 5, ± ovate, 3–4 mm long, margins tomentose-ciliate; petals usually white 5, elliptic-lanceolate, 0.6–0.7(–1) cm long; stamens numerous, filaments in bundles, 1.5 mm long, anthers ~1 mm long; ovary broadly conical, 2–3 mm long, stylar arms curled, 2–3 mm long. **Fruits** globose, 1.2–2 cm ∅; seeds ovoid, ~1 cm long, dark brown or black. **Distr:** Australia (Queensland, NSW, Norfolk Island, Lord Howe Island), New Caledonia, Loyauté Island. **Ecol:** In rainforests; 400–900 m. Fl. May–Sep. **Zones:** 9–12.

Codiaeum

Species ~15. **Shrubs** or **treelets**, monoecious, rarely dioecious; indumentum when present simple; latex present, clear and nontoxic. **Stipules** minute, caducous, sometimes obsolete. **Leaves** alternate, petiolate; lamina with margins entire, rarely lobed, venation pinnate. **Inflorescences** axillary, long, unbranched; ♂ flowers several per bract; ♀ flowers 1 per bract. **Flowers** usually unisexual; ♂ with calyx (3–)5(–6)-lobed, lobes imbricate, petals small, 5 or 6, rarely absent, disk divided into 5–15 free glands, stamens 15–100, rudimentary ovary absent; ♀ with calyx 5-lobed, petals absent, disk subentire or divided, ovary trilocular, ovules 1 per locule, styles 3, unlobed, rarely bifid. **Fruits** capsules; seeds caruncluate. **Distr:** Indonesia and Philippines to New Guinea, tropical Australia, and Pacific Islands. **Cult:** Propagate from cuttings. Requires a moist but well-drained soil in full sun.

C. variegatum (L.) Rumphius ex A. Jussieu, *Euphorb. Gen.* 80, 111, 1824. (Syn: *Codiaeum chrysosticton*, etc.). [Croton]. **Shrubs**, to 3 m tall; branches glabrous, with prominent leaf scars. **Leaves** very variable in shape and color; lamina linear, linear-lanceolate, oblong, elliptic, lanceolate, ovate, spathulate, or fiddle-shaped to obovate, sometimes divided to midvein into upper and lower parts, 5–30 × (0.3–)0.5–8 cm, base cuneate to obtuse, apex acute or acuminate to

Above: *Codiaeum variegatum* (male flowers), *Codiaeum variegatum* (female flowers), *Dimorphocalyx glabellus* (male flowers).

rounded, thinly leathery, green, purple-red, and/or yellow, sometimes with colored spots or stripes of spots; petiole 0.2–2.5 cm long. **Inflorescences** 8–30 cm long; ♂ flowers with pedicels delicate; ♀ flowers with pedicels slightly thickened. **Flowers** ♂ white, petals 5, smaller than sepals, disk glands 5, stamens 20–30; ♀ yellowish, sepals ovate-triangular, disk annular, styles recurved, entire. **Fruits** capsules, subglobose, slightly flattened, glabrous, ~9 mm Ø; seeds ~6 mm long. **Distr:** Malesia to SW Pacific. **Ecol:** Fl. Sep–Oct. Fr. Nov–Dec. **Zones:** 10–12.

Dimorphocalyx

Species ~15. **Trees** or **shrubs**, dioecious, rarely monoecious; much-branched; indumentum of simple hairs, sparse. **Leaves** alternate, petiolate; lamina simple, margins entire, venation pinnate, coriaceous, stipulate. **Inflorescences** terminal or axillary, racemelike thyrses or cymes, usually few-flowered, or ♀ solitary. **Flowers** ♂ with calyx cupular, rim 5-toothed or shortly 5-lobed, petals 5, free, disk glands 5, alternating with petals, stamens 10–15(–20) in 2 series, outer free, inner ones connate into column, pistillode absent; ♀ flowers with calyx 5-partite (almost to base, lobes oblong, petaloid), imbricate, accrescent, persistent, petals 5, shorter than calyx lobes, disk annular, ovary trilocular, ovules 1 per locule, styles 3, basally connate, each style bifid. **Fruits** capsules, trilocular; endocarp crustaceous; seeds ellipsoidal to ovoid, endosperm fleshy, cotyledon broad and flat. **Distr:** Mainly in tropical Asia (India and Sri Lanka to S China, Malaysia and Borneo), extending to Australia. **Cult:** Propagate from seeds or by air-layering. Requires a fertile, moist soil in shade or partial sun.

D. glabellus Thw., *Enum. P1. Zeyl.* 278, 1861. **Shrubs** or **small trees**, monoecious, 2–8(–12) m tall, much-branched; branches sometimes heavily lenticellate, or ridged with thick lines of cork. **Stipules** ~0.5–2 × 0.8–1 mm, broadly deltoid, acute, persistent. **Leaves** alternate, glabrous; lamina ovate to narrowly elliptic to lanceolate-elliptic, 5–9(–14) × 1.5–3.5(–5.5) cm, base rounded or cuneate, margins usually entire, apex acute or obtuse; petiole 0.5–1(–2) cm long. **Inflorescences** ♂ with pedicels 3–5 mm long, glabrous; ♀ with pedicels to 1.6 cm long. **Flowers** ♂ with calyx cupular, 1–1.5 mm deep, shortly pubescent, petals white, oblong, 0.6–1 × 0.2 cm, reflexed, stamens 10–11(–14), to 8 mm long, biseriate, usually 5 outer shorter and free, inner 5 connate most of their length, with one longer and centrally placed; ♀ with calyx lobes pale green, oblong, 0.6–1.3 × 0.2–0.3 cm, rounded or emarginate, petals smaller than sepals, ovary ~4 × 3 mm, conic, styles 5 mm long. **Fruits** capsules, ~1.3 × 1 cm; seeds oblong, 6–7 mm long, mottled. **Distr:** S India, Sri Lanka. **Ecol:** In hot, dry areas. **Zones:** 11–12.

Subfam. ACALYPHOIDEAE

Outer integument 3–6(–16) cells across, inner integument 3–24 cells across; (testa with vascular bundles) (Stevens 2001 onwards). **Distr:** Pantropical.

Tribe ACALYPHEAE

Macaranga

Species 240. **Trees** or **shrubs**, mostly dioecious (some forms of *M. andamanica* monoecious); indumentum usually of simple hairs and whitish to reddish glandular-scaly. **Stipules** small or enlarged; petiole insertion often peltate. **Leaves** alternate; lamina undivided or lobed, base with glands, abaxially glandular-scaly; venation palmate or pinnate. **Inflorescences** axillary, sometimes at old leafless nodes, unbranched, branched, or ♀ capitate or 1-flowered, unisexual, very rarely bisexual (*M. andamanica*); bracts sometimes with glands. **Flowers** 1 or few to many per bract. **Flowers** ♂ with calyx closed in bud, sepals 2–4, valvate, petals absent; disk absent, stamens (1–)2–22(–30), filaments free or shortly connate, anthers 3- or 4-locellate; pistillode absent; ♀ with calyx cup-shaped or urceolate, apex denticulate or lobed, sometimes truncate, caducous or persistent, petals absent, disk absent, ovary 1–6-locular, styles simple, free, rarely shortly connate, stigmatic surface papillose. **Fruits** capsules, (1 or)2- or 3(–6)-locular, smooth or spiny, often glandular scaly; seeds subglobose, smooth. **Distr:** Africa, tropical Asia, China, Australia, Madagascar, Pacific Islands. **Cult:** Propagate from seeds or cuttings. Requires moist but well-drained soil in full sun.

M. peltata (Roxb.) Müll.Arg., *Prodr.* 15(2): 1010, 1866. (Syn: *Mappa peltata*, etc.). **Trees**, small, to 15 m tall; branchlets stout, glabrous; younger parts rusty-tomentose. **Stipules** lanceolate-acuminate, 0.9–1.2 cm long, caducous. **Leaves** broadly ovate, 10–22(–32) × 7.5–18(–26) cm, margins entire, apex acute to caudate, shortly acuminate, adaxially mostly glabrous, abaxially pubescent, palmately veined with 7–9(–10) pairs of lateral veins above base; petiole 4–18 cm long. **Inflorescences** ♂ to 10 cm long, strict, shortly pubescent, flowers sessile, fascicled at intervals along axis, bracts ~2–2.5 × 4 mm, broadly ovate, toothed, densely rusty-pubescent; ♀ 3–4.5 cm long, shorter than ♂, densely pubescent, pedicels 2–3(–8) mm long, bract ovate, ~4 × 2.5–3 mm, densely pubescent. **Flowers** ♂ with stamens 2–5; ♀ with calyx lobes lacking, ovary ~2.5 mm Ø, globose, 1-celled, black-glandular, style 1. **Fruits** spherical, ~5 mm Ø, with persistent stigma on one side, black-glandular; seed ~3 mm Ø, globose, rough. **Distr:** Andaman Islands, India, Sri Lanka; Sikkim to Thailand. **Ecol:** In secondary forests; to 750 m. **Zones:** 10–12.

Acalypha

Species ~450. **Herbs**, **shrubs** or **trees**, mostly monoecious, sometimes dioecious; indumentum of simple or sometimes gland-tipped hairs. **Stipules** lanceolate or subulate, sometimes minute, deciduous. **Leaves** alternate; leaf lamina simple, margins crenate or dentate, rarely subentire; venation pinnate or palmate. **Inflorescences** axillary or terminal, mostly unbranched, bisexual or unisexual, sexes very diversely arranged, mostly bisexual with ♂ flowers sessile, in fascicles distally along slender axis, base with 1 to few ♀ flowers; ♀ flowers mostly enclosed within prominent

Below: *Dimorphocalyx glabellus* (female flowers), *Macaranga peltata*, *Acalypha chamaedrifolia*, *Acalypha hispida*.

leafy bracts, flowers 1–3 per bract, often sessile, bracts often dentate or lobed, accrescent in fruit. **Flowers** ♂ with sepals 4, valvate, membranous, petals absent, disk absent, stamens often 8, filaments free, anthers bilocular, anther-thecae pendulous and vermiform, pistillode absent; ♀ sepals 3–5, shortly connate, imbricate, ovary (2 or)trilocular, styles mostly free, laciniate. **Fruits** capsules, bi- or trilobed, small; seeds subglobose or ovoid, smooth, caruncle and aril absent. **Distr:** Widely distributed in tropical and subtropical regions. **Cult:** Propagate from cuttings. Needs well-drained moist soil in full sun.

A. chamaedrifolia (Lam.) Müll.Arg., *Prodr.* 15(2): 879, 1866. (Syn: *A. reptans*, etc.). [Red Cat Tails]. **Herbs**, suffrutescent, perennial or annual, much-branched from base; stems and branches, mostly prostrate, pubescent or glabrate, 5–30 cm long. **Leaves** broadly ovate to oblong-lanceolate, 5–7 × 3–4.5 cm, base truncate or slightly cordate, margins serrate or crenate-dentate, apex acute or cuspidate-acuminate, membranous, appressed-pubescent, flaccid, 5-nerved; petioles as long as lamina or longer. **Inflorescences** spikes, terminal, closely flowered, 5 cm long; ♀ bracts 1-flowered, ovate-triangular, 7–9 mm long, 3–5-lacerate to base, membranous. **Flowers** with calyx lobes ovate, acute; ovary hirsute above, styles elongated, delicate, attenuate, entire or shortly bifid, carpels slightly keeled. **Fruits** capsules, ~2 mm Ø; seeds ovoid, apiculate, brown, 1–3 mm long, testa very minutely pitted. **Distr:** S Florida to Caribbean. **Ecol:** Usually on rocky places and hillsides. **Zones:** 10–12.

A. hispida N.L. Burman, *Fl. Indica*, 303, 1768. [Chenile Plant, Foxtail, Red Hot Cat Tail]. **Shrubs**, dioecious, 1–3 m tall; branchlets grayish-tomentulose when young. **Stipules** triangular, 0.6–1 cm long, pilose. **Leaves** broadly ovate or ovate, 8–20 × 5–14 cm, base broadly cuneate, obtuse, or subcordate, margins coarsely serrate, apex acuminate or acute, papyraceous, basal veins 3–5; petiole 4–8 cm long. **Inflorescences** ♀ 3–7 per bract, fascicled, subsessile, spikes 15–30 cm long, axillary, pendulous, peduncle short, pubescent, bracts scattered, ovate-rhombic, ~1 mm long, entire. **Flowers** ♂ unknown; ♀ sepals (3 or)4, subovate, ~1 mm long, acute, puberulent, ovary subglobose, densely hirtellous, styles 3, 5–7 mm long, laciniate, crimson or red-purple. **Distr:** Cultigen of unknown origin, possibly Indonesia or W New Guinea. **Ecol:** Fl. Feb–Nov. **Zones:** 11–12. **Info:** Only ♀ plants are known, and cultivated (widely throughout the tropics). *A. hispida* var. *alba* has off-white or greenish-white flowers.

A. indica L., *Sp. Pl.* 2: 1003, 1753. [Copperleaf, Indian Acalypha, Indian Copperleaf, Indian Nettle, Ricinela]. **Herbs**, annual, 0.5–1 m tall, monoecious; branchlets adpressed-pubescent. **Stipules** triangular, ~1 mm long. **Leaves** often in apical rosettes; lamina rhombic-ovate or subovate, 2–3.5 × 1.5–2.5 cm, base cuneate, margins serrate, apex acute; petiole 1.5–3.5 cm long. **Inflorescences** axillary, unbranched, 2–7 cm long; ♀ bracts proximal, 3–7, 1- or 2-flowered, ovate-cordate, ~5 mm long; distal ♂ portion short, slender; ♂ flowers 5–7 per bract, pedicels ~0.5 mm long, bracts ovate-triangular or broadly triangular, ~0.5 mm long; allomorphic ♀ flower at apex, subsessile. **Flowers** ♂ with sepals 4, ovate, ~0.4 mm long, stamens 8; ♀ with sepals 3, triangular, ~0.5 mm long, ciliate, ovary pilose, styles 3, 2.5–3 mm long, 5-laciniate. **Fruits** capsules, trilocular, ~2 mm Ø, pilosulose; seeds ovoid, ~1.5 mm long, minutely puncticulate. **Distr:** Old World tropics and subtropics. **Ecol:** In grasslands, wastelands, roadsides; <100 m. Fl. & Fr. Mar–Oct. **Zones:** 11–12.

A. wilkesiana Müll. Arg., *Prodr.* 15(2): 817, 1866. [Copperleaf, Beefsteak Plant, Jacob's Coat]. **Shrubs**, 1–3 m tall, monoecious; young branchlets pubescent. **Stipules** narrowly triangular, ~8 mm long. **Leaves** broadly ovate, 10–18 × 6–12 cm, base obtuse, margins coarsely crenate, apex acuminate, often colored or red- or purple spotted, abaxially pilose along veins, basal veins 3–5; petiole 2–3 mm long, pilose. **Inflorescences** axillary, unbranched, unisexual; ♂ glomerate, 9–17-flowered, 10–20 cm long, pilose, bracts lax, ovate, ~1 mm long, pedicels ~1 cm long; ♀ 5–10 cm long, 1(or 2) flowers in each bract, sessile, peduncles ~2 cm long, bracts broadly ovate, ~5 × 8 mm, 7–11-serrate. **Flowers** ♂ with sepals 4, ovate, ~0.7 mm long, stamens 8; ♀ with sepals 3 or 4, ovate-lanceolate, 0.5–1 mm long, ciliate, ovary pubescent, styles 3, 6–7 mm long, 9–15-laciniate. **Fruits** capsules, ~4 mm Ø, sparsely villous; seeds globose, ~2 mm Ø, smooth. **Distr:** Malesia. **Ecol:** Fl. throughout the year. **Zones:** 11–12.

Ricinus

Species 1. *Generic description same as specific.* **Cult:** Propagate from seeds. Requires a fertile, moist, well-drained soil in full or partial sun.

R. communis L., *Sp. Pl.* 2: 1007. 1753. (Syn: *Croton spinosus*, etc.). [Castor Bean, Castor Oil Plant, Mole Bean, Ricin, Wonder Tree]. **Herbs** erect, variable (with numerous forms and varieties described); often single-stemmed, sometimes bush- or treelike, 2–5(–10) m tall; younger parts glaucous, whole plant often reddish or purplish. **Stipules** connate, 2–3 cm long. **Leaves** palmately 7–11-lobed, 0.3–0.5(–1) × 0.3–0.5(–1) m, margins serrate; petiole 20–40 cm long. **Inflorescences** to 30 cm long; ♂ flowers with pedicels 0.5–1.7 cm long; ♀ flowers with pedicels 0.5–1 cm long. **Flowers** ♂ with calyx lobes 5–8 × 3–5 mm, stamens 7–8 mm long; ♀ with sepals ~5 mm long, styles 2–5 mm long. **Fruits** capsules, ellipsoid or ovoid, 1.5–2.5 cm long, echinate, spines to ~5 mm long, sometimes smooth; fruiting pedicels to 4.5 cm long; seeds 0.7–1.2 cm, grayish, silvery, or beige with darker markings; caruncle depressed-conical, 2–3 mm across. **Distr:** NE tropical Africa; naturalized widely in tropical to temperate regions. **Ecol:** In various habitats; 0–2,000 m. Fl. Jun–Sep or Jan–Dec. **Zones:** 9–12. **Info:** Cultivated mainly for castor oil, which is used for medicinal and industrial purposes. All parts of the plant are extremely toxic.

TRIBE PLUKENETIEAE

Tragia

Species 170. **Herbs**, perennial, climbing or twining, hispid with stinging hairs, monoecious. **Leaves** alternate; lamina simple or palmately trilobed or tripartite. **Inflorescences** terminal, axillary, androgynous racemes; ♂ flowers many, superior in raceme; ♀ flowers few, inferior, bracts minute. **Flowers** apetalous; ♂ with calyx globose or obovoid, valvately trilobed, stamens 3, filaments short, distinct, anthers ovate or oblong; ♀ with sepals 6, imbricate in bud, entire or pinnatifid, stellately spreading in fruit or enclosing ripe capsule, ovary 3-celled, cells 1-ovuled, styles entire, united in column below. **Fruits** capsules of three 2-valved cocci; seeds globose. **Distr:** Tropical and warm temperate regions. **Cult:** Propagate from seeds. Requires a well-drained soil in full sun.

T. involucrata L., *Sp. Pl.*: 980, 1753. (Syn: *Croton urens*, etc.). [Climbing Nettle, Stinging Nettle]. **Twiners**, evergreen; stems densely hispid-pubescent throughout and scattered with stinging bristles. **Leaves** simple, narrowly lanceolate to narrowly ovate-lanceolate, 4–7.5 × 0.7–1.5 cm, base narrowly or broadly cuneate into petiole, margins serrate, apex acute-acuminate, adaxially slightly hispid-pubescent, abaxially densely so; petiole 2–4 mm long, densely hispid. **Inflorescences** racemes, 1.5–3 cm long; bracts small or minute. **Flowers** ♀ with sepals ~6 × 2 mm, pinnately linear-lobed when fruiting, hispid. **Fruits** ~7 mm Ø, hispid-pilose; seeds ~3 × 2.5 mm, subglobose, grayish-brown with slight mottling. **Distr:** India, Bangladesh, Sri Lanka. **Ecol:** Weedy in disturbed places. **Zones:** 11–12.

Subfam. EUPHORBIOIDEAE

Flowers ♂ with androecium not covered by perianth; disk usually absent; outer integument 3–6 or 8–22 cells across, inner integument 3–5(–22) cells across. **Seeds** carunculate or not. **Distr:** Pantropical, extending into temperate regions.

TRIBE EUPHORBIEAE

Euphorbia

Syn: *Poinsettia*, etc. [Fire-on-the-Mountain, Mole Plant, Snow-on-the-Mountain, Spurge, Wolfmilk].

Species over 2,000. **Herbs** (annual, biennial, or perennial), **shrubs**, or **trees**, very rarely subscandent, sometimes rhizomatous, monoecious or, rarely, dioecious; latex white, rarely yellow. **Stipules** present or not, sometimes modified into glands or prickles. **Leaves** alternate or opposite, rarely verticillate; lamina with margins usually entire, sometimes serrulate or dentate; petiole often ill-defined to absent. **Inflorescences** flowerlike cyathia, single or often several in terminal or axillary, dichasial or monochasial cymes often in a pseudo-umbel with a terminal ± sessile cyathium subtended by a whorl of pedunculate cymes, each subtended by an involucral leaf; cyathium consisting of a bowl-shaped to tubular involucre subtended by a pair of bracts, 'cyathophylls', enclosing several clusters of ♂ flowers and 1 central pedicellate ♀ flower, occasionally cyathia unisexual, involucre margin 4- or 5-lobed, cyathial glands (1–)4 or 5(–7), mostly alternating with involucral lobes, sometimes with petaloid appendages or a pair of horns. **Flowers** ♂ reduced to a single stamen, subtended by slender bracteoles; ♀ reduced to a single ovary, rarely subtended by a very reduced perianth, ovary triloculed, ovules 1 per locule, styles 3, usually free, stigma usually bilobed. **Fruits** capsules, breaking into three 2-valved cocci; seeds 1 per locule. **Distr:** Worldwide, particularly in drier areas of the tropics, especially diverse in Africa; one major group centred in temperate regions. **Cult:** Propagate from seeds or cuttings. Requires a fertile, well-drained soil in full sun.

E. antiquorum L., *Sp. Pl.* 1: 450, 1753. [Fleshy Spurge, Triangular Spurge, Square Milk Hedge, Square Spurge]. **Trees** or **shrubs**, 3–5(–8) m tall, glabrous except for cyathia; stems 5–7 cm thick, green, 3(or 4)-ribbed, branching from upper parts; ribs prominent, winglike, up to 1–2 cm wide, 3–5 mm thick, triangular-toothed. **Stipules** pricklelike; spine shields small, spines paired, 2–5 mm. **Leaves** alternate, apically clustered, caducous; lamina obovate or obovate-oblong, 2–5 × 1–2 cm, base attenuate, margins entire, apex rounded; petiole very short. **Cyathia** subterminal, single; peduncles 2–3 mm long; cyathophylls as long as involucre, membranous; involucre broadly campanulate, ~3 × 5 mm, lobes 5, rounded, denticulate; glands 5, entire; ♂ flowers many, bracts linear. **Flowers** ♀ with ovary pedicellate, exserted from involucre, smooth and glabrous, styles free, stigma slightly bilobed. **Fruits** capsules, trilobed, compressed, 3.4–4 × 4–5 mm; seeds subglobose, ~2 × 2 mm, brown-yellow, smooth, caruncle absent. **Distr:** Tropical Asia (exact origin unknown). **Ecol:** Fl. & Fr. throughout the year. **Zones:** 11–12.

E. burmanni (Klotzsch & Garcke) E. Mey. ex Boiss., *Prodr.* 15(2): 75, 1862. (Syn: *Arthrothamnus burmanni*, etc.). **Shrubs**, succulent, dioecious, to 70 cm high, or subscandent to 2 m, much-branched at base and above; branches ~4 mm Ø. **Stipules**

Top row: *Acalypha hispida* var. *alba*, *Acalypha indica*, *Acalypha wilkesiana*.
Bottom row: *Ricinus communis*, *Tragia involucrata*.

glandular, conspicuous, reddish, persistent. **Leaves** to 3 mm long, deciduous. **Inflorescences** cymes, terminal and axillary, in 3-rayed umbels, 2- to 3-forked, rays to 2.5 cm long; bracts broadly ovate, small. **Cyathia** minutely puberulous, to 3.5 mm ∅, minutely pubescent; pedicels ~3 mm long. **Fruits** capsules; seeds oblong, 2 mm long, ± 4-angled, tuberculate. **Distr:** S Namibia to Cape Province. **Ecol:** On mountainsides and under the shelter of bushes close to rivermouths; 0–120+ m. **Zones:** 9–11.

E. canariensis L., *Sp. Pl.*: 450, 1753. (Syn: *Torfosidis canariensis*, etc.). [Canary Island Spurge, Hercules Club]. **Shrubs**, to 3 m high, succulent, densely branched from base, sparsely above, forming compact clumps to 10 m across; branches parallel-erect, 4–6-angled, young parts dark green, winkled, angles with shallow teeth, ~1 cm apart. **Spine shield** obovate, ~8 × 6 mm, black, calloused. **Spines** divergent, ~4 mm long. **Inflorescences** cymes solitary, simple, reddish, subsessile. **Cyathia** ~8 mm ∅; nectar gland elliptic, dark red. **Fruits** deeply lobed, ~5 × 6 mm, deep red; pedicel shortly exserted, stout; seeds subglobose, smooth. **Distr:** Canary Islands. **Ecol:** On south-facing basalt cliffs and rocks; 120–800 m. **Zones:** 10–12.

E. cotinifolia* ssp. *cotinoides (Miquel) Christenhusz, *Harvard Pap. Bot.* 7: 3, 2002. (Syn: *E. cotinoides*). [Caracasana, Caribbean Copper Plant, Mexican Shrubby Spurge, Tropical Smoke Bush]. **Trees**, evergreen, 13–15(–19) m; trunk to 12(–17) cm thick; branches spreading, dark red. **Leaves** 3-whorled; lamina ovate-rounded, 2–6 × 2–4 cm, base subtruncate, margins entire, apex obtuse, surfaces red; petiole 2–9 cm long, less reddish. **Cyathia** numerous, peduncles ~2 cm long; involucre broadly campanulate, ~4 × 2.5–3 mm, lobes 4–6, triangular, pilose on margin; glands 4–6, dark green, rounded, appendages white, lobed; ♂ flowers numerous, bracts linear; ♀ flower exserted from involucre. **Flowers** ♀ with ovary 3-angular, with vertical furrows, conspicuous. **Fruits** capsules, 3-angular-ovoid, ~5 × 6 mm, smooth, glabrous; seeds subglobose, ~3 mm ∅, brown, adaxially dark-striate; caruncle absent. **Distr:** C and S America. **Ecol:** Fl. & Fr. Apr–Nov. **Zones:** 10–12. **Info:** Widely cultivated throughout the tropics.

E. cyathophora Murray, *Commentat. Soc. Regiae Sci. Gott.* 7: 81, 1786. (Syn: *E. heterophylla* var. *cyathophora*, etc.). [Fire on the Mountain, Painted Leaf, etc.]. **Herbs**, annual, shrubby, erect to 1 m tall; stems glabrous, hollow. **Stipules** brownish glands. **Leaves** alternate; lamina lanceolate in outline, violin-shaped, to 10 × 5 cm, base cuneate, margins shallowly and irregularly toothed, apex obtuse, abaxially sparsely pilose, glabrescent, adaxially glabrous; petiole to 1.5 cm long. **Inflorescences** terminal, cymes forking ~4x, cyathia densely clustered; basal bracts similar to leaves but with a bright orange-red blotch at base, upper bracts progressively smaller, more lanceolate, subsessile, entirely red; cyathia ~3.5 × 3 mm; involucre barrel-shaped, glabrous, lobes ~1.5 × 1.5 mm, rounded, margin deeply and bluntly toothed, glands 1(or 2), funnel-shaped, opening transversely elliptic, ~2 mm wide. **Flowers** ♂ with bracteoles ligulate, feathery, stamens ~4 mm long; ♀ with ovary pedicellate, glabrous, styles ~2 mm long, bifid almost to base. **Fruits** capsules, deeply trilobed, ~4 × 5 mm; seeds ovoid-conical, ~2.8 × 2.2 mm, acute, seed coat sharply warty, blackish-brown. **Distr:** C USA to NC Argentina. **Ecol:** Fl. & Fr. May–Nov. **Zones:** 9–11.

E. hirta L., *Sp. Pl.* 2: 454, 1753. (Syn: *Chamaesyce hirta*). [Garden Spurge, Hairy Spurge]. **Herbs**, annual, 30–60(–70) cm tall, usually few-branched; roots fibrous, 3–5 mm long; stems branching from middle or above, sometimes prostrate, hairy. **Stipules** membranous, triangular, 0.8–1.7 mm long, caducous. **Leaves** opposite; lamina lanceolate-oblong, long-elliptic, or ovate-lanceolate, 1–5 × 0.3–1.6 cm, base slightly oblique, margins entire or few serrulate below middle, apex acuminate or obtuse, adaxially green to red, sometimes with a purple blotch along midrib, abaxially gray-green, surfaces pilose; petiole 1–3.5 mm long. **Cyathia** dense, often headlike, pedunculate cymes at upper nodes; involucre campanulate, ~1 × 1 mm; glands 4, red, appendages white to reddish, narrowly elliptic to obdeltoid, to 0.3 × 0.2 mm; ♂ flowers 4 or 5; ♀ flowers shortly pedicellate, exserted from involucre. **Flowers** ♂ with anthers red; ♀ with ovary 3-angular, sparsely pilose, styles free, stigma slightly bilobed. **Fruits** capsules, 3-angular, 1–1.5 × 1–1.5 mm; seeds ± tetragonal, 0.7–0.9 × 0.4–0.5 mm, reddish. **Distr:** Tropical & subtropical America. **Ecol:** In various habitats, weedy. Fl. & Fr. Jun–Dec. **Zones:** 9–12.

E. leucocephala Lotsy, *Bot. Gaz.* 20: 350, 1895. [Snows of Kilimanjaro]. **Shrubs**, 1.5–4 m high, erect, deciduous; stems disarticulating at the nodes when dry, green, angulate, glabrous. **Leaves** mostly verticillate; lamina obovate to oblong, elliptic-oblong or oblong-

Top row: *Euphorbia antiquorum*, *Euphorbia burmanni*, *Euphorbia canariensis*, *Euphorbia cotinifolia* ssp. *cotinoides*.
Second row: *Euphorbia cyathophora*, *Euphorbia hirta*, *Euphorbia leucocephala*, *Euphorbia milii*.
Third row: *Euphorbia pulcherrima*, *Euphorbia tirucalli*, *Euphorbia tithymaloides*, *Euphorbia viguieri*.
Bottom row: *Hura crepitans* (female flower), *Hura crepitans* (male flower).

lanceolate, to 8 × 3.5 cm, base broadly cuneate, apex obtuse or rounded, apiculate, often red-tinged, adaxially glabrous, abaxially ± pilose; petioles almost filiform, slender, long. **Inflorescences** cymose, thinly pilose; cymes rather dense, umbellate at the ends of branches; bracts spathulate, 1–1.5 cm long, obtuse, white; involucres campanulate, almost sessile, pilose, lobes transverse-oblong, ciliate; glands 5, transverse-oblong, appendages white, petaloid, ovate, obtuse, entire, glabrous. **Fruits** capsules, 5 mm long, glabrous. **Distr:** Mexico (Oaxaca, Chiapas) to Honduras. **Ecol:** In moist or dry, often on rocky plains or hillsides; 600–2,000 m. **Zones:** 9–11.

E. milii Des Moulins, *Bull. Hist. Nat. Soc. Linn. Bordeaux.* 1: 27, 1826. [Crown of Thorn Euphorbia, Crown of Thorns]. **Shrubs**, scrambling, many-branched, 60–90 cm long; stems obscurely 3–5-angled, densely spiny. **Stipules** forming spines 1–1.3(–2) cm long. **Leaves** alternate, clustered at stem apex; lamina obovate to oblong-oblanceolate, 1.5–5 × 0.8–1.8 cm, base attenuate, margins entire, apex rounded; petiole absent or nearly so. **Cyathia** subapical, dichasial cymes; peduncles 4–7 cm long; cyathophylls 2, reniform-rounded, 0.8–1 × 1.2–1.4 cm, usually bright red (pink, white, or yellow in some cultivars), apex rounded; involucre campanulate, 3–4 × 3.5–4 mm, lobes 5, lanceolate; glands 5, reniform-rounded, ~1 × 2 mm, yellow-red; ♂ flowers many, bracteoles linear, apex pilose. **Flowers** ♀ with ovary smooth and glabrous, usually included with involucre, styles connate below middle, stigma bilobed. **Fruits** capsules, trilobed-ovoid, ~3.5 × 4 mm; seeds ovoid-terete, ~2.5 × 2 mm, gray-brown. **Distr:** Madagascar. **Ecol:** Fl. & Fr. throughout the year. **Zones:** 11–12.

E. pulcherrima Willdenow ex Klotzsch, *Allg. Gartenzeitung.* 2: 27, 1834. (Syn: *Poinsettia pulcherrima*). [Mexican Flame Leaf, Poinsettia]. **Shrubs** to **small trees**, 1–3(–4) m high, many-branched; stems glabrous. **Stipules** minute, membranous, caducous. **Leaves** alternate; lamina ovate-elliptic, oblong, or lanceolate, often shallowly lobed, 6–25 × 4–10 cm, base attenuate, apex acuminate or acute; petiole 2–5 cm long, glabrous. **Cyathia** congested into a 1-sided synflorescence, subtended by 5–7 petiolate leaflike bracts; bracts narrowly elliptic, 3–7 × 1–2 cm, usually entire, sometimes repand-lobed, scarlet, peduncles 3–4 mm long; involucre urceolate, light green, 7–9 × 6–8 mm, lobes 5, lacerate, triangular, glabrous; glands usually 1(2), yellow, compressed, 2-lipped, 4–5 × ~3 mm; ♂ flowers many, usually exserted from involucre, bracts linear, pilose. **Flowers** ♀ with ovary glabrous, pedicellate, exserted from involucre, smooth, styles connate below middle, stigma deeply bilobed. **Fruits** capsules, trilobed-globose, 1.5–2 × ~1.5 cm, smooth, glabrous; seeds ovoid, ~1 × 0.8–0.9 cm, grayish. **Distr:** Mexico to Guatemala. **Ecol:** Fl. & Fr. Oct–Apr. **Zones:** 9–12.

E. tirucalli L., *Sp. Pl.* 1: 452, 1753. (Syn: *Tirucalia indica*, etc.). [Indian Tree Spurge, Milk Bush, Petroleum Plant, Rubber Euphorbia]. **Trees** or **shrubs**, 2–6 m tall, dioecious, eventually forming trunk, 10–25 cm dbh; bark rugose, grayish; stems green, succulent, very finely longitudinally striate. **Stipules** very small, caducous. **Leaves** alternate, present only on new growth; lamina oblong-linear, 0.7–1.5 cm × 0.7–1.5 mm, base attenuate, margins entire, apex obtuse; petiole ± absent. **Cyathia** clustered at apex of branches, pedunculate, unisexual; involucral leaves minute, membranous, caducous; involucre turbinate, ~2 × 1.5 mm, shortly pubescent inside; glands 5, peltate-ovate or subrounded; ♂ flowers many, exserted from involucre. **Flowers** ♀ with ovary glabrous, exserted from involucre, styles connate below middle, stigma bilobed. **Fruits** capsules, trilobed, ~8 × 8 mm, smooth; seeds ovoid-globose, ~4 × 4 mm, smooth; caruncle small. **Distr:** Africa (Angola). **Ecol:** Fl. & Fr. Jul–Oct. **Zones:** 10–12. **Info:** The sap is toxic and can cause blindness on contact with eyes.

E. tithymaloides L., *Sp. Pl.*: 453, 1753. (Syn: *Pedilanthus tithymaloides*, etc.). [Christmas Candle, Devil's Backbone, Red Bird Cactus, Slipper Spurge]. **Subshrubs**, erect, 40–70 cm tall; stems stout, slightly fleshy, strongly flexuous, zigzagged with leaves emerging from nodes. **Stipules** small, with a rounded ~1 mm ∅ gland. **Leaves** distichous; lamina ovate or long-ovate, 3.5–8 × 2.5–5 cm, base rounded or obtuse, apex ± acuminate, fleshy, pubescent, glabrescent, midrib elevated abaxially, lateral veins 7–9 pairs. **Inflorescences** with cyathia in cymes clustered on leafless stems, terminal or axillary at upper part, each one with many ♂ flowers and 1 ♀ flower; involucres shoe-shaped, deep-red or purple-red, equilateral, ~1 cm long, glabrous, apex nearly lip-shaped, bifid, lobes small, oblong, apex triserrulate, other lobe boat-shaped, ~1 cm long, apex bipartite; ♂ flowers with pedicels 2.5–4 mm long, similar to filaments; ♀ flower inserted at center of involucres, oblique, exserted from involucre, pedicels 6–8 mm long. **Flowers** ♂ with anthers globose, slightly shorter than filaments; ♀ with ovary fusiform, styles usually connate, stigma 3, bifid. **Distr:** Florida, Mexico to tropical America. **Ecol:** Fl. Dec–Jun. **Zones:** 9–12. **Info:** Cultivated mainly for ornamental and medicinal purposes. The leaves are often eaten in the Maldives.

E. viguieri Denis, *Euphorb. Iles Austr. Afr.*: 73, 1921. **Shrubs**, succulent; stems to 1 m tall, usually simple, narrowed towards the base, to 3 cm ∅, 5–6-angled. **Leaves** clustered at stem apex; lamina ovate, to 10 × 3 cm, bright green, base red, abaxially with prominent white veins, subsessile, deciduous; leaf scars broadly elliptic, stipular. **Spines** stiff, ~2 cm long, expanded, divided and joined at base. **Inflorescences** cymes, ~4-forked on peduncles to 4 cm long; cyathophylls erect, enveloping the cyathia, bright red. **Cyathia** apparently unisexual, ~2.5 mm ∅; nectar glands minute. **Fruits** glabrous; seeds tuberculate. **Distr:** N & C Madagascar. **Ecol:** Usually on granite rocks. **Zones:** 10–12. **Info:** This species is variable across its range, with 5 described varieties.

TRIBE HUREAE

Hura

Species 2–3. **Trees**, monoecious; branching usually in whorls; trunk and branches spiny; indumentum of simple, multicellular hairs; exudate white latex. **Stipules** large, caducous. **Leaves** alternate; lamina with margins subentire or undulate-serrulate, pinnately veined; petiole biglandular at apex. **Inflorescences** terminal or axillary, unbranched, elongate; ♂ terminal, long-pedunculate catkin, bracts not glandular, united into membranous sheath around bud, afterward irregularly lobed; ♀ flowers solitary, axillary, pedicel long. **Flowers** ♂ with calyx membranous, shallowly cup-shaped, truncate or slightly serrulate, petals absent, disk absent, stamens (8–)10–20, several-seriate, filaments and connectives united into a stout column, theca free, extrorse, longitudinally dehiscent, pistillode absent; ♀ with calyx leathery, broadly cup-shaped, truncate at apex, petals absent, disk absent, ovary 5–20-loculed, ovules 1 per locule, styles fused into prominent, fleshy column, stigmas united into large umbrella-shaped lobed disk. **Fruits** oblate schizocarps, large, apex depressed, breaking up into laterally compressed woody mericarps; seeds laterally compressed, smooth, caruncle absent, endosperm fleshy, cotyledon rounded and flattened. **Distr:** Tropical America. **Cult:** Propagate from seeds or by air-layering. Requires a well-drained soil in full sun.

H. crepitans L., *Sp. Pl.* 2: 1008, 1753. [Hura Wood, Monkey Dinner-bell, Monkey Pistol, Possum Tree, Possum Wood, Sandbox Tree]. **Trees**, to 40 m tall; stem and main branches densely spiny; indumentum yellowish-brown; branches densely lenticellate, glabrous. **Stipules** lanceolate, 1–1.5 × 0.2–0.3 cm, pubescent, caducous. **Leaves** broadly ovate, 5–29 × 5–17 cm, base cordate, margins ± shallowly dentate-serrate, apex caudate-acuminate, mucronulate, papery, lateral veins 10–13(–16) per side; petiole 4–20 cm long, apex with 2 glands. **Inflorescences** ♂ ovoid-conical, (1.6–)4–5 × (0.8–)1.5–2 cm, mostly dark red, peduncles (1.2–)7–10 cm long, often with solitary ♀ flower at base, pedicel ~2 mm long; ♀ pedicel 1–1.7 cm long, after anthesis to 6 cm long. **Flowers** ♂ with calyx tube 2–3 mm long, staminal column 0.4–1.2 cm long, anthers in (1 or)2 or 3 whorls; ♀ with calyx tube truncate, 4–6(–8) mm long, enclosing ovary, style column 2–5 cm long, stigmatic disk dark purple, 1.5–2.5 cm ∅, with 11–14 obtuse lobes, spreading radially. **Fruits** oblate, 3–5 × 8–9 cm ∅, becoming reddish-brown, apex and base concave, longitudinally grooved; fruiting pedicels pendent, to 6 cm long; seeds ~2 cm ∅. **Distr:** Tropical America. **Ecol:** Fl. May. Fr. Aug. **Zones:** 10–12. **Info:** The milky sap can cause blindness on contact with eyes.

TRIBE HIPPOMANEAE

Sapium

Species 22. **Trees**, medium to big, with copious, white latex, usually only in twigs, flowers and fruits. **Leaves** spirally arranged; lamina sometimes with marginal glands; pedicels long, often with glands near leaf base. **Inflorescences** terminal or lateral, gland-bearing racemes or spikes, usually a few ♀ flowers at bottom of spikes, or separate ♂ and ♀ spikes. **Flowers** apetalous, without disk; calyx tiny, 2–trilobed, cup-shaped; ♂ with stamens 2–3, free or shortly connate; ♀ with ovary 2–triloculed, styles ± free, recurved and soon deciduous. **Fruits** capsules, woody or leathery, 2–trilobed, splitting into 2–3 bivalved segments, often leaving a central column; seeds sometimes with a fleshy covering. **Distr:** Tropical and subtropical America. **Cult:** Propagate from seeds or by air-layering. Requires a moist, well-drained soil in full sun.

S. stylare Müll. Arg., *Linnaea* 32: 119, 1863. (Syn: *Excoecaria stylaris*, *S. verum*, etc.). [Virgin Rubber]. **Trees**, 20–30 m tall; limbs ascending, slightly divaricate; crown rounded; floriferous branches thick. **Stipules** suboval, 3–4 mm long, margin ± fimbriate. **Leaves** clustered at ends of branchlets, rather large; lamina oblong or ovate-elliptic, 8–16(–20) × 3–5 cm, base ± rounded or cuneate, margins glandulose-denticulate, apex obtuse with a slightly inflexed, cucullate-glandulose acumen; petiole 2.5–4(–5) cm long. **Inflorescences** spikes, 14–15 cm long, in axils of upper leaves; ♂ flowers short-pedicellate, to 15 under each bractlet, mixed with small glanduliferous scales, glands ovate, ~4 mm long, bracts broadly triangular and subflabelliform, ~2.5 × 5 mm, thick at base, margin scarious, irregularly sinuate-denticulate. **Flowers** ♂ with perianth campanulate, attenuate at base, ~4 mm long, lobules rounded-sinuate, stamens half exserted, filaments thick, bulging at middle, anthers yellow. **Fruits** capsules, subglobose, ~1.2 × 1.5 cm, 3-celled; pedicels slender, ~4 mm long, crowned by persistent stylar column; seeds lenticular, 7–8 mm long, 5 mm thick. **Distr:** Costa Rica to Venezuela and Ecuador. **Ecol:** In montane forests; (1,000–)2,000–3,300. Fr. Aug–Sep. **Zones:** 10–12.

Excoecaria

Syn: *Commia*.

Species ~35. **Trees** or **shrubs**, monoecious or dioecious, glabrous; exudate milky. **Stipules** small, caducous. **Leaves** alternate or opposite, petiolate; lamina entire or serrulate, penninerved. **Inflorescences**

GL

Above: *Sapium stylare, Excoecaria cochinchinensis, Caryocar brasiliense, Caryocar nuciferum.*

racemelike thyrses, axillary or terminal. **Flowers** unisexual, apetalous, without disk; ♂ flowers (sub) sessile, sepals 3, small, imbricate, free, stamens 3, filaments free, anthers longitudinally dehiscent, without pistillode; ♀ flowers sessile to pedicellate, calyx trilobed or tripartite; ovary triloculed, smooth, ovules 1 per locule, stigmas extended or recurved, free to slightly basally connate, undivided, eglandular. **Fruits** capsules, dehiscent into 2-valved cocci, columella persistent, winged; seeds globose. **Distr:** Africa, Asia, Australia, Pacific Islands. **Cult:** Propagate from seeds or by air-layering. Requires a moist, well-drained soil in shade or full sun.

E. cochinchinensis Loureiro, *Fl. Cochinch.* 2: 612, 1790. [Blindness Tree, Chinese Croton, Firestorm]. **Shrubs**, evergreen, to 1 m tall; branches glabrous, lenticellate, often much-branched. **Stipules** ovate, ~1 mm long. **Leaves** opposite, rarely alternate or nearly 3-whorled; lamina narrowly elliptic or oblong, 6–14 × 2–4 cm, base attenuate, margins serrulate, apex long-acuminate, papery, glabrous, adaxially green, abaxially purple or dark red, or greenish abaxially when old, lateral veins 8–12; petiole 0.3–1.3 cm long, apex eglandular. **Inflorescences** dioecious, axillary or terminal racemes; ♂ 1–2 cm long, pedicels ~1.5 mm long, bracts broadly ovate, ~1.7 × 1.7 mm, serrulate inside base, biglandular, each bract 1-flowered, bractlets 2, linear, ~1.5 mm long, lacerate-serrulate on upper part, base biglandular; ♀ 3–5-flowered, pedicels 1.5–2 mm long, bract and bractlets as in ♂. **Flowers** ♂ with sepals 3, lanceolate, ~1.2 mm long, serrulate at apex, stamens exserted from calyx, anthers rounded, slightly shorter than filaments; ♀ with sepals 3, slightly basally connate, ovate, ~1.8 × 1.2 mm, ovary globose, glabrous, styles 3, free or ± basally connate, ~2.2 mm long. **Fruits** capsules, globose, ~8 mm ∅, base truncate, apex concave; seeds nearly globose, ~2.5 mm ∅. **Distr:** S China to peninsular Malaysia. **Ecol:** In forests, thickets, understory; <1,500 m. Fl. throughout most of the year. **Zones:** 9–12.

CARYOCARACEAE

Syn: Terstroemiaceae, etc.

Genera 2, **species** 23. **Trees** or rarely shrubs; stem erect, ligneous. **Stipules** 0–4, usually soon caducous. **Leaves** opposite or alternate, petiolate; lamina 3–5-foliolate; leaflets with margins usually serrate, dentate or crenate or rarely entire, often with stipels at base of leaflets. **Inflorescences** racemes, terminal, ebracteate. **Flowers** large, hermaphrodite, actinomorphic; sepals 5(–6), imbricate and large (in *Caryocar*) or small and reduced; petals 5(–6), imbricate, caducous, free or rarely slightly basally or apically connate forming a calyptra; stamens 55–750, filaments frequently caducous as a unit together with the petals, sterile filaments often with spirally arranged tubercules, anthers basifixed or medifixed; ovary superior, free, 4(–6)-loculed (in *Caryocar*) or 8–20-loculed, locules 1-ovuled, ovules basal, styles 4 or 8–20, apex stigmatose. **Fruits** drupes, large, 1–4-seeded (in *Caryocar*) or 8–20; seeds often reniform. **Distr:** Tropical America (Costa Rica to southern Brazil).

Caryocar

Syn: *Acantacarix, Pekea*, etc.

Species 15. **Trees**, large, rarely shrubs; branches opposite, frequently horizontal. **Stipules** absent or present and soon caducous. **Leaves** opposite, usually long-petiolate, trifoliolate; leaflets short-petiolate, base often with 2–4 stipels, margins usually serrate, crenate or dentate, pinnately nerved; stipels persistent or caducous, sometimes with 2 large and 2 small stipels. **Inflorescences** racemes, terminal; rachis short often rather corymbose; pedicels articulate at apex; bracteoles lateral, alternate, small. **Flowers** hermaphrodite, large; calyx distinctly 5(–6)-lobed, imbricate; petals 5(6), imbricate, basally fused together with filament bases, often caducous with filaments; stamens 57–750, exceeding petals, interior ones usually shorter and sterile; ovary usually 4(–6)-loculed, locules 1-ovuled, styles 4, long, filamentous. **Fruits** drupes, 4–6-loculed, 1–4 locules developing and dehiscing into 1-seeded cocci, mesocarp thick and fatty or fleshy, endocarp woody; seeds reniform or subreniform. **Distr:** Costa Rica to S Brazil, most abundant in the Guianas and Amazonia. **Ecol:** Predominantly in lowlands, rarely on forested uplands. **Cult:** Propagate from seeds or by air-layering. Requires a fertile, moist, well-drained, humus-rich soil in full sun.

C. brasiliense Cambessedes, *Fl. Bras. Mer.* 1: 322, t. 67bis, 1828. [Pekea Nut, Piquia-oil Plant, Souari Nut]. **Trees**, to 10 m tall, sometimes low shrubs or subshrubs; trunk gnarled, to 30 cm dbh. **Leaves** trifoliolate; leaflets ovate-elliptic, often markedly asymmetrical, base rounded, markedly unequal, margins coarsely crenate, apex rounded, serrate, terminal lamina 10–18 × 8–12 cm, lateral lamina equal or smaller, primary veins 10–13 pairs, usually slightly impressed; petiole 1–10.5 cm long, terminal petiolule 4–8 mm long, lateral petiolules 1–4 mm long. **Inflorescences** racemes, clustered; rachis 2.5–5.5 cm long; pedicels 1–5 cm long, with 2 membranous bracteoles; peduncles 4–10 cm long, usually densely tomentose. **Flowers** with calyx broadly cupuliform, 1–1.3 cm long, base often tomentose, lobes 5(–6), rounded; corolla 1.8–3 cm long, lobes 5–6, oblong, yellow or yellow within and orangish-red outside; stamens yellow, ~270–330, 3.5–5 cm long, innermost filaments 1.5–2 cm long, filaments shortly basally united into a caducous ring; ovary globose, usually 4(–6)-loculed, styles 4(–6), ~5 cm long. **Fruits** drupes, ovoid-globose, ~4–5 cm across, usually biloculed; stone ovoid ~2.5 cm across; endocarp with numerous fine spines 2–3 mm long. **Distr:** Bolivia, Brazil, Paraguay. **Ecol:** In the fringing areas of rainforests, also on plateaus and valleys. Fl. Jun–Oct. Fr. Nov–Mar. **Zones:** 10–12. **Info:** The pulp is edible.

C. nuciferum L., *Mantissa Plantarum* 2: 247, 1771. (Syn: *Amygdala guayanensis*, etc.). [Butter Nut, Butter Nut of Guiana, Pekea Nut, Souari Nut]. **Trees**, large, to 45 m tall. **Leaves** trifoliolate; leaflets elliptic, 12–30 × 6–18 cm, base rounded to subcuneate, margins entire to weakly crenate, apex acuminate, acumen 0.5–1.5 cm long, glabrous, primary veins 8–13 pairs; petiole 4–9(–15) cm long, petiolules 0.7–2 cm long, stipels absent. **Inflorescences** racemes, clustered; rachis 1–4.5 cm long; pedicels 4–6 cm long, ebracteolate; peduncles 6–10 cm long. **Flowers** very large; calyx campanulate, ~2 cm long, lobes 5, rounded; corolla ~6–10(–12.5) cm long, elliptic, deep red outside, paler inside; stamens >700, filaments basally united, caducous as a unit, and then dividing into fused groups before becoming free above, outer ones 7–10(–12.5) cm long, yellow, inner filaments many, shorter, 3.5+ cm long; ovary globose, 4-loculed, styles 4, filamentous, 8–9 cm long. **Fruits** subglobose to sublobate, to 15 cm long; stone ~7 cm across; seeds 1–2, subreniform. **Distr:** Brazil, Guianas, Venezuela. **Ecol:** In primary forests. Fl. Apr–May. **Zones:** 10–12.

C. villosum (Aublet) Persoon, *Syn. Pl.* 2: 84, 1806. (Syn: *C. butyrosum*, etc.). [Pekea]. **Trees**, large, to 40 m tall; trunk to 2.5 m dbh, buttressed. **Leaves** 3-foliolate; leaflets elliptic, 8–22 × 6–12 cm, laterals slightly smaller, base rounded to cordate, margins serrate to crenate, apex acuminate, acumen 0.3–1 cm long, primary veins 12–19 pairs; petiole 4–15 cm long, terminal petiolule 3–6 mm long, lateral petiolules 2–4 mm long. **Inflorescences** racemes, clustered; rachis 3–4 cm long; pedicels 1.8–3.5 cm long; bracteoles 2; peduncles 5–13 cm long. **Flowers** large; calyx campanulate-cupuliform, ~1.5 cm long, lobes 5, rounded; corolla ~2.5 cm long, lobes 5, oblong-elliptic, pale yellow; stamens numerous, ~300, filaments shortly united into a ring at base, but not into groups, subpersistent, of 2 distinct lengths with several of intermediate lengths, longest ~6.5–7 cm long, yellow, apical 1–3 mm tuberculate, shortest ~55, 1–1.5 cm long; ovary globose, 4-loculed, styles 4. **Fruits** drupes, oblong-globose, 6–7 × 7–8 cm, stone reniform, ~5 cm across, endocarp with numerous fine spines ~3 mm long. **Distr:** French Guiana and Amazonian Brazil. **Ecol:** In forests, on non-flooded areas. **Zones:** 10–12. **Info:** The flowers are night-blooming and fall by dawn, covering the ground completely.

VIOLACEAE

Genera 23, **species** 800. **Herbs** annual or perennial, **shrubs**, or **subshrubs**, sometimes scandent, rarely small trees. **Stipules** small or leaflike. **Leaves** usually alternate, sometimes opposite, petiolate; lamina simple,

Above: *Caryocar villosum, Viola odorata, Dovylas hebecarpa, Flacourtia indica* (fruits).

margins entire, serrate, or dissected. **Inflorescences** with flowers solitary or in axillary or terminal spikes, panicles, or racemes, bibracteolate, sometimes cleistogamous. **Flowers** bisexual or unisexual, rarely polygamous, actinomorphic or zygomorphic; sepals 5, equal or unequal, imbricate, persistent; petals 5, imbricate or convolute, unequal, anterior one usually larger than others, saccate, gibbous or spurred at base; stamens 5, anthers erect, free or connivent or connate, connectives often dilated into membranous appendages, filaments very short or absent, anterior 2 stamens with spurlike nectary at base; ovary superior, 1-loculed, 3–5-carpelled, syncarpous, with 3–5 parietal placentas each with 1 to many anatropous ovules, style simple, stigmas variously shaped. **Fruits** capsules, loculicidal, usually with elastic and abaxially carinate valves, rarely baccate; seeds often carunculate, testa hard, nitid, often with oily bodies. **Distr:** Worldwide.

Viola

Species 400–600. **Herbs** perennial or biennial, rarely subshrubs, rhizomatous; stem developed or absent, sometimes with creeping branches. **Stipules** small or large, leaflike, free or ± adnate to petioles. **Leaves** alternate or basal; lamina simple, margins entire, dentate, or dissected. **Inflorescences** with flowers solitary; pedicels axillary, bibracteolate. **Flowers** bisexual, zygomorphic, often dimorphic (cleistogamous flowers later than chasmogamous ones); sepals slightly equal, usually basally auriculate; petals unequal, anterior petal largest and basally spurred; filaments free, very short, anthers free or mostly connivent into a sheath around ovary, 2 anterior ones with spurlike or wartlike and nectariferous appendages at base, connectives produced apically into conspicuous, membranous appendages; ovary tricarpellate, with many ovuled parietal placentas, styles nearly erect or usually ± curved downward; style apex and stigma variously shaped. **Fruits** capsules, loculicidally and elastically 3-valved; seeds globose-ovoid, arillate or not, usually smooth; endosperm abundant. **Distr:** Worldwide. **Cult:** Propagate from seeds, basal cuttings or by division. Requires a moist, well-drained, humus-rich soil in shade or full sun.

V. odorata L., *Sp. Pl.* 2: 934. 1753. [Wild Violet]. **Herbs**, perennial, acaulescent, stoloniferous, 3–15 cm tall; rhizome erect or oblique, brownish, robust. **Leaves** basal; lamina orbicular or reniform to broadly ovate-cordate, smaller at anthesis, 1.5–2.5 × 1.5–2.5 cm, gradually accrescent, to 4.5 × 4.5 cm, base deeply cordate, margins crenate, apex rounded or ± acute. **Inflorescences** with flowers solitary; pedicels long, slender, puberulous or subglabrous, bibracteolate at or above middle. **Flowers** deep purple, large, fragrant; sepals oblong or oblong-ovate, base obtuse or shallowly dentate, apex obtuse, basal auricles 2–3 mm long; petals crenate, upper ones obovate, lateral ones inside shortly bearded, anterior one broadly obovate, 1.5–2 cm long; spur 2–4 mm long; spurs of 2 anterior stamens robust, ~4 mm long; ovary puberulous, styles thickened upward, slightly compressed, apically with a hooked beak. **Fruits** capsules, globose, densely puberulous. **Distr:** N Africa, C and SW Asia, and Europe. **Ecol:** Fl. Apr–May. Fr. Jun–Sep. **Zones:** 7–11.

SALICACEAE

Genera 55, **species** 1,010. **Trees** or **shrubs**, evergreen (deciduous). **Stipules** usually present. **Leaves** alternate, spiral or 2-ranked; lamina simple, margins serrate to dentate, teeth usually salicoid (vein expanding at tooth apex and associated with spherical, glandular setae). **Inflorescences** determinate to indeterminate, various in form, sometimes in erect to pendent catkins, sometimes reduced to a solitary flower, terminal or axillary. **Flowers** bisexual or unisexual (plants dioecious), radial, often reduced, subtended by a usually hairy bract; sepals usually 3–8, distinct to slightly connate, sometimes ± vestigial, forming a disk-shaped to cup-shaped structure; petals 3–8, distinct, or lacking; stamens 2 to numerous, filaments distinct to connate; carpels usually 2–4, connate, ovary superior to half-inferior, with parietal placentation, placentas sometimes deeply intruded and ovary then appearing axile, stigmas usually 2–4, ovules 1 to numerous on each placenta; nectary a disk or of separate glands, sometimes lacking. **Fruits** loculicidal capsules, berries, or drupes; seeds often arillate or with basal tuft of hairs. **Distr:** Pantropical, also temperate (but few in Australia, not in New Zealand) to Arctic.

Dovylas

Syn: *Aberia.*

Species 15. **Shrubs** or **trees**, dioecious or ♂ plant bearing occasionally a few bisexual flowers and fruits; trunk and branches sometimes with simple or compound spines; branchlets with simple axillary spines. **Leaves** alternate, sometimes fascicled on much-reduced lateral shoots, rarely deciduous, exstipulate, petiolate; lamina with margins entire, denticulate or crenate, nerves 1 or 2, basal or supra-basal. **Inflorescences** with flowers solitary or in short racemes, these generally reduced to few-flowered fascicles; pedicels with small basal bracts. **Flowers** greenish to yellowish, rarely white; calyx lobes (3–)4–6(–7), very rarely to 9, free almost to base, subimbricate to practically valvate, surfaces pubescent, sometimes with marginal glands; ♂ flowers with stamens 10–50(–80), rudimentary ovary absent; ♀ flowers with calyx lobes as many as (rarely to 12) and similar to those of ♂, staminodes rarely and stamens very rarely present, ovary unilocular or incompletely 2–4(–8+)-locular, placentas 2–4(–8+), each with 1 or 2(–6) ovules, styles 2–8(–40), divergent, stigmas ± lobed. papillose. **Fruits** berries, fleshy; seeds few, occasionally to 12, ellipsoid, flat. **Distr:** Africa (14 spp.), Sri Lanka (1 sp). **Cult:** Propagate from seeds or cuttings. Requires fertile, moist but well-drained soil in full sun. **Info:** Most species cultivated for the edible fruits or for ornamental purposes.

D. hebecarpa (Gardn.) Warb., Pflanzenfam. III, 6a: 44, 1893. (Syn: *Roumea hebecarpa*, etc.). [Ceylon Gooseberry]. **Trees** or **shrubs**, 2.5–6 m tall; stem unarmed; branchlets brownish-straw or purplish-brown, wrinkled, tips rusty-pubescent, sometimes with axillary spines to 3 cm long. **Leaves** ovate-oblong to oblong-lanceolate, 2.5–12 × 1.3–5 cm, base cuneate to rounded, margins entire or minutely glandular-dentate or -serrate, apex shortly acuminate, tip subacute; petiole 0.5–1.2 cm long, hairy. **Inflorescences** with ♂ flowers in umbellate clusters, sessile, 8–10(–20)-flowered, pedicels to 8 mm long; ♀ flowers solitary or 2–3 in fascicles at leafy or leafless axils, yellowish-tomentellous, pedicels 3–4 mm long. **Flowers** ♂ rusty- or yellowish-tomentellous, tepals 5(–7), ovate-acuminate, 3–4 × 1.5–2 mm, stamens 50–60; ♀ flowers with tepals as in ♂, though slightly larger, ovary densely spreading yellow-brown hairy, styles 5–7. **Fruits** purple to bronze-maroon, globular, 2.5 cm Ø, outside velvety; seeds ~10, flattened-ellipsoid, 7 × 5 mm, densely hairy. **Distr:** Sri Lanka. **Ecol:** In scrubs, open hillsides; 450–780 m. **Zones:** 10–12. **Info:** Cultivated for the edible fruits.

Flacourtia

Syn: *Stigmarota.*

Species 15–17. **Trees** or **shrubs**, dioecious or hermaphroditic, rarely polygamous, usually spiny. **Stipules** small, early caducous. **Leaves** alternate, petiolate; lamina usually pinnately veined, margins usually glandular-dentate. **Inflorescences** axillary, or terminal on lateral twigs, usually short, lax, racemose, or in form of small paniculate or umbellate clusters. **Flowers** hypogynous, unisexual or bisexual, small; sepals 4–7, imbricate, slightly connate at base, green, small; petals absent; disk fleshy, entire or comprised of distinct glands; ♂ with stamens many, exserted, filaments free, longitudinally dehiscent, connective not projected beyond thecae, disk extra-staminal, abortive ovary much reduced or absent; ♀ with disk surrounding base of ovary, ovary superior, globose, ovoid, or bottle-shaped, incompletely 2–8-loculed by false septa, placentas 2-ovuled, styles isomerous with placentas, columnar, stigmas slightly dilated, staminodes usually absent. **Fruits** berrylike indehiscent drupes with pyrenes twice as many as styles, globose, disk persistent at base, style or stigma remnants persistent at apex; seeds compressed-ellipsoid. **Distr:** Tropical Africa and Asia. **Cult:** Propagate from seeds, cuttings, by budding or separation of root suckers. Requires a rich, moist soil in full or partial sun.

F. indica (N.L. Burman) Merrill, *Interpr. Herb. Amboin.* 377, 1917. (Syn: *Gmelina indica, F. parvifolia*). [Governor's Plum, Gooseberry, Indian Plum,

Above: *Flacourtia ramontchi, Hydnocarpus heterophylla, Hydnocarpus kurzii, Hydnocarpus pentandrus.*

Madagascar Plum]. **Shrubs** or **small trees**, 2–4 m tall, deciduous; old branches usually not spiny; young branches with axillary, simple spines. **Leaves** obovate to oblong-obovate, 2–4 × 1.5–3 cm, base mostly acute to obtuse, margins serrulate above middle, apex rounded, sometimes retuse, adaxially deep green, rose-red when young, abaxially greenish, thickly papery, adaxially glabrous, abaxially ± pubescent, lateral veins 5–7 pairs; petiole red, 3–5 mm long. **Inflorescences** racemose, axillary or at ends of short lateral twigs, short; rachis 0.5–2 cm long; pedicels 3–5 mm long. **Flowers** with sepals 5 or 6, ovate, ~1.5 mm long; ♂ with stamen filaments 2–2.5 mm long, usually pubescent; ♀ with ovary globose, placentas 5 or 6, styles 5 or 6, united only at base, radiating, 1–2 mm long, slender. **Fruits** dull to blackish-red, globose, 0.8–1 cm Ø, longitudinally 5- or 6-angled, styles persistent; seeds 5 or 6. **Distr:** Asia and Malesia; tropical and subtropical Africa, Madagascar, Mascarenes, Seychelles. **Ecol:** In broad-leaved forests; 0–1,400 m. Fl. Jan–Mar. Fr. Mar–Jul. **Zones:** 9–12. **Info:** Cultivated widely for the edible fruits.

F. ramontchi L'Héritier, *Stirp. Nov.* 3: 59, 1786. [Batoko Plum, Ramontchi]. **Trees**, to 20 m tall; flowering and fruiting branches usually not spiny. **Leaves** broadly elliptic, elliptic, or elliptic-lanceolate, 4–10 × 2.5–6 cm, base cuneate, margins serrate, apex obtuse or acute, rarely retuse, adaxially deep green and shiny, abaxially greenish, surfaces glabrous, papery, lateral veins 4–6 pairs; petiole 4–8 mm long. **Inflorescences** terminal or axillary, racemose, 1–2 cm long, puberulous; sepals 5 or 6, ovate, ~1.5 mm across, outside glabrous, inside puberulous, margin ciliate, apex obtuse. **Flowers** ♂ with disk entire or shallowly lobed; ♀ with disk entire, ovary globose, placentas 5 or 6, each with 2 ovules, styles 5 or 6, free, stigmas bilobed. **Fruits** globose, 1.5–2.5 cm Ø, not longitudinally angled, with persistent styles; seeds 4–6. **Distr:** China, India, Malaysia, Philippines, Sri Lanka, Vietnam; Africa. **Ecol:** In evergreen broad-leaved forests; 200–1,700 m. Fl. Apr–May. Fr. Jun–Oct. **Zones:** 9–12.

Hydnocarpus

Syn: *Taraktogenos.*

Species 40. **Trees**, rarely shrubs, dioecious, rarely monoecious or polygamous. **Leaves** alternate. **Stipules** small, usually early caducous. **Leaves** usually petiolate, often thickened at apex; lamina pinnate-veined, margin entire or toothed. **Inflorescences** in axillary, ± branched cymes, sometimes reduced to fascicles or to a solitary flower, or rarely flowers in long racemelike panicles from trunk or older branches; bracts small to minute, sometimes persistent. **Flowers** hypogynous; sepals (3 or)4 or 5(or 7–11), imbricate, free or slightly joined at base, concave, becoming reflexed, caducous; petals 4 or 5(–14), free or slightly basally connate, each with usually hairy basal scale inside; ♂ with stamens 5 to many (>100), filaments free, sometimes very short, anthers longitudinally dehiscent, pistillode present or absent; ♀ with staminodes 5–many, resembling stamens but anthers mostly reduced or absent, ovary superior, 1-loculed, placentas 3–6, each with several ovules, styles 3–6, short or nearly absent, stigmas flattened. **Fruits** berries, baccate, globose, or ovoid; seeds several to many, angular-ovoid, packed in pulp. **Distr:** Tropical Asia to Malesia. **Cult:** Propagate from seeds, cuttings, by budding or separation of root suckers. Requires a rich, moist soil in full or partial sun.

H. heterophylla Blume, *Rumphia* 4: 22, t 178, 1848. (Syn: *Taraktogenos blumei, T. heterophylla*). **Trees**, to 20(–36) m, glabrous except inflorescences; bark nearly black; branches pendent. **Stipules** persistent, triangular, ~1 mm long. **Leaves** oblong-lanceolate, (10–)13–24 × 3.3–8 cm, base cuneate, ± unequal, margins entire or shallowly crenate, apex long acute-acuminate, subcoriaceous, ± shining, grayish-green when dry, nerves 6–7 pairs; petiole (0.5–)1–1.5 cm long. **Inflorescences** cymes, dichotomous, 4–8-flowered, ♀ cymes shorter than ♂ inflorescences; peduncles ~5 mm long, ± as long as its 2 branches, ± densely rusty-velvety. **Flowers** with sepals subrotund, concave, outer 5–6 mm long, inner ones 0.8–1 cm long, membranous, light-green or white; petals subrotund, ~5 mm Ø, greenish-white; scales ciliate; ♂ with stamens 20–32, pubescent, 5–7 mm long, anthers sagittate; ♀ with staminodes 15–24, ovary ovoid, rusty-sericeous, crowned by 4–5-lobed, radiate stigma. **Fruits** globose, 5.5–6.5(–8) cm Ø, blackish-velvety; seeds 15–20. **Distr:** Malaysia (S Sumatra), W Java, SE Borneo, N Sulawesi, Philippines. **Ecol:** In rainforests; 150–900 m. Fl. Jan–Jun. Fr. Apr–Dec. **Zones:** 11–12.

H. kurzii (King) Warb., *Nat. Pflanzenfam.* [Engler & Prantl] 3(6a): 21, 1893. (Syn: *Taraktogenos kurzii*). [Chalmoogra Tree, Kalaw Tree]. **Trees**, to 12–15 m tall; trunk and old branches gray-green. **Leaves** alternate; lamina lanceolate or oblong-lanceolate, sometimes elliptic, 16–30 × 6–12.5 cm, apex with a distinctly narrowed drip tip, shiny, subcoriaceous, primary veins 6–7 pairs. **Inflorescences** axillary cymes; ♂ many-flowered; ♀ few-flowered, pedicels 1.5 cm long. **Flowers** mostly dioecious, pale yellow, with unpleasant odor; ♂ with sepals 4, ovate-rotund, fulvous-pubescent outside, 5 × 4 mm, petals 8, broadly ovate, obtuse, ciliate, scales fleshy, stamens ~24, anthers elongate-cordate; ♀ with sepals and petals ± larger than ♂, staminodes 10–16, ovary elliptic, 6–7-sulcate. **Fruits** berries, globose or cucumber-shaped, rugose, 6–7.5 cm across; seeds many, ovoid, faceted, smooth, 2–3 × 1–1.5 cm, pulp brown-yellow. **Distr:** SE Asia. **Ecol:** In evergreen forests; 200–800 m. **Zones:** 10–12.

H. pentandrus (Buch.-Ham.) Oken, *Allg. Naturgesch.* 3(2): 1381, 1841. (Syn: *Chilmoria pentandra*, etc.). [Chalmoogra Nut, Jungli Almond (Wild Almond)]. **Trees**, 10–25 m tall; trunk often fluted; bark brownish-gray, smooth, sometimes white-mottled. **Leaves** alternate; lamina elliptic-lanceolate to elliptic-oblong, 5–20 × 3–8 cm, base acute or cuneate, margins toothed, apex acuminate, adaxially dark green, shiny, abaxially paler, lateral nerves 5–6 pairs; petiole to 2.2 cm long. **Inflorescences** ♂ clusters, 3–6-flowered; ♀ flowers solitary; pedicels ~8 mm long, brownish-hairy, enlarged in fruit to ~6–8 cm long, thick. **Flowers** ♂ green, ~6 mm across; ♀ ~1 cm across. **Fruits** berries, woody, globose or ovoid, 5–7.5(–10) cm Ø, apiculate, young fruit black-velvety maturing to brownish-tomentose, sometimes very shallowly 5-ridged, pericarp thick; seeds ~20, ~2 cm across, embedded in white pulp. **Distr:** India (Western Ghats). **Ecol:** In moist deciduous to semievergreen forests; <900 m. Fl. Nov–Jan. **Zones:** 10–12.

H. venenata Gaertn., *Fruct.* 1: 288, t. 6013, 1788. (Syn: *H. inebrians*). [Fish-poison Tree, Jungle Almond]. **Trees**, 3–25(–30) m tall; trunk usually fluted, 0.45–2.2 m dbh; bark grayish-white or green and gray patched. **Leaves** oblong-elliptic to narrowly oblong-lanceolate, 10–19(–22.5) × 3.5–7 cm, base attenuated, margins serrate, apex ± long acuminate, finely silky-pubescent when very young; petioles 6–7 cm long. **Inflorescences** ♂ several-flowered axillary cymes, peduncles ~1 cm long, pubescent, pedicels 5 mm long; ♀ with flowers solitary or paired, axillary, pedicels to 1.5 cm long. **Flowers** ♂ with sepals rounded, 2–3 mm long, ferruginous-pubescent outside, glabrous inside, petals white, slightly shorter than sepals, margins white-fimbriate, scale broadly ovate, 2 mm long, obtuse, densely golden-villous, stamens 5, rudimentary ovary absent; ♀ with staminodes 5 with narrow sterile anthers, ovary globose, densely golden-hairy, stigmas 3, sessile, bilobed. **Fruits** globose, 2.5–3 cm Ø, white or green, maturing to brownish-black, brownish velvety-tomentose; seeds 2–3(–6). **Distr:** Sri Lanka. **Ecol:** In wet evergreen gallery forests, dry deciduous forests; 300–800 m. **Zones:** 11–12. **Info:** Fruits often used as fish poison.

Oncoba

Species 4. **Shrubs** or **small trees**; stem and branches often spiny. **Stipules** caducous. **Leaves** deciduous, alternate; lamina with margins crenate-serrate, penninerved; petiole not apically thickened. **Inflorescences** with flowers solitary, axillary or terminal on short lateral shoots, pedunculate. **Flowers** bisexual or polygamous (andromonoecious), large, white, scented; sepals 3–5, ± united at base, ± imbricate, outer ones gradually larger than inner ones; petals (5–)8–15, imbricate in bud, subequal;

stamens numerous, filaments filiform, anthers linear to oblong, basifixed, longitudinally dehiscent, apiculate by the ± elongate connective; ovary (wanting in ♂ flowers) sessile, unilocular, placentas 6–8(–10) much protruding, multiovulate, style columnar, stigma either peltate, formed by short subglobular. **Fruits** globular, indehiscent, 1-locular, smooth or with slight longitudinal ridges, pericarp woody; seeds many, embedded in pulp, with a smooth bony testa. **Distr:** Tropical and subtropical Africa. **Cult:** Propagate from seeds or by air-layering. Requires a fertile, moist, well-drained soil in full or partial sun.

O. spinosa Forsk., *Fl. Aegypt.-Arab.*: CXIII, 103, 1775. [Fried Egg Tree, Snuffbox Tree]. **Shrubs** or **small trees**, glabrous, semideciduous, to ~5 m tall; branches with pale lenticels. **Spines** axillary, very sharp, spreading, to 5 cm long. **Leaves** elliptic, ovate-elliptic or oblong-elliptic, 3.5–12 × 2–10 cm, base abruptly cuneate, margins serrate or serrate-crenate except near base or serrations almost obsolete, apex acuminate, usually membranous, of lateral nerves 7–11 pairs; petiole to 8 mm long. **Inflorescences** with flowers solitary, terminal or lateral on short, axillary shoots. **Flowers** large, scented; calyx lobes ~1.5 × 1.2 cm, rounded; petals white, imbricate, oblong-spathulate, 8–10, ~3 × 2.5 cm, rounded; stamens numerous, with filaments to 6 mm long, anthers yellow, linear, 1.75 mm long; ovary globose, style as long as stamens, stigma 4 mm Ø. **Fruits** globose, ~5 cm Ø, brown, smooth, with a hard shell marked with ~8 rather faint longitudinal lines, calyx persistent at base; seeds many, ~6 × 4 mm, testa shiny, brown. **Distr:** Tropical Africa. **Ecol:** In warm, dry woodlands, sometimes forming thickets in river valleys. **Zones:** 9–12.

Pangium

Species 1. **Trees**, dioecious, more rarely andromonoecious. **Stipules** absent. **Leaves** spirally arranged, crowded at ends of branches; lamina trilobed in young trees, broadly ovate in old trees with base cordate to truncate, apex acute-acuminate, palmately nerved; petiole long. **Inflorescences** with flowers usually unisexual; ♂ flowers in few-flowered pseudo-racemes, terminal flowers sometimes bisexual; ♀ flowers mostly solitary in upper leaf axils; basal bracts 1 (in ♂ flowers) or several (in ♀ flowers), linear, caducous; pedicels elongate. **Flowers** with calyx closed before anthesis, nearly globose, splitting irregularly into 2–3(–4) caducous segments; petals (4–)5–8(–9), imbricate, each with a large, vaulted, thick, hairy scale inside base; ♂ flowers with stamens 20–25(–31), free, filaments thick, flat with a linear stripe of hairs ventrally, anthers minute, ovate, dorsifixed, rudiment of pistillode absent; ♀ flowers with staminodes as many as petals, rarely more, alternating with them, ovary thick-walled, with 2–4 many-ovuled placentas, stigma sessile, discoid. **Fruits** large, indehiscent; seeds many, large, transversal- or triangular-ovoid, in an oily and fleshy, sweet, aromatic, creamy-white or yellowish, edible pulp. **Distr:** Malaysia to Melanesia to Micronesia. **Cult:** Propagate from seeds. Requires moist but well-drained soil in full sun.

P. edule Reinw. ex Blume, *Syll. Ratisb.* 2: 13, 1825. (Syn: *Hydnocarpus polyandra*, etc.). [Black Nut, Football Fruit, Pangi, Raual]. **Trees**, polygamous-dioecious, medium-sized to lofty, to 40 m tall; crown dense; trunk to 1m dbh, buttressed. **Leaves** large, ovate-cordate, (10–)12–30(–60) × 8–20(–40) cm, margins entire, dark green, glabrous, adaxially shiny, abaxially dull, 5–7-nerved from base; petiole 7–30(–50) cm long, base much-swollen. **Inflorescences** ♂ racemes, 4–9-flowered, sometimes terminal flower (rarely 2) larger and hermaphrodite, others smaller and with stamens only, 6–24 cm long, including 5–12 cm long peduncle, pedicels 2.5–4 cm long, brown-pubescent, subtended by a bract 1–3 cm long; ♀ flowers mostly solitary, occasionally in few-flowered racemes (then flowers seemingly bisexual, but really ♂ with 20–25 empty anthers), pedicels 6–9(–12) cm long. **Flowers** ♂ with calyx rusty in bud, lobes reflexed at anthesis, glabrescent, 0.8–1.3(–2) cm long, petals ovate-oblong, somewhat fleshy, ciliate, patent, later reflexed, bright pale green, 1.5–2.5 cm long, each with an ovate or subrounded, densely hairy, stamens inserted on a flat villous disk, half as long as petals; ♀ with 5–8 anantherous staminodes, calyx and petals as in ♂, but petals somewhat larger. **Fruits** large, asymmetrical, mostly oblong-ovoid, 15–25(–30) × 7.5–12(–15) cm, brownish. **Distr:** Malaysia, Melanesia, Micronesia. **Ecol:** In rainforests; <300(–1,000) m. **Zones:** 11–12. **Info:** All parts of the plant are poisonous. The seeds are edible after washing or heating.

Salix

Syn: *Pleiariana.*

Species 450. **Trees** or **shrubs** (erect, ascending procumbent, creeping or cushion-forming), deciduous, rarely evergreen; branches terete. **Stipules** small, free, deciduous or persistent, developed mainly on vigorous branchlets. **Leaves** alternate, rarely subopposite or opposite; lamina variously shaped, often long and narrow; petiole short. **Inflorescences** catkins, upright or spreading, rarely pendulous; bracts entire, persistent or caducous; flowering precocious, coetaneous, or serotinous. **Flowers** entomophilous or anemophilous, each with 1 or 2 glands (1 abaxial or absent and 1 adaxial); ♂ with stamens 2–many, filaments free or partly to completely connate, usually exceeding bracts, anthers biloculed (rarely 4-loculed if filaments connate); ♀ with ovary biloculed, sessile or stipitate, style 1, short, slender, or absent, entire or 2-cleft, stigmas 1 or 2, lobed or entire. **Fruits** capsules, 2-valved; seeds mostly green or gray-green, small, surrounded by fine hairs. **Distr:** Mostly from cold and temperate regions of Northern Hemisphere, a few in Southern Hemisphere. **Cult:** Propagate from seeds, cuttings, or by air-layering. Requires a moist, well-drained soil in full sun.

S. babylonica L., *Sp. Pl.* 2: 1071, 1753. [Weeping Willow]. **Trees**, to 18 m tall; branchlets pendulous, brownish-yellow, or slightly purplish, slender, glabrous. **Stipules** lanceolate or ovate-orbicular, those of shoots obliquely lanceolate. **Leaves** narrowly lanceolate or linear-lanceolate, 9–16 × 0.5–1.5 cm, base cuneate, margins serrate, apex long-acuminate, adaxially green, abaxially light green, glabrous or slightly pilose; petiole (0.3–)0.5–1 cm long. **Inflorescences** with ♂ catkins 1.5–2(–3) cm long, peduncle short, bracts lanceolate, abaxially pilose; ♀ catkins 2–3(–6) cm long, with 3 or 4 leaves on the peduncle, bracts lanceolate, 1.8–2(–2.5) mm long, abaxially pilose. **Flowers** ♂ with glands 2, stamens 2, pubescent at base, anthers reddish-yellow; ♀ with adaxial gland, ovary ellipsoid, glabrous or minutely pubescent proximally, sessile or subsessile, style short, stigma 2–4-parted. **Fruits** capsules, slightly greenish-brown, 3–4 mm long. **Distr:** China, C Asia, Europe. **Ecol:** On riverbanks and sandbars, in disturbed woodlands and scrublands. Fl. Sep–Oct. **Zones:** 5–11.

Scolopia

Syn: *Aembilla, Phoberos.*

Species ~40. **Shrubs** or **small trees**; trunk and branches often spinose. **Stipules** small, caducous. **Leaves** alternate, usually petiolate; lamina leathery, pinnate-veined, sometimes 3-veined from base, margins entire or toothed (glandular). **Inflorescences** racemes, terminal or axillary, bracteate, sometimes in axillary fascicles or solitary; pedicels articulate at base. **Flowers** usually bisexual, hypogynous, small; sepals 4–6, imbricate, slightly basally united; calyx often opening early in bud to reveal closely packed anther tips; petals isomerous with and similar to sepals; disk extra-staminal, composed of a single row of 8–10, orange, short, thick glands, or rarely disk absent; stamens many, exserted, filaments free, filiform, anthers longitudinally dehiscent; ovary superior, sessile, 1-loculed, with 2–4 placentas, each few-ovuled, style 1, stigma entire or very shortly 2–4-lobed. **Fruits** berries, fleshy, drying blackish, with persistent perianth and stamens at base, and long slender persistent style at apex; seeds (1 or)2 or 3(–20). **Distr:** Tropical and subtropical regions of the Eastern Hemisphere. **Cult:** Propagate from seeds or half-hardened cuttings. Requires a moist, well-drained, humus-rich soil in full or partial sun.

S. crassipes Clos, *Ann. Sci. Nat. Bot.* IV (8): 251, 1857. (Syn: *Phoberos hookerianus, Scolopia crenata*). **Trees** or **large shrubs**, 6–13.5 m tall; trunk ~30 cm dbh; young branches and sometimes base of bole with spines. **Leaves** elliptic, rounded-elliptic, sometimes elliptic-lanceolate, 3–5 × 1.5–3 cm, base cuneate to ± rounded, margins entire to slightly crenate or undulate, apex obtuse to slightly emarginate or sometimes shortly acuminate, ± coriaceous, glabrous,

Below: *Hydnocarpus venenata, Oncoba spinosa, Pangium edule, Salix babylonica.*

Above: *Scolopia crassipes, Trichadenia zeylanica, Turnera diffusa, Turnera subulata.*

adaxially shiny, venation densely reticulate; petiole 2–4 mm long. **Inflorescences** axillary and terminal, few-flowered, gray-pubescent; rachis 1.5–4 cm long; pedicels 4–5 mm long. **Flowers** whitish or yellowish, sweet-scented, 4–6-merous; sepals ovate, 1.5–3 × 1.5 mm; petals similar, ovate-oblong, ~2.5 × 1 mm; disk lobes ~10–25; stamens 50–80, filaments to 6 mm long, connective elongate; ovary flask-shaped, 2–2.5 mm long, style 4–5 mm long, stigma obscurely 2–trilobed. **Fruits** ovoid-globose, to 1.8 × 1.3 cm, apiculate; pedicels 1.3 cm long. **Distr:** Sri Lanka. **Ecol:** In montane and submontane wet forests; 600–1,900 m. **Zones:** 11–12.

Trichadenia

Syn: *Leucocorema.*

Species 2. **Trees**, dioecious. **Stipules** foliaceous, deciduous. **Leaves** spirally arranged, petiolate; lamina pinnately nerved, margins subentire to serrate, often lobed when young. **Inflorescences** racemose panicles, axillary. **Flowers** with calyx globose, calyptriform, splitting irregularly or transversely; petals 5, imbricate, each with a lanceolate or narrowly oblong fleshy hairy scale adnate to inner side at base; ♂ with stamens 5, alternate with petals, anthers with broad connective; rudimentary ovary small or absent; ♀ with staminodes absent, ovary sessile, 1-locular with 3 placentas, each with 1(–2) ovules, styles 3, short, stigmas broad, reniform, crenate. **Fruits** globose, 1–3-seeded, pericarp crustaceous when dry; seed with bony testa, endosperm oily fleshy. **Distr:** Sri Lanka (1 endemic species); Philippines, E Borneo, New Guinea and Melanesia. **Cult:** Propagate from seeds or half-hardened cuttings. Requires a moist, well-drained, humus-rich soil in full or partial sun.

T. zeylanica Thw., *Hooker's J. Bot. Kew Gard. Misc.* 7. 197, t. 5, 1855. **Trees**, dioecious, large, 6–20 m tall; trunk 45 cm dbh, buttresses large; branches contorted, ferruginous-tomentose, leaf-scars prominent, apices with stipules. **Stipules** imbricate concave, ~7 mm long, tomentose. **Leaves** oblong, 6.5–32 × 3.5–24 cm, broadest towards apex, base rounded, margins sinuate (in ♂ trees) or distantly serrate (particularly in ♀ trees), or coarsely dentate, apex tapering-acuminate, abaxially densely tomentose; young leaves often large and deeply palmately 3–7-lobed, dentate and tomentose; petioles 2.5–17 cm long, tomentose. **Inflorescences** ~12 cm long. **Flowers** pale green or yellow; ♂ with calyx pubescent, petals thin, elliptic, glabrous, 5 × 3 mm, scale narrowly oblong, 4 × 1.5 mm, adnate except at apex, densely pilose, filaments linear-lanceolate, 2.5 mm long, flat, anthers 2 mm long, rudimentary ovary lacking; ♀ with ovary ovoid, densely pubescent. **Fruits** globose, 3.8–4.5 cm Ø; seeds 1–3, ovoid-globose, 2.5 × 1.7 cm. **Distr:** Sri Lanka. **Ecol:** In humid forests; 60–900 m. **Zones:** 11–12.

TURNERACEAE

Genera 7, **species** 80–100. **Herbs**, erect or ascending, annual or perennial, sometimes woody at base. **Leaves** alternate, petioled; lamina simple, serrate or serrate-dentate, biglandular at base or not. **Inflorescences** with flowers solitary in the leaf axils or in terminal racemes, hypanthoid. **Flowers** actinomorphic, hermaphrodite, ephemerous; sepals 5, borne on hypanthium, after anthesis circumsessile at base, segments imbricate in bud; petals 5, inserted in throat of hypanthium, contorted in bud, free, shortly clawed or subsessile, deciduous after anthesis; stamens 5, inserted on hypanthium, alternating with petals, filaments filiform or subulate, free, anthers introrse, 2-celled, opening longitudinally; ovary superior, sessile, 1-celled, with 3 placentas, placentation parietal, ovules numerous, anatropous, styles 3, terminal, free, slender, stigmas penicilliform. **Fruits** capsules, globose or ovoid, loculicidally 3-valved; seeds numerous, incompletely arillate, endosperm copious, horny or fleshy, embryo large, straight. **Distr:** Tropical America and Africa (including Madagascar).

Turnera

Species 122. **Herbs** annual or perennial, or **shrubs**, rarely trees, with simple or rarely stellate hairs. **Stipules** minute or absent. **Leaves** petiolate, alternate; lamina simple, margins entire or above base dentate-serrate, often with 2 marginal glands at base. **Inflorescences** usually with solitary flowers in the axils of the uppermost leaves, rarely cymose or capitate inflorescences, homo- or heterostylous; pedicels free or ± adnate to the petiole with 2 apical bracteoles; bracteoles sometimes foliaceous. **Flowers** with hypanthium often with thickenings above insertion point of stamens; calyx segments often very acute; petals on a shortly hairy claw, obovate, caducous after anthesis; stamens inserted above base of hypanthium and partially fused to hypanthium above; ovary globose or ovoid, stigmas multifid. **Fruits** capsules, globose or ovoid, obtuse, verruculose, 3-valved down to near base; valves with recurved tops; seeds globular to oblong, often curved with a raised scalariform reticulation and a unilateral aril. **Distr:** New World (from Texas to Argentina). **Cult:** Propagate from seeds. Requires a well-drained soil in full or partial sun.

T. diffusa Willd. ex Schult., *Syst.* 6: 679, 1820. (Syn: *T. aphrodisiaca*, etc.). [Damiana, Mexican Holly, Old-woman's Broom]. **Shrubs**, often densely branched, 0.2–2 m tall, usually densely pilose or pubescent throughout. **Leaves** very variable in shape and size, petiolate; lamina oblanceolate to spathulate or oblong-ovate, mostly 1–2 cm long, base cuneate, margins coarsely serrate or crenate, adaxially nerves impressed, abaxially often densely softly pilose. **Inflorescences** with peduncles very short; flowers often almost sessile. **Flowers** 0.8–1.2 cm long; calyx 5-dentate; petals spathulate, yellow, 4–8 mm long, much exceeding calyx teeth; ovaries superior. **Fruits** capsules, subglobose, 1.5–2 mm Ø. **Distr:** Baja, Mexico, C America, S America, West Indies. **Ecol:** Mostly in dry, rocky, open places; 500–1,950 m. **Zones:** 10–12.

T. subulata Sm., *Cyclop.* 36 (2), 1819. [White Alder]. **Herbs**, perennial, base woody, 30–70 cm high, densely pubescent. **Stipules** minute, erect, long-hairy. **Leaves** obovate or elliptic, 1–7 × 0.5–3 cm, base cuneate, margins entire basally, apically dentate-serrate, apex acute or obtuse; petiole 0.5–1.5 cm long with 2 large apical lateral glands. **Inflorescences** with flowers solitary, in upper leaf axils, heterodistylous, 2 types of flowers borne in separate plants; pedicel throughout its length adnate to petiole, 3–5 mm long; bracteoles linear-subulate, to 1.25 cm long, entire. **Flowers** with hypanthium densely pubescent, glabrous inside, tube abruptly widened in upper half, 0.5–1 cm long; calyx segments narrow, 1–1.5 cm long, acute; petals yellow, base purple, with dark purple streaks, claw 3–4 mm long, hairy, lamina obovate, 2–3 × 1.1–1.8 cm, base broadly rounded, overlapping, margins entire, apex light yellow; length of stamens and styles different in 2 types of flowers (stamens 1.2–1.8 cm long and styles 6–9 mm long, or stamens 6.5–9 mm long and styles 0.9–1.3 cm long); ovary mostly globose, sometimes ovoid, densely hairy. **Fruits** capsules, globose, finely verruculose, 5–7 mm long; seeds brownish-yellow, 2.5–3 mm long, aril white. **Distr:** Tropical America. **Ecol:** In edges of forests and savannas, usually on sandy or rocky, to clay areas; to 1,400 m. Fl. & Fr. Apr–Jan. **Zones:** 10–12.

T. ulmifolia L., *Sp. Pl.* 271, 1753. [Ram Goat Dash-along, Sage Rose, West Indian Holly, Yellow Alder]. **Herbs** or **small shrubs**, erect; branchlets usually densely pilose. **Leaves** elliptic to narrowly elliptic, sometimes almost linear, infrequently obovate, to 5 × 2 cm, base attenuate-cuneate, conspicuously biglandular, margins serrate-crenate especially above middle, thinly chartaceous, apex acute to rarely subacute; petiole to 1 cm long, usually densely puberulous. **Inflorescences** with flowers solitary, axillary, heterostylous; pedicels shorter than to equaling petioles of subtending leaves and adnate to them; bracteoles linear-subulate, to 1 cm long. **Flowers** with hypanthium ~5–7 mm long; sepals narrowly triangular, ~0.9–1.1 × 2 mm; petals inserted near throat of hypanthium, long-unguiculate, to 2.2 cm long, yellow, sometimes dark purple or brownish at base; stamens at base of hypanthium, filaments in short-styled flowers to 1.2 cm long, in long-styled ones to 8 mm long; ovary densely sericeous, style in short-styled flowers ~5 mm long, in long-styled ones, to 1 cm long. **Fruits** capsules, broadly ellipsoid, ~3–4 mm long; seeds arcuate, ~2.5 mm long. **Distr:** New World (from Texas to Argentina). **Ecol:** In hot and humid areas, under the shade of trees, coastal forest margins,

YK YK

Above: *Turnera ulmifolia*, *Passiflora amethystina*, *Passiflora aurantia*.

mountain slopes and cliffs, in wet or semideciduous tropical forests; 0–2,800 m. Fl. & Fr. throughout the year. **Zones:** 9–12.

PASSIFLORACEAE

Genera 16, **species** 705. **Climbers**, herbaceous or shrubby, rarely erect trees, with axillary tendrils. **Leaves** alternate, stipulate, petiolate; lamina simple, or compound, incised or entire. **Inflorescences** axillary, dichasial or monochasial cymes, with terminal or first flower modified into a tendril, or flowers solitary; flowers bracteate, or ebracteate. **Flowers** bisexual or unisexual, actinomorphic; hypanthium variously shaped, often hollow and usually with a central androphore or gynophore, hypogynous or perigynous; sepals (3–)5, free or basally connate, persistent, imbricate; petals (3–)5 or absent, free or basally united, imbricate; corona uni- or multiseriate, concave to cup-shaped, petaloid or staminodial in appearance, fleshy, rarely absent; stamens 3–5, rarely numerous, antipetalous, free or slightly basally united, anthers oblong, basifixed or versatile, dithecous, longitudinally dehiscent; carpels 3(–5), syncarpous, ovary unilocular, placentation parietal, ovules anatropous, styles 3–5, free or united, stigmas 3(–5), capitate, discoid, reniform or club-shaped. **Fruits** unilocular, loculicidally dehiscent capsules, rarely indehiscent berries; seeds ovoid or flattened, enclosed within a membranous or pulpy aril, endosperm fleshy. **Distr:** Tropics to warm temperate, especially Africa and America.

Passiflora

Species 525. **Vines**, herbaceous or woody, or scandent shrubs, usually climbing by tendrils, rarely small trees (one Peruvian species) or erect shrubs. **Stipules** setaceous or foliaceous or absent. **Leaves** simple, entire or palmately 3–7 lobed, rarely bilobed; petiole glandular. **Inflorescences** axillary, flowers solitary on articulated peduncles, or 2- to more-flowered small cymes, rarely racemose or fasciculate; flowers pedicellate. **Flowers** mostly bisexual, often large and showy; hypanthium cup-shaped or campanulate, perigynous; sepals 4–5, horned outside below apex, fleshy or membranous; petals 4–5, rarely absent; corona uni- or multiseriate, fleshy; stamens 4–5, filaments flat, basally connate into a tube and adnate to gynophore, separating below ovary, subreflexed, anthers versatile; ovary tricarpellate, syncarpous, stipitate on somewhat angular or cylindric gynophore, styles 3, free or somewhat basally united, stigmas 3, capitate. **Fruits** oblong to globose, pulpy berries, rarely somewhat 3-valved capsules; seeds flat, mostly ovate, testa minutely pitted. **Distr:** Mainly from tropical and subtropical America, New Zealand, Australia, Asia and 1 in Madagascar. **Cult:** Propagate from seeds, cuttings or by air-layering. Requires a moderately fertile, moist, well-drained soil in full or partial sun.

P. amethystina Mikan, *Delect. Fl. & Faun. Bras. Fasc.* 4: t. 20, 1825. (Syn: *P. onychina*, etc.). **Vines**, glabrous; stem slender, wiry, terete. **Stipules** semiovate-lanceolate, 0.5–1 × 0.2–0.4 cm, base rounded, apex aristate. **Leaves** trilobed to slightly below middle, to 6 cm along midvein, 10 cm between apices of lateral lobes; lobes oblong, to 3 cm wide, obtuse, sinuses minutely biglandular, base shallowly cordate, often subpeltate, 5-nerved, membranous, glabrous, abaxially glaucescent; petiole 2–4.5 cm long, glands 5–8, scattered, short-stipitate, ~0.5 mm long. **Inflorescences** with flowers solitary; peduncles 2.5–5 cm long; bracts narrowly lanceolate, glandular-serrulate, close flower. **Flowers** 6–8 cm Ø; calyx tube short-campanulate, green; sepals oblong, 5–6 mm wide, carinate, awn 4–5 mm long, bright blue inside, green outside; petals oblong, slightly longer, 6–8 mm wide, obtuse, bright blue; corona 4–5-ranked, filaments of 2 outer rows narrowly liguliform, 1+ mm wide, strongly compressed, 1-nerved, deep reddish-purple below, white, blue-maculate in middle, pale reddish-purple apically, successive filaments filiform, ~3 mm long, erect, dark purple; operculum membranous at base, filamentose, filaments 6–7 mm long, purple; nectar ring annular; limen cupuliform, laxly surrounding base of gynophore; ovary ovoid, tomentose. **Fruits** ellipsoid, 5–6 × 2–2.5 cm Ø, glabrescent. **Distr:** E Brazil (Minas Geraes to São Paulo). **Ecol:** In sandy soils in moist places on banks of watercourses or on high mountain slopes, and forest margins, often forming galleries; 0–2,500 m. Fl. & Fr. Oct–Jul. **Zones:** 10–12.

P. aurantia G. Forst., *Fl. Ins. Austral.* 62, 1786. (Syn: *Murucuia aurantia*, *P. glabra*, etc.). [Norfolk Island Passionfruit, Orange Passion Flower]. **Vines**, herbaceous, glabrous; stems subterete. **Stipules** minute, soon-deciduous. **Leaves**; lamina trilobed to about halfway, (2–)3–6(–9.5) × (2–)3.5–6.5(–10) cm, base rounded to subauriculate, margins entire; lobes rounded, often broadly so, sometimes slightly emarginate, or margins of lobes with ~3 rounded crenations in young plants; venation reticulate, with 1–6 laminal glands where reticulations from lobes meet; petiole glabrous, eglandular. **Inflorescences** with flowers solitary or paired; peduncles 1–5 cm long; bracts filiform, ~4 mm long, deciduous. **Flowers** creamy-white, darkening to orange-red with age, 5–8 cm Ø; calyx lobes 5, cream-yellow to orange-red, oblong-lanceolate, 2–5 × 1–1.5 cm, abaxially with fleshy keel; petals 5, cream to yellow to orange-red, oblong-lanceolate, 1.5–2.5 × 0.7–1 cm, membranous; corona 1-seriate, filaments red, filiform, 1–2 cm long; androgynophore 2.5–4 cm long; ovary subglobose to ellipsoid, glabrous. **Fruits** ellipsoid to subglobose, 3–5.5 × 2–4.5 cm, purplish when ripe; seeds reticulate. **Distr:** S & E Australia, Norfolk Islands, Fiji, Malaysia, New Guinea, and other nearby islands. **Ecol:** In forests. **Zones:** 9–12.

P. auriculata Kunth, *Nov. Gen. & Sp.* 2: 131, 1817. (Syn: *P. appendiculata*, etc.). Vines, glabrous to puberulent, or finely pilosulous; stem angulate. **Stipules** filiform, 2–4 mm long, soon-deciduous. **Leaves** lanceolate, ovate-lanceolate, or lance-oblong, 5–15 × 2–10 cm, trilobed, rarely unlobed, base subcordate or rounded, apex acuminate, (1)3–5-nerved, subcoriaceous, adaxially glabrous and lustrous, cinereous-pubescent, glabrate, abaxially dull or lustrous; petioles 0.5–2 cm long, puberulent, bearing near base 2 auriculate appendages ~2 mm wide. **Inflorescences** with flowers paired; peduncles 0.5–1 cm long; bracts setaceous, ~2 mm long, deciduous. **Flowers** 2–2.5 cm Ø; sepals narrowly oblong-lanceolate, 1–1.5 × 0.1–0.2 cm, acute, yellowish-green or pale greenish; petals linear, 5–7 × 0.5 mm, white; corona filaments biseriate, outer filiform, ~1 cm long, caudate, yellowish-green, base purple, the inner ~3 mm long, capitellate, white; operculum membranous, closely plicate, incurved, white; limen annular, fleshy, whitish; gynophore often swollen at base; ovary ovoid, pilosulous. **Fruits** globose, 1–1.5 cm Ø, densely or sparingly pilosulous, pale yellow; seeds obovate, 5 × 3 mm, flattened, testa with 6–10 undulating ridges. **Distr:** Nicaragua, southeastward to the Guianas and the Amazon Basin of Peru, Bolivia, and Brazil. **Ecol:** In tropical forests; <1,200 m. **Zones:** 10–12.

P. biflora Lam., *Encycl.* 3: 36, 1789. (Syn: *P. lunata*, etc.). [Two-flowered Passionflower]. **Climbers**, slender, glabrous; stems striate. **Stipules** minutely setose. **Leaves** simple, basifixed, subtrapezoid, 1–10 × 5–10 cm, base obtuse to broadly rounded, apically saliently and obtusely bilobed, occasionally with a smaller central lobe or mucro, conspicuously glandular-ocellate; petioles 0.5–3 cm long eglandular. **Inflorescences** paired, flowers solitary; tendril basal; peduncles ~1–1.5 cm long, jointed at about middle or somewhat below; bracts very inconspicuous, subsetaceous. **Flowers** greenish-white or yellow, 3–4 cm Ø; sepals ovate-lanceolate, ~1 cm long, not corniculate; petals oblong-lanceolate, slightly shorter than sepals; corona biseriate, outer about as long as petals. **Fruits** subglobose, 1–2 cm Ø, greenish-yellow. **Distr:** Mexico to Colombia and Venezuela; Bahamas. **Ecol:** 0–1,500 m. **Zones:** 11–12.

P. caerulea L., *Sp. Pl.* 2: 959, 1753. (Syn: *P. loureiroi*). [Blue Passion Flower]. **Vines**, herbaceous; stem terete. **Stipules** reniform, ~1.2 cm long, clasping. **Leaves** palmately (3–)5(–9)-lobed, 5–7 × 6–8 cm, base cordate, margins entire, lobes ovate-oblong, middle lobe slightly larger, papery, glabrous; petiole 2–3 cm long, with 2–4(–6) small glands. **Inflorescences** reduced cymes, central flower not developed, one lateral branch converted to a tendril, flower opposite tendril; bracts broadly ovate, 2–3 cm long, margins entire; pedicels 3–4 cm long. **Flowers**

6–8(–10) cm Ø; sepals light green outside, white inside, 3–4.5 cm long, awn 2–3 mm long; petals white to light green, 2.5–4 cm long; corona 3- or 4-seriate, filamentous, outer 2 series (0.6–)1–1.5 cm long, base dark purple, middle white, apex bright blue, inner 1 or 2 series 1–2 mm long, base light green, apex white and capitate; operculum fimbricate, lobes dark purple, with annular nectary at base; disk 1–2 mm high; androgynophore 0.8–1 cm tall; filaments ~1 cm long; ovary ovoid-globose, styles purple, 6–8 mm long. **Fruits** berries, orange-yellow or yellow, ovoid-globose or subglobose, ~6 cm across; seeds many, obcordate, ~5 mm long. **Distr:** N Argentina, S Brazil. **Ecol:** In understory of both dry forests, and wet edge of woods and forests, understory; 0–1,400 m. Fl. & Fr. throughout the year, mostly Sep–May. **Zones:** 9–12.

P. capsularis L., *Sp. Pl.* 957, 1753. (Syn: *P. hassleriana*, etc.). [Capsule-fruited Passion Flower]. **Vines**; stem 3–5-angled, striate, glabrate or pubescent. **Stipules** linear-subulate, slightly falcate, 5–7 mm long. **Leaves** bilobed, 2–7 cm along midvein, 4–10 cm along lateral nerves, cordate, trinerved, abaxially paler and densely pubescent; lobes lanceolate, occasionally asymmetrical, acute, rarely subobtuse, apiculate, sinus acute or occasionally truncate at base; petioles 1–3 cm long, glandless. **Inflorescences** with flowers solitary; peduncles 1–6 cm long, often very slender, articulate ~1 cm from apex; bracts none. **Flowers** fragrant, 2–6 cm Ø, greenish-white or pale yellow-green; sepals linear-lanceolate, 1–3 × 0.25–0.4 cm, acute, pilose out; petals narrowly oblong-lanceolate or subspathulate, 0.6–1.5 × 0.2–0.4 cm, obtuse; corona filaments 1- or 2-seriate, outer filiform, 1.2–1.5 cm long, united at base into a thin, transparent membrane, violet, base carinate, inner series capillary (sometimes lacking), <3 mm long, violet; operculum membranous, plicate, 1–1.2 mm high; limen annular; ovary narrowly ovoid or slightly obovoid, minutely puberulent or rarely glabrate, hexagonal. **Fruits** ellipsoid or fusiform, sharply hexagonal, 5–6 × 1.5–2 cm; seeds ovate, 3.5 mm long, transversely sulcate, with 5 or 6 smooth ridges. **Distr:** Greater Antilles, Guatemala to Costa Rica, Colombia, C Brazil to Paraguay. **Ecol:** <1,900 m. **Zones:** 10–12.

P. cincinnata Mast., *Gard. Chron.* 966, 1868. (Syn: *P. corumbaensis*). [Crato Passionfruit]. **Vines**, glabrous or softly pilose; stem terete. **Stipules** linear-subulate, 0.6–1 cm long. **Leaves** palmately 3–5-lobed, lobes oblong, 3–8 × 2.5–5cm, deeply divided or parted at base, adaxially dark glossy green; petiole 1.5–4 cm long, glands 2 or 4, 2mm Ø. **Inflorescences** with flowers solitary; peduncles to 6 cm long, terete; bracts ovate, to 3.5 × 2.5 cm, base glandular. **Flowers** pale pink to violet and, violet and blue, very fragrant, 7.5–12.5 cm across; sepals oblong-lanceolate, to 5 × 2 cm, green outside, pinkish-blue or violet inside, keeled with terminal awn; petals linear-lanceolate, to 3 × 1 cm, pinkish-blue and violet; corona filaments in several series, outer long, tapering, thin and twisted, 2–4 cm long, lower part deep purple, middle banded with pinkish-blue and pale blue, upper pale or deeper blue, other series 3 mm long, white or pale blue. **Fruits** globose or ovoid, 5 × 3 cm. **Distr:** Brazil, S Paraguay, Argentina, E Bolivia. **Ecol:** In wet places, floodprone areas, riparian forests and jungles, dry forests, and on hillsides; 400–1,300 m. Fl & Fr. throughout the year, mostly Nov–May in Argentina and Aug–Apr in Brazil. **Zones:** 9–12.

P. coccinea Aubl., *Pl. Guian.* 2: 828, p 324, 1775. (Syn: *P. fulgens*, *P. toxicaria*, *P. velutina*). [Red Granadilla, Red Passion Flower, Scarlet Passion Flower]. **Vines**, rufo-puberulent or rufo-tomentose nearly throughout. **Stipules** narrowly linear, entire or minutely glandular-serrulate. **Petiole glands** 2 or none at base, sometimes additional 2 or 3 in the middle. **Leaves** oblong, 6–14 × 3–7 cm, base subcordate, margins unlobed, duplicate-serrate or crenate, adaxially glabrous or sparingly pubescent, abaxially ferruginous or tomentose; petiole glandless, or biglandular at base. **Inflorescences** with peduncles to 8 cm long; bracts ovate, to 6 × 1–3.5 cm, free to base, crenate or sharply serrate, coriaceous, margins usually glandular, reddish. **Flowers** scarlet or red; calyx tube short, cylindric-campanulate, to 2 cm long; sepals linear-lanceolate, 3–5 cm long; petals linear; corona in 3 series, 2 outer consisting of subulate filaments ~1 cm long, purple above, pink or white below, inner rank tubular, filamentose only at margin, white; ovary yellowish-tomentose. **Fruits** subglobose or ovoid, ~5 cm Ø; seeds minutely reticulate. **Distr:** S America. **Ecol:** <2,000 m. **Zones:** 10–12.

P. coriacea Juss., *Ann. Mus. Hist. Nat.* 6: 109, pl. 39, f. 2, 1805. (Syn: *P. clypeata*, etc.). [Bat-leafed Passion Flower]. **Vines**, glabrous except stem and petioles; stem angulate. **Stipules** narrowly linear, 5 mm long, acute. **Leaves** 2- or obscurely trilobed, 2.5–7 cm long (along midvein), 7–30 cm wide, lateral lobes divergent, apex acute or rarely obtuse, peltate, 5-nerved, reticulate-veined, abaxially ocellate, coriaceous, margins cartilaginous, glabrous; petioles glabrate or puberulent, 2–4 cm long, glands 2 (very rarely 4), sessile or subsessile, ~1 mm Ø. **Inflorescences** with flowers solitary or paired in leaf axils, or in a terminal reduced racemes 4–6 cm long without leaves, but with stipules and bractlike biglandular appendages (petioles of abortive leaves) at base of pedicels. **Flowers** 2.5–3.5 cm Ø, yellowish-green; sepals oblong-lanceolate, 1–1.5 × 0.4–0.5 cm, obtuse; petals absent; corona filaments biseriate, outer series filiform, 7–8 mm long, inner linear, 2 mm long; operculum membranous, plicate, margin slightly incurved; limen annular, thick, <1 mm high, dark brown; ovary ovoid, glabrous. **Fruits** globose, 1–2 cm Ø, glabrous; seeds obcordate to obovate, ~4 × 2 mm, curved, beaked, faces coarsely reticulate. **Distr:** C & S America. **Ecol:** <2,000 m. **Zones:** 10–12.

P. cuprea L., *Sp. Pl.* 955, 1753. (Syn: *P. cavanillesii*, etc.). [Copper-colored Passion Flower]. **Vines**, glabrous; stem angulate, flattened, striate. **Stipules** setaceous, 2–3 mm long, early deciduous. **Leaves** oval or ovate-oblong, 2.5–7 × 1.5–5 cm, not lobed, apex rounded and usually mucronulate, rounded, often slightly emarginate, trinerved at base, sometimes ocellate, subcoriaceous or membranous; petioles 0.5–1 cm long, glandless. **Inflorescences** with flowers solitary or in pairs; peduncles to 2.5 cm long, enlarged at base of flower, articulate above middle; bracts setaceous, 0.5–1 mm long, soon-deciduous. **Flowers** red-brown; calyx bowl-shaped, 4–6 × 4–7 mm, usually narrowed to peduncle; sepals linear-oblong, 1.5–2 × 0.3–0.4 cm, obtuse; petals linear, 1–1.5 × 0.2–0.3 cm, obtuse; corona filamentose, filaments in a single series, 3–4 × 0.5–1 mm, yellowish; operculum membranous, scarcely 0.5 mm long, borne at throat of tube, deflexed, margin entire; limen none; ovary ovoid. **Fruits** globose, ~1 cm Ø; seeds obcuneate, ~4 × 2 mm, transversely 7–8-sulcate. **Distr:** Bahamas, E Cuba, Tortue Island, Haiti. **Ecol:** In coastal rainforests; near sea level. Fl. Jul. **Zones:** 11–12.

P. dispar Killip, *Journ. Wash. Acad. Sci.* 12: 330, 1922. **Vines**; stem terete, striate, glabrous. **Stipules** ovate, 2 × 0.6–0.8 cm, foliaceous, serrulate, aristulate. **Leaves** ovate, 11–13 × 6–7 cm, unlobed or occasionally with one lateral lobe, subpeltate ~2 mm above base, margins entire or remotely and minutely serrulate toward base, narrowed to an obtuse apex, adaxially glabrate, abaxially densely grayish-tomentulose, palmately 5- or 7-nerved with prominulous secondary veins, reticulate; petioles 3–3.5 cm long, bearing 4–6 stipitate glands 1.8 mm long. **Inflorescences** with flowers solitary; peduncles 2–2.5 cm long; bracts foliaceous, ovate-lanceolate, 1–1.5 × 0.5–0.6 cm, base cordate with unequal lobes, apex acute, closely callous-serrulate. **Flowers** 4.5–5.5 cm wide; sepals ovate-lanceolate, obtuse, 2–2.2 × 1 cm, short-horned just below apex; filaments of faucial corona in several series, outer filiform, equaling petals, successive series capillary, <3 mm high, middle corona membranous, erect, irregularly lacerate, basal corona closely surrounding base of gynophore, slightly folded, crenulate; gynophore 1 cm long, glabrous; anthers linear-oblong, 7 × 2 mm; ovary ellipsoid, glabrous, glaucous, styles filiform, 6–7 mm long, stigmas orbicular. **Distr:** Venezuela. **Ecol:** In forests; 10–100 m. **Zones:** 10–11. **Info:** This species has been confused with *P. oerstedii* in popular literature and trade.

P. edmundoi Sacco, *Sellowia* 18: 44, 1966. **Vines**, glabrous; stem terete, striate. **Stipules** reniform, 2–3 × 1–1.5 cm, entire. **Leaves** ovate, 5–9 × 5.5–8 cm, trilobed, margins entire; petiole 2.5–3.5 cm long, glands 2–6, mostly paired, elongated, 3–5 mm long. **Inflorescences** solitary or paired; peduncles 6.5–17 cm long; bracts ovate above, 4–6 × 2–3 mm, purplish, scattered along peduncle, entire, petiole 2–3 mm long. **Flowers** purplish-red, 6–7 cm Ø; floral tube cylindrical, 1–1.5 cm long, green outside; sepals purplish-red, oblong-lanceolate, 2.5–3 × 0.8 cm, keeled outside, keel terminating in an awn 4–5 mm long; corona 2- or triseriate, outer series to 6 mm long, bluish-violet, inner series a membrane 2–4 mm long; ovary ellipsoid, glabrous. **Fruits** ellipsoid, 6–9 × 2–3 cm. **Distr:** W Brazil. **Ecol:** ~1,000 m. **Zones:** 10–12.

P. edulis Sims, *Bot. Mag.* 45: t. 1989, 1818. (Syn: *P. middletoniana*, etc.) [Granadilla, Passionfruit, Purple Granadilla, Purple Passionfruit]. **Vines**, herbaceous, base woody, ~8 m long; stem slender-striate, glabrous. **Leaves** 6–13 × 8–13 cm, base cuneate or cordate, margins deeply trilobed, midlobe ovate, lateral lobes ovate-oblong, margins glandular-serrate, with 1 or 2 small cup-shaped glands near base of sinuses, glabrous, membranous. **Inflorescences** reduced cymes, central flower not developed, 1 lateral branch converted to a tendril, flower opposite tendril; bracts green, broadly ovate or rhombic, 1–1.2 cm long, margins irregularly serrulate; pedicel 4–4.5 cm long, apex biglandular. **Flowers** 4–7 cm Ø; hypanthium 0.8–1 × 1–1.2 cm; sepals green outside, light green or white inside, 2.5–4 × ~1.5 cm, awn 2–4 mm long; petals 2.5–3 × ~0.8 cm; corona in 4 or 5 series, outer 2 series filaments ligulate, distal half filiform, 2–2.5 cm long, base green, middle purple, apex white, inner 2 or 3 series filiform, 1–3 mm long, green and purple; operculum recurved, 1–1.2 mm; disk ~4 mm high; androgynophore 1–1.2 cm tall; trochlea just above disk; filaments 5–6 mm long, anthers yellow-green; ovary obovoid, ~8 mm long. **Fruits** purple, ovoid, 3–4 cm Ø, glabrous; seeds many, ovoid, 5–6 mm Ø. **Distr:** S America. **Ecol:** 100–1,900 m. Fl. Jun. Fr. Nov. **Zones:** 9–12. **Info:**

Top row: *Passiflora auriculata*, *Passiflora biflora*, *Passiflora caerulea*.
Second row: *Passiflora capsularis*, *Passiflora cincinnata*, *Passiflora coccinea*.
Third row: *Passiflora coriacea*, *Passiflora cuprea*, *Passiflora dispar*.
Bottom row: *Passiflora edmundoi*, *Passiflora edulis*, *Passiflora edulis* f. *flavicarpa*.
[All photos YK except *Passiflora edulis* and *Passiflora edulis* f. *flavicarpa* AF.]

The uncommon yellow form is ***P. edulis* f. *flavicarpa*** [Yellow Granadilla, Yellow Passion Fruit].

P. fernandezii L. Escobar, *Phytologia* 66(1): 80–81, 1989. **Vines**, pubescent; stems angulate, striate, gray-green. **Stipules** linear, ~1 × ~0.5 cm. **Leaves** oblong, (3.5–)5–9.4 × (1.4–)2.5–4.2 cm, base rounded, margins entire, apex obtuse then abruptly acuminate, coriaceous, lustrous, gray-green, mostly glabrous; petiole 1.2–1.8 cm long, nectaries 2, paired, dark brown, cup-shaped, ~1 mm Ø. **Inflorescences** with flowers solitary; peduncles 2.1–3 cm long; bracts 3, verticillate, ovate, 1.9–2.4 × 1–1.4 cm, base cuneate, margins irregularly crenate, with 4–6 scattered dark brown nectaries above middle, apex acute to rounded, densely pubescent; pedicel ~2 mm long, flowers pendent. **Flowers** campanulate, ~6 cm Ø, white; hypanthium funnelform, 1.6–2 × 1–2.9 cm, with irregular rows of filaments 0.5–2 mm long in lower half of inner surface; sepals ovate, 2.6–3.1 × 1.5–2.1 cm, apex rounded, softly coriaceous, with subapical corniculus ~2 mm long abaxially; petals oblong, ~1.4 × ~0.7 cm, base narrowed, apex rounded, delicately membranous; corona biseriate, filamentous, outer filaments ~6 × ~0.4 mm, inner ones 2.3–2.5 cm long, purple-banded; operculum horizontal, non-plicate, membranous; ovary ellipsoidal, densely pubescent. **Distr:** Bolivia. **Ecol:** In lowland areas; 0–500 m. **Zones:** 10–12.

P. foetida L., *Sp. Pl.* 959, 1753. (Syn: *Dysosmia foetida*, etc.). [Goat-scented Passion Fruit, Love in a Mist, Wild Passion Fruit, Wild Water Lemon]. **Vines**, herbaceous, highly variable, foul-smelling. **Stipules** partly clasping. **Leaves** broadly ovate to oblong-ovate, (3–)4.5–13 × (2.5–)4–12 cm, base cordate, trilobed or unlobed, margins irregularly undulate, apex acute, membranous, glandular-ciliate, adaxially sericeous, abaxially pubescent with small nectaries in distal part; petiole 2–6 cm long, without glands, pubescent. **Inflorescences** reduced to a single flower, opposite tendril; bracts 3, 2- or 3(or 4)-pinnatifid, lobes threadlike, glandular-pilose apically. **Flowers** white or light purple, with white spots or lines, 2–3 cm Ø; sepals ~1.5 cm long, awned; petals ~1.5 cm long; corona 3–5-seriate, filamentous, 2 outer series ~1 cm long, inner 3 series 1–2.5 mm long; operculum 1–1.5 mm long, membranous; disk cupular, 1–2 mm high; androgynophore 5–7 mm tall; stamens coherent at base, flat, anthers oblong, ~4 mm long; ovary shortly stipitate, ellipsoid, ~6 mm long, glabrous, styles 3(or 4), 5–6 mm long, stigmas capitate. **Fruits** berries, orange or orange-red, ovoid-globose, 2–3 cm Ø, glabrous, surrounded by persistent bracts; seeds many, light brown to black, elliptic, 3–4 mm long. **Distr:** West Indies, N South America. **Ecol:** Fl. throughout most of the year. **Zones:** 10–12. **Info:** Fruits edible with a sweet-sour flavor. ***P. foetida* var. *hirsutissima*** has larger and deep pink flowers. ***P. foetida* var. *oaxacana*** is densely villous throughout.

P. galbana Masters, *Gard. Chron.* ser. 3, 20: 555, f. 97, 1896. **Vines**, glabrous; stem terete, slightly flexuous. **Stipules** ovate-lanceolate, 1–2 × 1 cm, base oblique, subcordate, apex acute, mucronulate, coriaceous. **Leaves** oblong-lanceolate, 7–13 × 3–6.5 cm, base rounded or cordulate, apex obtuse, mucronulate, coriaceous; petioles 1–1.5 cm long, minutely biglandular toward apex. **Inflorescences** with flowers solitary; peduncles 5–9 cm long, terete; bracts oblong-lanceolate, 1–1.5 × 0.6–0.8 cm, acute, borne ~1 cm below base of flower. **Flowers** white or pale yellow; calyx tube broadly cylindric-campanulate, ~0.7 × 1 cm; sepals narrowly oblong, ~4 × 0.7 cm, dorsally broadly keeled, keel terminating in a foliaceous awn 5–6 mm long; petals similar in shape and size to sepals; corona filaments biseriate, outer liguliform, 0.8–1 cm long, inner filiform, 2–3 mm long; operculum borne at middle of calyx tube, filamentose; limen cupuliform; ovary ellipsoid, glabrous. **Fruits** narrowly ovoid, 6–7 × ~2 cm, narrowed to a blunt point, 5-angled; seeds obovate, 4–5 × 3–3.5 mm. **Distr:** E Brazil, Bahia to Rio de Janeiro. **Ecol:** In coastal rainforests and forest margins. Fl. & Fr. throughout the year, mostly Mar & Sep. **Zones:** 10–12.

P. gibertii N.E. Brown, *Trans. & Proc. Bot. Soc. Edinb.* 20: 58, 1896. **Vines**, glabrous; stem terete. **Stipules** semiovate, 2–5 × 0.7–3 cm, oblique, base rounded, margins subentire, apex acute, mucronate. **Leaves** 3(5)-lobes, 5–9(–18) × 5–12(–22) cm; lobes oblong-lanceolate, 2–2.5 cm wide, base cordulate, apex acute, sinuses 1–2-glandular, membranous; petioles 1–5 cm long, slender, glands (2–)4–5(–6), paired, subclavate, ~1 mm long. **Inflorescences** with flowers solitary; peduncles 4–8.5 cm long; bracts free, ovate, 1.5–3.3 × 1–2 cm, margins entire, apex acute, mucronulate. **Flowers** 6–9 cm Ø; calyx tube short-campanulate; sepals oblong-lanceolate, ~3 × 0.8 cm, dorsally awned near apex, awn 1–1.5 cm long, green outside, bluish inside; petals oblong, slightly shorter and broader than sepals, bluish; corona filaments filiform, 5–6-seriate, outer 2 series 0.6–2.8 cm long, successive series decreasing in length from 7 mm long to 1 mm long, greenish with purple; operculum membranous, 2–3 mm high; limen tubular, closely surrounding base of gynophore; ovary ellipsoid, glabrous. **Fruits** subglobose to ovoid, 4–6 × 3–6 cm, yellow to orange; seeds ~5 × 2–4 mm, reticulate. **Distr:** SC Brazil, Paraguay, NE Argentina. **Zones:** 9–11.

P. glandulosa Cav., *Diss.* 10: 453, pl. 281, 1790. (Syn: *Tacsonia glandulosa*, etc.). **Vines**; stem terete or subangulate, usually glabrous, purplish. **Stipules** linear-subulate or setaceous, soon deciduous. **Leaves** ovate-oblong to oblong-lanceolate, 6–15 × 4–10 cm, base rounded, or acute, margins entire, apex acute or acuminate, mucronulate, ± coriaceous, abaxially glaucous; petioles to 2.5 cm long, glands 2 below middle, sessile. **Inflorescences** with flowers solitary; peduncles to 8 cm long; bracts usually linear-lanceolate, 0.5–1 × 0.1–0.2 cm, margins glandular. **Flowers** red or scarlet; calyx tube cylindric, 1.5–2.5 × 0.8–1 cm, base ventricose; sepals oblong or linear-oblong, 2–5 × 0.6–1.3 cm, shortly awned near apex, erect, or reflexed with age; petals slightly shorter; corona biseriate, outer filaments free, awl-shaped, white or pale pink, ~1 cm long, inner a cylindric membrane, arising near base of tube, subadnate to tube, upper part free, apex lacerate-filamentose, cut portion 2–5 mm long; operculum borne ~3 mm above base of tube, pink, dependent, margin recurved, shortly filamentose; limen membranous, cupuliform, ~1 mm high, closely surrounding base of gynophore, minutely fimbrillate; ovary ellipsoidal, subangular, usually white-puberulent. **Fruits** ovoid, 5–6 × 2.5–3 cm; seeds obovate, ~8 × 4 mm, finely reticulate. **Distr:** Guianas to lower Amazon basin and E Brazil. **Ecol:** In tropical lowlands. Fl. & Fr. throughout year, mostly Sep–Dec. **Zones:** 10–12.

P. hahnii (Fourn.) Mast, *Fl. Bras.* 13, pt. 1: 569, 1872. (Syn: *P. cookii*). **Vines**, glabrous; stem wiry, terete, angulate above, striate. **Stipules** reniform, 1 × 1.2–1.7 cm, entire to crenate or minutely dentate. **Leaves** broadly ovate-lanceolate, 5–14.5 × 3.5–13 cm, margins entire or usually inconspicuously trilobed 1.5–3 cm from apex; lobes with base peltate, apex setosely tipped, 3–5-nerved, usually membranous; petioles 1.5–3 cm long, glandless. **Inflorescences** with flowers solitary or rarely paired; peduncles 2.5–7 cm long; bracts 2, cordate, 2.5–3 × 1.5–2 cm, denticulate toward base. **Flowers** 4–6 cm Ø; sepals and petals similar, oblong, 2–3 × 0.7–0.9 cm, obtuse, radiate, white or cream; corona filaments yellow, biseriate, outer to 1.5 cm long, apices dilated, spreading, inner to 5 mm long, clavate, erect; operculum membranous, closely plicate, 2.5–3 mm high, margin incurved; limen annular; ovary bluish-black, glaucous, subglobose or ovoid. **Fruits** globose, 3–3.5 cm Ø; seeds ovate-oblong, slightly flattened, curved, ~5 × 4 mm, faces reticulate. **Distr:** SE Mexico through C America to Colombia in NW South America. **Ecol:** <1,400 m. **Zones:** 10–12.

P. holosericea L., *Sp. Pl.* 958, 1753. (Syn: *P. reticulata*). **Vines**, usually densely pubescent; stem terete, striate, corky below. **Stipules** filiform, 6 mm long. **Leaves** trilobed, 5–10 × 4–7 cm; lobes rounded, mucronulate, middle one much longer, entire, bidentate at cordate base, trinerved, reticulate-veined, adaxially velvety-pubescent, abaxially densely and softly tomentose. **Inflorescences** solitary or paired, axillary; peduncles, (1)2–4-flowered; pedicels with 2 or 3 subulate bracteoles 2 mm long. **Flowers** 3–4 cm Ø; sepals ovate-lanceolate, 1.3–1.5 × 0.5 cm, obtuse, densely pubescent outside, glabrous inside, white, sparingly spotted with red; petals oblanceolate or spathulate, 1–1.3 × 0.6 cm, white, brown-streaked and red-mottled; corona filaments biseriate, lanceolate, 7 × 1 mm, apex yellow, purple toward base, inner capillary, 4–5 mm long, clavate; operculum membranous, closely plicate, margin incurved; limen annular, close to operculum; ovary obovoid, densely pilose. **Fruits** globose, 1.5 cm Ø, glabrous or softly pubescent; seeds obovate-obcordate, ~3 × 2 mm, coarsely reticulate. **Distr:** Mexico to Honduras, Cuba, Colombia and Venezuela. **Ecol:** <700 m. **Zones:** 10–12.

P. incarnata L., *Sp. Pl.* 959, 1753. (Syn: *P. edulis* var. *kerii*, etc.). [Apricot Vine, May Apple, May Pops, Wild Passion Flower]. **Vines**, herbaceous, glabrous or finely pilosulous; stem terete, younger parts angular. **Stipules** setaceous, 2–4 mm long, entire, soon deciduous. **Leaves** 3(5)-lobed, 6–15 × 7–17 cm, margins serrate, membranous. **Inflorescences** with flowers solitary; peduncles stout, to 10 cm long; bracts oblong, 4–8 × 2–4 mm, margins serrate, free, glands 2, basal. **Flowers** variable, white, pinkish, lavender, mauve or violet, 6–8(–9) cm Ø; sepals variable inside, green outside, oblong-lanceolate, ~3 × 1 cm, slightly keeled outside, keel terminating in an awn 2–4 mm long; petals slightly shorter than sepals; corona in several series, white, pinkish, lavender or violet, outer 2 series 1.5–2.5 cm long, wavy above, inner series 2–4 mm long; ovary ovoid, densely tomentose. **Fruits** ovoid or subglobose, 5–6 × 3.5–5 cm, greenish-yellow, edible; seeds obovate or nearly obcordate, 4–5 × 3–4 mm, reticulate. **Distr:** SE USA. **Ecol:** In sunny areas on moist, well-drained sandy soils and disturbed sites; 0–1,100 m. Fl. Jun–Sep. **Zones:** 8–12.

P. insignis (Mast.) Hook., *Bot. Mag.* 99: pl. 6069, 1873. (Syn: *P. callimorpha*). **Vines**, villous except adaxial leaf surfaces; stem terete. **Stipules** deeply bipinnatisect, ~2 × 1 cm, segments acute. **Leaves** ovate-lanceolate, 15–25 × 7–12 cm, base cordulate, margins denticulate, apex acute, 3–5-nerved, coriaceous, adaxially glabrous, lustrous, rugulose, abaxially densely ferruginous-lanate; petioles to 2 cm long, obscurely 2–4-glandular, densely lanate.

Top row: *Passiflora fernandezii, Passiflora foetida, Passiflora foetida* var. *hirsutissima*.
Second row: *Passiflora foetida* var. *oaxacana, Passiflora galbana, Passiflora gibertii*.
Third row: *Passiflora glandulosa, Passiflora hahnii, Passiflora holosericea*.
Bottom row: *Passiflora incarnata, Passiflora insignis*.
[All photos YK except *Passiflora foetida* AF and *Passiflora incarnata* MD.]

Top row: *Passiflora kermesina, Passiflora lancetillensis, Passiflora laurifolia*.
Bottom row: *Passiflora ligularis, Passiflora lobata, Passiflora maliformis, Passiflora mexicana*.
[All photos YK except *Passiflora ligularis* AF.]

Inflorescences solitary; peduncles 15–20 cm long, stout; bracts ovate-oblong, free to base, 3.5–4 × 1–1.5 cm, base narrowed, apex obtuse, shortly fimbriate, concave. **Flowers** with calyx tube cylindric, 3–4 × ~0.8 cm, enlarged at base to ~1.5 cm wide, slightly ferruginous-tomentose outside; sepals violet-crimson, oblong, 7–9 cm long, ~2 cm wide, obtuse, concave, dorsally keeled, keel terminating in an awn 1.5–2 cm long; petals oblong, 6–7 × ~1.5 cm, obtuse, rose-purplish; corona 1-seriate, filamentose, filaments erect, 0.8–1 cm long, blue and white; operculum attached ~1 cm above base of tube, dependent, ~7 mm long, shallowly lacerate, margin not recurved; ovary subglobose, obscurely 3-angled, densely lanate. **Distr:** N & NC Bolivia. **Ecol:** 1,300–3,000 m. **Zones:** 10–12.

P. kermesina Link & Otto, *Verh. Ver. Befoerd. Gartenb.* 2: 403, 1826. **Vines**, glabrous; stem terete, slender. **Stipules** subreniform, 1–2.5 × 0.5–1.2 cm. **Leaves** trilobed, 4–8 × 5–11.3 cm, margins entire; membranous, abaxially purplish-violet; petioles 2.8–4 cm long, glands 2–4, elongated, above middle, 1–3 mm long. **Inflorescences** with flowers solitary; peduncles 8–15 cm long; bracts minute, 2–2.5 cm below base of flower, soon-deciduous. **Flowers** deep pinkish-purple to pinkish-violet, (6–)8–9.5(–11) cm Ø; floral tube cylindrical-campanulate, 0.9–1 cm long, whitish-green, pink outside; sepals pale pinkish-purple outside, deep pinkish-purple to pinkish-violet inside, oblong-lanceolate, 3.5–5.6 × 0.5–1 cm; petals deep pinkish-purple to pinkish-violet, subequal to sepals; corona 5-seriate, coronal filaments thin, outer 2 series 1.4 cm long, violet with a small paler violet to white band beneath middle, 3rd series 8 mm long, 4th series 6 mm long, innermost series 0.9–1 cm long, base pale violet, deep violet above; ovary oblong-ovoid, green, glabrous. **Fruits** ellipsoid to oblong-ovoid, 7.5–9.5 × 2.2–3 cm, olive-green. **Distr:** Brazil. **Ecol:** In coastal forests on sandy soil near the sea. Fl. Apr–Sep. **Zones:** 10–12.

P. lancetillensis MacDougal & Meerman, *Novon* 13(4): 460–464, 2003. **Vines**, large, slightly pubescent; stem terete. **Stipules** minute, deciduous. **Leaves** unlobed, ovate to ovate-cordate, 8–22 × 7–19 cm, adaxially glabrous, abaxially pubescent; petioles stout, 3–5.5 cm long, glands 7–12. **Inflorescences** cymes, solitary, 1–6-flowered; peduncles minute to 3 cm long, terminating in a tendril; bracts to 2 mm long; pedicels 2.8–5 cm long. **Flowers** white to cream, 4.5–5 cm Ø; sepals green and pubescent outside, white to cream inside, narrowly ovate-oblong, 1.8–2.5 × 0.6–0.9 cm; petals white, oblong, 1.9–2.7 × 0.6–0.8 cm; corona in 3–4 series, outer series white to cream or very pale apricot becoming more intense near base, filiform, 1.5–2 cm long, slightly wavy, inner 2–3 series white, 3–6 mm long; ovary ellipsoid, densely pubescent. **Fruits** widely ellipsoid to globose, 6.5–8 ×4–4.8 cm, light yellowish-green; seeds 0.9–1.2 × 0.7–0.8 cm, reticulate, with 4 marginal wings. **Distr:** Belize, Guatemala, Honduras, El Salvador. **Ecol:** In the canopy of tropical evergreen or seasonal broad-leaved hill forests; 20–750 m. **Zones:** 11–12.

P. laurifolia L., *Sp. Pl.* 956, 1753. (Syn: *P. oblongifolia*, etc.). [Bell Apple, Water Lemon]. **Vines**, glabrous except ovary and bracts; stem terete. **Stipules** linear, 3–4 mm long, coriaceous. **Leaves** unlobed, ovate-oblong or oblong, 6–12 × 3.5–8 cm, margins entire, thickly coriaceous; petiole 0.5–1.5 cm long, glands 2, paired near apex. **Inflorescences** solitary; peduncles 2–3(–8) cm long; bracts free, ovate-oblong, 2.5–4 × 2–2.5 cm, glandular-serrate. **Flowers** purplish-white and white, 5–7 cm Ø; sepals white, mottled with purple spots inside, whitish-green outside, oblong, 2–2.5 × 1 cm; petals white, densely mottled with purple spots, oblong, 1.8–2.4 × 0.8–0.9 cm; corona 6-seriate, banded purple and white in lower third, banded violet and white above, outermost series ~2 cm long, 2nd series ligulate, 3–4 cm long, inner series ~1 mm long; operculum membranous, horizontal; limen cupuliform, closely surrounding base of gynophore; ovary ovoid, tomentose. **Fruits** ovoid, 5–8 × 4–5 cm, lemon-yellow to pale orange; seeds obcordate, 5–7 × 3–5 mm, finely reticulate. **Distr:** Guianas, West Indies, Trinidad, Venezuela, Peru, Brazil. **Ecol:** In forested slopes and thickets; 600–800 m. Fl. & Fr. throughout most of the year. **Zones:** 11–12. **Info:** The fruits are edible and are commercially cultivated.

P. ligularis Juss., *Ann. Mus. Hist. Nat. Paris* 6: 113, pi. 40., 1805. (Syn: *P. lowii*, *P. serratistipulata*). [Grenadia, Sweet Granadilla]. **Vines**, glabrous. **Stipules** ovate-lanceolate or oblong-lanceolate, to 2.5 cm long. **Leaves** broadly ovate, 8–15 × 6–13 cm, base deeply cordate, margins entire, apex abruptly acuminate; petiole to 10 cm long, bearing 4–6 scattered, liguliform or filiform glands 0.3–1 cm long. **Inflorescences** with flowers solitary or paired; peduncles 2–4 cm long; bracts ovate-lanceolate, basally connate. **Flowers** 6–9 cm Ø; sepals ovate-oblong, 2.5–3.5 × 1–1.5 cm, white inside, green outside; petals oblong, 3 × 1 cm, white or pinkish-white, with bluish or pinkish spots; corona 5–7-ranked, 2 outer rows of filaments equaling petals, radiate, blue at apex, banded with white and reddish-purple. **Fruits** ovoid, 6–8 × 4–5 cm Ø, pericarp parchmentlike, pulp white, edible; seeds narrowly obcordate, reticulate. **Distr:** Mexico to Venezuela, and from Peru to Bolivia. **Ecol:** 800–3,000 m. **Zones:** 9–12.

P. lobata (Killip) Hutchinson ex MacDougal, *Phytologia* 60: 446, 1986. (Syn: *Tetrastylis lobata*, etc.). **Vines**, mostly lightly pubescent; stem almost terete to slightly 3-keeled. **Stipules** foliaceous, ovate-auriculate, 0.4–1.8 × 0.2–1.1 cm. **Leaves** trilobed, 4–17 × 4–17 cm, base cordate, margins entire, bearing lamina nectaries near margins; petiole 2–6.5 cm long, glands 2(4). **Inflorescences** with flowers paired; peduncles 1.3–5.3 cm long; bracts setaceous, 1–4 mm long. **Flowers** white with purplish-red, ~4.5–6 cm Ø; sepals green outside, white inside, lanceolate-triangular, 2.2–3.5 × 0.7–1.1 cm, with a slight apical horn; petals white, lanceolate, 1.2–2.3 × 0.5–0.9 cm; corona filaments 1-seriate, 1–2.6 cm long, filiform, white or white with purplish-red bands, very rarely with a 2nd vestigial inner series; operculum usually dark red; ovary glabrous, ellipsoid. **Fruits** long-stipitate, narrowly obovoid, 11–18.5 × 3.5–6 cm, reddish or purplish-red, irregularly dehiscent; seeds ~7–9 × 4–6 mm, coarsely reticulate. **Distr:** Nicaragua, Costa Rica, Panama, Colombia. **Ecol:** In primary and secondary forests, premontane rainforests and mountain slopes; 10–1,500 m. Fl. & Fr. throughout most of the year. **Zones:** 11–12.

P. maliformis L., *Sp. Pl.* 956, 1753. (Syn: *P. caudata*, etc.). [Conch Apple, Sweet Calabash, Sweet Cup]. **Vines**, glabrous or finely pilosulous; stem terete or younger parts subangular. **Stipules** narrowly linear or lance-linear, 0.9–1.5 × 0.1–0.2 cm, entire or minutely serrulate. **Leaves** ± ovate, 6–12(–25) × 4–10(–15) cm, base rounded, truncate, or cordulate, margins undulate or finely serrulate, apex acute or abruptly acuminate, membranous; petioles 1.5–5 cm long, biglandular, glands to 1.5 mm wide, subsessile. **Inflorescences** with flowers solitary; peduncles to 5 cm long; bracts broadly ovate, 4–6 × 3.5–4.5 cm, united for ~1 cm from base, thinly membranous, completely enveloping the bud, lime-green. **Flowers** with calyx tube campanulate, 1 cm long, 1.2 cm wide at throat; sepals oblong or oblong-lanceolate, ~4 × 1.5 cm, apex cucullate, carinate, awn ~5 mm long, fleshy, green; petals linear-lanceolate, ~3 × 0.5 cm, green, densely mottled with dark red-purple inside; corona with filaments several-ranked, 2 outer rows white, violet-banded above middle, rose-purple near base, outermost terete, 1.5 cm long, second ligulate, 3 cm long, successive ranks consisting of minute tubercles, green, purple-tipped; operculum membranous, horizontal, cupuliform, 6 mm high, pale green, margin recurved, denticulate; nectar ring horizontal, margin entire; ovary oblong or subglobose, glabrous. **Fruits** globose, 3.5–4 cm Ø, green or orange-green, pericarp very hard; seeds oblong-obcordate, 5–6 × ~4 mm, finely reticulate, metallic-gray. **Distr:** West Indies, Venezuela, Colombia and N Ecuador. **Ecol:** In tropical regions; <1,700 m. **Zones:** 10–12. **Info:** The shell of the fruit is sometimes so hard that it must be broken with a hammer. The pulp has a grapelike flavor and is used in making cooling drinks (Killip 1938).

P. mexicana Juss., *Ann. Mus. Hist. Nat.* 6: 108, pl. 38, f. 2, 1805. (Syn: *P. contrayerva*, etc.). [Mexican Passion Flower]. **Vines**, glabrous; stem subquinquangular, grooved. **Stipules** setaceous or narrowly linear, 1.5–2 mm long. **Leaves** bilobed, 3–8 cm wide, base rounded or truncate, trinerved, abaxially ocellate, glabrous, adaxially dark green, abaxially paler; lobes oblong, occasionally linear, 0.5–4 cm wide, obtuse, rarely dilated near apex and slightly emarginate, suberect and nearly parallel or slightly divergent, length along midvein 0.5–4 cm, along lateral nerves 3–8 cm; petioles 1–2 cm long, glandless. **Inflorescences** with flowers solitary or paired; peduncles 2–3 cm long; bracts minute, setaceous, deciduous. **Flowers** 2.5–4 cm wide; calyx patelliform, about 1 cm wide, deep reddish-purple inside; sepals narrowly lanceolate, 1–1.5 × 0.4–0.5 cm, green to red outside, white inside; petals much shorter, recurved; corona filaments biseriate, filiform, outer ~1.5 cm long, rich pink to deep crimson, subreflexed, inner 2 mm long, erect; operculum closely plicate, strongly incurved, white, pink-tinged toward margin, minutely fimbrillate; limen annular, to 1.5 mm high, white; ovary globose, glabrous. **Fruits** globose, 0.6–1.2 cm Ø; seeds ovate, 3 × 2 mm, transversely sulcate with 7 or 8 irregular ridges. **Distr:** S Arizona to C Mexico. **Ecol:** Under trees in riparian areas and forests. Fl. Jul–Sep. **Zones:** 9–11.

P. miersii Mast, *Fl. Bras.* 13(1): 599, 1872. [Sjimio]. **Vines**, glabrous; stem terete, wiry, often golden-yellow. **Stipules** semiovate, 1–2.5 × 0.5–1.2 cm, margins ± crenulate, both ends rounded, apex mucronulate. **Leaves** lanceolate or ovate-lanceolate, 3–8 × 2–3 cm, base subpeltate, apex acute or rounded, 3–5-nerved, abaxially claret-colored; petioles to 1.5 cm long, biglandular. **Inflorescences** with flowers solitary; peduncles 3–5 cm long; bracts variable, narrowly ovate, oblanceolate, or linear-setaceous, to 6 × 2 mm. **Flowers** 4–5 cm Ø; calyx tube campanulate, base ventricose; sepals oblong, 1.5–2.5 × 0.5–0.7 cm, obtuse; petals subequal to sepals, obtuse, white; corona filaments filiform, in 4 series, outer 2 radiate, 1–1.5 cm long, whitish, base purple-barred, filaments of 3rd series ~1.5 mm long, violet, filaments of 4th series 5–7 mm long, erect, violet or deep purple; operculum membranous, erect below, bearing a row of filaments on outer surface, filaments 4–5 mm long; limen cupuliform; ovary ovoid, glabrous, glaucous. **Fruits** obovoid or ellipsoid, 3–4 × 1.5–2 cm; seeds obovate. **Distr:** E Brazil. **Ecol:** On mountains. **Zones:** 10–12.

P. miniata Vanderplank, *Curtis's Bot. Mag.* 23(3): 223–230, 2006. [Flaming Red Passion Flower, Monkey Guzzle, Red Granadilla]. **Vines**, robust, to 15 m long; stem deeply grooved, woody. **Stipules** narrowly linear, 3–6 mm long, nectar glands 1 or 2. **Tendrils** strong, 13–33 cm long. **Leaves** simple, oblong, 6–22 × 3–12 cm, base subcordate, margins double-crenulate, apex acute, lateral veins 5–7 per side; petiole 3–4 cm long, basal glands 2, nectariferous. **Inflorescences** with flowers solitary; peduncles 4–8 cm long; bracts 3, ovate, 3–5.5 × 1–2.4 cm, green or pink, red-veined, glands 6–16. **Flowers** bright red, 10–13 cm Ø; sepals and petals reflexed; hypanthium green, tinted pink or red; sepals linear-lanceolate, 4.7–6 × 1–1.9 cm, adaxially vivid scarlet, abaxially dull red, terminal awn yellowish, 3–5 mm long; petals linear-lanceolate, 4.5–5.7 × 1–1.8 cm, scarlet; corona filaments in 3 series; outer series 1.8–2.4 cm long, red or purple, free; central series, 6–8 mm long, white, free; inner series, 1.1 cm long, white, distally free; operculum white; androgynophore erect, 5.5–6 cm high, base white, olive-green red-speckled towards androecium; staminal filaments pale olive-green red-speckled, anthers green above; ovary ovoid, yellowish-green, style red, 1 cm long, stigma olive-green. **Fruits** spherical, 4.5–5.5 cm Ø, edible; seeds brown, ovate, ~6 mm long. **Distr:** Bolivia, Brazil, Peru and Colombia. **Ecol:** In tropical forests. **Zones:** 10–12.

P. misera HBK., *Nov. Gen. & Sp.* 2: 136, 1817. (Syn: *P. maximiliana*, etc.). **Vines**; stem angulate or strongly compressed, striate, glabrous or finely puberulous, sometimes scabrid. **Stipules** setaceous to narrowly linear, 2–3.5 mm long, falcate. **Leaves** bilobed (lobes widely divergent), 0.5–2.5 × 4–13 cm, base cordulate or subtruncate, trinerved, membranous; petioles 1–3.5 cm long, glands absent. **Inflorescences** with flowers solitary, rarely in pairs; peduncles slender, 1.5–10 cm long; bracts setaceous, 3–5 mm long. **Flowers** 2.5–4 cm wide; sepals lance-oblong to linear-oblong, 1–1.8 × 0.3–0.5 cm, obtuse, green, usually puberulous outside, white inside; petals linear-oblong, 0.8–1.3 × 0.2–0.4 cm, obtuse, white; corona filaments in 2 series, outer filiform, 1–1.5 cm long, purplish, inner linear, 3–4 mm long; operculum membranous, closely plicate, incurved, margin minutely fimbrillate; limen annular; ovary ovoid, glabrous. **Fruits** usually globose, 0.5–1.3 cm Ø; seeds narrowly ovate, ~3.5 × 2 mm. **Distr:** Panama, N & E South America to N Argentina. **Ecol:** In moist places of forest borders, gallery forests, mountains, savannahs, mountains, marshes and reservoirs, floodplains; 0–800 m. Fl. & Fr. Throughout the year, mostly Aug–Apr. **Zones:** 9–12.

P. mixta L. f., *Suppl. Pl.* 408, 1781. (Syn: *P. brachychlamys*, etc.). [Curuba, Curubita]. **Vines**, glabrous or pubescent; stem angulate, tip erect. **Stipules** reniform, 0.6–3.3 × 0.2–2 cm, glandular-

Below: *Passiflora miersii, Passiflora miniata, Passiflora misera, Passiflora mixta.*

MD

YK

MD

serrate. **Leaves** trilobed, 3.5–12 × 5–19 cm, margins serrate, adaxially glabrous; petiole 0.9–8 cm long, glands 4–10, subsessile or stipitate. **Inflorescences** with flowers solitary or rarely paired; peduncles 1–9 cm long, stout; bracts oblong to oblong-ovate, 1.2–7.5 × 0.5–2.5 cm, basally connate, entire. **Flowers** pink, occasionally pinkish-orange, rarely white, 5–9 cm Ø, 9.3–19.5 cm long; floral tube cylindrical, 7–15 × 0.5–1.5 cm, green; sepals pink, oblong to oblong-ovate, 2.3–5.5 × 0.9–2.1 cm, aristate, awn 2–5 mm long; petals pink, oblong to oblong-ovate, 2.5–5 × 1–2 cm; corona 1(2)-seriate, reduced to a purple ring with white teeth; ovary ellipsoid, densely pubescent or glabrous. **Fruits** oblong-ovate, 4–7.2 × 2–3.5 cm, green; seeds obcordate, 3–5 × 2–4 mm, reticulate, dark brown. **Distr:** W Venezuela to N Bolivia and Chile. **Ecol:** In cloud forests; 1,700–3,600 m. Fl. throughout most of the year. **Zones:** 9–12.

P. mollissima (HBK.) Bailey, *Rhodora* 18: 156, 1916. (Syn: *P. tomentosa*, etc.). [Banana Passionfruit, Curuba]. **Vines**, to 10 m long; stem terete, striate, densely or softly downy. **Stipules** denticulate, to 9 × 4 mm. **Leaves** trilobed for ~2/3 the length, to 12.5 × 15 cm, lobes ovate-oblong, 3–4 cm across, margins sharply serrated, softly pubescent; petiole to 3 cm long, petiole glands 8–12, sessile. **Inflorescences** with flowers solitary; peduncles to 6 cm long; bracts 2.5 cm long, united for 1/3–1/2 their length. **Flowers** pendent, large, showy, pink or coral-pink, 6–9 cm across, segments held horizontally or reflexed; calyx tube olive-green pink-tinged outside, white inside, to 8 × 1 cm; sepals pink, oblong, 2.5–3.5 × 1–1.5 cm; petals pink, slightly shorter than sepals; corona filaments a purple ring (warty rim) with pinkish tubercles. **Fruits** oblong-ovoid, resembling a small banana, to 10 × 3.5 cm, softly pubescent, green ripening to yellow, edible. **Distr:** S America; naturalized and often weedy in many countries. **Ecol:** 1,200–3,200 m. **Zones:** 9–11.

P. mucronata Lam., *Encycl.* 3: 33, 1789. (Syn: *P. albida*, etc.). [Sururu]. **Vines**, glabrous; stem terete, stout. **Stipules** ovate-lanceolate, 1.5–2.5 × 0.5–1.5 cm, base oblique, cordulate, apex subacute, mucronate, often glaucous, coriaceous. **Leaves** ovate-cordate, 4–12 × 2.5–6 cm, base subpeltate, apex rounded, 3–5-nerved, reticulate-veined, thickly coriaceous; petiole 1–2 cm long, obscurely 2–4-glandular near middle. **Inflorescences** with flowers solitary; peduncles to 8 cm long, articulate ~1 cm below apex, terete; bracts oblong-lanceolate, 2–2.5 × ~1 cm, base narrowed, apex acute, apiculate, sessile, thinly membranous, usually glaucous, borne ~1 cm below base of flower. **Flowers** 8–10 cm Ø, white; calyx tube campanulate; sepals linear, to 8 mm wide, obtuse, dorsally broadly keeled, keel terminating in an awn 2–3 mm long; petals linear, about as long as sepals, obtuse; corona filaments biseriate, outer very slender, ~1 cm long, inner filiform-clavate, 2–3 mm long; operculum borne at middle of tube, filamentose, filaments 2–3 mm long; limen cupuliform; ovary narrowly ovoid. **Fruits** ovoid, 4–5 × ~2.5 cm; seeds oblong-obcordate, ~5 × 4 mm, strongly flattened, reticulate. **Distr:** E Brazil, from Bahia to Rio de Janeiro. **Ecol:** In coastal rainforests and forest margins. Fl. Apr.–Sep. **Zones:** 10–12.

P. murucuja L., *Sp. Pl.* 957, 1753. [Virgin Island Passion Flower]. **Vines**, glabrous, slightly woody; stem angulate, deeply grooved, wiry. **Stipules** linear-setaceous, 2–4 mm long. **Leaves** bilobed, variable, 1–2(–3) × 2.5–4(–7) cm, margins entire, laminar nectaries present; petioles to 1.5 cm long, glands absent; tendrils simple. **Inflorescences** with flowers solitary or paired; peduncles 1–2.5 cm long; bracts setaceous, 2–4 mm long, borne below middle of peduncle, free. **Flowers** red to blood-red, 4–6 cm Ø; calyx 3–4 cm long, sepals red, lanceolate-oblong, 1.5–3 × 0.3–0.7 cm; floral tube 5–7 mm long, red outside; petals red, linear-oblong, 1–2 × 0.3–0.4 cm; corona in 1 series, cylindrical membrane 1–1.5 cm long, red or orange-red; ovary ovoid, glabrous. **Fruits** globose, 1–1.5 cm Ø, black-purple; seeds obovate, 2–3 × 1–2 mm, transversely 6–7-sulcate. **Distr:** Puerto Rico, Haiti, Dominican Republic. **Ecol:** In forests. Fl. & Fr. throughout the year. **Zones:** 11–12. **Info:** This species produces tiny greenish 'pseudo-flowers' or aborted flowers especially during winter (Ulmer & MacDougal 2004).

P. nephrodes Mast., *Bull. Torrey Club* 17: 282, 1890. **Vines**, densely hirsute; stem subangular. **Stipules** subreniform, 1.5–2 × 0.8–1 cm, aristate, coarsely serrate-dentate. **Leaves** trilobed, 6–13 × 5–9 cm, midlobe ovate or ovate-lanceolate, 1/2 to 2/3 the length of the leaf, 3–5 cm wide, base subcordate, margins denticulate, apex acute or subobtuse, lateral lobes deltoid-ovate; petioles 1–2 cm long, with 2 or 3 stipitate glands near middle, 2–3 mm long. **Inflorescences** with flowers solitary; peduncles 4–6 cm long, in fruit much longer; bracts borne ~3 mm below base of flower, ovate-lanceolate, 1–2 cm long, 0.7–1 cm wide, serrate. **Flowers** 6–8 cm wide; sepals oblong, 3–3.5 × 1–1.2 cm, obtuse, dorsally awned just below apex; petals linear-oblong, slightly shorter and narrower than sepals; corona filaments filiform, in several series, outermost 2.5–3 cm long, radiate, proximally purple, distally white, those of successive series 3–6 mm long, erect; operculum membranous at base, filamentose above, filaments 5–8 mm long; limen membranous, 3 mm high, closely surrounding base of gynophore; ovary ovoid, hirsute. **Fruits** ovoid, ~4 × 2 cm, base narrowed; seeds obovate, ~6 × 3 mm, reticulate. **Distr:** E Peru to W Bolivia and Amazonian Brazil. **Ecol:** <2,500 m. **Zones:** 10–12.

P. organensis Gardn., *London Journ. Bot.* 4: 104, 1845. (Syn: *P. maculifolia*, etc.). [Organ Mountain Passion Flower]. **Vines**, glabrous; stem subangular, compressed. **Stipules** linear-subulate, 2–3 mm long, subfalcate. **Leaves** bilobed (rarely trilobed), variable, lobes broadly ovate to lanceolate, 1.5–3 cm wide, base rounded, apex obtuse or acute, mucronulate, sinus wanting or lunate, 3–5-nerved (often white-fasciate along adaxial nerves), abaxially often glaucescent or reddish; petioles 1.5–3 cm long, slender, glands absent. **Inflorescences** with flowers paired; peduncles to 4 cm long; bracts setaceous, 2–3 mm long, borne near middle of peduncle. **Flowers** to 5 cm wide; calyx broadly patelliform, 1–1.5 cm wide; sepals oblong-lanceolate, ~1.5 × 0.5 cm, obtuse, reflexed, cream to purple; petals ovate-lanceolate, about half as long as sepals; corona filaments in 1 series, strongly dolabriform, ~5 × 1.5–2 mm, deep purple; operculum closely plicate, incurved, ~4 mm high, base pale, margin pink; limen a low ridge midway between operculum and gynophore; gynophore 5–7 mm long; ovary obovoid. **Fruits** globose, 1–1.5 cm Ø; seeds ovate, ~4.5 × 2.5 mm, transversely 7–9-sulcate. **Distr:** E Brazil. **Ecol:** On low mountains; <1,000 m. **Zones:** 10–12.

P. perfoliata L., *Sp. Pl.* 956, 1753. (Syn: *Murucuja perfoliata*, etc.). [Leafy Passion Flower]. **Vines**, nearly glabrous; stem angular, striate. **Stipules** linear-subulate, 1.5–3 mm long. **Leaves** deeply bilobed (lobes widely divergent), oblong or slightly lance-oblong, 2–6 cm along lateral nerve, 1.5–2.5 cm wide, base deeply cordate (basal lobes clasping), apex rounded, mucronulate, often emarginate, a small intermediate lobe sometimes present at sinus, trinerved, subcoriaceous; petioles to 5 mm long, glands absent. **Inflorescences** with flowers solitary or paired; peduncles 2–3 cm long; bracts setaceous, 2.5–3 mm long, borne at point of articulation. **Flowers** purple-red, 5 cm Ø; calyx tube 0.7–1.3 × 0.6–0.8 cm, ± 10-ribbed; sepals linear-subulate, 1–2 × 0.2–0.3 cm; petals oblanceolate, 5–7 mm wide, acute; corona filaments in 1 series, linear, 3–5 mm long, yellow; operculum membranous, 4–6 mm long; ovary ovoid, subangulate. **Fruits** globose, to 1.5 cm Ø; seeds ovate or obovate, ~3 × 1.5–2 mm, transversely 5- or 6-sulcate. **Distr:** Jamaica. **Ecol:** <1,000 m. **Zones:** 11–12.

P. punctata L., *Sp. Pl.* 957, 1753. (Syn: *Decaloba punctata*). [Dotted Passion Flower]. **Vines**, glabrous; stem subtriangular, compressed, striate. **Stipules** linear-falcate, 3–5 mm long. **Leaves** transversely oblong, 2–5 cm along midvein, 3–7 cm along lateral nerves, 6–12 cm between apices of lobes, apex truncate and very shallowly trilobed, usually trinerved, thinly membranous, abaxially glaucescent; lobes with base truncate or subcordate, apex broadly rounded and emarginate, rarely acute, minutely mucronulate; petiole 3–6 cm long, slender, glandless. **Inflorescences** solitary or in pairs; peduncles 5–8 cm long, very slender; bracts setaceous, 1–2 mm long, deciduous. **Flowers** 2.5–4 cm Ø, cream and purple; sepals greenish-cream outside, cream and often purplish inside base, oblong-lanceolate, 1.5–2.2 × 0.5–1 cm; petals cream, often base purplish, oblong-lanceolate,

Below: *Passiflora mollissima, Passiflora mucronata, Passiflora murucuja.*

MD YK

Top row: *Passiflora nephrodes, Passiflora organensis, Passiflora perfoliata.*
Bottom row: *Passiflora punctata, Passiflora quadriglandulosa, Passiflora quinquangularis.*
[All photos YK except *Passiflora organensis* and *Passiflora perfoliata* MD.]

0.8–1.2 × 0.2–0.6 cm; corona biseriate, outer series 0.7–1 cm long, base white, lower half purplish, cream-yellow above, dilated near middle, inner series 4–5 mm long, purplish; ovary narrowly ovoid to ellipsoid, puberulent or villosulous. **Fruits** subglobose to ellipsoid; seeds ovate, ~3–4 × 2 mm, transversely 6–7-sulcate, ridges rugulose. **Distr:** Panama, Colombia, Ecuador, Peru, Bolivia. **Ecol:** In lowland rainforests; 0–500 m. Fl. Jul. **Zones:** 10–12.

P. quadriglandulosa Rodschied, *Med. Bem. Kol. Rio Essequebo* 77, 1794. (Syn: *P. vitifolia* var. *minor*, *P. yacumensis*). [Five-angled-stemmed Passion Flower]. **Vines**, nearly glabrous to slightly pubescent; stem terete, occasionally angulate, striate. **Stipules** linear, 3–5 mm long. **Leaves** simple, occasionally trilobed or asymmetrically bilobed, 6–10 × 3–5 cm, margins serrate; petiole 1–2.5 cm long, glands 2, basal. **Inflorescences** with flowers solitary; peduncles 2.4–6 cm long; bracts free, linear or oblong-lanceolate, 1.1–1.7 × 0.3–0.6 cm, margins serrate, green to reddish-green, with 2–6 glands at base. **Flowers** bright red, 9–15 cm Ø; sepals red inside, red or greenish-red outside, oblong-lanceolate, 4.2–6(–8) × 0.8–1.4 cm, aristate, awn 2–5 mm long; petals red, oblong-lanceolate, 3.4–5.4(–7.6) × 0.9–1.5 cm; corona triseriate, outer series 1.2–1.8 cm long, slightly banded white and red, 2nd series 0.5–1 cm long, reddish, inner series 0.8–1.1 cm long, connate for about half its length; ovary ovoid to ellipsoid, tomentose. **Fruits** ellipsoid, 4–7.2 × 2.5–5 cm, yellowish-green, white-spotted; seeds oblong-ovoid, ~8 × 4 mm, reticulate, black. **Distr:** Guianas, Venezuela, Trinidad–Tobago. **Zones:** 10–12.

P. quinquangularis S. Calderon ex. J. MacDougal, *Novon* 14(4): 454, 2004. **Vines**, lightly pubescent; stem 5-angulate. **Stipules** setaceous, falcate, 4–7 mm long. **Leaves** bilobed to near middle, 4–14 × 4–11 cm, margins entire, laminar nectaries absent; petiole 1–2 cm long, glandless. **Inflorescences** solitary or rarely paired; peduncles 1.5–5.5 cm long, 1-flowered; bracts absent. **Flowers** whitish with white and purplish-red corona, ~3 cm Ø; sepals greenish-yellow outside, often red-flushed, whitish inside, lanceolate-triangular, 1.5–3 × 0.6–0.9 cm; petals white, lanceolate, 1.1–1.5 × 0.4–0.5 cm; corona biseriate, outer series1.1–1.8 cm long, filiform, white, basal half reddish-purple, inner series 3–5 mm long, capillary; ovary puberulent, ellipsoid. **Fruits** ellipsoid to widely fusiform, 6-keeled, 3–6 × 1.5–3 cm, yellowish-green, dehiscent; seeds ~3–4 × 2 mm, transversely grooved. **Distr:** Oaxaca, Mexico, to El Salvador. **Ecol:** In seasonally dry forests, brushy hillsides; 580–1,600 m. **Ecol:** In lowland rainforests; 0–1,000 m. Fl. Sep. **Zones:** 9–12.

P. reflexiflora Cav., *Icon. Pl.* 5: 15, pl. 425, 1799. (Syn: *Tacsonia reflexiflora*, etc.). **Vines**, herbaceous, glabrous, most parts membranous; stem slender, terete, striate. **Stipules** semiovate, 1–3 × 0.5–1.5 cm, base rounded, apex obtuse. **Leaves** trilobed, 3–5 × 4–6 cm; lobes obtuse, midlobe 1.5–3 cm wide, base cordate; petioles slender, to 1.5 cm, long, glands 4–6. **Inflorescences** solitary or paired; peduncles to 6.5 cm long; bracts ovate, 1–1.2 × 0.8–1 cm, base acute, apex rounded, sessile. **Flowers** crimson or red; calyx tube narrowly cylindric, 2–2.5 cm long, ~5 mm wide, base slightly enlarged, glabrous; sepals linear-lanceolate or lanceolate, 2.5–3 × 0.6–0.8 cm, cucullate, minutely corniculate, reflexed; petals linear, 2–2.5 × 0.4–0.5 cm, reflexed; corona triseriate, short-filamentose, outer filaments ~2 mm long, successively shorter; operculum 1.5–2 mm high, white; limen membranous, 5 mm high, adnate to gynophore; ovary ovoid, glabrous. **Fruits** ovoid, ~6 × 4.5 cm, green; seeds subturbinate, 6–7 × 3–4 mm, coarsely reticulate. **Distr:** Ecuador. **Ecol:** In coastal regions. **Zones:** 11–12.

P. serratifolia L., *Sp. Pl.* 955, 1753. (Syn: *P. denticulata*, etc.). [Sawtooth-leafed Passion Flower]. **Vines**; stem terete, striate, hirtellous or puberulous. **Stipules** linear-subulate, ~7 mm long, apically serrulate. **Leaves** ovate, 8–12 × 4–7 cm, base rounded, margins serrulate, apex acuminate, adaxially glabrescent, abaxially short-hirtellous; petioles 5–12 cm long, densely hirtellous, glands 6, clavate ~1 mm long. **Inflorescences** with flowers solitary; peduncles 5–7 cm long; bracts oblong-lanceolate, 2–3 × 0.8–1 cm, acute, finely pubescent, green. **Flowers** 4–6 cm across; calyx tube campanulate; sepals lanceolate, 2.5–3 × 0.5–0.8 cm, carinate, keel terminating in an awn ~2 mm long, purplish; petals oblong-lanceolate, 1.5–2 × 0.4–0.6 cm, obtuse, purplish; corona filaments in several series, outermost filiform, 2.5–3.5 cm long, radiate, base purple, white toward apex, next 3 or 4 series liguliform, 1–2 mm long, banded with purple and white, those of innermost series 0.8–1 cm long, basally connate; operculum membranous, ~3 mm high; limen cupuliform, ~4 mm high; ovary ellipsoid, glabrous. **Fruits** round, 5–9 × 3.5–5 cm, lemon-yellow, edible; seeds flattened, 6–7 × 4–5 mm. **Distr:** E Mexico to Costa Rica. **Ecol:** In moist, lowland, evergreen, seasonal broadleaved forests and montane forests; 60–2,000 m. Fl. & Fr. throughout the year. **Zones:** 10–12.

P. sprucei Mast, *Fl. Bras.* 13(1): 568, 1872. **Vines**, glabrous; stem terete, striate. **Stipules** semiovate, 1.5–3 × 0.5–1.2 cm, base rounded, margins undulate, apex mucronulate. **Leaves** 3(4)-lobed (lobes lanceolate, middle lobe produced, lateral lobes divergent), 2–8 × 5–12 cm, 5-nerved, abaxially green or glaucous; petioles to 3 cm long, glands 2–4. **Inflorescences** with flowers solitary; peduncles to 5 cm long; bracts oblong or lance-oblong, ~1.5 × 0.5–0.8 cm, acute. **Flowers** to 7 cm wide; calyx tube campanulate; sepals linear-oblong, 5–7 mm wide, obtuse, green outside, lilac, rose-purple, or violet inside, carinate, keel terminating in an awn 5–7 mm long; petals lanceolate, subequal to sepals, lilac, rose-purple, or violet; corona filaments in several series, outer 2 series filiform, transversely banded with lilac and purple, next 4 or 5 series 2–3 mm long; operculum with filaments to 1 cm long,

Above: *Passiflora reflexiflora, Passiflora serratifolia, Passiflora sprucei.*

banded with lilac and purple; nectar ring a low ridge; limen tubular, ~5 mm high; ovary narrowly ellipsoid. **Fruits** ovoid, 3–4.5 × 2–3 cm, greenish-yellow; seeds ovoid, 5 × 3 mm. **Distr:** Ecuador. **Ecol:** In coastal regions; to 2,300 m. **Zones:** 10–12.

P. standleyi Killip, *Journ. Wash. Acad. Sci.* 14: 110, 1924. **Vines**, glabrous; stem subquadrangular, striate, minutely pubescent above. **Stipules** narrowly linear-falcate, 2 × 0.3 mm. **Leaves** oblong, bilobed, length 2.5–5 cm along midvein, 6–12 cm along lateral nerves, 4–5 cm between apices of lobes; lobes lanceolate, 1.5–2 cm wide, base rounded or subcuneate, apex obtuse or acute, apiculate, trinerved, ocellate, reticulate-veined; petioles 1.5–2.6 cm long, glandless. **Inflorescences** with flowers solitary; peduncles slender, 2–3 cm long; bracts setaceous, 2–3 mm long, borne within 1 cm of apex of peduncle. **Flowers** 3–4 cm Ø, bluish-purple; sepals ovate-lanceolate, 1–1.5 × 0.4–5 cm, obtuse; petals half as long as sepals, obtuse; corona filaments capillary, biseriate, outer 4–7 mm long, base blue, white, apex blue-spotted, inner very numerous, 4–5 mm long, white; operculum membranous, closely plicate, minutely fimbrillate; limen annular, 1 mm high; ovary subglobose. **Fruits** globose, 1–1.5 cm Ø; seeds ovate or ovate-oblong, ~3 × 2.5 mm, transversely sulcate with ~7 straight, rugulose ridges. **Distr:** Salvador, Costa Rica. **Ecol:** On mountains; 1,000–2,000 m. **Zones:** 11–12.

P. stipulata Aubl., *Pl. Guian.* 830, pl. 325, 1775. (Syn: *P. glauca*). **Vines**, glabrous; stem terete. **Stipules** semiovate, 1–3 × 0.5–1 cm, oblique, base rounded, apex acute and aristulate. **Leaves** 5–8 × 7–10 cm, trilobed to middle (lobes broadly ovate, 3–5 cm wide, abruptly acuminate or acute, sinuses glandular), base subpeltate, cordate, 5-nerved, membranous, abaxially glaucous; petioles to 5 cm long, glands 2–5, sessile, minute. **Inflorescences** with flowers solitary; peduncles 2–5 cm long; bracts sessile, basal, lanceolate, ~1.5 × 0.6 cm, acute. **Flowers** 5–6 cm Ø; calyx tube turbinate-campanulate; sepals oblong-lanceolate, 2.5–3 × 1 cm, green outside, paler inside, keeled, keel terminating in a mucro ~2 mm long; petals as long as sepals, white; corona filaments filiform, in several series, 2 outer as long as petals, white, base violet, radiate, successive series much shorter, 2–3 mm long, erect; operculum ~1 cm high, filamentose nearly to base; limen tubular, closely surrounding base of gynophore; ovary ovoid, glabrous. **Distr:** West Indies and the Guianas. **Ecol:** In thickets. Fl. May–Jul. **Zones:** 11–12.

P. suberosa L., *Sp. Pl.* 2: 958, 1753. (Syn: *P. angustifolia*, etc.). [Corky Passionfruit, Ink Vine, Pap Bush]. **Vines**, herbaceous, 1–4(–10) m long; stem whitish, triangular, slender, striate. **Leaves** 5–5.5(–8) × 5–6(–11) cm, base cordate, margins trilobed, lobes ovate, apex acute, mucronate, abaxially gray-green, sparsely villous; petiole 2–4 cm long, white-strigose, with 2 glands at middle or slightly above. **Inflorescences** axillary, flowers solitary or paired. **Flowers** light green or white, 1.5–2.5 cm Ø; sepals oblong or lanceolate, 5–8 mm long; petals absent; corona biseriate, filamentous, outer series 3–7 mm long, inner series ~1 mm long; operculum plicate, 1–2 mm high; disk 0.5–1 mm across; androgynophore 2–4 mm tall; filaments flat, 1.5–4 mm long, free, anthers oblong, 2–3 mm long; ovary subglobose, glabrous, styles 3, almost threadlike, 7–8 mm long, stigma capitate. **Fruits** berries, turning bluish-black at maturity, subglobose, 1–1.2 cm Ø. **Distr:** West Indies, C & S America; naturalized and weedy throughout most of the tropics. **Ecol:** Fl. Aug–Sep. Fr. Sep–Nov. **Zones:** 10–12. **Info:** The fruits are used in the Maldives as a dye, among other uses.

P. subrotunda Mast, *Fl. Bras.* 13 (1): 601, 1872. (Syn: *P. barbosae*). **Vines**, glabrous; stem slender. **Stipules** semiovate or semiorbicular, 1–3 × 0.5–1.5 cm, base rounded, apex acute, aristulate, membranous. **Leaves** suborbicular or cordate, 2.5–4 × 2–4 cm, apex rounded and usually emarginate, membranous; petioles ~1 cm long, minutely biglandular just below apex. **Inflorescences** with flowers solitary; peduncles to 2 cm long, slender; bracts borne ~5 mm below flower base, ovate-lanceolate, 6–8 × 5–6 mm, both ends acute, thinly membranous. **Flowers** ~2.5 cm Ø; calyx tube campanulate, 5–7 mm wide at throat; sepals linear, 1.5–2.5 × 0.3–0.4 cm, obtuse, corniculate, margin deep blue, center pale blue; petals similar to sepals, slightly shorter and narrower, deep blue; corona filaments in 2 subequal series, filiform, 7–9 mm long, reddish-purple; operculum membranous, erect, finely crenulate; limen cupuliform, closely surrounding base of gynophore; ovary ovoid, stipitate. **Fruits** ovoid, 1.3–2 × 0.7–1.1 cm; seeds obovate, 3.5 × 2.5 mm, reticulate. **Distr:** NE Brazil. **Ecol:** In coastal areas. Fl. throughout the year. **Zones:** 10–12.

P. tarapotina Harms, *Verh. Bot. Vereins Prov. Brandenburg* 48: 185, 1907. **Vines**, herbaceous, glabrous, most parts membranous; stem terete, striate, slender. **Stipules** subreniform to ovate-oblong, 1.4–2.5 × 0.5–1 cm, margins entire to slightly serrate. **Leaves** trilobed, 4.7–8 × 5.5–10.7 cm, margins entire; petiole slender, 1.5–4.3 cm long, glands 2, stipitate, near middle. **Inflorescences** with flowers solitary; peduncles 8.5–17 cm long, slender, erect; bracts ovate-oblong to oblong-lanceolate, 1–1.5 × 0.3–0.6 cm, serrate, free. **Flowers** red, 4–5 cm Ø, ~5 cm long; floral tube cylindrical, 2–2.5 × 0.6 cm; sepals lanceolate, 2.5 × 0.5–0.6 cm, keeled, keel terminating in an awn 0.7–1 cm long; petals lanceolate, subequal to sepals; corona 3- or 4-seriate, violet, outer series filamentous, 1–2 mm long; ovary ovoid, glabrous. **Distr:** N Peru. **Ecol:** In gradual transitions between foothills and in lowlands; 300–750 m. Fl. throughout most of the year. **Zones:** 10–12.

P. tarminiana Coppens & V. Barney, *Novon* 11: 9, 2001. [Banana Passionfruit, Curuba]. **Vines**, pubescent; stem, terete, striate, tip erect. **Stipules** subreniform, 0.4–0.7 × 0.2–0.3 cm, margins serrate

Below: *Passiflora standleyi, Passiflora stipulata, Passiflora suberosa.*

YK YK

Above: *Passiflora subrotunda, Passiflora tarapotina.*

at base, apex with long arista, soon deciduous. **Leaves** trilobed, 5.5–11(–16) × 7–16(–29) cm, margins serrate, adaxially nearly glabrous; petiole 1.5–4 cm long, glands 2–8, mostly paired. **Inflorescences** solitary; peduncles 3–10 cm long; bracts ovate, 3–5 × 2–3 cm, connate for about half their length, margins entire. **Flowers** light pink, 8–11 cm Ø, 10–13 cm long; floral tube cylindrical, 6–8 × 0.7–1 cm, light green; sepals pink, oblong, 4.5–6 × 1.2–2.5 cm, aristate, awn 3–4 mm long; petals pink, oblong, 4–5.5 × 1.2–2 cm; corona 1 seriate, reduced to a purple ring with white teeth; ovary ellipsoid, pubescent. **Fruits** ellipsoid, 10–14 × 3.5–4.5 cm, yellow to orange-yellow; seeds asymmetrically ovoid, ~6 × 4 mm, reddish-brown, reticulate. **Distr:** Andes region, mostly in Ecuador and Colombia. **Ecol:** 2,000–2,700 m. **Zones:** 9–12. **Info:** Fruits are edible and widely cultivated.

P. trifasciata Lemaire, *Illus. Hort.* 15, 1868. **Vines**, glabrous, to 10 m long. **Stipules** minute, subulate, 2–4 mm long. **Leaves** trilobed, 5–10 × 4–10 cm, adaxially dark green with white, cream, yellow, pink or mauve markings along the main nerves, abaxially reddish-purple; petiole to 5 cm long, glands absent. **Inflorescences** with flowers solitary; peduncles 3 cm long; bracts setaceous, 2 mm long. **Flowers** 2.5–3.5 cm Ø; calyx tube broad, campanulate; sepals light green or cream, 1.5 × 0.5 cm; petals light green or cream, linear, 1 × 0.2–0.5 cm; corona filaments biseriate, white, sometimes pinkish, outer terete, 0.8–1.5 cm long, inner linear-clavate, ~3 mm long; operculum plicate, incurved, white, margin pink-tinged; limen thick, lobulate; ovary subglobose, glabrous. **Fruits** globose, 1.5–2.5 cm Ø; seeds narrowly oblong-ovoid, 4 × 2 mm, transversely 6-sulcate. **Distr:** Amazon regions and N Peru. **Ecol:** In dense forests; 200–840 m. Fl. Aug–Sep. **Zones:** 10–12. **Info:** Cultivated widely as a foliage plant.

P. triloba Ruíz & Pavón ex DC., Prodr. 3: 330, 1828. (Syn: *P. colubrina*, etc.). **Vines**, glabrous; stem terete, stout. **Stipules** ovate-lanceolate, 2–3 × 1.5–2.5 cm, base cordate, margins minutely serrulate-denticulate, apex acute or abruptly acuminate. **Leaves** unlobed or trilobed (lobes ovate or oblong-ovate, to 7 cm wide, obtuse), 8–17 × 9–20 cm, base cordate, lobes usually overlapping, margins minutely denticulate, thickly membranous, adaxially lustrous, abaxially glaucous; petioles to 10 cm long, glands 2, apical, saucer-shaped, 2–2.5 mm long, stipitate. **Inflorescences** solitary; peduncles to 5 cm long; bracts connate at very base, ovate, 4.5–7 × 3–5.5 cm, rounded or acute, glandular-serrulate, glabrous. **Flowers** fragrant, 7–10 cm Ø; calyx tube campanulate; sepals oblong-lanceolate, 3–4 × 1–1.8 cm, obtuse, green outside, densely spotted with deep purple inside; petals oblong-lanceolate, 2.5–3.5 × 1–1.4 cm, obtuse, purplish-red; corona triseriate, banded purple-violet and white at base, violet and white above, outer series 1–1.5 cm long, 2nd series 2.5–3 cm long; ovary ovoid, glabrous; operculum membranous; limen erect, adnate to gynophore; ovary ovoid, glabrous. **Distr:** Peru, Bolivia. **Ecol:** <1,000 m. **Zones:** 11–12. **Info:** This species is unique in having 6 sepals and 6 petals, giving the appearance of having 12 petals.

P. tulae Urban, *Symb. Antill.* 1: 374, 1899. **Vines**, slightly woody, tendriled, 4–5 m long; stem slender, angular, green or mulberry-colored, striate, glabrous. **Stipules** linear-subulate, 1–2 mm long, soon deciduous. **Leaves** alternate; lamina semicircular or semielliptic, 3–8 × 2.5–9 cm, base rounded, margins sinuate, apex ± truncate, with 2 or 3 rounded lobes, main veins 3 borne from the base, coriaceous, adaxially dark green, shiny, glabrous, abaxially pale green, dull, with a row of circular glands on each side of midvein; petioles cylindrical, 1.5–3 cm long, smooth, glands absent; tendrils axillary, simple, longer than the leaves. **Inflorescences** with flowers solitary or paired, axillary; peduncle 3–5 cm long. **Flowers** 5.5–7 cm Ø; sepals 5, oblong, 3–4 × 0.7 cm, pink-violet; petals oblong, ~3 cm long, same color as sepals; corona tubular, orange, ~1.5 cm long; gynophore ~3 cm long. **Fruits** berries, globose or ellipsoid, green, 1–2 cm long, sepals persistent at base. **Distr:** Puerto Rico. **Ecol:** On mountains. Fl. Dec–Apr. Fr. Feb–Jun. **Zones:** 11–12.

P. vellozii Gardn., *Lond. Journ. Bot.* 4: 103, 1845. (Syn: *Cieca vellozii*). **Vines**; stem subterete, sulcate, green or sometimes dark purplish, densely pilose. **Stipules** semiovate, 3–6 × 1.5–3 mm, lacerate-dentate, teeth aristate. **Leaves** 3–9 × 3–9 cm, trilobed, base shallowly cordate, 3–5-nerved, margins ciliate, minutely denticulate toward base (teeth often gland-tipped), membranous, surfaces appressed-pilose; lobes broadly ovate-lanceolate, acute or obtuse, apiculate, midlobe 2 or 3 times larger than lateral lobes; petioles 0.5–2.5 cm long, pilose. **Inflorescences** solitary; peduncles 0.5–2 cm long, slender; bracts lanceolate or oblong-lanceolate, 0.7–1.5 × 0.3–1 cm, lacerate-dentate or pinnatifid, membranous, pilose, ciliate. **Flowers** to 5 cm Ø; sepals oblong, ~2 × 0.6 cm, obtuse, pilose outside, dorsally awned just below apex, awn ciliate; petals linear, ~1.5 × 0.3 cm, obtuse, thinly membranous; corona filaments triseriate, the 2 outer series narrowly liguliform, ~2 cm long, inner clavate-filiform, 4 mm long; operculum 5 mm high, upper half filamentose; limen membranous, surrounding base of gynophore, 3 mm high; ovary ovoid, hirsute. **Fruits** ovoid, ~4 × 2.5 cm, sparingly hirsute; seeds cuneate, ~4 × 2 mm, reticulate. **Distr:** Minas Geraes and Rio de Janeiro, E Brazil. **Ecol:** ~900 m. **Zones:** 9–11.

P. vitifolia HBK., *Nov. Gen. & Sp.* 2: 138, 1817. (Syn: *P. sanguinea*, etc.). [Curuvito, Grape-leafed Passionfruit, Vined-leafed Passion Flower]. **Vines**; stem terete to subangulate, striate. **Stipules** linear, 4–9 mm long, usually with 2 or 3 glands. **Leaves** 7–15 × 8–18 cm, trilobed, base truncate to cordate, 3–5-nerved, margins dentate or crenate; petioles 2–5 cm long, biglandular, occasionally with 2 or

Below: *Passiflora tarmaniana, Passiflora trifasciata, Passiflora triloba.*

YK YK

MD

YK MD

Top row: *Passiflora tulae, Passiflora vellozii, Passiflora vitifolia.*
Bottom row: *Passiflora xiikzodz, Passiflora yucatanensis.* [Both YK.]

3 additional glands. **Inflorescences** with peduncles 4–9 cm long; bracts oblong, 1.5–2.5 × 0.4–0.8 cm, acuminate. **Flowers** scarlet, bright red, or vermilion; calyx tube cylindric, 1–1.8 × 1.5 cm; sepals lanceolate, 6–8 × 1–2 cm, obtuse, keel terminating in an awn ~1 cm long, inside magenta; petals linear-lanceolate, 4–6 × 0.8–1.5 cm, obtuse; corona 3-ranked, outermost filaments awl-shaped, 1.5–2 cm long, red or yellow, second rank ~1 cm long, pale red, third rank tubular, ~1 cm long, cleft into numerous filiform divisions above, pale red; operculum 0.7–1 cm long, fimbriate; ovary ellipsoidal, densely tomentulose. **Fruits** ovoid, ~5 × 3 cm, fragrant; seeds obcordate, 5 × 3.5 mm, reticulate. **Distr:** Nicaragua to E & S Venezuela, Colombia, E Ecuador and N Peru. **Ecol:** In lowland rainforests, secondary forests, montane forests and forest slopes; 0–2,000 m. Fl. & Fr. throughout the year. **Zones:** 10–12.

P. xiikzodz J. MacDougal, *Novon*, 2(4): 361, 1992. **Vines**, ~1–3(–6) m long, minutely puberulent, plant glabrescent below; stems subterete to obtusely subangular. **Stipules** ± linear, to 6.5 × 0.7 mm. **Leaves** obovate to narrowly elliptic, 2(–3)-lobed or truncate, (1.2–)1.5–7.5 × (3.5–)5–19 cm, base peltate, margins entire, usually strongly variegated along main veins as juveniles, 5(–7)-veined at base, lateral lobes (acuminate to) acute, central lobe retuse to obsolete (or obtuse); laminar nectaries (6–)8–16(–28); petioles 1–2(–3) cm long, apically biglandular, nectaries saucerlike, 0.7–1 × 1.3–2 mm, fleshy. **Inflorescences** (1)2 per node, on condensed terminal shoots with aborted laminas, to 20 cm long; peduncles 0.1–0.7(–1) cm long, 1-flowered; bracts absent; pedicels 1.2–3 cm long. **Flowers** light yellowish-green, corona dark violet or purple, apically yellowish; hypanthium ~6–9 mm Ø; sepals narrowly triangular to oblong-triangular, (0.8–)1–1.45 × 0.35–0.6 cm, greenish-yellow, or greenish outside and lightly flushed with reddish-purple, reflexed at anthesis; petals absent; coronal filaments ~5–7(–8)-seriate, outermost series 0.8–1.2 cm long, subequal to sepals, very slender-filiform, reflexed, inner ~3–7 series 0.5–3 mm long, 3 innermost series greatly apically dilated-capitate; operculum represented by vestigial fan-shaped denticles ~0.3 mm long; limen 5–6 mm Ø, floor convex, glossy-brown, dark purple, or reddish-purple; androgynophore when present light greenish-yellow, sometimes basally purplish; staminal filaments greenish-yellow, sometimes dark reddish-purple; ovary ellipsoid, 1.9–2.5 × 1.3–2 mm, glabrous, styles 1.5–5.5 mm long, greenish. **Fruits** ovoid. **Distr:** Yucatán Peninsula and adjacent Guatemala and Belize. **Ecol:** In thickets, forests, limestone outcrops; 100–300 m. **Zones:** 9–12.

P. yucatanensis Killip, *Publ. Field Columbian Mus.*, Bot. Ser. 8: 26, 1930. **Vines**, minutely puberulent to glabrescent throughout; stem angulate, grooved-striate. **Stipules** setaceous, falcate, 3–4 mm long. **Leaves** very shallowly bi- or trilobed with short middle lobe, or truncate, 2.3–5 × 4.8–8 cm, margins entire, surfaces bearing laminar nectaries; petiole 1.5–2 cm long, glands absent. **Inflorescences** paired, flowers solitary; peduncles 2–4 cm long; bracts setaceous, 2–3 mm long. **Flowers** white, corona short, yellow and reddish, ~3 cm Ø; sepals pale yellow-green outside, white inside, lanceolate-oblong, 1.8–2.5 × 0.6–0.8 cm; petals white, narrowly oblong, 1.3–2 × 0.5–0.7 cm; corona biseriate, outer series 0.7–1 cm long, lower half erect, red or purplish-red, curved and bright yellow above, inner series ~3–5 mm long, reddish-purple, capillary; ovary glabrous to slightly puberulent, ellipsoid. **Fruits** ellipsoid, purple. **Distr:** Mexico (Cozumel Island off the coast of the Yucatán Peninsula). **Ecol:** In coastal forests near the sea. Fl. throughout the year. **Zones:** 10–12.

CLUSIACEAE

Genera ~40, species ~1,200. **Trees, shrubs**, or sometimes herbs. **Leaves** simple, entire or rarely gland-fringed, opposite or sometimes whorled, nearly always estipulate. **Inflorescences** with flowers solitary or in cymes or thyrses; bracteoles often inserted just beneath calyx and not always easily distinguishable from sepals. **Flowers** bisexual or unisexual, regular, hypogynous; sepals (2–)4 or 5(or 6), imbricate or decussate or rarely wholly united in bud, inner ones sometimes petaloid; petals (3 or)4 or 5(or 6), free, imbricate or contorted in bud; stamens many to rarely few (9), in (3 or)4 or 5 bundles (fascicles) that are free and antipetalous or variously connate, filaments variously united or apparently free and then sometimes sterile (staminodes), anther longitudinally dehiscent; staminode bundles (fasciclodes) 3–5, free and antisepalous or variously connate or absent; ovary superior, with 2–5(–12) connate carpels, 1–1 biloculed, with axile to parietal or basal placentation, ovules 1 to many on each placenta, erect to pendulous, styles 1–5(–12), free or ± united or absent, stigmas 1–12, punctiform to peltate. **Fruits** capsules, berries or drupes, septicidal or septifragal, rarely loculicidal; seeds 1 to many, sometimes arillate. **Distr:** Mainly in tropical regions, few in temperate regions.

MS

Above: *Clusia rosea* (female flower), *Clusia rosea* (fruits; ripened woody flowerlike fruit with green unripened fruit at back), *Cratoxylum sumatranum*, *Garcinia dulcis*.

Clusia

Syn: *Decaphalangium*, *Havetia*, etc.
Species ~300. **Trees** or **shrubs**, free-standing, epiphytic, hemi-epiphytic or lithophytic, sometimes lianas; branchlets terete or tetragonal, sometimes winged; latex translucent, white, cream, yellow or orange-yellow. **Leaves** decussate, with or without conspicuous latex canals; petiole with small basal adaxial margined pits. **Inflorescences** terminal or lateral panicles of cymes, pyramidal or not, rarely flowers solitary. **Flowers** mostly petals 4–12, variously arranged, sometimes decussate or basally connate; stamens 4–numerous, free or basally connate, or united into a frequently resiniferous, globular, umbrella-shaped or annular synandrium, when annular surrounding a pistillode or a staminodial mass, filaments of varying shape, sometimes lacking, anthers variable, dehiscent by a variety of slits or pores; staminodes mostly smaller than stamens, with or without anthers, sometimes early caducous; pistil with 4–8(–21) carpels, styles short or rudimentary, stigmas free and radiate or connivent, large, 1–many ovules per carpel; pistillode somewhat similar to pistil or represented only by an undifferentiated central tissue mass producing waxy resin or absent. **Fruits** capsules, septifragally dehiscent, stigmas persistent; seeds with a reddish, non-vascularised aril. **Distr:** Tropics and subtropics of the New World. **Cult:** Propagate from seeds, root cuttings or by air-layering. Requires a fertile, moist, well-drained soil in full to partial sun.

C. rosea Jacq., *Enum. Syst. Pl.* 34, 1760. (Syn: *C. major*, *C. rosea* var. *colombiana*). [Autograph Tree, Balsam Apple, Fat Park Tree, Pitch Apple, Scotch Attorney]. **Trees** or **large shrubs**, free-standing or semiepiphytic shrubs or small trees, evergreen, to 30 m tall; produces masses of aerial roots; at least some populations apparently apomictic (only ♀ flowers known; stigmas 6 or 7); bark smooth. **Leaves** opposite; lamina obovate, 3.5 × 8 – 7 × 14 cm, base gradually tapered into petiole, margins slightly thickened, apex rounded and broadly obtusely acuminate, thickly coriaceous, adaxial surface with laterals hardly visible, midrib ± prominent, laterals ~25 pairs; petiole, 1–2 cm long, foveola distinct. **Inflorescences** axillary or pseudo-terminal, flowers solitary, sometimes 2 lateral, 3-flowered cymes; bracteoles 2–4, smaller than sepals. **Flowers** with sepals 4–6, to 2 cm long, persistent; petals 6–8, white or pink, obovate or suborbicular, 3–4 cm long; ♂ flowers with outer stamens fertile in several whorls, united at base into a cup or ring, inner stamens united into a solid resinous mass; ♀ flowers with staminodes united into a cup, ovary globose, stigma sessile, 6–9-rayed. **Fruits** capsules, septicidally dehiscent, resembling a woody flower, brownish outside, pale inside, axis angular, winged; seeds with thin, fleshy, soft, dark red arillode, valves thick, pointed, narrow. **Distr:** Tropical and subtropical America, West Indies. **Ecol:** In evergreen or semideciduous, lowland forests; 0–500 m. **Zones:** 9–12.

Cratoxylum

Species ~6. **Trees** or **shrubs**, deciduous to evergreen; branchlets with nodes sometimes flattened. **Leaves** opposite, sessile to petiolate; lamina with margins entire, abaxially often pruinose to waxy, often finely translucent-punctate between veins, secondary veins ± widely spaced, tertiary veins reticulate. **Inflorescences** cymose, terminal and/or axillary; bracteoles caducous, minute; flowers pedicellate. **Flowers** bisexual, homostylous or heterostylous; sepals 5, ± unequal, leathery, usually accrescent; petals 5, white to deep crimson or pink, sometimes tinged orange and green, often with punctiform or linear dark resinous glands, sometimes with ± adnate adaxial near-basal scalelike appendage; stamen fascicles apparently 3, with filaments united most of their length, anthers subpersistent, dorsifixed, sometimes with ± prominent resinous connective gland; fasciclodes 3, alternating with fascicles, scalelike, sometimes minute; ovary triloculed, locules (3–)5–16(–18)-ovuled, styles 3, free, often divergent, stigmas thickened. **Fruits** capsules, ± woody, ellipsoid to narrowly cylindric, loculicidally dehiscent; seeds obovoid to cylindric, with unilateral wing. **Distr:** China, Indo-China, SE Asia to Philippines. **Cult:** Propagate from seeds. Requires a moist, well-drained soil in full or partial sun.

C. sumatranum (Jack) Blume, *Mus. Bot. Lugd. Bat.* 2(1–8): 16, 1856. (Syn: *Elodea sumatrana*, etc.). **Trees**, 10–15 m tall; trunk to 90 cm dbh, latex white to yellow; bark reddish-brown, papery, scaly; inner bark orange-brown, with orange-red sap; young twigs slightly flattened. **Leaves** opposite; lamina simple, oblong, ovate-oblong to oblong-lanceolate, 5–18 × 2.5–5 cm, base rounded to obtuse, apex acuminate, subcoriaceous, glabrous, secondary nerves ~12 pairs, with short intermediate nerves, secondary nerves and reticulations prominent abaxially, adaxially faint; petiole 1.2–2.5 mm long. **Inflorescences** cymose panicles, many-flowered, axillary or terminal. **Flowers** ~8 mm Ø, reddish or reddish-purple; sepals and petals 5, persistent in fruit; stamens in 3 bundles; ovary triloculed. **Fruits** capsules, dehiscent, 7–8 × 4 mm, yellow to brownish-black; seeds 3–10 per valve, small, winged. **Distr:** Indo-China through peninsular Thailand to Malesia and Philippines. **Ecol:** In mixes dipterocarp forests, scrubs, disturbed open places; to 500 m. **Zones:** 11–12.

Garcinia

Syn: *Cambogia*, *Mangostana*, *Xanthochymus*, etc.
Species ~450. **Trees** or **shrubs**; functionally dioecious (sometimes apparently flowers bisexual or plant monoecious; polygamously dioecious); bark, branchlets and fruits with white or yellow, thick, sticky latex. **Leaves** opposite, or decussate; lamina with margins entire, leathery to papery, usually glabrous. **Inflorescences** terminal and/or axillary (or pseudoterminal) cymes (often thyrsiform), triads or fascicles, or flowers paired or solitary. **Flowers** with sepals (2 or 3 or)4 or 5, decussate or imbricate (quincuncial), usually free; petals (3 or)4 or 5(–8), contort, fascicles each with many to few stamens, anthers 1, 2, 4 or many-celled; fasciclodes (sterile stamen fascicles) 4 or 5; pistillode present or absent; ♂ with stamens indefinite, filaments connate into a central column or into phalanges as many as petals, anthers 2-celled; ♀ with staminode fascicles ± as in ♂, fasciclodes free as in ♂ but smaller or united in a ring at base of ovary or absent, stigmas usually papillate, ovary 2–12-celled, each uniovulate. **Fruits** berries, smooth or sulcate (or verrucose or rarely secondarily dehiscent), with leathery to thin exocarp and 1–5 or sometimes more seeds (–12), embedded in juicy, arillode, endocarpic pulp; seeds large. **Distr:** Tropical and S Africa, Madagascar, tropical Asia, NE Australia, W Polynesia, tropical America. **Cult:** Propagate from fresh seeds, cuttings, grafting or by air-layering. Requires fertile, moist soil in full sun.

G. dulcis Kurz, (Roxb.) Kurz, *J. Asiat. Soc. Bengal* 43(2): 88, 1874. (Syn: *Xanthochymus dulcis*). [Claudie Mangosteen]. **Trees**, medium-sized, 6–20 m tall; branchlets 4-angled. **Leaves** ovate-oblong, 10.5–25 × 4.4–12 cm, base rounded, apex shortly acuminate, deep green, coriaceous, nerves 10 pairs, not very prominent; petioles 1–1.5 cm long, stout. **Inflorescences** dense many-flowered fascicles, axillary, with unisexual or ♂ and bisexual mixed; pedicels 6–9 mm long. **Flowers** globular, hardly expanding ~6 mm across, greenish yellow; sepals 4–6, orbicular, outer 3 smaller; petals usually 5, larger; stamens 40 in pedicelled fan-shaped groups; disk in ♂ corrugated, fleshy, in ♀ 5-lobed; stigma in ♂ absent, in bisexuals, 5-rayed. **Fruits** drupes, globular, ovoid or pear-shaped, 6–7.5 cm long, peduncled, pulp yellow; seeds 1–5, oblong. **Distr:** Malay Islands, New Guinea, Australia. **Ecol:** In rainforests. **Zones:** 10–12. **Info:** Fruit eaten raw or cooked, often made into excellent jam.

G. humilis (Vahl) C.D.Adams, *Phytologia* 20: 312, 1970. (Syn: *Rheedia lateriflora*, *Mammea humilis*). **Shrubs** or **trees**, glabrous, to 10 m tall; sap yellow; twigs stout. **Leaves** ovate or elliptical, 15–28 × 6–12 cm, base obtuse or rounded, margins often slightly revolute, apex obtuse or acute, often umbonate, tips falcate, thickly coriaceous; petiole 1–2 cm long, stout. **Inflorescences** fasciculate, in axils of fallen leaves; pedicels 0.7–1 cm long. **Flowers** 0.9–1.5 cm across; sepals 2(–4), 1.5–2 mm long; petals white yellowish, or orangish, 4, 6–8 mm long, rotund, cucullate, slightly coriaceous, slightly unequal; stamens 12–25, filaments free, anthers minute; style short or obsolete, stigmas 3–4, covering young ovary. **Fruits** ellipsoid to globose, 3–4 cm long, yellow, acuminate-

Above: *Garcinia humilis*, *Garcinia mangostana*, *Garcinia morella*, *Garcinia quaesita*.

beaked at each end; seeds 1–3, ~2.5 cm long. **Distr:** The Antilles from Hispaniola to Trinidad. **Ecol:** 750–900 m. **Zones:** 11–12.

G. mangostana L., *Sp. Pl.* 1: 443, 1753. (Syn: *Mangostana garcinia*). [King's Fruit, Mangosteen]. **Trees**, dioecious, small, 12–20 m tall, glabrous; crown dense, pyramidal; branches many, dense, decussate; twigs distinctly angled; bark dark, with yellow, sticky latex; wood red. **Leaves** elliptic or elliptic-oblong, 14–25 x 5–15 cm, base broadly cuneate or subrounded, margins involute, apex shortly acuminate, thickly leathery, shiny, midvein raised on both surfaces, secondary veins 40–50 pairs; petiole robust, 2–2.5 cm long. **Inflorescences** with ♂ flowers rare, 2–9, clustered at apex of branchlets, pedicels short; ♀ flowers solitary or paired at apex of branchlets, pedicels ~1.2 cm long. **Flowers** large; ♂ with stamen fascicles 4, anthers 2-celled, cells longitudinally dehiscent, pistillode conic; ♀ slightly larger than ♂, 4.5–5 cm Ø, ovary 5–8-loculed, style nearly absent, stigma 5- or 6-lobed. **Fruits** berries, maturing to purple-red, sometimes yellow-brown spotted, globose, 5–8 cm Ø, smooth; seeds 4 or 5 or more, pulp white, juicy, fleshy, aromatic. **Distr:** Indonesia (Maluku). **Ecol:** Fl. Sep–Oct. Fr. Nov–Dec. **Zones:** 11–12.

G. morella (Gaertn.) Desr., Enc. 3 (2): 701, t. 405, f. 2, 1792. (Syn: *Cambogia gutta*, etc.). [Ceylon Camboge, Indian Camboge Tree]. **Trees**, glabrous, to 20 m tall, usually much smaller; trunk deeply fissured with age, to 50 cm dbh. **Leaves** obovate or oblanceolate, rarely elliptic, (2 x 5 –) 6 x 9 – 7 x 13 cm, base tapered, acute, apex broadly, shortly, obtusely, usually obscurely acuminate, thinly coriaceous to very stiffly coriaceous; petiole 0.5–1 cm long. **Inflorescences** in axils of fallen leaves; flowers sessile or nearly so; ♂ flowers 2–3 together; ♀ flowers usually solitary. **Flowers** with sepals 4, greenish-white, elliptic, concave, 4–6 mm long, rather thin; petals 4, white to pink, rather fleshy, veined, broadly elliptic, concave, 5–8 mm long; ♂ with stamens monadelphous, short filaments in a subquadrangular central column, anthers red, dehiscing transversely; ♀ similar to ♂, ovary greenish, globose, smooth, 4-celled, stigma peltate, first yellow, later brown-red, base of ovary surrounded by ~12 free stamens. **Fruits** globose, smooth, yellowish, to 2.5 cm Ø, pulp sweetly acidic, edible; seeds kidney-shaped, laterally compressed. **Distr:** S India. **Ecol:** In dry and wet; ~500 m. **Zones:** 11–12.

G. quaesita Pierre, *Fl. for. Cochinch.* 22, pl. 83 H, 1883. (Syn: *G. cambogia*). [Red Gourka, Red Goraka]. **Trees**, glabrous, to 20 m tall; trunk to 60 cm dbh. **Leaves** oblanceolate to subobovate or elliptic, 1.5–5 x 6–12 cm, base tapered, acute, slightly decurrent into petiole, apex obtuse or shortly acute or with a broad, very short acumen, chartaceous; petiole slender, 1–1.5 cm long. **Inflorescences** with ♂ flowers fascicled, axillary, pedicels to 2 cm long, slender; ♀ flowers solitary, pseudo-terminal, pedicel 2–5 mm long. **Flowers** ♂ white, outer 2 sepals spreading, concave, oblong-orbicular, 2 mm long, stiff, thick, inner 2 erect, oblong, 3–3.5 mm long, petals 4–5, fleshy, ovate to elliptic, 5–7 mm, base often narrowed, acute, stamens connate into a cylindrical, 1–1.5 mm long column, bearing ~10 anthers; ♀ with sepals to 5 mm long, stamens free, filaments slender, to 3 mm long, anthers 0.5 mm long, ovary with 6–9 grooves, crowned by a sessile, domelike stigma with rays of papillae. **Fruits** berries, globose to subovoid, to 5 cm Ø, red to orange-red, glossy with 6–9 very deep longitudinal grooves, sepals persistent. **Distr:** Sri Lanka. **Ecol:** In lowlands; to ~1,000 m. **Zones:** 11–12. **Info:** The arillode is thin, sweet-acid, edible.

G. xanthochymus J.D.Hooker ex T.Anderson, *Fl. Brit. India.* 1: 269, 1874. (Syn: *Xanthochymus pictorius*, etc.). [False Mangosteen, Mysore Camboge, Sour Mangosteen]. **Trees**, 8–15 m tall, 15–45 cm dbh; crown dense, pyramidal; latex bright yellow, thick; branches numerous, decussate, horizontal. **Leaves** elliptic or oblong to oblong-lanceolate, (14–)20–34 x (4–)6–12 cm, base ± broadly cuneate, margins involute, apex acute to obtuse (acuminate), thickly leathery, shiny, midvein robust, prominent, veins 35–40 pairs; petiole robust, 1.5–2.5 cm long, sometimes rose-colored. **Inflorescences** corymbose cymes, or fascicles, (2–)4–10(–14)-flowered, axillary; peduncles 0.6–1.2 cm long; pedicels 1.8–3cm; ♂ flowers 2–3 together; ♀ flowers usually solitary. **Flowers** 5-merous; ♂ with sepals (4)5, greenish-white, suborbicular, concave, margins fimbriate, 2 outer ones 4–6 mm long, inner ones thinner, to 7 mm long, petals 5, orbicular, shortly clawed, thin, white, to 9 mm long, stamens fascicled, to 1.3 cm long, anthers 3–10; ♀ with sepals and petals 3 large and 2 small, staminode fascicles 5, ~3 mm long, united below, ovary globose, usually 5-loculed, style short, ~1 mm long, stigma peltate, apex concave, (3–)5-cleft. **Fruits** berries, globose or ovoid, sometimes oblique, 3–5 cm Ø, maturing to yellow; seeds 1–4, oblong or ovoid, testa brown, smooth. **Distr:** Tropical Asia. **Ecol:** In dense humid forests; 100–1,400 m. Fl. Mar–May. Fr. Aug–Nov. **Zones:** 10–12.

Mammea

Syn: *Ochrocarpos*.

Species ~80. **Trees**, cryptically dioecious; apical bud not abortive; buds with scales. **Leaves** opposite, petiolate; lamina with margins entire, leathery; secondary veins many, nearly parallel, nearly perpendicular to midvein, tertiary venation densely and evenly reticulate, prominent, with translucent gland dots (or streaks) in areoles. **Inflorescences** axillary on older twigs, flowers solitary (or fasciculate). **Flowers** with calyx completely united in bud, splitting at anthesis into 2(or 3) sepals; petals (4 or 5 or)6(or 8), imbricate; filaments slender, basally connate, anthers erect, basifixed; ovary biloculed, each locule 2-ovuled (or incompletely 4- or 8-loculed, each locule 1-ovuled), style very short, stigma 2(–4)-lobed. **Fruits** berries, with (thick or) thin exocarp, fleshy mesocarp and 1–4(–8) seeds; seeds large, testa thin (to fibrous or woody). **Distr:** Tropical Asia, Madagascar, tropical Africa, Australasia, C America. **Cult:** Propagate from seeds or by air-layering. Requires a humus-rich, moist, well-

Below: *Garcinia xanthochymus*, *Mammea americana*, *Mesua ferrea*.

Above: *Mesua nagassarium, Mesua stylosa, Pentadesma butyracea.*

drained soil in full or partial sun.

M. americana L., *Sp. Pl.* 512, 1753. [Amarillo, South American Apricot, Mamey]. **Trees**, monoecious to polygamous, to 20 m tall; trunk to 50 cm dbh; bark ± smooth; sap yellow. **Leaves** oval to obovate, 8–16 x 5–10 cm, base rounded to obtuse, apex rounded or emarginate, thick, with pellucid dots, major lateral veins at nearly right angles to midrib, tertiary venation numerous, and as prominent as lateral veins; petiole to ~1 cm long. **Inflorescences** axillary, flowers solitary or in few-flowered clusters; peduncles short, stout; pedicels 1–1.5 cm long; buds globose. **Flowers** fragrant; sepals 2, nearly round, concave, ~1.5 cm long; petals white, 4–6 (usually 6), obovate, spreading, 2–2.5 cm long; ♂ with a cluster of yellow stamens ~2 cm Ø, ~1 cm long; ♀ with pistil 2–4-celled, style short, bilobed, stigmas ± reniform. **Fruits** ± globose, apiculate, to ~12 cm Ø, brownish, exocarp thick, mesocarp red to yellow, sap white; seeds 2–4, oblong, ~8 cm long, reddish-brown. **Distr:** West Indies. **Ecol:** In moist coastal forests, foothills and plains; <1,000(–1,500) m. Fl. May–Oct. **Zones:** 11–12. **Info:** Cultivated widely throughout the tropics for the edible fruits.

Mesua

Species ~5. **Trees**; apical bud abortive, axillary buds with scales. **Leaves** opposite, petiolate, exstipular; lamina leathery, usually with inconspicuous, translucent glands, midrib often partially channeled above, margins entire, secondary veins many, slender, tertiary veins scalariform, not prominent. **Inflorescences** axillary or pseudo-terminal, flowers solitary or sometimes in thyrses. **Flowers** bisexual (hermaphrodite), homostylous; sepals 4(or 5), imbricate, in 2 decussate pairs, outer ones smaller, thicker, persistent; petals 4(or 5), large, imbricate, alternate with sepals; stamens numerous, free (or basically slightly connate), not obviously fascicled, in continuous ring, filaments slender, anthers erect, basifixed, dehiscent by vertical slits; fasciclodes absent; ovary (1- or)biloculed, with 2 erect ovules per locule, anatropous, styles united, long, obscurely trilobed, stigma peltate. **Fruits** capsules (or capsular nuts), subwoody, septifragal, monolocular, dehiscent into 2–4 valves, septae persistent; seeds 1–4, erect, without appendage or arillode. **Distr:** India and Sri Lanka to peninsular Malaysia, possibly Java, and China, to as far as New Guinea. **Cult:** Propagate from seeds, half-hardened cuttings or by air-layering. Requires a moist, well-drained soil in shade, partial or full sun.

M. ferrea L., *Sp. Pl.* 1: 515, 1753. (Syn: *M. ferrea* ssp. *thwaitesii*, etc.). [Ceylon Ironwood, Indian Rose Chestnut, Ironwood Tree]. **Trees**, evergreen, 15–30 m tall, glabrous; trunk buttressed at base; exudate aromatic white resin. **Leaves** always pendulous; lamina reddish-yellow when young, becoming dark green, lanceolate or narrowly ovate-lanceolate to linear-lanceolate, (4–)6–10(–30) x (1–)2–11 cm, base cuneate, apex acuminate or long-acuminate to caudate, abaxially glaucous, white, thickly coriaceous; petiole 5–8 mm long. **Inflorescences** 1–3-flowered; peduncle short, in axils of apical leaves; each flower surrounded by 2–3 large, adpressed bracts; pedicels 0–5 mm long. **Flowers** bisexual, solitary, axillary, 5–11.5 cm across; sepals (outer 2 slightly larger) orbicular, convex, margin membranous, sometimes white-ciliate, sparsely pubescent; petals white, obovate-cuneate, 3–3.5 cm long; stamens with filaments filiform, 1.5–2 cm long, anthers golden-yellow, oblong; ovary conic, ~1.5 cm long, style 1–1.5 cm long. **Fruits** capsules, broadly ovoid or laterally depressed globose, ~3 x 2.5 cm, dry, longitudinally rugose, apex with persistent style, usually dehiscent by 2 valves, seeds 1–4, ± irregular, testa brown. **Distr:** Bangladesh, India, Indonesia (Java), Malaysia, Sri Lanka, Thailand. **Ecol:** In moist places; 500–600 m. Fl. Mar–May. Fr. Aug–Oct. **Zones:** 10–12.

M. nagassarium (Burm. f.) Kosterm., *Ceylon J. Sci., Biol. Sci.* 12 (1): 71, 1976. (Syn: *Calophyllum nagassarium*). [Ceylon Ironwood, Na Tree]. **Trees**, to 30 m tall; trunk to 70 cm Ø, buttresses sometimes to 2 m high; branchlets very slender, terete. **Leaves** very variable, linear-lanceolate, oblong-lanceolate, lanceolate or elliptic-oblong, 1 x 5 – 5 x 18 cm, usually 3 x 10 cm, base acute or obtuse, apex acute, acuminate, thinly to thickly coriaceous, abaxially waxy-white; petioles 0.5–1.2 cm long. **Inflorescences** with flowers usually solitary, in axils of upper leaves, rarely paired or ternate; peduncles short; pedicel 0.8–1.5 cm, densely rusty-tomentellous. **Flowers** 4–10 cm Ø, sweetly scented, white; sepals 4, orbicular, imbricate, fleshy, concave, 1.2–1.5 cm long, densely velvety-puberulous outside, persistent; petals 4, white, obovate or obcordate, cuneate, margins curled and erose, thin, very finely purplish-veined, caducous; stamens 4–5 mm long; ovary ovoid. **Fruits** ovoid to globose, apex conical, striate, 1-loculed, 1–4-seeded, persistent sepals enlarged, to 4 cm long, adpressed; seeds faceted, smooth, testa glossy, brown. **Distr:** India through Indo-China and Malay Peninsula to Sumatra. **Ecol:** <1,500 m. **Zones:** 9–12.

M. stylosa (Thw.) Kosterm., *Reinwardtia* 7: 430, 1969. (Syn: *Kayea cuspidata, K. stylosa*). **Trees**, to 5(–8) m tall, glabrous; trunk 40 cm dbh. **Leaves** ovate or ovate-oblong, 2.5 x 6–3 x 6–3 x 9 cm, base rounded, apex acuminate (more rarely acute or cuspidate), thinly coriaceous; petiole slender, 5–8 mm long, adaxially channeled. **Inflorescences** racemes, stalked, stiff, erect, 24-flowered, to 6 cm long, axillary, near upper part of branchlets; pedicels slender, thickened upward; bracteoles subulate, caducous. **Flowers** ~1–2 cm Ø; sepals 4, thick, concave, erect, persistent; petals 4, concave, longer than sepals, obtuse, white; stamens numerous, shorter than petals; ovary subglobose, 4-ovuled; style stout, longer than ovary, stigma small with subulate branches. **Fruits** mostly enveloped by enlarged, coriaceous, imbricate, sepals, depressed-globose, to 2.5 cm Ø, persistent style long; seeds 1–4, large, angular, dark reddish-brown. **Distr:** SW Sri Lanka (Kalutara, Pasdun Korale). **Ecol:** In moist lowlands. **Zones:** 11–12. **Info:** Very rare, not collected in the wild for over 50 years.

Pentadesma

Species 5. **Trees**, large, glabrous. **Leaves** coriaceous, penninerved. **Inflorescences** 3–15-flowered, terminal. **Flowers** hermaphrodite, large; sepals 5, imbricate, the inner larger; petals 5, broad, equaling to exceeding the inner sepals; stamens numerous, fascicled, 5-adelphous, filaments smooth, very shortly connate in each phalange, free above, anthers narrow, 0.7–1.6 cm long, locellate, anther cells extrorse; ovary 5-celled, narrowed into style, style divided at apex into 5 spreading linear lobes, ovules numerous per carpel. **Fruits** baccate, large, woody, pericarp thick, fleshy, 3–5-celled, each cell 1–2-seeded; seeds large, testa glossy. **Distr:** Tropical Africa (Mali to Zaire and Rwanda). **Cult:** Propagate from seeds or by air-layering. Requires any fertile soil in full sun.

P. butyracea Sabine, *Trans. Hort. Soc. London* 5: 457, 1824. (Syn: *P. grandifolia*, etc.). [African Butter Tree, Black Mango Tree, Butter Tree, Candle Butter Tree, Candle Tree, Tallow Tree]. **Trees**, evergreen, large, to 35 m tall; trunk cylindrical, 1–2.8 m dbh; exudate bright yellow. **Leaves** opposite, in dense terminal clusters; lamina obovate to oblong-oblanceolate, 12.5–25 x 4–6.5 cm, base rounded or cuneate, margins entire, apex shortly acuminate or obtuse, glossy; petiole 0.4–2.5 cm long. **Inflorescences** thyrses, 1–7-flowered, terminal or axillary; pedicels 1–4 cm long. **Flowers** bisexual, regular, 5-merous, white, cream or greenish-white, waxy, smelling like rancid butter; sepals ovate, 4–5 cm long; petals oblong to ovate, to 6 cm long, keeled; stamens numerous, in 5 bundles, opposite petals, 4–6.5 cm long, basally connate; ovary superior, ovoid-ellipsoid, 1–2 cm long, 5-celled, style elongate, apical lobes 5, linear, spreading, 5 mm long, style persistent. **Fruits** berries, ellipsoid to ovoid, 9–15 x 6.5–12.5 cm, woody; exudate copious, yellow; seeds 3–15, brown, pyramidal, 4–5 x 2.5–4 cm, embedded in yellowish pulp. **Distr:** West tropical Africa (Guinea Bissau to Cameroon, to DR Congo). **Ecol:** In primary forests; <1,000 m. Fr. Apr–May. **Zones:** 10–12. **Info:** Fruits edible when immature, but not palatable when ripened. The fruits yield edible fat and butter, used in cosmetics.

CALOPHYLLACEAE

Genera 13, **species** 460. **Trees** or **shrubs**; (vessel elements with scalariform perforations); schizogenous canals or cavities present. **Leaves** (spiral, 2-ranked), often flat (conduplicate; supervolute in *Kielmeyera*), (paired 'stipular glands' on the stem; colleters absent). **Flowers** 4–5-merous, corolla (contorted), (0–)4–5(–8); stamens not obviously fasciculate, many, (connate), anthers often with complex or simple glands; ovary 2–5-loculed, (placentation apical, basal), (1-few ovules/carpel), style (styles) present, usually long, stigmas much expanded to punctate, not papillate, wet. **Fruits** berries or drupes; seeds 1–many, embryo green or white (Stevens 2001 onwards). **Distr:** Throughout the tropics.

Calophyllum

Syn: *Apoterium*, *Augia*.

Species ~190. **Trees** or **shrubs**, glabrescent, buds finely tomentose, with clear (or milky or yellow) latex; branchlets quadrangular. **Stipules** absent. **Leaves** opposite, petiolate (rarely sessile); lamina margins entire, leathery, usually glabrous, secondary veins many, tertiary venation absent, translucent glandular canals present between veins. **Inflorescences** cymose or thyrsiform, terminal, pseudo-terminal or axillary; bracts caducous. **Flowers** bisexual (or rarely unisexual); sepals and petals together 4–12, 2- or 3-whorled, not always differentiated, outer (sepaloid) decussate, enveloping the bud, inner (petaloid) imbricate; stamens many, not obviously fascicled, filaments scarcely united or all free, slender, anthers erect, basifixed; ovary 1-loculed, glabrous (tomentose), with a single erect ovule, style elongate (usually as long as stamens), slender, often bent, stigma often peltate, concave with irregular margin. **Fruits** drupelike berries with thin exocarp, thin fleshy mesocarp and thin endocarp sometimes adherent to seed; seed 1, large, with thin (or thick) testa. **Distr:** Tropical regions, mainly in Asia; E Africa, Madagascar, the Mascarenes and Australasia; tropical America. **Cult:** Propagate from seeds or cuttings. Needs well-drained sandy soil in full sun and many species are salt-tolerant.

C. calaba L., *Sp. Pl.* 1: 514, 1753. (Syn: *C. apetalum*, etc.). [Calaba Oil Tree]. **Trees**, to 10 m tall; trunk 90 cm dbh; crown dense, round; branchlets slender, quadrangular. **Leaves** initially chartaceous, when mature rigidly coriaceous; lamina elliptic to ovate or obovate-elliptic, 3 × 2 – 8 × 6 – 9 × 5 cm, base very shortly cuneate, apex rounded or obtusely, obscurely acuminate to emarginate; petiole 0.4–1.2 cm long. **Inflorescences** racemes, axillary, 3–6-flowered; main peduncle densely rusty-pubescent, 0.5–1 cm long; pedicels filiform, 0.5–2 cm long. **Flowers** with sepals 4, oblong, concave, sometimes slightly pilose outside, 3–5 mm long, inner ones thinner, narrower, to 7 mm long; petals absent (rarely 4), spathulate or obovate, to 7 mm long; stamens 4 mm long; ovary globose, style slender, 3 mm long, stigma peltate, concave. **Fruits** ellipsoid to subglobose, bright orange or yellowish-orange, pendulous, 1 × 1.2 – 1.4 × 1.6 cm, apex obtuse, pleasantly sweet, acidic, edible; seed ovoid to ellipsoid. **Distr:** Sri Lanka; rare. **Ecol:** In dry savannas. **Zones:** 11–12.

C. inophyllum L., *Sp. Pl.* 1: 513, 1753. (Syn: *Balsamaria inophyllum*, etc.). [Alexander Laurelwood Tree, Borneo Mahogany]. **Trees**, 5–12 m tall; bark gray-brown or dark brown, thick, longitudinally fissured, exuding pellucid resin when wounded; young shoots striate. **Leaves** broadly elliptic or obovate-elliptic, rarely oblong, 8–15 × 4–8 cm, base rounded or broadly cuneate, apex rounded or emarginate, shiny, thickly leathery, midvein raised abaxially, impressed adaxially; petiole 1–2.5 cm long. **Inflorescences** thyrses, 7–11-flowered, in upper axils, to ~10 cm long; pedicels 1.5–4 cm long. **Flowers** scented, white, 2–2.5 cm ∅; sepals 4, outer 2 suborbicular, ~8 mm long, inner 2 obovate, petaloid; petals 4, oblanceolate to obovate, ~1.1 cm long, concave, apex subtruncate or rounded; ovary subglobose, stigma peltate. **Fruits** yellow, globose, ~2.5 cm ∅. **Distr:** Tropical Asia through Malesia to Pacific Islands. **Ecol:** On seashores, sandy wastelands; 0–500 m. Fl. Mar–Jun. Fr. Sep–Nov. **Zones:** 11–12.

C. tomentosum Wight, *Ic. Pl. Ind. Or* (*Illustrations of Indian Botany*) 1: t. 110, 1839. (Syn: *C. elatum*). **Trees**, to 30 m tall; trunk 90 cm dbh, sometimes with small buttresses. **Leaves** lanceolate to oblong-lanceolate, 1×5 – 2×7 – 5×15 cm, rarely narrower, both ends attenuated, apex sometimes acuminate with a blunt tip, almost white when young, mature leaves stiffly chartaceous; petiole slender, abaxially concave, glabrescent, 1.5–2.5 cm long. **Inflorescences** axillary, consisting of simple racemes, or once-branched, 2–5 cm long, 8–10-flowered, densely rusty-tomentose; bracts caducous; peduncles slender; pedicels very slender, white, 1–1.5 cm long, ± tomentose. **Flowers** white; sepals 4, concave, elliptic, outer ones reflexed, 3.3 mm long, inner ones 7.5 mm long, petaloid; petals 4, spathulate, elliptic, thin, to 8 mm long, inner ones smaller; stamens 4 mm long; ovary slender, style 2 mm long with peltate, concave, obscurely lobed stigma. **Infructescences** ~2 cm long, fruits pendulous; pedicels 1–2.5 cm long. **Fruits** globose, 2.5 cm ∅ (young fruits apiculate), pericarp thin. **Distr:** Sri Lanka; rare and endangered. **Ecol:** In lower hill zones and intermediate zones; 600–700 m. Fl. Aug–Nov. Fr. May–Jul. **Zones:** 11–12.

HYPERICACEAE

Genera 9, **species** 560. **Shrubs** to **trees**, or annual to perennial **herbs**, with clear or black resinous sap, in secretory cavities; hairs simple, multicellular. **Stipules** lacking. **Leaves** opposite or whorled; lamina simple, margins entire, venation pinnate, with pellucid or black dots. **Inflorescences** determinate, sometimes reduced to a single flower, usually terminal. **Flowers** bisexual, radial; petals usually 4–5, distinct, often asymmetrical, imbricate or convolute; stamens numerous, innermost developing before the outermost, often fascicled; carpels 3–5, connate, ovary superior, placentation axile, or sometimes parietal with deeply intruded placentas, stigmas minute, punctate, on 3–5 elongate, distinct to basally connate styles, ovules numerous per carpel; nectaries usually lacking. **Fruits** variously dehiscing capsules, berries or drupes; seeds not arillate, endosperm scanty (Judd et al. 2008; Stevens 2001 onwards). **Distr:** Worldwide.

Hypericum

Species ~400. **Trees**, **shrubs** or perennial to annual **herbs**, sometimes glabrous, with translucent red or black glands. **Leaves** opposite; lamina with margins entire or rarely gland-fringed. **Inflorescences** with flowers solitary or in terminal dichasial to monochasial cymes or panicles. **Flowers** with sepals 5, quincuncial or rarely 4, decussate, glandular like the leaves; petals (4)5, yellow, often tinged or veined red, deciduous or persistent; stamens fascicled, (4)5, antepetalous, free or variously united, deciduous or persistent, filaments slender, free or united, anthers dorsifixed, with an amber or black gland on the connective; ovary 1–5-locular, with (2)3–5 placentas, ovules 2 to numerous on each placenta, styles (2)3–5, free or ± united, slender. **Fruits** capsules, septicidal or rarely indehiscent; seeds sometimes winged or carinate, testa variously patterned. **Distr:** Worldwide excluding the polar regions. **Cult:** Propagate from seeds or softwood cuttings. Requires a moist, well-drained soil in shade or partial sun.

H. mysurense Wight & Arn., *Prod.* 1: 99, 1834. (Syn: *Norisca mysurensis*). [Mysore, St John's Wort]. **Shrubs**, much-branched, 1–2(–3) m tall; bark yellowish; young branches thin, glabrescent, ascending. **Leaves** very closely arranged in 4 ranks; lamina elliptic, lanceolate or oblanceolate, 2.8–3.8 × 0.3–1.5 cm, base attenuate, amplexicaul, apex acute, stiff, gland-dotted, pinnately veined, adaxially channeled, abaxially midrib prominent. **Inflorescences** terminal cymes, few-flowered; peduncle ~1 cm long. **Flowers** 4–5 cm across; sepals unequal in size, lanceolate, ~1.5 × 0.4 cm, acute, persistent; petals bright yellow, 2.7–3.4 × 1–1.5 cm, deciduous; stamens in 5 groups alternating with petals, filaments 1.5–1.7 cm long; ovary 5-lobed, tapering into 5 styles. **Fruits** capsules, crimson, 0.8–1.4 cm long; seeds cylindric, ~3 × 0.5 mm. **Distr:** S India, Sri Lanka. **Ecol:** In open bushy places and grasslands of upper montane regions. Fl. throughout most of the year. **Zones:** 10–12.

MALPIGHIACEAE

Subfamilies 2, **genera** ~65, **species** ~1,280. **Shrubs**, **small trees**, or **woody lianas** (or sometimes lianalike shrubs), monoecious or androdioecious, reddish-pubescent. **Stipules** free and deciduous, or connate and ± persistent, sometimes reduced or absent. **Leaves** usually opposite, rarely alternate or 3-whorled, petiolate; lamina simple, margins entire, paired glands often present either on petiole or on

Below: *Calophyllum calaba*, *Calophyllum inophyllum*, *Calophyllum tomentosum*, *Hypericum mysurense*.

abaxial surface of leaves. **Inflorescences** terminal or axillary, racemose, corymbose or umbellate, solitary or in panicles; pedicels articulate, bibracteolate. **Flowers** bisexual or rarely ♂, actinomorphic or weakly zygomorphic; perianth typically consisting of 2 whorls with 5 distinct segments in each; sepals 5, polysepalous or gamosepalous, imbricate, rarely valvate, one or more of the sepals usually have 1 or more conspicuous glands; petals 5, typically clawed, margin ciliate, dentate or fimbriate; disk inconspicuous; stamens 10, distinct or basally connate, 2-whorled, some or half of them are commonly reduced to staminodes; gynoecium consists of a single compound pistil of almost always 3 carpels, styles 3, distinct, ovary superior, triloculed, each 1-ovuled. **Fruits** schizocarps, carpels 3 or less, 1 seed per carpel; schizocarp splitting into winged samaras, indehiscent. **Distr:** Tropical and subtropical regions of the world, but mainly American.

Subfam. MALPIGHIOIDEAE

Genera 67, **species** 1,100. **Pollen** globally symmetrical (4-polyporate), style various; stigma usually not terminal. **Fruits** winged (bristly, unwinged) (Stevens 2001 onwards). **Distr:** Tropical and subtropical regions, especially the Americas.

Bunchosia

Species ~40. **Shrubs** or **small trees**, erect, pubescent. **Stipules** interpetiolar, mostly linear-lanceolate, distinct, acute, often very small. **Leaves** simple, entire, leathery; glands 2, large, at base of leaf lamina on opposite sides of midrib. **Inflorescences** racemes or clusters, axillary, short or usually elongate. **Flowers** yellow; calyx bearing 8–10 glands, these often confluent by pairs; petals and stamens glabrous; stamens 10, fertile, filaments winged, basally connate; styles free or connate, obliquely obtuse. **Fruits** drupaceous, fleshy, containing 2–3 nutlets, 2–3-seeded; nutlets smooth. **Distr:** Tropical America with a few additional species from southern C America. **Cult:** Propagate from seeds or by air-layering. Requires moist but well-drained soil in full to partial shade. **Info:** The generic name is said to be derived from an Arabic word, *bunchos*, signifying coffee. The fruit is often edible and highly nutritious, as with many of the species of this family.

B. armeniaca (Cav.) DC., *Prodr.* 1: 582, 1824. (Syn: *Malpighia armeniaca*, *B. nitida*). [Peanut Butter Plant]. **Shrubs** or **trees**, 4–12 m tall, younger parts sericeous. **Stipules** connivent, 0.5–1 mm long. **Leaves** narrowly ovate to oblong or ovate-lanceolate, to 24 × 11 cm, base obtuse or shortly narrowed, apex shortly to long-acuminate, scarcely revolute, chartaceous or subcoriaceous, glabrate, green, ± lustrous, primary nerves 6 or 7; petiole stout, 0.4–1 cm long. **Inflorescences** racemes, often geminate or ternate, to 40-flowered, 13 cm long; common peduncles 1–3.5 cm long, floriferous 1–5 mm long; pedicels 2–6 mm long, all enlarged; bracts ovate-lanceolate, 1.5–2.5 mm long, bractlets ovate, 1 mm long, very acute. **Flowers** 1.4 cm Ø; sepals ovate, ciliate, glands oblong-obovate; petal limb orbicular, ± glandular-ciliate; filaments basally connate, anthers subrotund, connective broad, slightly produced, brown; ovary 2-celled, sericeous, styles distinct (rarely connate). **Fruits** drupes, ovoid, 2.5 × 2 cm, subsericeous. **Distr:** Colombia to Bolivia and Brazil. **Ecol:** In flood-free, open woods. **Zones:** 10–12. **Info:** Fruits edible, with sweet dry pulp.

Galphimia

Species ~8. **Shrubs**, erect, glabrous or pubescent. **Leaves** with margins entire, thin; petiole slender. **Inflorescences** panicles, terminal, simple or branched; bracts and bractlets minute. **Flowers** yellow, subulate or lanceolate; calyx usually without glands, sepals 5, narrow, persistent; petals 5, entire or dentate, base abruptly unguiculate, persistent in fruit; stamens 10, all antheriferous, filaments glabrous, basally united, anthers glabrous; ovary trilobate, styles 3, slender, distinct, stigmas minute. **Fruits** capsules, small, trilobate. **Distr:** Tropical America. **Cult:** Propagate from seeds or cuttings. Requires a moderately fertile, well-drained soil in full or partial sun.

G. gracilis Bartl., *Linnaea* 13: 552, 1839. (Syn: *G. glauca*, etc.). [Spray of Gold, Thryallis]. **Shrubs**, densely branched, almost glabrous, 1–3 m high; young parts and inflorescences with sparse subappressed rufous hairs. **Leaves** ovate to oblong or oval, mostly 2–7.5 cm long, base obtuse, acute to rounded, membranous, adaxially glabrous, green, abaxially glaucous, with small glands at base of lamina or on abaxial surface of petiole; petiole slender. **Inflorescences** racemes, dense and many-flowered, terminal, 10–15 cm long; peduncles short, bracteolate; pedicels long. **Flowers** faintly fragrant; sepals oblong or lanceolate, 3–5 mm long; petals bright yellow, larger petals ovate, 0.7–1.2 cm long. **Fruits** capsules, 3.5–4.5 mm long, glabrous, almost smooth. **Distr:** Mexico, Guatemala, Honduras, Salvador, Nicaragua. **Ecol:** On brushy slopes or dry rocky hillsides; 900–2,100 m. **Zones:** 10–12.

Heteropterys

Syn: *Atopocarpus*, etc.
Species 120. **Vines** woody, **shrubs** or **small trees**. **Stipules** very small, distinct, triangular, persistent. **Leaves** opposite (very rarely alternate or whorled), usually with glands on petiole and/or lamina. **Inflorescences** umbels, corymbs, or pseudo-racemes, these single or grouped into racemes or panicles, axillary and/or terminal, pedunculate or not; bracteoles borne at or below apex of peduncle. **Flowers** with sepals biglandular or eglandular; corolla bilaterally symmetrical, posterior petal ± strongly differentiated from lateral 4 petals, petals light yellow, pink, white, bronze, or dark red, or combinations of some colors; lateral petals entire, denticulate, or erose; stamens 10, all fertile, filaments mostly proximally connate, anthers ± alike, posterior 3 sometimes smaller than others; ovary with 3 carpels proximally connate and distally distinct, all fertile, styles 3, anterior style different, apex of styles with a large, usually internal stigma. **Fruits** dry, breaking into 3 samaras or mericarps separating from a short pyramidal torus; samara with its largest wing dorsal, thickened on abaxial (lower) edge and mostly bent upward; much shorter winglets or crests present on sides of nut in some species; dorsal wing rudimentary or lost in a few species. **Distr:** Neotropics (N Mexico and West Indies to N Argentina and SE Brazil); 1 species in W Africa (from Senegal to Angola). **Cult:** Propagate from seeds or cuttings. Requires a moderately fertile, well-drained soil in full or partial sun.

H. angustifolia Griseb., *Linnaea* 13: 223, 1839. [Mariposa, Red Wing]. **Shrubs**, climbing; stems cylindrical, glabrous. **Leaves** alternate; lamina linear-lanceolate, 5–7.5 × 0.8–1 cm, base oblique, apex long-acuminate, tip acute, adaxially dark green, glabrous, shiny, abaxially paler, somewhat glaucous. **Inflorescences** racemes, 2.5–5 cm long, terminal or in upper leaf axils; peduncles opposite, verticillate, 1.5–2 cm long; bracts basal, obovate, ~2.5 mm long, glabrous. **Flowers** yellow; sepals oblong, ~5 mm long, base glandular, margins slightly ciliolate, apex obtuse; petals orbicular, ~1 cm long, margins crenulate, obtuse, penninerved; stamens 10, all fertile, ~4 mm long, glabrous, connective erect; ovary ovoid, ~2.5 mm long, setose, stigma linear, truncate, carpels 3, ovules pendulous. **Fruits** each composed of 3 samaras, 2 × 1 cm, glabrous, base green, wings pinkish-red, apices rounded, pericarp triangular. **Distr:** NW Argentina, S Brazil, Paraguay, Uruguay. **Ecol:** On the edges of natural rivers. Fl. Jan–Feb, Jul–Aug. Fr. Nov–Dec, Jun–Jul. **Zones:** 9–12.

Hiptage

Species ~30. **Lianas** woody, **lianoid shrubs**, or **small trees**. **Stipules** minute, glandlike, or absent. **Leaves** opposite; lamina with margins entire, leathery or subleathery, sometimes with 2 basal glands on abaxial surface. **Inflorescences** racemes, axillary or terminal; peduncles bracteate; pedicels articulate, bibracteolate at articulation point. **Flowers** white, sometimes pinkish, fragrant, bisexual, zygomorphic; calyx with 1 or 2 glands partially adnate to pedicel or eglandular; sepals 5, imbricate, basally connate; petals 5, clawed, ± unequal, silky-pubescent; stamens 10, unequal, one much larger than others, filaments basally connate; ovary shallowly trilobed, 3–5 appendages on abaxial surface, style 1, rarely 2, filiform, at first circinate, stigma capitate. **Fruits** samaras, 1–3, each 3-winged; abaxial wing long, erect; lateral wings short, spreading; seed angular-globose. **Distr:** Bangladesh, Cambodia, China, India, Indonesia, Laos, Malaysia, Mauritius, Philippines, Thailand. **Cult:** Propagate from seeds or cuttings. Requires a moderately fertile, well-drained soil in full or partial sun.

Below: *Bunchosia armeniaca*, *Galphimia gracilis*, *Heteropterys angustifolia*, *Heteropterys angustifolia* (fruits).

H. benghalensis (L.) Kurz, *J. Asiat. Soc. Bengal, Pt. 2, Nat. Hist.* 43: 136. 1874. [Helicopter Flower, Hiptage]. **Shrubs** or **lianas**, 3–10+ m long; branchlets and inflorescences densely pubescent. **Leaves** oblong, elliptic-oblong, or ovate-lanceolate, 9–18 × 3–7 cm, base broadly cuneate or rounded, apex acuminate, abaxially often biglandular, leathery, young leaves reddish, pubescent, becoming green and glabrous, lateral veins 6 or 7 pairs; petiole 0.5–1 cm long. **Inflorescences** racemes, axillary or terminal, 5–10 cm long, yellowish-brown pubescent, glabrate; pedicels articulate; bracteoles subulate-lanceolate. **Flowers** very fragrant; sepals broadly elliptic or ovate, 5–6 mm long, apex rounded, densely pubescent; gland thick, large, oblong, partially adnate on pedicel; petals white, base yellow maculate, or yellowish or pink, orbicular or broadly elliptic, 0.8–1.5 × 0.5–1 cm, base clawed, margins fimbriate, apex rounded, pubescent; stamens unequal, longest 0.8–1.2 cm long, others 3–5 mm long, anthers elliptic, 1–2 mm long; style ~1.2 cm, circinate. **Fruits** samaras, 3-winged; abaxial wing elliptic or obovate-lanceolate, 3.5–5(–7) × 1–1.6 cm, entire or retuse; lateral wings lanceolate-oblong, 1.5–3 cm long, base of wing with 1 triangular-crested appendage. **Distr:** Tropical Asia to Indonesia and Philippines. **Ecol:** In forests; (100–)200–1,900 m. Fl. Feb–Apr. Fr. Apr–May. **Zones:** 10–12.

Lophanthera

Species 5. **Shrubs** or **trees**; stems often containing white latex. **Stipules** intra- and slightly epipetiolar, ± completely connate. **Leaves** bearing glands on petiole or lamina, or both. **Inflorescences** terminal (rarely axillary in *L. spruceana*), thyrses composed of few-flowered cincinni or dichasia or a pseudo-raceme, bracts and bracteoles persistent, 1 or both bracteoles usually bearing a large gland. **Flowers** with receptacle glabrous or with a row of short, basifixed hairs between filaments and ovary; petals entire or at most minutely denticulate, deciduous, posterior petal moderately to strongly differentiated from the other 4; stamens 10, ~1-whorled, filaments distinct or slightly basally connate, glabrous, anthers similar, glabrous, deciduous, 4-locular, the outer locules bearing dark longitudinal wings, connective shorter than locules; ovary of 3 uniovulate carpels, styles 3, apical or subapical, subulate, stigmas minute, apical, glabrous, often persistent. **Fruits** schizocarpic, breaking apart into 3 (or fewer due to abortion) dry, unwinged, 1-seeded cocci; mericarps indehiscent or slightly dehiscent along keel. **Distr:** Tropical America. **Cult:** Propagate from seeds or cuttings. Requires a fertile, moist, well-drained soil in full or partial sun.

L. lactescens Ducke, *Arch. Jard. Bot. Rio de Janeiro* 4: 103, 1925. [Golden Chain]. **Trees**, to 15 m tall; stems lactiferous, very finely sericeous, soon glabrate and coarsely lenticellate. **Stipules** paired, triangular, 0.8–1.2 cm long. **Leaves** obovate, (12–)15–24 × (5–)9–12 cm, base decurrent, apex rounded or obtuse or apiculate, glabrate, abaxially glandular; petiole 1.5–2.5 cm long, eglandular or bearing 2 large flat glands above. **Inflorescences** thyrses, 30–40+ cm long, horizontal or reflexed, many-flowered, composed of 1–3-flowered dichasia or cincinni 2–4 cm long; primary peduncle 0.9–2 cm long; pedicels 4–7 mm long, sparsely velutinous. **Flowers** with sepals extending 1–2 mm beyond glands, triangular, apex acute, flat; petals yellow, 5–7 mm long, dorsally carinate, margin pilose; stamens with filaments 2–2.5 mm long, anthers 0.8–1 mm long, winged; ovary ~1.3 mm high, styles ~2.3 mm long. **Fruits** with cocci separating from a very short pyramidal base, ~5 × 3.5 mm, glabrous, obovoid, 1-seeded. **Distr:** Brazil (lower Rio Tapajos); very rare in the wild. **Ecol:** In rainforests. Fl. throughout most of the year. **Zones:** 10–12.

Malpighia

Species 40. **Shrubs** erect, or **trees**. **Stipules** slender, small and inconspicuous. **Leaves** mostly shortly petiolate; lamina entire to hollylike, coriaceous to membranous. **Inflorescences** corymbs or umbels, sometimes flowers solitary; pedicels equaling or usually much longer than peduncles. **Flowers** with calyx 6–10-glandular; petals unequal, glabrous, ± dentate or fimbriate, often ciliate, or some of them entire; stamens glabrous; ovary glabrous, styles free, obliquely obtuse. **Fruits** drupaceous, normally of 3 nutlets and 3-seeded, putamen of the nutlets longitudinally 3–5-cristate and transverse-costate. **Distr:** Tropical America. **Cult:** Propagate from seeds or cuttings. Requires any fertile, well-drained soil in full or partial sun.

M. coccigera L., *Sp. Pl.* ed. 2: 462, 1753. [Barbados Holly, Miniature Holly, Singapore Holly]. **Shrubs**, evergreen, often variously branched, to 1 m high; branches dark red and tomentose. **Leaves** opposite, mostly holly like; lamina ovate, 1.25–2.5 × 1–2 cm, rarely larger, both ends obtuse or orbicular, margins entire or serrate with sharply pointed lobes, leathery, adaxially dark green, abaxially light green, midrib elevated on both surfaces. **Inflorescences** axillary, flowers solitary or paired. **Flowers** pinkish to white; calyx lobes 5, lanceolate, with 2 glands on 2 lobes and a larger gland on the other 2 lobes, a reduced gland on the fifth lobe; petals 5, with a very narrow stalklike base, pale pink to mauve or almost white, margins fimbriate, crisped; stamens 10; styles 3, flexed. **Fruits** berries, glossy, red. **Distr:** West Indies. **Ecol:** On exposed rocky, dry hills, in dry thickets and lowland scrub-forests, usually as very low understory shrubs. Fl. throughout the year. **Zones:** 11–12.

M. emarginata DC., *Prodr.* (DC.) 1: 578, 1824. (Syn: *M. glabra*, etc.). [Barbados Cherry]. **Shrubs**, 2–3 m tall, sometimes a small tree to 5 m tall; young parts yellowish-sericeous, soon glabrate. **Leaves** ovate to ovate-lanceolate or oblong-lanceolate, mostly 8 × 3.5 cm, sometimes smaller, base acute or obtuse, apex acute or acuminate, chartaceous or subcoriaceous, becoming glabrous with age, bright green; petiole short. **Inflorescences** corymbs, 5–8-flowered, small, pedunculate; pedicels 0.6–1.5 cm long. **Flowers** pink or pale red, 1.2–1.6 cm across; sepals oblong or elliptic, glands oblong, 3 mm long or shorter; petals somewhat glandular on margins. **Fruits** drupes, cherry-red, trigonous-ovoid or subglobose, ~1 cm Ø. **Distr:** S USA (W Texas), Mexico, British Honduras to Salvador and Panama; West Indies; N South America. **Ecol:** In moist or dry forests or thickets; <1,800 m. **Zones:** 10–12. **Info:** Often cultivated in the tropics for the edible fruits.

Stigmaphyllon

Species ~55. **Vines**, woody. **Leaves** opposite, petiole long or short; lamina narrow or often very broad, margins entire, dentate, or lobate, often variable on the same plant, membranous or coriaceous. **Inflorescences** axillary, pedunculate, often umbelliform corymbs. **Flowers** with calyx with 8 glands; sepals 5, usually broad, persistent; corolla yellow, petals 5, unequal, erose, dentate, or fimbriate; stamens 10, usually only 4 of them fertile, filaments unequal, united to the middle or only at base, anthers short and thick; ovary trilobate, 3 styles distinct, unequal, hooklike or foliaceous at apex. **Fruits** samaras, 2–3, with a long broad wing, this thickened along the ventral (inner) edge, body smooth or cristate. **Distr:** Tropical America. **Cult:** Propagate from seeds, cuttings or by air-layering. Requires a well-drained soil in full sun.

S. ciliatum (Lam.) Juss., *Fl. Bras. Merid.* 3: 49, 1832. (Syn: *Banisteria ciliata*). [Golden Vine, Orchid Vine]. **Vines**, slender, small or large, to 5 m long; young branches sparsely sericeous. **Leaves** ovate or rounded-ovate, to 8 × 7 cm, base deeply cordate, basal lobes overlapping, apex acute or obtuse and mucronate, sparsely long-ciliate, glabrous, at least in age, abaxially paler; petiole long, slender, bearing 2 large discoid glands at apex. **Inflorescences** umbels, 4–7-flowered, terminating an axillary branch, 5–10 cm long, solitary, the pedicels 1–1.3 cm long. **Flowers** yellow, 4 cm across or larger; sepals ovate, glands oval, 1.5–2.5 mm long; petals with margins undulate, fimbriate; styles 3, widely divergent. **Fruits** samaras, ~2.5 cm long, wing 1.5 cm wide. **Distr:** S America. **Ecol:** In mangrove swamps at sea level. **Zones:** 10–12.

Tristellateia

Species ~20. **Lianas**, woody. **Stipules** connate with petiole. **Leaves** opposite or whorled; lamina with margins entire, glandular. **Inflorescences** racemes, axillary or terminal; pedicels bracteate. **Flowers** 5-merous, bright yellow, bisexual, actinomorphic; sepals with or without glands; petals clawed; stamens 10, unequal, filaments basally connate, articulate distally; ovary trilobed, globose, often with abaxial appendages, style 1, stigma minute. **Fruits** samaras, 3, each 3–10-winged; wings conic, of

Below: *Hiptage benghalensis*, *Lophanthera lactescens* (habit), *Lophanthera lactescens* (inflorescence), *Malpighia coccigera*.

Above: *Malpighia emarginata, Stigmaphyllon ciliatum, Tristellateia australasiae, Byrsonima crassifolia.*

unequal lengths, together thus star-shaped. **Distr:** Mainly in Madagascar, also in E Africa, India, China and Malaysia to Australia and Pacific Islands. **Cult:** Propagate from seeds or cuttings. Requires a moderately fertile, well-drained soil in full or partial sun.

T. australasiae A. Richard, *Voy. Astrolabe.* 2: 159, 1834. [Australian Golden Vine, Shower of Gold Climber, Vining Galphimia, Vining Milkweed]. **Lianas**, woody, ~10 m tall, mostly glabrous. **Stipules** 2, linear to lanceolate, ~1 mm long, apex acute. **Leaves** ovate, 6–12 × 4–7 cm, base cordate to rounded, papery or leathery, margins slightly reflexed, apex acute to acuminate, surfaces glabrous; petiole 1–1.5 cm long, with 2 glands adnate to petiole. **Inflorescences** terminal or axillary racemes; pedicels 1.5–3 cm long; bract 1, articulate distally; bracteoles 2. **Flowers** 2–2.5 cm across; sepals 5, triangular, ~3 × 1.5 mm, eglandular or with very minute glands; petals 5, yellow, sagittate-ovate, 0.8–1.3 × 0.5–0.6 cm, clawed, glabrous, entire, claw 2–3 mm long; stamens 3–4 mm long. **Fruits** schizocarps, stellate, 1–2 cm Ø. **Distr:** Malaysia, Thailand, S Vietnam; tropical Australia, Pacific Islands. **Ecol:** In forests near seashores. Fl. Aug. Fr. Oct. **Zones:** 10–12.

Subfam. BYRSONIMOIDEAE

Genus 1, **species** 150. **Style** subulate, stigma terminal (Stevens 2001 onwards). **Distr:** Tropical America.

Byrsonima

Species 150. **Shrubs** or **trees**, rarely almost herbaceous; young parts sericeous or tomentose. **Stipules** intrapetiolar, commonly connate. **Leaves** mostly shortly petiolate; lamina with margins entire. **Inflorescences** mostly many-flowered, simple, terminal racemes. **Flowers** yellow; sepals biglandular, rarely eglandular; petals long-unguiculate, glabrous, limb cordate or reniform, subentire; filaments scarcely connate, inserted on a hirsute torus, anthers about equaling filaments, glabrous or pilose; ovary glabrous or sericeous, styles subulate. **Fruits** drupes, with scant or rather copious flesh, stone 3-celled or by abortion 1–2-celled. **Distr:** Tropical America. **Cult:** Propagate from seeds or by air-layering. Requires moist but well-drained soil in full or partial sun.

B. crassifolia (L.) HBK., *Nov. Gen. & Sp.* 5: 149, 1822. (Syn: *B. cotinifolia*, etc.). [Craboo, Golden Spoon, Nance, Savanna Serette, Wild Craboo]. **Shrubs** or **small trees**, 5–10 m tall, often flowering ~1–2 m high; trunk straight or crooked, tall or short; crown rounded or spreading, sometimes rather tall and narrow; bark dark brown, rough; young branches densely or laxly tomentose. **Leaves** variable, mostly ovate to elliptic or oblong-elliptic, 8–15 × 4–7 cm, base acute or obtuse, apex acute or acuminate, sometimes rounded and apiculate, adaxially usually lustrous and glabrate, abaxially sparsely or densely tomentose with lax, rufous or grayish hairs, chartaceous; petiole ~0.8–1.5 cm long. **Inflorescences** racemes, equaling or longer than leaves, many-flowered, sparsely or densely rufous-tomentose. **Flowers** 1.5–2 cm across; petals yellow, turning dull red; ovary sparsely sericeous. **Fruits** drupes, 0.8–1.2 cm Ø, dull yellow or orange-tinged, flesh abundant. **Distr:** Mexico, C America, West Indies, N South America. **Ecol:** In moist or dry thickets or open forests; to 1,300 m. **Zones:** 10–12.

OCHNACEAE

Genera 27, **species** 495. **Trees** and **shrubs**, rarely herbs. **Stipules** entire or sometimes lacerate, persistent or caducous. **Leaves** alternate, petiolate; lamina simple, rarely pinnately compound, margins dentate, serrate or rarely entire, venation pinnate. **Inflorescences** terminal or axillary, cymose or racemose, rarely 1-flowered; pedicels articulated. **Flowers** generally bisexual, actinomorphic or more rarely zygomorphic, bracteate; sepals (2–4)5(10–15), free or more rarely united, imbricate or valvate; petals (3–)5(–8), mostly free, clawed to sessile, contorted or imbricate; stamens 5–10 or numerous, filaments generally free and persistent, or anthers sessile, anthers basifixed, longitudinally dehiscent or poricidal; staminodes present or not, sometimes persistent, awl-shaped, spathulate, or petaloid, sometimes connected into a tube; gynoecium (2 or)3–5(–15)-carpellate, gynophore present or rarely absent, ovary superior, entire or deeply lobed, generally long-styled, placentation basal, axile or parietal, rarely laminar, ovules 1 or 2 or numerous per locule, stigma entire or sometimes apex shortly divided. **Fruits** fleshy or non-fleshy, generally septicidal capsules, rarely a nut with accrescent sepals or drupes, or separating into up to 15 blackish drupelets on a brightly colored accrescent receptacle; seeds endospermic or non-endospermic, winged or not. **Distr:** Tropical regions worldwide, especially Brazil.

Ochna

Syn: *Diporidium*, etc.

Species ~85. **Shrubs** or **small trees**; usually glabrous. **Stipules** small, deciduous or caducous. **Leaves** petiolate; lamina with margins serrate to ciliate or rarely entire. **Inflorescences** paniculate to racemose or umbellate, branched, sometimes corymbose or reduced to a single flower, terminal or terminating short axillary shoots; bracts scalelike, caducous; pedicels articulated at or above base. **Flowers** with sepals (4 or)5, quincuncial in bud, green or yellow in flower, persistent, enlarging and becoming varying shades of red and coriaceous in fruit; petals 5–12, yellow or rarely orange or white, often unguiculate, deciduous; stamens numerous, 2 or more whorled, free, filaments ± slender, longer or shorter than anthers, persistent, anthers yellow; carpels (3)5–15, apparently free at base, 1 per locule; styles slender, gynobasic, united except sometimes towards apex, stigmas terminal, enlarged. **Fruits** 3–10(or 15), free, black, 1-seeded drupelets with fleshy mesocarp, inserted on enlarged red receptacle; seeds straight or ± curved or reniform, cylindric or somewhat flattened but not angular. **Distr:** Mainly in tropical Africa, a few species in tropical Asia. **Cult:** Propagate from seeds or cuttings. Requires well-drained soils in full sun or part-shade in tropical and subtropical climates.

O. integerrima (Loureiro) Merrill, *Trans. Amer. Philos. Soc.* n.s., 24(2): 11, 265, 1935. (Syn: *O. harmandii, Elaeocarpus integerrimus*). [Vietnamese Mickey Mouse Plant]. **Shrubs** or **small trees**, deciduous, 2–7 m tall; stem 6–16 cm dbh; branchlets gray-brown, glabrous. **Stipules** 2–7 mm long, soon deciduous. **Leaves** elliptic, obovate-oblong or obovate-lanceolate, 7–19 × 3–5.5 cm, base broadly cuneate, margins serrate, apex acute or obtuse, midvein prominent on both surfaces; petiole 2–5 mm long. **Inflorescences** corymbose, ~4 cm long, on short branchlets; pedicels 1.5–3 cm long. **Flowers** ~3 cm Ø; sepals oblong, 1–1.4 cm long, apex obtuse, reflexed during anthesis, red in fruit; petals 5(or 7), ovate, 1.3–2 cm long, apex obtuse or rounded; stamens 3-whorled, 0.9–1.2 cm long, filaments 5–8 mm long; ovary 10–12-locular, style terete, stigma slightly lobed. **Fruits** drupes, 1–1.2 × 0.6–0.7 cm, base slightly curved, apex obtuse. **Distr:** Cambodia, China, India, Laos, Malaysia, Myanmar, Pakistan, Thailand, Vietnam. **Ecol:** In deciduous forests, moist places; 300–1,400 m. Fl. Mar–Apr. Fr. May–Jun. **Zones:** 9–12.

O. jabotapita L., *Sp. Pl.* 513, 1753. (Syn: *Discladium koenigii, O. moonii, Polythecium moonii*, etc.). [Bird's Eye Bush]. **Shrubs** or **small trees**, to 5 m tall, much-branched; branchlets blackish, covered with lenticels. **Stipules** 4–5 × 1.5–2 mm. **Leaves** obovate to elliptic, 5–10 × 2–4 cm, base obtuse to acute, margins spinulose-denticulate, apex acute to acuminate, chartaceous; petiole 2.5–5 mm long. **Inflorescences** compound, many-flowered; main axis 0.6–2.5 cm long; branches of simple pedicels or to 3x branched, mono- or dichasial; pedicels 1.5–2 cm long, in fruit to 2.5 cm long; torus ~0.5 mm high. **Flowers** with sepals elliptic to obovate, 0.7–1 × 0.3–0.45 cm; petals 5, obovate to oblanceolate, 0.7–1 × 0.3–0.6 cm, ± tapering at base; stamens 12–25, filaments 1–2 mm long, anthers 3–6 × 0.4–0.6 mm; carpels 6–10, style 5–7 mm long. **Fruits** drupelets, mostly 1–2 per flower, to 8 × 6 mm. **Distr:** Sri Lanka. **Ecol:** In evergreen forests, grasslands; to 600 m. Fl. Feb–May. **Zones:** 11–12.

O. kirkii Oliv., *Flora of Tropical Africa* 1: 317, 1868. (Syn: *O. carvalhi, Polythecium kirkii*). [Bird's-eye Bush, Mickey Mouse Plant]. **Shrubs** or **small trees**, (1–)2–5(–6) m high, evergreen; branches ± spreading, reddish-purple or brownish with prominent lenticels at first, becoming grayish-white. **Leaves** elliptic or oblong-elliptic to narrowly obovate, (5–)6.5–11.2(–21) ×

Above: *Ochna intergerrima*, *Ochna jabotapita*, *Ochna kirkii*, *Ochna mossambicensis.*

2.6–5.1(–7) cm, base cuneate to cordate, margins entire or slightly undulate, usually beset with numerous equal dark-tipped cilia, apex obtuse to rounded and mucronate, ± coriaceous; petiole 4–5(–6) mm long, sometimes swollen. **Inflorescences** ± condensed panicles, 3–20-flowered; pedicels 1–2.5 cm long in fruit, articulated. **Flowers** with sepals 1–1.4 cm long, elliptic to elliptic-oblong, rounded, becoming red, 1.2–1.6 cm long, flat and spreading in fruit; petals bright yellow, sometimes dark-veined, obovate to broadly elliptic, 1.5–2.5 × 1–1.8 cm, clawed; stamens with anthers 4–5 mm long, often incurved; carpels (8)10–12, with styles free at apex and spreading, stigmas capitate. **Fruits** drupelets, 0.9–1 × 0.55–0.8 cm, flattened-cylindric, inserted near base. **Distr:** Mozambique. **Ecol:** In evergreen forests and scrubs; 0–600 m. **Zones:** 10–12.

O. mossambicensis Klotzsch, *Naturw. Reise Mossambique* 6(1): 88, t. 16, 1862. (Syn: *Diporidium purpureocostatum*, etc.). [African Mickey Mouse Plant]. **Shrubs** bushy, or **small trees**, to 3(–5) m tall, sometimes a rhizomatous shrublet ~5–40 cm high, evergreen, frequently galled. **Leaves** obovate to oblanceolate or rarely oblong, (5.5–)7–22.6 × (2–)3.4–7.8(–8.4) cm, base cuneate (rarely truncate), margins densely serrulate, apex obtuse to rounded (very rarely acute or acuminate) and sometimes shortly mucronate, coriaceous; petiole 1.5–8 mm long. **Inflorescences** panicles, ± laxly branched, terminating short lateral shoots; pedicels 1.5–3 cm long in fruit, articulated. **Flowers** with sepals elliptic to elliptic-oblong, 0.9–1.1(–1.25) cm long in flower, rounded, becoming red, 1.2–1.4 cm long, flat, and spreading in fruit; petals bright yellow, sometimes dark-veined, obovate or subcircular, 1–1.7(2.2) × 0.7–1.3(1.9) cm, shortly unguiculate; stamens with anthers 4–7 mm long; carpels (6–)8–10, with styles united almost to apex, free ends radially spreading, stigmas slightly enlarged. **Fruits** drupelets, 0.8–0.9(–1) × 0.6–0.8 cm, subglobose or flattened ovoid-cylindric, inserted near base. **Distr:** E Kenya, Mozambique, Tanzania. **Ecol:** In woodlands; 0–450(–800) m. **Zones:** 10–12.

PHYLLANTHACEAE

Genera 59, **species** 1,745. **Trees**, **shrubs**, or **herbs**, sometimes with dimorphic (horizontal or erect) branches or phylloclades; hairs simple. **Stipules** usually present. **Leaves** alternate, often 2-ranked; lamina usually simple, margins entire to serrate, venation pinnate. **Inflorescences** determinate, axillary, occasionally reduced to a single flower. **Flowers** unisexual (plants monoecious or dioecious), radial, ± inconspicuous; sepals usually 5, distinct or slightly connate; petals usually 0–5, distinct to slightly connate, imbricate, often reduced or lacking; stamens 3–8, occasionally few or numerous, filaments distinct to connate; carpels usually 3, connate, ovary superior, usually trilobed, styles usually 3, each usually bifid, occasionally entire, stigmas various, ovules 2 per locule; nectar disk usually present. **Fruits** usually schizocarps, segments elastically dehiscent from a persistent central column, sometimes berries or drupes; seeds not arillate. **Distr:** Pantropical, especially in Malesia, some temperate. **Info:** The classification followed here is according to Wurdack et al. (2004) and Stevens (2001 onwards).

Subfam. ANTIDESMATOIDEAE

Plants often dioecious, tanniniferous. **Leaves** spiral; lamina trifoliolate, margins with teeth deciduous; petiole pulvinate. **Flowers** often without corolla; ovary 1–5-loculed, stigma plumose. **Fruits** often indehiscent (Stevens 2001 onwards). **Distr:** Tropical and subtropical regions of the world.

Bischofia

Syn: *Microelus*, *Stylodiscus*.

Species 2. **Trees**, large, dioecious, rarely monoecious, with red or reddish latex; indumentum of simple hairs, often absent. **Stipules** small, falcate, caducous. **Leaves** alternate, usually crowded at stem apex; lamina palmately 3(–5)-foliolate; leaflet margins crenulate-serrate; petiole long. **Inflorescences** axillary or lateral, produced with young leaves, paniculate or racemose, pendent, flowers 1 per bract; ♂ flowers with pedicels articulated at base; ♀ flowers with pedicels articulated at middle. **Flowers** ♂ with sepals 5, free, valvate, initially curved around stamens, later reflexed, petals and disk absent, stamens 5, free, inserted at base of sepals, filaments short, anthers larger, thecae 2, parallel, introrse, longitudinally dehiscent, pistillode short and broad, peltate; ♀ with sepals imbricate, flat, free, petals and disk absent, staminodes sometimes present, soon lost, ovary 3(or 4)-locular, ovules 2 per locule, styles 3 or 4, long, stout, entire, erect or recurved. **Fruits** berries, small, baccate, globose, indehiscent, exocarp fleshy, endocarp papery to thinly woody, 3- or 4-locular; seeds 3–6, oblong-obovoid, crescent-shaped, smooth, testa crustaceous. **Distr:** Tropical and subtropical Asia to Pacific Islands. **Cult:** Propagate from seeds. Requires a deep, moist, well-drained soil in full or partial sun.

B. javanica Blume, *Bijdr.* 1168, 1826. (Syn: *Andrachne trifoliata*, etc.). [Javanese Bishopwood, Toog Tree]. **Trees**, dioecious, evergreen, to 40 m tall; trunk to 2.3 m dbh; latex red. **Stipules** lanceolate, ~8 mm long, membranous, caducous. **Leaves** palmately 3(–5)-foliolate; leaflets ovate, elliptic, obovate, or elliptic-ovate, 7–15 × 4–8 cm, base broadly cuneate to obtuse, margins with 2 or 3 teeth per cm, apex acute or caudate-acuminate, papery, sparsely pubescent only on nerves, glabrescent; petiole 8–20 cm long, terminal petiolule 2–5 cm long, lateral petiolules 0.5–2 cm long. **Inflorescences** axillary, paniculate; ♂ peduncles 8–13 cm long, puberulent to glabrous; ♀ peduncles 15–27 cm long, pendent. **Flowers** ♂ to 2.5 mm ∅, sepals membranous, semiorbicular, adaxially concave, ladle-shaped, abaxially puberulent outside, filaments short, pistillode small, peltate, pubescent; ♀ flowers with sepals similar to ♂ but oblong-ovate, margins membranous, ovary smooth, glabrous, 3- or 4-locular, styles 3 or 4, linear, entire. **Fruits** globose or subglobose, 0.6–1.3 cm ∅, brownish; seeds oblong, ~5 mm long. **Distr:** Tropical Asia; Australia, Pacific Islands (Polynesia). **Ecol:** In evergreen forests, open woodlands; <800 m. Fl. Apr–May. Fr. Aug–Oct. **Zones:** 9–12.

Baccaurea

Syn: *Gatnaia*, *Pierardia*.

Species 51. **Trees** or **shrubs**, dioecious; indumentum of simple hairs. **Stipules** caducous. **Leaves** alternate, usually clustered apically; lamina simple, margin entire or crenate-serrate, venation pinnate. **Inflorescences** axillary, often cauliflorous, compound spikelike or racemelike panicles, pendent, many-flowered. **Flowers** ♂ with sepals 4–8, usually unequal, imbricate, petals absent, disk absent or obscurely glandular and between stamens (sometimes interpreted as staminodes), stamens 4–8, as long as or longer than sepals, filaments free, anthers bilocular, introrse or extrorse, longitudinally dehiscent; pistillode usually enlarged at apex, depressed and bifid, usually pubescent; ♀ with sepals 4–8, larger than ♂, both surfaces pubescent, petals absent, disk absent, ovary shorter than sepals, 2- or 3(–5)-locular, ovules 2 per locule, styles 2–5, very short. **Fruits** berries or tardily loculicidally dehiscent fleshy capsules, ovoid, fusiform, or globose; seeds enclosed by fleshy edible aril, endosperm fleshy or thick. **Distr:** Tropical and subtropical Asia to Pacific Islands. **Cult:** Propagate from seeds or by air-layering. Requires a moist, well-drained soil in shade to full sun.

B. motleyana (Müll. Arg.) Müll. Arg., *Prodr.* 15(2): 461, 1866. (Syn: *Pierardia motleyana*). **Trees**, 6–10 m tall; branches terete, ferruginous-pubescent. **Stipules** lanceolate, ~8 mm long, usually caducous. **Leaves** elliptic or elliptic-lanceolate, 20–35 × 7.5–17 cm, base rounded to shallowly cordate, apex acute to shortly acuminate, papery, pubescent abaxially and on midvein adaxially, lateral veins 12–16 pairs, parallel, reticulate veins trapeziform; petiole 5–10 cm long, pubescent. **Inflorescences** axillary or cauliflorous, racemelike panicles, 13–35 cm long; bracts lanceolate. **Flowers** ♂ pubescent, sepals oblong-ovate, 1–1.5 mm long, apex acute, gray papillose-puberulent, stamens 4–6, pistillode retuse at apex, pubescent; ♀ with sepals oblong, 4–5 mm long, gray papillose-puberulent, ovary ovoid or globose, trilocular, sericeous. **Fruits** capsules, baccate, ovoid or globose, 2.5–3 cm across, indehiscent, pubescent;

Above: *Bischofia javanica, Baccaurea motleyana, Baccaurea racemosa, Baccaurea ramiflora.*

seed 1, ~1.5 cm across. **Distr:** Indonesia, peninsular Malaysia, Thailand. **Ecol:** In lowlands and hills, rarely in higher forests; to 900 m. Fl. May–Jul. Fr. Jul–Oct. **Zones:** 11–12. **Info:** The fruits are edible with sweet to acid aril, eaten raw, cooked, or preserved. The bark is rich in tannins and is used in dyeing.

B. racemosa (Reinw. ex Blume) Müll.Arg., *Prodr.* 15(2): 461, 1866. (Syn: *Coccomelia racemosa*, etc.). **Trees**, 15–25 m tall; trunk 25–70 cm dbh, often with buttresses; crown dense, irregular. **Stipules** ± triangular, 3–6 × ~2 mm, concave, acute. **Leaves** ovate-oblong or obovate, 7–18.5 × 3–7.5 cm, base broadly cuneate, margins entire, apex acuminate to obtuse, often with many glands which are distinct on abaxial surface, lateral nerves 6–12 pairs; petioles 0.5–4.5 cm long. **Inflorescences** ♂ racemes, 5–13 cm long, fascicled along old branches or burs on trunk, many-flowered, composed of numerous, short, 3-flowered cymes, densely hairy, bracts 3, central one largest; ♀ racemes, 10–20 cm long, solitary or fascicled on old branches. **Flowers** ♂ ~1–1.5 mm Ø, sepals 4–5, unequal, hairy, stamens 4–8, ~1 mm long, arising from small cavities at base of thick, subangular rudimentary ovary; ♀ relatively larger, 5–8 mm Ø, sepals mostly 5, unequal, apically incurved, puberulous, ovary 3–4-loculed, puberulous, stigmas sessile. **Infructescences** racemes, pendulous, 15–20 cm long, pedicels 1–1.5 cm long. **Fruits** yellowish-green, 2–2.4 cm Ø, indehiscent; seeds 1–3, pulp translucent white to pink. **Distr:** Peninsular Thailand to Malesia. **Ecol:** In lowland forests, and seasonal or shallow swamps. Fl. Mar–Apr. **Zones:** 11–12. **Info:** Fruits edible.

B. ramiflora Loureiro, *Fl. Cochinch.* 2: 661, 1790. (Syn: *Baccaurea cauliflora*, etc.). [Burmese Grape]. **Trees**, evergreen, to 20 m tall; trunk gray-brown, to 60 cm dbh; branchlets hispid, later glabrescent. **Leaves** obovate-oblong, oblanceolate, or oblong, 9–15 × 3–8 cm, base cuneate, margins entire or shallowly repand, apex shortly acuminate to acute, papery, adaxially green, abaxially yellowish-green, glabrous, lateral veins 5–7 pairs; petiole 3–5 cm long, glabrous. **Inflorescences** many-flowered, compound into racemelike panicles; bracts ovate-lanceolate, 2–3 mm long, chestnut-yellow, puberulent outside; ♂ densely papillose, often fascicled on branchlets as well as on trunk, to 15 cm long; ♀ to 35 cm long. **Flowers** small, dioecious, apetalous, puberulent outside; ♂ with sepals 4 or 5, oblong, 5–6 mm long, stamens 4–8, pistillode terete; ♀ with sepals 4–6, oblong-lanceolate, ~6 mm long, ovary ovoid or globose, 3-celled, densely hispid. **Fruits** capsules, baccate, ovoid or subglobose, 2–2.5 × 1.5–2 cm, red-yellow to purple; arils white; seeds flat-elliptic or rotund, 1–1.3 cm across. **Distr:** Bhutan, Cambodia, India, Laos, Malaysia (peninsular), Myanmar, Nepal, Thailand, Vietnam. **Ecol:** In evergreen forests; 100–1,300 m. Fl. May–Apr. Fr. Mar–Oct. **Zones:** 9–11. **Info:** The fruits are edible, and the wood is used for making furniture.

Antidesma

Syn: *Rhytis, Stilago.*

Species ~100. **Trees** or erect **shrubs**, dioecious; indumentum of simple hairs. **Stipules** small. **Leaves** alternate, eglandular; lamina simple, margins entire, venation pinnate; petiole usually short. **Inflorescences** spikes or racemes, sometimes much-branched, axillary, sometimes apparently terminal, or cauline. **Flowers** ♂ with calyx 3–5(–8)-lobed, cup-shaped, lobes imbricate, petals absent, disk extra-staminal, annular or cushion-shaped (enclosing bases of stamens and pistillode), or consisting of free lobes, stamens (1–)3–5(–7), inserted inside/within disk or sinus of disk lobes, filaments longer than sepals, anthers bilocular, connectives U-shaped, longitudinally dehiscent, pistillode usually present; ♀ with sepals as in ♂, disk annular, surrounding ovary, entire, ovary longer than sepals, usually 1-locular, locule 2-ovuled, pendent, styles 2–4, short, terminal to lateral, apex usually bifid. **Fruits** drupes, ovoid, ellipsoid, or lens-shaped, usually 1-seeded; seeds small, endosperm fleshy. **Distr:** Mainly tropical and subtropical Asia; Africa; Australia and Pacific Islands. **Cult:** Propagate from seeds of fully ripened fruits, cuttings or by air-layering. Requires a deep, well-drained soil in full or partial sun.

A. bunius (L.) Sprengel, *Syst. Veg.* 1: 826, 1824. [Bignay, Chinese Laurel]. **Trees**, rarely shrubs, to 30 m tall. **Stipules** linear, 4–6 × 1.5–2 mm, caducous. **Leaves** oblong, elliptic, or obovate, (5–)10–23(–32) × (2–)3–10 cm, leathery or thickly papery, lateral veins (5–)7–11 pairs; petiole 0.3–1(–1.7) cm long, glabrous to pubescent. **Inflorescences** axillary or terminal, axes glabrous to pubescent, robust; ♂ 6–17(–25) cm long, 3–8(–14)-branched; ♀ (4–)10–18 cm long, unbranched or more rarely up to 4-branched. **Flowers** ♂ with calyx 1–1.5 mm, cup-shaped, 3- or 4(or 5)-lobed, glabrous to pubescent outside, reddish-pubescent at base inside, margins fimbriate, apex of lobes obtuse to rounded, consisting of free lobes or enclosing stamens and pistillode, glabrous, stamens 3 or 4(or 5), 2–3 mm long, rudimentary ovary clavate to cylindric; ♀ with calyx trilobed, otherwise as in ♂, disk glabrous, ovary glabrous or pilose, stigmas 3 or 4(–6). **Fruits** drupes, ellipsoid, laterally compressed, 0.5–1.1(–1.8) × 0.4–0.7 cm, red to black. **Distr:** Tropical & subtropical Asia. **Ecol:** In forests; 200–1,800 m. Fl. Mar–May. Fr. Jun–Nov. **Zones:** 9–12. **Info:** Fruits are edible but sour. Most parts of plant used for medicinal purposes.

A. dallachyanum Baill., *Adansonia* 6: 337, 1866. [Herbert River Cherry]. **Trees**, bushy, to 7.5(–9) m tall; young shoots slightly hairy. **Leaves** ovate to lanceolate-elliptical, 5–15 cm long, almost hairless when mature, adaxially deep green, abaxially bright green, thick and leathery. **Inflorescences** ♂ spikes, hairy, generally in panicles, axillary, occasionally solitary, ± interrupted; ♀ flowers borne in racemes. **Flowers** odoriferous, ♀ greenish; calyx with 5 sepals, sepals ovate; stamens 3–5; ovary ± ovate, style basally constricted, apically trilobed. **Fruits** single or in clusters of 4–30, round to obovoid, to 2 cm across, rich-red when unripe, dark purple-red to nearly black when ripe, very acidic in taste. **Distr:** Australia (Queensland). **Ecol:** In rainforests near rivers and ponds. Fl. Dec–Feb. **Zones:** 9–12. **Info:** Fruits edible, often used to make jams and other foods.

Subfam. PHYLLANTHOIDEAE

Plants monoecious or dioecious; growth continuous. **Inflorescences** fasciculate; ovary 2–6(–15)-loculed, ovules hemitropous. **Fruits** indehiscent. **Distr:** Tropical to temperate regions.

Cleistanthus

Syn: *Leiopyxis, Paracleisthus.*

Species ~141. **Trees** or **shrubs**, monoecious or dioecious; indumentum of simple hairs. **Stipules** persistent or caducous. **Leaves** alternate, petiolate; lamina simple, margins entire, venation pinnate. **Inflorescences** axillary, glomerules or spikes, small, sessile or sometimes shortly pedicellate. **Flowers** ♂ with sepals 4–6, valvate, petals 4–6, small, scalelike, disk cupular or cushionlike, stamens 5, filaments connate to half their length, topped with rudimentary ovary, anthers usually ovoid, dorsifixed, extrorse, anther bilocular, longitudinally dehiscent; ♀ flowers with sepals and petals as in ♂, disk cup-shaped or conical, surrounding basal ovary, sometimes as long as ovary, tearing into lobes during fruit development, rarely absent, ovary 3(or 4)-celled, ovules 2 in each cell, styles 3, bifid at apex. **Fruits** capsules, subglobose, (2- or) 3-valvate when mature, sometimes stipitate (carpopodium), epicarp thin, endocarp bony, inside fibrous, columella and sepals persistent; seeds (1 or)2 in each cell, thinly to thickly fleshy, endosperm copious to submembranous. **Distr:** Tropical Africa, Asia, and Australia. **Cult:** Propagate from seeds. Requires a well-drained soil in shade to full sun.

C. collinus (Roxb.) Benth. ex Hook.f., *Fl. Brit. India* 5: 274, 1887. (Syn: *Amanoa collina*, etc.). **Trees**, dioecious, 5–7 m tall. **Stipules** 3–3.5 mm long, long-subulate, antrorse-strigose, ciliate. **Leaves** broadly elliptic to obovate-elliptic to suborbicular, 2–11 × 1.5–8 cm, apex obtuse, coriaceous, glabrous; petiole 4–9 mm long, sparsely hirsute to subglabrous. **Inflorescences** racemes, bracteate, short, borne at base of leafy branches of current season's growth; ♂ with buds rhomboid, tapering to stout pedicels ~2 mm long, buds and pedicels sparsely hirtellous; ♀ flowers short-pedicellate. **Flowers** ♂ with sepals lanceolate, ~4 mm long, base broad, petals reduced

Above: *Antidesma bunius*, *Antidesma dallachyanum*, *Cleistanthus collinus*, *Cleistanthus pallidus*.

to 1–4 linear segments ~1 mm long, between each pair of sepals, stamens glabrous, filaments united below, pistillode present; ♀ with sepals 6–7 mm long, triangular-acuminate, petals ~1 mm long, linear, ovary ~2.5 mm long, ellipsoid, glabrous, styles free, ~2 mm long, clavellate, spreading, lower half terete, upper half entire, expanded into broadened stigma. **Fruits** capsules, obovoid-globose, 2–2.5 cm long, obscurely trilobed, veined, glabrous, slightly glossy. **Distr:** India, Sri Lanka. **Ecol:** At low altitudes. **Zones:** 10–12.

C. pallidus (Thwaites) Müll.Arg., *Prodr.* 15(2): 508, 1866. (Syn: *Amanoa pallida*, etc.). **Trees**, 4–15 m tall, or occasionally shrubs as low as 2 m high, dioecious; twigs slender, glabrescent. **Stipules** (1.5–)2.5–5 mm long, subulate, glabrous. **Leaves** narrowly elliptic to lanceolate, oblanceolate, and oblanceolate-elliptic, 4–14.5 × 1.5–5 cm, base acute or obtuse, margins entire, often wavy, apex acute, or shortly acuminate; petiole 2–4.5(–6) mm long. **Inflorescences** glomerules, small, axillary, ♀ sometimes solitary; ♂ flowers with pedicels ~1–1.5 mm long, densely white-hirsute. **Flowers** ♂ with calyx ~3 mm long, sepals narrowly lanceolate, densely hirsute outside, glabrous inside, petals rotund, ~0.5 mm long, filaments 2.5 mm long, connate into staminal column 1.5 mm long, pistillode ~1.25 mm long, flask-shaped, slightly trilobed; ♀ with sepals as in ♂ flower, petals subflabellate, ovary ~1.25 mm ∅, densely hairy, styles spreading bifid, stigma minutely papillose. **Fruits** 5–6 × 6–7 mm, depressed-globose, trilobed, tawny-pubescent; seeds 3.5–4 mm long, deltoid-hemispheric, dull blackish. **Distr:** Sri Lanka. **Ecol:** In forests, open rocky hillsides; <800 m. **Zones:** 11–12.

Phyllanthus

Syn: *Cicca*, *Emblica*, etc.

Species 883. **Trees**, **shrubs**, or **herbs**, mostly monoecious, less often dioecious; branching often 'phyllanthoid', main stems with spiral phyllotaxy, ultimate branchlets sometimes clustered on short shoots, resembling pinnate leaves and often deciduous as a pinnate leaf, less often stems all similar with spiral or distichous phyllotaxy. **Stipules** small, deciduous or persistent. **Leaves** alternate, often reduced and scalelike on main stems, strongly distichous on leafy stems; lamina simple, margins entire, venation pinnate; petiole short. **Inflorescences** axillary, sometimes at leafless nodes, solitary or in fascicles, cymes, glomerules, racemes, or panicles; pedicels delicate. **Flowers** ♂ with sepals (2 or)3–6, in 1 or 2 series, free, imbricate, margin entire, eroded, denticulate or fimbriate, petals absent, disk glands 3–6, usually free, stamens 2–6, filaments free or connate, anthers bilocular, pistillode absent; ♀ with sepals as in ♂ or more, ovary bullate, or hairy, 3(–12)-locular, ovules 2 per locule, styles 3(–12), apex bilobed or 2-branched, rarely entire, erect, spreading, or recurved. **Fruits** usually capsules, globose or depressed-globose, smooth or warty, dehiscent into three 2-valved cocci; seeds without caruncle or aril. **Distr:** Tropical & subtropical regions worldwide. **Cult:** Propagate from seeds or cuttings. Requires a humus-rich, moist, well-drained soil in full sun.

P. acidus (L.) Skeels, Bull. *Bur. Pl. Industr. USDA* 148: 17, 1909. (Syn: *Averrhoa acida*, etc.). [Country Gooseberry, Otaheite Gooseberry, Star Gooseberry, Wild Plum, etc.]. **Shrubs** or **small trees**, 2–10 m tall; crown dense and bushy. **Leaves** borne at end of branchlets 15–30 cm long (giving the appearance of pinnate leaves), deciduous, 2-ranked; lamina ovate-lanceolate to broadly ovate, 2–7.5 × 1.2–3.8 cm, base rounded, apex acuminate, thin, pubescent, adaxially pale green and smooth, abaxially blue-green. **Inflorescences** cymose (or cymules) clusters of 5–12.5 cm long panicles, with ♂, ♀ and bisexual flowers, borne on leafless parts of main branches, in upper parts of the plant; ♀ flowers on stout pedicels. **Flowers** small, red or pinkish-red; floral whorls usually in 4s; ♂ 4-merous, filaments and anthers free; ♀ 4-merous, disk deeply lobed or split, styles connate, deeply bifid, staminodes present, ovary superior. **Fruits** produced in profusion, drupaceous berries, oblate, 1–1.5 × (1.2–)1.5–2(–2.5) cm, 6–8-ribbed, glossy, waxy, pale yellow or white when ripe, crispy and juicy; seeds 3–6, 3-angled. **Distr:** Brazil or perhaps Madagascar; cultivated or naturalized elsewhere. **Ecol:** In coastal forests. Fl. & Fr. throughout most of the year. **Zones:** 9–12. **Info:** Fruits edible but sour; sometimes used for medicinal purposes.

P. amarus Schumacher & Thonning, *Kongl. Danske Vidensk. Selsk. Skr.* 4: 195, 1829. [Black Catnip, Child Pick-a-back, Gulf Leaf-flower, Stone-breaker]. **Herbs**, annual or biennial, rarely perennial, monoecious, 0.1–1.7 m tall or long, erect or prostrate, glabrous throughout; stem simple. **Stipules** linear, green. **Leaves** distichous; lamina oblong or elliptic-oblong, 3–8 × 2–4.5 mm, base rounded, apex obtuse or rounded and often apiculate, ± membranous, lateral veins 4–7 pairs; petiole ~0.5 mm long. **Inflorescences** fascicles; ♂ usually along lower part of leafy shoots; bisexual usually in middle with 1 ♀ and 1 ♂ flower; ♀ often borne toward branchlet apex. **Flowers** ♂ with sepals 5, elliptic or ovate, ~0.5 × 0.2 mm, yellowish-green, disk glands 5, orbicular or obovate, or spathulate, stamens (2 or)3, filaments completely connate into a column; ♀ with sepals 5, obovate-oblong or ovate, 0.8–1 × 0.4–6 mm, disk flat or subulate, deeply 5-lobed, ovary globose-triangular, ~0.5 × 0.5 mm, smooth, styles free. **Fruits** capsules, smooth; seed sharply 3-angled, ~1 mm long, brownish. **Distr:** Pantropical; possibly originated from the Americas. **Ecol:** In dry places, weedy; 0–600 m. Fl. & Fr. throughout year. **Zones:** 9–12.

P. debilis Klein ex Willd., *Sp. Pl.* 4: 582, 1805. (Syn: *P. niruri* var. *debilis*). **Herbs**, annual, monoecious, glabrous, 10–80 cm high. **Stipules** on branchlets 0.9–1.5 mm long, truncate at base, acuminate, carinate. **Leaves** on main stems reduced to scales, deciduous branchlets mostly 4–10 cm long, with 15–35 leaves, sharply angled, smooth; lamina

Below: *Phyllanthus acidus*, *Phyllanthus acidus* (fruits), *Phyllanthus amarus*, *Phyllanthus debilis*.

TG

TG

NT

Above: *Phyllanthus emblica* (fruits), *Phyllanthus myrtifolius*, *Phyllanthus niruri*, *Phyllanthus pulcher*.

mostly 0.8–2 × 0.25–0.5 cm, base cuneate, apex acute. **Inflorescences** cymules, axillary on deciduous branchlets; ♂ flowers 3 or 4 per glomerule at proximal axils, pedicels <0.5 mm long; ♀ flowers solitary at distal axils, pedicel 1–1.6 mm long in fruit. **Flowers** ♂ with sepals 6, obovate to obcuneate, 0.5–0.6 mm long, entire, stamens 3; ♀ with sepals 6, obovate, 1–1.6 × 0.5–1 mm, rounded, margins whitish-scarious, ovary smooth, styles 0.2–0.3 mm long, free, bifid. **Fruits** capsular, oblate, 2–2.2 mm Ø, smooth; seeds trigonous, yellowish or pale brown, 1–1.1 mm long, with 6 or 7 longitudinal ribs. **Distr:** India, Sri Lanka. **Ecol:** In disturbed places; <1,000 m. **Zones:** 10–12.

P. emblica L., *Sp. Pl.* 2: 982, 1753. (Syn: *Emblica officinalis*, etc.). [Emblic Myrobalan, Myrobalan]. **Trees**, monoecious, deciduous, 3–8(–23) m tall; trunk to 50 cm dbh. **Leaves** distichous; lamina oblong or linear-oblong, 0.8–2.3 × 0.15–0.6 cm, base shallowly cordate and slightly oblique, margins narrowly revolute, apex truncate, rounded or obtuse, tip mucronate or retuse, papery to leathery, adaxially green, abaxially paler. **Inflorescences** fascicles with many ♂ flowers and sometimes 1 or 2 larger ♀ flowers. **Flowers** ♂ with sepals 6, membranous, yellow, obovate or spathulate, subequal, 1.2–2.5 × 0.5–1 mm, stamens 3; ♀ with sepals 6, oblong or spathulate, 1.6–2.5 × 0.7–1.3 mm, obtuse or rounded, thicker, margin membranous, ± lobate, ovary ovoid, ~1.5 mm, 3-celled, styles 3. **Fruits** drupes, globose, 1–1.3 cm Ø, pale green or yellowish-white; seeds reddish, 5–6 × 2–3 mm. **Distr:** Bhutan, Cambodia, China, India, Indonesia, Laos, Malaysia, Myanmar, Nepal, Philippines, Sri Lanka, Thailand. **Ecol:** In dry open sparse forests; 200–2,300 m. Fl. Apr–Jun. Fr. Jul–Sep. **Zones:** 9–12.

P. myrtifolius (Wight) Müll. Arg., *Prodr.* 15(2): 396, 1866. (Syn: *Macraea myrtifolia*, *P. myrtifolius*). [Myrteleaf Leaf-flower]. **Shrubs**, monoecious, ~0.5(–3) m tall; branches terete, upper parts puberulent; branchlets wing-angled, angles scabridulous or hispidulous. **Stipules** small, ovate, 0.5–0.7 mm long. **Leaves** oblanceolate, (1–)1.2–1.6(–2.5) × (0.2–)0.35–0.45(–0.8) cm, base shallowly cordate, apex acute or obtuse, leathery, glabrous, lateral veins numerous, subhorizontally spreading; petiole 0.7–1.7 mm long. **Inflorescences** fascicles, several-flowered, axillary; pedicels filamentous, unequal, 3–5 mm long; ♀ with pedicels 6–8 mm long. **Flowers** ♂ ~3 mm Ø, sepals 5, oblong, disk glands 5, orbicular, stamens 3, filaments connate at middle part to base; ♀ with sepals 6, 0.8–1 × 0.3–0.5 mm, disk cup-shaped, entire, ovary globose, 3-celled, styles 3, short, apex bifid, recurved. **Fruits** capsules, oblate, ~2 × 3 mm, glabrous, valves crustaceous; seed 3-angled (trivalvate), 1.7–2 mm long, brown, reticulate-striped. **Distr:** Sri Lanka. **Ecol:** Fl. May–Aug. Fr. Aug–Nov. **Zones:** 11–12.

P. niruri L., *Sp. Pl.*: 981, 1753. (Syn: *Nymphanthus niruri*, etc.). [Stone-breaker]. **Herbs** or **subshrubs**, base woody, 0.1–1.2 m tall, monoecious, usually with a single main stem and slender pinnatiform lateral branches 3–12 cm long with 16–36 leaves and internodes 3–6 mm long, leafy rachis glabrous. **Stipules** linear-subulate to filiform, 0.5–2.7 mm long, persistent. **Leaves** oblong or elliptic-oblong, usually 0.5–1.2 × 0.2–0.5 cm, base broadly cuneate or obtuse, apex broadly obtuse or rounded. **Inflorescences** axillary, unisexual, proximal nodes with 3–5(–7) ♂ flowers, distal nodes with solitary ♀ flowers. **Flowers** ♂ 3 mm across, sepals 5(6), lobes 0.5–1.5 mm long, disk with 5 rounded parts, stamens 3; ♀ with sepals 5, yellowish or marked with red, ovary smooth, styles ~0.5 mm long. **Fruits** oblate, ~3 × 3–4 mm, yellowish, smooth; seeds trigonous, ~2 mm long. **Distr:** Tropical & subtropical America. **Ecol:** In open sites of lower montane evergreen forests; 1,000–2,500 m. **Zones:** 10–12.

P. pulcher Wallich ex Müll. Arg., *Linnaea*. 32: 49, 1863. (Syn: *Diasperus pulcher*, etc.). [Tropical Leaf-flower]. **Shrubs**, monoecious, 0.5–1.5 m tall; glabrous except for puberulent or ciliate young branches and bracts; stem and branches terete. **Stipules** triangular-lanceolate, 3–4 × 1.5–2 mm, reddish-brown, entire or obscurely serrulate. **Leaves** distichous, 15–30 pairs along each branchlet; lamina obliquely oblong to ovate-oblong, 1.8–3 × 0.8–1.3 cm, adaxially green, abaxially gray-green, lateral veins 4–6 pairs, obscure; petiole 0.8–1.5 mm long. **Inflorescences** axillary fascicles, bisexual, usually with several ♂ and 1 ♀ flower; flowers with ciliate bracts at base. **Flowers** ♂ with sepals 4, ovate-triangular, 2–3 × 1–2 mm, dark red, margins lacerate, stamens 2; ♀ with sepals 6, ovate-triangular, 3.5–4 × ~1.5 mm, margins lacerate, fleshy, surrounding ovary at base, apex 6-lobed, ovary subglobose, smooth, 3-celled, styles 3. **Fruits** capsules, subglobose, ~3 mm Ø, brownish, smooth, sepals persistent. **Distr:** Cambodia, China, India, Indonesia, Laos, Malaysia, Myanmar, Vietnam. **Ecol:** In open forests and riverside scrubs; 700–1,800 m. Fl. May–Sep. Fr. Aug–Nov. **Zones:** 10–12.

Sauropus

Syn: *Diplomorpha*, *Synostemon*, etc.

Species ~56. **Shrubs**, erect, rarely herbs or climbers, monoecious or dioecious. **Stipules** small, paired. **Leaves** alternate, petiolate; lamina simple, margins entire, pinnately veined or rarely obscurely 3-veined from base. **Inflorescences** with ♂ flowers axillary or cauliflorous, clustered or solitary, rarely in racemelike thyrses or short cymes, pedicels usually bracteolate at base; ♀ flowers axillary, solitary or paired or with ♂, rarely inserted at base of ♂ inflorescences. **Flowers** ♂ smaller, apetalous, calyx discoid, cup-shaped, or turbinate, entire or 6-lobed, lobes obscurely biseriate, imbricate, erect, or spreading, margins slightly thick, with a whorl of scalelike adaxial lobes at bases of calyx lobes, rarely absent, petals absent, stamens 3, opposite outer sepals, filaments usually connate into a short cylinder, anthers free; ♀ with calyx usually 6-parted, sepals biseriate, imbricate, ovary ovoid or depressed-globose, apex truncate or retuse, trilocular, ovules 2 per locule, styles 3, free or basally connate, much-divergent, apex 2-branched, arms recurved. **Fruits** capsules, depressed or ovoid, breaking into three bivalved cocci; seeds not caruncular, endosperm fleshy. **Distr:** Mascarenes, India, and Sri Lanka to Indo-Malaysia, China, Philippines, Indonesia, and Australia. **Cult:** Propagate from seeds. Requires fertile, clay or acidic soil in shade.

S. androgynus (L.) Merrill, *Bull. Bur. Forest. Philipp. Islands* 1: 30, 1903. (Syn: *Aalius androgyna*, etc.). [Star Gooseberry, Sweetleaf Bush]. **Shrubs**, 1–3 m tall, erect, monoecious, glabrous throughout. **Stipules** lanceolate or linear-lanceolate, 1.5–3 mm long. **Leaves** ovate, oblong or lanceolate, 3–10 × 1.5–3.5 cm, base cuneate, rounded, or truncate, apex acuminate, submembranous or thinly papery, pinnately veined; petiole 2–4 mm long. **Inflorescences** axillary, 1- or 2-flowered, or several ♂ and ♀ per cluster; ♂ flowers with pedicels slender, 5–7.5 mm long; ♀ flowers usually solitary, axillary, pedicel 6–8 mm long. **Flowers** ♂ with calyx shallowly disk-shaped, 0.5–1.2 cm Ø, shallowly 6-fid, sepals obovate, disk segments 6, opposite to sepals, stamens 3, filaments connate, anthers extrorse; ♀ with calyx red, 6-lobed, sepals obovate or obovate-triangular, 5–6 × 3–5.5 mm, base attenuate into a short claw, disk absent, ovary depressed-globose, ~0.7 × 1.5 mm, trilocular, styles 3, bifid. **Fruits** capsules, white, depressed-globose or globose, ~1.2 × 1.7 cm, thinly crustaceous; seeds black, triquetrous, ~7 × 5 mm. **Distr:** Tropical Asia. **Ecol:** On slopes with brushwood, sunny forest margins; 100–400 m. Fl. Apr–Jul. Fr. Jul–Dec. **Zones:** 10–12.

Breynia

Syn: *Melanthesa*, *Melanthesopsis*.

Species 26–30. **Shrubs** or **trees**, monoecious; indumentum of simple hairs or absent. **Stipules** triangular, ± caducous. **Leaves** simple, alternate, distichous, petiolate; lamina entire, adaxially often drying to blackish, abaxially drying to brownish and often glaucous-papillate, pinnately veined. **Inflorescences** axillary; ♂ flowers in few-flowered clusters in proximal axils, sometimes in small bracteate racemes; ♀ flowers solitary in distal axils. **Flowers** ♂ with calyx turbinate, funnelform, or hemispheric, fleshy, apex usually shortly 6-lobed, lobes usually incurved and concealing stamens, petals absent, disk lobes

Above: *Sauropus androgynus, Breynia disticha, Mischodon zeylanicus, Chrysobalanus icaco.*

scalelike, at base of calyx lobes, stamens 3, filaments connate into cylindric column with parallel sessile anthers along sides, anthers bilocular, longitudinally dehiscent, pistillode absent; ♀ with calyx turbinate, hemispheric, campanulate, radial, or cup-shaped, 6-dentate to (5 or)6-fid, persistent, accrescent into saucer-shape or disk, ovary globose, truncate, or depressed, trilocular, styles 3, apex bifid, inserted within apical depression with lobed margin, locules 2-ovuled. **Fruits** usually berries, globose or depressed-globose, red, indehiscent, exocarp ± fleshy, with 3–6 trigonous cocci; seeds sharply 3-angled, yellow or reddish. **Distr:** Tropical Asia, Australia, Pacific Islands. **Cult:** Propagate from cuttings or suckers. Needs fertile and moist soil in full sun.

B. disticha J.R.Forst. & G.Forst., *Char. Gen. Pl.*: 73, 1775. (Syn: *B. disticha* var. *genuina*, etc.). [Foliage-flower, Leaf Flower, Snow Bush, Summer Snow, Sweetpea Bush]. **Shrubs** or **small trees**, monoecious, glabrous throughout, usually to 2 m tall. **Stipules** small. **Leaves** borne along main branches, spirally arranged and reduced, distichous along ultimate branchlets, shortly petiolate; lamina simple, ovate to broadly elliptic, margins entire. **Inflorescences** axillary, lower axils bearing fascicles of 3–many ♂ flowers, higher axils bearing solitary ♀ flowers. **Flowers** lacking petals and disk; ♂ flowers with calyx fleshy, obovoid-turbinate, shortly 6-lobed, stamens 3, filaments connate into a column bearing bilocular anthers apically, these not apiculate; ♀ with calyx campanulate to cupuliform, 6-lobed or toothed, persistent and sometimes accrescent, ovary trilocular, walls thickened distally, each locule with 2 ovules, stigmas small, subulate-dentiform, incurved. **Fruits** capsules, dehiscing into 2-seeded cocci; seeds with a somewhat fleshy testa. **Distr:** New Caledonia, Vanuatu. **Ecol:** At the edges of scrub woodlands near the beach. **Zones:** 9–12.

PICRODENDRACEAE

Genera 24, **species** 80. **Plants** dioecious; latex absent; hairs simple. **Stipules** petiolar, cauline or absent. **Leaves** spiral, opposite (2-ranked); lamina with secondary venation ± palmate or leaves palmate, margins entire or toothed, teeth with deciduous glands. **Flowers** petals absent; ♂ with perianth 3(–13), stamens 2–many, basally connate, often extrorse, pollen 4-brevicolporate, 5–6-porate to zoniaperturate or 6 colporoidate, echinate to verrucose, pistillode present or absent; ♀ with perianth (3–)4–8(–13), ovary (2–)3(–5)-loculed, ovules 2 per locule (usually both maturing), micropyle endostomal, stigmas stout, usually entire, dry (wet). **Fruits** indehiscent or explosively dehiscent; seeds carunculate; ant dispersal common (Stevens 2001 onwards; Sutter, Forster & Endress 2006). **Distr:** Tropical, especially New Guinea to Australia, New Caledonia; America; Africa, Madagascar.

Mischodon

Species 1. *Generic description same as specific.* **Cult:** Propagate from seeds. Requires a moist, well-drained soil in full or partial sun.

M. zeylanicus Thwaites, *Hooker's J. Bot. Kew Gard. Misc.* 6: 300, 1854. [Tammanaa Tree]. **Trees** or **large shrubs**, to 25 m tall, evergreen, much-branched; branchlets subquadrangular, pubescent. **Stipules** 2, minute. **Leaves** in whorls of 3–4; lamina narrowly elliptic to broadly elliptic to oblanceolate, 7–28 × (1.25–)2–7.5 cm, base narrowed-obtuse or subcordate, margins entire, apex rounded-obtuse to occasionally emarginate, coriaceous, red when young, glabrous at maturity, lateral veins 10–15 pairs; petiole 0.4–1.5(–5.75) cm long, densely pubescent to subglabrous. **Inflorescences** with ♂ flowers numerous, pedicellate, in bracteate, densely pubescent, axillary panicles to ~9 cm long, pedicels 2–3.5 mm long, densely pubescent; ♀ flowers very shortly pedicellate, in compact, densely pubescent, axillary fascicles or panicles to 1 cm long, pedicels ~1 mm long, vastly extending to 1.5(–2.5) cm in fruit. **Flowers** ♂ with sepals 4–5, lanceolate, 1.5 × 0.6 mm, acuminate, glabrous outside except for apical tuft of hairs, filaments 2–4.5(–6) mm long, slender, villous, anther ~1–1.25 × 0.6 mm, villous; ♀ with sepals as in ♂, ovary 2- or triloculed, locules 2-ovuled, stigmas short, simple or bifid. **Fruits** capsules, ~1 × 1 cm, trilobed, lobes bluntly keeled, dark green, glabrous; seeds ~7 × 5 mm, obscurely trigonous, shiny, dark brown. **Distr:** Sri Lanka, S India, Andaman Islands. **Ecol:** In dry, evergreen, primary and intermediate forests. **Zones:** 11–12.

CHRYSOBALANACEAE

Genera 16, **species** 450. **Trees**, **shrubs** or rhizomatous, geoxylic suffrutices. **Stipules** small and caducous to large and persistent. **Leaves** simple, entire, alternate, often coriaceous, usually with 2 glands at base of lamina or near apex of petiole. **Inflorescences** cymes, panicles or racemes. **Flowers** bisexual, actinomorphic to zygomorphic, strongly perigynous; receptacle tube short to elongate, straight or curved, often gibbous at base, lined with nectariferous tissue; throat with long hairs; sepals 5, free, imbricate, often unequal, ascending or reflexed; petals 5, rarely absent, sometimes unequal, imbricate, often caducous; stamens 2–100 or more, included or exserted; ovary superior, basically of 3 carpels but usually with only 1 fully developed carpel, each carpel 1-locular with 2 ovules, or bilocular, style filiform, stigma distinctly or indistinctly trilobed. **Fruits** drupes, dry or fleshy. **Distr:** Tropical regions worldwide.

Chyrsobalanus

Species 3. **Trees, small,** or **shrubs**. **Stipules** small, deciduous. **Leaves** alternate; lamina simple, entire, leathery, glabrous. **Inflorescences** cymes, axillary and terminal, pubescent. **Flowers** white, rather small; receptacle somewhat campanulate; stamens 10 to numerous, in a single whorl on upper margin of receptacle, mostly fertile; carpels solitary, ovary 1-celled, sessile at base of receptacle, style arising from one side near base, ovules 2, ascending. **Fruits** drupes, somewhat pulpy; seed 1, suberect, cotyledons fleshy, radicle inferior. **Distr:** Tropical and subtropical America, W Indies; tropical Africa. **Cult:** Propagate from seeds or by air-layering. Requires a moist, well-drained soil in shade or partial sun.

C. icaco L., *Sp. Pl.* 513, 1753. (Syn: *C. fruticosus*, *C. pellocarpus*). [Coco Plum, Pork-fat Apple]. **Trees** or **shrubs**, evergreen, 5–10 m tall. **Leaves** ovate-elliptical, obovate-roundish, or obovate, 3–10 × 2.5–7 cm, base rounded or wedge-shaped, apex rounded or emarginate. **Inflorescences** cymes, 3–6 cm long. **Flowers** white, bell-shaped, small; calyx triangular, ~2.5 mm long, tomentose outside and ± pubescent inside; petals obovate-spathulate, white, ~5 mm long; stamens exserted, filaments villose, anthers sometimes abortive on one side of the flower. **Fruits** drupes, 1 or 2 developing on each cyme, ellipsoidal, nearly spherical, or obovoid, 1.5–3 cm Ø, pericarp fleshy or thin. **Distr:** C America and N South America, West Indies. **Ecol:** In thickets, mostly near the coasts and in wet habitats; to ~1,000 m. **Zones:** 10–12. **Info:** Cultivated often for the edible fruits, and also for medicinal purposes.

Licania

Species ~75. **Trees** or **shrubs**, glabrous or pubescent. **Stipules** deciduous or persistent, free or connate, subulate or lanceolate. **Leaves** alternate; lamina simple, margins entire or nearly so, usually coriaceous, persistent, often tomentose; petiole often short, apex often with 2 glands. **Inflorescences** panicles, branched; flowers sessile or pedicellate, tribracteolate. **Flowers** small, perfect; hypanthium urceolate to globose or hemispheric, often hairy inside; sepals 5, small, imbricate or subvalvate; petals small, 5 or none; stamens mostly 3–10, inserted in the throat of hypanthium, included, sometimes unilateral, filaments short, subulate or complanate, often unequal, anthers small; ovary immersed at hypanthium base, globose, 1-celled, villous or strigose, style basilar, included in calyx, ovules 2, collateral, ascending from base of cell. **Fruits** small or large, fleshy, coriaceous, lignose, or crustaceous, often with abundant pulp, terete, obovoid to pyriform or globose; seed 1, erect, with membranous testa, cotyledons thick-fleshy, plano-convex. **Distr:** C & S America. **Cult:** Propagate from seeds or by air-layering. Requires a moist, well-drained soil in full or partial sun.

L. platypus (Hemsl.) Fritsch, *Ann. K. K. Naturhist. Hofmus.* 4: 53, 1889. [Monkey Apple, Sun Sapote, Wild Pear, Zapote]. **Trees**, 10–30(–50) m tall; trunk to 75 cm dbh, buttresses sometimes 1–2.5 m high; branches and leaves glabrous; branches often reddish.

JML

Above: *Licania platypus*, *Bauhinia acuminata*, *Bauhinia bidentata*, *Bauhinia bowkeri*.

Stipules ovate, ~2.5 mm long, stiff, adnate to petiole base, persistent. **Leaves** mostly narrowly oblong-elliptic, 10–30 × 3–9 cm, base acute to rounded, apex shortly acuminate, adaxially glossy, abaxially pale and ± glaucous; petiole ~1 cm long. **Inflorescences** panicles, terminal or upper-axillary, 10–25(–35) cm long, branches gray-tomentose, flattened at base; flowers sessile or short-pedicellate. **Flowers** minute, white, fragrant, gray-tomentose; hypanthium ± turbinate, ~1.5–2 mm Ø; sepals 5, triangular, ~1 × 1 mm, spreading; petals 5, obovate, 2–3 mm long; stamens 15, glabrous, filaments ~3 mm long, attached separately to disk; style 5–6 mm long. **Fruits** drupes, variable in size, to 20 cm long, to 14 cm Ø (weighing up to 900 g), green maturing to brown, mesocarp granular, yellow or orange, juicy, sweet, edible; seed usually 1, ovate-oblong, flattened, to 5 cm or longer. **Distr:** S Mexico to Colombia. **Ecol:** In forests. Fl. Feb–Apr. Fr. Jun–Aug. **Zones:** 10–12.

FABALES

FAMILIES 4, GENERA 754, SPECIES 20,055.

Vessel elements with single perforations. **Flowers** usually bisexual (less often unisexual), zygomorphic (less often actinomorphic); carpels free. **Fruits** mostly pods (legumes); seeds with embryo large, green. **Distr:** Worldwide. **Info:** The monophyly of Fabales is supported by molecular data (Chase et al. 1993; Soltis et al. 2000).

FABACEAE

Subfamilies 3, **genera** 730, **species** 19,400. **Herbs**, **vines**, **shrubs**, **trees**, and **lianas** (climbing by twining or tendrils); most species with high nitrogen metabolism (nitrogen-fixing). **Stipules** present. **Leaves** nearly always alternate, bipinnately or palmately compound to simple; petiole base usually enlarged into a pulvinus that functions in the orientation of the leaves. **Inflorescences** usually racemes, spikes, or heads. **Flowers** usually bisexual, actinomorphic to zygomorphic, slightly to strongly perigynous; hypanthium usually cup-shaped, short; perianth usually consisting of a calyx and a corolla of 5 segments each; androecium usually with 1–many stamens (usually 10), distinct or variously united, sometimes some of them reduced to staminodes; carpel 1 (rarely 2–16), distinct, elongate (occasionally shortened), gynophore short, ovary superior, placentation lateral, style 1, arching upward, sometimes hairy, stigma 1, small, ovules 1–numerous per carpel, borne in 2 rows. **Fruits** usually legumes, sometimes samaras, loments, follicles, indehiscent pods, achenes, drupes, or berries; seeds often with a hard coat with hourglass-shaped cells. **Distr:** Worldwide; mostly in the tropics. **Info:** The classification followed here is according to Lewis et al. (2005).

Subfam. CAESALPINIOIDEAE

Genera 160, **species** 1,930. **Trees**, **shrubs** or **climbers**, rarely herbs. **Stipules** often caducous; stipels present or absent. **Leaves** alternate; lamina pinnate or bipinnate, rarely simple (or simple leaflets). **Inflorescences** racemes or panicles, rarely spikes; bracteoles small or large, sepaloid; bracts reduced when bracteoles enclose flower bud. **Flowers** bisexual, rarely unisexual, usually ± zygomorphic, rarely actinomorphic; hypanthium very short or cup-shaped or elongated to tubular; sepals (4 or)5, free or connate on lower part; corolla usually showy, petals usually 5, rarely 1 or absent, as are sepals imbricate in flower bud, uppermost (adaxially) one smaller, covered by adjoining, sublateral 2; stamens 10 or fewer, rarely numerous, filaments free or connate, anthers 2-celled, usually opening by lateral slits, rarely by apical pores, pollen monads; ovary stalked or sessile, free or adnate to inner wall of hypanthium, ovules anatropous, 1–numerous, style slender, stigma terminal. **Fruits** pods, dehiscent or indehiscent, drupelike or samaralike; seeds sometimes arillate. **Distr:** Predominantly tropical (and subtropical) with a few genera in temperate regions.

TRIBE CERCIDEAE

Bauhinia

Species ~570. **Trees**, **shrubs** or **climbers**, sometimes armed, sometimes scandent with tendrils. **Stipules** usually small, caducous. **Leaves** alternate; lamina simple, usually bilobed or almost bifoliolate with midrib between the 2 leaflets produced as a small spur, less frequently entire, 3–many-nerved; petiole often aristate between leaflets (or lobes). **Inflorescences** with flowers arranged in simple or panicled, terminal or axillary racemes. **Flowers** showy; hypanthium sometimes long and cylindrical, sometimes short and turbinate; calyx variable, entire or spathaceous, or cleft into 2 or 5 teeth, limb at anthesis cleft; petals 5, erect or spreading, slightly unequal, narrowed at base into a claw, variously colored; stamens 10, or reduced to 5, 3 or 1 (some reduced to staminodia or absent), filaments free or shortly connate, filiform, anthers versatile, longitudinally dehiscent; ovary seated on a stalk (gynophore), ovules 2–many, style long or short and usually curved, stigma capitate. **Fruits** pods, oblong or linear, membranous to rigid, indehiscent or dehiscent (bivalvate); seeds compressed, albuminous, free or separated by tissue or septa, cotyledons plane, radicle often exserted. **Distr:** Tropics to warm temperate regions of both hemispheres. **Cult:** Propagate by air-layering, grafting, from softwood cuttings or seeds. Requires a fertile, moist but well-drained soil in partial to full sun.

B. acuminata L., *Sp. Pl.* 1:375, 1753. (Syn: *B. candida*, *B. grandiflora*, etc.). [Bell Bauhinia, Dwarf White Orchid Tree, Mountain Ebony, Snowy Orchid Tree, White Bauhinia]. **Shrubs** or **small trees**, 2.5–3 m tall; trunk short; branches spreading, bark gray-brown. **Stipules** semisagittate. **Leaves** bifarious; lamina bilobed, elliptic, 7.5–12.7 cm long, bifid nearly to middle, lobes acute or obtuse, 7–11-nerved, median nerve produced into a sharp point, adaxially glabrous, shiny, ± abaxially downy; petiole 7.5–10 cm long, swollen above. **Inflorescences** with flowers usually solitary or a few in short racemes; pedicels short; bracts and bracteoles minute. **Flowers** 6.5 cm Ø, pure white; calyx 3.8 cm long, spathaceous, narrowed into a hairy beak, tip lacerate; petals oblong, rounded; disk glands 5, globose; stamens 10, filaments alternately long and short, anthers equal; pistil geniculate, ovary linear, hairy, narrowed into a slender stipe and beak. **Fruits** pods, stipitate, linear-oblong, 10–13 cm long, flat, smooth, glabrous, margins keeled; seeds 8–11, turgidly ellipsoid, brown, shiny. **Distr:** SE Asia. **Ecol:** In dry dipterocarp forests, teak forests and brushwoods. Fl. throughout the year, mainly Apr–Jun. Fr. Jun–Aug. **Zones:** 10–12.

B. bidentata Jack, *Mal. Misc.* 2: 76, 1822. (Syn: *B. gracilipes*, etc.). [Orange Bauhinia, Vining Bauhinia]. **Climbers**, tendriled, large; young branches initially tomentose. **Stipules** broadly falcate to auriculate, ~5 mm long, early caducous. **Leaves** variable, to 14.5 × 12 cm, base ± cordate to truncate, margins entire, apex acute, emarginate or very shallowly bifid, nerves 4–7, adaxially usually ± glabrous, abaxially hairy; petiole 1–5 mm long. **Inflorescences** corymbose, to 10 cm long, ± hairy; pedicels slender; bracts ovate-triangular to linear-lanceolate, 0.2–1 cm long, early caducous; bracteoles smaller, 1–6 mm long. **Flowers** with hypanthium tubular, striate, 0.5–2.5 cm long; calyx splitting into 3(4) reflexed lobes; petals yellow turning red, unequal, obovate, shortly but distinctly clawed, 1.5–2.2 cm long; stamens 3, fertile, filaments glabrous or hairy, 2–9 mm long, base usually enlarged; staminodes 2–7; ovary 3–5 mm long, stipe ~3 mm long, hairy, style ~3 mm long, stigma capitate to peltate. **Fruits** pods, oblong, to 10 cm long, glabrous; seeds to 5, flat, ~1 cm across. **Distr:** Malesia. **Ecol:** In lowland rainforests to montane forests; 1,500 m. **Zones:** 11–12. **Info:** 7 varieties described.

B. bowkeri Harv., *Fl. Cap.* 2: 596, 1862. (Syn: *Pauletia bowkeri*). [Kei Bauhinia]. **Shrubs** or **small trees**, unarmed, 2–8 m tall; extremities puberulous or early glabrate. **Leaves** broadly obliquely oblong, ovate- or obovate-oblong, rounded above and below, 1.2–3.4 × 2–4 cm, base slightly cordate or truncate, leaflets connate from 2–5 mm to half their length, adaxially glabrous or sparsely pubescent and reticulate, abaxially paler or glaucous pubescent or glabrate. **Inflorescences** terminal and from upper axils,

JML KT

Above: *Bauhinia divaricata*, *Bauhinia ferruginea*.

usually in few-flowered leafy corymbs; pedicels very short. **Flowers** rather large, white; calyx puberulous or shortly tomentose, tube cylindrical elongate-sulcate, limb spathaceous, 2–3.2 cm long, shorter than or nearly equaling tube; petals oblanceolate, 3.2–5 cm long, narrowed into claw, with a strong midrib and conspicuously penniveined; stamens 10, all perfect, unequal, filaments filiform; ovary on a long gynophore, narrowed above into the slender pubescent style, stigma capitate. **Distr:** S Africa (in a broad coastal strip from Port Elizabeth to the Kei River). **Ecol:** Along riverbanks, in dense scrubby vegetation. **Zones:** 9–12.

B. divaricata L., *Sp. Pl.* 374, 1753. (Syn: *B. peninsularis*, etc.). [Bull Hoof, Butterfly Orchid Tree]. **Shrubs** or **small trees**, to 8 m tall. **Stipules** ovate to linear, 1–3 mm long, deciduous. **Leaves** suborbicular to narrowly ovate to lanceolate, emarginate to bilobate, lobes parallel or divaricate, 2–12 × 2–10 cm, base rounded to cordate, apex of lobes acuminate to rounded, margins smooth to slightly crisped, chartaceous to subcoriaceous, abaxially strigose, 5–9-nerved; petiole 0.5–4 cm long. **Inflorescences** racemose, terminal, or subterminal and axillary, (10–)20–50-flowered; bract lanceolate to linear, 1.5–4 mm long; bracteoles similar, but smaller; pedicels 0.5–2 cm long, strigose or tomentellous. **Flowers** with hypanthium cyathiform, 2–3 mm long; calyx spathaceous at anthesis; petals 5, subequal, white, turning pink with age, ovate to narrowly elliptic, glabrous, 1.5–2.5 cm long, claw glabrous; fertile stamen 1, filament glabrous, basally connate with staminodes, anther black; staminodes 9, subequal, 1–1.5 cm long, connate most its length, white, petaloid; gynoecium arcuate, shortly pilose to tomentose, stigma green. **Fruits** pods, linear, 2–2.5 cm long, style persistent, brown; seeds oblong to suborbicular, 0.8–1 × 0.6–0.8 cm, dark brown. **Distr:** USA (Baja California) to Costa Rica. **Ecol:** In moist deciduous forests to dry cactus scrubs; 0–1,500 m. **Zones:** 9–11.

B. ferruginea Roxb. [*Hort. Beng.* (1813) 90, nomen], *Fl. Ind., ed. Carey*, 2: 331, 1832. (Syn: *B. griffithiana*, etc.). **Climbers**, tendriled, large; branches glabrescent. **Stipules** foliaceous, orbicular to ovate-orbicular or reniform, 1–2.8 cm long. **Leaves** ovate-orbicular, 7–10 × 7–8 cm, base cordate, apex bifid to ⅓, lobe tips obtuse, adaxially glabrous, 9–11-nerved; petiole 2–4 cm long. **Inflorescences** corymbose to pyramidal racemes, terminal; rachis tomentose; pedicels 1–2 cm long; bracts variable, usually ovate or lanceolate, 0.7–1.5 cm long; bracteoles narrower, lanceolate. **Flowers** with calyx early splitting into 5 reflexed sepals; petals creamy-white turning yellowish, rarely pinkish-red, densely woolly tomentose outside, glabrous inside, petals becoming recurved, later whole lamina recurved; stamens 3, fertile, filaments pink, glabrous, 5–6 cm long; staminodes usually 2–3, 1.2–2 cm long, sometimes with minute anthers; whole gynoecium densely ferruginous-tomentose; ovary ~1 cm long, style 1–2.5 cm long, stigma peltate. **Fruits** pods, dehiscent, woody, 15 × 4–6 cm, glabrous; seeds few, 2–2.5 cm Ø. **Distr:** S Thailand, Malay Peninsula, E Sumatra. **Ecol:** In forest margins, peat swamp forests; <800 m. **Zones:** 11–12.

B. forficata Link, *Enum. Hort. Berol.* 1: 404, 1821. [White Orchid Tree]. **Trees**, to 10 m tall; stems long and straggling; branchlets pendulous, glabrous, usually aculeate with flat paired spines. **Leaves** broadly ovate, usually 7–10 cm long, base cordate, rounded or scarcely emarginate, usually divided below middle, lobes slightly divergent, ovate, obtuse or acute, glabrous or puberulent, membranous, typically 9-nerved. **Inflorescences** axillary, 1–3-flowered; peduncle short. **Flowers** variable in size; calyx pale green, sepals 5, linear-lanceolate, tube cylindrical, usually 1.2–2.5 cm long, lobes to 5 cm long, coherent in a spathe; corolla white, petals 5, spreading, about as long, lanceolate, broadly linear or oblong, ~7.5 cm long, somewhat obtuse; stamens about equal in length to corolla, ascending, lower ones the longest, filaments white, connate at base, anthers linear, yellow; style shorter than stamens, stigma thickened. **Fruits** pods, 15–25 × 2 cm, stipe to 5 cm long. **Distr:** Brazil. **Ecol:** In rainforests and other tropical areas; 0–950 m. Fl. Apr–Dec. **Zones:** 10–12.

B. galpinii N.E. Brown, *Gard. Chron.* 9: 728, 1891. (Syn. *B. punctata*, *Phanera punctata*). [Nasturtium Bauhinia, Pride of De Kaap, Red Bauhinia]. **Shrubs**, scrambling or subscandent, 2–8 m high, semideciduous, multistemmed, spiny; young branches hairy. **Leaves** alternate; lamina smaller than the flowers, orbicular, 3–7 × 2.5–6 cm, apex slightly bilobed to deeply notched for up to ¼ of lamina length, lobes rounded, coriaceous, venation prominent abaxially, nerves 11–13, slightly hairy; petiole 0.8–1.5 cm long. **Inflorescences** branched clusters or racemes; bracts linear to lanceolate. **Flowers** brick-red or scarlet to salmon-pink (rarely yellow), 8 cm across; petals 3–4 cm long, clawed; stamens 3, anthers ovate. **Fruits** pods, hard, 7–13 × 2 cm, hairless. **Distr:** S Africa. **Ecol:** In woodlands; along hillsides and streambanks in bushveld. Fl. Sep–Mar. **Zones:** 9–12.

B. grandidieri Baill., *Bull. Mens. Soc. Linn. Paris* 1(47): 373–374, 1883. [Blue Orchid Tree, Dwarf Orchid Tree]. **Trees**, small, evergreen, 2–4(–10) m tall; trunk branching near base; bark often flaking with small pieces. **Leaves** alternate; lamina bilobed (folded together before opening), lobes separate and bifoliate, shortly petiolate, oval, 7 × 5 cm, base oblique, apex truncate-rounded, margins entire, inner margins ± straight, coriaceous, palmately veined. **Inflorescences** with flowers solitary, axillary; pedicels 1.3–1.4 cm long. **Flowers** 3.5–4 cm across, showy, violet; calyx spathaceous, ~1 cm long; petals 2–2.4

Below: *Bauhinia forficata*, *Bauhinia galpinii*, *Bauhinia grandidieri*.

JML

Above: *Bauhinia herrerae*, *Bauhinia jenningsii*, *Bauhinia kockiana*.

× 1.1–1.5 cm; stamens 7, 2 fertile, anthers 2–3 mm long; ovary ~1 cm long, style 0.7–1.2 cm long. **Fruits** pods, sickle-shaped, 20 × 2 cm, flattened, somewhat broadened around the seeds; seeds ~8, irregularly oval, 1.2 × 1 cm, very flattened, brown. **Distr:** W Madagascar. **Ecol:** In thickets and woodlands; 0–1,000 m. **Zones:** 11–12.

B. herrerae (Britt. & Rose) Standl. & Steyerm. *Field Mus. Bot.* 23: 10, 1943. (Syn: *Schnella herrerae*). [Orchid Tree]. **Vines**, small or large, sometimes to 18 m long, provided with numerous tendrils; branches and inflorescences densely ferruginous-tomentulose. **Stipules** filiform, 6–8 mm long. **Leaves** long-petiolate; lamina mostly 3–8 cm long and about as wide, base shallowly or deeply cordate, bilobate to about middle, lobes with apex very obtuse or narrowly rounded, 9-nerved, adaxially glabrate, abaxially brownish-pilose with ± appressed hairs. **Inflorescences** racemes, many-flowered, terminal; pedicels 0.8–1.5 cm long; rachis thick and stout; buds ovoid. **Flowers** with calyx densely ferruginous-tomentose, campanulate, 8 mm long, teeth deltoid-ovate; petals 1.5–2 cm long, yellow; stamens glabrous. **Fruits** pods, 6–9 × 1.5 cm, shortly stipitate, strongly compressed and flat, densely appressed-pilose with ferruginous hairs, valves thin, apparently not elastic; seeds 1–4. **Distr:** Nayarit to Oaxaca and Yucatán; British Honduras. **Ecol:** In wet to dry thickets or forests; <300 m. **Zones:** 10–12.

B. jenningsii P. Wilson, *Bull. Torrey Bot. Club* 43: 463, 1916. (Syn: *B. castilloi*, *Casparia jenningsii*). [Cow's Tongue]. **Shrubs**, slender or **small trees**, to 6 m tall; branches glabrate. **Stipules** lanceolate, ~1.5 mm long, caducous. **Leaves** ovate to lanceolate, 7–14 × 2.5–5 cm, base truncate to rounded, apex acuminate to obtuse, subcoriaceous, adaxially bright green, glabrous, adaxially paler, strigulose to glabrate, 5-nerved; petiole 1.5–2 cm long. **Inflorescences** racemose, terminal or subterminal and axillary, 10–30-flowered, rachis reddish; bract lanceolate, ~0.5 mm long; pedicels 1–1.5 cm long, reddish. **Flowers** with hypanthium cyathiform, 2–3 mm long; calyx spathaceous; petals 5, subequal, white or cream, linear-lanceolate to oblanceolate, 1.2–2 cm long, ± equal to claw, claw glabrate; fertile stamen 1, slightly exceeding petals, filament 1.5–2 cm long, shortly connate with staminodes at base, anther 4–5 mm long, staminodes 1–1.5 cm long, alternate; ovary densely pilose. **Fruits** pods, linear, 5–10 × 1–2 cm, apiculate with persistent style, light brown; seeds oblong, 7–9 × 5–7 mm, dark brown. **Distr:** Cuba, Mexico, Belize, N Guatemala. **Ecol:** Usually in dry to mesic woods. Fl. Feb–Sep. **Zones:** 10–12.

B. kockiana Korth., *Kruidk. Verh. Nat. Gesch. Bot.* (1839–1842). 87: pl. 10, 1841 (Syn: *B. creaghii*, *Phanera kockiana*). [Climbing Bauhinia, Orange Bauhinia, Red Trailing Bauhinia]. **Lianas**, tendriled, large, to 30 m long; stems flattened; young shoots ± hairy, usually reddish-brown tomentose, later glabrous. **Stipules** auriculate, puberulous. **Leaves** narrowly ovate, elliptic to oblong, 7–13 × 4–6 cm, base obtuse, margins entire, apex acuminate, chartaceous to coriaceous, usually trinerved. **Inflorescences** corymbs or short racemes, lateral and terminal, axis usually reddish- or brownish-tomentose; pedicels 4–6(–7) cm long; bracts ovate-oblong, ~1 cm long, caducous; buds globose, 3–5 mm Ø. **Flowers** with hypanthium narrowly tubular, finely striate, 1.1–1.8 × 0.1–0.2 cm Ø; calyx splitting into 2 or 3 segments; petals yellow to orange turning red, short-clawed, broadly obovate, 2–2.5 cm long, margins crenulated, undulate or flat; fertile stamens 3, to 8 mm long; reduced stamens and staminodes 3–5; ovary shortly stipitate, 5–9 mm long, brownish- or reddish-hairy, style ~6 mm long. **Fruits** pods, narrowly elliptic, glabrous, ~15 cm long; seeds flat, broadly ovate, ~2 cm across. **Distr:** Malesia (from Malay Peninsula throughout Borneo to Lesser Sunda Islands). **Ecol:** Various habitats; 0–1,200 m. **Zones:** 11–12. **Info:** A very polymorphic species with 9 varieties described.

B. lunarioides A. Gray ex S. Wats., *Bibl. Ind. N. Amer. Bot.* 205, 1878. (Syn: *B. congesta*, etc.). [Anacacho Orchid Tree, Chihiahuan Orchid Tree, Pink/White Orchid Tree, Texas Plume]. **Shrubs**, to 4 m tall; branches loosely strigose when young, soon glabrate. **Stipules** broadly lanceolate, ~1 mm long, caducous. **Leaves** suborbicular, bilobate, usually bifoliolate, lobes suborbicular, slightly convex, 1(1–)1.5–2.5(–3) × (1.5–)2–2.5(–3) cm, base cordate or rarely truncate, apex of lobes rounded, margins ± crisped, adaxially glabrous, abaxially strigose to glabrate, (5–)11-nerved; petiole 0.5–1(–1.5) cm long. **Inflorescences** shortly racemose, terminal, or subterminal, 2–6(–10)-flowered, rachis strigose; pedicels slender, 0.5–1 cm long, strigose. **Flowers** with hypanthium cyathiform, 1–2 mm long; calyx spathaceous; petals 5, white or occasionally pink-tinged, elliptic-lanceolate to ovate, 1.5–2.5 cm long; fertile stamen 1, shortly connate with staminodes, anther 5–7 mm long; staminodes 9, 5–7 mm long, basally connate; stigma bilobate. **Fruits** pods, linear, 5–8 × 1–2 cm, apiculate with persistent style, light brown, glabrate; seeds 0.7–1 × 0.6–8 cm, dull, dark brown. **Distr:** Mexico, USA (Texas). **Ecol:** In desert scrubs, primarily on calcareous soils; 500–2,000 m. Fl. Mar–Jun. **Zones:** 9–12.

B. madagascariensis Desv., *J. Bot. Agric.* 1: 74, 1813. (Syn: *B. commersonii*). [Madagascar Orchid Tree]. **Shrubs**, much-branched, to ~4 m high, sometimes scrambling over other plants. **Leaves** alternate; lamina strongly bilobed, ~4 × 4.2 cm, base cordate to truncate, lobes with apices rounded, yellowish-green; petiole to 1.7 cm long. **Inflorescences** racemes, 2–5-flowered, terminal, sometimes appearing lateral on vigorously growing stems; peduncle 2–7 mm

Below: *Bauhinia lunarioides*, *Bauhinia madagascariensis*, *Bauhinia malabarica*.

Above: *Bauhinia mexicana, Bauhinia monandra, Bauhinia natalensis.*

long; pedicels ~5 mm long. **Flowers** with hypanthium ~1.5 cm long, green; calyx ~2.8 × 0.45 cm, green, split to base, spathelike, sepals tipped with green filiform appendage ~2.5 mm long; petals subequal, erect ~4.3 × 1.2 cm, limb yellow below, shading abruptly to red or orange-red with darker venation; upper petal clawed, limb oval, rounded, lower half yellow; lateral petals clawed, limb ovate, pointed, base yellow; fertile stamens 5, ascending, 2.5–3.3 cm long, filaments green, anthers gray, ~3.5 × 1 mm; staminodes 5, to 1.5 cm long, yellow, alternating with fertile stamens; gynoecium ~3.8 cm long, glabrous, ovary pale green, style and stigma cream-white. **Fruits** pods, pendent, to 12 × 1.4 cm, strongly flattened; seeds 6, flattened, ~7 × 5 mm, dark brown. **Distr:** Madagascar. **Ecol:** In thickets, woodlands, forests; 0–1,000 m. **Zones:** 11–12.

B. malabarica Roxb., Hort. Beng.: 31, 1814. (Syn: *Casparea castrata*, etc.). [Malabar Orchid Tree, Sour Tamarind]. **Trees**, dioecious, to 15 m tall; young branches pubescent to almost glabrous. **Stipules** linear-acute, 2–3 mm long, very early caducous. **Leaves**; ovate to rotund, 4–12 × 8–16 cm, base truncate to cordate, apex bifid, tip of lobes rounded, adaxially light green to subglaucous, sparsely pubescent, abaxially glaucous and glabrous to pubescent; petiole 2–4 cm long, glabrous to pubescent. **Inflorescences** racemes, short, simple or few-branched, axillary, to 5 cm long. **Flowers** with hypanthium turbinate, 3–5 mm long, longest in ♀; calyx splitting into 2–3(–5) lobes in upper part; petals white, oblong, ~1 cm long with short claw; ♂ with fertile stamens 10, shorter than petals, outer whorl longest; ♀ staminodes 10, ovary stipitate, 5–6 mm long, densely tomentose, style short, stigma peltate. **Fruits** pods, indehiscent or tardily dehiscent, strap-shaped, 20–25 × 1–1.5 cm; seeds 10–30, rounded-elliptic, 7 mm Ø, dark brown. **Distr:** Northern Hemisphere to the Philippines; and Malesia from Java to Timor; Australia. **Ecol:** In dry, deciduous, monsoon forests; <300(–500) m. **Zones:** 9–12.

B. mexicana Vog., *Linnaea* 13: 299, 1839. (Syn: *B. confusa*). [Mexican Orchid Tree, Pink Orchid Bush, White Orchid Tree]. **Shrubs** or **small trees**, semievergreen, 2–4.5(–6) m high, spreading to 3(–8) m wide; trunk sometimes to 12 cm dbh; bark gray, rough. **Leaves** 3–6.5 cm across, base cordate, bilobed to 1/3 the length, lobes divaricate, short, acute to acuminate, bright green, abaxially puberulent or glabrate, with numerous small glands, 7-nerved; petioles 1.5–2.3 cm long, pubescent-hirsute. **Inflorescences** racemes, terminal, pubescent-hirsute; bracts ~5 mm long; pedicels short. **Flowers** 2–2.5 cm long, white, often pink-tinged; calyx 5-lobed, lobes 1.2 cm long, pubescent; petals oblong-lanceolate, base clawed, apex acute to acuminate; stamens white, glabrous; ovary and style white, style long-exserted. **Fruits** pods, straight to falcate, to 7.5 × 1.2–1.5 cm, base subattenuate, stipitate, apiculate, subglabrous. **Distr:** S Mexico (Tamaulipas to Veracruz and Puebla). **Ecol:** Fl. Oct–Nov. **Zones:** 9–12.

B. monandra Kurz, *Journ. Asiatic Soc. Beng.* 42: 173, 1873. (Syn: *B. kappleri, B. krugii, Caspariopsis monandra*). [Butterfly Flower, Jerusalem Date, Mariposa, Napoleon's Plume, Pink Butterfly Tree]. **Shrubs**, ~5 m tall; young twigs pubescent. **Stipules** triangular to lanceolate, 3–6 mm long. **Leaves** 8–16 × 6–16 cm, base cordate to truncate, bilobed to ~1/3 the length, lobes obtuse, 11–11nerved, abaxially sparingly pubescent, especially at the nerves. **Inflorescences** racemes, few-flowered, opposite; pedicels ~3 cm long; receptacle cylindric, 3 cm long; buds attenuate at both ends, densely tomentose with 5 short dents at apex. **Flowers** with calyx limb laterally split, spathaceous, 2 cm long; petals pink, obovate, 4–5 × 2–2.5 cm; perfect stamens 1, 4 cm long, anther linear; staminodes very small, antheriferous, with free or partly connate filaments; ovary stipitate, tomentose, sometimes abortive, stipe adnate to the wand of the receptacle. **Fruits** pods, linear, 20–25 × 2 cm, dehiscent; seeds 10–20, oblong, ~1 cm Ø, flattened, black, shiny. **Distr:** Asia. **Ecol:** Natural habitat unknown. Fl. throughout most of the year. **Zones:** 9–12.

B. natalensis Oliv., *Bot. Mag.* vol. 100, ser. 3, 30: pl. 6086, 1874. [Camel's Foot, Dainty Bauhinia, Natal Bauhinia, Natal Orchid Tree]. **Shrubs**, small, bushy, glabrous, to 2.5 m high, spreading to ~3 m wide. **Stipules** subulate. **Leaves** alternate, somewhat distichous; lamina almost completely divided, somewhat free, leaflets obliquely obovate or subobovate-oblong, ~2.5 cm long, base of outer side rounded, apex rounded, adaxially dark green, abaxially paler; petiole 0.6–1.2 cm long, base swollen, ending in a subulate point between the leaflets. **Inflorescences** leaf-opposed, 1–2-flowered; bracts 2, minute, setaceous. **Flowers** pure white, ~4 cm Ø, often with a thin, faint, crimson streak along midrib of 3 smaller petals; calyx with a short turbinate tube, and a broad, spathaceous, green, apiculate limb, ~8 mm long, base truncate; petals erecto-patent, obovate-oblong, apex rounded, 3 upper petals smaller; stamens 10, filaments almost free, 5 *longer* stamens equaling the style, of which 2 have each a lateral spur above the base, 5 *shorter* stamens half as long, anthers yellow; ovary stipitate, free, style stout, stigma large, capitate. **Fruits** pods, 7.5–10 × 4–7 cm, flat, base contracted, apex acuminate, glabrous, 6-seeded. **Distr:** S Africa (KwaZulu-Natal coastal strip and the E Cape Region). **Ecol:** In valley bushveld and scrubs. Fr. Jan–Jun. **Zones:** 9–11.

B. petersiana Bolle, *Naturw. Reise Mossambique* [Peters] 6(Bot., 1): 24, 1861. (Syn: *Perlebia petersiana*). [Kalahari Orchid Tree, White Orchid Tree, Rag Bush]. **Shrubs** or **trees**, 2–8 m high, usually erect, sometimes scrambling; young branchlets ± densely brown-pubescent; tendrils occasionally present. **Leaves** often wider than long, (1.3–)2–8 × 2–10 cm, bilobed to ~1/3 the way, abaxially minutely puberulous, lobes rounded. **Inflorescences** racemes, 2–10-flowered, short, racemes often aggregated terminally; flower buds 3–5 cm long. **Flowers** with hypanthium (1.5–)2–5 cm long, longitudinally sulcate; petals white or pink-spotted, narrowly elliptic to oblanceolate-elliptic, 2.5–7.5 × 0.6–1.5(–2.7) cm, margins very crisped, midrib pubescent outside, with many small glands; fertile stamens 5; stigma 2.5–3 mm Ø, terminal, peltate, abruptly enlarged from the style. **Fruits** pods, woody, dehiscent, linear-oblong or oblanceolate-oblong, 10–24 × (2.3–)3–4.7 cm; seeds deep chestnut-purple, irregularly oblong to obovate or subcircular, 1.6–3 × 1.1–1.8 cm. **Distr:** DRC Congo, Mozambique, Tanzania, Zambia. **Ecol:** In woodlands and on escarpment slopes; 150–1,830 m. **Zones:** 10–12.

B. purpurea L., *Sp. Pl.* 375, 1753. [Orchid Tree, Fall Orchid Tree, Purple Orchid Tree]. **Trees**, medium-sized; bark ashy to dark brown, young parts pubescent. **Leaves** 7–18 cm long, longer than wide, 9–11-nerved, cleft about halfway down into 2 acute or rounded lobes, minutely pubescent below when young; petiole 2.5–5 cm long. **Inflorescences** panicles, few-flowered, borne at ends of branches; pedicels ~0.5–1.3 cm long; bract ~3 mm long; bracteole ~2 mm long. **Flowers** tomentose; hypanthium 0.7–1 cm long; calyx ~2.5–3 cm long, usually splitting into 2 reflexed segments, 1 emarginate the other 3 toothed; petals oblanceolate, 3.7–5 cm long, long-clawed, spreading, veined; stamens usually 3 fertile, others reduced to antherless filaments; ovary downy, long-stalked, style long, stigma oblique. **Fruits** pods 15–25 × ~1.5–2 cm; stalk ~2 cm long; seeds 12–15, almost round, ~1.2–1.3 cm Ø, brown, smooth. **Distr:** China, India, Myanmar, Pakistan, Sikkim, SE Asia, Sri Lanka. **Ecol:** Fl. Sep–Nov. **Zones:** 9–12.

B. racemosa, *Encycl. Meth.* 1: 390, 1784. (Syn: *Pilostigma racemosa*). [Mountain Ebony]. **Trees**, small, deciduous; trunk, short, unbranched; crown spreading; bark dark brown, rough with vertical cracks; young twigs hairy. **Leaves** 2–5 × 2.5–7.5 cm, divided halfway down into 2 lobes, base usually cordate, adaxially glabrous, abaxially hairy, 7–9-nerved; petiole 0.7–1.8 cm long. **Inflorescences** racemes, terminal or leaf-opposed, 7.5–12.5 cm long; pedicel 0.5–1 cm long, hairy, jointed near middle; bracts short, linear, acute. **Flowers** with hypanthium very short; calyx ~6–8 mm long, spathaceous, reflexed; petals narrow, oblanceolate, acute, white; stamens 10, all fertile, filaments hairy at base; ovary hairy, stigma sessile. **Fruits** pods, 12.5–25 × 1.7–2.5 cm, slightly inflated, beak long; stipe ~1.2–1.5 cm long, turgid, rigid, falcate; seeds 12–20, oblong, 7–8 mm long, compressed, black. **Distr:** China, India, Myanmar, Pakistan, Sri Lanka. **Ecol:** Fl. Mar–Jun. **Zones:** 10–12.

B. ramosissima Benth. ex Hemsl., *Diag. Pl. Nov.*

JML

Above: *Bauhinia petersiana, Bauhinia purpurea, Bauhinia racemosa, Bauhinia ramosissima.*

49, 1880. (Syn: *Casparia ramosissima*, etc.). **Shrubs**, densely branched, to 3 m tall. **Stipules** ovate to lanceolate, ~1.5 mm long, caducous. **Leaves** bifoliolate; leaflets oblong to suborbicular, (0.5–)1.5–2.5(–3) × (0.5–)1–1.5(–2) cm, inner margins ± parallel, base cuneate, margins slightly crisped, apex rounded, subcoriaceous, adaxially glabrous, abaxially strigose to glabrate, 2(–3)-nerved; petiole 0.3–1 cm long. **Inflorescences** shortly racemose or flowers solitary, terminal, or subterminal and axillary, 1–6-flowered, rachis strigose; bract lanceolate, 1–1.5 mm long; bracteoles similar to bract, but smaller; pedicels 3–6 mm long, strigose. **Flowers** with hypanthium cyathiform, 1–2 mm long; calyx spathaceous; petals 5, subequal, purple, elliptic, (1.5–)2–2.5(–3) cm long, claw tomentose; fertile stamen 1, filament slender, shortly connate with staminodes, anther oblong, 7–9 mm long; staminodes 9, 2–5(–7) mm long, basally connate tomentose, purple-tinged; gynoecium subequalling petals, stigma small. **Fruits** pods, linear, (5–)6–8(–9) × 0.5–1.5, apiculate with persistent style, light brown, strigose or glabrate; seeds oblong to suborbicular, 5–8 × 4–8 mm, chestnut-brown, dull. **Distr:** Mexico. **Ecol:** In desert shrub communities; 800–2,500 m. Fl. throughout most of the year. **Zones:** 9–12.

B. roxburghiana Roxb., *Fl. Ind.* 2: 322, 1832. (Syn: *Lasiobema retusa, Phanera retusa*). [Semla Gum]. **Trees**, medium-sized; bark dark brown. **Leaves** 7.5–15 cm long, slightly broader than long, base cordate or truncate, apex entire or notched, subcoriaceous, glabrous, 9–11-nerved; petiole 3.5–8.8 cm long, thickened at both ends. **Inflorescences** lax panicles, terminal, lower branches are axillary; pedicels 1.5–2 cm long, softly pubescent; bract linear, 2.5 mm long. **Flowers** with calyx 5–7.5 mm long, silky-pubescent, split almost to base into 2–3 segments; petals obovate, 1–1.5 cm long, clawed, hairy outside, white, 3 upper purple-mottled; stamens 3; ovary hairy on margins, stipitate, style elongated. **Fruits** pods, oblong or oblanceolate, 10–17.5 × 3–3.8 cm, flat, firm, late-dehiscent, shortly stalked; seeds 6–8, flat, suborbicular, dark-brown, smooth. **Distr:** W Pakistan, India. **Ecol:** Fl. Sept–Nov. **Zones:** 10–12.

B. scandens L., *Sp. Pl.* 1: 374, 1753. (Syn: *B. anguina*, etc.). [Snake Climber]. **Lianas**, large, woody; branches cylindric when young, flattened and ladderlike when old; tendrils in pairs, puberulent. **Stipules** caducous. **Leaves** ovate to broadly ovate, 5–9 × 4–8 cm, base shallowly cordate to truncate, margins entire on flowering branches, apex bifid to >1/2 in sterile or juvenile branches, lobes with obtuse or acuminate apices, papery, glabrous, primary veins usually 7–9; petiole 2–4 cm long. **Inflorescences** elongated racemes, 10–15 cm long, many-flowered, or several joined into panicles 15–25 cm across, terminal, puberulent; bracts and bracteoles linear; pedicels 3–4 mm long; flower buds ovoid, 1.8–2 mm Ø, apex open. **Flowers** with calyx lobes 5, triangular, outside pubescent; petals white, subequal, obovate to oblanceolate, ~3 mm long, shortly clawed; fertile stamens 3, filaments glabrous; staminodes 2; ovary shortly stalked, oblique, glabrous, style stout, stigma small. **Fruits** pods, rhombic to oblong, 1.8–3 × 1–1.6 cm, indehiscent or tardily dehiscent; valves thin, reticulate-veined; seeds 1 or 2(–4), ellipsoid to obovoid-orbicular, ~8 mm Ø. **Distr:** Cambodia, China (Hainan), India, Indonesia, Laos, Malaysia, Myanmar, Nepal, Thailand, Vietnam. **Ecol:** In forests, among shrubs in coastal habitats. Fl. Oct. Fr. Dec. **Zones:** 9–12.

B. thonningii Schum., *Beskr. Guin. Pl.* 203, 1827. (Syn: *Piliostigma thonningii*). [Camel's Foot, Monkey Bread, Rhodesian Orchid Tree, Wild Orchid Tree]. **Trees**, small, occasionally shrubs or scramblers, (2–)3–10 m tall, usually dioecious; trunk often crooked; bark rough, dark brown to gray or black; branchlets rusty-tomentellous or shortly rusty-tomentose when young. **Leaves** mostly 5–17 × 6–19 cm, bilobed almost to the middle, abaxially densely reticulate and rusty-puberulous or -pubescent, venation prominent abaxially. **Inflorescences** panicles, usually alternately leaf-opposed and axillary along branches; ♂ 5–19 cm long; ♀ 2–7 cm long. **Flowers** white to pinkish; calyx rusty-tomentose or -tomentellous, 1.3–2.3 cm long; petals obovate, 1.4–2.6 cm long, rugose or bullate, hairy on basal claw and outside of limb. **Fruits** pods, stipitate, woody, oblong or linear-oblong, mostly 13–26(–37) × 3–6.3 cm; seeds obovoid to ellipsoid, somewhat compressed, dark brown to blackish, ~7–9 × 5–7 mm, 3–4 mm thick. **Distr:** Tropical Africa (from Senegal and Abyssinia to Sudan, and southwards to SW Africa and the Transvaal). **Ecol:** In savanna regions; near sea-level to 1,830 m. **Zones:** 10–12.

B. tomentosa L., *Sp. Pl.* 375, 1753. [Bell Orchid Tree, Bush Neat's Foot, Yellow Bauhinia, Yellow Butterfly Tree]. **Shrubs** or **small trees**, to 6 m tall, multistemmed; twigs tomentose or hairy. **Stipules** lanceolate, ~3–8 mm long. **Leaves** alternate; lamina 2.5–5 × 4.5–6.5 cm, bilobed, lobes rounded at apex, 7–9 nerved at base, adaxially glabrous, abaxially hairy; petiole 2.5 cm long. **Inflorescences** opposite, axillary in pairs, sometimes clusters of 1–3 flowers, flowers drooping; pedicels 1.5 cm long with a pair of subulate persistent bracteoles ~3 mm long. **Flowers** large, yellow, occasionally with purplish markings at base of petals; calyx ~1.3–1.4 cm long, with 2 teeth at apex in bud, splitting down on one side when flower opens; petals 5, white or yellow, ~1.8–5 cm long, obovate, not clawed, conspicuously veined, one with a red blotch on the inner surface; stamens 10 (all fertile); ovary shortly stipitate, tomentose, dehiscent; style ~1.8–1.9 cm long. **Fruits** pods, linear-oblong, 10–12.5 × 1.5 cm, dehiscent, tomentose or glabrous; seeds 6–10. **Distr:** China, tropical Africa & Asia. **Ecol:** On rocky slopes in arid bushveld, often on dolomite. Fl. Aug–Oct. **Zones:** 9–12.

B. variegata L., *Sp. Pl.* 375, 1753. [Butterfly Tree, Camel's Foot, Mountain Ebony, Orchid Tree, Poor Man's Orchid]. **Trees**, medium–sized; bark dark brown, nearly smooth; young shoots pubescent. **Leaves** with petiole 2–3.8 cm long; lamina ~4.5–15 cm

Below: *Bauhinia roxburghiana, Bauhinia scandens, Bauhinia thonningii.*

JML

JML RF

Above: *Bauhinia tomentosa, Bauhinia variegata, Bauhinia viridecens, Lysiphyllum hookeri.*

long, as broad as or rather broader than long, with a medium cleft reaching from 1/4 to 1/3 the way down, lobes obtuse, base deeply heart-shaped, 9–15-nerved, pubescent beneath when young. **Inflorescences** few-flowered, pubescent racemes; pedicels short or absent; bracteoles minute. **Flowers** with hypanthium slender, ~1.2–2.5 cm long; calyx 2–2.7 cm long, tomentose, 5 toothed at apex; petals 5–6.3 cm long, obovate, with long rather broad claw, all white or 4 petals pale purple and fifth darker with purple veins; stamens 5, fertile, no staminodes; ovary hairy, stipe 1–1.7 cm long; style long, stigma capitate. **Fruits** pods, 15–30 × ~1–2.5 cm, hard, flat, dehiscent 10–15 seeded, stipe glabrous. **Distr:** China, India, Kashmir, Myanmar, Nepal, Pakistan. **Ecol:** Fl. Feb–Apr. **Zones:** 9–12.

B. viridescens Desvaux, *Ann. Sci. Nat.*, Sér. 1, 9: 429, 1826. (Syn: *B. laui, B. polycarpa, B. viridescens* var. *laui*). **Shrubs** or **small trees**, to 3 m tall, dioecious; young branches pubescent; older branches whitish. **Stipules** narrowly triangular and tapering, ~2 mm long. **Leaves** suborbicular, 5–9 × 6–10 cm, base truncate or shallowly cordate, apex bifid to halfway, lobes obtuse at apex, papery, adaxially glabrous, abaxially pubescent, primary veins 7–9; petiole 2–3.5 cm long. **Inflorescences** racemes, 10–15-flowered, opposite with leaves, 4–8 cm long; bracts 2–3 mm long, pubescent; flower buds narrowly fusiform, 0.6–1 cm long. **Flowers** with hypanthium turbinate, short; calyx spathaceous (open on one side); petals greenish-white, obovate-lanceolate, 0.7–1.2 cm long, shortly clawed; ♂ with fertile stamens 10, in 2 whorls, subequal, ~1 cm long, anthers oblong, 1–2 mm long; ♀ with reduced stamens present, ovary shortly stalked, 6–8 mm long, style short, stigma peltate, small. **Fruits** pods, straight or slightly curved, linear, 5–7 × 0.7–1 cm, compressed, beaked, valves leathery, glabrous; seeds 6–10, green, ellipsoid, ~3 mm Ø, slightly compressed. **Distr:** Cambodia, China (Hainan, Yunnan), Laos, Myanmar, Malaysia, Vietnam. **Ecol:** In open forests; low elevations. Fl. Mar–Jul. Fr. May–Jan. **Zones:** 9–11.

Lysiphyllum

Species ~8. **Trees**, **shrubs** (semiscandent) and **lianas** (tendriled in Malesia). **Leaves** bifoliolate or entire. **Flowers** with hypanthium tubular or campanulate; calyx split into 2, 3 or 5 lobes; nectariferous disk absent; stamens 10, anthers opening by longitudinal slits; ovary with peltate or capitate stigma. **Fruits** pods, woody, dehiscent or indehiscent; seeds with short funicular aril-lobes. **Distr:** Thailand, Malesia to N Australia. **Cult:** Propagation is usually by air-layering, softwood cuttings, grafting or by seeds. Needs fertile, moist but well drained soil in partial to full sun.

L. hookeri (F. Muell.) Pedley, *Austrobaileya* 1: 33, 1977. (Syn: *Bauhinia hookeri*). [Pegunny, Queensland Ebony]. **Shrub** or **trees**, to 12 m tall; crown spreading. **Stipules** curved-ovate, 1 mm long. **Leaves** bifoliolate, leaflets ovate to obliquely ovate, 1.5–5 × 1–3.2 cm, rounded, concolourous, glabrous; petiole 1.5–4 cm long, glabrous. **Inflorescences** short, terminal; rachis 1–5 cm long, glabrous; pedicels 0.5–1.3 cm long, puberulous; bracts ovate, 1.5–2 mm long, scarious. **Flowers** with calyx 2.3–3 cm long, lobes oblong-lanceolate, sparsely puberulous to almost glabrous; petals obovate, 3–4 cm long, white, pubescent outside, sparsely puberulous inside; stamens 10, all fertile, 3.7–5.2 cm long, base white, pink to red above, glabrous or sparsely puberulous, anthers 4–5 mm long, red; pistil 4.5–5.5 cm long, stipe pubescent, ovary and style glabrous, stigma peltate, green. **Fruits** pods, oblong, sometimes curved, 6–22 × 2.5–4.5 cm, glabrous; seeds to 12, oblong, 1.4–1.5 cm long, brown. **Distr:** Australia (E Queensland). **Ecol:** In monsoon forests, thickets, beach forests and open forests; 0–500 m. Fl. Jun–Nov. **Zones:** 10–11.

TRIBE DETARIEAE

Schotia

Species 4 or 5. **Shrubs** or **small trees**, much-branched, unarmed, deciduous before flowering. **Stipules** small, caducous. **Leaves** paripinnate; leaflets leathery, 3–18 pairs, lowermost usually smaller than upper pairs. **Inflorescences** dense panicles, short, showy, many-flowered, often borne from old wood; bracts and bracteoles small, ovate-oblong, membranous, caducous. **Flowers** bright red, pink or flesh-colored; calyx turbinate or cylindric, lobes 4, overlapping, not reflexed at flowering, sometimes pink, caducous; petals 5, ovate or oblong, subequal, longer than and inserted in mouth of calyx tube, subsessile, imbricate, uppermost petal innermost, sometimes small and scalelike; stamens 10, free or joined shortly at base, inserted with petals, anthers uniform, longitudinally dehiscent; ovary stalked, stalk attached to 1 side of calyx tube, several- to many-ovuled, style longer than stamens, stigma small, terminal. **Fruits** pods, linear, flat, oblong or falcate, leathery or woody, upper suture often thickened or winged, subdehiscent; seeds 1–few, globose, compressed, with or without a yellow, fleshy, cupular aril. **Distr:** E Zimbabwe, SW Mozambique, S Namibia, S Africa (excluding central regions) and Swaziland. **Ecol:** In forest margins, woodlands, scrub forests, bush lands, thickets, etc. **Cult:** Propagate from seeds or cuttings. Requires a well-drained soil in full sun, protected from strong wind.

S. brachypetala Sond., *Linnaea* 23: 39, 1850. (Syn: *S. latifolia, S. semireducta*). [African Walnut, Drunk Parrot Tree, Tree Fuchsia, Weeping Boerboon]. **Shrubs large** or **small trees**, glabrous, or nearly so. **Leaves** scattered, abruptly pinnate; leaflets 2–6 pairs, opposite, subsessile, ovate-oblong or elliptical, 3.7–6.5 × 2.5–3.8 cm, apex obtuse, mucronulate or muticous, reticulate with slender veins. **Inflorescences** panicles, much-branched, many-flowered, axillary, shorter than the leaves; bracts small, caducous; pedicels shorter than the calyx. **Flowers** with calyx richly colored, crimson, 1.5 cm long, with a turbinate tube and 4–5-parted limb; segments broadly elliptical, concave, imbricated, very obtuse; petals very minute, hidden in the base of the calyx tube, linear; stamens 10, exserted, combined at base into a campanulate tube, 5 opposite the petals, shorter than the rest; ovary on a long stipe, lanceolate, corrugated on the ventral margin, glabrous, style filiform, arched, stigma capitate. **Fruits** pods, flat, ~5 cm long, woody, upper suture winged, long-pedicelled; seeds 2–3, arillate. **Distr:** S Africa (Natal). **Ecol:** In sheltered valleys, where the soil is dry and rocky; ~600–900 m. **Zones:** 9–12.

Baikiaea

Species ~6. **Trees**, unarmed, evergreen (or rarely deciduous). **Leaves** simply pari- or imparipinnate; leaflets usually alternate, sometimes opposite, without pellucid gland-dots. **Inflorescences** racemes, terminal or axillary; flowers pedicellate, distichously arranged along the axes; bracteoles usually small, not enclosing flower buds, imbricate, almost valvate, fulvous villous-tomentose inside, soon falling off. **Flowers** hermaphrodite; sepals 4, one larger than the rest, very narrowly densely fuscous- or paler brown-tomentellous outside; petals 5, 4 of them equal, the fifth narrower and usually differently colored, all obovate, free, imbricate, crinkled around edges, villous along and near midrib; stamens 10, filaments glabrous or villous below, one of them free, the rest basally connate into a short tube, anthers dorsifixed, dehiscing by longitudinal slits; ovary stipitate, tomentose, ovules 1–many, style elongate, glabrous, stigma enlarged, peltate, depressed-subglobose. **Fruits** pods flattened, woody, dehiscing into 2 valves; seeds large, with thin and fragile or else hard testa. **Distr:** WC Africa and southern tropical Africa. **Ecol:** In tropical lowland rainforests, seasonally dry forests and woodlands on well-drained soils. **Cult:** Propagate from seeds or by air-layering. Requires a moist, well-drained soil in full or partial sun.

B. insignis Benth., *Trans. Linn. Soc*, Lond. 25: 314, 1865. **Trees**, 5–34 m tall; buttresses none or slight; bark gray or gray-brown, smooth or slightly fissured; branchlets and leaves glabrous (rarely minutely pubescent). **Leaves** pinnate; leaflets (2–)3–8(–10), coriaceous, ovate, elliptic, oblong-elliptic or oblong-lanceolate, to 40 × 17 cm, apex subacuminate to obtusely pointed; petiole with rachis to 23(–30) cm long, petiolules 0.6–1.5 cm long. **Inflorescences** racemes, 1.5–8.5(–12) cm long; bracts 3–4 × 4–5 mm; bracteoles 3–4.5 × 2–2.5 mm. **Flowers** large, fragrant; sepals to 11(–15) cm long; petals to 21 × 11 cm, white or cream, standard pale yellow;

Above: *Schotia brachypetala, Baikiaea insignis, Lysidice rhodostegia, Saraca asoca.*

filaments 6–16 cm long, glabrous or villous below, anthers 1.8–2(–3) cm long; ovary 2.5–3.8 cm long, style 5–9(–11) cm long. **Fruits** pods, 17–42(–60) × 5–12 cm, densely brown-velvety; seeds ellipsoid, 3–4.5 × 1.5–3 cm, compressed, dark red, thin-shelled. **Distr:** W Africa. **Ecol:** In evergreen rainforests in swampy areas. **Zones:** 10–12.

Lysidice

Species 2. **Shrubs** or **trees**. **Stipules** caducous or late deciduous, small, subulate or narrowly triangular. **Leaves** abruptly pinnate; leaflets 3–5 pairs, opposite, base slightly oblique, asymmetrical on two sides; petiolules short. **Inflorescences** panicles, terminal; bracts large, red or white at base of flowers; bracteoles small, in pairs inserted at apex or near apex on pedicels. **Flowers** showy, purplish-red or pink; calyx tubular, 4-lobed, lobes imbricate, reflexed after anthesis; petals 5, 3 at back large, obovate, long clawed, 2 in front small, reduced, scalelike or subulate; perfect stamens 2, free or slightly basally connate, filaments elongated, reflexed in bud, anthers elliptic, opening lengthwise; staminodes 3–8, unequal in length, subulate, without anthers or with 1–3 orbicular, small, abortive anthers; ovary oblong, compressed, stalked, stalk adnate to calyx tube, 6–14-ovuled, style filiform, long, stigma capitate, small. **Fruits** pods, oblong or obovoid-oblong, compressed, thickly leathery or woody, with stipe, 2-valved, valves flat or slightly curling or spirally twisted; seeds shiny, compressed, oblong, obliquely broadly elliptic to suborbicular. **Distr:** S and SW China, and Vietnam. **Cult:** Propagate from seeds and semihard cuttings under mist. Requires a warm climate with good soil.

L. rhodostegia Hance, *J. Bot.* 5: 299, 1867. [Casse]. **Shrubs** or **small trees**, 2–5(–10+) m tall. **Leaves** pinnate; leaflets oblong or ovate-lanceolate, 5–16 × 2–6.5 cm, base rounded, apex caudate-acuminate, papery, lateral veins subparallel, slender, evident on both surfaces; petiolules 2–3 mm long. **Inflorescences** panicles, 20–40 cm long; rachis, bracts, and bracteoles sparsely puberulous; bracts and bracteoles pink, ovate-oblong or elliptic; bracts 1.2–2.8 × 0.5–1.4 cm; bracteoles small, 2–5 mm long. **Flowers** with calyx tube 1.2–1.5 cm long, lobes oblong, 0.8–1 cm long, dark purplish-red; petals purplish-red, broadly obovate, ~1.2 cm long (including claw), apex rounded, emarginate; anthers ~4 mm long; staminodes usually 4, subulate; ovary hairy, 6–9-ovuled, style slender, hairy. **Fruits** pods, obovoid-oblong, 12–20 cm long, ventral suture longer and curved, dehiscent, valves often spirally twisted; seeds 2–7, brownish-red, oblong, 2.2–2.5 × 1.2–1.5 cm, margin not thickened, testa thinner and fragile, surface slightly wrinkled. **Distr:** China, Myanmar, Vietnam. **Ecol:** On mountain slopes, tropical forests; <500 m. Fl. Jun–Aug. Fr. Sep–Nov. **Info:** Most parts of this plant are somewhat toxic, and used medicinally to relieve pain.

Saraca

Species ~11. **Shrubs** or **trees**. **Stipules** axillary, enveloping bud, usually caducous. **Leaves** spirally arranged, paripinnate, 1–7-jugate, petiolate; leaflets opposite, sometimes with a pair of wartlike glands at base and/or at apex, nerves ± straight or slightly curved; petiolules very short. **Inflorescences** terminal or axillary, or on older branches, often rather dense, corymbose or paniculate, usually many-flowered; bracts caducous or persistent; bracteoles (sub) opposite, often colored and showy; pedicels very short. **Flowers** bisexual, or functionally ♂ (with short, small, sterile or rudimentary pistil), distinctly (or sometimes not) articulated at base; hypanthium cylindrical or tubular; calyx 4(–6)-lobed, subequal, petaloid; petals absent; disk absent; stamens (3 or)4–8(–10) fertile, in 1 whorl, filaments free or united at base into a thin rim or ringlike, ± equal in length, exserted, anthers ellipsoid, oblong, or broadly ovoid, dorsifixed, introrse; staminodes dentate or rarely subulate when present; pistil stipitate, excentric with stipe adnate to hypanthium, ovary oblong or lanceolate, 2- or (4–)6–8(–12)-ovuled, style filiform, stigma terminal, minute, obtuse. **Fruits** pods, suborbicular, broadly ovoid, ellipsoid, or obovoid, flattened, straight or distinctly curved, coriaceous or woody, dehiscent, 2-valved, valves twisting and enrolling during dehiscence, 1–8-seeded; seeds ovoid, ellipsoid, or subglobose, smooth, compressed or subterete, exarillate, exalbuminous. **Distr:** Bangladesh, Cambodia, China, India, Laos, Myanmar, Thailand, Vietnam; Malesia. **Cult:** Propagate from seeds or by air-layering. Requires a moist, well-drained soil rich in organic matter, in shade or full sun.

S. asoca (Roxb.) W.J. de Wilde, *Blumea.* 15: 393, 1967. (Syn: *Jonesia asoca, J. confusa, Saraca indica*). [Asoka, Ashoka]. **Trees**, to 10 m tall, sometimes shrubs, <4.5 m high; stem to 10 cm Ø. **Leaves** (1–)4–6-jugate, to 30 cm long, subsessile; leaflets elliptic-oblong or elliptic-lanceolate, or lanceolate, (3.5–)7–18(–26) × (1–)3–5(–9) cm, lowest pair usually smaller, base cuneate, obtuse, or rounded, symmetric, apex acuminate, chartaceous or subcoriaceous, glabrous; petiolules 2–5 mm long. **Inflorescences** 1.5–12(–20) cm across, glabrous, main branches to ~6 cm long; bracts 1–6 mm long; bracteoles embracing pedicel, 2–7 mm long; pedicels 0.8–1.8 cm long. **Flowers** yellow, orange, deep orange or orangish-red when older, fragrant at night; hypanthium 0.9–2 cm long; calyx lobes elliptic or obovate, 0.7–1.3 × 0.5–0.9 cm; stamens (5–)6–8(–10), filaments (1–)1.7–2.5 cm long, glabrous; staminodes 0–2; pistil stipitate, hairy, ovary 4–6 mm long, puberulous, 8–10(–12)-ovuled, style 1.5–2 cm long, glabrous. **Fruits** pods, oblong or elongate-oblong, 12.5–25 × 3.5–6.5 cm, leathery or woody, beaked, valves coiling; seeds (1–)4–6(+), oblong-ellipsoid, 2–3 × 1.25–2 cm. **Distr:** Bangladesh, India, Myanmar, Sri Lanka. **Ecol:** Fl. Jan–May. **Zones:** 10–12.

S. declinata (Jack) Miq., *Fl. Ind. Bat.* 1, 1: 84, 1855. (Syn: *Jonesia palembanica, S. triandra*, etc.). [Red Saraca]. **Trees**, to 10(–29) m tall, sometimes treelets 3–6 m tall; trunk 10(–60) cm dbh. **Leaves** (1–)3–5(–7)-jugate; leaflets usually lanceolate, 4–30(–50) × 1.5–11(–20) cm, base cuneate or rounded, symmetric, apex acuminate to obtuse, subcoriaceous; petiole and rachis (4.5–)8.5–56 cm long; petiolules to ~1 cm long. **Inflorescences** to 15(–30) cm across, main branches (1–)7–10(–30) cm long; bracts 0.3–1.2 × 0.2–0.7 cm; bracteoles 0.4–1.9 × 0.3–1 cm; pedicels 1–1.5(–3.5) cm long. **Flowers** yellow, orange, pink, pinkish-red, fragrant, articulated near base; hypanthium 0.7–1.5(–3.2) cm long; calyx lobes ovate, elliptic, or obovate, 0.6–1.5 × 0.3–0.7 cm; stamens (3 or)4(or 5), filaments 1.6–3.8 cm long, glabrous; staminodes absent; pistil stipitate, hairy, ovary 2.5–7 mm, puberulous, 6–8(–10)-ovuled, style 0.7–2 cm long. **Fruits** pods, dark brown, oblong or lanceolate, 7–31 × 3–7 cm, to ~1.2 cm thick, woody, beak to ~1 cm long; seeds 6–8, ovoid-oblong, ~4 × 2 cm. **Distr:** Indo-China to Malesia. **Ecol:** In primary and mixed dipterocarp forests; to 800 m. **Zones:** 10–12.

S. indica L., *Mant. Pl.* 1: 98, 1767. (Syn: *Jonesia minor*, etc.). [Asoka, Sorrowless Tree]. **Trees**, to 24 m tall; trunk to 34 cm dbh. **Leaves** (1–)2–4(–7)–jugate; leaflets elliptic or lanceolate, (5–)15–20(–30) × (1.5–)3.5–6(–11) cm, lower pairs usually smaller, base cuneate, obtuse or rounded, symmetric, glabrous, chartaceous or subcoriaceous, nerves 5–11 per side; petiole and rachis (0.5–)7–25(–35) cm long; petiolules (0–)0.4–1 cm long. **Inflorescences** to 10(–22) cm wide, main branches to ~14 cm long; pedicels (0.4–)1.2–2(–3.5) cm long, flowers articulated near base. **Flowers** orange, yellow, to deep orange or purple, or red, glabrous; hypanthium 0.9–1.2(–1.6) cm long; calyx lobes ovate, elliptic, or obovate, 0.5–1.2 × 0.2–0.8 cm; stamens (5 or)6–8(–10), filaments 1–3.5 cm long; staminodes absent; pistil stipitate, hairy, ovary 4–8 mm long, 6–8-ovuled, style 1.5–3.5 cm long, glabrous. **Fruits** pods, dark red, oblong, or narrowly oblong-lanceolate, 6–25 × 2–6 cm, 0.5–1.2 cm thick, leathery or woody; seeds 4–6(–8), ovoid-oblong, sometimes reniform, 4–5 × 2–2.5 cm. **Distr:** Thailand, Laos, S Vietnam; Malesia. **Ecol:** In forests; <900 m. Fl. & Fr. throughout the year. **Zones:** 10–12.

S. thaipingensis Cantley ex Prain, *J. As. Soc. Beng.* 66, ii 211, 490, 1897. (Syn: *S. cauliflora*). [Yellow Saraca]. **Trees**, to 24 m tall; trunk 38 cm dbh. **Leaves** (4–)7(or 8)-jugate; leaflets oblong-lanceolate, 7–32 × 3–9.5 cm, base cuneate, obtuse, or rounded,

Above: *Saraca declinata, Saraca indica.*

symmetric, apex acuminate or acute, chartaceous, adaxially densely pubescent, nerves 6–12 per side; petiole and rachis 15–56 cm long; petiolules distinct, (0.5–)1–1.5 cm long. **Inflorescences** (8–)15–35(–40) cm across, glabrous; bracts and bracteoles caducous; bracts oblanceolate, 1.2–3.5 × 0.7–2.2 cm; bracteoles oblanceolate or oblong-lanceolate, 0.6–1.7 × 0.2–0.5 cm; pedicels 0.7–1.5 cm long. **Flowers** opening yellow, changing through orange to red with age; calyx lobes ovate, elliptic or oblong, 0.5–1 × 0.3–0.5 cm; stamens (3 or) 4(to ~7), filaments 0.9–1.7 cm, glabrous; staminode 1, small, or absent; pistil stipitate, ovary 4–8 mm long, (6–)9–12-ovuled, style 1–1.5 cm long. **Fruits** pods, pink, purplish-red, narrowly oblong, curved, 15–45(–50) × 3.5–10 cm, woody, beaked, valves coiling; seeds 6–8, ellipsoid, 3.6–3.8 × 2–2.4 cm. **Distr:** Myanmar, Thailand; Malesia. **Ecol:** In forests, mostly along streams or rivers; <1,100 m. Fl. Jan–Nov. Fr. Jan–Feb, Apr–Jul, Sep. **Zones:** 10–12.

Cynometra

Species ~70. **Shrubs** to **large trees**, unarmed, usually evergreen. **Stipules** caducous. **Leaves** paripinnate, rarely unifoliate; leaflets nearly always opposite, in 1–many pairs, asymmetric, without gland dots. **Inflorescences** usually axillary, racemose or paniculate, sometimes corymbose, fasciculate, or cauliflorous, spirally arranged along the inflorescence axes; bracts scalelike, broadly ovate, imbricate, usually persistent; bracteoles present but caducous, small, not enclosing the flower buds. **Flowers** bisexual; sepals 4(–5), imbricate; petals usually 5, equal or subequal; stamens (8–)10(–15), sometimes with filaments (free or almost so) alternately longer or shorter, anthers uniform, dorsifixed, longitudinally dehiscent; ovary 1–4-ovuled, ± stipitate, style filiform, elongate, central, excentric or occasionally attached to wall of receptacle, stigma minute, terminal, truncate or capitate. **Fruits** capsules, usually oblique, recurved or subreniform, compressed or turgid, woody and elastically dehiscent or thickened and indehiscent, 2-valved; seeds 1 or 2(4), compressed or subglobular, funicle short, thin-walled. **Distr:** Pantropical. **Cult:** Propagate from seeds or by air-layering. Requires a moist soil rich in organic matter, in shade or partial sun.

C. cauliflora L., *Sp. Pl.* 382, 1753. (Syn: *C. acutiflora*, etc.). [Nam Nam]. **Trees**, 3–15 m tall. **Leaves** 1-jugate; leaflets ovate, ovate-oblong to ovate-lanceolate, ~5.5–16.5 × 1.6–5.6 cm, apex obtuse emarginate, or shortly acuminate, glabrous, nerves 7–10(–14) pairs; petiole 2–8 mm long, usually glabrous, petiolules very short, sometimes pubescent. **Inflorescences** with 4 or 5 small racemes fascicled on hard knots, cauliflorous, usually on trunk, sometimes on exposed roots; rachis 0.5–3 cm long, glabrous; bracts 0.1–1 cm long, ciliate; bracteoles 1.5 mm long; pedicels ~3–6.5 mm long. **Flowers** with hypanthium 1–2 mm long; sepals 4(–5), 2–4 mm long; petals white or pinkish, 3–4 mm long; stamens (8–)10, variable within a single plant, filaments rarely with 1 or 2 hairs; ovary densely hairy, stipe 0.8 mm long, style 5–6 mm long, hairy up to halfway. **Fruits** capsules, fleshy, kidney-shaped, 2.7–3 × 1.8–2 cm, 1 cm thick, rugose, glabrescent (green underneath). **Distr:** Malaysia. **Ecol:** In mountain forests; 1,300 m. Fl. Jun. **Zones:** 11–12. **Info:** Cultivated widely in the tropics. The edible fruits are sour. Seeds have medicinal uses.

C. iripa Kostel., *Allg. Med.-Pharm. Fl.* 4: 1341, 1835. (Syn: *C. ramiflora* var. *heterophylla*, etc.). [Wrinkled-pod Mangrove]. **Shrubs** to **straggly trees**, to ~8 m tall; branches with lenticels, glabrous or rusty-pubescent. **Leaves** pinnate with 2 pairs of leaflets, upper pair larger, rachis 0.9–2.7 cm long, shortly pubescent; leaflets obliquely obovate-oblong to oblong or elliptic, 1.5–10 × 0.9–5.5 cm, base asymmetrically acute, apex obtuse, emarginate or sometimes mucronate, coriaceous, glabrous, surfaces nitid; petiole to 4 mm long. **Inflorescences** corymbiform pseudo-racemes or clusters, short, axillary, borne on branchlets; pedicels to 9 mm long, pubescent. **Flowers** small; calyx with 4 or 5 sepals 2–4 mm long; petals white, ~4–5 mm long; ovary densely rusty-pubescent outside, ± appressed-pubescent inside, style slanting away from center of flower. **Fruits** obliquely subcircular or ellipsoid, ~2.5–3.6 × 2–3 cm, 2 cm thick, compressed, slightly pubescent or rugose, glabrescent, strongly rugose with a conspicuous lateral beak to ~6 mm long, indehiscent; seeds 1. **Distr:** NE Australia (Queensland), India, Malaysia, Sri Lanka. **Ecol:** In coastal areas, back of mangrove forests. **Zones:** 9–12.

C. zeylanica Kosterm., *Reinwardtia* 10: 63, 1982. (Syn: *C. longifolia* [misapplied], *C. ramiflora* [misapplied]). [Ceylon Nam Nam]. **Trees**, to ~25 m tall; young stems glabrous. **Leaves** with 1 pair of leaflets; leaflets sessile, obliquely oblong, ~5–13 × 2–4 cm, base asymmetrical, margins slightly entire, wavy, apex acute to subacuminate, tip rounded, glabrous, surfaces glossy, nitid, only midvein conspicuous, adaxially dark green, abaxially paler; petiole short, slightly curved and rotated. **Inflorescences** axillary; bracts caducous. **Flowers** small, white; calyx 3 mm long; petals 3–4 mm long; stamens 10. **Fruits** usually only a single one developed per inflorescence, subglobose-ovoid, ~2–2.5 × 1.5 cm, not conspicuously beaked, brown, verrucose and minutely and sparsely pilose, glabrescent, indehiscent, 1-seeded. **Distr:** Sri Lanka. **Ecol:** In gallery forests of the dry zone; 0–500 m. **Zones:** 11–12.

Tamarindus

Species 1. *Generic description same as specific.* **Cult:** Propagate from seed or softwood cuttings. Requires a sunny site in a tropical or subtropical climate, with any soil type.

T. indica L., *Sp. Pl.* 34, 1753. [Tamarind]. **Trees**, large, evergreen, 10–15(–30) m tall; trunk 30–50(–90) cm dbh; bark dark ashy, irregularly longitudinally

Below: *Saraca thaipingensis, Cynometra cauliflora, Cynometra iripa, Cynometra zeylanica.*

Above: *Tamarindus indica, Intsia bijuga, Humboldtiana laurifolia, Amherstia nobilis.*

splitting. **Stipules** linear, caducous. **Leaves** paripinnate, 3.5–15 cm long; leaflets 10–20 pairs, opposite, oblong, 1.2–1.8 cm × ~3.7–5 mm, apex obtuse. **Inflorescences** lax racemes, 10–15-flowered; pedicels 0.8–1 cm long. **Flowers** with calyx 0.7–1.2 cm long, tube turbinate, lobes lanceolate-oblong, lowest 2 connate; petals 3, obovate, subequal to calyx lobes, margins repand, curled, 1–1.5 cm long, pale yellow, red-veined, 2 lower reduced to scales; stamens monadelphous, 3 perfect, 1.2–1.5 cm long, pubescent near base, others reduced to bristles; ovary slightly incurved, terete, ~8 mm long, hairy. **Fruits** pods, 7.5–20 × 2–2.5 cm, slightly compressed, often irregularly constricted, indehiscent; seeds 3–14, dark brown or black, shiny. **Distr:** Tropical Africa. **Ecol:** 0–1,500 m. Fl. May–Jun. **Zones:** 10–12. **Info:** Cultivated widely in tropical Asia for the edible pulp of the fruits. The tree has other uses including medicinal. The wood is hard and often used for building houses and making furniture.

Intsia

Species ~3. **Trees**, unarmed, deciduous. **Stipules** basally connate. **Leaves** paripinnate; leaflets in few pairs, opposite, large. **Inflorescences** panicles, terminal; flowers pedicelled. **Flowers** large, white, pink or red; calyx with disk produced to the top of the elongated tube, lobes slightly unequal, imbricated; petals only 1 developed, orbicular, with a distinct claw, others absent or rudimentary; stamens 3, perfect, imperfect ones usually present, filaments long, pilose. **Fruits** pods, large, flat, oblong, somewhat woody; seeds 1–8, not arillate. **Distr:** SE Asia (Malesia and Papuasia). **Ecol:** In tropical humid coastal forests on sand, rainforests, mangrove fringes and tidal river mouths, occasionally inland on hills. **Cult:** Propagate from seeds or by air-layering. Requires a moist well-drained soil in full or partial sun.

I. bijuga (Colebr.) Kuntze, *Revis. Gen. Pl.* 1: 192, 1891. (Syn: *Afzelia bijuga, Intsia bijuga*, etc.). [Borneo Teak, Malacca Teak, Moluccan Ironwood]. **Trees**, erect, nearly glabrous, 5–25(–50) m tall; trunk 1.5–2 m dbh. **Leaves** even-pinnate; leaflets (2)4, opposite, 5–10 cm long, base somewhat oblique, apex retuse or rounded, glabrous. **Inflorescences** corymbose panicles, terminal or leaf-opposed, pubescent. **Flowers** with calyx pubescent, lobes ~8 mm long; corolla with petals somewhat >1 cm long, limb ~1 cm across, white or pink; fertile stamens 3; ovary stalked, pubescent on both sutures, exserted. **Fruits** pods, large, 10–30+ × 4–10 cm, thickly coriaceous; seeds 2–8, orbicular, 2–4 cm across, testa very hard (adapted to dispersal by sea). **Distr:** Madagascar, through Malaya to the Caroline Islands. **Ecol:** In coastal areas; 0–450 m. Fl. May–Jul. **Zones:** 11–12. **Info:** The wood is highly prized for timber and thus over-exploited.

Humboldtia

Species 6. **Trees**, unarmed. **Stipules** large, persistent; stipels lacking. **Leaves** paripinnate, shortly petiolate; rachis sometimes winged; leaflets sessile, 3–5 pairs, coriaceous. **Inflorescences** racemose or paniculate, many-flowered, dense, axillary or supra-axillary; bracts and bracteoles ovate to lanceolate. **Flowers** with calyx with 4 subequal lobes, imbricate in bud; petals 3 or 5, white or pinkish, long-clawed; stamens 5 perfect, exserted, equal in length, alternating with 5 staminodes, anthers dorsifixed; style long, filiform, stigma terminal, truncate, ovary ~3- or 4-ovuled, borne on a stipe adnate to disk. **Fruits** pods, dehiscent, 2-valved, elliptic-oblong, compressed, 1-seeded; seed transverse, ovate, exalbuminous. **Distr:** SW India (W Ghats), 1 sp. extending to Sri Lanka. **Ecol:** In tropical rainforests. **Cult:** Propagate from seeds or by air-layering. Requires a rich, moist, well-drained soil in shade or full sun.

H. laurifolia (Vahl) Vahl, *Symb. Bot.* 3: 106, 1794. (Syn: *Batschia laurifolia*). **Trees**, to ~10 m tall; young stems glabrous, often swollen and inhabited by ants. **Stipules** leaflike above the point of insertion, ~2–3.5 × 1–2 cm, acute, lower portion falcate. **Leaves** with 3–5 pairs of leaflets; rachis ~1–20 cm long, glabrous or nearly so; leaflets ovate to ovate-oblong or lanceolate-oblong, ~4–15 × 2–5.5 cm, base obliquely rounded to cuneate, apex acuminate, glabrous, adaxially nitid, abaxially puberulent, usually glabrous at maturity. **Inflorescences** axillary, racemose or paniculate, 8–12 cm long; bracts deltoid, ~4 mm long; bracteoles oblong-spathulate, 6–7 mm long; pedicels 5–6 mm long. **Flowers** with calyx usually reddish or orangish, tube ~3 mm long, segments 0.8–1 cm long, subequal; petals 5, white, 1–1.2 cm long; stamens exserted to ~1.2–2 cm long. **Fruits** puberulent or subglabrous when young, glabrous at maturity, ellipsoid, 5–10 × 2.5–3 cm, compressed, 3–4-seeded. **Distr:** India (W Ghats), Sri Lanka. **Ecol:** In damp places of lowland forests. Fl. Sep–Oct, Feb. **Zones:** 10–12.

Amherstia

Species 1. *Generic description same as specific.* **Cult:** Propagation is best by cuttings, since seeds are rarely formed in cultivation. Needs fertile, moist soil with partial shade to full sun.

A. nobilis Wall., *Phil. Mag.* n. s. 3: 223, 1828. [Orchid Tree, Pride of Burma, Queen of Flowering Trees]. **Trees**, erect, unarmed, to 12(–20) m tall. **Stipules** foliaceous, lanceolate, 2–4 cm long, caducous. **Leaves** paripinnate, (4–)6–8-jugate, 30–60 cm long; leaflets opposite or subopposite, oblong, 3(7–)14–34 × (4–)5.5–8.5, apex acuminate, subcoriaceous, glabrous. **Inflorescences** racemes, terminal, simple, pendent, 50–80 cm long; peduncles 12–30 cm long; pedicels red, 5–12 cm long; bracts caducous. **Flowers** with hypanthium red, 3.5–4.5 cm long; calyx lobes 4, petaloid, imbricate, narrowly oblong or -lanceolate, ~5 cm long, often reflexed, curved or coiled, rarely straight; petals 5, unequal, red blotched with a large yellow spot and a reddish-violet band; upper most petal 1, obcordate 5.5–7.5 × 5–5.5 cm; lateral petals 2, cuneiform, 5.5–7 × 1.75–2.25 cm; lowermost petals 2, minute, rudimentary, subulate, 6–7 mm long; stamens unequal, fertile, 10 (diadelphous) or 9 (monadelphous), alternately long and short, filaments 9, united into a sheath or slightly tubelike, anthers versatile, 0.4–1.1 cm long; ovary falcate, 1.2–2 cm long, compressed, densely hairy, 4–6-ovuled, style filiform, ~3 cm. **Fruits** pods, woody, oblong, 11–20 × 4–5 cm, dehiscent; seeds 4–6, transversely ovate-orbicular, 1.75–2.5 × 2–2.75 cm. **Distr:** Myanmar. **Ecol:** In dry lowland evergreen forests. Fl. & Fr. throughout the year. **Zones:** 9–11. **Info:** In lowland forests.

Brownea

Species 12. **Trees** or **shrubs**, glabrous, unarmed, evergreen, small to medium, 10–15(–20) m tall; stems often weak and clambering. **Stipules** foliaceous, sometimes colored, caducous. **Leaves** abruptly paripinnate; new leaves lax (resembling handkerchiefs), often bright pink or purplish, splashed with white, becoming stiff and green with maturity; leaflets 4–18 pairs, leathery or subcoriaceous, often large, usually caudate. **Inflorescences** tightly packed in globular heads, 10–30 cm Ø, borne on the trunk or at the tips of short branchlets ± concealed in the foliage, outermost flowers opening first, resembling a large rosette; bracts usually large, often colored, caducous; bracteoles usually 2, colored, ± connate into a floral sheath longer than calyx, cleft on 1 side. **Flowers** red, rose, orange, rarely creamish-white or white; calyx tubular-campanulate, lobes 4, petaloid, subequal, imbricate; petals (3–4)5, ovate to oblong, subequal, clawed; stamens 10–15(–20), exserted; ovary pubescent, adnate to calyx tube, many-ovuled, style filiform. **Fruits** pods, oblong or elongate, compressed, woody or leathery, 2-valved, upper suture often thickened, elastically dehiscent; seeds few, transversely arranged, compressed. **Distr:** W South America (Venezuela, Colombia, Peru, Ecuador, Brazil, the Guianas), C America (Panama, Costa Rica), the Caribbean. **Ecol:** Understory in tropical lowland rainforests. **Cult:** Propagate from seeds, cuttings (often difficult) or by air-layering. Requires moist, well-drained soil in full or partial sun.

B. ariza Benth. *Pl. Hartw.* 171, 1845. (Syn: *B. princeps, B. rosa-demonte, Hermesias ariza*). [Ariza, Mountain Rose, Rose of Venezuela]. **Trees**, evergreen, to 6–9(–20) m tall; crown spreading; young branchlets rusty-villous. **Leaves** pinnate, 5–6(7–10)-jugate, to 45 cm long; leaflets elliptic-oblong to oblong, rarely ovate, 4.5–11 × 2.5–3.5 cm, basal ones usually

Above: *Brownea ariza*, *Brownea grandiceps*, *Brownea macrophylla*, *Senna alata*.

smaller, base obliquely rounded, apex very narrowly acuminate, surfaces glabrous; petiolules to 6 mm long. **Inflorescences** 9–10 cm Ø; peduncles and rachises 2–2.5 cm long; bracts broad-obovate, 2.5–4.5 cm long, velvety-pubescent outside; bracteoles spathaceous, 1.7–2.5 cm long, densely velvety-pubescent; pedicels ~5 mm long. **Flowers** with hypanthium 0.9–1.8 cm long; calyx lobes 2 cm long; petals 3.5–4 cm long, orange-red or red; stamens 3–3.5 cm long, filaments united at lower 0.7–1.7 cm; ovary ~1 cm long, style ~4 cm long. **Fruits** pods, 10–18 × 4–5 cm, compressed, rusty-velvety; seeds 4, elliptic-ovate, 5 × 2.5 cm, depressed. **Distr:** Colombia, Venezuela, Ecuador, Peru. **Ecol:** <500 m. Fl. & Fr. Jul, Sep. **Zones:** 11–12.

B. grandiceps Jacq., *Collectanea* 3: 287, 1791. [Handkerchief Tree, Rose of Venezuela]. **Trees**, to 16 m tall; young branchlets densely brown-hairy. **Stipules** 7–16 cm long. **Leaves** 2–14-jugate; petiole and rachises (of lower ones) 25–45 cm long; leaflets ovate, ovate-oblong, or lanceolate, 2.5–19.5 × 1–6 cm, base cordate, rounded or obtuse, apex acuminate, abaxially midrib glabrous or hairy. **Inflorescences** 14–18 cm long; peduncles and rachises 3–4.5 cm long; bracteoles 2.25–3 cm long; pedicels 0.9–1.3 cm long. **Flowers** with hypanthium 1.2–1.5 cm long; calyx lobes 2–2.7 cm long; petals 4.5–5.25 cm long, bright red; stamens ~6 cm long, filaments united at lower 1.2–1.5 cm; ovary ~1 cm long, style 3–4.5 cm long. **Fruits** pods, 12–24 × 4.5–6 cm, densely shortly hairy. **Distr:** Venezuela. **Ecol:** Fl. throughout the year. **Zones:** 11–12.

B. macrophylla Linden, *Cat.* no. 18: 11, 1863. [Monterillo, Panama Flame Tree, Rose of Venezuela, Tiger Rosewood]. **Trees**, small, to 7(–10) m tall; young parts tomentulose. **Leaves** pinnate; leaflets in 5–7 pairs, ± elliptic, 10–25 × 2.5–6 cm, base asymmetric, acute to subcordate, apex long-acuminate, abaxially scattered-puberulent, glaucescent, bearing a single globose gland ~1 mm Ø at base abaxially; petiole swollen, ~1 cm long, to 10 cm long on non-flowering stems; petiolules to 5 mm long, swollen. **Inflorescences** axillary or nearly terminal, densely flowered, capitate to shortly racemose; pedicels ~5 mm long; bracteolar sheaths bilobed, 3–4 cm long, ± enclosing calyx. **Flowers** with sepals usually 4, somewhat petaloid, obovoid, ~3 × 0.6–3 cm; petals 5, orange, obovate and clawed, 3–3.5 × 1–1.5 cm, base rounded and oblique, claw narrow, ~1 cm long; stamens 10–15(20), 8 cm long, united at base into a tube 2.5 cm long, tube open on upper side, wrapping around pistil at base; pistil narrow, 10 cm long, ovary tomentose. **Fruits** pods, flat, slightly turgid, oblong, to 20 × 4.5 cm, beaked, brownish-tomentose when immature, with heavy ridges along both edges; seeds flat, rectangular, 3.5 × 2.5 cm, black. **Distr:** Panama and northern S America. **Ecol:** In tropical moist forests. Fl. Nov–May. **Zones:** 11–12.

TRIBE CASSIEAE

Senna

Species ~295–300. **Herbs**, **shrubs**, or **small trees**, rarely some monocarpic. **Leaves** abruptly pinnate; leaflets opposite; rachis and petiole with or without glands. **Inflorescences** racemes, axillary or terminal; bracteoles absent. **Flowers** with sepals 5; petals 5, subequal, usually yellow; stamens 10, filaments of all stamens straight, subequal, sometimes all fertile, sometimes 3 adaxial stamens staminodal, (6 or)7 fertile, anther thecae not ciliate along suture. **Fruits** pods, either indehiscent or inertly dehiscent through 1 or both sutures, in the last case not coiling, sometimes valves breaking up in 1-seeded joints; seeds numerous, funicle filiform. **Distr:** Pantropical. **Ecol:** In arid scrublands to tropical forests. **Cult:** Propagate from seeds or cuttings.

S. alata (L.) Roxb., *Fl. Ind.* (Roxburgh) 2: 349, 1832. (Syn: *Cassia alata*). [Candle Bush, Candle Stick, Ringworm Shrub]. **Shrubs**, 1–5 m tall; stems marked with leaf scars. **Stipules** deltoid, 0.6–2 cm long, base auriculate, apex acute to acuminate, persistent. **Leaves** eglandular, 30–60 cm long; leaflets 6–24 pairs, oblong to obovate-oblong, 3–15(–19) × 2–7(–12) cm, base rounded to subcordate, apex rounded to retuse; petiole 1.5–2.5 cm long. **Inflorescences** terminal or axillary, erect, spicate-racemose, many-flowered, 15–70 cm long; bracts petaloid, yellow or orange, 1–3 cm long, enclosing the bud, caducous; bracteoles lacking; pedicels 0.4–1 cm long. **Flowers** orangish-yellow; sepals 1–2 × 0.4–0.6 cm, puberulent; petals obovate to spathulate, 1.5–2 × 1 cm; stamens 9 or 10, fertile, variable in size. **Fruits** pods, linear, 10–15 × 1.5–2 cm, dehiscent, chartaceous, septate, valves with a longitudinal wing ~0.5–1 cm wide; seeds 40–60, light or dark brown, rhomboid, 5.5–7 × 4.5–5.5 mm, compressed, transverse, faces with an elliptic areole, 2–3 × 0.5–0.9 mm. **Distr:** Tropical S America. **Ecol:** Fl. Oct–Dec. **Zones:** 10–12.

S. auriculata (L.) Roxb., *Fl. Ind.* ii. 349, 1832. (Syn: *Cassia auriculata*). [Tanner's Cassia]. **Shrubs** or **small trees**, 2–7 m tall. **Stipules** foliaceous, reniform-rotund, 0.7–2.2 cm long, with long apicules on either side, produced into a filiform point to 7 mm long, persistent. **Leaves** (5–)10–12 cm long, leaflets 4–13 pairs, bearing erect linear glands on rachis usually between leaflets of each pair; leaflets elliptic-oblong, 1–3.5 × 1–2 cm, base rounded, mucronate, apex rounded, adaxially glabrous. **Inflorescences** corymbose, few-flowered, but aggregated into large terminal panicles, axillary; bracts linear, attenuate, 5–7 mm long; pedicels 1.2–3 cm long. **Flowers** with sepals yellow, 0.9–1.2 cm long, 2 outer much smaller than the other 3, apex rounded, base pubescent; petals 1.2–2.5 cm long, bright yellow, orange-veined; stamens 10, 3 with long filaments and anthers, 4 medium, 3 upper reduced to staminodes. **Fruits** pods, linear-oblong, 7.5–12.5 × 1–1.5 cm, compressed, flat, undulate between seeds, stipe ~5 mm long, persistent style 1.5 cm long; seeds 12–20, brown, ovate-oblong, 7–9 × 4–5 mm. **Distr:** India, Sri Lanka. **Ecol:** Fl. Jan–Jul. **Zones:** 10–12.

S. didymobotrya (Fresen.) H.S.Irwin & Barneby, *Memoirs of the New York Botanical Garden* 35: 467, 1982. (Syn: *Cassia didymobotrya*). [African Senna, Popcorn Cassia, Peanut Butter Senna]. **Shrubs**, large, to ~4.5(–9) m tall; young stems puberulent. **Stipules** cordate-ovate, acuminate, ~1.5–2.5 × 0.8–1.2 cm, pubescent or subglabrous, ciliate, persistent. **Leaves** 10–30(–36) cm long, rachis produced into a short spur; leaflets in 7–18 pairs, elliptic-oblong, 2–6.5 × 0.6–2.5 cm, base rounded, apex obtuse or subacute, mucronate, sparsely pubescent. **Inflorescences** racemose, subspicate, axillary, to ~50 cm long, pedunculate; bracts ovate, 0.9–2.7 × 0.5–1.4 cm; pedicels 5–7 mm long. **Flowers** with sepals puberulent, 1.2 cm long, narrowly obovate; petals yellow with darker veins, 1.5–3 mm long, claw 1 mm long; stamens 10, 2 lower anthers large, fertile, 5 medium, fertile, and 3 upper small, staminodal. **Fruits** pods, stipitate, oblong, 7.5–12 × 1.5–2.5 cm, compressed, dehiscent, septate; seeds 9–16, compressed, sublenticular, 8–9 × 4–5 mm, 2.5 mm thick, apiculate, faces with an oblong areole 4 × 1.5 mm. **Distr:** Tropical Africa. **Ecol:** In savanna thickets, deciduous bushlands, gallery forests, lake shores, riverbanks and other damp localities; 0–2,500 m. Fl. & Fr. throughout most of the year. **Zones:** 10–12.

S. fruticosa (Mill.) H.S. Irwin & Barneby, *Mem. New York Bot. Gard.* 35: 121, 1982. (Syn: *Cassia fruticosa*, etc.). **Shrubs** or **small trees**, 2–8 m tall; young stems zigzag. **Stipules** linear to linear-oblanceolate, straight or falcate, 0.4–2 cm long, acute. **Leaves** 12–30 cm long, with 2 pairs of leaflets, bearing an ovate-elliptic, conical, stout gland on rachis between lower (smaller) pair of leaflets; leaflets elliptic or ovate, (2.5–)4–16(–22) × 1.7–7.2(–11) cm, inequilateral, apex acute to subacuminate, sometimes obtuse. **Inflorescences** shortly racemose, 5–35-flowered, (1.5–)2–7(–8.5) cm long; bracts ovate to lanceolate, 1–3 mm long, caducous; pedicels 2–5.5 cm long. **Flowers** with sepals, pale green, 0.8–1.3 cm long, base clawed, apex rounded; petals pale or dull yellow, obovate, (1–)2–3.3 cm long, puberulent; stamens 9 or 10 with 2 or 3 large, 4 medium, 3 reduced to staminodes; ovary and style incurved. **Fruits** terete, (10–)15–27(–36) × 0.9–1.5 cm, slightly thickened along margins, dehiscent along one suture, apiculate; seeds many, dark brown, subreniform, 3–5.5 × 2–3.5 mm, faces with a faint areole, 2.6–4.4 × 1.7–2.4 mm, embedded in sticky pulp. **Distr:** Tropical America. **Ecol:** In dry thickets; 500–2,900 m. Fl. Jun–Sep. **Zones:** 10–12.

JML

Above: *Senna auriculata, Senna didymobotrya, Senna fruticosa, Senna nitida.*

S. nitida (L.C. Rich.) Irwin & Barneby, *Mem. New York Bot. Gard.* 35: 159, 1982. (Syn: *Cassia antillana*, etc.). [Hediondilla]. **Shrubs**, clambering or scandent, 2–5(–10) m high; stems to 2.5 cm Ø, grayish, almost cylindrical; branches green, striate. **Leaves** alternate, pinnate; leaflets 4, opposite, ovate or elliptical, 6–12 × 2–4.5 cm, distal pair larger, base unequal or rounded, margins slightly undulate, revolute, apex acuminate, adaxially shiny, glabrous or puberulent with midvein sunken, abaxially glabrous, yellowish-green, dull, punctate; petiolules swollen; rachis with a linear stipitate gland between leaflets; petioles longer than rachis, sulcate, base swollen. **Inflorescences** racemes, axillary, toward the end of the branches. **Flowers** with sepals green, oblong or ovate, ~5 mm long; petals yellow rounded, 1–2 cm long, base narrowed; stamens 10, 6 fertile, anthers cylindrical, ~5–7 mm long; ovary ~1 cm long, hairy, style curved, ~1.5 cm long, hairy, stigma small. **Fruits** pods, almost cylindrical, 10–18 × 1–1.5 cm, compressed along the suture, fleshy, producing a yellow exudate, oily and rancid; seeds elliptical, ~1 cm long, flattened. **Distr:** Caribbean. **Ecol:** In forests, grasslands and moist places. Fl. Oct–Dec. **Zones:** 11–12.

S. occidentalis (L.) Link, *Handbuch* 2: 140, 1833. (Syn: *Cassia occidentalis*, etc.). [Arsenic Bean, Coffee Senna, Coffee Weed, Fetid Cassia]. **Herbs** or **undershrubs**, to 3 m tall, sometimes an annual. **Stipules** linear-lanceolate, 4–8 mm long, caducous. **Leaves** pinnate, 12–25 cm long; leaflets 3–6 pairs, opposite, lanceolate or ovate-lanceolate or elliptic, (1–)2.5–9(–12) × 1.2–4 cm, base rounded, margins ciliolate, apex acute or attenuate, glabrous. **Inflorescences** corymbose racemes, forming few-flowered terminal panicles, axillary; bracts ovate, oblique, caducous; pedicels ~0.5–1.2 cm long. **Flowers** yellow, 1.2 cm across; calyx ~1 cm long, divided to base, glabrous, oblong, obtuse, white, pink-tinged; sepals 0.6–1 cm long, obtuse, ± glabrous; corolla 1.2 cm long, oblong-obovate, 1–1.5 cm long, ± unequal, obtuse; petals orange-yellow, tinged or veined with purple; stamens 9–10, 1 sterile or lacking, 2 large, 4 medium; staminodes 3. **Fruits** pods, linear, slightly falcate, 0.55–1.25 × 0.6–0.9 cm, compressed, glabrous; seeds 20–30, suborbicular, compressed, 4.5–6.5 × 3.5–5 mm, shiny, pale-brown. **Distr:** Pantropical (probably of South American origin). **Ecol:** Fl. Oct–Mar. **Zones:** 10–12.

S. siamea (Lam.) H.S. Irwin & Barneby, *Mem. New York Bot. Gard.* 35(1): 98, 1982. (Syn: *Cassia siamea*, etc.). [Iron Wood Tree, Kassod Tree, Thailand Shower]. **Trees**, 10–15 m tall; young branches ribbed. **Stipules** caducous, linear, minute. **Leaves** 20–30 cm long; rachis and petiole without glands, puberulous; leaflets 6–10(–15) pairs, oblong or ovate-oblong, 3–7 × 1.5–2.5 cm, base rounded, apex obtusely rounded, often emarginate, mucronate, leathery, adaxially smooth and glabrous, abaxially finely pubescent. **Inflorescences** racemes, in axils of apical leaves, often several forming a large terminal panicle; peduncle robust, 5–7 cm long; bracts linear, 5–6 mm long. **Flowers** with sepals suborbicular, unequal in size, outer smaller, inner larger, ~9 mm long, outside hairy; petals yellow, broadly obovate, 1.2–1.5 cm long, shortly clawed; stamens 10, 7 fertile, anthers opening by apical pores, 2 with filaments ~1 cm long, anthers 6–7 mm long, 4 or 5 with filaments 2–3 mm long, anthers 5–6 mm long, staminodes ~3 mm long; ovary sessile, densely white-pubescent. **Fruits** pods, 15–30 × 1–1.5 cm, flattened, suture thick, riblike, pubescent, purplish-brown; seeds 10–30, light brown, ovoid, 2–2.5 × 5–6 mm. **Distr:** Myanmar. **Ecol:** In lowland tropical regions with monsoon climates, and also in lowland semiarid conditions; 500–2,800 m. Fl. throughout the year. **Zones:** 10–12.

S. spectabilis (DC.) H.S. Irwin & Barneby, *Mem. New York Bot. Gard.* 35: 600, 1982. (Syn: *Cassia spectabilis*, etc.). [Candelillo, Carnaval, Golden Shower]. **Trees**, evergreen, small, 5–7(–10) m tall; branches long, spreading; young parts densely rusty-pubescent. **Leaves** 12–30 cm long, pinnate; leaflets (4–)8–15(–19) pairs, elliptic or oblong-lanceolate, 3–7 × 1–2 cm, lowermost pair usually much smaller and early caducous, base subrounded and slightly oblique, apex acute, cuspidate, adaxially glabrous or puberulous, abaxially densely yellowish-brown velutinous; rachis and petiole without glands. **Inflorescences** terminal panicles or axillary racemes; peduncles and pedicels densely yellowish-brown velutinous; bracts caducous, ovate, 4–5 mm long. **Flowers** fragrant; sepals 5, orange-yellow, unequal (outer 2, ovate, 5–7 mm long, pubescent, 3 inner suborbicular, 0.8–1.1 cm Ø); petals yellow, brownish-yellow dark-veined, unequal, obovate, shortly clawed; stamens 10, 7 of them fertile, staminodes 3, smaller; ovary linear, ~3 cm long, recurved, style and stigma inconspicuous. **Fruits** pods, pendulous, black, narrowly cylindric, 25–35 × 1–1.5 cm, slightly compressed, slightly contracted between seeds, annulate-septate, glabrous, tardily if at all dehiscent; seeds 50–70, suborbicular, ~5 mm Ø. **Distr:** Bolivia; Ecuador to C America and Trinidad. **Ecol:** In dry, coastal regions and secondary woodlands; 0–2,000 m. Fl. throughout the year. **Zones:** 10–12.

S. surattensis (Burm. f.) H.S. Irwin & Barneby, *Mem. New York Bot. Gard.* 35, 1982. (Syn: *Cassia surattensis*, etc.). [Glaucous Cassia, Glossy Shower, Scrambled-egg Tree]. **Shrubs** or **small trees**, to ~7 m tall; young stems puberulent or glabrous. **Stipules** linear-falcate, puberulent, 0.6–1.5 cm long, subpersistent. **Leaves** 14–30 cm long, bearing clavate glands between the leaflets of the 2–4 lowermost pairs; leaflets 4–7 pairs, ovate to ovate-oblong, 3–9 × 1.2–5 cm, base rounded to cuneate, apex emarginate, obtuse, adaxially glabrous, abaxially sparsely pubescent. **Inflorescences** racemose, 10–20-flowered, 5–13 cm long; bracts ovate-acute, 2.5–8 mm long, caducous or subpersistent; pedicels 2.5–4 cm long. **Flowers** with unequal sepals, to 8 mm long, ciliolate, otherwise glabrous; petals yellow, 2–3 cm long, puberulent outside; stamens 10, all fertile, 3 with longer filaments, 7 shorter. **Fruits** pods,

Below: *Senna occidentalis, Senna siamea, Senna spectabilis.*

Above: *Senna surattensis, Senna tora, Cassia brewsteri, Cassia fistula.*

15–20 cm long, compressed, glabrous, dehiscent; seeds 20–35, oblong-elliptic, 0.6–1 × ~0.4 cm, blackish, glossy. **Distr:** SE Asia. **Ecol:** In teak forests, open forests, marshy areas and waste places; to 300 m. Fl. throughout the year. **Zones:** 10–12.

S. tora (L.) Roxb., *Flora Indica* (ed. Carey) 2: 340, 1832. (Syn: *Cassia tora*, etc.). [Fetid Cassia, Sickle Senna]. **Herbs**, suffrutescent, erect, annual, 1–2 m high. **Stipules** linear, 1–1.5 cm long, caducous. **Leaves** 4–8 cm long, pinnate; leaflets (2)3 pairs, obovate or obovate-oblong, 2–6 × 1.5–2.5 cm, base cuneate to rounded, oblique, apex rounded, cuspidate, membranous, adaxially sparsely pubescent, abaxially pubescent; petiole without glands; rachis with a club-shaped gland between leaflets petiolules 1.5–2 mm long. **Inflorescences** racemes, 1- or 2(or 3)-flowered, axillary, short; peduncles 0.6–1 cm long; bracts linear, acute; pedicels 1–1.5 cm long. **Flowers** with sepals ovate or ovate-oblong, 5–8 mm long, membranous, outside pubescent; petals yellow, unequal, obovate, lower 2 slightly longer, 1.2–1.5 × 0.5–0.7 cm, shortly clawed; fertile stamens 7, nearly equal, filaments 1.5–2 mm, anthers opening by apical pores, ~4 mm long, 3 larger, 4 smaller, staminodes absent; ovary sessile, densely white-pubescent, style glabrous. **Fruits** pods, terete, subtetragonous, slender, 10–20 × 0.3–0.5 cm, both ends acuminate, valves membranous; seeds 20–30, glossy, rhomboid, ~5 × 3 mm, with an areole. **Distr:** Tropical America. **Ecol:** On mountain slopes, wastelands and riverbanks; 0–600 m. Fl. & Fr. throughout the year. **Zones:** 9–12.

Cassia

Species ~600. **Trees**, **shrubs**, less frequently **herbs**. **Stipules** of various shapes. **Leaves** paripinnate (abruptly pinnate) or rarely reduced to phyllodes; petiolar glands often present. **Inflorescences** axillary or terminal racemes, terminal panicles or flowers subsolitary axillary; bracts and bracteoles variable. **Flowers** with calyx 5, imbricate, tube short; petals 5, imbricate, subequal or lowermost larger; stamens 10, all perfect or subequal or the upper ones smaller or imperfect or only 5 stamens, anthers uniform or upper ones smaller, dehiscing by an apical pore or a short slit; ovary sessile or stipitate, ovules numerous, style with terminal stigma. **Fruits** variable, terete or compressed, membranous or woody, often 2 valved, sometimes indehiscent, rarely winged lengthwise; seeds mostly compressed, endospermous. **Distr:** Mostly in tropics and subtropics worldwide. Propagate from seeds or cuttings. Requires moist, but well-drained soil in full sun or partial shade.

C. brewsteri (F.Muell.) Benth., *Flora Australiensis* 2, 1864. [Cassia Pea, Leichardt Bean]. **Shrubs** or **trees**, deciduous, to 30 m tall. **Leaves** alternate, pinnately compound; leaflets 4–16, oblong to lanceolate, 2–11 × 1.5–3 cm, base wedge-shaped or acute, apex rounded or slightly notched, adaxially glabrous, abaxially glabrous or minutely hairy; rachis and petiole 10–25 cm long; petiolules 2–4 mm long. **Inflorescences** racemes, axillary or terminal, drooping or pendent. **Flowers** with calyx green or yellow; petals 5, yellow, pink or orange, 0.5–1.5 cm long. **Fruits** pods, indehiscent, woody, cylindrical, 15–45 × 1–3 cm, purplish or bluish to almost black, with reddish thickened sutures; seeds numerous, yellowish or brown, 0.8–1 cm long, within leathery valves. **Distr:** Australia (Queensland). **Ecol:** In rainforests, vine thickets and open forests; 200–600 m. Fl. Sep–Nov. Fr. Nov, Apr. **Zones:** 9–12.

C. fistula L., *Sp. Pl.* 377, 1753. [Golden Shower Tree, Pudding Pipe Tree, Indian Laburnum, Indian Senna]. **Trees**, to 20 m tall. **Stipules** deltoid, 1–2 mm long, acute, caducous. **Leaves** pinnately compound, 22.5–40 cm long; rachis 12–25 cm long, terete, glabrous; leaflets opposite, 3–8 pairs, 6–10(20) cm long, adaxially smooth, abaxially hairy. **Inflorescences** racemes, drooping, each raceme ~10–45 cm long; peduncle 2–10 cm long; pedicels 3.7–5.8 cm long, slender; bracts ovate, 0.8–1 cm long, acute, hairy. **Flowers** with calyx 5, green, folded backward on the stalk, hairy, ovate, 9 mm long; petals 5, obovate, blunt, distinctly veined; stamens 10, 3 longest stamens much curled and bear large oblong anthers, 4 smaller medium ones are quite straight, 3 remaining stamens are quite short, erect and sterile; ovary slender, thinly hairy, style sturdy, stigma punctiform. **Fruits** pods, terete, 40–60 × 1.5–2 cm, glabrous, indehiscent, glossy blackish-brown; seeds 40–100. **Distr:** W Pakistan, India, Myanmar, Sri Lanka. **Ecol:** In deciduous forests; to 1,200 m. Fl. Apr–Jun. **Zones:** 10–12.

C. grandis L.f., *Suppl.* 230, 1781. [Pink Shower, Pink Coral Shower, Horse Cassia, Stinking Toe] **Trees**, to ~30 m tall; young stems tomentulose. **Stipules** falcate or subulate, 0.5–1.5 mm long, caducous. **Leaves** shortly petiolate, eglandular, 15–25 cm long, pinnate; leaflets 8–20 pairs, deciduous, elliptic-oblong, 2.5–6 × 1.4–2.2 cm, ends rounded, abaxially puberulent or tomentulose, glabrescent. **Inflorescences** simple-racemose, many-flowered, lateral, 10–20 cm long, usually appearing when tree is ± leafless; bracts and bracteoles ovate, ~5 mm long, acute, tomentose, caducous; pedicels 1–2 cm long. **Flowers** with sepals obovate, 6–9 mm long, obtuse, tomentulose; petals pink or white, orbicular to obovate, ~1–1.5 cm long; stamens 10, unequal with 3 lower filaments ~2 cm long, sigmoid, others shorter, anthers pilosulous, the 3 larger opening by apical and basal pores or slits, 5 on shorter filaments opening by apical and basal pores, 2 smaller and abortive. **Fruits** pods, ± terete, (25–)45–80 × 2–5.5 cm, transversely rugose, indehiscent, 1 margin 1-ribbed, the other 2-ribbed; seeds 70–80, obovoid-ellipsoid, 1.4–1.6 × 0.9–1 cm, transverse, brown, lustrous, embedded in pulp. **Distr:** Tropical America. **Ecol:** In open, brushy or forested hillsides, or on thinly forested plains; to 900 m. Fl. & Fr. throughout the year. **Zones:** 10–12.

C. javanica L., *Sp. Pl.* 1: 379, 1753. [Apple Blossom Tree, Java Cassia, Pink Shower Tree, Rainbow

Below: *Cassia grandis, Cassia javanica, Cassia moschata, Cassia roxburghii.*

Above: *Cassia roxburghii* (flowers), *Haematoxylum campechianum*, *Caesalpinia bonduc*, *Caesalpinia coriaria*.

Shower Tree]. **Trees**, small to medium-sized, deciduous, to 12+ m tall; trunk to 30+ cm dbh; crown spreading; branchlets spine-tipped. **Leaves** alternate, even-pinnate, 20–42 cm long, usually in 2 rows; leaflets 6–15 pairs, regularly arranged, ± equal in size, ± glabrous, ovate to oval or oblong, 4–5.6 × 1.6–2.5 cm, apex obtuse or shortly acute, surfaces minutely hairy; adaxially dull green, abaxially gray-green; petiolules ~1.5 mm long. **Inflorescences** umbellate-corymbs, lateral, 12.5–23 cm long; pedicels finely hairy, 2.5–4 cm long, slightly curved upward. **Flowers** large, 5 cm Ø, pink, fragrant; sepals 5, concave, 6 mm long, pointed, dark red, green-tinged, finely hairy; petals 5, spreading, oblong, 2.5–3.2 cm long, base clawed, apex rounded, minutely hairy, pink, fading to white with age, reddish-veined; stamens 10, yellow, swollen beyond middle, 7 ~1.3 cm long; style slender, 3.2 cm long, curved, reddish. **Fruits** pods, terete, 30–60 × 1.6 cm, dark brown. **Distr:** N India through Indo-China, and Malesia to New Guinea. **Ecol:** Fl. May–Jul. **Zones:** 9–12. **Info:** This species is highly variable across its range, with a few named varieties.

C. moschata H.B.K., *Nov. Gen. & Sp.* 6: 338, 1824. [Orange Cassia]. **Trees**, to ~20 m tal1; young stems pilose, glabrescent. **Stipules** 1.5–4.5 mm long, auriculate below point of attachment, caducous. **Leaves** eglandular, pinnate; leaflets ~10–20 pairs, opposite or subopposite, deciduous, lance-oblong to ovate-oblong, 2.5–5 × 0.8–2 cm, base rounded to subcordate, margins somewhat revolute, apex obtuse, emarginate or subacute, puberulent. **Inflorescences** racemose, 25–70-flowered, terminal or axillary, ~(6)9–32 cm long; bracts and bracteoles lance-subulate, to ~3 mm long, caducous; pedicels (0.5–)0.6–1.2(–1.4) cm long. **Flowers** with sepals ovate-orbicular, obtuse, (5.5–)6–8(–8.5) mm long; petals yellow with red venation, 1–1.6 cm long; stamens 10, unequal with 3 filaments 1.5–2.5 cm long, sigmoid, anthers 3–4 mm long, other stamens with shorter filaments, 3 ± rudimentary. **Fruits** terete, 30–60 × 1.2–2 cm, septate and pulpy within, pulp with a musky odor; seeds ~75–100, ellipsoid, 7–8 × 5–6 mm, brown, lustrous. **Distr:** Tropical America. **Ecol:** In secondary forests; ~500 m. Fl. Feb–Sep. **Zones:** 10–12.

C. roxburghii DC., *Prodr.* 2: 489, 1825. (Syn: *C. marginata*). [Ceylon Cassia, Red Cassia]. **Trees**, 5–10 m tall; young stems tomentulose, glabrescent; branches slender, downward curving to pendent. **Stipules** falcate-reniform, ~5 mm long, pubescent, caducous. **Leaves** 13–35 cm long, eglandular; leaflets ~8–20 pairs, ovate to elliptic-oblong, 1.5–4.5 × 1–2 cm, base asymmetrically rounded, apex obtuse or retuse, adaxially puberulent or glabrate, abaxially appressed-pubescent. **Inflorescences** racemes or corymbs, axillary; peduncle 1–2 cm long; axis 5–9 cm long; bracts sericeous, ovate, ~3–8 × 4–5 mm, acute, caducous; bracteoles basal, oblong, 3–4 × 1 mm, acute; pedicels 1–2.5 cm long. **Flowers** with calyx hairy; sepals ovate, 3–7 × 3–4 mm, obtuse, puberulent; petals pink to salmon or rose, turning yellowish or orange, oblong-obovate, 1–1.5 × 0.5–0.8 cm, hairy outside; stamens 10, 3 lower filaments 2.5–3 cm long, others shorter, 3 stamens with reduced filaments and anthers, not swollen in middle. **Fruits** pods, cylindrical (terete), 20–32 × 2 cm, indehiscent, glabrous, black, many-seeded. **Distr:** S India, Sri Lanka. **Ecol:** Fl. May–Jun. **Zones:** 10–12.

TRIBE CAESALPINEAE

Haematoxylum

Species 3. **Shrubs** large, or **trees**; wood colored. **Stipules** either small and deciduous or spiny. **Leaves** paripinnate or bipinnate; leaflets in few pairs, obovate. **Inflorescences** racemes, axillary; bracts minute, bracteoles absent or caducous. **Flowers** with calyx segments 5, subequal, imbricate; petals 5, yellow, imbricate; stamens 10, free, filaments hairy at base, anthers uniform; ovary shortly stipitate, 2–3-ovuled, style filiform, stigma minute, terminal. **Fruits** pods, lanceolate, membranous, dehiscent, 2–3 seeded; seeds oblong. **Distr:** Mexico, through C America to Tropical S America; West Indies; S Africa. **Cult:** Propagate from seeds or cuttings. Requires a moist, moderately fertile soil in shade or full sun.

H. campechianum L., *Sp. Pl.* 384, 1753. [Bloodwood Tree, Logwood]. **Shrubs** or small **trees**, to 5 m tall; stem compressed and fluted; branches irregular; bark smooth, light gray, armed with stout spines. **Leaves** abruptly paripinnate, 2.5–7.5 cm long; leaflets 3- or 4-jugate, ~4 pairs, broadly obovate to wedge-shaped, 1.2–2.5 cm long, apex obtuse, emarginate, glabrous, nerves numerous, parallel. **Inflorescences** racemes, ~5–10 cm long. **Flowers** ~1.2 cm across, pale yellow; calyx 4 mm long; petals 5, oblong-obovate, 6 mm long; stamens free, filaments hairy at base; ovary shortly stiped. **Fruits** pods, ~5 × 0.8–1.2 cm, narrowed at both ends, dehiscent, membranous; seeds 2–3. **Distr:** C America, West Indies. **Ecol:** In lowland forests and thickets. **Zones:** 10–12. **Info:** The wood is the source of the Logwood dye.

Caesalpinia

Species ~280. **Trees**, **shrubs**, **herbs** or woody **lianas**, perennial; smooth or somewhat armed. **Stipules** if present, simple or pinnate. **Leaves** large, bipinnate; leaflets variable in number, size and texture. **Inflorescences** in lax racemes, borne in upper axils, or panicles at branchlets tips; bracts small or rarely large and membranous; bracteoles absent. **Flowers** usually yellow or red, often showy; calyx tube short, lower 5 imbricate or subvalvate or valvate segments often larger and concave or cymbiform, sometimes subequal; petals 5, orbicular-spathulate, rarely oblong, sometimes dissimilar; stamens 10, filaments often villous or glandular, anthers uniform, longitudinally dehiscent; ovary shortly stiped, free, style filiform. **Fruits** pods, usually compressed or flat but sometimes turgid, dehiscent or indehiscent, suture nerviform or thicker, albuminous or exalbuminous seeds often separated by tissue. **Distr:** Tropics and subtropics worldwide. **Cult:** Propagate from scarified seeds or cuttings. Requires any moist soil and is salt tolerant.

C. bonduc (L.) Roxb., *Fl. Ind.* 2: 362, 1832. (Syn: *C. crista, Guilandina bonduc, G. bonducella*). [Molucca Bean, Nickerbean, etc.]. **Shrubs**, climbing or scrambling, to over 5 m in height, slightly pubescent, very prickly all over; prickles on branchlets and leaves recurved, 1–6 mm long. **Stipules** apparently absent. **Leaves** pinnate, 20–40 cm long; leaflets 5–10 pairs, elliptic to oblong or ovate, 2–5 × 1–2.5 cm, apex acute to obtuse with a mucro, adaxially lustrous, abaxially puberulent; petiolules 0.5–1 cm long. **Inflorescences** racemes, narrow, elongate, often branched, 10–20 cm long, axillary, tomentulose; pedicels only a few mm long, jointed. **Flowers** yellowish; sepals 6 mm long, pubescent; petals spathulate, 1–1.2 cm long, yellowish; stamens shorter than petals, ~1 cm long, puberulent. **Fruits** pods, suborbicular, ovate or oblong-elliptic, 3–10 × 2.5–5 cm, compressed, usually bristly; seeds 1 or 2, ovoid to subglobose, gray or lead-colored. **Distr:** Tropical Asia to Australia. **Ecol:** Near waterways and seashores. **Zones:** 11–12.

C. coriaria (Jacq.) Willd., *Sp. Pl.* 2: 532, 1799. (Syn: *Libidibia coriaria, Poinciana coriaria*). [Ebano]. **Trees**, unarmed, to 10+ m tall; crown broad; branches crooked; branchlets puberulent or glabrous. **Leaves** small, bipinnate, 15 × 12 cm; pinnae 6–7 with an apical pinna, 5–6 cm long; leaflets 12–28, linear-oblong, 4–8 mm long, base cuneate, subcordate or truncate, apex obtuse to rounded, glabrous, adaxially green, abaxially paler, usually black-dotted; petiole and rachis tomentulose. **Inflorescences** racemes, few-flowered, shorter than leaves, 6–9 cm long; pedicels 2–4 mm long, glabrous. **Flowers** small, ~1 cm long; calyx glabrous, sepals 5, 5 mm long; petals white or yellowish, about equaling sepals, 4 petals equal, clawed, rounded; stamens 10, distinct, ~7–8 mm long, exserted to ~4 mm, filaments subulate, internally villous, anthers ovate; ovary compressed, glabrous, style slightly curved, colored, stigma obtuse. **Fruits** pods, twisted, very irregular in size and shape, 3–6 × 1–2 cm, thick and fleshy or coriaceous, indehiscent; seeds 8. **Distr:** W & S Mexico, Salvador to Panama; West Indies; Colombia and Venezuela. **Ecol:** On dry plains and hillsides; <900 m. **Zones:** 10–12.

C. decapetala (Roth) Alston, *Handb. Fl. Ceylon* 6: 89, 1931. (Syn: *C. sepiaria*, etc.). [Cat's Claw,

AC

Above: *Caesalpinia decapetala, Caesalpinia ferrea, Caesalpinia gilliesii, Caesalpinia pulcherrima.*

Mauritius Thorn, Mysore Thorn, Shoofly, Wait-a-bit]. **Shrubs** or **woody climbers**, to ~25 m tall; stems initially densely pubescent, armed with recurved prickles. **Stipules** recurved or obliquely semicordate, (0.6–)0.8–1.3 × 0.3–0.7 cm, acuminate, caducous. **Leaves** bipinnate, main rachis to ~47 cm long, armed with recurved prickles; pinnae 3–10 pairs, each pinna with 5–12 pairs of leaflets; leaflets elliptic to elliptic-oblong, essentially symmetrical with central costa, ~1–2.5 × 0.5–1 cm, base and apex rounded, finely puberulent. **Inflorescences** panicles, many-flowered, axillary and terminal; bracts ovate-deltoid to lanceolate, 4–8 × 2–2.5 mm, acuminate, pubescent, caducous; pedicels (1.5–)2–3(–3.5) cm long, tomentulose. **Flowers** 1.5–2 cm long; petals 1–1.5 cm long, bright yellow to yellowish-white; ovules 8–10. **Fruits** pods, sessile, elliptic-oblong, 6–12 × ~2–3 cm, 1–1.5 cm thick, beak 1–2 cm long, dehiscent, puberulent; seeds 4–9, black, or brown and tan-mottled, ellipsoid, ~0.8–1.2 × 0.6–0.8 cm, 3–4 mm thick. **Distr:** Tropical and subtropical Asia. **Ecol:** On mountain slopes, ravines, plains, hills and riversides; to 1800 m. Fl. & Fr. Apr–Oct. **Zones:** 10–12.

C. ferrea Mart., *Reise Bras.* (Spix & Mart.) ii. 611, 1828. [Brazilian Iron Wood, Leopard Tree]. **Trees**, deciduous, to 30 m tall; trunk 40–60 cm dbh; trunk 40–60 cm dbh; bark distinctively colored, smooth, fine and peeling, leaving irregular white and dark patches. **Leaves** bipinnate, 10–20 cm long, bright green; pinnae ~7, to 4.5 cm long; leaflets oblong, 1.5–2.5 × 0.6–1 cm, bases rounded or cuneate, often unequal, apices rounded, rarely notched, coriaceous, adaxially glossy dark green, abaxially paler. **Inflorescences** panicles, ~10 cm long, borne above the foliage; pedicels 5 mm long. **Flowers** golden-yellow, ~1 cm long; calyx cupular, 3 × ~3.2 mm; corolla 0.7 × ~1 cm; petals free, spreading or not, obovate, ~7 × 2.4 mm, rounded or sometimes notched; stamens ~7.5 mm long, exserted to ~4 mm. **Fruits** pods, indehiscent, dark-colored, hard, dry, 5–10 × ~2 cm, flattened, one margin wavy. **Distr:** Brazil. **Ecol:** Fl. Nov–Feb. Fr. Jul–Sep. **Zones:** 10–12.

C. gilliesii (Hook.) Wall. ex Hook. *Bot. Misc.* 1: 129, 1829. (Syn: *Erythrostemon gilliesii*, etc.). [Bird of Paradise, Paradise Caesalpina, Paradise Poinciana]. **Shrubs** or **trees**, 1–5 m tall; branches smooth or warty, prickles absent; branchlet tips, pedicels and calyces conspicuously glandular. **Leaves** bipinnate, 10–20 cm long; pinnae usually 7–14 pairs; leaflets usually 6–10 pairs per pinna, elliptic to elliptic-oblong, 0.5–1 × 0.2–0.3 cm, apices slightly acute, glabrous, membranous. **Inflorescences** racemes, simple, many-flowered, terminal, glandular-pubescent. **Flowers** with petals obovate, 2–4 cm long, clawed, light yellow with orange markings; stamens bright red, to 7.5 cm long, long-exserted. **Fruits** oblong, 6.5–12.5 × 1.2–2 cm, curved, flat, yellowish-tan, conspicuously dark-dotted, margins pubescent; fruiting pedicels to 2.5 cm long. **Distr:** Argentina, Uruguay. **Ecol:** On mountain slopes. Fl. throughout the year. **Zones:** 9–12.

C. pulcherrima (L.) Swartz, *Obs. Bot.* 166, 1791. (Syn: *Poinciana pulcherrima*). [Bird of Paradise Flower, Peacock Flower, Pride of Barbados]. **Shrubs** or **small trees**, to >4 m tall, sparsely prickly, on older stems, prickles to 1 cm long. **Leaves** bipinnate; pinnae 5–9 pairs, opposite; rachis and pinnular rachis bearing a minute, acicular, gland; leaflets usually 9–11 pairs per pinna, ± opposite, oblong or obovate, ~1.5(–2.3) × 0.7(–1) cm, base obtuse to rounded, apex rounded or emarginate; petiole to 5(–8) cm long. **Inflorescences** subcorymbose racemes, elongate, axillary, ± terminal; bracts caducous, subulate, to 3 mm long; pedicels 2–6 cm long. **Flowers** yellow, orange, red or pink; calyx tube 3–4 mm long, lobes 5, obovate, to 1.5 cm long; petals 5, obovate, clawed, spreading, ~2 cm long, 1 reduced, ~5 mm wide; stamens 10, ~4 cm long, unequal, filaments villous at base; style exserted, ~2 cm beyond stamens. **Fruits** pods, flat, 9–12 cm long, beaked, elastically dehiscent, brownish-black; seeds ~6, ovate, ~8 mm long, brown. **Distr:** Origin unknown, either from S America or tropical Asia. **Ecol:** Fl. Apr–Sep. **Zones:** 10–12.

C. sappan L., *Sp. Pl.* 1: 381, 1753. [Brazilwood, Sappanwood]. **Trees**, small, to 6 m tall, ± puberulous; branches densely lenticellate. **Leaves** pinnate, 30–45 cm long; pinnae 7–13 pairs, opposite, 8–12 cm long; leaflets 10–17 pairs, closely spaced, mostly oblong, 1–2 × 0.5–0.7 cm, base oblique, apex retuse or rounded, papery. **Inflorescences** panicles, terminal or axillary, about as long as leaves; bracts caducous, lanceolate, large; pedicels ~1.5 cm long, puberulous; receptacle shallowly campanulate. **Flowers** with sepals 5, slightly unequal, lower one larger than others, cucullate; petals yellow, broadly obovate, ~9 mm long, uppermost one entire at apex, base pink-tinged, clawed; stamens slightly exserted, filaments densely pubescent below; ovary grayish-velutinous, 3–6-ovuled, style hairy. **Fruits** pods, reddish-brown, shiny, suboblong to oblong-obovoid, ~7 × 3.5–4 cm, beaked; seeds 3 or 4, light brown, oblong. **Distr:** Cambodia, China, India, Laos, Malaysia, Myanmar, Sri Lanka, Taiwan, Vietnam; Africa; America. **Ecol:** Fl. May–Oct. Fr. Jul–Mar. **Zones:** 9–12.

C. spinosa (Molina) Kuntze, *Rev. Gen.* 3, pt. 2: 54, 1898. (Syn: *C. tinctoria, Poinciana spinosa*, etc.). [Chestnut, Spiny Holdback, Tara, Wattle]. **Shrubs**, sturdy, or **trees**; branchlets spreading, spinose, densely leafy. bark gray. **Leaves** pinnate, smooth or sparsely and shortly prickly; pinnae 2–3 pairs, usually 10 cm long; leaflets ~8 pairs, subsessile, oblong-elliptic, usually ~2.5 × 1 cm, base oblique, firm, glabrous, venation reticulate, abaxially obscurely puberulent, apex rounded; rachis puberulent. **Inflorescences** racemes, narrow, 8–12 cm long, pedicels 5 mm long, puberulent, articulate below the short calyx tube. **Flowers** reddish-yellow; calyx segments large, ~6 mm long, serrulate; petals ~1.2 cm long; stamens ~1.2 cm long. **Fruits** pods, often red or reddish. **Distr:** Lima. **Ecol:** Among rocks and irrigation ditches; at low altitudes. **Zones:** 10–12. **Info:** The dried fruits are used to dye fabric.

Moullava

Species 1. *Generic description same as specific.* **Cult:** Propagate from seeds or rooted cuttings. Requires a moist but well-drained soil in full or partial sun. A pergola or similar support is recommended.

M. spicata (Dalzell) Nicolson, *Bot. Hist. Hort. Malabaricus* 184, 1980. (Syn: *Caesalpinia mimosoides, Wagatea spicata*). [False Thorn, Rat Bean]. **Climbers**, lianas or scrambling shrubs, robust, spiny, armed (even on the inflorescences) with very sharp, small, curved prickles. **Leaves** bipinnate, 22–30 × 22–30 cm; pinnae 8–12, 15–22 cm long; leaflets 12–14, nearly sessile, ovate to oblong, 2.5–5 cm long, ± unequal, apex obtuse, or subacute, rigidly subcoriaceous, glabrous, venulose. **Inflorescences** subspicate racemes, dense, long, simple or panicled, 30–40 cm long; axis thick, furrowed, woody, finely downy; bracts minute, lanceolate-cuspidate. **Flowers** small, bright orange, never fully opening, apices of petals appearing slightly above the scarlet calyx; calyx and corolla tubular and connivent; calyx 1 cm long, orange to scarlet, 5-lobed, tubular to cup-shaped; corolla fleshy, petals 5, yellow to bright orange, equal in size; stamens 10, fertile, filaments subulate, base pilose; ovary tomentose, ovules 4–6, style filiform, stigma hollow, 2-lipped, fimbriate. **Fruits** pods, linear, coriaceous, margins thick, constricted around the seeds; seeds 3–4, obovate-oblong, not compressed, radicle obtuse. **Distr:** India (Western Ghats). **Ecol:** In seasonally dry tropical semievergreen forest margins. **Zones:** 10–12. **Info:** The seeds are enveloped in a transparent very tenacious gum, which can be easily drawn into long fine threads.

Schizolobium

Species 1. *Generic description same as specific.* **Cult:** Propagate from seeds or cuttings. Requires a well-drained soil in full sun.

S. parahyba (Vell.) S.F. Blake, *Contr. U.S. Natl. Herb.* 20: 240, 1919. (Syn: *Caesalpinia parahyba, Cassia parahyba*, etc.). [Brazilian Fern Tree, Tower Tree]. **Trees**, medium-sized to large, to 30 m tall, unarmed; trunk to 50+ cm dbh, often with high buttresses; crown wide, usually rather flat-topped. **Leaves** very large, bipinnate, fernlike, on young trees to >1 m long; pinnae variable, usually ~20; leaflets 15–20+ pairs, narrowly oblong, 2–3.5 cm long, base obtuse, apex obtuse or rounded, mucronate, adaxially green, abaxially paler, glabrous, lustrous,

MD

Above: *Caesalpinia pulcherrima* (yellow form), *Caesalpinia sappan*, *Caesalpinia spinosa*, *Moullava spicata*.

densely sericeous; petiole bearing a large conic gland below apex. **Inflorescences** panicles, very large, terminal; pedicels elongate, articulate above middle. **Flowers** pale or bright yellow; calyx sparsely strigillose with blackish hairs, 1 cm long, lobes obtuse; petals 1.5–2 cm long, narrow; stamens about equaling petals. **Fruits** pods, flat, 8–12 × 2.5–5 cm, base long-attenuate, apex rounded, glabrous, coarsely reticulate-veined, somewhat rugose; seed oval, 2 × 1.2 cm, strongly compressed, blackish-brown. **Distr:** Tropical America. **Ecol:** In wet to dry, mixed forests, on plains or hillsides, often in secondary growth; <900 m. **Zones:** 10–12.

Peltophorum

Species ~15. **Trees**, tall, evergreen and deciduous. **Stipules** caducous. **Leaves** glossy, bipinnate, to 45 cm long; leaflets 15 pairs, small, to 1.2 cm long. **Inflorescences** racemes or panicles, terminal; bract generally caducous. **Flowers** often fragrant, yellow; calyx segments 5, imbricate, subequal; petals 5, slightly unequal, imbricate; stamens 10, free, filaments hairy at base, anthers uniform; ovary sessile, 2–many-ovuled, style subulate, stigma peltate. **Fruits** oblong, flat, indehiscent, brownish-black; seeds 1–4, compressed. **Distr:** Africa, Asia, the Americas, N Australia. **Ecol:** Tropical savanna and coastal forests. **Cult:** Propagate by air-layering or from scarified seeds. Requires moist but well-drained soil in full sun.

P. pterocarpum (DC.) Backer ex K. Heyne, *Nutt. Pl. Ned.-Ind.* ed. 2. 2: 755, 1927. (Syn: *P. inerme*, *P. roxburghii*, etc.). [Copperpod, Flame Tree, Yellow Flamboyant Tree]. **Trees**, large, to 12–24 m tall; bark smooth, gray. **Leaves** abruptly bipinnate, deep green; main rachis 15–30 cm long; pinnae 6–20 pairs; leaflets 20–30 pairs per pinna, oblong, 2 cm long, apex notched and unequal sided. **Inflorescences** panicles or racemes, large, terminal, to 45 cm long. **Flowers** 4 cm across, golden-yellow; calyx coppery-red; petals 5, wavy or frilled, yellow with a central brownish-red mark, inversely oval, hairy at base and much crinkled near the margins; stamens 10, densely hairy at base, anthers golden-yellow; style long, threadlike. **Fruits** pods, copper-red, oblong, 5–10 × ~2.5 cm, flat, thin, hard, ends narrowed; seeds brown. **Distr:** Sri Lanka, Andaman Islands, SE Asia, Malayan Archipelago, New Guinea, N Australia. **Ecol:** Fl. Mar–Jun. **Zones:** 10–12.

Parkinsonia

Species 11–12. **Shrubs** or **trees**, armed; branches slender, often pendent. **Stipules** spinescent, persistent or caducous. **Leaves** crowded on short shoots, spiral, bipinnate; pinnae 2(4 or 6), opposite, long and flattened, phyllodial; leaflets numerous, small; rachis short, spine-tipped. **Inflorescences** racemes, long, axillary; bract small caducous. **Flowers** with calyx segments 5, slightly unequal, imbricate; petals 5, slightly unequal; stamens 10, free, filaments villous at base, anthers uniform; ovary shortly stipitate, many-ovuled, style filiform, stigma terminal. **Fruits** pods, linear, terete, torulose, finally dehiscing; seeds oblong, hilum near the tip. **Distr:** Tropical America, S Africa. **Cult:** Propagate from seeds. Requires fertile, moist but well-drained soil in partial to full sun.

P. aculeata L., *Sp. Pl.* 375, 1753. [Jerusalem Thorn]. **Shrubs** or **small trees**, to 4(–9) m tall, armed, glabrous. **Leaves** bipinnate; rachis short and ending into spines, often with a pair of stipular spines at its base; pinnae 2–6, ~15–40 cm long, often with much flattened rachis; leaflets 25–30 pairs, obovate-oblong, ~4–5 mm long; petiolule 1–2 mm long. **Inflorescences** racemes, axillary, shorter than leaves; pedicels ~0.7–1.5 cm long. **Flowers** with calyx ~7–8 mm long, segments distinct almost to base, oblong, apex obtuse; petals 5, yellow, 1.5–1.8 cm long, shortly clawed; stamens 10, free, flattened, base villous; ovary silky-hairy, ovules numerous. **Fruits** pods, linear, ~7.5–10 × ~0.6–1 cm, contracted between seeds, glabrous, 1–6 seeded; seeds oblong, smooth, mottled. **Distr:** Tropical America (SW USA to Mexico). **Ecol:** In ephemeral watercourses and dry areas; <100 m. Fl. Mar–May. **Zones:** 9–12.

Delonix

Species 11. **Trees**, deciduous, semievergreen or evergreen, unarmed; trunks often swollen, frequently fusiform (cigar-shaped, narrowed at base and apex); bark thin, usually smooth, often flaking or peeling-off in thin, papery strips, underbark green; wounds often producing a clear resin. **Leaves** bipinnate; pinnae 2–many pairs; leaflets few to numerous, opposite (paired) to alternate, somewhat unequal-sided, usually rounded and somewhat oblique basally and shallowly notched apically; stipulate. **Inflorescences** corymbose racemes, borne at end of branches; bracts small, caducous. **Flowers** showy, white, yellow, orange or red; calyx segments valvate, subequal; petals 5, orbicular, imbricate, subequal or uppermost dissimilar; stamens 10, free, declinate; ovary sessile, ovules numerous, style filiform, stigma truncate. **Fruits** elongated, compressed, hard, 2-valved; seeds large, spherical or ellipsoidal to fusiform-cylindrical. **Distr:** Africa, Madagascar (9 spp.), tropical Asia. **Cult:** Propagate from seeds, cuttings or by air-layering. Requires a moist but well-drained soil in full sun.

D. decaryi (R. Viguier) Capuron, *Adansonia*, n.s., 8: 16, 1968. (Syn: *Poinciana decaryi*, *P. adansonioides*). [Bottle Tree]. **Trees**, deciduous, 3–10 m tall, crown umbrella-shaped; trunk swollen, bottle-shaped, to ~80 cm dbh. **Leaves** pinnate; pinnae (1–)2–3 pairs; leaflets opposite, 14–26 per pinna, narrowly oblong, glabrous, adaxially matt, abaxially paler. **Inflorescences** racemes, combined into corymbs, axillary, held above the foliage. **Flowers** large and showy, white, aging to deep apricot-yellow, scented; calyx segments 1.5–2.3 cm long, spreading, outside glabrous, inside white-woolly; petals with broad overlapping laminas with obtuse tips, margins strongly crisped and frilled; stamens 4.5–7 cm long, exserted;

Below: *Schizolobium parahyba*, *Peltophorum pterocarpum*, *Parkinsonia aculeata*, *Delonix decaryi*.

TC

JML

JML

TG

TG

Top row: *Delonix elata, Delonix regia, Delonix floribunda.*
Bottom row: *Delonix floribunda* (flowers).

ovary densely white-woolly at the tip. **Fruits** long and slender, linear-oblong, compressed, dark blackish-brown; seeds ellipsoid. **Distr:** Madagascar. **Ecol:** In xerophytic or subarid shrublands and open woodlands; 0–500 m. **Zones:** 10–12.

D. elata (L.) Gamble, *Fl. Madras.* 1: 396, 1919. (Syn: *Poinciana elata*). [Creamy Peacock Flower, White Flamboyant Tree]. **Trees**, erect, 6–10(–15) m tall; bark ash-colored. **Leaves** 6–20 cm long, bipinnately compound; pinnae (2–)4–8(–12) pairs, 4–6 cm long; leaflets (8–)10–20(–25) pairs, ± sessile, linear-oblong, 0.7–1.5 × ~0.2–0.3 cm, base narrowed, apex obtuse, sometimes slightly mucronate. **Inflorescences** racemes, terminal; pedicels hairy, to 2.5(–3) cm long, lower flower with the longest pedicel. **Flowers** with calyx oblong, 2–3 cm long, fused at base, narrow and sharply pointed, leathery, silky-pubescent; corolla white, turning yellow with age, upper petal is smaller and darker in color, suborbicular, ~2.5 cm long, curled at the margins; stamens 5–10 cm long, hairy, dark-colored, thickened at base. **Fruits** pods, 12.5–25 × ~0.6–2.5 cm, ends narrowed, smooth; seeds 4–8. **Distr:** Abyssinia, Arabia. **Ecol:** Fl. Almost throughout the year. **Zones:** 11–12.

D. floribunda (Baill.) Capuron, *Adansonia*, s6r. 2, 8: 15, 1968. (Syn: *Aprevalia floribunda*, etc.). **Trees**, deciduous, umbrella-shaped, 3–12 (–15) m tall; trunk swollen, cigar-shaped, to 70+ cm dbh. **Leaves** pinnate; pinnae (1–)3–6 pairs, each with (8–)12–20 opposite leaflets; leaflets elliptic to oblong-elliptic, 1–3(–4.5) × 0.6–1.8(–2.5) cm, glabrous, thinly coriaceous. **Inflorescences** racemes, sometimes combined into a dense, flat-topped usually leafless corymbs; pedicels 0.8–2.5 cm long. **Flowers** held erect, yellow, nectariferous, petal(s) lemon-yellow, stamens and style greenish-yellow, anthers darker yellow; calyx segments 0.9–1.3 cm long, green outside, whitish and woolly inside; petals much-reduced, often only 1 petal present, sometimes with 1+ vestigial petals also present, the single petal small and narrow, 1.3–2.3 × 0.7–1.1 cm, without a distinctly broadened lamina, rolled into a narrow funnel, margins flat and entire; stamens 2.5–4.5 cm long, anthers ~4 mm long; ovary densely woolly. **Fruits** pods, linear-oblong, (13–)16–25(–31) × 2.5–3.5 cm, compressed, dark blackish-brown; seeds (12–)14–23, ellipsoid, 1–1.2 × 0.7–0.9 cm. **Distr:** SW & W Madagascar. **Ecol:** In xerophytic scrublands, deciduous woodlands. Fl. Sep–Dec. **Zones:** 10–12.

D. regia (Bojer) Rafin., *Fl. Tellur.* 2: 92, 1836. (Syn: *Poinciana regia*). [Flamboyant Tree, Flame of the Forest, Royal Poinciana]. **Trees**, deciduous, large, to 20 m tall; crown hemispherical. **Stipules** in lower part conspicuously pinnatipartite, in upper part setiform. **Leaves** 20–60 cm long; pinnae opposite, 15–20 pairs, 5–10 cm long; leaflets 25 pairs, crowded, opposite, oblong, 4–8 × 3–4 mm, base oblique, margins entire, apex obtuse, surfaces silky; petiole 7–12 cm long, base inflated, petiolules short. **Inflorescences** corymbose racemes, terminal or axillary; pedicels 4–10 cm long. **Flowers** bright red to orange-red, 7–10 cm Ø; receptacle discoid or shortly turbinate; sepals reddish inside, margin greenish-yellow; petals reflexed after anthesis, red, yellow-tinged and white-spotted, spathulate, 5–7 × 3.7–4 cm, claw long, ~2 mm long, slender; stamens curved upward, red, unequal in length, 3–6 cm long, filaments thick, woolly below, anthers red; ovary ~1.3 cm long. **Fruits** pods, dark reddish-brown, blackish-brown when mature, slightly curved, 30–60 × 3.5–5 cm, apex with persistent style; seeds 20–40, yellowish tinged with brown spots, ~1.5 × 0.7 cm, smooth, hard. **Distr:** Madagascar. **Ecol:** Fl. Jun–Jul. Fr. Aug–Oct. **Zones:** 10–12.

Colvillea

Species 1. *Generic description same as specific.* **Cult:** Propagate from seeds or cuttings. Requires a moist, well-drained soil in full or partial sun.

C. racemosa Boj., *Bot. Mag.* tt. 3325, 8326, 1834. [Colville Glory, Glory Covillea, Whip Tree]. **Trees**, unarmed, deciduous, to 15 m tall; trunk thick, straight; bark gray, horizontally ridged, with loose thin scales; branches thick, spreading, studded with corky nobs. **Stipules** minute, caducous. **Leaves** bipinnate, fernlike, long; pinnae opposite, 15–25 pairs; leaflets opposite, sometimes alternate, arranged in spiral rows, 18–30 pairs, small. **Inflorescences** 10–12, clustered at branch tips, racemes, long, densely crowded, cylindrical or cone-shaped, pendent; bracts membranous, colored, soon caducous. **Flowers** burnt-orange to scarlet, showy, 2 cm long; calyx tube short, 1 side swollen; sepals 5, rarely 4, thick, induplicate-valvate; petals 5, imbricate, uppermost petal inside, very broad, lateral ones smaller, obovate, lower ones outside, narrow; stamens 10, free, deflexed, filaments thick, hairy at base, anthers large, uniform; ovary subsessile, free, many-ovuled, style filiform, stigma small, terminal. **Fruits** pods, straight, elongate, turgid, 2-valved; seeds transverse, oblong, small. **Distr:** Madagascar. **Ecol:** In lowland areas of primary forests and wooded savannas. Fl. throughout most of the year. **Zones:** 10–12.

Subfam. MIMISOIDEAE

Genera 82, **species** 3,275. **Trees** or **shrubs**, sometimes lianas or rarely herbs. **Stipules** present or absent, sometimes developing into paired spines. **Leaves** alternate, usually bipinnate, rarely once pinnate or becoming phyllodes, scaly or absent; pinnae usually opposite; leaf rachis or petiole usually with glands; petiole with pulvinus. **Inflorescences** cylindric spikes or globose heads, rarely racemes or rearranged into panicles; bracts usually deciduous, small; bracteoles deciduous or absent. **Flowers** small, bisexual, sometimes unisexual, or some neuter and sterile, actinomorphic; calyx tubular (rarely sepals free), usually 5 (rarely 3-, 4-, 6-, or 7-)toothed; lobes valvate in bud (rarely imbricate); petals as many as calyx teeth, valvate in bud, free or connate into a tube; stamens as many as corolla lobes or numerous, or many, free or connate into a tube at base, free part of filaments usually exserted, anthers small, dithecal, opening by longitudinal slits, often with a deciduous gland at tip, pollen grains in monads or very often in tetrads or polyads; pistils usually 1, rarely 2–15, ovary superior, 1-loculed, ovules mostly numerous, style slender, stigma small. **Fruits** pods, straight or contorted, dehiscent or indehiscent, sometimes breaking into 1-seeded segments; seeds usually plano-compressed, in a few genera overgrown testa hard, usually with pleurogram. **Distr:** Tropical and warm temperate regions, especially in Africa and America.

TRIBE MIMOSEAE

Parkia

Species ~34. **Trees**, unarmed, to 50 m tall. **Leaves** alternate or opposite, bipinnate; pinnae numerous; leaflets numerous, opposite; petiole and rachis usually with extrafloral nectaries. **Inflorescences** compound, borne on a stout twig, to 50 cm long, bearing 2–10+ flexible peduncles, each to 50+ cm long, pendent; capitula pyriform, pyriform or biglobose, to 7 cm long; *fertile* flowers at apex of capitulum; *modified* structurally ♂ and nectar-secreting flowers in a constricted ring below; *neuter* flowers sometimes bearing elongated staminodia which form a fringe at base. **Flowers** *fertile* 5-merous, functionally hermaphrodite or ♂, then the gynoecium reduced or absent, bracts obdeltate-spathulate, slightly longer than calyx, pubescent outside, especially apically, calyx long-tubular or infundibuliform, gamosepalous, lobes 5, imbricate, 2 larger, corolla longer, lobes 5, narrowly spathulate or lingulate, stamens 10, all fertile, shortly exserted, filaments connate below into a tube, ovary ± sessile, style filiform, exserted, stigma terminal; *sterile* flowers with lower petals often free, staminodes 10, in a long bundle, free distally, filiform. **Infructescences** pendent, receptacle swollen, with 1–several pods and numerous rhomboidal scars where flowers have aborted. **Fruits** pods, linear, strap-shaped or oblong, sometimes corrugated over the seeds, straight or sometimes twisted; seeds to 38 per pod, ellipsoid. **Distr:** Pantropical. **Cult:** Propagate from seeds. Requires moist but well-drained organic soil in full sun.

P. biglandulosa Wight & Arnott, *Prod. Fl. Pen. Ind. Or.*: 279, 1834. (Syn: *Mimosa pedunculata, Parkia pedunculata*). [African Locust Tree]. **Trees**, ~18 m tall. **Leaves** alternate, rachis including petiole to 38 cm long with 2 adaxial glands on petiole, smaller glands on primary rachis, rachises white-pubescent; pinnae (13–)18–29 pairs, opposite or subopposite; leaflets 68–115(–130) pairs per pinna, opposite, linear, 5–8 × ~1 mm, base obliquely truncate, apex acute. **Inflorescences** capitula, clavate with a short cylindrical part at base and a spherical ball of fertile flowers ~4–5 cm Ø; peduncles alternate, to ~20 cm long. **Flowers** *hermaphrodite* with calyx 1.25–1.5 cm long including pseudo-pedicel to 3.5(–5) mm long, largest lobes 1–1.5 mm long, corolla 1.3–1.35 cm long, apical lobes 1–1.5 mm long, scarcely united below, separating readily, filaments to 1.45–1.9 cm long, united to ~1–1.5 cm, ovary ~3 mm long, style 5–6 mm long. **Fruits** pods, strap-shaped, dark brown, to 38 cm long, pulp mealy; seeds 10–12, ovate, ~1.9 × 1.2 cm, brown. **Distr:** Known only in cultivation (Hopkins 1994). **Ecol:** In tropical forests. Fl. Nov–Feb. **Zones:** 10–12.

Adenanthera

Species ~12. **Trees**; bark grayish-brown. **Leaves** bipinnate; leaflets many, alternate, paired, small, margins smooth. **Inflorescences** solitary or sometimes paired spiciform racemes, axillary; pedicels short. **Flowers** small, white or yellowish, usually bisexual; sepals 5, connate, campanulate; petals 5, almost distinct or basally connate, valvate, lanceolate; stamens 10, distinct, equaling or slightly longer than petals, anthers glandular at apex; ovary sessile, many-ovuled, style thin, stigma capitate. **Fruits** pods, linear, compressed, falcate or curved or spirally twisted, torulose, longitudinally dehiscent into 2 valves; seeds 10–100, small, thick, hard, red or red and black. **Distr:** Tropical and subtropical Asia to N Australia and Polynesia. **Cult:** Propagated easily from cuttings or seeds (after scarifying, scratching or boiling for 1 minute). Requires any well-drained soil (neutral to slightly acidic) in full sun; salt-tolerant.

A. microsperma Teijsm. & Binn., *Natuurk. Tijdschr. Ned.-Indië* 27: 58, 1864. (Syn: *A. microsperma* var. *luteosemiralis, A. pavonina* var. *microsperma*). [Bead Tree, Coral Wood, Curly Bean]. **Trees**, deciduous, 5–20 m tall; branchlets puberulent. **Leaves** bipinnate; pinnae 3–5 pairs; leaflets 4–7 pairs, alternate, oblong or ovate, 2.5–3.5 × 1.5–2.5 cm, ends rounded-obtuse, puberulent; petiolules short. **Inflorescences** racemes, simple, axillary or arranged in panicles at apices of branchlets, puberulent; pedicels short. **Flowers** white or yellow, small, fragrant; calyx <1 mm long, together with pedicel golden-yellow puberulous; petals oblong, 2.5–3.5 mm long, usually glabrous, slightly basally connate; stamens 3–4.5 mm long; ovary nearly sessile, villous, style filiform, stigma small. **Fruits** pods, narrowly oblong, 10–20 × 1.2–1.4 cm, valves contorted after dehiscence; seeds scarlet or yellow, suborbicular to ellipsoidal, 5–8 × 4.5–7 mm. **Distr:** Cambodia, China, Indonesia, Laos, Malaysia, Myanmar, Taiwan, Thailand, Vietnam. **Ecol:** In valleys, along streams, forests; 0–1,000 m. Fl. Apr–Jul. Fr. Jul–Oct. **Zones:** 9–12.

A. pavonina L., *Sp. Pl.* 384, 1753. [Red Bead Tree, Red Sandalwood Tree]. **Trees**, large, deciduous. **Leaves** bipinnate, to 24 cm long, rachis ~10–40 cm long; pinnae opposite, 2–6 pairs, ~7–12 cm long; leaflets 7–15 pairs, sessile, alternate, obovate, 1.5–4 × 0.5–2.5 cm, margins slightly curved backwards, base unequal; petiole ~5–10 cm long. **Inflorescences** racemes, axillary and terminal, ~7.5–20 cm long; pedicels 3.5–4 cm long. **Flowers** fragrant, opening creamish, changing to golden-yellow; calyx gamosepalous, 5-toothed, ~1 mm long; corolla pale yellow, ~2.5–3 mm long, petals basally united; stamens 10, free, hardly exserted, anthers gland-tipped. **Fruits** pods, curved, becoming coiled after dehiscence, 10–22.5 × ~0.7–1 cm; seeds many, glossy, bright red, beadlike, ~7–8 mm across. **Distr:** India, Sri Lanka, Maldives, SE China and Malaysia to the Moluccas. **Ecol:** Fl. Mar–May. **Zones:** 10–12. **Info:** The seeds are toxic when raw, but can be eaten after cooking. The seeds are also used as weights and in board games in the Maldives. The timber is of good quality, similar to sandalwood.

Tetrapleura

Species 2. **Trees**, unarmed. **Leaves** bipinnate, eglandular; pinnae with few–many alternate leaflets. **Inflorescences** racemes, solitary or rarely paired, axillary. **Flowers** hermaphrodite; calyx gamosepalous, 5-toothed; petals 5, free; stamens 10, fertile, anthers with a caducous apical gland. **Fruits** pods, straight or slightly curved, oblong, woody, indehiscent, valves with a thick winglike projection arising from the middle of the valve and running longitudinally the entire length (pod in cross-section thus ± cruciform), internally septate between seeds; seeds hard, dark brown, unwinged. **Distr:** Tropical Africa. **Cult:** Propagate from seeds or by air-layering. Requires a fertile, moist, well-drained soil in full or partial sun.

T. tetraptera (Schumach. & Thonn.) Taub., *Bot. Centralbl.* 47: 395, 1891. (Syn: *Adenanthera tetraptera, T. thonningii*). [Aidan/Aridan Tree]. **Trees**, 6–30 m tall. **Leaves** pinnate, rachis 7.5–23 cm long; pinnae 5–7(–10) pairs, opposite to alternate; leaflets 6–11(–13) pairs, oblong-elliptic to elliptic, (0.4–)0.65–2.1 × (0.3–)0.55–1.3 cm, apex rounded to emarginate, adaxially glabrous or almost so, abaxially appressed-puberulous; petiole 4–11 cm long; petiolules 0.5–0.7(–1) mm long. **Inflorescences** racemes, 4–14 cm long; peduncles 1.5–4 cm long; pedicels 1–3 mm long. **Flowers** yellowish to pinkish;

Below: *Colvillea racemosa, Parkia biglandulosa, Adenanthera microsperma, Adenanthera pavonina.*

TG

Above: *Tetrapleura tetraptera*, *Leucaena leucocephala* (fruits), *Mimosa pigra*, *Mimosa pudica*.

calyx ± puberulous, 0.5–1 mm long; petals 2–3.5 × 0.5–0.9 mm, usually puberulous outside towards apex; stamens with filaments 2.5–4.5 mm long. **Fruits** pods, 12–25 × 3.5–5.7(–6.5) cm, winged, dark brown, glossy, apex rounded, sometimes emarginate; seeds 9–9.5 × 7–8 mm, 3.5–4 mm thick. **Distr:** Kenya, Tanzania, Uganda. **Ecol:** To 1,200 m. **Zones:** 10–12.

Leucaena

Species 22. **Trees** or **shrubs**, small, erect, evergreen, unarmed. **Stipules** deciduous, small, setaceous. **Leaves** evenly bipinnate; leaflets opposite, numerous and small, or few and larger, oblique; petiole often glandular. **Inflorescences** heads, solitary or fasciculate, axillary, globose, pedunculate; bracts usually 2; flowers sessile. **Flowers** usually bisexual; calyx campanulate, 5-toothed; petals 5, free or nearly so; stamens 10, free, exserted, anthers apically eglandular; ovary funiculate, ovules numerous, style filiform. **Fruits** pods, stipitate, broadly linear, plano-compressed, leathery, 2-valved, continuous within; seeds transverse, ovoid, flat. **Distr:** Tropical America; 1 species (*L. leucocephala*) introduced and naturalized in the Old World. **Cult:** Propagate from seeds, cuttings, or by air-layering. Requires any well-drained soil in full or partial sun.

L. leucocephala (Lamarck) de Wit, *Taxon* 10: 54, 1961. (Syn: *L. glauca, Mimosa glauca, M. leucocephala*). [Ipil Ipil, Wild Tamarind]. **Shrubs** or **small trees**, 2–6 m tall, fast-growing. **Stipules** caducous, deltoid, very small. **Leaves** bipinnate; pinnae 4–8 pairs, 5–9(–16) cm long, rachis pubescent, with black glands near lowest pinnae; leaflets 5–15 pairs, linear-oblong, 0.7–1.3 × 0.15–0.3 cm, base cuneate, margins ciliate, apex acute. **Inflorescences** heads, usually 1 or 2, axillary, 2–3 cm Ø; peduncle 2–4 cm long; bracts deciduous, pubescent. **Flowers** white; calyx ~3 mm long, 5-toothed; petals narrowly oblanceolate, ~5 mm long, outside pubescent; stamens 10, ~7 mm long; ovary shortly funiculate, stigma cupulate. **Fruits** pods, straight, strap-shaped, 10–18 × 1.4–2 cm, flat, leathery, base attenuate, pedicel to 3 cm long, pubescent, beak acute, hard; seeds 6–25, brown, glossy, narrowly ovoid, flat, 6–9 × 3–4.5 mm. **Distr:** Tropical America; cultivated and naturalized in tropical and subtropical regions. **Ecol:** In dry places, somewhat salt-tolerant; 0–600 m. Fl. Apr–Jul. Fr. Aug–Oct. **Zones:** 10–12.

Mimosa

Species 510. **Shrubs** or **herbs**, rarely trees or climbers, usually armed. **Stipules** not spinescent, usually 2 per pinna, small, caducous. **Leaves** bipinnate, often sensitive. **Inflorescences** heads, globose or spikes cylindric, solitary or fasciculate, axillary, pedunculate; flowers sessile. **Flowers** small, bisexual or polygamous, usually 4-merous; calyx connate, valvate, campanulate, toothed; petals basally connate; stamens 4 or 8, free, exserted, anthers eglandular; ovary many-ovuled, style filiform. **Fruits** pods, oblong or linear, usually plano-compressed, membranous or leathery, composed of 1-seeded segments, separating from persistent margins; seeds elliptic or orbicular, flat. **Distr:** Mostly in tropical America; *M. pudica* widely naturalized and often weedy throughout the tropics. **Cult:** Propagate from seeds or cuttings. Requires any fertile but well-drained soil in full sun. Some species are salt-tolerant and are found in coastal regions.

M. pigra L., *Cent. Pl.* 1: 13, 1755. (Syn: *M. asperata*). [Cat-claw Mimosa, Giant Sensitive Plant, Thorny Sensitive Plant]. **Shrubs**, 0.6–3(–4.5) m high, sometimes scandent or rambling; stems armed with broad-based prickles to 7 mm long, also usually ± appressed-setose. **Leaves** sensitive; rachis 3.5–12(–18) cm long, with a prickle at the junction of each of the 6–14(–16) pairs of pinnae, sometimes with other prickles between the pairs; leaflets 20–42 pairs, linear-oblong, 0.3–0.8(–1.25) cm × 0.5–1.25(–2) mm, margins often setulose; petiole 0.3–1.5 cm long. **Inflorescences** heads, subglobose pedunculate, ~1 cm Ø, 1–2(–3) together in upper axils. **Flowers** mauve or pink; calyx minute, laciniate, 0.75–1 mm long; corolla ~2.25–3 mm long; stamens 8. **Fruits** pods, clustered, brown, densely bristly, 3–8 × 0.9–1.4 cm, breaking up transversely into segments 3–4 mm long, sutures persisting as an empty frame. **Distr:** Tropical Africa, Madagascar, Mauritius; tropical America. **Ecol:** In various habitats. **Zones:** 10–12.

M. pudica L., *Sp. Pl.* 1: 518, 1753. [Humble Plant, Sensitive Plant]. **Herbs**, diffuse, shrubby, to 1 m tall; stems cylindric, with bristles curved prickles. **Stipules** lanceolate, 0.5–1 cm long, bristly. **Leaves** bipinnate, sensitive; pinnae usually 2 pairs, digitate, 3–8 cm long; leaflets 10–20 pairs, linear-lanceolate, 0.6–1.5 × 0.15–0.3 cm, margins ciliate, apex acute, adaxially glabrous, abaxially slightly hispid. **Inflorescences** heads, globose, ~1 cm Ø, solitary or paired, axillary; peduncle long; bracts linear. **Flowers** numerous, pink, small; calyx minute; corolla campanulate, lobes pubescent outside; stamens 4, exserted; ovary shortly funiculate, glabrous, ovules 3 or 4, style filiform, stigma small. **Fruits** pods, arranged in a star, slightly recurved, flat, oblong, 1–2 × ~0.5 cm, consisting of 3–5 single-seeded segments which fall away from persistent, bristly sutures; seeds light brown, ovoid, ~3.5 mm Ø. **Distr:** Tropical America; naturalized widely in the tropics. **Ecol:** Usually in wastelands and disturbed places, also near beaches; 0–1,500 m. Fl. Mar–Oct. Fr. May–Nov. **Zones:** 10–12.

TRIBE ACACIEAE

Acacia

Species ~1,200. **Trees**, **shrubs**, or **lianas**, armed or unarmed. **Stipules** spinescent or not. **Leaves** bipinnate or modified into phyllodes by dilation of petiole and proximal part of rachis; extrafloral nectaries usually present on petiole and rachis; leaflets opposite, sessile or shortly stalked. **Inflorescences** consisting of pedunculate glomerules or spikes borne in axillary clusters or aggregated into terminal panicles. **Flowers** bisexual, or ♂ and bisexual, 4- or 5-merous, uniform; sepals connate, valvate; petals valvate, rarely absent; stamens numerous, exerted, filaments free or basally connate, anthers glandular or not; ovary sessile or stipitate, ovules many or in 2 rows, style filiform, stigma capitate, small. **Fruits** pods, extremely variable, dehiscent or indehiscent; endocarp not separated from exocarp, not forming envelope around each seed; seeds usually elliptic to oblong, ± flattened, with a hard black-brown testa; aril present only in some phyllodinous species. **Distr:** Tropical and subtropical regions, especially the Pacific region, and Africa. **Cult:** Propagate from scarified seeds or by air-layering. Requires a well-drained soil in full sun.

A. auriculiformis A. Cunningham ex Bentham, *London J. Bot.* 1: 377, 1842. (Syn: *A. auriculaeformis*, etc.). [Ear-Leaf Acacia, Earpod Wattle, Northern Black Wattle, etc.]. **Trees**, evergreen, to 20 m tall; bark gray-white, smooth; branches pendulous; branchlets angular, glabrous, with conspicuous lenticels. **Phyllodes** falcate-oblong, 10–20 × 1.5–4(–6) cm, conspicuous main veins 3 or 4, both ends attenuate. **Inflorescences** spikes, 1 to several, fasciculate, axillary or terminal, 3.5–8 cm long. **Flowers** orange-yellow, fragrant; calyx 0.5–1 mm long, shallowly dentate; petals oblong, 1.5–2 mm long; filaments 2.5–4 mm long; ovary densely puberulous. **Fruits** pods, contorted when mature, 5–8 × 0.8–1.2 cm, valves woody; seeds ~12, black, ~5 × 3.5 mm. **Distr:** N Australia and New Guinea. **Ecol:** In rainforests, gallery forests or along creeks and rivers; 0–500 m. Fl. throughout most of the year. **Zones:** 9–12.

A. chundra Willd., *Sp. Pl.* 4(2): 1078, 1806. (Syn: *Mimosa chundra*, etc.). [Red Cutch]. **Trees**, to 8 m tall; bark dark brown, rough, peeling off profusely. **Thorns** solitary below the stipules, small, hooked, ~3–4 mm long, large-based, or lacking. **Leaves** bipinnate; rachis slender, to 13 cm long, petiolar part 3–4 cm long with a large, oval, flat gland usually near base, and small, circular, concave glands between bases of 2 terminal pinnae, prickles absent; pinnae 9–18 pairs, ~3–4 cm long; leaflets subsessile, 20–30 pairs, overlapping, small, linear, ~3 × 1 mm, base oblique, apex obtuse, glabrous. **Inflorescences** spikes, erect, lax, axillary, shortly stalked, to 10 cm long; flowers sessile. **Flowers** glabrous, white; calyx campanulate, 1 mm long with short teeth; corolla 3× length of calyx, lobes linear-lanceolate. **Fruits** pods, strap-shaped, flat, thin, almost straight, tapering at both ends, stalked, sometimes with few indents between seeds, very dark reddish-brown; seeds 3–6, flat, greenish-brown, broadly oval. **Distr:** Peninsular India, Myanmar. **Ecol:** In dry places.

Above: *Acacia auriculiformis, Acacia chundra, Acacia farnesiana, Acacia podalyriifolia.*

Zones: 10–12. **Info:** The wood is very hard and used to make utensils.

A. farnesiana (L.) Willd., *Sp.Pl.* 4: 1083, 1806. (Syn: *Mimosa farnesiana*). [Cassie Flower, Farnese Wattle, Fragrant Acacia, Mimosa, Needle Bush, Sweet Acacia]. **Shrubs** or **small trees**; branches slightly zigzag, marked with gray or pale brown dots, young twigs glabrescent. **Spines** stipular, in pairs, straight, ~0.7–1.8 cm long. **Leaves** bipinnate; rachis 1.2–5.5 cm long, pilose; pinnae 2–8 pairs, ~1.2–2.5 cm long; leaflets 10–20 pairs, sessile, linear, oblong, acute, ~2.5–5.5 × 1–1.5 mm, base oblique, glabrous to subglabrous; petiole usually with a small gland near middle. **Inflorescences** axillary fascicles of globose heads; peduncles ~1.2–2.5 cm long, slender, pubescent; bracts whorled, at or near apex of peduncle. **Flowers** with calyx ~1.5–1.8 mm long, campanulate; corolla ~2.5 mm long. **Fruits** pods, 4.5–7.5 × ~1.2 cm, slightly curved, dark brown, mesocarp pulpy; seeds numerous, biseriate. **Distr:** Tropical America. **Ecol:** Fl. Nov–Mar. **Zones:** 10–12.

A. podalyriifolia A. Cunn. ex G. Don, *Gen. Syst.* 2: 405, 1832. [Queensland Silver Wattle]. **Shrubs** or **small trees**, 3–7 m tall, unarmed, spreading; young branchlets densely pubescent. **Leaves** alternate, spiral, simple (phyllodes), subsessile or sessile; lamina apparently simple, phyllodic, ovate to elliptic or elliptic-oblong, mostly 1.5–5.5 × 1–2.8 cm, base rounded, margins entire, apex mucronate, glaucous, softly ± pubescent; petiole 1–2 mm long. **Inflorescences** in heads arranged in axillary racemes, ± aggregated terminally; flowers sessile. **Flowers** regular, yellow; perianth 2-whorled; calyx 0.3–0.5 mm long, sepals 5, all joined; corolla 0.7–0.8 mm long, petals 5, all free; stamens numerous, free of the perianth and free from each other, anthers dehiscing via longitudinal slits; ovary monomerous, superior, 1-celled, ovules numerous, styles 1, simple. **Fruits** pods, to 12 × 1.5–2 cm, non-fleshy, dehiscent. **Distr:** Australia (Queensland, NSW). **Ecol:** Fl. May–Aug. **Zones:** 9–12.

TRIBE INGEAE

Calliandra

Species ~200. **Shrubs** or **small trees**, usually unarmed. **Stipules** often persistent, or sometimes spinescent, rarely absent. **Leaves** bipinnate, eglandular; pinnae 1 to several pairs; leaflets opposite, small and numerous, or larger and from few pairs to only 1. **Inflorescences** capitula, globose and axillary, or racemes terminal. **Flowers** (5- or)6-merous, polygamous; calyx campanulate, toothed; petals united to middle; *middle* flowers sometimes heteromorphic with an elongated tubular corolla; stamens numerous (to 100), red or white, ± united into a tube and long exerted, showy; anthers mostly glandular hairy; ovary sessile, ovules numerous, style filiform. **Fruits** pods, slightly falcate, strap-shaped, flat, rigidly leathery, often narrowed to base, margin thickened, 2-valved, valves elastically opening from apex to base, continuous inside; seeds obovoid or orbicular, compressed, testa hard, with pleurogram, without aril. **Distr:** Mainly in tropical and subtropical regions of the Americas, a few in India, Madagascar, Myanmar and China. **Cult:** Propagate from scarified seeds or cuttings. Needs fertile, moist but well-drained soil in full sun.

C. haematocephala Hasskarl, *Retzia* 1: 216, 1855. (Syn: *C. inequilatera*). [Red Powder-puff]. **Shrubs** or **small trees**, deciduous, 1–3 m high; branchlets brown, cylindric, rough. **Stipules** persistent, ovate-lanceolate. **Leaves** bipinnate; pinnae 1 pair, 8–13 cm long; leaflets 7–9 pairs, obliquely lanceolate, 2–4 × 0.7–1.5 cm, base oblique, margins sparsely pilose, apex obtuse, mucronate, midvein close to upper margin; petiole 1–2.5 cm long; petiolules ~1 mm long. **Inflorescences** heads, axillary, ~3 cm Ø (including filaments); peduncles 1–3.5 cm long. **Flowers** with calyx campanulate, ~2 mm long; corolla purplish, tube 3.5–5 mm long, lobes 5, reflexed, ~3 mm long, glabrous; stamens numerous, bright scarlet; staminal tube white, ~6 cm long, mouth inside with a subulate appendix; filaments deep red, ~2 cm long. **Fruits** pods, dull brown, linear-oblanceolate, 6–11 × 0.5–1.3 cm, valves elastically dehiscent from apex to base along sutures, reflexed; seeds 5 or 6, brown, oblong, 0.7–1 × ~0.4 cm. **Distr:** S America. **Ecol:** Fl. Aug–Sep. Fr. Oct–Nov. **Zones:** 10–12.

C. houstoniana (Mill.) Standl., *Contr. U.S. Natl. Herb.* 23: 386, 1922. (Syn: *Mimosa houstoniana*, etc.). [Angel Powderpuff, Red Angel's Hair, Tree Powderpuff]. **Shrubs** or **small trees**, unarmed, open, spreading, (1–)4–6(–12) m tall, pubescent or glabrous; stem to 30 cm dbh; bark reddish-brown. **Leaves** slender, feathery, alternate, bipinnate; pinnae (3–)6–20 pairs, (2–)4–7(–11) cm long; leaflets 19–60(–100) pairs, opposite, oblong, 4–8 × 1 mm, apex acute, adaxially dark green, glossy. **Inflorescences** umbels, few- to many-flowered, aggregated into spikelike racemes, 10–30(–40) cm long, terminal, erect. **Flowers** actinomorphic, showy, calyx 2 mm long; corolla 5–6 mm long, pale green; stamens numerous, 4–6 cm long, basally united, purplish-red. **Fruits** pods, linear-oblong, 7–11 × 1–1.5 cm, flattened, slightly tapering from apex to base, margins thickened and raised, densely brown-hirsute, elastically dehiscent from apex; seeds 3–15, ellipsoid, 5–7 mm long, flattened, dark brown mottled. **Distr:** Mexico to El Salvador. **Ecol:** In seasonally dry places; 0–1,300(–1,850) m. **Zones:** 9–12.

C. surinamensis Benth., *London J. Bot.* 3: 105, 1844. (Syn: *Acacia fasciculata*, etc.).[Pink Powder-puff, Pink Tassel Flower, Surinam Calliandra]. **Shrubs** or **treelets**, 3–6 m tall; young parts glabrous. **Stipules** linear-lanceolate, 5–8 mm long **Leaves** alternate, bipinnate; pinnae 1 pair, spreading; leaflets 7–12 pairs per pinna, oblong-lanceolate, 0.6–2.5 cm long, base oblique to falcate, apex obtuse to mucronate-acute, or acute; petioles 5–8 mm long. **Inflorescences** axillary, usually clustered, heads ± obconical, flowers many, crowded; bracts linear. **Flowers** with calyx slightly dentate; corolla funnel-shaped, greenish-white, sometimes red-tipped, lobes 5, partly fused, 6–8 mm long; stamens numerous, filaments joined in a tube for

Below: *Calliandra haematocephala, Calliandra houstoniana, Calliandra surinamensis.*

half their length, tube white, free above, red, or white with red tips. **Fruits** pods, oblanceolate, 7.5–10.8 cm long, flattened. **Distr:** Suriname. **Ecol:** Fl. throughout the year. **Zones:** 10–12.

Inga

Species ~300. **Trees** or **shrubs**, unarmed. **Leaves** pinnate, rachis ± winged. **Inflorescences** globosely umbellate, spicate or racemose; peduncles solitary or fascicled, axillary or panicled at branchlet tips. **Flowers** 5(6)-merous, mostly hermaphrodite, often showy; calyx tubular or campanulate, dentate or shortly lobed; corolla elongate, petals valvate, coalescent to the middle or higher; stamens many, long-exserted; ovary sessile, many-ovuled. **Fruits** pods, narrow, straight or somewhat incurved, rarely twisted, flat, 4-sided or subcylindric, margins usually enlarged, coriaceous or subfleshy, tardily if at all dehiscing; seeds sometimes surrounded with pulp. **Distr:** Neotropical (Mexico Uruguay); highest diversity in Peru, Ecuador, Colombia and southern C America. **Ecol:** In lowland and montane rainforests throughout the humid tropics, half the species on non-flooded land, the other half in riparian habitats on periodically flooded land, rarely in seasonally dry areas; 0–3,000 m. **Cult:** Propagate from seeds or half-hardened cuttings. Requires a moist soil in full or partial sun.

I. edulis Mart, *Flora* 20, Beibl. 113, 1837. [Ice Cream Bean]. **Trees**, to 20 m tall; trunk usually contorted, cylindrical, 30+ cm dbh, branching 1–2 m from base; crown broad, spreading, almost flat; branchlets ferruginous-tomentose. **Leaves** pinnate, 10–30 cm long; rachis narrowly or broadly winged, glands large, sessile; leaflets (3–)4–6 pairs, subcoriaceous, sessile, ovate to elliptic or lance-oblong, 5–19 × 2–10 cm, base rounded, apex acute or acuminate, adaxially scaberulous or velutinous, abaxially ± shortly pilose; lower leaflets smaller; terminal leaflet 15 × ~7.5 cm; petiole naked. **Inflorescences** spikes or heads, axillary, solitary or in fascicles; peduncles 2–6 cm long, usually dense; bracts ovate or lanceolate, ~5 mm long, deciduous. **Flowers** fragrant; calyx 5–8 mm long, grayish-pubescent; corolla sericeous-villous, 1.4–2 cm long; stamen tube shortly exserted. **Fruits** pods, subterete or somewhat tetragonous, 0.15–0.6(–2) m long, 1.5–3+ cm thick, flat or twisted, densely tomentulose, pulp white, sweet. **Distr:** C & S America. **Ecol:** In moist, wet, or sometimes rather dry forests, or in open habitats; 0–2,000 m. **Zones:** 10–12. **Info:** Cultivated for the edible fruit pulp.

I. velutina Willd., *Sp. Pl.* 4: 1014, 1806. (Syn: *I. expansa*, *Mimosa velutina*). **Trees** or **shrubs**, unarmed, to 20+ m tall, softly rusty-pubescent or shortly villous; crown spreading; trunk straight, columnar, 30–45 cm dbh, branching ~1 m from the ground. **Leaves** pinnate; rachis winged, circular glands conspicuous; leaflets usually 2–4 pairs, sessile or shortly petiolulate, elliptic, 12–15 × 6–7.5 cm, larger leaflet to 30 × 15 cm, base rounded or subcordate, apex obtuse or shortly acuminate, subcoriaceous, adaxially shortly pubescent, subscabrous. **Inflorescences** spikes, solitary or fasciculate in upper axils; peduncles to 6 cm long; bracts caducous before anthesis. **Flowers** with calyx tubular, 1.6–2 cm long, teeth nearly subulate, sometimes 3 mm long; corolla ~3 cm long, silky-villous, slightly ampliated toward apex, staminal tube often long-exserted. **Fruits** pods, plane, straight or curved, ~20 × 3 cm, margins only slightly enlarged, fleshy, densely reddish-hispid. **Distr:** Bolivia, Brazil, Peru. **Ecol:** In clearings or along margins of forests, free from periodical floods; to 500 m. **Zones:** 10–12. **Info:** Wood used for carpentry.

Zygia

Species ~45–50. **Trees** and **shrubs**, unarmed. **Leaves** bipinnate; pinnae 1 pair; leaflets 1–5 pairs, lower leaflets often alternate, larger; main and secondary rachises with glands; petiole short. **Inflorescences** fascicles of shortly pedunculate or subsessile heads or spikes, borne on bare parts of branches and trunk. **Flowers** similar to those of *Pithecellobium*. **Fruits** pods, flat, straight or falcate, margins ± undulated, indehiscent or tardily dehiscent, leathery to woody, stiff. **Distr:** Neotropics. **Ecol:** Mostly in tropical, riparian forests; to 900(–2,800) m. **Cult:** Propagate from seeds or half-hardened cuttings. Requires a moist soil in full or partial sun.

Z. cauliflora (Willd.) Killip, *Trop. Woods* 63: 6, 1940. (Syn: *Inga cauliflora*, etc.). [Hoopwood, Horsewood]. **Trees** small or tall **shrubs**; branches glabrous or slightly pubescent. **Stipules** deciduous or obsolete. **Leaves** bipinnate, rachis 1–7 cm long, puberulent; pinnae 2, each usually with 3–5 opposite leaflets, basal ones often much smaller; leaflets obliquely oblong or ovate-oblong, sometimes falcate, larger ones usually 7–11 × 3–4 cm, apex acuminate, tip rounded, subcoriaceous, lateral nerves unequal, distinct, strongly arcuate, ascending. **Inflorescences** several, fascicled in axils of older branchlets; peduncles 3–8 mm long. **Flowers** white; calyx 1–1.5 mm long, glabrous or slightly ciliate-puberulent; corolla 5–7 mm long, striate; staminal tube ± exserted, filaments red. **Fruits** pods, straight or curved (falcate), to 20 × 1.5 cm, coriaceous. **Distr:** Brazil to the Guianas. **Ecol:** In lowland tropical evergreen forests; to 900 m. Fl. Jun–Jul. **Zones:** 10–12.

Archidendron

Species ~35. **Trees** (to ~30 m tall) or **shrubs** (to ~2 m tall), unarmed. **Leaves** alternate, bipinnate; pinnae opposite, 1–4 pairs, rarely reduced to only 1; leaflets mostly opposite, or in some species with a few leaflets per pinna, at least some leaflets alternate; petiole and/or rachises with glands. **Inflorescences** heads, axillary, pedunculate, sometimes forming terminal panicles, sometimes cauliflorous or ramiflorous; heads few- to many-flowered. **Flowers** usually 5-merous, bisexual, all uniform; calyx large, campanulate, lobes short; corolla tubular, lobes spreading, usually longer than calyx; stamens numerous, filaments united into a tube about as long as corolla, free parts exserted, anthers small; ovaries 2–15 per flower, free, sessile, many-ovuled, styles long, stigmas inconspicuous or small, knob-shaped. **Fruits** pods, ± linear to oblong, often curved to ± circular, flattened, fleshy, leathery, wrinkled, valves twisted after dehiscence; seeds ovoid to ellipsoid, black, glossy, aril absent. **Distr:** India, Malesia, Australia. **Cult:** Propagate from seeds (soaked before sowing). Requires a humus-rich, moist, well-drained soil in full sun.

A. lucyi F.Muell., *Frag. Phyt. Australiae* 6: 201, 1868. (Syn: *Albizia lucyi*, etc.). [Scarlet Bean, Tassel Tree]. **Trees**, evergreen, to 20 m tall, glabrous. **Leaves** large, bi- to tri-pinnate, axes with sunken glands at insertion of pinnae and leaflets; leaflets 2- or 3-jugate, ovate, elliptic or obovate-elliptic, (5.2–)7.5–23 × 3.5–11.5 cm, base acute to acuminate, margins often undulate, apex obtuse; petiolules 2–7 mm long. **Inflorescences** clusters, usually cauliflorous or ramiflorous, to 26.5 cm long, pedunculate, 2–5 flowers per cluster; pedicels (2–)3.5–4.5 mm long. **Flowers** fragrant; calyx cylindric or campanulate, 5–9 mm long, greenish; corolla funnel-shaped, 1.7–2.5 cm long, white or greenish, lobes recurved; stamens numerous, white, exserted, 5 cm long. **Fruits** pods, 7–12 × 1.5–2.5 cm, slightly curved to once-coiled, deeply constricted on one margin between seeds, woody, reddish outside, orange inside, splitting open along one suture; seeds obovoid or ellipsoidal, ~1–1.6 × 1 cm, blue-black with white bloom. **Distr:** Malesia to Australia (N Queensland). **Ecol:** In coastal riverine rainforests or near waterways. **Zones:** 10–12.

Paraserianthes

Species 4. **Trees** or **shrubs**. **Stipules** minute or conspicuous. **Leaves** biparipinnate, with few to numerous paired pinnae along the rachis and each pinna with few to numerous, sessile paired pinnules; glands present on petiole or rachis. **Inflorescences** cylindric spikes or racemes, axillary, solitary or arranged in panicles; bracts and bracteoles usually small. **Flowers** actinomorphic, usually 5-merous, white to yellow, or yellow-green; calyx shortly toothed or lobed; petals connate to about middle; stamens numerous, united into a closed tube for part its length, exserted, anthers uniform, versatile; style filiform, stigma terminal. **Fruits** pods, 2-valved, straight, dehiscent; seeds without an aril. **Distr:** Malesia, Solomon Islands, Australia. **Cult:** Propagate from seeds. Requires a well-drained acidic soil in full sun.

P. lophantha (Willd.) J. Nielsen, *Bull. Mus. Nat. Hist.* (Paris) Sér. 4, Sect. B, 5: 326, 1983. (Syn: *Acacia lophantha*, *Albizia lophantha*). [Brush Wattle,

Below: *Inga edulis*, *Inga velutina*, *Zygia cauliflora*, *Archidendron lucyi*.

PL

Above: *Archidendron lucyi* (fruits), *Paraserianthes lophantha*, *Pithecellobium dulce*, *Pithecellobium dulce* (fruits).

Cape Leeuwin Wattle]. **Shrubs** to **small trees**, evergreen; twigs ribbed, densely hairy. **Stipules** minute. **Leaves** bipinnate, alternate, moderately hairy on rachis, pinnules sparsely hairy abaxially; pinnae in 8–15 pairs; pinnules in 20–40(–45) pairs, fairly closely set, oblong, 0.5–1 × 0.1–0.25 cm, margins entire, apex acute, asymmetric, one prominent vein closer to upper margin; petiole 3–8 cm long; usually a solitary gland present on petiole only, sometimes small glands present between 2 uppermost pairs of pinnules. **Inflorescences** simple racemes, flowers numerous, axillary, much shorter than leaves; flowers shortly pedicellate; bracts subtending raceme; bracteoles small, narrowly triangular, caducous. **Flowers** 5-merous, cream or greenish-yellow; petals inconspicuous; stamens long-exserted, prominent. **Fruits** pods, straight, 8–15 × 1.2–1.8 cm, glabrous; seeds dark brown or black, smooth, ellipsoid, ~7 mm long. **Distr:** W Australia. **Ecol:** In waste places, scrublands, riverbanks, coastal areas, often weedy. Fl. May–Aug. **Zones:** 9–11.

Pithecellobium

Species 18. **Trees** or **shrubs**, armed with spinescent stipules and stipels. **Leaves** bipinnate, not sensitive to touch; rachis and pinnae usually with extra-floral nectaries; leaflets 1 to several pairs per pinna. **Inflorescences** pedunculate glomerules or corymbs aggregated into terminal or axillary panicles. **Flowers** uniform, bisexual, 5-merous; calyx campanulate or tubular, shortly dentate; corolla tubular or infundibuliform, tube longer than lobes, gamopetalous, 5-lobed; stamens numerous, united into a tube at base; ovary sessile or stiped, many-ovuled. **Fruits** pods, straight or spirally contorted, in most species dark brown outside, reddish-orange inside, flat, dehiscing along both sutures; seeds brown or blackish, obovoid or asymmetric, testa hard, with a pleurogram, funicle developed into an aril. **Distr:** N and S America, including the Caribbean; 12 spp. restricted to Mexico, C America and the Greater Antilles (3 spp. extending north into subtropical Mexico, Bahamas and Florida). **Cult:** Propagate from seeds or greenwood cuttings. Requires any well-drained soil in full or partial sun.

P. dulce (Roxburgh) Bentham, *London J. Bot.* 3: 213, 1844. (Syn: *Mimosa dulcis*). [Sweet Tamarind]. **Trees**, evergreen; branches often pendulous; branchlets armed with spinescent stipules. **Leaves** bipinnate; pinnae 1 pair, with glands at junction of pinnae and leaflets; leaflets sessile, 1 pair per pinna, elliptic or obovate-elliptic, 2–5 × 0.2–2.5 cm, both surfaces glabrous, reticulate veins raised abaxially, base slightly oblique, apex obtuse or emarginate. **Inflorescences** pedunculate glomerules, aggregated in terminal panicles. **Flowers** white; calyx infundibular, 1–1.5 mm, tomentose; corolla ~6 mm; stamens numerous, connate into a tube at base. **Fruits** pods, blackish-brown, 10–18 cm long, curved into a circle, flat, 5–7 cm Ø; seeds dark brown, shiny, ovoid-ellipsoid, ~1.5 cm, hard, with pleurogram. **Distr:** C America to northern S America. **Ecol:** Fl. Mar–Jun. Fr. Jul. **Zones:** 9–12.

Albizia

Species ~145. **Trees** or **shrubs**, unarmed. **Stipules** setaceous, obsolete or rarely papery. **Leaves** bipinnate, petiolar and jugal glands ± conspicuous. **Inflorescences** pedunculate globose heads, racemosely arranged; flowers sessile or pedicellate. **Flowers** bisexual; calyx gamosepalous, dentate or shortly lobed; corolla gamopetalous, funnel-shaped, petals connate beyond middle; stamens generally indefinite, basally connate, or connate higher up into a tube, filaments elongated, white, rose or rarely purple, anthers small, eglandular; ovary shortly stipitate or sessile, style filiform, stigma capitate or minute. **Fruits** pods, broadly linear, thin, compressed, sometimes dehiscent, continuous inside the valves and without pulp; seeds ovate or orbicular, compressed, funicle filiform. **Distr:** Tropical and subtropical regions of the world. **Cult:** Propagate from seeds, cuttings or by air-layering. Requires well-drained soil in full sun.

A. adianthifolia (Schumach.) W.Wight, *USDA Bur. Pl. Industr. Bull.* 137, 12, 1909. (Syn: *Inga fastigiata*, etc.). [Flat Crown Albizia, Rough Bark Flat Crown, West African Albizia]. **Trees**, to 20 m tall; crown flat-topped; trunk single-stemmed; young branches densely hairy. **Leaves** alternate, 10–40 cm long, bipinnate; pinnae 4–8 pairs; leaflets 6–15 pairs per pinna, narrowing towards apex, oblong 0.7–2 × 0.4–0.8 cm, midrib running diagonally across leaflet, adaxially dark green, abaxially paler and densely hairy; petiole and rachis hairy, large gland borne above pulvinus; petiole 2.4–7 cm long. **Inflorescences** flowerheads, clustered, erect, pedunculate, 5–7 cm Ø; pedicels with indumentum. **Flowers** small, round, creamish-white; calyx and corolla brown. **Fruits** pods, flat, light brown. 7.5–16 × 1.9–3.4 cm, hairy; seeds swollen. **Distr:** Sudan, Ethiopia and Somalia southwards to Portuguese E Africa and S Rhodesia. **Ecol:** In forests and moist bushveld. Fl. Aug–Dec. Fr. Sep–Feb. **Zones:** 10–12. **Info:** The leaflets close when picked or after sunset.

A. anthelmintica Brongn., *Bull. Soc. Bot. Fr.* 7: 902, 1860. (Syn: *A. umbalusiana*, *Besenna anthelmintica*). [Worm-bark False-thorn, Worm-cure Albizia]. **Shrubs** or **trees**, 2–9(–12) m tall, deciduous; twigs often with short divaricate almost spinescent-tipped lateral branches. **Leaves** pinnate, pinnae 1–2(–4) pairs; leaflets 1–4(–5) pairs, obliquely obovate or elliptic to subcircular, (0.7–)1–3.6(–4.2) × (0.4–)0.6–3.1 cm, apex mucronate, venose, glabrous to ± sparsely shortly pubescent. **Inflorescences** heads; flowers usually on leafless twigs; pedicels 0.5–5.5 mm long. **Flowers** with calyx pale-greenish, (2–)3–5 mm long, glabrous to sparsely finely pubescent outside, apex irregularly denticulate, usually slit unilaterally to ~1–2.5 mm; corolla pale-greenish, 0.6–1.2 cm long, glabrous; staminal tube not or scarcely exserted beyond the corolla; filaments ~1.5–2.5 cm long, white. **Fruits** pods, oblong, (6–)7–18 × 1.5–2.9 cm, dehiscent, glabrous or sometimes puberulous, straw-colored; seeds 0.9–1.3 cm Ø, flattened, round. **Distr:** Sudan and Ethiopia (Eritrea) southwards to the Transvaal and Zululand. **Ecol:** In dry woodlands, wooded grasslands, thickets and bushland; ~30–1,070 m.

A. julibrissin Durazz., *Mag. Tosc.* 3(4): 11, 1772. [Mimosa, Persian Acacia, Pink Siris, Silk Tree]. **Trees**, medium-sized; bark dark-gray; young parts yellowish-brown hairy. **Stipules** linear, 7–8 mm long, caducous. **Leaves** alternate, bipinnate, rachis ~10–25 cm long with a large gland ~1–2.5 cm from base; pinnae 4–15 pairs, ± sessile, 7–15 cm long; leaflets 10–30 pairs, falcate-oblong, 1.2–1.8 × ~0.3–0.7 cm, oblique, apex acute, hairy. **Inflorescences** peduncled heads, solitary, or in fascicles of 2–3 arranged in terminal racemes; peduncles 3.5–7 cm long; bract ~3–6 mm long, linear; pedicels ~1–2 mm long. **Flowers** with calyx tubular, ~3–4 mm long, velvety or not, teeth short, triangular; corolla 7–8 mm long, hairy outside, lobes lanceolate, ~2–3 mm long, acute; stamens 2.5–3.2 cm long, staminal tubes as long as the corolla tube. **Fruits** pods, ~7.5–12.5 × ~1.5–2.5 cm, hairy till maturation, pale brown or yellowish; seeds 8–12. **Distr:** Asia minor, Caucasus, China, Cyperus, Japan, Kashmir, Persia, Russia. **Ecol:** Fl. Jun. **Zones:** 8–11. **Info:** 2 subspecies described.

A. lebbeck (L.) Benth., *Lond. J. Bot.*, 3, 1844. (Syn: *Acacia lebbeck*, etc.). [Black Siris, East Indian Walnut, Silk Tree, Siris Tree, Whistling Bean]. **Trees**, large, deciduous; young parts usually hairy. **Stipules** linear, ~3–4 mm long, caducous, tomentose. **Leaves** bipinnate, rachis 7.5–15 cm long, with a large gland 1.2–3.7 cm from base; pinnae 1–4 pairs, 5–20 cm long, often with glands, mostly between upper pairs; leaflets 3–9 pairs, lateral leaflets oblong, terminal ones obovate, obtuse or retuse; petiolule ~1 mm long. **Inflorescences** pedunculate heads, solitary or fascicled; peduncle 3.5–10 cm long; pedicel hairy, ~2–3 mm long, bracteate; bract linear, 5 mm long, caducous. **Flowers** greenish-white, very fragrant; calyx campanulate 3–4 mm long, hairy, teeth short, teeth deltoid-acute; corolla funnel-shaped, 7–8 mm long, lobes ovate, ~2 mm long, acute, hairy outside; stamens 2.5–3.8 cm long, staminal tube slightly shorter than corolla tube, anthers minute. **Fruits** pods, 15–30 × ~2.5–5 cm, thin, pale straw-colored; seeds 6–12 compressed, pale brown, surfaces foveolate. **Distr:** Tropical Africa and Asia, N Australia. **Ecol:** Fl. Apr–May. **Zones:** 10–12.

A. odoratissima (L.f.) Benth., *Lond. J. Bot.*

PL

Top row: *Albizia adianthifolia, Albizia anthelmintica.*
Bottom row: *Albizia julibrissin, Albizia lebbeck, Albizia odoratissima, Albizia saman.*

3:88, 1844. (Syn: *Acacia odoratissima, Mimosa odoratissima*). [Black Siris, Ceylon Rose Wood, Fragrant Albizia]. **Trees**, almost evergreen; stem erect; bark grayish with patches; young shoots brown-hairy. **Leaves** alternate, rachis 10–20 cm long, base with a large black gland, another gland between the uppermost pair, minute gland below the last 1–2 pairs of leaflets; pinnae 3–8 pairs, 4.5–9 cm long; leaflets 8–20 pairs, oblong, 1.7–2.5 × ~0.5–1 cm, oblique, usually obtuse or acute, abaxially hairy. **Inflorescences** peduncled heads solitary or in panicles, brownish-pubescent; peduncle 0.7–1.2 cm long; flowers sessile, bracteate; bract ovate, ~2–3 mm long, acute, hairy. **Flowers** pale yellow, fragrant; calyx ~1–2 mm long, densely pubescent, teeth subobsolete; corolla to 8 mm long, gray-hairy outside, lobes lanceolate, 2–3 mm long, acute; stamens ~20, 1.5 cm long, staminal tube ~2–3 mm long; style filiform, as long as the filaments, green. **Fruits** pods, 10–30 × ~1.7–3.5 cm, reddish-brown, thin, shiny, hairy when young, shortly stalked; seeds 8–12. **Distr:** India, Myanmar, W Pakistan, Sikkim, Sri Lanka. **Ecol:** Fl. Apr–Jun. **Zones:** 9–12.

A. saman (Jacq.) Merr., *Sel. Pl. Indust. Cult.* 12, 1891. (Syn: *Samanea saman*, etc.). [East Indian Walnut, Monkey Pod Tree, Rain Tree, Saman]. **Trees**, large, unarmed; old trees with pendent branches. **Stipules** caducous. **Leaves** large, bipinnate; pinnae 2–6 pairs; leaflets 2–8 pairs per pinnae; primary rachis 4–6 cm long; axis of pinna ± like primary rachis; leaflets asymmetrically ovate, oblong or obovate, to ~4 × 2 cm, base obliquely obtuse or rounded, apex obtuse or rounded, adaxially puberulent or subglabrous and lustrous, abaxially yellow-tomentulose; petiole mostly 4–6 cm long. **Inflorescences** umbels, 1 to few, congested, axillary and subterminal on branchlets; peduncles 4–8 cm long, tomentose; pedicels mostly 1–3 mm long; bracts lanceolate, ~3 mm long. **Flowers** whitish or pinkish, dimorphic; *central* (sessile) flower of umbel larger, persistent, calyx campanulate, ~5 × 3–4 mm, teeth 10, staminal column gross and thick, stamens numerous, relatively short, ovary ovulate, this flower surrounded by 5 oblanceolate bracts ~5 mm long; *outer* (pedicellate) flowers first to fall, calyx funnelform, ~5 × 1–2 mm, 5-toothed, yellow-tomentose, corolla funnelform, ~1 cm long, valvate, apex only yellow-pubescent, stamens ~3.5 cm long, staminal tube included, ovaries sessile, subglabrous, style base pubescent, becoming apically ± glabrous. **Fruits** pods, linear or linear-oblong, to 25 × 2 cm, valves thick; seeds hard. **Distr:** Mexico, C America. **Ecol:** In sandy, coastal areas; 0–1,800 m. Fl. & Fr. Aug–Apr. **Zones:** 10–12.

Enterolobium

Species 5. **Trees**, large, erect, unarmed. **Stipules** inconspicuous. **Leaves** evenly bipinnate; pinnae and leaflets in numerous pairs; petiole with gland. **Inflorescences** capitula, globose, solitary or in racemes, axillary. **Flowers** dimorphic or uniform, bisexual, 5-merous; calyx campanulate, shortly 5-dentate; corolla funnel-shaped, 5-lobed; stamens monadelphous, numerous, basally connate into a tube, anthers eglandular; ovary sessile, many-ovuled, style filiform. **Fruits** pods, broadly circinate or incurved-reniform, thick, hard, indehiscent; mesocarp spongy, at length hardened; seeds transverse, compressed, with pleurogram, funicle filiform. **Distr:** Tropical America. **Cult:** Propagate from seeds, cuttings or by air-layering. Requires well-drained soil in full sun.

E. cyclocarpum (Jacquin) Griseb, *Fl. Brit. W. Ind. Isl.* 226, 1860. (Syn: *Mimosa cyclocarpa*). [Earpod Tree, Elephant Ear, Monkey Soap]. **Trees**, unarmed, deciduous, large, 10–20 m tall; crown spreading, thin; young parts white-pubescent. **Stipules** caducous, small. **Leaves** bipinnate; pinnae (3 or)4–9 pairs; leaflets 12–25(–30) pairs, subsessile, falcate-lanceolate, 0.8–1.4 × 0.3–0.6 cm, base truncate, apex mucronate, sparsely pubescent, midvein close to upper margin; petiole and rachis with glands. **Inflorescences** heads, globose, 1–1.5 cm Ø, fasciculate or in racemes. **Flowers** greenish or white; calyx ~3 mm long, pubescent; corolla ~6 mm long, pubescent; stamens numerous, basally connate into a tube. **Fruits** pods, black-brown, curved, auriculate-reniform, 5–7 cm Ø, fleshy, ends rounded, indehiscent; seeds dark brown, 10–20, arranged in 2 rows, shiny, narrowly ellipsoidal, ~1.5 cm long, hard. **Distr:** C & S America. **Ecol:** Fl. Apr–Jun. Fr. Oct–Dec. **Zones:** 10–12.

Subfam. FABOIDEAE

Syn: PAPILIONOIDEAE.
Genera 476, **species** 13,855. **Herbs**, **shrubs** , or **trees**. **Leaves** pinnate or palmate to trifoliolate or apparently simple. **Flowers** with corolla usually showy, zygomorphic; petals imbricate, posterior (upper or banner) petal outermost in bud; stamens 10 or 9+1 (diadelphous), not showy; pollen released in monads. **Seeds** with U-shaped line (pleurogram) lacking. **Distr:** Worldwide.

Above: *Enterolobium cyclocarpum, Swartzia pinnata, Castanospermum australe, Myroxylon balsamum.*

TRIBE SWARTZIEAE

Swartzia

[Bastard Rosewood].
Species ~180. **Trees**, small (rarely large shrubs, sarmentose trees or vines), glabrous, unarmed; smaller branches usually conspicuously lenticellate. **Stipules** usually small or very small, caducous. **Leaves** alternate; lamina imparipinnate or pinnately trifoliate or unifoliate; leaflets opposite or more rarely alternate, without pellucid dots, ovate to elliptic, membranous to coriaceous, prominently veined, principal lateral vein anastomosing towards margin; petiole and rachis (if present) usually obsagittate-alate or angled; petiolules short, terete. **Inflorescences** racemes, few- to many-flowered, lateral, short, fascicled at older nodes, sometimes panicled, rarely solitary; bracts caducous, bracteoles caducous or rarely persistent. **Flowers** showy, hermaphrodite; calyx tube disclike or obsolete, limb before anthesis entire, obovoid, ellipsoid or globose, afterwards variously ruptured (usually 3–5 irregular sections) or rarely cyathiform and erose-dentate; petals 1 (the standard), rarely 2 reduced lateral petals or entirely absent, crinkled, usually large, suborbicular, clawed; stamens numerous (>30), arranged in several rows (of 2 types), declinate and incurved-ascending; ovary stipitate, many-ovuled. **Fruits** pods, stipitate, ovoid or elongate, turgid, cylindrical or subterete, coriaceous, woody or fleshy; seeds 1–many, reniform, ovoid or globose. **Distr:** Tropical America; 2 species from tropical Africa. **Cult:** Propagate from seeds or by air-layering. Requires a humus-rich, moist, well-drained soil in shade or partial sun.

S. pinnata (Vahl) Willdenow, *Sp. Pl.* 2: 1220, 1799. (Syn: *Rittera pinnata*, etc.). **Trees**, 7–15 m tall; trunk ~30 cm dbh. **Stipules** triangular, 0.6–0.7 mm long, caducous. **Leaves** pinnate; leaflets (1- or) bijugate, rarely trijugate, lamina of the lowermost pair elliptic and smaller, (8.5–)11(–15) × (4.5–)5–7(–8) cm, other pairs elliptic to oblanceolate-elliptic, (13–)20–25(–30) × (5–)8–10(–13.5) cm, bases rounded-obtuse, apices obtuse, acute, or acuminate; petiole terete, 3–8 cm long; petiolules (3–)5(–7) cm long. **Inflorescences** 10–35(–44) cm long, often fasciculate, racemose; bracts triangular, (0.3–)0.7–1(–1.5) mm long, persistent; pedicels slender, 1.5–2(–2.7) cm long. **Flowers** with calyx segments 4, deciduous; petal yellow, rounded, 1.2–1.5 cm Ø, ± cuneate, claw 1.5–4 mm long; larger stamens 2, smaller stamens 0.9–1.8 cm long; ovary arcuate-linear, 0.7–1.5 cm × 1–1.5 mm, style 1–5 mm long, gynophore 5.5–7.5 mm long. **Fruits** pods, moniliform, seeds enclosed in an oval or elliptic section 2–4.5 × 1.5–2 cm; seeds 1–10, ovate, 2–2.8 × 1.5–2 cm, smooth, dark brown. **Distr:** Trinidad and coastal states of N Venezuela. **Ecol:** In forests; 50–800 m. **Zones:** 10–12.

TRIBE SOPHOREAE

Castanospermum

Species 2. **Trees**. **Leaves** alternate, imparipinnate; leaflets exstipellate, large, coriaceous. **Inflorescences** racemes, short, borne on one year old branches to old wood; bracts minute; bracteoles absent. **Flowers** large, yellow to orange; calyx colored, teeth short and broad; vexillum obovate-orbicular, larger than other petals; stamens 10, free, anthers versatile; ovary stipitate, style curved, stigma terminal. **Fruits** large, subfalcate, woody coriaceous, spongy between seeds; seeds subglobose, large. **Distr:** Subtropical Australia and New Caledonia. **Ecol:** In rainforests. **Cult:** Propagate from seeds or by air-layering. Requires humus-rich, moist but well-drained soil in full or partial sun.

C. australe A. Cunn. & Fraser, *Bot. Misc.* 1: 241, t.51, 1830. [Black Bean Tree, Moreton Bay Chestnut, Queensland Black Bean]. **Trees**, evergreen, ~15 m tall; crown open, spreading; bark dark brown. **Leaves** imparipinnate, 30–45 cm long, deep green; leaflets 9–17, elliptic-oblong, ~7.5–15 × 5–6 cm, slightly curved, margins wavy, apex acuminate, glabrous. **Inflorescences** racemes, lateral, arising from the old wood; pedicels ~1.6–2.2 cm long. **Flowers** yellow, aging to bright orange-red, 3–4 cm long; calyx cup-shaped, teeth obtuse; vexillum ~3 cm long. **Fruits** pods, ~20–30 × 5 cm (sometimes smaller in cultivation), turgid, woody, spongy inside between the seeds; seeds 1–5, dark brown to black. **Distr:** Australia (Queensland & NSW). **Ecol:** In rainforests, gallery forests, along creeks and rivers; 0–1,000 m. Fl. Oct–Mar. Fr. Feb–Apr. **Zones:** 10–12. **Info:** Cultivated in the tropics for ornamental purposes. The seeds are poisonous, containing substances of medical interest.

Myroxylon

Syn: *Toluifera*.
Species 2. **Trees**, unarmed; wood aromatic. **Leaves** alternate, imparipinnate; leaflets 5–15, alternate, estipulate, pellucid-lineate or punctate. **Inflorescences** paniculate racemes, terminal and axillary; bracts and bracteoles caducous; flowers pedicellate. **Flowers** fragrant, white, small; calyx turbinate-campanulate, 5-lobed, subequal, irregularly dentate; standard broadly orbicular, the 4 lower petals subequal, narrow; stamens 10, free, deciduous with the petals, free or nearly; anthers uniform, acuminate; ovary long-stiped, 1(–2)-ovuled at apex, style short with a small terminal stigma. **Fruits** samaras, compressed, indurate and 1-seeded at tip, the sterile lower part 2-winged, like an inverted samara; seed reniform. **Distr:** Mexico, C America, W & N South America to N Argentina. **Ecol:** In tropical lowland forests to seasonally dry forests and woodlands. **Cult:** Propagate from seeds and by air-layering. Requires a moist, well-drained soil in full or partial sun.

M. balsamum (L.) Harms, *Notizbl. Bot. Gart. Berlin* 5: 94, 1908. (Syn: *M. toluifera, Toluifera balsamum*). [Balsam, Balsam of Peru, Tolu Balsam]. **Trees**, to 40 m tall, usually glabrous except the inflorescences; trunk ~1 m dbh. **Leaves** pinnate; leaflets alternate, 5–10, ovate or ovate-oblong, usually 6–9 × 3–4 cm, apex acuminate, often obtusely, ± subcoriaceous, venation reticulate, lustrous; rachis and petiolules pubescent, terete; rachis 5–15 cm long. **Inflorescences** racemes, tomentulose; pedicels 1.5 cm long, flowers borne oblique on the pedicel. **Flowers** white; calyx campanulate, 1.5–4.5(–6) mm long, finely ribbed, lobes ~1.5 mm long; standard orbicular, ~9 mm Ø, cordate basally, claw ~1 mm long; wings elliptic to narrowly spathulate, ~1 × 0.4 cm, claw 2 mm long; keel subelliptic, ~8.5 ×3 mm, claw ~1.5 mm long; ovary shortly stipitate for ~2 mm, villosulose, style subulate. **Fruits** samaras, usually ~8 × 2.5 cm, style-tip noticeably situated below the curved upper edge, seed portion with 2 globose balsam pits between the corky mesocarp and coriaceous endocarp, leaving the seeds smooth, testa if present dry and thin. **Distr:** Brazil to Venezuela and C America. **Ecol:** In wet to dry, lowland forests; <300 m. Fl. May–Jun. **Zones:** 10–12.

Ormosia

Species ~130. **Trees** or **shrubs**, very tall to small and shrubby. **Stipules** small and inconspicuous; stipels minute or none. **Leaves** odd-pinnate; leaflets coriaceous. **Inflorescences** terminal panicles, or rarely axillary racemes or panicles; bracts and bractlets small or minute, linear. **Flowers** white to lilac or purple, upper 2 calyx campanulate, hypanthoid, lobes (deltoid teeth) 5, subconnate, usually broader and incurved; corolla papilionoid, twice as long as calyx, composed of 5 different petals, yellow to blackish-purple; standard suborbicular, sometimes with a contrasting whitish or reddish spot; wings oblique, oval-oblong; keel similar to the others, more incurved, free; stamens 10, free, unequal, all perfect or 1–2 without anthers, anthers versatile; ovary subsessile, 2-many-ovuled, style filiform, involute at apex, stigma introrse, lateral. **Fruits** pods, oblong or rarely elongate, compressed or turgid about the seeds, coriaceous or ligneous, 2-valvate, continuous or septate within; seeds obovate or oblong, thick, lustrous, scarlet or bicolored. **Distr:** Tropical America, Asia. **Ecol:** In tropical rainforests, seasonally dry forests, xerophytic woodlands, etc. **Cult:** Propagate from seeds or by air-layering. Requires a well-drained soil in full sun.

O. monosperma (Sw.) Urb., *Ant.* 1: 321, 1899. (Syn: *O. dasycarpa*, etc.). [Snakewood]. **Trees**, to ~17 m tall; young stems fulvo- to ferrugino-tomentulose. **Stipules** linear-deltoid, ~3 mm long. **Leaves** (5–)7–11-foliolate, axis ~6–16 cm long, tomentulose, glabrescent; leaflets sometimes subopposite, ovate to oblong, 4–20 × 1.5–6 cm, base rounded to acute, apex acute to acuminate, acumen sometimes to 2 cm long, subcoriaceous, adaxially glabrous, subnitid, abaxially tomentulose along major veins, secondary veins ~12–15 pairs; petiole 2–4 cm long, petiolules 3–4 mm long. **Inflorescences** with axes ferrugino-tomentulose; bracts linear-deltoid, ~3–5 × 1 mm; bracteoles linear, 2 mm long. **Flowers** 1.5–2 cm long; calyx ferrugino-tomentulose, 0.8–1 cm long, tube 4–5 × 5 mm, teeth deltoid, 3–5 mm long; corolla dark purple, standard with a white spot. **Fruits** dehiscent, 2–6 × (2–)2.5–3.5 cm, ~2 cm thick, lignous or sublignous, densely fulvo- to ferrugino-velutinous, valves 2–3 mm thick; seeds 1–3, red and black, (1.3–)1.5–1.7 × (1–)1.5–1.7 cm. **Distr:** Lesser Antilles, Trinidad, and NE Venezuela. **Ecol:** In primary forests, on hillsides; to ~1,600 m. **Zones:** 10–12.

Sophora

Species ~50. **Trees**, **shrubs**, **subshrubs** or perennial **herbs**, deciduous or evergreen, rarely twining. **Stipules** present or absent, rarely with stipels. **Leaves** imparipinnate; leaflets many, margins entire. **Inflorescences** racemes, terminal or axillary; bracts small, linear or absent. **Flowers** white, yellow or purple; calyx campanulate or cup-shaped, 5-lobed, equal, upper bilobed nearly bilabiate; banner orbicular, oblong, elliptic, obovate or oblanceolate; wings asymmetrical or symmetrical; keels similar to wings; stamens 10, free or joined at base, anthers versatile; ovary short-stalked or absent, many-ovuled, stigma small, terminal, rounded. **Fruits** pods, cylindric, moniliform, fleshy or leathery, sometimes winged, indehiscent or tardily dehiscent; seeds 1 or many, ovate, elliptic or ± spherical, black, dark brown or light red. **Distr:** SE Europe to W, C & E Asia and south through tropical regions to Australasia and the Pacific. **Ecol:** In seasonally dry tropical to warm temperate lowland and highland forests, and also in arid regions. **Cult:** Propagate from seeds or cuttings. Requires a well-drained soil in full or partial sun.

S. microphylla Ait., *Hort. Kew.* 2: 43, 1789. (Syn: *Edwardsia grandiflora* var. *microphylla*, etc.). [Kowhai]. **Trees**, to ~10 m tall; trunk to 60 cm dbh; branches spreading or drooping, young branchlets pale to fulvellous tomentose. **Leaves** to 15+ cm long, with 20–40 pairs of leaflets; leaflets obovate-oblong, 0.5–0.7(–1) cm long, surfaces silky-hairy, usually becoming glabrous, midrib distinct abaxially. **Inflorescences** racemes, 4–10-flowered. **Flowers** to 4.5 cm long, pale to golden-yellow; calyx silky-tomentose, 1+ cm long; standard to 3 cm long, wings equal in size or slightly larger than the standard; keel ~4 cm long. **Fruits** pods, fulvellous-tomentose when young, broadly winged, to 15+ cm long; seeds 6+, brownish to golden-yellow, 5–8 mm long. Distr: New Zealand, Chatham Islands. **Ecol:** In lowland and lower montane regions. **Zones:** 8–11.

S. prostrata Buchan., *Trans. NZ. Inst.* 16: 395, t. 36, 1884. (Syn: *Edwardsia prostrata*, *S. tetraptera* var. *prostrata*). [Prostrate Kowhai]. **Shrubs**, bushy, prostrate, to ~2 m tall; branches densely divaricating, mature branchlets brown, glabrous. **Leaves** to 2.5 cm long; leaflets to 8 pairs, oblong, ~4 mm long, glabrous when mature except on midrib abaxially, veins prominent. **Inflorescences** with flowers solitary, paired, or 3 together; peduncles slender, silky-hairy. **Flowers** orange to brownish-yellow, rarely bright yellow; calyx appressed-hairy, ~1 cm long; standard to 1.5 cm long, slightly or distinctly shorter than wings; keel ~2 cm long. **Fruits** pods, to 3(–5) cm long, including beak, silky-hairy when young, very narrowly winged; seeds up to 5, dark to reddish-brown, ~7 mm long. **Distr:** New Zealand. **Ecol:** In grasslands and rocky places, of lowland to montane regions. **Zones:** 8–11.

S. tetraptera J. Mill., *Ic. Pl.*, t.1, 1780. (Syn: *Edwardsia grandiflora*, *E. tetraptera*). [Kowhai]. **Trees**, to ~12 m tall; trunk to 60 cm dbh; branches spreading, young branchlets covered with fulvellous tomentum. **Leaves** to 15+ cm long; leaflets 10–20 pairs, sometimes more, ovate to elliptic-oblong, 1.5–3.5 × 0.5–0.8 cm, surfaces covered with silky appressed hairs, midrib distinct abaxially. **Inflorescences** racemes, 4–10-flowered. **Flowers** to 5 cm long, golden-yellow; calyx 1–1.5 cm long, covered in appressed silky hairs; standard ~3 cm long; wings ~3.5 cm long; keel ~4.5 cm long. **Fruits** pods, to 20 cm long, broadly winged; seeds 6 or more, yellowish, ~7 mm long. **Distr:** New Zealand. **Ecol:** Along streamsides, forest margins, in lowland regions. **Zones:** 8–11.

S. tomentosa L., *Sp. Pl.* 1: 373, 1753. [Coast Sophora, Sea-coast Laburnum, Yellow Necklacepod, Yellow Sophora]. **Shrubs** or **small trees**, 2–4 m high; branches shortly tomentose. **Stipules** absent. **Leaves** 12–18 cm long; leaflets 11–15(–19), oval or rounded, rarely ovate, 2.5–5 × 2–3.5 cm, nearly leathery, adaxially glabrous, abaxially densely gray-white tomentose, lateral veins obscure, base rounded, apex rounded or retuse. **Inflorescences** dense racemes, terminal, 10–20 cm long, gray-white tomentose; pedicel equal to calyx, 1.5–1.7 cm long; bracts linear. **Flowers** with calyx campanulate, 5–6 mm long, shortly tomentose; corolla yellow or creamy-white; standard broadly ovate, ~1.7 × 1 cm, claw ~3 mm long; wings long-elliptic, 1-auriculate, claw slender, ~5 mm long; keel similar to wings, slightly shorter; stamens 10, free; ovary densely

Top row: *Ormosia monosperma*, *Sophora prostrata*.
Bottom row: *Sophora microphylla*, *Sophora tetraptera*, *Sophora tomentosa*.

TG

Above: *Camoensia maxima, Baphia nitida, Baphia racemosa, Dipteryx odorata.*

pubescent, style short, <2 mm long. **Fruits** pods, moniliform, 7–10 × 0.9–1 cm, shortly hairy; seeds many, brown, globose, lustrous. **Distr:** Throughout the tropics. **Ecol:** On sand dunes, seashores. Fl. Aug–Oct. Fr. Sep–Dec. **Zones:** 10–12.

Camoensia

Species 2. **Shrubs**, climbing. **Leaves** digitately trifoliate. **Inflorescences** racemes, axillary. **Flowers** with calyx turbinate, lobes imbricate; petals papilionaceous with long claws; standard orbicular, others narrower, all free; stamens free, anthers uniform, linear, versatile; ovary stipitate, multiovulate, style filiform, involute in the unexpanded flower, stigma terminal, small, capitate. **Fruits** pods, broad-linear, flattened, coriaceous, 2-valved. **Distr:** West tropical Africa. **Cult:** Propagate from seeds, cuttings or by air-layering. Requires a fertile, moist soil in full or partial sun. Needs support for climbing.

C. maxima Welw. ex Benth., *Trans. Linn. Soc. London* 25(2): 302, t. 36, 1865. (Syn: *C. scandens, Giganthemum scandens*). [Climbing Camoensia]. **Climbers**, woody, lofty; branches terete, glabrous. **Stipules** rigid and subspinescent. **Leaves** digitately trifoliate; leaflets with very short petiolules and linear stipels, obovate-oblong, 12–15 cm long, narrowed below middle, cuspidate, subcoriaceous, glabrous; petiole 5–6.3 cm long. **Inflorescences** racemes, 6–8-flowered, axillary; pedicels 2.5–5 cm long, downy; bracts and bracteoles lanceolate. **Flowers** large, to 28 cm long, extremely fragrant in the evening; calyx densely reddish woolly tomentose, tubular 15–17.5 cm long; stamens inserted halfway up, slit into 5 lanceolate segments 2.5–3.8 cm deep, which are finally recurved; petals cream-colored, with deeper colored veins and a golden border, edges crisped and crenulated, standard protruding 10 cm beyond calyx, limb suborbicular, 7.5–10 cm across, other petals shorter and <2.5 cm wide; stamens 11, united below. **Fruits** pods, 15–20 × 2.5–3.8 cm, nearly straight, ferruginous woolly tomentose; seeds 3–4. **Distr:** Angola, Zaire and Equatorial Guinea. **Ecol:** In forests. **Zones:** 10–12.

Baphia

Species 47. **Trees** (<45 m tall, sometimes buttressed), **shrubs** and **lianas** (<80 m long). **Stipules** usually soon caducous. **Leaves** alternate, unifoliolate; lamina entire, variable, base usually rounded, cuneate, apex usually acuminate, coriaceous to chartaceous; petiole short. **Inflorescences** lax racemes or pseudo-racemes, sometimes grouped into pseudo-panicles, sometimes reduced to paired or solitary flowers, terminal or axillary; bracts usually soon caducous; bracteoles 2, linear to suborbicular. **Flowers** usually fragrant, erect or deflexed; calyx splitting by 1 or 2 longitudinal fissures at anthesis forming a sort of spathe or bilobed calyx respectively; standard usually clawed, broadly ovate to suborbicular, usually white with a yellow or orange blotch at base, sometimes cream, purplish-pink or pinkish-red, glabrous; wings usually basally pouched, white sometimes purplish-pink or purplish-red; keel oblong incompletely fused at dorsal margin, color as in standard and wing; stamens 10, filaments free; ovary subsessile, style subulate, ovules 2–8. **Fruits** pods, linear-oblong, curved, usually laterally compressed, woody, brown or purplish-black, dehiscing into 2 valves; seeds 1–5, brown or blackish. **Distr:** Tropical Africa, S Africa, Madagascar. **Ecol:** In tropical lowland rainforests and seasonal forests. **Cult:** Propagate from seeds or by air-layering. Requires a fertile, moist, well-drained soil in full or partial sun.

B. nitida Lodd., *Bot. Cab.* t. 367, 1825. (Syn: *B. angolensis*, etc.). [African Sandalwood, Barwood, Camwood]. **Shrubs** or **small trees**, many-stemmed, erect, to 9 m tall; branchlets glabrous or densely pubescent. **Leaves** very variable, usually ovate or oblong-elliptic, 5–21 × 3–9 cm, base rounded or cuneate, apex acuminate, acumen pointed, papyraceous to subcoriaceous, glabrous, lateral veins 5–8 pairs; petiole 1–4 cm long, pulvini discrete. **Inflorescences** fascicles, 2–5-flowered or flowers solitary, axillary; pedicel 0.3–1.7 cm long; bracteoles under calyx, opposite, free at base, broadly ovate, suborbicular or reniform, 1.5–2.5 × 1.5–3.2 mm, margins ciliate or sparsely hairy. **Flowers** with calyx spathaceous, 0.8–1 cm long, with tufts of brown hairs at apex; standard 1.3–2 × 1.3–2 cm; wings and keel with a pocket near base; stamens with filaments 3.5–7 mm long; ovary glabrous or with a row of long silvery-gray hairs along dorsal margin. **Fruits** pods, 8–16.5 × 1–1.5 cm; seeds 1–4, 1–1.5 × 1–1.5 cm, brown. **Distr:** Tropical Africa. **Ecol:** In rainforests of coastal regions, secondary forests; 0–600 m. Fl. & Fr. throughout the year. **Zones:** 10–12.

B. racemosa (Hochst.) Baker, *Fl. Trop. Afr.* 2: 248, 1871. (Syn: *Bracteolaria racemosa*). [Natal Camwood, Tree Violet, Violet Pea]. **Shrubs** or **small trees**, to 8 m tall; branchlets grayish-brown tomentose. **Leaves** ovate to lanceolate, sometimes elliptic, 4–11 × 1.5–5.5 cm, base rounded or cuneate, apex acute to acuminate or obtuse, subcoriaceous or chartaceous, glabrescent, major lateral veins 5–8 pairs; petiole 1–3.5 cm long, glabrescent, pulvini discrete. **Inflorescences** axillary racemes, rarely terminal racemes or pseudo-panicles, 3–4.5 cm long, brown-tomentose; pedicels 6–8 mm long, grayish-brown tomentose; bracteoles under calyx, opposite, free at base, linear-lanceolate, 2–3.5 × 0.7–1 mm, tomentose. **Flowers** with calyx bilobed, 5–7 mm long, toothed, grayish-brown tomentose; standard 1–1.25 × 1–1.25 cm; wings and keels with a pocket near base; filaments 4–7.5 mm long, anthers 1–1.3 mm; ovary densely villous. **Fruits** pods, 4.5–6.5 × 1.3–1.6 cm, straw-colored; seeds 1–2, 1.3–1.7 × 0.9–1.2 cm, brown. **Distr:** S Africa (Natal Province). **Ecol:** In the subcanopy layer of lowland rainforests. Fl. Sept–Jan. Fr. Mar–Apr, Aug–Oct. **Zones:** 9–12.

TRIBE DIPTERYGEAE

Dipteryx

Species ~12. **Trees**, unarmed, small to very large; bark and sapwood of some species coumarin-scented. **Leaves** opposite alternate, paripinnate; leaflets opposite or alternate, 3–14, often gland-dotted; rachis slightly winged or flattened, apex projected beyond leaflets. **Inflorescences** panicles, terminal; bracts and bracteoles small or later absent. **Flowers** violet or rose, showy, fragrant in some species; calyx tube short, campanulate, pinkish, coriaceous, pubescent, glandular, short-toothed, or with 2 upper lobes very large, winglike or petaloid, membranous, lower 3 joined into a small entire or 3-toothed lip; petals clawed, very unequal; standard obovate, emarginate; wings falcate or oblique-oblong, free, often emarginate; keel petals similar to wings, smaller, free or joined dorsally; stamens 10, all joined into a sheath split above, 5 alternative ones shorter or abortive; ovary linear, glabrous, 1-ovuled, style slender, stigma terminal. **Fruits** pods, drupaceous, elliptic, somewhat compressed, endocarp woody, with or without an oily, aromatic pericarp, indehiscent; seed 1, flat, oblong, strongly coumarin-scented or odorless. **Distr:** C & S America. **Ecol:** Usually in tropical rainforests (riverine). **Cult:** Propagate from seeds or by air-layering. Requires a well-drained soil in full sun.

D. odorata (Aubl.) Willd., *Sp. Pl.* 3: 910, 1803. (Syn: *Coumarouna odorata*). [Cumarú, Tonka Bean]. **Trees**, tall (at least in virgin forests); crown spreading; stems and branchlets with a pealike smell; bark yellowish-brown, smooth but scaly. **Leaves** pinnate; rachis winged, tomentose, tip long-produced; leaflets 4–6(8), oblong, 10–15 × 4–6 cm or larger, apex acuminate, coriaceous, glandular, not pellucid-punctate; leaflets alternate or opposite, as many as 8 on young shoots. **Inflorescences** panicles, ample, loosely flowered, rusty-tomentulose. **Flowers** very fragrant, except the whitish banner, rose-colored, 1.2 cm long; calyx wings uniformly rusty-tomentose, coriaceous, 1–1.2 cm long, lower calyx tube 3-toothed; ovary glabrous. **Fruits** drupes, with oily and aromatic pericarp, inedible. **Distr:** S America. **Ecol:** In lowlands and rainforests; 0–1,000 m. **Zones:** 10–12. **Info:** A substance called coumarin (with a vanillalike fragrance) is extracted from the fruits, which is used in the perfume industry. The fruits also produce solid fat or tonka butter used to flavor food.

Above: *Crotalaria retusa*, *Crotalaria verrucosa*, *Spartium junceum*, *Ulex europaeus*.

TRIBE CROTALARIEAE

Crotalaria

Species ~690. **Herbs** or **shrubs**. **Leaves** simple or palmately trifoliolate or rarely up to 7-foliolate, generally stipulate. **Stipules** free from petiole, sometimes decurrent on stems, often small or none. **Inflorescences** racemes, terminal or leaf-opposed, or occasionally flowers solitary; bracts and bracteoles small or rarely absent, bracteoles inserted on pedicel or calyx. **Flowers** usually yellow, rarely blue or purplish; calyx lobes free or calyx rarely bilabiate, teeth linear, lanceolate, subequal, free or upper 2 fused, forming a lip; wings obovate or oblong, shorter than standard; standard orbicular, shortly clawed; keel incurved, rostrate, beaked; vexillum orbicular, rarely ovate; stamens monadelphous (connate into a cleft sheath), anthers dimorphic, alternate, small and versatile, others long and basifixed; ovary generally sessile, 2–many-ovuled, style incurved or abruptly inflexed above ovary, bearded or barbellate above along inner side, stigma small, rarely bilobed. **Fruits** globose or linear-oblong, turgid or inflated, bivalvate, 2–many-seeded, continuous within; seeds on filiform funicles. **Distr:** Worldwide, especially in the Southern Hemisphere with the highest diversity in Africa and Madagascar. **Cult:** Propagate from seeds. Requires any soil in full sun and many species are salt-tolerant.

C. retusa L., *Sp. Pl.* 715, 1753. [Rattle Weed, Sun Hemp]. **Undershrubs**, ~0.6–1.2 m tall; branches glabrous or appressed-pubescent. **Stipules** subulate. **Leaves** simple, oblanceolate-oblong, ~3.7–9 × 1–2.2 cm, apex obtuse, retuse or mucronate, adaxially glabrous, abaxially silky-pubescent; petiole very short. **Inflorescences** erect, terminal racemes, ~15–30 cm long; pedicel ~4–8 mm long; bracts 2–6 mm long; bracteoles 2, ~1–2 mm long. **Flowers** with calyx ~1–1.4 cm long, ± glabrous, upper teeth ovate, acute, 3 lower narrower; corolla ~2.5–2.6 cm long, yellow, purple-tinged. **Fruits** capsular, linear-oblong, ~2.5–4.8 cm long, stipitate, glabrous, 15–20-seeded. **Distr:** Pantropical. **Ecol:** In montane grasslands and seashore sandy areas. Fl. & Fr. throughout the year. **Zones:** 10–12. **Info:** The seed pods when dry rattle in the wind.

C. verrucosa L., *Sp. Pl.* 715, 1753. [Virgin Flower]. **Herbs**, annual, to ~1 m tall; stems much-branched, erect or spreading, striate, 4-angled, pubescent, sometimes glabrescent. **Stipules** foliaceous, ovate to falcate, 0.5–2 × 0.4–1.3 cm, acuminate. **Leaves** simple, ovate to lanceolate, rhomboid, or elliptic, ~1–12 × 0.7–8 cm, base acute, apex obtuse or acute, appressed-pubescent; petiole 2–3 mm long. **Inflorescences** racemose, leaf-opposed, pseudo-terminal, longer than leaves, many-flowered; bracts lanceolate, caudate, ~1.5 mm long; bracteoles filiform to lanceolate, 1–4 mm long; pedicels 3–5 mm long. **Flowers** 1.1–1.6 cm long; calyx appressed-pubescent, 0.7–1.3 cm long, lobes deltoid, acute, longer than tube; petals blue, purplish, or white, glabrous. **Fruits** pods, capsular, sessile, inflated, oblong-cylindrical, pubescent, ~2.5–3(–5) × 1–1.2 cm; seeds 10–16, yellowish to brown, obliquely cordiform, 3–4.5 × 3–4 mm. **Distr:** Asia; naturalized widely. **Ecol:** In grasslands and sparse forests; 100–200 m. Fl. Aug–Oct. Fr. Nov–Dec. **Zones:** 9–12.

TRIBE GENISTEAE

Spartium

Species 1. *Generic description same as specific.* **Cult:** Propagate from seeds or cuttings. Requires a well-drained soil in full sun.

S. junceum L., *Sp. Pl.* 708, 1753. [Spanish Broom, Weaver's Broom]. **Shrubs**, evergreen, 1–3 m tall; crown rounded; stems erect, terete, ridged, glabrous, branchlets slender, young twigs green; bark gray. **Leaves** simple, narrowly elliptic to linear-lanceolate, 1–4 × 0.5–1.7 cm, base tapering, apex rounded, papery, adaxially glabrous, abaxially sparsely puberulent, midrib prominent, lamina caducous. **Inflorescences** racemes, 5–20-flowered, terminal; bract and bracteole minute, caducous; bracts ovate-lanceolate, 4 mm long; pedicels 2–3 mm long. **Flowers** with calyx spathaceous, membranous, 6–7.5 mm long; corolla golden-yellow; vexillum largest, 2–2.5(–3) cm long; wings shortest; keel acuminate, beak 2 mm long. **Fruits** pods, linear, 4–10 × 0.5–0.8 cm, ends acute, dark brown, pubescent when young; seeds 10–20, ellipsoid, 4 × 3 mm, reddish-brown or black, shiny. **Distr:** Around the Mediterranean. **Ecol:** In shrublands, cliffs, ravines, sometimes near waterways and lakes. Fl. Apr–Jul. **Zones:** 8–10.

Ulex

Species 10–20. **Shrubs**; branchlets sharply spiny. **Stipules** absent. **Leaves** usually alternate, trifoliolate on seedlings, on mature plants reduced to scalelike or narrow, usually spinelike phyllodes; petioles spiny. **Inflorescences** with flowers solitary or 2–3 in racemes, axillary; bracts small; bracteoles ovate, adnate to base of calyx. **Flowers** with calyx 2-lipped, membranous; corolla yellow, petals subequal, shortly clawed; stamens monadelphous, anthers dimorphic; ovary sessile, ovules numerous. **Fruits** pods, ovoid, oblong or shortly linear, compressed or turgid, tomentose, usually embraced in marcescent calyx, 2-valved, continuous within; seeds 2–4, strophiolate. **Distr:** Europe, N Africa. **Ecol:** In shrublands, woodlands and exposed places. **Cult:** Propagate from seeds. Requires a moist, well-drained, sandy soil in full sun.

U. europaeus L., *Sp. Pl.* 741, 1753. [Gorse]. **Shrubs**, biennial or perennial, densely branched, densely spiny, 0.5–1.5(–6) m high; stems terete, longitudinally ridged, densely villose, tips spiny; fascicular petioles becoming spines, 0.5–1.5 cm long, deeply furrowed. **Leaves** alternate; lamina ~2.5 cm long, reduced to dorsally flattened spines. **Inflorescences** racemes, 1–3-flowered, borne at the tip of the stems; bracts ovate, 2 mm long; pedicels 3–9 mm long; bracteoles 2 mm long. **Flowers** fragrant; calyx yellow-brown, 1.2–1.4 cm long, membranous, all densely tomentose; corolla golden-yellow, 1.3–1.5 cm long, glabrous, distal margin of keel tomentose. **Fruits** pods, narrowly ovoid, 1.1–2 cm long, blackened, densely tomentose; seeds 2–4, ovoid, 3 × 2 mm, dark brown, shiny. **Distr:** Europe; naturalized widely in many parts of the tropics and subtropics, regarded as a noxious weed. **Ecol:** In tropical and subtropical montane habitats. **Zones:** 6–11.

TRIBE DALBERGIEAE

Andira

Species 29. **Trees**, usually large trees, unarmed. **Leaves** alternate, odd-pinnate; leaflets generally opposite; stipels setaceous or none. **Inflorescences** panicles, terminal; often densely crowded, subsessile; bracts and bractlets small, caducous. **Flowers** mostly violaceous; calyx truncate, teeth short or obsolete; standard suborbicular, not appendaged; wings almost straight, oblong, obtuse, free; keel petals similar to the wings, not connate; vexillar stamen free or rarely connate with the others to form a sheath, anthers versatile; ovary stipitate or rarely subsessile, 2–4-ovuled or sometimes 1-ovuled, style short, incurved, stigma small, terminal. **Fruits** drupaceous, ovoid, indehiscent, endocarp ligneous; seed 1, pendulous. **Distr:** Tropical America and Africa. **Cult:** Propagate from seeds or by air-layering. Requires a fertile, moist, well-drained soil in full or partial sun.

A. inermis (W.Wright) DC., *Prodr.* 2: 475, 1825. (Syn: *A. excelsa*, etc.). [Barley Wood, Bat Seed, Partridge Wood, Wild Olive, etc.]. **Trees**, large, to 10–15(–35) m tall; trunk to 1 m dbh. **Stipules** linear-subulate, to 2 cm long, deciduous. **Leaves** alternate, large, imparipinnate, ± glabrous; leaflets usually (5)7–15 pairs, opposite to subopposite along rachis, oblong to ovate, 4–9.5(–13) × 1.5–4.5(–5) cm long, base obtuse or rounded, apex acute or acuminate, subcoriaceous. **Inflorescences** panicles, many-flowered, 15–30 cm long, terminal, tomentose; flowers subsessile. **Flowers** red-violet or purplish, ~1 cm long, sweetly fragrant; calyx campanulate, brown or dark purple, puberulent, 3 mm long, teeth short ± equal; standard 1 cm long, darker inside in a broad submarginal band, white at center below; wings slightly longer than the keel; stamens diadelphous, staminal sheath open, glabrous, filaments alternating

Above: *Andira inermis, Brya ebenus, Pterocarpus indicus, Pterocarpus santalinus.*

long and short, curved inward; stigma small, brushlike. **Fruits** drupaceous pods, oval, sometimes subglobose, 2.5–5+ cm long, 1-seeded. **Distr:** British Honduras to Salvador and Panama, Guatemala, S Mexico, S America, West Indies, W Africa. **Ecol:** In wet to dry forests; <900 m. Fl. Mar–Apr. **Zones:** 10–12.

Brya

Species 4. **Shrubs** or **small trees**, erect, prickly, deciduous, to ~8 m tall; trunk ~20 cm dbh. **Stipules** spinescent and persistent, or inconspicuous and caducous; stipels absent. **Leaves** imparipinnate; leaflets 1–many, but usually trifoliolate, fascicled or clustered with scarcely a common petiole. **Inflorescences** cymes or fascicles, few-flowered, short, axillary or terminal; bracts and bracteoles small, persistent. **Flowers** usually yellow; calyx lobes subequal or lower lobe longer; standard suborbicular, clawed; wings falcate-oblong; keel incurved, blunt; stamens all connate into a sheath split above, anthers uniform; ovary sessile, or shortly stalked, 2- or few-ovuled, style filiform, incurved, stigma terminal, very small. **Fruits** pods, sessile or shortly stalked, with 1–2 broad flat indehiscent joints, upper joint usually sterile, lower suture curved; seeds reniform. **Distr:** Caribbean (Greater Antilles). **Cult:** Propagate from scarified seeds. Requires a well-drained soil in full sun.

B. ebenus (L.) DC., *Prodr.* 2: 421, 1825. [Cocuswood, Jamaican Ebony, Jamaican Rain Tree, West Indian Ebony]. **Shrubs** or **small trees**, evergreen or deciduous, 2.5–7(–12) m tall; branchlets armed with short, sharp, subulate spines. **Leaves** imparipinnate; leaflets 1–many, sessile, obovate or spathulate, base cuneate, apex usually emarginate, leathery. **Inflorescences** fascicles, 1–3-flowered, axillary. **Flowers** yellow to orangish-yellow; calyx bell-shaped, pubescent, obscurely 2-lipped, upper lip bipartite, lower tripartite, lowest one spreading, the rest erect; standard glabrous, lobes ovate, acute; corolla bright orangish-yellow; vexillum ± rounded, with deep purple streaks at the center; stamens 10, monadelphous, anthers subglobose; pistil hairy, ovary oblong, of 2 joints. **Fruits** pods, <2.5 cm long, compressed, hairy. **Distr:** Cuba. **Ecol:** In hot and dry habitats. **Zones:** 11–12. **Info:** The wood is highly valued in the Caribbean. The heartwood is almost black and are often used in making musical instruments and furniture.

Pterocarpus

Species 20. **Trees** or **shrubs**, unarmed. **Stipules** usually deciduous. **Leaves** alternate, compound, odd-pinnate; leaflets mostly alternate, not stipellate. **Inflorescences** racemes or panicles, axillary, often appearing before leaves; bracts and bractlets small, caducous. **Flowers** yellow to orange, sometimes whitish or violet-tinged, often fragrant; calyx turbinate, often incurved, upper 2 teeth or lobes ± connate; petals glabrous; standard orbicular or broadly ovate; wings obliquely obovate or oblong; keel petals similar to wings or shorter, free or shortly connate dorsally; stamens monadelphous, or the vexillar one free, anthers versatile; ovary sessile or stipitate, few-ovulate, style filiform, slightly incurved, stigma small, terminal. **Fruits** pods, orbicular to ovate or oval-oblong, compressed, somewhat oblique or sickle-shaped, indehiscent, margins winged or carinate; seeds 1–3, separated by septa, oblong or subreniform. **Distr:** Pantropical. **Cult:** Propagate from seeds, cuttings or by air-layering. Requires a moist, well-drained soil in full sun.

P. indicus Willd., *Sp. Pl.* 3: 904, 1800. [Bloodwood, Burmese Rosewood]. **Trees**, deciduous, to 30 m tall; branches arching to almost pendent, glabrous; bark grayish-brown, peeling. **Leaves** 15–30 cm long; leaflets 7–11, ovate to elliptic, 10–20 cm long, base rounded or deltoid, apex narrowed into a point, moderately firm, glabrous; petiole 1–4 cm long; petiolules 6–8 mm long. **Inflorescences** panicles, axillary, ± crowded; bracteoles caducous; pedicels as long as calyx. **Flowers** fragrant, ~1.25 cm long, short-lived; calyx top-shaped, finely brown-silky, teeth rounded, upper 2 much larger; corolla yellow-orange, exceeding calyx; standard 0.9–1.3 cm wide. **Fruits** pods, orbicular, 4–5 cm Ø, silky-veined against the seed, wing 0.8–1.3 cm wide, thicker in middle, wrinkled, style a considerable distance above base, pointing outward; seeds 1(2). **Distr:** S Myanmar across SE Asia to Indonesia, Philippines and W Pacific Islands. **Ecol:** Open forests on slopes and evergreen rainforests; to 1,300 m. Fl. Mar–Jun. Fr. Apr–Jul. **Zones:** 11–12. **Info:** A popular street tree throughout the tropics.

P. santalinus L. f., *Suppl. Pl.* 318, 1781. (Syn: *Lingoum santalinum*). [Red Sandalwood]. **Trees**, to 11 m tall; branches obscurely gray-downy; bark blackish brown. **Leaves** compound, ternate; leaflets 3(5), subrotund to ovate, 5–10 cm long, base rounded, apex obtuse or rounded, slightly emarginate, adaxially smooth, abaxially with adpressed gray hairs, venation fine. **Inflorescences** racemes, axillary, shorter and less copious than those of *P. indicus*; pedicels shorter than the calyx. **Flowers** with stamens 2–3-adelphous; calyx 5–6 mm long, teeth deltoid, minute; standard with limb not longer than the calyx. **Fruits** pods (similar to those of *P. indicus* but the center more turgid, wing narrower and style brought down to the basal corner), silky at first, beak at the basal corner; fruiting pedicel much exceeding the calyx. **Distr:** S India (Western Peninsula). **Ecol:** In dry deciduous forests; ~750 m. Fl. Jul–Aug. **Zones:** 10–12.

Arachis

Species 19. **Herbs**, annual or perennial, erect or prostrate. **Stipules** adnate at base to petiole, partly adnate to leaves; stipels absent. **Leaves** paripinnately 3–4-foliolate. **Inflorescences** with flowers solitary or in 2–7-flowered spikes; bracts membranous, primary ones biapiculate. **Flowers** yellow, sometimes red-striped; calyx membranous, filiform, 5-lobed, 4 upper lobes united, lower ± free; vexillum, rounded, shortly narrowed at base, wings free, keel beaked, incurved; stamens 8–10, monadelphous, anthers 4–5, elongate, alternating with 4–5 short ones; ovary subsessile, linear, 1–7-ovuled, style very long, soon deciduous, stigma minute, terminal. **Fruits** slightly constricted, functionally indehiscent, thick-walled, developing below the soil, having been pushed beneath by lengthening of the gynophore; seeds 1–6-seeded, ovoid or oblong, cotyledons rich in oil. **Distr:** S America (mainly S Brazil). **Cult:** Propagate from seeds. Requires any moist soil in full or partial sun.

A. hypogaea L., *Sp. Pl.* 741, 1753. [Earth Nut, Ground Nut, Peanut]. **Herbs**, annual, erect or straggling, ~30 cm long. **Stipules** 1.5–4 cm long, ciliate. **Leaves** with free part of petiole 1.5–7 cm long; leaflets 4, obovate or elliptic, 1–7 × 0.7–3.2 cm, apex rounded or emarginate and mucronate, abaxially glabrous or sparsely pilose. **Inflorescences** axillary, flowers solitary, pedicellate; primary bracts ovate-lanceolate, 1–1.4 × 0.4–0.5 cm, biapiculate; secondary bracts similar but bifid. **Flowers** yellow, red-nerved; hypanthium 2–4 mm long, pubescent; corolla 0.7–1.3 cm long; stamens 8–9. **Fruits** pods, 2–6 × 1–1.5 cm, gynophore becoming 1–20 cm long; seeds subovoid, 1–2 cm long. **Distr:** S America (probably originated from Brazil). **Ecol:** Fl. Jul–Aug. **Zones:** 8–12. **Info:** Cultivated since ancient times, and widely for commercial peanuts.

Dalbergia

Species ~300. **Trees** and **shrubs**, sometimes ± scandent, unarmed. **Leaves** alternate, imparipinnate, rarely unifoliolate; lateral leaflets alternate, exstipellate. **Inflorescences** terminal or axillary, racemes, cymes or panicles; bract and bracteoles small. **Flowers** mostly small, white or purplish; calyx campanulate, teeth unequal; vexillum ovate or orbicular; keel petals united at the tip; stamens 9 or 10, monadelphous or diadelphous; ovary stipitate, ovules 1 to few, style incurved, glabrous, stigma capitate. **Fruits** samaroid, indehiscent, not winged; seeds 1–4, reniform, compressed. **Distr:** Tropical and subtropical regions worldwide. **Cult:** Propagate from seeds or cuttings. Requires moist but well-drained soil in full sun.

D. melanoxylon Guill. & Perr., *Fl. Seneg. Tent.*: 227, t. 53, 1832. (Syn: *Amerimnon melanoxylon*, etc.). [African Blackwood, African Ebony, Senegal/Sudan Ebony]. **Shrubs** or **trees**, spiny, to 5–12(–30) m tall, typically several-stemmed and much-branched; branchlets clustered at nodes, sometimes spine-tipped, initially hairy. **Stipules** oblong-ovate or oblong,

Above: *Arachis hypogaea, Dalbergia melanoxylon, Indigofera decora, Indigofera tinctoria.*

2–6 mm long, fugacious. **Leaves** imparipinnate, 6–22 cm long; lateral leaflets 4–6(–8), oblong-obovate to obovate, 1–5(–5.5) × 0.7–3(–5) cm, base broadly cuneate to rounded or slightly cordate, apex truncate or emarginate, primary lateral nerves 8–12 per side. **Inflorescences** panicles, appearing with new flush of leaves, ~3–12 cm long, many-flowered; bracts linear-lanceolate to ovate, concave, 0.5–1 mm long, caducous. **Flowers** with calyx 2–3(–4) mm long; corolla 4–6 mm long, white, fragrant; standard elliptic-obovate or obovate, narrowed to claw; wings slightly shorter than standard and slightly exceeding keel; stamens usually 9, united or variously divided. **Fruits** elliptic-oblong or irregularly oblong, 3–7 cm long, bluntly pointed, papery, glabrous, 1–2-seeded. **Distr:** Senegal to Eritrea and N Ethiopia, south to Angola and the Transvaal; W peninsular India. **Ecol:** 0–1,350 m. **Zones:** 10–12.

TRIBE INDIGOFERAE

Indigofera

Species >300. **Shrubs** or **herbs**, generally pubescent. **Stipules** subulate-setaceous, slightly adnate to petiole. **Leaves** pinnate or imparipinnate or palmately trifoliolate or unifoliolate or simple; leaflets entire. **Inflorescences** racemes, spikes or flowers solitary, axillary; bract generally caducous, bracteoles absent. **Flowers** small, usually pinkish or purplish; calyx lobes or teeth subequal or lowermost longer; vexillum sessile or clawed; wing slightly cohering with keel; keel gibbous on each side and often spurred, keel petals obliquely lanceolate, apex rounded to acute, rarely produced into a beak; stamens 9+1, diadelphous, anthers uniform, apiculate; ovary sessile, ovules numerous or 1–2, style glabrous, stigma capitate, often penicillate. **Fruits** pods, linear or globose, terete or compressed, septate between seeds; seeds 3–4, globose to cylindric, truncate, compressed or quadrate. **Distr:** Warm tropical regions of the world. **Cult:** Propagate from seeds or half-hardened cuttings. Requires a moist, well-drained soil in full sun.

I. decora Lindley, *J. Hort. Soc. London* 1: 68, 1846. [Chinese Indigo]. **Shrubs**, 0.4–2 m tall. **Stipules** caducous; stipels subulate, ~1.5 mm long. **Leaves** 8–25 cm long, 5–15(–23)-foliolate; leaflets opposite or rarely alternate, usually ovate or lanceolate, 2–7.5(–10) × 1–3.5 cm, base cuneate to broadly cuneate, apex usually acuminate, papery, abaxially with white trichomes; petiole 1–2.5 cm long; petiolules ~2 mm long. **Inflorescences** racemes, 13–21 cm long; peduncle 2–4 cm long; bracts linear-lanceolate, ~3 mm long, caducous; pedicel 3–6 mm long. **Flowers** with calyx cup-shaped, 2.5–3.5 mm long, tube 1.5–2 mm long, teeth triangular, ~1 mm long; corolla light purple, pink, or rarely white; standard elliptic, 1.2–1.8 × ~0.7 cm, outside white-pubescent, sometimes brown-hairy; wings 1.2–1.4 cm long, margin ciliate; keel as long as wings, spur ~1 mm long; stamens (1.1–)1.3–1.4 cm long, anthers ovoid, hairy; ovary glabrous, >10-ovuled. **Fruits** pods, brown, cylindric, 2.5–6.5(–8) cm long, glabrous, endocarp blotched; seeds 7 or 8, ellipsoid, 4–4.5 mm long. **Distr:** China, Japan. **Ecol:** In woods, near streams; 200–1,800 m. Fl. Apr–Jun. Fr. Jun–Oct. **Zones:** 6–11.

I. tinctoria L., *Sp.Pl.* 751, 1753. [Bengal Indigo, Black Henna, Ceylon Indigo, Commercial Indigo, Indian Indigo]. **Shrubs**, slender, 0.6–2 m high; stems terete or subangulate, strigose; branches pubescent. **Stipules** subulate-setaceous, 2–3 mm long. **Leaves** 1.2–12.5 cm long; leaflets 7–15, elliptic or oblong, ~0.5–2.6 cm long, base obtuse or acute, apex rounded and mucronate, adaxially ± glabrous, abaxially strigose- or appressed-pubescent; petiolule ~1 mm long. **Inflorescences** racemes, subsessile, ~5–12 cm long, mostly shorter than the leaves. **Flowers** with calyx ~1–1.5 mm long, pubescent, teeth as long as the tube; corolla pink; vexillum ~4–4.5 mm long. **Fruits** pods, 1.8–3.8 cm long, ~2.5 mm broad, straight or slightly curved, glabrous; seeds 8–12, 2 mm long. **Distr:** Ceylon, India, Indonesia, Indo-China, Malaya, Myanmar, Pakistan, Philippines, Siam, Tropical Africa. **Ecol:** Fl. Nov–Dec. **Zones:** 10–12. **Info:** Cultivated for commercial indigo and medicinal purposes.

TRIBE MILLETTIEAE

Wisteria

Species ~6. **Lianas**, deciduous; scales 3–5. **Stipules** present; stipels present. **Leaves** odd-pinnate; leaflets opposite, margins entire. **Inflorescences** racemes, terminal, elongate, pendulous; nodes scattered on rachis, each with 1 flower. **Flowers** 1.5–2.5 cm long; corolla standard orbicular, large, reflexed, glabrous or outside with some trichomes at base, with 2 basal calluses; wings free from keel; stamens diadelphous, vexillary stamen distinct from other 9 or slightly connate at middle of sheath; ovary stipitate, with trichomes, 5–16-ovuled. **Fruits** pods, linear to oblanceolate, leathery, convex over seeds; seeds 1–3(+), reniform to lenticular. **Distr:** E Asia, N America. **Cult:** Propagate from seeds, cuttings, by air-layering or grafting. Requires a humus-rich, moist, well-drained soil in full sun.

W. sinensis (Sims) Sweet, *Hort. Brit.* 121, 1826. (Syn: *W. alba*, etc.). [Chinese Wisteria, Primavera]. **Lianas**, to 25 m long; stems twined leftwards, white-villous when young, soon glabrescent. **Leaves** odd-pinnate; rachis 15–25 cm long (including petiole 3–5 cm long); leaflets 7–13, elliptic-ovate to lanceolate-ovate, 5–8 × 2–4 cm, basal pair smallest and becoming larger apically, base rounded to cuneate and somewhat asymmetric, apex attenuate to caudate, surfaces appressed-pubescent when young. **Inflorescences** racemes, terminal or axillary from branchlets of previous year, 15–30 × 8–10 cm, white-villous; pedicels 2–3 cm long, slender. **Flowers** 2–2.5 cm long, fragrant; calyx with adaxially tooth longer than others; corolla purple or occasionally white; standard orbicular, sometimes retuse, glabrous, apex truncate; ovary tomentose, 6–8-ovuled. **Fruits** pods, oblanceolate, 10–15 × 1.5–2 cm, tomentose,

Below: *Wisteria sinensis, Mundulea sericea, Derris elliptica.*

JML

hanging on branches persistently; seeds 1–3, brown, thickly lenticular, ~1.5 cm Ø, shiny. **Distr:** China, Japan. **Ecol:** In montane forests; 500–1,800 m. Fl. Apr–May. Fr. May–Aug. **Zones:** 5–11.

Mundulea

Species 12. **Trees small** or **shrubs**, to ~7 m tall. **Stipules** minute; stipels absent. **Leaves** alternate, imparipinnate; leaflets opposite or subopposite. **Inflorescences** racemose, axillary, terminal, or pseudo-terminal; bracts minute; bracteoles lacking. **Flowers** with calyx campanulate, teeth or lobes 5, short, upper (vexillar) often connate; petals bluish to pink or purple, usually pubescent; wings slightly adherent to keel; keel petals connate; stamens 10 with the vexillar filament sharply bent, free at base, united with the others above; filaments alternately dilated toward apex, pubescent above, anthers uniform, elliptic; ovary sessile, many-ovuled, style bent above, hardened, glabrous, stigma terminal, minutely capitate. **Fruits** pods, indehiscent or tardily dehiscent, linear-oblong, compressed, many-seeded; seeds oblong to reniform, brown, non-arillate, hilum lateral. **Distr:** Madagascar; 1 species (*M. sericea*) widespread from Africa to Sri Lanka and India. **Cult:** Propagate from seeds or by air-layering. Requires a moist, well-drained soil in full sun.

M. sericea (Willd.) A. Chevalier, *Compt. Rend. Hebd. Séances Acad. Sci.* 180: 1521, 1925. (Syn: *Cytisus sericeus*, etc.). [Cork Bush, Silver Bush]. **Shrubs** or **small trees**, to ~7 m tall; young stems velutinous, glabrescent. **Stipules** deltoid, 1–3 mm long. **Leaves** 9–17-foliolate; axis ~4–10 cm long, pubescent; leaflets opposite to subopposite, ovate to oblong or lanceolate, 1–5 × 0.4–1.5(–1.7) cm, base cuneate or rounded, sometimes attenuate, apex usually acute or acuminate, adaxially glabrous or subsericeous, abaxially ± sericeous. **Inflorescences** racemose, terminal or axillary, ~5–10 cm long, few- to many-flowered, flowers 2 per node; bracts deltoid, 0.5–2 mm long. **Flowers** 1.7–2.5 cm long; calyx ~2 mm long, pubescent; petals pinkish or reddish-violet to purple; vexillum sericeous outside. **Fruits** pods, linear-oblong, ~6–8(–10) × 0.6–0.8 cm, sericeous to velutinous, thickened along both sutures; seeds 6–8, ± reniform, dark green to brown, sublustrous, 4–5 × 3 mm. **Distr:** Madagascar, Tropical and S Africa; Sri Lanka, India. **Ecol:** On dry rocky hills. **Zones:** 10–12.

Derris

Species ~70. **Climbers** or **scandent shrubs**, woody, sometimes trees. **Stipules** small and caducous; stipels present or absent. **Leaves** imparipinnate, leaflets opposite. **Inflorescences** racemes or panicles; axillary and terminal; bract and bracteoles small. **Flowers** with calyx cupulate and subtruncate or with very short teeth; corolla white, pink or purple, much-exserted; vexillum ovate to obovate or suborbicular, usually with auricles or calluses at base; wing adhering to keel petal; stamens usually monadelphous, vexillary (upper) stamen free at base only; ovary sessile or often subsessile, few-ovulate, style glabrous above, stigma terminal. **Fruits** oblong or strap-shaped, flat, indehiscent, rather rigid, upper or both sutures winged; seed reniform or orbicular. **Distr:** Mainly in tropical regions of the world. **Cult:** Propagate from seeds, cuttings or by air-layering. Requires any moist, well-drained soil in full sun.

D. elliptica (Wall.) Benth., *J. Linn. Soc. Bot.* 4, suppl.: 111, 1860. (Syn: *Pongamia elliptica*, etc.). [Derris, Jewel Vine, Oil Tree, Poison Vine, Tuba Root]. **Lianas** or **scandent shrubs**, robust, 5–12 m long, densely velvety brown-pubescent; shoot apex often leafless for several metres. **Leaves** pinnate, 15–35 cm long, mostly with 11–15 leaflets; leaflets narrowly oblong-obovate, 4–25 cm long, base rounded, margins entire, apex rounded to shortly acuminate, rusty-velvety, somewhat glabrescent, abaxially glaucous; petiole 4–8 cm long. **Inflorescences** pseudo-racemes, composed of stalked clusters of 3, 15–25 cm long, rusty-pubescent. **Flowers** ~2 cm long; calyx shallowly cup-shaped, 6–8 mm long, toothed; corolla pink to whitish, 1.5–1.8 cm long; standard suborbicular, 1.2–1.7 cm Ø, with 2 auricles at base, softly ferruginous-hairy outside; wings and keel with interlocking longitudinal folds. **Fruits** elliptic to oblong-elliptic, 3.5–7 cm long, flat, leathery, narrowly winged; seeds 1–4, reniform, flat. **Distr:** India to Malesia. **Ecol:** Fl. Apr–May. Fr. Jun. **Zones:** 10–12. **Info:** Often cultivated for use as an insecticide or fish poison.

D. scandens (Roxb.) Benth., *J. Linn. Soc.* 4. Suppl. 103, 1860. (Syn: *Dalbergia scandens*). [Hog Creeper, Malay Jewelvine]. **Shrubs**, climbing, woody, 3–6 m high; branches sparsely pubescent. **Leaves** imparipinnate, 7–15 cm long; leaflets 9–19, lower pairs smaller, oblong or obovate-oblong, 2–6 × 1–2.5 cm, apex obtuse or acute, adaxially ± glabrous, abaxially pubescent; petiolules short. **Inflorescences** racemes, few- to many-flowered, axillary, twice as long as the leaves; pedicels ~4–6.5 mm long. **Flowers** with calyx dull purple, somewhat oblique, 3–3.5 mm long, silky-pubescent, teeth obscure; corolla pale rose to white, 1 cm long; vexillum ~1–1.1 cm long. **Fruits** pods, lanceolate or oblong-lanceolate, ~2.5–7.5 × 0.9–1.2 cm, flattened, narrowed at both ends, glabrous; seeds 1–4, wings not more than 1.5 mm wide. **Distr:** India, E Pakistan, Sri Lanka; SE Asia, China; N Australia. **Ecol:** In woods, usually near the sea, and forests, mostly in open places; 0–500 m. Fl. Jan–May. Fr. Apr–Jul. **Zones:** 10–12.

Millettia

Species ~150. **Trees**, **shrubs** or **woody climbers**. **Stipules** small. **Leaves** imparipinnately compound; leaflets mostly stipellate and opposite. **Inflorescences** racemes or panicles, terminal; bracts and bracteoles often deciduous before flowering. **Flowers** white, rose or purple; calyx teeth almost absent or short; vexillum large, bare or callose above the claw or rarely appendaged with auricles; wings free from the keel; keel incurved, obtuse; stamens monadelphous or diadelphous, vexillary stamen free at base but ± connate above or quite free, anthers uniform; ovary sessile or rarely stipitate, style glabrous, stigma terminal, ovules numerous; annular disk surrounding the ovary usually present. **Fruits** pods, compressed, flat, thick or woody, 2-valved or tardily dehiscent; seeds orbicular or reniform. **Distr:** Tropical and subtropical regions worldwide. **Info:** The genus *Pongamia* is now merged with *Millettia*. **Cult:** Propagate from fresh seeds (soaked overnight in hot water). Requires a rich, moist, well-drained soil in full sun.

M. pinnata (L.) Panigrahi, *Fl. Bilaspur District*, 1: 210, 1989. (Syn: *Pongamia pinnata*, etc.). [Indian Beech]. **Trees**, to ~30 m tall, sometimes scandent; stems glabrous. **Stipules** (bud scales) cucullate, 3.5–4 × 3 mm, striate, puberulent, caducous. **Leaves** 5–9-foliolate; axis 6–17 cm long, glabrous or nearly so; leaflets ovate to elliptic-ovate, (2–)3–10 × 2–5 cm, base rounded, apex acuminate. **Inflorescences** racemose, axillary; axes ± glabrous; flower buds nutant on pedicels ~3–5 mm long; bracts linear-deltoid, 3–4 × 0.5–1 mm, pubescent, caducous; bracteoles at base of calyx, minute. **Flowers** 1.2–1.4 cm long; calyx campanulate, sericeous, 3–4 × 4–5 mm, ± truncate, teeth or lobes deltoid, minute; petals white to pinkish; vexillum ± sericeous, essentially straight; keel petals cohering at tip; anthers elliptic, to 1 mm long; ovary subsessile, usually 2-ovuled, puberulent. **Fruits** pods, obliquely oblong or elliptic, 5–8 × 2.5–3 cm, ~0.8–1 cm thick, indehiscent, glabrous, somewhat compressed, essentially sessile; seeds 1 or 2, reniform, ~2 × 1 cm, compressed, brown, lustrous. **Distr:** India, Sri Lanka, Seychelles; Malaysia; N Australia, Polynesia. **Ecol:** In lowland forests and along tidal streams and rivers; 0–1,200 m. Fl. Dec–Apr. Fr. Feb–Jun. **Zones:** 10–12.

Tephrosia

Species >400. **Herbs**, perennial or rarely annual, often suffrutescent, usually sericeous. **Stipules** caducous; stipels absent. **Leaves** odd-pinnate; leaflet opposite, margins entire, abaxially often sericeous, secondary veins to ~30 per side, closely parallel. **Inflorescences** pseudo-racemes, terminal or axillary, sometimes opposite a leaf; bracts usually caducous; bracteoles absent. **Flowers** with corolla white, cream-colored, or mauve, occasionally orange or red; standard suborbicular, reflexed, outside villous or sericeous; stamens monadelphous; vexillary stamen somewhat distinct from the other 9; ovary sessile, with trichomes, ovules numerous. **Fruits** pods, flat, occasionally inflated, dehiscent, apex often beaked, valves twisted; seeds 5–16, oblong, ellipsoid, or occasionally globular. **Distr:** Tropical and subtropical regions worldwide. **Cult:** Propagate from seeds, scarified or soaked in hot water. Requires any well-drained soil in full sun.

T. purpurea (L.) Persoon, *Syn. Pl.* 2: 329, 1807.

Below: *Derris scandens*, *Millettia pinnata*, *Tephrosia purpurea*.

TG

Above: *Abrus fruticulosus*, *Abrus fruticulosus* (fruits), *Abrus precatorius*, *Abrus precatorius* (fruits).

[Faux Indigo, Fish Poison, Wild Indigo]. **Herbs**, perennial, 0.3–0.6(–1.5) cm tall, many-branched, puberulous, densely villous, or glabrescent; stems nearly erect to spreading, base woody, ridged. **Leaves** odd-pinnate; rachis 7–15 cm long (including petiole ~1 cm long); leaflets 9–17(–21), oblong-elliptic, oblanceolate-elliptic or obovate-elliptic, 1.5–3.5(–4.5) × 0.4–1.5 cm, base narrowly rounded, apex obtuse, truncate, or retuse and cuspidate, secondary veins 7–12 per side. **Inflorescences** pseudo-racemes, terminal, opposite to a leaf or axillary near apex of branchlets, ~2 or 10–15 cm long; pedicel 2–4 mm long, slender, slightly elongate in fruit. **Flowers** ~8 mm long; calyx 2–4 × ~3 mm, teeth equal; corolla mauve; standard orbicular, white-puberulous; ovary with trichomes, 5–8-ovuled. **Fruits** pods, linear, 3–5 × 0.35–0.4(–0.6) cm, with sparse trichomes, apically slightly curved; seeds ~6, grayish-brown, ellipsoid, ~3 × 1.5 mm. **Distr:** Cambodia, China, India, Indonesia, Laos, Malaysia, Nepal, Sri Lanka, Taiwan, Thailand, Vietnam. **Ecol:** On open grasslands, often in coastal areas; 0–700 m. Fl. Mar–Oct. Fr. Sep–Dec. **Zones:** 9–11.

TRIBE ABREAE

Abrus

Species ~15. **Shrubs** or **shrublets**, often climbing **vines**, woody. **Leaves** pinnate; leaflets many-paired; rachis bristle-tipped. **Inflorescences** racemes, lateral. **Flowers** with calyx truncate, teeth very short, upper 2 subconnate; vexillum ovate, clawed; wing narrowly falcate, oblong; keel larger than wings; stamens 9, adnate into a sheath, vexillary stamen absent, anthers uniform; ovary subsessile, many-ovuled. **Fruits** pods, plano-compressed, 2-valved, subseptate between seeds; seeds subglobose or ellipsoid, black or white, sometimes red and black. **Distr:** Pantropical. **Cult:** Propagate from scarified seeds. Requires a humid environment and any fertile, well-drained soil in full or partial sun.

A. fruticulosus Wight & Arnott, *Prod. Floræ Peninsulæ Indiæ Orientalis*, vol. 1: 236, 1834. [White Crab's Eye Vine]. **Climbers** or climbing undershrubs; ultimate branches, leaves and racemes pubescent. **Leaves** alternate, even-pinnate, distichous; leaflets 5–16 pairs, oblong or obovate, 1.3–2 × 0.6–0.9 cm, apex mucronate, abaxially appressed-pubescent. **Inflorescences** racemes, few-flowered, terminal or axillary. **Flowers** white, calyx 4-toothed, teeth triangular-acute, upper one bifid at apex; claw of the vexillum very short, cohering with the stamen sheath only at the very base. **Fruits** pods, flat, 5 cm long; seeds ~4, white, subglobose. **Distr:** Malabar. **Ecol:** In tropical forests and forest margins. Fl. Jul–Aug. Fr. Aug–Nov. **Zones:** 10–12. **Info:** The leaves contain a sweetening substance called abrusoside which is up to 100 times sweeter than sugar (Saralamp *et al* 1996).

A. precatorius L., *Syst. Nat.* ed. 12: 472, 1767. (Syn: *Glycine abrus*). [American Pea, Coral Pea, Crab's Eye Vine, Indian Liquorice, Precatory Bean, Rosary Pea]. **Climbers**, perennial, to 10 m long; branches slender. **Stipules** linear, 3.5–6 mm long, deciduous. **Leaves** pinnate, 5–10 cm long; leaflets 10–20 pairs, opposite, oblong, 0.8–2.2 × 3.5–6 cm, apex rounded, apiculate, adaxially glabrous, abaxially sparsely hairy; petiole 0.6–1.2 cm long; petiolule <1 mm long. **Inflorescences** racemes, axillary, pedunculate, shorter or equaling the leaves; pedicels short. **Flowers** rose-pink to purple; calyx 2.5 mm long, glabrous or sparsely silky, teeth very short; corolla pink or white with a pink tinge; vexillum 0.9–1 cm long. **Fruits** pods, inflated, oblong, 2.5–5 × 1–1.3 cm, beaked; seeds 3–6, ovoid, 7–8 mm long, scarlet, with a large black spot near the hilum. **Distr:** India, Pakistan, Sri Lanka, tropical Africa. **Ecol:** In moist or wet thickets, forests; 0–500 m. Fl. Jul–Oct. **Zones:** 10–12. **Info:** The seeds are highly toxic, containing abrin (related to ricin) and other substances, and are used medicinally. The stems, roots and leaves, contain glycyrrhizin (principle compound in commercial liquorice).

TRIBE PHASEOLEAE

Canavalia

Species ~60. **Lianas** to slender **vines**, woody or herbaceous, perennial. **Stipules** present; stipels small, deciduous. **Leaves** pinnately trifoliolate. **Inflorescences** racemose or flowers in groups along the rachis, axillary; bract and bracteoles small. **Flowers** with calyx 2-lipped, upper lip larger, truncate or bilobed, lower entire or trifid; vexillum rounded, reflexed; keel incurved, obtuse or beaked, often twisted; stamens generally monadelphous, vexillary stamen free at base, rarely quite free, anthers uniform; ovary multiovulate, style incurved, stigma terminal. **Fruits** linear or oblong, with a wing or rib near the upper suture, sometimes thinly filled within between the seeds. **Distr:** Pantropical. **Ecol:** In tropical flooded or swamp forests, river banks, seaside beaches, etc. **Cult:** Propagate from seeds. Requires a moist soil in full or partial sun.

C. maritima (Aubl.) Thouars, *Journ. Bot. Desv.* 1: 80, 1813. (Syn: *Dolichos maritimus*). [Bay Bean, Beach Bean, Mackenzie Bean]. **Climbers**, twining; stems coarse, herbaceous, often greatly elongate and ropelike, prostrate or sometimes climbing. **Leaves** trifoliate; leaflets, oval to ovate or orbicular, 5–8 cm long, apex obtuse to rounded or retuse and apiculate, sparsely strigillose when young but soon glabrate; petiole about equaling leaflets. **Inflorescences** racemes, 6–30-flowered; peduncles stout, somewhat longer than leaves. **Flowers** with calyx campanulate, sparsely strigillose, upper lip wide, emarginate, almost equaling tube, lower lip tridentate, lobes triangular, obtuse; corolla pink; standard 1–1.5 cm long. **Fruits** pods, linear-oblong, almost straight, 7–15 × 2–2.5 cm, not much compressed, sparsely strigillose, valves with 3 longitudinal costae, 2 of them close to the margins, the third 3–5 mm from the ventral suture; seeds 4–9, ovoid or subglobose, 1.2–1.6 × 0.7–1 cm, marbled with brown, and fulvous. **Distr:** Pantropical. **Ecol:** Usually found near beaches, and often climbing on other vegetation. **Zones:** 11–12.

C. rosea (Sw.) DC., *Prod.* 2: 404, 1825. (Syn: *C. lineata*, etc.). [Beach Bean]. **Vines**, coarse, trailing or climbing. **Stipules** lanceolate, ~4 mm long, deciduous; stipels small, deciduous. **Leaves** trifoliate; rachis 1.5–4 cm long; leaflets coriaceous, elliptic to suborbicular, to ~12 × 7 cm, base broadly cuneate, rounded to truncate, apex obtuse or emarginate, adaxially glabrous or sparsely pubescent, abaxially glabrous to densely pubescent, primary lateral veins in ~7 pairs. **Inflorescences** several-flowered, erect or pendent; rachis 4–18 cm long, tuberculate; peduncles 10–20 cm long; pedicels 3 mm long. **Flowers** ~3 cm long, red, bright rose, mauve to reddish-purple, rarely white; calyx tube ~8 mm long, pubescent, upper lobes 4–5 mm long; vexillary petal somewhat orbicular to elliptic, ~2.5 × 2 cm; ovary many-ovulate, style slender, upcurved. **Fruits** oblong, compressed, 6.5–15 × 2.2–3 cm, shortly beaked, dehiscent, exocarp smooth or wrinkled; seeds ~7, ellipsoid or ovoid, 1.5–2 × 1–1.4 cm, slightly compressed, dark, smooth, glossy, hilum linear, ~8 mm long. **Distr:** Pantropical. **Ecol:** On beaches and with littoral vegetation. Fl. throughout the year. **Zones:** 11–12.

Clitoria

Species ~30. **Herbs**, climbing or erect, or shrubs, rarely trees. **Stipules** and stipels present. **Leaves** mostly pinnately 3–9-foliolate. **Inflorescences** racemose or axillary solitary or paired; bracts stipulelike; bracteoles large. **Flowers** with calyx 5 toothed, 2 upper teeth subconnate; corolla white, red or blue; vexillum without appendages, larger than other petals; stamens monadelphous or diadelphous, 9+1, vexillary stamen free, anthers uniform or 5 dorsifixed alternating with 5 subbasifixed; ovary stipitate, 2- to many-ovuled, style ± flattened, bearded inside, stigma terminal. **Fruits** linear-oblong, compressed or turgid. **Distr:** Tropical and subtropical regions worldwide. **Cult:** Propagate from seeds soaked overnight in water or from cuttings. Requires any well-drained soil in full sun.

C. ternatea L., *Sp.Pl.* 753, 1753. [Asian Pigeon Wings, Blue Butterfly Pea, Blue Pea, Butterfly Pea]. **Climbers**, woody, over 6 m long. **Stipules** linear, 0.4–1 cm long, apex acute. **Leaves** pinnate, 5–9.5 cm long; leaflets usually 5–7, oblong, 1.4–6.5 × 1–4 cm, apex slightly emarginate or obtuse, shortly stalked; petiolules to 2 mm long. **Inflorescences** with flowers solitary, axillary; bracts 4–5 mm long; bracteoles orbicular, ~7 mm long; pedicel 6–9 mm long. **Flowers** with calyx 1.2–1.6 cm long, 2 upper teeth joined near the base; corolla rich blue with greenish veins on the vexillum; vexillum 2.5–5 × 3.5 cm; stamens 10, diadelphous, 9+1, dimorphic, vexillary stamen free; ovary hairy, style simple, hairy along the inner side. **Fruits** pods, 5–12 cm long, beaked; seeds 8–10, flat, round, with marbled patterns. **Ecol:** Fl. Jun–Jan. **Distr:** Tropical Africa. **Zones:** 10–12. **Info:** The white-flowered form (White Butterfly Pea) has edible petals, which tastes like young coconut. The double-corolla form often has white markings on the petals.

Centrosema

Species 30. **Herbs**, scandent subshrubs or vines, mostly scandent. **Stipules** persistent, striate. **Leaves** pinnately trifoliolate, sometimes 1-foliolate; leaflets entire, stipellate. **Inflorescences** solitary or geminate, 1–several-flowered; peduncles axillary; lowest bracts stipuliform, geminate; pedicels 1–2 at each bract; bractlets appressed to calyx, striate, larger than bracts. **Flowers** large, showy, violaceous, bluish, pink, or whitish; calyx short-campanulate, lobes or teeth subequal, or upper 2 connate; standard broad, explanate, somewhat calcarate dorsally near base; wings falcate-obovate; keel incurved, scarcely shorter than the wings; vexillar stamen free or ± connate with others, anthers uniform; ovary subsessile, many-ovulate, style incurved, ± dilated at apex, barbellate about the terminal stigma. **Fruits** pods, subsessile, linear, compressed, flat, 2-valvate, sutures thickened; seeds transverse-oblong, thick or compressed, not strophiolate, hilum small. **Distr:** Tropical to temperate America. **Ecol:** In seasonal forests, grasslands, etc. **Cult:** Propagate from seeds soaked overnight in water or from cuttings. Requires any well-drained soil in full sun.

C. plumieri (Turp.) Benth., *Ann. Wien. Mus.* 2: 118, 1837. [Mariposa, Marmelada]. **Vines** herbaceous, small or rather large; stems glabrous or sparsely pilose. **Stipules** ovate, 4–7 mm long, acute, striate. **Leaves** trifoliate; leaflets ovate or rhombic-ovate, 5–12 cm long, base obtuse or rounded, apex shortly acuminate, often abruptly so, glabrous or nearly so. **Inflorescences** 2–6-flowered, mostly shorter than the petiole; bractlets broadly ovate, obtuse, multistriate, 2–3 times as long as calyx and concealing it; calyx 6–7 mm long, teeth much shorter than the tube; standard white with a dark purple center, 4–5 × 4–5 cm. **Fruits** pods, 10–15 × 1 cm, margins of the valves much-thickened, very long-rostrate. **Distr:** British Honduras to Salvador and Panama, S Mexico, S America, West Indies. **Ecol:** Moist or wet thickets; ~900 m. **Zones:** 10–12.

C. pubescens Benth., *Ann. Wien. Mus.* 2: 119, 1837. (Syn: *Bradburya pubescens*). [Butterfly Pea, Campanilla, Snake Plant]. **Vines**, small, perennial; stems slender, procumbent or twining, usually hispidulous. **Stipules** ovate, acute, 2–3 mm long. **Leaves** long-petiolate, trifoliate; leaflets ovate to oblong, mostly 4–7.5 cm long, base rounded, apex acute, obtuse, or abruptly short-acuminate, conspicuously reticulate-veined, adaxially hispidulous or glabrate, abaxially usually densely and softly pubescent. **Inflorescences** few-flowered; peduncles mostly longer than petioles, bractlets ovate, about equaling the calyx, densely whitish-sericeous with long hairs. **Flowers** with standard 2–4 cm wide, dull purple or rose-purple, sometimes whitish. **Fruits** pods, 10–20 × 0.5–0.7 cm, pubescent or glabrate, tapering to a very long slender beak. **Distr:** Costa Rica, Guatemala, Panama, S Mexico, Salvador, West Indies, S America. **Ecol:** In moist or dry thickets or open, often rocky forest; 250–2,100 m. **Zones:** 10–12.

Mucuna

Species ~100. **Lianas**, **herbs**, scandent or erect **shrubs**, woody-stemmed, evergreen, annual or perennial. **Stipules** and often stipels caducous. **Leaves** pinnately trifoliolate, large; lateral leaflets ± asymmetrical. **Inflorescences** fasciculate-racemose on axillary peduncles or subcymose at apex of peduncle; peduncles often greatly elongate, pendent, cordlike; bracts small or caducous. **Flowers** large, showy, mostly dark purple, red or greenish-yellow; calyx tube campanulate, upper 2 calyx teeth wholly connate, lowest tooth longer than middle ones; corolla much exserted; standard complicate, shorter than wings, base with inflexed auricles; wings oblong or ovate, incurved, often adherent to keel; keel equaling or longer than wings, incurved at apex, acute or cartilaginous-rostrate; stamens diadelphous, vexillar stamen free, others connate, anthers dimorphous, alternate anthers longer and subbasifixed, others shorter, versatile, usually barbate; ovary sessile, villous, few-ovulate, style filiform, not barbate, stigma small, terminal. **Fruits** pods, thick, ovate to oblong or linear, smooth or transversely ridged, often covered with stinging hairs, 2-valvate; seeds rounded or transverse-oblong. **Distr:** Tropics of both hemispheres. **Cult:** Propagate from seeds or by air-layering. Requires a humus-rich soil, moist, well-drained soil in shade or partial sun.

M. bennettii F.Muell., *Papuan Plants* 1: 63, 1876. (Syn: *M. warburgii*, *Stizolobium bennettii*). [New Guinea Creeper, Red Jade Vine]. **Lianas**, woody, to 30+ m long. **Leaves** trifoliate; leaflets lanceolate-ovate, to 7.5–10 × 4 cm, glabrous; petioles 2.5–5 cm long; petiolules very short. **Inflorescences** racemes, pendent, to 45 cm long, usually shorter; peduncles

Top row: *Canavalia maritima*, *Canavalia rosea*, *Clitoria ternatea*, *Centrosema plumieri*.
Bottom row: *Centrosema pubescens*, *Mucuna bennettii*, *Mucuna novoguianensis*, *Mucuna pruriens*.

ZS

Above: *Mucuna sempervirens*, *Butea monosperma*, *Cajanus cajan*.

with rachis scarcely exceeding 2.5 cm, silky; bracts soon deciduous; pedicels crowded, ~2.5 cm long, silky. **Flowers** rich scarlet; calyx densely hairy, tube ~1 cm long, upper lip as long as the tube, lower lip 1.25 cm long, narrow and acuminate; standard ~4 cm long, almost deltoid above middle, 2-toothed at apex; keel petals nearly 7.5 × 0.8–1 cm, tips curved, subulate; wing petals nearly as long and much broader; stamens 9, basally united, the 4 lower ones ~2.5 mm long; style within the staminal sheath with distant bristlelike hairs, exserted portion almost glabrous. **Distr:** New Guinea. **Ecol:** In tropical forests; 0–920 m. **Zones:** 11–12. **Info:** *M. bennetti* and *M. novo-guineensis* are very similar, but the latter has much longer inflorescences.

M. novo-guineensis Sheff., *Ann. Jard. Bot. Buitenzorg* 1: 18, 1876. [Flame of the Forest, New Guinea Creeper, Scarlet Jade Vine]. **Lianas**, high climbing; mature stems glabrous. **Stipels** setaceous, 1–2 mm long. **Leaves** trifoliolate; rachis ~2 cm long; leaflets base mostly rounded, apex acute to long-acuminate, thin, abaxially glabrous or glabrescent, terminal leaflet elliptic, laterals ovate-elliptic, inequilateral. **Inflorescences** pendent, pseudo-racemose, to ~75-flowered, each tubercle 2–4-flowered, mostly cauliflorous, to 60(–90+) cm long, florate to base, glabrous or glabrescent; pedicels ~1 cm long. **Flowers** falcate, to ~7 cm long; calyx broadly cup-shaped, ~8 mm long, with long ferruginous irritant hairs basally, margin usually truncate; petals red, shortly clawed; vexillum spreading, ovate, ~3 cm long; wings narrow, ~5.5 cm long; keels narrow, to ~7 cm long, beak narrow, elongate; ovary several-ovulate, stiffly hairy. **Fruits** pods, linear-oblong, to ~27 × 5.5 cm, exocarp with shallowly ribbed; seeds 5, 4–5.5 × 4–5.5 cm. **Distr:** New Guinea, New Ireland. **Ecol:** In primary and secondary forests; 0–1,800 m. Fl. Jul–Aug. **Zones:** 11–12.

M. pruriens (L.) DC., *Prodr.* 2: 405, 1825. (Syn: *Dolichos pruriens*, etc.). [Velvet Bean, Nescafé, Cow Itch]. **Climbers**, semiwoody, often large, to 4 m long. **Stipules** 3–4 × 1 mm, hairy, deciduous; stipels terete, 4–5 mm long. **Leaves** variable in size, to 46 cm long; terminal leaflet elliptic or rarely rhombic-ovate, (8.5–)14–16 × (4.5–)8–10 cm, base rounded, apex rounded, ± acute, or narrowing into a 5 mm long acumen, extreme tip often mucronate; lateral leaflets asymmetrical, 7–19 cm long; petiole 26 cm long; rachis 1.2–2 cm long; petiolules 4–6 mm long. **Inflorescences** 2–3-flowered, 3–4 cm long, axillary, branches 2–13. **Flowers** showy; calyx densely hairy, tube broadly cup-shaped, 1 × 0.5 cm; corolla purple or white; standard 1.6–2.5 cm long; wings ~1.2 cm long, clawed; keel 2.8–4.2(–4.5) cm long, clawed; staminal tube 2–2.7 cm long. **Fruits** pods, fleshy, oblong-linear, slightly curved and narrowing towards tip, 5–9 × 0.8–2 cm. **Distr:** Tropical Asia. **Ecol:** In thickets, forests; 250–1,100 m. Fl. Sep–Mar. **Zones:** 10–12.

M. sempervirens Hemsl., *J. Linn. Soc. Bot.* 23: 190, 1887. (Syn: *M. japonica*, *M. mairei*). [Cow and Horse Creeper]. **Climbers**, to 25 m, usually woody. **Stipules** very deciduous, to 3 × <1 mm, densely hairy. **Leaves** 21–39 cm long; leaflets with lateral veins 4–7; *terminal* leaflet mostly elliptic, 8–15(–19) × 3.5–9 cm, base cuneate, apex with acumen to 1–5 cm long; *lateral* leaflets asymmetrical, 9–14 × 2–5 cm; petiole 7–16.5 cm long. **Inflorescences** 10–36 cm long, lateral branches 5–7, distant, each 3-flowered, often in groups, borne from very old wood; pedicels 1–2.5 cm long. **Flowers** variable; calyx densely hairy, tube broadly cupular, 0.8–1.2 × (1.2–)1.8–2.5 cm; corolla dark purple or reddish; standard 3.2–4 cm long, rounded; wings (4.5–)4.8–6 × 1.8–2 cm, base clawed; keel (5.3–)6–7 cm long, base clawed; staminal tube ~4 cm long; style and ovary hairy. **Fruits** pods, woody, linear, 30–53 × 3–3.5 × 1–1.3 cm, laterally flattened, brown-hairy; seeds 4–12, reddish, brownish or black, oblong or discoid, 2.8–3 × 2–2.2 × 1 cm. **Distr:** China, Myanmar. **Ecol:** In subtropical forests; 300–3,000 m. **Zones:** 9–11.

Butea

Species 3. **Trees** or climbing shrubs, tomentose. **Stipules** small caducous. **Leaves** pinnately trifoliolate. **Inflorescences** densely fascicled, fascicles racemose or paniculate; bracts and bracteoles narrow, caducous. **Flowers** large, showy, orange or flame-colored; calyx broadly campanulate, teeth shortly deltoid; corolla much-exserted; vexillum ovate, acute, recurved, not appendaged; wings falcate, adherent to the keel; keel much-incurved, acute; stamens diadelphous, vexillary stamen free, others connate, anthers uniform; ovary 2-ovuled, style incurved, stigma terminal. **Fruits** oblong or broadly linear, 2-valved, 1-seeded at the top. **Distr:** India, Myanmar, Pakistan, Sri Lanka. **Cult:** Propagate from seeds or by air-layering. Requires a moist, well-drained soil in full sun.

B. monosperma (Lain.) Taubert, *Pflanzenfam.* 3(3): 366, 1894. (Syn: *B. frondosa*, *Erythrina monosperma*). [Flame of the Forest]. **Trees**, deciduous; young branches tomentose. **Stipules** small, deciduous; stipels small. **Leaves** pinnately trifoliolate; *terminal* leaflet broadly ovate, 12.5–20 × 11–17.5 cm, apex obtuse or retuse, glabrescent; *lateral* leaflets obliquely ovate, apex obtuse, slightly smaller than terminal, glabrescent; petiole 10–22.5 cm long; petiolules 5–7.5 mm long. **Inflorescences** racemes, 10–17.5 cm long, flowers fascicled at nodes, axillary or terminal; bracts and bracteoles small, velvety; pedicels 1.2–2.5 cm long, velvety. **Flowers** with calyx ~1.2 cm long, velvety, teeth deltoid, upper 2 teeth united; corolla bright red, orange-tinged, silky-pubescent outside; vexillum 3.7–5 cm long. **Fruits** pods, 10–20 × 2.5–5 cm, silky, 1 seed at apex; stalk 1.2–1.8 cm long. **Distr:** India; Pakistan, Sri Lanka. **Ecol:** Fl. Mar–Apr. **Zones:** 10–12.

Cajanus

Species ~34. **Shrubs** or **suffrutescent herbs**, erect, abundantly pubescent, branched. **Stipules** caducous. **Leaves** trifoliolate, exstipellate; leaflets entire, resinous-punctate beneath. **Inflorescences** racemes, axillary; scattered along the short axis of peduncles; bracts caducous; bractlets none. **Flowers** rather large, yellow and purple; calyx tube campanulate, lobes acute or acuminate, 2 upper ones connate to form a bidentate lobe; corolla much-exserted, petals equal in length; standard orbicular, reflexed, appendaged at base with 2 reflexed auricles; wings obliquely obovate; keel incurved at apex, obtuse or truncate; stamens diadelphous; vexillar stamen free, others connate, anthers uniform; ovary subsessile, many-ovulate, style long, filiform, thickened above middle, not barbate, slightly dilated below the oblique terminal stigma. **Fruits** pods, linear, narrowed at both ends, obliquely acute, compressed, bivalvate, valves transversely impressed between the seeds, scarcely septate within; seeds 3–5, subcompressed, hilum lateral, oblong. **Distr:** Old World tropics. **Cult:** Propagate from seeds. Requires a moderately fertile, well-drained soil in full sun.

C. cajan (L.) Millsp., *Publ. Field Mus. Nat. Hist. Bot.* Ser. 2: 53, 1900. (Syn: *C. indicus*, *Cytisus cajan*). [Pigeon Pea]. **Shrubs**, erect, 1.5–4 m tall; stem pubescent. **Leaves** pinnately trifoliolate; leaflets elliptic to lanceolate, 2.5–10 × 1.5–3.5 cm, apex acute, adaxially velvety-pubescent, abaxially pilose, both surfaces glandular; petiole 1–5.5 cm long; petiolules present; rachis 0.8–1.3 cm long. **Inflorescences** panicles, terminal; peduncles 2–7 cm long; bracts ~8 mm long; pedicels 1–1.7 cm long, pubescent. **Flowers** with calyx velvety-pubescent and glandular, tube 5 mm long, teeth 4–7 mm long; corolla bright yellow with reddish-brown or crimson lines; vexillum 1.2–1.7 cm Ø. **Fruits** 4.5–10 × 0.8–1.4 cm, pubescent and glandular. **Distr:** Tropical Africa; cultivated widely. **Ecol:** Fl. Jul–Aug. **Zones:** 10–12.

Flemingia

Species ~30. **Herbs** or **subshrubs**. **Stipules** deciduous, stipels absent. **Leaves** palmately trifoliolate or 1-foliolate; leaflets adaxially glandular punctate. **Inflorescences** terminal or axillary, racemes or panicles, or in small cymes, each cyme hidden by a large folded bract, the bracts distichously arranged in zigzag racemes. **Flowers** with calyx lobes longer than the tube, usually glandular; corolla greenish or yellowish; vexillum oblong or elliptic, with inflexed basal auricles; wings narrow; stamens diadelphous, 9+1, vexillary stamen free, anthers uniform; ovary subsessile, 2-ovuled, style enlarged above, stigma

TG

Above: *Flemingia strobilifera, Flemingia wightiana, Erythrina abyssinica.*

small. **Fruits** oblong, inflated, 2-seeded. **Distr:** Tropical Asia, Africa, Malaya, Philippines and Australia. **Cult:** Propagate from seeds or cuttings. Requires a well-drained soil in full or partial sun.

F. strobilifera (L.) R.Br., *Ait. Hort. Kew. ed.* 2(4): 350, 1812. (Syn: *Moghania strobilifera*, etc.). [Luck Plant, Wild Hops]. **Shrubs**, erect, 1.2–3 m tall; young branches pubescent. **Stipules** 6–9 mm long, caducous; stipels absent. **Leaves** 1-foliolate; leaflet ovate-oblong, 8.5–13.5 × 4–5 cm, apex acute, adaxially glabrous to subglabrous, abaxially silky-pubescent, gland-dotted; petiole ~0.6–2.5 cm long. **Inflorescences** usually simple, racemes or panicles, 7.5–15 cm long, axillary and terminal; bracts large, membranous, 1.2–2.5 cm long, usually broader than long, base cordate; each bract folded and enclosing 2 or more small flowers. **Flowers** with calyx 6–6.5 mm long, pubescent, teeth longer than the tube; corolla 8 mm long, white or yellowish-green, purple-tinged; vexillum 0.9–1 cm long, auricled. **Fruits** oblong, 0.9–1 × 0.45–0.5 cm, densely pubescent, completely concealed by bracts, 2-seeded. **Distr:** Indo-China, SE Asia. **Ecol:** In thickets and open dry places. Fl. Oct–Feb. **Zones:** 10–12.

F. wightiana Grah. ex Wight & Arn., *Prod.* 1: 242, 1834. (Syn: *F. congesta* var. *wightiana*, etc.). **Shrubs**, erect, ~0.5–1 m high; young stems somewhat 4-angled, tomentose or somewhat puberulent. **Stipules** ~5 mm long, acuminate, scarious, caducous. **Leaves** trifoliolate; leaflets ovate-oblong or lanceolate, 7.5–12 × 3–7 cm, base cuneate, apex acute, obtuse, adaxially glabrous, somewhat rugose, abaxially tomentose or densely pubescent (golden-silky), black gland-dotted; venation prominent; petioles 3–5 cm long. **Inflorescences** racemose, to ~5 cm long, numerous-flowered, flowers crowded on the rachis, axillary; bracts ovate, acute, not exceeding the buds, caducous. **Flowers** ~6 mm long, subsessile; calyx tube ~3 mm long, hirsute, lobes long-acuminate, greatly exceeding the tube; petals pinkish-white. **Fruits** somewhat oblong, ~9 × 6 mm, light brown, apex apiculate, without glands; seeds 2, somewhat orbicular, ~3 × 2 mm, dark, smooth. **Distr:** S India, Sri Lanka. **Ecol:** In open, dry places. <500 m. Fl. Mar. **Zones:** 10–12.

Erythrina

Species ~108. **Trees, shrubs** or **undershrubs**, rarely herbs; branches often prickly; young branches thick, aculeate. **Stipules** small; stipels glandular. **Leaves** pinnately trifoliolate; leaflets large and broad. **Inflorescences** racemes, axillary or terminal, flowers usually clustered; bracts and bractlets small or absent. **Flowers** usually scarlet, sometimes orange, showy; calyx with an oblique mouth, splitting down to base or campanulate, 2-lipped; vexillum much longer than the wing and the keel; stamens mono- or diadelphous, vexillar stamen free nearly to base or connate with others, anthers uniform; ovary stipitate, many-ovuled, style incurved, stigma small, terminal. **Fruits** stipitate, torulose, falcate. **Distr:** Tropical and subtropical regions worldwide. **Cult:** Propagate from seeds, cuttings or by air-layering, herbaceous species by division. Requires moist but well-drained sandy soil in full sun.

E. abyssinica Lam. ex DC., *Prodr.* 2: 413, 1825. (Syn: *E. tomentosa*, etc.). [Coral Tree, Erythrina, Red Hot Poker Tree]. **Trees**, rarely shrubs, 3–10 m tall. **Stipules** linear-triangular, 3–7 mm long. **Leaves** trifoliate; *terminal* leaflet broadly elliptic to subrotund, 6–14 × 6–16 cm; *lateral* leaflets broadly ovate-rhombic, 4.5–14 × 3–11 cm, base asymmetric; petiole 8–17 cm long, armed with prickles. **Inflorescences** pseudo-racemes, terminal; peduncle 4–16 cm long, densely tomentose; rachis 5–20 cm long; bracts oblong, 6–8 × 2–3 mm, caducous; pedicels 2–5 mm long. **Flowers** with calyx spathaceous, densely tomentose, tube 0.6–1.8 cm long, lobes variable, filiform to obovate, or turbinate, 0.6–4.5 cm long; corolla red, glabrous; standard elliptic-oblong, 2.8–4 × 1.1–1.5 cm; wings 0.7–1.1 × 0.4–0.6 cm; keel petals free, 5–6 × 3–4 mm; vexillary stamen joined only at base, others basally fused into a tube; ovary narrowly cylindrical-oblong, 1.5–2.5 cm long. **Fruits** pods, brown to black, moniliform, 4.8–10.2 cm long, curved, woody, pubescent; seeds 1–10, red, 6–1.2 × 0.5–0.7 cm. **Distr:** E Africa. **Ecol:** In wooded grasslands, rocky hillsides; 200–1,400 m. **Zones:** 10–12.

E. berteroana Urban, *Symb. Ant.* 5: 370, 1908. (Syn: *E. neglecta*). [Brucal, Mother Tree]. **Trees**; branchlets lustrous, armed, to 6 mm long. **Leaves** trifoliolate; leaflets obliquely ovate or ovate; terminal leaflet subrhomboidal, 5–17 × 5–20 cm, base broadly cuneate, margins slightly revolute, apex shortly acuminate, thickly chartaceous, adaxially glabrous, abaxially minutely puberulent, main veins 5–6; petiole 4.5–15 cm long. **Inflorescences** to 50 cm long; pedicels to 1 cm long; bracts ovate, ~1 cm long; bracteoles narrowly ovate, ~1.2 cm long, acute. **Flowers** with calyx narrowly campanulate, ~1.6 × 0.4 cm; standard elliptic, ~8 × 2 cm, petaloid, glabrous; wings oblong, ~1 × 0.25 cm, claw ~2 mm long; keel petals subcircular, ~8.5 × 6.5 mm; odd stamen free at base, anthers ~3 mm long; ovary slender, linear, ~1.7 cm × 1.5 mm, ferruginous-puberulent, style ~3 cm long, stigma capitate, ~0.3 mm long. **Fruits** pods, moniliform, to 15 × 1.5 cm. **Distr:** West Indies, Mexico to Colombia and Venezuela. **Ecol:** In tropical moist forests; 0–1,800 m. **Zones:** 10–12.

E. caffra Thunb., *Prodr. Pl. Cap.*, part 2: 121, 1800. (Syn: *E. constantiana*, etc.). [Coastal Erythrina, Coral Tree, Kaffir Boom, Lucky Bean]. **Trees**, (3.5–)7–18 m tall; branches armed with prickles. **Stipules** minute, caducous. **Leaves** scattered or clustered at ends of twigs, unarmed, glabrous; *terminal* leaflet broadly ovate, 8–16 × 8–18 cm, acuminate; *lateral* leaflets asymmetric, ovate to elliptic, 7.5–13 × 4.5–12(–13.5) cm; petiole 6–16 cm long; rachis 2.8–7.8 cm long. **Inflorescences** dense pseudo-racemes, terminal, appearing before the leaves; peduncle 19–38 cm long, glabrous; bracts membranous, ovate, caducous; pedicels 2–6 mm long. **Flowers** with calyx 1–1.9 cm long, 2-lipped, tomentose; corolla red, glabrous; standard obovate or elliptic, 3.5–5.5 × 1.4–2.8 cm, strongly reflexed; wings oblong, 0.9–2.4 × 0.3–1.1 cm; keel oblong, 0.8–2.1 × 0.4–1.4 cm, petals fused along their lower margin; vexillary stamen free or basally joined, others fused into a sheath; ovary narrowly cylindrical, 1.4–2.7 cm long. **Fruits** pods, black, (10–)12–20 cm long, irregularly constricted between seeds; seeds ~4–6, red, 6–8 × 4–6 mm. **Distr:** Eastern coast of S Africa. **Ecol:** In coastal forests, wooded stream banks; <200 m. **Zones:** 9–11.

E. corallodendrum L., *Sp. pl.* 2: 706, 1753. [Jamaica Coral Tree]. **Trees** or **shrubs**, to 8 m tall. **Leaves** trifoliate; leaflets rhombic-ovate to rhombic-orbicular, broader than long, 5–15 × 4–12 cm, base rounded or obtuse, apex acute or acuminate; petiole slender, 6–15 cm long. **Inflorescences** racemes, 10–30 cm long; pedicels short, 3–5 mm long. **Flowers** with calyx campanulate, 0.7–1.2 cm long, truncate; corolla coral red, 5–6 cm long; standard narrowly oblong or oblong-oblanceolate; wings scarcely exceeding the calyx; keel petals distinct, shorter than the calyx. **Fruits** pods, moniliform, slightly flattened, 10–15 cm long, stipe 1.5–2 cm long, beak to 2 cm long; seeds 3–9, scarlet. **Distr:** Carribean (Haiti, Jamaica). **Ecol:** In coastal open forests, thickets and rocky hillsides; 30–900 m. Fl. Feb–Apr. **Zones:** 11–12.

E. crista-galli L., *Mant.* 99: 228, 1767. [Cock's Comb, Cockspur Coral Tree, Coral Tree]. **Trees**, evergreen, 6–9 m tall; bark thick, corky; branches with stout spines. **Leaves** trifoliate; leaflets ovate-oblong to broadly elliptic, 5–9 × 1.9–5.7 cm, coriaceous, glabrous; petioles to 20(–30) cm long, often spiny. **Inflorescences** appearing with leaves, terminal, drooping racemes, to 60 cm long, or axillary clusters of 2 or 3. **Flowers** with calyx bell-shaped, to 2 cm long, entire; standard reflexed, elliptic to broadly elliptic, 3.8–5 cm long, crimson to blood red, to 5 cm long. **Fruits** cylindrical, curved, 7.5–20 × 1.2–1.5 cm; seeds 1.6 cm long, dark brown. **Distr:** SE Brazil, Paragual, Uruguay, and N Argentina. **Ecol:** In gallery forests, along watercourses, in swamps and wetlands; to 1,500 m. Fl. Dec–Apr. **Zones:** 9–11.

E. fusca Lour., *Fl. Cochinch.* 427, 1790. (Syn: *E. ovalifolia*). [Cape Kaffirboom, Kaffirboom Coral Tree, Lucky Bean Tree]. **Trees**, large, to ~20 m tall; trunk short; branches armed. **Stipels** ~1 mm long, persistent. **Leaves** present at flowering; leaflets mostly elliptic or ovate, ~15 × 7 cm, apices rounded or obtuse, coriaceous, adaxially glabrous, lateral veins 7–9 pairs; petioles to ~19 cm long. **Inflorescences** erect, ferruginous hairy, soon glabrescent, 1–2(–5) per node, ~5 cm long; bracts and bracteoles ovate, to ~2.5 × 2 mm; pedicels 0.5–2 cm long. **Flowers** with calyx broadly campanulate, asymmetrically split, to ~1.7 cm long, spur to ~1.4 cm long on the vexillum side, lacerate, ferruginous pubescent; petals orange or dark red; vexillum rounded-rhomboid, 3–5 cm long, glabrous, claw ~1.5 cm long; wings obliquely obovate, ~2.5 × 1.5 cm; keels ovate-falcate, ~3 × 1.5 cm; ovary long-stipitate, several–many-ovulate. **Fruits** linear, torulose, ~30 × 1.5 cm. **Distr:** West Indies, C America, tropical S America; India, Malaysia, Indonesia, Philippines, Pacific Islands. **Ecol:** Near rivers and streams. Fl. Jan–Apr. **Zones:** 10–12.

E. herbacea L., *Sp.Pl.* 706, 1753. [Coral Bean]. **Shrubs** or **small trees**, 1–3(–6) m tall; prickles few; stems often clustered, armed with prickles; bark pale, thick. **Leaves** trifoliolate, 15–20 cm long; leaflets orbicular or ovate, 3.5–10 × 3–5 cm, shallowly trilobed, apex acute or acuminate, glabrous, light to mid green, abaxially with prickles on the midribs; petiole 5–8 cm long; petiolules ~4 mm long. **Inflorescences** racemes, terminal, 30–60 cm long, leafy below, flowers distant. **Flowers** tubular, reddish; calyx ~1–1.25 cm long, slightly 2-lipped, glabrous; corolla scarlet; vexillum 4.5–5 × ~2 cm. **Fruits** pods, drooping, 7.5–12.5 cm long, shortly beaked, constricted between the seeds; seeds 5–10, bright red or scarlet, with a black spot, 1 cm long. **Distr:** S USA, Mexico. **Ecol:** Fl. Mar–Jun. **Zones:** 8–11.

E. latissima E. Mey., *Comm. Pl. Afr. Austr.* 151, 1836. (Syn: *Chirocalyx mollissimus*, *E. gibbsae*, etc.). [Broad-leaved Coral Tree, Lucky-bean Tree]. **Trees**, 3–8(–10) m tall; stems and branches armed with large conical spines to 1 cm long. **Stipules** ovate, 6–8 × 4–5 mm. **Leaves** pinnately trifoliolate; leaflets with prickles along nerves; *terminal* leaflet deltate to oblate, 7–25 × 6–30 cm; *lateral* leaflets broadly ovate to deltate, 6–22 × 7–22 cm, base often asymmetric; petiole 12–30 cm long. **Inflorescences** pseudo-racemes, terminal; peduncle 15–24(–30) cm long, densely tomentose; bracts linear-obovate, 1.2–2 cm long. **Flowers** with calyx spathaceous, 5-lobed, tube 1.3–3 cm long, lobes ± filiform, 0.7–1.8 cm long; corolla scarlet, wing margins white; standard oblong-elliptic, 4.5–5.5 × 2.8–3.3 cm; wings 2.2–2.5 × 1–1.2 cm; keel petals free, 1.7–2 × 1.3–1.5 cm; stamens basally fused; ovary ± filiform, 1.5–2 cm long. **Fruits** pods, falcate, 18–25 × 2–3 cm; seeds (1–)3–4(–6), red, reniform, 1.1–1.5 × 0.9–1 cm. **Distr:** Zimbabwe, Botswana, Mozambique, S Africa, Swaziland. **Ecol:** In woods, grasslands; 1,000–1,750 m. **Zones:** 10–12.

E. livingstoniana Baker, *Fl. Trop. Africa* 2: 182, 1871. [Aloe Coral Tree, Mozambique Coral Tree]. **Trees**, (2)5–25 m tall; spines to 2.5 cm long. **Stipules** triangular, 8 × 2 mm; stipels ovate, 3 mm long, caducous. **Leaves** trifoliate; leaflets palmately lobed or rarely broadly ovate, 4–16.5 × 5–17.2 cm, apex acuminate, terminal lobe triangular, lateral lobes truncate, sometimes emarginate, bases truncate; petiole 5–19.5 cm long, often with scattered prickles; rachis 2.5–8 cm long, armed with scattered prickles. **Inflorescences** pseudo-racemes, flowers in fascicles; peduncles (5–)8–15(–30) cm long, ferruginous-tomentose to glabrescent. **Flowers** with calyx spathelike, 1.5–2.8 cm long, split ventrally, glabrous, lobes obscure, to 3 mm long; standard scarlet, elliptic to ovate, 3.8–5.8 × 2.4–4.2 cm; wings 1.2–1.5 × 0.8–1 cm; keel petals free, 1–1.4 × 0.7–1.1 cm. **Fruits** pods, straw-colored, moniliform, 11–39 × to 3.5 cm; seeds 2–13, orange-red, reniform, 1.3–1.5 × 0.5–0.7 cm. **Distr:** Malawi, Mozambique, Zambia, Zimbabwe. **Ecol:** On rocky hillsides, river banks, scrubs; 95–630 m. **Zones:** 11–12.

E. lysistemon Hutch., *Bull. Misc. Inform.*, Kew 1933: 422, 1933. (Syn: *E. caffra* [misapplied], *E. humei*, etc.). [Coral Tree, Sacred Coral Tree]. **Trees**, 2–12(–20) m tall; branches and bark armed with prickles. **Stipules** ovate, 2–3 mm long, caducous. **Leaves** with terminal leaflet broadly ovate or deltate, 4.5–12(–13) × 3.5–10(–12) cm, apex acuminate; lateral leaflets broadly ovate, 4–11 × 2.5–7 cm, base asymmetric; petiole 3.5–8.5 cm long, with 0–2 prickles, rachis 8.5 cm long. **Inflorescences** pseudo-racemes, dense, terminal; peduncle (4–)6–22 cm long, ferruginous-hairy; rachis 2–10 cm long. **Flowers** with calyx 1–1.4 cm long, 2-lipped, tomentose; corolla bright red, glabrous; standard elliptic to obovate, 2.4–5.5(–6.5) × 1–2.1 cm; wings oblong, 0.9–1.1 × 0.3–0.4 cm; keel narrowly oblong, 0.7–1 × 0.3–0.5 cm, petals fused along lower margin; vexillary stamen free, others fused into a sheath; ovary narrowly cylindrical, 1–1.7 cm long. **Fruits** pods, black, 12–30 × 1.2–1.8 cm; seeds (1–)3–8, red, 6–8 × 4–6 mm. **Distr:** Botswana, Malawi, Mozambique, S Africa, Swaziland, Tanzania, Zimbabwe. **Ecol:** In dry habitats; 0–1,600 m. **Zones:** 9–12.

E. sandwicensis O. Degener, *Fl. Hawaiiensis Fam.* 169c, 1932. (Syn: *E. monosperma*). [Wiliwili]. **Trees**, deciduous, 5–15(–16.8) m tall; trunk, crooked or gnarled, 0.3–0.9(–3.8) m dbh. **Leaves** alternate, 13–30 cm long; leaflets 3, broadly ovate to diamond-shaped or circular, 4–10 × 6–15 cm, coriaceous, abaxially ± densely white-hairy. **Inflorescences** racemes, 7.5–15 cm long, ± horizontal, flowers crowded at apex. **Flowers** densely felty-hairy outside; calyx splitting along upper side; corolla orange, salmon, yellowish, greenish, or off-white; standard reflexed, ovate to elliptic, 4–5 cm long; lower 4 petals 1.5–2 cm long; stamens protruding from keel. **Fruits** pods, 10 × 13 cm, curved, constricted; seeds usually 1–2(–4), 1.3–1.5 cm long, bright red to yellowish orange. **Distr:** Hawai'i. **Ecol:** Fl. Sep–Nov. **Zones:** 10–12.

E. variegata L., *Dis. Herb. Amboin.* (L.) 10, 1754. (Syn: *E. indica*, etc.). [Coral Tree, Indian Coral Tree, Tiger's Claw]. **Trees**, large, evergreen, 9–25 m tall; trunk and branches often prickled; bark grayish-green. **Leaves** trifoliate; leaflets ovate to broadly elliptic or flattened-circular, 7.5–20 × 10–22 cm, base rounded or truncate, apex cuspidate, terminal one subrounded, membranous, glossy green or variegated with white and yellow, ± glabrous; petiole 10–20 cm long. **Inflorescences** pseudo-racemes, erect, 14–16 cm long, crowded. **Flowers** grayish felty-hairy; calyx to 4 cm long, splitting on one side; standard spreading; standard paler, orange-red or scarlet (rarely white), 5.7–7 × 3.3 cm; other petals much shorter and dark red; stamens long-protruding from keel; ovary felty-hairy. **Fruits** pods, distinctly torulose, 15–40 cm long, blackish, reticulate, slightly constricted; seeds usually 4–13, kidney-shaped, 1.5–2 cm long. **Distr:** Tropical E Africa, through tropical Asia to Polynesia. **Ecol:** 0–50(–500) m. **Zones:** 11–12.

E. velutina Willd., *Ges. Naturf. Fr. Neue Schr.* 3: 426, 1801. (Syn: *E. aurantiaca*, etc.). [Coral Tree, Red Bean Tree]. **Trees**, large, spreading; branches armed. **Stipules** linear, rigid, ~5 mm long, deciduous; stipels ~1 mm long, persistent. **Leaves** present at flowering, densely tomentose, soon glabrescent; leaflets ovate-rhomboid to triangular, to ~11 × 12 cm, base broadly cuneate to truncate, apex rounded to obtuse, chartaceous, primary lateral veins ~6 pairs; petioles to ~20 cm long. **Inflorescences** erect, stellate-tomentose, soon glabrescent, to ~20 cm long, flowers numerous, in fascicles of 2 or 3; bracts ovate, ~2.5 mm long, caducous; pedicels ~1 cm long. **Flowers** to ~6 cm long; calyx splitting to form spathe, 2–3 cm long; petals orange-red to vermilion, vexillum ovate, obovate or oblong-elliptic, emarginate, inequilateral, glabrous, usually strongly reflexed, 3–5.5 × 1.5–3 cm, claw ~6 mm long, wings obliquely obovate or -elliptic, slightly exceeding keels, keels free, obovate; stamens monadelphous; ovary long-stipitate, few- to several-ovuled, pubescent, style glabrous. **Fruits** indehiscent, woody, linear to narrowly oblong, 7.5–14 × 1.2–1.7 cm, beak ~1 cm long; seeds 1–4, oblong, to 1.7 × 1.1 cm, red. **Distr:** West Indies, tropical S America. **Ecol:** In lowland dry regions. **Zones:** 10–12.

Psophocarpus

Species ~10. **Herbs climbing** or **subshrubs**. **Stipules** and **stipels** persistent; stipules produced below insertion. **Leaves** 1-foliolate or pinnately trifoliolate. **Inflorescences** axillary, fasciculate-racemose, tuberculate, few- to several-flowered; bracteate and bracteolate. **Flowers** with calyx 4–5-lobed, upper lobes connate, entire or emarginate; petals blue or purple, vexillum suborbicular, biauriculate, with appendages, glabrous, wings free, somewhat ovate, keels obliquely oblong, somewhat beaked; vexillary stamen free or fused to staminal sheath, anthers uniform; ovary linear, ridged, ovules few to numerous, style geniculate, upper style bearded, bent at apex, stigma penicillate. **Fruits** oblong to linear, dehiscent, 4-winged, somewhat septate; seeds few to numerous, ovoid or oblong, hilum short, oval or elliptic. **Distr:** Africa and Madagascar. **Ecol:** In seasonally dry tropical forests, secondary vegetation, moist grasslands. **Cult:** Propagate from seeds. Requires a moist, well-drained soil in shade or full sun.

P. tetragonolobus (L.) DC., *Prod.* 2: 403, 1825. (Syn: *Dolichos ovatus*, etc.). [Princess Bean, Winged Bean]. **Vines** climbing, twining; stems terete, glabrous. **Stipules** lanceolate, ~5 mm long above insertion, 1–4 mm below, striate, glabrous; stipels lanceolate, 3–5 mm long. **Leaves** pinnately trifoliolate; rachis 2–5 cm long; leaflets ovate, 6–14.5 × 3–9 cm, base usually broadly cuneate, apex acuminate, thin, glabrous; terminal leaflet often rhombic or deltoid; petiole 6–20 cm long. **Inflorescences** ~15 cm long, each tubercle 1–2-flowered; bracts lanceolate, ovate, ~2 mm long, striate, glabrous, persistent; bracteoles ovate-orbicular, striate, ~3 mm long; pedicels 0.8–1.2 cm long. **Flowers** ~3 cm long; calyx tube to ~8 mm long, glabrous, upper lobe shallowly emarginate, obtuse, laterals acute, acuminate, ~5 mm long; vexillary petal reflexed, to ~3 cm long, wings to ~1.8 cm long, with an auricle almost equaling the claw; vexillary stamen usually slightly fused; ovary few- to many-ovulate. **Fruits** linear, to ~18 × 1 cm, glabrous, each wing 0.5–1.5 cm wide, margins crenate; seeds ~14, ovoid, dark, ~6 × 5 mm. **Distr:** Origin unknown; cultivated worldwide. **Ecol:** Natural habitat unknown. Fl. Mar–July Fr. Oct–Nov. **Zones:** 10–12.

Strongylodon

Species 15–20. **Climbers**, perennial, evergreen or

Top row: *Erythrina berteroana*, *Erythrina caffra*, *Erythrina corallodendrum*, *Erythrina crista-galli*.
Second row: *Erythrina fusca*, *Erythrina herbacea*, *Erythrina latissima*, *Erythrina livingstonia*.
Third row: *Erythrina lysistemon*, *Erythrina sandwicensis*, *Erythrina sandwicensis* (off-white form).
Bottom row: *Erythrina variegata*, *Erythrina velutina*.

F&KS

F&KS

F&KS

F&KS

TG

F&KS

F&KS

TG

MS

Above: *Psophocarpus tetragonolobus, Strongylodon elmeri, Strongylodon macrobotrys.*

deciduous, vigorous; stems woody. **Stipules** small. **Leaves** pinnate, petiolate; leaflets 3. **Inflorescences** racemes, pendent, borne in leaf axils or leaf scars on bare stems, flowers clustered at nodes. **Flowers** with calyx truncate or broadly toothed; corolla with petals clawed; standard petal 2-appendaged above claw; keel petal joined, incurved; stamens 9, joined, upper 1 free. **Fruits** pods, woody, indehiscent, wrinkled; seeds few, rounded, hilum extending halfway around. **Distr:** Madagascar, through Asia to Polynesia, with the highest diversity in Philippines with 8 eight endemic species. **Cult:** Propagate from cuttings. Requires any fertile, well-drained soil in full sun.

S. elmeri Merr., *Philipp. Gov. Lab. Bur. Bull.* 29: 20, 1905. **Vines**, woody. **Leaves** 20–25 cm long, trifoliate; leaflets oblong, 11–17 × 4–6.5 cm, apex abruptly acuminate, acumen short, blunt, nerves 5 or 6 per side; *terminal* leaflets with base slightly rounded or subacute; *lateral* leaflets with base strongly inequilateral, rounded or truncate. petiole glabrous, 8–9 cm long; petiolules 5–7 mm long. **Inflorescences** racemes, many-flowered, flowers borne from each protuberance, forming a rather dense, subglobose inflorescence, axillary, axis to 20 cm long; peduncles glabrous, 15 cm long, flower-bearing portion 5–6 cm long; pedicels slender, ~3 cm long, glabrous. **Flowers** dark greenish-blue (jade), 2.5–3 cm long; calyx tubular, slightly inflated, 6–9 × 5–6 mm, glabrous, teeth very short, rounded, bracts, if present, caducous; corolla glabrous; standard 2 × 1.5 cm, strongly reflexed, acute, wings 1.5 cm long, claw 5 mm long, lamina obtuse, 1 × 0.6 cm; keel 2.5–3 cm long, strongly curved, clawed; ovary pedicellate, densely and minutely pubescent, ovules 5 or 6. **Fruits** pods fleshy, oblong-lanceolate, 5–7 × 2 cm, compressed, apex with a short, curved apicula. **Distr:** Philippines (Luzon). **Ecol:** In thickets and forests; to 1,600 m. Fl. Feb–Mar. **Zones:** 11–12.

S. lucidus (Forster f.) Seem., *Fl. Vit.* 61. 1865. [Hawaiian Jadevine]. **Climbers**, large; stems glabrous, thin hollow but firm. **Stipules** triangular, ~6 mm long, membranous; stipels lanceolate, ~4 mm long, membranous. **Leaves** 20 cm long; leaflets 3, ovate, 10 × 6.5 cm, base subequal, apex shortly bluntly cuspidate, membranous, bright green, glabrous, adaxially shiny, trinerved at base, central nerve with 2–3 pairs of arching lateral nerves; *terminal* leaflet with base rounded and petiolule 2.5 cm long; *lateral* leaflets shortly petiolulate, base obliquely cuneate; petiole 5 cm long, glabrous. **Inflorescences** racemes, axillary, composed of 30–40 1–3-flowered fascicles, flowers pedicellate arising from small produced nodes with obsolete bracts; peduncles 10–15 cm long; pedicels slender, spreading, 1.2–2 cm long, 2-bracteolate; bracteoles caducous. **Flowers** with calyx tubular, ~8 mm long, glabrous, teeth very short, rounded; corolla bright red 2.5 cm long; standard lanceolate, acute; wings short, obtuse, intimately attached to lamina of narrowly lanceolate curved keel. **Fruits** pods, oblong, 7.5 cm long, obliquely pointed, distinctly stalked, 2-seeded. **Distr:** Africa, Tropical Asia, Australasia and Pacific Islands. **Ecol:** In coastal areas, forests and thickets along streams. Fl. Aug–Feb.**Zones:** 11–12.

S. macrobotrys A.Gray, *U.S. Expl. Exped., Phan.* 15: 448, t. 49, 1854. [Blue Jade Vine, Emerald Creeper, Jade Vine, Turquoise Jade Vine]. **Climbers**, robust, woody, 9–15 m long; stems twisted, rope-like. **Leaves** pinnately–compound, 3–foliate; leaflets elliptic to oblong, 8.9–17 × 4–7 cm, apex abruptly tapering, membranous, glossy green, new flush pinkish–bronze in colour; petioles 3.2–7.5 cm long. **Inflorescences** pendent, 0.6–3 m long, from new or old wood, flowers 4–8 per node. **Flowers** blue-green, 4–9 cm long; calyx truncate, 4–5–dentate; standard petal ~$1/3$ as long as keel, reflexed, wing petal about half as long as keel; free, keel petal sickle–shaped, apex beaked; ovary 4–6–ovuled. **Fruits** pods, ± cylindrical, woody, 1.3–2 m long, glabrous; seeds 5–12, oblong, 3.5–5 cm long. **Distr:** Philippines (Luzon and Mindanao). **Ecol:** In humid, montane forests, and in damp forested ravines; 650–1,000 m. Fl. Jan–May. **Zones:** 11–12.

S. siderospermus Cordemoy, *Fl. Ile Reunion* 392, 1895, 1979. (Syn: *S. lucidum*, etc.). [Australian Pea]. **Vines**, climbing, to 10 m long. **Stipules** acute, 3–4 mm long, persistent; stipels linear, 4–5 mm long, persistent. **Leaves** trifoliolate; rachis 2–3 cm long; leaflets to ~10 × 6 cm, base cuneate to rounded, apex acute or acuminate, glabrous, basal veins 3, prominent. **Inflorescences** 1–3-flowered, borne on nodes; bracts acute, ~2 mm long; bracteoles ovate, ~2 × 1 mm; pedicels 1–1.5 cm long. **Flowers** 2–2.5 cm long, calyx tube ~6 mm long, glabrous, lobes ciliate, <1 mm long; petals reddish-pink; vexillum ovate, ~2.2 × 0.9 cm, shortly clawed, inflexed biauriculate basally, appendages 2, glabrous; wings somewhat oblong, ~9 × 4 mm; keels slightly falcate, ~1.5 × 0.4 cm, acute; wings and keels long-clawed; ovary 2-ovuled. **Fruits** pods, slightly compressed, ~8 × 4 cm, ~1.5 cm thick, beaked; seeds 1–2, ~2.5 × 1.8 cm, black, smooth, hilum 3–4 cm long, arils red. **Distr:** W Indian Ocean (Reunion Island, Mascarenes, Madagascar), Sri Lanka, Andaman Islands, Christmas Island, N Australia, New Guinea, New Caledonia. **Ecol:** A littoral species from humid lowlands. Fl. Jun. **Zones:** 10–12.

Vigna

Species ~104. **Herbs**, erect or scandent. **Stipules** sessile or sometimes produced below point of insertion. **Leaves** pinnately trifoliolate, stipellate. **Inflorescences** sometimes umbelliform racemes, short, borne at apex of peduncle, axillary; peduncle elongate; bracts and bractlets small, caducous. **Flowers** generally yellow, upper 2 calyx lobes or teeth connate or distinct; standard orbicular, with inflexed basal auricles; wings falcate-obovate, slightly shorter than the standard; keel equaling the wings, incurved, not rostrate or with a short incurved beak; vexillar stamen free, others connate, anthers uniform; ovary sessile, many-ovulate, style filiform or thickened above, often barbate along inner side, stigma oblique. **Fruits** pods, linear, straight or nearly so, subterete, 2-valvate, interrupted within between the seeds; seeds reniform or subquadrate,

Below: *Strongylodon lucidus, Strongylodon siderospermus, Vigna caracalla, Phaseolus lunatus.*

TR

JS

hilum lateral, short, not strophiolate. **Distr:** Tropics and subtropics; Old World (~90 spp.), New World (~22 spp.). **Ecol:** In seasonally dry tropical woodlands, grasslands. **Cult:** Propagate from seeds. Requires a fertile, moist, well-drained soil in full sun.

V. caracalla (L.) Verdc., *Kew Bull.* 24: 552, 1970. (Syn: *Phaseolus bertonii*, *P. caracalla*). [Bertoni Bean, Corkscrew Flower, Snail Bean, Snail Flower]. **Vines**, short-lived perennials, twining, 3–6 m long; stems sparsely pubescent, glabrescent. **Stipules** deltoid, striate, ciliate; stipels ovate, glabrous, striate, 2–4 mm long. **Leaves** pinnately trifoliolate; leaflets ovate, mostly 8–15 × 4–9 cm, base obtuse, truncate or rounded, apex shortly acuminate, glabrate; petioles stout, 8–15 cm long; petiolules stout, 3–6 mm long, tomentose. **Inflorescences** 5–30 cm long, enlarged nodes distributed along upper portion of peduncle; bracteoles ovate, 2–2.5 mm long, striate; bracts ovate, striate, caducous; pedicels slender, to 9 mm long, glabrous. **Flowers** sweetly fragrant, pale lavender or purple, sometimes partly white; calyx broadly campanulate, basally cupular, to 1.3 cm long, lobes short, rounded; standard to 5 mm long, curled back; keel in 4 spirals. **Fruits** pods, linear oblong, falcate, 8–15 × ~1 cm, compressed, buff-colored; seeds compressed-globose, ~8 mm across, black, hilum white. **Distr:** Tropical S America. **Ecol:** In cooler humid areas; 0–1,500 m. Fl. Feb–Apr. **Zones:** 10–11.

Phaseolus

Species ~60–65. **Herbs** or **subshrubs**, erect, prostrate or climbing. **Stipules** persistent. **Leaves** not prolonged below point of insertion, pinnately trifoliolate, rarely unifoliolate. **Inflorescences** pseudo-racemose, axillary, flowers fascicled along the rachis; bracts and bracteoles present. **Flowers** with calyx 5-lobed, 2-lipped; upper lip emarginate, bifid, lower lip trilobed; vexillum round, auriculate, with 2 appendages or with only slight appendages at side of the claw; wings often following the spiral of the keel; keel often narrow, elongated, apex beaked and forming the spiral of 1–5 complete turns; stamens diadelphous, 9+1, anthers subuliform or 5 dorsifixed alternating with 5 basifixed; ovary 2–many-ovuled, style tenuous below, thickened above, curved through 360°. **Fruits** pods, linear or oblong, compressed or subcylindrical, sometimes beaked; seeds 3–many, oblong or reniform. **Distr:** Tropical to warm temperate regions of America. **Cult:** Propagate from seeds. Requires a humus-rich, well-drained soil in full sun.

P. lunatus L., *Sp. Pl.* 724, 1753. [Butter Bean, Lima Bean]. **Climbers**, perennial, stem glabrous or pubescent. **Stipules** 2–3.5 mm long. **Leaves** trifoliolate; leaflets 3–15 × 1.2–10 cm, lateral leaflets oblique, ovate to lanceolate or narrowly rhombic, acuminate; petiole 1.5–19 cm long; petiolules 3–5 mm long. **Inflorescences** racemes; peduncles 1.5–30 cm long; bracts 1.5 mm long; bracteoles 1.5–2 mm long; pedicel 0.5–1 mm long. **Flowers** with calyx pubescent, tube 1.5–2.5 mm long, teeth 5–8 mm long, upper 2 joined to form an emarginate lip; corolla white, yellowish or pale rose; vexillum 5–7 mm long; keel 1–1.4 cm long, spirally twisted for 1½ turns. **Fruits** oblong-falcate or oblong, 5–10.5 × 1.2–2.5 cm, 3–4-seeded. **Distr:** Tropical America; widely cultivated throughout the tropics. **Ecol:** In extremely dry coastal areas, tropical areas and temperate areas; to 2,500 m. **Zones:** 8–12. **Info:** The seeds are edible, often eaten toasted.

TRIBE SESBANIEAE

Sesbania

Species ~60. **Herbs**, **shrubs** or **trees**. **Stipules** small, deciduous. **Leaves** even-pinnate; rachis with a setaceous tip; leaflets numerous, mostly oblong, apex obtuse or rounded and mucronate. **Inflorescences** racemes, axillary; bracts and bractlets small, caducous. **Flowers** with calyx broadly campanulate, turbinate at base, as broad as long, teeth short, deltoid, acute or acuminate; corolla yellow or white; standard longer than the other petals, suborbicular, reflexed, often appendaged at base, shortly unguiculate; keel petals lunate, apex obtuse or rounded; stamens diadelphous. **Fruits** pods, linear, slender, terete or somewhat compressed, rostrate, 2-valvate; seeds many, cylindric-oblong, brown, smooth. **Distr:** Tropical and warmer regions of both hemispheres. **Cult:** Propagate from seeds or hardwood cuttings. Requires fertile, well-drained soil in full sun and some species are salt-tolerant.

S. grandiflora (L.) Poiret, *Encycl. Meth.* 7:127, 1806. (Syn: *Agati grandiflora*, etc.). [Baby Booties, Hummingbird Tree, Scarlet Wisteria Tree, Vegetable Hummingbird]. **Trees**, small, soft-wooded, fast-growing, 6–10(–15) m tall; stem cylindrical, 25 cm Ø. **Stipules** minute, deciduous. **Leaves** borne at ends of branches, alternate; lamina pinnately compound, ~13–30 cm long; leaflets 24–60, ~1.8–4 × 0.5–1 cm, oblong, apex obtuse, glabrous. **Inflorescences** racemes, axillary, 2–4-flowered, flowers pendulous. **Flowers** 5–10 cm long; calyx ~2.5 cm long, glabrous, bell-shaped, oblique, somewhat 2-lipped; corolla white, cream, pink, or red; vexillum ~7–10 cm long. **Fruits** pods, pendent, ~30–50 × ~0.8 cm, suture much-thickened, 15–40-seeded. **Distr:** Cultivated and naturalized widely in the tropics; origin obscure, perhaps Indonesia. **Ecol:** Fl. Oct–Dec. **Zones:** 10–12. **Info:** The leaves, flowers (excluding stamens and calyx) and pods are edible.

TRIBE ROBINEAE

Gliricidia

Species 5. **Trees** or **shrubs**, unarmed. **Leaves** not stipellate, odd-pinnate. **Inflorescences** axillary, racemes; bracts small, deciduous; bractlets none. **Flowers** large, showy, mostly pink, often appearing before leaves; calyx short-campanulate, as broad as long, truncate, teeth 5, short and broad or obsolete; standard orbicular, reflexed, often with 2 callosities at base of lamina, shortly unguiculate; wings oblong-oblanceolate, erect, free, keel petals strongly arcuate above middle, obtuse, united at apex; stamens diadelphous, anthers uniform; ovary stipitate, 7–12-ovulate, style glabrous, inflexed at almost a right angle, stigma capitate, papillose. **Fruits** pods, linear, shortly stipitate, compressed, 2-valvate; seeds suborbicular, compressed. **Distr:** Tropical America. **Cult:** Propagate from seeds or cuttings. Requires a moist, well-drained soil in full sun.

G. sepium (Jacq.) Steud., *Nom. Bot.* ed. 2. 1: 688, 1841. (Syn: *Robinia sepium*, etc.). [Mother of Coffee]. **Trees**, deciduous, glabrous, 3–10 m tall; trunk to 30 cm dbh. **Stipules** ovate to lanceolate, 2 mm long. **Leaves** odd-pinnate; leaflets 7–17, lanceolate-oblong to ovate or elliptic, 3–7 × 2–3 cm, base acute, apex acute or obtuse-acuminate, surfaces initially sparsely sericeous, glabrate with age, adaxially green, abaxially faintly blotched with pale purplish. **Inflorescences** racemes, 5–10 cm long, many-flowered and often dense; bracts ovate, 1 mm long; pedicels 5 mm long. **Flowers** with calyx puberulent or almost glabrous, 4–5 mm long; corolla 1.5–2 cm long, bright rose-pink to almost white. **Fruits** pods, 10–15 × 1–1.5 cm, short-stipitate, glabrous, valves thick and somewhat ligneous; seeds lenticular, ~1 cm long, dark brown. **Distr:** Mexico, C America, West Indies. **Ecol:** On hillsides, thickets or in forests; <1,600 m. Fl. Dec–Apr. **Zones:** 10–12.

Robinia

Species 4. **Trees** or **large shrubs**, all parts (excluding corolla) sometimes with glandular bristles; buds only axillary. **Stipules** bristlelike or spinelike. **Leaves** odd-pinnate; leaflets petiolulate and stipellate, margins entire. **Inflorescences** racemes, axillary, pendulous; bracts membranous, caducous. **Flowers** with calyx campanulate, 5-lobed, adaxially 2 lobes subconnate into an apical lip; corolla white, pink or rosy; petal stipitate; standard large, retroflexed; wings curved; keel incurved, blunt; stamens diadelphous, single one distinct from other 9 connate ones, anthers homomorphic, biloculed, longitudinally dehiscent; ovary stipitate, many-ovuled, style subulate, apically piliferous, stigma small, terminal. **Fruits** pods, compressed, narrowly winged along ventral suture, segments often with appressed bristles; seeds oblong or obliquely reniform, strophiole absent. **Distr:** USA and NW Mexico. **Cult:** Propagate from scarified seeds, suckers or by air-layering. Requires a moist, well-drained soil in full sun.

R. pseudoacacia L., *Sp. Pl.* 2: 722, 1753. (Syn: *R. pseudoacacia* var. *inermis*, etc.). [Black Locust, False Acacia, Yellow Locust]. **Trees**, 10–25 m tall, deciduous; **Stipulate spine** to 2 cm long; stipel acicular. **Leaves** 10–25(–40) cm long; leaflets 2–12

Below: *Sesbania grandiflora*, *Gliricidia sepium*, *Robinia pseudoacacia*.

Top row: *Clianthus puniceus, Carmichaelia aligera, Carmichaelia odorata.*
Bottom row: *Carmichaelia williamsii, Chordospartium stevensonii.*

pairs, ± opposite; leaflets elliptic, oblong, or ovate, 2–5 × 1.5–2.2 cm, base rounded to broadly cuneate, margins entire, apex rounded, retuse, and apiculate, adaxially green, abaxially grayish-green, pubescent when young; petiolule 1–3 mm long. **Inflorescences** racemes, axillary, 10–20 cm long, pendulous, many-flowered; bracts caducous; pedicels 7–8 mm long. **Flowers** fragrant; calyx obliquely campanulate, 7–9 mm long, teeth 5, triangular to ovate-triangular, densely pubescent; corolla white, stipitate; standard suborbicular, ~1.6 × 1.9 cm, retroflexed, inside yellow-spotted, base rounded, apex with a sinus; wings obliquely obovate, ~1.6 cm long; keel falcate, triangular, connate at leading edge, obtuse; stamens diadelphous, one opposite to standard free; ovary linear, ~1.2 cm long, glabrous, style subulate, ~8 mm long. **Fruits** pods, brown or with reddish-brown stripes, linear-oblong, 5–12 × 1–1.3(–1.7) cm, compressed; seeds 2–15, brown. **Distr:** E North America. **Ecol:** Fl. Apr–Jun. Fr. Aug–Sep. **Zones:** 4–10.

TRIBE GALEGEAE

Clianthus

Species 2. **Shrubs** softly woody or **small trees**. **Stipules** broad, awl-shaped, herbaceous; stipels absent. **Leaves** imparipinnate; leaflets numerous, ovate, margins entire. **Inflorescences** umbellate racemes, axillary and terminal; bracts long-acuminate; bracteoles subpersistent. **Flowers** large, red or white; calyx campanulate, 5-toothed, usually hairy; petals shortly clawed; standard acuminate, reflexed; wings shorter; keel equal in size or larger than the standard; stamens diadelphous, anthers uniform, dorsifixed; ovary stipitate, glabrous, ovules numerous, style subulate, bearded, stigma minute. **Fruits** pods, terete, turgid, beaked, 2-valved; seeds numerous, small, reniform. **Distr:** New Zealand (North Island). **Ecol:** In subtropical thickets and lowland forests. **Cult:** Propagate from seeds or half-hardened cuttings. Requires a moist, well-drained soil in full or partial sun.

C. puniceus (G. Don) Sol. ex Lindl., *Bot. Reg.* 18, 1835, t. 1775. (Syn: *Donia punicea*). [Kaka's Beak, Lobster Claw, New Zealand Parrot's Bill, Parrot-beaked Kowhai, Red Kowhai]. **Shrubs**, soft-wooded, 1–4 m high; branches spreading; branchlets subherbaceous, silky-pubescent to glabrate. **Leaves** imparipinnate, to 15 cm long; leaflets to 15 pairs, subsessile, linear-oblong, 1.5–2.5 cm long, apex retuse to obtuse. **Inflorescences** racemes, to 15-flowered, pendent. **Flowers** scarlet, to 8 cm long; calyx campanulate, teeth subulate, erect, 1.5–3 mm long; standard ovate-acuminate, ~6 cm long; wings lanceolate-falcate, ~3 cm long, acute; keel ~6+ cm long, falcate-acuminate. **Fruits** pods, to 8 cm long, turgid; seeds numerous. **Distr:** New Zealand. **Ecol:** In scrubby forest margins, on cliffs, near lakes and river margins. Fl. Sep–Dec. **Zones:** 8–11. **Info:** This species is now endangered, and extinct in some localities. A white-flowered form is common in cultivation.

Carmichaelia

Species 55. **Shrubs** or **small trees**; branches flattened, grooved; leafless, except in very young plants. **Inflorescences** racemes or fascicles, simple or branched, or flowers solitary. **Flowers** with calyx ± campanulate, 5-toothed; standard ± orbicular, clawed; keel incurved, clawed, auricled; wings ± falcate, auricled; stamens diadelphous; ovary with slender style (persisting as a 'beak') and minute capitate stigma, ovules usually few. **Fruits** pods, dry, margins thickened, persistent; valves 2, hardened; seeds 1–12, ± reniform. **Distr:** Predominantly in New Zealand; Australia (Lord Howe Island). **Cult:** Propagate from seeds or half-hardened cuttings. Requires a well-drained soil in full or partial sun.

C. aligera G.Simpson, *Trans. & Proc. Roy. Soc. NZ* lxxv. 250, 1945. [North Island Broom]. **Shrubs** erect, or **trees**, to ~10 m tall; trunk to 20 cm dbh; branchlets much-flattened, finely grooved, (4–)6–7 mm wide, ± glabrous. **Leaves** on young plants glabrous, (1–)5–7-foliolate; leaflets obcordate- to obovate-cuneate, to 1.5 × 1.2 cm. **Inflorescences** of 1–3 racemes, 8–12-flowered; peduncles to ~1 cm long, pubescent; pedicels ± 1 mm long. **Flowers** ~4 × 3 mm; calyx 1.5–2 × 1.5–2 mm, glabrous, teeth broadly triangular; standard, keel and wings white, purple-veined and purple-flushed; auricles rounded or bluntly pointed; ovary glabrous. **Fruits** pods, black, subturgid, obliquely oblong, ~0.8–1 × 0.4–0.5 cm, beak 1–2 mm long, subulate; seeds (1–)2(–3), orangish-red, ± mottled with black, ~2–3 × 1–2 mm. **Distr:** New Zealand (North Auckland Peninsula and adjacent islets to Taranaki). **Ecol:** In coastal to lowland forest margins and shrublands. **Zones:** 9–11.

C. odorata Col. ex Hook. f., *Fl. NZ.* 1: 50, 1852. [Scented Broom]. **Shrubs**, bushy, deciduous, to 2 m tall; branches spreading; branchlets drooping, flanged, much-compressed, striate to grooved, ± pilose, to 20 cm × 2 mm. **Leaves** 5–7-foliolate; leaflets subsessile, obcordate to obovate, 3 × 2 mm, ± pilose. **Inflorescences** racemes, simple, 5–10(–15)-flowered,

SZ

Above: *Parochetus africanus, Parochetus communis, Suriana maritima.*

ascending to erect; peduncles ± pubescent, ~1 cm long; pedicels ~1 mm long, pubescent. **Flowers** 3–4 × 3 mm, calyx ~1 × 1 mm, ciliate, teeth minute, broadly triangular; standard, keel and wings all white, purple-veined, auricles rounded; ovary ± glabrous. **Fruits** pods, broadly ovate to suborbicular, ~7 × 4 mm, ± compressed, stramineous, beak rather stout, straight, subulate, almost pungent, 2–3 mm long; seeds 1–2, ~3 × 2 mm, somewhat flattened, pale brown to straw-colored, usually with a dark blotch. **Distr:** New Zealand (North Island). **Ecol:** In lowland to montane streamsides and forest margins. **Zones:** 8–11.

C. williamsii Kirk, *Trans. NZ. Inst.* 12: 394, 1880. [Broom]. **Shrubs** or **small trees**, erect, to ± 4 m tall; trunk to ~10 cm Ø; branchlets ~0.8–1.2 cm wide, much-compressed, closely grooved, glabrous. **Leaves** of young plants 1–3-foliolate; leaflets narrowly obcordate to obovate. **Inflorescences** of 1–2 racemes per notch, each 1–5-flowered; peduncles ~1 cm long, ± pubescent; pedicels stout, 0.5–1 cm long, ± pubescent. **Flowers** falcate, 2–2.5(–3) cm long; calyx 5–6 × 4–5 mm, campanulate, teeth narrowly triangular, acute; standard, keel and wings yellowish, ± purple-veined and purple-blotched, auricles rounded or bluntly pointed; ovary glabrous. **Fruits** pods, ± elliptic-oblong, ~2–3 × 0.6–0.7 cm, very dark brown to black; seeds 6–15, to 5 × 4 mm, red, black-mottled or not. **Distr:** New Zealand (North Island and adjacent small islands). **Ecol:** Mostly in coastal areas. **Zones:** 9–11.

Chordospartium

Species 1. *Generic description same as specific.* **Cult:** Propagate from seeds or half-hardened cuttings. Requires a well-drained soil in full or partial sun.

C. stevensonii Cheesem., *Trans. NZ. Ins.* 43: 175, 1911. [Cord Broom, Weeping Broom]. **Trees**, small, leafless, 3.6–8 m tall; trunk to ~30 cm dbh; branches pendulous, branchlets slender, deeply grooved, nearly terete, ~1.75 mm Ø. **Inflorescences** racemes, solitary or fascicled, to 8 cm long; peduncle densely pilose; pedicels very short. **Flowers** ~9 mm long, pale lavender; calyx cup-shaped, pilose, minutely 5-toothed; standard orbicular, reflexed; wings dolabriform, falcate, obtuse, smaller than the keel; keel incurved, obtuse, about equal in size to standard; stamens diadelphous; ovary sessile, to 5-ovuled, silky-pubescent, style long, slender, incurved bearded on ventral side, stigma minute, terminal. **Fruits** pods, rhomboid to suborbicular, ~5 mm long, turgid, indehiscent; seed usually solitary. **Distr:** New Zealand (South Island). **Ecol:** In moist places, near streams; 450–760 m. **Zones:** 8–11.

TRIBE TRIFOLIEAE

Parochetus

Species 2. **Herbs**, prostrate, rooting at nodes. **Stipules** present, free or somewhat adnate to petiole; stipels lacking. **Leaves** digitately trifoliolate leaflets obcordate, margins entire or shallowly lobed or shallowly dentate; petioles long. **Inflorescence** umbellate, 2–4-flowered, or sometimes flowers solitary, axillary; bracts stipulelike; bracteoles lacking. **Flowers** with calyx campanulate, 5-lobed, vexillary lobes 2, fused almost to apex; petals blue or purplish, glabrous, deciduous; stamens free from petals, diadelphous 9:1, vexillary filament free, anthers essentially uniform; ovary glabrous, many-ovuled, style glabrous, stigma minute, capitate. **Fruits** linear to linear-oblong, acute, somewhat compressed; seeds many, reniform, hilum lateral. **Distr:** Tropical Asia (1 sp.); E Africa (1 sp.). **Cult:** Propagate from seeds or by division of the rooted runners. Requires a moist but well-drained soil in partial sun.

P. africanus Polhill, *Kew Mag.* 8: 56, 1991. (Syn: *P. communis* var. *grossecrenatus*). [African Shamrock Pea]. **Herbs**, creeping, prostrate; stems rooting at nodes. **Stipules** ovate, 0.6–1 cm long, acute, scarious. **Leaves** digitately trifoliolate; leaflets obovate, to 4 × 4 cm, base cuneate, margins usually coarsely toothed towards apex, apex emarginate, green with white central blotches; petiole 3–30 cm long, pilose. **Inflorescences** 1–2(3)-flowered; peduncle 1.5–25 cm long, ascending; bracts similar to stipules but smaller; pedicels 1.5–6 cm long, reflexed in fruit. **Flowers** sweetly scented; calyx ~7 mm long, 2 upper lobes connate almost to their tips, lateral lobes triangular-lanceolate, lowest lobe longer than the 2 lateral; corolla pale blue (rarely white), glabrous; standard 1.2–1.8 cm long, broadly ovate-elliptic, emarginate, base shortly clawed; wings oblong-ovate, ~1.3 cm long, auricle 0.5–1.5 mm long; keel ~1 cm long, acute, tip darker blue; stamens glabrous; ovary glabrous, style slender, glabrous. **Fruits** pods, narrowly oblong, 1.5–2.5 × 0.4–0.5 cm, shortly beaked, tardily dehiscent or sometimes forced open by seeds germinating inside; seeds 2–2.5 × 1.8–2 mm, slightly reniform, brown, sometimes mottled darker. **Distr:** Tropical Africa. **Ecol:** In montane forests; 1,500–2,000 m. **Zones:** 9–12.

P. communis Buch.-Ham. ex D. Don, *Fl. Nepal.* 240, 1825. (Syn: *Cosmiusa repens*, etc.). [Blue Oxalis, Shamrock Pea]. **Herbs**, perennial, prostrate, creeping, 10–20 cm long. **Stipules** scarious, deltoid-ovate, 0.4–1 cm long, entire, acute, membranous, glabrous. **Leaves** digitally trifoliolate; leaflets obcordate, 0.8–2 × 1–2 cm, base narrowly cuneate, margins entire, or sometimes undulate, lateral veins 4–5 pairs; petiole 8–15 cm long; petiolule 1 mm long. **Inflorescences** umbellate, 1–3-flowered, erect; peduncles long, axillary; bracts 2–4; pedicels 0.5–1 cm long. **Flowers** with calyx 6–9 mm long, densely puberulous, teeth subequal to tube; corolla blue to purple, rarely white or pale red, 2 cm long; vexillum obovate, tapered, shortly clawed; wings falcate-oblong, keel broadly triangular-falcate, apex abruptly inflexed, acute, auricle 2.5–4 mm long; stamens diadelphous, filaments not dilated; ovary sessile, glabrous, ovules numerous. **Fruits** pods, linear, 2–2.5 × 0.3–0.4 cm, apex obliquely acute; seeds 8–12, reniform, 2 × 1 mm, brown, sometimes spotted, hilum small, round and lateral. **Distr:** Tropical Asia. **Ecol:** In thickets, grasslands; 0–3,000 m. Fl. & Fr. Apr–Nov. **Zones:** 9–11.

SURIANACEAE

Genera 5, **species** 8. **Trees** or **shrubs**. **Stipules** small or absent. **Leaves** alternate, simple or pinnately compound. **Inflorescences** cymose or paniculate, terminal or axillary, or solitary and axillary, rarely cauliflorous. **Flowers** bisexual or sometimes unisexual, hypogynous, actinomorphic; sepals 5(–7), distinct, imbricate or quincuncial, often persistent, sometimes accrescent; petals 5 or rarely deciduous or lacking, imbricate, free; stamens 10, inner cycle sometimes rudimentary, anthers oblong, basifixed, opening by a longitudinal slit; gynoecium superior, of 1–5 distinct carpels with adaxial-basal style, ovary with (1 or)2(–5) ovules, ovule basal-marginal or marginal, style gynobasic, stigma clavate to capitate. **Fruits** 1–5 drupelike or nutlike monocarps per flower; endosperm present. **Distr:** Pantropical, but mostly in Australia.

Suriana

Species 1. *Generic description same as specific.* **Cult:** Propagate from seeds or cuttings. Requires a well-drained sandy soil in full sun.

S. maritima L., *Sp. Pl.* 1: 284, 1753. [Bay Cedar, Tassel Plant]. **Shrubs** or **small trees**, 1–3 m tall; branches pubescent when young. **Stipules** absent. **Leaves** often fascicled at ends of branchlets, shortly petiolate; lamina narrowly oblanceolate, 2.5–3.5 × ~0.5 cm, base gradually narrowed, margins entire, apex obtuse, somewhat fleshy. **Inflorescences** cymes, 2–4-flowered, axillary; bracts lanceolate, 4–9 × 1–1.5 mm, pubescent; pedicels ~1 cm long, pubescent. **Flowers** with sepals ovate-lanceolate or ovate-oblong, 0.5–1 × 0.2–0.4 cm, pubescent; petals yellow, imbricate, obovate, oblong, or rounded, shortly clawed, deciduous; filaments ~5 mm long, basally villous; carpels hairy, obovoid to globose, ~5 mm long, styles glabrous, stigma small. **Fruits** drupes, pubescent, subglobose, ~3.5 mm Ø, with persistent basifixed style. **Distr:** Pantropical. **Ecol:** Along the shorelines, and sandy places. Fl. Jun–Jul. Fr. Aug–Oct. **Zones:** 10–12.

POLYGALACEAE

Genera 13–17, **species** ~1,000. **Herbs** (perennial or annual), **shrubs** or **trees**, rarely small herbs (the latter sometimes saprophytic). **Stipules** absent, sometimes spiniform or scalelike appendages present. **Leaves** alternate, opposite, or whorled, rarely reduced and scalelike; lamina simple, margins entire, papery or leathery, veins pinnate. **Inflorescences** racemes, panicles, or spikes, axillary or terminal; flowers pedicellate or sessile; bracts and usually bracteoles present. **Flowers** bisexual, zygomorphic, white, yellow, or purple-red; calyx persistent or caducous; sepals 5, free or basally connate, outer 3 small, inner 2 large, petal-like, or all 5 nearly equal; petals 3 or 5, often basally connate, lower (median or 'keel') usually inflexed, carinate, sometimes with fimbriate or lamellate or papilionaceous apical appendages; stamens 8, 7, 5, or 4, filaments free, or variously united and forming a sheath open on upper side and troughlike, anthers basifixed, usually dehiscing by a single apical pore; disk usually absent, if present annular or glandular; ovary superior, 1- or biloculed, ovule 1 per locule, anatropous, pendulous, rarely ovules numerous and placentas parietal, style 1, erect or curved; stigmas 1 or 2, capitate. **Fruits** biloculed capsules, dehiscing by valves, or 1-loculed samaras or berrylike drupes, seeds 2, or 1 with 1 sterile locule, yellow-brown. **Distr:** Worldwide, especially in tropical and subtropical regions of both hemispheres.

Monnina

Species ~85. **Herbs** or **shrubs**, rarely somewhat scandent. **Leaves** alternate, estipulate or with stipular glands; lamina with margins entire. **Inflorescences** paniculate racemes, terminal and axillary. **Flowers** small, usually blue or lilac; sepals 5, deciduous, the 3 outer ones herbaceous, free or the 2 lower ones united, the 2 inner ones (wings) much larger, petaloid; petals 3, the lower one (keel) boat-shaped, not unguiculate, not appendaged, free or nearly so; 2 upper petals usually oblong or liguliform, united below to the staminal sheath; stamens 8 or 6, filaments united nearly to apex into a sheath, anthers confluently 1-celled, opening by a large introrse apical pore; ovary 1-celled, style sickle-shaped, stigma lobes dissimilar, ovules solitary, pendulous; disk usually reduced to a gland at base of ovary. **Fruits** samaroid, narrowly and almost equally winged, or more often drupaceous, surface rugose; seed glabrous, not arillate, testa thin, endosperm thick. **Distr:** SW USA to Chile and Argentina. **Cult:** Propagate from seeds or cuttings. Requires a moist, well-drained soil in shade or full sun.

M. xalapensis HBK., *Nov. Gen. & Sp.* 5: 414, 1823. [Pica]. **Shrubs**, 1–3.5 m high, often much-branched; branches erect or strongly ascending, strigillose. **Leaves** petiolate; lamina oblanceolate or narrowly obovate, 4–8.5 cm long, base attenuate, apex acuminate or acute, thin, sparsely strigillose throughout, abaxially paler. **Inflorescences** racemes, dense or interrupted, mostly 10 cm long or shorter; peduncles axillary and terminal; flowers shortly pedicellate; bracts triangular-ovate, acute or obtuse, deciduous. **Flowers** violet or purple and yellow; sepals triangular-ovate or oblong-ovate, 2–3 mm long, acute or obtuse; wings rounded-ovate or ovate, 5–6 mm long; ovary oblong 1-celled, style thick, bent, terminating in a dilated truncate stigma. **Fruits** drupes, ellipsoid-ovoid, rugose-reticulate, 6–8.5 mm long, initially red, becoming purple or purplish-black when ripe, very juicy. **Distr:** S Mexico to C America. **Ecol:** In wet to dry thickets and forests, often in pine forests, common in secondary growth; 550–3,500 m. **Zones:** 8–12.

Polygala

Species ~500. **Herbs**, annual or perennial, or **shrubs** or **small trees**, rarely small, woody climbers. **Leaves** alternate, rarely opposite or whorled, petiolate; lamina simple, margins entire, papery or subleathery, glabrous or pilose. **Inflorescences** racemes, terminal, axillary, or extra-axillary; bracts 1–3, persistent or caducous; bracteoles present. **Flowers** bisexual, zygomorphic; sepals 5, unequal, persistent or caducous, 2-ranked, outer 3 small, inner 2 petaloid, large; petals 3, connate in lower half, white, yellow, or purple-red, keel boat-shaped, cucullate, or helmet-shaped enclosing stamens and style, apex with often highly divided appendage; stamens 8, filaments united at least at base, forming a trough-shaped staminal sheath, open on upper side, adnate with petals; ovary biloculed, compressed, locules 1-ovuled, anatropous, pendulous, style 1, erect or curved, stigmas 1 or 2. **Fruits** capsular, compressed, often winged; seeds 2, usually black, ovoid, rounded or shortly cuneate, pubescent or glabrous. **Distr:** Worldwide. **Cult:** Propagate from seeds, cuttings or by air-layering. Requires a well-drained soil in shade, partial sun or full sun.

P. javana DC., *Prod.* 1, 2: 327, 1824. **Herbs**, sometimes shrubs to 1 m tall; stem divaricate, base woody. **Leaves** oblanceolate, 1–3 × 0.5–1 cm, base attenuate, margins entire, apex mucronate, apiculate, abaxially more puberulous; petiole 1–2 mm long, puberulous. **Inflorescences** racemes, 2–8 cm long, 5–15-flowered, bracteate; bracts 3, ovate, *larger* one ~1.5 × 0.7 mm, 2 *smaller* 1.2 × 0.5 mm, puberulous; rachis puberulous; bracteoles 3, ovate, *larger* one ~1.5 × 0.8 mm, 2 *smaller* ~1.2 × 0.5 mm; pedicels to 6 mm long, puberulous. **Flowers** pale yellow, beard purple; sepals 3, ovate, *larger* 4 × 2.4 mm, 2 *smaller*, 3 × 1.4 mm, *petaloid* sepals very pale yellow or cream, ovate, to 9 × 7 mm, enclosing corolla and capsule, usually persistent; lateral petals elliptic, to 3 mm long, slightly twisted, keel petal to 5 mm long with beard to 2 mm long, highly branched; stamina 1 sheath to 4 mm long, terminal free filaments to 0.5 mm long, anthers oblong, to 0.2 mm long; ovary obcordate, 0.2 × 0.2 mm, margin ciliate, style to 0.4 mm long. **Fruits** capsules; seeds obovate, to 4 × 2.5 mm; caruncle trifid, 2 arms to 3.5 × 1.5 mm, short arm ~1 mm long, beaklike. **Distr:** S India, Sri Lanka, Java. **Ecol:** In dry open areas. **Zones:** 11–12.

P. myrtifolia L., *Sp. Pl.* 2: 703, 1753. [Parrot Bush, September Bush]. **Shrubs**, sprawling or erect, evergreen, 0.6–2(–4) m high, glabrous or pubescent (especially on young stems), densely branched. **Leaves** alternate, petiolate, erect or spreading, dense, variable; lamina flat, elliptic-oblong, obovate, lanceolate or sublinear, to 5 × 2 cm, base tapered, apex obtuse, acute or acuminate. **Inflorescences** racemes, few-flowered, to 5 cm long, sessile, subterminal; pedicels much longer than bracts; alae ovate-cordate, oblique, mucronulate. **Flowers** ~1.2–2 cm long, purple and white; anterior sepals ovate (sometimes sharply keeled); lateral petals bilobed, posterior lobe ear-shaped and reflexed, anterior acute, deltoid or cuspidate. **Fruits** capsules, obcordate, almost circular-flat, 1 cm across; seeds 2, ovoid-cylindrical, ~4 mm long, with scattered hairs. **Distr:** S Africa (S, SW & E Cape Region). **Ecol:** On rocky slopes. Fl. Jul–Dec. **Zones:** 9–11.

ROSALES

FAMILIES 9, GENERA 261, SPECIES 7,725.

Roots diarch (lateral roots 4-ranked). **Leaves** with margins dentate. **Inflorescence(s)** cymose; hypanthium present. **Flowers** nectariferous; calyx valvate; corolla clawed; ovary with 1 apotropous ovule per carpel, styles present, stigma dry. **Fruits** with calyx and/or hypanthium persistent. **Distr:** Worldwide.

ROSACEAE

Genera 90, **species** 2,520. **Trees**, **shrubs**, or **herbs**, deciduous or evergreen; stems erect, scandent, arching, prostrate, or creeping, armed or unarmed; buds usually with several exposed scales, sometimes with only 2. **Stipules** paired, rarely absent, persistent or deciduous. **Leaves** alternate, rarely opposite, simple or compound; lamina with margins often serrate, rarely entire; petiole usually biglandular apically. **Inflorescences** variable, from solitary flowers to umbellate, corymbose, racemose or cymose-paniculate. **Flowers** usually actinomorphic, bisexual, rarely unisexual (plants dioecious); hypanthium (formed from basal parts of sepals, petals, and stamens) free from or adnate to ovary, short or elongate; sepals usually 5, rarely fewer or more, imbricate; epicalyx segments sometimes also present; petals as many as sepals, inserted below margin of disk, free, imbricate, sometimes absent; disk lining hypanthium, usually entire, rarely lobed; stamens usually numerous, rarely few, always in a complete ring, filaments usually free, very rarely connate, anthers small, bilocular; carpels 1–many, free, or ± connate, ovary inferior, semi-inferior, or superior, ovules usually 2 per carpel, rarely 1 or several, styles as many as carpels, terminal, lateral, or basal, free or sometimes connate. **Fruits**

Below: *Monnina xalapensis*, *Polygala javana*, *Polygala javana* (flower close-up), *Polygala myrtifolia*.

Above: *Duchesnea indica*, *Eriobotrya japonica*, *Prunus walkeri*, *Raphiolepis indica*.

follicles, pomes, achenes, or drupes, rarely capsules, naked or enclosed in persistent hypanthium and sometimes also by sepals; seeds sometimes winged. **Distr:** Worldwide, mainly the Northern temperate zone.

Duchesnea

Species 2. **Herbs** perennial; rhizome short; stolons procumbent, long, filiform, bearing adventitious roots and forming plantlets at nodes. **Stipules** paired, adnate to base of petiole, persistent. **Leaves** *radical* several; *stolon* leaves alternate, long-petiolate; lamina trifoliolate, leaflets with margins dentate. **Inflorescences** axillary, ebracteate, flowers solitary; sepals 5, persistent; epicalyx segments 5, larger than and alternating with sepals, margin incised; petals 5, yellow, obovate; stamens numerous, anthers subglobose; carpels numerous, free, inserted on convex receptacle, style adaxial or subterminal, deciduous, stigma entire. **Fruits** aggregates of achenes formed from enlarged receptacle, hemispheric or turbinate, fleshy; achenes seated on surface of aggregate fruit, compressed-ovoid, minute; seeds reniform, smooth. **Distr:** Europe through Asia to Japan and SE Asia. **Cult:** Propagate from seeds or by division of runners. Requires a well-drained soil in full sun.

D. indica (Andrews) Focke, *Nat. Pflanzenfam.* 3(3): 33, 1888. [False Strawberry, Indian Mock-Strawberry, Indian Strawberry, Mock Strawberry]. **Herbs**, perennial, densely villous; rhizome robust, short; stolons 0.3–1 m long. **Stipules** narrowly ovate to broadly lanceolate, 5–8 mm long. **Leaves** trifoliate; leaflets petiolulate, obovate to rhombic-oblong, 1–5 × 1–3 cm, margin obtusely serrate, apex rounded, surfaces usually hairy; petiole 1–5 cm long. **Inflorescences** with flowers solitary, axillary, ebracteate; pedicel 0.2–6 cm long, pilose or densely villous. **Flowers** 1–2.5 cm Ø; epicalyx segments obovate, longer than sepals, apex usually 3–5-serrate; sepals ovate, apex acute; petals with apices rounded; stamens 20–30; carpels numerous, free. **Fruits** aggregates of achenes, ripening to red, shining, 1–2 cm Ø, spongy; achenes shiny when fresh, ovoid, ~1.5 mm long. **Distr:** Asia; naturalized in Africa, Europe, and N America. **Ecol:** On mountain slopes, open grasslands; <3,100 m. Fl. Jun–Aug. Fr. Aug–Oct. **Zones:** 6–11.

Eriobotrya

Species ~30. **Trees** small, or large **shrubs**, evergreen. **Stipules** free, rarely intrapetiolarly connate. **Leaves** usually petiolate; lamina simple, margins serrate or entire, penninerved. **Inflorescences** panicles or compound racemes, many-flowered, terminal. **Flowers** with hypanthium cupular or obconoid, open at top; sepals 5, persistent; petals 5, white or yellow, obovate or orbicular, base clawed; stamens 15–20(–40); ovary inferior, carpels 2–5, ventrally and laterally connate, ovules 2 per locule, styles 2–5, dorsally adnate to hypanthium, apex often pubescent and exposed; stylodia usually basally connate, hairy, stigmas capitate. **Fruits** pomes, yellow, orange or red, with persistent incurved sepals, fleshy or dry, fleshy mostly of hypanthial origin, endocarp membranous; seeds 1–3, large, testa thin but firm. **Distr:** Himalayas through continental SE Asia to Japan and the islands of W Malesia. **Ecol:** In montane or tropical lowland forests. **Cult:** Propagate from seeds. Requires a moist, well-drained soil in full sun.

E. japonica (Thunberg) Lindley, *Trans. Linn. Soc. London*, Bot. 13: 102, 1822. (Syn: *Mespilus japonica*, *Crataegus bibas*). [Japanese Plum, Japanese Loquat, Loquat]. **Trees**, small, to 10 m tall. **Stipules** subulate, 1–1.5 cm long, apex acuminate, pubescent. **Leaves** lanceolate, oblanceolate, obovate, or elliptic-oblong, 12–30 × 3–9 cm, base cuneate, margins basally entire, apically remotely serrate, apex acute or acuminate, leathery, lateral veins 11 or 12 pairs, adaxially lustrous, rugose, abaxially densely tomentose; petiole ~0.6–1 cm long. **Inflorescences** panicles, many-flowered, 10–19 cm long; peduncle densely tomentose; bracts subulate, 2–5 mm long, densely tomentose; pedicels 2–8 mm long, rusty-tomentose. **Flowers** fragrant, 1.2–2 cm Ø; hypanthium shallowly cupular, abaxially rusty-tomentose; sepals triangular-ovate, 2–3 mm long, obtuse, abaxially densely tomentose; petals white, oblong or ovate, 5–9 × 4–6 mm, obtuse or emarginate; stamens 20; ovary rusty-pubescent apically, 5-loculed, with 2 ovules per locule, styles 5, free. **Fruits** edible, yellow or orangish-yellow, globose or obovate, 1–5 cm Ø, rusty-tomentose, soon glabrescent. **Distr:** China, Taiwan. **Ecol:** Fl. Jun. Fr. Jul–Aug. **Zones:** 8–12.

Prunus

Species ~30. **Trees** or **shrubs**, usually deciduous; branchlets sometimes spine-tipped; axillary winter bud solitary, ovoid. **Stipules** membranous, soon caducous. **Leaves** alternate, convolute (or rarely conduplicate) when young, petiolate or sessile; lamina simple, margins variously crenate or coarsely serrate; petiole apex or base of lamina margins with or without nectaries. **Inflorescences** with flowers solitary or in 3-flowered fascicles, sometimes in dense racemes, apparently axillary; bracts small, soon caducous; flowers usually opening before or at the same time as leaves. **Flowers** with hypanthium campanulate; sepals 5, imbricate; petals 5, white, sometimes purple-veined, rarely greenish, inserted on rim of hypanthium, imbricate; stamens 20–30, 2-whorled, filaments unequal; carpel 1, ovary superior, 1-loculed, glabrous or sometimes villous, ovules 2, collateral, pendulous, style terminal, elongated. **Fruits** drupes, glabrous, often glaucous, usually longitudinally grooved. **Distr:** Asia, Europe, N America. **Cult:** Propagate from seeds, cuttings, or by grafting. Requires a fertile, humus-rich, moist, well-drained soil in full or partial sun.

P. walkeri (Wight) Kalkman, *Blumea* 13: 55, 1965. (Syn: *Polydontia walkeri*, etc.). **Trees**, evergreen, 20–40 m tall. **Stipules** linear, 3–6 mm long, densely tomentose, deciduous. **Leaves** ovate to ovate-oblong, 12–19 × 6–9 cm, base obtuse to rounded, margins entire, apex acute to slightly acuminate, membranous, adaxially usually glabrous, abaxially pubescent; young leaves densely pubescent (golden-tomentose); basal glands 2(–5), spherical, flattened, light green; petiole 1 cm long, lanate-tomentose. **Inflorescences** racemes, many-flowered, usually solitary, axillary, 2.5–7 cm long, pubescent; bracts pubescent, deciduous; pedicels 2–3 mm long. **Flowers** small, white; calyx tube obconic, lobes usually 6, subulate, pilose; petals 6, hardly differentiated from sepals, to 1 mm long; stamens 12–17; ovary densely tomentose. **Fruits** drupes, transversely ellipsoid, 1.5–1.6 × 2.2–2.4 cm, woolly pubescent; young fruits globose, rufous-tomentose, tipped with remnant of style; endocarp bony, glabrous. **Distr:** Sri Lanka. **Ecol:** In fringes of secondary forests, heavily shaded forests; <1,000 m. Fl. Mar–Apr. **Zones:** 11–12.

Rhaphiolepis

Species ~15. **Shrubs** or **small trees**, evergreen. **Stipules** caducous, subulate. **Leaves** shortly petiolate; lamina simple, margins serrate or entire, leathery, secondary veins curved toward the margins without forming loops. **Inflorescences** racemes or panicles, terminal. **Flowers** with hypanthium campanulate to tubular; sepals 5, erect or reflexed; petals 5, white or pink, base shortly clawed; stamens 15–20; ovary inferior, biloculed, with 2 erect ovules per locule, styles 2 or 3, basally connate. **Fruits** pomes, drupelike, purplish-black or bluish, subglobose, succulent, with early caducous sepals leaving an annular ring; seeds 1 or 2, subglobose, large, testa thin. **Distr:** E Asia. **Cult:** Propagate from seeds, cuttings or by air-layering. Requires a well-drained, sandy soil in full sun.

R. indica (L.) Lindley, *Bot. Reg.* 6: t. 468, 1820. [Indian Hawthorn]. **Shrubs**, rarely small trees, to 4 m high; branchlets purplish-brown when young. **Stipules** lanceolate, acuminate, caducous. **Leaves** usually ovate or oblong, (2–)4–8 × 1.5–4 cm, base attenuate, margins crenulate, serrate, or obtusely serrate, apex obtuse, acute, acuminate, or long-caudate, leathery, adaxially lustrous, glabrous, abaxially pale; petiole 0.5–1.8 cm long. **Inflorescences** panicles or racemes, many- or few-flowered, terminal; peduncle and pedicels rusty-tomentose; bracts and bractlets caducous, lanceolate or narrowly lanceolate, 2–7 mm long, acuminate; pedicel 0.5–1.5 cm long. **Flowers** 1–1.3(–1.5) cm Ø; hypanthium tubular; sepals triangular-lanceolate or linear, 4.5–6 mm long, acute; petals white or pinkish, obovate or lanceolate, 5–7 × 4–5 mm, base pubescent, apex obtuse; stamens 15, about as long as or shorter than petals; ovary glabrous, 2- or triloculed, with 2 ovules per locule, styles 2 or

3, basally connate, subglabrous. **Fruits** purplish-black, globose, 5–8 mm Ø, glabrous. **Distr:** Cambodia, China, Japan, Laos, Taiwan, Thailand, Vietnam. **Ecol:** On slopes, thickets; 700–1600 m. Fl. Apr. Fr. Jul–Aug. **Zones:** 8–12.

Rubus

Species 700. **Shrubs** or low **undershrubs**, often straggling or scrambling; branches mostly sarmentose, mostly armed with straight or hooked prickles. **Leaves** simple or compound, stipulate. **Inflorescences** axillary and terminal, panicles, corymbose-cymes, or axillary glomerules or clusters, rarely flowers solitary. **Flowers** bisexual, white or pink; calyx expanded, tube short, often shallowly campanulate, lobes 5, imbricate, persistent; petals 5, mostly unguiculate, white or pink; stamens indefinite, mostly biseriate; carpels numerous, inserted on a ± prominent convex receptacle, ovules 2 per ovary, styles terminal. **Fruits** an aggregate of few or numerous drupelets, arranged on a rather dry or spongy receptacle; stones cartilaginous, rugose; seeds without endosperm. **Distr:** Worldwide, mostly in temperate regions of Northern Hemisphere, a few species extending into Southern Hemisphere. **Cult:** Propagate by dividing suckering species, or from seeds or half-hardened cuttings. Requires a fertile, humus-rich, moist, well-drained soil in shade or partial sun.

R. niveus Thunberg, *Diss. Bot.-Med. de Rubo* 9, 1813. (Syn: *R. bonatii*, etc.). [Ceylon Raspberry, Hill Raspberry, Mysore Raspberry, Snowpeaks Raspberry, Mysore Black Raspberry]. **Shrubs**, erect or semiscandent, much-branched at base; stem glabrous, glaucous; young branches villous; prickles numerous, scattered, straight or slightly recurved. **Stipules** lanceolate, to 7 mm long, adnate to base of petiole, persistent. **Leaves** compound, imparipinnate; leaflets 5–9, membranous; lateral leaflets sessile, ovate, oblong or elliptic, 2.5–5.5 × 1–2.2 cm, base cordate or truncate, margins distantly serrate, serratures mucronate, apex acute, tip mucronate, adaxially slightly pilose, abaxially grayish-white tomentose; terminal leaflet long-petiolulate, broadly ovate, often lobed on one or both sides; rachis and petiole 5–15 cm long, slender. **Inflorescences** cymose panicles, many-flowered, chiefly terminal, sometimes reduced to 1- or 2-flowered cymes in uppermost leaf axils, densely villous; pedicels 1 cm long. **Flowers** to 1 cm Ø; calyx lobes distant, triangular, acute-aristate, densely villous, not prickly; petals broadly obovate or suborbicular, shorter than calyx, hairy at base inside, pink; carpels ~50, borne on a rather low, villous receptacle. **Fruits** subglobose, <1 cm across, greenish-red, dark purple when ripe; drupelets flattened, dry or slightly fleshy, white-tomentose, tipped with persistent style. **Distr:** Tropical and subtropical Asia to Malesia, Philippines and Taiwan. **Ecol:** 500–2,800 m. Fl. May–Jul. **Zones:** 9–12.

RHAMNACEAE

Genera ~50, **species** >900. **Trees** or **shrubs**, rarely climbers, evergreen or deciduous. **Leaves** alternate or subopposite, generally stipulate; lamina simple, coriaceous, 3–8-nerved, sometimes modified into spines. **Inflorescences** cymes, corymbs, racemes, spikes, panicles or fascicles, axillary. **Flowers** hermaphrodite, rarely unisexual, actinomorphic, small, perigynous; calyx tubular, 5(4 or 6)-lobed; petals 5, rarely 4 or 6, sometimes absent, usually clawed and cucullate; stamens as many as petals, inserted opposite the petals, often enclosed within their folds; disk fleshy, intra-staminal, filling and covering the hypanthium, glabrous, rarely tomentose; ovary superior to inferior, (1–)2–3 (-4)-locular, locules 1-ovuled, style 2–4-fid. **Fruits** often drupaceous, pyrenes 1–3(–4), 1–2-seeded, or capsules, rarely samaroid; seeds with a large erect embryo. **Distr:** Subtropical to tropical regions worldwide.

Alphitonia

Species 10–20. **Trees** or **shrubs**. **Leaves** alternate; lamina coriaceous, pinnately nerved, abaxially usually pale to white or reddish-brown tomentose. **Inflorescences** thyrses, branched, relatively open. **Flowers** perfect; sepals 5; petals 5; nectary disk annular, filling the floral tube; stamens 5; ovary partly inferior, 2–triloculed, style short, 2–trilobed. **Fruits** partly enclosed by the adherent nectary disk and floral tube, exocarp spongy, somewhat brittle, endocarp crustaceous, usually dehiscing along the ventral suture; seeds with a thin reddish aril, often completely enclosing the seed. **Distr:** Malesia, E Australia, Pacific Islands. **Cult:** Propagate from seeds. Requires a fertile, moist, well-drained soil in full sun.

A. ponderosa Hillebr., *Fl. Haw. Isl.* 81, 1888. [Hawai'i Kauila Tree]. **Trees**, 4–25 m tall. **Leaves** ovate, elliptic-ovate, or oblong-ovate to lanceolate, (6–)8–15(–22.5) × (2.5–)3.5–7(–9.6) cm, margins entire, apex usually acuminate, sometimes acute or obtuse, ± coriaceous, adaxially dark green, somewhat glossy, glabrate, abaxially densely rusty-tomentose; petiole 1.6–5 cm long, ± tomentulose. **Inflorescences** thyrses, usually 6–8 cm long, rusty-tomentose; peduncles 0.6–1 cm long; pedicels 3–6 mm long; bracteoles small. **Flowers** with sepals deltate, 1.5–2.5 mm long, rusty- to whitish-tomentose, inside glabrous; petals spathulate, ~1.5 mm long, each partly enclosing a stamen; nectary disk pentagonal. **Fruits** black, subglobose, 1.4–1.8 cm Ø, indehiscent to tardily dehiscent, surface wrinkled; seeds brown, oblong, ~6 mm long, aril red. **Distr:** Hawai'i. **Ecol:** In dry, coastal mesic and mixed mesic forests; 240–1,250 m. Fl. Aug–Nov. Fr. Apr–Jun. **Zones:** 10–12.

Colletia

Species 5. **Shrubs**, often leafless; branches decussately opposite, spinose, notably compressed and sometimes enlarged at the nodes, these not articulate. **Spines** opposite, very prominent. **Inflorescences** with flowers solitary or fascicled below the spines, solitary flowers sessile. **Flowers** 4–6-merous; calyx membranous, tube produced above the adnate disk; petals often lacking; ovary 3-celled. **Fruits** drupes, finally capsular, coriaceous with 3 crustaceous bivalved cocci. **Distr:** Southern S America. **Cult:** Propagate from seeds or half-hardened cuttings. Requires a well-drained, sandy soil in full sun.

C. paradoxa (Spreng.) Escal., *Boletín de la Sociedad Argentina de Botánica* 1: 219, 1946. (Syn: *C. cruciata*, etc.). [Anchor Plant]. **Shrubs**, rather slow-growing, 0.5–3.5 m tall, copiously branched; branchlets transformed into formidable flat triangular spines. **Leaves** sparsely on the branches, minute, soon deciduous; lamina triangular, 5–8 mm long. **Inflorescences** with flowers solitary or fascicles of 2–4(–10) drooping flowers, borne below the spines. **Flowers** yellowish-white, base green-tinged; calyx tube cylindrical, 6–8 mm long, base slightly swollen, limb 5-lobed, lobes narrowly ovate, apices hooked; disk ringlike; ovary small, half-sunk into base of perianth, triloculed, each locule 1-ovuled, style terete, as long as the perianth tube, stigma trilobed. **Fruits** drupaceous-capsular, 5 mm Ø. **Distr:** N Argentina, Uruguay. **Ecol:** Fl. Mar–Apr. **Zones:** 8–11.

Colubrina

Species ~23. **Trees**, **shrubs** (sometimes scandent), deciduous or evergreen, unarmed. **Stipules** minute, early deciduous. **Leaves** alternate (opposite); lamina margins entire, serrate or crenate, papery to subleathery, venation pinnate or 3-veined from base. **Inflorescences** cymes or small thyrses, mainly few-flowered, axillary, sessile or shortly pedunculate. **Flowers** bisexual, 5-merous; calyx tube hemispherical, sepals 5, triangular, adaxially conspicuously keeled, abaxially densely pubescent, deciduous; petals 5, erect to spreading, clawed; stamens 5, about as long as petals, anthers dorsifixed, introrsely dehiscent; disk massive, fleshy, adnate to calyx tube; ovary semi-inferior, 3(or 4)-loculed, locules 1-ovuled, style trilobed to trifid. **Fruits** capsules, subglobose, mesocarp thin, explosively splitting, or slowly into 3 ventrally dehiscent 1-seeded endocarpids; seeds shiny, broadly obovoid. **Distr:** Tropical Africa, S Asia, China; Australia, Pacific Islands; S America. **Cult:** Propagate from seeds. Requires any soil in full sun; salt tolerant.

C. asiatica (L.) Brongniart, *Mém. Fam. Rhamnées.* 62, 1826. (Syn: *Ceanothus asiaticus*). [Asian Nakedwood, Asian Snakewood, Latherleaf, Wild Coffee]. **Shrubs**, scandent, evergreen; young branchlets glabrous. **Leaves** alternate; lamina ovate, broadly ovate or cordate, 4–8 × 2–5 cm, base rounded or subcordate, margins coarsely crenate, apex acuminate, emarginate, thinly papery, surfaces ± glabrous, venation pinnate, secondary veins 2 or

Below: *Rubus niveus, Alphitonia ponderosa, Colletia paradoxa, Colubrina asiatica.*

FK&S

Above: *Gouania hillebrandii, Phylica plumosa, Zizyphus mauritiana, Zizyphus xylopyrus.*

3 pairs, prominent; petiole 1–1.6 cm long, pilose. **Inflorescences** thyrses, axillary; pedicels 2–3 mm long. **Flowers** yellow, 5-merous; calyx tube hemispherical, sepals 5, triangular, adaxially distinctly keeled; petals 5, obovate, cucullate, about as long as stamens, clawed; stamens 5; disk fleshy; disk stout; ovary 2- or triloculed, style ± deeply bifid or trifid. **Fruits** capsules, globose, 7–9 mm Ø, loculicidally dehiscent, locules 1-seeded; seed grayish-brown. **Distr:** Pantropical. **Ecol:** In forests, thickets, along beaches; sea level. Fl. Jun–Sep. Fr. Sep–Dec. **Zones:** 10–12.

Gouania

Species 50–70. **Shrubs** or **lianas**. **Stipules** usually caducous. **Leaves** alternate; lamina with margins entire to dentate or crenate, pinnately veined, with 3–5 conspicuous nerves spreading from base. **Inflorescences** cymes or fascicles, these sometimes racemose, terminal or axillary, flowering branches ± with a tendril arising from the leaf axil. **Flowers** perfect or functionally unisexual; floral tube cuplike, surrounding ovary; sepals 5, small, deltate; petals 5, clawed, embracing and usually concealing a stamen; nectary disk flattened, surrounding ovary, with an oblong lobe opposite each sepal; stamens 5; ovary inferior, 3(4)-loculed, style trilobed. **Fruits** septicidal schizocarps, longitudinally 2–3(4)-winged; seeds obovoid to ellipsoid. **Distr:** Pantropical. **Cult:** Propagate from seeds or cuttings. Requires a moist, well-drained soil in full sun.

G. hillebrandii Oliv. ex Hillebr., *Fl. Hawaiian Isl.* 83, 1888. (Syn: *G. cucullata*, etc.). [Hairyfruit Chewstick]. **Shrubs**, erect to sprawling, clump-forming; stems 0.2–2 m long. **Stipules** linear-lanceolate, 2–5(–8) mm long, caducous to persistent. **Leaves** usually elliptic, 1.4–6.5 × 0.8–4 cm, base rounded to broadly cuneate, margins entire, apex shortly acuminate, acute, or obtuse to rounded, chartaceous to coriaceous, adaxially glabrate, abaxially ± pilose; petioles (0.4–)0.7–4 cm long, white-pilose. **Inflorescences** axillary cymes, 1.5–5 cm long, without tendrils; ♂ flowers borne laterally to the central perfect ones, pilose; peduncles (0.2–)0.6–4 cm long; pedicels 1–7 mm long, bracts deltate, 2(–3) mm long. **Flowers** sweetly fragrant, perfect and ♂; sepals 1.5–2.5 mm long, densely pilose; petals white, ~0.6–1.5 mm long, each one enclosing a stamen; ovary abortive in ♂ flowers; style trilobed. **Fruits** 0.4–1.1 cm long, densely puberulent, 3-winged; seeds narrowly ellipsoid, 3.5–4 mm long, brown, glossy. **Distr:** Hawai'i. **Ecol:** In dry forests; 240–500 m. Fl. Feb–Jun. **Zones:** 10–12.

Phylica

Species 150. **Shrubs** or **small trees**, much-branched, unarmed. **Leaves** alternate, crowded; lamina simple, ovate, lanceolate or linear, margins entire, usually revolute. **Inflorescences** thyrses or capitula. **Flowers** small; calyx obconical, urceolate or cylindrical; sepals 5; petals (0)5, cucullate, unguiculate, bristlelike; stamens 5, very short, anthers ovate or kidney-shaped, 1- or 2-celled; disk ± distinct; ovary inferior or adnate to the tube of the calyx, trilocular, style simple, stigma trilobed or 3-toothed, in some entire. **Fruits** capsules, usually crowned by persistent calyx, tricoccous; seeds solitary, erect. **Distr:** S Africa, Madagascar. **Cult:** Propagate from seeds or half-hardened cuttings of vegetative stems. Requires gritty, slightly acidic, well-drained soil in full sun.

P. plumosa E.Mey. ex Sond., *Fl. Cap.* (Harvey) i. 486, 1860. [Cape Myrtle]. **Shrubs**, evergreen, hairy, 0.3–2 m tall; branches erect or semiscandent, pubescent; twigs pubescent. **Leaves** linear-lanceolate, awl-shaped, 1.5–2.5 × 0.25–0.5 cm, margins revolute, adaxially smooth, tomentose, abaxially hairy; young leaves hairy; petiole short. **Floral leaves** villous, or hirsute, spreading, horizontal, 1.5–2.3 cm long, much longer than flowers. **Inflorescences** spikes, oblong or roundish, 2.5 cm long; bracteoles villous, half as long as the calyx. **Flowers** yellow; calyx tubular, minutely pilose, 6 mm long, segments erect, at length spreading; corolla tube reversed, appressed, lobes oblong to ovate-lanceolate, erect, half as long as the tube. **Fruits** capsules, ~5 mm Ø, glossy. **Distr:** S Africa (Cape Province). **Ecol:** On lower mountain slopes; 0–1,200 m. Fl. Feb–Jul. **Zones:** 9–11.

Zizyphus

Species ~100. **Shrubs** or **trees**, erect or straggling, often climbing, evergreen or deciduous, often spinose. **Stipules** usually transformed into 1 or 2, erect or ± recurved spines. **Leaves** alternate, petiolate; lamina distinctly triplinerved or rarely pinnately veined. **Inflorescences** axillary corymblike cymes, or axillary or terminal thyrses. **Flowers** yellow-green, small, bisexual; calyx tube shallow, patelliform to hemispherical; sepals ovate-triangular or triangular, adaxially ± keeled; petals clawed, obovate or spathulate, rarely absent; disk shallow, fleshy, 5–10-lobed; ovary superior, globose, 2- or 3(or 4)-loculed, style ± deeply branched, 2(–4)-fid. **Fruits** drupes, single-stoned, (1)2- or triloculed, globose or oblong, locules 1-seeded. **Distr:** Mainly in tropical and subtropical areas of Asia and the Americas, a few species in Africa and temperate regions. **Cult:** Propagate from seeds, root cuttings, by air-layering or grafting. Requires a moist, well-drained soil in full sun.

Z. mauritiana Lamarck, *Encycl.* 3: 319, 1789. (Syn: *Z. jujuba*, etc.). [Indian Jujube, Stone Apple]. **Trees** or **shrubs**, evergreen, to 15 m tall; stipular spines 2, 1 oblique, hooklike, recurved. **Leaves** ovate or oblong-elliptic, rarely subrounded, 2.5–6 × 1.5–4.5 cm, base subrounded, slightly oblique, margins serrulate, apex rounded, rarely acute, papery, adaxially dark green, shiny, abaxially tomentose, 3-veined from base; petiole 0.5–1.3 cm long. **Inflorescences** dichotomous cymes, few- to 10-flowered, axillary, subsessile or shortly pedunculate; pedicels 2–4 mm long, tomentose. **Flowers** green-yellow; sepals ovate-triangular, acute, abaxially hairy; petals oblong-spathulate, base clawed; stamens subequalling petals; disk thick, fleshy, 10-lobed, concave at middle; ovary globose, glabrous; style bifid or branched to middle. **Fruits** drupes, yellow, orange or red, maturing to brown or black, oblong or globose, 1–1.2 × 1 cm; seeds 1–2, red-brown, shiny. **Distr:** Asia; Africa, Australia. **Ecol:** In humid forests; 0–1,800 m. Fl. Aug–Nov. Fr. Sep–Dec. **Zones:** 10–12.

Z. xylopyrus (Retz.) Willd., *Sp. Pl.* 1(2): 1104, 1798. (Syn: *Rhamnus xylopyrus*). [Ghont, Woody-fruited Jujube]. **Trees**, small or large straggling shrubs, 7.5–12 m tall; young branches rusty-tomentose; spines in pairs, one straight and the other curved. **Leaves** broadly elliptic or ovate-oblong, base obliquely subcordate, margins serrate, apex obtuse with a mucro, adaxially glabrous, abaxially hairy, basally 3–4 nerved; petioles ~1.2 cm long, fulvous-tomentose. **Inflorescences** dichotomous cymes, dense, axillary; peduncles to 1.5 cm long; pedicels 3–4 mm long, tomentose. **Flowers** 4–6 mm Ø, yellowish-green, densely pubescent; calyx lobes 5, triangular, 2–2.5 mm long, keeled to middle, pubescent outside, glabrous inside; petals 5, spathulate, 1.5–2 mm long, yellowish-green; stamens equal to petals; disk 10-lobed, rarely 5-lobed, pitted, glabrous; ovary 3-celled, globose, ovule solitary, style 3-cleft to middle. **Fruits** drupes, globose, 1.6–2.5 cm Ø, woody, slightly rugose, stone hard; seeds 3, oblong, compressed, black. **Distr:** India; Nepal; Sri Lanka. **Ecol:** In lowland areas. Fl. Apr–Aug. Fr. May–Jan. **Zones:** 10–12.

ELAEAGNACEAE

Genera 3, **species** ~90. **Trees** or **shrubs**, deciduous or evergreen, most parts silvery or brownish scales and/or stellate hairs; branches sometimes spine-tipped. **Leaves** alternate, opposite, or whorled, exstipulate, usually petiolate; lamina simple, margins entire or subentire, often leathery, abaxially densely hairy or peltate-scaly, pinnately veined. **Inflorescences** with flowers solitary or in clusters or short racemes. **Flowers** actinomorphic, bisexual, or unisexual (plants dioecious); calyx in bisexual and ♀ flowers tubular, 2–6(–8)-lobed; petals absent; stamens 4–8, free, adnate to calyx tube, in ♂ flowers twice as many as lobes, in bisexual flowers as many as lobes and alternate with them; ovary superior but tightly enclosed in differentiated basal part of calyx and apparently inferior, 1-loculed, 1-ovuled, style elongate, stigma lateral. **Fruits** drupelike, indehiscent, enclosed in base of calyx tube, 1-seeded. **Distr:** N temperate and tropical regions.

Above: *Elaeagnus latifolia, Holoptelea integrifolia, Artocarpus communis.*

Elaeagnus

Species ~90. **Shrubs**, sometimes climbing, or **small trees**, deciduous or evergreen, sometimes spiny. **Leaves** alternate, petiolate, lamina margins usually entire. **Inflorescences** with flowers clustered on short axillary shoots or sometimes solitary. **Flowers** bisexual; calyx tubular, 4-lobed, constricted above ovary and breaking at constriction as fruit develops, lobes usually spreading, deciduous, white or yellow inside; stamens 4, inserted in mouth of calyx tube, alternate with lobes; style linear, not exserted. **Fruits** drupes, globose or ellipsoid, rarely longitudinally winged; stone usually 8-ribbed. **Distr:** Asia, S Europe, N America. **Cult:** Propagate from fresh seeds or cuttings. Requires a moist, well-drained soil in full or partial sun.

E. latifolia L., *Sp. Pl.* 1: 121, 1753. (Syn: *E. thwaitesii*). [Silver Berry]. **Shrubs**, ~3 m tall, ± scandent, branches often thorny; thorns to ~3 cm long. **Leaves** very variable, usually elliptic, 1–12 × 9 cm, base acute or obtuse, margins entire, sometimes undulate, apex obtuse or rounded or abruptly acute; petiole 0.5–1.5(–1.8) cm long. **Inflorescences** fascicles, 2–15-flowered, axillary; bracts filiform, ~2 mm long; pedicels 1–3 mm long. **Flowers** fragrant, silvery-white to greenish-orange, scaly, lower part tubular, 1.5–2 mm long, upper part campanulate, 3.5–5.5 mm long, ± 4-sided; perianth lobes spreading, broadly ovate, 2–4 mm long, acute, scaly; filaments 0.3–1 mm long; style 5–7 mm long. **Fruits** berrylike, ovoid-oblong, 1–2.5 cm long, reddish-pink to bright red to reddish-brown; seeds narrowly ovoid, ~1.2 cm long, brown. **Distr:** China, Myanmar, Sri Lanka, Himalayas; Malesia. **Ecol:** In forest margins. Fl. Apr–May. **Zones:** 9–12. **Info:** Fruits edible, but slightly sour.

ULMACEAE

Genera 6, **species** 35. **Trees** or **shrubs**, evergreen or deciduous. **Stipules** usually membranous, caducous. **Leaves** simple, alternate or rarely opposite, usually distichous, petiolate; lamina pinnately veined, basally 3(or 5)-veined, margins entire or serrate. **Inflorescences** axillary. **Flowers** monochlamydeous, bisexual, or rarely unisexual or polygamous; perianth lobes 4–9, imbricate (rarely valvate); stamens usually equal in number to and opposite perianth lobes, basally adnate to tepals, filaments distinct, anthers 2-celled, longitudinally fissured; pistil bicarpellate, ovary superior, 1(or 2)-loculed, ovule 1, suspended, anatropous, integuments 2, style very short, stigmas 2, linear. **Fruits** samaras, drupes, or winged nutlets, stigmas persistent stigmas; seeds flattened. **Distr:** Mostly N temperate, especially Asian, scattered elsewhere in tropical areas, except Australia and the Pacific.

Holoptelea

Species 2. **Trees**, deciduous. **Stipules** small, lateral, scarious, caducous. **Leaves** alternate, distichous; lamina entire, penninerved. **Inflorescences** fasciculate on previous year's leafless shoots. **Flowers** polygamous or bisexual; perianth simple, calycine, lobes 4–8, imbricate, often unequal; stamens as many as tepals or sometimes more, filaments erect, anthers hairy; ovary stipitate, compressed, ovule solitary, pendulous, style short, bifid, arms stigmatose. **Fruits** samaras, indehiscent; wings membranous or chartaceous, usually reticulate; seed flat. **Distr:** Tropical Africa, India, Indo-China, Myanmar and Sri Lanka. **Cult:** Propagate from seeds or by air-layering. Requires a moist, well-drained soil in full or partial sun.

H. integrifolia (Roxb.) Planch., *Ann. Sci. Nat. Bot.* ser. 3, 10: 266, 1848. (Syn: *Ulmus integrifolia*). [Indian Elm, Jungle Cork Tree]. **Trees**, deciduous, 15–25 m tall; bark malodorous. **Leaves** elliptic-oblong to elliptic-ovate, 7.5–13 × 4–8.5 cm, base unequally rounded to subcordate, margins entire, apex acute to cuspidately acuminate, subcoriaceous, abaxially pubescent; lateral nerves 5–7 pairs; petioles 7–9 mm long. **Inflorescences** fascicles, ♂ and bisexual flowers mixed, borne on scars of fallen leaves on old shoots. **Flowers** ~4 mm ∅; perianth segments usually 4, free, concave, 1.5 mm long, imbricate, pubescent; stamens 7–8, subbiseriate, filaments 1.6 mm long; ovary ovate, compressed, 2 × 1.5 mm, style with 2 arms. **Fruits** compressed samaras, broadly ovate, 1.2–1.5 × 1 cm, notched at the top, lobes incurved; wings membranous, reticulated; seed flat. **Distr:** India, Indo-China, Myanmar, Sri Lanka. **Ecol:** In forests of the dry regions. Fl. Mar–Apr. Fr. Apr–Jun. **Zones:** 10–12.

MORACEAE

Genera 38, **species** 1,100. **Trees**, **shrubs**, **vines**, or rarely **herbs**, frequently with milky or watery latex, sometimes spiny. **Stipules** present, frequently caducous. **Leaves** alternate, rarely opposite; lamina simple, sometimes with cystoliths, margins entire or palmately lobed, venation pinnate or palmate; petiole often well-defined. **Inflorescences** axillary, frequently paired, racemose, spicate, capitate, or rarely cymose, sometimes a fig or syconium with flowers completely enclosed within a hollow receptacle. **Flowers** unisexual (plants monoecious or dioecious), small to very small; calyx lobes (1 or)2–4(–8), free or connate, imbricate or valvate; corolla absent; ♂ with stamens as many as and opposite to calyx lobes (except in *Artocarpus*), straight or inflexed in bud, anthers 1- or biloculed, crescent-shaped to top-shaped, pistillode (rudimentary sterile pistil) often present; ♀ with calyx lobes usually 4, ovary superior, ± inferior, 1(or 2)-loculed, locules 1-ovuled, anatropous or campylotropous, style branches 1 or 2, stigmas usually filiform. **Fruits** usually drupes, rarely achenes, enveloped by an enlarged calyx and/or immersed in a fleshy receptacle, often joined into a syncarp; seed solitary. **Distr:** Widespread in tropical and subtropical regions, less common in temperate regions.

Artocarpus

Species ~50. **Trees**, evergreen or deciduous, with latex; monoecious. **Stipules** free. **Leaves** spirally arranged or distichous; lamina simple to pinnatifid, rarely pinnate, leathery, margins entire. **Inflorescences** unisexual, capitate, many-flowered, sometimes borne on main branches or trunk; ♂ flowers free, surrounded by peltate to clavate interfloral bracts; ♀ flowers at least partially adnate to each other and/or to interfloral bracts. **Flowers** ♂ with calyx tubular, slightly bilobed or 2–4-lobed, lobes imbricate or valvate, stamen 1, slightly to conspicuously exserted from calyx, anthers globose to oblong, biloculed, pistillode absent; ♀ with calyx tubular, basally thin-walled, apically thick walled and either completely fused or not, ovary free, style central or ± lateral, stigmas 1 or 2. **Fruits** syncarps (formed by the lateral fusion of flowers and bracts), fleshy throughout or at least at basal portions of calyx, sometimes very large, flowers and bracts fused at their tips to form an areolate surface or free and forming variously shaped processes on surface; seeds without endosperm. **Distr:** Tropical and subtropical Asia, Pacific Islands. **Cult:** Propagate seeds cuttings, suckers or by air-layering. Requires a humus-rich, moist well-drained soil in full sun.

A. communis J. R. Forster & G. Forster, *Char. Gen. Pl.* 51, 1775. (Syn: *A. altilis, A. incisus*, etc.). [Breadfruit, Breadnut]. **Trees**, 10–15 m tall, evergreen. **Stipules** amplexicaul, lanceolate, 10–25 cm long, pubescent. **Leaves** spirally arranged; lamina mostly ovate, 10–50 cm long, margins entire, apex acuminate, thickly leathery, glabrous, adaxially dark green and shiny, abaxially pale green, secondary veins 10 per side; leaves on mature trees pinnately lobed or pinnatipartite, lobes or segments 3–8, lanceolate; petiole 8–12 cm long. **Inflorescences** axillary, solitary; ♂ yellow, narrowly cylindric, narrowly ellipsoid, or clavate, 7–30(–40) cm long. **Flowers** inconspicuous; ♂ flowers with calyx tubular, apically bilobed, pubescent, lobes lanceolate; ♀ flowers with calyx tubular, ovary ovoid, style long, apically 2-branched. **Fruits** syncarps, green to yellow, brown to black when mature, obovoid to ± globose, 15–30 × 8–15 cm, tuberculate, pericarp soft; mesocarp of milky-white fleshy calyx; drupes ellipsoid to conic, ~2.5 cm ∅; seeds ~2.5 cm long, dark brown to black, glossy (rarely produced in cultivation). **Distr:** Origin probably from tropical Asia, naturalized and cultivated throughout the tropics. **Ecol:** In primary rainforests and coastal forests; 0–500+ m. Fl. & Fr. throughout most of the year. **Zones:** 11–12.

A. heterophyllus Lamarck, *Encycl.* 3: 209. 1789. (syn. *A. integer, A. integrifolia*). [Jack Fruit]. **Trees**, 10–20 m tall, evergreen, trunk 30–50 cm dbh. **Stipules** amplexicaul, ovate, 1.5–8 cm, caducous, scar annular and conspicuous. **Leaves** spirally arranged; lamina elliptic to obovate, 7–15(+) × 3–7 cm, base cuneate, margins of mature leaves entire, apex blunt to

Above: *Artocarpus heterophyllus, Artocarpus integer.*

acuminate, leathery, adaxially dark green, glabrous, and shiny, abaxially pale green, secondary veins 6–8 per side; leaves on mature trees entire, lobed on new growth of young trees; petiole 1–3 cm long. **Inflorescences** on old stems or brachyblasts; ♂ inflorescences axillary on apical branchlets, sometimes axillary on axillary brachyblasts, cylindric to conic-ellipsoid, 2–7 cm long, many-flowered but some sterile; peduncle 1–5 cm long; ♀ inflorescences with a globose fleshy rachis. **Flowers** inconspicuous; ♂ with calyx tubular, apically bilobed, 1–1.5 mm long; ♀ with calyx tubular, apically lobed, ovary 1-celled. **Fruits** syncarps, pale yellow, yellowish-brown when mature, ellipsoid, globose, or irregularly shaped, 30–90+ × 25–50 cm, covered with stiff hexagonal tubercles and thick hairs; drupes narrowly elliptic, ~3 × 1.5–2 cm; seeds 3–6 × 2–3 cm, glossy. **Distr:** India. **Ecol:** Fl. Feb–Mar. **Zones:** 11–12. **Info:** Cultivated widely for the edible fruits. The seeds are also eaten roasted.

A. integer (Thunb.) Merr., *Interpr. Rumph. Herb. Amb.* 190, 1917. (Syn: *A. champeden*, etc.). [Champedak, Small Jackfruit]. **Trees**, evergreen, to 20 m tall. **Stipules** ovate, 1.5–9 cm long, acute. **Leaves** variable, usually obovate or elliptic, 5–25 × 2.5–12 cm, base cuneate or rounded, margins entire, apex acuminate, adaxially dull midgreen; juvenile leaves elongate, or with 1–2 pairs of lateral lobes; petiole 0.8–3 cm long. **Inflorescences** solitary, axillary, cauliflorous and ramiflorous; ♂ cylindrical, to 3–5.5 × 1–1.25 cm, pale greenish or yellowish, clavate, smooth, peduncles 2.5–6 cm long; ♀ without green ring at the top of the stalk. **Flowers** ♂ with perianth tubular, 0.7–1 mm long, shortly bilobed above, stamen 1–1.3 mm long, filament cylindric, anthers ellipsoid; ♀ with styles filiform, exserted to 1.5 mm. **Fruits** syncarps, cylindric, 20–35 × 10–15 cm, yellow, covered with closely set, firm, tapering, obtuse, processes, 2–4 × 3 mm, odor similar to that of durian; seeds embedded in slimy opaque pulp. **Distr:** Malay Archipelago. **Ecol:** In evergreen forests; 450–1,200 m. **Zones:** 11–12. **Info:** Fruits and seeds are edible with a flavor similar to those of *A. heterophyllus*.

A. lakoocha Roxburgh, *Fl. Ind.*, ed. 1832. 3: 524, 1832. (Syn: *A. ficifolius, A. yunnanensis*). [Monkey Jack]. **Trees**, 10–15 m tall; branchlets 3–6 mm Ø, densely pubescent. **Stipules** ovate-lanceolate, 4–5 cm long, membranous, densely pubescent. **Leaves** distichous; lamina elliptic, 25–30 × 15–20 cm, sometimes pinnately lobed, base broadly cuneate to obtuse, margins entire or with small teeth, apex obtuse, abaxially with yellow bristles, adaxially coarse, secondary veins (9–)10–12(–18) pairs; petiole 2–3 cm long, densely covered with yellow bristles. **Inflorescences** ♂ ovoid to ellipsoid, 1–3.5(–4) × 1.5–2 cm, pedunculate, bracts shield-shaped, margin ciliate. **Flowers** ♂ with calyx deeply bilobed. **Fruits** syncarps, reddish-brown when dry, ± globose, ~7 cm Ø, bristles bent. **Distr:** China, India, Indonesia, Laos, Myanmar, Nepal, Sikkim, Vietnam. **Ecol:** In forests on limestone mountains; 100–1,300(–1,800) m. **Zones:** 9–12.

A. nitidus* ssp. *lingnanensis (Merrill) F. m. Jarrett, *J. Arnold Arbor.* 41: 124, 1960. (Syn: *A. lingnanensis*). **Trees**, to 17 m tall. **Leaves** oblong-elliptic to obovate-elliptic, 7–15 × 3–7 cm, base cuneate to ± rounded, margins entire or irregularly shallowly toothed, apex mucronate to caudate, thinly leathery, secondary veins 6–10 per side, adaxially prominent; petiole 0.5–1.5 cm long. **Inflorescences** ♂ capitate, obovoid, or oblong, 0.25–1.2 × 0.27–0.7 cm; ♀ with peduncle 1.5–5 mm long. **Flowers** ♂ with calyx lobes 2–4, basally connate for 0.5–0.7 mm. **Fruits** syncarps, red when mature, ± globose, coarsely pubescent. **Distr:** China. **Ecol:** In evergreen and mixed forests, and in savannas; to 900 m. Fl. & Fr. throughout most of the year. **Zones:** 9–12.

A. nobilis Thw., *Enum. Pl. Zeyl.* 262, 1861. [Ceylon Breadfruit, Wild Breadfruit]. **Trees**, evergreen, to 25 m tall; bark gray with orange scale scars. **Stipules** 3–12 cm long, hairy. **Leaves** usually obovate, 14–32 × 8–23 cm, base cuneate, margins strongly sinuate and in-rolled, apex shortly acuminate, surfaces smooth, stiff, lateral veins 10–13 pairs; petiole 2–3 cm long; sapling leaves pinnately lobed. **Inflorescences** solitary or paired, axillary; ♂ inflorescences cylindric, 7–13 × 1.5 cm, densely covered with flowers and bracts; ♀ inflorescences stouter, covered with peltate bracts; peduncles 3–7 cm long. **Flowers** ♂ with perianth bilobed, stamen 1.8 mm long; ♀ with stigma simple, exserted to 1 mm. **Fruits** syncarps, cylindric, to 20 × 10 cm, covered with short, close, rigid, subcylindric processes 1 × 1–1.5 mm, with persistent peltate bracts; fruiting peduncles 10–15 cm long; fruiting perianths ± fleshy; seeds 8 × 7 mm. **Distr:** Sri Lanka. **Ecol:** In wet zone forests; <700 m. **Zones:** 11–12.

Brosimum

Species 13. **Trees** small or **shrubs**, monoecious or dioecious, unarmed; trunk sometimes buttressed; exudate white or yellowish latex. **Stipules** free and paired or united and solitary, fully encircling stem or not, usually caducous. **Leaves** distichous, shortly petiolate; lamina entire, sometimes dentate or repand, abaxially hairy. **Inflorescences** bisexual or unisexual, (sub)globose, hemispherical, turbinate, or convexly discoid, paired or solitary, axillary, pedunculate; ♂ flowers several to numerous; ♀ flowers 1 to several. **Flowers** ♂ with perianth well-developed, (2–)4-lobed to (2–)4-parted, vestigial or lacking, stamens 1–4, pistillode absent; ♀ with thickened part of testa sometimes vascularised, perianth vestigial and united with ovary beneath, stigmas 2. **Fruits** drupes, borne within enlarged and pulpy yellowish or red receptacle, with a hard stone; seeds large. **Distr:** Mexico and the Greater Antilles, Cuba, and Jamaica, to S Brazil. **Cult:** Propagate from seeds or by air-layering. Requires a fertile, moist, well-drained soil in full or partial sun.

B. guianense (Aublet) Huber, *Bol. Mus. Emilio Goeldi* 5: 337, 1909. (Syn: *B. aubletii*, etc.) [Bastard Breadnut, Leopardwood, Mountain Breadnut]. **Shrubs** or **trees**, to 30 m tall, bisexual; latex white to yellow. **Stipules** 2–5 mm long, puberulent. **Leaves** usually

Below: *Artocarpus lakoocha, Artocarpus nitidus* ssp. *lingnanensis, Artocarpus nobilis, Brosimum guianense.*

FK&S

Above: *Broussonetia papyrifera*, *Broussonetia zeylanica*, *Castilla elastica*.

elliptic-oblong to obovate-oblong, (2–)4–10(–15) × (1–)2–4.5(–6) cm, base acute to obtuse, often asymmetric, margins entire, apex abruptly narrowed, acute or shortly acuminate, adaxially lustrous, abaxially puberulent, secondaries 6–14 pairs; petiole 2–6 mm long, hairy. **Inflorescences** usually bisexual, solitary, capitula hemispheric to broadly turbinate or discoid, often lobed, 0.3–1.2 cm Ø; peduncles to 2(–3) cm long; bracts peltate, many, 0.4–1 mm across; ♂ flowers few to many; ♀ flowers 1 to several. **Flowers** ♂ with perianth 3- or 4-lobed, stamen 1; ♀ with style 0.4–1 mm long. **Fruits** syncarps, succulent, globose or subturbinate ~1.2 cm Ø, or lobed and to 1.6 cm Ø, when containing 2 or more fruits, reddish at maturity; seeds ~8 × 5 mm. **Distr:** Tropical America, West Indies. **Ecol:** In evergreen forests; 0–1,000 m. Fl. throughout the year. **Zones:** 10–12.

Broussonetia

Species 4. **Trees**, **shrubs**, or **climbers**, deciduous, monoecious or dioecious, with latex. **Stipules** free, lateral, ovate-lanceolate, caducous. **Leaves** alternate, spirally arranged or distichous; lamina simple to palmately lobed, margins toothed, primary veins 3–5 and plinerved, secondary veins pinnate. **Inflorescences** ♂ axillary, spicate or ± capitate, many-flowered; ♀ densely capitate to ± elongated, bracts clavate, persistent. **Flowers** ♂ with calyx (3 or) 4-lobed, valvate, stamens inflexed in bud; ♀ with calyx lobes connate into a tube, apically entire or 3- or 4-lobed, persistent, ovary enclosed, style lateral, stigma linear, often with vestigial second basal lobe. **Fruits** densely aggregated into enlarged globose syncarps. **Distr:** E Asia, Pacific Islands. **Cult:** Propagate from seeds or cuttings. Requires a moist soil in full sun.

B. papyrifera (L.) L'Héritier ex Ventenat, *Tabl. Règn. Vég.* 3: 547, 1799. (Syn: *Morus papyrifera*, *Smithiodendron artocarpioideum*). [Paper Mulberry]. **Trees**, 10–20 m tall, dioecious. **Stipules** ovate, 1.5–2 × 0.8–1 cm, attenuate. **Leaves** spirally arranged; lamina ± ovate, simple or 3–5-lobed on young trees, 6–18 × 5–9 cm, base cordate, asymmetric, margins coarsely serrate, apex acuminate, adaxially sparsely pubescent, abaxially densely pubescent, secondary veins 6 or 7 per side; petiole 2.3–8 cm long. **Inflorescences** usually on leafy stems; ♂ inflorescences long-spicate, 3–8 cm long, bracts lanceolate, pubescent; ♀ inflorescences globose, bracts clavate, apically pubescent. **Flowers** inconspicuous; ♂ flowers with calyx lobes 4, triangular-ovate, pubescent, anthers globose; ♀ flowers with calyx pipelike, lobes apically connate with style, ovary ovoid, stigma linear, pubescent. **Fruits** syncarps, orange-red when mature, 1.5–3 cm Ø, mostly pubescent, fleshy; drupelets equal in length to peduncle, with 2 rows of small verruca. **Distr:** Asia; Pacific Islands. **Ecol:** In dry or humid woodlands. Fl. Apr–May. Fr. Jun–Jul. **Zones:** 6–12.

B. zeylanica (Thw.) Corner, *Gard. Bull. Singapore* 19: 235, 1962. (Syn: *Allaeanthus zeylanicus*). [Ceylon Paper Mulberry]. **Trees**, deciduous, to 18 m tall. **Stipules** ovate-lanceolate, 6–7 mm long, denticulate, membranous. **Leaves** distichous; ovate to somewhat elliptic or lanceolate, 4–10.5 × 1.5–4 cm, base cordate to rounded, margins acutely serrate, apex acuminate, acumen to 1.5 cm long, abaxially hirsute along main veins, lateral veins 7–15 pairs; petiole 0.8–1 cm long. **Inflorescences** ♂ spikes 2.5–6 cm long, 1–2 axillary on short shoots from long twigs, yellow, peduncle 4–5 mm long, bracts ovate, 1–1.3 mm long, acute, hairy; ♀ heads 5–8 mm wide, peduncle 2–5 mm long, bracts 1–1.3 mm long, hairy. **Flowers** ♀ with perianth saccate, lobes 4, short, style 1(2), 6–8 mm long, puberulous. **Fruits** syncarps, 2 cm wide; seeds 2.2–2.6 mm long, smooth. **Distr:** Sri Lanka. **Ecol:** In moist regions, mostly by rivers. Fl. Feb, Aug. **Zones:** 11–12.

Castilla

Species 3. **Trees**, monoecious or dioecious; trunk buttressed; twigs self-pruning. **Stipules** connate, fully amplexicaul, with distinct parallel veins, caducous. **Leaves** chartaceous, adaxially ± scabrous, most tertiary veins parallel. **Inflorescences** ♂ pedunculate, flabellate, bivalvate, the complemental ones ± infundibuliform to cyathiform, entire to bilobed, stamens solitary or paired along radiating and branching ridges of the receptacle bearing membranous, free or connate interstaminal bracts; ♀ solitary, sessile, discoid to cupuliform or subglobose, many- to several-flowered, ovary free, partly adnate to perianth, or immersed in the receptacle, stigmas vitiform, short. **Distr:** C America, NW South America and the Amazon Basin. **Cult:** Propagate from seeds, cuttings or by air-layering. Requires a fertile, moist soil in full or partial sun.

C. elastica Sessé, *Supl. Gaz. Lit.* 7, fig. (1–35), 1794. [Panama Rubber]. **Trees**, to 30 m high, most parts golden-hairy; trunk buttressed; latex copious, white; vegetative branches wide-spreading; flowering branches short, caducous, reaching 1–3 m long before abscission, small-leaved. **Stipules** 3–6 cm long. **Leaves** borne on long branches, mostly oblong, 20–60+ × 1–25+ cm, base cordate, often unequal, apex acuminate, stiffly membranous to coriaceous, ciliate-denticulate to subentire, lateral veins 16–22 pairs, basal veins 2–3 pairs, short; petiole 0.7–2.5 cm long, stout. **Inflorescences** ♂ heads, usually bigeminate in groups of 4, turnip-shaped, compressed, conduplicate-flabellate, 2–3.7 cm wide, peduncle 0.2–1.5 cm long; ♀ heads smaller at anthesis than the ♂, subsessile. **Fruits** syncarps, 5–9 cm wide, 1.5–2.5 cm thick, orange-red, pulpy; seeds 0.8–1 × 0.6–0.7 cm, pale straw colored. **Distr:** S Mexico to Colombia. **Ecol:** In open areas of moist forests, and on rainforest margins; to 1,000 m. Fl. Mar–Apr. **Zones:** 10–12. **Info:** Latex formerly used for commercial rubber.

Chlorophora

Species ~3. **Trees**, dioecious, with white latex, often armed with spines. **Stipules** lateral, caducous. **Leaves** alternate, petiolate; lamina with margins entire or dentate, penninerved. **Inflorescences** ♂ dense catkinlike spikes, long, slender, bracts small; ♀ capitate, globose or oblong, bracts similar to perianth segments and of equal length. **Flowers** ♂ with perianth 4-parted, segments broad, obtuse, slightly imbricate, stamens 4, filaments inflexed in bud, porrect and exserted at anthesis; ♀ with perianth 4-parted, segments concave and apically thickened, ovary included, oblique, style sublateral, filiform, usually simple, ovule laterally affixed, descending. **Fruits** with fruiting perianths fleshy, forming globose or oblong syncarps; achenes equaling perianth or somewhat exserted, ovate, compressed, apically oblique, pericarp coriaceous; seed with membranous testa. **Distr:** Africa; Mexico to British Honduras and Panama, West Indies, S America. **Cult:** Propagate from seeds, rooted cuttings or by air-layering. Requires a moist, well-drained soil in full sun.

C. tinctoria (L.) Gaud. ex Benth., *Gen. Pl.* 3: 363, 1880. (Syn: *Morus tinctoria*). [Mora]. **Trees**, sometimes shrubs, to 20 m tall; trunk 30–60 cm dbh, sometimes buttressed; bark light brown with numerous lighter excrescences; branches often armed with stout sharp axillary spines. **Leaves** deciduous, shortly petiolate; lamina ovate to ovate-oblong, 5–10 × 3–4 cm, base obtuse to subcordate, margins entire or serrate, often deeply lobate on young branches, apex usually cuspidate-acuminate, membranous, glabrous or nearly so. **Inflorescences** ♂ spikes cylindric, 4–12 × ~0.4 cm, whitish or greenish, shortly pedunculate; ♀ capitula spherical, 0.6–1 cm Ø, styles very long, threadlike. **Fruits** syncarps, globose, 1–1.5 cm Ø. **Distr:** S Mexico to British Honduras and Panama, West Indies, S America. **Ecol:** In thickets or forests in the tierra caliente, and the plains and lowlands of the Oriente and the Pacific coast; <1,200 m. **Zones:** 10–12.

Dorstenia

Species 105. **Herbs**, often succulent or pachycaulous, unarmed, monoecious. **Stipules** paired, sometimes minute. **Leaves** spirally arranged; lamina pinnately or palmately lobed to simple. **Inflorescences** bisexual, pedunculate, dilated into a circular, oblong, linear, or angled disk; ♂ flowers numerous, sessile, superficial or slightly immersed; ♀ flowers numerous, generally surrounded by 3–4 ♂ flowers. **Flowers** ♂ with perianth bilobed or -toothed, stamens 1–3, mostly 2, anthers extrorse, filaments inflexed in bud, pistillode absent; ♀ with perianth slight or none, ovary immersed in a hollow in the receptacle, style short, with 2 short stigmatic arms. **Fruits** syncarps, fleshy, extruding the small crustaceous seeds. **Distr:** Tropical parts of Africa, America, Arabia, India. **Cult:** Propagate from seeds or

Above: *Chlorophora tinctoria, Dorstenia foetida, Dorstenia indica, Ficus auriculata.*

by division. Requires humus-rich, moist but well-drained compost with partial sun.

D. foetida (Forssk.) Schweinf., *Bull. Herb. Boiss.* 4, App. 2: 120, 1896. (Syn: *D. crispa*, etc.). [Shield Flower]. **Undershrubs**, to 15(–30) cm high; stems succulent, erect; scars conspicuous and prominent. **Stipules** subulate, 1–4 mm long. **Leaves** spirally arranged, subrosulate; lamina oblong to narrowly lanceolate or elliptic, 3.5–14 × 0.3–2 cm, base attenuate, margins ± undulate and crenate to dentate or subentire, apex acute, surfaces minutely puberulous, lateral veins 5–15 pairs; petiole 0.1–2.5 cm long. **Inflorescences** solitary; peduncle 1–3 cm long; receptacle discoid, ~0.5–1 cm Ø, flowering face substellate, margins almost lacking, appendages in 2 rows, inner (marginal) row several, toothlike, to 1 mm long, outer (submarginal) row 5–10, band-shaped to filiform, 0.2–1.2 cm long; ♂ flowers ± spaced; ♀ flowers numerous. **Flowers** ♂ with perianth lobes 2; ♀ with perianth shortly tubular, stigma 1. **Fruits** with endocarp body tetrahedral, ~1.5 mm Ø, tuberculate. **Distr:** Kenya, Tanzania, Ethiopia, Somalia, Saudi Arabia, Yemen, Oman. **Ecol:** In rock outcrops; 100–1,000 m. **Zones:** 10–12.

D. indica Wight, *Ic. Pl. Ind. Or.* 6: 1964, 1853. [Indian Shield Flower]. **Herbs**, tufted; stems fleshy, tapering, curved-ascending, to 45 cm high, unbranched but proliferating from base, rooting from underside; young parts shortly hairy; latex white. **Stipules** absent. **Leaves** spirally arranged; lamina elliptic to sublanceolate, 2.5–12.5 × 1.3–4.5 cm, base subrotund or cuneate, margins dentate to sinuate, apex subacute, membranous, lateral veins 5–9 pairs; petiole 0.6–2 cm long. **Inflorescences** solitary, axillary, often somewhat decurved, green, set with 5–12 linear marginal lobes to 7 mm long; peduncle 2–3.5 cm long, dilating into disk 0.8–1.6 cm Ø; ♂ flowers numerous; ♀ immersed in disk, opening before ♂. **Flowers** ♂ 0.5 mm high, perianth with 2 slight lobes, puberulous, stamens 1–2(–3); ♀ perianth vaguely bilobed, ovary stalked, stigma bifid. **Fruits** with seeds 2 mm across, crustaceous, minutely papillate. **Distr:** S India, Sri Lanka. **Ecol:** In shady places of lowland woods; <2,000 m. Fl. Jan–Apr. **Zones:** 10–12.

Ficus

Species ~1,000. **Trees**, **shrubs**, **climbers**, **stranglers**, or sometimes woody **epiphytes**, evergreen or deciduous, with latex; monoecious species with ♂, gall (sterile ♀), and ♀ flowers in each fig or dioecious with either ♂ and gall flowers or only ♀ flowers in each fig. **Stipules** often connate, scar ringlike. **Leaves** usually alternate, rarely opposite or ± verticillate; lamina simple to lobed, rarely palmate, abaxially usually with 'wax glands' at base, or in axil of secondary veins, margins entire or toothed, veins pinnate to ± palmate. **Inflorescences** axillary or on specialized cauliflorous branches, syconium (or 'fig') with many minute flowers inserted on inner wall of hollow receptacle communicating with outside through an apical pore, sometimes closed by scalelike bracts; involucral bract usually 3 at base of fig; lateral bracts sometimes present on side of fig, scalelike. **Flowers** ♂ with calyx lobes 2–6, stamens 1–3(+), straight in bud, pistillode present or absent, *gall* flowers similar to ♀ flowers but never producing seeds and usually occupied by a fig wasp (from family the Agaonidae which are specialized symbiotic pollinators in the wild; ♀ is a normal winged insect, while the ♂ are mostly wingless); ♀ flowers with calyx lobes 0–6, ovary free, straight or oblique, styles 1 or 2 and unequal, apical or lateral. **Fruits** seedlike achenes, usually enclosed within syncarp formed from enlarged hollow fleshy receptacle; seeds very small. **Distr:** Mainly in tropics and subtropics, with the highest diversity in SE Asia. **Ecol:** Usually, initially epiphytic or epilithic (rarely terrestrial). **Cult:** Propagate from seeds, cuttings or by air-layering. Requires a humus-rich, moist but well-drained soil in full sun. **Info:** Several forest species display the 'strangler' habit in which the seedling develop high up in the tree, canopy eventually strangling the host tree which rots away leaving only the *Ficus* tree. Numerous species display the 'banyan' habit in which long aerial adventitious roots form dense curtains which support the tree canopy and eventually becoming robust trunks.

F. auriculata Loureiro, *Fl. Cochinch.* 2: 666, 1790. [Elephant Ear Fig, Roxburgh Fig]. **Trees**, dioecious, 4–10 m tall; trunk 10–15(–30) cm dbh. **Stipules** reddish-purple, triangular-ovate, 1.5–2 cm long, adaxially shortly pubescent. **Leaves** alternate; lamina broadly ovate-cordate, 15–55 × (10–)15–27 cm, base cordate to occasionally rounded, margins regularly shallowly dentate, apex obtuse and mucronate; petiole 5–8 cm long. **Figs** on specialized leafless, serpentine branchlets at base of trunk and main branches, reddish-brown, pear-shaped, depressed-globose, or top-shaped, with 8–12 conspicuous longitudinal ridges, 3–5(–6) cm Ø; peduncles (2–)4–6 cm long; involucral bracts triangular-ovate; apical bracts in 4 or 5 rows, broadly triangular-ovate, imbricate, rosulate; ♂ flowers sessile. **Flowers** ♂ with calyx lobes 3, transparent, spathulate, thinly membranous, stamens 2; *gall* flowers with calyx lobes 3, style lateral, hairy; ♀ with calyx lobes 3, ovary ovoid, style lateral, hairy. **Fruits** achenes, with adherent fluid. **Distr:** India through China to Indo-China and Thailand. **Ecol:** In moist forests; 100–1,700(–2,100) m. Fl. Aug–Mar. Fr. May–Aug. **Zones:** 9–12. **Info:** The figs are edible.

F. benghalensis L., *Sp. Pl.* 1059. (Syn: *F. indica*, etc.). [Banyan Tree, Bengal Fig, Indian Fig]. **Trees**, large, evergreen to deciduous, to 20(–50) m tall; crown wide, leafy; branches spreading to 100(–200) m with pillarlike aerial prop roots and accessory trunks. **Stipules** ± bulging, coriaceous, stout, 1.5–2.5 cm long, acute. **Leaves** ovate or obovate to elliptic, (8–)10–20(–25) × (6–)8–15(20) cm, base cordate or rounded, margins entire, apex obtuse, adaxially glabrous, abaxially finely pubescent, coriaceous, basal veins (2–)3–4 pairs, lateral nerves 4–7 pairs; petiole (1.5–)2–6(–8) cm long. **Figs** paired, sessile, orange or red, depressed-globose,

Below: *Ficus benghalensis, Ficus benghalensis* var. *krishnae, Ficus benjamina* var. *benjamina.*

Above: *Ficus benjamina* var. *nuda, Ficus carica, Ficus dammaropsis, Ficus deltoidea.*

1.5–2.5 cm Ø, pubescent; subtending bracts ovate, 3–7 mm long; ostiole closed by 3 flat or ± umbonate apical bracts 3–4 mm across, internal bristles absent; ♂ flowers numerous; ♀ flowers sessile, mixed with gall flowers; *gall* flowers numerous, pedicellate. **Flowers** ♂ ostiolar, sepals 2–3, stamen solitary, anther shortly mucronate; ♀ with sepals 3–4, small, style elongate; *gall* with sepals as in ♀, style short. **Fruits** achenes. **Distr:** S Asia. **Ecol:** In disturbed thickets at low elevations. Fl. throughout the year. **Zones:** 11–12. **Info:** Among the varieties, the most unusual is ***F. benghalensis* var. *krishnae*** (Syn: *F. krishnae*) [Krishna's Cup] with most of its leaves having its lower half of the lamina formed into a cup; vegetative propagation is needed to maintain this leaf form.

F. benjamina L., *Syst. Nat.*, ed. 12. 2: 681, 1767. [Weeping Fig]. **Trees**, to 20 m tall; crown spreading to >60 m with the help of prop roots; main trunk 0.3–1(–2.5) m dbh. **Stipules** caducous, lanceolate, 0.6–1.5 cm long, membranous, glabrous. **Leaves** ovate to broadly elliptic, 4–8(–14) × 2–4(–8) cm, base rounded to cuneate, margins entire, apex shortly acuminate, ± leathery, glabrous, secondary veins 8–10 per side, parallel; petiole 1–2 cm long, adaxially sulcate. **Figs** axillary on leafy branchlets, paired or solitary, purple, red, or yellow (sometimes red with white dots), globose to depressed-globose or sometimes pear-shaped, 0.8–2 cm Ø, base attenuate into stalk, sessile; involucral bracts inconspicuous, triangular-ovate, persistent; ♂ few, shortly pedicellate; ♀ flowers sessile; *gall* flowers many. **Flowers** ♂ with calyx lobes (3)4, broadly ovate, stamen 1; *gall* with calyx lobes (3 or)4 or 5, narrowly spathulate, ovary ovoid, style ± lateral; ♀ with calyx lobes 3, shortly spathulate, style ± lateral, short. **Fruits** achenes, ovoid-reniform. **Distr:** Asia to Malesia; N Australia, Pacific Islands. **Ecol:** In moist mixed forests; 400–800 m. Fl. Aug–Nov. **Zones:** 9–12. **Info:** This species is highly variable. ***F. benjamina* var. *benjamina*** are trees, to 20 m tall, trunks 30–50 cm dbh, figs 0.8–1.5 cm Ø, glabrous. ***F. benjamina* var. *nuda*** are trees, trunks 18–40 cm dbh, figs 1.8–2 cm Ø, glabrous or pubescent.

F. carica L., *Sp. Pl.* 2: 1059, 1753. [Calimyrna Fig, Fig]. **Shrubs**, 3–10 m tall; branches many. **Stipules** red, ovate-lanceolate, ~1 cm long. **Leaves** alternate; lamina broadly ovate, usually with 3–5 ovate lobes, 10–20 × 10–20 cm, base ± cordate, margins irregularly toothed, thickly papery, adaxially scabrous, abaxially densely covered with small cystoliths, shortly pubescent, basal lateral veins 2–4, secondary veins 5–7 per side; petiole 2–5 cm long. **Figs** axillary on normal leafy shoots, sessile, solitary, purplish-red to yellow, pear-shaped, 3–5 cm Ø, apical pore concave; involucral bracts ovate; ♂ flowers near apical pore. **Flowers** ♂ with calyx lobes 4 or 5, stamens (1 or)3(–5); *gall* flowers with style lateral; ♀ flowers with calyx lobes 4 or 5, ovary ovoid, smooth, style lateral, stigma 2-branched, linear. **Fruits** achenes, lenslike. **Distr:** Mediterranean region eastward to Afghanistan. **Ecol:** Fl. & Fr. May–Jul. **Zones:** 9–12.

F. dammaropsis Diels, *Bot. Jahrb. Syst.* 67: 204, 1935. (Syn: *Dammaropsis kingiana*). [Dinner Plate Fig, Highland Breadfruit, Shuwat Fig Tree]. **Trees** small, or pachycaulous **shrubs**, 3–10 m tall; trunk slender; branches straggling. **Stipules** conical, to 26 cm long. **Leaves** spirally arranged at ends of branches; lamina ovate, to 90 × 60 cm, broader when mature, base ± cordate, margins undulate, apex rounded, heavily veined, leathery, deeply corrugated, adaxially deep green, shiny, abaxially paler, basal veins 5, ribs reddish when young; petiole red. **Figs** axillary, dioecious, sessile, solitary or paired, subglobose, covered entirely by large lateral bracts (to 13 cm long), rose-red to deep purple (variable, seed figs are usually red while gall figs are usually greenish), very large when mature (resembling a pineapple), to 8–15(–18) cm Ø; flowers are borne on lobing processes on the inner surface (unusual in the genus); ♂ flowers near orifice; *gall* flowers with ostiolar ♂ flowers. **Flowers** with perianth red, entire, saccate, ± covering ovary; ovary white, no internal bristles or bracteoles except between ♂ flowers. **Fruits** achenes. **Distr:** Papua New Guinea. **Ecol:** In montane regions; (0–)800–2,750(–2,800). **Zones:** 11–12. **Info:** Cultivated in PNG for the edible leaves, bark and fruits.

F. deltoidea Jack., *Mal. Misc.* 7: 71, 1822. (Syn: *F. diversifolia, F. ovoidea*). [Mistletoe Fig, Rusty-leaved Bush Fig]. **Shrubs**, evergreen, glabrous, to 2 m high. **Stipules** linear-lanceolate, 0.8–1.2 cm long, convolute. **Leaves** polymorphic, broadly obovate, spathulate or elliptic to deltoid, 2.5–7.5 × 1.5–5 cm, base cuneate-rounded, apex truncate, rounded-obtuse or notched, coriaceous, abaxially rusty to pale-olive, midrib forked or simple and penninerved, abaxial glands (black spots) present in forks or in axils of nerves; petiole to 8 mm long. **Figs** solitary or paired, axillary, ovoid, globose or pyriform, 0.5–1.5 cm Ø, green (yellow to orange-red), glabrous, subtended by 3, broad, puberulous basal bracts, apical orifice covered by minute, ± triangular bracts; peduncles 0.5–2 cm long; ♂ dispersed among all flowers, long-pedicellate; ♀ flowers ± sessile. **Flowers** ♂ with sepals 4, obovate; *gall* with sepals 3, linear-ovate; ♀ with sepals 3, ovate, ovary globose with long subterminal style. **Fruits** achenes. **Distr:** Malayan Archipelago. **Ecol:** Fl. & Fr. Jul–Oct. **Zones:** 11–12. **Info:** Variable species.

F. elastica Roxburgh, *Fl. Ind.*, ed. 1832. 3: 541, 1832. (Syn: *F. cordata*, etc.). [Assam Rubber, Indian Rubber]. **Trees**, 20–30 m tall, epiphytic when young; trunk 0.3–1 m dbh. **Stipules** dark red, ~10 cm long, membranous. **Leaves** oblong to elliptic, 8–30 × 7–10 cm, base broadly cuneate, margins entire, apex acute, thickly leathery, adaxially dark green, shiny, abaxially pale green, secondary veins many, closely parallel, inconspicuous; petiole robust, 2–5 cm long. **Figs** axillary on leafless branchlets, paired, subsessile, yellowish-green, ovoid-ellipsoid, ~1 × 0.5–0.8 cm; involucral bracts hoodlike, caducous; ♂ scattered among other flowers, pedicellate; ♀ flowers sessile. **Flowers** ♂ with calyx lobes 4, ovate, stamen 1, filament absent, anther ovoid-ellipsoid; *gall* with sepals 4, ovary ovoid, smooth, style subapical, curved; ♀

Below: *Ficus elastica* (variegated form), *Ficus lyrata, Ficus macrophylla, Ficus nymphaeifolia.*

PL

Top row: *Ficus pseudopalma, Ficus religiosa, Ficus pumila.*
Bottom row: *Ficus racemosa, Ficus sur.*

style persistent, long, stigma enlarged, ± capitate. **Fruits** achenes, ovoid, tuberculate. **Distr:** Bhutan, China, N India, Indonesia, Malaysia, Myanmar, Nepal, Sikkim. **Ecol:** 800–1,500 m. **Zones:** 10–12.

F. lyrata Warb., *Bot. Jahrb.* 20: 172, 1984. (Syn: *F. pandurata*). [Banjo Fig, Fiddle Leaf Fig]. **Trees**, large, to 9–12(–30) m tall. **Stipules** narrowly triangular-lanceolate, 4–5 cm long, acuminate, glabrous, persistent. **Leaves** fiddle-shaped or pandurate, 15–30 × 12–20 cm, base cordate, margins sinuate-entire, apex almost truncate, shortly acuminate, thick, glabrous, lateral nerves 4–5 pairs, bulging beneath; petiole stout, 2–5 cm long, glabrous, grooved. **Figs** sessile, in axillary pairs, globose, 2.5–4.5 cm Ø, finely puberulous, subtended by 3 triangular, appressed basal bracts, apical orifice somewhat depressed, ± 2-lipped, closed by lanceolate inwardly directed bracts. **Flowers ♂** with sepals 2–3, ovate-lanceolate, stamen solitary, anther ovoid; ♀ flowers with sepals 3, ovate, obtuse, style short, stigma papillate. **Distr:** Tropical Africa. **Ecol:** In rainforests. **Zones:** 9–12.

F. macrophylla Desf. ex Pen., *Syn. Pl.* 2: 609, 1806. (Syn: *F. macrocarpa*). [Australian Banyan, Black Fig, Figwood, Moreton Bay Fig]. **Trees**, large, to 20+ m tall. **Stipules** showy, 5–10 cm long, acute. **Leaves** ovate-elliptic to oblong, 15–25 × 7.5–12 cm, base rounded or ± cuneate, margins entire, apex obtuse-acute, thick, adaxially glabrous and glossy, abaxially silvery to rusty dotted, distinct lateral nerves 10–20 pairs, parallel, intercostals present; petiole to 8 cm long. **Figs** axillary group of 3–4, globular or ± pyriform (globular when mature), 1.5–2 cm Ø (2.5 cm Ø, dark reddish-brown to purple, with pale spots when mature), subtended by 3 basal bracts; peduncles short, thick. **Flowers ♂** ostiolar and mixed with ♀, sepals 3, stamen solitary; ♀ flowers with sepals 4–5, ovary ovoid, style lateral, stigma elongate-acute. **Fruits** achenes. **Distr:** Australia. **Ecol:** Fl. Oct–Dec. **Zones:** 9–11.

F. nymphaeifolia Mill., *Gard. Dict.*, ed. 8. n. 9, 1768. (Syn: *Urostigma nymphaeifolium*, etc.). [Black Fig]. **Trees**, 15–35 m tall, becoming massive with age. **Stipules** lanceolate or deltoid, to 4 cm long. **Leaves** ovate to ovate-round, 9–23 × 6.5–17 cm, base cordate, apex acute, broadly acute, rounded or sometimes shortly acuminate, lateral veins (4–)5–11, intercostals inconspicuous; petiole 3–12.5 cm long. **Figs** globose, minutely puberulent or glaucous, borne among leaves, 1.5–2.5 cm Ø, greenish, purple-spotted; peduncles ~0–5 mm long; bracts deltoid or ovate, 0.4–1.9 cm long, connate, minutely puberulent or glaucous; ostiole 2–3 mm Ø; ♂ flowers stalked; bracts present between flowers. **Flowers ♂** ~2 mm long, tepals 3, hooded; ♀ ~2 mm long, tepals 3, hooded. **Fruits** achenes. **Distr:** Panama, Colombia and N South America to Brazil. **Ecol:** In coastal plains and savanna belt. **Zones:** 10–12.

F. pseudopalma Blanco, *Fl. Filip.*, ed. 2: 473, 1845. [Dracaena Fig, Palm Fig, Philippine Fig]. **Shrubs** or **small trees**, erect, glabrous, 2–6(–9) m tall; stem unbranched, terete, 4–6(–7.5) cm Ø, with prominent leaf scars (heart-shaped). **Stipules** persistent, lanceolate, 5–7 cm long. **Leaves** alternate, crowded at the top of the stem, giving the appearance of a palm; lamina oblanceolate, to 0.8(–1) m long, base slightly cordate, narrowed towards petiole, margins coarsely and irregularly sinuate-toothed, apex acute or acuminate, coriaceous; petiole subsessile, stout. **Figs** usually paired and crowded, axillary, dark-green (maturing to purplish), ovoid, 2.5–4 cm Ø, obscurely angular, with raised white flecks outside. **Distr:** Philippines. **Ecol:** In thickets and grasslands; to ~200 m. Fl. throughout the year. **Zones:** 11–12. **Info:** This species is known to grow unbranched in the wild and multistemmed in cultivation. The figs are edible, and in Philippines, young sprouts are used as a vegetable.

F. pumila L., *Sp. Pl.* 2: 1060, 1753. (Syn: *F. repens*, etc.). [Climbing Fig, Creeping Fig]. **Vines** woody, or **shrubs**, sprawling; vines closely appressed to substrate with adventitious roots at nodes; shrubs loosely ascending, evergreen; branches glabrous with age. **Stipules** 3–8 mm long. **Leaves** dimorphic, spreading, *appressed stems* with distichous and smaller leaves, those of *flowering stems* loose, extended; lamina oblong to ovate-elliptic or obovate, 4–10 × 2.5–4.5 cm, base obtuse to rounded, margins recurved, apex obtuse to nearly acute, ± glabrous, basal veins 1 pair, lateral veins 3–6 pairs; petiole 1.5–2 cm long. **Figs** solitary, pedunculate, green, oblong, obovoid, pyriform, or nearly globose, 3–4(–8) × 3–4 cm, slightly pubescent, becoming glabrescent with age; peduncles 0.8–1.5 cm long; subtending bracts ovate, 5–7 mm long; ostiole closed by 3 bracts, umbonate. **Fruits** achenes. **Distr:** SE Asia. **Ecol:** In disturbed thickets at low altitudes. Fl. throughout the year. **Zones:** 9–11. **Info:** Needs pruning to maintain the creeping habit, else it will become erect robust shrubs.

F. racemosa L., *Sp. Pl.* 2: 1060, 1753. (Syn: *F. glomerata*). [Cluster Fig, Red River Fig]. **Trees**, monoecious, 25–30 m tall; trunk 0.6–1 m

Above: *Morus nigra*, *Streblus asper*, *Streblus asper* (fruit), *Streblus elongatus*.

dbh. **Stipules** ovate-lanceolate, 1.5–2 cm long, membranous. **Leaves** alternate; lamina ± elliptic, 10–14 x 3–4.5(–7) cm, base cuneate to obtuse, margins entire, apex acuminate to obtuse, ± leathery, adaxially dark green, glabrous, abaxially pale green, basal lateral veins 2, secondary veins 4–8 per side; petiole 2–3 cm long. **Figs** clustered on short branchlets on trunk, occasionally axillary on leafy shoots or on older leafless branchlets, paired, reddish-orange, pear-shaped, 2–2.5 cm Ø, base attenuated into a stalk, apical pore navel-like, flat; peduncles ~1 cm long; involucral bracts triangular-ovate; ♂, gall, and ♀ flowers within same the fig; ♂ flowers near apical pore, sessile; *gall* and ♀ flowers pedicellate. **Flowers** ♂ with calyx lobes 3 or 4, stamens 2; *gall* and ♀ with calyx lobes linear, apex 3- or 4-toothed, style lateral, stigma clavate. **Distr:** Asia through Malesia to Australia. **Ecol:** Near waterways, occasionally in streams; 100–1,700 m. Fl. May–Jul. **Zones:** 19–12.

F. religiosa L., *Sp. Pl.* 2: 1059, 1753. (Syn: *Urostigma religiosum*). [Bo Tree, Peepul Tree, Pipal Tree, Sacred Fig]. **Trees**, epiphytic when young, 15–25 m tall; trunk 30–50 cm dbh. **Stipules** ovate, small, acute. **Leaves** triangular-ovate, 9–17 x 8–12 cm, base broadly cuneate to ± cordate, margins entire or undulate, apex acute to caudate, cauda (drip tip) (0.5–)2–5(10) cm long, leathery, adaxially dark green, shiny, abaxially green, basal lateral veins 2, secondary veins 5–7 pairs; petiole slender, ~8–12 cm long. **Figs** axillary on leafy branchlets, paired or solitary, red when mature, globose, 1–1.5 cm Ø, smooth; peduncles 4–9 mm long; involucral bracts ovate; ♂, gall, and ♀ flowers within same fig; ♂ flowers few, near apical pore, sessile; *gall* flowers pedicellate; ♀ flowers sessile. **Flowers** ♂ with calyx bi- or trilobed, margins revolute, stamen 1, filament short; *gall* with calyx 3- or 4-lobed, ovary globose, smooth, style short, stigma enlarged, bilobed; ♀ with calyx 4-lobed, broadly lanceolate, ovary globose, smooth, style thin, stigma narrow. **Distr:** N India, Nepal, Pakistan. **Ecol:** Fl. Mar–Apr. Fr. May–Jun. **Zones:** 9–12.

F. sur Forssk., *Fl. Aegypt.-Arab.* 124: 180, 1775. (Syn: *F. capensis*, *F. mallotocarpa*). [Broom-Cluster Fig, Sur]. **Trees**, to 12 m tall (in woodlands) or 25 m tall (in forests); crown rounded; trunk sometimes buttressed; all parts exudate milky latex when cut. **Leaves** alternate, spiraled; lamina ovate to elliptic, 5.5–20 x 2–13 cm, base rounded to cordate, margins widely and irregularly dentate, apex rounded to acute, leathery, mid to dark green, abaxially paler, usually glabrous; petiole 1.3–9 cm long. **Figs** borne on much-branched leafless branches or trusses on the main branches, trunk or roots (thus the fruit borne just below soil level), pear-shaped, 2.8–4 cm Ø, green ripening to pink or red, spotted with cream spots, glabrous or pubescent. **Distr:** S Africa (Cape Region) northwards throughout Africa to Saudi Arabia and W Africa. **Ecol:** In open woodlands, riverine and semideciduous forests, rocky hills. **Zones:** 10–12.

Morus

Species ~16. **Trees** or **shrubs**, monoecious or dioecious, deciduous, with latex. **Stipules** free, sublateral, caducous. **Leaves** alternate; lamina simple to deeply palmately lobed, margins toothed, primary veins 3–5 from base, secondary veins pinnate. **Inflorescences** ♂ axillary, spicate, many-flowered, shortly pedunculate; ♀ shortly spicate to capitate, sessile. **Flowers** ♂ with calyx lobes 4, imbricate, stamens inflexed in bud, pistillode top-shaped; ♀ with calyx lobes 4, imbricate, fleshy in fruit, ovary 1-loculed, style present or not, stigma 2-branched, abaxially pubescent or papillose. **Fruits** aggregated syncarps, calyces enlarged, succulent, juicy; syncarps with achenes enclosed by enlarged and succulent calyx; seed ± globose. **Distr:** Widespread in temperate and tropical regions, mostly in N America, Europe, and Asia. **Cult:** Propagate from cuttings or seeds. Requires a fertile, moist, well-drained soil in full or partial sun.

M. nigra L., *Sp. Pl.* 2: 986, 1753. [Black Mulberry]. **Trees**, monoecious or dioecious, to 10 m tall; branchlets pale brown pubescent; bark dark brown. **Stipules** lanceolate, membranous, brown-pubescent. **Leaves** broadly ovate, unlobed, 6–12(–20) x 7–11 cm, base cordate, margins regularly and coarsely serrate, apex acute to shortly acuminate, thick, abaxially pale green, shortly pubescent, tomentose, adaxially dark green, coarse; petiole 1.5–2.5 cm long, pubescent. **Inflorescences** ♂ catkins, cylindric, 2–4 cm long, pubescent; ♀ catkins, ellipsoid, 2–2.5 cm long, peduncle short. **Flowers** ♀ with style inconspicuous, stigmas without mastoidlike protuberance, 2-branched and pubescent. **Fruits** syncarps, blackish-purple when mature, elliptic, 2–2.5 x 1.5–2.5 cm. **Distr:** W Iran; cultivated worldwide. **Ecol:** Natural habitat unknown; widely cultivated in the tropics and subtropics. Fl. Mar–Jul. **Zones:** 5–12.

Streblus

Species 22. **Trees** or **shrubs**, occasionally scandent shrubs, monoecious or dioecious, with latex, sometimes spiny (spines axillary). **Stipules** free or connate, lateral to ± amplexicaul, caducous. **Leaves** alternate, distichous, shortly petiolate; lamina with margins entire or toothed and eglandular, secondary veins pinnate. **Inflorescences** ♂ axillary, sometimes bisexual with a single central ♀ flower, racemose, spicate, or ± capitate, pedunculate; ♀ often 1-flowered. **Flowers** ♂ with calyx lobes (3 or)4(or 5), free or basally adnate with pistillode, imbricate or valvate, stamens inflexed in bud, anthers small, reniform, extrorse; ♀ with calyx lobes 4, opposite, free or ± connate, imbricate, ovary superior, ovoid to oblique, base thick and fleshy, enclosed by persistent calyx lobes or not, style 2-branched. **Fruits** drupes, free (not aggregated into a syncarp), globose, to 1.5 cm Ø, dehiscent or indehiscent; seed large, globose. **Distr:** Tropical and subtropical Asia. **Cult:** Propagate from seeds or by air-layering. Requires a fertile, moist, well-drained soil in full or partial sun.

S. asper Loureiro, *Fl. Cochinch.* 2: 615, 1790. (Syn: *Achymus pallens*, etc.). [Sandpaper Tree, Siamese Rough Bush, Toothbrush Tree]. **Trees** or **shrubs**, monoecious or dioecious, to 14 m tall. **Stipules** small, caducous. **Leaves** shortly petiolate or sessile; lamina elliptic-obovate to elliptic, 2.5–6 x 2–3.5 cm, base obtuse to ± cordate, margins entire or irregularly crenate, apex blunt to shortly acuminate, leathery, scabrous, secondary veins 4–7 pairs. **Inflorescences** *bisexual* capitate with 1 central sessile ♀ flower surrounded by ♂ flowers; ♂ solitary or paired, capitate, peduncle 0.8–1 cm long, pubescent, bracts few, bracteoles 2, basal; ♀ pedunculate, bracts 1 or 2, basal, minute, bracteoles basal. **Flowers** ♂ subsessile, pistillode conic to cylindric, apically tuberculate; ♀ with calyx lobes pubescent, ovary globose, style apically branched, 0.6–1.2 cm long in fruit. **Fruits** drupes, yellow, globose, ~6 mm Ø, indehiscent, enclosed by enlarged calyx lobes when immature, without a fleshy base. **Distr:** Indo-China, SE Asia. **Ecol:** In forests, monsoon regions; 200–1,000 m. Fl. Feb–Apr. Fr. May–Jun. **Zones:** 9–12. **Info:** The fruits are edible, sweet.

S. elongatus (Miq.) Corner, *Gard. Bull. Singapore* 19: 227, 1962. (Syn: *Sloetia elongata*, etc.). [Tampines Tree]. **Trees**, to 33 m tall; trunk to 2.5 m dbh, often slightly fluted, without large buttresses. **Leaves** oblong-elliptic, 10–37.5 x 2.5–7.5 cm, base unequal, apex shortly pointed, secondary veins 10–20 pairs, ± sunken above; petiole 0.25–1.25 cm long. **Inflorescences** catkins, hanging, solitary or 2–4 together, axillary, one side without flowers; ♂ flowers numerous; ♀ flowers 1–3. **Flowers** ♂ minute, tepals 3, stamens 3; ♀ with tepals 4, styles 2, long. **Fruits** drupes, 1.25 cm Ø, enclosed in swollen, fleshy, bilobed white base of flower; seed 1, 6 mm Ø, pale brown. **Distr:** Malay Peninsula to Sumatra. **Ecol:** Scattered in lowland forests. **Zones:** 11–12. **Info:** Timber is hard and durable, thus used for making knife handles, etc.

S. taxoides (Heyne) Kurz, *For. Fl. Burma* 2: 465, 1877. (Syn: *Epicarpurus spinosus*, etc.). [Fig Lime]. **Trees**, small, evergreen, 2–3 m tall; twigs with short spines; branchlets curved, one side yellow-hairy; spines woody, 1–1.5 cm long; latex white. **Stipules** lanceolate, 2–5 mm long, acute or acuminate, often sharply pointed, caducous. **Leaves** variable, elliptic obovate, often asymmetric, (2–)5–10(–18) x 0.7–6.5 cm, base cuneate or slightly rounded, margins dentate, especially distally, apex acute to long-acuminate, often tridentate, smooth, coriaceous; petiole 2–3 mm long. **Inflorescences** ♂ 1–4, axillary, capitate, 2–5 mm Ø, 8–14-flowered, flowers sessile except central ones, or

SZ

Above: *Streblus taxoides*, *Treculia africana*, *Boehmeria nivea*, *Cecropia insignis*.

shortly racemose and flowers with pedicels to 1.5 mm long, shortly pedunculate; ♀ flowers solitary, peduncles 3–6 mm long. **Flowers** ♀ with tepals 2–5 mm long, style 1 mm long, stigmatic arms 0.5–1.5 mm long. **Fruits** drupes, globose or ellipsoid, 0.5–1 cm Ø, asymmetric. **Distr:** India, Indonesia, Malaysia, Philippines, Sri Lanka, Thailand, Vietnam. **Ecol:** In lowland forests. **Zones:** 11–12.

Treculia

Species 7. **Trees** or **shrubs**, glabrous or hairy. **Stipules** lanceolate, caducous. **Leaves** alternate, shortly petiolate; lamina margins entire, coriaceous or chartaceous, penninerved. **Inflorescences** capitate, arranged on the outside and densely surrounded by bracts of the fleshy receptacle, axillary or at leafless nodes, sessile or shortly pedunculate; bracts below the head few, small, subbiseriate, not forming a true involucre; bracteoles on capitula surrounding flowers peltate or rounded. **Flowers** dioecious; ♂ with perianth tubular, membranous, apex 2–4-fid or dentate, base thickened, stamens 2–4, often 3, sometimes 1 imperfectly developed, anthers oblong or ovate, erect, exserted, rudimentary ovary absent; ♀ with perianth absent, ovary ovoid, enclosed by the fleshy receptacle or bracts, style subulate, with stigmatose exserted filiform branches, ovule pendulous from apex. **Fruits** syncarps, globose, sometimes very large, bristly with bracts surrounding the flowers; achenes in an indefinite series in the fleshy part of syncarps; seeds ovoid or ellipsoid, testa membranous. **Distr:** Tropical Africa (6 spp. endemic), Madagascar (1 sp.). **Cult:** Propagate from seeds or by air-layering. Requires a fertile, moist, well-drained soil in full or partial sun.

T. africana Decne. ex Trécul, *Ann. Sci. Nat.*, Bot. sér. 3, 8: 109, 1847. (Syn: *Ficus welwitschii*, etc.). [Bread Tree]. **Trees**, to 24 m tall; latex whitish; trunk straight, 30–60 cm dbh. **Stipules** caducous, lanceolate, 1.2–2 cm long, acute, submembranous, often minutely puberulous outside. **Leaves** elliptic or ovate-elliptic, usually ~7.5–20 × 3.2–10 cm, base rounded, oblique, apex shortly and obtusely pointed, coriaceous; petiole 0.6–1.2 cm long. **Inflorescences** ♂ heads, axillary, very shortly pedunculate, globose or slightly ellipsoid, 4–5.7 cm Ø. **Flowers** dioecious; perianth usually bifid, lobes erect, ciliolate, closely embracing filaments; stamens 2–4, filaments stout, glabrous, anthers 2.5–3 mm long. **Fruits** syncarps, ± globose, very large, 30–45 cm Ø, yellowish outside when ripe, very rough with the knoblike tops of perianth-segments; seeds numerous, embedded in syncarp. **Distr:** Tropical Africa. **Ecol:** In dense tropical forests. **Zones:** 11–12.

URTICACEAE

Genera ~45, **species** ~800. **Herbs**, **subshrubs**, **shrubs**, **trees** or **lianas**, monoecious or dioecious, without stinging hairs, with linear cystoliths. **Stipules** interpetiolar. **Leaves** opposite, petiolate; lamina margins entire or dentate, 3-costate. **Inflorescences** axillary, compact, globose or paniculate, sessile or peduncled cymose clusters; flowers sessile, bracteate. **Flowers** minute, green, unisexual; calyx (2–3–)4-lobed or truncate and cupular in ♂, lobes horned outside below apex, unequally trilobed in ♀, median lobe largest, cucullate or gibbous; stamens (2–3)4; staminodes scalelike, inflexed in ♀; ovary straight, style absent, stigma penicillate; pistillode conical or oblong in ♂. **Fruits** achenes, flattened, ovate or orbiculate, not or partly enveloped by calyx. **Distr:** Worldwide, mainly tropical and subtropical regions.

Boehmeria

[False Nettle].

Species 50. **Herbs**, **subshrubs**, or **shrubs**, perennial, rhizomatous, sparsely to ± densely pubescent or tomentose with hooked, curved, or straight, non-stinging hairs; stems simple to freely branched, erect. **Stipules** present. **Leaves** opposite or nearly opposite, or alternate. **Leaves** lanceolate to broadly ovate, base rounded to cordate, sometimes cuneate, margins dentate or serrate, apex acuminate; cystoliths rounded. **Inflorescences** axillary, spikelike or paniculate; ♂ and ♀ on same plant, rarely on different plants, in remote or crowded clusters in same or separate inflorescence; bracts linear. **Flowers** unisexual; ♂ with tepals 4, stamens 4, pistillode globose; ♀ with tepals 4, connate, adnate to ovary, staminodes absent, style persistent, elongate, stigma linear, straight or hooked. **Fruits** achenes, sessile, laterally compressed, ovoid, nearly orbicular, or ellipsoid, tightly enclosed by persistent perianth. **Distr:** Tropical and subtropical regions worldwide. **Cult:** Propagate from seeds. Requires a moist soil in shade or full sun.

B. nivea (L.) Gaudichaud Beaupré, *Voy. Uranie*, Bot. 12: 499, 1830. (Syn: *Urtica nivea*, *Ramium niveum*). [Chinese Grass]. **Shrubs** or **subshrubs**, 2(–3) m high. **Leaves** alternate; lamina broadly ovate to nearly orbiculate, 8–15 × 5–12 cm, adaxially slightly scabrous, abaxially densely white-tomentose. **Inflorescences** panicles of moniliform (beaded) clusters, branches not leafy at apex; ♂ flowers in proximal leaf axils, sessile; ♀ flowers in distal axils. **Flowers** ♂ 4-merous, perianth lobes connate to middle, ~1.5 mm long, pubescent; ♀ rhomboid-ellipsoid, 0.6–0.8 mm long, stigma ~1 mm long. **Fruits** achenes, compressed or lenticular, ovoid or ellipsoid, ~1.5 × ~0.9 mm, pubescent, smooth; fruiting perianth rhomboid-obovoid, compressed, ~1 mm long, strigose on shoulder, base constricted and stalklike, apex almost without neck, 2–3-toothed. **Distr:** Tropical and subtropical Asia to Malesia. **Ecol:** In forest margins, thickets and moist places along streams; 0–1,700 m. Fl. May–Aug. Fr. Sep–Nov. **Zones:** 9–12.

Cecropia

Species 30–40. **Trees**; trunk and branches hollow with septa at intervals; exudate milky. **Stipules** large, connate, intrapetiolar, enclosing the terminal bud, deciduous. **Leaves** peltate, lobes prominently penniveined; petiole long. **Inflorescences** spikes, dense, cylindrical, several, borne at apex of peduncles; peduncles within a spathaceous caducous bract, ♂ slender, ♀ thick; flowers numerous, sessile. **Flowers** dioecious, very small; ♂ with perianth tubular, stamens 2, filaments straight, scarcely exserted, ovary absent; ♀ with perianth tubular with a small opening at apex, ovary included, style very short, stigma exserted, penicillate, ovule, erect, orthotropous. **Fruits** oblong, enclosed in the very thin perianth. **Distr:** Tropical America (West Indies and Mexico to Brazil). **Cult:** Propagate from seeds or soft branch cuttings. Requires a fertile, well-drained soil in full sun.

C. insignis Liebm., *Kong. Danske Vidensk. Selsk. Skrift.*, ser. 5, 2: 318, 1851. (Syn: *C. eximia*, etc.). [White Guarumo]. **Trees**, dioecious, to ~40 m tall; trunk 70 cm dbh, often with stilt-roots. **Stipules** red when falling, mostly 17–26(–40) × 5–8 cm, with 2 ridges. **Leaves** clustered at branch ends; lamina to 90 cm Ø, deeply divided to middle, lobes 7–9, 22 cm across, base narrowed, margins held upward, apex acute to obtuse, adaxially shiny; petiole stout, basal pulvinus brown. **Inflorescences** ♂ with spathes oblong, 10–12(–16) × 5–6.5 cm, rounded, tip long-apiculate, spadices held erect in clusters of usually 6 or 7, 12–15 × 0.8–1 cm, whitish or green, stipes purplish, 1.2–1.6 cm long, peduncle 5–10 cm long; ♀ with spathes similar to ♂, spadices in clusters of 4–7, pale yellow, 8–10 × ~0.8 cm, becoming bright yellow in fruit (to 15 × 1.3 cm), stipes ~1 cm long, peduncle 8–13 cm long. **Fruits** achenes, ovate-elliptic, flattened, 1.7–2.3 × ~1 mm, muricate, reddish-brown. **Distr:** Nicaragua to Colombia. **Ecol:** In forests. Fl. Dec–Apr. Fr. Apr–Aug. **Zones:** 10–12.

Pilea

Species ~400. **Herbs**, **shrubs**, or **subshrubs**, annual or perennial, glabrous; stems simple or branched, erect, ascending, or repent. **Stipules** present. **Leaves** opposite, paired, equal or unequal; lamina ovate, margins dentate or entire; cystoliths linear, ± conspicuous. **Inflorescences** cymes, axillary, compact to lax; ♂ and ♀ flowers in the same cyme; bracts deltate to linear. **Flowers** unisexual; ♂ with tepals 4, stamens 4, pistillode conic; ♀ with tepals 3, equal or sometimes 1 tepal enlarged and hoodlike, staminodes 3, opposite tepals, under tension and ejecting mature achene, style and tufted stigma deciduous. **Fruits** achenes, sessile, laterally compressed, ovoid to teardrop-shaped, free from perianth at maturity, partly covered by hoodlike tepal. **Distr:** Tropical, subtropical regions worldwide. **Cult:** Propagate perennials from stem cuttings and annuals from seeds. Requires a

Above: *Pilea cadierei, Pilea involucrata, Pilea microphylla, Pilea nummulariifolia.*

moist, well-drained soil in shade or partial sun.

P. cadierei Gagnepain & Guillemin, *Bull. Mus. Natl. Hist. Nat.*, sér. 2. 10: 629, 1939. [Aluminum Plant, Watermelon Plant]. **Herbs perennial**, or **subshrubs**, rhizomatous, glabrous, dioecious, most parts covered with cystoliths and leaves densely; stems erect, 15–40 cm high, somewhat succulent, base woody. **Stipules** caducous, green, oblong, 1–1.3 cm long, papery, longitudinally 2-ribbed. **Leaves** subequal, obovate, 2.5–6 × 1.5–3 cm, base broadly cuneate or subrounded, margins obscurely dentate or erose, apex mucronate, papery, 3-veined, adaxially with 2 interrupted white grooves; petiole subequal, 0.7–1.5 cm long. **Inflorescences** paired; ♂ capitula with peduncle 1.5–4 cm long, glomerules 0.6–1 cm Ø, bracts broadly ovate, ~3 mm long, pedicel 2–3 mm long; ♀ flowers subsessile. **Flowers** ♂ ~2.5 mm long, perianth lobes 4, cymbiform, stamens 4, rudimentary ovary conic; ♀ with persistent perianth lobes 0.5–0.7 mm long, staminodes oblong. **Fruits** achenes, ovoid, ~1.5 mm long, compressed. **Distr:** China, Vietnam. **Ecol:** In moist forests; 500–1,500 m. Fl. Sep–Nov. Fr. Nov–Dec. **Zones:** 9–12.

P. involucrata (Sims) Urb., *Symb. Antill.* (Urban). 1: 298, 1899. [Friendship Plant]. **Herbs**, bisexual or unisexual; stems branched, erect to trailing, hairy, 5–30 cm long. **Stipules** 3–4 mm long, rounded. **Leaves** ± sessile; lamina obovate to elliptic-obovate, to 2–6 × 1.3–3.8 cm, base obtuse to slightly rounded, margins toothed, apex obtuse to rounded, surface often puckered, reddish, with 3 prominent veins; cystoliths linear to fusiform and usually confined to adaxial margin. **Inflorescences** usually unisexual, subsessile, borne in axils of distal leaves near apex of stems, 0.5–3 cm long, much-branched. **Flowers** minute. **Fruits** ovoid or slightly flattened, 0.5–0.4 mm long, yellowish-brown. **Distr:** C North America, S America. **Ecol:** In moist evergreen places; to 1,500 m. **Zones:** 10–12.

P. microphylla (L.) Liebm., *Kong. Danske Vidensk. Selsk. Skrift.*, ser. 5. 5(2): 302, 1851. (Syn: *Parietaria microphylla*). [Artillery Fern, Military Fern, Rock Weed]. **Herbs**, weak, glabrous, monoecious; stems erect or ascending, simple or branched, 3–17 cm high, succulent, cystoliths dense. **Stipules** persistent, triangular, ~0.5 mm long, membranous. **Leaves** obovate or spathulate, unequal, 2–7 × 1.5–3 mm, base cuneate or attenuate, margins entire, recurved, apex obtuse, succulent, adaxially green, abaxially paler; petiole slender, 1–4 mm long. **Inflorescences** often androgynous, compactly cymose-capitate; peduncle 1.5–6 mm long, sometimes sessile; glomerules few-flowered; ♂ flowers pedicellate. **Flowers** ♂ ~0.7 mm Ø, perianth lobes ovate, subapically corniculate, rudimentary ovary minute, conic; ♀ with perianth lobes subequal, oblong. **Fruits** achenes, ovoid, ~0.4 mm across, compressed, smooth, enclosed by persistent perianth. **Distr:** Tropical S America; naturalized worldwide. **Ecol:** Fl. Jun–Aug. Fr. Sep–Oct. **Zones:** 10–12.

P. nummulariifolia Wedd., *Ann. Sci. Nat.*, Bot. sér. 3, 18: 225, 1852. [Creeping Charlie]. **Herbs**, creeping, rooting freely from nodes; stems reddish, decumbent, to 20 cm high. **Stipules** ovate, ~3 mm long, obtuse and sometimes ciliate. **Leaves** paired, subequal; lamina broadly elliptic to circular, 0.8–5 cm long, base rounded to cordate, margins scalloped to toothed, apex rounded, bright green, surface puckered; petiole 0.4–2.5 cm long. **Inflorescences** ♂ and ♀ separate, borne in upper leaf axils; ♂ sessile, partly concealed by leaves, pedicels elongate, hairy; ♀ with peduncle ~8 mm long, glabrous. **Flowers** minute. **Distr:** West Indies, S America (Panama to Peru). **Ecol:** In moist low elevation areas; to 1,500 m. **Zones:** 10–12.

CUCURBITALES

FAMILIES 7, GENERA 129, SPECIES 2,295.

Stems with separate vascular bundles. **Flowers** unisexual; calyx or perianth valvate; ovary inferior, placentation parietal, stigmas usually forked, micropyle bistomal (Judd et al. 2008; Stevens 2001 onwards). **Info:** The monophyly of the Cucurbitales is very strongly supported by molecular data (Zhang et al. 2006).

CORYNOCARPACEAE

Genus 1, **species** 6. **Trees**. **Stipules** absent. **Leaves** alternate; lamina simple, margins entire. **Inflorescences** panicles, terminal. **Flowers** perfect, regular, 5-merous; sepals and petals imbricate; petals adnate to base of sepals; stamens perigynous, opposite petals, alternating with 5 subpetaloid staminodes, anthers 2-celled, opening lengthwise; disk-glands prominent; ovary superior, 1–2-celled, styles 1–2, free, ovule solitary, pendulous. **Fruits** drupes; seed lacking endosperm. **Distr:** New Guinea to New Zealand.

Corynocarpus

Species 6. *Generic description same as familial.* **Cult:** Propagate from seeds or cuttings. Requires fertile, moist, well-drained soil in full or partial sun.

C. laevigatus J.R. et G. Forst., *Char. Gen. Pl.* 32, t. 16, 1776. [New Zealand Laurel]. **Trees**, of canopy, to ~15 m tall; trunk to ~60 cm dbh; branches stout. **Leaves** elliptic- to obovate-oblong, (5–)10–15(–20) × (3–)5–7 cm, margins recurved, thickly coriaceous, dark green, glossy; petioles stout, ~1–1.5 cm long. **Inflorescences** panicles, stiff, stout, 10–20 cm long; peduncles and pedicels short. **Flowers** 4–5 mm Ø; sepals suborbicular; petals greenish-yellow, ~5 mm long, minutely crenately toothed, obovate-spathulate; staminodes spathulate, minutely toothed; style ~2 mm long, thick, stigma capitate. **Fruits** drupes, ellipsoid to ovoid, 2.5–4 cm long, orange; seed with distinct testa-veins, cotyledons plano-convex, fleshy. **Distr:** New Zealand. **Ecol:** In coastal and lowland forests. Fl. Aug–Nov. Fr. Jan–Apr. **Zones:** 9–11. **Info:** Weedy in Hawai'i.

CORIARIACEAE

Genus 1, **species** 5. *Familial description same as generic.*

Coriaria

Species 5. **Subshrubs**, **shrubs** decumbent, or sometimes **small trees**, polygamous or with perfect flowers; branchlets quadrangular or ribbed; buds usually perulate. **Stipules** small. **Leaves** whorled, opposite or exstipulate; lamina simple, ± flat. **Inflorescences** racemes; bracteoles absent. **Flowers** 6-merous; sepals imbricate, persistent; petals keeled on inside; stamens 10, free, or the petal-opposed ones adnate, anthers large; ovaries superior, carpels free, 5–10, placentation apical, 1-celled, carpels 1-ovuled, ovules pendent; receptacles conical, styles free, elongate. **Fruits** capsules (or pseudo-drupes) of 5–10 individual cocci surrounded by the enlarged, succulent petals; seeds compressed. **Distr:** Pacific region to China and Himalayas, Mediterranean. **Cult:** Propagate from seeds, cuttings or by division.

C. arborea Lindsay, *Contr. NZ. Bot.* 84, 1868. (Syn: *C. sarmentosa*). [Tutu]. **Shrubs**, stout, much-branched from base, to ~6 m tall; trunk short, inclined, to ~30 cm dbh, rarely a tree with erect cylindric trunk to 3 m tall. **Leaves** opposite, or occasionally in whorls of 3; lamina ovate to ovate-oblong, (3–)5–8(–10) × (2.5–)4–5(–6) cm, base subcordate to rounded, apex abruptly acuminate or acute, subcoriaceous, adaxially dark shiny green, abaxially paler, glabrous, lateral veins 2–4; petiole stout, 1–3 mm long. **Inflorescences** racemes (10–)15–20(–30) cm long, slender, drooping, pubescent, flowers numerous, distant to closely set; pedicels patent, very slender, to ~5 mm long. **Flowers** with sepals broadly ovate to ovate-oblong, 3–4 mm long, ± crenulate; petals ovate-oblong, ~1.5 mm long. **Fruits** with ripe carpels 5, surrounded by enlarged, succulent, dark purple petals ~4 mm long. **Distr:** New Zealand. **Ecol:** In coastal to montane forests. **Zones:** 9–11. **Info:** All parts of the plant are highly toxic.

CUCURBITACEAE

Genera 118, **species** 845. **Herbs** annual or perennial, or **undershrubs**, prostrate or climbing; tendrils simple or branched. **Leaves** exstipulate, petiolate, alternate; lamina entire, palmately , pedately or pinnately divided, rarely compound. **Inflorescences** paniculate, racemose, umbellate or flowers solitary. **Flowers** unisexual (plants then monoecious or dioecious), very rarely bisexual, actinomorphic,

Above: *Corynocarpus laevigatus, Coriaria arborea, Benincasa hispida, Citrullus lanatus.*

yellow or white; sepals mostly 5, united into a rotate, campanulate, saucer-shaped or tubular calyx, adnate to and often produced beyond the ovary in ♀ flowers, with 5, imbricate or open lobes; petals mostly 5, free or united into a rotate or campanulate corolla, with 5, imbricate or induplicate-valvate, entire or rarely fimbriate lobes; stamens basically 5, of which 4 mostly connate in pairs and appearing as 1–3 stamens; staminodes often present in ♀ flowers; carpels (2–)3(–5), syncarpous, ovary inferior to semi-inferior, mostly unilocular, rarely more loculed, ovules numerous, style simple, rarely (2–)3, free, stigmas bilobed, thick. **Fruits** berries or pepos, elongated or globose, rarely capsules, indehiscent or variously dehiscent, smooth or covered with tubercles or prickles; seeds mostly flattened, occasionally winged, exalbuminous. **Distr:** Mainly tropical and subtropical, especially in drier parts of Africa.

Benincasa

Species 1. *Generic description same as specific.* **Cult:** Propagate from seeds. Requires a moist, well-drained soil in full or partial sun.

B. hispida (Thunb.) Cogn., *Monogr. Phan.* 3: 513, 1881. (Syn: *B. cerifera, Cucurbita hispida*). [Chinese Squash, Wax Gourd, White Pumpkin, etc.]. **Climbers**, herbaceous, annual, large, softly hairy, monoecious, much-branched. **Tendrils** slender, usually 2–4-fid. **Probracts** opposite the tendril at base of petiole, reniform-ovate, entire. **Leaves** cordate, reniform-orbicular, 10–25 × 10–25 cm, palmately ± 5–7-lobed, lobes ovate-triangular, base deeply cordate, margins sinuate, dentate or crenulate, adaxially pilose or scabrous, abaxially shortly hispid; petiole eglandular, long-villous. **Inflorescences** axillary, flowers solitary; ♂ 5–15 cm long; ♀ 2–4 cm long; peduncles hirsute. **Flowers** large, yellow; ♂ with calyx tube broadly campanulate, 1–1.5 cm long, 5-lobed, lobes leaflike, lanceolate, 0.6–1.2 cm long, serrate, acute, densely hairy; corolla subcampanulate, petals oblong-ovate, 3–5 × 2–4 cm, obtuse-mucronate, stamens 3; ♀ with ovary oblong-ovoid, 2–4 cm long, many-ovuled, style 2–3 mm long. **Fruits** large, fleshy, globose or oblong, baccate, 25–50 × 10–25 cm, densely hairy when young, waxy when mature, indehiscent; seeds many, ovoid-oblong, ~1 × 0.6 cm, compressed, white. **Distr:** India, Pakistan; China, Japan; Malaya, E Australia, Polynesia. **Ecol:** Natural habitat unknown. Fl. & Fr. throughout most of the year. **Zones:** 9–12.

Citrullus

Species 3. **Herbs**, trailing or climbing, annual or perennial, monoecious. **Tendrils** simple, or proximally 2–3(–4)-fid. **Leaves** simple, usually deeply bipinnately lobed. **Inflorescences** with flowers solitary. **Flowers** yellow; calyx short, campanulate, lobes moderately large; petals 5, united below middle; stamens inserted on the tube, 3, two 2-thecous, one 1-thecous, free, connective broad, thecae flexuous; ovary hairy, subglobose, many-ovuled, horizontal; staminodes present; stigma trilobed. **Fruits** medium to large, globose or subspherical, greenish, pericarp firm, mesocarp fleshy, indehiscent; seeds many, usually ovate, compressed, smooth or slightly rough. **Distr:** Africa, Asia. **Cult:** Propagate from seeds, cuttings or by layering. Requires a fertile, moist, well-drained soil in full sun.

C. lanatus (Thunb.) Mats. & Nakai, *Cat. Sem. Spor. Hort. Bot. Univ. Imp. Tokyo.* 30, 1916. (Syn: *C. vulgaris*, etc.). [Water Melon]. **Climbers** or **trailers**, annual, monoecious; stem herbaceous, villous. **Tendrils** robust, pubescent, bifid to trifid. **Probracts** obovate-spathulate, 0.8–1 cm long. **Leaves** ovate, deeply 3–5-lobed, lobes elongate-ovate, obtuse, ± pinnately sinuate-lobulate, margins minutely denticulate, adaxially smooth, abaxially hirsute; petiole long-villous. **Inflorescences** with flowers solitary; ♂ flowers with pedicels ~2(–4) cm long; ♀ flowers with pedicels ~6 cm long. **Flowers** ♂ with calyx broadly campanulate, ~7 mm long, lobes lanceolate, ~5(3–5) mm long, petals yellow, obovate-oblong, (0.7–)1.2(–1.6) × ~0.4 cm; ♀ with calyx and corolla as in ♂, ovary ellipsoid or subglobose, hairy. **Fruits** large, ellipsoid or subspherical, to >30 cm Ø, hairy, green mottled with paler green and yellowish ± longitudinal stripes; mesocarp fleshy, red or yellow, rarely white; indehiscent; seeds ovate, flat, ~1 × 0.5 cm, black, rarely red or of different shades, smooth or slightly verrucose. **Distr:** Africa (Kalahari). **Ecol:** In very dry arid regions. Fl. Jan–May. **Zones:** 10–12.

Coccinia

Species ~30. **Climbers** and **trailers**, dioecious; stem glabrous or scabrous. **Tendrils** simple, rarely bifid. **Leaves** petiolate; lamina simple, very variable, 5-angular or 5-lobed, deltoid or rounded-cordate, adaxially glabrous or punctate, abaxially rarely glandular. **Inflorescences** ♂ with flowers solitary, clustered or shortly racemose, rarely in short spikes; ♀ flowers solitary, rarely racemose. **Flowers** medium to large, white or brownish-yellow, often green-veined; ♂ with calyx tube short, campanulate or turbinate, teeth 5, corolla large, campanulate, shortly or acutely 5-lobed, stamens 3, inserted at base of calyx tube, filaments connate into a central column, rarely free, pistillode absent; ♀ with calyx and corolla as in ♂, staminodes 3, oblong or subulate, ovary ovoid, oblong to linear, triplacentiferous, smooth, style slender, stigmas 3. **Fruits** small to large, globose to elongate, indehiscent; seeds many, small, ± compressed, ovate, testa fibrillose. **Distr:** Tropical Africa, S Africa; Tropical India and Malaysia. **Cult:** Propagate from seeds. Requires a moist, well-drained soil in full sun.

C. grandis (L.) Voigt, *Hort. Suburb. Calc.* 59, 1845. (Syn: *Bryonia grandis*, etc.). [Ivy Gourd, Scarlet-fruited Gourd]. **Climbers**, perennial, widely spreading, sometimes prostrate; stem much-branched, cylindrical, glabrous or white scaly. **Probracts** small. **Leaves** 5–10 × 5–10 cm, base cordate, margins usually entire to 5-angular, with few glistering glands on both sides of the midrib towards base; petiole slender, striate, 2–5 cm long. **Inflorescences** with flowers pedicellate, solitary, ♂ and ♀ on different plants. **Flowers** large, white, ~4 cm across. **Fruits** cylindrical or fusiform, 2.5–5 × 1.2–2.5 cm, bright scarlet when ripe, slightly beaked; seeds many, asymmetrically pyriform in outline, ~6 × 4 mm, compressed. **Distr:** India, Malaya, Pakistan, tropical Africa. **Ecol:** Fl. Dec–Jun. **Zones:** 10–12.

Cucumis

Species ~25. **Herbs**, annual or perennial, climbing or trailing, monoecious, rarely dioecious; stems and leaves mostly scabrid or hispid. **Tendrils** simple. **Leaves** variously 3–7-lobed, margins dentate or serrate. **Inflorescences** with ♀ flowers solitary; ♂ flowers mostly fascicled. **Flowers** with calyx campanulate, lobes 5, subulate; corolla yellow, turbinate, rotate or subcampanulate, 5-lobed, lobes mostly oblong or ovate, acute; stamens 3, free, inserted at calyx base, 1–2-celled, filaments short, connectives produced, bifid; pistillode glandular; staminodes setiform or ligulate; ovary ovoid or globose, ovules horizontal, numerous, style short, simple, surrounded by annular disk, stigmas 3(–5), obtuse, spherical or obovate. **Fruits** smooth, echinate or spinous; seeds smooth, compressed. **Distr:** Mostly in tropical Africa. **Cult:** Propagate from seeds. Requires a moist, well-drained soil in full or partial sun.

C. melo L., *Sp. Pl.* 2: 1011, 1753. [Cantaloupe, Honeydew, Melon, Muskmelon]. **Herbs**, annual, creeping; stems and branches scabrous, puberulent or hispid. **Tendrils** filiform. **Leaves** suborbicular or reniform, 6–15 × 6–15 cm, undivided or 3–7-lobed, dentate, sinus truncate or rounded, papery, adaxially scabrous, abaxially hispid; petioles 7–12 cm long, setose. **Inflorescences** ♂ fasciculate, 1–several-flowered, pedicels filiform, 0.5–4 cm, pubescent; ♀ flowers solitary. **Flowers** monoecious or bisexual; ♂ with calyx tube narrowly campanulate, densely hispidulous to white-villous, subulate or linear, 3–8 mm long, corolla yellow, tube 1–2 mm long, segments ovate-oblong to broadly ovate or obovate, 0.3–2.4 × 0.25–2 cm, obtuse, stamens 3; ♀ with ovary ellipsoid or fusiform, densely white-lanate, villous or puberulent, 0.4–1.1 × 0.2–0.5 cm, stigmas 3. **Fruits** variable; seeds numerous, yellow-white, ovate-oblong, 4–8 × 2.4–4 mm. **Distr:** Tropical and subtropical Old World. **Ecol:** Fl. & Fr. May–Sep. **Zones:** 9–12.

C. sativus L., *Sp. Pl.* 1012, 1753. [Cucumber, Gherkin, Sikkim Apple, Snake Cucumber]. **Herbs**, trailing or climbing, annual; stem angular, with scabridulous hairs. **Leaves** ovate, 12–18 cm long,

Above: *Coccinia grandis, Cucumis melo, Cucumis sativus, Cucurbita maxima.*

3–8-lobed, lobes acute, hispidulous or scabridulous on both surfaces; petiole elongate, 0.8–2 cm long. **Inflorescences** ♂ with flowers mostly fascicled; ♀ flowers solitary, rarely fascicled. **Flowers** with sepals spreading; corolla ~2.5 cm long, lobes oblong-lanceolate; anthers 3–4 mm long; ovary fusiform, muricate. **Fruits** oblong and obscurely trigonous, or cylindric, when young sparsely tuberculate, otherwise smooth and glabrous. **Distr:** Origin unknown; cultivated widely throughout the tropics and subtropics. **Ecol:** Fl. throughout most of the year. **Zones:** 9–11.

Cucurbita

Species ~20. **Herbs**, annual or perennial, monoecious; stem angular, hispid or hairy; tendrils branched, mostly 2–4-fid. **Leaves** petiolate; lamina ovate, generally 5-lobed, base cordate, hirsute or scabrescent. **Inflorescences** ♂ flowers rarely fascicled; ♀ flowers solitary, shortly pedunculate. **Flowers** large, yellow; ♂ with calyx campanulate, lobes 5, subulate, spathulate or leafy, corolla campanulate, shallowly 5-lobed, lobes usually recurved, stamens 3, inserted low on calyx tube, filaments free, anthers linear, connate, 1 monothecous, 2 dithecous, pistillode absent; ♀ with ovary oblong, style simple, short, stigmas 3(5), papillose, ovules many, horizontal. **Fruits** fleshy, indehiscent; seeds many, ovate or oblong, smooth, marginate or emarginate. **Distr:** Tropical America. **Cult:** Propagate from seeds, cuttings or by air-layering. Requires a fertile, moist, well-drained soil in full sun.

C. maxima Duch. ex Lam., *Encycl. Meth. Bot.* 2: 151, 1786. [Pumpkin, Red Gourd, Squash]. **Herbs**, climbing or sprawling, somewhat robust, annual. **Leaves** shallowly 5-lobed or subentire, coarsely hairy on both surfaces with hispidulous soft hairs, margins densely and minutely dentate; petiole hairy, not pungent. **Inflorescences** with flowers mostly solitary. **Flowers** faintly scented; calyx lobes linear; corolla campanulate, ~10+ cm across, yellow, petals ruffled. **Fruits** generally oval or rounded, bluntly ribbed, large, to ~2 m Ø; flesh yellow or reddish when ripe; seeds white, brown or bronze, strongly marginate; fruiting peduncles circular, striated, corky or spongy, not expanded at the attachment. **Distr:** Probably from S America; cultivated widely in the tropics and subtropics. **Ecol:** Natural habitat unknown. Fl. & Fr. throughout most of the year. **Zones:** 8–11.

Lagenaria

Species 6. **Herbs**, annual or perennial, scandent or trailing, monoecious or dioecious. **Tendrils** bifid. **Leaves** simple, palmately 3–5-lobed, ovate, cordate, suborbicular; petiole with 2 glands at apex. **Inflorescences** ♂ peduncled, flowers solitary or racemose; ♀ flowers solitary. **Flowers** large, white; ♂ with calyx tube campanulate or funnel-shaped, lobes 5, small, distant, petals 5, free, obovate or oblong-obovate, retuse, stamens 3, anthers included, connate or free, oblong, greenish-gray, oblong; ♀ with calyx tube cup-shaped, petals similar to ♂, staminodes 3, obsolete, ovary ovoid or cylindric, triplacentiferous, style thick, short, stigmas 3, bilobed, thick, ovules numerous, horizontal. **Fruits** large, epicarp hard, mesocarp fleshy, indehiscent, polymorphous; seeds many, oblong, ovate-oblong, compressed, margins grooved. **Distr:** Mostly tropical Africa with 1 sp. pantropical. **Cult:** Propagate from seeds. Requires a moist, well-drained soil in full sun.

L. siceraria (Molina) Standley, *Publ. Field Mus. Nat. Hist. Chicago, Bot. Ser.* 3: 435, 1930. (Syn: *Cucurbita lagenaria*, etc.). [Bottle Gourd, Calabash]. **Climbers** or **trailers**, annual; stem densely hairy. **Tendrils** bifid. **Leaves** palmate, broadly ovate, reniform or suborbicular, cordate, obscurely 3–5-lobed, lobes rounded, 3–23 × 4–23 cm; petiole 2.7–13 cm long with a pair of small apical lateral conical glands. **Inflorescences** with flowers solitary; ♂ peduncles as long as petiole; ♀ peduncles shorter. **Flowers** with corolla white; ♂ with petals obovate, apiculate, 2.7–4.5 × 1.8–3.7 cm, staminal filaments 3–4 mm long, anthers oblong, coherent, included; ♀ with petals ~3 × 2.4 cm, ovary ovoid, 1.1–1.7 × 0.6–0.8 cm, villous. **Fruits** large, subglobose, ellipsoid, lageniform, sometimes biventricose, greenish, hairy, indehiscent; seeds white, oblong, compressed, slightly 2-horned, 0.7–2 cm long. **Distr:** Pantropical, possibly of Asian and African origin; cultivated widely. **Ecol:** Fl. Mainly Mar–May. **Zones:** 10–12.

Luffa

Species ~6. **Herbs**, annual, scandent, climbing or trailing, monoecious or dioecious, scabrous or hairy. **Tendrils** proximally (2)3–6-fid. **Leaves** simple, palmately 5–7-lobed, rarely subentire, sometimes gland-dotted; petiole eglandular. **Probracts** axillary, glandular. **Inflorescences** ♂ racemose; ♀ flowers solitary. **Flowers** medium to large, yellow; ♂ with calyx tube shortly campanulate, lobes relatively large, enclosing petals in bud, entire, petals 5, free, stamens 3(5), all monothecous or one 1-locular, others bilocular, filaments inserted on the tube, free, anthers free, connective broad, thecae marginal, much-convoluted; ♀ with calyx tube short, ovary smooth, ribbed, tuberculate or spiny, stigmas 3, ovules many, horizontal. **Fruits** medium-sized to large, subglobose to elongate-cylindrical, beaked, smooth, ribbed or sparsely to densely spiny with short to long spines, brownish, dry and fibrous, dehiscent by an apical operculum; seeds oblong-elliptic, compressed. **Distr:** Tropical regions with unknown origin. **Cult:** Propagate from seeds. Requires a moist, well-drained soil in full sun.

L. acutangula (L.) Roxb., *Flora Indica* 3, 1832. [Angled Loofah, Ribbed Loofah]. **Climbers**, annual; stem 5-angled, scabrous along ribs. **Tendrils** mostly trifid, subscabrous. **Leaves** roundish, palmately 5–7-lobed, 15–20 cm across, glabrous, pale green; petiole 8–12 cm long. **Probracts** fleshy with 3–10, glistening glands on upper surface, 3–7 × 2–4 mm. **Inflorescences** ♂ racemes, 10–20-flowered, erect, 10–15 cm long, axillary; ♀ with flowers solitary in the same axil as ♂; flowers pedicellate. **Flowers**

Below: *Lagenaria siceraria, Luffa acutangula* (fruits), *Luffa cylindrica, Momordica charantia.*

PL

Above: *Momordica charantia* (fruits), *Momordica foetida*, *Sechium edule* (flowers), *Sechium edule* (fruit).

pale yellow, ~4.5 cm Ø; calyx tube pentagonal, slightly hairy, lobes equal to slightly longer than tube; petals obcordate, ~2 × 2–2.5 cm, emarginate or not; stamens 3, filaments ~4 mm long; ovary 10-angled. **Fruits** (10–)15–30(–45) × 6–10 cm, acutely 10-angled; seeds black, 1–1.2 × 0.6–0.8 cm, rugose, emarginate. **Distr:** Origin unknown; cultivated widely. **Ecol:** In warm tropical areas. Fl. & Fr. throughout most of the year. **Zones:** 10–12.

L. cylindrica (L.) Roem., *Syn. Monogr.* 2: 63, 1846. (Syn: *L. aegyptiaca*, etc.). [Loofah, Sponge Gourd]. **Climbers** or **trailers**, annual; stem 5-angled. **Tendrils** 3–6-fid, slightly pubescent. **Leaves** palmately 5-lobed, orbicular-cordate, 8–25 cm across, lobes triangular, lanceolate, margins entire or sinuate, apex acute-apiculate, dark green, scabrous; petiole 5–15 cm long. **Probracts** fleshy, ovate, with 3–7 glistening glands on upper surface. **Inflorescences** ♂ racemose, 15–20-flowered, axillary, 12–25 cm long; ♀ flowers in the same axil as ♂; flowers pedicellate. **Flowers** bright yellow, 5–6 cm Ø; calyx tube short, broadly campanulate, slightly pubescent, lobes triangular-lanceolate, longer than tube; petals obovate-cuneiform, 2.5–3.5 × 1–2.5 cm, obtuse; stamens 3–5, filaments 6–8 mm long; ovary cylindrical, finely hairy. **Fruits** cylindrical, fusiform, 20–50 × 6–10 cm, smooth; seeds dull black, elliptic-ovoid, ~1–1.2 × 0.6–0.8 cm, margin ~1 mm wide. **Distr:** Asia and Africa. **Ecol:** Fl. Apr–Oct. **Zones:** 9–12.

Momordica

Species ~45. **Climbers** or **trailers**, annual or perennial, monoecious or dioecious. **Tendrils** simple or bifid. **Leaves** simple, palmately lobed or 3–9-pedately foliate, margins mucronate-dentate or lobulate-dentate or undulated, remotely denticulate. **Inflorescences** ♂ flowers solitary, corymbose, racemose or umbellate; ♀ flowers solitary; peduncle mostly with large foliaceous bracts. **Flowers** small or large, yellow, rarely white; ♂ with calyx tube closed at bottom by 2–3 incurved oblong scales, lobes ovate-lanceolate, entire, corolla rotate or broadly campanulate, deeply 5-partite, lobes not fimbriate, obovate or oblong or petals free, stamens 3, rarely 2, inserted at mouth of calyx tube, one 1-thecous, others bithecous, pistillode absent or glandular; ♀ with perianths as in ♂, staminode absent or glandular, ovary oblong or fusiform, ovules many, usually horizontal. **Fruits** oblong-fusiform, cylindric, baccate, tuberculate or muriculate, indehiscent or dehiscent by 3-valves or irregularly exposing seeds enveloped in scarlet pulp; seeds turgid or flattened, smooth or sculptured, grooved at margins. **Distr:** Paleotropical (mostly in Africa). **Cult:** Propagate from seeds. Requires a moist, well-drained soil in full sun.

M. charantia L., *Sp. Pl.* 1009, 1753. (Syn: *M. muricata*, etc.). [Balsam Apple, Balsam Pear, Bitter Gourd]. **Climbers** or **trailers**, annual, herbaceous, monoecious; stem glabrous or hairy. **Tendrils** unbranched. **Leaves** suborbicular to orbicular, 5–10(–12) cm Ø, mucronate-dentate, deeply 5–7-lobed, lobes ovate-oblong or ovate-elliptic, base constricted, apex acute, glabrous or pubescent; petiole 1.5–3.5(–6) cm long, villous. **Inflorescences** with flowers solitary; ♂ peduncles ~7 cm long, bearing bracts ± about middle. **Flowers** yellow, ~3 cm Ø; calyx pubescent; corolla slightly zygomorphic, lobes obovate, obtuse; ovary fusiform, muricate. **Fruits** oblong-fusiform or ovoid, 7–25 cm long, dehiscent; seeds oblong, 1–1.5 × 0.6–0.9 cm, embedded in red pulp, sculptured. **Distr:** Tropical & S Africa, SE & Far East Asia to Australia. **Ecol:** Fl. Apr–Jul. **Zones:** 9–12.

M. foetida Schumach., *Beskr. Guin. Pl.*: 426, 1827. (Syn: *M. schimperiana*, etc.). [Wild Cucumber]. **Herbs**, perennial, dioecious, fetid, prostrate or scandent, to 7 m long; rootstock tuberous; stems and petioles green-flecked. **Tendrils** simple or bifid. **Leaves** ovate-triangular to broadly ovate, 1.3–19 × 1.3–18 cm, base cordate, margins minutely denticulate to strongly sinuate-denticulate, unlobed, apex acuminate, apiculate; petiole 1.3–17 cm long. **Inflorescences** ♂ subumbelliform, 1–8-flowered, peduncle 2–23 cm long, bract 0.3–3 cm long, pedicels 0.2–7 cm long; ♀ flowers solitary, peduncle 0.2–6.5 cm long, bract 0.3–2 cm long, pedicel 0.2–18.6 cm long. **Flowers** ♂ with calyx lobes usually ovate, 0.5–1 cm long, blackish, petals 1.7–3.5 cm long, pale yellow to orange, base deeper orange to black, obovate, rounded; ♀ with calyx lobes usually ovate, 0.2–1.1 cm long, petals 1.5–3.7 cm long, ovary ellipsoid-rostrate, 0.8–2.6 × 0.3–1.1 cm, densely muricate. **Fruits** ovoid or ellipsoid, 3.5–7.5 × 2.5–5 cm, softly spiny, bright orange, dehiscent into 3 valves; seeds ovate-oblong or elliptic-oblong, 0.7–1.2 × 0.5–0.75 cm, enveloped in bright red pulp. **Distr:** Tropical Africa, S Africa. **Ecol:** In rainforests, grasslands; 350–2,250 m. **Zones:** 10–12.

Sechium

Species ~5. **Herbs**, perennial scandent; roots tuberous. **Tendrils** 3–5-fid. **Leaves** cordate, membranous. **Inflorescences** ♂ racemes; ♀ flowers solitary or paired, usually in same axil as ♂. **Flowers** monoecious, white; ♂ with calyx tube semiglobose, segments 5, corolla rotate, 5-partite, segments ovate-lanceolate, acute, stamens 3, inserted on base of calyx tube, filaments connate, anthers free, one 1-locular, two bilocular, locules flexuous, pistillode absent; ♀ with calyx and corolla as in ♂, ovary fusiform, 1-locular, setose, style short, stigma 5-lobed, lobes reflexed, with only one pendulous ovule. **Fruits** fleshy, obovate, sulcate; seed 1, usually large, ovate, compressed, woody. **Distr:** Mexico, C America; one species cultivated worldwide. **Cult:** Propagate from seeds. Requires a moist, well-drained soil in full sun.

S. edule (Jacquin) Swartz, *Fl. Ind. Occ.* 2: 1150, 1800. (Syn: *Sicyos edulis*). [Spiny Fruited Chayote, Vegetable Pear]. **Herbs**, perennial, scandent; stem angular-sulcate. **Leaves** spirally arranged; lamina nearly orbicular, 7–25 cm Ø, base cordate, margins dentate, apex acuminate, membranous, adaxially slightly scabrous, abaxially puberulent; petioles slender, 5–15 cm, glabrous. **Inflorescences** ♂ 10–30-flowered, peduncle 8–30 cm long, pedicels 1–6 mm long; ♀ flowers usually solitary, pedicel 1–1.5 cm long. **Flowers** ♂ with calyx tube short, segments explanate, 5–7 mm long, corolla rotate, 1.2–1.7 cm long, segments ovate-lanceolate; ♀ with calyx and corolla as in ♂, ovary obovate, 5-angled, style 2–3 mm long, stigma ~2 mm long. **Fruits** pale green, obovate, slightly hispid, 8–15 × 6–8 cm, 5-sulcate in upper part; seed large, ovate, ~10 × 7 cm, compressed. **Distr:** Mexico. **Ecol:** Fl. Jul–Sep. Fr. Aug–Oct. **Zones:** 9–12.

Trichosanthes

Species ~40. **Herbs**, annual or perennial, climbing, dioecious or monoecious. **Tendrils** bifid–trifid or unbranched. **Leaves** palmately 3–5(–9)-lobed, base cordate, margins entire or denticulate. **Inflorescences** ♂ racemose, rarely solitary, often bracteate; ♀ flowers solitary. **Flowers** white; ♂ with calyx tube oblong-cylindrical, expanded above, lobes 5, lanceolate, entire or serrate, corolla deeply 5-lobed, rotate, lobes oblong or lanceolate, fimbriate, stamens 3, inserted on calyx tube, filaments short, anthers syngenesious, rarely free, staminodes absent, style slender, stigmas 3, pistillodes 3, filiform; ♀ with ovary oblong-ovoid, fusiform, globose, ovules many, horizontal or semipendulous. **Fruits** oblong, ovoid-globose or fusiform, large, indehiscent, glabrous, smooth; seeds many, ellipsoid or oblong, packed in pulp. **Distr:** C Asia eastward to Indo-Malaysia and Australia. **Cult:** Propagate from seeds. Requires a moist, well-drained soil in full sun.

T. cucumerina L., *Sp. Pl.* 2: 1008, 1753. (Syn: *T. anguina*). [Club Gourd, Serpent Cucumber, Snake Gourd]. **Climbers**, stout, annual; stem slightly pubescent. **Tendrils** (2)3–fid, puberulous. **Leaves** orbicular-reniform or broadly ovate, 7–10 × 8–12 cm, 5(–7)-lobed, lobes acute or obtuse, denticulate; petiole 2–7 cm long, pubescent. **Inflorescences** ♂ racemose, peduncles paired, one 1-flowered, others many-flowered; ♀ flowers solitary; flowers pedicellate. **Flowers** white; calyx tube broadened at apex, ~3 mm across, lobes minute; petals lanceolate-oblong; staminal filaments ~2 mm long; ovary oblong, style 1.5–1.8 cm long. **Fruits** ovoid-fusiform, 5–6 × 3.5–4 cm, pointed at both ends, white-striped when young, yellow when mature; seeds ovate-oblong, 1.1–1.2 × 0.6–0.7 cm, enclosed within red pulp. **Distr:** Bangladesh, India, Malaysia, Myanmar, N Australia, Pakistan, Philippines, Sri Lanka. **Ecol:** Fl. Jun–Oct. **Zones:** 10–12.

BEGONIACEAE

Genera 2, species 1,401. **Herbs**, perennial, succulent, very rarely subshrubs; stem erect, frequently rhizomatous, or plants tuberous and either acaulescent or shortly stemmed, rarely lianoid or climbing with adventitious roots, or stoloniferous. **Stipules** usually deciduous **Leaves** alternate or all basal, petiolate; lamina simple, rarely palmately compound, often oblique and asymmetric, rarely symmetric, margins irregularly serrate and divided, occasionally entire, venation usually palmate. **Inflorescences** (1 or)2–4 to several, rarely numerous in dichotomous cymes, sometimes in panicles, with pedicels and bracts. **Flowers** unisexual, plants monoecious, rarely dioecious; ♂ with tepals 2 or 4 and decussate, usually outer ones larger, inner ones smaller, stamens usually numerous, filaments free or basally connate, anthers 2-celled, apical or lateral; ♀ with tepals 2–5(–10), usually free, rarely basally connate, ovary nodding, pendulous, or ascending, 1–3(4–8)-loculed, placentation axile or parietal, styles 2 or 3(or more), free or basally fused, forked once or more, stigma turgid, spirally twisted-tortuous or U-shaped, capitate or reniform and setose-papillose. **Fruits** capsules, dry, sometimes berrylike, unequally or subequally 3-winged, rarely wingless and 3- or 4-horned; seeds numerous, minute, oblong, testa pale brown, reticulate. **Distr:** Tropical and subtropical regions of the world.

Begonia

Species >1,400. **Herbs**, rarely subshrubs, perennial, succulent; stem erect, frequently rhizomatous, or plants tuberous and either acaulescent or shortly stemmed, rarely lianoid or climbing with adventitious roots, or stoloniferous. **Stipules** membranous, usually deciduous. **Leaves** alternate or all basal; lamina simple, rarely palmately compound, often oblique and asymmetric, rarely symmetric, margins often irregularly serrate and divided, occasionally entire, venation usually palmate; petiole long, weak. **Inflorescences** with flowers (1 or)2–4 to several, rarely numerous in dichotomous cymes. **Flowers** unisexual, plants monoecious, rarely dioecious; ♂ with tepals 2 or 4 and decussate, filaments connectives extended at apex, sometimes apiculate; ♀ with tepals 2–5(–10), pistil 2–5(–7)-carpellate, ovary inferior, 1–3(–7)-loculed, placentation axile or parietal, styles 2 or 3(or more), free or basally fused, forked once or more. **Fruits** capsules, dry, sometimes berrylike, unequally or subequally 3-winged, rarely wingless and 3- or 4-horned; seeds numerous. **Distr:** Tropical and subtropical regions worldwide; highest diversity in tropical America. **Cult:** Propagate from seeds, leaf cuttings, or by division of offsets. Requires a fertile, humus-rich, moist soil in partial sun or shade.

B. bowerae Ziesenhenne, *Begonian* 17: 78, 1950. [Eyelash Begonia]. **Herbs**, perennial, creeping; rhizome green to purple, branched. **Stipules** persistent, ovate to triangular-ovate, 0.5–1.2 × 0.5–0.9 cm. **Leaves** asymmetric, ovate, 2–11.5 × 1.5–10 cm, base cordate, margins wavy, angular-lobed, minutely saw-toothed, with several long white cilia, apex acuminate, adaxially green or green with separate purple to almost black markings around margins, sometimes next to main veins, hairless, abaxially paler, sparsely hairy, venation palmate; petiole pale green with purple lenticels, sometimes purplish throughout, 4–26 cm long, hairy. **Inflorescences** cymose, few-flowered, bisexual; bracts persistent, ovate to broadly elliptic, 0.15–1.3 × 0.1–1.4 cm, margins ciliate; ♀ flowers with 2 bracts beneath ovary, persistent, elliptic, 1.2–3 × 0.5–1 mm. **Flowers** ♂ with tepals 2, white, sometimes red-spotted, obovate to almost circular, 0.8–1.1 × 0.8–1.1 cm, stamens ~10–15; ♀ with tepals 2, white, obovate to almost circular, 5–9 × 0.5–8.5 mm, ovary green, sometimes wings pink-tinged, broadly ellipsoid to spherical. 4.5–5.5 × 4–5 mm, slightly unequally 3-winged, trilocular. **Distr:** Mexico. **Ecol:** In understory of tropical rainforests. Fl. & Fr. throughout most of the year. **Zones:** 9–12. **Info:** The variety ***B. bowerae* var. *nigramarga*** has small green leaves with black bands bordering the main veins.

B. coccinea W. J. Hooker, *Curtis's Botanical Magazine* 69: pl. 3990, 1843. (Syn: *B. rubra*). [Angel Wing Begonia, Cane Begonia, Scarlet Begonia]. **Herbs**, erect, perennial, to ~3 m tall; stem green, sometimes reddish-tinged, glabrous. **Stipules** soon caducous, ovate-lanceolate to elliptic, strongly concave, 2.5–3.4 × 1.1–1.4 cm. **Leaves** distichous; lamina asymmetric, ovate-elliptic, 6.5–14 × 3–7 cm, base shallowly cordate, lowermost lobe extended for 4–5.5 cm, margins wavy-toothed, apex acute to acuminate, venation palmate-pinnate, glabrous, adaxially green, abaxially paler; petiole green or yellowish-green, reddish-tinged, 1–4 cm long. **Inflorescences** cymose, 14–48-flowered, axillary; peduncle, rachis and pedicels reddish; bracts soon caducous, ovate or elliptic, 6–9 × 4–9 mm, concave. **Flowers** ♂ with tepals 4, red, *outer* pair ovate-cordate, 1.5–2 × 1.3–2.1 cm, *inner* pair ovate-elliptic to elliptic, concave, 0.8–1.7 × 0.4–0.5 cm, stamens 20–35; ♀ bracteate, tepals 5 or 6, ovate to oblong-ovate, ± equal, 0.6–1.6 × 0.3–1.7 cm, ovary red, ellipsoid or obovoid, 0.8–1.7 × 0.4–0.7 cm, wings 3, almost rounded. **Distr:** Brazil. **Ecol:** In coastal forested mountains, along streams, often growing as an epiphyte; 1,800–2,500 m. Fl. & Fr. throughout most of the year. **Zones:** 10–12.

B. listada L.B. Smith & Wasshausen, *Begonian* 49: 155, 1981. [Striped Begonia]. **Subshrubs**, erect, branched, non-rhizomatous, to 1 m tall; stem green to pinkish-purple, hairy. **Stipules** persistent, of 2 distinct sizes, one side ovate-elliptic, 1.8–2.3 × 1.6–2 cm, keeled, the other side ovate, 0.8–1.2 × 0.6–0.8 cm, without a keel. **Leaves** usually angular-ovate, occasionally hastate, 6–10 × 3.5–5 cm, base obliquely cordate, longer lobe acute, projecting, shorter lobe rounded, margins shallowly toothed, ciliate, apex acuminate, adaxially bronze with a pale yellowish-green band 0.5–1 cm wide along main vein, densely hairy, abaxially purple with 2 paler green veins, densely reddish-hairy; petiole pale green to pink, shortly hairy, 4.5–15 cm long. **Inflorescences** cymose, bisexual, few-flowered, axillary; bracts soon caducous, ovate to elliptic, 6–7 × 3–4 mm. **Flowers** ♂ with tepals 4, white, outer pair red-hairy, ovate-cordate to almost circular, 1.1–1.5 × 1.1–1.5 cm, inner pair narrowly obovate, 1–1.2 × 0.3–0.4 cm, stamens ~25–35, symmetrically arranged, anther connectives projecting; ♀ ebracteolate, tepals 5, white, sometimes red-tinged, elliptic, slightly unequal, 0.7–1.5 × 0.4–1 cm, ovary green pink-tinged, ellipsoid, 0.5–1 cm long, unequally 3-winged. **Distr:** Brazil. **Ecol:** Epiphytic in rainforests. **Zones:** 10–12.

B. masoniana Irmscher ex Ziesenhenne, *Begonian*. 38: 52, 1971. [Iron-cross Begonia]. **Herbs**; rhizomes creeping, 1.5–2.3 cm Ø. **Stipules** triangular, keeled. **Leaves** all basal; lamina obliquely broadly ovate to suborbicular, 10–25 × 9–20 cm, base deeply cordate, margin densely denticulate, apex acuminate or shortly caudate, adaxially with broad blackish-brown (red abaxially) bands along main veins, with many hairy-tipped pustules, sparsely setose, abaxially pale green, basally 5–7-veined; petiole 10–30 cm long. **Inflorescences** dichasial-cymes, axillary, 20–100-flowered; peduncles 10–54 cm long, sparsely pilose; pedicels 0.5–1.4 cm long; bracts and bracteoles ovate or oblong to lanceolate, 3–5 × 2–3 mm, membranous, margins ciliate, acute, caducous. **Flowers** ♂ with tepals 4, outer 2 broadly ovate or suborbicular, 0.5–1.1 × 0.6–1.2 cm, inner 2 oblong, 0.5–0.8 × 0.2–0.45 cm, obtuse; ♀ with tepals 3, outer 2 broadly ovate to suborbicular, (0.6–)0.8–1.2 × (0.4–)0.7–1.3 cm, inner 1 oblong or obovate, 5–7.5 × 2–4.5 mm, ovary red, styles 3. **Fruits** capsules, nodding, oblong to ellipsoid, 0.9–1.3 × 0.5–0.6 cm, reddish, glandular-hispid, unequally 3-winged. **Distr:** China, Vietnam. **Ecol:** On rocky limestone slopes; 100–300 m. Fl. Mar–Sep. Fr. Jun–Sep. **Zones:** 9–11.

B. scharffii J.D. Hooker, *Curtis's Botanical Magazine* 114: pl. 7028, 1888. (Syn: *B. haageana*). **Herbs**, erect, non-rhizomatous, perennial, to ~50 cm tall, densely hairy; stem branched, pale green with short purple bands above each node. **Stipules** persistent, ovate, 1.2–2.8 × 0.9–1.9 cm. **Leaves** asymmetric, ovate, 8–16 × 6–14 cm, base cordate, margins wavy, crenately lobed, irregularly toothed, ciliate, veins palmate-pinnate, adaxially green, abaxially paler, pink-tinged; petiole pale green to pale pink, 2.5–15 cm long. **Inflorescences** cymose, many-flowered, bisexual, in upper leaf axils; bracts deciduous, ovate 0.3–1.6 cm × 0.5–7.5 mm. **Flowers** ♂ with tepals 4, pale pink, broadly ovate to almost circular, 1.4–3 × 1.4–3 cm, inner pair obovate, 1.4–1.7 × 0.4–0.5 cm, stamens numerous, symmetrically arranged; ♀ bracteate, tepals 5, pale pink, ± oblong, 0.5–2.9 × 0.3–2.8 cm, inner 3

Below: *Trichosanthes cucumerina* (flowers), *Trichosanthes cucumerina* (fruits), *Begonia bowerae* var. *nigramarga*, *Begonia coccinea*.

Above: *Begonia listada, Begonia masoniana, Begonia scharffii, Castanopsis javanica.*
Right: *Casuarina equisetifolia.*

elliptic, 0.6–1.4 × 0.5–0.9 cm, ovary pinkish-white red-hairy, almost spherical, 3–7 × 2–5 mm, unequally 3-winged. **Distr:** Brazil. **Ecol:** In understory of rainforests. **Zones:** 10–12.

FAGALES

FAMILIES 8, GENERA 55, SPECIES 1,877.

Flowers unisexual; tepals much reduced or lacking; ovary usually inferior, ovules 1-integumented, 1 or 2 per locule; nectaries absent. **Fruits** indehiscent, 1-seeded (Judd et al. 2008). **Distr:** Cosmopolitan. **Info:** Monophyly of Fagales is supported by molecular data (Chase et al. 1993; Soltis et al. 2000).

FAGACEAE

Genera 7, **species** 670. **Trees**, rarely shrubs, monoecious, evergreen or deciduous. **Stipules** usually early deciduous. **Leaves** alternate, sometimes in pseudowhorls. **Inflorescences** unisexual or androgynous with ♀ cupules at base of an otherwise ♂ inflorescence; ♂ pendulous head, or erect or pendulous catkin, sometimes branched, flowers in dense cymules; ♀ with 1–7+ flowers subtended individually or collectively by a cupule formed from numerous fused bracts, arranged individually or in small groups along an axis or at base of an androgynous inflorescence or on a separate axis. **Flowers** ♂ with sepals 4–6(–9), scalelike, connate or distinct, petals absent, filaments filiform, anthers dorsifixed or versatile, opening by longitudinal slits, with or without a rudimentary pistil; ♀ with pistil 1, ovary inferior, 3–6(–9)-loculed, style and carpels as many as locules, placentation axile, locules 2-ovuled. **Fruits** nuts; seed usually solitary by abortion, endosperm absent, embryo large. **Distr:** Worldwide.

Castanopsis

Species 110. **Trees**, evergreen. **Stipules** extrapetiolar. **Leaves** alternate, distichous, or sometimes spirally arranged. **Inflorescences** usually unisexual, erect, spicate or paniculate; ♂ fascicles, 3–7-flowered, rarely flowers solitary and scattered; ♀ flowers solitary or in clusters of 3–5(–7) per cupule. **Flowers** ♂ with perianth 5- or 6(–8)-lobed; stamens (8 or)9–12, rudimentary pistil very small, densely hairy; ♀ with staminodes (when present) opposite perianth lobes, ovary trilaculed, styles (2 or)3(or 4), stigmas punctiform or shallow terminal pores, cupules solitary on rachis, actinomorphic or zygomorphic, rarely indehiscent, completely or partially enclosing nut, bracts sparsely to densely covering outside of cupule, spinelike or rarely scalelike or tubercles. **Fruits** nuts, 1–3 per cupule; abortive ovule apical. **Distr:** Tropical and subtropical Asia. **Cult:** Propagate from seeds or half-hardened cuttings. Requires a humus-rich, moist, well-drained soil in full sun.

C. javanica (Blume) A. DC., *J. Bot.* 1: 182, 1863. (Syn: *Castanea javanica, Quercus discocarpa*, etc.). [Turkey Oak]. **Trees**, to 36 m tall; buttresses frequent, equal to steep, to 1 m thick. **Leaves** oblong-elliptic, 7.5 × 2.5 – 19 × 8 cm, base sometimes unequal, apex long-tipped, to 2.5 cm long, adaxially pale gray-green, abaxially beige and scattered hairy, thinly leathery, secondary nerves 7–10 pairs bending towards the margins, prominent below, tertiary nerves ladderlike; petiole 0.5–1.25 cm long, slightly hairy, brown. **Inflorescences** ♂ terminal and axillary, hardly branched, golden-hairy; ♀ stout, unbranched usually with few fruits. **Flowers** ♂ white to pale yellow; cupule variable, often conical, 2–3.8 cm Ø, variably, sometimes sparsely spiny, spines chocolate-brown, sharp, branched, inside golden-hairy. **Fruits** triangular, chestnut-brown, often hairy. **Distr:** S Thailand to W Malesia. **Ecol:** In montane regions, to 1,500 m. **Zones:** 10–12.

CASUARINACEAE

Genera 4, **species** 95. **Trees**, rarely shrubs, evergreen; photosynthetic branchlets slender, wiry, with several very short basal segments, and 1 to numerous elongate segments; segments terete (quadrangular), with as many longitudinal ridges as leaves, ridges separated by furrows containing stomates. **Leaves** reduced to small teeth, arranged in whorls of (4–)6–17 at apex of each segment of photosynthetic branchlets. **Inflorescences** with flowers in alternating whorls, each flower subtended by toothlike bract and 2 bracteoles; bracteoles usually persistent, lateral, scalelike; ♂ and ♀ on same or different plants; ♂ catkinlike spikes, short to elongate; ♀ heads, globular to ovoid. **Flowers** unisexual; ♂ with sepals deciduous at anthesis, 1–2, hooded, scalelike, stamen 1, anthers basifixed, bilocular; ♀ without perianth, pistil 1, compound, bicarpellate, 1 fertile, the other usually reduced or absent, ovules 2, an additional 2 abortive ovules in reduced carpel, styles 2-branched, reddish. **Infructescences** ± woody, cylindric, conelike; floral bracteoles 2, enlarged as lateral valves. **Fruits** winged nuts (samaras), compressed; seeds 1 in each samara. **Distr:** Tropical and subtropical dry regions, warm temperate regions.

Casuarina

[She-oak].

Species 17. **Trees**, evergreen; young persistent branchlets distinguished from deciduous branchlets by shorter segments and differences in shape or size of leaves; furrows deep and closed, concealing stomates. **Leaves** reduced to small teeth. **Inflorescences** of alternating whorls of flowers, each flower subtended by toothlike bract and 2 bracteoles; bracteoles usually persistent, lateral, scalelike. **Flowers** unisexual. **Infructescences** pedunculate, pubescent at least when immature; bracts thin in exposed portion, not vertically expanded; bracteoles ± protruding from surface of infructescence, never greatly thickened, always lacking dorsal protuberance. **Fruits** samaras, pale yellow-brown or grayish, dull, glabrous. **Distr:** Tropical, dry subtropical and warm temperate regions. **Cult:** Propagate from seeds or cuttings. Requires moist or dry soils in full sun. Suitable for coastal regions as they tolerate salty, and strong windy conditions.

C. equisetifolia L., *Amoen. Acad.* 4: 143, 1759. [Australian Pine, Beach She-Oak, Beef Wood, Common Iron Wood]. **Trees**, monoecious, to 35 m tall, not suckering from roots; trunk straight, to 70 cm dbh; crown conical; bark scaly, adaxially red-brown on old trees; ultimate branchlets usually pendulous, grayish-green, 10–27 cm long; articles (2.5–)4–9 mm long. **Leaves** erect and appressed to branchlets, (6 or)7(or 8) per whorl; lamina lanceolate or triangular, 1–3 mm long. **Inflorescences** ♂ 1–4 cm long. **Infructescences** ellipsoid, 1.2–2.5 cm long, grayish-green or yellowish-brown tomentose when young, glabrous at maturity, base and apex truncate to obtuse; apex of bracteoles slightly obtuse or acute. **Fruits** samaras, 5–8 mm long (including the wing). **Distr:** Indonesia, Malaysia, Maldives, Myanmar, Papua New Guinea, Philippines, Thailand, Vietnam; Australia, Pacific Islands. **Ecol:** Fl. Apr–May. Fr. Jul–Oct. **Zones:** 10–12.

11 ROSIDS II [MALVIDAE]

ORDERS 9, FAMILIES 72, GENERA 1,763, SPECIES 30,108.

Info: The monophyly of Rosids II (Malvidae) is very strongly supported by molecular data (Wang et al. 2009).

GERANIALES

FAMILIES 5, GENERA 17, SPECIES 836.

Leaves with margins gland-toothed. Inflorescences cymose. **Flowers** with nectary outside androecium; gynoecium opposite petals, (outer integument largely dermal in origin), stigma dry. **Seeds** testal (Stevens 2001 onwards). **Distr:** Predominantly Temperate and warm temperate, especially in southern Africa. **Info:** The monophyly of the Geraniales is very strongly supported by molecular data (Wang et al. 2009).

MELIANTHACEAE

Genera 3, **species** 11. **Trees** or **shrubs**, with indumentum of simple hairs. **Stipules** present, usually large, either in pairs at base of petiole or fused and intrapetiolar. **Leaves** alternate, imparipinnate or 3-foliolate. **Inflorescences** racemes, terminal or axillary. **Flowers** bisexual in appearance but often functionally unisexual and then apparently dioecious or polygamous, slightly to markedly zygomorphic, usually 4–5-merous; sepals and petals dissimilar; sepals 4–5, shorter or longer than petals, basally united, imbricate; petals 4–5, free, imbricate, unequal, unguiculate; disk extrastaminal, annular-pentagonal or variously unilateral; stamens 4–5(8), free or basally connate, anthers 2-thecous, longitudinally dehiscent; ovary superior, 4–5-locular, placentation basal or axile, locules 1–4-ovuled, style 1, stigma capitate. **Fruits** capsules, papery or woody, loculicidal; seeds large, endosperm copious, aril present or absent. **Distr:** Africa (south of the Equator).

Greyia

Species 3. **Shrubs** or **small trees**, deciduous or evergreen; bark thick and rough. **Leaves** crowded on thick terminal branches; lamina simple, margins irregularly crenate, shallowly lobed, glabrescent to densely villous, venation palmate; petiole sheathing. **Inflorescences** racemes, terminal. **Flowers** bisexual, polysymmetric; sepals 5, much shorter than petals, imbricate, basally fused; petals 5, imbricate, red, rarely white, glandular-ciliate, fugacious; stamens 10, at maturity diplostemonous, filaments subulate, free, anthers basifixed, much-exserted from corolla; disk extrastaminal, annular, with 10 peltate or lacerate-dentate lobes; ovary 5-locular, deeply 5-grooved, style simple, ovules numerous, in 2 rows on each axile placenta. **Fruits** capsules, narrowly oblong, septicidal; seeds numerous, minute. **Distr:** Southern Africa (± below the equator). **Cult:** Propagate from seeds or half-hardened cuttings. Requires a fertile, well-drained soil in full sun.

G. radlkoferi Szyszył., *Polypet. Discifl. Rehmann.* 49, 1888. [Woolly Bottlebrush]. **Trees**, small, often gnarled, to 6 m high; branchlets green to yellowish and velvety; bark dark brown to black, rough, flaky. **Leaves** alternate, clustered toward ends of branchlets; lamina simple, round, 5–15 cm Ø, base cordate to slightly tapering, margins shallowly lobed and serrated, apex rounded, thick, ± fleshy, adaxially glossy green, seasonally turning reddish or yellow, strongly multinerved from base, adaxially sparsely hairy, abaxially densely velvety; petiole clasping stem, ~5 cm long, hairy. **Inflorescences** dense clusters, ~10 cm Ø, at end of branches; pedicels ~1 cm long. **Flowers** bright red, rarely white; petals with base tapered. **Fruits** 4–5 follicles (cylindrical tubes), initially held together, later flaring open, tubes split open to release the seeds. **Distr:** S Africa (N KwaZulu-Natal, Swaziland and Limpopo Province). **Ecol:** On cliffs, among rocks in grasslands, forest margins and along rivers. Fl. Jul–Nov. Fr. Sep–Jan. **Zones:** 9–11.

Melianthus

Species 8. **Shrubs** or **subshrubs**, sometimes short-lived, multistemmed; stems often hollow, branching near ground, often strongly scented. **Stipules** paired or fused and intrapetiolar, persistent. **Leaves** imparipinnate; leaflets 5–27, margins coarsely serrated to rarely sinuous or entire; rachis with serrated or entire wings. **Inflorescences** racemose, terminal or axillary, flowers alternate or whorled, apical flowers sometimes sterile forming a red crown. **Flowers** monosymmetric with ± well-developed mentum, resupinate, 5-merous; sepals 5, greenish or reddish, trimorphic, outer sepals ovate, oblong-ovate, ovate-lanceolate or linear, odd sepal saccate, pouched or flat, ovate to lorate, simple to trilobed, acute to acuminate; petals 4, seated on the mentum, reddish; stamens 4, dimorphic, basally fused; disk unilateral, cup-shaped, dark-colored, with copious blackish nectar; ovary oblong, 4-locular, ovules axile, 2–6 per locule, style persistent, penicillate. **Fruits** capsules, membranous, 4-winged, valves rounded or acute, opening at apex; seeds small, globose, subglobose or pear-shaped, glossy, black. **Distr:** Southern Africa. **Cult:** Propagate from seeds, rooted suckers or hardwood cuttings. Requires an average to rich, evenly moist but well-drained soil full or partial sun.

M. major L., *Sp. Pl.* 2: 639, 1753. [Cape Honey Flower, Honey Bush, Turkey Bush]. **Shrubs**, glabrous, to 2–3 m high, clump-forming, with a strong unpleasant smell when crushed; stems bamboolike; branches sprawling. **Stipules** foliaceous, large, 2.5–5 cm long. **Leaves** alternate, imparipinnate, 20–75 cm long; leaflets sessile, 7–13(–17), oblong, 5–10 cm long, base decurrent on petiole, margins coarsely serrate, bluish-green. **Inflorescences** racemes, dense, ~30 cm long, 2–4 flowers per node; peduncle long. **Flowers** dark reddish-brown, ~2.5 cm across; sepals maroon to greenish; petals reddish, shorter than sepals; ovary glabrous. **Fruits** capsules, 4-lobed, 2.5–4 cm long, membranous, glabrous; seeds 2 per chamber, glossy, black. **Distr:** S Africa (Cape Province). **Ecol:** Along rocky watercourses, especially in semi-arid areas. Fl. throughout year. **Zones:** 9–11.

MYRTALES

FAMILIES 11, GENERA 2,380, SPECIES 11,027.

Plants with vessel elements with vestured pits; stems with internal phloem; bark flaky. **Stipules** absent or present as small lateral or axillary structures. **Leaves** often opposite; lamina simple, margins usually entire. **Flowers** with short to elongate hypanthium, stamens incurved in bud (straight in Onagraceae); style single, carpels completely connate. **Distr:** Worldwide. **Info:** Monophyly of Myrtales is strongly supported by molecular data (Chase et al. 1993; Hilu et al. 2003; Soltis et al. 2000; Wang et al. 2009).

COMBRETACEAE

Genera 14, **species** 500. **Trees**, **shrubs**, or **woody lianas**, evergreen or deciduous, rarely subherbaceous. **Leaves** opposite, subopposite, whorled, spiraled, or alternate, usually petiolate, estipulate; lamina simple, margins entire or subentire, sometimes toothed, glands often present between crenations of proximal margin and at base or on petiole; petiole sometimes persistent and thornlike. **Inflorescences** spikes, branched spikes, racemes, panicles, or sometimes capitula, terminal, axillary, or extra-axillary, bracteate. **Flowers** usually regular, rarely slightly zygomorphic, usually bisexual, sometimes unisexual and ♂ flowers present in the same inflorescence; receptacle surrounding and adnate to ovary and extended into a short or long calyx tube dilated distally, lobes 4 or 5(–8), valvate in bud, persistent or deciduous, sometimes almost absent; petals 4 or 5, inserted near mouth of calyx tube, imbricate or valvate in bud, conspicuous or not, or absent; stamens usually twice as many as calyx lobes, 2-seriate, inserted inside distal part of calyx tube, included in or exserted from calyx tube, filaments incurved in bud, anthers dorsifixed, usually versatile, longitudinally dehiscent; disk usually present,

MD

Above: *Greyia radlkoferi, Melianthus major, Combretum constrictum, Combretum grandiflorum.*

intrastaminal, hairy or glabrous; ovary inferior, 1-loculed, ovules 2(–6), pendulous, anatropous, usually only 1 developing, style 1, simple, usually free from distal part of calyx tube, subulate to filiform; stigma capitate or inconspicuous. **Fruits** pseudocarps, variable, fleshy or dry, usually indehiscent, often longitudinally 2–5-winged, -ridged, or -angled; seed 1, endosperm absent. **Distr:** Pantropical.

Combretum

Species ~250. **Lianas** woody, or **shrubs** when lacking climbing support, rarely non-climbing shrubs, trees, or subherbaceous plants. **Leaves** opposite, whorled, or rarely alternate; lamina variable, generally elliptic or oblong-elliptic to broadly ovate, hairy or glabrous; petiole sometimes persistent, thornlike. **Inflorescences** simple or branched spikes, racemes, or panicles, terminal, axillary, or extra-axillary. **Flowers** with calyx tube usually shorter <2 cm long, proximally ellipsoid or fusiform, slightly contracted above ovary, distally narrowly funnelform to saucer-shaped, lobes 4 or 5, rarely more, deltoid to subulate, sometimes almost absent; petals 4 or 5, white, yellow, orange, red, or purple, small and inconspicuous or showy and exceeding calyx lobes; stamens usually 8 or 10, usually exserted from calyx tube; style not adnate to inside of calyx tube. **Fruits** often shortly stipitate, dry, rarely fleshy, longitudinally 4- or 5-winged, -ridged, or -angled, wings papery, transversely striate, endocarp not sclerenchymatous. **Distr:** Mostly in tropical and S Africa, also in tropics of America and Asia, and Madagascar. **Cult:** Propagate from seeds or half-hardened cuttings. Requires any well-drained soil in full or partial sun.

C. constrictum Benth. Laws., *F.T.A.* 2: 423, 1871. (Syn: *C. bussei*, etc.). [Thailand Powderpuff]. **Shrubs** or **climbers**. **Leaves** subopposite; lamina mostly oblong, to 6(–12) × 3(–6.5) cm, base rounded, apex rounded or acuminate, chartaceous to subcoriaceous, ± glabrous, lateral nerves 6–9 pairs; petiole 4–6 mm long, forming a slightly hooked spine at base with age. **Inflorescences** spikes, subcapitate, terminal or axillary, to 6 cm long; bracts narrowly elliptic, 6 × 2 mm; lower receptacle 5–6 mm long; upper receptacle 1–1.8 × to 7 mm, constricted in middle, broadly infundibuliform at apex. **Flowers** red or pinkish, 5-merous; sepals triangular, 4 × 3 mm; petals narrowly elliptic, 7–8 × 2·5 mm, unguiculate, pilose; stamens with filaments 1.7(–2.5) cm long, anthers 0–8 mm long; disk 2 mm Ø; style 2.5 cm long. **Fruits** 5-angled, sessile, ellipsoid, ~2.5 × 1.2–1.5 cm, glabrous, apical peg absent. **Distr:** Kenya, Mozambique, Nigeria, Somalia, Tanzania. **Ecol:** In edges of mangrove, seasonal swamps. **Zones:** 10–12.

C. grandiflorum G. Don, *Edinburgh Philos. J.* 346, 1824. [Large-flowered/Showy Powderpuff]. **Lianas**, twining, 3–10 m long; branches cylindrical, somewhat compressed at nodes, pith usually hollow. **Leaves** opposite; lamina elliptical, obovate or ovate, 7.5–14.5(–25) × 3.5–7(–12) cm, base obtuse, slightly asymmetrical, margins entire, apex obtuse or rounded, chartaceous or coriaceous, adaxially glabrous; petiole 1–1.3 cm long, geniculate at base, pubescent. **Inflorescences** racemes, axillary, horizontal, flowers numerous, ascending, 6–12 cm long; pedicels 3–6 mm long; hypanthium infundibuliform, 0.8–1.2 cm long, glabrous, red-purple, puberulous. **Flowers** 4–5-merous; sepals triangular, ~3 mm long; petals red-purple, spathulate, 1–1.2 cm long; stamens exserted, filaments unequal; style simple, free, exserted. **Fruits** ovoid, 3–4 cm long, base stipitate, apex retuse, 5-winged, pale yellow or stramineous; seed 1, pentagonal, 0.7–1 cm long. **Distr:** C Africa (Ghana, Guinea-Bissau, Ivory Coast, Liberia, Senegal, Sierra Leone). **Ecol:** Fl. Sep–Nov. Fr. Feb. **Zones:** 10–12.

C. microphyllum Klotzsch, *Reise Mossamb. Bot.* 1: 74, 1861. (Syn: *C. lomuense*, *C. paniculatum* ssp. *microphyllum*). [Burning Bush Combretum, Flame Climbing Bushwillow, Riverine Flame Creeper]. **Shrubs**, scandent or ± prostrate, deciduous. **Leaves** opposite; lamina usually to ~4.5 × 4 cm and subcircular, sometimes to 11 × 6 cm and ± oblong, base usually subcordate, apex usually rounded, often apiculate, chartaceous, lateral nerves 5–6 pairs; petiole to 1.6 cm long forming a spine. **Inflorescences** panicles, terminal and axillary, large, leafy or leafless; rachis tomentose; bracts ~2 mm long, often transitional to small inflorescence leaves which are in turn transitional to normal foliage; lower receptacle 3.5–4.5 mm long, densely pubescent; upper receptacle 2.5–3.5 × 1.5–3 mm, usually sericeous; flowers ± sessile. **Flowers** 4-merous, red (due to red petals and stamens); sepals to 0.9 × 1.2 mm; petals red, subcircular, 2–2.5 × 2 mm, usually overlapping and connivent, glabrous; stamens with filaments 1–1.3 cm long; style 0.8–1 cm long. **Fruits** capsules, broadly winged (samaralike), to ~5 × 3.3 cm. **Distr:** Botswana, Malawi, Mozambique, S Africa, Tanzania, Zambia, Zimbabwe. **Ecol:** In woodlands or savannas; 10–500 m. **Zones:** 10–12.

C. ovalifolium Roxb., *Fl. Ind.* ed. 2, 2: 226, 1832. (Syn: *C. roxburghii*, *C. wightianum*, etc.). **Shrubs**, straggling or climbing, large. **Leaves** opposite, broadly ovate to lanceolate-elliptic, (6–)9.5–11(–13) × 3.5–6.5(–8) cm, base broadly shortly cuneate, or subtruncate, apex shortly acuminate, densely scaly, veins 5–6 pairs; petiole 1–1.5(+) cm long, densely scaly. **Inflorescences** spikes, elongate racemes, to 15 cm long, densely flowered, axillary; bracts minute, early dehiscent; buds globose, apiculate; lower receptacles 3–4(–5.5) mm long, narrowing at top before expanding sharply into upper receptacle; upper receptacle cup-shaped, 1.5–2 × 2 mm. **Flowers** yellow or white; calyx lobes 4, triangular-acute, 1.2–1.8 × 1.25–1.8 mm, glabrous to densely glandular, subreflexed; petals 4, obovate-oblong, ~1–1.5 × 0.3–0.5 mm, obtuse or slightly emarginate, shortly clawed; filaments longer than petals; style 7+ mm long, glabrous, stigma simple. **Fruits** 2.5 cm

Below: *Combretum microphyllum, Combretum ovalifolium, Combretum racemosum, Lumnitzera littorea.*

RT

MS

HL

PL

Above: *Lumnitzera racemosa, Quisqualis conferta, Quisqualis falcata, Quisqualis indica.*

long, dark brown, 4-winged, wings ~1.1–1.2 cm deep. **Distr:** India, Sri Lanka. **Ecol:** In dry regions. Fl. Apr–Jul. **Zones:** 10–12.

C. racemosum P. Beauv., *Fl. Owar.* 2: 90, t.118, 1820. [False Bougainvillea]. **Lianas**, sometimes scandent shrubs. **Leaves** opposite or subopposite, sometimes ternate; lamina oblong-ovate to ovate, to 16 × 6 cm, base rounded or subcordate, apex shortly acuminate, lateral nerves 6–10 pairs; petiole to 5 mm long. **Inflorescences** contracted racemes, terminal and axillary; rachises rufous-pubescent; floral leaves to 6 × 2–5 cm, similar to foliage leaves but white or pink; lower receptacle 3–4.5 mm long, somewhat constricted above and below ovary; upper receptacle campanulate, 5–6 × 3 mm, glabrous or nearly so. **Flowers** crimson, 4-merous; sepals shortly triangular, acuminate, 0.5 mm long; petals oblong-elliptic, 2–3 × 0.1–2 mm, densely pubescent outside; stamens with filaments 1.2 cm long; disk inconspicuous, pilose, without a free margin. **Fruits** obtriangular to obovate, 1.8–2.5 × 1.7–2.5 cm, wings membranous, glabrous, apical peg absent or minute, stipe 3 mm long. **Distr:** W Africa from Senegal southwards to Angola and east to S Sudan. **Ecol:** 1,100–2,200 m. **Zones:** 10–12.

Lumnitzera

Species 2. **Trees small** or **shrubs**, evergreen. **Leaves** spiraled, densely crowded at apices of branchlets; lamina spathulate to narrowly oblanceolate, gradually narrowed into a short petiole, somewhat fleshy, glossy and glabrous when mature, lateral veins inconspicuous or obscure. **Inflorescences** spikes or racemes, several-flowered, axillary or terminal, short. **Flowers** with calyx tube cylindric or cylindric-ellipsoid, not differentiated into proximal and distal parts, bearing 2 deltoid bracteoles minutely pilose at margin, lobes 5, persistent, deltoid or broadly triangular, margin minutely pilose; petals 5, red or white, rarely pink or yellow; stamens 5–10; style persistent. **Fruits** fusiform or ellipsoid, obtusely angled, dry, ± woody, nearly smooth or longitudinally wrinkled, apex bearing persistent calyx lobes and style. **Distr:** Asia; E Africa, Madagascar; N Australia, Pacific Islands. **Ecol:** In mangrove forests, banks of tidal creeks. **Cult:** Propagate from seeds. Requires a slightly moist, well-drained, sandy soil mixed with clay, in full sun.

L. littorea (Jack) Voigt, *Hort. Suburb. Calcutt.* 39, 1845. (Syn: *Bruguiera littorea*, etc.). [Red Teruntum, Red-flowered Mangrove]. **Trees**, to 7(–25) m tall; trunk to 0.5 m Ø; bark dark blackish-brown, deeply longitudinally fissured; branchlets reddish or green when young, glabrous. **Leaves** oblanceolate or narrowly oblong-oblanceolate, (2–)4–8 × 1–3 cm, apex retuse or rounded, adaxially dark green, lateral veins in 4 or 5 pairs. **Inflorescences** terminal, 3–4.5 cm long; bracteoles 1.5–2 mm long. **Flowers** with calyx tube 1.2–1.8 cm long, lobes broadly triangular, 1–1.5 mm long, obtuse; petals bright red, oblong-elliptic, 5–6 mm long, obtuse or acuminate; stamens 5–10 (usually 7), ~1 cm long, about twice as long as petals; style ~1 cm long. **Fruits** blackish-brown, fusiform, 1.6–2 × 0.4–0.5 cm excluding stipe, longitudinally striate; stipe ~5 mm long. **Distr:** Asia through Malesia, N Australia and Pacific Islands. **Ecol:** In open remnant mangrove forests along sea shores. Fl. Nov–Dec, May. Fr. Jun–Aug. **Zones:** 10–12.

L. racemosa Willdenow, *Ges. Nat. Freun. Berlin Neue Schriften.* 4: 187, 1803. (Syn: *Bruguiera madagascariensis*, etc.). [Black Mangrove]. **Shrubs** or **small trees**, to 8 m tall; trunk to 30 cm Ø; branchlets red or grayish-black. **Leaves** spathulate to oblanceolate or obovate, 3.5–7 × 1–2.5 cm, apex retuse or rounded, adaxially pale green, lateral veins usually in 3 or 4 pairs. **Inflorescences** axillary, 2–6 cm long; bracteoles ~1 mm long. **Flowers** fragrant; calyx tube ~9 mm long, lobes deltoid, 1–2 mm long, apex obtuse or shortly mucronate; petals white, elliptic, 4.5–5 mm long, apex obtuse; stamens 5 or 10, 4–5 mm long; style ~4 mm long. **Fruits** blackish-brown when ripe, ellipsoid or ovoid, slightly compressed on one side, 1–2 × 0.5–0.8 cm excluding stipe, 2- or 3-ridged, smooth proximally, longitudinally striate distally; stipe ~1 mm long. **Distr:** Asia to Malesia; E Africa, Madagascar; N Australia, Pacific Islands. **Ecol:** Open remnant mangrove forests along sea shores. Fl. Nov–Aug. Fr. Aug–Apr. **Zones:** 10–12.

Quisqualis

Species ~17. **Lianas** woody. **Leaves** opposite or subopposite; petiole persistent and thornlike; lamina ± elliptic, glabrous or hairy. **Inflorescences** simple or sometimes compound spikes, terminal or axillary. **Flowers** with calyx tube (1.7–)5–9 cm long, ± uniformly narrowly tubular except funnelform at apex, deciduous above ovary, hairy or subglabrous, lobes 5, deltoid or triangular-lanceolate, small, apex sometimes cuspidate; petals 5, white or red, larger (often much more so) than calyx lobes; stamens 10, not or scarcely exserted from calyx tube; style partly adnate to inside of calyx tube. **Fruits** fusiform to subglobose or ovoid, longitudinally 5-ridged or -winged, dry, leathery. **Distr:** Tropical Africa and Asia. **Cult:** Propagate from cuttings, air-layering or from suckers. Plants propagated from suckers will produce flowers after 1–2 years. Requires any moist but well-drained fertile soil in full sun.

Q. conferta (Jack) Exell, *J. Bot.* 69: 122, 1931. (Syn: *Q. densiflora, Sphalanthus confertus*). [Vientiane Creeper]. **Lianas**; branchlets brown-pilose. **Leaves** oblong, 5–13 × 2–5.5 cm, base rounded, apex acuminate, adaxially glabrous and finely verruculose, abaxially glabrous except sparsely pilose on midvein and with denser hairs in axils of lateral veins; petiole 3–7 mm long, with an inflated joint near base, brown-pilose. **Inflorescences** dense; bracts leaflike, lanceolate, 0.5–1.2 × 0.2–0.4 mm, brown pilose, becoming sparsely so. **Flowers** with calyx tube 1.7–2.4 cm long, brown-pilose; lobes triangular-lanceolate, ~2 mm long, apex cuspidate, cusp filiform, 1–3 mm long, often recurved; petals opening dark red or reddish, oblong-elliptic, ~3.5 × 2 mm. **Fruits** glossy black when ripe, ovoid, conspicuously 5-ridged, ~2.5 cm long, glabrous. **Distr:** Cambodia, China, Indonesia (Sumatra), Malaysia, Thailand, S Vietnam. **Ecol:** In dense forests and wetlands; 400–1,100 m. Fl. Jan. **Zones:** 9–12.

Q. falcata Welw. ex Hiern, *Cat. Afr. Pl.* (Hiern) i. 355, 1855. (Syn: *Combretum falcatum*). [Wild Poinsettia]. **Shrubs**, erect or occasionally scandent, giving the appearance of a liana, 1–3 m tall. **Leaves** opposite, ovate-oblong, 6–11.5 × 3–5 cm, base rounded or cordate, margins entire, apex shortly acuminate, flat, membranous, papery, glabrous or nearly so, adaxially dark green, abaxially paler and whitish-green, lateral veins ~5 per side; petiole 5–6 mm long, jointed about middle. **Inflorescences** racemes, many-flowered, axillary and terminal; peduncles 1.2–3.8(–7.5) cm long; bracts narrowly lanceolate, 0.8–1.6 cm long, ± pubescent, petiolate, soon falcate-recurved, usually yellowish-green, persisting for some time after flowers fall; pedicels 2.5–5 mm long, tawny-pubescent. **Flowers** ~1.6–2.5 cm long, ± erect; calyx tube narrowly obconical, pilose-hairy, calyx teeth linear-subulate from an ovate base, equaling petals, promptly recurved; petals oval, ~2 mm long. **Fruits** drupaceous, 2.5 cm long, whitish-rose colored, at length broadly winged making the fruit ~2 cm Ø. **Distr:** Tropical Africa (Angola). **Ecol:** In secondary thickets. Fl. Jan, Jun. Fr. Mar, Aug. **Zones:** 10–12.

Q. indica L., *Sp. Pl.*, ed. 2. 1: 556, 1762. (Syn: *Combretum indicum*, etc.). [Drunken Sailor, Rangoon Creeper]. **Lianas**, to 15+ m long; branchlets pubescent. **Leaves** mostly oblong-elliptic or elliptic, 5–18 × 2.5–7 cm, base obtuse, apex acuminate to shortly caudate, adaxially ± glabrous, abaxially sometimes brown-pilose, lateral veins in 7 or 8 pairs; petiole 5–9 mm, without an inflated joint near base, densely pilose when young. **Inflorescences** large leafy panicles, many-flowered; bracts deciduous, filiform-linear to ovate, 0.3–1.2 cm long, brown-pilose. **Flowers** fragrant; calyx tube 5–9 cm long, yellow-pilose, lobes deltoid, 2–3 mm long, acute or shortly acuminate, not cuspidate; petals opening white, adaxially darkening through all shades of pink, later turning reddish (usually on the third day), yellowish or whitish abaxially, obovate to oblanceolate, 1–2.4 × 0.4–1 cm, rounded to obtuse. **Fruits** red when young, greenish-black or brown when ripe, fusiform or narrowly ovoid, sharply 5-ridged, 2.7–4 × 1.2–2.3 cm, apex mucronate, glabrous. **Distr:** Coastal

Above: *Quisqualis indica* (double form), *Terminalia arjuna*, *Terminalia catappa* (flowers), *Terminalia catappa* (fruit).

E Africa, Asia, Indian Ocean islands, to Malesia and Pacific Islands. **Ecol:** In rainforests; <1,500 m. Fl. Mar–Nov. Fr. Jun–Nov. **Zones:** 10–12. **Info:** A form with double petals is commonly cultivated.

Terminalia

Species ~150. **Trees**, often very large, buttressed, rarely shrubs; branches often in tiers. **Leaves** spiraled, alternate, subopposite, or opposite, often crowded into pseudowhorls at apices of branchlets; lamina oblong, elliptic, obovate, or orbicular, hairy or glabrous, often minutely verruculose and translucent-dotted, often with domatia and with 2 or more glands at or above lamina base or on petiole. **Inflorescences** spikes or racemes, sometimes panicles, axillary or terminal, with bisexual flowers toward base of inflorescence and ♂ flowers toward apex. **Flowers** with calyx tube proximally broadly cylindric to ellipsoid or ovoid, distally cupular or sometimes scarcely developed, lobes 4 or 5, deltoid or ovate; petals absent; stamens 8 or 10. **Fruits** drupelike often fleshy, sometimes dry and leathery or corky, often longitudinally 2–5-winged, or -ridged, sometimes weakly so. **Distr:** Tropical Africa, America, and Asia, extending to S Africa, Australia, and Pacific Islands. **Cult:** Propagate from seeds, cuttings or by air-layering. Requires fertile, well-drained soil in full sun.

T. arjuna (Roxb. ex DC.) Wight and Arn, *Prodr.* 314, 1834. (Syn: *T. glabra*, etc.). [Arjun Tree, White Marudah]. **Trees**, large, to 6–15(–25) m tall; stems buttressed, often fluted; bark smooth, pale greenish to whitish-gray. **Leaves** usually subopposite; lamina oblong-elliptic or somewhat suborbicular, 7–18(–25) × 4–6 cm, base rounded or cordate, margins entire with somewhat crenate or serrate in upper half or throughout, apex obtuse, rarely subacute, adaxially ± glabrous, abaxially partially pubescent; petiole 0.5–1 cm long, with (1–)2 rounded glands at apex. **Inflorescences** paniculate spikes, axillary or terminal, 3–6 × 1 cm; bracteole small, deciduous, linear-lanceolate. **Flowers** yellowish-white, sessile; hypanthium broadly campanulate, 4–5 mm long, teeth triangular ~1.5 cm long, glabrous; stamens much-exserted; ovary glabrous; disk barbate. **Fruits** ovoid-oblong, 2.5–5 cm long, brown with 5 projecting wings, wings striated with 5 ascending veins. **Distr:** India, Sri Lanka. **Ecol:** Fl. Apr–May. **Zones:** 10–12. **Info:** Used for medicinal purposes.

T. catappa L., *Syst. Nat.*, ed. 12. 2: 674, 1767. (Syn: *Myrobalanus catappa*, etc.). [Country Almond, Indian Almond, Sea Almond]. **Trees**, deciduous, to 20(–30) m tall; trunk to 2 m Ø; branches spreading, forming tiers. **Leaves** alternate, crowded into pseudowhorls at branch tips; lamina obovate to oblanceolate, narrowed in proximal half, 12–30 × 8–15 cm, base narrowed, cordate or truncate, apex obtuse or mucronate, surfaces glabrous or abaxially sparsely hairy when young, lateral veins in 10–12 pairs; leaves turn bronze or scarlet red before falling; petiole 0.5–2 cm long, stout, tomentose. **Inflorescences** spikes, numerous-flowered, simple, slender, axillary, 15–20 cm long; axis shortly tomentose. **Flowers** fragrant; calyx tube distally cupular, 7–8 mm long, abaxially white-tomentose, densely so on ovary, sparsely so on cupular part, adaxially glabrous, lobes 5; stamens 10, 2–3 mm long, exserted. **Fruits** not stipitate, yellow, red or blackish-green when ripe, ellipsoid, slightly to strongly compressed, strongly 2-ridged to narrowly 2-winged (wings to 3 mm wide), 3–5.5 × 2–3.5 cm, glabrous, pericarp woody, rigid; seed 1, narrowly ellipsoid, ~2.5–3 cm long, one end rounded while the other acute. **Distr:** Old World tropics. **Ecol:** On sandy seashores. Fl. Mar–Jun, Oct. Fr. May, Jul–Sep. **Zones:** 10–12.

T. kaernbachii Warb., *Bot. Jahrb. Syst.* 18: 201, 1894. (Syn: *T. okari*). [Brown Terminalia, Okari Nut]. **Trees**, 20–30 m tall; trunk straight. **Leaves** clustered toward the tips of branches; obovate or narrowly obovate, 15–28 × 6–13 cm, slightly flattened, base tapering, apex usually acute, abaxial surface and petiole with persistent reddish-brown hairs; petiole to 2 cm long. **Inflorescences** ± erect spikes; buds globular, 0.8–1 cm long. **Flowers** with calyx lobes triangular, to 2 mm long, densely hairy; style to 2 cm long. **Fruits** drupes, large, ellipsoid, slightly flattened, to 9–11 × 6–8 × 5–6 cm, coated with short reddish-brown hairs when young, red, fleshy, and glabrous when ripe, containing a massive woody seed. **Distr:** Irian Jaya, Aru Islands, Papua New Guinea, Solomon Islands. **Ecol:** In lowland rainforests, riverine forests; <1,000 m. **Zones:** 10–12. **Info:** Cultivated in SE Asia and Malesia for the edible nuts (kernels).

T. melanocarpa F.Muell., *Fragm. Phytogr. Austral.* 3: 92, 1862. (Syn: *Myrobalanus melanocarpa*). **Trees**, deciduous, to 15 m tall. **Leaves** alternate or opposite; lamina simple, spathulate, 4.5–12 × 3–7.5 cm, surfaces glabrous or sparsely hairy, domatia often present as tufts of hair; petiole 0.5–2 cm long. **Inflorescences** spikes or racemes, axillary, usually toward the ends of the branchlets; ♂ flowers toward the apex, bisexual flowers near the base. **Flowers** with sepals 5, white or cream, 5–6 mm long; petals absent; stamens exserted. **Fruits** drupes, 2.3–3.3 × 2–2.4 cm, ripening from green to black, beaked, green immature fruits winged, becoming less distinctly winged in ripe fruits; seed 1, 2.2–2.5 cm long, brown. **Distr:** Australia. **Ecol:** In rainforests, littoral rainforests and vine thickets; 0–300 m. Fl. Jul–Dec. Fr. Feb–May. **Zones:** 10–12.

T. spinosa Engl., *Pflanzenw. Ost-Afrikas C.* 294, 1895. [Small-leaved Terminalia, Spiny Cluster-leaf]. **Trees**, to 20 m tall; branchlets grayish or reddish-brown; long shoots zigzag with occasional spur shoots; lateral shoots irregularly arranged, ending in a spur shoot, with 0–2 lateral spur shoots; spines 2(–3), frequently at base of spur shoots, to 1.5(–2) cm long. **Leaves** borne on spur shoots; lamina broadly obovate, to 4 × 3 cm, base cuneate, apex rounded to subtruncate, emarginate, glabrous to sparsely pubescent, lateral nerves 4–5 pairs; petiole to 5 mm long. **Inflorescences** spikes, to 6 cm long, borne on spur shoots; peduncle sparsely to densely pubescent; flowers shortly pedicellate. **Flowers** small, white or purplish, glabrous outside. **Fruits** reddish-purple, oblong-elliptic, 2–3 × 1–2 cm, apex emarginate, glabrous, broadly winged, membranous; stipe 3 mm long. **Distr:** Kenya, Somalia, Sudan, Tanzania, Uganda. **Ecol:** Savannas, woodlands; 0–1,770 m. **Zones:** 10–12.

LYTHRACEAE

Genera ~31, species 625–650. **Herbs, shrubs,** or **trees**; young stems often quadrangular. **Stipules** vestigial or absent. **Leaves** opposite, often decussate, or whorled, rarely ± alternate; lamina simple, entire, pinnately veined, secondary veins typically joined in a series of intramarginal arches. **Inflorescences** racemes, cymes, or panicles, axillary or terminal. **Flowers** usually 4-, 6- or 8-merous, sometimes 3- or 5-merous, bisexual, regular or irregular; floral tube perigynous, hemi-epigynous, or epigynous, persistent, membranous to leathery, often 6–12-ribbed; sepals valvate, equal to or much shorter than floral tube, membranous to thickly leathery, persistent; epicalyx alternating with sepals or absent; petals inserted at rim of floral tube, alternating with sepals, crinkled, sometimes clawed, frequently caducous, rarely absent; stamens usually biseriate and twice as many as sepals, sometimes uniseriate, inserted near base of floral tube or higher, or numerous and multiseriate, with at least some inserted at floral rim just below sepals, anthers usually versatile; ovary superior or inferior, 2–6- or multiloculed, locules many-ovuled, style simple, stigma capitate, conic-peltate, or punctiform, placentation axile. **Fruits** loculicidally dehiscent or irregularly dehiscent capsules, infrequently indehiscent, or berrylike, partly or completely surrounded by persistent floral tube, leathery; seeds usually numerous. **Distr:** Predominantly in the tropics, less common in temperate regions.

Cuphea

Species ~200. **Herbs**, woody, annual or perennial, rarely shrubs. **Leaves** opposite or whorled. **Inflorescences** with flowers solitary or in racemes, axillary; bractlets 2 or none. **Flowers** zygomorphic, calyx tubular, tube ridged, 6-toothed, spurred or appendaged at base, with 6 accessory lobes; corolla 6, rarely 2 or none, very rarely 4; stamens 4–11 (rarely 6 or 9), filaments short; ovary superior, sessile, usually with a basal, dorsal or rarely cupuliform disk, or

Top row: *Terminalia kaernbachii*, *Terminalia kaernbachii* (fruits), *Terminalia melanocarpa*, *Terminalia spinosa*.
Bottom row: *Cuphea hyssopifolia*, *Cuphea ignea*, *Cuphea micropetala*.

disk sometimes absent, 2-celled, with few to numerous ovules, often 3 (very rarely 2), stigma bifid. **Fruits** capsules, dehiscent; seeds lentiform, usually narrowly winged. **Distr:** Tropical America. **Cult:** Propagate from seeds or cuttings. Requires a moist, well-drained soil in full or partial sun.

C. hyssopifolia HBK., *Nov. Gen. and Sp.* 6: 199, 1823. (Syn: *C. hyssopifolia* f. *subrevoluta*). [False Heather, Mexican Heather]. **Shrubs**, erect, 25–50 cm high; branches usually numerous, suberect, slender; stems pubescent and sometimes appressed-hispidulous. **Leaves** crowded; lamina linear or nearly so, 1–3 cm long, suberect or spreading, glabrous or with a few scattered hairs on the costa, 1-nerved. **Inflorescences** racemes, small, axillary; floral leaves not reduced, subsessile; pedicels 2–7 mm long, bracteolate at apex. **Flowers** with calyx glabrous or with a few scattered short hairs, 6–8 mm long; petals pale purple or white; ovules 5–8. **Distr:** C America (Mexico, Honduras). **Ecol:** Usually on large rocks along the edges of usually swift streams, often submerged seasonally, also in wet forests; <1,500 m. **Zones:** 10–12.

C. ignea A. DC., *Fl. Serres Jard. Eur.* 5: t. 500c, 1849. (Syn: *C. platycentra* [misapplied], *Parsonsia ignea*). [Cigar Flower, Cigar Plant, Cigarette Plant, Firecracker Plant]. **Subshrubs** or spreading perennials, 1 m high, glabrous; stems usually dark red. **Leaves** opposite; lamina ovate to narrowly elliptic, 2.5–4.4 × 0.8–1.9 cm, both ends tapering, base sometimes decurrent onto petiole; petiole 6 mm long. **Inflorescences** with flowers solitary (1 per node), axillary; pedicels 0.6–1 cm long. **Flowers** tubular, glabrous or sparsely hairy but not sticky-glandular; calyx 2–3.8 cm long, bright red or reddish-orange with violet and white or yellow at apices; petals absent; stamens projecting from tube; basal spur saccate. **Fruits** capsules, ellipsoid, enclosed within the hypanthium. **Distr:** Mexico (Chiapas and Oaxaca). **Ecol:** Along streams and in open woods. Fl. throughout the year. **Zones:** 10–12.

C. micropetala H.B.K., *Nov. Gen. and Sp.* 6: 209, pl. 551, 1823. (Syn: *C. eminens*, *Parsonsia micropetala*). [Candy Corn Plant, Tall Cigar Plant]. **Shrubs**, sometimes herbaceous, small, rounded, to 0.75–1 m high; stems ± branched; branchlets densely leafy, flexible, puberulent, sometimes hirtellous. **Leaves** opposite; lamina oblong-lanceolate to narrowly lanceolate, 5–16 cm long, base narrowed and petiolelike to 1 cm long, apex attenuate, tip acute, bright green, scaberulous. **Inflorescences** solitary, borne slightly above the upper leaf axils, appearing as leafy racemes, with the flowers on one side, usually nodding. **Flowers** narrowly tubular, golden-yellow becoming orange-red with age, tips greenish-yellow; calyx tube 2–2.35 cm long, scabrous, base unequal, calyx teeth 12; petals 6, minute, white; stamens 11, unequal, often exserted beyond calyx teeth, filaments filiform, red, anthers linear-oblong; ovary superior. **Fruits** capsules, many-seeded. **Distr:** Mexico (Colima to Morelos and Oaxaca). **Ecol:** Along streams, in open woods and pastures. Fl. throughout the year. **Zones:** 9–11. **Info:** Flowers pollinated by hummingbirds.

Duabanga

Species 2–3. **Trees**, 8–45 m tall, buttressed; branchlets 4-angled, becoming terete with age; ultimate branches pendulous. **Leaves** opposite, shortly petiolate; lamina oblong to ovate or lanceolate, base rounded to cordate, apex acute to acuminate, leathery to papery, glabrous, secondary vein pairs numerous, joined in a series of intramarginal arches. **Inflorescences** corymbs, terminal, laxly few- to many-flowered. **Flowers** actinomorphic, bisexual, usually 4- or 6-merous, but variably 4–8-merous on same or different plants; floral tube obconic or broadly campanulate, persistent in fruit; sepals 4–8, triangular-ovate, thick; petals 4–8, broad, crinkled, shortly clawed; stamens 12 or 24–50+, filaments filiform-subulate, base broadened, long, anthers recurved or replicate over one end of connective; ovary semi-inferior, ovules numerous, style long-exserted, slender, stigma capitate to slightly 4-lobed. **Fruits** indurate capsules, partially surrounded by persistent floral tube, oblong to ovoid-globose, loculicidally deeply dehiscent, 4–9-valved; seeds numerous, filiform, fragile, 2-tailed. **Distr:** SE Asia. **Ecol:** In evergreen lowland rainforests. **Cult:** Propagate from seeds or by air-layering. Requires a fertile, moist, well-drained soil in full sun.

D. grandiflora (Roxburgh ex Candolle) Walpers, *Repert. Bot. Syst.* 2: 114, 1843. (Syn: *D. sonneratioides*, *Lagerstroemia grandiflora*). [Lampati]. **Trees**, large, 30(–40) m tall, glabrous. **Leaves** distichous; lamina ovate to oblong, 12–15 × 5–7 cm, base cordate, apex shortly acuminate, rigid, adaxially dark green, abaxially pale green, abaxially glaucous, lateral veins (12–)20–24 pairs; petiole to 1.2 cm long. **Inflorescences** corymbs, 3–20-flowered, ± pendent at ends of branches; pedicels 3–4 cm long, stout. **Flowers** with an unpleasant smell; floral tube broadly campanulate, 1.6–2.5 × 1.8–3 cm; sepals (5 or)6, green, thick, enlarged in fruit; petals 6, white, obovate,

2.5–3 × 1.5–2 cm; stamens numerous, ~50+, white, surpassing petals. **Fruits** capsules, subglobose, 3–4 × (3–)4–4.5 cm, 6–9-valved; seeds 4–6 mm long. **Distr:** Andaman Islands, Cambodia, China, E India, Laos, Malaysia, Myanmar, Thailand, Vietnam. **Ecol:** In valley forests, open places, especially on riverbanks; 900–1,500 m. Fl. Mar–May, Jul–Sep. **Zones:** 10–12.

Lagerstroemia

Species ~55. **Trees** or **shrubs**; young stems terete to frequently 4-angled or subulate, glabrous to tomentose. **Stipules** minute, caducous. **Leaves** variably subopposite to ± alternate, rarely opposite or truly alternate, subsessile or petiolate. **Inflorescences** paniculate cymes, terminal and axillary. **Flowers** actinomorphic, campanulate to turbinate, flexibly 5–7-merous; floral tube leathery, smooth walled with 6–12(–14) veins visible as dark lines, or with broad to narrow ribs or ridges; sepals deltate, sometimes subtended within by a narrow annulus; epicalyx segments alternating with sepals, small to prominent, or absent; petals 6(–12), rose, purple, or white, crinkled, slenderly clawed; stamens (6–)12–100+, inserted at or near base of floral tube, exserted, dimorphic in most species, 6 solitary in front of sepals with thick filaments and large anthers, 12–100+ in clusters in front of petals with thin filaments and small anthers, rarely monomorphic, then all filaments thin, subequal; ovary globose to oblong, 3–6-loculed; style long-exserted; stigma capitate. **Fruits** capsules, dry, indurate, surrounded at base by persistent floral tube, loculicidally dehiscent, 3–6-valved; seeds numerous, obpyramidal, unilaterally winged from raphe. **Distr:** Tropical and subtropical Asia to Australia, north to Japan. **Cult:** Propagate from seeds, cuttings or by air-layering. Requires any well-drained soil in full sun.

L. indica L., *Sp. Pl.*, ed. 2. 1: 734, 1762. (Syn: *L. chinensis, Murtughas indica*). [Crepe Myrtle]. **Shrubs** or **small trees**, to 7 m tall; branchlets slender, 4-angled or subulate. **Leaves** elliptic, oblong, obovate, or suborbicular, typically at least some suborbicular to obovate and mucronate, 2.5–7(–10) × 1.5–4 cm, base broadly cuneate to rounded, apex acute, obtuse with a small mucro, or retuse, papery to slightly leathery, ± glabrous, lateral veins 3–7 pairs; petiole to ~2 mm long. **Inflorescences** panicles, densely flowered, subpyramidal, 7–20 cm. **Flowers** 6-merous; floral tube, 0.7–1.1 cm long, smooth-walled or ± 6-ribbed, glabrous; sepals 3.5–5.5 mm long, adaxially glabrous; annulus present; epicalyx absent; petals purple, fuchsia, pink, or white, orbicular, 1.2–2 cm long including claw 6–9 mm long; stamens 36–42, dimorphic; ovary glabrous. **Fruits** capsules, ellipsoidal, 1–1.3 × 0.7–1.2 cm, 4–6-valved; seeds including wing ~8 mm. **Distr:** Asia to Malesia. **Ecol:** Fl. Jun–Sep. Fr. Sep–Nov. **Zones:** 7–12.

L. speciosa (Linn.) Pers., *Syn.* 2: 72, 1807. (Syn: *L. regina*, etc.). [Pride of India, Queen's Crepe Myrtle]. **Trees**, partly deciduous, to 15(–20) m tall; bark flaky and peeling. **Leaves** opposite; lamina oblong to elliptic-ovate, 8–30 × 3–14 cm, base rounded or somewhat narrowed, apex obtuse, acute or acuminate, leathery, glabrous, often reddish; petiole 0.5–1.2 cm long, reddish. **Inflorescences** panicles, 14–40 × 5–20 cm, erect, rusty-hairy, later glabrous. **Flowers** 3.8–10 cm Ø; calyx tube 0.8–1 × 1.1–1.3 cm, 12-ribbed, tomentose; sepals 5–8 × 2.5–5 mm, thin or ± thickened at margins, erect or spreading or reflexed in fruit; petals ± circular or obovate, 3.2–3.5 × 2.1–3.5 cm, margins wavy, purple, lavender, mauve or rarely white, claw 5–6 mm long; stamens 120–200, all similar, anthers yellow; pistil simple, ovary superior, subglobose, 6 mm Ø, style simple, ± straight, to 5 cm long, purple, stigma dark green. **Fruits** capsules, ellipsoid, 1.5–2.5 × 1.2–2.3 cm, rugose. **Distr:** Assam, S India. **Ecol:** Fl. May–July. **Zones:** 10–12.

L. subcostata Koehne, *Bot. Jahrb. Syst.* 4: 20, 1883. (Syn: *L. subcostata* var. *hirtella, L. unguiculosa*). [Chinese Crepe Myrtle]. **Trees** or **shrubs**, to 14 m tall. **Leaves** oblong, ovate-lanceolate, elliptic, obovate-elliptic, rarely obovate, 2–9(–11) × 1–5 cm, base broadly cuneate to subrounded, apex acuminate, papery, abaxially glabrous or slightly pubescent, lateral veins 3–10 pairs; petiole 2–4 mm long. **Inflorescences** panicles, densely flowered, pyramidal, 7–16(–30) cm long, brownish-pubescent. **Flowers** with floral tube 6-merous, 3.5–4.5 mm long, with (10 or)12(or 14) dark veins or obscurely (10 or)12(or 14)-ribbed, glabrous to densely pubescent; sepals 1–2 mm long, adaxially glabrous; annulus thin or apparently absent; epicalyx absent; petals white, pink, or purple, broadly ovate, 0.7–1 cm long including claw 3.5–6 mm long, base obtuse; stamens 15–30, dimorphic; ovary glabrous. **Fruits** capsules, globose to mostly oblong, 0.6–0.9(–1.1) cm long, 3–6-valved; seeds including wing ~4 mm long. **Distr:** China, Japan, Philippines, Taiwan. **Ecol:** In forest margins, streamsides; 0–500 m. Fl. Jun–Aug. Fr. Jul–Oct. **Zones:** 9–12.

L. tomentosa C. Presl, *Bot. Bemerk.* 142, 1844. (Syn: *Murtughas tomentosa*, etc.). [White Crepe Myrtle]. **Trees**, 20–30(–35) m tall; bark fibrous, to 1 cm thick; stem, petiole, leaves, floral tubes, and ovaries densely golden-tomentose. **Leaves** oblong-lanceolate or ovate, 8–18 × 4–6.5 cm, base rounded, apex acute to shortly acuminate, abaxially

Top row: *Duabanga grandiflora, Lagerstroemia indica, Lagerstroemia speciosa, Lagerstroemia subcostata.*
Bottom row: *Lagerstroemia tomentosa, Lagerstroemia villosa, Lawsonia inermis* (red-flowered form).

Above: *Lawsonia inermis, Pemphis acidula, Punica granatum.*

tomentose, glabrescent except on midvein, lateral veins 9–11 pairs; petiole 4–8 mm long. **Inflorescences** panicles, 8–20 cm long, relatively open. **Flowers** 6-merous, floral tube, 0.8–1 cm long, 12-ribbed; sepals 3–4 mm long, adaxially completely glabrous; annulus absent; epicalyx absent; petals white, pale pink, or purple, ovate, 1–1.6 cm long including claw 3–5 mm long, base obtuse; stamens 24–70, dimorphic. **Fruits** capsules, globose to ovoid-oblong, 1–1.5 x 0.8–1.1 cm, 6-valved, apex tomentose; seeds 6–7 mm long including wing. **Distr:** Laos, Myanmar, Thailand, Vietnam. **Ecol:** In mixed forests; 600–1,200 m. Fl. May. Fr. Aug–Nov. **Zones:** 11–12.

L. villosa Wallich ex Kurz, *J. Asiat. Soc. Bengal*, Pt. 2, Nat. Hist. 42: 234, 1873. (Syn: *Murtughas villosa*). **Trees**, 10–15(–30) m tall; branchlets, leaves and inflorescences white-pubescent. **Leaves** ovate to broadly lanceolate or elliptic, 6–10 x 2.5–4 cm, base subacute to rounded, apex acute to shortly acuminate, sometimes mucronate, papery to leathery, lateral veins (5)6(–8) pairs; petiole 2–4(–6) mm long. **Inflorescences** panicles, densely flowered, nearly globose, 3(–8) cm long. **Flowers** 5- or 6-merous, floral tube 4–5 mm long, with 5 or 6 prominent undulate winged ribs; wings attenuate on pedicellate floral base; sepals 1.5–2.5 mm long; annulus absent; epicalyx absent; petals white, lanceolate, 1.5–3 mm long, not crumpled, sometimes missing, not or scarcely clawed; stamens ~25, dimorphic; ovary glabrous. **Fruits** capsules, subglobose to narrowly ovoid-oblong, 1.5(–2.2) x ~1.1 cm, 3- or 4-valved; seeds including wing (0.3–)0.9–1 cm long. **Distr:** Myanmar, Sri Lanka, Thailand. **Ecol:** In forests; 700–1,000 m. Fl. & Fr. Sep–Feb. **Zones:** 10–12.

Lawsonia

Species 1. *Generic description same as specific.* **Cult:** Propagate from seeds, cuttings or by air-layering. Requires a well-drained soil in full sun.

L. inermis L., *Sp. Pl.* 349, 1753. (Syn: *L. alba, L. spinosa*). [Henna]. **Shrubs**, glabrous, fragrant, to 2.5 m high; old stems spinescent. **Leaves** elliptic, ovate or obovate, 0.8–4.4 x 0.2–2 cm, base narrowed into a short petiole, apex acute, obtuse or subacuminate. **Inflorescences** panicles, terminal, 3–22 cm long; pedicels 2–3.5 mm long. **Flowers** fragrant, 4-merous, white or yellowish-white; hypanthium cupular, much shorter than calyx, peltate in fruit; epicalyx lacking; petals corrugate, obovate, obtuse; sepals ovate; petals 3–4 x 4–5 mm; stamens 8, in pairs, inserted at the top of hypanthium, exserted, all of equal length, filaments 4 mm long; ovary subglobose or obovate, narrowed below, 4-locular, stigma simple. **Fruits** capsules, globose, (3–)4–8(–9) mm Ø, wall veined, much exceeding hypanthium, 1- or incompletely 4-locular, irregularly dehiscent; seeds pyramidal. **Distr:** Africa and Asia. **Ecol:** In dry regions. Fl. Oct–Nov. **Zones:** 10–12. **Info:** Cultivated widely in the Old World for medicinal purposes and as a very popular dye used for staining skin. A red-flowered form is also commonly cultivated.

Pemphis

Species 1. *Generic description same as specific.* **Cult:** Propagate from seeds, cuttings or by air-layering. Requires any soil in full sun.

P. acidula J.R. Forster and G. Forster, *Char. Gen. Pl.* 34, 1775. [Iron Wood, Small-leafed Mangrove]. **Shrubs** to **trees**, densely branched, spreading, maritime, small, to 11 m tall, often ~1 m high; branchlets, young leaves, and inflorescences densely silky-strigose. **Leaves** opposite, ± sessile; lamina narrowly elliptic to lanceolate, 1–3 x 0.5–1.5 cm, base attenuate, apex often terminating in a minute circular hydathode, thick, succulent; petiole to 2 mm long. **Inflorescences** with flowers solitary or paired, axillary; pedicels 0.5–1.3 cm long. **Flowers** 6-merous, actinomorphic, heteromorphic, distylous; floral tube 4–5 mm long, turbinate, 12-ribbed, thick-walled; sepals short, ~1 mm long; epicalyx segments equaling sepals or ~½ as long; petals 6, white or pale pink, obovate to suborbicular, 3–5 mm long; stamens 12, scarcely biseriate, 6 exserted in short-styled thrum flowers, all included in long-styled pin flowers; ovary vestigially 3- or 4-loculed, appearing 1-loculed, ovules ascending. **Fruits** capsules, dry, circumscissile, obovoid, ~6 mm long, slightly exserted at maturity; seeds ~20, irregularly obpyramidal, not winged, 2–3 mm long. **Distr:** E Africa west across Indian Ocean (including Maldives), to Marshall Islands in Pacific Ocean, north to Japan (Ryukyu Islands), Taiwan. **Ecol:** On tropical sandy shores, inlets in mangrove forests. **Zones:** 11–12.

Punica

Species 2. **Shrubs** or **small trees**; branches often terminating in spines. **Leaves** ± opposite, sometimes crowded on short lateral shoots, estipulate; lamina simple, entire. **Inflorescences** 1–5-flowered clusters, or flowers solitary. **Flowers** actinomorphic, bisexual; floral tube thick, leathery, adnate to ovary and produced above it; sepals thick, valvate, persistent; petals showy, red (or white), strongly crumpled; stamens numerous, covering inner surface of floral tube from rim to ovary; ovary inferior, multiloculed; style exserted; stigma capitate. **Fruits** berrylike, with leathery rind, retaining a crown of sepals; seeds many, with translucent, clear to dark pink or rose, juicy sarcotesta; cotyledons spirally rolled. **Distr:** 1 species endemic to Indian Ocean island of Socotra and the other species of uncertain origin (perhaps the Mediterranean region to C and SW Asia). **Cult:** Propagate from seeds, cuttings or by air-layering. Requires a moist, well-drained compost in full or partial sun.

P. granatum L., *Sp. Pl.* 1: 472, 1753. [Pomegranate]. **Shrubs** or **small trees**, 2–3 m high, glabrous; branches and branchlets 4-angled, becoming terete with age, often terminating as indurate spines. **Leaves** lanceolate, elliptic-oblanceolate or oblong, 2–9 x 1–2 cm, base attenuate, apex obtuse or mucronate, adaxially shiny; petiole 0.2–1 cm long. **Inflorescences** 1–5-flowered. **Flowers** campanulate-urceolate, floral tube red-orange or pale yellow, 2–3 x 1–1.5 cm; sepals 5–9, erect, deltate; petals 5–9, bright reddish-orange (or white), obovate, 1.5–3 x 1–2 cm, rounded or obtuse; stamens numerous, included to exserted; ovary 8–13-loculed, in 2 or 3 superposed layers, lower locules with axile placentation, upper ones with apparent parietal placentation. **Fruits** globose, leathery berries, variable in color, red to yellow-green or red-brown, 5–12 cm Ø, crowned by persistent sepals, irregularly dehiscent; seeds obpyramidal within juicy sarcotestal layer, ruby-red, pink, or yellowish-white. Fl. Mar–Jul. **Distr:** Origin unknown; cultivated and naturalized worldwide, especially in the tropics and subtropics. **Ecol:** On dry stony or rocky slopes, cliffs, open forests, thickets and riverbanks; 400–1,400 m. Fl. Apr–July. **Zones:** 9–12.

Sonneratia

Species 9. **Trees**, evergreen, glabrous, columnar or spreading, not buttressed, surrounded by pencil-like pneumatophores arising from long, shallowly buried, horizontal roots. **Leaves** opposite, borne in a single plane; lamina simple, margins entire, apex often with a minute hydathode forming a thick mucro, leathery. **Inflorescences** 1–3-flowered, clustered at apex of ultimate, pendulous branchlets. **Flowers** actinomorphic, 4–8-merous, opening at night; floral tube shallowly campanulate, at fruit maturity persistent, partly surrounding fruit or expanded and saucer-shaped with fruit fully exposed; sepals 4–8, adaxially green or red, thick, leathery; petals red or white, lanceolate-linear, caducous, vestigial, or absent; stamens numerous, filaments red or white, anthers reniform; ovary superior to partly inferior, depressed, 10–20-carpellate. **Fruits** globose, berrylike, leathery, indehiscent, crowned by style base; seeds numerous, embedded in foul-smelling pulp. **Distr:** E Africa through Maldives to Indo-Malesia, China, Australia, New Guinea, and the W Pacific Islands. **Ecol:** In mangroves. **Cult:** Propagate from cuttings or seeds. Requires brackish mangrovelike conditions in full sun.

S. alba Smith, *Cycl.* 33: Sonneratia no. 2, 1816. (Syn: *Chiratia leucantha*, etc.). [Mangrove, Mangrove Apple, Perepat Tree]. **Trees**, 3–15(–30) m tall, spreading; crown broad. **Pneumatophores** 0.3(–1) m high, thick, blunt. **Leaves** elliptic to ovate or obovate, 5–11 x 4–8 cm, base rounded, apex broad, rounded, with broad recurved mucro, adaxially pale; petiole 0.5–1.5 cm long. **Flowers** 5–8-merous; floral tube shiny, 3–3.5 cm long, smooth, often 6-ribbed; sepals adaxially strongly red-tinged, 1.3–2 cm long, ± erect at anthesis, recurved in fruit; petals white,

RT

MM

Above: *Sonneratia alba, Sonneratia caseolaris, Woodfordia fruticosa, Bellucia pentamera.*

linear, 1.3–2 × ~0.1 cm, or variably semipetalous to absent; staminal filaments white. **Fruits** 2–4.5 cm Ø, about equal to width of floral tube; seeds falcate. **Distr:** China, India, Malaysia, Maldives, Myanmar, Sri Lanka, Thailand, Vietnam; tropical E Africa, Madagascar, Seychelles; N Australia, W Pacific Islands. **Ecol:** In shallow parts of calm seas and seashores, tidal creeks. Fl. Oct–Nov. Fr. Feb. **Zones:** 10–12.

S. caseolaris (L.) Engler, *Nat. Pflanz., Nachtr.* 1: 261, 1897. (Syn: *Sonneratia acida*, etc.). [Crab Apple Mangrove, Mangrove Apple]. **Trees**, columnar, 5–20 m tall. **Pneumatophores** ~1 m high, thin, pointed; ultimate branchlets pendulous, 4-angled. **Leaves** elliptic to broadly elliptic or broadly oblong, 4–11 × 2–7 cm, base rounded, narrowly to broadly attenuate on petiole, apex rounded with minute thick mucro, midrib prominent with inconspicuous lateral veins; petiole 2–9 mm long. **Flowers** 5–7-merous; floral tube 2.5–3 cm long, smooth, flattened in mature fruit; sepals adaxially green, often red-streaked, 1.4–1.9 cm long; petals red, linear, 1.8–2.9 × 0.1–0.3 cm; staminal filaments red, sometimes white distally. **Fruits** (2–)4–5 cm Ø, broader than floral tube; seeds irregularly angular, ~7 mm long. **Distr:** Cambodia, China, India, Indonesia, Malaysia, Maldives, N New Guinea, Sri Lanka, Thailand, Vietnam; N Australia, Pacific Islands. **Ecol:** In coastal mangrove communities, tidal creeks. **Zones:** 11–12.

Woodfordia

Species 2. **Shrubs** or **small trees**; stems irregularly branching; branches pendulous. **Leaves** opposite, sessile or subsessile; lamina pubescent to tomentose, abaxially orange to black glandular-punctate. **Inflorescences** condensed axillary clusters, rarely flowers solitary. **Flowers** 6-merous, slightly zygomorphic; sepals very short, alternating at sinus with minute, thickened epicalyx segments; floral tube red-orange, cyathiform, slightly constricted at level of stamen insertion; petals red, pink or white, small; stamens 12, 2-whorled, alternating in length; ovary ellipsoidal, sessile or shortly stipitate, style thicker than staminal filaments, ultimately long-exserted, stigma punctiform. **Fruits** capsules, thin-walled, translucent at maturity, irregularly dehiscent, rarely loculicidally dehiscent; seeds many, narrowly obpyramidal, small. **Distr:** Africa, the Arabian peninsula (1 sp.); SE Asia including China (1 sp.). **Cult:** Propagate from cuttings, seeds or by air-layering. Requires a well-drained soil in shade or full sun.

W. fruticosa (L.) Kurz, *J. Asiat. Soc. Bengal*, Pt. 2, Nat. Hist. 40: 56, 1871. (Syn: *Grislea punctata*, etc.). [Fire Flame Bush, Red Bell Bush]. **Shrubs**, 1–5 m high; stems and branches pendulous, long, pubescent when young, becoming glabrous. **Leaves** lanceolate or ovate-lanceolate, 3–14 × 1–4 cm, base rounded to subcordate, apex acuminate, leathery, adaxially glabrous, abaxially sparsely to densely tomentose and orange to black glandular punctate. **Inflorescences** condensed axillary clusters, 1–15-flowered. **Flowers** with floral tube light red, red-orange, or deep red, greenish basally, narrowly cyathiform, 0.9–1.5 cm long; sepals oblong-ovate or deltate, 2–3 mm long; epicalyx segments scarcely present; petals 6, thin, linear-lanceolate, 1–5 mm long; stamens 12, inserted above ovary base, long-exserted; ovary 2-loculed, ovules 100+. **Fruits** capsules, elongate, elliptic; seeds reddish-brown, ~1.5 mm long. **Distr:** Bhutan, China, India, Indonesia, Laos, Myanmar, Nepal, Pakistan, Thailand. **Ecol:** In forests and on open slopes. Fl. Jan–May (mainly Mar–Apr). Fr. Apr–May. **Zones:** 10–12.

MELASTOMATACEAE

Genera 182, species 4,572. **Herbs**, **shrubs**, or **trees**, to 20 m tall, erect, climbing, or rarely epiphytic. **Stipules** absent. **Leaves** commonly opposite and decussate with one of a pair slightly smaller than other, rarely verticillate or alternate by abortion of one of a pair; lamina simple, usually 1–4(or 5) veins per side, originating at or near base and anastomosing apically, tertiary veins numerous, parallel, and connecting secondary veins and midvein. **Inflorescences** cymose, umbellate, corymbose, in paniculate clusters, or cincinni, rarely flowers solitary, fascicled, or borne on a spike; bracts sometimes conspicuous and persistent; bracteoles opposite, usually caducous. **Flowers** bisexual, actinomorphic, androecium often slightly zygomorphic, usually (3 or)4- or 5(or 6)-merous, perianth biseriate, perigynous; hypanthium funnel-shaped, campanulate, cyathiform, or urceolate; calyx lobes (3–)5(or 6), valvate; petals (3–)5(or 6), equal to number of sepals, distinct, imbricate; stamens usually twice as many as petals, 2-whorled, filaments distinct, anthers 2-celled, introrse, basifixed, dehiscent by 1 or 2 apical pores or by short longitudinal slits, connective often variously appendaged; pistil and style 1, stigma minute, capitate or truncate, ovary commonly inferior or semi-inferior. **Fruits** dry or fleshy capsules or berries, loculicidally dehiscent or indehiscent; seeds usually small. **Distr:** Predominantly tropical and subtropical regions worldwide; greatest diversity in S America and the Caribbean.

Bellucia

Species 8. **Shrubs** or **trees**. **Leaves** arranged in terminal clusters, petiolate; lamina coriaceous, 3–5-plinerved. **Inflorescences** with flowers solitary, fascicled or in short cymes, often axillary or sometimes borne on older wood below terminal cluster of leaves. **Flowers** large or small; calyx at anthesis splitting into several regular or irregular lobes or rarely circumscissile; stamens similar, 10–16, filaments short, stout or slender, anthers often somewhat coherent, 4-celled, minutely 1- or 2-pored, connective ± elevated but not prolonged, smooth; ovary inferior, 5- or 8–15-celled, style short with a capitate or ovoid stigma. **Fruits** baccate, many-seeded, edible. **Distr:** C America to S America. **Cult:** Propagate from seeds, rooted cuttings or by air-layering. Requires a fertile, humus-rich, moist, well-drained soil in shade or partial sun.

B. pentamera Naud., *Ann. Sci. Nat.* sér. 3. 16: 105, 1851. (Syn: *B. axinanthera, B. costaricensis*). [Coronillo]. **Shrubs** or **trees**, glabrous, small or medium-sized, 4–10 m tall; branchlets and petioles grooved. **Leaves** opposite; lamina simple, ovate to elliptic, to 20–40 × 6–28 cm, base acute, apex acute or shortly acuminate, adaxially lustrous, abaxially somewhat glaucous, light greenish-blue, secondary veins prominent, running parallel to central vein, tertiary veins smaller, running perpendicular to prominent parallel veins in a ladderlike manner; petioles 1–2(–3.5) cm long. **Inflorescences** with flowers solitary or in cymes of 3–9 flowers; pedicels 0.5–1.5 cm long. **Flowers** white or pale rose-colored, 1.8–3 cm across; calyx 1.5–2 cm wide, strongly lobed, the 5 equal or subequal, triangular-acute divisions slightly shorter than the tube; style 8–9 mm long, stigma slightly dilated, deeply multisulcate. **Fruits** yellow, 4 cm Ø; seeds numerous. **Distr:** Mexico to Tropical S America. **Ecol:** In forests. **Zones:** 10–12.

Blakea

Species ~100. **Shrubs** or **small trees**, erect or occasionally scandent and rooting, or apparently sometimes epiphytic, glabrous or variously pubescent; branches usually terete. **Leaves** petiolate or subsessile; lamina large or small, broad, entire or nearly so, mostly coriaceous, 3–5-nerved, striolate with numerous transverse, parallel veins. **Inflorescences** fasciculate or flowers solitary, axillary; pedicels long or short. **Flowers** usually large, pink, purple or white, surrounded by 4 or rarely 6 subcoriaceous bracts, these opposite by pairs. **Flowers** 6-parted; calyx tube broadly campanulate, limb simple, short or elongate, 6-lobate, persistent; petals oblong or obovate, glabrous; stamens 12, equal, filaments thick, glabrous, anthers thick, subhorizontal or pendulous, laterally compressed, short, obtuse and minutely 2-pored at apex, connective thickened and unappendaged or basally dilated into a dorsal, deflexed, horizontal or upturned spur; ovary inferior, 4–6-locular, stigma punctiform to capitate. **Fruits** berries, baccate, juicy; seeds ovoid, clavate, or narrowly pyriform, testa smooth. **Distr:** S Mexico, West Indies to Bolivia and Brazil. **Ecol:** In wet montane regions. **Cult:** Propagate from seeds, rooted cuttings or by air-layering. Requires a fertile, humus-rich, moist, well-drained soil in shade or partial sun.

B. trinervia L., *Syst.* ed. 10, 1044, 1759. (Syn: *B. fruticosa, B. trinervia* var. *normanii*). [Jamaica Rose]. **Shrubs**, climbing on trees; young branchlets, petiole, nerves of leaves beneath, and peduncles with thick brown hairs or minute scales. **Leaves** narrowly elliptical, elliptical to roundish-elliptical, 10–15 × 5–8.5 cm, base rounded or obtuse, apex shortly and abruptly acuminate, 3-plinerved or 3-nerved with a marginal pair, glabrous except abaxial nerves; petiole 1–3 cm long. **Inflorescences** with flowers solitary, in upper leaf axils; peduncles 3–6 cm long; bracteoles roundish, with powdery-stellate scales, puberulous, or glabrate, margin above middle tomentose, with many longitudinal nerves, rigid, 1.5–2 cm long. **Flowers** large; calyx green, lobes tomentose near apex, 2–4 mm long, obtuse; petals 2–3.5 cm long, purple; filaments white, anthers creamish-white, 1.5 cm long. **Distr:** Jamaica. **Ecol:** In moist places, often growing under shade of trees; 500–1,200 m. Fl. throughout the year. **Zones:** 11–12.

Clidemia

Species 120. **Shrubs**, commonly setose, villous, or tomentose. **Leaves** mostly petiolate, sometimes bearing inflated appendages (usually housing ants); lamina 3–5-nerved, entire or denticulate. **Inflorescences** axillary panicles or fascicles, sometimes congested capitula. **Flowers** 4–5-parted, usually small, white, pink, or purple; calyx variously pubescent, tube campanulate to urceolate, limb often produced beyond ovary, truncate or usually shortly lobed, lobes often appendaged with elongate teeth; petals obtuse or retuse, usually glabrous; stamens equal or slightly unequal, twice as many as petals, filaments glabrous, subulate or filiform, anthers linear-elongate or filiform, straight or curved, connective commonly not produced below cells, not appendaged at base or sometimes gibbous or subcalcarate; ovary 3–5-celled, glabrous or more often setulose. **Fruits** baccate, generally globose, small, crowned with the persistent calyx limb; seeds usually minute, smooth or granulate. **Distr:** C America. **Cult:** Propagate from seeds. Requires a moist, well-drained soil in full sun to shade.

C. hirta (L.) D. Don., *Mem. Wern. Soc.* 4, 1823. [Koster's Curse, Soap Bush]. **Shrubs**, slender, densely branching forming dense monotypic stands, to 3(–4) m high; branches densely long-hirsute and ± stellate-puberulent. **Stipules** absent. **Leaves** opposite; lamina simple, ovate or oblong-ovate, 4–15 cm long, base rounded or cordate, often unequal, margins crenulate, denticulate or subentire, ciliate, apex acute or shortly acuminate, 5–7-nerved, hirsute, papery. **Inflorescences** cymes, few-flowered, short, 3–5 cm across, axillary; flowers shortly pedicellate. **Flowers** small, ~1.2 cm Ø; calyx tube 5 mm long, setulose and stellate-puberulent, the outer teeth 2–4 mm long; petals white or pink, 0.8–1 cm long. **Fruits** berries, blackish, dark purple, bluish or red, 5–7 mm Ø, fleshy; seeds >100. **Distr:** C and S America; naturalized pantropically. **Ecol:** In disturbed habitats; <900 m. Fl. & Fr. throughout the year. **Info:** A pantropical weed.

Dissotis

Species ~140. **Herbs** annual or perennial, **shrubs** or **small trees**, usually ± pilose, pubescent, lanate or tomentose, rarely completely glabrous. **Leaves** opposite or verticillate, usually petiolate. **Inflorescences** usually heads or cymes, frequently paniculate, rarely solitary. **Flowers** 4–5(6)-merous; receptacle with various indumentum or rarely glabrous; sepals persistent or caducous; intersepalar segments present, rarely absent; petals obovate, rose, mauve, purple or violet, rarely white or orange; stamens 8–10(12), very unequal, sometimes subequal, anthers of longer stamens lanceolate-subulate, rostrate, 1-porose, with introrse pore and undulate thecae, generally purple, with a long pedoconnective provided at base with an anterior bilobed or bifid appendage like a spur, those of shorter stamens usually yellow, shorter, with a short pedoconnective and smaller appendage; ovary 4–5-locular, adnate to receptacle by 8–10(12) longitudinal septa, usually bristly above, style filiform, curved, slightly dilated toward apex, stigma punctiform. **Fruits** capsules, 4–5-valved, coriaceous, included in receptacle; seeds minute, cochleate, numerous. **Distr:** Tropical and subtropical southern Africa. **Cult:** Propagate from seeds or cuttings. Requires a moist, well-drained soil in full or partial sun.

D. rotundifolia (Sm.) Triana, *Trans. Linn. Soc.* 28: 58, 1872. (Syn: *D. plumosa*, etc.). [Dwarf Tibouchina, Pinklady, Spanish Shawl, Trailing Tibouchina]. **Herbs**, decumbent, rooting at nodes; branches, erect, to 16+ cm high. **Leaves** ovate-suborbicular to oblong-ovate or ovate-lanceolate, 1.5–7 × 0.8–4 cm, base truncate to shortly attenuate, margins slightly crenulate-ciliate, apex acute, ± densely pilose, basal longitudinal nerves 1(–2) pairs; petiole 0.5–2.5 cm long. **Inflorescences** with flowers solitary, rarely 2–4-flowered cymes. **Flowers** 5-merous; calyx tube cylindric-campanulate, 5–7 × 3–4 mm, ± pilose, lobes subulate-lanceolate, 6 mm long, margin ciliate, apex setose; intersepalar appendages ligulate, 2–2.5 mm long, apex setose; petals pink, obovate, 2 × 1.5 cm; stamens 10, unequal, *large* stamens with anthers 7 mm long, connective produced to 6 mm, bilobed appendage 1.5 mm long, filament 6 mm long, *small* stamens with anthers 5.5 mm long, connective produced to 1 mm, bilobed appendage 0.5 mm long, filament 5 mm long. **Distr:** Tropical Africa. **Ecol:** In moist places; 0–1,900 m. **Zones:** 10–12.

Medinilla

Species 400. **Trees** small, or **shrubs** (erect or scandent), terrestrial, epiphytic, or rarely climbing; stems 4-sided or terete, sometimes succulent or corky. **Leaves** opposite or verticillate, petiolate or sessile; lamina usually glabrous, margins entire, crenate or dentate, 3–5-nerved, lateral nerves joining midrib ± above base. **Inflorescences** umbellate, cymose, paniculate, or flowers solitary, terminal or axillary, inserted on leafless stems or at nodes of rootstock; bracts small, caducous; bracteoles 2, small at pedicel base. **Flowers** 4(–6)-merous; hypanthium cup-shaped, funnel-shaped, campanulate or tubular; calyx lobes conspicuous or inconspicuous, apiculate; petals obovate, ovate or suborbicular, sometimes oblique; stamens twice as many as petals, whorls equal or slightly unequal; anthers linear, lanceolate, or oblong, base tuberculate or appendaged, apex beaked, dehiscent by a single pore, connective slightly decurrent forming a spur; ovary inferior, 4-locular, ovate, apex truncate or with a membranous crown, sometimes septate, style filiform. **Fruits** berries, globular, ovate, or often urceolate, apex indehiscent, soft, juicy; seeds numerous, obovate to shortly cuneate, small, glabrous or tuberculate. **Distr:** Tropical Africa, Asia; Pacific Islands. **Cult:** Propagate from rooted cuttings or by air-layering. Requires a fertile, moist, well-drained soil in shade or partial sun.

M. albiflora Merr., *Philipp. J. Sci.*, vol. 8: 385–386, 1913. **Shrubs**, epiphytic, ~2 m tall; branches terete; growing branchlets very obscurely and thickly 4-winged or merely sulcate. **Leaves** opposite, sessile; lamina ovate-elliptic, 25–30 × 15–17 cm, prominently narrowed below, the very base ~3 cm wide, apex shortly acuminate, coriaceous, abaxially slightly paler, 5-plinerved, interior pair reaching apex, exterior pair evanescent at middle or upper two-thirds. **Inflorescences** panicles, few-flowered, terminal, pedunculate, ~12 cm long, without bracts or bracteoles. **Flowers** white, 5-merous, rather large; calyx somewhat funnel-shaped or distinctly urceolate, truncate, ~6 mm long; petals obliquely obovate, 1.8 cm long; stamens 10, 5 somewhat longer than others, longer ones with filaments 1.1 cm long and anthers 1.2 cm long, shorter ones with filaments ~1 cm long and anthers 0.9–1 cm long, anthers lanceolate, acuminate, somewhat curved, longer ones with a very minute dorsal spur, shorter ones with dorsal spurs nearly 1 mm long; ovary 5-celled. **Distr:** Philippines (Leyte). **Ecol:** In forests. **Zones:** 11–12.

M. fuchsioides Gardn., *Calcutta J. Nat. Hist.* 8: 12, 1847. **Shrubs**, generally epiphytic, often pseudo-dichotomously branched, to ~3 m high; branchlets terete or subquadrangular, stout. **Leaves** sessile or with a very short stout petiole; lamina elliptic, 3.5–9.5 × 1.5–4.5 cm, base acute in outline but rounded to subcordate at the very base, margins entire or shallowly crenate, apex obtuse or subacute, 5-nerved. **Inflorescences** clusters, few-flowered, or flowers solitary; flowers ± pendent on a 0.5–1 cm long peduncle in axils of fallen leaves; pedicels 3–7 mm long. **Flowers** with hypanthium 3–6 mm wide, pink to red; petals much overlapping, rotund, 1.3–1.6 cm long, waxy and translucent, pink; anthers narrowly ovate, ~4 mm long, acute to obtuse, connective appendage small but distinct, straight; 2 basal theca projections very small, filaments about twice as long as anthers. **Fruits** subglobose to campanulate, red and becoming dark purple. **Distr:** Sri Lanka. **Ecol:** In

Below: *Blakea trinervia, Clidemia hirta, Dissotis rotundifolia, Medinilla albiflora.*

montane moss forests such as in Horton Plains. **Zones:** 10–12.

M. magnifica Lindl., *Paxt. Flow. Gard.* i. 55. t. 12, 1850. (Syn: *M. bracteata*). [Chandelier Tree, Red Medinilla, Rose Grape, Showy Medinilla]. **Shrubs**, evergreen, erect, glabrous; branches compressed, 4-winged, bearing tufts of bristles at joints. **Leaves** opposite; lamina broadly oblong, very large, base cordate and semiamplexicaul, coriaceous, triple-nerved below middle, and the rest penninerved, adaxially dark green, abaxially paler. **Inflorescences** panicles, terminal, >45 cm long, flowers numerous, when in bud wholly clothed with densely imbricated, large, rose-colored bracts, these are gradually deciduous, the lower and larger ones remaining and becoming reflexed, lowermost green-tinged; branches of panicle whorled, much-divided; bracts to 10 cm long. **Flowers** deep rose colored; calyx cuplike, pink; corolla with petals 5, 0.8–1.5 cm long, borne atop the calyx, surrounding the stamens; stamens 10, yellowish, curved, anthers linear, affixed to yellow, conspicuously bent filaments. **Fruits** berries, cuplike, purplish-pink. **Distr:** Philippines (Manila). **Ecol:** Fl. Apr–Jun. **Zones:** 11–12.

M. myriantha Merrill, *Philipp. J. Sci.* 1 (Suppl. 3): 215, 1906. [Malaysian Grapes, Malaysian Orchid, Rose Grape Medinilla]. **Shrubs**, 4–5 m tall, glabrous; branches terete, rather slender, younger ones obscurely angled, nodes densely barbellate. **Leaves** opposite, sessile; lamina elliptic-ovate, 9–13 × 5.5–7 cm, base cordate, often broadly so, apex abruptly shortly acuminate, subcoriaceous, nerves 7–9, prominent, sometimes with an additional pair of outer short nerves at base. **Inflorescences** panicles, many-flowered, terminal, 10–20 cm long, pink, branches verticillate, lower ones ~9 cm long, gradually shorter upward; bracts minute, subulate. **Flowers** 4-merous, pink; calyx urceolate-cylindrical, 4 mm long, truncate; petals 4, oblique, ~8 mm long; stamens 8, subequal; filaments 4 mm long, anthers ~5 mm long, spur <1 mm long. **Distr:** Phillipines (Luzon). **Ecol:** In mossy forests; 900–1,100 m. **Zones:** 11–12.

M. scortechinii King, *J. Asiat. Soc. Bengal*, Pt. 2, Nat. Hist. 69: 62, 1900. [Coral Medinilla, Orange Medinilla, Orange Spike]. **Shrubs**, epiphytic or semiterrestrial, 30–60(–90) cm high; stems woody at base; rootstock often caudiciform, caudex to >6 cm thick. **Leaves** opposite, sessile, ascending; lamina ovate, 20–25 × 12–15 cm, base truncate, slightly clasping stem (basal lobes of opposite pairs often overlapping), margins entire, sometimes wavy, apex acute to acuminate, thickly coriaceous, adaxially bright to dark green, glossy, abaxially much paler, 3-nerved (with an additional very faint marginal pair), midvein relatively thick. **Inflorescences** panicles, lax, terminal and axillary, 25–32 × 22–27 cm, all parts orange, glabrous, branching to third order; pedicels ~0.8–1.2 cm long. **Flowers** orange, ~5 cm ∅, somewhat fleshy; calyx ~1.5 cm long, base urceolate ~9 mm ∅ at widest point, narrowed below limb, lobes acute; petals free, ovate, truncate to rounded; stamens with anthers white. **Distr:** Peninsular Malaysia, throughout Indonesia to Philippines. **Ecol:** Epiphytic or semiepiphytic on trees in dense forest; 900–1,200 m. Fl. Feb–Mar, Aug–Sep. **Zones:** 11–12.

M. speciosa Hook., *Bot. Mag.* 73: t. 4321, 1847. (Syn: *M. eximium*). [Showy Medinilla]. **Shrubs**, 1.5 m tall, erect, glabrous; branches di- or trichotomous, tri- or tetragonous, angles winged, leaf scars with a circle of harsh wiry hairs around the stem. **Leaves** mostly in whorls of 3–4, rarely opposite, nearly sessile; lamina oval or oval-oblong, large, base usually obtuse, margins quite entire, apex acuminate, surfaces often wavy, 5- or 7-nerved, nerves abaxially and often margins red. **Inflorescences** panicles, terminal, branched almost to base, drooping, main peduncle and branches red; secondary branches whorled; ultimate pedicels short. **Flowers** with calyx, urceolate, pale red, truncate, deepest colored toward the margin; petals 4–5, ovate, acute, spreading, delicate rose; stamens 8–10, filaments subulate, anther longer than the filament, curved, subulate, with 2 blunt anterior horns at base and a short spur behind; style filiform, shorter than stamens, curved. **Distr:** Java. **Ecol:** On rocks in hill forests; ~900 m. Fl. Feb–Jun. **Zones:** 11–12.

M. venosa (Blume) Blume, *Flora* 14: 518, 1831. [Holdtight]. **Shrubs**, epiphytic or terrestrial, erect 2–3 m tall. **Leaves** somewhat unequal in each pair; lamina elliptic, elliptic-oblong, or elliptic-lanceolate, 7–18 × 2.5–9 cm, base of larger leaves rounded to subauriculate, margins entire, apex acute to acuminate, venation pinnate, with 3–4 pairs diverging from central nerve, adaxially glabrous, abaxially hairy; petioles 0.4–4 cm long. **Inflorescences** 8–11 cm long; bracts and bracteoles ovate to elliptic-ovate or lanceolate, 0.5–1.5 × 0.2–0.8(–1.4) cm, 3–5-nerved, deciduous; pedicels 6–8 mm long in fruit. **Flowers** with hypanthium ~4 mm long, ± stellate-pubescent; calyx lobes 4, erect, broadly rounded, ~2 mm long; petals 4, pink, obliquely obovate, 0.9–1.1 × 0.5–0.6 cm; stamens 8, ± equal, filaments 4.5–5 mm long, anthers 3.5–4 mm long, dorsal spur truncate, ~0.25 mm long, deflexed, ventral appendages ~1 mm long; ovary densely bristly at apex. **Fruits** berries, white, 4-celled,

Top row: *Medinilla fuchsioides, Medinilla magnifica, Medinilla myriantha, Medinilla scortechinii.*
Bottom row: *Medinilla speciosa, Medinilla venosa, Melastoma malabathricum, Miconia calvescens.*

F&KS

Above: *Osbeckia aspera, Osbeckia octandra, Tibouchina urvilleana, Memecylon floribundum.*

5–7 mm long; seeds papillose, ~0.5 mm long. **Distr:** Moluccas, Philippines, Sulawesi. **Ecol:** In wet coastal forests. **Zones:** 11–12.

Melastoma

Species ~70. **Shrubs. Leaves** entire, pubescent, 5-nerved. **Inflorescences** cymes, few-flowered, bracteate; flowers pedicelled. **Flowers** 5-merous, hypanthium cyathiform-urceolate, adnate to ovary base, covered with hairlike scales; intersepalar emergences alternating with sepals present, small, subulate; petals large, ± ciliate apically at margin; stamens 10, dimorphic, opening by an apical pore, 5 *longer* ones red-violet with a connective basally projected into a curve, at the end bifurcate stalk about the same length as the thecae, 5 *shorter* ones with yellow anthers and a connective forming 2 small lobes basally at insertion of the filament; style filiform, apically slightly curved; ovary apically with hairs, 5-locular. **Fruits** capsules, coriaceous or fleshy, borne inside the persistent hypanthium; seeds numerous. **Distr:** Sri Lanka, India to SE Asia, N Australia, and the Pacific. **Cult:** Propagate from seeds or half-hardened cuttings. Requires a fertile, humus-rich, moist, well-drained soil in full sun.

M. malabathricum L., *Sp. Pl.* 390, 1753. (Syn: *M. royenii, M. ellipticum*). [Blackmouth, Blue Tongue, Malabar Melastome, Indian/Straits Rhododendron]. **Shrubs** to 3 m tall; branchlets covered with fimbriate brownish scales. **Leaves** elliptic, 5–20 × 1.5–7 cm, base acute to obtuse, apex acute or shortly acuminate, 5-nerved, surfaces hairy and with scales similar to those on stems; petiole 0.3–1.5 cm long. **Inflorescences** clusters, dense, but few-flowered; bracts widely elliptic, ventrally glabrous, dorsally scaly, soon deciduous. **Flowers** with hypanthium 5–9 mm wide, wider and ovoid in fruit, becoming shallowly 5-toothed after loss of sepals and intersepalar emergences, densely scaly; intersepalar emergences subulate, with similar scales; sepals ovate-triangular, ventrally glabrous, dorsally with scales as on hypanthium; petals 2–3 cm long, pink to mauve and purple; anthers narrowly ovate, 5–8.5 mm long excluding the basally projected connective. **Distr:** Sri Lanka, India; SE Asia. **Ecol:** In wet montane regions, thickets, riverbanks. **Zones:** 10–12. **Info:** The fruit pulp is edible, and stains black.

Miconia

Species 1,000. **Shrubs** or **trees. Leaves** usually petiolate; lamina entire or dentate. **Inflorescences** panicles or corymbs, rarely glomerate or spicate, terminal or very rarely lateral; flowers shortly pedicellate or sessile, sometimes secund on branches. **Flowers** mostly white, small, 4–8-parted, normally 5-parted; calyx tube urceolate, campanulate, or globose, limb truncate or lobate, lobes short, not appendaged externally or appendages punctiform or minute; petals obovate or oblong, obtuse or retuse; stamens equal or nearly so, filaments subulate or filiform, anthers polymorphous, short or elongate, commonly linear, oblong, or subulate, 1–4-pored at apex, connective usually shortly produced, anteriorly biauriculate or bituberculate; ovary 2–5-celled. **Fruits** baccate, usually globose; seeds pyramidal or angulate, rarely ovoid, generally smooth and lustrous. **Distr:** C America. **Ecol:** In forest undergrowth. **Cult:** Propagate from seeds or rooted cuttings. Requires a fertile, moist, well-drained soil in full or partial sun.

M. calvescens DC., *Prod.* 3: 185, 1828. (Syn: *Cyanophyllum magnificum, M. magnifica*). [Bush Currant, Purple Plague, Velvet Tree]. **Shrubs** or **small trees**, to 8 m tall; branchlets ± tomentose with small stellate hairs, soon becoming glabrous. **Leaves** elliptic, elliptic-obovate or obovate-oblong, >50 × 20 cm, base rounded or subcordate, apex shortly acuminate or sometimes obtuse to rounded, 5-nerved with many distinct lateral veins, adaxially glabrous, abaxially ± purple and puberulous, especially on veins, hairs stellate; petiole 4–15 cm long. **Inflorescences** panicles, flowers numerous, pyramidal, large, 20–50 cm long, ± erect, branched, terminal, puberulous to tomentose. **Flowers** small; hypanthium 2–2.5 mm wide, puberulous, hairs stellate; petals obovate, ~2.5 mm long, white; anthers narrowly ovate-oblong, 2.3–3 mm long. **Fruits** globose, 10-costate, 2–4 mm Ø, each crowned with persistent hypanthial rim. **Distr:** C and S America. **Ecol:** In moist areas, as understory or near forest margins; <650 m. **Zones:** 10–12.

Osbeckia

Species ~50. **Herbs, shrublets,** or **shrubs**, erect; stems (3 or)4-sided, pubescent. **Leaves** opposite or sometimes 3-whorled, petiolate or sessile; lamina with margins entire, usually pubescent, secondary veins 1–3 per side, tertiary veins numerous and parallel. **Inflorescences** capitate or paniculate, terminal; flowers usually bracteate. **Flowers** 4- or 5-merous, purple or white; hypanthium urceolate or flask-shaped, often setiform-tuberculate; calyx lobes linear, lanceolate, or ovate-lanceolate, margin ciliate; petals obovate or broadly ovate, margin ciliate or not; stamens twice as many as perianth segments, isomorphic, ± equal, filaments shorter or as long as anthers, anthers oblong or oblong-ovate, beak long or short, connective decurrent, base abaxially slightly inflated or shortly spurred and sometimes with 1 or 2 setose, adaxially lengthened tubercles; ovary half inferior, 4- or 5-celled, ± adherent to hypanthium, apex with a setose ring, style filiform. **Fruits** capsules, ribbed, contracted in the middle, dehiscing at truncate apex by 4 or 5 pores; seeds numerous, curved (cochleate), small, densely granulate. **Distr:** Tropical W Africa, tropical and subtropical Asia. **Cult:** Propagate from seeds or half-hardened cuttings. Requires a moist, well-drained, sandy soil in full sun.

O. aspera (L.) Blume, *Bijdr.* 6: 220, 1831. (Syn: *Melastoma asperum*, etc.). [Bowitiya]. **Shrubs**, much-branched, to 3 m tall; branchlets ± pubescent. **Leaves** elliptic or elliptic-ovate, 1.5–11 × 0.8–5 cm, base rounded to acute, apex acute or sometimes shortly acuminate, 3–5-nerved, ± pubescent; petiole 0.1–1.2 cm long. **Inflorescences** elongated clusters, rather dense, few- to 10-flowered, pedicelled; hypanthium 4–9 mm wide, with scattered short bristlelike hairs. **Flowers** 5-merous; sepals triangular-ovate, 2–4 mm long, rounded to truncate, deciduous in fruit; petals 1.4–2.5 cm long, pink to mauve and purple; anthers very narrowly ovate, 5.5–9 mm long, twisted; ovary apically ± covered with erect hairs. **Distr:** S India, Sri Lanka. **Ecol:** On grasslands and in open forests. **Zones:** 10–12.

O. octandra (L.) DC., *Prod.* 3: 142, 1828. (Syn: *Asterostoma octandrum*, etc.). [Eight-stamen Osbeckia, Heen Bowitiya]. **Shrublets**, much-branched, to 2 m tall; branchlets sparsely pubescent. **Leaves** elliptic, 1.5–6 × 0.5–2 cm, base rounded to acute, apex acute, 3-nerved, ± sparsely pubescent; petiole 1–5 mm long. **Inflorescences** loose clusters, few- to 10-flowered; flowers pedicelled. **Flowers** 5-merous; hypanthium 3–6 mm wide, often glabrous or with scattered emergences and occasional bristles; intersepalar emergences generally with 1–few bristles, whitish, sometimes stellate with several bristles, purplish; sepals triangular, 2–4 mm long, apically with bristles as in intersepalar emergences, deciduous in fruit; petals 1.5–1.9 cm long, pink to mauve and purple; anthers very narrowly ovate, 6–8.5 mm long, twisted; ovary apically crowned by several basally connate bristles. **Distr:** Sri Lanka. **Ecol:** In montane regions, on grasslands and open places. Fl. Oct–Dec. **Zones:** 11–12.

Tibouchina

Species 245. **Trees, shrubs,** or **herbs. Leaves** petiolate; lamina with margins entire, membranous to coriaceous, 3–7-nerved. **Inflorescences** with flowers solitary or paniculate, sometimes subtended by large bracts. **Flowers** typically 5-merous, small or large and showy, mostly pink or purple; calyx tube oblong to campanulate, lobes subulate to ovate, equaling or exceeding tube, persistent or deciduous; petals obovate, sometimes retuse; stamens twice as many as petals, subequal or alternately unequal, filaments glabrous, glandular, or pilose, anthers alike, mostly linear-subulate, 1-pored, arcuate or sigmoid, connective ± produced below cells, anteriorly bilobate or bituberculate; ovary commonly 4-celled. **Fruits** capsular; seeds cochleate, minutely tuberculate. **Distr:** C America. **Cult:** Propagate from seeds or cuttings. Requires a moist, well-drained soil in full sun.

T. urvilleana (DC.) Cogn., *Fl. Bras.* 14 (3): 359, 1885. (Syn: *Lasiandra urvilleana, T. semidecandra* [misapplied]). [Glory Flower, Princess Flower, Purple Glorybush]. **Shrubs**, large, to 6 m tall; branchlets densely pubescent. **Leaves** elliptic or elliptic-ovate,

Above: *Memecylon macrocarpum, Memecylon rostratum, Memecylon umbellatum, Mouriri guianensis* (flowers).

4–11 × 1.5–5.5 cm, base rounded, apex shortly acuminate, 5–7-nerved, surfaces densely pubescent; petiole 0.6–1.8 cm long. **Inflorescences** several-flowered, branched, pubescent; bracts elliptic, 1.5–3 cm long, dorsally pubescent and ventrally glabrous. **Flowers** with hypanthium ~5 × 1.5 mm, wider and ellipsoid in fruit, densely pubescent; sepals narrowly triangular, dorsally pubescent; petals 2–3 cm long, deep purple; anthers very narrowly ovate, ~7 mm long excluding the basally projected connective. **Distr:** S America. **Ecol:** Usually in thickets of moist areas. **Zones:** 10–12.

MEMECYLACEAE

Genera 6, species 435. **Trees** or **shrubs**, indument absent or of uniseriate hairs; with nodes swollen; young branchlets cylindrical. **Stipules** present. **Leaves** opposite; lamina lanceolate or oblong, or ovate, margins entire, pinnately veined (with a single midrib), cross-venulate. **Inflorescences** aggregated forming cymes, axillary; pedicels articulated. **Flowers** regular, usually small, often bluish, 4–5-merous; calyx valvate; stamens 4–5, or 8, or 10, anthers versatile, dehiscing via longitudinal slits; ovaries inferior. **Fruits** berries, fleshy, indehiscent; seeds 1–5. **Distr:** Tropical regions, especially S America and SE Asia. **Ecol:** Usually understory plants of the forest. **Info:** The monophyly of this family is strongly supported by molecular data (Clausing and Renner 2001).

Memecylon

Species 200–300. **Shrubs** or small to medium-sized **trees**. **Leaves** entire, glabrous, midrib prominent, with distinct to obscure intramarginal and lateral veins. **Inflorescences** cymes, compound, bracteate, usually thyrsoid to umbel-shaped, often condensed to sessile fascicles of flowers or few-flowered heads at the tips of peduncles (primary axes); flowers sessile or pedicelled. **Flowers** tetramerous; calyx campanulate to cyathiform or urceolate, glabrous (occasionally papillose), smooth or shallowly striate to deeply furrowed with 8 radiating wings inside, truncate or ± 4-lobed; calyx lobes if present valvate in bud; petals rather thin, easily dropped at anthesis; stamens 8, equal, filaments filiform, anthers with thecae opening by slits and with a somewhat axe-shaped, curved or straight connective, with or without a dorsal disk-shaped gland; ovary unilocular, ovules several, placentation axile, style filiform. **Fruits** globose or occasionally ellipsoid, pulpy or juicy; seeds 1(2), large. **Distr:** Old World tropics. **Cult:** Propagate from seeds or by air-layering. Requires a humus-rich, moist, well-drained soil in full or partial sun.

M. floribundum Blume, *Mus. Bot.* 1: 361, 1851. **Shrubs**. **Leaves** ovate to oblong-elliptic, 6–18 × 2.5–8 cm, base obtuse or rounded, rarely cuneate, margins entire, apex acute, obtuse or sometimes emarginate, mucronate, mucro easily damaged, thickly coriaceous, secondary venation obscure on both sides, basal nerves absent, midrib obvious; petiole 3–7 mm long. **Inflorescences** umbels, simple or compound, 8–15-flowered; peduncles 2–5 mm long; pedicels articulate near middle, thick, 3–5 mm long, bibracteolate; bracteoles persistent, curved, ~1 mm long. **Flowers** hermaphrodite; hypanthium broadly campanulate, 4 mm long, upper rim minutely 4-dentate; petals purple or blue inside, outside reddish, obovate-orbicular, 3–4 mm long; ovules 7–10, often 8, style 4–5 mm long. **Fruits** berries, obovoid or ellipsoid, with an unequal base, 1–1.5 cm long. **Distr:** Cambodia, China, Indonesia, Vietnam. **Ecol:** In forests and forest margins. **Zones:** 10–12.

M. macrocarpum Thw., *Enum. Pl. Zeyl.* 110, 1859. **Trees**, small to medium-sized; branchlets stout, subterete. **Leaves** elliptic, 6–15 × 3.5–9 cm, base cuneate and narrowed into a short but distinct stout petiole, margins somewhat revolute, apex subacute to obtuse or rounded, occasionally shortly caudate, with indistinct intramarginal and lateral veins, very thick, adaxially ± shiny; foliar sclereids narrowly filiform. **Inflorescences** condensed or sometimes thyrsoid to umbel-shaped, borne at the nodes on the branches below leaves; pedicels at least as long as the calyx; buds acute to obtuse and apiculate, with exposed petals. **Flowers** pale blue to pinkish-red; calyx ~3 mm wide, deeply furrowed with radiating wings inside, shallowly 4-lobed with a somewhat scarious margin, inside enlarged and semiglobose in fruit; anthers ~2.2 mm long, connective with a gland. **Fruits** globose, constricted to a neck below calyx remnants with persistent radiating wings inside, 1.5–2 cm Ø. **Distr:** Sri Lanka; very rare. **Ecol:** In montane regions. **Zones:** 10–12.

M. rostratum Thw., *Enum. Pl. Zeyl.* 111, 1859. (Syn: *M. elegantulum*). [Heen Kuretiya]. **Shrubs** or small to medium-sized **trees**; branchlets terete to subquadrangular. **Leaves** elliptic or elliptic-ovate, 3–8 × 1–3 cm, base cuneate and narrowed into an up to 5 mm long petiole, apex distinctly caudate, obtuse to rounded at the very apex, with indistinct intramarginal and lateral veins, membranous, adaxially shiny. **Inflorescences** thyrsoid to umbellate, 0.5–1.2 cm Ø, sessile or with up to 3 mm long primary axes in leaf axils and at nodes on branches just below leaves; pedicels 1–3 times the calyx length; buds acute to obtuse and often apiculate at apex, with exposed petals. **Flowers** bluish to white; calyx 1.8–2.8 mm wide, smooth or shallowly striate inside, shallowly 4-lobed; anthers 1–1.4 mm long; connective with a gland. **Fruits** globose. **Distr:** Sri Lanka. **Ecol:** In wet and moist forests at low and medium elevations. **Zones:** 11–12.

M. umbellatum Burm. f., *Fl. Ind.* 87, 1768. (Syn: *M. rubro-caeruleum, M. tinctorium*). [Blue Mist]. **Shrubs large** or **small trees**, to ~5 m tall; branchlets subterete, whitish. **Leaves** elliptic, often canaliculate, 2.5–8 × 1.2–4.5 cm, base cuneate and narrowed into petiole, apex obtuse to rounded and often notched at, occasionally subacute, coriaceous, surfaces dull; foliar sclereids narrowly filiform; petiole to 6 mm long. **Inflorescences** thyrsoid to umbellate, many-flowered, 1–3 cm Ø, sessile or with up to 1 cm long primary axes at nodes on branches below leaves or occasionally axillary; pedicels 1–4 times the calyx length; buds acute to obtuse, with exposed petals. **Flowers** with calyx 1.8–3.4 mm wide, pale bluish or pinkish, sometimes pale yellowish-green, smooth or shallowly striate inside, truncate or shallowly 4-lobed; petals elliptic-ovate to roughly orbicular, blue, occasionally pink; filaments and style blue, occasionally pink, anthers 1.2–1.7 mm long, connective blue, occasionally pink, with a gland, pollen sacs yellow. **Fruits** globose, initially yellowish-green, maturing to bluish-black, ~7 mm Ø, edible. **Distr:** India, Sri Lanka. **Ecol:** In dry lowland forests. **Zones:** 10–12.

Mouriri

Species ~40–51. **Shrubs** or **small trees**, glabrous. **Leaves** sessile or shortly petioled; lamina entire, coriaceous, 1-nerved or faintly pinnately veined or veinless, midnerve prominent. **Inflorescences** axillary fascicles or flowers solitary at nodes of twigs. **Flowers** small, commonly 5-merous; calyx limb cupular, lobate or subtruncate, tube turbinate to hemispheric, smooth or in some species scurfy; petals acute or acuminate; stamens twice as many as petals and equal in length, exserted, filaments filiform, anthers 2-rimose; ovary inferior, 1–5-celled, style slender, stigma small, locules 2–several-ovuled. **Fruits** berries, globose, baccate, drupelike; seeds 1–4. **Distr:** Tropical S America. **Cult:** Propagate from seeds or by air-layering. Requires a fertile, humus-rich, moist, well-drained soil in full or partial sun. **Info:** Often misspelled as *Mouriria*.

M. guianensis Aublet, *Pl. Guian.* 1: 452, t. 180, 1775. (Syn: *M. polyantha*, etc.). **Trees**, often shrublike, to 10(–20) m tall; trunk to 20(–50) cm Ø; young twigs sometimes 4-winged. **Leaves** dimorphic, those of *dorsiventral* shoots usually ovate, 3–13 × 1.5–5.3 cm, base rounded or rarely shallowly cordate, apex acute to acuminate; leaves of *leader* shoots shorter, broader, base cordate when dimorphy is strong. **Inflorescences** axillary or at upper leafless nodes, 1 or 2 per side, each (1–)3–17-flowered, to 2.6 cm long; bracts triangular to ovate, 1–2.6 mm long, acute; pedicels to 1.1 cm long. **Flowers** with calyx yellowish or brown, campanulate, 3.6–5.2 mm long, lobes triangular, 0.7–1.3 × 1.6–2.8 mm, acute to broadly rounded; petals white to pink, violet, or sometimes yellow, squarish or roundish to broadly ovate to trullate, 4–7 × 2–6 mm, base clawed, margin erose, apex acute; *antesepalous* filaments 4–7.5 mm long, *antepetalous* filaments ~6–9 mm long, anthers yellow, purple or white; ovary 2–5-locular, ovules 9–33, style 1–1.3 cm

Above: *Mouriri guianensis* (fruits), *Acca sellowiana*, *Callistemon viminalis*, *Eucalyptus deglupta.*

long. **Fruits** orange to red, sometimes purple-black, globose, 0.65–1.1 cm Ø, edible; seeds 1 or 2(–3). **Distr:** SE Venezuela, Trinidad, the Guianas, south through Brazil to Rio de Janeiro. **Ecol:** In moist forests; 0–300 m. Fl. Nov. Fr. Jan. **Zones:** 10–12.

MYRTACEAE

Subfamilies 2 , **genera** 131, **species** 4,620. **Trees, shrubs**, or rhizomatous geoxylic **suffrutices. Stipules** absent. **Leaves** usually opposite or subopposite, rarely alternate, ternate, or in fours; lamina simple, entire, often coriaceous, pellucid-punctate. **Inflorescences** paniculate, thyrsoid, umbellate, cymose, racemose, or flowers solitary or fasciculate, axillary or terminal. **Flowers** mostly bisexual, sometimes unisexual by abortion, actinomorphic, partly or completely epigynous; sepals (0)4–5, often persistent, rarely accrescent, sometimes fused to form an operculum, often with punctate glands; petals (0)4–5, imbricate, free or coherent to form a calyptra or fused to form an operculum, often with punctate glands; stamens usually numerous, free or basally connate, filaments often coiled or folded in bud, anthers basifixed or dorsifixed, connective sometimes with an apical gland; ovary inferior or half-inferior, (1)2–5(10)-loculed, septa sometimes incomplete, placentation axile or parietal, locules or placenta 1–numerous-ovuled, style 1, punctiform, capitate, funnel-shaped or shallowly 2–trilobed at apex. **Fruits** berries, drupes or loculicidal capsules (dehiscing near apex), rarely nutlike and indehiscent; seeds 1–many per locule. **Distr:** Tropical and subtropical.

Acca

Species 3. **Shrubs** or **trees**, to ~4 m tall, usually partially covered with dense indumentum. **Leaves** simple, abaxially gray-downy. **Inflorescences** with flowers solitary or 3-flowered dichasia, sometimes aggregated in bracteate shoots; bracteoles persistent or caducous before anthesis. **Flowers** 4-merous; calyx lobes suborbicular to oblong; petals dark red to light purple or pink, somewhat fleshy or waxy, strongly glandular; stamens ~20–90, short and erect in young bud, filaments at first shorter than anthers, then lengthening rapidly at anthesis to exceed petals, stiff, dark red, anthers ~1 mm long; ovary 2–4-locular, locules usually not completely fused at anthesis, placenta peltate, sometimes appearing parietal because of lack of fusion in locule walls, locules 6–60-ovuled. **Fruits** berries, globose or ellipsoid; seeds several to many, flattened horizontally, sometimes packed in vertical rows, ± round. **Distr:** S America. **Cult:** Propagate from seeds or cuttings. Requires a fertile, well-drained soil in full sun.

A. sellowiana (Berg) Burret, *Repert. Spec. Nov. Regni Veg.* 50: 59, 1941. (Syn: *Feijoa sellowiana*, etc.). [Feijoa, Pineapple Guava]. **Shrubs** or **small trees**, to 4 m tall; young growth whitish or rusty-lanate. **Leaves** obovate, elliptic, 3–7 × 1.5–4 cm, base acute, cuneate, or rounded, apex acute to rounded or sometimes abruptly acuminate or emarginate, lateral veins 6–10 pairs; petiole shallowly channeled, 2–8 mm long. **Inflorescences** with solitary flowers; peduncles densely lanate, 0.5–3 cm long, terete to ± flattened; bracteoles narrowly triangular to linear, ~2.5–3 mm long, early caducous. **Flowers** with calyx lobes suborbicular or oblong, 4–9 × 4–8 mm, densely lanate outside, glabrous and reddish inside, inner lobes longer than outer; petals suborbicular to elliptic, ~1.5 cm long, reddish to pink; hypanthium sometimes ellipsoid, 0.7–1.2 cm long, densely tomentose; stamens ~60–90, 1.3–2.4 cm long, stiff, dark red, anthers ~1 mm long; ovary 3–4-locular, locules ~36–60-ovuled, style glabrous, 1.7–2.7 cm long, dark red. **Fruits** berries, ellipsoid to obovoid, ~2.5–8 cm long; seeds numerous, ~2 mm long. **Distr:** SE Brazil, Paraguay, Uruguay. **Ecol:** In forests and dry mountainous regions. **Zones:** 8–11.

Callistemon

Species 37. **Shrubs** or **small trees**; young parts villous. **Leaves** alternate, ± sessile, aromatic; lamina terete to lanceolate or oblanceolate, margins entire, apex ± acute, sometimes pungent. **Inflorescences** spicate, oblong to cylindric, at first terminal, then the axis usually growing into a leafy shoot. **Flowers** 5-merous; hypanthium urceolate to campanulate; sepals circular; petals circular, exceeding sepals; stamens numerous, exceeding petals, free or rarely shortly basally united, filaments usually glabrous, anthers versatile; ovary half-inferior, 3- or 4-locular, style with capitate stigma. **Fruits** capsules, woody, opening loculicidally, 3- or 4-valved, often persistent on stem and enlarging for many years; seeds numerous. **Distr:** Australia. **Cult:** Propagate from seeds, cuttings or by air-layering. Requires a dry, moderately fertile, well-drained soil in full sun.

C. viminalis (Sol. ex. Gaertn.) G.Don ex Loudon, *Hort. Brit.* [Loudon] 197, 1830. (Syn: *Melaleuca viminalis*). [Weeping Bottlebrush]. **Shrubs** or **small trees**, weeping, to 9 m tall; branches mostly pendent; bark dark, furrowed. **Leaves** linear, narrowly elliptic to oblong with one side straighter than the other, 3–7 × 0.3–0.7 cm, base wedge-shaped, apex acute or apiculate, silky-hairy when young, becoming glabrous with age, lateral veins obscure. **Inflorescences** spikes, many-flowered, 4–12.5 × 3–6 cm. **Flowers** with filaments fused into a ring at base and shed as a unit, 1.5–2.5 cm long, bright red to pink. **Fruits** capsules, cup-shaped with a wide orifice, 5–6 mm Ø. **Distr:** Australia. **Ecol:** Along watercourses, in sandstone or granite areas. Fl. throughout most of the year. **Zones:** 9–12.

Eucalyptus

Species 749. **Trees** or **shrubs**; bark smooth, fibrous or tessellated. **Leaves** often aromatic with oil glands, usually polymorphic with different juvenile and mature forms and sometimes with intermediate forms; *juvenile* leaves opposite, 3–several pairs, shortly petiolate or sessile, sometimes persisting throughout the life of plant, lamina often glaucous or with glandular trichomes; *mature* leaves alternate, petiolate, lamina usually leathery, secondary veins numerous, with intramarginal veins. **Inflorescences** axillary or clustered into terminal or axillary panicles, consisting of umbelliform condensed dichasia. **Flowers** bisexual; hypanthium campanulate, obconic, or semiglobose, stipitate or not, apex usually truncate; sepals rarely distinct; petals connate, either adnate to sepals into a 1-layered calyptra or not adnate and then with connate sepals forming a 2-layered calyptra; calyptra deciduous at anthesis; stamens numerous, usually distinct, in several whorls with outer whorl usually sterile, anthers 2-celled, parallel or oblique, elliptic, ovate, cordate, or bifurcate, dehiscing longitudinally or occasionally poricidally; ovary adnate to hypanthium, 2–7-loculed, ovules numerous, style persistent. **Fruits** capsules, woody, mostly included in expanded hypanthium, opening at apex; seeds numerous, many sterile and undeveloped, developed seeds ovate or angular, testa rigid, sometimes developed into wings. **Distr:** Philippines to Australia. **Cult:** Propagate from seeds. Requires a fertile, neutral to slightly acidic, moist, well-drained soil in full sun.

E. deglupta Blume, *Mus. Bot.* 1: 83, 1850. [Mindanao Gum, Painted Gum Tree, Rainbow Gum]. **Trees**, very large, to ~60 m tall; crown columnar diffused; bark smooth grayish to greenish and coppery mottled, flaky. **Leaves** opposite; *mature* leaves with lamina ovate-lanceolate, 5–14 × 2–7 cm, apex acuminate, pale olive-green, petiolate; *juvenile* leaves petiolate, lamina ovate to oblong-lanceolate, ~5 × 4 cm, apex acuminate, chartaceous. **Inflorescences** corymbs with compound umbels, large, terminal, 3–7-flowered; flower buds ~5 × 4 mm, clavate to cylindrical, acute; flowers pedicellate. **Flowers** with hypanthium 3–5 mm Ø, operculum acutely conical, as long as calyx tube; corolla absent; stamens white, 0.5–1 cm long, anthers reniform, with a small terminal glands. **Fruits** capsules, woody, ovoid-clavate to globose, ~5 × 5 mm, disk thin, valves as long as fruiting calyx, protruding. **Distr:** Philippines to New Guinea (New Britain, New Ireland). **Ecol:** In lowland wet and dry forests, mixed rainforests, and montane forests; 250–1,200 m. Fl. Nov–Dec. **Zones:** 10–12.

E. grandis W. Hill ex Maiden, *Cat. Timbers Queensland* (London Exhib.). 25, 1862. [Flooded Gum]. **Trees**, large; bark at base of trunk rough and flaky, apically silvery white, smooth, and gradually exfoliating; branchlets grayish-white, ridged. **Leaves** *mature* with petiole ~2 cm long, lanceolate, 13–20 × 2–2.5 cm, surfaces with small glands, adaxially dark green and slightly glossy; *juvenile* leaves shortly petiolate, broadly lanceolate to ovate, thinly leathery.

Inflorescences umbels, simple, 3–10-flowered, axillary; peduncles 1-1.5 cm long, compressed; flower buds narrowly obovate, 0.8–1 × ~0.5 cm. **Flowers** with hypanthium obconic to campanulate; stipe 0–3 mm long; calyptra semiglobose to conic, as long or slightly shorter than hypanthium, apex acute to apiculate; stamens 0.8–1 cm long, anthers oblong, nearly basifixed, longitudinally dehiscent, glands abaxial; style shorter than stamens. **Fruits** capsules, gray, pyriform to awl-shaped, 7–8 × 6–8 mm, disk narrow, valves 4 or 5(or 6), exserted from hypanthium. **Distr:** E and NE Australia. **Ecol:** On flats or lower slopes of deep, fertile valleys or often fringing rainforests; 0–600 m. Fl. Apr–May. **Zones:** 9–12.

Eugenia

Species 1,010. **Trees**, **shrubs**, or rhizomatous, geoxylic **suffrutices**. **Leaves** opposite, petiolate; lamina pinnately veined. **Inflorescences** flowers solitary or in fascicles, or in very short racemes, rarely in 2–3-flowered cymules, axillary or often lateral below leaves; bracteoles conspicuous and persistent at base of receptacle; receptacle sharply differentiated from usually elongate and slender pedicel, in bisexual flowers completely fused to inferior ovary, in ♂ flowers usually deeply concave; upper receptacle absent or poorly developed; flower buds subglobose. **Flowers** bisexual, or ♂ by abortion; calyx of 4–5 well-developed, free, ciliolate sepals; petals 4–5, ± ovate, ciliolate, free, persistent; stamens numerous, in ♂ flowers inserted on rim of receptacle, filaments free, shorter than or scarcely longer than petals; ovary 2-locular, placentation axile, ovules numerous, style as long as or longer than filaments, punctiform, funnel-shaped, capitate or shortly bilobed at apex, in ♂ flowers rudimentary and <1 mm long. **Fruits** berries; seeds 1–3. **Distr:** Tropical and subtropical regions worldwide. **Cult:** Propagate from seeds or half-hardened cuttings. Requires a moist, well-drained soil in full or partial sun.

E. brasiliensis Lam., *Encycl.* 3: 203, 1789. (Syn: *E. bracteolaris*, etc.). [Brazilian Cherry, Spanish Cherry]. **Trees**, slender, to 7.5–15 m tall; young shoots reddish. **Leaves** ovate or oblong-obovate, 6.3–16 × 2.5–6.3 cm, base cuneate, apex obtuse, coriaceous; petiole 1 cm long. **Inflorescences** with solitary flowers, or in terminal clusters on branches; pedicels 2.5–5 cm long, with 2 basal narrowly elliptic densely glandular-pustulate bracts, 1.3 cm long. **Flowers** with calyx green, sepals 4; petals white, 1.5 cm long; stamens white, to 100, anthers pale yellow. **Fruits** green, turning bright red to dark purple or nearly black, edible, oblate, 1.2–2 cm Ø, with persistent sepals 1.25 cm long; seeds 1(± round)–3(hemispherical) seeds. **Distr:** E and S Brazil. **Ecol:** In coastal areas, on exposed hills and in thickets; 0–330 m. Fl. Apr, Jul–Dec or Nov–Feb. Fr. May, Jul–Dec or Nov–Feb. **Zones:** 10–12.

E. rufofulva Thwaites, *Enum. Pl. Zeyl.* 416, 1864. (Syn: *E. xanthocarpa* [illegal name]). **Trees**, medium-sized. **Leaves** elliptic-lanceolate, (4–)5–12 × (1–)1.5–3.5 cm, base attenuate-cuneate, apex with up to 3 cm long caudate acumen, abaxially ± concave, nerves ~10 pairs; petiole 0.7–1.2 cm long, slender, deeply channeled above. **Inflorescences** axillary or terminal, usually 2-flowered; peduncles to 3 mm long, very slender, subtended by bracts; bracts linear, to 1 cm long; flower bud to 4 × 3 mm. **Flowers** white, 8 mm Ø; calyx urceolate at anthesis, segments 4–5, linear, to 3 mm long; petals elliptic, to 6 × 3 mm, exceeding sepals; stamens to 6 mm long; disk prominent, polygonal. **Fruits** subglobose, to 4 mm Ø, with calyx segments ± persisting in a ring, ~4 mm Ø seeds 2–4. **Distr:** Sri Lanka; rare. **Ecol:** In highland forests and forests margins; ~500–700 m. Fl. Dec–Jan. Fr. Mar–Apr. **Zones:** 11–12. **Info:** Fruits edible, with a mediocre flavor.

E. uniflora L., *Sp. Pl.* 1: 470, 1753. (Syn: *Myrtus brasiliana*, etc.). [Florida Cherry, Pitanga, Surinam Cherry]. **Shrubs** or **trees**, to 5 m tall, generally glabrous. **Leaves** ovate to elliptic, 3.2–4.2 × 2.3–3 cm, base rounded, slightly cordate, or cuneate, apex acuminate, mucronate, or obtuse, papery, surfaces dark green, glabrous, with numerous pellucid glands, secondary veins ~5 per side, slightly conspicuous, and at an angle of ~45° from midvein, intramarginal veins ~2 mm from margin; petiole ~1.5 mm long. **Inflorescences** clusters or flowers solitary. **Flowers** white, slightly fragrant; calyx lobes 4, long-elliptic, 3–4 mm long, reflexed; petals 4, white, thin, fugacious, obovate, ~1 cm long; stamens many (~50), 8 mm long; ovary slightly ridged. **Fruits** berries, dark red when ripe, globose, 1–2 cm Ø, 8-ridged, 1–2-seeded. **Distr:** South America (Brazil). **Ecol:** Fl. Aug–Oct. Fr. Sep–Dec. **Zones:** 10–12. **Info:** The fruits are edible with a unique flavor and high in Vitamin C.

Leptospermum

Species 87. **Shrubs** or **trees**; bark smooth, fibrous or papery. **Leaves** alternate, sessile to shortly petiolate; lamina usually small, ± rigid, margins entire, often aromatic when crushed, occasionally lemon-scented. **Inflorescences** with flowers solitary or in clusters of 2–several; flowers sessile in axils of bracts on condensed shoots. **Flowers** 5-merous; hypanthium

Top row: *Eucalyptus grandis*, *Eugenia brasiliensis*, *Eugenia rufofulva*.
Bottom row: *Eugenia uniflora* (flowers), *Eugenia uniflora* (fruits).

Top row: *Leptospermum brachyandrum, Leptospermum scoparium, Lophostemon confertus, Melaleuca leucadendra.*
Bottom row: *Melaleuca linariifolia, Melaleuca quinquenervia, Metrosideros carminea, Metrosideros excelsa.*

obtuse or tapered at base; sepals imbricate, sometimes persistent; stamens numerous, usually shorter than petals, in 5 bundles opposite petals but appearing free, anthers opening by slits, with a gland near connective; ovary half-inferior, mostly 3–5-locular, placentation axile, ovules few to numerous, style simple, stigma entire or ± lobed. **Fruits** capsules, firm or woody, usually 3–5(–10)-valved; seeds ± ovoid, reticulate and occasionally ridged or winged, or irregularly linear and striate. **Distr:** Indo-China to New Zealand. **Cult:** Propagate from seeds or half-hardened cuttings. Requires a moist, well-drained soil in full sun.

L. brachyandrum (F.Muell.) Druce, *Rep. Bot. Exch. Club Brit. Isles* 1916: 632, 1917. [Weeping Tea Tree]. **Shrubs** or **small trees**, to 6 m tall; bark seasonally flaking; younger stems pubescent, glabrescent. **Leaves** linear to narrowly elliptic, 2–5 × 0.2–0.4 cm, base somewhat tapering, apex acuminate, infolded, short point usually pungent, ultimately glabrous; petiole ± absent. **Inflorescences** with flowers 3–7 together. **Flowers** ~3–7 mm ∅, white; hypanthium ~3 mm long, usually glabrous; sepals ~1 mm long, usually ciliate; petals 2–3 mm long; stamens 1–1.3 mm long; ovary 3-locular. **Fruits** capsules, ~4 mm ∅, glabrous, valves not exserted, later raised slightly. **Distr:** Australia (Queensland to NE NSW). **Ecol:** In shrubby dry sclerophyll forests along rocky creeks. Fl. Nov–Jan. **Zones:** 9–11.

L. scoparium J.R.Forst. and G.Forst., *Char. Gen. Pl.*: 36, 1775. (Syn: *Melaleuca scoparia*, etc.). [Manuka, Tea Tree]. **Shrubs** or **small trees**, to ~4 m tall; bark brown, shedding in long strips; branchlets and young leaves ± silky-hairy. **Leaves** subsessile, of 2 main forms on different plants, erect to patent; narrowly lanceolate or ovate, ~0.4–1.2 × 0.1–0.4 cm, apex acute, rigid, pungent. **Inflorescences** with flowers usually solitary, axillary or occasionally terminal on branchlets; flowers ± sessile. **Flowers** to 1.2+ cm ∅, white to various shades of pink or rose; receptacle broadly turbinate, adnate to ovary; sepals suborbicular, soon deciduous; petals suborbicular, with short claw. **Fruits** capsules, woody, brownish, long-persistent, distinctly exserted beyond the rim of the receptacle, 5-valved, dehiscing usually after a year. **Distr:** SE Australia, New Zealand, Chatham Islands. **Ecol:** In lowland to subalpine regions, often forming extensive thickets. Fl. throughout most of the year. Fr. Sep–Jun. **Zones:** 8–11.

Lophostemon

Species 4. **Trees** or rarely shrubs; terminal buds covered with scales; stems and petiole exude a milky oil when cut. **Leaves** alternate on mature branchlets, or pseudo-whorled, clustered at apex of branchlets, rarely opposite; lamina with venation conspicuous. **Inflorescences** cymose dichasia, axillary; bracts caducous or absent. **Flowers** bisexual, 5-merous, white; hypanthium ovate or obconiform, pubescent; sepals 5, imbricate, persistent in fruit; petals 5, white or yellow, pubescent; stamens numerous, filaments usually basally connate into 5 bundles (fascicles) opposite petals, inflexed, anthers dorsifixed, locules parallel, longitudinally dehiscent; ovary semi-inferior, 3-loculed, ovules numerous, style shorter than stamens, stigma capitate or slightly ampliate. **Fruits** loculicidal capsules, semiglobose or cup-shaped, apically flatted truncate, splitting into 3 parts but still surrounded by hypanthium; seeds few, linear, sometimes winged. **Distr:** Australia, S New Guinea. **Cult:** Propagate from seeds, by air-layering or grafting. Requires a fertile, moist, well-drained soil in full sun.

L. confertus (R. Brown) Peter G. Wilson and J. T. Waterhouse, *Austral. J. Bot.* 30: 424, 1982. (Syn: *Tristania conferta*). [Brush Box]. **Trees**, large to very large, to 20 m tall; trunk to 50 cm ∅. **Leaves** clustered at apex of branches, pseudo-whorled; lamina oblong to ovate-lanceolate, 7–15 × 3–7 cm, base cuneate, apex acuminate to sharply pointed, leathery, adaxially with subdermal pellucid-punctate oily glands, abaxially sometimes grayish, secondary veins 12–18 per side, 3–6 mm apart; petiole 1–2 cm, flattened. **Inflorescences** cymes, 3–7-flowered, axillary, 2–3 cm long; peduncle 0.6–1.5 cm long; pedicels 3–6 mm long. **Flowers** with hypanthium obconic, 4–5 mm long, grayish-white sericeous; sepals triangular, 4–5 mm long, apex sharply pointed; petals obovate-rounded, ~6 mm long, outside pubescent; stamen bundles 1–1.2 cm long, filaments partly distinct, anthers ~0.5 mm long. **Fruits** capsules, semiglobose to turbinate, 0.8–1.2 cm ∅, enclosed in hypanthium, apex truncate; seeds 2–3 mm long. **Distr:** Australia. Fl. Oct–Dec. **Zones:** 10–12.

Melaleuca

Species 246. **Trees** or **shrubs**. **Leaves** alternate or opposite-decussate, petiolate or sessile; lamina leathery, secondary veins basal and parallel to leaf axis or pinnate. **Inflorescences** spikes, arranged in spikes or heads and pseudoterminal or lateral. **Flowers** bisexual; hypanthium subglobose or campanulate; sepals 5, caducous or persistent; petals 5; stamens numerous, greenish-white, filaments basally connate into 5 bundles opposite petals, anthers dorsifixed,

TR

Above: *Metrosideros excelsa* (yellow form), *Metrosideros perforata*, *Metrosideros polymorpha* (yellow form), *Metrosideros robusta*.

longitudinally dehiscent; ovary slightly adnate to hypanthium, 3-loculed, apex prominent, ovules numerous, style linear, stigma ± enlarged. **Fruits** capsules, semiglobose to globose, apically dehiscent; seeds obovoid-oblong to obovoid, testa thin. **Distr:** Indo-China to Australia, New Caledonia, Society Islands. **Cult:** Propagate from seeds or cuttings. Requires an acidic, moist, well-drained soil in full or partial sun.

M. leucadendra (L.) L., *Mant. Pl.* 1: 105, 1767. [Broad-leaved Teatree, Cajeput, Weeping Paperbark]. **Trees**, spreading, 22–32(–43) m tall; trunk to ~0.6(–1.5) m Ø; branches pendulous; bark white to brown, smooth, slippery, papery. **Leaves** alternate, shortly petiolate, pendulous, aromatic; lamina ± lanceolate, 10–20 × 1–2(–3.5) cm, thinly coriaceous, longitudinal veins 5 (outer pair often inconspicuous). **Inflorescences** usually spikes, 1–3 together, terminal, or solitary in upper leaf axils, glabrous; spikes cylindrical, mostly 6–20 × 2.2–3 cm. **Flowers** whitish to creamy-white; sepals 5, deciduous, small, triangular to rounded; petals 5, ± orbicular, much longer than sepals; stamens 15–many, long-exserted. **Fruits** capsules, sessile, persistent, ± cylindrical, ~3–4 × 3.5–4.5 mm, thin-walled, brown, opening very widely and valves often near rim level; seeds numerous. **Distr:** Maluku to N Australia. **Ecol:** Along watercourses, river flats, coastal plains or seasonal swamps; 5–500 m. Fl. Mar–Oct. **Zones:** 10–12.

M. linariifolia Sm., *Trans. Linn. Soc. London* 3: 278, 1797. (Syn: *M. hyssopifolia*, etc.). [Flax-leaved Paperbark, Snow in Summer]. **Shrubs** or **small trees**, to 10 m tall; bark papery. **Leaves** ± opposite; lamina narrowly elliptic to linear-lanceolate, 2–4.5 × 0.2–0.35 cm, apex acute, glabrous; petiole ~1 mm long. **Inflorescences** spikes, many-flowered, to 4 cm long; rachis glabrous or sparsely pubescent; flowers solitary within each bract. **Flowers** white; sepals tardily caducous; petals obovate to broadly elliptic, ~3 mm long; stamens 30–60 per bundle; claw 0.8–1.5 cm long. **Fruits** capsules, cylindrical to subglobose, 3–4 mm Ø, orifice ~2 mm Ø, valves level with the orifice. **Distr:** E Australia. **Ecol:** In heath and dry sclerophyll forests, in moist or swampy ground, also in coastal areas. **Zones:** 8–12.

M. quinquenervia (Cav.) S.T. Blake, *Proc. Roy. Soc. Queensland* 69: 76, 1958. (Syn: *M. leucadendra* var. *angustifolia*, etc.). [Broad-leaved Paperbark, Cajeput Tree, Paperbark, Punk Tree]. **Trees**, usually 10–15 m tall; bark papery. **Leaves** alternate; lamina lanceolate to elliptic, 3–9 × 0.6–2.5 cm, apex obtuse, with 5 prominent longitudinal veins (sometimes 3 or 7), mostly hairy; petiole 0.4–10 cm long. **Inflorescences** spikes, many-flowered, usually 2–5 cm long, flowers in 3s within each bract; rachis densely hairy to almost glabrous. **Flowers** white or cream, sometimes greenish; sepals not persistent in fruit, floral tube cup-shaped; petals obovate, 2–4 mm long; stamens 0.8–1.2 cm long, 6–10 per bundle, claw 1.5–2 mm long. **Fruits** capsules, broadly cylindrical to ± globose, flat-topped, 4–5 mm Ø, orifice 3–4 mm Ø. **Distr:** New Guinea, New Caledonia, E Australia. **Ecol:** In coastal swamps and around lake margins, sometimes in dry places. **Zones:** 10–12.

Metrosideros

Species 54. **Trees**, **shrubs**, or **woody lianas**, often epiphytic; branches sometimes with hanging roots. **Leaves** opposite, aromatic when bruised, exstipulate; lamina simple, gland-dotted. **Inflorescences** usually cymes or racemes, axillary or terminal; receptacle adnate to ovary; flowers pedicelled. **Flowers** red, pink, white or yellow, usually 5-parted; sepals 5, free, overlapping or imbricate; petals 5, free; stamens numerous, much-exserted, filaments free, filiform, anthers versatile; ovary inferior, partly inferior, or ± superior, 3-celled, style filiform, ovules numerous. **Fruits** capsules, 3-valved or opening irregularly, enclosed in leathery floral tube, loculicidally 3-valved or irregular-dehiscent; seeds numerous, minute. **Distr:** Cape Province; Ogasawara Islands to Malesia and Pacific. **Cult:** Propagate from seeds, half-hardened cuttings or by air-layering. Requires a fertile, well-drained soil in full sun.

M. carminea W.R.B. Oliver, T.N.Z.I. 59, 1928, 420. (Syn: *Melaleuca diffusa*, *Metrosideros diffusa*). [Carmine Rata, Crimson Rata]. **Lianas**, over 15 m tall, with branchlets. **Leaves** elliptic- to ovate-oblong to broadly ovate, 1.5–3.5 × 0.7–1(–2) cm, apex obtuse to subacute, coriaceous; petiole ~1 mm long. **Inflorescences** compound cymes, flowers numerous, usually terminal; axes and receptacles setose to pubescent; pedicels ~5 mm long. **Flowers** very bright carmine; sepals broadly oblong, ~2 mm long; petals suborbicular, ~5 × 4 mm, base shortly clawed, margins usually toothed; stamens 1–1.5 cm long; ovary adnate to and distinctly exceeded by receptacle. **Fruits** capsules, subglobose, 6–7(–9) mm Ø. **Distr:** New Zealand (North Island). **Ecol:** In lowland forests. **Zones:** 8–11.

M. excelsa Sol. ex Gaertn., *Fruct.* 1, 1788, 172, t. 34, f. 8. (Syn: *M. tomentosa*). [New Zealand Christmas Tree, Pohutukawa]. **Trees**, to 20 m tall; trunk to 2 m Ø; branches spreading; branchlets stout, tomentose. **Leaves** elliptic to oblong, (2.5–)5–10 × 2.5–3.5 cm, apex acute or obtuse, coriaceous, thick, abaxially clad in white tomentum (young plants occasionally glabrous); petiole short, stout. **Inflorescences** broad compound cymes, flowers numerous; pedicels stout, tomentose; receptacle obconic. **Flowers** with sepals deltoid; petals crimson or yellow, oblong; stamens numerous, crimson or yellow, 3–4 cm long; ovary adnate to receptacle. **Fruits** capsules, 7–9 mm long, tomentose, distinctly exserted, loculicidally 3-valved. **Distr:** New Zealand. **Ecol:** In coastal forests and shores of lakes. Fl. Dec–Jan. Fr. Jan–Feb. **Zones:** 8–11. **Info:** Weedy in some places, such as the Fynbos area of S Africa.

M. perforata (J.R. et G. Forst.) A. Rich., *Essai Fl. NZ* 1832, 334. (Syn: *Metrosideros scandens*, etc.). [Climbing Rata]. **Lianas**, rather slender, >15 m tall; branchlets terete, ± setose. **Leaves** subsessile, closely set; lamina broadly ovate to broadly oblong to suborbicular on the same plant, ~0.6–1.2 × 0.5–0.9 cm, margins recurved, apex obtuse, coriaceous, adaxially glabrous, abaxially paler and ± setose. **Inflorescences** cymes, few-flowered, crowded toward apex of branchlets, axillary; peduncles and pedicels pubescent to setose; receptacle broad-turbinate. **Flowers** with sepals broad, obtuse; petals suborbicular, white or pink; stamens 0.8–1 cm long, white to pink; ovary adnate to receptacle. **Fruits** capsules, subglobose, 4–5 mm Ø, exserted, loculicidally 3-valved. **Distr:** New Zealand (Three Kings Is, N and S coastal to Banks Peninsula on east and Martin's Bay on west). **Ecol:** In lowland forests and margins. Fl. Jan–Mar. Fr. Feb–Jul. **Zones:** 8–11.

M. polymorpha Gaudich., *Voy. Uranie*: 482, 1830. (Syn: *M. collina* ssp. *polymorpha*, *Nania polymorpha*). [Ohi'a Lehua]. **Shrubs** prostrate to tall **trees**, evergreen, 20–25(–30) m tall; bark rough, flaky. **Leaves** obovate to circular or elliptic, 1–8.2 × 1–5.6 cm, margins flat to rolled under, leathery, adaxially glabrous, abaxially hairy to glabrous; petiole 0.1–1.6 cm long. **Inflorescences** compound cymes, axillary or terminal; peduncle 1–1.8 cm long, glabrous or woolly hairy. **Flowers** red, orange or yellow, floral tube ± globose; sepals rounded to triangular; petals obovate to circular. **Fruits** capsules, ± enclosed in floral tube, 2.5–7.5 mm long. **Distr:** Hawai'ian Islands. **Ecol:** In dry forests, seasonally wet forests and cloud forests; 400–2,500 m. **Zones:** 11–12.

M. robusta A. Cunn., *Ann. Nat. Hist.* 3, 1839, 112. (Syn: *Melaleuca florida*, *Metrosideros florida*). [Northern Rata, Rata]. **Trees**, >25 m tall; trunk >2 m Ø; branchlets 4-angled, puberulous. **Leaves** elliptic- to ovate-oblong, 2.5–5 × 1.5–2 cm, apex obtuse, coriaceous, glabrous; petiole stout ~2 mm long. **Inflorescences** cymes, many-flowered, terminal; pedicels to ~5 mm long; receptacle obconic, to 9 mm long. **Flowers** with sepals broadly triangular; petals ± oblong, ~3 mm long; stamens ~3 cm long, red; ovary adnate to receptacle. **Fruits** capsules, oblong, 6–9 mm long, distinctly exserted, loculicidally 3-valved. **Distr:** New Zealand. **Ecol:** In lower montane forests. **Zones:** 8–11.

Myrcianthes

Species 26. **Trees**, small or medium-sized. **Leaves** opposite or ternate. **Inflorescences** dichasia, 1–7(15–31)-flowered; dichasia usually solitary and axillary, sometimes numerous in upper axils forming compound inflorescences, or modified into terminal myrcioid panicles, central flowers of the dichasia usually sessile. **Flowers** with calyx lobes and petals 4 (in some species regularly 5), or an occasional flower 5-merous; hypanthium not prolonged beyond the summit of the ovary; petals conspicuous; ovary usually 2-locular, ovules numerous (usually 8–20), placentation axile. **Fruits** capsules; seeds 1(–4), plano-convex. **Distr:** S Florida to Tropical America. **Cult:** Propagate from seeds or by air-layering. Requires a moist, well-drained soil in full sun.

M. fragrans (Sw.) McVaugh, *Fieldiana, Bot.* 29: 485, 1963. (Syn: *Eugenia fragrans*, etc.). [Simpson Stopper, Twinberry Stopper]. **Shrubs** or **small trees**, evergreen, to ~8 m tall; bark reddish, smooth, flaking. **Leaves** opposite, shortly petiolate, aromatic; lamina simple, elliptic to ovate or obovate, 1–8 × 1–4 cm long, margins entire, leathery, nearly glabrous, surfaces covered with tiny, often blackish dots, glossy, light green. **Inflorescences** dichasia, ± appressed-pubescent; dichasia 3- or often 7-flowered; peduncles 2–4(–6) cm long, 1.2–1.5 mm wide at apex; hypanthium often densely white-hairy and contrasting with the green glabrous calyx lobes; flower buds 3–4.5 mm long. **Flowers** white, small, calyx lobes 2.5–3.5 mm wide at base; stamens numerous, spreading; disk 3–3.5 mm wide; styles to 6.5 mm long. **Fruits** berries, rounded, 0.6–1 cm Ø, red, reddish-brown or orange. **Distr:** S Florida to Tropical America. **Ecol:** In hammocks. **Zones:** 9–12.

Pimenta

Species 15. **Shrubs** or **trees**, to 20 m tall, usually strongly glandular, strongly aromatic with a spicy, lemony fragrance; hairs whitish, yellowish, or reddish-brown. **Leaves** simple, submembranous to coriaceous, persistent. **Inflorescences** myrcioid panicles, dichasia or panicles, 3–100-flowered; central axis, and those of the principal branches of panicles prolonged and bearing several pairs of lateral branches; bracts and bracteoles rarely present (caducous at or before anthesis). **Flowers** small, 4- or 5-merous; calyx lobes 4 or 5 (if developed), distinct, imbricate, persistent in fruit; hypanthium not or scarcely produced beyond the summit of the ovary; petals 4, whitish, suborbicular; stamens ~16–170, whitish, folded toward the center, filaments much longer than anthers in young buds, not stiff in mature flowers; ovary 1–2(or 3)-locular, locules 1–2(or ~9)-ovuled. **Fruits** drupes or berries, small, fleshy, verrucose, globose; seeds usually 1 or 2, ± round, testa leathery or membranous, embryo involute-cyclic, forming a double spiral (pimentoid). **Distr:** C America, West Indies and Caribbean islands, SE Brazil. **Cult:** Propagate from seeds or by air-layering. Requires a moist, well-drained soil in full or partial sun.

P. dioica (L.) Merr., *Contr. Gray Herb.* 165: 37, 1947. (Syn: *P. officinalis*, etc.). [Allspice, Jamaican Pepper, etc.]. **Trees**, to 20 m tall, aromatic; trunk 30 cm Ø; bark peeling; branchlets flattened, 4-angled. **Leaves** variable, ovate or elliptic, 9–20 × 3–9 cm, base acute, rounded or cuneate, apex acute, very tip obtuse or subacuminate, coriaceous, lateral veins 9–12 pairs; petiole 1.5–2(–3) cm long. **Inflorescences** myrcioid panicles, many-flowered, 3 to 4 times compound, 6–12 cm long; peduncles 2–4.5(–7) cm long; flowers sessile, lateral ones on short branches, appearing pedicellate; bracts subtending flower clusters 1–2 mm long, deciduous; buds turbinate, 2.5–3.5 mm long. **Flowers** small, cream or tan; calyx lobes 4, broadly rounded, concave, densely tomentose inside, ~2 mm long; petals white, suborbicular, ~5 mm long; disk cup-shaped, 2.5 mm wide; stamens ~150, to 5 mm long; ovary bilocular, locules 1(–2)-ovuled, style 4–5 mm long, capitate, stigma thicker. **Fruits** subglobose or oblate to pyriform, (4–)6–8 mm Ø, with convex oil-bearing glands; seeds usually 2, laterally compressed, suborbicular. **Distr:** Cuba, El Salvador, Guatemala, Honduras, Jamaica, Mexico, Nicaragua. **Ecol:** In moist or wet forests; <700 m. Fl. Mar–May. **Zones:** 10–12.

P. racemosa (Mill.) J.W. Moore, *Bull. Bernice P. Bishop Mus.* 102: 33, 1933. (Syn: *Caryophyllus racemosus*). [Bay Tree, Bay-Rum Tree]. **Trees**, to ~12 m tall; young twigs usually 4-angled. **Leaves** ± elliptic, oblong, obovate, oblanceolate, or suborbicular, 3–15 × 1.2–6.5 cm, base and apex acute, acuminate, obtuse, or broadly rounded, margins often slightly revolute, adaxially glabrous, abaxially densely strigose, lateral veins ~10–20 pairs; petiole channeled, 0.3–1 cm long. **Inflorescences** panicles of (5–)15 to ~100 flowers, 2.5–12 cm long, pubescent or strigose, peduncles 1–7 cm long; bracteoles and bracts narrowly triangular to linear, 1–2 mm long. **Flowers** small; calyx lobes triangular, 0.5–2 mm long, pubescent inside; petals suborbicular, ~2 mm Ø; hypanthium 1.5–5 mm long, tearing between calyx lobes; disk ~3–4 mm Ø; stamens 50–170, 4–5 mm long; ovary 2-locular, locules 4–9-ovuled, style 4–5 mm long. **Fruits** subglobose, ~0.6–1.2 cm Ø; seeds 1–4 per, ~4 mm long, embryo once-spiraled. **Distr:** Surinam, West Indies. **Ecol:** In mountain forests, at moderate elevations. Fl. Apr–Nov. Fr. Feb–Mar. **Zones:** 8–12.

Psidium

Species ~150. **Shrubs** or **small trees**; bark gray, smooth; branchlets pubescent. **Leaves** opposite, petiolate; lamina pinnately veined. **Inflorescences** axillary, usually 1 or 2 per axil; bracts 2. **Flowers** large, hypanthium campanulate or urceolate; calyx lobes 4 or 5, unequal; petals 4 or 5, white; stamens many, separate, in many whorls, anthers ellipsoid, basifixed, locules parallel, longitudinally dehiscent; ovary inferior, adnate to hypanthium, 4- or 5-loculed or more, ovules numerous, style linear, stigma expanded. **Fruits** berries, globose to pyriform, fleshy, apex with persistent calyx lobes; seeds many, testa hard. **Distr:** Tropical America. **Cult:** Propagate from seeds, cuttings or by air-layering. Requires a moist, well-drained soil in full sun.

P. cattleianum Afzel. ex Sabine, *Trans. Hort. Soc. London.* 4: 317, 1821. (Syn: *Guajava cattleyana*, etc.). [Cherry Guava, Strawberry Guava]. **Shrubs** or **trees**, evergreen, to 2–8 m tall; bark grayish-brown, smooth; branchlets terete. **Leaves** narrowly to broadly obovate-elliptic, 5–10 × 2–4 cm, base cuneate, apex acute, thickly leathery, surfaces glabrous, secondary veins obscure. **Inflorescences** flowers solitary, axillary; hypanthium obconiform. **Flowers** with calyx lobes 4 or 5, long-ovate, 2 × 3 mm, glabrous inside; petals white, obovate, ~0.5–1 cm long; stamens shorter than petals, ~5 mm long; ovary adnate to hypanthium, 4-loculed, style slender, 5 mm long, distinctly capitate at apex, stigma peltate. **Fruits** berries, dark purple-red or violet when mature (yellow in var. *lucidum*), pyriform to globose or obovoid, 2.5–4 cm long; flesh white, yellow, or reddish. **Distr:** S America (Brazil). **Ecol:** Usually in dense thickets near the beach, and at low elevations, invasive in places; 0–1,500 m. Fl. & Fr. several times a year. **Zones:** 9–12.

P. guajava L., *Sp. Pl.* 1: 470, 1753. (Syn: *Guajava pyrifera*, etc.). [Guava]. **Trees**, to 13 m tall; bark gray, smooth, peeling in strips; branchlets angular, pubescent. **Leaves** oblong to elliptic, 6–12 × 3.5–6 cm, base rounded, apex acute to obtuse, adaxially slightly rough, abaxially pubescent, leathery, secondary veins 12–15 per side, reticulate veins obvious; petiole ~5 mm long. **Inflorescences** cymes, 2–3-flowered, or flowers solitary. **Flowers** with hypanthium campanulate, ~5 mm long, pubescent; calyx cap nearly rounded, 7–8 mm long, irregularly opening; petals white, 1–1.4 cm long; stamens 6–9 mm long; ovary adnate to hypanthium, style as long as stamens. **Fruits** berries, globose, ovoid, or pyriform, 3–8 cm long, with persistent calyx lobes at apex; flesh white or yellow; seeds many. **Distr:** Tropical America. **Ecol:** In coastal and inland regions. Fl. throughout most of the year. **Zones:** 10–12.

P. montanum Sw., *Prodr.*: 77, 1788. (Syn: *Guajava montana*, *P. wrightii*). [Jamaican Guava, Mountain Guava]. **Trees**, to 9–15(–30) m tall, glabrous; young branchlets compressed and 4-angled; bark peeling, distinctly smooth, with pink and gray patches. **Leaves** opposite; lamina simple, elliptic, oblong-elliptic, or elliptic-lanceolate, 3–9.5 × 1–3.5(–4) cm, base cuneate to rounded, apex obtuse to acuminate, gland-dotted, silky hairy when young. **Inflorescences** with flowers solitary, sometimes appearing to be in pseudo-racemes due to falling or non-development of leaves. **Flowers** berries, round, green, to 2.5 cm Ø, smelling of bitter almonds. **Fruits** green, subglobose, ~2 cm Ø, very fragrant. **Distr:** Jamaica. **Ecol:** On wooded hillsides; 550–1,220 m. Fl. May, Aug. Fr. Jan, Apr, Jun. **Zones:** 10–12. **Info:** Fruits edible, with a mangolike flavor.

Rhodamnia

Species ~20. **Shrubs** or **small trees**. **Leaves** opposite, petiolate; lamina 3-veined, abaxially usually white

Below: *Myrcianthes fragrans*, *Pimenta dioica*, *Pimenta racemosa*, *Psidium cattleianum*.

glaucous or tomentose. **Inflorescences** axillary clusters, cymes or racemes; bracteoles small, caducous. **Flowers** small, hypanthium subglobose, adnate to ovary; calyx lobes 4, persistent; petals 4, larger than calyx lobes; stamens numerous, arranged in many whorls, curved in bud, anthers dorsifixed, versatile, longitudinally dehiscent; ovary inferior, 1-loculed, with 2 parietal placentation, ovules many, style linear, stigma peltate. **Fruits** berries, globose, apex with persistent calyx lobes; seeds globose or depressed-globose, testa hard, embryo horseshoe-shaped, hypocotyl long, cotyledon short. **Distr:** Tropical Asia, Australia, China, New Caledonia. **Cult:** Propagate from seeds or by air-layering. Requires a fertile, moist, well-drained soil in full or partial sun.

R. trinerva Blume, *Museum Botanicum Lugduno-Batavum* 1(6), 1849. (Syn: *Eugenia dumetorum*, etc.). [Brown Mallet-wood]. **Shrubs** or **small trees**, to 25 m tall, bark reddish-brown, fissured; young stems densely tomentose. **Leaves** ovate to elliptic, 5–10 × 2–4.5 cm, base cuneate to rounded, apex shortly acuminate, adaxially green and sparsely hairy, abaxially paler and sparsely to densely hairy with erect hairs, strongly 3-veined from base, lateral veins transverse; oil glands distinct, moderately dense; petiole 4–9 mm long. **Inflorescences** 1–3 per axil, each usually 3-flowered; peduncle 0.5–2.2 cm long. **Flowers** with hypanthium sparsely pubescent; sepals 2–3 mm long, caducous; petals 4–6 mm Ø, white; stamens 3–5 mm long; style 4–5 mm long. **Fruits** globose, 5–8 mm Ø, maturing through red to black. **Distr:** Myanmar, India, Andaman Is, Malacca to Philippines, through Malaya to Australia. **Ecol:** 1,500–1,800 m. **Zones:** 10–12. **Info:** The fruits are edible and have medicinal properties.

Rhodomyrtus

Syn: *Myrtus* sect. Rhodomyrtus.

Species ~18. **Shrubs** or **trees**. **Leaves** opposite, petiolate; lamina 3-veined. **Inflorescences** 1(–3)-flowered, axillary. **Flowers** with hypanthium ovoid to subglobose; calyx lobes 4–5, leathery, persistent; petals 4 or 5, larger than calyx lobes; stamens many, separate, in many whorls, usually shorter than petals, anthers dorsifixed to nearly basifixed, versatile, longitudinally dehiscent; ovary inferior, adnate to hypanthium, 3(or 4)-loculed, ovules arranged in 2 lines in each locule, style linear, stigma capitate or peltate. **Fruits** berries, ovoid, urceolate, or globose; seeds many, flat, reniform, separated by longitudinal septa and a false transverse septum; seed coat hard. **Distr:** Tropical Asia, Australia, China, SW Pacific Islands. **Cult:** Propagate from seeds or by air-layering. Requires a moist, well-drained soil in full sun.

R. tomentosa (Aiton) Hasskarl, *Flora*. 25(Beibl.): 35, 1842. (Syn: *Myrtus tomentosa*, *M. canescens*). [Ceylon Hill Gooseberry, Downy/Rose Myrtle, Hill Gooseberry]. **Shrubs**, 1–2 m high; branchlets grayish-tomentose. **Leaves** opposite; lamina elliptic to obovate, 3–8 × 1–4 cm, base broadly cuneate, apex usually rounded to obtuse, leathery, pubescent when young, abaxially gray-tomentose, secondary veins 1 per side, originating near lamina base and meeting at apex, tertiary veins 4–6 per side, intramarginal veins 3–4 mm from margin; petiole 4–7 mm long. **Inflorescences** 1(–3)-flowered; flowers pedicellate. **Flowers** 2–4 cm Ø; hypanthium obovoid, ~6 mm long, gray-tomentose; calyx lobes 5, subrounded, 4–5 mm long, persistent; petals 5, violet, obovate, 1.3–2 cm long; stamens red, 7–8 mm long; ovary 3(or 4)-loculed, style ~1 cm long. **Fruits** berries, edible, purplish-black, urceolate, 1.5–2 × 1–1.5 cm, tomentose; seeds numerous. **Distr:** Cambodia, China, India, Indonesia, Japan, Laos, Malaysia, Myanmar, Philippines, Sri Lanka, Vietnam. **Ecol:** On mountain slopes, grasslands. Fl. Apr–May. **Zones:** 10–12.

Syzygium

Species ~1,200. **Trees** or **shrubs**; branchlets sometimes 2–4-ridged, usually glabrous. **Leaves** opposite or sometimes whorled, petiolate to subsessile; lamina densely to sometimes sparsely pinnately veined. **Inflorescences** terminal or axillary, usually panicles of cymes, 3- to many-flowered; bracts small, caducous after flowering. **Flowers** with hypanthium obconic or sometimes clavate; calyx lobes 4 or 5, rarely more, usually short, caducous or persistent, apex usually obtuse, rarely connate and then calyptrate; petals 4 or 5, rarely more, distinct and then expanding separately or coherent and then caducous as a unit; stamens numerous, distinct but occasionally slightly adhering at base, anthers minute, versatile, 2-celled; ovary inferior, 2 or 3-loculed, ovules many per locule, style linear. **Fruits** drupaceous, 1(or 2)-seeded; seeds sometimes with or without a testa. **Distr:** Tropical and subtropical Old World to Pacific. **Cult:** Propagate from cuttings, fresh seeds or by air-layering. Requires a fertile, moist, well-drained soil in full or partial sun. Some species are moderately salt-tolerant.

S. aqueum (Burm. f.) Alston, *Ann. R. Bot. Gard. Perad.* 11: 204, 1929. (Syn: *Eugenia aquea*, etc.). [Water Apple, Water Rose Apple]. **Trees**, to 20 m tall, glabrous; trunk crooked, to 4 m Ø. **Leaves** elliptic-oblong, 4.5–23 × 1.5–11 cm, base ± broadly cuneate to narrowly subcordate, apex obtuse, subretuse or shortly broadly acuminate, coriaceous, drying dark reddish brown abaxially, main nerves ~9 pairs, distinctly elevated and ± channeled adaxially; petiole 1–5 mm long, stout, drying to black; young leaves brilliant crimson. **Inflorescences** cymes, short, to 10-flowered, terminal or subterminal axillary; flowers subsessile or shortly pedicellate. **Flowers** with calyx funnel-shaped, tube 1.5–3 cm long, segments ovate, to 6 × 6 mm, obtuse, becoming rotate, forming a ~1.5 cm Ø ring at anthesis; petals oblong, ~1.2 × 0.8 cm, obtuse, concave, pale pinkish-white; stamens ~1.5 cm long, folded in bud, filaments brilliant pink, anthers white. **Fruits** ~2–5 cm Ø, globose, red, crowned with a persistent calyx ring to 8 mm Ø. **Distr:** Tropical Asia to N Queensland. **Ecol:** In lowland wet forests; <1,500 m. **Zones:** 10–12.

S. aromaticum (L.) Merr. and L.M.Perry., *Mem. Amer. Acad. Arts* 18: 196, 1939. (Syn: *Eugenia aromatica*, etc.). [Clove]. **Trees**, to 15 m tall, glabrous; bark smooth, pale brown; twigs terete, becoming pale gray-brown. **Leaves** aromatic, pinkish-red when young; lamina oblanceolate, 7–12 × 3–5 cm, base narrowly attenuate-cuneate, apex shortly broadly acuminate, thinly coriaceous, lustrous, abaxially densely punctate, nerves many, parallel, obscure, intramarginal nerve ~1 mm within margin; petiole 1–2 cm long, slender. **Inflorescences** cymes, ~4 cm long. **Flowers** with calyx tubular, ~1.5 × 0.6 cm, verruculose, with 4 hooklike involute ascending segments; petals elliptic, calyptrate, ~1 × 0.5 cm. **Fruits** berries, elongate. **Distr:** Moluccas. **Ecol:** In woodlands and rainforests; 0–1,000 m. **Zones:** 11–12. **Info:** Cultivated widely and commercially for the cloves (dried flower buds).

S. caryophyllatum (L.) Alston, *Handb. Fl. Ceylon* 6 (Suppl.): 116, 1931. (Syn: *Myrtus caryophyllatus*, etc.). [South Indian Plum]. **Trees**, small; branches ascending; bark pale grayish-brown; branchlets rather straight, terete, rufous-brown, scurfy. **Leaves** frequently subopposite; lamina obovate or oblanceolate, 3.5–10 × 2–5.5 cm, base cuneate, margins flat or narrowly revolute, apex obtuse or shortly subacuminate, surfaces densely punctate, drying dark reddish-brown, nerves many, adaxially narrowly channeled, intramarginal nerve straight, ~1 mm within margin; petiole 4–7 mm long. **Inflorescences** cymes, dense, much-branched, to 6 cm long, ascending, terminal or subterminal axillary. **Flowers** white; calyx urceolate, small, ~2 × ~2 mm, shallowly 4-lobed; petals suborbicular, calyptrate, to 2 × 2 mm; stamens ~4 mm long, very slender. **Fruits** subglobose, to 8 mm Ø, with prominent, ~2 mm Ø unlobed apical collar, ripening to dark purple, succulent. **Distr:** S India, Sri Lanka. **Ecol:** In secondary forests. **Zones:** 10–12.

S. cumini (L.) Skeels, *Bull. Bur. Pl. Industr. U.S.D.A.* 248: 25, 1912. (Syn: *Eugenia cumini*, *E. jambolana*, etc.). [Black Plum, Indian Blackberry, Jambolan, Java Plum]. **Trees**, 6–20 m tall; branchlets grayish-white when dry. **Leaves** elliptic, 6–12 × 3.5–7 cm, base broadly cuneate to rarely rounded, apex rounded to obtuse, with a short cusp, leathery, adaxially brownish-green to blackish-brown, surfaces with small glands, secondary veins numerous; petiole 1–2 cm long. **Inflorescences** axillary on flowering branches or occasionally terminal, paniculate cymes, to 11 cm long; hypanthium obconic or long pyriform, ~4–8 mm long. **Flowers** with calyx lobes inconspicuous, 0.3–0.7 mm long; petals 4, white or light purple, coherent, ovate and slightly rounded, ~2.5 mm long; stamens 3–4 mm long; style as long as stamens. **Fruits** red to black, ellipsoid to pot-shaped, 1–2 cm long, 1-seeded, persistent calyx tube 1–1.5 mm long. **Distr:** Tropical and subtropical Asia to Australia (N Queensland).

Below: *Psidium guajava*, *Psidium montanum*, *Rhodamnia trinerva*, *Rhodomyrtus tomentosa*.

Top row: *Syzygium aqueum, Syzygium aromaticum, Syzygium caryophyllatum, Syzygium cumini.*
Bottom row: *Syzygium floribundum, Syzygium grande, Syzygium hemisphericum, Syzygium jambos.*

Ecol: In secondary forests, streamsides; ~100–1,200 m. Fl. Feb–Mar or Apr–May. Fr. Jun–Sep. **Zones:** 10–12.

*S. **floribundum*** F. Muell., *Fragm.* 4: 58, 1864. (Syn: *Metrosideros floribunda*, etc.). [Weeping Lilly Pilly]. **Trees**, small to large; bark dark gray, ± fissured, somewhat flaky. **Leaves** lanceolate to narrowly elliptic, 5.5–16 × 1.5–5 cm, base cuneate, margins undulate, apex acuminate, glabrous, adaxially dark green and glossy, abaxially dull and paler, lateral and intramarginal veins distinct; oil glands numerous, fine; petiole 2–7 mm long. **Inflorescences** panicles, many-flowered; pedicels to 3 mm long. **Flowers** white; sepals inconspicuous; petals cohering, shed together, 1.5–2 mm long; stamens 3–7 mm long. **Fruits** ± globose, 1.5–2 cm Ø, green but developing a reddish tinge at maturity, crowned by persistent hypanthium, 1–2 mm long. **Distr:** E Australia (NSW, Queensland). **Ecol:** In riverine rainforests, often lining streambanks. **Zones:** 10–12.

*S. **grande*** (Wight) Walp., *Repert. Bot. Syst.* 2: 180, 1843. (Syn: *Eugenia grandis*, etc.). [Sea Apple]. **Trees**, large, to 25 m tall, glabrous; trunk to 3 m Ø, pale orange-brown, fluted; bark thinly flaked, inner bark purplish; buttresses spreading, concave, stout; twigs stout, terete, pale cream brown, becoming minutely striated. **Leaves** broadly elliptic to obovate, 10–17 × 5.5–11 cm, base broadly cuneate, apex retuse, obtuse or acuminate, thickly coriaceous, lustrous, midrib stout, prominent abaxially; petiole 1.2–2 cm long, stout; young leaves coppery-brown. **Inflorescences** cymes, 3-flowered, remotely branched, terminal or axillary, 10 cm long; flowers subsessile. **Flowers** pale creamish-yellow; calyx funnel-shaped, 1.4 cm Ø, segments 4–5, ovate, obtuse; petals to 6 × 5 mm, concave, caducous; stamens to 2 cm long, many, spreading. **Fruits** subglobose, to 2 cm Ø, green, with ~1 cm Ø prominent crown of persistent calyx segments. **Distr:** Seychelles to W Malesia. **Ecol:** In wet lowland forests. **Zones:** 10–12.

*S. **hemisphericum*** (Walp.) Alston, *Handb. Fl. Ceylon* 6: 115, 1931. (Syn: *Eugenia hemispherica*, etc.). [Rose Apple]. **Trees**, medium-sized, glabrous. **Leaves** elliptic, 5–11 × 2–6 cm, base narrowly cuneate, margins slightly revolute, apex acuminate, acumen to 1.5 cm long, thinly coriaceous, nerves ~12 pairs, ascending, midrib prominent abaxially; petiole 1–1.5 cm long, drying to black; young leaves mauve. **Inflorescences** cymes, many-flowered, to 9 cm long, terminal or axillary; peduncles rather stout; pedicels to 1.5 cm long, slender. **Flowers** large, white; calyx shortly funnel-shaped, tube ~6 mm long, ~4 mm Ø; sepals 4, suborbicular, cupped, rotate, to 5 × 4–5 mm, prominent; petals to 9 × 7 mm, suborbicular, concave; stamens many, 1.4 cm long. **Fruits** globose, to 1.5 cm Ø, crowned with apical disk ~8 mm Ø, bearing persistent calyx segments. **Distr:** S India, Sri Lanka, Thailand. **Ecol:** In montane regions; 500–1,500 m. **Zones:** 10–12.

*S **jambos*** (L.) Alston, *Handb. Fl. Ceylon.* 6: 115, 1931. [Malabar Plum, Rose Apple]. **Trees**, to 10 m tall; stems very short, broadly branched. **Leaves** lanceolate, ovate-lanceolate, oblong, or linear, 8–26 × 2–4.5 cm, base narrow to broadly cuneate, apex acuminate to long-acuminate, leathery or stiffly papery, surfaces with pellucid glands, secondary veins 8–25 per side, 0.7–1.5 cm apart; petiole 0.5–1 cm long. **Inflorescences** cymes, terminal, several-flowered, sometimes axillary and flowers solitary; peduncle 1–3.5 cm long; hypanthium obconic, 0.8–1.5 cm long. **Flowers** white or pink, 3–4 cm Ø; calyx lobes 4, semiorbicular or triangular-ovate, 5–8 × 6–9 mm; petals distinct, broadly ovate, 1.4–1.5 cm long; stamens 1.5–2.8 cm long, anthers ~1.5 mm long; style 2–3.5 cm long. **Fruits** pale yellow or red when ripe, globose or ellipsoid, 2.5–5 cm Ø, with oil glands, pericarp fleshy; seeds 1–2, embryos numerous. **Distr:** Philippines. **Ecol:** In mixed forests; <100–1,500 m. Fl. Mar–Apr. Fr. May–Jun or Nov–Dec. **Zones:** 11–12.

*S. **malaccense*** (L.) Merrill and L.M. Perry, *J. Arnold Arbor.* 19: 215, 1938. [Malay Apple]. (Syn: *Eugenia malaccensis*, etc.). **Trees**, to 15 m tall; branchlets grayish-brown when dry, stout, terete. **Leaves** narrowly elliptic to elliptic, 16–24 × 6–8 cm, base cuneate, apex acute, leathery, adaxially dark green, secondary veins 11–14 per side, 1–1.5 cm apart; petiole ~1 cm long. **Inflorescences** lateral cymes on older leafless branches or often on the trunk, 4–9-flowered clusters; peduncle very short. **Flowers** red, ~2.5 cm Ø, stout, ridged; hypanthium broadly obconic, ~1 × 1 cm; calyx lobes 4, suborbicular, 5–6 × 7–8 mm, rounded; petals rounded, ~1 × 1 cm, distinct; stamens completely distinct, 1–1.3 cm long; style as long as stamens. **Fruits** ovoid to pot-shaped, ~4 cm long, 1-seeded. **Distr:** Probably native to Malaysia. **Ecol:** In mixed forests. Fl. May or Jan–Feb. Fr. Apr–May. **Zones:** 10–12. **Info:** Widely cultivated species for the aromatic and sweet fruits. The stamens are also edible.

*S. **oleosum*** (F. Muell.) B. Hyland, *Austral. J. Bot., Suppl.* Ser. 9: 107, 1983. (Syn: *Eugenia coolminiana*, etc.). [Blue Lilly Pilly]. **Shrubs** or **small**

trees; bark flaky. **Leaves** lanceolate to elliptic, 3–12 × 1–4 cm, base cuneate, apex usually long-acuminate, glabrous, adaxially glossy green, abaxially paler, lateral veins faint in fresh material, more distinct when dried, intramarginal vein distinct; petiole 0.2–1 cm long. **Inflorescences** panicles, borne in upper axils. **Flowers** white; petals 2–3.5 mm long, free and spreading; stamens 0.6–1.5 cm long. **Fruits** globose to ovoid, 1–2.5 cm Ø, bluish to purple or magenta; seed solitary, embryo solitary with smooth cotyledons. **Distr:** E Australia (NSW, Queensland). **Ecol:** In littoral, subtropical and warm-temperate rainforests. Fl. Nov–Aug. **Zones:** 9–12.

S. paniculatum Gaertn., *Fruct. Sem. Pl.* 1: 167, 1788. (Syn: *Eugenia paniculata*, etc.). [Magenta Cherry, Magenta Lilly Pilly, Scrub Cherry]. **Shrubs** or **small trees**; bark flaky. **Leaves** lanceolate to obovate, 4.5–10 × 1.5–3 cm, base cuneate, apex acuminate, glabrous, adaxially surface green and glossy, abaxially paler, lateral veins numerous, intramarginal vein usually discernible, oil glands small, rather scattered, distinct; petiole 0.2–1 cm long. **Inflorescences** cymose or paniculate, terminal and in upper axils. **Flowers** white; petals 4–5 mm long, free and spreading; stamens 0.6–1.5 cm long. **Fruits** globose to ovoid, 1.5–2.5 cm Ø, magenta; seed solitary, polyembryonic, cotyledons smooth. **Distr:** E Australia. **Ecol:** In subtropical and littoral rainforests on sandy soils or stabilized coastal dunes. **Zones:** 9–12.

S. samarangense (Blume) Merrill and L.M. Perry, *J. Arnold Arbor.* 19: 115, 1938. (Syn: *Eugenia javanica*, etc.). [Java/Lady/Rose/Water Apple, Makopa, Wax Jambu]. **Trees**, to 12 m tall; branchlets compressed. **Leaves** elliptic to oblong, 10–22 × 5–8 cm, base narrow, rounded, or slightly cordate, apex obtuse to slightly acute, thinly leathery, adaxially turning yellowish-brown when dry, abaxially with numerous small glands, secondary veins 14–19 per side, 0.6–1 cm apart; petiole <4 mm long or absent. **Inflorescences** cymes, several-flowered, 5–6 cm long, terminal or axillary; hypanthium obconic, 7–8 × 6–7 mm. **Flowers** white; calyx lobes 4, semiorbicular, ~4 × 4 mm or larger; petals 4, distinct, 1–1.3 cm long; stamens numerous, ~1.5 cm long; style 2.5–3 cm long. **Fruits** dark red, pyriform to conic, 4–5 cm long, fleshy, glossy, apex impressed, persistent sepals fleshy; seed 1. **Distr:** Indonesia, Malaysia, Papua New Guinea, and Thailand. **Ecol:** Fl. Mar–Apr. Fr. May–Jun. **Zones:** 11–12.

S. tierneyanum (F. Muell.) T.G. Hartley and L.M. Perry, *J. Arnold Arbor.* 54: 200, 1973. (Syn: *Eugenia tierneyana*, etc.). [River Cherry, Water Cherry]. **Trees**, to 25 m tall; trunk sometimes buttressed. **Leaves** alternate or opposite; lamina simple, elliptic, 9–19 × 3–7.8 cm, base acute, margins entire, apex acuminate, acumen tip rounded, glabrous, covered with numerous minute dull-orange oil glands, intramarginal veins 1 or 2; petiole 0.3–1.2 cm long. **Inflorescences** panicles, axillary or ramiflorous. **Flowers** with calyx green, white, cream or yellow; petals 4, white or cream, 5–9 mm across; stamens numerous, exserted. **Fruits** berries, white, cream, pink or red, 1.5–2 × 1.5–2 cm, persistent calyx lobes protruding, sometimes incurved; seed 1, 5–8 mm Ø, cut seed purple or reddish-purple. **Distr:** Papuasia to Australia (N and NE Queensland). **Ecol:** In rainforests and galley forests; 0–460 m. Fl. Oct–Feb. Fr. Jan–May. **Zones:** 9–12.

Tristaniopsis

Species 40. **Trees** or **shrubs**. **Leaves** alternate and often crowded toward the end of branches, petiolate; lamina with venation obscure. **Inflorescences** with flowers in triads or dichasia, axillary. **Flowers** 5-merous, yellow; sepals persistent; petals ± circular, free; stamens usually numerous, fused into fascicles opposite the petals, inflexed; ovary half-inferior, usually 3-locular, ovules usually pendent from the top of the axis, stigma capitate. **Fruits** loculicidal capsules, strongly exserted from the hypanthium, 3-valved, only splitting as far as the hypanthium; seeds mostly winged. **Distr:** Australia, Malesia, Myanmar, New Caledonia, Thailand. **Cult:** Propagate from seeds or by air-layering. Requires a moist, well-drained soil in full or partial sun.

T. laurina (Sm.) Peter G. Wilson and J.T. Waterh., *Austral. J. Bot.* 30(4): 435, 1982. (Syn: *Tristania laurina*). [Water Gum, Kanooka, Kanuka]. **Trees**, to 30 m tall; bark smooth, pale, shedding in strips. **Leaves** oblanceolate to narrow-oblanceolate, usually 6.5–12 × 1.5–3 cm, apex obtuse to acute, adaxially dark green, glabrous, abaxially grayish-white, pilose, oil glands scattered; petiole 3–8 mm long. **Inflorescences** dichasia, 7–15-flowered. **Flowers** with sepals triangular; petals obovate, 4–5 × 3–4 mm, yellow; stamens mostly 12–20 per fascicle, shorter than petals. **Fruits** capsules, 0.6–1 × 0.4–0.6 cm; seeds winged, 5–6 mm long. **Distr:** Australia. **Ecol:** In rainforests and sclerophyll forests, usually along banks of streams. Fl. Dec–Jan. **Zones:** 10–12.

Xanthostemon

Species 47. **Trees** or **shrubs**. **Leaves** alternate; lamina penninerved. **Inflorescences** cymes, dense, terminal or axillary; bracts and bracteoles usually very small or none. **Flowers** with calyx tube broadly campanulate or open, adnate to ovary at base, lobes 4 or 5, slightly imbricate, often unequal; petals 4 or 5, spreading; stamens indefinite, free or slightly basally united, in one or more series much exceeding petals, filaments often rigid, anthers versatile, longitudinally dehiscent; ovary enclosed in calyx tube, half-inferior or free except the broad base, 2–6-loculed, ovules numerous per locule, closely packed in a single ring round a clavate or peltate placenta, style filiform, stigma small. **Fruits** capsules, seated on the expanded calyx, or half-enclosed in the cup-shaped calyx tube, opening loculicidally in 2–6 valves; seeds flat or angular. **Distr:** C Malesia to New Caledonia. **Cult:** Propagate from seeds or cuttings. Requires a moist or well-drained soil in full sun.

X. chrysanthus (F. Muell.) Benth., *Fl. Austral.* 3: 268, 1867. (Syn: *Metrosideros chrysantha*, *Nania chrysantha*). [Black Penda, Golden Penda, Johnstone River Penda]. **Trees**, to 25 m tall; trunk sometimes buttressed. **Leaves** alternate or whorled; lamina simple, 7–22 × 2–9.5 cm, glabrous, dotted with rust-colored oil glands; petiole 0.5–1.2 cm long, base swollen. **Inflorescences** panicles, axillary or terminal. **Flowers** with calyx green, cream or yellow, spotted with reddish oil glands, broadly campanulate, ~7.5–9 mm Ø, lobes ovate-triangular; petals 4 or 5, cream, yellow or greenish-yellow, 7–9 mm long; stamens 20–25, yellow, uniseriate, longest ~2.5 cm long, anthers oblong, versatile; ovary usually 3-celled, ovules numerous, flat, closely packed, style very long. **Fruits** capsules, green to brown, woody, ~1–1.4 × 1–1.4 cm, dotted with reddish glands, 2–4-valved, partially enclosed in the persistent calyx; seeds several, brownish, flat, 3–4 mm long. **Distr:** Australia (N and NE Queensland). **Ecol:** In coastal rainforests, along waterways; 0–1,000 m. Fl. Feb–Nov. Fr. Aug–Feb, May. **Zones:** 9–12.

ONAGRACEAE

Genera 22, **species** 656. **Herbs** annual or perennial, **shrubs**, rarely trees, to 30 m tall. **Stipules** present and usually caducous, or absent. **Leaves** simple, spirally arranged, opposite, or occasionally whorled; lamina with margins entire or toothed to pinnatifid. **Inflorescences** axillary, leafy spikes or racemes, or flowers solitary or occasionally in panicles. **Flowers** perfect and hermaphroditic or occasionally unisexual, actinomorphic or zygomorphic, (2–)4(–7)-merous, all but *Ludwigia* with a distinct floral tube, nectariferous within; sepals green or colored, valvate; petals as many as sepals or rarely absent, variously colored, imbricate or convolute and occasionally clawed; stamens as many as sepals in one series or twice as many as sepals in 2 series, anthers versatile or basifixed, dehiscing by longitudinal slits; ovary inferior, with as many carpels and locules as sepals, placentation axile or parietal, locule 1–many-ovuled, in 1 or several rows or clustered, style 1, stigma with as many lobes as sepals or clavate to globose. **Fruits** capsules, loculicidal or indehiscent nuts or berries; seeds small, smooth or variously sculptured, sometimes with a coma (or wing). **Distr:** Worldwide.

Fuchsia

Species ~100. **Shrubs** or **small trees**, pubescent or glabrate. **Stipules** small, caducous; **Leaves** opposite, alternate, verticillate; lamina entire or usually dentate. **Inflorescences** flowers solitary or clustered, rarely racemose or paniculate, axillary; flowers long-pedunculate, nutant or pendulous. **Flowers** pink, red, purple, or violet, small or often large and showy,

Below: *Syzygium malaccense*, *Syzygium oleosum*, *Syzygium paniculatum*, *Syzygium samarangense*.

Above: *Syzygium tierneyanum*, *Tristaniopsis laurina*, *Xanthostemon chrysanthus*, *Fuchsia arborescens*.

sometimes polygamous; calyx tube globose or ovoid at base, produced above ovary into a 4-lobate, deciduous, campanulate or funnelform limb; petals 4 or rarely none, convolute, spreading, or reflexed; stamens 8, filaments filiform, anthers linear or oblong; ovary 4-celled, style elongate, stigma capitate or clavate, entire or 4-lobate, ovules numerous, multiseriate. **Fruits** baccate, 4-celled, few–many-seeded; seeds angulate or reniform, testa membranous. **Distr:** Predominantly in tropical America (from Mexico to Patagonia); New Zealand, Tahiti. **Ecol:** In cool montane regions. **Cult:** Propagate from seeds, softwood or half-hardened cuttings. Requires a fertile, moist, well-drained soil in full or partial sun.

F. arborescens Sims, *Bot. Mag.* 53: t. 2620, 1825. (Syn: *F. arborescens* var. *megalantha*). [Lilac/ Tree Fuchsia]. **Shrubs** or **small trees**, 1.5–8 m tall, glabrous throughout or nearly so; trunk short and thick, crown dense and rounded. **Leaves** opposite or ternate, oblanceolate or oblanceolate-oblong, 7–20 cm long, base attenuate, margins entire or serrulate, apex acute or acuminate, paler beneath; petiole slender. **Inflorescences** dense panicles, large, terminal; flower buds rounded or very obtuse at apex. **Flowers** rose-red or red-purple, ~1.2 cm long; calyx lobes about equaling tube, linear-oblong, obtuse, reflexed in flower; petals shorter than sepals; stamens usually conspicuously exserted. **Fruits** berries, subglobose, to 1 cm Ø, dark purple or purple-black. **Distr:** C and S America (Costa Rica, Guatemala, Honduras, Panama, S Mexico). **Ecol:** In moist or wet, mixed forests, forests margins, open meadows, thickets; 1,300–2,900 m. Fl. throughout most of the year. **Zones:** 10–12.

F. boliviana Carrière, *Rev. Hort.* (Paris). 48: 150, pl. 1876. (Syn: *F. boliviana* var. *luxurians*, etc.). [Bolivian Fuchsia]. **Shrubs**, 6–7 m tall, most parts densely hairy. **Leaves** opposite or 3-whorled; lamina elliptical to oblong-elliptical, 7–18 cm long, base blunt, often oblique and unequal-sided, margins entire, minutely dentate, apex tapered, adaxially sparsely puberulous, nerves 14–24 per side; petiole 1–4(–6) cm long. **Inflorescences** with flowers solitary, in axils of much-reduced leaves at ends of branches forming terminal corymbiform racemes or panicles, flowers pendulous; peduncles 1–1.5 cm long; flower buds mucronate at apex with cusps of calyx lobes. **Flowers** with calyx dark red, outside sparsely puberulous, lobes lanceolate, 1.7–2.1 × 0.4 cm, acuminate, cuspidate, tube 4.5–6.5 cm long, apex 7 mm Ø, puberulous inside; petals oblong, 1.5 cm long, acute, light red; stamens 1.4 cm long, filaments light red, anthers creamy-white; ovary 0.8–1.3 cm long, style with stigma ultimately as long as calyx lobes. **Fruits** berries, linear-oblong, 1.8 × 0.4 cm. **Distr:** Argentina, Bolivia, Colombia, Guatemala, Peru, Venezuela. **Ecol:** On moist slopes; 1,200–1,500 m. Fl. throughout most of the year. **Zones:** 9–11.

F. excorticata (J.R. and G. Forster) L. f., *Suppl. Pl.* 217, 1781. [New Zealand Fuchsia, Tree Fuchsia]. **Trees** or **large shrubs**, 2–13 m tall, deciduous; trunk twisted, to 70 cm Ø; bark coppery, papery, peeling off readily. **Stipules** lanceolate, purplish, caducous. **Leaves** alternate; lamina lanceolate to ovate-elliptic, (3.5–)4.5–13 × (1.5–)2.5–5.5 cm, base acute to rounded, margins entire to remotely serrulate, apex acute to acuminate, strongly bicolored, adaxially green, abaxially silvery-white, occasionally red-suffused; petiole 0.9–4 cm long. **Inflorescences** with *perfect* flowers solitary or ± tightly grouped, sometimes appearing cauliflorous, pedicels 0.5–1.5 cm long; ♀ flowers with pedicels 4.5–8.5 mm long. **Flowers** *perfect* with dull purple streaks, changing abruptly to red-crimson, sepals narrow-triangular, 0.8–1.9 × 0.3–0.6 cm, floral tube 1.1–2.2 cm long, petals dark purple, elliptic or lanceolate, 1.8–5 × 0.7–2.1 mm, filaments 0.6–1.5 cm long, light to dark purple, ovary 4.5–8 × 1.7–3 mm, green, style glabrous, 2.2–4 cm long, exserted part purple, greenish below, stigma yellow-green; ♀ similar to perfect ones, sepals 5.5–8 × 2.2–3.5 mm, floral tube 0.6–1.3 cm long, petals 1.2–2.6 × 0.35–1 mm, style 1.8–2.3 cm long. **Fruits** berries, edible, oblong-ellipsoid, 0.9–1.3 cm long, dark purple; seeds 320–670, oblong-triangular, ~1 mm long, tan. **Distr:** New Zealand. **Ecol:** In lowland to upper montane forests, 0–1,000 m. Fl. Jun–Jan. **Zones:** 8–11.

F. microphylla HBK., *Nov. Gen. and Sp.* 6: 103, t. 534, 1823. (Syn: *F. aprica*, *F. microphylla* var. *aprica*). [Small-leaved Fuchsia]. **Shrubs**, erect, 1–2.5 m high, sometimes low and prostrate; branches brown or ferruginous, puberulent. **Leaves** opposite, petiolate; lamina small, lance-oblong to oblong-elliptic or obovate, mostly 1–2.5 cm long, base acute, margins remotely denticulate, apex generally obtuse, membranous or rather thick, ± glabrous. **Inflorescences** with flowers solitary, axillary; peduncles short. **Flowers** rose-red to red or purple-red, ~1.5 cm long, glabrous; calyx tube deeply constricted above ovary, somewhat dilated above; sepals ovate-lanceolate, acuminate, slightly shorter than the tube; petals shorter than sepals, pale; stamens shorter than petals. **Fruits** berries, 5–6 mm Ø, maturing to black, very juicy. **Distr:** S Mexico. **Ecol:** In dense, moist or wet, mixed forests, sometimes in low thickets or in exposed rocky places; 1,200–3,800 m. **Zones:** 9–11.

F. procumbens R. Cunningham ex A. Cunningham, *Ann. Nat. Hist.* 3: 31, 1839. (Syn: *F. kirkii*). [Creeping/Trailing Fuchsia]. **Subshrubs**, subdioecious, procumbent, to 2 m long. **Stipules** narrowly lanceolate, 0.3–0.9 × ~0.2 mm, caducous. **Leaves** alternate; lamina suborbicular to broadly ovate, 0.5–2.8 × 0.3–2.7 cm, base subcordate to truncate, margins sinuate to serrulate, apex obtuse to rounded, pale green to purple-flushed, abaxially paler; petiole 0.5–5 cm long. **Inflorescences** with flowers solitary, erect, axillary; pedicels 0.3–1 cm long. **Flowers** ♂ (or perfect) with sepals lanceolate, 7–9 × 2.7–3.8 mm, acute, reflexed, purple, base green, petals absent, floral tube cylindrical, 0.7–1.1 cm long, orange-yellow, becoming darker, stamens with filaments to 5 mm long,

Below: *Fuchsia boliviana*, *Fuchsia excorticata*, *Fuchsia microphylla*, *Fuchsia procumbens*.

Above: *Fuchsia splendens, Ludwigia sedoides, Canarium bengalense, Canarium commune.*

pink to red or yellow near base, anthers purple, ovary cylindrical, 3–6 mm long, green, 4-locular, locules 15–55-ovuled, ± biseriate, style 0.6–0.9(–1.4) cm long, pale pink, stigma yellow-green or dull brown, nectary yellow-orange; ♀ flowers similar to ♂, sepals 4.1–6.4 × 2.2–2.8 mm, floral tube 5.5–9 mm long, filaments red-purple, anthers pink, ovary 5–8 mm long, style 0.7–1.5 cm long, stigma yellow-green, ovules 50–230 per ovary, biseriate. **Fruits** berries, ± ovoid, 1–2.5 × 0.8–1.6 cm, bright red to magenta, purple-wax; seeds 20–80. **Distr:** New Zealand. **Ecol:** In coastal forests; 0–150 m. Fl. Oct–Mar. **Zones:** 9–11.

F. splendens Zucc., *Flora* 1882, pt. 2: Beibl. 102, 1832. [Fuchsia]. **Shrubs**, terrestrial, stout and erect, or epiphytic and pendent, to 2.5 m tall; bark of older branches brown, exfoliating. **Leaves** ovate, 5–12 cm long, base usually deeply cordate, sometimes rounded, apex acute or acuminate, membranous, adaxially puberulent, abaxially somewhat paler, densely or sparsely villous-hirsute; petioles long, slender. **Inflorescences** with flowers solitary, axillary near ends of branches; flowers pendent; peduncles long, slender, 3–4 cm long, sparsely or densely pilose. **Flowers** large and showy; calyx tube broad and thick, dilated upward, 0.8–1 cm wide at apex, bright red; sepals ovate, acuminate, green, half as long as the tube; petals pale green or yellow, half as long as the sepals; stamens yellow, much longer than petals. **Distr:** Mexico (Chiapas), Costa Rica. **Ecol:** In moist or wet, mixed, dense forests; 2,000–3,500 m. **Zones:** 8–11.

Ludwigia

Species 82. **Herbs**, slender, erect to prostrate and rooting at nodes, or shrubs or rarely small trees; underwater parts often swollen and spongy or with inflated white spongy pneumatophores. **Stipules** present, reduced and/or deciduous. **Leaves** alternate (or opposite); lamina margins usually entire. **Inflorescences** in upper leaf axils or in spikes, racemes, or clusters; bracteoles 2, at or near base of ovary, or absent. **Flowers** perfect, actinomorphic, floral tube not prolonged beyond ovary; sepals (3 or)4 or 5(–7), green, persistent after anthesis; petals as many as sepals or absent, yellow or white, caducous; stamens as many as or twice as many as sepals, anthers versatile or sometimes basifixed; ovary with as many locules as sepals, rarely more, often with a depressed nectary surrounding base of each epipetalous stamen, stigma capitate or hemispheric, entire or lobed. **Fruits** capsules, obovoid to cylindric, irregularly dehiscent or by a terminal pore or by flaps separating from valvelike apex; seeds numerous, in 1 to several rows per locule. **Distr:** Cosmopolitan. **Cult:** Propagate from seeds or by division. Requires shallow water in full sun.

L. sedoides (Bonpl.) H. Hara, *Journal of Japanese Botany* 28(10): 294, 1953. (Syn: *Jussiaea sedoides*). [False Loosestrife, Mosaic Plant]. **Herbs**, aquatic, free-floating or with the lower stem rooted; branches ending in floating rosettes of leaves. **Leaves** arranged alternately forming a symmetrical rosette 7.5–25 cm across; lamina diamond-shaped or ovate-rhomboid, 1 × 1 cm, base acute, margins coarsely toothed above middle, apex obtuse, green, reddish-flushed (especially older leaves); petiole usually reddish, 3–3.5 cm long. **Inflorescences** with flowers solitary, borne above water, erect or suberect; pedicels glabrous, 2–2.5 cm long; bracteoles 2. **Flowers** bright yellow, 2–2.5(–5) cm Ø; calyx tube tetragonal, 0.8–1 cm long, lobes 4, ovate, 4–6 × 3–5 mm, acute to obtuse, glabrous, 5-nerved; petals 4, rotund-obovate, 1–1.2× 0.8–1 cm, base clawed, apex emarginate; stamens 8. **Fruits** capsules, obconic. **Distr:** Brazil and Venezuela. **Ecol:** In rainforests, wetlands, ponds, riverine areas that are free of currents, and in stagnant waters. Fl. throughout most of the year. **Zones:** 9–12. **Info:** The leaf rosettes contract at night and open up in the day, leaving gaps as the rosettes are formed by alternate leaves that gradually differ in the length of their petioles (Weberling 1998).

SAPINDALES

FAMILIES 9, GENERA 460, SPECIES 5,670.

Leaves spirally arranged; lamina pinnately compound (occasionally becoming palmately compound, trifoliate, or unifoliate), conduplicate. **Flowers** 4–5-merous; perianth parts imbricate; nectar disk distinct (Judd et al. 2008; Stevens 2001 onwards). **Distr:** Worldwide, mainly in tropical regions. **Info:** The monophyly of Sapindales is strongly supported by molecular data (Chase et al. 1993; Hilu et al. 2003; Soltis et al. 2000).

BURSERACEAE

Genera ~16, species ~550. **Trees** or **shrubs**, resiniferous. **Leaves** alternate or rarely opposite, odd-pinnate, 3-foliolate (or rarely 1-foliolate), stipulate or exstipulate. **Inflorescences** racemose or paniculate. **Flowers** regular, small, unisexual or bisexual, often polygamous; calyx 3–6-lobed, imbricate or valvate; petals 3–6, free or rarely connate, imbricate or valvate; disk annular or cupular, usually conspicuous; stamens as many as or twice as many as petals, inserted at base or margin of disk, equal or unequal, filaments free, rarely basally connate, anthers dorsifixed or rarely adnate, 2-thecate, longitudinally dehiscent; ovary superior, usually 2–5-loculed, locules (1)2-ovuled, usually anatropous, axile, pendulous or rarely ascending, style simple, stigma undivided or 2–5-lobed. **Fruits** drupaceous, indehiscent, containing 2–5 pyrenes, or rarely pseudo-capsular and dehiscent; seeds solitary, exalbuminous, testa membranous, cotyledons usually membranous, rarely fleshy and plano-convex, radicle superior. **Distr:** Tropical regions of both hemispheres.

Canarium

Species ~75. **Trees**, evergreen; branchlet pith usually with vascular strands. **Leaves** alternate, odd-pinnate, stipulate or exstipulate. **Inflorescences** panicles, terminal or axillary. **Flowers** unisexual; calyx cupular or campanulate, trilobed or -fid, valvate; petals 3, imbricate or valvate, usually exceeding calyx; stamens 6, free to connate, inserted on margin or outside of disk; ovary 2- or 3-loculed, locules 2-ovuled, style short to long, stigma capitate. **Fruits** drupes, usually ellipsoid, ± distinctly trigonous, with a 1–3-celled, 1–3-seeded pyrene; seeds with cotyledons often partite, contortuplicate. **Distr:** Tropical Africa, Asia to NE Oceania and Pacific Islands. **Cult:** Propagate from seeds or by air-layering. Requires a deep, moist, well-drained soil in partial sun.

C. bengalense Roxburgh, *Fl. Ind.* 136, 1832. [East Indian Copal]. **Trees**, to 25 m tall; trunk ~1.2 m Ø. **Stipules** subulate, pubescent, early deciduous. **Leaves** pinnate; leaflets 6–8(–10) pairs, oblong or obovate-lanceolate, 10–20 × 4.5–6 cm, base rounded, margins sinuate or entire, apex acuminate, acumen 1–1.5 cm long, adaxially glabrous, abaxially pubescent, lateral veins 18–25 pairs. **Inflorescences** cymose panicles in ♂ plants, 30–40 cm long, axillary, nearly glabrous, branches 3–4 cm long, **Flowers** with calyx ~2 mm long; stamens glabrous, filaments united below; disk tubular, 1.5–1.8 mm high, margin and abaxial surface hirsute in ♂ flowers, annular, trifid, fimbriate in ♀ flowers. **Infructescences** extra-axillary or axillary, 5–8 cm long, 1–3-fruited, persistent calyx disk-shaped, shallowly trilobed, ~1 cm Ø; drupes green, spindle-shaped, 3-ribbed, 4.5–5 × 1.8–2 cm, glabrous, or obovoid and 3- or 4-ribbed, apex acute; cross-section of pyrene acutely triangular or rounded. **Distr:** China, India, Laos, Myanmar, Thailand. **Ecol:** In mixed forests; 400–1,300 m. Fr. Jul–Oct. **Zones:** 10–12.

C. commune L., *Mant. Pl.* 127, 1767. (Syn: *C. mehenbethene*, etc.). [Canary Nut, Java Almond, Pili Nut]. **Trees**, large, to ~40 m tall; buttresses, narrow winglike, to >2 m high; exudate resinous, strongly scented. **Stipules** embracing the stem, ovate-round, 2.5 × 1.3 cm, dentate or fringed. **Leaves** unequally pinnate; leaflets 4 or 5 pairs, opposite, ovate, 12.5–18 × 5–7.5 cm, base tapered, margins entire, coriaceous, smooth; common petiole woody; petiolules ~1.25 cm long. **Inflorescences** panicles, terminal, branches subdivided, divaricate, 3–4-angular, compressed, tomentose. **Flowers** white, small; calyx urceolate, coriaceous, lobes 3, rounded, teeth

Above: *Canarium zeylanicum*, *Campnosperma auriculata*, *Campnosperma zeylanicum* (flowers), *Campnosperma zeylanicum* (fruits).

upright; petals 3, perigynous, alternate with calyx segments, ovate-acuminate, erect, converging, upper half carinate, margins undulated; stamens 6, filaments 4–6 mm long, basally connate, in a ring surrounding pistil, anthers ovate, erect; ovary superior, round-oblong, ~5 mm Ø, prolonged into a short, thick style, stigma thick, globular, with 3 furrows. **Fruits** ovoid, slightly 3-angled; seeds 1–3, elongate, edible. **Distr:** Indonesia. **Ecol:** In forests. **Zones:** 11–12.

C. zeylanicum (Retz.) Blume, *Mus. Bot.* 1: 218, 1850. (Syn: *Amyris zeylanica*, *C. balsamiferum*). **Trees**, large, much-branched, 25–30 m tall; trunk buttressed; young branches densely pubescent; exudate clear, resinous. **Stipules** rotund-sagittate, deciduous. **Leaves** imparipinnate; rachis ~25 cm long, pilose; leaflets 5–9, broadly elliptic or oblong-oval, 10–15 x 3–6 cm, base subcordate, margins entire, reflexed, apex shortly acuminate, glabrous, subcoriaceous, veins yellow, very broad; petiolules 6–9 mm long. **Inflorescences** panicles, terminal; ♂ with flowers numerous, crowded; ♀ few-flowered; pedicels short, stout; buds subglobose; bracts rotund, brown-tomentose, deciduous. **Flowers** green; ♂ 5–8 mm Ø, pilose outside, calyx cup-shaped, trilobed, petals 3, imbricate, erect, acute, stamens 6, erect, inserted on disk, glabrous, ovary rudimentary; ♀ 1–1.3 cm long, calyx and petals as in ♂, stamens 6, rudimentary, ovary 3-locular, glabrous, incrassate above, style short, stigma large, capitate, trilobed. **Fruits** drupes, ellipsoid or oblong-ovoid, 2.5–4 cm long, obscurely trigonous, plum-colored, stone bony; seed large, oblong, edible. **Distr:** Sri Lanka. **Ecol:** In wet lowlands. Fl. & Fr. Apr–Sep. **Zones:** 11–12.

ANACARDIACEAE

[Poison Ivy, Sumac].

Genera ~77, **species** 600. **Trees** or **shrubs**, woody climbers or perennial herbs, hermaphroditic, polygamodioecious or dioecious; bark and foliage with resiniferous secretory ducts, plants turpentine-smelling, blackening when wounded. **Leaves** often clustered distally, alternate, exstipulate, simple, trifoliolate or imparipinnate. **Inflorescences** thyrsoids or panicles, terminal or axillary; floral bracts small, sometimes large, membranous, fused to pedicel. **Flowers** small, actinomorphic, 3–5-merous, bisexual to unisexual; receptacle sometimes elongate and barrel-shaped; perianth usually double; sepals fused basally and lobed, imbricate or valvate in bud, caducous or persistent; petals free or adnate basally to extended receptacle, imbricate or valvate, deciduous to persistent; stamens 1- or 2-whorled, 1, several, or all fertile, filaments slender, sometimes basally connate, anthers ovoid or oblong; disk fleshy, stipe-shaped or 5–10-notched; ovary superior, sometimes half inferior or inferior, either (a) 1-carpellate and 1-locular, (b) syncarpous and 2–5-locular (rarely more), (c) 4–6-carpellate and apocarpous, or (d) 5-carpellate and incompletely connate; stigmas 1–5 (rarely more), ± distinct, locules 1-ovuled, usually one carpel developing to maturity. **Fruits** drupaceous, or dry and indehiscent, sometimes borne on enlarged fleshy hypocarp formed by pedicel and receptacle or fused to membranous accrescent floral subtending bract, composed of usually 1–5 cells; cells 1-seeded; epicarp thin; mesocarp usually fleshy, fibrous and resinous; endocarp crustaceous to bony. **Distr:** Mainly in tropical, subtropical, and temperate areas, with the center of diversity in the Malesian region.

Subfam. SPONDIADOIDEAE

Genera 10, **species** 115. **Ovule** pendulous. **Distr:** Tropical regions worldwide.

Campnosperma

Species 10. **Trees**; twigs with terminalioid branching. **Leaves** alternate, spirally arranged; lamina simple, margins entire, very coriaceous. **Inflorescences** racemes, axillary; ♀ less branched than ♂; flowers sessile. **Flowers** very small, greenish-yellow, physiologically unisexual; calyx 3–5-partite, segments erect, persistent, imbricate; petals 3–6, suborbicular, erect, imbricate; disk urceolar; stamens 6–10, inserted at base of disk, sterile in ♀ flowers; ovary 1-celled (or 2-celled), free, sessile, ovoid, style very short, stigma sessile, discoid or capitate, lobes large, dilated, ovule 1, pendulous. **Fruits** drupes, ovoid, fleshy, stone bony, almost 2-celled by a vertical plate; seed curved, pendulous. **Distr:** Mostly in Malesia, a few in tropical Africa and Asia (India, Sri Lanka), Madagascar (1 sp.), Seychelles (1 sp.), tropical America (1 sp.). **Cult:** Propagate from seeds or by air-layering. Requires a moist, well-drained soil in full sun.

C. auriculata Hook. f., *Fl. Brit. India* 2: 41, 1876. (Syn: *Buchanania auriculata*). **Trees**. **Leaves** obovate-spathulate, 15–25 x 5–7.5 cm in mature plants, base gradually decurrent along petiole, apex rounded or retuse (lobes rounded), quite glabrous, dark green, surfaces reticulated, midvein and secondary veins prominent, raised, pale green, nerves 12–18 pairs; petiole 1.25–2.5 cm long, auricled at the very base, ± clasping the stem, auricles small; lamina of juvenile plants to 1.5 m x 30–45 cm. **Inflorescences** panicles, conical, pubescent, very slender, laxly branching at base. **Flowers** small, 4-merous. **Fruits** globose, covered with white spots; infructescence rachis thickened, held ± horizontal. **Distr:** Singapore, Sumatra, Borneo. **Ecol:** In mixed dipterocarp and freshwater swamp forests, and in secondary forests; to 1,000+ m. **Zones:** 11–12. **Info:** The sap may cause severe skin irritation to some people.

C. zeylanicum Thw., *Kew J. Bot.* 6: 65, 1854. **Trees**, moderate to large-sized; twigs with conspicuous leaf-scars; young parts with dark brown scales. **Leaves** crowded near apex of stem; lamina elliptic-oblong, 10–20 x 3–9 cm, base gradually narrowed and decurrent along petiole, apex rounded or slightly retuse, coriaceous, adaxially shinier than abaxially, abaxially papillose and covered with scattered rufous scales, lateral nerves 20–25 pairs, crowded near base of lamina; petiole ~1 cm long. **Inflorescences** spicate, 4–6 cm long (♀), axillary; peduncles 2–3 cm long; axes smooth; pedicels ~1–2 mm long. **Flowers** with calyx lobes triangular, ~1 mm long, persistent around the young fruit; corolla lobes tongue-shaped; stamens shorter than petals, rather long-persistent, visible around young fruits. **Fruits** ~1.5 cm long, apiculate, pulpy, purple; young fruits inside a low-rimmed disk. **Distr:** Sri Lanka. **Ecol:** In lowland dipterocarp forests and the wet zone. Fl. Mar–Apr. **Zones:** 11–12.

Dracontomelon

Species 8. **Trees**. **Leaves** imparipinnately compound; leaflets entire, rarely serrulate. **Inflorescences** paniculate, axillary or subterminal; flowers pedicellate. **Flowers** 5-merous, bisexual; stamens 10, equal to petals in length, filaments linear-subulate; disk dish-shaped, ± lobed; ovary 5-loculed, locules 1-ovulate, styles 5, joined in upper part. **Fruits** drupes, subglobose; mesocarp fleshy; endocarp pentagonous, slightly compressed in cross-section, 5-loculed; seed elliptic, 3-ribbed. **Distr:** Tropical Asia. **Cult:** Propagate from seeds or cuttings. Requires any well-drained soil in full sun.

D. dao (Blanco) Merr. and Rolfe, *Philipp. J. Sci.*, C 3: 108, 1908. (Syn: *D. edule*, etc.). [Argus Pheasant-tree]. **Trees**, to 20 m tall; branches thickened, terete, ~1 cm Ø, grayish-brown, marked with large petiolar scars. **Leaves** crowded at the apices of branchlets, ~40 cm long; petiole and rachis somewhat cinerous-puberulent, becoming glabrous; leaflets ~11, oblong, 12–17 x 4–5.5 cm, base acute to rounded, somewhat inequilateral, margins entire, apex sharply acuminate, chartaceous, adaxially shiny, glabrous, abaxially slightly paler, glabrous except for the bearded glands in axils, lateral nerves ~12 per side; petiolules ~2 mm long. **Inflorescences** panicles, 50 cm long, much-branched, flowers numerous, borne in upper axils, puberulent; flowers pedicellate. **Flowers** white, bisexual; calyx 5-lobed; corolla urceolate, petals 5, imbricate apically; disk hairy; stamens 10, filaments subulate; ovary superior, 5-carpellate, 5-loculed, 1–4 abortive, styles 5. **Fruits** globose, 5-loculed, green, ripening to yellowish, 3 cm Ø, glabrous, smooth; mesocarp fleshy, edible but acidic; stone depressed, ~1.8 cm Ø, 8 mm thick. **Distr:** Tropical Asia to Malesia. **Ecol:** In forests behind mangroves, and on steep slopes; <200 m. **Zones:** 11–12.

Lannea

Species ~70. **Trees**, dioecious. **Leaves** imparipinnately compound; leaflets opposite, margins entire. **Inflorescences** panicles or racemes, terminal. **Flowers** 4-merous; stamens 8, anthers ovoid, reduced and sterile in ♀; disk intrastaminal; ovary (rudimentary in ♂ flower) 4-locular, with 1 apical pendulous ovule per locule, styles 3 or 4, short, stigma subglobose. **Fruits** drupes, small, subreniform, compressed; mesocarp thin, endocarp woody, 1–4-locular with operculum. **Distr:** Tropical Africa, S and SE Asia. **Cult:** Propagate from seeds or by air-layering. Requires a well-drained soil in full sun.

L. coromandelica (Houttuyn) Merrill, *J. Arnold Arbor.* 19: 353, 1938. (Syn: *L. grandis*, etc.). [Indian Ash Tree]. **Trees**, deciduous, 5–10 m tall; branchlets densely hairy. **Leaves** imparipinnately compound, 10–33 cm long; leaflets (5–)7–9(–11), ovate or oblong-ovate, 5.5–9 × 2.5–4 cm, base oblique, subrounded, margins entire, apex long-acuminate or caudate-acuminate, membranous or papery, adaxially glabrous, abaxially pubescent, ferruginous, lateral veins 6–10 pairs; petiolule 1–3 mm long, ferruginous-hairy. **Inflorescences** paniculate or racemose, appearing before leaves, on branch apices or on short shoots in leaf-scar axils; ♂ 15–30 cm long; ♀ smaller; bracts 1–2 mm long, margins ciliate. **Flowers** small, yellow or purplish; calyx lobes ovate, ~1 mm long, glabrous, margins ciliate; petals ovate-oblong, ~2.7 × 1.5 mm, recurved; stamens 8, subequal to petals in ♂, reduced and sterile in ♀; disk annular; ovary glabrous, ovoid, 4-locular, usually only 1 ovule fertile. **Fruits** drupes, ovoid to slightly reniform, purplish-red, 0.6–1 × 0.5–1 cm. **Distr:** Bhutan, China, India, Myanmar, Nepal, Sri Lanka. **Ecol:** In lowland and hill forests; 100–1,800 m. Fl. Mar. Fr. Apr–Jun. **Zones:** 10–12.

Nothopegia

Species 6. **Trees**, small to medium-sized; exudate drying black. **Leaves** spirally arranged or ± in 2 rows or subopposite; lamina coriaceous. **Inflorescences** racemes or panicles, ± spicate, axillary. **Flowers** polygamous; calyx 4–5-lobed, persistent, lobes imbricate; petals 4–5, imbricate, spreading; stamens 4–5, inserted under the margin of a 4–5-lobed disk, filaments free, hairy, anthers introrse, longitudinally dehiscent; ovary abortive in ♂ flowers, sessile in ♀, ovoid, 1-celled, style short, stigma capitate. **Fruits** drupes, globose, acute, depressed, tipped with the style; seed pendulous. **Distr:** India, Sri Lanka (1 sp.). **Cult:** Propagate from seeds or by air-layering. Requires a moist, well-drained soil in full sun.

N. beddomei Gamble, *Kew Bull.* 1918: 227, 1918. (Syn: *Glycycarpus racemosa*, *N. colebrookiana*). **Trees**, small to medium-sized; bark closely fissured; exudate white latex, slowly darkening; twigs and leaves glabrous. **Leaves** oblong-lanceolate, ~10–15 cm long, base cuneate, slightly decurrent toward petiole, apex acuminate, adaxially midrib sunken, abaxially keeled and slightly furrowed, lateral nerves to 20 pairs; petiole ~1 cm long, partly twisted, bringing all the leaves into 1 plane. **Inflorescences** bisexual, spicate, 2–3 cm long, in upper axils, lower ones longer; ♂ paniculate, 7–8 cm long, lowest side branches 1.5 cm long, branches short patent-hairy. **Flowers** in bud to 1 mm long; calyx shallowly lobed; petals oblong, obtuse; stamens with filaments hairy (villous); ovary glabrous. **Fruits** drupes, ~1.5 cm long; about twice as wide as long, figlike, longitudinally striate, purple, pulp copious. **Distr:** W peninsular India, Sri Lanka. **Ecol:** In wet lowlands and virgin forests. Fl. Mar–Jul. **Zones:** 11–12.

Rhodosphaera

Species 1. *Generic description same as specific.* **Cult:** Propagate from scarified seeds. Requires a moist, well-drained soil in full to partial sun.

R. rhodanthema (F. Muell.) Engl., *Botanische Jahrbucher fur Systematik* 1: 424, 1880. [Yellow Cedar, Tulip Satinwood]. **Trees**, evergreen, dioecious, to 20 m tall; bark brown to gray, scaly; branchlets reddish-lenticellate. **Leaves** alternate, imparipinnate, mostly 8–20 cm long; leaflets 3–11, ± opposite, oblong-ovate to narrowly elliptic, 3–10 × 1–2.5 cm, base obtuse, asymmetric with midvein curved, margins entire or ± lobed in juveniles, apex shortly acuminate, ± glabrous except for small red-brown-hairy domatia on abaxial surface; petiole 3–8 cm long, petiolules 3–5 mm long. **Inflorescences** panicles, mostly 10–20 cm long, terminal and axillary; pedicels 1–2 mm long, bracteate. **Flowers** with sepals 5, 1.8–2 mm long; petals 5, ~3 mm long, pinkish-red; ♂ with 10 stamens inserted below margin of disk, disk lobed between filaments, sterile ovary absent; ♀ with 10 staminodes, ovary seated on disk, 1-loculed. **Fruits** drupes, borne in clusters that persist for many months, ± globose, 0.8–1.1 cm Ø, glossy-brown; exocarp papery; mesocarp thick, woody; endocarp bony, compressed. **Distr:** Australia. **Ecol:** In subtropical rainforests and drier rainforests; ~400 m. Fl. Sep–Oct. Fr. Feb–Jun. **Zones:** 9–11.

Spondias

Species ~11. **Trees**, entirely or partly deciduous. **Leaves** alternate, spirally arranged, petiolate, imparipinnately compound; leaflets with margins serrate or entire. **Inflorescences** paniculate, terminal or axillary. **Flowers** 4- or 5-merous, bisexual or functionally unisexual; stamens 8–10, filaments subulate to filiform, equal in length; disk intrastaminal, 10-crenate to crenulate; ovary 4- or 5-locular, locules 1-ovules, styles 4 or 5, free, or style 1. **Fruits** drupaceous, mesocarp juicy, endocarp woody or bony, covered by a fibrous matrix. **Distr:** Tropical America and tropical Asia. **Cult:** Propagate from seeds, cuttings or by air-layering. Requires a moist, well-drained soil in full or partial sun.

S. dulcis Soland. ex. Parkinson, *Voy. South Seas* 39, 1773. (Syn: *S. Mangifera*, etc.). [Golden Apple, Otaheite Apple, Polynesian Plum]. **Trees**, to 6–12 m tall; trunk 45 cm Ø; branches few, stout; twigs stout; bark smooth, greenish; exudate resinous. **Leaves** alternate, pinnate, 20–30 cm long, paired except the terminal one; leaflets 11–23, lanceolate or oblong, 3.8–6.4 cm long, base acute, margins toothed, apex long-acuminate, thin; petiole <3 mm long. **Inflorescences** panicles, 20–30 cm long, flowers numerous; pedicels ~1.5 mm long. **Flowers** ♂, ♀ or bisexual, whitish, fragrant, 5-merous, 6 mm Ø; calyx small, 5-lobed; petals 5, whitish, <3 mm long, spreading, bent downward; stamens 10; pistil on a disk, ovary with 5 styles. **Fruits** 2–10, in drooping clusters, drupes, 5–10 × 5 cm, pericarp thick; seeds, hard 2.5–4 cm long, covered with stiff spinelike projections, embedded in juicy yellowish pulp. **Distr:** Society Islands. **Ecol:** Fl. & Fr. throughout the year. **Zones:** 9–12. **Info:** The edible fruits are sweet and sour.

Schinus

Species 30. **Trees** or **shrubs**. **Leaves** alternate, simple or odd-pinnate (rarely paripinnate); leaflets sessile, margins often crenate-dentate, especially toward the tips. **Inflorescences** terminal panicles or axillary pseudo-racemes, bracteate. **Flowers** polygamodioecious, small, whitish; calyx lobes 5, rounded; petals 5, oblong, imbricate; stamens 10, inserted at base of thick annular disk, finally reflexed, filaments subulate, anthers oval; ovary free, sessile, 1-celled, ovule suspended from near apex, style terminal, trifid or simple, with 3 minute capitellate stigmas or simple and 3-lobate. **Fruits** drupes, globose, small, putamen coriaceous or osseous, usually oily. **Distr:** Mexico to Argentina. **Cult:** Propagate from seeds or cuttings. Requires any well-drained soil in full sun.

S. molle L., *Sp. Pl.* 388, 1753. [California Pepper, Pepper Tree, Peruvian Mastic, Pink Peppercorns, etc.]. **Trees** or large **shrubs**, 6–9 m tall, abundantly but loosely leafy and gracefully branched, slender upper branches and branchlets ± pendent. **Leaves** imparipinnate, to 25 cm long, feathery (fernlike), pendent; leaflets 17–35, linear-lanceolate, 1.5–5 × 0.2–1 cm, margins subentire or serrate, apex acute or subobtuse, light green to grayish-green. **Inflorescences** compound panicles, conical, terminal, 8–25 cm long. **Flowers** small, greenish- or yellowish-white, 5-merous; sepals deltate, 0.4 mm long; petals oblong or narrowly ovate, 1.5–2 mm long; styles 3, stigmas 3, free. **Fruits** drupes, 5–6 mm Ø, globose, pink. **Distr:** Tropical America. **Ecol:** Fl. Feb–Mar. **Zones:** 8–12.

S. terebinthifolius Reddi, *Mem. Mat. As. Soc. Ital. Sci.* 18:399, 1820. [Brazilian Pepper Tree, Christmas Berry, Florida Holly]. **Shrubs** or small trees, dioecious, mounding, to 2.5–6 m tall. **Leaves** imparipinnate, to 13 × 1.2–3 cm, spreading; leaflets ± sessile, (7–)9(–13), elliptic to narrowly obovate or elliptic-oblong, 3.8–9 × 1.3–2.3 cm, terminal leaflet largest, tapering at both ends, margins subserrate, adaxially dark glossy green, abaxially paler. **Inflorescences** panicles or subracemose, 1–11.5 cm long, compact, axillary. **Flowers** greenish to yellow; sepals broadly ovate, ~0.8 mm long, margins ciliate; petals 5, ovate,

Below: *Dracontomelon dao, Lannea coromandelica, Nothopegia beddomei, Rhodosphaera rhodanthema.*

TG

Above: *Spondias dulcis, Schinus molle, Schinus terebinthifolius, Sclerocarya caffra.*

1.2–2.5 mm long, obtuse; ♂ with a pistillode and ♀ with staminodes; ovary subglobose ~1 mm long. **Fruits** drupes, red, ± globose, 5 mm Ø. **Distr:** S America (from Venezuela to Argentina). **Ecol:** Fl. Aug–Nov. **Zones:** 8–12. **Info:** As with several species in this family, the plant produces harmful toxins.

Sclerocarya

Species 2. **Trees**, small or medium-sized. **Leaves** alternate, imparipinnate, crowded at apex of branchlets, petiolate; leaflets petiolulate. **Inflorescences** axillary or terminal, spikelike (♀ ones sometimes reduced), simple or with a few basal branches; flowers shortly pedicellate. **Flowers** unisexual (sometimes bisexual), sepals and petals 4–5, imbricate; ♂ with stamens (8–10)15–25(–30), filaments subulate, anthers oblong, dorsifixed, disk patelliform, pistillode absent; ♀ slightly larger than ♂, sterile stamens ± developed, ovary subglobose, immersed in disk, 2–3(4)-locular, styles 2–3, short, lateral, stigmas capitate. **Fruits** drupes, obovoid-subglobose, mesocarp fleshy, endocarp thick, bony, 3-operculate at apex, opercules ovate; seeds 2–3, obclavate, brown. **Distr:** Africa, Madagascar. **Cult:** Propagate from seeds or cuttings. Requires any well-drained soil in full sun.

S. caffra Sond., *Linnaea*, 23: 26, 1850. (Syn: *S. birrea*, etc.). [Cat Thorn, Cider Tree, Jelly Plum, Maroola Nut, Moroola Plum, Marula]. **Trees**, to 18 m tall; trunk thick; branches spreading. **Leaves** 7–13(–17)-foliolate; leaflets ovate or elliptic, 3–9.5(–11) × 1.5–3.5(–6) cm, base asymmetric, margins usually entire, apex acuminate, acumen to 1 cm long, glabrous; petiole and rachises 15–30 cm long, lateral petiolules 0.5–3 cm long. **Inflorescences** ♂ axillary and terminal, 7–17 cm long, bracts ovate, 2 × 1.5 mm, obtuse; ♀ subterminal, shorter, flowers fewer. **Flowers** ♂ with sepals ~2 × 2 mm, petals yellow to purplish-pink (red in bud), oblong-ovate, 4–6 × 3–4 mm, obtuse, filaments ~3 mm long, anthers 1–1.5 mm long; ♀ with sepals and petals ± as in ♂, staminodes 15–25, ovary subglobose. **Fruits** drupes, yellow, 3–3–5 cm Ø, mesocarp very juicy, stone obovoid, 2–3 × 2.5 cm, thick; seed ~1·5–2 × 0.4–0.8 cm. **Distr:** Africa, Madagascar. **Ecol:** In woodlands and savannas. **Zones:** 11–12.

Subfam. ANACARDIOIDEAE

Genera 60, **species** 485. **Trees**, rarely shrubs, bisexual, polygamous or dioecious. **Leaves** simple (compound), margins entire to slightly serrate. **Flowers** with 1 or 2 whorls of stamens, equal or unequal in size; ovary composed of 4–6 free carpels, each with a short style or composed of a single carpel, with an eccentric to lateral style. **Distr:** Mainly in tropical regions, also in temperate regions.

Anacardium

Species ~10–11. **Trees** or **shrubs**, evergreen, andromonoecious. **Leaves** petiolate, simple; lamina margin entire. **Inflorescences** with functionally ♂ and bisexual flowers together, panicles, terminal, frondo-bracteate, branches numerous; pedicel enlarging after anthesis into a fleshy hypocarp bearing fruit. **Flowers** with calyx 5-parted, cup-shaped, lobes pointed; petals 5, recurved at anthesis; stamens 7–10, usually 1(or 2) noticeably larger, filament basally fused, anther broadly ellipsoid; disk lacking; ovary obovoid, asymmetrical, 1-locular and 1-ovulate, style 1, lateral, linear. **Fruits** drupes, nutlike, glabrous with bony endocarp. **Distr:** Tropical America. **Cult:** Propagate from seeds, cuttings, grafting or by air-layering. Requires a moderately fertile, well-drained sandy soil in full sun.

A. excelsum Skeels, *U.S.D.A. Bur. Pl. Industr. Bull.* 242: 36, 1912. (Syn: *A. rhinocarpus, Rhinocarpus excelsa*). [Espavé, Espavel, Wild Cashew]. **Trees**, very large, to 20–35 m tall, almost glabrous; trunk clear for (3–4–)10–18 m, 1–2.5(–3) m Ø; roots massive, exposed or partly buried, to 50 cm thick. **Leaves** obovate-oblong, 15–28 × 6–12 cm, broadest well above middle, base acute, apex rounded or sometimes emarginate, glabrous, veins 12–22 pairs, strong, parallel. **Inflorescences** panicles, much longer than the leaves, to 30 cm long, brownish-pubescent. **Flowers** small, stamens 10, 4 of them much longer than the others. **Fruits** drupes, nutlike, asymmetric, biconvex, elongate, 3–3.5 × 2 cm, ~1 cm thick; pedicel, thickened, accrescent, fleshy, sigmoid, 1.5–2 cm long, 2–3 mm thick. **Distr:** C America (Costa Rica to the Guianas). **Ecol:** In forests, often along streams; <1,200 m. Fl. Feb. Fr. Mar–May. **Zones:** 10–12. **Info:** One of the largest trees in C America. Fruits are edible and the bark is used as fish poison.

A. occidentale L., *Sp. Pl.* 1: 383, 1753. [Cashew Apple, Cashew Nut]. **Trees** or **shrubs**, 4–10 m tall; branchlets ± glabrous. **Leaves** obovate, 8–11 × 6–8.5 cm, base broadly cuneate, margins entire, apex rounded, truncate to retuse, leathery, glabrous, lateral veins ~12 pairs; petiole 1–1.5 cm long. **Inflorescences** paniculate, 10–20 cm long, glabrous to densely sericeous; floral bracts ovate-lanceolate, 0.5–1 cm long, keeled; flowers ± sessile. **Flowers** with calyx abaxially grayish-sericeous, ~4 × 1.5 mm; petals 5, greenish-yellow to red, linear-lanceolate, 7–9 × ~1.2 mm; stamens 7–10, larger one 8–9 mm long in ♂, 5–6 mm long in bisexual flowers, sterile stamens 3–4 mm long; ovary ~2 mm long, glabrous; style 4–5 mm long. **Fruits** fleshy hypocarps, 3–7 × 4–5 cm, purplish-red; drupe reniform, 2–2.5 × ~1.5 cm. **Distr:** Tropical America. **Ecol:** Fl. Mar–Apr. Fr. Jul–Aug. **Info:** Widely cultivated for the edible nuts (seeds). The fruit, often referred to as the Cashew Apple, is edible and sweet.

Mangifera

Species 69. **Trees**, evergreen, andromonoecious. **Leaves** petiolate, simple; leaf lamina margin entire. **Inflorescences** with functionally ♂ and bisexual flowers together, pleiothyrsoids, terminal. **Flowers** small, 4- or 5-merous, with imbricate aestivation; sepals sometimes basally united; petals with 1–5 prominent veins

Below: *Anacardium excelsum, Anacardium occidentale, Anacardium occidentale* (fruit), *Mangifera indica.*

Above: *Mangifera zeylanica*, *Semecarpus anacardium*, *Semecarpus walkeri*, *Aesculus hippocastanum*.

adaxially; stamens 5, free or basally united with disk, usually 1(or 2) noticeably larger; disk of 5 separate extrastaminal glands; ovary glabrous, 1-locular, 1-ovulate; style 1, terminal. **Fruits** drupaceous; mesocarp fleshy or fibrous; endocarp thick and bony, compressed, polyembryonic. **Distr:** Tropical Asia. **Cult:** Propagate from seeds, by air-layering or grafting. Requires a deep, well-drained soil in full sun. Most of the species are intolerant to salty or coastal habitats.

M. indica L., *Sp. Pl.* 1: 200, 1753. (Syn: *M. austroyunnanensis*). [Mango]. **Trees**, 10–20 m tall; branchlets brown, glabrous. **Leaves** oblong to oblong-lanceolate, 12–30 × 3.5–6.5 cm, base cuneate to obtuse, margin entire, undulate, apex acute to long-acuminate, leathery, glabrous, adaxially deep green, abaxially paler, lateral veins 20–25 pairs; petiole 2–6 cm long, base inflated. **Inflorescences** paniculate, terminal, 20–35 cm long, glabrous to tomentose-pilose; bracts lanceolate, ~1.5 mm long, pubescent; pedicels 1.5–3 mm long, articulate. **Flowers** with sepals ovate-lanceolate, 2.5–3 × ~1.5 mm, acuminate; petals light yellow with prominent red tree-shaped pattern adaxially, oblong or oblong-lanceolate, 3.5–4 × ~1.5 mm, glabrous, recurved; fertile stamen 1, ~2.5 mm long, anther ovate; staminodes 4, 0.7–1 mm long; disk inflated, fleshy, 5-lobed; ovary oblique, ovate, ~1.5 mm Ø; style ~2.5 mm long, eccentric. **Fruits** drupes, oblong to subreniform, greenish-yellow to red, 5–10 × 3–4.5 cm; mesocarp fleshy, bright yellow; endocarp ± compressed. **Distr:** Continental SE Asia. **Ecol:** 200–1,400 m. Fl. Mar–Apr. Fr. May–Jul. **Zones:** 11–12.

M. zeylanica (Blume) Hook. f., *Fl. Br. Ind.* 2: 16, 1876. (Syn: *Buchanania zeylanica*). [Sri Lankan Mango]. **Trees**, large; bark rather rough, slightly cracked-fissured, brownish-gray, with dispersed corky lenticels; inner bark orange-brown, sapwood white; twigs glabrous, angular when young. **Leaves** lanceolate or ovate-oblong, 7–13 cm long, in saplings to 20 cm long, base tapered along petiole, apex rounded or retuse (in saplings acute), coriaceous, midrib raised on both faces, lateral nerves 7–10 pairs, prominently raised abaxially; petiole 1–2 cm long. **Inflorescences** panicles, terminal, glabrous. **Flowers** with sepals ovate, obtuse, glabrous; petals twice as long as sepals, obtuse, clawed, reflexed; sterile stamens reduced to short tabulate processes. **Fruits** ovoid, 3–4 cm long. **Distr:** Sri Lanka. **Ecol:** In lowlands of moist and dry regions, to ~1,000 m. Fl. Feb–Mar, Nov. **Zones:** 11–12.

Semecarpus

Species ~50. **Trees**, rarely shrubs; inner bark with black-drying juice which causes skin irritation. **Leaves** spirally arranged; lamina simple, coriaceous, margins entire, abaxially often lighter and dull-colored than adaxially, tertiary nerves near midrib often partly parallel with lateral nerves as intermediate nerves. **Inflorescences** paniculate, terminal or axillary. **Flowers** small, polygamous or dioecious, ♂ usually smaller; calyx 5-lobed, lobes deciduous; petals 5, imbricate; stamens 5, inserted at base of disk, in ♀ flowers smaller, filaments linear, anthers medifixed, ovate-cordate, versatile, in ♀ flowers empty; disk broad, annular, faintly lobed; ovary (absent in ♂ flowers) superior to semi-inferior, 1-celled with 1 pendulous ovule, styles 3, free or basally united, stigmas lobed. **Fruits** drupes, ± compressed, obliquely ovoid, with a thick fleshy receptacle; seed testa not fused with endocarp. **Distr:** Tropical Asia to Oceania. **Cult:** Propagate from seeds or cuttings. Requires any well-drained soil in full sun.

S. anacardium L. f., *Suppl. Plant.*, 182, 1782. [Marking Nut Tree]. **Trees**, medium- to large-sized, 15–25 m tall; trunk large, straight; bark gray, exfoliating in small regular flakes; exudate white, soft, gumlike. **Leaves** alternate, borne at end of branchlets; lamina simple, obovate-oblong, 22–45 × 10–20 cm, margins entire, apex rounded, coriaceous, adaxially glabrous, abaxially ± pubescent, main nerves 15–25 pairs; petiole 3.8–5 cm long. **Inflorescences** fascicled panicles, composed of many simple spikes, pubescent, terminal, very large (on ♂ trees spikes are more slender); ♂ flowers on separate trees; bracts many, deciduous. **Flowers** greenish-white, numerous, small; ♂ with calyx and corolla similar to bisexual flowers, filaments 5, as long as petals, anthers larger than in bisexual flowers, pistil absent or rudimentary. **Fruits** drupes, obliquely ovoid or oblong, 2.5 cm long, black when ripe; receptacle fleshy, yellow when ripe; seed 1, smooth, black, shiny. **Distr:** India. **Ecol:** In semievergreen and moist deciduous forests. Fl. Jul–Aug. Fr. Jan–Feb. **Zones:** 11–12. **Info:** The fruits are used for medicinal purposes.

S. walkeri Hook. f., *Fl. Br. Ind.* 2: 33, 1876. **Trees**, medium-sized; trunk without buttresses, shallowly fissured; branchlets glabrous. **Leaves** oblong-lanceolate, ovate-elliptic in saplings, to 38 cm long, base cuneate, almost decurrent, apex acuminate, often twisted, stiff, coriaceous, lateral veins 8–14 pairs, almost at right angles with midrib, 4–5 intermediate nerves reaching halfway toward the margin between adjacent laterals, anastomosing at the margins; petiole 2.5–8 cm long, base corky, swollen. **Inflorescences** panicles, terminal and axillary, (2–)10–15 cm long, slightly branched or almost simple, slender, cauliflorous. **Flowers** small. **Fruits** drupes, >1.2 cm long, ovoid, not oblique; receptacle cup-shaped, enclosing the drupe halfway or slightly more. **Distr:** Sri Lanka. **Ecol:** In lowlands of the wet Zones:; <600 m. Fl. Feb–Mar. Fr. Jun–Jul. **Zones:** 11–12.

SAPINDACEAE

Genera 135, species 1,580. **Trees, shrubs,** or **woody vines** with tendrils (in *Cardiospermum* and allied genera), rarely herbaceous climbers. **Leaves** alternate, usually estipulate, pinnate or digitate, rarely simple; leaflets alternate to opposite, entire or dentate to serrate. **Inflorescences** thyrses, terminal or axillary; bracts and bracteoles small. **Flowers** unisexual, rarely polygamous or bisexual, actinomorphic or zygomorphic, usually small; sepals 4 or 5(or 6), equal or unequal, free or basally connate, imbricate or valvate; petals 4 or 5(or 6), sometimes absent, free, imbricate, usually clawed, often with scales or hair-tufted basal appendages; disk conspicuous, fleshy, complete or interrupted, lobed or annular, rarely absent; stamens 5–10(–74), usually 8, rarely numerous, variously inserted but usually within the disk, often exserted in ♂ flowers, filaments free, rarely connate, anthers dorsifixed, longitudinally dehiscent, introrse; staminodes sometimes present in ♀ flowers, but filaments shorter and anthers with a thick wall, indehiscent; ovary superior, (1–)3(or 4)-loculed, ovules 1 or 2(or several) per locule, placentation usually axile, style usually apical, semigynobasic in some genera, stigma entire or 2 or 3(or 4)-lobed, usually rudimentary in ♂ flowers. **Fruits** loculicidal capsules, berries, or drupes, or consisting of 2 or 3 samaras, often 1-seeded and 1-loculed by abortion; seeds 1(or 2 or more) per locule, testa black or brown, hard, often with a conspicuous fleshy aril or sarcotesta. **Distr:** Worldwide.

Aesculus

Species 12. **Trees** or **shrubs**, deciduous. **Leaves** 5–11-foliolate; leaflets with margins crenate to serrate. **Inflorescences** thyrses, cylindric or conic, often large and showy; branches simple; bracts absent. **Flowers** with sepals connate to form a tubular to campanulate calyx tube; petals often unequal, base clawed, limb obovate, oblong, oblanceolate, or spathulate; ovary without a gynophore, style long, slender, stigma depressed-globose, entire or obscurely lobed. **Fruits** capsules, depressed-globose to pyriform, without a long gynophore, often 1-seeded, pericarp usually smooth, often dotted, rarely verrucose or prickly; seeds depressed-globose to pyriform, large, 2–7 cm Ø, testa brown, hilum large, pale, occupying up to half of the seed. **Distr:** Mainly in N America and Asia; Europe. **Cult:** Propagate from fresh seeds or by grafting. Requires a deep, fertile, moist soil in full sun.

A. hippocastanum L., *Sp. Pl.* 1: 344, 1753. (Syn: *Hippocastanum vulgare*). [Horse Chestnut]. **Trees**, to 30 m tall; trunk to 2(–5) m Ø. **Leaves** 5–7-foliolate; leaflets sessile, abaxially green, oblanceolate, 10–25 × 5–12 cm, base cuneate, margins crenate-serrate, teeth crenulate-serrulate, apex acuminate to subcaudate, lateral veins 18–25 pairs; petiole 8–20 cm long. **Inflorescences** thyrses conic or cylindric-conic, 10–30 × 6–10 cm; branches 2.5–

Above: *Allophylus africanus, Allophylus africanus* (fruits), *Allophylus cobbe.*

5 cm long, 4–12-flowered; peduncles 2.5–5 cm long; pedicels 3–6 mm long. **Flowers** with calyx 5–6 mm long; petals 4 or 5, white, with red spots and a yellow (later brown) claw, equal, ~1.1 cm long; stamens 5–8, 1–2 cm long, filament villous; style apically glabrous. **Fruits** capsules, brown, ± globose, prickly, 2.5–4 × 2.5–6 cm (excluding prickles); prickles subulate, slightly curved, to 1 cm long; seeds 1–3(–6), glossy chestnut-brown, subglobose, 2–4 cm Ø; hilum pale brown. **Distr:** SE Europe. **Ecol:** Fl. May–Jun. Fr. Sep. **Zones:** 7–10.

Allophylus

Species ~200. **Shrubs**, rarely trees (very rarely woody climbers), monoecious or dioecious. **Stipules** absent. **Leaves** digitate; leaflets 1–5, petiolate or subsessile, margins usually serrate, sometimes entire. **Inflorescences** thyrses, racemose, axillary; bracts and bracteoles subulate or lanceolate, small. **Flowers** unisexual, zygomorphic, closed, small; sepals 4, concurved, outer 2 elliptic, slightly smaller, inner 2 suborbicular; petals 4, in 2 pairs; disk 4-lobed, lobes glandlike; stamens (in ♂ flowers) 8, sometimes fewer, exserted, filaments free or connate below middle; ovary (in ♀ flowers) 2(or 3)-partite, 2-loculed, lobes subglobose or obovate, locules 1-ovuled, style basilar, entire or bi- or trilobed nearly to base, stigma recurved. **Fruits** parted into 2 or 3 schizocarps, often only 1 developed, berrylike, subglobose or obovoid, base with rudimentary style and small, globose, sterile schizocarps, exocarp fleshy, juicy, endocarp crustaceous; seeds nearly same as schizocarp in shape. **Distr:** Widely distributed in tropical and subtropical regions. **Cult:** Propagate from seeds or by air-layering. Requires a moist, well-drained soil in full or partial sun.

A. africanus P. Beauv., *Fl. Owar.* 2: 75, t. 107, 1819. (Syn: *Schmidelia africana*). [African False Currant]. **Shrubs** or **trees**, (2–)4.5–10 m tall. **Leaves** trifoliolate, typically ± glabrous, when glabrous dark and shiny adaxially; leaflets narrowly elliptic to elliptic, 2–14.5 × 1–7 cm, base attenuate to cuneate, margins subentire to crenate or dentate, apex rounded to acuminate, subcoriaceous, sometimes with hairy domatia abaxially; petiole (1–)2–7.5 cm long. **Inflorescences** 4.5–15 cm long, unbranched or 1–6-branched, usually fulvous-pubescent; cymules subsessile; pedicels (0.5–)1.5–2 mm long, often glabrous. **Flowers** white, greenish-white or yellowish-green, scented; outer sepals broadly elliptic, ~1 × 1 mm, inner sepals oblate, 1 × 1–1.5 mm, ± glabrous; petal spathulate, 1 × 0.5 mm; filaments ~2 mm long, base hairy; style 1 mm long, hairy, stigma 0.5 mm long. **Fruits** ellipsoid, 5–7 × 4–4.5 mm, very sparsely hairy; seed ellipsoid, 5.5–6 × 4.5 mm. **Distr:** Tropical Africa. **Ecol:** In forest margins, streambanks of savanna regions. **Zones:** 11–12.

A. cobbe (L.) Raeusch., *Nomencl.* ed. 3, 108, 1779. (Syn: *A. rheedii*, etc.). [Wild Berry, Tit Berry]. **Shrubs** or **small trees**, much-branched. **Leaves** 3-foliolate, central leaflet larger; leaflets ovate, elliptic or elliptic-ovate, 3–13.5 × 2–7.5 cm, base rounded, obtuse, truncate or cuneate, margins irregularly crenulate, serrulate, apex acute-acuminate, usually softly pubescent, lateral nerves 7–10 pairs; petiole 2–8 cm long; petiolules 1–8 mm long. **Inflorescences** panicles, axillary, 6–12 cm long, spiciform, with scattered clusters of 2–5-flowered, subsessile cymules; pedicels 1–2 mm long. **Flowers** ~2.5 mm Ø, greenish-white; sepals elliptic to suborbicular, 1–2 × ~1 mm, concave, membranous; petals about as long as sepals, spathulate, long-clawed, apex emarginate or dentate; stamina1 filaments filiform, exserted, 1–2 mm long; ovary bilobed, lobes suborbicular, hairy, style twice as long, stigmas recurved. **Fruits** with ripe carpels usually solitary (1 abortive), subglobose, ovoid to ellipsoid, 5–6 × 4–5 mm, bright red, glabrescent. **Distr:** Pantropical. **Ecol:** In forests of dry regions, coastal regions. Fl. Mar–May, Sept–Oct. **Zones:** 11–12.

Blighia

Species ~6–7. **Trees**, large, polygamous. **Leaves** large, sparse, petiolate, even-pinnate; leaflets entire, pinnately veined. **Inflorescences** axillary, racemose; bracteoles small, buds subconical; flowers pedicellate; bracts small; buds subconic. **Flowers** polygamous, small; calyx 5-parted, segments ovate-lanceolate, narrowly imbricate; petals 5, linear-lanceolate, squamate at base and saccate; disk swollen, somewhat 8-lobed; stamens 8, inserted within the disk. **Fruits** 3-celled, large, trigonous-pyriform; seeds surrounded by a large fleshy aril. **Distr:** Africa. **Cult:** Propagate from seeds or by air-layering. Requires a moist, well-drained soil in full or partial sun.

B. sapida Koenig, *Ann. Bot.* 2: 571, tab. 16–17, 1806. [Akee, Vegetable Marrow]. **Trees**, to 8–15(–50) m tall. **Leaves** even-pinnate; leaflets in (3–)4(–5) pairs, cuneate-obovate to elliptic or oblong, 4–21 × 2.5–8.5 cm, base obtuse to acute, apex obtuse, rounded or shortly acuminate; petiolules 3–7 mm long. **Inflorescences** racemes, in terminal axils, simple or compound, tomentose; pedicels 3–5 mm long. **Flowers** cream to yellowish-green, fragrant; calyx 2.5–3 mm long, calyx deeply 5(6)-lobed, lobes narrowly ovate, tomentose; petals linear-lanceolate to ovate-oblong, 3.5–4.5 mm long, acute, villous, ciliate, basal scale wider than petals; disk annular, flattened, tomentose or densely villous; stamens 8–10; ovary shortly stalked, densely pubescent; ♂ flowers with stamens long-exserted, 4–6 mm long, filaments villous, ovary to ~1 mm long; *bisexual* flowers with stamens to ~3 mm long, ovary ovoid, narrowly tapered to style, style and stigmas together to 7 mm long. **Fruits** capsules, oblong-obovate, obtusely 3-angled, valves 3, dehiscent, 6–10 cm long, yellowish to red, aril fleshy, white, large; seeds usually 3, large, globose, to 1.8 cm Ø, blackish-brown, shiny, partially embedded in aril. **Distr:** W Africa. **Ecol:** In evergreen forests, riverine forests and grasslands; 0–1,200 m. Fl. Mar–Apr. Fr. Dec–Jan. **Zones:** 10–12. **Info:** The arils with a nutty flavor are only edible immediately after opening (ripening) naturally on the tree. Any earlier or later, the arils are highly toxic. The fruit was introduced to Jamaica, where the arils are consumed as a delicacy, by Captain Bligh on the *Bounty*.

Cardiospermum

Species ~12. **Climbers**, herbaceous or woody, rarely shrubs. **Stipules** small, deciduous. **Leaves** usually biternate or bi-trilobate; leaflets pinnately lobed or dentate, usually with pellucid glands. **Inflorescences** panicles, axillary; peduncles long, first pair of branches forming tendrils or spines; bracts and bracteoles subulate. **Flowers** unisexual, zygomorphic; pedicels slender, noded; sepals 4 or 5, imbricate, outer 2 smaller; petals 4, with a large scale inside and slightly above base; scales of abaxial petals with broadly winged appendage, scales of adaxial petals simple; disk lobed into 2 large glandular lobes at base of petals; stamens (in ♂ flowers) 8, slightly longer than petals; ovary (in ♀ flowers) ellipsoid, trigonous, 3-loculed, ovules 1 per locule, style short, stigma trilobed. **Fruits** capsules, inflated, saclike, 3-loculed; pericarp membranous or papery, veined; seeds 1 per locule, subglobose. **Distr:** Mostly in tropical and subtropical America. **Cult:** Propagate from seeds or stem cuttings. Requires fertile, moist, well-drained soil in full or partial sun.

C. grandiflorum Sw., *Nov. Gen. and Sp. Pl.* 64, 1788. (Syn: *C. hirsutum, C. grandiflorum* f. *hirsutum*). [Balloon Vine, Heart Seed, Love-in-a-puff]. **Herbs**, climbing; stems, rachis and peduncles ± hirsute. **Leaves** biternate; pinnae 3–9.5 × 2–11.5 cm; pinnules elliptic-ovate, base acute or abruptly narrowing into petiolules, apex acute to acuminate, abaxially puberulent; petiole 7.5–15.5 cm long, petiolules 1.5–4.5 cm long, terete; rachises 1.5–3 cm long. **Inflorescences** panicles corymbiform, 2.5–5 × 2.5–5 cm; peduncles 5–13.5 cm long, with a pair of recurved tendrils at base; secondary peduncles with many subulate bracts, tip with ~7-flowered circinnate racemes; pedicels ~5 mm long. **Flowers** 0.75–1 cm long, white; inner sepals longer, 8.5 × 8 mm, outer ones 2 × 2 mm; petals 0.85–1.1 × ~0.5 cm, glabrous; scales oblong cucullate, recurved; crests broad, truncate; disk bilobed; filaments glabrous, 3–5.5 mm long; ovary pubescent. **Fruits** capsules, ellipsoid, 4–7.5 × 2.5–4 cm, 6-ribbed, puberulent; seeds patelliform, aril white. **Distr:** Tropical America. **Ecol:** Along forest margins. **Zones:** 9–12.

C. halicacabum L., *Sp. Pl.* 1: 366, 1753. (Syn:

Above: *Blighia sapida* (flowers), *Blighia sapida* (fruit), *Cardiospermum grandiflorum* (flowers), *Cardiospermum grandiflorum* (fruits).

C. halicacabum var. *microcarpum*, *C. microcarpum*). [Small Balloon Vine, Heart Pea, Heartseed Vine, Love-in-a-puff]. **Climbers**, herbaceous, 1–1.5 m; stems and branches green, 5- or 6-sulcate, glabrous or sparsely hairy. **Leaves** biternate, triangular in outline; leaflets subsessile, thinly papery, margins sparsely serrate or pinnately parted; *terminal* leaflet obliquely lanceolate or subrhombic, 3–8 × 1.5–2.5 cm, apex acuminate; *lateral* leaflets slightly smaller, ovate or narrowly elliptic; petiole 3–4 cm long. **Inflorescences** panicles, few-flowered; peduncles straight, 4–8 cm long, tendrils spiraled. **Flowers** small; sepals 4, ciliate, outer 2 ovate, 0.8–1 cm long, inner 2 narrowly elliptic, twice as long as outer ones; petals milky-white, obovate; stamens (in ♂ flowers) subequal to petals; ovary (in ♀ flowers) obovoid or sometimes subglobose, pubescent. **Fruits** capsules, brown, pearlike (base wider), turbinate-obtriangular or sometimes nearly ellipsoid, 1.5–3 × 2–4 cm, pubescent; seeds black, shiny, ~5 mm Ø. **Distr:** Tropical and subtropical regions. **Ecol:** In forest margins, grasslands. **Zones:** 9–12.

Dimocarpus

Species ~7. **Trees** or **shrubs**, usually monoecious. **Leaves** paripinnate; leaflets opposite or subopposite, margins entire or serrate. **Inflorescences** thyrses, usually large, terminal, or fascicles, stellate or tomentose; bracts and bracteoles subulate, small. **Flowers** unisexual, actinomorphic; calyx cupular; sepals 5, imbricate, stellate or tomentose; petals 5 or 1–4, sometimes absent, usually spathulate or lanceolate, without scales; disk acetabuliform; stamens (in ♂) often 8, exserted, filaments hirsute, anthers ellipsoid; ovary (in ♀) obcordiform, bi- or trilobed and 2- or 3-loculed, densely tuberculate, tufted-stellate and tomentose on tubercles, locules 1-ovuled, style between ovary lobes, stigma bi- or trilobed. **Fruits** deeply lobed into 2 or 3 schizocarps, usually 1 or 2 developed, fertile schizocarps berrylike, subglobose, with small sterile schizocarps at base; seeds subglobose or ellipsoid, testa leathery, smooth. **Distr:** S and SE Asia, from Sri Lanka and India to E Malesia; Australia. **Cult:** Propagate from seeds or by air-layering. Requires a humus-rich, moist, well-drained soil in full or partial sun.

D. longan Loureiro, *Fl. Cochinch.* 1: 233, 1790. (Syn: *Euphoria longan*, etc.). [Cat's Eye, Dragon's Eye, Longan]. **Trees**, evergreen, usually ~10(–40) m tall; trunk to ~1 m Ø. **Leaves** with leaflets (3)4 or 5(6) pairs, oblong-elliptic to oblong-lanceolate, 6–15 × 2.5–5 cm, base strongly asymmetrical, cuneate at side toward leaf base, broadly cuneate to truncate at side toward leaf apex, apex acute, sometimes slightly obtuse, thinly leathery, adaxially deep green, shiny, abaxially powdery green, surfaces glabrous, lateral veins 12–15 pairs; petiole 15–30+ cm; petiolules <5 mm long. **Inflorescences** large, many-branched, terminal or axillary near apex; pedicels short. **Flowers** with sepals triangular-ovate, ~2.5 mm long, subleathery, tomentose; petals 5, milky-white, lanceolate, nearly as long as sepals, abaxially pilosulose; filaments hirsute. **Fruits** usually yellowish-brown, sometimes grayish-yellow, subglobose, 1.2–2.5 cm Ø, abaxially slightly rugose, or with few slightly prominent tubercles; seeds brown, nitid, with fleshy arillode. **Distr:** Asia to Malesia. **Ecol:** In forests of dry and wet regions; <700 m. Fl. & Fr. Apr–Aug. **Zones:** 11–12.

Dodonaea

Species 70. **Shrubs** or **small trees**, evergreen, sometimes with glandular resinous exudate on leaves and inflorescences. **Leaves** simple or pinnate, estipulate. **Inflorescences** with flowers solitary and axillary, or arranged in terminal and axillary racemes, corymbs or panicles. **Flowers** bisexual, or unisexual and plants dioecious, actinomorphic; sepals (3–)5(–7), valvate or sometimes imbricate, deciduous; petals absent; disk inconspicuous, absent in ♂; stamens (in ♂) 5–8, filaments short, anthers ellipsoid; ovary (in ♀) ellipsoid, obcordiform, or transversely ellipsoid, 2- or 3(or 5 or 6)-gonous, 2- or 3(or 5 or 6)-loculed, locules 2-ovuled, 1 ascending, 1 pendulous, style apical, much longer than ovary, often twisted, deciduous, stigma 2–6-lobed. **Fruits** capsules, samaroid, 2- or 3(–6)-loculed, septifragal, boatlike; seeds 1 or 2 per locule, obovoid, lenslike, or subglobose. **Distr:** Tropical to temperate regions. **Cult:** Propagate from seeds or tip cuttings. Requires moderately fertile, moist, well-drained soil in full sun. Some species tolerates drought and salt.

D. viscosa Jacquin, *Enum. Syst. Pl.* 19, 1760. (Syn: *Ptelea viscosa*). [Candlewood, Hopbush, Hopseed, Hopwood, Soapwood]. **Shrubs** or **small trees**, 1–3+ m tall; branches flat, narrowly winged or ridged; branches and leaves with sticky juice. **Leaves** shortly petiolate or subsessile; lamina simple, variable, linear, linear-spathulate, linear-lanceolate, or oblong, 5–12 × 0.5–4 cm, margins usually entire, apex acute, obtuse, or rounded, papery, glabrous. **Inflorescences** densely flowered, terminal or axillary near apices, shorter than leaves, rachis and branches ridged; pedicels 0.2–0.5(–1) cm long, slender. **Flowers** small; sepals 4, lanceolate or narrowly elliptic, ~3 mm long, obtuse; stamens 7 or 8, filaments <1 mm long, anthers incurved, ~2.5 mm long, glandular; ovary ellipsoid, abaxially with sticky juice, 2- or 3-loculed, style ~6 mm long, apex bi- or trilobed. **Fruits** capsules, obcordiform or compressed-globose, 2- or 3-winged, 1.5–2.2 cm long, wings 1.8–2.5 cm wide; testa membranous or papery, veined; seeds 1 or 2 per locule, black, lenslike. **Distr:** Tropical and subtropical regions. **Ecol:** In forest margins, savannas, coastal vegetation, sandy beaches. **Zones:** 9–12.

Glenniea

Species ~8. **Trees**, monoecious or dioecious. **Leaves** spirally arranged or partly decussate, unifoliate or paripinnate, 1–6-jugate; leaflets opposite to alternate, exstipulate, margins entire or wavy-repand, coriaceous. **Inflorescences** racemes or panicles, with few spreading branches, terminal, usually in

Below: *Cardiospermum halicacabum*, *Dimocarpus longan*, *Dodonaea viscosa* (flowers), *Glenniea unijuga*.

TG

Above: *Harpullia arborea*, *Harpullia pendula*, *Lepisanthes tetraphylla*, *Majidea zanguebarica*.

upper leaf axils. **Flowers** regular, polygamous; sepals 4- or 5-lobed, basally connate, valvate to narrowly imbricate, equal persistent and recoiled after anthesis; petals 5 (absent in Malaysian species), minute, broader than long, shortly clawed, concave, woolly ciliate, with lateral bilobed scale, caducous; disk complete, bilobed, adnate to base of calyx, broad and flat, glabrous; stamens 5–8, equal, exserted, glabrous, filaments subulate, anthers short; ovary sessile, ovoid, contracted at base, 3-loculed (2-loculed in Malaysian species), thinly hairy, style short; stigma grooved or bi–trilobed, cells 1-ovuled. **Fruits** variable, depressed-globose, indehiscent, fleshy or hard, sometimes bilobed; seeds with a thin crustaceous testa, without aril. **Distr:** Tropical Africa, Madagascar, Sri Lanka, Malaysia. **Cult:** Propagate from seeds or by air-layering. Requires a fertile, moist, well-drained soil in full or partial sun.

G. unijuga Hook.f., *Gen. Pl.* 1(1): 404, 1862. (Syn: *G. zeylanica*, etc.). **Trees**, medium-sized, deciduous. **Leaves** pinnate, 10–20 cm long; leaflets 3–4 pairs, sometimes unifoliolate, ovate to lanceolate-oblong, 6–10 cm long, base cuneate, sometimes oblique, margins entire or undulate, apex acute to rounded, often slightly emarginate, coriaceous, glabrous, lateral nerves 8–10 pairs; petiole 2–5.5 cm long, base swollen; petiolules <5 mm long. **Inflorescences** to 25 cm long; ♂ trees with terminal panicles, much exceeding leaves; *bisexual* trees with inflorescences mostly axillary, simple, shorter than leaves; pedicels 2 mm long. **Flowers** polygamous; calyx lobes 5, basally connate, ovate, 2 × 1.5 mm, acute, pubescent; petals 5, broadly deltoid, 1 × 1.5–2 mm, clawed; stamens (5–)8, filaments subulate, exserted, spreading and exceeding calyx in ♂, short in *bisexual*; ovary sessile, ovoid, ~1.5 mm long, 3-loculed, base hairy, locules 1-ovuled, style a grooved knob. **Fruits** depressed-globose, 2.5–4 cm Ø, indehiscent, bi–trilobed, green; seeds globose, 1.2–1.5 cm Ø, without aril. **Distr:** Sri Lanka. **Ecol:** In dry forests; <400 m. Fl. Sep–Dec. **Zones:** 11–12.

Harpullia

Species ~26. **Shrubs** or **trees**, dioecious. **Leaves** paripinnate, estipulate; rachis and petiole sometimes winged; leaflets with margins often entire. **Inflorescences** thyrses, compound racemose or sometimes racemose, terminal or axillary; bracts and bracteoles small. **Flowers** functionally unisexual, actinomorphic; sepals 5, imbricate, persistent or deciduous; petals 5, subcuneate, longer than calyx, slightly fleshy, scales absent, apex reflexed, or petals subspathulate, distinctly clawed, adaxially with 2 earlike, thin scales; disk small; stamens (in ♂) 5–8, subequal to petals; ovary (in ♀) globose or ovoid, sessile or shortly stalked, bilaterally compressed, 2(–4)-loculed, locules 1- or 2-ovuled, pendulous, style short or long, twisted. **Fruits** capsules, usually compressed, 2(–4)-loculed, grooved between locules, pericarp papery or crustaceous; seeds 1 or 2 per locule, ellipsoid or ovoid, testa shiny, thinly crustaceous, often with white or orange fleshy aril. **Distr:** Sri Lanka and India through SE China and Malesia to Australia. **Cult:** Propagate from fresh seeds or by air-layering. Requires a moist, well-drained soil, rich in organic matter, in full or partial sun.

H. arborea (Blanco) Radlk., *Sitz. Königl. Akad. Wiss.* 16: 404, 1886. (Syn: *Otonychium imbricatum*, etc.). [Cooktown Tulipwood]. **Trees**, to 45 m tall; trunk often buttressed. **Leaves** pinnately compound; leaflets 6–12, 7–19 × 4–8 cm, margins entire or toothed; rachis and petiole 4–22 cm long, basal pulvinus present; petiolules 3–6 mm long. **Inflorescences** panicles, axillary or ramiflorous, to 18 cm long; axis hairy; flowers of mixed sexes or totally ♂, ♀ flowers when present terminal; bracts linear-lanceolate. **Flowers** 1–1.6 cm Ø; calyx cream or greenish; sepals ± ovate, equal, ~0.5–1 × 0.35 cm, obtuse; petals 5, green, white, cream, yellow or pink, obovate-oblong or oblong-spathulate, 0.8–1.6 × 0.3–0.8 cm, obtuse, claw 3–6 mm long; stamens 5; ovary 2(4)-loculed, compressed, densely pubescent, style 1.4–1.6 cm long, locules 1- or 2-ovuled. **Fruits** capsules, orange or yellow, 1.3–2.5 × 2.5–5 cm, 2- or rarely 3-valved, ribbed, inside red, glabrous; seeds 1–3, black, 1.2–1.5 cm long, aril orange. **Distr:** Australia, India, Malesia, New Guinea, Pacific Islands, SE Asia, Sri Lanka. Fl. Nov. **Ecol:** In rainforests, monsoon forests; 0–500 m. Fr. Jan–Apr. **Zones:** 9–12.

H. pendula Planch. ex F. Muell., *Trans. Phil. Inst. Vict.* 3: 26, 1859. [Australian Tulipwood, Moreton Bay Tulipwood]. **Trees**, evergreen, 7–25 m tall; trunk often buttressed; crown 4–6 m wide; branches weeping; bark rough. **Leaves** alternate, pinnately compound; leaflets 4–8(–10), alternate or opposite, 4–11 × 2–4 cm, adaxially glabrous, glossy, abaxially glabrous or sparsely hairy along midrib; rachis and petiole 15–60 cm long, basal pulvinus present; petiolules absent or to 6 mm long. **Inflorescences** panicles, axillary, pendulous. **Flowers** 0.5–1.2 cm Ø; calyx green, cream or red; petals green, cream or yellow. **Fruits** capsules, bladderlike (hollow inside), woody, 1.3–2.5 × 1.5–3 cm, yellow, pink, orange or red, ± glabrous, 1- or mostly 2-valved, valves yellow and glabrous inside; seeds usually 2, glossy, black, aril absent or minute, yellow. **Distr:** Australia (NSW, Queensland). **Ecol:** In rainforests, monsoon forests; 0–850 m. Fl. Dec–Jan. Fr. throughout the year. **Zones:** 9–12.

Lepisanthes

Species ~24. **Trees** or **shrubs**, monoecious. **Leaves** usually paripinnate, alternate, estipulate, usually petiolate; leaflets 2 to several pairs, opposite or alternate, usually entire. **Inflorescences** thyrses, axillary, above axils or on old branches, solitary or several in fascicles. **Flowers** unisexual, actinomorphic or zygomorphic; sepals 5, leathery, concave, imbricate, outer 2 smaller, orbicular, inner ones usually broadly ovate or elliptic; petals 4 or 5, often spoon-shaped, longer than sepals, base clawed, apex with adaxial scale; disk acetabuliform or semilunar, entire or lobed; stamens (in ♂) 8, rarely more or fewer, disk longer than petals, filaments flat, usually hairy, anthers ellipsoid; ovary (in ♀) 2- or 3-loculed, usually with grooves between locules, locules 1-ovuled, style short, apex inflated, entire or bi- or trilobed. **Fruits** ellipsoid or subglobose, 2- or 3-loculed; pericarp leathery or slightly fleshy; seeds ellipsoid, testa brown. **Distr:** Tropical Africa, Madagascar; S and SE Asia; NW Australia. **Cult:** Propagate from seeds or by air-layering. Requires a fertile, moist, well-drained soil, rich in organic matter, in full or partial sun.

L. tetraphylla (Vahl) Radlk., *Sitz. Math.-Phys.* 8: 276, 1878. (Syn: *Sapindus tetraphylla*, etc.). **Trees**, 8–12(–18) m tall; young parts brownish-hairy. **Leaves** even-pinnate, 15–25 × 10–16 cm; leaflets 2–4 pairs, elliptic-oblong or lanceolate-oblong, 6–8(–20) × 2–4(–6) cm, base acute or rounded, sometimes slightly oblique, margins entire, apex obtuse or rounded, tapering to abruptly acuminate, lateral nerves 8–10 pairs; petiole 4–5 cm long; petiolules 2–4 mm long. **Inflorescences** panicles, from axils of fallen leaves, to 25 cm long; flowers numerous, laxly fasciculate; pedicels to 2–3 mm long; bracts triangular to subulate, minute. **Flowers** with sepals 5, outer 2 ovate-elliptic to orbicular, 1.5 mm long, inner ~4 mm long, densely ciliolate; petals usually 4 (in ♂), 4 or 5 (in ♀), obovate, 6 × 4 mm, obtuse, shortly clawed, claws woolly; scale ~3 mm long, inner side with an appendage; disk 4–5-lobed; stamens 8, ~6 mm long; ovary globose, 2-loculed, gibbous, locules 1-ovuled. **Fruits** drupes, oblong-ovoid to 3-gonous, 1.5–2 cm Ø, densely hairy, yellow; seeds usually solitary, 1.5 × 0.8 cm, brown, smooth. **Distr:** India, Myanmar, Sri Lanka; Malaysia to New Guinea. **Ecol:** Fl. Mar–May. **Zones:** 10–12.

Majidea

Species 1. **Trees**, tall, monoecious, variable; bark grayish; twigs, leaves and inflorescences initially densely hairy. **Leaves** with 4–11 pairs of leaflets; rachis 4-angled, with lines of hairs on angles; leaflets ovate to lanceolate, slightly asymmetric, margins entire, glabrescent. **Inflorescences** paniculate, terminal; flowers subtended by herbaceous bracts. **Flowers** regular; sepals 5, free, concave, herbaceous, hairy outside; petals 4, caducous, without claw or scale; disk a complete circle; stamens 7–8(–10), filaments folded in bud, glabrous, anthers sagittate; ovary trilobed, 3-loculed, locules 2-ovuled, style long, hairy, tip channeled, glabrous. **Fruits** capsules, dehiscent, trigonous, opening to reveal 3 bright pink valves; seeds 1–2 per valve, ovoid, without aril, softly velvety, blue-black; cotyledons spirally arranged. **Distr:**

Madagascar and tropical Africa. **Cult:** Propagate from fresh seeds or by air-layering. Requires a moist, well-drained soil, in full or partial sun.

M. zanguebarica Kirk ex Oliv., *Icon. Pl.* 11: 78, 1871. (Syn: *Harpullia zanguebarica, M. zanquebarica* etc.). [Black Pearl Tree, Mgambo Tree, Velvet Seed Tree]. **Shrubs** or **trees**, 2–22 m tall. **Leaves** with leaflets 5–7 pairs, drooping when young, ovate-lanceolate, 5–7 × 2–3 cm, base asymmetrically cuneate, apex shortly acuminate or acute, glabrous and shiny, lateral veins 7–15 pairs; rachis 10–30 cm long; petiolules 1 mm long. **Inflorescences** with main axes 8–15(–30) cm long, densely pubescent; bracts ± elliptic, 1.3–1.7 × 0.6–0.8 cm; pedicels 3–6 mm long. **Flowers** scented; sepals ovate, 5–7 mm long, outermost larger; petals 4, yellow-green to red, white-hairy, elliptic, 6 × 2 mm, soon deciduous; disk hardly visible in ♀, pink-tinged, conspicuous in ♂; stamens exserted, filaments yellow-green to orange-red, 6 mm long, staminodes very short; ♂ flowers without a rudimentary ovary. **Fruits** capsules, trigonous, apiculate, brownish-yellow, internally bright pink, 3–3.5 × 3.5–4.5 cm; seeds bluish-black, ovoid, 1.2 cm long, finely silky-pubescent, ± persistent; mature fruits and ♂ flowers usually occur together in the inflorescence. **Distr:** Madagascar and tropical Africa. **Ecol:** In tropical forests; 0–1,300 m. Fl. Feb–Jun. Fr. Apr–Aug. **Zones:** 11–12.

Nephelium

Species ~22. **Trees**, rarely shrubs, monoecious or dioecious. **Leaves** paripinnate, alternate, petiolate; leaflets entire. **Inflorescences** thyrses, terminal or axillary; bracts and bracteoles small. **Flowers** unisexual, actinomorphic, small; calyx cupular, 5- or 6-lobed, sepals small, valvate or imbricate, often opening early; petals absent or 5 or 6; disk annular, entire or lobed; stamens (in ♂) 6–8, exserted, filaments villous; ovary (in ♀) obcordiform, 2(or 3)-lobed, 2(or 3)-loculed, densely tubercled, locules 1-ovuled, style inserted between ovary lobes, stigma bi- or trilobed. **Fruits** capsules, deeply parted into 2 or 3 schizocarps, usually 1 developed, ellipsoid, pericarp leathery, with soft spines; seeds same as schizocarps in shape, arillode fleshy, adnate to testa, thoroughly covering seeds. **Distr:** SE Asia. **Cult:** Propagate from seeds, by budding, grafting or air-layering. Requires a deep, fertile, well-drained soil in full sun to partial shade.

N. chryseum Blume, *Rumphia.* 3: 105, 1847. [Golden-yellow Rambutan, Usau]. **Trees**, evergreen, 10–20+ m tall; branches striate, grayish-brown when dry, ferruginous-pubescent when young. **Leaves** paripinnate; leaflets (2–)4 pairs, 6–18 × 2.5–7.5 cm, base and apex nearly acute, thinly leathery, glaucous, abaxially pilose, lateral veins 9–14 pairs or more, prominent and nitid abaxially, smooth or slightly concave adaxially; petiole 20–40 cm long, petiolules 5–8 mm long. **Inflorescences** many-branched, ♂ ones nearly as long as leaves, ♀ ones shorter. **Flowers** with calyx ~1.5 mm long, densely pilose; disk pilose; stamens 7 or 8, filaments ~3 mm long, villous; ovary bilobed, 2-loculed, pilose. **Fruits** red, ellipsoid, 4–5 × 3–4 cm including spines; spines ~1 cm or longer, bilaterally compressed, base broad, apex acute, hooklike. **Distr:** China, Vietnam, Borneo, Philippines. **Ecol:** In dense forests; 500–1,500 m. **Zones:** 10–12. **Info:** The fruits are edible.

N. lappaceum L., *Mant. Pl.* 125, 1767. (Syn: *Dimocarpus crinitus*, etc.). [Hairy Lychee, Rambutan]. **Trees**, evergreen, ~10(–25) m tall; branches grayish-brown, terete, rugose, ferruginous-pilosulose when young. **Leaves** with leaflets (1 or)2 or 3(or 4) pairs, elliptic or obovate, 6–18 × 4–7.5 cm, base cuneate, margins entire, apex obtuse or slightly rounded, sometimes nearly acute, thinly leathery, glabrous, lateral veins 7–9 pairs; petiole 15–45 cm long; petiolules ~5 mm long. **Inflorescences** many-branched, subequal to leaves, ferruginous-tomentose; pedicels short. **Flowers** with calyx ~2 mm long, leathery; sepals ovate, tomentose; petals absent; stamens ~3 mm long. **Fruits** capsules, yellow to red, broadly ellipsoid, ~5 × 4.5 cm including spines, spines ~1 cm long. **Distr:** Indonesia, Malaysia, Philippines, Thailand. **Ecol:** In lower or middle story forests, mixed dipterocarp forests, swamp forests and submontane forests; to 1,200 m. Fl. Mar–Apr. Fr. Aug–Sep. **Zones:** 10–12. **Info:** Cultivated widely in tropical Asia for the edible fruits (arils).

N. ramboutan-ake (Labill.) Leenh., *Blumea* 31(2): 415, 1986. (Syn: *Litchi ramboutan-ake, N. mutabile*, etc.). **Trees**, to 10 m tall; trunk 80 cm Ø; buttresses to 2.4 m high. **Leaves** 1-foliate or each with 1–7 pairs of leaflets; leaflets narrowly elliptic, 4–20 × 1.75–11 cm, base acute or in lower leaflets obtuse to rounded, attenuate, margins mostly curved, apex mostly acuminate, thinly coriaceous; petioles 0.75–11.5 cm long, petiolules 3–8 mm long, grooved. **Inflorescences** axillary or pseudo-terminal. **Flowers** with sepals slightly or up to halfway connate, 1–2.8 mm long; petals absent; disk glabrous; stamens 5–8; ovary 2(–3)-loculed. **Fruits** ellipsoid to subglobose, 4–6.5 × 2.5–5 cm, glabrous; pericarp coriaceous, to 7 mm thick, coarsely spiny, spines to 1.5 cm long, base bulbous and often confluent, or often with short tongue-shaped knobs. **Distr:** Assam, Myanmar, peninsular Malaysia, Sumatra, Borneo, Philippines. **Ecol:** In primary mixed dipterocarp forests; 0–1,950 m. **Zones:** 10–12.

Pometia

Species 1. **Trees**, large, monoecious; trunk usually buttressed; bark pink-brown, peeling off in flakes; exudate red. **Leaves** paripinnate, alternate, sessile; leaflets usually in many pairs, first pair (basal) small, stipulelike, others larger, 12–30 × 4–10 cm, margins usually serrate. **Inflorescences** thyrses, terminal or axillary, stiff or pendent, large, ~15–60 cm long, with basal auricles. **Flowers** small, regular, polygamous or unisexual, actinomorphic; calyx cupular, 2–3 mm Ø; sepals half-connate, valvate, lobes 5; petals 5, usually broadly obovate or subtriangular, short, 0.5–1.5 × 0.5–1 mm; disk annular, fleshy, shallowly 5-lobed, glabrous; stamens (in ♂) 5, long exserted, filaments glabrous or base hairy, anthers small; ovary (in ♀) obcordiform, bilobed, 2-loculed, lobes subglobose, glabrous or tomentose, locules 1-ovuled, style filiform, very long, apex twisted. **Fruits** deeply parted into 2 schizocarps, usually only 1 developed, ellipsoid; pericarp thick, spongy in middle, smooth adaxially; seeds same shape as schizocarps, seeds fully covered by a mucilaginous aril. **Distr:** Tropical Asia, Pacific Islands. **Ecol:** In tropical rainforests. **Cult:** Propagate from seeds, by grafting or air-layering. Requires a moist, well-drained soil in full sun.

P. pinnata J.R. Forster and G. Forster, *Char. Gen. Pl.* 110, 1775. (Syn: *P. eximia*, etc.). [Island Lychee, Oceanic Lychee]. **Trees**, large, ~20–50 m tall; trunk buttressed; branches grooved, sometimes hirsute. **Leaves** large, axis and leaflets subglabrous to tomentose; leaflets dense, 5–9 (–15) pairs, subopposite; petiolules short, inflated; *first pair* orbicular, small, base cordate, stipulelike, *others* oblong or upper ones nearly cuneate, 15–40 × 5–10 cm, margins regularly serrate, apex acute or acuminate; petiole to 1.5 m long. **Inflorescences** thyrses, terminal or axillary; rachis and branches strong, straight, 30–50 cm long, pilosulose; pedicels ~6 mm long, base jointed. **Flowers** small; sepals ~1 mm long, pilosulose; petals obovate-triangular, ~2 × 2 mm; stamens ~5 mm long. **Fruits** shiny, ellipsoid, sometimes subglobose, ~3 × 2 cm, glabrous. **Distr:** China, Taiwan, Andaman and Nicobar Islands, Indonesia, Malaysia, New Guinea, Philippines, Sri Lanka, Thailand, Vietnam; Pacific Islands. **Ecol:** In moist forests; <900 m. Fl. & Fr. May–Sep. **Zones:** 10–12.

Sapindus

Species ~13. **Trees**, rarely shrubs, monoecious. **Leaves** paripinnate, rarely simple, alternate, estipulate; leaflets opposite or alternate, margins entire. **Inflorescences** thyrses, large, many-branched, terminal or fascicled at branchlet apices; bracts and bracteoles subulate, small. **Flowers** unisexual, actinomorphic or zygomorphic; sepals 5 or sometimes 4, imbricate, outer 2 smaller; petals 5, clawed, with 2 earlike scales at base or margins thickened adaxially, or 4, without claw, with 1 large scale at base adaxially; disk acetabuliform or semilunar, fleshy, sometimes lobed; stamens (in ♂) usually 8, exserted, filaments hairy below; ovary (in ♀) obovoid or turbinate, usually trilobed, 3-loculed, locules 1-ovuled, ascending, style terminal. **Fruits** parted into 3 schizocarps, usually 1 or 2 developed, developed schizocarps subglobose or obovoid, abaxially slightly flat, adaxially with 1 or 2 semilunar sterile schizocarps; schizocarps free, broadly ellipsoid or subglobose, containing abundant saponin. **Distr:** Warm regions of Asia, Australia, and N and

Below: *Nephelium chryseum, Nephelium lappaceum, Nephelium ramboutan-ake, Pometia pinnata.*

Above: *Sapindus emarginatus, Sapindus trifoliatus, Schleichera oleosa, Picrasma excelsa.*

S America. **Cult:** Propagate from seeds or cuttings. Requires any well-drained soil in full or partial sun.

S. emarginatus Vahl, *Symb. Bot.* 3: 54, 1794. [Soap Berry, Soapnut Tree]. **Trees**, medium-sized. **Leaves** 12–17 cm long, pinnate; leaflets 2–3 pairs, opposite, elliptic-ovate, 6–11 × 4.5–9 cm, terminal pair largest, apex obtuse to emarginate, abaxially glabrous or pubescent; petiolule ~5 mm long, pubescent. **Inflorescences** panicles, terminal, pubescent; bract 1; bracteoles 2, ~1 mm long, subulate, pubescent, persistent; pedicels 3–4 mm long. **Flowers** with sepals 5, ovate, clawed, ciliate, with woolly scales above claw; disk concave, 5-lobed, hirsute; stamens 8, free; unequal, filaments 3–4 mm long, hairy, anthers 1.5 mm long, apiculate; pistillode present in ♂; ovary ovoid, ~3 mm long, 3-loculed, hairy, style 2 mm long, stigma trifid; staminodes (in ♀) 8. **Fruits** berries, trilobed, with 1 or 2 undeveloped cocci, obovoid, 1.2–1.5 × 1.5 cm, rusty-pubescent; seeds black, globose, ~6 mm Ø. **Distr:** Myanmar, Deccan Peninsula, Sri Lanka. **Ecol:** In dry deciduous forests; <500 m. Fl. Nov–Jan. **Zones:** 10–12.

S. trifoliatus L., *Sp. Pl.* 367, 1753. (Syn: *S. laurifolius*). [Soapnut Tree]. **Trees**, large. **Leaves** 15–30 cm long; leaflets subsessile, 2–3 pairs, elliptic-lanceolate, 8–18 × 5–7.5 cm, terminal pair longest, base slightly oblique, margins entire, apex acute, glabrous. **Inflorescences** panicles, terminal, slightly pubescent; bract 1; bracteoles 2, 1 mm long, subulate, pubescent; pedicels 3 mm long, pubescent. **Flowers** greenish-white; sepals 5, slightly basally connate, ovate-oblong, 4–5 mm long, pubescent; petals 5, free, lanceolate-ovate, 5–6 mm long, clawed, hispid; disk 5-lobed, concave hirsute; stamens 8, free, filaments 2–3 mm long, villous, anthers 1.5–2 mm long, basifixed, apiculate; ovary 3-loculed, trilobed, ovoid, ~3 mm long, tomentose, locules 1-ovuled, style 1.5 mm long. **Fruits** 2–trilobed, 1.3–2 cm long, pubescent when young, hard and glabrous when mature; seed 1 per cell, 6–9 mm long, black. **Distr:** Bengal, S India, Sri Lanka. **Ecol:** In moist forests; <500 m. Fl. Jul–Nov. Fr. Jan–Feb. **Zones:** 11–12.

Schleichera

Species 1. **Trees**; young parts glandular-hairy. **Leaves** even-pinnate; leaflets opposite or subopposite, sessile, usually 3 pairs, base cuneate, apex obtuse. **Inflorescences** fascicles, in interrupted, slender racemes or panicles; pedicels slender. **Flowers** regular, polygamodioecious, small; calyx cupular, lobes (4–)5(4), usually valvate; petals absent; disk flat, margins undulate; stamens 5–8, inserted within the disk, filament hairy; ovary ovoid, 3(–4)-loculed in *bisexual* flowers, erect, locules 1-ovuled, stigma 3–4-lobed, in ♂ a villous pistillode. **Fruits** drupes, ovoid, 1-locular by abortion, pointed at apex, tipped with style, indehiscent, smooth or sometimes 1-echinate; seeds 1–2, erect with a fleshy aril. **Distr:** Tropical SE Asia. **Cult:** Propagate from seeds, root suckers or by air-layering. Requires a moist, well-drained soil in full or partial sun.

S. oleosa (Lour.) Oken, *Allg. Natur. Bot.* 3(2): 1341, 1841. (Syn: *Pistacia oleosa, S. tijuga*). [Ceylon Oak, Macassar Oil-tree]. **Trees**, deciduous, to 25 m tall. **Leaves** even-pinnate, to 20 cm long; leaflets 3 pairs, ± opposite, oblong-elliptic or obovate, 4.5–15(–20) × 2.5–5(–8) cm, base cuneate-subcordate, often oblique, margins entire to repand, apex usually obtuse, abaxially paler, lateral nerves 6–10 pairs; leaflets bright red when young; petiole pulvinate (2–)6–7 cm long. **Inflorescences** panicles, lax, to 15 cm long, pendulous below new leaves. **Flowers** to 2–3.5 mm Ø, regular, yellow; ♂ with calyx lobes 5, triangular, ~1 mm long; petals absent; stamens 7 or 8, filaments 1.5 mm long; pistillode small, hairy; *bisexual* flowers with ovary 1.5–2 mm long, 3-loculed, hairy, locules 1-ovuled, style to 2 mm long, glabrous, stigma bi–trilobed. **Fruits** drupes, ovoid, 2–2.5 × 1.5 cm, sometimes spiny; seeds 1 or 2, ± ovoid or subglobular, 1–1.5 cm long, aril fleshy, pulpy, with a pleasant acidic taste. **Distr:** Tropical Himalayas, India, Myanmar to E Malaysia and Sri Lanka. **Ecol:** In forests; <700 m. Fl. Mar–Apr. Fr. Apr–Aug. **Zones:** 9–12.

SIMAROUBACEAE

Genera 19, **species** 95. **Trees** or **shrubs**; bark usually bitter. **Stipules** present or absent. **Leaves** alternate, sometimes opposite; lamina pinnate to unifoliolate, rarely simple. **Inflorescences** axillary, racemose, paniculate, cymose, rarely spicate. **Flowers** usually small, bisexual or unisexual (andromonoecious, functionally dioecious, or distinctly dioecious), often with rudiments of opposite sex, actinomorphic; calyx of 4 or 5(–8) sepals, imbricate or valvate; petals usually 4 or 5(–8), free, imbricate or valvate; stamens same number as petals and alternate with them, or twice as many as petals (or more), filaments free, base often with an appendage, anthers oblong, versatile, 2-celled, with a longitudinal slit; disk usually present, sometimes a gynophore or androgynophore; gynoecium weakly basally united, or virtually free, usually 2–5-carpellate, ovary superior, placentation axile, locules 1- or 2-ovuled, style connate or free, stigmas 2–5. **Fruits** drupes or samaras, actually 1–5 fruitlike monocarps produced per flower, each a druparium or a samarium. **Distr:** Mainly in tropical and subtropical regions, some species in temperate regions.

Picrasma

Species ~9. **Trees**; branches with a pith, glabrous. **Stipules** early deciduous or persistent. **Leaves** odd-pinnate, base of petiole or petiolule often dilated into a pulvinus; leaflets opposite or nearly so, margins entire or serrate. **Inflorescences** cymose panicles, axillary; bracts small or early deciduous; pedicel with lower half articulated. **Flowers** unisexual or polygamous, 4- or 5-merous; sepals small, free or lower half connate, persistent; petals ± valvate in bud, apex shortly incurved, cuspidate, longer than sepals, persistent in ♀; stamens 4 or 5, inserted at base of disk; disk slightly thickened, entire or shallowly 4- or 5-lobed, sometimes dilated in fruit; carpels 2–5, free, degenerate or rudimentary in ♂, each with a single ovule, style basally connate, apex free, stigma free. **Fruits** comprising of one or more drupelike monocarps (each a druparium); seeds with hilum broad, testa membranous. **Distr:** Tropical and subtropical America and Asia. **Cult:** Propagate from seeds. Requires a fertile, well-drained soil in full or partial sun.

P. excelsa Planch., *London J. Bot.* 5: 574, 1846. (Syn: *Simaruba excelsa*, etc.). [Bitter Wood, Jamaica Quassia]. **Trees**, 12–18(–24) m tall; trunk 0.6–1 m Ø; bark grayish-brown, smooth, wrinkled. **Leaves** imparipinnate; leaflets 4–5 pairs, narrowly elliptic or oblong-elliptic, 5–12 cm long, base unequal-sided, mostly cuneate or acute, when young covered with fulvous down. **Inflorescences** panicles; peduncles long. **Flowers** small, yellowish-green, polygamous; sepals 0.6–0.7 mm long; petals of ♂ flowers oblong-elliptical, 2 mm long, those of *hermaphrodite* flowers oblong-lanceolate, 3 mm long; stamens ~4 mm long in ♂, 2–2.5 mm long in *hermaphrodite* flowers. **Fruits** drupes, globular or obovoid-globular, black, 6–7 mm long. **Distr:** Cuba, Jamaica, Puerto Rico. **Ecol:** Fl. Jun–Aug, Oct–Nov. Fr. Sep–Oct, Dec–Jan. **Zones:** 11–12.

Quassia

Species ~40. **Trees** or **shrubs**, monoecious or bisexual. **Leaves** alternate, petiolate, exstipulate; lamina simple or pinnately compound, usually with pitted glands on lower surface. **Inflorescences** simple or branched racemes, panicles or umbels, axillary or terminal. **Flowers** 4–6-merous; calyx lobed, imbricate; petals imbricate or contorted, longer than calyx; stamens twice as many as petals, appendiculate; disk ± cylindrical or subglobose, varying in size; carpels free or basally coherent, styles fused above, free at base, stigma stellate to capitate, ovule 1 per locule, pendulous. **Fruits** drupaceous, often compressed; seed solitary, testa membranous. **Distr:** Pantropical and subtropical. **Cult:** Propagate from seeds, cuttings or by air-layering. Requires a moist, well-drained soil in full or partial sun.

Q. amara L., *Sp. Pl.*, ed. 2. 1: 553, 1762. [Bitterwood, Quassia Wood]. **Shrubs** or **small trees**, 2–5(–9) m tall. **Leaves** alternate, pinnately compound, ~30 × 20 cm, midrib and petiole broadly winged; leaflets 3 or 5, sessile, 3 near the tip widely separated from the 2 near base by winged midrib, elliptic-oblong, 5–16 × 3–6 cm, both ends tapered, adaxially dark green, abaxially paler. **Inflorescences** racemes, 8–20 cm long, terminal, pinkish or pale reddish; peduncles and pedicels reddish. **Flowers** tubular, constricted apically and base widened, bright red or pinkish-red, white inside, radially symmetrical, 5-merous; corolla 2.5–4.5 cm long; stamens exserted, anthers yellow. **Fruits** fleshy, black, oval, 1.5 cm Ø, in groups of 4–5 on a broad red base. **Distr:** Mexico to Venezuela. **Ecol:** In moist to wet lowland forests, usually understory; 0–700 m. Fl. & Fr. throughout the year. **Zones:** 10–12.

Q. indica (Gaertn.) Nooteb., *Blumea* 11: 517, 1962. (Syn: *Samadera indica*, etc.). [Indian Bitterwood]. **Shrubs** or **small trees**, evergreen, to 5–12 m tall. **Leaves** large, elliptic-oblong to lanceolate, 14–28.5 × 4–12 cm, base ± acute, margins entire, apex usually acuminate, coriaceous, glabrous, shiny; petiole 1–2 cm long, curved. **Inflorescences** umbels, few- to many-flowered, axillary; peduncles 2–18 cm long; bracts minute. **Flowers** 4-merous; calyx 4-lobed, lobes semiorbicular, ± as long as the tube, 2 mm wide, thick, persistent; petals 4, free, oblanceolate, 1.2–1.8 × 0.3–0.4 cm, base narrowed, apex obtuse, cream-green, purple, white or pale yellow; stamens 8, 1–1.5 cm long, filaments 1–1.5 cm long, with a basal appendage; disk ± glabrous; carpels 4, 2 × 2 mm, free, ± puberulous, styles filiform, to 2 cm long. **Fruits** drupes, 1–4 together, flattened, semicircular, 4–6 × 2.5–4.5 cm; seed large. **Distr:** Cambodia, India, Indonesia, Madagascar, Malaysia, Myanmar, Philippines, Solomon Islands, Sri Lanka, Thailand, Vietnam. **Ecol:** In lowland moist forests. Fl. & Fr. Mar–May. **Zones:** 10–12.

MELIACEAE

Genera ~50, **species** ~650. **Trees**, **shrubs**, or sometimes shrublets, monopodial or sympodial, usually dioecious, less often monoecious or polygamodioecious. **Stipules** absent. **Leaves** in spirals, very rarely opposite, usually pinnate; leaflets opposite, subopposite, or alternate, lamina with base oblique, margins usually entire or rarely lobed or serrate. **Inflorescences** usually in axillary thyrses, rarely racemose or spicate. **Flowers** usually small; calyx small, 3–6-lobed or with distinct sepals, usually cup-shaped or tubular; corolla sometimes quincuncial; petals (3 or)4 or 5(or 6+), sometimes adnate to staminal tube and then valvate; stamens 3–10+, hypogynous, mostly with a staminal tube (distinct stamens in *Cedrela* and *Toona*), anthers usually sessile on staminal tube, erect, included or exserted, 2-celled, longitudinally dehiscent; disk tubular, annular, obsolete, or absent, free or adnate to ovary; ovary usually free, 2–5(or more)-loculed, with 1 to many collateral or superposed ovules per locule, style single or rarely absent, stigma disciform or capitate. **Fruits** berries (dry but indehiscent, with arillate seeds in some *Aglaia* sect. *Aglaia*), capsules, or rarely drupes; seeds winged or with a fleshy aril or sarcotesta wholly or partly covering seed. **Distr:** Tropical, subtropical, occasionally in warm temperate regions of both hemispheres.

Aglaia

Species ~120. **Trees** or **shrubs**, dioecious, young parts usually lepidote or stellately pubescent. **Leaves** alternate to subopposite, odd-pinnate, 3-foliolate, or rarely simple; leaflets with margins entire. **Inflorescences** thyrses, axillary. **Flowers** small, usually globose; calyx slightly or deeply 3–5-lobed; petals 3–5, short, concave, quincuncial or imbricate in bud, distinct or rarely basally connate and adnate to staminal tube; stamens as many as or more than petals, staminal tube usually subglobose, obovoid, or cup-shaped with apex incurved, apical margin entire, crenate, or shallowly lobed, anthers 5 or 6(–12), included, slightly exserted, or rarely semiexserted; disk absent; ovary 1–3(or 4)-locular, locules 1- or 2-ovuled, style short or absent, stigma ovoid or shortly cylindric. **Fruits** with fibrous pericarp, indehiscent with 1 or 2 locules, or loculicidally dehiscent with 3 locules, locules empty or 1-seeded, pericarp often containing latex; seeds usually surrounded by a colloidal and fleshy aril. **Distr:** Tropical and subtropical Asia, tropical Australia, Pacific Islands. **Cult:** Propagate from fresh seeds or cuttings. Requires a moist, well-drained soil in full or partial sun.

A. elaeagnoidea (A. Juss.) Benth., *Fl. Austral.* 1: 383, 1863. (Syn: *A. roxburghiana*, etc.). [Coastal Boodyarra, Droppy Leaf]. **Trees small** or **shrubs**, 1.5–10(–15) m tall; trunk 25(–40) cm Ø. **Leaves** imparipinnate, obovate 6–29 × 7–21 cm; leaflets (1–)3–7, elliptic or obovate, 2–13(–16) × 1–5(–6) cm, base cuneate, asymmetric, apex rounded or acuminate, acumen obtuse, 0.2–0.5(–2) cm long, subcoriaceous, adaxially shiny, lateral veins 5–10 pairs; petiole 2.5–10.5 cm long; petiolules 0.5–1.5 cm long, terminal one to 2 cm long. **Inflorescences** ♂ (3–)9–34 × (1–)2.5–25 cm; ♀ 12.5 × 10 cm; peduncle to 6 cm long. **Flowers** ± globose, to 2 × 2 mm, sweetly fragrant; corolla aestivation quincuncial, petals usually 5, white or yellow; staminal tube, subglobose, yellow, anthers 5; ovary subglobose, 1-ovuled (usually absent in ♂), stigma ovoid, apical lobes 2. **Infructescences** to 12 × 10 cm; peduncle to 5 cm long. **Fruits** subglobose, ellipsoid or obovoid, to 1.1–2 × 1.3–1.5 cm, orange, yellow, brown or red, indehiscent; seeds 0–1, ~1 × 0.6 cm. **Distr:** Tropical Asia to Pacific. **Ecol:** In dense and moist regions; 0–1,500 m. **Zones:** 10–12.

A. odorata Loureiro, *Fl. Cochinch.* 1: 173, 1790. (Syn: *A. odorata* var. *microphyllina*). [Chinese Perfume Plant, Chinese Rice Flower, Mock Lime, Ricegrain Flower]. **Shrubs** or **small trees**, much-branched; young branches apically with trichomes. **Leaves** 5–12(–16) cm long; leaflets 3–7(or 9), opposite, usually obovate, sometimes elliptic, 1–7(–11) × 0.5–3.5(–5) cm, base cuneate, apex obtuse, apical one largest, glabrous, secondary veins 8 per side; petiole and rachises narrowly winged. **Inflorescences** thyrses, 5–10 cm long, axillary, lax; ♂ pedicels 1.5–3 mm long; ♀ pedicels short, thick. **Flowers** fragrant, ~2 mm Ø; calyx lobes 5, round; petals 5, yellow, oblong to suborbicular, 1.5–2 mm long, rounded to truncate; staminal tube slightly shorter, obovoid to subcampanulate, apical margins entire or lobed, anthers 5, included; ovary ovoid, densely covered with yellow trichomes. **Fruits** indehiscent, ovoid to subglobose, 1–1.2 cm Ø; seeds with aril fleshy. **Distr:** Cambodia, China, Laos, Thailand, Vietnam. **Ecol:** In lowland sparse forests, thickets. Fl. May–Dec. Fr. Jul–Mar. **Zones:** 10–12.

Aphanamixis

Species 3. **Trees** or **shrubs**, polygamodioecious. **Leaves** odd-pinnate; leaflets opposite, base frequently oblique, margins entire. **Inflorescences** ♂ panicles; ♀ or *bisexual* racemes; flowers sessile. **Flowers** globose; sepals 5, distinct or basally connate, imbricate; petals 3, concave, imbricate in bud; staminal tube nearly globose, slightly shorter than petals; anthers 3–6, included; disk extremely small or absent; ovary 3-loculed, locules (1- or)2-ovuled, style absent, stigma large, pointed or conic. **Fruits** capsules, septicidal with 3 valves, segments leathery; seeds arillate. **Distr:** Tropical Asia, Pacific Islands. **Cult:** Propagate from seeds or by air-layering. Requires a humus-rich, moist, well-drained soil in full or partial sun.

A. polystachya (Wallich) R. Parker, *Indian Forester.* 57: 486, 1931. (Syn: *Amoora rohituka*, etc.). **Trees** or **shrubs**, (2–)20–30 m tall. **Leaves** odd- or even-pinnate, 30–60(–90) cm long; leaflets (5–)9–21, opposite, oblong-elliptic, elliptic, or ovate, (7–)17–26 × 4–10 cm, basal pair smallest, base oblique, cuneate to broadly cuneate, sometimes one side rounded, margins entire, apex caudate-acuminate to obtuse, membranous when young, ± leathery when mature, surfaces glabrous, secondary veins (8–)11–20 per side; petiolules (0.2–)0.6–1.2 cm long. **Inflorescences** axillary, <30 cm long. **Flowers** 6–7 mm Ø, with 3 bracteoles; sepals 5, suborbicular, 1–1.5 mm Ø, margin sometimes ciliate; petals 3–7 mm Ø, concave; staminal tube globose, glabrous, anthers 5 or 6, oblong; ovary 3-loculed, with thick trichomes. **Fruits** capsules, spherical-pyriform to nearly ovoid, 2–2.5 × 2.5–3 cm, orangish when mature; seeds grayish-brown, oblate, 1.3–1.5 × 1–1.2 cm. **Distr:** Tropical Asia to Malesia, Solomon Islands. **Ecol:** In mixed evergreen and deciduous forests. Fl. May–Sep. Fr. Oct–Apr. **Zones:** 11–12.

Azadirachta

Species 2. **Trees**, tall, evergreen; exudate colorless. **Leaves** spirally arranged, pinnately compound, with or without a terminal leaflet; leaflets opposite and of the same type, oblique, margins smooth or serrate (with small to large teeth), venation pinnate. **Inflorescences** panicles, axillary, with ♂ and ♀ flowers on the same plant. **Flowers** bisexual (with some unisexual); calyx 5-partite; petals 5, free; stamens 10, monadelphous, anthers included, sessile on staminal tube; disk absent;

Below: *Quassia amara*, *Quassia indica*, *Aglaia elaeagnoidea*.

Above: *Aglaia odorata, Aphanamixis polystachya, Azadirachta indica, Carapa guianensis.*

ovary trilocular, locules 2-ovuled, stigma trifid. **Fruits** drupes, fleshy, not splitting; seeds 1–2. **Distr:** Sri Lanka, India and Indo-Malaya, Java. **Cult:** Propagate from freshly collected seeds, or cuttings. Requires fertile, moist but well-drained soil in full sun.

A. indica Adr. Juss., *Mem. Mus. Par.* 19. 221, 1830. (Syn: *Melia azadirachta, M. indica*). [Margosa Tree, Neem Tree]. **Trees**, to 15 m tall; branches glabrous. **Leaves** imparipinnate, pulvinus at base; leaflets alternate to opposite, subsessile, ovate, 2.5–7 × 1.5–4 cm, apex acuminate, bright green, thin. **Flowers** small, white, sweetly scented; sepals obovate, 1.5 mm long, puberulous, imbricate; petals obovate to oblong, 6 mm long, white, margin ciliate; staminal tube ~5 mm long, puberulous, 10-striate, 10-toothed, teeth bilobed, anthers oblong, basifixed; ovary subglobose, style linear, ~2.5 mm long, stigma trifid. **Fruits** drupes, oblong, 1.3–2 cm long, greenish-yellow, 1-seeded. **Distr:** China, India. **Ecol:** Fl. Apr–May. **Zones:** 10–12. **Info:** Plant parts used as an insecticide, and have numerous medicinal uses.

Carapa

Species ~10. **Tree**, glabrous. **Leaves** impari- or abruptly pinnate; leaflets with margins entire, epunctate. **Inflorescences** racemiform panicles, borne at the tips of leafy bracteate branchlets, or terminal. **Flowers** small, hermaphrodite; calyx parts as petals 4 or 5, latter alternate with the former, free, contorted in bud; stamens 8–10, connate into an urceolate dentate tube, anthers sessile, included; ovary on a fleshy disk, 4–5-loculed, ovules 4–8, biseriate in each locule, style short, stigma discoid. **Fruits** capsules, loculicidal, ligneous, ± warty; seeds large, somewhat angled, exalbuminous. **Distr:** Tropical America, W Africa. **Cult:** Propagate from seeds or by air-layering. Requires a moist, well-drained soil, rich in organic matter, in full or partial sun.

C. guianensis Aublet, *Pl. Guian. Suppl.* 32. pl. 387, 1775. [Bastard Mahogany, Brazilian Mahogany, Crab Wood]. **Trees**, large or medium-sized, glabrous almost throughout, to 30(–50) m tall; trunk to 2 m Ø. **Leaves** very large, 30–50 cm long; leaflets mostly 10–14 or more, opposite or subopposite, petiolulate, oblong to oblong-elliptic, mostly 15–25 × 6–10 cm, base subacute to rounded, apex acute or obtuse and mucronate, coriaceous, adaxially somewhat lustrous, abaxially slightly paler; petiole and rachis terete; petiole long. **Inflorescences** panicles, axillary, glabrous, branches 20–40+ cm long, flowers ± sessile. **Flowers** 5 mm long, glabrous; sepals 4, semiorbicular, 2 inner larger; petals 4, broadly obovate to elliptic, 5–5.5 mm long, apex obtuse or rounded; staminal tube cupular, 3 mm long; ovary 4-loculed, glabrous, style 1 mm long, stigma 1 mm across. **Fruits** obtusely 4-angulate, 7–10 cm across; seeds 12, large. **Distr:** C America to Brazil. **Ecol:** In forests of the Atlantic coast; 600–1,450 m. **Zones:** 10–12.

Chukrasia

Species 1. **Trees**, deciduous. **Leaves** in spirals, usually even-pinnate or sometimes odd-pinnate; leaflets usually alternate, margins entire. **Inflorescences** thyrses, axillary to subterminal. **Flowers** bisexual, oblong; calyx short, 4- or 5-lobed; petals 4 or 5, distinct, contorted; staminal tube cylindric, slightly shorter than petals, apical margins entire or 10-crenate, lobes each bearing one anther, anthers exserted; disk absent to narrowly cushion-shaped; ovary 3–5-loculed, ovules numerous per locule, style thick, stigma capitate. **Fruits** septicidal capsules, 3–5-locular, with 60–100 seeds per locule; seeds flat, terminally winged, cotyledons leaflike. **Distr:** Tropical and subtropical Asia. **Cult:** Propagate from seeds or by air-layering. Requires a fertile, moist, well-drained soil in full sun.

C. tabularis A. Jussieu, *Bull. Sci. Nat. Géol.* 23: 241, 1830. (Syn: *Dysoxylum esquirolii*, etc.). [Chittagong Wood, Indian Redwood]. **Trees**, to 25 m tall. **Leaves** usually 30–50 cm long; leaflets 10–16, ovate to oblong-lanceolate, 7–12 × 3–5 cm, base oblique, margins entire, apex acute to acuminate, papery, glabrous or abaxially pubescent, secondary veins 10–15 per side; petiole 4.5–7 cm long; petiolules 4–8 mm long. **Inflorescences** thyrses, lax, about half as long as leaves, branches ± glabrous; peduncle short; bracts linear, caducous; pedicels short, jointed. **Flowers** 1.2–1.5 cm long, fragrant; calyx ~2 mm long; petals cream-colored to ± lavender, linear-oblong to spathulate, 1.2–1.5 × 0.5–0.6 cm; staminal tube cylindric, anthers 10, oblong; ovary on a short disk, elongate, style short. **Fruits** capsules, yellowish-gray to brown, subglobose to oblong, ~4.5 × 3.5–4 cm, usually 3-valved, woody; seeds many, flat, oblong, ~3 × 0.5 mm, broadly winged. **Distr:** Bhutan, China, India, Indonesia, Laos, Malaysia, Nepal, Sri Lanka, Thailand, Vietnam. **Ecol:** In evergreen and deciduous forests; 300–1,600 m. Fl. Apr–May. Fr. Jul–Jan. **Zones:** 9–12.

Lansium

Species 3. **Trees**. **Leaves** odd-pinnate; leaflets few, coriaceous. **Inflorescences** ♂ often panicles; ♀ spikes or racemes, axillary or cauline. **Flowers** small, subglobose, bisexual or polygamodioecious; sepals and petals 5, rounded, imbricate; stamens united in a tube; anthers 10, included, mostly biseriate; ovary globose, 3–5-loculed, locules 2-ovuled, style very short, thick, stigma 3–5-lobed. **Fruits** baccate, fleshy, 1–5-loculed, locules 1–2-seeded; seeds oblong, surrounded by soft, fleshy pulp. **Distr:** India to Malaya, Philippines. **Cult:** Propagate from seeds or by air-layering. Requires a fertile, humus-rich, moist, well-drained soil in full or partial sun.

L. domesticum Corrêa, *Ann. Mus. Natl. Hist. Nat.* 10: 157–158, 1807. [Langsat]. **Trees**, 4–8+ m tall, slightly pubescent or nearly glabrous; bark pale. **Leaves** alternate, pinnate, 20–40 cm long; leaflets 5–7, oblong or elliptic-oblong, 7–18 × 7–10 cm, base narrowed, apex acuminate or abruptly shortly acuminate, adaxially shiny green, glabrous, nerves 10 pairs, prominent abaxially; petiolules ~1.2 cm long. **Inflorescences** spikes, pubescent, solitary or fascicled

Below: *Chukrasia tabularis, Lansium domesticum, Melia azedarach.*

Above: *Sandoricum koetjape, Swietenia macrophylla, Swietenia mahagoni, Toona ciliata.*

on the trunk and larger branches, much shorter than leaves; flowers sessile. **Flowers** small, perfect, yellow; calyx fleshy, puberulous, lobes 5, round; petals glabrous; staminal tube subglobose, anthers in 1 row; ovary tomentose 5-loculed. **Fruits** berries, oblong-ovoid or ellipsoid, usually 2.5–3.2 cm long, pubescent, pericarp tough; seeds 1 or 2, surrounded by translucent pulp, as are the remaining 3 or 4 aborted seeds. **Distr:** Malay Peninsula and Archipelago. **Ecol:** Fl. Jun. **Zones:** 11–12. **Info:** Often cultivated in SE Asia for the edible fruits (arils).

Melia

Species 3. **Trees** or **shrubs**, usually deciduous; young parts usually covered with stellate trichomes; branchlets with leaf scars and lenticels. **Leaves** in spirals, 2- or 3-pinnate; leaflets petiolulate, margins usually dentate, sometimes entire. **Inflorescences** thyrses, axillary, much-branched, consisting of several dichasia. **Flowers** bisexual; calyx 5- or 6-parted, imbricate; petals 5 or 6, white or purple, distinct, linear-spathulate, expanding; staminal tube cylindric, 10–12-ribbed, mouth expanding, margins 10–1bilobed, anthers 10–12, inserted between filament tube lobes, included or partly exserted; disk annular; ovary subglobose, 3–6-locular, with 2 superposed ovules per locule, stigma capitate, 3–6-lobed. **Fruits** drupes, subfleshy, stone bony; seeds pendulous, exotesta hard chitinous, endosperm fleshy, thin or absent, cotyledons leaflike. **Distr:** S tropical Africa, tropical to temperate Asia. **Cult:** Propagate from seeds or cuttings. Needs a well-drained soil in full sun.

M. azedarach L., *Sp. Pl.* 1: 384, 1753. (Syn: *Melia azedarach* subvar. *intermedia*, etc.). [Bead Tree, Chinaberry, Persian Lilac, Texas Umbrella Tree, White Cedar]. **Trees**, to 12 m tall, deciduous. **Leaves** odd-pinnate, 2-pinnate or 3-pinnate, 20–40 cm long; leaflets opposite, ovate, elliptic, or lanceolate, 3–7 × 2–3 cm, base ± oblique, cuneate, margins crenate or sometimes entire, apex shortly acuminate, terminal one usually larger, secondary veins 12–16 per side. **Inflorescences** thyrses, ± half as long as leaves. **Flowers** fragrant; calyx 5-parted, sepals ovate to oblong-ovate, acute; petals lilac-colored, obovate-spathulate, 0.9–1.3 cm long; staminal tube purple, 7–8 mm long, longitudinally striped, apical margin with 10 narrow lobes, lobes conic, anthers 10, alternate to lobes, narrowly elliptic; ovary spherical, glabrous, 5–8-locular, locules 2-ovuled, style acerose, stigma not included within filament tube, apex 5-dentate. **Fruits** drupes, orange, globose to ellipsoid, 1–3 × 0.8–1.5 cm; seed 1, ellipsoid. **Distr:** Asia to Australia and Pacific Islands. **Ecol:** In mixed evergreen and deciduous forests; 500–2,100 m. Fl. Mar–May. Fr. Oct–Dec. **Zones:** 8–12.

Sandoricum

Species 5. **Trees**; indumentum of simple hairs. **Leaves** trifoliolate. **Inflorescences** thyrses, axillary. **Flowers** with calyx ± truncate to shallowly 4- or 5-lobed; petals (4)5, free, imbricate; staminal tube cylindrical, ribbed distally, margin with 5 or 10 short lobes; anthers 10, glabrous, included; disk tubular, free, margin coarsely toothed; ovary somewhat sunk in receptacle, 4- or 5-locular, each locule with 2 collateral ovules, style head with 4- or 5-lobed stigma. **Fruits** drupes, 1–5-locular; pyrenes 1(2)-seeded; seeds kidney-shaped, laterally compressed, with thin sarcotesta. **Distr:** Predominantly Malesian; 1 sp. widely naturalized throughout tropical Asia from Mascarenes through Indian Subcontinent and Indo-China to Malesia. **Cult:** Propagate from seeds or cuttings. Needs well-drained soil in full or partial sun.

S. koetjape (Burm. f.) Men., *Philipp. J. Sci.* 7: 237, 1912. (Syn: *S. indicum, Trichilia nervosa*, etc.). [Lolly Fruit, Red Santol, Santol]. **Trees**, to 45(–50) m; trunk to 1 m Ø, sometimes with buttresses to 3 m high. **Leaves** 18–40 cm long; leaflets ovate, base acute to rounded, laterals ± asymmetric, apex acuminate, veins 7–14(–20) per side, apical leaflet 8–20 × 5–14 cm, laterals smaller; petiole 7.5–16 cm long, flattened; lateral petiolules 4–9 mm long, apical petiolule 3–5.5 cm long. **Inflorescences** thyrses, 2.5–24 cm long, fragrant, in apical axils, ± erect; secondary branches 1–5-flowered glomerules; bracts narrowly triangular, ~7 mm long; pedicels ~3–5 mm long. **Flowers** with calyx ~3.5 mm long, campanulate to cupular, lobes 5, irregular, yellow-green; petals (4)5, linear-lanceolate to oblanceolate, 6–9 mm long, reflexed, yellowish-green or pinkish; staminal tube yellow to orangish, margin 10-lobed, anthers (8)10, 1–1.5 mm long; disk ~1.5 mm long. **Fruits** drupes, 5–8 cm Ø, flattened-globose, velvety, yellow or brownish, pericarp with milky latex; mesocarp white, translucent, juicy, sweet to sour; seeds 1 or 2, 2–3.5 × 1.2–2.1 cm. **Distr:** Malesia; naturalized in tropical Asia. **Ecol:** In forests; to 1,200+ m. **Zones:** 10–12.

Swietenia

Species 3. **Trees**, deciduous; wood red. **Leaves** in spirals, even-pinnate, glabrous; leaflets opposite to subopposite. **Inflorescences** thyrses, axillary or subterminal. **Flowers** small; calyx small, 5-parted to middle, imbricate; petals (4 or)5, distinct, broad, imbricate; staminal tube cup-shaped, apically (8 or)10-lobed, anthers (8 or)10, inserted on inside of tube throat, alternate with lobes; disk annular; ovary ovoid, (4 or)5(or 6)-locular, ovules 9–16 per locule, pendulous, style cylindric, stigma disciform with (4 or)5 lobes. **Fruits** capsules, oblong or ovoid, 5-locular, woody; seeds 9–16 per locule, winged, hanging by wing-end from distal part of columella. **Distr:** Tropical W Africa, tropical and subtropical America. **Cult:** Propagate from seeds or cuttings. Requires a well-drained soil in full sun; salt-tolerant.

S. macrophylla G. King, *Icon. ser.* 3. 6: pi. 1550, 1886. (Syn: *S. belizensis, S. tessmannii*). [American Mahogany, Broadleaf Mahogany]. **Trees**, large, to 35–50 m tall; trunk clear of limbs for 18–20 m, 75–1.5 m Ø; buttresses sometimes 3.5–4.5 m high. **Leaves** large, petiolate, glabrous; leaflets mostly 8–12, slender-petiolulate, obliquely lanceolate, usually 8–15 × 2.5–7 cm, base acute and very oblique; apex acuminate to long-acuminate. **Inflorescences** panicles, 10–20+ cm long, glabrous. **Flowers** with calyx 2–2.5 mm Ø, lobes short, rounded; petals obovate, white, 5–6 mm long; stamen tube cylindric, teeth acute or acuminate. **Fruits** capsules, ovoid, usually 12–15 × 7 cm, apex pointed; seeds winged samaras, 7.5–8.5 cm long, wing reddish-brown to 3 cm wide. **Distr:** British Honduras, along the Atlantic coast to Panama, and probably southward locally to the Amazon Valley of Brazil and Peru. **Ecol:** Scattered in wet, lowland or hillside, mixed forests; <400 m. **Zones:** 10–12.

S. mahagoni (L.) Jacquin, *Enum. Syst. Pl.* 20, 1760. (Syn: *Cedrela mahagoni*). [Mahogany, Spanish Mahogany, West Indian Mahogany]. **Trees**, to 45 m tall; trunk to 4 m Ø, base buttressed. **Leaves** alternate, ~35 cm long; leaflets 8–12, ovate to lanceolate, 10–16 × 4–6 cm, base oblique, margins entire or with 1 or 2 serrations, apex long-acuminate, leathery, adaxially dark green, abaxially lighter, secondary veins 10 per side; petiole 3–6 mm long, base slightly swollen. **Inflorescences** thyrses, axillary, 6–15 cm long, glabrous; pedicels ~3 mm long. **Flowers** small; calyx cup-shaped, lobes 5, short, truncate, apex rounded; petals greenish-white, obovate, 3–4 mm long, glabrous; staminal tube subcylindric, glabrous, anthers 10; disk annular; ovary conic to ovoid, longer than disk, locules 12-ovuled, style longer than ovary, glabrous, stigma disciform. **Fruits** capsules, brown, ovoid, ~8 cm Ø, apically splitting into 5 parts on maturity, releasing seeds; seeds many, apically winged samaras, ~7+ cm long including wing. **Distr:** Tropical America. **Ecol:** Fl. May–Jun. Fr. Oct–Nov. **Zones:** 11–12.

Toona

Species ~5. **Trees**, to 50 m tall, monoecious, deciduous or semideciduous. **Leaves** spirally arranged, even-pinnate or occasionally odd-pinnate, leaflets usually >8 per side; leaflets with margins entire, serrate, or dentate. **Inflorescences** thyrses, much-branched, pendent, often >1 m long. **Flowers** 5-merous, unisexual with well-developed vestiges of opposite sex present, rarely hermaphrodite, small; calyx 5(6)-lobed or with 5(6) distinct sepals; sepals imbricate or cup-shaped in bud, margins always ciliate; petals 5(6), white, cream-colored, or pink, distinct, longer than calyx in bud, imbricate, basally adnate to a short pulvinate androgynophore (disk); stamens 5(6), distinct, arising from androgynophore, sometimes alternating with 1–5 filamentous staminodes, anthers in ♂ yellow; antherodes in ♀ often sagittate, brown; ovary 5-locular, locules 6–10-ovuled, vestigial in ♂, style short in ♀; pistillodes long and slender in ♂; style head discoid with stigmatic papillae, usually 5-rayed. **Fruits** capsules, ellipsoid or obovoid, pendulous, septifragal, 5-valved; seeds numerous per locule, winged. **Distr:** Asia, Australia, Malesia, New Guinea. **Cult:** Propagate from seeds or suckers. Requires a deep, fertile, moist, well-drained soil in full sun.

T. ciliata M. Roemer, *Fam. Nat. Syn. Monogr.* 1: 139, 1846. (Syn: *Cedrela toona*, etc.). [Cedar, Indian Mahogany, Red Cedar]. **Trees**, to 30 m tall; trunk to 22 m tall, to 1.5 m Ø, buttresses to 3.5 m or absent. **Leaves** pinnate, (15–)26–69 cm long; leaflets usually (5–)9–15 pairs, lanceolate to ovate-lanceolate, (7–)9–12.8(–16) × (2.2–)3.2–5(–6) cm, base asymmetric, margins entire, apex acute to acuminate; petiole 6–11 cm long; rachis often reddish; petiolules 0.2–1(–1.4) cm long. **Inflorescences** to 55 cm long, pendent; rachis pilose to pilose-villous; pedicels 0.5–1 mm long. **Flowers** 3.5–5(–6) mm long, sweetly scented; calyx 0.7–1.3 mm long, lobes imbricate; sepals spathulate, (0.4–)0.7–1 × (0.5–)0.7–1.3 mm; petals white to creamy-white, 3.5–5.8 × 1.3–3.1 mm; ovary 1.2–1.8 mm Ø, locules to 8-ovuled. **Fruits** capsules, 1.5–2(–2.5) cm long; columella 1.5–2(–2.4) × 0.5–0.7(–1) cm; valves red to reddish-brown; seeds 1.1–1.9 cm long, both ends winged. **Distr:** Australia (NSW, Queensland), New Guinea, Malesia, Asia. **Ecol:** In tropical rainforests; 0–1,000 m. Fl. Jan–Jun. Fr. Feb–Nov. **Zones:** 9–12.

Turraea

Species 60. **Shrubs** or **small trees**, sometimes scrambling. **Leaves** alternate, sometimes fasciculate; lamina simple, margins entire, rarely shallowly lobed. **Inflorescences** with flowers solitary or fasciculate, or in axillary or terminal cymes or pseudo-racemes. **Flowers** bisexual, 4–5-merous; calyx cupuliform, margins with (4)5(6) teeth or lobes, sometimes almost entire, persistent in fruit; petals (4)5, much longer than calyx, linear-spathulate to linear; staminal tube cylindric, sometimes expanded distally, terminated by simple or bilobed free or partly or completely fused appendages opposite to or alternating with anthers, anthers (8–)10(–20), apiculate; disk always present, sometimes vestigial; ovary small, (4)5–10(–20)-loculed, locules 2-ovuled, style elongate, distally expanded. **Fruits** capsules, small, leathery or woody, loculicidal with (4)5–10(–20) valves; seeds large, with a large or small, orange or red aril, testa black or red, shiny. **Distr:** Africa, Madagascar; tropical Asia to Australia. **Cult:** Propagate from seeds or half-hardened cuttings. Requires a moderately fertile, moist, well-drained soil in full or partial sun.

T. floribunda Hochst., *Flora*, 27: 297, 1844. (Syn: *T. heterophylla*). [Honeysuckle Tree, Splendid Honeysuckle]. **Shrubs** (sometimes scrambling) or **small trees**, deciduous, to 10(–13) m tall; first-year branchlets shortly pubescent, second-year purple-brown and glabrous. **Leaves** ovate to lanceolate, to 14 × 7 cm, base subtruncate, rounded or broadly cuneate, apex acuminate, densely setose when young; petiole to 1.5 cm long. **Inflorescences** pseudo-racemes or flowers subfasciculate, 2–7-flowered; peduncles 3–8 mm long; bracts 2 mm long, squamiform; pedicels 0.5–1.7 cm long. **Flowers** large, creamy to greenish-white, sweetly fragrant; with calyx 2.5–5 mm long, densely puberulous; petals linear-spathulate, (3.8–)4.5–5.2 × 0.3 cm, greenish-white; staminal tube cylindrical, (2.5)3.6–4.5 cm long, pure white, glabrous inside; appendages in pairs alternating with anthers, 3 mm long; ovary 10-locular, puberulous, style (4)4.8–5.5 cm long, style head globose. **Fruits** capsules, obovoid-cylindric, globose or depressed-globose, 1.3–2.4 × 2 cm, deeply ribbed and sulcate, woody, glabrous; aril covering about half the seed. **Distr:** Tropical Africa. **Ecol:** In evergreen forests; 15–1,800 m. Fl. Oct–Feb. **Zones:** 10–12.

Walsura

Species ~16. **Trees**, sometimes small. **Leaves** in spirals, odd-pinnate or occasionally a single leaflet (sometimes in *W. pinnata*); leaflets opposite, margins entire, abaxially pale; petiolule bases inflated and jointlike. **Inflorescences** thyrses, axillary or subterminal. **Flowers** bisexual or occasionally only ♂, small; calyx short, 5-lobed or with 5 distinct sepals, imbricate in bud; petals 5, much longer than calyx, distinct, broad and expanding, valvate or imbricate in bud; stamens 10; filaments flat, broad, usually basally connate into a tube or sometimes distinct, shorter than petals; anthers introrse, inserted on apex or between 2 lobes of filament; disk annular, fleshy; ovary short, 2- or 3-locular, locules 2-ovuled, style almost as long as ovary, stigma disciform or conic, tip 2- or 3-cleft. **Fruits** berries (rarely capsules), usually 1(or 2)-locular; seeds 1 or 2 per locule, arillate. **Distr:** Tropical Asia. **Cult:** Propagate from seeds or by air-layering. Requires a moist, well-drained soil in full sun.

W. gardneri Thw., *Enum. P1. Zeyl.* 61, 1858. **Treelets**, to 4 m tall; trunk to 15 cm Ø. **Leaves** undivided, all along twigs; lamina ± elliptic, 8.5–14 × 3.8–6.4 cm, base acute to shortly attenuate, usually weakly asymmetric, apex acute to subacuminate, subcoriaceous, abaxially glaucous, secondary veins 6 or 7 per side; petiole 0.8–2(–2.5) cm long. **Inflorescences** thyrses, 2.5 cm long, in axils of undeveloped leaves, branched to second (or third) order with primary branches to 2 cm long, hairy. **Flowers** with calyx ~1 mm long, lobes to ~0.5 mm long; petals ~2.5–3 × 1 mm, imbricate, creamish, apices acute; staminal tube lobed to ~¾ the length, lobes long-hairy except at base and apically bifid, anthers ~0.5–0.7 mm long, apiculate, long-hairy; style obconical, glabrous, style head apically and shortly bilobed. **Fruits** berries, ovoid-ellipsoid, 2.2–2.5 × 1.2–1.5 cm, reddish, densely tomentellous; pericarp to 1.5 mm thick; seed 1, ± enveloped in colorless sweet aril. **Distr:** Sri Lanka; rare and endangered. **Ecol:** In forests on mountain slopes; 175–600 m. Fl. Jan–Mar. **Zones:** 11–12.

W. trifoliolata (A. Juss.) Harms, *Nat. Pflanz.*, ed. 2 19(b1): 119, 1940. (Syn: *Heynea trifoliolata*, etc.). **Trees**, to 15(–18) m tall; trunk to 50 cm Ø, sometimes with buttresses to 50 cm high. **Leaves** trifoliolate, 9–18 cm long; leaflets ovate-elliptic to subobovate, 5–12.8 × 2.5–4.5 cm, base attenuate, apex acuminate to obtuse or even emarginate, subcoriaceous, abaxially glaucous, sparsely hairy; petiole 1.4–4.4 cm long, terete; petiolules 0.2–1 cm long. **Inflorescences** thyrses, 2.8–10.5(–14) cm long, axillary at apex, branched to third order, primary branches to 5(–7) cm long, minutely hairy. **Flowers** with calyx ~1 mm long, lobes to 0.8 mm long; petals 2.5–3.5 × ~1 mm, white, ± imbricate, apices ± acute; staminal tube lobed, hairy within, lobes apically, bifid, anthers ~0.7 mm long, apiculate; style obconical, subglabrous, apically shortly bilobed. **Fruits** berries, subglobose to ellipsoid, 1.1–1.9 × 0.8–1.8 cm, pale yellow- or orange-brown, minutely tomentose; seed 1(2), enveloped in succulent, whitish, sweet aril. **Distr:** India, Sri Lanka. **Ecol:** In moist forests. **Zones:** 11–12.

Xylocarpus

Species 3. **Trees** or **shrubs**, semievergreen. **Leaves** in spirals, even-pinnate; leaflets (2–)4(–8) pairs; leaflets with margins entire. **Inflorescences** thyrses, axillary, consisting of sparse cymes. **Flowers** appearing bisexual; calyx short, lobes 4, oblong, contorted in bud; staminal tube kettlelike, apically narrowed, apex 8-lobed, anthers 8, sessile, inserted between and alternating with lobes, oblong, included; disk thick, hemispheric, fleshy, adnate with ovary base; ovary small, globose, 4-locular, with 3 or 4 superposed ovules per locule, style cylindric, stigma disciform. **Fruits** capsules, globose; pericarp fleshy, dehiscing into 4 segments; seeds 5–20, large and thick, irregularly tetrahedral to pyramidal, sarcotesta spongy. **Distr:** Tropical E Africa, Asia, W Pacific Islands. **Ecol:** Mostly in mangroves. **Cult:** Propagate from seeds or by air-layering. Requires a moist soil in full or partial sun; tolerates brackish water.

X. granatum J. Koenig, *Naturforscher (Halle)*. 20: 2, 1784. (Syn: *Carapa obovata*). [Cannonball Mangrove]. **Trees** or **shrubs**, to 5 m tall. **Leaves** ~15 cm long; leaflets usually 4, opposite; leaflets elliptic to obovate-oblong, 4–9 × 2.5–5 cm, base cuneate, margin entire, apex rounded, subleathery, surfaces glabrous, usually pale, secondary veins 8–10 per side; petiole and rachises cylindric, glabrous; petiole 3–5 cm long; petiolules ~4 mm long, base inflated. **Inflorescences** cymes, 1–3-flowered, many cymes forming thyrses, glabrous; pedicels >1 cm long. **Flowers** small; calyx lobes rounded; petals white, obovate-oblong, ~6 mm long, leathery; staminal tube lobes suborbicular and weakly 2-cleft, anthers

Below: *Turraea floribunda, Walsura gardneri, Walsura trifoliolata, Xylocarpus granatum.*

RT

Above: *Xylocarpus granatum* (fruit), *Xylocarpus moluccensis*, *Acronychia pedunculata*.

ellipsoid; style subquadrigonal, glabrous, stigma ± as long as staminal tube. **Fruits** capsules, 10–12 cm Ø, stipitate; seeds 8–12. Fl. & Fr. Apr–Nov. **Distr:** China, India, Indonesia, Malaysia, Papua New Guinea, Philippines, Sri Lanka, Thailand, Vietnam; E Africa, W Pacific Islands. **Ecol:** In mangrove forests. **Zones:** 10–12.

X. moluccensis (Lam.) M. Roem., *Fam. Nat. Syn. Monogr.* 1: 124, 1846. (Syn: *Carapa moluccensis*, *X. australasicus*). [Marugas (Death Tree), Puzzlenut Tree]. **Trees**, evergreen, 4–15 m tall, much-branched; surface roots laterally compressed and forming a spreading network of ribbonlike pneumatophores (cone-shaped), upper edges protruding above mud and suggesting a mass of snakes. **Leaves** paripinnate; leaflets opposite, 1–2(3)-jugate, elliptic, oblong-elliptic or obovate-elliptic, to 12 x 5 cm, base cuneate, apex rounded, rarely obtuse or emarginate, glabrous, coriaceous; petiole and rachises to 8.5 cm long; petiolules 2–5 mm long. **Inflorescences** lax racemes of (2)3-flowered cymes; peduncle and rachis together 4–7 cm long; bracts minute, usually caducous. **Flowers** white to creamish or pale pink; calyx ~3 mm long, glabrous, lobed to middle, lobes rounded; petals 5–6.5 x 2.5 mm, glabrous; staminal tube 4–5 mm long, glabrous; ovary <1 mm Ø, style 1.5 mm long; disk fused to lower half of ovary. **Fruits** capsules, large, to 20 cm Ø, obscurely 4-sulcate; seeds, many, 4–8 cm long. **Distr:** Mostly Old World tropics to Australia and Pacific Islands. **Ecol:** Mostly in tidal mud of mangrove swamps to further inland. Fl. May–Sep. **Zones:** 11–12.

RUTACEAE

Genera ~155, species ~1,600. **Shrubs**, **trees**, or sometimes **herbs**, sometimes scrambling or scandent, sometimes armed, with aromatic volatile oils contained in glands visible at surface of at least leaves, young branchlets, inflorescences, flower parts, fruit, or cotyledons in seed. **Stipules** absent (or stipular excrescences rarely present). **Leaves** alternate, opposite (or whorled), simple (petiole neither apically swollen nor articulate with lamina), 1-foliolate (in individual specimens at least some 1-foliolate leaves with petiole apically swollen and/or articulate with lamina), or variously compound. **Inflorescences** solitary, axillary or terminal cymes, panicles, rarely simple racemes. **Flowers** bisexual or unisexual, usually 3–5-merous, actinomorphic or rarely zygomorphic, hypogynous (or rarely perigynous); perianth biseriate, with clearly differentiated calyx and corolla or sometimes in 1 or 2 irregular series, with ± undifferentiated tepals; sepals distinct or connate to entire length; petals distinct (or rarely coherent or connate for part of their length); stamens usually as many as or twice as many as petals or sometimes more, filaments distinct or ± connate; disk (rarely lacking), nectariferous; gynoecium of 1–5 distinct 1-loculed carpels or 2–many partially to completely connate carpels, placentation usually axile, locules 1–many-ovuled. **Fruits** of 2–5 follicles (drupes or samaras), or a single follicle, capsule, or berry (or samara); seeds oblong or reniform with relatively large embryo, endosperm present and fleshy or lacking. **Distr:** Nearly cosmopolitan, mainly tropical and subtropical regions of SE Asia, Mediterranean countries, N America, Australia and S Africa.

Acronychia

Species ~48. **Shrubs** or **trees**, evergreen. **Leaves** opposite, 1-foliolate (or digitately 3-foliolate). **Inflorescences** cymulose to thyrsiform (or reduced to solitary flowers), axillary or basal to leaves. **Flowers** bisexual; sepals 4, distinct or basally connate; petals 4, valvate in bud; stamens 8, distinct, ones opposite sepals nearly as long as petals, filaments with indumentum toward base (or rarely glabrous), gradually tapering from ± broad base to subulate apex, becoming reflexed; anthers obtuse or obtusely mucronulate; disk pulvinate; gynoecium 4-loculed, ± completely syncarpous (or grading to subapocarpous), ovules 2 per locule; stigma punctiform or capitellate. **Fruits** ± completely syncarpous 4-loculed, drupaceous berries (or grading to subapocarpous with 4 basally connate drupaceous berries); seeds ellipsoid to pyriform. **Distr:** S and SE Asia, Australia, SW Pacific Islands. **Cult:** Propagate from seeds or cuttings. Requires a moist, well-drained soil in full or partial sun.

A. pedunculata (L.) Miquel, *Fl. Ned. Ind., Eerste Bijv.* 532, 1861. (Syn: *Clausena simplicifolia*, etc.). [Heartwood, Rosewood]. **Shrubs** or **trees**, to 28 m tall. **Leaves** with leaflets usually elliptic to elliptic-oblong but grading to obovate, oblanceolate, or nearly oblong, 3.5–24.5 x 2–8.5 cm, base cuneate or sometimes rounded or attenuate, apex obtusely acuminate, acumen to 2 cm long, or sometimes obtuse or rounded; petiole 0.5–5 cm long. **Inflorescences** 2–25 cm long, few- to many-flowered; pedicels 0.2–1.2 cm long. **Flowers** with sepals 0.6–1.5 mm long; petals 0.4–1.2 cm long; ovary pubescent, rarely only apically pubescent, with or without apical septicidal fissures; style glabrous, base pubescent. **Fruits** subglobose to ellipsoid, pyriform, or broadly conic, 0.5–1.5 cm Ø; seeds reddish-black to black, 3–7 mm long. **Distr:** Asia to Malesia. **Ecol:** In secondary forests, woods or thickets on lower hills; 0–900 m. Fl. Apr–Aug. Fr. Aug–Dec. **Zones:** 10–12.

Aegle

Species 1. *Generic description same as specific.* **Cult:** Propagate from seeds or ripe cuttings (if leafy, will root in sand under glass, in heat). Requires a moist, well-drained, rich, loamy soil in full sun.

A. marmelos (L.) Corrêa, *Trans. Linn. Soc. London* 5: 223, 1800. (Syn: *Bilacus marmelos*, etc.). [Bael Fruit, Elephant Apple, Golden Apple, Wood Apple]. **Trees**, deciduous, to 10 m tall; spines straight, axillary, to 3 cm long. **Leaves** alternate, odd-pinnate, (1 or)3(or 5)-foliolate; leaflets ovate to elliptic, 4–12 x 2–6 cm, base rounded to narrowly cuneate, margins crenulate, apex acuminate or sometimes acute. **Inflorescences** loosely fasciculate or racemose and few-flowered or flowers solitary, terminal and axillary. **Flowers** bisexual, fragrant; calyx cup-shaped, lobes 4- or 5, ~1 mm long; petals 4 or 5, imbricate in bud, ~1 cm long, white; stamens 30–50+, nearly as long as petals, filaments short, subulate, distinct or irregularly coherent at base, anthers linear-lanceolate; gynoecium ~6 mm long; disk columnar or bell-shaped; gynoecium 8–20-loculed, syncarpous, locules many-ovuled, in 2 rows, style short, thick, stigma capitate, cylindric, or bluntly conic, longitudinally grooved. **Fruits** edible, greenish-yellow, berries, globose to ellipsoid to pyriform, 10–12 x 6–8 cm; exocarp thin, parenchymatous; mesocarp woody; endocarp fleshy, soft and pulpy, becoming hard and reddish-orange when dry; seeds depressed-ovoid, ~8 mm long. **Distr:** India. **Ecol:** Fr. Oct. **Zones:** 10–12.

Atalantia

Species ~17. **Shrubs** or **trees**, unarmed or with straight axillary spines, rust-colored villosulous indumentum on terminal and axillary buds or young inflorescences. **Leaves** alternate, 1-foliolate or simple; petiole 0.2–1.3 cm long, usually not bent, twisted, or swollen. **Inflorescences** fasciculate, racemose or paniculate, terminal and/or axillary. **Flowers** bisexual; sepals 3–5, basally connate to entirely connate, or calyx irregularly splitting into 2 or 4 segments; petals 3–5, imbricate in bud; stamens 6–10, equal or alternately unequal in length, distinct, monadelphous, filaments coherent in phalanges, or filaments irregularly coherent; disk annular, cup-shaped, or columnar; gynoecium 2–5-loculed, syncarpous, ovules 1 or 2 per locule, style to 1.5x as long as ovary, deciduous in fruit. **Fruits** berries, usually with pulp vesicles and with or without mucilaginous pulp; endocarp membranous; seeds with membranous testa. **Distr:** S and SE Asia. **Cult:** Propagate from seeds, cuttings or by air-layering. Requires a fertile, moist, well-drained soil in full sun.

A. ceylanica (Arn.) Oliv., *J. Linn. Soc. Bot.* 5, Suppl. 2: 25, 1861. (Syn: *Limonia monophylla*, etc.). [Ceylon Atalantia, Yakinaran]. **Shrubs**, to 2.5 m tall, densely branched, armed with sharp slender spines to 2.5 cm long solitary in axils. **Leaves** ovate to ovate-lanceolate or elliptic, rarely somewhat oblong-elliptic to obscurely obovate, mostly 4–12(–15) x 2–6 cm, margins entire, adaxially very dark dull green, abaxially pale or yellowish-green, lateral nerves 5–10 pairs; petiole mostly 3–6 mm long. **Inflorescences** cymose or racemose, sometimes fascicled, ~1–2 cm long, several-flowered, axillary; pedicels 2–3 mm long. **Flowers** calyx saucer-shaped, sepals 4, broadly deltoid, 1.5 mm long, glabrous, margins minutely ciliolate; petals 4 (rarely 3 or 5), ovate-oblong, white,

PL

Top row: *Aegle marmelos, Atalantia ceylanica, Calodendrum capense*.
Bottom row: *Casimiroa edulis* (flowers), *Casimiroa edulis* (fruits), *Citrus* x *aurantiifolia, Citrus maxima*.

rounded, 5 × 3 mm, 3–5-veined, glandular, shortly broadly clawed; stamens usually 8, free, longer ones subequal to petals, filaments glabrous, usually ~3 mm long; ovary sessile, glabrous, 2.5–3 mm long, 2-loculed, locules 2-ovuled, stigma truncate. **Fruits** subglobose, dry, with very few vesicles; seeds 2–4, plump. **Distr:** Sri Lanka, S India. **Ecol:** In lowland areas. Fl. Mar–Oct. **Zones:** 11–12.

Calodendrum

Species 2. **Trees**, tall; branches opposite. **Leaves** opposite, exstipulate, petiolate; lamina simple, dotted with pellucid glands. **Inflorescences** panicles, terminal. **Flowers** bisexual, large; sepals 5, ovate, small; petals 5, imbricate; stamens 5, opposite sepals, anthers with an apical gland; staminodes 5, petaloid, opposite petals, inserted at base of a cupuliform disk; ovary on an elongated gynophore, verrucose-glandular, 5-lobed, 5-loculed, locules 2-ovuled, style filiform, long, stigma small. **Fruits** capsules, globose, woody, verrucose, stipitate, 5-loculed, locules 2-seeded, dehiscence valvate; seeds large, angular, black, shiny. **Cult:** Propagate from seeds or by air-layering. Requires a fertile, moist but well-drained soil in full sun.

C. capense (L. f.) Thunb., *Nov. Gen. Pl.* 2: 41, 43, 1782. (Syn: *Dictamnus capensis*). [Cape Chestnut]. **Trees**, to 20 m tall; young branches, leaves and inflorescences densely tomentose. **Leaves** elliptic, 6–13(–18) × 3.5–7.5(–11) cm, base cuneate or obtuse or slightly cordate, apex acute or slightly apiculate or rounded; petiole 0.2–1 cm long. **Inflorescences** to 15 cm long, with opposite cymose branches; bracteoles linear, 1.3–1.5 mm long, caducous; pedicels to 1.5 cm long, stout. **Flowers** erect, mauve; sepals 3.2 × 1.8 mm, free or slightly basally united, ovate, minutely tomentose; petals 3–3.5 × 0.6–0.7 cm, narrowly oblanceolate, whitish or mauve, tomentose; stamens as long as petals, inserted at base of an annular disk, filaments filiform, anthers 1.3–1.5 mm long; staminodes longer than petals, narrowly oblanceolate, caudate, dotted with bright crimson glands; gynophore 5 mm long, glandular, ovary 2–3 mm Ø, globose, apiculate, papillose, style 1.8–2 cm long, stigma capitate. **Fruits** capsules, 5-lobed, strongly rugose, warty, septicidal, 3.5 cm Ø. **Distr:** Tropical Africa. **Ecol:** In forests, often forming part of the canopy, in evergreen montane forests, in riverine forests, and also in coastal forests and scrubs; 1,200–2,300 m. Fl. throughout the year, mostly Oct–Dec. **Zones:** 9–11.

Casimiroa

Species ~6. **Shrubs** or **trees**. **Leaves** alternate, deciduous or persistent, petiolate, digitately 1- to several-foliolate; leaflets entire or obscurely crenate, pellucid-punctate, lateral ones sessile or short-petiolulate. **Inflorescences** panicles. **Flowers** small, whitish, perfect or sometimes with an abortive ovary; sepals 4–6, usually 5; petals generally 5; stamens as many as petals, inserted on base of disk, filaments subulate or linear-lanceolate, anthers elliptic or oval; ovary 2–8-loculed, commonly 4–5-loculed, locules 1-ovuled, usually stigma 5-lobate or entire. **Fruits** resembling apples, drupaceous, green or yellow, 2–5-celled; seeds 2–5, large seeds, these oblong to obovate or oval. **Distr:** Mexico, C America. **Cult:** Propagate from seeds, by grafting and air-layering. Requires a moist, well-drained soil in full or partial sun.

C. edulis Llave and Lex., *Nov. Veg. Descr.* 2: 9, 1825. [White Sapote]. **Trees**, medium-sized or sometimes large, to 15 m tall; trunk often thick, pale; crown 4.5 m wide; branches drooping. **Leaves** palmate, long-petiolate; leaflets 3 or mostly 5, long- or shortly petiolulate, elliptic to oval or broadly ovate, mostly 10–18 cm long, base acute to rounded, margins entire or undulate, apex acute or acuminate, thin or rather thick. **Inflorescences** panicles, mostly shorter than the petiole, branches puberulent or glabrous. **Flowers** fragrant, greenish-yellow or whitish, small; calyx small, lobes acute or subacute; petals 3.5–4 mm long. **Fruits** variable in size and shape, usually green or pale yellow and resembling an apple or a persimmon, commonly 6–10 cm across. **Distr:** Costa Rica, Mexico, Salvador. **Ecol:** In wet to dry forests or thickets; 600–2,700 m. **Zones:** 10–12.

Citrus

Species 20–25. **Shrubs** or **small trees**, evergreen, rarely deciduous; young branches often flat and angled, usually with solitary (rarely paired) spines at axils. **Leaves** 1-foliolate, rarely 3-foliolate or simple; petiole usually articulated with base of leaf lamina, usually conspicuously winged; lamina with margins crenulate or rarely entire, subleathery to leathery, with dense pellucid fragrant oil dots. **Inflorescences** axillary, solitary or in small fascicles. **Flowers** fragrant, hermaphrodite or ♂; calyx cup-shaped; lobes 3–5, subglabrous; petals (3 or)4 or 5(–8), white or outside pinkish-red, imbricate, thick; stamens usually 4(–10) times as many as petals, free or basally coherent; disk annular or short, with nectary glands; ovary (3–)5–14(–18)-loculed, locules 2–8(or more)-ovuled, stigma large. **Fruits** berries, sarcocarp segments of pulp vesicles and adaxially attached seeds; seed coat smooth or ridged. **Distr:** Asia, Australia, SW Pacific Islands. **Cult:** Propagate from seeds, cuttings, air-layering or by grafting. Requires a moist but well-drained soil in full sun.

C. × ***aurantiifolia*** (Christmann) Swingle, *J. Wash. Acad. Sci.* 3: 465, 1913. (Syn: *C.* × *aurantiifolia* ssp. *murgetana*, *C.* × *lima*, etc.). [Lime, Mexican Lime]. **Trees** small; branchlets numerous and irregular, with short stout spines. **Leaves** broadly ovate to elliptic, 5–8 × 2–4 cm, base rounded, margins crenulate, apex obtuse and sometimes mucronate, slightly stiff; petiole short. **Inflorescences** ~7-flowered, rarely flowers solitary. **Flowers** white; calyx cup-shaped; lobes 4 or 5; petals (4 or)5, 1–1.2 cm long, rather thick; stamens 20–25; ovary globose; style short, ± as long as stigma, stigma large. **Fruits** greenish-yellow, globose, ellipsoid, or obovoid, usually 4–5 cm Ø, smooth, with prominent oil glands, apex with a papilla; pericarp thin; sarcocarp with 9–12 segments, very acidic; seeds few, ovoid, testa smooth, cotyledons milky-white. **Distr:** Cultivated widely in the tropics and subtropics. **Ecol:** At low elevations. Fl. Apr–May. Fr. Sep–Dec. **Zones:** 11–12. **Info:** Possibly a hybrid involving *C. medica* and *C. hystrix*.

C. ***maxima*** (Burman) Merrill, *Interpr. Herb. Amboin.* 296, 1917. (Syn: *Aurantium maximum*, etc.). [Grapefruit, Pomelo, Shaddock]. **Trees**, small to medium; young parts pilose; branches usually purplish, flat, ridged, angular when young. **Leaves** broadly ovate to elliptic, 9–16 × 4–8 cm or larger, base rounded, margins subentire to shallowly crenulate, apex rounded to obtuse and sometimes mucronate, thick, dark green; petiole obcordately winged, 2–4 × 0.5–3 cm, wing to a third as long as the lamina. **Inflorescences** clusters or racemes, few-flowered or flowers solitary, axillary; flower buds purplish or rarely milky-white. **Flowers** large, 5-merous, puberulous; calyx 3–5-lobed; petals white, 1.5–2 cm long; stamens 20–35 (some undeveloped), anthers linear, large; ovary large, globose, style caducous, long and thick, locules numerous (>10). **Fruits** large to very large, globose, oblate, pyriform or broadly obconic, >10 cm Ø; pericarp, spongy, pale yellow to yellowish-green, with large prominent oil dots; sarcocarp with 10–15(–19) segments, firm, pale green, white, pink, reddish, or rarely milky-yellow; seeds 0–200, large, monoembryonic. **Distr:** Malesia, SE Asia. **Ecol:** In coastal areas. Fl. & Fr. several times a year. **Zones:** 10–12.

C. ***medica*** L., *Sp. Pl.* 2: 782, 1753. (Syn: *Aurantium medicum*, etc.). [Citron]. **Shrubs** or **small trees**; branches, leaf buds, and flower buds purplish when young; branches with ~4 cm long spines. **Leaves** simple or rarely 1-foliolate; lamina elliptic to ovate-elliptic, 6–12 × 3–6 cm or larger, margins serrate, apex rounded, obtuse, rarely mucronate; petiole short, not winged. **Inflorescences** ~12-flowered, sometimes flowers solitary, axillary. **Flowers** bisexual, sometimes ♂ by ± complete abortion of pistil; petals 5, 1.5–2 cm long; stamens 30–50; ovary cylindric, style long and thick, stigma clavate. **Fruits** pale yellow, elliptic to subglobose, to 2 kg in weight, surface coarse; pericarp white to pale yellow, soft within, thicker than sarcocarp; sarcocarp with 10–15 segments, colorless, nearly pellucid to pale milky-yellow, acidic to slightly sweet, fragrant; seeds small, smooth. **Distr:** NE India and possibly Myanmar. **Ecol:** Fl. Apr–May. Fr. Oct–Nov. **Info:** The Buddha-hand Citron, with separated segments ± surrounded by pericarp, is treated as a cultivar of *C. medica*.

C. ***reticulata*** Blanco, *Fl. Filip.* 610, 1837. (Syn: *C. deliciosa*, etc.). [Clementine, Mandarin, Satsuma, Tangerine]. **Trees**, small; branchlets numerous, spines few. **Leaves** 1-foliolate; leaflets lanceolate, elliptic, or broadly ovate, basal articulated part to leaf lamina usually narrow or only a remnant, midvein furcate near apex, margins apically obtusely crenulate or rarely entire, apex emarginate. **Inflorescences** fascicles of 3, or flowers solitary. **Flowers** fragrant; calyx irregularly 3–5-lobed; petals usually to 1.5 cm long; stamens 20–25; style long, slender, stigma clavate. **Fruits** pale yellow, orange, red, or carmine, oblate to subglobose, smooth or coarse; pericarp very thin to thick, easily removed; sarcocarp with 7–14 segments or rarely more, sweet to acidic, sometimes bitter; seeds few–many, rarely seedless, usually ovoid, base rounded, apex narrow and acute; pulp vesicles plump, short, rarely slender and long. **Distr:** SE China and/or S Japan. **Ecol:** On hillside forests; low elevations. Fl. Apr–May. Fr. Oct–Dec. **Zones:** 9–11.

Clausena

Species 15–30. **Shrubs** or **trees**, unarmed; rust-colored villosulous indumentum on terminal and axillary buds or young inflorescences. **Leaves** alternate, odd-pinnate. **Inflorescences** paniculate or in loose racemes, terminal or axillary. **Flowers** bisexual or very rarely ♀, globose to pyriform or rarely ovoid in bud; sepals 4 or 5, connate at base or to their entire length; petals 4 or 5, imbricate in bud; stamens 8 or 10, distinct, alternately unequal in length, filaments ± abruptly dilated toward base, ± straight or geniculate; disk columnar, conic, bell-shaped, or ± hourglass-shaped; gynoecium (2–)4- or 5-loculed, syncarpous, locules 2-ovuled, style to 2.5 times as long as ovary, deciduous in fruit. **Fruits** berries, with neither pulp nor pulp vesicles, endocarp membranous; seeds with membranous testa. **Distr:** Africa, Asia, NE Australia, SW Pacific Islands. **Cult:** Propagate from seeds or cuttings. Requires a moist, well-drained soil in full sun.

C. ***indica*** (Dalz.) Oliver, *J. Linn. Soc. Bot.* 5 Suppl. 2: 36, 1861. (Syn: *Bergera nitida*, *Piptostylis indica*). **Shrubs** or **small trees**, to 5 m tall. **Leaves** to 30 cm long; rachis gray-green; leaflets 7–13; *basal* leaflets smaller and broader; *lateral* leaflets ovate-lanceolate, mostly 3–8 cm long, base strongly asymmetric, margins subentire or minutely and irregularly subcrenulate, apex obtusely acuminate, adaxially dark glossy, abaxially paler, nerves 4 or 5 pairs; petiole base swollen, 3 mm thick; petiolules 3–4 mm long. **Inflorescences** paniculate, axillary and terminal; bracteoles minute, deltoid, dorsally puberulent; pedicels ~1 mm long; flower buds pale green. **Flowers** obovoid-rounded, 2–3 mm long; sepals 5, ovate-deltoid, concavo-convex, ~1 × 1 mm; petals greenish or white, elliptic, ~3 × 1.4 mm, rounded, sparsely glandular; stamens 10, 1–2 mm long, apex narrowed, white, glabrous, anthers with a large dorso-apical gland; disk obsolete; ovary green, subglobose-obovoid, 1.4 mm long, 2–3-loculed, locules 2-ovuled, white, style 0.6 mm long, smooth, green, truncate, stigma obscure. **Fruits** subglobose, cream to salmon-colored, to 1.3 cm Ø, smooth, glandular, pulp copious; seeds green, 1 or 2. **Distr:** India, Sri Lanka. **Ecol:** In dry areas; 200–700 m. Fl. May–Jun. **Zones:** 11–12.

Erythrochiton

Species ~6. **Shrubs** or **trees**, 0.5–12 m tall. **Leaves** alternate, 1-foliolate, often clustered at apex of trunk or branch; lamina with margins entire, chartaceous, glabrous; petiole ± swollen and articulate at apex. **Inflorescences** lateral, pedunculate dichasia, branched, elongate, cauline; flowers pedicellate in epiphyllous cymes. **Flowers** with sepals 5, valvate, connate, forming a campanulate to tubular calyx, 1.5–4.6 cm long, red, white, or green, usually persistent; corolla 3.8–7.9 cm long, white, glabrous; petals 5, connate forming a ± straight tube, lobes imbricate, arranged actinomorphic or zygomorphic; androecium of (4)5 fertile stamens or of 2 fertile stamens flanking innermost corolla lobe and (1)3–5 linear staminodes, filaments connate or free, anthers usually connate; ovary 5-carpellate, carpels 2-ovuled, style 1, stigma capitate. **Fruits** of 5 free mericarps (rarely 1 aborting) rarely base connate, each dehiscing adaxially; seeds 1 or 2 per carpel, ovoid, reniform, or cuboidal. **Distr:** Costa Rica across northern S America and south to Bolivia and Brazil. **Ecol:** Understory of humid tropical forests; <1,200 m. **Cult:** Propagate from seeds or cuttings. Requires a fertile, moist, well-drained soil in shade or partial sun.

E. ***brasiliensis*** Nees and C. Martius, *Nova Acta Phys.-Med. Acad.* 11: 166, t. 18C and 22, 1823. (Syn: *Pentamorpha graveolens*). [Chiquita Blanca, Cola de Pava]. **Shrubs** or **trees**, often unbranched, 0.5–12 m tall. **Leaves** oblanceolate to lanceolate, 15–94 × 3.5–14.7 cm, base long-cuneate to long-decurrent, apex acuminate to acute, secondary veins 15–34 per side; petiole 0.7–7.6 cm long. **Inflorescences** lateral, umbelliform, ± racemose, (6–)17–95 cm long; peduncle and rachis 3–4-angled; bracts linear to narrowly elliptic, 1.5–10 × 0.2–1.5cm; pedicels 0.7–2.2 cm long. **Flowers** with calyx lobed, 1.5–4.3 × 0.6–1.4 cm, red to pink-red; corolla actinomorphic, 4.3–7.5 cm long, tube straight,

Below: *Citrus medica*, *Citrus reticulata*, *Clausena indica*, *Erythrochiton brasiliensis*.

Above: *Flindersia australis, Glycosmis pentaphylla, Limonia acidissima, Murraya koenigii.*

2–4.2 cm long, lobes oblanceolate to widely obovate, 1.5–4 × 0.6–1.9 cm; stamens 5, fertile, filaments connate into a tube, 3–8 mm long; carpels basally connate, style 3.3–4.2 cm long. **Fruits** mericarps, 1.1–1.7 × 0.7–1.3 cm, red when fresh; seeds 6–8 × 3.5–5 mm, brown or tan. **Distr:** Venezuela, the Guianas, fringes of Amazon Basin, E Brazil. **Ecol:** In moist tropical forests; <400 m. Fl. & Fr. throughout most the year. **Zones:** 10–12.

Flindersia

Species 17. **Trees. Leaves** alternate or more frequently opposite; lamina pinnate or rarely simple, marked with pellucid dots. **Inflorescences** panicles, terminal. **Flowers** with calyx small, 5-lobed; petals 5, imbricate in bud, spreading; disk broad, concave; stamens 5, inserted on outside of disk, with as many or fewer staminodia alternating with them, sometimes absent, filaments subulate, anthers versatile; ovary 5-loculed, 5-lobed, locules 4–6-ovuled, style short, thick, inserted between lobes, stigma capitate. **Fruits** capsules, oblong, hard, tuberculate or muricate, opening septicidally in 5 boat-shaped valves or cocci, without any persistent axis; seeds flat, winged, 2 or 3 on each side of a flat placenta, which almost divides each locule into 2. **Distr:** Australia, Malesia, New Caledonia, New Guinea. **Cult:** Propagate from seeds or by air-layering. Requires a fertile, moist, well-drained soil in full sun.

F. australis R. Br., *Voyage to Terra Australis* 2, App. III, 595, t. 1, 1814. [Crow's Ash, Australian Teak]. **Trees**, to 40 m tall; larger trees moderately buttressed; exudate gumlike. **Leaves** alternate, rarely opposite, crowded toward end of branches, usually imparipinnate; leaflets 3–13, usually 5–9, elliptic or narrowly ovate, 2.4–13 × 0.8–4.3 cm, base obtuse to cuneate, often asymmetric, apex acute to obtuse, surfaces glossy green, abaxially paler, oil dots prominent; rachis and petiole 5–35 cm long; lateral petiolules 0–3 mm long, terminal petiolule 0.3–3 cm long. **Inflorescences** panicles, terminal or in upper axils, to 15 cm long. **Flowers** with calyx green, sepals 2.2–2.5 mm long, densely hairy; petals 5–7 mm long, white to cream, densely hairy outside except for the margins. **Fruits** capsules, woody, 5–10 cm long, 5-valved, remaining united at base after splitting; seeds 3.4–5 cm long, apically winged. **Distr:** Australia (NSW, Queensland). **Ecol:** In dry and subtropical rainforests; 0–750 m. Fl. Aug–Feb. Fr. Mar–Jan. **Zones:** 9–12.

Glycosmis

Species ~50. **Shrubs** or **trees**, unarmed. **Leaves** alternate (rarely opposite), odd-pinnate, 1-foliolate, or simple. **Inflorescences** paniculate, compound-racemose, or reduced to 1 or a few flowers, terminal and/or axillary. **Flowers** bisexual, globose to ellipsoid in bud; sepals 4 or 5, basally connate; petals 4 or 5, imbricate in bud; stamens 8 or 10, distinct, alternately ± unequal in length, filaments ± straight; disk annular, pulvinate, columnar, conic, or bell-shaped; gynoecium 2–5-loculed, syncarpous, radial walls of locules straight, locules 1(or 2)-ovuled, style to nearly as long as ovary, persistent in fruit. **Fruits** berries, with mucilaginous pulp or dry, without pulp vesicles, endocarp membranous; seeds with testa membranous. **Distr:** Asia, Australia. **Cult:** Propagate from seeds or by air-layering. Requires a fertile, moist, well-drained soil in full or partial sun.

G. pentaphylla (Retzius) Candolle, *Prodr.* 1: 538, 1924. (Syn: *G. arborea*, etc.). [Gin/Rum Berry, Jamaican Mandarin, Orangeberry]. **Trees**, to 5 m tall. **Leaves** (3- or)5-foliolate; leaflets oblong, 10–25 × 3–7 cm, base cuneate, margins serrate, apex mucronate, papery; petiolules 0.2–1 cm long. **Inflorescences** paniculate, axillary or terminal. **Flowers** globose in bud; sepals broadly ovate, <1 mm long; petals white or pale yellow, 3–4 mm long, caducous; stamens 10; ovary globose to broadly ovoid, style extremely short, stigma slightly expanded. **Fruits** reddish, subglobose, 0.8–1 cm Ø. **Distr:** Bhutan, Cambodia, China, NE India, Indonesia, Laos, Malaysia, Myanmar, Nepal, Pakistan, Philippines, Sri Lanka, Thailand, NW Vietnam. **Ecol:** On hillsides and valley woods; 600–1,200 m. Fl. Jul–Oct. Fr. Jan–Mar. **Zones:** 9–12.

Limonia

Species 1. *Generic description same as specific.* **Cult:** Propagate from seeds, ripe cuttings or by air-layering. Requires a well-drained soil in full sun.

L. acidissima L., *Sp. Pl.* ed. 2, 1: 554, 1762. (Syn: *Feronia elephantum*, etc.). [Elephant Apple, Wood Apple]. **Trees small**, to 7 m tall; branchlets with straight spines to ~4 cm long. **Leaves** to 10 cm long; leaflets opposite, usually 2 or 3 pairs and a terminal, *lateral* leaflets sessile, obovate, to 3.8 cm long, apex obtuse, obovate, glabrous; rachis and petiole narrowly winged or marginate; petiole shorter than leaflet. **Inflorescences** cymose-paniculate, axillary, small, spreading, often from leafless nodes, puberulent, flowers numerous; ♂ and perfect flowers usually together in the same inflorescence. **Flowers** small, white, green or reddish-purplish; sepals <0.5 mm long, deltoid, minutely puberulent, ciliolate especially near apex; petals lance-ovate, 4–6 mm long, acute, glabrous; stamens 7–12, filaments excavate and woolly inside at base, ~3 mm long, anthers 3–5 mm long, with 1 small apical gland; ovary glabrous, rotund, incompletely 4–6-loculed, style very short, stigma 5-lobulate. **Fruits** edible, woody, hard-shelled, globose, pale, with sticky pulp; seeds numerous. **Distr:** India, Pakistan, Java, Sri Lanka. **Ecol:** Usually in dry areas. **Zones:** 10–12.

Murraya

Species ~12. **Shrubs** or **trees**, unarmed. **Leaves** alternate, odd-pinnate (occasional leaves even-pinnate or 1-foliolate). **Inflorescences** paniculate, or reduced to cymes or few- to several-flowered racemes, terminal, or terminal and axillary. **Flowers** bisexual, ellipsoid to obovoid, or narrowly so or subcylindric in bud; sepals 4 or 5, connate at base or to ± half their length; petals 4 or 5, imbricate in bud; stamens (5–)8 or 10, distinct, alternately unequal in length, filaments ± linear, ± straight; gynoecium 2–5-loculed, syncarpous, radial walls of locules straight, locules 1- or 2-ovuled, style 3–7× as long as ovary, deciduous in fruit or sometimes basal portion persistent. **Fruits** berries, the pulp mucilaginous, without pulp vesicles; endocarp fleshy; seeds with membranous or fleshy testa. **Distr:** Asia, Australia, China, SW Pacific Islands. **Ecol:** In thickets, from near sea level to montane forests. **Cult:** Propagate from seeds, cuttings, suckers or by air-layering. Requires a fertile, moist, well-drained soil in full sun.

M. koenigii (L.) Sprengel, *Syst. Veg.* 2: 315, 1817. (Syn: *Bergera koenigii, Chalcas koenigii*). [Curry Leaf Tree, Curry Tree]. **Shrubs** or **trees**, to 4(–5) m tall. **Leaves** 17–31-foliolate; leaflets ovate, 2–5 × 0.5–2 cm, base obtuse to rounded and oblique, margins entire or crenulate. **Inflorescences** paniculate, many-flowered, terminal. **Flowers** 5-merous, ellipsoid in bud, fragrant; sepals ovate, <1 mm long; petals white or greenish-white, oblanceolate to oblong, 5–7 mm long; stamens 10; stigma capitate. **Fruits** bluish-black, ovoid to oblong, 1–1.5 cm long; seeds 1 or 2, testa membranous. **Distr:** Bhutan, China, India, Laos, Nepal, Pakistan, Sri Lanka, Thailand, Vietnam. **Ecol:** In moist forests; 500–1,600 m. Fl. Mar–Apr. Fr. Jul–Aug. **Zones:** 10–12. **Info:** This species is widely cultivated throughout the tropics for the edible leaves and for medicinal purposes. In the Maldives, the flowers are used to keep shrews away.

M. paniculata (L.) Jack, *Malayan Misc.* 1: 31, 1820. (Syn: *Chalcas exotica*, etc.). [Jasmine Orange, Satin Wood]. **Shrubs** or **trees**, 1.8–12 m tall; older branchlets grayish-white to pale yellowish-gray; bark papery and peeling with age. **Leaves** 2–5-foliolate; leaflets mostly suborbicular to ovate to elliptic, 2–9 × 1.5–6 cm, margins entire or crenulate, apex rounded to acuminate; petiolules <1 cm long. **Inflorescences** terminal, or terminal and axillary. **Flowers** 5-merous, very fragrant; sepals ovate to lanceolate, to 2 mm long, persistent in fruit; petals white, narrowly elliptic to oblanceolate, to 2 cm long; stamens 10. **Fruits** orange to vermilion, narrowly ellipsoid or rarely ovoid, 1–2 × 0.5–1.4 cm; seeds villous. **Distr:** Asia through Malesia to Australia and SW Pacific Islands. **Ecol:** In montane forests; 0–1,300 m. Fl. Apr–Oct. Fr. Apr–Feb. **Zones:** 10–12. **Info:** The extremely fragrant flowers last for only a day, petals usually fall by morning.

Above: *Murraya paniculata, Pilocarpus racemosus, Ravenia spectabilis, Ruta graveolens.*

Pilocarpus

Species 16+. **Shrubs** or **small trees. Leaves** alternate to subverticillate by crowding at tips of branchlets, simple or divided; leaflets with apices mostly emarginate. **Inflorescences** racemes; peduncles rudimentary or lacking; pedicels alternate, adaxially incurved when in bud. **Flowers** 5-merous to 4-merous; calyx toothed or lobed; petals deciduous, cohering with their hooked tips when in bud, reflexed at anthesis, ovate, inner margin thickened and (slightly) induplicate, apex acute and uncinately inflexed; filaments deciduous; disk annular to somewhat cup-shaped; carpels basally connate, free distally, carpels 1- or 2-ovuled, style inserted with 5 strands, pentangular, stigma 5-lobed with a gland in each lobe. **Fruits** divided into mericarps, 1–4 of which are usually rudimentary; seeds usually reniform, slightly keeled dorsally or sometimes double keeled. **Distr:** Mexico to Argentina, rare in the Amazon basin. **Cult:** Propagate from seeds, rooted cuttings or by air-layering. Requires a moist, well-drained soil in shade or partial sun.

P. racemosus Vahl, *Eclog.* 1: 29, t. 10, 1796. [Guadeloupe Jaborandi]. **Shrub** or **small trees**, 2–8(–20) m tall. **Leaves** alternate, often crowded and subopposite at ends of branches, pinnate, 1–2(–3)-jugate, or 1–3-foliolate, occasionally simple, 8–23(–32) cm long; leaflets usually elliptic to slightly obovate, 4–16(–30) × 1.5–9.5(–11.5) cm, base shortly attenuate, base of lateral leaflets ± unequal, apex obtuse or rounded, sometimes slightly acuminate; petiole often slightly winged, 0.5–6(–11) cm long; petiolules 0–0.6(–1.1) cm long. **Inflorescences** racemes, usually solitary, subterminal, erect, 5–50 × 3.5–4(–5) cm, usually 10–25 cm long, flowers numerous; bracts small; pedicels 0.8–2.1 cm long. **Flowers** 7.5–9(–9.2) mm Ø, red; calyx 5-lobed, ~0.6–0.7 mm high, lobes depressed-triangular or semicircular, 0.2–0.6 × 1–1.7 mm, rounded; petals subvalvate, 3.2–4(–4.5) × 1.8–2.8 mm, thickly coriaceous, keel ~1–2 mm long; filaments glabrous, anthers pale yellow; carpels 2-ovuled, style and stigma clavate. **Fruits** mericarps, subobovate, 0.9–1.2 × 0.8–0.9 cm, dehiscent halfway; seed 1 per mericarp, reniform, 6–9 × 5–7 mm, testa black-brown. **Distr:** Caribbean. **Ecol:** In lowland and montane forests, often as understory shrubs; ~300–1,500 m. Fl. Feb–Mar, Aug–Oct. **Zones:** 11–12.

Ravenia

Species ~14. **Shrubs** or **trees. Leaves** opposite, 3-foliolate or simple. **Inflorescences** axillary, few-flowered, pedunculate. **Flowers** with sepals 4 or 5, 2 outer larger, enclosing the bud; corolla tube straight, limb oblique with 4 or 5 somewhat unequal segments; disk cup-shaped, crenate, 4–5-lobed, enclosing ovary; stamens 4 or 2, filaments adherent to corolla tube, anthers 2–4 at corolla throat, staminodes 3–1, linear, exserted; ovary 4–5-carpellate, immersed in the disk, ovules 2 per cell, superposed, style 1, slender, stigma 4–5-lobed. **Fruits** capsules, surrounded by the persistent calyx, 1–5-loculed, locules 2-valved, endocarp hard and brittle; seeds 1 or 2 per locule, endosperm in a thin layer. **Distr:** West Indies and tropical S America. **Cult:** Propagate from seeds or cuttings. Requires a moist, well-drained soil in full or partial sun.

R. spectabilis Planch. ex Griseb., *Mem. Amer. Acad.* n. s. 8: 170, 1860. (Syn: *Limonia spectabilis*). [Lemonia, Limonia, Pink Ravenia]. **Shrubs. Leaves** digitately 3-foliolate, glabrous, glandular-dotted, dots ± pellucid; leaflets narrowly elliptical to oblong, 3–5 cm long; petiole 1–2 cm long. **Inflorescences** 1–3-flowered; peduncles 2–6 cm long. **Flowers** markedly protandrous; sepals 5, glandular-dotted, 2 outer ovate, obtuse, 1.1–1.3 × 0.7–1 cm, 3 inner much shorter, roundish, adpressed to corolla tube; corolla purplish-red or pinkish, glandular-dotted, fleshy, tube 1.3 cm long, lobes somewhat unequal, upper being the shorter; stamens glandular-dotted, 3 lower linear, sterile, the middle one trilobed, the 2 lateral and each branch ending in a gland. **Fruits** berries; seeds ellipsoid, muriculate, ~3.5 mm long. **Distr:** Cuba, Hispaniola, Guadeloupe. **Ecol:** In lowland tropical forests. Fl. throughout most of the year. **Zones:** 11–12.

Ruta

Species ~50. **Herbs**; all parts strongly scented. **Leaves** alternate; lamina compound or simple. **Inflorescences** terminal, cymose-corymbose or paniculate. **Flowers** with calyx 4–5-merous, with distinct, persistent sepals; corolla 4–5-merous; petals greenish or yellowish, imbricate, concave, often toothed or ciliate; stamens 8 or 10, free, unequal; disk glandular, 8–10-lobulate; ovary 4–5-lobed, 3–5-loculed, style central, ovules pendulous, axial. **Fruits** capsules, 3–5-lobed, indehiscent or splitting open at apex; seeds angular, testa pitted. **Distr:** Mainly Mediterranean, Europe, and W Asia. **Cult:** Propagate from seeds or half-hardened cuttings. Requires a moist, well-drained soil in full sun.

R. graveolens L., *Sp. Pl.* 1: 383, 1753. [Garden Rue, Herb of Grace, Rue Bitterwort]. **Herbs**, perennial, glaucous, 60–90 cm high, spreading to ~90 cm wide; stems often woody at base, but herbaceous apically. **Leaves** fetid-scented, petiolate, 6–13 cm long, decompound (2–3 times subdivided); segments cuneate or spathulate-oblong, sea-green and often covered with a powdery bloom. **Inflorescences** corymbose or paniculate clusters, divaricate, held above the foliage, often produced in profusion; pedicels (in fruit) longer than the capsules; bracts lanceolate. **Flowers** small, yellow, ~1.3 cm Ø; sepals deltoid, acute; petals 4, oblong-obovate, clawed, concave, margins pectinate, apex notched. **Fruits** capsules, obtuse with rounded lobes; seeds angulate. **Distr:** Iran and adjacent areas, extending westward. **Ecol:** In open, rocky limestone areas, and on cliffs; 0–2,300 m. Fl. Dec–Feb. **Zones:** 6–11. **Info:** The leaves are used for medicinal purposes.

Triphasia

Species 3. **Shrubs**, erect, branched, armed with spines. **Leaves** persistent, alternate, 3-foliolate, sometimes with only 1–2 leaflets. **Inflorescences** with flowers solitary or in axillary cymes (usually 3-flowered). **Flowers** perfect; calyx cupular, 3–4-lobed; petals 3–4, free, imbricate; stamens 6, inserted on a fleshy disk, filaments compressed, dilated toward the base, anthers oblong; ovary ovoid or obovoid, 3–4-loculed, locules 1-ovuled, ovary narrowed above into the style, style deciduous, stigma obtuse or capitate, 3–4-lobate. **Fruits** berries, baccate, small, ovoid, 1–3-loculed; seeds 1–3, ovoid or subglobose, testa fleshy. **Distr:** SE Asia to Malesia. **Cult:** Propagate from seeds. Requires a very moist and fertile soil in shade or full sun.

T. trifolia (Burm. f.) P. Wilson, *Torreya* 9: 33, 1909. (Syn: *Limonia trifolia*, etc.). [Limeberry]. **Shrubs**, 1–4 m high, glabrous, spiny; spines axillary, in pairs, slender, straight. **Leaves** trifoliate, aromatic when crushed; leaflets ovate to oblong-ovate, margins crenate, apex obtuse or retuse, glabrous, dark green, terminal leaflet 1–4 cm long, lateral leaflets smaller; petiole very short. **Inflorescences** with flowers very shortly pedicellate. **Flowers** white, fragrant, ~1 cm long; calyx puberulent, lobes broadly triangular, margins ciliate, apex acute; petals oblong, 1.2–1.6 cm long. **Fruits** berries, ellipsoid, ovoid or globular-ellipsoid, 1–1.5 cm long, fleshy, maturing to red, gland-dotted, aromatic. **Distr:** SE to Malesia; widely cultivated and naturalized in SE Asia and to a lesser extent pantropically. **Ecol:** In thickets, often as understory. Fl. & Fr. throughout most of the year. **Zones:** 11–12. **Info:** The fruits are edible, with a mediocre flavor. The fruits contain a clear, gumlike mucilaginous substance often used as glue.

MALVALES

FAMILIES 10, GENERA 338, SPECIES 6,005.

Plants predominantly woody; usually stellate-pubescent; tissues producing mucilage. **Leaves** pulvinate, venation palmate. **Flowers** usually bisexual and actinomorphic; perianth mostly 5-merous; calyx valvate in bud; stamens usually numerous or in more than 1 whorl; ovary multicarpellate, placentation usually axile and derived from a parietal condition by a deep intrusion of placentas. **Fruits** often capsules. **Distr:** Worldwide, mainly tropical and subtropical.

F&KS

Above: *Triphasia trifolia, Dais cotinifolia, Wikstroemia indica, Wikstroemia uva-ursi.*

THYMELAEACEAE

Genera 50, **species** ~720. **Shrubs** or **small trees**, rarely herbs, evergreen or deciduous, plants mostly bisexual, sometimes dioecious. **Leaves** opposite or alternate, rarely some ternate, estipulate; lamina simple, entire, pinnately veined, articulate at base. **Inflorescences** basically racemose, sometimes capitate, spicate, umbelliform, or fascicled, terminal or subterminal, less often axillary, sometimes on brachyblasts, sessile or pedunculate; flowers bracteate (sometimes bracts forming an involucre) or ebracteate, sessile or pedicellate. **Flowers** usually bisexual or unisexual (plants then mostly dioecious), actinomorphic; calyx tubular, campanulate, or infundibuliform, usually corollalike, 4- or 5(or 6)-merous, mostly caducous, sometimes circumscissile, or persistent, lobes imbricate; petals absent or represented by 4–12 scales, inserted at or near throat of calyx tube; stamens 2 to many; ovary superior, 1- or 2-loculed, ovules solitary, pendulous, style filiform, caducous. **Fruits** mostly indehiscent, dry or fleshy, sometimes a loculicidal capsule. **Distr:** widely distributed in both hemispheres.

Dais

Species 2. **Shrubs**, evergreen or semideciduous. **Leaves** scattered, alternate or opposite, large, flat; lamina obovate to oblong-elliptic. **Inflorescences** terminal heads, peduncled, surrounded by 4-leaved involucre. **Flowers** fragrant, small, pink; calyx colored, funnel-shaped, tube slender, continuous or at length expanded above ovary, pubescent inside; corolla tubular, limb 5-lobed, spreading, regular, throat without scales; stamens (8)10, inserted in the throat in 2 rows, filaments conspicuous, setaceous, the alternate shorter, upper row or all exserted, anthers oblong; ovary with a membranous cup beneath, style lateral, capillary, exserted, stigma capitate. **Fruits** berries, fleshy or dry, enclosed in base of calyx. **Distr:** Tropical and S Africa, Madagascar. **Cult:** Propagate from seeds or semiripe cuttings. Requires a fertile, well-drained compost in full or partial sun.

D. cotinifolia L., *Sp. Pl.*, ed. 2. 1: 556, 1762. (Syn: *D. laurifolia*). [Pincushion Tree, Pompom Tree]. **Shrubs** or small semideciduous **trees**, 3–8 m tall; stem single or sometimes multiple, slender, upright; branches tough, difficult to break; bark stripping off in tough strips. **Leaves** opposite, often crowded at ends of branches; lamina oblong-elliptic to obovate, 3–12 × 2.5–6.5 cm, base and apex acute, thinly textured, waxy, hairless, adaxially olive-green to bluish-green, abaxially paler, veins translucent; petiole 5 mm long. **Inflorescences** globose heads or umbellate clusters, dense, terminal, showy, 4–6(–10) cm across, produced in profusion; involucral bracts 4, green, conspicuous. **Flowers** pink to mauve, drying to reddish-brown, fragrant; corolla narrowly funnelform, to 3 cm long, lobes 5, spreading; anthers prominent, golden-yellow. **Fruits** nutlets, small, reddish-brown, enclosed in persistent calyx tube and dry bracts. **Distr:** S Africa. **Ecol:** Along rivers, forest edges and in riverine thickets. Fl. Nov–Mar. Fr. Jan–Apr. **Zones:** 9–11.

Wikstroemia

Species ~70. **Shrubs** or **subshrubs**, occasionally small trees or rarely herbaceous perennials, evergreen or deciduous. **Leaves** opposite or alternate. **Inflorescences** fascicled or solitary, spicate, racemose, umbelliform or capitate, sometimes in compound terminal panicles, often without involucre, usually terminal and subterminal, rarely axillary; peduncle long or short; flowers subsessile or pedicellate; pedicels articulate. **Flowers** bisexual or unisexual, 4- or 5-merous; calyx tube yellow or green, less often purplish, red or white, cylindric or tubular, sometimes slightly funnel-shaped, often caducous, lobes 4 or 5, spreading; petaloid appendages absent; stamens twice as many as calyx lobes, biseriate, filaments very short, anthers oblong; disk with 2 or 4, very rarely 1 or 5, scales; ovary usually sessile, ellipsoid, 1-loculed, style terminal, short, stigma large, globose. **Fruits** berries, succulent or dry, endosperm scanty or absent. **Distr:** Sri Lanka to S China, Malaysia, Australia, Pacific Islands; Hawai'i (where many species are dioecious). **Cult:** Propagate from seeds or cuttings. Requires a moist, well-drained soil in full or partial sun.

W. indica (L.) C.A. Meyer, *Bull. Cl. Phys.-Math. Acad. Imp. Sci. Saint-Pétersbourg*. 1: 357, 1843. (Syn: *Capura purpurata*, etc.). [Tiebush]. **Shrubs**, 0.5–2+ m tall; branches reddish-brown, glabrous. **Leaves** opposite; lamina obovate, elliptic-oblong, or lanceolate, 2–5 × 0.5–1.5 cm, base cuneate, apex obtuse or acute, papery to thinly leathery, surfaces glabrous, green or bluish-green, glaucous, reddish-brown when dried; petiole ~1 mm long. **Inflorescences** terminal, capitate, several-flowered; peduncles 0.5–1 cm long, glabrous; pedicels 1–2 mm long. **Flowers** with calyx yellowish-green, 0.7–1.2 cm long, outside glabrescent, lobes 4, broadly ovate to oblong, ~3 mm long, apex acute or obtuse; stamens 8; disk scales often 2 or 4; ovary obovoid or ellipsoid, style very short, stigma capitate. **Fruits** drupes, red to dark purple, ellipsoid, 7–8 mm long. **Distr:** China, India, Malaysia, Myanmar, Philippines, Taiwan, Thailand, Vietnam; Australia, Pacific Islands east to Fiji; Mauritius. **Ecol:** In forests, rocky shrubby slopes; <1,500 m. Fl. & Fr. Jun–Aug. **Zones:** 9–12.

W. uva-ursi A. Gray, *J. Bot.* 3: 304, 1865. (Syn: *Diplomorpha uva-ursi*, etc.). [Fish Poison Plant, Hillside False Oheio]. **Shrubs**, densely branching, prostrate or sprawling, to 1.5 m high; young branches pubescent. **Leaves** elliptic, ovate to rounded, (0.9–)1.7–3(–5) × (0.5–)1.2–2.8(–3.8) cm, base cuneate to rounded, apex rounded to acute, adaxially dark to grayish-green, abaxially lighter, sometimes glaucous; petiole 1–2.5(–4) mm long. **Inflorescences** several-flowered; peduncle 1–6 mm long; pedicels 0.8–3 mm long. **Flowers** unisexual, sometimes perfect, yellow, glabrous or puberulent; ♂ with calyx tube 0.5–0.9(–1.2) cm long; ♂ with calyx 3.5–8 mm long; perfect flowers with calyx tube 5–9 mm long. **Fruits** bright red to dark crimson, ellipsoid 0.8–1.1 × 0.5–1 cm. **Distr:** Hawai'i. **Ecol:** On dry ridges, open hillsides, etc.; <670 m. **Zones:** 10–12.

BIXACEAE

Genera 4, **species** 21. **Shrubs** or **small trees**; young branches and leaves with peltate scales. **Stipules** small, sheathing bud, fugacious. **Leaves** alternate; lamina simple, palmately veined; petiole swollen at base and apex. **Inflorescences** panicles, terminal. **Flowers** bisexual, actinomorphic; sepals 5, free, imbricate, with basal abaxial glands, fugacious; petals 5, imbricate, large and conspicuous; stamens many, free or slightly united at base, anthers oblong, with 2 inverted, U-shaped thecae, dehiscing by short, apical slits on the bend; ovary superior, 1-loculed, ovules many, on 2 parietal placentas, style slender, stigmas bilobed. **Fruits** capsules, loculicidally 2-valved, usually spiny; seeds many, testa red, slightly fleshy. **Distr:** Pantropical; mostly in tropical America.

Bixa

Species 5. *Generic description same as familial.* **Cult:** Propagate from cuttings or seeds. Requires a fertile, moist, well-drained soil in full or partial shade.

B. orellana L., *Sp. Pl.* 1: 512, 1753. [Annatto, Lipstick Plant]. **Shrubs** or **small trees**, evergreen, 2–5(–10) m tall; branches densely hairy. **Leaves** cordate-ovate or triangular-ovate, (5–)10–25 × (3.3–)5–13(–16.5) cm, base rounded or subtruncate, sometimes slightly cordate, margins entire, apex acuminate, adaxially deep green, abaxially pale green, with resinlike gland dots, palmately 5-veined, glabrous; petiole erect, 2.5–5 cm long, glabrous. **Inflorescences** panicles, robust, often flat-topped, 5–10 cm long, densely scaly and glandular-hairy; bracts caducous, leaving scalelike scars; pedicel 0.4–1.2 cm long. **Flowers** 4–5 cm Ø; sepals obovate, 0.8–1 × ~0.7 cm, densely red-brown scaly, with glands at base; petals bright pink, mauve, or white with pale red veins, obovate, (1–)1.5–3 × 0.8–2 cm; stamens many, anthers yellow, apically dehiscent. **Fruits** capsules, subglobose or ovoid, slightly laterally compressed, (1.4–)2–4.5 cm Ø, usually densely purple-brown spiny, rarely smooth; spines 1–2 cm long; seeds numerous, red-brown, obovoid-angular, 4–5 mm long. **Distr:** Tropical America. **Ecol:** <2,200 m. **Zones:** 10–12. **Info:** The red dye obtained from

Above: *Bixa orellana, Cochlospermum vitifolium, Dipterocarpus zeylanicus* (flowers), *Dipterocarpus zeylanicus* (flowering branch).

the seeds are used for food coloration and fabric dyeing.

Cochlospermum

Species 12. **Shrubs** or **trees**. **Stipules** deciduous. **Leaves** petiolate; lamina simple, palmately lobed, or compound with 5–9 leaflets. **Inflorescences** racemes or panicles, terminal. **Flowers** showy; sepals 5, outer 2 lanceolate, smaller, inner 3 broadly ovate, much larger; petals yellow, obovate, base tapering, apex ± notched; stamens numerous, equal, free or filament bases slightly connate, anthers narrow; ovary 1-loculed, style slender. **Fruits** 3–5-valved, woody; seeds kidney- or ear-shaped, long silky-hairy. **Distr:** Tropical America, Africa and Asia. **Cult:** Propagate from seeds or semiripe cuttings. Requires a moderately fertile soil in full sun.

C. vitifolium (Willd.) Spreng., *Syst.* 2: 596, 1825. (Syn: *Bombax vitifolium*, *C. hibiscoides*). [Brazilian Rose, Buttercup Tree, Wild Cotton]. **Trees**, deciduous, glabrous. **Leaves** deeply palmately 3-, 5- or 7-lobed or compound, 14–30 × 15–30 cm, base cordate, lobes oblong-lanceolate, margins repand-serrulate or serrate, apex acute or acuminate, nearly glabrous; petiole 10–20 cm long. **Inflorescences** panicles or racemes, few- to several-flowered, terminal; flowers ± pendent or suberect. **Flowers** showy, 7.5–12.5 cm Ø; sepals 5, persistent, brownish, puberulent, ~1.5 cm long, outer 2 lanceolate, smaller, inner 3 broadly ovate, much larger; petals 5 (or numerous in the double form), bright yellow or golden, finely red-lined, obovate; stamens with filaments 0.5–2 cm long, anthers yellow, curved; ovary ≤ 2.5 mm Ø. **Fruits** capsules, obovate, 5–8.2 × 4–6 cm, velvety-hairy; seeds reniform, 5–6 mm long, brownish, white-hairy. **Distr:** N Mexico to Peru, Venezuela, N Brazil, and Bolivia. **Ecol:** In seasonally dry habitats. **Zones:** 10–12.

DIPTEROCARPACEAE

Genera 17, **species** ~680. **Trees**, often large, sometimes shrubs, evergreen or semievergreen, rarely deciduous; xylem with aromatic resin. **Stipules** persistent or caducous, large or small. **Leaves** alternate, petiolate; lamina simple, margins entire or sinuate-crenate, lateral veins pinnate; in African species always with an extra-floral nectary at base of midrib adaxially. **Inflorescences** racemes or panicles, few- or many-flowered, terminal or axillary; flowers usually sweetly scented; bracts usually fugacious and minute, rarely persistent and large. **Flowers** bisexual, actinomorphic, contorted, most parts hairy; calyx lobes 5, free or basally united, imbricate in bud if not united; petals 5, adnate or basally connate; stamens (10–)15 to many, free from or connate to petals, filaments usually dilated at base, anthers 2-celled, with 2 pollen sacs per cell; ovary superior, rarely semi-inferior, slightly immersed in torus, usually 3-loculed, locules 2-ovuled, rarely many-ovuled, ovules pendulous, lateral or anatropous. **Fruits** usually nutlike, sometimes capsular and 3-valved, 1(to many)-seeded, with persistent, variously accrescent calyx of which 2 or more lobes are usually developed into lorate wings (always 5 in African spp.); seed exalbuminous. **Distr:** Tropical Africa, Asia (most species and genera in NW Borneo), S America.

Dipterocarpus

Species ~70. **Trees**, lofty, emergent; buttresses stout; bark lenticellate bark; exudate aromatic, oily, white, resinous. **Stipules** large, enclosing terminal bud, finally caducous and leaving an annular scar. **Leaves** leathery, plicate in bud and ± corrugate when opened, lateral veins pinnate, straight, tertiary veins subscalariform, conspicuous, margins entire or sinuate-crenate. **Inflorescences** racemes, 3–9-flowered, hardly branched. **Flowers** large, sweetly scented; calyx with urceolate or cup-shaped free basal tube; sepals valvate, unequal; petals white or with a reddish median stripe, pubescent, especially on parts exposed in bud; anthers yellow, linear, equivalved, connective appendages aristate or filiform; ovary narrowly ovoid, pubescent, style filiform, stigma slightly dilated. **Fruits** samaras, nutlike, enclosed in accrescent calyx tube; winglike calyx lobes 2, erect; seed adnate to base of pericarp. **Distr:** Cambodia, China, India, W Indonesia, Laos, Malaysia, Myanmar, Philippines, Sri Lanka, Thailand, Vietnam. **Cult:** Propagate from seeds, cuttings or by air-layering. Requires a fertile, moist, well-drained soil in full or partial sun.

D. zeylanicus Thw., *Enum. P. Zeyl.* 33, 1858. (Syn: *D. turbinatus*). [Hora]. **Trees**, large, evergreen, to 40 m tall; trunk 4 m Ø, buttresses low, rounded; crown dense, hemispherical. **Stipules** spathulate, 13 × 3 cm, obtuse. **Leaves** clustered around twig apices; lamina ovate to elliptic, 10–25 × 7.5–14 cm, base obtuse to subcordate, apex with acumen 6 mm long, thickly coriaceous, nerves 15–18 pairs, prominent abaxially; petiole 2–5 cm long; sapling leaves elliptic, to 45 × 16 cm. **Inflorescences** racemes, to 14 cm long, axillary, hanging, in dense pendent masses near ends of twigs. **Flowers** with stamens 15, appendages shorter than anthers; style and stylopodium columnar, pubescent in basal ²/₃. **Fruits** samaras; fruiting calyx tube to 2.5 cm Ø, subglobose, with 5 ± prominent ribs or becoming smooth on drying, 2 longer lobes to 13 × 3 cm, lorate, obtuse, tapering to ~8 mm wide at subauriculate base, 3 shorter lobes oblong, to 3 × 2 cm, obtuse. **Distr:** Sri Lanka. **Ecol:** In mixed dipterocarp forests; <1,000 m. Fl. Feb. **Zones:** 11–12.

Hopea

Species ~100. **Trees**, evergreen, of main canopy or understory, white resinous. **Stipules** caducous, small. **Leaves** with margins entire, thinly leathery, lateral veins pinnate, tertiary veins scalariform. **Inflorescences** lax panicles of unilateral racemes; flowers sessile or shortly pedicellate; bracts early caducous, linear. **Flowers** with calyx segments imbricate; petals with exposed parts usually pubescent in bud; stamens (10–)15, slightly connate and adnate to petals, anthers ovoid, small, subequal, connectival appendages aristate or filiform; ovary ovoid, with prominent filiform style, or hourglass-shaped, with short tapering style. **Fruits** ovoid or globose, pericarp usually thin, waxy, 1-seeded, closely surrounded by thickened base of calyx segments; 2 fruiting sepals winglike, linear, or all orbicular or ovate, subequal. **Distr:** Cambodia, China, S India, Indonesia, Laos, Malaysia, Myanmar, Papua New Guinea, Philippines, Sri Lanka, Thailand, Vietnam. **Cult:** Propagate from seeds, cuttings or by air-layering. Requires a fertile, moist, well-drained soil in full or partial sun.

H. jucunda Thw., *Enum. Pl. Zeyl.* 403, 1864. [Rat/Red Beraliya]. **Trees**, to 25 m tall; trunk to 2 m Ø, buttresses low; crown dense. **Leaves** ovate, 3.5–11 × 1.5–5.5 cm, base broadly cuneate, subequal, apex acuminate with acumen slender, caudate, to 2 cm long, nerves 3–4 pairs, arched, ascending, slender but ± prominent abaxially, with prominent pubescent axillary domatia; petiole 0.5–1.2 cm long, relatively short and slender; sapling leaves similar to adult leaves. **Inflorescences** panicles, to 11 cm long, branches to 1.5 cm long, to 4-flowered. **Flowers** with sepals broadly ovate, subacute; corolla at anthesis to 1 cm Ø, petals lanceolate; stamens 15, connectival appendage long, slender; ovary ovoid, with ± prominent columnar style. **Fruits** nuts, ovoid, to 1.5 × 1 cm, apiculate, often resin-coated; fruiting sepals with 2 *longer* (to 8 × 2.5 cm), obtuse, tapering to saccate base, thickened, 3 *shorter* lobes broadly ovate, to 6 × 6 mm, mucronate. **Distr:** Sri Lanka. **Ecol:** On wet lowland hills; <1,000 m. Fl. April. **Zones:** 11–12.

Shorea

Species 360. **Trees**, usually large; trunk buttressed. **Stipules** caducous, large or small. **Leaves** with margins entire, ± leathery, tertiary veins parallel. **Inflorescences** cymose panicles, axillary or terminal; bracts persistent, caducous, or absent. **Flowers** with sepals with 3 outer larger than 2 inner; petals white, yellow, or pink, usually pubescent; stamens (12–)15 or 20–100, anthers ovoid, oblong, or panduriform, connective subulate-cuspidate or stout, club-shaped, valves equal or outer one slightly larger; ovary ovoid, pubescent, style subulate, stigma entire or 3-toothed. **Fruits** usually 1-seeded, closely surrounded by thickened bases of accrescent calyx segments; sepals developed into lorate wings, outer 3 much larger than 2 inner. **Distr:** Bhutan, Cambodia, China, NE India, Indonesia, Laos, Malaysia, Myanmar, Nepal, Philippines, Thailand,

KT

Above: *Hopea jucunda, Shorea lissophylla, Vateria copallifera, Vatica diospyroides.*

Vietnam. **Cult:** Propagate from seeds, cuttings or by air-layering. Requires a fertile, moist, well-drained soil in full or partial sun.

S. lissophylla Thw., *Enum. Pl. Zeyl.* 402, 1864. (Syn: *Isoptera lissophylla*). [Malmora]. **Trees** small, to 20 m tall; trunk to 1 m Ø, often branching from base, buttresses low, concave. **Stipules** oblong, to 8 × 5 mm, acute, fugacious. **Leaves** broadly ovate to suborbicular, 6–17 × 4–10 cm, base obtuse to cordate, margins ± revolute, apex obtuse, acuminate or mucronate, mucro to 1.2 cm long, thickly coriaceous, lustrous, dark green, nerves 5–12 pairs, abaxially midrib prominent; petiole 0.6–1.2 cm long, stout. **Inflorescences** panicles, to 10 cm long, terminal or axillary to ramiflorous, 1-branched, branchlets to 3 cm long, bearing up to 5 secund flowers. **Flowers** with sepals ovate, 3 outer acute, 2 inner subacuminate; corolla to 1 cm Ø, hardly contorted, petals linear-lanceolate, narrowly obtuse, cream; stamens ~47–52; ovary broadly ovoid, style short, glabrous. **Fruits** nuts, ovoid, to 1 × 0.9 cm, with stylopodium to 4 mm long; fruiting sepals ovate, to 1 × 0.7 cm, subequal, acuminate, loosely clasping. **Distr:** Sri Lanka. **Ecol:** In humid forests, often hanging over cliffs. Fl. Apr–Jun. **Zones:** 11–12.

Vateria

Species 3. **Trees**, large; trunk usually somewhat crooked; branchlets stout, rugulose. **Stipules** hastate, subpersistent, leaving conspicuous pale horizontal scars. **Leaves** ± oblong, large, coriaceous, venation abaxially prominent, intercostals scalariform; petiole long, stout, geniculate. **Inflorescences** paniculate, lax, subterminal, axillary, ascending, twice-branched, branches long, bearing many secund pedicellate flowers and conspicuous ovate acute bracts. **Flowers** with sepals slightly imbricate; petals oblong; stamens many, subequal, filaments very short, long linear 2-celled anthers the abaxial sacs of which are prolonged apically into recurved awns, and short connectival appendages; ovary ovoid, short, tomentose, crowned by a slender columnar glabrous style many times its length. **Fruits** ovoid, large, acute, with short equal ± reflexed sepals. **Distr:** S India, Sri Lanka. **Cult:** Propagate from seeds, cuttings or by air-layering. Requires a fertile, moist, well-drained soil in full or partial sun.

V. copallifera (Retz.) Alston, *Handb. Fl. Ceylon* 6: 26, 1931. (Syn: *V. acuminata*, etc.). **Trees**, large, to 40 m tall; trunk to 4 m Ø in the wild; branches dense, ascending; twigs and panicles densely tomentose, twigs ~5 mm Ø, stout, much-branched. **Stipules** narrowly hastate, to 2.5 × 0.6 cm. **Leaves** oblong, 11–50 × 5–18 cm, base obtuse to cordate, apex acuminate, acumen to 8 mm long, thickly coriaceous, nerves 18–25 pairs, prominent abaxially, spreading; petiole 1.8–5.5 cm long, usually long, stout, geniculate. **Inflorescences** panicles, to 25 cm long, branchlets to 12 cm long, bearing to 8 ± secund flowers; bracts ovate, concave, to 1 × 0.6 cm, acute. **Flowers** with sepals hastate, subequal, imbricate at base only, subacute; petals cream, oblong; stamens 45–55, anthers yellow. **Fruits** nuts, very large, to 11 × 7 cm, ovoid, apiculate, pericarp to 2.5 cm thick, fibrous, spongy; fruiting calyx to 1.5 × 0.7 cm, lanceolate, subacute, reflexed, subequal. **Distr:** Sri Lanka. **Ecol:** In lowland mixed dipterocarp forests; <1,000 m. Fl. Apr. **Zones:** 10–12.

Vatica

Species ~65. **Trees**, of mainly canopy and understory; trunk unbuttressed; exudate white, resinous; bark gray, smooth, often with annular striations. **Stipules** caducous. **Leaves** small, margins entire, ± leathery, pinnately veined, tertiary reticulate veins conspicuous. **Inflorescences** cymose panicles, axillary or terminal, usually pubescent. **Flowers** with calyx tube short; sepals narrowly imbricate, ± equal; petals white, often with a mauve suffusion, lorate, large, falling separately; stamens 15, filaments unequal, stout, anthers ellipsoid, connective appendages short, obtuse; ovary ovoid, pubescent, style cylindric, short, stigma capitate or conic, prominent, entire or dentate. **Fruits** nuts, globose or ellipsoid, 1- or 2-seeded; calyx segments equal or unequal, with 2 developed into long wings. **Distr:** Cambodia, China, E and S India, Indonesia, Laos, Malaysia, Myanmar, Philippines, Sri Lanka, Thailand, Vietnam. **Cult:** Propagate from seeds, cuttings or by air-layering. Requires a fertile, moist, well-drained soil in full or partial sun.

V. diospyroides Symington, *Gard. Bull. Straits Settlements*, 9: 347, 1938. (Syn: *Synaptea cinerea, V. cinerea*). **Trees**, 6–30 m tall, slow-growing; trunk relatively straight, >1 m Ø; bark smooth. **Leaves** ovate to obovate, 14–20 × 4–7 cm, base cuneate, margins entire, apex acute, coriaceous, dark green, midrib and lateral veins prominent abaxially, lateral veins prominent, surface with brownish, glandular trichomes; petioles ~1.5 cm long, pubescent, stout. **Inflorescences** cymose panicles, 2.5–3 cm across, axillary; pedicels short or flowers sessile, pubescent. **Flowers** pleasantly fragrant, white, creamish, ivory or rarely pinkish; calyx brownish-tomentose outside, lobes leathery, united into a cup and persistent in fruit, not adherent to the nut; corolla with limb 1.2–1.5 cm Ø, lobes free, rotate, recurved, slightly overlapping. **Fruits** nuts, globose, indehiscent. **Distr:** Peninsular Thailand; rare in the wild. **Ecol:** In humid rainforests; 0–100 m. Fl. Nov–Jul. Fr. Feb–Aug. **Zones:** 11–12.

V. rassak (Korth.) Blume, *Mus. Lugd.-Bat.* 2: 31, 1852. (Syn: *Vateria rassak*, etc.). **Trees**, canopy, large, to 30 m tall; trunk to 70 cm Ø; twigs stout, crooked, ribbed, becoming rugose, flaky, stipule scars prominent. **Stipules** lorate, 1.4 × 0.4 cm, subacute, subpersistent. Leaves oblong to narrowly elliptic, 13–32 × 5–11 cm, base broadly cuneate to subcordate, apex with tapering acumen to 1.5 cm long, thickly coriaceous, midrib prominent abaxially, flat adaxially, lateral veins 16–20 pairs; petiole 2–2.5 cm long, not drying to black. **Inflorescences** terminal or axillary; rachis ribbed, irregularly branched mostly from the base, appearing fascicled, to 14 cm long; pedicels to 3 mm long. **Flowers** with buds spindle-shaped, to 1.4 × 0.3 cm; calyx densely shortly pale buff pubescent, lobes free, subequal, deltoid, to 1.2 × 0.7 cm, acute, incrassate, reflexed, recurved. **Fruits** oblong to ovoid, to 5 × 3.5 cm, very variable, obtuse to tapering-apiculate, often asymmetric, verrucose with thick corky pericarp, prominently 3-furrowed. **Distr:** Borneo, Sulawesi, Maluku, New Guinea, Sudest Islands. **Ecol:** On riverbanks and in lowland mixed forests. **Zones:** 11–12.

MUNTINGIACEAE

Genera 3, **species** 3. **Trees** (small to medium-sized) or **shrubs**, pubescent, especially on young growth; stipulelike appendages dimorphic at least on plagiotropic branches, filiform or foliaceous and peltate. **Leaves** alternate, petiolate, distichous on plagiotropic branches; lamina with base asymmetrically cordate, margins serrate, palminerved. **Inflorescences** supra-axillary, flowers solitary or in few-flowered clusters, pedicellate. **Flowers** relatively large, actinomorphic, usually hermaphrodite; sepals (4)5(–7), basally ± fused, forming a saucer- to cuplike tube, caducous or persistent; petals (4)5(–7), crumpled in bud, imbricate, longer than calyx, thin, caducous; stamens numerous, filaments filiform, anthers basifixed, dehiscing by longitudinal slits; ovary superior to inferior, syncarpous, usually 5- to multiloculed, ovules numerous, style thick, ± short or almost lacking, stigma thick. **Fruits** berries, indehiscent, baccate, sometimes juicy and sweet; seeds numerous, small. **Distr:** Tropical America.

Muntingia

Species 1. *Generic description same as specific.* **Cult:** Propagate from seeds; fruiting starts usually within the first 2 years. Difficult to propagate from cuttings. Requires any soil in full sun and is salt-tolerant.

M. calabura L., *Sp. Pl.* 509, 1753. [Calabura, Capulin, Jamaican Cherry, Jam Tree, Japanese Cherry, Strawberry Tree]. **Trees small** or **shrubs**, terrestrial or epiphytic (then usually dwarfed), 5–8 m tall; bark with very strong fibers; branchlets densely hairy. **Stipules** 2, hairy, deciduous. **Leaves** herbaceous, distichous; lamina simple, oblong-ovate, 4.5–9 × 1.5–3.5 cm, base with one side cordate, the other acute, margins acutely dentate, apex acuminate, oblique, densely hairy, 3–5-nerved, adaxially green, abaxially grayish-white; petiole 2–6 mm long, hairy, base nodose. **Inflorescences** with flowers solitary or in pairs, supra-axillary; peduncles absent; pedicels hairy, 1.5–1.6 cm long; buds ovoid, 9 × 5 mm. **Flowers** white, 2 cm Ø; calyx valvate, free; sepals 5, green, lanceolate,

Above: *Vatica rassak*, *Muntingia calabura*, *Corchorus aestuans*, *Entelea arborescens*.

1–1.2 × 0.3 cm, base truncate, apex tailed, hairy outside, glabrous inside, reflexed; petals 5, broadly obovate, glabrous, crumpled, 1–1.1 × 0.9 cm; stamens numerous, filaments glabrous, 5–6.5 mm long; ovary 2.5 mm thick, 5(–6)-ridged and grooved, stigma 5–6-sulcate or lobed. **Fruits** baccate, fleshy, globose, 0.75–1.5 cm Ø, 5–6-celled, smooth, red, sweet and edible; seeds numerous, seeds minute. **Distr:** Tropical regions of the New World; widely cultivated and naturalized in tropical Asia. **Ecol:** On low dry hills. Fl & Fr. throughout the year. **Zones:** 10–12.

MALVACEAE

Syn: Bombacaceae, Byttneriaceae, Sterculiaceae, Tiliaceae, Triplochitonaceae.
Subfamilies 9, **Genera** ~204, **species** ~2,330. **Trees**, **shrubs**, **lianas** or **herbs**; hairs various but mostly stellate or peltate scales; mucilage canals and often mucilage cavities present. **Leaves** alternate, spiraled or 2-ranked; lamina simple, often palmately lobed or palmately compound, venation palmate or sometimes pinnate, margins entire to serrate, teeth malvoid (with the major vein unexpanded and ending at tooth apex). **Stipules** present. **Inflorescences** indeterminate, mixed or determinate, sometimes reduced to a single flower, axillary, with determinate basic repeating unit and bearing 3 bracts, one of which is always sterile, while the others subtend lateral cymes or single flowers. **Flowers** bisexual or functionally unisexual, usually actinomorphic or radial, often associated with conspicuous bracts that form an epicalyx; perianth consists of usually 5, valvate, distinct or basally connate sepals; petals usually 5 distinct, imbricate, convolute or valvate (sometimes lacking), often basally adnate to androecium; androecium consisting of 5 to numerous stamens, these distinct, fascicled, or monadelphous with apically divergent filaments, anthers 2-loculed or unilocular (due to developmental modifications, anther appearing as a half anther); staminodes sometimes present, these sometimes elongate and alternating with stamens or groups of stamens; gynoecium is a single compound pistil of 2–many (sometimes loosely united) carpels with styles and stigmas equaling carpels in number and distinct, or variously united; ovary superior with 2–many locules, each bearing 1–numerous, usually axile ovules. **Fruits** loculicidal capsules, schizocarps, nuts, indehiscent pods, aggregates of follicles, drupes, berries, or samaras. **Distr:** Worldwide. **Info:** Malvaceae is monophyletic and was traditionally classified into 4 large families Bombacaceae, Malvaceae, Sterculiaceae and Tiliaceae (these 4 families not being monophyletic). Recent phylogenetic analysis in the order Malvales has led to the suggestion that the core Malvales (Sterculiaceae, Tiliaceae, Bombacaceae, and Malvaceae) be considered an enlarged family, Malvaceae *s.l.*, and that 9 subfamilies (Bombacoideae, Brownlowioideae, Byttnerioideae, Dombeyoideae, Grewioideae, Helicterioideae, Malvoideae, Sterculioideae and Tilioideae) be recognized (Baum et al. 1998, 2004; Bayer *et al.* 1999). The larger Malvoideae is further divided into 4 tribes (Gossypieae, Hibisceae, Kydieae, Malveae), with several molecular studies conducted on these (Tate et al. 2005).

Subfam. GREWIOIDEAE

Syn: Grewiaceae, Sparmanniaceae.
Genera 25, **species** 770. **Shrubs** or **trees**, rarely herbs. **Stipules** present. **Leaves** alternate; lamina simple, deciduous, base asymmetric, pubescent. **Inflorescences** cymose, usually in small clusters, axillary. **Flowers** bisexual, rarely unisexual, actinomorphic, hypogynous; calyx with 3–5 sepals, ± free, valvate, nectary absent; corolla with 3–5 petals, free, imbricate or valvate, sometimes with glandular (nectariferous) hairs at their bases, rarely absent; androgynophore present (rarely absent), with many stamens, sometimes only 5, filaments usually free or united into groups of 5 or 10, adnate to base of petals, anthers bithecous, dehiscence longitudinal, or by apical pores; staminodes absent; gynoecium with 2–many united carpels, multilocular (locules as many as carpels) with many ovules, placentation axile, ovary superior, style single, stigma lobed or capitate. **Fruits** capsules or fleshy; seeds 1–many. **Distr:** Pantropical, few in warm temperate regions.

Corchorus

Species 40–100. **Herbs** or **subshrubs**, frequently hirsute. **Stipules** filiform. **Leaves** petiolate; lamina with margins serrate, serrulate, or crenate, papery, basal veins usually 3, usually with linear appendages at or near base. **Inflorescences** with flowers solitary or several arranged in cymes, axillary or extra-axillary. **Flowers** bisexual, yellow; sepals 4 or 5; petals 4 or 5; glands absent; stamens 15 to many, on androgynophore, free; staminodes absent; ovary 2–5-loculed, locules many-ovuled, style short, stigma peltate or disk-shaped, entire or lobed. **Fruits** capsules, elongate-linear, cylindrical or globose, sometimes angled, 2–5-valved, loculicidally dehiscent, sometimes septate between the numerous pendulous or horizontal seeds. **Distr:** Mostly in tropical regions worldwide. **Cult:** Propagate from seeds or cuttings. Requires a moist, well-drained soil in full sun; some species salt-tolerant **Info:** Some Asian species (*C. capsularis* and *C. clitoritis*) are used for the production of fiber (jute).

C. aestuans L., *Syst. Nat.*, ed. 10. 2: 1079, 1759. [Children's Clove, Jute]. **Herbs**, annual, to 1 m high; stem red-brown. **Leaves** ovate or broadly ovate, 4.5–6.5 × 3–4 cm, base rounded, margins serrate, basal pair of teeth usually elongating into filiform or caudate appendages, apex shortly acuminate or acute, surfaces sparsely pilose, basal veins 5–7, lateral veins 3–6 pairs; petiole 0.9–1.6 cm long, hairy. **Inflorescences** with flowers solitary or several together in cymes, axillary or leaf-opposed; peduncles and pedicels very short or absent. **Flowers** with sepals 5, purple-red abaxially, narrowly oblong, ~5 mm long, semifoveolate, boat-shaped distally, apex awned; petals 5, yellow, nearly as long as sepals, obovate; stamens many, yellow, ~3 mm long; ovary 3–5-loculed, long-cylindrical, puberulent, style rodlike, stigma beaklike, 5-toothed. **Fruits** capsules, cylindrical, angled, 3–5-valved, 1.2–2.5 × ~0.5 cm, awns bifurcate; seeds separated by transverse septum. **Distr:** Tropical Asia, Indo-China, Malesia; tropical Africa; Australia; C America, West Indies. **Ecol:** In waste areas, plains and hills; 0–2,000 m. Fl. Aug–Oct. **Zones:** 10–12.

Entelea

Species 1. *Generic description same as specific.* **Cult:** Propagate from seeds or cuttings. Requires a deep, fertile, moist, well-drained soil in full or partial sun.

E. arborescens R. Br., *Bot. Mag.* t. 2480, 1824. (Syn: *Apeiba australis*). [Whau]. **Shrubs** or **small trees**, 3–6 m tall; trunk to 25 cm Ø; branchlets, leaves, petiole and inflorescences densely whitish-hairy; bark gray; leaf scars ± oval. **Stipules** linear-acuminate, ± persistent. **Leaves** alternate; lamina oblique, broadly ovate, (5–)10–15(–25) × (5–)10–15(–20) cm, base cordate, margins doubly crenate-serrate, sometimes obscurely subpalmately 5–7-lobed, apex abruptly acuminate; petiole to 20 cm long. **Inflorescences** subumbellate cymes, near branch tips. **Flowers** (3–)4–5-merous; sepals free, acuminate; petals white, crumpled, (3–)4–5; stamens numerous, free or united only at very base, anthers versatile; ovary hispid, 4–7-loculed, ovules numerous, style simple, stigma fringed or toothed. **Fruits** capsules, loculicidally dehiscent, subglobose, ~2 cm Ø, covered with rigid bristles ~1.5–2.5 cm long; seeds numerous. **Distr:** New Zealand. **Ecol:** In coastal to lowland forests. **Zones:** 9–11. **Info:** The wood is one of the lightest known, less dense than cork (Hogan 2003).

Grewia

Species 290. **Trees** or **shrubs**. **Stipules** minute, caducous. **Leaves** alternate; lamina with margins serrate or rarely lobed, basal veins 3–5; petiole short. **Inflorescences** usually arranged in 3- to many-flowered cymes, solitary or fascicled, axillary or opposite leaves, rarely terminal; peduncles and pedicels usually hairy; bracts caducous. **Flowers** bisexual, polygamous, or unisexual (plants monoecious); sepals 5, yellow, white, or lilac, rarely brown, free, valvate, abaxially hairy, adaxially usually glabrous, rarely hairy; petals 5, mostly yellow or white, shorter than sepals; glands usually scalelike, borne on base of petals, margins usually villous; androgynophore short, usually glabrous; stamens many, free, irregular in length, anthers globose, dorsifixed, longitudinally dehiscent; ovary

Top row: *Grewia asiatica, Grewia occidentalis, Grewia orientalis.*
Bottom row: *Luehea divaricata, Luehea speciosa, Triumfetta rhomboidea.*

2–4-loculed, locules 2–8-ovuled, style simple, stigma swollen, peltate, entire or 2–4-lobed. **Fruits** drupes, 1–4-lobed, usually with 2 or 4 drupelets, mesocarp somewhat fleshy or fibrous, endocarp hard and woody, endosperm copious. **Distr:** Old World tropics. **Cult:** Propagate from seeds or by air-layering. Requires a fertile, moist, well-drained soil in full sun.

G. asiatica Linn., *Mant.* 1: 122, 1767. (Syn: *G. hainesiana*, etc.). [False Berry]. **Shrubs** or trees, to 8 m tall; young shoots tomentose. **Stipules** narrowly oblique-lanceolate or falcate, 1–1.3 cm long, hairy. **Leaves** broadly ovate to almost orbicular, 5–15 × 5–15 cm, 5(–6)-costate, base obliquely shallowly cordate, margins serrate, apex acute to obtuse, adaxially scabrous, abaxially grayish-tomentose; petiole 1–1.5 cm long, hairy, clavate. **Inflorescences** cymes, mostly 3-flowered (dichasial), 2–6(–10) in axillary clusters, peduncles 2–3.5(–5) cm long, densely hairy; pedicels 1–1.3(–1.5) cm long, densely hairy; bracts linear-lanceolate, 3–4 mm long. **Flowers** orange-yellow, ~2 cm Ø; sepals oblong, ~1.2 × 0.25–0.3(–0.5) cm, acute, hairy outside; petals oblong, claw ~2 mm long, limb 4–5 × ~3 mm, apex irregularly lobed, orange-yellow; stamens numerous, filaments 4–6 mm long, orange-yellow, turning purplish; ovary globose, strigose, style ~5 mm long, stigma indistinctly 4-lobed. **Fruits** drupes, dark purple, ± globose, ~0.5–1.2 cm Ø, entire or obscurely bilobed, hairy, mesocarp fibrous, acidic. **Distr:** S India. **Ecol:** Fl. Mar–Sept. **Zones:** 10–12. **Info:** Cultivated in India for the sweet and sour acidic fruits.

G. occidentalis L., *Sp. Pl.* 2: 964, 1753. [Kaffir Hemp, Lavender Star Flower]. **Shrubs** scrambling, or **small trees**, to 4 m tall; multistemmed; branchlets purple to green. **Leaves** alternate; lamina simple, elliptic to slightly ovate, 2–14 × 1–10 cm, base tapering to rounded, symmetric, margins finely toothed, serrated to scalloped, apex usually tapering, sometimes rounded, 3-veined from base, thin, adaxially glossy green, abaxially paler; petiole 0.3–2.5 cm long. **Inflorescences** opposite, flowers solitary or 1–3(–6)-flowered small clusters, opening one at a time. **Flowers** usually purple, mauve or pink, rarely white, ~3 cm Ø; sepals narrowly lanceolate, similarly colored, to 2.5 cm long; corolla with 5 petals, oblanceolate, 1–1.5 cm long; stamens numerous, filaments red-violet, anthers golden-yellow. **Fruits** drupes, usually 4-lobed, to 2.5 cm Ø, reddish-brown to purple. **Distr:** S Africa to Senegal, Ethiopia; Middle East. **Ecol:** In forest margins, grasslands. Fl. Oct–Jan. Fr. Dec–May. **Zones:** 9–11. **Info:** This species is often confused with *G. caffra*, which has yellow flowers.

G. orientalis L., *Sp. Pl.* 964, 1753. (Syn: *G. columnaris*). **Shrubs**, to 3–4 m high. **Stipules** subulate, fugacious. **Leaves** ovate to elliptic, to 8 × 4 cm, base subcordate to obtuse, margins bluntly and irregularly crenulate, apex acute to subobtuse, sometimes acuminate, membranous to chartaceous, adaxially glabrescent, 3-nerved from base; petiole to 8 mm long. **Inflorescences** umbels, rarely 3-flowered, axillary, supra-axillary or opposite; peduncles 0.6–1.5 cm long; pedicels 1–1.5 cm long; bracts ovate to narrowly so, to 5 mm long. **Flowers** to 1.8 cm long; buds ovate-oblong, tawny when dry; sepals recurved, narrowly ovate, to 7 × 3.5 mm, acute, membranous, glabrous, densely tomentellous; stamens indefinite, filaments free, unequal, 0.7–1.2 cm long, glabrous; gynophore to 2.5–3 mm long, 5-angulate-lobulate, upper half minutely tomentellous; ovary ovoid, ~2.5 mm long, densely villous, style 1–1.2 cm long, glabrous, stigma dilated, lobulate. **Fruits** depressed-globose, ± 4-lobulate, to 1.3 cm Ø. **Distr:** India, Sri Lanka; East tropical Africa to Java. **Ecol:** In forests, scrubs, or grasslands. **Zones:** 10–12.

Luehea

Species ~25. **Trees** or **tall shrubs**. **Leaves** with margins usually dentate, abaxially stellate-tomentose. **Inflorescences** axillary cymes or terminal panicles, or flowers solitary, multibracteolate; bractlets somewhat coalescent but never permanently united, caducous. **Flowers** showy, white or rose-colored, epicalyx bracts 5–9, free; petals 5, glandular-thickened at base; androgynophore absent; stamens many, obscurely or very shortly coalescent (in one species completely connate) into 5 bundles, integrating into outer antherless filaments, sometimes fused and petaloid, basally lanate or pubescent, anthers fixed near base of connective, in bud introrse and straight, at anthesis turned outward and becoming hippocrepiform; ovary 5-locular, ovules numerous, ascending, style simple, stigma capitate, discoid or lobed. **Fruits** capsules, woody, imperfectly loculicidally 5-valvate; seeds numerous, winged. **Distr:** Mexico to Argentina. **Ecol:** In deciduous forests or secondary vegetation. **Cult:** Propagate from seeds or rooted cuttings. Requires a fertile, moist, well-drained soil in full or partial sun.

L. divaricata Mart., *Nov. Gen. Sp. Pl.* 1: 101, 1824. (Syn: *Thespesia brasiliensis*, etc.). **Trees**, 6–20(–30) m tall; crown rounded, dense; trunk short, crooked, bark smooth, gray, fissured with age, buttressed; branches ashy-tomentose; vigorous regrowth by sprouting from the trunk, new branches woody. **Leaves** simple, alternate; lamina oblong, rarely elliptic or oblong-lanceolate, 7–10 × 3–5 cm, margins serrated, ribs brown, abaxially glabrescent, white-pubescent, with prominent trichomes. **Inflorescences** panicles, small, few-flowered, terminal. **Flowers** white, pale pink to rose-colored, 3–4 cm long, ~2.5 cm Ø; petals reflexed; staminal column ~1.5 cm long. **Fruits** capsules, dehiscent, ellipsoid, ~3 cm long, opening at one end into 5 parts when mature. **Distr:** Argentina, Brazil, Paraguay, Uruguay. **Ecol:** Along riverbanks in forests. Jan–Jul. Fr. Jun–Aug. **Zones:** 10–12. **Info:** Drought tolerant and has many medicinal uses.

L. speciosa Willd., *Ges. Naturf. Freund. Berlin Neue Schrift.* 3: 410, 1801. (Syn: *L. alternifolia*, *L. platypetala*). [Broadleaf Bay Cedar, Mountain Mono]. **Trees** or **large shrubs**, to 25 m tall; trunk 45–60 cm Ø, often buttressed. **Leaves** shortly petiolate, deciduous; lamina elliptic to oval or elliptic-ovate, 8–24 × 6–12 cm, base rounded or cordate, margins serrate, apex abruptly acuminate, often thick and firm, adaxially green, sparsely scabrous, abaxially densely tomentose. **Inflorescences** branched clusters at the end of the twigs; bractlets commonly ~1.5 cm long. **Flowers** white, cream or pale yellow, radially symmetric, 5-merous, showy, odor unpleasant; calyx densely tomentose, generally 1.5 cm long or somewhat larger; petals white, 2.5–4 × 1.5 cm long, apex irregular. **Fruits** capsules, narrowly elliptic, 3–4 × 1.5 cm, woody, rusty-hairy, opens into 5 sections to release the seeds; seeds small, winged, 1 × 0.4 cm. **Distr:** Guatemala, Mexico, British Honduras to Salvador and Panama, Cuba, S America. **Ecol:** In forests or thickets; 100–1,100(–1,400) m. Fl. Nov–Jan. **Zones:** 10–12.

Triumfetta

Species 150. **Herbs**, annual or perennial, shrublets or shrubs, erect or procumbent; rootstock woody. **Stipules** lateral. **Leaves** alternate, petiolate; lamina simple or digitate, often lobed, margins serrate or crenate, often several-nerved from base. **Inflorescences** cymes, terminal, borne from nodes. **Flowers** actinomorphic; sepals 5, usually linear and with a short horn just behind the apex, usually hairy outside; petals 5, yellow or orange, linear to obovate, narrowed to base, often hairy near base; stamens 4–40, raised on a short glabrous androgynophore; androgynophore with apex produced into a ciliate or pubescent or villous disk; ovary often tubercled, each tubercle with 1 or more minute bristles, 2–5-locular with 2 pendulous collateral ovules per locule, style terete, about as long as the stamens, stigma entire or very shortly 2–5-lobed. **Fruits** capsules, usually globose, sometimes ovoid, echinate or setose, dividing into 3–5 valves, 1–2 seeds per locule or indehiscent; seeds obovoid or subreniform. **Distr:** Tropical and subtropical regions. **Cult:** Propagate from seeds or by air-layering. Requires a fertile, moist, well-drained soil in full sun.

T. rhomboidea Jacquin, *Enum. Syst. Pl.* 22, 1760. (Syn: *Bartramia indica*, etc.). [Burr Bush, Diamond Burbank]. **Subshrubs** or **herbs**, usually annual, sometimes perennial, highly polymorphic, to 2 m tall; branchlets almost glabrous to tomentose. **Leaves** *lower* broadly ovate-orbicular, rhomboid, or broadly ovate, trilobed, 2.5–15 × 2–10 cm, base broadly cuneate or rounded, margins irregularly bluntly serrate, apex acute, adaxially sparsely hairy, abaxially stellate-pilose, basal veins 3–5, 2 lateral ones reaching the tip of lobes; *upper* leaves oblong-lanceolate, not lobed; petiole 1–5 cm long. **Inflorescences** cymes, 3–5 per axil, crowded at the nodes; peduncle to 2 mm long; flower buds cylindrical, ~4 mm long, apex slightly wider; pedicels <1 mm long. **Flowers** with sepals narrowly oblong, 4.5–5.5 mm long, villous, appendaged at apices; petals yellow, slightly shorter than sepals, hairy along the margins; stamens 10–15; ovary spiny, 2–3-loculed, closely setulose or echinulate. **Fruits** capsules, globose or ovoid-globose, 3–5 mm Ø, spiny, indehiscent; spines ~2 mm long, gray-yellow puberulent, tip hooked; seeds 2–6. **Distr:** Throughout the tropics. **Ecol:** In forest margins, open areas, wastelands; 100–1,500 m. Fl. Aug–Nov. **Zones:** 10–12.

Subfam. BYTTNERIOIDEAE

Genera 26, **species** 650. **Trees**, usually small, or **shrubs**, rarely herbs or lianas. **Stipules** sometimes large and leaflike, rarely reduced. **Leaves** spiraled or distichous, on plagiotropic branches; lamina usually simple, sometimes palmate; petiolar bundle ± incurved-arcuate. **Inflorescences** many- to few-flowered, rarely flowers solitary, often terminal and forming sympodia with leaf-opposed flowers. **Flowers** with calyx segments connate, rarely petaloid, sometimes persistent; petals broad at base, often with apical appendage, rarely flat or reduced to a concave to scalelike structure; androgynophore rarely present, androecium epipetalous, stamens antepetalous, solitary or in groups of 2–3(–6), filaments fused within groups or rarely almost distinct, anthers dithecal or exceptionally trithecal; staminodes antesepalous, petaloid to inconspicuous, usually fused with stamens to form a tube, or sometimes absent; ovary syncarpous, usually 5-locular, more rarely 3- or rarely 2- or 1-locular, style usually simple, apically ± branched, stylar branches usually slender. **Fruits** capsules or indehiscent, sometimes separating into mericarps, locules 1- to many-seeded; seeds sometimes arillate or strophiolate, rarely pubescent. **Distr:** Pantropical, especially S America.

Abroma

Species ~10. **Trees** or **shrubs**, evergreen, fast-growing. **Leaves** cordate or ovate-elliptic, usually large, margins entire or serrate, sometimes palmately lobed. **Inflorescences** leaf-opposed or terminal, several-flowered. **Flowers** bisexual; sepals 5, divided to near base; petals 5, red-purple, abruptly narrowed below middle, lower part deeply concave, glandular, basally connate with staminodes, upper part narrower, spathulate; stamens 15, filaments connate into a tube enclosing pistil, anthers in 5 groups of 3 on outer part of filament tube alternating with staminodes; staminodes 5, margin ciliate, apex obtuse; ovary sessile, 5-grooved, 5-loculed, locules many-ovuled, style 5-lobed. **Fruits** capsules, membranous, 5-angular or longitudinally 5-winged, apex truncate, loculicidally dehiscent; seeds many. **Distr:** Tropical Asia to Australia. **Cult:** Propagate from seeds. Requires any well-drained soil in full sun.

A. augustum (L.) L. f., *Suppl. Pl.* 341, 1782. (Syn: *A. augusta*, etc.). [Devil's Cotton, Indian Flax]. **Shrubs**, 1–4 m tall; branchlets densely stellate-velutinous when young. **Stipules** linear, 0.5–1 cm long, caducous. **Leaves** cordate or ovate-cordate, sometimes 3–5-lobed, 10–22 × 9–18 cm, base cordate or obliquely cordate, apex acute or acuminate, adaxially densely puberulent, basal veins 3–7; petiole 1–10 cm long. **Inflorescences** cymes, 1–5-flowered; flowers pendulous. **Flowers** ~5 cm Ø; sepals lanceolate, 1.5–1.8 cm long, densely puberulent; petals pendent, dark reddish-purple, ~2.5 cm long, upper part elliptic-spathulate, acute or obtuse; staminodes nearly spathulate, hairy; ovary oblong, ~1.5 mm long, slightly hairy, 5-grooved, style triangular, tongue-shaped. **Fruits** capsules, erect, obconic, 3–6 cm Ø, stellate-hairy, 5-winged, cuplike after dehiscence, margins villous; seeds oblong, ~2 mm long, black. **Distr:** Bhutan, China, India, Indonesia, Malaysia, Nepal, Philippines, Thailand, Vietnam; Australia, Pacific Islands. **Ecol:** In forest margins. Fl. & Fr. May–Jul. **Zones:** 10–12. **Info:** The dried fruits have extremely irritant hairs.

Guazuma

Species 4. **Trees** or **shrubs**, stellate-hairy. **Leaves** alternate; lamina simple, dentate, serrate, often oblique. **Inflorescences** cymes, axillary, pedunculate. **Flowers** white, pink or yellow, bisexual, actinomorphic, bracteate; calyx 2–5-lobed or partite, lobes reflexed; petals 5, cucullate, apex with a long, deeply bifid liguliform appendage, valvate; stamens 15, monadelphous, staminal column split above into triangular or lanceolate staminodes and 5 lobes, each with 3 functional anthers, antipetalous; carpels 5, syncarpous, ovary ellipsoid-globose, 5-loculed, ovules biseriate, style filiform. **Fruits** capsules, subglobose, ligneous, woody, muricate, tuberculate-echinate or setose; seeds small, endospermic. **Distr:** Tropical America. **Cult:** Propagate from seeds. Requires a moist, well-drained soil in full or partial sun.

G. ulmifolia Lam., *Encycl. Meth. Bot.* 3: 52, 1789. (Syn: *G. tomentosa*, *Theobroma guazuma*). [Bastard Cedar, Bay Cedar, Pigeon Wood, West Indian Elm]. **Trees**, medium-sized, 8–30 m tall; trunk 30–40 cm Ø. **Leaves** ovate, or ovate-oblong or lanceolate, 5–10 × 2–6 cm, base oblique, apex acuminate, adaxially

Below: *Abroma augustum*, *Abroma augustum* (fruits), *Guazuma ulmifolia*, *Herrania mariae*.

somewhat scabrous, abaxially velvety; petiole 0.5–2 cm long. **Inflorescences** thyrsiform cymes, many-flowered, axillary or terminal; pedicels 3–6 mm long. **Flowers** small, ~8 mm Ø, yellow; sepals 3–5, elliptic-lanceolate or ovate, basally connate, 2–3 × ~1.5 mm, tomentose outside, nearly glabrous inside, reflexed; petals obovate, cucullate, ~3 × 2 mm, appendage 3–4 mm long; stamens 2–2.5 mm long, concealed in petals, anthers ~15, orange or reddish, stamen column around pistil; ovary ~1 mm long, light-green 5-celled, style, stigmas 5, united. **Fruits** capsules, globose or oblong-ellipsoid, 1–2 × 1–1.5 cm, black, tuberculate; seeds many, egg-shaped, 3 mm long, gray. **Distr:** Tropical America, Java. **Ecol:** In secondary forests, sometimes along streams and rivers. Fl. Mar–May, Sep–Dec. **Zones:** 10–12. **Info:** The sweet fruit pulp is edible.

Herrania

Species 17. **Trees** or **treelets**, small, unbranched. **Leaves** unusually large, digitate, palmately compound (4–9-foliolate). **Inflorescences** with flowers fasciculate, cauline. **Flowers** with calyx 3–5-fid; sepals ± divided, valvate in bud; corolla 5-parted; petals hypogynous, cucullate-concave, apically inflexed, with a linear or filiform ligule which, before anthesis, is circinnate-involute; androphore 5-fid, fleshy, glabrous; sterile segments alternate with petals, enlarged above into erect or reflexed, ovate-lanceolate or linear, petaloid staminodes; fertile segments 1–4(mostly 3)-antheriferous; anthers short-stipitate, with diverging, didymous locules; ovary sessile, 5-locular, the lower with many anatropous ovules, style filiform, cylindrical, stigmas 5, terete, obtuse. **Fruits** ovate-oblong, costate, apex usually ± acuminate, toughly coriaceous, indehiscent; seeds enclosed in pulp, flattened-ovate, angulate, testa papery. **Distr:** C America to S America. **Cult:** Requires a fertile, moist, well-drained soil. Propagate from fresh seeds or by air-layering.

H. mariae (Mart.) Decaisne ex Goudot, *Ann. Sci. Nat.* III. 2: 233, 1844. (Syn: *Abroma mariae*, *Theobroma mariae*). **Trees**, to 9(–20) m tall; trunks often 5 or 6, erect, columnar, 7–30 cm Ø; bark brownish-black. **Stipules** caducous, linear, 2.5–4 cm long, acute, tomentose. **Leaves** large, digitate, long-petiolate, 6–9(usually 7)-foliolate; leaflets 27–54 × 7–19 cm, base decurrent, margins usually entire, armed with spinelike prolongations of veins to 1 mm long, apex acuminate, acumen ~2 cm long, adaxially dark green; petiole 30–50 cm, base swollen, hairy. **Inflorescences** fasciculate, in contracted racemes, 10–15(sometimes 80–90)-flowered, cauliflorous on trunk; flower buds subglobose, 1.7 cm Ø, densely hispid; pedicels long; bract 3 mm long. **Flowers** with calyx subcymbiform; sepals 3–5, broadly elliptic-oblong, 1.2–1.5 × 0.6–1 cm, subacute; petals 4 or 5, concave, rotund-obovate, 7–9 × 6–8 mm, very strongly cucullate, 5-nerved, pale purplish-red with black or purple veins, apically ligulate; ligules filiform, pendulous, membranous, pale yellow or white, with purple venation, minutely granulose, tails 7.5–10 cm long, slightly coiled apically; staminal tube 5-parted; staminodia petaloid; ovary subglobose, pentagonal, rose-colored or yellow-white, style short, subcylindric-pentagonal, reddish, stigmas 5, filiform, rose-colored. **Fruits** baccate, elliptic-ovoid, 10–12 × 6–7 cm, acuminate, yellow; seeds 30–40(–60), obtusely rhomboid, flattened, ~10 × 0.9 cm, pulp white. **Distr:** Amazon basin. **Ecol:** In moist lowland forests and forest margins. Fl. Oct–Dec. **Zones:** 10–12.

Kleinhovia

Species 1. *Generic description same as specific.* **Cult:** Propagate from seeds or cuttings. Requires a moist, well-drained soil in full sun.

K. hospita L., *Sp. Pl.*, ed. 2. 2: 1365, 1763. [Bataria Teak, Guest Tree, Timanga Tree]. **Trees**, to 12 m tall; bark gray; branchlets gray-green, sparsely pilose. **Leaves** broadly ovate or ovate, 5.5–18 × 5.5–18 cm, base cordate or subcordate, margins entire or sparsely dentate in upper part, apex acuminate or acute, adaxially glabrous, abaxially puberulent when young; petiole 3–5.5 cm long. **Inflorescences** thyrses, dense, to 50 cm long, hairy. **Flowers** with sepals pink, petaloid, ~6 mm long; petals pink, shorter than sepals, 1 cucullate, apex yellow, shorter than others; androgynophore long, hairy; ovary globose, hairy, usually only 1 ovule developed in each locule, style glabrous. **Fruits** capsules, broadly pyriform to almost globose, 5-angled, 1–1.7 cm long, greenish-pink when mature, pendulous; seeds globose, 1.5–2 mm Ø, black or dark brown. **Distr:** Throughout tropical Africa, Asia, Australia. **Ecol:** In hilly or montane forests. Fl. Mar–Jul. **Zones:** 10–12.

Theobroma

Species ~22. **Trees**; branches spreading. **Stipules** small or medium-sized, geminate, lateral, deciduous. **Leaves** alternate, distichous, nodding or pendulous; lamina simple, large, margins entire, coriaceous. **Inflorescences** 1-flowered or cymose, usually on trunk or major branches. **Flowers** bisexual, small, perfect; calyx 5-lobed, divided nearly to base; petals 5, lower part deeply concave, middle part becoming narrow, upper part spathulate; stamens in 5 groups of 1–3, alternate with staminodes, filaments basally connate into a tube; staminodes 5; ovary sessile, 5-loculed, locules many-ovuled, stigma 5-lobed. **Fruits** large, drupaceous; seeds many, embedded in pulp, cotyledons fleshy, endosperm absent. **Distr:** Tropical America. **Cult:** Requires a fertile, moist, well-drained soil. Propagate from fresh seeds or by air-layering. **Info:** The name 'theobroma' derived from Greek means 'food of the gods'.

T. cacao L., *Sp. Pl.* 2: 782, 1753. [Cocoa]. **Trees**, evergreen, to 12 m tall; bark thick, dark gray-brown; branchlets brown, puberulent. **Stipules** linear, caducous. **Leaves** narrowly ovate- to obovate-elliptic, 20–30 × 7–10 cm, base rounded to shallowly cordate, apex long-acuminate, surfaces glabrous or sparsely puberulent. **Inflorescences** cymose, small and delicate; pedicels ~1.2 cm long. **Flowers** ~1.8 cm Ø; calyx pink, lobes narrowly lanceolate, persistent, margin hairy; petals 5, yellowish, slightly longer than calyx, lower part helmet-shaped and abruptly narrowed, reflexed, apex acute; staminodes linear; ovary obovoid, slightly 5-angular, 5-loculed, ovules 14–16 per locule, in 2 rows, style cylindrical. **Fruits** drupes, ellipsoid, 15–20 × ~7 cm, longitudinally 10-grooved; endocarp thick, fleshy, hard and woody when dried, 4–8 mm thick; seeds 12–14 per cell, ovoid, slightly flattened, ~2.5 × 1.5 cm. **Distr:** S America. **Ecol:** Fl. throughout the year. **Zones:** 11–12. **Info:** Widely cultivated throughout tropics for commercial cocoa and chocolate.

Subfam. STERCULIOIDEAE

Genera 12, **species** 430. **Trees**, often tall, monoecious. **Leaves** spirally arranged; lamina simple or lobed to digitate, sometimes palmately compound; petiole bundle annular, with medullary bundle. **Inflorescences** axillary, paniculate, sometimes reduced to dense clusters, rarely cauliflorous; epicalyx absent. **Flowers** usually gamosepalous, apetalous, usually functionally unisexual; androgynophore present, in ♀ occasionally reduced; sepals petaloid; stamens (4)5–many, filaments connate, free parts of filaments short or anthers ± sessile, thecae distinct, sometimes forming a densely packed whorl or head; staminodes absent; carpels (2–)5(–60), free, micropyle endostomal. **Fruits** composed of distinct and dehiscent (follicles) or indehiscent (nuts) carpels, sometimes winged, endocarp sometimes pubescent; seeds with or without endosperm. **Distr:** Pantropical.

Brachychiton

Species 31. **Trees** or **shrubs**, deciduous or evergreen, monoecious; trunk sometimes greatly swollen at base. **Leaves** alternate, often deciduous in flower; lamina simple, margins entire or palmately lobed. **Inflorescences** solitary racemes or panicles, axillary or terminal. **Flowers** actinomorphic, unisexual or bisexual or both on the same plant; calyx lobes (4)5(–8), usually spreading, petaloid, reddish or pale green to white, sometimes with laminar adaxial appendages; petals absent; stamens 10–30, irregularly crowded; carpels 5, syncarpous, free at maturity, ovary 5-loculed, ovules numerous per locule, stigma peltate, radiate or lobed. **Fruits** follicles, usually 5 (or fewer by abortion) stalked, woody, endocarp and seeds pubescent; seeds numerous, biseriate, enclosed within thick, honey-comb like endocarp. **Distr:** Australia, Papua New Guinea. **Cult:** Propagate from fresh seeds or by grafting. Requires a moist, well-drained soil in full sun.

B. acerifolius (A. Cunn. ex G. Don) Macarthur and Moore, *Cat. Coll. Bois Indig.* 6, 1855. (Syn: *Sterculia acerifolium*). [Illawarra Flame Tree]. **Trees**, to 35 m tall; trunk cylindrical, to 70 cm Ø, base sometimes slightly flanged. **Leaves** alternate, coriaceous, glossy; lamina variable on *younger trees*, palmate, deeply 5–7-lobed, deciduous at flowering, 10–20 cm long; lamina on *mature trees* either rhomboid or ovate, margins entire, venation palmate; petiole to 14 cm long, slender. **Inflorescences** lax panicles, axillary, to ~40 cm long; pedicels 0.5–2 cm long, bright coral-red or scarlet. **Flowers** unisexual, ♂ and ♀ frequently on the same tree, campanulate, 2–3 cm long; calyx bright coral-red or scarlet, petaloid, tubular below, lobes 5 or 6, shortly triangular, 6–8 mm long, glabrous; petals absent; ♂ with staminal column ~5–9 mm long, lower half globular, 5-lobed, anthers 15, yellow, in an apical globular mass; ♀ with carpels 5, yellow, 3–4 mm long, staminodes 15, in 5 clusters. **Fruits** clusters of 5 follicles; follicles boat-shaped, 10–20 cm long; seeds numerous, ovate, exotesta bristly. **Distr:** E Australia. **Ecol:** In coastal lowlands, subtropical rainforests; 0–800 m. Fl. Nov–Jan. Fr. Dec–Apr. **Zones:** 9–12.

Cola

Species ~50. **Trees**, usually deciduous, rarely unbranched, sometimes lepidote. **Leaves** simple, entire, palmati-lobed or digitately compound, sometimes with sacciform domatia; petiole often swollen at base and apex. **Inflorescences** racemes, panicles or fascicles, axillary, sometimes cauliflorous or ramiflorous; pedicels usually articulate. **Flowers** unisexual or polygamous; calyx (3–)5(–7)-lobed, petaloid, campanulate to cup-shaped, margins often induplicate and corrugate-membranous; petals absent; ♂ with stamens (5–)10–12(–20) in 1 or apparently 2 rings at apex of column with vestigial carpels often projecting from apex; ♀ with carpels (3–)5–10(–12), coherent or ± free, carpels 2–20-ovuled (carpels dehiscent in fruit), styles short, as many as carpels, filiform or dilated, often coherent. **Fruits** groups of 1–several follicles; follicles leathery or woody; seeds 1–many. **Distr:** Tropical Africa. **Cult:** Propagate from seeds (germination is slow, taking 2–3 months in most species) or preferably by cuttings and air-layering. Requires a well-drained soil in shade and humid conditions. Some species can withstand long dry seasons.

C. acuminata (Beauv.) Schott and Endl., *Metelm. Bot.* 33, 1832. (Syn: *C. ledermannii*, etc.). [Bissy Nut, Cola Nut, Cola Tree, Goora Nut]. **Trees**, evergreen, to 8 m tall; trunk to 20 cm Ø; branches smooth, cylindrical; branchlets sometimes bearing hooked spines 3–4 mm long. **Leaves** oblong to ovate, or elliptic, 10–20 × 4–9 cm, base narrowed,

Top row: *Kleinhovia hospita, Theobroma cacao* (flowers), *Brachychiton acerifolius, Cola acuminata.*
Bottom row: *Cola lateritia, Cola millenii, Cola millenii* (fruits).

margins entire, apex acuminate, surfaces glabrous. **Inflorescences** racemes, to 10 cm long, several- to many-flowered. **Flowers** 2–3 cm Ø; calyx lobes 5, spreading, yellowish or whitish with reddish or purplish markings toward base; stamens ~10; pistil with 5 or fewer locules and styles. **Fruits** clusters of follicles, <5 follicles, 10–13 cm long, obovate, warty, gray-green; seeds 6–12 per follicle, white or bright red, drying to brown, 2.5–3 cm long. **Distr:** Tropical Africa. **Ecol:** 0–900 m. Fl. Feb–Aug. **Zones:** 10–12.

C. lateritia K.Schum., *Notizbl. Königl. Bot. Gart. Berlin* 2: 307, 1899. (Syn: *Sterculia lateritia*, etc.). **Trees**, to 10–15(–40) m tall; trunk 30–45 cm Ø, twisted, buttresses winged. **Leaves** simple, broadly ovate 17–25 × 15–17 cm, base ± truncate, apex acuminate, coriaceous, adaxially dark green, abaxially paler with lighter midrib; petiole ~15 cm long. **Inflorescences** panicles, loose, axillary; pedicels white-pubescent. **Flowers** fragrant, uniformly dull creamy-white inside, darker yellowish, reddish-tinged outside, aging to orange, rich pink and red; calyx urceolate, 0.9–1 cm long, lobes 5, spreading. **Fruits** clusters of follicles (carpels); follicles light tawny-brown or red outside, pinkish inside, ovoid, curved, narrowed at both ends. **Distr:** Tropical Africa (Guinea to Nigeria, Cameroon, Gabon, Liberia). **Ecol:** In deciduous forests and secondary jungles; <800 m. Fl. Jun–Aug. Fr. Jul–Aug. **Zones:** 10–12.

C. millenii K. Schum., *Bot. Jahrb.* 33: 313, 1903. (Syn: *C. heterophylla, C. togoensis*). [Monkey Cola]. **Trees** or **large shrubs**, deciduous, to 12(–20) m tall. **Leaves** palmately lobed, to 30 × 35 cm, lobes usually (3 or)5, obovate, dissected ± halfway, apical lobe larger and broader, basal lobes smaller, apices acuminate, glabrous, adaxially dark green, abaxially paler; petiole slender, to 30 cm long. **Inflorescences** clusters, small, axillary, ± hemispheric, ~5–75 cm Ø. **Flowers** small, ~2.5 cm Ø, greenish outside (varying to dull purple), red-brown to purplish inside (varying to purple and greenish-yellow), rarely flowers wholly pale yellow; petals absent; ♂ with rudimentary pistil at apex of androgynophore; ♀ with stigma ~8-lobed apically, ovary 5-loculed, ovules numerous, stamens rudimentary at base of ovary, ringlike. **Fruits** clusters of follicles, usually arranged in a single plane resembling a large open flower; follicles rich pink to bright orange-red, obovoid, ridged on one side, base narrowed. **Distr:** Tropical Africa (Côte d'Ivoire to S Nigeria). **Ecol:** In dry forests. Fr. Apr–Aug. **Zones:** 10–12.

Heritiera

Species 35. **Trees**; trunk usually buttressed. **Leaves** alternate; lamina simple or palmately compound, abaxially usually scaly. **Inflorescences** cymose, paniculate, axillary, many-flowered, hairy or scaly. **Flowers** unisexual, usually small; calyx campanulate or urceolate, 4–6-lobed; petals absent; ♂ with androgynophore short, anthers 4–15, in 1 or 2 rings at apex of androgynophore, pistillode present; ♀ with ovary consisting of 3–5 coherent carpels, base with sterile anthers, ovules 1 per carpel, style very short, stigma very small. **Fruits** woody or leathery, with keel-like prominence or wing, indehiscent; seeds without endosperm. **Distr:** Tropical Africa, Asia, Australasia. **Cult:** Propagate from seeds or by air-layering. Requires a moist soil in full sun. Mangrove species can be grown in shallow brackish water.

H. littoralis Aiton, *Hort. Kew.* 3: 546, 1789. [Looking-glass Mangrove, Looking-glass Tree]. **Trees**, evergreen, to 10(–15) m tall. **Stipules** lanceolate, caducous. **Leaves** oblong-lanceolate, elliptic or ovate, 10–20 × 5–10 cm, base obtuse, apex acute or obtuse, leathery, abaxially densely silver-white scurfy-scaly, adaxially glabrous or nearly so; petiole 1–2 cm long. **Inflorescences** paniculate, ~8 cm long, axillary, densely hairy or scaly. **Flowers** with calyx red-brown, campanulate, 4–6 mm long, surfaces hairy, lobes triangular, ~2 mm long; ♂ with disk thin, papillate, androgynophore short and glabrous, anthers 4 or 5, in 1 ring; ♀ with carpels 4 or 5, stigmas as many as carpels, short and curved downward. **Fruits** nutlike, woody, drying yellow-brown to dark brown, nearly ellipsoid, ~6 × 3.5 cm, keeled on back, glabrous, often glossy; seeds ovoid, ~2 cm long. **Distr:** Australia, Cambodia, China, E Africa, India, Indonesia, Malaysia, Maldives, Philippines, Sri Lanka, Taiwan, Vietnam. **Ecol:** In mangrove forests; salt tolerant. **Zones:** 10–12.

Pterygota

Species ~20. **Trees. Leaves** cordate, usually entire, but lobed when very young. **Inflorescences** racemose or paniculate, axillary. **Flowers** unisexual; calyx campanulate, 5-lobed to base, lobes reflexed at apex; petals absent; ♂ with androgynophore cylindrical, enclosed by calyx, apex cup-shaped, filaments clustered into 5 groups, staminodes usually present, anthers sessile; ♀ androgynophore very short, with 5 undeveloped staminodes, carpels almost free, ovules many per carpel, stigma swollen, radiate. **Fruits** follicles, woody, subglobose, long-stiped; seeds many, with long and wide apical wing. **Distr:** Africa, tropical Asia. **Cult:** Propagate from seeds or by air-layering. Requires a fertile, moist, well-drained soil in full sun.

P. alata (Roxburgh) R. Brown, *Pl. Jav. Rar.* 234,

Top row: *Heritiera littoralis* (flowers), *Heritiera littoralis* (fruits), *Pterygota alata*, *Pterygota thwaitesii*.
Bottom row: *Sterculia balanghas*, *Sterculia balanghas* (fruits), *Sterculia foetida*, *Sterculia lanceolata*.

1844. (Syn: *Sterculia alata*, *Pterygota roxburghii* [misapplied]). [Buddha Coconut]. **Trees**, large, to 30 m tall. **Stipules** subulate, caducous. **Leaves** cordate or broadly ovate, 13–35 × 10–17 cm, base truncate, cordate or rounded, margins nearly entire, apex acute or obtuse, surfaces glabrescent; petiole 5–15 cm long. **Inflorescences** paniculate, shorter than petiole, axillary; pedicels nearly absent. **Flowers** sparse, red; calyx campanulate, 1.7–2 cm long, lobes linear-lanceolate, densely puberulent; ♂ with androgynophore cylindric cone-shaped, half as long as calyx, puberulent, anthers ~20, 3–5 grouped into fascicles on top of androgynophore, undeveloped carpels apparent; ♀ with androgynophore very short, ovary globose and puberulent, ovules 40–50 per carpel, in 3 rows, styles 5, curved, pubescent. **Fruits** follicles, woody, compressed-globose, ~12 cm Ø, puberulent outside, adaxially corky; seeds many, oblong, flat, ~7 cm long including long and wide wing. **Distr:** Bangladesh, Bhutan, China, India, Malaysia, Myanmar, Philippines, Thailand, Vietnam. **Ecol:** In open forests. Fr. Dec. **Zones:** 10–12.

P. thwaitesii (Masters) Alston., *Handb. Fl. Ceylon* 6: 31, 1931. (Syn: *Sterculia alata*, *S. thwaitesii*). **Trees**, 10–50 m tall; trunk cylindrical, 2.8 m Ø, fluted or buttressed, buttresses thin, to 2 m high. **Leaves** crowded at ends of branchlets; lamina oblong, 6–29 × 2.5–21 cm, base rounded, truncate or cordate, apex acute or shortly acuminate, 5-nerved at base (3 stronger); petiole to 15 cm long, ends swollen. **Inflorescences** velvety-tomentose, 5 cm long; pedicels 3 mm long. **Flowers** ♂ brown outside, green to red-brown inside, calyx tube 1 mm long, lobes 3 × 1.5 mm, staminal column reddish-brown, 1 mm long, anthers yellow, 1 mm long; ♀ similar, calyx tube 2–3 mm long, lobes elliptic or ovate-elliptic, 6 × 2.5 mm, carpels ~2 mm long, ferruginous-tomentose, styles 1 mm long, recurved. **Fruits** follicles, woody, oblong-obovoid, 8–11 × 5–9 cm, strongly convex along suture, shortly stipitate; seeds compressed-ellipsoid, 1.5 × 0.8 cm, wing oblong, 2.7 × 1.5 cm. **Distr:** Sri Lanka. **Ecol:** In deciduous forests; 100–500 m. **Zones:** 11–12.

Sterculia

Species 100–150. **Trees** or **shrubs**. **Leaves** simple, entire or palmately lobed, rarely palmately compound, margins entire or dentate. **Inflorescences** usually axillary, usually paniculate, rarely racemose. **Flowers** unisexual; calyx 5-lobed or 5-partite; petals absent; ♂ with anthers clustered at top of androgynophore, enclosing undeveloped carpels; ♀ with androgynophore very short, staminodes at top of androgynophore in whorl around base of carpels, carpels 5, 2- to many-ovuled, styles basally connate, stigmas as many as carpels, free. **Fruits** clusters of follicles; follicles usually leathery, less often woody, dehiscent when mature; seeds 1 to many per follicle. **Distr:** Tropics and subtropics of both hemispheres, predominantly in Asia. **Cult:** Propagate from seeds or by air-layering. Requires a fertile, moist, well-drained soil in full sun.

S. balanghas L., *Sp. Pl.* 1007, 1753. **Trees**, 4–20 m tall; trunk buttressed. **Leaves** crowded at branch ends; lamina simple, ovate to elliptic or ovate-oblong, 8–17.5 × 5–11 cm, base rounded to subcordate, margins entire, apex rounded to shortly acuminate, abaxially hairy to velvety, rarely ± glabrous; petiole 14.5 cm long, ends swollen. **Inflorescences** paniculate, drooping, axillary, slender, 7–13 cm long; axes hairy; pedicels 1.2–3 cm long. **Flowers** scented, greenish, crimson-hairy, crimson inside; calyx tube obconic, 1–2 × 2–3 mm; lobes lanceolate, 5–7 mm long, densely hairy outside, densely velutinous inside, apices coherent; ♂ with column 2–3 mm long, anthers 15, bright yellow forming a globose head 1 mm Ø; *hermaphrodite* with column 0.7 mm long, glabrous, ovary globose, 2.5 × 3 mm, densely hairy, stigmas 5, oblong, slender, ~1 mm long. **Fruits** follicles ~5, bright orange-scarlet, spreading, 6.5–8 × 3.5–4.5 cm, brownish-tomentose; stipe ~1 cm long; seeds black, base yellow, ovoid, 1.5–2 × 0.9–1.5 cm, glossy. **Distr:** India, Sri Lanka, Myanmar, Malay Peninsula, Langkawi and Timor. **Ecol:** In monsoon and evergreen forests; 0–900 m. **Zones:** 11–12.

S. foetida L., *Sp. Pl.* 2: 1008, 1753. [Peon, Hazel/Stinky Sterculia, Indian/Wild Almond, Java Olive, Skunk Tree]. **Trees**; branches verticillate and spreading. **Stipules** arrow-shaped, caducous. **Leaves** apically clustered; lamina palmately 7–9-foliolate; leaflets elliptic-lanceolate, 10–15 × 3–5 cm, base cuneate, margins entire, apex long-acuminate or caudate, initially pilose; petiole 10–20 cm long. **Inflorescences** apical on branchlets, paniculate, erect, many-flowered; pedicels shorter than flowers. **Flowers** with epicalyx lobes minute; calyx purple-red, ~1.2 cm long, divided nearly to base, lobes elliptic-lanceolate, abaxially yellowish brown pubescent, adaxially upper half white-villous; ♂ with stamens 12–15, capitate; ♀ with carpels 5, hairy, style curved, stigma 5-divided. **Fruits** clusters of follicles; follicles ellipsoid,

Top row: *Sterculia lanceolata* (fruits), *Sterculia urens* (fruits), *Sterculia urens* (habit).
Bottom row: *Dombeya burgessiae*, *Dombeya wallichii*.

boat-shaped, 5–8 cm long, woody, nearly glabrous, apex acute, beaked; seeds 10–15, black, ellipsoid, ~1.5 cm long, smooth. **Distr:** Cambodia, India, Indonesia, Malaysia, Myanmar, Philippines, Sri Lanka, Thailand, Vietnam. **Ecol:** Fl. Apr–May. **Zones:** 10–12.

S. lanceolata Cavanilles, *Diss.* 5: 287, 1788, (Syn: *S. balansae*). [Lance-leaved Sterculia, Scarlet Sterculia]. **Trees**, usually to ~8 m tall; crown spreading; branchlets at first pilose. **Leaves** elliptic, lanceolate, or elliptic-lanceolate, 9–20 × 3.5–8 cm, base obtuse or nearly rounded, apex acute, surfaces ± glabrous, lateral veins 7–9 per side, curved upward, connected near margin; petiole 2.5–3.5 cm. **Inflorescences** paniculate, 4–10 cm, densely many-branched. **Flowers** small; calyx reddish, divided almost to base, pubescent abaxially, lobes oblong-lanceolate or oblong-elliptic, 4–6 mm, stellately spreading, margins ciliate, apex obtuse or minutely mucronate; ♂ with androgynophore 2–3 mm, curved, anthers ~10; ♀ with ovary globose, hairy, style curved, stigma minutely 5-lobed. **Fruits** clusters of follicles, red when fresh, narrowly ovoid or ellipsoid, 5–7 × 2–2.5 cm, base attenuate, apex beaked, densely puberulent; seeds 2–4, black-brown, ellipsoid-ovoid, glossy, ~1 cm long. **Distr:** China, Laos, Myanmar, Thailand, Vietnam. **Ecol:** Usually near streams. Fl. Apr–Jun. **Zones:** 10–12. **Info:** The bark fiber is used for making paper.

S. urens Roxb., *Pl. Corom.* 1: 25, t. 24, 1795. [Ghost Tree, Gum Karaya, Indian Gum Tragacanth]. **Trees**, fairly large, soft-wooded, deciduous, 10–20 m tall; pith with red resin canals. **Leaves** borne at ends of branches; lamina palmately (3–)5-lobed, 15–25(–35) × 15–25(–35) cm, base cordate and 7-nerved, adaxially slightly pubescent, abaxially velvety, lobes entire, acuminate; petiole 12–25(–30) cm long. **Inflorescences** panicles, dense, terminal, many-flowered, to 25 cm long, few hermaphrodite flowers mixed with many ♂ flowers; peduncles 5 cm long, axes sticky-glandular, tomentose; pedicels 1.5–4 mm long; bracts 6 mm long. **Flowers** malodorous, yellow or green, center purple or maroon, small, ~6–8 mm Ø; ♂ with calyx tube campanulate, 2 mm long, lobes 5, rusty-green, 4 mm long; staminal column 1.5 mm; stamens 10; *hermaphrodite* with ± no column, anthers 10; ovary globose many-ovuled, style reduced. **Fruits** clusters of follicles, red or yellowish-brown, bristly pubescent with irritant yellow hairs; seeds 3–6, black, oblong. **Distr:** India, Sri Lanka. **Ecol:** On exposed hills, grasslands; 10–450 m. **Zones:** 11–12.

Subfam. DOMBEYOIDEAE

Syn: Dombeyaceae.
Genera 21, **species** 381. **Shrubs**, rarely herbs or small trees. **Leaves** spirally arranged. **Inflorescences** usually axillary cymes or flowers solitary; epicalyx present. **Flowers** with calyx basally connate to free, often persistent, flat, not cucullate, apical appendage absent, nectary present at base; stamens 10 or more, usually connate, rarely free, pollen often porate, spinulose; staminodes elongated, antesepalous, staminodes forming a short tube, rarely staminodes absent. **Distr:** Old World tropics (St Helena, with highest diversity in Madagascar and Mascarenes). **Ecol:** Mostly in xeric habitats, savannas and gallery forests.

Dombeya

Species ~350. **Trees** or **shrubs**. **Leaves** 5-costate, cordate, usually palmately lobed. **Inflorescences** pedunculate cymes, axillary or terminal, much-branched. **Flowers** bisexual, actinomorphic; bracts 3, deciduous or absent. **Flowers** with sepals 5, basally united, eventually reflexed; petals 5, persistent, imbricate; stamens 10–20, alternating with 5 strap-shaped staminodes, both monadelphous at base, anthers parallel; carpels 2–5, syncarpous, ovary sessile, 2–5-loculed, locules 2–3-ovuled, style simple, stigmas 2–5, patent. **Fruits** capsules, loculicidally dehiscent, as many loculed as ovary, locules 1–2-seeded; seeds endospermic. **Distr:** Mainly in Africa and Madagascar. **Cult:** Propagate from seeds, cuttings or by air-layering. Requires a fertile, well-drained soil in full or partial sun. Regular watering is very important for pot-bound specimens.

D. burgessiae Gerr. ex Harv. and Sond., *Fl. Cap.* (Harvey) 2: 590, 1862. (Syn: *D. elegans*, etc.). [Pink Wild Pear, Pink Dombeya]. **Shrubs** or **small trees**, to 5 m tall; bark reddish-brown, smooth with some warts. **Leaves** alternate; lamina simple, broadly ovate to rounded, 3- or 5-lobed, 6–30 × 5–18 cm, base cordate, margins serrated, apex very broadly tapering, adaxially pale green, abaxially paler, surfaces hairy, 5–7-veined from base; petiole 2.5–12 cm long. **Inflorescences** capitula, axillary, many-flowered; peduncles terminally branched, ~7.5 cm long. **Flowers** cup-shaped, uniformly deep to pale pink or white, ~3–5 cm Ø, persistent; sepals thin, ~1.2 cm long, showy in flower. **Fruits** capsules, small, dark brown when mature, ~8 mm Ø, hairy, splitting open to reveal brown seeds, the persistent perianth dries forming papery wings, aiding in wind dispersal. **Distr:** KwaZulu-Natal, Swaziland, Mozambique, Zimbabwe,

Above: *Pterospermum acerifolium, Pterospermum semisagittatum.*

Malawi, Zambia, Uganda, Tanzania, Kenya. **Ecol:** Along streams, forest margins, on rocky outcrops, in high rainfall areas; 1,370–1,870 m. Fl. Mar–Jul. Fr. June–Oct. **Zones:** 9–12.

D. wallichii (Lindl.) Bailey, *Stand. Cycl. Hort.* 1: 1066, 1947. (Syn: *Astrapaca wallichii*). [Tropical Hydrangea]. **Shrubs** or **small trees**. **Leaves** cordate, 20–30 cm long, margins serrate. **Inflorescences** umbellate cymes, many-flowered, axillary, drooping, 12–15 cm across; peduncles >20 cm long, hairy; pedicels ~4 cm long, hairy. **Flowers** pink, ~3 cm Ø; sepals linear-oblong, ~1.5 cm × ~3–4 mm, densely hairy outside; petals obovate, oblique, 2–2.5 × 1–1.5 cm; fertile stamens ~1 cm long; staminodes ~1.5 cm long; carpels 5, ovary ovate-oblong, densely villous, stigma exserted. **Fruits** capsules, pentagonal, ovate-oblong, densely beset with rusty hairs, tipped by ~2 cm long, persistent style. **Distr:** Madagascar. **Ecol:** Fl. Apr–May. **Zones:** 9–12.

Pterospermum

Species 18–40 **Trees** or **shrubs**, velutinous or scurfy and scaly. **Stipules** linear to palmately fimbriate, sometimes forming nectaries, caducous. **Leaves** simple; lamina lobed or not, base often oblique, margins entire or serrate, leathery, apex sometimes truncate; juvenile leaves sometimes much larger, petiole long, lamina often with peltate base and prominently palmately lobed; petiole short. **Inflorescences** with flowers solitary or 1–5-flowered cymes, axillary. **Flowers** bisexual; epicalyx lobes usually 3, usually distant from calyx, entire, fimbriate, or palmately lobed, rarely absent, caducous; calyx 5(or 6)-lobed, divided nearly to base, usually spreading to ± reflexed, caducous; petals 5, white or yellow, often erect; androgynophore present, very short, glabrous; stamens 15, in 5 groups of 3 alternating with staminodes, filaments connate into tube, free at top; staminodes 5, filiform, longer and thicker than filaments; ovary 5-locular, locules 4–22-ovuled, style clavate or filiform, stigma longitudinally 5-grooved. **Fruits** schizocarps, cylindric or ovoid, angular or rounded, usually woody, sometimes leathery, dehiscent into 5 loculicidal mericarps; seeds 2–many per locule, with long, oblong, membranous wing. **Distr:** Tropical and subtropical Asia. **Cult:** Propagate from seeds, cuttings or by air-layering. Requires a moist, well-drained soil in full or partial sun.

P. acerifolium (L.) Willdenow, *Sp. Pl.* 3(1): 729, 1800. [Maple-leaf Bayur]. **Trees**, large; young shoots densely velutinous when young. **Stipules** linear. **Leaves** nearly orbicular or oblong, sometimes ± lobed, 24–34 × 14–29 cm, base cordate, margins entire or crenate, apex truncate, nearly rounded, or pointed, leathery, adaxially sparsely hairy or glabrous, abaxially densely brown-velutinous; juvenile leaves palmately lobed, peltate; petiole robust, striate. **Inflorescences** with flowers solitary. **Flowers** fragrant; epicalyx lobes fimbriate or palmately divided; sepals linear-oblong, ~9 × 0.7 cm, adaxially white villous, abaxially densely brown-velutinous; petals white, linear-oblong, slightly cuneate, slightly shorter than sepals, glabrous; staminodes clavate, hairy; ovary oblong, 5-angular, locules many-ovuled. **Fruits** capsules, woody, cylindrical, stipitate, obscurely 5-grooved, 10–15 × 5–5.5 cm, brown-velutinous, glabrescent, base tapering, apex rounded; seeds many per locule, obliquely ovate, flat, wing large and thin, brown, smooth. **Distr:** Bangladesh, Bhutan, China, India, Laos, Malaysia, Myanmar, Nepal, Thailand. **Ecol:** On forested slopes; 1,200–1,700 m. **Zones:** 10–12.

P. semisagittatum Buch.-Ham. ex Roxb., *Fl. Ind.* ed. 1832, 3: 160, 1832. **Trees**, deciduous or evergreen, 5–15 m tall; young shoots downy. **Stipules** finely pinnatifid. **Leaves** alternate, bifarious, shortly petioled, spreading; lamina lanceolate, to 15 × 5 cm, base of inferior half protruded into a lobe, generally longer than the petiole, margins entire, apical margins sometimes strongly toothed, adaxially smooth, abaxially whitish-hairy. **Inflorescences** with flowers solitary, terminal and axillary; peduncles short, downy, rust-colored; involucral bracts usually 3-leaved, 3–5-parted, divisions variously laciniate; pedicels >0.5 cm long. **Flowers** large, pure white; calyx 5-cleft, lobes recurvate, linear longer than corolla, inside white and villous, outside brownish-hairy; corolla subcampanulate; petals 5, obliquely obovate-cuneate; filaments 15; staminodes 5, not basally united; ovary ovate, downy, style single, filiform, stigma clavate. **Fruits** capsules, oblong-elliptic, to 8 cm long, woody, splitting into 5 sections, densely pubescent; seeds ~10 per cell, obliquely oval, compressed, with a large thin, brown wing. **Distr:** Myanmar, India, Indo-China, Laos, Sri Lanka, Thailand. **Ecol:** Scattered in secondary and primary evergreen forests. **Zones:** 9–12. **Info:** The wood is heavy and durable, and used for construction.

Schoutenia

Species 9. **Trees**, small to medium; trunk prominently fluted; wood white to yellow-brown; twigs slender, pubescent, with obscure stipular scars. **Leaves** alternate; lamina simple, base usually unequal, 3-nerved. **Inflorescences** racemose or paniculate, axillary or terminal, usually few-flowered; pedicels jointed; bracts usually present below the joint. **Flowers** with calyx membranous, white to yellow, usually bell-shaped, 5-lobed, lobes with conspicuous nerves; petals absent in Malesian spp.; disk usually present as a small, circular strip, adnate to calyx base; stamens ~20–125, usually basally connate in 5 bundles; ovary 5-loculed, style stout, cylindric, stigmas 5, usually recurved. **Fruits** capsules, globular, 5-chambered, usually 1-seeded, densely rusty-pilose, with enlarged papery, persistent calyx, style and stigma. **Distr:** SE Asia (Cambodia and Thailand to Myanmar and Java). **Cult:** Propagate from seeds or by air-layering. Requires a moist, well-drained soil in full or partial sun.

S. accrescens (Mast.) Curtis, *J. Str. Br. Roy. Asiat. Soc.* 25: 84, 1894. (Syn: *Chartacalyx accrescens*, etc.). **Trees**, briefly deciduous, to 33 m tall; trunk prominently fluted, to 2.4 m Ø; crown coppery; limbs twisted; twigs slender, densely yellow-brown to reddish pubescent; bark brownish, deeply and closely fissured. **Leaves** narrowly ovate to obovate, (2 × 1 –) 5 × 1.5

Below: *Schoutenia accrescens, Berrya cordifolia, Cullenia ceylanica, Durio zibethinus.*

Above: *Helicteres angustifolia, Helicteres isora.*

(– 11.5 × 4.5) cm, base unequal, obtuse to wedge-shaped, apex pointed, adaxially yellowish-brown, densely hairy, secondary nerves 4–5 pairs, basal pair continuing to halfway along lamina. **Inflorescences** racemes, terminal or axillary, few-flowered, produced in profusion, covering the crown. **Flowers** with calyx bell-shaped, prominent. **Fruits** 7 mm Ø, with persistent pale brown calyx. **Distr:** Borneo, Sumatra. **Ecol:** In lowlands to montane forests. **Zones:** 11–12. **Info:** A very attractive tree when in bloom, crown changes color from white (flowering) to golden brown (fruiting).

Subfam. BROWNLOWIOIDEAE

Syn: Berryaceae, Brownlowiaceae.
Genera 8, **species** 68. **Trees**, usually tall; indumentum often lepidote. **Inflorescences** axillary. **Flowers** with calyx connate, campanulate, irregularly splitting into 2–3 lobes; stamens ~30, in antepetalous bundles, anther thecae sagittate, often broadly so; staminodes antesepalous, rarely absent, petaloid; carpels 2-ovuled, style or styles present (Stevens 2001 onwards). **Fruits** with persistent calyx. **Distr:** Tropical regions, especially Old World. **Ecol:** Usually in rainforests.

Berrya

Species 6. **Trees**. **Leaves** alternate; lamina with base cordate, margins entire, basal veins 5–7. **Inflorescences** paniculate, terminal or axillary. **Flowers** many, minute; calyx campanulate, 3–5-divided; petals 5, white or pink, spathulate or filiform; stamens many, free; staminodes absent; ovary 3(or 4)-loculed, stigma trilobed. **Fruits** capsules, globose, 3(or 4)-valved, loculicidally dehiscent, each valve with 2 erect wings; seeds 1 or 2(–4) per cell, villous, endosperm fleshy. **Distr:** Primarily in tropical Asia and Polynesia. **Cult:** Propagate from seeds or by air-layering. Requires a moist, well-drained soil in full sun.

B. cordifolia (Willdenow) Burret, *Notizbl. Bot. Gart. Berlin-Dahlem.* 9: 606, 1926. (Syn: *B. ammonilla, Espera cordifolia*). [Trincomalee Wood]. **Trees**, 4–6(–35) m tall; trunk 10–20 cm Ø. **Stipules** filiform, 1–2 cm long, glabrous. **Leaves** ovate to ovate-oblong, 10–20(–25) × 6–8(–14) cm, base cordate, apex acuminate, thinly leathery, very sparsely pubescent, soon glabrous, lateral veins 4 or 5 pairs; petiole 3.5–10 cm long. **Inflorescences** panicles, to ~20 cm long, hairy; pedicel ~1 cm long. **Flowers** with calyx 3–5-lobed, lobes triangular-ovate, 3–5 mm long, hairy; petals white or pink, filiform, 6–8 mm long, becoming reflexed; stamens 5–6 mm long; ovary hairy, style short. **Fruits** capsules (samaras), with persistent calyx, globose, 1–1.3 cm Ø, pubescent, with 6(or 8) horizontally spreading thin wings, wings 2.5–3 × 0.7–1 cm; seeds ~6 × 3 mm, with brown or yellow, caducous bristles. **Distr:** Cambodia, India, Indonesia, Laos, Malaysia, Myanmar, Philippines, Sri Lanka, Taiwan, Thailand, Vietnam. **Ecol:** In dry open places; 0–600 m. **Zones:** 11–12.

Subfam. HELICTEROIDEAE

Syn: Durionaceae, Helicteraceae, Triplochitonaceae.
Genera 8–10, **species** 95. **Shrubs** or **trees**; indumentum lepidote. **Leaves** with secondary veins pinnate. **Flowers** with calyx connate; petals often with lateral constrictions; androgynophore present; androecium usually with short tube and/or fascicles, pollen baculate or microverrucate to suprareticulate; gynoecium sometimes apocarpous. **Seeds** arillate (Stevens 2001 onwards). **Distr:** Tropical, especially SE Asia and W Malesia.

Cullenia

Species 3. **Trees**, tall, evergreen; branchlets lepidote. **Leaves** petiolate; lamina simple, margins entire, adaxially glabrous, abaxially densely scaly. **Inflorescences** densely fasciculate on protuberances of old wood (ramiflorous); pedicels articulate near middle, densely lepidote. **Flowers** densely lepidote outside; epicalyx, irregularly 3–4-lobed at apex, splitting to base on one side and deciduous; calyx ± irregularly 5-dentate apically, carnose, deciduous; corolla absent; staminal tube 5-dentate to 5-lobate, lobes with 7–11 stamens along margins; stamens with short filaments, connectives irregularly spherical to club-shaped; ovary sessile, densely scaly, 5(6)-celled, with 2–6 superposed ovules per cell, style longer than staminal tube, sometimes long-exserted, stigma capitellate and densely papillate. **Fruits** capsules, ± globose, densely covered with stout spines, 5-valved; seeds 1–few per locule, sometimes reduced to 1 per capsule, completely to incompletely covered by a fleshy aril or exarillate. **Distr:** India (one species, *C. exarillata*) and Sri Lanka (2 species). **Ecol:** In rainforests. **Cult:** Propagate from seeds or by air-layering. Requires a moist, well-drained soil in full sun.

C. ceylanica (Gardn.) K. Schum., *Pflanzenfam.* 3 (6): 68, fig. 35 D., 1890. (Syn: *C. excelsa, Durio ceylanicus*). [Wild Durian]. **Trees**, to 20–36 m tall; trunk straight, to >1.8 m Ø. **Leaves**; lamina usually narrowly elliptic, 8.5–20 × 2.5–5.5 cm, base rounded, margins somewhat undulated, apex long-acuminate, acumen to 2 cm long, subcoriaceous, adaxially deep green, shiny; petiole 1.2–2.5 cm long. **Inflorescences** ramiflorous; pedicels thick, 1.5–1.8 cm long. **Flowers** with epicalyx shortly tubular, 1–1.2 cm long; calyx oblong-tubular, 2.5–3 × 0.6–0.9 cm, base slightly swollen, apex 5-dentate, densely scaly; staminal tube white at base and apex, reddish at middle, to 3.5 cm long, apex 5-dentate, lobes ~2–3 mm long, with 7–9 stamens along their margins; ovary densely lepidote, locules 2-ovuled, style long-exserted, to 4 cm long, white-hairy, stigma orange, to 2.3 mm Ø. **Fruits** capsules, globular, to 10 cm Ø, dehiscent on tree, spines conical, 0.6–1.5 cm long; seeds 1–5, chestnutlike, to 3.8 cm long with orange-brown aril. **Distr:** Sri Lanka. **Ecol:** In dense humid evergreen forests; 0–900 m. **Zones:** 11–12.

Durio

Species ~28. **Trees**, medium to large; trunk straight, buttresses rounded. **Stipules** caducous. **Leaves** alternative; lamina usually oblong to elliptic or lanceolate, midrib depressed adaxially, secondary venation generally not prominent abaxially; petiole slender. **Inflorescences** cymes, few-flowered, borne on young branchlets, older branches and trunk, sometimes at base of the trunk; peduncles reduced, ± unbranched; bracts around flowers, scaly, sometimes large and persistent. **Flowers** with epicalyx splitting into 2 lobes; calyx 5-lobed, sometimes splitting, lobes becoming free at anthesis, base becoming saclike; petals (4)5(6), larger than sepals, white, yellow, pink to red, falling separately; stamens numerous, longer than ovary, free or united into 5 groups, groups themselves sometimes basally united; ovary sessile, ovoid or ribbed (3–)5(–6)-chambered, chambers 2(or more)-ovuled, style long, slender, stigma capitate. **Fruits** capsules, globular to ellipsoid, woody, with long spines, usually 5-chambered, splitting from tip to base; seeds usually with prominent aril. **Distr:** Malesia. **Ecol:** In primary forests; to ~900 m. **Cult:** Propagate from seeds or by air-layering. Requires a fertile, moist, well-drained soil in full sun.

D. zibethinus Murray, *Syst. Nat. Veg.*, ed. 13: 581, 1774. [Civet Cat Fruit, Durian]. Trees, large, to >30 m tall; trunk ~3 m Ø, buttresses steep. Stipules to 2 cm long, scaly, caducous. Leaves elliptic or lanceolate-elliptic, 10–15 × 3–4.5 cm, base pointed or blunt, apex pointed, abaxially densely silvery- or golden-scaly, secondary nerves ~15 pairs, faintly visible on both surfaces; petiole 1–1.5 cm long. Inflorescences clusters or fascicles, borne on older branches. Flowers 5–6 cm long; calyx colored, 5-lobed, girded by a bilobed concave involucrum; petals 5, white, spathulate, base narrowed to a conspicuous claw, joined together at base into a tube, longer than the calyx; stamens in 5 free bundles, anthers anfractuous; ovary 5-ribbed, stigma roundish. Fruits green to yellow, globose, ovoid or ellipsoid, to 15–25 cm Ø, dropping unopened, spines broadly pyramidal, 1 cm long; seeds, shiny, ovate-oblong, completely covered by soft, white or yellowish, sweet but malodorous aril. **Distr:** Sumatra, Borneo. **Ecol:** In lowland tropical rainforests, where the temperature does not drop below 22°C. Fl. Mar–Jun. Fr. Sep–Dec. **Zones:** 11–12.

Helicteres

Species ~60. **Trees** or **shrubs**. **Leaves** simple, margins entire or serrate. **Inflorescences** with flowers solitary or cymose, axillary or rarely terminate. **Flowers** bisexual; epicalyx lobes minute; calyx tubular, 5-lobed, lobes usually unequal and 2-lipped; petals 5, equal to each other or 2-lipped, clawed and usually with auriculate appendage; filaments ± connate, enclosing pistil, anthers 10, on top of elongate androgynophore; staminodes 5, within stamens; ovary 5-loculed,

Above: *Abelmoschus esculentus, Abelmoschus manihot, Abelmoschus moschatus, Abelmoschus moschatus* ssp. *tuberosus.*

5-angular, ovules numerous in each locule, styles 5, linear, apically slightly swollen. **Fruits** capsules, when mature straight or spirally twisted, usually densely hairy, often with soft setae which are themselves hairy; seeds verrucose. **Distr:** Tropical America and Asia. **Cult:** Propagate from seeds or by cuttings. Requires a moist, well-drained soil in shade or partial sun.

H. angustifolia L., *Sp. Pl.* 2: 963, 1753. [Narrow-leaved Screwtree]. **Shrubs**, small, to 1 m tall; branchlets gray-greenish puberulent. **Leaves** narrowly oblong or linear-lanceolate, 3.5–5 × 1.5–2.5 cm, base rounded, apex obtuse or acute, adaxially glabrous or nearly so, abaxially gray or yellowish-tomentose, sometimes hirsute; petiole 5–7 mm long. **Inflorescences** cymose, 2- or many-flowered; pedicels usually with 4 pointed bracteoles. **Flowers** with calyx tubular, ~6 mm long, puberulent, 5-lobed, lobes triangular; petals 5, unequal in length, pale blue, pink, or purplish, darker at base of limb, slightly longer than calyx, reflexed at anthesis, uniformly pubescent, claws 2-auriculate; staminodes linear, very short; ovary 5-loculed, hairy, slightly shorter than style, locules ~10-ovuled. **Fruits** capsules, ovoid-oblong, 1.2–2 × 0.7–0.8 cm, densely setose, setae relatively slender, sparsely hairy, apex acute; seeds small, brown, with elliptic spots. **Distr:** Cambodia, China, Indonesia, Japan, Laos, Malaysia, Myanmar, Philippines, Taiwan, Thailand, Vietnam; Australia. **Ecol:** Usually in dry grasslands. Fl. throughout most of the year. **Zones:** 9–12.

H. isora L., *Sp. Pl.* 2: 963, 1753. (Syn: *H. chrysocalyx, H. roxburghii*). [Spiral Bush, East Indian Screw Tree]. **Shrubs** or **small trees**, to 2(–8) m tall; branchlets puberulent. **Stipules** linear, 0.7–1 cm long, caducous. **Leaves** usually broadly oblong-obovate, 10–12(–23) × 7–9(–17) cm, base rounded or obliquely cordate, margins serrate, apex ± truncate, tip shortly acuminate, thinly leathery, abaxially densely puberulent to tomentose; petiole 0.8–2.5 cm long. **Inflorescences** axillary, usually 2 or 3 per axil, flowers densely clustered. **Flowers** 3.5–4 cm Ø; epicalyx lobes subulate, to 7 mm long; sepals 1.7–2 cm long, usually 4–5-lobed, lobes triangular, ± 2-lipped, densely hairy; petals red or purple, upper 2 larger, 1.2–1.5 cm long, obliquely falcate; androgynophore 3–4 cm long; stamens 10; staminodes 5; ovary spirally twisted after pollination. **Fruits** capsules, black, cylindric, spirally twisted, 4–8 × (0.5–)0.7–0.9(–1) cm, apex acute, beak to 1 cm long, densely puberulent at first; seeds many, to 2 mm long. **Distr:** Asia through Malesia to N Australia. **Ecol:** In dry deciduous forests; <100–600 m. Fl. Apr–Oct. Fr. Oct–Jan. **Zones:** 9–12.

Subfam. MALVOIDEAE

Genera 78, **species** 1,670. **Herbs**, **shrubs** or **trees**; indumentum of stellate, lepidote, simple, and unbranched glandular hairs; sometimes spiny. **Leaves** mostly stipulate; lamina simple or lobed, exceptionally digitate, sometimes with nectaries on primary veins at base abaxially. **Inflorescences** with flowers solitary or in cymes, collectively forming spikes, panicles or heads; epicalyx present or absent. **Flowers** regular or somewhat irregular, perfect or sometimes unisexual; calyx gamosepalous, lobes valvate in bud, usually nectariferous inside at base; corolla usually regular, petals 5, convolute in bud, distinct but adnate at base to staminal column; stamens monadelphous, filaments few to many, terminating the staminal column or borne below 5 sterile teeth; ovary superior, carpels (1–)3 to many, borne in a single whorl, each with 1–9 or many pendulous ovules, style exceeding the staminal column, single, stigmas lobed decurrent or divergent, hairy, as many as carpels, or branched with as many or twice as many stylar branches as carpels and terminal (capitate) or decurrent stigmas. **Fruits** schizocarps; seeds hairy, embryo curved, cotyledons folded. **Distr:** Temperate to tropical regions worldwide.

Abelmoschus

Species ~15. **Herbs**, annual, biennial, or perennial, often hispid or tomentose, hairs often mostly simple. **Leaves** entire or palmately lobed. **Inflorescences** with flowers solitary, axillary. **Flowers** yellow or red; epicalyx lobes 5–15, filiform, very rarely lanceolate, persistent; calyx spathaceous, splitting along 1 side at anthesis, apex 5-toothed, caducous with corolla; corolla with a dark red center, funnel-shaped; petals 5; staminal column shorter than corolla, 5-toothed at apex, anthers at base; ovary 5-loculed, locules many-ovuled, style single with 5 sessile capitate stigmas. **Fruits** capsules, elongate, loculicidal, pubescent or hispid; seeds reniform or globose, many, glabrous, smooth. **Distr:** tropical and subtropical regions in E Hemisphere. **Cult:** Propagate from seeds. Requires a fertile, well-drained soil in full sun.

A. esculentus (L.) Moench, *Methodus.* 1: 617, 1794. (Syn: *Hibiscus esculentus, H. longifolius*). [Gumbo, Lady's Finger, Okra]. **Herbs**, annual, 1–2 m high, most parts with very sparse prickly hairs; stem often hollow. **Stipules** filiform, 0.7–1 cm long, sparsely hirsute. **Leaves** palmately 3–7-lobed, (5–)10–30 cm Ø; lobes broad to narrow, surfaces sparsely hirsute, margins sparsely dentate, apex emarginate; petiole 7–15(–35) cm long, hirsute. **Inflorescences** with flowers solitary, axillary; pedicels (0.5–)1–2(–5) cm long, sparsely strigose. **Flowers** epicalyx lobes 7–10(–12), filiform, 0.5–1.8 cm long, sparsely hirsute; calyx campanulate, 2–3 cm long, densely puberulent; corolla yellow or white with a dark purple center, 5–7 cm Ø; petals obovate, 3.5–5 × 3–4 cm; filament tube 2–2.5 cm long. **Fruits** capsules, cylindric to tower-shaped, 10–25 × 1.5–2(–3) cm, long-beaked, sparsely strigose; seeds dark brown or gray, globose to reniform, 5–15 per locule, (3–)4–5(–6) mm long, striate, minutely warty. **Distr:** India. **Ecol:** Fl. May–Sep. **Zones:** 8–12.

A. manihot (L.) Medikus, *Malvenfam.* 46, 1787. (Syn: *Hibiscus manihot*). [Aibika, Manihot, Manihot Mallow]. **Herbs**, annual or perennial, 1–2 m high, hispid and minutely pubescent. **Stipules** linear-lanceolate, sometimes 2 per petiole, 1–1.5 cm long. **Leaves** palmately 5–9-lobed, 15–30 cm Ø, lobes oblong-lanceolate, sometimes ± pinnately lobed, 8–18 × 1–6 cm, margins obtusely serrate, surfaces sparsely hispid; petiole 6–18 cm long. **Inflorescences** with flowers solitary, subapical, forming terminal racemes; pedicels 1.5–4 cm long. **Flowers** with epicalyx lobes 4 or 5, ovate-lanceolate, 1.5–2.5 × 0.4–0.5 cm, sparsely hispid; calyx spathaceous, minutely 5-lobed, nearly entire, longer than epicalyx, puberulent only, not hispid, caducous at fruiting; corolla yellow with a purple center, ~12 cm Ø; staminal column 1.5–2.5 cm long, anthers nearly sessile; stigma purple-black, spathulate-disk-shaped. **Fruits** capsules, ovoid-ellipsoid, 4–5 × 2.5–3 cm, densely hirsute; seeds many, reniform, with several lines of hairs. **Distr:** China, India, Nepal, Philippines, Taiwan, N Thailand. **Ecol:** In grasslands, streamsides; 1,000–2,100 m. Fl. Aug–Oct. **Zones:** 10–12. **Info:** 2 varieties described (var. *manihot* and var. *pungens*).

A. moschatus Medikus, *Malvenfam.* 46, 1787. (Syn: *Hibiscus abelmoschus*, etc.). [Ambrette, Musk Mallow]. **Herbs**, annual or perennial, (0.25–)1–2 m high, uniformly yellow hispid/setose, rarely glabrous; taproot tapered, not swollen. **Stipules** filiform, 7–8 mm long. **Leaves** 6–15 cm Ø, very variable in form, usually palmately (3–)5–7-lobed, lobes lanceolate to triangular, base cordate, margins irregularly serrate, lamina on distal part of stem narrower, surfaces sparsely hirsute; petiole 7–15 cm long. **Inflorescences** with flowers solitary, axillary; pedicels 2–3 cm long, hirsute. **Flowers** with epicalyx lobes 6–10, linear to narrowly oblong, (0.8–)1–1.3 cm × 1.5–2 mm, incurved, appressed to capsule; calyx spathulate, much longer than epicalyx, 2–3 cm long, 5-parted, usually caducous, uniformly densely pubescent; corolla 7–12 cm Ø, yellow with a dark purple center; petals obovate; staminal column ~2.5 cm long, glabrous; ovary hairy, style hairy, branches 5, stigma disk-shaped. **Fruits** capsules, oblong, 5–6 cm long, apex acute, uniformly yellow-hirsute; seeds black-brown, reniform, concentrically ribbed, glandular-reticulate, with musk smell. **Distr:** Cambodia, India, Laos, Thailand, Vietnam. **Ecol:** In valleys, streamsides. Fl. Jun–Oct. **Zones:** 10–12. **Info:** 3 subspecies described. The often cultivated ***A. moschatus* ssp. *tuberosus*** [Red Musk Mallow] has red flowers with a white center.

Abutilon

Species ~200. **Herbs**, **subshrubs**, **shrubs**, or **small trees**, evergreen, semievergreen or deciduous. **Stipules** usually caducous. **Leaves** usually entire (lobed in *A. pictum*), palmately veined, base cordate, margins crenate or serrate. **Inflorescences** axillary or subterminal, flowers solitary, paired or in small cymes, often aggregated into terminal panicles. **Flowers** without epicalyx; calyx campanulate, lobes 5; corolla mostly yellow or orange (red in *A. roseum*), often with a dark center, campanulate to wheel-shaped, rarely ± tubular (*A. pictum*), petals 5, basally connate and adnate to filament tube; anthers many, clustered at filament tube apex; ovary (5–)7–20-loculed, locules (carpels) 2–9-ovuled, style branches as many as carpels. **Fruits** schizocarps, often blackish when mature, subglobose to hemispherical, often covered with irritant hairs, mericarps (5–)7–20, eventually dehiscent, apex rounded or acute, sometimes 2-awned, pericarp leathery; seeds reniform, glabrous or slightly pubescent. **Distr:** Mostly in tropic and subtropical regions worldwide. **Cult:** Propagate from seeds, by softwood cuttings or by air-layering in larger species, which root quite readily. Requires a moist, well-drained soil in full sun.

A. indicum (L.) Sweet, *Hort. Brit.* 1: 54, 1826. (Syn: *A. asiaticum*, etc.). [Country Mallow, Monkey Bush]. **Herbs**, subshrublike, annual or perennial, erect, many-branched, 1–2.5 m high, gray-puberulent. **Stipules** subulate, 1–2 mm long, curved outward. **Leaves** ovate-orbicular or nearly orbicular, 3–9 × 2.5–7 cm, base cordate, margins irregularly serrate, apex acute or acuminate, densely puberulent; petiole 2–4 cm long, gray-puberulent, sparsely hairy, hairs ~1 mm long. **Inflorescences** with flowers solitary, axillary; pedicel ~4 cm long, articulate near apex, gray-puberulent. **Flowers** 2–2.5 cm Ø; calyx green, disk-shaped, 0.6–1 cm Ø, densely puberulent, lobes 5, broadly ovate, apex acute; corolla uniformly yellow, petals 7–8 mm long; staminal column scabrous; ovary 15–20-loculed. **Fruits** capsules, hispid, black, flat-topped, ~1.5 cm Ø; mericarps 15–20, apex acute, slightly awned, long-scabrous; seeds reniform. **Distr:** Bhutan, Cambodia, China, India, Indonesia, Laos, Maldives, Myanmar, Nepal, Sri Lanka, Taiwan, Thailand, Vietnam. **Ecol:** In disturbed sites on sandy soils; <800(–1,500) m. Fl. Jul–Oct. **Zones:** 9–12.

Alcea

Species ~60. **Herbs**, annual, biennial, or perennial, usually erect, unbranched, most parts pubescent, sometimes mixed with long hairs. **Leaves** long-petiolate; lamina ovate to suborbicular, angled, weakly lobed, or deeply palmatipartite, margins crenate or dentate, apex acute to obtuse. **Inflorescences** with flowers solitary or fascicled, often arranged into terminal racemes, axillary. **Flowers** with epicalyx lobes 6 or 7, basally connate; calyx 5-lobed, ± pubescent; petals pink, white, purple, or yellow, usually >3 cm wide, apex notched; staminal column glabrous with anthers clustered at apex, anthers yellow and compact; ovary 15(or more)-loculed, locules 1-ovuled, erect, styles as many as locules, stigmas decurrent, filiform. **Fruits** schizocarps, disk-shaped; mericarps >15, laterally compressed, circular with a prominent ventral notch, glabrous or pubescent, 2-celled, proximal cell 1-seeded, distal cell sterile; seeds glabrous or pustulose. **Distr:** C and SW Asia, E and S Europe. **Cult:** Propagate from seeds. Requires a rich, moist, well-drained soil in full sun.

A. rosea L., *Sp. Pl.* 2: 687, 1753. (Syn: *A. rosea*, etc.). [Hollyhock]. **Herbs**, biennial, erect, to 2(–3) m tall; stem densely hirsute. **Stipules** ovate, ~8 mm long, apically trilobed. **Leaves** nearly orbicular, palmately 5–7-lobed or crenate-angled, 6–16 cm Ø, papery, abaxially long-hirsute or stellate-tomentose, adaxially sparsely stellate-pilose, lobes triangular or rounded, central lobe ~3 × 4–6 cm; petiole 5–15 cm long, hirsute. **Inflorescences** with flowers solitary or fascicled, aggregated into a terminal, spikelike inflorescence; bracts foliaceous; pedicel ~5 mm long, hirsute. **Flowers** with epicalyx cup-shaped, usually 6- or 7-lobed, 0.8–1 cm long, densely hirsute, lobes ovate-lanceolate; calyx campanulate, 2–3 cm Ø, lobes ovate-triangular, 1.2–1.5 cm long, densely hirsute; corolla red, purple, white, pink, yellow, or black-purple, 6–10 cm Ø, sometimes double; petals obovate-triangular, ~4 cm long, base attenuate, claw tipped with long thin hairs, apex emarginate; staminal column glabrous, ~2 cm long, filaments ~2 mm long, style branches many, puberulent. **Fruits** schizocarps, disk-shaped, ~2 cm Ø, puberulent; mericarps many, nearly orbicular, longitudinally grooved. Fl. Feb–Aug. **Distr:** Mediterranean region. **Ecol:** In dry open areas. Fl. Feb–Aug. **Zones:** 4–11.

Gossypium

Species ~20. **Herbs**, annual or perennial, sometimes shrubs; all parts irregularly dotted with dark oil glands. **Leaves** usually palmately 3–9-lobed, rarely entire. **Inflorescences** subterminal, flowers solitary; pedicels not articulated, usually with glands below insertion of epicalyx lobes. **Flowers** with epicalyx lobes 3(–7), foliaceous, glandular, free or basally connate, entire or toothed to deeply laciniate; calyx cup-shaped, nearly truncate to 5-lobed; corolla white or yellow, sometimes with a purple center, large, petals 5, apex rounded; staminal column with many anthers along entire length, apex truncate; ovary 3–5-loculed, locules 2–many-ovuled, style short, rod-shaped, stigma clavate, 5-grooved. **Fruits** capsules, globose or ellipsoid, loculicidally dehiscent; seeds globose, densely white long-woolly, mixed with short hairs or without short hairs. **Distr:** Tropical and subtropical regions worldwide. **Cult:** Propagate from seeds or cuttings. Requires a fertile, moist, well-drained soil in full sun.

G. herbaceum L., *Sp. Pl.* 2: 683, 1753. (Syn: *G. zaitzevii*). [Arabian/Levant/Syrian Cotton]. **Herbs** or **subshrubs**, annual, to 1.5 m high, pilose. **Stipules** filiform, 0.5–1 cm long, caducous. **Leaves** usually 5-lobed, 5–10 cm Ø, base cordate, apex acute, lobes broadly ovate, abaxially minutely puberulent, villous on veins, adaxially hirsute; petiole 2.5–8 cm long, villous. **Inflorescences** with flowers solitary, axillary; pedicels 1–2 cm long, villous. **Flowers** with epicalyx lobes basally connate, broadly triangular, 2–3 cm long, sparsely villous on veins, 6–8-toothed; calyx cup-shaped, 5-lobed; corolla yellow, purple at center, 5–7 cm Ø; petals 2.5–3 × 3–4 cm; filaments uniform in length. **Fruits** capsules, usually 3–4-celled, ovoid, ~3 cm long, apex beaked; seeds free, obliquely cone-shaped, ~1 cm long, with white wool and persistent short fuzz. **Distr:** SW Asia. **Ecol:** Fl. Jul–Sep. **Zones:** 8–12.

G. tomentosum Nutt. ex Seemann., *Fl. Vit.* 22, 1865. (Syn: *G. sandvicense*). [Hawaiian Cotton]. **Shrubs**, 0.6–2 m high, spreading to ~3 m wide. **Stipules** awl-shaped. **Leaves** petiolate; lamina usually 3(–5)-lobed, ~10 cm across, wider than long, margins wavy, gray-green, densely white-hairy, giving the leaves a silvery, gray-green appearance. **Flowers** with epicalyx bracts to 2.5 cm long, toothed; corolla ~5–7.5 cm Ø, bright sulfur yellow, center not dark. **Fruits** ovoid, 1.2–1.8 cm long, 3-loculed; seed 6–12, hairs short, reddish-brown. **Distr:** Hawai'i. **Ecol:** In arid, windy, rocky areas near the ocean. Fl. throughout the year. **Zones:** 10–12.

Hibiscus

Species ~200. **Shrubs**, **subshrubs** (or undershrubs), **trees**, or **herbs** (annuals or perennials), usually stellate-pubescent. **Leaves** palmately lobed or entire, basal veins 3 or more. **Inflorescences** usually axillary solitary flowers, sometimes subterminal and ± congested into terminal racemes. **Flowers** 5-merous, bisexual; epicalyx lobes 5–many, free or basally connate, rarely very short (*H. schizopetalus*) or absent (*H. lobatus*); calyx campanulate, rarely shallowly cup-shaped or tubular, 5-lobed or 5-dentate, persistent; corolla usually large and showy, variously colored, often with a dark center; petals adnate at base to staminal tube; filament tube well-developed, apex truncate or 5-dentate, anthers throughout or only on upper half of tube; ovary 5-loculed or, as a result of false partitions, 10-loculed, locules 3–many-ovuled, style branches 5, stigmas capitate. **Fruits** capsules, cylindrical to globose, valves 5, loculicidally dehiscent and sometimes partially septicidal or indehiscent (*H. vitifolius*); seeds reniform, hairy or glandular verrucose. **Distr:** Tropical and subtropical regions worldwide. **Cult:** Propagate annuals from seed, and perennials from seed, cuttings, grafting or by air-layering. Requires a fertile, moist, well-drained soil in full sun.

H. acetosella Welw. ex Hiern, *Cat. Afr. Pl. Welw.* 1: 73, 1896. [African Rose Mallow, False Roselle, Red-leaved Hibiscus]. **Herbs**, annual or perennial, sometimes suffruticose, most parts red; stems crisped-pubescent or nearly glabrous. **Stipules** linear, to 1.5 cm long. **Leaves** ovate to suborbicular in outline, to 8 × 8 cm, almost entire to irregularly 3–5-lobed or palmatipartite, slightly fleshy, somewhat glaucous; petiole to 10–11 cm long, minutely pubescent. **Inflorescences** solitary, axillary; peduncle to 0.8–1 cm

Below: *Abutilon indicum*, *Alcea rosea* (white form), *Alcea rosea* (pink form), *Gossypium herbaceum*.

Above: *Gossypium tomentosum, Hibiscus acetosella, Hibiscus arnottianus* ssp. *arnottianus.*

long, articulated near middle. **Flowers** to 10 cm Ø, purple-red or sometimes lemon-yellow, with a purple center; epicalyx segments 9–10, to 1.5 cm long, forked, inner branch linear, outer branch spathulate; calyx 2 cm long, 10-nerved, nearly glabrous, lobes ovate-lanceolate to triangular, rigid, sharp-pointed; petals obovate, to 5.5 × 3.5 cm, glabrous; staminal tube 1 cm long, free parts of filaments 1 mm long; style branches 1.5 mm long. **Fruits** capsules, ovoid-acuminate, 1.7 × 0.9 cm, hispid; seeds subreniform, 3 × 2.5 mm, ± glabrous. **Distr:** Tropical Africa, Mauritius. **Ecol:** In disturbed lowland areas, riverine areas, open woodlands, plantations, moist thickets and on roadsides; to 1,200 m. Fl. & Fr. Jun–Dec. **Zones:** 10–12.

H. arnottianus A. Gray, *U.S. Expl. Exped.*, Phan. 15: 176, 1854. [White Hibiscus]. **Shrubs** or **treelets**, evergreen, to 3(–9) m tall; stems 7.5+ cm Ø. **Stipules** with apex long-pointed, early caducous. **Leaves** ovate, 5–10 × 4–7.5 cm, base rounded, margins finely wavy or straight, apex blunt, thin, surfaces green and slightly shiny, abaxially often finely hairy, main veins 5; petiole 2–4 cm long. **Inflorescences** with flowers solitary, borne at 1 or 2 uppermost leaves; epicalyx bracts 5–7. **Flowers** showy, large, fragrant, funnel-shaped, 13–15 cm Ø; calyx tubular, cylindrical, light green, 2 cm long, 5-toothed and split on 1 side; petals 5, spreading, white, sometimes pinkish, elliptic, 7.5–11 cm long, basally united into a narrow tube 2.5 cm long; stamens many, along upper part of long dark red or white staminal column; staminal column 8–10 cm long; ovary cylindrical, 5-loculed, locules several-ovuled, style slender, with 5 exposed erect dark red branches, stigmas brown. **Fruits** capsules, oblong, 2.5 cm long, thin-walled, 5-loculed, enclosed by calyx; seeds 5 mm long, brown hairy. **Distr:** Hawai'i. **Ecol:** In wet forests; 300–915 m. **Zones:** 10–12. **Info:** ***H. arnottianus* ssp. *arnottianus*** (Syn: *H. arnottianus* f. *parviflorus*) has a reddish staminal column and a pale pink flush to the petals. ***H. arnottianus* ssp. *immaculatus*** (Syn: *H. immaculatus*) has pure white flowers.

H. brackenridgei A. Gray, *U.S. Expl. Exped., Phan.* 15: 175, 1854. [Hawai'i Yellow Hibiscus]. **Shrubs** or **treelets**, 1–5 m tall; new growth stellate-hairy. **Leaves** circular, 5–15 cm across, usually deeply 3-, 5- or 7-lobed, margins serrate. **Inflorescences** with flowers solitary or in short terminal racemes. **Flowers** with epicalyx bracts 7–10, awl-shaped to linear-lanceolate, 1–2.8 cm long; calyx 1.5–4 cm long, principal veins raised; corolla with petals flaring, sulfur yellow, base usually maroon-blotched, 3.5–9 cm long; staminal column ± included, anther-bearing, ± throughout. **Fruits** capsules, 1.2–2 cm long, beaked; seeds ~5 mm long, puberulent. **Distr:** Hawai'i; rare. **Ecol:** In dry forests and scrubs; 40–750 m. **Zones:** 10–12. **Info:** ***H. brackenridgei* ssp. *brackenridgei*** has epicalyx bracts half as long as or equaling the calyx; calyx 1.5–2.5 cm long; petal bases with or without maroon blotches.

H. campylosiphon* var. *glabrescens (Warb. ex Perkins) Borss. Waalk., *Blumea* 14(1): 56, 1966. (Syn: *Bombycidendron glabrescens, B. vidalianum, H. vidalianus*, etc.). [Cotton Tree, Tree Hibiscus]. **Trees**, to 15+ m tall, densely pubescent; trunk straight, terete, glabrous. **Leaves** oblong, 10–15 × 4–5 cm, apex acuminate, adaxially glabrous, abaxially paler, sparsely tomentulose, veins ~7, curved; petiole 1–2 cm long. **Inflorescences** with flowers solitary; pedicel 2 cm long. **Flowers** large, showy, pink; epicalyx lobes lanceolate, 1 cm long, acute; calyx glabrous, villous below. **Fruits** capsules, lignous, ovoid, 2.2–3 cm long, 5-loculed, locules cuspidate, many-seeded; seeds reniform. **Distr:** Philippines. **Ecol:** In forested ravines and secondary forests; to 1,200 m. Fl. Oct–Dec. **Zones:** 11–12.

H. cannabinus L., *Syst. Nat.*, ed. 10. 2: 1149, 1759. (Syn: *Abelmoschus verrucosus*, etc.). [Bastard Jute, Deccan/Indian Hemp, Wild Kenaf]. **Herbs**, annual or perennial, erect, robust, to 3 m tall, glabrous, sparsely sharply spiny. **Stipules** silklike, 6–8 mm long. **Leaves** dimorphic, lamina on *proximal* part of stem cordate, those on *distal* part of stem palmately 3–7-lobed, lobes lanceolate, 2–12 × 0.6–2 cm, base cordate or nearly rounded, margins serrate, apex acuminate, glabrous, basal veins 5–7, glandular near base of midrib; petiole 6–20 cm, sparsely spiny. **Inflorescences** with flowers solitary, axillary, nearly sessile. **Flowers** with epicalyx lobes 7–10, not red, filiform, 6–8 mm long, free, sparsely spiny; calyx nearly campanulate, connate for about half its length, ~3 cm long, spiny and white-tomentose, lobes 5, lanceolate, 1–2 cm long, long-caudate; corolla yellow with a red center, petals oblong-obovate, ~6 cm long; staminal column 1.5–2 cm long, glabrous; styles 5, glabrous. **Fruits** capsules, globose, ~1.5 cm Ø, densely spiny, apex shortly beaked; seeds reniform, subglabrous. **Distr:** Africa, India.

H. clayi O. Deg. and I. Deg., *Fl. Hawaiiensis* 221, 1959. (Syn: *H. newhousei*). [Red Kauai Rosemallow]. **Shrubs** or **trees**, 4–8 m tall; stems bearing sparse hairs at branch tips. **Leaves** ovate or elliptical, usually 2.5–7.5 × 1.5–3.5 cm, margins entire or toothed toward the apex, adaxially glabrous, abaxially slightly hairy. **Inflorescences** with flowers solitary, borne at ends of the branches. **Flowers** with epicalyx segments 5 or 6; calyx green, tubular or urn-shaped, usually 1.5–2.5 cm long; petals flared, dark red, 4.5–6 × 1–1.8 cm. **Fruits** capsules, pale brown, 1.2–1.4 cm long; seeds ~10, ovoid, ~4 mm long, brownish-black. **Distr:** Hawai'i (Kaua'i). **Ecol:** In lowland, dry forests, often on slopes; 230–350 m. **Zones:** 10–12.

H. furcatus Roxb., *Hort. Beng.* 51, 1814. **Herbs**, shrubby, perennial or annual, scrambling to suberect; stem softly stellate-pubescent or glabrous, with scattered, slender, recurved prickles with thickened bases. **Stipules** oblong- to linear-lanceolate, 3.5–7(–8) × 0.75–1.5 mm, entire, variously ciliate-hispid. **Leaves** orbicular in outline, 3–7.5(–9) × 4–9(–12) cm, entire or palmately 3–5-lobed (lower leaves sometimes

Below: *Hibiscus arnottianus* ssp. *immaculatus, Hibiscus brackenridgei* ssp. *brackenridgei, Hibiscus campylosiphon* var. *glabrescens, Hibiscus cannabinus.*

TR

F&KS

Above: *Hibiscus clayi, Hibiscus furcatus, Hibiscus indicus, Hibiscus kokio.*

subentire), lobes acute, serrate, pubescent, especially abaxially, prickly along adaxial veins; petiole 2.5–5(–11.5) cm long, stout, very prickly. **Inflorescences** with flowers solitary, axillary; pedicels 1.5–6.5(–10) cm long, softly pubescent, prickly. **Flowers** ~8–10 cm Ø, yellow with dark crimson center; epicalyx segments 10–12, linear, spathulate above, 0.7–0.8(–1) cm long, bristly; sepals lanceolate-ovate, 1.5–1.8 cm long, connate to half its length, acuminate, strongly veined, margins and major veins bristly ciliate, accrescent in fruit to 2.75 × 1.24 cm; corolla ~7.5–10 cm Ø, yellow with a crimson center. **Fruits** capsules, ovoid, pointed, 1.2–1.5 cm long, yellow-setose, prominently reticulate-veined, enclosed in calyx; seeds triangular-reniform, ~3 × 2 mm, longitudinally compressed, blackish-brown, papillose. **Distr:** Tropical Asia and Africa, S Africa. **Ecol:** On coastal hills; to ~1,000 m. Fl. Jan–Mar. **Zones:** 10–12.

H. indicus (N.L. Burman) Hochreutiner, *Mém. Soc. Hist. Nat. Afrique* N. 2: 163, 1949. [Indian Hibiscus]. **Shrubs**, deciduous, erect, to 3 m tall, densely puberulent. **Stipules** lanceolate, ~5 mm long, caducous. **Leaves** cordate, 8–12 × 10–15 cm, usually 7-lobed proximally, usually 3-5-lobed distally, lobe margins irregularly dentate or entire, apices acuminate or obtuse, papery, palmately 5–11-veined; petiole terete, 6–11 cm long. **Inflorescences** with flowers solitary, axillary on upper branchlets; pedicel 6–15 cm long, articulate at apex. **Flowers** with epicalyx lobes 4 or 5, ovate, basally connate, ~2 × 0.8–1.2 cm, apex obtuse or acute, densely tomentose; calyx cup-shaped, ~2.5 cm long, lobes 5, ovate, connate for a third its length, densely tomentose; corolla pink to white, 7–10 cm Ø; petals obovate, ~6.5 cm long, barbate at base, abaxially hairy; staminal column 3.5–4 cm long; style sparsely villous. **Fruits** capsules, nearly globose, ~3 cm Ø, hirsute; mericarps 5–6; seeds reniform, ~3 mm Ø, densely ferruginous-puberulent. **Distr:** China, Taiwan. **Ecol:** In scrubs; 500–2,000 m. Fl. Jul–Dec. **Zones:** 9–12.

H. kokio Hillebrand ex Wawra, *Flora* 56: 174, 1873. (Syn: *H. arnottianus* var. *kokio*). [Clay Hibiscus, Red Hibiscus]. **Shrubs** or **treelets**, 2–6 m tall. **Leaves** elliptic, ovate or obovate, 3.8–10(–20) cm long, margins usually toothed above middle, ± coriaceous, glossy. **Inflorescences** with flowers solitary. **Flowers** with epicalyx bracts 6–8, awl-shaped to lanceolate-linear, 0.8–1.8 cm long; calyx tubular, usually 2.5–3.3 cm long, smooth; corolla wheel-shaped, petals reflexed, ± oblong, 4.5–7.5 cm long, red to orange, rarely yellow; staminal column projecting, 3.8–6.5 cm long; filaments clustered below apex; style branches spreading. **Fruits** 1.5–2 cm long, sparsely pubescent; seeds to ~6 mm long, sparsely pubescent. **Distr:** Hawai'i. **Ecol:** In dry or wet forests; 130–750 m. **Zones:** 10–12.

H. moscheutos L., *Sp. Pl.* 2: 693, 1753. [Crimson-eyed Rose-mallow]. **Herbs** perennial, erect, 1–2.5 m tall; stem pubescent or ± glabrous. **Stipules** silklike, caducous. **Leaves** ovate to ovate-lanceolate, sometimes with 2 lateral lobes, 10–18 × 4–8 cm, base cuneate or ± rounded, margins bluntly dentate, apex caudate, adaxially ± glabrous, abaxially whitish-tomentose; petiole 4–10 cm long, puberulent. **Inflorescences** with flowers solitary, axillary; pedicel 4–8 cm long, sparsely puberulent, apically articulate. **Flowers** with epicalyx lobes 10–12, filiform, ~ 1.8 cm long, densely puberulent; calyx half as long as epicalyx, lobes rounded, connate for more than half its length; corolla white, pink, red, or purple, with a dark red center, 10–14 cm Ø, petals obovate, ~10 cm long, adaxially bearded on margins, abaxially sparsely pilose; staminal column ~4 cm long; ovary glabrous, style branches 5, sparsely scabrous. **Fruits** capsules, conic-ovoid, 2.5–3 cm long; mericarps 5; seeds reniform, 2–3 mm Ø, pointed. **Distr:** SE USA. **Ecol:** In moist places. Fl. Jul–Sep. **Zones:** 6–11.

H. mutabilis L., *Sp. Pl.* 2: 694, 1753. (Syn: *Abelmoschus mutabilis, H. sinensis*, etc.). [Confederate Rose, Cotton Rose]. **Shrubs** or **small trees**, erect, 2–5 m tall, deciduous; branchlets, petiole, pedicel, epicalyx, and calyx densely woolly pubescent. **Stipules** lanceolate, 5–8 mm long, usually caducous. **Leaves** broadly ovate to round-ovate or cordate, 5–7-lobed, 10–15 cm Ø, margins obtusely serrate, apex acuminate, papery, adaxially sparsely hairy, abaxially densely tomentose, lobes triangular, basal veins 7–11; petiole 5–20 cm long. **Inflorescences** with flowers solitary, axillary on upper branches; pedicels 5–8 cm long, articulate near end. **Flowers** with epicalyx lobes 8, filiform, basally connate, 1–1.6 × ~0.2 cm; calyx campanulate, 2.5–3 cm long, lobes 5, ovate, acuminate; corolla white or reddish, becoming dark red, ~8 cm Ø; petals nearly orbicular, 4–5 cm Ø, hairy abaxially, barbate at base; staminal column 2.5–3 cm long, glabrous; styles 5, pilose. **Fruits** capsules, flattened globose, ~2.5 cm Ø, yellowish-hispid and woolly, mericarps 5; seeds reniform, villous abaxially. **Distr:** SE China, Taiwan. **Ecol:** In thickets along streams. Fl. Aug–Oct. **Zones:** 8–12. **Info:** The form with double flowers, *H. mutabilis* f. *plenus*, is commonly cultivated.

H. ovalifolius (Forsskål) Vahl, *Symb.* 1.50. 1790. (Syn: *H. calyphyllus, H. rockii, H. wildii*). [Rock's Hibiscus]. **Shrubs** or **perennial herbs**, 1–3 m. high; stems tomentose or pubescent when young. **Stipules** filiform, to 1.5 cm long. **Leaves** suborbicular in outline, obscurely or distinctly 3–5-lobed, to 12 × 12 cm, base cordate or subcordate, margins serrate, apex acute, adaxially pubescent or pilose, abaxially pubescent or tomentose; petiole to 9(–18) cm long, pilose. **Inflorescences** with flowers solitary, axillary; peduncle ~0.7(–1) cm long, pubescent. **Flowers** to 12 cm Ø, yellow usually with a brownish-maroon to maroon center; epicalyx of 5 bracts, pubescent; bracts to 1.8 cm long; calyx to 1.6 cm long, tomentellous, lobes ovate to ovate-lanceolate, joined to nearly halfway, usually 3-nerved; petals obliquely obovate, to 6 × 4·5 cm, pubescent outside, glabrous inside; staminal tube to 1.5 cm long, filaments free for 1.5–

Below: *Hibiscus moscheutos, Hibiscus mutabilis, Hibiscus mutabilis* f. *plenus.*

F&KS

TG

3 mm; style branches 4–5 mm long. **Fruits** capsules, ellipsoid, 1.8–2.5 × 1.5 cm; valves scarious, aristate, pubescent; seeds subreniform, 3 × 2.5 mm, tomentose. **Distr:** Tropical and southern Africa, Madagascar, Mascarene Islands. **Ecol:** In rainforests, riverine forests, thickets and grasslands. **Zones:** 10–12.

H. radiatus Cav., *Diss.* 3: 150, t. 54, fig. 2., 1787. [Monarch Rose Mallow]. **Herbs**, annual, 1–3(–5) m tall; stem simple or branched, sparsely aculeolate, prickles pustular-based and upturned, otherwise ± glabrous. **Stipules** filiform. **Leaves** polymorphic, simple, ± subcircular to usually 3–5(–7)-palmatifid, lobes elliptic to linear-ovate, to 15 cm long, margins serrate, apices acute or acuminate, glabrate or with few minute prickles on veins; petioles slender, elongate. **Inflorescences** with flowers solitary or in terminal pseudo-racemes (by reduction of uppermost leaves), axillary; flowers subsessile or shortly pedicellate, pedicels stout, to 5 mm long, aculeolate. **Flowers** with epicalyx segments 7–10, linear, basally united and adnate to base of calyx, flattened and slightly dilated at apex, usually with a small toothlike appendage on inner surface below apex, 0.8–1.5 cm long, setose; calyx cupuliform, lobed to below middle, 1.5–2.5 cm long, scarious, greenish or whitish, setose, lobes long-acuminate; petals cuneate-obovate, 3.5–8 cm long, rounded, yellow or red, with crimson basal spots; staminal tube about half as long as corolla, antheriferous most of its length, filaments mostly paired; style about as long as the staminal tube, stigmas densely fimbrillate. **Fruits** capsules, surrounded by the persistent epicalyx and calyx, ovoid, 1.5–2 cm long, acuminate, densely appressed-setose, valves ligneous; seeds subreniform-angulate, 3–5 mm long. **Distr:** SE Asia. **Ecol:** In lowland dry areas, grasslands and wastelands; 0–2,000 m. Fl. & Fr. throughout most of the year. **Zones:** 10–12. **Info:** This species is vegetatively very similar to *H. cannabinus* (with yellow flowers) and also superficially to *Cannabis sativa* (family Cannabaceae).

H. rosa-sinensis L., *Sp. Pl.* 2: 649, 1753. [China Rose, Chinese Hibiscus, Hawai'ian Hibiscus, Hibiscus, Rose of China, Shoe Black, Shoe Flower]. **Shrubs**, evergreen, 1–3 m high. **Stipules** filiform, 0.5–1.2 cm long, hairy. **Leaves** broadly or narrowly ovate, not lobed, 4–9 × 2–5 cm, base rounded or cuneate, margins dentate or lobed, apex acuminate, papery, pilose on veins abaxially only; petiole 0.5–2 cm long, villous. **Inflorescences** with flowers solitary, axillary, on upper branches, sometimes pendulous; pedicels 3–7 cm long, sparsely pilose or nearly glabrous, articulate near apex. **Flowers** epicalyx lobes 6–7, filiform, 0.8–1.5 cm long, basally connate, apex obtuse or acute, sparsely pilose; calyx campanulate, ~2 cm long, puberulent, lobes 5, ovate to lanceolate; corolla rosy-red, reddish, or orange-yellow, funnel-shaped, 6–10 cm Ø, often double; petals obovate, apex rounded, abaxially pilose; staminal column 4–8 cm long, glabrous; style branches 5. **Fruits** capsules, ovoid, ~2.5 cm long, glabrous, apex beaked. **Distr:** Origin obscure (perhaps China). **Ecol:** Fl. throughout the year. **Zones:** 9–12. **Info:** Plants with double flowers are referred to as *R. rosa-sinensis* var. *rubroplenus*.

H. sabdariffa L., *Sp. Pl.* 695, 1753. (Syn: *H. cannabinus*). [Roselle]. **Herbs**, annual or perennial, simple or branched; stems with sparse, simple, bulbous, spiny hairs. **Stipules** filiform, 6–8 mm long. **Leaves** ovate, unlobed in the lowermost part, 3–7 partite in upper part, lobes elliptic-lanceolate, coarsely serrate, surfaces nearly glabrous; petiole 4–15 cm long, almost spiny near apex. **Inflorescences** with flowers solitary, axillary, subsessile. **Flowers** with epicalyx segments 7–8, linear, 0.6–1 cm long, sparsely prickly; calyx fused at base, 1–2.5 cm long, woolly, also setose or prickly, lobes long acuminate-aristate, with a swollen, linear gland on central nerve at base; corolla yellow with a crimson center, 3–5 cm Ø; petals obovate, 4–6 × 2–4 cm; staminal column inserted. **Fruits** capsules 1.5–2 × ~1 cm, conical, beaked, appressed-setose; seeds many, 2–3 mm long, brown. **Distr:** Origin obscure. Common in tropical and subtropical regions of the world. **Ecol:** In lowland and medium altitude disturbed areas, open woodlands; to 1,200 m. Fl. & Fr. Dec–Apr. **Zones:** 9–12.

H. schizopetalus (Dyer ex Masters) J. D. Hooker, *Bot. Mag.* 106: t. 6524, 1880. (Syn: *H. rosa-sinensis* var. *schizopetalus*). [Coral Hibiscus, Fringed Hibiscus, Japanese Hibiscus, Japanese Lantern]. **Shrubs**, evergreen, erect, to 3 m tall; branchlets slender, usually pendulous, glabrous. **Stipules** subulate, ~2 mm long, usually caducous. **Leaves** elliptic or oblong, not lobed, 4–7 × 1.5–4 cm, papery, glabrous, base obtuse or broadly cuneate, margins dentate, apex acute or shortly acuminate; petiole 1–2 cm long. **Inflorescences** with flowers solitary, axillary, on upper branchlets, flowers pendulous; pedicels slender, 8–14 cm long, glabrous or slightly hairy, articulate in middle. **Flowers** with epicalyx lobes 5, lanceolate, 1–2 mm long, ciliate, apex obtuse or acute; calyx tubular, ~1.5 cm long, sparsely hairy, 5-lobed, usually dehiscent on 1 side; petals 5, red, ~5 cm long, deeply pinnatifid, strongly reflexed; staminal column longer than corolla, 9–10 cm long, glabrous, curved upward toward tip, anthers restricted to upper part; style branches 5,

Top row: *Hibiscus ovalifolius, Hibiscus radiatus, Hibiscus rosa-sinensis.*
Bottom row: *Hibiscus sabdariffa, Hibiscus schizopetalus, Hibiscus surattensis, Hibiscus tiliaceus.*

F&KS

F&KS

F&KS

Above: *Hibiscus tiliaceus* (variegated form), *Hibiscus trionum*, *Hibiscus waimeae*, *Kokia cookei*.

glabrous. **Fruits** capsules, oblong-cylindric, ~4 × 1 cm. **Distr:** E Africa. **Ecol:** Fl. throughout the year. **Zones:** 10–12.

H. surattensis L., *Sp. Pl.* 2: 696, 1753. (Syn: *Furcaria surattensis*). [Wild Sour, Bush Sorrel]. **Herbs** annual, subshrublike, usually procumbent, sometimes scandent, 0.5–2(–4) m tall or long, most parts sparsely villous and aculeate, prickles retrorsely curved. **Stipules** auriculate, foliaceous, ~5 mm long, sparsely long-hirsute. **Leaves** palmately 3–5-lobed, 5–10 × 5–11 cm, lobes ovate-lanceolate, 3–7 × 1.5–3 cm, margins irregularly serrate, sparsely strigose, basal veins 5, sparsely aculeate; petiole 2–7 cm long, ± hirsute. **Inflorescences** with flowers solitary, axillary; pedicel 1–5 cm long, villous. **Flowers** with epicalyx lobes 8–10, base terete, linear-lanceolate to filiform, 1–2 cm long, aculeate, outer branch/abaxial appendage ovate to elliptic, 0.7–1 cm long; calyx turning reddish in fruit, shallowly cup-shaped, lobes 5, ovate-lanceolate, ~2.5 cm long, apex long-acuminate, aculeate; corolla pale yellow with a dark red center; petals ~3.5 cm long. **Fruits** capsules, ovate-globose, ~ 1.2 × 1 cm, densely long-hairy, shortly beaked; seeds reniform, sparsely white-pubescent. **Distr:** Old World tropics. **Ecol:** In forest margins, open slopes; 300–1,200 m. Fl. Sep–Mar. **Zones:** 10–12.

H. tiliaceus L., *Sp. Pl.* 2: 694, 1753. (Syn: *Talipariti tiliaceum*, etc.). [Beach Hibiscus, Tree Hibiscus]. **Shrubs** or **trees**, evergreen 4–10 m tall; trunk to 60 cm Ø. **Stipules** foliaceous, oblong, ~2 × 1.2 cm, stellate pilose, apex rounded, caducous. **Leaves** orbicular to broadly ovate, 8–15 × 8–15 cm, base cordate, margins entire or obscurely crenate, apex abruptly acuminate, leathery, green, abaxially densely puberulent, adaxially sparsely scaly, glabrescent, basal veins 7 or 9; petiole 3–8 cm long. **Inflorescences** racemelike cymes, 1- to few-flowered, terminal or axillary; peduncles 4–5 cm long; pedicels 1–3 cm long, with 1 pair of stipulelike bracteoles at base. **Flowers** with epicalyx lobes 7–10, joined for about half its length, free lobes 2–2.5 mm long, triangular-acuminate with slightly rounded sinuses, densely puberulent; calyx 1.5–2.5 cm long, proximally connate, lobes 5, lanceolate, puberulent, persistent; corolla yellow with a dark purple center, campanulate, 6–7.5 cm Ø; petals obovate, 4–4.5 cm long, abaxially yellow-puberulent; filament tube ~3 cm long, glabrous; style branches 5, slender. **Fruits** capsules, subglobose to ovoid, ~2 cm Ø, obscurely beaked, densely hirsute, valves 5, woody; seeds reniform, smooth, glabrous. **Distr:** Pantropical. **Ecol:** On seashores, along streams, sandy soils; 0–300 m. Fl. Jun–Aug. Zones: 10–12. **Info:** A form with variegated leaves is commonly cultivated.

H. trionum L., *Sp. Pl.* 2: 697, 1753. (Syn: *H. africanus*, *Trionum annuum*, etc.). [Flower-of-an-hour, New Zealand Hibiscus, Venice Mallow]. **Herbs** annual, erect or procumbent, 25–70 cm tall; stems slender, white-hirsute. **Stipules** filiform, ~7 mm long, coarsely hirsute. **Leaves** 3–6 cm across, dimorphic; lamina on *proximal* parts of stem orbicular, those on *distal* part of stem palmately 3–5-lobed, central lobe longer, lateral lobes shorter, lobes obovate to oblong, usually pinnate, adaxially sparsely hirsute or glabrous, abaxially sparsely hairy; petiole 2–4 cm long, hirsute and puberulent. **Inflorescences** with flowers solitary, axillary; pedicel ~2.5 cm long, hirsute. **Flowers** with epicalyx lobes 12, filiform, basally connate, ~8 mm long, hirsute; calyx greenish, campanulate, swollen, connate for about half its length, 1.5–2 cm long, membranous, hirsute, lobes 5, triangular, longitudinally purple-veined; corolla very pale yellow with a purple center, 2–3 cm Ø, petals 5, obovate, ~2 cm long, abaxially ± puberulent; staminal column ~5 mm long; filaments slender, free for ~3 mm, anthers yellow; styles 5, glabrous. **Fruits** capsules, oblong-globose, ~1 cm Ø, hirsute; mericarps 5, endocarp black, thin, exocarp papery; seeds black, reniform. **Distr:** Old World (from S Europe to S Africa, Madagascar, Asia and Australia, New Zealand). **Ecol:** Fl. Jul–Oct. **Zones:** 9–12.

H. waimeae A. Heller, *Minnesota Bot. Stud.* 1(9): 851, t. 53, 1897. [Hawaiian/Kauai White Hibiscus]. **Trees**, 6–10 m tall; branchlets densely pubescent. **Stipules** filiform, 5–8 mm long, early deciduous. **Leaves** orbicular to broadly ovate-elliptic, 5–18(–25) × 3–13(–18) cm, base obtuse to truncate, margins ± crenate or crenate-serrate, apex obtuse to acute, adaxially pubescent or glabrous, abaxially grayish velvety-tomentose; petioles pubescent, about half as long as lamina. **Inflorescences** with flowers solitary, borne near branch ends; pedicels articulate, 2–3 cm long, densely pubescent; involucral bracts 7–8, linear-lanceolate, 0.8–1 cm long. **Flowers** strongly fragrant; calyx tubular, (2–)3–4.5 cm long, lobes 0.8–1.5 cm long, tube usually longitudinally splitting in fruit; petals flaring, opening white in the morning, fading to pinkish by afternoon, usually broadly obovate, (4–)9–13 cm long, basally adnate to staminal column forming a tube (1.5–)3–6 cm long; staminal column exserted, reddish to crimson apically, to 15 cm long, filaments in upper half of staminal column, spreading, to 2.5 cm long. **Fruits** capsules, cartilaginous, obovoid, to 1.8–2.5 cm long, apiculate, glabrous; seeds angular-reniform, ~4 mm long, brownish-tomentose. **Distr:** Hawai'i (Kaua'i). **Ecol:** In coastal mesic, mixed mesic, and moist/wet forests; 250–1,200 m. Fl. throughout most of the year. **Zones:** 10–12.

Kokia

Species 4. **Trees**, deciduous, glandular-punctate, glabrous, to 10 m tall; branches with prominent petiole scars. **Stipules** caducous. **Leaves** orbicular in outline, 5-, 7-, or 9(11)-lobed with shallow, angulate or deltate divisions, base cordate to subtruncate, margins ± entire, adaxially glabrous, abaxially densely pubescent at base, sometimes with glands on major veins; petioles deciduous with prominent scars. **Inflorescences** with flowers solitary, clustered near ends of branches, axillary, pedicels stout, articulate near middle and sometimes with a persistent bract, terminal nectaries absent; involucral bracts 3, foliaceous, distinct or basally connate, entire or lobed. **Flowers** with calyx cylindrical, 2–5-lobed, lobes usually falling away leaving a truncate tube; corolla zygomorphic, orangish-red to scarlet or brick-red, curved in bud, densely yellowish silky-pubescent, petals reflexed above at anthesis, twisted at base; staminal column exserted, curved, antheriferous above, terminated by 5 teeth; ovary 5-locules, ovules 1 per locule, style clavate, unbranched, stigma 5-lobed, sulcate. **Fruits** capsules, ovoid-acuminate to globose-apiculate, woody, loculicidally dehiscent; seeds obovoid, 1–1.5 cm long, densely reddish-lanate. **Distr:** Hawai'i. **Cult:** Propagate annuals from seed, and perennials from seed, cuttings, grafting or by air-layering. Requires a fertile, moist, well-drained soil in full sun.

K. cookei O. Deg., *Fl. Hawaiiensis* 221, 1934. [Hawaiian Tree Cotton, Molokai Tree Cotton]. **Trees**, deciduous, 3–5 m tall. **Leaves** 5–13 cm across, 5-lobed or 7-lobed, base cordate to subcordate with an open sinus, abaxially tufted-pubescent in axils of major veins near base. **Inflorescences** with flowers solitary; pedicels 3–7 cm long, bracts to 1.5 cm long, sometimes persistent; involucral bracts broadly ovate, 2.5–3.5 × 1.5–3 cm, shallowly to deeply 3–5-lobed, apex acute to obtuse. **Flowers** with calyx tube ~1 cm long; petals orangish-red, ~8 × 6 cm; staminal column ~6.5 cm long. **Fruits** capsules, depressed-globose to globose, ~to 3 cm Ø, apex apiculate; seeds 1–1.3 cm long. **Distr:** Hawai'i (Moloka'i). **Ecol:** In lowland dry forests; ~200 m. Fl. & Fr. May–Jun. **Zones:** 10–12.

K. drynarioides (Seem.) Lewton, *Smithsonian Misc. Collect.* 60(5): 3, t. 5, 1912. (Syn: *Gossypium drynarioides*, *Kokia rockii*). [Hawaiian Tree Cotton]. **Trees**, to 8 m tall. **Leaves** 8–20 cm across, (5)7- or 9-lobed, base cordate with a usually closed sinus, abaxially pubescent near base and at apex of petioles, sometimes with nectaries on major veins. **Inflorescences** with flowers solitary; pedicels 5–10 cm long; involucral bracts broadly ovate to suborbicular, 4–7 cm long, usually sinuately 3–5-lobed, apex obtuse. **Flowers** with calyx irregularly 2–5-lobed, ~2.2–3 cm long, lobes deciduous or sometimes persistent, tube 1.5 cm long, densely yellowish-pubescent, later glabrate except at base; petals scarlet, 9–11 cm long; staminal column to 11 cm long. **Fruits** capsules, ovoid, ~3 cm long, apex acuminate; seeds 1.2–1.5 × ~0.8 cm, pubescent, hairs to 3 mm long. **Distr:** Hawai'i. **Ecol:** In dry forests; 450–1,900 m. Fl. & Fr. Jan–Jun. **Zones:** 10–12.

Malva

Species ~30. **Herbs**, annual or perennial, or **undershrubs**, ascending or erect, glabrous or hairy. **Stipules** sessile, usually ciliate, persistent. **Leaves** alternate; lamina palmately lobed or sometimes deeply dissected. **Inflorescences** with flowers solitary or fascicled, axillary; involucellar bracts usually 3, linear or foliaceous, usually free. **Flowers** with calyx cup-shaped, 5-lobed, often accrescent and spreading in fruit; petals 5, usually purple, sometimes white, rose to dark red, apex usually emarginate or with a prominent notch; staminal column included, apex antheriferous; ovary with 8–15 pistils, ovule 1 per locule, erect, style branches as many as pistils, adaxial surface stigmatic. **Fruits** schizocarps, oblate, pubescent or glabrous; mericarps 9–15, indehiscent, mature carpels without spines; seeds 1 per mericarp. **Distr:** N Africa, Europe, Asia. **Cult:** Propagate herbaceous species from seeds, and shrubby seeds from softwood cuttings. Requires a rich, well-drained soil in full to partial sun.

M. wigandii (Alefeld) M.F. Ray, *Novon* 8(3): 293–294, 1998. (Syn: *Axolopha wigandii, Lavatera africana, L. maritima*). [Sea Mallow]. **Shrubs**, evergreen, perennial, branched, to 0.5–2 m high, with indumentum of short hairs; stems brownish-green, becoming gray and woody; branches stiff. **Stipules** green, erect, triangular, 4–6 mm long. **Leaves** alternate; lamina suborbicular, to 5 × 4 cm, palmately 5-lobed, base cordate, margins crenate, 7–9 nerved at base, surfaces tomentose; petiole green, to 3 cm long, sparsely hairy. **Inflorescences** with flowers solitary, slightly nodding, axillary; pedicel 3 cm long, shortly hairy, articulated ~5 mm below epicalyx; bracteoles 3, nearly free, elliptic-lanceolate to ovate, ~1 × 0. 5 cm, acute. **Flowers** with epicalyx valvate; calyx valvate; sepals 5, triangular-ovate, 1.2–1.5 cm long, acuminate, basally connate; corolla convolute in aestivation, ~8 cm Ø; petals 5, spirally arranged, hypogynous, obovate, ~4 cm long, retuse, purplish-white with a magenta claw and base, feathered with ~15 magenta veins; staminal column ~2 cm long, purplish-red, lower half with short white hairs, upper half antheriferous, anthers pink, pollen yellow; ovary 12-loculed, style branches filiform, purple, as many as locules. **Fruits** schizocarps, mericarps 12, 5–6 mm long, pale brown, glabrous, angles acute, denticulate; seeds midbrown, reniform. **Distr:** W Mediterranean regions. **Ecol:** In dry rocky places, usually near the sea. Fl. Sep–Nov. **Zones:** 8–11.

Malvaviscus

Species 3. **Shrubs** or **small trees**, sometimes scandent, pubescent or glabrous. **Stipules** subulate, caducous. **Leaves** simple, elliptic or ovate, entire or 3–5-lobed, base truncate or cordate, margins crenate to dentate; foliar nectaries lacking. **Inflorescences** with flowers solitary, axillary, or sometimes in terminal cymelike clusters. **Flowers** with epicalyx lobes 5–10(–12), linear, lanceolate, or spathulate; calyx campanulate or tubular, 5-lobed; corolla tubular; petals 5, red or crimson-red, bases auriculate, remaining adpressed to staminal column; filament tube longer than corolla, apex 5-toothed, anthers below teeth, on outside of staminal column only; ovary 5-loculed, locules 1-ovuled, style branches 10, stigmas capitate. **Fruits** schizocarps, fleshy, usually red (sometimes white), broader than long, ultimately drying and separating into mericarps; mericarps 5, indehiscent; seeds 1 per carpel. **Distr:** Tropical America. **Cult:** Propagate from seeds or cuttings. Requires a moist, humus-rich, well-drained soil in full sun.

M. arboreus Cavanilles, *Diss.* 3: 131, 1787. (Syn: *Hibiscus malvaviscus*, etc.). [Sleeping Hibiscus, Turk's Cap Mallow, Wax Mallow]. **Shrubs**, ~1 m high; branchlets sparsely villous to glabrate, rarely glabrous. **Stipules** filiform, ~4 mm long, usually caducous. **Leaves** broadly cordate to ovate-cordate, usually trilobed, sometimes entire, 6–12 × 2.5–10 cm, base broadly cuneate to nearly rounded or cordate, margins crenate, sometimes irregularly so, apex acuminate, surfaces nearly glabrous or stellate-pilose, basal veins 3 or 5; petiole 2–5 cm long, puberulent. **Inflorescences** with flowers solitary, axillary, pendulous; pedicels 0.3–1.5 cm, villous or puberulent; epicalyx lobes spathulate, 0.8–1.5 cm long, basally connate, hairy. **Flowers** tube-shaped, slightly expanding only at top, 2.3–5 cm long; calyx campanulate, ~1 cm Ø, lobes 5, slightly longer or shorter than bracteoles, hirsute; petals 5, scarlet-red, 2.5–5 cm long; staminal column 5–7 cm long, exceeding corolla tube; style branches 10. **Fruits** bright red, usually 3- or 4-seeded. **Distr:** C and N America (SE United States); widely cultivated in the tropics. **Ecol:** In scrublands, wet ravines, forests and thickets; to 2,100 m. Fl. throughout most of the year. **Zones:** 9–12.

M. penduliflorus Candolle, *Prodr.* 1: 445, 1824. (Syn: *M. arboreus* ssp. [var.] *penduliflorus*). [Cardinal's Hat, Sleeping Hibiscus, Turk's Cap Mallow]. **Shrubs**, to 2 m high; branchlets villous to glabrate. **Stipules** filiform, ~4 mm long, caducous. **Leaves** lanceolate to narrowly ovate, 6–12 × 2.5–6 cm, surfaces nearly glabrous or stellate-pilose, basal veins 3, base broadly cuneate to nearly rounded, margins crenate-serrate; petiole 1–2 cm long, villous. **Inflorescences** with flowers solitary, axillary, pendulous; pedicels ~1.5 cm long, villous. **Flowers** with epicalyx lobes ~8, spathulate, 1–1.5 cm long, margins ciliate; calyx slightly longer than epicalyx, hirsute; corolla tubular, slightly expanded apically only, ~5–10 cm long; petals red; staminal column ~7 cm long, exserted. **Fruits** capsules, rarely formed in cultivation. **Distr:** Tropical America; widely cultivated in the tropics. **Ecol:** In moist or wet thickets, or forests; below 2,500 m. Fl. throughout the year. **Zones:** 11–12.

Pavonia

Species ~180. **Herbs** prostrate perennial, erect subshrubs, or shrubs (rarely arborescent), stellately pubescent, sometimes glabrate or viscid. **Leaves** simple, margins serrate, entire or angulate-lobed often pellucid-punctate, sometimes asymmetrical. **Inflorescences** with flowers solitary or paired, axillary or aggregated into clusters at ends of branches; epicalyx (bractlets) (4)5–22(–many), distinct or ± basally connate with calyx. **Flowers** with calyx 5-lobed; petals white, yellow, lavender, purple, sometimes with a dark basal spot; androecium included or exserted; carpels obovoid-trigonous, never glochidiate but sometimes echinulate and often 1–3-aristate or -rostrate, often separating septicidally (sometimes carpels early or tardily dehiscent), stylar branches 10, stigmas capitate. **Fruits** minutely pubescent or glabrous, often barbed, adapted for dispersal by animals. **Distr:** Tropical and warm temperate regions. **Cult:** Propagate from seeds or half-hardened cuttings. Requires a moist, well-drained soil in full or partial sun.

P. bahamensis Hitchcock, *Rep. Missouri Bot. Gard.* 4: 63, t. 11, 1893. (Syn: *Malache bahamensis, Malvaviscus cokeri*). [Bahama Marsh/Swamp Bush]. **Shrubs**, 2–4 m high, stems minutely sparsely pubescent. **Stipules** linear, 5–6 mm long. **Leaves** ovate, mostly 4–6 × 2.5–5 cm, base cordate to subtruncate, margins entire, acute, palmately 5–7-nerved, surfaces minutely and sparsely pubescent, densely bearded in axils of principal nerves beneath; petiole 1–3 cm long. **Inflorescences** with flowers solitary, axillary; pedicels 1–3 cm long, slender; involucral bracts 5–6(–10), ligulate or lanceolate, 1–1.2 cm × 1.5–3 mm, briefly connate, minutely pubescent basally, subglabrous distally. **Flowers** with calyx subequal to involucel, sparsely pubescent to glabrate outside, densely puberulent inside; corolla tubular, greenish-yellow, pubescent outside, 1.5–2 cm long; androecium exserted, 3 cm long. **Fruits** ~1 cm Ø, glabrous, subglobose and 5-winged; mericarps 6–7 mm long, with broad lateral wings. **Distr:** Bahamas Archipelago. **Ecol:** Growing on rocky coasts or near mangroves; near sea level. **Zones:** 11–12.

P. multiflora St.-Hilaire, *Fl. Bras. Merid.* 1: 239, 1827. (Syn: *Malvaviscus multiflorus*, etc.). [Brazilian Candles]. **Shrubs**, 1.5–2.5 m high. **Stipules** subulate, erect, 0.9–1.3 cm × 0.5–1 mm. **Leaves** narrowly elliptic, mostly 12–26 × 2–8 cm, base truncate to cuneate, margins ± serrate, apex acuminate, hairy to ± glabrous, penninerved (6–8 secondary nerves per side); petiole 2.5–12 cm long, sometimes hirsute, usually. **Inflorescences** usually axillary 2–3-flowered fascicles, or sometimes flowers solitary; flowers often crowded apically because of shortened internodes and deciduous leaves; pedicels to 7 cm long; involucellar bracts 20–24, seemingly biseriate, outer spreading to reflexed, shorter than the arcuate to erect inner bracts, all narrowly ligulate, pubescent and ciliate, sometimes reddish, longer bracts 1.2–2(–3) cm long. **Flowers** with calyx 1–2 cm long, deeply divided, with stellate and glandular-hairy, sometimes basally setose; corolla tubular, 2–2.5 cm long, barely exceeding calyx to twice calyx length, pubescent outside, yellow (reddish in bud) or dark purple in some cultivated forms; androecium exserted, about twice corolla length, pallid

Below: *Kokia drynarioides, Malva wigandii, Malvaviscus arboreus, Malvaviscus penduliflorus.*

SZ

CS

Top row: *Pavonia bahamensis, Pavonia multiflora.*
Bottom row: *Sida acuta, Sida fallax, Sida mysorensis, Sida rhombifolia.*

or purplish, filaments 3–5 mm long; styles exserted, slender, stigmas villous. **Fruits** 7–9 mm Ø, short-beaked, glabrous, yellowish; mericarps 7–9 mm long, smooth. **Distr:** Atlantic coast of Brazil. **Ecol:** In primary forests. Fl. Nov–Mar. **Zones:** 10–12.

Sida

Species 200. **Herbs** perennial or annual, **subshrubs** or **shrubs**, to 2 m high, most parts glandular-hairy. **Stipules** threadlike to narrowly lanceolate. **Leaves** simple, entire, entire, sometimes lobed, margins usually dentate, without foliar nectaries. **Inflorescences** with flowers solitary or paired, axillary or subterminal, often in axillary or terminal racemes or panicles, rarely in umbels or glomerules; epicalyx absent. **Flowers** with calyx campanulate or cup-shaped, 5-lobed, often 10-ribbed basally and plicate in bud; corolla mostly yellow, rarely white or ± orange (or rose or purplish), sometimes with a dark center; petals 5, free, basally connate; staminal tube pubescent or glabrous, anthers many; ovary 5–10-loculed, ovules 1 per locule, pendulous, style branches as many as carpels, stigma capitate. **Fruits** schizocarps, ± disk-shaped or globose; mericarps (4–)5–10(–14), sculptured or smooth, sometimes partly membranous, mostly beaked, often with 1 or 2 apical awns, often minutely puberulent, dehiscent or indehiscent; seeds 1 per mericarp, smooth, glabrous except sometimes for minute hairs around hilum. **Distr:** Africa, Asia, Australia, Pacific Islands; N and S America. **Cult:** Propagate from seeds. Requires any well-drained soil in full or partial sun.

S. acuta N.L. Burman, *Fl. Indica.* 147, 1768. (Syn: *Malvastrum carpinifolium*, etc.). [Wireweed]. **Subshrubs** or **herbs**, erect, 1–2 m tall; branchlets pilose or subglabrous. **Stipules** filiform, 4–6 mm long. **Leaves** ± distichous; lamina ovate, oblong, lanceolate, or linear-lanceolate, 2–5 x 0.4–1 cm, base obtuse, margins dentate, sometimes basally entire, apex acute or acuminate, glabrous or sparsely pilose; petiole 4–6 mm long. **Inflorescences** with flowers solitary or paired, axillary, sometimes congested at stem apex; pedicel 0.4–1.2 cm long. **Flowers** with calyx shallowly cup-shaped, ~6 mm across, basally connate, margins often ciliate, lobes 5, caudate; corolla yellow, sometimes white or yellow-orange, 0.8–1 cm Ø; petals obovate, 6–7 mm long, ciliate, base attenuate, apex rounded; filament tube ~4 mm long. **Fruits** schizocarps, nearly globose; mericarps (4–)6(–9), tetrahedral, ~3.5 mm long, basally transversely ridged, apex beaked, extending into 2 awns, apically dehiscent; seeds trigonous, ~2 mm long. **Distr:** SE Asia, China, Indo-China. **Ecol:** In scrubs, roadsides and wastelands. Fl. Oct–Feb. **Zones:** 10–12.

S. fallax Walp., *Nov. Actorum Acad. Caes. Leop.-Carol. Nat. Cur.* 19(suppl. 1): 306, 1843. (Syn: *S. meyeniana*, etc.). **Shrubs**, prostrate or diffuse to erect, 0.2–1.5(–3) m high, ± velvety-tomentose. **Stipules** filiform to linear, 0.3–1 cm long. **Leaves** usually ovate to subcircular, 1–12 cm long, base obtuse to subcordate, margins serrate or crenate, apex acute to acuminate or obtuse, discolorous, adaxially bright green, glabrous or ± pubescent, abaxially densely pubescent; petiole 0.5–5.5 cm long. **Inflorescences** with flowers solitary or 2–7 clustered per node, axillary, borne especially at branch ends; pedicels articulate, 0.5–2 cm long. **Flowers** with calyx green or dark purple, campanulate, ~6–9 mm long, tube strongly ribbed, lobes triangular to ovate, acute to acuminate; corolla yellow to orangish-yellow, sometimes dark maroon at base, rotate, 1.2–2.5 cm Ø, petals broadly obovate, 0.6–1.5 cm long, apex asymmetrically bilobed, claw ciliate; staminal column included, yellow, 2.5–4 mm long, glabrous or apically puberulent. **Fruits** capsules, ~ 6 mm long, mericarps (5)6–9, pale brown to black, ± trigonous, 2.8–4 mm long; seeds brown or black, trigonous-obovoid, 1.7–2.2 mm long. **Distr:** Widespread across Pacific Islands to China. **Ecol:** In dry sandy areas and rocky coastlines, often near the ocean, and also in montane dry shrublands; to 600 m. Fl. throughout the year. **Zones:** 10–12.

S. mysorensis Wight and Arnott, *Prodr. Fl. Ind. Orient.* 1: 59, 1834. (Syn: *S. glutinosa*, etc.). [Mysore Wildweed]. **Herbs**, shrubby, erect, to 1 m tall; stem hairy. **Stipule** filiform, ~5 mm long. **Leaves** ovate-cordate, 3–6 x 2.5–4.5 cm, base cordate, margin crenate, apex acuminate, surfaces viscid-stellate; petiole 1–3 cm long. **Inflorescences** with flowers solitary or paired, axillary or subterminal, often on congested reduced axillary shoots; pedicel 0.2–0.6(–1.5) cm long. **Flowers** with calyx widely campanulate, 6–8 mm long, sparsely pilose, lobes 2.5–3 x ~2.5 mm, acute or acuminate; corolla yellow, 1–1.2 cm Ø; petals obtriangular, glabrous; filament tube strigose. **Fruits** schizocarps, nearly globose, 3–4 mm Ø; mericarps 5, ovoid-tetrahedral, ~2.5 mm long, apex not awned, enclosed in persistent calyx; seeds ovoid, slightly 3-sided, ~2 mm long, glabrous. **Distr:** SE Asia, China, Indo-China, Thailand, Indonesia, Philippines, Taiwan. **Ecol:** In forest margins. **Zones:** 10–12.

S. rhombifolia L., *Sp. Pl.* 2: 684, 1753. (Syn: *Malva rhombifolia*, etc.). [Arrowleaf Sida, Cuban

Above: *Thespesia grandiflora, Thespesia populnea, Urena lobata, Wissadula periplocifolia.*

Jute]. **Subshrubs**, erect or prostrate, many-branched, to ~1 m high. **Stipules** spinelike, 3–5 mm long. **Leaves** rhombic to oblong-lanceolate or obovate, rarely linear-lanceolate, 1–4.5 × 0.6–2 cm, base broadly cuneate, margins dentate, apex obtuse to acute, adaxially sparsely pilose to subglabrous, abaxially gray-white pilose; petiole 2–5(–8) mm long, puberulent. **Inflorescences** flowers solitary, axillary; pedicel 1–2.5 cm long, densely tomentose, articulate above middle. **Flowers** with calyx cup-shaped, 4–5 mm long, abaxially pubescent, lobes triangular, apices acute; corolla ~1 cm Ø, petals yellow, obovate, ~8 mm long, base attenuate, apex rounded; filament tube 4–5 mm long, glabrous; style branches 8–10. **Fruits** schizocarps, semiglobose to broadly turbinate, 6–7 mm Ø, mericarps 7–10, 2.5–3 mm excluding awn, shallowly grooved to near base, eventually dehiscent, side walls usually thin, not veined, puberulent, apex usually (1 or)2-awned, awns to 1.5 mm long; seeds reniform, ~2 mm long, blackish. **Distr:** Bhutan, Cambodia, China, India, Laos, Nepal, Taiwan, Thailand, Vietnam; pantropical. **Ecol:** In scrubs, on open slopes, and streamsides. **Zones:** 9–12.

Thespesia

Species about 17. **Trees** or **shrubs**, glabrous or pubescent, with an indumentum of scales or hairs. **Stipules** very slender. **Leaves** simple, ovate, entire or lobed, margins entire, often with abaxial foliar nectaries. **Inflorescences** with flowers solitary, rarely in small cymes, axillary. **Flowers** large and showy; epicalyx lobes 3–5, minute, caducous after flowering, sometimes subtended by 3-merous nectaries; calyx cup-shaped, truncate to 5-lobed; corolla campanulate; petals 5, yellow (or white or pink), with or without a dark purple basal spot; staminal column antheriferous throughout, apex 5-toothed, usually included; ovary 5-loculed; locules several-ovuled, style rod-shaped, 5-grooved, stigma 3–5-sulcate or rarely 3–5-lobed, decurrent. **Fruits** capsules, 3–5-locular, globose or pyriform, leathery or woody, dehiscent or indehiscent, sometimes slightly fleshy; seeds 3–many per locule, obovoid, glabrous or hairy. **Distr:** Tropical Africa, America, Asia, Australia. **Cult:** Propagate from seeds, cuttings or by air-layering. Requires a moist, well-drained soil in full or partial sun.

T. grandiflora DC., *Prodr.* 1: 456, 1824. (Syn: *Montezuma speciosissima*, etc.). [Maga, Maga Wood]. **Trees**, to ~15 m tall; trunk to 30 cm Ø, bark gray or brown, rough and furrowed; twigs lepidote when young. **Stipules** 3–4 mm long. **Leaves** cordate, 5–20 cm long, base cordate to subtruncate, margins entire, apex acute or acuminate, 5–7-nerved from base, abaxially lepidote; petiole long. **Inflorescences** flowers solitary, axillary; peduncles stout 10–14 cm long; involucre bracts linear, 1–2 cm long, caducous. **Flowers** with calyx cup-shaped, ~2 cm long, coriaceous, lepidote, shedding as a ring after flowering, with 3 narrow bracts 1 cm long at base; petals broader on one side, 7–9 cm long, opening to deep rose, aging to crimson; anthers in 5 double ranks, bright yellow. **Fruits** fleshy or leathery, subglobose, 3–5 cm Ø; seeds obovoid, brown, 1.2 cm long. **Distr:** Puerto Rico. **Ecol:** In woods and on hillsides at low elevations. **Zones:** 11–12.

T. populnea (L.) Solander ex Corrêa, *Ann. Mus. Natl. Hist. Nat.* 9: 290, 1807. (Syn: *Hibiscus populneus*, etc.). [Milo, Portia Tree, Tulip Tree]. **Trees** or **shrubs**, evergreen, 3–6 m tall; branchlets with minute-scaly, sometimes dense. **Stipules** filiform-lanceolate, 2–7 mm long, usually caducous. **Leaves** ovate-cordate to triangular, 7–18 × 4.5–12 cm, base truncate to cordate, margins entire, apex long-caudate, adaxially glabrous or with scales, abaxially scaly; petiole 4–10 cm long, scaly. **Inflorescences** with flowers solitary, axillary; pedicels 2.5–6 cm long, densely scaly. **Flowers** with epicalyx lobes 3–4, filiform-lanceolate, 0.8–1 cm long, usually caducous, scaly; calyx cup-shaped, 1–1.5 cm Ø, subtruncate with 5 teeth, each ~0.5 mm long, densely scaly; corolla yellow, adaxially purple at base, campanulate, ~5 cm long; staminal column ~2.5 cm long. **Fruits** capsules, globose to pyriform, ~5 × 2 cm Ø, stipe 6–8 cm long, slightly fleshy, ± indehiscent; seeds triangular-ovoid, 8–9 mm long, brown-hairy or glabrous, veined. **Distr:** Africa, Cambodia, India, Japan, Maldives, Philippines, Sri Lanka, Thailand, Vietnam. **Ecol:** Common on sea coasts. Fl. throughout the year. **Zones:** 11–12.

Urena

Species ~6. **Herbs** perennial, or **shrubs**, stellate. **Leaves** alternate; lamina orbicular or ovate, palmately lobed or sinuate, with 1 or more prominent foliar nectaries on abaxial surface. **Inflorescences** with flowers solitary or fascicled, rarely racemose, axillary or rarely aggregated on twig tips; epicalyx campanulate, 5-lobed. **Flowers** with calyx 5-parted; petals 5, abaxially puberulent; staminal column truncate or slightly incised, anthers numerous, on outside of staminal column only, nearly sessile; ovary 5-loculed, locules 1-ovuled, style branches 10, reflexed, stigma discoid, apically ciliate. **Fruits** schizocarps, subglobose; mericarps 5, ovoid, usually with spines, spines with a cluster of short barbs at tips; seed 1, obovoid-trigonous or reniform, glabrous. **Distr:** Tropical and subtropical regions. **Cult:** Propagate from seeds. Requires a moist, well-drained soil in full or partial sun.

U. lobata L., *Sp. Pl.* 2: 692, 1753. [African/Congo Jute, Caesar Weed, Hibiscus Burr]. **Herbs**, subshrublike, erect, to 1 m high; branchlets tomentose. **Stipules** filiform, ~2 mm long, caducous. **Leaves** on *proximal* part of stem nearly orbicular, 4–5 × 5–6 cm, base rounded or nearly cordate, margins serrate, apex trilobed, leaves on *middle* part of stem ovate, 5–7 × 3–6.5 cm, leaves on *distal* part of stem oblong to lanceolate, 4–7 × 1.5–3 cm, adaxially puberulent, abaxially gray-puberulent; petiole 1–4 cm long. **Inflorescences** with flowers solitary or slightly aggregated, axillary; pedicels ~3 mm long, woolly. **Flowers** with epicalyx lobes basally connate, ~6 mm long, puberulent; calyx cup-shaped, lobes 5, slightly shorter than bracteoles, puberulent, caducous; corolla reddish, ~1.5 cm Ø, petals 5, obovate, ~1.5 cm, abaxially puberulent; staminal column ~1.5 cm long, glabrous; style branches 10, hirsute. **Fruits** flattened-globose, ~1 cm Ø, mericarps puberulent and spiny with hooked spines. **Distr:** Pantropical. **Ecol:** In grasslands, forest margins; 500–2,200 m. Fl. Jul–Oct. **Zones:** 10–12. **Info:** Variable species with numerous infraspecific taxa described.

Wissadula

Species 25–30. **Herbs** or **subshrubs**, annual or perennial, usually erect, hairy or sometimes glabrate, sometimes with long simple hairs. **Leaves** simple, ovate to narrowly triangular, without foliar nectaries, base cordate, margins entire or crenate-dentate, apex acuminate. **Inflorescences** axillary or subterminal, usually grouped into terminal panicles; pedicels articulate. **Flowers** without epicalyx; calyx shallowly cup-shaped to campanulate, lobes 5, longer than tube; corolla yellow, cream, or white, small, ± rotate; petals 5, basally connate, adnate to filament tube, filament tube shorter than free portions of filaments, anthers all terminal, not exserted; ovary 3–5-loculed, ovules 3 per locule, usually erect, style arms as many as locules, filiform, longer than filaments, stigmas capitate. **Fruits** schizocarps, obconical; mericarps 3–5, eventually dehiscent; seeds 1–3 per mericarp. **Distr:** Predominantly in tropical America, few in tropical Asia and Africa. **Cult:** Propagate from seeds. Requires a moist, well-drained soil in full or partial sun.

W. periplocifolia (L.) C. Presl ex Thwaites, *Enum. Pl. Zeyl.* 27, 1858. (Syn: *Sida periplocifolia*, etc.). [White Velvetleaf]. **Subshrubs**, ~1 m high; stems densely fawn-puberulent. **Stipules** subulate, ~3 mm long. **Leaves** long-triangular, ovate, or ovate-lanceolate, 3–7 × 1.5–2.5 cm, base truncate to subcordate, margins entire or shallowly undulate, apex long-acuminate to acute, abaxially tomentose, adaxially minutely pubescent; petiole 0.3–2 cm long, tomentose. **Inflorescences** with *lower* flowers solitary, axillary; *upper* flowers in lax, terminal panicles to 80 cm long; pedicels 1–2 cm long, to 4 cm in fruit, puberulent, articulate near tip. **Fruits** with calyx cup-shaped, ~3 mm long, lobes ovate to triangular, acute; corolla pale yellow, ~7 mm Ø; petals obovate, ~4 mm long; staminal column glabrous, free filaments many, 1–2 mm long; style short. **Fruits** pseudo capsules, obconic, ~1 cm Ø, apically truncate; mericarps with a short mucro, abaxially rounded, subglabrous; seeds black, ~3 mm long, single

HS

Above: *Adansonia digitata*, *Bombax ceiba*, *Cavanillesia platanifolia* (flowers), *Cavanillesia platanifolia* (fruits).

proximal seed densely hairy, 2 distal seeds hairy. **Distr:** Cambodia, China, India, Indonesia, Laos, Sri Lanka, Thailand. **Ecol:** In thickets near sea level, dry slopes. Fl. Sep–Feb. **Zones:** 10–12. **Info:** Cultivated for medicinal purposes.

Subfam. BOMBACOIDEAE

Syn: Bombacaceae.
Genera 16, **species** 120. **Trees**, often massive, rarely shrubs; trunk often stout and bottle-shaped, with parenchymatous water-storage tissue; prickles stout; bark thin, often green. **Leaves** palmately lobed. **Inflorescences** with flowers solitary or 1–2 together, axillary. **Flowers** often very large; carpels 2–5; filaments usually fasciculate, anthers bisporangiate/monothecal, pollen ± flattened; staminodes usually absent. **Fruits** capsules, rarely indehiscent; endocarp often pubescent, sometimes with abundant floss (kapok) or sparse hairs; seeds usually glabrous. **Distr:** Tropical, especially America and Africa.

Adansonia

Species 8. **Trees**, deciduous (during dry seasons), long-lived (Swart 1963), 5–30 m tall; trunks unarmed, massive, cylindrical, tapering from bottom to top or bottle-shaped; crowns usually compact; trunk of some extant specimens are >23 m Ø, and are estimated to be 1,000–2,000 years old (Llamas 2003); primary branches massive, unarmed (rarely with conical spines), horizontal or ascending; bark smooth or foliaceous. **Stipules** to (1–1.5)2 mm long, usually caducous. **Leaves** palmately compound (juvenile leaves simple), spirally arranged; leaflets 5–11, elliptic, lanceolate or oblanceolate, lateral leaflets smaller than medial, sessile or petiolulate, margins entire or toothed, apex acute (rarely obtuse) to apiculate, membranous to coriaceous; petiole bipulvinate. **Inflorescences** with flowers solitary, rarely paired, axillary; peduncles and pedicels ± distinct; bracteoles usually 3, caducous. **Flowers** perfect, hypogynous; calyx synsepalous, enclosing flower until at least 6 hours before anthesis, lobes 5, reflexed or contorted at base of flower, sericeous and cream, pink, or red within; corolla actinomorphic or slightly zygomorphic, petals 5, free, convolute, inserted below staminal tube, creamy-white, yellow, or red; staminal tube cylindrical or slightly tapering, filaments numerous, free, usually glabrous, white or yellowish; ovary superior, conical to oblong, tomentose, ovules numerous, stigma usually with 5–10 short lobes, yellowish or dark red. **Fruits** berries, large, dry, globose, ovoid, or oblong-cylindrical, tomentose, usually indehiscent; seeds many, pulp white or cream, powdery or spongy; seeds many, reniform, 1–1.5(–2) cm long. **Distr:** Tropical Africa (1 sp.), Australia (2 spp.), Madagascar (6 spp.). **Cult:** Propagate from seeds or cuttings. Requires a fertile, well-drained, sandy soil in full sun.

A. digitata L., *Syst. Nat.* ed. 10, 2: 1144, 1759. (Syn: *A. baobab*, etc.). [Baobab Tree, Monkey Bread Tree, Sour Gourd Tree, Upside-Down Tree]. **Trees**, usually massive, deciduous, to 30 m tall; trunk to 10 m Ø, single or multiple, cylindrical or fluted, often buttressed; crowns spreading, rounded. **Stipules** caducous. **Leaves** 5–7(–9)-foliolate; leaflets ± sessile, variable, medial leaflet usually elliptic-obovate, 5–15 × 3–7 cm, base decurrent, margins entire, apex acuminate. **Inflorescences** with flowers solitary, rarely paired, axillary; pedicels 15–90 cm long; flowers pendulous. **Flowers** with calyx lobes (3–)5, triangular, 5–9 × 3–5 cm, green and tomentose outside, cream and villous inside, reflexed; petals white, broadly obovate, 4–8 × 4–8 cm, base narrowed; staminal tube cylindrical or tapering, 3–6 cm long, filaments apical, 720–1,600, free; ovary conic-ovoid or globose, style white, bent over at right angles, rarely straight, stigma white, irregularly lobed. **Fruits** variable, globose to cylindrical, 20–25 cm long, velvety hairy; seeds reniform, laterally flattened, 1–1.3 × 0.8–1 cm. **Distr:** Africa (Southern Africa to S Sudan and Ethiopia). **Ecol:** In semiarid subSaharan regions, around margins of seasonal pools and along rivers. **Zones:** 11–12.

Bombax

Species ~50. **Trees**, deciduous, very large; young trunk usually spiny. **Leaves** palmately compound; leaflets 5–9, sometimes petiolulate, margins entire. **Inflorescences** with flowers solitary or fascicled, axillary or terminal; pedicels to 10 cm long; bracteoles absent. **Flowers** bisexual, large, produced before leaf flush; calyx tubular, campanulate, or cup-shaped, apex truncate to deeply lobed, sometimes with abaxial glands, leathery, falling with petals and stamens; petals 5, usually red, sometimes yellow, orange, or white, obovate or obovate-lanceolate, asymmetrical, sometimes reflexed; stamens 70–900, bases connate into a short tube, filaments connate into 5–10 distinct phalanges, alternating with petals, anthers reniform; ovary syncarpous, 5-locular, locules many-ovuled, style filiform, longer than stamens, stigma stellately lobed. **Fruits** capsules, loculicidally dehiscent into 5 valves, valves woody or leathery, with silky wool inside; seeds small, black, enclosed by wool. **Distr:** Tropical America, tropical Africa, Asia, and Australasia. **Cult:** Propagate from seeds or cuttings. Requires a fertile, moist, well-drained soil in full sun.

B. ceiba L., *Sp. Pl.* 1: 511, 1753. (Syn: *Gossampinus malabarica*, etc.). [Malabar Simal, Red Cotton Tree, Red Kapok Tree, Silk Cotton Tree]. **Trees**, very large, deciduous, to 25 m tall; trunk buttressed, usually very spiny on young trees. **Stipules** minute. **Leaves** palmately compound; leaflets 5–7, oblong to oblong-lanceolate, 10–16 × 3.5–5.5 cm, base broad or tapering, apex acuminate, glabrous, lateral veins 15–17 per side; petiole 10–20 cm long; petiolules 1.5–4 cm long. **Inflorescences** terminal, flowers solitary. **Flowers** ~10 cm Ø; calyx cup-shaped, 2–3(–4.5) cm Ø, abaxially glabrous, adaxially densely sericeous, lobes 3–5, semiorbicular, ~1.5 × 2.3 cm; petals usually red, sometimes orange-red, obovate-oblong, 8–10 × 3–4 cm, fleshy, surfaces puberulent; filament tube short, filaments with outer series in 5 fascicles, each with >10 stamens, inner series bifid, central series with 10 stamens. **Fruits** capsules, ellipsoid, 10–15 × 4.5–5 cm, densely white villous, puberulent; seeds many, obovate, smooth. **Distr:** Tropical Asia to Papua New Guinea. **Ecol:** In dry areas; <1,400 m. Fl. Mar–Apr. **Zones:** 10–12.

Cavanillesia

Species 5. **Trees**, large, tall, deciduous; trunk often enormously swollen, pachycaulous, sometimes enlarged in the middle. **Leaves** appearing after anthesis of flowers; lamina simple, palmately 5–7-lobed, palmately nerved. **Inflorescences** umbellate cymes, appearing before leaves. **Flowers** relatively small; calyx 5-lobed; petals 5, base glandular inside; stamens many, basally united in a column, contracted above base, finally separating into 5 bundles, filaments each with 1 reniform anther; ovary 3–5-celled, ovules geminate and erect from the cell base, style 1. **Fruits** capsules, relatively large, indehiscent with 5 large membranous wings much enlarged laterally, often with only 1 narrow seed. **Distr:** Tropical America (Panama to Brazil). **Cult:** Propagate from seeds (which germinates readily when placed on a moist compost), or from cuttings. Requires a fertile, moist, well-drained soil in full sun.

C. platanifolia (Humb. and Bonpl.) HBK., *Nov. Gen. and Sp.* 5: 306, 1823. (Syn: *Pourettia platanifolia*). [Cuipo Tree, Quipo Tree]. **Trees**, tall, massive, >30 m tall, deciduous; trunk smooth, bottle-shaped, narrowed at base and then widened to a great height before gradually narrowing; bark pale, porous; crown rather open, subglobose. **Leaves** palmately 3-, 5- or 7-lobed, base sinuate-cordate, nerves 5–7, reticulate veins prominent abaxially, surfaces somewhat puberulent; younger leaves cordate-ovate, margins entire or 1–bilobed; petiole to 13 cm long. **Inflorescences** large; flowering precociously, usually before the new flush. **Flowers** with calyx finely ferruginous-tomentose, campanulate, coriaceous, lobes ovate, spreading, unequal; petals ligulate, oblong-obtuse, membranous, to 2 cm long, flesh-colored outside, reddish-puberulent inside; stamens red, staminal tube about half as long as the petals; ovary pubescent, 5-loculed, stigma 5-parted. **Fruits** spherical in outline, to 15 cm Ø, with large semicircular wings, reticulate-veined, to 7.5 cm wide, membranous; locules 1-seeded; seeds oblong, both ends acute. **Distr:** C America. **Ecol:** In moist forests, sometimes near lakes or rivers; 30–100 m. Fl. Mar–Apr. Fr. May–Jul. **Zones:** 11–12.

Top row: *Cavanillesia platanifolia* (habit), *Ceiba pentandra*, *Ceiba schottii*, *Ceiba speciosa*.
Bottom row: *Chiranthodendron pentadactylon*, *Lagunaria patersonia*, *Ochroma pyramidale*.

Ceiba

Species 17. **Trees**, deciduous; trunk swollen and/or buttressed, trunk and branches usually spiny. **Leaves** long-petiolate, palmately compound; leaflets 3–5(–9), shortly petiolulate, base attenuate, margins serrate or entire, surfaces glabrous, abaxially gray-white. **Inflorescences** with flowers solitary, or 2–15-flowered fascicles axillary or subterminal; flowers pendulous, opening before or simultaneously with leaf flush. **Flowers** actinomorphic, rarely zygomorphic; calyx campanulate, truncate or irregularly 3–5(–12)-lobed, thickly fleshy, persistent; petals 5, pink or yellow-white, basally connate and coherent to staminal tube, falling with stamens and style; stamens (3–)5–15, filament tube short, sometimes thickened or with sterile appendages, free parts of filament long to absent, each bearing 2(or 3) straight to strongly twisted thecae; ovary syncarpous, 5-locular, locules many-ovuled, style filiform; stigma capitate to lobulate. **Fruits** capsules, oblong or nearly obovate, pendulous, woody or leathery, loculicidally dehiscent into 5(or 6) valves, inner walls densely woolly; seeds many, enclosed in wool. **Distr:** W Africa (1 sp.), tropical America. **Cult:** Propagate from fresh seeds or semiripe cuttings. Requires a fertile, humus-rich, moist, well-drained soil in full sun.

C. pentandra (L.) Gaertner, *Fruct. Sem. Pl.* 2: 244, 1791. (Syn: *Bombax pentandrum*, etc.). [Cotton Tree, Silk Cotton Tree, Kapok Tree]. **Trees**, to 70 m tall; trunk often sparsely spiny; buttresses small or absent; main branches verticillate, spreading horizontally; young branches spiny. **Leaves** palmately compound; leaflets 5–9, oblong to lanceolate, 5–20 × 1.5–6.5 cm, base acuminate, margins entire or very sparsely and minutely toothed near apex, apex shortly acuminate, thinly leathery, glabrous; petiole 7–14(–25) cm long; petiolules 0.3–0.4(–1) cm long. **Inflorescences** with flowers solitary, up to 15-flowered fascicles, subterminal; flowers produced before or with new leaves; pedicels (1.8–)2.5–5 cm long. **Flowers** with calyx (0.9–)1.2–2 cm long, adaxially glabrous; petals pink or white, obovate-oblong, 2.5–4 × 0.7–1.5 cm, abaxially densely white-villous, adaxially glabrous; filaments on staminal tube uneequal, anthers reniform; ovary glabrous, style 2.5–3.5 cm long, stigma rod-shaped, 5-lobed. **Fruits** capsules, oblong, tapering toward tip, 7.5–15(–26) × 3–5(–11) cm; fruiting pedicel 7–25 cm long; endocarp leathery, smooth; seeds globose, ~6 mm Ø. **Distr:** Tropical America, W Africa. **Ecol:** Fl. Mar–Apr. **Zones:** 11–12. **Info:** The cotton (kapok) was earlier commercially harvested, and used for pillows and other stuffing.

C. schottii Britten and Baker f., *J. Bot.* 34: 173, 1896. [Kapok Tree]. **Trees**, ~8 m tall; trunk and branches aculeate, spines on younger branches ~0.5 mm long, black, only slightly curved. **Leaves** 3–7-foliolate; leaflets oblanceolate to elliptical, 5–10 × 1.9–3 cm, margins entire, apex acute, with a small mucro, coriaceous, glabrous; petiole ~6 cm long, petiolules 0.6–1 cm long. **Inflorescences** with flowers borne solitary or in fascicles, axillary; pedicels to 4 mm long, stout. **Flowers** with calyx cylindrical, 2.5–4 × 0.8–1 cm, glabrous; petals narrowly oblong, 17–19 × ~1.5 cm, white, densely hispid-hairy outside, scabrous-glabrous inside; staminal tube 8–10 cm long, slender, white silky-hairy, which are prominent in fresh flowers, filaments ~7 cm long, anthers versatile; ovary pyriform, style slender, glabrous, stigma globose at about same level as anthers. **Fruits** capsules, elongate to ellipsoidal, ~10 × 5 cm. **Distr:** SW Mexico, Guatemala. **Ecol:** In dry woodlands. Fl. (Jun)Aug–Feb. **Zones:** 10–12.

C. speciosa (A.St.-Hil., A. Juss. and Cambess.) Ravenna, *Onira* 3(15): 46, 1998. (Syn: *Chorisia speciosa*). [Pink Floss Silk Tree, Silk Floss Tree]. **Trees**, mainly evergreen, briefly deciduous (during flowering), 9–18 m tall; trunk stout, green, often armed with thick, large spines, base often enlarged (bottle-shaped), to 2 m Ø; branches often, horizontal, covered with prickles. **Leaves** palmately 5–7-foliate, long petiolate; leaflets oblong, margins serrate, apex acute, 6–10 cm long, pale green. **Flowers** violet-red, 10–15 cm Ø, creamish-white at center, pink toward petal tips; petals 5; staminal column usually fused together along its entire length (sometimes filaments partially or completely separated), stamens 5. **Fruits** capsules, pear-shaped, woody, 15–20 cm long; seeds black, surrounded by a mass of fibrous, fluffy, cotton- or silklike coma. **Distr:** S America (Argentina, Brazil and Paraguay). **Ecol:** In tropical and subtropical forests; to 600 m. **Zones:** 9–12.

Chiranthodendron

Species 1. *Generic description same as specific.* **Cult:** Propagate from seeds or cuttings. Requires a moist, well-drained soil in full sun. **Info:** The placement of *Chiranthodendron* and *Fremontodendron* (not treated here) needs to be further investigated. There is very weak support by molecular data for the placement of these 2 genera in Malvoideae (Baum et al. 2004).

C. pentadactylon Larreategui, *Descr. Chiranthod.* 17, 1805. (Syn: *Cheirostemon platanoides*,

Above: *Pachira aquatica*, *Pachira insignis*, *Tropaeolum majus*, *Tropaeolum majus* (yellow form).

Chiranthodendron platanoides). [Mexican Hat Plant, Monkey Hand Tree]. **Trees**, to 12–30 m tall or even larger; trunk frequently 1–2 m Ø. **Leaves** long-petiolate; lamina ± ovate-rounded in outline, 12–30 cm long, base deeply and narrowly cordate, margins shallowly lobate or subentire, apex subacute to acuminate, adaxially deep green and glabrate, abaxially densely and loosely brownish-tomentose, palmately nerved. **Inflorescences** with flowers solitary, opposite; peduncles shorter than calyx. **Flowers** large; calyx 3.5–5 cm long, brown-tomentose outside, glabrous and dark lurid red inside, with a large nectariferous pit inside base of each lobe; staminal column equaling calyx, bearing at its apex 5 long curved branches, these prolonged into very slender, tapering tips. **Fruits** capsules, oblong-ellipsoid, very hard and woody, 10–15 cm long, deeply 5-lobate, angles narrow and blunt-edged; aril of seeds orange. **Distr:** Mexico. **Ecol:** In montane wet mixed forests; 2,000–3,000 m. **Zones:** 9–12. **Info:** Its closest relatives are in the genus *Fremontodendron*, with which many hybrids have been produced.

Lagunaria

Species 1. *Generic description same as specific.* **Cult:** Propagate from seeds. Requires a fertile, moist, well-drained soil in full sun.

L. patersonia (Andrews) G. Don, *Gen. Hist.* 1: 485, 1831. [Norfolk Island Hibiscus, Sally Wood, White Oak]. **Trees**, to 20 m tall, lepidote. **Stipules** caducous. **Leaves** alternate, spiraled; lamina simple, elliptic or ovate, 6.5–10 × 3.5–6 cm, undissected, base tapering, margins entire, apex acute or obtuse, scaly, younger leaves with scurfy scales becoming glabrous with age; petiole 1.5–2 cm long. **Inflorescences** with flowers solitary (occasionally paired), axillary; pedicel 2–2.3 cm long. **Flowers** predominantly pink, regular; perianth 2-whorled; epicalyx fused, enclosing floral bud, ± caducous, leaving an annular scar; calyx 1.2–1.5 cm long, sepals 5, basally connate; corolla 4–4.5 cm long; petals 5, free, cream to pink; stamens numerous, adnate to perianth at base, coherent to each other (filaments fused in a column surrounding the style), anthers dehiscing via longitudinal slits; ovary syncarpous, superior, 5-loculed, locules 2-ovuled, style 1, branches 5, joined at base, free near tip. **Fruits** capsules, stinging-pubescent inside, dehiscent, not fleshy, pentagonal, 3–4 × ~2.5 cm; seeds reniform, glabrous. **Distr:** Australia, Norfolk and Lord Howe Islands. **Ecol:** On cliff tops and disturbed natural vegetation. Fl. Nov–Feb. **Zones:** 10–11.

Ochroma

Species 1. *Generic description same as specific.* **Cult:** Propagate from fresh seeds (needs heat to germinate); flowering as early as 3–5 years. Requires a moderately fertile, moist, well-drained soil in full sun.

O. pyramidale (Cav. ex Lam.) Urban., *Repert. Spec. Nov. Regni Veg. Beih.* 5: 123, 1920. (Syn: *O. lagopus*, etc.). [Balsa, Down Tree, Silk Cotton Tree]. **Trees**, unarmed, evergreen, to 20–30(–40) m tall; trunk to 1–1.5 m Ø, often buttressed. **Stipules** ovate to suborbicular, caducous. **Leaves** large, broadly ovate or ovate-orbicular, to 15–50 cm long, nearly as wide, base cordate, sinus deep, margins entire, ± shallowly angulately lobed or undulate, membranous, adaxially green, glabrous, abaxially ferruginous or tawny-white, mostly 7-nerved, nerves densely brown-puberulent abaxially, adaxially ± glabrate; petiole 8–25 cm long. **Flowers** 10–15 cm long, not opening fully, musky scented, bat-pollinated; calyx tubular-campanulate, lobes 5, 2 acute or triangular ovate, the other 3 obtuse to truncate; petals 5, fleshy, white, much longer than calyx, adnate to staminal column at base; staminal column large, twisted, ± enclosed by corolla, with 5 short but broad lobes, apically densely covered with stamens; ovary 5-loculed, style sulcate, stigmas exserted slightly beyond staminal tube. **Fruits** capsules, cylindrical, 5-ribbed, brown, to 18 cm long; seeds many, round, with pale brown, abundant silky floss. **Distr:** Mexico to Bolivia; West Indies. **Ecol:** In wet lowland forests, often near streams, but drought tolerant; 0–2,000 m. **Zones:** 10–12.

Pachira

Species ~10–20. **Trees**, sometimes deciduous, spiny and/or buttressed. **Leaves** palmately compound; leaflets 3–11, base jointed, sometimes petiolulate, margins entire or serrate. **Inflorescences** with flowers solitary, or 2–3-flowered fascicles, axillary; pedicels to 10 cm long; bracteoles 2 or 3. **Flowers** bisexual; calyx cup-shaped to tubular, adaxially glabrous, truncate to lobed, often with glands abaxially, persistent, sometimes accrescent; petals spathulate to linear, yellowish-green, white, or reddish, abaxially tomentose; stamens 90–1,000 in fascicles of 7–10, basally connate into tube, anthers reniform; ovary 5-locular, ovules many, style exserted; stigma 5-lobed. **Fruits** nearly oblong or nearly pyriform, woody or leathery, loculicidally dehiscent into 5 valves, inner surface long-woolly; seeds irregularly scariform-cuneate, to 2.5 cm long, glabrous, seed coat fragile, smooth. **Distr:** Tropical America. **Cult:** Propagate from seeds or cuttings. Requires any moist but well-drained soil in full sun.

P. aquatica Aublet, *Hist. Pl. Guiane* 2: 726, 1775. (Syn: *Bombax aquatica*, etc.). [Guiana Chestnut, Provision Tree, Shaving Brush Tree, Water Chestnut]. **Trees**, 4–12(–22) m tall. **Leaves** leaflets 5–11, ± subsessile, oblong to obovate-oblong, base cuneate, margins entire, apex acuminate, abaxially ferruginous-tomentose, adaxially glabrous, central leaflet 13–24 × 4.5–8 cm, size decreasing outward, lateral veins 16–20 per side; petiole 11–15 cm long, ferruginous-tomentose. **Inflorescences** with flowers solitary, axillary near twig tips; pedicels robust, ~2 cm long, yellow-tomentose, glabrescent. **Flowers** with calyx cup-shaped, nearly leathery, ~1.5 × 1.3 cm in Ø, abaxially sparsely pilose, adaxially glabrous, truncate or obscurely 3–6-toothed, persistent, with 2–3 globose glands at base; petals yellowish-green, narrowly lanceolate or filiform, to 15 cm long, distal half reflexed; staminal tube short, proximally yellow, distally red, 13–15 cm long including filaments; anthers 2–3 mm long; style dark red, longer than stamens. **Fruits** capsules, nearly pyriform, 9–10 × 4–6 cm, endocarp yellow-brown, woody, adaxially densely woolly; seeds many per cell, dark brown, white-spiraled, 2–2.5 × 1–1.5 cm, edible. **Distr:** Tropical America. **Ecol:** Often in swamps, sometimes near brackish water; 0–300 m. Fl. May–Nov. **Zones:** 10–12.

P. insignis (Sw.) Savigny, *Encycl. Méthe. Bot.* 4: 690, 1798. (Syn: *Bombax insignis*, etc.). [Shaving Brush Tree, Wild Chestnut]. **Trees**, evergreen, unarmed, to 30 m tall; trunk gray to green, 50–60 cm Ø. **Leaves** grouped at apex of twigs, palmate; leaflets 5–9, obovate-oblong, to 35 × 15 cm, base cuneate, apex emarginate and apiculate, glabrous, abaxially sometimes brown-scaly; petiole ~25 cm long, base swollen; petiolules 0.5–1 cm long. **Inflorescences** borne at ends of twigs, axillary, flowers 1 or 2 per axil; pedicels 1–7 cm long. **Flowers** very large, showy; calyx dark red, tubular-campanulate, 3–3.5 cm long, both sides finely woolly; petals 5, narrowly strap-shaped, to 23–35 × 2 cm, both sides pink; stamens numerous, completely white or red and white, ~20 cm long including the column; style to 30 cm long. **Fruits** capsules, large, woody, globose or ellipsoid, 15–25 × 9 cm, base ± rounded, finely brown-tomentose, valves 1 cm thick; seeds 2.5 cm long, 4-angled, brownish, edible. **Distr:** S America (Colombia to Peru). **Ecol:** In wet forests; 0–5,000 m. **Zones:** 10–12.

BRASSICALES

FAMILIES 17, GENERA 398, SPECIES 4,450.

Stipules small. **Leaves** spirally arranged. **Inflorescences** racemose. **Flowers** with petals clawed; ovules in 1 or 2 rows. **Embryo** often green. **Distr:** Worldwide. **Info:** The order is generally united in having glucosinolate secondary compounds (Simpson 2006; Stevens 2001 onwards).

TROPAEOLACEAE

Genus 1, **species** 95. *Familial description same as generic.*

Tropaeolum

Species 95. **Herbs**, scandent, creeping or twining,

Above: *Tropaeolum pentaphyllum*, *Tropaeolum speciosum*, *Moringa oleifera*, *Carica papaya* (female flowers).

annual or perennial, usually ± glabrous; roots usually tuberous. **Stipules** minute. **Leaves** alternate, usually long-petiolate; lamina simple, angulate or palmately compound, base peltate, margins entire or lobate. **Inflorescences** with flowers solitary, or less frequently umbels, axillary. **Flowers** bisexual, irregular, 5-merous, yellow or red; hypanthium produced posteriorly into a spur; calyx deeply lobate; sepals 5, imbricate or valvate, basally connate; petals free, 5 or rarely by abortion fewer, imbricate, upper ones exterior and ± different to lower ones; stamens 8, free, 2-whorled, filaments unequal, declinate, anthers dehiscent by longitudinal sutures; ovary superior, of 3 connate carpels, 3-loculed, placentation axile-apical, ovule 1 per carpel, pendulous from apex of locule, style 1, filiform, with 3 short stigmatic branches, introrsely stigmatose. **Fruits** schizocarps, separating from the axis into 3 fleshy mericarps, each 1-seeded; seeds without endosperm. **Distr:** New World. **Cult:** Propagate from basal cuttings, seeds or by division. Requires a moist, well-drained soil in full or partial sun.

T. majus L., *Sp. Pl.* 1: 345, 1753. [Nasturtium]. **Vines**, herbaceous, scandent or creeping, to 2–3 m long; stems cylindrical, glabrous, fleshy. **Leaves** alternate, ascendant, simple; lamina ± circular, peltate, 5–8 cm ∅, membranous, glabrous, adaxially surface dark green, dull, with pale venation, abaxially lower surface pale green; petiole glabrous, 5.5–15(–25) cm long. **Inflorescences** axillary, flowers solitary, ascendant; pedicels glabrous, as long as or longer than petiole. **Flowers** with calyx yellow or pale orange, sepals unequal, ovate, oblong, 2–2.3 cm long; spur straight or slightly curved, to 3 cm long; petals orange or red, obovate, long-unguiculate, 3–4 cm long, 3 of them long-ciliate on margin adjacent to claw; anthers lanceolate, filaments slender, unequal; ovary trilobate, glabrous. **Fruits** tricoccal, ~1 cm long. **Distr:** Andean cordillera. **Ecol:** In moist forests. Fl. Feb. **Zones:** 9–11.

T. pentaphyllum Lam., *Encycl. Method.* 1(2): 612, 1785. [Digitate-leaf Nasturtium]. **Climbers**, vigorous; roots tuberous, tubers large, oblong; stem slender, branched. **Leaves** digitate, ~5 cm across; leaflets 5, oblong, margins entire; petiole 5 cm long, twisted in form of a tendril, forming the chief support of the stem. **Inflorescences** with flowers solitary, axillary; peduncle 10 cm long, curved, purple, glabrous, thickened above, flowers held horizontal. **Flowers** dusky coral-red to scarlet, tubular; calyx 3.2 cm long, persistent, spurred, limb 5-parted, green, brighter and spotted or streaked with deep purple inside, segments ovate-acute, uppermost narrowest; spur ± horizontal, fleshy, dull purple outside, yellow inside, nectariferous, conical; petals 2, small, subrotund, clawed, reflexed, bright vermillion, inserted into calyx throat; stamens 8, longer than calyx segments, filaments subulate, streaked or spotted with purple, anthers 4-sided, oblong; style yellow, 3-sided, stigmas 3, acute, diverging. **Fruits** 3-loculed, glabrous. **Distr:** S America. **Ecol:** In dry scrubs and grasslands, near the coast; to 2,500 m. Fl. Aug–Nov. **Zones:** 8–11.

T. speciosum Poepp. and Endl., *Nov. Sp. pl.* 1: 22, 1836. [Chilean Flame Creeper, Flame Flower, Flame Nasturtium]. **Vines**, perennial, vigorous, rhizomatous, creeping or climbing, to 6(–10) m long; rhizomes stoloniferous; stems thin, wiry. **Tendrils** to 7 mm long, coiling and attaching to supporting vegetation. **Leaves** deeply 5–6-lobed, almost palmate, ~10 cm across; leaflets obovate, ~5 cm long, base narrowed, margins undulated, apices rounded, shallowly notched, adaxially bright green to bluish green, abaxially paler, thin, fleshy, venation prominent abaxially; petiole very long, wiry, hairy. **Inflorescences** with flowers solitary; pedicels reddish, very long. **Flowers** bisexual, irregular, 5-merous, scarlet, tubular, ~1.5–2.5 cm across; petals 5, spur ~2–3 cm long; ovary 3-carpellate. **Fruits** berries, usually 1–3 together, bluish-black, somewhat glaucous, fleshy, ~1 cm ∅, with persistent floral segments. **Distr:** Chile. **Ecol:** In thickets, clambering through bushes and climbing on to trees; 400–1,200 m. Fl. Nov–Apr. Fr. Dec–Mar. **Zones:** 8–11.

MORINGACEAE

Genus 1, species 12. **Trees**, **shrubs**, or **herbs**, deciduous, massive to slender, tuberous when young and sometimes as adults; bark and pith with gum canals. **Stipules** absent, but stalked glands present at base of petiole and leaflets. **Leaves** alternate, 1–3(or 4)-imparipinnate; leaflets opposite, margins entire, sometimes caducous, glabrous to puberulent. **Inflorescences** panicles, axillary. **Flowers** bisexual, white to yellow or red, actinomorphic to zygomorphic, with or without hypanthium; sepals 5, often reflexed at anthesis; petals 5, equal or unequal, often reflexed; stamens 2-seriate, 5 perfect, 5 reduced to staminodes, filaments free; pistil 1, ovary superior, borne on short gynophore, 1-loculed, ovules numerous, style 1, stigma minute. **Fruits** capsules, 3–12-ribbed, sometimes with elongated beak; valves 3, dehiscent; seeds numerous, 3-winged or wingless, without endosperm. **Distr:** India to Africa, Madagascar.

Moringa

Species ~13. *Generic description same as familial.* **Cult:** Propagate from cuttings or seeds. Requires any soil in full sun.

M. oleifera Lamarck, *Encycl.* 1: 398, 1785. (Syn: *M. pterygosperma*). [Drumstick Tree, Horseradish Tree]. **Trees**, to 12 m tall. **Leaves** petiolate, 3-pinnate, 25–60 cm long, stalked glands often exuding clear or amber liquid at base of petiole and leaflets; leaflets in 4–6 pairs, ovate, elliptic, or oblong, 1–2 × 0.5–1.2 cm, base rounded to cuneate, apex rounded to emarginate, puberulous when young, glabrous at maturity; petiolules slender, 1–2 mm long. **Inflorescences** panicles, 10–30 cm long; bracts linear, ~1 mm long; each flower borne on false pedicels 0.7–1.5 cm long, true pedicels 1–2 mm long. **Flowers** white to cream, fragrant, with 2 dorsal sepals and 1 dorsal petal usually remaining unreflexed and forming a projecting 'keel' while the rest of the perianth reflexes down to form a 'banner' at right angles to the 'keel', sepals lanceolate to linear-lanceolate, 0.7–1.4 mm long, usually puberulent; petals spathulate, 1–2 cm long; stamens hairy at base; ovary hairy. **Fruits** capsules, long, podlike, 3-valved, 20–50 cm long, 1–3 cm ∅, dehiscent; seeds subglobose, 3-angled, 0.8–1.5 cm ∅, wings 0.5–1 cm wide, rarely absent. **Distr:** India. **Ecol:** Fl. throughout most of the year. Fr. Jun–Dec. **Zones:** 10–12.

CARICACEAE

Genera 6, species 34. **Trees** small, palmlike, or shrubs, rarely vines, often prickly, monoecious, dioecious, andromonoecious, gynomonoecious, or polygamomonoecious; stem stout, unbranched, rarely branched, with a terminal cluster of leaves; exudate flowing, latexlike. **Stipules** when present, spiny. **Leaves** alternate, long-petiolate; lamina palmate or palmatifid, rarely entire or pinnatifid, large; sap of leaves milky, sometimes orange-colored. **Inflorescences** axillary; ♂ flowers aggregated in cymose panicles; ♀ flowers usually solitary or aggregated in corymbose cymes, large. **Flowers** with calyx 5-lobed; lobes small, basally connate; corolla 5-lobed (the 5 petals soon caducous), tube long in ♂ flowers, short in ♀ flowers; stamens 5 or 10, 1- or 2-whorled, inserted in throat of corolla tube, filaments free, basally connate, anthers introrse, tetrasporangiate, dehiscing via longitudinal slits; gynoecium in ♂ flowers vestigial, or absent; in ♀ flowers syncarpous, synovarious to synstylovarious; ovary superior, 1- or 5-loculed, placentation when 1-loculed parietal (placentas ± deeply intruded) or laminar-dispersed, when 5-loculed axile, ovules numerous, anatropous, bitegmic, styles 1 or 5, free to partly joined, apical; stigmas 5, papillate, dry. **Fruits** berries, large, fleshy, indehiscent; seeds numerous, surrounded by mucilage. **Distr:** Predominantly C and S America. One genus with 2 species (*Cylicomorpha*) in tropical Africa.

Carica

Species 23. **Trees** small, or shrubs. **Leaves** subpeltate; lamina palmatifid, rarely entire. **Inflorescences** ♂ cymose panicles; ♀ flowers solitary. **Flowers** unisexual or bisexual; ♂ with corolla lobes oblong or linear, valvate or contorted, stamens 10, 5 sessile or subsessile, alternating with corolla lobes, 5 inserted in the throat on short filaments, ovary sterile, subulate; ♀ with corolla lobes linear-oblong, stamens absent, ovary 1-loculed, ovules several to numerous, placentation parietal, style wanting or short, rarely elongate, stigmas 5, inflated or linear, entire to variously divided. **Fruits** large, succulent; seeds numerous, ovoid or slightly

Above: *Carica papaya* (male flowers), *Carica pentagona*, *Capparis brevispina*, *Capparis floribunda*.

compressed, arillate. **Distr:** C and S America; unknown in the wild. **Cult:** Propagate from seeds, or cuttings in case of sterile hybrids. Needs a fertile, moist, well-drained compost in full sun.

C. papaya L., *Sp. Pl.* 2: 1036, 1753. (Syn: *Papaya carica*). [Papaya, Pawpaw]. **Trees** or **shrubs**, 8–10 m tall; stem simple (rarely branched), with helically arranged stipulate scars. **Leaves** usually 5–9-palmatifid, ~60 cm long, lobes pinnatifid; petiole hollow, 0.6–1 m long. **Inflorescences** ♂ pendulous, to 1+ m long, pedicels absent; ♀ flowers solitary or aggregated in corymbose cymes, pedicels short or nearly absent. **Flowers** ♂ with corolla tube creamy-yellow, 1.6–2.5 cm long, lobes lanceolate, ~1.8 × 0.45 cm, stamens 5 longer and 5 shorter, shorter ones almost without filaments, filaments white, white-tomentose; ♀ with calyx lobes ~1 cm long, corolla lobes creamy-yellow, oblong or lanceolate, 5–6.2 × 1.2–2 cm, ovary ovoid, stigmas partite, nearly fimbriate; *bisexual* with corolla tube 1.9–2.5 cm long, lobes oblong, ~2.8 × 0.9 cm, stamens 5 or 10 in 1 or 2 whorls, ovary smaller than in ♀ flowers. **Fruits** orange-yellow or yellow at maturity, cylindric, ovoid-cylindric, or subglobose, 10–30 cm long, sarcocarp soft with a mild, pleasant flavor; seeds numerous, black at maturity, ovoid. **Distr:** Tropical S America. **Ecol:** On dry mountain slopes, rainforests and semievergreen seasonal forests; to ~2,000 m. Fl. & Fr. throughout the year. **Zones:** 11–12.

C. pentagona Heilborn, *Ark. Bot.* 17(12): 2, 1922. (Syn: *C. heilbornii* [notho]var. *pentagona*). [Babaco, Mountain Papaya]. **Shrubs**, to 3(–4) m tall; stem usually rather slender. **Leaves** palmatifid, lobes fewer and broader than other species, surfaces glabrous. **Inflorescences** with flowers solitary, axillary; pedicels thick, ~1–2 cm long. **Flowers** ♀ only, fleshy; corolla lobes linear, 2.5–4 cm long, apex recurved, outside pinkish, aging to salmon, inside white, tinged pale pink; ovary cylindrical, shorter than corolla, stigma 5-lobed. **Fruits** cylindrical, conspicuously 5-angled, ~30 × 7.5–12.5 cm, base truncate, apex sharply acute, ripening to yellow or orange, flesh ~1 cm thick, nearly white, distinctly fragrant with an acidic flavor, the hollow inside has large quantities of cottonlike unpalatable substance; seeds absent or few (if pollinated by other *Carica* spp.). **Distr:** Throughout the Andean region; known in cultivation only. **Ecol:** In various habitats; 1,800–3,000 m. **Info:** Only purely ♀ flowers are known. Widely cultivated in the native range and other parts of the tropics.

CAPPARACEAE

Genera 16, **species** 480. **Herbs**, annual or perennial, subshrubs, shrubs or small trees, sometimes climbing, scrambling or rarely lianaceous. **Stipules** absent or present, rarely developed into spines. **Leaves** alternate, petiolate, persistent or seasonally deciduous, rarely minute and caducous; lamina simple or palmately compound, (1–)3–7(–13)-foliolate; petiole not sheathing, rarely remaining after the lamina has been shed as a phylodium; petiolules short. **Inflorescences** racemose, these often corymbose or subumbellate, or flowers solitary, or fascicled on short shoots; pedicels bracteate or ebracteate. **Flowers** actinomorphic or bilaterally symmetric to zygomorphic, bisexual or rarely unisexual; sepals (3)4 or rarely more, 1-seriate and valvate, reduplicate or induplicate, 2-seriate and decussate; petals none, (2–)4 or rarely more, free, usually grouped in pairs and ± dimorphic; androecium exclusively of fertile stamens or including staminodes; stamens usually 6 to numerous, rarely (1)2 or 4, filaments filiform; staminodes ~4–10, only in zygomorphic flowers; carpels 2(–8), ovary superior, ovules 1–many, style present or absent, stigma punctate to capitate, usually entire. **Fruits** usually capsules, rarely baccate, pepos, nutlike, schizocarps or samaras, dehiscent or indehiscent by 2 or rarely 4 valves; seeds mostly reniform or rounded. **Distr:** Tropical regions worldwide.

Capparis

Species 250. **Shrubs**, **small trees**, or **vines**, evergreen, erect, climbing, or sometimes prostrate. **Stipular spines** straight or curved, sometimes absent. **Leaves** alternate, spirally arranged or distichous (or rarely secund), petiolate, sometimes subsessile; lamina simple, herbaceous to leathery, margins entire. **Inflorescences** racemes, corymbs, umbels, or panicles, sometimes reduced to a single flower, but usually 2–10-flowered, supra-axillary, axillary, or terminal; bract usually present at base of pedicels, often caducous. **Flowers** with sepals 4, 2-whorled; sepals of outer whorl often thick, often inwardly concave or becoming navicular; sepals of inner whorl often thin, ± equal; petals 4, imbricate, sometimes all 4 ± equal, distinct, often 2 pairs ± similar, *anterior* pair distinct, sometimes clawed, *posterior* pair with lateral margins adhering, base asymmetric, margin revolute or rarely inner lateral margin involute from base to near middle; stamens 7–120; gynophore ± as long as filaments, ovary 1-loculed, placentas 2–6(–8), ovules few–many, style short or obsolete, stigma sessile. **Fruits** baccate, globose or ellipsoid, usually not dehiscent; seeds 1–numerous, reniform to nearly polygonal. **Distr:** Tropical and subtropical regions worldwide, few in temperate regions. **Cult:** Propagate from seeds or half-hardened cuttings. Requires a fertile, well-drained soil in full or partial sun.

C. brevispina DC., *Prod.* 1: 246, 1824. (Syn: *C. zeylanica*). **Shrubs** or **slender trees**; stems pubescent, soon glabrescent. **Thorns** 1–3 mm long. **Leaves** variable; lamina usually lanceolate to obovate, 3–8.5 × 14.5 cm, base obtuse or rounded, margins entire, apex acute or obtuse, often with stiff mucro to 1.5 mm long, coriaceous, glabrous, major veins prominent; petiole 3–7 mm long, deeply channeled above. **Inflorescences** with flowers solitary, axillary, at times on short lateral branchlets; pedicels 1.5–3.5 cm long, shortly pubescent, soon glabrescent, or subglabrous. **Flowers** with sepals ovate, 0.5–1 × 0.3–0.5 cm, outer pair glabrous outside, margin finely pubescent, inside tomentose to villous, inner pair slightly longer than outer, pubescent outside, subglabrous inside; petals obovate, ~1.5–3 × 0.5–1.25 cm, white, base yellow turning maroon; stamens to 40–50, ~2.5 cm long, slender; gynophore 1.5–3 cm long, glabrous; ovary ovoid, ~5 × 2 mm, beaked; seeds globose-reniform, yellow. **Distr:** S India, Sri Lanka. **Ecol:** In dry country. **Zones:** 10–12.

C. floribunda Wight, *Ill. Ind. Bot.* 1: 35, 1838. (Syn: *C. oligostema*). **Shrubs** or **climbers**, 2–4 m high, without stipular spines or with reflexed small spines. **Leaves** oblong-ovate, (4–)6–10(–13) × (1.5–)2–4(–6) cm, base rounded to cuneate, margins often revolute, apex obtuse to slightly mucronate, secondary veins 7–9 per side; petiole 0.5–1.5 cm long. **Inflorescences** terminal panicles of many subumbels, to 15 cm long, often also with axillary small panicles; peduncle 1–2 cm long; pedicel 1–1.5 cm long; sepals elliptic to nearly orbicular, 2–4 × 1.5–2.5 mm, ± equal, spreading, caducous; petals white, oblong to ovate, 3–5 × 1.5–2 mm, extremely thin; stamens 7–9(–12), filaments 0.8–1 cm long; gynophore 4–6 mm long, glabrous, ovary ovoid, 1.5–2 mm long, glabrous. **Fruits** orangish-red, globose, 1.5–2(–2.5) cm ∅, smooth; seeds 1–3 per fruit, 1–1.2 × ~0.6 cm. **Distr:** Indonesia, Myanmar, Philippines, Sri Lanka, C and S Thailand, S Vietnam, S Taiwan; Indian Ocean islands. **Ecol:** In seasonal forests, thickets; <200 m. Fl. May. Fr. Jan. **Zones:** 10–12.

C. sandwichiana DC., *Prodr.* 1: 245, 1824. (Syn: *C. spinosa* var. *mariana*). [Hawaiian Caper]. **Shrubs**; stems usually decumbent or prostrate, 1–5 m long. **Leaves** ovate, elliptic, or broadly elliptic, 2.4–6 × 1.9–5 cm, apex obtuse or emarginate, fleshy, often glaucous, fetid when crushed; petioles 1–4 cm long. **Inflorescences** with flowers solitary, axillary. **Flowers** opening after sunset, sweetly fragrant; calyx zygomorphic, outer sepal strongly saccate, 2–4 cm long, others 2–2.5 cm long; petals white, turning pink with age, upper pair rhombic, 2–5.5 cm long, densely pilose along the line of union, lower pair broadly obovate to suborbicular, 3–4.5 cm long; stamens ~120–180, white, turning pink with age, filaments 3–7 cm long; gynophore 1–1.5 cm long at anthesis, elongating to 5–6 cm long in fruit. **Fruits** berries, orange, ellipsoid, somewhat cucumber-shaped, 4–5 × 1.7–3 cm, usually 6-ridged; seeds dark reddish-brown to gray, asymmetrically reniform, 2.5–5 mm long, embedded in fetid orange pulp. **Distr:** Hawai'i. **Ecol:** In hot places exposed to salt spray such as coastal rocky cliffs and coral flats. **Zones:** 11–12.

C. zeylanica L., *Sp. Pl.* ed. 2. 720, 1762. (Syn: *C. horrida*). [Ceylon Caper]. **Shrubs**, 0.75–10 m tall, erect or straggling, much-branched. **Thorns** 2–4 mm long, recurved to occasionally patent. **Leaves** ovate-elliptic or ovate-lanceolate, 3.8–11 × (1.25–)1.75–4.5 cm, base rounded, apex acute or rounded-obtuse, subcoriaceous, adaxially pubescent, soon glabrescent, becoming glossy; petiole 3–8 mm long, glabrous. **Inflorescences** with flowers solitary or paired, axillary; pedicels (0.5–)2.5–3.5 cm long. **Flowers** with sepals ovate, 0.75–1 × 0.3 cm, densely pubescent outside; petals white, upper pair yellow at base, becoming reddish-violet with age, rounded-oblong, 2–2.5 × 0.8–1.2 cm, subglabrous; stamens 25–62, 1.5–3.5 cm long, anthers ~2 × 1 mm; gynophore 2.5–4.5 cm long, ovary 4–5 × 1–1.5 mm, narrowly pyriform to ellipsoid, minutely pubescent, placentas 4. **Fruits** ~5 × 3 cm, ripening to red or orange; seeds 12–15, blackish-brown, enclosed in white aril, embedded in white gelatinous pulp. **Distr:** India, Sri Lanka, Pakistan, eastward to Malesia and Indo-China. **Ecol:** In dry scrublands, sandy beaches and dry hillsides. Fl. May–Jul. **Zones:** 10–12.

Crateva

Species ~8. **Trees** or sometimes **shrubs**, evergreen or deciduous, glabrous. **Stipules** triangular, small, caducous. **Leaves** alternate, palmately compound; leaflets 3, disarticulating from petiole; lateral leaflets with bases asymmetric; petiole long, distal portion near rachis often with glands; petiolules short, thin when young, becoming thick in maturity. **Inflorescences** at tip of new branches, corymbose racemes; rachis either with arrested growth after anthesis or growing through a leafy twig, often with obvious scars after pedicels fall; bract basal, caducous; pedicel long. **Flowers** bisexual or unisexual (by failure of one sex to develop); receptacle disklike, with nectary; sepals 4, greenish, equal, smaller than petals, deciduous; petals 4, white, cream-colored, or yellow (aging to darker colors), equal, clawed, ovate to rhomboid with 4–6 secondary veins per side; stamens (8–)12–50, filaments basally connate to form a 1–4 mm long androgynophore; gynophore 2–8 cm long, degenerate in ♂ flowers; ovary 1-locular, ovules many, style short or absent, stigma inconspicuous. **Fruits** berries, globose or ellipsoid, drooping; seeds 25–50, embedded in creamy fetid or pungent mesocarp. **Distr:** Tropical and subtropical regions worldwide. **Cult:** Propagate from seeds or by air-layering. Requires a moist, well-drained soil in full sun.

C. adansonii* ssp. *odora (Buch.-Ham.) Jacobs, *Blumea* 12, 2: 198, 1964. (Syn: *C. roxburghii*, etc.). [Garlic Pear Tree, Three-leaf Caper]. **Trees**, 3–12(–20) m tall, deciduous (before flowering). **Leaves** 5–8(–12) × 6.5–17(–20) cm, *terminal* leaflet broadly elliptic to lanceolate, 5–12 × 2–6 cm, acute, acuminate, *laterals* 3.5–7 × 1–10 cm, base asymmetric, acute, apex acuminate; petiole 3–7(–13) cm long, usually with small glands adaxially; petiolules 0.4–1 cm long. **Inflorescences** 17–30-flowered, on small twigs, floriferous portion to 10 cm long; pedicels 1.5–4 cm long. **Flowers** white or cream, aging to yellow or golden-yellow, sweetly scented; torus 4–5 mm Ø; sepals 3.5–6 × 2–2.5 mm, elliptic; petals with limb 1–2.5 × 0.6–1.2 cm, broadly elliptic, claw 2–5(–7) mm long; stamens 18–22+, to 5 cm long; gynophore 3–5 cm long; ovary ellipsoid, 2.5–3 × ~2 mm. **Fruits** globose, 1.8–2.5 cm Ø; seeds ~5 × 2 mm, deeply reniform or horseshoe-shaped. **Distr:** Myanmar, India, Sri Lanka. **Ecol:** In lowland dry woodlands and scrubs. **Zones:** 10–12.

C. religiosa G. Forster, *Diss. Pl. Esc.* 45, 1786. (Syn: *C. membranifolia*). [Sacred Garlic Pear, Spider Tree, Temple Plant]. **Trees**, 3–15 m tall; trunk to 40 cm Ø. **Leaves** palmately compound; leaflets (4–)5.5–7(–10) × (2–)3–4 cm, apex acuminate to abruptly acuminate, thin and leathery, adaxially dull green, abaxially gray, midvein reddish, secondary veins 5–10 per side; petiolules 3–5(–7) mm long; petiole (5–)6–7(–10) cm, adaxially with minute triangular glands near rachis. **Inflorescences** racemes or corymbs, 10–25-flowered, branchlets just below flowers having fewer leaves; axis 3–7 cm long, after flowering slightly elongated; bracts 0.8–1.5 cm long, leaflike or slender, caducous; pedicels 2–5+ cm long. **Flowers** fragrant, opening with new flush; sepals ovate, 2–4.5 × 1.5–3 mm, acuminate; petals white to yellow aging to golden-yellow, claw 3.5–5 mm long, lamina 1.5–2.2 cm long; stamens 16–22(–30), filaments 3–6 cm long, anthers 2–3 mm long; gynophore 3.5–6.5 cm long; ovary ovoid to subcylindric, 3–4 × 1–2 mm. **Fruits** ovoid to obovoid, 1.8–2.6(–3.5) cm long; seeds 25–30, dark brown, 1.2–1.8 cm long, tuberculate. **Distr:** Asia through Malesia to Pacific Islands. **Ecol:** <200 m. Fl. Mar–May. Fr. Jul–Aug(–Oct). **Zones:** 10–12.

Steriphoma

Species 8. **Shrubs** or **small trees**, orange-pubescent. **Leaves** alternate, exstipulate; lamina simple. **Inflorescences** racemose, bracteate. **Flowers** zygomorphic, hermaphroditic; receptacle with 4 scaly appendages adnate to calyx base; calyx cylindric-campanulate, minutely 2- or 4-lobed, sepals 4, connate, concrescent, rupturing into 2–3 irregular lobes at anthesis, inside base glandular; petals 4, ± unequal; stamens 5–8, filaments attached to a short androphore; pistil borne on elongate gynophore, superficially 2-celled, ovules numerous. **Fruits** fleshy, dehiscent or

Top row: *Capparis sandwichiana*, *Capparis zeylanica*, *Crateva adansonii* ssp. *odora*.
Bottom row: *Crateva religiosa*, *Steriphoma paradoxum*.

FK&S

SZ

Top row: *Steriphoma paradoxum* (fruits), *Cleome gynandra*, *Cleome rutidosperma*. Bottom row: *Cleome spinosa*, *Cleome spinosa* (white form).

indehiscent, silique; seeds long, enclosed in a thin, hyaline, slimy pulp, sometimes extended to form a very delicate marginally wavy wing around the seed. **Distr:** Trinidad and Guatemala to Peru. **Ecol:** In forests. **Cult:** Propagate from seeds, cuttings or by air-layering. Requires a moist, well-drained soil in shade to full sun.

S. paradoxum (Jacq.) Endlicher, *Flora* 15(2): 396, 1832. (Syn: *Capparis paradoxa*, etc.). **Shrubs** or **small trees**, to 3(–5) m tall, evergreen; branches orangish-brown pubescent. **Leaves** alternate, crowded toward ends of branches, spreading and (abruptly) deflexed (at base of lamina); lamina oblong to oblong-lanceolate, 10–20 × 3–6 cm, base rounded, margins undulate, apex acuminate; petiole slender, swollen at both ends, to 12(–18) cm long. **Inflorescences** racemes, densely flowered, short, stout, terminal, flowers downward pointing; rachis furfuraceous, tumid; pedicels orange-yellow, 2.5–4 cm long, furfuraceous, often curved. **Flowers** imbricating, abruptly deflexed from apex of pedicels; calyx bright orange, oblong-cylindric, 1–1.7 cm long, splitting irregularly for half its length; petals 2–3.5 cm long, linear-oblong, pale yellow; stamens 5–7, green to pale yellow, 7–10 cm long, horizontal, slightly upcurved; ovary cylindric, on a curved gynophore, style pale green. **Fruits** capsules, indehiscent, furfuraceous, to ~12 × 2 cm Ø, thicker in the middle. **Distr:** S America (Caracas and New Grenada). **Ecol:** Mostly in shaded understorys. Fr. Jun–Aug. **Zones:** 10–12. **Info:** The specific name is often written as '*paradoxa*'.

CLEOMACEAE

Genera 10, **species** 300. **Herbs** or **shrubs**, monoecious; root hairs absent. **Leaves** palmately compound; petiole bundle(s) arcuate. **Inflorescence** racemose. **Flowers** monosymmetric; petals 4, clawed; stamens 6, filaments long, anthers linear, coiled at dehiscence; gynoecium with gynophore, carpels 2. **Fruits** dehiscent; placenta thin, looplike, woody, persistent, remaining on plant after the fruit valves have fallen off; seeds arillate (Stevens 2001 onwards). **Distr:** Tropical and warm temperate, especially America.

Cleome

Species ~20. **Herbs**, annual, sparsely branched, glandular pubescent or glabrous, lacking spines. **Stipules** absent or scalelike and caducous. **Leaves** alternate, spirally arranged, pinnately compound; leaflets (1 or)3–7(–11), lanceolate to oblanceolate, margins entire or serrulate; petiole with a pulvinus at basal or distal end; petiolule bases connate, forming a pulvinar disk. **Inflorescences** racemose, sometimes flat-topped or elongated, terminal, or axillary from apical leaves; bract present (or absent), basal. **Flowers** slightly zygomorphic; sepals 4, equal, connate for half their length, each often subtending a basal nectary; petals 4, distinct, equal; stamens (4–)6, distinct; filaments inserted on a discoid or conical receptacle (androgynophore); gynophore slender, elongating and recurving in fruit or sometimes obsolete; carpel 1, style 0.2–0.8 mm long, thick, stigma 1, capitate. **Fruits** capsules, oblong, dehiscent; seeds 10–40 per capsule, reniform, arillate or not. **Distr:** Warm-temperate and tropical areas of the Old World. **Cult:** Propagate from seeds. Requires a fertile, moist, well-drained soil in full sun.

C. gynandra L., *Sp. Pl.* 671, 1753. (Syn: *C. pentaphylla*, etc.). [Cat's Whiskers, Spider Wisp, Wild Spider Flower]. **Herbs**, annual, 0.3–1 m high, erect, widely branched; stems glandular-pubescent to long, shaggy-pilose, unarmed. **Leaves** long-petiolate, palmately (3–)5(–7)-foliolate; leaflets lanceolate to obovate-lanceolate or rhombic, 1.2–7.5(–10) × 0.8–3 cm, base acute to subobtuse, margins ciliate to denticulate, apex acute, shortly acuminate, thinly herbaceous; petiole 2.5–10 cm long, densely pubescent, occasionally with short prickles; petiolules 1–3 mm long, webbed at base, pubescent. **Inflorescences** long corymbose racemes, bracteate; bracts composed of 3 palmately arranged bractlets; pedicels 1.5–2 cm long, slender. **Flowers** white, yellowish, pink or pale purplish; sepals linear, 2.5–3(–5) × ~0.5 mm, acute, glandular-pubescent; petals rotund, 0.9–1.6 × 0.2–0.4 cm, slender-clawed; stamens 6; gynophore slender, 1–4 mm long, glabrous, to 2–2.5 cm long in fruit; ovary linear-oblong-cylindric, ~4 × 0.5 mm, glandular-pubescent. **Fruits** linear-cylindric, 6.5–8 × 0.3–0.35 cm, compressed, slightly curved; seeds ~1.3 mm Ø, helicoid-reniform, black, rough. **Distr:** Africa to Seychelles and Madagascar; tropical Asia; America. **Ecol:** In disturbed areas. **Zones:** 10–12.

C. rutidosperma Candolle, *Prodr.* 1: 241, 1824. (Syn: *C. aspera*, etc.). [Fringed Spider Flower, Spider Flower]. **Herbs**, usually annual, 0.3–1 m high, branched; branches often decumbent. **Stipules** ~0.5 mm long, scalelike or absent. **Leaves** pinnately compound; leaflets 3, oblanceolate to rhomboid-elliptic, 1–3.5 × 0.5–1.7 cm, margins entire or serrulate-ciliate, apex acute, obtuse, or sometimes acuminate; petiole 0.5–3.5 cm long, proximally winged. **Inflorescences** 3–6-flowered, 2–4 cm long, terminal and axillary; bracts leaflike, 3-foliate, 1–3.5 cm long; pedicels 1.1–2.1 cm long. **Flowers** pink to pale purple, or white; sepals yellow, narrowly lanceolate, 2.5–4 × 0.2–0.3 mm, ± persistent, margins ciliate, glabrous; petals white or purple-speckled, oblong to narrowly ovate, 0.7–1 cm × 1.5–2.3 mm, central 2 petals outside with a yellow transverse band; filaments yellow, 5–7 mm long, anthers purplish-brown, 1–2 mm long; pistil 2–3 mm long, glabrous, style 0.5–1.4 mm long. **Fruits** capsules, 4–7 × 0.3–0.4 cm, striate; seeds 4–25, reddish-brown to black with white funicular aril, 1–1.5 mm long. **Distr:** Tropical Africa. **Ecol:** In disturbed places; 0–200 m. Fl. & Fr. Jun–Sep. **Zones:** 10–12.

C. spinosa Jacq., *Pl. Carib.* 26, 1760. (Syn: *C. hassleriana*, *C. sesquiorygalis*). [Spider Flower]. **Herbs**, annual or biennial, 0.6–0.9(–1.2) m tall, viscid, pubescent, strongly scented. **Leaves** large, 5–7-foliolate; leaflets oblong-lanceolate, to 10 × ~2–3 cm, margins subentire to finely or slightly serrulate; petiole in lower leaves usually longer than the leaflets, with 2 small spiny stipules at base. **Inflorescences** racemes, many-flowered, conspicuously bracteate. **Flowers** large, white or pinkish, ~2–2.5 cm across; pedicels long, but shorter than the gynophore, both elongated and thickened in fruit; sepals 5–8 mm long; petals ~2–3 times as long as sepals, limb obovate, clawed, ~1 cm across with an equally long claw; stamens 6, about twice length of petals; gynophore ~6–8 cm long in fruit. **Fruits** cylindric-linear, 3–4.5 × ~0.3 cm; seeds many, glabrous. **Distr:** Tropical S America. **Ecol:** On open slopes, in sandy thickets, often in sand/gravel along streams, weedy in wastelands; below 1,000 m. Fl. throughout most of the year. **Zones:** 10–12.

C. viscosa L., *Sp. Pl.* 672, 1753. (Syn: *C. icosandra*, etc.). [Dog Mustard, Tick Weed, Wild Mustard]. **Herbs**, annual, erect, to 0.4–1.5 m tall, widely branched, unarmed; stems yellowish-hirsute, viscid, fetid. **Leaves** 3–5-foliolate; leaflets obovate, elliptic-oblong, very variable in size, often 2–4 × 1.5–2.5 cm, middle one largest; petiole to 5 cm long. **Inflorescences** racemes, to 30 cm long, with corymbose flowers at the top and elongated mature fruits below, bracteate; pedicels 0.6–2 cm long; bracts foliaceous. **Flowers** 1–1.5 cm Ø, whitish or yellowish; sepals oblong-lanceolate, 3–4(–6) × 1–2 mm, glandular-pubescent; petals oblong-obovate, 0.8–1.5 × 0.2–0.4 cm; stamens 10–12(–20), not exceeding petals; gynophore absent. **Fruits** linear-oblong, 3–7.5 × 0.3–0.5 cm, erect, obliquely striated, ends tapered, glandular-pubescent, slender; style 2–5 mm long; seeds many, 1–1.4 mm Ø, dark brown. **Distr:** Tropical regions worldwide. **Ecol:** In rocky and sandy places. **Zones:** 10–12.

Podandrogyne

Herbs, erect or ascending, suffrutescent or suffruticose. **Leaves** alternate, exstipulate; lamina simple or palmately compound. **Inflorescences** racemose or corymbose, terminal, several- to many-flowered, bracteate or ebracteate; flowers monoecious, rarely andromonoecious or dioecious through abortion, lower flowers ♀, upper ♂. **Flowers** with calyx ± deeply 4-parted, persistent or deciduous, sometimes ± petalaccous; petals 4, ± unequal, usually unguiculate; disk symmetric or eccentric; fertile stamens 6, inserted on a short or moderately elongate, concentric or eccentric gynophore, filaments somewhat unequal and declinate, anthers dorsifixed near base, accompanied by an abortive pistillode; fertile ovary borne upon a concentric or eccentric gynophore, stigma capitate, sessile or stipitate, ovules numerous, accompanying staminodia greatly reduced, sagittate, borne upon an androgynophore. **Fruits** siliques, dry, terete or compressed, usually irregularly dehiscent, replum finally separating at apex and irregularly contorted; seeds cochleate-reniform, with a conspicuous lamellate, funicular aril. **Distr:** Tropical C and S America. **Cult:** Propagate from seeds or cuttings. Requires a moist, well-drained soil in shade or partial sun.

P. decipiens (Triana and Planch.) Woodson, *Ann. Missouri Bot. Gard.* 35: 142, 1948. (Syn: *Cleome decipiens*, etc.). **Shrubs** or **shrubby herbs**, 1–3 m high. **Leaves** alternate; lamina 3-foliate, terminal leaflet larger, elliptic, 10–25 × 7–13 cm, veins pinnate, adaxially deeply depressed, shiny, yellow-green; petiole to 20 cm long. **Inflorescences** umbellate clusters, borne at ends of stems; ♂ and ♀ flowers separate, with ♂ toward the top of the cluster, and ♀ below. **Flowers** pale coral to orange, radially to bilaterally symmetrical; sepals fused; petals 4, free, spathulate, 1.5–2.7 × 0.5–1 cm; stamens 6, anthers dark blue; pistil long and thin. **Fruits** pods, pendent, at first pinkish-red, ripening to green, somewhat beanlike, 5–21 × ~0.5 cm, straight to curved, glabrous, often formed with flowers, and hanging beneath the inflorescences, splitting open spirally to expose the seeds; seeds many, small, dark brown, 3 mm long, embedded in whitish aril. **Distr:** Guatemala to Panama, Colombia and Peru. **Ecol:** In rocky areas near creeks, moist, open or partly shaded areas, streamsides; 200–2,700 m. Fl. & Fr. throughout the year. **Zones:** 10–12.

BRASSICACEAE

Genera 338, **species** 3,710. **Trees**, **shrubs** or **herbs**; roots lacking mycorrhizae; hairs variously furcate. **Stipules** absent. **Leaves** usually alternate and spiral, sometimes in basal rosettes; lamina simple, often pinnately lobed, margins entire to serrate, venation palmate or pinnate. **Inflorescences** indeterminate, occasionally reduced to a solitary flower, terminal or axillary; bracts absent (foliaceous). **Flowers** usually bisexual, radial or bilateral, often lacking subtending bracts; receptacle prolonged, forming an elongate or shortened gynophore (or androgynophore); sepals 4, distinct; petals 4, distinct, often forming a cross, often with an elongate claw and abruptly spreading limb, imbricate or convolute; stamens (2–)6, or numerous, ± similar in length or the 2 outer shorter than the 4 inner, filaments elongate to rather short, distinct or connate in pairs; carpels usually 2, connate, ovary superior, placentation parietal, ovules 1 to numerous per placenta; nectar disk or gland usually present. **Fruits** berries or capsules, usually dehiscing into 2 valves; seeds usually numerous, small. **Distr:** Worldwide, especially N temperate regions.

Brassica

Species ~40. **Herbs**, annual, biennial or perennial, rarely subshrubs or shrubs, often glaucous; stems erect or ascending, simple or branched, rarely leafless. **Leaves** *basal* petiolate, rosulate or not, simple, margins entire, dentate, lyrate-pinnatifid, or pinnatisect; *cauline* leaves petiolate or sessile, base cuneate, attenuate, auriculate, sagittate, or amplexicaul, margins entire, dentate, or lobed. **Inflorescences** racemes, ebracteate, elongated in fruit. **Flowers** with sepals ovate or oblong, erect, ascending, or rarely spreading, base of lateral pair saccate or not; petals usually yellow, rarely white or pink, obovate, spathulate, or rarely oblanceolate, apex obtuse or emarginate, claw distinct; stamens 6, tetradynamous, anthers ovate or oblong, obtuse at apex; nectar glands 4, median and lateral, rarely 2 and lateral; ovules 4–50 per ovary, style obsolete or distinct, stigma capitate, entire or bilobed. **Fruits** dehiscent siliques, linear or rarely oblong, terete, 4-angled; seeds few to numerous, uniseriate or rarely biseriate, wingless, globose or rarely oblong. **Distr:** Mediterranean region, especially SW Europe and NW Africa; few spp. in Asia. **Cult:** Propagate from seeds or cuttings. Requires a fertile, moist, well-drained soil in full sun.

B. oleracea L., *Sp. Pl.* 2: 667, 1753. [Broccoli, Brussels Sprouts, Cabbage, Cauliflower, Kale]. **Herbs**, biennial or perennial, rarely annual, (0.3–)0.6–1.5(–3) m high, glabrous, glaucous; stems erect or decumbent, branched at or above middle. **Leaves** *basal* and *lowermost cauline* leaves long-petiolate, sometimes strongly overlapping and forming a head, petiole to 30 cm, lamina ovate, oblong, or lanceolate, to 40 × 15 cm, margins entire, repand, or dentate, sometimes pinnatifid or pinnatisect, lobes 1–13 per side, oblong or ovate; *upper cauline* leaves oblanceolate, ovate, or oblong, to 10 × 4 cm, base amplexicaul, auriculate, margins entire, repand. **Inflorescences** racemes, sometimes fleshy and condensed into a head. **Flowers** with sepals oblong, 0.8–1.5 cm × 1.5–2.7 mm, erect; petals creamy-yellow or rarely white, ovate or elliptic, (1.5–)1.8–2.5(–3) × (0.6–)0.8–1.2 cm, rounded, claw 0.7–1.5 cm long; filaments 0.8–1.2 cm long. **Fruits** linear, terete (2.5–)4–8(–10) × (0.2–)0.3–0.4(–0.5) cm, 10–20-seeded per locule; seeds dark brown or blackish, globose, 1.5–2.5 mm Ø. **Distr:** W Europe; cultivated worldwide. **Ecol:** Fl. Mar–Jun. Fr. Apr–Jul. **Zones:** 8–12.

Below: *Cleome viscosa*, *Podandrogyne decipiens*, *Brassica oleracea*.

SZ

12 SANTALALES AND CARYOPHYLLALES

The placement of these orders is yet to be determined in relation to Rosids and Asterids. Molecular studies based on *rbcL* gene sequences by Savolainen et al. (2000) places Caryophyllales sister to Dilleniaceae and Vitaceae. The same study places Santalales (together with Saxifragales) at a higher level to both Rosids and Malvids. Molecular studies based on 81 genes from 64 plastid genomes by Jansen et al. (2007) places Caryophyllales sister to Asterids (Euasterids) with very good support. Wilkström et al. (2001) places Caryophyllales, Santalales and Dilleniaceae (Dilleniales) in a distinct clade and sister to Asterids. Studies based on mitochondrial *matR* sequences by Zhu et al. (2007) places Caryophyllales and Santales sister to Asterids. It seems that ultimately Caryophyllales, Santalales and Dilleniaceae (Dilleniales) will form a unified clade sister to Asterids, or perhaps integrated into Asterids, in future revisions.

SANTALALES

FAMILIES 8, GENERA 151, SPECIES 1,985.

Roots lacking root hairs. **Leaves** with lamina entire. **Inflorescences** cymose. **Flowers** with calyx often reduced; corolla valvate; stamens opposite petals in most species, epipetalous, anthers basifixed, pollen colpate or porate; disk present; gynoecium with carpels 3, connate, 1 pendulous apotropous tenuinucellate ovule/carpel, micropyle endostomal, ovary inferior in most species, style present, stigma small, globose. **Fruits** drupes, indehiscent, 1-seeded, calyx persistent; testa crushed, embryo minute, green (Judd et al. 2008; Stevens 2001 onwards). **Distr:** Pantropical. **Info:** The monophyly of Santalales is supported by molecular data (Hilu et al. 2003; Soltis et al. 2000, 2003).

BALANOPHORACEAE

Genera 17, **species** 50. **Herbs**, monoecious or dioecious, fleshy, parasitic on roots or rhizomes of various hosts; rhizome usually branched, usually with scales or warts and/or lenticels; flowering shoots endogenously arising from rhizome; scapes with or without leaves, unbranched. **Leaves** scaly, opposite or alternate and distichous or spiraled, sometimes whorled, rarely contorted or clustered, without stomata. **Inflorescences** unisexual or androgynous, terminal, spadix or spadixlike structure covered with minute branches; branches frequently subtended by variously modified bract; flowers pedicellate or sessile; ♀ congested on branches or basally on spadicles and/or shoot axis. **Flowers** unisexual; ♂ larger than ♀, 3(or 4 or more)-merous, perianth apically lobed or dentate, sometimes absent, lobes valvate, stamens 1 or 2 when perianth absent or usually as numerous as and opposite to perianth lobes when perianth present, filaments free or connate into a synandrium, anthers free or connate, 2-loculed or more, dehiscent by slits; ♀ with perianth absent or reduced and bilobed to irregularly lobed, adnate to ovary; ovary inferior, styles 1 or 2, stigmas slightly capitellate. **Fruits** achenes, 1-seeded. **Distr:** Mostly in tropical and subtropical regions.

Balanophora

Species 15. **Herbs**, monoecious or dioecious; rhizome containing sticky wax, with small scaly warts and/or stellate lenticels. **Leaves** opposite, alternate and distichous or spiral, or whorled, sessile; lamina fleshy or scalelike. **Inflorescences** spadixlike, cylindric, ellipsoid, ovoid-globose, or globose, enlarged after anthesis; ♂ flowers pedicellate or sessile, subtended by U-shaped or variously reduced bracts, spadicles minute, subclavate to clavate, very numerous, congested on ♀; ♀ flowers congested on main axis or also on basal stipe of spadicles. **Flowers** ♂ with perianth 3–6-lobed, lobes ovate, lanceolate, or orbicular, concave, isomorphic or heteromorphic, valvate, reflexed, stamens in a hemispheric or ± elongated synandrium, anthers bilocular, usually as numerous as perianth lobes, straight or sometimes U-shaped, bended near apex of synandrium, occasionally variously fused or interrupted into locelli, pollen white; ♀ composed of a pistil, ovary ellipsoid to fusiform, 1-loculed, attenuate toward both ends, ovules anatropous, shortly stiped, style elongated, persistent. **Fruits** small, exocarp crustaceous. **Distr:** Mainly in tropical Africa and Australia; temperate to tropical Asia; Pacific Islands. **Cult:** Propagate from seeds. The seeds can be placed on newly formed roots of the host plant (such as members of Fabaceae), the seeds then form a germ tube or haustorium that penetrates the root and links with the vascular tissues of the host. Requirements similar to host, but in shade or partial sun.

B. fungosa J.R. Forster and G. Forster, *Char. Gen. Pl.* 50, 1775. [Fungus Root]. **Herbs**, monoecious, rarely dioecious; rhizome yellowish-brown, unbranched or in a mass, surface with granular warts and scattered yellow stellate lenticels; branches subglobose, depressed, ~1.5 × 2.5 cm. **Scapes** pink, reddish-orange, or occasionally yellowish, 3–8 × 0.6–1.5 cm at anthesis, to 12 cm long in fruit. **Leaves** 15–30, spirally arranged, rarely opposite, imbricate; lamina broadly ovate, 1–1.5 × 1.5–1.7 cm, apex obtuse. **Inflorescences** ellipsoid, ovoid, or conically ovoid, 1–3 × 1–2 cm; ♂ flowers basally on androgynous inflorescences, actinomorphic, subtended by short rudimentary bracts, pedicels 4–5 mm long; ♀ flowers basally on spadicles and main axis. **Flowers** ♂ with perianth lobes 4 or 5, ovate-elliptic, apex acute, synandria subglobose, stipe 0.5–1 mm long, anthers 4 or 5, U-shaped, spadicles subclavate, apical ½ ellipsoid; ♀ yellowish. **Distr:** Indonesia, Ryukyu Islands, New Guinea, Philippines, Taiwan; Australia, Pacific Islands. **Ecol:** In coastal forests; 0–900 m. Fl. Dec. **Zones:** 9–12.

B. laxiflora Hemsley, *J. Linn. Soc.*, Bot. 26: 410, 1894. (Syn: *B. formosana*, etc.). **Herbs**, parasitic, dioecious, red to dark red, sometimes purplish; rhizome branched; branches subspherical, 1–3 × 1–2.5 cm, surface densely covered with scabrous speckles and yellowish stellate warts. **Leaves** 8–14, alternate; lamina elliptic-oblong, 1.5–2.5 × 1–1.5 cm, apex obtuse. **Inflorescences** with scapes 5–10 cm long; ♂ cylindric, 3–18 × 0.5–2 cm, apex acuminate; ♀ ovoid-spheroid to oblong-ellipsoid, 2–6 × 0.8–2 cm, apex acuminate, spadicles subclavate, to 1.5 mm long, apically obovate; ♂ flowers subsessile, subtended by 2 slender rudimentary or indistinct bracts; ♀ flowers on basal stipes of spadicles and main axis. **Flowers** ♂ zygomorphic, perianth lobes 4–6, suborbicular to ovate, 2–3 mm long, apex acute to obtuse, synandria sessile, subdiscoid or sometimes expanded laterally, 4.5–6 mm ∅, anthers dehiscent by short slits. **Distr:** China, Laos, Taiwan, Thailand, Vietnam. **Ecol:** In dense forests; (200–)600–1,700 m. Fl. Sep–Nov. **Zones:** 9–12.

B. wrightii Makino, *Bot. Mag. Tokyo* 24: 292, 1910. (Syn: *B. torbiracola*). **Herbs**, monoecious, creamy-yellow to reddish, 4–5.5 cm high; rhizome branched, surface scabrous, densely covered with speckles and stellate lenticels; branches subspheroid to flat-spheroid, 1.5–2.2 cm wide. **Leaves** scalelike, 5–10; lamina ovate, 0.8–1.4 × 0.5–0.8 cm, apex obtuse. **Inflorescences** androgynous, conic, ovoid-ellipsoid, or ovoid, 1.8–4 × 1–2 cm; scapes yellowish, 1–5.5 × 0.4–0.8 cm; ♂ flowers scattered between spadicles and ♀ flowers, shortly pedicellate, spadicles subglobose, shortly stiped, apical cells smooth; ♀ flowers only on main axis of inflorescences. **Flowers** ♂ 2–3 mm ∅, perianth lobes 3, spreading, ovate to subrounded, synandria subdiscoid, anthers 3, transversely dehiscent. **Distr:** Japan, Ryukyu Islands, Taiwan. **Ecol:** In lowland forests, usually parasitic on *Pittosporum* and *Rhaphiolepis* species. Fl. Aug–Dec. **Zones:** 9–11.

LORANTHACEAE

Genera 68, **species** 950. **Shrubs**, usually aerial hemiparasites on other seed plants (rarely on other members of Loranthaceae), often spreading along host by runners (epicortical roots), more rarely terrestrial root-parasitic shrubs or trees, nodes not articulated, glabrous or hairy. **Stipules** absent. **Leaves** opposite

Top row: *Balanophora fungosa, Balanophora laxiflora, Balanophora wrightii, Amyema miraculosum.*
Bottom row: *Amylotheca dictyophleba, Dendrophthoe falcata, Dendrophthoe suborbicularis, Helixanthera parasitica.*

or alternate; lamina simple, usually pinnately veined, margins entire; petiole often indistinct. **Inflorescences** racemes, spikes, or umbels (sometimes condensed into heads), terminal or axillary; bracts usually inconspicuous, sometimes forming conspicuous involucre. **Flowers** usually bisexual, rarely unisexual (plants dioecious), 4–6-merous, actinomorphic or zygomorphic; calyx adnate to ovary, limb annular to cupular, entire or shortly toothed, persistent; petals usually 4–6, free or connate, valvate; disk usually inconspicuous to ± absent; stamens as many as petals, opposite and adnate to them, anthers mostly basifixed, sometimes dorsifixed, 2–4-loculed, longitudinally dehiscent; ovary inferior, 1- or 3- or 4-loculed, without true ovules, embryo sacs originating from a central column or at ovary base, integument absent, style simple, stigma small. **Fruits** berries (rarely drupes or capsules), with a viscin layer (sticky mucilaginous tissue) outside the vascular bundles; seed 1, testa absent, endosperm copious, embryo large. **Distr:** Primarily in tropical and subtropical regions.

Amyema

Species ~90. **Shrubs**, parasitic; aerial stems with or without external runners. **Leaves** opposite or whorled, rarely scattered; lamina penniveined or with curved longitudinal venation. **Inflorescences** usually axillary, sometimes arising between the nodes or from runners, primarily a pedunculate, many-rayed umbel of triads, often variously reduced; bract single beneath each flower. **Flowers** with corolla free; usually 4-, 5-, or 6-merous; anthers basifixed, immobile; styles usually articulate at base. **Distr:** Malesia, Australia. **Cult:** Propagate from seeds. Requires a suitable host plant.

A. ***miraculosum*** (Miq.) Tiegh., *Australian Journal of Botany* 14(3): 476, 1966. (Syn: *Loranthus miraculosus*, etc.). [Fleshy Mistletoe]. **Shrubs**, parasitic, erect, spreading or pendent, glabrous except rarely the ovary and young shoots with a sparse indumentum; external runners absent. **Leaves** lanceolate to narrowly elliptic or obovate, 3–11 × 0.5–2.5 cm, base attenuate or contracted, apex acute to rounded; petiole obscure, 0.2–1 cm long. **Inflorescences** pedunculate umbels of triads, central flower sessile, laterals pedicellate; peduncle 0.6–3 cm long; rays of umbel 2 or 3, 0.3–1 cm long; pedicels usually 1.5–7 mm long; bracts 0.7–1.5 mm long. **Flowers** with corolla 0.8–2 cm long, usually 5-merous, clavate, longitudinally ridged or winged, crimson. **Fruits** narrowly ellipsoid, ~0.8–1 cm long, apex constricted. **Distr:** E Australia. **Ecol:** Parasitic on various hosts including other *Amyema* species. Fl. Jan–Apr. **Zones:** 10–11.

Amylotheca

Species 5. **Shrubs**, parasitic; aerial stem erect to pendent; external runners present. **Leaves** opposite; lamina penniveined. **Inflorescences** axillary, racemes of opposite triads with the central flower sessile and lateral flowers pedicellate, sometimes reduced to simple racemes or umbels; bract 1 under each flower. **Flowers** with corolla 6-merous, regular; petals united to the middle or higher into a slightly inflated tube; stamens equal, anthers dorsifixed, immobile, introrse, 4-loculed. **Fruits** berries. **Distr:** New Guinea, Melanesia, Australia. **Cult:** Propagate from seeds. Requires a suitable host plant.

A. ***dictyophleba*** (F. Muell.) Tiegh., *Bulletin de la Societe Botanique de France* 41: 262, 1894. (Syn: *Loranthus dictyophlebus*, etc.). [Brush Mistletoe, Red Mistletoe]. **Shrubs**, parasitic, spreading to pendent, glabrous; runners present. **Leaves** opposite, rarely 3-whorled; lamina simple, lanceolate to elliptic, 5–14 × 1.5–6 cm, base attenuate or contracted, apex acute to rounded, adaxially usually shiny, abaxially dull, reticulate venation distinct; petiole 2–8 mm long. **Inflorescences** racemes, with 1–6 decussate pairs of triads or single flowers; peduncle usually 4–8 mm long; axis slender, 0.5–1.4(–2.8) cm long; pedicels of lateral flowers if present 2–4 mm long; bracts 1 mm long. **Flowers** with corolla in mature bud 2.8–3.8 cm long, inflated in middle, acute; calyx green or red; petals 6, pink or red with green tips or yellow with red bases, rarely completely yellow, 1.8–6 cm long, erect; style articulate 1–3 mm above base. **Fruits** globose, 0.5–1 cm Ø, red or purple. **Distr:** Australia (NSW, Queensland). **Ecol:** Parasitic on many species of rainforest trees; 0–1,270 m. Fl. & Fr. any time of the year. **Zones:** 10–11.

Dendrophthoe

Species ~30. **Shrubs**, parasitic, some parts hairy. **Leaves** alternate or subopposite; lamina pinnately veined. **Inflorescences** axillary, sometimes at leafless nodes, racemes or spikes; 1 bract subtending each flower, shorter than calyx. **Flowers** bisexual, 5-merous, actinomorphic or zygomorphic by the presence of a single lateral split of varying length; mature flower bud tubular; calyx ovoid or urceolate, limb annular or cupular, persistent; corolla sympetalous, tube inflated, tip clavate, lobes reflexed or twisted with age; stamens inserted at base of corolla lobes, filaments short, flattened, anthers 4-loculed; ovary 1-loculed, placentation basal, style filiform, 5-angled, stigma capitate. **Fruits** berries, ovoid, exocarp leathery. **Distr:** Tropical Africa, Asia, and Australia. **Cult:** Propagate from seeds; the sticky seeds can be removed from the fruit and placed on a suitable branch of a host tree. Requires a suitable host.

D. falcata (L. f.) Ethingsh., *Denkschr. Kaiserl. Akad. Wiss., Math.-Naturwiss. Kl.* 32: 52, 1872. (Syn: *L. amplexifolius*, etc.). [Honeysuckle Mistletoe]. **Shrubs**, parasitic. **Leaves** sometimes amplexicaul, subopposite, highly variable; lamina mostly ovate-oblong, to 20 × 8 cm, base and apex obtuse, margins minutely calloused, glabrous; petioles (0)2–3 mm long. **Inflorescences** racemes, solitary, to 50-flowered, curved, to ~7 cm long, axillary; peduncle ~2–3 × 5 mm; pedicels ~3 mm long; bracts ~1 mm long, broadly rounded. **Flowers** with mature floral buds white to pink or bright red to constriction, with a dark green band just below the constriction, clavate portion bright green; calyx rudimentary, ~1 mm high, minutely lobed; corolla tube at anthesis colored as buds, 2.5–5 cm long, sinuses ~2 mm deep, tube split ~1.3 cm deep, margin of sinuses and split tuberculate, lobes white to red, margins green, sublinear, ~9 × 1 mm, acute; filaments bright red, erect, free, anthers yellow; style light reddish-brown, exserted, ~6 cm long, slightly enlarged below stigma for ~5 mm, stigma dark reddish-brown. **Fruits** bright pink, oblong, ~1.7 × 0.6 cm. **Distr:** India and Sri Lanka to Australia. **Ecol:** In dry regions. Fl. throughout the year. **Zones:** 10–12.

D. suborbicularis (Thw.) Danser, *Bull. Jard. Bot. Buitenzorg Ser.* 3. 10: 311, 1929. (Syn: *Loranthus suborbicularis*). **Shrubs**, parasitic. **Leaves** alternate; lamina obovate-orbicular, ~5–7 cm ∅, base and apex usually obtuse-rounded, brownish-tomentose when young, often 5-veined; petiole ~8 mm long. **Inflorescences** racemose, brownish-tomentose, in axillary clusters, to ~20-flowered; rachis ~3–4 cm long; peduncles ~4 mm long; bracts ovate-acute, ~1 mm long. **Flowers** with mature floral buds greenish-yellow to constriction, dark brownish-red at clavate portion, ~3 cm long, with 2 prominent lateral swellings at constriction, slightly curved; calyx lobed, ~1 mm long; corolla tube yellow, ~2.5 cm long, lobes yellow, linear, ~7 × 1 mm, acute, all spreading in the same direction and opposite the split; filaments maroon where adnate to corolla, sinuses between lobes ~3 mm deep, tube split ~1.4 cm deep, free portion of filament erect, bright red, ~3 mm long, anthers light yellow, oblong, ~4 × <1 mm, exserted ~7 mm above tube; style light green, reddish-orange just below stigma, ± terete, curved, ~3 cm long, stigma dark red, subglobose, exserted to ~3 mm. **Fruits** brownish, oblong, ~1 cm long, with persistent calyx, puberulent. **Distr:** Sri Lanka. **Ecol:** In high montane regions. Fl. Apr–Aug. **Zones:** 11–12.

Helixanthera

Species ~50. **Shrubs**, parasitic, glabrous or hairy and glabrescent. **Leaves** opposite or alternate, rarely ± whorled; lamina pinnately veined. **Inflorescences** racemes or spikes, axillary, rarely terminal; 1 bract subtending each flower; flowers sessile or pedicellate. **Flowers** bisexual, 4–6-merous, actinomorphic; mature flower bud straight, basal half usually slightly inflated and angled, distal portion usually subclavate; calyx limb 4–6-denticulate, persistent; corolla red, pink, orange, or yellowish, petals free, straight, spreading; stamens usually inserted at middle of the petals, filaments short, anthers 2–4-loculed, sometimes multilocellate, ellipsoid; ovary 1-loculed, placentation basal, style subcylindric, 4–6-angled, stigma capitate or truncate. **Fruits** berries, ovoid or ellipsoid, exocarp leathery, smooth or pubescent. **Distr:** Tropical and subtropical Africa and Asia. **Cult:** Propagate from seeds; the sticky seeds can be placed on a suitable branch of a host tree. Requires a suitable host.

H. parasitica Loureiro, *Fl. Cochinch.* 1: 142, 1790. (Syn: *Helicia parasitica*, etc.). **Shrubs**, parasitic, 1–1.5 m high, glabrous; branches elongated. **Leaves** opposite (or subopposite); lamina ovate to ovate-lanceolate, 5–12(–18) × (1.5–)3–4.5(–6) cm, base broadly cuneate to rounded, apex acute to acuminate, papery to thinly leathery; petiole 0.5–1.5(–2) cm long. **Inflorescences** racemes, solitary or paired, axillary, sometimes at leafless node, 40–60-flowered, 5–10(–30) cm long, sometimes verticillate, dull brown or gray papillose; bracts subovate or ovate-triangular, (0.7–)1–1.5 mm long; pedicel 1–2 mm long. **Flowers** 5-merous; mature bud with basal half abruptly expanded into 5 prominent semicircular keels, distally clavate; calyx ellipsoid, 1.5–2 mm long, limb annular, ~0.5 mm long, entire or 5-denticulate; corolla red, pink, or yellowish, petals (4–)6–8 mm long, reflexed from above basal keels; filaments 1–2.5 mm long, anthers 1–1.5(–2.5) mm long; style cylindric, 5-angled, constricted in middle, 3–6 mm long. **Fruits** berries, red, ellipsoid, ~6 × 4 mm, papillose. **Distr:** NE India through SE Asia to Philippines. **Ecol:** Mostly in forests; 100–1,500(–1,800) m. Fl. Jan–Jul. Fr. May–Aug. **Zones:** 9–12.

Macrosolen

Species ~40. **Shrubs**, parasitic, base often with epicortical roots, glabrous. **Leaves** opposite; lamina pinnately veined. **Inflorescences** racemes or spikes, rarely umbels, axillary; 1 bract and 2 bracteoles subtending each flower; bract small, shorter than calyx; bracteoles often connate. **Flowers** bisexual, 6-merous, actinomorphic or sometimes zygomorphic by the presence of a single split; mature flower bud tubular; calyx ovoid to ellipsoid, limb annular or cupular, persistent; corolla sympetalous, tube gradually dilated, usually 6-keeled in middle portion, then constricted abruptly to a neck and expanded to a clavate tip, lobes reflexed; filaments short, anthers 4-loculed, sometimes multilocellate; ovary at first 3-loculed, then 1-loculed, placentation free, central, style filiform, base usually articulated, stigma capitate. **Fruits** berries, ovoid or ellipsoid. **Distr:** Tropical S and SE Asia; China. **Cult:** Propagate from seeds; the sticky seeds can be placed on a suitable branch of a host tree. Requires a suitable host.

M. albicaulis Wiens, *Ceylon J. Sci.* 9 (2): 45, 1971. **Shrubs**, parasitic, without haustoria-bearing surface runners, usually forming large, pendulous masses to ~4 m long; branchlets usually twisted 90º. **Leaves** lanceolate to ovate-lanceolate, ~7–10 × 3–4 cm, base usually obtuse, sometimes slightly oblique, apex usually acute and sometimes attenuated, pale green; petiole ~1–1.5 cm long. **Inflorescences** spicate, usually with 2–3 pairs of decussate apically crowded flowers; peduncle ~1 cm long; bracts and bracteoles usually rounded, ~2 mm long. **Flowers** with mature floral bud pink at base, becoming pale pink at point of maximum dilation, constricted section green, upper clavate portion dull pink, ~6 cm long, base often strongly curved; calyx cylindrical, ~2 mm long, truncate, entire; corolla tube pink at base, pale toward the throat, ~4.5 cm long, sinuses ~4 mm deep, lobes when reflexed silvery against a dark red background, linear-oblanceolate, ~1.4 cm long, acute, usually twisting with age; filaments reddish-pink, thickened, free for ~8 mm, anthers bright yellow, linear, ~5 mm long; style pink, terete, curved, ~6 cm long, stigma bright red. **Fruits** green, ovate-elliptic, ~8 × 4 mm, calyx persistent. **Distr:** Sri Lanka. **Ecol:** In montane regions. Fl. Apr, Jun–Sep. **Zones:** 11–12.

OLACACEAE

Genera 14, **species** 103. **Shrubs**, **scandent shrubs**, **trees**, or **lianas**, sometimes hemiparasitic, dioecious. **Stipules** absent. **Leaves** alternate, rarely opposite, spiral or 2-ranked; lamina conduplicate, simple, margins entire, veins usually pinnate, sometimes palmate and 3–5-veined; petiole well-developed. **Inflorescences** axillary, rarely also terminal, cymose, fasciculate, racemose, spicate, or umbellate, mostly small; bracts small, inconspicuous. **Flowers** small, usually bisexual, actinomorphic, 3–6-merous, sometimes heterostylous; hypanthium present; calyx small at anthesis, cupular, discoid, or lobed; petals free or basally connate, valvate; disk cuplike or annular; stamens 3–15, rarely in part staminodes, epipetalous, anthers basi- or medifixed, usually 2-loculed, longitudinally dehiscent; ovary usually superior, rarely semisuperior, sometimes apparently inferior through fusion with calyx, 2–5-loculed or 1-loculed, locules 1-ovuled, pendent from apex of free central placenta, usually anatropous, integuments 1 or 2, style terminal, simple, stigma 2–5-lobed. **Fruits** drupes, sometimes with an accrescent calyx; seed solitary, testa thin, endosperm abundant. **Distr:** Pantropical.

Ximenia

Species 8. **Shrubs** or **trees**; branchlets often spine-tipped; often with very short shoots (brachyblasts) bearing leaves and inflorescences. **Leaves** alternate or fascicled on brachyblasts; lamina pinnately veined. **Inflorescences** axillary fascicles, pedunculate cymes, or sometimes flowers solitary. **Flowers** rarely functionally unisexual; calyx cupular, 4- or 5-dentate, persistent, not accrescent; petals 4 or 5, free, densely villous inside, apical half revolute; stamens 8 or 10, filaments filiform, anthers linear, longitudinally dehiscent; disk absent; ovary superior, (3 or)4-loculed, placentation axile, ovules 3 or 4, style ± slender, stigma capitate. **Fruits** drupes, ovoid to globose, 1-seeded. **Distr:** Mostly in tropical America, one species restricted to Africa; one species ± pantropical. **Cult:** Propagate from fresh seeds or cuttings. Requires any moist, well-drained soil in full sun.

X. americana L., *Sp. Pl.* 2: 1193, 1753. [False Sandalwood, Hog Plum, Yellow Plum]. **Shrubs** or **trees**, 2–7(–11) m tall, often straggling, glabrous (except the flowers); branchlets usually spiny. **Leaves** elliptic, ovate, obovate, almost round or lanceolate, 1.3–10 × 0.8–6 cm, base obtuse, apex obtuse, apiculate, mucronulate, sometimes emarginate, secondary veins 3–5 per side; petiole 0.2–1.2 cm long. **Inflorescences** cymes or racemes, often umbellate, 2–10-flowered or flowers solitary, 1.5–2.5 cm long. **Flowers** fragrant, usually hermaphrodite; calyx cupular, ~1 mm long; sepals 4(–5), 0.5–1.4 mm long, sometimes ciliate; petals 4 or 5, white or greenish, oblong, 0.5–0.9(–1.2) × 0.1–0.25 cm, apiculate, densely white-barbate inside to 1–3 mm below apex, recurved when mature; stamens 8(–10), filaments 1–4 mm long; ovary ovoid-conic, 2.4–3.7 mm long, style to 4.3(–5.5) mm long, rarely absent. **Fruits** drupes, yellow, sometimes orange to red, edible, ovoid or ± globose, 1–3.5 × 1–3 cm; seeds ellipsoid, 1.5–2.5 × 1–2 cm. **Distr:** Asia to Malesia, Africa, tropical America, Australia, Pacific Islands. **Ecol:** In coastal areas; 0–100 m. Fl. & Fr. Mar–Jun. **Zones:** 10–12.

SANTALACEAE

Genera 44, **species** 990. **Herbs** or **shrubs**, rarely trees, usually root hemiparasites, occasionally aerial hemiparasites, rarely endophytic. **Stipules** absent. **Leaves** usually alternate, spiraled to opposite, sometimes scalelike; lamina simple, margins entire, rarely spiny, usually pinnately veined, sometimes palmately 3–9-veined; petiole often indistinct. **Inflorescences** cymose, umbellate, paniculate, racemelike, spicate, or clustered, sometimes 1-flowered, mostly axillary, occasionally terminal; bracts scalelike, small, sometimes forming involucre;

RT

NT

Above: *Macrosolen albicaulis, Ximenia americana, Phacellaria tonkinensis, Santalum album.*

bracteoles sometimes present, paired. **Flowers** bisexual or unisexual (plants usually dioecious, rarely monoecious), actinomorphic, 3–6(–8)-merous, very small; perianth lobes 3–6(–8), slightly fleshy; ♂ with perianth lobes sparsely hairy or with tongue-shaped appendage at insertion of stamens, stamens as many as and opposite to perianth lobes, usually on base of lobes, filaments filiform; ♀ and *bisexual* with perianth tube usually longer than ♂, ovary inferior or half-inferior, 1- or 5–12-loculed, ovules 1–3(–5), style 1. **Fruits** drupes or nuts, exocarp usually fleshy; seed 1. **Distr:** Tropical and temperate regions.

Phacellaria

Species 8. **Subshrubs** or **herbs**, usually epiparasitic. **Leaves** alternate, reduced to rudimentary scales. **Inflorescences** axillary, 1-flowered or flowers densely clustered. **Flowers** bisexual or unisexual (then plants dioecious, monoecious, or polygamous), sessile, minute; ♂ with perianth tube short, lobes 3–6, valvate, stamens as many as and opposite to perianth lobes, filaments short, compressed, anthers cordate, 2-celled, longitudinally dehiscent, disk flat or slightly conic; ♀ with perianth lobes 4–8, valvate, ovary inferior, 1-loculed, ovules 2 or 3; *bisexual flowers* similar to ♀ ones, but with stamens. **Fruits** drupes, ellipsoid, with persistent lobes and disks at apex, exocarp fleshy, endocarp bony, 5- or 6-furrowed; seed 1, endosperm copious. **Distr:** Tropical and subtropical SE Asia. **Cult:** Propagate from seeds. Requires a compatible host. **Info:** All species are epiparasites growing on *Dendrotrophe* and species of the family Loranthaceae (Xia and Gilbert 2009).

P. tonkinensis Lecomte, *Bull. Mus. Natl. Hist. Nat.* 20: 399, 1914. **Subshrubs**, 10–30 cm high; stems in clusters, unbranched or with a few branches, slender, terete, 1–1.5 mm thick, glabrous. **Leaves** alternate, reduced to rudimentary scales. **Inflorescences** elongated clusters, dense, axillary; bracts imbricate when young, ± orbicular, small, ~0.5 mm long, apex sometimes acute; bracteoles inconspicuous or absent; involucre absent. **Flowers** mostly unisexual, less often bisexual (then plants dioecious or monoecious); ♂ with perianth tube whitish, lobes 5, triangular, distinct from disks, stamens 5, filaments very short, anthers small, disks nearly flat, shallowly divided; ♀ and bisexual with perianth ellipsoid, lobes 5(or 6), triangular, 1.5–2 mm long, style shortly terete, stigmas truncate or slightly oblong-scutiform. **Fruits** drupes, ovoid-ellipsoid, 8–9 × 2.5–3 mm, widest at base, gradually narrowing to apex, endocarp 5- or 6-furrowed; seed ellipsoid. **Distr:** China, Vietnam. **Ecol:** In forests; ~1,000 m. Fl. Jun–Aug. Fr. Oct–Feb. **Zones:** 9–11.

Santalum

Species 8–9. **Trees** or **shrubs**, glabrous, parasitic on grasses, herbs, shrubs or trees. **Leaves** opposite or subopposite; lamina coriaceous. **Inflorescences** axillary or terminal, paniculate cymes; bracts minute. **Flowers** bisexual; perianth 4- or rarely 5-lobed, lobes with tuft of hairs, tube adnate to base of ovary; stamens 4, or more usually 5, short, united at base of lobes; ovary semi-inferior, ovules 2–3, inserted below top of central free column. **Fruits** subglobose; seed subglobose. **Distr:** Tropical or subtropical regions (Indo-Malaysia to Australia and Hawai'i; Juan Fernandez). **Cult:** Propagate from seeds or by grafting. Requires any well-drained soil in full sun.

S. album L., *Sp. Pl.* 349, 1753. [Sandal Tree, Sandalwood]. **Trees**, evergreen, semiparasitic, evergreen, glabrous, to ~4 m tall; branches slender, drooping. **Leaves** elliptic-ovate or ovate-lanceolate, 3.5–7 × 1.5–3.25 cm, base acute, apex acute or subacute, membranous, drying pale-brown, abaxially glaucous, lateral venation indistinct; petiole 0.75–2 cm long, very slender. **Inflorescences** paniculate, terminal and lateral; pedicels opposite, 1–1.5 mm long, subequaling perianth tube. **Flowers** 4–5 mm ∅, pale fawn to red or maroon, usually with orange center, lobes ~1.5 mm long, broadly triangular. **Fruits** ~8 mm ∅ when dry, globose, ripening to purplish-black, juicy; seed 1, globose to obovoid. **Distr:** India. **Ecol:** In coastal dry forests, on dunes and cliff tops; to 700 m. Fl. May–Sep. **Zones:** 11–12. **Info:** Cultivated commercially for the sandalwood oil and wood.

S. haleakalae Hillebr., *Fl. Hawaiian Isl.* 390, 1888. [Haleakala Sandalwood]. **Trees** small, 2–4 m tall; trunk to 20 cm ∅; crown 2–4 m wide. **Leaves** ovate, obovate or orbicular, 2.5–7.5 × 2–6 cm, stiff to coriaceous, olive green, often tinged purple, usually glaucous; petiole 2–7 mm long. **Inflorescences** cymes, compound, terminal, sometimes few borne in extra axillary clusters, flowers congested; pedicels 2–4 mm long; bracts persistent until anthesis. **Flowers** reddish to cream in bud, slightly fragrant; corolla 0.8–1.7 cm long, outside dark red, glaucous, turning to same color inside after anthesis, tube cylindrical, white, turning dark red with age; ovary partly inferior. **Fruits** drupes, 1–1.5 cm long, black or purplish black, with a distinctive subapical receptacular ring. **Distr:** Hawai'i (Maui). **Ecol:** In subalpine shrublands; 1,900–2,700 m. **Zones:** 11–12.

CARYOPHYLLALES

FAMILIES 33, GENERA 692, SPECIES 11,155.

Plants not mycorrhizal. **Leaves** with margins entire. **Flowers** with anther outer parietal cells developing directly into endothecium, pollen colpate; style branch long. **Fruits** loculicidal capsules; seed testal (Stevens 2001 onwards). **Distr:** Worldwide.

DROSERACEAE

Genera 3, **species** 115. **Herbs**, perennial or annual, mostly terrestrial or rarely aquatic, carnivorous or insectivorous; stem with much reduced leaf laminas that function as rhizoids below ground, with or without tubers or rhizomes. **Leaves** basal and rosulate, or alternate, rarely whorled, stipulate or exstipulate; lamina with sticky, glandular hairs, or with sensitive hairs that trigger closing of lamina to trap small animals, juvenile laminas circinate. **Inflorescences** monochasial cymes, axillary, lateral or terminal, or flowers solitary. **Flowers** bisexual, actinomorphic, 4- or 5-merous; sepals 4 or 5(or 6–8), persistent; petals as many as sepals or more; stamens 4 or 5 (to 15 in *Dionaea*), hypogynous, free, alternate with petals, anthers 2-loculed, dehiscing by longitudinal slits; ovary superior or nearly so, globose or ovoid, 2–5-carpellate, 1–3-loculed, placentation parietal or basal, styles (2 or)3–5(or 6), simple or branched, stigma simple or multifid, expanded, papillate. **Fruits** capsules, dehiscent or indehiscent; seeds few to numerous, endosperm fleshy. **Distr:** Worldwide.

Dionaea

Species 1. *Generic description same as specific.* **Cult:** Propagate by division. Best grown in pots containing peat moss, in full or partial sun.

D. muscipula J. Ellis, *St. James Chron. Brit. Eve. Post* No. 1172: 4, 1768. [Venus Flytrap]. **Herbs**, to 10 cm high, rosetted, insectivorous; rosette 10–25 cm ∅. **Leaves** consisting of 2 parts, a winged basal petiole and a terminal highly modified lamina, the 'trap'. **Trap** resembling and acting like a very small 'bear trap', hinged in the middle, with stiff bristles along the margins, inside surface usually purplish-red, with 3 minute trigger hairs, which are triggered by prey (usually wandering insects); the trap is activated when the hairs are triggered in a specific sequence, to avoid false triggering other than by insects, the trap is reset after the prey has been digested, and each trap is functional for a few times only. **Inflorescence** an umbellate cluster, few-flowered, erect, 7.5–15(–30) cm tall, unbranched; bracts lanceolate, acute, green, glabrous; floral bract minute, triangular. **Flowers** white, ~2.5 cm ∅; calyx 5-lobed, lobes free, narrowly lanceolate, ~5 mm long, acute, green; corolla with petals 5, spreading, free, pure white, narrowly obovate, ~1.2 cm long, margins often ruffled, apices rounded, prominently veined; stamens many, free, anthers globular; ovary, ovoid, style long. **Fruits** capsules, ~6 mm long, opening irregularly; seeds, minute, black. **Distr:** USA (N and S Carolina). **Ecol:** In savannas and bogs, usually in nutrient-deficient areas. **Zones:** 5–11.

NEPENTHACEAE

Genus 1, species ~87. **Climbers**, **shrubs** or

Above: *Santalum haleakalae, Dionaea muscipula, Nepenthes alata.*

subshrubs, woody or subwoody, carnivorous, to 20 m long; stem simple or branched, erect, climbing, or procumbent, cylindric or triangular in cross-section. **Leaves** alternate, petiolate or sessile; lamina with midvein prolonged into a tendril, apex of tendril usually developing into a dimorphic pitcher with a lid usually projecting over mouth and a spur inserted at base of lid. **Lower pitchers** (toward the base of the plant) ovoid, facing inward, with 2 fringed wings, tendril ventral, uncoiled. **Upper pitchers** (toward the summit of the plant) cylindric or funnelform, facing outward, wings generally unfringed or reduced to ridges, tendril dorsal, coiled. **Inflorescences** terminal (appearing leaf opposed due to axillary growth), racemose, or with secondary branches 2-flowered or cincinni, usually sparsely bracteate. **Flowers** regular; perianth in 1 whorl of (3 or)4 tepals; ♂ with 4–24 anthers, filaments united into a column, pistils inconspicuous; ♀ with anthers inconspicuous, pistils (3 or)4, ovary superior, usually ovoid-globose or tetragonal-terete, (3 or)4-loculed, placentation axile, ovules numerous, in many rows, style very short to obsolete, stigma disklike, (3 or)4-lobed. **Fruits** capsules, (3 or)4-valved, leathery or woody, loculicidally dehiscent; seeds numerous, usually hairlike. **Distr:** Madagascar, through SE Asia to Malesia, N Australia and Pacific Islands.

Nepenthes

Species ~87. *Generic description same as familial.* **Cult:** Propagate from seeds or stem cuttings. Requires a fertile, loose soil rich in organic matter, high humidity and full to partial sun.

N. alata Blanco., *Fl. Filip.*, ed. 1: 805, 1837. (Syn: *N. graciliflora*, etc.). [Winged Nepenthes]. **Climbers**, terrestrial, to 4 m long; climbing stems terete or rounded-triangular, 4–8 mm Ø. **Leaves** narrowly elliptic or oblanceolate-spathulate, coriaceous to chartaceous; *rosette* leaves to 13 × 3 cm; *cauline* leaves 10.5–25 × 1.6–5.2 cm, base cuneate or distinctly petiolate, apex acute; petiole winged 2.5–5 × 0.3–1 cm, clasping stem; pitchers usually greenish, sometimes red-flushed, with dark purple peristome, with mauve bloom inside. **Lower pitchers** ± ellipsoid, slightly constricted toward subcylindrical upper half, 9–17 × 2.8–3.5 cm, mouth ovate; lid (suborbicular-) elliptic(-ovate), (1.7–)2.2–3.8(–5.5) × (1.6–)2–3(–4) cm, base rounded, apex rounded to retuse; spur simple, 6–8 mm long. **Upper pitchers** as lower, but slightly ventricose, abruptly constricted and cylindrical above, (7.5–)12.5–19.5 × (2–)2.7–4.5 cm. **Inflorescences** ♂ racemes 28–43 × 3–5 cm, peduncle 1–1.3 cm long, partial peduncles 1(–2)-flowered, pedicels 1.4–1.9(–2.7) cm long. **Flowers** greenish-yellow; ♂ with tepals elliptic, 3–4.5 × 1.5–3 mm, androphore (3.5–)4–5 mm long. **Fruits** with valves 1.8–2.8 × 0.2–0.3 cm; seeds filiform, 1–1.4 cm × 0.3 mm. **Distr:** Philippines (Luzon to Mindanao). **Ecol:** In mossy forests; 400–2,400 m. **Zones:** 11–12.

N. ampullaria Jack, *Camp. Bot. Mag.*: 271, 1835. (Syn: *N. ampullaceae*). [Flask-shaped Pitcher Plant]. **Climbers**, terrestrial, to 15 m long, with many terrestrial and some aerial rosettes; stem cylindrical, 1–1.5 cm Ø. **Leaves** ± sessile, lanceolate to spathulate, base attenuate, clasping stem, apex usually acute, thickly chartaceous, longitudinal nerves 3–5 per side; *rosette* leaves 2–5 × 0.5 cm; *cauline* leaves ~25 × 6 cm; pitchers usually green, deeply maroon-flecked, rarely entirely red, sometimes almost whitish-yellow, pink-flecked. **Lower pitchers** obliquely urceolate, dorsal side semicircular, flat ventrally, to 10 × 9 cm, with 2 fringed wings to 1.5 cm wide, fringe elements 0.5–1 cm long; mouth oval, almost horizontal, straight; peristome flattened, to 1.5 cm wide; lid narrowly oblanceolate, to 4 × 1.5 cm; spur to ~1 cm long. **Upper pitchers** rudimentary, broadly infundibuliform, ~2 × 2 cm. **Inflorescences** ♂ panicles to 40 × 4–5 cm; peduncle 2.5 cm long; partial peduncles 8–12(–50) cm long, fasciculate at apex, (1–)3–6(–10)-flowered; bracts foliose, spathulate, 1.2–1.4 × 0.4–0.5 cm; pedicels 7–8 mm long. **Flowers** ♂ with tepals broadly elliptic, 4–5 × 3–5 mm, androphore 3–5 mm long; anther head 2 × 1.5 mm. **Distr:** Malesia. **Ecol:** In damp, shady forests; 0–200(–2,100) m. **Zones:** 11–12.

N. burbidgeae Hook. f. ex Burb, *Gard. Chron.* 1: 56, 1882. (Syn: *N. phyllamphora*). [Painted Pitcher Plant].**Climbers**, terrestrial, to 10 m long. **Leaves** of *rosette* oblong, 44–50 × 9–11.5 cm, base attenuate, apex acute, petiole 10–17 cm long, margins sheathing and clasping stem; *cauline* leaves elliptic-lanceolate, 13–25 × 3.8–7.5 cm, base acute to rounded, apex acute, petiole 6–11 cm long, ± winged, clasping stem, decurrent into 2 wings. **Lower pitchers** pale green, streaked and blotched red or purple, peristome with numerous red stripes, ellipsoid or ovoid, 19–25 × 8.5–10 cm; mouth obliquely concave; peristome broad, subcylindrical to slightly flattened, ~2–3 mm wide; lid ovate, 7–9.5 × 6.4–9 cm, base cordate, margin straight, apex round; spur ~1.5 cm long, unbranched. **Upper pitchers** translucent-white or yellowish, speckled or blotched with red or purple; shortly infundibuliform, 7.2–11 × 5–8 cm, arising abruptly from the tendril; mouth round, horizontal, ~1 cm long; peristome cylindrical. (1–)2.5(–4) mm wide, ribbed; lid ovate, 2.5–6.5 × 2.6–5.2 cm, base cordate, margin slightly sinuate, apex rounded; spur ~1 cm long, unbranched or shortly bifurcate. **Inflorescences** 25–35 cm long, partial peduncles 2-flowered; pedicels 1–1.5 cm long. **Distr:** Borneo (Sabah). **Ecol:** In open moss forests; 1,200–2,250 m. **Zones:** 11–12.

N. clipeata Danser., *Bull. Jard. Bot. Buitenzorg* III. 9: 281, 1928. [Shield-leaved Pitcher Plant]. **Scramblers** terrestrial, or prostrate **herbs**, ~30 cm high; stem terete or slightly flattened. **Leaves** orbicular to ovate, 7–20 × 6–16 cm, base obtuse to cordate, apex rounded, peltate halfway from apex, thickly coriaceous; petiole 5–10.5 cm long, lower 3 cm clasping, longitudinal nerves 3–5 per side; tendril robust, curved downward, 5–12 cm long. **Upper pitchers** white, with a few red flecks, 10–30 cm long, lower third globose to obovoid, leathery-woody, 8–11.5 × 6–8.5 cm, abruptly constricted above, narrowly infundibuliform, 2.8–5 cm wide at base, chartaceous, flaring gradually to 5.2–9 cm at mouth, ridges and wings absent except the smallest pitchers with 2 thin fringed wings 1.5 mm wide; mouth ± circular, horizontal to slightly oblique; peristome white, heavily red-streaked, subcylindrical, 0.1–1.2 cm wide, ribs 0.3–0.7 mm apart, inner edge toothed; lid ovate, vaulted, concave below, 3–8 × 2–5 cm, base cordate, apex rounded, lower surface with a laterally flattened basal appendage to 8 mm tall, nectar glands craterlike, densely scattered; spur simple, ~5 mm long. **Distr:** Borneo (W Kalimantan). **Ecol:** In crevices of granite walls; 600–800 m. **Zones:** 11–12.

N. distillatoria L., *Sp. Pl.* 955, 1753. (Syn: *N. indica, N. zeylanica*). [Ceylon Pitcher Plant]. **Climbers**, woody, robust, lianalike, several meters long, reaching tops of tall trees, sometimes shrubby or semishrubby, woody at base. **Leaves** alternate; lamina lanceolate, 5.5–35.5 × 2.5–9.5 cm, base tapered, midrib prominent abaxially, apex constricted, coiled, to 4.5–17.5 cm long. **Pitchers** pendent, usually pale green, sometimes tinged purple-red, urn-shaped or ± cylindric, 2.5–20.5 cm long, curved upward, flattened on one side between 2 narrow longitudinal, ciliate wings or ridges; mouth with rim fluted, horizontal; lid erect, puberulous outside, orbicular, 2–4.5 cm long, with a false midrib. **Inflorescences** dioecious, panicles, leaf-opposed, >40 cm long, narrow; peduncles pubescent. **Flowers** ♂ >1.5 cm, tepals 4, densely pubescent outside, glabrous inside, filaments monadelphous, androphore with anthers in a whorl and in an apical group; ♀ >1.5 cm long, tepals 4, persistent, densely pubescent outside, glabrous inside, ovary silky-pubescent, stigma discoid, 4-lobed, sessile. **Fruits** capsules, oblong, 0.8–2.5 cm long, blunt and pilose. **Distr:** W and SW Sri Lanka. **Ecol:** In lowland rainforests. **Zones:** 11–12.

N. maxima Reinw. ex Nees., *Ann. Sci. Nat.* 3: 369, 1824. (Syn: *N. boschiana*, etc.). [Great Pitcher Plant]. **Shrubs** epiphytic or terrestrial, or low climbers, to 4 m long; stems terete to triangular, often 2-winged, occasionally 4-marked. **Leaves** obovate to lanceolate, 15–30 × 2.5–7 cm, base attenuate, apex obtuse to acute, chartaceous; petiole narrowly winged, to 7 cm long; pitchers greenish-white or yellowish-green, usually red-mottled, occasionally 'waisted'; peristome crimson. **Lower pitchers** narrowly ovoid, becoming cylindrical above, 8–20 × 2.2–5 cm, with 2 fringed wings;

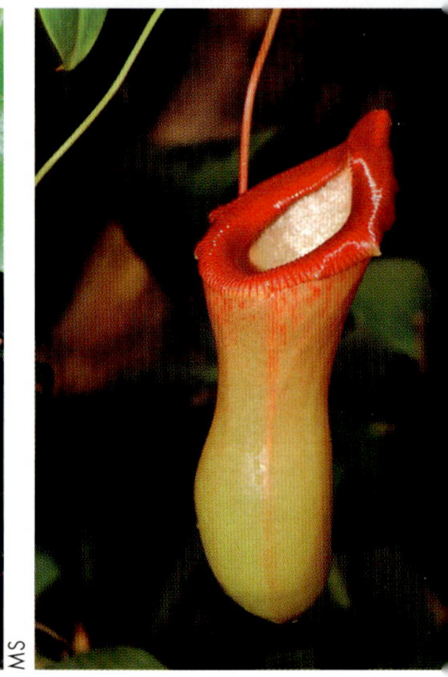

Top row: *Nepenthes ampullaria, Nepenthes burbidgeae, Nepenthes clipeata, Nepenthes distillatoria.*
Bottom row: *Nepenthes maxima* (inflorescense), *Nepenthes rafflesiana, Nepenthes truncata, Nepenthes ventricosa.*

mouth ovate, concave; peristome subcylindrical at front, 2–5 mm wide; lid ovate, ~2.5 × 2.5 cm, base cordate, apex rounded; spur filiform, to 6 mm long, entire. **Upper pitchers** narrowly infundibuliform, to 19(–30) × 6(–8) cm, with 2 prominent ribs; mouth ovate, forming a column to 3 cm tall; peristome flattened, to 2 cm wide near lid, ribbed; lid elliptic to ovate, 2–5 × 3–6 cm, base ± cordate, apex obtuse; spur entire or shortly bifid, 4–8 mm long. **Inflorescences** ♂ to 19(–40) × 3(–5) cm, densely hairy; peduncles 5.5–8(–14) cm long; partial peduncles 2-flowered; pedicels to 0.9(–1.6) cm long. **Flowers** dark red; tepals elliptic, to 4.5 × 2.5(–3.5) mm; androphore 2.5–4 mm long, anther head to 1.5 × 2 mm. **Distr:** Sulawesi, Moluccas, New Guinea. **Ecol:** In mossy forests or swamp grasslands; (600–)1,200–2,500 m. **Zones:** 10–12.

N. rafflesiana Jack, *Camp. Bot. Mag.* I: 270, 1835. (Syn: *N. hookeriana*, etc.). **Climbers**, terrestrial, 2–6 m long; stems terete. **Leaves** papery; *rosette* leaves narrowly oblong, to 37(–45) × 7(–11) cm, base abruptly acute or attenuate, apex usually acute; *cauline* leaves to 20(–30) × 5.5(–7) cm, longitudinal nerves 3 or 4 (or 5) per side; petiole 4–17 cm long, base sheathing. **Lower pitchers** white, strongly mottled with purplish-red, rarely pure white, or white flushed with pink, broadly ovoid to subglobose, or ovoid or narrowly ovoid, 8.5–25 × 5.5–9 cm, with 2 fringed wings, 1–3 cm wide; mouth strongly concave; peristome subcylindrical, to 0.8(–1) cm long; lid orbicular, ovate to long-elliptic, 4–9.5 × 3.2–7.5 cm, base shallowly cordate or rounded, apex rounded, truncate or retuse; spur to 2.5(–3) cm long. **Upper pitchers** pale green, infundibulate, rarely subcylindrical, 9–34 × 3–8 cm, with 2 ridges; mouth oblique; peristome striped yellow, green and red, cylindrical, 2–8 mm wide; lid ovate, to 7(–9) × 7 cm, base obtuse to slightly cordate, apex usually rounded; spur to 1.2(–2.6) cm long. **Inflorescences** ♂ to 49 × 4(–5.5) cm. **Flowers** ♂ with tepals red or brown-red, elliptic, (5–)6–7 × 5 mm, androphore 4.5–6 mm long, anther head 1.2–1.5 × 2–2.5 mm. **Fruits** (15–)30–50; seeds filiform, to 1.8 cm long. **Distr:** Sumatra, peninsular Malaysia, Borneo. **Ecol:** In forests; 0–300(–1,000) m. **Zones:** 11–12.

N. truncata Macfarl., *Trans. and Proc. Bot. Soc. Pennsylv.* 3, 3: 209, 1911. **Shrubs**, terrestrial, to 1 m high, probably not climbing; stems terete, 1.5–3 cm Ø. **Leaves** thickly coriaceous; *rosette* leaves subrectangular to obtriangular, to 45 × 22 cm, base truncate to shallowly cordate, margin ± undulate, apex ± deeply emarginate or retuse, longitudinal nerves 5 per side; petiole to 27 cm long, wings 1.8 cm wide, clasping stem completely; pitchers green, peristome red, green-striped, ferruginously villose. **Lower pitchers** with 2 fringed wings. **Upper pitchers** cylindrical, to 30 × 10 cm, wings absent; mouth ovate, oblique; peristome flattened, to 6 cm wide, with coarse ribs 1 mm apart, inner surface dentate; lid ovate to subtriangular, to ~9.5 × 6.5 cm, base cordate, apex usually acute, basal appendage ~1.3 × 1.2 cm, midline keeled, nectar glands sparsely scattered, longitudinally elliptic, bordered, mixed with sessile red glands; spur 1–1.2 cm long. **Inflorescences** with peduncle to 30 cm long, partial peduncles 2-flowered, 7 mm long; bracts absent; pedicels 1.8–3 cm long. **Flowers** with tepals elliptic-ovate, 5 × 3 mm; androphore 6 mm long, anther head 1.25 mm wide. **Fruits** with valves 2.5–3.8 cm long. **Distr:** Philippines (Mindanao). **Ecol:** On open mountainsides; 230–600 m. **Zones:** 11–12.

N. ventricosa Blanco, *Fl. Filip.*, ed. 1: 807, 1837. **Climbers**, sometimes epiphytic, to ~2 m long; stems winged to strongly triangular. **Leaves** *cauline* oblanceolate or narrowly oblong, 15–25 × 2–3 cm, base slightly peltate, gradually tapered, clasping stem, decurrent, winged, apex acute, chartaceous to thinly coriaceous, sessile, longitudinal nerves 3–6 per side. **Lower pitchers** rare, same as upper but with tendril in front. **Upper pitchers** creamish-white, hourglass-shaped, 9–16 × 3–8 cm, basal half obliquely ellipsoid, apical half broadly infundibular, center constricted, fringed wings absent; mouth circular to shortly elliptic; peristome red, inside yellowish, broad, subcylindrical, 1–2.5 cm wide, ribs coarse, 1–2 mm apart; lid membranous, narrowly elliptic, 4–6 × 1.8–2.25 cm, base and apex rounded, lower surface lacking appendages; spur flattened, 9 mm long, entire. **Inflorescences** ♂ 28–60 cm long; peduncle 11–21 cm long; partial peduncles 70–180, 1-flowered; bracts absent; pedicels 0.5–1 cm long. **Flowers** with tepals elliptic, 2–3.5 × 1.5 mm; androphore 3–3.5 mm long, anther head 1.5 mm wide. **Fruits** with valves 2.5–3.2 cm long. **Distr:** Philippines (Luzon). **Ecol:** In low mossy oak forests; 1,200–1,500 m. **Zones:** 10–12.

TAMARICACEAE

Genera *5*, species *90*. **Shrubs**, **subshrubs**, or **trees**. **Stipules** absent. **Leaves** small, mostly scalelike, alternate, usually sessile, mostly with salt-secreting glands. **Inflorescences** racemes or panicles, rarely flowers solitary. **Flowers** usually hermaphroditic, regular; calyx 4- or 5-fid, persistent; petals 4 or 5, free, deciduous after anthesis or sometimes persistent; disk inferior, usually thick, nectarylike; stamens 4, 5, or more, usually free, inserted on disk, rarely united into a fascicle at base, or united up to half the length into a tube, anthers 2-thecate, longitudinally dehiscent; pistil 1, consisting of 2–5 carpels, ovary superior, 1-loculed, placentation parietal, rarely septate, or basal, ovules numerous, (rarely few), styles short, usually 2–5, free, sometimes united. **Fruits** capsules, conic, abaxially dehiscent; seeds numerous, hairy throughout or awned at apex; awns puberulous from base or from middle. **Distr:** Eurasia and Africa, especially the Mediterranean region to C Asia.

Tamarix

Species 55. **Shrubs** or **trees**, 2–10(–12) m tall, densely branched; root system deep; bark variously colored, mostly brownish to blackish-brown; young branches usually glabrous, dimorphic, *ligneous growing* branches (not deciduous in winter) and *green vegetative* branchlets (deciduous in winter). **Leaves** alternate, sessile, amplexicaul, vaginate or auriculate; lamina small, scalelike, usually 0.5–7 mm long, glabrous, rarely hairy, mostly with salt-secreting glands. **Inflorescences** simple or compound racemes or panicles, 1–15 cm long, borne on younger and older branches; flowers usually pedicellate; bract 1. **Flowers** small, mostly bisexual, rarely unisexual, 4 or 5(or 6)-merous; calyx 4 or 5-lobed, subequal or equal with imbricate aestivation, mostly ovate or orbicular, denticulate or entire, obtuse or acute; petals 4 or 5, free, ovate, oblong, obovate or elliptical, apex obtuse, slightly notched or somewhat cuneate, pink, violet, purple or white, deciduous or persistent; disk nectariferous, fleshy; stamens 4–10(–12); ovary pyramidal or bottle-shaped, stigmas (2)3–4(5). **Fruits** capsules, conical or pyramidal, 3–5-angled, dehiscing via 3 valves; seeds numerous, minute, compressed, glabrous, hairy from apex, coma sessile. **Distr:** Africa, Asia, Europe. **Cult:** Propagate from ripe seeds, hardwood cuttings or semiripe cuttings. Requires a well-drained soil in a sunny position (in coastal areas) or a slightly moist soil (in inland areas).

T. indica Willd., *Ges. Naturf. Freunde Berlin Neue Schr.* 4: 214, 1803. (Syn: *T. bengalensis*, etc.). **Shrubs** tall, or **small trees**, 2–5(–8) m tall; bark reddish-brown, glabrous. **Leaves** sessile, amplexicaul or auriculate, clasping at base; lamina scalelike, ovate, deltoid, 1–2 mm long, without coherent margins, apex acute to acuminate, glandular-punctate. **Inflorescences** racemes, 3–12(–15) cm long, in drooping panicles; bract triangular, sometimes slightly keeled, ~1 mm long, margins somewhat irregularly denticulate, apex acuminate; pedicel ~0.5 mm long. **Flowers** pink to white; sepals 5, fused at base, inner somewhat broader than the outer, outer more acute, ovate or trullate-ovate to orbicular, 0.5–1 × ~0.5 mm, margin somewhat membranous, apex deeply denticulate; petals 5, free, elliptic or elliptic-obovate, obtuse 1–2 × 0.75–1 mm; disk minute, horizontal, hypogynous, lobed; stamens 5, exserted, filaments filiform; ovary bottle-shaped, trigonous, styles 3, clavate, separate to base, stigmas irregularly dentate. **Fruits** capsules, conical, triangular or lanceolate, 3–6 mm long, maturing from bright to chestnut-brown color; seeds numerous, beaked, rostrum densely covered with long brown hairs. **Distr:** Afghanistan, Bangladesh, India, Pakistan, Sri Lanka. **Ecol:** In dry coastal areas. Fl. throughout most of the year. Fr. Oct–Feb.

PLUMBAGINACEAE

Genera 27, **species** 836. **Herbs** or **shrubs**, rarely lianas, perennial, rarely annual, tap-rooted or rhizomatous; stems woody erect to prostrate, nodes swollen; indument of simple hairs, capitate glands or multicelled glandlike structures. **Stipules** absent. **Leaves** often basal, alternate, spiraled; lamina linear to broadly obovate, ovate, or round, margins entire or lobed; petiole present or absent. **Inflorescences** cymes, panicles, racemes, or corymbs, or solitary heads, terminal or axillary; bracts herbaceous, scarious, sometimes absent; involucral bracteoles usually present; flowers sessile or shortly pedicellate. **Flowers** bisexual, radially symmetric; perianth and androecium hypogynous; sepals persistent in mature fruits, 5, connate into 5- or 10-ribbed tube, mostly dry and membranous, sometimes petaloid, toothed or with distinct simple or lobed limbs; petals 5, nearly distinct, basally connate (corolla salverform), lamina sometimes clawed, margins entire; stamens 5, filaments adnate to bases of petals or free; ovary superior, 1-locular, 1-ovuled, anatropous, styles 1 with apically lobed stigma, or 5, each with linear stigma. **Fruits** utricles, achenes, or capsules, 1-seeded. **Distr:** Predominantly Mediterranean to C Asia, scattered elsewhere, especially maritime areas.

Plumbago

Species 12. **Shrubs** perennial, or suffrutescent **herbs**; stems erect, prostrate, or climbing, ribbed. **Leaves** cauline, sessile or shortly petiolate (petiole usually <1.5 cm long); lamina elliptic to oblanceolate or spathulate, base narrowed, margins entire, apex acute, acuminate, or obtuse, membranous. **Inflorescences** spikelike racemes or panicles, terminal or axillary; bracts absent; pedicels 2-bracteolate, short. **Flowers** sometimes heterostylous; calyx persistent, 5-ribbed, tubular, with stalked, capitate-glandular protuberances along ribs; lobes triangular, 1–2 mm long; corolla salverform, evenly to somewhat unevenly 5-lobed, lobes spreading, obovate, round, or truncate, mucronate; stamens included or exserted, free from corolla; style 1 included or exserted, stigmas 5, linear. **Fruits** capsules, included, brownish, long-beaked; valves coherent at apex. **Distr:** Tropical and subtropical regions (America, Europe, Asia, Africa). **Cult:** Propagate from seeds, half-hardened cuttings or by air-layering. Requires any moist, well-drained soil in full or partial sun.

P. auriculata Lam., *Encycl.* (Lamarck) 2(1): 270, 1786. (Syn: *P. capensis*).[Cape Leadwort, Plumbago]. **Shrubs**, semiclimbing, 1–2 m high; often diffusely branched, glabrous except the inflorescence. **Stipules** ear-shaped, leaflike. **Leaves** borne in fascicles in alternating clusters on the stem; lamina elliptic to elliptic-spathulate, oblanceolate or elliptic-ovate, 3–7 cm long, base tapering into a short petiole, margins undulate, apex obtuse and shortly mucronate. **Inflorescences** panicles, compact, many-flowered, to 15 cm across; bracts ovate-mucronate. **Flowers** 2–2.5 cm long, sky-blue or white; calyx tubular, 2–3 times shorter than corolla, finely pubescent with stalked glands on upper part, limb 5-fid; corolla gamopetalous, tubular, tube 2 cm long, limb usually 5-lobed, 2.5 cm Ø. **Distr:** S Africa. **Ecol:** Fl. Jul–Oct. **Zones:** 9–12.

P. indica L., *Herb. Amb.* 24, 1754. (Syn: *P. rosea*, *Thela coccinea*). [Crimson Leadwort, Scarlet Leadwort]. **Herbs**, perennial, evergreen, 0.5–2 m high; stems pliable, often lianoid, simple or branched from base. **Leaves** narrowly ovate to elliptic-ovate, to 9.5(–13) × 4(–6) cm, base rounded to obtuse, apex acute, papery; petiole bases without auricles. **Inflorescences** spicate-racemose, (20–)35–90-flowered; peduncles 1–3 cm long, not glandular; rachis to 40(–50) cm long, not glandular; bracts ovate, 2–3 × 1.5–2 mm, acuminate. **Flowers** heterostylous; calyx 7.5–9.5 mm long, glandular almost throughout, tube ~2 mm Ø; corolla purple to dark red, tube 2–2.5 cm long, limb ~2 cm Ø, lobes obovate, ~1.2 × 0.7 cm, apex rounded, mucronate; anthers blue, 1.5–2 mm long; ovary ellipsoid-ovoid, indistinctly angular; style basally pilose; *short-styled* form with style arms only partly exserted, stigmatic glands without enlarged apex; *long-styled* form with style arms completely exserted from corolla throat, stigmatic glands capitate. **Fruits** capsules, rarely formed in cultivation. **Distr:** Old World tropics. **Ecol:** In wet grasslands and open dry grasslands; 0–300 m. Fl. Nov–Apr. **Zones:** 10–12.

P. zeylanica L., *Sp. Pl.* 1: 151, 1753. (Syn: *P. scandens*). [Doctorbush]. **Herbs**, perennial or shrubs, evergreen, 1–3 m high, prostrate, climbing, or erect, glabrous. **Leaves** sessile or with petiole to 1.5 cm long, petiole bases sometimes auriculate; lamina ovate, lanceolate-elliptic, or spathulate to oblanceolate, (3–)5–9(–15) × (1–)2.5–4(–7) cm, base attenuate, apex acute, acuminate, or obtuse. **Inflorescences** spicate-racemose, (3–)5–70-flowered, 3–15(–30) cm long, rachises glandular, viscid; floral bracts lanceolate, 3–7 × 1–2 mm. **Flowers** heterostylous; calyx 0.7–1.1(–1.3) cm long, tube glabrous, with stalked glands along length of ribs; corolla white, 1.7–3.3 cm long, tube 1.25–2.8 cm long, lobes 0.5–1.2 × ~0.35 cm; stamens included, anthers blue, ~2 mm long; ovary ellipsoid, 5-angular, style glabrous. **Fruits** capsules, pale yellow-brown, oblong, ~8 mm long; seeds reddish-brown to dark brown, 5–6 mm long. **Distr:** S USA, Mexico, C and S America; Africa; Asia; Pacific Islands. **Ecol:** In thickets; 0–1,600 m. Fl.

Below: *Tamarix indica*, *Tamarix indica* (fruits), *Plumbago auriculata* (white form), *Plumbago auriculata* (blue form).

Top row: *Plumbago indica, Plumbago zeylanica, Antigonon guatemalense, Antigonon leptopus.*
Bottom row: *Antigonon leptopus* var. *alba, Coccoloba uvifera, Muehlenbeckia astonii, Muehlenbeckia platyclada.*

throughout the year. **Zones:** 10–12. **Info:** This species is referred to as *P. zeylanica* in the Old World, and *P. scandens* in the New World.

POLYGONACEAE

Genera 43, species 1,110. **Herbs, shrubs,** or **small trees,** sometimes monoecious or dioecious; stems erect, prostrate, twining, or scandent, often with swollen nodes, striate, grooved, or prickly. **Stipules** often united to a sheath (ochrea). **Leaves** alternate, rarely opposite or whorled, petiolate or subsessile; lamina simple. **Inflorescences** terminal or axillary, spicate, racemose, paniculate, or capitate; pedicels occasionally articulate. **Flowers** small, actinomorphic, bisexual, rarely unisexual; perianth 3–6-merous, in 1 or 2 series, herbaceous, often enlarged in fruit or inner tepals enlarged, with wings, tubercles, or spines; stamens usually (3–)6–9, rarely more, filaments free or basally united; anthers 2-loculed, opening lengthwise; disk annular (often lobed); ovary superior, 1-loculed, styles 2 or 3, rarely 4, free or connate at lower part. **Fruits** achenes, trigonous, biconvex, or biconcave; seed with straight or curved embryo and copious endosperm. **Distr:** Worldwide.

Antigonon

Species ~6. **Vines,** perennial, rarely annual; roots tuberous; stems scandent, tendril-bearing, pubescent or glabrous; tendrils terminal and axillary, branched. **Leaves** deciduous or persistent, cauline, alternate, petiolate; ochrea usually deciduous, chartaceous; lamina broadly ovate to deltate or truncate, margins entire, sometimes undulate. **Inflorescences** with flowers (1–)2–5 per ochreate fascicle, terminal and axillary, often clustered near tips of stems, racemelike, pedunculate; pedicels present. **Flowers** bisexual, base stipelike; perianth accrescent in fruit, pink to purple or, rarely, white or yellowish, campanulate, glabrous; tepals 5, proximally connate, petaloid, dimorphic, outer 3 broader than inner 2; stamens (7–)8(–9), filaments connate for about half their length, forming a staminal tube, adnate to perianth tube, glandular-pubescent, anthers yellow to reddish, ovate to elliptic; styles 3, recurved, distinct, stigmas reniform-capitate. **Fruits** achenes, enclosed in membranous perianth, brown, unwinged, subglobose to bluntly 3-gonous proximally, 3-gonous distally. **Distr:** Subtropical and tropical regions of the New World. **Cult:** Propagate from tubers, semiripe cuttings or seeds. Requires a fertile and well-drained soil in full sun.

A. guatemalense Meisn., *Prodr.* 14: 184, 1856. (Syn: *A. grandiflorum*, etc.). [Bellisima, Chain of Love]. **Climbers,** to 6 m long; internodes leafy, 1.5–15 cm long, densely brownish-puberulent. **Leaves** broadly ovate to ovate-triangular, 6–15 × 3.5–10 cm, base cordate to cordulate, basal sinus usually narrow and obscuring petiole and node, margins entire, apex tapering gradually to an acute tip, venation pinnate; petiole 0.2–1 cm long, densely brownish-tomentulose, often obscured by lamina base. **Inflorescences** solitary and axillary with a single axis but occasionally terminal with several axes in a paniculate arrangement, to ~25 cm long, alternating flower fascicles 0.5–2.5 cm apart on rachis; floral bracts broadly ovate, 0.5 mm long, apex cuspidate, puberulent; pedicels 0.8–2.5 cm long, densely puberulent. **Flowers** with perianth lobes ~4 mm long before anthesis; tepals with lobes spreading, edges ruffled, white or pink; staminal tube relatively short and thick. **Fruits** achenes, narrowly ovoid, faintly 3-angled, enclosed within the persisting perianth, 3.5 cm long. **Distr:** Guatemala. **Ecol:** In thickets, growing on small trees and shrubs; 0–1,500 m. **Zones:** 10–12.

A. leptopus Hooker & Arnott, *Bot. Beechy Voy.* 308, pl. 69, 1838. [Chain of Love, Confederate Vine, Mexican Creeper, Mountain Rose, Coral Vine]. **Herbs,** base sometimes woody; stems climbing or sprawling by tendrils, branched, angular, to 15 m long, brownish- or reddish-pubescent or glabrous; leafy internodes 1–10 cm long, puberulent. **Leaves** with ochrea 0.2–2 mm long; 5–14 × (2–)4–10 cm, base usually cordate, margins ciliate, apex acute to acuminate, glabrous or pubescent; petiole often winged distally, 1–2.5(–5) cm long. **Inflorescences** 4–20 cm long; peduncle angular, 1–5 cm long; pedicels articulated proximally, 0.3–0.5(–1) cm long. **Flowers** with tepals ovate to elliptic, 4–8 × 2–6 mm, 0.8–2 × 0.4–1.5 cm in fruit, margins entire, apex acute. **Fruits** achenes, 0.8–1.2 × 0.4–0.7 cm, shiny, short-angled, conical nut surrounded by enlarged, heart-shaped sepals. **Distr:** C America (probably Mexico). **Ecol:** In wet forest margins, disturbed places; 0–600 m. Fl. throughout the year. **Zones:** 10–12. **Info:** Cultivated and naturalized throughout the tropics. The tubers are edible. ***A. leptopus* var. *alba*** is white-flowered, common in cultivation.

Coccoloba

Species 120. **Trees** or **shrubs,** evergreen; roots woody; stems erect or spreading. **Leaves** persistent, cauline, alternate, petiolate; ochrea often deciduous, membranous to coriaceous; lamina lanceolate to round or transversely elliptic, margins entire. **Inflorescences** racemelike, terminal, pedunculate; pedicels present; ♂ flowers 1–7 per ochreate fascicle; ♀ flowers 1 per ochreate fascicle. **Flowers** functionally unisexual, some plants having only ♂ flowers, others with only ♀ flowers, base stipelike; perianth white or greenish-white, campanulate, glabrous; tepals 5, proximally connate, sepaloid, monomorphic; ♂ with perianth non-accrescent, stamens 8, filaments basally connate, adnate to perianth, glabrous, anthers white or bluish-white, elliptic to round, pistil rudimentary; ♀ with perianth accrescent and fleshy in fruit, stamens rudimentary, styles 3, erect, distinct, stigmas capitate. **Fruits** achenes, usually included in fleshy perianth tube, brown to black, unwinged, bluntly trigonous, glabrous. **Distr:** Tropical parts of New World. **Cult:** Propagate from seeds, cuttings or by air-layering. Requires any well-drained soil in full sun; plants salt-tolerant.

C. uvifera (L.) L., *Syst. Nat.* ed. 10. 2: 1007, 1759. (Syn: *Polygonum uvifera*). [Sea Grape, Seaside Grape]. **Trees**, briefly deciduous, 2–7(–15) m tall; bark peeling. **Stipules** forming ochrea, cylindric to funnelform, 3–8 mm long. **Leaves** of adventitious or juvenile shoots often much larger and of different shapes; lamina round to transversely elliptic, (6–)10–20(–27) × 6–20(–27) cm, base cordate, margins sometimes revolute, apex rounded to blunt or emarginate, coriaceous, adaxially green to bluish-green, abaxially paler; petiole 0.5–1.5 cm long. **Inflorescences** 10–30 cm long, pendent in fruit; ♂ flowers 1–7 per ochreate fascicle; peduncles 1–5 cm long, glabrous; pedicels 1–4 mm long, glabrous. **Flowers** with tepals round to broadly elliptic, margins entire, apex obtuse; ♀ with tube obpyriform, 1.2–2 × 0.8–1.2 cm, becoming fleshy. **Fruits** achenes, 0.8–1.1 × 0.8–1 cm, shiny. **Distr:** S USA (Florida), Mexico, West Indies, C and S America. **Ecol:** In sandy or rocky coastal hummocks, sand dunes; 0–10 m. Fl. throughout the year. **Zones:** 10–12.

Muehlenbeckia

Species 23. **Shrubs**, vinelike, perennial; rhizomatous; stems suberect, prostrate, or scandent. **Leaves** deciduous, cauline, alternate; ochrea usually deciduous, chartaceous; lamina linear to orbiculate, panduriform, or triangular-lanceolate, margins entire or irregularly wavy; petiole base articulated, extrafloral nectaries present. **Inflorescences** spikelike, terminal and axillary, essentially not pedunculate; flowers pedicellate; bisexual and unisexual, with ♂, ♀, or sometimes both sexes occurring with bisexual flowers on the same plant, 1–2(–5) per ochreate fascicle. **Flowers** with perianth accrescent, white to greenish-white, campanulate, glabrous, base stipelike; tepals 5, proximally connate, sepaloid, dimorphic, outer slightly larger than inner; ♂ with stamens 8 (9), filaments distinct, adnate to base of perianth tube, glabrous, anthers yellow or pink to purple, ovate to elliptic, pistil rudimentary; ♀ with tube white or reddish-purple to black in fruit, becoming fleshy, stamens rudimentary, styles 3, spreading, proximally connate, stigmas fimbriate. **Fruits** achenes, completely or partly included in fleshy perianth, black or dark brown, unwinged, trigonous to subglobose, glabrous. **Distr:** C and S America, Pacific Islands, New Zealand, Australia.

M. astonii Petrie, *Trans. New Zealand Inst.* 43: 257, 1911. [Shrubby Tororaro, Wiggy-wig Bush, Zig-zag Bush]. **Shrubs**, erect, to 2.4 m high; stems many, to 1 cm Ø, much-branched; branchlets slender, flexuous, divaricate and ± interlacing; canelike shoots often developed from near base. **Leaves** in fascicles of 2–3 on short branchlets, or distributed along elongate branchlets; lamina usually broadly obcordate, 0.3–1 cm Ø, base cuneately narrowed to short slender petiole, submembranous, usually entire, occasionally incised at apex. **Inflorescences** fascicles, small, axillary; peduncles very short. **Flowers** minute; ♀ ~2.5 mm Ø; tepals green to white or pinkish; staminodes usually 8; ovary bluntly trigonous, faces grooved; stigmas 3, fimbriate. **Fruits** 2–2.5 × 1.25 mm, dull, rugulose; persistent tepals succulent or dry. **Distr:** New Zealand. **Ecol:** In coastal to lowland riverbeds and terraces. Fr. Dec–Feb. **Zones:** 8–11.

M. platyclada (F. Muell.) Meisner, *Bot. Zeitung* 23: 313 (as '*platydados*'), 1865. (Syn: *Homalocladum platycladum, Polygonum platycladum*). [Ribbon Bush, Tapeworm Plant, Wirevine]. **Shrubs**, unisexual, 0.5–3 m high; branches many, closely crowded, glabrous, basal and primary stems terete, distal stems flat and straplike, to 1 mm thick and 0.5–1.5 cm wide, with distinct transverse lines, distant and demarking the nodes, surface with closely parallel ribs 0.2–0.5 mm apart. **Stipules** (ochrea) represented by a transverse line across the flattened stem. **Leaves** usually absent, sessile; lamina 2–5 × 0.5–1.5 cm, base cuneate to hastate, narrowly lanceolate distally, acute at apex, glabrous, soon caducous. **Inflorescences** fascicles, borne alternately along edges of stems just above the transverse lines; pedicels ~1 mm long. **Flowers** small, unisexual; perianth ~2 mm long. **Fruits** achenes, ~3 mm long, embedded within the red, succulent perianth, ~5 mm long. **Distr:** SW Pacific. **Ecol:** In thickets and grasslands. Fl. Jul–Aug. **Zones:** 11–12.

Polygonum

Species ~230. **Herbs**, **shrubs** or **subshrubs**, usually annual, homophyllous or heterophyllous, sometimes heterocarpic, rarely dioecious; stems prostrate to erect. **Leaves** simple, cauline, usually alternate; ochrea with distal part persistent, often hyaline, white or silvery, bilobed, chartaceous, glabrous; lamina linear, lanceolate, elliptic, ovate, or subround, margins entire; petiole base articulated with ochrea or not. **Inflorescences** spikelike, flowers 1–7(–10) per ochreate fascicle, or flowers solitary, axillary and/or terminal. **Flowers** bisexual, rarely unisexual; perianth non-accrescent, persistent, white or greenish-white to pink, campanulate to urceolate, glabrous; tepals (4 or)5, petaloid or sepaloid, monomorphic or, rarely, dimorphic, inner usually flat, outer flat or sometimes keeled and cucullate distally, sometimes of different length than inner; stamens (4 or) 7 or 8 (some reduced to staminodes), filaments distinct, free or adnate to perianth tube, glabrous, anthers whitish-yellow, pink to purple or orange-pink, elliptic to oblong, styles 2 or 3, mostly spreading, elongate, distinct or proximally connate, deciduous, stigmas 2–3, capitate. **Fruits** achenes, included or exserted, yellow-green, brown, or black, unwinged, (2–)3-gonous, rarely biconcave, glabrous. **Distr:** Nearly worldwide. **Cult:** Propagate from seeds or by division, and climbing species from semihardened cuttings.

P. capitatum Buch.-Ham. ex D. Don, *Prodr. Fl. Nepal.* 73, 1825. (Syn: *Cephalophilon capitatum, Persicaria capitata*). [Knotweed, Pink Fleece Flower, Smartweed]. **Herbs**, perennial; stems creeping, tufted, base ligneous, much-branched, rooting from nodes. **Leaves** sometimes with a large blackish spot adaxially, ovate or elliptic, 1.5–3 × 1–2.5 cm, base cuneate, margins entire, glandular-hairy, apex acute, surfaces glandular-hairy; ochrea tubular, 5–8 mm long, membranous, apex ciliate; petiole 2–3 mm long, base sometimes auriculate. **Inflorescences** capitate, solitary or geminate, 0.6–1 cm Ø, terminal; peduncle glandular-hairy; bracts narrowly ovate, membranous; pedicel very short. **Flowers** with perianth pinkish, 5-parted; tepals elliptic, 2–3 mm long; stamens 8, included; styles 3, connate to below middle, equaling perianth, stigmas capitate. **Fruits** achenes, included in persistent perianth, black-brown, slightly shiny, narrowly ovoid, trigonous, 1.5–2 mm long, densely punctate. **Distr:** Tropical and subtropical Asia. **Ecol:** On mountain slopes; 600–3,500 m. Fl. Jun–Sep. Fr. Aug–Oct. **Zones:** 8–12.

P. chinense L., *Sp. Pl.* 1: 363, 1753. [Chinese Knotweed]. **Herbs**, perennial; rhizomes stout; stems erect, 0.7–1 m high, base ligneous, much-branched, striate, glabrous or retrorsely hispid. **Leaves** ovate, elliptic, or lanceolate, 4–16 × 1.5–8 cm, base truncate or broadly cordate, margins entire, apex shortly acuminate, surfaces glabrous or hispid, abaxially sometimes pubescent along veins; ochrea tubular, 1.5–2.5 cm long, membranous, glabrous, much-veined, apex oblique, not ciliate; petiole 1–2 cm long, base usually auriculate, upper leaves subsessile. **Inflorescence** capitate, 3–5 mm, terminal or axillary, usually several capitula aggregated and paniclelike; peduncle densely glandular-hairy; bracts broadly ovate, each 1–3-flowered. **Flowers** with perianth white or pinkish, 5-parted; tepals ovate, accrescent in fruit, becoming blue-black, fleshy; stamens 8, included; styles 3, connate to below middle. **Fruits** achenes, included in persistent perianth, black, opaque, broadly

Below: *Polygonum capitatum, Polygonum chinense, Polygonum glabrum, Triplaris weigeltiana.*

RF

Above: *Achyranthes aspera, Achyranthes bidentata, Aerva lanata, Aerva lanata* (inflorescences).

ovoid, trigonous, 3–4 mm long. **Distr:** Bhutan, China, India, Indonesia, Japan, Malaysia, Myanmar, Nepal, Philippines, Sikkim, Taiwan, Thailand, Vietnam. **Ecol:** In mixed forests, thickets; to 3,000 m. Fl. Jul–Nov. Fr. Jul–Dec. **Zones:** 9–12.

P. glabrum Willdenow, *Sp. Pl.* 2: 447, 1799. (Syn: *Persicaria glabra*, etc.). [Denseflower Knotweed]. **Herbs**, annual, glabrous throughout; stems erect, 0.7–1 m high, nodes dilated, rarely branched. **Leaves** lanceolate or oblong-lanceolate, 8–18 × 1.5–3 cm, base narrowly cuneate, margins entire, not ciliate, apex narrowly acuminate; ochrea tubular, 1–3 cm long, membranous, veins numerous; petiole 0.8–1 cm long, robust. **Inflorescences** spicate, 4–12 cm long, usually with several dense spikes aggregated and paniclelike, terminal; bracts funnel-shaped, not ciliate, usually each 3- or 4-flowered; pedicel large, longer than bracts, articulate at apex. **Flowers** with perianth white or pinkish, 5-parted; tepals elliptic, 3–4 mm long, veins slender, apex furcate, not curved downward; stamens 6–8; styles 2, connate to below middle. **Fruits** achenes included in persistent perianth, black-brown, shiny, ovoid, biconvex, 2.5–3 mm long. **Distr:** Asia; Africa, Australia, N and S America, Pacific Islands. **Ecol:** On riverbanks, streamsides, marshy areas; 0–100 m. Fl. Jun–Aug. Fr. Jul–Sep. **Zones:** 10–12.

Triplaris

Species 18. **Trees**, dioecious; bark often peeling off; twigs striate, hollow and inhabited by stinging ants. **Leaves** large, short-petiolate; lamina often with 3–6 longitudinal distant lines per side. **Inflorescences** thyrses, axillary or terminal; ♂ flowers 3–5(–7) in each partial inflorescence, sessile or subsessile; ♀ flowers 1 in each partial inflorescence, pedicellate. **Flowers** with outer perianth segments of ♀ enlarged into red to purple wings in fruit; perianth tube at fruiting longer than the achene. **Fruits** samaras, achenes trigonous; seed ovoid-trigonous. **Distr:** Mexico (Oaxaca) to Brazil (Paraná). **Cult:** Propagate from seeds, cuttings or by grafting. Requires a fertile, moist, well-drained soil in full or partial sun.

T. weigeltiana (Reichb.) Kuntze, *Revis. Gen. Pl.* 3(3): 270, 1898. (Syn: *Blochmannia weigeltiana, T. surinamensis*). [Ant Tree, Long Jack, Long John, Mulato Tree]. **Trees**, dioecious, to over 5 m tall. **Leaves** subsessile; lamina 10–22 × 4–6 cm, base cuneate, margins entire, apex acute, longitudinally nerved, with numerous pairs of prominent lateral nerves. **Inflorescences** spikes, 6–10 cm long; bracts numerous, ovate to ovate-lanceolate, hirtellous. **Flowers** ♀ with perianth accrescent and enclosing fruit, 3 of the sepals much enlarged, 2.8–3.2 × 0.4–0.5 cm, suffused red-purple, showy. **Fruits** samaras, achenes ~1 cm long including the style, styles red-purple. **Distr:** S America. **Ecol:** In lowland forests, seasonally flooded palm marshes, along rivers and in swamp forests. Fl. May–Sep. **Zones:** 10–12.

AMARANTHACEAE

Genera 174, **species** 2,050. **Herbs** (annual or perennial), clambering **subshrubs**, **shrubs**, **lianas**, rarely trees hermaphrodite, dioecious, monoecious, or polygamous. **Leaves** alternate or opposite, exstipulate; lamina with margins entire. **Inflorescences** elongated or condensed spikes (heads), racemes, or thyrsoid structures of varying complexity, flowers solitary or aggregated in cymes; bracteoles membranous or scarious; bract 1 membranous; bracteoles 2. **Flowers** small, bisexual or unisexual, or sterile and reduced, tepals 3–5, membranous, 1-, 3-, 5-, or 7(–23)-veined; stamens as many as tepals and opposite these, rarely fewer than tepals; filaments free, united into a cup at base or ± entirely into a tube, filament lobes present or absent, pseudo-staminodes present or absent; anthers (1)2-loculed, dorsifixed, introrsely dehiscent; ovary superior, 1-loculed, ovules 1–many, style persistent, short and indistinct or long and slender, stigma capitate, penicillate, bilobed or forming 2 filiform branches. **Fruits** dry utricles, or fleshy capsules, indehiscent, irregularly bursting, or circumscissile; seeds lenticular, reniform, subglobose, or shortly cylindric, smooth or verruculose. **Distr:** Worldwide; mostly in the tropics.

Achyranthes

Species ~15. **Herbs** or **subshrubs**; stem with conspicuous nodes; branches opposite. **Leaves** opposite, petiolate; lamina with margins entire. **Inflorescences** terminal (on the stem and branches) or axillary spikes, erect; flowers with 3 bracts; bracteoles with midvein spiny, long, excurrent. **Flowers** reflexed or spreading, perfect; tepals 4 or 5, membranous, apex awn-pointed, rigid in fruit; stamens 2–5, rarely 4 or 2, shorter than perianth, filaments basally connate into a short cup, anthers 2-loculed; pseudo-staminodes fringed; ovary oblong, style linear, persistent, stigma capitate. **Fruits** utricles, ovoid-oblong, ovoid, or subglobose, falling off with tepals and bracteoles; seed oblong, lenticular. **Distr:** Tropical and subtropical regions worldwide. **Cult:** Propagate from seeds. Requires a moist, well-drained soil in full sun.

A. aspera L., *Sp. Pl.* 1: 204, 1753. [Prickly Chaff Flower]. **Herbs**, perennial, 0.2–1.2 m high; stem quadrangular, pubescent; nodes slightly inflated; branches opposite. **Leaves** broadly obovate or elliptic-oblong, 1.5–7 × 0.4–4 mm, base cuneate or rounded, margins entire or undulate, apex obtuse, with a mucro, papery, surfaces hairy; petiole 0.5–1.5 cm long, somewhat hairy. **Inflorescences** spikes terminal, erect, reflexed after anthesis, 10–30 cm; rachis angular, stout, densely hairy; bracts lanceolate, 3–4 mm long, acuminate; bracteoles spiny, shiny, 2.5–4.5 mm long, rigid, base 2-winged; wings 1.5–2 mm wide, membranous, margin entire. **Flowers** with tepals lanceolate, 3.5–5 mm long, with a vein; stamens 2.5–3.5 mm long; pseudo-staminodes truncate or crenate at apex, fimbriate and ciliate. **Fruits** utricles, ovoid, 2.5–3 mm long; seeds brown, ovoid, ~2 mm long. **Distr:** Asia to Europe and Africa. **Ecol:** On hillsides, dry scrubs, dry forest margins, sandy sea shores; 0–2,300 m. Fl. Jun.–Aug. Fr. Oct. **Zones:** 8–12.

A. bidentata Blume, *Bijdr. Fl. Ned. Ind.* 545, 1826. [Ox Knee]. **Herbs**, perennial, 0.7–1.2 m high; stem green or purple-tinged, angulate or quadrangular, pubescent or subglabrous; branches opposite. **Leaves** elliptic or elliptic-lanceolate, rarely oblanceolate, 4.5–12 × 2–7.5 cm, base cuneate, caudate, surfaces pubescent; petiole 0.5–3 cm long, hairy. **Inflorescences** spikes, 3–5 cm long, densely flowered, terminal or axillary; rachis 1–2 cm long, white-hairy; bracts reflexed after anthesis, broadly ovate, 2–3 mm long, acuminate; bracteoles 2.5–3 mm long, spiny, base 2-parted, apex curved. **Flowers** ~5 mm long; tepals shiny, lanceolate, 3–5 mm long, with a midvein, apex acute; stamens 2–2.5 mm long; pseudo-staminodes slightly serrulate, apex rounded. **Fruits** utricles, yellowish-brown, shiny, oblong, 2–2.5 mm long, smooth; seeds light brown, oblong, ~1 mm long. **Distr:** Russia and Asia to Malesia. **Ecol:** On hillsides; 200–1,800 m. Fl. Jul–Sep. Fr. Sep–Oct. **Zones:** 8–12.

Aerva

Species 10. **Herbs** or **subshrubs**, perennial; stem erect, stoloniferous or climbing. **Leaves** alternate or opposite; lamina with margins entire. **Inflorescences** spikes, terminal or axillary, simple or in complex thyrsoid structures. **Flowers** perfect, unisexual or dioecious, sometimes polygamous, small or very small; bracts and bracteoles membranous, persistent or bracteoles falling off with perianth in fruit; tepals 4 or 5, ovate or oblong, membranous or papery, lanose, with only 1 vein; stamens 4 or 5, filaments subulate, unequal, basally united into a short cup, alternating with pseudo-staminodes, anthers 2-loculed; pseudo-staminodes subulate to oblong; ovary obovoid or subglobose, glabrous, style persistent, stigmas 2, capitate. **Fruits** utricles, ovoid, compressed, membranous, indehiscent or irregularly dehiscent (bursting), falling off with perianth; seeds thin-walled, reniform-orbicular, lenticular, compressed, black. **Distr:** Africa, Asia. **Cult:** Propagate from seeds. Requires a moist, well-drained soil in full or partial sun.

A. lanata (L.) Juss. ex Schult., *Syst. Veg.* ed. 15, 5: 564, 1819. (Syn: *Achyranthes lanata*, etc.). [Mountain Knot Grass]. **Herbs**, perennial, frequently woody and suffrutescent below, prostrate to decumbent or erect, stiff or weak and straggling, (0.1–)0.3–2 m high. **Leaves** alternate; lamina suborbicular to lanceolate-elliptic or ovate-elliptic, lamina 1–5 × 0.5–3.5 cm, leaves of upper parts reduced and often smaller, base cuneate, apex rounded and apiculate to acute; petiole 0–2 cm long. **Inflorescences** sessile spikes, solitary or in axillary clusters, 0.4–1.5(–2)

Above: *Alternanthera sessilis, Alternanthera tenella, Amaranthus caudatus.*

× 0.3–0.4 cm, divergent, cylindrical, silky, white to creamy; bracts to 1 mm long; bracteoles similar or slightly smaller, persistent. **Flowers** hermaphrodite; outer 2 tepals hyaline, oval-oblong, ~1.2 mm long, tip abruptly contracted; all tepals densely lanate dorsally; stamens delicate; style and 2 short, divergent stigmas subequaling ovary. **Fruits** utricles, rotund, ~1 mm long, compressed; seed reniform, ~0.6–0.8 mm long, black, shiny. **Distr:** Widespread in the Old World tropics and subtropics. **Ecol:** In dry and disturbed places. **Zones:** 10–12.

Alternanthera

Species ~200. **Herbs**, annual or perennial, ascending or creeping; stem much-branched. **Leaves** opposite; lamina with margins entire. **Inflorescences** sessile or peduncled heads, solitary in axils of bracts; bracts and bracteoles persistent, membranous. **Flowers** perfect; tepals 5, often unequal, membranous; stamens 2–5, filaments connate into a tube or a cup at base, anthers 1-loculed; pseudo-staminodes entire, dentate, or laciniate; ovary globose or ovoid, ovule 1, pendulous, style short or long, stigma capitate. **Fruits** utricles, globose or ovoid, indehiscent, falling off with perianth; seeds lenticular, vertical. **Distr:** Mainly in N and S America, widely naturalized elsewhere. **Cult:** Propagate from seeds, cuttings or by division. Requires a rich, moist, well-drained soil in full or partial sun.

A. sessilis (L.) R. Brown ex Candolle, *Cat. Pl. Horti Monsp.* 77, 1813. (Syn: *A. denticulata*, etc.). [Sessile Joyweed]. **Herbs**, perennial, 10–45 cm high; stem ascending or creeping, green or purple-tinged, striped, hairy. **Leaves** linear-lanceolate, oblong-obovate, or ovate-oblong, 1–8 × 0.2–2 cm, base attenuate, margins entire or slightly serrate, apex acute or obtuse, glabrous or pilose; petiole 1–4 mm long. **Inflorescences** capitula, 1–4, axillary, sessile, at first globose, later cylindric, 3–6 mm ∅, flowers dense; rachis densely white-hairy; bracts and bracteoles white, glabrous, apex acuminate; bracts ovate-lanceolate, ~1 mm long; bracteoles subulate, 1–1.5 mm long. **Flowers** with tepals white, ovate, 2–3 mm long, acuminate or acute, glabrous; stamens 3, filaments ~0.7 mm long, connate into a cup at base, anthers oblong; pseudo-staminodes subulate, shorter than stamens; style very short, stigma shortly parted. **Fruits** utricles, enclosed in perianth, dark brown, obovoid, 2–2.5 mm long; seeds ovoid. **Distr:** China, Indo-China, SE Asia. **Ecol:** In disturbed places. Fl. May–Jul. Fr. Jul–Sep. **Zones:** 10–12.

A. tenella Colla, *Mem. R. Acad. Sc. Torino* 33: 131, 1849. (Syn: *Achyranthes ficoidea*, etc.). [Calico Plant, Joy Weed, Jacob's Coat, Parrot Leaf]. **Herbs**, erect or clambering; stems to 1 m long, densely puberulent at first. **Leaves** opposite; lamina elliptic to ovate-elliptic, (1.5–)2–9 × (0.5–)1–4 cm, base gradually tapered, obtuse or acute and decurrent into petiole, margins entire, apex obtuse to bluntly acute, smooth, becoming glabrous, major secondary veins 3–6 pairs; petioles 1–8 mm long. **Inflorescences** capitate or very short spikes, 0.4–1 × 0.3–0.6 cm, whitish, solitary and sessile, axillary; floral bracts ~2 mm long, tip sharp, aristate, margins translucent; flowers sessile. **Flowers** with outer perianth parts ~4 mm long, stiff, with a sharp tip, with 3(5) strongly raised longitudinal ridges on back and short stiff hairs near base. **Fruits** utricles, suborbicular, within persisting perianth; seed cochleate-oblongoid, 1 mm long, lustrous dark reddish-brown. **Distr:** Tropical America. **Ecol:** In moist evergreen and seasonally deciduous areas; 0–500 m. Fl. Nov–Feb. **Zones:** 10–12.

Amaranthus

Species ~40. **Herbs**, annual; stem erect or decumbent. **Leaves** alternate, petiolate; lamina with margins entire. **Inflorescences** clustered at axils and at apex, in solitary or complex thyrsoid structures; flowers sessile, each with 1 bract and 2 bracteoles. **Flowers** unisexual, both sexes on the same plant or not; tepals (1–4)5, erect or obliquely spreading, sometimes indurate at base after anthesis, green, equal or subequal, membranous; stamens (1–4)5, filaments free, filiform, anthers 2-loculed; pseudo-staminodes absent; ovary with 1 erect ovule, style short or absent, stigmas 2 or 3, persistent, subulate or linear. **Fruits** utricles, globose or ovoid, laterally compressed, membranous, dehiscent via lid, irregularly dehiscent or falling off together with perianth and indehiscent; seeds black or brown, globose or lenticular, without aril. **Distr:** Worldwide. **Cult:** Propagate from seeds. Needs a fertile, moist, well-drained soil in full sun.

A. caudatus L., *Sp. Pl.* 2: 990, 1753. [Love Lies Bleeding, Tassel Flower]. **Herbs**, annual, ~1.5 m high; stem erect, green, strong, angular, slightly branched or not, glabrous when mature. **Leaves** green or red, rhombic-ovate or rhombic-lanceolate, 4–15 × 2–8 cm, base cuneate, margins entire or undulate, apex acuminate or obtuse, surfaces glabrous; petiole 1–15 cm long, green, pilose. **Inflorescences** complex thyrsoid structures, terminal, pendulous, much-branched, composed of many spikes; bracts red, lanceolate, ~3 mm long, transparent, with a midvein abaxially, apex apiculate. **Flowers** with tepals red, shorter than fruit, 2–2.5 mm long, transparent, overlapping at margins, with a midvein; ♂ segments oblong; ♀ segments oblong-lanceolate; stamens 5; stigmas 3, <1 mm long. **Fruits** utricles, red above, subglobose, ~3 mm ∅, longer than perianth, circumscissile; seeds light yellowish-brown, subglobose. **Distr:** Neotropical. **Ecol:** Fl. Jul–Aug. Fr. Sep–Oct. **Zones:** 8–12.

A. spinosus L., *Sp. Pl.* 2: 991, 1753. [Pig Weed, Spiny Amaranth]. **Herbs**, erect, 0.3–1 m high; stem green or somewhat purple-tinged, terete or obtusely angulate, much-branched, glabrous or slightly pubescent. **Leaves** ovate-rhombic or ovate-lanceolate, 3–12 × 1–6 cm, base cuneate, margins entire, apex obtuse, with a mucro, glabrous or slightly pubescent along veins when young; petiole 1–8 cm long, glabrous, 2-armed at base. **Inflorescences** complex thyrsoid structures, terminal or axillary, 8–25 cm long; terminal spike usually with all ♂ flowers at or toward apex; bracts becoming very sharply spiny in proximal part of spike. **Flowers** with tepals green, margins

Below: *Amaranthus spinosus, Amaranthus tricolor, Celosia argentea.*

Above: *Celosia cristata, Gomphrena globosa, Iresine diffusa, Oscularia deltoides.*

transparent, with a green or purple median band, apex acute, with a mucro; ♂ oblong, 2–2.5 mm long; ♀ oblong-spathulate, ~1.5 mm long; filaments subequal to perianth; stigmas 3(or 2). **Fruits** utricles, included in perianth, oblong, 1–1.2 mm long, circumscissile slightly below middle; seeds brownish-black, subglobose, ~1 mm Ø. **Distr:** Neotropical; naturalized elsewhere. **Ecol:** In disturbed and waste places. Fl. & Fr. Jul–Nov. **Zones:** 9–12.

A. tricolor L., *Sp. Pl.* 2: 989, 1753. (Syn: *A. gangeticus*, etc.). [Joseph's Coat, Jacob's Coat, Chinese Spinach]. **Herbs**, annual; stem green or red, 0.8–1.5 m high, stout, often branched. **Leaves** green, red, purple, or yellow, ovate, ovate-rhombic, or lanceolate, 4–10 × 2–7 cm, base cuneate, margins entire or undulate, apex obtuse or notched, with a mucro, glabrous; petiole green or red, 2–6 cm long. **Inflorescences** dense clusters at leaf axils or spikes at apex; ♂ and ♀ flowers in the same inflorescence; bracts and bracteoles ovate-lanceolate, 2.5–3 mm long, transparent, apex long pointed; stamens 3; stigmas 3. **Fruits** utricles, included in perianth, ovate-oblong, 2–2.5 mm long, circumscissile; seeds brownish-black, subglobose or obovoid, ~1 mm Ø. **Distr:** Tropical Asia. **Ecol:** Fl. May–Aug. Fr. Jul–Sep. **Zones:** 8–12.

Celosia

Species ~60. **Shrubs**, **subshrubs** or **herbs**, annual or perennial; sometimes rather woody at the base, occasionally scandent. **Leaves** alternate, petiolate; lamina ovate to linear, margins entire or nearly so. **Inflorescences** spicate, terminal or axillary, uppermost part frequently becoming a panicle; flowers in axils of bracts, subtended by 2 bracteoles. **Flowers** perfect; tepals 5, persistent, erect, spreading, colored, scarious, glabrous; stamens 5, filaments basally connate into a cup, apical part free, linear; pseudo-staminodes absent; ovary 1-loculed, ovules 2–many, style 1, persistent, stigma capitate, or bi- or trilobed. **Fruits** utricles, ovoid or globose, thinly walled, dehiscent by lid; seeds black, lenticular. **Distr:** In subtropical and temperate regions of Africa, N and S America, and Asia. **Cult:** Propagate from seeds. Requires a fertile, moist, well-drained soil in full sun.

C. argentea L., *Sp. Pl.* 1: 205, 1753. (Syn: *C. swinhoei*). [Cockscomb]. **Herbs**, annual, 0.3–1 m high; stem erect, green or red, glabrous, often branched. **Leaves** green, often red-tinged, oblong-lanceolate, lanceolate, or lanceolate-linear, rarely ovate-oblong, 5–8 × 1–3 cm, base attenuate, apex acuminate or acute; petiole 0–1.5 cm long. **Inflorescences** spikes, densely flowered, narrowly cylindric or with a conic apex, 3–10 cm long, not branched; flowers with white bracts and bracteoles, shiny, lanceolate, 3–4 mm long, with midvein, apex acuminate. **Flowers** with tepals white, with a pink tip or nearly pink, then white, oblong-lanceolate, 0.6–1 cm long, with midvein, apex acuminate; filaments 5–6 mm, free part 2.5–3 mm long, anthers purple; ovary shortly stalked, style purple, 3–5 mm long. **Fruits** utricles, ovoid, 3–3.5 mm long, enveloped in persistent perianth; seeds compressed-reniform, ~1.5 mm Ø. **Distr:** Tropical Asia and Africa. **Ecol:** On hillsides, often weedy. Fl. May–Aug. Fr. Jun–Oct. **Zones:** 11–12.

C. cristata L., *Sp. Pl.* 1: 205, 1753. (Syn: *C. argentea* f. *cristata*, *C. argentea* var. *cristata*). [Cockscomb, Crested Celosia, Feather Amaranth]. **Herbs**, annual, upright, 0.15–1 m high, glabrous; stem erect. **Leaves** unlobed, variable, mostly ovate, ovate-lanceolate, or lanceolate, 8–15 × 1–8 cm, base tapered, apex long-acuminate; petiole 1–3 cm long. **Inflorescences** spikes, often cristate, convolute, or feathery, relatively large; some small branches pyramidal-oblong; bracts lanceolate, with midvein, apex acuminate. **Flowers** with tepals red, purple, yellow, or orange, faintly 3-veined at base, 5–8 mm long, scarious; style elongate, 3–4 mm long, stigmas 3. **Fruits** utricles, ovoid, 3–4 mm long, enveloped in persistent perianth; seeds 6–10, compressed-reniform, 1.5 mm Ø, faintly reticulate, shiny. **Distr:** Cosmopolitan; origin unknown, some regard this species as of garden origin. **Ecol:** Fl. & Fr. Jul–Sep. **Zones:** 10–12.

Gomphrena

Species ~100. **Herbs** or **subshrubs**, annual or occasionally perennial. **Leaves** opposite, rarely alternate. **Inflorescences** clustered heads; bracteoles with distal crest. **Flowers** perfect; tepals 5, equal or not, long-hairy or glabrous; stamens 5, filaments united into a tube or cup at base, lateral appendages present, 3-parted at apex, anthers 1-loculed; pseudo-staminodes absent; ovary 1-loculed, with a pendulous ovule, stigmas 2 or 3, linear, or stigma 2-parted. **Fruits** utricles, globose or oblong, compressed, indehiscent; seeds shiny, lenticular. **Distr:** Mainly in N and S America, Australia, Pacific Islands; naturalized elsewhere. **Cult:** Propagate from seeds. Requires a humus-rich, moist, well-drained soil in full sun.

G. globosa L., *Sp. Pl.* 1: 224, 1753. [Bachelor's Buttons, Globe Amaranth]. **Herbs**, annual, erect, 20–60 cm high; stem stout, branched; branches slightly quadrangular, gray strigose, slightly inflated at nodes. **Leaves** oblong or oblong-obovate, 3.5–13 × 1.5–5 cm, narrowing toward base, margins undulate, apex acute or obtuse, papery, long white hairy and ciliate; petiole 1–1.5 cm long, gray-hairy. **Inflorescences** capitula, terminal, 1–3, usually purple, sometimes light purple or white, 2–2.5 cm Ø; bracts 2, opposite, green, ovate or cordate, 1–1.5 cm long, gray-hairy; bracts white, apex purple, ovate, 3–5 mm long; bracteoles purple, triangular-lanceolate, 1–1.2 cm long. **Flowers** with tepals not rigid after anthesis, lanceolate, 5–6 mm long, acuminate, outside densely white-lanose; filaments connate into a tube, 5-parted at apex; styles linear, shorter than staminal tube, stigmas 2, furcate. **Fruits** utricles, subglobose, 2–2.5 mm Ø; seeds brown, shiny, reniform. **Distr:** Neotropical. **Ecol:** Fl. & Fr. Jun–Sep. **Zones:** 8–12.

Iresine

Species ~70. **Herbs**, **shrubs** or **vines**, annual or perennial, polygamous or dioecious; stems erect to prostrate, often trailing. **Leaves** opposite or alternate, petiolate; lamina elliptic, ovate, or lanceolate to lance-oblong, margins entire or serrulate, glabrous or pubescent. **Inflorescences** terminal, diffuse, open panicles of spikes and axillary spikes; bracts and bracteoles membranous. **Flowers** unisexual; tepals 5, basally connate or distinct, 1–3-veined; stamens mostly 5, filaments basally connate, anthers 2-locular; pseudo-staminodes absent or short; ovary compressed, style very short or absent, stigmas 2, rarely 3, subulate, ovule 1, pendulous. **Fruits** utricles, globose, compressed, indehiscent or irregularly dehiscent; seeds shiny, lenticular or reniform. **Distr:** Tropical Asia, N and S America, Pacific Islands. **Cult:** Propagate from seeds, cuttings or by division. Requires a moist, well-drained soil in full sun.

I. diffusa Humb. & Bonpl. ex Willd., *Sp. Pl.* 4: 765, 1805. (Syn: *I. celosia*, etc.). [Blood Leaf]. **Herbs** or **subshrubs**, clambering or less frequently creeping, 1–3 m long, much-branched from woody base; stems glabrous or pubescent, with nodes compressed, usually pilose. **Leaves** opposite; lamina lanceolate or ovate, chartaceous, 3–9(–11) × 1–5 cm, base obtuse, acute, or truncate, margins slightly undulate and revolute, apex acute or obtuse, surfaces dull, venation prominent; petiole slender, 0.5–3 cm long. **Inflorescences** panicles, terminal or axillary, pendulous, 7–15 cm long; bracts and bracteoles not keeled, apex acute or short-acuminate, bracteoles smaller than tepals. **Flowers** unisexual; tepals cream-colored or whitish, ovate or oblong, with a tuft of undulate hairs at base, 0.5–0.8 mm long in ♂ flowers, oblong, 0.9–1.1 mm long in ♂ flowers; stamens 5, filaments unequal. **Fruits** capsules, ellipsoid, ~1 mm long, membranous; seeds reddish-brown, shiny. **Distr:** C America, Lesser Antilles; Colombia, Venezuela. **Ecol:** Fl. & Fr. Sep–May. **Zones:** 9–12.

AIZOACEAE

Subfamilies 4, **genera** 123, **species** ~2,020. **Herbs** annual or perennial, **subshrubs**, or **shrubs**, succulent; stems erect or prostrate. **Stipules** absent, sometimes a stipulelike sheath present at base of petiole. **Leaves** mostly opposite, sometimes alternate; lamina simple, rarely pinnate, in many species fleshy, margins entire, rarely with teeth. **Inflorescences** cymes, or flowers solitary, terminal or apparently axillary. **Flowers**

bisexual, rarely unisexual, actinomorphic, perigynous or epigynous; tepals (4 or)5(–8), connate below into a tube; petals absent or present; stamens 3–many, free or basally connate, outermost often as filamentous staminodes, anthers dehiscing by longitudinal slits; ovary inferior, syncarpous, carpels 2–many, ovules 1–many, on long funicles, mostly campylotropous, placentation axile or parietal, sometimes basal-parietal, stigmas as many as carpels; nectaries separate or in a ring around ovary. **Fruits** capsules, hygroscopic or circumscissile, more rarely berries or nuts; seeds rarely arillate, endosperm scanty or absent. **Distr:** Mainly in southern Africa; some species pantropical. **Ecol:** Mainly in tropical, subtropical and arid regions.

Oscularia

Species ~25. **Shrubs** or **shrublets**, dense; stems spreading to decumbent, with reddish or reddish-brown indument, shiny. **Leaves** opposite; lamina triangular to 3-angled, short, to ~2 cm long, mostly with a toothed keel and margins, convex, thick, very light gray distinct layer of crystal sand in the outer epidermal wall and a thick wax cover of densely placed platelets. **Inflorescences** terminal cymes, often with additional cymes below, forming larger aggregate inflorescences, or flowers solitary and arranged in a cluster of 3. **Flowers** to ~1.5 cm Ø, almond-scented, white to deep pink, open day and night; calyx 5-lobed; petals in ~1 series; filamentous staminodes variable in number; stamens and staminodes in a cone; ovary surrounded by 5 distinct nectary glands, often broad and touching each other. **Fruits** 5-segmented, base funnel-shaped, gray to brown. **Distr:** S Africa, mainly on the mountains along the West Coast. **Cult:** Propagate from seeds or cuttings. Requires a well-drained soil in full sun; plants tolerate xeric conditions.

O. deltoides Schwantes, *Mollers Deutsch. Gartner-Zeit.* xlii. 187, 1927. (Syn: *Lampranthus deltoides*, etc.). **Shrubs** or **shrublets**, sprawling or rounded, to 20 cm high; branches shiny, with reddish or deep maroon indumentum. **Leaves** deltoid or triangular, 0.5–1.8 × 0.3–1 cm, broadest above middle, keel and margins toothed, mostly 2–3 on the keel and 3–4 on the margins, grayish. **Inflorescences** cymes, flowers crowded, terminal and several additional cymes, flowers 1–3 together. **Flowers** 1–2 cm Ø, pink to deep pink (appearing reddish), fragrant; calyx 5-lobed; petals in ± 1 series; filamentous staminodes variable in number; stamens and staminodes in a cone; ovary surrounded by 5 distinct nectary glands, often broad and touching each other. **Fruits** 5-segmented, base funnel-shaped, gray to brown. **Distr:** S Africa (Boland Mountains in the Fynbos Region). **Ecol:** On sandstone rocks. Fl. Oct–Dec. **Zones:** 8–11.

PHYTOLACCACEAE

Subfamilies 3, genera 18, species 65. **Herbs** annual or perennial, **vines**, **subshrubs**, **shrubs**, or rarely **small trees**. **Stipules** minute or absent. **Leaves** alternate, opposite, or appearing whorled; lamina conduplicate, pinnately veined, margins entire, often undulate. **Inflorescences** ± racemose, panicles, or compound dichasia, axillary, terminal, or leaf-opposed spikes; bract 1; bracteoles 2. **Flowers** hypogynous (rarely epigynous), radially symmetric; perianth 4–5(–10)-merous; sepals usually persistent, imbricate in bud, often unequal; petals absent; stamens twice in number as perianth segments, usually 4–30(–100), sometimes 1–2-whorled on a hypogynous disk; gynoecium 1–12(–18)-carpellate, carpels connate or distinct, ovaries superior (rarely inferior), 1–12(–18)-locular, placentation basal, ovules 1 per locule, styles distinct (rarely connate). **Fruits** achenes, berries, or cypselas, with persistent perianth; seeds with perisperm present, embryo white. **Distr:** Tropical and warm temperate regions, especially America.

Phytolacca

Species 25. **Herbs** perennial (somewhat fleshy), **shrubs** or pachycaulous **trees**. **Leaves** alternate, petiolate; lamina with margins entire. **Inflorescences** racemes or spikes, 5–100-flowered, axillary, terminal, or leaf-opposed, proximal pedicels sometimes bearing more than 1 flower. **Flowers** unisexual or bisexual; sepals 5–8; stamens (5–)8–25(–30), sometimes 1–2-whorled; carpels 6–12(–30), distinct or connate at least proximally, if connate, ovary 6–12-loculed, style and stigma 1 per carpel, style free. **Fruits** either oblate berries with persistent style or group of lenticular achenes; seeds 6–12(–30) per berry or 1 per achene. **Distr:** N America to S America, West Indies; Eurasia; Africa; Pacific Islands, Australia. **Cult:** Propagate from seeds or cuttings. Requires a fertile, moist, well-drained soil in full sun.

P. dioica L., *Sp. Pl.* 632, 1753. (Syn: *Pircunia dioica*). [Bella Sombra]. **Trees**, erect, dioecious, 7–10 m tall. **Leaves** long-petiolate; lamina broadly elliptic to oblong-ovate, 4.5–8(–14) × 2.25–4.5(–8.5) cm, rounded-obtuse at base with sides at times unequal, apex shortly acuminate-mucronate, mucro ~3.5 mm long, recurved, leathery, glabrous, 5–8 (–10) pairs of lateral veins, prominent beneath with midrib markedly so; petiole 1.5–4(–5) cm long. **Inflorescences** 5–12 cm long, terminal or axillary, shortly pedunculate, flowers sublaxly arranged, spreading, becoming subreflexed at maturity; rachis shortly pedunculate, flexuous, pubescent; bracts 1–1.5 mm long, narrowly acute to subulate, glabrous; pedicels 6–7 mm long, puberulous, thickened toward apex. **Flowers** with calyx lobes orbicular-ovate, 3 × 2 mm, obtuse, shortly laciniate; stamens 20–30, unequal, filaments filiform, ~5 mm long, anthers oblong, 1 mm long. **Fruits** aggregate of 7–10 carpels, 0.7–1 cm Ø, ripening black. **Distr:** Tropical S America. **Ecol:** In forests. Fl. Dec–Jan. Fr. Mar–Apr. **Zones:** 10–11.

Rivina

Species 1. *Generic description same as specific.* **Cult:** Propagate from seeds. Requires a moist, well-drained soil in shade to full sun.

R. humilis L., *Sp. Pl.* 1: 121, 1753. (Syn: *R. laevis*, *R. portulaccoides*). [Coral Berry, Pigeon Berry, Rouge Plant]. **Herbs** or **subshrubs**, erect, straggling, or vinelike, 0.4–2 m high, ± glabrous or densely pubescent. **Leaves** lanceolate, elliptic, or oblong to deltate or ovate, to 15 × 9 cm, base cuneate or rounded to truncate or cordate, apex acuminate or acute to obtuse or emarginate; petiole 1–11 cm long. **Inflorescences** racemes, 4–15 cm long; peduncle 1–5 cm long; pedicel 2–8 mm long. **Flowers** with sepals white or green to pink or purplish, elliptic or oblong to oblanceolate or obovate, 1.5–3.5 mm long; style often curved. **Fruits** berries, 2.5–5 mm Ø; seeds lenticular, 2–3 mm, enclosed in thin densely pubescent membranes. **Distr:** N America, Mexico, West Indies, C and S America, Asia, Pacific Islands, Australia. **Ecol:** In woods, thickets, disturbed areas, often weedy; 0–1,700 m. Fl. throughout the year. **Zones:** 10–12.

NYCTAGINACEAE

Genera 30, species 395. **Herbs**, **shrubs**, often weak-stemmed **trees**, or sometimes spiny vines or lianas. **Stipules** absent. **Leaves** opposite, alternate, or ± whorled; lamina simple, herbaceous or slightly fleshy, margins entire; petiole usually present, well-defined. **Inflorescences** cymes, umbels, or verticils, sometimes 1-flowered or fasciculate, often grouped into panicles, mostly terminal, less often axillary; bracts often inconspicuous, sometimes forming calyxlike involucre, or large and brightly colored. **Flowers** bisexual, rarely unisexual or polygamous, actinomorphic; perianth connate, petaloid, lobes induplicate, constricted beyond the ovary, base persistent, closely enclosing ovary which appears inferior, limb petaloid beyond constriction, tubular, funnelform, or campanulate, apex 5–10-lobed, lobes plicate or valvate in bud, persistent or caducous, nectary on receptacle; disk absent; stamens (1–)3–5(–many), hypogynous, free or basally connate, involute in bud, anthers 2-loculed, longitudinally dehiscent; ovary superior, 1-loculed, ovule 1, style 1, stigma globose. **Fruits** achenelike anthocarps or nutlets, enclosed by persistent perianth, ribbed or winged, often glandular; seed 1, endosperm present, embryo green. **Distr:** Tropical to warm temperate.

Boerhavia

Species 20–40. **Herbs**, annual or perennial, erect, ascending, or procumbent; stems sometimes glandular. **Leaves** opposite, often unequal, petiolate; lamina with margins entire or sinuate. **Inflorescences** of small umbels, often irregular, rarely 1-flowered, grouped into cymose panicles, usually terminal, sometimes apparently axillary; bracteoles minute, often deciduous.

Below: *Phytolacca dioica*, *Rivinia humilis*, *Boerhavia diffusa*.

Above: *Bougainvillea glabra, Bougainvillea spectabilis, Mirabilis jalapa, Pisonia brunoniana.*

Flowers bisexual, small; perianth constricted beyond ovary, limb campanulate, apex 5-lobed, truncate, or plicate, caducous; stamens 1–5, included or shortly exserted, filaments basally connate; ovary oblique, stipitate, stigma peltate or capitate. **Fruits** obovoid, turbinate, clavate, or terete, small, 5-ribbed or deeply 5-angled, often rough, with sessile sticky glands; seed with embryo curved. **Distr:** Widespread in tropical and subtropical regions. **Cult:** Propagate from seeds. Requires any well-drained soil in full sun.

B. diffusa L., *Sp. Pl.* 1: 3, 1753. (Syn: *B. diffusa* var. *mutabilis*, *B. repens* var. *diffusa*). [Hog Weed, Horse Purslane, Pig Weed, Spreading Hogweed, Tar Vine]. **Herbs**, perennial; stems trailing, to 2 m long; roots thick, fleshy. **Leaves** ovate, 1–5 × 1–4 cm, base rounded or cuneate, margins undulate, apex obtuse or acute, with stout, muticellular hairs, surfaces sparsely pubescent, abaxially gray-yellow; petiole 0.4–2 cm long. **Inflorescences** capitate-cymose panicles, terminal; peduncle slender; pedicels short to almost absent; bracts small, lanceolate, pubescent. **Flowers** with perianth limb bright purple or purple-red, 1.5–2 mm long; stamens 1–3(–5), slightly exserted or included. **Fruits** anthocarps, clavate, 3–3.5 mm long, 5-ribbed, with viscid glands and sparse pubescence, apex rounded. **Distr:** Tropical regions worldwide. **Ecol:** In open places near sea, usually weedy; 100–1,900 m. **Zones:** 10–12.

Bougainvillea

Species ~18. **Shrubs** or **small trees**, sometimes climbing; branches spiny. **Leaves** alternate, petiolate; lamina ovate or elliptic-lanceolate. **Inflorescences** 3-flowered cymes, each flower subtended by a persistent, often brightly colored, ovate bract adnate to the pedicel, often grouped into terminal panicles, axillary, pedunculate; pedicel adnate to midvein or the bract. **Flowers** bisexual; perianth connate, limb rose or yellow, funnelform, 5–6-lobed, lobes short; stamens 5–10, included, filaments shortly basally united; ovary fusiform, stipitate, style lateral, shortly linear, stigma fimbriate. **Fruits** anthocarps, cylindric or clavate, 5-ribbed, without sticky glands; seed with testa thin. **Distr:** S America. **Cult:** Propagate from cuttings or by air-layering. Requires a fertile, moist, well-drained soil in full sun.

B. glabra Choisy, *Prodr.* 13(2): 437, 1849. (Syn: *B. spectabilis* var. *glabra*). [Bougainvillea, Paper Flower]. **Shrubs**, vinelike, evergreen, vigorous, woody, climbing to 3–6 m high; stems thick, branches pendent. **Spines** axillary, 0.5–1.5 cm long. **Leaves** elliptic, ovate or ovate-lanceolate, 5–13 × 3–6 cm, chartaceous, adaxially glabrous, abaxially sparsely pubescent; petiole 1 cm long. **Inflorescences** with bracts foliaceous, purple or magenta, oblong or elliptic, 2.5–3.5 × 2 cm, chartaceous. **Flowers** with perianth tube greenish, distinctly angled, ~2 cm long, sparsely pubescent, ribbed, apex 5-lobed, hairs short, curved away from pedicel; stamens 6–8. **Fruits** anthocarps, 1-seeded. **Distr:** S America (Brazil). **Ecol:** Fl. throughout most of the year. **Zones:** 10–12.

B. spectabilis Willdenow, *Sp. Pl.* 2: 348, 1799. (Syn: *B. brasiliensis*). [Bougainvillea, Paper Flower]. **Climbers**, vigorous, thorny, woody, climbing to 5–6 m high; branches and leaves densely pubescent. **Spines** recurved. **Leaves** petiolate; lamina elliptic or ovate, base rounded, green. **Inflorescences** with bracts dark red or light purple-red, elliptic-ovate, 2.5–6.5 × 1.5–4 cm, base rotund to cordate. **Flowers** with perianth tube green, narrowly tubular, rounded, 1.6–2.4 cm long, densely pubescent, apex 5–6-lobed, lobes spreading, yellow, 3.5–5 mm long, hairs copious, spreading, to 1 mm long; stamens 8; ovary stipitate. **Fruits** anthocarps, 1–1.5 cm long, densely hairy, 1-seeded. **Distr:** Tropical America. **Ecol:** Fl. throughout most of the year. **Zones:** 10–12.

Mirabilis

Species 45. **Herbs**, perennial (rarely annual), often suffrutescent; taproots slender to swollen and often tuberous; stems erect to decumbent, unarmed, noded. **Leaves** petiolate or sessile, ± equal in each pair; lamina with base ± symmetric, thin to thick, fleshy. **Inflorescences** compound dichasia, usually cymose, sometimes consisting of a single involucre in leaf axil, terminal and axillary, terminating in pedunculate involucres subtending 1–3(–16+) flowers; bracts persistent, 5, connate, or rarely distinct, ± ovate, forming accrescent, herbaceous to papery involucre. **Flowers** bisexual; perianth (4–)5(–8)-lobed, sometimes enlarged and membranous in fruit; stamens 2–6, usually exserted; styles exserted beyond stamens, stigmas capitate. **Fruits** anthocarps, radially symmetric, with (4–)5 round or angular ribs separated by usually broad sulci, or shallow ribs, obovoid, ellipsoid, or nearly globose, smooth or tuberculate, stiffly coriaceous, without glands. **Distr:** Mostly in C and W USA, C America, N South America; 1 species in W Himalayas. **Cult:** Propagate annual species from seed, and tuberous perennial species by division or from seeds. Requires a fertile, moist, well-drained soil in full or partial sun.

M. jalapa L., *Sp. Pl.* 1: 177, 1753. (Syn: *Nyctago jalapa*). [Beauty of the Night, Four O'Clock Flower, Flower of Abbas, Marvel of Peru]. **Herbs**, annual, 0.3–1(–1.5) m high; roots tuberous; stems erect, much-branched, cylindric, glabrous or slightly pubescent, nodes inflated. **Leaves** ovate or ovate-triangular, 3–15 × 2–9 cm, base truncate or (sub) cordate, margins entire, apex acuminate; petiole 1–4 cm long. **Inflorescences** usually several clustered at apex of branches; pedicels 1–2 mm long. **Flowers** fragrant; involucre campanulate, ~1 cm long, 5-lobed, lobes triangular-ovate, acuminate, glabrous, persistent; perianth purple, red, yellow, white, or variegated, tube 2–6 cm long, limb 2.5–3 cm Ø, opening in late afternoon, closing the next morning; stamens 5, filaments slender, exserted, anther globose. **Fruits** black, globose, 5–8 mm Ø, coriaceous, ribbed and plicate, endosperm white mealy. **Distr:** Tropical America. **Ecol:** Fl. Jun–Oct. Fr. Aug–Nov. **Zones:** 8–12.

Pisonia

Species >35. **Shrubs**, **trees**, or **vines**; stems sometimes armed. **Leaves** opposite or alternate, petiolate; lamina with margins entire. **Inflorescences** many-flowered cymes or panicles, axillary (or terminal); bracts absent or inconspicuous, caducous; bracteoles 2–4. **Flowers** unisexual (plants dioecious), bisexual, or polygamous; ♂ with perianth limb campanulate or funnelform, stamens 6–10; ♀ with perianth ovate-tubular or cylindric, 5–10-lobed, ovary sessile, oblique, stigma capitate or lobed. **Fruits** oblong or clavate, ribbed, with sticky glands; seed oblong, with a deep longitudinal furrow. **Distr:** Tropical and subtropical regions, with the highest diversity in SE Asia and tropical America. **Cult:** Propagate from cuttings or fresh seeds. Requires a moist, well-drained soil in full or partial sun.

P. brunoniana Endl., *Prodr. Fl.* norf. 43, 1833. (Syn: *Heimerliodendron brunonianum*, etc.). [Bird-catcher Tree]. **Shrubs** or **trees**, to 6+ m tall; trunk to 60 cm Ø; branches brittle. **Leaves** opposite to subwhorled; lamina oblong to obovate-oblong, 10–40 × 5–15 cm, margins entire, apex obtuse, glabrous, submembranous; petiole to 3 cm long. **Inflorescences** panicles, many-flowered; bracts foliaceous; pedicels ± pubescent. **Flowers** with perianth to 1 cm long, greenish, pubescent to glabrate; stamens 6–8, anthers not or very slightly exserted. **Fruits** elongate, narrow, constricted below mouth, ribs 5, papillose, very viscid. **Distr:** New Zealand. **Ecol:** In coastal forests and dry to mesic areas. Fl. & Fr. Aug–Dec. **Zones:** 8–12. **Info:** Small birds often get caught on the sticky seeds. This species is often confused with the Australasian species *P. umbellifera*.

P. grandis R. Brown, *Prodr.* 1: 422, 1810. (Syn: *Ceodes grandis*, etc.). [Lettuce Tree]. **Trees**, to 14(–30) m tall; trunk 30–50(–70) cm Ø; branches unarmed. **Leaves** elliptic, oblong, or ovate, (7–)10–20(–30) × (4–)8–15(–20) cm, base rounded or slightly cordate, mostly oblique, margins entire, apex acute to acuminate, papery or membranous, lateral veins 8–10 pairs; petiole 1–8 cm long. **Inflorescences** cymes, terminal, 1–4 × 3–5 cm; peduncles ~1.5 cm long, brown-hairy; pedicels 1–1.5 mm long, apex with 2–4 oblong bracteoles. **Flowers** bisexual; perianth tube funnelform, ~4 mm long, 5-lobed, with 5 rows of black glands; stamens 6–10, exserted; stigma fimbriate, included. **Fruits** clavate, ~1.2 cm × 2.5 mm, 5-ribbed, with sessile glands, without persistent perianth, rib with a row of viscid prickles, hairy between the ribs; seed 0.9–1 cm × 1.5–2 mm. **Distr:** India, Indonesia, Malaysia, Sri Lanka; NE Australia, Madagascar,

Above: *Pisonia grandis, Basella alba, Talinum paniculatum, Portulaca grandiflora.*

Maldives, Pacific Islands. **Ecol:** On seashores to further inland in forests. Fl. Feb–Mar. Fr. Jul–Sep. **Zones:** 11–12.

BASELLACEAE

Genera 4, **species** 20. **Vines**, herbaceous, perennial, succulent, mucilaginous, or shrubs, glabrous. **Leaves** alternate, sessile or petiolate; lamina with margins entire, sometimes toothed. **Stipules** absent. **Inflorescences** racemose, sometimes spicate, or paniculate, axillary or terminal, bracteate; flowers sessile or pedicellate, subtended by 1–2 bracts. **Flowers** usually bisexual; perianth hypogynous, hypanthium present; sepals 2, distinct or basally connate; petals (4–)5(–13), basally connate; nectary annular; stamens (4–)5(–9), epipetalous, basally connate, filaments erect or reflexed in bud, broadened at base, anthers versatile or sometimes basifixed, 4-locular, dehiscence longitudinal, rarely terminal; ovule 1, styles 3, basally connate, stigmas slender, bifid to clavate or capitate (or style 1, with stigma capitate to trilobed). **Fruits** utricles, surrounded by dry or fleshy perianth, winged or not; seeds usually rust-colored, testa membranous, endosperm copious. **Distr:** Tropical America; Eurasia, Africa; Australia.

Basella

Species 5. **Herbs**, twining, creeping or decumbent, annual or biannual, succulent; stems fleshy, much-branched. **Stipules** absent. **Leaves** alternate, succulent, petiolate; lamina simple. **Inflorescences** simple or branched spikes, axillary; rachis long, thick; bracts minute; bracteoles and urceolate perianth adnate, ovoid or oblong, opening very little at anthesis, fleshy and enclosing fruit after flowering; flowers sessile. **Flowers** actinomorphic, bisexual; perianth segments short, fleshy, blunt, ridged on back, hardly opening at anthesis; calyx of 2 white or pale green sepals, free to base, much shorter than corolla, united to petals at base; corolla tubular, urceolate, 5-lobed, succulent; stamens 5, included, filaments short, inserted near apex of perianth tube, very short, erect in bud, anthers dorsifixed, or versatile, dehiscent by longitudinal sutures; ovary superior, globose or pyriform, of 3 carpels, uniovulate, styles 3, free to base, stigmas linear. **Fruits** utricles, globose, fleshy, completely covered by persistent or accrescent corolla. **Distr:** Africa, Madagascar; 1 species pantropical. **Cult:** Propagate from seeds or cuttings. Requires a moist or well-drained, fertile soil rich in organic matter, in full or partial sun.

B. alba L., *Sp. Pl.* 1: 272, 1753. (Syn: *B. rubra*). [Ceylon Spinach]. **Vines**, herbaceous, twining, climbing or creeping, annual, to 10 m long, glabrous; stems greenish, reddish or purplish, succulent, cylindrical, to 2.5 cm Ø. **Leaves** alternate; lamina green, reddish, or purple, ovate, orbicular, or oblong, 3–15 × 1.2–14 cm, base acute, subtruncate, rounded, or cordiform, decurrent on petiole, margins entire, undulate, apex obtuse, rounded, or acute, fleshy, adaxially dull, venation slightly conspicuous, abaxially dull, venation slightly prominent; petiole green or reddish, 0.5–5 cm long, slightly winged. **Inflorescences** spikes, simple or branched, 3–15(–20) cm long, erect, axillary; bracteoles oblong. **Flowers** with perianth reddish or purplish, sepals 2, ovate-oblong, 3–4 mm long, base pale, margins entire, apex blunt; corolla fleshy, white, with a reddish or violet tinge toward apex, 3–4 mm long; filaments white, anthers yellowish. **Fruits** utricles, red to dark red or black, 4–6 mm Ø, fleshy, 4-lobate, covered by accrescent corolla, 5–7 mm long, containing violet juice (photosensitive). **Distr:** Pantropical. **Ecol:** In humid or dry areas; 0–500 m. Fl. May–Sep. Fr. Jul–Oct. **Zones:** 10–12.

TALINACEAE

Genera 2, **species** 22. **Herbs** annual or perennial, or **shrubs**, underground parts often tuberous; stems erect, glabrous. **Leaves** with paired axillary scales. **Flowers** with ovary rarely septate. **Fruits** capsules, rarely indehiscent and mucilaginous (Stevens 2001 onwards). **Distr:** America and Africa.

Talinum

Species 10. **Herbs**, annual or perennial, or subshrubs, often with thick roots; stems erect, glabrous; axils naked. **Leaves** alternate or partly opposite, sessile or shortly petiolate; lamina flat. **Inflorescences** terminal or rarely axillary, thyrses or panicles, very rarely flowers solitary. **Flowers** with sepals free or shortly basally connate, caducous; petals 5(–10), free; stamens 5–30, usually adnate to base of petals; ovary superior, stigma (bi- or)trilobed. **Fruits** capsules, 3-valved, often recurved, globose, ovoid, or ellipsoid; seeds glossy blackish-brown or black, subglobose or oblate, orbicular-reniform, tuberculate or ribbed, with pale caruncle. **Distr:** Tropical Americas to Africa and Arabia. **Cult:** Propagate from seeds. Requires a well-drained, sandy soil in full sun.

T. paniculatum (Jacquin) Gaertner, *Fruct. Sem. Pl.* 2: 219, 1791. (Syn: *Portulaca paniculata*, etc.). [Jewels of Opar]. **Herbs**, annual or perennial, 0.3–1 m tall; roots few-branched, obconic, thick, epidermis blackish-brown, flesh milky white; stems branched basally, semiwoody. **Leaves** obovate or obovate-lanceolate, 5–10 × 2.5–5 cm, base narrowly cuneate, apex acute, sometimes retuse, mucronate. **Inflorescences** panicles, terminal or axillary, large, often dichotomous, long-peduncled; bracts lanceolate, ~1 mm long, apex acute, membranous; pedicels 0.5–1 cm long. **Flowers** small, 0.6–1 cm wide; sepals purplish-red, ovate, caducous; petals pink or reddish-purple, obovate or elliptic, 0.6–1.2 cm long, rounded, rarely retuse; stamens (10–)15–20, shorter than petals; ovary ovoid, ~2 mm long, style ~2 mm long, stigma trilobed. **Fruits** capsules, subglobose, ~4 mm Ø, papery; seeds oblate, ~1 mm long. **Distr:** Tropical America; naturalized elsewhere in the tropics. **Ecol:** In moist and disturbed places. Fl. Jun–Aug. **Zones:** 9–12.

PORTULACACEAE

Genus 1, **species** 40–100. *Familial description same as generic.*

Portulaca

Species 40–100. **Herbs**, succulent, annual or perennial; roots tuberous, fleshy; stems erect to prostrate, branched, fleshy or suffrutescent. **Leaves** alternate or subopposite, congested and involucrelike immediately proximal to inflorescence; lamina ± terete, subterete, or flattened. **Inflorescences** heads, terminal, or axillary on short branches; flowers sessile or subsessile with involucre. **Flowers** with sepals broadly clasping at base, herbaceous to scarious; petals ephemeral, 5–7, usually distinct, margins usually entire; stamens (4–)6–40(–100); ovary half inferior to inferior, plurilocular proximally to 1-locular distally, placentation free-central, style 1, short, stigmas 3–8(–18). **Fruits** capsules, membranous, chartaceous, dehiscence circumscissile, pericarp undifferentiated; seeds many, brown to black or gray, reniform to cochleate; seed with testa smooth or variously sculptured, granular to stellate-tuberculate or spiny. **Distr:** Worldwide, primarily tropical and subtropical N and S America. **Cult:** Propagate from cuttings or seeds. Requires a moist, well-drained, sandy soil in full sun.

P. grandiflora Hooker, *Bot. Mag.* 56: pl. 2885, 1829. [Japanese Rose, Moss Rose, Sun Plant]. **Herbs**, annual; roots fibrous; stems prostrate to suberect; trichomes conspicuous at nodes and in inflorescence; branches to 30 cm long. **Leaves** linear to lanceolate, terete to hemispheric, 0.5–3 cm × 1–5 mm, apex acute or subacute; involucrelike leaves 8–9(–14). **Inflorescences** terminal in clusters, or axillary on short branches. **Flowers** 2.5–5.5 cm Ø; petals pink, red, purple, yellow, bronze, or white, obovate, 1.5–2.5 × 1.5–2 cm; stamens 40 or more; stigmas 5–8. **Fruits** capsules, ovoid, (3.5–)4–6.5 mm Ø; seeds grayish, often iridescent, orbiculate or elongate, flattened, ~1 mm diam. **Distr:** S America; cultivated and naturalized worldwide. **Ecol:** In sandy soils and rocky areas; 0–2,000 m. **Zones:** 8–12.

CACTACEAE

Subfamilies 4 Genera 111, **species** ~1,500. **Shrubs**, **trees** or **vines**, fleshy perennials, terrestrial or epiphytic; stems jointed, terete, globose, flattened, or fluted, mostly leafless and variously spiny. **Leaves** alternate; lamina flat or subulate to terete, vestigial, or entirely absent; spines, glochids (easily detached, small, bristlelike spines), and flowers always arising from cushionlike, axillary areoles (modified short shoots). **Inflorescences** with flowers solitary, sessile, rarely

Above: *Pereskia bleo, Pereskia grandifolia, Opuntia cochinellifera, Opuntia ficus-indica.*

clustered and stalked (in *Pereskia*). **Flowers** bisexual, rarely unisexual, actinomorphic or occasionally zygomorphic; receptacle tube (hypanthium or perianth tube) absent or short to elongate, naked or invested with leaflike bracts, scales, areoles, and hairs, bristles, or spines; perianth segments usually numerous, in a sepaloid to petaloid series; stamens numerous, variously inserted in throat and tube, anthers 2-loculed, longitudinally dehiscent; ovary (pericarpel) inferior, rarely superior, 1-loculed, with 3 to many parietal (rarely basal) placentas, ovules usually numerous, style 1, stigmas 2 to numerous, papillate, rarely bifid. **Fruits** juicy or dry, naked, scaly, hairy, bristly, or spiny, indehiscent or dehiscent, when juicy then pulp derived from often deliquescent funicles (except *Pereskia*); seeds usually numerous, often arillate or strophiolate. **Distr:** Predominantly in tropical to temperate regions of N America, West Indies, C America, S America, including Galápagos Islands; 1 species in Africa, Madagascar, Sri Lanka, and other Indian Ocean islands.

Subfam. PERESKIOIDEAE

Genus 1, **species** 9. *Subfamilial description same as generic.*

Pereskia

Species 9. **Trees** or **shrubs**, to 15 m tall; roots, tuberous; stems straight or zigzag; spines 1–12 per areole, binate or fasciculate in the leaf axils; glochids none. **Leaves** alternating with spines, deciduous with age; lamina green, margins entire, fleshy. **Inflorescences** with flowers solitary, corymbose or paniculate. **Flowers** from areoles of new growth, fragrant or not; outer tepals often greenish, colored near margins, rotate; inner tepals white, yellow, orange-red, red, pink to purplish; stamens 50–100 (in small-flowered species), to 300 (in large-flowered species), filaments colorless near base, in some species pigmented distally, color either matching inner tepals or contrasting with them; carpel primordia 5 or more, originating separately on the convex floral apex, styles shorter to longer than stamens, ovary ± superior, stigma lobes 3–20, linear. **Fruits** often bearing smaller leaves, fleshy and juicy; seeds black, glossy. **Distr:** S South America. **Cult:** Propagate from seeds or cuttings. Requires a dry and well-drained soil in full or partial sun.

P. bleo (HBK.) DC., *Prodr.* 3: 475, 1828. (Syn: *P. corrugata*, etc.). [Desert Rose, Orange Rose Cactus]. **Shrubs** or **small trees**, to 4–7 m tall; branches terete, rather stout, armed with few to several slender straight spines usually ~1 cm long. **Leaves** long-petiolate; lamina usually elliptic-obovate, 4–20 × 2–7 cm, base cuneate, apex acuminate; petiole 0.5–3.5 cm long. **Inflorescences** with flowers usually few or solitary, in terminal or axillary panicles shorter than the subtending leaves; pedicels ~5 mm long. **Flowers** with perianth pink to deep red, rotate, ~5–7 cm Ø, segments numerous, inner broadly obovate, outer oblong-linear; stamens numerous, about half as long as perianth segments; style stout, slightly longer than stamens. **Fruits** berries, turbinate, 5–6 × 5–6 cm, with several ± persistent bracts, yellow when ripe. **Distr:** Panama and NW South America. **Ecol:** Along streams and rivers in secondary forests, sometimes in thickets; 0–500 m. Fl. Sep–May. Fr. Jan–Aug. **Zones:** 9–12.

P. grandifolia Haworth, *Suppl. Pl. Succ.* 85. 1819. [Rose Cactus, Wax Rose]. **Shrubs** or **trees**, erect, 2–10 m tall; stems to 20 cm Ø, very spiny, areoles to 1.2 cm Ø on basal portion of stem. **Leaves** elliptic-oblong, 6–30 × 3–9 cm, 0.5 mm thick; spines all alike, 1–6 per areole, straight, acicular, 1–6.5 cm long. **Inflorescences** densely cymose-paniculate aggregations, 10–50-flowered; lateral pedicels 1–3 cm long; terminal flowers without pedicels. **Flowers**, not fragrant, 1.5–3.2 × 3–7 cm; tepals epigynous; scales and areoles in distal half of receptacles; perianth pink. **Fruits** reddish-green to yellowish-green, pyriform or turbinate, often nearly as broad as long, 5–10 × 3–7 cm, often proliferating; seeds obovoid-ellipsoid, 5–7 × (3.6–)4.5–5(–5.4) mm, glossy. **Distr:** S America (Brazil). **Ecol:** In lowland forests. Fl. throughout most of the year. **Zones:** 9–12.

Subfam. OPUNTIOIDEAE

Genera 16, **species** 220–350. **Stems** succulent, articulated, roots often tuberous. **Seeds** ± covered by bony funicular aril. **Distr:** Canada, to Patagonia; widespread. **Info:** The monophyly of Opuntioideae is strongly supported by molecular data, but many currently accepted genera may be para- or polyphyletic (Griffith and Porter 2009).

Opuntia

Species 200. **Shrubs** or **small trees**, some low and cespitose; stems fleshy, segmented, usually many-branched; segments terete, club-shaped, subglobose, ± laterally compressed, sometimes tuberculate, rarely ribbed; areoles with glochids and usually 1–many spines; spines acicular, subulate or papery, usually 1–many, rarely absent, sometimes sheathed. **Leaves** conic to terete, usually small, caducous. **Flowers** solitary, lateral or subterminal, rarely terminal, sessile; receptacle obovoid, truncate and depressed at apex; perianth rotate, spreading, rarely erect or slightly zygomorphic, inserted at rim of receptacle tube, yellow, pink, red or off-white; segments numerous, outer ones sepaloid, inner ones petaloid; stamens inserted in perianth throat, usually touch-sensitive; ovary (pericarpel) inferior, placentas parietal, style often ± expanded at base. **Fruits** fleshy or dry, umbilicate, globose or ovoid, with areoles, glochids, and sometimes spines; seeds circular to broadly ovate, 3–9 mm long, encased in a white, bony, rarely hairy or fleshy aril. **Distr:** S Canada to Patagonia, Galápagos Islands, West Indies; widespread. **Cult:** Propagate from soaked seeds or rooting stem segments. Requires a sandy, fertile, humus-rich, well-drained soil in full sun.

O. cochenillifera (L.) Miller, *Gard.* dict. ed. 8: Opuntia no. 6. 1768. (Syn: *Cactus cochenillifera, Nopalea cochenillifera*). [Velvet Opuntia, Wooly Joint Prickly Pear]. **Shrubs** or **small trees**, 2–4 m tall; trunk terete, woody. **Large joints** green, elliptic to narrowly obovate, 8–40(–50) × 5–7.5(–15) cm, ends rounded, margins entire. **Areoles** ~2 mm Ø. **Spines** usually absent, when present, 1–3 per areole, spreading, grayish-tan, acicular, 3–9 mm long; glochids early deciduous. **Leaves** conic, 3–4 mm long, early deciduous. **Flowers** 1.2–1.5 cm Ø, erect; sepaloids with brilliant red or green midrib, largest ones ovate-deltoid, 0.5–1.2 × 0.6–0.9 cm, margin entire, apex acute; petaloids bright red, ovate to obovate, 1.3–1.5 × 0.6–1 cm, margin entire or undulate, apex rounded or acute; filaments pink, 3–4 cm long, anthers pink, ~1.5 mm long; style pink, 4–4.5 cm long, stigmas 6–8, greenish, ~3 mm long. **Fruits** red, ellipsoid, 3–5 × 2.5–3 cm; seeds gray or tan, thickened-discoid, ~3 mm Ø. **Distr:** Mexico. **Ecol:** In dry rocky places. Fl. Jul–Feb. **Zones:** 9–12.

O. ficus-indica (L.) Miller, *Gard. Dict.*, ed. 8. Opuntia No. 2, 1768. (Syn: *Cactus ficus-indica*, etc.). [Barbary Fig, Cactus Pear, Indian-fig Opuntia, Mission Cactus, Prickly Pear]. **Shrubs** erect, or **small trees**, 1.5–5 m tall; trunk terete, woody. **Joints** dull greenish, broadly to narrowly obovate, elliptic, or oblong, (20–)25–60 × 7–20 cm, thick. **Areoles** usually narrowly elliptic, 2–4.5 mm long. **Spines** usually absent, sometimes 1–6 per areole, bristlelike or acicular, 0.3–3.2 cm long, basally flattened; glochids yellow. **Leaves** conic, 3–4 mm long, early deciduous. **Flowers** 5–8 cm Ø; sepaloids yellow with reddish or green center, broadly ovate or obovate, to 2 cm long; petaloids spreading, yellow to orange, obovate to oblong-obovate, 2.5–3.5 × 1.5–2 cm, margins entire or erose; filaments yellowish, ~6 mm long, anthers yellow; style greenish, ~1.5 cm, stigmas (6–)7–10, 3–4 mm long, cream. **Fruits** yellow, orange, or purplish, 5–10 × 4–9 cm, umbilicus low and concave; seeds elliptic-orbicular, 4–5 × 3.5–4 mm. **Distr:** Mexico. **Ecol:** In hot dry valleys, rocks; 600–2,900 m. Fl. May–Jun. **Zones:** 9–12.

Subfam. MAIHUENIOIDEAE

Genus 1, **species** 2. *Subfamilial description same as generic.*

Maihuenia

Species 2. **Shrubs**, cespitose, low-growing as mats or cushions to several meters across; stems segmented or not, succulent, globose to cylindrical, olive-green, becoming brownish with age, sometimes bearing spur shoots; photosynthetic parenchyma at base of areolar

MD

RF

Above: *Maihuenia patagonica, Cleistocactus icosagonus, Hylocereus undatus, Rhipsalis baccifera.*

crypts. **Areoles** numerous, small, silky-hairy. **Spines** usually 3, white to brown, tips darker; central spine 1, sometimes absent, stronger than the laterals, straight or slightly curved, sometimes flattened; lateral spines 2, rarely absent, sometimes inconspicuous, pointing outward to lying against the surface, straight to slightly curved, sometimes flexible. **Leaves** persistent, small, conical to linear, round in cross-section. **Inflorescences** with flowers solitary, borne terminally on main branches or spur shoots. **Flowers** with perianth segments spreading, white or yellow. **Fruits** subglobose, oblong or club-shaped, with numerous fleshy bracts; seeds almost circular, shiny black, smooth to slightly tuberculate. **Distr:** Argentina, Chile. **Cult:** Propagate from seeds or cuttings that have been dried for 1–2 weeks. Requires a rich, well-drained soil in full sun.

M. patagonica (Philippi) Britton and Rose, *Cactaceae* 1: 41, 1919. (Syn: *Opuntia patagonica*, etc.). **Shrubs**, to 40 cm high, forming cushions 3 m across; taproots to 40 cm long; stems lax, not segmented, elongated, to 40 cm long, 1–2 cm Ø; spur shoots numerous, knoblike to cylindrical. **Spines** with central spine 1, rigid, 3–7.5 cm long; lateral spines (0–1)2, sometimes inconspicuous, 2–8 mm long, sometimes flattened against the surface. **Leaves** numerous, conical, ovoid to linear, round or elliptical in cross-section, green, 2–6 mm long, 1.5–3 mm Ø, tips acute. **Inflorescences** with flowers solitary, usually borne terminally on spur shoots. **Flowers** white, 2.8–3 cm long. **Fruits** oblong to subglobose to club-shaped, 2–4 cm long. **Distr:** S Argentina, Patagonia. **Ecol:** <2,400 m. **Zones:** 9–11.

Subfam. CACTOIDEAE

Genera 92, **species** 1,250. **Shrubs** or **trees**, columnar, cespitose, scandent, globular or epiphytic; roots usually fibrous, sometimes tuberous; stems usually unsegmented, ribbed or tuberculate, depressed to globular or cylindric; glochids absent. **Leaves** vestigial or lacking. **Flowers** sessile, diurnal or nocturnal; pericarpel usually with scales and/or areoles, or naked, tube short to elongate, with few to numerous scales and/or areoles; floral areoles with wool and spines, bristles, hair-spines or hairs, sometimes naked or rudimentary; tepals usually decurrent into tune; stamens usually numerous, inserted in the tube; ovary inferior. **Fruits** globose to oblong or ovoid, small to large; pericarp ± leathery, fleshy or dry; seeds variable, blackish to brownish. **Distr:** New World, S Canada to S USA southwards; few (*Rhipsalis*) in Africa and Madagascar to some Indian Ocean Islands.

Cleistocactus

Species 40–50. **Shrubs**; stems erect, prostrate or pendulous, usually slender, narrowly cylindric; ribs 5–30, low. **Areoles** with and without flowers, mostly similar; spines variable, numerous and densely arranged to almost absent. **Flowers** narrow, tubular, red, orange, yellow or green; perianth tube straight, bent or S-shaped, and together with the pericarpel, covered with scales that are woolly or hairy in the axils, petals tend to remain tightly clustered with many of the species with flowers almost completely closed; stamens in 2 groups, the lower ones on a rim over the nectar chamber. **Fruits** small, spherical, 1–4 cm Ø, fleshy, sometimes with hairs and usually with persistent remnants of the perianth. **Distr:** C Peru to Bolivia, N Argentina, Paraguay, Uruguay. **Cult:** Propagate from seeds, or cuttings which needs to be dried for a week before planting. Requires a rich, well-drained soil in full sun.

C. icosagonus (Kunth) F.A.C. Weber, *Bull. Mens. Soc. Centr. Agric. Hort. Acclim. Nice Alpes-Marit.* 44: 34, 1904. (Syn: *Binghamia humboldtii*, etc.). **Shrubs**, forming large, low clumps; stems prostrate or ascending, light green, 20–60 cm long, 3–5 cm Ø; ribs 8–21, low, rounded, forming into protuberances separated by grooves. **Areoles** closely set; spines 25–60, needle- or bristlelike, golden-yellow, 1–1.5 cm long; flowering areoles with many, fine, long bristles. **Flowers** pink to scarlet to orange, ± straight, cylindrical, 7–8 cm long, widening toward the tip; perianth parts oblique; pericarpels with small scales bearing a few long or short hairs. **Fruits** globose, 2–4 cm Ø, green to yellow, usually with some hairs. **Distr:** S Ecuador, N Peru. **Ecol:** In tropical alpine areas such as grasslands, woodland patches, scrubs and wetlands. Fl. Mar–Aug. **Zones:** 10–12.

Hylocereus

Species ~16. **Shrubs**, scandent, scrambling, or epiphytic; stems branched, to 5 m long, usually 3-winged or -angled, segmented, margins often horny, often producing numerous aerial roots; areoles spaced far apart in notches along wings or angles with short spines or rarely spineless. **Leaves** absent. **Flowers** usually very large, nocturnal, white or rarely red, funnelform; receptacle tube (pericarpel) usually elongate, stout, with usually broad, leaflike scales; areoles of pericarpel and tube usually naked or nearly so; stamens numerous, inserted in receptacle tube and perianth throat, shorter than perianth, in a continuous series; placentas parietal, stigmas numerous, sometimes bifid or flabellate. **Fruits** large, globose, ellipsoid, or ovoid, fleshy, spineless, with broad scales; seeds numerous, ovate-reniform, testa glossy black, smooth or minutely spotted. **Distr:** Mexico to C America, West Indies, N South America. **Cult:** Propagate from cuttings. Requires a moist, well-drained compost in full or partial sun.

H. undatus (Haworth) Britton and Rose, *Fl. Bermuda*, 256, 1918. (Syn: *Cereus undatus*). [Dragon Fruit]. **Shrubs**, epiphytic, clambering or sprawling; stems deep green, 3–15 m long; joints 20–50 × 3–8(–12) cm, winged or angled, margins of wings or angles coarsely crenate, horny; areoles 3–6 cm apart, 2–5 mm Ø. **Spines** 1–3(–6) per areole, spreading in various directions, gray-brown, conic to subulate, 0.2–0.5(–1) cm. **Flowers** night-blooming, 25–30 × 15–34 cm; receptacle tube funnelform; sepaloids with greenish midrib and mostly white margin, lanceolate-linear to linear, 10–15 × 1–1.5 cm, margins entire, apex acuminate, reflexed; petaloids erect to spreading, white, narrowly oblanceolate, margin entire or erose; filaments cream, 5–7.5 cm long, anthers linear, ~4.5 mm long; style cream, 14–20 cm long, stout; stigmas 20–24, exserted, 2–2.3 mm long, slender. **Fruits** red, globose to ellipsoid, 7–12.5 × 6–12 cm, with numerous triangular scales, umbilicus small; pulp white; seeds obovate, ~2 × 1 mm. **Distr:** Mexico to C America. **Ecol:** On trees, lithophytic on rocks, and in maritime scrubs; 0–500 m. Fl. Jul–Dec. **Zones:** 10–12.

Rhipsalis

Species 50. **Small shrubs**, mostly epiphytic or saxicolous; stems cylindrical, ribbed, angled, winged or flat, succulent, segmented, segments with small areoles; areoles prominent in younger plants or branches. **Leaves** absent or represented by small scales. **Flowers** solitary, small, sessile, lateral, sometimes terminal; perianth lobes 5–12; stamens rather few, often separated from style by nectariferous collar; ovary smooth, stigmas 3 or more. **Fruits** berries, globose, smooth, white or colored, viscidly juicy; seeds oblong, smooth or minutely wrinkled. **Distr:** Throughout tropical America, especially Brazil; 1 species (*R. baccifera*) extending to tropical and subtropical Africa, Madagascar and Sri Lanka. **Cult:** Propagate from seeds or by cuttings. Requires a moist, well-drained compost in shade or partial sun; best grown in hanging baskets.

R. baccifera (J.S. Mueller) Stearn, *Cact. J. (Croydon)* 7: 107, 1939. (Syn: *Cassytha baccifera*, etc.). [Mistletoe Cactus]. **Shrubs**, epiphytic or epilithic, perennial, small, pendent, ~0.5–3(–6)+ m long, slightly woody at base; stem succulent, green, cylindrical, slender, dichotomously branched, with a fibrous root system. **Leaves** very minute, often absent in mature plants. **Inflorescences** with flowers solitary, sessile. **Flowers** small, white, buds greenish-white; perianth lobes 9–12, outer unequal, acute, inner oblong, ~2–3 mm long, obtuse, spreading; stamens many, shorter than to subequaling perianth lobes, inserted at base of lobes; ovary glabrous, style thick, longer than stamens, stigma 3–5-branched, ovules 3–4. **Fruits** small, globose, 0.5–1 cm Ø, white, translucent; seeds oblong to pyriform, ~1 mm long, black, obscurely striate, in juicy pulp. **Distr:** Tropical America; tropical Africa (Côte d'Ivoire to S Africa); Madagascar, Comoros, Mascarene Islands, Sri Lanka. **Ecol:** In moist habitats. **Zones:** 10–12.

13 BASAL ASTERIDS

ERICALES

FAMILIES 25, GENERA 346, SPECIES 11,515.

Trees, **shrubs**, **lianas**, **herbs**, chlorophyll-deficient **mycoheterotrophs** and **carnivorous plants**. **Leaves** spirally arranged, vascular bundles collateral; lamina simple. **Flowers** with twice or more as many stamens as the number of petals (Judd et al. 2002; Stevens 2001 onwards). **Info:** Morphologically, Ericales is very diverse and finding common characters is difficult. The monophyly of Ericales is strongly supported by molecular data (Bremer et al. 2002; Hilu et al. 2003; Soltis et al. 2000).

BALSAMINACEAE

Genera 2, **species** 1,001. **Herbs**, ± succulent, annual or perennial, rarely epiphytic or shrubs; stems usually translucent, erect or procumbent, often rooting at lower nodes. **Leaves** alternate, opposite or verticillate, not stipulate or with stipular glands at base of petiole, hydathodes are commonly present, hairs usually absent; lamina simple, margins entire, crenate or serrate. **Inflorescences** cymes or umbels, mostly flowers solitary. **Flowers** bisexual, 5-merous; corolla with 5 petals often appearing to be 3 by fusion of 2 lateral petals; anterior organs appear posterior and a large spurred posterior sepal appear anterior in mature flowers; stamens 5, alternating with petals. **Fruits** of *Hydrocera* are fleshy, indehiscent berrylike drupes (pseudoberries), while those of *Impatiens* are non-fleshy, elastically dehiscent, loculicidal capsules (explosive). **Distr:** Mostly Old World, Africa (especially diverse in Madagascar) to montane regions of SE Asia. **Info:** *Impatiens* is widely distributed across the range of the family, while *Hydrocera* is restricted to the lowlands of Indo-Malesia (Yuan *et al.*, 2004). The plants have developed very elaborate pollination mechanisms with irregular, zygomorphic and resupinate flowers.

Impatiens

Species >1,000. **Herbs**, terrestrial, occasionally epiphytic, annuals or perennials, or suffrutescent perennials; stems succulent, rarely woody below, upright or procumbent, often rooting at lower nodes, occasionally acaulous. **Leaves** spirally arranged, opposite or in verticils. **Inflorescences** racemes or fascicles, or flowers solitary, axillary. **Flowers** with lateral sepals 2, lower sepal with a short or long, broad or filiform spur; petals 5, the 4 lateral ones united into lateral pairs. **Fruits** capsules, elastically dehiscent; seeds usually many. **Distr:** Throughout tropical and subtropical montane regions of eastern hemisphere, including temperate Asia, Europe and N America. **Cult:** Propagate from seeds or rooted cuttings. Requires a moist, well-drained soil in shade or partial sun.

I. balsamina L., *Sp. Pl.* 2: 938, 1753. (Syn: *B. hortensis*). [Balsalm, Garden Balsalm, Rose Balsalm, Spotted Snapweed, Touch-me-not]. **Herbs**, annuals, erect to 1 m high, erect, robust, succulent, simple or sparsely branched, glabrous when mature, lower nodes swollen. **Leaves** alternate; lamina lanceolate-elliptical, 4–15 × 1.5–3 cm, base cuneate, margins deeply serrate, apex acuminate, surfaces glabrous or sparsely pubescent; petiole usually 3 cm long, with a pair of sessile black glands toward base. **Inflorescences** with flowers solitary or 2–3-flowered fascicles, axillary; peduncles absent; pedicels 2–2.5 cm long, densely pubescent. **Flowers** white, pink to purple or sometimes red, sometimes with double petals; lateral sepals 2, ovate or ovate-lanceolate, lower sepal deeply navicular, pubescent, abruptly narrowed into an incurved spur 1–2.5 cm long, slender; upper petal orbicular, apex retuse; lateral united petals shortly clawed to 2.5 cm long, bilobed; basal lobes obovate-oblong, distal lobes suborbicular, apex retuse; ovary fusiform, densely pubescent. **Fruits** capsules, broadly fusiform, 1–2 cm long, densely tomentose, narrowed at both ends; seeds many, black-brown, globose, tuberculate. **Distr:** SE Asia. **Ecol:** Fl. Jul–Oct. **Zones:** 10–12.

I. grandis Heyne ex Wall. in Roxb., *Fl. Ind.* 2: 465. 1824. (Syn: *I. hookerana, I. biglandulosa*). **Herbs** perennial, to 2 m tall. **Leaves** spirally arranged; lamina broadly elliptical to ovate, 8–20 × 3–8 cm, base with a pair of dark glands, margins shallowly crenate, apex acuminate, adaxially bright green, abaxially pale whitish-green, lateral veins 8–10 pairs; petiole to 11 cm long. **Inflorescences** subumbellate racemes, 3–6-flowered, in upper leaf axils; peduncle erect, to 15 cm long; pedicels to 6 cm long. **Flowers** flat-faced, pure white, some or all petals dotted and lined or flushed with purplish-red or magenta; lateral sepals ovate, lower sepal funnel-shaped, narrowing into a thick filiform spur to 7.5 cm long; dorsal petal orbicular, erect, to 2.5 × 3.8 cm, margin undulate, with a stout crest behind; lateral united petals 2–3.8 cm long, margins undulate; lower petal longer and wider, unevenly and shallowly bilobed; ovary glabrous. **Fruits** capsules, ellipsoid, to 1.8 × 0.7 cm. **Distr:** Sri Lanka, S India (W Ghats). **Ecol:** In lower montane rainforests; 350–1,550 m. Fl. Mar. **Zones:** 10–12.

I. leptopoda Arnott, *Companion Bot. Mag.* 1: 321, 1836. (Syn: *I. brevicornu, I. gibbosa*). **Herbs**, perennial, glabrous, 6 cm high to 0.7(–1) m tall; stems much-branched, rarely simple, green or reddish. **Leaves** alternate; lamina lanceolate to elliptical or ovate-lanceolate, 2–10 × 0.7–3.2(–4) cm, base attenuate, with several extra-floral nectary glands ~1 mm long, margins crenulate or almost serrulate, apex acuminate or slightly cuspidate, adaxially dark green, abaxially paler; petiole 0.4–3.4 cm long. **Inflorescences** 2–3-flowered fascicles or flowers solitary, axillary; pedicels 1.8–4.3 cm long. **Flowers** pink, mauve-pink or whitish with a dark magenta spot toward base of each pair of lateral united petals; lateral sepals linear-lanceolate, pointed, lower sepal shallowly boat-shaped, constricting into a curved, filiform spur, 2.5 cm long; dorsal petal obcordate, 1 cm across, lateral petals 0.7–1 cm long; upper petals almost covering the much smaller lower petals; ovary glabrous. **Fruits** capsules, fusiform, ~1.3 cm long, glabrous. **Distr:** Sri Lanka. **Ecol:** In montane rainforests; 1,200–3,150 m. Fl. Apr–Jul. **Zones:** 10–12. **Info:** Flowers variable in size and color, spur length varies from to almost absent, all forms frequently growing together (Dassanayake 1985).

I. niamniamensis Gilg., *Bot. Jahrb. Syst.* 43: 104, 1909. [Congo Cockatoo, Parrot Plant]. **Herbs**, erect, evergreen, perennial, to 60 cm high; stems thick, succulent, densely foliated, becoming woody with age. **Leaves** spirally arranged; lamina broadly ovate-oblong or elliptical, 6–20 × 4–8 cm, margins crenate, crenations overlapping; petiole 1–7 cm long, often with fimbriae near base. **Inflorescences** of 2–6-flowered fascicles, sometimes solitary, often flowering directly from bare stem. **Flowers** deeply pouched; spur quite variable in shape and color, from red and yellow to red, magenta, and pink; lateral sepals minute, 2 mm long, reflexed, lower sepal 2–3 cm long, deeply saccate, constricted into a sharply incurving, often bilobed spur; dorsal petal small, 1 × 0.5 cm, hooded; lateral united petals 1–1.5 cm long, upper pair larger than the lower. **Fruits** capsules, fusiform, 1.5 cm long. **Distr:** Tanzania, C Africa to Cameroon, Fernando Po Island. **Ecol:** 0–1,342 m. Fl. Jan.–Dec. **Zones:** 10–12.

I. repens Moon, *Cat.* 18, 1824. [Ceylon Balsalm, Ceylon Jewelweed, Golden Dragon, Rock Balsalm, Yellow Impatiens]. **Herbs** perennial, trailing, procumbent or creeping with thick and fleshy stems to 1 m long, pinkish-red, rooting at some of the nodes, somewhat angled. **Leaves** alternate; lamina broadly reniform or heart-shaped, occasionally suborbicular, 0.7–1.2 × 1–1.7 cm, base subcordate with several filiform appendages on the shallowly crenate margins, apex obtuse, often apiculate, green red-tinged, adaxially sparsely hairy, abaxially glabrous, lateral veins 2–3 pairs; slender petiole to 2 cm long. **Inflorescences** with flowers solitary, axillary. **Flowers** bright yellow; lower sepal with reddish-brown netted veins inside, saccate, narrowing to a short thin spur, tip swollen; dorsal petal large, hooded; lateral petals united, to 2.5 cm long; upper pair ovate, lower pair longer and wider; ovary densely pubescent. **Fruits** capsules, fusiform, 1.1 × 0.25 cm, pilose. **Distr:** Sri Lanka. **Ecol:** Among moist rocks in submontane rainforests; <950 m. Fl. Jun–Sep. **Zones:** 10–12. **Info:**

Top row: *Impatiens balsamina, Impatiens grandis, Impatiens niamniamensis.*
Bottom row: *Impatiens leptopoda, Impatiens repens, Impatiens scabrida, Impatiens sodenii.*

This plant is very rare in the wild, and has not been collected during the past 70 years and may well be extinct in the wild.

I. scabrida DC, *Prodr.* 1: 687, 1824. (Syn: *I. praetermissa*). [Himalayan Jewelweed]. **Herbs**, annual, habit variable, 0.3–1.2 m tall; stem erect, thick, green or purple-tinged basally, much-branched, pubescent or subglabrous basally. **Leaves** alternate, sessile or subsessile; lamina ovate, 5–15 × 1.5–2 cm, base subrounded, apices of teeth with marginal glands, apex acuminate, margins sharply serrate, adaxially sparsely strigose, abaxially pubescent; petiole with 2 globose basal glands. **Inflorescences** axillary, 1–5-flowered, shortly pedunculate. **Flowers** variable, golden-yellow, yellow or pale yellow, purple-spotted, ~3.5 cm deep; lateral sepals 2, ovate, pale greenish-brown; lower sepal large, 3–3.5 cm long, abruptly narrowed into a curved spur ~1 cm long, saccate, pale yellow, with reddish-brown markings in throat; dorsal petal large, orbicular, 1.5–2 cm Ø, with keeled dorsal crest; lateral united petals 2.2 cm long; upper petal broadly obovate, 1 cm Ø, lower narrowly oblong, 1–1.2 cm long. **Fruits** capsules, linear, 3.5–5 cm long, glabrous or sparsely hairy. **Distr:** Himalayas, Bhutan, Nepal, N India, and Pakistan. **Ecol:** In forest understory, thickets. Fl. Jul–Sep. **Zones:** 9–11.

I. sodenii Engl. and Warb., *Abh. Königl. Akad. Wiss. Berlin* 46: 55, 1894. (Syn: *I. oliveri, I. thomsonii*). [Shrub Balsalm, Oliver's Touch-me-not, Poor Man's Rhododendron]. **Herbs**, somewhat shrubby, perennial, robust, erect, to 3 m high; stems thick, succulent, moderately branched, glabrous, often becoming slightly woody near the base. **Leaves** arranged in dense whorls of 6–10, becoming congested toward the apex of the stem; lamina oblanceolate, rarely oblong, 15–18 × 2–5 cm, margins shallowly serrate or serrulate with a few fimbriae near base. **Inflorescences** 1- or 2-flowered; peduncle 2.5–6 cm long; pedicel 2–4.5 cm long. **Flowers** lilac or white with central magenta markings, or sometimes pure white; lateral sepals ovate to elliptic-lanceolate, 6–8 mm long; lower sepal narrowly navicular, 1.4–1.8 cm long, abruptly constricted into a filiform 6–10 cm long spur; dorsal petal broadly ovate to suborbicular, 1.8–2.3 × 2.5–3 cm, emarginate, with a narrow dorsal crest; lateral united petals 3–3.7 cm long; upper petal oblong or suborbicular, 2.1–2.8 × 1.7–2.7 cm, emarginate; lower petal oblong, 2–3.3 × 1.4–2.1 cm, slightly emarginate. **Fruits** capsules, fusiform, 2.5 cm long. **Distr:** Kenya, N and E Tanzania. **Ecol:** In exposed rocky areas. **Zones:** 10–12.

I. tinctoria A. Richard, *Tent. Fl. Abyss.* 1: 120, 1847. [Perennial Impatiens]. **Herbs**, perennial, tuberous, to ~2+ m high; stems succulent, hollow; rootstock tuberous, to 30 × 10 cm. **Leaves** spirally arranged, petiolate; lamina ovate to oblong-lanceolate, ~5–32 × 2.5–10 cm, base cuneate, margins serrate, apex usually acuminate; petiole to 9 cm long, with glands near apex. **Inflorescences** racemes, 3–9-flowered; peduncle to 30 cm long; pedicel to 4 cm long. **Flowers** 5+ cm across, fragrant, white with central pinkish-magenta markings; lateral sepals 4, ovate or oblong, 0.6–1 × 0.4–0.7 cm; lower sepal funnel-shaped, gradually constricted into a filiform spur 3–13 cm long; dorsal petal hoodlike, 1.7–2.2 × 1–1.4 cm, dorsally crested; lateral petals 4, united in pairs, upper pair small, 0.6–1.5 × 0.2–0.5 cm, lower one larger, 1.6–5 × 1.6–4.6 cm; stamens 5, connate in a ring; ovary superior, 5-loculed, style short. **Fruits** clavate or club-shaped, to 4 cm long. **Distr:** Tropical Africa. **Ecol:** In damp and shady places of upland rainforests, often near streams; 300–3,600 m. **Zones:** 10–12.

I. walleriana Hook. f., *Fl. Trop. Afr.* 1: 302, 1868. (Syn: *I. holstii, I. sultani*). **Herbs**, perennial, succulent, to 30–50(–80) cm high. **Leaves** spirally arranged, variable; lamina broadly elliptic to ovate, (3–)4.5–10(–15) × (2–)2.5–5(–7.5) cm, base cuneate, margins crenate-denticulate, apex acute, acuminate or cuspidate, subsucculent, translucent, glabrous, surfaces green to reddish-green; petiole to 4.5(–6.5) cm long. **Inflorescences** racemes, few-flowered, axillary and/or terminal, rarely flowers solitary; bracts lanceolate-triangular, to 6 mm long; pedicels 1–2.5(–3) cm long. **Flowers** variable in color and size, dark carmine, red, pink, orange, violet, rarely white; lateral sepals ovate-lanceolate or lanceolate-triangular, (3–)4–7 × 2.25–3 mm, acute; posterior sepal to 1.5 × ~0·5 cm, shallowly navicular, blunt-keeled, abruptly constricted into the spur; spur to 4 cm long, slender, slightly curved; anterior petal very broadly ovate or obcordate, to 1.2 ×1.8 cm, retuse to emarginate; lateral united petals deeply bilobed, lobes obovate, 1–2 × 0.7–1.5 cm, rounded or subtruncate; ovary glabrous. **Fruits** capsules, to 1.8 cm long, broadly oblique-fusiform, glabrous; seeds numerous, ~1.75 × 1.5 mm, pyriform. **Distr:** Tropical Africa. **Ecol:** In rainforests; 100–1,200 m. **Zones:** 10–12.

SZ

Above: *Impatiens tinctoria, Impatiens walleriana, Cantua bicolor, Cantua buxifolia.*

POLEMONIACEAE

Subfamilies 3, **genera** ~18, **species** 385. **Herbs** annuals, perennials and sometimes subshrubs, shrubs or lianas; plants have an unpleasant smell. **Leaves** alternate or opposite, sessile to petiolate; lamina simple, pinnatifid to palmately lobed, or compound. **Inflorescences** determinate, corymbose, paniculate, or capitate, or flowers solitary. **Flowers** (4- or) 5-merous; calyx is ± connate, cylindric to campanulate, consisting of a green central portion and the rest being whitish; corolla strongly sympetalous, actinomorphic or zygomorphic, rotate to funnelform or salverform, lobes overlapping in bud; stamens 5, exserted to inserted, frequently inserted at same or different levels on corolla at base, alternate with corolla lobes; ovary superior, trilocular, 1–many-ovuled, style 1, usually trilobed, stigma (2 or)trilobed. **Fruits** capsules; seeds ovoid or fusiform, sometimes winged, often sticky and mucilaginous when wet. **Distr:** N temperate, W North America, S America.

Cantua

Species ~11. **Shrubs** or **small trees**, evergreen or semievergreen; branches pubescent. **Leaves** alternate; lamina with margins toothed, subcoriaceous; petiole short, pubescent. **Inflorescences** corymbs, 1–many-flowered, terminal; pendent or semipendent; pedicels thin and weak. **Flowers** perfect, trumpet-shaped, zygomorphic; calyx purplish to greenish, lobes 5, 3-veined, pubescent; corolla tubular, 2–4 times longer than the calyx, white, yellow, greenish-yellow or purple, lobes 5, flared; stamens 5, slightly exserted, partially adnate to corolla; ovary superior, carpels 3, many-ovuled. **Fruits** capsules, coriaceous, valves 3; seeds winged. **Distr:** Andes of S America (Ecuador, Bolivia and Peru), with the highest diversity in Peru. **Cult:** Propagate from tip cuttings or fresh seeds. Requires a moist, humus-rich, well-drained soil in full sun.

C. bicolor Lem., *Fl. Serres Jard. Eur.* 3: 242(Verso), 1847. [Cantua]. **Shrubs**, erect, 1.3–2.5 m high; branches and young foliage pubescent. **Leaves** shortly petiolate; lamina small, obovate, ~2.5 cm long, base tapered or cuneate, margins entire, apex apiculate, dark green, hairy, penninerved. **Inflorescences** corymbs of solitary flowers, terminal or subterminal, usually massed at branch tips, flowers curved downward, subpendent or drooping. **Flowers** tubular, large, to 5 cm long; calyx monophyllous, campanulate to cylindrical, membranous, veined, slightly reticulate, 5-dentate, teeth triangular, erect; corolla infundibuliform, tube yellow, 6 cm long, pinkish red at the slightly constricted neck, lobes 5, flared, cream, obcordate; stamens exserted, filaments yellow, unequal, anthers nearly black; disk large, glandular; ovary ovoid, triloculed, glabrous, style longer than the stamens, much-exserted, stigma with 3 linear lobes. **Fruits** capsules; seeds many. **Distr:** Bolivia. **Ecol:** In rocky places and scrubs; to 4,000 m. Fl. Aug–Oct. **Zones:** 10–12.

C. buxifolia Juss. ex Lam., *Encycl. Bot.* 1: 603, 1785. (Syn: *C. ovata*, etc.). [Magic Flower, Magic Tree, Sacred Flower of the Incas]. **Shrubs** or **small trees**, erect, branching, to 5 m tall, ± pubescent. **Leaves** sessile to shortly petiolate; lamina usually elliptic, sometimes obovate, oblanceolate, or linear-oblanceolate, 0.6–4 cm long, base tapered, margins usually entire, rarely irregularly dentate toward apex, apex acute, fleshy. **Inflorescences** usually lax terminal corymbs, rarely reduced to solitary flowers, few- to many-flowered; pedicels varying in length. **Flowers** with calyx tubular or tubular-campanulate, 1.6–3.2 cm long, lobes unequal, triangular, acute, rarely acuminate, nearly glabrous to heavily pubescent; corolla funnelform, usually rose to purple, sometimes white, yellow, or tube yellow with lobes white or rose, 5.5–8 cm long, lobes bilobate, retuse-apiculate or rounded and irregularly dentate; stamens exserted or included, adnate to tube at a point ~1–1.5 cm from base; ovary pyriform to oblong, to 1.2 cm long, style usually exserted, to 9 cm long. **Fruits** capsules, oblong, to 2.5 cm long; seeds usually 20–30, winged, 0.8–1.2 cm long. **Distr:** Peru, Bolivia and N Chile. **Ecol:** On mountains. **Zones:** 9–12.

Cobaea

Species 20. **Lianas**, usually somewhat pubescent, tendriled, fast growing. **Leaves** alternate, pinnately compound; leaflets entire, lateral leaflets broad, shortly petiolate, terminal leaflet usually modified into a tendril. **Inflorescences** with flowers solitary, axillary, borne along the stems; peduncle long. **Flowers** with calyx persistent, disklike, prominent, herbaceous; sepals 5, divided almost to base, regular; corolla green, greenish-white, yellow, pink, violet or purple (rarely striped), large, campanulate, lobes orbicular to ovate, sometimes caudate; stamens inserted near base of corolla tube, filaments slender, anthers versatile, sometimes appearing sagittate; style included or exserted, stigma trilobed or branched; disk 5-lobed; ovary superior, glabrous, elliptic-ovate, tricarpellate, trilocular. **Fruits** capsules, exceeding calyx, elliptic or elliptic-ovoid, septicidally dehiscent, each locule containing 2–several, large, flat, winged seeds. **Distr:** Mexico to Venezuela and in the Andes from Colombia to N Chile. **Ecol:** In humid, montane forests. **Cult:** Propagate from seeds or cuttings. Requires a rich, moist, well-drained soil in partial sun. **Info:** The flowers are pollinated by bats and nocturnal insects.

C. scandens Cav., *Icones et Descriptiones Plantarum* 1: 11, t. 16, 17, 1791. (Syn: *Rosenbergia scandens*). [Cathedral Bells, Cup and Saucer Vine]. **Lianas**, vigorous, to 4.5 m long; stems sparsely pubescent. **Leaves** sessile, pinnately compound, upper 2 pairs short-petioluled, lower pair usually auriculate, sessile or subsessile; leaflets elliptic to obovate, 7–8(–10) x 3–3.5(–5) cm, base oblique, margins entire, apex acuminate, usually glabrous; petiole pubescent. **Inflorescences** with flowers solitary; peduncles 15–22 cm long. **Flowers** with calyx broadly campanulate, glaucous outside, minutely pubescent inside, lobes 2.5–3.5 x 2–3 cm, margins undulate, apex mucronate; corolla campanulate, green, greenish-white, pink, rose, purple, or variously striped, 5–6 cm long, sometimes sparsely pubescent outside of tube beneath the lobes, lobes orbicular to obovate; stamens exserted, filaments woolly above point; disk with lobes deeply bilobate; ovary elliptic-ovoid, glabrous, style slightly exceeding or about equaling corolla tube. **Fruits** capsules, elliptic-ovoid, 5–7 cm long, glabrous; seeds light brown, ~1.6 x 1 cm, broadly winged. **Distr:** Mexico to Western S America, Venezuela to N Chile. **Ecol:** Fl. Jan.–Dec. **Zones:** 9–11. **Info:** The flowers emit a rather unpleasant odor in the evenings that diminishes with age.

LECYTHIDACEAE

Subfamilies 5, **Genera** 20, **species** ~450. **Shrubs** or **trees**. **Leaves** alternate, often large, simple, shortly petioled, arranged toward the end of the (usually bare) branchlets. **Inflorescences** racemes, spikes, sometimes on specialized serpentine branches from the trunk (cauliflorous) or flowers borne solitary. **Flowers** often large, showy, bisexual, actinomorphic or zygomorphic; calyx with 4–6 lobes, wide campanulate tube adnate to the ovary; corolla with 4–6 lobes, free, usually thick, rarely absent; stamens numerous (some of which are often sterile) are united at the base arranged in several whorls, either monadelphous and equally arranged around the disk (which is sometimes lobed) or diadelphous in 2 unequal bundles, with the outermost being staminodial, anthers basifixed, bilocular with narrow longitudinal openings; ovary inferior or semi-inferior, 2–6 locules with 1 or more ovules per locule. **Fruits** berries, indehiscent, usually very large, or operculate capsules, with persistent calyx lobes; seeds 1 to many, endosperm absent. **Distr:** Tropical, especially America and W Africa, with the greatest diversity found in the Amazon rainforests. **Info:** Many species are grown for ornamental purposes while some are grown commercially for the edible nuts (seeds). Most species of this family possess numerous flower and fruit features that reflect adaptations for various pollination and dispersal methods. Among the 5 subfamilies, Napoleonaeoideae and Fetidioideae are restricted to tropical Africa. Scytopetalodideae is predominantly from tropical W Africa with some representation in S America. Lecythidoideae is restricted to the neotropics.

Subfam. NAPOLEONAEOIDEAE

Genera 2, **species** 11. **Secondary** xylem with crystal

Above: *Cobaea scandens, Napoleonaea heudelotii, Napoleonaea imperialis, Napoleonaea vogelli.*

chains. **Leaves** distichous; lamina often with glands abaxially at the base of the lamina, margins serrate. **Flowers** with staminodial corona present; anthers 10, long, extrorse, monothecal; gynoecium opposite petals, style absent, stigma broad, pentagonal. **Fruits** drupes, 1- to many-seeded; seeds with embryo curved (Stevens 2001 onwards). **Distr:** Western tropical Africa.

Napoleonaea

Species ~6. **Trees** or **shrubs**, rarely lianas (*N. lutea* and *N. reptans*). **Stipules** absent. **Leaves** alternate; lamina elliptic-oblong to ovate or obovate, 5–32 × 1.5–11 cm; petiole 3–9 mm long. **Inflorescences** with flowers solitary, axillary or cauliflorous, panicles; flowers subsessile to sessile. **Flowers** with calyx tube turbinate, limb valvate, 5-fid; petals, staminodia, and stamens in 4 concentric rows, and ± adnate at base, those of the exterior row, or petals, connate, forming an orbicular petaloid membrane, with 20–40 folds, those of the second row numerous, free or nearly so, narrowly ligulate and antherless, those of the third row connate more than the lower half and forming a cup 20–40-lobed (pseudocorolla), lobes inflexed, antherless, those of the inner row connate at the base both with each other and the neighboring row, free portions broadly ligulate, mostly bearing anthers, anthers oblong, adnate, 1-locular; ovary 5-locular, ovules 2–6 per locule, style short, expanded into a very broad peltate disk. **Fruits** drupes, fleshy, globose; seeds few. **Distr:** W tropical Africa. **Cult:** Propagate from seeds, cuttings or by air-layering. Requires a moist, well-drained soil in partial or full sun.

N. heudelotii A. Juss., *Ann. Sc. Nat.*, ser. 3, Bot., 2: 227, tab. 4, 1844. (Syn: *N. whitfieldii, N. imperialis*). [Chimpanzee's Chair, Napoleon's Button]. **Trees** small, evergreen; young shoots with uppermost node swollen with whorled branches. **Leaves** elliptic to obovate, 10–17 × 3.5–8 cm, base rounded, apex obtuse to acuminate, secondary veins 7–9 pairs; petiole 3–5(–8) mm long. **Inflorescences** fascicles or flowers solitary, axillary or cauliflorous. **Flowers** large, pale reddish, red and yellow, double-cupped; calyx 2–2.2 cm Ø, glabrous; corolla ~4 cm Ø, purple; stamens many. **Fruits** ellipsoid, 2.2 cm long and 2.8 cm Ø, ripening through dark orange to reddish-brown, skin warty. **Distr:** Western Africa (Guinea, Sierra Leone). **Ecol:** In forests, often in forest regeneration. **Zones:** 1–12. **Info:** The mucilage of the fruit is edible, leaves (regarded as poisonous) and other plant parts are used for medicinal purposes.

N. imperialis P. Beauv., *Fl. Oware Benin*, 2: 30, tab. 78, 1810. (Syn: *Belvisia caerulea*, etc.). [Napoleon's Hat]. **Trees**, to 7 m tall, low-branching; crown dense. **Leaves** elliptic to obovate, 13–24 × 4.5–9 cm, base acuminate, apex rounded; petiole 3–5 mm long. **Inflorescences** fascicles or solitary flowers, in axils of older leaves or cauliflorous on old wood; flowers subsessile. **Flowers** variable. usually cream at the circumference with red at the center, but varying to apricot; calyx 1.6–2 cm Ø, glabrous; corolla 3.5–4 cm Ø. **Fruits** 3–3.5 cm long, 4–4.5 cm Ø. **Distr:** Congo to Angola. **Ecol:** Mostly in secondary forests and understory of closed forests. **Zones:** 11–12. **Info:** This species is often cultivated for ornamental purposes. The fruit pulp is edible and other parts of the plant are used for medicinal purposes.

N. vogelii Hook. and Planch., Hook., *Ic. Pl.*, 8: 799–800, 1848. (Syn: *N. angolensis*, etc.). **Trees** small, or shrubs. **Leaves** elliptic-oblong to ovate or obovate, 7.5–18 × 2.5–7 cm, base cuneate or rounded, apex acuminate, rarely acute; petiole ~3–5 mm long. **Inflorescences** fascicles, or flowers solitary, borne axillary or cauliflorous, sessile, usually hanging below leaves. **Flowers** with calyx 1.2–2.2 cm Ø, greenish-yellow, glabrous, lobes 5, ovate; corolla, 4–6 cm Ø, vaguely 5-lobed, margin ragged, crinkled, pinkish-cream, with ~40 maroon-crimson pleated ribs, corolla usually bent over calyx; outer staminodes threadlike, to 5 mm long, cream-colored, strongly reflexed; inner staminodes forming a shallow cup ~3 cm Ø, 1.5 cm high, margin with ~40 crimson teeth; cup pinkish-cream below, base crimson; stamens forming cream-colored cup ~2 cm Ø, 1 cm high, with 20 flattened filaments strongly curved inward, filaments arranged in pairs, one pair with an anther, alternating with a pair without anthers, anthers curved, hidden by stigma; disk ~1 cm Ø, forming a shallow fleshy cup; stigma 5-lobed, starlike, ~1 cm Ø, style short, fat. **Fruits** 2.2-5 × 3.5–7 cm; seeds 2.5×1.5 cm. **Distr:** Tropical W Africa. **Ecol:** In riverine forests and near seashores. **Zones:** 11–12.

Subfam. PLANCHONIOIDEAE

Genera 6, **species** 58. **Secondary xylem** without crystal chains. **Leaves** spiral spirally arranged. **Flowers** with androecium with a basal ring; disk annular; style long. **Fruits** usually 1-seeded; embryo hypocotylar or with long radicle and leaflike cotyledons (Stevens 2001 onwards). **Distr:** Restricted to Old World Tropics.

Barringtonia

Species ~40–56. **Shrubs** or **trees**, seldom with buttresses. **Leaves** spirally arranged, at ends of branchlets; lamina simple, all of the same type, base always cuneate, margins entire to serrate-crenulate, apex usually acute or acuminate, veins prominent abaxially. **Inflorescences** racemes, terminal or lateral on twigs, or cauliflorous, generally pendulous and long, very rarely erect and short; peduncle often with a tuft of cataphylls at base. **Flowers** with globular buds; calyx of 4 or 5 persistent, imbricate lobes inserted on rim of receptacle, or breaking at anthesis into 2–5 persistent pseudo-lobes, or circumscissile leaving a cup-shaped rim; petals (3–5) free; sepals free, adnate to staminal tube, outer ones smaller than inner; stamens numerous, basally connate in 3–8 whorls; style long, terete, filiform, folded in bud, persistent. **Fruits** obovoid to ellipsoid to fusiform, terete or angled to winged, fleshy, fibrous; seed large, testa brownish (Payens 1967). **Distr:** Tropical and subtropical regions of E Africa, Madagascar, Maldives and other islands of the Indian Ocean, South Asia, China, Malesia, Australia and the Pacific Islands. **Cult:** Propagate from seeds or large hardwood cuttings. Requires a very moist soil in full sun. Most species are salt and flood tolerant.

B. acutangula (L.) Gaertn., *Fruct.* 2: 97, Pl. 101. 1791. (Syn: *Butonica terrestris*, etc.). [Cut Nut, Freshwater Mangrove, Indian Oak]. **Shrubs** or **small trees**, to 25 m tall, most parts usually glabrous; trunk 20–90 cm dbh. **Leaves** elliptical or obovate-oblong; 5–25 × 2–8 cm, base cuneate, margins serrate or serrate-crenulate, apex acute, acuminate, cuspidate, rounded, truncate or retuse, tip emarginate, chartaceous; petiole 0.1–1.5 cm long. **Inflorescences** spikes or racemes, pendulous, densely flowered (up to 90 flowers), 20–80 cm long, terminal (with several terminal inflorescences or in upper axils). **Flowers** pinkish or reddish, scented, 2–3 cm Ø; hypanthium funnel-shaped, 1–5 mm long; sepals 4–5, semiorbicular or oblong, 1–4 × 1–3 mm; petals 4–5, suborbicular, oblong or spathulate, 0.5–1 × 0.2–0.6 cm, concave; petal and sepal rim hyaline minutely dentate, adhering to staminal tube; stamens 3-whorled, filaments 1–2 cm long; ovary biloculed, locules 2–5-ovuled, style 2–4 cm long. **Fruits** globose, ovoid, obovate or cylindrical, 1–7 × 1–5 cm, with 4–8 ribs or wings; winged when young. **Distr:** Widespread from Afghanistan to N Australia. **Ecol:** Mostly along rivers, swamps and fresh water mangroves, often submerged for several weeks during the wet monsoon season. **Zones:** 10–12.

B. asiatica (L.) Kurz, *Prelim. Rep. Forest and other Vegetation of Pegu*, App. A: 65, 1875. (Syn: *Barringtonia speciosa*, etc.). [Fish Poison Tree, Freshwater Mangrove, Sea Putat]. **Trees**, to 7–30 m tall, trunk fissured; twigs with large leaf scars. **Leaves** crowded at ends of branchlets; lamina large, oblong-obovate, 15–40 × 10–20 cm, base cuneate, tapering to petiole, margins entire, apex obtuse or broadly rounded, leathery, surfaces glossy green, nerves 6–10 pairs, marginal often pale or reddish; petiole to 1 cm long. **Inflorescences** racemes, to 5–15 cm long, terminal (rarely several terminal inflorescences) or occasionally lateral, 5–10(–20)-flowered; cataphylls oblong or ovate, 1.5–3 × 0.7–1 cm; bracts ovate; bracteoles triangular, 1.5–5 mm long; pedicels 5–10 cm long. **Flowers** nocturnal, 10–15 cm Ø; calyx 2–4 cm Ø, tube rupturing into 2 or 3 unequal, segments or lobes; petals 4, white, ovate or elliptic, convex, 5–6 cm long; staminal tube with 6 whorls, inner one staminodial, stamens red-tipped, filaments to

Above: *Barringtonia acutangula, Barringtonia asiatica, Barringtonia racemosa, Planchonia careya.*

10 cm long; ovary 4-loculed, tetragonous, locules 4- or 5-ovuled, style 11–13 cm long. **Fruits** large, 10–15 × 8–10 cm, quadrangular (pyramidal), smooth, bluntly pointed, ovoid, crowned by persistent calyx, pericarp very thick, fibrous and spongy; seed oblong, >5 cm long. **Distr:** Mainly Old World Tropics. **Ecol:** Fl. throughout the year.

B. racemosa (L.) Sprengel, *Syst. Veg.* 3: 127, 1826. (Syn: *B. ceylanica*, etc.). [Freshwater Mangrove, Powderpuff Tree, Small-leaved Barringtonia]. **Shrubs** or **trees**, to 27 m tall. **Leaves** obovate-oblong, 20–35 × 6–14 cm, base cuneate, margins serrate-crenulate, apex acute or acuminate; petiole 0.2–1.5 cm long, winged. **Inflorescences** racemes usually terminal or in axils of fallen leaves, pendulous, many-flowered; bracts triangular, 5–6 mm long; bracteoles triangular, 1.5–2 mm long; pedicel to 2.5 cm long. **Flowers** with calyx 2–4-lobed, tube ~2.5 mm long; petals 4, green or tinged red or yellow, oblong, 1–1.3 × 0.5–0.8 cm; stamens 5- or 6-whorled, innermost sterile, tube 3.5–6 mm long, filaments red, proximally white or pink, 3–3.5 cm long; staminodes 1–1.3 cm long; ovary 2–4-loculed, 1.5–3.5 mm long, locules 2- or 3-ovuled, style 4–6 cm long. **Fruits** ovoid-cylindric, 4-angled, 5–9 × 3–4 cm; pericarp fibrous but somewhat fleshy; seeds ovoid, 2–4 cm Ø. **Distr:** Mozambique, Madagascar, India, Sri Lanka, Thailand, Taiwan, Laos, Malesia, Japan (Ryukyu) and N Australia. **Ecol:** Usually on seashores, along tidal rivers and estuaries. Fl. & Fr. throughout the year. **Zones:** 10–12.

Planchonia

Species ~15. **Trees**, to 20 m tall. **Leaves** spirally arranged, crowded at tips of branchlets. **Inflorescences** spikes, crowded, erect. **Flowers** with calyx tube adnate to ovary, and not produced beyond it, segments 4, imbricate; petals 4, imbricate; stamens in many whorls, numerous, connate at the base; ovary inferior, 4-loculed, with numerous ovules in 2 rows, style long, simple. **Fruits** large, fleshy, indehiscent, with numerous seeds immersed in the flesh. **Distr:** Afghanistan through India to Sri Lanka and east to Thailand. **Cult:** Propagate from cuttings or by air-layering. Requires a moist soil in full or partial sun.

P. careya R. Kunth, *Pflanzenr. IV.* 219a (Heft 105): 56, 1939. (Syn: *Careya arborea*, etc.). [Cocky Apple, Patana Oak]. **Trees**, to 20 m tall; branchlets with prominent leaf scars. **Leaves** usually sessile; lamina large, broadly ovate, 15–30 × 5–15 cm, base tapered, margins denticulate or crenate, apex obtuse or rounded, thick, surfaces smooth, shiny, pale green. **Inflorescences** spikes, swollen, terminal, flowers crowded on a thick rachis, each with a central oval bract and 2 lateral linear ones; flowers sessile. **Flowers** large, 10 cm Ø; calyx tube campanulate, ~2–3 cm long, glabrous, segments rounded, stiff, erect; petals ovate, 5–7 cm long, margin often revolute, obtuse, or acute; filaments about as long as petals, spreading. **Fruits** ~8 cm Ø, green, glabrous, crowned with persistent calyx segments and style, solid; seeds many, immersed in flesh. **Distr:** From Afghanistan through India to Sri Lanka and east to Thailand. **Ecol:** On exposed montane grasslands and other hilly areas. Fl. Mar–Apr. Fr. May–Jun. **Zones:** 10–12. **Info:** These trees are fire resistant.

Subfam. LECYTHIDOIDEAE

Genera 10, **species** 215. **Stems** with bark fibrous; secondary xylem with crystal chains. **Leaves** 2-ranked or spirally arranged; lamina involute. **Flowers** monosymmetric via androecium; androecium with a basal ring, filaments contracted at the apex; style short (rarely long). **Fruits** operculate (indehiscent); seeds with swollen funicle, or aril (wings), or neither, embryo curved or not (Stevens 2001 onwards). **Distr:** Neotropical.

Bertholletia

Species 1. **Trees**, emergent to canopy. **Leaves** scattered along branches; lamina 17–36 cm long, glabrous. **Inflorescences** terminal or axillary, spicate or once-branched panicles, growth of rachises determinate; flowers present with leaves. **Flowers** zygomorphic, fragrant; sepals 2 at anthesis; petals 6, flat at apices; staminal lip present, yellow; androecial hood swept inward but not coiled, bearing vestigial stamens; stamens not reflexed, anthers laterally dehiscent; ovary usually 4-locular, locules 16–25-ovuled, attached at base of septum, style relatively long, geniculate, without stylar collar. **Fruits** globose, dehiscent, but the opening is smaller in Ø than the seeds, thus dropping from tree at maturity with seeds inside, pericarp thick (fruit size and shape are variable between populations); seeds not winged, testa hard, glabrous. **Distr:** Amazonian Colombia, Venezuela, Guyana, Surinam, Peru, Brazil, and Bolivia. **Cult:** Propagate from seeds, by grafting or from rooted cuttings (air-layered). Seedlings take over 15 years to mature and produce fruits. Requires a rich, moist, well drained soil in full sun.

B. excelsa Bonpl. *Pl. Aequinoct.* 1: 122–127, t 36, 1808. (Syn: *B. nobilis*). [Brazil Nut, Cream Nut, Para Nut]. **Trees** to 60 m tall, unbuttressed. **Leaves** scattered at ends of branches; lamina oblong, 16–40 × 5–16 cm, base rounded, margins entire to slightly crenulate, undulate, apex apiculate, adaxially glabrous, secondary veins 25–45 pairs; petiole 2–3.5 cm long. **Inflorescences** spikes or panicles, terminal or axillary; pedicels not well-defined. **Flowers** zygomorphic, fragrant; ~3 cm Ø; calyx lobes 2 at anthesis, margins entire; petals 6, pale yellow to white; androecium prolonged on one side into a hood with appendages swept inward but not forming a full coil, white with yellow at entrance into flower; stamens with anthers laterally dehiscent; ovary inferior, 3–6-loculed. **Fruits** dehiscent, the opening smaller than seeds, falling to ground with seeds inside at maturity, 10–25 cm Ø, usually globose; pericarp very thick; seeds with bony and wrinkled testa, 12–18, crescent-shaped. **Distr:** Guyana and Amazonian Colombia, Venezuela, Peru, Brazil, and Bolivia. **Ecol:** In non-flooded forests and grows only in areas with a 3–5-month long dry season (Müller 1981). Fl. Sep–Feb. **Zones:** 11–12. **Info:** In the wild the only animal equipped to open the Brazil Nuts is the agouti (*Dasyprocta punctata*). The brazil nuts are economically important in its native range with exports second only to rubber.

Couroupita

Species 3. **Trees** large, canopy. **Leaves** in apical whorls of 7–40; lamina 8–30 cm long, usually with hirsute pubescence in axils of secondary veins. **Inflorescences** cauliflorous on specialized serpentine branches on the trunk or rarely ramiflorous on oldest branches, usually racemes or rarely panicles, growth of rachis indeterminate. **Flowers** zygomorphic; sepals 6; petals 6; androecial hood absent or flat, staminodes anther-bearing; ovary 6-loculed, ovules numerous. **Fruits** indehiscent, dropping from tree at maturity, globose or subglobose, pericarp thin, often cracking when fruit hits ground; seeds numerous, ovate to lenticular, 1–1.5 cm long, embedded in pulp which oxidizes to bluish-green when exposed to air, dry pulp breaks into 6 segments, these crescent shaped when viewed laterally, wedge-shaped in cross section. **Distr:** C America (El Salvador into eastern Panama) and S America (Amazonia and the Guianas). **Cult:** Propagate from seeds or by air-layering. Requires a fertile, moist, well-drained soil in full sun.

C. guianensis Aubl. *Hist. Pl. Guiane* 2: 708, 1775. (Syn: *C. surinamensis*, etc.). [Cannonball Tree]. **Trees**, to 35 m tall, unbuttressed; bark not fissured. **Leaves** in clusters at ends of branches; lamina usually narrowly obovate to obovate, sometimes elliptic, to (6–)8–31(–57) × 3–10 cm, base cuneate, margins entire, apex generally acute or acuminate; petiole to 3 cm long. **Inflorescences** cauliflorous on specialized serpentine branches along the trunk, usually unbranched racemes, sometimes branched and paniculate, sometimes covering the entire lower trunk; pedicels 1.2–6 cm long. **Flowers** zygomorphic, 5–6 cm Ø; calyx lobes 6, imbricate, margins ciliate; petals 6, mostly yellow toward apices and pink to red toward bases abaxially, pink to red except for white bases adaxially; androecium prolonged on one side into a flat hood, with stamen free area between staminal ring and hood; hood white or pale yellow tinged with pink outside; *ring stamens* with filaments white, anthers pale yellow; *hood staminodes* often white at very base, pink for most its length, yellow at

Top row: *Bertholettia excelsa* (flowers), *Bertholettia excelsa* (fruits), *Couroupita guianensis*, *Gustavia augusta*.
Bottom row: *Gustavia superba*, *Lecythis lanceolata*, *Lecythis ollaria* (empty fruit).

apex, sometimes white for most its length and yellow at apex, stamens laterally dehiscent; ovary inferior, 6-loculed. **Fruits** indehiscent, globose, 12–25 cm Ø, falling from the tree at maturity; seeds embedded in pulp; pulp oxidizing to bluish-green when exposed to air. **Distr:** S America. **Ecol:** In wet areas of lowland forests. Fl. Nov.–May. Fr. Oct–Apr. **Zones:** 10–12.

Gustavia

Species ~40. **Trees**, evergreen, slow-growing, growth form pachycaulous or leptocaulous. **Leaves** scattered or clustered at branch ends on slender or stout twigs; lamina 10 cm to 1.2 m long, glabrous, secondary venation reticulate. **Inflorescences** terminal or axillary, racemose, growth of rachises determinate; flowers present with leaves. **Flowers** actinomorphic, not markedly fleshy; sepals absent (then calyx entire) or 4–6; petals 6–8 (12 or 18 in *G. romeroi*), flat at apices; androecium a wide tube, the stamens at the margin of tube, curved inward, but not reflexed as in *Allantoma*, ovary 4–6(10)-loculed, ovules 7–100 per locule, attached on septum toward apex of locule, style short, without stylar collar. **Fruits** capsules, usually globose, indehiscent, remaining on tree or dropping to ground at maturity, infrequently cylindric (e.g., *G. brachycarpa*), the pericarp somewhat soft at maturity, thin or thick; seeds not winged, rarely 1, usually 2–5 per fruit, not flattened, not embedded in pulp, testa glabrous. **Distr:** Panama into NW South America. **Ecol:** Understory (often shrubs) to canopy. **Cult:** Propagate from seeds, cuttings or by air-layering. Requires a moist, well-drained, organic soil in full sun.

G. augusta L., *Pl. Surin.* 12: 17, 1775. (Syn: *Japarandiba augusta*, etc.). [Gustavia, Janiparindiba]. **Trees**, understory, to 22 m tall, the trunk not buttressed. **Leaves** clustered at ends of branches; lamina narrowly obovate to oblanceolate, 16–48 × 4–13 cm, base acute to cuneate, margins entire to serrulate, rarely serrate, apex acute or acuminate, glabrous, secondary veins in 14–22 pairs; petiole 0–4 cm long. **Inflorescences** usually terminal, rarely axillary or cauline, racemose; pedicels 1.5–7.5 cm long. **Flowers** actinomorphic, 9–15 cm Ø; calyx with lobes nearly absent or of 4 broadly rounded or triangular lobes, not imbricate; petals (6)8(9), white, often with tinges of pink; androecium with stamens arising from apex of rim, filaments white throughout or white toward base and pink toward apex, the anthers with poricidal dehiscence; ovary inferior, 4–7-locular. **Fruits** capsules, indehiscent, 3–7 × 3–8 cm, globose, truncate at apex; seeds not winged, with yellow, contorted funicle, the testa without trichomes, black. **Distr:** Tropical America, widespread. **Ecol:** Mostly in riverine habitats. Fl. Sep–Dec. Fr. Jan–May. **Zones:** 10–12.

G. superba (Kunth) O. Berg, *Linnaea* 27: 444, 1854. (Syn: *Pirigara superba*, etc.). [Stinkwood, Wild Mango]. **Trees**, to 20 m tall; trunk to 25 cm dbh. **Leaves** tightly compacted at branch ends; lamina oblanceolate, 0.25–1.28 m × 6–25 cm, base long tapered to acute or cuneate, margins serrate, apex acuminate, glabrous, chartaceous, secondary veins 16–36 pairs; petiole to 15 cm long. **Inflorescences** cauline, racemose, puberulous, 3–12-flowered; rachis 1–6 cm long; pedicels 2.9–8.5 cm long, subtended by a single ovate to oblong basal bract and 2 ovate, cucullate bracteoles, the latter often fused forming a cuplike structure. **Flowers** 10–15 cm Ø; calyx entire or very slightly 4-lobed; petals (7)8(9), obovate, oblanceolate, or oblong, 3.3–6.7 × 1.2–3.7 cm, white, pink-speckled; androecial base 0.8–1.5 cm high; outermost filaments 1–2.3 cm long, base yellow, apex pink, anthers 2.5–4 mm long, yellow; ovary smooth, puberulous (5)6-locular, locules 18–30-ovuled, style 1–2 mm long. **Fruits** ± globose, 3–9 × 4–10 cm, opercular region sunken below fruit surface; seeds usually angled, 1.2–3.5 × 1.1–2.2 cm. **Distr:** C Panama to NW Colombia. **Ecol:** In lowland moist forests. Fl. May–Jun, Dec. Fr. Apr–Aug. **Zones:** 10–12.

Lecythis

Species 26. **Trees** large, emergent, understory or canopy. **Leaves** scattered along branches; lamina 2.5–40 cm long, glabrous, secondary venation reticulate. **Inflorescences** ramiflorous, terminal, or axillary, racemose or spicate or once-branched paniculate arrangements of racemes or spikes, growth of rachises determinate; flowers present with leaves but sometimes flowering as new leaves are just about to open, depending on species. **Flowers** zygomorphic; sepals 6; petals 6, flat at apices; staminal lip present; androecial hood variable; stamens not reflexed, anthers laterally dehiscent; ovary 3–6-locular, ovules 2–30 in each locule, attached to lower part of septum, style often obliquely oriented or geniculate, stylar collar absent. **Fruits** dehiscent or less frequently indehiscent, remaining on tree or dropping to ground at maturity, generally globose, pericarp thick; seeds not winged, one to numerous. **Distr:** Nicaragua into the Andean valleys and Pacific coast of Colombia and Ecuador, throughout Amazonia and the Guianas, and the Atlantic coastal forests of Brazil. **Cult:** Propagate from seeds or half-hardened cuttings. Requires a moist, well-drained soil in full sun.

L. lanceolata Poir., *Encyclopédie Méthodique Botanique* 6: 27, 1804. (Syn: *L. pohlii*, etc.). **Trees**, deciduous, to 30 m tall, with very low branches, glabrous. **Leaves** sparse; lamina lanceolate, 2–9 × 1.2–4 cm, base cuneate, margins serrated, apex acuminate, acumen 1–2 cm long, glossy green, midrib prominent abaxially, minutely puberulous; petioles 2.5–9 mm long, puberulous. **Inflorescences** axillary racemes; flowers appearing with the new flush of leaves; pedicels 2.5–5 mm long; bracteoles caducous. **Flowers** 5 cm across, bluish purple (aging to white, before falling), very fragrant; calyx flat, ~1 cm Ø; sepals 6, obovate, 2.5 mm long, minutely puberulous; petals obtuse, 0.7–1.2 cm long; stamens depressed-globose, ~7 mm Ø, anthers subrotund; styles long. **Fruits** capsules, large with an operculum, over 20 cm Ø. **Distr:** S America. **Ecol:** Fl. Dec–Jan. **Zones:** 10–12.

L. ollaria Loefl., *Iter Hispanicum* 189, 1758. (Syn: *L. cordata, Eschweilera cordata*). [Monkey Pot Tree, Venezuela Coco de Mono]. **Trees**, evergreen, to 25 m tall. **Leaves** sessile or subsessile, alternate; simple, ovate to oblong-ovate, 5.2–9 × 2.5–5 cm, base rounded, cordate or subcordate, margins subserrate, apex acute to obtuse, glossy, bright green, chartaceous. **Inflorescences** racemes, axillary, toward the end of the branchlets, or terminal; rachis thickened like a branchlet at fruit maturity; bracts ovate, deciduous. **Flowers** variable, mauve; sepals 6, oblong, uneven with margins rounded, concave, persistent; petals 6, larger than the sepals, spathulate, subequal, oblong, to subrounded, concave, with margins reflexed, white. **Fruits** capsules, spherical or urn-shaped, to 30 cm Ø, with an operculum; operculum falls off to release the seeds, pericarp woody; seeds brown. **Distr:** Restricted to the east of the Andes in the Venezuelan states of Anzoátegui, Apure, Aragua, Guarico, Lara, and Portuguesa. **Ecol:** In dry forests. Fl. Apr–Nov. **Zones:** 10–12. **Info:** The seeds are poisonous as with some of the other species in this genus. Its bark separates into thin sheets like paper, and used by the natives as cigarette wrappers.

SAPOTACEAE

Genera 53, **species** 1,100. **Trees** or **shrubs**, usually producing latex. **Stipules** early deciduous or absent **Leaves** spirally arranged or alternate and distichous, rarely ± opposite, sometimes crowded at apex of branchlets; lamina with margins entire. **Inflorescences** usually in sessile axillary clusters, rarely solitary flowers, cluster pedunculate or in a racemelike inflorescence, bracteolate. **Flowers** bisexual or unisexual,. calyx 1 whorl of usually 4–6 sepals, or 2 whorls each with 2–4 sepals; corolla lobes 1 or twice as many as sepals, usually entire, rarely with 2 lacerate or lobular appendages; stamens inserted at corolla base or at throat of corolla tube, as many as and opposite corolla lobes to many and in 2 or 3 whorls; staminodes when present alternate with stamens, scaly to petal-like; ovary superior, 4- or 5-locular, placentation axile; ovules 1 per locule, anatropous; style 1, often apically lobed. **Fruits** berries or drupes, 1- to many-seeded; seeds with testa hard, shiny, rich in tannin, endosperm usually oily, seed scar lateral and linear to oblong or basal and rounded. **Distr:** Throughout the tropics.

Chrysophyllum

Species 70. **Shrubs** or **trees**. **Leaves** alternate; stipules absent. **Inflorescences** 2 to many, axillary fascicles; flowers pedicellate or sessile. **Flowers** small, sepals 5(or 6), usually glabrous or inside tomentose; corolla tube campanulate, usually extended, (4 or)5–11-lobed; stamens (4 or)5–10 in 1 whorl, inserted in throat, opposite corolla lobes; staminodes absent; ovary 1–10-locular, villous or glabrous; style longer to shorter than ovary. **Fruit** 1–8-seeded, exocarp thick to very thin; seed coat papery to crusty; scar narrow to wide, lateral, sometimes almost covering entire seed. **Distr:** Tropical and subtropical America; tropical Africa and Madagascar; Asia. **Cult:** Propagate from seeds or cuttings. Requires a rich, well-drained, in full sun.

C. argenteum Jacq., *Enum. Syst. Pl.*: 15, 1760. (Syn: *C. glabrum*). [Milky Iron, Star Apple, Wild Kaimit]. **Trees** or **shrubs**, 5–13(–20) m tall; crown low; trunk 10–20 cm dbh; branchlets greenish, finely hairy, becoming brown or gray; sap milky. **Leaves** alternate; lamina elliptic or oblong-elliptic, 4.5–12 × 2–5.2 cm, base cuneate, apex acuminate, abaxially silvery greenish or ferruginous pubescent, becoming glabrous; petiole 0.8–1.2 cm long. **Inflorescences** fascicles 5–8-flowered, axillary; pedicels ~1 cm long. **Flowers** with calyx ~2 mm long, pubescent; sepals 5, rounded, brownish-green, ~3 mm long, finely brown-hairy; corolla bell shaped, greenish yellow, 3–4 mm long, lobes 5–6, rounded, shorter than tube, sparsely sericeous. **Fruits** dark blue, ellipsoid, subglobose or ovoid, 1.5–2 cm long, 1-seeded; seed large, brown, shiny. **Distr:** Panama, West Indies, Venezuela. **Ecol:** In limestone and lower mountain forests, woodlands and hillsides. 1,100 m. Fl. & Fr. throughout most of the year. **Zones:** 10–12.

C. cainito L., *Sp. Pl.* 1: 192, 1753. (Syn: *C. bicolor*, etc.). [Caimito, Cainito, Golden Leaf, Satin Leaf, Star Apple]. **Trees**, hermaphrodite; crown open, 12(–20) m tall; trunk to 60 cm dbh. **Leaves** alternate, distichous or spirally arranged; lamina elliptic-oblong, 6–15 × 2.5–6 cm, apex and base acute, adaxially dark green, abaxially deep reddish silky; petiole 1.3–1.6 cm long, reddish-brown hairy. **Inflorescences** fascicles, 5–35-flowered, axillary, ramiflorous or cauliflorous; pedicels 0.6–1.6 cm long. **Flowers** unisexual or bisexual; calyx in aa single whorl; sepals 4–6; corolla 5-lobed, globose, campanulate or cylindrical, tube shorter to longer than the lobes, lobes (4)5(–8); stamens 4–8, usually adnate to corolla tube; ovary (4–)5–7(–12)-loculed, stigma 7–11-crested. **Fruits** globose to broadly ellipsoid, 7.5–10 cm Ø, soft when ripe, skin green or purplish, smooth, pulp whitish, sweet, mucilaginous; seeds 4–8(–10), compressed, blackish, drying to brown. **Distr:** C America. **Ecol:** In dry hot areas (*tierra caliente*), and lowland moist to wet forests; <1,000 m. Fl. Aug–Sep. Fr. Feb–Mar. **Zones:** 11–12.

C. oliviforme L., Syst. Nat. ed. 10, 2: 937, 1759. (Syn: *Chrysophyllum oliviforme* var. *typicum*, etc.). [Damson Plum, Satin Leaf, Wild Star Apple]. **Trees**, slender, 4–6(–15) m tall; crown open and ± layered; bark scaly, grayish. **Leaves** ovate to broadly elliptic, 4–12.5 × 3–6 cm, adaxially deep green, abaxially tawny or reddish-golden hairy. **Inflorescences** fascicles of 5–10 flowers; pedicels 6–8 mm long. **Flowers** with corolla 3.8–5 mm long, tube longer than lobes; ovary 5-loculed, stigma 5-crested. **Fruits** spindle-shaped, 2.5 cm long and 1.25 cm Ø, purplish black, smooth, inedible, produces latex; seed 1, compressed, brownish. **Distr:** Bahamas Islands, extreme southern Florida, and the Greater Antilles (Cuba, Hispaniola, Jamaica and Puerto Rico). **Ecol:** Near sea coasts, but not salt-tolerant. **Zones:** 10–12.

Madhuca

Species 100–125. **Trees**, often very large, producing latex. **Stipules** small to very large, usually early deciduous. **Leaves** spirally arranged, often closely clustered at end of branchlets; lamina ± leathery. **Inflorescences** with flowers solitary, clustered in axils or sometimes terminal, usually with a long pedicel. **Flowers** with sepals 4(–6), 2-whorled, inner ones often with margin membranous and ciliate, outer sepals usually hirsute; corolla cylindric, with hairs at throat; lobes (5–)8(–18), usually woolly pubescent between stamens, appendages absent; stamens 1–3-whorled, inserted in the corolla tube and alternating with lobes, glabrous or tomentose, sessile or subsessile; staminodes absent; ovary usually pubescent, 6–8(–12)-locular, style subulate, persistent. **Fruits** berries, globose to ellipsoid, with enlarged persistent brown calyx; seeds 1–4. **Distr:** Asia through Malesia to Australia. **Cult:** Propagate from seeds, rooted cuttings or by air-layering. Requires a moist, well-drained soil in full or partial sun.

M. longifolia (Koenig) Macbride, *Contr. Gray Herb. Harv. Uni.* NS. 53: 17, 1918. (Syn: *Bassia longifolia, Madhuca indica*). [Butter-Nut Tree]. **Trees** up to 16 m tall, fast growing, deciduous, stem glabrous, young growth tomentose. **Leaves** scattered or clustered at the end of the branches, 7–25 × 5–11 cm, glabrate, young leaves tomentose, broadly elliptic-lanceolate, varying in shape, 12–15-nerved; new flush bronze. **Inflorescences** terminal or axillary clusters; pedicel tomentose, 2.5–5 cm long. **Flowers** fragrant, sepals ovate, acuminate, pubescent; corolla white, lobes ovate-acuminate; stamens 20 or more; style exserted. **Fruits** berries, fleshy, brown tomentose. **Distr:** Sri Lanka, India, Myanmar. **Ecol:** In open plains

Below: *Chrysophyllum argenteum, Chrysophyllum cainito, Chrysophyllum cainito* (fruit), *Chrysophyllum oliviforme.*

Above: *Chrysophyllum oliviforme* (fruit), *Madhuca longifolia*, *Madhuca neriifolia*, *Manilkara hexandra.*

and forests. **Zones:** 10–12. **Info:** The fragrant flowers are edible and are largely used in the preparation of distilled liquors and vinegar. Flowers are regarded as a cooling, tonic and demulcent also used for coughs, colds and bronchitis. The honey from flowers is edible and used to treat diseases of the eye. A low quality of oil is extracted from the seeds which is edible and also used in making soap. The bark is used medicinally in treating leprosy.

M. neriifolia (Moon) H.J. Lam., *Bull. Jard. Bot. Buitenzorg Ser.* 3, 7: 182, 265, 1925. (Syn: *Bassia malabarica*, etc.). [Illipe Butter Tree]. **Trees**, 5–20 m tall, sometimes ± shrubby to 4 m high; trunk to 25 cm dbh. **Leaves** scattered; lamina oblong-elliptic to narrowly oblong-lanceolate or linear-oblong, 2.8–13(–21) x 0.8–4 cm, base tapered and decurrent into petiole, apex obtuse to subacute to narrowly rounded, thin, glabrous. **Inflorescences** fascicles 2–6-flowered, axillary; pedicels 1.5–2(–2.5) cm long, glabrous. **Flowers** with a strong disagreeable smell; sepals ovate-lanceolate, 0.8–1 cm long; corolla green, white or greenish white, ~1.7 cm long, tube tomentose outside, woolly inside, lobes 6–8, ovate-oblong, 8 mm long; anthers 12–16, 1- or 2-whorled; style 2.5 cm long, much-exserted. **Fruit** narrowly ellipsoid, 2–3 x 1–1.5 cm, beaked, 1-seeded, with white sticky latex; seed pale brown, shiny, narrowly ellipsoid, 2.5 cm long, compressed, acute at both ends, hilum extending the whole length. **Distr:** Sri Lanka, India. **Ecol:** In lowland rainforests, usually by streams and riverbanks. **Zones:** 10–12.

Manilkara

Species ~65. **Trees**, evergreen. **Stipules** early deciduous. **Leaves** often obovate, apex rounded, leathery, densely lateral-veined. **Inflorescences** fascicles, or flowers solitary; flowers pedicellate. **Flowers** bisexual; sepals 6 in 2 whorls of 3; corolla lobes 6, simple or with 2 appendages; stamens 12, 6 fertile, inserted on petal, 6 sterile and petal-like, alternating with petals; ovary 6–14-loculed, locules 1-ovuled. **Fruits** berries, dry, indehiscent; seeds compressed, hilum elongate. **Distr:** Tropical regions of the world. **Cult:** Propagate from seeds or by air-layering. Requires a well-drained soil in full sun.

M. hexandra (Roxb.) Dubarb, *Revis. Gen. Pl.* 406, 1891. (Syn: *M. emarginata*, etc.). [Kimi, Milky Tree]. **Shrubs** or **trees**, 3–12 m tall; bark gray; branchlets glabrous. **Leaves** alternate, often closely clustered at ends of branchlets, with conspicuous scars; lamina obovate to obovate-elliptic, 5–10 x 3–7 cm, base broadly cuneate to obtuse, apex retuse, surfaces glabrous, midrib raised abaxially, lateral veins slender and parallel, veinlets dense; petiole 0.8–2 cm long. **Inflorescences** fascicles, axillary; pedicel thick, 1–1.8 cm long. **Flowers** with sepals ovate-triangular, 3–4 mm long, outside yellowish-gray tomentose; corolla white or light yellow, ~4 mm long; lobes oblong, ~3 mm long; stamens ~5 mm long; staminodes 2-parted, lobes linear, ~3 mm long; ovary ovoid, ~2 mm long, 6-locular, tomentose. **Fruits** berries, obovoid-oblong to ellipsoid, 1–1.5 cm long; seeds 1 or 2, ~0.8–1 cm long. **Distr:** Cambodia, India, Sri Lanka, Thailand, Vietnam. **Ecol:** In thickets and forests of dry and arid regions, mangroves, scrub forests, sandy flats and rocky outcrops; 0–270 m. Fl. Aug–Mar. Fr. July. **Zones:** 10–12. **Info:** Seeds contain 25% edible oil and are of medicinal value.

M. zapota (L.) P. Royen, *Select. Stirp. Amer. Hist.* 57 1763. (Syn: *Achras sapota*, etc.). [Chicle, Noseberry, Sapodilla, Sapote]. **Trees** 15–30 m tall; crown conical. **Leaves** clustered at branch ends; lamina elliptic to oblong-elliptic, 3.8–15.5 x 1.9–5.7 cm, base wedge-shaped, apex rounded to acuminate, abaxially often reddish-tomentose when young, glabrous at maturity. **Inflorescences** with flowers solitary, axillary; pedicels 2.5 cm long. **Flowers** with corolla white or yellowish, tube longer than lobes, lobes 6, entire, toothed or with 2 appendages; sterile stamens petal-like; ovary hairy, ± sunken in fleshy disk. **Fruits** berries, fleshy, ellipsoid or flattened-globose, to 10 cm Ø, skin brown, scurfy, flesh gritty, sweet; seeds 2–12, ~2 cm long, compressed, dark brownish-black, shiny. **Distr:** W Indies and tropical America. **Ecol:** Throughout the year. **Zones:** 10–12. **Info:** The edible fruits are cultivated widely. The coagulated resinous latex derived from the bark is used commercially for making chewing gum.

Mimusops

Species ~40–57. **Trees**, evergreen, with milky juice. **Stipules** very small. **Leaves** often elliptic or obovate, usually glabrous, leathery, lateral nerves 15–20(–30) pairs, parallel. **Inflorescence** fascicles or flowers solitary, axillary or on leafless twigs; flowers pedicellate. **Flowers** with calyx 6–8-lobed, lobes biseriate (3–4 in each), outer valvate, inner imbricate; corolla tube short, 18–24-lobed, 2–3-seriate; stamens 8, opposite, inserted at base of corolla and opposite the interior lobes, filaments short, anthers lanceolate, cells divergent at base; staminodes as many as stamens and alternating with them, variously toothed or entire, hairy; ovary hirsute, 6–8-loculed, style slightly longer than corolla. **Fruits** ovoid, seeds 1–6, compressed, ellipsoid. **Distr:** Tropical Africa, Madagascar, Indian Ocean Islands and tropical Asia. **Cult:** Propagate from bare (hard shell removed) seeds, cuttings or by air-layering. Requires a fertile soil in full or partial sun; plants are usually drought- and salt-resistant.

M. coriacea (A. DC.) Miq., *Fl. Bras.* 7: 44, 1863. (Syn: *M. hexandra*, etc.). [Monkey Apple]. **Shrubs** (to 6 m tall) or **small trees** (to 15–20 m tall), tortuous; bark rough, fissured, brown; very young shoots ferruginous-tomentose. **Leaves** born at apices of shoots, persistent; lamina obovate, elliptic, obovate-elliptic or oblong-elliptic, 7–15 x 3.5–8.5 cm, base obtuse, margins slightly revolute, apex obtuse or rounded, coriaceous, pale green; petiole 1–1.5 cm long, initially pubescent. **Inflorescences** few-flowered; pedicels 5–8 cm long.

Below: *Manilkara zapota*, *Mimusops coriacea* (fruits), *Mimusops elengi*, *Palaquium maingayi.*

Flowers ~1.5 cm Ø; calyx with 8 sepals, 1.1–1.2 cm, outside hairy; corolla ~1.2 cm long, lobes 0.9–1 cm long, each with 2 appendages 1.1 cm long, tube 2 mm long; stamens short, 3–3.5 mm long, hairy at base, anthers 4–4.5 mm long; staminodes 5–5.5 mm long, hairy dorsally and inside the edges; ovary pubescent, style glabrous. **Fruits** yellow, ± round, 3–4 cm Ø, with pink, sweet and edible flesh. **Distr:** Comoros Islands, Madagascar. **Ecol:** In coastal forests. **Zones:** 11–12.

M. elengi L., *Sp. Pl.* 349, 1753. (Syn: *M. elengi* var. *parvifolia*). [Bulletwood Tree, Indian Medlar, Spanish Cherry]. **Trees**, large, evergreen, to 25 m tall; trunk cylindrical. **Leaves** alternate; lamina broadly ovate, 3.5–5 × 7–10.5 cm, base cuneate, margins entire or slightly wavy, apex acuminate, glabrous, petiole 2–3.5 cm long; new flush bronze-colored. **Inflorescences** fascicles, or flowers solitary, axillary; pedicel ~2 cm long, hairy. **Flowers** very fragrant, creamish-white, ~1.5 cm Ø; calyx 6-lobed, lobes free, lanceolate, tomentose; corolla tube very short, lobes 24, biseriate; stamens 8, anthers subsessile, staminodes hairy; ovary pubescent. **Fruits** ovoid, 2.5–4 cm long, orange, pubescent when young, with persistent calyx; pulp yellowish-orange with a starchy texture and bland or slightly sweet taste. **Distr:** India, Sri Lanka, Malay Peninsula. **Ecol:** In wet evergreen forests near coastal areas to submontane and montane areas, disturbed forests and secondary bushes near streams; 0–800 m. Fl. & Fr. throughout most of the year. **Zones:** 11–12. **Info:** The fruit is eaten in the Maldives but now thought to contain a carcinogenic or an astringent substance. The hard seed coat is often chewed. The bark, fruits and flowers are used medicinally in Maldives and other parts of the Asia.

Palaquium

Species ~110. **Trees**, producing latex; branchlets with terminal scaly bud. **Stipules** small to very large, usually early deciduous. **Leaves** often clustered; lamina leathery. **Inflorescences** with flowers solitary, or fascicled, sometimes in short terminal fascicles, axillary; bracts conspicuous at peduncle base. **Flowers** with sepals (4–)6(or 7), 2-whorled; corolla 4–6-lobed, imbricate or twisted; stamens (8–)12–18(–36), in 2 or 3 whorls, inserted in corolla throat, filaments long; staminodes absent; ovary (5–)6(–11)-locular, villous, ovules pendulous. **Fruits** berries, often edible, oblong to ellipsoid; seeds 1–4, with scar covering half the surface. **Distr:** Cambodia, China, India, Indonesia, Malaysia, New Guinea, Pacific Islands, Philippines, Sri Lanka, Thailand and Vietnam. **Cult:** Propagate from seeds or by air-layering. Requires any well-drained soil in full sun.

P. maingayi (C.B. Clarke) Engl., *Bot. Jahrb. Syst.* xii. 511, 1890. (Syn: *Croixia maingayi*. *Dichopsis maingayi*). **Trees**, to 18 m tall; trunk 25–38 cm dbh; young branchlets densely shortly ferruginous-pubescent. **Leaves** obovate, 13–33 × 7.5–18 cm, base cuneate to slightly rounded, often ± auricled, apex obtuse, adaxially glabrous, abaxially densely velvety with thick ferruginous pubescence, lateral nerves 12–20 pairs, drying to reddish beneath; petiole 2.5–3.8 cm long, thickly rusty-pubescent. **Inflorescences** fascicles, 5–10-flowered, axillary mostly on leaf-scars on thick branches; pedicels 1–1.8 cm long, ferruginous-pubescent. **Flowers** with sepals broadly ovate-semicircular, 3–4 mm long, outer 3 densely pubescent; corolla tube short, 2.5 mm long, lobes ovate-oblong, 6–9 mm long, glabrous. **Fruits** shiny, green, ~1.3 cm long; seeds 2, dark brown, oblong. **Distr:** Peninsular Malaysia and Thailand. **Ecol:** In lowland and hill forests; to 1,100 m. **Zones:** 10–12.

Pouteria

Species ~50. **Trees** or **shrubs**; branchlets glabrous or pubescent but glabrescent. **Leaves** usually alternate, not stipulate; lamina papery to leathery, surfaces usually pubescent when young, glabrescent when mature, lateral veins conspicuous. **Inflorescences** axillary clusters, sometimes with 2–4 bractlets. **Flowers** with 4 (rarely 5 or 6) sepals, outside pubescent, inside glabrous or sericeous, early deciduous or persistent in fruit; corolla tubular to campanulate, 4- or 5–8-lobed; stamens 5 (rarely 4 or 5–8), inserted in throat of corolla; staminodes less than 5 or 5–8, opposite sepals, lanceolate to subulate, sometimes scalelike to petal-like; ovary conical, sometimes with cup-shaped disk at base, 5(or 6)-locular, ± villous. **Fruits** globose, glabrous or tomentose, sometimes setose; pericarp thin to thick, sometimes very hard when dry; seed with testa shiny, scar oblong to broadly ovate, covering half to nearly whole length of seed. **Distr:** Tropical regions of mostly America, and China. **Cult:** Propagate from seeds or by air-layering. Requires a moist, well-drained soil in full sun; moderately salt-tolerant.

P. campechiana (Kunth) Baehni, *Candollea* 9: 398, 1942. (Syn: *Lucuma campechiana*, etc.). [Canistel, Eggfruit, Yellow Sapote]. **Trees**, 7.5–25 m tall; trunk to 50 cm dbh. **Leaves** clustered, spirally arranged; lamina narrowly elliptic to oblanceolate, 8–25(–33) × 2.7–7(–15) cm, base cuneate to attenuate, apex round to attenuate, glabrous, lateral nerves 9–18 pairs; petiole 1–2.5(–4.5) cm long. **Inflorescences** fascicles, 2–3-flowered, axillary and below leaves; pedicels 0.8–1.1 cm long. **Flowers** scented; sepals 5–6, broadly ovate to ± round, 0.45–1.1 cm long, apex obtuse or rounded, silky outside; corolla, pale green, 0.8–1.4 cm long, lobes 5–6, oblong, apex rounded or truncate, shorter than the tube. **Fruits** orange, yellow or green, ellipsoid, conical or globose, 2.5–10(–12.5) × 6.5–8 cm, apex slightly mammiform; seeds 1–6, shiny, brown, ellipsoid or crescent-shaped, 2–3.8 × 1.6–2 cm, hilum 3 × 1.5 cm. **Distr:** Mexico through Guatemala to Panama; widely cultivated in the Old World tropics. **Ecol:** In moist or wet mixed forests, sometimes in pine forests; 0–1,400 m. Fl. May–Jul. Fr. Nov–Jan. **Zones:** 10–12. **Info:** The fruits are edible with a sweet taste and a starchy texture.

EBENACEAE

Genera 3, **species** ~500. **Trees** or **shrubs**, erect; branchlets occasionally spine-tipped. **Leaves** alternate, rarely opposite; margins entire. **Inflorescences** with ♂ flowers often in cymes, sometimes in clusters or solitary; ♀ flowers often solitary, axillary. **Flowers** actinomorphic, usually unisexual, dioecious, or polygamous, rarely bisexual; ♂ with pistil rudimentary or absent; ♀ imperfect or without stamens; calyx 3–7-lobed, persistent and often becoming enlarged in ♀ or bisexual flowers, lobes valvate or overlapping in bud; corolla 3–7-lobed, lobes convolute, rarely overlapping or valvate; stamens hypogynous or at base of corolla, 2–4 times as many as corolla lobes, rarely as many as corolla lobes and alternate with them, filaments free or united in pairs; ovary superior, 2–16-locular, locules 1- or 2-ovuled, styles 2–8, free or basally united, stigmas entire or bilobed. **Fruits** berries, fleshy, few- to several-seeded; seeds usually oblong; endosperm sometimes ruminate. **Distr:** Tropical and subtropical regions of the world.

Diospyros

Species ~475–485. **Shrubs** and **trees**, evergreen and deciduous; branchlet tips sometimes spinelike. **Leaves** alternate; lamina simple, often colorful, occasionally with minute translucent glands. **Inflorescences** with ♂ flowers are borne in axillary cymes, usually on basal part of the current year's growth; ♀ flowers axillary are usually solitary. **Flowers** dioecious or polygamous; ♂ with stamens 4, often paired and forming 2 whorls; ♀ with 1–16 staminodes or absent, stigma often 2-cleft; calyx usually 3–7 lobed, sometimes truncate; corolla urn-shaped, campanulate or tubular, 3–7-lobed, sometimes truncate. **Fruits** ranging from small fleshy berries to leathery pearlike persimmons, usually with the persistent calyx; seeds 1–10 or more, laterally compressed. **Distr:** Tropical and temperate regions of the world. **Cult:** Propagate from seeds, root cuttings, grafting or by air-layering. Requires a moist, well-drained soil in full sun. Some species are relatively salt-tolerant. **Info:** Cultivated commercially for timber, edible fruits, medicinal purposes and ornamental purposes. *Diospyros* is the source of ebony wood, of which two kinds are known, the evenly pitch-black one (*D. ebenum*) and the variegated one called Calamander (*D. quaesita*) which is almost extinct in the wild (Dassanayake 1981).

D. atrata Alston, *Handb. Fl. Ceylon*, 6: 180, 1931. (Syn. *D. embryopteris* var. *atrata*). **Trees**, to 25 m tall; trunk 70 cm dbh; bark smooth, black, live bark pink, inside dirty white. **Leaves** oblong, to 7 × 17 cm, base rounded, apex broadly acuminate, chartaceous to subcoriaceous, glabrous; petiole ~1 cm long. **Inflorescences** axillary, 3–8 (♂) or less (♀) flowers on short pedicels. **Flowers** ♂ with a cup-shaped calyx,

Below: *Pouteria campechiana* (fruits), *Diospyros atrata*, *Diospyros atrata* (fruits), *Diospyros ebenum* (female flowers).

Top row: *Diospyros kaki, Diospyros malabarica, Diospyros mollis.*
Bottom row: *Diospyros philippensis, Diospyros philippensis* (fruits), *Diospyros racemosa, Diospyros walkeri.*

lobes fleshy, corolla pinkish-white or white, fleshy, 0.8–1.2 × 1–1.2 cm, tube very short, fleshy, lobes 4, ovate, 4 mm long; ♀ about same size, ovary globose, styles 4. **Fruits** globose, to 7 cm Ø, rusty-tomentose, easily rubbed off; calyx persistent, lobes 0.4–1 cm long, black-hairy, reflexed, stiffly coriaceous; seeds 2.5 × 1.5 cm. **Distr:** Sri Lanka (restricted to the Gannoruwa forest between Peradeniya and Kandy, considered rare). **Ecol:** In humid forests; ~500 m. Fl. Mar–Apr. Fr. May–Sep. **Zones:** 11–12.

D. ebenum Koenig ex Retz., *Physiogr. Salsk. Handl. (Lund)* 1: 176, 1781. (Syn: *D. glaberrima*, etc.). [Ceylon Ebony, Ceylon Persimmon, Ebony]. **Trees**, to 30 m tall, dioecious, polygamous, occasionally monoecious; trunk straight, 90 cm dbh, buttresses to 2 m high, 30 cm wide. **Leaves** alternate; lamina oblong-oval to oblong-lanceolate, 2 × 5–6 × 13 cm, base contracted into petiole, apex short, obtuse, thinly coriaceous, glabrous; petiole 4–8 mm long. **Inflorescences** mostly polygamous (♂ and hermaphrodite); ♂ dichasial compressed cymes, 3–15-flowered, axillary, peduncle 2–5 mm long; ♀ flowers solitary, axillary, subsessile; bracts small. **Flowers** ♂ with calyx funnel-shaped,~4 mm long, lobes 4, green or white-speckled, margin ciliate, apex obtuse, corolla glabrous, tubular, yellow to yellowish-white, throat constricted, ~1 cm long, lobes (3)4, rotate, narrow, fleshy; ♀ with calyx cuplike, lobes patent or reflexed, 3–4, acute, glabrous, 9 mm long, corolla 1.5 cm long, mouth wide, lobes 3–4, rotate, broad, obtuse, reflexed, 9 mm long, staminodes 8, ovary 8-loculed, stigma 4–5-lobed. **Fruits** berries, ± globose, to 1.5 cm Ø, with persistent calyx, glabrous; seeds 2–8, flattened, brown or black. **Distr:** S India. **Ecol:** In dry regions. **Zones:** 11–12.

D. kaki Thunberg, *Nova Acta Regiae Soc. Sci. Upsal.* 3: 208, 1780. (Syn: *D. discolor, D. philippensis*). [Chinese Persimmon, Sharon Fruit]. **Trees**, to 27 m tall, deciduous; young branchlets densely pubescent to glabrous. **Leaves** lanceolate, elliptic or ovate, occasionally obovate, 5–18 × 2.6–9 cm, papery, pubescent when young; petiole 0.8–2 cm long. **Inflorescences** ♂ 3–5-flowered cymes; ♀ flowers solitary. **Flowers** ♂ small, with calyx ± as long as corolla, both sides hairy, lobes 4, corolla white, yellowish-white, or red, 0.6–1 cm long, stamens (14–)16–24; ♀ with calyx 3+ cm Ø, lobes 4, corolla usually yellowish-white, campanulate, (0.9–)1–1.6 cm long, lobes recurved and ovate, staminodes 8(–16), ovary glabrous or pubescent. **Fruits** berries, with persistent calyx to 8.5 cm Ø, yellow to orange, flattened-globose to ovoid but usually globose, 8-locular, glabrescent; seeds dark brown. **Distr:** China; widely cultivated elsewhere. **Ecol:** Fl. May–Jun. Fr. Sep–Oct. **Zones:** 8–11.

D. malabarica (Desr.) Kostel., *Allg. Med.-Pharm. Fl.* 3: 109, 1834. (Syn. *Garcinia malabarica*, etc.). [Gaub Persimmon]. **Trees**, dioecious and polygamous, to 14 m tall; trunk 70 cm dbh; bark black; branchlets yellowish-tomentose. **Leaves** alternate; lamina oblong to lanceolate-oblong, to 5 × 22 cm, base cuneate, apex obtuse or broadly acuminate, glabrous, coriaceous, adaxially dark green, glossy; petiole 0.5–2 cm long. **Inflorescences** ♂ 3–7-flowered; ♀ 1–5 flowered. **Flowers** ♂ 1–1.5 cm long, calyx cup-shaped, silky, lobes 4–5, corolla widely tubular, 0.8–1.3 cm long, throat silky, lobes 4–5, yellow, ovate, rounded; ♀ to 2 × 3 cm, calyx cup-shaped, lobes ovate, tapered, 1.5 × 1.5 cm, glabrescent, corolla widely tubular, lobes to 1.8 cm long, tip recurved, ovary globose, apiculate in bud, yellowish-green, densely pilose. **Fruits** subsessile, ± globose, to 6 cm Ø, pulp transparent, sweet, astringent; seeds 4–8, elliptical wedge-shaped, to 2.5 cm long, brown. **Distr:** S and C East India, Siam, Indo-China, W Malesia. **Ecol:** In dry and intermediate zones, but always near rivers or waterways. **Zones:** 9–12. **Info:** The heartwood is variegated black and brown and much desired, though inferior in quality to calamander (*D. quaesita*) (Dassanayake 1981).

D. mollis Griff., *Journ. Agr. Hort. Soc. Ind.* 3: 145, 1844. [Maklua]. **Shrubs** or medium-sized **trees**; branchlets ± greenish. **Leaves** alternate; lamina ovate, ovate-oblong or obovate-lanceolate, 5.5–13 × 2.5–4 cm, base rounded, sometimes slightly unequal, margins ± entire, apex ± obtusely acuminate, initially pubescent on both surfaces, later adaxially glossy dark green, glabrescent, abaxially dull and paler. **Inflorescences** cymes, few-flowered, axillary, or flowers solitary. **Flowers** with corolla golden-yellow. **Fruits** globose, 1.5–2 cm Ø, glossy pale yellowish-green, with persistent green calyx which ± clasps the stem; seeds 1–1.1 × 0.7 cm. **Distr:** Tropical Asia. **Ecol:** In natural and mixed deciduous forests. Fl. Apr–Jun. Fr. Aug–Sep. **Zones:** 9–12. **Info:** A black dye known as 'shan black dye' is made from the fruits and leaves of this species, mainly used for dyeing silk. The plant has toxic properties, and are used for medical purposes.

D. philippensis (Desr.) Gurke, *Nat. Pflanzenfam.* 4, 1: 164, 1891. (Syn: *Cavanillea philippensis, D. discolor, D. utilis*). [Velvet Apple]. **Trees**, to 20 m

Above: *Clavija spathulata, Jacquina macrocarpa, Ardisia crenata, Ardisia elliptica.*

tall, evergreen. **Leaves** oblong to oblong-elliptic, (7.5–)20–30 × (3.5–)7–11 cm, base rounded to shallowly cordate, apex acuminate, acute, or rarely obtuse, leathery, glabrous, abaxially glaucous, lateral veins 10–14+ per side; petiole 0.5–1.7 cm long, robust, densely pubescent. **Inflorescences** ♂ cymes or racemes, (1–)3–7-flowered, sericeous, fragrant, pedicels 1–7 mm long, densely villous; ♀ with flowers solitary, similar to ♂, subsessile. **Flowers** ♂ with calyx densely villous, lobes 4, oblong, corolla white, urn-shaped, outside slightly hairy, lobes 4, reflexed, stamens 24; ♀ with staminodes apparently absent, style apparently 3-cleft. **Fruits** berries, edible, red to pink, globose, ~8 cm Ø, 8–10-locular, densely rusty, yellowish, or gray-villose, sessile; fruiting calyx 1.4–2.8 cm Ø; seeds dark brown, 3-sided, ~2.7 × 1.4 cm. **Distr:** Indonesia, Philippines Taiwan. **Ecol:** In coastal forests; 0–200 m. Fl. Mar–May. Fr. Nov. **Zones:** 10–12.

D. racemosa Roxb., *Fl. Ind.* 2: 536, 1824. (Syn: *D. toposia, Embryopterus racemosa*). **Trees**, dioecious, to 20 m tall; trunk to 40 cm dbh. **Leaves** alternate; lamina mostly elliptic to oblong, 2 × 10 – 5 × 3 cm, base acute or rounded, apex acuminate, thinly coriaceous, glossy, lateral nerves 8–9(–17) pairs; petiole 0.5–1.2 cm long. **Inflorescences** ♂ usually 3-flowered cymes, in more complex ones the dichotomy is repeated, to 12 flowers, axillary; ♀ flowers solitary or cymose, on lower part of young shoots, axillary, pedicel 0.6–1 cm long, hairy. **Flowers** ♂ yellow, 1 cm long, fragrant, lobes 2–3, small, thick, corolla yellow, urceolate, throat narrow, lobes 4, stamens 18–33, pistil none or a small disk; ♀ remaining closed for a long time as in ♂, calyx 4-lobed, corolla yellowish-white, ± urceolate, lobes 4, triangular, staminodes 12–16, anthers broad, ovary densely pubescent, 4–6-celled, style absent, stigmas 3–4. **Fruits** ovoid-ellipsoid to subglobose, 2.5–6 × 2–4 cm, yellowish, brown-hairy; persistent calyx 1.5–2.5 cm Ø, lobes triangular-ovate, 0.5–1 cm long. **Distr:** Sri Lanka, NE India, Bangladesh. **Ecol:** In moist regions; <1,000 m. **Zones:** 10–12.

D. walkeri (Wight) Guerke, *Pflanzenfam.* 4 (1): 162, 1897. (Syn. *Patonia walkeri*, etc.). **Trees**, dioecious and polygamous, to 6(–15) m tall; trunk 10(–20) cm dbh; bark black. **Leaves** oblong, ovate-oblong to lanceolate-oblong, to 9 × 22 cm, apex acuminate, chartaceous to thinly coriaceous, glabrous, glossy; petiole 0.7–2 cm long. **Inflorescences** cymes, axillary; ♂ and hermaphrodite in the same or separate inflorescences; ♂ cymes, ± sessile, pubescent; ♀ flowers solitary, axillary, ± sessile. **Flowers** ♂ with calyx green, campanulate, base with neck, lobes 4, deltoid, acuminate, corolla yellow or white, 1–1.3 cm long, tube inflated, lobes 4; ♀ with calyx green, widely campanulate, thick, lobes (3–)4, 1 cm long, margin slightly recurved, corolla white or yellow, tube 1.5 cm long, lobes strongly reflexed or patent, 0.7–1 cm long. **Fruits** yellowish, glabrescent, depressed-globose, apiculate, to 3.5 cm Ø, glossy; calyx persistent; seeds 8, thin, to 1.8 cm long, brown, smooth. **Distr:** Sri Lanka. **Ecol:** In wet lowland evergreen forests. Fl. Apr–May. Fr. May–Jun. **Zones:** 11–12.

THEOPHRASTACEAE

Genera 5, **species** ~105. **Shrubs** or **trees**, rarely acaulescent, sometimes dioecious, wood often brittle and breaking along enlarged rays. **Leaves** opposite; lamina simple, margins entire or serrate, mostly coriaceous, sometime spinose, venation pinnate, all species with characteristic linear sclereids running under the epidermis oriented without reference to the venation. **Inflorescences** racemes, short or long, mostly with minute bracts subtending the short pedicels. **Flowers** white, orange or green, 3–5-merous; sepals imbricate; corolla rotate or campanulate, basally united into a short tube, lobes mostly rotund; staminodes inserted on corolla near base of tube or at mouth, often resembling petals; stamens free or united; ovary 1-locular. **Fruits** berries or drupes, pericarp usually coriaceous, indehiscent; seeds few or numerous, often in gelatinous pulp. **Distr:** Mostly from tropical and warm temperate regions of the New World and the Old World.

Clavija

Species ~50. **Shrubs** or **trees**, erect, mostly unbranched. **Leaves** alternate, clustered at the tip of the stem; lamina simple, mostly elongate, margins entire or serrate, mostly coriaceous and glabrous, with numerous linear sclereids running beneath the epidermis. **Inflorescences** racemes, clustered among leaves, dioecious, mostly cauliflorous; peduncles slender; pedicels subtended by minute bracts. **Flowers** 3–5-merous; sepals small, rotate, persistent; corolla with petals united into a short tube, rotate, lobes round, fleshy, bearing staminodes as thickened appendages at corolla mouth; ♂ with filaments united into a stipelike tube, anthers forming a peltate disk at mouth of corolla, dehiscing on lower surface; ♀ with stamens free, reduced and sterile, ovary glabrous, stigma subsessile. **Fruits** berries, mostly globose, coriaceous; pericarp becoming hard, brittle; seeds several. **Distr:** Tropical America. **Ecol:** In humid forests. **Cult:** Propagate from seeds or by air-layering. Requires a moist, well-drained soil in partial sun.

C. spathulata Ruiz and Pavón, *Syst. Veg. Fl. Peru. Chil.* 1: 285, 1798 (non Hooker, 1844). (Syn: *Theophrasta longifolia*, etc.). **Trees**, to 10 m tall; stems with leafy portion 5–7 mm Ø, **Leaves** very large, narrowly obovate, 40 × 7–9 cm, base cuneate and winged toward petiole, margins entire or inconspicuously serrate, apex acute or acuminate, stiff, coriaceous, glabrous, adaxially dark green, abaxially paler; petiole 1–2 cm long. **Inflorescences** short, becoming stout and pendulous, few-flowered, 36 cm long, mostly borne below the terminal leaves. **Flowers** 4-merous, small, bright orange; sepals basally united, suborbicular, 1–1.2 mm long, fimbriate; corolla 6 mm long, ~1 cm Ø, orange, lobes round; ovary of ♂ flowers spikelike. **Fruits** berries, globose, coriaceous. **Distr:** Panama. **Ecol:** In forests; 0–100 m. Fl. Jul–Aug. **Zones:** 10–12.

Jacquinia

Species ~35. **Shrubs** or **small trees**, glabrate; wood brittle. **Leaves** mostly alternate or pseudo-verticillate, shortly petiolate; lamina with margins entire, apex with a sharp spine, mostly coriaceous and rigid. **Inflorescences** racemes, short, rarely flowers solitary, mostly terminal. **Flowers** mostly perfect, 5-merous, white or orange; calyx with sepals free or basally connate, imbricate, round, sometimes ciliate or crenate; corolla salverform, tube straight, campanulate to urceolate, lobes imbricate, round, thick, margins thin; staminodes 5, resembling corolla lobes, connate to top of corolla tube but free above, sometimes dorsally appendaged; filaments flat, narrowing upward, basally connate into a tube which is basally adnate to corolla tube, anthers ovate, pointed, exserted; ovary conical unilocular, ovules numerous, style not demarcated from ovary, stigma peltate, entire or minutely lobed. **Fruits** berries, ovoid, elliptical or globose, sometimes cuspidate, coriaceous; pericarp thick; seed large, single, embedded in spongy or mucilaginous pulp. **Distr:** Tropical America. **Cult:** Propagate from seeds or by air-layering. Requires a poor, sandy, well-drained, alkaline soil in shade or partial sun.

J. macrocarpa Cav., *Icon.* 5: 55, tab. 483, 1799. (Syn: *J. aurantiaca, J. panamensis*). [Barbasco]. **Shrubs** or **small trees**, to 8 m tall; branches brittle; stems drying to dark reddish-brown. ; **Stipules** linear to lanceolate. **Leaves** mostly elliptic, often narrow, mostly 5–7 × 1.5–2.5 cm, base obtuse or rounded, margins often revolute, apex obtuse or acute with a sharp blackish spine, coriaceous. **Inflorescences** racemes, round-topped, short, condensed, terminal or axillary; peduncle and rachis 1–3 cm long; pedicels ~1 mm long, drying to orange, subtended by cucullate, lanceolate, ciliolate bracts 3–7 mm long. **Flowers** bright orange; calyx cupular, 5–6 mm long, lobes 5, ciliolate imbricate sepals; corolla salverform, tube ~5 × 3 mm, petals 5, round ~5 mm long; staminodes resembling petals but narrower with 4 longitudinal ridges (appendages); stamens 5, filaments flat, basally fused into a tube narrowing upward, anthers yellowish-white, ovoid, acute, converging around style; ovary conical, style not demarcated, stigma peltate. **Fruits** berries, ovoid or ellipsoidal, 3–4 cm Ø, coriaceous, brown, rugose; pericarp ~5 mm thick; seeds 1–many, ~2 cm long, embedded in a soft yellowish matrix.

Above: *Ardisia solanacea, Ardisia willisii, Elingamita johnsonii.*

Distr: The Pacific coast of Panama. **Ecol:** Fl. Nov–Apr. **Zones:** 11–12.

MYRSINACEAE

Genera ~30, species 900–1,000. **Trees, shrubs** or **lianas. Leaves** exstipulate, alternate, often clustered at branch ends; lamina simple, margins entire or toothed, always with schizogenous resiniferous dots or lines, but these varying from very obvious to obscure. **Inflorescences** racemose, paniculate, umbellate or fasciculate, lateral or terminal. **Flowers** usually small, individually inconspicuous, 5- or 4-merous, actinomorphic, hermaphrodite or plants very often dioecious; calyx with free or connate sepals, often ciliate, often with dark spots, aestivation valvate, imbricate or contorted; petals usually white or pink, rarely purple or yellow, free or more usually connate, often with dark dots or stripes, often papillose; stamens as many as and opposite petals, filaments long or short, adnate to corolla, sometimes almost free; ovary globose, ovoid or clavate, usually superior, number of carpels obscure, probably 3–4, style long and slender or short and thick or rarely absent, stigma punctiform, capitate, discoid or lobed, ovules few to many, in 1 or more rows. **Fruits** indehiscent berries or drupes; seed 1, in copious, often ruminate, endosperm. **Distr:** Throughout the tropics.

Ardisia

Species ~400–500. **Trees, shrubs**, suffrutescent (or rarely herbs). **Leaves** alternate or pseudo-verticillate; lamina usually punctate or punctate-lineate. **Inflorescences** paniculate, cymose, corymbose, or umbellate, rarely racemose. **Flowers** bisexual, often punctate, (4)5-merous; calyx campanulate or cupular, sepals free or barely basally united, imbricate or quincuncial, usually punctate or punctate-lineate; corolla campanulate, often punctuate, lobes basally united, overlapping to right or very rarely to left, imbricate, or quincuncial, often conical in bud; stamens attached at base or middle of corolla tube, filaments very short, base broad, anthers dehiscing longitudinally or by apical pores; ovary ovoid or subglobose, as long as or longer than petals, ovules 3–many, style base persistent, stigma minute, apiculate. **Fruits** drupaceous, 1-seeded, punctate, sometimes longitudinally ribbed, with somewhat fleshy exocarp and crusty or slightly bony endocarp; seeds covered by membranous remnants of placenta. **Distr:** Tropical E and SE Asia, China, Americas, Australia, and the Pacific Islands. **Cult:** Propagate from semiripe cuttings or seeds. Requires a fertile, moist, well-drained, humus-rich compost in partial sun.

A. crenata Sims, *Bot. Mag.* 45: pl. 1950, 1817. (Syn: *A. bicolor*, etc.). [Coral Berry, Hilo Holly]. **Shrubs**, 1–1.5(–3) m high; branchlets minutely reddish glandular-papillate. **Leaves** elliptic, narrowly lanceolate, or oblanceolate, 7–15 × 2–4 cm, base cuneate, margins subrevolute, crenate, or undulate, with large vascularised marginal nodules, apex acute or acuminate, leathery or papery, punctate, lateral veins 12–18 per side; petiole narrowly marginate, 0.6–1 cm long. **Inflorescences** umbellate or cymose, terminal, on specialized, 2- or 3-leaved lateral branches 4–16(–40) cm long; pedicel 7–10 cm long. **Flowers** fragrant, membranous, white or rarely pinkish, 4–6 mm long; sepals oblong-ovate, 1–1.5(–2.5) mm long, margin entire, apex rounded or obtuse, glabrous, punctate; petals nearly free, ovate, acute, punctate, adaxially glandular-papillose near base; stamens shorter than petals, filaments nearly obsolete, anthers triangular-lanceolate; pistil subequaling petals, ovary glabrous, punctate, ovules ~5, uniseriate. **Fruits** red, globose, 6–8 mm Ø, punctate. **Distr:** China, SW India, Japan, Malaysia, Philippines, Vietnam. **Ecol:** In forests, hillsides; 100–2,400 m. Fl. May–Jun. Fr. Oct–Dec. **Zones:** 7–11.

A. elliptica Thunb., *Nov. Gen.* 8: 119, 1795. (Syn: *A. littoralis, A. polycephala, A. humilis*). [Shoebutton Ardisia]. **Shrubs**, much-branched, 2 m high; branches stout, with large scars of fallen leaves, glabrous; bark dark-gray, vertically cracked. **Leaves** obovate or oblong or elliptic-sublanceolate, 7.5–12 × 2.5–4 cm, base tapering to petiole or cuneate, margins entire, apex usually obtuse, sometimes shortly subacuminate, chartaceous or subcoriaceous, glabrous, abaxially pale, punctate with small pellucid glands, lateral veins inconspicuous; petiole 2–5 mm long. **Inflorescences** corymbose racemes, dense, many-flowered, pedunculate, axillary; pedicels 1–1.8 cm long, stout. **Flowers** pink; calyx pubescent, lobes large, semiorbicular to rotund, 3–4 mm Ø, with a narrow membranous margin, punctate; petals shortly basally connate, lobes obliquely ovate-acuminate, 0.8–1 cm long, gland-dotted; stamens much shorter than petals; ovary subglobose, 2 mm Ø, style small. **Fruits** berries, depressed-globose, 5–6 mm Ø, dark-purple. **Distr:** Maldives, Sri Lanka to S India, Malaysia, Philippines. **Ecol:** In lowland areas, especially in coastal areas. Fl. & Fr. Sep–Apr. **Zones:** 10–12.

A. solanacea Roxburgh, *Pl. Coromandel.* 1: 27, 1795. [Baludan, China Shrub, Jet Berry]. **Shrubs** or **trees**, to 6 m tall, glabrous; branchlets prominently angular, 5–7 mm Ø, **Leaves** elliptic or oblanceolate, 12–20 × 4–7 cm, base cuneate or narrowly decurrent on petiole, margins subrevolute, entire, apex acute, papery, black-punctate, lateral veins ~20 per side, raised on both surfaces, marginal vein absent; petiole 1–2 cm long, canaliculate. **Inflorescences** paniculate with racemose or rarely corymbose branches, at bases of new shoots, 3–8 cm long. **Flowers** leathery, pink, ~1 cm Ø; sepals broadly ovate to reniform, ~3 mm long, base subauriculate, margin subentire or crenulate, ciliate, apex rounded, densely black-punctate, scarious; petals nearly free, broadly ovate, ~9 mm long, punctate, margins entire, hyaline, apex obtuse or acute; stamens subequaling petals, anthers linear-lanceolate, densely punctate dorsally; pistil subequaling petals, ovary globose, densely punctuate, ovules numerous, multiseriate. **Fruits** purplish-red or blackish, oblate, 7–9 mm Ø, densely black-punctate. **Distr:** China, India, Nepal, Singapore, Sri Lanka. **Ecol:** In mixed forests, mountains or hillsides; 400–1,600 m. Fl. Feb–Mar. Fr. Aug–Nov. **Zones:** 9–12.

A. willisii Mez, *Pflanzen.* 9 (IV-236): 140. 1902. (*A. humilis, A. moonii* var. *subsessilis*). [Lunudan]. **Shrubs**, to 2 m high; stems cylindrical, glabrous. **Leaves** clustered at ends of branches; lamina elliptic or obovate-elliptic, 10–12 × 5–7 cm, base subacute to cuneate, margins entire, apex obtuse or subacute, often twisted, thick, shiny, glabrous, lateral veins oblique, not conspicuous. **Inflorescences** simple or compound panicles, 5–10 cm long, (sub)terminal, glabrous; bracts small; pedicels ~7 mm long at anthesis. **Flowers** ~6 mm long, pink; sepals slightly basally connate, broadly ovate, obtuse or subacute, puberulous, margin ciliolate; petals asymmetric, elliptic, 5–6 mm long, acute, twisted, brown-punctate; stamens much shorter than petals, filaments small, anthers dorsally punctate; ovary globose, 4–5 mm Ø, style small. **Fruits** drupes, globose, 0.8–1 cm Ø, with fleshy, copious pulp, bright scarlet; seeds with a fibrous coat. **Distr:** Sri Lanka. **Ecol:** In forests. Fl. & Fr. Dec–May. **Zones:** 11–12. **Info:** The fruits are edible and have medicinal uses.

Elingamita

Species 1. *Generic description same as specific.* **Cult:** Propagate from semiripe cuttings or seeds. Requires a fertile, moist, well-drained soil in full partial sun.

E. johnsonii G.T.S. Baylis, *Rec. Auck. Inst. Mus.* 4: 100, 1951. **Trees**, glabrous; bark smooth. **Stipules** absent. **Leaves** obovate to broad-obovate, ~10–18 × 4–9 cm, margins entire, very coriaceous, gland-dotted; petiole stout, ~1 cm long. **Inflorescences** panicles, ~5 cm Ø, terminal, enlarged in fruit; bracts broad, caducous; pedicels ~5 mm long. **Flowers** bisexual, yellowish, 4–6-partite; sepals valvate, free, ovate, ~1.5 mm long, acute; corolla shorter than sepals, tubular, ~1 × ~2.5 mm, limb truncate or obscurely lobed; stamens ~3·5 mm long, filaments almost wholly free from corolla, in mature flowers much exceeding, anthers dorsifixed, splitting down their whole length; pistil ~2.5 mm long, ovary ovoid, narrowed into a rather stout style with a stigmatic pit at apex, ovules 2–4, at one level in upper half of placenta. **Fruits** drupes, bright red, globose to subglobose, ~1.7 cm Ø, crowned with persistent style, 1-seeded, endocarp brittle. **Distr:** New Zealand (Three Kings Islands). **Ecol:** Fr. Dec–Feb. **Zones:** 8–11.

Above: *Synardisia venosa, Camellia sinensis, Gordonia axillaris, Schima wallichii.*

Synardisia

Species 1. *Generic description same as specific.* **Cult:** Propagate from seeds or cuttings. Requires a moist, well-drained soil in full or partial sun.

S. venosa (Mast.) Lundell, *Wrightia* 3(5): 90, 1963. (Syn: *Ardisia venosa*). **Shrubs** or **trees**, sometimes to 20 m tall, rarely epiphytic. **Leaves** large, mostly oblanceolate, 15–25 × 5–9 cm, base decurrent, attenuated, margins entire, apex acute or broadly acuminate, acumen obtuse, surface punctate, glabrous; petioles short, broadly marginate, to 2 cm long. **Inflorescences** subpyramidal, 3–4-pinnate-paniculate, many-flowered, terminal, often exceeding leaves, to 45 cm long, flowers corymbose; branches densely pubescent; basal bracts leafy, oblanceolate, to 10 cm long, glabrous; pedicels slender, to 2 cm long, densely pubescent. **Flowers** 0.6–1.2 cm long; sepals almost free, narrowly lanceolate, 2.5–5 mm long, caudate-acuminate, punctate, pubescent; corolla campanulate or suburceolate, thin, semitransparent, pink, 0.5–1.1 cm long; petals connate for most their length, subovate, obtuse-rounded; stamens small, included, attached at base of corolla, to 3.5 mm long, filaments united into a shallow ring; ovary subglobose, glabrous, style included, 2.5–4 mm long, ovules 8+, pluri-seriate. **Fruits** depressed-globose, to 1 cm Ø, purplish. **Distr:** Mexico, El Salvador, Honduras. **Ecol:** In humid forests; 1,300–3,200 m. **Zones:** 10–12.

THEACEAE

Genera ~7, **species** 195. **Shrubs** or **trees**, evergreen, rarely deciduous, usually bisexual, rarely dioecious. **Stipules** absent. **Leaves** alternate, petiolate or rarely sessile; lamina simple, margins usually serrate, rarely entire, secondary veins pinnate. **Inflorescences** wit flowers solitary or sometimes to 3(–10+) in clusters or racemes, axillary or subterminal; flowers pedicellate or subsessile; bracteoles 2–8 or rarely more, persistent or caducous, sometimes undifferentiated from sepals. **Flowers** with sepals 5(or 6) or rarely more, persistent; corolla white, red, or yellow; petals 5 or rarely more, basally connate or rarely distinct, adnate to androecium; stamens numerous, in 1–6 whorls, outer filaments ± basally connate, anthers dorsifixed or basifixed, biloculed, laterally and longitudinally dehiscent; gynoecium 3–5-carpellate, carpels connate or rarely incompletely connate to nearly distinct, ovary superior, rarely half inferior, 3–5-loculed, locules 2–5(to ~100+)-ovuled, styles distinct to basally connate, rarely completely united. **Fruits** loculicidal capsules, or indehiscent and drupaceous or baccate, seeds 1–many per locule, pericarp woody, leathery, or fleshy, columella persistent or ± degenerating; seeds globose, semiglobose, compressed oblong, ovoid, or reniform, winged or wingless, testa bony, leathery, or sometimes with a fleshy red outer layer or sarcotesta. **Distr:** Mostly SE Asia to Malesia; SE USA to tropical S America.

Camellia

Species 120. **Shrubs** or **small trees**, rarely large trees, evergreen. **Leaves** petiolate or rarely sessile and amplexicaul; lamina with margins serrate, serrulate, or rarely entire, leathery to thinly leathery. **Inflorescences** with flowers solitary or rarely to 3-flowered clusters, axillary or subterminal; flowers ± pedicellate or sessile; bracteoles differentiated from sepals or not, 2–10, spirally arranged, persistent or caducous. **Flowers** with sepals 5(or 6), persistent, distinct or basally connate; petals 5–8(–12), white, red, or yellow, ± basally connate; stamens numerous, 2–6-whorled, outer filament whorl ± basally connate into a tube and adnate to petals, anthers dorsifixed, biloculed; ovary superior, 3–5-loculed, placentation axile. **Fruits** capsules, globose or oblate, (1–2)3–5-loculed; seeds ± globose. **Distr:** Bhutan, Cambodia, China, NE India, Indonesia, S Japan, S Korea, Laos, Malaysia, Myanmar, Nepal, Philippines, Thailand, Vietnam. **Cult:** Propagate from seeds, cuttings or by grafting. Requires an acidic to neutral, moist, well-drained soil in shade to partial sun.

C. sinensis (L.) Kuntze, *Trudy Imp. S.-Peterburgsk. Bot. Sada.* 10: 195, 1887. [Tea]. **Shrubs** or **trees**, 1–5(–9) m tall. **Leaves** elliptic, oblong-elliptic, or oblong, 5–14 × 2–7.5 cm, base cuneate, margins serrate to serrulate, apex bluntly acute to acuminate, adaxially dark green, shiny, abaxially paler, leathery; petiole 4–7 mm long, pubescent. **Inflorescences** flowers solitary or to 3-flowered clusters, axillary; pedicel 0.5–1 cm long, recurved; bracteoles 2, caducous, ovate, ~2 mm long. **Flowers** 2.5–3.5 cm Ø; sepals 5, persistent, broadly ovate to suborbicular, 3–5 mm long, outside glabrous or white-pubescent; petals 6–8, white, outer 1–3 sepaloid, inner ones obovate, 1.5–2 × 1.2–2 cm; stamens numerous, 0.8–1.3 cm long, outer filament whorl basally connate; ovary globose, triloculed, style ~1 cm long, apically trilobed. **Fruits** capsules, oblate, bicoccal, or rarely globose, 1–1.5 × 1.5–3 cm, 1- or biloculed, locules 1-seeded; seeds brown subglobose, 1–1.4 cm Ø, **Distr:** China, NE India, S Japan, S Korea, Laos, Myanmar, Taiwan, Thailand, Vietnam. **Ecol:** In evergreen forests, thickets; 100–2,200 m. Fl. Oct–Feb. Fr. Aug–Oct. **Zones:** 10–12.

Gordonia

Species ~70. **Trees. Leaves** spirally arranged; lamina coriaceous. **Inflorescences** with flowers solitary, axillary; pedicels with 4–7 bracteoles. **Flowers** with sepals 5, imbricate, unequal, persistent; petals 5–6, unequal, basally connate, stamens numerous, basally connate and adhering to petals, anthers inserted dorsally in basal incision; ovary 3–5-loculed, ovules 4–8 per locule, pendulous, style slender, stigmas 3–6-lobed. **Fruits** capsules, woody, oblong, loculicidally dehiscent, with a solid persistent columella; seeds several, compressed, obliquely winged. **Zones:** Tropical and subtropical Asia and N America. **Cult:** Propagate from seeds or half-hardened cuttings. Requires a fertile, humus-rich, moist, well-drained soil in full or partial sun.

G. axillaris (Roxb.) Dietr., *Syn. Pl.* 4: 863, 1847. (Syn. *Camellia axillaris*, etc.). [Fried Egg Plant]. **Shrubs** or **small trees**; branchlets and terminal buds densely pubescent. **Leaves** oblong, oblong-lanceolate or oblanceolate, 8–12 × 2–4 cm, base acute or cuneate, margins remotely crenate in upper part, apex retuse, sometimes obtuse or rounded, coriaceous, green, adaxially glabrous, abaxially slightly pubescent when young, midrib impressed above, veins obscure on both surfaces; petiole 0.5–1 cm long. **Inflorescences** with flowers solitary or paired, axillary or subterminal; flowers nearly sessile; bracts deciduous, subrounded, densely pubescent outside. **Flowers** with sepals 5, obovate, 1.5–2 cm long, densely pubescent outside; petals 5, white, obovate, 3–4.5 cm long, emarginate, pubescent outside; stamens numerous, unequal, filaments 1.5–2 cm long; ovary pubescent, style 1.5–2 cm long. **Fruits** capsules, oblong or oblong-ovoid, 2–3.5 cm long, pilose; seeds compressed, oblong, 1.5–2 cm long. **Distr:** Indo-China, China, Taiwan. **Ecol:** In broad-leaved forests; 0–2,300 m. **Zones:** 8–11.

Schima

Species ~20. **Trees**, evergreen; branchlets with white lenticels. **Leaves** petiolate; lamina with margins entire or serrate. **Inflorescences** with flowers solitary, or rarely 3–5-flowerd short racemes, axillary; pedicels ± slender, apically thickened and recurved; bracteoles usually 2(or more), apically on pedicel, caducous. **Flowers** bisexual; sepals 5, imbricate, persistent, slightly basally connate; petals 5, white, basally connate and adnate with distinct filaments; stamens numerous, anther dorsifixed; ovary densely tomentose, 5-loculed with 2 or 3 ovules per locule, placentation axile, style stout, glabrous, stigma capitate, 5-lobed. **Fruits** capsules, globose or depressed-globose, pericarp woody, splitting for half its length into 5 valves, columella persistent; seeds small, reniform, flat, with a marginal membranous wing. **Distr:** Bhutan, Cambodia, China, NE India, Indonesia, Japan (Ryukyu Islands), Laos, Malaysia, Myanmar, Nepal, Thailand, Vietnam. **Cult:** Propagate from seeds or half-hardened cuttings. Requires a moist, well-drained, humus-rich soil in full sun.

S. wallichii (Candolle) Korthals, *Verh. Nat. Gesch. Ned. Bezitt. Bot.* 143, 1842. (Syn: *Gordonia chilaunia*, etc.). [Needlewood Tree]. **Trees**, 10–15(–20) m tall. **Leaves** elliptic, 8–17.5 × 4–7.5 cm, base broadly cuneate, margins entire, apex bluntly acute, adaxially dark green and glabrous, abaxially paler, sparsely pubescent, leathery, secondary veins

SA/SAJF

TG

Above: *Bejaria aestuans, Epacris longiflora, Epacris purpurascens, Rhododendron hippophaeoides.*

8–12 per side; petiole 1–2 cm long, yellowish-villous. **Inflorescences** flowers solitary or 3-flowered clusters, axillary; pedicel 1–1.5 cm, white-lenticellate; bracteoles 2, caducous. **Flowers** white, fragrant, 3–4 cm Ø; sepals lunate to semiorbicular, 1.5–2 × 3–3.5 mm, ciliolate; petals white, broadly obovate, 1.5–2 × ~1.5 cm, rounded; stamens 0.8–1 cm long, filaments many, basally adnate with petals; ovary globose, style ~7 mm long, glabrous, stigmas 5, capitate. **Fruits** capsules, brown, subglobose, 1.5–2 cm Ø, 5-loculed, locules 2-seeded, pericarp woody, splitting into 5 valves, white-lenticellate; seeds reniform, winged, 0.8–1 × 0.5–1 cm. **Distr:** Bhutan, China, N India, Laos, Myanmar, Nepal, Thailand, Vietnam. **Ecol:** In forests; (300–)800–1,800(–2700) m. Fl. Apr–May. Fr. Nov–Dec. **Zones:** 9–11.

ERICACEAE

Subfamilies 8, **Genera** 126, **species** 3,995. **Trees, shrubs, herbs**, or sometimes herbs lacking chlorophyll. **Leaves** spirally arranged, sometimes decussate; lamina with margins often toothed. **Inflorescences** racemose; bracteoles paired and basal. **Flowers** are 4–5-merous; calyx imbricate with a connate and imbricate corolla; stamens 10, sometimes with spurs or awns, dehiscing by pores, pollen in tetrads, rarely single; ovary superior or inferior, placentation axile, rarely parietal, locules often many-ovuled, style ± as long as corolla, slender. **Fruits** capsules or berries, rarely drupes, with persistent calyx. **Distr:** Temperate and subarctic regions and at high elevations in tropical regions.

Subfam. ERICOIDEAE

Genera 19, **species** ~1,790. **Shrubs** to **trees**, evergreen or deciduous. **Leaves** spirally arranged, flat (convolute). **Inflorescences** racemose, spicate, corymbose, umbellate, fasciculate, capitate, or flowers solitary, terminal or axillary; bracts and bracteoles extremely various; flowers articulated or continuous with pedicel. **Flowers** usually 4–7(–9)-merous, actinomorphic or slightly zygomorphic, held erect to ± pendulous; corolla free; anthers lacking appendages, (pollen with viscin threads). **Fruits** capsules, septicidal to loculicidal, indehiscent pods, or drupes; seeds with variable testa cells. **Distr:** Worldwide.

Bejaria

Species 15–25. **Shrubs** or small slender **trees**, glabrous to densely glandular-setose. **Leaves** alternate, petiolate; lamina flat to tightly revolute, margins entire, rarely denticulate, pinnately veined. **Inflorescences** racemose to corymbose or rarely paniculate, few- to many-flowered, terminal or axillary; floral bract small and inconspicuous; pedicels continuous with calyx; bracteoles 2. **Flowers** 5–8-merous, sometimes mildly fragrant; calyx synsepalous, persistent in fruit, deeply lobed, lobes much longer than tube, separate or imbricate at anthesis; corolla polypetalous, aestivation imbricate; petals showy, suberect, imbricate or spreading at anthesis, soon deciduous; stamens 10–20 (twice as many as petals) about as long as petals. **Fruits** capsules, depressed-obovoid or depressed-globose, woody, 5–7-valvate; seeds numerous, winged or tailed. **Distr:** S USA to Cuba, and Mexico south into Bolivia, and east to Guyana. **Cult:** Propagate from cuttings or seeds. Requires a humus-rich, moist, well-drained soil in filtered sun.

B. aestuans L., *Mont. Pl.* 242, 1771. (Syn: *B. glauca*). **Shrubs** or **small trees**, 5–6 m tall; twigs terete, coarsely ribbed. **Leaves** flat or slightly revolute, elliptic to lanceolate or oblanceolate, lance-elliptic or rarely ovate, 2–5 × 0.8–1.5 cm, base long-cuneate, apex acute to short-acuminate; petiole subterete, slightly winged, 0.5–1 cm long. **Inflorescences** short-racemose or subcorymbose, 10–20-flowered, rachis 1.5–2.5 cm; pedicels 1.5–2.4 cm long. **Flowers** erect, 7-merous; calyx 2–3 mm long; corolla spreading campanulate, petals obovate, 1.5–2 cm long, white to dark rose-pink. **Fruit** capsules, depressed-ovoid, 7–8 mm Ø, **Distr:** WC Mexico almost continuously to Bolivia. Very rare in Nicaragua and Panama. Considered endangered in Panama. **Ecol:** In pine woods, cloud forests, dry montane forests, shrub páramo and grasslands; 300–3,500 m. Fl. Jan, May, Jul. Fr. Jan, Jul–Aug. **Zones:** 9–11.

Epacris

Species 40. **Shrubs**, usually erect, often rigid; branchlets usually pubescent; branches with leaf scars differing in prominence and shape. **Leaves** scattered or imbricate, sometimes stem-clasping but not sheathing, sessile or petiolate; lamina usually concolorous, flat to concave or keeled. **Inflorescences** with flowers solitary, axillary in upper leaf axils or extending along branches to form leafy inflorescences; bracts numerous, imbricate, grading in size to sepals. **Flowers** with sepals 5, usually glabrous, margins ciliolate; corolla tube cylindrical or campanulate, shorter or much longer than sepals, glabrous, lobes spreading or erect, shorter or longer than tube, glabrous, usually ± obtuse, imbricate in bud; stamens 5, filaments short, inserted in throat, anthers included in tube or half-exserted, versatile; nectary of 5 scales, rarely joined; ovary 5-locled, locules several-ovuled, placentation axile, style inserted in deep depression in ovary, short, or long and exserted. **Fruits** loculicidal capsules. **Distr:** Australia, New Caledonia, New Zealand. **Cult:** Propagate from seeds or cuttings. Requires a moist, well-drained soil in partial sun.

E. longiflora Cav., *Icones et Descr. Plant.* 4(1): 25, t. 344, 1797. [Fuchsia Heath]. **Shrubs**, erect to spreading, straggly, 0.5–2 m high; stems with prominent short broad leaf scars; branchlets villous. **Leaves** spreading to ± reflexed; lamina ovate, 0.5–1.7 × 0.3–0.7 cm, base obtuse or cordate, margins minutely toothed, apex acuminate, thin, flat, adaxially sometimes scabrous. **Inflorescences** with flowers solitary, axillary, extending down branches, initially erect, later pendent; peduncle to 2 mm long; bracts usually long-acuminate. **Flowers** narrowly tubular, 1.2–2 cm long, 5–6 mm Ø, corolla tube red, lobes white, or all white; sepals 4.5–6 mm long; corolla tube 1.2–2.7 cm long, lobes erect to spreading, 2.4–4.4 mm long; anthers hidden in tube. **Fruits** capsules, 3–4 mm long. **Distr:** Australia (NSW, Queensland). **Ecol:** In sandstone ledges and cliffs, heath and open forests, mostly in moist places; 0–760 m. Fl. throughout most of the year. **Zones:** 9–11.

E. purpurascens R. Br., *Prodr.* 550, 1810. [Port Jackson Heath]. **Shrubs**, erect, 0.5–1.5 m high; stems with prominent short, broad, leaf scars; branchlets villous. **Leaves** appressed to stem near base, spreading and recurving above, ovate or cordate, 0.7–2.1 × 0.4–0.9 cm, base obtuse or cordate, margins entire or fimbriate toward base, apex acuminate, thin, concave; petiole 0.8–1.5 mm long. Inflorescences with flowers solitary, along much of the branchlets; peduncle 1–2 mm long; bracts acuminate. **Flowers** usually 0.7–1 cm Ø, white or red-tinged; sepals 4.3–6.5 mm long; corolla tube 4.3–7.7 mm long, lobes 3.6–5 mm long; anthers half-exserted. **Fruits** capsules, 2–2.5 mm long. **Distr:** Australia (NSW). **Ecol:** In dry sclerophyll forests and scrubs, and near creeks and swamps on sandstone. Fl. Jul–Nov. **Zones:** 8–11.

Rhododendron

Subgenera 5, **species** 850. **Trees** or **shrubs**, terrestrial or epiphytic, variously hairy, hairs usually tangled and coming away as a layer. **Leaves** evergreen, deciduous or semideciduous, alternate, sometimes clustered at stem apex; lamina with margins entire, very rarely crenulate. **Inflorescences** racemes or corymbs, mostly terminal, sometimes lateral, few- to many-flowered, sometimes reduced to a single flower. **Flowers** with calyx persistent, 5–8-lobed, sometimes reduced to a rim; corolla funnelform, campanulate, tubular, rotate or hypocrateriform, regular or slightly zygomorphic, 5(–8)-lobed, lobes imbricate in bud; stamens 5–10(–27), inserted at base of corolla, usually declinate, filaments linear to filiform, anthers without appendages, opening by terminal or oblique pores; disk usually thick, 5–10(–14)-lobed; ovary 5(–18)-loculed, style straight or declinate to deflexed, persistent, stigma capitate-discoid, crenate to lobed. **Fruits** capsules, cylindrical, coniform, or ovoid, sometimes curved, dehiscent from top, septicidal; valves thick or thin, straight or twisted; seeds numerous, minute, fusiform, always winged, or both ends with appendages or threadlike tails. **Distr:** Asia, Malesia, Australia; Europe; N America. The tropical species (~300 spp., also known as vireyas) are predominantly from Malesia, with the greatest diversity in New Guinea. **Cult:** Propagate from seeds, air-layering or from rooted cuttings. Requires a moist, humus-rich, acidic soil, in full or partial sun. **Info:** The classification followed here is that of Craven et al. (2008).

Subgenus Rhododendron

Section Rhododendron

Corolla campanulate to funnel-shaped or tubular; scales entire to crenulate. **Seeds** without distinct tails (Craven et al. 2008).

R. hippophaeoides I.B. Balfour and W.W. Smith, *Notes Roy. Bot. Gard. Edinburgh*. 9: 236, 1916. **Shrubs**, erect, 0.25–1(–1.5) m high; branchlets virgate; bud scales deciduous. **Leaves** oblong, elliptic, oblong-lanceolate to oblong-ovate, (0.8–)1.2–2.5(–4) × (0.4–)0.5–1(–1.7) cm, base broadly cuneate, apex rounded or obtuse, adaxially pale green, scaly, abaxially yellowish-buff, uniformly creamy-yellow; petiole 2–5 mm long, scaly. **Inflorescences** 4–7(–8)-flowered; pedicels 2.5–7 cm long. **Flowers** with calyx reddish, lobes 1–2 mm long, rounded, unequal, persistent in fruit, scaly, margin ciliate; corolla broadly funnelform, bright rose or lavender-blue to bluish-purple, rarely white, 1–1.3(–1.5) cm long, tube ~4 mm long, outer surface not scaly, inner surface pubescent; stamens (8–)10, shorter than corolla, filaments pubescent toward base; ovary 1–2.5 mm long, scaly, style 0.4–1.6 cm long, usually as long as stamens, glabrous or occasionally pubescent at base. **Fruits** capsules, narrowly ovoid, 5–6 mm long, densely scaly. **Distr:** China (Sichuan, Yunnan). **Ecol:** In forests, thickets; 2,400–4,800 m. Fl. May–Jun. Fr. Oct. **Zones:** 9–11.

R. leptothrium I.B. Balfour and Forrest, *Notes Roy. Bot. Gard. Edinburgh*. 11: 84, 1919. (Syn: *R. australe*). **Shrubs** or **small trees**, 3–4(–6) m tall; young shoots densely white-pubescent. **Leaves** lanceolate or oblong-lanceolate, 4–12 × 1.8–3.5 cm, base cuneate, margins shallowly undulate, slightly revolute, apex tapered, truncate, subcoriaceous, adaxially pubescent only on midrib; petiole 1–1.8 cm long, densely pubescent or sparsely hairy. **Inflorescences** subapical, usually in clusters of 2–4, 1-flowered; pedicel 1.5–2.5 cm long, densely setose. **Flowers** with calyx lobes elliptic or oblong-ovate, 6–8 mm long, hairy at base outside; corolla rotate, pale rose to magenta-purple, with dark flecks on upper lobes, lobes obovate, 2–2.5 cm long; stamens 5, unequal, shorter than corolla, filaments flat, pubescent below middle; ovary ovoid, ~3 mm long, apex hairy, style ~2.2 cm long, longer than stamens, exserted, glabrous. **Fruits** capsules, ovoid, ~6 mm Ø, with persistent calyx lobes. **Distr:** China, Myanmar. **Ecol:** In thickets; 1,700–3,200 m. Fl. May–Jun. Fr. Sep–Nov. **Zones:** 9–12.

R. spinuliferum Franchet, *J. Bot.* (Morot). 9: 399, 1895. **Shrubs**, 0.5–1(–3.5) m tall; old branches brownish-red. **Leaves** obovate, elliptic to lanceolate, 3–10.5 × 1.8–3.8 cm, base cuneate, apex acuminate, rarely acute, apiculate, papery, adaxially yellowish-green, glabrous or pubescent, abaxially paler, glabrous; petiole 3–6 mm long, ± pubescent, setose or scaly. **Inflorescences** subterminal, 2–4-flowered; bud scales white-pubescent and densely scaly outside; pedicels 1–2 cm long, densely pubescent and scaly. **Flowers** with calyx shallowly cup-shaped, without conspicuous lobes, small; corolla tubular, narrowed at both ends, bright or orange red, 1.5–2.5 cm long, outside glabrous, rarely pubescent at lobe base, lobes 5, erect, ovate; stamens 10, unequal, slightly exserted from corolla, filaments glabrous, anthers purplish-black; ovary 5-locular, scaly, densely hairy, style exserted from corolla, usually glabrous, rarely pubescent at base. **Fruits** capsules, cylindric, 1–1.4 cm, scaly, pubescent. **Distr:** China. **Ecol:** In mixed forests, thickets; 1,900–2,500 m. Fl. Feb–Jun. **Zones:** 8–11.

Section Vireya

Scales sessile or sometimes stalked, lobed to deeply incised or sometimes entire. **Flowers** with corolla campanulate, trumpetlike, salver-shaped, tubular or funnel-shaped; stamens (5–)10(–16), exserted to included, staminal filaments glabrous or hairy from the base. **Fruits** capsules, valves twisting after dehiscence; seeds with a distinct tail at each end (Craven et al. 2008).

Subsection Euvireya

Scales sessile or stalked, scattered or dense, not of 2 obviously different size classes, lobed to deeply incised (or sometimes entire) and the center not dark-colored (Craven et al. 2008).

R. alborugosum Argent and J. Dransf., *Notes RGB Edinb*. 46(1): 27, 1989. **Shrubs**, to 3 m tall; twigs smooth; internodes 5–12 cm long. **Leaves** in pseudowhorls of 3–5; lamina elliptic, narrowly ovate or obovate, 4–8 × 2.3–3.7 cm, base broadly tapering to rounded, margins entire, apex broadly acute, obtuse or rounded, adaxially laxly silvery-scaly, soon glabrescent and dark green, abaxially persistently laxly scaly and paler; petiole 0.8–1.2 cm long, red. **Inflorescences** umbels, 6–10-flowered, open or one-sided, flowers semierect to horizontal; pedicels red, 1.2–2.6 cm long. **Flowers** white or pale pink, fragrant; calyx shallowly lobed, scaly and hairy, ringlike, 1–1.5 mm high; corolla 4–4.6 × 3.5–4.5 cm, tube 3.5–4 × 0.9 cm, straight, scaly outside, lobes 1.8 × 1.4–1.8 cm; stamens included, filaments white or pale pink, anthers brown; disk lobed; ovary ovoid to conical, 5.5 × 3.5 mm, ends tapered, densely scaly, hairy, style 3.7 cm long, stigma irregularly lobed, ~1 mm, cream, eventually exserted to 6 mm. **Fruits** narrowly ellipsoid, 1.7–2.5 × 0.4–0.7 cm, longitudinally grooved; seeds 4–4.5 mm long. **Distr:** Indonesia (Borneo). **Ecol:** In upper montane, mossy forests; 1,400–1,800 m. **Zones:** 11–12.

R. archboldianum Sleumer, *Reinwardtia* 5: 121, 1960. **Shrubs**, tall, epiphytic, occasionally terrestrial; internodes 6–16 cm long. **Leaves** in pseudowhorls of 5–7, 3–4 smaller; lamina ovate to elliptic-ovate, 4–11 × 2–6 cm, often asymmetric, base subtruncate-rounded or broadly tapered, margins entire, apex shortly acuminate, subacute; petiole 0.4–1.2 cm long. **Inflorescences** umbels, 8–15-flowered; pedicels 0.8–1.1 cm long, densely scaly; bracts ovate to broadly obovate, 1 × 0.7 cm; bracteoles narrowly linear, to 1.5 cm long. **Flowers** with calyx small, rimlike; corolla pink, trumpet-shaped, tube cylindrical, 4.5–5 × 0.45 cm, slightly curved, lobes broadly spathulate to subcircular, 1–1.2 × 0.8–1.2 cm; stamens exserted, filaments linear below, filiform above, anthers oblong, ~4 mm long; ovary cylindrical, 7–8 × ~2 mm, style as long as corolla tube, stigma thick, obconic. **Distr:** Papua New Guinea. **Ecol:** In ridge forests. **Zones:** 11–12.

R. arfakianum Becc., *Malesia* I: 201, 1878. (Syn: *R. undulaticalyx*). [Arfak Mountain Vireya]. **Shrubs**, epiphytic, rarely terrestrial, to 2.5 m high; young parts sparsely hairy, becoming glabrescent. **Leaves** in pseudowhorls of 3–5; lamina obovate-elliptic, 4–10 × 1.5–4.5 cm, base tapering, margins entire, slightly revolute, apex obtusely tapering or nearly rounded; petiole 0.6–1.2 cm long. **Inflorescences** umbels, 4–10-flowered; *outer bracts* ovate, 1.5 × 0.8 cm, *inner ones* oblong-obovate; bracteoles filiform, to 2.3 cm long; pedicels 0.8–1.2 cm long. **Flowers** with calyx disk-shaped, ~3 mm Ø, ciliate; corolla deep pink, tubular below, expanded upward, oblique, 2.5–3.5 cm long, tube 1.5–2 cm long, lobes suberect, or somewhat spreading, obovate-spathulate, 0.7–1 × 0.5–0.8 cm; stamens spreading all around the mouth, unequal, anthers obovate, 2 mm long; disk shortly hairy; ovary subcylindrical, ~4 × 2 mm, densely hairy, style slender, stigma club-shaped, obconical. **Fruits** fusiform, ~2 × 0.4 cm, densely hairy. **Distr:** Indonesia (W New Guinea). **Ecol:** At edge of forests; 1,200–2,135 m. Fl. Jan–Dec. **Zones:** 11–12.

R. burtii P. Woods, *Notes RBG Edinb*. 37(1): 57, 1978. **Shrubs**, terrestrial, to 80 cm high. **Leaves** in tight pseudowhorls of 5–8, with smaller, reduced, scale leaves; lamina obovate, 1.8–2.8 × 0.9–1.2 cm, base tapering, margins entire, slightly recurved, apex broadly pointed to rounded; petiole 4–5 mm long, hairy, scaly. **Inflorescences** 1–2(–4)-flowered, lateral and terminal, flowers pendent; pedicels 1.7 cm long, red, white-hairy. **Flowers** with calyx disklike, green; corolla bright red, 2.3 × 2.6 cm, tube ~1.7 cm long, swollen just above the base, then narrowed slightly, lobes spreading, 1 × 0.6 cm; stamens spreading, filaments 1.8 cm long, basally white to pink distally, anthers brown; ovary shortly cylindrical, ~4 × 3 mm, densely white-hairy, style ~1.2 cm long, stigma rounded, bright red, becoming exserted to ~8 mm. **Fruits** obovoid-cylindric, 1.2 × 0.6 cm, dark purple, densely hairy; seeds 1.8 mm long. **Distr:** Borneo (Sabah: Mt Kinabalu). **Ecol:** In summit shrubberies; 3,160–3,650 m. Fl. Jul–Dec. Fr. Aug. **Zones:** 11–12.

R. caliginis Kores, *Blumea* 30(1): 45, 1984. **Shrubs**, small, to 50 cm high, straggling, much-branched. **Leaves** in tight pseudowhorls of 4–7; lamina linear or narrowly ovate or elliptic, 1–7 × 0.1–0.6 cm, base broadly tapering or rounded, margins entire or slightly revolute, apex obtuse, densely reddish-brown scaly initially; petioles 2–5 mm long. **Inflorescences** with flowers solitary or 2–3 together; pedicels 0.6–1.2 cm long, densely scaly, flowers pendent; outer bracts subulate, inner bracts ovate. **Flowers** with calyx ~2.5 mm Ø, small, disklike; corolla 3–3.5 cm long, white to cream, or pale pink, tubular below, slightly curved, zygomorphic, densely scaly, lobes broadly obovate to subcircular, 7–9 × 6–9 mm, spreading horizontally; stamens clustered on upper side, unequal, exserted to ~1 cm, filaments white, linear, ± glabrous, anthers brown; ovary subovoid, gradually tapering to style, densely scaly, 5-ribbed, style greenish-white, on upper side of tube, stigma white or green, club-shaped to globose, obscurely 5-lobed. **Fruits** fusiform. **Distr:** Indonesia, Papua New Guinea. **Ecol:** In open submontane shrubberies; 2,400–2,900 m. **Zones:** 11–12.

R. celebicum (Blume) DC., *Bijdr.* 855, 1826. **Shrubs**, epiphytic, to 1.5 m high; roots sometimes swollen and tuberous. **Leaves** in loose pseudowhorls of 3–7; lamina elliptic to narrowly ovate-elliptic, 3–7 × 1.5–3 cm, base broadly tapering, margins entire,

Top row: *Rhododendron leptothrium, Rhododendron spinuliferum, Rhododendron alborugosum.*
Second row: *Rhododendron archboldianum, Rhododendron arfakianum, Rhododendron burtii.*
Third row: *Rhododendron caliginis* (white form), *Rhododendron celebicum, Rhododendron christianae.*
Bottom row: *Rhododendron christi, Rhododendron christi* (Mt Miap form), *Rhododendron citrinum, Rhododendron crassifolium.*

narrowly cartilaginous, flat, apex shortly acuminate; petiole 0.4–1.2 cm long. **Inflorescences** umbels, 3–8-flowered; bracts ovate to obovate, 0.5–1.2 × 0.4–0.7 cm; pedicels 1.5–2 cm long. **Flowers** with calyx disk-shaped, 2.5 mm Ø; corolla pink to red, tubular, 3.5–4.6 × 2–2.5 cm, lobes subcircular, 0.8–1.1 × 0.6–1 cm; stamens irregularly spread around mouth or exserted, anthers dark brown; disk green, glabrous; ovary pink, subcylindrical, 5–6 × 2 mm, very densely silvery-scaly, stigma thick-globose. **Fruits** fusiform-cylindrical, 2.5–3.8 × 3.5 cm, scaly; seeds 7–8 mm long. **Distr:** Indonesia (Sulawesi). **Ecol:** In mossy forests; 1,800–2,600 m. Fl. Jan, Jun–Jul, Nov. **Zones:** 11–12.

R. christianae Sleumer, *Reinwardtia* 5: 211, 1960. **Shrubs**, terrestrial, to 3 m tall; internodes 2–12 cm long. **Leaves** in tight pseudowhorls of 3–5; lamina broadly elliptic or subobovate-elliptic, 4–7.5 × 3–4.5 cm, base tapering, margins entire, apex acuminate, obtuse or slightly emarginate. **Inflorescences** umbels, 2–5-flowered; pedicels 1–2.7 cm long, densely hairy. **Flowers** with calyx oblique, disk-shaped, ~4 mm Ø, lobes 5, obtuse, scaly and hairy outside; corolla yellow, aging to dark orange, especially lobe margins, tubular, 2.5–3.5 cm long, lobes broadly obovate or subcircular, 1.5–2.5 × ~2 cm; stamens unequal, arranged around mouth, longer ones exserted to ~1.2 cm, filaments linear, yellow, densely hairy, anthers blackish-purple; ovary subcylindrical, abruptly contracted distally, ~6 × 2.5 mm, white-hairy, style 3–4 cm long, green, densely hairy, stigma, thick, 5-lobed. **Fruits** fusiform, 3–3.4 × ~0.6 cm, longitudinally grooved. **Distr:** Papua New Guinea. **Ecol:** On rock faces or cliffs; 600–1,525 m. Fl. May–Oct. **Zones:** 11–12.

R. christi Foerster, *Fedde Rep.* 13: 222, 1914. (Syn: *R. christi* var. *loniceroides*, *R.* aff. *wentianum*). **Shrubs**, mostly epiphytic, to 1.2 m high; branchlets rounded, initially laxly scaly. **Leaves** in tight pseudowhorls of 2–4, at upper 2–3 nodes, subsessile; lamina ovate, 4–11 × 2.5–6.5 cm, base cordate or rounded, margins entire, flat, apex acuminate, stiff, adaxially glabrescent, abaxially persistently scaly; petiole to 2 mm long. **Inflorescences** umbels, 2–5-flowered. **Flowers** with mouth facing downward; corolla, 3–4.5 cm long, tubular below, expanded and oblique at lobes, bicolored with a yellow tube and orange lobes, densely long-hairy; stamens exserted; ovary cylindrical, to 7 mm long. **Fruits** capsules, subcylindrical, becoming twisted. **Distr:** Papua New Guinea, Indonesia. **Ecol:** In mossy forests and on grassland tree-ferns, or terrestrial in open forests or rock faces on cliffs; 1,200–3,000 m. Fl. May–Jun. **Zones:** 10–12. **Info:** The form from Mt Miap differs in habit (stems somewhat pendent) and leaf morphology.

R. citrinum (Hassk.) Hassk., *Cat. Hort. Bog.* 161, 1844. (Syn: *Azalea citrina*, etc.). **Shrubs**, spreading, to 2 m tall; internodes 1–5 cm long. **Leaves** in pseudowhorls of 4–5, unequal; lamina usually elliptic, 2–6 × 1–3 cm, base tapered, margins flat, apex broadly acute, obtuse to rounded, sometimes with a small terminal gland; petiole 0.5–1.2 cm long, scaly. **Inflorescences** 1–4-flowered, flowers ± hanging; pedicels 1.5–2 cm long, red, scaly, hairy; bracteoles filiform, to 8 mm long. **Flowers** with calyx oblique, disk-shaped, 5-lobed, 3–4 mm Ø; corolla campanulate, 1.5–1.9 × 1–1.4 cm, lobed nearly to halfway, yellowish-white, lobes obovate-spathulate, ~0.9 × 1 cm, rounded or slightly retuse; stamens 5, loosely clustered on lower side of corolla, filaments 1.2–1.5 cm long. glabrous, pale-yellow, red or orange, anthers yellow or orange; ovary obliquely oblong-ellipsoid, 3–4 × ~1.5 mm, style 6–8 mm long, red or orange, glabrous, stigma rounded-subglobose. **Fruits** obliquely cylindrical, 1.5–2.7 cm long, deeply 5-grooved; seeds 5.5–6 mm long. **Distr:** Indonesia (Java). **Ecol:** In humid primary forests; 1,000–2,900 m. Fl. Sep–Mar. **Zones:** 10–12.

R. crassifolium Stapf, *Trans. Linn. Soc. London*, II, Bot. 4: 195, 1894. (Syn: *R. brevitubum*, *R. crassinervium*). **Shrubs**, epiphytic, to 2.5 m high; internodes 1.5–15 cm long. **Leaves** spirally arranged or in lax pseudowhorls of 3–4; lamina usually elliptic, 8–14 × 4–8 cm, base broadly tapering, usually rounded or truncate, margins entire, apex usually obtuse or rounded; petiole 0.7–1.5 cm long. **Inflorescences** umbels, semierect to hanging, 8–20-flowered; bracts ovate-acuminate, to 2.5 × 1 cm, glabrous; bracteoles linear, to 2 cm long. **Flowers** with calyx shortly funnelform, as a fleshy ring 3–4 mm Ø, glabrous; corolla shortly funnelform, pink, red or more rarely orange or white, 2.5–3 × 2.5–4.5 cm, distinctly 5-pouched at base, lobes obovate-elliptic, 1.2–2.2 ×1.3–1.9 cm, not overlapping, margins revolute; stamens dimorphic, 1.2–2 cm long, regularly arranged around mouth, filaments filiform, anthers 2.5–3 mm long; ovary 5–6 mm long, glabrous, style 3–4 mm long, stigma dark red, 2–3 mm Ø, 5-lobed. **Fruits** 1.8–2.5 × 0.5–0.6 cm; seeds 0.8 mm long with tails to 2.8 mm long. **Distr:** Borneo. **Ecol:** In deep shade of mossy forests, also terrestrial on open landslides; 1,200–2,200 m. Fl. Jan–Dec. **Zones:** 11–12. **Info:** ***R. × planecostatum*** is a natural hybrid between *R. crassifolium* and *R. stenophyllum*.

R. gardenia Schltr., *Bot.Jahr.* 55: 158, 1918. [Gardenia Vireya]. **Shrubs**, epiphytic, to 2.5 m high; internodes 15–20 cm long. **Leaves** in pseudowhorls of 3–5, at upper nodes; lamina elliptic, rarely subobovate-elliptic, 8–15 × 5.5–9 cm, base broadly tapering to rounded, margins revolute, apex obtuse or rounded to slightly retuse, stiffly leathery, densely scaly, soon glabrescent; petiole 2–4 cm long, brown-scaly. **Inflorescences** umbels, 3–4-flowered, flowers ± half-hanging; pedicels 1–3.5 cm long. **Flowers** strongly fragrant; calyx 0.8–1 cm Ø, oblique, scaly and hairy outside, white ciliate along margins; corolla white or suffused with cream, sometimes green on the tube when young, sometimes pink spots at the base of the lobes, tubular-funnelform, 9–11 × 11–14 cm, fleshy; stamens 10–16, slightly unequal, clustered on lower side of mouth, anthers pale brown; stigma 5–8-lobed, style 6–11 cm long. **Fruits** capsules, ellipsoid, 7.5–8.5 × ~1.5 cm. **Distr:** New Guinea. **Ecol:** In forests of mountain ridges, 1,400–2,300 m. Fl. Jun–Jul, Oct–Nov. **Zones:** 10–12. **Info:** ***R. gardenia* 'Odyssey'** is a strongly scented form, regarded as a hybrid.

R. goodenoughii Sleumer, *Reinwardtia* 5: 131, 1960. **Shrubs**, to 2 m high; internodes 4–15 cm long. **Leaves** in tight pseudowhorls of 4–6; lamina broadly elliptic to obovate, 5–10 × 3–6 cm, base broadly tapering, margins entire, apex usually shortly acuminate, adaxially glabrescent, abaxially scaly; petiole 1–1.5 cm long. **Inflorescences** umbels, 8–22-flowered, initially held erect in a tight cluster by the soon caducous bracts; bracts ovate, to 2.5 × 1.8 cm; bracteoles narrowly linear, to 1.5 cm long, hairy. **Flowers** sweetly scented; calyx oblique, ~2.5 mm Ø; corollas white, trumpet-shaped, 6.5–7.5 × 3–3.5 cm, slightly curved, lobes obliquely spreading, 1.5–1.7 × 1.2–1.5 cm, ± glabrous; stamens exserted to 1.8 cm in a tight group on upper side of mouth, later irregularly spreading, filaments linear, white, anthers 4–6 mm long, brown; ovary elongate-cylindrical, 0.8–1 × 0.2–0.3 cm, densely white-hairy, style ~5 cm long, on upper side of corolla, green, densely scaly, stigma obconical, green, initially deep inside corolla tube, later exserted to ~1.5 cm. **Distr:** Papua New Guinea (Goodenough Islands). **Zones:** 10–12.

R. gracilentum F. Muell., *Trans. R. Soc. Vict.* n.s. 1(2): 22, 1889. **Shrubs**, terrestrial or epiphytic, erect or prostrate; branches spreading, to 60 cm long. **Leaves** in tight whorls of 2–7, often recurved; lamina narrowly ovate or -elliptic, 0.6–1.8 × 0.3–0.6 cm, base acutely tapering, margins entire, apex usually shortly acuminate, adaxially glossy, dark green, abaxially paler; petiole 1.5–3 mm long. **Inflorescences** with flowers solitary, rarely paired, flowers hanging; bracts ovate, to 8 × 4 mm; pedicels 0.8–1.6 cm long, pink, densely scaly. **Flowers** with calyx membranous, 5-lobed; corolla red or pink, cylindrical, 2.6–3.3 × 2–2.5 cm, lobes semierect to spreading, obovate to subcircular, 0.8–1.1 × 0.8–1 cm; stamens unequal, clustered on lower side of mouth, exserted, filaments glabrous, pink, anthers purplish-brown; disk glabrous, green; ovary oblong-cylindrical, 4–7 × 1.5–2 cm, densely scaly, style on lower side of mouth, exserted, white-hairy, stigma club-shaped. **Fruits** erect, oblique, oblong, 1.5–2.2 × 0.3–0.4 cm, slightly curved. **Distr:** Papua New Guinea. **Ecol:** In forests; 2,000–2,745 m. Fl. May–Oct. **Zones:** 11–12.

R. herzogii Warb., *Bot. Jahr.* 16: 25, 1892. **Shrubs**, epiphytic or terrestrial, erect, to 2 m high. **Leaves** in tight pseudowhorls of 4–6, aromatic; lamina elliptic, 4–6 × 2–4 cm, base broadly tapering to

Below: *Rhododendron gardenia* 'Odyssey', *Rhododendron gardenia*, *Rhododendron goodenoughii*.

Above: *Rhododendron gracilentum* (Papua New Guinea form), *Rhododendron herzogii, Rhododendron jasminiflorum* ssp. *jasminiflorum, Rhododendron jasminiflorum* ssp. *oblongifolium.*

subrounded, margins slightly basally revolute, apex usually acuminate, lateral veins 6–10(–12) per side; petiole 0.4–1 cm long, scaly, pink or green. **Inflorescences** umbels, 5–10-flowered, flowers semierect; pedicels 4–8 mm long; bracts ovate, to 1.5 × 1 cm. **Flowers** fragrant, ± erect, trumpet-shaped; calyx very small, oblique, disk-shaped; corolla 4–9 cm long, straight or slightly curved, pure white, or slightly pinkish outside, densely scaly outside, lobes circular to elliptic-obovate, 8–9 × 5–6 mm, spreading to somewhat reflexed; stamens exserted in a tight cluster, filaments white, anthers orange, oblong; disk sparsely hairy or glabrous; narrowly cylindrical, ovary 6–7 × 1.5 mm, densely scaly, style exserted, stigma green. **Fruits** cylindrical, straight, 3–5 × 0.4–0.5 cm, reddish, scaly. **Distr:** Indonesia, Papua New Guinea. **Ecol:** On cliff faces, and in montane forests; 1,530–2,500 m. Fl. & Fr. throughout the year. **Zones:** 11–12.

R. jasminiflorum Hook., *Bot. Mag.*, t: 4525, 1850. (Syn: *R. elongatum, R. jasminiflorum* var. *punctatum*, etc.). [Jasmine Vireya]. **Shrubs**, epiphytic or terrestrial, to 2.5 m high; internodes 2–9 cm long. **Leaves** in tight pseudowhorls of 3–5; lamina broadly elliptic, sometimes subcircular, 2.5–6 × 1–3.6 cm, base usually subtruncate, margins entire, apex broadly obtuse, acuminate or rounded, adaxially dark green, glossy, abaxially paler, dull; petiole 2–4 mm long, scaly. **Inflorescences** umbels, 5–15-flowered; flowers suberect; bracts ovate to obovate-spathulate, obtuse, glabrous; bracteoles linear, to 1 cm long, glabrous. **Flowers** white or pink-flushed, especially when young, scented; calyx ~2 mm Ø, obscurely 5-lobed; corolla trumpet-shaped, 4–5 × 1–1.4 cm, straight, laxly scaly, lobes broadly obovate, 0.8–1.5 × 0.6–1.1 cm, horizontally spreading, glabrous or laxly scaly outside; stamens exserted to 4 mm, filaments linear, anthers ~1.8 mm long; disk densely hairy; ovary 5 × 2 mm, shortly cylindrical, densely hairy and scaly proximally, stigma shortly obconical, later becoming exserted to ~1 cm. **Fruits** narrowly ellipsoid, 2.8 × 0.4–0.5 cm. **Distr:** Peninsular Malaysia. **Ecol:** In ridge forests, shrubberies and in open places on rocks; 1,100–1,500 m. Fl. Dec. **Zones:** 11–12. **Info:** 6 subspecies have been described. ***R. jasminiflorum* ssp. *oblongifolium*** (Syn: *R. verticillatum*) have narrowly elliptic leaves and pedicels >1.5 cm long.

R. konori* var. *konori Becc., *Malesia* 1: 200, 1878. **Shrubs** or **small trees**, epiphytic or terrestrial, to 6 m tall; internodes 6–20 cm long. **Leaves** scattered in loose pseudowhorls, or tightly crowded, 3–5 together; lamina usually elliptic, 7–20 × 5–10 cm, base broadly tapering to rounded, margins entire, flat or slightly revolute; apex obtuse or rounded; petiole 1–3.5 cm long. **Inflorescences** umbels, 4–10-flowered, ± held horizontally; pedicels 0.4–1.5 cm long. **Flowers** sweetly and strongly scented; calyx oblique, 0.6–1 cm Ø; corolla 8–19 × 7–15 cm, tubular below, expanded, funnel-shaped toward lobes, fleshy, pure white or pink, base often pink-spotted; stamens 12 or 14, exserted, filaments white, anthers yellow to orange; style 4–8.5 cm long, white to reddish, stigma yellow. **Fruits** fusiform, 4–12 × 1.2–1.8 cm; seeds flattened. **Distr:** New Guinea, Fergusson Islands, New Britain, New Ireland. **Ecol:** In montane forests; 750–2,500 m. Fl. Jan.–Dec. **Zones:** 10–12.

R. lanceolatum Ridl., *J. Str. Br. As. Soc.* 63: 60, 1912. (Syn: *R. partitum*). **Shrubs**, epiphytic, to 1.2 m high; internodes 3.5–8 cm long. **Leaves** in pseudowhorls of 3–7, with several slender, scale leaves; lamina narrowly ovate to elliptic, 7–12 × 2.5–4.5 cm, base obtusely tapering to rounded, margins entire, apex acuminate, acute, surfaces densely scaly when young, abaxially scaly; petiole 2–3 mm long, scaly, reddish. **Inflorescences** umbels, 4–10-flowered, flowers ± horizontal; bracteoles subulate-linear, to 1.5 cm long, scaly; pedicels 2–3 cm long, densely scaly, white-hairy. **Flowers** white, sometimes scented; calyx disk-shaped, ~4 mm Ø, shallowly lobed; corolla subcampanulate, deeply lobed, densely scaly outside, lobes erect, concave, subobovate, 1.2–1.6 × 0.8–1.3 cm; stamens spreading around mouth, filaments white, hairy, ~1–1.2 cm long, apically curved, anthers white, brown-tinged; disk 10-lobed, lobes linear, greenish-yellow; ovary green, conical, 4–5 × 3–3.5 mm, white-hairy, style ~5 mm long, pale greenish-yellow; stigma greenish-yellow. **Fruits** ~2.5 × 0.5 cm. **Distr:** Borneo. **Ecol:** In wet mossy forests; ~1,000–1,600 m. **Zones:** 11–12.

R. leptobrachion Sleumer, *Reinwardtia* 5: 203, 1960. **Shrubs** or **small trees**, to 3 m high; internodes 3.5–12 cm long. **Leaves** in tight pseudowhorls of 4–7, in upper nodes; lamina narrowly elliptic, 5.5–9 × 1.8–2.8 cm, base acutely tapering, margins entire, slightly revolute, apex shortly acuminate, acute; petiole 1–2 cm long, scaly. **Inflorescences** umbels, 4–10-flowered, flowers ± hanging; pedicels 1–2.4 cm long, pink; bracteoles linear, 1.5–2 cm long. **Flowers** with calyx obliquely disk-shaped, shallowly 5-lobed, densely scaly outside; corolla bright glossy red, tubular, 3–4 × ~3 cm, straight, lobes spreading, broadly obovate or suborbicular, 1–1.4 × 0.9–1.1 cm; stamens irregularly arranged around mouth, filaments linear, pink, densely hairy, anthers obovate-oblong; disk prominent, glabrous; ovary subovoid to cylindrical, 4–6 × 2–3 mm, densely hairy, scaly, style slender, pale pink, stigma globose. **Fruits** cylindrical, 2–3 × 0.5–0.6 cm, slightly curved; seeds 3.8 mm long. **Distr:** Indonesia (Sulawesi). **Ecol:** In mossy forests; 2,000–3,000 m. Fl. May–Jun. **Zones:** 11–12.

R. lochiae F. Muell., *Vict. Nat.* 3: 157, 1887. (Syn: *Azalea lochae, R. notiale*). [Australian Rhododendron]. **Shrubs** or **small trees**, terrestrial, to 6.5 m tall. **Leaves** in tight pseudowhorls of 3–5; lamina usually broadly elliptic, 5–9 × 3–5 cm, base broadly tapering to rounded, margins entire, apex usually acute to rounded; petiole 1–2 cm long. **Inflorescences** umbels, 2–6-flowered; pedicels 1–4 cm long. **Flowers** with calyx disk-shaped or 5-lobed, to 8 × 3.5 mm, pink; corolla red or pink, 2.5–5.5 × 3.5–4.5 cm, lobes spreading horizontally

Below: *Rhododendron konori* var. *konori* (pink form), *Rhododendron lanceolatum, Rhododendron leptobrachion.*

Top row: *Rhododendron lochiae, Rhododendron loranthiflorum, Rhododendron luraluense, Rhododendron macgregoriae* (yellow form).
Bottom row: *Rhododendron macgregoriae* (orange form), *Rhododendron macgregoriae* (large form), *Rhododendron majus.*

to slightly reflexed, 1.5–2 × 1.5–2.2 cm; stamens loosely clustered on upper side of mouth, not exserted, filaments 22–31, 2.1–3 cm long, anthers oblong, pale brown; disk green; ovary subcylindrical, 3.5–5 × 2–3 mm, style on upper side of tube, stigma rounded. **Fruits** fusiform, 2–2.5 × 0.4–0.5 cm; seeds 2.5–3 mm long. **Distr:** Australia (Bellenden Ker Range). **Ecol:** In montane forests, shrubberies among rocks; 1,200–1,520 m. Fl. Oct–Nov. **Zones:** 10–12.

R. loranthiflorum Sleumer, *Notizbl. Berl.-Dahl*, 12: 485, 1935. (Syn: *R. subpacificum*). **Shrubs**, epiphytic, much-branched; internodes 1.5–12 cm long. **Leaves** in tight pseudowhorls of 3–6; lamina obovate, 2.5–6 × 1.5–2.5 cm, base tapering, margins entire, apex broadly obtuse or rounded, initially scaly; petiole 3–8 mm long, scaly. **Inflorescences** loose umbels, 3–7-flowered, ± hanging; bracts triangular, appressed, brownish-green; bracteoles filiform, to 6 mm long; pedicels 1–2.2 cm long. **Flowers** white, faintly scented; calyx 3–4 mm Ø, green, disk-shaped; corolla tubular trumpet-shaped, tube cylindrical, 2.5–3.5 × 0.5–0.9 cm, straight or slightly curved, lobes broadly elliptic, obovate or rounded, 1–1.2 × 0.9–1 cm, spreading to weakly reflexed; stamens initially exserted, filaments filiform, curving; disk low, glabrous; ovary subcylindrical, ~6 × 1.5 mm, slightly 5-ribbed, style 3.5–4 cm long, stigma ~2 mm Ø, **Fruits** cylindrical, 4–4.5 × ~0.3 cm, slender, curved, apically tapering, 5-ribbed, brown-scaly. **Distr:** Papua New Guinea, Solomon Islands. **Ecol:** In high rainforests; 180–1,500 m. Fl. Dec. **Zones:** 10–12.

R. luraluense Sleumer, *Notizbl. Berl.-Dahl.* 12: 485, 1935. **Shrubs** or **trees**, epiphytic, to 9 m tall. **Leaves** in tight pseudowhorls of 4–7; lamina obovate-elliptic to elliptic, 3.5–6 × 2–3 cm, base broadly tapering, margins narrowly recurved, apex broadly acute to obtuse, or occasionally rounded or apiculate; petiole 3–5 mm long. **Inflorescences** umbels, 3–6-flowered; bracts ovate-acuminate, to 1.2 × 0.6 cm,, *outer bracts* ovate-acuminate, *inner bracts* obovate to spathulate; bracteoles filiform, ~1 cm long; pedicels ~1.3 cm long. **Flowers** with calyx disk-shaped, ~3 mm Ø; corolla white, funnel-shaped, 3.5–4.5 × ~5 cm, lobes spreading, obovate-subcircular, 1–1.5 × 1–1.4 cm; stamens exserted, filaments filiform, anthers oblong, ~2.5 mm; disk glabrous; ovary subcylindrical, ~4 × 1.8 mm, densely white-hairy and scaly, style slightly shorter than stamens, stigma rounded. **Distr:** Papua New Guinea (Bougainville Island). **Ecol:** In rainforests; 900–1,500 m. Fl. Aug, Oct. Zones 10–12.

R. macgregoriae F. Muell., *J. Bot.* 29: 177, 1891. (Syn: *R. calceolaroides*, etc.). **Shrubs** or **trees**, terrestrial, sometimes epiphytic, to 15 m tall; internodes 2–10 cm long. **Leaves** in tight pseudowhorls of 3–7; lamina elliptic, 4–14 × 2.5–5 cm, base tapering, margins entire, narrowly revolute, apex shortly acuminate, acute or obtuse to rounded; petiole 0.5–1.2 cm long, scaly. **Inflorescences** umbels, 5–15-flowered; bracteoles linear, to 1.5 cm long; pedicels slender, 2.5–5 cm long. **Flowers** usually scentless; calyx oblique, rounded, ~4 mm Ø; corolla variable in size and color, tube and/or center light yellow to orange, or yellow, lobes reddish-orange, shortly tubular or tubular-campanulate, 1–2.5 × 2–3 cm, straight, 5-angular, lobes broadly obovate-spathulate, 1.2–2 × 0.8–1.4 cm; stamens strongly exserted to 1.5 cm, unequal, spread around mouth, filaments linear, densely hairy, anthers 2.5–3.2 mm long; ovary elongate-conical, ~4 × 1.8 mm, densely hairy, style 1–1.5 cm long, stigma rounded to lobed. **Fruits** fusiform, 3.5–4.5 × 0.3–0.5 cm, strongly 5-ribbed, hairy and scaly; seeds ~4 mm long. **Distr:** New Guinea. **Ecol:** In primary mossy forests; 120–3,000 m. Fl. throughout the year. **Zones:** 9–12. **Info:** This variable species readily hybridizes in the wild.

R. majus (J.J. Sm.) Sleumer, *Reinwardtia*, 5: 120 ('*maius*'), 1960. (Syn: *R. carringtoniae* var. *majus*). **Shrubs**, terrestrial or epiphytic, to 2 m high; internodes 6–14 cm long. **Leaves** in well-marked pseudowhorls of 6–10; lamina elliptic, 5–9 × 2.2–4 cm, base broadly tapering to rounded, margins slightly revolute, apex broadly tapering, obtuse, abaxially scaly; petiole 0.4–1 cm long, scaly. **Inflorescences** umbels, (5)12–15-flowered, flowers horizontal; pedicels 1–2.5 cm long. **Flowers** carnation-scented; calyx ~6 mm Ø; corolla trumpet-shaped, 7–9 × 2.5–3 cm, white, or white with a pink tube, base sometimes darkening to red, slightly curved, lobes obovate to circular, 1.2–1.4 × 1–1.7 cm, margins irregularly undulate; stamens initially clustered on upper side, later spreading, exserted, filaments linear, anthers narrowly oblong, 3.5 mm long; ovary cylindrical or elongate-conical, 0.9–1.2 × 0.2 cm, style 5–5.5 cm long, stigma globose-conical, crenulate. **Distr:** New Guinea. **Ecol:** In mossy forests; 2,770–3,000 m. Fl. Feb–Oct. **Zones:** 10–12.

R. maxwellii Gibbs, *J. Linn. Soc. Bot.* 42: 103, 1914. **Shrubs**, to 2 m high. **Leaves** pseudowhorled, 3–5 large, few smaller; lamina obovate or elliptic,

6.5–10 × 3–5 cm, base broadly tapering, margins slightly recurved, apex rounded or retuse, glabrescent; petiole 0.7–1.2 cm long. **Inflorescences** umbels, 11–14-flowered, flowers ± horizontal; pedicels green or bright red, densely scaly. **Flowers** very strongly and sweetly scented; calyx a low ring, scaly; corolla shortly funnelform, ~2.6 × 3 cm, pale yellow, tube often pink-flushed outside; stamens evenly arranged in a circle around mouth; stigma green. **Fruits** ovoid-cylindrical, ~1.5 × 0.7 cm. **Distr:** Borneo (Sabah: Mt Kinabalu). **Ecol:** In mossy submontane forests; 2,300–2,900 m. Fl. Feb–Jun. **Zones:** 10–12. **Zones:** 10–12.

R. multicolor Miq., *Fl. Ind. Bat.* Suppl. 1: 251, 586, 1860. (Syn: *Azalea multicolor*, etc.). **Shrubs**, terrestrial, to 1.5 m high; internodes 1.5–7 cm long. **Leaves** in pseudowhorls of 3–6; lamina narrowly elliptic, 3–7 × 0.7–2 cm, base narrowly tapering, margins entire, apex acute or shortly acuminate, veins 6–8 per side; petiole 2–6 mm long. **Inflorescences** umbels, 3–6-flowered, ± hanging; pedicels 1.5–2.5 cm long, green; bracteoles linear, 1.5 × 0.1 cm, glabrous. **Flowers** sometimes sweetly scented; calyx oblique, rimlike, ~4 mm Ø, shortly lobed; corolla whitish-cream to yellow, pink or red, broadly funnelform, 2–2.5 × 2.5–3.2 cm, lobed to halfway, lobes spreading, subcircular, 1.3–1.5 × 1.1–1.3 cm; stamens arranged around mouth, exserted to 1 cm, filaments linear, densely hairy, anthers yellow; disk glabrous; ovary obovoid-cylindrical, ~4 × 2 mm, style 0.7–1 cm long, glabrous, stigma obconical. **Fruits** subcylindrical, 1.8–2.2 × ~0.5 cm. **Distr:** Indonesia (Sumatra). **Ecol:** In open places; 900–2,100 m. Fl. Jan–Dec. **Zones:** 11–12.

R. pauciflorum King and Gamble, *J. As. Soc. Beng.* 74(2): 75, 1905. (Syn: *R. elegans*, etc.). **Shrubs**, epiphytic or terrestrial, to 2 m high; roots swollen. **Leaves** in pseudowhorls of 3–6, at upper 2–3 nodes; lamina obovate to subcircular, 1.2–3.2 × 1–2.5 cm, base broadly tapering, margins entire, cartilaginous, apex usually rounded, dark green, subglossy, abaxially paler; petiole 1–2 mm long. **Inflorescences** with flowers solitary, occasionally paired, half-hanging to pendent; outermost bracts triangular; bracteoles filiform, to 1 cm long; pedicels 1–1.5 cm long, pink, densely hairy. **Flowers** with calyx obliquely cup-shaped, ~2.5 mm Ø; corolla deep pink or red, broadly tubular or tubular-campanulate, 2–2 × 1.5–1.9 cm, lobes broadly obovate, 4–5 × 8–9 mm, reflexed; stamens regularly spread around mouth, filaments pale pink, dimorphic, white-hairy, anthers brown; ovary ovoid-conical, 3–4 × 2–2.5 mm, silvery-scaly, style 3–8 mm long, pale green or pink, stigma green. **Fruits** subcylindrical, 1–1.8 × ~0.4 cm, pink or green, hairy, scaly; seeds 3.5 mm long. **Distr:** Peninsula Malaysia. **Ecol:** in mossy forests; 1,370–2,135 m. Fl. Feb–Nov. **Zones:** 11–12.

R. polyanthemum Sleumer, *Blumea*, 12: 111, 1963. **Trees** or **shrubs**, to 7 m tall. **Leaves** in pseudowhorls of 5–7, with additional much-reduced leaves; lamina elliptic, 7–10 × 4–9 cm, base usually broadly tapering, margins entire, apex usually gradually tapering; petiole 2–4 cm long, green, densely scaly. **Inflorescences** umbels, 25–30-flowered, flowers ± horizontal; bracteoles linear to filiform, ~2 cm long; pedicels orange-red, 6–8 cm long. **Flowers** brilliant pink, or orange with a yellow center, strongly fragrant; calyx circular, oblique, 4–5 mm Ø; corolla funnelform, 3–3.5 × 4–8 cm, strongly angled, lobes broadly obovate-spathulate, 1.8–2 × 1.5–1.8 cm; stamens irregularly arranged in 2 groups on either side of flower, filaments 2.5 cm long, base yellow, distally pink to red, anthers pale brown; ovary 5–6 × 3–4 mm, green, style green, hairy, base yellow, stigma pinkish-brown. **Fruits** fusiform, 2.1–2.8 × 0.7–0.9 cm; seeds 3 mm long. **Distr:** Borneo (Sabah). **Ecol:** In dense mossy forests, hanging from cliffs; 1,300–2,300 m. Fl. Jan–Feb, May, Sep. **Zones:** 11–12.

R. praetervisum Sleumer, *Blumea* 21(2): 376, 1973. **Shrubs**, epiphytic, to 2 m high. **Leaves** in pseudowhorls of 4–6 large and few small ones; lamina ovate or elliptic, 5–7.2 × 1.8–3.2 cm, base tapering, margins entire, broadly revolute, apex rounded or retuse, adaxially dark green, shiny, abaxially paler, dull; petiole 2–6 mm long, often purplish-red. **Inflorescences** umbels, 3–7-flowered, flowers pendent; bracts broadly ovate, rounded; bracteoles filiform; pedicels 2–3.7 cm long. **Flowers** with calyx circular, disklike; corolla pink or pinkish-violet, 5–8 × 2.7–3 cm, slightly curved, proximally strongly fluted, lobes ~2 × 1–1.2 cm, spreading; stamens distributed around lower side of mouth, filaments 6 cm long, glabrous, proximally white, distally becoming pink, anthers 2.5 mm long, pale brown; ovary green, 1.2 × 0.35 cm, hairy, style 5 cm long, green, white or pink-flushed, stigma circular, ~2 mm Ø, **Fruits** 3–3.2 × 0.5–0.6 cm, valves curving back on opening; seeds 3.2 mm long. **Distr:** Malaysia (Borneo). **Ecol:** In montane forests; 1,100–1,800 m. Fl. Mar–Apr. **Zones:** 11–12.

R. rarilepidotum J.J. Sm., *Contr. Arn. Arb.* 8: 126, 1934. **Shrubs**, terrestrial, rarely epiphytic, to 4 m tall; internodes 4–11 cm long. **Leaves** in loose pseudowhorls of 3–7; lamina usually elliptic, 5–9 × 2–3.5 cm, base tapering, margins entire, apex shortly acuminate or acute, initially densely scaly, later ± glossy; petiole 0.7–1.8 cm long. **Inflorescences** umbels, ± dense, 6–12-flowered, flowers ± horizontal; bracteoles linear, ~2 cm long; pedicels 1.5–2 cm long, ± scaly, reddish-orange. **Flowers** sometimes scented; calyx oblique, rimlike, ~4 mm Ø, wavy; corolla reddish-orange, darker toward center, or flowers wholly yellow, broadly funnelform, 2–2.5 × 3.5–4 cm, lobes obovate, 1.3–1.9 × 1.2–1.7 cm, spreading; stamens exserted to ~1 cm, filaments linear, glabrous, reddish (yellow in yellow form), anthers ± globose, reddish (orange in yellow form); disk prominent, glabrous; ovary subcylindrical, 4–6 × 2–2.5 mm, green, style 1.2–1.8 cm long, apically curved, reddish, with reddish stigma (yellowish with greenish stigma in yellow form). **Fruits** fusiform, slightly curved, 2.5–3 × 0.5–0.7 cm, glabrous. **Distr:** Indonesia (Sumatra). **Ecol:** In subalpine shrubberies; 1,000–2,500 m. Fl. Jan–Dec. **Zones:** 11–12.

R. rarum Schltr., *Bot. Jahr.* 55: 150, 1918. **Shrubs**, epiphytic, occasionally terrestrial, slender, much-branched, to 1.5 m high. **Leaves** in pseudowhorls of 2–5, of variable sizes; lamina usually narrowly elliptic, 2.5–6 × 0.6–1.5 cm, base rounded or broadly tapering, margins entire or slightly sinuate, apex long-acuminate, acute; petiole 2–4 mm long, densely scaly. **Inflorescences** with flowers solitary or paired, rarely more, appearing umbellate, flowers ± hanging; bracts to 1 × 0.6 cm; bracteoles linear, to 1 cm long; pedicels 0.7–1 cm long, densely scaly. **Flowers** with calyx disklike, ~2 mm Ø; corolla pink to dark red, zygomorphic, 2.5–3.5 × 2–2.5 cm, curved, densely scaly outside, lobes broadly obovate to subcircular, 0.8–1.2 × 0.8–1.3 cm; stamens clustered on upper side of corolla, exserted, filaments pink to red, anthers oblong, brown to purplish-black; ovary subcylindrical, ~6 × 2–2.5 mm, densely scaly, style held on upper side of tube, pink, stigma club-shaped, purple. **Fruits** narrowly oblong, 3–4 × 0.4–0.5 cm; seeds ~4 mm long. **Distr:** Papua New Guinea. **Ecol:** In lower moss of forests; 1,585–3,350 m. Fl. Apr–Nov. **Zones:** 10–12.

R. retivenium Sleumer, *Reinwardtia* 5: 222, 1960. **Shrubs** straggling, or **small trees**, terrestrial or epiphytic, to 6 m tall. **Leaves** spirally arranged or in loose pseudowhorls; lamina usually lanceolate, 11–16 × 3–5 cm, base cuneate, margins entire, apex shortly acuminate, coriaceous, glabrous, nerves 12–18 pairs; petiole 1–2 cm long, green or red. **Inflorescences** umbels, 4–7(–8)-flowered; outer perulae ovate, inner ones broadly ovate-oblong, innermost ones spathulate, to 3 × 1.2 cm, obtuse, membranous, glabrous; bracteoles linear, ~2 cm long, glabrous; pedicels 2.2–3.2 cm long, red or green. **Flowers** sweetly scented; calyx obliquely disk-shaped, 4–5 mm Ø; corolla rich yellow or orange, funnelform, 3–6.4 × 3.5–7.5 cm, limb to 7.5 cm Ø, lobes broadly obovate to suborbicular, 2.5–3 cm Ø; stamens 10, unequal, slightly exserted, filaments linear, base densely hairy, anthers oblong; ovary cylindric, 0.8–1 × ~0.2 cm, style slender, stigma capitate, 5-lobed. **Fruits** subfusiform, 5.2 × 0.9 cm; seeds 5 mm long. **Distr:** Borneo (Mt Kinabalu). **Ecol:** In jungle ridges or on wet rocks near waterfalls; 1,220–2,985 m. Fl. Jan–Dec. **Zones:** 11–12.

R. robinsonii Ridl., *J. Fed. Mal. St. Mus.* 4: 44, 1909. **Shrubs**, to 2.5 m; internodes 4–10 cm long. **Leaves** in loose pseudowhorls of 3–5; lamina elliptic, 8–16 × 3.5–6 cm, base narrowly tapering, sometimes slightly decurrent, margins entire, slightly revolute, apex shortly acuminate, acute, rarely obtuse, adaxially minutely scaly; petiole 1–2.5 cm long. **Inflorescences** umbels, complete or slightly open, 5–12-flowered. **Flowers** with calyx oblique, disk-shaped or rimlike, 3–4 mm Ø; corolla yellow often flushed with orange or red, funnelform, 3–3.5× 3–3.5 cm; stamens to 2.5 cm long, exserted; stigma, knoblike, thickly 5-lobed. **Fruits** fusiform, ~3 × 0.6 cm, somewhat oblique and curved. **Distr:** Malaysia (Perak to Selangor). **Ecol:** Epiphytic on tall trees, especially those overhanging rivers, also terrestrial on rocks by streams, occasionally on wooded hillsides or swamps. Fl. & Fr. Jan–Dec. **Zones:** 11–12.

R. rubineiflorum Craven, *Notes RBG Edinb.* 38(1): 141, f.1, 1980. (Syn: *R. anagalliflorum*). **Shrubs**, epiphytic, to 20 cm high; internodes 1–4 cm long. **Leaves** in tight pseudowhorls of 4–7, with fragile cataphylls along internodes; lamina elliptic to ovate, 0.5–1 × 0.2–0.6 cm, base usually broadly tapering, margins entire, apex broadly acute; petiole 1–2 cm long, white-hairy. **Inflorescences** with flowers solitary, occasionally paired, ± semipendent; bracteoles linear, 4–5 mm long; pedicels slender, 0.7–1.2 cm long, densely scaly. **Flowers** with calyx obliquely cupular, obtusely 5-lobed, ~2 mm Ø, densely scaly; corolla red or pink, broadly campanulate, 1.5–2 × 2.5–3.5 cm, densely scaly outside, lobes elliptic, 1.3–1.5 × 1–1.3 cm; stamens dimorphic, regularly arranged around mouth, filaments pink, 8 and 6.5 mm long, anthers dark purplish-brown, 0.7–1.2 mm long; ovary broadly subovoid-cylindrical, ~3 × 2 mm, hairy, style 4–4.5 mm long, stigma green or red, globose ~2 mm Ø, **Fruits** pink, fusiform, 1.2–1.5 × ~0.2 cm, hairy, scaly; seeds 3–4 mm long. **Distr:** New Guinea. **Ecol:** In alpine shrubberies, upper montane forests; 2,600–3,400 m. Fl. Jan, Apr–Jul. **Zones:** 10–12.

R. rugosum Low ex Hook. *f.*, *Ic. Pl.* t.885, 1852. **Shrubs** or **small trees**, epiphytic or terrestrial; internodes 2–8 cm long. **Leaves** in tight pseudowhorls of 3–6; lamina mostly elliptic, 6–12 × 2.5–5.5 cm, often variable in the same pseudowhorl, , base broadly tapering, margins entire or weakly crenulate, broadly revolute, sometimes strongly revolute, apex shortly acuminate, broadly acute, extreme point deflexed downward, surfaces with dense indumentum, adaxially quickly glabrescent; petiole 7–25 mm long, dark red, initially densely scaly. **Inflorescences** umbels, 8–14-flowered, flowers ± hanging. **Flowers** with calyx thick, disk-shaped, shortly 5-lobed; corolla 2.5–3 × ~2.5 cm, pink or purplish-pink; stamens irregularly arranged around flower, predominantly on lower side, exserted, filaments ~2 cm long; ovary 4–5 × 2–2.5 mm, subcylindrical, gray-green, hairy. **Fruits**

Top row: *Rhododendron maxwellii, Rhododendron multicolor, Rhododendron pauciflorum, Rhododendron polyanthemum.*
Second row: *Rhododendron praetervisum, Rhododendron rarilepidotum, Rhododendron rarilepidotum* (yellow form).
Third row: *Rhododendron rarum, Rhododendron retivenium, Rhododendron robinsonii, Rhododendron rugosum* (Borneo form).
Bottom row: *Rhododendron rubineiflorum, Rhododendron saxifragoides, Rhododendron sessilifolium.*

1.5–2 × 0.5–0.6 cm, densely hairy, scaly. **Distr:** Malaysia (Borneo: Sabah and Sarawak). **Ecol:** In low mossy forests and open exposed ridges; 2,000–3,500 m. Fl. & Fr. Jan–Dec. **Zones:** 11–12.

R. saxifragoides J.J. Sm., *Med. Rijksherb.* n. 25: 3, 1915. (Syn: *R.* aff. *fuchsioides*). **Shrubs,** dwarf, forming tussocks, large cushions or dense mats, 0.3–1 m Ø, 5–15(–30) cm high, divaricately branched. **Leaves** in pseudowhorls of 6–10; lamina linear-lanceolate or oblanceolate, 1.6–4(–5.5) × 0.3–0.7 cm, base gradually attenuate, apex usually shortly acuminate, stiffly coriaceous, adaxially slightly glossy, glabrous; petiole 3–8 mm long. **Inflorescences** with flowers solitary, rarely paired, nodding at first, erect and held above foliage later; perulae ovate, subulate-acuminate, to 1.5 × 0.5 cm; bracteoles linear, to 1 cm long, fimbriate; pedicels erect, reddish. **Flowers** with calyx oblique-discoid, wavy, obtusely 5-lobed, ~4 mm Ø; corolla blood-red, pale scarlet or pinkish-red, obliquely tubular below, 2.5–3 cm long, densely lepidote outside, lobes broadly spathulate or suborbicular, ~1 × 0.7–0.9 cm; stamens 10, subequal, purple, filaments linear, ~2.5 mm long; ovary conical, 5–6 × 2.5–3 mm, style stout, base hairy, stigma large, oblique. **Fruits** erect, thickly fusiform, 1.5–2 × (0.5–)0.7–0.8 cm, shortly hairy, splitting into 5 valves, starlike; pedicel 6–11 cm long; seeds ~2.5 mm long. **Distr:** New Guinea. **Ecol:** In alpine grasslands; 3,225–4,000 m. Fl. & Fr. Aug–Dec. **Zones:** 10–12.

R. sessilifolium J.J. Sm., *Contr. Arn. Arb.* 8: 125, 1934. **Shrub,** terrestrial, rarely epiphytic, to 3 m tall; twigs red-brown, glabrous. **Leaves** in tight pseudowhorls of 4–6, ± sessile; lamina elliptic, 8–17.5 × 2.4–5 cm, base broadly tapering to rounded or subtruncate, occasionally slightly cordate, often with conspicuous folds at base, margins entire, flat, often red, apex gradually or abruptly acuminate, acute, adaxially sparsely scaly and quickly glabrescent, abaxially persistently and more densely scaly. **Inflorescences** open umbels, 5–10-flowered, flowers semierect to horizontal. **Flowers** pale yellow, without scent; calyx ~4 mm Ø; corolla 2.5–5 × 4–5 cm, broadly funnelform, tube 2–2.2 cm long, 1.5–2.2 cm Ø; stamens exserted to ~1.5 cm; ovary cylindrical. **Fruits** subcylindrical, 3.5–4 × 0.8–1 cm. **Distr:** Indonesia (Sumatra). **Ecol:** In open primary or secondary forests; 1,100–1,980 m. Fl. Jan–Dec. **Zones:** 11–12.

R. stapfianum Hemsl. Ex Prain, *Bot. Mag.* sub t. 8372, 1911. **Shrubs,** epiphytic; internodes 1.5–3(–4) cm long. **Leaves** scattered, mostly toward apex of internodes, sometimes in (3–)4–7-merous pseudowhorls; lamina linear-oblong, to 5(–5.6) × 1.5(–1.7) cm, base truncate, margins strongly revolute, apex ± obtuse, ± coriaceous; petiole 1–2 mm long. **Inflorescences** umbels, 5–7-flowered; perulae ovate to spathulate-obovate, to 2 × 1.5 cm, sericeous outside; bracteoles linear to linear-subspathulate, to 3 cm long; pedicels to 6(–8) mm long, densely tomentulose. **Flowers** with calyx minute, shortly cupular, densely hairy, lobes obsolete; corolla salverform, to 5.5 × 2.5 cm, white, or slightly suffused with violet or pink, tube ± straight, lobes obovate to oblong-elliptic, 1–1.5 × 0.6–0.7 cm, ± spreading horizontally; stamens 10, exserted, filaments linear, ± glabrous, anthers broadly oblong, ~1.5 mm long; disk hairy; ovary cylindric, 4(–5) cm long, stigma shortly obconical, lobulate. **Fruits** capsules, reddish, oblong-fusiform, 3–3.5 × 0.4 cm, slightly 10-ribbed. **Distr:** Borneo (Mt Kinabalu, Mt Palimasan). **Ecol:** In mossy ridge forests; (800–) 1,065–1,525 m. Fl. Jan–Dec. **Zones:** 11–12.

R. stenophyllum* ssp. *stenophyllum Hook. *f. ex* Stapf, *Trans. Linn. Soc. London*, II, Bot. 4(2): 196, 1894. **Shrubs,** to 3m. **Leaves** in pseudowhorls of 10–15, gradually passing to deciduous scale leaves on elongate parts of stem; lamina linear, 4–7 cm × 2.5–5 mm, base narrowly tapering, margins entire, flat, apex acute, adaxially glabrescent, abaxially sparsely scaly; petiole ~3 mm long. **Inflorescences** 1–3-flowered, flowers held stiffly horizontal or slightly angled downward; bracts to 2.4 cm long, glabrous; bracteoles filiform, to ~1 cm long; pedicels 0.8–1 cm long, red, scaly, occasionally hairy. **Flowers** without scent; calyx a low scaly disk; corolla bell-shaped, 3.2 × 4.5 cm, bright waxy orange to red, tube 2.2 cm long, lobes 2 × 2.2 cm, strongly overlapping; stamens arranged around mouth of flower, filaments 1.2–1.5 cm long, orange or red, base hairy, anthers cream; stigma cream, slightly lobed, style glabrous. **Fruits** capsules, subcylindrical, 1.8 × 0.5 cm, hairy. **Distr:** Borneo (Sabah: Mt Kinabalu). **Ecol:** In mossy submontane forests, mainly terrestrial in open situations, 2,700–2,800 m. Fl. Jan–Dec. **Zones:** 11–12.

R. stenophyllum* ssp. *angustifolium Argent, A. Lamb and Phillipps, *Notes RBG Edinb.* 42(1): 115, 1984. (Syn: *R. stenophyllum* var. *angustifolium*). Differing from *R. stenophyllum* ssp. *stenophyllum* in the longer, narrower **leaves,** 6–12 cm × 1.4–2.2 mm and **pedicels** always softly hairy besides being scaly, 0.8–1.8 cm long. **Distr:** Borneo (Sabah, Sarawak, Brunei, Kalimantan). **Ecol:** In mossy submontane forests; 1,500–2,400 m. **Zones:** 11–12.

R. stevensianum Sleumer, *Blumea* 21(2): 371, 1973. **Shrubs,** epiphytic, sometimes terrestrial, to 60 cm high. **Leaves** in tight pseudowhorls of 2–3, sometimes with 1 or 2 smaller ones; lamina usually ovate, 1.5–3.5 × 1–2.5 cm, base cordate, margins entire or minutely crenulate, apex obtuse to rounded, sometimes minutely emarginate, adaxially green, sparsely scaly, soon glabrescent, abaxially sparsely scaly, lateral veins 4–6 per side; petiole 2–3 mm long. **Inflorescences** with flowers solitary, rarely paired, pendent; pedicels 0.8–1.2 cm long, pink, densely hairy; outer bracts broadly ovate, scaly; inner bracts ovate, glabrous. **Flowers** with calyx green, densely hairy; corolla 2.8–3.3 × 2.2–2.8 cm, dull purple, strongly curved, glabrous, lobes ~1–1.2 × 0.9–1.1 cm, ± spreading; stamens tightly clustered on upper side of mouth, sometimes exserted, filaments purplish-red, glabrous, anthers brown; disk green, subglabrous to densely hairy; ovary conical, 4 × 2.5 mm, green, silvery-scaly, style 0.7–1 cm long, green or red, glabrous. **Fruits** 1.6–2.1 × ~0.5 cm, green, with a persistent purple style; seeds 4–5 mm long. **Distr:** Papua New Guinea. **Ecol:** 2,000–2,300 m. **Zones:** 11–12.

R. suaveolens Sleumer, *Bot. Jahr.* 71: 147, 1940. **Shrubs** or **trees,** epiphytic and terrestrial, to 3 m tall; internodes 2–10 cm long. **Leaves** variably sized in tight pseudowhorls of 3–6; lamina elliptic, 6–10 × 3.5–7 cm, base usually cordate, margins smooth, with a narrow cartilaginous flat edge, apex usually rounded; petiole 1–2 mm long. **Inflorescences** umbels, 12–25-flowered, flowers held erect to half-hanging; pedicels 1.5–2.5 cm long, pedicels bright red. **Flowers** strongly and sweetly scented; calyx irregular, ringlike; corolla 3.5–7 × 2–2.5 cm, pure white or sometimes slightly pink-flushed especially on tube, distinctly narrowed, lobes 1.3 × 1 cm, horizontal or irregularly reflexed against the tube; stamens irregularly clustered at or below mouth, inserted, filaments slender, white, anthers 2 mm long, pale-brown; disk green; ovary cylindrical, 9 × 2 mm, stigma 1.5–2 mm wide, irregularly 5-lobed. **Fruits** cylindrical, 3–4.5 × 0.3–0.5 cm, curved, narrowed to persistent style, longitudinally ribbed; seeds 5–5.5 mm long. **Distr:** Borneo (Sabah). **Ecol:** On trees and ridges; 1,200–1,700 m. Fl. Dec–Mar. **Zones:** 11–12.

R. superbum* var. *superbum Sleumer, *Reinwardtia* 5: 76, 1960. **Shrubs,** robust, to 6 m tall; internodes 5–10 cm long. **Leaves** in pseudowhorls 4, or spirally arranged; lamina broadly ovate or broadly elliptic, 7.5–12 ×4.5–8 cm, base usually broadly tapering, margins entire, flat, apex broadly tapering, usually shortly acuminate, young leaves golden-scaly, glabrescent with age; petiole 1.5–3 cm long. **Inflorescences** umbels, 3–6-flowered; pedicels 1.2–2.4 cm long, flowers horizontal to half-hanging.

Below: *Rhododendron stapfianum, Rhododendron stenophyllum* ssp. *stenophyllum, Rhododendron stenophyllum* ssp. *angustifolium, Rhododendron stevensianum.*

Top row: *Rhododendron suaveolens, Rhododendron superbum* var. *superbum, Rhododendron taxifolium.*
Bottom row: *Rhododendron truncicola, Rhododendron vanvurnii, Rhododendron viriosum.*

Flowers strongly scented; calyx cupular, 7–9 mm Ø, scaly outside, lobes pointed; corolla funnelform or widely trumpet-shaped, 5.5–14 × 7–12 cm, fleshy or waxy, white to cream or very light yellow, sometimes faintly pink-flushed, tube 6.5–7.5 cm long; stamens 12 or 14, exserted. **Fruits** capsules, cylindrical, ~5 × 1.3 cm, ends shortly narrowed; seeds 7–8 mm long. **Distr:** Papua New Guinea. **Ecol:** In mossy forests and on edge of cloud forests; 1,525–3,050 m. Fl. & Fr. Apr.–Sept. **Zones:** 10–12.

R. taxifolium Merr., *Philip. J. Sc.* 30: 419, 1926. **Shrubs**, epiphytic, to ~1 m high; internodes 1–3(–5) cm long. **Leaves** crowded to densely pseudowhorled, in groups of 20+; lamina linear, (1.8–)2–3.5 cm × 1–1.5 mm, base attenuate into petiole, margins shallowly crenulate, apex obtuse, rigidly coriaceous, adaxially glabrous, abaxially laxly lepidote; petiole 0–2 mm long. **Inflorescences** 3–5-flowered, occasionally 1-flowered; perulae membranous, broadly ovate, to 1.2 × 0.6 cm; bracteoles linear, ~6 mm long; pedicels slender, ~1 cm long, pubescent, laxly lepidote. **Flowers** with calyx disk-shaped, lepidote and pubescent outside, obtusely and shortly 5-lobed, 3–4 mm Ø; corolla white, tubular-subcampanulate, 2–2.2 cm long, thin, lobes broadly obovate, ~1 × 0.8 cm, rounded; stamens 10, subequal, filaments linear, anthers oblong; ovary subcylindrical, ~3 × 1.5 mm, densely hairy, style columnar, hairy below, ~5 mm long, stigma subcapitate. **Fruits** ~1 cm long. **Distr:** Philippines (Luzon, Mt Pulog). **Ecol:** In trees of mossy forests; ~2,700 m. Fl. Feb–Mar. **Zones:** 11–12.

R. truncicola Sleumer, *Reinwardtia* 5: 91, 1960. **Shrubs**, small. **Leaves** in pseudowhorls of 3–4; lamina elliptic, 3–4.5 × 1.6–2.7 cm, base broadly tapering to truncate or rounded, margins entire, narrowly revolute, apex obtuse to rounded, occasionally minutely retuse, adaxially glabrescent, abaxially scaly; petiole 0.7–1 cm long, pale cream, densely scaly. **Inflorescences** umbels, ~3-flowered, flowers half-hanging, unless held ± horizontal by the leaves below; bracts ovate to spathulate, to 1.5 × 0.6 cm; bracteoles filiform, to 1.5 cm long; pedicels 1–1.5 cm long, pink, densely scaly. **Flowers** with calyx disklike, ~2.5 mm Ø; corolla pink, zygomorphic, tubular, 3–3.5 × 2.9 cm, curved, lobes broadly obovate-spathulate, 1–1.5 × 1–1.5 cm; stamens initially tightly clustered on upper side of corolla, later irregularly arranged, ~12, to 1 cm long, exserted; ovary cylindrical, ~4 × 1.8 mm, densely scaly, style on upper side of tube, stigma conical-globose. **Distr:** Papua New Guinea; ~2,135 m. Fl. Jun. **Zones:** 11–12.

R. vanvurnii J.J. Sm., *Bull. Jard. Bot. Buit.* III, 1: 399, t. 48, 1920. **Shrubs** or **small trees**, terrestrial, to 4 m tall; internodes 3–15 cm long. **Leaves** usually spirally arranged, sometimes in loose pseudowhorls; lamina elliptic or ovate-elliptic, 5–14 × 1.5–4 cm, base broadly tapering to rounded, margins entire, apex shortly acuminate or subacute; petiole 0.5–1 cm long, densely scaly. **Inflorescences** umbels, 5–15-flowered; pedicels 0.9–2.5 cm long, ± hairy. **Flowers** with calyx oblique, ~5 mm Ø, shortly 5-lobed; corolla orange to yellow, occasionally white or pink, funnelform, 4–6 × 5–5.5 cm, hairy inside, lobes broadly obovate, 2–2.5 × 1.5–2 cm; stamens subequal, to ~1–1.2 cm long, exserted, filaments linear, filiform; disk densely hairy marginally; ovary subcylindrical, 5–9 × 2–2.5 mm, densely hairy, scaly, style slender, stigma rounded, reddish. **Fruits** subfusiform to cylindrical, 5.5–7 × 0.6–0.7 cm, ± hairy and scaly. **Distr:** Indonesia (Sulawesi). **Ecol:** In dry forests, shrubberies and bare slopes; 600–1,600 m. Fl. Jun–Aug, Dec. **Zones:** 11–12.

R. viriosum Craven., *Edinb. J. Bot.* 59(3): 448, 2002. [Australian Rhododendron]. **Shrubs**, terrestrial or epiphytic, to 3 m tall. **Leaves** in tight pseudowhorls of 2–6; lamina elliptic or obovate, 2.5–11 × 1–7 cm, base tapering, margins entire, narrowly revolute or flat, apex shortly acuminate to obtuse; petiole 0.5–2 cm long. **Inflorescences** umbels, open, 2–7-flowered; pedicels 1.8–3.5 cm long. **Flowers** with calyx an angular disk, ~4 mm Ø; corolla red or deep pink, unscented, funnelform, straight, 5–5.8 × 5–5.8, lobes semierect to spreading, obovate to ± circular, 2–2.7 × 2–2.7 cm, emarginate, overlapping; stamens irregularly spread around mouth, dimorphic, slightly exserted, filaments 23–30, 2–2.8 cm long, anthers dark red; disk 10-lobed, hairy; ovary subcylindrical, 3–6 × 2–2.5 mm, densely hairy, style 1.4–2.3 cm long, appressed to lower side of tube, stigma rounded ~1.5 mm Ø, **Fruits** ellipsoid, 1.5–3 × ~0.5 cm. **Distr:** Australia (Queensland: Mt Finnigan). **Ecol:** In rainforests; 910–1,330 m. Fl. Aug–Mar. **Zones:** 10–12.

R. womersleyi Sleumer, *Reinwardtia* 5: 136, 1960. (Syn: *R.* aff. *inconspicuum*). **Shrubs**, terrestrial, rarely epiphytic, stiffly erect, to 2 m high. **Leaves** densely spirally arranged; lamina ovate to broadly elliptic or subcircular, 0.7–1.4 × 0.4–0.9 cm, base broadly tapering to rounded, margins entire or weakly crenulate, apex usually shortly acuminate; petiole 1–2 mm long densely hairy. **Inflorescences** 1–4-flowered, flowers hanging; outer bracts ovate-acuminate; bracteoles filiform, ~5 mm long; pedicels 1.5–2 cm long, densely scaly. **Flowers** with calyx discoid, ~2.5 mm Ø, lobes broadly triangular, ~1 mm long, obtuse, scaly, hairy; corolla bright or deep red, tubular, 2–2.5 × 1.6–2 cm, slightly dilated at mouth, outside white-hairy, lobes subcircular, ~8 × 9 mm, overlapping; stamens arranged around lower half, exserted, filaments filiform, deep pink, glabrous, anthers oblong, purplish-black, pollen gray; ovary subcylindrical, 3–4 × 1–1.5 mm, densely hairy, style on lower side of mouth, white-hairy below, stigma ± globose. **Fruits** fusiform, 1.5–2 × 0.3–0.4 cm, hairy, grayish or purplish; seeds 4 mm long. **Distr:** Papua New Guinea. **Ecol:** In mossy forests or grasslands; 3,200–3,960 m. Fl. Feb–Sep. **Zones:** 11–12.

R. yelliotii Warb., *Bot. Jahr.* 16: 25, 1892. (Syn: *R. saruwagedicum*, *R. saruwagedicum* var. *alpinum*). **Shrubs** compact, or **trees**, erect, much-branched, to 12 m tall. **Leaves** regularly spirally arranged; lamina usually elliptic, 0.7–2.5 × 0.5–1.5 cm, base usually rounded, margins crenulate, apex mostly broadly obtuse, adaxially grayish-green, shiny, abaxially brownish, silvery or yellowish; petiole 1–3 mm long. **Inflorescences** open umbels, 3–5-flowered, flowers ± hanging; outer bracts ovate, to 6 × 5 mm, subacute; pedicels 0.8–1.3 cm long, densely scaly. **Flowers** with calyx disk-shaped, oblique, 2.5–3.5 mm Ø, densely scaly; corolla deep red, zygomorphic, cylindrical, 1.2–1.6 × 0.8–1 cm, ± straight, densely scaly outside, lobes suberect, subcircular, 3–5 mm Ø, overlapping; stamens irregularly arranged, slightly exserted, filaments glabrous; ovary ovoid-conical, ~5 × 2.5 mm, densely scaly, stigma 5-lobed. **Fruits** 1.2–1.5 × 0.5–0.6 cm; seeds 2 mm long. **Distr:** Papua New Guinea. **Ecol:** Epiphytic in montane forests, terrestrial in forest margins or grasslands; 1,300–3,680 m. Fl. & Fr. throughout the year. **Zones:** 10–12.

R. zollingeri J.J. Sm., *Ic. Bog.* 4: 73, 1910. (Syn: *R. tubiflorum*, etc.). **Shrubs** or **trees**, erect, much-branched, to 4 m tall. **Leaves** in tight pseudowhorls of 4–6; lamina narrowly ovate to obovate or elliptic, 3–7 × 1–2.2 cm, base tapering, margins entire, flat, apex usually shortly acuminate, adaxially glabrescent, abaxially scaly; petiole 2.5–5 mm long. **Inflorescences** umbels, 3–8-flowered, terminal, some axillary, flowers half-hanging; bracts ovate to obovate, 8 × 4 mm, obtuse; bracteoles filiform, to 6 mm long; pedicels 1–1.5 cm long, densely scaly. **Flowers** with calyx, obliquely cupular, ~3 mm Ø, 5-lobed; corolla red or orange, tubular, 1.4–1.5 × 0.9 cm, lobes subcircular, ~4 × 4 mm, scaly outside; stamens ~1 cm long, filaments glabrous, anthers obovate; ovary conical, 2–4 × 2–2.5 mm, 5-ribbed, densely scaly, style glabrous, stigma 5-lobed. **Fruits** fusiform or subcylindrical, 1.5–1.8 × 0.5–0.6 cm, thick, densely scaly, style persistent; seeds 2.7–3 mm long. **Distr:** Indonesia (Java, Lesser Sunda Islands, Sulawesi), Philippines (Luzon). **Ecol:** On bare open slopes or ridges; 1,200–3,090 m. Fl. Apr–Oct. **Zones:** 11–12.

Subsection Malayovireya

Scales sessile, dense, of 2 obviously different size classes, lobed and the center dark-colored (Craven et al. 2008).

R. fallacinum Sleumer, *Reinwardtia* 5: 99, 1960. **Shrubs** or **small trees**, terrestrial, sometimes epiphytic, to 6 m tall; internodes 2–10 cm long. **Leaves** subsessile, in tight pseudowhorls of 3–6, with small, slender, early deciduous scale leaves arranged along the stems; lamina ovate, 8–16 × 3–5.5 cm, base rounded to auriculate, margins irregularly wavy, flat, apex usually acuminate or subcaudate, adaxially densely silvery-scaly, abaxially densely brown-scaly; petiole ~4 mm long. **Inflorescences** umbels, 15–35-flowered, flowers erect to horizontal; pedicels 1.5–3.5 cm long. **Flowers** with calyx a low, scaly ring; corolla 1.8–2.5 × 3–3.7 cm, bright orange, densely scaly outside; stamens first erect then reflexing back away from stigma against corolla lobes, filaments 1.2–1.6 cm long; ovary 5.5–7 × 1.5–2 mm, densely scaly. **Fruits** to 3.5 × 0.4 cm; seeds 6 mm long. **Distr:** Borneo (Sabah and Sarawak). **Ecol:** In damp, shady places of mossy forests; 1,220–2,600 m. Fl. & Fr. Jan–Dec. **Zones:** 11–12.

R. himantodes Sleumer, *Bot. Jahr.* 71: 145, 1940. **Shrubs**, epiphytic, to 2 m high. **Leaves** in pseudowhorls of 12–25, or regularly spirally arranged; lamina linear, 1.5–12 × 0.2–0.6 cm, base tapering, margins irregular, usually entire, slightly reflexed, apex usually acute, adaxially densely scaly, becoming usually glabrescent, abaxially scaly; petiole 2–6 mm long, densely scaly. **Inflorescences** umbels, 8–15-flowered, flowers ± horizontal; pedicels 2–3 cm long, pale green, densely scaly; bracts ovate to obovate-spathulate, brown-spotted, margins finely white-ciliate; bracteoles narrowly linear, to 8 mm long, ± glabrous. **Flowers** with calyx discoid or minutely 5-lobed, densely scaly; corolla 0.7– weakly funnelform, 1.2 × 2–2.5 cm, deeply longitudinally grooved, white, brown-scaly outside, densely hairy inside, lobes 0.8–1 × 0.7–1 cm, wide-spreading, white, conspicuously brown-speckled outside, sometimes with distinctive but irregular radial lines inside; stamens wide-spreading, irregularly arranged around corolla, filaments 5 mm long, white, hairy, anthers dark brown, disk glabrous; ovary conical, 3.5–5 mm long, scaly, style green, glabrous, stigma green. **Fruits** narrowly fusiform, 1.5–2 × 0.4 cm, scaly; seeds 6–8 mm long. **Distr:** Borneo (Sabah: Mt Kinabalu). **Ecol:** On large trees in mossy montane forests; 1,300–2,000 m. **Zones:** 11–12.

R. malayanum Jack, *Mal. Misc.* 2: 17, 1822. (Syn: *R. malayanum* var. *axillare*, *Vireya tubiflora*). **Shrubs** or **small trees**, erect, to 5 m tall. **Leaves** in tight pseudowhorls; lamina variable, elliptic or narrowly ovate, 4–19 × 1–4 cm, base acutely tapering, margins often slightly revolute, apex shortly acuminate, tip glandular, adaxially densely scaly, abaxially brown; petiole 1–2.2 cm long, densely scaly. **Inflorescences** umbels, 1–5-flowered, flowers hanging; bracts ovate-triangular, ovate to spathulate, densely scaly outside, glabrous inside; bracteoles filiform, to 1.5 cm long; pedicels slender, 0.6–1 cm long. **Flowers** with calyx oblique, 5-lobed, densely scaly; corolla waxy, glossy, bright red or purplish, to pink, or greenish-white, 1.5–3 × 1.7–2.2 cm, straight or slightly curved, laterally compressed, lobes 0.7–1 × 0.5 cm; stamens exserted, unequal, loosely clustered on lower side of flower, filaments glabrous, anthers 1.2–1.6 mm long; disk scaly; ovary oblong-conical, 4–5 × 1.5–2 mm, densely scaly, style 1–2 cm long, on lower side of corolla, green or white, stigma club-shaped. **Fruits** fusiform, 2.2–3.3 × 3–4 cm, brown-scaly; seeds 7.5 mm long. **Distr:** Indonesia, Malaysia, Thailand. **Zones:** 10–12.

R. micromalayanum Sleumer, *Blumea* 21: 364, 1973. **Shrubs**, erect to spreading, to 1.2 m high. **Leaves** in tight pseudowhorls of 5–7, 2–6 cm apart; lamina narrowly elliptic, 2.5–4 × 0.7–1.2 cm, base tapering, margins entire, apex obtuse to rounded, adaxially soon glabrescent, abaxially ± scaly; petiole 5–7 mm long. **Inflorescences** 1–4-flowered, flowers pendent; *basal* bracts brown-scaly; *upper* bracts glabrous, fringed with white hairs; pedicels 1.5 cm long, densely brown-scaly. **Flowers** with calyx brown, scaly, disklike; corolla purplish-pink, pale pink or greenish-white, 2.6 × 1.8 cm, tube cylindrical, fluted, base slightly swollen, laterally flattened, outside brown-scaly, inside glabrous, lobes wide-spreading, 1 × 0.7 cm; stamens irregularly arranged with more on lower side of corolla, filaments 2.4 cm long, glabrous, anthers pale orange-brown; ovary 5 × 2 mm, densely scaly, style 1.7 cm long, stigma green, 1.5 mm Ø, **Fruits** fluted, slightly curved, 2–3.7 × 0.4 cm, with persistent stigma; seeds 5–6 mm long. **Distr:** Borneo (Sarawak, Sabah). **Ecol:** Epiphytic in mossy forests, terrestrial on open ridges; 800–2,00 m. Fl. throughout the year. **Zones:** 11–12.

Section Pseudovireya

Corolla campanulate; staminal filaments glabrous proximally and distally, and hairy in the middle region (Craven et al. 2008).

R. santapaui Sastry, Kataki, P.A.Cox, E.P.Cox and Hutchison, *J. Bombay Nat. Hist. Soc.* 65: 744, 1969. **Shrubs**, epiphytic, to 1.5 m high. **Leaves** in tight pseudowhorls of 5–9; lamina elliptic to obovate, 1.5–3 × 0.5–1.2 cm, base tapering, margins entire, apex acute to subobtuse, shortly apiculate, adaxially densely scaly, soon glabrescent, dark green, abaxially pale green, brown-scaly, lateral veins 3–4 per side; petiole 2–3 mm long. **Inflorescences** umbels, 2–6-flowered; pedicels 1–2 cm long, densely silvery-scaly; bracts ovate, to 5–8 × 3–6 mm, mucronate. **Flowers** with calyx lobes triangular, 1 × 1.5 mm, densely brownish-scaly outside; corolla campanulate, ~1 × 1.5–2.5 cm, white, fleshy, lobes; stamens ~1.2 cm long, exserted to 7 mm, filaments white, hairy in middle; disk circular,

Below: *Rhododendron womersleyi*, *Rhododendron yelliotti*, *Rhododendron zollingeri*, *Rhododendron fallacinum*.

Above: *Rhododendron himantodes, Rhododendron malayanum, Rhododendron micromalayanum, Rhododendron santapaui.*

10-lobed, glabrous; ovary ovoid, 5-ridged, ~5 × 2.5 mm, densely scaly, style ~5 mm long, stigma truncate. **Fruits** spindle-shaped, 2–2.7 × 0.4–0.5 cm, brown-scales; seeds to 1.1 cm long. **Distr:** India (Arunachal Pradesh). **Ecol:** On large tree trunks; ~900–1,600 m. **Zones:** 10–12.

Section Discovireya

Corolla tubular-cylindric; staminal filaments glabrous or base hairy (Craven et al. 2008).

R. adinophyllum Merr., *Notes Natl. Acad. Nat. Sci. Philad.* 47: 3, 1940. **Shrubs**, to 1.5 m high. **Leaves** densely spirally arranged; lamina elliptic to narrowly elliptic, 0.7–1.2 × 0.2–0.5 cm, base tapering into petiole, margins flat or slightly recurved, apex acute to obtuse, gradually acuminate, stiff, adaxially shiny dark green, abaxially paler, subdensely scaly; petiole to 2.5 mm long, curved to a right angle. **Inflorescences** umbels, (4–)6–12-flowered; flowers ± pendent. **Flowers** without scent, tubular, 1.8–2.1 × ~0.9 cm, bright orange-red; corolla covered with white patent hairs and small scales outside, glabrous inside, lobes short, straight, not reflexed, overlapping; stamens exserted, loosely clustered at lower side of mouth; ovary ovoid-conical, densely scaly, hairless. **Fruits** elongate-ovoid, to 1.1 cm long, 5-ribbed, with persistent style, lepidote. **Distr:** Indonesia, N Sumatra. **Ecol:** In open summit heathlands; 2,600–3,460 m. Fl. Jan–Apr. **Zones:** 10–12.

R. borneense (J.J. Sm.) Argent, A. Lamb and Phillips, *Rhododendrons of Sabah, Sabah Parks Publ.* 8: 74, 1988. (Syn: *R. quadrasianum* var. *borneense*). **Shrubs**, epiphytic, to 2 m high. **Leaves** spirally arranged, mostly at the very ends of twigs; lamina obovate or spathulate, 1.4–1.8 × 0.5–0.8 cm, base tapering, margins entire, slightly revolute, apex rounded or slightly retuse, adaxially scaly, quickly glabrescent, abaxially sparsely scaly; petiole 2–3 cm long. **Inflorescences** with flowers solitary, occasionally 2 or 3 together, horizontal to semipendent; pedicels 6–8 mm long, hairy, scaly, pink. **Flowers** without scent; calyx a low disk, hairy, scaly; corolla 2 × 1 cm, red or yellow, tube straight, lobed near base, densely scaly, white-hairy at least proximally outside, glabrous inside, lobes 4 × 5 mm, erect or slightly spreading, overlapping, scaly outside; stamens loosely arranged on lower side of mouth, exserted, filaments ~1.8 cm long, glabrous, anthers 7 mm long, brown; ovary ~2 × 1 mm, densely hairy, style 1.3–1.6 cm long, glabrous, or proximal hairy, distally slightly swollen, lying on lower side of tube, stigma exserted, club-shaped, purple. **Fruits** ~8 × 3 mm; seeds 1.6 mm long. **Distr:** Borneo. **Ecol:** In montane forests. **Info:** ***R. borneense* ssp. *villosum*** has leaves narrowly obovate, <2 mm wide, and ovary densely white-hairy.

R. ericoides Low ex Hook. f., *Ic. Pl.*: t. 887, 1852. **Shrubs**, erect, (1–)2–3 m high, sometimes dwarfed (~4,000 m altitude) to ~15 cm high, occasionally epiphytic; branchlets densely covered with wartlike leaf-cushions in lower defoliated parts. **Leaves** spirally arranged, very dense, ± imbricate in upper part of branchlets, ericoid, suberect; lamina linear-lanceolate, 0.4–0.7(–1) cm × ~1 mm, base attenuate, apex with an obtuse glandular point, coriaceous, adaxially glabrous, abaxially laxly lepidote; petiole 0.5 mm long. **Inflorescences** with flowers solitary or paired, rarely 3-flowered umbels, flowers nodding; bracteoles linear-spathulate, long-ciliate upward, ~3 × 0.5 mm, early caducous; pedicels slender, 5–8 mm long, densely lepidote. **Flowers** with calyx lobes 5, erect, narrowly subovate-triangular, ~2 × 1–1.3 mm, obtuse; corolla bright scarlet or purplish-red, tubular, (1–)1.5 cm long, subdensely greenish-lepidote outside, lobes broadly ovate to rounded, 2–3 mm Ø, suberect or ± spreading, margin erose-denticulate; stamens 10, ± equaling corolla or slightly exserted, filaments filiform, glabrous; ovary conical, green, densely lepidote, ~1.8 mm long, gradually attenuate to style, style red, to 1.4 cm long, exserted, stigma club-shaped. **Fruits** oblong-ovoid, often ± oblique and curved, 6–7 × 2.5–3 mm. **Distr:** N Borneo (Mt Kinabalu). **Ecol:** In primary mossy forests; >2,440 m. Fl. & Fr. Mar–Nov. **Zones:** 10–12.

R. quadrasianum S. Vidal, *Rev. Pl. Vasc. Filip.* 170, 1886. (Syn: *R. retusum*). **Shrubs**, occasionally treelets, erect and terrestrial, rarely epiphytic, 0.3–1(–3) m high. **Leaves** usually in pseudowhorls of 4–8 at upper nodes; lamina obovate, to 3 × 1.3 cm, base cuneate, margins ± revolute, apex mostly broadly attenuate or rounded, ± coriaceous, adaxially dark green, glossy, abaxially dull, nerves 2–4 pairs; petiole to 1.5 mm long. **Inflorescences** with flowers solitary or in 2(3–4)-flowered umbels; pedicels ~1(–1.5) cm, densely pubescent. **Flowers** with calyx small, discoid, 5-lobed, lepidote, ± hairy, lobes broadly ovate, ~0.5 mm long, obtuse; corolla red, tubular, slightly oblique, subcampanulate, 1.2–1.8 × 1.3–1.5 cm, shortly 5-lobed, ± glossy, lobes irregular, broadly elliptic-obovate or suborbicular, 3–5 × 2.5–5 mm, apex finely erose or entire, stamens 10, filaments linear, glabrous; ovary obliquely cylindrical, ~2 × 1.5 mm, 5-ribbed, gradually attenuate into style, densely lepidote, style ~8 mm long, glabrous, stigma subcapitate, irregularly lobed. **Fruits** obliquely cylindrical, to 1.2(–1.5) cm long, densely lepidote; seeds ~2 mm long. **Distr:** Philippines (Luzon, Mindoro). **Ecol:** On rocky open slopes, 1,330–2,100 m. Fl. & Fr. May–Jun, Sep. **Zones:** 10–12. **Info:** ***R. quadrasianum* var. *rosmarinifolium*** (Syn: *R. rosmarinifolium*, etc.) grows to 2(–6) m tall, with leaves narrowly elongate, 1–2.2 × 0.3–0.6 cm, and corolla elongate-campanulate, 1–1.6 × 0.9–1.1 cm, red to orange, tube straight.

R. retusum (Blume) Benn., *Pl. Jav. Rar.* 86 in text, 88 t.20, 1838. (Syn. *Vireya retusa, Azalea retusa*). **Shrubs**, mostly terrestrial, erect or occasionally creeping; branchlets with younger parts red. **Leaves** in pseudowhorls of 4–7 at upper nodes; lamina obovate to oblong-obovate, to 4(–6.5) × 1.8(–3) cm, base cuneate, margins ± recurved, apex rounded

Below: *Rhododendron adinophyllum, Rhododendron borneense, Rhododendron borneense* ssp. *villosum, Rhododendron ericoides.*

Above: *Rhododendron quadrasianum, Rhododendron quadrasianum* var. *rosmarinifolium, Rhododendron retusum, Rhododendron indicum.*

or retuse, ± coriaceous, adaxially dark green, shiny, abaxially paler to almost whitish-green, ± dull, nerves 4–6(–8) pairs; petiole 1–5 mm long. **Inflorescences** umbels, (2–)4–10-flowered. **Flowers** with calyx small, oblique, shortly 5-lobed or -dentate, ~4 mm Ø, densely lepidote outside, lobes ~1 mm long, ciliate; corolla tubular, light red to intense scarlet, or orange-red, 2–2.5(–3) cm long, tube cylindric, base slightly 5-saccate, or dilated into a ring, 4–6 mm Ø at base, to 8 mm Ø at apex, ± densely yellow-lepidote, short whitish-hairy outside, glabrous inside, lobes erecto-patent or spreading, broadly oblong to suborbicular, 4–6 × 3–5 mm; stamens 10. **Fruits** capsules, erect, oblong, fusiform, 1–1.5(–1.8) × 0.4–0.5 cm, lepidote; seeds oblong, ~2 mm long. **Distr:** Indonesia. **Ecol:** In subalpine mossy forests or shrubberies; 1,350–3,400 m. Fl. throughout the year. **Zones:** 10–12.

Subgenus Azaleastrum

R. indicum (L.) Sweet, *Hort. Brit.*, ed. 2: 343, 1830. (Syn: *Azalea indica*, etc.). [Indian Azalea, Indian Rhododendron]. **Shrubs**, semievergreen, 1–2 m high; young shoots densely rusty-strigose, glabrescent. **Leaves** narrowly lanceolate or oblanceolate, 1.7–3.2(–4.5) × ~0.6 cm, base narrowly cuneate, margins sparsely and finely crenate-serrate, apex obtusely pointed, papery, adaxially shiny-strigose, abaxially sparsely brownish-strigose; petiole 2–4 mm long, coarsely brownish-strigose. **Inflorescences** 1–3-flowered; pedicels 0.6–1.2 cm long, coarsely white-strigose. **Flowers** with calyx lobes elliptic-ovate or suborbicular, 2–3 × 1.5–2 mm, pale green, softly white-hairy; corolla broadly funnelform, bright red, occasionally rose-red, with dark red flecks, 3–4 × 3.7(–6) cm, lobes broadly elliptic, 1.7–2 × ~1.6 cm; stamens 5, unequal, 1.6–2.2 cm long, inserted, filaments puberulent at base; ovary ~3 mm long, densely strigose, style 2.3(–4.5) cm long, longer than stamens, glabrous. **Fruits** oblong-ovoid, 6–8 mm long, densely reddish-strigose. **Distr:** Japan. **Ecol:** On wooded riverbanks. Fl. May–Jun. **Zones:** 9–12.

Subfam. VACCINIOIDEAE

Genera ~50, **species** 1,580. **Shrubs**, **trees**, or **lianas**, often epiphytic, evergreen or deciduous. **Leaves** alternate; lamina with margins entire to serrulate or serrate. **Inflorescences** racemose, paniculate, or fasciculate, or flowers solitary, terminal or axillary; bracts present; bracteoles usually paired, variable in position; flowers articulated or continuous with pedicel, usually pendulous. **Flowers** usually 4- or 5-merous, actinomorphic; calyx lobes usually 4 or 5, usually persistent, sometimes fleshy; corolla usually sympetalous, urceolate to cylindrical or campanulate, deciduous; stamens usually 5–10, usually included, anthers spurred; ovary usually 4- or 5-locular, or 10-locular due to secondary division of each locule, placentation axile, superior to inferior, style usually impressed, stigma truncate. **Fruits** loculicidal capsules, berries or drupes. **Distr:** Worldwide.

Agapetes

Species 80. **Shrubs** or **small trees**, epiphytic, rarely terrestrial, evergreen; rootstock usually swollen. **Leaves** spirally arranged, subopposite, or pseudo-verticillate, subsessile or shortly petiolate; lamina with margins entire or serrate. **Inflorescences** racemes, corymbs, or fascicles, or flowers solitary, axillary; pedicels articulate, sometimes expanded and forming a cup apically. **Flowers** 5-merous; calyx some-times 5-winged or -angled; limb 5-fid or deeply lobed; corolla red, rarely white or yellow, ± plain-colored or with dark-colored zigzag horizontal bands, tubular, rarely globose, urceolate, or campanulate, often >1 cm long, usually shortly 5-lobed; stamens 10, as long as corolla; ovary inferior, carpels many-ovuled, stigma small, truncate or capitate. **Fruits** berries, globose, succulent or nearly dry, with persistent calyx; seeds several, testa hard or mucilaginous. **Distr:** E Himalayas through SW China and Indo-China to SE Asia. **Cult:** Propagate from seeds, cuttings or by air-layering. Requires an acidic, humus-rich, moist, well-drained soil in shade or partial sun. **Info:** The majority of species of *Agapetes* is part of the same lineage as many SE Asian-Malesian species of *Vaccinium*. Species of *Agapetes* from New Guinea and the SW Pacific are now placed in the genus *Paphia* (Brach 1996).

A. buxifolia Nuttall ex J.D. Hooker, *Bot. Mag.* 83: t. 5012, 1857. **Shrubs**, epiphytic, 1.2–1.5 m high; branchlets angled, 2–3 mm thick, densely glandular-hispid. **Leaves** crowded; lamina obovate or elliptic, 1.5–2.8 × 1–1.2 cm, base cuneate, margins entire, crenate above middle, apex acute, leathery, glabrous, adaxially bright green, abaxially pale green, secondary veins ~3 pairs, inconspicuous; petiole ~2 mm long. **Inflorescences** axillary, 1- or 2-flowered; pedicels slender, ~8 mm long. **Flowers** tubular; calyx tube green, narrowly obconic-campanulate, villous, limb divided to base, lobes triangular-lanceolate, 3–4 mm long; corolla bright red, tubular, ~2.5 cm long, glabrous, lobes spreading, pale or greenish, ovate-lanceolate, 4–6 mm long; filaments ~4 mm long, glabrous, anthers ~1.8 cm long, thecae shortly caudate at base, tubules ~4× as long as thecae, without spurs. **Fruits** berries. **Distr:** NE India (Assam). **Ecol:** Epiphytic on trees of damp forests; 600–900 m. Fl. Apr–May. **Zones:** 10–11.

A. hosseana Diels, *Rep. Nov. Spec. Regni Veg.* 1, 16, 1905. **Shrubs**, epiphytic; stems reddish-hairy. **Leaves** sessile or shortly petiolate; lamina ovate, ovate-oblong or oblong, 1.8–4.5 × 1–1.5 cm, base cuneate, margins slightly recurved, apex acute, adaxially glossy midgreen, abaxially paler, coriaceous, veins 5–6 per side. **Inflorescences** 1–2(–4)-flowered, axillary; peduncles 1–1. 5 cm long; pedicels 1.3 cm long, glabrous. **Flowers** 2.5 cm long, calyx green, corolla bright red, constricted near mouth and blackish, corolla lobes greenish; calyx ~5 mm long, pilose; corolla cylindrical, tubular, smooth, red, 1.5–2 cm

Below: *Agapetes buxifolia, Agapetes hosseana, Agapetes serpens, Agapetes smithiana.*

YR

Above: *Cavendishia bracteata, Ceratostema silvicola, Dimorphanthera alpina, Dimorphanthera kempteriana.*

long, ~5 mm Ø, lobes short; anthers 10, 1.3–1.7 cm long, red, with horns at the back, 6–8 mm long. **Distr:** S China, Myanmar, Thailand, Vietnam. **Ecol:** On forest trees in high mountain areas; 800–2,400(–2,550) m. **Zones:** 9–11. **Info:** In China the roots are used for treating fractures.

A. serpens (Wight) Sleumer, *Bot. Jahrb. Syst.* 70: 105, 1939. (Syn. *Vaccinium serpens*, etc.). [Creeping Agapetes, Flame Heath, Himalayan Lantern]. **Shrubs**, epiphytic, 40–60 cm high; rootstock tuberous. **Leaves** crowded; lamina narrowly ovate or ovate-oblong, 1.2–2 × 0.5–0.8 cm, margins slightly recurved, serrate above middle, apex acute or obtuse, mucronulate, leathery, adaxially deep green, abaxially pale green; petiole ~1 mm long, glabrous. **Inflorescences** fasciculate, 1–3-flowered; pedicel erect or pendulous, 0.7–2.5 cm long. **Flowers** tubular; calyx tube 3.5–4 mm long, 5-winged, lobes ovate-triangular; corolla bright red, orange, or pinkish-white, with dark red zigzag bands, tubular, 1.2–2.8 cm long, 5-angled, lobes recurved, triangular, 2–3(–6) mm long. **Fruits** berries, obovoid, ~6 mm Ø, 5-winged, with persistent calyx lobes enlarged. **Distr:** Bhutan, NE India, E Nepal. **Ecol:** In evergreen forests, on mossy tree trunks and rocks; 1,200–2,400 m. Fl. May–Jun. Fr. May–Nov. **Zones:** 9–11.

A. smithiana Sleumer, *Bot. Jahrb.* 70: 106, 1939. (Syn: *Pentapterygium sikkimense*). **Shrubs**, terrestrial or epiphytic. **Leaves** 1.4–2.8 × 0.8–1.4 cm, base narrowly tapering to the almost sessile petiole, with a pair of marginal hydathodes a short distance above base of lamina (inconspicuous to conspicuous), margins serrate, apex acute (with sharp tip), midrib paler in color, prominent. **Inflorescences** axillary cymes, 1–8(or more)-flowered, flowers pendent; pedicel hairy, yellowish-green. **Flowers** with calyx 5-winged, yellowish-green, lobes triangular; corolla cylindrical, 1.7–2.1 cm long, yellow to greenish-yellow, 7–9 mm Ø, ribbed, lobes recurved. **Distr:** E Himalayas (E Nepal to Bhutan, Sikkim). **Ecol:** Grows as an understory shrub or as an epiphyte, 2,000–2,800 m. **Zones:** 9–10. **Info:** There are two described varieties for this species.

Cavendishia

Species ~130–150. **Shrubs**, epiphytic or terrestrial, evergreen. **Leaves** alternate; lamina simple, margins entire and revolute, surfaces usually with multicellular, reddish or blackish, slightly sunken, glandular trichomes. **Inflorescences** racemose or subfasciculate, axillary, usually solitary, encircled at base by a progression of bracts, the smallest bracts minute and triangular, the largest conforming to the shape and size of individual floral bracts; inflorescence bracts few to numerous; floral bracts solitary at base of pedicel, caducous or usually persistent, usually large and showy. **Flowers** 5-merous, without scent; corolla tubular, 5-lobed, usually flaring at anthesis; stamens 10, subequal, anthers alternately unequal; stigma minutely 5-lobed, truncate but slightly flaring at anthesis. **Fruits** berries, juicy, spherical, dark blue-black; seeds many, 0.5–1 mm long. **Distr:** Mexico (Oaxaca), south through montane C America to Bolivia, and east through the Guianas to Brazil (Amapa). **Cult:** Propagate from cuttings, seeds or by air-layering. Requires peaty, humus-rich soil, in humid conditions and filtered sun.

C. bracteata (Ruiz and Pavón ex Jaume Saint-Hilaire) Hoerold, *Bot. Jahrb. Syst.* 42: 280, 1909. (Syn. *Thibaudia cordifolia*, etc.). **Shrubs**, terrestrial or epiphytic, 1–4 m tall, rarely scandent and lianoid or arborescent (to 15 m tall with stem base to 30 cm Ø). **Leaves** oblong, elliptic, ovate, or lanceolate, (2.5–)4–15(–22) × (1–)1.5–5(–11) cm, base rounded, cordate, cuneate or subtruncate, apex acute to acuminate, often abruptly. **Inflorescences** 2–40-flowered, surrounded at base by bracts, 0.5–8 cm long, green or red, floral bract smooth or muricate, oblong, ovate, elliptic or oblanceolate, 1.1–4.0 × 0.5–2.7 cm, pink to dark red, sometimes ciliate along margins or at distal tip; pedicel 1.5–2 cm long. **Flowers** with calyx 3.5–9 mm long, lobes triangular, rarely ovate or oblong, 0.5–3 mm long; corolla cylindric to bottle-shaped, 1–2.8 cm long, narrowed to throat, red, pale pinkish or white at base, otherwise tube dark pink to red, limb and lobes yellowish-green, lobes deltate, widely flaring, 1–2(–3) mm long; stamens to ~2 cm long, filaments sometimes exserted. **Fruits** berries, 0.8–1.4 cm Ø, **Distr:** Mexico through C America, south from Andes to Bolivia, and east through Cordillera de la Costa of Venezuela. **Ecol:** In tropical wet forests to montane cloud forests; (300–) 1,000–3,200(–4,000) m. **Zones:** 9–12.

Ceratostema

Species 33. **Shrubs**, terrestrial or epiphytic; to 3.5 m tall, sometimes scandent, often arising from lignotubers. **Leaves** alternate, petiolate; lamina coriaceous, plinerved or pinnately veined. **Inflorescences** axillary, solitary, fasciculate, or racemose; floral bract small to large and showy; bracteoles 2; flowers pedicellate. **Flowers** (4–)5(–6)-merous, without odor, aestivation valvate; calyx articulate with pedicel; corolla thick-carnose, often large, subcylindric but often ventricose at base, terete or more commonly bluntly angled, usually deeply lobed, lobes usually narrowly elongate; stamens (8–)10, equal or alternately slightly unequal, nearly as long as corolla, slightly exserted or appearing exserted when corolla lobes become reflexed, filaments distinct or connate; stigma truncate. **Fruits** berries, spherical, usually dark blue-black, rarely translucent, greenish-white; seeds many. **Distr:** S Colombia to N Peru (with a disjunct species in Guyana). **Cult:** Propagate from cuttings or seeds. Requires a moist, well-drained soil in partial sun.

C. silvicola A.C. Smith, *Contr. U.S. Natl. Herb.* 29(8): 363, 1950. **Shrubs**, epiphytic; bark peeling; twigs with sterile bracts (almost miniature leaves) to 6 mm long scattered along length. **Leaves** subsessile; lamina ovate-oblong, 9–16 × 5–8.5 cm, base cordate with a sinus 1–1.5 cm deep and rounded auricles which are separate or imbricate, margins entire and slightly recurved, apex acuminate, glabrous, coriaceous, midrib thickened and slightly raised in proximal 1–5 cm long. **Inflorescences** racemose, usually 3–8-flowered, axillary at end of branchlets; floral bract membranous, ovate, ~2 mm long, puberulent, persistent after anthesis; pedicel striate, 2.2–3.5(–4) cm long. **Flowers** with calyx articulate, 5 mm long, densely pilose with white hairs; corolla thickly carnose when fresh, subcylindric-urceolate, terete to bluntly and broadly 5-angled, 2.7–3.8 × 1–1.7 cm, crimson, shortly pilose, deeply 5-lobed to near middle; corolla lobes reflexed, ~1.5 cm long, dark blue-black; stamens 10, 2.4–3.3 cm long; style 2.8–3.6 cm long, glabrous. **Fruits** berries. **Distr:** Ecuador. **Ecol:** In tropical wet forests to premontane moist forests; 450–650 m. **Zones:** 10–12.

Dimorphanthera

Species ~85. **Lianas**, terrestrial or epiphytic, often robust, to >20 m long; branches are usually pendent and may reach over 3 m in length. **Leaves** elliptic to ovate-lanceolate, base subcordate, margins entire, apex acute, adaxially dark green, abaxially paler, leathery, glabrous; petiole short. **Inflorescences** axillary, at end of branchlets, and leafy or leafless stems; flowers shortly stalked. **Flowers** usually large, tubular, tomentose with white hairs; calyx adnate to ovary, hemispherical, green or reddish; corolla fleshy, lobes small, rounded, slightly flared at anthesis (incurved in some species). **Fruits** berries, fleshy, subglobose. **Distr:** New Guinea. **Cult:** Propagate from cuttings or seeds. Requires high humidity and a moist, free-draining soil in shade to full sun. **Info:** The flowers are collected by the native people for making leis. Some species of birds of paradise collect them to decorate their nesting sites. The branches of mature specimens reach >7 m into high light areas to attract nectar feeders and potential pollinators.

D. alpina J.J. Sm., *Nova Guinea* 7: 92, 1961. [Alpine Heath]. **Lianas**, robust; branches pendent. **Leaves** spirally arranged; lamina lanceolate to oblong, to 15 × 5 cm, base subcordate, margins entire, apex acute, thickly coriaceous, glossy green, secondary veins arising from base; petiole short. **Inflorescences** axillary, few-flowered; pedicels thick. **Flowers** with calyx cup-shaped, green; corolla fleshy, tubular, to 10 cm long, constricted at neck, pale pink (fading to almost white before falling), sparsely pubescent, lobes small, incurved. **Fruits** berries, subglobose, to 2.5 cm Ø, **Distr:** Papua New Guinea. **Ecol:** In high humidity areas such as near lakes; 2,800 m. **Zones:** 9–12. **Info:** 2 varieties have been described for this species

Above: *Gaultheria leschenaultii, Vaccinium acrobracteatum, Vaccinium auriculifolium, Vaccinium leschenaultii.*

(var. *alpina* and var. *pubigera*). The branchlets of var. *alpina* are glabrescent while that of var. *pubigera* are pubescent. Flowers are usually produced on branches exposed to the sun.

D. kempteriana Schltr., *Bot. Jahrb.* 55: 187, 1918. (Syn: *D. intermedia*, etc.). [Kempter's Heath]. **Lianas** or **small shrubs**, robust; branches ± densely pubescent. **Leaves** spirally arranged; lamina elliptic or oblong to ovate-lanceolate, 7–15(–24) x 2.5–6(–8) cm, base subcordate, margins entire, apex acute, abaxially glabrous or pubescent, secondary veins 5, arising from base; new flush pinkish-brown; petiole 0.6–1(–1.5) cm long. **Inflorescences** racemes or clusters, (5–)8–15-flowered, axillary, toward end of branches; pedicels reddish, 1.5–2(–3) cm long; bracteoles below middle. **Flowers** tubular; calyx cup-shaped, limb spreading, shallowly 5-dentate; corolla deep red to rose-pink, fleshy, campanulate, 1.5–2 cm long, 5-lobed, limb ~1.5 cm Ø, grayish-puberulous; stamens (8–)9–10, (6–)7–8 mm long; style slender, ~2.5 cm long, ± exserted. **Fruits** berries, subglobose, 7–8 mm Ø, **Distr:** New Guinea. **Ecol:** In dense primary and secondary forests; 1,000–2,300(–3,000) m. Fl. throughout the year. **Zones:** 10–12. **Info:** 2 varieties (var. *kempteriana* and var. *breviflos*) have been described for this species. Florally they are both identical.

Gaultheria

Species 35. **Shrubs** or **small trees**, deciduous or evergreen; stems erect, creeping or procumbent, often from underground woody burl or producing thickened horizontal underground rhizomes. **Leaves** spirally arranged, petiolate, persistent; lamina serrate, rarely entire. **Inflorescences** racemes or panicles, or flowers solitary, axillary or terminal; bracteoles variable in position. **Flowers** (4–)5(–8)-merous; calyx with variously aestivated lobes, usually valvate in bud; corolla white, rarely red, tubular or urceolate, lobes short; stamens 10, inserted, filaments flattened, geniculate, with or without 1 pair of spurs at anther-filament junction; ovary superior, locule many-ovuled, stigma truncate. **Fruits** capsules (appearing like fleshy berries, completely enclosed in enlarged succulent calyx), loculicidally dehiscent, with pale ± thickened sutures sometimes separating from valves at dehiscence; seeds oblong-ovoid or spindle-shaped to shortly linear, minute, ends often truncated, without wings. **Distr:** E and S Asia; SE Australia, Tasmania, New Zealand; America. **Cult:** Propagate from seeds, cuttings or by division of suckers. Requires an acidic, humus-rich, well-drained soil in partial shade. **Info:** Methyl salicylate (oil of wintergreen) occurs in many species in all their parts (Dassanayake 1999).

G. leschenaultii DC., *Prod.* 7: 593, 1839. (Syn: *G. fragrantissima, Gaultheria rudis*). [Asian Snowberry]. **Shrubs**; twigs stiff, erect, much-branched, spreading; bark orange-brown; young stems purplish-pink, darkly pubescent, compressed; buds scaly. **Leaves** elliptic to ovate, 1.5–8 x 1–4.3 cm, base rounded to acute, margins reflexed, serrated, apex obtuse or acute and bluntly apiculate, stiffly coriaceous, glabrous, adaxially dark green, abaxially light green, glaucous, abaxially with scattered dark brown sunken glands, adaxially veins impressed and lighter than lamina; new flush reddish; petiole to ~7 mm long, red. **Inflorescences** racemes, axillary toward end of branches, ~5 cm long, to 15-flowered; pedicels ~5 mm long; bract basal, ~3 mm long; flowers pendent. **Flowers** with calyx greenish-white, segments deltoid, 3–4 mm long, acute, ciliate; corolla urceolate, 5–7 x 4–5 mm, white, with scattered hairs inside, lobes 5, reflexed; stamens 10, filaments ~1.5 mm long, white; stigma shortly 5-lobed, hollowed at apex. **Fruits** capsules, ~5 mm Ø, white-hairy, completely enclosed in fleshy enlarged calyx, the whole pseudo-berry 0.8–1 cm Ø, deep purplish-blue, stalk red; seeds angular, ovoid. **Distr:** S India (Nilgiri mountains), Sri Lanka (highlands such as Horton Plains). **Ecol:** In upper montane regions. Fl. Apr–Jul. **Zones:** 10–12. **Info:** Flowers slightly fragrant.

Vaccinium

Species ~450. **Shrubs** creeping, or **small trees**, terrestrial or epiphytic, evergreen or deciduous. **Leaves** spirally arranged, usually with a marginal gland on each side close to base, petiolate; lamina with margins entire or serrate. **Inflorescences** racemes or fascicles, or solitary flowers, terminal or axillary; pedicels usually with 2 bracteoles. **Flowers** with calyx limb lobed or toothed; corolla urceolate, campanulate or tubular, lobes 4 or 5; stamens mostly 8 or 10, epigynous, inserted at outer margin of disk or base of corolla tube; ovary inferior, 4- or 5-loculed, locules few- to many-ovuled. **Fruits** berries, crowned by the disk and persistent calyx; seeds, many, small, ellipsoid, irregularly compressed. **Distr:** C and S America; Europe; Africa, Madagascar; Asia. **Cult:** Propagate by division, from cuttings or seeds. Requires an acidic, humus-rich, moist, well-drained soil in full to partial sun.

V. acrobracteatum K. Sch., Nachtr. 339, 1905. (Syn: *V. torricellense*, etc.). **Shrubs** or **small trees**, terrestrial or mostly epiphytic, erect or pendulous, much-branched, (0.5–)1.5–2(–5) m high. **Leaves** mostly obovate, (2.4–)4–9 x (1–)2–5.5 cm, base ± cuneate, slightly decurrent on petiole, margins entire, apex broadly attenuate, coriaceous, glabrous, shiny, new leaves red; petiole red, flattened, to 5(–7) mm long. **Inflorescences** racemes, axillary, suberect, laxly subsecundly (3–)5–8(–10)-flowered; rachis (1–)4–5(–7) cm long; perulae few, ovate to suborbicular, 3–4 x 3 mm; pedicels to 1.5(–1.8) cm long; bract ovate-oblong, to 1(–2) x 0.5 cm, green, red-suffused. **Flowers** with calyx tube broadly cup-shaped, 1.5–2 mm long, base subtruncate, limb ± expanded, 1.5 mm long, lobed halfway; corolla broadly urceolate, fleshy, reddish, pink or nearly white, lobes darker, 0.6–0.8(–1) x 0.3–0.4(–0.5) mm, lobes suborbicular, 1–1.5 mm long, finally reflexed; filaments linear, 2.5–3 mm long; style columnar, 5–8 mm long. **Fruits** depressed-globose, 6–7 mm Ø, reddish to dull purplish. **Distr:** New Guinea. **Ecol:** In lower and montane forests, 200–2,400 m. Fl. & Fr. Jan–Dec. **Zones:** 10–12.

V. auriculifolium Sleumer, *Blumea* 12, 131, 1963. **Shrubs**, 1.2–3 m high, sometimes scrambling; stems shortly branched distally. **Leaves** imbricate, ovate or elliptic-ovate, 1–1.5 cm Ø, base deeply auriculate-cordate, apex attenuate or subacuminate, acute or obtuse, coriaceous, (2.5–)3–6 x (2–)2.5–4.5(–5) cm, 7–9-plinerved, new leaves pink; petiole 1.5–3 mm long. **Inflorescences** racemes, from upper 1–2 axils, condensed, almost umbelliform, (3–)4–8(–12)-flowered; rachis to 1 cm long; pedicels 3–5 mm long; bract foliaceous, persistent, broadly ovate to ovate-elliptic, 0.5–1 x 0.5–1 cm. **Flowers** with calyx tube cup-shaped, 1–1.5 mm long, lobes erect, deltoid, 1–1.5 mm long, apical gland small, glabrous; corolla subglobose-urceolate, depressed, ~4 x 5–6 mm, opening ~2 mm Ø, fleshy, pale to intense pink, hairy outside, lobes reflexed, 0.7 mm long; filaments subulate, 1.5 mm long, villous; disk prominent, hairy; style 3 x 1 mm, glabrous. **Fruits** subglobose, 7–8 mm Ø, apex truncate, blackish-purple. **Distr:** New Guinea (W and S Highlands). **Ecol:** In forests; 1980–2860 m. Fl. & Fr. Jul–Sep. **Zones:** 10–12.

V. leschenaultii Wight, *Ic. P1. Ind. Or.* t.1188, 1848. (Syn: *V. leschenaultii* var. *zeylanica*, *V. symplocifolium*). **Shrubs** large, or **small trees**, ~5 m tall, much-branched. **Leaves** elliptic, ovate, lanceolate or rotund, 1.5–9 x 0.8–4 cm, base acute, obtuse or rounded, margins serrate, apex acuminate or rounded, stiffly coriaceous, glabrous; petiole 1–5 mm long. **Inflorescences** racemes, axillary or terminal; peduncle 0.2–1 cm long, slightly pubescent; pedicels 0.6–1 cm long; flowers nodding. **Flowers** regular; calyx cup-shaped, lobes triangular, ~1.5 mm long, acute, ciliate; corolla tubular-urceolate, 0.8–1.4 x 0.4–0.8 cm, 5-angled, white to bright dark pink, lobes 1–3 mm long, recurved; filaments 2.5–3 mm long, pubescent, white or pink below, yellow-brown distally, anthers yellow-brown; ovary with a yellow disk at apex, 5-carpellate and 5-loculed, locules several-ovuled, style 0.8–1 cm long, greenish distally, persistent, stigma subcapitate. **Fruits** berries, subglobose, to ~1.2 cm Ø, smooth, glabrous, red; seeds few, ovoid, compressed, brown, reticulated. **Distr:** Sri Lanka, S India (Nilgiri mountains). **Ecol:** In montane forests. Fl. Feb–Apr, Sept. **Zones:** 10–12. **Info:** The berries are edible but acidic in taste.

14 ASTERIDS I

ORDERS 4, FAMILIES 40, GENERA 2,521, SPECIES 46,917.

Flowers with late sympetaly (petals connate later in their development); gynoecium bicarpellate, carpels superposed. **Distr:** Worldwide. **Info:** Boraginaceae is yet to be placed into a suitable order and is treated here as an individual family.

BORAGINACEAE

Subfamilies 6, **genera** 148, **species** 2,740. **Herbs**, perennial, biennial, or annual, less often lianas, shrubs, or trees, usually bristly or scabrous-pubescent. **Leaves** exstipulate, alternate, rarely opposite; lamina simple, margins entire or serrate. **Inflorescences** often double-scorpioid cymes, rarely flowers solitary; bracts present or absent. **Flowers** bisexual, actinomorphic, rarely zygomorphic; calyx usually 5-parted or -lobed, usually persistent in fruit, lobes free; corolla tubular, campanulate, rotate, funnelform, or salverform, tube appendages 5, rarely more, mostly trapeziform, rarely absent, sometimes a ring of hairs present, limb usually 5-parted, lobes overlapping, rarely twisted in bud; stamens 5, inserted on corolla tube or rarely at throat, included or rarely exserted, anthers introrse, biloculed, usually basally dorsifixed, sometimes medifixed, longitudinally dehiscent; nectaries at corolla base or on disk below ovary; ovary superior, bicarpellate, biloculed, locules 2-ovuled, or 4-loculed and locules 1-ovuled, style usually terminal or gynobasic, branched or not; gynobase flat, fastigiate, or subulate. **Fruits** drupes or nutlets (mericarps), 1–4-seeded; nutlets mostly dry, often ornamented with wings, prickles and/or glochids; seeds vertical or oblique, testa membranous. **Distr:** Temperate and tropical regions worldwide.

Argusia

Species ~150. **Shrubs**, **climbing shrubs**, **trees**, or **herbs**. **Leaves** alternate, petiolate; lamina with margins entire. **Inflorescences** cymes, terminal or axillary, corymbose, ebracteate. **Flowers** with calyx 4- or 5-parted, not changing in fruit, lobes narrow; corolla white or pale green, usually funnelform, tube pubescent outside, usually exceeding calyx, throat unappendaged, lobes 4 or 5, overlapping or valvate in bud, spreading at anthesis; filaments short, inserted on corolla tube, anthers ovate to oblong, apex mucronate or obtuse; disk slightly convex or sometimes subcupular; ovary 4-loculed, locules 1-ovuled, ovules pendulous, style terminal, stigma entire or 2-cleft, base fleshy, ringlike, inflated. **Fruits** drupes, with watery, sticky, or corky mesocarp, endocarp divided at maturity into two 2-seeded or four 1-seeded mericarps, sometimes 1 seed sterile; seeds oblique. **Distr:** Tropical and subtropical regions worldwide. **Cult:** Propagate from seeds or cuttings. Requires a moist, well-drained, sandy soil in full sun.

A. argentea (L. f.) Heine, *Fl. Nouv.-Calédonie and Dépend.* 7: 109, 1976. (Syn: *Tournefortia argentea*, etc.). [Beach Heliotrope, Velvet-leafed Soldier Bush]. **Shrubs** or **small trees**, 1–5 m tall; branchlets stout, densely rusty- or white-pubescent. **Leaves** terminating branches; lamina oblanceolate to obovate, 7–13 × 2–4 cm, base attenuate, apex obtuse to rounded, densely whitish-hairy. **Inflorescences** corymbose, 5–10 cm Ø, terminal, densely rusty-pubescent; cymes scorpioid. **Flowers** with calyx fleshy, sessile, 1.5–2 mm long, 5-parted, lobes oblong or obovate to rounded, about half as long as corolla, densely rusty-pubescent outside, pubescent only at base or subglabrous inside; corolla white, tubular, 2.5–3 mm long, lobes spreading, ovate, ~2 mm long, longer than tube; stamens ± exserted, filaments inserted ~4 mm above base of corolla tube, anthers ovate-oblong, ~1 mm long; ovary subglobose, glabrous, style obscure, stigma 2-cleft, base ringlike, inflated. **Fruits** drupes, subglobose, ~5 mm Ø, glabrous. **Distr:** China, Indonesia, Japan, Maldives, Pacific Islands (New Caledonia, Polynesia), Philippines, Sri Lanka, Taiwan, Vietnam. **Ecol:** On sandy beaches, salt-tolerant. Fl. & Fr. Apr–Jun. **Zones:** 10–12.

Carmona

Species 1. *Generic description same as specific.* **Cult:** Propagate from seeds or cuttings. Requires a moist, well-drained soil in full sun.

C. retusa (Vahl) Masam., *Trans. Nat. Hist. Soc. Taiwan* 30: 61, 1940. (Syn: *C. microphylla*, *Cordia retusa*, *Ehretia microphylla*). [Fukien/Philippine Tea]. **Shrubs** or **small trees**, 1–3 m tall, much-branched; internodes 1–2 cm long. **Leaves** alternate on annual twigs, fascicled on dwarf shoots; lamina obovate to spathulate, 1.5–3.5 × 1–2 cm, base attenuate, margins coarsely crenate, apex rounded to acute, leathery, abaxially subglabrous, adaxially hirtellous or spotted; petiole short. **Inflorescences** widely spaced glomerules, usually 2–6-flowered, mostly solitary, axillary, spreading, 0.5–1.5 cm wide; peduncles 1–1.5 cm long, pubescent; pedicels 1–1.5 mm long, or ± obsolete. **Flowers** with calyx 4–6 mm long, 5-parted nearly to base, lobes spreading, usually linear, 0.5–0.8 mm long, attenuate below middle, hirtellous, densely pubescent inside; corolla white or slightly reddish, campanulate, 4–6 mm long, tube short, lobes spreading, oblong, longer than tube; filaments 3–4 mm long, inserted near base of corolla, anthers oblong, 1.5–1.8 mm long, exserted; style deeply 2-cleft, subequal to corolla, persistent, beaked in fruit, 4–6 mm long, glabrous, stigmas 2, capitate. **Fruits** drupes, red or yellow, subglobose, 3–4 mm Ø, endocarp bony, 4-seeded, not separating at maturity, globose, reticulate-wrinkled, 2–3 mm Ø, apex shortly beaked. **Distr:** Australia, China, Indonesia, Japan (Ryukyu Islands). **Ecol:** In thickets. **Zones:** 9–12.

Cordia

Species ~325. **Trees** or **shrubs**. **Leaves** alternate, rarely opposite, usually petiolate; lamina with margins entire or serrate, rarely lobed. **Inflorescences** cymes,

Below: *Argusia argentea*, *Carmona retusa*, *Cordia africana*, *Cordia africana* (flower).

mostly corymbose, ebracteate. **Flowers** bisexual, frequently heterostylous or ± functionally unisexual; calyx tubular or campanulate, enlarged after anthesis, persistent; corolla white, yellow or orange-red, campanulate to funnelform, usually (4- or)5(–8)-lobed, lobes antrorse or recurved; stamens usually well-developed, filaments often pubescent at base; ovary 4-loculed, glabrous, locules 1-ovuled, style twice 2-cleft, each branch with a spathulate or capitate stigma. **Fruits** drupes, ovoid, globose, or ellipsoid, frequently with watery or sticky fleshy mesocarp and bony endocarp, rarely with corky mesocarp or nutlike without fleshy mesocarp; seeds 1–4, without endosperm. **Distr:** Predominantly in tropical America; fewer species in Africa and Asia. **Cult:** Propagate from cuttings or seeds. Requires a moist, well-drained soil in full sun. Most species are salt-tolerant.

C. africana Lam., *Tab. Encycl. Meth. Bot.* 1: 420, 1792. (Syn: *C. abyssinica*, *C. holstii*). [East African Cordia, Large-leafed Cordia, Sudan Teak]. **Trees**, deciduous, small to medium sized, to 24 m tall; trunk usually forking a few meters from base. **Leaves** alternate or sometimes almost ternate; lamina ovate or broadly ovate, sometimes subcircular or elliptic, 6–21 × 4–16.5 cm, base subacute to cordate, often asymmetrical, margins entire or shallowly crenate, apex usually rounded and abruptly shortly acuminate, sometimes acute, usually stiffly coriaceous, secondary nerves 5–7 per side; petiole (1.5–)2.5–10.5 cm long. **Inflorescences** cymes, arranged in terminal leafy usually ample panicles; rachis and branches minutely tomentose; pedicels to 1 mm long. **Flowers** hermaphrodite; calyx tubular, 7–9 × 4–7 mm, 10–12-ridged, coriaceous, minutely tomentose outside, glabrous inside, splitting into 3–5 teeth, sometimes bilabiate; corolla funnel-shaped, 1.7–2(–2.4) cm long, white, tube 1.5–1.9(–2.2) cm long, lobes 1.5–3 × 1–1.3(–1.5) cm, retuse, tip ~1 mm long, pubescent, margin undulate; stamens enclosed, filaments 0.7–1 cm long; ovary ovoid, ~2 × 1.5 mm, glabrous, style 1.3–1.8 cm long. **Fruits** ovoid, ellipsoid or obovoid, ~1.2 × 0.8 cm, apiculate, glabrous, dark brown, surrounded by enlarged cup-shaped calyx; pyrene subquadrangular in cross-section; seeds 1–2(3–4). **Distr:** Tropical Africa; Arabian Peninsula (Yemen). **Ecol:** In dry or evergreen forests, rocky mountain slopes; 750–1,700 m. **Zones:** 10–12.

C. dichotoma G. Forster, *Fl. Ins. Austr.* 18. 1786. (Syn: *C. myxa* [misapplied], *Varronia sinensis*). [Bird Lime Tree, Cordia Plum, Sebesten Plum]. **Trees**, 3–4(–20) m tall. **Leaves** ovate to oblong or elliptic, 6–13 × 4–9 cm, base rounded to broadly cuneate, margins usually subundulate to undulate-dentate, rarely entire, apex obtuse to mucronate, sparsely pubescent or glabrous; petiole 2–5 cm long. **Inflorescences** terminating leafy lateral branches, dichotomously branched into corymbose cymes, widely spaced, 5–8 cm apart; flowers sessile. **Flowers** small, dimorphic; calyx campanulate, 5–6 mm long, 5-lobed, lobes unequal, triangular; corolla white, about as long as calyx, lobes shorter than tube, margin somewhat undulate; filaments of ♀ flowers ~3.5 mm long, filaments of bisexual flowers 1–2 mm long; rudimentary pistil globose, style basally united, stigma spathulate. **Fruits** drupes, yellow or reddish, glaucous, subglobose, 1–2.5 cm Ø, with sticky mesocarp, surrounded by persistent calyx. **Distr:** Tropical Asia through Malesia to Australia and Pacific Islands. **Ecol:** In open woods, mountain streamsides. Fl. Feb–Apr. Fr. Jun–Aug. **Zones:** 11–12. **Info:** The fruits are edible, and young fruits are often pickled. Consumption of ripe fruits in large numbers causes drowsiness.

C. leucosebestena Griseb., *Cat. Pl. Cub.* 208, 1866. (Syn: *Tabebuia setulosa*, *Tecoma setulosa*). [Anacagüita, Pale Cordia]. **Trees**, small, 3–6 m tall; branchlets white-tomentose, strigulose; bark thick, gray, ridged. **Leaves** deciduous; lamina ovate, 5–12.5 cm long, base rounded-obtuse, margins repand, apex retuse, adaxially pale-strigulose, abaxially white-tomentose. **Inflorescences** cymes, corymbose; flowers long-pedicellate. **Flowers** white, throat yellow, large, trumpet-shaped; calyx tubular-campanulate, ~2 cm long, tomentose, 2–5-dentate; corolla tubular, lobes 5, ovate, obtuse, tube glabrescent. **Fruits** drupes, ovoid-globose, fleshy, white, calyces persistent. **Distr:** Greater Antilles (Cuba). **Ecol:** In coastal scrubs; 0–500 m. Fl. Feb–Apr. **Zones:** 11–12. **Info:** Fruits edible in moderate quantities, and often used for medicinal purposes.

C. macrophylla Vahl ex Griseb., *Fl. Brit. W. Ind.* 479, 1861. [Broad-leaved Cherry]. **Trees**, large, 6–20 m tall; trunk to 40 cm dbh; branchlets rusty-villous. **Leaves** alternate, deflexed; lamina large, ovate or ovate-oblong, 15–33 × 7.5–18 cm, base subcordate, margins repand-entire, apex pointed, adaxially scabrous-pubescent, abaxially villous-pubescent and reticulate-veined; petiole short, terete, channeled. **Inflorescences** panicles, expanded; flowers shortly gyrate, approximate, sessile. **Flowers** with calyx obovate-globose, rusty-villous, smooth, 3–5-dentate; corolla white, tube cylindrical, villous, lobes 5, ovate-oblong, blunt, 1/3 as long as the tube; stamens exserted; ovary superior, ovoid, style filiform. **Fruits** drupes, ovate-conical, pointed. **Distr:** Jamaica. **Ecol:** In woodlands. **Zones:** 11–12.

C. sebestena L., *Sp. Pl.* 190, 1753. [Anaconda, Broadleaf Cordia, Geiger Tree]. **Shrubs** or **small trees**, evergreen, to 3–8 m tall. **Leaves** ovate, 8–20 × 4.5–15 cm, margins entire (toothed near the tip), undulate, stiffly leathery, coriaceous, adaxially dark green and rough like sandpaper, abaxially paler, glabrous, minutely pubescent on nerves; petiole 4.5 cm long. **Inflorescences** cymes, several-flowered, terminal. **Flowers** showy, large, 2–5 cm Ø, bright orange-red; calyx tubular to narrowly bell-shaped, 1.2–1.5 cm long, not ribbed, ± hairy; corolla funnelform, 2.5–5.8 cm long, limb 5–7-lobed, crinkly, lobes crenulate, papery (crepelike); stamens usually as many as lobes. **Fruits** drupes, ovoid, turbinate, 2.4–4 cm long, enclosed in fleshy white calyx, waxy; seeds 2–4. **Distr:** W Indies, SW USA, Venezuela. **Ecol:** In dry habitats, often near sea coasts on sand or limestone soils. Fl. throughout most of the year. **Zones:** 10–12. **Info:** The fruits are edible.

C. subcordata Lamarck, *Tabl. Encycl.* 1: 421, 1792. [Sea Trumpet]. **Trees**, to 3–9(–15) m tall; bark yellow-brown; branchlets glabrous. **Leaves** ovate to narrowly ovate, 8–18 × 6–13 cm, base obtuse to rounded, rarely cordate, margins entire to subundulate, apex acuminate to acute, abaxially densely cottony in vein axils, adaxially ± spotted; petiole 3–6 cm long, glabrous. **Inflorescences** cymes, opposite leaves, ~12 cm across at anthesis; pedicels 3–6 mm long. **Flowers** with calyx green, cylindric, ~1.3 × 0.8 cm, leathery, lobes irregular, short; corolla orange, funnelform, 3.4–4.5 cm long, limb ~4 cm Ø, lobes divaricate, orbicular. **Fruits** drupes, ovoid or obovoid, ~2.5 cm long, mesocarp corky, enclosed by enlarged persistent calyx. **Distr:** Africa (E coast), China, India, Indonesia, Maldives, Pacific Islands, Thailand, Vietnam. **Ecol:** In sandy, open woodlands, often near sea coasts. Fl. Jun. **Zones:** 10–12. **Info:** Grown mainly for ornamental purposes or for the attractively grained, light golden wood with darker markings.

Echium

Species >40. **Herbs**, annual, biennial, or perennial, strigose. **Leaves** lanceolate. **Inflorescences** cymes, terminating stems and branches, or in panicles; flowers often pedicellate. **Flowers** with calyx 5-parted, lobes lanceolate-linear, sometimes broadly lanceolate, slightly enlarged, connivent in fruit, 2 adaxial lobes usually smaller; corolla blue, purple, or pink, campanulate or tubular, distinctly zygomorphic, usually pubescent outside, tube usually short, with a pubescent or glabrous protective ring inside base, throat unappendaged, limb oblique, lobes unequal; stamens unequal, filaments thin, long, attached at different levels in corolla tube, appearing spirally arranged, often exserted; ovary 4-divided, style filiform, exserted, appressed-pubescent, apex 2-cleft, each branch with a small capitate stigma; gynobase flat. **Fruits** nutlets, usually light brown, straight, ovoid or narrowly ovoid, tuberculate or smooth. **Distr:** Africa, W Asia, Europe. **Cult:** Propagate from seeds or cuttings. Requires a moderately fertile, moist, well-drained soil in full or partial sun.

E. candicans L. f., *Suppl. Pl.* 131, 1782. (Syn: *E. fastuosum*). [Pride of Madeira]. **Shrubs**, biennial, monocarpic, woody, strigose, 1.8–2.5 m high; branches stout, leafy near apex; bark whitish, papery. **Leaves** lanceolate to oval-lanceolate, to 25 × 4 cm, densely silvery-hairy, abaxially veins prominent. **Inflorescences** spikelike panicles, dense, clusters 8-flowered, terminal, erect, cylindrical, to 25–30 cm tall; pedicels very short. **Flowers** with calyx 4–5 mm long, to 7 mm long in fruit, lobes lanceolate, pointed;

Below: *Cordia dichotoma* (flowers), *Cordia dichotoma* (fruits), *Cordia leucosebestena*, *Cordia macrophylla*.

Above: *Cordia sebestena*, *Cordia subcordata*, *Echium candicans*, *Heliotropium indicum*.

corolla funnel-shaped, 0.9–1.1 cm long, white, or blue streaked white, red-purple in bud, limb to 2 mm across, hairy along midvein, with a ring of 10 conical lobes at base of corolla; stamens exserted from corolla, filaments pink, glabrous; style exserted, apically 2-divided. **Fruits** nutlets, 2 mm long, with small elongate papillae. **Distr:** Madeira and Canary Islands. **Ecol:** In lower laurel forests. Fl. May–Aug. **Zones:** 9–11.

Heliotropium

Species ~250. **Herbs**, annual or perennial, rarely subshrubs, pubescent or strigose, rarely scabrous. **Leaves** alternate, less often opposite, sessile or petiolate. **Inflorescences** cymes, terminal, rarely axillary, unilateral, scorpioid, bracteate or not. **Flowers** small; calyx 5-parted; corolla white or light bluish-purple, less often yellow, cylindric or funnelform, strigose outside, glabrous and rarely appressed pubescent inside, throat frequently pubescent, limb 5-parted, lobes orbicular, sometimes linear, margin with folds or undulate; filaments extremely short, anthers included; ovary completely or incompletely divided into 4 lobes, ovules 4, style terminal, stigma conical or ringlike. **Fruits** drupes, dry, nutletlike, without evident mesocarp at maturity, endocarp bony, dividing into 4 1-seeded or 2 2-seeded mericarps; seeds straight or curved, usually with a thin endosperm. **Distr:** Tropical and temperate regions worldwide. **Cult:** Propagate from cuttings or seeds. Requires a fertile, moist, well-drained soil in full sun.

H. indicum L., *Sp. Pl.* 1: 130, 1753. [Indian Heliotrope]. **Herbs**, annual, 0.2–1 m high; stems erect, stout, much-branched, strigose. **Leaves** alternate or subopposite; lamina 3–9 × 2–4 cm, pubescent or strigose, base rounded or truncate, decurrent to petiole, margins undulate, apex acute, lateral veins 5–7 pairs; petiole 2–5 cm long. **Inflorescences** cymes, solitary, scorpioid, 5–15 cm long, ebracteate; flowers sessile, crowded. **Flowers** small; calyx lobes lanceolate, 1.5–2 mm long, strigose; corolla light blue or blue-purple, salverform, 3–4 mm long, base ~1 mm wide, throat contracted to ~0.5 mm wide, limb 2–2.5 mm wide, lobes rotund, ~1 mm wide, margin crispate; anthers narrowly ovate, ~0.5 mm long, inserted ~1 mm above base; ovary glabrous, style ~0.5 mm long, stigma conical, pubescent. **Fruits** 3–3.5 mm long, ribbed, glabrous or nearly so, deeply 2-cleft into mericarps, each divided into two 1-seeded pyrenes, mericarps longitudinally ribbed. **Distr:** Asia, Malesia, Pacific Islands; Africa; N and S America. **Ecol:** On hills, riverbanks; 0–700 m. Fl. & Fr. Apr–Oct. **Zones:** 10–12.

GENTIANALES

FAMILIES 5, GENERA 1,118, SPECIES 16,637.

Glandular hairs absent. **Vestured pits** present. **Stipules** sometimes reduced to stipular line, and thick glandular hairs (colleters) on adaxial surface or on base of petioles; the colleters produce mucilage and help to protect the shoot apex. **Leaves** opposite. **Flowers** with corolla convolute in bud, swollen at apex; carpels many-ovuled (Bremer and Struwe 1992; Judd et al. 2008; Nicholas and Baijnath 1994; Stevens 2001 onwards; Struwe et al. 1994; Wagenitz 1959, 1992). **Distr:** Worldwide, especially tropical regions. **Info:** The monophyly of the Gentianales and its placement within the Core Asterid group is strongly supported by molecular data (Bremer et al. 1994, 2002; Chase et al. 1999; Hilu et al. 2003; Soltis et al. 2000).

RUBIACEAE

Subfamilies 4, **Genera** 611, **species** 13,150. **Trees**, **shrubs**, annual or perennial **herbs**, **vines**, or **lianas**, terrestrial or sometimes epiphytic, sometimes armed with spines; young stems tetragonal. **Stipules** persistent or caducous. **Leaves** opposite, sometimes pseudo-verticillate; lamina simple, margins entire, or rarely serrate, pinnately veined or 1-veined. **Inflorescences** in various compound cymes, thyrsiform, cymose, compound-cymose, or occasionally heads, racemes, spicate, corymbs, or with flowers solitary, terminal or axillary, sometimes hypanthia of individual flowers in a fused inflorescences. **Flowers** usually hermaphroditic, actinomorphic, rarely zygomorphic; calyx gamosepalous and fused in hypanthium portion to inferior ovary, limb cupular, tubular, or occasionally obsolete, truncate to (2–)4- or 5(–9)-lobed or sometimes spathaceous, lobes often small, ± unequal, one sometimes enlarged into a petaloid calycophyll; corolla gamopetalous, actinomorphic, rarely zygomorphic, occasionally bilabiate, funnelform, salverform, tubular, campanulate, white, yellow, orange, red, blue, purple, or pale green or with colors in various combinations, lobes (3 or)4 or 5(or 10); stamens adnate to corolla, usually isomerous and alternate with corolla lobes, occasionally 2, inserted variously in corolla throat, corolla tube, or near corolla base and appearing free, anthers included or exserted, free, introrse, dorsifixed to basifixed, dehiscing by lateral slits, connective sometimes shortly prolonged; ovary usually inferior, (1 or)2(–10)-loculed, crowned by a fleshy, pulvinate, hemispheric, or conic disk, ovules 1 or 2 to numerous per locule, style 2–10-lobed, rarely 2-parted to base, stigmatic portion capitate, filiform, linear, or spathulate. **Fruits** capsular, dehiscing loculicidally, septicidally, circumscissile, or by an apical lid, or fleshy and baccate, drupaceous or schizocarpous; mericarps indehiscent, sometimes dehiscing into 2 halves; pyrenes when present papery to hard, 1–10-loculed; seeds 1–numerous, sometimes variously winged or appendaged. **Distr:** Worldwide; especially diverse in the tropics.

Alberta

Species 6. **Shrubs small** to **trees**. **Stipules** interpetiolar, cup-shaped and acuminate, deciduous. **Leaves** decussate, petiolate; lamina leathery, without domatia. **Inflorescences** thyrsic or pleiothyrsic, terminal on lateral branches; bracts small to minute. **Flowers** 5-merous; calyx lobes free, either very unequal (1 or 2 lobes enlarged) or subequal (all lobes enlarged); enlarged calyx lobes brightly colored; corolla zygomorphic to ± actinomorphic, tubes curved or ± straight, cylindrical and gradually widening above, lobes small, ± acuminate, contorted to left in bud; stamens subsessile, inserted in upper part of corolla tube, anthers linear, sagittate, with an apical appendage, dorsally hairy to subglabrous; ovary 2(3)-locular, each locule with one anatropous ovule, pendulous from an apical placenta, style ± curved, united, long-exserted, glabrous, stigmas bifid, minute, filiform to elliptic. **Fruits** crowned by persistent calyx lobes (enlarged lobes winglike), dry, indehiscent in *A. magna*, in other species dehiscent into 2 mericarps; mericarps with elongated, ventrally concave seed. **Distr:** Madagascar (5 endemic spp.), SE Africa (1 endemic sp.). **Cult:** Propagation is difficult, air-layering thus preferred. Requires a fertile, moist, well-drained soil, in full sun.

A. magna E. Mey., *Linnaea* 12: 258, 1838. [Natal Flame Bush]. **Trees**, to ~13 m tall, occasionally shrubby. **Leaves** narrowly elliptic to obovate, 6–12.5 × 1.3–5 cm, base cuneate, apex rounded to acute, leathery, glabrous; petiole ~2–5 mm long. **Inflorescences** paniculate, subpyramidal, to 20 cm long; axes stout; peduncles and pedicels ~0.4–1 cm long; bracts narrowly triangular, 1–2 mm long. **Flowers** with calyx lobes subequal, 2–4 × 0.5–1 mm, 2 obovate (greatly enlarging postflorally), the other 3 triangular; corolla curved, tube 2–2.5(–3) cm long, ± cylindrical below, 2–3 mm Ø, gradually widening above to 6 mm Ø, bright red, lobes ± trapezoidal to rounded-acuminate, 1–3 × 1.5–2.5 mm; anthers basally bearded; ovary 2–3 mm long, slightly pubescent, style 2.5–3.5 cm long. **Fruits** indehiscent, hard, ovoid to ellipsoid, longitudinally ribbed, 4–6 × 3 mm, crowned by calyx lobes, 2 much-enlarged, subsessile or with petioles to 2 mm long, obovate, 1.5–2 × 4–7 cm, bright red at first and brownish and papery later, reticulately veined. **Distr:** SE Africa (Natal and eastern part of the Transkei). **Ecol:** In forests and valleys; to 1,800 m. Fl. Mar–May. Fr. Apr–Jul. **Zones:** 9–12.

Carphalea

Species 6. **Shrubs**, small, with erect branched stems. **Stipules** with (1–)3–5 linear or filiform colleter-tipped setae from a short base. **Leaves** paired or in whorls of 3, petiolate. **Inflorescences** densely few- to many-flowered, terminal. **Flowers** hermaphrodite, dimorphic, some flowers with anthers included and style exserted and others completely vice versa; calyx tube narrowly obconic or turbinate, ribbed, limb variously deeply 4–5-lobed or eccentrically elliptic, tube ± placed at one of the foci, sometimes shallowly 3–4-lobed; corolla tube narrowly cylindrical, lobes 4–5, oblong or ovate, throat densely hairy; ovary 2–trilocular, each locule with a slender basal placenta bearing ~(3–)4–6 ovules, style filiform, stigma divided into 2(–4) filiform lobes. **Fruits** obconic, sometimes curved, bony, strongly ribbed, ribs running into the strongly nervose accrescent calyx limb or lobes, indehiscent; seeds 1–2, narrowly oblong-obconic. **Distr:** Madagascar. **Cult:** Propagate from cuttings or seeds. Requires a fertile, moist, well-drained soil in full sun.

C. kirondron Baill., *Bull. Mens. Soc. Linn. Paris* 1: 188, 1878. [Flaming Beauty, Flaming Red Beauty]. **Shrubs**, >3 m tall. **Stipules** connate below, free parts 3–5-fimbriate. **Leaves** decussate (occasionally 3-whorled), petiolate; lamina membranous. **Inflorescences** corymblike thyrses, many-flowered, terminal. **Flowers** mostly 4-merous, heterostylous; calyx lobes enlarged, unequal (1 much larger than the others), bright red; corolla tube narrowly cylindrical, reddish outside, lobes valvate in bud, spreading-recurved in open flowers, white or creamish-white; stamens inserted below or around throat, anthers on long filaments, exserted in short-styled, included in long-styled; ovary biloculed, each locule with a few ovules on a peltate placenta, style with bilobed stigma included in short-styled, exserted in long-styled. **Fruits** nutlike, indehiscent, obconic, crowned by the persistent enlarged calyx lobes; typically only 1 fertile seed in each locule. **Distr:** Madagascar. **Ecol:** Mostly at forest edges, in clearings or along rivers in dry, deciduous forests; 0–1,500(–1,600) m. Fl & Fr. throughout the year. **Zones:** 10–12.

Catunaregam

Species 12. **Shrubs** or **small trees**, erect; branches with spines; spines solitary, paired or ternate. **Stipules** interpetiolar. **Leaves** usually borne on short shoots. **Inflorescences** terminal on short shoots. **Flowers** bisexual; corolla tube inside with a band of hairs at base, lobes 5–10; anthers and style exserted, stigmatic lobes adherent; ovary bilocular, ovules numerous. **Fruits** globose to obpyriform, fruit wall woody or fleshy and chartaceous; seeds immersed in waxy matrix. **Distr:** Tropical and S Africa, Madagascar; tropical and subtropical Asia; Malesia. **Cult:** Propagate from seeds or cuttings. Requires an acidic, humus-rich, moist, well-drained soil in full sun.

C. spinosa (Thunb.) Tirveng., *Bull. Mus. Natl. Hist. Nat.*, Sér. 3, Bot. 35: 13, 1978. (Syn: *Canthium coronatum*, etc.). [Emetic Nut, False Guava, Mountain Pomegranate, Spiny Randia, Thorny Bone-apple]. **Shrubs**, 0.5–5 m tall; branches armed with spines to 2 cm long. **Stipules** triangular, 2–3 mm long. **Leaves** very variable generally either obovate-spathulate to orbicular, 0.5–2 × 0.5–1.5 cm, or on older branches obovate to obovate-lanceolate, 3–6 × 1–2 cm, base obtuse or cuneate to attenuate, apex obtuse to abruptly acute, veins slightly pubescent on both surfaces, lateral nerves 5–8 pairs; petiole to 0.5 cm long. **Flowers** 5-merous; calyx cupular, lobes obovate, to 8 mm long; corolla infundibular, white turning yellow, tube 4–6 mm long, outside densely sericeous above, glabrous below, lobes orbicular 0.6–0.8(–1) cm long, inside slightly pubescent, outside densely sericeous; anthers 4–5 mm long; style 6–9 mm long, stigma obovoid. **Fruits** 1–2 cm Ø, crowned with persistent calyx lobes. **Distr:** Tropical E Africa, Mozambique and S Africa (Natal); India, Sri Lanka, continental SE Asia and Malesia. **Ecol:** In dry scrubs. **Zones:** 10–12.

Cinchona

Species 23. **Shrubs** or **trees**, evergreen; bark with a bitter taste. **Stipules** interpetiolar, caducous. **Leaves** opposite, petiolate. **Inflorescences** cymes, paniculate, axillary or terminal, with bracts and bracteoles. **Flowers** bisexual, usually fragrant; calyx tube turbinate, abaxially softly hairy, 5-dentate, persistent; corolla salverform or trumpet-shaped, abaxially softly hairy, tube cylindrical, often slightly 5-ridged, glabrous or hairy in throat, lobes 5, shorter than corolla tube, valvate in bud, margin ciliate; stamens 5, inserted at corolla tube, filaments short or long, glabrous, anthers included or slightly exserted, dorsifixed, longitudinally dehiscent, linear; disk pulvinate; ovary inferior, biloculed, locules many-ovuled, placenta linear, adnate to septum, style glabrous, heterostylous, stigma bilobed, included or slightly exserted. **Fruits** capsules, biloculed, usually ovate, conical or subcylindrical, septicidally dehiscent into 2 valves; seeds many, slightly flattened, surrounded by a membranous wing. **Distr:** Costa Rica to W South America. **Cult:** Propagate from seeds or by air-layering. Requires a fertile, moist, well-drained soil in full sun.

C. succirubra Pav. ex Klotzsch, *Abh. Akad. Berlin* 60, 1857. [Quinine]. **Trees**, evergreen, usually 3–12 m tall; bark grayish-brown, sometimes white-laciniate or longitudinally fissured. **Leaves** ovate, ovate-elliptic or oblong, 10–24.5 × 5.5–17 cm, base broadly cuneate or rounded, usually decurrent, apex obtuse, papery or thinly leathery, abaxially pale red when fresh, adaxially glabrous, abaxially pubescent, lateral veins 6–11 pairs; petiole 1.5–4 cm long, glabrous or pubescent. **Inflorescences** axillary and terminal, to 23 × 23 cm, yellowish-pubescent; bracts and bracteoles lanceolate, ~2 mm long; pedicels 1–3 mm long. **Flowers** with calyx ~3 mm long, limb slightly expanded, lobes triangular, 0.5 mm long; corolla white or pink, 1–1.2 cm long, tube cylindrical, 2–3 mm wide, slightly 5-ridged, lobes ovate-lanceolate, ~4 mm long, yellowish-villose at margin adaxially. **Fruits** subcylindrical or conical, 1–1.8 × 0.3–0.6 cm, pubescent; seeds elliptic, 4–6 mm long, surrounded by wing. **Distr:** Ecuador, Peru. **Ecol:** Fl. & Fr. Jun–Feb. **Zones:** 10–12.

Coffea

Species 102. **Shrubs** or **trees**; branches slightly terete, slightly compressed apically. **Stipules** broad, interpetiolar, persistent. **Leaves** opposite, rarely 3-whorled, sessile or petiolate; lamina membranous or thinly leathery. **Inflorescences** clusters or few-flowered cymes, occasionally flowers solitary, axillary; flowers sessile or shortly pedicellate; bracts usually connate. **Flowers** usually fragrant; calyx tube short, subtubulate or turbinate, apex truncate or 4–6-toothed, adaxially often glandulose, persistent; corolla white or pale yellow, rarely rosy, salverform or funnelform, glabrous or villose at throat, apex (4 or)5–9-lobed, lobes patent or spreading, contorted in bud; stamens 4–8, inserted at corolla throat, filaments absent or short, anthers dorsifixed near base, linear, exserted or included; disk swollen; ovary biloculed, locules 1-ovuled, style linear, slightly thick, stigma bilobed. **Fruits** berries, globose or oblong, dry or fleshy, with 2 pyrenes; pyrenes leathery or fleshy, convex abaxially, abaxially longitudinally grooved when leathery, or not grooved when membranous or fleshy; seeds abaxially longitudinally grooved or depressed. **Distr:** Tropical and S Africa, W Indian Ocean. **Cult:** Propagate from seed or cuttings. Requires a moist, humus-rich soil in partial shade.

C. arabica L., *Sp. Pl.* 172, 1753. [Arabian Coffee, Coffee]. **Trees small** or **large shrubs**, 5–8 m tall, usually many-branched at base; old branches enlarged at nodes. **Stipules** broadly triangular, subulate long pointed, 3–6 mm long. **Leaves** ovate-lanceolate or lanceolate, 6–14 × 3.5–5 cm, base cuneate or slightly obtuse, rarely rounded, margins entire or undulate, apex long-acuminate, acumen 1–1.5 cm long, glabrous, thinly leathery, midribs prominent, lateral veins 7–13 pairs; petiole 0.8–1.5 cm long. **Inflorescences** cymes, several fascicled in leaf axils, each 2–5-flowered, sessile or shortly pedunculate; pedicels 0.5–1 mm long; bracts ± basally connate, dimorphic, foliaceous. **Flowers** fragrant; calyx tubular, 2.5–3 mm long, limb truncate or 5-denticulate; corolla white, length varying in varieties, usually 1–1.8 cm long, apex 5-lobed, rarely 4- or 6-lobed, lobes usually longer than corolla tube, apex often obtuse; anthers exserted to 6–8 mm; style 1.2–1.4 cm long, stigma bilobed, 3–4 mm long. **Fruits** berries, broadly elliptic at maturity, red, 1.2–1.6 × 1–1.2 cm, epicarp rigid membranous, mesocarp fleshy, sweet; seeds convex abaxially, flat adaxially, longitudinally grooved, 0.8–1 × 0.5–0.7 cm. **Distr:** Ethiopia and Arabia. **Ecol:** Fl. Mar–Apr. **Zones:** 10–12.

C. canephora Pierre ex Froehn., *Notizbl. Bot. Gart. Berlin* 1: 237, 1897. (Syn: *C. robusta*). [Coffee, Robust Coffee]. **Trees small** or **shrubs**, 4–8 m tall; lateral branches long and pendulous. **Stipules**

Below: *Alberta magna, Carphalea kirondron, Catunaregam spinosa, Cinchona succirubra.*

triangular, 7 mm long, basally connate. **Leaves** elliptic, ovate-oblong or lanceolate, 15–30 × 6–12 cm, base cuneate, sometimes slightly obtuse, margins entire or undulate, apex acute, tip 1–1.8 cm long, thickly papery, glabrous, lateral veins 10–12 pairs; petiole 1–2 cm long. **Inflorescences** cymes, 1–3 fascicled in leaf axils, each 3–6-flowered; peduncles very short; bracts slightly connate, dimorphic, 2 broadly triangular, 2 lanceolate or oblong. **Flowers** with calyx shortly tubular, apex truncate or indistinctly denticulate; corolla white, rarely reddish, 2–2.6 cm long, tube shorter in bud, elongated at anthesis, apex (4)5–7(8)-lobed; filaments short, anthers 0.8–1 cm long, exserted; style 1.2–1.7 cm long, exserted, stigma bilobed, ~6 mm long. **Fruits** berries, subglobose, 1–1.2 cm long; seeds abaxially convex, 0.9–1.1 × 0.7–0.9 cm. **Distr:** Tropical Africa. **Ecol:** Fl. Apr–Jun. **Zones:** 10–12.

Ernodea

Species ~6. **Shrubs**, vinelike, glabrous or somewhat pubescent; branches erect, decumbent or trailing. **Stipules** connate into a sheath. **Leaves** opposite, nearly sessile; lamina linear to lanceolate, leathery. **Inflorescences** with flowers solitary and sessile, axillary. **Flowers** small; calyx tube short, limb 4–6-parted, lobes triangular to linear or subulate, persistent; corolla white to pink, tube nearly cylindric, lobes 4–6, narrow, revolute, valvate; ovary biloculed, locules 1-ovuled, style slender, stigma subcapitate. **Fruits** drupes, fleshy, grooved, containing 2 cartilaginous, 1-seeded pyrenes. **Distr:** S USA to C America and West Indies. **Cult:** Propagate from seeds or by air-layering. Requires a moist, well-drained soil in full to partial sun.

E. littoralis Sw., *Prodr. Veg. Ind. Occ.* 29, 1788. [Beach Creeper, Coughbush, Golden Creeper]. **Shrubs**, glabrous or glandular-puberulent, 0.3–1.6 m high; erect or prostrate; branches 4-angled, usually densely leafy. **Stipules** 1.5–2 mm long. **Leaves** sessile; lamina oblong, elliptic, linear-oblong or oblanceolate, 2–3.5 × 0.4–1 cm, margins entire or glandular-serrulate, apex acute, 3–5-nerved, shiny, somewhat fleshy, glabrous or nearly so. **Inflorescences** axillary, flowers sessile. **Flowers** small, white or pink; calyx lobes linear-lanceolate, longer than tube; corolla white or pink, tube 1–1.5 cm long, lobes 4–6, relatively long; stamens 4–6, filaments adnate to top of corolla tube, anthers narrow. **Fruits** subglobose, yellow, 4–6 mm Ø, about as long as or somewhat shorter than calyx lobes, thinly fleshy. **Distr:** S USA to C America and West Indies. **Ecol:** On seashores, coastal sands and rocks. Fl. throughout the year. **Zones:** 9–12.

Euclinia

Syn: *Gardenia* (Sect. Rothmannia), *Randia* (Sect. Euclinia).

Species ~8. **Shrubs** or **trees**, erect, unarmed. **Stipules** short, intra-petiolar, free or connate. **Leaves** in equal pairs, opposite or with one often arrested. **Inflorescences** fascicled or flowers solitary, axillary or leaf-opposed. **Flowers** white or yellowish; calyx tube various, lobes very long, limb often tubular; lobes short, long or leafy; corolla large, funnelform, salverform or bell-shaped, tube 5–25 cm long, throat glabrous or hairy; lobes 5, rarely more, short or long, twisted in bud; stamens 5; anthers subsessile, narrow; disk annular or cushion-shaped; ovary 2(3–4)-loculed, style short or slender, stigma usually fusiform, entire or bifid, ovules usually numerous, sunk in placentas attached to septum. **Fruits** berries, globose, ellipsoid or ovoid, biloculed; seeds many, often immersed in pulp, angled, testa thin. **Distr:** Tropical Africa. **Cult:** Propagate from cuttings, rooted easily in water. Requires a moist, well-drained soil in full or partial sun.

E. longiflora Salisb., *Parad. Lond.* t. 93, 1808. *Randia macrantha*, etc.). [Angel's Trumpets, Tree Gardenia]. **Shrubs** or **small trees**, to 6 m tall, deciduous; mostly glabrous. **Stipules** persistent, glumaceous, ovate or lanceolate, 0.9–1.6 cm long, obtuse. **Leaves** opposite, crowded near the extremities of branches; lamina membranous, oval or obovate-oblong, 6–20 × 2–9 cm, base narrow, apex cuspidate to caudate, lateral veins 8–9 pairs, inconspicuous; petiole 0.8–2.5 cm long. **Inflorescences** with flowers solitary, erect, terminal or often at end of short branchlets, subsessile. **Flowers** fragrant, 5-merous; calyx tube oblong, lobes subfoliaceous, linear or linear-spathulate, spreading 1.3 cm long, persistent; corolla white, turning to cream with red markings, tube slender, cylindrical, 15–25 cm long, pale green or yellowish, limb whitish, 5–6.5 cm Ø, lobes oval, obtuse, dextrorsely contorted in bud; anthers and style partly exserted. **Fruits** globose or somewhat pear-shaped, not ribbed, 2.5–4 cm Ø. **Distr:** Tropical Africa. **Ecol:** In forests or secondary jungles. **Zones:** 10–12.

Gardenia

Species ~250. **Shrubs** to medium-sized **trees**, sometimes spiny; young parts often resin-coated. **Stipules** conspicuous, entire, connate into a tube, sheathing, buds often conspicuously resinous. **Leaves** opposite or whorled, petiolate; lamina sometimes with domatia in axils of secondary veins abaxially. **Inflorescences** with flowers solitary (appearing axillary) or clustered (or fascicled). **Flowers** fragrant, often large and showy; calyx tube truncate to 5–6-toothed (or -lobed), or spathelike and splitting along one side, sometimes winged; corolla white, often aging to yellow, golden or orange, corolla tube long or short, cylindric or funnelform to salverform, lobes 5–9(–12), spreading, contorted (convoluted) in bud; stamens as many as corolla lobes, anthers linear, dorsifixed, attached in upper part of tube, ± sessile, included or partially exserted; ovary 1-loculed, ovules numerous on 2–6 parietal placentas, style stout, stigma clavate, fusiform, tapered, or 2-cleft. **Fruits** ovoid, globose or ellipsoid, ribbed or winged, usually splitting in 5–8 parts, exposing orange mucilaginous mass; seeds many, immersed in pulp. **Distr:** Old World tropics. **Cult:** Propagate from cuttings, rooted easily in water. Requires a fertile, humus-rich, moist, well-drained soil in full or partial sun.

G. brighamii H. Mann, *Proc. Amer. Acad. Arts* 7: 171, 1867. [Hawai'ian Gardenia]. **Trees**, small, to ~5 m tall; branches quadrangular. **Stipules** connate into a collar 2.5–4 mm long. **Leaves** few, clustered toward branch tips; lamina elliptic to elliptic-ovate, 2.2–10.5 × 1.5–5.5 cm, firm-chartaceous, lateral veins 6–11 pairs, adaxially somewhat viscid, abaxially dull, glabrous; petiole 2–7 mm long. **Inflorescences** with flowers solitary, terminal, flanked by 2 lateral vegetative buds; pedicels 3–5 mm long. **Flowers** fragrant; hypanthium narrowly funnelform, 4–5 mm long; calyx narrowly funnelform, 5–8 mm long, sparsely puberulent, apically produced into 4–5 terminal spurs, these coriaceous, linear to obliquely linear-spathulate, 0.8–1.6 cm × 1.2–2.5 mm, puberulent; corolla white, thick, 6(7)-lobed, tube 1.5–2 cm long, lobes 0.8–1.8 cm long; stamens scarcely exserted, anthers 5–6 mm long. **Fruits** green, whitish-spotted, subglobose or broadly ellipsoid, 2.5–3.5 cm Ø; seeds 4–5 mm long. **Distr:** Hawai'ian Islands (W Maui, Hawai'i). **Ecol:** In low elevation dry forests, rocky gulches, on plateau lands with deep soils, and in old forested lava flows; ~300–600 m. Fl. & Fr. throughout most of the year. **Zones:** 10–12.

G. crameri Tirveng., *Nordic J. Bot.* 3: 458, 1983. (Syn: *G. latifolia*). [Ceylon/Indian Boxwood]. **Shrubs** (2–4 m high) or **trees** (8–10 m tall), dichotomously branching at base. **Stipules** tubular, to 7 mm long. **Leaves** ovate to broadly elliptic or obovate, 10–15(–25) × 5.5–9 cm, base cuneate, sometimes slightly decurrent, apex rounded or obtuse, tip shortly acuminate; petiole 0.5–1.5 cm long. **Inflorescences** with flowers solitary; pedicels to 1 cm long. **Flowers** with hypanthium obconical, to 7 mm long; calyx tube infundibular, 1–2 cm long, longitudinally ridged, ridges extending to calyx segments as short lateral wings; corolla tube 2.5–7 cm long, apex dilated, lobes 7–9, obovate to rhomboid, obtuse, 3–5 × 1.5–2.5 cm, pale lemon-yellow; anthers linear, 1.2 cm long, sessile, inserted toward apex of tube, slightly exserted; ovary 1-locular with 3 parietal placentas, style about equal to corolla tube, stigmatic lobes 3, oblong, 7 × 3 mm, exserted. **Fruits** globular, 2–2.5 cm Ø; seeds numerous, compressed, 3.5 × 5 mm. **Distr:** Sri Lanka. **Ecol:** In lowland forests. **Zones:** 11–12. **Info:** Flowers are larger than those of *G. fosbergii*, but produced in relatively smaller numbers.

G. fosbergii Tirveng., *Nordic J. Bot.* 3: 460, 1983. (Syn: *G. carinata*, *G. coronaria*). [Golden Gardenia, Malaysian Tree Gardenia]. **Shrubs**, 1–3(–5) m high; buds usually with yellow resin. **Stipules** shortly tubular, truncate or with 3(–5) lobes to 3.5 mm long. **Leaves** opposite, on short lateral shoots, elliptic to elliptic-oblong, 10–23 × 3–8 cm, base obtuse to

Below: *Coffea arabica*, *Coffea canephora*, *Coffea canephora* (fruits), *Ernodea littoralis*.

F&KS

KT

F&KS

Top row: *Euclinia longiflora, Gardenia brighamii, Gardenia crameri, Gardenia fosbergii.*
Bottom row: *Gardenia jasminoides* (double form), *Gardenia sootepensis, Gardenia taitensis, Gardenia ternifolia* ssp. *jovis-tonantis.*

slightly cuneate, apex obtuse, adaxially ± glabrous, shiny, abaxially glabrous to densely pubescent; petiole 2–5 mm long. **Inflorescences** with flowers solitary, produced in profusion; peduncles 0.5–1 cm long. **Flowers** fragrant; hypanthium 5 mm long, 6-angled; calyx tube campanulate, to 2.5 cm long, 6-angled; corolla tube to 8 cm long, throat hairy inside, lobes 6–7, obovate to spathulate, 3–4.5 × 2.5 cm, yellow, darkening to deep golden-orange; anthers 1.5 cm long, slightly exserted, filaments short, inserted at about middle of tube; ovary 1-loculed, style ± equal to corolla tube, stigmatic lobes 3, spathulate, ~8 × 3 mm, exserted. **Fruits** ellipsoid, 3–4.5 × 2–3 cm, glabrous, 6–7 ribbed; seeds numerous, 5 × 4 mm. **Distr:** Sri Lanka. **Ecol:** In dry regions. **Zones:** 11–12.

G. jasminoides J. Ellis, *Philos. Trans.* 51(2): 935, 1761. (Syn: *G. augusta*, etc.). [Cape Jasmine, Gardenia]. **Shrubs** or **small trees** (many-branched), to 5 m tall. **Stipules** 0.5–1 cm long, bractlike, with a short tube 2–4 mm long, persistent. **Leaves** 2(–3) per node; mostly broadly elliptic, lamina 3–12 × 1.5–5 cm, base acute and decurrent on petiole, apex acuminate, glabrous, veins 6–9 pairs; petiole 1.4 cm long, slightly differentiated from leaf base. **Inflorescences** often of 3 terminal flowers, or flowers solitary, axillary (subterminal); pedicels ~1 mm long. **Flowers** large, 6–10 cm long, sweetly fragrant; hypanthium ~1 cm long; calyx lobes 0.8–3 cm long, narrowly oblong, spurlike, persistent; corolla white, tube 2–5 cm long, corolla lobes usually 6 or 7 (some varieties with >2 series), obovate, ~2.5 × 1.8 cm; anthers linear, ~1.8 cm long, projecting. **Fruits** obovoid to oblong, 2.5 cm Ø, ribbed, orange. **Distr:** Indo-China to S Japan. **Ecol:** Fl. Jul–Dec. **Zones:** 10–12.

G. sootepensis Hutch., *Bull. Misc. Inform. Kew* 1911: 392, 1911. (Syn: *G. massieana*). [Golden Gardenia]. **Trees**, deciduous, to 15 m tall; trunk smooth, to 30 cm dbh; buds usually with yellow resin. **Stipules** interpetiolar, triangular, connate into a tube, ~1 cm long. **Leaves** opposite, crowded at ends of branchlets; lamina narrowly elliptic, 9–15 × 3–6 cm, base acute, margins entire, apex acuminate, papery to slightly coriaceous, surfaces sparsely hairy, nerves parallel, 10–14 pairs. **Inflorescences** with flowers solitary, axillary; pedicel >5 mm long. **Flowers** opening to pure white or creamish-white, ageing to golden-yellow, sweetly fragrant; receptacle subcylindrical, 5 mm long; calyx tubular, 1.3–1.5 cm long, pubescent; corolla tube cylindrical, 5–6 × 0.3–0.4 cm, outside minutely puberulous, inside glabrous, lobes 5, spreading, broadly obovate, ~5 × 3 cm, glabrous; anthers 1.5 cm long, exserted to ~2 mm; ovary with 2 placentas, style 6 cm long, sulcate, subappressed pubescent, stigma clavate. **Fruits** berrylike, oblong or oblong-elliptic, 4 × ~2–4 cm, dehiscent, exposing seeds and orange pulp; seeds numerous, foveolate. **Distr:** Indo-China. **Ecol:** In open dry dipterocarp forests; 500–600 m. **Zones:** 9–12.

G. taitensis DC., *Prodr.* 4: 380, 1830. (Syn: *G. weissichii*). [Star of Tahiti, Tahitian Gardenia]. **Shrubs** or **trees**, 2–6 m tall. **Stipules** forming papery, bilobed sheath. **Leaves** usually subsessile, opposite; lamina ovate, obovate to broadly elliptic, 5–23 × 5–11.5 cm, apex rounded or toothed, glossy green, leathery. **Inflorescences** with flowers solitary, borne at end of branches. **Flowers** white, fragrant, waxy, 7.5 cm Ø; calyx lobes usually 4–5, often sickle-shaped, to 3.2 cm long; corolla salverform, tube 2–4.5 cm long, limb spreading, lobes 6–8, ± oblong to elliptic, usually 2–4.5 cm long; ovary ~1.2 cm long, 4- or 5-angled. **Fruits** drupes, obovoid or ellipsoid, 5 × 4 cm, 3–5-ribbed or -winged, with persistent lanceolate calyx lobes, pulp orange. **Distr:** SW Pacific. **Ecol:** Fl. throughout the year. **Zones:** 10–12.

G. ternifolia* ssp. *jovis-tonantis (Welw.) Verdc., *Kew Bull.* 34: 354, 1979. (Syn: *G. jovis-tonantis*, etc.). [Large-leaved Gardenia]. **Trees**, small, shrubby, 3–4 m high; branches always in 3s, flattened toward base, thorny; bark often reddish-brown and breaking down to a powdery surface; young shoots not heterophyllous. **Leaves** elliptic or oblong to oblanceolate-obovate, 0.7–18 × 0.7–11 cm, subcoriaceous, discolorous, lateral nerves arcuate. **Inflorescences** with flowers solitary. **Flowers** 8–9(–13)-merous, sweetly fragrant, white, waxy, opening at night, turning yellow next day; calyx forming an irregular frill, lobes 8–9, subulate, to 1.7 cm long; corolla lobes 8–9(–13), elliptic or oblong, 2–5.5 cm long. **Fruits** ellipsoid, 4–10 × 1.6–6 cm, yellowish or reddish-brown to gray-green, long-persistent, ± conspicuously lenticellate; seeds smaller, 3–4(–4.5) mm long. **Distr:** Senegal to Cameroon, Sudan, tropical Africa. **Ecol:** In savannas, often extending into moist regions; 0–2,100 m. **Zones:** 10–12.

G. tubifera Wall. ex Roxb., *Fl. Ind.* 2: 562, 1824. [Golden Gardenia, Water Gardenia]. **Shrubs** or **small trees**, to 3–6(–20) m tall, unarmed; young parts resinous. **Leaves** in closely approximate pairs, petiolate; lamina cuneate-oblong or oblanceolate, 8–24 × 4–6 cm, ribbed, base much-narrowed, apex acuminate, acumen tapering downward, adaxially slightly scabrous, abaxially pubescent, nerves 15–18 pairs; petiole 2.5 cm long. **Inflorescences** flowers solitary, terminal. **Flowers** strongly fragrant, creamy-white, turning yellow, then golden (or orange); calyx 1.2 cm long; corolla tube 10–12 cm long, lobes 6–7, ovate to oblanceolate, 3 × 1 cm, margins often revolute, apex rounded; stamens exserted. **Fruits** drupes, globose, 3–5 cm Ø, uneven, crowned with a very long truncate, calycine tube, divisible into 8 valves. **Distr:** Malay Peninsula, Sumatra, Borneo. **Ecol:** Along riverbanks in mud. **Zones:** 11–12.

Genipa

Species 3. **Trees. Stipules** interpetiolar and intrapetiolar, connate to form a short tube, caducous or deciduous with leaves. **Leaves** opposite and decussate, subsessile or petiolate; lamina often large, pinnately veined, chartaceous to coriaceous, without domatia. **Inflorescences** with flowers solitary or in few-flowered cymes, or subcapitate, terminal, subterminal or axillary; pedicels present and continuous with hypanthium base. **Flowers** radially symmetric, usually bisexual, 5- or 6-parted; hypanthium turbinate to campanulate; calyx tube truncate and entire or with 5–6 short lobes; corolla salverform to funnelform, carnose, white to yellowish-white, tube barbate inside throat and at base of lobes, corolla lobes 5–6, thick, spreading, convolute in bud; stamens 5–6, inserted in upper part of tube, anthers subsessile, dorsifixed, linear, partly exserted; ovary 1-locular or becoming bilocular, style thick, stigmas fusiform, placentation parietal, ovules many. **Fruits** baccate, large, ovoid to subglobose or obovoid, calyx tube persistent at apex of fruit; seeds many, large, compressed. **Distr:** S USA (Florida) and Mexico through C America into tropical S America. **Cult:** Propagate from seeds or half-hardened cuttings. Requires a humus-rich, fertile, moist, well-drained soil in full sun.

G. americana L., *Syst. Nat.* ed. 10, 2: 931, 1759. (Syn: *Gardenia genipa*, etc.). [Jagua]. **Trees**, 4–27 m tall; trunk to 50 cm dbh. **Stipules** triangular, 1–2.5 cm long, basal sheathing tube 1–3 mm long, acute, deciduous with leaves. **Leaves** obovate to elliptic-obovate or broadly oblanceolate, 12–42 × (4–)6–19 cm, base cuneate, gradually narrowed, apex usually acuminate, lustrous, abaxially sometimes densely pubescent; petiole 2–1.3 cm long. **Inflorescences** cymes, 1–9-flowered, 4–10 cm long, ± terminal; peduncles to 2.5 cm long; pedicels 0.4–1.2 cm long. **Flowers** apparently bisexual, functionally unisexual; hypanthium with calyx tube 0.7–1.7 cm long; calyx glabrous; corolla salverform, ~4 cm Ø, densely sericeous, white or creamish-white, darkening with age, tube 0.5–1.5 cm long, lobes 5–6, obovate, 1.1–2.8 × 0.5–1.2 cm, rounded; anthers 0.6–1.4 cm long, becoming recurved between lobes; stigmas ~5 mm long. **Fruits** obovoid to subglobose, 4–11 × 3–11 cm, brownish; seeds 0.6–1.2 × 0.4–0.7 cm. **Distr:** Tropical America. **Ecol:** In wet evergreen rainforests and seasonally dry deciduous forests; 0–900 m. Fl. Mar–Aug. Fr. throughout the year. **Zones:** 10–12. **Info:** The fruits are used for making dyes. The color lasts for about 3 weeks on skin, similar to henna (*Lawsonia inermis*).

Geophila

Species 27. **Herbs**, with white raphide bundles, prostrate to somewhat ascending stems; branches sometimes initially ascending, later prostrate. **Stipules** broad, membranous, becoming firm, subpersistent. **Leaves** petiolate; lamina cordate. **Inflorescences** with flowers solitary or in subcapitate clusters at ends of peduncles, subtended by bracts, terminal, often on short branches; bracts sometimes trifid or incised, 5-merous. **Flowers** with calyx basally united; corolla tube cylindric with 5 spreading or recurving lobes; anthers included; style with stigma bilobed. **Fruits** soft drupes, with 2 pyrenes, each with a single seed, dorsally rounded, ventrally flattened; seed with endosperm. **Distr:** Pantropical. **Cult:** Propagate from seeds. Requires a moist, humus-rich soil in shade or partial sun.

G. repens (L.) Johnston, *Sargentia* 8: 281, 1949. (Syn: *Cephaëlis reniformis*, etc.). [Rat/Snake Pennywort]. **Herbs**, prostrate, creeping, pubescent. **Stipules** orbicular to reniform on main stems, triangular on flowering branches. **Leaves** orbicular to reniform, or ovate-cordate, 1.2–3.6(–5) × 1.3–4.3(–5.5) cm, apex rounded, base emarginate-cordate, adaxially glabrous, abaxially glabrous or pubescent; petiole 0.4–11.5 cm long. **Inflorescences** with flowers solitary or several in condensed cymes or heads, at ends of peduncles, terminal on short erect branches; peduncles subtended by lanceolate or trifid acuminate bracts. **Flowers** white; calyx lobes 5, lanceolate, caudate-acuminate, 1.2–2 mm long, apices filiform; corolla salverform, tube 0.5–1.3 cm long, lobes 5, oblong, 4–9 × 2–5 mm, acute; anthers 5, narrowly sagittate, inserted low in corolla tube. **Fruits** soft drupes, red, with 2 plano-convex pyrenes. **Distr:** Pantropical. **Ecol:** In shaded parts of moist or wet forests. **Zones:** 10–12.

Guettarda

Species ~100. **Trees** or **shrubs**; bark grayish-brown. **Stipules** interpetiolar, deciduous or persistent. **Leaves** opposite or ternate. **Inflorescences** axillary, usually cymose and cincinnoid. **Flowers** bisexual or polygamodioecious; calyx cupuliform, truncate or 2–9-dentate; corolla hypocrateriform to infundibular, lobes 4–9, imbricate, stamens 4–9, inserted at throat of corolla, anthers dorsifixed, included; ovary inferior, 2–9-locular, 1 pendulous ovule per locule, style filiform, stigma capitate. **Fruits** drupaceous, hard, subglobose to oblong, often angled, mesocarp fibrous, with a hard 4–9-chambered putamen. **Distr:** Predominantly American with 1 widespread species in the Old World. **Cult:** Propagate from seeds or cuttings. Requires a well-drained soil in full sun.

G. speciosa L., *Sp. Pl*, 991, 1753. (Syn: *Cadamba jasminiflora*, etc.). [Beach Gardenia, Indian Lavendar]. **Shrubs** to **small trees**, evergreen, to 10 m tall. **Stipules** triangular, 1–1.5 cm long, entire, caducous. **Leaves** broadly obovate, 9–24 × 6–20 cm, base rounded to subcordate, apex obtuse, coriaceous, surfaces hairy. **Inflorescences** cymose, axillary, flowers sessile on upper sides of the inflorescence branches, branches subtended by linear-lanceolate bracts; peduncles (3–)6–9(–10) cm long. **Flowers** bisexual, or variably polygamodioecious, fragrant; calyx cupular, truncate, 5–6 mm long; corolla hypocrateriform (trumpet-shaped), tube 1.5–3.5 cm long, outside pallidly hairy, inside throat hairy, lobes ovate, 4–9 mm long, anthers dorsifixed, sessile, inserted in upper part of corolla tube, included, style 1.5–3.5 cm long, stigma capitate, entire, slightly exserted; ovary 4–9-loculed, locules 1-ovuled, ovule pendulous. **Fruits** depressed-globose, 1.5–2.5 cm Ø, pyrenes 4–9, fused into a hard stone. **Distr:** Widespread from E Africa to Malesia, Micronesia and S Pacific. **Ecol:** On seashores and beach forests. **Info:** Flowers open shortly after sunset and fall the next morning, often pollinated by fruit bats.

Hamelia

Species 17. **Shrubs** or **herbs**. **Stipules** interpetiolar, multifid or bristlelike, usually caducous. **Leaves** opposite or in whorls of 3 or 4, petiolate. **Inflorescences** cymes, terminal, dichotomous or trichotomous, branches scorpioid, with flowers often arranged on 1 side of the branches; flowers short-pedicellate or subsessile; bracteoles usually small. **Flowers** with calyx tube ovoid or turbinate, lobes 4–6, usually short and erect, persistent; corolla tubulate or campanulate, tube 4–6-ribbed, constricted at base, villose in throat, lobes 4–6, imbricate in bud; stamens 4–6, inserted at base of corolla tube, filaments short, anthers with appendages at apex of connectives, basifixed, included or apically exserted; disk swollen; ovary 5-loculed, locules many-ovuled, style filiform, stigma clavate, slightly twisted. **Fruits** berries, small, 5-loculed, many-seeded, crowned with swollen disk; seeds small, testa membranous, reticulate. **Distr:** Tropical and subtropical America. **Cult:** Propagate from cuttings. Requires a moist, well-drained soil in full or partial sun.

H. patens Jacq., *Select. Am.* 72, 1763. (Syn: *H. coccinea*, *H. erecta*). [Fire Bush, Hummingbird Bush, Scarlet Bush, Redhead]. **Shrubs**, deciduous, 1.5–4 m high, much-branched; young parts grayish-pubescent; branchlets, petiole and peduncles reddish. **Stipules** subulate, ~6 mm long. **Leaves** usually 3- or 4-whorled; lamina elliptic to oblanceolate, 7–15(–20) × 4–6 cm, base cuneate, apex acute, lateral nerves 7–9 pairs; petiole 1.5–4.5 cm long. **Inflorescences** cymes, terminal, with 3–5 radiate branches; primary branches 2.5–4 cm long, secondary branches 1.5–2 cm long, flowers arranged along one side of ultimate branches; pedicels 1–2 mm long. **Flowers** with calyx campanulate, ~3 mm long, lobes 5, triangular, ~1 mm

Below: *Gardenia tubifera*, *Genipa americana*, *Geophila repens*, *Guettarda speciosa*.

MIW

PL

Above: *Hamelia patens, Hedyotis verticillata, Hoffmannia refulgens, Ixora brachypoda.*

long; corolla scarlet-orange, narrowly cylindrical, 1.8–2.5 cm long, glabrous, tube 5-ribbed, lobes 5, ovate-triangular, ~2 mm long; stamens 5, as long as corolla tube, filaments 5–6 mm long, anthers linear, 0.9–1 cm long, connective protruded/projected; style 1.1–1.2 cm long, stigma clavate. **Fruits** berries, ovoid, 6–7 mm Ø, dark red or purple. **Distr:** C America. **Ecol:** In disturbed sites. Fl. May–Dec. **Zones:** 10–12.

Hedyotis

Species 118. **Herbs**, **undershrubs** or **shrubs**, erect or climbing. **Stipules** free or connate into sheath or adherent with petiole in sheath. **Leaves** opposite, rarely whorled or cespitose. **Inflorescences** usually cymes or further arranged into panicles, capitula, umbels or corymbs, rarely other types of inflorescences or reduced into solitary flowers, terminal or axillary; flowers with or without bracts, pedicellate or sessile. **Flowers** with calyx tube usually turbinate, lobes usually (2 or 3)4(5); corolla tubulate, funnelform or rotate, lobes (2 or 3)4 or 5, valvate in bud; stamens usually as many as corolla lobes, exserted or included, filaments hairy or glabrous, anthers dorsifixed; disk usually small, 4-lobate or shallowly 4-lobed; ovary biloculed, ovules few to numerous, very rarely solitary, style linear, included or exserted, stigma bilobed, rarely undivided. **Fruits** small, membranous, rarely leathery, septicidally or loculicidally dehiscent, or indehiscent; seeds (1)2 to many, small. **Distr:** Tropical and subtropical Asia to NW Pacific. **Cult:** Propagate from seeds. Requires a moist, well-drained soil in full sun.

H. verticillata (L.) Lam., *Tabl. Encycl.* 1: 271, 1792. (Syn: *Oldenlandia verticillata*, etc.). **Herbs**, annual, diffuse, 25–30 cm high; branches often prostrate. **Stipules** slightly hairy, basally connate with petioles into a sheath, apically lacerate into several bristles 3–4 mm long. **Leaves** opposite, ± sessile; lamina elliptic or lanceolate, 2.5–5 × 0.6–2 cm, base cuneate or obtuse, apex acute or acuminate, papery or thinly leathery. **Inflorescences** glomerules, axillary, without peduncles; bracts lanceolate, 3–4 mm long; flowers sessile. **Flowers** with calyx tube obconical, ~1 mm long, hirsute, lobes 4, lanceolate, 1–1.5 mm long; corolla white, subfunnelform, 3.8–4 mm long, glabrous except lobes bearded at apex, tube ~2 mm long, lobes 4, lanceolate, 1.8–2 mm long; stamens inserted at throat, filaments ~2 mm long, anthers ~1 mm long, exserted; style 4–4.5 mm long, apex inflated. **Fruits** capsules, ovoid, 1.5–2.5 × 1.5–2 mm; seeds many. **Distr:** China, Hong Kong, India, Indonesia, Malaysia, Nepal, Sri Lanka, Vietnam. **Ecol:** In thickets. Fl. Mar–Nov. **Zones:** 9–12.

Hoffmannia

Species 115. **Herbs**, herbaceous **subshrubs**, or slender **shrubs**; usually with few or no lateral branches. **Stipules** interpetiolar, usually small, triangular, often succulent, divergent, caducous. **Leaves** usually opposite and decussate; lamina with base often decurrent on petiole, margins entire, pinnately veined; petioles with inflated chambers (ant vesicles) in a few species. **Inflorescences** few- to many-flowered, usually cymose or dichasial and often with helicoid branches, fasciculate to capitate in a few species, axillary; flowers usually pedicellate. **Flowers** bisexual, monomorphic; hypanthium often longitudinally ribbed; calyx very short, lobes 4(5), short, triangular; corolla salverform to funnelform or rotate, white, yellow, orange, or rose to deep red or purple, lobes 4(5), usually acute; stamens (3)4(5), inserted, filaments very short, anthers linear to narrowly oblong, usually white; ovary 2(3 or 4)-locular, locules many-ovuled, placentation axile, style slender, stigmas bilobate or clavate. **Fruits** baccate, usually small, 2(3 or 4)-locular, often becoming enlarged and spongy; seeds many, minute. **Distr:** Mexico to tropical America. **Cult:** Propagate from seeds or cuttings. Requires a humus-rich, moist, well-drained soil in shade or partial sun.

H. refulgens (Hook.) Hemsl., *Biol. Cent.-Amer.*, Bot. 2: 37, 1881. (Syn: *Campylobotrys argyroneura*, etc.). [Quilt Plant, Strawberry Splash, Taffeta Plant]. **Herbs** or **subshrubs**, simple, usually <40 cm high; stems ferruginous-villous or glabrate, densely leafy above. **Leaves** opposite; lamina obovate to oblong-oblanceolate, 9–25 × 4–11 cm, base acute to long-attenuate, apex rounded or obtuse and abruptly apiculate, adaxially glabrous or sparsely villous, abaxially often purple or reddish, ferruginous-villous; petioles stout, to 3.5 cm long. **Inflorescences** subumbellate or subracemose, sometimes secund; peduncles slender, 2.5–12 cm long; flowers shortly pedicellate. **Flowers** with hypanthium 3–4 mm long, glabrous; calyx 3–4 mm long, lobes to 2 mm long, acute, sparsely villosulous; corolla red or pale red, 1 cm long, glabrous, lobes 2–3 times as long as the tube. **Fruits** bright red. **Distr:** SE Mexico to C America. **Ecol:** In deeply shaded parts of humid forests; 600–900 m. **Zones:** 10–12.

Ixora

Species 530. **Shrubs** or **small trees**, evergreen. **Stipules** interpetiolar, base broad, usually basally fused into a sheath, apex prolonged or aristate. **Leaves** opposite, rarely 3-whorled, petiolate or sessile. **Inflorescences** terminal cymes, dense or spreading, corymbose or trichotomous; flowers pedicellate or sessile; bracts and bracteoles present. **Flowers** with calyx tube usually ovoid, limb with 4(or 5) persistent lobes; corolla salverform, glabrous or bearded at throat, apically 4(or 5)-lobed, lobes shorter than corolla tube, spreading or reflexed, contorted in bud; stamens as many as corolla lobes, inserted at throat, filaments short or absent, anthers dorsifixed, 2-celled, ± exserted; disk fleshy, swollen; ovary biloculed, locules 1-ovuled, style filiform, stigma 2, short, recurved at anthesis. **Fruits** drupes, globose or slightly compressed, bisulcate, leathery or fleshy, with 2 pyrenes; pyrenes leathery, plano-convex or concave adaxially; seeds same shape as the pyrene, testa membranous. **Distr:** Tropical and subtropical regions worldwide. **Cult:** Propagate from seeds, half-hardened cuttings or by air-layering. Requires a humus-rich, moist, well-drained soil in full or partial sun.

I. brachypoda DC., *Prodr.* 4: 488, 1830. (Syn: *I. radiata*, etc.). **Shrubs** or **small trees**, 1.8–8 m tall, glabrous. **Stipules** ~5 mm long, basally connate, strongly keeled. **Leaves** narrowly elliptic-oblong or oblong-ovate, 8–37 × 4.5–13.5 cm, base usually obtuse-rounded, apex usually acute, coriaceous, adaxially dull, lateral nerves 7–14(–18) pairs; petiole 0–7 mm long. **Inflorescences** many-flowered, contracted-trichotomous, ~10–12 cm Ø across corollas; peduncle 1–9.5 cm long, sometimes with lobed involucral bract of stipules and reduced leaves, above leaves subtending the inflorescence; bracts ovate or triangular, 3 mm long; bracteoles minute; pedicels of lateral flowers of triads to 2 mm long, central flowers subsessile. **Flowers** fragrant; calyx tube 1.5 mm long, lobes triangular, to 1.5 mm long; corolla white, pink-tinged or lobe apices pink, tube 3.7–6 × 0.1–0.2 cm, lobes elliptic-oblong, 8–9 × 2–3.2 mm, sometimes hairy inside; anthers 6 mm long, pollen presenter arms 2–3 mm long; style exserted to 5 mm. **Fruits** red or orange-red, often becoming blackish with age, ~8 mm Ø. **Distr:** Tropical Africa. **Ecol:** In gallery and swamp forests; 1,200–1,500 m. **Zones:** 10–12.

I. chinensis Lam., *Encycl.* 3: 344, 1789. (Syn: *I. stricta*, etc.). [Chinese Ixora]. **Shrubs**, 0.8–2 m high, glabrous. **Stipules** base broad, fused into sheath, 5–7 mm long, long-acuminate. **Leaves** opposite, sometimes 4-whorled as internodes shortened, very shortly and thickly petiolate or sessile; lamina lanceolate, oblong-lanceolate to oblong-oblanceolate, 6–13 × 3–4 cm, apex obtuse or rounded. **Inflorescences** many-flowered, terminal; peduncle 0.5–1.5 cm long, together with branches red, ± softly hairy, often with 2 small basal leaves; bracts and bracteoles minute. **Flowers** with calyx tube 1.5–2 mm long, limb 4-lobed, lobes 0.8 mm long, acute or obtuse; corolla red or reddish-yellow, 2.5–3 cm long, apically 4-lobed, lobes obovate or subrotund, spreading or reflexed, 5–7 × 4–5 mm, obtuse or rotund; filaments very short, anthers oblong, ~2 mm long, base bilobed; style shortly exserted, stigmas 2, connivent at first, furcate at anthesis, slightly decurved. **Fruits** subglobose, binate, grooved at middle, reddish-black; seeds 4–4.5 mm long. **Distr:** China, Hong Kong, Indonesia, Malaysia, Philippines, Vietnam. **Ecol:** In thickets or sparse forests; 200–800 m. Fl. May–Jul. **Zones:** 10–12.

I. coccinea L., *Sp. Pl.* 110, 1753. [Flame of the

ZS

Above: *Ixora chinensis, Ixora coccinea, Ixora finlaysoniana, Ixora hookeri.*

Woods, Jungle Flame, Jungle Geranium]. **Shrubs** or **small trees**. **Stipules** 0.4–1.2 cm long, cusp 2–8 mm long. **Leaves** elliptic, oblong, or obovate, 3–14 × 1.5–6 cm, base cuneate to cordate, apex obtuse to acute or mucronate, glabrous, lateral nerves 8–14 pairs; petiole 0–5 mm long. **Inflorescences** cymose panicles, usually with 3 main branches, flowers usually in triads, central flower sessile and ebracteolate, others with pedicel to 4 mm long; peduncles very short or absent, puberulous; bracts linear, 0.1–1.2 cm long. **Flowers** variable in color, but not white; hypanthium 1–2 mm long; calyx lobes triangular, 0.5–2 mm long, glabrous or pubescent; corolla tube 1.5–4.5 cm long, pubescent, lobes lanceolate to ovate, 0.75–2 × 0.3–1 cm; filaments 0.7–2 mm long, anthers 3–5 mm long; style 2–4.5 cm long, stigmatic arms 1.5–3 mm long. **Fruits** globose, 5–7 × 5–6 mm. **Distr:** W and S India, Sri Lanka, Indo-China. **Ecol:** In lowland coastal forests and forest margins; 0–300 m. Fl. throughout the year. **Zones:** 10–12.

I. finlaysoniana Wall. ex G. Don, *Gen. Syst.* 3: 572, 1834. [Fragrant Ixora, White Ixora]. **Shrubs** or **small trees**, 5–6 m tall, glabrous. **Stipules** broadly ovate, ~5 × 4–6 mm, aristate. **Leaves** oblong, elliptic or oblong-lanceolate, 10–17(–20) × 3–6(–9) cm, base cuneate, apex short or obtuse acuminate; petiole 4 mm long. **Inflorescences** cymes, arranged in trichotomous corymbs, first order branches longest, third order branches very short, forming many-flowered capitulum, terminal; flowers sessile, or lateral ones shortly pedicellate; bracts lanceolate, 0.8–1 cm long. **Flowers** fragrant; calyx tube subglobular, 1.5–2 mm Ø, lobes elliptic, 4–5 × 2–3 mm, acute; corolla white, tube ~3 mm long, lobes 4, obovate, 5–6 × 3 mm; filaments very short, anthers ~3 mm long, exserted; style slightly longer than corolla tube, stigmas 2, connivent at first, furcate at blooming. **Fruits** subglobose, 6–8 mm Ø, apex with persistent calyx lobes. **Distr:** China, India (Assam), Indo-China, Philippines. **Ecol:** In lowland sparse forests. Fl. Apr–Oct. **Zones:** 9–12.

I. hookeri (Oudem.) Bremek., *Repert. Spec. Nov. Regni Veg.* 37: 194, 1934. (Syn: *I. odorata* [misapplied], *Pavetta hookeri*). [Fragrant Ixora, Sweet Ixora]. **Shrubs**, to 3 m high, or **small trees** with multiple trunks to 6 m tall; branches purplish. **Stipules** broadly ovate, 3.5–5 mm long, pressed against the twig, bases briefly united. **Leaves** usually opposite; lamina ovate to elliptic-lanceolate, to 2.5–7.5(–30) × 1–4(–7.5) cm, base wedge-shaped, apex acute to acuminate, coriaceous, lateral veins 7–9 pairs; petiole 3–4 mm long. **Inflorescences** many-flowered, often 25–30 cm across, terminal. **Flowers** fragrant, white to pink, especially toward base; calyx lobes ovate, to ~1 cm long, acute; corolla white, 6–10(–12.5) cm long, lobes white, apices often rose-colored, aging to yellowish, oblong, 5–7 × 1–3 mm, narrowed toward base. **Fruits** globose, 0.6–1 cm Ø. **Distr:** Madagascar, Mozambique. **Ecol:** In humid forests. Fl. May–Aug. **Zones:** 10–12.

I. jucunda Thw., *Enum. P1. Zeyl.* 155, 1859. **Trees**, small, 2–6(–12) m tall; bark fissured; branchlets glabrous. **Stipules** 3–5 mm long, outside slightly pubescent when young, cusp short to 1 mm long. **Leaves** elliptic to oblanceolate, (9–)12–25 × (3.5–)5–9(–12) cm, base acute, apex acuminate, glabrous, coriaceous, lateral nerves 6–13 pairs; petiole (0.7–)1–2 cm long. **Inflorescences** often with 3 main branches, usually in triads, central flower sessile and ebracteolate, others with pedicel to 3 mm long; hypanthium 1–1.5 mm long, very finely pubescent; peduncles 0.5–1.5 cm long, minutely pubescent; bracts linear, 1–2 mm long. **Flowers** pink to pinkish-white; calyx lobes ovate, 0.4–0.6 mm long, glabrous; corolla tube 2.5–3.5 cm long, glabrous, lobes ovate-oblong, 5–6 × 2–3 mm; filaments 1–1.5 mm long, anthers 4–5 mm long; style 2.8–4 cm long, stigmatic arms 1–3 mm long. **Fruits** globose. **Distr:** Sri Lanka. **Ecol:** In wet evergreen forests. **Zones:** 11–12.

I. pavetta Andr., *Bot. Rep.* t. 78, 1799. (Syn: *I. arborea*, etc.). [Small-flowered Ixora, Sweet Ixora]. **Shrubs** or **small trees**; branchlets glabrous. **Stipules** 3.5–5 mm long, outside hairy, cusp 2–3 mm long. **Leaves** ovate to obovate, 4–18 × 2.5–7.5 cm, base rounded to cordate, apex acute to obtuse, sometimes apiculate, glabrous, lateral nerves 9–12 pairs; petiole 2–5 mm long. **Inflorescences** racemes of corymbs with 3–5 pairs of short branches, generally in triads, central flower sessile and ebracteolate; peduncles 0.3–4 cm long, pubescent, purplish; bracts 1–5 mm long; pedicels to 2.5 mm long. **Flowers** white; hypanthium 0.75–1.5 mm long, glabrous or sparsely pubescent at base; calyx lobes triangular, 0.25–0.75 mm long, glabrous; corolla tube 0.5–1.2 cm long, glabrous, lobes linear-oblong, 2–3 × 1–2 mm; filaments 0.5–0.7 mm long, anthers 2–3 mm long; style 0.6–1.1 cm long, stigmatic arms 1–2 mm long. **Fruits** globose, 5–7 × 5 mm, red-violet to black. **Distr:** Bangladesh, India, Sri Lanka. **Ecol:** In dry deciduous forests. **Zones:** 10–12.

I. philippinensis Merr., *Philip. Journ. Sci. Bot.* 5: 238, 1910. (Syn: *I. hayatae*). [Santan]. **Shrubs** or **small trees**, glabrous; branchlets terete, grayish-green to reddish-brown. **Stipules** lanceolate, 3–5 mm long, long-acuminate. **Leaves** elliptic or oblong, 4.5–6 × 2–2.5 cm, base subrotund or cordate, sometimes amplexicaul at upper parts, apex obtuse, usually minutely apiculate by excurrent midrib (protruding out of tip), thinly papery, lateral veins 11 pairs; petioles <3 mm long. **Inflorescences** cymes, dense, terminal, solitary; peduncle 1–1.5 cm long, usually subtended by a pair of reduced, ovate or subovate, cordate leaves 1–3 cm long; bracts ovate-lanceolate, acuminate, 1.5–1.8 mm long; flowers ± sessile. **Flowers** with calyx ~2 mm long, limb 4-lobed, lobes dentiform, minute, apiculate; corolla white to pale-pink or reddish, tube 1.8–2 cm long, apically 4-lobed, lobes elliptic-oblong, ~7 × 3.5 mm, rounded or obtuse, anther 3.5 mm long, acute, base bilobed; stigma 2 mm long. **Fruits** ovoid, somewhat fleshy, binate, slightly compressed, 1 cm long, red. **Distr:** Taiwan (Lü Toa) to Malesia. **Ecol:** In broad-leaved forests, thickets. Fl. Jun–Aug, Oct–Jan. **Zones:** 9–12.

Below: *Ixora jucunda, Ixora pavetta, Ixora philippinensis.*

MS

Keetia

Species 32. **Climbers** or **scandent shrubs**; stems glabrous or frequently pubescent. **Leaves** petiolate, not restricted to new growth at apex of branches. **Stipules** interpetiolar, lanceolate to ovate or triangular at base and acuminate to linear above, never keeled, never white silky-hairy inside. **Leaves** subtending lateral branches smaller and broader than main leaves or not, chartaceous or occasionally coriaceous. **Inflorescences** branched cymes, pedunculate; bracts and bracteoles conspicuous or not. **Flowers** 4–6-merous; calyx tube ellipsoid to ovoid, limb repand to dentate or less often lobed; corolla white, cream or yellow, tube ± cylindrical, typically with a ring of deflexed hairs inside, lobes reflexed; stamens set at throat of corolla, anthers fully or partly exserted but usually not reflexed, narrowly ovate or oblong; ovary bilocular, each locule 1-ovuled, style long, slender, stigmatic-club cylindrical, longer than wide. **Fruits** drupes; seeds 2, or often 1 aborted, ovoid. **Distr:** Tropical and southern Africa. **Cult:** Propagate from seeds. Requires a moist, well-drained soil in partial sun.

K. gueinzii (Sond.) Bridson, *Kew Bull.* 41: 970, 1986. (Syn: *Canthium gueinzii*, etc.). [Climbing Canthium]. **Climbers**, robust, or **scrambling shrubs**, 4–10(–20) m long; branchlets finely hairy. **Stipules** lanceolate to ovate, 0.9–1.3 × 0.6 cm. **Leaves** opposite; lamina ovate to oblong-elliptic 4–10 × 3–4.5 cm, base rounded, margins entire, apex tapering, thin, adaxially mid to dull deep green, abaxially paler, densely hairy, with hair-tufted pits (domatia); petiole ~7 mm long, hairy. **Inflorescences** dense clusters, axillary, 20–50-flowered, erect. **Flowers** sweetly fragrant, creamy-white to yellowish; calyx tube 1 mm long, usually densely hairy, limb 1.25–1.5 mm long, deeply divided, usually ciliate; corolla creamy-white, tube 2.2–4 mm long, hairy inside below throat, lobes oblong-lanceolate to ovate, 2.5–4 × 1.2–2.2 mm, apex acute, thickened; anthers fully exserted, seldom reflexed, pollen presenter 1.2–2.2 mm long; disk pubescent; style 0.5–1 cm long, glabrous. **Fruits** clustered, ± round, 0.8–1.4 cm Ø, 1–bilobed, blackish, glabrous. **Distr:** Cameroon to Ethiopia and S Africa. **Ecol:** In evergreen forests, riverine forests, montane grasslands. **Zones:** 9–12.

Morinda

Species 129. **Lianas**, climbing or erect **shrubs**, or **small trees**. **Stipules** interpetiolar or intrapetiolar, free or connate forming a sheath, appressed, membranous or papery. **Leaves** opposite, rarely 3-whorled. **Inflorescences** capitula, mulberrylike or subglobose, few- to many-flowered, solitary in leaf axils or opposite a leaf in woody species, several arranged in umbels terminal on branches, pedunculate or rarely sessile; flowers sessile, bisexual, (3–)5(–7)-merous. **Flowers** with calyx semiglobose or conical, fused below, annular above, apex truncate or 1–3-dentate; corolla white, funnelform, salverform or campanulate, tube as long as or much longer than limb, throat densely hairy or glabrous; stamens equal in number as lobes, inserted at throat or lateral base of lobes, filaments short, anthers dorsifixed, exserted or included; ovary 2–4-loculed, locules 1-ovuled. **Fruits** multiple drupes, developed from (1 or)2 to many fused flowers and receptacles, ovoid, mulberry-shaped or subglobose, each drupe with 2–4 pyrenes; pyrenes subtrigonal, curved abaxially, flat or laterally sulcate, 1-seeded; seeds subtrigonal or oblong. **Distr:** Tropical and subtropical regions. **Cult:** Propagate from cuttings or seeds. Requires any soil in full sun.

M. citrifolia L., *Sp. Pl.* 176, 1753. (Syn: *M. tinctoria*, etc.). [Awl Tree, Cheese Fruit, Indian Mulberry, Limburger Tree, Noni]. **Shrubs** or **small trees**, evergreen, 1–5(–8) m tall; branches ± tetragonal. **Stipules** interpetiolar, opposite, widely ovate. **Leaves** decussate; lamina oblong, elliptic or ovate, 10–25 × 5–13 cm, base and apex acute or acuminate, margins entire, fleshy, glossy, glabrous, lateral veins 5–7 pairs; petioles 0.5–2 cm long. **Inflorescences** dense head, many-flowered; rachis/peduncle 1–1.5 cm long; flowers sessile. **Flowers** with calyx tube somewhat mutually connate, limb subtruncate; corolla white, funnelform, ~1.5 cm long, throat densely villose, lobes 5, ovate-lanceolate, ~6 mm long; stamens (4 or)5(or 6), inserted at throat, filaments ~3 mm long; ovary 4-loculed, sometimes 1 or 2 locules infertile, style subequal to corolla tube. **Fruits** compound drupes with numerous blunt protuberances, green, 4–7.5 cm long, ripening to soft straw or pale grayish-yellow, malodorous. **Distr:** Tropical and subtropical Asia to N Australia. **Ecol:** On seashores, sparse forests. Fl. & Fr. throughout the year. **Zones:** 10–12.

M. umbellata L., *Sp. Pl.* 176, 1753. (Syn: *Guttenbergia umbellata*). [Indian Mulberry]. **Lianas**, climbing or twining, sometimes shrubby. **Stipules** tubulate, scarious, 4–6 mm long. **Leaves** ovate, obovate-lanceolate or obovate-oblong, 6–9 × 2–3.5 cm, base attenuate or cuneate, margins entire, apex acuminate or mucronulate, adaxially shiny, abaxially pale brown or straw-colored, lateral veins 4 or 5 pairs; petioles 4–6 mm long. **Inflorescences** capitula, 0.1–1 cm Ø, 6–12-flowered, 3–11 arranged in umbels, terminal on branches; peduncles 0.4–1.1 cm long; flowers sessile. **Flowers** 4- or 5-merous; calyx basally connate, annular above; corolla white, campanulate, ~4 mm long, lobes 4–5, oblong, apex hooked-inflexed, adaxially densely bearded above, tube ~2 × 2 mm; stamens as many as corolla lobes, at base of corolla lobes; ovary 2–4-loculed, locules 1-ovuled, style often absent, stigma conical. **Fruits** compound drupes, red, ± globose, 0.7–1.2 mm Ø. **Fruits** drupes, with 2–4 pyrenes; pyrenes 1-seeded. **Distr:** China, India, Sri Lanka, Taiwan. **Ecol:** In montane forests, dense thickets; 300–1,200 m. Fl. Jun–Jul. Fr. Oct–Nov. **Zones:** 9–12.

Mussaenda

Species ~200. **Shrubs** (erect or scandent), or **climbers**. **Stipules** interpetiolar, ovate to triangular, entire or bifurcate, sometimes connate around stem and circumscissilly detaching. **Leaves** opposite, rarely ternate. **Inflorescences** terminal. **Flowers** 5-merous, bisexual, heterostylous; calyx tube oblong to turbinate, lobes triangular to obovate, usually 1(or more) enlarged into a petaloid appendage (often mistaken as bracts), white (or colored); corolla salverform, cylindrical or infundibular, lobes valvate; stamens 5, sessile or with short filaments, inserted toward the middle or the top of the tube; ovary bilocular, placenta axile, peltate, ovules numerous, style short or long, stigma bilobed, lobes divergent or appressed. **Fruits** berries. **Distr:** Tropical Africa and Asia; Pacific Islands. **Cult:** Propagate from cuttings. Requires a thick layer of mulch to protect the roots and retain moisture, in full or partial sun.

M. erythrophylla Schum. and Thonn., *Beskr. Guin. Pl.*: 116, 1827. (Syn: *M. fulgens* [misapplied], *M. splendida*). [Ashanti Blood, Red Flag Bush, Tropical Dogwood]. **Shrubs** to 3 m tall, or scandent and climbing to 8 m long. **Stipules** bifid. **Leaves** elliptic to broadly ovate, ~7–12 × 5–8 cm, base rounded, subcordate to cuneate, apex acute to acuminate, hairy; petiole to 4 cm long. **Inflorescences** many-flowered, terminal; axes densely red-hairy. **Flowers** 5-merous, heterodistylous; calyx lobes of most flowers small, lanceolate, ~1–1.5 mm long, red-hairy, but 2–6 flowers per inflorescence with a single calyx lobe enlarged, foliaceous and colored bright red, elliptic to round, to ~10 × 8 cm; corolla tube red-hairy outside, to ~3 mm long, lobes white to creamy-yellow, round, to ~1 cm long, a ring of dark red hairs surrounding the throat; anthers and style and stigmas included in corolla tube both in long- and short-styled forms. **Fruits** berrylike, ellipsoid, to ~2 cm long; seeds numerous, minute. **Distr:** Western tropical Africa. **Ecol:** In deciduous forests or jungles. Fl. throughout most of the year. **Zones:** 11–12.

M. frondosa L., *Sp. Pl.* 1: 177, 1753. (Syn: *M. frondosa* var. *zeylanica*). [Flag Bush]. **Shrubs**, scandent; branchlets hirsute. **Stipules** ovately triangular, 3.5–6.5(–9.5) × 2.5–5 mm, bifurcate most its length, densely hairy. **Leaves** variable, ovate to lanceolate, elliptic to orbicular or obovate, 4.5–13 × 2–8.5 cm, above hirsute, below densely tomentose, base cuneate to obtuse, apex acute, lateral nerves 6–10 pairs. **Inflorescences** cymes, few-flowered. **Flowers** with hypanthium obconical to turbinate, 3–4 mm long, outside hairy; calyx lobes narrowly linear, 0.6–1.5 cm × 0.5–1.5 mm, outside pubescent, petaloid sepal creamy-white, ovate(-oblong), 3.5–12 × 2–9 cm; corolla tube 2–2.7 mm long, outside densely hairy, inside densely hairy above, lobes 3–4.5 × 4–5 mm; stamens adnate to corolla tube halfway in long-styled forms, 3/4 in short-styled forms, filaments short, anthers 4.5–5.5 mm long; style and stigma lobes 1.6–2.5 cm

Below: *Keetia gueinzii, Morinda citrifolia, Morinda umbellata.*

PL

Above: *Mussaenda erythrophylla, Mussaenda frondosa, Nauclea orientalis.*

and 2–5.5 mm respectively in long-styled forms, 1–1.5 cm and 2.5–4 mm in short-styled forms. **Fruits** globose, ~1 cm Ø, sparsely hirsute. **Distr:** SE Asia, Indo-Malaysia. **Ecol:** In secondary forests. **Zones:** 10–12.

Nauclea

Species 6. **Trees**, medium to large; terminal vegetative bud strongly flattened. **Stipules** ovate, elliptic or obovate, flattened to strongly keeled, adpressed. **Leaves** opposite on all axes. **Inflorescences** terminal, or terminal and lateral heads; flowering axes unbranched with a node bearing reduced leaves and stipules, these not surrounding the young flowering heads. **Flowers** 4- or 5-merous; hypanthia mutually connate; calyx lobes triangular, obtuse to oblong or (sub)spathulate, persistent; corolla infundibular, lobes oblong, imbricate; stamens inserted in upper part of tube, filaments short, glabrous, anthers basifixed, introrse, protruding from throat; ovary bilocular, ovules numerous per locule, mostly pendulous, some erect or horizontal but these mostly abortive, style exserted, stigma spindle-shaped. **Fruits** indehiscent syncarps of connate fruitlets; seeds ovoid to ellipsoid, sometimes bilaterally compressed. **Distr:** Continental SE Asia, through Malesia to New Guinea, N and NE Australia; continental Africa (3 spp.). **Cult:** Propagate from cuttings, seeds or by air-layering. Requires a moist, well-drained soil, rich in organic matter, in full sun.

N. orientalis (L.) L., *Sp. Pl.* ed. 2, 1: 243, 1762. (Syn: *Bancalus orientalis*, etc.). [Cheesewood, Leichhardt Tree]. **Trees**, to 30 m tall, usually smaller; bark dark brown, thick, fissured and flaky. **Stipules** broadly ovate to orbicular or obovate, 2–4 × 1–2 cm, keeled, glabrous or keel slightly pubescent. **Leaves** broadly ovate to orbicular, (8–)15–30 × (6–)7.5–15 cm, base obtuse to cordate, sometimes decurrent into petiole, apex obtuse, coriaceous, adaxially glabrous, abaxially glabrous to pubescent; petiole to 4 cm long. **Inflorescences** usually solitary, flowering axis 2–5 cm long, mature flowerheads 1.5–2.5(–3) cm Ø. **Flowers** with calyx lobes clavate, 3 mm long, pubescent; corolla tube 6–9 mm long, glabrous, lobes ovate, 4–5 mm long; anthers ellipsoid, 1 mm long; style 1.2–1.4 cm long, stigma spindle-shaped. **Fruits** woody syncarps, 2.5–3 cm Ø; seeds ovoid. **Distr:** Sri Lanka, SE Asia, Malesia to New Guinea, N and NE Australia. **Ecol:** In dry deciduous forests and secondary forests in moist regions. **Zones:** 10–12.

Nertera

Species ~15. **Herbs**, perennial, slender, creeping, branching, slightly fetid, rooting at nodes. **Stipules** interpetiolar. **Leaves** opposite, small. **Inflorescences** with flowers solitary, axillary or terminal, sessile or very shortly stalked. **Flowers** 4–5-merous, perfect; calyx truncate, sometimes very obscurely toothed; corolla tube funnelform, lobes valvate; stamens inserted at base of tube, filaments filiform, exserted, anthers large, usually pendent; ovary biloculed, locules 1-ovuled, styles 2, filiform, hairy. **Fruits** drupes fleshy, rarely dry; pyrenes 2. **Distr:** Australia, New Zealand; Malaya; C and S America. **Cult:** Propagate from seeds. Requires a moist, humus-rich soil in shade or partial sun.

N. depressa Banks et Sol. ex Gaertn., *Fruct.* 1: 124 (t. 26), 1788. (Syn: *N. montana*). [Bead Plant, Fruiting Duckweed]. **Herbs**, forming small to large patches to ~3 m Ø, usually glabrous, ± fetid when bruised; stems to ~30 cm or in some forms 1 m long, usually much-branched. **Stipules** small, usually thick, triangular. **Leaves** ovate to occasionally suborbicular or deltoid-ovate, to 0.8(–1.5) × 0.5(–1) cm, base rounded, truncate or subcordate, abruptly narrowed, margins thickened, recurved, sometimes slightly wavy, apex acute to obtuse, usually minutely apiculate, submembranous to thick; petiole (1–)2–3(–4) mm long, sometimes to 1 cm long in shade. **Inflorescences** with flowers solitary, axillary and terminal, sessile. **Flowers** very small; calyx truncate; corolla subcampanulate, (1–)2–3 mm long, tube very short, lobes 4, about a third the length of the tube; styles slightly exserted. **Fruits** drupes, glabrous, succulent, bright to dark red, globose, ~4 mm Ø. **Distr:** New Zealand. **Ecol:** In coastal to subalpine damp forests, boggy grounds. Fl. Nov–Feb. Fr. Dec–May. **Zones:** 9–11.

Oldenlandia

Species 248. **Herbs**, annual or perennial, rarely subshrubs; stems erect to prostrate. **Stipules** fimbriate to weakly pectinate with 1–several bristles from a short base which is often adnate to leaf base. **Leaves** opposite. **Inflorescences** terminal or axillary, flowers sometimes solitary or fasciculate at nodes. **Flowers** small, mostly 4-merous, hermaphrodite, heterostylous or isostylous; calyx tube short, lobes triangular to ovate; corolla tube short, infundibular to hypocrateriform, throat hairy or glabrous; stamens inserted or exserted; ovary biloculed, ovules numerous, usually on peltate placentas, style included or exserted, stigma mostly bilobed, lobes linear to subglobose. **Fruits** capsules, subglobose to ovoid, usually loculicidally dehiscent at apex. **Distr:** Tropical and subtropical regions worldwide. **Cult:** Propagate from seeds. Requires a moist, well-drained soil in shade or full sun.

O. auricularia (L.) K. Schum., *Nat. Pflanzenfam.* 4(4): 25, 1891. (Syn: *Hedyotis auricularia*, etc.). **Herbs**; rootstock woody, nodes with adventitious roots, stem grooved, ± pubescent. **Stipules** adnate to petiole, pectinate, basal part deltoid, to 5 mm long, with 3–5 pilose bristles. **Leaves** ovate to elliptic or lanceolate, 1–10 × 1–3.5 cm, base rounded to cuneate, apex acute to acuminate, adaxially glabrous to sparsely hairy, abaxially ± pubescent, lateral nerves 3–7 pairs; petiole 0–6 mm long. **Inflorescences** condensed cymes, axillary; flowers shortly pedicelled. **Flowers** heterostylous; hypanthium 0.8–1 mm long, ± pubescent, calyx tube short, lobes narrowly triangular, 0.8–1(–1.5) × 0.2–0.5 mm, margins ciliate; corolla infundibular, tube (0.3–)0.7–1.5(–2) mm long, outside glabrous, inside densely pubescent, lobes ovate-

Below: *Nertera depressa, Oldenlandia auricularia, Pavetta agrostiphylla, Pavetta indica.*

Above: *Pavetta lanceolata, Pentas lanceolata, Posoqueria latifolia, Posoqueria longiflora.*

oblong, 1–1.5 mm long, outside sparsely hairy, inside minutely pubescent; filaments *short* 0.2–0.5 mm long, included, *long* 1–1.5 mm long, exserted, anthers oblong; style *short* ~0.3–0.5 mm long, *long* 2–3 mm long, stigma bilobed. **Fruits** capsules, subglobose, 1.5–2.5 × 1.5–2 mm, indehiscent; seeds 5–10 per placenta, blackish, slightly reticulate. **Distr:** Tropical and subtropical Asia. **Ecol:** In secondary vegetation and open places, weedy in many places; 0–800 m. Fl. throughout the year. **Zones:** 9–12.

Pavetta

Species 357. **Shrubs** or **small trees**; branches terete. **Stipules** usually ± united into a short sheath. **Leaves** opposite. **Inflorescences** corymbiform cymes or panicles, usually open, branched, terminal; flowers numerous, pedicelled. **Flowers** 4-merous; calyx tube ovoid, limb 4-toothed, persistent; corolla tube long, slender, lobes 4, spreading; stamens inserted on mouth of corolla; ovary biloculed, locules 1-ovuled, style slender, long-exserted, usually about twice as long as corolla tube, stigma slender, entire, rarely notched. **Fruits** globose or ovoid, small, each with two 1-seeded pyrenes. **Distr:** Tropical Old World to S Pacific. **Cult:** Propagate from seeds, cuttings or by air-layering. Requires a humus-rich, well-drained soil in full sun.

P. agrostiphylla Bremek., *Repert. Spec. Nov. Regni Veg.* 37: 89, 1934. (Syn: *P. angustifolia* [misapplied], *P. hispidula* var. *angustifolia*). **Shrubs** or **treelets**, to 1 m high; branchlets glabrous. **Stipules** triangular to oblong, 0.5–1 cm long, acute, outside glabrous, inside sparsely pubescent. **Leaves** linear, (5–)7–14 × (0.4–)1–2 cm, base cuneate to attenuate, apex acute to obtuse, surfaces glabrous, lateral nerves 7–10 pairs, bacterial nodules present; petiole to 2 cm long. **Inflorescences** cymose, terminal, or terminal and axillary, rarely axillary, ~10–30-flowered; peduncles to 3 cm long, glabrous or minutely puberulous; pedicels to 5 mm long.. **Flowers** 4-merous, white; calyx lobes triangular to oblong, 0.3–1 mm long, outside glabrous or minutely pubescent; corolla tube 1.5–2.5 cm long, outside glabrous, inside pilose in the upper third, lobes oblong, 1–1.2 × 0.2–0.4 cm, acute; filaments 1 mm, anthers 4–5 mm long; style 4.5–7 cm long. **Fruits** globose, 5–8 mm Ø, black. **Distr:** Sri Lanka (Adam's Peak). **Ecol:** In wet evergreen forests. Fl. May–Jun. **Zones:** 11–12.

P. indica L., *Sp. Pl.*: 110, 1753. (Syn: *Ixora indica*). [Bride's Bush, Indian Pavetta, Indian Pellet Shrub]. **Shrubs** or **small trees**, 2–4 m high. **Stipules** cupular to broadly triangular, 3–6 mm long, long-cuspidate, cusp 2–3 mm long, glabrous. **Leaves** elliptic, (5–)7–12(–16) × (3–)4–6(–9) cm, base acute, sometimes slightly decurrent, apex obtuse or subacute, usually glabrous, lateral nerves 8–12 pairs, bacterial nodules present; petiole to 2 cm long. **Inflorescences** cymose, flowers (20–)40+, terminal, or terminal and axillary, rarely axillary; peduncle to 3 cm long, usually glabrous; pedicels to 8 mm long. **Flowers** 4-merous, white; hypanthium 0.3–0.8 mm long, glabrous; calyx slender, ~1 mm Ø, lobes dentate to triangular, to 0.5 mm long, glabrous; corolla tube 0.8–1.2 cm long, outside glabrous, inside pilose above, lobes oblong, 4–6 × 2 mm, acute to mucronate; filaments 1 mm long, anthers 3–4 mm long; style 2–3 cm long. **Fruits** 6–8 mm Ø. **Distr:** Indian Subcontinent to S China and peninsular Malaysia. **Ecol:** In lowland dry scrubs, secondary vegetation. **Zones:** 10–12.

P. lanceolata Eckl., *S. African Quart. J.* 1830: 374, 1830. (Syn: *P. tristis*, etc.). [Forest Bride's Bush, Weeping Bride's Bush]. **Shrubs** or **small trees**, evergreen, to 4 m high; main stem single, much-branched; bark gray, finely vertically striated. **Leaves** opposite; lamina narrowly elliptic, 5–7.5 × 0.9–1.5 cm, base and apex narrowly tapering, margins entire, flat to wavy, adaxially dull to glossy deep green, glabrous, small hair-fringed pits present in vein axils on abaxially, nodules round to elongate, scattered over the whole leaf; petiole 2–6 mm long. **Inflorescences** clusters, dense, large, terminal. **Flowers** strongly fragrant, white; calyx lobes 2 mm long, acute; corolla tube 5–8.5 mm long, lobes ~4 mm long; stamens exserted, placed between corolla lobes, greenish, turning yellow, finally black; style long-exserted, greenish, turning black with age. **Fruits** berries, black. **Distr:** Mozambique to S Africa. **Ecol:** On forest margins and rocky hillsides, also in lowland riverine bush. Fl. Nov–Jan. **Zones:** 10–12.

Pentas

Species 16. **Herbs** or **shrubs**, mostly perennial (rarely biennial); stems erect or straggling. **Stipules** divided into 2 to many filiform colleter-tipped segments. **Leaves** paired or 3–5-whorled. **Inflorescences** cymose, much-branched, terminal, complicated, individual branches often becoming spicate in fruit. **Flowers** small to very large, hermaphrodite, mono-, di- or trimorphic; calyx tube ovoid or globose, sometimes with a free annular part above, lobes usually 5, equal or unequal, 1–3 larger than others or sometimes foliaceous; corolla tube shortly cylindrical to narrowly tubular, short to long, throat hairy, lobes ovate or oblong; *monomorphic* flowers with stamens enclosed in an abrupt apical dilation of tube, style exserted; *dimorphic* flowers with tube gradually dilated at apex in short-styled flowers and abruptly dilated in long-styled flowers; *trimorphic* rarely, with both stamens and style included in tube; ovary bilocular, ovules numerous per locule. **Fruits** capsules, obtriangular or ovoid, ribbed, beaked, opening at apex, beak splitting into 4 valves, sometimes into 2 cocci; seeds minute, brownish, irregularly globose or tetrahedral. **Distr:** Tropical and S Africa, Madagascar, Arabian Peninsula. **Cult:** Propagate from seeds or cuttings. Requires a moist, well-drained, fertile soil in full sun.

P. lanceolata (Forssk.) Deflers, *Voy. Yemen*: 142, 1889. (Syn: *Manettia lanceolata*, etc.). [Egyptian Star Cluster]. **Herbs**, perennial, slightly woody at base, to ~50 cm high. **Leaves** ovate to lanceolate, 4–6 × 2.5–3 cm, base narrowed into petiole, apex acute to acuminate, thin, pubescent to puberulous; petiole to ~1 cm long. **Inflorescences** umbellate, many-flowered, terminal. **Flowers** 5-merous, mostly heterodistylous; corolla, white, shades of pink, reddish-purplish or lilac or mauve, sometimes 2-toned (dark pink with white stripes, etc.), tube narrowly cylindrical, to ~2 cm long, lobes ovate to elliptic, ~0.5–1 cm long; *short-styled forms* with filaments and anthers exserted, styles and stigmas included; *long-styled forms* with upper part of style and 2 filiform stigmas exserted. **Fruits** capsular, dehiscing into 2 valves, to ~5 mm long; seeds numerous, minute. **Distr:** Arabian Peninsula, Comoros, Ethiopia to Mozambique. **Ecol:** In woodlands and grasslands; 0–700 m. Fl. throughout most of the year, mostly Mar–May. **Zones:** 10–12. **Info:** A variable species across its range.

Posoqueria

Species ~15. **Trees** or **shrubs**. **Stipules** free, subulate to broad, or foliaceous, glandular inside. **Leaves** opposite, petiolate; lamina usually large, often coriaceous. **Inflorescences** cymose-corymbose to umbellate, terminal; bracts minute. **Flowers** large and showy, zygomorphic, 5-merous, buds curved; calyx 5-lobed, segments ± imbricate, base auriculate, glandular on back; corolla funnelform or salverform (infundibuliform or hypocrateriform), lobes unequal, ascending, spreading or reflexed, imbricate, tube much-elongate and slender, glabrous inside, glandular-papillose outside; stamens unequal, inserted in mouth of corolla, anthers linear, slender, apiculate, base sagittate; ovary biloculed or incompletely 1-loculed, ovules numerous, style elongate. **Fruits** large, baccate, biloculed; seeds numerous, large, irregular, embedded in gelatinous pulp. **Distr:** Mexico, C and S America, West Indies. **Cult:** Propagate from half-hardened cuttings or by air-layering. Requires a humus-rich, fertile, well-drained soil in shade to full sun.

P. latifolia (Rudge) R. and S., *Syst. Veg.* 5: 227, 1819. (Syn: *P. decora*, etc.). [Brazilian Oak, Monkey Apple]. **Trees** or **shrubs**, 5–15 m tall, glabrous; branchlets terete, nodes 3–15 cm apart. **Stipules** triangular, 0.5–1.3 cm long, sometimes foliaceous and much larger, apex obtuse. **Leaves** ovate-oblong to broadly ovate, rarely rotund, 8–32 × 3.5–22 cm, base usually obtuse, cuneate, acute to subcordate, margins usually revolute, apex usually shortly acuminate, coriaceous, adaxially shiny, abaxially glabrous to finely puberulent, lateral veins 6–8; petiole 0.3–2.5 cm long. **Inflorescences** corymbiform or cymose, few- or many-flowered, dense, terminal, glabrous, drying to black; peduncle 0.5–2.2 cm long; pedicels ± sessile. **Flowers** fragrant; hypanthium oblong or turbinate, ~8 mm long; calyx lobes ovate-triangular, to 0.5 mm long, acute or subobtuse; corolla white, aging to creamish-yellow, usually 4–17 cm long, 2–4 mm ∅, lobes oblanceolate, 1.2–2 cm long, obtuse; stamens 4, filaments attached at mouth, shorter than corolla lobes, anthers exserted, ~4 mm long; style filamentous, as long as corolla tube, stigmas ~2 mm long. **Fruits** globose, ~3–7 cm ∅, yellow, glabrous; seeds numerous, white to greenish, pulp greenish. **Distr:** S Mexico, south to Panama, Colombia, the Guianas, Venezuela, Amazonian Brazil, Peru. **Ecol:** In or at edge of forests. **Zones:** 10–12.

P. longiflora Aubl., *Pl. Guian.* 134. pi. 51. 1775. (Syn: *Kyrtanthus longiflorus, Solena longiflora*). [Needle Flower Tree]. **Shrubs** or **small trees**, glabrous, 1.5–2 m tall; trunk 7–10 cm dbh. **Stipules** 7–9 mm long, glabrous, glandulose, base subulate, apex acute. **Leaves** oblong or narrowly oblong, (10–)15–19(–22) × (3.5–)4–8(–10) cm, base usually acute, apex long-acuminate, veins 8–9; petiole 1.5–2 cm long, trigonous, adaxially channeled. **Inflorescences** umbel-like, 4–6-flowered; peduncle 1–2 cm long; bracts lanceolate to suborbicular; flowers pendent. **Flowers** ± pendent; calyx lobes triangular-ovate, 2 mm long, acute; corolla pure white, 24–27(–32) cm long, lobes linear-lanceolate, 3–4(–5) cm long, acute or subobtuse, reflexed; stamens 0.8–2.2 cm long, filament subulate, anther 0.7–0.8(–1.1) cm long; ovary ~4 mm long, style 20–22 cm long. **Fruits** yellow, subglobose, 6–8 cm ∅; seeds round, 1.2 × 0.8 cm, testa thin, reddish. **Distr:** Brazil, Colombia, the Guianas, Peru. **Ecol:** Fl. Nov. Fr. Jan. **Zones:** 10–12.

Pseudomussaenda

Species 6. **Shrubs**. **Stipules** with apex lobed. **Leaves** decussate. **Inflorescences** cymes, few- to several-flowered. **Flowers** 5-merous, tubular; calyx lobes filiform, some flowers in each inflorescence with enlarged foliaceous lobes; stamens inserted in upper half of tube; ovary biloculed, locules with numerous ovules, style with stigma bilobed. **Fruits** capsular. **Distr:** Tropical Africa. **Cult:** Propagate from cuttings. Requires a thick layer of mulch to protect the roots and retain moisture, in full or partial sun. **Info:** This genus is closely related to *Mussaenda*, differing mainly in the fruits. The fruits of *Mussaenda* are berries while those of *Pseudomussaenda* are dehiscent capsules.

P. flava Verdc., *Kew Bull.* 6: 378, 1951. (Syn: *Mussaenda flava*, etc.). [Dwarf Mussaenda, Yellow Mussaenda]. **Shrubs**, to ~3 m high. **Stipules** with 2 apical filiform fimbriae. **Leaves** opposite, decussate, shortly petiolate; lamina elliptic to oblong-elliptic, ~4–8 × 2–4 cm, base narrowed, apex acute to acuminate, softly pubescent. **Inflorescences** cymes, several-flowered, terminal. **Flowers** 5-merous, heterodistylous; calyx lobes of most flowers small, filiform, ~2–5 mm long, but 1–2 flowers per inflorescence with a single, enlarged foliaceous white to creamy-yellow calyx lobe, enlarged lobe oblong, ovate or elliptic, ~2–5 × 1.5–4 cm; corolla yellow or greenish, infundibular, tube cylindrical, 2–3 cm long, swollen where the subsessile anthers are located, throat often with a ring of lemon- to orange-yellow hairs, lobes ovate, ~5–8 mm long; stamens inserted in upper half of tube, below the throat in short-styled forms; ovary biloculed, locules with numerous ovules, style with stigma bilobed, included in both long- and short-styled forms, but relatively longer in long-styled forms. **Fruits** capsular, dry, oblong, to ~5 mm long, loculicidally splitting into 2 halves; seeds numerous, small. **Distr:** Nigeria to Ethiopia and Kenya. **Ecol:** In woodlands and open forests. Fl. throughout the year. **Zones:** 10–12.

Psilanthus

Species 18. **Shrubs** or **small trees**. **Stipules** triangular to ovate or truncate. **Leaves** opposite. **Inflorescences** terminal or axillary, brachyblasts bearing 1 or 2 pairs of bracts and one pair of bracteoles. **Flowers** bisexual, (4–)5-merous; calyx tube ovate to campanulate, subequaling disk, 5-lobed or irregularly toothed; corolla tube cylindrical, widening at apex, lobes contorted; anthers attached above middle of corolla tube, sessile, linear, partly exserted or included; ovary bilocular, placentas small, attached to middle or just above middle of septum, ovule 1, style short, stigma included, bifid. **Fruits** drupes, bilobed. **Distr:** Old World tropics. **Cult:** Propagate from seeds, cuttings or by air-layering. Requires a moist, well-drained soil rich in organic matter, in full or partial sun.

P. bengalensis (Roxb. ex Schult.) J.-F.Leroy, *Bull. Mus. Natl. Hist. Nat., B, Adansonia* 3: 252, 1981. (Syn: *Coffea bengalensis*, etc.). [Bengal Coffee, Paracoffea]. **Shrubs**, to 2–3 m high, sometimes (briefly) deciduous. **Stipules** triangular-aristate, 5–6 mm long. **Leaves** broadly ovate or elliptic, 3–8(–10) × 4–7 cm, base rounded or slightly cuneate, apex acuminate, subcoriaceous, lateral veins 3–5 pairs; petiole 0.3–0.5(–1) cm long, pubescent. **Inflorescences** 1–5-flowered, terminal or axillary short shoots; bracts 2 pairs, lanceolate-acuminate, 1–1.4 × 0.3–0.4 cm, pubescent outside; bracteoles paired, sometimes developing into 'normal' leaves. **Flowers** fragrant, 5-merous; calyx with a short basal tubular part, <1 mm long, lobes 5–7, minute, teethlike; corolla white, tube 1.7–2.2 cm long, lobes ovate or elliptic, 1–1.5 × 0.5–0.6 cm, acute or obtuse; stamens sessile, inserted, anthers 1–1.2 mm long; ovary ~3–4 mm high, glabrous, style to 3.5 mm long, glabrous, stigma lobes 1.5 mm long. **Fruits** drupes, obovoid or ellipsoid, 1–1.3 × 0.5–8 cm, crowned by persistent calyx, shiny, dark bluish-black; seeds oblong or ellipsoid, ventrally longitudinally grooved. **Distr:** Bangladesh, Bhutan, India, Myanmar, Nepal, Thailand. **Ecol:** In evergreen forests, often along streams; 300–500 m. **Zones:** 10–12.

Psychotria

Species 1,902. **Shrubs**, **small trees** or **climbers**. **Stipules** intrapetiolar, usually connate, apex entire or bilobed, caducous or persistent. **Leaves** opposite, rarely 3–4-whorled. **Inflorescences** cymes or headlike, terminal, rarely axillary. **Flowers** greenish-white, bisexual or unisexual; corolla-tube short, cuplike, (4–)5(–6)-lobed, valvate in bud, mouth of throat pilose; stamens as many as the lobes, exserted or included, filaments glabrous or pilose, anthers 2-celled, longitudinally dehiscent; ovary biloculed, locules 1-ovuled, styles cylindrical, stigma bilobed, papillose; *bisexual* flowers dimorphic, heterostylous; *unisexual* flowers dioecious; ♂ flowers with rudimentary pistil, included, stamens long exserted; ♀ flowers with rudimentary empty stamens in mouth of corolla, styles exserted. **Fruits** berries or drupes, red, black or white; seeds 2-ridged. **Distr:** Tropical and subtropical regions worldwide. **Cult:** Propagate from seeds or cuttings. Requires a moist, well-drained soil, rich in organic matter, in shade or partial sun.

P. gardenioides (Scheidw.) Standl., *J. Wash. Acad. Sci.* 17: 342, 1927. (Syn: *Palicourea gardenioides*, etc.). **Shrubs**, erect, glabrous, to ~2 m high; branches opposite, cylindric, articulate. **Stipules** basal, acute. **Leaves** petiolate, opposite; lamina ovate-oblong, to 6 × 3.2 cm, margins entire, apex acuminate, glabrous, bright green, nerves alternate. **Inflorescences** cymes, terminal, bracteate. **Flowers** with calyx creamish, reddish-tinged, tube cylindrical, ~4 mm long, tribracteolate, connate to ovary, lobes recurved; corolla funnelform, tube long-cylindric, 2–3 cm long, 2–3 mm ∅, glabrous, limb 5-lobed, 1.75 cm ∅, lobes lanceolate, 9 × 4 mm, reflexed, mucronate, tips usually pinkish; stamens 5, inserted in tube, filaments short, anthers linear, medifixed, base sagittate; disk epigynous, carnose; ovary inferior, biloculed, locules 1-ovuled, style simple, stigma linear. **Distr:** SE Mexico (Veracruz to Orizaba). **Ecol:** In rainforests and moist forest margins. Fl. Nov–Feb. **Zones:** 10–12.

P. psychotrioides (DC.) Roberty, *Bull. Inst. Franç. Afrique Noire* 16: 62, 1954. (Syn: *Cephaëlis congensis*, etc.). **Shrubs** or **small trees**, 3–4(–6) m tall, branching from base, glabrous. **Stipules** triangular, 6–8 mm long, obtuse, caducous. **Leaves** arranged toward end of stems, opposite; lamina very variable in shape, narrowly or broadly elliptic, obovate to suborbicular, 10–20 × 4–14 cm, base cuneate, apex obtuse, thick, glossy, dark green, glabrous, nerves pinnate; petiole 1.5–2(–5) cm long, glabrous. **Inflorescences** dense capitula, 1–1.5 cm ∅, axillary; flowers sessile. **Flowers** white, fragrant; calyx short, tubular; corolla 5-lobed, tubular, 2–4 mm ∅; style bottlebrush-shaped, exserted. **Fruits** berries, ovoid, 1 × 0.8 cm, glabrous, ± glossy, red, ± bumpy when ripe, topped with remnants of persistent calyx. **Distr:** W Tropical Africa to Sudan. **Ecol:** In closed and fringing forests, savanna woodlands, beside streams. **Zones:** 10–12.

Below: *Pseudomussaenda flava, Psilanthus bengalensis, Psychotria gardenioides, Psychotria psychotrioides.*

PL

Above: *Rondeletia odorata, Rothmannia globosa, Rothmannia longiflora, Rothmannia urcelliformis.*

Rondeletia

Species 267. **Shrubs** or **trees**, usually pubescent; branchlets terete or angulate. **Stipules** interpetiolar, usually broad and sometimes foliaceous, obtuse to cuspidate, usually persistent. **Leaves** opposite or verticillate, sessile or petiolate; lamina membranous or coriaceous. **Inflorescences** terminal or axillary, usually cymose or paniculate. **Flowers** with hypanthium commonly subglobose; calyx 4–5-lobate, lobes narrow or wide, often unequal, persistent; corolla funnelform or salverform, white, yellowish, pink, or red, tube usually slender and elongate, throat often ampliate, often annular-thickened, sometimes barbate, lobes 4–5, spreading, broad, obtuse, imbricate in bud, 1 or 2 of them exterior; stamens 4–5, inserted in corolla throat, included or exserted, filaments short, anthers dorsifixed, narrowly oblong or broader, obtuse, erect; ovary biloculed, style filiform, obtuse or shortly bifid, ovules numerous. **Fruits** capsules, usually small, globose, chartaceous or coriaceous, biloculed, loculicidally or septicidally bivalvate, valves commonly bipartite; seeds minute, compressed or angulate, often winged or appendaged, testa thin. **Distr:** Mexico to tropical America. **Cult:** Propagate from half-hardened cuttings or by air-layering. Requires a fertile, moist, well-drained soil in shade to full sun.

R. odorata Jacq., *Enum. Pl. Carib.*: 16, 1760. [Panama Rose]. **Shrubs**, ~1–2 m high. **Stipules** triangular or deltoid, 5–6 mm long, persistent. **Leaves** decussate, opposite; lamina bullate, ovate to elliptic, 3–10 × 1.5–5 cm, base obtuse to subcordate, margins entire, apex rounded or acute, thickly leathery; petiole subobsolete, to 2 mm long. **Inflorescences** cymes, united into a semiglobose 10–30-flowered thyrses, terminal; peduncle 3–7 mm long; bracteoles 4–5 mm long; pedicels 4–5 mm long. **Flowers** 5-merous, bisexual, 2–3 × 1–2 cm; calyx lobes 6, narrowly oblong, 4–5 mm long, hairy; corolla bright red or orange-red, center with distinct yellowish-orange 'eye', tube cylindrical, lobes 5, ovate, ~4 mm long, obtuse, imbricate in bud and spreading; stamens inserted in upper third of tube, anthers subsessile, included; ovary biloculed, locules with numerous ovules on a large placenta attached to septum, style with stigma bilobed, barely exserted from throat. **Fruits** capsular, small, globose, splitting into 2 valves; seeds numerous, minute. **Distr:** Northern C America (S Mexico) southward to NW South America; the Caribbean. **Ecol:** In dry forests, seasonally deciduous woodlands, dry scrubs and mesic forests, at low altitudes. Fl. throughout most of the year. **Zones:** 11–12.

Rothmannia

Species ~30. **Shrubs** or **small trees**, 8–12 m tall, monopodial with lateral branches. **Stipules** deltoid, ± acuminate, deciduous. **Leaves** opposite or sometimes borne at base of flower or a vegetative node; lamina elliptic, base cuneate sometimes cordate, margins entire, apex acuminate or rarely acute, lateral veins arcuate; domatia sometimes present and tufts of hair at axils of secondary veins; petiole to 2 cm long. **Inflorescences** with flowers solitary, rarely 2–3-flowered, terminal or axillary. **Flowers** large, often fragrant; calyx tube gradually funnelform or cylindric; corolla funnelform or campanulate, lobes 5; ovary 1-locular, with 2 parietal placentas. **Fruits** spherical or ellipsoid; seeds numerous, fused into a pulpy solid mass. **Distr:** S Africa, Seychelles, Malay Peninsula. **Cult:** Propagate from seeds, half-hardened cuttings or by air-layering. Requires a fertile, humus-rich, moist, well-drained soil in shade to full sun.

R. globosa (Hochst.) Keay, *Bulletin du Jardin botanique de l'État a Bruxelles*, 28: 56, 1958. (Syn: *Gardenia globosa*). [Bell Gardenia, September Bells, Star of Africa]. **Shrubs** or **small trees**, 4–7(–12) m tall, evergreen, sometimes briefly deciduous, much-branched, with opposite, spineless branches. **Stipules** small, subulate, base broad. **Leaves** opposite; lamina broadly lanceolate, 7.5–10 × 2.5–3 cm, base and apex acute or subobtuse, membranous, coriaceous, glossy, penninerved. **Inflorescences** cymes, 1–3(–4)-flowered, terminal and axillary; pedicels very short, articulated with calyx. **Flowers** fragrant; calyx campanulate, base bibracteolate, sharply 2–5-dentate, pubescent; corolla white to creamy-white, villous, campanulate, to 4 × 3.5 cm, base sharply contracted, lobes 4–5, broadly ovate, ~1.5 cm long, throat sometimes pink-speckled; stamens 5, linear, inserted; disk fleshy; style inserted, club-shaped, gradually passing into stigma, stigma shortly bifid, apex reflexed; ovary oblong. **Fruits** mostly globose, 2.5 cm Ø, green, turning brown. **Distr:** S Africa (Swaziland, Transvaal and Natal). **Ecol:** Near sea coast, along riverbanks, and in evergreen forests; <1,200 m. Fl. Aug–Dec. Fr. Jan. **Zones:** 9–12. **Info:** The fruits are eaten by birds, monkeys and baboons.

R. longiflora Salisb., *Parad. Lond.*, 1(2): t. 65, 1807. (Syn: *Randia maculata*). [Spotted Randia]. **Shrubs**, 1–9 m tall; often with climbing stems. **Leaves** elliptic, 6–18 × 2–8 cm, base cuneate, apex acuminate, pinnately veined, lateral veins 4–6 pairs, with axillary domatia, surfaces hairless; petiole 0.3–1.5 cm long. **Inflorescences** with flowers solitary, terminal on axillary branches; pedicels to 1 cm long; bracteoles scalelike. **Flowers** bisexual, regular, 5-merous, sweetly scented; calyx tubular, 1–2 cm long, lobes triangular, 1–4 mm long; corolla trumpet-shaped, to 24 × 4 cm, lobes ovate 1–4 × 1–1.5 cm, purple-red to green outside, whitish, purplish-spotted inside; stamens inserted in corolla tube, included or with tips of anthers exserted; ovary inferior, biloculed, style 12–21 cm long, stigma bilobed, to 3 cm long, partly exserted. **Fruits** berries, globose to ellipsoid, 3.5–7 × 5–6 cm, greenish-black, 10-ribbed, with persistent calyx, glabrous; seeds numerous, lens-shaped, 6–8 × 5–6 mm, brownish-red. **Distr:** Gambia east to Sudan and Kenya, and south to Tanzania and Angola. **Ecol:** In tropical primary and secondary forests, sometimes on edges of scrubby creeks; 200–700 m. Fl. throughout most of the year. **Zones:** 10–12.

R. urcelliformis (Hiern) Bullock ex Robyns, *Fl. Spermatophyt. Parc Nat. Albert* 2: 340, 1947. (Syn: *Gardenia urcelliformis*, etc.). **Shrubs** or **small trees**, (1–)4–8.5(–15) m tall. **Stipules** triangular, 0.2–1.2(–1.5) cm long, long-acuminate, caducous. **Leaves** elliptic to obovate-elliptic, 5.7–18 × 1.2–10 cm, base cuneate, apex distinctly acuminate, adaxially usually glabrous, abaxially usually glabrescent, nerves densely hairy on the nerves; domatia present; petiole 0.4–1.1 cm long, glabrescent to pubescent. **Inflorescences** with flowers solitary; peduncle very abbreviated; pedicel 3–7 mm long; bracteoles filiform. **Flowers** with calyx sparsely pubescent to pubescent, tube 0.3–0.6(–1.1) cm long, 5-angled, lobes linear, 0.3–2(–2.7) cm long; corolla white to cream, with dark brownish-red blotches, pubescent outside, narrowly cylindrical below, funnel-shaped above, tube 3–7.7 cm long, lobes lanceolate, 1.2–4.5 × 0.8–1.8 cm, acuminate, shortly hairy; anthers 0.6–2.3 cm long, exserted, pollen presenter 1.2–2.6 mm long, exserted; style 5–7.5 cm long, widening apically. **Fruits** spherical to ellipsoid, 2.5–7.6(–10.2) cm long; seeds to 8 × 6 mm, blackish. **Distr:** Tropical Africa. **Ecol:** In evergreen forests, gallery woodlands; 200–1,300 m. **Zones:** 10–12.

Sarcocephalus

Species 2. **Trees** or **scandent shrubs**; branchlets subterete or obtusely quadrangular. **Stipules** interpetiolar, caducous. **Leaves** opposite, very rarely 3-whorled; lamina subcoriaceous. **Inflorescences** with flowers crowded on globose common receptacles, forming compact pedunculate globose heads, without intervening bracteoles, terminal and axillary. **Flowers** whitish, pale pink or yellowish; calyx tubes cohering, teeth 5–6, hairy, tips jagged or in some species with as many alternating appendages; corolla narrowly funnelform, fleshy, glabrous, 5-lobed, imbricate in aestivation, caducous; anthers 5, subsessile, inserted at mouth of corolla, ovate-oblong; disk inconspicuous; ovary biloculed, ovules numerous, anatropous, style filiform, exserted, caducous, stigma oblong or fusiform, thicker than style, glabrous, emarginate or bifid. **Fruits** syncarpous, fleshy; seeds small, not winged. **Distr:** Tropical Africa. **Cult:** Propagate from seeds or cuttings. Requires a moist, well-drained soil in full or partial sun.

S. latifolius (Sm.) E.A. Bruce, *Kew Bull.* 2: 31, 1947. (Syn: *Cephalina esculenta, Nauclea latifolia*, etc.). [African Peach]. **Trees** or **shrubs**, ± scandent,

Above: *Sarcocephalus latifolius, Tarenna wallichii, Vangueria madagascariensis, Warszewiczia coccinea.*

4–5(–9) m tall; trunk twisted, to 30 cm dbh; branches entangled, initially erect, then drooping. **Stipules** paired between leaves, 3–7 mm long, thin, forming a triangular collar, apex pointed or rounded. **Leaves** opposite; lamina broadly elliptic or suborbicular, 10–22 × 7–15 cm, base broadly cuneate, rounded or subcordate, apex shortly acuminate, mostly glabrous, adaxially bright green, greasy, lateral nerves pinnate, 6–8(–9) pairs; petiole 1–2 cm long. **Inflorescences** heads, dense, spherical, 3.5–4.5 cm Ø, arranged at end of branches, flowers numerous. **Flowers** fragrant, white or ± pinkish, small; corolla tubular, lobes 4, ± ciliate; style club-shaped, exserted to 5–8 mm. **Fruits** berries, fleshy, irregularly globose, 3–5(–8) cm Ø, red to deep chestnut, skin covered with shallow cavities; seeds numerous, embedded in pinkish flesh, with strawberrylike scent. **Distr:** W tropical Africa to Ethiopia and NW Angola. **Ecol:** In fringing forests, near waterways, savannas. **Zones:** 10–12.

Tarenna

Species ~180. **Shrubs** or **small trees**; branchlets glabrous or pubescent. **Stipules** interpetiolar, ovate to broadly triangular, usually slightly basally connate, apex usually cuspidate. **Leaves** opposite, often with domatia in axils of secondary veins abaxially. **Inflorescences** corymbose cymes or panicles, terminal, short and erect (or pendulous). **Flowers** 5-merous, bisexual; calyx 5-lobed; corolla hypocrateriform to infundibular, lobes 5, lanceolate, contorted in bud; filaments subsessile, anthers linear, longitudinally dehiscent, dorsifixed, inserted in throat of corolla; disk conspicuous; ovary bilocular, ovules 2 to several, placentation axile, styles cylindrical, stigma capitate, linear, rarely bifid, lobes coherent for all or most its length. **Fruits** berries, globose to ovoid; seeds 2 to several, angular. **Distr:** Old World tropics. **Cult:** Propagate from cuttings or by air-layering. Requires a fertile, moist, well-drained soil in full or partial sun.

T. wallichii (Hook.f.) Ridl., *Fl. Malay Penins.* 2: 106, 1923. (Syn: *Ixora wallichii*, etc.). **Shrubs**, glabrous; branches slender. **Stipules** triangular, cuspidate. **Leaves** variable, elliptic-lanceolate, 12–18 × 3.5–5 cm, base acute, apex caudate-acuminate, membranous, adaxially usually glossy, abaxially paler; lamina often folded to ~90º along midrib; petiole slender, 1.3–2.5 cm long. **Inflorescences** cymes, 5–8 cm across, not brachiate, branches with sometimes a line of pubescence on one side, lower suberect; bracts linear-lanceolate, persistent; bracteoles triangular, acute. **Flowers** opening creamish-white, aging to golden-yellow; corolla tube 2 cm long, lobes 5, linear, margins sometimes ciliate, apex truncate, spreading; style villous, long-exserted, stigma narrowly fusiform. **Fruits** berries, ~5 mm Ø; seeds plano-convex. **Distr:** India, Cambodia, peninsular Malaysia, Singapore, Borneo. **Ecol:** In lowland evergreen and secondary forests; <450 m. Fl. & Fr. Sep–Jun. **Zones:** 10–12.

Vangueria

Species 59. **Shrubs** or **small trees**, deciduous, unarmed or rarely spiny. **Stipules** basally connate into a short sheath, apex produced into a linear or subulate appendage, hairy inside, deciduous or ± persistent. **Leaves** paired, never whorled; lamina often quite large, glabrous to densely velvety. **Inflorescences** cymes, usually many-flowered, divaricately branched, axillary, opposite nodes, often at leafless nodes, sometimes flowering before leaves are fully developed; branches ± elongate; flowers regularly spaced, shortly petiolate. **Flowers** with calyx tube hemispherical or depressed-campanulate, limb divided into 5(6) triangular, oblong, linear or ligulate lobes; corolla small, usually greenish-yellow, tube shortly cylindrical or campanulate, ± glabrous outside, densely hairy at throat, lobes 5, narrowly triangular, mostly reflexed, anthers apiculate, shortly exserted; disk annular; ovary 5-locular, locules 1-ovuled, ovules pendulous, style filiform-cylindrical, glabrous, shortly exserted. **Fruits** large, indehiscent, fleshy, ± globose, (4)5(6)-locular, containing (4)5(6) pyrenes, sometimes bearing remnants of calyx lobes, glabrous, often edible. **Distr:** Tropical and S Africa, Madagascar. **Cult:** Propagate from seeds, cuttings or by air-layering. Requires a fertile, moist, well-drained soil rich in organic matter, in full or partial sun.

V. madagascariensis J.F. Gmel., *Syst. Nat.*: 367, 1791. (Syn: *V. edulis, Vavanga chinensis, V. edulis*). [Spanish Tamarind, Wild Medlar]. **Shrubs** or **small trees**, 1.5–15 m tall, often multistemmed. **Stipules** with base broad, 3–5 mm long, apex narrowed, 0.4–1.8 cm long. **Leaves** opposite, petiolate; lamina elliptic to rotund, 7–28 × 2–15 cm, base usually cuneate to rounded, apex acute to shortly acuminate, membranous, usually glabrous, adaxially dark green, abaxially paler; petiole 0.8–1.8 cm long. **Inflorescences** dichasial cymes, axillary, paired, branches 1–4.5 cm long, 7–10-flowered; peduncle ~1 cm long; pedicels ~2–4 mm long. **Flowers** 5-merous, bisexual, small, fragrant; calyx tube 1.2–3 mm long, lobes triangular-oblong to narrowly oblong, 0.5–1.5 mm long, spreading or recurved; corolla greenish-yellow, yellowish or creamy-white, tube wide, broadly funnelform, 3–4.5 mm long, inside hairy, lobes mostly shorter, acute; stamens inserted at throat, anthers partly exserted; ovary 5-locular, locules 1-ovuled, style with a ± cylindrical stigmatic knob. **Fruits** drupaceous, globose, 3–5 cm Ø, with 4–5 woody pyrenes. **Distr:** Tropical and S Africa, Madagascar. **Ecol:** In grasslands, sand banks at edge of lakes and beaches; 90–470 m. Fl. Jul–Sep. **Zones:** 10–12. **Info:** The fruits are edible and sour.

Warszewiczia

Species 8. **Trees** or **shrubs**, ± pubescent. **Stipules** large, base glandular inside, persistent. **Leaves** opposite, petiolate; lamina large, membranous or coriaceous. **Inflorescences** cymose, terminal; cymes arranged in racemiform panicles. **Flowers** small, 5-merous; calyx lobes persistent, one of them dilated into a large, foliaceous, petiolate, brightly colored limb; corolla shortly funnelform, throat villous, lobes obtuse, imbricate in bud; stamens and anthers exserted; ovary biloculed, ovules numerous, stigmas 2, short. **Fruits** capsules, small, oblong or globose, septicidally bivalvate; seeds numerous, minute, horizontal, marginate, reticulate. **Distr:** Mexico, C America, and tropical S America. **Cult:** Propagate from seeds or softwood cuttings. Requires a moist, well-drained soil in full or partial sun.

W. coccinea (Vahl) Klotzsch, *Ber. Bek. Verh. Königl. Berlin* 1853: 497, 1853. (Syn: *Calycophyllum coccineum*, etc.). [Sanguenaria, Wild Poinsettia]. **Shrubs** or **trees**, to 12 m tall; twigs stout, angular, ± pubescent. **Stipules** narrowly triangular-oblong, 1.5–2 cm long, acuminate or attenuate. **Leaves** oblanceolate-oblong to oval-obovate, usually (5–)20–50(–60) × <20 cm, apex acute or short-acuminate, adaxially sparsely pubescent or almost glabrous, abaxially pubescent; petiole ~2.5 cm long. **Inflorescences** 30–60+ cm long, narrow, flower clusters 1.3–6.5 cm apart. **Flowers** with calyx lobes 5, rounded; enlarged lobe bright red, elliptic 12 × 4 cm, glabrous, long-stalked; corolla yellow or orange, 5–8 mm long, tube flaring above, lobes 5; style projecting, bilobed. **Fruits** capsules, 4–5 mm long, pubescent. **Distr:** SE Caribbean to C and S tropical America. **Info:** The often cultivated variety 'David Auyong' has more than one enlarged calyx lobes.

GENTIANACEAE

Genera 87, **species** ~1,665. **Herbs** or **shrubs**, rarely small trees, annual, biennial, or perennial, mycorrhizal (saprophytic), glabrous; stems ascending, erect, or twining. **Stipules** sometimes present. **Leaves** opposite, sometimes alternate or whorled, usually basally connate, sessile; lamina simple, secondary venation ± palmate (pinnate). **Inflorescences** simple or complex cymes, sometimes reduced to sessile clusters, often thyrses or flowers solitary. **Flowers** bisexual, rarely unisexual, 4–5(–16)-merous; calyx tubular, obconic, campanulate, or rotate, lobes joined at least basally; 'disklike' structure between calyx and corolla (Stevens 2001 onwards); corolla tubular, obconic, salverform, funnelform, campanulate, or rotate, rarely with basal spurs, lobes overlapping to right or rarely valvate in bud; stamens inserted on corolla tube or occasionally at sinus between corolla lobes, alternate with lobes, anthers basifixed or dorsifixed, bilocular; nectaries absent or attached to ovary base or corolla; ovary

F&KS

Above: *Exacum walkeri, Exacum trinervium, Fagraea berteroana, Fagraea ceilanica.*

usually 1(2)-locular at least apically. **Fruits** septicidal capsules, rarely berries, 2-valved, often with persistent calyx; seeds many or rarely few, small, endosperm abundant (scanty in saprophytic species), embryo white or green. **Distr:** Worldwide, especially temperate regions.

Exacum

Species ~40. **Herbs**, annual or perennial, glabrous; stems branched, ± 4-angular. **Leaves** opposite, sessile or shortly petiolate; lamina ovate to elliptic or lanceolate, margins entire, 3–5-nerved at base. **Inflorescences** cymes, paniculate, axillary or terminal. **Flowers** 4- or 5-merous; calyx lobed nearly to base; corolla rotate, lobes longer than tube; stamens inserted at throat of corolla tube just below sinus between corolla lobes, anthers cylindric, bilocular, dehiscing by apical pores; ovary bilocular. **Fruits** capsules, 2-valved; seeds many, testa warty. **Distr:** Tropical Africa, tropical and subtropical Asia, Madagascar. **Cult:** Propagate from seeds. Requires a moist, well-drained soil mix in full or partial sun.

E. trinervium (L.) Druce, *Bot. Exch. Club Brit. Isles Rep.* 3 (5): 418, 1914. (Syn: *Chironia trinervia*, etc.). [Binara]. **Herbs**, annual, 0.5–1 m high; stem sharply 4-angular, angles often ribbed. **Leaves** sessile or subsessile; lamina narrowly oval-oblong to oval-lanceolate, 5.5–9 × 1.2–2 cm, base cuneate to tapering, margins faintly undulate, apex acute to acuminate, trinerved at base. **Inflorescences** cymes, lax, terminal or in upper leaf axils; bracts foliaceous, to 2.6 × 0.7 cm, acuminate; pedicels 1–2.6 cm long. **Flowers** with calyx tube broadly campanulate, 4–4.5 mm long, lobes 5, lanceolate, 7–9 × 2–3 mm, acuminate; corolla tube 4–7 mm long, pale yellowish-green inside at throat, lobes 5, obovate, 1.3–2.4 × 0.9–1 cm, obtuse, concave, pale to dark blue; anthers 0.8–1.3 cm long, straight, gradually attenuate to apex. **Fruits** capsules, ovoid, 1–1.3 × 0.4–0.45 cm, tipped with persistent style; fruiting calyx with wings 2–4 mm wide, truncate to cordate at base; seeds irregularly rhomboid, 0.5 mm long, angular. **Distr:** Sri Lanka. **Ecol:** In wet lowlands; 0–190 m. Fl. Jul–Apr. **Zones:** 11–12.

E. walkeri Arn., *Ann. Sci. Nat. Bot.* 2 (11): 176, 1839. [White Binara]. **Herbs**, annual, to 60 cm high; stem dichotomously branched, ± 4-angular. **Leaves** broadly ovate to suborbicular, base rounded or truncate, apex subacute to acute, 3(–5)-nerved; petiole 2–4 mm long. **Inflorescences** with flowers solitary, axillary and in forks of branches, occasionally terminal dichasial cymes; pedicels slightly nodding, 1.5–2.6 cm long; bracts ovate-lanceolate, 0.2–1.1 × 0.12–0.6 cm, foliaceous above. **Flowers** with calyx lobes 5, lanceolate, to 7 × 1.75 mm, long-acuminate; corolla white, lobes 5, obliquely erect, oblong-elliptic, 1.2–1.6 × 0.6–0.8 cm, acute; anthers uniformly broad, 4–5 mm long. **Fruits** capsules, ovoid to elliptic, 0.8–1 cm long; fruiting calyx winged, wings 2–3 mm wide, rounded at base. **Distr:** Sri Lanka. **Ecol:** In moist places and edges of disturbed forests of the highlands; ~900–2,500 m. Fl. throughout the year. **Zones:** 10–12.

LOGANIACEAE

Genera 13, **species**. **Trees**, **shrubs**, woody **climbers**, or annual **herbs**, sometimes epiphytic, sometimes with axillary spines or tendrils. **Stipules** usually present, often reduced to lines connecting petiole bases, sometimes reduced to petiolar ochrea or a connate leaf-sheath. **Leaves** opposite, occasionally alternate, rarely verticillate, fasciculate, or whorled; lamina ± flat, margins usually entire, secondary veins pinnate or 3–7 from base. **Inflorescences** cymes, often grouped into thyrses; cymes sometimes umbellate, scorpioid or reduced to a single flower; bracts usually small. **Flowers** usually bisexual, 4- or 5-merous; calyx lobes free or basally connate, mostly persistent, imbricate or valvate; corolla sympetalous, lobes valvate, imbricate or contorted in bud; stamens inserted on corolla, included to exserted, usually equal in number to corolla lobes and alternating with them, filaments free, anthers basifixed, 2–4-locular; disk absent or poorly developed; ovary usually superior, (1 or)2(–4)-loculed, locules 1- to many-ovuled, style simple, terminal, persistent or deciduous, stigma usually capitate, entire or shortly 2–4-cleft. **Fruits** capsules, drupes, or berries; seeds 1–many, sometimes winged, embedded in pulp. **Distr:** Pantropical, especially Australia and New Caledonia.

Fagraea

Species ~35. **Trees** or **shrubs**, sometimes scandent, epiphytic or semiepiphytic. **Stipules** connate into an ochrea which usually splits early into 2 axillary scales. **Leaves** petiolate or sometimes sessile; lamina with margins entire or rarely crenulate, veins often inconspicuous. **Inflorescences** with flowers solitary or in cymes forming racemes, corymbs, or panicles, terminal or axillary. **Flowers** 5-merous, large; calyx deeply divided, lobes broad, thick, imbricate, base inside with colleters; corolla funnelform to salverform, tube long, lobes shorter than tube, contorted, overlapping to right in bud; stamens inserted at or near corolla mouth, often exserted, filaments filiform, anthers introrse; ovary 1-locular with parietal placentation or bilocular with axillary placentation, locules many-ovuled, style filiform to thick, stigma capitate, obconical, peltate, or 2-cleft. **Fruits** berries, globose to ellipsoid, 1- or bilocular, many-seeded. **Distr:** SE Asia, Australia, Pacific Islands. **Cult:** Propagate from seeds, cuttings or by air-layering. Requires a humus-rich, moist, well-drained soil in full or partial sun.

F. berteroana A. Gray ex Benth., *J. Proc. Linn. Soc., Bot.* 1: 98 *'berteriana'*, 1857. **Shrubs** or **trees**, 5–15 m tall. **Stipules** forming a fleshy, notched structure (ochrea) in leaf axil. **Leaves** elliptic-oblong or obovate, 8–22 × 4.5–12 cm, base tapered, margins entire, apex rounded or obtuse, ± leathery; petiole 1.5–5 cm long. **Inflorescences** lax panicles or cymes, many-flowered (mostly 9–25-flowered), terminal; rachis to 11 cm long, with small paired ovate bracts at nodes; pedicels to 1 cm long. **Flowers** 4–8.5 cm long; calyx bell- or urn-shaped, 0.9–1.6 cm long, deeply divided into 5 ovate sepals; corolla creamy, turning orange-yellow, trumpet-shaped, tube ± cylindrical, 4–11 cm long, limb lobes 5, oblong, contorted, 1.7–3.6 cm long; stamens 5, epipetalous, attached on thickened ring inside tube, anthers linear, long, included above throat; ovary superior, style reaching corolla mouth, stigma bilobed. **Fruits** berries, ± ellipsoid, 3.2–5.6 × 2.5–4.5 cm, orange, skin sap milky. **Distr:** Malesia, Australia, Pacific Islands. **Ecol:** In highlands, mangroves, on beaches, forest margins, cloud forests; 0–1,000 m. Fl. & Fr. throughout the year. **Zones:** 10–12.

F. ceilanica Thunberg, *Kongl. Vetensk. Acad. Nya Handl.* 3: 132, 1782. (Syn: *F. chinensis*, etc.). [Perfume Flower Tree]. **Shrubs**, scandent **epiphytes**, or **trees** to 15 m tall; branchlets stout, with prominent leaf scars. **Stipules** scalelike, ~1 × 4 mm. **Leaves** elliptic, ovate, obovate, or narrowly obovate, 5–25 × 2–10 cm, margins entire, fleshy to leathery, glabrous, lateral veins 4–10 pairs, inconspicuous; petiole 1–5 cm long. **Inflorescences** with flowers solitary or in dichasial cymes, terminal. **Flowers** large, fragrant; calyx broadly campanulate, lobes ovate to rounded, ~1 cm long; corolla white, funnelform, tube 3–3.5 cm long, lobes obovate, 2.5–3 × ~2 cm; stamens included, anthers oblong to narrowly ovate, 5–7 mm long; ovary ellipsoid to ovoid, glabrous, bilocular, style filiform, stigma obconical, slightly cup-shaped to peltate. **Fruits** berries, ovoid, ellipsoid, or subglobose, 3–5 × 2–4 cm, apex beaked. **Distr:** Cambodia, China, India, Indonesia, Laos, Malaysia, Myanmar, Philippines, Sri Lanka, Taiwan, Thailand, Vietnam. **Ecol:** Mostly in dense primary and secondary forests; 0–1,800 m. Fl. Apr–Aug. **Zones:** 10–12.

F. fragrans Roxb., *Fl. Ind.* ed. Wall. 2: 32, 1824. (Syn: *Cyrtophyllum lanceolatum*, etc.). [Iron Wood]. **Trees**, medium-sized, to 30 m tall; trunk 1.75 m dbh; crown light green, finely leafed, conical when young, without buttresses; bark dark brown, deeply irregularly fissured; twigs slender. **Stipules** forming small cups in leaf axils. **Leaves** elliptic, 4 × 1.5 – 15 × 6 cm, base acute, apex pointed, leathery, secondary nerves 8–10 pairs, clear below, anastomosing near margins, tertiary nerves and reticulations inconspicuous; petiole 1.25–2 cm long, apparently clawed due to the axillary scale. **Inflorescences** usually (1–)3- to many-flowered, in upper leaf axils. **Flowers** fragrant, 2 cm

NT

Above: *Fagraea fragrans, Strychnos nux-vomica, Gelsemium elegans, Gelsemium sempervirens.*

Ø, creamish-white, aging to golden-yellow; calyx campanulate, 2.5–8 mm long, deeply divided; corolla tube narrowly funnelform, 0.8–2.2 cm long; stamens exserted, anthers oblong, 1.8–3 mm long; style 1.2–6 cm long, stigma small, capitate or obconical, faintly bilobed. **Fruits** berries, broadly ellipsoid, 0.8–1 cm Ø, apex pointed, orange or scarlet; seeds minute, numerous. **Distr:** Throughout Malaya and Singapore; Borneo, Philippines, Siam, Sumatra, Tenasserim. **Ecol:** In open and swampy lowland areas; 0–800 m. Fl. Apr–Jun. Fr. Jul–Nov. **Zones:** 10–12.

Strychnos

Species ~190. **Shrubs**, **trees**, or **lianas**, armed or unarmed; lianas with axillary simple or double-curled tendrils, sometimes with axillary thorns. **Stipules** often reduced to a straight ciliate ridge connecting petiole bases. **Leaves** with margins entire, basal veins 3–7, secondary veins 1–3 distinct pairs. **Inflorescences** thyrsoid, terminal and/or axillary; bracts scalelike to sepal-like; flowers pedicellate or sessile. **Flowers** 4- or 5-merous; corolla rotate to salverform, lobes valvate in bud, spreading to reflexed when open; stamens inserted at corolla throat to middle of corolla tube, exserted to included, filaments long to short, mostly filiform, anthers orbicular to narrowly oblong, base mostly slightly 2-cleft, introrse, bilocular and separate; ovary (1- or)bilocular, locules few- to many-ovuled, style cylindrical, stigma capitate or faintly 2-cleft. **Fruits** berries, orange or red, usually globose to ellipsoid, thin- to thick-walled, glabrous, pulp fleshy, usually orange; seeds 1–15. **Distr:** Tropical and subtropical regions worldwide. **Cult:** Propagate from cuttings or seeds. Requires a well-drained, acidic soil in full or partial sun. **Info:** Many species are poisonous, but used for medicinal purposes.

S. nux-vomica L., *Sp. Pl.* 1: 189, 1753. (Syn: *S. nux-vomica* var. *oligosperma*, *S. spireana*). [Nux-Vomica Tree, Strychnine Tree]. **Trees**, to 25 m tall. **Leaves** suborbicular, broadly elliptic, or ovate, 5–18 × 4–12.5 cm, base rounded to cordate, apex shortly acuminate to acute and often mucronulate, papery, abaxially minutely hairy, adaxially glabrous and shiny, basal veins 3–5; petiole 0.5–1.5 cm long. **Inflorescences** thyrses, 3–6 cm long, axillary; peduncle puberulent; bracteoles pubescent; pedicels puberulent. **Flowers** 5-merous; calyx lobes ovate, outside densely pubescent; corolla greenish-white to white, salverform, ~1.3 cm long, tube longer than lobes, outside glabrous, inside sparsely villous at base, lobes narrowly ovate, ~3 mm long, margin thickened and puberulent; stamens inserted at mouth or corolla tube, filaments very short, anthers elliptic, ~1.7 mm long, apex exserted; pistil 1–1.2 cm long, ovary ovoid, glabrous, style to 1.1 cm long, glabrous, stigma capitate. **Fruits** berries, orange when ripe, globose, 2–4 cm Ø, glabrous; seeds 1–4, orbicular to elliptic, 2–4 cm wide, densely sericeous. **Distr:** India. **Ecol:** In deciduous forests; 180–600 m. Fl. May–Jun. **Zones:** 10–12.

GELSEMIACEAE

Genera 2, species 11. **Shrubs**, **subshrubs** or **lianas**, with simple hairs or almost glabrous. **Leaves** opposite or whorled, stipulate or estipulate; lamina entire or obscurely sinuate-dentate. **Inflorescences** cymes or flowers solitary, axillary or terminal, bracteate. **Flowers** nearly always heterostylous, actinomorphic or nearly so, bisexual; perianth and androecium (4)5-merous; sepals free or connate to halfway, equal or unequal, imbricate, persistent; corolla funnelform, tube much longer than the lobes, lobes imbricate, equal or subequal; stamens 4(5), inserted near base of corolla tube, or midway down the corolla tube, equal or unequal, filament very short, anthers dorsifixed, extrorse or latrorse; gynoecium bicarpellate, style 1–2, free to partially connate, stigmas 2 or 4 (by bifurcation), ovary superior, bilocular, glabrous or hairy, ovules anatropous, 2 or 2–8 per locule. **Fruits** capsules, septicidal or loculicidal; seeds rather large, flattened, winged or wingless, testa thin (Takhtajan 2009). **Distr:** Tropical and subtropical parts of SE Asia, Africa and America.

Gelsemium

Species 3. **Shrubs**, or straggling to twining **climbers**. **Stipules** reduced to an interpetiolar line. **Leaves** opposite, rarely verticillate; lamina with margins entire. **Inflorescences** thyrses, many-flowered, or few-flowered fascicles to flowers solitary, axillary and terminal. **Flowers** 5-merous; sepals imbricate; corolla funnelform, lobes imbricate; stamens inserted at base to middle of corolla tube, filaments strap-shaped to filiform, anthers narrowly ovate to narrowly oblong, included to exserted, bilocular, base sagittate; ovary bilocular, ovules several per locule, style filiform, stigma 4-cleft. **Fruits** capsules, ovoid to narrowly ellipsoid, septicidal, 4-valvate; seeds several, winged. **Distr:** N America (2 spp.); E and SE Asia (1 sp.). **Cult:** Propagate from seeds or half-hardened cuttings. Requires a fertile, moist, well-drained soil in full sun. **Info:** All parts of these plants are poisonous.

G. elegans (Gardner and Champion) Bentham, *J. Linn. Soc.*, Bot. 1: 90, 1857. (Syn: *Medicia elegans*). **Climbers**, twining, to 12 m long. **Leaves** ovate to narrowly ovate, 5–12 × 2–6 cm, base broadly cuneate to subrounded, apex acuminate, membranous to papery, lateral veins 5–7 pairs; petiole 0.6–1.2 cm long. **Inflorescences** thyrses, main nodes with 2 lateral cymes each; bracts acutely triangular to awl-shaped, to 4 mm long, ciliate; pedicels 3–8 mm long. **Flowers** with corolla yellow to orange, 1.2–1.9 cm long, tube 0.7–1 cm long, outside brownish, mouth reddish-spotted, lobes ovate, 5–9 mm long, apex blunt to acute; stamens inserted at middle of corolla tube, filaments 3.5–4 mm long, anthers narrowly ovate, 1.5–2 mm long, exserted; ovary narrowly ovoid, 2–2.5 mm long, style 0.8–1.2 cm long, stigma pubescent. **Fruits** capsules, ellipsoid to ovoid, 1–1.4 × 0.6–0.8 cm, smooth, glabrous; seeds elliptic to reniform, ~5.5 mm Ø, winged all around. **Distr:** China, India, Indonesia, Laos, Malaysia, N Myanmar, Taiwan, N Thailand, Vietnam. **Ecol:** In scrubby forests, thickets; 200–2,000 m. Fl. May–Nov. **Zones:** 9–12.

G. sempervirens (L.) J. St.-Hil., *Expos. Fam. Nat.* 1: 338, 1805. (Syn: *Bignonia sempervirens*). [Carolina Jasmine (Jessamine), Yellow Jessamine]. **Climbers**, twining, perennial, glabrous, to 6 m long; stems wiry, reddish-brown. **Leaves** arranged in opposite pairs; lamina simple, lanceolate to ovate-lanceolate, or narrowly oblong, 2.5–10 × 1–4 cm, margins entire, apex blunt or rounded, adaxially glossy, green; petiole short. **Inflorescences** cymes, axillary or terminal. **Flowers** sweetly fragrant, bright yellow, bell-shaped; sepals ovate, apices blunt or rounded, minutely ciliate; corolla 2–4 cm long, tube with throat orangish-yellow, limb flared, lobes 5, spreading; stamens included, adnate near base of tube; ovary oblong. **Fruits** capsules, oblong, 1.2–2.5 × 0.8–1.2 cm, beaked, greenish-brown, prominently veined; seeds ~7 per locule, apically winged. **Distr:** SE USA to Guatemala. **Ecol:** In well-drained woodlands. Fl. Dec–Apr. **Zones:** 8–11.

APOCYNACEAE

Subfamilies 5, **genera** 415, **species** 4,555. **Trees**, **shrubs** or **vines**, rarely subshrubs or herbs, with latex, or rarely with watery juice. **Stipules** absent or rarely present. **Leaves** opposite, rarely whorled or alternate; lamina simple, pinnately veined. **Inflorescences** cymose, terminal or axillary, bracteolate. **Flowers** bisexual, actinomorphic, (4 or)5-merous; calyx (4)5-partite, quincuncial, basal glands usually present; corolla (4)5-lobed, salverform, funnelform, urceolate, or rarely rotate, lobes overlapping to right or left, rarely valvate; stamens (4)5, filaments short, anthers mostly sagittate, free or connivent into a cone adherent to pistil head, longitudinally dehiscent, base rounded, cordate, sagittate, or prolonged into an empty spur, pollen granular; disk ringlike or cup-shaped, 2–5-lobed, or absent; ovaries superior, rarely half-inferior, connate or distinct, 1- or bilocular, ovules (1- or)2 to numerous per locule, style 1, pistil head capitate, conical, or lampshade-shaped, base stigmatic, apex 2-cleft and not stigmatic. **Fruits** berries, drupes, capsules, or often follicles; seeds with or without coma. **Distr:** Tropical to warm temperate regions.

Subfam. RAUVOLFIODEAE

Genera 84, species 980. **Anthers** connivent; gynoecium apocarpous or not, placentation then axile or parietal. **Fruits** berries, drupes or follicles; seeds with coma or arils (Stevens 2001 onwards). **Distr:** Tropical regions worldwide.

Allamanda

Species 14. **Shrubs**, erect or trailing. **Leaves** whorled, with axillary glands. **Inflorescences** terminal or seemingly axillary corymbs. **Flowers** large, showy, sometimes fragrant; calyx deeply divided, without basal glands; corolla yellow, funnelform, narrow part with fringed scales, limb campanulate, lobes overlapping to left; stamens inserted at narrow part of corolla, filaments very short, anthers narrowly oblong, base rounded, coherent with pistil head; disk cup-shaped, obscurely 5-lobed or entire, fleshy, thick; ovary 1-loculed, with 2 parietal placentae, ovules numerous, style filiform, pistil head thick, with a membranous, reflexed rim, apex conical, shortly bilobed. **Fruits** capsules, globose, glabrous, sharply spiny, 2-valved; seeds numerous, imbricate, compressed, winged or with a membranous margin, radicle short. **Distr:** Tropical America. **Cult:** Propagation from cuttings, which roots readily in water. Requires a fertile, moist, well-drained soil in full sun.

A. blanchetii A.DC., *Prodr.* 8: 319, 1844. (Syn: *A. violacea*). [Purple Allamanda]. **Shrubs**, vinelike, evergreen, to 2–3(–6) m high; stems bristly hispid, red-tinged; sap milky. **Leaves** 3- or 4-whorled, sometimes opposite; lamina simple, oblanceolate to elliptic, usually 4–11 × 1.5–3 cm, light green, surfaces hispid; petiole very short. **Inflorescences** clusters, few-flowered, terminal; pedicels erect, rigid, 0.7–1 cm long. **Flowers** mauve-purple, throat dark red; calyx short, densely hispid, lobes ovate-oblong to linear-lanceolate, 1.8 × 0.6–0.7 cm; corolla fused, bell-shaped, 6–10 cm long, tube 1.5–3 cm long, base abruptly narrowed, cylindrical above, limb 5–6.5 cm Ø, lobes 5, rounded, ~4–5 cm long, reddish-purple to rose-pink. **Fruits** capsules, spiny, subglobose. **Distr:** NE Brazil. **Ecol:** In moist and wet areas. Fl. throughout most of the year. **Zones:** 11–12.

A. cathartica L., *Mant. Pl.* 2: 214, 1771. (Syn: *A. hendersonii*, *A. cathartica* var. *hendersonii*). [Climbing Allamanda, Common Allamanda, Golden Trumpet]. **Shrubs**, trailing, to 4 m long, with white latex; stems glabrous. **Leaves** opposite or 3–5-whorled; lamina obovate, narrowly obovate or oblong, 6–15 × 4–5 cm, glabrous or villous along veins on abaxial surface, lateral veins flattened; petiole ~5 mm long. **Inflorescences** cymes, few-flowered, axillary; peduncles short. **Flowers** 7–14 cm long; calyx greenish, sepals 5, lanceolate, 1.2–1.8 cm long; corolla yellow, funnelform, tube 4–8 cm long, cylindric in proximal half, campanulate in distal half, limb 8–14 cm Ø, lobes obovate-truncate to orbicular, revolute. **Fruits** capsules, subglobose to ellipsoid, 3–7 × 3–5 cm, with spines to 1 cm long; seeds compressed, 1.2–1.5 cm long, winged or with a membranous margin. **Distr:** S America. **Ecol:** Fl. throughout most of the year. Fr. Apr–May. **Zones:** 10–12.

A. schottii Pohl, *Pl. Bras. Icon. Descr.* 1: 73, 1827. (Syn: *A. neriifolia*). [Bush Allamanda]. **Shrubs**, erect, to 2 m high, with clear sap. **Leaves** 3–5-whorled; lamina elliptic or narrowly obovate, 5–14 × 2–4 cm, minutely hispid along veins, lateral veins 10–15 pairs, elevated abaxially; petiole 3–7 mm long. **Inflorescences** cymes, few-flowered, axillary at ends of twigs, pubescent; peduncle to 1 cm long; bracts linear, 1 mm long; pedicels 3–5 mm long. **Flowers** 4–6 cm Ø; sepals ovate, glabrous; corolla tube narrowly funnelform, ~3 cm long, distinctly swollen at base, proximal tube 1.5 × 0.3 cm, distal tube cylindric-campanulate, 3–1.5–2 cm, limb ~4 cm Ø, lobes pale yellow, ovate or orbicular, 1.5–2 × 1–1.2 cm, obtuse; ovary 1.5 mm long; disk 5-partite. **Fruits** capsules, globose, 2.5–3 cm Ø, long-spiny, spines 1 cm long; seeds 1–2 × 1–1.5 cm. **Distr:** Brazil. **Ecol:** Fl. May–Aug. Fr. Oct–Dec. **Zones:** 11–12.

Alstonia

Species ~60. **Trees** or **shrubs**, latex present; branches whorled, mostly 4 or 5 together. **Leaves** whorled (usually of 7), rarely opposite, at ends of branchlets; lamina glossy, lateral veins numerous, ending in a marginal vein. **Inflorescences** cymes, terminal, usually 1–5 in thyrses or compound umbels, terminal. **Flowers** white, yellow or pink; calyx without glands inside, lobes basally connate; corolla salverform, tube cylindric, dilated in distal half, pubescent inside, lobes overlapping to right or left; stamens included, inserted near or above middle of corolla tube, anthers ovate, free from pistil head, not caudate; disk absent or of scales; ovaries 2, distinct or connate, ovules numerous. **Fruits** paired follicles, free or connate; seeds oblong or linear, ends long-bearded. **Distr:** Tropical Asia, Africa, C America, N Australia, Pacific Islands. **Cult:** Propagate from seeds or cuttings. Requires a fertile, well-drained soil in full sun. **Info:** All parts of these plants produce milky, corrosive, poisonous sap.

A. macrophylla Wallich ex G. Don, *Gen. Hist.* 4: 87, 1837. [Batino, Devil Tree, Hard Alstonia]. **Trees**, to 20 m tall; branches nearly 4-angled; bark smooth. **Leaves** 3- or 4-whorled; narrowly obovate or narrowly elliptic, 10–53 × 4–19 cm, apex usually acuminate, leathery, abaxially pubescent, lateral veins 16–33 pairs at 60–70° to midvein; petiole 1–4 cm long. **Inflorescences** cymes, 3-branched, terminal, 5–9 together, pubescent; peduncles 4–6 cm long; pedicels 4–5 mm long. **Flowers** small, white; calyx campanulate, lobes short, obtuse; corolla tube slightly longer than lobes, 4.5–6 mm long, lobes overlapping to right, ciliate; disk absent; ovaries distinct, glabrous, ovoid, acute. **Fruits** follicles, linear, 30–60 cm × 2–5 mm; seeds pubescent, ends with deltoid wings, with long stiff hairs all around. **Distr:** China (S Yunnan, S Guangdong) to tropical Asia. **Ecol:** Fl. Oct–Nov. **Zones:** 10–12. **Info:** Cultivated for medicinal and ornamental purposes; can become weedy in tropical regions.

A. scholaris (L.) R. Brown, *Asclepiadeae*: 64, 1810. (Syn: *Echites scholaris, Pala scholaris*). [Devil Tree, Milky Pine, White Cheesewood]. **Trees**, to 40 m tall, glabrous; bark gray; branchlets copiously lenticellate. **Leaves** 3–10-whorled; lamina narrowly obovate to very narrowly spathulate, 7–28 × 2–11 cm, base cuneate, apex usually rounded, leathery, lateral veins 25–50 pairs; petiole 1–3 cm long. **Inflorescences** cymes, dense, pubescent; peduncles 4–7 cm long; pedicels usually as long as or shorter than the calyces. **Flowers** with corolla white, tube 0.6–1 cm long, lobes broadly ovate or broadly obovate, 2–4.5 mm long, overlapping to left; ovaries distinct, pubescent. **Fruits** follicles, distinct, linear, to 57 cm × 2–5 mm; seeds oblong, margins ciliate, ends with tufts of hairs 1.5–2 cm long. **Distr:** Tropical and subtropical Asia to Solomon Islands. **Ecol:** In mixed forests; 200–1,000 m. Fl. Jun–Nov. Fr. Oct–Dec. **Zones:** 10–12.

A. spatulata Blume, *Bijdr.*: 1037, 1826. (Syn: *A. cochinchensis, A. cuneata*). [Hard Milkwood, Siamese Balsa]. **Trees**, to 12–24 m tall, glabrous; trunk 1.2–1.6 m dbh, bole 6 m tall; crown pagodalike; bark grayish; latex milky. **Leaves** exstipulate, 3–5-whorled, sessile or shortly petiolate; lamina simple, obovate-cuneate to obovate-oblong, 3.8–10 × 2.3–2.7 cm, base acute, apex rounded, thickly coriaceous, adaxially shiny, abaxially concolorous, nerves 20–30 pairs, horizontal. **Inflorescences** cymes, few-flowered, small, umbellately branched; peduncled; bracts short; pedicels to 1.3 cm long. **Flowers** with calyx lobes orbicular; corolla salverform, tube 0.6–1.2 cm long, lobes 5, elliptic, throat villous; disk small; ovary ovoid, glabrous. **Fruits** paired follicles, 12.5–18 cm × 2 mm, slender, greenish-blue. **Distr:** Indo-China to New Guinea. **Ecol:** In marshes. **Zones:** 10–12.

Carissa

Species 7. **Shrubs**, **climbers** or **small trees**, mostly spiny; branches dichotomous. **Leaves** opposite; petiole 2–3 mm long. **Inflorescences** cymes, dichotomous, usually many-flowered, pedunculate, terminal or axillary. **Flowers** (4 or)5-merous; calyx without glands or rarely with many basal glands inside; corolla salverform, tube cylindric, dilated at staminal insertion, lobes overlapping to left or to right; stamens included in throat; anthers lanceolate, obtuse or apiculate, base not appendaged; disk absent; ovary biloculed, locules 1–4-ovuled, rarely numerous, biseriate, style filiform, pistil head narrowly oblong or fusiform, apex shortly 2-cleft. **Fruits** berries 1- or biloculed; seeds 2 or more, peltate; endosperm fleshy. **Distr:** Africa to Indo-China, Australia to New Caledonia. **Cult:** Propagate from seeds or cuttings. Requires a moist, well-drained soil in full sun.

C. carandas L., *Mant. Pl.* 1: 52, 1767. (Syn: *Arduina carandas, Damna-canthus esquirolii*). [Small Num-Num]. **Shrubs**, **small trees**, or **climbers**, to 5(–8) m tall; spines simple or forked, to 5 cm long. **Leaves** broadly ovate to oblong, 3–7 × 1.5–4 cm, base broadly cuneate to rounded, apex shortly apiculate,

Below: *Allamanda blanchetii, Allamanda cathartica, Allamanda schottii, Alstonia macrophylla.*

Above: *Alstonia scholaris, Alstonia spatulata, Carissa carandas, Carissa macrocarpa.*

lateral veins ~8 pairs, ascending, convergent, anastomosing near margin. **Inflorescences** cymes, usually 3-flowered, terminal; peduncles 1.5–2.5 cm long; bracteoles minute; pedicels about as long as calyx or slightly longer. **Flowers** fragrant; sepals 2.5–7 mm long, with many basal glands inside; corolla white or pale rose, tube to 2 cm long, puberulent inside, lobes lanceolate, ~1 cm long, acute, overlapping to right, puberulent, ciliate; ovules numerous in each locule. **Fruits** berries, reddish-purple, ellipsoid, 1.5–2.5 × 1–2 cm. **Distr:** India, SE Bangladesh. **Ecol:** Fl. Mar–Jun. Fr. Jul–Dec. **Zones:** 10–12. **Info:** Cultivated for the edible fruits and also as an ornamental.

C. macrocarpa (Ecklon) A. de Candolle, *Prodr.* 8: 336, 1844. (Syn: *Arduina grandiflora*, etc.). [Natal Plum]. **Shrubs** or **small trees**, dense, 3–5 m high; spines 1- or 2-forked, 2–4 cm long, strong. **Leaves** broadly ovate, 2.5–7.5 × 2–5 cm, base rounded to obtuse, apex mucronate, thickly leathery, glabrous, dark green, lateral veins obscure. **Inflorescences** cymes, (1–)3-flowered, terminal; pedicels 2–3 mm long. **Flowers** fragrant, 5 cm Ø; sepals very narrowly ovate, 3–6 mm long; corolla white or pink, tube 1.1–1.8 cm long, pubescent inside, lobes oblong, 0.9–2.4 cm long, overlapping to left; ovules numerous. **Fruits** berries, bright red to violet, ovoid to ellipsoid, 2–5 cm long; seeds 6–16. **Distr:** Kenya to S Africa (SE Cape Province). **Ecol:** In dry open places. Fl. Aug. **Zones:** 10–12. **Info:** This species is often cultivated for its edible fruits, and also planted as a seaside wind-breaker.

Catharanthus

Species 8. **Herbs**, erect, perennial and often woody at base, juice watery. **Leaves** opposite; lamina herbaceous to somewhat leathery, entire; petiole short, intra- and interpetiolar glands present. **Inflorescences** with flowers solitary, rarely 2- or 3-flowered cymes, terminal and axillary. **Flowers** with sepals small, narrowly oblong, awl-shaped, without glands; corolla purple, red, pink, or white, salverform, tube glabrous or sparsely puberulent, throat constricted, woolly to velvety, lobes spreading, obliquely obovate, overlapping to left, apex apiculate; stamens inserted in widened portion of corolla tube, anthers free, oblong, base obtuse; disk of 2 glands; ovaries 2, ovules numerous, style filiform, pistil head with a cylindric base and reflexed hyaline frill. **Fruits** paired follicles, cylindric, apex acute; seeds black, oblong, testa rugose. **Distr:** Madagascar (8 spp.); India and Sri Lanka (1 sp.). **Cult:** Propagate from seeds or cuttings, often grown as an annual. Requires a moist, well-drained soil in full or partial shade.

C. roseus (L.) G. Don, *Gen. Hist.* 4: 95, 1837. (Syn: *Vinca rosea*, etc.). [Madagascar Periwinkle, Old Maid, Vinca, Rose Periwinkle]. **Subshrubs** or perennial **herbs**, to 1 m high, erect or decumbent; young stems puberulent. **Leaves** obovate or elliptic, 2.5–9 × 1–3.5 cm, apex minutely apiculate, herbaceous, glossy, deep green, midrib paler, lateral veins 7–11 pairs; petiole 1–5 mm long. **Inflorescences** with flowers solitary. **Flowers** with calyx (2.5–)5 mm long; corolla red to pink or white and then mostly with a pink or less often yellow eye, tube (1.5–)2.5–3 cm long, pilose inside, throat villous, lobes broadly obovate, (0.5–)1–2 × (0.3–)0.6–0.8 cm; anthers 2.5 mm long, inserted; ovary 2 mm long. **Fruits** pair of follicles (mericarps), cylindrical (podlike), 2–3.8 × 0.3–0.4 cm, with a prominent longitudinal groove; mericarps divergent; seeds numerous, 2.5–3 mm long. **Distr:** E and S Madagascar; naturalized in the tropics. **Ecol:** Fl. throughout most of the year. **Zones:** 11–12. **Info:** Cultivated for ornamental and medicinal purposes.

Cerbera

Species 6. **Trees**, with latex; branches stout. **Leaves** alternate; lamina with lateral veins 20–30 pairs, diverging almost at right angles to midvein. **Inflorescences** cymes, terminal, long-pedunculate. **Flowers** with calyx deeply divided, without glands inside; corolla white, funnelform, throat slightly dilated, ribbed, or with 5 pubescent scales, lobes broad, overlapping to left; stamens inserted at corolla throat, anthers narrowly oblong, apiculate, free from pistil head, lobes rounded at base; disk absent; ovaries 2, free, locules 4-ovuled, style filiform, dilated distally, pistil head shortly 2-cleft. **Fruits** drupes, single or paired, large, ellipsoid or globose, 1- or biloculed, endocarp thick, woody-fibrous; seeds 1 or 2 per locule, wingless, not comose, without endosperm. **Distr:** Africa, tropical Asia, Australia, Pacific Islands. **Cult:** Propagate from cuttings or seeds. Requires any well-drained soil in full sun. **Info:** The leaves and fruits contain the extremely poisonous glycoside 'cerberin'.

C. manghas L., *Sp. Pl.* 1: 208, 1753. (Syn: *Odollamia manghas, Tanghinia manghas*). [Madagascar Ordeal Bean, Pink-eyed Cerbera, Sea Mango]. **Trees**, to 8(–12) m tall; bark gray-brown; branches whorled, marked with leaf scars. **Leaves** narrowly ovate, 6–37 × 2.3–7.8 cm, base cuneate, apex acuminate; petiole 2.5–6 cm long. **Inflorescences** panicles, terminal; peduncles stout, 5–21 cm long; pedicel terete, 1–2 cm long. **Flowers** 4–7 cm Ø; corolla white, center pinkish, tube 2.5–4 cm long, villous inside, lobes ovate, falcate, 1.5–2.5 cm long; stamens 5; ovary superior. **Fruits** drupes, smooth, 5–8(–10) × 4–6 cm, bright red at maturity, smooth, pericarp fibrous; seeds usually single. **Distr:** Tanzania (Pemba), W Indian Ocean to Pacific. **Ecol:** On seashores, tidal riverbanks. Fl. Mar–Oct. Fr. Jul–Dec. **Zones:** 10–12.

C. odollam Gaertn., *Fruct.* 2: 193, t. 124, fig. 1, 1791. (Syn: *Tanghinia odollam*). [Pong Pong Tree, Yellow-eyed Cerbera]. **Trees**, small, 3–6 m tall; branchlets whorled with leaves crowded at end of a year's growth. **Leaves** elliptic to lanceolate, 12–30 × 3–7 cm, widest above middle, base much-tapered, apex abruptly acuminate and almost acute, slightly coriaceous; petiole 2–5 cm long. **Inflorescences** panicles or cymes, 4–6-flowered, terminal; peduncles ± as long the calyx. **Flowers** white with a yellow

Below: *Catharanthus roseus, Cerbera manghas, Cerbera odollam, Kopsia arborea.*

PL

Above: *Kopsia flavida, Kopsia fruticosa, Kopsia singapurensis, Landolphia buchananii.*

throat, 5–7.5 cm Ø; calyx lobes 1–2 cm long, glabrous; corolla tube swollen just above middle, 1.5–2 cm long, lobes 2–3.5 cm long. **Fruits** mericarps, drupelike, 6–10 cm Ø, glabrous, green, often pink-tinged when ripe. **Distr:** S India to Pacific. **Ecol:** On sea coasts, tidal riverbanks and in mangroves. **Zones:** 11–12.

Kopsia

Species 24. **Trees** or **shrubs**, with white latex. **Leaves** opposite. **Inflorescences** cymes, 3- to many-flowered, bracteate, terminal; peduncle long or short; bracteoles small or large. **Flowers** white or rose-colored, 5-merous; calyx small, deeply divided, without glands; corolla salverform, tube 2.3–5 cm long, narrow, dilated at or below apex, throat without scales, pilose inside, lobes overlapping to right; stamens inserted above middle of corolla tube, filaments very short, anthers narrowly oblong or ovate, included, free from pistil head, base rounded; disk scales alternate with ovaries; ovary biloculed, distinct, ovules 2 per locule, style filiform, pistil head thickened, with a short basal collar and an apiculate apex. **Fruits** drupes, single or paired, ellipsoid, 1- or 2-seeded; seeds oblong, testa membranous, not comose. **Distr:** S China to W Pacific. **Cult:** Propagate from seeds or half-hardened cuttings. Requires a rich, moist, well-drained soil in full sun.

K. arborea Blume, *Catalogus.* 13, 1823. (Syn: *K. longiflora*, etc.). **Trees**, to 15 m tall; branches greenish, terete, slightly compressed, puberulent when young. **Leaves** elliptic, narrowly elliptic or narrowly ovate, 8–24 × 3.5–8.5 cm, apex acute or shortly acuminate, glabrous, lateral veins 10–20 pairs; petiole 5–15 cm long. **Inflorescences** corymbose cymes, many-flowered; peduncle to 14 cm long, puberulent or glabrous; bracteoles narrowly oblong, puberulent or glabrous; pedicels 3–4 mm long. **Flowers** with sepals narrowly oblong, 4–6 mm long, puberulent or glabrous; corolla white, tube ~2.5 cm long, lobes narrowly oblong, 1.5–2 cm long; disk scales narrowly oblong, longer than ovary, fleshy; ovary ovoid, puberulent. **Fruits** drupes, black or blue-black, ellipsoid, 2.5–3.5 × 1.5–2 cm. **Distr:** China (S Yunnan to S Guangdong) to Malesia, Australia (N Queensland). **Ecol:** In montane forests, often along moist ravines; 400–1,000 m. Fl. Apr–Sep. Fr. Jul–Dec. **Zones:** 10–12.

K. flavida Blume, *Rumphia* 4: 28, 1849. (Syn: *Calpicarpum albiflorum, K. albiflora*, etc.). **Trees**, to 8 m tall. **Leaves** elliptic to oblong-lanceolate, 12.5–20 × 2.5–7 cm, base acute, margins entire, slightly undulate, apex acuminate, dark green, semiglossy, midvein raised and prominent abaxially, lateral veins, raised, prominent, horizontal (right angles to midvein), slightly angled toward apex; petiole ~8 mm long. **Inflorescences** cymes, few-flowered. **Flowers** ~3–4 cm Ø, white, 'eye' at center yellowish; corolla tube 2 cm long, ~2 mm Ø, narrowed at apex, lobes 5, narrowly oblong, ~1.5 × 0.5 cm, outer margin revolute or involute, apex rounded. **Fruits** follicles, drupelike, oblong-ovoid, apex round, black when ripe. **Distr:** Philippines to W Pacific. **Ecol:** In swamp forests to lowland evergreen forests; 0–270 m. Fl. Jul–Oct. **Zones:** 11–12.

K. fruticosa (Ker Gawler) A. de Candolle, *Prodr.* 8: 352, 1844. (Syn: *Cerbera fruticosa*, etc.). [Pink Gardenia, Pink Kopsia, Shrub Vinca]. **Shrubs**, evergreen, to 4 m high, glabrous except for the inflorescences. **Leaves** narrowly elliptic or narrowly oblong, 10–23 × 2.5–9 cm, apex acute or obtusely caudate, lateral veins 10–15 pairs; petiole ~1 cm long. **Inflorescences** few-flowered, puberulent; peduncles to 1 cm long; bracteoles to 1.5 mm long, pubescent; pedicels 5–7 mm long. **Flowers** with sepals ovate, 1.5–2.5 mm long, apex obtuse, pubescent; corolla pink, tube 3–5 cm long, throat pubescent, lobes oblong, 1.5–2.5 cm long; disk scales sublinear, as long as or shorter than ovary; ovary tomentose, style 3–4 cm long. **Fruits** drupes, ellipsoid, usually solitary, 1-seeded, to 2.5 × 2 cm, pubescent. **Distr:** S Myanmar. **Ecol:** In lowland tropical forests; ~80 m. Fl. throughout the year. **Zones:** 10–12. **Info:** Cultivated for medicinal and ornamental purposes. Fruits are rarely formed in cultivation.

K. singapurensis Ridl., *Fl. Malay Penins.* 2: 336, 1923. [White Kopsia]. **Trees**, small, to 5 m tall; latex white. **Leaves** oblong-ovate, 14–26 × 7–12.5 cm, base rounded, margin ± entire, slightly undulate, apex acuminate, acumen ~1 cm long, coriaceous, adaxially dark green, semiglossy, nerves to ~16 per side, ± paired, joining to a marginal vein, nerves and midvein prominent abaxially. **Inflorescences** cymes, few- to many-flowered, terminal. **Flowers** fragrant, 3.5 cm Ø; corolla white (sometimes pale pink), with a central crimson eye, tube to 2 cm long, ~2 mm Ø, constricted at apex, lobes 5, 1.5 × 0.8 cm, apex obtuse, often mucronate, one side of the margin (outer margin) involute, unequal. **Fruits** drupes, rhomboid, 2.5 cm long, dull red. **Distr:** S Malaya, Singapore (only in the Nee Soon Swamp Forest). **Ecol:** In swampy forests of the lowlands. Fl. Twice a year in the wild, but free-flowering in cultivation. **Zones:** 11–12. **Info:** This species is threatened in the wild.

Landolphia

Species 60. **Lianas**, rarely sarmentose shrubs with long curled tendrils, branched at the very tip; latex present in all parts, often abundant. **Stipules** absent. **Leaves** opposite; petiole often alternating with 1 or several pairs of bracts. **Inflorescence** cymes, lax to dense, terminal and/or axillary, sometimes tendril-like. **Flowers** 5-merous, fragrant; sepals imbricate, free or shortly base-united; corolla variously colored, tube colored, cylindrical, slightly enlarged at the level of anthers, thickened above anthers, lobes broadly elliptic to narrowly linear, contorted, overlapping to the left; stamens included, free from each other and the pistil head, filaments very short; ovary glabrous or hairy, ovules numerous, style terete, stigma bifid, subtended by a clavuncula positioned at the level of anther bases. **Fruits** berries, indehiscent, with or without a sclerified pericarp, pulp sweet and acid; seeds many, endospermous. **Distr:** Tropical and S Africa (50 spp.), Madagascar (13 spp.). **Cult:** Propagate from seeds, cuttings or by air-layering. Requires a humus-rich, moist, well-drained soil in full sun.

L. buchananii (Hallier f.) Stapf, *Fl. Trop. Afr.* 4(1): 35, 1902. (Syn: *Clitandra buchananii*, etc.). [East African Gum Vine].**Shrubs** scandent, or **lianas** to 40 m long. **Leaves** ovate-or elliptic-oblong, (4–)5–12(–14) × 1–7.5 cm, base acute to rounded, apex cuspidate-attenuate, acumen short, round-tipped, membranous, adaxially glabrous, abaxially glabrous or midrib pubescent; petiole 3–8 mm long, deeply channeled and often folded into a tube; tendrils common. **Inflorescences** cymes or racemes, few- to several-flowered forming dense clusters or frequently elongating and becoming tendrillous, terminal and axillary; flowers subtended by 2 sepal-like bracteoles. **Flowers** strongly jasmine-scented; calyx with 5 strongly imbricate ovate sepals 1.5–2 mm long, basally united; corolla white, tube and outside of lobes tinged pink or yellow, tube 7.5–9 mm long, lobes narrowly elliptic, ± as long as tube, glabrous; stamens inserted at the middle of corolla tube; anthers ~1.6 mm long; ovary conical, 1 mm long, glabrous, style and stigma to 2 mm long. **Fruits** globose, to 10 cm Ø; seeds 0.8–1.7 cm long. **Distr:** Tropical Africa. **Ecol:** In montane evergreen forests, riverine forests. **Zones:** 11–12.

L. mannii Dyer, *Rep. Progr. Condition Roy. Bot. Gard. Kew* 1880: 39, 1881. (Syn: *L. foretii, L. klainei*). [African Rubber Plant]. **Lianas**, robust; branches and branchlets covered with lenticels. **Leaves** alternate; lamina lanceolate, 10–15.5 × 2.9–5.2 cm, base acute to rounded, sometimes slightly unequal, margins entire, apex acuminate, acumen ~1.5 cm long, coriaceous, adaxially glossy dark green, abaxially paler, venation depressed adaxially, prominent abaxially, slightly arcuate; petiole ~6.5 mm long. **Inflorescences** cymes, dense, ± hemispherical, terminal, bracteate; peduncle very short. **Flowers** white, ~7 mm across; calyx ~3.3 mm long; corolla rotate, tube ~7.5 mm long, lobes ~3.7 mm long, twisted. **Fruits** berries, indehiscent. **Distr:** Nigeria to Angola. **Ecol:** In swamp forests, gallery forests; <450 m. Fl. & Fr. throughout the year. **Zones:** 11–12. **Info:** The seeds are usually dispersed by animals. The synonym *L. klainei* is often written as *L. klanii* and is still used in many botanical collections.

Ochrosia

Species 42. **Trees**, with latex; branches stout, often whorled. **Leaves** 3–5-whorled, rarely opposite; lamina with lateral veins numerous, subparallel, almost at

Above: *Landolphia mannii, Ochrosia oppositifolia, Picralima nitida, Pleiocarpa mutica.*

right angles to the midvein. **Inflorescences** cymes, subterminal, pedunculate. **Flowers** with calyx deeply divided, usually without glands; corolla salverform, tube slightly dilated above middle, to 1 cm long, throat without scales, lobes overlapping to right; stamens inserted in the widening of corolla tube, anthers free from pistil head, narrowly oblong, rounded at base; disk absent; ovaries 2, free or basally connate, ovules 2–6, biseriate on each side of a prominent placenta, style filiform, pistil head shortly 2-cleft at apex. **Fruits** drupes, 1 or 2, smooth, endocarp thick, hard; seeds 2–4 per locule, flat, not comose, endosperm none. **Distr:** Malaysia, Indian Ocean Islands, W Pacific Islands. **Cult:** Propagate from fresh seeds or cuttings. Requires a fertile, moist, well-drained soil in full sun; some species are salt-tolerant.

O. oppositifolia (Lam.) K.Schum., *Nat. Pflanzenfam.* 4(2): 156, 1895. (Syn: *Cerbera oppositifolia*, etc.). [Corkwood]. **Trees**, small, to 10 m tall; bark smooth; branches glaucous green with leaves often crowded at the end of a year's growth. **Leaves** obovate to obovate-elliptic-oblong, 13–20 x 4–10 cm, widest above middle, base tapered, apex rounded or faintly acuminate and obtuse, leathery, glossy green, aging to yellow, midrib impressed adaxially, prominent abaxially, lateral veins faintly sunk adaxially, prominent abaxially; petiole 2–3.5 cm long. **Inflorescences** clusters, 7 cm across; pedicels as long as or shorter than the calyx. **Flowers** strongly fragrant, 1.2 cm Ø; calyx lobes 1–1.5 mm long, glabrous; corolla tube 5 mm long. **Fruits** paired mericarps; mericarps oblong, 6–7 x 4 cm, glabrous, ripening to bright yellow. **Distr:** Seychelles to Pacific. **Ecol:** In coastal regions (on sandy beaches, near seashores) to dry open areas and in tropical forest margins; salt tolerant. Fl. & Fr. throughout most of the year. **Zones:** 11–12.

Picralima

Species 1. **Trees** or **shrubs**, glabrous, except inside the corolla tube; white latex present in all parts. **Leaves** opposite. **Inflorescences** compound umbellate cymes, pedunculate, terminal, rarely axillary. **Flowers** fragrant or not; sepals imbricate, with 2–4 rows of colleters at the extreme base; corolla tube thick and fleshy, almost cylindrical, lobes overlapping to the left, entire, spreading and later recurved; stamens included, filaments short; ovary of 2 separate carpels, united at the extreme base by a disklike thickening, ovules numerous in each carpel, pistil head of a stigmatic oblong basal part and a stigmoid apex to 1.5 mm long. **Fruits** paired mericarps; mericarps to 80-seeded. **Distr:** W tropical Africa to Uganda. **Cult:** Propagate from seeds, cuttings or by air-layering. Requires a humus-rich, moist, well-drained soil in full or partial sun.

P. nitida (Stapf) T. Durand and H. Durand, *Syll. Fl. Congol.*: 338, 1909. (Syn: *P. klaineana, Tabernaemontana nitida*, etc.). **Trees** or **shrubs**, 4–35 m tall; crown dense; trunk cylindrical, 5–60 cm dbh; wood very hard. **Leaves** elliptic to oblong, 10–27 x 1.9–13 cm, base rounded to obtuse, apex abruptly acuminate, acumen 0.6–1.4 cm long; petiole 1–2.2 cm long. **Inflorescences** 12–36-flowered, 6–10 cm long; peduncle 0.2–3.7 cm long; bracts small; pedicels 0.2–2 cm long. **Flowers** with sepals broadly ovate to suborbicular, 3.5–7.5 x 3.5–5 mm, obtuse to rounded; corolla tube often greenish, 1.45–2.1 cm long, lobes white to yellow, ovate, 1.35–3 x 0.55–1 cm, obtuse; stamens inserted at 0.95–1.5 cm from base; pistil 0.97–1.25 cm long, style 5–7.3 mm long, ovules 120–150 per carpel. **Fruits** yellow to orange, smooth, irregularly pyriform or obovoid to ellipsoid, 8–15 x 13–14 cm, apex rounded. **Distr:** W tropical Africa to Uganda. **Ecol:** In forests; 800–1,200 m. **Zones:** 11–12.

Pleiocarpa

Species 7. **Shrubs** or **small trees**, rarely lianas, glabrous except inside the corolla tube; white latex present in all parts; spines and tendrils absent. **Stipules** absent. **Leaves** opposite or 3–5-whorled. **Inflorescences** fasciculate, 1–40-flowered, predominantly axillary, sometimes also terminal, ramiflorous, sessile; bracts very small, scalelike; pedicels 1–2 mm long. **Flowers** very fragrant; calyx glabrous, lobes free to base, herbaceous, imbricate, not revolute, without scales on inner surface; corolla hypocrateriform, glabrous outside, pubescent inside, pure white to yellow, tube greenish in bud, tube often with 5 splits, lobes contorted, overlapping to left in bud; stamens inserted near mouth of corolla tube, filaments short, anthers without carina; ovary 2–5-carpellate, apocarpous, carpels 1–4-ovuled, clavuncle present, pistil head ellipsoid or oblong, stigma reduced or as a rudimentary apiculus. **Fruits** compound berries of 2–5 separate mericarps; mericarps yellow to bright orange, smooth to rugose, apex rounded to hooked (not beaked); seeds brown, sometimes angular. **Distr:** Tropical Africa. **Cult:** Propagate from seeds, cuttings or by air-layering. Requires a moist, well-drained soil in shade to full sun.

P. mutica Benth., *Hooker's Icon. Pl.* 12: t. 1181, 1876. (Syn: *Hunteria pleiocarpa*, etc.). [Jasmine Bush]. **Shrubs** or **small trees**, glabrous, 0.6–8 m tall, usually smaller; branches tawny. **Leaves** opposite; lamina elliptic or oblong, 7.5–15 x 3.8–5.7 cm, base cuneate, apex abruptly acuminate, thinly coriaceous, adaxially shiny, nerves 11–14 per side; petiole ~8 mm long. **Inflorescences** globose clusters, many-flowered, axillary, ramiflorous, often flowering all over the plant, even near the very base of the stem close to the exposed roots; flowers shortly pedicellate. **Flowers** white, very fragrant; sepals ovate, obtuse or subacute, ~2.5 mm long; corolla pure white, tube cylindric, 1.5–2 cm long, lobes ovate or oblong, obtuse, ~8–9 mm long; carpels 5, each 1-ovuled. **Fruits** berrylike (consisting of 2 divaricate mericarps), ovoid-globose, ~1 cm long, obtuse or apiculate, orange, rugose. **Distr:** W and WC tropical Africa (Sierra Leone to W Cameroon, and into E Cameroon and the Congo basin). **Ecol:** In the undergrowth of evergreen and deciduous forests. **Zones:** 11–12.

Plumeria

Species 7. **Trees**, with copious latex; branchlets 2–3 cm thick, nearly fleshy. **Leaves** alternate or spirally arranged at the ends of branchlets, long petiolate; lamina with margins usually entire, rarely lobed, thickly coriaceous. **Inflorescences** cymes, 2- or 3-branched, pedunculate, terminal; bracts usually large, deciduous before anthesis. **Flowers** fragrant, waxy; calyx small, without glands; corolla white, yellowish, pink-red, or rose-purple, funnelform, tube narrow, hairy inside, faucal scales absent, lobes overlapping to left; stamens inserted at or near base of corolla tube, anthers free from pistil head, oblong, rounded at base; disk absent; ovaries 2, distinct, ovules numerous, multiseriate on each placenta, style short, pistil head with obtusely 2-cleft apex. **Fruits** paired follicles, cylindrical, usually elongate; seeds many, flat proximally, with a membranous wing, endosperm fleshy. **Distr:** Tropical America. **Cult:** Propagate from seeds or cuttings; ends of cuttings must be dried before planting in moist compost. Requires a well-drained soil in full sun.

P. alba L., *Sp. Pl.*: 210, 1753. (Syn: *P. revolutifolia*). [White Frangipani]. **Trees**, small, sometimes shrubby, 3–5 m tall, evergreen, with white latex; trunk to 10 cm dbh; branchlets with ± round leaf scars. **Leaves** alternate; lamina narrowly lanceolate, 15–40 x 1.2–5 cm, base and apex long-pointed, thickly leathery, adaxially shiny, green, abaxially densely minutely pubescent; petiole 2–4 cm long. **Inflorescences** cymes, several to many-flowered, compact, flattened, borne among the apical leaf clusters; peduncle 8–20 cm long; flowers shortly pedicellate. **Flowers** white, with a central yellow eye, waxy, very fragrant, 4–5 x 4–5 cm; calyx ~3 mm long, 5-lobed; corolla with tube narrow, lobes 5, spreading, overlapping, elliptic; stamens 5, inserted near base of corolla tube; pistil bicarpellate, carpels ~3 mm long, style short. **Fruits** paired follicles, 10–15 cm long, 1.2 cm Ø, long-pointed; seeds many, flat, winged. **Distr:** Puerto Rico to Windward Islands. **Ecol:** In coastal thickets, on rocky plains or mountain slopes. **Zones:** 10–12.

P. obtusa L., *Sp. Pl.* 1: 210, 1753. [Frangipani, Plumeria, Temple Flower]. **Trees**, 2–6(–8) m tall; branchlets pale green, thick, fleshy. **Leaves** obovate to obovate-oblong, variable in outline, 3.5–18 x 1–8.5 cm, base obtuse or rounded to acutely cuneate, apex rounded, emarginate, or very shortly acuminate-submucronulate, coriaceous or subcoriaceous, rarely firmly membranous, adaxially dark green, ± lustrous, glabrous, secondary veins anastomosing distally to form a well-developed marginal vein; petiole 1–4 cm

DM

Top row: *Plumeria alba, Plumeria obtusa, Plumeria pudica.*
Bottom row: *Plumeria rubra, Plumeria x stenopetala, Rauvolfia serpentina, Rauvolfia verticillata.*

long, puberulent; stamens wholly included; carpels 2, many-ovuled. **Inflorescences** subumbellate, usually congested, secondary peduncles usually greatly shortened, bearing numerous flowers; pedicels 0.7–1 cm long. **Flowers** white with a yellow center; calyx lobes ovate-subdeltoid, 1–1.5 mm long, usually rounded to truncate; corolla salverform, tube 0.9–2 cm long, lobes usually ovate-oblong, rarely obovate, 1.5–4.5 cm long, obtuse or rounded, spreading, slightly recurved. **Fruits** follicles, stout, 6–24 x 1–2 cm. **Distr:** USA (Florida Keys), Caribbean, SE Mexico to Belize. **Ecol:** In dry open forests near the coast, lowland broad-leaved moist forests, coastal thickets and lowland savannahs; 0–900 m. Fl. throughout the year. **Zones:** 10–12.

P. pudica Jacq., *Enum. Syst. Pl.*: 13, 1760. (Syn: *P. caracasana, P. cochleata*). [Bashful Frangipani, Bridal Bouquet, Wild Plumeria]. **Trees small**, or **shrubs**, 2–3 m high. **Leaves** obovate-oblong, ± cochleate or pandurate, 7–30 x 3–15 cm, base obtusely to acutely cuneate, apex obtuse to shortly acuminate, delicately membranous, adaxially glabrous, abaxially ± pilose to essentially glabrate; petiole scarcely differentiated. **Inflorescences** subumbellate, several-flowered, congested; secondary peduncles very short; pedicels 0.8–1.25 cm long. **Flowers** showy, pure white or pale yellowish-white; calyx lobes deltoid-subquadrate, rounded to truncate, 1.5–2.5 mm long; corolla subinfundibuliform, tube 1.5–2.5 cm long, ~2 mm Ø at base, gradually dilated above the insertion of stamens to an orifice ~4 mm Ø, lobes obovate to obovate-oblong, 2–3.5 cm long, about half-convolute in bud, longitudinal in aestivation or scarcely spiral. **Fruits** follicles, 15–17 x 2 cm. **Distr:** Panama to Venezuela. **Ecol:** On coastal hills; 0–450 m. Fl. Jul–Oct. **Zones:** 10–12.

P. rubra L., *Sp. Pl.* 1: 209, 1753. (Syn: *P. acuminata*, etc.). [Frangipani, Mexican Plumeria, Nosegay, Pagoda Tree, Red Plumeria]. **Trees**, 1.5–8 m tall; bark pale green, smooth, thin. **Leaves** broadly elliptic, obovate to oblong-lanceolate, or elliptic to very narrowly so, 12–30(–50) x 3.5–8(–15) cm, apex obtuse, acute or acuminate, base obtuse to acutely cuneate, adaxially glabrous, glaucous, abaxially glabrous to densely pubescent, lateral veins 30–40 pairs, slightly elevated abaxially; petiole 1.5–11 cm long. **Inflorescences** corymbose cymes, several-flowered, terminal, glabrous to ± pubescent; secondary peduncles well-developed, 2–15 cm long; pedicels 1–2 cm long. **Flowers** white, yellow, rose or multicolored; calyx lobes ovate-quadrate to ovate-deltoid, 1–2 mm long, usually obtuse to truncate; corolla salverform, 4–6 cm Ø, tuber 1–2.5 cm long, lobes usually broadly obovate, 2.5–6 x 1.5–2.5 cm, rounded to obtuse, obliquely spreading. **Fruits** follicles oblong, 9–30 x 1.2–4 cm. **Distr:** Mexico to Venezuela. **Ecol:** In dense forests. Fl. Mar–Sep. Fr. Jun–Dec. **Zones:** 10–12.

P. × stenopetala Urb., *Symb. Antill.* 3: 335, 1902. [Weeping Frangipani]. **Trees**, 6–9 m tall, evergreen. **Leaves** elliptic-oblong to oblong-spathulate, 5–13 x 2–4 cm, base acute, apex rounded to shortly acuminate, plicate, chartaceous, adaxially glabrous, abaxially shortly pubescent, midrib pale, canaliculate-impressed, lateral nerves 20–25 pairs; petiole 0.8–2 cm long. **Inflorescences** subcapitate-cymose, 3–7 cm long; peduncle terete, 2–3 mm long, pilose; pedicels 5–8 mm long. **Flowers** strongly fragrant, white, to 10 cm across; calyx lobes suborbicular, 1 x 1.2–2 mm long, subtruncate and apiculate, coriaceous; corolla white, 4.8–5 cm long, outside glabrous, inside pubescent, tube 1.2 cm long, lobes lanceolate to linear-lanceolate, 4–5 mm wide; filaments short, anthers linear-lanceolate, 2 mm long; style 2.5 mm long. **Distr:** Hispaniola. **Ecol:** In rocky places; 100 m. Fl. throughout most of the year. **Zones:** 10–12. **Info:** The parents of this natural hybrid are *P. obtusa* and *P. subsessilis*.

Rauvolfia

Species 110. **Trees** or **shrubs**, with latex. **Leaves** whorled, rarely opposite, with glands axillary and sometimes on petiole. **Inflorescences** cymes, pedunculate, terminal or axillary; each end of branch with as many inflorescences as leaves. **Flowers** with calyx deeply divided, without glands; corolla white, yellow, green, or pink, rarely with a red tube, salverform or campanulate, tube cylindric, swollen on 1 side at or above middle, villous inside distal half, throat not scaly, often with long hairs inside, lobes overlapping to left; stamens inserted in widening of corolla tube, filaments very short, anthers ovate, free from pistil head, base rounded; disk ringlike or cup-shaped, entire or lobed at apex; ovaries 2, free or connate, style filiform, pistil head drum shaped, with a pendulous ring, apex shortly 2-cleft. **Fruits** paired drupes, distinct or connate; seed 1, without coma. **Distr:** Africa, Asia, America. **Cult:** Propagate from seeds or cuttings. Requires a moist, well-drained soil in full sun.

R. serpentina (L.) Bentham ex Kurz, *Forest Fl. Myanmar.* 2: 171, 1877. (Syn: *Ophioxylon serpentinum*, etc.). [Devil Pepper, Serpentine Wood, Snake Wood]. **Shrubs**, to 1 m high, erect, glabrous; stems usually unbranched, slender, straw-colored.

Leaves grouped near stem apex, 3–5-whorled; lamina narrowly elliptic or obovate, 7–17 × 2–9 cm, base cuneate, apex acuminate or rarely obtuse, membranous, lateral veins 7–15 pairs; petiole 1–1.5 cm long. **Inflorescences** cymes, congested; peduncles 5–13 cm long, red or reddish; pedicels reddish. **Flowers** with calyx red or reddish; corolla white, tube cylindrical, 1–1.8 cm long, inflated at middle and pilose inside in distal half, lobes obliquely suborbicular, 1.5–3.5 mm long; stamens inserted at middle of corolla tube; ovaries connate in basal half. **Fruits** drupes, ellipsoid, ~8 mm long, connate for half their length. **Distr:** Himalayas to W Malesia. **Ecol:** In montane forests; 800–1,500 m. Fl. Feb–Oct. Fr. May–Dec. **Zones:** 11–12. **Info:** This plant has numerous medicinal uses including the treatment for snake and scorpion bites.

R. verticillata (Lour.) Baill., *Hist. Pl.* 13: 170, 1895. (Syn: *Dissolaena verticillata*, etc.). [Common Devil Pepper]. **Shrubs**, to 3 m tall, erect, glabrous; branchlets pale gray, lenticellate. **Leaves** *lower* opposite; *terminal* leaves in 3- or 4-whorled, narrowly to broadly ovate or oblong, 3.5–25 × 5–13 cm, nearly papery to membranous, lateral veins 6 or 7 pairs; petiole 0.5–1.5 cm long. **Inflorescences** cymes, rather lax, 3–9 together; peduncles 2–15 cm long; pedicels 3–6 mm long. **Flowers** with corolla white, tube cylindrical, 1–1.8 cm long, inflated and villous from middle to throat, lobes broadly elliptic or ovate, 1–4.5 mm long; stamens inserted at middle of corolla tube; ovaries distinct. **Fruits** drupes, ellipsoid or ovoid, distinct, ~1 × 0.5 cm, 1-seeded. **Distr:** India to Taiwan and W and C Malesia. **Ecol:** In lowlands, montane rainforests, monsoon forests, riverbanks, seashores; 0–2,200 m. Fl. Feb–Oct. Fr. Apr–Dec. **Zones:** 11–12. **Info:** Used medicinally, especially to treat snake bites.

Saba

Species 3. **Lianas**, with curled tendrils and all parts with white latex. **Leaves** opposite, with some colleters in axils. **Inflorescences** terminal, in forks or on tendrils. **Flowers** sweetly scented; corolla white or creamy with orange or yellow throat, tube inside with a pilose belt around the anther, lobes overlapping to the left; stamens inserted below the middle of the tube, anthers glabrous, acute at apex, cordate at base; ovary glabrous or with a setose belt just below the truncate apex, pistil head oblong or ellipsoid. **Fruits** berries, with a glabrous, smooth or rough sclerified pericarp and edible pulp, many-seeded. **Distr:** Tropical Africa, Comoros, Madagascar. **Cult:** Propagate from seeds, cuttings or by air-layering. Requires a humus-rich, moist, well-drained soil in full sun.

S. comorensis (Bojer ex A.DC.) Pichon, *Mém. Inst. Franç. Afrique Noire* 35: 303, 1953. (Syn: *Landolphia florida*, etc.). [Rubber Vine]. **Lianas**, vigorous, climbing to 30 m long; tendrils common. **Leaves** elliptic or ovate, 7–19.5 × 4–9.5 cm, base rounded to subcordate, apex rounded or obtuse, membranous, not strongly discolorous, glabrous, midrib ± impressed adaxially, raised abaxially; petiole 0.8–1.4 cm long, glabrous. **Inflorescences** paniculate cymes, many-flowered, usually forming large dense heads. **Flowers** large, white, showy, very sweetly scented; calyx 2.5–3 mm long, lobes strongly imbricate, free to base, apex oblong, rounded, dorsally glabrous or tomentose, minutely ciliate; corolla tube 1.8–2.4 cm long, slender, lobes oblong-elliptic or obovate-elliptic, 1.6–3 × 0.3–0.9 cm, apex rounded, irregularly ciliate; stamens inserted just below the middle of the corolla tube; ovary subglobose, ~1 mm long, with a ring of stiff erect hairs at apex, style, clavuncle and stigma 4–5 mm long. **Fruits** globose, to 6 cm Ø, yellow or orange with white or orange pulp, edible. **Distr:** Tropical Africa, Comoros, Madagascar. **Ecol:** 0–2,000 m. **Zones:** 10–12.

Tabernaemontana

Species 99. **Shrubs** or **small trees**; latex white; stems repeatedly dichotomously branched. **Leaves** opposite; petiole with adaxial surface often with a basal semicircular or semiamplexicaul ochrea. **Inflorescences** cymes, corymbose or umbellate, at branch forks, many-flowered or rarely flowers solitary. **Flowers** with calyx divided halfway down or deeper, with few to many basal glands inside; corolla salverform, widened at or near middle, lobes sharply overlapping to left (or right); stamens inserted in widened part of corolla tube, filaments short or almost none, anthers oblong or narrowly triangular, free from pistil head, base sagittate or deeply cordate and not spurred; disk absent; ovaries 2, free, ovules numerous, style filiform, pistil head with a subglobose or lampshade-shaped basal part, and stigmoid 2-cleft apical part. **Fruits** paired follicles, divaricate; seeds with a red or orange fleshy aril, coma absent. **Distr:** Tropical and subtropical regions worldwide. **Cult:** Propagate from seeds, cuttings or by air-layering. Requires a moist, well-drained soil in full sun.

T. crassa Benth., *Niger Fl.*: 447, 1849. (Syn: *Conopharyngia crassa*, etc.). [Adam's Apple]. **Trees**, small to medium-sized (sometimes large shrubs); latex sticky. **Leaves** opposite or verticillate; lamina obovate or oblong, 13–40 × 6–20 cm, margins slightly wavy, apex shortly acuminate, thickly leathery, glabrous, adaxially glossy dark green, lateral nerves 6–12 pairs. **Inflorescences** axillary at the end of branches; pedicels 10 cm long. **Flowers** sweetly fragrant, (3–)6–10 cm long; calyx 2.5–8 mm long; corolla tube thin, 3–10 cm long, broadened and twisted at base, lobes 5, white with a yellow throat, curved; anthers with long-tipped, inserted 0.5–1.2 cm above base of corolla tube. **Fruits** (single or) paired follicles, free or united at base; follicles large, globose-ovoid, green, smooth, to 15 cm Ø, fleshy; seeds numerous. **Distr:** W Tropical Africa to Angola. **Ecol:** In closed forests, and in secondary vegetation; 0–2,000 m. **Zones:** 11–12. **Info:** Plant parts contain many poisonous substances, and are used for medicinal purposes. The sap is used as arrow poison.

T. dichotoma Roxb. ex Wall., *Edwards's Bot. Reg.* 15: t. 1273, 1829. (Syn: *Rejoua dichotoma*, etc.). **Trees**, small; young parts covered with a glossy resinous coat. **Leaves** 8–21 × 2.5–9 cm, broadest at or above middle, base tapered, apex rounded or very shortly acuminate, coriaceous, lateral veins faintly impressed adaxially, reticulate venation scarcely noticeable; petiole 2–4 cm long, bearing at base on upper side a semicircular rim of rudimentary coalescent stipules. **Inflorescences** clusters, few-flowered; pedicels much longer than calyx. **Flowers** white, short-lived, opening in the evening and falling early in the morning next day; calyx lobes 3–4 mm long, glabrous; tube 1.8–2.5 cm long. **Fruits** mericarps, pendulous, paired or solitary by abortion, 3.5–7 × 3–5 cm, glabrous, orange to scarlet when ripe; seeds with a coat of crimson pulp. **Distr:** Sri Lanka, W and S Deccan Peninsula. **Ecol:** In secondary forests of moist regions; to ~1,200 m. **Zones:** 11–12. **Info:** The white latex on contact can cause inflammation of the eyes.

T. divaricata (L.) R. Brown ex Roemer and Schultes, *Syst.* 4: 427, 1819. (Syn: *Ervatamia coronaria*, etc.). [Butterfly Gardenia, Ceylon Jessamine, Crepe Gardenia, Crepe Jessamine, Pinwheel Flower]. **Shrubs** or **small trees**, 0.5–5 m high, glabrous. **Leaves** elliptic or narrowly so, 3–18 × 1–6 cm, apex acuminate, lateral veins 5–17 pairs; petiole 0.3–1 cm long. **Inflorescences** cymes, dichotomous, 1–8-flowered; bracts scalelike. **Flowers** with flower buds ovoid-headed, tip acute or obtuse; calyx lobes often ciliate; corolla white, tube 1.5–2.7 cm long, lobes simple or double, obovate or broadly so, 1.5–2.7 × 0.8–2 cm; stamens inserted at basal third of corolla tube. **Fruits** follicles, obliquely and narrowly ellipsoid, 2–7 × 0.6–1.5 cm. **Distr:** Himalayas to China (S Yunnan) and Indo-China. **Ecol:** In montane brushwoods, sparse forests; 100–1,600 m. Fl. Apr–Sep. Fr. Jul–Nov. **Zones:** 10–12. **Info:** All parts of the plant are poisonous, often used for medical purposes.

T. persicariifolia Jacq., *Collectanea* 4: 139, 1791. (Syn: *Pagiantha persicariifolia*, etc.). [Milk Wood]. **Shrubs** or **small trees**, 5–10 m tall, dichotomously branched; latex abundant, white; trunk to 40 cm dbh. **Leaves** opposite, ochrea conspicuous, widened into stipules in axils of petioles; lamina simple, elliptic, 4–18 × 1–9 cm, base cuneate or rounded, margins entire, apex obtuse, acute to acuminate, leathery, glossy green, lateral veins 7–20 pairs; petiole 0.5–2.2 cm long. **Inflorescences** irregularly corymbiform, 3–20-flowered, 3–6 cm long, paired in forks of branches; peduncle 5–25 cm long; pedicel 0.3–1 cm long; bracteoles 1–2. **Flowers** bisexual, 5-merous, white to yellowish, sweetly fragrant, densely pubescent; sepals suborbicular to broadly elliptic, 2–3.5 mm long; corolla tube cylindrical, 0.9–2.3 cm long, inside hairy, base slightly twisted, lobes obliquely elliptic, 0.9–1.6 × 0.3–0.7 cm, rounded, undulate, spreading; ovary

Below: *Saba comorensis*, *Tabernaemontana crassa*, *Tabernaemontana dichotoma*, *Tabernaemontana divaricata.*

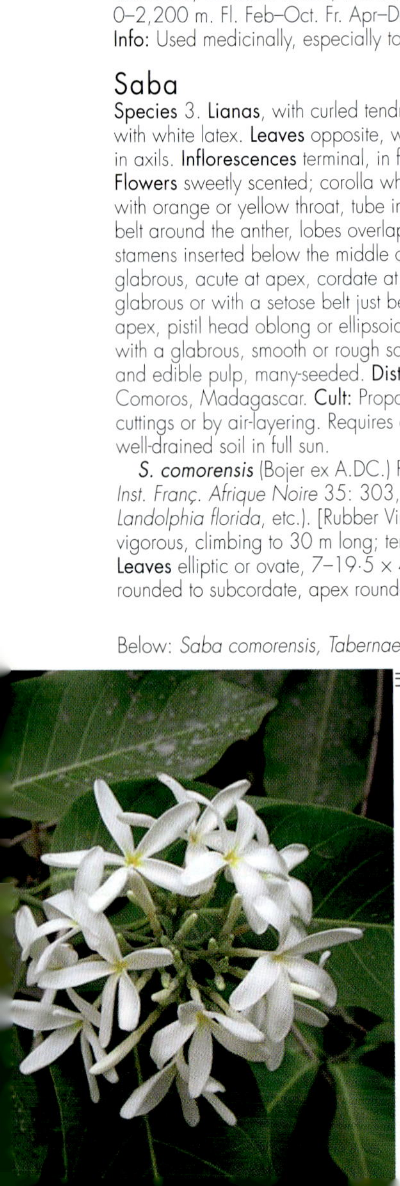

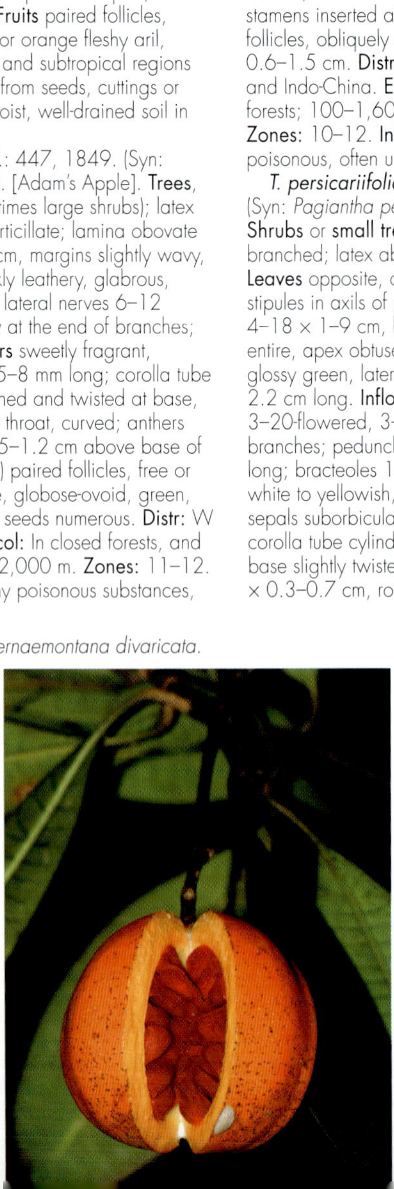

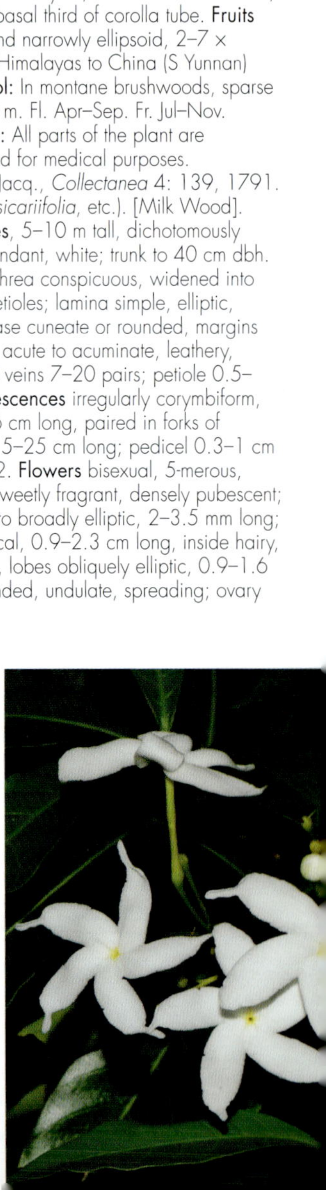

TG

Above: *Tabernaemontana persicariifolia, Tabernaemontana ventricosa, Thevetia peruviana* (yellow form).

superior, ovoid, carpels 2, 0.6–1.5 cm long. **Fruits** composed of 2 mericarps; mericarps ovoid, 3–6.5 cm Ø, smooth, green, dehiscent; seeds many, large, brown, aril red. **Distr:** Mascarenes (Réunion and Mauritius). **Ecol:** In semidry forests, humid forests; <1,200 m. Fl. Dec–Mar. **Zones:** 11–12.

T. ventricosa Hochst. ex A.DC., *Prodr.* 8: 366, 1844. (Syn: *Conopharyngia ventricosa*, etc.). [Forest Toad Tree]. **Trees**, 4–8 m high, glabrous; sap milky. **Stipules** united into a short tubular sheath 2 mm long. **Leaves** opposite; lamina oblong to oblong-oblanceolate, 9–17 × 3.5–6 cm, base cuneate to rounded, apex acute to acuminate, adaxially glossy dark green, abaxially paler; petiole 0.5–1.2 cm long. **Inflorescences** corymbose, dense; peduncle stout, 3–7 cm long; bracts ovate 4–5 mm long, acute; pedicels 3–6 mm long. **Flowers** white, fragrant, to 2–5 cm Ø; calyx cup-shaped, 4–5 mm long, sepals 5, rounded; corolla salverform, tube spindle-shaped to urceolate, 0.8–1 cm long, lobes ovate-oblong, 1–1.2 cm long, base auriculate, overlapping to the left, margins crisped; stamens 5; ovary of 2 free carpels, style columnar, compressed, 4–5 mm long. **Fruits** of 2 mericarps spreading at 180°; mericarps baccate, ellipsoid to oblong, 4.5–5.5 × 2.5–3 cm, pulp fleshy; seeds numerous, ovoid-triquetrous, 0.9–1.1 cm long, brown. **Distr:** Tropical and S Africa. **Ecol:** In coastal forests. **Zones:** 10–12.

Thevetia

Species 8. **Trees** or much-branched **shrubs**, evergreen, erect; latex white. **Leaves** alternate, densely together on slender branchlets. **Inflorescences** cymes, terminal and leaf-opposed. **Flowers** with calyx deeply divided, with many basal glands inside; corolla yellow, funnelform, lobes overlapping to left, throat with 5, narrow, long-hairy scales; stamens inserted in distal narrow part of corolla tube, filaments very short, anthers narrowly oblong, small, free from pistil head; disk absent; ovary biloculed, placenta prominent, style filiform, pistil head disklike, thick, dilated, apex shortly 2-cleft. **Fruits** follicles (drupelike), depressed-globose, endocarp hard, woody or fleshy; seeds 2 per locule, wingless, not comose, without endosperm. **Distr:** Mexico to S tropical America, Cuba. **Cult:** Propagate from seeds or cuttings. Requires a well-drained soil in full sun.

T. peruviana (Persoon) K. Schumann, *Nat. Pflanzenfam.* 4(2): 159, 1895. (Syn: *Cascabela thevetia, Cerbera thevetia*, etc.). [Lucky Nut, Yellow Oleander]. **Trees** or **large shrubs**, 4–6 m tall; bark chocolate-brown, lenticellate; lower branches pendulous, young branches greenish-gray. **Leaves** linear to narrowly oblong or narrowly lanceolate, 10–15 × 0.5–1.2 cm, apex acuminate, adaxially lustrous green, somewhat leathery, glabrous, abaxially light green, lateral veins obscure; petiole ~3 mm long. **Inflorescences** cymes, few-flowered; pedicels 2.5–5 cm long. **Flowers** fragrant, funnelform; sepals green, narrowly triangular, apex acuminate; corolla 6–7 × 4.5–5.5 cm, tube 4–5 cm long, shorter than lobes; corona scales present, connected by a transverse row of long white hairs, lobes obliquely obovate. **Fruits** drupes, fleshy, compressed triangular-globose, 2.5–4 cm Ø; seeds light gray, lenticular, ~2 × 3.5 cm. **Distr:** Mexico to tropical America. **Ecol:** Fl. May–Dec. **Zones:** 10–12. **Info:** All parts of this plant are extremely poisonous.

Vinca

Species 5. **Herbs**, erect or trailing, perennial, with stolons and watery juice. **Leaves** opposite, shortly petiolate, intra- and interpetiolar glands present; lamina with margins entire. **Inflorescences** with flowers solitary or rarely in 2-flowered cymes, axillary. **Flowers** with calyx small, without glands, lobes 5, usually equal; corolla violet, funnelform or infundibuliform, not appendaged inside, tube cylindric, hairy or with scales at throat, lobes 5, equal, obliquely obovate, spreading, shorter than tube, overlapping to left; stamens inserted just below middle of corolla tube, anthers not connivent, connective produced into a relatively large apical appendage; disk glands 2, ligulate, alternating with ovaries; ovary apocarpous, ovules 6–many, style filiform, pistil head ringlike, apex densely hairy. **Fruits** paired follicles, erect or spreading, cylindric, striate, relatively slender; seeds numerous, naked, subcompressed, glabrous. **Distr:** Europe, Macaronesia to C Asia. **Cult:** Propagate by division or by air-layering. Requires a moist, well-drained soil rich in organic matter, in full or partial sun.

V. major L., *Sp. Pl.* 1: 209, 1753. (Syn: *Pervinca major*). [Bigleaf Periwinkle, Blue Buttons, Blue Periwinkle, Greater Periwinkle]. **Herbs**, perennial, to 1 m high; flowering stems to 30 cm long. **Leaves** elliptic, ovate, or broadly ovate, 2–9 × 2–6 cm, base truncate or subcordate, margins ciliate with hairs 0.1–1 mm long, apex acute to acuminate, dark green, glossy, lateral veins to 5 pairs. **Inflorescences** with pedicels 3–5 cm. **Flowers** with sepals narrowly triangular, ~9 mm long, densely ciliate; corolla bluish-purple, tube 1.2–1.5 cm long, limb 3–5 cm Ø, lobes obliquely truncate; anthers short, applanate, apex puberulent. **Fruits** follicles, spreading, ~5 cm long. **Distr:** S Europe to E Mediterranean. **Ecol:** Fl. Mar–May. **Zones:** 7–11. **Info:** Cultivated widely throughout the tropics and subtropics, sometimes for medicinal purposes.

V. minor L., *Sp. Pl.* 1: 209, 1753. (Syn: *Pervinca minor*, etc.). [Common Periwinkle, Creeping Myrtle, Dwarf Periwinkle, Myrtle, Running Myrtle]. **Herbs**, perennial, creeping or trailing, evergreen; flowering stems to 20 cm long. **Leaves** oblong, ovate, or elliptic, 1–6 × 0.5–2.5 cm, base rounded or somewhat obtusely cuneate to broadly acute, margins not ciliate, apex obtuse to broadly acute, firmly membranous to subcoriaceous, glabrous, adaxially lustrous; petiole 1–2 mm long, usually minutely glandular-puberulent. **Inflorescences** with flowers solitary, borne in alternate leaf-axils, usually on suberect branches; pedicels 1–3.5 cm long, glabrous. **Flowers** with calyx lobes ovate-lanceolate or narrowly elliptic, 1.5–5 mm long, acute, glabrous; corolla infundibuliform, bright-blue, lilac-blue, rarely white, tube 0.3–1.1 cm long, throat

Below: *Vinca major, Vinca minor, Voacanga africana, Adenium obesum.*

PL

conic to conic-campanulate, 5–7 mm long, limb 2.5–3 cm Ø, lobes obliquely truncate; filaments longer than anthers. **Fruits** follicles, erect, slender, 2–7 cm long. **Distr:** Europe to Caucasus. **Ecol:** Fl. May. **Zones:** 5–11.

Voacanga

Species 13. **Shrubs** or **trees**, repeatedly dichotomously branched; bark and branchlets with white latex. **Leaves** opposite, petiole or leaf bases of a pair connate into a short ochrea, with a single row of colleters in the axils. **Inflorescences** cymose, paired, at forks of branches; bracts usually deciduous, leaving conspicuous scars. **Flowers** often fragrant, sometimes malodorous; calyx green, usually shed with corolla, basally united, inside with colleters; corolla creamy or yellow, tube shorter or only slightly longer than calyx, twisted, lobes in bud overlapping to left; stamens exserted or included, anthers introrse, sessile, narrowly triangular, glabrous; disk present; ovary broadly ovoid, carpels 2, free or basally connate, rarely entirely fused, pistil head with a ring at base, obovoid, coherent with the connectives of the anthers, stigmoid apex short, style and stigma shed with corolla. **Fruits** composed of 2 fleshy mericarps, free, sometimes partly or wholly united; seeds numerous, surrounded by a yellow or orange pulpy aril, endosperm copious. **Distr:** Tropical and S Africa, Madagascar, Malesia to Queensland. **Cult:** Propagate from seeds or cuttings. Requires a moist, well-drained soil in full or partial sun.

V. africana Stapf ex Scott-Elliot, *J. Linn. Soc.*, Bot. 30: 87, 1894. (Syn: *V. glabra*, etc.). [Small-fruited Wild Frangipani]. **Shrubs** or **small trees**, 1–10(–25) m tall; trunk terete, 2–30(–40) cm dbh; latex white. **Leaves** variable, elliptic, 7–41.5 × 3–20 cm, base cuneate or decurrent into petiole, apex acuminate, acute or obtuse, usually pale green becoming darker, surfaces glabrous to pubescent abaxially, secondary veins 8–22 per side; petiole 0–2 cm long. **Inflorescences** usually many-flowered, 6–25 × 4–15 cm; peduncle long, pale green, glabrous; pedicels 0.3–2 cm long; bracts soon deciduous. **Flowers** malodorous; calyx pale green, 0.7–1.9 cm long, tube cyathiform, 3.5–9 mm long, lobes erect, broadly ovate to oblong, 3.5–8 × 3.5–8 mm; corolla creamy, greenish-creamy, yellow, rarely white, tube subcylindrical, 0.7–1.5 cm long, basally twisted, lobes obovate or elliptic, 1.2–3.7 × 0.7–1.6 cm; stamens exserted to 1.2 mm; pistil glabrous, 0.7–1.25 cm long, ovary 1.6–2.5 × 1.7–3 mm, carpels 2, ~200-ovuled, style basally split. **Fruits** paired mericarps, often only 1 developing; mericarps dark green, with paler spots, obliquely subglobose, 3–8 × 3–8 × 2.5–7 cm, 2-valved; seeds many, dark brown, obliquely ellipsoid, to 5 mm long, aril orange. **Distr:** Tropical Africa. **Ecol:** In open woodlands, light forests, riverine forests and moist areas of savannahs; 0–1,200 m. Fl. Oct–Dec. **Zones:** 11–12.

Subfam. APOCYNOIDEAE

Genera 77, **species** 860. **Corolla** left-contorted; anthers inserted well below bases of corolla lobes (Stevens 2001 onwards). **Distr:** Mainly tropical regions worldwide.

Adenium

Species 5. **Shrubs** or **trees**, succulent, 0.2–6 m high, with clear or white latex; trunk sometimes to 2 m Ø at base. **Stipules** minute or absent. **Leaves** alternate, (sub)sessile, confined to apices of branchlets, with colleters in the axils; lamina variable in shape and color, entire, usually somewhat fleshy. **Inflorescences** thyrsoid, lax; peduncle very short or absent. **Flowers** slightly zygomorphic; corolla tube infundibuliform to salverform, widened at throat, lobes obovate, overlapping to right in bud, with an obcordate corona at base of and in between lobes; stamens included to exserted, anthers with long, filiform, usually coherent and twisted appendages at apex; ovary of 2 globose carpels, basally connate, style cylindrical, basally split, pistil head subcylindrical, stigmoid apex short, bifid. **Fruits** paired follicles; follicles divergent, basally united, oblong, tapering at both ends, woody, pubescent; seeds many, with tufts of dirty white to light brown hairs at both ends. **Distr:** Tropical and S Africa, Arabian Peninsula. **Cult:** Propagate from cuttings, seeds or by grafting (which is useful in producing different colored flowers on one plant). Requires a well-drained sandy soil in full sun.

A. obesum (Forssk.) Roem. and Schult., *Syst. Veg.* 4: 411, 1819. (Syn: *A. arabicum*, etc.). [Desert Rose, Impala Lily, Mock Azalea]. **Shrubs** or **small trees**, succulent, 2- to many-branched, 0.4–6 m high; trunk sometimes bulbous at extreme base, to 2 m Ø. **Leaves** sessile to subpetiolate; lamina obovate to linear, 3–12(–17) × 0.2–6 cm, base cuneate, apex rounded and mucronate or emarginate, pubescent to glabrous; petiole 0–4 mm long. **Inflorescences** dense cymes, terminal; bracts linear to narrowly oblong, 3–8 mm long; pedicels 5–9 mm long. **Flowers** with sepals 0.5–1.1 mm long, pubescent; corolla pink to red, tube reddish-pink to white suffused with pink, sometimes red-striped inside throat, 2–5 × 0.9–1.7 cm, inside glandular-hairy on main veins, lobes 0.9–3 × 0.5–2.5 mm, apiculate to mucronate; corona 3–5 mm long; stamens ± exserted; pistil 1.1–2 cm long, pistil head 1–1.5 mm long. **Fruits** paired follicles; follicles brownish, sometimes pink-fringed, 11–22 × 0.9–2 cm; seeds 1–1.4 × 0.2–0.4 cm, with tuft of hairs 2.5–3.5 cm long. **Distr:** W tropical Africa to Tanzania, Arabian Peninsula. **Ecol:** In savannas, on rocky or compacted soil between dunes; 0–1,200 m. Fl. throughout most of the year. **Zones:** 10–12.

Anodendron

Species 17. **Lianas**, with white latex. **Leaves** opposite; lateral veins usually wrinkled above. **Inflorescences** cymes, paniculate, terminal or axillary. **Flowers** small; calyx deeply divided, with basal glands inside; corolla salverform, tube cylindric, slightly dilated at staminal insertion, throat constricted, faucal scales absent, lobes overlapping and twisted to right; stamens included, inserted just below middle of corolla tube, filaments short, anthers sagittate, connivent, adherent to pistil head, cells spurred at base; disk ringlike or cup-shaped, apex truncate or shortly 5-lobed; ovaries 2, distinct, slightly higher than disk, ovules numerous in each ovary, style short, pistil head thick, base with a ringlike membrane. **Fruits** follicles, divaricate, thick, narrowly ovoid, acuminate; seeds compressed, ovate or oblong, beak with a long apical coma. **Distr:** Tropical and subtropical Asia. **Cult:** Propagate from seeds, cuttings or by air-layering. Requires a fertile, moist, well-drained soil in full sun.

A. paniculatum A.DC., *Prodr.* 8: 444, 1844. (Syn: *Echites manubriata*, etc.). [Twin Net]. **Lianas**, woody, large, often >20 m long. **Leaves** elliptic to elliptic-oblong, 3.5–17 × 1.2–9 cm, base cuneate to rounded, apex acute to short-acuminate, thinly coriaceous, lateral veins strongly prominent abaxially; petiole 0.5–2 cm long. **Inflorescences** dichasial cymes, to 8 cm long; peduncle 0.6–3 cm long; pedicels 0.5–3 mm long; bracts lanceolate, 1.3–2.2 × 0.5–0.8 mm. **Flowers** hypocrateriform, 0.5–1.1 × 0.9–1.4 cm, malodorous; calyx lobes 0.6–1 mm long, glabrous; corolla cream-yellow, tube 1–2.4 × 0.6–1.5 mm, glabrous, lobes lanceolate, slightly falcate, 4–9 × 0.6–1.8 mm, twisted to right in bud; stamens inserted at base of tube, anthers lanceolate-ovate, 0.8–1 mm long, filaments ± lacking, style much-reduced, stigma-head ~0.8 × 0.8 mm. **Fruits** follicles, woody, fusiform, 10–14 × 2–4 cm, glabrous; seeds narrowly oblong, ~0.9–2 cm long (including beak to 1 cm long), coma 8–9 mm long, brown. **Distr:** Tropical Asia. **Ecol:** In moist forests; <700 m. **Zones:** 10–12.

Beaumontia

Species 9. **Lianas**, woody, robust, with latex. **Leaves** opposite, usually with petiolar glands. **Inflorescences** cymes, terminal or axillary; bracteoles leaflike, large. **Flowers** large, fragrant; sepals free, bearing many basal glands inside, lobes leaflike, large; corolla white, funnelform, tube short or long, limb widely campanulate, throat not scaly, lobes overlapping to right; stamens inserted at distal narrow portion of corolla tube, filaments arcuate, long, thickened distally, anthers sagittate, usually exserted, connivent, adherent to pistil head, cells spurred at base; disk ringlike, shallowly 5-lobed; ovaries 2, connate, surrounded by disk, ovules numerous in each locule, style long, pistil head fusiform. **Fruits** follicles, elongated, thick, hard; seeds compressed, apex attenuate, coma silky. **Distr:** E and SE Asia. **Cult:** Propagate from seeds or semiripe cuttings. Requires a fertile, moist, well-drained soil in full to partial sun.

B. grandiflora Wallich, *Tent. Fl. Napal.* 15, t. 7, 1824. (Syn: *B. longiflora*, *Echites grandiflora*).

Below: *Anodendron paniculatum*, *Beaumontia grandiflora*, *Beaumontia multiflora*, *Cameraria latifolia*.

MD

SZ

Above: *Chonemorpha fragrans, Funtumia africana, Holarrhena floribunda, Holarrhena mitis.*

[Easter Lily Vine, Herald's Trumpet]. **Lianas**, to 20 m long; young branches rusty-pubescent. **Leaves** narrowly obovate or elliptic, 6–30 × 3.5–15 cm, pubescent (reddish-hairy) when young, glabrous when older, lateral veins 8–20 pairs; petiole to 3 cm long. **Inflorescences** cymes, 3–19-flowered, 12–25 cm long, pubescent; peduncle 2.5–9 cm long; bracts leaflike, pale green; pedicels 2.5–4.5 cm long. **Flowers** large; calyx reddish- or rust-hairy, sepals pale green, 3–6 mm long; corolla white, creamy, or pale yellow, base pale green, tube funnelform, 6.5–13 cm long, glabrous inside, limb ~10 cm Ø, base attenuate, pubescent outside, glabrous inside, lobes suborbicular to broadly ovate, 1.7–4 cm long, acuminate; stamens white, filaments 3.2–6 cm long, anthers 1.5–1.7 cm long, included; disk ringlike, apex sparsely puberulent; ovary tomentose, style 7–9 cm long. **Fruits** follicles, usually narrowly ellipsoid, 22–31 × 5–6 cm; seeds 1.5–2.5 cm, coma 4–7 cm long. **Distr:** Nepal to China (S Yunnan, SW Guangxi) and Indo-China. **Ecol:** In humid montane forests; 300–1,500 m. **Zones:** 10–12.

B. multiflora Teijsm. and Binn., *Natuurw. Tijdschr. Ned.-Indië* 4: 395, 1853. [Easter Lily Vine, Herald's Trumpet]. **Lianas** or **vinelike shrubs**, often climbing into trees, woody, evergreen, robust, to 10+ m long; sap milky. **Leaves** opposite; lamina simple, elliptic to oblanceolate, usually 10–25 cm long, base acute, margins wavy, apex acuminate. **Inflorescences** terminal clusters (at the end of branches), 13–15-flowered; bracteoles caducous; pedicels rubiginose-pilose. **Flowers** white, green-tinged, mildly fragrant, 15–20 cm Ø; calyx reddish-hairy; corolla of fused petals, funnelform, 5–7 cm long, rotate, 8–20 cm across, divided about halfway into 5 broad lobes, white, green-tinged in the center. **Fruits** paired follicles. **Distr:** Peninsular Malaysia to Lesser Sunda Islands (Bali). **Ecol:** Fl. Sep–Mar. **Zones:** 10–12.

Cameraria

Species 7. **Shrubs** or **small trees**, glabrous. **Leaves** opposite, eglandular. **Inflorescences** cymose, few-flowered, terminal. **Flowers** with calyx without squamellae, 5-cleft almost to the receptacle, lobes subequal; corolla salverform, white, small, not appendaged inside; anthers not connivent, connective not enlarged; ovary of 2 distinct carpels, without a nectary, 1–4-ovulate. **Fruits** follicles 2, indehiscent, samaroid, with a broad inequilateral wing, usually containing a single naked seed. **Distr:** Caribbean, C America. **Cult:** Propagate from cuttings or seeds. Requires a loamy and peaty soil, in full or partial sun.

C. latifolia L., *Sp. Pl.*: 210, 1753. (Syn: *Neriandra havanensis*, etc.). [White Poisonwood]. **Shrubs** or **small trees**; bark ashy-gray bark. **Leaves** mostly oblong-ovate or oblong-oval, 1.5–4 cm long, base obtuse or rounded, apex obtuse or rounded and emarginate, very lustrous above; firmly membranous to subcoriaceous; petioles 2–6 cm long. **Inflorescences** usually 2–4-flowered, terminal, about equaling the subtending leaves; pedicels 5–8 mm long. **Flowers** with calyx lobes imbricate, ovate, 1–1.5 mm long, acute or obtuse; corolla white, tube 5–8 mm long, somewhat dilated above, lobes broadly obovate, 0.6–1.5 cm long; stamens included, sessile, anthers short, obtuse, apex of connective produced into a long, free filament. **Fruits** 4–4.5 × 1.5–2 cm. **Distr:** Belize to Greater Antilles. **Ecol:** In lowland or coastal mixed forests. **Zones:** 10–12. **Info:** The plant is highly poisonous, the latex causes serious swelling and inflammation of skin.

Chonemorpha

Species ~15. **Lianas**, stout, woody, with latex. **Leaves** opposite, interpetiolar lines and colleters present; lamina large. **Inflorescences** cymes, lax, paniculate or racemose, terminal or subaxillary. **Flowers** large; calyx tubular, shortly 5-toothed or 5-partite, basal glands large, denticulate; corolla white or reddish, funnelform, tube cylindric, throat not scaly, lobes overlapping to right; stamens inserted near base or middle of corolla tube, anthers sagittate, connivent, adherent to pistil head, cells spurred at base; disk ringlike, fleshy, shorter than ovary, apex 5-cleft; ovaries 2, free, ovules numerous in each ovary, style filiform, pistil head club-shaped, slightly thickened, apex 2-cleft. **Fruits** paired follicles, elongated, cylindric; seeds ovate-oblong, flat, shortly beaked, beak with a long coma. **Distr:** Tropical and subtropical Asia. **Cult:** Propagate from seeds, cuttings or by air-layering. Requires a moderately moist, well-drained soil in full sun.

C. fragrans (Moon) Alston, *Ann. Roy. Bot. Gard.* (Peradeniya). 11: 203, 1929. (Syn: *Echites fragrans*, etc.). [Frangipani Vine]. **Lianas**, to 30 m long, hirsute except for the flowers. **Leaves** suborbicular to broadly ovate, 15–45 × 13–45 cm, base cordate, apex acute or rounded, cuspidate, papery, lateral veins 10–12 pairs; petiole to 2 cm long. **Inflorescences** cymes, lax, few-flowered. **Flowers** large, fragrant, white or cream-white, darkening with age; calyx tubular, ~1 cm long, apex toothed, glabrescent outside; corolla white, tube 3.5–4.5 cm long, dilated at base, glabrous outside, limb to 8 cm Ø, throat villous, lobes obliquely obovate, ~3.5 cm long; filaments shortly pubescent. **Fruits** paired follicles, cylindric to fusiform, to 30 × 2 cm; seeds oblong, coma ~5 cm long. **Distr:** Himalayas to C Malesia. **Ecol:** In dense montane forests, often growing on trees. Fl. May–Jul. **Zones:** 10–12.

Funtumia

Species 2. **Trees** or **shrubs**, evergreen, with white latex; trunk mostly straight, cylindrical. **Leaves** opposite, those of a pair connate into a short ochrea, with many small colleters in 2 or 3 rows in the axils; lamina with domatia abaxially, secondary veins prominent abaxially. **Inflorescences** cymose, congested, terminal and axillary; bracts ovate or elliptic with colleters inside; peduncle short. **Flowers** 5-merous, actinomorphic, fleshy, fragrant; sepals free, thick, with colleters inside; corolla tube ventricose at middle, throat thickened, lobes overlapping to right; stamens included, filaments short or absent, anthers 2-celled, introrse, connective coherent with pistil head; disk 5-lobed; ovary bicarpellate, carpels almost free, basally united with style, carpels 200–350-ovuled, ovules pendulous, pistil head ovoid, stigma conical. **Fruits** paired follicles, separate, woody, basally connate, flattened adaxially and dehiscent; seeds slender, fusiform, with a hairy coma at apex. **Distr:** Egypt, tropical Africa. **Cult:** Propagate from seeds or by air-layering. Requires a moist, well-drained soil in full sun.

F. africana (Benth.) Stapf, *Hooker's Icon. Pl.* 27: t. 2696, 1901. (Syn: *F. latifolia*, etc.). [Bush Rubber, Lagos Silk Rubber Tree, West African Rubber Tree]. **Trees**, 8–30 m tall; trunk to 50 cm dbh; latex not easily coagulating. **Leaves** ovate or elliptic, 5–32 × 1.7–17 cm, base cuneate, apex shortly acuminate, often puberulous on midrib; petiole 0.3–1.5 cm long. **Inflorescences** 3–40-flowered, 2–2.5 cm long; bracts 0.9–2 mm long, obtuse or acute; pedicels 0.3–1.5 mm long. **Flowers** with sepals triangular to broadly ovate, 1.5–4 mm long, with 5–50 colleters in a single flower; corolla tube very pale green to creamy, 0.58–1 cm long, lobes creamy, obliquely ovate to narrowly oblong, sometimes auriculate, 0.5–1.7 × 0.2–0.5 cm, obtuse or acute; disk 5-lobed; pistil 4–7 mm long, ovary pubescent at apex, style 1.2–3 mm long. **Fruits** gray-brown, fusiform, 8.4–32 × 1.4–2 cm; seed 3.5–7.5 × 0.2–0.6 cm, coma 1.6–6 cm long, glabrous basal portion of beak 0.1–1.1 cm long, hairs 3.2–9 cm long. **Distr:** Tropical Africa. **Ecol:** In forests; 30–1,600 m. **Zones:** 11–12.

Holarrhena

Species 5. **Trees** or **shrubs**. **Leaves** opposite; lamina mostly ovate or ovate-lanceolate, margins entire, membranous. **Inflorescences** corymbose cymes, many-flowered, terminal or subaxillary. **Flowers** white; calyx 5-lobed, glandular inside, lobes small; corolla salverform, tube narrow, cylindrical, throat contracted, naked, lobes oblong, overlapping to left in bud; stamens 5, included, inserted near base of tube, filament short, anthers not connivent with stigma; disk absent; carpels 2, many-ovuled, style short, filiform, stigma oblong or fusiform, tips entire or somewhat 2 lobed. **Fruits** paired follicles, elongate, incurved, circular; seeds many, linear or oblong, compressed and concave, coma deciduous. **Distr:** Tropical and S Africa, Indian Subcontinent to peninsular Malaysia. **Cult:** Propagate from seeds or by air-layering. Requires a moist, well-drained soil in full sun.

H. floribunda (G. Don) T. Durand and Schinz, *Études Fl. Congo*: 190, 1896. (Syn: *H. africana*,

Above: *Mandevilla laxa, Mascarenhasia arborescens, Nerium oleander, Odontadenia macrantha.*

etc.). [False Rubber Tree]. **Shrubs** to medium-sized **trees**, 1.5–17(–26) m tall; trunk to 1(–2) m dbh; bark gray, smooth; latex copious. **Leaves** ovate to ovate-lanceolate, 8–12 × 4–6 cm, adaxially dark green, glossy, abaxially glabrous, lateral nerves 8–12 pairs, prominent. **Inflorescences** umbellate or corymbiform, axillary; flowers produced in profusion. **Flowers** white, fragrant, 1.3 × 1.67 cm; calyx short, lobes with apices triangular, green, glossy; corolla rotate, white, slightly overlapping at base, tube ~1 cm long, lobes 5, obovate, ~1 cm long, rounded. **Fruits** paired follicles, linear or narrowly cylindrical, 30–60 cm long; seeds numerous, flattened, with brownish silky hairs tufted at apex, ~5 cm long. **Distr:** Tropical Africa. **Ecol:** In deciduous forests, savanna woodlands and in secondary growths. **Zones:** 11–12. **Info:** All parts of the plant are toxic, often used for medicinal purposes.

H. mitis (Vahl) Roem. and Schult., *Syst. Veg.* 4: 400, 1819. (Syn: *Carissa mitis, Echites lanceolata*). [Ceylon Rubber Tree]. **Trees**; bark of trunk whitish, smooth; branchlets slender, drooping, dark purplish-brown; lenticels inconspicuous. **Leaves** 5–12 × 1–3 cm, widest at or below middle, base abruptly tapering to almost rounded at base, apex acuminate, obtuse, lateral veins slightly raised or inconspicuous on both surfaces, reticulate venation not sculptured; petiole 4–6 mm long. **Inflorescences** cymes; pedicels much longer than calyx. **Flowers** fragrant, white; calyx lobes ~2 mm long, puberulous; corolla tube 0.8–1.2 cm long, lobes free. **Fruits** mericarps, 30–45 × 0.3–0.4 cm, glabrous. **Distr:** Sri Lanka. **Ecol:** In moist forests; to 650 m. Fl. Apr–May. **Zones:** 11–12. **Info:** Plant parts used for medicinal purposes.

Mandevilla

Species 185. **Lianas**, usually glabrous, with latex. **Stipules** interpetiolar, reduced to many linear segments. **Leaves** opposite; lamina elliptic to lanceolate, large, apex acuminate, deep green. **Inflorescences** racemes, few-flowered, axillary. **Flowers** large; calyx deeply divided, with many basal glands inside; corolla funnelform, tube narrow, >2 cm long, limb campanulate, faucal scales absent, lobes overlapping to right; stamens inserted and included in widened part of corolla tube, filaments short, anthers oblong, adherent at middle to pistil head, cells obtusely caudate; disk 5-cleft; ovaries free, glabrous, ovules numerous, style glabrous, pistil head with a 2-cleft mucro. **Fruits** follicles, long, slender; seeds narrowly oblong, not beaked, comose. **Distr:** Mexico to tropical America. **Cult:** Propagate from seeds or cuttings. Requires a humus-rich, moist, well-drained soil in full or partial sun.

M. laxa (Ruiz and Pavon) Woodson, *Ann. Missouri Bot. Gard.* 19: 68, 1932. (Syn: *Echites laxus, Mandevilla suaveolens*, etc.). [Chilean Jasmine]. **Climbers**, woody; branchlets glabrous; older stems with many prominent lenticels. **Stipules** several, short, filamentous. **Leaves** ovate to heart-shaped or narrowly cordate, 5–18 × 1.5–5 cm, base cordate, apex acuminate, herbaceous, adaxially glabrous, with numerous hairy domatia and 3–6 small raised glands at base of midvein, abaxially glaucous, pale green and bearded at vein axils, venation finely reticulate; petiole slender, 2.5–7.5 cm long, reddish. **Inflorescences** racemes, long pedunculate, nodding, simple, many-flowered, axillary. **Flowers** fragrant, white or pale pink, 4–7 cm Ø; calyx ~1.3 cm long, sepals narrow, 0.7–1 cm long, acute; corolla white, ivory-white or pale pink, funnel-shaped, tube 2.5–3.5 cm long, hairy inside, limb 10-folded inside, lobes oblong, broadly ovate or almost round, margins undulate, apiculate; filaments broad, pubescent, much shorter than anthers, anthers glabrous, yellow; disk apex obtuse or truncate, glabrous. **Fruits** follicles, cylindrical, 25–40 cm long, curved and cohering at the tips. **Distr:** Argentina. **Ecol:** In tropical forests, climbing to the tops of tall trees. Fl. May–Jun, Dec–Jan. **Zones:** 9–11.

Mascarenhasia

Species 8. **Shrubs** or **small trees**, with milky latex; spines and tendrils absent. **Leaves** opposite, exstipulate, without domatia, with petiolar glands. **Inflorescences** with flowers solitary or in few-flowered fasciculate cymes, terminal. **Flowers** often large, conspicuous; calyx lobes imbricate, free to base, with many glandular scales inside at base; corolla hypocrateriform, tube divided into 2 distinct regions of varying shape and proportions, lobes 5, with induplicate aestivation, straight in bud or twisted to right; stamens 5, sessile, inserted at base of upper region of corolla tube, anthers conniving in a cone around the gynoecium; receptacular disk cupular or represented by separate scales, these sometimes united in pairs; ovary of 2 free multiovulate carpels, glabrous or hairy, style pubescent, clavuncle swollen, stigma pointed or flared at apex, not bifid. **Fruits** paired follicular mericarps; seeds linear-compressed with an apical coma. **Distr:** Kenya to Mozambique, W Indian Ocean. **Cult:** Propagate from seeds or by air-layering. Requires a moist, well-drained soil in full sun.

M. arborescens A.DC., *Prodr.* 8: 488, 1844. (Syn: *Echites arborescens, M. elastica*, etc.). [Mgoa Rubber]. **Trees** or **shrubs**, 1.5–15(–20) m tall, with white latex; trunk 30–40 cm dbh. **Leaves** petiolate; lamina elliptic or obovate, 4–18 × 2–6.5 cm, base cuneate, margins revolute, apex acuminate to retuse, glabrous; petiole 1–8 mm long, glabrous. **Inflorescences** fasciculate, 1–15-flowered, 2–4 × 2–6 cm; peduncles 2–5 mm long, glabrous; pedicels 0.7–1.2 cm long. **Flowers** with sepals ovate or triangular, 1.5–5.5 × 1.5–4 mm, with a row of 7–10 colleters inside; corolla white, yellow-streaked inside, pubescent inside on lobes, tube 0.5–1.4 cm long, lobes ovate, 0.4–1.5 × 0.2–1.4 cm, apiculate, acuminate or acute, densely hairy; stamens inserted 5–8 mm from corolla base; disk (1–)2–3 mm Ø; ovary cylindrical, pubescent at apex, rarely entirely glabrous, carpels 30–60-ovuled, style pubescent, pistil head almost conical. **Fruits** follicles, dark brown, 6–17.5 × 0.6–2 cm; seeds 1–1.4 cm long, dark brown, coma 1–2 cm long. **Distr:** Kenya to Mozambique, W Indian Ocean. **Ecol:** In riverine forests; 0–750 m. Fl. Jan–Feb. **Zones:** 11–12.

Nerium

Species 1. *Generic description same as specific.* **Cult:** Propagate from cuttings (roots readily in water), seeds or by air-layering. Requires a fertile, moist, well-drained soil in full sun; moderately salt-tolerant.

N. oleander L., *Sp. Pl.* 1: 209, 1753. (Syn: *N. indicum, N. odorum*). [Oleander, Rose Bay]. **Trees** or **shrubs**, evergreen, to 6 m high, often much-branched from base, juice watery. **Leaves** 3-whorled; lamina very narrowly elliptic, 5–21 × 1–3.5 cm, base cuneate or decurrent into petiole, apex acuminate or acute, leathery, midvein prominent abaxially, lateral veins numerous, parallel. **Inflorescences** corymbose, terminal. **Flowers** showy, fragrant; sepals narrowly triangular to narrowly ovate, 0.3–1 cm long; corolla purplish-red, pink, white, salmon, or yellow, funnelform, tube cylindric near base, 1.2–2.2 cm long, throat open, lobes 1.3–3 cm long, single or double, overlapping to right; corona segments 5, petal-like, large, fringed; stamens inserted at apex of corolla tube, anthers sagittate, with a bristly, filiform apical appendage, pilose, connivent, adherent to pistil head; disk absent; ovaries 2, distinct, ovules numerous in each locule. **Fruits** paired follicles, united until just before dehiscence, cylindrical, 12–23 cm long; seeds numerous, oblong, densely pilose, coma at truncate apex, 0.9–1.2 cm long, golden-brown. **Distr:** Mediterranean to SC China. **Ecol:** Along watercourses, places with gravel, and in damp ravines. Fl. throughout most of the year. **Zones:** 8–12. **Info:** All parts of the plant are extremely toxic.

Odontadenia

Species 20. **Vines**, woody, large, lactescent, often robust; branches of the twining stems sometimes alternate above. **Leaves** opposite, eglandular, stipulate or exstipulate. **Inflorescences** thyrsoid to simply scorpioid, few- to many-flowered, lateral or terminal. **Flowers** often showy, infundibuliform, rarely salverform; calyx lobes sometimes very unequal, the squamellae inside 5–many, alternate or scattered; corolla tube straight or slightly gibbous, not appendaged inside; filaments short, usually pilosulous; carpels apically united by the fusiform, subcapitate or 5-digitate stigma; nectaries 5, usually concrescent. **Fruits** paired follicles, distinct, terete or dorsally compressed; seeds numerous,

Above: *Pachypodium decaryi, Pachypodium lamerei, Parsonsia heterophylla, Pentalinon luteum.*

truncate, comose. **Distr:** Tropical America. **Cult:** Propagate from cuttings or by air-layering. Requires a moist, well-drained soil in full sun. Best grown on an arch or pergola.

O. macrantha (Roem. and Schult.) Mgf., *Fl. Surinam.* 4, pt. 1: 461, 1937. (Syn: *Echites macrantha*, etc.). [Climbing Frangipani]. **Lianas**, glabrous; stems rather stout, not or obscurely lenticellate. **Stipules** obsolete. **Leaves** broadly elliptic to obovate-lanceolate, 15–25 × 5–12.5 cm, base acute to obtuse, usually shortly subcaudate-acuminate, mostly membranous or chartaceous; petiole to 2 cm long. **Inflorescences** few- to many-flowered, usually lateral, dichasially cymose, often much-reduced; pedicels 2–2.5 cm long; bracts minute, ovate, persistent. **Flowers** with calyx lobes ± equal, broadly ovate, 4–6 mm long, obtuse, squamellae solitary or sometimes geminate; corollas broadly funnelform with tube much shorter than throat, deep yellow, tinged with red or orange, tube 0.5–1 × 0.3–0.5 cm, strongly constricted at insertion of stamens (base of conical throat 2.5–3.5 cm long, 1.25–1.75 cm Ø at orifice), lobes broadly acuminate, 2–3.25 cm long; anthers elliptic-sagittate, densely hirtellous, nectary annulate, deeply multifid, half as long as ovary; ovary glabrous, ovoid, stigma fusiform. **Fruits** follicles, somewhat falcate, 15–30 × 1–2 cm; seeds 3–4 cm long, coma yellowish, about as long. **Distr:** Tropical America. **Ecol:** In wet forests, on coastal slopes, and in disturbed areas; 0–800 m. Fl. Feb–Jul. Fr. Jan–Oct. **Zones:** 10–12.

Pachypodium

Species 20. **Shrubs** or **small trees**; trunk and stems thick, succulent; latex present; conspicuous ternate spines (the central one often much-reduced or absent) present at each node. **Leaves** alternate or spirally arranged. **Inflorescences** cymes, few- to many-flowered, terminal, sessile or pedunculate. **Flowers** usually showy, pink, purple, yellow or partly white; calyx lobes imbricate, free almost to base; corolla hypocrateriform to subcampanulate, tube constricted near base, tubular to campanulate above, lobes contorted, overlapping to right; stamens inserted above the constriction, anthers subsessile, conniving in a cone; disk cupular, lobed or with 5 scalelike processes; ovary bicarpellate, carpels free, each containing numerous ovules, style slender, cylindric, bilobed. **Fruits** paired follicles; seeds numerous, ovate to oblong with an apical coma. **Distr:** S tropical and S Africa, Madagascar. **Cult:** Propagate from seeds or stem tip cuttings. Requires a fertile, well-drained soil in full sun.

P. decaryi Poiss., *Bull. Trimestriel Acad. Malgache*, n.s., 3: 235, 1917. **Shrubs**, to 4 m high; trunk without thorns, base bottle-shaped; branches with few spines grouped in 3s. **Leaves** deciduous; lamina elliptic, 7–10 × 3.5–4.5 cm, apex acuminate, abaxially slightly hairy, lateral veins 20–25 pairs; petiole 0.5–1 cm long. **Inflorescences** ~10-flowered; peduncle 0–6 mm long; pedicels 1–1.5 cm long. **Flowers** with sepals linear, expanded at base, 4 mm long; corolla white, hypocrateriform, inferior tube 1 × 0.3 cm, superior tube 2.5 × 0.6 cm, narrowed to throat, lobes obliquely obovate, 4 × 2 cm; anthers 1.5 cm long; disk cupuliform; ovary globose, finely hairy, 3 mm long. **Fruits** paired follicles, smooth, 22 × 8 cm; seeds 1 × 0.3 cm, pappus 3–3.5 cm long. **Distr:** N Madagascar. **Ecol:** Among strongly eroded limestone rocks and sunny places in dry forests; 200–400 m. **Zones:** 9–11.

P. lamerei Drake, *Bull. Mus. Hist. Nat.* (Paris) 5: 308, 1899. (Syn: *P. champenoisianum*, etc.). [Clubfoot]. **Trees**, succulent, to 6 m tall; trunk thick, conical; crown with a few thick ascending branches; stem covered in low spiraled tubercles, each with 2 diverging lateral spines 2.5 cm long, and 1 shorter central spine. **Leaves** in terminal rosettes; lamina ± linear, 13–36 × 2.5–9 cm, olive-green, glossy, lateral nerves numerous; petiole short, running into midrib, 3–5 cm long. **Inflorescences** clusters, terminal; peduncle 5–9 cm long; pedicels 0.5–1 cm long. **Flowers** showy, white, hypocrateriform, 6–12 cm Ø, fragrant; sepals triangular, 4 × 2 mm; corolla white, center yellowish, inferior tube 3 × 2 mm, superior tube 2.5–4.5 × 0.5–0.7 cm, lobes obliquely obovate, 2.5–6 × 1.5–3 cm, obtuse to truncate; anthers 1.7 cm long; ovary glabrous, 3 mm long. **Fruits** paired follicles, paired, 25–30 × 1.5 cm; seeds 1.5 × 5 cm, pappus 4 cm long. **Distr:** SC and S Madagascar. **Ecol:** In deciduous seasonal forests and on xerophytic bare limestone rocks; 700–800 m. Fl. Nov–Feb, Apr–Jun, Aug-Sep. Fr. Oct–Dec. **Zones:** 9–11.

Parsonsia

Species 85. **Lianas**, woody; latex white. **Leaves** opposite. **Inflorescences** cymes, corymbose or paniculate, dichotomous, terminal or axillary, pedunculate. **Flowers** small; calyx with basal glands inside or 5-scaled; corolla salverform, tube short, hairy inside distally, faucal scales absent, lobes overlapping to right; stamens inserted at middle of corolla tube or at throat, filaments long, strongly inter-twisted or geniculate, anthers narrowly sagittate, exserted, glutinous, connivent into a subcylindric cone, adherent to middle of pistil head, cells with an empty tail; disk 5-lobed or 5-scaled; ovaries 2, ovules numerous per carpel, style filiform, pistil head thickened, apex entire or 2-cleft. **Fruits** paired follicles, terete, parallel or divergent; seeds linear or oblong, apex crowned with a coma. **Distr:** Tropical and subtropical Asia to SW Pacific. **Cult:** Propagate from seeds. Requires a moist, well-drained soil in full or partial sun.

P. heterophylla A. Cunn., *Ann. Nat. Hist.* 2: 46, 1839. (Syn: *P. albiflora, P. macrocarpa*). [New Zealand Jasmine]. **Lianas**, soft-wooded, stem-twining, to 10+ m long; main stems to ~10 cm Ø. **Leaves** of adults on stout petioles to ~1(–2) cm long; lamina highly variable (within and among plants), linear, lanceolate, elliptic, rhomboid, broadly oblong to broad-obovate, ~3–7(–10) × (0.5–)1–4 cm, apex acute, obtuse to broadly rounded, often apiculate, coriaceous, dark green, adaxially ± glossy, abaxially paler. **Inflorescences** paniculate cymes, to ~10 cm long, axillary and terminal, bracts often foliaceous. **Flowers** fragrant, to 0.8(–1) cm long; calyx pubescent, 3–4 mm long, deeply lobed, lobes bluntly narrowly triangular; corolla white to yellow, tube cylindric to slightly flaring, ~7–8 mm long, ± pubescent, lobes oblong, 4–5 mm long, acute to obtuse, recurved; anthers sagittate, ~2 mm long, included; style ~4 mm long, stigmas 2, erect, connivent. **Fruits** paired follicles, narrow, subterete, 7–15 cm long, glabrous, persistent; seeds linear, ~5 mm long, coma of white tufted hairs to 1 cm long. **Distr:** New Zealand. **Ecol:** In coastal to lowland and lower montane forests. **Zones:** 8–11.

Pentalinon

Species 2. **Vines** twining, or scandent **shrubs**, with abundant milky latex. **Stipules** glandular, intrapetiolar. **Leaves** opposite, petiolate; lamina simple, without glands. **Inflorescences** cymes, axillary. **Flowers** with calyx campanulate, sepals 5, elongate, with digitiform glandular hairs (colleters) at base; corolla infundibuliform; stamens 5, inserted, anthers concrescent around stigma, with apical appendages, filaments very short; ovary superior, with 5 glandular nectaries at base, carpels 2, free or united by a common style. **Fruits** paired follicles, cylindrical, apically connate; seeds numerous, narrowly elliptic, with a tuft of long hairs at apex. **Distr:** Tropical America. **Cult:** Propagate from cuttings or seeds. Requires a fertile, moist, well-drained soil in full sun.

P. luteum (L.) Hansen and Wunderlin, *Taxon* 35: 167, 1986. (Syn: *Urechites lutea, Vinca lutea*). [Wild Allamanda]. **Vines**, woody, twining, to 5 m long, much-branched from base, with abundant milky latex; stems smooth, cylindrical, pubescent when young, lenticellate; bark thin, copper-colored. **Leaves** opposite; lamina elliptic or obovate, 3–7 × 1.5–3.7 cm, base rounded to obtuse, margins entire, revolute, apex rounded and mucronate, coriaceous, glabrous, adaxially shiny with midvein yellowish, abaxially dull, with a prominent midvein, puberulous; petiole ~5 mm long, pubescent. **Inflorescences** cymes, few-flowered, axillary; pedicels 1.5–2 cm long, pubescent. **Flowers** with calyx light green, pubescent, sepals 5, lanceolate, ~1.2 cm long; corolla yellow, infundibuliform, ~3–5 cm long, with cardinal-red lines inside, limb 3–6 cm Ø, lobes 5, rounded. **Fruits** paired follicles, woody, linear, divaricate; seeds numerous, with long silky hairs. **Distr:** S Florida to Caribbean. **Ecol:** In

MS

TT

Above: *Strophanthus caudatus, Strophanthus divaricatus, Strophanthus gratus, Strophanthus preussii.*

coastal forests and thickets. Fl. & Fr. throughout the year. **Zones:** 10–12.

Strophanthus

Species 39. **Lianas**, or erect stolon-bearing **climbing shrubs**, rarely trees, with clear, white or yellow latex; dichotomously or trichotomously branched or with single or opposite lateral branches, unarmed, lenticellate. **Leaves** opposite and decussate or ternate, rarely 3–4-whorled. **Inflorescences** cymes, 1- to many-flowered, mostly dichasial, terminal or in forks of branches, pedunculate or sessile. **Flowers** large; sepals free or basally connate, imbricate or quincuncial, basal glands 5–many; corolla in shades of white and red initially, later turning yellow and purple, funnelform, usually turning darker and dark streaked at anthesis, tube short, lobes overlapping, mostly twisted to right, distal portions mostly forming filiform, involute long, linear tails, overlapping to right; corona 10-lobed with lobes arranged in pairs between corolla lobes, inserted at base of corolla lobes; stamens inserted at apex of corolla tube, filaments short, anthers sagittate, connivent into a cone at apex of style, spurred at base; disk absent; ovaries 2, ± basally connate, ovules numerous per locule, style filiform. **Fruits** paired follicles, divergent, rigid or woody, tapering; seeds numerous, beaked with an apical tuft of hairs. **Distr:** Tropical and S Africa to C Malesia. **Cult:** Propagate from cuttings, seeds or by air-layering. Requires a moist, well-drained soil in full or partial sun.

S. caudatus (L.) Kurz, *J. Asiat. Soc. Bengal*, Pt. 2, Nat. Hist. 46(2): 257, 1877. (Syn: *Echites caudatus*, etc.). **Climbers**, shrubby, erect, glabrous, to 1–2 m high; branches brown or reddish-brown, lenticellate. **Leaves** elliptical-ovate to oblong-ovate, 8–13 × 3–6 cm, base usually rounded to acute, apex shortly acuminate, glabrous, adaxially ± dull, abaxially paler, nerves 5 or 6 per side; petiole to 1 cm long. **Inflorescences** cymes, few-flowered, terminal, glabrous; bracts lanceolate, long-acuminate. **Flowers** with calyx lobes lanceolate, 7–9 mm long, long-acuminate; corolla tube white, slightly yellow inside, 2 × ~0.8 cm, slightly constricted at middle, enlarged above, lobes ~8 mm wide at base, attenuate into long, slender, drooping, yellowish tips, ~13 cm long; appendages ~1 cm long, cleft into 2, long, slender, acuminate parts; anthers 3 mm long, base sagittate, filaments very short, somewhat hirsute, apical awn very slender, ~1.5 cm long; ovary glabrous, style cylindrical, ~1.2 cm long. **Fruits** paired follicles, woody, oblong-ovoid, ~16 × 0.7 cm; seeds lanceolate, 1.5–2 cm long. **Distr:** China (S Guangxi), Indo-China to New Guinea. **Ecol:** In forests and thickets; 500–900 m. Fl. Apr–Jun. **Zones:** 9–12.

S. divaricatus (Loureiro) Hooker and Arnott, *Bot. Beechey Voy.* 199, 1837. (Syn: *Pergularia divaricata*, etc.). [Goat Horns]. **Lianas** or **sarmentose shrubs**, stolon bearing when pruned, to 4.5 m high, latex clear or yellowish; stem to 4 cm Ø; branches densely lenticellate; branchlets reddish-brown. **Leaves** elliptic or slightly obovate, 3–10 × 1.5–5 cm, lateral veins 4–9 pairs; petiole 0.5–1 cm long. **Inflorescences** cymes, 3–15-flowered; peduncles to 1.5 cm long; bracts linear or narrowly ovate, deciduous; pedicels to 1 cm long. **Flowers** with sepals narrowly triangular, 0.4–1.1 cm long; corolla yellow, tube 0.9–1.6 cm long, lobes with a red basal spot inside, ovate, to 10 cm long, abruptly narrowed into pendulous tails to 1 mm wide; corona lobes 10, greenish-yellow, triangular or awl-shaped, 0.9–3 mm long; anthers included, connective tail to 0.6 mm long; ovary glabrous. **Fruits** paired follicles, ellipsoid-oblong, 9–15 × 2–3.5 cm, hard, woody, divergent at 180–250°; seeds fusiform, 1.3–2 cm long, beak 1.2–3.4 cm long, coma 3.5–5.5 cm long. **Distr:** S China to N Vietnam. **Ecol:** In forests, thickets; 100–1,000 m. Fl. Mar–Jul. **Zones:** 9–12.

S. gratus (Wallich and Hooker) Baillon, *Hist. Pl.* 10: 171, 1888. (Syn: *Roupellia grata*, etc.). [Climbing Oleander, Indian Rubber Vine]. **Lianas**, woody, to 25 m long, latex white; main stem to 10 cm Ø; branches dark or purplish-brown, densely lenticellate. **Leaves** ovate, elliptic, or obovate, 5–18 × 2–9 cm, leathery, glabrous, lateral veins 5–11 pairs; petiole 0.5–3.2 mm long. **Inflorescences** dichasial cymes, 3–32-flowered, terminal, glabrous; peduncle 0–0.6(–1.5) cm long; bracts ovate or triangular, 2–9 mm long. **Flowers** fragrant; sepals obovate or broadly so, 0.7–1.8 cm long; corolla white, turning yellow at base, reddish or purple above, throat reddish, tube 2.5–4.5 cm long, inflated in distal half, lobes orbicular or nearly so, 1.5–3.5 × 1.5–3.2 cm, apex rounded; corona lobes pink, awl-shaped, 0.5–1.5 cm long, exserted, filaments pubescent, connective glabrous, extending into a tail 0.6–1.2 cm long. **Fruits** paired follicles, divergent at 180°, cylindric, 23–41 × 3–4.3 cm; seeds to 2 cm long, beak 1.6–6.2 cm long, coma to 5 cm long. **Distr:** W and WC Africa. **Ecol:** Fl. Feb. **Zones:** 10–12.

S. preussii Engler and Pax, *Bot. Jahrb. Syst.* 15: 369, 1892. (Syn: *S. bracteatus*, etc.). [Poison Arrow Vine]. **Shrubs** sarmentose (0.7–5 m high) or lianas (1–12 m long), evergreen; branches reddish or purple-brown, lenticellate; latex clear or white. **Leaves** opposite; lamina elliptic or ovate to obovate, 2–19 × 1.5–7.5 cm, base cuneate, rounded or rarely subcordate, apex acuminate, glabrous or rarely scabrid; petiole 0.2–1.4 cm long. **Inflorescences** laxly 1–48-flowered, terminal or in forks, sessile or pedunculate; pedicels 0.4–2.5 cm long. **Flowers** fragrant, white, turning cream, yellow, later orange, red-suffused near mouth, spotted and streaked with red inside; sepals unequal, outer ovate, inner linear, 0.4–2.5 cm long, acute or obtuse; corolla tube 1.2–2.6 cm long, with pendulous tails >20 cm long; corona lobes 1–2.5 mm long. **Fruits** paired mericarps, ± opposite-divergent, ± cylindrical, 15–29 × 1–3 cm, glabrous; seeds 1.2–2 cm long, densely puberulous to pubescent, shortly beaked, coma 6–10 cm long. **Distr:** W tropical Africa to Tanzania and Angola. **Ecol:** In moist forests, swamp forests; 750–1,200 m. **Zones:** 10–12.

S. sarmentosus de Candolle, *Bull. Soc. Philom. Paris.* 3: 123, 1802. [Poison Arrow Vine, Spider Tresses]. **Shrubs**, trailing, stolon-bearing, to 40 m long, deciduous, latex clear or white; branches densely

Below: *Strophanthus sarmentosus, Strophanthus speciosus, Trachelospermum jasminoides, Vallaris solanacea.*

AC

lenticellate; branchlets dark or reddish-brown. **Leaves** elliptic or ovate, 2–15 × 1.5–7 cm, apex acuminate, acumen 0.2–2 cm, papery or thinly leathery, lateral veins 3–6 pairs, glabrous; petiole 0.2–2.1 cm long. **Inflorescences** cymes, few-flowered, terminal; peduncle 0–0.6(–1) cm long; bracts ovate, 0.5–1.4 cm long, sepal-like; pedicels to 1.2 cm long, puberulent. **Flowers** with sepals ovate or elliptic, 0.5–2 cm long, puberulent; corolla yellow-white outside, purple-yellow inside, tube 1.7–4 cm long, glabrous outside, lobes including tail to 13.5 cm long, pendulous; anthers included; ovary puberulent. **Fruits** paired follicles, 10–28 × 2.2–4.4 cm, apex obtuse; seeds 0.8–2 cm long, beak to 8 cm long, coma 2.5–10.5 cm long. **Distr:** W tropical Africa to Uganda and NW Angola. **Ecol:** Fl. Dec–May. **Zones:** 10–12. **Info:** This species is very similar to *S. petersianus*.

S. speciosus (Ward and Harvey) Reber, *Der Fortschr.* 3: 299, 1887. (Syn: *Christya speciosa, S. capensis*). [Forest Poison Rope]. **Shrubs**, to 1–4 m high, or **lianas** to 16 m long, evergreen; stems trichotomously branching; latex clear or white. **Leaves** usually ternate; lamina narrowly elliptic or slightly obovate, 2–12 × 0.6–3.5 cm, base decurrent into petiole, apex ± acuminate, acumen 1–6 mm long, adaxially glossy, dark green, abaxially dull, paler, coriaceous, veins 6–19 pairs; petiole 0.2–1.2 cm long. **Inflorescences** 3–26-flowered, congested, pedunculate; pedicels 0.6–2 cm long; bracts sepal-like. **Flowers** with sepals subequal, narrowly ovate, to 1.5 × ~0.2 cm, acute; corolla tube to 1.4 cm long, yellow, lobes ovate, 2–6 × 3–5 mm, gradually narrowed into spreading tails, lobes including tails 1.9–5 cm long; corona lobes subulate, 1.8–5 × 1–2.2 mm; stamens included, filaments short; ovary 1.8 × 2.1 mm, style 3.9–5.5 mm long. **Fruits** single (or paired) follicles, (7–)10–22 × 0.7–1.5 cm, divergent; seeds 1.3–2.2 cm long, coma 2.2–4.3 cm long. **Distr:** Zimbabwe to S Africa. **Ecol:** Mostly in evergreen forests; 300–1,800 m. Fl. Sep–Oct. **Zones:** 10–12.

Trachelospermum

Species 11. **Lianas**, woody, latex white. **Leaves** opposite; lamina ovate, glossy. **Inflorescences** cymes, lax, terminal, pseudo-axillary or axillary. **Flowers** white or purplish, 5-merous; calyx small, deeply divided, basal glands 5–10, apex usually denticulate; corolla salverform, tube cylindric, 5-angled, dilated at staminal insertion, throat constricted, lobes sharply overlapping to right; stamens inserted at lower third of corolla tube, anthers sagittate, connivent, adherent to pistil head, anther tips included or exserted, cells spurred at base; disk scales 5, free; ovaries 2, free, usually longer than disk, ovules numerous in each ovary, style short, pistil head conical. **Fruits** paired follicles, linear or fusiform, divergent or parallel; seeds linear-oblong, not beaked, coma silky-white. **Distr:** Tropical and subtropical Asia. **Cult:** Propagate from semiripe cuttings. Requires a moist, well-drained soil in full or partial sun.

T. jasminoides (Lindley) Lemaire, *Jard. Fleur.* 1: t. 61, 1851. (Syn: *Rhynchospermum jasminoides*, etc.). [Confederate Jasmine, Star Jasmine]. **Lianas**, woody, to 10 m long; stem brownish, lenticellate; young branchlets pubescent, glabrous when older. **Leaves** ovate to obovate, or narrowly elliptic, 2–10 × 1–4.5 cm, papery, glabrous, abaxially sometimes sparsely pubescent; petiole 0.3–1.2 cm long. **Inflorescences** cymes, paniculate, terminal and axillary; peduncles 2–6 cm long. **Flowers** very fragrant; sepals narrowly oblong, 2–5 mm long, spreading or reflexed, pubescent outside, ciliate, apex obtuse or acute; corolla white, tube dilated at middle, 0.5–1 cm long, throat glabrous or pilose facing stamens; lobes obovate, as long as tube; stamens included, inserted at middle of corolla tube; ovary glabrous. **Fruits** paired follicles, linear, 10–25 × 0.3–1 cm; seeds oblong, 1.5–2 cm long, coma 1.5–4 cm long. **Distr:** WS and S Japan to Indo-China. **Ecol:** On sunny forest margins; 200–1,300 m. Fl. Mar–Aug. Fr. Jun–Dec. **Zones:** 9–11. **Info:** Whole plant poisonous, often used for medicinal purposes. Flowers yield perfumed oil.

Vallaris

Species 3. **Shrubs**, trailing or scandent, latex white. **Leaves** opposite; lamina elliptic, oblong to lanceolate, glabrous, minutely dotted. **Inflorescences** cymes, umbellate or compound corymbose, dichotomously branched, axillary or terminal. **Flowers** white; calyx 5-partite, glandular or eglandular inside, lobes narrowly ovate-lanceolate, acute; sepals free, with or without basal glands inside; corolla subrotate, limb spreading, throat without scales, lobes broad, overlapping to right; stamens inserted at apex or middle of corolla tube, filaments short, with a large subapical dorsal gland, anthers sagittate, partially or completely exserted, connivent, adherent to pistil head, cells with rigid basal spurs; disk ringlike or cup-shaped, 5-lobed, lobes sometimes free, ciliate; ovaries bicarpellate, carpels distinct, many-ovuled, style filiform, pubescent, stigma somewhat annulate, thick. **Fruits** paired follicles, free, oblong, acuminate, splitting from below into 2 distinct follicles along ventral suture; seeds numerous, biseriate, apex crowned with coma, endosperm starchy. **Distr:** S China, tropical Asia. **Cult:** Propagate from seeds or cuttings. Requires a moist, well-drained soil in full sun. Best grown on a fence, arch or pergola.

V. solanacea (Roth) Kuntze, *Revis. Gen. Pl.* 2: 417, 1891. (Syn: *Peltanthera solanacea*). [Bread Flower]. **Shrubs**, climbing, often twining, to 10 m; bark dirty whitish-gray; flowering branchlets, slender, grayish-pubescent. **Leaves** elliptic to narrowly so, 2–15 × 0.8–6 cm, base cuneate or rounded, densely pubescent on both surfaces, lateral veins 5–12 pairs; petiole 0.2–2 cm long. **Inflorescences** cymes, di- or trichasial; peduncles 0.5–3 cm long; pedicels 0.5–2.5 cm long. **Flowers** fragrant; sepals ovate or narrowly elliptic, 2–7 mm long; corolla white or pale yellow, limb 1.4–2.5 cm Ø, tube 0.5–1 cm long, lobes rounded at apex; staminal glands yellow, globose; disk shorter than ovary, apex pilose. **Fruits** follicles, oblong, 8–14 × 1.5–3.5 cm; seeds ellipsoid, 0.9–1 cm long, coma 3–4 cm long. **Distr:** Indian Subcontinent to China (Hainan). **Ecol:** In forests, thickets, streambanks; 0–2,700 m. Fl. Mar–Jul. **Zone** 9–12.

Wrightia

Species 29. **Trees** or **shrubs**, with latex. **Leaves** opposite, petiolate, glands axillary. **Inflorescences** cymes, terminal or subterminal, dichasial, few- to many-flowered. **Flowers** with sepals quincuncial, with 5–10 basal scalelike glands inside; corolla salverform, funnelform, subrotate or rotate, tube cylindric to campanulate, lobes overlapping to left; corona ligulate, fringed, or cup-shaped, apex entire or subentire, shallowly or deeply divided, sometimes absent; stamens inserted at middle, apex, or rarely base of corolla tube, anthers sagittate, connivent and adherent to pistil head, exserted, spurred at base; disk absent; ovary bicarpellate, carpels distinct or connate, ovules numerous per carpel, style filiform, pistil head ovoid, usually dilated at base. **Fruits** paired follicles, connate or divaricate; seeds narrowly fusiform, with an apical coma directed toward fruit base, beakless. **Distr:** Ethiopia to S Africa, tropical and subtropical Asia. **Cult:** Propagate from cuttings, seeds or by-air-layering. Requires a moist, well-drained soil in full sun.

W. antidysenterica (L.) R.Br., *Mem. Wern. Nat. Hist. Soc.* 1: 74, 1811. (Syn: *Nerium antidysentericum*, etc.). [Arctic Snow, Milky Way, Snowflake, Winter Cherry Tree]. **Treelets**, evergreen, 0.7–2 m high. **Leaves** broadly ovate, 4–11 × 1–4 cm, widest at ± middle, base cuneate or rarely almost rounded, apex slightly to markedly acuminate, firmly membranous, dark green, lateral veins flat or faintly elevated above, more clearly prominent beneath, reticulate venation not sculptured on adaxially, often slightly raised abaxially; petiole 1–6 mm long. **Inflorescences** cymes, few-flowered; pedicels much longer than calyx; pedicels subequal to corolla tube, ~2.5 cm long. **Flowers** white; calyx lobes 1.5–3 mm long, glabrous, green; corolla tube 1.7–2.8 cm long, lobes obovate, ~2.5 cm long, rounded to acute; corona minutely puberulous, frilled. **Fruits** paired follicles, 9–15 × ~0.5 cm, glabrous. **Distr:** Sri Lanka. **Ecol:** In moist lowland regions, secondary scrubs, forest clearings; <600 m. **Zones:** 10–12.

W. dubia (Sims) Spreng., *Syst. Veg.* 1: 638, 1824. (Syn: *Cameraria dubia*, etc.). [Red Wrightia]. **Shrubs**, to 3 m high. **Leaves** elliptic or obovate, 9–20 × 2.5–8 cm, base acute to obtuse, apex acuminate, secondary veins 8–10 pairs; petiole 3–5 mm long. **Inflorescences** monochasial, few-flowered, condensed, terminal; peduncles ~5 mm long; bracts ovate, ~3 mm long; pedicels ~7 mm long. **Flowers** pink, yellow, orange to dull red inside, white to yellowish or greenish outside, fragrant; calyx lobes ovate, 2.5–6 mm long, with 5 orbicular to ovate squamellae; corolla

Below: *Wrightia antidysenterica, Wrightia dubia, Wrightia pubescens, Wrightia religiosa.*

HS

Above: *Wrightia tinctoria, Cryptostegia grandiflora, Hemidesmus indicus, Araujia sericifera.*

subinfundibuliform, tube suburceolate to campanulate, 0.6–1.2 cm long, lobes obliquely ovate and acute, to ovate-lanceolate and long-acuminate, 1.6–2.2 cm long; corona inconspicuous, antepetalous segments reduced to a thickened, triangular ridge, 2–3 mm long, alternipetalous segments entire, obtuse, ~1 mm long; stamens 5, inserted near base, anthers ~6 mm long, usually included, acumen barbate, filament inflexed; carpels 2, free, glabrous, ~1.5 mm long, common style gradually dilated near conical stigma. **Fruits** paired follicles, free, usually apically coherent, ~32 cm long; seeds linear-fusiform, ~1.5 cm long, coma white, ~4 cm long. **Distr:** Indo-China to Malaya. **Ecol:** In bamboo forests, evergreen forests. Fl. Nov–Aug. **Zones:** 9–12.

W. pubescens R. Brown, *Mem. Wern. Nat. Hist. Soc.* 1: 73, 1811. **Trees**, to 35 m tall; trunk to 60 cm dbh; bark yellowish-brown; young branchlets yellowish-pubescent, glabrate with age, densely lenticellate. **Leaves** narrowly oblong, ovate, or narrowly ovate, 5–10 x 3–6 cm, papery, adaxially puberulent to glabrous, abaxially densely pubescent to puberulent or glabrescent except along veins, veins 8–15 pairs; petiole ~1 cm long. **Inflorescences** cymes, ~5 cm Ø, pubescent. **Flowers** with sepals ovate to broadly ovate, 2–5 mm long; corolla white or pinkish-white, funnelform, tube 5–6.5 mm long, lobes oblong, 1–2 cm long; corona fringed, scales 10, as long as or longer than anthers, puberulent inside; stamens pubescent, inserted at mouth of corolla, anthers exserted; ovaries connate, glabrous. **Fruits** paired follicles, connate, sublinear, 15–30 x 1–2 cm, not lenticellate; seeds narrowly fusiform, coma to 3.5 cm long. **Distr:** S China, Indo-China to N Australia. **Ecol:** In secondary rainforests, dry woods; ~400 m. Fl. Apr–Aug. Fr. Aug–Dec. **Zones:** 10–12.

W. religiosa (Teijsm. and Binn.) Kurz, *Forest Fl. Myanmar:* 194, 1877. (Syn: *Echites religiosus*). [Sacred Buddhist, Water Jasmine, Wild Water Plum, Wondrous Wrightia]. **Shrubs**, to 3 m high; branchlets thin, terete, often with many lateral short branchlets, minutely puberulent. **Leaves** elliptic, ovate or narrowly oblong, 2.5–7.5 x 1.5–3 cm, pubescent along midvein, lateral veins 5–7 pairs; petiole 2–4 mm long. **Inflorescences** cymes, 1–13-flowered, often on short, few-leaved branches, shortly pedunculate; pedicels 1.5–2 cm long, thin, finely hairy. **Flowers** fragrant; sepals ovate, ~1.5 mm long; corolla white, subrotate, tube 3–4 mm long, glabrous, lobes ovate, ~7 mm long, surfaces densely pubescent; corona obsolete; stamens inserted at mouth of corolla tube; ovaries free. **Fruits** follicles, free, linear, 12–17 cm long; seeds narrowly fusiform, ~8 mm long, coma to 3.5 cm long. **Distr:** China (S Guangdong) to peninsular Malaysia. **Ecol:** Fl. throughout the year. **Zones:** 9–12. **Info:** Cultivated for medicinal purposes.

W. tinctoria R.Br., *Asclepiadeae*: 59, 1810. [Pala Indigo Plant, Dyers's Oleander]. **Trees**, small, deciduous; branchlets glabrous or puberulous. **Leaves** elliptic-oblong or obovate, 7.5–15 x 2.5–5.5 cm, base acute or rounded, apex acute or acuminate, abaxially glabrous or puberulous, lateral nerves 8–14 pairs; petiole 2.5–5 mm long. **Inflorescences** lax cymes, terminal; pedicels 2–1.2 cm long; bracts ovate, ~2.5 mm long. **Flowers** white or yellowish, fragrant, 2.5 cm Ø; calyx ~2.5 mm long, deeply divided, pubescent or glabrous, lobes oblong, obtuse, glandular inside, margin scarious; corolla tube ~5 mm long, lobes narrow, oblong, obtuse; corona of many linear scales, inserted on filaments and corolla lobes; anthers white or yellowish, forming a cone. **Fruits** paired follicles, pendulous, cylindrical, 2.5–5 x 0.6–0.8 cm, apically united when young, glabrous, striate; seeds 1.25–2 cm long, pale brown, coma 2.5–4 cm long. **Distr:** India. **Ecol:** In lower montane areas. Fl. Jul–Sep. **Zones:** 10–12.

Subfam. PERIPLOCOIDEAE

Genera 31, **species** 180. **Flowers** with corolla valvate, tube formation intermediate; anthers inserted at top of corolla tube, (staminal feet erect, connate, forming tube around ovary), anthers without guide rails, pollen in tetrads (Stevens 2001 onwards). **Distr:** Tropical to dry temperate regions of the Old World.

Cryptostegia

Species 2. **Lianas** twining, or scandent **shrubs**, glabrous; stems cylindrical, lenticels conspicuous; sap milky, abundant. **Stipules** interpetiolar, minute. **Leaves** opposite, shortly petiolate; lamina simple, coriaceous, without glands. **Inflorescences** cymes, several-flowered, terminal. **Flowers** with calyx campanulate, split nearly to base, lobes 5, ovate to lanceolate, acute; corolla whitish-pink or lavender, funnelform, lobes longer than tube; true corona absent, 5 corolline appendages present on tube; stamens 5, short, inserted at base of corolla, filaments short, free, distinct, anthers lanceolate, adnate to stigma; ovary superior, bicarpellate, stigma capitate (globular), large. **Fruits** paired follicles, divergent, boat-shaped, woody with ribs or wings; seeds numerous, small, ovate, crowned with a tuft of long hairs. **Distr:** Madagascar. **Cult:** Propagate from cuttings or seeds. Requires a moist, well-drained soil in full or partial sun.

C. grandiflora R. Br., *Bot. Reg.* tab. 435, 1820. [Purple Allamanda, Rubber Vine]. **Vines**, woody, twining, to 6 m long; stems cylindrical, glabrous or pilose at nodes, dark brown, with numerous minute lenticels; latex milky, abundant. **Stipules** interpetiolar and intrapetiolar. **Leaves** opposite; lamina elliptic to rounded, 5–10 x 2–6.3 cm, base obtuse to rounded, decurrent on petiole, margins entire, apex acute or obtuse, membranous, glabrous, venation pinnate, secondary veins 11–13 pairs, adaxially dull dark green, abaxially paler, with obscure venation; petiole 1–1.7 cm long, glabrous. **Inflorescences** arranged in dichasial cymes, 6–9-flowered, pedunculate; bracts foliaceous, ovate, 1–1.2 cm long. **Flowers** with calyx green, campanulate, sepals ovate, 1–1.5 cm long, pubescent; corolla infundibuliform, 5–7 cm long, violet, whitish inside, tube ~2 cm long; corona lobes lanceolate, ~1.5 cm long, bifid. **Fruits** paired follicles, divergent, brown when mature, 10–15.4 cm long, boat-shaped (keeled), woody; seeds reddish-brown, ovate-lanceolate, 3–4 mm long, with long white silky hairs. **Distr:** Madagascar. **Ecol:** In coastal regions; 0–500 m. Fl. & Fr. throughout the year. **Zones:** 10–12. **Info:** All parts of the plant are poisonous.

Hemidesmus

Species 1. *Generic description same as specific.* **Cult:** Propagate from seeds or cuttings. Requires a moist, well-drained soil in full sun.

H. indicus (L.) R. Br., *Hort. Kew*, ed. 2.2: 75, 1811. (Syn: *Periploca indica*). [Indian Sarsaparilla]. **Shrubs**, ± herbaceous, perennial, slender, laticiferous, twining, sometimes prostrate or semierect; roots woody, aromatic; stems numerous, slender, terete, thickened at nodes; branches terete, glabrous or pubescent. **Leaves** elliptic-oblong to linear-lanceolate, 5–10 x 0.8–1.8 cm, apex acute to obtuse, apiculate, glabrous, sometimes pubescent abaxially; petiole 3–5 mm long. **Inflorescences** crowded in subsessile axillary cymes; pedicels short, covered by numerous imbricating bracts. **Flowers** with calyx ~2.5 mm long, lobes ~2 mm long; corolla 5–6.5 mm long, greenish outside, purple inside, lobes 4–5 mm long. **Fruits** paired follicles, cylindrical, 10–15 x ~0.6 cm, tapering to a point; seeds ovate, oblong, 6–8.5 mm long, flattened, black, coma white, 2.5 cm long. **Distr:** India, Malaysia, Pakistan, Sri Lanka. **Ecol:** Fl. throughout the year. **Zones:** 10–12. **Info:** Cultivated for medicinal purposes.

Subfam. ASCLEPIADOIDEAE

Genera 214, **species** 2,365. **Leaves** spirally arranged. **Flowers** with corolla valvate; anthers bisporangiate (1 sporangium from each theca is lost), pollinia 2, caudicle present (absent in Fockeeae), anther secreting wall around pollinium, (pollen in tetrads, no pollinium wall in *Fockea*; orbicules ± present in *Riocreuxia*), granular layer of exine thin (Stevens 2001 onwards). **Distr:** Tropical to temperate regions worldwide, especially diverse in drier parts of Africa.

Araujia

Species 2–4. **Shrubs**, woody, evergreen; stems twining; exudate white latex. **Leaves** opposite; lamina with margins entire, grayish-green. **Inflorescences** cymes, axillary. **Flowers** fragrant; calyx 5-partite, sepals ovate; corolla slightly contorted in bud, hypocrateriform, lobes 5, ± erect at anthesis; corona single, segments 5, free, without awns. **Fruits** paired follicles, narrowly ovoid or pear-shaped, not spiny; seeds numerous, coma of silky hairs at one end. **Distr:** Tropical S America. **Cult:** Propagate from seeds or cuttings. Requires a fertile, well-drained soil in full or partial sun.

A. sericifera Brotero, *Trans. Linn. Soc. London* 12: 69, t. 4–5 1817. (Syn: *A. sericofera* ['*sericofera*' is an orthographic error], etc.). [Bladder Vine, Cruel Plant, Moth Catcher]. **Vines**, evergreen, sometimes partially deciduous, perennial, usually to 12 m long. **Leaves** opposite, spaced 7–18 cm along stems; lamina narrowly triangular, sagittate at maturity, 5–12 × 2–6 cm, base truncate to slightly lobed, apex acuminate, adaxially glabrous, glossy dark green, abaxially gray-green, minutely pitted, densely hairy; petiole 1.5–3 cm long. **Inflorescences** cymes or clusters, usually 2-flowered, or dichotomously branched and 4–8-flowered, to 10 cm long, below leaf axils; peduncles 0.8–2 cm long; pedicels 1–1.7 cm long. **Flowers** fragrant, waxy, white and pink-striped, to pink, 1–2 × 2–3 cm; sepals 5, basally fused, leaflike; corolla tubular, base inflated, 5-lobed, bell-shaped to funnelform, lobes spreading, triangular-ovate, 5–7(–9) × 4–6 mm, glabrous; stamens fused into a filament column; styles fused into a pistil head with 2 erect, elongate lobes. **Fruits** single (paired) follicles, narrowly ovoid, 8–15 × 4–5 cm, pendent; seeds numerous, 5–6 mm long, with numerous silky white, deciduous hairs at apex to 5 cm long. **Distr:** Central S America. **Ecol:** Fl. Dec–Feb. **Zones:** 9–11. **Info:** Pollinated mostly by insects. The pollen in the flowers are very sticky, sometimes trapping the pollinators, temporarily or permanently. All parts of the plant are poisonous, and sap causes skin irritation.

Asclepias

Species ~120. **Herbs**, base frequently woody. **Leaves** opposite or whorled, shortly petiolate. **Inflorescences** cymes, umbel-like, many-flowered, terminal and extra-axillary, erect. **Flowers** with calyx glands 5–10; corolla rotate, deeply parted, lobes reflexed, valvate, rarely overlapping to right; corona lobes 5, inserted on gynostegium, erect, apex hooded, with an incurved, ligular hornlike adaxial appendage; stamens inserted at base of corolla tube, filaments connate into a tube, anther appendages incurved; pollinia 2 per pollinarium, pendulous; stigma head conical. **Fruits** single or paired, follicles or capsules, fusiform, apex acuminate; seeds flat, with a long coma. **Distr:** Tropical America. **Cult:** Propagate from seeds. Requires a fertile, well-drained soil in full sun.

A. curassavica L., *Sp. Pl.* 1: 215, 1753. [Blood Flower, Indian Root, Swallow Root]. **Shrubs**, to 1–2 m high; stems pale gray, puberulent to glabrous. **Leaves** opposite; lamina lanceolate or oblong-lanceolate, 6–15 × 1–4 cm, base decurrent into petiole, apex acuminate or acute, glabrous or abaxially puberulent along veins; petiole to 1 cm long. **Inflorescences** cymes, 10–20-flowered, as long as leaves, mostly terminal; peduncle 3.5–6 cm long, pubescent; pedicels 1.2–2.5 cm long, pubescent. **Flowers** with sepals lanceolate, ~3 × 1 mm, pubescent; corolla purple or red, 7–9 mm long, lobes oblong, 5–8 × ~3 mm, reflexed; corona lobes yellow or orange, 3.5–4 mm long; gynostegium 2.5–3 mm long. **Fruits** follicles, fusiform, 5–10 × 1–1.5 cm; seeds ovate, 6–7 × ~3 mm, coma 2–4 cm long. Fl. Almost all year. **Distr:** Tropical America. **Ecol:** In moist or wet thickets, fields, open places, on hillsides, river banks and often along roadsides; <1,900 m. Fl. throughout most of the year. **Zones:** 8–12.

A. physocarpa Schltr., *Bot. Jahrb. Syst.* 21 (Beibl. 54): 8, 1896. (Syn: *Gomphocarpus physocarpus*). [Balloon Plant, Family Jewels Milkweed Tree, Swan Plant]. **Shrubs**, 1–2 m high, few-stemmed; young stems and inflorescences pubescent. **Leaves** narrowly lanceolate, 5–10 × 0.6–1.5 cm, adaxially sparsely pubescent, abaxially hairy along midvein, both ends tapering or acute; petiole ~1 cm long. **Inflorescences** cymes; peduncles 2–4.5 cm long, shortly hairy; pedicels 1.5–2.5 cm long. **Flowers** with sepals lanceolate; corolla white, 1.4–2 cm Ø; lobes ovate, 0.8–1 cm long, reflexed, margin densely bearded; corona lobes white, inner margin of hoodlike apex with 2, short, recurved or straight cusps, with a large adaxial nectary. **Fruits** follicles, capsulelike, relatively large, obliquely ovoid to subglobose, 6–8 × 2.5–5 cm, base oblique, apex rounded, beakless, pericarp with soft bristles or spines, minutely tomentose when young, glabrescent when ripe; seeds ovate, ~5 mm long, coma shiny white, ~5 mm long. **Distr:** Africa. **Ecol:** On mountains, and also on roadsides. Fl. throughout most of the year. **Zones:** 9–12.

A. tuberosa L., *Sp. Pl.* 1: 217, 1753. [Butterfly Weed, Flame Weed]. **Shrubs**; stems relatively stout, (36–)43–90 cm high; stems branching at the top, usually into 2–5 parts, these sometimes much-elongated and spreading, and each bearing inflorescences, rough-hairy, without milky latex. **Leaves** crowded, usually alternate, shortly petiolate or sessile; lamina variable, usually obovate, lanceolate-oblong to linear-lanceolate, 5.5–11.5 × 0.8–3.4 cm, base cuneate to cordate, abaxially hirsutulous; petiole to 5 mm long. **Inflorescences** of 1–5 umbels, mostly cymose, borne on terminal branches, shortly pedunculate. **Flowers** with corolla in various shades of orange, 4–8 mm long, lobes oblong, purplish or slightly greenish-orange; column short; hoods 4–6 mm long, lanceolate or narrowly oblong, erect, deep bright orange, much surpassing anthers, nearly equalled by the subulate horns; horns narrowly acicular. **Fruits** capsules, erect, 8.5–12.2 cm long, minutely pilose. **Distr:** S USA to Mexico. **Ecol:** In dry and sandy soils. Fl. Jun–Sep. Fr. Jul–Oct. **Zones:** 4–10.

Calotropis

Species 3. **Shrubs**, erect, canescent; stems succulent. **Leaves** opposite, almost sessile; lamina broad, subsessile. **Inflorescences** cymes, umbel-like, many-flowered, extra-axillary or terminal, long-pedunculate. **Flowers** usually fleshy; calyx with basal glands, deeply divided into 5 lobes, tube short or almost absent; corolla campanulate or bowl-shaped to subrotate, divided to the middle, lobes valvate or overlapping to right, reflexed; corona lobes 5, arranged edge to edge, adnate to gynostegium, fleshy, laterally compressed, apex with a tubercle on each side, with abaxial, basal, revolute spur; filaments connate, anther appendages incurved; pollinia 2 per pollinarium, oblong, pendulous; styles long, stigma head slightly convex. **Fruits** paired follicles, ovoid, subglobose, or oblong-lanceolate, mesocarp inflated. **Distr:** N Africa, Arabia, and tropical Asia. **Cult:** Propagate from cuttings or seeds. Requires a well-drained soil in full sun.

C. gigantea (L.) W.T. Aiton, *Hortus Kew.* ed. 2. 2: 78, 1811. (Syn: *Asclepias gigantea*, etc.). [Bowstring Hemp, Crown Flower]. **Shrubs**, 1–5 m high. **Leaves** obovate-oblong or oblong, 7–30 × 3–15 cm, base cordate, apex obtuse, cottony-tomentose when young, frequently glabrescent and glaucous green; lateral veins 4–8 pairs; petiole 1–4 mm long. **Inflorescences** umbel-like cymes, woolly hairy; peduncle robust, 5–12 cm long; pedicels thick, 2–5 cm long. **Flowers** with calyx almost flat, 1.2–1.5 cm Ø; corolla usually purplish, lilac, pink or pale greenish-white, base paler green, 2.5–3.5 cm Ø, fleshy, glabrous, lobes ovate, 1–1.5 × 0.6–1 cm, spreading or reflexed, margin revolute; corona shorter than gynostegium. **Fruits** paired follicles, large, obliquely elliptic to oblong-lanceolate, 5–12 × 2.5–4 cm, ends incurved, surface often ridged when mature; seeds broadly ovate, 5–7 × 3–4 mm, coma 2.5–4 cm long. **Distr:** Tropical Asia and Africa. **Ecol:** On streambanks and seashores; 0–1,400 m. Fl. Almost throughout the year. **Zones:** 10–12. **Info:** Semiripe fruits are eaten by an indigenous species of lizard in the Maldives. The salt-tolerant plant has many medicinal and other uses.

C. procera (Aiton) W.T. Aiton, *Hortus Kew.* ed. 2. 2: 78, 1811. (Syn: *Asclepias procera*, *A. gigantea*). [Sodom's Apple]. **Shrubs** or **small trees**, to 6 m high. **Leaves** ovate to obovate, 10–30 × 5–15 cm, base cordate, apex abruptly pointed, grayish-green or hoary-white; petiole very short. **Inflorescences** cymes, umbel-like, 3–10-flowered, finely woolly; peduncles 2–5.5 cm long; pedicels 1.5–2.5 cm long; flower buds subglobose. **Flowers** with sepals elliptic or ovate, ~5 × 3 mm; corolla white outside, pinkish inside, 1.5–2 cm Ø, lobes with purple-brown apices, broadly ovate or ovate-triangular, 0.7–1 × 0.6–1 cm, spreading; corona as long as gynostegium. **Fruits** paired follicles, subglobose to obliquely ovoid, inflated, 6–10 × 3–7 cm, pericarp thick, spongy; seeds ~6 × 4 mm, coma 3.5–5 cm long. **Distr:** Afghanistan, China, India, Myanmar, Nepal, Pakistan, Thailand, Vietnam; Africa, SW Asia. **Ecol:** Often near waterways. Fl. May–Dec. **Zones:** 9–12.

Below: *Asclepias curassavica*, *Asclepias physocarpa*, *Asclepias physocarpa* (fruits), *Asclepias tuberosa*.

PL

Above: *Calotropis gigantea*, *Calotropis gigantea* (white form), *Calotropis procera*, *Ceropegia haygarthii*.

Ceropegia

Species ~170. **Herbs**, perennial, erect or twining, sap clear or cloudy, rarely milky; rootstock often a cluster of fusiform roots or a subglobose tuber, sometimes a rhizome (or with fibrous roots only); stems herbaceous (to very succulent). **Leaves** relatively small or often absent. **Inflorescences** mostly umbel-like, less often racemelike and sometimes branched, extra-axillary (rarely terminal). **Flowers** usually large; calyx deeply 5-parted; basal glands many, small; corolla tubular, base swollen, often asymmetrically, upper part often funnelform; lobes usually slender, apically coherent; corona double, outer lobes 5, joined to form a cup, entire to deeply bilobed so that outer corona is 10-toothed; inner lobes 5, subulate to narrowly spathulate, basally incumbent on anthers, apical part usually long, erect; filaments connate into a very short tube, anthers without apical appendages; pollinia 2 per pollinarium; stigma head convex or impressed. **Fruits** follicles, linear, narrowly fusiform, or cylindric, rough or smooth. **Distr:** Africa, extending through tropical Asia to Oceania. **Cult:** Propagate from seeds, cuttings or by division of tubers. Requires a well-drained gritty soil in full or partial sun.

C. haygarthii Schltr., *Bot. Jahrb.* 38: 46, f. 7A, 1905. (Syn: *C. tristis*, etc.). [Wine-glass Vine]. **Climbers**; stems twining; roots fibrous. **Leaves** ovate to broadly elliptic, ~7.5 × 5 cm, margins entire, apex acute, fleshy, adaxially green, abaxially pale grayish-green. **Inflorescences** 1- to few-flowered; pedicels 1–1.5 cm long. **Flowers** to 10 cm long; corolla 3–5 cm long and sharply bent at right angles near the somewhat inflated base, tube 2–2.5+ × 0.45–0.6 cm, widening at mouth to 1–2.2 cm Ø, lobes with 3 distinct portions (5 quadrate, platelike extensions of the keels formed from the replication within the tube; a single slender almost filiform erect central column made by the junction of all 5 lobes; and an apical cagelike or flanged structure formed from the expanded lobes with ciliate margins), central column of corolla lobes is usually 1–1.5 cm long, sometimes only 2–3 mm long. **Distr:** S Africa. **Ecol:** In open scrub bush, sometimes on slopes near water courses; 0–1,300+ m. **Zones:** 9–12. **Info:** This species is often confused with the related species *C. distincta*.

C. linearis **ssp.** ***debilis*** (N.E. Brown) H. Huber, *Mem. Soc. Brot.* 12: 118, 1957. (Syn: *C. debilis*). [Chain of Hearts]. This species differs from *C. linearis* ssp. *woodii* with the following features: **Leaves** ± sessile; lamina narrowly linear, succulent, semicircular in cross-section. **Flowers** with corolla lobes narrowly spathulate, apically connate, lamina folded back along midrib, inside black-purple with violet; corona glabrous. **Distr:** Malawi, Zimbabwe. **Ecol:** In coastal sandhills and bushes, rocky outcrops, and twining or hanging on cliffs on rock ledges; 0–30 m. Fl. Feb–Apr. **Zones:** 10–12.

C. linearis **ssp.** ***woodii*** (Schlechter) H. Huber, *Mem. Soc. Brot.* 12: 118, 1957. (Syn: *C. euryacme*, etc.). [Chain of Hearts]. **Climbers**, succulent; rootstock tuberous, tubers usually globose, 2.5–5(–10) cm Ø. **Leaves** variable, usually broadly cordate, ovate to lanceolate, 0.6–1.8 × 0.3–1.6 cm, apex often acuminate, succulent, adaxially flat, abaxially convex, pale green, often red-suffused; petiole 0.3–1 cm long. **Inflorescences** 2–3-flowered, borne from nodes, flowers held upright. **Flowers** pallid greenish with purplish vertical stripes; corolla 1.8–2.5 cm long, pale green-white, whitish-rose, often striped with dark violet, lobes dark purple, rarely greenish, inflated base globose, ~4 × 4 mm, inside purple, merging ± gradually into the tube, tube 2–3 mm Ø, base purple, apex whitish, inside hairy, lobes narrowly spathulate, basally folded back along midrib, erect, 6–7 mm long, forming an elongated cage with a blunt (rarely cone-shaped) tip, inside purple-hairy; corona white, sessile or shortly stipitate, ~2.5–3 × 2 mm; interstaminal corona lobes with upper margin often setulose; staminal corona 2–2.5 mm long. **Distr:** Kenya, Zimbabwe, S Africa, Swaziland. **Ecol:** In coastal sandhills and bushes, rocky outcrops, and twining or hanging on cliffs on rock ledges; 0–30 m. Fl. Feb–Apr. **Zones:** 10–12.

C. nilotica Kotschy, *Sitzb. Acad. Wien, Math.-Nat.* 51: 356, 1865. **Climbers**, tuberous, glabrous; stem twining, slender, fleshy; internodes 5 cm long. **Leaves** ovate-lanceolate, 2.5 × 1.7 cm, narrowed at both ends, apex acuminate, denticulate toward apex with callose cuspidate teeth, somewhat fleshy. **Inflorescences** 2-flowered; peduncles lateral, quadrangular. **Flowers** with sepals subulate, ~3 mm long, spreading; corolla 4 cm long, tube clavate, base slightly inflated, dilated and 1.25 cm wide at mouth, glabrous, lobes deltoid, 6 mm long, apically connate, blackish-purple and pilose within, marked with a yellow spot at base; outer coronal lobes horizontal, short, dilated-truncate, entire, glabrous; inner coronal lobes linear-spathulate, straight, 3× as long as the outer corona, glabrous, yellowish. **Distr:** Tropical Africa; widespread. **Ecol:** In tropical and subtropical areas; <900 m. Fl. Nov–Jan. **Zones:** 10–12.

C. radicans Schlechter, *Bot. Jahrb. Syst.* xviii. Beibl. n. 45: 12, 1894. **Creepers**, sparingly branched, rooting from nodes; stems pliant and fleshy; roots clustered, fleshy. **Leaves** ovate to oblong, to 4 × 2.5 cm, succulent, hairless. **Inflorescences** 1(–3)-flowered; pedicels to 3 cm long. **Flowers** with sepals linear-lanceolate, 4–6 mm long; corolla 5.5–8 cm long, base inflated, 1.3–1.7 × 0.5–0.7 cm, weakly 5-angled, whitish-green, tube 4–5.5 × ~0.3 cm, basally narrowed to ~3 mm Ø, cream, inside base hairy, maroon-flecked, widening gradually to a funnelform mouth, 1.5 cm wide, purple-blotched, lobes 5, elongated and forming a conical crown, obtusely lanceolate to triangular, 1.5–3 × ~0.15 cm, broadly keeled, thinly hairy, base rich maroon, middle white, apex green to yellow, tips fused, folded back along midrib, basal angles projecting, inside basally black-purple with a white and then a black-purple transverse stripe; corona shortly stipitate, fused into a cup with upper margin ± straight and slightly revolute; interstaminal corona lobes pouchlike; staminal corona linear, erect. **Distr:** S Africa (E Cape). **Ecol:** Usually creeping below shrubs. **Zones:** 10–12.

C. sandersonii Decne. ex Hook., *Bot. Mag.* 95: t. 5792, 1869. (Syn: *C. fimbriata*, etc.). [Parachute Plant, Umbrella Flower]. **Climbers**, twining, lofty, glabrous; stems green, stout, succulent, sparingly branched. **Leaves** relatively small and distant, shortly stoutly petioled; lamina ovate-cordate, 3.8–6.5 cm long, obtuse, thick, succulent, nerveless, deep green. **Inflorescences** 3–4-flowered, axillary; peduncles short, curved, bright green; bracteoles small, subulate; pedicels short. **Flowers** with calyx lobes subulate, 6 mm long, green; corolla 6.5 cm long, curved at base, 5 cm Ø at top of the tube, tube slightly inflated and green at base, expanding into a funnel-shaped, 5-angled, transparent limb, with opaque green reticulated veins, presenting 5 short distant lobes on its margin, bearing 5 horizontal appendages forming the umbrellalike cap of the flower, cap bright green, surface pitted, formed of 5 confluent convex lobes with a conical central papilla, blackish-mottled, tips darker, lobes with outer margin bilobed, margins turned up bearing a series hairs; staminal corona yellow, terminating in 5 erect strap-shaped connivent processes, apices recurved. **Distr:** KwaZulu Natal, Mozambique and Swaziland. **Ecol:** Usually climbing on shrubs, in bushes on stream banks, often in stony soils. Fl. Nov–Feb. **Zones:** 10–12.

Dischidia

Species ~80. **Herbs**, epiphytic or rarely epilithic; stems and branches fleshy, usually climbing by adventitious roots, sometimes also twining, or hanging; latex white. **Leaves** opposite or partly 3–4-whorled, fleshy and pitcher-shaped, these often inhabited by ants, sometimes leafless. **Inflorescences** usually umbel-like groups of flowers, small, extra-axillary; rachis thick, often branched, eventually elongated. **Flowers** mostly very small; calyx with 5 basal glands; corolla white, red, or violet, ovoid to urceolate, with a narrow mouth, fleshy, lobes valvate in bud, short, often hairy; corona lobes 5, slender, inserted on gynostegium, erect or ascending, apex entire, notched, 2-cleft, or with recurved arms; anthers erect, with apical membrane overlying stigma head, pollinia 2 per pollinarium, margin translucent, caudicles enlarged at apex, erect. **Fruits** follicles, lanceolate or cylindric; seeds ovate, flat, brown, with a tuft of white coma at the microphylar end. **Distr:** Tropical and subtropical Asia and Oceania. **Cult:** Propagate from seeds or stem cuttings. Requires a compost containing a mixture of moist peat and perlite, in partial sun.

SZ

Top row: *Ceropegia linearis* ssp. *debilis*, *Ceropegia linearis* ssp. *woodii*, *Ceropegia nilotica*, *Ceropegia radicans*.
Bottom row: *Ceropegia sandersonii*, *Dischidia major*, *Dregea volubilis*, *Edithcolea grandis*.

D. major (Vahl) Merrill, *Interpr. Herb. Amboin.*, 437, 1917. (Syn: *D. rafflesiana*, etc.). [Ant Plant, Malayan Urn Vine]. **Herbs**, epiphytic; rooting at nodes and adventitious, those at nodes often growing into the pitcherlike leaves; stems thick, 3–4 mm Ø, glabrous. **Leaves** of 2 types: (1) with petiole 2–3 mm long, lamina circular, 2.5–2.7 × 1.6–1.8 cm, base cuneate, apex acuminate, flat, (2) with petiole 8 mm long, lamina pitcher-shaped or tubular, 8–10 × 3–4 cm, inside dark purple, tips obtuse, one side flattened. **Inflorescences** clusters, few-flowered; peduncles 1.5 cm long; pedicels 3.5–4 mm long. **Flowers** with sepals ovate, 1–1.5 × 1 mm, margins minutely denticulate; corolla 3–4 × 2.5–3 mm, yellow-green, or striped with yellow and green, lobes lanceolate, 1.5 × 0.5 mm, acuminate, revolute, tips reddish, petaloid; corona flat; anthers ovate; style head obtuse, 0.25 mm long, pollinia 0.5–0.6 × 0.1–0.2 mm. **Fruits** fusiform, 4.5–5 × 0.3 cm, reniform in cross-section, yellow-green, glabrous; seeds 2–2.5 × 1 mm, with a tuft of hairs 1.3–1.5 mm long. **Distr:** Tropical Asia through Malesia to Australia. **Ecol:** On trees bordering rainforests, and also on trees in dry areas. Fl. May–Aug. **Zones:** 10–12.

Dregea

Species ~12. **Lianas. Leaves** ovate, rather firm, but not fleshy. **Inflorescences** umbel-like clusters of cymules, solitary, borne at nodes, flowers numerous; peduncle and pedicels long and slender. **Flowers** with sepals overlapping, with 5 basal glands, triangular-ovate, obtuse to almost acute; corolla rotate to shallowly bowl-shaped, deeply 5-divided, lobes overlapping to right, usually green; corona lobes 5, ± spreading, thickly fleshy, adnate to stamens, outer angle obtuse or rectangular, interior angle produced into an acute tooth incumbent to anthers; anthers with membranous apical appendages, pollinia 2 per pollinarium, oblong, erect; stigma head umbonate or thickly conical. **Fruits** paired follicles, widely divergent, thick, finely longitudinally ribbed or corrugate, rarely winged; seeds comose. **Distr:** Africa, S Asia. **Cult:** Propagate from seeds or cuttings. Requires a moist, well-drained soil in full sun.

D. volubilis (L. f.) Bentham ex J. D. Hooker, *Fl. Brit. India*. 4: 46, 1883. (Syn: *Tylophora macrantha*, *Wattakaka volubilis*, etc.). [Cottom Milk Plant, Green Wax Flower, Sneeze Wort, Sneezing Silk]. **Lianas**, to 12 m long. **Leaves** broadly ovate or suborbicular, 7–18 × 4–17 cm, base shallowly cordate, apex acute or shortly acuminate, glabrous or softly pubescent, lateral veins ~4 pairs; petiole 2.5–6 cm long. **Inflorescences** many-flowered, pendent; peduncles 2–6 cm long, slender, puberulent; pedicels 2–2.5 cm long. **Flowers** green or yellowish-green, fragrant; sepals ovate-oblong, 2.5–3 mm long, pubescent, ciliate; corolla glabrous, lobes broadly ovate, 0.6–1.2 × 0.5–1.2 cm, obtuse, ciliate; corona yellowish-green, 4–4.5 mm Ø; anther appendages white, pollinia oblong; ovaries pilose. **Fruits** follicles, narrowly ovoid, 10–15 × 3–4 cm, longitudinally wrinkled-striate or irregularly ribbed; seeds ovate, ~1.2 × 0.6 cm, flattened, marginate; coma ~4.5 cm long. **Distr:** Tropical Asia through Indo-China and Malaysia to Taiwan and Philippines. **Ecol:** In montane forests. Fl. Apr–Sep. Fr. Jul–Dec. **Zones:** 9–12. **Info:** Young leaves are edible; leaves and other parts used for medicinal purposes.

Edithcolea

Species 1. *Generic description same as specific.* **Cult:** Propagate from stem cuttings or seeds. Requires a well-drained, gritty, potting mix.

E. grandis N.E. Brown, *Bull. Misc. Inform. Kew* 220, 1895. (Syn: *E. sordida*, etc.). [Persian Carpet Flower]. Succulents, perennial; stems prostrate, much-branched, ~10 × 1–1.5 cm, 4–5-angled, glabrous, ribs sometimes spirally twisted, tubercles conical, apically with brown acute hard persistent spines. **Inflorescences** mostly 1-flowered, borne on upper parts of stems; pedicels 1.8–2.3 cm long, glabrous. **Flowers** with sepals ovate-lanceolate to subulate, 5–9 × 1.5–3 mm; corolla ± flat, 8–12.5 cm Ø, incised halfway, outside olive-green, inside pale yellow with scattered to densely velvety purplish-brown spots, spots sometimes raised, tube saucer-shaped, 6 × 6 mm, mouth raised slightly like an annulus, inside finely papillate with narrow raised purplish-brown concentric rings, lobes ovate-tailed, 3.5–5 × 2.5–3 cm, upper half dark brown to olive-green, smooth, glabrous, tube margin with 5 narrow radial bands with clavate purplish-brown hairs, to 1 cm long; corona biseriate, yellowish; anther rectangular, pollinia ~0.5 × 0.2 mm. **Fruits** fusiform, spreading; seeds broadly ovate, ~6 × 4 mm, broadly winged, pale brown, hair crown 2.5–3 cm long. **Distr:** Ethiopia, Kenya, Socotra, Somalia, Tanzania, Uganda. **Ecol:** In dry rocky places; 200–1,500 m. Fl. throughout most of the year. **Zones:** 11–12.

Hoya

Species ~100. **Subshrubs** or **lianas**, epiphytic or epilithic, often twining or climbing by adventitious roots. **Leaves** opposite; lamina fleshy, papery or membranous. **Inflorescences** umbel-like, producing a succession of globose or flat-topped clusters, extra-axillary or sometimes terminal; old rachis thick, with contiguous pedicel scars. **Flowers** with calyx small, with basal glands; corolla fleshy, rotate, reflexed (or campanulate), lobes valvate, often densely hairy or scurfy inside; corona lobes 5, depressed-patent or vertically adnate to gynostegium, fleshy, inner angle a tooth incumbent on anthers, outer angle rounded or acute, margin revolute, basally longitudinally grooved; stamens shortly connate, apical membrane of anthers appressed against stigma head; pollinia 2 per pollinarium, oblong, erect, with raised, translucent margin; stigma head discoid, rounded or subapiculate. **Fruits** follicles, often solitary, cylindric-fusiform. **Distr:** SE Asia to Oceania. **Cult:** Propagate from stem or leaf cuttings. Requires a well-drained potting mix in full or partial sun.

H. bella Hook., *Bot. Mag.* 74: t. 4402, 1848. [Miniature Wax Plant, Wax Flower]. **Shrubs**, scandent; stems branching, weak; branches somewhat climbing, copiously leafy. **Leaves** opposite; lamina ovate, base broadly cuneate, 4.5 × 1.6 cm, widest near base, margins entire, apex shortly acuminate, fleshy, 1-nerved, midvein depressed above, prominent below, adaxially dark green, abaxially paler; petioles short, stout, curved, pale reddish. **Inflorescences** corymbs, 8–10-flowered, ~6 cm Ø; peduncles lateral; pedicels 2 cm long, hairy. **Flowers** scented, to 2 cm Ø; calyx slightly downy, sepals 5, spreading, elliptic; corolla rotate, convex, nearly white, lobes 5, obovate-triangular, 1.1 × 0.7 cm, acute, hairy outside; corona alternating with corolla, ovate or rather cymbiform, ~9 mm Ø, concave, adaxially deep-purple, paler below, lobes 4.5 × 2.4 mm. **Distr:** Myanmar (Taung Kola Mountain, Moulmein). **Ecol:** In cool montane forests or moist highland rainforests. Fl. throughout most of the year. **Zones:** 10–12.

H. bilobata Schltr., *Philipp. J. Sci.* 1 (Suppl. 4): 301, 1906. **Epiphytes**; stems elongate, terete, puberulous, foliate. **Leaves** elliptic to suborbicular, 1.7–2.2 × 1.3–1.8 cm, apex obtuse, glabrous, coriaceous; petiole 3 mm long, carnose. **Inflorescences** umbellate cymes, ~20-flowered, upright; peduncles terete, 1–3 cm long, glabrous; pedicels filiform, glabrous, usually to 8 mm long. **Flowers** with calyx lobes oblong, ~1 mm long, obtuse, glabrous; corolla rotate, lobes 5, decurved, ovate, 1.5 mm long, obtuse, outside glabrous, inside shortly and densely papillose; corona scales produced at back into 2 rather large lobes; pollinia oblong. **Distr:** Philippines (Mindanao). **Ecol:** On trees at low altitudes. Fl. throughout most of the year. **Zones:** 11–12.

H. carnosa (L. f.) R. Brown, *Prodr.* 460, 1810. [Wax Plant]. **Shrubs**, epiphytic, climbing, glabrous except for inflorescences; stems robust, to 6 m long, pale gray, smooth. **Leaves** broadly ovate-cordate to ovate-oblong or elliptic, 3.5–13 × 3–5 cm, base rounded to shallowly cordate, apex obtuse or shortly acuminate, lateral veins ~4 pairs, obscure; petiole 1–1.5 cm long. **Inflorescences** pseudo-umbels, ~30-flowered, globose, extra-axillary, pubescent; peduncles ~4 cm long; pedicels 2–4 cm long. **Flowers** with corolla white, sometimes with a pink center, rotate, 1.5–2 cm Ø, lobes triangular, densely papillate inside, margin recurved, apex reflexed; corona lobes stellate-spreading, outer angle acute, middle ridge prominent, margin strongly reflexed and enclosing a hollow space at base, inner angle acute, incumbent on anthers; stigma head obtuse or obscurely apiculate. **Fruits** follicles, linear-lanceolate, 6–10 × 0.5–1.5 cm; seeds ~5 × 1 mm, coma ~2.5 cm long. **Distr:** China, India, Malaysia, Taiwan, Vietnam. **Ecol:** In montane forests, usually on trees; 200–1,200 m. **Zones:** 9–12.

H. crassicaulis Elmer ex Kloppenb., *Fraterna*, 1995(3): 10, 1995. **Climbers**, to 7 m long. **Leaves** deciduous, opposite or in clusters; lamina oblong, sometimes ovate-oblong, 10–17 × 5–8 cm, base obtusely rounded to broadly cuneate or rounded, margins entire, apex gradually tapered, adaxially yellowish-green, abaxially paler; petiole 1.5–5 cm long. **Inflorescences** umbels, spherical, upright, borne from terminal leaf axils; peduncles 5 cm long; pedicels filiform, 1.5–2.5 cm long, yellowish-brown. **Flowers** fragrant, dull, perianth tips purplish-brown; calyx segments narrowly oblong, 1.5 mm long, subhyaline, tips acute, reddish-brown, with 5 outer lobes, and a pair of inner ones; corolla rotate, basally united, also hyaline, glabrous, membranous, 5 × 5 mm, upper half strongly inflexed, apex acute, prominent; corona with 5 horns alternating with corolla lobes, wheel-like, ovately oblong and with a pair of pockets in their sinuses. **Fruits** follicles, dry, dehiscing along upper side, terete, linear, 12 cm long; seeds 3–5 mm long, compressed, pale brown, base obtusely rounded, apex truncate with white coma. **Distr:** Philippines (Luzon). **Ecol:** At low elevations. **Zones:** 11–12.

H. imbricata Decne., *Prodr.* (DC.) 8: 637, 1844. (Syn: *Concophyllym imbricatum*, etc.). **Climbers**; stem wiry, forming rootlets at nodes; stems and leaves densely appressed to host tree. **Leaves** alternate, imbricate, shortly petiolate; lamina orbicular, 8–10 cm Ø, base rounded to cordate, margins reflexed, apex obtuse to rounded, platelike, very thick and rigid, adaxially mostly dull glaucous green, abaxially purplish and concave with a rigid root system, providing housing for ants that live in a symbiotic relationship with the plant. **Inflorescences** many-flowered, erect, concave; peduncles ~10 cm long; pedicels variable in length, green. **Flowers** odorless; sepals narrowly elliptic, acute, glabrous; corolla 0.8–1 cm Ø, cream-colored, lobes much-revolute, triangular, velvety; staminal corona lobes narrowly rounded, inner appendages elliptic, steeply ascending, white; anther appendages extending beyond corona, acuminate; stigma apiculate. **Fruits** follicles, fusiform; seeds 7 mm long, brown with a tuft of hairs to 1.4 cm long. **Distr:** Indonesia (Sulawesi), Philippines. **Ecol:** Usually epiphytic on medium-sized trees; 300 m. **Zones:** 11–12.

H. kerrii Craib, *Bull. Misc. Inform. Kew.* 1911: 418, 1911. (Syn: *H. obovata* var. *kerrii*). [Heart Leaf, Sweetheart Hoya, Valentine Hoya, Wax Hearts]. **Climbers**, to 3 m long; stems stout, to 1 m long, glabrous, pale. **Leaves** broadly obovate to orbicular or circular heart-shaped, 4–10 × 5–10 cm, base rounded to broadly cuneate, margins revolute, hairy, apex retuse, sinus to 1 cm wide, glabrous, dark green, leathery, lateral veins obscure; petiole 0.5–2 cm long, robust. **Inflorescences** umbels or pseudo-umbels, to ~20-flowered, 4–5 cm Ø, pendent; peduncle 2–3 cm long; pedicel slender, 1.3–1.8 cm long, puberulent. **Flowers** ~1.3 cm Ø, densely hairy inside; sepals ~2.5 mm long, puberulent; corolla creamy-white, papillate-puberulent, reflexed, lobes ovate-triangular, 4–5 × ~4 mm; corona lobes reddish or rose-purple, furrowed on back, outer angle rounded, apical part concave, ~2.5 mm long. **Distr:** Laos, Malaysia, Thailand, Vietnam. **Ecol:** In lowland deciduous dipterocarp-oak forests, dry forests and seasonal hardwood forests; 450–1,300 m. Fl. May–Aug. **Zones:** 10–12.

H. lacunosa Blume, *Bijdr.* 1063, 1826. (Syn: *Otostemma lacunosum*). [Wax Plant]. **Subshrubs**, epiphytic; stems climbing to 1.5 m by adventitious roots, glabrous or sparsely pubescent. **Leaves** ovate or ovate-lanceolate, 2–5 × 1.2–3 cm, base cuneate or rounded, apex acute, glabrous or subglabrous, lateral veins 2–4 pairs, obscure; petiole 0.3–1.2 cm long, subglabrous. **Inflorescences** umbels, pendent, ~30-flowered; peduncle 2.5–7 cm long, robust; pedicels 0.5–2.5 cm long. **Flowers** with sepals obtuse; corolla yellowish-white with dense soft white hairs, 6–7 mm Ø, tube conspicuously white-villous inside, lobes suborbicular, much longer than tube; corona yellow, lobes ovate, constricted at middle, outer angle rounded, inner angle acute, surface with a thin keel; apical anther membranes triangular, exceeding stigma head; pollinia subobovate. **Fruits** follicles, lanceolate, 5–7 cm long. **Distr:** Thailand, Malaysia, Indonesia. **Ecol:** In lowland forests. Fl. Jan–Aug. **Zones:** 11–12.

H. linearis Wallich ex D. Don, *Prodr. Fl. Nepal.*, 130, 1825. (Syn: *H. linearis* var. *nepalensis*, etc.). [Porcelain Flower]. **Shrubs**, epiphytic; stems pendent, to 1.5 m long, thin, soft, hirsute, puberulent, grayish-green, ± villose. **Leaves** linear, 2.5–5 × ~0.5 cm, base cuneate, apex mucronate, lateral veins obscure, abaxially canaliculate; petiole ~2 mm long. **Inflorescences** pseudo-umbels, terminal, sessile, many-flowered; pedicel ~2.5 cm long. **Flowers** with sepals linear-lanceolate, villose; corolla ± white, 0.8–1.2 cm Ø, inside glabrous or papillose, margin ciliate; corona lobes stellate, spreading, outer angle acute, inner angle acute, base long-spurred. **Fruits** follicles, fusiform, ~6.5 × 0.5 cm; seeds 3 mm long, coma white, ~2 cm long. **Distr:** China (Yunnan), India (Sikkim), Myanmar, Nepal. **Ecol:** In evergreen forests; 1,500–2,000 m. Fl. Sep. **Zones:** 8–11.

H. mindorensis Schltr., *Philipp. J. Sci.* 1(Suppl. 4): 303, 1906. [Aloha Hoya]. **Shrubs**, epiphytic; stems terete, glabrous, laxly foliated. **Leaves** oblanceolate-elliptic, 9–12 × 3.5–4.5 cm, apex shortly acuminate, glabrous, coriaceous; petioles terete, 2–2.5 cm long, carnose. **Inflorescences** cymes, umbelliform, many-flowered; peduncle terete, ~2 cm long, glabrous; pedicels slender, filiform, ~1.5 cm long, glabrous. **Flowers** small; calyx segments ovate, ~1.5 mm long, obtuse, minutely ciliate; corolla campanulate or reflexed, ~9 mm Ø, recurved, usually 5-lobed, glabrous outside, slightly puberulous inside, lobes sparsely hispid, ovate, base obtuse, apex obtuse, auricles 2 between the corolla lobes; corona horizontal, lobes thick, elliptic, margins laterally compressed, anterior apex acuminate, posterior apex acute, exterior angles rising upward, inner angles toothlike incumbent upon the anthers; pollinaria ovoid or oblongoid, laterally compressed. **Distr:** Philippines (Mindoro). **Ecol:** In shaded areas of tropical forests. Fl. anytime during the year, often Mar–May. **Zones:** 11–12.

H. multiflora Blume, *Catalogus.* 49, 1823. (Syn: *Centrostemma multiflora*, *C. platypetalum*). [Shooting Star Hoya]. **Shrubs**, glabrous; stems erect or decumbent, to 2.5 m high, leaf scars persistent. **Leaves** oblong-lanceolate, 8–18 × 2–6 cm, base cuneate, apex obtusely acuminate; petiole 1–2 cm long. **Inflorescences** many-flowered, hemispherical, extra-axillary, subterminal, or terminal; peduncles stout,

Top row: *Hoya bella, Hoya bilobata, Hoya carnosa, Hoya crassicaulis.*
Second row: *Hoya imbricata* (habit), *Hoya imbricata* (flowers), *Hoya kerrii, Hoya lacunosa.*
Third row: *Hoya linearis, Hoya mindorensis* (red form), *Hoya mindorensis* (yellow form), *Hoya multiflora.*
Bottom row: *Hoya myrmecopa, Hoya obovata, Hoya obscura, Hoya pubicalyx.*

MS

MS

MS

MS

SZ

MS

MS

MS

MS

SZ

MS

Above: *Marsdenia floribunda*, *Orbea semota*, *Oxystelma esculentum*, *Pergularia daemia*.

1.5–3 cm long; pedicels 3.5–7 cm long. **Flowers** with sepals ovate, ~2.5 × 1.7 mm, glands many, linear; corolla yellowish-white, lobe apices orange, 1.6–1.8 cm long, strongly reflexed from base, limb ~6 mm Ø, throat white-villous, lobes oblong-triangular, ~1.2 × 0.8 cm; corona on distinct stalk, lobes yellow, narrowly lanceolate, 8–9 mm long, outer angles extended into acuminate spurs, inner angles acuminate, higher than stigma head. **Fruits** follicles, linear-lanceolate in outline, 12–18 cm long; seeds ovate, ~4 × 2 mm, coma to 5 cm long. **Distr:** China, Indonesia, Laos, Malaysia, Myanmar, Philippines, Thailand, Vietnam. **Ecol:** In open forests, bushlands; 500–1,200 m. Fl. May–Jul. Fr. Sep–Dec. **Zones:** 10–12.

H. myrmecopa Kleijn and Van Donkelaar, *Blumea* 46(3): 477, 2001. [Ant Hoya]. **Shrubs**, epiphytic, small, rarely twining; stem clustered, pendent. **Leaves** ovate to elliptic, 2.5–7.5 cm long, apex acute to acuminate, venation obscure; petioles 0.5–1.5 cm long. **Inflorescences** umbels, convex, negatively geotropic, 6–16-flowered; peduncles 1–2 cm long, thin, generally upturned. **Flowers** small, ~5–8 mm across; corolla sparsely hirsute, strongly recurved, only marginally wider than the coronal lobes, corolla and corona greenish or yellowish-white; corona ~2–3 mm high, inner coronal lobe acuminate, outer coronal lobe with bilobed extensions that are an integral part of the entire lateral sides of lobe, inner lobe tips much elevated above outer lobe tips, outer lobe tips elevated above corolla; pollinaria caudicles with small wings. **Distr:** C Sulawesi. **Ecol:** In open rainforests on limestone ridges, usually growing from cavities in tree branches, often with ant communities; ~600 m. **Zones:** 11–12.

H. obovata Decne., *Prodr.* (DC.) 8: 635, 1844. [Porcelain Flower]. **Climbers**, epiphytic; stems twining. **Leaves** elliptic or obovate, 6.5–10 × 10 cm, base tapered to petiole, apex obtuse or rarely notched, thick, fleshy. **Inflorescences** umbels, 15–30-flowered, pendent, 6 cm across, convex; peduncle thick, ~4 mm Ø; pedicels 3.3 cm long, green to purplish, hairy; floral bracts small, ~1.3 mm long. **Flowers** with corolla ~1.5 cm across, fleshy, white, papillose inside, lobes 9 × 7 mm, obtuse, sinuses rounded; corona purple or reddish-purple, 9.5 mm Ø, 2 mm high, lobes obovate, 4.5 × 2.5 mm, with white tips. **Fruits** paired follicles. **Distr:** Indonesia (Sulawesi). **Ecol:** In lowland forests. Fl. anytime during the year. **Zones:** 11–12.

H. obscura Elmer ex Merr., *Enum. Philipp. Fl. Pl.* 3: 352, 1923. [Porcelain Flower]. **Climbers**, epiphytic; stems tangles, branched from near root cluster, woody at base. **Leaves** scattered along all branches, opposite; lamina flat, elliptic to obovate-oblong and subelliptic, 8(–10) × 2–4 cm, base broadly obtuse or cuneate, margins entire, apex tapered, acute to subacuminate, adaxially glabrous, smooth and glossy, midrib shallowly depressed adaxially, straight, lateral nerves faint, 5–8 per side; petiole 0.5–1(–1.5) cm long, glabrous. **Inflorescences** cymes, flowers numerous, axillary; peduncle stout, coriaceous, winglike, ridged, 5 cm long, thickened flower-bearing tubercle at apex; pedicels glabrous, striate, filiform, mostly curved; bracteoles 1 cm long, deciduous. **Flowers** with calyx lobes 5, round, hyaline, to 1.5 × 1.5 mm; corolla adnate and united toward base, subhyaline, upper or inner surface slightly fuzzy-gray, ovate, tips acute, recurved; corona horns 5, suberect, small, with minute appendages; stamens surrounding the 2 flask-shaped, erect pistils. **Distr:** Philippines (Luzon). **Ecol:** In primary forests on ridges extending into mossy forests; ~1,000 m. Fl. Jan–Jul. **Zones:** 11–12.

H. pubicalyx Merr., *Philipp. J. Sci.*, C 13: 331, 1918. [Harlequin Wax Plant]. **Shrubs**, scandent, glabrous; internodes to 20 cm long. **Leaves** opposite; lamina fleshy, oblong to oblong-obovate, 10–14 × 3–5 cm, base obtuse, margins entire, apex acuminate, pale, shiny, ± wrinkled, lateral nerves ~6 per side; petiole ~1 cm long. **Inflorescences** umbels, many-flowered, 8–9 cm Ø; rachis thickened, above; pedicels spreading, ~8.5 cm long, sparingly pubescent. **Flowers** fragrant, ~1.8 cm Ø; calyx segments oblong-ovate to ovate-lanceolate, acute or subacute, externally pubescent, ~4 mm long; corolla lobes broadly triangular, ~6 × 6 mm, apex acuminate, somewhat recurved, spreading, rotate, glabrous outside, inside densely papillose; corona 1–1.2 cm Ø, stellate, lobes spreading, shining, oblong-ovate to ovate-lanceolate, ~5 mm long, acuminate, coriaceous, brown, external tip slightly retuse, internal one ascending, adaxially plano-convex and somewhat keeled in the median portion. **Distr:** Philippines (Luzon). **Ecol:** On large trees along streams in damp forests. **Zones:** 11–12.

Marsdenia

Species ~125. **Lianas**, twining, rarely erect shrubs. **Leaves** opposite; lamina cordate. **Inflorescences** umbel-like, paniculate, or racemelike, terminal or extra-axillary, pedunculate. **Flowers** with calyx 5-lobed, sepals overlapping, basal glands or scales often present; corolla urceolate or campanulate, throat often hairy, lobes twisted to right; corona lobes 5, erect, acuminate, fleshy, attached to gynostegium; filaments connate into a tube, anthers erect, appendages inflexed; pollinia 2 per pollinarium, often oblong, erect; stigma head flat, convex, or beaked. **Fruits** follicles, mostly thick, acuminate, smooth or longitudinally winged; seeds comose. **Distr:** Asia, America, tropical Africa. **Cult:** Propagate from seeds, semiripe cuttings or by air-layering. Requires a moderately fertile, moist, well-drained soil in partial sun. In full sun the growth may be stunted. **Info:** The popular *Stephanotis floribunda* is now included in this genus, as the genera *Stephanotis* and *Marsdenia* are merged.

M. floribunda (Brongn.) Schltr., *Symb. Antill.* 1: 275, 1899. (Syn: *Stephanotis floribunda*, *S. jasminoides*). [Bridal Bouquet, Bridal Wreath, Madagascar Jasmine, Stephanotis]. **Climbers**, evergreen, glabrous, 2.5–10 m long; older stems whitish-warty; sap milky. **Leaves** ovate to oblong-elliptic, 5–10(–15) × 2.8–4.6 cm, base cordate, apex acute, thickly coriaceous, adaxially gland-dotted at base, dull dark green; petiole 0.8–2.5 cm long. **Inflorescences** crowded clusters; pedicels 0.5–1.9 cm long. **Flowers** sweetly fragrant, strongest at night; sepals green, ovate, spreading; corolla waxy, white to ivory, salverform, 2.5–3 cm long, tube ± cylindrical, base swollen, limb 3–4 cm Ø, 5-lobed; corona with 5 erect scales. **Fruits** large, ovoid, 10–15 × 6–10 cm, thick, leathery, glaucous; seeds flat, brownish, comose. **Distr:** Madagascar. **Ecol:** In humid forests; 0–500 m. Fl. throughout the year. **Zones:** 10–12.

Orbea

Species 56. **Succulents**, spineless, often rhizomatous, forming compact to diffuse clumps, occasionally to 3 m Ø; stems decumbent to prostrate, fleshy and firm, glabrous, green to gray-green, purplish-mottled; tubercles deltoid to conical, spreading, often laterally flattened, joined into 4 wings along stem, each tipped with a leaf rudiment, usually subtended by 2 stipular denticles. **Inflorescences** glabrous, 1–3(–5) per stem, 1–40-flowered; bracts 1–several, lanceolate, with lateral teeth; pedicel mostly spreading. **Flowers** with sepals 5, lanceolate, acuminate; corolla usually rotate to campanulate, often strongly thickened into a prominent annulus around mouth, mostly deeply lobed, smooth, often purple-streaked, inside mostly deeply rugulose, tube mostly shallowly bowl-shaped to ± absent, lobes spreading to reflexed, deltate, acute, margins often with clavate cilia; corona biseriate, outer lobes with 5 discrete spreading to erect ± subquadrate lobes, inner lobes adpressed to anthers. **Fruits** follicles, erect, divergent, terete-fusiform, slender, with 2 horns, glabrous. **Distr:** Africa, SW Arabia. **Cult:** Propagate from cuttings. Requires a well-drained compost in full or partial sun.

O. semota (N.E. Brown) Leach, *Kirkia* 10: 290, 1975. (Syn: *Stapelia semota*). [Yellow Star Huernia]. **Succulents**, small, clump-forming, 6 cm to 1 m Ø; stems 3–10 × 0.6–0.8 cm, greenish, purplish-flecked; tubercles 0.6–1.2 cm long, in 4 obtuse rows. **Inflorescences** 1 per stem, 1–3-flowered, flowers facing upward; peduncle absent; bracts concave, ovate, ~2 ×1.5 mm, with lateral teeth; pedicel 1.5–4 cm long. **Flowers** with sepals lanceolate, 5–7 × 2–3 mm, acuminate; corolla rotate, 2–5.5 cm Ø, deeply lobed, cream to green, with 5 darker lines along lobes, inside bright yellow, to yellow and maroon-spotted to wholly dark maroon, slightly yellow-mottled near apices, lobes spreading, ovate-lanceolate, 0.7–2.5 × 0.6–1 cm, acuminate, margins with dense white (in yellow form) to maroon marginal cilia 1.5–3.5 mm long; corona ~0.3 × 0.6–1 cm, cream, ~1.5 mm long; outer lobes subquadrate, ~2 × 2 mm, maroon, becoming yellow near base, inner lobes ovate, 0.7–1.5 mm long, maroon with yellow markings, to yellow. **Distr:** Kenya, Rwanda, Tanzania. **Ecol:** 1,000–1,800 m. **Zones:** 11–12.

Oxystelma

Species 2. **Lianas**, base woody. **Leaves** usually linear or lanceolate. **Inflorescences** cymes, lax, racemelike or umbel-like, rarely flowers solitary, pedunculate, extra-axillary; flower buds globose. **Flowers** with calyx glands 5 or many; corolla bowl-shaped, lobes basally valvate and apically overlapping to right; corona biseriate: *outer* corona at base of corolla tube, annular, truncate, margin membranous; *inner* series of 5 ovate-lanceolate lobes attached to gynostegium and exceeding anthers; filaments short, connate into a tube, anthers with appendages; pollinia 2 per pollinarium, pendulous, on long translator arms; stigma head convex. **Fruits** solitary or paired follicles, oblong or ovate-lanceolate in outline, smooth. **Distr:** Tropical and subtropical Asia; tropical Africa. **Cult:** Propagate from seeds or cuttings. Requires a moist, well-drained soil in full sun.

O. esculentum (L. f.) Schult., *Syst. Veg.*, ed. 15 bis [Roemer and Schultes] 6: 89, 1820. (Syn: *Sarcostemma esculentum*, etc.). [Rosy Milkweed Vine]. **Lianas**, to over 4 m long, glabrous except for flowers; stems slender. **Leaves** linear or linear-lanceolate, 6–11 × 0.7–2 cm, base rounded, membranous, lateral veins 9–12 pairs, marginal vein present; petiole 1–1.5 cm long. **Inflorescences** (1- or)2–4-flowered, longer than leaves; flower buds 8–9 mm Ø. **Flowers** relatively large; sepals ovate-lanceolate, ~3.5 × 1 mm; corolla white with purple veins or blotches, 2–3 cm Ø, limb 1.5–2 cm Ø, lobes triangular, 1–1.5 cm long, densely ciliate; outer corona densely pubescent; pollinia ~1.5 mm long; ovaries glabrous. **Fruits** follicles, lanceolate in outline, ~5 × 1.5 cm, apex subacute; seeds ovate, ~2 × 1.5 mm, coma 1.5–2 cm long. **Distr:** Asia, NE Africa. **Ecol:** In moist bushlands, streambanks. Fl. Jul–Sep. Fr. Oct–Dec. **Zones:** 10–12.

Pergularia

Species 3–5. **Herbs**, twining; branches pubescent, hispid or prickly. **Leaves** opposite; lamina cordate, surfaces. **Inflorescences** umbellate to racemose cymes, solitary at nodes. **Flowers** with calyx lobes 5, ovate-lanceolate, acute; corolla yellowish or greenish-white, campanulate lobes imbricate in bud; corona biseriate, outer glabrous, a 5-lobed annulus at base of the staminal column, the inner of 5 fleshy lobes, each with 2 spurs, one directed outward and downward and the other upward with the tip over the gynostegium; anthers with prolonged membranous connectives; pollinium solitary in each loculus, pendulous, hyaline margin narrow. **Fruits** follicles, lanceolate, usually echinate; seeds comose. **Distr:** Africa, S Asia. **Ecol:** In dry areas. **Cult:** Propagate from seeds or cuttings. Requires a moist, well-drained soil in full sun.

P. daemia (Forssk.) Chiov., *Result. Sc. Miss. Stefanini Paoli Somal. Ital.* 1: 115, 1916. (Syn: *Cynanchum extensum*, etc.). **Herbs**, perennial, twining, juice milky; branches with spreading hairs. **Leaves** broadly ovate or suborbicular, 5–15 × 3.8–9.5 cm, base cordate, apex acute to acuminate, glabrous or shortly pubescent above, usually velvety-pubescent below; petiole 1.8–9 cm long. **Inflorescences** cymes, lateral; pedicels 1.8–3.5 cm long, pubescent; bracts minute, lanceolate. **Flowers** with sepals ovate-lanceolate, ~2–3 mm long, generally hairy; corolla greenish-yellow or dull white, tube ~3.5 mm long, lobes ~7 mm long, spreading, ovate-oblong, pubescent; outer corona membranous, truncate, inner corona lobes 5–8 mm long including horns. **Fruits** follicles, lanceolate, 5–8 × 12–13 cm, beaked, echinate, puberulous; seeds ~8 × 4 mm, ovate, tip truncate, velvety-pubescent; seeds with coma 2.5–3 cm long. **Distr:** Arabia to S and tropical Africa; Afghanistan, W. Pakistan, India, Myanmar, Sri Lanka; Malaya. **Ecol:** Fl. Jan–Apr. **Zones:** 10–12.

Sarcostemma

Species ~10. **Lianas**, **shrubs** or **subshrubs**, scrambling or twining; stems succulent, base woody, cylindrical; branches pendulous. **Leaves** reduced to minute scales, soon deciduous. **Inflorescences** cymes, umbel-like, sessile, terminal or extra-axillary. **Flowers** with calyx with or without 5 basal glands; corolla rotate or nearly so, deeply lobed, lobes overlapping to right; corona double, *outer* corona inserted at junction of corolla tube and gynostegium, ringlike or cupular, membranous, 5-angled, truncate or shallowly lobed, *inner* corona lobes 5, inserted at base of gynostegium, fleshy, basally dilated, 2-keeled inside; anther appendages membranous; pollinia 2 per pollinarium, waxy, pendulous, caudicle slender; stigma head short conical or oblong-fusiform. **Fruits** follicles, thick, smooth. **Distr:** Africa, tropical Asia, Australia. **Cult:** Propagate from seeds or stem cuttings. Requires a moist, well-drained soil in full sun.

S. viminale (L.) R. Br., *Mem. Wern. Soc.* 1: 51, 1810. (Syn: *S. brunonianum*). [Rapunzel Plant]. **Shrubs** or **subshrubs**, twining; stem cylindrical, 3–7 mm Ø, green, glabrous, internodes 10–20 cm long. **Inflorescences** umbels, many-flowered, terminal, sessile; pedicels 0.6–1.2 cm long, pubescent; bracts minute, linear-lanceolate. **Flowers** sweetly scented, cream or greenish; sepals 5, 1–1.5 mm long, subacute; corolla campanulate, pale greenish-white, divided almost to base, lobes ovate-oblong, ~5 mm long, subacute; outer corona crenately 10-lobed, lobes of inner corona thick, fleshy, obtuse; staminal column very short; pollen-masses compressed, clavate. **Fruits** follicles, lanceolate, 10–13 × 0.8 cm, ends tapered, slightly divergent; seeds 3–6 mm long, flattened-ovate, coma 1.8 cm long. **Distr:** Tropical Africa, S India, Myanmar. **Ecol:** In dry savannas and semiarid regions, forming dense scrambling bushes with intricately interlaced branches. Fl. Jun–Nov. **Zones:** 10–12. **Info:** The milky juice is reputedly consumed by travellers. Plant parts used for medicinal purposes.

Stapelia

Species 29. **Succulents**, spineless, rarely rhizomatous, forming clumps to 2 m Ø; stems 2–30 cm long; tubercles 1–5 mm long, laterally flattened into 4(5 or 6) obtuse often winglike angles, each bearing apically a caducous leaf-rudiment 0.1–1.1 cm long. **Inflorescences** 1–5-flowered, 1–3(–6) per stem; peduncles <1(–3) cm long; bracts several, deltoid, 1–2(–5) mm long; pedicel to 8(–12) cm long. **Flowers** with sepals lanceolate to ovate, to 1(–1.5) cm long, acute; corolla rotate to shallowly cupular, rarely campanulate, to 20(–40) cm Ø, mostly deeply lobed, usually with darker veins on lobes, inside usually transversely rugulose, hairy, tube usually <5 mm long, lobes deltate or ovate to narrowly lanceolate, to 10 × 3.5 cm, margins strongly recurved; corona to 1.9 × 1.5 cm, biseriate, outer lobes simple, inner lobes adpressed to anthers; anthers horizontal on top of style head. **Fruits** follicles, erect, diverging at 30–60º, terete-fusiform, obclavate, 2-horned, longitudinally mottled with purple stripes. **Distr:** S Africa. **Cult:** Propagate from seeds or stem cuttings. Requires a moist, well-drained soil in full sun.

S. leendertziae N.E. Brown, *Ann. Transvaal Mus.* 2: 168, 1910. (Syn: *Gonostemon leendertziae*, *S. wilmaniae*). [Black Bells, Carrion Flower, Lizard Plant, Star Flower]. **Succulents**, forming lax clumps, to 50 cm Ø; stems 8–20 cm long, 1–2.5 cm Ø, green, purple-mottled apically; tubercles 1–2 mm long; leaf rudiments 2–3 mm long. **Inflorescence** 1, basal, 1–3-flowered; peduncle <1 cm long; bracts few, ovate, <2 mm long; pedicel 2–3 cm long. **Flowers** with sepals ovate to lanceolate, 0.6–1.4 × 0.3–0.5 cm, acute; campanulate to urceolate, corolla 7–10 × 9–10 cm, dark purple to reddish-purple with darker longitudinal veins, inside with purple hairs to 6 mm long, tube 5.5–8 × 7–8 cm, cupular, often narrowing toward mouth, lobes spreading, deltate, 4–5 × 3.5–4.5 cm, acuminate, margins finely hairy; corona ~1.3–1.5 × 1 cm, glabrous, stipe pinkish to yellow, 2–3 mm tall, outer lobes ~3 mm long, deeply bifid, reddish, inner lobes linear, 0.9–1 cm long, dark purple. **Distr:** S Africa (KwaZulu-Natal) to Swaziland,. **Ecol:** On rocky ground, slopes or flats. **Zones:** 10–12.

Telosma

Species ~10. **Lianas**. **Leaves** opposite, long petiolate. **Inflorescences** cymes, extra-axillary, pendulous; rachis laxly branched. **Flowers** with calyx with 5 glands at base; corolla tube cylindric, with enlarged base, throat sometimes constricted, lobes patent, twisted to right; corona lobes 5, adnate to base of staminal tube, margin and apex free, erect, oblong, obtuse, with ligulate inner lobes adnate to near apex, or

Below: *Sarcostemma viminale, Stapelia leendertziae, Telosma cordata, Tweedia caerulea.*

Above: *Jasminum grandiflorum, Jasminum humile, Jasminum laurifolium, Jasminum mesnyi.*

corona lobes notched; filaments connate into a tube; membranous anther appendages incurved; pollinia 2 per pollinarium, oblong, erect; stigma head convex or conical. **Fruits** follicles, thick, acuminate, smooth; seeds comose. **Distr:** Asia, tropical Africa, Oceania. **Cult:** Propagate from seeds or cuttings. Requires a moist, well-drained soil in full sun.

T. cordata (N.L. Burman) Merrill, *Philipp. J. Sci.* 19: 372, 1921. (Syn: *Pergularia minor*, etc.). [Pakalana Vine, Tonkin Jasmine, Cowslip Creeper]. **Lianas**, stems 1–10 m long; yellowish-green. **Leaves** ovate, 4–12 × 3–10 cm, base deeply cordate with a narrow sinus, apex acuminate, basal veins 3, lateral veins to 6 pairs; petiole 1.5–5 cm long. **Inflorescences** cymes, umbel-like, 15–30-flowered; peduncles 0.5–1.5 cm long, puberulent. **Flowers** very fragrant; sepals oblong-lanceolate, puberulent; corolla yellowish-green, tube 0.6–1 × 0.4–0.6 cm, lobes oblong, 0.6–1.2 × 0.3–0.6 cm, ciliate; corona lobes slightly fleshy, basal part ovate, apex acuminate, often notched to deeply lobed; pollinia oblong or reniform; stigma head capitate. **Fruits** follicles, lanceolate, 7–13 × 2–3.5 cm, glabrous, obtusely 4-angled; seeds broadly ovate, ~1 × 1 cm, coma 3–4 cm long. **Distr:** China, India, Kashmir, Myanmar, Pakistan, Vietnam; Europe, N and S America. **Ecol:** In open woods, bushlands. Fl. May–Oct. Fr. Oct–Dec. **Zones:** 10–12.

Tweedia

Species 1. *Generic description same as specific.* **Cult:** Propagate species from seeds or cuttings. Requires a moderately fertile, well-drained soil in full sun.

T. caerulea D. Don ex Sweet, *Brit. Fl. Gard.* [Sweet] Ser. II. t. 407. (Syn: *Oxypetalum coeruleum*, etc.). [Blue-flowered Milkweed, Heavenly blue Tweedia, Light Blue-flowered Tweedia, Southern Star]. **Shrubs** or **subshrubs**, climbing, perennial, herbaceous, densely hairs, 30–90 cm high; stems twining, woody at base; latex opaque, milky. **Leaves** opposite; lamina oblong-lanceolate to ± ovate, 3–6 × 1–3 cm, base cordate, margins entire, apex obtuse, mucronate, posterior lobes rounded, connivent; petiole 0.2–1.2 cm long, woolly, base glandular. **Inflorescences** umbels or open cymes, 3–5-flowered, axillary and terminal; pedicels densely hairy. **Flowers** 2–2.5 cm ∅, sky-blue; calyx lobes 5, lanceolate, acuminate; corolla rotate, tube 4–5 mm long, lobes 5, blue, elliptic oblong, ~2 cm long, obtuse, densely hairy beneath, base with nectariferous cavities; corona simple, lobes 5, ligulate, obtuse, fleshy, revolute at apex; stamens 5, monadelphous, filaments white, anthers yellow; ovaries 2, style head conical, apex bifid, stigmas compressed, acute. **Fruits** paired follicles, erect, narrowly ovoid, 9–13 cm long, pubescent; seeds with long coma. **Distr:** S America (Brazil, Uruguay). **Ecol:** Fl. May–Jul. **Zones:** 10–12.

LAMIALES

FAMILIES 23, GENERA 1,059, SPECIES 23,275.

Plants with gland-headed hairs. **Leaves** opposite. **Flowers** with parenchyma tissue extending from anther connective into locules. **Fruits** capsules, septicidally dehiscent (Stevens 2001 onwards). **Distr:** Worldwide.

OLEACEAE

Tribes 5, **genera** 24, **species** 615. **Trees** or erect or scandent **shrubs**; branches and branchlets lenticellate. **Stipules** present. **Leaves** opposite, rarely alternate or whorled; lamina simple, trifoliolate, or pinnately compound, venation pinnate or palmate. **Inflorescences** terminal or axillary, in cymes, panicles, racemes, umbels, or fascicles. **Flowers** actinomorphic, bisexual, rarely unisexual or polygamous and plants monoecious, dioecious, or polygamodioecious; calyx 4(–16)-lobed or -parted, rarely absent; corolla 4(–16)-lobed, sometimes almost free to base, rarely absent, lobes sometimes united in pairs at base or into a very short tube; stamens 2(–4), inserted on corolla tube or hypogynous, anthers longitudinally dehiscent; ovary superior, biloculed, ovules 2 per locule, sometimes 1 or numerous, style 1 or absent, stigma bilobed or capitate. **Fruits** drupes, berries, capsules, or samaras; seeds without endosperm. **Distr:** Worldwide, especially Asia.

Jasminum

Species >200. **Trees**, or erect or scandent shrubs, evergreen or deciduous; branchlets terete or angular and grooved. **Leaves** opposite or alternate, rarely whorled, simple, trifoliolate, or odd-pinnate; petiole usually articulated. **Inflorescences** basically cymose, in panicles, racemes, corymbs, umbels, or heads; bracts subulate or linear, sometimes leafy. **Flowers** bisexual, usually heterostylous, usually fragrant; calyx campanulate, cupular, or funnelform, 4–16-lobed; corolla white or yellow, rarely red or purple, salverform or funnelform, lobes 4–16, imbricate in bud, sometimes doubled in cultivation; stamens 2, included, inserted about middle of corolla tube, filaments short; anthers dorsifixed, introrse; ovules 1 or 2 per locule, style filiform, stigma capitate or bilobed. **Fruits** berries, didymous or one half aborted. **Distr:** Africa, Asia, Australia, South Pacific Islands, Mediterranean region (1 sp.). **Cult:** Propagate from seeds, cuttings or by air-layering. Requires humus-rich, moist but well-drained soil in full or partial sun.

J. grandiflorum L., *Sp. Pl.*, ed. 2. 1: 9, 1762. (Syn: *J. officinale* var. *grandiflorum*, etc.). [Jasmine, Royal Jasmine]. **Shrubs**, scandent, 2–4 m high; branchlets terete, angular or grooved. **Leaves** opposite, pinnatipartite or compound; leaflets 5–9, ovate or narrowly ovate (terminal one usually narrowly rhomboid), 0.7–3.8 × 0.5–1.5 cm, base cuneate or blunt, apex acute, acuminate, or blunt, sometimes mucronate; petiole 0.5–4 cm long. **Inflorescences** cymes, 2–9-flowered, terminal or axillary; bracts linear, 2–3 mm long; pedicel 0.5–2.5 cm long, middle pedicel of cymes conspicuously shorter. **Flowers** with calyx glabrous, lobes subulate-linear, (0.3–)0.5–1 cm long; corolla white, salverform, tube 1.3–2.5 cm long, lobes often 5, oblong, 1.3–2.2 cm long. **Distr:** Arabia. **Ecol:** Fl. Aug–Oct. **Zones:** 10–12. **Info:** Cultivated in warm and tropical regions worldwide for the fragrant flowers.

J. humile L., *Sp. Pl.* 1: 7, 1753. [Italian Jasmine]. **Shrubs** or **small trees**, sometimes scandent, 0.3–3 m high; branchlets angular. **Leaves** alternate, compound or simple; leaflets 3–9(–13), usually 5, ovate to lanceolate, rarely obovate, terminal leaflet 0.6–6 × 0.2–2 cm, lateral leaflets 0.2–4.5 × 0.2–2 cm, base rounded or cuneate, margins sometimes spiny-ciliate, apex acute to caudate, primary veins 2–4 pairs, leathery or papery, glabrous or adaxially shortly setose, abaxially pubescent along veins; petiole 0.5–2 cm long. **Inflorescences** cymes, in umbels, corymbs, or panicles, 1–10(–15)-flowered, terminal, rarely with linear bracts 2–4 mm long; pedicels 0.2–3 cm long. **Flowers** with calyx lobes deltate to subulate, shorter than tube; corolla yellow, nearly funnelform, tube 0.8–1.6 cm long; lobes orbicular or ovate, 3–7 mm long, apex often rounded. **Fruits** berries, ellipsoid or globose, 0.6–1.1 × 0.4–1 cm, purple-black. **Distr:** Afghanistan, SW Asia, Tajikistan, India, China. **Ecol:** In woods, thickets; 1,100–3,800 m. **Zones:** 8–11.

J. laurifolium Roxb., *Hort. Bengal.* 3, 1814. (Syn: *J. angustifolium*). [Angel-wing Jasmine, Confederate Jasmine, Shining Jasmine, Star Jasmine, Windmill Jasmine]. **Vines**, or scandent **shrubs**, evergreen, twining, 0.5–5 m high, glabrous; branchlets terete. **Leaves** articulate at base; lamina simple, oblong or narrowly elliptic, 6.5–10 × 1.5–2 cm, base obtuse or rounded, apex acuminate, coriaceous to ± membranous, usually trinerved, nerves anastomosing near the margin; petiole 4–6 mm long. **Inflorescences** cymes, laxly 3–5-flowered, occasionally flowers solitary, terminal or long-peduncled and axillary; bracts linear, 5 mm long; pedicels 0.8–2 cm long, not much thickened. **Flowers** white; calyx teeth linear, 6 mm long; corolla salverform, tube 2 cm long, lobes long, narrow, 2 × 0.3 cm. **Fruits** berries, oblong-ovoid, 0.8–2.2 × 0.4–1.1 cm, black and shiny. **Distr:** India (Khasia Mountains), China. **Ecol:** In thickets. **Zones:** 10–12.

J. mesnyi Hance, *J. Bot.* 20: 37, 1882. (Syn: *J. primulinum*). [Primrose Jasmine, Yellow Jasmine].

Top row: *Jasminum nobile* ssp. *rex*, *Jasminum multiflorum*, *Jasminum officinale*.
Bottom row: *Jasminum polyanthum*, *Jasminum sambac*.

Subshrubs, erect, evergreen, 0.5–5 m high; branchlets 4-angled, glabrous. **Leaves** opposite, trifoliolate, or simple at base of branchlets; leaflets narrowly ovate or ovate-lanceolate to narrowly elliptic, base cuneate, apex blunt and mucronulate, *terminal* one 2.5–6.5 × 0.5–2.2 cm, basally decurrent into a short petiolule, *lateral* ones sessile, 1.5–4 × 0.6–2 cm; petiole 0.5–1.5 cm long. **Inflorescences** with flowers usually solitary, axillary or rarely terminal; bracts leafy, obovate or lanceolate, 0.5–1 cm long; pedicel 3–8 mm long. **Flowers** with calyx campanulate, lobes 5–8, leafy, lanceolate, 4–7 mm long; corolla yellow, funnelform, 2–4.5 cm Ø, tube 1–1.5 cm long, lobes 6–8, doubled in cultivation, broadly obovate or oblong, 1.1–1.8 cm long. **Fruits** berries, ellipsoid, 6–8 mm Ø. **Distr:** China. **Ecol:** In ravines, woods; 500–2,600 m. Fl. Nov–Aug. Fr. Mar–May. **Zones:** 9–11.

J. multiflorum (N.L. Burman) Andrews, *Bot. Repos.* 8: t. 496, 1807. (Syn: *J. pubescens, Nyctanthes multiflora*). [Angel-hair Jasmine, Downy Jasmine, Star Jasmine]. **Shrubs**, scrambling or weakly climbing, to 3 m high; branchlets terete, densely pilose. **Leaves** opposite; lamina simple, ovate-cordate, often broadly so, 3–8 × 1.5–5 cm, base cordate, apex acute to sometimes slightly acuminate, mucronulate, papery, ± pilose, glabrescent, except on midrib and veins, primary veins 3 or 4 on per side; petiole 0.5–1 cm long, densely pilose. **Inflorescences** many-flowered, congested, terminal on side shoots; bracts leafy, basal ovate, 1.5–2 cm long, upper ones linear, 3–5 mm long; pedicels 0–2 mm long. **Flowers** sweetly scented; calyx densely pilose; tube ~1 mm long, lobes 6–9, filiform, (2–)5–7 mm long; corolla white, tube 1.2–1.5 cm long, lobes 7–9, 1–1.5 cm long, acute. **Distr:** India. **Ecol:** Fl. Oct–Apr. **Zones:** 10–12.

J. nobile C. B. Clarke, *Fl. Brit. Ind.* 3: 597, 1882. **Climbers**, woody; young stems glabrous. **Leaves** ovate, 5–7 × 2–3.5 cm, base obtuse to rounded, apex manifestly acuminate, ± coriaceous, glabrous, venation ± obscure, especially adaxially, 2 main veins arising from base of midrib, becoming submarginal nerves, 2–3 lateral nerves per side, without domatia. **Inflorescences** cymose, 3–7-flowered, terminal on side shoots, glabrous; bracts linear, 2–5 mm long; pedicels 0.5–1 cm long. **Flowers** with calyx tube 2–3 mm long, lobes subulate, 0.5–1 cm long, glabrous; corolla white, tube 2.2–2.8 cm long, lobes 8, 1.8–2.2 × 0.6–0.9 cm. **Fruits** unknown. **Distr:** Myanmar. **Ecol:** 1,219–1,523 m. **Zones:** 10–12. **Info:** *J. nobile* ssp. *rex* from Cambodia and Thailand, is similar except: **Leaves** oblong-elliptic or ovate to broadly lanceolate, 5–14 × 2.5–5 cm, base often acute, apex only somewhat acuminate. **Inflorescences** 1(–3)-flowered; pedicels 2–4 cm long. **Flowers** with corolla tube 2.5–3.5 cm long, lobes 2.3–3 × 1–1.8 cm.

J. officinale L., *Sp. Pl.* 1: 7, 1753. [Common Jasmine]. **Shrubs**, scandent, 0.4–6(–8) m high; branchlets angular or grooved. **Leaves** opposite, pinnatipartite or pinnately compound, often simple at base of branchlets; leaflets 3–9, glabrous or sparsely pubescent, *terminal* leaflet ovate to narrowly elliptic, 0.5–4.5 × 0.2–2 cm, base cuneate, apex acute or acuminate, rarely obtuse, *lateral* leaflets ovate to elliptic or suborbicular, 0.3–3 × 0.2–1.3 cm, base rounded or cuneate, apex acute or obtuse; petiole 0.4–4 cm long, glabrous or appressed-hairy. **Inflorescences** cymes, umbellate or subumbellate, 1–10-flowered, terminal or rarely axillary; bracts linear, 0.1–1 cm long; pedicels 0.4–2.5 cm long. **Flowers** heterostylous; calyx cupular, 1–3 mm long, lobes subulate-linear, (0.3–)0.5–1 cm long; corolla white, sometimes reddish or pinkish outside, tube 1–1.5(–2) cm long, lobes 5, narrowly ovate to oblong, 0.6–1.2 cm long. **Fruits** berries, globose or ellipsoid, 0.7–1 × 0.5–0.9 cm, ripening through dark red to purple. **Distr:** Bhutan, China, India, Kashmir, Nepal, Tajikistan. **Ecol:** In ravines, thickets, woods, along rivers; 1,800–4,000 m. **Zones:** 7–12.

J. polyanthum Franchet, *Rev. Hort.* 1891: 270, 1891. (Syn: *J. blinii, J. delafieldii*). [Climbing Jasmine, Pink/White Jasmine]. **Vines**, woody, twining, 1–10 m long; branchlets terete or angular, glabrous. **Leaves** opposite, pinnatipartite or pinnately compound; leaflets 5–7, papery or thinly leathery, glabrous or with tufts of hairs in vein axils abaxially, *terminal* leaflet lanceolate or ovate, (1.5–)2.5–9.5 × (0.6–)1–3.5 cm, base cuneate or rounded, apex acute to caudate-acuminate, *lateral* leaflets sessile or shortly petiolulate, ovate, (1–)1.5–8.5 × (0.5–)1–2.7 cm, base rounded or subcordate, apex obtuse or acute; petiole 0.4–2 cm long; terminal petiolule 0–2 cm long. **Inflorescences** racemes or panicles, 5–50-flowered, terminal or axillary; bracts subulate, 1–6 mm long; pedicels 0.5–2.5 cm long. **Flowers** heterostylous; calyx glabrous or puberulent, tube 1–2 mm long, lobes 5, deltate or subulate-linear, <2 mm long; corolla white, red outside and in bud, tube 1.3–2.5 cm long, lobes 5, oblong or narrowly ovate, 0.9–1.5 cm long. **Fruits** berries, black, subglobose, 0.6–1.1 cm Ø. **Distr:** SW China. **Ecol:** In valleys, thickets, woods; 1,400–3,000 m. Fl. Feb–Aug. Fr. Nov. **Zones:** 7–11.

J. sambac (L.) Ait., *Hort. Kew.* 1: 8, 1789. (Syn: *Nyctanthes sambac*). [Arabian Jasmine]. **Shrubs**, erect, suberect or scandent, to 3 m high; branchlets terete or slightly compressed, sometimes hollow, sparsely pubescent. **Leaves** opposite, simple; lamina orbicular to elliptic or obovate, 4–12.5 × 2–7.5 cm, both ends blunt, sometimes base subcordate, papery, glabrous except for tufted hairs at vein axils abaxially, primary veins 4–6 pairs; petiole 2–6 mm long, articulate, pubescent. **Inflorescences** cymes, (1 or)3(or 5)-flowered, terminal; bracts subulate, 4–8 mm long; pedicels 0.3–2 cm long. **Flowers** very fragrant; calyx glabrous or sparsely pubescent, lobes 8–9, linear, 5–7 mm long; corolla white, tube 0.7–1.5 cm long, lobes oblong to suborbicular, 5–9 mm across. **Fruits** berries, globose, ~1 cm Ø, purple-black. **Distr:** India. **Ecol:** Fl. May–Aug. Fr. Jul–Sep. **Zones:** 10–12. **Info:** Oil from the flowers are used in perfumes.

Ligustrum

Species ~45. **Shrubs** or **small trees**, deciduous or evergreen. **Leaves** opposite, shortly petiolate; lamina simple, margins entire. **Inflorescences** panicles of cymes, terminal, rarely lateral; flowers sessile or pedicellate. **Flowers** bisexual; calyx campanulate, truncate or 4-toothed, persistent; corolla white, rotate, funnelform, or salverform, lobes 4, about as long as or shorter than corolla tube, valvate in bud; stamens 2, inserted at mouth of corolla tube, included or exserted, anthers yellow or sometimes purple, oblong; ovary ovoid, bilocular, locules 2-ovuled, ovules pendulous, style shorter than stamens, stigma 2-cleft. **Fruits** drupes, berrylike, with membranous or papery endocarp, rarely drupaceous or loculicidal; seeds 1–4, endosperm fleshy. **Distr:** Asia, Australia, Europe. **Cult:** Propagate from seeds or cuttings. Requires a moist, well-drained soil in full or partial sun.

L. robustum* ssp. *walkeri (Decne.) P.S. Green, *Kew Bull.* 40: 130, 1985. (Syn: *L. ceylanicum*, etc.). [Sri Lankan Privet, Wild Privet]. **Shrubs** or **small trees**, evergreen to 10 m tall. **Leaves** lanceolate, 3–9 × 1.5–2.5 cm, base rounded, obtuse, or attenuate into petiole, margins entire, apex acute-attenuate, glabrous, venation somewhat obscure, 5–7 primary veins per side; petiole glabrous. **Inflorescences** paniculate, many-flowered, pyramidal, 7–20 cm long, terminal, glabrate or minutely puberulous. **Flowers** white, scented; calyx campanulate, 1 mm long, ± entire or with 4 very shallow triangular teeth; corolla tube 1 mm long, lobes 4, valvate, 2 mm long, ± reflexed at anthesis; stamens 2, exserted, filaments inserted at top of corolla tube, 1.5 mm long, anthers narrowly oblong, 1 mm long; ovary rounded, 0.5 mm long, style 1.5 mm long, stigma terminal, ± fused, elongate. **Fruits** berries, ellipsoid, 0.7–1 × 0.4–0.5 cm, purplish-blue. **Distr:** Peninsular India. **Ecol:** In secondary montane forests, often near streams, 450–2,000 m. **Zones:** 10–12.

Nestegis

Species 5. **Shrubs** or **trees**. **Leaves** opposite; lamina simple, margins ± thickened, entire, coriaceous. **Inflorescences** axillary, decussate or terminal panicles; bracts small and slender. **Flowers** perfect or unisexual (and then plants polygamous or dioecious); calyx short, 4-toothed; corolla 4-lobed, lobes imbricate in bud, elliptic, tube short to long, or corolla absent; stamens 2 or 4, subsessile; ovary biloculed, ovules 2 per locule, stigma bilobed. **Fruits** drupes, ellipsoid or globose; seed 1. **Distr:** Norfolk Island, New Zealand, Hawai'i. **Cult:** Propagate from seeds or cuttings. requires a moist, well-drained soil in full sun.

N. sandwicensis (A. Gray) Degener, I. Degener and L. Johnson, *Flora Hawaiensis* 300: s.p., 1958. (Syn: *Gymnelaea sandwicensis*, etc.). [Hawai'i Olive]. **Trees**, to 8(–25) m tall; bark thick, corrugated. **Leaves** lanceolate to elliptic-lanceolate or narrowly elliptic, 7–22 × (1.5–)2.2–6(–7) cm, base rounded to cuneate, margins entire, apex acute to acuminate (obtuse), adaxially glossy, abaxially paler, glabrous to sparsely puberulent, coriaceous; petiole 0.6–1.3 cm long, base thickened. **Inflorescences** decussate, 2–4 cm long, axillary, 7–11-flowered; bracts lanceolate to ovate, ~1.5 mm long, caducous; pedicels 3–4 mm long. **Flowers** with calyx teeth ~0.6–1.1 mm long, entire to suberose; corolla rotate, ~2–3 mm long; stamens 4; stigma bilobed. **Fruits** drupes, purplish to black, ovoid to ovoid-conical, 1.6–2.2 cm long. **Distr:** Hawai'i. **Ecol:** In dry to mesic forests; 30–1,300 m. **Zones:** 10–12.

Nyctanthes

Species 2. **Shrubs** or **small trees**; branches tetragonal. **Leaves** opposite, petiolate; lamina ovate, margins entire or with a few large, unequal, distant teeth, apex acuminate, adaxially papillose-scabrous, abaxially pubescent. **Inflorescences** small, bracteate heads disposed in terminal trichotomous cymes or fascicles of 3–5 flowers. **Flowers** fragrant; calyx gamosepalous, ovoid-cylindric, tubular or campanulate; corolla actinomorphic or subactinomorphic, hypocrateriform, tube cylindric, yellow or orange, lobes 4–8, white, crenate or emarginate; stamens 2, inserted included, anthers 2; ovary biloculed, ovules 1 per locule, style simple, cylindric, apex very shortly bifid, stigma capitate. **Fruits** capsules, chartaceous, orbicular or obcordate, septicidally splitting into 2 discoid 1-seeded cells (carpels); seeds erect, orbicular, flattened. **Distr:** Tropical Asia. **Cult:** Propagate from seeds, softwood cuttings or by air-layering. Requires a moist, well-drained soil in full or partial sun.

N. arbor-tristis L., *Sp. Pl.* ed. 1: 6, 1753. (Syn: *Bruschia macrocarpa*, etc.). [Coral Jasmine, Tree of Sorrow]. **Shrubs** or **small trees**, to 10 m tall; branches quadrangular. **Leaves** decussate-opposite; lamina ovate or those near flowers sometimes oblong, 3.5–13 × 2–9 cm, base rounded or cordate to cuneate, margins entire to very coarsely and unequally glandular-dentate or serrate on lower portions, angular-lobed, apex acute or acuminate, coriaceous, adaxially whitish-hairy; petiole 0.5–2 cm long. **Inflorescences** axillary or forming large terminal, leafy, decussately branched panicles, composed of small terminal involucrate corymbs or headlike umbels, each 2–7-flowered; involucre 4-leaved; bracts sessile, opposite, broadly ovate; flowers sessile, subtended by 2 broadly ovate or elliptic bracts. **Flowers** strongly fragrant, opening at sunset and mostly falling before sunrise; calyx campanulate, cylindric, subturbinate, ~5 mm long; corolla hypocrateriform, tube yellowish, orange or reddish-orange, cylindric, 0.5–1.5 cm long, mouth orange, limb white, spreading, (5–)6–7(–8)-parted, ± contorted, obliquely triangular or elliptic, 5–6 mm wide; stamens 2; ovary superior, stigma capitate or peltate. **Fruits** capsules, cordate or suborbicular, ~2 × 0.3–0.4 cm, compressed, seeds 1 per cell. **Distr:** India, Indonesia, Thailand. **Ecol:** 0–1,000 m. Fl. throughout most of the year. **Zones:** 10–12.

Olea

Species >40. **Trees** or **shrubs**, evergreen, sometimes polygamodioecious. **Leaves** opposite, petiolate; lamina simple, margins entire or serrate. **Inflorescences** panicles, sometimes racemes or umbels, axillary or terminal. **Flowers** small, bisexual or unisexual; calyx small, campanulate, 4-lobed to subtruncate, lobes deltate or ovate, usually ciliate; corolla with a short tube, 4-lobed, lobes shorter or longer than tube, usually valvate in bud, cucullate; stamens 2(–4), inserted at base of corolla tube, included, filaments short or absent, anthers ovate, elliptic, or subrounded; ovules 2 per locule, pendulous, style short or absent, stigma capitate or slightly 2-cleft. **Fruits** drupes, endocarp thick and hard, sometimes papery; seeds often 1, endosperm fleshy or bony. **Distr:** Africa, Asia, Europe, Australia and Pacific Islands. **Cult:** Propagate from seeds, cuttings or suckers. Requires a fertile, well-drained soil in full or partial sun.

O. europaea L., *Sp. Pl.* 1: 8, 1753. [Olive]. **Trees** or **shrubs**, evergreen, to 10 m tall; branchlets angular, along with petiole and leaf lamina densely silvery-gray or rusty lepidote. **Leaves** narrowly lanceolate to elliptic, rarely narrowly ovate, 1.5–10 × 0.5–2 cm, base cuneate or attenuate, margins entire, apex acute to acuminate and mucronate to cuspidate, leathery, glabrous; primary veins 5–11 per side, obscure, somewhat raised adaxially; petiole 2–5 mm long. **Inflorescences** panicles, 2–4 cm long, axillary or terminal; flowers subsessile. **Flowers** bisexual or functionally unisexual; calyx 1–1.5 mm long; corolla white, 2.5–4 mm long, tube ~1 mm long, lobes elliptic-oblong, 1.5–3 mm long, involute. **Fruits** drupes, ellipsoid or subglobose, 0.7–4 cm long. **Distr:** Mediterranean region or Middle East. **Ecol:** In semiarid to subhumid regions, widely cultivated in warm-temperate regions of the world. Fl. Oct–Nov. Fr. Nov–Feb. **Zones:** 8–12. **Info:** Cultivated for thousands of years for the edible fruits and oil.

CALCEOLARIACEAE

Genera 2, species 260. **Herbs** or **shrubs**. **Leaves** with lamina toothed (rarely entire). **Inflorescences** thyrsic, with accessory flowers, paired, each composed of a terminal flower and the flower from its reduced prophyll. **Flowers** 4-merous; calyx valvate; corolla with short tube, abaxial corolla lobe saccate, shoelike, elaiophores as pads of hairs on inside of abaxial lip; stamens 2 or 3, thecae separate at apex; stamens absent; ovary bilocular. **Fruits** capsules, both septicidal and loculicidal. **Distr:** Tropical and W temperate S America, Brazil, New Zealand.

Calceolaria

Species ~300–400. **Herbs** annual, or **shrubs**; stem erect. **Leaves** opposite or verticillate. **Inflorescences** compound cymes, axillary or terminal. **Flowers** with calyx 4-partite; corolla yellow, white or red, tube very short, limb bilabiate, upper lip outside in bud, hooded, entire or obscurely bilobed, much smaller than the saccate lower lip; stamens 2, inserted near base of

Below: *Ligustrum robustum* ssp. *walkeri*, *Nestegis sandwicensis*, *Nyctanthes arbor-tristis*.

Above: *Olea europaea*, *Calceolaria chelidonioides*, *Achimenes grandiflora*, *Achimenes longiflora*.

corolla tube, anthers 2-celled, united by an elongated connective, posterior cell fertile, anterior one sterile or fertile; style not thickened at apex, stigma small. **Fruits** capsules, septicidally 2–4-valved; seeds numerous, testa covered with rows of tubercles. **Distr:** C and S America. **Cult:** Propagate from seeds or by division. Requires a moderately fertile, acidic soil in full to partial sun. The alpine species require moist but well-drained, gritty soil in partial sun.

C. chelidonioides HBK., *Nov. Gen. and Sp.* 2: 378, 1818. [Bedding Slipperwort]. **Herbs**; stem profusely rooting at lower nodes, ascending to 45 cm high, occasionally decumbent at base, quadrangular, glandular-pilose, reddish-purple. **Leaves** in unequal pairs; lamina ovate in outline, 2.5–7 × 2–5.5 cm, pinnatipartite to pinnatisect; *lower* leaves 7-lobed, *intermediate* ones 5-lobed, *uppermost* ones trilobed; lobes serrate, apex acute, adaxially scabrid and dark green; petioles 0.8–4.2 mm long. **Inflorescences** with flowers solitary or in dichasial cymes, axillary; pedicels 1–2.2 cm long, slender, glandular-hispid. **Flowers** with calyx lobes broadly ovate, to 5 × 3 mm, acute, glandular-pilose outside; corolla chrome-yellow, tube 1.5 mm long, glandular-hispid outside, upper lip oblong-spathulate, 3–5 mm long, finely pubescent outside, lower one orbicular, 1–1.4 × 1–1.3 cm, with a small orifice; style short. **Fruits** capsules, broadly ellipsoid, 5–6 mm long, ± equal to calyx, tipped with withered style, glandular-hirsute, 4-valved; seeds ellipsoid, 0.8 mm long, obtusely 4–5-angular, striate, orange-yellow. **Distr:** Ecuador; naturalized in the Old World. **Ecol:** In grasslands; >1,500 m. Fl. Dec–Aug. **Zones:** 10–12.

GESNERIACEAE

Genera over 147, **species** 3,200. **Herbs**, **shrubs**, or rarely weak-stemmed trees; hairs of stalked glands, or with thickened terminal cells. **Leaves** opposite or rarely alternate, whorled or basal, rosette-forming, exstipulate; lamina usually simple, rarely shallowly to deeply lobed, margins toothed or entire, pinnately veined, rarely palmately veined, rather fleshy, usually hairy. **Inflorescences** usually cymes, rarely racemes, axillary, often near apex and appearing terminal, usually pedunculate. **Flowers** perfect, zygomorphic, seldom actinomorphic; calyx actinomorphic, rarely zygomorphic, connate, usually (4 or)5-divided; corolla gamopetalous, zygomorphic, rarely actinomorphic, usually 2-lipped; fertile stamens 2 or 4, then often didynamous, rarely 5, epipetalous, anthers free or coherent, thecae 2, parallel, divergent, or divaricate; staminodes 1–3 or absent; disk ringlike to cupular, rarely absent; ovary 1(2)-loculed, placentas (1 or)2, parietal, placentation axile, ovules numerous, anatropous, style 1, stigmas 1 or 2, broadly bilobed to trumpet-shaped, wet or dry. **Fruits** usually capsules, loculicidal, septicidal, or circumscissile, rarely indehiscent berries; seeds numerous, fusiform to ellipsoid or ovoid, minute. **Distr:** Mainly tropical.

Achimenes

Species ~25. **Herbs**, perennial, terrestrial; rootstock scaly, rhizomatous; stems erect to decumbent, usually pubescent to pilose, unbranched. **Leaves** opposite or 3-whorled, pair subequal to very unequal; lamina with margins variously toothed, membranous; petiole short. **Inflorescences** axillary, flowers solitary or 2- or more-flowered short cymes. **Flowers** often showy, floral tube turbinate; calyx lobes 5, free, entire; corolla erect or oblique in calyx, either tubular with a flat spreading limb, or obliquely funnelform, often swollen or spurred on adaxially at base, limb 5-lobed, nearly regular to 2-lipped, lobes broad, spreading or nearly erect, entire to fringed, lower 3 lobes usually larger; stamens 4, included, didynamous, filaments adnate to base of corolla tube, slender, anthers oblong or quadrate, apically united; disk annular, cuplike, entire; ovary semi to completely inferior, 1-celled with 2 parietal placentas, style curved, stigma bilobed. **Fruits** capsules, bivalved, apex conic. **Distr:** C and S America, Jamaica. **Cult:** Propagate from rhizomes or cuttings. Requires a fertile, humus-rich, moist, well-drained soil in full or partial sun.

A. grandiflora DC., *Prodr.* 7(2): 536, 1839. [Cupid's Bower]. **Herbs**, erect, perennial, 40–60 cm high; stems deep maroon, evenly white-hairy. **Leaves** equal, in opposite pairs; lamina ovate, to 14 × 8 cm, base unequal, obtuse, tapered or somewhat cordate, margins toothed and edges densely white-hairy, apex acute to acuminate, tip pointed, 2–9 mm long, adaxially copper-colored or dark green, white-hairy, abaxially deep maroon to scarlet-red, white-hairy on veins, lateral veins 6–10 pairs; petiole 1–2.5 cm long. **Inflorescences** axillary, 1–3-flowered; common peduncle obsolete, or if present <1 cm long; pedicels solitary, bracteate, usually shorter than leaves. **Flowers** 2–2.5 cm Ø; calyx tube villous or pilose, often densely so, lobes linear, acute, mostly 0.5–1.2 cm long, ± villous or sometimes scabrous; corolla tubular, 2–5 × 5 cm, obliquely placed in calyx, bright bluish-purple to red or pinkish-purple, base of throat yellow, mottled with scarlet or purple, with white markings near mouth, tube spurred, curved slightly upward, broadening toward mouth; style pilose or puberulent, stigma bilobate. **Distr:** Mexico. **Ecol:** In mountain forests, damp or wet, mixed forests or pine forests, often on bluffs or rocky cliffs; 700–2,400 m. Fl. Aug–Dec. **Zones:** 10–12.

A. longiflora DC., *Prodr.* 7: 536, 1839. (Syn: *Gesneria longiflora*, etc.). [Cupid's Bower, Hot Water Plant, Magic Flower, Monkey-faced Pansy]. **Herbs**; stems erect or ascending, 10–40 cm high, green or reddish, villous; branches few. **Leaves** opposite or 3 at a node, pair subequal; lamina lanceolate to ovate, 1.9–5.9 × 0.8–2.7 cm, base cuneate to acute, margins serrate, apex acute, adaxially green and villous, abaxially pale green or more usually reddish between veins; petiole short. **Inflorescences** with flowers solitary, axillary; pedicels ~5 mm long, densely pilose. **Flowers** variable in color, showy; floral tube 2–5 mm long, densely pilose; calyx lobes linear lanceolate, 0.7–1.2 cm long, glabrescent, long-ciliate; corolla blue or lavender, erect in calyx, lacking a spur, tube narrow, 3.5–6.5 × ~0.5 cm, spotted or lined, glabrous, limb oblique, broad, flat, 3.9–6.5 cm Ø, nearly regular, lobes entire; stamens included, subequal to corolla tube, filaments adnate to base of corolla tube, glabrous; style puberulous, stigma bilobed. **Distr:** Mexico to Panama. **Ecol:** In moist forests, near streams, on rock outcrops. **Zones:** 10–12.

Aeschynanthus

Species ~185. **Herbs** perennial, or **subshrubs**, epiphytic, rarely terrestrial, woody or lignescent; stem terete, rarely ± quadrangular, erect to pendulous, rooting at nodes. **Leaves** opposite, often in a single plane by twisting of the internodes, rarely whorled; lamina ± lanceolate, ovate, or cordate, rarely suborbicular, margins usually entire, fleshy or leathery, with a water-storing hypodermis just below adaxial epidermis; petiole short. **Inflorescences** few-flowered cymes, or flowers solitary, axillary, or emerging from scaly bracts and forming pseudoterminal inflorescences; bracteoles rarely large and colored, mostly deciduous. **Flowers** with sepals free, usually narrow, partially or entirely connate, forming a cup-shaped or cylindrical, often showy (colored) calyx with small rounded lobes; corolla usually bright red, rarely orange, yellow or greenish, tube usually long, arcuate, limb often with dark spots or stripes, bilabiate, lobes rounded; stamens 4; ovary stipitate, linear, style slender, stigma capitate or peltate. **Fruits** capsules, linear, to 50 cm long, bivalvate, loculicidally dehiscent; seeds very small, numerous. **Distr:** S China, N and S India, throughout Malesia to New Guinea and the Solomon Islands. **Ecol:** In lowland or montane rainforests. **Cult:** Propagate from cuttings or seeds. Requires a moist, well-drained compost in partial shade.

A. radicans Jack, *Trans. Linn. Soc. London* 14: 43, 1823. (Syn: *A. javanicus*, etc.). [Basket Vine, Lipstick Plant, Lipstick Vine]. **Subshrubs**; stems pendent, 0.3–1.2 m long, rooting at the nodes, glabrous to finely pubescent. **Leaves** opposite, spreading; lamina 2.5–5 × 2–3 cm, margins entire to subentire, fleshy-coriaceous, glabrous to pubescent, green, with reddish-flushed; petiole 3–8 mm long. **Inflorescences** clusters or fascicles, 2–6-flowered, axillary and terminal; pedicels 0.5–1.5 cm long, glabrescent. **Flowers** with calyx tubular, 2–2.5 cm long, 5-dentate, reddish-green to black-purple, glossy or sometimes finely pubescent; corolla bright red, 4–5 cm long, tube terete and decurved, limb bilabiate, 3 lower segments ~8 mm long, upper one bilobed; anthers extending to apex of

Above: *Aeschynanthus radicans, Aeschynanthus speciosus, Alsobia dianthiflora.*

upper corolla lobe or slightly exserted; ovary downy, style pubescent. **Fruits** 20–30 cm long. **Distr:** Malayan Peninsula, Borneo, Java, Philippines. **Ecol:** Epiphytic in the angles of branches and limbs of rainforest trees, also grows on rocks; to 1,750 m. Fl. throughout most of the year. **Zones:** 11–12.

A. speciosus Hook., *Bot. Mag.* t. 4320, 1847. **Subshrubs**, woody, epiphytic; stems erect to scandent, to 80 cm long. **Leaves** 3–5-whorls, shortly petiolate; lamina fleshy, glabrous, elliptic to narrowly elliptic, 5–14.5 × 1.1–5 cm, base acute to cuneate, margins usually crenulate, apex acute to acuminate, midrib impressed adaxially, prominent abaxially, other venation obscure. **Inflorescences** 2- to many-flowered, subterminal, erect; pedicels to 1.2 cm long. **Flowers** with calyx pale green, lobes linear, dissected to base, 0.7–1.9 cm long; corolla tubular, arcuate, 7.5–10 cm long, tube straight for lower half and then flaring gradually toward throat, lobes rounded, to 1 cm long, minutely pubescent outside, tube base pale green to greenish-orange, turning bright red toward apex and lobes with a darker red line on inside of lobes; stamens 4, exserted, filaments red to violet, minutely pubescent; style stout, pale red to pale yellow, minutely pubescent. **Fruits** to 40 cm long. **Distr:** Peninsular Malaysia, Thailand; Borneo, Java and Sumatra; rare in the wild. **Ecol:** Fl. Feb–Mar, Oct–Nov. **Zones:** 11–12.

Alsobia

Species 2. **Herbs**, epiphytic, perennial; stems 20–80 cm long, creeping and rooting, producing stolons. **Leaves** opposite; lamina elliptic to ovate, 1.5–10 × 1–6 cm, margins crenate to serrate, apex acute or obtuse, ± fleshy, adaxially ± appressed-pilose with veins green to reddish, abaxially paler; petiole 1–4 cm long. **Inflorescences** with flowers solitary, axillary; pedicels 0.3–1.2 cm long. **Flowers** with calyx lobes 5, equal or unequal, 0.9–1.9 × 0.1–0.5 cm, margins entire, pilose and ciliate; corolla funnelform, lobes white or red spotted inside, tube ~3 cm long, spur and tube white or violet-purple spotted inside, limb 5-lobed, 0.7–1.2 × 1–1.4 cm, spreading, fringed with fine teeth 2–5 mm long; stamens 4, filaments white, anthers coherent; ovary superior, style white, stigma mouth-shaped; nectary a bilobed dorsal gland. **Fruits** capsules. **Distr:** Mexico, Guatemala, Costa Rica. **Cult:** Propagate from cuttings or roots. Requires a moist, well-drained compost in partial sun or shade. Best grown in hanging baskets.

A. dianthiflora (H.E. Moore and R.G. Wilson) Wiehler, *Selbyana* 5(1): 28, 1978. (Syn: *Episcia dianthiflora*). [Lace-flower Vine]. **Herbs**, to 20 cm high; stems creeping and rooting, 20–60 cm long, producing densely pilose green stolons. **Leaves** clustered in 4–5 pairs; lamina elliptic to ovate, 1.5–4 × 1–3 cm, margins crenate, apex obtuse, thinly fleshy, densely appressed-pilose; petiole 3–4 cm long. **Inflorescences** with flowers solitary, pedicels 6–8 mm long, reddish. **Flowers** with calyx lobes linear, subequal, 0.9–1.1 cm × 1–2 mm, green, pilose and ciliate; corolla funnel-shaped, white, tube 3 cm long, 2–3 mm Ø at base to 7–9 mm Ø at throat, spur and tube violet-purple spotted inside, limb 5-lobed, spreading, 1.2–1.4 cm Ø, conspicuously fringed with fine teeth 3–5 mm long; style glabrous. **Fruits** green, ~1.5 × 1 cm. **Distr:** S Mexico, Costa Rica. **Ecol:** Epiphytic on tree trunks, in tropical forests and rocky habitats. Fl. Apr–Oct. **Zones:** 10–12.

Chirita

Species ~100. **Herbs**, annual or perennial, sometimes slightly woody at base; stems fleshy. stout; vertical and horizontal rhizomes and stolons present in perennial species. **Leaves** opposite, those in each pair unequal, rarely whorled, usually densely downy. **Inflorescences** with flowers solitary or in many-flowered cymes; peduncle fused to the short petioles. **Flowers** with calyx tubular or divided to base, with 2, 3 or 5 lobes; corolla tube narrowly tubular to broadly funnelform, limb spreading, 5-lobed; fertile stamens 2, fused at apex; staminodes 2 or 3; ovary superior, stigma bilobed; nectary low, ringlike. **Fruits** capsules, straight, rarely oblique in relation to pedicel, linear to very narrowly ovoid, valves 2 or 4, straight, not twisted; seeds unappendaged. **Distr:** Bhutan, Borneo, China, India, Indonesia, Laos, Malaysia, Myanmar, Nepal, Thailand, Vietnam. **Cult:** Propagate from cuttings or seeds. Requires a moist compost in shade or partial sun.

C. lavandulacea Stapf, *Bot. Mag.* 150: t. 9047, 1925. (Syn: *C. horsfieldii, Didymocarpus horsfieldii*). [Lavender Chirita]. **Herbs**, branching, annual or perennial, upright, to 1 m high; stems fleshy. **Leaves** opposite; lamina elliptic-oblong, to 20 × 12.5 cm, concave, base cordate, margins serrated, apex acute to acuminate and recurved, light green, hairy; petiole short. **Inflorescence** 1–8-flowered, arranged around stem, in axils of leaves, flowers held ± horizontal; pedicels glandular, long. **Flowers** with sepals narrowly elliptic, ~6.5 × 1.5 mm, green; corolla 3–3.5 cm long, lower side bulging, off-white, lobes mauve-blue, almost hairless inside and outside; anthers hairy. **Fruits** capsules, 4–5 × 2–2.5 cm, sparsely hairy. **Distr:** Indo-Malaya. **Ecol:** On moist rock surfaces near waterfalls. Fl. Aug–Dec. **Zones:** 10–12.

C. sinensis Lindley, *Edward's Bot. Reg.* 30: pl. 59, 1844. (Syn: *C. dryas*, etc.). [Chinese Chirita]. **Herbs**, perennial, stemless. **Leaves** basal, opposite or ternate; lamina usually ovate to elliptic, 4–20 × 3.5–6.5(–9) cm, base attenuate to cordate, margins usually repand to crenate, apex acuminate to rounded, papery, pubescent; petiole 2–8(–19) cm long. **Inflorescences** cymes, 1–12-flowered; peduncle 12–32 cm long; bracts 2, free, ovate to lanceolate, 1–4 × 0.4–1.6 cm; pedicel 0.8–2.5 cm long. **Flowers** with calyx 5-lobed, lobes 3–6 mm long; corolla purple to white, inside purple-tinged, with 2 yellow longitudinal lines below the tube, (2–)2.7–4.6 cm long, tube tubular to broadly funnelform, often gibbous abaxially, (1.5–)2–3.5 × 1.1–1.5 cm, adaxial lip 4–6 mm long, abaxial lip 0.7–1.5 cm long; filaments 0.7–1.4 mm long; staminodes 2; pistil 1.8–2.5 cm long, ovary 1–1.8 cm long, stigma bilobed. **Fruits** capsules,

Below: *Chirita lavandulacea, Chirita sinensis, Chrysothemis friedrichsthaliana, Chrysothemis pulchella.*

4–7.5 cm long. **Distr:** China (W Guangdong). **Ecol:** In moist, shady places near streams; 100–500 m. Fl. May–Feb. Fr. Aug–Feb. **Zones:** 9–12.

Chrysothemis

Species 7. **Herbs**, terrestrial or epiphytic, growing from perennial tubers; stems short-lived, weak, succulent, erect or sometimes decumbent, subquadrangular, green, pubescent. **Leaves** opposite, in equal or subequal pairs; lamina lanceolate, elliptic, ovate, or oblanceolate, margins dentate or crenate, membranous, adaxially usually dark green, sparsely pubescent, abaxially light green or purplish to roseate and pubescent, lateral veins 7–8; petiole short, pubescent. **Inflorescences** cymose or umbellate, 1–9-flowered, axillary; bracts small, leaflike; peduncles and pedicels green or reddish, pubescent. **Flowers** somewhat zygomorphic; calyx of 5 connate sepals, winged or not, inflated or campanulate to urceolate, yellow-green, orange or red; corolla of 5 connate petals, limb flaring, 5-lobed, yellow to orange, with orange or reddish lines, tube cylindric; stamens 4, didynamous, included, anthers orbicular; nectary usually bilobed; ovary superior, spherical, conic or ovoid, stigma bilobed, ovules numerous. **Fruits** capsules, fleshy, globose to ovoid, bivalved, with persistent calyx; seeds dark brown, shiny. **Distr:** C and northern S America. **Ecol:** At low elevations. **Cult:** Propagate from cuttings or bulbils. Requires a moist soil or compost, moderately drained in partial shade.

C. friedrichsthaliana (Hanst.) H.E. Moore, *Baileya* 2: 87, 1954. (Syn: *Tussacia friedrichsthaliana*). **Herbs**, terrestrial, rarely epiphytic, to 60 cm high, ± pilose; stem erect or trailing, green; tuber 1–3 cm Ø. **Leaves** lanceolate, elliptic, ovate or oblanceolate, 7.5–37.5 × 2.5–13.5 cm, base long-decurrent, occasionally oblique, margins crenate to dentate, apex acute to acuminate, adaxially dark green, abaxially light green to purple; petiole to 1 cm long, green. **Inflorescences** with flowers solitary or 2–9-flowered cymes or umbels, axillary; peduncles <4 cm long, green; bracts leaflike, to 1.5 × 0.3 cm, green to purple; pedicels to 3 cm long, green. **Flowers** calyx of connate sepals, inflated cuplike, winged, 5-angled, tube 0.6–1.5 cm long, green to yellow; corolla greenish-yellow to orange or salmon-red, reddish-lines, tube 1.4–2.7 cm long, limb flared, 1.2–2 cm Ø; stamens not exserted, filaments coiling or recurved at apex after dehiscence; nectary yellow; ovary superior, style <1.5 cm long. **Fruits** globose, 0.7–1 cm Ø; seeds minute. **Distr:** Nicaragua south to Ecuador. **Ecol:** Mostly in shaded wet forests; <200 m. **Zones:** 10–12.

C. pulchella (Donn ex Sims) Decne., *Rev. Hort.* 21, sér. 3, 3: 242, 1849. (Syn: *Besleria pulchella*, etc.). [Black Flamingo, Copper Leaf, Square Stem, Sunset Bells]. **Herbs**, terrestrial, rarely epiphytic, to 90 cm high, ± pilose; stems decumbent with age, green or purplish; tuber spheroid, 2–3 cm Ø. **Leaves** lanceolate, elliptic to ovate, 9–30 × 3–12 cm, base long-decurrent, margins crenate or dentate, apex acute, adaxially dark green, shiny, subbullate, abaxially lighter green to pink purple along veins or generally; petiole <1 cm long, green. **Inflorescences** solitary, erect, umbellate, 3–9-flowered; peduncles 1.6–2.8+ cm, green or red; bracts 2, linear, leaflike, 0.4–1.5 cm long; pedicels 1–2.5 cm long, green or red. **Flowers** with calyx tubular-campanulate, 5-angled or -winged, 1–1.8 cm long, orange to red; corolla orange or yellow, with reddish-orange stripes, 1.6–3 cm long; stamens 4, filaments ~1 cm long; ovary conic, superior, style shorter than filaments, stigma deeply bilobed. **Fruits** ovoid, ~6 mm Ø; seeds ellipsoid. **Distr:** Panama, West Indies, northern S America from Surinam to W Colombia and Brazil. **Ecol:** In lowland damp forests. **Zones:** 10–12.

Cobananthus

Species 1. *Generic description same as specific.* **Cult:** Propagate from cuttings or seeds. Requires a moist, well-drained compost in shade or partial sun.

C. calochlamys (Donn. Sm.) Wiehler, *Selbyana* 2(1): 94, t. 288, 1977. (Syn: *Alloplectus calochlamys*, etc.). [Red Lantern]. **Herbs**, suffrutescent, terrestrial or epiphytic, usually erect, 30–60 cm high; stems simple, densely pilose. **Leaves** opposite, in subequal or unequal pairs; lamina oblong-lanceolate, mostly 6–20 × 1.5–6 cm, base oblique, attenuated, margins serrate, apex acute, surfaces velutinous-sericeous, thick; petiole 0.5–3 cm long. **Inflorescences** cymes, 1–3-flowered, axillary; bracts linear, mostly 1–1.5 cm long; pedicels 0.5–4 cm long. **Flowers** with calyx 1.5–2 cm long, densely sericeous, limb star-shaped, segments ± ovate, entire, obtuse or acute, usually red or orange; corolla red, densely sericeous and obscuring base color, tube cylindrical, slightly gibbous at base, slightly ventricose, appearing straight due to dense indument, usually 2–2.5 cm long, narrowed below the short lobes; stamens 4, included; staminode often present; disk gland solitary, bifid; ovary villous, stigma obliquely bilobed. **Fruits** capsules, subglobose, papyraceous, bivalved. **Distr:** Guatemala, Honduras. **Ecol:** Terrestrial or facultatively epiphytic, in forests. **Zones:** 10–12.

Codonanthe

Species ~20. **Subshrubs**, epiphytic, 0.2–0.8(–2) m high; stems climbing or trailing, rooting at nodes, usually glabrous; internodes 0.5–7 cm long. **Leaves** decussate, pairs subequal to unequal; lamina ovate to elliptic, 0.8–8 × 0.5–4 cm, margins entire to finely serrate, somewhat succulent, adaxially green to reddish, abaxially sometimes with reddish extrafloral; petiole 0.1–1.5 cm long, green. **Inflorescences** 1–4-flowered, axillary; pedicels to 1.5(–2.5) cm long. **Flowers** 5-merous; calyx green or reddish, lobes linear, elliptic or ovate, dorsal lobe sometimes recurved around corolla spur; corolla funnelform to subcampanulate and ± ventricose, 1–3 cm long, white, yellowish or pinkish, throat often tinged with brown or maroon, shortly tubular to prominently spurred at base, lobes spreading, unequal; stamens 4, included; ovary superior, style white, stigma mouth-shaped. **Fruits** berries or fleshy capsules, orange, pink or purplish-red; seeds ellipsoid, brown to reddish, 1–3 mm long. **Distr:** Mexico, Lesser Antilles, C and S America to Peru, Bolivia and Brazil. **Ecol:** On tree trunks and branches, sometimes on ant nests. **Cult:** Propagate from seeds or cuttings. Requires a well-drained soil mix, with high humidity and light.

C. gracilis (Martius) Hanstein, *Linnaea* 26: 209, 1854. (Syn: *Hypocyrta gracilis*, etc.). **Subshrubs**, epiphytic or lithophytic; stems glabrous, creeping or trailing, 30–80 cm long; internodes 1.5–6 cm long. **Leaves** paired, subequal; lamina narrowly to broadly ovate, 2.5–6 × 1–3 cm, margins entire, often with reddish punctation, succulent, adaxially glossy green, abaxially paler or rarely reddish; petiole 1–3 mm long. **Inflorescences** 1–3-flowered, axillary; pedicels 0.3–1.5 cm long. **Flowers** glabrescent; calyx 0.7–1.5 × 0.3–0.8 cm, lobes elliptic to ovate, margin entire, green, often with reddish veinlets; corolla tubular-ventricose, 2–2.5 cm long, white, or sometimes brown-spotted outside, throat yellow, brown-streaked, lobes subequal, 5–7 × 6–8 mm. **Fruits** berries, ovoid, 7–9 mm long, bright orange; seeds 1.2 mm long, brown, striate. **Distr:** Brazil. **Ecol:** Often on humus-covered rocks. **Zones:** 10–12.

Columnea

Species >75. **Shrubs**, small, usually epiphytic, sometimes terrestrial, often subscandent or pendent. **Leaves** opposite, in subequal to unequal pairs, usually shortly petiolate; lamina with margins entire or somewhat serrate. **Inflorescences** cymes, 1- to few-flowered, axillary, sometimes borne below the foliage; flowers pedicellate, usually ebracteate. **Flowers** mostly large and showy; calyx lobes 5, almost distinct, often colored, entire, dentate, or incised; corolla erect in calyx, tube saccate at base but ecalcarate, gradually ampliate upward, limb bilabiate, upper lip trilobate, with midlobe hooded, 2 lateral lobes spreading from it, lower lip entire; stamens 4, didynamous, filaments basally connate, free above, anthers coherent, dehiscent; staminode usually present; disk gland large, posterior, usually solitary, sometimes with 2–4 smaller glands; ovary superior, stigma usually bilobate. **Fruits** ± globose, fleshy and berrylike; seeds minute, ± ellipsoid, shiny, striate. **Distr:** C and S America, West Indies. **Cult:** Propagate from cuttings. Requires a moist, well-drained compost in partial sun. Best grown in hanging baskets.

C. gloriosa Sprague, *Bot. Mag.* 137: pl. 8378, 1911. [Goldfish Plant]. **Shrubs**, epiphytic, perennial, 30 cm high; stems wiry, to 1 m long, purplish to hirsute, rooting from nodes, eventually becoming pendulous. **Leaves** in subequal pairs; lamina ovate to ovate-oblong, 2–3.5 × 1–1.5 cm, base rounded,

Below: *Cobananthus calochlamys, Codonanthe gracilis, Columnea gloriosa, Episcia cupreata.*

Above: *Gloxinia perennis*, *Gloxinia sylvatica*, *Kohleria hirsuta*, *Mitraria coccinea*.

margins entire, apex obtuse, adaxially hirsute and nerves impressed, abaxially hirsute and nerves elevated, hairs brownish-red. **Inflorescences** with flowers solitary; pedicel 2–2.5 cm long, densely hairy. **Flowers** large, fiery-red, hooded; calyx lobes spreading, broadly ovate, ~1.5 cm long, margin entire, densely villous; corolla 6.5–8 cm long, scarlet, sparsely villous, tube funnelform, uniformly widening to throat, throat open, yellow, limb 2-lipped, lower lip reflexed, ~2.5 cm long; stamens exserted; ovary silky-hairy, style glandular, yellow, stigma bilobed. **Fruits** white, ovoid, ~1.5 cm Ø; seeds yellowish-brown. **Distr:** Panama, Costa Rica. **Ecol:** Epiphytic in shaded areas cloud forests; ~2,000 m. Fl. Jan–Jul(–Oct). **Zones:** 11–12.

Episcia

Species 6. **Herbs**, evergreen, terrestrial or epiphytic, creeping and mat-forming with stolons arising in pairs at nodes. **Leaves** opposite, petiolate to sessile, in equal or unequal pairs, smaller of the pair often deciduous; lamina puckered, often colorful. **Inflorescences** with flowers solitary or in clusters or racemes, axillary. **Flowers** with sepals 5 or calyx 5-lobed, acute or acuminate; corolla tube funnelform, spurred at base and contracted above spur and at throat, limb with 5 spreading, ± equal lobes; stamens 4, anthers joined in pairs; ovary superior; nectary of a single gland. **Fruits** capsules, ± spherical; seeds oblong. **Distr:** S America. **Cult:** Propagate from cuttings or seeds. Requires a fertile, moist. well-drained soil in shade.

E. cupreata (W. J. Hooker) Hanstein, *Linnaea* 34: 340, 1865. (Syn: *Achimenes cupreata*, etc.). [Carpet Plant, Flame Violet]. **Herbs**, perennial, creeping, mat-forming, stoloniferous, to 20 cm long. **Leaves** elliptic, 5.5–15 × 3.5–7 cm, margins toothed, surfaces usually wrinkled or puckered, green or often bronzy-red, adaxially usually silver- or purple-veined or -banded, abaxially purple, midrib and veins prominent; petiole to 7.5 cm long, coarsely hairy. **Inflorescences** clusters; pedicels to 4.5 cm long. **Flowers** with calyx green, base pinkish-flushed, to 1.2 × 0.3 cm, lobes toothed near apex; corolla 6 × 0.9 cm, wider at throat, reddish above, yellow with reddish markings or spots below and inside tube, densely covered with white or translucent papillae, shaggily hairy, 2 lower lobes reflexed. **Distr:** Brazil, Colombia, Venezuela. **Ecol:** In moist forests, and along lake margins; 1,500–1,600 m. Fl. throughout most of the year. **Zones:** 10–12.

Gloxinia

Species 15. **Herbs**, perennial; rhizomes scaly, occasionally aerial; stem usually erect. **Leaves** ± isophyllous, petiolate. **Inflorescences** with flowers usually solitary and ebracteolate in leaf or bract axils. **Flowers** showy, purple, orange, red, rarely white or yellow; sepals free; corolla tubular, funnelform or campanulate, erect or obliquely placed in calyx, limb with lobes subequal; stamens included, filaments often coiling after anthesis, anthers apically coherent; nectary annular or reduced, rarely absent; ovary semi-inferior to inferior. **Fruits** capsules, usually dry, rarely fleshy, 2-valved, rostrate. **Distr:** C and S America. **Cult:** Propagate from seeds or cuttings. Requires a humus-rich, moist, well-drained soil in shade or partial sun.

G. perennis (L.) Fritsch, *Nat. Pflanzenfam.* 4, Abt. 3b: 174, 1894. (Syn: *G. maculata*). [Canterbury Bells]. **Herbs**, perennial; stems to 75 cm high, simple or branched, ± glabrous. **Leaves** opposite; lamina ovate, to 20 × 15 cm, base rounded to cordate, margins scalloped, adaxially deep green and slightly hairy, abaxially sometimes red-flushed. **Inflorescences** many-flowered racemes in upper parts, and flowers solitary in lower parts; pedicels to 2.5 cm long, glabrous. **Flowers** relatively large; calyx tube to 1 cm long, 10-ribbed, lobes unequal, spreading; corolla horizontal in calyx, campanulate, to 4 cm long, purple, hairy, lobes equal, lower incurved with toothed or fringed margin; stigma mouth-shaped; disk pale. **Distr:** Colombia to Peru. **Ecol:** In moist forests, and along margins of waterways. Fl. Mar–Aug. **Zones:** 10–12.

G. sylvatica (Kunth) Wiehler, *Selbyana* 1: 33, 1975. (Syn: *Gesneria sylvatica*, etc.). [Bolivian Sunset]. **Herbs**, perennial, deciduous. **Leaves** opposite; lamina ovate-elliptic, 11–17 × 2.5–3 cm, base gradually tapered, apex acute, adaxially dark glossy green, abaxially pale green, surfaces with scattered. **Inflorescences** with flowers solitary or 2–3 in leaf-axils; pedicels to 16 cm long, dark red to purple, hairy. **Flowers** with calyx lobes green, linear; corolla tubular, to 2 cm long, swollen above middle, bright red, outside densely shortly hairy, inside yellow, spotted with reddish-brown, short, thick hairs, lobes 5, deep red, triangular; disk ringlike, indistinctly 5-lobed; style protruding, stigma club-shaped. **Fruits** capsules, ends blunt, with straight hairs. **Distr:** Bolivia, Brazil, Ecuador, Peru. **Ecol:** In coastal forests, often on mountain slopes. Fl. throughout most of the year. **Zones:** 10–12.

Kohleria

Species 17. **Shrubs**, **subshrubs** or **perennial herbs**, terrestrial or epilithic; rhizomes scaly; stem usually erect, sometimes decumbent, terete. **Leaves** opposite or in whorls of 3–4, ± isophyllous, petiolate; lamina pilose or hirsute. **Inflorescences** few-flowered, fasciculate to umbellate cymes, or flowers solitary, borne from leaf or bract axils. **Flowers** with sepals erect or recurved, equaling or longer than floral tube; corolla somewhat zygomorphic, red or greenish, tube usually somewhat inflated or ventricose or campanulate above a narrow base, hirsute outside; limb bilabiate or regular, usually spotted or striped; stamens 4, adnate to base of corolla tube, anthers at first coherent, later often becoming free, thecae dehiscent by longitudinal slits; nectary deeply lobate or of 5 separate glands, the 2 dorsal ones usually united; ovary more than half inferior, stigma bilobed. **Fruits** capsules, dry, dehiscing into 2 valves, apex conical or rostrate. **Distr:** Western Cordillera from Peru to Mexico, Venezuela, Trinidad, the Guianas. **Cult:** Propagate by division or from seeds. Requires a moist, well-drained soil in full or partial sun.

K. hirsuta (Kunth) Regel, *Flora* 31: 250, 1848. **Herbs** or **shrubs**, often epilithic; rhizomes scaly; stems villous to 1.5(–4) m long. **Leaves** usually opposite; lamina elliptic, sometimes ovate to elliptic, 4–12(–18) × 2–6(–10) cm, green, margins purplish, abaxially pale green to red or purplish; petioles to 4(–8) cm long. **Inflorescences** with partial inflorescences 1–4(–6)-flowered; peduncles to 2(–4) cm long; pedicels 2–6(–10) cm long, villous, yellow-green to purple; bracts usually scaly, sometimes lanceolate, <2 cm long. **Flowers** with floral tube villous; calyx (0.4–)0.7–1.2(–2) cm long, lobes erect, lanceolate to subulate, rarely triangular, to 1(–1.5) × 0.6 cm, green or purplish; corolla tubular, occasionally subventricose, to 4(–5) cm long, tube to 3.8(–4.5) cm long, red, sometimes yellowish to orangish, purple-dotted, base usually saccate, limb 0.6–2.2 cm wide, lobes 3–6 × 4–7 mm, yellow, purple-dotted; filaments 1.8–3.5 cm long; ovary half-inferior, 5–7 × 3–4 mm, style to 3.3 cm long. **Fruits** capsules, ovoid, 1–1.4 × 0.5–0.9 cm, 2-valved; seeds ellipsoid, ~0.5 mm long. **Distr:** Venezuela, Guyana, Surinam, Trinidad, Colombia. **Ecol:** On exposed, dry, rocky slopes; 0–2,600+ m. **Zones:** 10–12.

Mitraria

Species 1. *Generic description same as specific.* **Cult:** Propagate from seeds, cuttings or by air-layering. Requires a moist, humus-rich, peaty soil in a cool situation and partial sun.

M. coccinea Cav., *Anal. Cienc. Nat.* 3: 281, 1801. [Chilean Mitre Flower, Little Bottles, Patridge Berry]. **Herbs** perennial, small **shrubs** or **subshrubs**; stems wiry, climbing or straggling, to 2 m long, glabrous or shaggily hairy. **Leaves** opposite, pairs equal; lamina elliptic to ovate, 1–3 cm long, margins toothed, apex acute, leathery, glossy, dark green, hairy. **Inflorescences** with flowers solitary, axillary; pedicels 2–5 cm long; bract large, bilobed. **Flowers** scarlet or orangish-red; calyx lobes 5, free, 0.7–1.2 cm long; corolla tubular, much longer than calyx, to 5 cm long, tube inflated around middle, narrowing just below limb, limb (4–)5-lobed; stamens 4, inserted at base of corolla tube, anthers united in pairs; ovary superior, unilocular; nectary poorly developed, ringlike. **Fruits** capsules, berrylike, 1.2 cm across, green, red-flushed. **Distr:** Argentina, Chile. **Ecol:** In cloud forests, growing on moss-covered tree trunks and rocks in deep shade; 400–1,200 m. Fl. May–Aug or Dec–Mar. **Zones:** 9–11.

Above: *Nematanthus gregarius, Rhabdothamnus solandri, Rhynchoglossum gardneri.*

Nematanthus

Species ~26. **Shrubs**, epiphytic or climbers, 0.3–2 m high; stems often woody at base and rooting at nodes, glabrous to densely hairy. **Leaves** opposite, pairs ± equal; lamina with apex acute, abaxially often purple-tinged. **Inflorescences** axillary, usually ± pendent; pedicels usually long. **Flowers** with calyx lobes 5, much longer than tube, often reflexed; corolla usually orange, red, yellow, pink or creamish-white, tube usually conspicuously pouched on lower side, mouth narrow, or ± widening to throat but laterally compressed; stamens 4, anthers united, forming a rectangle; ovary superior, hairy or not; nectary of a single gland. **Fruits** capsules, opening widely, exposing the fleshy mass of seeds and funicles. **Distr:** Brazil. **Ecol:** In tropical forests. **Cult:** Propagate from seeds or stem cuttings. Requires a moisture-retentive potting mixture, high humidity and ample lighting.

N. gregarius D.L.Denham, *Baileya* 19(3): 126, 1974. (Syn: *Hypocyrta radicans, N. radicans*). [Clog Plant, Goldfish Plant, Guppy Plant]. **Herbs** or **subshrubs**; stems climbing or pendent, to 80 cm long. **Leaves** broadly elliptic to ovate, to 1.5–3 × 1.2–1.7 cm, base and apex ± acute, margins entire, succulent, glabrous, adaxially very glossy, dark green, abaxially paler; petiole 1–3 mm long. **Inflorescences** 1–3-flowered, axillary, flowers ± horizontally oriented, not resupinate, glabrous; pedicels 0.6–1.5 cm long. **Flowers** glossy, appearing waxy; calyx 1–1.5 cm long, green with orange marks at apex, lobes ovate, 1–1.5 cm long, margins entire, somewhat crassulescent; corolla tubular, strongly ventricose, 1.8–2.5 cm long, bright orange with a purple-brown stripe leading to each lobe, lobes 2 × 2 mm, margins not rolled under, apex obtuse. **Fruits** capsules, 1 × 0.8 cm, yellow-orange, flesh creamish. **Distr:** SE Brazil. **Ecol:** Epiphytic on tree trunks, in tropical rainforests. Fl. Jul–Feb. **Zones:** 10–12.

Rhabdothamnus

Species 1. *Generic description same as specific.* **Cult:** Propagate from seeds or cuttings. Requires a moist, well-drained soil in shade or partial sun.

R. solandri A. Cunn., *Ann. Nat. Hist.* 1: 460, 1838. (Syn: *Columnea scabrosa, R. scabrosus*). [New Zealand Gloxinia]. **Shrubs**, much-branched, scabrous, to ~2 m high; branches and branchlets opposite. **Leaves** broadly ovate to suborbicular, to 3(–5) × 3(–5) cm, margins coarsely serrate-dentate, appearing gray-green due to indumentum; petiole slender, to ~1 cm long. **Inflorescences** axillary, flowers solitary and nodding; pedicels slender, to 2(–4) cm long. **Flowers** variously colored; calyx free from ovary, deeply lobed, ~1 cm long, pilose, lobes ovate- to triangular-acuminate; corolla usually orange, red-veined, subcampanulate, tube 2–2.5 cm long, limb obscurely 2-lipped, lobes spreading, ciliolate, to ~2 cm long; stamens 4, staminode present, inserted on corolla tube, filaments long, slender, incurved above, anthers cruciately paired; disk small, annular, obscure; ovary superior, ovoid, style filiform, curved at apex, ~1.5 cm long, stigma small, obscurely bilobed. **Fruits** capsules, ovoid, ~0.7–1 cm long, with persistent calyx, tapering to apex, placentas thick, bilobed; seeds numerous, ovoid. **Distr:** New Zealand. **Ecol:** In coastal to lowland forests. **Zones:** 8–11.

Rhynchoglossum

Species ~12. **Herbs**, perennial or annual, terrestrial. **Leaves** few, along stem, alternate; lamina with base sometimes strongly oblique, cuneate to cordate. **Inflorescences** racemose, lax, later appearing scorpioid, pseudo-terminal and/or axillary, few- to many-flowered cymes; bract 1, opposite each flower, alternate (or absent). **Flowers** with calyx actinomorphic, 5-lobed, sometimes winged; corolla blue to purple, zygomorphic, tube cylindrical, 1.5–4 mm Ø, limb 2-lipped, upper lip bilobed, abaxial lip longer, trilobed, seldom undivided, rounded and mucronulate or acute to obtuse; stamens 2 or 4, if 2 on abaxial side, included, coherent in pairs, longitudinally dehiscent; staminodes 2, 3, or absent, adnate to adaxial or both sides of corolla tube; disk cupular; ovary ovoid, 1-loculed, stigma 1, undivided. **Fruits** capsules, ovoid, shorter than calyx, loculicidally dehiscent, 2-valved; seeds unappendaged. **Distr:** China, SE Asia, Pacific Islands. **Cult:** Propagate from seeds or cuttings. Requires a moist, well-drained soil in shade to partial sun.

R. gardneri Theobald and Grupe, *Ceylon J. Sci., Biol. Sci.* 10: 70, 1972. (Syn: *Klugia zeylanica*). **Herbs**, erect or prostrate, 20–60 cm high. **Leaves** ovate to ovate-oblong, 4–9.5 × 2.5–6.5 cm, base oblique, one side acute to obtuse, the other usually incurved and cordate, margins entire, sinuate, apex acuminate, adaxially bright green, abaxially paler; petiole 0.3–3.5 cm long. **Inflorescences** racemose; peduncles 1–3 cm long; bracts linear-filiform, 1–3 mm long; pedicels 2–5 mm long. **Flowers** with calyx 6–9 mm long, tube 5-angled, angles equal, subwinged, lobes 5, lanceolate-acuminate, 4–6 mm long; corolla tube white, 0.8–1 × 0.3–0.4 cm, upper lip white, ~2–3 × ~3 mm, lower lip rich-blue with a yellow spot at base, 0.8–1.4 × 0.9–1.4 cm, glabrous; stamens 4, filaments paired, ~5 mm long, the other ~4 mm long; disk cupular; ovary glabrous, style ~7 mm long. **Fruits** capsules, ovate, membranous, 4–5 × 3–4 mm. **Distr:** Sri Lanka. **Ecol:** In moist lowlands; <1,100 m. Fl. Apr, Jul–Aug. **Zones:** 11–12.

Streptocarpus

Species ~130. **Herbs**, rarely subshrubs, caulescent, erect or creeping, or acaulescent, annual, perennial or monocarpic, rosulate or unifoliate. **Leaves** opposite and petiolate in *caulescent* species (rarely alternate and sessile, and then the flowering axis with a very large basal leaf); in *acaulescent* species leaf show continued growth from a basal meristem; lamina abaxially sometimes white-dotted. **Inflorescences** cymose, branching, sometimes ± 1-sided, flowers paired, rarely reduced to 1 or 2, axillary or apparently from base of lamina or from petiole. **Flowers** with calyx usually divided to base into 5 segments, rarely with a distinct tube and 5 teeth; corolla gamopetalous, 5-lobed, often bilabiate; stamens arising at various levels on corolla tube; ovary usually unilocular, sometimes apparently bilocular by fusion. **Fruits** capsules, ± cylindrical, usually very long and slender, spirally twisted, sometimes splitting into 4 valves; seeds numerous, small. **Distr:** Africa, Madagascar, Mascarene Islands. **Cult:** Propagate by division, or from seeds or cuttings. Requires a fertile, humus-rich, moist, well-drained soil in shade.

S. saxorum Engl., *Bot. Jahrb. Syst.* 19: 154, 1894. [Cape Primose, False African Violet]. **Herbs**, perennial, succulent. **Leaves** opposite, crowded in upper parts of stems, pairs equal; lamina ovate-elliptic, 1.2–3(–3.8) × 0.7–1.2(–2) cm, base obtuse to rounded, shortly attenuate in largest leaves, margins revolute, entire, apex obtuse, fleshy, pubescent; petiole 0.2–1 cm long. **Inflorescences** axillary, solitary, 1–2-flowered; peduncles red-tinged, 7.5–13.5(–17) cm long; pedicels to 1.5 cm long. **Flowers** with calyx lobes oblong-lanceolate, 2–3.5 mm long; corolla white, lobes violet, ~2.7–4 cm long, base saccate, mouth laterally compressed, limb bilabiate, oblique, upper lip reflexed, bilobed, rounded, 0.5–1 cm Ø, lower lip protruding, to 2.3 cm long, lateral lobes to 1.2 × 1.35 cm, rounded, median lobe to 1.4 × 1.6 cm; stamens with filaments 1.7–2 mm long; staminodes 1–2 mm long; ovary narrowly conical, 6–8 mm long, densely pubescent, style to 1 mm long. **Fruits** capsules, 2.8–5 × ~2.5–0.4 cm, pubescent, style persistent; seeds ~0.4 mm long. **Distr:** Kenya, Tanzania. **Ecol:** On rocks and cliffs; 500–2,000 m. **Zones:** 10–12.

PLANTAGINACEAE

Genera ~90, species 1,700. **Herbs**, rarely shrubs or rooted aquatic herbs. **Stipules** absent. **Leaves** spirally arranged to opposite (rarely whorled), usually in a basal rosette; lamina simple to compound, veins (1- or)3–11, arcuate. **Inflorescences** spikes, rarely racemes or 1-flowered, pedunculate, each flower 1-bracteate. **Flowers** bisexual, small, rarely polygamous or unisexual; calyx usually 4-segmented, persistent, segments imbricate; corolla scarious or membranous, gamopetalous, actinomorphic or rarely zygomorphic, (3- or)4-lobed, lobes imbricate, often bilabiate, rarely spurred; stamens (1 or 2)4, inserted on corolla tube, alternate with lobes, anthers free, versatile, with an apical appendage; pistil 2-carpellate, ovary superior, (1)biloculed, axile or rarely basal, style 1, filiform, mostly longitudinally stigmatic. **Fruits** septicidal (loculicidal) capsules, rarely

Above: *Streptocarpus saxorum, Bacopa monnieri, Campylanthus salsoloides, Digitalis purpurea.*

nutlets; seeds 1 to numerous, minute. **Distr:** Worldwide, mostly temperate.

Bacopa

Species 55. **Herbs**, erect or creeping, 5–10 cm high, spreading to 60 cm Ø. **Leaves** opposite; lamina small, rounded, fleshy, often fragrant when crushed. **Inflorescences** with flowers solitary, axillary, or in terminal racemes; bracteoles 1 or 2 or absent. **Flowers** with sepals (4 or)5, entirely free, imbricate, upper 1 largest, lower 2 next large, lateral 2 innermost and narrowest; corolla tubular, limb patent, conspicuously or obscurely 2-lipped, lower lip trilobed, upper lip bilobed; stamens 4, didynamous or equal in length, very rarely 5, anther locules parallel, free; stigma dilated, capitate or bilobed. **Fruits** capsules, ovoid or globose, 2-grooved, 4-valved; seeds numerous, minute. **Distr:** Tropics and subtropics, mainly in America. **Cult:** Propagate from seeds or runners. Requires a damp place or shallow water, in full or partial sun.

B. monnieri (L.) Pennell, *Proc. Acad. Nat. Sci. Philadelphia*. 98: 94, 1946. (Syn: *Lysimachia monnieri*, etc.). [Water Hyssop]. **Herbs**, mat-forming; stems creeping, rooting at nodes, succulent, glabrous. **Leaves** opposite, each pair at right angles to the next, sessile; lamina oblong-oblanceolate, 0.8–2 × 0.3–0.6 cm, margins entire or rarely dentate, apex rounded, glossy, succulent, gland-dotted. **Inflorescences** with flowers solitary, axillary; pedicel 0.5–3.5 cm long; bracteoles 2, linear, below calyx. **Flowers** with sepals 5, ~5 mm long, lower and upper sepals ovate-lanceolate, lateral 2 sepals lanceolate to linear; corolla blue, purple, or white, 0.8–1 cm long, obscurely 2-lipped; stamens didynamous; stigma capitate. **Fruits** capsules, narrowly ovoid, enveloped in persistent calyx, apex acute; seeds minute, yellow-brown, ellipsoid, truncate at one end, longitudinally striate. **Distr:** Tropical and subtropical regions worldwide. **Ecol:** In brackish areas, sandy beaches; 0–1,100 m. Fl. May–Oct. **Zones:** 9–12.

Campylanthus

Species 15. **Shrubs**, **subshrubs** or woody-based **perennial herbs**; stems terete, erect to decumbent, often much-branched, sometimes with spine-tipped branches. **Leaves** alternate, sessile to subsessile; lamina simple, subulate to flattened, often succulent, entire. **Flowers** racemes, few- to many-flowered, terminal; flowers pedicellate or subsessile, bracts usually leaflike, similar to ordinary leaves; bracteoles 2, inserted at base of pedicel. **Flowers** white to purplish or pinkish-brown; calyx 5-lobed; corolla hypocrateriform to infundibuliform, lobes 5, subequal, tube cylindric with 2 geniculations; stamens 2, anterior, filaments filiform, inserted in corolla tube at upper geniculation, glabrous, anther longitudinally dehiscent; ovary ovoid, style filiform, glabrous, stigma capitate, oblique. **Fruits** capsules, orbicular to ovate, laterally compressed, septicidally dehiscent; seeds numerous, discoid, margins winged. **Distr:** Canary Islands, Cape Verde Islands, NE tropical Africa, southern Arabia, Pakistan. **Cult:** Propagate from cuttings or seeds. Requires a well-drained soil, containing sandy loam and peat, in full sun.

C. salsoloides (L. f.) Roth, *Nov. Pl. Sp.* 4, 1821. (Syn: *Eranthemum salsoloides*, etc.). **Shrubs**, to 2 m high; stems erect, branched; young branches leafy, old ones with persistent leaf bases. **Leaves** linear, ± terete, 1–5 cm long, apex acute, succulent, glabrous. **Inflorescences** racemes, ~10–30-flowered, simple or branched, to 20 cm long, hirsute; pedicels 3.5–6.5 mm long; bracts 2.5–6 mm long; bracteoles 1.3–2.5 mm long, hirsute. **Flowers** hirsute outside; calyx lobes, narrowly ovate to triangular, 4–6 × 1.5–2 mm, acute, ciliate; corolla hypocrateriform to infundibuliform, pink to mauve, or sometimes white, tube 7–9 × 1.5–2 mm, geniculate at level of calyx mouth, hairy below insertion of stamens inside, lobes elliptic to broadly obovate, 3.5–7.5 × 3–4 mm, obtuse to acute; anthers 2.2–2.5 mm long; ovary ovoid, abruptly tapered, style 1–1.5 mm long. **Fruits** capsules, broadly elliptic to ovate, 4–6.5 × 3.5–4 mm, brown to black; seeds 0.8 × 0.9–1 mm, reddish-brown, margins winged. **Distr:** Canary Islands. **Ecol:** On cliffs, rocky slopes; 10–660 m. **Zones:** 10–12.

Digitalis

Species 25. **Herbs**, rarely basally woody; stems simple or branched at base. **Leaves** alternate, often crowded below and rosulate; lamina large, margins entire or toothed. **Inflorescences** often secund racemes, terminal. **Flowers** with calyx, lobes 5, imbricate; corolla oblique, tubular-campanulate, often constricted beyond ovary, somewhat ventricose, limb slightly 2-lipped, lower lip trilobed, midlobe longer than lateral lobes, upper lip lobes much-reduced, emarginate; stamens 4, didynamous, included, anthers touching in pairs, locules divaricate, apically confluent; style bilobed. **Fruits** with capsules, ovoid, septicidal; seeds numerous, oblong to subovoid, small, ribbed, testa scrobiculate-reticulate. **Distr:** Asia, Europe. **Cult:** Propagate from seeds or basal offsets (in perennial species). Requires a humus-rich, moist, well-drained soil in full or partial sun.

D. purpurea L., *Sp. Pl.* 2: 622, 1753. [Foxglove, Lady's Glove, Purple Foxglove]. **Herbs**, annual or perennial, 0.6–1.2 cm high, gray-white pubescent, glandular-hairy, except for corolla and sometimes for subglabrous stems; stems 1 or few and cespitose. **Leaves** dimorphic; *basal* leaves mostly rosulate with petiole narrowly winged, to 15 cm long, lamina ovate to oblong-elliptic, 5–15 cm long, base tapering, margins crenate and rarely serrate, apex acuminate to obtuse; *cauline* leaves similar to basal leaves, decreasing in size upward, sessile or shortly petiolate and forming bracts. **Inflorescences** racemes, terminal. **Flowers** with calyx campanulate, ~1 cm long, lobes free, oblong-ovate; corolla purple to white, 3–4.5 cm long, inside spotted, lobes with apices white-pubescent. **Fruits** capsules, ~1.5 cm long; seeds clavate, puberulent. **Distr:** Europe; naturalized widely. **Ecol:** Fl. May–Jun. **Zones:** 4–11. **Info:** All parts of the plant are highly poisonous.

Isoplexis

Species 4. **Shrubs**, glabrous or glandular-pubescent to densely villose; stems erect, terete. **Leaves** usually congested at end of branches, alternate, subsessile; lamina lanceolate, base attenuate, margins serrate or serrulate, apex acuminate. **Inflorescences** spikelike racemes; flowers subsessile. **Flowers** tubular, 2-lipped; calyx deeply 5-lobed, campanulate; corolla cylindrical to campanulate or inflated-globose, usually constricted toward base, orange-yellow, purple-veined, limb 2-lipped; stamens 4, slightly exserted; ovary ovoid to conical, stigma capitate. **Fruits** capsules; seeds numerous, testa reticulate. **Distr:** Canary Islands, Madeira. **Cult:** Propagate from seeds, cuttings or basal suckers. Requires a fertile, moderately fertile soil in full or partial sun. **Info:** All the species are endangered in the wild. The flowers are mostly pollinated by birds.

I. canariensis (L.) Lindl., *Encycl.* pl. 528, 1829. (Syn: *Callianassa canariensis, Digitalis canariensis*). [Canary Island Foxglove, Cresta de Gallo, Tenerife Shrub Foxglove]. **Shrubs** or **subshrubs**, herbaceous, evergreen, 0.6–2 m high, with candelabra-shaped growth. **Leaves** whorled; lamina lanceolate to elliptic, 15–30 cm long, margins sharply dentate, dark green, stiffly coriaceous, pilose, abaxially felty; petiole erect. **Inflorescences** dense racemes, spikelike, sometimes up to 7 per plant, 16–40-flowered. **Flowers** unisexual and hermaphroditic (protandrous), tubular, bright orange or reddish-orange, large, 2.5–3 cm long, 2–2.5 cm Ø; corolla bilabiate, dorsal lobes oblong, acute, shortly bilobed; stamens subincluded in corolla tube. **Fruits** capsules, 0.4–1.4 × 0.2–0.8 cm, persistent; seeds numerous, minute. **Distr:** Canary Islands. **Ecol:** Mainly in laurel forests. Fl. May–Aug. **Zones:** 9–11. **Info:** Plant parts poisonous, containing cardiac glycosides.

Otacanthus

Species 4. **Subshrubs**, rarely **herbs**, covered with glands (3 types); stems terete to quadrangular, base often woody. **Leaves** opposite; lamina lanceolate to obovate, margins entire to clearly divided, pinnately nerved. **Inflorescences** spikes or flowers solitary, terminal or in axils of upper bracteate leaves; bracteoles sometimes present; pedicels <2 mm long. **Flowers** showy; calyx green, persistent, segments 5, imbricate, dorsal sepal ovate, blunt, foliaceous, larger, which are lanceolate; corolla violet to blue with a

IK

Above: *Isoplexis canariensis, Otacanthus azureus, Plantago ovata, Rhodochiton atrosanguineum.*

white spot at base of lower lip, hypocrateriform, tube long, cylindrical, villous outside, bilabiate, lips equal, lower lip sometimes slightly trilobed, swollen with a convex palate closing the throat; stamens 4, inserted at 2 different heights in corolla tube, 2 posterior sterile, 2 anterior fertile and longer, filaments filiform, sterile anthers reniform, fertile anthers coherent; ovary ovoid, biloculed, style filiform, stigma funnelform, bilobed. **Fruits** capsules, ovoid, septicidal, 2-valved; seeds numerous, truncate to nearly ovoid, small, brownish-black. **Distr:** Brazil. **Cult:** Propagate from half-hardened cuttings. Requires a moderately fertile, moist, well-drained soil in full to partial sun.

O. azureus (Linden) Ronse, *Brittonia*, 53(1): 144–145, 2001. (Syn: *O. coeruleus*, etc.). [Brazilian Snapdragon, Little Blue-boy]. **Subshrubs**, to 1–2 m high, branched at base, stems and leaves sticky, glandular-hairy; stems terete. **Leaves** broadly elliptic, to 12 × 4.4 cm, margins subcrenate to serrate, apex triangular to acute; petioles of opposite leaves connate or touching at base, winged. **Inflorescences** 6–17 cm long; pedicels to 2 mm long. **Flowers** with calyx glabrous, minutely glandular, lower segments narrowly triangular to subulate, 1.5 × 0.1 cm, upper segment broadly ovate, to 1.1–2 × 1 cm, apex subrounded; corolla violet-blue to sky-blue with a white spot on lower lip at throat, tube 2.5–4 cm long, subcylindrical, apically slightly curved downward, mostly glabrous, upper lip reniform, 1.5–2 × 1.6–2.5 cm, lower lip slightly trilobed, smaller; ovary conical, 3 × 2 mm, bilocular, style 3 cm long. **Fruits** capsules, 8 × 4 mm, 2-valved; seeds minute. **Distr:** Brazil (Rio de Janeiro and Espírito Santo). **Ecol:** On rocks, and in forests. **Zones:** 10–12.

Plantago

Species ~200. **Herbs**, or rarely small shrubs, annual, biennial, or perennial, acaulescent or rarely stemmed, with a taproot or only numerous fibrous roots. **Leaves** simple, ovate, elliptic, oblong, lanceolate, linear, or subulate, margins entire, repand, erose, or toothed, rarely pinnately or palmately cleft; petiole base vaginate. **Inflorescences** spikes, narrowly cylindric, capitate, or rarely 1-flowered, arising from rosette or leaf axil of stem. **Flowers** with corolla salverform or tubular, persistent, tube cylindric or constricted at throat, as long as or slightly longer than calyx, limb 4-parted; stamens (2–)4, anthers ovoid, suborbicular, elliptic, or oblong; ovary biloculed, sometimes with an apical third compartment, placentation axile, with 1 to numerous ovules per locule. **Fruits** pyxides, with 1 to numerous seeds; seeds peltately inserted, testa mucilaginous when wet. **Distr:** Cosmopolitan; weedy. **Cult:** Propagate from seeds. Requires any fertile, moist soil in full sun.

P. ovata Forsskal, *Fl. Aegypt.-Arab.* 31, 1775. (Syn: *P. ispaghula*). [Desert Indian-wheat, Indian Plantago, Indian Plantain, Ispaghul(a), Spogel Seeds]. **Herbs**, annual; taproot tenuous, long; stem short, branching at base, internodes 0.2–1.5 cm long, densely lanate. **Leaves** spirally alternate; lamina linear-lanceolate, 3–15 × 0.1–0.6 cm, base attenuate, decurrent, margins entire or remotely denticulate, apex with 1.5–5 mm subulate tip, papery, white-lanate, veins 3; petiole tenuous, 2–5 cm long. **Inflorescences** spikes, densely flowered, ovoid, 0.5–2 cm long; peduncle 3–22 cm long, villous-lanate; bracts broadly ovate to orbicular, 3–4 mm long, keeled. **Flowers** with sepals 2.5–3 mm long, keeled, anterior sepals obovate-elliptic, posterior sepals elliptic; corolla white, lobes orbicular-ovate, 2.5–2.8 mm long, acute; stamens exserted, anthers 0.9–1.3 mm long. **Fruits** pyxides, ellipsoid, 4–5 mm long, borne in narrowly ovoid to shortly cylindric spikes to 5 cm long; seeds 2, ellipsoid, 2.5–3 mm long. **Distr:** Mediterranean region; cultivated and naturalized worldwide. **Ecol:** Fl. May–Jun. Fr. Jun–Jul. **Zones:** 9–12. **Info:** Seeds are used for medicinal purposes and also in foods and beverages.

Rhodochiton

Species 3. **Herbs**, suffrutescent, perennial, glandular-pubescent to glabrescent; stems climbing using twining petioles. **Leaves** cauline, alternate, petiolate; lamina cordate to deltate, margins irregularly mucronate-dentate or subentire, teeth minute; pedicels pendent, not cirrhous. **Flowers** with calyx shallowly 5-lobed, broadly infundibuliform or campanulate, lobes equal, entire; corolla blackish-purple or deep pinkish-purple, subactinomorphic, subequally bilabiate, tube broad, terete or somewhat 5-sided, with 4 shallow longitudinal folds, base entire abaxially, without basal palate; ovary oblong-globose, locules equal. Fruits capsules; seeds numerous, somewhat dorsiventrally subsymmetrical, tuberculate, with 2 longitudinal wings. **Distr:** Mexico. **Cult:** Propagate from seeds or cuttings. Requires a rich, well-drained soil in full or partial shade.

R. atrosanguineum (Zucc.) Rothm., *Repert. Spec. Nov. Regni Veg.* 52: 25, 1943. (Syn: *Lophospermum atrosanguineum*, etc.). [Purple Bell Vine]. **Climbers**, perennial, evergreen or semievergreen, climbing by twining its slender stems or leaf petioles, to 3+ m high; stems often forming adventitious roots. **Leaves** heart-shaped, 2.6–7.5 cm, 2.1–4.3 cm at widest point, margins shallowly crenate; petiole sensitive, 2.6–5.5 cm long. **Inflorescences** with flowers solitary; peduncle 6–12 cm long; pedicels slender, flowers pendent. Flowers protogynous, long-lasting, somewhat fuchsialike; calyx large and showy, pink to purple or fuchsia-red, 4-lobed, lobes 2.5–3 cm long, connate in proximal 1.2–1.5 cm; corolla deep purple-red to almost blackish-purple, subtubular, 4–4.5 cm long, lobes 1–1.2 cm long, subequal, all recurved; stamens subequal, filaments straight, 3.6–4 cm long, connective expanded; style 3.8–4.3 cm long, slightly exserted, receptive stigma conical and lobed, straight. **Fruits** capsules, 1.5–2 cm long. **Distr:** Mexico. **Ecol:** In montane cloud forests; 1,400–3,500 m. Fl. & Fr. Mostly Jun–Jan. **Zones:** 9–11.

Russelia

Species 50. **Shrubs**, or herbaceous **subshrubs** from a woody base, perennial, erect scandent. **Leaves** opposite or whorled (caducous in *R. equisetiformis*), subsessile or shortly petiolate; lamina usually ovate to lanceolate, margins entire or dentate to deeply incised, glandular scales often present, pinnately veined. **Inflorescences** cymes, dichotomous, 1–3, axillary or subtended by bractlike leaves, often verticillate; flowers subsessile to long-pedunculate; bracts linear, subtending the slender pedicels. **Flowers** mostly small, 1–2 cm long; calyx deeply 5-lobed or with 5 sepals, ovate to lanceolate; corolla tubular to funnelform, slightly 2-lipped, red to pink (white), lobes 5, subequal, 2 adaxial lobes outside in bud, lower lip trilobed; stamens 4, in 2 unequal pairs, included, filaments in lower half of the tube; staminode usually present, short; ovary bilocular, many-ovuled, style slender, stigma slightly thickened. **Fruits** capsules, dry, ovoid to subglobose, glabrous, dehiscence septicidal (loculicidal); seeds small, slightly rugulose. **Distr:** Mexico, Guatemala; 1 sp. extending to Cuba and Colombia. **Cult:** Propagate from cuttings or rooted layers. Requires a well-drained, gritty or sandy soil in full sun.

R. equisetiformis Schltdl. and Cham., *Linnaea* 6: 377, 1831. (Syn: *R. juncea*). [Coral Plant, Firecracker Fern, Firecracker Plant, Fountain Bush]. **Subshrubs**, erect, to 1 m high, much-branched, branches 4–8-whorled at lower nodes, distal nodes with 2 branches, distal stems usually leafless. **Leaves** opposite or verticillate, usually caducous; lower larger lamina ovate to elliptic, 0.8–1.5 × 0.6–0.9 cm; petiole 3–4 mm long, often remaining appressed to stems when lamina fall. **Inflorescences** usually with 2–4 peduncles/node, each peduncle with 1 or 2 flowers, often resembling open racemose panicles, 2–16 cm long; bracteoles linear, 1–2 mm long; pedicels 0.6–1.5 cm long. **Flowers** glabrous outside; calyx 2–3 mm long, imbricate, acute or acuminate; corolla bright red, tubular-funnelform, 1.5–3 × 0.2–0.4 cm; stamens 18, 2 cm long, anthers near mouth, staminode present; ovary ovoid, 2 mm long, style 1.5 cm long, stigma minute. **Fruits** capsules, globose, 3–6 mm Ø; seeds light brown, among white hairs. **Distr:** Mexico. **Ecol:** In open sunny areas. Fl. throughout most of the year. **Zones:** 9–12.

R. sarmentosa Jacq., *Enum. Pl. Syst.* 6: 25, 1760. (Syn: *R. colombiana*, etc.). [Coral Plant, Leafy Coral Blow, Red Firecracker Plant]. **Herbs**, or weakly

Top row: *Russelia equisetiformis, Russelia sarmentosa, Buddleja davidii, Buddleja madagascariensis.*
Bottom row: *Phygelius capensis, Halleria lucida, Anisomeles indica, Callicarpa nudiflora.*

stemmed **shrubs**, 0.5–2 m high, base woody, leafy stems strongly 4- or 6-angled. **Leaves** 2 or 3 per node, subsessile; lamina ovate or ovate-triangular, to 7(–9) × 5(–6) cm, base obtuse or truncated, margins dentate, apex acute. **Inflorescences** congested verticels of cymes, 12–20-flowered, 2–3 cm long, in distal leaf axils; peduncles to 1.8 cm long, stemlike; bracts linear, 3–4 mm long; pedicels 5–8 mm long. **Flowers** with calyx 3–4 mm long, lobes 2–3 mm long; corolla deep red, tubular, 1–1.4(–1.6) cm long, lobes subequal, 1.3–3 mm long; stamens ~8 and 6 mm long, inserted near middle of tube, anthers white; ovary 1–2 mm long, style 4–8 mm long. **Fruits** rounded-ovoid to subglobose, 4–5 × 4–5 mm; seeds minute, dark, among yellowish hairs. **Distr:** Mexico to Cuba and Colombia. **Ecol:** In forest margins; 0–1,300 m. Fl. & Fr. throughout most of the year. **Zones:** 10–12.

SCROPHULARIACEAE

Genera 65, **species** 1,700. **Herbs**, sometimes **shrubs**, mostly autotrophic, less often hemiparasitic or parasitic, rarely trees. **Stipules** usually absent. **Leaves** alternate, opposite, whorled, or basally opposite and apically alternate; lamina simple or sometimes pinnately dissected. **Inflorescences** racemes, spikes, or thyrsoid panicles, determinate or indeterminate, or flowers solitary. **Flowers** perfect, usually zygomorphic, rarely actinomorphic; calyx often persistent, (2)4 or 5-lobed or -parted, variously connate; corolla sympetalous, limb (3 or)4 or 5-lobed, often 2-lipped; stamens mostly 4, didynamous, sometimes 1 or 2 staminodes present, less often 2 or 5 stamens, anther locules 1 or 2, equal or subequal, free or confluent; nectary often present at base of ovary, ringlike, cupular, or reduced to a gland; ovary superior, biloculed, rarely apically 1-loculed, ovules numerous, rarely 2 per locule, on axile placentas, anatropous or hemitropous, style simple, stigma capitate, bilobed, or bilamellate. **Fruits** capsules, septicidal, loculicidal, or septifragal, sometimes opening by pores or irregularly dehiscent, rarely berries; seeds minute or rarely conspicuous, sometimes winged, testa often reticulate. **Distr:** Cosmopolitan.

Buddleja

Species ~100. **Shrubs**, less often trees, lianas, or suffrutescent herbs; branches terete, 4-angled, or 4-winged. **Stipules** usually leafy, suborbicular and auriculate or reduced to a transverse line. **Leaves** opposite, rarely alternate, often shortly petiolate; lamina with margins entire, crenate, or dentate. **Inflorescences** usually many-flowered, terminal and/or axillary; bracts mostly leafy; bracteoles resembling sepals. **Flowers** 4-merous, bisexual or unisexual; calyx usually ± campanulate, tube usually longer than lobes; corolla campanulate, cup-shaped, salverform, or funnelform, tube cylindrical, straight to curved, usually longer than lobes, lobes imbricate, rarely valvate; stamens inserted on corolla tube, usually included, alternating with corolla lobes, anthers bilocular; ovary 2(–4)-locular, locules several- to many-ovuled, style short to long, stigma often large, clavate, capitate, or less often bilobed. **Fruits** capsules, septicidally 2-valved or berries; seeds many, small, often winged. **Distr:** Tropics and subtropics of America, Africa, and Asia. **Cult:** Propagate from half-hardened cuttings. Requires a fertile, moist, well-drained soil in full sun.

B. davidii Franchet, *Nouv. Arch. Mus. Hist. Nat.*, sér. 2. 10: 65, 1887. (Syn: *B. davidii*, etc.). [Butterfly Bush, Summer Lilac]. **Shrubs**, 0.5–5 m high, most parts white-tomentose or -pubescent; branchlets nearly 4-angled. **Stipules** suborbicular to ovate, 1–6 mm long. **Leaves** narrowly ovate or narrowly elliptic, 4–20 × 0.3–7.5 cm, base cuneate, margins serrate, apex acuminate, adaxially dark green, ± glabrous, lateral veins 9–14 pairs; petiole 1–5 mm long. **Inflorescences** seemingly racemose or thyrsoid cymes, 4–30 × 2–5 cm, terminal; lower bracts leafy, others small, linear. **Flowers** with calyx campanulate, 2–3.5 mm long, lobes narrowly triangular, 0.5–2 mm long; corolla violet to dark purple, sometimes white, throat orange-yellow, 0.8–1.4 cm long, tube ± cylindrical, 0.6–1.2 cm long, lobes suborbicular, 1.5–3 mm Ø; anthers <1.2 mm long; ovary ovoid, 1.2–2 mm long. **Fruits** capsules, brown, narrowly ellipsoid to narrowly ovoid, 5–9 × 1.5–2 mm; seeds ellipsoid, 2–4 × ~0.5 mm, ends winged. **Distr:** S China. **Ecol:** In thickets; 500–3,000 m. Fl. May–Oct. **Zones:** 4–11.

B. madagascariensis Lam., *Tabl. Encycl.* 1: 291, 1792. (Syn: *Adenoplea madagascariensis*, etc.). [Butterfly Bush, Smoke Bush]. **Shrubs**, often scrambling, 2–6(–10) m high, most parts densely tomentose; branchlets terete. **Leaves** narrowly ovate to elliptic, 2–14 × 1.5–7 cm, base rounded, acute, cuneate, or rarely subcordate, margins entire, apex acuminate, adaxially ± glabrous, lateral veins 6–12 pairs; petiole 0.5–2 cm long. **Inflorescences** paniculate or thyrsoid, 5–25 × 2–15 cm, terminal. **Flowers** with calyx campanulate to urceolate, 2–3.5 cm long, lobes broadly triangular, ~1 mm long; corolla dark yellow, orange, or salmon, tube 0.7–1 cm long, outside tomentose, lobes ovate to suborbicular, 2–4 ×

2–3 mm; stamens inserted just below corolla mouth, anthers 1–1.4 mm long; ovary subglobose, 1–1.5 mm long, apex tomentose, 4-locular, style base tomentose, stigma 1.2–1.7 mm long. **Fruits** berries, white at first, purple-blue, globose, 2.5–5 mm Ø; seeds ellipsoid to obliquely ovoid, ~1 × 0.5 mm, unwinged. **Distr:** Madagascar. **Ecol:** 100–300 m. Fl. Apr–Jun. Fr. Aug–Nov. **Zones:** 9–11.

Phygelius

Species 2. **Shrubs** or **undershrubs**, robust, evergreen, ± glabrous, erect; shoots 4-angled, woody at base, ± herbaceous above. **Leaves** opposite; lamina with margins dentate. **Inflorescences** secund cymose panicles, terminal, flowers pendent, numerous; branches mainly opposite, alternate toward apex. **Flowers** with calyx thickly chartaceous, 5-lobed, lobes imbricate, persistent, not accrescent in fruit; corolla trumpet-shaped, tube long, incurved or nearly straight, gradually dilated upward, limb strongly oblique or subequal, lobes 5, short, obtuse, subequal, ± spreading; stamens 4, didynamous, glabrous, filaments thickly filiform, declinate, inserted below corolla mouth, exserted, anthers oblong, 2-celled, connective thick; staminode obsolete; ovary 2-celled, style thickly filiform, exserted beyond stamens, glabrous, stigma small, minutely bifid, ovules numerous. **Fruits** capsules, ovoid, ± oblique, septicidal; valves entire; seeds irregularly ovoid, bluntly angular, subreticulately veined, testa thickly spongy. **Distr:** S Africa. **Cult:** Propagate from cuttings. Requires a humus-rich, moist, well-drained soil in full sun.

P. capensis E. Meyer ex Benth., *Companion to the Botanical Magazine* 2: 53, 1836. [Cape Figwort, Cape Fuchsia, Orange Cape Fuchsia]. **Undershrubs**, erect, loosely branched, ± glabrous, 60–90 cm high, base woody, subherbaceous above; branches narrowly 4-winged. **Leaves** opposite; lamina ovate or ovate-lanceolate, 2.5–12.5 × 1.2–6.5 cm, base usually subtruncate, margins crenulate-serrulate, apex obtuse, coriaceous or chartaceous; petiole clasping, narrowly decurrent, base shortly biauriculate, 0.8–6.5 cm long. **Inflorescence** loosely cymose, compound, secund, pyramidal, 15–45 cm long, drooping, terminal, bracteate; ultimate pedicels 1.2–2.5 cm long. **Flowers** with calyx lobes ovate or oblong-lanceolate, ~6 mm long, ± obtuse or; corolla deep scarlet, ~4 cm long, tube shortly contracted above ovary, dilated above, oblique at apex, lobes ovate-rounded, ~3–5 mm long; stamens and style shortly exserted. **Fruits** capsules, ovoid, 1–1.6 × 0.6–0.9 cm, oblique, bisulcate, smooth, glaucous. **Distr:** S Africa. **Ecol:** On stream banks; 1,300–2,200 m. Fl. Oct–Dec. **Zones:** 8–11.

STILBACEAE

Genera 11, species 39. **Trees**, **shrubs** or **herbs**, with ericoid habit; shrubs single-or multistemmed. **Leaves** densely crowded, verticillate, opposite, rarely alternate; lamina with glandular trichomes, narrowly lanceolate to linear, margins recurved. **Inflorescences** terminal, rarely on axillary branches; flowers subtended by a leafylike bract and 2 bracteoles. **Flowers** bisexual; calyx with 4–5 sepals, usually tubular basally, lobes 4–5, free, imbricate to valvate; corolla actinomorphic or zygomorphic, tube narrowly funnelform, lobes (4)5, imbricate; stamens usually 4 or 5, equal, filaments free, short to long, slender or stout, inserted in corolla throat or between lobes, anthers bilocular, dorsifixed; gynoecium bicarpellate, ovary superior, bilocular, often becoming 1-locular by abortion or less of septum, locules 1–4(–6)-ovuled, style terminal, filiform, erect, stigma capitate or minutely bilobed. **Fruits** capsules, loculicidal or valvular; seeds reticulate. **Distr:** S Africa (Cape Province), tropical Africa, Madagascar, Mascarenes, S Arabia.

Halleria

Species 10. **Shrubs** or **small trees**, glabrous. **Leaves** opposite, petiolate; lamina ovate or elliptic, margins dentate or subentire. **Inflorescences** with flowers subfascicled in axillary clusters; pedicels bibracteolate. **Flowers** with calyx 3–5-lobed, subrotate or cup-shaped, persistent; corolla 4–5-lobed, bilabiate, tube curved or straight, inflated above to funnelform, upper lip bilobed, lower lip shortly trilobed, lobes subequal, all much shorter than tube; stamens 4, didynamous, subequal, usually exserted, filaments filiform, inserted subcentrally on corolla tube, anther cells diverging, at length divaricate; staminode absent; ovary ovoid, bilocular, ovules numerous, style filiform, persistent, usually exserted. **Fruits** berries; seeds compressed, narrowly winged. **Distr:** Tropical Southern Africa, Madagascar. **Cult:** Propagate from seeds or cuttings. Requires a fertile, moist, well-drained soil in full or partial sun.

H. lucida L., *Sp. Pl.*: 625, 1753. (Syn: *H. abyssinica*). [African Honeysuckle, Tree Fuchsia]. **Trees**, small or erect to straggling **shrubs**, to 12 m tall; branches subquadrangular. **Leaves** broadly ovate, 4.5–10 × 2–6.2 cm, base rounded to broadly cuneate, margins shortly serrate to serrate-crenate, apex acuminate to caudate-acuminate, coriaceous, glabrous, abaxially minutely glandular-punctate; petiole 0.4–1.2 cm long. **Inflorescences** with flowers subfascicled in axillary clusters; bracteoles lanceolate, 1–1.8 × 0.4–0.5 mm; pedicels 1–1.4 cm long with a pair of bracteoles below middle, sparsely hirsute. **Flowers** with calyx 2.4–3.5 mm long, lobes 3.8–4.2 mm wide, joined to beyond halfway, rounded; corolla orange-yellow to brownish-red, 2.5–3.3 cm long, tube curved, widening above to 0.6–1 cm Ø, base slightly gibbous, mouth oblique, unequally 4-lobed; stamens long-exserted; style 2–3.8 cm long. **Fruits** berries, ovoid to subglobose, 1.2–1.8 × 0.8–1.4 cm, blackish-purple. **Distr:** Ethiopia southwards to S Africa; Angola, Yemen. **Ecol:** In montane forests. Fl. May–Jan. **Zones:** 8–11.

LAMIACEAE

Subfamilies 7, genera 236, species 7,173. **Herbs**, **subshrubs** or **shrubs**, annual or perennial, rarely trees or vines, usually aromatic; stems and branches usually 4-angled; hairs eglandular, usually uniseriate. **Stipules** absent. **Leaves** opposite, rarely whorled or alternate; lamina simple to pinnately dissected or palmately compound, margins toothed. **Inflorescences** usually compound, sometimes flowers solitary and axillary, verticillasters 2- to many-flowered, subtended by leaves or bracts; flowers bracteolate or not. **Flowers** bisexual, zygomorphic, rarely subactinomorphic; calyx persistent, 5-toothed, 2-lipped, upper lip 3-toothed or entire, lower lip 2- or 4-toothed; corolla with limb usually 2-lipped, upper lip bilobed, rarely entire, and lower lip 3(4)-lobed, tube hairy annulate inside; stamens epipetalous, 4 or 2, free, rarely filaments connate, sometimes 1 staminodial, anther 1- or 2-celled, usually longitudinally dehiscent; disk persistent; ovary superior, biloculed, locules 2-ovuled, style subterminal, or ovary 4-parted and each lobe 1-ovuled and style gynobasic (from bases of ovary lobes) with apex 2-cleft, stigma inconspicuous. **Fruits** schizocarps, berries or drupes, calyx persistent or accrescent; seeds with or without endosperm. **Distr:** Worldwide.

Anisomeles

Species 5. **Herbs**, erect, robust. **Leaves** with margins dentate. **Verticillasters** many-flowered, in dense or interrupted, long, terminal spikes; upper floral leaves reduced, bractlike; bracts linear, minute. **Flowers** with calyx campanulate, inconspicuously 10-veined, apex netted, teeth 5, equal, straight, slightly dilated in fruit; corolla 2-lipped, tube as long as calyx, finely pilose annulate inside, upper lip erect, margin entire, slightly concave, lower lip longer, spreading, trilobed, midlobe largest, emarginate or bilobulate; stamens 4, exserted, didynamous, anterior 2 usually longer, anterior anthers transversely 2-celled, posterior anthers 1-celled; style apex 2-cleft, lobes subulate, subequal; disk cupulate, crenate. **Fruits** nutlets, black, subglobose, shiny. **Distr:** W Indian Ocean, tropical and subtropical Asia to N Australia. **Cult:** Propagate from cuttings or seeds. Requires a rich, well-drained soil in full sun.

A. indica (L.) Kuntze, *Revis. Gen. Pl.* 2: 512, 1891. (Syn: *A. ovata*, etc.). [Cat Mint]. **Herbs**; stems erect, 1–2 m high, branched, densely white-pubescent. **Leaves** broadly ovate, 4–9 × 2.5–6.5 cm, base broadly truncate-cuneate, margins irregularly dentate, apex acute or shortly acuminate, adaxially finely strigose, abaxially densely tomentose; petiole 1–4.5 cm long. **Inflorescences** spikes, ~2.5 cm Ø; floral leaves shortly petiolate or ± sessile, longer than verticillasters; bracts 3–4 mm long. **Flowers** with calyx ~6 mm long, hirsute, glandular-pubescent, yellow-glandular, teeth purple-red, triangular-lanceolate, ~2.7 mm long, ciliate; corolla purplish, ~1.3 cm long, glabrous outside, tube funnelform, mouth to 3.5 mm wide, upper lip oblong, 4.5–5 × 3 mm, margin entire, lower lip subhorizontally spreading, ~9 × 5 mm, midlobe obcordate, ~3 × 4.5 mm, slightly undulate, bearded, lateral lobes ovate; ovary glabrous. **Fruits** nutlets, ~1.5 mm Ø. **Distr:** Tropical and subtropical E Asia, to Philippines, Taiwan. **Ecol:** In forest margins; 0–2,400 m. Fl. Aug–Sep. Fr. Sep–Nov. **Zones:** 10–12.

Callicarpa

Species ~140. **Shrubs** or **trees**, erect or rarely climbing, mealy tomentose, hairs stellate, verticillately branched, dendritic, rarely simple and hooked; branchlets terete or 4-angled. **Leaves** opposite or in 3s. **Inflorescences** cymes, sessile or pedunculate, axillary; bracts linear. **Flowers** with calyx campanulate, truncate or minutely 4-dentate, unaltered in fruit; corolla actinomorphic, campanulate or tubular, lobes 4; stamens 4, inserted in corolla tube, filaments slender, often exserted, anthers ovate or oblong, dehiscing by longitudinal slits or circular pores; ovary imperfectly bilocular, ovules 2 per locule, attached to middle or distal part of ovary, style usually longer than stamens, stigma usually dilated. **Fruits** drupes, globose, small, endocarp bony, mesocarp fleshy, exocarp thin; seeds small, oblong, testa membranous. **Distr:** Mainly tropical and subtropical Asia, few in tropical America and Africa, and very few in temperate Asia and N America. **Cult:** Propagate from half-hardened cuttings or seeds. Requires a moist, well-drained soil in partial sun.

C. nudiflora Vahl, *Bot. Beechey Voy.* 206, 1836. (Syn: *C. reevesii*, etc.). [Beauty Bush]. **Shrubs** or **small trees**; young branchlets, petiole, abaxial leaf surfaces, and inflorescences unevenly gray-brown stellate tomentose. **Leaves** ovate-elliptic to lanceolate, 12–22 × 4–7 cm, base obtuse to subrounded, margins serrate, slightly undulate, or subentire, adaxially dark green-black and glabrescent except for tomentose veins, abaxially densely gray-brown unevenly tomentose. **Inflorescences** cymes, 8–13 cm across; peduncle 3–8 cm long; bracts linear-lanceolate; pedicel ~1 mm long. **Flowers** with calyx usually glabrous, truncate to obscurely 4-dentate; corolla purple to pink, ~2 mm long, glabrous; stamens 2–3 times as long as corolla; ovary glabrous. **Fruits** ~2 mm Ø. Fl. Jun–Aug. Fr. Aug–Dec. **Distr:** India to SE China and peninsular Malaysia. **Ecol:** In mixed forests; ~1,000 m. **Zones:** 10–12.

C. pedunculata R. Br., *Prod. Fl. Nov. Holl.* 1: 513, 1810. (Syn: *C. cuspidata*, etc.). [Beauty Berry, Wild Heliotrope]. **Shrubs**, most parts densely tomentose;

MD

Above: *Callicarpa pedunculata, Clerodendrum bungei, Clerodendrum calamitosum, Clerodendrum incisum.*

branches obtusely tetragonal. **Leaves** decussate-opposite; lamina oblong or oblong-elliptic, 5.5–14.5 × 2.5–5.8 cm, base usually acute or rounded, margins sharply serrate in the middle, apex usually long-acuminate, secondaries 5–13 per side; petiole 0.3–1 cm long. **Inflorescences** axillary or supra-axillary, cymes, solitary, densely many-flowered, 2–4 × 1.5–4.5 cm, decussate, dichotomous, densely tomentose; peduncles 1–2.5 cm long; pedicels absent or 1 mm long; bractlets 1–3 mm long. **Flowers** with calyx campanulate, 1.2–1.3 mm Ø, densely pubescent, rim toothed; corolla infundibular or hypocrateriform, pale-purple, ~2 mm long, lobes ~0.8 mm long; stamens exserted, filaments filiform, ~3.5 mm long, glabrous; pistil exserted, surpassing stamens, style 5–5.2 mm long, ovary depressed-globose, ~0.5 mm long. **Fruits** depressed-subglobose, deep lilac or purple, small; fruiting-calyx ~2 mm Ø, densely pubescent. **Distr:** Tropical and subtropical Asia to Solomon Islands. **Ecol:** In and around tropical and subtropical rainforests. Fl. Apr–Jun or Nov–Mar. **Zones:** 10–12.

Clerodendrum

Species 325. **Shrubs** or **small trees**, rarely subshrubs or herbs, erect or rarely climbing, deciduous or evergreen; branchlets usually 4-angled when young. **Leaves** opposite, rarely whorled; lamina simple. **Inflorescences** loosely cymose or capitate, in terminal or rarely axillary paniculate thyrses. **Flowers** with calyx campanulate or cup-shaped, ~5-dentate or deeply 5-lobed; corolla tube slender, lobes 5, spreading; stamens 4, ± exserted; ovary 4-locular, ovules pendulous or laterally attached, style with 2 acute stigmatic lobes. **Fruits** drupes, with four 1-seeded pyrenes, sometimes separating into two biloculed or four 1-locular mericarps; fruiting calyx ± inflated and often showy. **Distr:** Tropical and subtropical regions worldwide. **Cult:** Propagate from seeds, cuttings or suckers. Requires a moist, well-drained, humus-rich soil in full or partial sun; some species are salt-tolerant.

C. bungei Steudel, *Nomencl. Bot.*, ed. 2. 1: 382, 1840. (Syn: *C. fragrans* var. *foetidum, Volkameria bungei*). [Glory Flower]. **Shrubs**, 1–2 m high; branchlets subterete, lenticellate. **Leaves** ovate, 8–20 × 5–15 cm, base cuneate, truncate, or cordate, margins serrate, apex acuminate to acute, papery, adaxially sparsely pubescent, abaxially sparsely glandular-pubescent to subglabrous, with several peltate glands near base, veins 4–6 pairs; petiole 4–17 cm long, densely brownish or purplish pubescent when young, glabrescent. **Inflorescences** cymes, flat-topped, dense, capitate, terminal; bracts lanceolate to ovate-lanceolate, ~3 cm long, deciduous; bractlets lanceolate, ~1.8 cm long. **Flowers** with calyx campanulate, tube 2–6 mm long, pubescent, with several peltate glands, lobes narrowly triangular to triangular, 0.1–1 cm long; corolla pinkish, red, or purple, tube 2–3 cm long, lobes obovate, 5–8 mm long; stamens and style exserted. **Fruits** drupes, blue-black, subglobose, 0.6–1.2 cm Ø. **Distr:** China, Taiwan, Vietnam. **Ecol:** 1,100–2,500 m. Fl. & Fr. May–Nov. **Zones:** 8–11.

C. calamitosum L., *Mant. Pl.* 1: 90, 1767. (Syn: *Volkameria alternifolia*). [White Butterfly]. **Shrubs**, small to medium-sized. **Leaves** opposite, decussate; lamina ovate, rarely some leaves shallowly trilobed, 3.5–12.5 × 2.5–7.5 cm, base broadly tapered, margins irregularly dentate, adaxially bright green, abaxially paler, surface somewhat puckered, venation prominent, more so abaxially, surfaces softly pubescent; petiole 0.3–1.5 cm long. **Inflorescences** cymes, 5–10 cm long, axillary; peduncles 1.5–3 cm long; pedicels ~6 mm long. **Flowers** pure white, mildly fragrant, 2.5–3 cm long; calyx green or pale reddish, 0.7–1 cm long; corolla salverform, tube 2–3 cm long, lobes 5, oblanceolate, to 1.5 × 0.5 cm, rounded; stamens exserted to 2 cm long, filaments glabrous, white, anthers orangish. **Distr:** Indo-China to W and C Malesia. **Ecol:** Natural habitat unknown, naturalized widely and cultivated in moist or dry situations. Fl. throughout most of the year. **Zones:** 10–12.

C. chinense (Osbeck) Mabb., *Plant-book, Repr.*: 707, 1989. (Syn: *C. philippinum*, etc.). [Cashmere Bouquet, Chinese Glory Bower]. **Shrubs**, 0.5–1.2 m high; branchlets nearly 4-angled to terete, pubescent when young, later glabrescent. **Leaves** broadly ovate to subcordate, 9–22 × 8–21 cm, base truncate, broadly truncate or subcordate, margins sparsely irregularly dentate, apex acuminate, abaxially pubescent especially on veins and with several large glands near base, adaxially strigose; petiole 3–17 cm long, pubescent to tomentose. **Inflorescences** dense corymbose cymes, terminal; peduncle tomentose; bracts lanceolate, 1.5–3 cm long; pubescent, with several large glands. **Flowers** single or double, very fragrant; calyx 1.5–2.5 cm long, deeply 5-lobed, pubescent, with several large glands, lobes linear-lanceolate or lanceolate, 0.7–1.6 cm long; corolla white, pinkish or reddish, lobes elliptic or ovate, ~1 cm long, rounded. **Fruits** globose, ~5 mm Ø. **Distr:** S China to Tropical Asia. **Zones:** 10–12.

C. incisum Klotzsch in W.C.H. Peters, *Naturw. Reise Mossambique*: 257, 1861. (Syn: *C. bernieri*, etc.). [Glory Bower]. **Shrubs** or **undershrubs**, erect, 0.5–2 m high, strongly branched; young branches subtetragonal, ± pubescent. **Leaves** decussate-opposite or ternate, usually with small ones in their axils; lamina elliptic to obovate, 3–12.5 × 1–5 cm, base tapering, margins entire or coarsely dentate, apex acute to long-acuminate, chartaceous, dark green, abaxially densely puberulent, lateral veins 5–8 per side; petioles 0–1.5 cm long, densely pubescent. **Inflorescences** cymes, subcapitate or corymbose, few- or many-flowered, axillary and terminal, subsessile. **Flowers** large; calyx 5–6 mm long, deeply split; corolla short-lived, white or yellowish-white, tubular, 10–13 cm long, limb ~3 cm Ø, shallowly 5-lobed, lobes often directed forward, median one longest; stamens red or purplish, 6–7 cm long. **Fruits** drupes, obovate, ~7 × 9 mm, dark purple, 4–lobed, usually 2-seeded; fruiting calyx coriaceous, campanulate, ~5 × 5 mm, sharply 5-lobed, lobes narrowly triangular, spreading. **Distr:** S Somalia to Mozambique, Madagascar. **Ecol:** On dry gravel hills, sand dunes, rocks, on sandstones, and also on periodically burned grasslands; to 1000 m. Fl. Oct–Jun. Fr. Apr. **Zones:** 11–12.

C. indicum (L.) Kuntze, *Revis. Gen. Pl.* 2: 506, 1891. (Syn: *Siphonanthus indica*, etc.). [Champagne Clerodendrum, Sky Rocket, Turks Turban]. **Subshrubs** to **shrubs**, 1–2 m high; branchlets purplish, channeled. **Leaves** 3–5-whorled or opposite, ± sessile; lamina lanceolate, 10–21 × 1.3–2.5 cm, base attenuate, margins entire or sinuate, apex shortly acuminate, membranous, glabrous, veins 10–12 pairs. **Inflorescences** leafy thyrses, 20–45 × 10–15 cm, terminal; cymes few-flowered; peduncle to 3 cm long; bracts lanceolate, 1–2 cm long; bractlets awl-shaped. **Flowers** with calyx 1–1.5 cm long, deeply divided, densely glandular, lobes ovate-lanceolate, 0.8–1.5 × 0.3–0.6 cm, acute; corolla white, becoming cream-colored, tube funnelform, 5–9 cm long, curved, lobes lanceolate, elliptic or ovate-oblong, 0.8–1.5 × 0.3–0.6 cm, obtuse; stamens long-exserted; ovary glabrous, style longer than stamens. **Fruits** drupes, dark blue, ~1.2 cm Ø; fruiting calyx crimson, to 3.8 cm Ø, leathery. **Distr:** S China, tropical Asia to SW Pacific. **Ecol:** On mountain slopes; 500–1,000 m. Fl. & Fr. Aug–Nov. **Zones:** 10–12.

C. inerme (L.) Gaertner, *Fruct. Sem. Pl.* 1: 271, 1788. (Syn. *Volkameria neriifolia*, etc.). [Garden Quinine]. **Shrubs**; branches and branchlets pubescent. **Leaves** elliptic-lanceolate, ovate-lanceolate, or elliptic, 3–8 × 1.5–4 cm, base obtuse, margins entire, apex obtuse, thickly papery, adaxially subglabrous and glandular, abaxially pubescent; petiole to ~1 cm long. **Inflorescences** usually 3-flowered, pubescent; peduncle 2–4 mm long; bracts linear, ~2 cm long. **Flowers** with calyx subtruncate, teeth minute, rudimentary, outside pubescent, inside glabrous, sparsely glandular; corolla white, tube 2–3 cm long, ~2 mm Ø at throat, lobes elliptic, ~7 mm long; stamens exserted, pubescent at base of filaments; style and ovary glabrous. **Fruits** drupes, gray-yellow, obovoid to subglobose, 0.6–1.1 cm Ø. **Distr:** S and SE Asia, Maldives, Australia, Pacific Islands. **Ecol:** In coastal habitats, salt tolerant; 100–200 m. Fl. & Fr. Mar–Dec. **Zones:** 10–12.

C. infortunatum L., *Sp. Pl.*: 637, 1753. (Syn: *C. calycinum*). **Shrubs**, branching, 1–2 m high, or erect **trees** to 5 m tall; branchlets obtusely tetragonal, densely pubescent. **Leaves** decussate-opposite; lamina ovate, 7–24 × 4.5–15 cm, base ± truncate, sometimes cordate, margins entire, apex abruptly acuminate, dark green, adaxially strigillose; petioles 1.5–13 cm long, densely pubescent. **Inflorescences** paniculate, to 22 × 22 cm, pubescent, composed of 5–8 pairs of decussate-opposite and divaricate many-flowered cymes, terminal; peduncles 1.5–9 cm long; sympodia 2–5 cm long; pedicles to 2.2 cm long, green, densely

Top row: *Clerodendrum chinense* (double form), *Clerodendrum indicum* (fruits), *Clerodendrum inerme*.
Bottom row: *Clerodendrum infortunatum*, *Clerodendrum minahassae*, *Clerodendrum paniculatum*, *Clerodendrum paniculatum* var. *alba*.

pubescent or reddish-hairy; bracts few, foliaceous, ovate, to 6 × 3.2 cm. **Flowers** fragrant; calyx green, 5-fid, lobes lanceolate, acuminate; corolla white, hypocrateriform, tube 2–3x as long as calyx, white-hairy inside, outside ferruginous-villous. **Fruits** drupes, subglobose, black or violet-black, shiny; fruiting pedicels scarlet; fruiting calyx accrescent, persistent, dark maroon to scarlet. **Distr:** S India to Indo-China, peninsular Malaysia. **Ecol:** In damp forests; <1,500 m. **Zones:** 10–12.

C. minahassae Teijsm. and Binn., *Tijdschr. Nederl. Ind.* 25: 409, 1863. (Syn: *C. calycinum*, *Siphobaea commersonii*). [Fountain Clerodendrum, Starfish Clerodendrum]. **Shrubs**, 1–3 m high, or **small trees** to 6 m tall; trunk to 7.5 cm dbh; branchlets obtusely tetragonal. **Leaves** decussate-opposite; lamina elliptic or oblong, 7.5–27.5 × 3–13 cm, base usually obtuse, margins entire, apex usually acuminate, chartaceous, adaxially dark green, abaxially lighter; petiole 0.5–7.5 cm long. **Inflorescences** cymose, 1- to many-flowered, aggregated into terminal panicles, 5–12 × 4.5–5 cm; pedicels 0.8–2.5 cm long. **Flowers** fragrant, large; calyx in bud filled with fluid, fleshy, green or yellowish-green, tubular, 1.5–4 × 0.8–1 cm, deeply incised, apex often reddish; corolla hypocrateriform, very long, tube narrowly cylindrical, creamish, glabrous, 5–11 cm long, lobes white or pink-streaked, 2.5–4 × 1.3 cm; filaments and style white, pink or purple, becoming whitish basally. **Fruits** drupes, blue-green to turquoise-blue or purple; fruiting calyx fleshy, accrescent, maroon to red or blood-red, lobes sharply acute or acuminate becoming widely divaricate, stellate, often to 5 cm Ø. **Distr:** W and C Malesia. **Ecol:** In thickets along waterways, forest patches and secondary forests at low and medium altitudes. Fl. May–Aug. Fr. Aug–Sep. **Zones:** 11–12.

C. paniculatum L., *Mant. Pl.* 90, 1767. (Syn: *C. pyramidale*, etc.). [Lindley's Clerodendrum, Orange Tower Flower, Pagoda Flower]. **Herbs** or **shrubs**, perennial, erect, to 3 m high, base often woody; stems sometimes hollow, obtusely tetragonal. **Leaves** decussate-opposite, gradually smaller upward; lamina ovate, 4–40 × 7–38 cm, base cordate, 3–7-lobed, margins denticulate to entire, apex acute or apiculate, adaxially bright green, shiny, abaxially paler, dull; petiole 1.2–3.5 cm long. **Inflorescences** cymes, 9–13 × 3–7 cm, many-flowered, axillary and terminal, 4–16 pairs of cymes arranged into large thyrsoid terminal panicles, to 45 × 45 cm; peduncles 1.5–12 cm long, reddish; pedicels rose-colored, 0.4–1.5 cm long; bracts foliaceous, large. **Flowers** with calyx red or orange-red, campanulate, 3–5 mm long, deeply divided, segments spreading; corolla hypocrateriform, orange-red to scarlet, rarely white, tube 1.2–2 cm long, lobes 7–8 mm long; filaments 2.5–3.5 cm long, exserted. **Fruits** drupaceous, greenish-blue to black. **Distr:** Tropical and subtropical Asia. **Ecol:** Weedy in disturbed areas. **Zones:** 9–12. **Info:** ***C. paniculatum* var. *alba*** is a form with creamish-white flowers.

C. quadriloculare (Blanco) Merr., *Philipp. Gov. Lab. Bur. Bull.* 35: 63, 1905. (Syn: *C. blancoanum*, etc.). [Bronze-leaved Clerodendrum, Philippines Fire Works, Philippine Glorybower]. **Shrubs** or **small trees**, erect, 2–5 m high, ± glabrous; suckering from roots. **Leaves** paired, petiolate; lamina elliptic to oblong, 12.5–25 × 3.8–12.5 cm, base rounded, apex shortly acuminate, adaxially deep dull green, sometimes purplish-red, abaxially usually uniformly dark purple, sometimes dull green. **Inflorescences** cymes, arranged in panicles, terminal, usually densely many-flowered; bracts inconspicuous. **Flowers** very showy; calyx purple, urceolate or bell-shaped, 0.6–1 × ~0.2 cm, 5-toothed; corolla salverform, tube 8.9–10 cm long, pinkish, limb 5-lobed, lobes recurved, ~1.5 cm long, white; stamens projecting, anthers purplish-black. **Fruits** drupes, ellipsoid, 1–1.5 cm long, purplish, with persistent calyx; fruiting calyx red, 1–1.5 cm long. **Distr:** C and E Malesia. **Ecol:** In tropical forests. Fl. Dec–Jul. Fr. Apr. **Zones:** 11–12.

C. schweinfurthii Gürke, *Bot. Jahrb. Syst.* 18: 177, 1894. (Syn: *C. bakeri*, etc.). **Shrubs**, highly variable, to 2 m high, branching from base; rhizome creeping; branches erect or ascending, simple or few-branched; branchlets glabrous. **Leaves** decussate-opposite, early deciduous from lower portions of stems and branches; lamina oblong to elliptic or obovate, 10–15 cm long, base deltoid or rounded, margins irregularly sharply serrate, teeth antrorse, apex cuspidate, membranous, green, glabrous; petioles short. **Inflorescences** cauliflorous, borne on main stems, lower leafless parts of branches or directly from the rhizome, in dense, long-pedunculate, globose, many-flowered heads. **Flowers** white, sweetly fragrant; calyx campanulate, ~4 mm long, glabrous, deeply lobed, lobes ovate, shorter than the oblong tube; corolla hypocrateriform, white, tube elongate, 3–3.6 cm long, outside glabrous, limb 4-lobed, lobes obovate, subequal, ~4 mm long; stamens 1.2–2 cm long. **Fruits** drupes, ellipsoid, green, lower half embedded in the persistent enlarged calyx; fruiting calyx white, fleshy, glossy, ± ovoid. **Distr:** Tropical Africa. **Ecol:** On lower mountain slopes; ~700 m. Fl. Apr–Jun. Fr. Jun–Jul. **Zones:** 10–12.

C. speciosissimum Paxton, *Mag. Bot.* iii. 217 and 271, 1837. (Syn: *C. pulchrum*, etc.). [Flaming Glory Bower, Pagoda Flower, Red Honeysuckle]. **Shrubs**, to 4 m tall; internodes 1.5–7.5 cm long. **Leaves** decussate-opposite; lamina broadly ovate or ovate-rotund, 10–35 × 8.5–26 cm, base cordate, margins entire or ± repand-denticulate, apex acute, membranous, pubescent, adaxially dark-green, abaxially paler; petiole 1.4–21 cm long.

Above: *Clerodendrum quadriloculare, Clerodendrum schweinfurthii, Clerodendrum schweinfurthii* (fruits), *Clerodendrum speciosissimum.*

Inflorescences panicles, laxly many-flowered, 15–45 × 14–25 cm, composed of 5–13 decussate-opposite pairs of cymes, often bracteate, terminal; peduncles 5–7 cm long; pedicels 5–9 mm long; bracts large, foliaceous; bractlets linear, 3–5 mm long. **Flowers** with calyx red, campanulate, 5–9 × 3–4 mm, lobes 5, ovate or triangular-ovate to deltoid, ~3 mm long, long-acuminate; corolla hypocrateriform, red to scarlet, tube 2–2.5 cm long, limb wide-spreading, lobes obovate, 1.5–1.8 × 0.7 cm; stamens and pistil exserted to ~3 cm, stamens ~4–6 cm long; style 6–7 cm long. **Fruits** drupes, bright-red or scarlet, depressed-quadrangular, ~5 × 7 mm, deeply 4-lobed, sulcate, separating into 4 distinct pyrenes; fruiting calyx patelliform, bright-red, ~1 cm wide, firm. **Distr:** Malesia to Solomon Islands. **Ecol:** Forest clearings of low mountains. Fl. throughout most of the year. **Zones:** 11–12.

C. splendens G. Don ex James, *Edinburgh New Philos. Journ.* 11: 349, 1824. (Syn: *Siphonanthus splendens*, etc.). [Glory Bower Vine]. **Shrubs**, or twining woody **vines**, to 5+ m long; branchlets ± acutely tetragonal, pubescent; internodes 3–7 cm long. **Leaves** decussate-opposite; lamina variable, elliptic, ovate, orbicular or narrowly lanceolate, 5–18 × 3–8 cm, base usually cordate, margins entire or undulate, apex acuminate, glabrous, usually thin, adaxially dark green, shiny, abaxially paler, bright green, base with patelliform extra-floral nectaries; petiole 0.8–2 cm long, sometimes spinescent at base. **Inflorescences** cymes, corymbose, 7–11 × 6–8 cm, many-flowered, usually dense, supra-axillary, terminal; peduncles 1.5–3.5 cm long; pedicels 0.3–1 cm long. **Flowers** with calyx elongate-campanulate, ~1 cm long, ± glabrate, deeply divided, lobes 4 or 5, broadly triangular, long-acuminate; corolla hypocrateriform, deep-red to crimson, tube ~2 cm long, lobes oblong-elliptic, 1–1.5 × 0.6–0.7 cm, rounded; stamens exserted to ~3 cm, filaments yellow or pink, glabrous, anthers bluish. **Fruits** drupes, black, shiny; seeds black, enclosed in orange pulp; fruiting calyx accrescent, red. **Distr:** W Tropical Africa to Angola. **Ecol:** In secondary forests; 0–900 m. Fl. throughout the year. **Zones:** 10–12.

C. thomsoniae Balf. f., *Gard. Chron.* 1862: 70, 1862. (Syn: *C. balfourii, C. speciosum*). [Bleeding Heart Vine, Bag Flower]. **Vines**, twining, slightly woody, to 3–7 m long; stems obtusely quadrangular, puberulent. **Leaves** opposite; lamina elliptic or lanceolate, 5.2–14 × 2.7–7 cm, base obtuse or rounded, margins entire or remotely sinuate, apex acuminate, chartaceous, adaxially puberulent, dark green, dull, abaxially puberulent, pale green, with numerous dots, veins prominent; petiole 1–2.8 cm long, sulcate, puberulent. **Inflorescences** dichasial cymes, axillary; bracts minute, subulate. **Flowers** with calyx ± urceolate, 1.5–2.5 cm long, white, puberulent, sepals lanceolate or elliptic-lanceolate, basally connate, apices acuminate; corolla red or crimson, hypocrateriform, ~2 cm long, tube quite narrow, lobes rounded; filaments pink, twice as long as corolla; style pink, as long as the filaments. **Distr:** W Tropical Africa to Cameroon. **Ecol:** Fl. Dec. **Zones:** 10–12.

C. wallichii Merrill, *J. Arnold Arbor.* 33: 220, 1952. (Syn: *C. nutans, C. penduliflorum* [both names illegal]). [Bridal Veil, Nodding/Pendent Glorybower]. **Shrubs** or **small trees**, 2–4(–6) m tall, erect; branchlets 4-angled, ± winged, glabrous. **Leaves** oblong to oblong-lanceolate, 11–18 × 2.5–4 cm, base narrowly cuneate, margins entire, apex acuminate to acute, subleathery, glabrous, veins 7 or 8 pairs, adaxially slightly distinct; petiole ~1 cm long. **Inflorescences** thyrses, 20–33 cm long, pendent, glabrous; axis and peduncle 4-angled or winged; bracts small, linear to awl-shaped. **Flowers** with calyx red to purple, ~1 cm long, tube very short, lobes ovate-lanceolate, 7–8 mm long; corolla white, tube ~1.1 cm long, lobes ovate, 1.1–1.5 cm long; stamens and style exserted. **Fruits** drupes, yellow-green when young, black and shiny at maturity, globose, 1–1.3 cm Ø; fruiting calyx red to purple, inflated, thickened. **Distr:** Pakistan to W Malesia. **Ecol:** On open forests of mountain slopes; 100–1,200 m. Fl. & Fr. Oct–Apr. **Zones:** 10–12.

Colebrookea

Species 1. *Generic description same as specific.* **Cult:** Propagate from seeds. Requires a moist, well-drained soil in full sun.

C. oppositifolia Smith, *Exot. Bot.* 2: 111, 1806. (Syn: *Elsholtzia oppositifolia*, etc.). [Indian Squirrel Tail]. **Shrubs**, 1–3 m high, much-branched. **Leaves** sometimes 3-whorled; lamina 10–20 × 3–7 cm, base broadly cuneate to rounded, margins crenulate-serrulate, apex long-acuminate, adaxially rugulose, puberulent, abaxially densely tomentose; petiole 0.8–2.5 cm long. **Inflorescences** panicles, 10–15 cm long, branches 4–7 cm long; verticillasters 10–18-flowered, globose; bracteoles ~1 mm long, densely tomentose outside. **Flowers** ~2 mm; ♀ with calyx campanulate, ~1.5 mm long, tube very short, becoming ribbed, teeth subulate, later spinescent, purplish, corolla with lower lip slightly longer than upper, midlobe ovate, stamens inserted on apical part of tube, included, style twice as long as corolla; *bisexual* with calyx ~0.6 mm long, corolla to 3 mm long, upper lip ovate-oblong, ~0.5 mm long, straight, emarginate, lower lip elongate, spreading, ~1.5 mm long, midlobe ovate-oblong, twice as long as ovate lateral lobes, style erect. **Fruits** nutlets, obovoid, ~1 mm long, yellow-brown. **Distr:** Pakistan to China (Yunnan). **Ecol:** In savanna forests, thickets; 200–2,200 m. Fl. Jan–Mar. Fr. Mar–Apr. **Zones:** 9–12.

Congea

Species 12. **Shrubs**, climbing; branchlets subterete, tomentose with stellate and/or simple hairs. **Leaves** opposite. **Inflorescences** cymes, capitate, 3–9-flowered, arranged in panicles, pedunculate; involucral bracts 4 or by fusion reduced to 3 with third bract deeply emarginate. **Flowers** with calyx funnelform or campanulate, 5-dentate, slightly enlarged in fruit; corolla tube slender, outside glabrous, pubescent inside throat, 2-lipped, lower lip 3-cleft, upper lip 2-cleft; stamens 4, didynamous, inserted at corolla throat, anthers subglobose; ovary obovate, apex glandular,

Below: *Clerodendrum splendens, Clerodendrum thomsoniae, Clerodendrum wallichii, Colebrookea oppositifolia.*

HS

RF

Above: *Congea griffithiana, Congea tomentosa, Faradaya splendida, Gmelina arborea.*

imperfectly bilocular, locules 2-ovuled, stigma capitate or shortly 2-cleft. **Fruits** drupes, obovate, indehiscent, enclosed in enlarged calyx forming samaras. **Distr:** India to SC China and W Malesia. **Cult:** Propagate from seeds or cuttings. Requires a moist, well-drained soil, rich in organic matter, in full sun.

C. griffithiana Munir, *Gard. Bull. Singapore* 21: 285, 1966. (Syn: *C. griffithiana* var. *elliptica*). [Congea, Pink Sandpaper Vine]. **Shrubs**, climbing; branchlets cylindric, tawny-tomentose. **Leaves** elliptic, to 12 × 6 cm, base cuneate, margins entire, apex acuminate, subcoriaceous to chartaceous, adaxially glabrous, abaxially ferruginous-pubescent, lateral veins 4 pairs; petioles 0.5–1 cm long, pubescent. **Inflorescences** panicles, to 45 cm long; rachis tawny-tomentose, axillary and terminal; cymes 5-flowered; involucral bracts 4, free, spathulate to oblanceolate, to 3 × 1.3 cm, base much-narrowed, violet, adaxially thickly tomentose, abaxially pubescent; peduncles ~1 cm long, pubescent. **Flowers** with calyx campanulate, 4–5 mm long, 5-lobed, densely canescent-pilose outside, appressed-pilose inside, lobes ~1 mm long, margins crisped, apex acute; corolla exserted, white, tube slightly longer than calyx, glabrous except for a villous band in throat, lobes almost rounded; stamens 4, exserted, filaments elongate, to 1+ cm long, anthers almost orbicular; ovary ± obovoid, glabrous, apex glandular, style longer than filaments, stigma indistinctly bilobed. **Fruits** drupes, enclosed in enlarged calyx. **Distr:** Indo-China to peninsular Malaysia. **Ecol:** In hedges, brushwoods, bamboo and evergreen forests, and near streams; 100–1,000 m. Fl. Jan–Apr, Oct. **Zones:** 10–12.

C. tomentosa Roxburgh, *Pl. Coromandel.* 3: 90, 1820. (Syn: *C. tomentosa* var. *oblongifolia*). [Shower of Orchids, Woolly Congea]. **Shrubs**, climbing; branchlets densely yellowish-tomentose when young, becoming grayish. **Leaves** elliptic, broadly elliptic or ovate, 6–16 × 3–9.5 cm, base rounded to subcordate, margins entire, apex acuminate to acute, papery, adaxially densely pilose when young, becoming sparsely pilose to subglabrous, abaxially densely pilose, veins 5 or 6, abaxially prominent; petiole 0.5–1.3 cm long, densely tomentose. **Inflorescences** cymes, (5–)7(–9)-flowered, densely white-pilose; involucral bracts 3 or 4, purple, oblong, obovate-oblong, or broadly elliptic, 2–3 × 0.8–1.2 cm, pubescent, base pilose. **Flowers** with calyx funnelform, ~7 mm long, lobes half as long as tube, outside densely pilose, inside pubescent; corolla, glabrous except pilose in throat, tube longer than calyx; stamens long-exserted; ovary glabrous, style exserted. **Fruits** drupes, enclosed in enlarged calyx. **Distr:** India to peninsular Malaysia and China (SW Yunnan). **Ecol:** In mixed forests; 600–1,200 m. **Zones:** 10–12.

Faradaya

Species 7. **Trees**, **shrubs** or **climbers**. **Leaves** decussate-opposite to subopposite or verticillate in 3's, mostly petiolate, exstipulate, deciduous; lamina simple, margins entire, sometimes with basal glands or scales abaxially, usually glabrous, chartaceous or coriaceous. **Inflorescences** cymose, cymes usually many-flowered, aggregated in terminal, often large, loosely corymbose panicles or sessile in leaf axils, rarely cauliflorous. **Flowers** large, showy; calyx gamosepalous, campanulate, during anthesis unequally split into 2–4 short, valvate, often recurved lobes, corolla gamopetalous, zygomorphic, large, usually white, hypocrateriform or infundibular, tube cylindric, straight, exserted, apically ampliate, limb wide-spreading, 4(5)-fid, lobes imbricate, subequal or posterior one wider and apically entire or emarginate, others smaller and subequal; stamens 4(5), 2 inserted in upper part and 2 near base of corolla tube or sometimes all inserted near apex, middle, or base of tube, sometimes long-exserted, filaments often hairy; pistil single, compound, bicarpellate, style sunken between terminal ovary lobes, stigma subulate, ovary superior, composed of two biloculed carpels, apically shortly to deeply 4-lobed. **Fruits** drupaceous, 4-lobed, sometimes so deeply lobed appearing as 2–4 nearly separate 1-seeded pyrenes; endocarp hard, 1–4-seeded; seed 1 per locule. **Distr:** Malesia to SW Pacific. **Cult:** Propagate from seeds or cuttings. Requires a moist, humus-rich soil in full or partial sun.

F. splendida F. Muell., *Fragm.* 5: 21, 1865. (Syn: *F. papuana*, etc.). [October Glory, Potato Vine]. **Climbers**, woody, robust, to 15 m long. **Leaves** decussate-opposite; lamina simple, ovate, 9–39 × 4.5–14.5 cm, base rounded to truncate or deeply cordate, apex acute or acuminate, secondaries 5–7 pairs; petiole 1.5–6 cm long. **Inflorescences** cymose, mostly lax, forming large panicles, terminal; peduncles 2.5–6 cm long; pedicels 0.3–1.4 cm long. **Flowers** large, fragrant, tubular, 5 cm long, 3.5–6 cm across; calyx green, 1.5–2.5 cm long, usually bilobed to about middle; corolla white or greenish, 4-lobed, glabrous, tube infundibular, 2–4 cm long, limb 4(5)-lobed, occasionally 2-lipped, lobes to ~2 × 1.5 cm; stamens 4(5), inserted near base of corolla tube, exserted, filaments white, ~6.5 cm long, base long-pilose, anthers cream or pale-brown; style white, ~6.5 cm long, stigma shortly bifid. **Fruits** drupes, white or cream, ovoid, 4.8–9 × 4–6 cm, usually with (1–3)4 basally connate 1-seeded pyrenes, flesh white, turning yellowish, smelling of bleach; seeds 4.5 cm long, brown. **Distr:** Malesia to Solomon Islands, Australia (NE Queensland). **Ecol:** In rainforests; 0–1,000 m. Fl. Jul–Nov. Fr. Aug–Apr. **Zones:** 11–12.

Gmelina

Species ~35. **Trees**, tall **shrubs**, rarely **subshrubs**, often climbing when young; branchlets tomentose, often spiny. **Leaves** opposite; lamina simple, sometimes lobed, usually with large gland patches near base, often abaxially gray-mealy. **Inflorescences** usually cymes or panicles, terminal, sometimes 1-flowered in leaf axils; cymules axillary, decussate, few-flowered, short; bracts leaflike. **Flowers** with calyx campanulate, persistent and enlarged in fruit, truncate, 4- or 5-dentate or lobed, often unequal or oblique, usually with large glands; corolla zygomorphic, ventricose-funnelform, tube narrow at base, throat wide, limb oblique, ~2-lipped, upper lip bilobed or entire, lower lip trilobed with midlobe larger; stamens 4, didynamous, inserted on basal part of corolla tube, usually included, sometimes slightly exserted, anthers divaricate, opening by longitudinal slits; ovary (2–)4-locular, ovules pendulous or laterally attached, often with a central cavity, style slender, usually unequally bilobed or stigma awl-shaped. **Fruits** drupes, succulent, endocarp bony, mesocarp fleshy; seeds 4, or by abortion 2 or 3. **Distr:** Pakistan to China (S Yunnan) and W and NC Malesia. **Cult:** Propagate from seeds, cuttings or by air-layering. Requires a moist, well-drained soil in full or partial sun.

G. arborea Roxburgh, *Pl. Coromandel.* 3: 41, 1815. (Syn: *G. sinuata*, etc.). [Beechwood, Gray/White Teak, Kashmir Tree, Malay Beechwood]. **Trees**, ~15 m tall; bark grayish-brown; branchlets, petiole and inflorescences densely yellow-brown tomentose; branchlets slightly 4-angled when young, becoming terete, lenticellate, leaf scars prominent. **Leaves** broadly ovate, 8–19 × 4.5–15 cm, base broadly cuneate to subcordate, apex acuminate, thickly papery, veins 3–5 pairs, abaxially prominent; petiole terete, 3.5–10 cm long. **Inflorescences** narrow thyrses, terminal; peduncle 15–30 cm long. **Flowers** with calyx 3–5 mm long, with several black discoid gland patches, teeth 5, sharply triangular; corolla yellow, 3–4 cm long, 2-lipped, sparsely glandular; lower lip trilobed, outside yellowish-brown puberulent, inside glabrous, upper lip entire or slightly 2-cleft; ovary glabrous, glandular, stigma unequally 2-cleft. **Fruits** drupes, yellow when ripe, black when dry, ellipsoid to obovoid-ellipsoid, 1.5–2 cm Ø. **Distr:** Pakistan to China (S Yunnan) and W and NC Malesia. **Ecol:** In open forests; <1,500 m. Fl. Apr–May. Fr. May–Jul. **Zones:** 10–12.

G. asiatica L., *Sp. Pl.* 2: 626, 1753. (Syn: *G. lobata*, etc.). [Asian Bush-beech]. **Shrubs**, climbing, 1–3(–10) m tall, usually with spines and minute yellowish-brown hairs on young shoots. **Leaves** ovate to obovate, 3–5(–9) × 2.2–3(–8.5) cm, base cuneate, margins entire or ~3–5-lobed, apex acuminate, papery, adaxially glabrescent, abaxially brown-villous, glandular, veins 3 or 4 pairs; petiole 0.5–4.5 cm long. **Inflorescences** few to many-flowered, terminal, usually pendulous; primary floral branches very short, 1–5-flowered; bracts leaflike, small to large; flowers pendulous. **Flowers** with calyx 3–6 × 2.5–4 mm, outside densely brown-pubescent and with 2–several disk-shaped glands, inside glabrous; corolla yellow, 2–5 cm long, 2-lipped and 4-lobed, lower lip trilobed,

Above: *Gmelina asiatica, Gmelina philippensis, Holmskioldia sanguinea, Leonotis leonurus.*

upper lip entire, outside scattered-hairy, inside densely glandular-hairy; filaments densely glandular-hairy; ovary 4-locular, glabrous, style slender, hardly exserted, apically curved, stigma unequally 2-cleft. **Fruits** drupes, yellow, ovoid to obovoid, glabrous. **Distr:** Mascarenes, India to Malesia. **Ecol:** In mixed forests; 400–800 m. **Zones:** 10–12. **Info:** This species differs from other species in its climbing habit.

G. philippensis Cham., *Linnaea.* 7: 109, 1832. (Syn: *G. hystrix*, etc.). [Hedgehog, Wild Sage]. **Shrubs** or **small trees**, to 7 m tall, climbing or scandent, sometimes prostrate; stems lenticellate; dwarfed branchlets spinescent, spines to 1.5 cm long. **Leaves** ovate or rhomboid-elliptic, 1.5–10 × 1.5–6 cm, base acute, margins entire or slightly lobed to slightly dentate, apex obtuse or subacute, glabrous, glaucous, submembranous, abaxially smooth and glandular; petiole 2–2.5 cm long. **Inflorescences** terminal cymes, 6–8 cm long, paniculately arranged (10–20 cm long) in axils of foliaceous, large, petaloid bracts; bracts reddish, maroon, purple or green, broadly ovate, 1.5–4 × 1–3.5 cm long, apex mucronate, shortly apiculate, membranous. **Flowers** slightly fragrant; calyx green, ~5 mm long, with 2–4 external glands; corolla yellow, 4–5.5 cm long, tube curiously inflated upward; longer stamens 2, often with gland-tipped hairs. **Fruits** drupes, pendulous, obovoid, ~1.2 cm long, yellow; pericarp soft and watery. **Distr:** Indo-China to Philippines. **Ecol:** Widely scattered on open slopes and thickets; <100 m. Fl. Apr–Nov. **Zones:** 10–12.

Holmskioldia

Species 1. *Generic description same as specific.* **Cult:** Propagate from seeds or half-hardened cuttings. Requires a moist, well-drained soil in full or partial sun.

H. sanguinea Retz., *Observ. Bot.* 6: 31, 1791. [Chinese Hat Plant, Cup and Saucer, Mandarin's Hat, Parasol Flower]. **Shrubs**, large, large, evergreen, scandent, to 3–9 m high; young branches quadrangular. **Leaves** ovate or elliptic-ovate, 3–12 × 1.5–8.5 cm, margins entire to coarsely dentate, apex shortly acuminate, pubescent to glabrous, abaxially gland-dotted; petiole 1.5–3 cm long. **Inflorescences** cymes or racemes, rarely panicles, dense, to 5 cm long, pedunculate, axillary or terminal; pedicels ~5 mm long, slender, glandular-pubescent; bracts lanceolate, 4–5 mm long. **Flowers** brick-red to orange, ~6 mm Ø; calyx petaloid, 1.5–2.2 cm Ø, margins entire, with minute dots; corolla tube curved, broadening upward, hypocrateriform, crimson, 1.5–2.5 cm long, limb subbilabiate, unequally 5-lobed; stamens slightly exserted, filaments pink; ovary glabrous, style as long as the filaments. **Fruits** drupes, obovoid to globose, 5–8 mm long, 4-lobed, subfleshy, enclosed by persistent, enlarged, colored calyx. **Distr:** Himalayas to Myanmar. **Ecol:** Fl. Oct–April. **Zones:** 10–12.

Leonotis

Species 9. **Shrubs** or **annual herbs**. **Leaves** usually petiolate; lamina with margins dentate. **Inflorescences** of verticillasters, usually few or solitary near apex of stem or branches, dense, many-flowered; floral leaves similar to foliage leaves but smaller. **Flowers** with calyx tubular, 10-nerved, often incurved toward apex; limb 8–10-lobed, mouth oblique, lobes rigid, often spinescent, ± unequal, posterior usually large and broader than others; corolla tubular, tube usually exserted, cylindric or slightly enlarged, faintly curved, glabrous or imperfectly pilose-annulate inside, limb bilabiate, upper lip erect, elongated, concave, densely villous, lower lip spreading, trilobed, median lobe scarcely longer than lateral ones; stamens didynamous, ascending under upper lip, anthers 2-celled, cells divaricate, filaments villous at base; style subulate at apex, bilobed. **Fruits** nutlets, ovoid, triquetrous, apex obtuse or truncate, glabrous. **Distr:** Tropical and S Africa, Madagascar. **Cult:** Propagate from seeds or cuttings. Requires a moderately fertile, moist, well-drained soil in full sun.

L. leonurus (L.) R. Br., *Hortus Kew.* 3: 410, 1811. (Syn: *Phlomis leonurus*, etc.). [Lion's Ears, Lion's Tail, Wild Dagga]. **Shrubs**, perennial, to 2 m high; branches densely hairy. **Leaves** lanceolate to narrowly elliptic, 5–8 × 1–1.5 cm, base tapering into petiole, margins irregularly toothed or slightly scalloped, apex subacute, green, sparsely hairy, strongly aromatic; petiole 0.3–1 cm long. **Verticillasters** dense clusters, >12-flowered, ± globose, axillary, arranged in whorls at nodes along apical part of stems. **Flowers** with calyx tube 1–1.5 cm long, densely shortly hairy, lobes 10, spreading, to 1 mm long, mucronate; corolla 3–5.5 cm long, bright orange. **Distr:** Angola to S Africa (Cape Region). **Ecol:** In forest margins, rough grasslands. Fl. throughout most of the year. **Zones:** 9–11.

L. nepetifolia (L.) R. Br., *Hortus Kew.* 3: 409, 1811. (Syn: *Phlomis nepetifolia*, etc.). [Annual Wild Dagga]. **Herbs**, large, annual, to 2.6 m high, woody at base; stem obtusely 4-angular, tomentose. **Leaves** *upper* oval-lanceolate, 5.6–8.5 × 1.5–2.6 cm, base acute; *lower* ovate to broadly ovate, 4–20 × 2.9–15 cm, base broadly rounded or truncate; margins coarsely crenate-serrate, apex acute to subacute; petiole 1.5–12 cm long. **Verticillasters** 2–8 per stem, distant, globose, 3–7 cm Ø; floral leaves linear-lanceolate, 1.2 × 0.2 cm, spine-tipped. **Flowers** with calyx 1.6–2.3 cm long; tube 1–1.4 cm long, faintly curved, pubescent below, villous-hirsute above; posterior lobe broadly ovate, 6–8 × 4 mm, acuminate, spine-tipped, anterior one 5–6 mm long, 7-toothed, teeth unequal; corolla 1.9–2.8 cm long, orange-red, tube 1–1.4 cm long, base glabrous, shaggily villous above, inside hairy, upper lip 1.2–1.5 cm long, lower lip 7–8 mm long; filaments flattened; style branches unequal, linear. **Fruits** nutlets trigonous, 3–4 mm long, black, densely covered with white glands. **Distr:** Africa, Indian Subcontinent. **Ecol:** In waste places; 0–500 m. Fl. Nov–Jun. **Zones:** 10–12.

Mentha

Species 23. **Herbs**, annual or perennial, aromatic,

Below: *Leonotis nepetifolia, Mentha spicata, Ocimum basilicum, Ocimum tenuiflorum.*

often rhizomatous or stoloniferous. **Leaves** upper sessile or subsessile, margins dentate, serrate, or crenate. **Verticillasters** (rarely 2–6)- to many-flowered; floral leaves similar to stem leaves or reduced; bracts lanceolate to linear, ± distinct. **Flowers** bisexual or ♀; calyx funnelform to campanulate, 10–13-veined, throat glabrous or hairy, limb equally 5-toothed or 2-lipped, upper lip 3-toothed, lower lip 2-toothed; corolla funnelform, ± regular or slightly irregular, tube generally included, throat slightly dilated or saccate in front, limb 4-lobed, lobes equal, entire, upper lobe slightly wider, emarginate or bilobulate; stamens 4, subequal, divaricate, erect, exserted in bisexual flowers, included and often rudimentary in ♀ flowers, posterior 2 slightly longer than anterior 2, filaments glabrous, anther cells 2, parallel; style exserted, apex equally 2-cleft. **Fruits** nutlets, ovoid, dry, smooth or slightly tuberculate, apex rounded, rarely hairy. **Distr:** Cosmopolitan, mainly in northern temperate regions. **Cult:** Propagate by division, cuttings or from seeds. Requires a fertile, moist soil in full or partial sun.

M. spicata L., *Sp. Pl.* 2: 576, 1753. [Garden Mint, Menthol Mint, Silver Mint, Spearmint]. **Herbs**, stoloniferous; stems erect, 0.4–1.3 m long, glabrous or subglabrous, striate. **Leaves** sessile or subsessile; lamina ovate-oblong to oblong-lanceolate, 3–7 × 1–2 cm, base broadly cuneate to rounded, margins irregularly serrate, apex acute. **Verticillasters** in cylindric terminal spikes 4–10 cm long, spikes interrupted at base; bracteoles linear, 5–8 mm long; pedicels ~2 mm long. **Flowers** with calyx campanulate, ~2 mm long, glandular, obscurely 5-veined, teeth triangular-lanceolate, ~1 mm long; corolla purplish, ~4 mm long, glabrous, tube ~2 mm long, lobes subequal, apex emarginate; ovary brown, glabrous. **Distr:** Europe to China. **Ecol:** In wet sandy areas. Fl. Jul–Sep. **Zones:** 3–12. **Info:** Cultivated for culinary, medicinal and commercial purposes.

Ocimum

Species ~100–150. **Herbs**, **subshrubs**, or **shrubs**, aromatic. **Leaves** petiolate; lamina with margins dentate. **Verticillasters** 6(–10)-flowered, in terminal, pedunculate thyrses or panicles; bracts early deciduous, entire, rarely longer than flowers; pedicel straight, apex recurved. **Flowers** white; calyx ovoid to campanulate, limb 2-lipped, upper lip 3-toothed, middle tooth circular to obovate, margin winged, decurrent, lateral teeth shorter, lower lip bidentate, teeth narrower, apex acuminate to spinescent, sometimes approximate; corolla tube not hairy annulate inside, dilated, obliquely campanulate at throat, limb 2-lipped, upper lip subequally (3- or) 4-lobed, lower lip declined, margin entire; stamens 4, exserted, declined on lower corolla lip, anterior 2 longer, filaments free or anterior 2 basally connate, anthers ovoid-reniform, 1-locellate; style longer than stamens, 2-cleft at apex, lobes subequal, subulate or flat. **Fruits** nutlets, ovoid or subglobose, smooth or glandular-foveolate, viscid when moist, with a white basal areola. **Distr:** Tropical and subtropics regions worldwide. **Cult:** Propagate from seeds (annuals) or cuttings (perennials). Requires a moist, well-drained soil in full sun.

O. basilicum L., *Sp. Pl.* 2: 597, 1753. [Basil, Sweet Basil]. **Herbs**, annual; stems erect, 20–80 cm high, sometimes red-tinged, much-branched. **Leaves** ovate to oblong, 2.5–5 × 1–2.5 cm, base attenuate, margins irregularly dentate or subentire, apex subobtuse to acute, subglabrous, abaxially glandular, lateral veins 3- or 4-paired; petiole ~1.5 cm long, ± narrowly winged apically. **Inflorescences** thyrses, 10–20 cm long; verticillasters puberulent or densely pilose, approximate apically; bracts sessile, oblanceolate, 5–8 mm long, colored; pedicel ~3 mm long. **Flowers** with calyx campanulate, ~4 × 3.5 mm, pubescent outside, pilose at throat inside, tube ~2 mm long; middle tooth of upper lip, subcircular, ~2 × 1 mm, concave, mucronate; corolla purplish or with upper lip white, limb puberulent outside, tube ~3 mm long, throat ± dilated, upper lip wide, ~3 × 4.5 mm, 4-lobed, ± flat, lower lip purple, ~6 mm long; stamens free, slightly exserted, posterior bidentate, base puberulent. **Fruits** nutlets, dark brown, ovoid, ~2.5 × 1 mm, glandular-foveolate; fruiting calyx persistent, veined. **Distr:** Africa, Asia. **Ecol:** Fl. Jul–Sep. Fr. Sep–Dec. **Zones:** 10–12.

O. tenuiflorum L., *Sp. Pl.*: 597, 1753. (Syn: *Lumnitzera tenuiflora*, *O. sanctum*, etc.). [Holy Basil, Sacred Basil]. **Subshrubs**, much-branched, to 1 m tall; stems erect, base woody, spreading pilose. **Leaves** oblong, 2.5–5.5 × 1–3 cm, base cuneate to rounded, margin shallowly undulate-serrate, apex obtuse, puberulent, glandular, pilose on veins; petiole 1–2.5 cm long. **Verticillasters** 6-flowered, in pedunculate, terminal thyrses or panicles, 6–8 cm long; bracts sessile, cordate, ~ 1.5 × 1.5 mm, apex acute; peduncles 1–1.5 cm; pedicels ~2.5 mm long. **Flowers** with calyx campanulate, ~2.5 mm long, villous, tube ~1.5 mm long; corolla white to reddish, ~3 mm long, sparsely puberulent, tube ~2 mm long, upper lip <1 × 2.5 mm, lobes ovate, lower lip oblong, ~1 × 0.6 mm, flat; stamens free. **Fruits** nutlets, brown, ovoid, ~1 × 0.7 mm; fruiting calyx to 6 × 4 mm, conspicuously veined. **Distr:** Tropical and subtropical Asia to N Australia. **Ecol:** In dry sandy areas; 0–1,800 m. Fl. Feb–Jun. **Zones:** 9–12. **Info:** Cultivated for culinary and medicinal purposes.

Orthosiphon

Species 39. **Herbs** perennial, or **subshrubs**; roots often thickened, woody. **Leaves** with margins dentate. **Verticillasters** (4–)6-flowered separate long thyrses; bracts shorter than pedicels, circular to oblate, margin entire. **Flowers** with calyx tubular to broadly tubular, tinged with brilliant colors, 2-lipped, upper lip ovate to oblate, scarious, decurrent into tube, margin reflexed, lower lip 4-toothed, teeth awned to needlelike, anterior teeth longer than lateral ones; corolla white or reddish to purple, 2-lipped, tube exserted, straight or incurved, obconical, upper lip 3- or 4-lobed, lower lip entire, concave; stamens 4, anterior 2 longer, declined, included or slightly exserted, filaments separate, edentate, anthers 1-locellate; style globose, apex entire or 2-cleft; disk produced, fingerlike in front. **Fruits** nutlets, ovoid or subglobose, minutely tuberculate, glabrous; fruiting calyx enlarged, 10-veined. **Distr:** Tropical and S Africa to NE Australia. **Cult:** Propagate from seeds or cuttings. Requires a moderately fertile, moist, well-drained soil in shade to full sun.

O. aristatus (Blume) Miq., *Fl. Ned. Ind.* 2: 943, 1858. (Syn: *Clerodendranthus spicatus*, etc.). [Cat's Moustache/Whiskers]. **Herbs**, perennial, 1–1.5 m high; stems retrorse-pubescent. **Leaves** rhombic- to oblong-ovate, (1.2–)2–5.5 × (0.8–)1.3–3.5 cm, base broadly cuneate to truncate-cuneate, margins mucronate to dentate, apex acute, papery, pubescent, sparsely glandular, lateral veins 4- or 5-paired; petiole to 1.5 cm long. **Verticillasters** 6-flowered pedunculate thyrses, 8–12 cm long; rachis densely pubescent; bracts ~3.5 × 3 mm, longitudinally parallel-veined; pedicels to 5 mm long, densely pubescent. **Flowers** with calyx 5–6 × ~2.5 mm, rusty-glandular outside, upper lip ~2.5 × 2.5 mm, teeth of lower lip triangular, awned, margin ciliolate, upper lip reflexed, lower lip projected; corolla purplish or white, sparsely rusty-glandular on upper lip, tube 0.9–1.9 cm long, lower lip oblong, ~5 mm long; filaments edentate. **Fruits** nutlets, dark brown, ovoid, ~2 mm long, corrugate; fruiting calyx to 1.1 cm long. **Distr:** Tropical and subtropical Asia to N Australia. **Ecol:** In wet forests and plains; 0–1,500 m. Fl. & Fr. May–Nov. **Zones:** 10–12.

Oxera

Species 20. **Lianas**, **shrubs** or **trees** (sometimes pachycaulous). **Leaves** decussate, spirally arranged or whorled, petiolate or subsessile; lamina simple, margins entire, thickened, reticulate-veined, main veins abaxially prominent. **Inflorescences** determinate or indeterminate thyrses, axillary or cauliflorous. **Flowers** large, showy, bisexual, hypogynous; calyx 4- or 5-lobed, persistent, campanulate, cup-shaped but split during flowering into 2–5 lobes; corolla hypocrateriform or bucket-shaped, 4-lobed, zygomorphic, deciduous; stamens 2(4), didynamous, antisepalous, epipetalous, exserted, usually borne between lateral and anterior lobes, anthers extrorse, dorsifixed; staminodes 2, usually not exserted, usually borne between lateral and posterior lobes; ovary deeply lobed, bilocular (4 by secondary partition), seated on a disk, ovule 1 per locule, pendulous, style filiform, terete, gynobasic, with 2 short stigmatic lobes; disk unlobed. **Fruits** compound; mericarps 1–4, exocarp fleshy, rather thin, endocarp stony, 1-seeded. **Distr:** SW Pacific. **Cult:** Propagate from seeds or cuttings. Requires a fertile, moist, well-drained soil in full to partial sun.

Below: *Orthosiphon aristatus*, *Oxera pulchella*, *Petraeovitex bambusetorum*, *Plectranthus ecklonii*.

Above: *Plectranthus hadiensis* var. *tomentosus*, *Plectranthus scutellarioides*, *Rotheca myricoides*, *Salvia africana-lutea*.

O. pulchella Labill., *Sert. Austro-Caledon.* 1: 23, 1824. (Syn: *Oncoma pulchellum*). [Royal Climber]. **Climbers**, woody, glabrous; branches slender, terete, smooth. **Leaves** opposite, petioled; leaves on *upper* branches oblong, apex obtuse or subacute, leaves on *lower* branches oblong-lanceolate, 5–12.5 cm long, base rounded or acute, margins entire or with broad shallowly crenate, apex obtusely acuminate, bright green, abaxially finely reticulate, nerves 6–8 pairs; petiole 1.2–2 cm long. **Inflorescences** cymes, many-flowered, sessile, axillary, branched, pendulous; bracts minute. **Flowers** large; sepals 4, free or connate, elliptic, 1.2–2 cm long, membranous, trinerved, subacute, pale green, punctate; corolla 5 cm long, funnel- to bell-shaped, pale yellowish or greenish-white, tube curved, terete, lobes broadly oblong; stamens 2, inserted about middle of tube, filaments exserted for half their length, anthers small, bright yellow; staminodes between stamens short, filiform; ovary deeply 4-lobed, seated on a broad fleshy disk, style filiform, stigma minute. **Distr:** New Caledonia. **Ecol:** In dry or sclerophyll forests; 5–50 m. Fl. several times a year, peaking Aug–Sep. **Zones:** 10–12.

Petraeovitex

Species 8. **Shrubs**, usually climbing, sometimes creeping or scrambling; branches and branchlets tetragonal. **Leaves** decussate-opposite, exstipulate, deciduous, petiolate, usually compound, trifoliolate, sometimes twice trifoliolate, rarely unifoliolate; leaflets usually 3–6. **Inflorescences** terminal, compound and pyramidal-paniculate, loose (flowers long-pedicellate) to densely congested (flowers shortly pedicellate), or axillary and cymose; bracts and bractlets usually small or subulate, sometimes colored, sometimes later becoming normal green leaves. **Flowers** inconspicuous to showy; calyx very small, shortly tubular or campanulate, 5-dentate, lobes much-enlarged; corolla gamopetalous, zygomorphic, obliquely hypocrateriform, posteriorly split, limb expanded, ± distinctly bilabiate, lobes 5, subequal, obtuse, incurved; stamens 4, epipetalous, alternate with and longer than corolla lobes; pistil bicarpellate, ovary obovoid, imperfectly biloculed, biovulate, style single, stigma bilobed or bifid. **Fruits** capsular, subcylindric or 4-lobed, usually longitudinally striate, apically conic or truncate; seeds 1–4. **Distr:** Indo-China to SW Pacific. **Cult:** Propagate from seeds or cuttings. Requires a fertile, moist, well-drained soil in full to partial sun.

P. bambusetorum King and Gamble, *Bull. Misc. Inform. Kew* 1908: 113, 1908. (Syn: *P. bambusetorum* f. *simplicifolia*). [Nong Nooch Vine]. **Climbers**, long, woody; branches fulvous. **Leaves** decussate-opposite, 1–3-foliolate; leaflets ovate or ovate-oblong, base rounded or shortly cuneate, margins entire, apex acuminate, glabrous, secondary veins 4–6(7 or 8), central leaflet 7–10 × 3–6 cm, lateral leaflets 6–9 × 2.5–4.3 cm; petioles 2.5–5 cm long. **Inflorescences** lax cymose-thyrsoid panicles, to 30 cm long, axillary and terminal; bracts foliaceous, lanceolate, long-acuminate, early deciduous; bracteoles minute; pedicels ~4 mm long. **Flowers** with calyx tube 3 mm long, lobes 5, ~2 mm long, puberulent; corolla white or greenish-yellow, zygomorphic, bilabiate, ~6 mm long, 5-lobed, upper lip bilobed, lower lip trilobed, lobes elliptic-ovate, ~4 mm long, incurved; stamens 4, ~4 mm long; ovary orbicular-ellipsoid, style elongate. **Fruits** wedge-shaped, ~7 mm long; fruiting calyx 2–2.6 cm long, tube 6–8 mm long, faintly ribbed, lobes membranous, oblong, 3-veined, 1.3–1.9 × 0.3–0.5 cm. **Distr:** Peninsular Malaysia (Perak), Borneo. **Ecol:** In dense bamboo forests; 160–200 m. Fl. throughout most of the year. **Zones:** 11–12.

Plectranthus

Species ~150. **Herbs** or **shrubs**, perennial, rarely annual, to 3.5 m high, evergreen, aromatic, glabrous or hairy, often glandular, succulent to semisucculent; stems usually 4-angled. **Leaves** opposite; lamina undivided, usually toothed or scalloped or rarely more deeply divided, petioled. **Inflorescences** racemes, terminal, rarely axillary, often branched and paniculate, flowers whorled, arranged in 1- to many-flowered cymes or dichasia; bracts soon deciduous. **Flowers** bisexual, white, purple or yellow; calyx 2-lipped, persistent in fruit, 5-lobed, tube short, upper lobe larger, ovate, 4 lower lobes equal; corolla 2-lipped, tube curved or abruptly bent or dilated, rarely straight, upper lip 4-lobed, 2 larger lobes notched with 2 smaller lateral lobes, lower lip boat-shaped; stamens 4, rarely 2 abortive, fused to mouth of corolla, style lying on lower lip of corolla. **Fruits** nutlets, 4, oblong to rounded, smooth. **Distr:** Tropical and subtropical Old World. **Cult:** Propagate from stem cuttings or seeds. Requires a moist well-drained soil in full or partial sun.

P. ecklonii Benth., *Prodr.* 12: 64, 1848. [Blue/Large Spur Flower, Swedish Ivy]. **Herbs**, erect, glabrous or nearly so; stems striate. **Leaves** ovate, 9–15 × 4–7.5 cm, base cuneate, usually unequally, margins crenate-serrate, glabrous, apex acute or acuminate; petioles 2.5–7.5 cm long. **Inflorescences** panicles, branched, 5–23 cm long, formed of several opposite lax racemes of various lengths; verticils laxly 6-flowered; peduncles scarcely any; pedicels to 10 cm long, slender; bracts small, oblong, deciduous. **Flowers** with calyx campanulate, 5 mm long, upper tooth ovate, 3 × 2 mm, acute, 4 other teeth linear-subulate, ciliate, 2 lateral 2 mm long, 2 lower 3 mm long; corolla 2 mm long, tube straight, funnel-shaped at month, 1.4 cm long, upper lip 5–6 × 5–6 mm, 2 terminal lobes rounded, lower lip boat-shaped, 6 × 2.5 mm; stamens very long, much-exserted, longer pair to 2 cm long; style shorter than stamens. **Fruits** with fruiting calyx to 1.2 cm long, tubular-campanulate, ribbed, glabrous, upper tooth erect; nutlets ellipsoid, 3 × 2 mm, smooth, brown. **Distr:** S Africa. **Ecol:** In moist coastal forests and forest margins. Fl. Mar–May. **Zones:** 9–11.

P. hadiensis var. ***tomentosus*** (Benth. ex E. Mey.) Codd, *Fl. S. Afr.* 28: 153, 1985. (Syn: *P. tomentosus*, etc.). **Herbs**, tuberous, suffruticose, to 1.2 m high; stems stout, erect, tomentose or villous. **Leaves** broadly ovate to suborbicular, 2.6–12.5 × 2.2–12.5 cm, base rounded to truncate, margins coarsely crenate-serrate, apex obtuse or rounded, thick, abaxially gland-dotted; petiole <4 cm long. **Inflorescences** panicles, terminal, 12–42 cm long; verticils lax; peduncles slender, hirtellous; floral leaves sublunar, to 5 × 7 mm, concave, subacute; pedicels 2–3 mm long. **Flowers** with calyx to 4 mm long, tube 1.25 mm long, 10-nerved, gland-dotted, upper lip broadly ovate, to 2 × 2.5 mm, abruptly acute, subdecurrent on tube, later deflexed over calyx mouth, lower lip 2.25 mm long, lateral lobes triangular-ovate, median ones lanceolate, longer than laterals, acuminate; corolla finely pubescent, 0.8–1 cm long, pale blue, tube 4–5 mm long, whitish, upper lip 4 mm long, shallowly lobed, back gland-dotted, lower lip orbicular, 4–5 mm long; style glabrous. **Fruits** nutlets, globose, 0.4 mm Ø. **Distr:** S Africa, Sri Lanka. **Ecol:** In moist forest margins; 0–600 m. **Zones:** 10–12.

P. scutellarioides (L.) R. Br., *Prodr.*: 506, 1810. (Syn: *Coleus scutellarioides*, *Solenostemon scutellarioides*, etc.). [Coleus, Flame Nettle, Painted Nettle]. **Herbs**, perennial, aromatic, 0.5–1.5 m high; stems semisucculent, finely pubescent. **Leaves** variable, diamond-shaped to ovate, 1–15 × 1–10 cm, base usually cordate, margins variously toothed, apex usually acuminate, brightly colored or blotched, membranous; petiole 1–5(–8) cm long. **Inflorescences** verticillasters or irregularly branched cymes composed of simple or branched thyrses 5–10(–25) × 3–5(–8) cm, terminal; bracts ovate, 2–3 mm long, long-acuminate; pedicels 3–4 mm long. **Flowers** with calyx obliquely campanulate, 2–2.5 mm long, 5-toothed; corolla boat-shaped, blue or violet, tube whitish, upper lip often paler than lower one, 0.8–1.3(–1.8) cm long, lower lip exceeding upper, blue-violet; stamens in 2 pairs, basally united. **Fruits** nutlets, broadly ovate or orbicular, 1–1.2 mm long, brown, shiny. **Distr:** SE Asia, Malaysia. **Ecol:** On streamsides, hills, forests and open areas. Fl. anytime during the year, mainly Jul. **Zones:** 10–12.

Rotheca

Species 32. **Shrubs**, **subshrubs** or **herbs**, sometimes geoxylic, more rarely scandent or decumbent, or even small trees. **Leaves** opposite or verticillate, sessile or shortly petiolate; lamina with margins often crenate or coarsely serrate, sometimes entire, acrid when crushed.

Inflorescences terminal thyrses or axillary cymes, usually lax, occasionally reduced to solitary axillary flowers. **Flowers** non-resupinate; calyx campanulate to subspherical, mostly red or purple, lobes (4)5(6), acute to rounded; corolla mostly blue to whitish, subactinomorphic to unilabiate (all lobes on lower lip due to cleft between posterior lobes), tube curved or straight, often anteriorly saccate, lobes 5, anterior lobe often much larger and differently colored; stamens parallel, long-exserted, usually arched; ovary glabrous, glandular or hairy. **Fruits** drupes or drupaceous schizocarps, 4-lobed, lobes with a 1-seeded pyrene. **Distr:** Tropical and S Africa to Tropical Asia. **Cult:** Propagate from seeds, cuttings or suckers. Requires a moist, well-drained, soil in full or partial sun.

R. myricoides (Hochst.) Steane and Mabb., *Novon* 8: 205, 1998. (Syn: *Clerodendrum myricoides*, etc.). [Blue Butterfly Bush, Blue-flowered Clerodendrum, Oxford and Cambridge Bush, Rough-leaved Cat's-whiskers]. **Shrubs**, to 3 m high; crown open; stems slender. **Leaves** whorled; lamina elliptic to narrowly obovate, 8.9–15 × 3.2–7.5 cm, base wedge-shaped, margins serrate, apex pointed to obtuse, surfaces glabrous. **Inflorescences** panicles, terminal, lower nodes with leaflike bracts; flowers widely spaced. **Flowers** showy, odorless; calyx bell-shaped, shallowly 5-lobed; corolla blue, tube ~6 mm long, lobes 5, spreading, 4 pale, 1 darker and larger; stamens projecting forward, curved. **Fruits** 4-lobed, 1.2 cm Ø, black, fleshy. **Distr:** Ethiopia to S Africa. **Ecol:** Among shrubs near water; ~900–2,200 m. Fl. Apr–Jul, Oct–Dec. **Zones:** 9–12.

Salvia

Species >900. **Herbs**, **subshrubs** or **shrubs**. **Leaves** simple or pinnately compound. **Inflorescences** racemes, panicles or spikes, composed of 2- to many-flowered verticillasters, rarely solitary, axillary; bracteoles minute. **Flowers** with calyx tubular to campanulate, 2-lipped, upper lip entire, 2-toothed, or trimucronate, lower lip 2-toothed; corolla 2-lipped, tube straight, horizontal, curved upward or adaxially dilated, upper lip folded, straight or falcate, entire or emarginate, lower lip spreading, trilobed, midlobe widest, margin entire, apex emarginate, fringed, or bilobulate, lateral lobes oblong or circular; stamens 2, filaments short; staminodes 2, small or absent; style 2-cleft, lobes subulate, linear or circular, equal, anterior lobe larger or posterior lobe indistinct; disk slightly swollen in front or ringlike. **Fruits** nutlets, triquetrous, ovoid or oblong, glabrous, smooth. **Distr:** Tropical to temperate zones of the Old and New World. **Cult:** Propagate from seeds or cuttings (shrubby species). Requires a well-drained soil in full sun.

S. africana-lutea L., *Sp. Pl.*: 26, 1753. (Syn: *Crolocos aurea*, *S lutea*, etc.). [Beach Sage, Golden Sage, Wild Sage]. **Herbs**, shrubby, evergreen, to 2 m high; stems numerous, densely leafy, woody, much-branched, with minute orange-red dots. **Leaves** elliptic to paddle-shaped, to 3.5 cm long, margins sometimes toothed, gray-green, aromatic, gray-hairy, thickly textured. **Inflorescences** clusters, erect, each comprising of several 2-flowered verticillasters. **Flowers** rusty-orange or golden-brown, 3–5 cm long; calyx large, funnelform, persistent, enlarged in fruit, dark rusty-colored on the side exposed to light and olive-green on the opposite side, shortly hairy, gland-dotted; corolla golden-brown, reddish-brown or rarely purplish-brown, upper lip 1.2–2.5 cm long. **Distr:** S Africa (Cape Province: Namaqualand to E Cape). **Ecol:** In coastal dunes, fynbos and scrubs. Fl. Jun–Dec. **Zones:** 9–11. **Info:** The leaves are used to flavor food and also for medicinal purposes.

S. confertiflora Pohl, *Pl. Bras. Icon. Descr.* 2: 134, 1833. (Syn: *S. confertiflora* var. *brachyantha*, *S. rufa*, etc.). [Harvest Sage, Sabra Spike Sage]. **Herbs**, perennial, 1.2–1.5(–2.5) m high, base woody, ± hirsute-pubescent; stem quadrangular, angles thickened and reddish. **Leaves** petiolate; lamina ovate, to 20 cm long, lower leaves larger, margins scalloped or coarsely serrated, apex acuminate, yellowish-green, abaxially paler, wrinkled, adaxially reddish-brown hairy, abaxially densely so, with an unpleasant smell when crushed. **Inflorescences** spikes, to 30(–60) cm long, terminal, densely hairy; flowers arranged in few-flowered verticillasters, subsessile. **Flowers** downy, orangish-brown or reddish, 1 cm long; calyx deeply red-tinged, densely velvety-hairy or tomentose, upper lip obtuse; corolla clavate, twice as long as the calyx, 1shortly 2-lipped, obtuse, lips subequal, concave, upper lip entire, lower lip trilobed, with the midlobe longest and entire; stamens with anthers clubbed at apex. **Distr:** SE Brazil. **Ecol:** 650–1,350 m. **Zones:** 9–11.

S. involucrata Cav., *Icon.* 2: 3, 1793. (Syn: *Belospis laevigata*, *S. laevigata*, etc.). [Roseleaf Sage]. **Herbs**, perennial, suffrutescent or shrubby, to 1.5 m high, woody at base; stems sparsely branched. **Leaves** ovate or oblong, 6–12 cm long, margins crenate-serrate, apex acuminate, purplish-green, abaxially paler. **Inflorescences** racemes, dense, elongate; pedicels short. **Flowers** with calyx tubular, ~1.5 cm long, reddish-green, red or purplish, ribbed, slightly pubescent, 2-lipped, upper lip 1-toothed, lower lip 2-toothed; corolla rosy-magenta, tubular, to 5 cm long, somewhat inflated or ventricose, especially on the underside, upper lip very hairy, entire, laterally compressed, lower lip reflexed, trilobed; stamens with transverse appendage elongated; style scarcely exserted, stigma bifid. **Distr:** Mexico to Guatemala. **Ecol:** In shady places, such as at the edge of forests. Fl. Jul–Nov. **Zones:** 9–12.

S. leucantha Cav., *Icones* 1: 16, t. 24, 1791. (Syn: *S. bicolor* [illegal name], *S. leucantha f. iobaphes*). [Mexican Bush Sage, Velvet Sage]. **Herbs** perennial, or **subshrubs**, spreading; stems upright, base woody with age, 0.7–1.3 m high, covered with woolly hairs. **Leaves** oblong-lanceolate, 6–15 × 0.8–1.6 cm, margins with fine teeth pointing forwards, soft-textured, adaxially green, densely gray-woolly and finely blistered, abaxially densely white-woolly. **Inflorescences** dense spikes, to 50 cm long, white-woolly, verticillasters 3–6-flowered. **Flowers** with calyx ~0.7–1.3 cm long, densely purple-, pink- or white-woolly; corolla white, tubular, slightly curved, 1.3–1.5 cm long, lips short, upper lip to 8 mm long, lower lip bilobed, lobes sickle-shaped. **Distr:** Mexico. **Ecol:** In subtropical and tropical conifer forests. Fl. throughout most of the year, mainly Oct–Mar. **Zones:** 9–12.

S. splendens Sellow ex Roem. and Schult., *Mant.* 1: 185, 1822. (Syn: *Fenixanthes splendens*, etc.). [Bonfire Salvia, Red/Scarlet Sage]. **Herbs**, suffruticose, to 90 cm high. **Leaves** ovate to triangular-ovate, 2.5–7 × 2–4.5 cm, base truncate or ± rounded, margins serrate, apex acuminate, glabrous, abaxially glandular; petiole 3–4.5 cm long. **Inflorescences** racemes, to 20 cm long, composed of 2–6-flowered verticillasters; bracts ovate, red, enveloping flowers in bud, apex caudate-acuminate; pedicels 4–7 mm long, red glandular-villous. **Flowers** with calyx red, campanulate, ~1.6 cm long, dilated to 2 cm after anthesis, red-glandular, veins villous, shallowly 2-lipped, upper lip triangular-ovate, 0.5–0.6 × 1 cm, mucronate, lower lip slightly longer, deeply 2-toothed, teeth triangular; corolla scarlet (purple or white in some cultivars), 4–4.2 cm long, pubescent, tube slightly dilated at throat, upper lip straight, somewhat concave, oblong, 8–9 × ~4 mm, lower lip shorter; filaments ~5 mm long. **Fruits** nutlets, dark brown, ellipsoid, ~3.5 mm long, narrowly winged. **Distr:** Brazil (Rio de Janeiro). **Ecol:** 500–1,000 m. Fl. Mar–Oct. **Zones:** 9–12.

Tectona

Species 3. **Trees**, tall, deciduous, indumentum stellate. **Leaves** large, opposite or whorled, petiolate. **Inflorescences** panicles, terminal, with many dichotomous cymes; bracts small, narrow, early deciduous. **Flowers** with calyx campanulate, tube short, lobes spreading; stamens 5 or 6, inserted on apical part of tube, exserted, anthers dehiscing by longitudinal slits; ovary 4-locular, locules 1-ovuled, stigma 2-cleft. **Fruits** enclosed in enlarged persistent calyx, endocarp bony, exocarp thin; seeds oblong. **Distr:** Tropical Asia. **Cult:** Propagate from seeds, cuttings or by air-layering. Requires a moist, well-drained soil in full sun.

T. grandis L. f., *Suppl. Pl.* 151, 1781. [Teak]. **Trees**, to 40 m tall; branchlets gray to grayish-brown, 4-angled, yellowish to grayish-brown stellate-tomentose. **Leaves** ovate-elliptic to ovate, 15–45(–70) × 8–23(–37) cm, base cuneate and decurrent, margins entire, apex acuminate to obtuse, papery, abaxially densely grayish-brown to yellowish-brown minutely tomentose, adaxially scabrous and puberulent along veins, veins 7–12 pairs; petiole robust, 2–4 cm long.

Below: *Salvia confertiflora*, *Salvia involucrata*, *Salvia leucantha*, *Salvia splendens*.

Above: *Tectona grandis, Vitex agnus-castus, Vitex capitata, Vitex leucoxylon.*

Inflorescences panicles 25–40 × 30 cm. **Flowers** fragrant; calyx tube 2.5–3 mm long, with white stellate hairs; corolla white, tube 2.5–3 mm long, outside glandular-puberulent, lobes ~2 mm long, obtuse; ovary strigose, style 3–4 mm long. **Fruits** globose, 1.2–1.8 cm Ø, minutely tomentose. **Distr:** India to Indo-China. **Ecol:** 0–900 m. Fl. Jun–Aug. Fr. Sep–Dec. **Zones:** 10–12. **Info:** Cultivated in the tropics for the valuable timber.

Vitex

Species ~250. **Trees** or **shrubs**; branches glabrous or sparsely pubescent. **Leaves** opposite, palmately (1–)3–8-foliolate; leaflets petiolulate, margin entire, dentate, serrate, or incised. **Inflorescences** cymes, thyrses, or panicles, terminal or axillary; bracts usually small, often early deciduous. **Flowers** with calyx campanulate, tubular, or funnelform, sometimes 2-lipped, usually truncate or shortly 5-dentate; corolla blue, white or yellow, 2-lipped, lower lip trilobed with midlobe greatly elongated, upper lip usually bilobed; stamens 4, didynamous, sometimes exserted; ovary 2–4-locular, locules 1- or 2-ovuled, style filiform, stigma 2-cleft. **Fruits** drupes, subtended by enlarged calyx, globose, ovoid, or obovoid, normally 4-locular and 4-seeded, often some locules suppressed and base of pyrene forming a hollowed cavity, endocarp a bony pyrene, mesocarp generally fleshy; seeds obovoid or oblong. **Distr:** Tropical and subtropical regions worldwide. **Cult:** Propagate from seeds or cuttings. Requires a fertile, moist, well-drained soil in full sun.

V. agnus-castus L., *Sp. Pl.* 638, 1753. [Chaste Tree]. **Shrubs**, erect, 1–2(–4) m high, whitish-tomentose; branches obtusely quadrangular. **Leaves** palmately 5–7-foliolate, 2 extreme leaflets (especially in 7-foliolate leaves) much shorter and smaller than the others; leaflets subsessile to petiolulate, usually lanceolate, (2–)5–10(–12) × (0.4–)1–1.5(–2) cm, margins usually entire, apex ± acuminate, abaxially whitish-tomentose. **Inflorescences** terminal, subcylindrical, 10–20 cm long, interrupted; cymes forming verticillasters, ± compact, sessile or subsessile, each usually 0.8–1.3 cm long. **Flowers** lilac or deep violet, ~4 mm across; calyx campanulate, 3 × 2 mm, truncate, triangulate to obscurely toothed, persistent; corolla 0.7–1 cm long, exceeding calyx; lower lobes glabrous to slightly pubescent at base. **Fruits** drupes, globose, 2–2.5 mm Ø, exserted halfway above the slightly enlarged and expanded, persistent calyx, glabrous, 4-loculed, locules 1-seeded. **Distr:** E Mediterranean, SW and C Asia. **Ecol:** On riverbanks and in cool coastal areas, forming dense thickets. Fl. Jun–Nov. **Zones:** 8–11.

V. capitata Vahl, *Eclog. Amer.* 2: 50, 1798. (Syn: *V. bignonioides*, etc.). [Five-leaf Fiddlewood]. **Shrubs**, or **small trees**, to 18 m tall; trunk short, cylindric, to 40 cm dbh. **Leaves** decussate-opposite or subopposite, 3- or 5-foliolate, sometimes 2 lowermost much smaller; leaflets with central one usually narrowly oblong or elliptic, 3.5–14.5 × 2–3.5 cm, margins entire, base acute or acuminate, apex long-acuminate or caudate, shiny, adaxially dark green, abaxially lighter; petioles 2–6.5 cm long. **Inflorescences** ± capitate, rarely subumbelliform, 5–12 × 1–3.5 cm, usually densely many-flowered, axillary; peduncle 4–9.5 cm long; pedicels 1–2 mm long; bractlets 1–3 mm long. **Flowers** with calyx campanulate, 1–2 × 1–2 mm, often repand-denticulate or obsoletely 5-dentate; corolla hypocrateriform, blue to lavender or violet, 1–1.2 cm long, densely puberulent, tube cylindric, limb 2-lipped, upper lip bilobed, lower lip trilobed, midlobe largest, ovate; stamens purple, exserted, anthers dark-purple; style glabrous. **Fruits** drupaceous, fleshy, black, oblong, 1.2–1.4 × 0.8–0.9 cm; fruiting-calyx enlarged and incrassate, to 0.3 × 1 cm. **Distr:** Trinidad–Tobago to S Tropical America. **Ecol:** On dry rocky slopes; 600 m. Fl. & Fr. Mar–May. **Zones:** 10–12.

V. leucoxylon L. f., *Sp. Pl.* 293, 1782. (Syn: *Wallrothia leucoxylon*, etc.). **Trees**, 4–20 m tall; trunk to 1.3 m dbh. **Leaves** 3–5-foliolate; leaflets usually oblong to elliptic, *terminal* 3.5–14.5 × 1–4.5 cm, base and apex acute or ± attenuate-subacuminate, *lateral* 5.5–9.5 × 3–4.5 cm, base mostly inequilateral, ± coriaceous, bright green; petiole 2.5–6.5 cm long; terminal petiolule 1.4–2.5 cm long, lateral ones 1–2 cm long. **Inflorescences** cymose, 3–17 × 3–17 cm, lax, loosely many-flowered, axillary; peduncle 1.8–7.5 cm long. **Flowers** with calyx cupuliform, tube ~2 mm long, rim 5-toothed; corolla irregular, white or whitish, purple-pubescent in throat, lower lip with purplish-blue hairs toward base above, occasionally wholly white or yellowish-tinged, tube ~4.5 mm long, densely villous inside, upper lip bilobed, lower lip trilobed, lateral lobes rounded; stamens inserted ~1 mm below mouth of corolla tube, slightly exserted, anthers purple; ovary globose, stigma shortly bifid. **Fruits** drupes, obovoid or elliptic, 1.2–2 cm long, mostly 1-seeded, black or purple-black, base usually covered by fruiting calyx. **Distr:** Pakistan to peninsular Malaysia. **Ecol:** In wet evergreen forests. **Zones:** 10–12.

V. lucens Kirk, *Trans. and Proc. New Zealand Inst.* 29: 525, 1897. [New Zealand Oak]. **Trees**, to ~20 m tall; trunk to ~1.5 m dbh; branches stout, spreading; branchlets 4-angled. **Leaves** opposite, palmately lobed; leaflets 3–5, basal one or pair usually much smaller than the terminal 3, digitate; laminas of 3 main leaflets elliptic-oblong to obovate, ~5–12.5 × 3–5 cm, margins entire, ± undulate, apex abruptly acute to subacuminate, glabrous, coriaceous, dark green, glossy, domatia present at axils of costa and main veins; petiole to ~10 cm long. **Inflorescences** panicles, (4–)10–15-flowered, dichotomous, axillary. **Flowers** with calyx cupular, minutely 5-toothed; corolla dull red, pubescent, 2-lipped, ~2.5–3·5 cm long, upper lip entire or bifid, lower deflexed, trilobed; style slender, bifid, ~2.5 cm long. **Fruits** drupes, subglobose, bright red, ~2 cm Ø. **Distr:** New Zealand (North Island). **Ecol:** In lowland forests. Fl. Jun–Oct, Feb–Apr. Fr. Feb–Sep. **Zones:** 9–11.

Below: *Vitex leucoxylon* (fruits), *Vitex lucens, Vitex negundo, Vitex trifolia.*

MS

PM MS

Above: *Paulownia tomentosa, Aeginetia indica, Brandisia swinglei, Lindenbergia philippinensis.*

V. negundo L., *Sp. Pl.* 2: 638, 1753. [Five-leaved Chaste Tree]. (Syn: *Agnus-castus negundo, V. agnus-castus* var. *negundo*). **Shrubs** or **small trees**; branchlets densely tomentose, tetragonal. **Leaves** mostly 5-foliate, occasionally 3–7-foliolate; central leaflets elliptic-ovate, lanceolate, or narrowly lanceolate, 3–11 × 0.5–3 cm, base cuneate, acute or obtuse, margins entire or irregularly dentate apically, apex attenuate, long-caudate or acuminate, adaxially green, densely pubescent, abaxially very densely so, gray, central leaflet largest; petiolules 0.5–1.1 cm long, densely pubescent; petiole 2–4.5 cm long. **Inflorescences** thyrses, 12–20 cm long, densely pubescent, terminal. **Flowers** light purplish-blue; calyx campanulate, ~3 mm long, teeth 5, triangular, gray-pubescent outside, slightly ridged; corolla 5–6 mm long, 5-lobed, one lobe twice as long as the others, shortly pubescent outside, villous inside; stamens exserted, ~2.5 mm long, slightly villous on lower part; ovary glabrous. **Fruits** obconic. **Distr:** S Somalia to Mozambique and tropical and subtropical Asia, Marianas. **Ecol:** 0–3,200 m. **Zones:** 7–11.

V. trifolia L., *Sp. Pl.* 2: 638, 1753. (Syn: *V. agnus-castus* var. *trifolia*). [Arabian Lilac, Three-leaf Chaste Tree]. **Shrubs** or **small trees**, 1.5–5 m high, erect; branchlets densely pubescent. **Leaves** 1–3(–5)-foliolate; leaflets sessile, oblong, lanceolate, or obovate, base cuneate, margins entire, apex obtuse, adaxially green and glabrous or subglabrous, abaxially densely tomentose, veins ~8 pairs and slightly prominent on both surfaces; central or single leaflet 2.5–9 × 1.7–3 cm; petiole 1–3 cm long. **Inflorescences** panicles, 3–15 cm long; peduncles densely gray-tomentose. **Flowers** with calyx slightly 5-dentate, outside gray-pubescent, inside glabrous; corolla purplish to bluish-purple, 0.6–1 cm long, outside scaly white, pubescent at filament bases and on inside of lower lobe; stamens exserted; ovary glabrous, with or without glands, style glabrous. **Fruits** black, subglobose, ~5 mm Ø. **Distr:** Somalia to Mozambique and Pacific. **Ecol:** 100–1,700 m. Fl. Apr–Aug. Fr. Aug–Nov. **Zones:** 9–11.

PAULOWNIACEAE

Genus 1, **species** 7. *Familial description same as generic.*

Paulownia

Species 7. **Trees**, deciduous, evergreen in the tropics; bark smooth with conspicuous lenticels when young, longitudinally splitting with age; branches opposite, without terminal buds. **Leaves** opposite, occasionally 3-whorled; lamina entire or shallowly 3–5-lobed, margins undulate and often serrate when young; petiole long. **Inflorescences** thyrses, pyramidal to cylindric, large; cymes (1–)3–5(–8)-flowered, pedunculate or not. **Flowers** with calyx campanulate or obconical, hairy, lobes 5, subequal, upper 1 enlarged; corolla purple or white, funnelform-campanulate to tubular-funnelform, tube base constricted and slightly curved, limb 2-lipped, lower lip elongated and trilobed, upper lip yellow, bilobed; stamens 4, didynamous, filaments included, twisted near base, anthers divergent; ovary biloculed, style nearly as long as stamens. **Fruits** capsules, loculicidal, 2-valved or incompletely 4-valved, pericarp thin or thick and woody; seeds small, numerous, membranous-winged. **Distr:** Mostly in China with one species extending to Laos and Vietnam. **Cult:** Propagate from seeds, root cuttings. Requires a moist, well-drained soil in full sun.

P. tomentosa (Thunberg) Steudel, *Nomencl. Bot.* 2: 278, 1841. [Princess Tree]. **Trees**, to 20 m tall; crown broad, umbelliform; bark brown-gray; twigs conspicuously lenticellate, viscid-glandular when young. **Leaves** cordate, ~40 cm long, apex acute, abaxially densely to sparsely hairy, adaxially sparsely hairy. **Inflorescences** thyrses, pyramidal to narrowly conical, to 50 cm long; cymes 3- or 4-flowered; peduncles 1–2 cm long, nearly as long as pedicels. **Flowers** with calyx shallowly campanulate, ~1.5 cm long, outside tomentose, lobes to half or more calyx length, ovate-oblong; corolla purple, funnelform-campanulate, 5–7.5 cm long, ventrally ridged, outside glandular, inside glabrous, limb ~4.5 cm Ø; stamens to 2.5 cm long; ovary ovoid, glandular, style shorter than stamens. **Fruits** capsules, ovoid, 3–4.5 cm long, densely viscid-glandular hairy, pericarp ~1 mm thick, persistent calyx lobes flat; seeds 2.5–4 mm long including wing. **Distr:** China. **Ecol:** To 1,800 m. Fl. Apr–May. Fr. Aug–Sep. **Zones:** 6–11.

OROBANCHACEAE

Genera 99, **species** 2,060. **Herbs**, annual, biennial, or perennial parasites, without chlorophyll; stems unbranched or sometimes branched. **Leaves** scalelike, spirally or subimbricately arranged. **Inflorescences** racemose, spicate, or subcapitate, rarely 1-flowered; bract 1, usually similar to leaves; bractlets 2, adnate to base of calyx or pedicel; flowers subsessile or pedicelled. **Flowers** bisexual; calyx tubular, cupular, or campanulate, (3 or)4–6-lobed, 2–6-parted, 6-toothed, or spathelike, sometimes absent or of 3 free sepals; corolla bilabiate, usually curved, sometimes tubular-campanulate or funnelform with 5 subequal lobes; upper lip entire, emarginate, or bilobed; lower lip trilobed; stamens 4, didynamous, inserted at base of corolla tube, filaments slender, anthers 2-celled, longitudinally dehiscent, sometimes 1 cell fertile, the other sterile or reduced to a spur; pistil 2- or 3-carpellate, ovary superior, placentas 2–4 or 6(–10), parietal or sometimes axile at ovary base, ovules 2–4 or numerous, anatropous, style long, stigma inflated, discoid, peltate, or 2–4-lobed. **Fruits** capsules, usually loculicidally dehiscent by 2 or 3(or 4) valves; seeds minute, testa pitted or reticulate. **Distr:** Worldwide, especially N (warm) temperate and Africa-Madagascar.

Aeginetia

Species 4. **Herbs**, fleshy, parasitic; stems very short, branched or not. **Leaves** scalelike. **Inflorescence** with flowers solitary or clustered at stem apex; bractlets absent; pedicels very long, erect. **Flowers** large; calyx spathelike, acute or obtuse-rounded; corolla tubular or campanulate, indistinctly bilabiate, upper lip bilobed, lower lip trilobed, lobes subrounded; stamens 4, included, anthers adherent, 1 cell fertile, another reduced into a spur; carpels 2, ovary usually 1-loculed, parietal placentas 2 or 4, occasionally imperfectly biloculed and placentas axile, style slightly curved, stigma fleshy, peltate. **Fruits** capsules, dehiscing by 2 valves; seeds numerous, testa reticulate. **Distr:** Tropical Asia to Indo-Malaya and Philippines. **Cult:** Propagate from seeds. Requires a moist soil rich in organic matter, in shade or partial sun.

A. indica L., *Sp. Pl.* 2: 632, 1753. (Syn: *Orobanche aeginetia*, etc.). [Ghost Flower]. **Herbs**, parasitic, 15–40(–50) cm tall; root slightly fleshy, with small branches; stems unbranched or branched from near base. **Leaves** red, ovate-lanceolate or lanceolate, 0.5–1 × 0.3–0.4 cm, glabrous. **Inflorescence** with flowers usually solitary; pedicel usually erect, 10–30(–40) cm long, ~3 mm Ø. **Flowers** with calyx apex acute or acuminate; corolla purple-red striate, indistinctly bilabiate, tubular-campanulate, 2–4.5 cm long, tube slightly curved, lobes subentire; filaments purple, 7–9 mm long, glabrous, anthers yellow; ovary 1-loculed, parietal placentas 4, style 1–1.5 cm long, stigma pale yellow. **Fruits** capsules, conical, or long ovoid-globose, 2–3 cm long; seeds yellow, ellipsoid, minute. **Distr:** Tropical and subtropical Asia, China, Indo-China to Indonesia, Philippines and Taiwan. **Ecol:** On slopes, forest margins, parasitic on *Miscanthus* and *Saccharum* spp.; 200–1,800 m. Fl. Apr–Aug. Fr. Aug–Oct. **Zones:** 9–12.

Brandisia

Species ~11. **Shrubs**, erect or scandent, occasionally parasitic, usually stellate-tomentose. **Leaves** ± opposite, shortly petioled; lamina with margins entire or denticulate. **Inflorescences** with flowers solitary, paired, or in racemes, axillary; bracteoles 2, inserted on pedicels. **Flowers** with calyx campanulate, rarely cylindric-ovate, outside stellate-hairy, regular or somewhat 2-lipped, lobes 2 or 5; corolla tube somewhat incurved, limb 2-lipped, lower lip as long as or shorter than upper lip, trilobed, patent, upper lip bilobed, concave; stamens 4, somewhat inserted at base of corolla tube, filaments glabrous, anthers biloculed; staminode absent; ovary ovoid, hairy, biloculed, ovules numerous, linear-oblong, style elongated, stigma simple. **Fruits** capsules, ovoid to

Above: *Citharexylum spinosum, Duranta erecta, Lantana camara, Lantana montevidensis.*

ovoid-globose, loculicidal; seeds linear, testa reticulate, membranous-winged. **Distr:** Subtropical E Asia. **Cult:** Propagate from seeds. Requires a humus-rich, moist compost in shade. **Info:** The placement of this genus is still unclear, but some studies shows that they are closely related to Orobanchaceae (Bennett and Mathews 2006).

B. swinglei Merrill, *Philipp. J. Sci.* 13: 157, 1918. **Shrubs**, erect to scandent, to 2 m high, densely gray-brown stellate-tomentose; branches glabrescent. **Leaves** ovate, rarely ovate-oblong, 3–11 × 1–5.5 cm, base broadly cuneate to subcordate, margins entire or irregularly sparsely serrate, apex acute to caudate-acute, densely gray-brown stellate-tomentose, adaxially glabrescent; petiole to 8 mm long. **Inflorescences** with flowers solitary or sometimes paired; pedicel to 8 mm long; bracteoles gray stellate-tomentose, inserted at pedicel apex. **Flowers** with calyx campanulate, to 1.5 cm long, gray stellate-tomentose, inside sericeous, outside 10-veined, lobes 5, narrowly triangular-ovate, ~8 mm long, acuminate and apiculate; corolla yellow, ~2.5 cm long, gray-brown stellate-tomentose except for tube base, lateral lobes of lower lip oblong-ovate, smaller than middle lobe, upper lobes obliquely ovate; stamens included; ovary base and style stellate-tomentose. **Fruits** capsules, shorter than calyx; seeds ~4.5 mm long. **Distr:** China. **Ecol:** On mountain slopes; 500–1,000 m. Fl. Jun–Nov. **Zones:** 9–11.

Lindenbergia

Species 12. **Herbs**, annual or perennial, basally woody, erect or decumbent, much branched, hairy, rooting from lower nodes. **Leaves** opposite, or upper alternate. **Inflorescences** with flowers solitary or in terminal spikes or racemes; bracts foliaceous; bracteoles absent. **Flowers** with calyx campanulate, 5-lobed, parted to about middle, hairy; corolla 2-lipped, tube tubular, often puberulent, lower lip large, trilobed, often plicate, upper lip short, wide, apex truncate, subemarginate, or bilobed; stamens 4, didynamous, included, inserted below middle of corolla tube, filaments usually glabrous, anther locules distinct, connective often forming a stalk; ovary hairy or glabrous, stigma not lobed. **Fruits** capsules, generally enclosed by persistent calyx, 2-grooved, loculicidal, valves entire, separating from placental column, style persistent; seeds numerous, minute. **Distr:** Tropical and subtropical Africa and Asia. **Cult:** Propagate from seeds. Requires a moist soil rich in organic matter, in shade or partial sun.

L. philippensis (Chamisso and Schlechtendal) Bentham, *Prodr.* 10: 377, 1846. (Syn: *Stemodia philippensis*, etc.). **Herbs**, perennial, to 1 m tall, stout, erect, straight, much-branched, glandular-hairy; stems terete. **Leaves** ovate to ovate-lanceolate, 2–8 cm long, base cuneate, margins serrate, apex acute to acuminate, papery; petiole 0.6–1.2 cm long. **Inflorescences** terminal, spicate-racemose, dense, 6–20 cm long; bracts narrowly lanceolate, shorter than calyx; flowers subsessile. **Flowers** with calyx 5–6 mm long, 5-veined, lobes subulate-triangular, as long as tube, apex acute; corolla yellow, outside purple-patched, sparsely hairy, tube about twice as long as calyx, lower lip long, plicate, upper lip apically subtruncate, sometimes emarginate; anthers long-stalked; style basally and apically hairy. **Fruits** capsules, narrowly ovoid, 5–6 mm long, apically densely brown-hirsute; seeds yellow, ~0.5 mm long, scabrous. **Distr:** Cambodia, China, India, Laos, Myanmar, Philippines, Thailand, Vietnam. **Ecol:** On dry mountain sides, rocky crevices; 1,200–2,600 m. Fl. & Fr. Nov–Mar. **Zones:** 9–12.

VERBENACEAE

Genera 34, **species** 1,175. **Herbs**, **shrubs** and **trees**, rarely woody climbers; branches and twigs tetragonal, not prominently nodose or articulated. **Leaves** opposite, decussate, rarely whorled or alternate, exstipulate; lamina simple, rarely compound, margins toothed to deeply lobed. **Inflorescences** definite and centrifugal or indefinite and centripetal, spicate, racemiform or panicled to umbellate or a head, very rarely solitary and axillary. **Flowers** usually zygomorphic, bisexual, hypogynous; calyx gamosepalous, mostly tubular or campanulate, (2–)4(–7)-lobed or toothed to subentire, persistent, sometimes enlarging and concealing the fruit; corolla gamopetalous, often with as many lobes as the calyx and somewhat bilabiate, with unequal lobes above and tubular below; stamens mostly 4 and didynamous, inserted on corolla tube, alternating with lobes, staminodes often present, anthers dithecous; gynoecium 2(4–5)-carpellate, syncarpous, somewhat 4-lobed and 2–4-loculed, locules usually 1-ovuled; style simple, terminal. **Fruits** usually schizocarpic or drupaceous, endocarp ± hard, indehiscent or dehiscent into 1–2(–10)-celled pyrenes; seeds 1–few. **Distr:** Mostly tropical and subtropical regions worldwide.

Citharexylum

Species 110. **Shrubs** or **trees**, rarely climbing; branches and branchlets usually tetragonal, sometimes spiny. **Leaves** deciduous, opposite-decussate, ternate or verticillate, usually with a pair of glands at base of lamina, petiolate or sessile, petiole rarely myrmecophilous at apex, exstipulate; lamina with margins entire. **Inflorescences** spicate racemes, terminal or axillary, indeterminate, occasionally sparsely branched, mostly elongate and many-flowered, rarely few-flowered. **Flowers** small; calyx tubular or cyathiform, regular or somewhat zygomorphic, thin, accrescent, apex 5-toothed or lobed to truncate; corolla infundibuliform or hypocrateriform, mostly yellow or white, varying to blue, violet, or lilac, tube narrowly cylindrical and regular, throats ± pubescent lobes (4)5(6), patent, spreading; stamens 4–5, lower 2 forming staminodes, not exserted; ovary bicarpellate, perfectly or imperfectly 4-loculed, locules 1-ovuled, style terminal, stigma bifid. **Fruits** fleshy drupes (2 pyrenes, each 2-seeded, often separated by a median fissure), partly enclosed by enlarged, persistent calyx. **Distr:** Tropical and subtropical America. **Cult:** Propagate from seeds, cuttings or by air-layering. Requires a moist, well-drained soil in full sun.

C. spinosum L., *Sp. Pl.* ed. 1, 2: 625, 1753. (Syn: *C. caudatum* [illegal name], *C. cinereum* [name superfluous], *C. fruticosum* [superfluous name], *C. quadrangulare*, etc.). [Fiddlewood, Juniper Berry, Savanna Wattle]. **Shrubs** or **trees**, to 16 m tall; trunk to 35 cm Ø; crown narrow. **Leaves** decussate-opposite; lamina elliptic or elliptic-oblong, 3.5–29 × 1.3–11.3 cm, base acute or subacuminate, margins entire (rarely irregularly and coarsely dentate), apex obtuse (rarely emarginate) to acute to shortly acuminate, adaxially usually dull rich-green, secondaries 5–8 pairs; petiole 0.7–2.4 cm long, orange. **Inflorescences** racemes, axillary and mostly terminal, simple or compound with 1–5 pairs of basal branches, 2.5–35 × 2 cm, loosely many-flowered; peduncle 1.5–2.5 cm long; pedicels 1.5–4 mm long; bracts large, foliaceous; bractlets linear, to 1 cm long, upper 1–2 mm long. **Flowers** very fragrant; calyx cyathiform, 3-4 mm long, pale-green, glabrous, rim ciliate, minutely 5-dentate; corolla white, creamish or reddish-white, ± hypocrateriform to infundibular, tube 4–6 mm long, limb 5-lobed, lobes suborbicular-lingulate, 2–3 mm long, sparsely ciliate, rounded. **Fruits** drupaceous, obovoid-oblong, ~8 × 6 mm, fleshy, brownish- or reddish-black to black, shiny. **Distr:** Caribbean, Panama to N South America. **Ecol:** In forests and on slopes. Fl. Aug–Nov. **Zones:** 10–12.

Duranta

Species 21. **Shrubs**. **Leaves** opposite or whorled; lamina simple, margins entire or serrate-dentate. **Inflorescences** racemes, axillary or terminal; bracts small. **Flowers** with calyx 5-veined, 5-dentate, persistent; corolla 2-lipped, lower lip trilobed and largest, upper lip bilobed; stamens 4, short, inserted above middle of cylindric corolla tube, obscurely didynamous; ovary 8-locular, ovules 1 per locule, pendulous, style club-shaped, stigma nearly oblique. **Fruits** drupes, equaling or shorter than calyx, with 2-seeded pyrenes. **Distr:** Tropical and subtropical America. **Cult:** Propagate from seeds or cuttings. Requires a moderately fertile, moist, well-drained soil in full sun. Some species are salt- and drought-tolerant. **Info:** The colorful fruits are usually poisonous.

D. erecta L., *Sp. Pl.* 2: 637, 1753. (Syn: *D. inermis, D. plumieri, D. repens*, etc.). [Golden Bead

Tree, Golden Dew Drop, Pigeon Berry, Sky Flower]. **Shrubs**, often climbing, 1.5–3 m high; branches unarmed or spiny, pubescent when young. **Leaves** opposite; lamina oblong or ovate to lanceolate, 2–6.5 × 1.5–3.5 cm, base cuneate, margins entire to distally crenate, apex acute or obtuse, papery, veins 6 pairs; petiole ~1 cm long, pubescent. **Inflorescences** 5–12 racemes, arranged in panicles or clusters, drooping, terminal or lateral, 10–15 cm long, to 30-flowered per raceme; bracts linear. **Flowers** lavender-blue or lilac-blue, rarely white; calyx 3–4 mm long, pubescent; corolla tube ~7 mm long, limb expanded, 6–8 mm Ø, lobes suborbicular, pubescent inside; stamens included; ovary glabrous. **Fruits** drupes, yellow, ~5 mm Ø, shorter than calyx, shiny, glabrous. **Distr:** S USA to tropical America. **Ecol:** 0–200 m. Fl. & Fr. May–Oct. **Zones:** 9–12.

Lantana

Species 122. **Shrubs**, climbing, aromatic, pubescent or glabrous; branches 4-angled, sometimes prickly. **Leaves** petiolate; lamina simple, margins crenate, often rugose. **Inflorescences** dense capitula, pedunculate; bracts exceeding calyx. **Flowers** with calyx small, membranous, truncate or sinuate dentate; corolla nearly actinomorphic or slightly 2-lipped, tube slender, lobes 4 or 5, spreading; stamens 4, didynamous, included; ovary bilocular, style shorter than corolla tube, stigma obliquely subcapitate. **Infructescences** globose, dense. **Fruits** drupes, with two 1-seeded pyrenes. **Distr:** Tropical and subtropical America, Africa to Thailand. **Cult:** Propagate from seeds or cuttings. Requires a fertile, well-drained soil in full sun.

L. camara L., *Sp. Pl.* 2: 627, 1753. [Angle Lips, Red & Yellow Sage, Shrub Verbena, Spanish Flag]. **Shrubs**; branches long, weak, armed with stout recurved prickles, pubescent. **Leaves** ovate to oblong, 3–8.5 × 1.5–5 cm, base rounded to subcordate, margins crenate, papery, wrinkled, very rough, with short stiff hairs, aromatic when crushed, lateral veins 5 pairs, very prominent, elevated; petiole 1–2 cm long, pubescent. **Inflorescences** terminal heads, 1.5–2.5(–4) cm Ø. **Flowers** with calyx small, inconspicuous, ~3 mm long, thinly textured; corolla hypocrateriform, mostly opening yellow and turning to rose, rose-pink, or pink, sometimes orange-yellow to red or scarlet with age, tube ~1 cm long, barely enlarged above middle, slightly curved, puberulent, limb (4–)6–8 mm Ø; ovary glabrous. **Fruits** drupes, deep purple, globose, ~4 mm Ø. **Distr:** Mexico to tropical America. **Ecol:** In open waste areas and near coastal areas; 100–1,500 m. **Zones:** 9–12.

L. montevidensis (Spreng.) Briq., *Annuare Conserv. Jard. Bot. Géneve* 7–8: 301, 1904. (Syn: *Lippia montevidensis*, etc.). [Polecat Geranium, Trailing Lantana, Weeping Lantana]. **Herbs** suffrutescent, or low **shrubs**, unpleasantly aromatic, scrambling to 4 m long. **Leaves** ovate, subrhomboid-oblong or lanceolate, mostly 1.2–3.5 ×0.8–1.6 cm, 6 × 3 mm on dwarf forms, base narrowed, acute to cuneate-attenuate, margins coarsely toothed, apex subacute or obtuse, abaxially paler; petiole ~4 mm long. **Inflorescences** initially capitate and hemispheric, >2.5 cm Ø, later elongating and finally oblong in fruit; peduncles long, filiform; bractlets thinly herbaceous, concave, ciliate, margins deflexed apically; flowers sessile. **Flowers** with calyx ~2 mm long, obsoletely 4- or 5-dentate; corolla infundibular, slightly irregular, varying from pink, lilac, or lavender to violet, rose, purple or magenta, paler outside, tube slender, limb ~6 mm Ø, 4- or 5-lobed. **Fruits** drupes, dark-violet, globose, ~4 mm Ø, fleshy, endocarp woody, bilocular, splitting into 2 halves; seeds incrassate dorsally. **Distr:** Tropical America. **Ecol:** In various habitats of subtropical and tropical regions. Fl. throughout the year. **Zones:** 9–12.

L. rugosa Thunb., *Prodr. Pl. Cap.*: 98, 1800. (Syn: *L. violacea*, etc.). [Bird's Brandy]. **Shrubs** or **small trees**, erect, sparse, to 5 m high; main stem single- or multistemmed, densely branched; branches rough, quadrangular, unarmed; young parts tomentellous, glabrescent. **Leaves** opposite or 3–4-whorled; lamina ovate, lanceolate or narrowly lanceolate, 1.5–10 × 1–2.5 cm, base rounded or cordate, margins toothed or crenulate, apex shortly attenuate and acute, surfaces green, wrinkled, venation reticulate; petiole ~3 mm long. **Inflorescences** compound racemes of dense clusters of cymules; lateral racemes to 6 × 2.5 cm; pedicels obscure. **Flowers** small, ~5 mm Ø; corolla lilac to mauve or violet, with darker or orange centers, lobes very broadly oblong, apex rounded; anthers subsessile, inserted below throat; ovary almost spherical, densely hirsute above. **Fruits** drupes, globose, ~5 mm Ø, lilac, closely set in a conical infructescence. **Distr:** Ethiopia to S Africa, Arabian Peninsula. **Ecol:** In thickets in bushveld or on grassy hillsides. **Zones:** 10–12.

Petrea

Species 12. **Lianas**, **shrubs** or **trees**. **Leaves** opposite-decussate or verticillate; lamina with margins entire, coriaceous. **Inflorescences** racemes, terminal or axillary; flowers pedicellate. **Flowers** slightly zygomorphic, violet, blue or white; calyx tube campanulate, 5-lobed, with 5 scales in throat, lobes colored, longer than tube; corolla tube short, cylindric, urceolate or infundibuliform, limb 5-lobed with unequal, brightly colored lobes; stamens 4, didynamous, inserted at middle of corolla tube, not exserted; ovary 1-carpellate by abortion, bilocular, locules 1-ovuled, style simple, stigma capitate, bilobed. **Fruits** drupes, enclosed within persistent, hardened calyx, forming a 5-winged samara. **Distr:** Florida to tropical America. **Cult:** Propagate from seeds or cuttings. Requires a fertile, well-drained soil in full sun; some species are drought tolerant.

P. volubilis L., *Sp. Pl.* 626, 1753. (Syn: *P. arborea*, etc.). [Petrea, Purple Wreath, Queen's Wreath, Sandpaper Leaf, Sandpaper Vine]. **Shrubs**, twining to suberect, to 6(–8) m high, woody, hairy. **Leaves** elliptic, ovate-elliptic to oblong, 5–20 × 2.5–10 cm, base often subcordate, margins entire or wavy, apex obtuse or usually acute, coriaceous, rough, sandpaperlike; petiole 1–1.5 cm long. **Inflorescences** racemes, many-flowered, to 30 cm long, pendent, terminal; pedicels 7–8 mm long. **Flowers** blue or pale violet; calyx petaloid, tube 2–3 mm long, lobes oblong, 1.2–1.8 × 0.4–0.6 cm, throat scales triangular, acute; corolla salverform, bright blue, tube infundibuliform, pubescent above and densely hairy near stamen insertion inside, anterior lobe largest. **Fruits** drupes, obovate-oblong, enclosed within the hardened calyx forming a 5-winged samara. **Distr:** Florida, Mexico to Tropical America. **Ecol:** In dry thickets, dry forests and on rocky slopes; 0–1,800 m. Fl. Mar–May. **Zones:** 10–12.

Phyla

Species 11. **Herbs**; branches usually acutely 4-angled, sometimes spiny, hirsute, glandular. **Leaves** opposite or clustered; lamina hairy. **Inflorescences** spicate or capitate, terminal or axillary, solitary or paired, elongated in fruit; bracts obovate. **Flowers** small; calyx membranous, 2-lipped, 4-lobed; corolla sometimes 2-lipped, lobes 5, spreading; stamens 4, didynamous, inserted on distal part of corolla tube, included; ovary bilocular, locules 1-ovuled, style short, stigma capitate. **Fruits** capsules, small, splitting into two 1-seeded cocci. **Distr:** Tropical and subtropical regions worldwide. **Cult:** Propagate from seeds or by division. Requires a moist, well-drained soil in full or partial sun. Some species are drought- and salt-tolerant.

P. nodiflora (L.) E.L. Greene, *Pittonia*. 4: 46, 1899. (Syn: *Verbena nodiflora*, *Lippia nodiflora*, etc.). [Capeweed, Matgrass, Turkey Tangle]. **Herbs**, perennial; branches many, creeping, rooting at distal nodes, minutely strigose. **Leaves** subsessile; lamina spathulate, 0.5–4 × 0.4–2 cm, base cuneate, margins distally sharply serrate, papery, pubescent, veins 4-paired. **Inflorescences** cylindric to ovate heads, 1–4.5 × 0.6–0.8 cm; peduncle 1–7 cm long; bracts imbricate, ~2 mm long, mucronate or acuminate. **Flowers** very small, ~3 mm long; calyx flattened, shorter than bracts, hyaline-membranous, deeply dissected, lobes lanceolate, pubescent; corolla pinkish-purple or white, slightly exceeding bracts, unequally 4-lobed, lobes spreading, glabrous. **Fruits** ovate, ~1.6 mm long, subcompressed, enclosed by persistent calyx, separating at maturity into two 1-seeded pyrenes. **Distr:** Tropical and subtropical regions worldwide. **Ecol:** Weedy on streambanks and in grassy places; 0–2,300 m. **Zones:** 9–12.

Stachytarpheta

Species 122. **Herbs** or **shrubs**; branches usually forked, hairy. **Leaves** opposite, petiolate; lamina simple, margins dentate or crenate, often rugose. **Inflorescences** spikes, usually long, lax, terminal;

Below: *Lantana rugosa*, *Lantana rugosa* (fruits), *Petrea volubilis*.

Above: *Phyla nodiflora, Stachytarpheta jamaicensis, Stachytarpheta mutabilis, Stachytarpheta urticifolia.*

rachis often angular, excavated; flowers ± immersed in excavated rachis when young. **Flowers** with calyx membranous, 4- or 5-dentate, becoming variously slit; corolla slender, cylindric, tube often curved, ± villous in throat, lobes 5, spreading; stamens 2, fertile, inserted on distal part of corolla tube, anthers divaricate; ovary supported by a short ringlike disk, bilocular, locules 2-ovuled, stigma capitate. **Fruits** capsules, splitting into two 1-seeded cocci. **Distr:** Tropical and subtropical America, tropical Africa. **Cult:** Propagate from seeds. Requires any well-drained soil in full sun.

S. jamaicensis (L.) Vahl, *Enum. Pl.* 1: 206, 1804. (Syn: *Verbena jamaicensis*, etc.). [Blue Porterweed, Devil's Coachwhip, Jamaican Vervain]. **Herbs** or **subshrubs**, perennial, robust, 0.6–2 m high; young branches 4-angled. **Leaves** opposite or alternate; lamina elliptic to oblong-ovate, 2.4–8 × 4.5 cm, base elongated, margins entire, serrate or dentate, apex acuminate, papery, subglabrous, veins 5 or 6 pairs; petiole winged. **Inflorescences** spikes, to 20+ cm long; flowers sessile or immersed into rachis, bracteate; bracts ~5 mm long, membranous. **Flowers** small; calyx ~6 mm long, glabrous, tube narrow, 5-lobed or toothed, persistent, not enlarged in fruit; corolla dark blue, tube cylindrical, 0.7–1.2 cm long, slightly curved, limb 5-lobed; stamens 4, the 2 anterior ones fertile and with divergent anther cells, included; ovary bilocular, locules 1-ovuled, style filiform, stigma ± capitate. **Fruits** capsules, included in persistent calyx, dry, linear-oblong, splitting into two 1-seeded pyrenes; seeds linear. **Distr:** SE USA to tropical America; widely introduced elsewhere. **Ecol:** In thickets of the Atlantic tierra caliente, often growing on seashores; 0–600 m. **Zones:** 10–12.

S. mutabilis (Jacq.) Vahl, *Enum. Pl.* 1: 208, 1804. (Syn: *Verbena mutabilis*, etc.). [Pink Porterweed, Snakeweed]. **Herbs**, often suffrutescent, erect, 1–2(–5) m high, stout, copiously pubescent; stems often woody at base; branches tetragonal, rigid, hairy-tomentose. **Leaves** opposite; lamina simple, broadly ovate to oblong-ovate or spathulate-ovate, 3–20 cm long, base abruptly contracted below and decurrent on petiole, margins serrate, apex rounded to acuminate, pubescent or hirsute; petiole winged. **Inflorescences** spikes, 10–30(–70) cm long, terminal, stout, pubescent; bracts scalelike. **Flowers** small, pink to magenta, sometimes red or purple; corolla with petals fused, tube curved, 1.2–2.2 cm long, throat cream or pink with a red border, limb salverform 1–1.7 cm Ø, short-lived, lobes 5, rounded, opening sequentially. **Fruits** schizocarps, 2-seeded, embedded in the rachis. **Distr:** Mexico to Tropical America; introduced elsewhere. **Ecol:** Fl. throughout the year. **Zones:** 10–12.

S. urticifolia (Salisb.) Sims, *Bot. Mag.* 43: t. 1848, 1816. (Syn: *Cymburus urticifolius, Zappania urticifolia, S. urticaefolia* f. *albiflora*). [Blue Porter Weed, Nettle-leaf Velvetberry]. **Herbs** coarse, or **subshrubs**, 0.8–2 m high, with a long tap-root; internodes often quite elongate. **Leaves** decussate-opposite; lamina oval or ovate to broadly elliptic, 4–8 × 2–4.5 cm, base cuneately acuminate into petiole, margins regularly sharply serrate, apex abruptly acute, glabrous, midrib sometimes purple when fresh, membranous, adaxially usually dark glossy green bullate-rugose when fresh, abaxially paler; petiole 0.5–2 cm long, cuneately winged, indistinct from lamina base. **Inflorescences** spikes, elongate, many-flowered, mostly terminal, rarely a few in upper leaf axils, erect or nutant, 14–40 cm long, slender; peduncles 1–2 cm long; rachis slender, ~2.5 mm Ø, often weak and flexible during anthesis, stiffer in fruit; flowers imbricate, opening centripetally a few at a time, each lasting a day; bracts lanceolate, conspicuous, 7 × 2 mm. **Flowers** with calyx cylindric, ~7 × 2 mm, closely appressed to rachis, 5-costate, rim shortly 5-toothed; corolla hypocrateriform, pure white (usually dark purplish-blue with a lighter or white 'eye'); stamens and style white; stigma yellowish-white. **Distr:** Caribbean to N Brazil, widely introduced elsewhere. **Ecol:** In disturbed places. Fl. throughout the year. **Zones:** 10–12.

PEDALIACEAE

Genera 13, **species** 70. **Herbs**, annual or perennial, rarely shrubs or deciduous trees, sometimes with swollen stems. **Stipules** absent. **Leaves** opposite, sometimes alternate on upper part of stem; lamina densely covered with 4-merous mucilaginous glands. **Inflorescences** with flowers solitary, axillary, rarely in few-flowered cymes; pedicel with extrafloral nectaries. **Flowers** zygomorphic; calyx (4 or)5-parted; corolla bilabiate or obscurely so, lobes 5, imbricate; stamens 4 and didynamous, sometimes 2, anthers bilocular, longitudinally dehiscent; staminodes usually 1; disk fleshy; ovary superior, rarely inferior, 2- or 4-locular, placentation axile, ovules 2 to numerous per locule, anatropous, style filiform, stigma with 2 broad lobes, wet, often sensitive. **Fruits** dehiscent or indehiscent, winged or usually with horns or hooks; seeds 1 to many per locule. **Distr:** Mostly Old World tropics. **Ecol:** In coastal or arid habitats.

Sesamum

Species 19. **Herbs** annual or perennial, rarely shrubs, terrestrial. **Leaves** opposite or alternate; lamina simple or divided, margins entire or digitate. **Inflorescences** with flowers solitary; pedicels short. **Flowers** bell-shaped, similar to foxgloves (*Digitalis*); calyx deciduous or persistent; corolla white, pink or purple, obliquely campanulate, obscurely bilabiate, lowest lobe longest; disk glandular; ovary bilocular, divided by a false septum almost to apex, ovules numerous, uniseriate in each locule. **Fruits** capsules, longitudinally dehiscent, oblong to obconical, 4-sulcate, beaked; seeds obovate, compressed, winged, fringed, or rarely smooth, testa smooth or rugose. **Distr:** Primarily in Africa, India, and Sri Lanka; China. **Cult:** Propagate from seeds. Requires a rich, moist, well-drained soil in full sun.

S. angolense Welw., *Apont. Phyto-Geogr.*: 588, 1859. (Syn: *S. macranthum*). **Herbs**, erect, annual or perennial, simple or branched, 0.8–3 m

Below: *Sesamum angolense, Sesamum indicum, Martynia annua, Martynia annua* (fruits).

PL

Above: *Torenia fournieri, Torenia travancorica, Campsis grandiflora.*

high, malodorous. **Leaves** subsessile or shortly petiolate; lamina narrowly oblong, narrowly elliptic or oblanceolate, 2–11 × 0.4–4 cm, base cuneate, truncate, margins entire, ± involute, apex retuse or rarely acute, and usually mucronate, abaxially white-tomentose. **Inflorescences** with flowers solitary. **Flowers** reddish, pink, or pale mauve with darker markings; calyx lobes narrowly lanceolate or narrowly ovate, 0.5–1 cm long, base ~2 mm wide, apex acute or acuminate, pubescent; corolla large, 3.5–7 cm long, 2–3 cm Ø at the throat. **Fruits** capsules, narrowly oblong, subquadrangular, 4-sulcate, 1.8–2.8 × 0.5–0.7 cm, gradually narrowed into a flattened broad and short beak, rather densely pubescent; seeds with a double fringe, ~2 × 1.5 mm, faintly rugose. **Distr:** Kenya, Tanzania, Uganda. **Ecol:** In grasslands and open woodlands. 350–2,300 m. **Zones:** 10–12.

S. indicum L., *Sp. Pl.* 2: 634, 1753. (Syn: *S. orientale*). [Benne, Gingelly, Sesame]. **Herbs**, annual, erect, to 1.8 m high; stems branched or unbranched, 4-angled, finely pubescent to glabrescent. **Leaves** opposite or alternate; lamina lanceolate to ovate, variously 3-parted, 4–20 × 2–10 cm, upper cauline leaves oblong to linear-lanceolate, 0.5–2.5 cm wide, base cuneate, margins entire; petiole 3–11 cm on basal leaves. **Inflorescences** with flowers solitary, axillary; pedicels mostly 2–5 mm long. **Flowers** white, cream, pink or mauve-pink with darker markings; calyx persistent, parted nearly to base, lobes linear to linear-lanceolate, 5–8 mm long, villous outside; corolla 1.5–3.3 cm long, ± villous outside, throat 1–1.5 cm wide. **Fruits** capsules, explosively dehiscent, narrowly oblong, 1.5–3 × 0.6–0.7 cm, base rounded, densely pubescent, beak broad, short; seeds horizontally arranged, compressed, ~3 mm long, double fringe conspicuous, testa white, brown, or black, smooth. **Distr:** Origin unknown, probably S Asia. **Ecol:** Original habitat unknown. Fl. Jun–Oct. **Zones:** 10–12. **Info:** Widely cultivated for the edible seeds.

MARTYNIACEAE

Genera 5, **species** 16. **Herbs**, annual or perennial, usually viscid-hairy. **Stipules** absent. **Leaves** alternate or opposite, petiolate; lamina simple, margins entire to undulate, dentate or somewhat lobed. **Inflorescences** racemose, terminal or axillary; flowers pedicellate, often subtended by 1 or 2 large bracteoles, sometimes becoming thick and fleshy with age. **Flowers** bisexual; calyx 5-parted or 5-lobed and split down 1 side, glandular-puberulent outside; corolla bilaterally symmetrical, somewhat 2-lipped, base tubular and expanded to campanulate or funnelform distally, 5-lobed with 2 upper lobes exterior in bud; stamens 4 or 2, filaments equal or of 2 lengths, bithecous; staminodes 1–3; disk annular; ovary superior, unilocular, with 2 intruding parietal placentas and appearing 2- or 4-locular by intrusion of placentas, few- to many-ovuled, style slender, stigma bilobed. **Fruits** woody capsules, usually with distal recurved horns, exocarp fleshy and deciduous, endocarp hard and woody; seeds few to many, sculptured, black, compressed or oblong. **Distr:** Tropical and subtropical America.

Martynia

Species ~3. **Herbs**, coarse, diffusely branched, glandular-pubescent and viscid, strongly scented. **Leaves** opposite or alternate, long-petiolate; lamina subcordate, margins entire or lobed. **Inflorescences** racemes, short, terminal or axillary; flowers 1–bibracteolate at base. **Flowers** large, violet, purple, whitish or mottled; calyx campanulate, inflated, unequally 5-cleft, deciduous; corolla funnelform-campanulate, oblique, decurved, 5 lobes nearly equal, spreading; stamens 4, fertile stamens 2, anthers gland-tipped, their sacs divergent; ovary 1-celled, the 2 parietal placentas intruded and expanded in the center of the cavity into broad surfaces bearing the ovules in 1 or 2 rows. **Fruits** capsules, incurved, beaked, loculicidally 2-valved, exocarp somewhat fleshy, endocarp fibrous, woody, crested below or also above, 4-celled by the extension of placentae; seeds numerous, tuberculate. **Distr:** C America. **Cult:** Propagate from seeds. Requires a moist, well-drained soil in full or partial sun.

M. annua L., *Sp. Pl.* 618, 1753. (Syn: *M. diandra*). [Devil's Claw, Ice Plant, Small-fruit Devil's Claw, Tiger's Claw]. **Herbs**, branched, to 1.5 m high; stem succulent, glandular-tomentose. **Leaves** opposite; lamina subcordate, 7–15 × 7–20 cm, distantly dentate with veins ending in denticles at margin, glandular; petiole slightly shorter than or as long as lamina, glandular. **Inflorescences** 5–15(–20)-flowered, 3–6 cm long, terminal or axillary; bracteoles 1–3, unequal, oblong, 1.2–2.3 cm long, petaloid or lilac; pedicles 1.2–2.8 cm long. **Flowers** showy, lilac, mottled red, blue or yellow inside, ± glandular; sepals 5, free, unequal, oblong-lanceolate, 1–1.5 cm long; corolla tube ~2.7 cm long, gibbous, 5-lobed, lobes purple-tinged, obtuse; fertile stamens 2, filaments ~1.3 cm long; staminodes 2, ~4 mm long; ovary globose, 2 mm Ø, glabrous, style ~2 cm long, stigmas 2, sensitive. **Fruits** capsules, ± ovoid, ~3 cm long, 2-valved, with persistent style splitting into 2 curved, lignified and spiny processes. **Distr:** Mexico to NE Costa Rica. **Ecol:** In deciduous forests; <300(–2,400) m. Fl. Jul–Aug. Fr. Aug–Feb. **Zones:** 10–12.

LINDERNIACEAE

Genera 13, **species** 195. **Herbs**, ephemerals to suffruticose perennials; stems quadrangular. **Leaves** opposite, rarely basally connate; lamina with margins entire or serrate, venation palmate. **Inflorescences** racemose or axillary racemes; bracteoles absent. **Flowers** monosymmetric; corolla strongly 2-lipped, glandular-hairy inside; stamens 4, filaments of abaxial pair geniculate, Z-shaped or with spurlike appendages or 2, abaxial pair reduced to filiform or geniculate staminodes, anthers in pairs, with 2 equal and spreading thecae, unequal mucronate thecae or monothecous anthers; stigma bilobed, sensitive. **Fruits** capsules, septicidal; seeds with endosperm ruminate. **Distr:** Pantropical to warm temperate, mostly New World.

Torenia

Species 40. **Herbs**, glabrous or villous, rarely hispidulous. **Leaves** cauline, opposite, usually petiolate; lamina lanceolate-ovate to orbicular, margins dentate to crenate, apex acuminate or obtuse. **Inflorescences** racemose or in axillary umbellate fascicles, rarely dichotomous and terminal flowers reduced, or 1-flowered; flowers pedicellate. **Flowers** with calyx ribbed or winged, often 2-lipped and short toothed, usually 5-lobed; corolla 2-lipped, lower lip trilobed, lobes subequal, upper lip erect, apex entire, emarginate, or bilobed; stamens 4, anterior filaments appendaged or less commonly unappendaged; posterior stamens included, anthers connivent in pairs, locules often apically confluent; ovary apically usually with short coarse hairs, style apically bilamellate. **Fruits** capsules, oblong, included in persistent calyx, septicidal; seeds numerous, scrobiculate. **Distr:** Mostly in tropical Africa and Asia. **Cult:** Propagate from seeds. Requires a fertile, moist, well-drained soil in partial to full sun.

T. fournieri Linden ex Fournier, *Ill. Hort.* 23: 129, 1876. [Bluewings, Wishbone Flower]. **Herbs**, 15–50 cm high; stems erect, quadrangular, simple or branched above middle, subglabrous. **Leaves** oblong-ovate to ovate, 3–5 × 1.5–2.5 cm, margins coarsely serrate, subglabrous; petiole 1–2 cm long. **Inflorescences** racemes, often terminal; bracts linear, 2–5 mm long; pedicels 1–2 cm long. **Flowers** with calyx ellipsoid, 1.3–1.9 × ~0.8 cm, green or purple-red at apex and margin, 5-winged, wings decurrent, ~2 mm wide, becoming ~3 mm wide in fruit, lips subtriangular, 1.5–1.7 cm long, sometimes apically lobed; corolla 2.5–4 cm long, exceeding calyx by 1–2.3 cm, tube pale violet, upper side yellow, lower lip lobes purple-blue, midlobe with a yellow patch near base, oblong to suborbicular, ~1 × 0.8 cm, subequal, upper lip pale blue, erect, broadly obovate, 1–1.2 × 1.2–1.5 cm, emarginate; stamens unappendaged. **Fruits** capsules, narrowly ellipsoid, ~1.2 cm long; seeds yellow. **Distr:** Cambodia, China, Laos, Taiwan, Thailand, Vietnam. **Ecol:** <1,200 m. Fl. & Fr. Jun–Dec. **Zones:** 9–12.

T. travancorica Gamble, *Fl. Pres. Madras* 957, 1923. (Syn: *T. asiatica* [misapplied], *T. leucosiphon*). **Herbs**; stem decumbent, rooting at lower nodes,

SZ

GHIR

Top row: *Campsis radicans, Catalpa bignonioides, Catalpa longissima, Clytostoma callistegioides.*
Bottom row: *Crescentia alata, Crescentia cujete, Deplanchea tetraphylla.*

diffusely branched, quadrangular, glabrous, often purple-tinged at upper nodes; internodes to 6.5 cm long. **Leaves** ovate to ovate-deltoid, 1.5–5 × 1–3.2 cm, base subcordate, margins coarsely crenate-serrate, apex usually acute, penninerved, scabrous; petiole 0.8–2.2 cm long. **Inflorescences** axillary, flowers solitary; bracts linear, 3–3.5 mm long; pedicel 2.2–3.2 cm long. **Flowers** with calyx tube 0.9–1.1 cm long, lobes lanceolate, 6–8 mm long; corolla 3–3.7 cm long, lower half orange-yellow, upper half pale purple, glabrous, upper lip oblong-orbicular, 1–1.4 cm long, dark purple, lower lip 1.6–1.8 cm across, pale bluish-purple; posterior filaments 4 mm long, anterior ones longer, appendages 5–5.5 mm long; hypogynous disk pale yellow; style glabrous. **Fruits** capsules, linear-cylindric, 0.8–1.5 × 0.2 cm, apex tapered, glabrous, pale brown; seeds ellipsoid, 0.6 mm long, pitted. **Distr:** India, Sri Lanka. **Ecol:** In lowland wet forests. **Zones:** 11–12.

BIGNONIACEAE

Genera 110, **species** 800. **Trees, shrubs**, or more often **lianas** (rarely herbs or twining herbs), often climbing with the aid of twisting stems or tendrils, tendrils simple or distally trifid rarely with sticky pads, often becoming woody; stems terete or angulate, glabrous or pubescent, often with interpetiolar ridges or gland fields at nodes. **Stipules** absent, pseudo-stipules present or absent. **Leaves** opposite or whorled, rarely alternate, simple, bifoliolate, trifoliolate, pinnately compound or palmately compound, occasionally twice compound, petiolate; distal leaflet often replaced by a tendril in climbing species; lamina usually entire, sometimes serrate (not lobed), glabrous to densely puberulent, venation palmate or pinnate. **Inflorescences** cymose, paniculate, racemose, fasciculate, or flowers solitary, terminal, axillary, or cauliflorous; bracts and bracteoles often present, small and caducous; pedicels usually well-developed. **Flowers** bisexual, mostly large and showy; calyx united and tubular to campanulate, usually with 5 lobes or teeth, rarely entire, bilabiate, calyptrate, or spathelike; corolla united and funnelform to campanulate with a basal tube, usually 2-lipped and bilaterally symmetrical, rarely radially symmetrical, lobes 5; stamens usually (2 or)4(or 5), alternate with corolla lobes, filaments usually of 2 unequal pairs, borne on lower half of tube and free, anthers (1)bithecous, free; staminode usually present; disk usually present, annular; ovary superior, bilocular with 2 axile placentas or unilocular with 2–4 intruded parietal placentas, rarely 4-locular, style simple, stigma bilobed, rounded. **Fruits** capsules, mostly 2-valved, septicidally or loculicidally dehiscent, rarely fleshy or hard and indehiscent; seeds mostly flat and winged in capsular fruits, often embedded in pulp or mucilage in indehiscent fruits. **Distr:** Tropical regions worldwide, especially S America.

Campsis

Species 2. **Vines**, woody, deciduous, climbing by aerial roots. **Leaves** opposite, pinnately compound; leaflets with margins serrate. **Inflorescences** shortly paniculate, terminal, or flowers clustered. **Flowers** with calyx campanulate, subleathery, irregularly 5-lobed; corolla red or orange-red, campanulate-funnelform, slightly bilabiate; lobes spreading, semirounded; stamens 4, didynamous, curved, included; ovary bilocular, base surrounded by a large disk. **Fruits** capsules, loculicidally dehiscent; seeds numerous, compressed, membranous, with semitransparent wings. **Distr:** N America (1 sp.); China and Japan (1 sp.). **Cult:** Propagate from cuttings, seeds or by air-layering. Requires a well-drained soil in full sun.

C. grandiflora (Thunberg) Schumann, *Nat. Pflanzenfam.* 4(3b): 230, 1894. (Syn: *Bignonia grandiflora*, etc.). [Chinese Trumpet Creeper, Chinese Trumpet Vine]. **Vines**, robust; stems woody, with few aerial roots. **Leaves** pinnate; rachis 4–13 cm long; leaflets 7–9, ovate to ovate-lanceolate, 3–6(–9) × 1.5–3(–5) cm, base broadly cuneate, margins serrate, apex caudate-acuminate, glabrous, lateral veins 6 or 7 per side; petiolules 0.5(–1) cm long. **Inflorescences** shortly paniculate, large, to 50 cm long, lax, terminal; peduncle 15–20 cm long. **Flowers** trumpet-shaped, 8–10 cm long; calyx campanulate, ~3 cm long, divided halfway, lobes lanceolate, ~1.5 cm long; corolla red adaxially, orange-red abaxially, ~5 cm long; lobes semirounded; stamens inserted at base of corolla tube; filaments linear, 2–2.5 cm long, anthers divergent, yellow; style linear, ~3 cm long, stigma compressed, bilobed. **Fruits** capsules, apex obtuse. **Ecol:** Fl. May–Aug. **Distr:** China, India, Japan, Pakistan, Vietnam. **Zones:** 7–11.

C. radicans (L.) Seem., *J. Bot.* 5: 372, 1867. (Syn: *Bignonia radicans, Tecoma radicans*). [Cow Itch/Vine, Trumpet Creeper/Vine]. **Vines**, vigorous, woody, climbing with self-clinging aerial roots, sometimes growing with a shrubby habit and trailing branches, to 15 m long. **Leaves** pinnate; leaflets 7–11, oval, ovate or elliptic, 2–10 cm long, base gradually narrowed, margins coarsely serrate, apex acuminate or sometimes acute, softly pubescent to glabrate, adaxially green, abaxially paler; petiolules short. **Inflorescences** racemes or corymbs, few-flowered, terminal; pedicels stout, 0.5–2 cm long. **Flowers** trumpet-shaped, smaller than those of *C. grandiflora*; calyx tubular-campanulate, 1.4–1.8 cm long, margins denticulate, lobes triangular or triangular-ovate, much shorter than the tube; corolla red and orange, or yellow inside, tubular-funnelform, 5–9 cm long, limb 3–5 cm Ø, oblique, lobes 5, spreading, suborbicular or reniform. **Fruits** capsules, spindle-shaped or cylindric, oblong, 10–20 cm long, keeled, with 2 ridges running lengthwise. **Distr:** N America. **Ecol:** In open woods, swamps and thickets. Fl. May–Sep. **Zones:** 5–12.

Catalpa

Species ~13. **Trees**, deciduous. **Leaves** simple, purple glandular-punctate at vein axils abaxially. **Inflorescences** paniculate, corymbose, or racemose, terminal. **Flowers** bell-shaped; calyx bilabiate or irregularly divided, globose in buds; corolla campanulate, bilabiate, upper lip bilobed, lower lip trilobed; fertile stamens 2, included, inserted at base of corolla tube; staminodes 3; disk conspicuous; ovary bilocular, ovules several. **Fruits** capsules, loculicidally dehiscent, long-terete, to 75 cm long, pendent; valves thin, septum slender; seeds in several rows, rounded, thinly membranous, with hair tufts at both ends. **Distr:** N America and E Asia. **Cult:** Propagate from seeds or softwood cuttings. Requires a rich, moist, well-drained soil in full or partial sun.

C. bignonioides Walter, *Fl. Carol.* 64, 1788. [Bean Tree, Indian Bean Tree, Southern Catalpa]. **Trees**, broadly conical, 9–18 m tall; branches wide-spreading; bark pinkish-brown or gray, flaky or ridged. **Leaves** large, often 3-whorled; lamina broadly ovate, 10–20 cm, base cordate, margins entire or shallowly angularly lobed, apex abruptly acuminate, abaxially downy, eventually hairless, unpleasant odor when crushed; petiole 7–15 cm long. **Inflorescences** panicles, trichotomous, many-flowered, large, 15–20 cm long, erect. **Flowers** bell-shaped; calyx purple; corolla white with purple markings and yellow-spotted, throat yellow, limb oblique, 4–5.3 cm Ø, lobes undulate, lower lobe entire. **Fruits** capsules, beanlike, slender, elongated, 15–38 cm long, pendulous; seeds ~2.5 cm long, with a papery wing. **Distr:** USA (Georgia). **Ecol:** Near riverbanks. Fl. Apr–Jun. **Zones:** 5–11.

C. longissima Sims, *Bot. Mag.* 27: sub t. 1094, 1808. (Syn: *Bignonia longissima, Macrocatalpa longissima*). [French Oak, Indian Bean, West Indian Catalpa, Yokewood]. **Trees**, evergreen, fast-growing, 3–15 m tall; bark grayish-brown, vertically furrowed, ± corky. **Leaves** opposite or 3-whorled; lamina simple, usually elliptic-lanceolate or oblong lanceolate, 2.5–15 × 1.2–4.6 cm, base acute, margins undulate, apex acuminate, glabrous or sparsely hairy; petiole slender. **Inflorescences** panicles, few-flowered, terminal or subterminal. **Flowers** with calyx deeply bilobed; corolla of fused petals, bell-shaped, 2.5–3.5 cm long, white or pinkish-white with yellow and sometimes purple markings in the throat, somewhat 2-lipped, upper lip bilobed, lower lip trilobed, lobes rounded, white to pale pink with purple longitudinal lines running toward the throat, reflexed, margins ruffled; stamens inserted. **Fruits** cylindrical, 30–60 × 0.6–0.8 cm; seeds numerous. **Distr:** West Indies (Jamaica and Hispaniola, Lesser Antilles). **Ecol:** Fl. throughout the year. **Zones:** 9–11.

Clytostoma

Species 9. **Lianas**, climbing with simple tendrils; stems 4-angled or terete, nodes lacking gland fields. **Pseudostipules** usually a small condensed axis of subulate cataphylls. **Leaves** trifoliolate (rarely simple), with or without a terminal simple tendril (by modification of the terminal leaflet), petiolate; leaflets with margins entire, venation pinnate. **Inflorescences** few-flowered fascicles, cymes, panicles, or flowers solitary, terminal or axillary, flowers subtended by bracts similar to the cataphylls. **Flowers** with calyx cupular or campanulate, truncated distally and minutely 5-denticulate or with 5 linear lobes, subglabrous to minutely lepidote or puberulent; corolla tubular-funnelform, white to purple, glandular-puberulent or minutely hairy outside; stamens 4, included, filaments of 2 lengths, anthers glabrous, thecae straight and divaricate; staminode present; disk absent; ovary bilocular, ovules 2(–4)-seriate per locule. **Fruits** capsules, ellipsoid to suborbicular, valves convex or flattened parallel with the septum, woody and echinate with curved spines; seeds transverse-oblong, corky, flattened-winged or wingless. **Distr:** Mexico to Brazil. **Cult:** Propagate from seeds or cuttings. Requires a moist, well-drained soil in full or partial sun.

C. callistegioides (Cham.) Baill., *Hist. Pl.* (Baillon) 10: 34, 1888. (Syn: *Bignonia speciosa*). [Argentine Trumpet Vine, Love Charm, Painted Trumpet Vine]. **Vines**, slender, to 4 m long; stems and leaves of new growth finely soft-hairy, eventually glabrescent. **Leaves** trifoliate, terminal leaflet often modified into a tendril, thus appearing 1–2-foliate; leaflets oblong, obovate-oblong or oblong-elliptic, 4–7 × 1.8–2.8 cm, base somewhat oblique, subacute, margins wavy, apex acute or shortly acuminate, adaxially glabrous, shiny, abaxially with netted venation. **Inflorescences** with flowers solitary or paired; pedicels long. **Flowers** funnelform; calyx green, 6–9 mm long, teeth 5, filiform or subulate, 4–5 mm long; corolla pale purple or pale lilac-purple inside, outside white or tinged with the interior color, 5–7 cm long, minutely puberulent outside, tube laterally flattened, lobes 5, ± spreading, unequal, margins ruffled, apices somewhat rounded, with dark purple lines emanating from the middle of the lobes down to the throat, throat yellow-tinged. **Fruits** capsules, flattened, 7–8 × 4–4.5 cm; seeds ovate, flat, winged. **Distr:** Argentina, Brazil, Paraguay. **Ecol:** In margins of forests. Fl. Mar–Aug. Fr. Nov. **Zones:** 10–11.

Crescentia

Species 6. **Trees**, small to medium-sized; branches many, thick, twisted; leafy stems becoming terete with enlarged nodes. **Leaves** alternate, simple or (1 or 2)3-foliolate; leaflets with margins entire, subglabrous or sparsely puberulent, venation pinnate; petiole and rachis often broadly winged and leaflike. **Inflorescences** fascicles, 1–3-flowered, borne on older thick branches or cauliflorous on trunks; peduncles absent; pedicels short, thick. **Flowers** with calyx large, tubular, usually splitting into 2 halves at maturity, thick, glabrous or minutely puberulent; corolla broadly campanulate, bilabiate, narrowed only at base, thick, outside usually with minute peltate hairs, tube with a transverse fold midway across the lower side, distally 5-lobed or variously erose along the margin; stamens 4, subexserted, filaments of 2 lengths; disk large; staminode present; ovary lepidote, 1-locular, ovules multiseriate on 4 parietal placentas, stigma flat, expanded. **Fruits** large pepos (a gourdlike calabash), pericarp hard, corky, subglobose to ovoid, pulpy inside; seeds angular, wingless, <1 cm long, embedded in pulp. **Distr:** Mexico, West Indies, C America, Amazonia. **Cult:** Propagate from seeds, cuttings or by air-layering. Requires a moist, well-drained soil in full sun; some species are moderately drought- and salt-tolerant.

C. alata Kunth, *Nov. Gen. Sp.* 3: 158, 1819. (Syn: *Parmentiera alata, C. ternata*). [Mexican Calabash Tree]. **Trees,** to 10 m tall; trunks to 25 cm dbh. **Leaves** mostly in fascicles of 3–9, simple or (1 or 2)3-foliolate, simple leaves sessile, compound leaves with winged petiole 2.5–1 1 × 0.4–1.1 cm, petiolules absent; lamina oblanceolate to narrowly obovate (small leaves often rounded), 4–15 × 0.5–3.5 cm; leaflets oblanceolate, 1.5–10 × 0.4–2.3 cm, base acute or cuneate, gradually narrowed, apex rounded or emarginate, glabrous. **Inflorescences** 1–3-flowered, borne directly on larger stems and trunks; pedicels 0.5–1 cm long, subtended by bracts, 1.5 mm long. **Flowers** malodorous; calyx 1.4–2.8 × 0.9–1.4 cm, usually splitting into 2 subequal halves, greenish; corolla 4–6 cm long, tube narrowed near base, 2.2–3 cm Ø. greenish-purple to reddish-white, 5-lobed or margin irregularly toothed; filaments ~3, 1.8 cm long, thecae 5–6 mm long. **Fruits** globose, 7–10 cm Ø, greenish, smooth; seeds 6–7 × 7–9 mm. **Distr:** Mexico, along the Pacific slope to NW Costa Rica. **Ecol:** In seasonally dry deciduous forests; <800 m. Fl. throughout the year. **Zones:** 10–12.

C. cujete L., *Sp. Pl.* 2: 626, 1753. [Calabash Tree, Gourd Tree, Rum Plant]. **Trees**, to 10 m tall; crown dense; trunks to 40 cm dbh; branches thick, often crooked. **Leaves** usually fasciculate on elevated alternate nodes, 1–15 and varying in size at the same node, unifoliolate, sessile; lamina oblanceolate to narrowly obovate, 3–22(–26) × 1–6(–7.5) cm, base gradually narrowed, acute or cuneate, apex acute to acuminate or rounded (emarginate), ± glabrous, veins 5–14 per side. **Inflorescences** of 1 or 2 flowers, borne on older branches and trunks, pedicels 0.9–3 cm long, ± glabrous. **Flowers** malodorous; calyx 1.8–3.4 cm long, splitting into 2 subequal lips 1.2–2.4 cm wide; corolla 5.5–7 cm long, dull white to yellowish-white, with purple lines, tube 1.8–2.5 cm Ø, lobes triangular, tip narrow, ± subglabrous outside; filaments ~30, 1 cm long, anthers 6–9 mm long, slightly divergent. **Fruits** subglobose or ovoid, to 30 × 13–25 cm, smooth, subglabrous; seeds 7–8 × 4–6 mm. **Ecol:** 1,200 m. Fl. Irregularly throughout the year. **Distr:** SE Mexico. **Ecol:** In both deciduous and wet evergreen forests; to 1,200 m. Fl. irregularly throughout the year. **Zones:** 10–12.

Deplanchea

Species 5. **Trees**; stems glabrous to pubescent. **Leaves** 3–4-whorled, ± tufted at end of branches; lamina simple, margins entire. **Inflorescences** thyrses, few-flowered, terminal, horizontal to erect. **Flowers** erect, showy, yellow; calyx campanulate, 5-dentate; corolla yellow, tubular to campanulate; stamens 4, exserted, anthers glabrous, thecae divergent; staminode rarely present; disk annular; ovary ovoid-elliptic, glabrous, ovules numerous. **Fruits** capsules, ellipsoid, not subtended by calyx, valves glabrous; seeds with a hyaline wing all around. **Distr:** Australia, Malesia, New Caledonia, New Guinea. **Cult:** Propagate from seeds, cuttings or by air-layering. Requires a moist, well-drained soil in full sun.

D. tetraphylla (R. Br.) F. Muell., *Second Syst. Cens. of Austr. Pl.* 167, 1889. [Burmese Pagoda Tree, Golden Bouquet Tree]. **Trees**, 4–25 m tall. **Leaves** opposite or 3–4-whorled, clustered at ends of branches; lamina simple, usually obovate, 10–35(–60) × 5.5–18(–30) cm, base shallowly cordate to truncate, apex rounded, tip slightly retuse, ± hairy; petiole 1.2–5 cm long. **Inflorescences** racemes or panicles, axillary or terminal, indeterminate, with an apical ± flat 'platform'; peduncle 4–12 cm long; rachis 3–9 cm long; branches 2–7.5 cm long. **Flowers** with calyx green or yellow, 1.2–1.4 cm long; corolla strongly zygomorphic, trumpet-shaped, tube short, mouth laterally compressed, closed, petals 5, yellow, 2–3.5 cm long; stamens and style long-exserted, ± erect; stamens 4; staminode absent; nectar dark brown to almost black, exposed in the spoon-shaped lowermost corolla lobe, acting as a visual cue for pollinators (Weber and Vogel 1986). **Fruits** capsules, brown, flat, 8–11(–15) × ~2.6 cm; seeds several,

RT

TNR

Above: *Dolichandrone spathacea*, *Eccremocarpus scaber*, *Fernandoa adenophylla*, *Fernandoa magnifica*.

obovate, ~1.4 cm long, brown, surrounded by a papery, ± transparent wing. **Distr:** Australia, New Guinea. **Ecol:** In rainforests, monsoon forests, littoral rainforests; 0–670 m. Fl. May–Nov. Fr. Nov–Dec. **Zones:** 10–12.

Dolichandrone

Species 9. **Trees**. **Leaves** usually opposite, pinnate, rarely simple; leaflets with margins entire, rarely serrulate. **Inflorescences** racemes or panicles, few-flowered, terminal. **Flowers** large, usually >10 cm long; calyx spathaceous, curved, caducous; corolla opening toward evening, lower portion of corolla tube very long, cylindrical, much exceeding calyx, upper portion inflated and funnelform, lobes 5, subequal, rounded, crisped; stamens 4, didynamous, equally inserted at base of swollen portion of tube, anthers very large, bilocular, locules divergent, glabrous; staminode small; disk large, annular, cushionlike; ovary linear-oblong, bilocular with 2 false septa, appearing 4-locular, style slender, stigma 2-lipped. **Fruits** capsules, linear, curved, subcylindrical or compressed, loculicidally 2-valved, valves perpendicular to septum; seeds numerous, corky or membranous-winged. **Distr:** E Africa, India, SE Asia, Malesia, New Caledonia, New Guinea, Australia. **Cult:** Propagate from seeds, cuttings or by air-layering. Requires a moist, well-drained soil in full sun.

D. spathacea (L. f.) K. Sch., *Fl. Kais. Wilh. Land*: 123, 1889. (Syn: *D. alternifolia*, *D. rheedii*). [Mangrove Trumpet Tree]. **Trees**, evergreen, sometimes deciduous, to 24 m tall, usually much smaller. **Leaves** opposite or alternate, pinnate, with a terminal leaflet; leaflets 4–9, elliptic, 5–15 × 2.5–6.25 cm, base asymmetric, apex pointed, glabrous, abaxially with domatia as foveoles with a tuft of hairs; rachis and petiole 11–30 cm long, pulvinus at base; lateral petiolules 0.5–1.2 cm long. **Inflorescences** clusters, short, terminal; opening 1 flower at a time. **Flowers** white, very fragrant, 12.5–17.5 cm long; calyx conical and closed in bud, filled with water and apex recurved, splitting open along one side; corolla tube 12–18 cm long, bell-shaped near mouth, limb 7.5–12.5 cm Ø, lobes 5, margins frilled; stamens 4, not exserted; style exserted. **Fruits** capsules, hanging in clusters, flattened-cylindrical, 39–49 × 1.8–2.5 cm, ± arcuate o or twisted; seeds numerous, with thin corky wing, 1.2–1.8 × 0.6–0.8 cm. **Distr:** Australia, New Caledonia, New Guinea, Malesia, SE Asia, India. **Ecol:** In rainforests and mangroves. Fl. May–Nov. Fr. Apr–Nov. **Zones:** 10–12.

Eccremocarpus

Species 6. **Climbers** or **scandent shrubs**, evergreen or deciduous; stems sulcate-angulate. **Leaves** bipinnate or bipinnately lobed, with terminal tendrils. **Inflorescences** racemes, laxly few-flowered, terminal, pendulous, opposite. **Flowers** with calyx campanulate, 5-lobed; corolla tubular, entire or 2-lipped, scarlet, orange or yellow, tube long and narrow, widening above, throat swollen but abruptly contracted at mouth; stamens 4, subexserted, anthers glabrous, thecae straight; ovary ovoid, contracted at base, lepidote, ovules multiseriate on 2 parietal placentas; disk annular. **Fruits** capsules, ovoid to ellipsoid, flattened, valves remaining apically joined, splitting below, woody. **Distr:** Peru, S Colombia, Chile, Argentina. **Cult:** Propagate from seeds or cuttings. Requires a fertile, well-drained soil in full sun.

E. scaber Ruiz and Pav., *Syst. Veg. Fl. Peruv. Chil.* 157, 1798. [Chilean Glory Flower]. **Climbers**, to 3 m long; stems quadrate, sulcate. **Leaves** bipinnate or bipinnately lobed, to 7 cm long, pulverulent; leaflets or lobes alternate, ovate, to 1–2 × 1–2 cm, bases obliquely cordate, margins entire, toothed or serrated, apices acute, glabrous or hairy; tendrils 2–3-parted. **Inflorescences** racemes, spreading-ascending, 10–15 cm long; bract basal, linear, acute; pedicels 2–3.5 cm long. **Flowers** turning to one side, 2.5–4 cm long; corolla 2–3 cm long, scarlet to orange or yellow, ventrally enlarged, contracted below, tube glandular as anthers and calyx, limb 5-lobed, short, revolute; anthers with divaricating lobes, connective glandular; ovary oblong, style slender, stigmas 2, diverging. **Fruits** capsules, 4–5 cm long, inflated-membranous, attenuate to stipe, pendulous, wrinkled, 2-valved; seeds with wings lustrous, margins sinuate. **Distr:** Chile. **Ecol:** In coastal mountains, valleys; 500–2,000(–3,200) m. Fl. throughout most of the year. **Zones:** 9–11.

Fernandoa

Species ~14. **Trees**, medium-sized. **Leaves** large, pinnately compound; leaflets 2–5(or 6) pairs, margins entire. **Inflorescences** panicles, cymose, terminal or axillary, pubescent. **Flowers** large; calyx large, not spathaceous, glandular, tubular or campanulate, usually 2–5-lobed, subzygomorphic or regular, persistent; corolla funnelform or campanulate, lobes 5, subrounded, sinuate or crenate; stamens 4, didynamous, anthers 2-celled, slightly divergent, included or subexserted; staminodes small; disk large, annular, occasionally dentate; ovary appressed-hairy, terete, glabrous or pubescent, bilocular, ovules numerous, biseriate. **Fruits** capsules, cylindrical, loculicidally dehiscent, 4-angular; valves thin leathery; septum thick, smooth, vertical with fruiting valve; seeds in 2 rows, numerous, discoid, nearly square, narrowly membranous-winged at both ends. **Distr:** Tropical Africa, Madagascar, China, SE Asia. **Cult:** Propagate from seeds, cuttings or by air-layering. Requires a moist, well-drained soil in full or partial sun.

F. adenophylla (Wall. ex G. Don) V. Steenis, *Blumea*. 23. no.1: 135, 1976. (Syn: *Bignonia adenophylla*, etc.). **Trees**, deciduous, 10–15 m tall. **Leaves** opposite, pinnate, 20–30 cm long; pinnae rusty-tomentose; leaflets 5–7, lateral, elliptic-oblong, 6–18 × 6–15 cm, margins entire, apex acute, rusty-tomentose (especially abaxially). **Inflorescences** large, panicles, laxly few-flowered, erect, brown-tomentose. **Flowers** large; calyx 3–5-lobed, 2–2.5 cm long, rusty-tomentose; corolla tube 3.5 cm long, tomentose outside, lobes subequal, obtuse; style 3.5 cm long, stigma elliptic, 5 mm long. **Fruits** capsules, very long, often curved. **Distr:** Myanmar, the Andaman Islands, E Bengal and Assam. **Ecol:** Fl. Jul–Nov. **Zones:** 10–12. **Info:** The flowers open after dusk and fall before midday, and are bat-pollinated.

F. magnifica Seem., *Journ. Bot.* 8: 280, 1870. (Syn: *Heterophragma longipes*). **Shrubs** or **trees**, 2–20 m tall. **Leaves** pinnate, 18–30 × 12–23 cm; rachis 5–25 cm long; leaflets (4–)5–7, ovate to oblong or lanceolate, 3–14(–16) × 2–6 cm, to 19.5 × 8 cm in saplings, base broadly cuneate to rounded, margins usually entire, apex attenuate to long-acuminate, ± glabrous; petiole (1–)2.5–9 cm long; terminal petiolule to 2.8 cm long, lateral leaflets ± sessile. **Inflorescences** racemose cymes, 6–10 cm long; peduncles (4–)5–17(–21) cm long, thick; pedicels 5–7(–10) cm long, bent around below calyx. **Flowers** large, showy; calyx campanulate, 1.3–2.3 cm long, irregularly 3–5-lobed; lobes broadly ovate-triangular; corolla orange-red or crimson, throat yellow, ± crimson-lined, 5–11 cm long, 5-lobed, lobes unequal, broadly rounded; filaments to 6.5 cm long, adnate to corolla for ~8 mm, thecae 0.9–1.2 cm long; staminode 2.5 mm long; ovary cylindrical, 0.6–1.8 cm long, glabrous, style 4–8 cm long, stigma lobes elliptic, 2–3 mm long. **Fruits** terete, 33–54 × 1–2 cm, spirally twisted, curved; seeds 0.7–1.2 × 1.8–3.6 cm. **Distr:** Kenya, Malawi, Mozambique, Tanzania, Zimbabwe. **Ecol:** In dry forests, woodlands; 50–450 m. Fl. Aug–Oct. **Zones:** 11–12.

Jacaranda

Species 49. **Trees** or **shrubs**; nodes without gland fields or interpetiolar lines. **Pseudostipules** absent. **Leaves** opposite, usually bipinnate, sometimes simple, often large, petiolate; lateral pinnae opposite or alternate; leaflets ± sessile, often strongly asymmetric, margins usually entire, venation pinnate. **Inflorescences** panicles or thyrses (racemes), many-flowered, terminal, axillary or from older leafless nodes, pedunculate; pedicels subtended by enlarged ebracteate nodes. **Flowers** showy, calyx broadly cupulate to tubular-campanulate, short, 5-lobed to denticulate or subentire; corolla campanulate above a narrowed base, slightly bilabiate, blue or lavender to magenta (white), tube contracted near base; stamens 4, filaments of 2 lengths, anthers glabrous; staminode usually exceeding stamens, glandular-hairy; disk pulviniform; ovary bilocular, ovules ~8-seriate in each locule. **Fruits** capsules, oblong or rounded, strongly flattened; seeds thin, flat, with translucent or brown circumferential wings. **Distr:** Mexico to Argentina. **Cult:** Propagate

Above: *Jacaranda mimosifolia, Kigelia pinnata, Macfadyena unguis-cati, Mansoa hymenaea.*

from seeds or half-hardened cuttings. Requires a fertile, moist, well-drained soil in full or partial sun.

J. mimosifolia D. Don, *Bot. Reg.* 8: tab. 631, 1822. (Syn: *J. ovalifolia*). [Blue Haze Tree, Brazilian Rosewood, Fern Tree]. **Trees**, 8–15 m tall. **Leaves** bipinnate, 13–34 cm long, lateral pinnae 3–10 cm long, narrowly winged; leaflets 13–41, opposite or alternate, subsessile, elliptic to elliptic-obovate, 0.5–1.4 × 0.1–0.4 cm, terminal leaflets to 2.2 × 0.6 cm, base obtuse or cuneate, slightly asymmetric, apex acute to acuminate, glabrous, veins 3–7 per side; petiole to 4 cm long. **Inflorescences** open panicles, 9–38 cm long, terminal, branches well-separated, opposite; peduncle 3.4–8.8 cm long; pedicels 1–2 mm long. **Flowers** with calyx cup-shaped, 1–2 × ~2 mm, with 5 triangular or acute teeth to 1 mm long; corolla tubular-campanulate, 3–4.5 cm long, bluish-lavender or lavender, base densely puberulent, tube 0.9–1.3 cm Ø near the mouth, lobes 3–8 mm long; filaments ~16, 1.3 cm long; staminode 2–2.5 cm long. **Fruits** orbicular to oblong-rounded, 3–7 × 3.7–6 cm, lenticular, edges flattened; seeds 0.9–1.3 × 1.1–1.9 cm, differentiated from the transparent circumferential wing. **Distr:** N Argentina, Bolivia. **Ecol:** Fl. Jan–Jul. **Zones:** 10–12.

Kigelia

Species 1. *Generic description same as specific.* **Cult:** Propagate from seeds. Requires a rich, moist, well-drained soil in full or partial sun.

K. pinnata (Jacq.) DC., *Rev. Bign. (Bibl. Univ. Geneve)* 24, 1838. (Syn: *Crescentia pinnata, K. africana*). [Sausage Tree]. **Trees**, 6–12(–20) m tall; twigs often hollow. **Leaves** pinnate, to 50 cm long; leaflets usually 7 or 9; lamina elliptic-oblong to elliptic-obovate, (2–)5–16 × (1.5–)3–7 cm, base asymmetric in lateral leaflets, rarely rounded/cuneate, margins entire or dentate distally, apex obtuse to shortly acuminate, secondary veins 7–11 per side; petiole 6–14 cm long; petiolules of lateral leaflets 1–4 mm long. **Inflorescences** to 40 cm long, terminal, pendulous; peduncle ~20 cm long; lateral branches to 9 cm long; pedicels 1–1.4 cm long, flowers held horizontally. **Flowers** large; calyx 2–3 × 1.2–2 cm, base truncate, lobes to 1.2 cm long, coriaceous; corolla 6–10 cm long, tube ~1.4 cm Ø at base, narrowing to 1.1 cm and then opening to 4–6 cm Ø, dark maroon or wine-red; anthers ~8 mm long; ovary 1–1.1 × 0.3 cm. **Fruits** tubular-oblong, 25–50(–80) × 5–12 cm, pendulous, glabrous. **Info:** Flowers are malodorous, night-blooming and are pollinated by bats, thus a few flowers remain open and intact till morning.

Macfadyena

Species 4. **Lianas**, tendrils with hardened curved trifid tips. **Pseudostipules** small, lanceolate to ovate. **Leaves** dimorphic, juvenile form smaller, opposite, bifoliolate, petiolate, often with a tendril; lamina with margins entire, venation pinnate. **Inflorescences** with flowers 1 or 2, or 3–15 on short cymes or reduced panicles, axillary or terminal on short-shoots. **Flowers** variable; calyx tubular or campanulate, lobed or variously split to spathelike, usually thin; corolla tubular-campanulate or tubular-funnelform, 2-lipped, bright yellow, glabrous outside, lobes 5, rounded; stamens 4, filaments of 2 lengths, anthers glabrous, thecae straight, divaricate, staminode present; disk annular-pulvinate; ovary biloculed, ovules 2–4-seriate in each locule. **Fruits** capsules, elongate-linear, valves parallel to septum, flat, smooth; seeds flat, transversely oblong, central area not clearly differentiated from the 2 thin lateral wings. **Distr:** Mexico and West Indies to Uruguay and Argentina. **Cult:** Propagate seeds or semiripe cuttings. Requires a moderately fertile, moist, well-drained soil in full sun. **Info:** The plants climb up tree trunks using their short, 3-clawed tendrils as grappling hooks; adventitious roots also aid in adhering to their support.

M. unguis-cati (L.) A. Gentry, *Brittonia* 25: 236, 1973. (Syn: *Bignonia unguis-cati*, etc.). [Cats' Claw]. **Lianas**, deciduous, to 30 m tall, 7 cm Ø; tendrils 2–5 cm long, with 3 terminal 'claws' to 1.4 cm long. **Pseudostipules** lanceolate to ovate. **Leaves** dimorphic, bifoliolate; lamina narrowly ovate or elliptic, 5–16 × 1–7 cm, base rounded to acute, apex acuminate, tip narrowed to 2 cm long, glabrous, veins 4–8 per side; petiole 1.1–2.7(–4.7) cm long; petiolules to 1.6(–2.5) cm long. **Inflorescences** panicles, to 15-flowered, terminal on short-shoots or axillary to leafless nodes; peduncles to 2.8 cm long; pedicels to 1 cm long. **Flowers** with calyx 0.6–1.5 × 0.7–1.2 cm, cupular to campanulate, ± glabrous; corolla tubular-campanulate or funnelform, 4.2–8 cm long, deep yellow or yellow-orange, glabrous outside, mouth 1.2–1.8 cm wide, lobes 0.9–1.8 cm long; filaments ~20, 1.4 cm long; staminode rarely forming a fifth stamen. **Fruits** capsules, linear, 15–45+ × 1.1–1.9 cm, ends tapered; seeds 0.9–1.4 × 4–6.5 cm. **Distr:** Mexico and West Indies to Argentina. **Ecol:** In seasonally dry deciduous forests; 10–900(–1,300) m. Fl. Feb–Jun. **Zones:** 10–12.

Mansoa

Species 15. **Lianas**, climbing with distally trifid tendrils or simple tendrils with a terminal peltate disk. **Pseudostipules** well-developed to inconspicuous. **Leaves** opposite, 2- or 3-foliolate; lamina margins entire, glabrous or puberulent, venation pinnate or subpalmate; petiole sometimes with a glandular field at apex. **Inflorescences** panicles, racemes or small corymbs, usually few-flowered, axillary to distal leaves or leafless nodes (terminal); pedicels subtended by minute bracts or ebracteate. **Flowers** showy; calyx cupulate or tubular-campanulate, 5-lobed or margin truncate, often with plate-shaped glands on distal surface; corolla narrowly tubular at base, distally tubular-funnelform to tubular-campanulate, white to reddish-purple; stamens 4, filaments of 2 lengths; staminode present; disk annular-pulvinate; ovary cylindric, lepidote to papillate, bilocular, ovules 2–4-seriate in each locule. **Fruits** capsules, linear-oblong, woody, thin, flat to thick and terete, surfaces smooth or with elevated spines; seeds flat with 2 thin wings, or corky wingless. **Distr:** S America. **Cult:** Propagate from seeds, cuttings or by air-layering (roots easily in water). **Info:** This genus now includes species previously placed in *Pseudocalymma* and *Pachyptera.* Plant parts of *M. hymnaea* and *M. standleyi* has a garlic smell when crushed.

M. hymenaea (DC.) A. Gentry, *Ann. Missouri Bot. Gard.* 66: 782, 1979. (Syn: *Bignonia hymenaea*, etc.). [Garlic Vine]. **Lianas**; tendrils 10–15 cm long. **Pseudostipules** ~3 mm long. **Leaves** 2-foliolate; leaflets ovate to ovate-oblong or suborbicular, 4–14 × 2–8 cm, base obtuse to rounded-truncate (subcordate), apex shortly acuminate to bluntly obtuse, glabrous, veins 3–10 per side; petiole to 3.8 cm long; petiolules to 2.6 cm long. **Inflorescences** panicles, axillary or terminal, 4–10 cm long; peduncles 1–5 cm long; pedicels to 2 cm long. **Flowers** with calyx cupulate, 4–7 × 4–6 mm, margin subentire or minutely 5-lobed, ciliolate; corolla tubular-campanulate (slightly 2-lipped), 3.7–6.5 cm long, white to lavender or dark pink, tube to 1.4 cm Ø, lobes to 1.4 cm long; filaments ~18, 1.3 cm long, thecae ~3 mm long. **Fruits** capsules, linear-oblong, 15–34 × 1.5–2.3 cm, flat, smooth; seeds 1.1–1.5 × 3.4–6.2 cm, not clearly differentiated, wings translucent. **Distr:** N Mexico to SE Brazil. **Ecol:** Usually in seasonally dry deciduous forests; 10–1,300 m. Fl. Dec–Mar. **Zones:** 10–12.

Markhamia

Species ~10. **Trees**, evergreen. **Leaves** opposite, pinnately compound, large; leaflets few, oblong to lanceolate. **Inflorescences** racemes, showy, terminal, borne at end of branches. **Flowers** yellow or yellow-red, trumpet- to bell-shaped; calyx closed in bud, laterally divided to base, spathelike at anthesis, floccose to lanate-woolly; corolla lobes 5, subequal, rounded, tube short; stamens 4, didynamous. **Fruits** capsules, loculicidally dehiscent, long, terete, compressed, yellow-brown floccose to lanate-woolly, septum woody, compressed, with 1 concave midrib; seeds long-ellipsoid, 2-rowed in each locule, both ends with transparent and membranous wings. **Distr:** Mostly in tropical Africa, few in Asia. **Cult:** Propagate from seeds or cuttings. Requires a moist, well-drained, humus-rich soil in full or partial sun.

M. lutea (Bentham) K. Schumann, *Naturl. Pflanzenfam.* iv . 3b: 342, 1895. (Syn: *M. hildebrandtii*, etc.). [Nile Tulip Tree]. **Shrubs** or **trees**, 5–15 m tall; trunk to 1 m dbh; crown narrow, irregular; branchlets finely scaly, becoming glabrous; bark reddish-brown, finely cracked. **Leaves** pinnate, 30–50 cm long, often in bunches; leaflets 7–13,

Above: *Markhamia lutea, Markhamia obtusifolia, Millingtonia hortensis, Oroxylum indicum.*

variable in size, to 10 cm long, base narrowed, apex wider, veins prominent. **Inflorescences** corymbose panicles or clusters, terminal and axillary; buds furry, splitting on one side. **Flowers** bright yellow marked with purple, >5 cm long; calyx incompletely spathelike; corolla trumpet-shaped, with 5 frilly orange-red stripes in the throat. **Fruits** capsules, falcate, narrow, 40–90 cm long, hanging in spiral clusters, splitting while on the tree; seeds 3 cm long, including the flat hyaline wing at each end. **Distr:** Tropical Africa. **Ecol:** In fringing and savanna forests; 900–2,000 m. Fl. Mar, Nov–Dec. Fr. Feb, Nov–Dec. **Zones:** 11–12.

M. obtusifolia (Baker) Sprague, *Kew Bull.* 1919: 312, 1919. (Syn: *Dolichandrone obtusifolia*, etc.). [Golden Bean Tree]. **Shrubs** or **small trees**, 1.5–15 m tall; branchlets initially golden-brown velvety-tomentose. **Pseudostipules** 0.5–1.3 cm long. **Leaves** pinnate, 18–56 cm long; leaflets (1)3–5(6), sessile, unequal, elliptic to oblong, ovate or obovate, (4.5–)6.5–17(–23) × 2.5–9.5(–13.5) cm, base usually obtuse, margins usually entire, apex usually obtuse to rounded, adaxially pubescent, abaxially densely woolly; petiole 2–9.5 cm long. **Inflorescences** panicles, densely branched, many-flowered, 10–30(–40) cm long, terminal, tomentose; pedicels 1–2.3(–2.8) cm long; bracts to 1(–1.5) cm long. **Flowers** with calyx 1.9–3.3 × 1.5–2.5 cm, deeply split, densely tomentose; corolla bright yellow, funnelform to campanulate, lobes spreading, subcircular, 1.7–3 cm Ø, 3 lower ones streaked brown-reddish, with glands near mouth, tube 2.5–4.5 cm long; filaments 1.5–2 cm long; disk 4–5 mm Ø; ovary 5–8 mm long; style 1.7–2.7 cm long. **Fruits** capsules, dehiscent, 20 × 1.8–2.5 cm, ± straight, flattened, velvety-tomentose; seeds 0.7–1.2 × 3.4 cm, winged. **Distr:** Angola, Namibia, Zaire and Kenya to S Africa. **Ecol:** In woodlands, xerophytic forests, savannas; <2,400 m. **Zones:** 11–12.

Millingtonia

Species 1. *Generic description same as specific.* **Cult:** Propagate from suckers, cuttings or air-layering. Requires a fertile, moist, well-drained soil in full sun.

M. hortensis L. f., *Sp. Pl.* 291, 1782. [Cork Tree, Tree Jasmine]. **Trees**, 8–25 m tall; trunk very straight. **Leaves** opposite, 2- or 3-pinnately compound, 0.4–1 m log; leaflets elliptic, ovate or ovate-oblong, (2–)5–7 × 1.5–4 cm, base rounded, oblique, margins entire, apex acuminate, glabrous; petiolule ~1 cm long. **Inflorescences** cymose-paniculate, terminal, ~25 cm across; peduncles and pedicels pubescent; bracts and bractlets deciduous; pedicels ~1 cm long. **Flowers** fragrant; calyx small, cupular, apex subtruncate, 2–4 × 2–4 mm, sinuolate-lobed, lobes slightly reflexed; corolla white, bilabiate, upper lip bilobed, lower lip trilobed, tube 3–7 × 0.2–0.3 cm, lobes ovate-lanceolate, 1–2 cm long; stamens 4, inserted high in corolla tube; disk annular-cupular; ovary glabrous, sessile, ovoid, ovules numerous, 4-rowed, style long, stigma bilobed. **Fruits** capsules, septicidally dehiscent, linear, 30–35 × 1–1.5 cm, compressed; seeds discoid-oblong, compressed, 1.5–3.5 × 1–1.5 cm, surrounded by a hyaline wing. **Distr:** Cambodia, China, Laos, Myanmar, Thailand, Vietnam. **Ecol:** On slopes; 500–1,200 m. Fl. Sep–Dec. **Zones:** 10–12. **Info:** This tree has a suckering habit, with small plants developing as far as 10 m from trunk base.

Oroxylum

Species 1. *Generic description same as specific.* **Cult:** Propagate from seeds, cuttings or by air-layering. Requires a moist, well-drained soil in full or partial sun.

O. indicum (L.) Bentham ex Kurz, *Forest Fl. Myanmar.* 2: 237, 1877. (Syn: *Bignonia indica*, etc.). [Broken Bones Tree, Indian Trumpet Tree]. **Trees**, 6–10 m tall, few-branched; trunk 15–20 cm dbh. **Leaves** opposite, 2(–4)-pinnately compound, borne nearly at stem apex, 0.6–1.3 m long; leaflets triangular-ovate, 5–13 × 3–10 cm, base subrounded or cordate, oblique, margins entire, apex shortly acuminate, glabrous, lateral veins 5 or 6 per side. **Inflorescences** racemose, erect, 4–1.5 m long, terminal; pedicels 3–7 cm long. **Flowers** usually open at night, malodorous, bat-pollinated; calyx large, fleshy, purple, broadly campanulate, 2.2–4.5 × 2–3 cm, subtruncate, glabrous; corolla purple-red, campanulate, tube fleshy, 3–9 × 1–1.5 cm, limb 5.5–8 cm Ø, lobes spreading, slightly reflexed, rounded, margin sinuous, upper lip bilobed, lower lip trilobed; stamens 5, filaments ~4 cm long, anthers ellipsoid, 2-celled, to 1 cm long; disk ~1.5 cm Ø, 5-lobed; style filiform, 5–7 cm long, stigma lingulate, 2-parted, ~7 × 5 mm. **Fruits** capsules, woody, long-lanceolate, to 1.2 m × 5–9 cm, ~1 cm thick, septicidally dehiscent; seeds rounded, in several rows, very thin, 6–7 × 3.5–4 cm, wing transparent. **Distr:** Asia to Malesia. **Ecol:** In tropical and subtropical forests; 500–900 m. Fl. Sep–Dec. **Zones:** 10–12.

Pandorea

Species 6. **Climbers**, rarely scandent; stems woody, terete, without gland patches between petiole. **Leaves** imparipinnate, tendrils absent. **Inflorescences** thyrses, terminal, also in upper axils, sometimes appearing racemose. **Flowers** with calyx cup- or bell-shaped, shallowly lobed; corolla tubular to funnelform, lobes imbricate, slightly 2-lipped; stamens 4, didynamous, usually included in corolla tube; ovary 2-celled. **Fruits** capsules, oblong-ovoid, stipitate, beaked, opening perpendicular to septum; seeds flat, ± circular, surrounded by a hyaline wing. **Distr:** Australia, Malesia, New Caledonia. **Cult:** Propagate from cuttings, seeds or by air-layering. Requires a moist, well-drained soil in full sun.

P. jasminoides (Lindl.) K. Schum., *Die Nat. Pflanz.*, 4(3b), 1891. (Syn: *Bignonia jasminoides*). [Bower of Beauty, Bower Plant/Vine]. **Climbers**, long, woody. **Leaves** mostly opposite, sometimes 3-whorled, 12–17 cm long; leaflets 4–7, ovate to ± lanceolate, 4.5–6 × 1.5–3 cm, base cuneate to rounded, margins entire, apex acuminate, basal leaflets asymmetric, glabrous; petiole 2–4 cm long; petiolules 2–4 mm long. **Inflorescences** 6–12 cm long. **Flowers** with calyx 5–7 mm long; corolla 4–6 cm long, white or pink, with throat red, hairy inside, minutely hairy outside, tube ± straight, 1–2 cm Ø, lobes 2–3 cm long. **Fruits** capsules, oblong-ovoid, 4–6 × 1–2 cm, leathery, beaked; seeds winged, 1–1.5 cm Ø. **Distr:** NE Australia. **Ecol:** In dry rainforests. **Zones:** 9–12.

P. pandorana (Andrews) Steenis, *Bull. du Jard. Bot. de Buit.*, Series 3: 10, 1928. (Syn: *Bignonia pandorana*, etc.). [Wonga Wonga Vine]. **Climbers** or **scramblers**, woody, ± twining branches, glabrous. **Leaves** mostly opposite, strongly dimorphic; *juvenile* leaves 2–8 cm long, leaflets 8–17, small, bluntly toothed; *adult* leaves 8–16 cm long, mostly 3–9-foliate, leaflets linear to ± ovate, 2.5–8 × 0.2–3 cm, base rounded to cuneate, margins usually entire, apex acuminate with a mucro, glabrous; petiole to 1–4.5 cm long; lateral petiolules 1–2 mm long. **Inflorescences** 5–15 cm long, many-flowered. **Flowers** with calyx 3–5 mm long; corolla 1–2.5 cm long, whitish or yellow, often with purple blotches or stripes in throat, bearded inside, glabrous outside, tube ± straight, 0.5–1 cm Ø, lobes 3–6 mm long. **Fruits** capsules, oblong-ovoid, 4–6 × 1–2 cm, beaked; seeds winged, 1–1.5 cm Ø. **Distr:** Australia, Malesia, New Guinea. **Ecol:** In moist gullies, sclerophyll forests and woodlands. **Zones:** 9–12.

Parmentiera

Species 10. **Trees** or **shrubs**; stems terete, leaves subtended by simple spines. **Pseudostipules** absent. **Leaves** opposite or subopposite (alternate), sometimes fasciculate, usually 3- or 5-foliolate (rarely simple or 2- or 4-foliolate); leaflets with margins entire or with few distal teeth, mostly glabrous, venation pinnate; petiole sometimes with narrow lateral wings. **Inflorescences** 1–3-flowered, terminal or often on leafless short shoots or cauliflorous on trunk and older stems; bracts minute; pedicels drying black. **Flowers** glabrous, drying black, usually bat-pollinated; calyx tubular, entire distally but split abaxially and spathelike, subglabrous; corolla white to greenish-white, tubular-campanulate, often bent, 2-lipped, 2 upper lobes partly united, thick, glabrous outside; stamens 4; staminode present; disk annular; ovary cylindric, bilocular, ovules multiseriate. **Fruits** fleshy and indehiscent, elongate-oblong to linear, terete or longitudinally ridged; seeds borne on a fibrous-fleshy central core, with or without a narrow circumferential wing. **Distr:** Mexico to Panama, Colombia. **Cult:** Propagate from seeds, cuttings or by

Above: *Pandorea jasminoides, Pandorea pandorana, Parmentiera aculeata, Parmentiera cereifera.*

air-layering. Requires a fertile, moist, well-drained soil in full or partial shade.

P. aculeata (Kunth) Seem., *Bot. Voy. Herald* 183, 1854. (Syn: *Crescentia aculeata*, etc.). [Cow Okra, Cuajilote]. **Shrubs** to **small trees**, 2–6 m tall; leaf bases subtended by a forward-pointing sharp-tipped conical spine 3–8 mm long. **Leaves** opposite or occasionally alternate, trifoliolate or simple, 2–4 per node; leaflets elliptic to rhombic-elliptic, 2.5–9 × 0.6–4 cm, base cuneate, decurrent on petiole, apex obtuse, adaxially glabrous, veins 3–7 per side; petiole 1.1–4.7 cm long, with narrow wings 0.1–0.5 mm wide; petiolules 0–1.5 cm long. **Inflorescences** of 1–4 terminal flowers, or 1–2 flowers in axils of near-terminal leaves, or fasciculate on older stems; pedicels 1.2–2.1 cm long, with few minute peltate trichomes. **Flowers** with calyx 2.5–3.5 cm long, spathelike or split more than once, glabrous; corolla campanulate, 4–6.5 cm long, white or greenish with purple lines, glabrous outside, tube 2–2.5 cm wide at mouth; stamens with thecae 8–9 mm long. **Fruits** linear-cylindric, 9–17 × <3 cm, ends narrowed, usually slightly curved, with thick longitudinal ridges. **Distr:** Mexico to Honduras. **Ecol:** In hot lowland areas. Fl. Jan–Apr. **Zones:** 10–12. **Info:** Fruits edible, but fibrous.

P. cereifera Seem, *J. Bot.* 3: 302, 1841. [Candle Tree]. **Trees**, 7–15 m tall; trunk to 20–30 cm dbh; main branches strongly ascending from base. **Leaves** opposite, often with a smaller leaf in the axil of a larger (4 per node), trifoliolate; leaflets elliptic to elliptic-rhombic, terminal leaflets larger, to 7.5(–9.5) × 3(–4) cm, base acute, decurrent on petiole, apex acute to acuminate, adaxially glabrous, veins 3–6 per side; petiole 1–4(–6) cm long, to 3 mm wide with narrow greenish wings. **Inflorescences** with flowers solitary, or 2–3-flowered on short-shoots, borne on trunks and larger branches; peduncles ~3 mm long; pedicels 0.9–1.6 cm long, ± glabrous. **Flowers** with calyx 2–5 × 0.9–2.5 cm, split and spathelike, glabrous; corolla, tubular-campanulate, 3.7–6.4 cm long, with a transverse fold on lower side, white, glabrous outside, limb 1.8–2.9 cm wide, lobes ~1 cm long; filaments 2.6–3.5 cm long. **Fruits** linear-cylindric and slightly curved, 30–54 × 1–2.5 cm, pendulous, yellow, ± glabrous, fleshy; seeds 3–4 × 3–4 mm, wing narrow, mucilaginous. **Distr:** C America. **Ecol:** On mountain slopes. Fl. Jan–Jul. Fr. Jun–Dec. **Zones:** 10–12.

Podranea

Species 1. *Generic description same as specific.* **Cult:** Propagate from seeds or cuttings. Requires a fertile, moist, well-drained soil in full sun.

P. ricasoliana Sprague, *Fl. Cap.* (Harvey) 4. (2): 450, 1904. (Syn: *Tecoma ricasoliana*, etc.). [Pink Trumpet Vine]. **Lianas** or **clambering shrubs**, 3–4 m high; stems cylindrical, smooth. **Pseudostipules** absent. **Leaves** opposite, imparipinnate, 7–9-foliolate, without tendrils; leaflets ovate, 2.5–3.8 × 1.5–2 cm, base truncate, cuneate or attenuate, margins serrate, apex obtuse or shortly acuminate, adaxially dull, abaxially light green, punctate, dull; rachis glabrous, marginate; petiole 2.5–5 cm long, glabrous; petiolules 4–8 mm long, glabrous. **Inflorescences** panicles, terminal; pedicels 1.5–2 cm long. **Flowers** with calyx green, pink-tinged, campanulate, 1.5–2 cm long, with 5 apiculate lobes; corolla violet-pink, campanulate, 5–7 cm long, tube white with reddish-purple lines, limb ~5 cm Ø, with 5 rounded lobes, unequal, stamens inserted; ovary glabrous. **Fruits** capsules, linear-cylindrical, 25–30 cm long, coriaceous; seeds numerous, ~1.5 cm long, brown, with 2 marginal wings ~5 mm long. **Distr:** S Africa. **Ecol:** 0–2,000 m. Fl. Aug–May. Fr. Feb. **Zones:** 9–12.

Pyrostegia

Species 3 or 4. **Vines** or **lianas**, climbing with the aid of coiling petiole-borne tendrils; tendrils usually trifid near the tip; stems longitudinally ribbed, lacking interpetiolar gland fields. **Pseudostipules** small or inconspicuous. **Leaves** ± opposite, petiolate, 2- or trifoliolate, petiolules present; lamina with margins entire, venation pinnate. **Inflorescences** panicles, terminal or axillary, often much-branched; pedicels subtended by small paired bracts. **Flowers** with calyx cupular, minutely glandular-lepidote or subglabrous, distal margin truncated with 5 small teeth; corolla tubular or narrowly funnelform, orange or reddish-orange, lobes 5, valvate in bud; stamens 4, exserted, anthers glabrous, thecae straight, parallel or slightly divergent; disk annular; ovary narrowly tubular, stigma bifid. **Fruits** capsules, dry, valves parallel to septum, smooth and leathery; seeds thin, 2-winged, transversely oblong, wings smooth, margins hyaline. **Distr:** S America. **Cult:** Propagate from cuttings or by air-layering. Requires a fertile, well-drained soil in full sun.

P. venusta (Ker-Gawler) Miers, *Proc. R. Hort. Soc.* N.S. 3: 188, 1863. (Syn: *Bignonia ignea*, *B. venusta*, *P. ignea*). [Flame Vine, Golden Shower, Orange Trumpet Creeper]. **Vines**, 6–9 m long, climbing with tendrils to 15 cm long; leafy stems with 6–8 longitudinal ribs. **Leaves** trifoliolate, or bifoliolate and often with a tendril; leaflets ovate-elliptic to ovate-oblong, 3–9 × 1.5–6 cm, base usually obtuse, apex acute to caudate-acuminate, glabrous, abaxially minutely punctate, veins 3–5 per side; petiole to 3.5 cm long; petiolules to 1.8 cm long, distal one longer. **Inflorescences** dense panicles, usually terminal on short shoots, often pendulous, 8–25-flowered; rachis to 9 cm long; peduncle woody, to 5 cm long; pedicels 1.2–2 cm long, glabrous. **Flowers** with calyx 4–5 × 4–6 mm, distal margin minutely ciliolate, subentire or minutely lobed; corolla 6–9 cm long, 2–3 mm Ø at base, 0.8–1.2 cm Ø at mouth, orange, tubular to narrowly tubular-funnelform, ± straight, lobes oblong, 0.6–1.8 cm long, apex blunt, becoming reflexed; filaments 3–5 cm long. **Fruits** capsules, 20–30 × 1–1.6 cm; seeds 3–4 cm long. **Distr:** Argentina, Bolivia, Brazil, Paraguay. **Ecol:** In rainforests. Fl. throughout most of the year. **Zones:** 8–12.

Below: *Podranea ricasoliana, Pyrostegia venusta.*

HL

MS

Top row: *Radermachera eberhardtii, Radermachera pinnata, Radermachera sinica, Saritaea magnifica.*
Bottom row: *Spathodea campanulata, Stereospermum colais, Stereospermum kunthianum, Stereospermum suaveolens.*

Radermachera

Species ~16. Trees. **Leaves** opposite, 1–3-pinnately compound; leaflets petiolate, margins entire. **Inflorescences** paniculate or reduced almost to fascicles, terminal or lateral; bracts and bractlets linear or leaflike. **Flowers** with calyx campanulate, apex 5-lobed or truncate; corolla funnelform-campanulate or salverform, tube short or long, limb slightly bilabiate, lobes rounded, spreading; stamens 4(or 5), didynamous or subequal; staminode usually present; disk annular, slightly fleshy; ovary terete, ovules numerous, 2-rowed in each locule, style included, stigma ligulate, bilobed. **Fruits** capsules, loculicidally dehiscent, long, terete, 2-angular, septum compressed, woody; seeds compressed, both ends winged, wings transparent, membranous. **Distr:** Tropical Asia. **Cult:** Propagate from seeds, cuttings or by air-layering. Requires a fertile, moist, well-drained soil in full sun.

R. eberhardtii Dop, *Bull. Mus. Hist. Nat.* (Paris) 32: 233, 1926. (Syn: *R. microcalyx*). **Trees**, evergreen, 7–15 m tall. **Leaves** pinnate, 25–30 cm long; rachis minutely glandular; leaflets 5–7, elliptic to ovate, (10–)14–20(–25) × (3.5–)4.5–6(–11) cm, base cuneate, oblique, margins entire, apex acute, acuminate or caudate, ± chartaceous, glabrous, veins 6–10 pairs; petiolules 1–2 cm long, terminal one to 6 cm long. **Inflorescences** panicles, ~20 cm long, terminal; pedicels 0.5–1 cm long. **Flowers** with calyx green, 6–7 mm long, tube 4–5 mm long, lobes 5, irregular, ~2 mm long, acute; corolla yellow, 3–3.5 cm long, tube 2.5–3 cm long, lobes ovate to orbicular, 0.6–1 cm long, margin fimbriate; stamens 4, didynamous, filaments glabrous, anthers 1.5–2 mm long; staminode 1, ~2 mm long; disk ~0.5 mm high; ovary cylindric, 2–2.5 mm long, style 0.7–1 cm long, glabrous, stigma bilobed. **Fruits** capsules, terete, curved, 10–25 cm long, dehiscing by 2 valves, valves 4–6 mm wide; seeds flat, 0.8–1.5 × 0.1–0.2 cm including wing. **Distr:** China, Thailand, Vietnam. **Ecol:** In evergreen forests; 300–950 m. Fl. Mar–Apr. Fr. Apr–Dec. **Zones:** 10–12.

R. pinnata (Blanco) Seem, *Journ. Bot.* 8: 147, 1870. (Syn: *Millingtonia pinnata*, etc.). **Trees**, to 20 m tall, glabrous; trunk 15–40 cm dbh. **Leaves** pinnate; leaflets elliptic, (3–)5–16 × (1.5–)2–5(–8) cm, base acute, apex acute to shortly acuminate, adaxially dull green. **Inflorescences** thyrses or panicles, 35–45 × 25–30 cm; primary branches round, gray, woody and sparsely pubescent. **Flowers** 2.5–3(–4.5) cm long, pink or pale purple; calyx pinkish-brown, cup-shaped, 1–1.5 cm long, deeply split into 2 lobes; corolla pink to pale purplish, with yellow markings in the throat, 2–3.5(–4.5) cm long, tube funnelform, flattened, bent at the end of the calyx, lobes 5, pink, with an orange blotch at base, recurved, margins ruffled, ciliate; stamens spring-loaded, curved, inserted in corolla tube, filaments white, anthers dark brown; pistil green, ovary and style glabrous, 1.2–1.7(–2.2) cm long. **Fruits** capsules, (6–)30–50 cm long; seeds compressed, (0.7–)1.3–1.5 × 0.2–0.3 cm, winged. **Distr:** Philippines (Luzon). **Ecol:** In lowland and montane primary and secondary forests; to 600 m. Fl. Aug–May. Fr. Feb–Nov. **Zones:** 11–12.

R. sinica (Hance) Hemsley, *Icon. Pl.* 28: pl. 2728, 1902. (Syn: *Stereospermum sinicum, R. tonkinensis*). [China Doll, Emerald Tree, Serpent Tree]. **Trees**, ~10 m tall; leaves, petiole, leaf axis, and inflorescences glabrous. **Leaves** 2(or 3)-pinnately compound; leaf rachis ~30 cm long; lateral petiolules <5 mm long, terminal one 1–2 cm long; leaflets ovate to ovate-lanceolate, 4–7 × 2–3.5 cm, base broadly cuneate, margins entire, apex caudate-acuminate, glabrous, lateral veins 5 or 6 per side. **Inflorescences** paniculate, erect, 25–35 cm long, terminal; bracts linear-lanceolate, ~10 cm long, deciduous; bractlets linear, 4–6 cm long. **Flowers** with calyx teeth 5, ovate-lanceolate, ~1.2 cm long; corolla white to pale yellow, campanulate-funnelform, 6–8 cm long, lobes rounded, ~2.5 cm long; stamens 4, didynamous; staminode present, filiform; ovules in 2 rows, style exserted, stigma bilobed. **Fruits** capsules, terete, nodding, angular, ~85 × 1 cm; seeds ellipsoid, including wing ~2 × 0.5 cm. **Distr:** Bhutan, China, India, N Myanmar, Taiwan, Vietnam. **Ecol:** On slopes and in forests; 300–800 m. Fl. May–Sep or Dec–Apr. Fr. Oct–Dec or May–Jul. **Zones:** 10–12.

Saritaea

Species 1. *Generic description same as specific.* **Cult:** Propagate from seeds, semiripe cuttings or by air-layering. Requires a moist, fertile, well-drained soil in full sun.

S. magnifica (W. Bull) Dugand, *Caldasia* 3: 263, 1945.(Syn: *Arrabidaea magnifica, Bignonia magnifica*). [Purple Bignonia, Saritaea]. **Lianas**, climbing using tendrils, 3–7 m long. **Pseudostipules** foliaceous, to 4.2 cm long. **Leaves** opposite, bifoliolate, sometimes with a simple tendril; leaflets obovate, 4.2–11.5 × 3.1–6.4 cm, base cuneate or decurrent, margins entire, apex obtuse, chartaceous, dull; petiole 1.7–2.8 cm long, lepidote; petiolules 0.3–1.6 cm long, lepidote. **Inflorescences**

Above: *Stereospermum personatum, Stereospermum xylocarpum, Tabebuia aurea.*

corymbiform panicles, few-flowered, terminal; pedicels 4–6 mm long. **Flowers** with calyx yellowish-green, simple, crateriform, 7–8 mm long, truncate, lepidote; corolla purple-pink, tubular-campanulate, 8–9 cm long, glabrous or minutely glandular, tube pubescent inside, throat white inside, with purple-pink lines, lobes unequal, 2.2–3.1 cm long; stamens 4, didynamous, inserted; disk hypocrateriform-pulviniform, 1 mm high; ovary linear, ~4 mm long, glandular-lepidote, biloculed, ovules biseriate per locule. **Fruits** capsules, linear, 20–25 × 1–1.2 cm, compressed, brown, coriaceous; seeds numerous, oblong, 2-winged, wings hyaline, membranous. **Distr:** Colombia, Ecuador. **Ecol:** Fl. Sep–Dec. **Zones:** 10–12.

Spathodea

Species 1. *Generic description same as specific.* **Cult:** Propagate from seeds, runners, cuttings or by air-layering. Requires a fertile, moist soil in full sun.

S. campanulata Beauvois, *Fl. Oware* 1: 47, tab. 27, 1805. (Syn: *S. nilotica*). [African Tulip Tree]. **Trees**, to 25(–30) m tall; trunk to 40(–60) cm dbh. **Leaves** pinnate, 19–40 cm long; leaflets, 7–13(–17), elliptic to ovate-elliptic or ovate-oblong, 5–14 × 2–6 cm, base acute to obtuse, or rounded on 1 side and asymmetric, apex acute to acuminate; petiole to 20 cm long; petiolules 1–3 mm long, terminal one to 1.5 cm long. **Inflorescences** corymbose racemes, compact, with a closely placed spiral of 6–35 flowers, ~7 × 15 cm (not including corollas), terminal, pedicels to 6 cm long. **Flowers** with calyx 2–6 cm long, split on 1 side, spathelike, recurved and acuminate, densely sericeous, yellowish-brown; corolla 6–15 × 5–9 cm, widely campanulate and curved from a shortly tubular base, deep red to orange-red, lobes 2–3 cm long, obtuse, distal edge of corolla crinkled and yellowish; stamens 4; ovary bilocular, narrowly oblong, ovules multiseriate per locule, stigma flat and bifid. **Fruits** capsules, dehiscent (each half canoe-shaped), narrowly ellipsoid-oblong, 17–27 × 3–7 cm, 1–2 cm thick; seeds ~1.5 × 2 cm, wing thinly membranous. **Distr:** Tropical C Africa. **Ecol:** In evergreen forests; <1,500 m. **Zones:** 11–12.

Stereospermum

Species ~15. **Trees** or **shrubs**, deciduous. **Leaves** opposite, pinnately compound; leaflets elliptic or elliptic-lanceolate, margins entire. **Inflorescences** cymose-paniculate, large, terminal. **Flowers** with calyx campanulate, 5-dentate, sometimes unequal; corolla lobes subequal, rounded or lacerate, tube small, yellow or pale red; stamens 4, didynamous, included, anthers divergent; disk cushionlike; ovary sessile, ovules numerous, in 1 to several rows. **Fruits** capsules, loculicidally dehiscent, terete, slender, 4-angular, septum terete, woody; seeds inserted in septum, transparent and membranous winged at both ends. **Distr:** Tropical Asia and Africa. **Cult:** Propagate from seeds or by air-layering. Requires a fertile, moist, well-drained soil in full sun.

S. colais (Buchanan-Hamilton ex Dillwyn) Mabberley, *Taxon* 27: 553, 1978. (Syn: *Bignonia colais*, etc.). [Trumpet Flower Tree, Yellow Snake Tree]. **Trees**, 15–20(–35) m tall; trunk 15–25(–80) cm dbh. **Leaves** pinnate, 25–50 cm long; leaflets 3–6 per side, long-elliptic, 8–14 × 2.5–6 cm, pubescent or glabrous. **Inflorescences** paniculate, 20–40 cm long, terminal, many-flowered; peduncles pubescent or glabrous; bract and bractlets deciduous; pedicels 3–4 mm long, pubescent or glabrous. **Flowers** with calyx campanulate, purple, pubescent or glabrous, 4–5 × 4–5 mm, 3–5-lobed; corolla pale yellow, slightly curved, ~2 cm long, upper lip bilobed, lower lip trilobed, tomentose at mouth, tube terete; filaments ~1 cm long, glabrous; stigma bilobed. **Fruits** capsules, terete, 4-angular, slightly curved, 30–70 × ~1 cm, pericarp thick, subwoody, septum 4–7 mm Ø; seeds ovate, ~2.8 × ~5 cm, apices winged, wings white, membranous. **Distr:** Tropical Asia. **Ecol:** In forests, slopes; 400–1,800 m. Fl. May–Jul. Fr. Sep–Nov. **Zones:** 10–12.

S. kunthianum Cham., *Linnaea* 7: 720, 1833. [Pink Jacaranda]. **Shrubs** or **trees**, to 10–12 m tall; crown open, rounded. **Leaves** opposite, imparipinnate, to 32 cm long; leaflets ± opposite, 3–4 pairs, elliptic or oblong, 5–10 × 3.5–5 cm, base asymmetrical, cuneate or rounded, apex apiculate, pointed or obtuse, nerves pinnate, prominent, with 5–8 pairs of lateral nerves. **Inflorescences** lax panicles, to 25 cm long, terminal; flowers appearing with new flush of leaves. **Flowers** pink, with crimson stripes; corolla tubular, 2.5–5 cm long, lobes 3–4 cm across, margins wavy; stamens with filaments 0.6–1.5 cm long, anthers 2 mm long; staminode 1.5 mm long; disk 1.5 mm long; ovary 5–6 mm long, glabrous, reddish-brown. **Fruits** capsules, cylindrical, 40–60 × 0.7–0.8 cm, twisted when mature; seeds flat, 2.5–3 cm long, with papery wings at each end. **Distr:** E and tropical Africa. **Ecol:** On rocky hillsides, in open woodlands, along margins of riverine forests; <1,500 m. Fl. Aug–Oct. **Zones:** 11–12.

S. personatum (Hassk.) Chatterjee, *Bull. Bot. Soc. Bengal* 2: 70, 1948. (Syn: *Dipterosperma personatum*, etc.). **Trees**, to 25 m tall. **Leaves** odd-pinnate, broadly obovate to obovate-oblong; leaflets 5–9, elliptical-oblong, 3–11 × 2.5–5 cm, base acute, obtuse, or oblique, margins entire, apex narrowly caudate, glabrous. **Inflorescences** panicles; peduncles 1–7 cm long; bracts minute, caducous. **Flowers** with calyx yellowish-purple, 5–8 × ~5 mm, lobes obtuse, glabrous; corolla tube 1.5–1.8 × 0.6–0.8 cm, upper portion campanulate-ventricose, abruptly inflated and sharply geniculate, 1–1.3 × 0.6–0.8 cm, dull pink to yellow outside, with numerous deep reddish-purple flecks and stripes, ± pubescent, 2 pubescent ridges evident on lower surface inside, upper 2 lobes partially laterally fused, reflexed, ~8 mm long, lower 3 distinct, ~0.7–1 × 0.6–0.8 cm, each with 2 prominent ridges; filaments ~8 mm long; disk 1.5 mm Ø; style ~1 cm long, stigma ~1 × 1 mm. **Fruits** capsules, curved or spirally twisted, to ~50 cm long, ~5 mm wide; seeds 4 × 4 mm, wings ~6 mm long. **Distr:** Sri Lanka, India, eastwards to Borneo. **Ecol:** From coastal regions to moist low montane regions; <650 m. Fl. May–Aug. **Zones:** 11–12.

S. suaveolens DC., *Bibl. Univ. Genev.* 17: 124, 1838. (Syn: *Bignonia suaveolens*, etc.). **Trees**, 9–18 m tall; young parts viscous-hairy. **Leaves** pinnately compound, 30–45 cm long; leaflets 7–9, broadly elliptic, 14 × 7.5 cm, margins entire or serrulate, apex acuminate or acute, hairy when young; petiolules to ~2.5 mm long. **Inflorescences** compound panicles, many-flowered, viscous-hairy. **Flowers** with calyx 8 mm long, hairy, lobes 3–5, very short, broad; corolla pale or dark purple, 2.5–3.2 cm long, puberulous outside, hairy in the throat, lobes rounded, crisped-crenate. **Fruits** capsules, linear, terete, 45 × 1.7 cm, woody, slightly rough with tubercles, obscurely 4-ribbed, glabrous; seeds 3.2 × 0.6 cm, deeply notched in the middle, subtrigonous, embedded in notches of the septum. **Distr:** India (from the Himalayan Terrain to Travancore and Tenasserim). **Ecol:** In moist regions. **Info:** According to Thwaites the accessions of this species found in Sri Lanka (including those at the Royal Botanical Gardens, Peradeniya) are presumably cultivated. **Zones:** 10–12.

S. xylocarpum (Roxb.) Wight ex C.B.Clarke, *Fl. Brit. India* [J.D. Hooker] 4(11): 383, 1884. (Syn: *Bignonia xylocarpa*, etc). **Trees**, 10–20 m tall; innovations pubescent. **Leaves** very large, 0.3–1.3 m long, 2–3-pinnate; leaflets elliptic to ovate, 5–15 × 3.8–7.5 cm, margins entire or coarsely serrate, acute to shortly acuminate, glabrous, abaxially pale or silvery; subsessile. **Inflorescences** panicles, ovoid, compound, 7.5–18 cm long, densely pubescent, somewhat rigid, erect; pedicels subcapitately corymbose on branches; bracts caducous. **Flowers** large, 4–6.5 cm long, white with yellow or pink, sweetly scented; calyx 1.25–1.5 cm long, pubescent, glabrate when mature, lobes 3–5, very short, broad; corolla 3.8–5 cm long, white, yellow-tinged, ventricose from near base, subglabrous, lobes 3–5, very short, broad, round, crisped; filaments hairy below. **Fruits** capsules, stout, rugged, cylindric, 0.3–1 m × 2.5–4 cm, glabrous, tubercular; seeds (including wings) 3.2–5 × 1.3 cm, thinly discoid. **Distr:** Deccan Peninsula. **Ecol:** In moist areas of forests, often near rivers. Fl. Apr–May (during the deciduous phase Mar–May). Fr. Jul–Apr. **Zones:** 11–12.

Tabebuia

Species ~100. **Trees** or **shrubs**, often deciduous. **Pseudostipules** absent. **Leaves** opposite, palmately 3–7(–9)-foliolate or less often simple or 1-foliolate,

Above: *Tabebuia chrysantha, Tabebuia guayacan, Tabebuia impetiginosa, Tabebuia heterophylla.*

petiolate, basal leaflets usually smaller and with shorter petiolules than the terminal ones, venation pinnate. **Inflorescences** terminal, short, open or congested panicles, few-flowered racemes or fascicles of 1 to few flowers; bracts subtending the pedicels; some species flowering when the tree is leafless. **Flowers** with calyx tubular to cupular or campanulate, margin 2–5-lobed, irregular or truncate, surface hairy; corolla tubular-funnelform to tubular-campanulate, white, yellow, rose, purple or red, lobes 5, rounded; stamens 4, exserted or included, filaments of 2 lengths, anthers glabrous; staminode present; disk cushion-shaped; ovary oblong, bilocular, ovules 2–many-seriate in each locule. **Fruits** capsules, linear-cylindric to oblong-cylindric, subterete, valves dehiscing approximately perpendicular to the septum, smooth to verrucose-muricate; seeds thin, transversely oblong with 2 thin lateral wings. **Distr:** Neotropical. **Cult:** Propagate from seeds, cuttings or by air-layering. Requires any well-drained soil in full sun.

T. aurea (Manso) Bentham and Hooker f. ex S. Moore, *Trans. Linn. Soc.* 2, Bot. 4: 423, 1895. (Syn: *Bignonia aurea*, etc.). [Caribbean/Paraguayan Trumpet Tree, Silver Trumpet Tree, Tree of Gold]. **Trees**, to ~16 m tall. **Leaves** palmately 5–7-foliolate, lepidote; leaflets oblong-elliptic to narrowly oblong-lanceolate, to 13 × 9.5 cm, lateral leaflets smaller, base rounded to subcordate, margins entire, apex rounded or retuse, coriaceous; petiole to 14 cm long; petiolules to 5 cm long. **Inflorescences** panicles, terminal; branches densely lepidote. **Flowers** with calyx campanulate, irregularly bilabiate, 0.8–1.6 × 1–1.7 cm, densely lepidote; corolla yellow, tubular-infundibuliform, 5.5–9 × 1.2–2.5 cm, tube 4.2–6.6 cm long, lobes 1.2–2.2 cm long, glabrous outside; stamens didynamous; pistil 2.9–3.2 cm long, ovary linear-oblong, 4 × 1 mm, densely lepidote, ovules biseriate in each locule; disk pulvinate, 1 × 3 mm. **Fruits** capsules, oblong, 8.5–15 × 1.7–3 cm, ends narrowed, grayish, densely lepidote; seeds thin, large, ~2 × 4.5–5.5 cm, wings hyaline-membranous. **Distr:** Brazil, Argentina, west to Bolivia and disjunct in S Surinam. **Ecol:** In seasonally dry forests and savannas; 0–1,500 m. **Zones:** 10–12.

T. chrysantha (Jacq.) G. Nichols., *Ill. Dict. Gard.* 4: 1, 1887. (Syn: *Bignonia chrysantha*, etc.). [Golden Goddess, Golden Trumpet Tree]. **Trees**, to 30 m tall; trunk to 1 m dbh. **Leaves** usually 5-foliolate (3-, 4-, 7-foliolate), leaflets elliptic or elliptic-obovate, (4–)7–21(–26) × (2–)3–12 cm, basal leaflets smaller, base obtuse to rounded, slightly truncate, margins entire or sometimes bluntly serrate, apex acuminate to caudate-acuminate, secondary veins 6–12 per side; petiole (5–)8–19(–28) cm long, densely puberulent; petiolules 1–9 cm long. **Inflorescences** condensed panicles, 7–25-flowered, terminal; peduncles and pedicels to 2 cm long. **Flowers** with calyx cupular-funnelform or campanulate, to 1.4(–1.9) × 1(–1.3) cm, densely pubescent, lobes 2–3 mm long, broadly triangular; corolla tubular-funnelform, 5.2–9.5 cm long, bright yellow with narrow reddish lines inside, glabrous outside, lobes to 2.8 cm long; filaments 16–22, 1–1.5 cm long. **Fruits** linear-cylindric, 20–50(–90) × 1.4–2.4 cm, ends narrowed, smooth; seeds 0.6–1.2 × 1.4–3.8 cm, wings transparent. **Distr:** Mexico to Nicaragua, Venezuela and Peru. **Ecol:** In wet evergreen lowland rainforests; 20–600 m. **Zones:** 11–12.

T. guayacan (Seem.) Hemsley, *Biol. Centr. Am. Bot.* 2: 495, 1882. (Syn: *Tecoma guayacan*). [Brazilian Walnut, Guayacan Trumpet Tree]. **Trees**, to 50 m tall; trunk to 1.7 m dbh. **Leaves** 5- or 7-foliolate; leaflets ovate-elliptic to elliptic or lanceolate, 5–20(–30) × 3–11(–15) cm, terminal one larger than laterals, base rounded to obtuse (lateral leaflets usually asymmetric), margins entire (serrate in juveniles), apex acute to acuminate, secondary veins 6–9 per side; petiole 8–23 cm long; petiolules 1–7 cm long, basal ones 1–3 cm long. **Inflorescences** panicles, compact, flowers in diads or triads, opening together; peduncles to 1.4 cm long; pedicels 1–1.8 cm long. **Flowers** with calyx campanulate, 0.9–1.5 × 0.4–0.9 cm, minutely scurfy-hairy, margin split or with 2–5 rounded or obtuse lobes; corolla tubular-funnelform, 6–10 cm long, yellow with brown lines inside, glabrous outside, 1.3–2.2 cm wide at mouth, lobes 2.2–3.5 cm long; filaments 15–21, 1.1–1.6 cm long. **Fruits** capsules, linear, 30–60 × 1.2–2.8 cm, subglabrous to thickly puberulent; seeds 0.6–1.1 × 2.3–4 cm, wings transparent. **Distr:** S America. **Ecol:** In lowland evergreen rainforests; 5–600 m. Fl. Feb–May. Fr. May–July. **Zones:** 10–12.

T. heterophylla Britton, *Ann. Missouri Bot. Gard.* 2: 48, 1915. (Syn: *Bignonia pentaphylla*, etc.). [Pink Tecoma, Pink Trumpet Tree, White Cedar]. **Shrubs** or **trees**, evergreen or semideciduous, to 18 m tall; crown pyramidal or dome-shaped. **Leaves** pinnate; leaflets usually (1)3–5, variable in size, shape and texture, often obovate-elliptic or oblong-elliptic, 1.25–16.5 cm long, base and apex obtuse to rounded, both sides densely scaly. **Inflorescences** panicles, usually few-flowered. **Flowers** with calyx cup-shaped, 0.8–1.25 cm long, irregularly lipped, leathery, scaly, persistent; corolla 3.5–7 cm long, lavender or pale magenta to ± white, throat yellow, fading to white; stamens included. **Fruits** linear-cylindrical, 5–20 cm long, densely scaly. **Distr:** West Indies. **Ecol:** In woodlands and thickets. **Zones:** 10–12.

T. impetiginosa (Mart. ex DC.) Standl., *Publ. Field Mus. Nat. Hist.*, Bot. Ser. 11: 176, 1936. (Syn: *T. palmeri*, etc.). **Trees**, deciduous, to 30 m tall; trunk to 70 cm dbh. **Leaves** 3- or 5-foliolate; leaflets elliptic to ovate-elliptic, (3–)11–19 × (2–)4–8 cm, basal leaflets much smaller, base acute to obtuse or rounded, margins serrate in juvenile plants, apex short- to long-acuminate, secondary veins 9–12 per side; petiole 4–14 cm long, longitudinally striate; petiolules 0.7–5 cm long. **Inflorescences** compact-paniculate (congested-fasciculate), terminal, peduncles to 1.2 cm long, densely pubescent; pedicels 2–6 mm long. **Flowers** with calyx tubular-campanulate, 5–8 × 3–6 mm, ± puberulent, margin truncate or slightly 5-lobed; corolla tubular-funnelform, 3.5–7.5 cm long, purple to red-purple (yellow inside), ± puberulent outside, 1.5–2.5 cm wide at mouth, lobes 1–1.7 cm long; filaments 16–23, 1–1.6 cm long. **Fruits** 15–56 × 1.5–2.5 cm;

Below: *Tabebuia pallida, Tabebuia rosea.*

Above: *Tabebuia serratifolia, Tecoma capensis, Tecoma guarume, Tecoma stans.*

seeds to 1.3 × 3–4.8(–8) cm, central area 1.7–1.8 cm wide, wings translucent, clearly differentiated. **Distr:** Mexico to Argentina. **Ecol:** In seasonally dry forests; 10–300(–1,400) m. Fl. Nov–Feb. Fr. Jan–Feb. **Zones:** 10–12.

T. pallida Miers, *Proc. Roy. Hort. Soc.* 3: 199, 1863. (Syn: *Bignonia leucoxylon*, etc.).[White Cedar]. **Trees**, evergreen or semideciduous, 5–20 m tall; bark dark gray, shallowly fissured; foliage and inflorescences lepidote. **Leaves** 1–5-foliate (in some places mostly 1-foliate); leaflets elliptic to oblong or oblong-elliptic, to 20 × 12 cm, apex acute or obtuse, smooth, glossy, subcoriaceous, somewhat glossy; petiole to ~10 cm long; petiolules short, sometimes absent in lower leaflets. **Inflorescences** clusters, 1–few-flowered, terminal; pedicels slender, to ~2 cm long. **Flowers** pale pink to nearly white, large, 5–8 cm long; calyx ~1 cm long; corolla 5–7 cm long. **Fruits** capsules, 10–20 cm long, ~6 mm thick. **Distr:** C America, Lesser Antilles, Venezuela. **Ecol:** In woodlands, hillsides and thickets. **Zones:** 11–12.

T. rosea (Bertol.) DC., *Prodr.* 9: 215, 1845. (Syn: *Bignonia pentaphylla*, etc.). [Pink Poui, White Oak]. **Trees**, 5–30 m tall; trunk to 1 m dbh. **Leaves** 5-foliolate, basal leaflets often smaller; leaflets elliptic to elliptic-oblong, 5–18(–35) × 3–10(–18) cm, base cuneate or rounded, margins entire, apex acute to acuminate, veins 7–12 per side; petiole 6–17(–32) cm long; petiolules of distal leaflets 2–7(–11) cm long. **Inflorescences** panicles, often dichotomously branched, 2–20 cm long; peduncles to 2.5 cm long; bracts 1–2 mm long; pedicels to 2 cm long. **Flowers** with calyx cupular-funnelform, 1.1–2.1 × 0.6–1.1 cm, lepidote, lobes 2 or 3, rounded, sinuses 1–3 mm deep; corolla tubular-funnelform, 5.3–9.7 cm long, pink to purple or white, lobes pinkish, often yellowish inside, glabrous outside, lobes 1.8–2.6 cm long; filaments 14–20, 1–1.5 cm long. **Fruits** capsules, linear-cylindric, 21–38 × 1–1.8 cm, densely hairy, apex narrowly acuminate; seeds 0.8–1.2 × 2.8–4.5 cm, wings transparent, sharply differentiated. **Distr:** Mexico to Venezuela and W Ecuador. **Ecol:** In deciduous and lowland evergreen rainforests; 0–1,100 m. Fl. Jan–Mar. Fr. Apr–May. **Zones:** 10–12.

T. serratifolia (Vahl) Nichols., *Dict. Gard.* 4: 1, 1887. (Syn: *Bignonia serratifolia*, etc.). [Yellow Poui, Yellow Trumpet Tree]. **Trees**, deciduous, tall; branchlets terete, finely ribbed, when young minutely scurfy, glabrous with age. **Leaves** usually 5-foliate, sometimes trifoliate; leaflets very variable in size and shape, usually elliptic, lanceolate to elliptic-ovate, base rounded or obtuse, apex acuminate, terminal ones >20 × ~10 cm, margins ± wavy and crenate-serrate, sometimes entire, membranous, abaxially hairy in axils of main nerves, otherwise glabrous. **Inflorescences** usually on leafless branchlets, resembling a fascicle of pedicels, brownish scurfy-tomentellous with minute stellate hairs. **Flowers** with calyx campanulate, irregularly lobed, 0.8–1.8 cm long, stellate-pubescent; corolla bright yellow, funnelform, 6–8 cm long, limb 4–9 cm Ø, frilled, ciliate. **Fruits** capsules, linear, to 45 × 2 cm, valves glabrous or sparingly scurfy. **Distr:** Southern Antilles, Guiana to Brazil and Bolivia. **Ecol:** In rainforests, evergreen forest, often forming pure stands in some areas, also on ridges and in swampy areas. Fl. Jul–Oct. **Zones:** 10–12.

Tecoma

Species 14. **Shrubs** or **small trees**. **Leaves** opposite, simple, trifoliolate or pinnately compound with opposite lateral leaflets and a terminal leaflet, petiolate; leaflets very variable in form, margins serrate, venation pinnate. **Inflorescences** racemes or racemose panicles, terminal, bracts small and inconspicuous; flowers pedicellate. **Flowers** showy, calyx cupulate, glands often present on distal surface, margin with 5 deltoid lobes; corolla tubular-campanulate to tubular-funnelform, radially symmetric, glabrous outside, yellow to orange, 5-lobed; stamens 4, exserted or included, filaments of 2 lengths, anthers glabrous or puberulent, thecae divergent; staminode present; disk cupular; ovary lepidote, bilocular, ovules biseriate in each locule, stigma flat. **Fruits** capsules, long, linear, compressed parallel to septum but dehiscing perpendicular to septum, valves smooth and glabrous; seeds thin, with 2 lateral membranous wings sharply differentiated from seed. **Distr:** Africa (2 spp.), Neotropics (12 spp.). **Cult:** Propagate from fresh seeds, suckers, cuttings or by air-layering. Requires a fertile, moist, well-drained soil in full or partial sun.

T. capensis (Thunberg) Lindley, *Bot. Reg.* 13: t. 1117, 1828. (Syn: *Bignonia capensis*, etc.). [Cape Honeysuckle, Cock-a-doodle-doo]. **Shrubs** or woody-based **herbs**, sometimes climbing, to 3 m high. **Leaves** opposite, compound; leaflets usually 7–11, elliptic or almost circular, ~1,5 × 1 cm, apex rounded or obtuse; terminal leaflet often with base abruptly tapered, apex acute, sessile. **Inflorescences** racemes or racemelike panicles, many-flowered. **Flowers** with calyx 5–7 × 4–5 mm, persistent, ± downy; corolla 3.5–5 cm long, 6–7 mm wide at mouth, tubular, slightly curved, orange-red, mostly hairless, lobes with marginal hairs, lower lobes reflexed, much shorter than tube; stamens ± equal, anthers protruding. **Fruits** capsules, 7–12 × 0.7–1.2 cm, wrinkled. **Distr:** S Africa, S Mozambique; widely cultivated throughout the tropics and subtropics. **Ecol:** In thickets and near rivers; 300–900 m. Fl. throughout most of the year. **Zones:** 9–12.

T. guarume A.P. de Candolle, *Prodr.* 9: 224, 1845. (Syn: *Bignonia alata*, etc.). [Orange Bells, Orange Trumpet Flower]. **Shrubs**, 1.5–3 m high. **Leaves** pinnate, (1)5–11-foliolate; leaflets elliptic to oblanceolate, *laterals* 0.8–3 × 0.4–1.4 cm, *terminal* to 4 × 2 cm, base cuneate, margins shallowly serrate, apex mostly obtuse, chartaceous; rachis narrowly winged; petiole mostly 1–2 cm long; lateral petiolules poorly defined, mostly 1–3 mm long. **Inflorescences** 1–several racemes, terminal, or from axils of terminal leaves, sometimes ± paniculate. **Flowers** with calyx cupular, evenly 5-denticulate; corolla tube red to red-orange, lobes orange to yellow-orange, throat usually yellow, broadly salverform-tubular, 5–6.5 ×

Below: *Tecomanthe dendrophila, Tecomanthe hilli.*

Top row: Tecomanthe montana, Tecomanthe speciosa, Acanthus ebracteatus.
Bottom row: Acanthus ilicifolius, Acanthus mollis, Acanthus montanus, Acanthus pubescens.

0.8–1.1 cm, lobes 4–5 mm long, glabrous, lobe margins ciliate; stamens held in mouth of tube, thecae divaricate, 3 mm long; pistil 5–6 cm long, ovary linear-oblong to narrowly ovate, 3 × 1–1.2 mm, somewhat keeled; disk annular-pulvinate, 0.5 × 1.2 mm. **Fruits**, capsules, linear, 5.5–7 × 0.6–0.8 cm, ends tapered, with persistent calyx; seeds thin, 0.4–0.6 × 0.9–1.5 cm, wings membranous. **Distr:** Peru. **Ecol:** In sandy areas, along rivers; 450–1,200 m. **Zones:** 10–12.

T. stans (L.) Juss. ex H.B.K., *Nov. Gen. Sp.* 3: 144, 1818. (Syn: *Bignonia stans*, etc.). [Yellow Bells, Yellow Elder]. **Shrubs**, to 2.5(–6) m high, glabrous. **Leaves** opposite, pinnate; leaflets (1–)3–7(–9), ovate-lanceolate, elliptic to elliptic-ovate, 2.5–10 × 0.8–3 cm, base acute or obliquely acute, margins sharply dentate, apex acuminate. **Inflorescences** panicles, ± erect, 5–15 cm long, terminal. **Flowers** bright yellow with reddish lines in throat, fragrant; calyx tubular-campanulate, 5–7 mm long, lobes triangular, acuminate, ciliolate; corolla tube 3.2–5 cm long, base abruptly contracted, lobes subequal, 3 cm long; stamens included, longer filaments 2 cm long, pubescent; staminodes 5–6, filiform, 5 mm long; disk shallowly cupular, 5-lobed, fleshy; ovary ~3.5 mm long, elongated; style 1.8–2 cm long, stigma bilobed, lobes oval to ovoid, ~1.5 mm long. **Fruits** capsules, linear-oblong, 10–22 × 0.7–2 cm; seeds orbicular-compressed, 2 × 0.6 cm, wing 1.6–1.8 cm wide. **Distr:** Tropical S America. **Ecol:** In deciduous and semideciduous forests; 0–1,300(–1,800) m. Fl. throughout most of the year. **Zones:** 10–12.

Tecomanthe

Species ~5. **Lianas**, climbing or creeping, rarely erect; stems glabrous or pubescent. **Leaves** opposite, pinnate; leaflets 1–7 pairs. **Inflorescences** racemes, few-flowered, short, pendent, cauliflorous (mostly on old wood, near leaf scars) or axillary. **Flowers** showy; calyx campanulate to suburceolate, equally 5-lobed; corolla pink, campanulate, lobes unequal; stamens 4, included, anthers glabrous, thecae divaricate, staminode present; disk annular; ovary cylindrical, lepidote, ovules multiseriate on 2 placentas per locule. **Fruits** capsules, linear-terete, sometimes subtended by persistent calyx, flattened, stipitate, valves coriaceous or woody, loculicidally dehiscent; seeds orbicular, winged. **Distr:** Malesia, Australia, New Zealand. **Cult:** Propagate from seeds or cuttings. Requires a fertile, moist, well-drained soil in full sun.

T. dendrophila K. Schum., *Fl. Schutzgeb. Südsee*: 539, 1900. (Syn: *T. venusta*). [New Guinea Trumpet Vine]. **Climbers** or **twiners**, vigorous, evergreen, woody, to 15 m long; young stems brown with white lenticels, second year bark whitish. **Leaves** pinnate; leaflets 3–7, elliptic, obovate, ovate or oblong-lanceolate, 4–15 cm long, margins entire or apically few-toothed, apex mucronate, dull green. **Inflorescences** racemes or dense clusters, usually below leaves, on old wood; pedicels short; buds angular, dark pink, becoming light pink at anthesis. **Flowers** with calyx bell-shaped, 1.25–3.8 cm long; corolla 7.5–12 cm long, outside dark pink to magenta, inside cream and rose-pink, lobes slightly flaring, throat pale pink to white. **Fruits** capsules, ± cylindrical, 7–12 cm long, stalked. **Distr:** Moluccas to New Guinea, Solomon Islands and other adjacent small islands. **Ecol:** In various types of forests, including swampy mangrove forests to freshwater riverine forests and inland rainforests; 0–1,500 m. Fl. Jul–Aug. **Zones:** 10–12.

T. hillii Steenis, *Recueil Trav. Bot. Neerl.* ♀iv. 874, 894, 1927. (Syn: *Tecoma hillii*). [Fraser Island Creeper, Pink Trumpet Vine]. **Climbers**, rampant, woody, twining, slender, evergreen; secondary branches often pendent. **Leaves** pinnate; leaflets narrowly ovate, 2.5–8 × 1.5–3.5 cm, base rounded, ± oblique, margins entire, apex acuminate, glossy dark green, glabrous. **Inflorescences** racemes or dense clusters, to 10-flowered, usually borne on old and mature wood, sometimes from newer growth later in the season; pedicels 1–1.5 cm long. **Flowers** showy; calyx 2–2.5 cm long; corolla deep pink to reddish-pink, tubular-campanulate, 5–8.5 cm long, paler inside with purplish stripes, lobes with minute hairs, otherwise glabrous; stamens shorter than the style. **Fruits** capsules, cylindrical, 5.5 × 2 cm; seeds flattened, winged. **Distr:** Australia (Queensland, Fraser Island); rare. **Ecol:** In lowland and coastal rainforests, on sandy soil along banks of freshwater streams, often in swampy areas associated with *Melaleuca* species. **Zones:** 10–12.

T. montana Diels, *Bot. Jahrb. Syst.* 57: 497, 1922. [Mountain Trumpet Vine]. **Climbers**, woody, usually to 4 m long; stems scandent, wiry. **Leaves** pinnate; leaflets 3–9, oblong to obovate, 3–6 × 2.5–4.5 cm, terminal leaflet much larger, margins usually entire, apex acute, tip blunt, adaxially glossy green, abaxially paler. **Inflorescences** clusters, few-flowered. **Flowers** showy; calyx green, campanulate-infundibuliform, 6 mm Ø, teeth triangular, 4.5 mm long; corolla oblique-campanulate, 2–2.5 × 0.8–0.9 cm, throat creamish-white, tube slightly flattened, curved, outside pink (darker at base), inside creamish-white with long dark pink lines from base to corolla mouth, lobes creamish-white, triangular, 5–6 × 4–5 mm, pubescent; stamens 4, symmetrically arranged with 2 stamens on either side of stigma, filament dark pink, anthers white; style creamish-white, translucent with an apical stigma. **Fruits** capsules; seeds winged. **Distr:** New Guinea. **Ecol:** In montane regions; 1,000–4,000 m. Oct–Nov. **Zones:** 10–12.

T. speciosa W.R.B. Oliver, *Rec. Auck. Inst. Mus.* 3, 233, 1948. [New Zealand Bignonia, New Zealand Trumpet Vine]. **Lianas**, robust, woody, glabrous, >10 m long. **Leaves** imparipinnately 3–5-foliolate, coriaceous, adaxially dark green, abaxially paler; *lateral* leaflets broadly elliptic to broadly obovate, ~5–9(–13) × 3–7(–8) cm, base obliquely cuneate, margins entire, apex apiculate to retuse; *terminal* leaflet, broadly elliptic to broadly obovate, 8–12(–15) × 5–9 cm, base cuneately narrowed, apex retuse to apiculate; petiole to 6 cm long; petiolules *lateral* ~5 mm long, *terminal* to 2 cm long. **Inflorescences** corymbose cymes, to 30-flowered; peduncle 2–3 cm long; pedicels ~1 cm long. **Flowers** with green calyces, tubular, to ~2.5 × 1 cm, (3–)4(–5)-lobed; corolla creamy-white, green-flushed, tube cylindric, ~3 × 1.5 cm, limb 2-lipped, lower lip 2 × 1 cm, entire, upper lip 4-lobed, ~4 × 5 cm, subacute, recurved; stamens usually 4, to 5–6 cm long, 2 outer, 2 inner, anthers orange, ~1 cm long; staminode filiform; ovary bilocular, placentas 2 per locule, stigma truncately bilobed. **Fruits** capsules, woody, subcylindrical, to 19 × 3·5 cm, loculicidally splitting into 2 boat-shaped valves; seeds numerous, densely packed, flattened, ~1 × 2.5 cm, wing broad, membranous. **Distr:** New Zealand (Three Kings Island). **Ecol:** On a lowland stony stream bed within a mixed coastal forest. Fl. Jun–Jul. Fr. Feb–May (following year). **Zones:** 9–11. **Info:** Extinct in the wild except for a single known plant, from which all presently cultivated plants originated.

ACANTHACEAE

Genera ~202, species 3,520. **Herbs** or **shrubs**, rarely climbers or small trees, perennial sometimes annual, suffrutescent, armed or unarmed. **Leaves** exstipulate, opposite-decussate; lamina usually simple, margins entire or undulate-dentate, rarely crenate-dentate or lobed, adaxially mostly with cystoliths. **Inflorescences** commonly in dichasial cymes becoming monochasial, often in dense to lax spikes, racemes, panicles, sometimes fasciculate or verticillate, rarely solitary; bracts 1, 2 or more, often colored, frequently large, imbricate, persistent, rarely absent. **Flowers** bisexual, hypogynous, usually zygomorphic; calyx persistent, lobes 4–5, valvate or imbricate, sometimes spinescent or reduced to truncate or 10–15-dentate minute tube; corolla tube cylindrical or ventricose, limb 5-lobed with lobes sub-equal, spreading, or strongly 2-lipped, upper lip ± erect, usually entire or bifid, lower lip more spreading, trifid or trilobed, lobes imbricate or contorted; stamens 4 and didynamous or 2, longitudinally dehiscent; staminodes 0–3; ovary conical or oblong-cylindric, bicarpellate, ovules 2–10(–many) per locule, in 1 or 2 vertical rows, style simple, stigmas 1 or 2, filiform, minute. **Fruits** capsules, bivalved, biloculed, clavate to elliptic-ovoid or oblong-cylindric, terete or compressed, usually ± stipitate, elastically dehiscent from apex downwards, rarely drupaceous; seeds mostly flattened and lenticular or discoid, rarely globose or angular, borne on retinacula or papillae. **Distr:** Widely distributed from tropical to warm temperate regions.

Acanthus

Species 50. **Shrubs**, erect, spiny or unarmed. **Leaves** opposite; lamina pinnatifid, margins spinous-toothed, rarely entire, cystoliths absent. **Inflorescences** uninterrupted spikes, terminal; bracts imbricate, ovate, large; bracteoles in 2 pairs, oblong. **Flowers** with calyx, 3- or 4-merous, lobes imbricate, 2 outer ones larger; corolla tube short, horny, bilabiate, lower one elongate, obtusely trilobed; stamens 4, didynamous, filaments stout, attached to corolla throat, anther theca 1, linear-oblong, bearded along one margin; ovary bilocular, locules 2-ovuled, style slender, stigma bifid. **Fruits** capsules, erect, oblong to ellipsoid, sometimes elastically dehiscent; seeds 4, orbicular, muricate, glandular. **Distr:** Tropics and subtropics of the Old World; S Europe. **Cult:** Propagate from seeds or by division. Requires a fertile, well-drained soil in full or partial sun.

A. ebracteatus Vahl, *Symb.* 2: 75, 1791. (Syn: *Dilivaria ebracteata*). [Holly Leaf Mangrove, Sea Holly]. **Herbs**, shrubby, erect, multistemmed, to 1–2(–3) m high; stems terete, glabrous. **Leaves** hollylike, excretes salt; lamina oblong or elliptic, 6–15(–23) × 2.5–6.5(–7.5) cm, margins deeply lobed, toothed or pinnatifid, with sharp spines, rigid, glabrous, dark green; petiole short, base spinous or not. **Inflorescences** spikes, 18–30-flowered, elongate, somewhat lax, terminal; bracts 6 mm long, caducous, before anthesis; bracteoles 3–4 × 1–2 mm. **Flowers** white; calyx 8 mm long, sepals 4, fused, 0.7–1.5 cm long; corolla 1.8 cm long, white, turning brownish with age; ovary biloculed, locules 2-ovuled. **Fruits** capsules, rather square, slightly flattened, ~2 cm long, elastically dehiscent when ripe, projecting the seeds away from the plant; seeds 2 or more. **Distr:** SE Asia to Australia. **Ecol:** In saline tidal muds. Fl. Oct–Feb. **Zones:** 10–12. **Info:** This species is very similar to *A. ilicifolius*, differing mainly in the size and color of the flowers.

A. ilicifolius L., *Sp. Pl.* 639, 1753. (Syn: *Dilivana ilicifolia*). [Holly Mangrove, Sea Holly]. **Shrubs**, prickly, semiaquatic; often with small prop roots; stem to 1.25 m high, glabrous. **Leaves** oblong, 6.4–11 × 3–4.8 cm, pinnatifid, base acute, margins with few large spines, glossy, stiff; petiole 0.8–1 cm long, with 2 stipulelike spines at base. **Inflorescences** spikes, flowers crowded, erect, to 16.5 cm long; bracts lanceolate, 1 × 0.5 cm, mucronate; bracteoles in 2 pairs, oblong-lanceolate, 3–8 × 2–3 mm, mucronate. **Flowers** with calyx lobes obovate-oblong, 2 outer ones larger, 2 × 1 cm, 2 inner to 1.8 × 0.8 cm, all mucronate, scarious at margins, ciliolate; corolla pale to bright blue; disk annular, penicillate; stamens unequal, filaments stiff, anther thecae linear-oblong; style glabrous, stigma bifid. **Fruits** capsules, ovoid, 1.6–3.4 cm long, glabrous, elastically dehiscent; seeds orbicular, to 1.5 cm wide, rugulose or corrugated. **Distr:** S India through Sri Lanka to Myanmar, Malaysia, Philippine Islands and N Australia. **Ecol:** Along banks of estuaries and lagoons, and in marshy areas close to seashore. **Zones:** 10–12.

A. mollis L., *Sp. Pl.* 639, 1753. [Bear's Breeches]. **Herbs**, erect, to 70 cm high. **Leaves** mostly basal; lamina to 60 × 30 cm, base cordate, margins sinuate pinnatifid-dentate; petiole to 21 cm long, subquadrangular, 3 mm wide. **Inflorescences** spikes, large, terminal, to 65 cm tall; peduncles quadrangular, to 35 cm long; floral bracts ovate, to 3.5 × 1.5 cm, margins with 6–8 spiny teeth to 5 mm long, apex acute, spine-tipped; bracteoles linear, 1.5–2 cm long, spine-tipped. **Flowers** with upper calyx segment obovate, 4.5 × 2 cm, apically bilobed, lobes 2.5 cm long, acuminate, lower segment obovate, 3.5 × 1.2 cm, apex bilobed, lobes 2.2 cm long, acuminate, lateral lobes orbicular, 5 cm Ø; corolla whitish, rose or lilac, 1-lipped, to 4 cm long, lip trilobed, lobes suborbicular, 4 mm long, midlobe to 1.9 cm wide, lateral lobes 1.4 cm wide, tube 4 mm long; stamens included inside upper calyx lobe, filaments sigmoid, to ~2 cm long, indurate, anthers 1 cm long. **Fruits** capsules, oblong-cylindrical, ~3–4 cm long; seeds 4. **Distr:** Italy. **Ecol:** In coastal forests, riparian areas, and disturbed scrublands or shrublands. Fl. Oct–Apr. Fr. Nov–May. **Zones:** 6–10.

A. montanus T. Anders., *Journ. Linn. Soc.* 7: 37, 1863. (Syn: *A. barteri*, *Cheilopsis montana*). [Alligator Plant, False Thistle, Mountain Thistle]. **Undershrubs**, herbaceous, glabrous, 1–2 m high; stem woody or semiwoody at base, sparsely branched or unbranched. **Leaves** ovate, obovate or oblong-lanceolate, to 30 × 15 cm, base caudate, margins deeply ± pinnatifid, irregularly lobed, lobes acute and spine-tipped, apex acuminate, papery, adaxially very dark green, glossy, midrib lighter and prominent, abaxially pale green. **Inflorescences** spikes, compact, 20–30 cm tall, terminal, glabrous; bracts green, leaflike, obovate-oblong, spiny, incised-dentate, teeth setaceous, apex acute; scabrid bracts and calyx lobes whitish with green lines; bracteoles linear-lanceolate, dentate. **Flowers** white or pale pinkish-white, marked with pale purple lines; calyx 4-parted, to 2.5 cm long, spinose; corolla to 3.25 cm long, rose or white, mauve-tinted, 1-lipped, tube short, lobes 3–5; stamens 4, curved, anthers unilocular, ciliate-barbate, erect. **Fruits** capsules, biloculed, compressed; seeds compressed. **Distr:** Tropical W Africa. **Ecol:** In thickets and montane forest understory. **Zones:** 10–12. **Info:** This species like *A. pubescens*, lacks the basal rosette of leaves.

A. pubescens (Oliv.) Engl., *Fl. Trop. Afr.* 2: 390, 1892. (Syn: *A. arboreus* [misapplied], etc.). [Prickly/Spiny Acanthus]. **Herbs**, large, shrublike, erect, robust,

Below: *Andrographis paniculata, Aphelandra aurantiaca, Aphelandra deppeana, Aphelandra scabra.*

Above: *Aphelandra sinclairiana*, *Aphelandra squarrosa*, *Asystasia gangetica*.

1–3(–6) m high; stems quadrangular. **Leaves** prickly, oblong-elliptic to elliptic-lanceolate, usually to 20 × 9 cm, base cordate to rounded, margins irregularly lobed, lobes spinescent, coarsely undulate-dentate, apex acute, spinescent, secondary nerves usually 14 pairs; petiole 1–2 cm long. **Inflorescences** spikes, many-flowered, terminal, erect, ~20 cm tall; bracts ovate, usually to 2 × 1.5 mm, margins spinescent, apex acute to acuminate, densely puberulous; bracteoles linear-lanceolate, 1.5 × 0.2–0.3 cm, acuminate. **Flowers** bright pink or magenta; sepals 4, obtuse, slightly ciliate, *anterior* oblong-ovate, 1.5 × 0.9 cm, denticulate, *posterior* ovate, usually 1.5 × 1.1 cm, *lateral* elliptic-ovate, 1 × 0.7 cm; corolla tube 7 mm long, interior densely pubescent, limb 5-lobed, usually 3.5 × 3.5 cm; stamens 4, filaments 2 cm long, glabrous, anthers 5 mm long, densely ciliate; ovary 2 × 1 mm, glabrous, style 2.5 cm long, shortly bifid. **Distr:** E Africa. **Ecol:** In grasslands; <1,500 m. **Zones:** 10–12. **Info:** Confused with *A. arboreus* (Turrill 1913).

Andrographis

Species 20. **Herbs** or **undershrubs**; stem quadrangular, occasionally narrowly winged or acutely quadrangular. **Leaves** shortly petiolate; lamina with margins entire, marked with cystoliths. **Inflorescences** lax racemes or panicles, axillary or terminal; bracts and bracteoles linear-lanceolate. **Flowers** with calyx 5-lobed, lobes narrow, subequal; corolla tube narrow, limb 2-lipped, lobes imbricate in bud, upper lip erect, bifid, lower one trilobed; stamens 2, exserted, filaments setulose, broadened at base, anther thecae equal, each bearded at base with a tuft of hairs; ovary bilocular. **Fruits** capsules, linear-oblong, dorsally compressed; seeds 4–12, oval to subquadrate, rugose or pitted, glabrous. **Distr:** Tropical Asia. **Cult:** Propagate from seeds. Requires a well-drained soil in full sun.

A. paniculata (Burm. f.) Wall. ex Nees, *Pl. As. Rar.* 3: 116, 1832 (Syn: *Justicia paniculata*). **Herbs**, to 75 cm high; stems sharply quadrangular, glabrous. **Leaves** ovate-elliptic to elliptic-lanceolate, 3.5–8.5 × 1.5–2.3 cm, base tapered, glabrous; petiole 3–4 mm long. **Inflorescences** racemes, terminal and axillary, forming pyramidal panicles, 5–20 cm long; peduncles slender, glabrous; bracts linear-lanceolate, 1.5 × 0.5 mm; pedicels 5–20 cm long. **Flowers** with calyx lobes linear-lanceolate, 3 mm long, glandular-bristly outside; corolla white, pink-tinged, tube 6–9 mm long, glandular-bristly outside, lower lip of limb to 8 mm across lateral lobes, violet-spotted, filaments flattened, bearded down one side, anther thecae with a tuft of hairs at base; style sparsely bristly. **Fruits** capsules, erect, linear-oblong, 1.8–2 × 0.4 cm, compressed, furrowed on opposite sides, mucronate, sparsely bristly; seeds 12, orbicular. **Distr:** India, Sri Lanka, Malay Peninsula, China, Indo-China, Thailand. **Ecol:** In shaded areas of moist and dry lowlands. Fl. Jan–Apr, Sep. **Zones:** 10–12.

Aphelandra

Species ~200. **Shrubs** or **herbs**, basally woody, erect. **Leaves** petiolate; lamina, oblong to elliptic, usually large, margins entire, spinose-dentate, crenate or sometimes sinuately lobed, cystoliths absent. **Inflorescences** spikes, terminal or axillary; bracts large, showy, ovate to lanceolate, entire, dentate or ciliate, imbricate; bracteoles akin to calyx lobes or reduced; flowers sessile. **Flowers** large, red, yellow, orange or white; calyx deeply 5-lobed, lobes ± equal or the posterior lobe broader, lanceolate; corolla tube straight or incurved, limb 2-lipped, upper lip erect, shallowly bilobed or entire, lower lip reflexed or patent, trilobed; stamens 4, all fertile, ± exserted or shorter than upper lip; anthers monothecous, dorsifixed, apex hairy, muticous; ovary bilocular, locules 2-ovuled, style long, stigma obtuse or shallowly bifid. **Fruits** capsules, oblong-clavate, base slightly contracted; seeds 4, compressed, ± orbicular, silky-hairy. **Distr:** Tropical America. **Cult:** Propagate from seeds or cuttings. Requires a moist, well-drained soil in partial sun.

A. aurantiaca (Scheidw.) Lindl., *Bot. Reg.* 31: pl. 12, 1845. (Syn: *A. simplex*, *Hemisandra aurantiaca*). [Fiery Spike]. **Herbs** or **shrubs**, erect, unbranched, perennial, to 1.5 m high; young stems quadrangular, glabrous, older stems terete. **Leaves** elliptic to ovate-elliptic, 8–30 × 3–12.5 cm, base attenuate, occasionally obliquely decurrent, margins undulate to entire, apex acuminate, glabrous; petiole 1 cm long. **Inflorescences** quadrifarious spikes, to 15 × 2 cm (excluding corollas), mostly terminal; rachis puberulous, nodes flattened; flowers sessile; floral bract imbricate, elliptic or ovate, to 3 × 1 cm, margins ciliolate, serrate, teeth 1.5 mm long; bracteoles 2 per flower, narrowly lanceolate, to 1 cm long, striate-ridged. **Flowers** with calyx 5-merous, 1 cm long, posterior segment oblong, 2.5 mm wide, apex 3-toothed; corolla red, orange-red or scarlet, tube to 4 × 0.3 cm, bilabiate, upper lip oblong-obovate, to 2 × 1 cm, apiculate, forming a partial hood over anthers, lower lip trilobed, elliptic, midlobe 2 × 1.2 cm, acute, lateral lobes elliptic-ovate, to 1.2 × 0.8 cm, acute; style to 4.6 cm long. **Fruits** 1.6 cm long; seeds hispidulous. **Distr:** Mexico to Bolivia. **Ecol:** In forests, forested slopes. **Zones:** 11–12.

A. deppeana Schlecht. and Cham., *Linnaea* 5: 96, 1830. (Syn: *A. pulcherrima*, etc.). **Shrubs**, to 4 m high; young stems subquadrangular to terete, glabrous to velutinous; internodes 3–10 cm long. **Leaves** elliptic, to 22 × 8 cm, base attenuate, margins entire to undulate, apex acuminate, adaxially glabrous to scabridulous, abaxially strigose to velutinous; petiole to 5 mm long, strigose. **Inflorescences** dense spikes, to 19 × ~1 cm, terminal; bracts 1 per flower, elliptic, imbricate, to 1.5 × 0.6 cm, acuminate, strigose; bracteoles 2 per flower, lanceolate, to 6 × 1.5 mm, keeled; flowers sessile. **Flowers** with calyx segments lanceolate, ~8 mm long, 2 *anterior* segments ~1.5 mm wide, 2 *lateral* segments ~1 mm wide, *posterior* segment ~3 mm wide; corolla pink, purplish-red, scarlet or red, bilabiate, to 3 cm long, mouth to 4 mm wide, lips ovate, upper lip bilobed, ~1 cm long, lobes 3 mm long, acuminate, lower lip slightly longer, acuminate; stamens with filaments hispid, exserted. **Fruits** elliptic, to 1.5 × 0.6 cm, acute; seeds 2 per locule, flattened, irregular, ~3 mm across, dark brown. **Distr:** S Mexico to N South America; West Indies. **Ecol:** In rainforests; <1,600 m. **Zones:** 10–12.

A. scabra (Vahl) Sm., *Cyclopaedia* 39, Aphelandra n. 3, 1818. (Syn: *Justicia scabra*, etc.). **Shrubs**, 1–3.5 m high; young stems subquadrangular to terete, glabrous to velutinous, internodes 3–10 cm long. **Leaves** elliptic, to 25 × 8 cm, base attenuate, margins entire to undulate, apex acuminate; petiole to 5 mm long. **Inflorescences** dense spikes, to 19 cm long, 1 cm Ø excluding corollas, terminal; peduncle subsessile; rachis pilose; bracts imbricate to 1.5 × 0.6 cm, base cuneate, apex acuminate, margins ciliate, teeth 4–8; bracteoles lanceolate, to 6 × 1.5 mm, keeled. **Flowers** to 3 cm long; calyx segments lanceolate, to ~8 mm long, anterior segments ~1.5 mm wide, lateral segments ~1 mm wide, posterior segment ~3 mm wide; corolla pink, purplish-red, scarlet or red, bilabiate, to 3 cm long, mouth 4 mm wide, lips ovate, upper lip bilobed, to 1 cm long, lobes 3 mm long, acuminate, lower lip slightly longer, acuminate; stamens with filaments hispid, exserted. **Fruits** elliptic, to 1.5 × 0.6 cm, acute; seeds 2 per locule, flattened, irregular, ~3 mm across, dark brown. **Distr:** Mexico to northern S America. **Ecol:** In tropical dry forests <1,100 m. Fl. throughout the year. **Zones:** 10–12.

A. sinclairiana Nees, *Bot. Voy. Sulfur*, 146: pl. 47, 1844. [Orange Shrimp Plant, Coral Aphelandra, Panama Queen]. **Shrubs** or **small trees**, to 6 m high; younger stems terete to subquadrangular, velutinous. **Leaves** narrowly elliptic to suboblanceolate, to 38 × 10 cm, base attenuate, margins entire to shallowly crenate, apex acuminate; petiole 1–2 cm long. **Inflorescences** spikes, in clusters of up to 10, terminal and subterminal, spikes to 15 × 2 cm; rachis hispid to woolly; bracts orange-red, densely imbricate, obovate-rhombic, to 2 × 1.5 cm; flowers falling at maturity when bracts are expanded. **Flowers** with calyx 5-merous, segments free, lanceolate, *posterior* segment to 9 × 3 mm, other segments to 9 × 2 mm, striate; corolla purplish-red, bilabiate, tube funnelform, to 4.5 × 0.35–0.65 cm, upper lip erect, to 1.7 × 0.9 cm wide, bilobed, lobes to 7.5 mm long, acuminate, keeled, lower lip spreading, midlobe lanceolate, to 2.5 × 0.8 cm, keeled, lateral lobes triangular, one side attached to upper lip, acute; stamens exserted, just below upper lip. **Fruits** clavate, to 1.8 × 0.3–0.6 cm, obtuse; seeds 4, flattened, ~3.5 mm Ø, **Distr:** Costa Rica, Panama. **Ecol:** Mostly in rainforests. Fl. Sep–Apr. **Zones:** 10–12.

A. squarrosa Nees, *Fl. Bras.* (Martius) 9: 89,

Above: *Avicennia alba, Barleria cristata, Barleria micans.*

1847. [Saffron Spike, Zebra Plant]. **Shrubs**, ± succulent, evergreen, 30–60(–90+) cm high. **Leaves** crowded; lamina ovate to ovate-elliptic, 25–30 × 7–10 cm, base narrowed into petiole, margins slightly undulate, thick, waxy, adaxially deep glossy green with white venation, abaxially paler; petiole 2.5–3.3 cm long, apically winged. **Inflorescences** spikes, 1–3, erect, terminal, 10–20 cm long, central spike ~11 × 4 cm and without flowers; bracts overlapping, ovate, 2.5–3.8 × 1.8–2.5 cm, waxy yellow to orange. **Flowers** with calyx 0.8–1 cm long; corolla yellow, 3.8–5 cm long, protruding well beyond subtending bracts, glabrous, tube ± cylindrical, 3–3.5 cm long, lip trilobed, lobes slightly recurved, hairy inside, midlobe ~7 mm long, lateral lobes shorter; stamens equal to, or protruding slightly beyond upper lip, anthers ~5.5 mm long. **Fruits** capsules, ~8 mm long, rarely formed in cultivation. **Distr:** SE Brazil. **Ecol:** In lower montane rainforests; 60–800 m. Fl. Oct–Feb. **Zones:** 11–12.

Asystasia

Species ~40. **Herbs** or **undershrubs**, erect or ± scandent. **Leaves** petiolate; lamina simple, margins entire to crenulate, apex acute to acuminate, thin, adaxially with numerous cystoliths. **Inflorescences** simple or compound racemes or spikes, loose or compact; bracts lanceolate, not 4-ranked, acute; bracteoles linear, shorter than calyx; flowers shortly pedicellate. **Flowers** variable in color; calyx deeply 5-lobed, lobes equal, linear-lanceolate, hairy, acute or acuminate; corolla infundibuliform, tube narrow or ventricose, limb subequally 5-lobed, lobes ± orbicular, obtuse, imbricate, patent; stamens 4, all fertile, didynamous, included, anthers dithecous, oblong, unequal, each lobe muticous or basally minutely beaked; ovary oblong-cylindrical, hairy, bilocular, locules 2-ovuled, style basally hairy, stigma minutely bifid or ± capitate. **Fruits** capsules, stipitate, elliptic-oblong, flattened, pubescent; seeds 2–4, ± orbicular or angled, compressed, glabrous, on hooked funicles. **Distr:** Old World tropics. **Cult:** Propagate from seeds or cuttings. Requires a well-drained soil in full sun.

A. gangetica (L.) T. Anders., *Enum. Pl. Zeyl.* 235, 1860. (Syn: *A. coromandeliana, Justicia gangetica*). [Chinese Violet, Creeping Foxglove, Ganges Primrose]. **Subshrubs**, suberect to climbing; twigs 4-angled, pilose to glabrescent. **Leaves** elliptic-ovate to deltoid, 2–5 × 1.5–3.5(–4) cm, gray-pubescent to glabrate, base truncate to rounded, margins crenulate to entire, apex acute or shortly acuminate; petiole to ~2 cm long. **Inflorescences** racemes, 1-sided, lax, terminal; bracts lanceolate, 2.5–3 mm long, pilose. **Flowers** with calyx lobes lanceolate, (5–)7–9 mm long, sparsely pubescent; corolla yellow to creamy-white, throat purple-streaked, tube to 2.5 cm long, narrowed below, widened to ~1 cm at the throat, hairy outside, lobes ± orbicular, to 1.5 cm long; anthers oblong, 2.5–3 mm long; style to 2 cm long. **Fruits** capsules, explosively dehiscent, oblong, to 2.8 cm long including stipe, pubescent, 4-seeded. **Distr:** India, Sri Lanka, SE Asia, Tropical Africa. **Ecol:** Fl. Aug–Feb. **Zones:** 10–12. **Info:** *A. gangetica* ssp. *micrantha* (Syn: *A. intrusa*, etc.) differs in having pure white corolla with a large violet spot on lower lip, tube 1.4–1.8 cm long, ~1 cm Ø at the mouth, dorsal and lateral lobes reflexed, lower strongly pleated.

Avicennia

Species 8. **Shrubs** or **trees**, maritime; branches terete, sometimes 4-ridged when young. **Leaves** opposite. **Inflorescences** small spikes or capitula; bracts and bractlets ovate, shorter than calyx, persistent; flowers opposite, sessile. **Flowers** small; calyx cup-shaped, deeply 5-lobed, lobes overlapping, persistent; corolla subactinomorphic, campanulate, shortly inserted on an inconspicuous disk, lobes 4 or 5, upper lobe often broader than others; stamens 4, adnate to apical part of corolla tube; ovary imperfectly 4-locular, with a free central winged placenta, ovules pendulous. **Fruits** capsules, subtended by persistent calyx, dehiscent into 2 leathery valves. **Distr:** Tropics and subtropics of both hemispheres. **Ecol:** In maritime habitats. **Cult:** Propagate from seeds or seedlings. Requires a wet area in full sun; salt-tolerant. **Info:** Genus previously placed in the monogeneric family Avicenniaceae.

A. alba Blume, *Bijdr.* 821, 1826. [White Mangrove]. **Shrubs** or **small trees**, maritime, fast-growing, to 10–18 m tall; branches pale. **Leaves** lanceolate, 5–7.5 × 2–2.5 cm, base acute, apex very acute, adaxially waxy, adaxially glossy dark green, abaxially whitish-tomentose; petioles ~3 mm long. **Inflorescences** heads, 3–4, terminal or solitary and axillary; peduncle usually stout, angular; flowers sessile. **Flowers** yellow, ~4 mm Ø; calyx 5-partite almost to base, sepals broadly ovate or suborbicular, 3–4 mm long, obtuse, concave, ciliate; corolla 1.2 cm Ø, lobes 4, subequal, ovate, acute; stamens included; ovary villous, style villous, shorter than the ovary, 2-cleft. **Fruits** capsules, ovoid, 1.6–2 cm long, compressed, apiculate, bright-yellow and smooth when ripe; seed viviparous. **Distr:** India, SE Asia, Malaya, W Pacific, N Australia. **Ecol:** In maritime habitats such as mangroves and tidal creeks. Fl. Mar–May. **Zones:** 10–12. **Info:** This species have commercial and medicinal uses.

Barleria

Species ~230. **Herbs** or **shrubs**, armed or unarmed. **Leaves** petiolate to almost sessile; lamina simple, margins entire, surfaces with cystoliths. **Inflorescences** spikes, axillary or terminal; bracts entire or serrate, often spinescent or ciliate, rarely absent; bracteoles linear or absent. **Flowers** often large and showy, yellow, blue-purple or white; calyx deeply 4-lobed, lobes imbricate, unequal, narrow; corolla infundibuliform or salverform, tube slender, straight or incurved, limb subequally 5-lobed or obscurely 2-lipped, upper lip 4-lobed, lobes patent, imbricate, rounded or ovate; stamens 4, all fertile and ± exserted, or 2 posterior ones staminodial and partly sterile or absent, anthers dithecous; ovary oblong-conical, 2 or 4-ovuled; style thickened at apex, stigma minutely bifid or entire. **Fruits** capsules, ovoid-oblong to elliptic-obovoid, flattened, ± stipitate, pointed above, seminiferous below middle; seeds 2 or 4, flat, ovate or orbiculate, covered with hygroscopic hairs, rarely

Below: *Barleria lupulina, Barleria prionitis, Barleria repens.*

NT

Above: *Barleria strigosa, Blepharis integrifolia, Clinacanthus nutans, Crossandra infundibuliformis.*

glabrous, on hooked funicles. **Distr:** Tropical and subtropical regions. **Cult:** Propagate from seeds or cuttings. Requires a fertile, moist, well-drained soil in full or partial sun.

B. cristata L., *Sp. Pl.* 636, 1753. (Syn: *B. ciliata*, etc.). [Blue Bell, Crested Philippine Violet]. **Shrublets**, erect, to 2 m high, much-branched, unarmed; twigs hairy, densely at nodes. **Leaves** elliptic-oblong to lanceolate, 3–12.5 × 1–3.5 cm, base attenuate, margins entire, apex acute-acuminate, hairy; petiole 1–2 cm long, hairy. **Inflorescences** spikes, 1–5-flowered, short, axillary or terminal; bracts absent or lanceolate, 0.8–1.2 cm long, margins bristly, scabrous, acute; bracteoles absent. **Flowers** purple, blue, pink or white, 4.5–5 cm long; calyx deeply 4-cleft, outer 2 lobes much larger than the inner pair, ovate-lanceolate, 1.5–2.5 × 0.6–0.8(–1) cm, acuminate, margins bristly toothed, hairy or glabrescent, inner 2-lobes linear, 7–8 mm long, pointed; corolla glandular-pubescent outside, tube 3.5–4 cm long, infundibuliform, lobes oblong-obovate, unequal, to 2 cm long; stamens 2, filaments hairy, anthers oblong, ~3 mm long; staminodes 2, anthers sterile; ovary oblong-conical, 4-ovuled, style ~4 cm long. **Fruits** ellipsoid, 1.5–2 cm long, glabrous, ends pointed; seeds 4, orbicular, ~4 mm Ø, hairy. **Distr:** Bangladesh, China, India, Myanmar, Nepal, Pakistan, Philippines, Sri Lanka. **Ecol:** Fl. Nov–Feb. **Zones:** 10–12.

B. lupulina Lindl., *Bot. Reg.* t. 1483, 1833. [Hop-headed Barleria, Hophead Philippine Violet]. **Shrubs**, to ~1.5 m high, much-branched; branches purplish, armed at the nodes with slender, 1–2 cm long spines. **Leaves** linear-oblong to linear-lanceolate, 3–12 × 0.9–1.4 cm, base cuneate, apex obtuse, mucronulate, glabrous, midrib usually reddish or yellowish; petiole to 6 mm long. **Inflorescences** dense spikes, erect, hoplike, 5–9 × 2 cm, terminal; bracts erect, green with brown tips, broadly ovate, 1.5–2 × 1.2 cm, mucronulate, ciliolate, on the back at base with cupular glands; bracteoles lanceolate, to 4 × 1 mm, mucronulate. **Flowers** with calyx lobes broadly ovate, unequal, 6–9 × 1–4 mm, shortly aristate; corolla trumpet-shaped, yellow, tube to 4 cm long, bent at base, expanded above; stamens 2, exserted; ovary glabrous, 1-ovuled, style filiform. **Fruits** ovoid, 1.2–1.3 cm long, glabrous; seeds ovate. **Distr:** Mauritius. **Ecol:** Fl. throughout the year. **Zones:** 10–12.

B. micans Nees, *Bot. Voy. Sulfur* 146, 1844. (Syn: *B. discolor*). [Giant Yellow Shrimp]. **Herbs**, erect, to 1.5 m high; stems strigose. **Leaves** lanceolate to oblanceolate, to 25 × 7.5 cm, base attenuate, apex acuminate, adaxially ± strigose; petiole to 4 cm long. **Inflorescences** spikes, dense, cylindrical, to 7 cm long, terminal; bracts ovate, to 3.5 × 1.5 cm, margins with spiny hairs to 2 mm long, apex acuminate, chartaceous; bracteoles 2, oblanceolate, to 2 cm long, ciliate. **Flowers** yellow, blue-purple when dry, sessile; calyx segments unequal, *anterior* segment elliptic, to 2.5 × 1 cm, marginal spines to 2 mm long, apex bidentate, *lateral* segments lanceolate, to 2 × 0.2 cm, aristate; corolla bilabiate, tube to 3 cm long, lobes obovate, ~1.5 cm long; *perfect* stamens 2, filaments glabrous, to 4 cm long, anthers ~5 mm long, *imperfect* stamens staminodal, ~0.2 mm long, with a central spur; ovary pandurate, ~5 mm long, style ~4.5 cm long, stigma cylindrical, to 1 mm long. **Fruits** elliptic, to 2 × 1.6 cm, acuminate; seeds 4, ~5 mm long. **Distr:** Mexico, British Honduras to El Salvador and Panama, Colombia. **Ecol:** In thickets and forests; <1,300 m. **Zones:** 10–12.

B. prionitis L., *Sp. Pl.* 636, 1753. [Porcupine Flower]. **Shrubs**, much-branched, armed; spines (1–)2–4, axillary, grayish-white, 1–2 cm long, sharp. **Leaves** elliptic or ovate-elliptic to obovate, (4–)6–12(–17) × 2–5(–6) cm, base attenuate, margins entire, ciliate, apex bristly, acuminate, adaxially scabrid-lineolate, abaxially sparsely pubescent, petiole 0–2.5 cm long. **Inflorescences** short terminal spikes, or often sessile and solitary flowers on lower leaf axils; bracts ± oblong, 1–2.5 × 0.3–0.5 cm, keeled, spine-tipped; bracteoles linear-subulate, 1–2 cm long, spinescent. **Flowers** large, golden-yellow, to 7 cm long; calyx lobes spine-tipped, outer 2 lobes broadly oblanceolate, 1.5–1.7 cm long, acuminate, glabrous, inner ones narrowly ovate-lanceolate, 1–1.2 cm long, mucronate; corolla ~1.5 cm Ø, tube 2–2.5 cm long, limb nearly as long as tube, lobes oblong-ovate, obtuse; fertile stamens 2, filaments exserted, anthers 2–6 mm long; staminodes 2, filaments short, included, hairy at insertion; ovary ~3.5 mm long, style to 4 cm long. **Fruits** ovoid-conical, 1.5–2 × 0.8–1 cm, beaked; seeds 2, ± orbicular, ~6 × 5 mm, hairy. **Distr:** Tropical Africa, eastwards to SE Asia through Arabia. **Ecol:** In thickets. Fl. Oct–Apr. **Zones:** 10–12.

B. repens Nees, *Prodr.* (DC.) 11: 230, 1847. [Coral Creeper, Creeping Barleria, Pink Ruellia, Small Bush Violet]. **Herbs**, 0.3–0.7(–2) m high, ± prostrate to almost creeping (spreading to ~1 m Ø, rooting where the branches touch the ground), evergreen. **Leaves** elliptic-ovate, 2.5–5 cm long, base gradually tapered into petiole, apex rounded to acute, adaxially glossy green; petiole very short. **Inflorescences** with flowers solitary, axillary; bracts 2, ovate, erect, below flower, persistent, enclosing the fruit, turning semitransparent and papery with dark prominent venation. **Flowers** relatively larger than the leaves, ~5 cm long and almost as wide; corolla trumpet-shaped, purple-mauve, salmon-pink or almost red, limb deeply 5-lobed, 2.5–4 cm Ø; stamens and style inserted near the mouth of corolla tube. **Fruits** slender, oblong, ~2 cm long, explosively dehiscent; seeds 4, flattened, disk-shaped, ~6 mm Ø, **Distr:** S Africa. **Ecol:** Fl. Feb–Apr. Fr. Mar–May. **Zones:** 10–12.

B. strigosa Willd., *Sp. Pl.* 3: 379, 1803. (Syn: *B. coerulea*). **Herbs**, shrubby, to 80 cm high, much-branched; stem subterete, coarsely fulvous-strigose. **Leaves** ovate, 11–18 × 4.5–9 cm, base cuneate, long-decurrent into petiole; petiole to 2.5 cm long, subulate, coarsely fulvous-strigose. **Inflorescences** spikes, dense, to 5 cm long, terminal and axillary; bracts oblong to elliptic-oblong, sharply acute, shortly strigose on nerves outside, pectinate-ciliate; bracteoles ovate, to 2 × 1 cm, strigose on nerves outside, pectinate-ciliate. **Flowers** with *outer* calyx lobes compacted, rhomboid-ovoid, 2.4–3 × 1.6–1.8 cm, acute, 8–11-nerved from base, pectinate-ciliate, *inner* calyx lobes linear-lanceolate, to 1 × 0.2 cm, long-acuminate, densely strigose; corolla tube linear-cylindric, to 3.5 cm long, expanded above, white, limb 2-lipped, upper lip 4.5 cm across, lobes obovate-oblong, 2 × 1–1.5 cm, pale blue, lower lip similar, pale blue; stamens slightly exserted; staminodes included at base of expanded part of corolla tube; disk cupular; style exserted. **Fruits** ellipsoid, 1.4–1.8 × 0.5–0.6 cm, glabrous, shiny. **Distr:** India, Indonesia, Myanmar, Sri Lanka. **Ecol:** In moist places, often near waterways. Fl. Sep. **Zones:** 10–12.

Blepharis

Species ~100. **Herbs**, perennial, prostrate-erect, or undershrubs. **Leaves** sessile or petiolate, opposite or pseudo-whorled; lamina simple, margins entire or spinescent-toothed. **Inflorescences** axillary clusters, strobilate spikes, or flowers solitary; bracts 4-ranked, closely imbricated, spinose, persistent; bracteole lanceolate to spathulate, hairy. **Flowers** with calyx unequally deeply 4-lobed, 2 inner lobes shorter than the 2 outer, imbricate, persistent; corolla tube shorter than the limb, indurate, limb not 2-lipped, upper lip absent or rudimentary, lower lip almost flat, usually trilobed, imbricate; stamens 4, didynamous, all fertile, exserted, filaments inserted at base of corolla tube, thickened, anthers monothecous, white-hairy, the other cell sterile or absent; ovary bilocular, 4 or 2-ovuled, style bifid. **Fruits** capsules, oblong or ellipsoid, compressed, 2 or 4-seeded; seeds flattened. **Distr:** Mainly in the Old World tropics, Mediterranean, S Africa and Madagascar. **Cult:** Propagate from seeds. Requires a well-drained soil in full sun.

B. integrifolia (L. f.) E. Meyer ex Krauss, *Flora* 28: 70, 1840. (Syn: *Acanthus integrifolius*, etc.). **Herbs**, procumbent; stem widely branched from base, sometimes mat-forming, not rooting at nodes, sparsely puberulous to pubescent, hairs downward-pointed. **Leaves** oblong-linear, 1–2 × 0.4–0.5 cm, or obovate to obovate-spathulate, 0.8–1.4 × 0.4–0.6 cm, margins reflexed, apex mucronate, glabrous. **Inflorescences** clusters, axillary; bracteoles ovate-oblong, to 1.5 × 0.5 mm, mucronate; bracts concave, spinous; basal bracts orbicular, 2.5–3 × 1 mm, lateral

Above: *Eranthemum pulchelum, Fittonia albivenis, Graptophyllum pictum, Gymnostachyum ceylanicum.*

ones orbicular-spathulate, 4 × 2 mm, inner ones obovate-oblong, 7–8 × 2 mm, mucronate. **Flowers** with outer calyx lobes linear-oblong, 8 × 1.5 mm, 1-nerved, long-mucronate, inner lobes slightly unequal, 1 × 0.25 cm, one lobe of the pair bifid at apex, the other mucronate, both lobes trinerved; corolla bright violet. **Fruits** capsules, broadly ovoid, 6 × 3 mm, glabrous, black, shiny. **Distr:** S Africa, India, Sri Lanka. **Ecol:** Often in coastal areas of lowlands. Fl. Jan–Mar. **Zones:** 10–12.

Clinacanthus

Species 2. **Herbs**, tall. **Leaves** with margins entire or toothed. **Inflorescences** pendent cymes, in large terminal clusters on ends of branches; bracts small, linear-lanceolate. **Flowers** large and showy; pedicels short; calyx deeply 5-parted, segments subequal; corolla tube narrow, incurved near base, funnelform, limb 2-lipped, abaxial lip slightly spreading, trilobed, lobes imbricate, adaxial lip erect, narrow, bilobed; stamens 2, inserted at throat, as long or shorter than limb, anther cell 1, narrowly oblong, spurless, base rounded. **Fruits** capsules, clavate, stalked; seeds 2 per cell, supported by a retinaculum. **Distr:** Cambodia, China, Indonesia, Laos, Malaysia, Vietnam. **Cult:** Propagate from seeds. Requires a moist, well-drained soil in full sun.

C. nutans (N.L. Burman) Lindau, *Bot. Jahrb. Syst.* 18: 63, 1894. (Syn: *Clinacanthus burmanni* [misapplied], *Justicia nutans*). [Snake Plant]. **Herbs**, tall, erect, sometimes rambling shrubs; stems cylindric, yellow when dry, densely striate, subglabrous. **Leaves** lanceolate to ovate-lanceolate, 5–11 × 1–4 cm, base oblique, margins subentire, apex caudate-acuminate, papery, secondary veins 5 or 6 per side and convex on both surfaces when dry; petiole 5–7+ cm long. **Inflorescences** ~1.5 cm long, glandular; bracts linear, ~8 mm long, acute. **Flowers** with calyx segments ~8 mm long, acuminate; corolla dark red, 4 cm long, pubescent; stamens and pistil glabrous. **Distr:** Cambodia, China, Indonesia, Laos, Malaysia, Vietnam. **Ecol:** In forests and thickets; low elevations. **Zones:** 10–12.

Crossandra

Species ~50. **Shrubs** or **herbs**, glabrous or hairy, unarmed. **Leaves** opposite or whorled; lamina with margins entire or ± crenate. **Inflorescences** mostly subterminal spikes, long-pedunculate; bracts 4-ranked, large, ovate-lanceolate, imbricate; bracteoles linear-lanceolate, shorter than calyx, hairy. **Flowers** showy, yellow or red; calyx deeply 5-lobed, lobes unequal; corolla tube narrow, longer than the limb, incurved, limb deeply divided to throat forming a 5-lobed lip, lobes unequal, imbricate; stamens 4, all fertile, included, anthers longer than filaments, monothecous, sterile theca bearded throughout, muticous, connective mucronate; ovary bilocular, locules 2-ovuled, style thickened upward, stigma minutely bifid. **Fruits** capsules, oblong, flattened, ± quadrangular; seeds 4, ellipsoid-ovate, beset with fringed scales or hairs. **Distr:** Tropical Africa, Madagascar, Arabia and Indo-Pakistan subcontinent. **Cult:** Propagate from seeds or cuttings. Requires a humus-rich, well-drained soil in full sun or partial shade.

C. infundibuliformis (L.) Nees, *Pl. Asiat. Rar.* 3: 98, 1832. (Syn: *C. axillaris*, etc.). [Firecracker Flower, Orange Tropic Flame]. **Undershrubs**, small, glabrous, evergreen, to 1(–1.5) m high; branchlets terete. **Leaves** crowded near the inflorescence; lamina ovate-elliptic to lanceolate-oblong, 5–10 × 2–5 cm, base attenuate, margins undulate-crenate, apex acute, glabrous; petiole 1–4 cm long. **Inflorescences** dense spikes, terminal and axillary, to 15 cm long; bracts lanceolate, 1.2–1.5 × 0.4–0.5 cm, mucronate, glandular-pubescent to glabrescent; bracteoles linear-lanceolate, hairy, mucronate. **Flowers** bright orange-scarlet, throat yellow; calyx lobes hairy, lanceolate, 7–9 mm long, outer 2 obtuse or bifid, inner smaller; corolla tube ~2.5 cm long, limb ~1.5 cm long; stamens inserted at the middle of tube; ovary 2.5–3 mm long, style filiform, to 2 cm long, glabrous. **Fruits** oblong, 1–1.2 cm long, glabrous; seeds densely clothed with fringed scales. **Distr:** India, Nepal, Sri Lanka. **Ecol:** Fl. Dec–Mar. **Zones:** 10–12.

Eranthemum

Species ~45. **Shrubs** or **herbs**, suffrutescent, perennial. **Leaves** petiolate; lamina simple, margins usually entire, surfaces lineolate. **Inflorescences** dense spikes, simple or branched, terminal or axillary; bracts large, longer than calyx, neither ciliate nor spinescent; bracteoles smaller than sepals. **Flowers** blue, purple, violet or purplish-white; calyx deeply 5-lobed, lobes equal, valvate, almost scarious; corolla tube cylindrical, more than twice as long as the ± equal patent limb lobes, imbricate; stamens 4, inserted at throat, 2 fertile with dithecous anthers, parallel; staminodes 2, small, shortly connate to filaments; ovary bilocular, locules 2-ovuled, style long, ± hairy. **Fruits** capsules, oblong-clavate; seeds 4, mucilaginous hairy, supported on a retinaculum. **Distr:** Bangladesh, India, Myanmar, Nepal, Sri Lanka. **Cult:** Propagate from cuttings or seeds. Requires a rich, moist soil in partial sun.

E. pulchelum Andrews, *Bot. Repos.* 2: t. 88, 1800. (Syn: *Daedalacanthus nervosa*, etc.). [Blue Sage, Blue Eranthemum]. **Shrubs**, large glabrous, well-branched, to 1.5 m high. **Leaves** elliptic or ovate-elliptic, 10–24 × 4.5–8 cm, base attenuate, margins entire or crenulate, apex acute-acuminate, prominently nerved, lateral nerves ~10 pairs, abaxially glabrous or ± hairy on nerves; petiole 1.5–2 cm long. **Inflorescences** dense spikes, 2–4 cm long, forming close terminal panicles; bracts leafy, ovate-elliptic to obovate, 1.2–1.5 × 0.6–0.8 cm, shortly acuminate, glabrous or ± puberulous; bracteoles narrowly lanceolate, 4–5 mm long, acute. **Flowers** blue, ~2.5–3 cm long; calyx lobes subequal, linear-lanceolate, ~6 mm long, minutely pubescent; corolla narrow, ~2 cm long, glabrous, dilated near the top, limb oblique, lobes ± equal, patent, ~1 cm long, obtuse; fertile stamens 2, anthers narrowly oblong, slightly exserted. **Fruits** ~1.2–1.5 cm long. **Distr:** Indian Subcontinent. **Ecol:** Fl. Feb–Mar. **Zones:** 10–12.

Fittonia

Species 1. *Generic description same as specific.* **Cult:** Propagate from rooted cuttings. Requires a moist, well-drained soil in full or partial sun.

F. albivenis (Veitch) Brummitt, *Bot. Mag.* 182: 165, 1979. (Syn: *F. verschaffeltii*). [Mosaic Plant, Painted Net Leaf, Silver Net/Nerve Plant]. **Herbs**, creeping, highly variable; stems rooting at nodes, tips upturned. **Leaves** broadly ovate-elliptic, 2–12 × 1.2–2 cm, base shallowly cordate, margins entire, apex rounded, venation conspicuously reticulate (netlike), white or reddish. **Inflorescences** spikes, erect; peduncle green, hairy; bracts wedge-shaped to ± circular, 0.5–1.3 cm long, margins ciliate, outside sparsely hairy; flowers sessile. **Flowers** with calyx green, shorter than floral bracts; corolla 1–1.3 cm long, tube cylindrical, yellowish, outside sparsely hairy, limb lemon-yellow, purplish-tinged, upper lip bilobed, lower lip trilobed, recurved, with a darker yellow line along midlobe. **Fruits** capsules, ~1 cm long; seeds 2, disk-shaped, minutely warty. **Distr:** Tropical S America. **Ecol:** In shady areas of rainforests. Fl. intermittently throughout the year. **Zones:** 10–12. **Info:** There is still much confusion with regard to the taxonomy of this species; see Llamas (2003) and Staples and Herbst (2005) for a further discussion.

Graptophyllum

Species ~10. **Shrubs**, evergreen, glabrous. **Leaves** opposite; lamina often colored or variegated, margins entire or dentate. **Inflorescences** panicle of subsessile cymes, short, terminal, or solitary and axillary; bracts shorter than calyx, not 4-ranked; bracteoles very small; flowers pedicellate. **Flowers** red or purple; calyx small, shallowly 5-lobed, lobes equal, linear-lanceolate; corolla cylindrical-tubular, tube curved, limb 2-lipped, lobes imbricate, upper lip shortly bilobed, lower one trilobed; stamens 2, fertile, anthers oblong, bithecous, thecae parallel, muticous; staminodes 2, small; ovary bilocular, locules 2-ovuled, style filiform. **Fruits** capsules, stipitate, clavate, 2–4-seeded; seeds compressed, orbicular. **Distr:** West Africa, New Guinea, Australia and Polynesia. **Cult:** Propagate from cuttings. Requires any well-drained soil in full or partial sun.

G. pictum (L.) Griff., *Not. Pl. Asiat.* 4: 139, 1854. (Syn: *G. hortense, Justicia picta*). [Caricature Plant].

Above: *Hemigraphis alternata*, *Hygrophila auriculata*, *Hypoestes phyllostachya*, *Justicia adhathoda*.

Shrubs, erect, to 2.5(–3) m high; stems glabrous. **Leaves** shortly petiolate; lamina ovate-elliptic, 10–20 × 5–12 cm, base attenuate, margins entire, apex acuminate-acute, glabrous, purplish or green mottled with yellow. **Inflorescences** panicles of cymes, terminal; peduncles sparingly puberulous; bracts and bracteoles inconspicuous; pedicels ~5 mm long. **Flowers** dark purple or crimson, ~4.5–5 cm long; calyx lobes linear, ~3–4 mm long; corolla glabrous outside, glandular-pubescent inside, ~2 cm across, tube narrowly cylindrical, ~1.5 cm long, upper lip erect, lower lip reflexed; stamens exserted, attached at throat of tube, filaments ~5 mm long; staminodes linear, ~1.5 mm long; ovary oblong, ~2.5 mm long, style ~2.5 cm long. **Fruits** capsules, clavate; seeds 2–4. **Distr:** Probably native to New Guinea; introduced and cultivated elsewhere. **Ecol:** Fl. throughout the year. **Zones:** 10–12.

Gymnostachyum

Species ~30. **Herbs** or **undershrubs**. **Leaves** cauline or subradical. **Inflorescences** cymes, axillary or in terminal panicles, often spicate or racemose, sometimes small and axillary; bracts and bracteoles very small. **Flowers** with calyx small, 5-partite, segments subequal, linear-lanceolate; corolla bilabiate, tube cylindric, slender, limb small, 2-lipped, upper lip narrowly bidentate, lower lip tridentate, 5-lobed, imbricate; stamens 2, inserted at middle and below, nearly as long as corolla, included; anthers 2-celled, cells parallel, 1 or 2 mucronate at base, anthers without mucro soon becoming 1-celled; staminode absent; ovary biloculed, locules 3- to many-ovuled, stigma bilobed, lobes compressed. **Fruits** capsules, linear; seeds many, compressed, ovoid, white-hairy, very shortly spreading when moistened. **Distr:** Tropical parts of Asia, India, Sri Lanka, Indo-China peninsula to C Malaya, Java, Philippines. **Cult:** Propagate from seeds. Requires a moist, well-drained soil in shade or partial sun.

G. ceylanicum Arn. and Nees, *Prod.* 11: 93, 1847. **Undershrubs**; stem rhizomatous, slender, oblique, with long roots. **Leaves** oblong-oval, 5.5–12.5 × 2–4.7 cm, base cuneate, tapering into petiole, apex obtuse, adaxially glabrous with copious cystoliths, abaxially tomentose, with prominent nerves; petiole to 5.5 cm long. **Inflorescences** cymes, in erect panicles, 7–16 cm long; peduncles to 7 cm long; pedicels 2 mm long; bracts and bracteoles linear-lanceolate, 2–4 mm long. **Flowers** with external floral parts glandular-hirtellous; calyx lobes linear lanceolate, to 1.3 cm long; corolla greenish-yellow to lemon-yellow; tube 1–1.2 cm long, purplish inside, upper lip of limb 3.6 mm long, bilobed; lower lip to 6 mm across lateral lobes, midlobe broader than laterals; stamens slightly exserted, filaments flattened, glabrous. **Fruits** linear, to 1.3 cm long, grooved on opposite sides, glandular-hirtellous; seeds 12–14, suborbicular, 1.25 mm Ø. **Distr:** Sri Lanka. **Ecol:** In forests, often in rocky ground, under shade. Fl. Mar–Aug. **Zones:** 11–12.

Hemigraphis

Species ~100. **Herbs** or **shrubs**, erect or ascending. **Leaves** opposite; lamina with margins entire, undulate or ± crenate. **Inflorescences** short spikes or heads, rarely flowers solitary, terminal or axillary; bracts large, ovate or lanceolate, imbricate, 4-ranked; bracteoles present or absent. **Flowers** blue or yellow, infundibuliform; calyx deeply 5-lobed, lobes linear-lanceolate, ± equal; corolla tube narrow, straight or slightly incurved, limb subequally 5-lobed, lobes obtuse, twisted in bud; stamens 4, all fertile, didynamous, almost included, anthers dithecous, muticous; staminodes absent; ovary biloculed, locules 3–10-ovuled, style filiform, stigma unequally bilobed. **Fruits** capsules, linear or linear-oblong, stipe absent; seeds 6–20, compressed, discoid, brown, margins white. **Distr:** S China, India, eastwards to Malaysia; tropical Australia, Pacific Islands. **Cult:** Propagate from cuttings or seeds. Requires a moist, well-drained soil in partial sun.

H. alternata (Burm. f.) T. Anderson, *J. Linn. Soc. Bot.* 7: 114, 1864. (Syn: *Blechum cordatum*, etc.). [Red Ivy, Red Flame Ivy, Waffle Plant]. **Herbs**; stems ascending to 35 cm high, subquadrangular, sparingly puberulous, roots emerging from lower nodes. **Leaves** cordate, to 6 × 4 cm, base cordate, margins crenate and sparingly ciliate, apex obtuse, surfaces sparingly pilosulous, portions turning indigo when dry; petiole to 2.5 cm long, pilose. **Inflorescences** spikes, axillary, to 1.5 × 1 cm; peduncles pilose; flowers generally 2 per bract; bracts ovate-oblong, to 1.3 × 0.5 cm, base and apex acute, surfaces puberulous. **Flowers** with calyx segments slightly unequal, linear subulate, to 9 mm long, peeled, ciliate; corolla to 1.5 cm long, strigose at throat adjacent to stamens, minutely pubescent; ovary ciliate at terminus, style ciliate for entire length. **Fruits** capsules. **Distr:** Malaysia. **Ecol:** In moist evergreen forests. Fl. throughout most of the year. **Zones:** 10–12.

Hygrophila

Species ~80. **Herbs**, erect or diffuse, glabrous or villose, ± suffrutescent, occasionally armed. **Leaves** opposite, sessile to subsessile; lamina with margins entire. **Inflorescences** cymose fascicles or terminal dense heads or spikes; bracts leafy, elliptic or linear-lanceolate; bracteoles linear-subulate; flowers sessile. **Flowers** bluish-purple or white; calyx deeply 4–5-lobed, lobes linear or lanceolate, subequal, valvate; corolla glabrous, tube cylindrical, ampliate near mouth, limb bilabiate, upper lip erect, concave, shorter than lower, shallowly bilobed or emarginate, lower lip patent, shallowly trilobed; stamens 4, didynamous, all fertile or 2 fertile and 2 staminodes, attached at throat, exserted, filaments basally dilated and connate in pairs, anthers oblong, dithecous; ovary oblong-cylindric, biloculed, locules (2–)4–8(–18)-ovuled, style filiform, tip hairy, recurved, stigma simple. **Fruits** capsules, oblong-cylindric, not stipitate; seeds (4–)8–16(–36), compressed, oblique-ovate or orbicular, on hooklike funicles, hairy. **Distr:** In tropical and subtropical regions of both New and Old World. **Ecol:** In wet or marshy habitats. **Cult:** Propagate from seeds or by division. Requires a fertile, well-drained soil in full or partial sun.

H. auriculata (Schumacher) Heine, *Kew Bull.* 16: 172, 1962. (Syn: *H. spinosa*, etc.). [Marsh Barbel]. **Herbs**, stout, suffrutescent, strigose-hispid, to 80 cm high, armed with 6 axillary, 2–3.5 cm long, straight or curved thorns at each node. **Leaves** in pseudowhorls of 6, outer 2 larger, sessile; lamina lanceolate or oblong-lanceolate, (4–)5–12(–15) × 1.5–2.5 cm, base cuneate, margins inconspicuously dentate, apex acute, surfaces white-hairy. **Inflorescences** whorled fascicles, axillary; bracts lanceolate, 2–2.5 cm long, hairy; bracteoles linear, 6–8 mm long, pilose. **Flowers** purplish-blue, ~2.5 cm long; calyx 4-lobed, lobes unequal, lanceolate, acute, outer ones longest, 2–2.2 cm long, limb 2-lipped, lips subequal, upper lip bilobed, lower lip trilobed, lobes obtuse; stamens 4, all fertile, anthers oblong, ~2.5 mm long; ovary oblong, 8-ovuled, tip hairy, style filiform, to 2.5 cm long. **Fruits** oblong, shorter than calyx, glabrous; seeds 4–8, orbicular, ~3 mm Ø, **Distr:** Bangladesh, India, Indo-China, Myanmar, Nepal, Pakistan, Sri Lanka; Tropical Africa. **Ecol:** Fl. Aug–Mar. **Zones:** 10–12. **Info:** Plants used for medicinal purposes.

Hypoestes

Species ~90–100. **Herbs**, ± rosette-forming; stems ascending and rooting at nodes, partly decumbent, rarely erect. **Leaves** isophyllous. **Inflorescences** spiciform, variable in length, terminal or axillary on upper branchlets; bracts persistent, 1-nerved or pinninerved, rarely trinerved from base; bracteole usually absent. **Inflorescences** with flowers solitary or 2–3 superposed in axils of bracts. **Flowers** with calyx 5-fid, segments ± equal, usually triangular or sublinear, usually ± ciliate, incision between anticous and lateral segments extended into the lower half as 2-lipped, upper lip trifid, lower lip bifid; corolla blue, pale violet or white, sometimes with violet striated dots on the helmet-shaped lobe, tube straight and non-resupinate, campanulate at throat, hairy inside, lobes equal, ovate, rounded or obcordate, often recurved at anthesis; stamens 4, included; ovary ± hairy or pubescent, locule 3–8-ovuled. **Fruits** capsules, elongated; seeds 6–18, areola gray, with brown and annular hairs around the areola. **Distr:** India to Malaysia; tropical Australia, Pacific Islands. **Cult:** Propagate from seeds or cuttings. Requires a fertile, moist, well-drained soil in partial sun.

Top row: *Justicia aurea, Justicia betonica, Justicia brandegeana, Justicia candicans.*
Bottom row: *Justicia carnea, Justicia floribunda, Justicia fulvicoma, Justicia procumbens.*

H. phyllostachya Baker, *J. Linn. Soc., Bot.* 22: 511, 1887. (Syn: *H. sanguinolenta*). [Freckle Face, Polka-dot Plant]. **Herbs**, to 70 cm high; young basal shoots especially velutinous; internodes between leaf-bearing nodes 2–6 cm long, subquadrangular. **Leaves** ovate, 2–4.5 × 1.5–2.5 cm, base obtuse to rounded, margins entire, apex acute, surfaces dotted with numerous pink spots to ~4 mm Ø, and cystoliths; petiole to 2.5 cm long. **Inflorescences** with flowers solitary or in terminal and axillary spikes, often with a gradual transition from solitary flowers to spikes; rachises glandular-puberulous; bracts 2 per flower, unequal, oblong, 1.2–1.5 cm long, acute to obtuse; bracteoles subulate, ~1 cm long, tapered; flowers sessile. **Flowers** with calyx 5-merous, 6–7 mm long, segments equal, subulate; corolla magenta, ~2.5 × ~0.3–0.7 cm, puberulous, upper lip ~1.25 cm long, trilobed, lobes rounded, ~1 × 3.5 mm, lower lip ~1 cm long; stamens exserted, anthers 1-celled. **Fruits** clavate, ~1 cm long; seeds 4, oval, ~1.7 mm long, papillate. **Distr:** Tropical Africa. **Ecol:** In cloud forests; 1,500–2,000 m. Fl. Apr, Dec. **Zones:** 10–12.

Justicia

Species >400. **Herbs** or **shrubs**, erect, rarely procumbent or ascending. **Leaves** petiolate; lamina ovate to elliptic, margins entire, rarely crenate or crenulate, often with cystoliths. **Inflorescences** congested spikes or panicles, sometimes solitary or in thyrsoid cymes, terminal or axillary; bracts linear or subulate, foliaceous and imbricate; bracteoles 2, linear-lanceolate, occasionally absent. **Flowers** white, pink or purple; calyx deeply 4–5-lobed, lobes usually narrow, equal or ± unequal, narrowly scarious, obtuse to acuminate; corolla bilabiate, tube often narrow, funnelform, straight or incurved; upper lip narrow, ± galeate, entire to bifid, lower lip broad, ± patent or recurved, trilobed, imbricate; stamens 2, filaments filiform, ± basally dilated, attached near middle of tube, slightly exserted, anthers dithecous; staminodes absent; ovary hairy, stigma minutely unequally bilobed. **Fruits** capsules, oblong, ovoid or obovoid-clavate; seeds usually 4, on curved retinaculum, compressed, usually orbicular-lenticular. **Distr:** Mainly in tropical and subtropical regions of both hemispheres. **Cult:** Propagate from seeds or cuttings. Requires a moist, well-drained soil in full or partial sun.

J. adhatoda L., *Sp. Pl.* 15, 1753. (Syn: *Adhatoda vasica, A. zeylanica*). [Adhatoda, Malabar Nut Tree, Physic Nut]. **Shrubs**, evergreen, erect, much-branched, gregarious, to 2(–2.5) m high; stem ± quadrangular to subterete, young shoots pubescent. **Leaves** elliptic-lanceolate, 10–20 × 3.5–8 cm, base attenuate, margins entire, apex acuminate, glabrous; petiole to 3.5(–4.5) cm long. **Inflorescences** spikes, to 10 × 2.5–3 cm, terminal and axillary; flowers subsessile, in leafy bracts; bracts broadly elliptic, to 2.5 × 1.5(–1.8) cm, pubescent; bracteoles elliptic-oblong to lanceolate, to 1.5(–2) cm long, acute. **Flowers** white, ~3 cm long; calyx 5-lobed, lobes linear-lanceolate, 0.6–1 cm long, acute, puberulous, imbricate; corolla pale-white, tube 1.2–1.5 cm long, pubescent outside, throat villous, limb 2-lipped, upper lip erect, shortly bifid, galeate, lower lip trilobed, lobes elliptic, obtuse; stamens 2, filaments 1–1.5 cm long, anthers oblong; ovary oblong, ~3 mm long, style 2–2.5 cm long. **Fruits** broadly clavate, ~2.5 cm long, pubescent, stipitate; seeds 4, ± orbicular, 2–3 mm Ø, glabrous. **Ecol:** Fl. throughout the year. **Distr:** Tropical Asia. **Zones:** 10–12.

J. aurea Schlecht., *Linnaea* 7: 393, 1832. (Syn: *J. umbrosa*, etc.). [Yellow Justicia]. **Shrubs**, to 3.5 m high; stem subquadrangular when young, tomentose. **Leaves** ovate to ovate elliptic, 15–40 × 8–17 cm, base attenuate, margins entire, undulate or crenulate, apex acuminate, adaxially glabrous to sparingly strigose, abaxially glabrous to strigose; petiole 1–9 cm long, tomentose. **Inflorescences** densely thyrsoid, subterminal, 8–33 × ~5 cm excluding flowers; rachises tomentose; bracts linear-lanceolate, mostly 1.1 × 0.1–0.2 cm; bracteoles similar but smaller. **Flowers** with calyx 5-merous, segments lanceolate, to 7 × 1.5 mm, puberulous; corolla yellow, bilabiate, cylindrical, to 5.5 × 0.3 cm, puberulous outside, glabrous inside, throat bearded; lips ~2.5 cm long, upper lip erect, 7 mm wide, acute, lower lip 6 mm wide, trilobed, lobes ~1 mm long, rounded and slightly hooded, especially the midlobe; stamens glabrous, exserted to tip of upper lip, anthers 3 mm long, filaments stout; ovary 2 × 1 mm, glabrous. **Fruits** clavate, 2–2.5 × ~0.5 cm, puberulous; seeds 4, flattened, suborbicular, ~2 mm Ø, surface roughened. **Distr:** S Mexico, C America. **Ecol:** In moist habitats along streambanks; <2,000 m. **Zones:** 9–12.

J. betonica L., *Sp. Pl.* 15, 1753. (Syn: *Nicoteba betonica*). [Paper Plume, Squirrel Tail, White Shrimp Plant]. **Shrubs**, glabrous, to ~2(–3) m high; stem terete, tumid and purple above nodes. **Leaves** oval-oblong to oblong-lanceolate, 6–17.5 × 2.2–4 cm, base acute, tapered into petiole, apex subacuminate. **Inflorescences** spikes, terminal, 6.5–23.5 cm long, usually simple; peduncles 3–8.5 cm long; bracts

Above: *Mackaya bella, Megaskepasma erythrochlamys, Odontonema callistachyum* (lavender form), *Odontonema cuspidatum.*

4-ranked, imbricate, broadly ovate, to 2 × 1.2 cm, margins usually scarious, apex acute, prominently trinerved; bracteoles elliptic, 1.2–1.6 × 0.4–0.6 cm. **Flowers** fragrant; calyx lobes linear-lanceolate, 3–5 × 1–1.5 mm, densely bristly outside; corolla infundibular, tube 5–8 mm long, greenish-white, limb pinkish-mauve, upper lip broadly ovate, 6–7 mm long, slightly hooded, lower lip ovate-oblong, 4–5.5 × 7–9 mm long, lobes oval-oblong, spur of lower anther thecae 1–1.25 mm long. **Fruits** ovoid-clavate, 1–1.6 × ~0.6 cm, bristly, with a short, solid base; seeds orbicular, 4–8 mm Ø, subcompressed, rugulose-tuberculate. **Distr:** Throughout the Old World tropics. **Ecol:** In disturbed areas, forest margins; 0–2,000 m. Fl. throughout the year. **Zones:** 10–12.

J. brandegeana Wash. and Smith, *Fl. Illust. Catarinense* (Acanthaceae). 102, 1969. (Syn: *Beloperone guttata, Drejerella guttata*). [False Hops, Red Shrimp Plant]. **Shrublets**, small, to 1.5 m high, much-branched from base; young twigs with patent hairs. **Leaves** ovate or lanceolate-ovate, 2.5–9 × 2–5 cm, base attenuate, margins entire, apex acute, sparsely hirsute; petiole 1–3 cm long, slender, strigose. **Inflorescences** spikes, strobilate, to 10 cm long, reddish-brown, terminal or axillary; bracts broadly ovate, 1–2 × 0.7–1.5 cm, obtuse or ± acute, pilose and densely ciliate, reddish-brown, closely imbricate; bracteoles oblong-lanceolate to ovate, ~1 cm long, ciliate. **Flowers** white or pale-lilac, 2–3.5 cm long; calyx lobes narrowly lanceolate, 5–7 mm long, ciliate; corolla tube 1.5–2.5 cm long, softly hairy outside, upper lip smaller than lower, ± entire, lower lip purple-spotted inside, shallowly trilobed; ovary glabrous with filiform, style hairy. **Fruits** oblong, 1–1.2 cm long, puberulent; seeds ± ovate, ~3 mm long, smooth. **Distr:** Mexico. **Ecol:** In damp forests; to 1,800 m. Fl. throughout most of the year. **Zones:** 9–12.

J. candicans (Nees) V.A.W. Graham, *Kew Bull.* 43: 611, 1988. (Syn: *Adhatoda candicans, Jacobina ovata*, etc.). [Arizona Water-willow, Hummingbird Bush, Red Justicia]. **Shrubs**, evergreen or semievergreen, 1–1.8 m high; stems slender, and arching, quadrangular when young. **Leaves** ovate or broadly ovate, 1.5–8 × 0.7–3.7 cm, base rounded to truncate, margins entire to subsinuate, apex acute or acuminate, thin, surfaces pubescent or glabrate; petiole to 1.3 cm long. **Inflorescences** dichasia, few-flowered, dense, sessile or subsessile, axillary; dichasial spikes to 1.8 cm long; bracts subfoliose, ± elliptic, 0.9–1.5 cm long; bracteoles ± linear, 0.8–1.2 cm long; flowers ± subsessile. **Flowers** with calyx 5-lobed, 6–8 mm long, lobes subulate, 5–7 mm long; corolla red with white markings on lower lip, 3.2–4 cm long, tube ~2 cm long, throat hairy, 2-lipped, lower lip trilobed, 1.3–1.5 cm long, lobes 7–9.5 mm long; stamens inserted near apex of corolla tube, 1.2–1.5 cm long; style 3–3.5 cm long. **Fruits** 1.1–1.5 cm long, glabrous; seeds 4, reddish, compressed, ~3.5 mm long. **Distr:** Mexico (Oaxaca), S USA (Arizona). **Ecol:** 500–1,000 m. Fl. & Fr. Apr. **Zones:** 8–11.

J. carnea Lindley, *Edward's Bot. Reg.* 17: pl. 1397, 1831. (Syn: *Jacobina carnea, J. magnifica*). [Brazilian Plume, Flamingo Flower, King's Crown]. **Shrubs** or **herbs**, evergreen, 1.5–2 m high; stems 4-angled or with prominent grooves. **Leaves** petiolate; lamina oblong-ovate, 25 × 7 cm, base acute to rounded, decurrent on petiole, apex acuminate, membranous. **Inflorescences** thyrses, dense, to 20 × 8 cm, solitary, terminal, sessile or subsessile; rachis puberulous, lowermost flowers subtended by elliptic bracts 2 × 1 cm, obtuse or rounded, sparingly ciliate; bracts subtending flowers oblong, 2 × 1.5 cm, rounded, glabrous, ciliate; bracteoles lanceolate, 1.5 cm long. **Flowers** with calyx lobes 5, lanceolate, 1 × 0.2 cm, subhyaline, ciliate near tip; corollas lilac, red or pink, sparingly puberulous, 6.5 cm long, tube 2.5 cm long, throat 5 mm wide, upper lip erect, arched, tip bilobed, lobes 1 mm long, lower lip spreading or recurved, trilobed, cuneate, 8 mm wide at base of lobes, lobes ovate, 3 × 2 mm; anther thecae slightly superposed. **Fruits** capsules, rarely formed in cultivation, 4-seeded. **Distr:** Colombia, Venezuela to Brazil. **Ecol:** In moist forests, in clearings and edges of trails, often in lowland areas near watercourses. Fl. mainly Nov–May. **Zones:** 10–12.

J. floribunda (C. Koch) Wassh., *Darwiniana* 35: 151, 1998. (Syn: *J. rizzinii*, etc.). [Firefly, Water Willow]. **Shrubs**, branching, 0.5–1(–2) m high. **Leaves** petiolate; lamina ovate, elliptic or obovate, very variable in size, 1.5–7 × 0.8–2.5 cm, base cuneate, margins entire or slightly serrate, apex obtuse, discolored. **Inflorescences** spikes, few-flowered, frequently reduced to solitary flowers, in axils of upper leaves; peduncles slender, to 3 cm long; bracts and bracteoles triangular-lanceolate, 1 × 0.5 mm, glabrous to shortly ciliolate; calyx 5-lobed, 5 mm long, lobes lanceolate, to 3.5 mm long, puberulous; corolla red, ± upper half yellow, 2–3 cm long, tube to 2 cm long, throat 5 mm Ø, puberulous outside, glabrous inside, lips oblong, ~5 mm long, upper lip erect, obtuse or emarginate, lower lip erect or spreading, trilobed, lobes 2 mm long, triangular; anther thecae slightly superposed, oblique. **Fruits** claviform, 1.2–1.7 × 0.4 cm; seeds flattened, suborbicular, ~2 mm Ø, brownish, somewhat rugulose. **Distr:** Argentina, Brazil, Paraguay. **Ecol:** In forests and wet forests, especially on periodically flooded soils; 0–1,400 m. Fl. May–Aug. **Zones:** 9–11.

J. fulvicoma Schlecht., *Linnaea* 6: 369, 1831. (syn: *Beloperone blechioides*, etc.). [Mexican Plume]. **Herbs** or **shrublets**, branching; stems 40+ cm long, pilose. **Leaves** ovate or lance-ovate, mostly 4–9 × 2–5 cm, base acute or rounded, margins entire, apex acute or acuminate, pilose; petiole slender, usually 1–3 cm long. **Inflorescences** spicate, mostly 2–10 cm long, terminal or axillary; bracts closely imbricate, broadly ovate, obtuse or acute, first green, becoming reddish with age, 1–2 × 0.7–1.5 cm, pilose and densely ciliate; bracteoles oblong-lanceolate, lance-ovate, or ovate, ~1 cm long, ciliate. **Flowers** with calyx segments 5, narrowly lanceolate, ~6 mm long, green, ciliate; corollas 3–3.5 cm long, pubescent outside, white or pale lilac, anterior lip purple-spotted, lips 1–1.5 cm long, ciliate, posterior lip entire or minutely bifid, anterior one shallowly trilobate; stamens subequal to upper lip, thecae 1.5–2 mm long, hirsute or glabrate, superposed, calcarate; style pubescent, stigma minutely bifid. **Fruits** 1–1.2 cm long, puberulent; seeds 2–4, compressed, ovate, smooth. **Distr:** Mexico, Honduras. **Ecol:** In damp, shaded slopes, thickets or forests; 600–1,800 m. **Zones:** 9–12.

J. procumbens L., *Sp. Pl.* 15, 1753. (Syn: *Rostellaria procumbens, Rostellularia procumbens*). [Water Willow].**Herbs**, annual, diffusely branched, to 30 cm high; stems procumbent to ascending strigose-hispid; branches sulcate. **Leaves** lanceolate-elliptic, 1–2(–3) × 0.5–1(–1.2) cm, base and apex acute, margins entire, adaxially scabrid, abaxially glabrescent, strigose-hairy; petiole ~5 mm long. **Inflorescences** spikes, to 2.5 cm long, terminal; bracts obovate, ~4 mm long, midrib broad, scarious-ciliate on margins; bracteoles lanceolate, margins narrowly scarious-ciliate, acute. **Flowers** pink, ~1 cm long; calyx 4-lobed, rarely a 5th rudimentary lobe present, lobes lanceolate, ~4 mm long, subequal, margins scarious-ciliate, acute; corolla tube ~3.5 mm long, narrow, upper lip nearly as long as tube, slightly notched, lower lip ± orbicular, ~6 mm Ø, puberulous outside, trilobed; staminal filaments ~3 mm long, anthers oblong, ~0.7 mm long; ovary oblong, pubescent, style filiform, ~4 mm long. **Fruits** oblong, as long as style, sparsely hairy, pointed; seeds ± ovate, ~1 mm long. **Distr:** Australia, India, Malaysia, Pakistan, Sri Lanka. **Ecol:** Fl. Dec–Mar. **Zones:** 10–12.

Mackaya

Species 1. *Generic description same as specific.* **Cult:** Propagate from seeds or semiripe cuttings. Requires a moderately fertile, moist, well-drained soil in full or partial sun.

M. bella Harv., *Thes. Cap.* i. 8. t. 13, 1859. (Syn: *Asystasia bella*). [Forest Bell Bush, River Bells]. **Shrubs**, usually multistemmed, sub glabrous, 1–2 m high. **Leaves** elliptic to ovate-oblong, 8–12 cm long, margins irregularly scalloped or widely and coarsely toothed, apex tapering to a drip tip, adaxially glossy dark green, abaxially dull green with midrib and lateral veins raised, lateral veins with reddish-purple hair-tuft domatia in their axils; petiole short. **Inflorescences** racemes, unilateral (cymoid), terminal, lax; bracts opposite, minute, subulate; pedicels 5–8 mm long, cernuous. **Flowers** large and showy, 5 cm Ø, mauve to white, usually marked with fine purple-pink lines;

Above: *Pachystachys lutea, Pachystachys spicata, Peristrophe paniculata, Pseuderanthemum carruthersii.*

calyx continuous with pedicel, equally 5-lobed, lobes narrowly subulate; corolla ~5 cm long, pale-lilac, thin, membranous-veiny, tubular below, campanulate upward, with a deeply 5-lobed, subequal, erect limb, lobes oblong, blunt; stamens inserted at the top of the narrow part of corolla, shorter than limb, anthers, 2 perfect, sagittate, equal-sided, 2-celled, 2 filiform, nearly as long as the others, without anthers; ovary bilocular, locules 2-ovuled, style filiform, stigma minute, bifid. **Fruits** capsules, narrowly cup-shaped, splitting into 2 valves to explosively eject the seed. **Ecol:** In evergreen forests. **Zones:** 9–11.

Megaskepasma

Species 1. **Shrubs**. *Generic description same as specific.* **Cult:** Propagate from cuttings or seeds. Requires a moderately fertile, humus-rich, moist, well-drained soil in full sun.

M. erythrochlamys Lindau, *Bull. Herb. Boissier* 5: 666, 1897. [Brazilian Plume, Brazilian Red Cloak, Red Justicia]. **Shrubs**, to 4 m high; stems subquadrangular. **Leaves** elliptic, to 24 × 10.5 cm, base acute, margins entire to crenulate, apex acuminate, tip rounded, apiculate, cystoliths absent; petiole to 2 cm long. **Inflorescences** loosely clustered spikes, terminal; spikes to 19 × 4 cm; peduncles brown puberulous, to 3.5 cm long; bracts and bractlets large, colorful; bracts solitary, ovate to elliptic, to 3.5 × 2 cm, base acute, margins entire, apex acute, deep red; bractlets elliptic, to 2 × 0.5 cm, base attenuate, apex acute. **Flowers** with calyx of 5 equal segments, lanceolate, to 1 × 0.2 cm, densely strigose; corolla white, pilosulous, tube cylindrical, to 2.5 × 0.2 cm, salverform, bilabiate, upper lip lanceolate, bifid, to 3.5 × 0.5 cm, bilobed, lobes subulate, 1.5 mm long, lower lip elliptic, 3.5 × 0.8 cm, trilobed, lobes lanceolate, 7 mm long; stamens 2, exserted to ~3 cm, filaments glabrous, anthers 2-celled; ovary 2 mm long, glabrous, stigma subcapitate. **Fruits** capsules, clavate, to 3 × 0.8 cm, apex attenuate, mucronate, glabrous; seeds 4, suborbiculate, flattened, ~6 mm Ø, glabrous, borne on hooklike funicles. **Distr:** Venezuela. **Ecol:** In pluvial forests and rainforests. Fl. throughout most of the year. **Zones:** 10–12.

Odontonema

Species ~40. **Herbs** or **shrubs**. **Leaves** often large with cystoliths visible on both surfaces, sessile to petiolate. **Inflorescences** panicles or thyrses, with inconspicuous bracts, terminal. **Flowers** white, red or yellow, 5-merous; calyx with equal, subulate segments; corolla bilabiate, upper lip bilobed, lower trilobed; stamens 2, exserted or extending to mouth, anthers 2-celled; staminodes 2. **Fruits** capsules, clavate; seeds 4. **Distr:** New World tropics. **Cult:** Propagate from seeds or cuttings. Requires a fertile, humus rich, moist compost in partial sun.

O. callistachyum (Schlecht. and Cham.) Kuntze, *Rev. Gen. Pl.* 2: 494, 1891. (Syn: *Justicia callistachya, Thyrsacanthus callistachyus*). [Fire Spike]. **Herbs**, suffrutescent, to 2 m high; stems subquadrangular. **Leaves** elliptic to elliptic-ovate, 15–30 × 4–12 cm, base attenuate, margins entire to crenulate, apex acuminate, tip often curved to one side, adaxially glabrous, abaxially puberulous along midrib and major costa, surfaces with cystoliths; petiole 0–1 cm long. **Inflorescences** thyrses, to 30 cm long, terminal; lateral peduncles to ~5 mm long, flowers fascicled; fascicle bract subulate, keeled, 2–3 mm long, flowers mostly borne in dichasial cymes; floral bracts similar, slightly smaller, more triangular; rachises ± tomentose; pedicels 3–6 mm long. **Flowers** puberulous; calyx 5-merous, segments equal, subulate, keeled, 2–3 mm long; corolla lavender to deep magenta, bilabiate, funnelform, often curved to one side, to 3 × 0.2–0.5 cm, tube upper lip 6–7 × 4.5 mm, bilobed, lobes ovate, 4.5 × 2.5 mm, obtuse, lower lip trilobed, lobes 7–8 × 2 mm, obtuse; stamens near upper lip, filaments villous; staminodes to 3.5 mm long. **Fruits** 2 cm long, glabrous. **Distr:** Mexico and C America. **Ecol:** In lowland rainforests. **Zones:** 10–12.

O. cuspidatum (Nees) Kuntze, *Rev. Gen. Pl.* 2: 493, 1891. (Syn: *Thyrsacanthus cuspidatus*). [Cardinal's Crest, Scarlet Fire Spike]. **Shrubs**, 1 m high; stems quadrangular, pilose, older stems glabrate. **Leaves** petiolate; lamina elliptic to ovate, 9.5–15 × 4–6 cm, margins entire to undulate, glabrous. **Inflorescences** racemes or panicles, with 3- to many-flowered fascicles, terminal; pedicels 5–7 mm long, puberulent; bracts subulate, 1.5–4 × 0.5–1 mm, margin entire, ciliolate; bracteoles subulate, 1–1.5 × 0.5 mm, margin ciliolate. **Flowers** with calyx 5-lobed, lobes subulate, 1.5–2 mm long, margin ciliolate; corolla red, tubular, 2.5–3.5 cm, glabrous, weakly bilabiate, lower lip trilobed, lobes elliptic elliptic-ovate, ciliolate, upper lip bilobed, lobes elliptic-ovate, rounded; stamens 2, included, anthers bithecous, thecae blunt at base; staminodes 2; ovary conical, ~3 × 1 mm long, glabrous. **Fruits** clavate, glabrous; seeds 4. **Distr:** Mexico. **Ecol:** Fl. throughout most of the year. **Zones:** 10–12. **Info:** The inflorescences are sometimes crested, similar to those sometimes observed in *Coccoloba uvifera*.

Pachystachys

Species 12. **Herbs** or **subshrubs**. **Leaves** large, petioled. **Inflorescences** dense spikes, with 3- or 4-flowered clusters, terminal; bracts conspicuous, herbaceous; bractlets small or absent. **Flowers** with calyx 5-lobed, lobes relatively short; corolla ringent, slenderly obconic, curved, 2-lipped, lower lip trilobed, lobes subequal, oblong or ovate, upper lip narrow, bilobed; stamens 2, about as long as corolla, attached near base of tube, anthers deeply sagittate, basal thecae muticous, equal; staminodes, if present, rudimentary. **Fruits** capsules, 4-seeded. **Distr:** Neotropical (from Wes Indies southward to Peru and Bolivia and eastward to French Guiana and Amazonian Brazil). **Ecol:** Along edges of lowland rainforests; <1,000 m. **Cult:** Propagate from cuttings or seeds. Requires a fertile, moist, well-drained soil in partial shade.

P. lutea Nees, *Prodr*, 11: 320, 1847. (Syn: *Justicia lutea*). [Candle Plant, Golden Candles, Lollipop Plant, Yellow Candles, Yellow Shrimp Plant]. **Shrubs**, evergreen, erect or sprawling, lax, 0.5–2.5 m high; stems angular. **Leaves** opposite, ± sessile; lamina oblong-lanceolate, 5–12 × 3–4 cm, both ends tapered, margins entire, membranous, glabrous, shiny. **Inflorescences** spikes, 4-sided, erect, 10–20 cm long, solitary, terminal; bracts densely imbricate, in 4-rows, bright yellow or orange-yellow, ovate to broadly elliptic, 1–2 cm long, glandular-pilose along the entire margins, apex obtuse to rounded. **Flowers** with corolla white, bilaterally symmetric, tubular, 3.5–5.5 cm long, pilose and glandular-punctate, limb 2-lipped, upper lip erect, projected forward, lanceolate, minutely bilobed, lower lip spreading, ovate; stamens 2, borne just inside upper lip, protruding, anthers green, dithecous. **Fruits** club-shaped, dry; seeds brown. **Distr:** Peru. **Ecol:** In lowlands. **Zones:** 10–12.

P. spicata (Ruiz and Pav.) Wassh., *Proc. Biol. Soc. Wash.* 99(1): 175, 1986. (Syn: *P. coccinea*, etc.). [Black Stick, Cardinal's Guard]. **Shrubs** or **subshrubs**, to 2(–3) m high. **Leaves** shortly petiolate; lamina oblong, elliptic or broadly oblanceolate, 10–20 × 5–11.5 cm, margins entire, surface ± puckered. **Inflorescences** spikes, ± 4-sided, erect, 25 cm long, solitary, terminal; bracts green, imbricate, ovate or ovate-lanceolate, to 2.5 cm long, puberulous, margins entire; flowers numerous, projecting well beyond bracts. **Flowers** with calyx campanulate, 4.5 mm long, segments narrowly triangular, 3 × 1 mm; corolla scarlet, becoming reddish-orange with age, 5–7.5 cm long, glabrous to hirtellous, tube curved, limb 2-lipped, upper lip projected forward, erect, minutely bilobed, lower lip spreading, recurved; stamens projecting, anthers deeply arrowhead-shaped; staminodes rudimentary, puberulous; ovary glabrous. **Fruits** elastically, loculicidally dehiscent. **Distr:** Widespread in tropical America from West Indies, south to N Bolivia and Amazonian Brazil. **Ecol:** Along edges of lowland rainforests and riverbanks; 150–700 m. Fl. Jul–Sep. **Zones:** 10–12.

Peristrophe

Species ~30. **Herbs** or **undershrubs**, erect to spreading. **Leaves** opposite; lamina with margins entire. **Inflorescences** with flowers solitary or in lax trichotomously branched paniculate cymes; bracts linear or ovate, longer than calyx; bracteoles paired, linear-lanceolate, shorter than bracts. **Flowers** showy; calyx deeply 5-lobed, lobes linear-lanceolate, all equal; corolla pink or rose-colored, hairy, tube slender,

Above: *Pseuderanthemum carruthersii* var. *atropurpureum*, *Pseuderanthemum carruthersii* var. *reticulatum*, *Rhinacanthus nasutus*, *Ruellia brevifolia*.

twisted, ± equal to the 2-lipped limb, lobes imbricate, upper lip shorter than lower, subentire, lower lip shortly trilobed; stamens 2, attached to corolla throat, free, equal, well-exserted, filaments hairy below, anthers dithecous, muticous, lobes rotund, unequal, superposed; staminodes absent; ovary bilocular, locules 2-ovuled, style filiform, stigma ± bilobed. **Fruits** capsules, oblong-ellipsoid, stiped, ± pubescent; seeds 4, compressed, ovoid-orbicular, papillate, glochidiate or smooth, without hygroscopic hairs. **Distr:** E Malaysia to Tropical Africa. **Cult:** Propagate from seeds. Requires a well-drained soil in full sun.

P. paniculata (Forssk.) Brummitt, *Kew Bull.* 38: 451, 1983. (Syn: *Dianthera bicalyculata*, etc.). **Herbs**, erect to spreading, hairy, to 1 m high; stems sharply 4–6-angled. **Leaves** ovate-elliptic to lanceolate, 4–6 × 1.5–4 cm, base rounded to acute, apex acute to acuminate, densely lineolate, ± pubescent especially on nerves beneath; petiole 1–1.7 cm long. **Inflorescences** lax panicles of cymes, terminal or axillary; bracts 2, linear-spathulate, much unequal, 0.7–1.5 cm long, acute; bracteoles in 2 valvate pairs, unequal to subequal, margins scarious, apex acute; flowers pedicellate. **Flowers** pink or purple; calyx lobes linear, ~3.5–4 mm long, acute, valvate, hairy; corolla tube cylindrical, 4–5 mm long, lips subequal, puberulous, upper lip elliptic-oblong, 5–7 mm long, entire or notched, lower lip slightly longer, deflexed, slightly crested with 3 acute lobes; staminal filaments ~5 mm long, hairy; ovary oblong, apically pubescent, style ~1 cm long. **Fruits** ellipsoid, 1–1.2 cm long, basally solid and hairy, acuminate; seeds orbicular, ~2.5 mm Ø, minutely papillate. **Distr:** Tropical Africa, India, Indo-China, Malaya, Myabnmar, Pakistan. **Ecol:** In wastelands; 0–600 m. Fl. Oct–Feb. Zones 10–12.

Pseuderanthemum

Species 120. **Herbs** or **small shrubs**. **Leaves** ovate or elliptic. **Inflorescences** spikes, racemes or panicles, terminal or axillary; 1 bract and 2 bracteoles subtending each flower or flower cluster. **Flowers** with calyx segments 4 or 5, subulate; corolla white, blue, purplish or pink, tube narrow, limb spreading, lobes 5, subequal; stamens 2, included or exserted, anther cells basally obtuse or acute; staminodes 2, minute. **Fruits** capsules, long-stipitate; seeds 2 or 4. **Distr:** Old and New World tropics. **Cult:** Propagate from seeds, stem cuttings or leaf cuttings (which root very easily in moist compost). Requires a well-drained soil in shade or partial sun.

P. carruthersii (Seem.) Guillaumin, *Ann. Mus. Colon. Marseille* VI, 5–6: 48, 1948. (Syn: *Eranthemum atropurpureum*, etc.). [False Eranthemum]. **Shrubs**, evergreen, 1–2(–4) m high; branches quadrangular. **Leaves** mostly elliptic but variable, to 20 × 10 cm, base attenuate, margins unevenly wavy (sinuate), apex acuminate, cuspidate or obtuse, green, variously colored or variegated. **Inflorescences** to 30 cm long, branches cymose; bracts ovate, 2–4 mm long; pedicels 1–3 cm long. **Flowers** salverform; calyx lobes lanceolate, 3–4 mm long, pubescent; corolla white to violet or spotted around the throat, pubescent, 2 posterior lobes slightly wider than the others, elliptic to 1 × 0.8 cm, obtuse; stamens exserted, anthers muticous. **Fruits** clavate; seeds 4. **Distr:** New Caledonia, Vanuatu (New Hebrides); naturalized widely in the tropics. **Ecol:** In lowland valleys and slopes; 0–250 m. Fl. throughout most of the year. **Zones:** 11–12. **Info:** Cultivated for ornamental purposes or for the edible leaves. ***P. carruthersii* var. *atropurpureum*** (Syn: *P. atropurpureum*) have mottled silver-gray to purplish variegation. ***P. carruthersii* var. *reticulatum*** (Syn: *P. reticulatum*), [Purple False Eranthemum] have yellowish young leaves, maturing to green with golden-yellow venation, to full green.

Rhinacanthus

Species ~6. **Shrubs** or perennial erect **herbs**, laxly branched, sometimes subscandent. **Leaves** entire or undulate. **Inflorescences** 1- to few-flowered subsessile fascicles, lax axillary or terminal cymes, combined into paniculate thyrses; bracts and bracteoles small, linear-lanceolate. **Flowers** with calyx 5-partite, lobes narrow, acuminate; corolla tubular, tube elongate, slender, lobes imbricate in bud, limb 2-lipped, upper lip linear, erect, lower lip broadly trilobed; stamens 2, inserted on throat of corolla, anther thecae superposed, muticous at base; disk annular; ovary with locules 2-ovuled, style filiform. **Fruits** capsules, clavate with a long solid base; seeds 4, orbicular, compressed. **Distr:** Tropical Africa, Madagascar, India, Sri Lanka, Malesia. **Cult:** Propagate from seeds. Requires a well-drained soil in full sun.

R. nasutus (L.) Kurz, *J. Asiat. Soc. Bengal*, Pt. 2: 79, 1870. (Syn: *Justicia nasuta*, *Rhinacanthus communis*). [Dainty Spurs]. **Herbs**, perennial, erect, to 1.5 m high; stem, subquadrangular, faintly striate, glabrescent. **Leaves** oblong-elliptic to oval-elliptic, 6.3–11 × 2–4.2 cm, acute at both ends, glabrous, marked with copious cystoliths; petiole to 5 mm long. **Inflorescences** lax panicles, terminal and axillary, to 55 cm long; flowers subsessile, discrete at ends of divaricate branches; bracts lanceolate, to 2 × 0.5 mm; bracteoles smaller. **Flowers** with calyx lobes lanceolate, to 4 × 0.75 mm, unequal, glandular-bristly outside; corolla greenish-white, tube slender, to 2.5 cm long, curved from middle upward, hirsutulous, upper lip erect, 6 mm long, lower lip to 1.2 cm across lateral lobes, hispidulous outside; stamens exserted. **Fruits** to 2 × 0.3 cm, with a long solid base, bristly; seeds orbicular-helicoid, to 2.25 mm wide, papillose. **Distr:** China, India, Madagascar, Malesia, S Africa, Sri Lanka, Taiwan. **Ecol:** In secondary forests of dry lowlands. Fl. Dec–Jun. **Zones:** 9–12.

Ruellia

Species ~250. **Herbs** perennial, erect or prostrate to decumbent or shrubs. **Leaves** petiolate or sessile; lamina elliptic, ovate-lanceolate, oblong-spathulate, margins entire, apex acute-acuminate or obtuse. **Inflorescences** with flowers solitary or in cymes or panicles, rarely spicate, axillary or terminal; bracts inconspicuous, usually linear and small or

Below: *Ruellia chartacea*, *Ruellia macrantha*, *Ruellia makoyana*, *Ruellia tuberosa*.

absent; bracteoles 2 or absent flowers pedicellate or sessile. **Flowers** usually large, red, yellow, purple or white, showy and chasmogamous, or small and cleistogamous; calyx deeply 5-partite, lobes linear or lanceolate, ± equal; corolla infundibuliform or salverform, tube straight, incurved or ventricose, limb spreading, subequally 5-lobed, lobes ovate, obtuse, contorted in bud; stamens 4, didynamous, all fertile, filaments inserted below throat, basally united in pairs, anthers dorsifixed, dithecous; ovary bilocular, locules 2-ovuled, style slender, stigma unequally bilobed. **Fruits** capsules, oblong-cylindrical or clavate-obovate or ellipsoid; seeds 4–20, compressed, flat, ovate or orbicular, on acute, strong retinacula. **Distr:** Tropical and temperate regions of both hemispheres. **Cult:** Propagate from seeds or cuttings. Requires a fertile, moist, well-drained soil in full sun.

R. brevifolia (Pohl) C. Ezcurra, *Darwiniana* 29: 278, 1989. (Syn: *R. ventricosa*, etc.). [Red Spray Ruellia]. **Herbs**, suffruticose or subshrubs, to 1 m high; stems numerous, erect, glabrous. **Leaves** ovate, 4–15 × 1.5–8 cm, base cuneate, margins entire or shallowly crenate, apex acuminate, usually glabrous, green, abaxially paler; petiole 1–5 cm long. **Inflorescences** cymes of compound dichasia, lax, multiflowered, axillary, long-pedunculate; bracteoles ovate-lanceolate, foliaceous, exceeding calyx lobes. **Flowers** red; calyx puberulous, 0.5–1 cm long, lobes linear, ~4 mm long, acute; corollas curved, scarlet, to 3 cm long, basal tube 0.5–1 cm long, opening into an anteriorly inflated, ventricose throat 1–1.5 × 0.5–1 cm, narrowed toward orifice, lobes suberect, small, rounded, 2–3 mm long, retuse; stamens subexserted, filaments ~9 mm long, anthers 3 mm long, laterally connate, basally apiculate; ovary puberulous, style exserted. **Fruits** clavate, 1–1.7 cm long, solid-stipitate; seeds 8–12, suborbicular, 2mm Ø, **Distr:** Tropical and subtropical S America, south to N Argentina. **Ecol:** In seasonally dry forests; <1,500 m. Fl. Apr–Oct. **Zones:** 10–12.

R. chartacea (T. Anderson) Wassh., *Opera Bot.* 92: 265, 1987. (Syn: *Aphelandra chartacea*, *R. colorata* [misapplied]). [Red Shrimp Plant]. **Herbs**, suffrutescent, evergreen, erect, 1–1.5 m high; stems glabrous, quadrangular, base woody. **Leaves** oblong-elliptic, 10–30 × 4–15 cm, both ends ± abruptly acuminate, base narrowed and decurrent on petiole, margins undulate or crenate, apex cuspidate, green, often with red venation, lateral veins 8–9 pairs. **Inflorescences** spikes, 12 × 8 cm, terminal, flowers subtended by relatively large bracts; bracts ovate, scarlet or red, to 4–8 × 2–4.5 cm, acute to obtuse, persistent; bractlets linear, 1.5 cm long, red; peduncles to 1 cm long. **Flowers** tubular, golden to pinkish-orange, 7.5–9 cm long; calyx 1 cm long, red, pubescent, segments lanceolate, 8–9 mm long; corolla red, tube slightly curved, 6 mm wide at base, narrowed to 2 mm at 1 cm wide above base, enlarged to 1.2 cm wide at mouth, lobes reflexed, oblong, to 1.9 cm long, rounded; stamens and styles exserted; ovary glabrous. **Distr:** S Colombia, Ecuador, Peru. **Ecol:** In rainforests and riverside forests. Fl. throughout most of the year. **Zones:** 10–12.

R. macrantha Mart. ex Nees, *Fl. Bras.* (Martius) 9: 37, 1847. [Christmas Pride]. **Herbs**, branched, often woody at base, 1–2 m high; stems erect, angular, glabrous. **Leaves** lanceolate to ovate, 15 × 5–8 cm, abaxially dull dark green, rough with veins impressed, sparsely hairy; upper leaves shortly petiolate; lower leaves with petiole to 2 cm long. **Inflorescences** with flowers solitary. **Flowers** large, showy; calyx to 2.5 cm long, sparsely hairy; corolla funnelform, to 8 cm long, sparsely hairy, lavender to pink, throat paler, sometimes yellow, with dark red venation, lobes to 5 cm wide, emarginate, spreading, recurved, strongly trinerved; stamens 4, filaments subequal, slender, basally connate in pairs, anthers included, linear-oblong; ovary narrow, terete, pubescent, style slender glabrous, tip unequally bifid. **Fruits** cylindrical. **Distr:** Brazil. **Ecol:** In tropical savannas (cerrados). Fl. Aug–Feb. **Zones:** 10–12.

R. makoyana Hort. Makoy ex Closon, *Rev. Hort. Belg.* 109, 1895. (Syn: *Dipteracanthus devosianus*, *R. devosiana*) [Monkey Flower, Monkey Plant, Trailing Velvet Plant]. **Herbs** or **subshrubs**, sprawling or low-spreading, to 60 cm high, evergreen, perennial, hairy; stems weak. **Leaves** opposite, petiolate; lamina broadly lanceolate to ovate, 5–7.5 cm long, margins wavy, velvety-hairy, adaxially olive green, dull, red-flushed, with silvery-gray along veins, abaxially rich purple. **Inflorescences** with flowers solitary, axillary; flowers flanked by 2 leaflike bracts; pedicels absent. **Flowers** trumpet-shaped; corolla to 5 × 5 cm, magenta or rose-pink, tube slender, limb abruptly flattened and expanded, swollen upward, shallowly 5-lobed, lobes unequally bifid; stamens 4, included, ± near the mouth; stamens and style ± equal, usually at upper side of mouth. **Fruits** club-shaped, 1.2–2 cm long. **Distr:** Brazil. **Ecol:** Fl. throughout most of the year. **Zones:** 10–12.

R. tuberosa L., *Sp. Pl.* 635, 1753. [Cracker Plant, Water Cannon]. **Herbs**, erect, to 60(–70) cm tall; stems stout, 4-angled, nodes swollen, purplish, tender parts sparsely hairy; roots slender, elongated, tuberous. **Leaves** oblong-obovate to oblanceolate, 5–9(–10) × 2–4 cm, base cuneate to attenuate, margins entire to ± undulate, apex obtuse to acute, glossy, often mucronulate; petiole to 2 cm long. **Inflorescences** with flowers solitary or in 1–3-flowered cymes, terminal or axillary; bracts and bracteoles linear-lanceolate, ~5 mm long; flowers shortly pedicellate. **Flowers** blue-violet, showy, 5–5.5 × ~3.5 cm; calyx lobes ± equal, linear, 1.3–2 cm long, ciliate, acute; corolla pubescent outside, tube ~3.5 cm long, base abruptly narrowed, limb 5-lobed, lobes ± equal, ovate, ~1.5 cm long, patent; stamens with oblong-sagittate, ~4 mm long, puberulous anthers; ovary oblong, ~4 mm long, glabrous, style 2.3–2.5 cm long. **Fruits** oblong, 2.2–3 cm long, tip minutely beaked, explosively dehiscent; seeds 24–28, flattened, orbicular, ~2 mm Ø, hairy. **Distr:** Tropical America. **Ecol:** In disturbed places. Fl. throughout most of the year. **Zones:** 10–12.

Ruttya

Species ~3. **Shrubs**. **Leaves** ovate, apex acuminate. **Inflorescences** spikes, dense, terminal; bracts subulate. **Flowers** bright red; calyx 5-lobed, lobes linear-attenuate, very long; corolla 2-lipped, upper lip erect, semibifid, flat, lower lip deflexed, equally 3-parted, tube wide, shorter than calyx, inflated on one side; stamens inserted in the throat, exserted, fertile 2, 1-celled, anthers oblique, base mucronate, sterile stamen toothlike, very short; ovary sessile, biloculed, locules 2-ovuled, style slender, pubescent, stigma bifid. **Fruits** capsules, clavate, base empty, 4-seeded above, elastically dehiscent. **Distr:** Tropical Africa. **Cult:** Propagate from seeds or cuttings. Requires a fertile, well-drained soil in full or partial sun.

R. fruticosa Lindau, *Bot. Jahrb. Syst.* 20: 45, 1894. [Hummingbird Plant, Jammy Mouth, Orange Bird, Orange Rabbit Ears]. **Shrubs**, evergreen, spreading, to 1.5–4 m high; stems terete, glabrous, woody, slender, twigs tightly grouped, often suckering. **Leaves** opposite, petiolate; lamina elliptic to ovate, 5–7.5 cm across, base tapered, apex tapered or acuminate, wrinkled, smooth, adaxially midgreen, abaxially paler, venation sparsely pilose. **Inflorescences** spikes, 1–4-flowered, terminal. **Flowers** with corolla ~5 cm Ø, yellow, orange-red, 2-lipped, lower lip deflexed with a dark brown blotch at its center, upper lip erect; stamens 2; nectar copious. **Fruits** elongate, woody, stipitate, glabrous, explosively dehiscent; seeds 2, flattened. **Distr:** Tropical E and NE Africa, Arabian Peninsula. **Ecol:** On rocky slopes; 420–2,000 m. Fl. throughout most of the year. **Zones:** 10–12.

Sanchezia

Species ~30. **Herbs** or **shrubs**, erect, mostly glabrous. **Leaves** with numerous cystoliths on both surfaces, glabrous. **Inflorescences** with flowers solitary or fascicled; fascicled flowers subtended by bracts which are often large, partly connate and showy. **Flowers** usually large and showy, yellow, orange or red; calyx 5-merous; corolla tube cylindric, limb 5-lobed, lobes equal; stamens 2, usually exserted, anthers 2-celled, base mucronulate, puberulous; staminodes 2. **Fruits** capsules, oblong; seeds 6–8, orbicular. **Distr:** Tropical America, mainly in northern Andes. **Ecol:** In wet forests. **Cult:** Propagate from cuttings. Requires a fertile, moist, humus-rich, well-drained soil in full sun.

S. parvibracteata Sprague and Hutchinson, *Bull. Misc. Inform.* 253, 1908. (Syn: *S. glaucophylla* [hort.]) [Fire Fingers, Small-bracted Sanchezia, Zebra Plant]. **Shrubs**, to 2.5 m high; stems subquadrangular, glabrous; internodes 4–6 cm long. **Leaves** elliptic to obovate, 12–24(–35) × 5–11(–15) cm, base attenuate, margins undulate-serrate, apex abruptly acuminate, glabrous, costa and veins edged in

Below: *Ruttya fruticosa*, *Sanchezia parvibracteata*, *Sanchezia speciosa*, *Strobilanthes dyerianus*.

yellow, surfaces with numerous cystoliths; petiole to 3 cm long, glabrous. **Inflorescences** sparingly branched panicles, to 20 cm long, terminal; bracts loosely imbricate, ovate, 1.5–2 × 0.75–1 cm, obtuse, glabrous; bracteoles oblong-ovate, ~1.5 × 4 mm, rounded, glabrous; peduncles to 5 cm long, glabrous; rachis glabrous; flowers 3–5 per pair of bracts, sessile. **Flowers** with calyx segments oblong, 2–2.5 × 0.3–0.5 cm, obtuse, glandular-puberulous inside; corolla red or ochre, puberulous, hairs yellowish, 4–5 × 0.6–0.7 cm at throat, 5 mm wide just above ovary, lobes oblong, reflexed, 5–7 × 5 mm; stamens exserted to 8 mm beyond corolla mouth, filaments sparsely villous, anthers 6–7 mm long; staminodes 1.5 cm long. **Distr:** C America. **Ecol:** In thickets; <1,500 m. **Zones:** 10–12.

S. speciosa Leonard, *J. Wash. Acad. Sci.* 16: 490, 1926. (Syn: *S. nobilis* [misapplied], *S. spectabilis* [hort.].). [Large-bracted Sanchezia, Zebra Plant]. **Shrubs**, to 3 m high; stems 4-angled. **Leaves** opposite; lamina simple, oblong to elliptic, 10–45 × 5–15 cm, base long-decurrent on petiole, green or with yellow veins; petiole 1–4 cm long, winged nearly to base. **Inflorescences** spikes, terminal, erect, flowers several, clustered, sometimes divided at base; bract below each flower cluster, ovate, orange to red, 2.5–5.5 cm long. **Flowers** with sepals oblanceolate, orange ~2.5 cm long; corolla of fused petals, tubular, rich yellow to orange, 4.5–5.5 cm long, lobes 5, short, rounded, yellow to orange; stamens 2, exserted, rich yellow; style yellow. **Fruits** narrowly cylindrical. **Distr:** N South America. **Ecol:** In wet forests, forming thickets; 0–800 m. Fl. Through the year. **Zones:** 10–12.

Strobilanthes

Species 250. **Shrubs** erect or **herbs**; branchlets terete or ± quadrangular, slightly 4-winged. **Leaves** opposite or sometimes pseudo-alternate, pair occasionally dissimilar; lamina with margins entire, crenate or toothed, apex acute to shortly acuminate, thickly coriaceous. **Inflorescences** with flowers solitary or in axillary and/or terminal, lax or condensed, interrupted spikes or panicles or capitula; bracts solitary, foliaceous or small, persistent or caducous; bracteoles 2, small or absent. **Flowers** white, purplish-pink or blue and white; calyx deeply 5-lobed, lobes equal or unequal, valvate; corolla campanulate, tube cylindric below, straight or curved, ventricose above, limb subequally 5-lobed; stamens usually (2)4, included or exserted, staminodes 2 or absent; ovary oblong-globular, bilocular, 4–8-ovuled, style filiform, stigma with one obsolete lobe. **Fruits** capsules, oblong, apically pubescent or glabrous; seeds 4–8, glabrous or hairy, much-compressed. **Distr:** Madagascar and Tropical Asia. **Cult:** Propagate from seeds or cuttings. Requires a humus-rich, well-drained soil in full or partial sun.

S. dyerianus Mast., *Gard. Chron.* i. 442, 1893. (Syn: *Perilepta dyeriana*). [Persian Shield]. **Undershrubs**, erect, branched, 1–2 m high; stems hairy when young, later glabrescent. **Leaves** paired, unequal, smaller one about two-third the larger one; lamina ovate or elliptic, 8–31 × 2–11.5 cm, gradually narrowed toward base, then abruptly widened to a broad, clasping base, margins toothed, apex acute, adaxially sparsely covered with bristly white hairs, both surfaces variegated with rose-purple, white and green. **Inflorescences** spikes, dense, glandular-hairy, terminal on small axillary branches, flowers overlapping; bracts obovate, 4–5 × 4–5 mm, acute, persistent; bracteoles absent. **Flowers** with calyx 4–6 mm long, glandular-hairy, obscurely 2-lipped, lobes oblong, obtuse, 1 slightly longer than the others; corolla 2.5–3 cm long, pale violet, hairy, tube narrow for ~1 cm then widened and slightly bent; ovary hairless. **Distr:** Myanmar. **Ecol:** Fl. Aug–Feb. **Zones:** 10–12.

S. gossypinus T. Anderson, *J. Linn. Soc.*, Bot. 9: 466, 1867. [Persian Shield]. **Shrubs**, perennials or flowering for a few years and then dying, all green parts covered with dense yellowish wool. **Leaves** ovate, 10 × 5 cm, base broadly rhomboid, margins entire, apex acute, subcoriaceous, adaxially woolly at first, soon glabrescent, abaxially woolly, nerves 13 pairs, approximate, parallel; petiole 2 cm long. **Inflorescences** spikes, linear-oblong, 2.5–7.5 cm long, compound and (by reduction of upper leaves) running into terminal panicles, woolly; bracts elliptic, 0.8–1.2 cm long, obtuse; bracteoles linear-oblong, 8 mm long. **Flowers** with calyx ~1.2 cm long, divided about halfway down, segments narrowly lanceolate, subobtuse; corolla nearly straight, 3.2 cm long, glabrous, ventricose portion of the tube much longer than the cylindrical basal portion; filaments hairy; ovary glabrous, with a few long hairs near apex, style glabrous, ovules 4. **Distr:** Mountains of the S Deccan Peninsula. **Ecol:** Often on nearly vertical rocks; 1,200–1,500 m. **Zones:** 10–12.

S. hamiltoniana (Steud.) Bosser and Heine, *Bull. Mus. Natl. Hist. Nat.*, B, Adansonia 10: 148, 1988. (Syn: *Goldfussia crinita*, *Ruellia hamiltoniana*). [Indian Blue Bells]. **Shrubs**, to 1.5 m high; stems glabrous. **Leaves** isophyllous to weakly anisophyllous; lamina ovate to lanceolate, 5–15(–19) × 1.6–8(–9) cm, base usually symmetrical, margins sharply serrate, sinuses hairy, apex acuminate, lateral veins 3–6(–7) per side; petioles 0–2.5(–8.5) cm long. **Inflorescences** thyrses, much-branched, diffuse; partial inflorescences to 30 cm long; bracts leaflike, narrowly ovate to narrowly elliptic, to 3.5 × 1.3 cm; pedicels slender, 1.5–5 mm long; floral bracts green, widely obovate, 3–3.5 mm long; bracteoles green, oblanceolate to narrowly elliptic, 3–4 long. **Flowers** glabrous; calyx green, membranous, lobes basally connate, subequal, narrowly elliptic, 4.5–7 × 1.5–2.5 mm; corolla pink or purple, tube white, 3.7–3.9 cm long, subventricose above, tube 1.3 cm long, throat to 1.4–1.8 cm Ø, lobes 4–4.5 × 6 mm, rounded or retuse; stamens included, to 9 mm long, anthers white; ovary 2.6 mm long, style white, 3.8 cm long. **Fruits** dark purple, fusiform, 1–2.2 × 0.3–0.6 cm, glabrous or glandular-hairy; seeds 4, 3 × 2 mm. **Distr:** E Himalayas from E Nepal to N Thailand and SW China. **Ecol:** In dry and shaded forests; 75–2,130 m. Fl. throughout most of the year. **Zones:** 9–12.

Thunbergia

Species ~200. **Climbers**, herbaceous or shrubby, pubescent or glabrous, rarely erect. **Leaves** petiolate, opposite; lamina simple, ovate to lanceolate, base cordate or hastate, apex acute to acuminate. **Inflorescences** with flowers solitary and axillary, or in terminal racemes; bracts 2, large, foliaceous, ovate or lanceolate; bracteoles 2, spathelike, enveloping calyx, mostly persistent; flowers pedicellate. **Flowers** mostly large and showy, usually blue, yellow, white, orange or red; calyx cuplike, with 10–16 subulate teeth or reduced to a rim; corolla infundibuliform, tube incurved or oblique, widened above, limb patent, 5-lobed, lobes subequal, obtuse or truncate, contorted in bud; stamens 4, all fertile, didynamous, inserted near base of corolla tube, anthers bithecous; disk shortly annular or pulvinate; ovary fleshy, globose, bilocular, locules with 2 collateral ovules, style long, stout, stigma bifid, entire or fringed. **Fruits** capsules, globose below, abruptly beaked in upper part, 4-seeded, loculicidally dehiscent; seeds semiglobose to ovoid, glabrous, funicle papillalike. **Distr:** C and S Africa, Madagascar; Tropical Asia. **Cult:** Propagate from seeds, cuttings or by air-layering. Requires a humus-rich, moist, well-drained soil in shade to full sun.

T. alata Bojer ex Sims, *Bot. Mag.* 52, t. 2591. 1825. [Black-eyed Susan, Clock Vine]. **Vines**, herbaceous, creeping or climbing, twining, 2–3 m long; stems slender, puberulous. **Leaves** opposite; lamina 4.5–10.5 × 3.2–6 cm, ovate, lobed, chartaceous, base subcordiform, apex acute, adaxially dark green, dull, pubescent, abaxially pale green, dull, with prominent venation; petiole 4–8 cm long, winged, pubescent. **Inflorescences** with flowers solitary, axillary; pedicels 4–5 cm long, pubescent; bracts green, ovate, pubescent, 1.5 cm long, covering calyx and corolla tube. **Flowers** with calyx yellowish-green, with 12 filiform lobes, ~4 mm long; corolla orange, pale yellow, or sometimes whitish, infundibuliform, with 5 lobes, tube ~2.5 cm long, base narrow, dark violet inside, lobes ~2.5 cm long, truncate, limb ~5 cm Ø; stamens with glandular hairs on basal portion. **Fruits** depressed-globose to 4-lobed at base, ~4 mm long, upper half beaked, dehiscent by 2 valves; seeds 2 or 4, semicircular, 1.2–1.5 mm long, reticulate. **Distr:** E Africa. **Ecol:** In moist areas, forest margins and slopes; 0–700 m. Fl. throughout most of the year. **Zones:** 10–12.

T. battiscombei Turrill, *Ic. Pl.* 31: t. 3041, 1915. (Syn: *T. adjumaensis*). [Blue Glory Vine, Scrambling Sky Flower]. **Herbs**, perennial; rootstock woody; stems erect, to 75 cm long (rarely twining to 3 m long). **Leaves** ovate to elliptic or slightly obovate, 8–17.5 × 5.5–9.5 cm, base decurrent, margins entire or with a few large teeth, apex acuminate to broadly rounded, apiculate, slightly fleshy, palmately veined; petiole 1–4 cm long. **Inflorescences** racemes, many-flowered, axillary, solitary (rarely 2 racemes per axil or 1 raceme with a solitary flower); peduncle and axis 2.5–30 cm long; bracts caducous, lower foliaceous, upper bracteolate; pedicels 0.8–3.5 cm long; bracteoles pale green, with raised dark green reticulation, ovate to elliptic, 2–2.7 × 0.8–1.2 cm, acute to obtuse, apiculate. **Flowers** with calyx puberulous, 2–4 mm long, lobes broadly triangular; corolla limb and upper part of tube purple to royal blue, lower part whitish, throat yellow, tube (2.5–)3–4.5 cm long, lobes 1–1.5 × 1.5–2 cm long; filaments 1–1.5 cm long, glabrous, anthers 2–3 mm long. **Fruits** 0.8–1 cm Ø, beak 1.6–2 cm long; seed pale brown, ~7 Ø, surface reticulate. **Distr:** Kenya, Uganda. **Ecol:** 750–2,300 m. **Zones:** 10–12. **Info:** In zones 8–9 the plant dies to the ground and comes up during warmer months.

T. coccinea Wall. ex D. Don, *Prodr. Fl. Nap.* 120. 1825. (Syn: *Hexacentris coccinea*). [Scarlet Clock Vine]. **Climbers**, large, woody; twigs subglabrous, sharply 4-angled, pendent. **Leaves** ovate to broadly ovate, (6–)8–15(–18) × 5–12 cm, base hastate, truncate-cordate, margins entire in upper half, sinuate-dentate to shallowly lobed in basal half, 3–5-costate, thick, abaxially ± glaucous; petiole 1–5 cm long, uppermost leaves sessile. **Inflorescences** racemes, pendulous, to 45 cm long, terminal and axillary; bracts ovate-oblong to lanceolate, 1–3.5 cm long, acute-acuminate; bracteoles ovate, subequal to bracts, enclosing the corolla. **Flowers** scarlet-red, center orange-yellow, 2.5–3 cm long; calyx minutely 12-toothed; corolla tube 5–6 mm long, constricted between tube and throat, lobes round, 5–7 mm Ø, obtuse, reflexed; anthers of anterior stamens with a single spur and those of posterior 2-spurred. **Fruits** ± orbicular, 2.5–3.5 cm long, puberulous, beak ensiform, ~2.5 cm long. **Distr:** W China, Nepal, India eastwards to Myanmar, Indo-China and Java. **Ecol:** In forests of mountain slopes. Fl. Dec–Mar. **Zones:** 9–12.

T. erecta (Benth.) T. Anderson, *J. Proc. Linn. Soc. Bot.* 7: 18, 1863. (Syn: *Meyenia erecta*). [Bush Clock Vine, King's Mantle]. **Undershrubs**, upright, to 1.25(–3) m high, sometimes twining shrubs, glabrous. **Leaves** lanceolate-ovate to elliptic-ovate to obovate, 3–8 × 2–4 cm, base rounded, margins distantly serrate, apex acute, apiculate. **Inflorescences** usually with flowers solitary, axillary; bracts 1.5 × 1 cm, caducous. **Flowers** purple-blue to dark blue, rarely white, to ~6 × 4 cm; calyx teeth ~15, subulate, glandular-hairy; corolla tube ± curved yellowish-white or white, glandular-

Top row: *Strobilanthes hamiltoniana, Strobilanthes gossypinus, Thunbergia alata, Thunbergia battiscombei.*
Middle row: *Thunbergia coccinea, Thunbergia erecta, Thunbergia fragrans, Thunbergia grandiflora.*
Bottom row: *Thunbergia grandiflora* (white form), *Thunbergia gregorii, Thunbergia kirkii, Thunbergia mysorensis.*

hairy outside, lobes subequal, ~1.5 cm long, obtuse; staminal filaments glandular-hairy, anthers oblong, ~4 mm long, bearded, mucronate; style ~3.5 cm long, stigma bilobed, upper lobe upright, lower patent. **Fruits** ± globose, beak ~1.5 cm long. **Distr:** W Tropical Africa. **Ecol:** Fl. Nov–Mar. **Zones:** 10–12.

T. fragrans Roxb., *Pl. Coromand.* 1: 47. t. 67, 1795. (Syn: *T. volubilis*). [Angel Wings, White Thunbergia]. **Climbers** or **twiners**, herbaceous, slender; young shoots quadrangular, strigose. **Leaves** broadly elliptic to ovate or lanceolate-ovate, (3–)4–10 × (1.5)2–5(–6) cm, margins entire or few-toothed, sparsely strigose on both sides near the truncate to cordate base, apex acute-acuminate, palmately 3–5-nerved; petiole 2–4.5 cm long, slender. **Inflorescences** with flowers solitary, axillary; pedicels 2–7 cm long, stout; bracts connate and spathaceous, oblong-lanceolate, 1.5–2 × ~0.6–0.8 cm, ± acute, hairy. **Flowers** white, 4–5 × ~5 cm; calyx cupular, teeth 12–16, subulate, 1–3 mm long, glandular-hairy; corolla tube cylindrical, 2–2.5 cm long; lobes 5, subequal, nearly as long as tube, puberulous outside; staminal filaments glandular-hairy below, anthers oblong, slightly beaked above, without spur at base; style ~2 cm long. **Fruits** depressed-globose, ~8 mm Ø, scabrid, puberulent, beak 1–1.5 cm long; seeds reticulate. **Distr:** Indo-China, Myanmar, Nepal, India, N. Australia and W. China. **Ecol:** Fl. Oct–Dec. **Zones:** 10–12.

T. grandiflora (Roxb. ex Rottler) Roxb., *Bot.*

Above: *Utricularia aurea, Argyreia nervosa, Cuscuta chinensis, Evolvulus alsinoides.*

Reg. 6. t. 495, 1820. [Bengal Clock Vine, Bengal Trumpet, Blue Trumpet Vine, Sky Vine, Sky Flower]. (Syn: *Flemingia grandiflora, T. cordifolia*). **Climbers** or **twiners**, large, woody; twigs quadrangular, strigose. **Leaves** ovate or lanceolate-ovate, (6–)8–15(–16) × (3–)4–12 cm, base ± cordate, 3–5-costate, margins entire or coarsely toothed in lower half, apex acute to acuminate, surfaces hairy except the scabrous costa and veins; petiole 2–5(–6) cm long, scabrous. **Inflorescences** racemes, drooping, usually axillary or terminal; flowers pedicellate; bracts free or ± connate, ovate-oblong, 3–3.5 × 1.5–2 cm, acute-acuminate. **Flowers** blue or violet with yellow centers, 6–8 × ~6 cm; calyx truncate, rim velvety, purplish; corolla glabrous, tube 3–4.5 cm long, abruptly contracted near base, lobes ± equal, patent, 3–4 cm long, obtuse; stamens inserted at throat, filaments flattened, anthers oblong, bithecous, anterior anthers with a single spur, posterior ones with 2, sharply curved spurs; ovary glabrous. **Fruits** globular, contracted into a stout, ~2 cm long flat beak. **Distr:** China, India, Myanmar. **Ecol:** Fl. Jun–Aug. **Zones:** 10–12.

T. gregorii S. Moore, *J. Bot.* 32: 130, 1894. (Syn: *T. gibsonii*). [Golden Glory Vine/Climber, Orange Clock Vine]. **Herbs**, climbing or trailing, evergreen, perennial, to 4 m long, often forming a ground cover; stems slender, woody below, twining. **Leaves** ovate-triangular, 7–8 cm long, margins toothed, hairy, apex shortly acuminate to mucronate, venation prominent; petiole winged. **Inflorescences** with flowers solitary, rarely a few together, flowers erect on long peduncles; peduncles slender, erect, green, hairy; bracts roughly hairy. **Flowers** trumpet-shaped; calyx 10–11-lobed, long-hairy, brownish with green prominent ridges; corolla ~4.5 cm long, tube slender with an abruptly expanded, flattened limb, bright orange with a dark throat, lobes broadly obovate, margins ± entire, slightly ruffled, apices broad, bilobed, lobes rounded, sinus rounded, venation prominent; stamens inserted. **Distr:** Tropical E Africa. **Ecol:** In montane grasslands; (900–)1,350–2,300(–2,450) m. Fl. throughout most of the year. **Zones:** 10–12.

T. kirkii Hook. f., *Bot. Mag.* 109: t. 6677, 1883. (Syn: *T. hookeriana*). [Blue Sky Shrub]. **Shrubs**, erect, small, to ~1.5 m high; stems and branches slender, rigid, divaricating, acutely 4-angled, often forming a tangled mass. **Leaves** in opposite pairs, shortly petiolate; lamina variable, narrowly lanceolate, 4–7.5 × 2 cm, base tapered, margins entire or with each side dilated around halfway into an obtuse lobe, making the lamina appear rhombic or somewhat trilobed, apex subacute or obtuse, apiculate, coriaceous and rigid, adaxially dark glossy green, abaxially paler, midrib excurrent, trinerved. **Inflorescences** cymes, short, usually 2-flowered; peduncle and pedicels short, stiff; bracteoles oblong, 8 mm long, subacute. **Flowers** violet blue; calyx very short, irregularly obtusely toothed, cupular; corolla 2.5–3.2 cm long, tube ± equal to lobes, constricted near apex of calyx, above somewhat inflated, then narrowed again near mouth, lobes spreading, ± rectangular with truncate apices, throat yellowish; stamens on upper side of throat, bases slightly hairy; ovary glabrous, stigma shortly 2-lipped. **Fruits** loculicidally dehiscent, beaked. **Distr:** Tropical E Africa. **Ecol:** Fl. throughout most of the year. **Zones:** 11–12.

T. mysorensis (Wight) T. Anderson ex Bedd., *Trans. Linn. Soc. London,* 25: 222, 1865. (Syn: *Hexacentris mysorensis*). [Mysore Clockvine]. **Vines**, woody; internodes 2–8.5 cm long, subquadrangular, glabrous. **Leaves** narrowly elliptic to oblong-hastate or lanceolate-oblong, 5–15 × 1.2–7.5 cm, base obtuse to subcordate, margins entire to irregularly lobed or angled, often undulate, apex attenuate, veins 3, surfaces glabrous; petiole 1.6 cm long, glabrous. **Inflorescences** racemes, terminal and axillary, pendent, 2–3 m long, nodes 2-flowered, opposite to each other; pedicels to 2.5 cm long, glabrous; bracts brick-red to purplish, ovate, 2–2.5 × 0.8–1 cm, acute, glabrous. **Flowers** laterally compressed; calyx annular, much-reduced, at base of corolla; corolla bilaterally symmetrical, tube 4–5.5 cm long, yellow inside, pale cream outside, throat opening obliquely, limb 5.7–8 cm across, lobes orange-red to deep red-brown or scarlet, lower lobes reflexed and 2 upper lobes fused for most of their length to partially enclosing the stamens and style; stamens curved upward, filaments hairy, anthers ~8 mm long, basally long-spurred, spurs ~4 mm long; nectar copious, often overflowing from the sides of the corolla mouth. **Distr:** W India (Nilgiri Hills). **Ecol:** In tropical rainforests and evergreen forests, often near waterfalls; to ~2,000 m. Fl. throughout most of the year. **Zones:** 9–12.

LENTIBULARIACEAE

Genera 3, **species** 320. **Herbs**, annual or perennial, insectivorous, rosette-forming, terrestrial, epiphytic or aquatic; roots present or absent; stem (stolons) short and delicate in terrestrial and epiphytic species while usually more robust and long in aquatic species. **Leaves** alternate, opposite or verticillate, sometimes rosulate; lamina entire or divided, sometimes polymorphic, covered with glandular hairs (in *Pinguicula*) or with bladders (in *Utricularia*). **Inflorescence** racemose, simple or sparingly branched or in some aquatic species supported on the water by a whorl of modified spongy leaves (floats). **Flowers** showy, hermaphrodite and zygomorphic; calyx 2–5-lobed, persistent; corolla gamopetalous, 2-lipped usually spurred or saccate; stamens 2, attached at base of corolla throat, filaments often winged, short, usually curved, anthers 2-celled, dorsifixed, cells diverging or confluent; ovary superior, 1-loculed, carpels 2, style usually short, persistent, stigma ± 2-lipped, ovules many, on free central or free basal placenta. **Fruits** capsules, indehiscent or opening via 2–4 valves, pores or circumscissile; seeds minute, without endosperm. **Distr:** Worldwide. **Ecol:** Usually in lakes and other damp places.

Utricularia

[Bladderworts].

Species 220. **Herbs**, annual or perennial, insectivorous, terrestrial, epiphytic or aquatic; stem (stolons) monomorphic or dimorphic, filiform, terete and branched. **Leaves** rosulate at base of inflorescence or alternate, opposite or verticillate on the stolons, ± dichotomously divided into capillary segments. **Bladders** (traps) hollow, ovoid to globose, stalked with lateral, terminal or basal mouth. **Inflorescence** few- to many-flowered; peduncle usually simple, sometimes branched, filiform, erect or twining; bracts persistent, basifixed, medifixed or variously produced below the point of attachment; sterile bracts (scales) present or absent; pedicel usually short, terete, often reflexed or recurved in fruit. **Flowers** with sepals bilobed, ± equal, free or ± united at the base, persistent; upper lobe usually entire, lower lobe usually emarginate; corolla yellow, violet or purple variously shaded or white or rarely blue or red, 2-lipped; upper lip 2 or more lobed, entire; lower lip 1–2 or 5-lobed, spurred and with gibbous palate; stamens 2, filaments usually curved; ovary ovoid or globose, style short, stigma unequally 2-lipped, lower lip usually much larger than the upper which may be obscure or obsolete. **Fruits** capsules, ovoid or ± globose, indehiscent or dehiscent by longitudinal slits, dorsiventral or lateral valves, pores or circumscissile; seeds 1 to many, small, variously shaped and sculptured. **Distr:** Worldwide, especially in the tropics. **Cult:** Propagate from leaf cuttings or seeds. Requires a still and stagnant water, such as in lakes or ponds, in full or partial sun.

U. aurea Lour., *Fl. Cochinch.* 1: 26, 1790. (Syn: *U. confervifolia*, etc.). [Golden Bladderwort]. **Herbs**, aquatic, floating submerged; stolons (stems) filiform, to 1 m long. **Leaves** very numerous, filiform, 1–8 cm long, primary segments 3 or more, each repeatedly divided into numerous segments, ultimate segments capillary, setulose. **Bladders** numerous, dimorphic, lateral on secondary and tertiary segments, and also at base of primary segments, ovoid-subglobose, 1–4 mm long, with branched antennae. **Inflorescence** a raceme, erect, elongating with age, 5–10-flowered, 5–25 cm long, held above water; peduncle terete, sometimes with a whorl of narrowly fusiform floats; bracts basifixed, broadly ovate, 1–2 mm long; pedicels erect, filiform, 0.4–2 cm long. **Flowers** with calyx lobes subequal, ovate, 2–3 mm long at anthesis, becoming much larger and reflexed in fruit; corolla yellow, 1–1.5 cm long, outside ± pubescent; upper lip broadly ovate; lower lip larger, transversely elliptic, apex rounded or emarginate; spur narrowly conical, about as long as the limb; filaments curved, 1–1.5 mm long, anther thecae ± confluent; ovary ovoid, style distinct, stigma lower lip semicircular, upper lip ± obsolete.

Above: *Evolvulus glomeratus, Hewittia malabarica, Ipomoea alba, Ipomoea asarifolia.*

Fruits globose, to 5 mm Ø, style greatly enlarged, persistent; seeds disk-shaped, angular, 1.5–2 mm wide, angles narrowly winged. **Distr:** India, SE Asia, from Malaysia, eastwards to China, Japan and N Australia. **Ecol:** In still and stagnant water of lakes, ponds and swamps. Fl. Dec–Apr. **Zones:** 9–12.

SOLANALES

FAMILIES 5, GENERA 165, SPECIES 4,080.

Nodes unilacunar. **Stipules** absent. **Leaves** alternate, spirally arranged; lamina simple. **Flowers** radially symmetric; calyx persistent in fruit; corolla plicate, sympetalous; stamens equal in number to petals; ovary with 4 or more ovules (Judd et al. 2008). **Distr:** Worldwide.

CONVOLVULACEAE

Genera *57*, species 1,625. **Herbs** or **shrubs**, usually with twining or climbing stems or erect, laticiferous. **Leaves** alternate, absent in parasitic species; lamina simple, margins entire, dissected, or compound. **Inflorescences** with flowers solitary and axillary, or in cymes, racemes, panicles, umbels or heads. **Flowers** bisexual, actinomorphic, usually 5-merous, often showy; calyx quincuncial, large, free; sepals free, often persistent, sometimes enlarged in fruit; corolla sympetalous, funnelform, campanulate, salverform or urceolate, limb subentire or deeply lobed; stamens alternating with corolla lobes, adnate to corolla, filaments filiform, equal or unequal in length, anthers introrse, dehiscing laterally and longitudinally; disk ringlike or cupular; ovary superior, mostly bicarpellate, 1- or 2(3 or 4)-loculed, ovules apotropous, basal, erect, styles 1 or 2, terminal (gynobasic) or very short or absent, stigma dry, entire or 2(3)-lobed, rarely peltate. **Fruits** capsules, dehiscing by valves, circumscissile, or irregularly shattering, sometimes berries or nutlike; seeds usually trigonous, smooth or pubescent. **Distr:** Worldwide.

Argyreia

Species ~90. **Shrubs**, scandent or lianas; stems woody. **Leaves** petiolate; lamina with margins entire, abaxially sometimes silvery sericeous. **Inflorescences** cymes, or loose or compact capitula, few- to many-flowered, axillary, rarely terminal; bracts persistent or early deciduous, small or large. **Flowers** with sepals persistent, herbaceous or ± leathery, abaxially pubescent, adaxially glabrous and often red, ± enlarged and reflexed in fruit; corolla purple, red, pink, or white, campanulate, funnelform, or tubular; limb subentire to deeply 5-lobed, with 5 distinct, mostly pubescent midpetaline bands; stamens inserted near base of corolla, filaments filiform, base dilated, often pubescent or glandular; disk ringlike or cupular, margin entire or shallowly 5-lobed; ovary 2- or 4-loculed, 4-ovuled, style 1, filiform, stigma capitate, bilobed or biglobular. **Fruits** berries, red, purplish, orange, or yellowish, globose or ellipsoid, fleshy, mealy, or leathery; seeds 4 or fewer. **Distr:** Predominantly in tropical Asia; 1 sp. in Australia (Queensland). **Cult:** Propagate from seeds or softwood cuttings. Requires a moderately fertile, moist, well-drained soil in full sun.

A. nervosa (Burm. f.) Bojer, *Hort. Maurit.* 224, 1837. (Syn: *A. speciosa*, etc.). [Elephant Creeper, Woolly Morning Glory]. **Lianas**, to 8(–10) m high; stems herbaceous toward the tips, woody at base; younger branches densely white-pubescent, glabrate with age. **Leaves** ovate to orbicular, 10–30 × 8–25 cm, base deeply cordate, margins entire, apex acute to shortly cuspidate, adaxially glabrous, abaxially densely white-pubescent. **Inflorescences** capitate cymes; peduncle to 20 cm long, white-tomentose; pedicels short, angular; bracts ovate to oblong, 3.5–5 cm long. **Flowers** with sepals densely white-tomentose outside, outer 2 sepals broadly elliptic, 1.5 cm long, 3 inner sepals broadly elliptic to orbicular, ~1–1.2 cm long; corolla lavender or pink-purple, throat darker, tubular to funnelform, ~6 cm long, limb shallowly lobed, midpetaline bands and tube densely lanate outside; stamens and style included; ovary glabrous, 4-loculed. **Fruits** indehiscent, dry-baccate, ± globose, ~2 cm Ø, yellowish-brown; seeds brown, glabrous. **Distr:** India (from Assam and Bengal to Belgaum and Mysore). **Ecol:** Usually climbing on other vegetation and trees. **Zones:** 10–12.

Cuscuta

Species ~170. **Herbs**, parasitic, yellow or reddish, glabrous; stems twining, filiform, obtaining nourishment from hosts by haustoria. **Leaves** reduced to minute scales. **Inflorescences** with flowers sessile or shortly pedicellate, mostly in globular, spicate, racemose or cymose clusters; bracts minute or absent. **Flowers** 4- or 5-merous; calyx gamosepalous, ± deeply lobed, or sepals free; corolla white, pinkish, or cream-colored, urceolate, tubular, globose or campanulate, inside with fimbriate or crenulate, membranous, infrastaminal scales at base of tube; stamens as many as corolla lobes, inserted on corolla above scales, alternating with corolla lobes; ovary biloculed, locules 2-ovuled, styles 1 or 2, stigmas 2, subglobose or elongated, sometimes united. **Fruits** capsules, ovoid or globose, dry or sometimes fleshy, circumscissile or opening irregularly; seeds 1–4, glabrous. **Distr:** Mainly in N and S America, several in Asia and Europe. **Cult:** Propagate from seeds. Requires a suitable host, in full or partial sun.

C. chinensis Lamarck, *Encycl. Meth. Bot.* 2: 229, 1786. [Chinese Dodder]. **Herbs**, parasitic; stems yellow, thin, ~1 mm Ø. **Inflorescences** compact cymose glomerules, few- to many-flowered, lateral, subsessile; bracts and bracteoles scalelike. **Inflorescences** dense glomerules, or very rarely flowers arranged in small umbels, 2–4 mm long; pedicels ~1 mm long; bracts narrowly triangular, keeled, entire, nearly transparent. **Flowers** with calyx cupular, sepals triangular, ~1.5 mm long, obtuse, partly thickened; corolla white, urceolate, ~3 mm long, lobes persistent, triangular-ovate, apex acute or obtuse, reflexed; stamens inserted at throat, scales oblong, reaching stamens, long-fimbriate; ovary subglobose, styles 2, equal or unequal in length, stigma globose. **Fruits** capsules, enclosed by withered corolla, globose, ~3 mm Ø, circumscissile; seeds 2–4, pale brown, ovoid, ~1 mm long, scabrous. **Distr:** Africa, Asia, Australia. **Ecol:** On open mountain slopes, thickets, sandy beaches, usually twining on other herbs or shrubs; 200–3,000 m. **Zones:** 10–12.

Evolvulus

Species ~100. **Herbs**, **subshrubs**, or **shrubs**; stems never twining. **Leaves** petiolate or sessile; lamina with margins entire. **Inflorescences** spikes or capitula, terminal, 1- to several-flowered, axillary; flowers pedicellate or sessile; bracts tiny. **Flowers** with sepals free, equal or subequal; corolla rotate, funnelform, or salverform, limb ± entire to 5-lobed, midpetaline bands often pilose outside; stamens included or exserted; disk cupular or absent; ovary glabrous or pilose, biloculed, locules 2-ovuled, styles 2, filiform, free or basally united, each 2-cleft, stigma filiform, terete or slightly clavate. **Fruits** capsules, globose or ovoid, usually 4-valved; seeds 1–4, smooth or minutely tuberculate, glabrous. **Distr:** N and S America. **Cult:** Propagate from seeds, cuttings or by root division. Requires a well-drained soil in full sun.

E. alsinoides (L.) L., *Sp. Pl.*, ed. 2. 392, 1762. [Dwarf/Slender Morning Glory]. **Herbs**, perennial; stems few to numerous, prostrate or ascending, slender, hairy. **Leaves** petiolate or subsessile; oblong , elliptic or spathulate, 0.7–2(–2.5) × 0.3–1 cm, base rounded or attenuate into petiole, apex obtuse or shallowly emarginate and mucronulate, usually ± densely pilose. **Inflorescences** cymes, 1- to few-flowered; peduncle filiform, 2.5–3.5 cm long; bracts linear-subulate to linear-lanceolate, 1.5–4 mm long; pedicels usually longer than calyx. **Flowers** with sepals lanceolate, 3–4 mm long, acute or acuminate, villous; corolla pale-blue or white, rotate, 0.5–0.7(–1) cm Ø; stamens included, filaments filiform, adnate to base of corolla tube; ovary glabrous, styles 2, free. **Fruits** globose, 4-valved, glabrous, 4-valved; seeds 4 or fewer, black, smooth. **Distr:** Asia to Philippines, Australia, Pacific Islands; Africa; N and S America. **Ecol:** On sandy soils, dry slopes, maritime areas, grasslands, thickets; 0–1,800 m. Fl. & Fr. throughout most of the year. **Zones:** 10–12.

E. glomeratus Nees & C. Mart., *Nov. Actorum Acad. Caes. Leop.-Carol. Nat. Cur.* 11(1): 81, 1823.

Above: *Ipomoea cairica, Ipomoea hederifolia, Ipomoea horsfalliae, Ipomoea indica.*

(Syn: *E. pilosus* [misapplied]). [Blue Daze, Dwarf Morning Glory]. **Subshrubs**, perennial, evergreen, 25–40 cm high, spreading to ~1 m Ø. **Leaves** dense, alternate; lamina simple, oblanceolate to oblong 2.5–5 × 1–2.5 cm, base attenuate, bluish-green, white downy-pubescent on both surfaces. **Inflorescences** clusters, small, few-flowered, in upper leaf axils; bracts linear. **Flowers** rotate, cup-shaped, blue with a central white eye; sepals filiform, acute, persistent; corolla with petals fused, petals 1.5–2 cm long, limb spreading, 1.8–2.5 cm Ø; stamens white, filiform, exserted beyond corolla mouth to ~1 cm. **Fruits** globose 5 cm Ø, glabrous, 4-seeded. **Distr:** Brazil, Paraguay. **Ecol:** Fl. throughout the year. **Zones:** 10–12.

Hewittia

Species 1. *Generic description same as specific.* **Cult:** Propagate from seeds or cuttings. Requires a moist (for aquatic species) or well-drained soil in full sun.

H. malabarica (L.) Suresh, *An Inter. of Van Rheede's Hort. Malabar.* 88, 1988. (Syn: *Convolvulus malabaricus, Ipomoea malabarica*, etc.). **Herbs** perennial; stems herbaceous; axial parts ± densely pubescent, twining or prostrate, 1–2 m long, occasionally rooting at nodes. **Leaves** ovate, 3–10 × 3–8 cm, base cordate, hastate, or truncate, margins entire or trilobed, ciliate, apex acuminate, acute or mucronate, pilose or subglabrous; petiole 1–6 cm long. **Inflorescences** often 1-flowered, axillary; peduncle 1.5–10 cm long; bracts oblong-lanceolate, 0.7–1.5 cm long, acuminate, pubescent; pedicels 2–4 mm long. **Flowers** with sepals unequal, abaxially pubescent, outer 3 broadly ovate, 0.9–1.5 × 0.6–0.9 cm, to 1.7 cm long in fruit, inner 2 oblong-lanceolate, smaller; corolla pale yellow or whitish, with a purple center, 2–2.5 cm Ø, midpetaline bands densely pilose; stamens ~9 mm long, filaments basally minutely papillate, anthers ovoid-deltoid; disk ringlike; pistil included, ovary villous, style 1, filiform, stigmas 2, ovate-oblong, complanate. **Fruits** capsules, enclosed by persistent calyx, depressed-globose, 0.8–1 cm Ø, pilose; seeds dull black, ovoid-trigonous, 4–6 mm long, glabrous, hilum pubescent. **Distr:** Africa; Asia to Malesia, Pacific Islands. **Ecol:** In thickets, riverbanks; 0–600 m. Fl. & Fr. throughout the year. **Zones:** 10–12.

Ipomoea

Species ~500. **Herbs** or **shrubs**, often twining, sometimes prostrate, erect or floating. **Leaves** petiolate; lamina with margins entire, lobed, or divided. **Inflorescences** cymes, 1- to many-flowered, rarely panicles, usually axillary; bracts various. **Flowers** small to large; sepals persistent, equal to unequal, ± enlarged in fruit; corolla variously colored, rarely yellow, funnelform, campanulate or salverform, limb 5-lobed to entire, midpetaline bands well defined; stamens included or exserted, filaments filiform, usually unequal, dilated and pubescent basally, anthers ovate or linear, longitudinally dehiscent, not twisted; disk ringlike; ovary 2–4-loculed, 4- or 6-ovuled, style 1, filiform, included or exserted, stigma capitate, or 2- or triglobulose. **Fruits** capsules, globose or ovoid, 4- or 6-valved; seeds 4(–6) or fewer, glabrous or pubescent. **Distr:** Tropical to warm temperate regions, worldwide, especially of N and S America. **Cult:** Propagate from seeds or cuttings. Requires a moist (for aquatic species) or well-drained soil in full sun.

I. alba L., *Sp. Pl.* 1: 161, 1753. (Syn: *Calonyction aculeatum*, etc.). [Giant Moonflower]. **Herbs**, annual or perennial, twining, usually glabrous; stems to 10 m long, terete, smooth or with soft prickles, sap milky. **Leaves** ovate to ± circular, 10–20 × 5–16 cm, base cordate, margins entire, angular to trilobed, apex acuminate, mucronulate; petiole 5–20 cm long. **Inflorescences** helicoid cymes, rarely dichasial, 1- to several-flowered; peduncle stout, terete, 1–24 cm long; bracts early deciduous, small; pedicels 7–15 cm long. **Flowers** nocturnal, fragrant, large; sepals elliptic to ovate, ± leathery, glabrous, outer 3 sepals to 1.2 cm long, apex with a stout spreading awn 4–9 mm long, inner 2 sepals to 1.5 cm long, mucronate; corolla white, with greenish bands, salverform, tube 7–12 cm long, ~5 mm Ø, limb 7–12 cm Ø, shallowly 5-undulate; stamens exserted, filaments glabrous; pistil exserted, ovary narrowly conical, glabrous, stigma bilobed. **Fruits** ovoid, 2.5–3 cm long, apiculate; seeds white, brown, or black, ~1 cm long, glabrous. **Distr:** N and S America. **Ecol:** In wet forests, watercourses; 0–700 m. **Info:** Often used treating snake bites.

I. asarifolia (Desr.) Roem. and Schultes, *Syst. Veg.* 4: 251, 1819. (Syn: *Convolvulus asarifolius*, etc.). [Ginger-leaf Morning Glory]. **Vines**; stems herbaceous, decumbent, tips twining, rooting at nodes, glabrous. **Leaves** rounded-cordate to subreniform, 4–8 cm long, base cordate, apex rounded, glabrescent. **Inflorescences** with flowers solitary, or in axillary or terminal simple or compound cymes, glabrous or puberulent; pedicels 1.4–2.4 cm long. **Flowers** with sepals unequal, outer 5–6 mm long, inner elliptic to ovate, 1–1.2 cm long, apex rounded, mucronate, coriaceous, glabrous, verrucose; corolla lavender to purple, rarely white, funnelform, 6–8 cm long. **Fruits** tardily dehiscent, subglobose, 1–1.2 cm long, brown, glabrous; seeds brown to dark brown, 6–7 mm long, minutely gray-pubescent. **Distr:** Tropical America; naturalized pantropically. **Ecol:** Mostly in wetlands, near streams, canals and lagoons. **Zones:** 10–12.

I. cairica (L.) Sweet, *Hort. Brit.*, ed. 1. 287, 1827. [Coast Morning Glory]. **Herbs**, perennial, twining; rootstock tuberous; axial parts ± glabrous; stems to 5 m long, thinly angular, ± tuberculate or smooth. **Leaves** palmately 5-parted to base; lobes with margins entire or minutely undulate, apex acute or obtuse, mucronulate, basal pair usually again lobed or parted, midlobe larger, ovate, ovate-lanceolate, or elliptic, (2.5–)4–5 × (0.5–)2–2.5 cm; petiole 2–8 cm long, base with leafy pseudo-stipules. **Inflorescences** 1- or several-flowered; peduncle 2–8 cm long; bracts and bracteoles early deciduous, squamiform, small; pedicels 0.5–2 cm long, sometimes verruculose. **Flowers** with sepals unequal, outer 2 sepals 4–6.5 mm long, inner ones 5–9 mm long, glabrous, abaxially ± verruculose, margin paler, scarious; corolla pink, purple, or reddish-purple, with a darker center, rarely white, funnelform, (2.5–)5–7 cm long; stamens

Below: *Ipomoea mauritiana, Ipomoea nil, Ipomoea pes-caprae, Ipomoea pes-tigridis.*

included, unequal; ovary glabrous, stigma bilobed. **Fruits** ± globose, ~1 cm Ø; seeds black, ~5 mm long, densely tomentose, margin long-hairy. **Distr:** Origin unknown; cultivated and naturalized throughout the tropics. **Ecol:** In lowland thickets, waste places and open meadows; 100–600 m. Fl. Jun–Feb. **Zones:** 10–12.

I. hederifolia L., *Syst. Nat.* ed. 10: 925, 1759. (Syn: *I. angulata*, *I. coccinea*, *I. phoenicea*). [Cardinal's Flower, Red Ipomoea, Scarlet Creeper]. **Vines**; stems slender and herbaceous; annuals, glabrous to sparsely pubescent. **Leaves** ovate to suborbicular, 2–15 cm long, base cordate, margins entire, dentate, trilobate or less commonly with 5 or 7 lobes, apex acute to acuminate, glabrous or sparsely pubescent. **Inflorescences** few- to several-flowered cymes or flowers solitary; pedicels 5–7 mm or longer. **Flowers** with sepals oblong to elliptic 1.5–3 mm long, obtuse or truncate, outer sepals with a 1.6–6 mm long, subterminal arista, glabrous; corolla red or red-yellow, 2.5–4.5 cm long, tube 1–2 mm Ø, throat often yellow, limb 1.8–2.5 cm Ø. **Fruits** subglobose, 6–8 mm long; seeds dark brown or black, pyriform, with usually 2 lines of short dark trichomes on dorsal surface. **Distr:** Tropical America; naturalized in many tropical regions. **Ecol:** Usually in disturbed places. **Zones:** 10–12.

I. horsfalliae Hook., *Bot. Mag.* t. 3315, 1834. [Cardinal's Creeper]. **Vines**, perennial; stems twining, glabrous, becoming woody with age, lenticellate. **Leaves** orbicular, 5–20 × 5–20 cm, deeply palmately lobed to beyond middle or to base into 3–5 segments, middle segment usually larger than laterals, ovate, elliptic or elliptic-oblong, mostly attenuate toward both ends, apex acuminate with an acute or obtuse, mucronulate point, lateral segments ovate-lanceolate to linear-lanceolate, margins of segments slightly crisped, entire or coarsely dentate to crenate. **Inflorescences** few- to several-flowered, widely cymosely branched, axillary; peduncle 1.5–14 cm long; pedicels 0.8–1.5 cm long. **Flowers** with sepals subequal or the outer slightly shorter, elliptic or ovate-elliptic, 0.7–1 cm long, obtuse, concave; corolla red or red-purple, salverform, tube ~4 cm long, limb 4–4.5 cm Ø; stamens and style exserted. **Fruits** rarely produced in cultivation. **Distr:** Jamaica, Puerto Rico. **Ecol:** In moist forests. Fl. Oct–Jan. **Zones:** 10–12.

I. indica (J. Burman) Merrill, *Interpr. Herb. Amboin.* 445, 1917. (Syn: *Ipomoea acuminata*, etc.). [Dawn Flower, Ocean-blue Morning Glory]. **Herbs**, twining or sometimes prostrate; axial parts ± densely retrorse-pilose; stems 3–6 m long, sometimes rooting at nodes. **Leaves** ovate or circular, 5–15 × 3.5–14 cm, base cordate, margins entire or ± trilobed, apex acuminate or abruptly acuminate, adaxially ± sparsely pubescent, abaxially densely pubescent; petiole 2–18 cm long. **Inflorescences** dense umbellate cymes, several-flowered; peduncle 4–20 cm long; bracts linear, sometimes lanceolate; pedicels 2–5(–8) mm long. **Flowers** with sepals subequal, 1.4–2.2 cm long, gradually linear-acuminate apically, glabrous to appressed-pilose, outer 3 lanceolate to broadly lanceolate, inner 2 narrowly lanceolate; corolla bright blue or bluish-purple, aging to reddish-purple or red, with a paler center, funnelform, 5–8 cm long, glabrous; stamens included; pistil included, ovary glabrous, stigma trilobed. **Fruits** ± globose, 1–1.3 cm Ø; seeds ~5 mm long. **Distr:** S America; naturalized pantropically. **Ecol:** In coastal habitats, moist forests, weedy; 0–500 m. **Zones:** 10–12.

I. mauritiana Jacquin, *Collectanea.* 4: 216, 1791. [Giant Potato]. **Herbs**, perennial, twining; axial parts glabrous or minutely muricate; roots tuberous; stems to 10 m long, thinly angular. **Leaves** circular, 7–18 × 7–22 cm, usually palmately 5–7-divided to or beyond middle, rarely entire or shallowly lobed, segments lanceolate or elliptic, glabrous or sparsely pubescent along midvein, margins entire or irregularly undulate, apex acuminate or acute, mucronulate; petiole 3–11 cm long. **Inflorescences** few- to many-flowered; peduncle 2.5–20 cm long; bracts early deciduous; pedicels 0.9–2.2 cm long. **Flowers** with sepals ± circular, oblong to broadly elliptic, concave, equal or outer 2 shorter, 0.7–1.2 cm long, obtuse, glabrous; corolla pink or reddish-purple, center darker, funnelform, 5–6 cm long, limb 5–7 cm Ø, undulate; stamens included; pistil included, ovary glabrous, stigma bilobed. **Fruits** ovoid, 1.2–1.4 cm long; seeds dark brown, ~6 mm long, woolly sericeous with long, easily detached hairs. **Distr:** Old World tropics; thought to be originated from tropical America. **Ecol:** In dwarf forests near seashores, thickets, montane forests, streamsides; 0–1,100 m. **Zones:** 10–12.

I. nil (L.) Roth, *Catal. Bot.* 1: 36, 1797. (Syn: *I. hederacea*, etc.). [Blue Morning Glory, Japanese Morning Glory]. **Herbs**, annual, twining; axial parts retrorsely hirsute; stems 2–5 m long. **Leaves** broadly ovate or subcircular, 4–15 × 4.5–14 cm, base cordate, margins entire or ± 3- (or 5)-lobed, apex acuminate, hirtellous; petiole 2–15 cm long. **Inflorescences** axillary, 1- to few-flowered; peduncle 1.5–18.5 cm long; bracts linear or filiform, 5–8 mm long, hirtellous; pedicels 2–7 mm long. **Flowers** with sepals lanceolate, ± equal, 1–2.5 cm long, with a linear acumen, abaxially hirsute, apically subglabrous, hairs with swollen bases; corolla pale to bright blue, tube whitish, fading to pinkish with age, funnelform, 5–6(–8) cm long, glabrous; stamens included, unequal; pistil included, ovary glabrous, triloculed; stigma trilobed. **Fruits** straw-colored, ovoid to ± globose, 0.8–1 cm Ø, glabrous; seeds black, ovoid-trigonous, 5–6 mm long, gray-puberulent. **Distr:** S America; naturalized pantropically. **Ecol:** In thickets on mountain slopes; 0–1,600 m. **Zones:** 9–12.

I. pes-caprae (L.) R. Brown, *Narr. Exped. Zaire.* 477, 1818. (Syn: *Convolvulus pescaprae*, etc.). [Beach Morning Glory, Goat's-foot Creeper]. **Herbs**, perennial, glabrous; tap root thick; stems 5–30 m long, prostrate, sometimes twining, rooting at nodes. **Leaves** ovate, elliptic, circular, reniform or ± quadrate to oblong, 3.5–9 × 3–10 cm, base broadly cuneate, truncate, or shallowly cordate, margins entire, apex emarginate or deeply bilobed, mucronulate, thick, abaxially biglandular; petiole 2–10 cm long. **Inflorescences** 1- to several-flowered; peduncle stout, 4–14 cm long; bracts early deciduous, broadly deltate, 3–3.5 mm long; pedicel 2–2.5 cm long. **Flowers** with sepals unequal, ± leathery, glabrous, apex obtuse, mucronulate, outer 2 ovate to elliptic, 5–8 mm long, wider, inner 3 subcircular and concave, 0.7–1.1 mm long; corolla purple or reddish-purple, with a darker center, funnelform, 4–5 cm long; stamens included; pistil included, ovary glabrous, stigma bilobed. **Fruits** ± globular, 1.1–1.7 cm Ø, glabrous, leathery; seeds black, trigonous-globose, 7–8 mm long, densely brownish-tomentose. **Distr:** Pantropical. **Ecol:** Usually on sandy seashores, salt-tolerant; 0–100 m. **Zones:** 11–12.

I. pes-tigridis L., *Sp. Pl.* 1: 162, 1753. (Syn: *Convolvulus pes-tigridis*, etc.). [Tiger-foot Morning Glory]. **Herbs**, annual, twining; axial parts spreading-hispid; stems 0.5–3 m long. **Leaves** circular or transversely elliptic in outline, 2–10 × 3–13 cm, palmately deeply divided; segments (3–)5–9, elliptic or oblong, both ends tapered, densely pubescent, apex mucronate; petiole 2–8 cm long. **Inflorescences** capitate, few-flowered; peduncle 4–11 cm long; bracts hirsute; outer involucral bracts oblong to linear-oblong, 2–2.5 cm long; inner bracts smaller; pedicels obsolete. **Flowers** with sepals lanceolate, slightly unequal, 1–1.4 cm long, surfaces hirsute; corolla white, funnelform, 3–4 cm long, midpetaline bands sparsely pubescent; stamens included, filaments glabrous; pistil included, ovary biloculed, glabrous, stigma bilobed. **Fruits** ovoid, ~7 mm long, 4-valved; seeds ellipsoid, ~4 mm long, gray-tomentellous. **Distr:** Old World tropics. **Ecol:** In thickets, roadsides, seashores; 0–400 m. **Zones:** 9–12.

I. quamoclit L. *Sp. Pl.* 159, 1753. (Syn: *Quamoclit pinnata*, etc.). [Cypress Vine, Sweet Willy, Star Morning Glory]. **Herbs**, annual, twining, glabrous; stems slender. **Protostipules** present, 1.5 cm long. **Leaves** finely pinnatisect, usually 3–4 cm long, divided to midrib, segments 8–15 pairs, parallel, linear to filiform, often 1–1.5+ cm long, apex acute; petiole usually 1.2–3 cm long. **Inflorescences** 1- to several-flowered, axillary; peduncle 4–15 cm long; bracts minute, caducous; pedicels to 2 cm long. **Flowers** with sepals ovate-lanceolate or oblong, 5 × 1.5 mm, obtuse mucronate, subequal, ~6 mm long in fruit, inner sepals ovate-lanceolate, 6 × 2 mm; corolla tubular-cylindric or hypocrateriform, 2.5–3 cm long, scarlet or white, limb with spreading subacute lobes; stamens and style exserted; stamens 1.9 cm long, inserted ~4 mm from base of tube, pubescent, anthers sagittate, 2 mm long; ovary ellipsoid, to 1.5 mm long, biloculed, locules 2-ovuled, style 2 cm long, stigma globular. **Fruits** obpyriform, to ~8 mm Ø, with persistent sepals, dehiscent into 4 valves; seeds 4, obovoid 6 × 2 mm, pubescent. **Distr:** Pantropical. **Ecol:** In disturbed places. Fl. Aug–Feb. Fr. Feb–Apr. **Zones:** 10–12.

Below: *Ipomoea quamoclit*, *Ipomoea triloba*, *Jacquemontia paniculata*, *Merremia tridentata*.

Above: *Merremia tuberosa*, *Merremia tuberosa* (fruits), *Porana volubilis*, *Poranopsis paniculata*.

I. triloba L., *Sp. Pl.* 1: 161, 1753. (Syn: *I. blancoi*, etc.). [Little Bells, Pink Convolvulus, Trilobed Morning Glory]. **Herbs**, annual; stems twining or prostrate, glabrous or nodes sparsely pubescent. **Leaves** broadly ovate to circular in outline, 2.5–7 × 2–6 cm, base cordate, margins entire or coarsely dentate to deeply trilobed, glabrous or sparsely pilose; petiole 2.5–6 cm long, glabrous, sometimes tuberculate. **Inflorescences** dense umbellate cymes, 1- to several-flowered; peduncle 2.5–5.5 cm long, glabrous, angular, distally verruculose; bracts lanceolate-oblong, minute; pedicels 5–7 mm long, ± angular, glabrous, verruculose. **Flowers** with sepals slightly unequal, 5–8 mm long, glabrous or sparsely pilose abaxially, margin fimbriate, apex obtuse or acute, mucronulate, outer 2 oblong, slightly shorter, inner 3 elliptic-oblong; corolla pink or pale purple, funnelform, 1.5–2 cm long, glabrous, limb obtusely 5-lobed; stamens included; pistil included, ovary pubescent, stigma bilobed. **Fruits** ± globular, 5–6 mm Ø, bristly pubescent, apiculate, biloculed, 4-ovuled; seeds dark brown, ~3.5 mm long, glabrous. **Distr:** West Indies. **Ecol:** In disturbed places; 0–800 m. **Zones:** 10–12.

Jacquemontia

Species ~120. **Herbs** or **woody twiners**, rarely erect, hairs usually stellate, rarely glabrous. **Leaves** petiolate; lamina often cordate, margins entire, rarely dentate or lobed. **Inflorescences** axillary, often umbelliform or capitate cymes, less often scorpioid cymes, dense terminal spikes, capitula, or flowers solitary; bracts linear, subulate or leaflike. **Flowers** with sepals equal or outer 3 often larger; corolla blue, lilac, or pink, rarely white, funnelform or campanulate, with 5 distinct midpetaline bands, limb 5-toothed or subentire; stamens included, filaments basally adnate to corolla, anthers ellipsoid; disk small or absent; pistil included, ovary biloculed, locules 2-ovuled, style 1, filiform, stigmas 2, elliptic or oblong, flat, rarely linear or globose. **Fruits** capsules, globose, 4- or 8-valved; seeds 4 or fewer, smooth or minutely papillate, glabrous or velutinous, abaxial edges often with a narrow scarious wing. **Distr:** N and S America, with a few species also in Africa and Asia. **Cult:** Propagate from seeds. Requires a moist, well-drained soil in full sun.

J. paniculata (N.L. Burman) H. Hallier, *Bot. Jahrb. Syst.* 16: 541, 1893. (Syn: *Convolvulus parviflorus*, *Ipomoea paniculata*). **Herbs**, twining; stems terete, to 2(–4) m long, pubescent when young, glabrescent. **Leaves** lanceolate, ovate or ovate-oblong, 1.5–8 × 0.5–5 cm, base cordate or rounded to cuneate, apex acuminate, mucronulate, adaxially ± villous or glabrous, lateral veins 5–8 pairs; petiole 1–6 cm long, pubescent. **Inflorescences** umbelliform-cymose, loose to dense; peduncles 0.3–4 cm long; bracts subulate; pedicels 3–5 mm long, pubescent. **Flowers** with sepals unequal, apex ± acuminate, abaxially pilose to glabrous, outer 3 ovate to ovate-lanceolate, 5–7 × ~4 mm, inner 2 smaller; corolla lilac, pale purple, pink, or white, funnelform, 0.8–1.2 cm long, limb shallowly 5-lobed; filaments subequal, base pubescent, anthers oblong; ovary glabrous, style ~8 mm long, stigmas oblong. **Fruits** globose, 3–4 mm Ø, 8-valved; seeds brownish-yellow to purplish-black, 1.5–2 mm long, minutely tuberculate, glabrous, edges narrowly winged. **Distr:** Asia to Malesia, Australia, Pacific Islands; Africa. **Ecol:** In thickets, grassy areas; 0–600 m. **Zones:** 10–12.

Merremia

Species ~80. **Herbs** or **shrubs**, often twining, sometimes prostrate. **Leaves** usually petiolate, rarely sessile; lamina with margins entire, dentate, palmately lobed, pedately lobed or compound. **Inflorescences** with flowers solitary or in few- to many-flowered, variously branched cymes, axillary; bracts usually small. **Flowers** with sepals variable in shape, often convex, subequal or outer 2 smaller, persistent, often enlarged in fruit; corolla often yellow or white, sometimes with a darker center, funnelform or campanulate, usually glabrous, or midpetaline bands ± sericeous, sometimes only at apex, limb entire or ± 5-angled; stamens included, often unequal, anthers often spirally twisted, filaments basally dilated, distally filiform; disk ringlike; pistil included, ovary 2–4-loculed, 4-ovuled, style 1, filiform, stigmas biglobular. **Fruits** capsules, 1–4-loculed, usually 4-valved or ± irregularly dehiscent; seeds 4 or fewer, glabrous or pubescent to villous. **Distr:** Tropical regions of Africa, Asia, Australia, N and S America. **Cult:** Propagate from seeds or cuttings. Requires a moist (for aquatic species) or well-drained soil in full sun.

M. tridentata (L.) Hall. f., *Bot. Jahrb. Syst.* 16: 552, 1893. (Syn: *Ipomoea tridentata*, etc.). [Arrow-leaf Morning Glory]. **Vines**, perennial, highly variable; stems prostrate to twining, glabrous or pubescent. **Leaves** linear or lanceolate to oblong, 2.5–10 cm long, base truncate, hastate or auriculate, lobes often 1- to several-toothed, apex acuminate to emarginate and mucronulate, glabrous or pubescent. **Inflorescences** with flowers often solitary, sometimes few-flowered and cymose, axillary; peduncle slender, 2.5–4 cm long. **Flowers** with sepals oblong, ovate-oblong or lanceolate, 0.4–1 cm long, outer 2 usually shorter, obtuse, attenuate-acuminate or cuspidate, sometimes winged and crisped; corolla pale yellow, cream, or white, often with a reddish or purplish eye, campanulate, 1–2 cm long. **Fruits** globose, 4–9 mm long; seeds 2–3 mm long, black, glabrous. **Distr:** Tropical and South Africa, Mascarene Islands, tropical Asia to Formosa, Malaysia, Australia and Micronesia. **Ecol:** In disturbed areas including roadsides, grasslands. **Zones:** 10–12.

M. tuberosa (L.) Rendle, *Fl. Trop. Africa* 4 (2): 104, 1905. (Syn: *I. tuberosa*, *Operculina tuberosa*). [Wood Rose]. **Lianas**; stems herbaceous toward the tips, basally woody, glabrous. **Leaves** orbicular in outline, usually 7-lobed, sometimes almost to base, at least to below middle of lamina, lobes lanceolate to elliptic, 0.6–1.6 cm long, margins entire, apex acuminate, glabrous. **Inflorescences** axillary, few- to several-flowered, cymose, occasionally flowers solitary; peduncles 4–20 cm long. **Flowers** with sepals unequal, outer oblong, 2.5–3 cm long, obtuse, mucronate, inner 1.2–2 cm long, acute; corolla yellow, campanulate, 5–6 cm long. **Fruits** irregularly dehiscent, subglobose, 3–3.5 cm long, sepals somewhat accrescent and subtending the fruits; seeds black, densely tomentose, 1.6–1.7 cm long. **Distr:** Tropical America; naturalized in the Old World Tropics. **Zones:** 10–12. **Ecol:** In margins of tropical forests; 0–600 m. Fl. Jun–Aug. Fr. Oct onwards, and persistent for over a year. **Info:** The flowers and dry fruits are highly ornamental.

Porana

Species 31. **Twiners**, large, woody or herbaceous. **Leaves** herbaceous, petiolate; lamina ovate, base mostly cordate, margins entire, palmately nerved, rarely penninerved. **Inflorescences** racemose or paniculate, rarely flowers solitary; bracts leaflike, or minute and subulate, or none. **Flowers** with sepals 5, small in flower, 3 outer ones or all in fruit much-accrescent, scarious, reticulately veined, spreading, often spathulate, falling off with the fruit; corolla regular, white, small, campanulate or funnelform, rarely larger and funnelform or salverform, limb subentire or 5-lobed; stamens and style included, rarely exserted; stamens 5, filaments adnate to corolla, filiform, glabrous, or glandular or pubescent at base; disk annular or none; ovary mostly glabrous, 1-loculed, 2-ovuled, or 1–biloculed and 4-ovuled, style 1, simple, or bifid with unequal branches, stigma globose or bilobed, solitary, or one on each branch. **Fruits** capsules, small, subglobose to oblong, 2-valved, or indehiscent; seed usually 1, glabrous. **Distr:** Tropical and subtropical Asia, Africa and Australia. **Cult:** Propagate from seeds or cuttings. Requires a moist, well-drained soil in full sun. Best grown over an arch.

P. volubilis Burm., *Fl. Ind.* Vol. 1: 465, 1832. [Horse-tail Creeper]. **Twiners**, very large, woody, shrubby, glabrous except the shoots; branches warted. **Leaves** alternate; lamina ovate, 3–9 cm long, base subcordate, margins entire, apex shortly acuminate; petiole 1.7 cm long. **Inflorescences** paniculate cymes, large, subebracteate, minutely fulvous-pubescent, axillary and terminal, drooping, flowers numerous; pedicels 0–6 mm long; bracts small, villous, several at base of each flower, 1 longer. **Flowers** fragrant, small, white; sepals oblong, 5 mm long; corolla bell-shaped or campanulate, limb deeply 5-lobed, 0.8–1 cm long, white, segments oblong, obtuse; ovary ovoid, 1-loculed, 4-ovuled, placentation basal, style bipartite,

Above: *Stictocardia beraviensis, Brugmansia arborea, Brugmansia sanguinea, Brugmansia suaveolens.*

stigma small, capitate, somewhat bilobed. **Fruits** pea-sized, surrounded by persistent sepals, these all equal, subspathulate, 8.5 mm long, scarious, with 5–7 longitudinal nerves. **Distr:** Myanmar to Indo-China and Indonesia. **Ecol:** In tropical forests and forest margins. Fl. Jul–Sep. **Zones:** 10–12. **Info:** This species is very similar to *Poranopsis paniculata* and often confused in literature. The distinctly and deeply lobed corolla of *P. volubilis* can be used to differentiate the 2 species.

Poranopsis

Species 3. **Lianas**, dull yellow or grayish villous or sericeous, glabrescent. **Leaves** petiolate; lamina simple, cordate-ovate, papery, rugulose, abaxially densely pubescent, venation pedate or subpalmate. **Inflorescences** panicles, flowers fascicled, axillary (or terminal) bracteate; bracteoles 2, scalelike, basal to calyx; pedicels filiform. **Flowers** tiny, often fragrant; sepals free, quincuncial, unequally enlarged, outer 3 greatly enlarged, inner 2 slightly so, midvein 1, secondary veins reticulate; corolla white, ± funnelform, <8 mm long, limb 5-lobed, outside villous apically, inside glabrous; stamens included or exserted, anthers ellipsoid, longitudinally dehiscent; disk ringlike or absent; pistil included, ovary unilocular, ovules 4, style simple, stigma biglobose. **Fruits** papery, indehiscent; seed 1, smooth. **Distr:** Bhutan, China, India, Laos, Myanmar, Nepal, Pakistan, N Thailand, Vietnam. **Cult:** Propagate from seeds or cuttings. Requires a moist, well-drained soil in full sun. Best grown over an arch.

P. paniculata (Roxburgh) Roberty, *Candollea*. 14: 26, 1952. (Syn: *Dinetus paniculatus, Porana paniculata*). [Christmas Vine, Snow Creeper]. **Climbers**. **Leaves** cordate-circular, 7.5–16.5 × 5.3–15 cm, smooth to rugulose, abaxially silvery villous; petiole 2.6–10.8 cm long. **Inflorescences** panicles, crowded; bracts cordate; pedicels 2–4 mm long, elongating in fruit. **Flowers** (4–)5–6(–7) mm long; sepals lanceolate-linear, flat or concave, equal, 1–2 mm long, abaxially tomentose-villous, outer 3 sepals elliptic-oblong to narrowly ovate, 1.6–2.2(–2.4) × 0.7–0.9 cm, puberulent, margins free; corolla white to cream, narrowly funnelform, limb (3–)5–7 mm Ø, 5-lobed; stamens included, ± equal, filaments glabrous; disk absent or ringlike; ovary glabrous, style obsolete, stigma subsessile. **Fruits** tan to brownish with darker lines, globose-ellipsoid, 5–6(–7) × 4–5 mm, pubescent, apiculate; fruiting calyx tan, reddish, or pale brown, loosely clasping; seeds dark brown, globose-ellipsoid, 4–6 × 3–5 mm, glabrous. **Distr:** Bhutan, China, India, Myanmar, Nepal, Pakistan. **Ecol:** In forests, thickets, scrubs, open plains; 0–2,000 m. Fl. Oct–Dec. Fr. Mar–Apr. **Zones:** 10–12.

Stictocardia

Species ~12. **Twiners**, woody or herbaceous. **Leaves** petiolate; lamina ovate to circular, base usually cordate, margins entire, abaxially spotted with minute glands. **Inflorescences** cymes, 1- to many-flowered, axillary, peduncled; bracts early deciduous, small. **Flowers** with sepals ovate to elliptic or circular, equal or slightly unequal, much-enlarged in fruit, ± leathery, abaxially glandular punctate, margin membranous, apex obtuse to emarginate; corolla funnelform, midpetaline bands ± pilose and with minute glands; stamens and pistil included; filaments filiform, inserted near corolla base, pollen globular; disk ringlike; ovary glabrous, 4-loculed, locules 1-ovuled, style 1, stigma biglobular. **Fruits** capsules, enclosed by enlarged calyx, globose, pericarp thin, eroding irregularly between septa; seeds 4 or fewer, pubescent. **Distr:** Africa, Asia. **Cult:** Propagate from seeds or cuttings. Requires a moist (for aquatic species) or well-drained soil in full sun.

S. beraviensis Hallier f., *Bot. Jahrb. Syst.* 18: 159, 1894. (Syn: *Argyreia bagshawei*, etc.). [Braveheart Vine, Hawaiian Bell, Hawaiian Sunset Vine]. **Twiners**, strong, woody; young stems densely pubescent; older stems glabrescent, yellowish. **Leaves** ovate, to 16(–23) × 15(–20.5) cm, base shallowly cordate or truncate, apex mucronate, tip acute or blunt, adaxially glabrous or very sparingly pilose, abaxially very densely gray-velvety to glabrous or glabrescent, lateral nerves closely parallel; petiole to 17 cm long. **Inflorescences** many-flowered, axillary; peduncle 0.8–1(–2.5) cm long; pedicels 0.6–1 cm long; buds finely pubescent at apices. **Flowers** with calyx lobes elliptic or suborbicular, 0.7–1(–1.3) × 0.5–0.8(–1.2) cm, blunt or emarginate, mucronulate, basally hairy, glabrescent above; corolla very bright crimson with base of tube orange-yellow, funnelform, not markedly narrowed basally, 4.5–5.5 cm long, with tufts of hairs at the tips of midpetaline areas. **Fruits** globose, 1.4 × 1.3 cm; seeds black, ± granular, 6 × 4 mm. **Distr:** Ethiopia to W Africa and Madagascar. **Ecol:** In riverine forests; 900–1,800 m. **Zones:** 10–12.

SOLANACEAE

Subfamilies 8, **genera** 102, **species** 2,460. **Herbs**, **shrubs**, **small trees**, or **climbers**; stems sometimes prickly, rarely thorny; hairs simple, branched, or stellate, sometimes glandular. **Stipules** absent. **Leaves** alternate, solitary or paired, simple or pinnately compound; lamina or leaflets with margins entire, dentate, lobed, or divided. **Inflorescences** terminal, overtopped by continuing axes, appearing axillary, extra-axillary, or leaf opposed, often apparently umbellate, racemose, paniculate, clustered, or flowers solitary, rarely true cymes, sometimes bracteate. **Flowers** mostly bisexual, usually regular, (4)5(6–9)-merous; calyx mostly lobed; petals united; stamens as many as and alternate with corolla lobes, inserted within corolla, all alike or 1 or more reduced, anthers dehiscing longitudinally or by apical pores; ovary 2–5-locular, placentation mostly axile, ovules usually numerous, style 1. **Fruits** berries or capsules; fruiting calyx often becoming enlarged, mostly persistent; seeds with copious endosperm. **Distr:** Worldwide, highly diverse in tropical America.

Brugmansia

Species 6–10. **Shrubs** or **small trees**; pubescence of simple hairs. **Leaves** petiolate; lamina mostly ovate, margins subentire. **Inflorescences** with flowers solitary, at leaf nodes near branch ends. **Flowers** very large, usually fragrant, perfect, actinomorphic, 5-merous; calyx tubular, 5-lobed or spathaceous; corolla white, pink, red, orange, greenish, or yellowish, trumpet-shaped, mostly with 5 pointed teeth; stamens inserted near middle of corolla tube, sometimes geniculate, equal, pubescent or not, anthers linear, basifixed, longitudinally dehiscent; ovary bilocular. **Fruits** capsules, unarmed, elongate or rotund, woody, indehiscent. **Distr:** S America. **Cult:** Propagate from cuttings or seeds. Requires a moderately fertile, moist, well-drained soil in shade to full sun.

B. arborea (L.) Lagerh., *Bot. Jahrb. Syst.* 20: 663, 1895. (Syn: *B. candida*, etc.). [Angel's Trumpet, Peruvian Trumpets, Tree Datura]. **Shrubs** or **small trees**, much-branched, to 6 m high; twigs stout, densely pubescent; young parts velvet-hairy. **Leaves** ovate to oblong or ovate-lanceolate, 15–25 cm long, base usually unequal, rounded or acute, margins entire or sinuate-dentate, apex acute to acuminate, light green, surfaces pubescent; petioles 2.5–7.5 cm long, pubescent. **Inflorescences** with flowers solitary, flowers pendent; peduncles 2.5–5 cm long; pedicels white-hairy. **Flowers** fragrant; calyx spathelike, 2-cleft to about middle, 7.5–10 cm long, subulate, rarely divided, smooth, basally adpressed to corolla, deciduous; corolla white, trumpet-shaped, 12–25 cm long, inflated beyond calyx, often double, united for most of the length, lobes 2.5–5 cm long, veins softly hairy, mouth and lobes recurved, distinct notches between lobes; anthers free; stigma capitate. **Fruits** capsules, spherical or ovoid, 4–7.5 cm long, light green, with short, velvet-hairy; seeds ~0.7–1.2 cm long, angled, corky. **Distr:** S America. **Ecol:** Often in disturbed and waste places. **Zones:** 10–12.

B. sanguinea (Ruiz & Pav.) D.Don, *Brit. Flow. Gard. Ser.* II. t. 272, . (Syn: *Datura sanguinea*, etc.). [Red Angel's Trumpet, Red Floripontio, Scarlet Datura]. **Shrubs** or **small trees**, to 10 m tall; young parts sparsely to densely white-hairy. **Leaves** variable, pale to very dark green, densely white-hairy; *lower leaves* ovate, margins wavy to dentate; *upper leaves* ovate to lanceolate, margins wavy or sinuate (with lobes usually angular) to entire. **Inflorescences** with flowers solitary, axillary or terminal, pendent. **Flowers** large, not fragrant; calyx tubular, ribbed, inflated, 6–9 cm long, lobes 2–5, acuminate; corolla yellow, red, orange-red apically, and greenish or yellowish basally, or variously combined, tubular to funnelform, gradually

Top row: *Brunfelsia americana, Brunfelsia australis, Brunfelsia densifolia, Brunfelsia grandiflora.*
Bottom row: *Brunfelsia pauciflora, Capsicum annuum, Capsicum frutescens, Cestrum aurantiacum.*

expanded toward mouth and recurved, 15–25 cm long, united for almost its entire length, ribbed, lobes 5, long-acuminate, recurved with margins incurved, 1–2 cm long; stamens reaching near mouth but not exserted, free, anthers creamish-white; style equal in length to stamens, stigma club-shaped. **Fruits** ovoid, 7–9 × 4–6 cm; seeds ~4 × 7 mm, wrinkled, partially covered by persistent calyx, hairy. **Distr:** W South America. **Ecol:** In cool subtropical montane areas; 3,000–4,000 m. Fl. throughout most of the year. **Zones:** 9–11.

B. suaveolens (Willd.) Bercht. and Presl, *Prir. Rostlin.* 1 (Solanac.): 45, 1823. (Syn: *Datura suaveolens*). [Angel's Tears, White Angel's Trumpet]. **Shrubs** or **trees**, to 5 m high; twigs finely pubescent with simple hairs, soon glabrescent. **Leaves** ovate, 15–30 × 8–15 cm, base rounded or obtuse, margins subentire, apex acute or acuminate, pubescent; petiole 2–8 cm long. **Inflorescences** with flowers solitary, pendulous, borne at leaf nodes near end of branches; pedicels 2–5 cm long. **Flowers** large, showy, fragrant; calyx broadly tubular, inflated in bud and flower, 9–12 × 3–4 cm, lobes 5, 1.5–2 cm long, acute or obtuse; corolla white or yellowish, 25–30 cm long, pendent, tubiform, tube less than half as wide as calyx mouth, abruptly expanded above calyx, limb sinuate-lobed with 5 narrowly lanceolate spreading teeth 1–1.5 cm long; filaments inserted just below middle of tube, sometimes geniculate, pilose near point of insertion, anthers 2.5–3.5 cm long, cohering into a cylindrical tube 4–6 mm Ø. **Fruits** fusiform, 20 cm long; seeds 0.8–1.2 cm Ø, rugose. **Distr:** S America. **Ecol:** In lowland areas, dense forests, forest margins and often along streams. Fl. Aug–Nov. **Zones:** 10–12.

Brunfelsia

Species ~46. **Shrubs** or **small trees**, evergreen, to 10 m tall. **Leaves** alternate, shortly petiolate; lamina usually elliptic, oblong, lanceolate, obovate to oblanceolate, base narrowed, cuneate or blunt, margins entire, apex acuminate to rounded, ± glabrous, lateral nerves pinnate. **Inflorescences** cincinnal cymes or reduced to solitary flowers, terminal or axillary, 1- to many-flowered; bracts 1–3 per flower, linear to lanceolate, caducous. **Flowers** white turning yellow, reddish-purple or violet-purple fading to pale violet or white, with or without fragrance; calyx gamosepalous, tubular to campanulate, green or purplish, 5-dentate, teeth persistent, enclosing capsule or basal; corolla gamopetalous, hypocrateriform, slightly zygomorphic bilaterally symmetrical, tube cylindric, ± straight; stamens 4, didynamous, included or upper pair briefly exserted, filaments with lower pair shorter, anthers all fertile, longitudinally dehiscent; ovary superior, bilocular, ovules numerous, anatropous, style filamentous, stigma included between anthers. **Fruits** capsular or subbaccate, bilocular or unilocular, valves ± indehiscent; seeds few to numerous, reddish-brown. **Distr:** C and S America, West Indies. **Cult:** Propagate from seeds or cuttings. Requires a fertile, well-drained soil in partial to full sun. **Info:** All species contain toxic alkaloids.

B. americana L., *Sp. Pl.* 1: 191, 1753. (Syn: *B. undulata, Brunfelsiopsis americana*). [Jamaican Rain Tree, Lady of the Night]. **Shrubs**, erect, subglabrous or puberulent, 1–3(–5) m high. **Leaves** somewhat crowded on ultimate twigs; lamina elliptic to obovate or oblong-ovate, 5–10 cm long, base cuneate, tapered toward petiole, margins entire, apex acute or obtuse, papery to somewhat leathery, often yellowish-green; petiole to 6 mm long. **Inflorescences** with flowers solitary or in pairs, terminal (at branch tips); pedicels to 1.3 cm long. **Flowers** very fragrant; calyx green, bell-shaped, 6 mm long, 5-lobed, lobes narrow; corolla white, turning yellow with age, tube ~5–6.5 cm long, slender, somewhat straw-colored, limb 5–10 cm Ø, oblique, subequally 5-lobed, lobes rounded. **Fruits** globose, 1–2 cm Ø, pericarp yellow, somewhat fleshy, 2-valved; seeds numerous. **Distr:** Tropical America. **Ecol:** In woodlands at high elevations. Fl. throughout the year. **Zones:** 10–12.

B. australis Benth., *Prodr.* 10: 200, 1846. (Syn: *B. hopeana* var. *australis*, etc.). [Paraguay Jessamine, Yesterday-today-and-tomorrow]. **Shrubs** to **small trees**, 2–4 m high; trunk to 8 cm Ø. **Leaves** scattered along branchlets; lamina broadly elliptic to obovate, rarely orbicular or rhombic, 4–12.5 × 2.5–6 cm, base cuneate, apex usually blunt to rounded, adaxially dull dark green, abaxially paler; petiole 3–8 mm long. **Inflorescences** 1–4-flowered, terminal at branchlet tips, sessile; bracts lanceolate, 0–2 mm long, concave, ciliolate, early caducous; pedicels stout, 4–7 mm long. **Flowers** showy, violet, fading to pure white with age, fragrant; calyx campanulate, 0.7–1.2 × 0.4–1 cm, ± inflated, glabrous; corolla tube 2.5–3.2 cm long, glabrous, limb 3–4 cm Ø, flat, spreading, thickening at mouth prominent, round, white or yellowish, lobes equal, broadly rounded, laterally overlapping, base abruptly narrowed; stamens included; ovary ovoid-conical, 2.5 × 1 mm, style 2.2–2.8 cm long. **Fruits** subglobose to ovoid, 1–2.2 × 1–2 cm, glabrous, dark green, indehiscent or tardily dehiscent; seeds 20–40, oblong-ellipsoid, 4–6 × 3–4 mm, angular, dark brown. **Distr:** Brazil, Paraguay, Argentina, Uruguay. **Ecol:** In understory of lowland forests near rivers, also in upland

Above: *Cestrum diurnum, Cestrum elegans, Cestrum nocturnum, Cestrum parqui.*

areas and swampy areas; 0–450 m. Fl. Aug–Nov. **Zones:** 9–12.

B. densifolia Krug and Urban, *Not. Berl. Bot. Gart.* 1: 324, 1897. [Serpentine Hill Raintree]. **Trees**, 6–10 m tall, glabrous; branches densely leafy. **Leaves** linear-oblong to linear-oblanceolate, 5–11 × 1 cm, base long-attenuate, margins entire, apex acute or obtuse, coriaceous, glossy, venation nearly parallel; petiole 4–9 mm long. **Inflorescences** with flowers solitary, terminal; peduncle to 1 cm long. **Flowers** with calyx narrowly campanulate, ~7 mm long, lobes short, rounded; corolla white, aging to yellow, tube slender 10–14 cm long, lobes short, obtuse or rounded, 1.2 cm long; stamens unilaterally affixed, 2 fertile, 2 sterile. **Fruits** ovoid, 2.5 cm Ø, rounded. **Distr:** Puerto Rico. **Ecol:** On serpentine hillsides at high elevations. **Zones:** 10–12.

B. grandiflora D. Don, *Edinburgh New Philos. J.*: 86, 1829. (Syn: *Franciscea grandiflora*). [Brazilian Rain Tree, Yesterday-today-and-tomorrow]. **Shrubs** or **small trees**, 1–6(–10) m high; stem to 7 cm Ø near base, much-branched. **Leaves** ± 2-ranked, scattered along branchlets; lamina lanceolate to oblong, 6–23 × 2–8 cm, base cuneate to narrowed, apex acuminate, acumen subfalcate, ± glabrous, adaxially dark green, abaxially paler; petiole 2–6 mm long. **Inflorescences** simple or branched, 5- to many-flowered, ± terminal; axis to 4.5 cm long; bracts 1–3 per flower, lanceolate to ovate, to 0.5(–1) cm long, ciliolate, caducous; pedicels to 0.5(–1) cm long. **Flowers** scentless, violet, fading to white, with rounded white ring at mouth; calyx tubular-campanulate, 0.5–1.3 × 0.3–0.7 cm, narrowed toward base, yellow-green to gray-green, teeth 1–5 mm long; corolla tube 1.5–4 cm long, limb 2–5 cm Ø, spreading, lobes 0.7–1.5 cm long, upper lobe larger, rounded or emarginate; stamens included; ovary conical to ovoid, 1.5–2 mm long, yellowish, stigma ~1 mm long, briefly bifid, unequal. **Fruits** ovoid to subglobose, 0.8–2 cm long, obtuse or apiculate, smooth, brownish; seeds 10–20, ellipsoid to oblong, 5–7 mm long, dark reddish-brown. **Distr:** Bolivia, Brazil, Colombia, Ecuador, Peru, Venezuela. **Ecol:** In humid, montane rainforests, often in valleys; 650–2,000 m. Fl. sporadically throughout the year. **Zones:** 10–12.

B. pauciflora (Cham. & Schltdl.) Benth., *Prodr.* 10: 199, 1846. (Syn: *Brunfelsia calycina*, etc.). [Brazil Rain Tree, Morning-noon-and-night, Yesterday-today-and-tomorrow]. **Shrubs**, 1–3(–5) m high. **Leaves** scattered or crowded near branch tips; lamina usually oblong to oblong-lanceolate, 6.5–16 × 2–6.5 cm, base cuneate to attenuate, apex acute to shortly acuminate, glabrous, adaxially dark green, abaxially paler, lateral nerves 5–11; petiole 0.5–1.2 cm long. **Inflorescences** 1–11-flowered, terminal, ± sessile; bracts small, 1–3 per flower, linear-lanceolate, 1–8 mm long, caducous; pedicels 1.1–2.5(–3.5) cm long, stout. **Flowers** reddish-purple, fading to pale lavender or white, with white ring at mouth sometimes edged with violet; calyx 1.8–3.2 × 0.6–1 cm, green; corolla tube 2.8–3.6 cm long, lobes broadly rounded to subelliptic, 1.5–3 cm long; stamens included filaments white, upper pair 4–6 mm long, lower pair 2–4 mm long, anthers greenish-brown; ovary conical-ovoid, 2–3 × 1.5 mm, pale green, style 2.5–3 cm long, filamentous, lavender, stigma 1 mm long, briefly bifid, lobes subequal, white. **Fruits** capsules, enclosed in calyx, subglobose to ovoid, 2–2.2 × 1.5–1.8 cm, green; seeds 12–30, ovoid or oblong, 5–6 × 2.5–3 mm, angular, brownish. **Distr:** Brazil. **Ecol:** In coastal lowland rainforests, in shade along small streams and ravines, also in humid primary and secondary forests; 0–1,500 m. Fl. Sep–Dec. Fr. Jan–Mar. **Zones:** 10–12.

Capsicum

Species ~25. **Shrubs**, or annual or perennial **herbs**, pubescent; stems branched. **Leaves** solitary or paired, petiolate; lamina simple, margins entire or sinuate. **Inflorescences** solitary or few-flowered clusters at branch forks or leaf axils; peduncle absent; pedicels erect or nodding. **Flowers** actinomorphic; calyx broadly campanulate to cup-shaped, denticulate, sometimes slightly enlarged; corolla white, blue, or violet, campanulate or rotate, divided halfway or more; stamens inserted near distal end of corolla tube, filaments slender, anthers yellow or purplish, ovoid, longitudinally dehiscent; ovary 2(or 3)-locular, ovules numerous, style slender, stigma small, capitate. **Fruits** berries, variously shaped, erect, nodding, or reflexed; seeds yellowish, discoid. **Distr:** S America. **Cult:** Propagate from seeds, grown as annuals. Requires a moist, well-drained soil in full sun.

C. annuum L., *Sp. Pl.* 1: 188, 1753. (Syn: *C. frutescens* var. *conoides*, etc.). [Bell Pepper, Capsicum, Chilli, Paprika, Pepper]. **Shrubs**, or annual or perennial **herbs**, 20–80 cm high; stems glabrescent. **Leaves** solitary or paired; lamina oblong-ovate, ovate or ovate-lanceolate, 4–13 × 1.5–4 cm, base narrowed, margins entire, apex shortly acuminate or acute, surfaces glabrescent; petiole 4–7 cm long. **Inflorescences** with flowers solitary or few-flowered clusters; pedicel bent at apex, 1–2 cm long. **Flowers** with calyx cup-shaped, undulate, 2–3 × 3 mm; corolla white, ~1 cm long; anthers purplish, 1.8–2 mm long. **Fruits** berries, mostly red, variously shaped (and colored in cultivars), to 15+ cm long; seeds pale yellow, discoid or reniform, 3–5 mm Ø, flattened. **Distr:** Mexico to S America. **Ecol:** Fl. May–Aug. Fr. Jul–Nov. **Zones:** 6–12. **Info:** Cultivated widely and commercially for the edible fruits, used as a spice.

C. frutescens L., *Sp. Pl.* 1: 189, 1753. (Syn: *C. minimum*). [Bird Pepper, Chilli, Hot Pepper, Spur Pepper]. **Herbs** or **shrublets**, 0.2–1.5 m high, densely branched; branches glabrous or sparsely pubescent. **Leaves** solitary or often paired, petiolate; lamina ovate to ovate-lanceolate, 4–12 × 1–4.5 cm, base attenuate, apex long-acuminate. **Inflorescences** (1)2- or more-flowered, borne at nodes; flowers held erect. **Flowers** with calyx truncate, or minutely lobed, prominent constriction lacking between base of calyx and pedicel; corolla greenish-white or yellowish-white. **Fruits** berries, held erect, intensely pungent, red, ellipsoid-lanceolate or lanceolate, 1.5–3 cm long, 0.5–1.2 cmØ; seeds cream to yellow. **Distr:** Tropical America. **Ecol:** In lowland forests. Fl. & Fr. throughout most of the year. **Zones:** 9–12.

Cestrum

Species ~175. **Shrubs** or **trees**, rarely vines, glabrous or pubescent, unarmed; hairs simple, branched or stellate, sometimes glandular. **Leaves** solitary, petiolate; lamina simple, margins entire, pinnately nerved. **Inflorescences** racemes, spikes or cymes, often large and appearing paniculate, sometimes clustered in leaf axils, few- to many-flowered, pseudoterminal or axillary, often bracteate or bracteolate; flowers pedicellate. **Flowers** diurnal or nocturnal, mostly fragrant, mostly 5-merous, calyx campanulate or tubular; corolla long-tubular, tube sometimes expanded or contracted around anthers, sometimes pubescent abaxially, limb lobed, usually spreading; stamens inserted at various levels in corolla tube, filaments sometimes pubescent or appendaged at or below point of insertion, anthers longitudinally dehiscent; disk mostly evident; ovary bilocular, ovules few to several, rarely to 20, style slender, stigma entire or bilobed, rarely exserted. **Fruits** berries, mostly white or blackish, globose, ovoid, or oblong, often juicy; seeds 1 or several, oblong. **Distr:** N to S America. **Cult:** Propagate from cuttings or seeds. Requires a fertile, moist, well-drained soil in full or partial sun.

C. aurantiacum Lindley, *Bot. Reg.* 30 (Misc.): 71, 1844. [Orange Cestrum]. **Shrubs**, to 2–3 m high; young growth pubescent, glabrescent. **Leaves** ovate or elliptic, 4–7 × 2–4 cm, base acute, margins entire, slightly undulate, apex acute, surfaces glossy green, with an unpleasant odor when crushed; petiole 1–1.4 cm long. **Inflorescences** racemose panicles, many-flowered, erect, terminal or axillary; bracts deciduous, leaflike, to 7 mm long; pedicels ~1 mm long. **Flowers** with calyx green, campanulate, ~6 mm long, sometimes with 5 prominent longitudinal ribs, teeth ~1 mm long, often somewhat unequal; corolla golden-yellow, 1.4–2 cm long, gradually expanded upward, glabrous, lobes 3–4 mm long; filaments denticulate, pubescent at point of insertion. **Fruits** berries, white, ovoid, 0.7–1.1 cm long; persistent calyx often splitting; seeds ~4. **Distr:** C America. **Ecol:** In moist thickets or forests, often in pine-oak forests; 1,000–2,600 m. Fl. throughout most of the year. **Zones:** 10–12.

C. diurnum L., *Sp. Pl.* 191, 1753. (Syn: *C. fastigiatum*, etc.). [Chinese Ink Berry, Day Cestrum/ Jessamine]. **Shrubs** or **small trees**, 2–6 m high. **Leaves**

Above: *Cyphomandra betacea*, *Datura metel*, *Datura metel* (double form), *Datura stramonium*.

ovate or elliptic, 4.5–12.5 × 1.5–4.5 cm, base narrowly acute to rounded, slightly asymmetrical, apex obtuse or acute, subcoriaceous, glabrous, adaxially glossy green, abaxially paler, veins 6–7 per side; petiole 0.5–1 cm long; minor leaves rotund, 1–1.2 × 0.4–0.7 cm, sessile, caducous. **Inflorescences** congested spicate cymes, racemes, or umbels, axillary and terminal; peduncle to 9 cm long; bracts foliaceous, 3 × 1.5 mm, pubescent; pedicels obsolete. **Flowers** fragrant day and night, 1.2–1.7 cm long, often 6-merous; calyx light green, cupular, 2.5–3.5 × 1.5–2 mm, costate, tube 2–3 mm long, glabrous; corolla funnelform, obconical, 1.1–1.6 cm long, lobes 5–6, rounded, 1.5 mm long, recurved; stamens 0.8–1.1 cm long; ovary oblong, 1 mm long, style white, 0.8–1 cm long, stigma dark green, capitate, slightly exserted. **Fruits** purple-black, subglobose, 1.1–1.2 × 0.8 cm; seeds 13–14, dark brown, 2.5–3.5 mm long. **Distr:** West Indies. **Ecol:** Fl. Jan–Jun and Aug–Oct. **Zones:** 10–12. **Info:** The fruits are moderately poisonous.

C. elegans (Brongniart) Schlechtendal, *Linnaea*. 19: 261, 1847. (Syn: *C. purpureum*, *Habrothamnus elegans*). [Butterfly Flower, Pink/Purple/Red Cestrum]. **Shrubs**; stem copiously pubescent; branches arching. **Leaves** ovate, elliptic, ovate-oblong to lanceolate, ~8 × 3 cm, base acute, margins entire, apex acuminate, tip rounded, surfaces sparsely pubescent, olive-green, gives of a disagreeable odor when crushed; petiole 0.6–1.2 cm long, pubescent. **Inflorescences** erect, terminal or axillary, congested racemose panicles; bracts ovate, minutely tomentose, 4–5 mm long; pedicels ~1 mm long. **Flowers** scentless; calyx narrowly campanulate, 6–8 mm long, glabrous, unribbed; teeth ~3 mm long, ciliate; corolla red, pink, or violet, ~2 cm long, expanded upward, abruptly contracted at throat, glabrous, lobes deltoid, ~2 mm long; filaments glabrous, unappendaged, slightly bent at point of insertion. **Fruits** berries, dark pink, globose, 0.8–1.3 cm Ø; fruiting calyx not splitting; seeds ~8. **Distr:** Mexico. **Ecol:** In understory of cloud forests. Fl. Aug–Mar. **Zones:** 10–12.

C. nocturnum L., *Sp. Pl.* 1: 191, 1753. (Syn: *C. suberosum*). [Night Jessamine, Night-scented Jessamine, Queen of the Night]. **Shrubs**, erect or sprawling, 1–5 m high; young growth puberulent, glabrescent; branches slender. **Leaves** oblong-ovate or oblong-lanceolate, 6–15 × 2–4.5 cm, margins entire; petiole 0.8–2 cm long. **Inflorescences** many-flowered, axillary or terminal racemose panicles, drooping, 7–10 cm long; pedicels 1–5 mm long. **Flowers** strongly scented at night; calyx campanulate, 2–3 × 1–1.5 mm, lobes deltoid, 0.5–0.8 mm long; corolla greenish or whitish-yellow, 1.5–2.5 cm long, tube slightly contracted at throat, lobes 3–4 mm long; filaments with dentate appendages, usually puberulent at point of insertion; ovary ~1 mm long. **Fruits** berries, white, juicy, oblong or globose, 0.6–1 × 0.4–1 cm; seeds 1–5, ellipsoid, 3.5–4.5 mm long. **Distr:** Tropical America. **Ecol:** In lowland moist or wet thickets or forests, sometimes in open places; below 1,800 m. Fl. Feb–Mar. **Zones:** 9–12. **Info:** Widely cultivated throughout the tropics for the fragrant flowers.

C. parqui L'Heritier, *Stirpes Novae aut minus cognitae*, 4: 73, tab. 36, 1788. [Chilean Cestrum, Green Cestrum, Willow Leaf Jessamine, Willow-leaved Cestrum]. **Shrubs**, suckering, evergreen or deciduous, to 3–12 m tall, unarmed, glabrous or pubescent. **Leaves** subsessile, unpleasant odor when crushed; lamina linear, lanceolate, elliptic-lanceolate to elliptic, 5–14(–32) × 2.5–7(–15) cm, base acute, margins entire, apex acute, pubescent. **Inflorescences** cymes or panicles, many-flowered, axillary or terminal; flowers pendent or erect. **Flowers** greenish-yellow to orange, 2.5 cm long; calyx lobes short; corolla infundibuliform, lobes 5, ovate-acute, ciliate; stamens attached near throat of corolla, anthers oblong. **Fruits** berries, blackish; seeds few. **Distr:** Chile. **Ecol:** In coastal mountains, forest margins; 0–2,000 m. Fl. Mostly throughout the year. **Zones:** 9–11. **Info:** The fragrant flowers bloom at night. All parts are poisonous.

Cyphomandra

Species ~60. **Shrubs** or **small trees**, unarmed. **Leaves** alternate, long-petiolate; lamina with margins entire, trilobed, or pinnatifid, often large, sometimes dimorphic. **Inflorescences** scorpioid cymes, mostly axillary; rachis nodose; flowers pedicellate. **Flowers** with calyx campanulate, 5-lobed; corolla subrotate, pink to purplish, or greenish-purple, limb deeply 5-parted, lobes valvate in bud; stamens 5, exserted, filaments short and broad, anthers oblong or attenuate, dehiscent by 2 apical pores or slits, these often continued as longitudinal introrse slits halfway down the thecae, connective prominent, adaxially thickened; disk annular; ovary bilocular, ovules numerous; entire or dentate, style thick or elongate and filiform, stigma scarcely or strongly dilated. **Fruits** berries, often large, ovoid, oblong, or elliptic; seeds numerous, compressed. **Distr:** Tropical America. **Cult:** Propagate from seeds or cuttings. Requires a well-drained soil in full sun.

C. betacea (Cavanilles) Sendtner, *Flora* 28: 172, 1845. (Syn: *C. crassifolia*, etc.). [Tamarillo, Tree Tomato]. **Shrubs**, herbaceous, 1–5(–7.5) m high, perennial, fast-growing. **Leaves** at ends of branches; lamina broadly cordate-ovate, 15–30 × 10–20 cm, shiny, hairy, pungent smelling. **Inflorescences** 10–50-flowered, 2.5–15 cm long, usually branched, puberulent; peduncles 1.5–9 cm long; rachises 2–8 cm long; pedicels 1–2 cm long. **Flowers** with calyx fleshy, puberulent, 3–5 mm long, lobes 1–2 × 2–3 mm, obtuse to truncate, apiculate; corolla pinkish-white, subcoriaceous to fleshy, stellate, 1–1.5 cm long, tube 2–3 mm long, lobes narrowly triangular, 0.7–1.2 × 0.2–0.4 cm, glabrous, margin tomentose, apex acute; anther thecae yellowish, lanceolate, 5–6 × 2–2.5 mm; ovary glabrous, style glabrous, cylindrical, not dilated distally, 5–6 mm long, exserted. **Fruits** berries, egg-shaped, purple-red or orangish-yellow, to 5 cm long, glabrous; seeds numerous, embedded in pulp. **Distr:** Andean regions of S America, from Peru to Argentina. **Ecol:** Fl. Jan. **Zones:** 9–11. **Info:** Cultivated widely in the tropics and subtropics for the edible (acidic) fruits.

Datura

Species ~11. **Shrubs**, or annual or perennial **herbs**; pubescence of simple, sometimes glandular hairs. **Leaves** petiolate; lamina simple, margins entire or sinuate-dentate. **Inflorescences** with flowers solitary, in leaf axils or in branch forks; peduncle, bracts and bracteoles absent; pedicel often stout. **Flowers** actinomorphic, large; calyx long tubular or cylindric, often circumscissile near base; corolla elongated, funnelform, lobes sometimes cuspidate; anthers mostly elongated, longitudinally dehiscent, included; ovary 2–4-locular. **Fruits** dry capsules, 4-valved or irregularly dehiscent, prickly or unarmed, often subtended by remnants of persistent calyx; seeds numerous, laterally compressed. **Distr:** N and S America. **Cult:** Propagate from seeds, cuttings or root divisions. Requires a moist, well-drained soil in full to partial sun. **Info:** All species are toxic, but possess numerous medicinal properties.

D. metel L., *Sp. Pl.* 1: 179, 1753. (Syn: *D. alba*, *D. fastuosa*, *D. fastuosa* var. *alba*). [Horn of Plenty, Thorn Apple]. **Herbs**, annual, 0.5–1.5 m high, glabrescent; stems often dark violet. **Leaves** ovate or broadly ovate, 5–20 × 4–15 cm, base truncate or cuneate, asymmetrical, margins irregularly sinuate-dentate, lobed, or entire, apex acuminate, membranous, glabrescent, veins 4–6 pairs; petiole 2–6 cm long. **Inflorescences** with flowers solitary, erect; pedicels ~1 cm long. **Flowers** with calyx tubular, 4–9 cm long; corolla white, yellowish, or pale purple, funnelform, sometimes doubled or tripled, 14–20 cm long, limb 6–10 cm Ø, lobes elongate; anthers 1–1.2 cm long. **Fruits** deflexed, subglobose, ~3 cm Ø, tuberculate, irregularly 4-valved, subtended by remnants of persistent calyx; seeds pale brown, reniform-discoid, ~3 mm Ø. **Distr:** Tropical America. **Ecol:** 1,200–2,100 m. Fl. & Fr. Mar–Dec. **Zones:** 9–12.

D. stramonium L., *Sp. Pl.* 1: 179, 1753. (Syn: *D. stramonium* var. *tatula*, *D. tatula*). [Thorn Apple, Jamestown Weed, Nightshade]. **Herbs** or **subshrubs**, sometimes robust, 0.5–1.5 m high, glabrescent. **Leaves** broadly ovate, 8–17 × 4–14 cm, base asymmetric, cuneate, irregularly sinuous or dentate-lobed, apex acuminate, membranous, glabrescent, veins 3–5 pairs; petiole 3–5.5 cm long. **Inflorescences** with flowers solitary; flowers erect; pedicel 0.5–1.2 cm long. **Flowers** with calyx tubular, 5-angular, 3–5 cm long; corolla white or pale purple, greenish at base, sometimes purple distally, funnelform, limb 3–5 cm Ø, lobes 6–10 cm long, apex mucronate; filaments ~3 cm

Above: *Goetzea elegans, Iochroma australe, Iochroma cyaneum, Iochroma fuchsioides.*

long, anthers 3–4 mm long. **Fruits** erect, globose or ovoid, 3–4.5 × 2–4 cm, with copious prickles, rarely smooth, dehiscent by 4 equal valves, subtended by remnants of persistent calyx; seeds black, ovate or discoid, ~4 mm Ø. **Distr:** Mexico. **Ecol:** 600–1,600 m. Fl. Jun–Oct. Fr. Jul–Nov. **Zones:** 7–12.

Goetzea

Species 2. **Trees. Leaves** alternate, petiolate; lamina coriaceous, pinnately many-veined. **Inflorescences** cymes, small. **Flowers** relatively large, orange; calyx campanulate, 6-cleft; corolla funnelform, limb 6-lobed, lobes reflexed; stamens 6, exserted, borne near base of corolla tube, filaments filiform, anthers oblong, ovary sessile, biloculed, ovules few, pendulous, villous, style filiform, stigma bilobed. **Fruits** berries, obovoid, coriaceous. **Distr:** Puerto Rico (1 endemic species) and Hispaniola (1 endemic species) (Santiago-Valentin and Olmstead, 2003). **Cult:** Propagate from seeds or cuttings. Requires a moist, well-drained soil in full sun.

G. elegans Wydler, *Linnaea* 5: 423, t. 8, 1830. (Syn: *Ilex exandra*). [Manzanilla]. **Trees**, 8–15 m tall, evergreen; trunk 25 cm dbh. **Leaves** alternate, exstipulate; lamina simple, ovate to elliptic, 2–10 × 3–5 cm, base acute, margins entire, apex acute to shortly acuminate, adaxially dark green, shiny, glabrous, abaxially pale green; petiole 3–8 mm long. **Inflorescences** cymes or usually flowers solitary, terminal or axillary; peduncle curved, 0.9–1.1 cm long; pedicels 0.6–1.2 cm long. **Flowers** actinomorphic, perfect; calyx, campanulate, 4–6 mm long, lobes 6, narrow, acute, free apices 1–2 mm long; corolla yellowish, funnelform, 1.6–2.5 × 1–1.2 cm, lobes 6, ovate, 6–8 mm long, acute, reflexed; stamens 6, epipetalous at base of corolla and alternate with corolla lobes, free for the rest of their length and long-exserted, anthers versatile, ~2.5 mm long; ovary superior, ~1 mm Ø, biloculed, hairy at base, style somewhat shorter than stamens but exserted, stigma bilobed, appearing globose. **Fruits** berries, orange, subglobose, ~2 cm Ø; seeds ~6 mm long. **Distr:** Puerto Rico; rare. **Ecol:** Understory trees of woodlands and limestone forests. **Zones:** 10–12.

Iochroma

Species ~20. **Shrubs** or **small trees**, sometimes thorny; young branches and leaves woolly or downy. **Leaves** petiolate; lamina ovate to lanceolate, margins entire. **Inflorescences** umbels or clusters; *primary* near stem apices, to 120-flowered; *secondary* borne from nodes of lower leaves or along branches, 1–15-flowered; flowers long-pedicellate, usually pendent. **Flowers** with calyx cup-shaped, urceolate, campanulate or inflated, with small teeth or large lobes or 2-cleft, expanded in fruit; corolla funnelform, campanulate or narrowly tubular, often slightly curved and inflated toward middle, with 5 or 10 teeth or short lobes, blue to purple, red to orange, greenish-yellow or white; stamens 5, inserted near base of tube, sometimes exserted, anthers 2-celled, splitting lengthwise; style thin, longer than stamens, stigma lobed or capitate. **Fruits** berries, fleshy, spherical or conical-ellipsoid, partly or completely enclosed by the expanded calyx; seeds many, reniform. **Distr:** Argentina, Bolivia, Colombia, Ecuador, Galapagos, Peru. **Cult:** Propagate from seeds or cuttings. Requires a moist, well-drained soil in full sun.

I. australe Griseb., *Abh. Königl. Ges. Wiss. Göttingen* 19: 218, 1874. (Syn: *Acnistus australe*). [Blue Angel Trumpet]. **Shrubs**, deciduous, to 5 m high; bark silvery-gray, with numerous lenticels and fissures, corky; new shoots borne from the center of the plant. **Leaves** lanceolate, to 8 × 2.5 cm, adaxially dark green, sparsely hairy, abaxially light green with prominent venation; petiole to 2 cm long. **Inflorescences** umbels, 1–5-flowered; peduncle to 3 cm long, finely hairy. **Flowers** with calyx, cupular, 5–6 mm long, lobes 5, shallow, mucronate, ~1 mm long, ribs prominent; corolla funnelform to campanulate, deep blue, purple or white with a pink tinge, ~2.5 cm long, outside velvety, inside ± hairless, lobes usually 5, margins hairy to papillate; stamens with filaments basally flattened, elongating after anthesis from 1–2.5 cm long, anthers elliptic, pollen yellow; ovary conical, style to 3 cm long, initially curved, later straight, stigma bilobed. **Fruits** berries, globular, ripening from green to yellow, style bases persistent; seeds numerous, flattened, ~1.5 × 2 mm. **Distr:** Bolivia to Argentina. **Ecol:** In tropical forests. Fl. anytime during the year. **Zones:** 9–11.

I. cyaneum (J. Lindley) M.L. Green, *Stand.-sp. of nom. cons.* 54: 63, 1926. (Syn: *Habrothamnus cyaneus*, etc.). [Blue Cestrum]. **Shrubs**, to 2 m high, or taller in cultivation. **Leaves** elliptic, 10–23 × 5–10 cm, adaxially pale green, slightly hairy but more on midrib, abaxially pale green with dense light brown hairs when young becoming sparse with age; petiole 2–9 cm long. **Inflorescences** primary umbels, ~60-flowered, secondary umbels ~10-flowered, terminal and supra-axillary. **Flowers** usually blue, or dark blue to light purplish-blue, pale pink, apricot and white; calyx usually somewhat inflated, green (rarely dark blue), sometimes tinged blue-purple; corolla tube 2.5–4 cm long, ~0.8–1 cm wide at mouth at anthesis, lobes 5; gynoecium with stamens 5, ± exerted, anthers, 3–3.5 mm long, yellowish, usually bluish-tinged before anthesis. **Fruits** conical, ~2 × 1.5 cm, partly enclosed in calyx, pale greenish-yellow to purplish when ripe. **Distr:** Colombia. **Ecol:** Near streams in shady riverine forests; <1,500 m. **Zones:** 9–11.

I. fuchsioides (Bonpl.) Miers, *London J. Bot.* 7: 345, 1848. (Syn: *Chaenesthes fuchsioides*, etc.). [Red Iochroma]. **Shrubs**, to 4 m high, ± glabrous, unarmed. **Leaves** often fascicled; lamina obovate, ovate or oblong, base cuneate, tapered into the short petiole, margins entire, apex very obtuse; petiole 1.2–1.5 cm long. **Inflorescences** with flowers solitary, axillary, supra-axillary or terminal; peduncles aggregated forming pseudo-umbels. **Flowers** drooping, large, showy, ~4 cm long; calyx subcampanulate, ~1.3 cm long, 5-toothed, often splitting on 1 side, or 2 sides forming 2 unequal lobes; corolla tubular, ~4 cm long, orange-scarlet, tube elongated, nearly straight, limb moderately spreading, 5-toothed or angled, with a smaller intermediate tooth; stamens 5, inserted near base of corolla, filaments included, downy at base; ovary pyramidal or conical, obscurely 5-lobed, style as long as corolla, stigma capitate, bilobed, lobes acute. **Fruits** berries, ovoid, acuminate, with persistent cleft calyx. **Distr:** Ecuador, Peru. **Ecol:** 400 m. **Zones:** 9–12.

I. gesnerioides (Kunth) Miers, *London J. Bot.* 7: 346, 1848. (Syn: *I. umbrosum, Lycium gesnerioides*). **Shrubs** or **small trees**, 1–6 m high; young parts felted or densely woolly hairy. **Leaves** lanceolate to broadly elliptic, sometimes ovate, 7–18 × 3.5–7.5 cm, base wedge-shaped or unequal, tapered into petiole, apex acute to acuminate, rarely long-acuminate, adaxially sparsely hairy, abaxially usually densely hairy; petiole 1.5–3 cm long. **Inflorescences** in dense clusters, to 120-flowered, borne near apices of branchlets; pedicels 1–4 cm long. **Flowers** with calyx cupular, 3–5 mm long, with small teeth, outside densely hairy; corolla tubular, 2.5–4 cm long, bright red, reddish-pink or reddish-orange, outside hairy, inside basally hairy, lobes 5, sometimes recurved; stamens with filaments basally adnate to corolla tube, anthers ~4 mm long. **Fruits** berries, ellipsoid to conical, partially enclosed in the persistent, split calyx. **Distr:** Colombia to Ecuador. **Ecol:** In lowland areas. Fl. Sep–Dec. **Zones:** 9–12.

I. grandiflora Benth., *Edwards's Bot. Reg.* 31: sub t. 20., 1845. [Giant Iochroma]. **Trees small** or **shrubs**, to 5 m high; young growth mucilaginous and long-hairy; branches grayish-brown with numerous lenticels. **Leaves** alternate; lamina ovate-elliptic to broadly ovate, 7–14 × 4–8 cm, base rounded or gradually tapered into petiole, apex acute rarely mucronate, adaxially glandular-hairy, abaxially prominent venation densely hairy; petiole 1–2.5 cm long, adaxially channeled. **Inflorescences** with flowers solitary or few together, borne from nodes of new growth, or flowers in clusters on older growth; peduncle to 5 cm long; pedicels 1–8 cm long. **Flowers** with calyx 0.7–1.2 × 0.4–1.1 cm, slightly inflated, ribbed, lobes 5, small, mucronate, densely glandular-hairy, enlarged in fruit; corolla tubular to trumpet-shaped, blue to violet-purple, 4.5–7 × 0.5–0.7 cm, sparsely glandular-hairy, limb 2.3–3 cm Ø, lobes 5, 0.7–1 cm long, apices hairy; stamens 5, filaments inserted at base of corolla, anthers pale yellow; style inserted or slightly exserted, stigma bilobed. **Fruits** berries, ellipsoid, with persistent, split calyx; seeds reniform. **Distr:** Ecuador, Peru. **Ecol:** In cloud forests. Fl. Nov–Mar. **Zones:** 10–12.

Top row: *Iochroma gesnerioides, Iochroma grandiflora, Nicandra physalodes, Nicotiana tabacum.*
Bottom row: *Physalis peruviana, Solandra grandiflora, Solandra longiflora, Solandra maxima.*

Nicandra

Species 1. *Generic description same as specific.* **Cult:** Propagate from seeds. Requires a rich, moist, well-drained soil in full sun.

N. physalodes (L.) Gaertner, *Fruct. Sem. Pl.* 2: 237, 1791. (Syn: *Atropa physalodes*). [Apple of Peru, Shoo Fly]. **Herbs**; stems erect, angular, 0.4–1.5 m high, glabrescent or sparsely pubescent; hairs simple and glandular. **Leaves** simple, ovate or elliptic, 4–20 × 2–13 cm, base cuneate, margins lobed or coarsely sinuate-dentate, apex acute or shortly acuminate, papery, sparsely pubescent; petiole 1.5–6 cm long. **Inflorescences** with flowers solitary, axillary or in branch forks; pedicels 1.5–4 cm long. **Flowers** actinomorphic; calyx campanulate, 0.8–3 cm long, deeply 5-lobed, lobes broadly ovate, 2.5–4 cm Ø, acute; corolla pale blue, blue-purple, or bluish with a white center, campanulate, 2.5–4 cm Ø, shallowly lobed, lobes ovate; stamens included, inserted on corolla tube, filaments slender, anthers elliptic, longitudinally dehiscent; ovary 3–5-locular, ovules numerous, style filiform, stigma subcapitate, 3–5-lobed. **Fruits** berries, globose, dry, brown or yellow, 1–2 cm Ø; fruiting calyx conspicuously enlarged, persistent, enclosing fruit, lobes cordate-sagittate, each with 2 basal auriculate lobes; seeds reniform-discoid, compressed, 1–1.2 mm Ø, pitted, pale brown. **Distr:** Peru. **Ecol:** On hillsides and wastelands; 800–2,600 m. **Zones:** 9–10.

Nicotiana

Species ~95. **Herbs**, **shrubs** or **small trees**, pubescent; hairs simple and glandular. **Leaves** petiolate or sessile; lamina with margins entire or subentire. **Inflorescences** paniculate, racemose, or reduced to solitary flowers; peduncle mostly erect; bracts mostly present; flowers pedicellate. **Flowers** sometimes showy, mostly fragrant, zygomorphic, 4- or 5-merous; calyx regular or irregular, tubular or tubular-campanulate, lobed; corolla tubular, funnelform, or salverform, tube lobed to subentire; stamens inserted below middle of corolla tube, filaments slender, anthers longitudinally dehiscent; disk ringlike, nectariferous; ovary bilocular, stigma bilobed. **Fruits** capsules, dry, dehiscing by apical valves; fruiting calyx persistent, slightly inflated, partially or completely enveloping fruit; seeds numerous, minute. **Distr:** Africa, Americas, Australia, and Oceania. **Cult:** Propagate from seeds. Requires a fertile, moist well-drained soil in full sun. **Info:** Contact with the leaves may cause skin irritation.

N. tabacum L., *Sp. Pl.* 1: 180, 1753. (Syn: *N. chinensis*). [Tobacco]. **Herbs**, viscid, annual or short-lived perennial, 0.7–2 m high, glandular-hairy. **Leaves** decurrent, winging stems; lamina ovate to elliptic or lanceolate, 30(–70) × 8–15(–30) cm, base narrowed, nearly half-clasping, apex acuminate, membranous, glandular-hairy. **Inflorescences** panicles, many-flowered, much-branched; pedicels 0.5–2 cm long. **Flowers** with calyx tubular or tubular-campanulate, 2–2.5 cm long, lobes deltate, acuminate, unequal; corolla base yellowish, distally yellowish, greenish, red or pink, funnelform, 3.5–5 cm long, limb 1–1.5 cm Ø, lobes unequal, acute; stamens unequal, sometimes partly exserted. **Fruits** ellipsoid or ovoid; seeds brown, rounded, ~0.5 mm Ø. **Distr:** S America. **Ecol:** On old terraces and in archaeological sites. Fl. Mar–Sep. **Zones:** 8–12. **Info:** Cultivated widely for medicinal, commercial and ornamental purposes.

Physalis

Species ~75. **Herbs**, annual or perennial, glabrous or pubescent; hairs simple. **Leaves** mostly equal, sometimes paired, petiolate; lamina simple, margins entire, denticulate or sinuate, membranous. **Inflorescences** with flowers solitary, borne in leaf axils or branch forks. **Flowers** actinomorphic, 5-merous; calyx campanulate; corolla yellow or white, often with a contrasting eye, rotate or rotate-campanulate, subentire or lobed; stamens shorter than corolla, inserted at base of corolla tube, filaments sometimes expanded at base; anthers longitudinally dehiscent; ovary bilocular, ovules numerous, stigma minutely bilobed. **Fruits** berries, juicy, globose; fruiting calyx becoming enlarged, inflated, entirely enveloping fruit, membranous or leathery, with 5 or 10 longitudinal ribs, base often invaginated; seeds numerous, compressed, discoid or reniform, minutely pitted. **Distr:** Predominantly in the Americas, a few in Asia and Europe. **Cult:** Propagate perennials by division of rhizomes or from cuttings, and annuals by seeds. Requires a fertile, well-drained soil in full sun.

P. peruviana L., *Sp. Pl.*, ed. 2. 2: 1670, 1763. (Syn: *P. edulis*). [Cape Gooseberry, Golden Berry]. **Herbs**, perennial, 45–90 cm high; stems erect, sparingly branched, densely pubescent. **Leaves** broadly ovate to cordate, 6–15 × 4–10 cm, base cordate, margins entire or with a few indistinct teeth, apex shortly acuminate, densely pubescent; petiole 2–5 cm long. **Inflorescences** with flowers solitary; pedicels ~1.5 cm long. **Flowers** with calyx broadly campanulate, 7–9 mm long; corolla yellow, throat spotted inside, 1.2–1.5 × 1.2–2 cm; filaments and anthers bluish-purple, anthers 3–4 mm long. **Fruits** berries, yellow, 1–1.5 cm Ø; fruiting calyx green, ovoid, with 5–10 weak angles, 2.5–4 cm long, pubescent; seeds yellow, ~2 mm Ø. **Distr:** S America. **Ecol:** In disturbed areas;

Above: *Solanum aviculare, Solanum indicum, Solanum jasminoides, Solanum laciniatum.*

0–2,000 m. **Zones:** 8–12. **Info:** The fruits are edible.

Solandra

Species ~8. **Scandent shrubs, high-climbing hemi-epiphytes** or large stemmed **lianas**, unarmed; stems mostly glabrous, stout, to 5 cm Ø, with no apparent pith, sometimes with flaking bark, often with aerial roots; twigs stout with a large pith. **Stipules** absent. **Leaves** alternate; lamina simple, entire, membranous, coriaceous or somewhat succulent, glabrous or pubescent. **Inflorescences** with flowers solitary or few-flowered clusters, terminal, at ends of twigs or on short shoots; pedicels stout, rarely elongate. **Flowers** with calyx mostly tubular, sometimes ellipsoidal or campanulate, angled and sometimes inflated, deeply splitting in flower into 2–5 irregular lobes; corolla imbricate, often very large, fleshy, mostly glabrous, campanulate, crateriform or funnelform, often slightly zygomorphic, lobes entire to fimbriate and often recurved; stamens 5, equal, mostly exserted, filaments inserted at base of tube, often pubescent at point of insertion, upper portion ± straight, anthers 4-loculed, longitudinally dehiscent; ovary conical, sometimes apically flattened, 4-loculed, sometimes becoming partly biloculed, ovules many, style sometimes much-exceeding anthers and persistent for a time on fruit. **Fruits** berries, leathery, loosely enclosed by the ruptured calyx; seeds many, reniform. **Distr:** Mexico and the Antilles to Brazil. **Cult:** Propagate from seeds, cuttings or by air-layering. Requires a fertile, well-drained soil in full or partial sun.

S. grandiflora Sw., *Kongl. Vet. Acad. Handl.* 8: 300–306, Tab. 11, 1787. (Syn: *S. hirsuta*, etc.). [Cup of Gold, Showy Chalice Vine]. **Hemi-epiphytes**, high climbing, to over 10 m; stems often with aerial roots. **Leaves** broadly elliptic, 10–22 × 4.5–10 cm, ends mostly obtuse, or apex shortly acuminate, coriaceous, glabrous; petiole slender, 3–5 cm long. **Inflorescences** with flowers solitary; pedicels stout, sometimes tuberculate. **Flowers** showy, fragrant; calyx tubular or campanulate, 4–9 cm long, angled and inflated, splitting into 3–5 lobes, mostly apiculate, glabrous, somewhat succulent; corolla fleshy, crateriform, yellow or golden, becoming orange-brown with age, inside with 10 purplish ribs (5 darker alternating with 5 lighter), 19–25 cm long, tube slender, 16–20 cm long, included or partly exserted from calyx, abruptly expanded into a large limb, goblet-shaped, 6–12 cm Ø, the lobes rotund, 3.5–5 cm long, margin entire to erose, glabrous outside; stamens inserted near middle of tube, filaments ± straight, glabrous, anthers 0.6–1.2 cm long; ovary 4-locular, style exserted, 1.8–2.1 cm long. **Fruits** berries, ovoid-globose, to 5 × 5 cm, white, glossy; seeds numerous, 5 mm long. **Distr:** C to S America, West Indies. **Ecol:** Growing as a woody liana, over tops of large trees, in woodlands and forests; 600–1,000 m. **Zones:** 9–12.

S. longiflora Tussac, *Fl. Antill.* 2: 49–51, t. 12, 1818. (Syn: *S. grandiflora* var. *macrantha*, etc.). [Chalice Vine, Cup of Gold]. **Climbers**, or terrestrially with a semiscandent habit, woody, evergreen, to over 9 m high. **Leaves** elliptic to ovate or obovate, to 15 cm long, base cuneate, decurrent into the short petiole, apex acuminate, glabrous. **Inflorescences** with flowers solitary, axillary or terminal; pedicels very short or flowers sessile. **Flowers** showy, fragrant; calyx green, white-pubescent, 6.5–11 cm long, usually trilobed, lobes triangular or apices acute; corolla opening white or very pale cream-white, becoming golden-yellow with age, 23–32 × 5–7.5 cm, cylindrical portion much longer than the calyx and the expanded upper portion, expanded gradually into a trumpet shape, lobes 5, base narrowed, margins ruffled, inside with 5 reddish-purple guides, colors diffused into the nearby venation; stamens 12–18.5 cm long, inserted at base of corolla, anthers 5–7 mm long. **Fruits** capsules, subglobose. **Distr:** Ecuador, Venezuela, Suriname, West Indies. **Ecol:** In shaded areas of forests. Fl. Sep–Feb. **Zones:** 9–12.

S. maxima (Sessé & Moc.) P.S. Green, *Bot. Mag.* 176: Tab. 506, 1967. (Syn: *Datura maxima, S. hartwegii, S. nitida*). [Golden Chalice Vine, Cup of Gold]. **Hemi-epiphytes**, high climbing, to >12 m high. **Leaves** broadly elliptic, to 14 cm long, base mostly obtuse, apex acuminate, mostly glabrous; petioles slender, 2–5 cm long. **Inflorescences** with flowers solitary or several in a cluster, terminal; peduncle short, stout; pedicels stout, 1–1.5 cm long, often with tubercles. **Flowers** showy, fragrant; calyx tubular, 5–8 cm long, angled, sometimes inflated, deeply splitting into 3–5 lobes, lobes slender or broad, apiculate, glabrous outside, sometimes fleshy; corolla fleshy, crateriform, yellow, becoming golden, 5 conspicuous greenish ribs outside and 5 or 10 purplish ribs inside the tube and limb, tube slender, 8–13 cm long, abruptly expanded into a shallowly cup-shaped limb, 8–15 cm Ø, lobes round, margins crenate to entire, glabrous outside; stamens 10–13 cm long, filaments geniculate at point of insertion, glabrous; ovary with center depressed, style sometimes apically purplish, stigma green. **Fruits** ovoid, smooth, glossy, ~7 cm Ø, with persistent calyx. **Distr:** Mexico, C America, Colombia, Peru, Venezuela. **Ecol:** 1,500–2,000 m. **Zones:** 10–12.

Solanum

Species ~1,200. **Herbs, shrubs, climbers**, or **small trees**, sometimes prickly; hairs simple, branched or stellate, sometimes glandular. **Leaves** solitary or paired, mostly petiolate; lamina simple or pinnately compound; lamina with margins entire, dentate, lobed, or parted. **Inflorescences** mostly racemose, paniculate, umbellate, fasciculate, or flowers solitary, axillary, extra-axillary, or leaf-opposed, ebracteate; peduncle branched or not, sometimes obsolete. **Flowers** bisexual or andromonoecious, mostly actinomorphic, often 5-merous; calyx mostly lobed and splitting further at sinuses; corolla mostly rotate or stellate; stamens inserted high in corolla tube, anthers often connivent or connate around style, dehiscing by apical pores or sometimes longitudinally; ovary 2–5-locular, with enlarged placentas, ovules axile, numerous, stigma small. **Fruits** berries; calyx persistent, sometimes enlarged and enclosing fruit; seeds discoid or lenticular. **Distr:** Tropical and subtropical regions with the highest diversity in tropical America. **Cult:** Propagate from seeds, cuttings or by division. Requires a fertile, moist, well-drained soil in full sun or partial shade.

S. aviculare Forst. f., *Pl. Esc.* 42, 1786. [Kangaroo Apple]. **Shrubs** or **small trees**, glabrous, to ± 3 m high; stems soft-wooded, branching, green or purplish. **Leaves** alternate; lamina lanceolate or linear-lanceolate, (4–)15–30(–40) cm long, margins entire or irregularly pinnatifid (usually on the same plant) with ~1–3 lanceolate acute lobes, dark green, membranous to subcoriaceous; petioles decurrent. **Inflorescences** 1–3, cymes, few- to many-flowered, axillary. **Flowers** with calyx lobes short, broad, spreading; corolla broadly campanulate to rotate, to 3.5 cm Ø, lobes ± as long as tube, lanceolate, lavender, becoming paler toward the margins, or white; anthers oblong, spreading, opening by apical slits. **Fruits** berries, broadly ovoid, 2–2.5 cm long, drooping, purple to yellowish or orange; seeds to 2 mm long. **Distr:** New Guinea, E Australia, New Zealand. **Ecol:** In lowland forest margins and shrublands. Fl. throughout most of the year. **Zones:** 9–11.

S. indicum L., *Sp. Pl.*: 187, 1753. [Bush Tomato, Indian Nightshade]. **Undershrubs**, much-branched, 0.3–1.5 m high, very prickly; prickles large, with a long compressed base, sharp, often slightly recurved; stem stout, often purple; branches covered with minutely hairy. **Leaves** ovate in outline, 5–15 × 2.5–7.5 cm, base cordate, cuneate or truncate, often unequal-sided, margins subentire or with a few large triangular-ovate subacute lobes, apex acute, sparsely prickly, adaxially sparsely hairy, abaxially densely hairy; petioles 1.2–2.5 cm long, prickly. **Inflorescences** racemose cymes, extra-axillary; peduncles short; pedicels 0.6–1.2 cm long, hairy and prickly. **Flowers** with calyx 3 mm long, hairy, teeth triangular, 1.5 mm long; corolla 8 mm long, pale-purple, with darker purple hairs outside, lobes deltoid-ovate, 5 mm long, acute; filaments very short, or inconspicuous, anthers oblong-lanceolate, opening by small pores; ovary often hairy at apex, style hairy, apex curved. **Fruits** berries, globose, ~8 cm Ø, dark yellow when ripe, glabrous or sometimes apex sparsely hairy; seeds ~4 mm Ø, minutely pitted. **Distr:** Tropical India, Sri Lanka, China, Malaya, Philippines. **Ecol:** In tropical lowland open areas; to 1,500 m. Fl. throughout most of the year. **Zones:** 10–12.

S. jasminoides Paxt., *Mag. Bot.* 8. t. 5, 1841. [Jasmine Nightshade, Potato Vine]. **Climbers**, shrubby, semievergreen, unarmed, glabrate, rooting by runners.

Above: *Solanum lycopersicum, Solanum melongena, Solanum nigrum, Solanum pseudocapsicum.*

Leaves ovate to ovate-lanceolate, 3.5–7.5 × 1.5–3 cm, margins both entire and basally lobed on the same plant, glabrous, green, sometimes purplish-tinged; petiole 1–2 cm long. **Inflorescences** open paniculate cymes, many-flowered, slender, short, extra-axillary, axillary or terminal; pedicels ~1 mm long. **Flowers** bluish-white to white; calyx lobes obtuse, mucronate, sparsely pubescent, apices ciliate with umbonate teeth, later splitting briefly into 2 mm long lobes; corolla rotate, 2 cm Ø, lobes spreading, narrowly obtuse with a prominently angled midvein; stamens subequal and alike, filaments 1 mm long, anthers 4 mm long, stout with large pores and longitudinal slits; ovary glabrous, style basally glandular-villous. **Distr:** Argentina, Brazil, Paraguay, Uruguay. **Ecol:** Fl. throughout most of the year. **Zones:** 10–12.

S. laciniatum Aiton, *Hort. Kew.* 1: 247, 1789. [Large Kangaroo Apple]. **Shrubs**, to 3 m high, pubescent, soon glabrescent; hairs simple; stems glabrous, young parts minutely pubescent. **Leaves** paired, unequal; lamina with margins lobed and entire (often on the same plant); *lobed* leaves broadly ovate, deeply pinnately 3–5-parted, 12–17 × 6–13 cm; *entire* leaves lanceolate, 10–20 × 1.5–3 cm; petiole of lobed leaves ~1–1.5 cm long, that of entire leaves 0.5–1 cm long. **Inflorescences** scorpioid-racemose, ~8 cm long, terminal, axillary, or extra-axillary; peduncle short or obsolete; pedicels 1.5–3 cm long. **Flowers** with calyx 5–7 mm long, lobes deltate, 2–3 mm long, apex mucronulate, leathery; corolla blue-purple, rotate, 2–2.5 cm long, lobes ± rounded, notched; filaments subulate, anthers oblong, 3–4 mm long; style ~8 mm long. **Fruits** berries, orange-yellow, ovoid or ellipsoid, ~3 × 2.5 cm, pendulous; seeds subovate, ~3 × 2 mm, compressed. **Distr:** S Australia and New Zealand. Fl. Apr–Jun. Fr. Jul–Sep. **Zones:** 9–11.

S. lycopersicum L., *Sp. Pl.* 1: 185, 1753. (Syn: *Lycopersicon esculentum*, etc.). [Love Apple, Tomato]. **Herbs**, annual, sprawling, 0.6–2 m high, viscid pubescent, odorous. **Leaves** mostly pinnately compound or divided, sometimes entire, 10–40 cm long, base oblique, cuneate, apex obtuse; leaflets mostly 5–9, sessile or petiolulate, unequal, ovate or oblong, 5–7 cm long, margins entire or irregularly dentate, sparingly glandular-pilose; petiole 2–5 cm long. **Inflorescences** often 3–7-flowered; peduncle 2–5 cm long, ± unbranched; pedicels 1–1.5 cm long. **Flowers** with calyx rotate-campanulate, lobes lanceolate; corolla 2–2.5 cm Ø, lobes narrowly oblong, 0.8–1 cm long, often reflexed; filaments ~1 mm long, anthers 0.6–1 cm long; style ~1.2 cm long. **Fruits** berries, red or orange-yellow, subglobose, 1.2–1.8 cm Ø (cultivars much larger), fleshy, juicy, shiny; seeds straw-colored, 2–4 mm long, pilose. **Ecol:** Fl. May–Sep. Fr. Sep–Nov. **Zones:** 9–12. **Info:** Cultivated widely for the edible fruits.

S. melongena L., *Sp. Pl.* 1: 186, 1753. (Syn: *S. esculentum*, etc.). [Aubergine, Brinjal, Egg Plant, Jew's Apple]. **Herbs** or **subshrubs**, to 60 cm high, sparingly armed, pubescent; stems and branches minutely tomentose, sometimes with stout recurved prickles. **Leaves** ovate to oblong-ovate, 6–18 × 5–11 cm, base oblique, margins sinuately lobed, apex obtuse, surfaces stellate-tomentose or sometimes with a few fine prickles, abaxially denser; petiole 2–4.5 cm long. **Inflorescences** mostly with flowers solitary, rarely reduced racemes; peduncle obsolete; pedicel 1–1.8 cm long. **Flowers** andromonoecious; calyx stellate-tomentose, abaxially often with ~3 mm long prickles; lobes lanceolate; corolla purplish or violet, rotate, 3(–5) cm long, lobes deltate, ~1 cm long; filaments ~2.5 mm long; style 4–7 mm long. **Fruits** berries, highly variable (especially in cultivation), black, purple, pink, brown, or yellow, usually >6 cm Ø, mesocarp and septal region thick, spongy, whitish; seeds lenticular, yellowish, 2.8–3.9 × 2.5–3.5 mm. **Distr:** SE Asia. **Ecol:** Natural habitat unknown. Fl. Jul–Sep. **Zones:** 9–12.

S. nigrum L., *Sp. Pl.* 1: 186, 1753. (Syn: *S. nigrum* var. *atriplicifolium*). [Black Nightshade, Poisonberry]. **Herbs**, annual, erect, 0.2–1 m high, pubescent, unarmed; hairs simple; stems often angular, sparsely pubescent. **Leaves** ovate, 4–10 × 3–7 cm, base cuneate, decurrent, margins entire or coarsely dentate, apex obtuse, pubescent or glabrescent; petiole 2–5 cm long. **Inflorescences** umbels, extra-axillary; peduncle 2–4 cm long; pedicels 0.8–1.2 cm long. **Flowers** with calyx cup-shaped, 2–3 × 2–3 mm, lobes subdeltate, 0.5–1 × 1–1.5 mm, abaxially pubescent, ciliate; corolla white, 0.8–1 cm long, lobes ovate-oblong, 4–5 × 3–3.5 mm, abaxially pubescent, ciliate, spreading; filaments 1–1.5 mm long, anthers oblong, 2.5–3.5 mm long; style 5–6 mm long. **Fruits** berries, dull black, globose, 0.8–1 cm Ø; fruiting pedicel strongly deflexed; fruiting calyx adnate to fruit; seeds discoid, ~2 mm Ø. **Distr:** China, India, Japan, Taiwan; SW Asia, Europe. **Ecol:** In disturbed places, usually weedy; 500–3,000 m. Fr. May–Aug. Fr. Jul–Nov. **Zones:** 8–12.

S. pseudocapsicum L., *Sp. Pl.* 1: 184, 1753. [Jerusalem Cherry]. **Shrubs**, evergreen, erect, to 1.2 m high, branched, unarmed, pubescent. **Leaves** narrowly oblong to lanceolate, 1–6 × 0.5–1.5 cm, base cuneate, decurrent, margins entire or sinuate, apex acute or obtuse, pubescent or glabrous, dark green; petiole 2–5 mm long. **Inflorescences** with flowers solitary, or rarely paired or in short racemes, leaf-opposed or extra-axillary; peduncle short or obsolete; pedicels 3–4 mm long. **Flowers** small; calyx green, ~4 mm Ø, prominently lobed, lobes ~1.5 mm long; corolla white or mauve, 0.8–1.5 cm Ø, lobed to about half way; filaments <1 mm long, glabrous, anthers ~2 mm long; style ~2 mm long. **Fruits** berries, orange or red, showy, fleshy, spherical, 1.2–1.5(–2) cm Ø, glossy; fruiting pedicel ~1 cm long, expanded at apex; seeds discoid, 2–3 mm Ø. **Distr:** S America. **Ecol:** Natural habitat unknown. Fl. Feb–Apr. Fr. May–Aug. **Zones:** 9–12. **Info:** Often cultivated for ornamental purposes. Fruits often eaten by birds, but poisonous to humans.

S. seaforthianum Andrews, *Bot. Repos.* 8: t. 504, 1808. (Syn: *S. kerrii*). [Brazilian Nightshade, Italian Jasmine, St. Vincent Lilac, Star Potato Vine]. **Vines**, slender, woody, 1–5 m high, glabrescent, pubescent; hairs simple. **Leaves** simple or pinnately compound;

Below: *Solanum seaforthianum, Solanum trilobatum, Solanum tuberosum, Solanum violaceum.*

Above: *Solanum virginianum*, *Solanum wendlandii*, *Solanum wrightii*, *Streptosolen jamesonii*.

lamina 4–10 × 4–7 cm, base acute, obtuse, truncate, or cordate, puberulent along veins, usually pinnately (5–)7(–9)-parted; lobes ovate to oblong or lanceolate, unequal, 1.5–5 × 0.5–2.5 cm, apex acute or acuminate; petiole 2–5 cm long. **Inflorescences** open panicles, many-flowered, showy, leaf-opposed or appearing terminal; peduncle to 4 cm long, branched; pedicels 0.5–1 cm long, abscising nearly at base. **Flowers** with calyx cup-shaped, 1–1.5 × 2–3 mm, truncate or with minute, shortly deltate lobes, glabrous; corolla usually purple, rarely white, 0.9–1.3 cm long, lobes ovate-oblong, 5–8 × 2.5–4.5 mm, glabrous; filaments 2–3 mm long, anthers free, 3–4 mm long; style ~1 cm long. **Fruits** berries, red, juicy, globose, 1–2 cm long; seeds discoid, 2–3.5 mm Ø. **Distr:** S America. **Ecol:** In disturbed areas. Fl. Jan–Feb, Jun. **Zones:** 10–12.

S. trilobatum L., *Sp. Pl.* 188, 1753. [Purple-fruited Pea Eggplant]. **Shrubs**, scrambling, slender, prickly; prickles curved, base broadened, yellowish; stems with numerous prickles along the stems, otherwise subglabrous. **Leaves** rounded-ovate in outline, obtusely trilobed, 2–7 × 1–4 cm, glabrous or slightly stellate, with a few prickles along petiole and midrib. **Inflorescences** 3–9-flowered, extra-axillary; peduncle short; pedicels 1–2 cm long, widely divergent. **Flowers** with calyx 3 mm long, teeth narrow; corolla 1.2–1.6 cm long, deeply lobed, stellate-pubescent outside, purple, reflexed; stamens protruding, anthers 7 mm long, yellow. **Fruits** globose, ~1.5 cm Ø, scarlet; seeds compressed, 3 mm Ø. **Distr:** Sri Lanka, India to Java. **Ecol:** In thickets of dry lowlands, often in coastal areas. Fl. Dec–Jan. **Zones:** 10–12.

S. tuberosum L., *Sp. Pl.* 1: 185, 1753. [Potato]. **Herbs**, erect or sprawling, tuberous, 30–80 cm high, glabrous or sparsely pubescent; stolons bearing underground tubers; tubers white, red, or purplish, globose, oblate, or elliptic, 3–10 cm Ø, fleshy. **Leaves** interruptedly odd-pinnate, with 6–8 pairs of leaflets and smaller, unequal interstitial leaflets; lamina ovate or oblong, mostly sparingly pilose; petiole 2.5–5 cm long. **Inflorescences** panicles, many-flowered, appearing terminal, leaf-opposed, or axillary, sparingly branched; pedicels articulate near middle, 1–2 cm long. **Flowers** with calyx sparsely pubescent, lobes lanceolate; corolla white, pink, or blue-purple, sometimes all on 1 plant, rotate, 2.5–3 cm Ø, lobes deltate, ~5 mm long; filaments ~1 mm long, anthers 5–6 mm long; ovary glabrous, style ~8 mm long. **Fruits** berries, green or yellowish-green, often striped, globose, ~1.5 cm Ø, smooth. **Distr:** S America. **Ecol:** Natural habitat unknown. Fl. May–Sep. **Zones:** 7–12. **Info:** The potato is widely cultivated for the edible tubers.

S. violaceum Ortega, *Nov. Pl. Descr. Dec.* 56, 1798. (Syn: *S. chinense*, etc.). [Asian/Indian Nightshade, Poisonberry]. **Shrubs**, 0.5–1.5(–2) m high, much-branched, densely pubescent; stems and branches with pale yellow, recurved prickles 0.4–1 cm long, gray-tomentose. **Leaves** paired, unequal; lamina ovate, 5–8(–11) × 2–7(–8.5) cm, base cordate or truncate, margins sinuately 5–7-lobed, apex obtuse or acute, adaxially tomentose, abaxially densely hairy, surfaces with straight needlelike prickles; petiole 1.5–4 cm long. **Inflorescences** scorpioid racemes, 2–6 cm long, extra-axillary, sparingly branched; peduncle ~1.5 cm long, prickly or not; pedicels 0.4–1.5 cm long. **Flowers** with calyx 4–7 mm long, lobes lanceolate, 3–5 mm long, unequal, pubescent and prickly; corolla blue-purple, sometimes white, rotate, 1–1.2 × 1–2 cm, lobes ovate or ovate-lanceolate, 5–8 × 2–5 mm, obtuse; filaments ~1 mm long, anthers oblong, 5–6 mm long; style 0.8–1 cm long, pubescent. **Fruits** berries, glossy orange, globose, 0.8–1.3 cm Ø; pubescent, prickly; persistent calyx reflexed, pubescent, prickly; seeds subdiscoid, ~2 mm Ø. **Distr:** Tropical Asia. **Ecol:** In forests, dry thickets, wastelands; 100–2,700 m. Fl. & Fr. throughout the year. **Zones:** 10–12.

S. virginianum L., *Sp. Pl.* 1: 187, 1753. (Syn: *S. xanthocarpum*, etc.). [Indian/Thorny Nightshade, Thai Eggplant]. **Herbs**, erect or creeping, sometimes woody at base, 50–70 cm high, copiously armed with sturdy, sharp, needlelike, broad-based prickles 0.5–2 cm long, pubescent. **Leaves** in unequal pairs; lamina ovate-oblong, 4–9 × 2–4.5 cm, base subcordate or unequal, margins usually 5–9-lobed or pinnately parted, lobes unequal, sinuate, apex acute, pubescent and prickly along veins, glabrescent; petiole 2–3.5 cm long, prickly, hairy. **Inflorescences** racemes, 4–7 cm long; peduncle unbranched, copiously armed; pedicels ~1 cm long. **Flowers** with calyx campanulate, ~1 cm Ø; lobes oblong, pubescent, prickly; corolla blue-purple, rotate, 1.4–1.6 × 2.5 cm, lobes ovate-deltate, 6–8 mm long, densely pubescent; filaments ~1 mm long, anthers ~8 mm long; style ~1 cm long. **Fruits** berries, pale yellow, 1.3–2.2 cm Ø; fruiting pedicel 2–3.6 cm long, prickly, sparsely hairy; fruiting calyx prickly, sparsely pubescent; seeds subreniform, ~1.5 mm Ø. **Distr:** Old World tropics and subtropics. **Ecol:** On sandy river beaches; 100–1,300 m. Fl. Nov–May. Fr. Jun–Sep. **Zones:** 9–12.

S. wendlandii Hook. f., *Bot. Mag.* t. 6914. 1887. [Giant Potato Climber, Paradise Flower, Potato Vine]. **Vines**, slightly woody, scandent, attaining several meters long; stems cylindrical, glabrous, unarmed or less frequently with some minute recurved spines. **Leaves** alternate; lamina deeply (3)5–7-pinnatilobate or sometimes simple, 4.5–17 × 4.6–12 cm, elliptic when simple, 4.5–14 × 4.6–7.5 cm, base obtuse to almost rounded, apex shortly acuminate, segments or lobes elliptic or oblong-elliptic, base usually asymmetric, margins entire, apex acute or acuminate, membranous, glabrous, with numerous minute dots, abaxially dull, primary vein prominent, sometimes with recurved spines; petiole glabrous, 3–6.5 cm long, usually narrowly winged, sometimes with recurved spines. **Inflorescences** panicles, pseudo-dichotomous, 8–20 cm long, terminal, glabrous; pedicels 1–1.7 cm long, glabrous. **Flowers** unisexual; ♂ flowers with calyx green, cyathiform, 5–7 mm long, glabrous, lobes 5, ovate-deltate or ovate, 1–3 mm long, apex apiculate; corolla lavender, rotate, pentagonal, 3.5–6 cm Ø, lobes apiculate, ciliate; anthers oblong, 0.9–1 cm long. **Fruits** ovoid, to 10 cm long, yellowish **Distr:** Costa Rica. **Ecol:** In moist areas of the cordilleras. Fl. Apr–Oct. **Zones:** 10–12.

S. wrightii Bentham, *Fl. Hongk.* 243, 1861. [Giant Star Potato]. **Trees**, to 12 m tall, sometimes armed with straight prickles, pubescent. **Leaves** broadly ovate, 30 × 15–20 cm, base oblique, margins entire or sinuately lobed, apex obtuse or acute, adaxially hairy, abaxially sparsely to densely hairy and prickly, often with straight, yellow or green prickles; petiole 5–10 cm long. **Inflorescences** racemes, many-flowered, axillary or extra-axillary; peduncle mostly unbranched; pedicel 0.6–1.5 cm long, often leaving a conspicuous scar, with dense, bristlelike hairs. **Flowers** showy, andromonoecious; calyx 1–1.5 cm long, deeply lobed, lobes lanceolate, hispid with mostly simple hairs; corolla opening deep purple, fading through blue to white over several days, 6–9 cm Ø, shallowly lobed, stellate-hairy abaxially; filaments 2–3 mm long, glabrous, anthers oblong, apex narrowed, beaklike, 1.2–1.5 cm long; style dimorphic. **Fruits** berries, red, sometimes remaining green, globose, 3–4 cm Ø; seeds lenticular, 2–3 mm Ø. **Distr:** Bolivia, Brazil. **Ecol:** Natural habitat unknown. Fl. & Fr. throughout most of the year. **Zones:** 10–12.

Streptosolen

Species 1. *Generic description same as specific.* **Cult:** Propagate from cuttings. Requires a fertile, moist, well-drained soil in full sun.

S. jamesonii (Benth.) Miers, *Ann. Mag. Nat. Hist.* ser. 2, 5: 209, 1850. (Syn: *Browallia jamesonii*). [Marmalade Bush, Orange Browallia]. **Shrubs**, climbing or sprawling, evergreen or semievergreen, 1–4 m high; stem densely branched, finely bristly hairy. **Leaves** alternate; lamina ovate to elliptic, (1–)1.5–5 × 0.9–1.5 cm, apex acute or obtuse, rugose, surface wrinkled with prominent venation; petiole ~1 cm long. **Inflorescences** panicles or subcorymbose cymes, terminal, strigillose or hirsutulous. **Flowers** with calyx tubular-campanulate, ~1 cm long, lobes 5, short; corolla 2.5–3 cm long, tube narrow, spirally twisted at base, yellow-orange to orange-red, basally greenish, limb abruptly expanded, lobes 5, broad, obtuse, darker in color, appearing 2-lipped; fertile stamens 4, in 2 pairs, inserted at base of corolla, exserted, 1 pair shorter, anthers opening via longitudinal slits; ovary biloculed, shortly stalked, style threadlike, stigma bilobed. **Fruits** capsules, leathery, dehiscing by 2 bifid valves. **Distr:** W South America. **Ecol:** 2,100 m. **Zones:** 9–12.

15 ASTERIDS II

ORDERS 4, FAMILIES 32–34 (4 UNPLACED), GENERA 2,235, SPECIES 33,151.

Vessel elements with scalariform perforation plates. **Flowers** rather small; style short. **Seeds** with endosperm copious, embryo short or very short (Stevens 2001 onwards). **Distr:** Worldwide.

AQUIFOLIALES

FAMILIES 5, GENERA 21, SPECIES 536.

Trees or **shrubs**, sometimes woody lianas. **Stipules** absent or rarely very small and caducous. **Leaves** alternate or rarely opposite; lamina simple; petiole bundles arcuate. **Inflorescences** cymose or racemose, rarely flowers solitary. **Flowers** small, greenish or whitish, unisexual or bisexual, actinomorphic, usually 4- or 5-merous; sepals free or ± connate; corolla valvate or imbricate, free; stamens free from perianth or corolla, anthers longitudinally dehiscent; gynoecium mostly of 2–5 united carpels, ovules usually anatropous, 1–2 per carpel. **Fruits** drupes, baccate, or samaras, stones 1-seeded (Stevens 2001 onwards; Takhtajan 2009). **Distr:** ± Worldwide, especially diverse in the tropics.

CARDIOPTERIDACEAE

Genera 5 or 6, **species** 43–45. **Herbs**, climbing, with abundant milky juice. **Leaves** spirally arranged, estipulate. **Inflorescences** cymes, axillary, branched, ± scorpioid, ebracteate. **Flowers** very small; sepals (4 or)5, basally coherent; petals (4 or)5, basally coherent; stamens (4 or)5, epipetalous; disk absent; ovary superior, 1-loculed, apically with 2 pendulous ovules, styles 2, dissimilar, 1 longer and thicker, cylindric or subclavate, persistent on fruit, other style shorter and thinner, with capitate stigma. **Fruits** samaras, 2-winged, indehiscent, flat, 2-seeded. **Distr:** Tropical regions worldwide.

Gonocaryum

Species 9–10. **Shrubs** or **small trees**. **Leaves** alternate; lamina with margins entire, leathery, glabrous. **Inflorescences** dense or interrupted spikes or racemes, 1 to several together, axillary. **Flowers** bisexual or plants polygamodioecious; calyx free at least apically, sepals 5 or 6, imbricate; corolla tubular, lobes 5, valvate; stamens 5, rudimentary in ♀ plants, filaments glabrous, 3–5 times as long as anthers, rarely very short, adnate on corolla tube, alternate to petals, anthers ovoid or narrowly ellipsoid; ovary conical, reduced and sterile in ♂ flowers, style subulate or terete, stigma thickly peltate. **Fruits** drupes, ellipsoid, apex subtruncate, exocarp thick, corky-spongy, endocarp thin, woody. **Distr:** S and SE Asia. **Cult:** Propagate from seeds, cuttings or by air-layering. Requires a fertile, moist, well-drained soil in full or partial sun.

G. litorale (Blume) Sleumer, *Notizbl. Bot. Gart. Berlin-Dahlem* 15: 233, 1940. (Syn: *G. pyriforme*, etc.). [Aria, Balula, Karo]. **Trees**, (2–)6–20(–30) m tall. **Leaves** alternate; lamina usually ovate or ovate-elliptic, (10–)12–20(–30) x 6–12(–19) cm, base acute, margins entire, apex acute to shortly acuminate, adaxially glossy green, abaxially paler; petiole thick, bright orange, to 0.4–1(–3) cm long, ~6 mm Ø. **Inflorescences** ♂ spikes 1–4(–8), (3–)5–15(–30) cm long, flowers numerous; flowers sessile; ♀ spikes generally shorter than ♂. **Flowers** ♂ with calyx lobes ovate, ciliate, greenish-brown, 1.5(–2) mm long, petals narrowly ovate-acuminate, united below for halfway, cuplike, expanding at anthesis, slightly keeled, white to cream or greenish, sometimes pinkish, fragrant, (4–)5–6 x 2–2.5 mm, filaments basally dilated, filaments 1.5–2 mm long, rudimentary ovary hairy; ♀ with calyx less deeply lobed, (2–)3 mm long, petals 6–7 x 2–2.5 mm, anther cells 1 mm long, devoid of pollen, ovary ovoid, densely fulvous-hairy, stigma rather large, padlike. **Fruits** drupes, ellipsoid or obovoid, sometimes globular, (3–)4–5(–6.5) x 2–3(–4.5) cm, base rounded or truncate with a small tip, orange to reddish; seed pear-shaped or ellipsoid. **Distr:** Moluccas. **Ecol:** In rainforests; to 1,650 m. **Zones:** 11–12.

AQUIFOLIACEAE

Genus 1, **species** 405. **Trees** or **shrubs**, dioecious, evergreen or deciduous. **Stipules** minute, persistent or caducous, scar callose. **Leaves** alternate, rarely opposite; lamina with margins entire, serrate, or spinose, leathery, papery, or membranous; petiole present, rarely absent. **Inflorescences** cymes, usually of order 1, 2, or 3(–5), respectively composed of up to 1, 3, or 7(–31) developed flowers, axillary, solitary on current year's branchlets or fasciculate on second year's branchlets. **Flowers** hypogynous, regular, small, unisexual (with stamens abortive in ♀ or ovary abortive in ♂); calyx persistent; corolla often white or cream, rarely green, yellow, pink, or red; petals imbricate, mostly basally connate; ♂ with calyx 4–8-lobed, petals 4–8, stamens isomerous, alternating with petals, epipetalous, anthers oblong-ovoid, introrse, longitudinally dehiscent; ovary rudimentary, subglobose or pulvinate, rostrate; ♀ with calyx 4–8-lobed, petals 4–8, staminodes sagittate or cordate, isomerous, alternating with petals, epipetalous, ovary superior, ovoid, 4–8(–10)-loculed, rarely pubescent, style rarely developed, stigma capitate, discoid, or columnar. **Fruits** drupes, red, brown, or black (rarely green), usually globose; exocarp membranous or papery; mesocarp fleshy; pyrenes (1–)4–6(–23); endocarp smooth, leathery, woody, or stony. **Distr:** Tropical regions worldwide.

Ilex

Species 405. *Generic description same as familial.* **Cult:** Propagate from seeds or half-hardened cuttings. Requires a fertile, humus-rich, moist, well-drained soil in full or partial sun.

I. paraguariensis A. St.-Hil., *Mém. Mus. Hist. Nat.* 9: 351, 1822. [Brazilian Tea, Mate (Yerba Maté), Paraguayan Tea]. **Trees**, dioecious, evergreen, to 15 m tall; trunk grayish, 20–40 cm dbh. **Leaves** alternate; lamina ovate to subovate, 5–11 x 3–5.5 cm, base attenuate, margins somewhat dentate or serrated, particularly in abaxial half, apex slightly obtuse, adaxially dark green, abaxially paler; petiole ~6 mm long, pubescent. **Inflorescences** cymes, borne in upper leaf axils; axes pubescent, pedicels ~0.7–1.5 cm long, sparsely pubescent. **Flowers** small, white, unisexual (♀ with non-functional stamens; ♂ with non-functional and abortive pistils); calyx with 4 sepals, green; corolla white or greenish-white, petals 4, concave; stamens 4, wide, filaments white, anthers yellowish-orange; ovary greenish, ovoid, glabrous. **Fruits** drupes, globose, 4–8 mm Ø, reddish-brown to black; seeds 4, small exocarp hard, rough, yellowish or brown. **Distr:** S America. **Ecol:** Fl. Sep–Dec. Fr. Jan–Mar. **Zones:** 9–11. **Info:** The leaves are widely used in S America for the preparation of a coffeelike beverage known as Yerba Maté, which is high in caffeine.

STEMONURACEAE

Genera 12, **species** 95. **Plants**, woody, dioecious. **Leaves** with petiole bundles arcuate, exstipulate; lamina with margins entire. **Inflorescences** cymose; pedicels articulated. **Flowers** rather small, perfect, 4–5(–7)-merous; calyx ± connate; corolla rarely connate and keeled; petals free, apex inflexed; nectary (semi)annular (rarely absent); ♂ flowers with filaments often stout and with club-shaped hairs, pistillode present; ♀ flowers with staminodes present or absent, style absent, stigma broad, sessile. **Fruits** drupes, dorsiventrally compressed (rounded), the 2 sides very different and 1 with a fleshy 'appendage', endocarp with ridges and grooves on the side of the appendage, when present; seeds with testa thick (Stevens 2001 onwards). **Distr:** Tropics, especially Indo-Malesia to Australia (Queensland).

Stemonurus

Species 12. **Trees**, sometimes buttressed, glabrous except for some parts of the inflorescences. **Leaves** spirally arranged; lamina with margins entire, ± coriaceous and thick; petiole thick. **Inflorescences**

Top row: *Gonocaryum litorale* (flowers), *Gonocaryum litorale* (fruits), *Ilex paraguariensis*.
Bottom row: *Stemonurus apicalis, Canarina canariensis.*

axillary, usually solitary, umbel-like or heads supported by ± conspicuous often persistent bracts; flowers sessile. **Flowers** hermaphrodite; calyx cupular, truncate or ± 5-lobed; petals (4–)5, valvate, joined at base, apex inflexed; stamens (4–)5, filaments fleshy, flattened, shortly bearded below anthers on ventral side, long-penicillate dorsally, hairs inflexed in bud but erect and exserted in open flowers; disk rim- or cuplike; ovary ovoid-conic, unilocular with 2 pendulous ovules, style short, slender, stigma punctiform. **Fruits** drupes, ovoid, ellipsoid or fusiform-oblong, 1-seeded, usually bicolored, lower part red or purple, upper part cream or greenish; endocarp coriaceous to woody, outer part fibrous. **Distr:** From Sri Lanka (1 sp.), Indo-China and Malesia to the Solomon and Palau Islands. **Cult:** Propagate from seeds or cuttings. Requires a moist, well-drained soil in full or partial sun.

S. apicalis (Thw.) Miers, *Trans. Linn. Soc. London* 22: 110, 1856. (Syn: *Lasianthera apicalis, Urandra apicalis*). **Trees**, 6–20(–70) m tall; crown dense pyramidal; trunk 10–40(–45) cm dbh; young parts glabrous, dark, varnished. **Leaves** oblong-elliptic to obovate-oblong, 5–18 × 2–7.5 cm, base ± cuneate, margins entire, apex ± obtuse to abruptly shortly acuminate, thickly coriaceous; petiole 0.7–2 cm long. **Inflorescences** with flowers ± crowded into heads, 0.5–1 cm long; peduncle stout, 0.7–1 cm long; bracts 4, ovate, 2.5–4 × 2–2.5 mm, pubescent. **Flowers** green or reddish in bud; calyx ~3 mm long, lobes ± acute or obtuse, slightly pubescent; petals reddish or white, oblong, 4–5 × 1.5 mm; stamens with hairs completely concealing anthers; disk slightly 5-lobed. **Fruits** drupes, at first with upper part gray-green and lower dark green becoming yellow-green, finally white above and red-brown or purplish-green below, the 2 colors sharply demarcated, oblong-ovoid, 4–5 × ~2.3 cm. **Distr:** Sri Lanka. **Ecol:** In primary wet evergreen forests; 45–700 m. **Zones:** 11–12. **Info:** The resinous waxy exudate from the bark of old trees is inflammable and slightly scented when burned (Meijer and Verdcourt 1996).

ASTERALES

FAMILIES 11, GENERA 1,649, SPECIES 25,790.

Herbs perennial or annual, **subshrubs**, less often shrubs, lianas, small trees or medium-sized trees. **Leaves** spirally arranged. **Flowers** with corolla valvate; petals valvate, apices inflexed; stamens rarely basifixed, free from corolla, pollen grains trinucleate; nectary disk present; ovary with ovules many per carpel, style long (Stevens 2001 onwards). **Distr:** Worldwide. **Info:** Monophyly of Asterales is strongly supported by molecular data (Chase et al. 1993; Hilu et al. 2003; Lundberg and Bremer 2003; Soltis et al. 2000).

CAMPANULACEAE

Genera ~40, **species** ~600. **Herbs**, annual or perennial, sometimes lactiferous; roots and rhizomes often thickened and tuberous; stems erect, decumbent, trailing or climbing. **Leaves** usually alternate, sometimes opposite, exstipulate; lamina simple. **Inflorescences** racemose, terminal. **Flowers** usually hermaphrodite, actinomorphic, regular, sometimes cleistogamous; calyx tube adnate to ovary; sepals with linear to triangular or lanceolate segments; corolla united, campanulate, infundibuliform or tubular, 5-lobed, sometimes the segments split to base; stamens 5, filaments flattened; ovary inferior, (2–)3–5-locular, ovules numerous, styles thick, stigmas 3, sometimes 2. **Fruits** capsules, dehiscing by valves or pores. **Distr:** Mainly in the northern hemisphere.

Canarina

Species 3. **Herbs**, perennial, terrestrial or epiphytic, glabrous, with abundant white latex; roots thickened, fleshy; stems herbaceous, terete and hollow, dichotomously or trichotomously branched from leaf axils; most leaf axils also producing small, usually rudimentary, accessory shoots. **Leaves** opposite or ternate, petiolate. **Inflorescences** with flowers solitary, borne in dichasial forks or terminal. **Flowers** large, pendent, (5–)6(–7)-merous throughout; calyx lobes entire or sometimes dentate, erect, spreading or reflexed; corolla funnelform or campanulate, lobes short; filament bases almost linear to broad, shieldlike; ovary inferior, style shorter than corolla, markedly thickened toward apex, style lobes short. **Fruits** baccate, with persistent calyx; seeds numerous, testa finely pitted or striate. **Distr:** Tropical Africa (2 spp.), Canary Islands (1 sp.). **Cult:** Propagate from seeds or cuttings. Requires a fertile, moist, well-drained soil in full or partial sun.

C. canariensis (L.) Vatke, *Linnaea* 38: 700, 1874. (Syn: *C. campanula*). [Canary Bellflower]. **Herbs**, perennial, erect, climbing or trailing, to 1.5 m high; rootstock thick, fusiform; stems to 3 m long, fleshy, usually trichotomously branched in lower parts, dichotomously branched in upper parts. **Leaves** opposite, spreading, petiolate; lamina ovate, linear to narrowly triangular, 4–10 cm long, base hastate to heart-shaped, margins coarsely toothed or finely serrated, occasionally scalloped; petiole about half as long as the lamina, concave. **Inflorescences** with flowers solitary, pendent, borne in upper leaf axils. **Flowers** orange-red, with darker veins, usually 6-partite, 3–6 cm long; calyx lobes free, erect, spreading or reflexed, margins often toothed; corolla campanulate, to 6 cm long, lobes recurved, orange-yellow, longitudinally striped with reddish-brown to maroon or purplish; filament bases almost linear; ovary obconic, 6-angled. **Fruits** berries, black. **Distr:** Canary Islands. **Ecol:** In coastal areas, scrambling through scrub, or hanging down from cliffs. Fl. & Fr. throughout most of the year. **Zones:** 9–11.

C. eminii Asch. & Schweinf., *Sitzb. Ges. Naturf. Fr.*: 173, 1892. (Syn: *C. elegantissima, C. eminii* var. *elgonensis*). **Herbs**, epiphytic or terrestrial, usually glaucous; stems erect and scandent or pendent, to several meters long, dichotomously branched, usually with a fine purplish mottling. **Leaves** ± triangular to ovate, 2.5–10 × 1.5–9 cm, base cordate to truncate, margins obtusely to sharply dentate to doubly dentate or doubly serrate, apex acute; petiole less than half the length of lamina, not coiled. **Inflorescences** with flowers solitary. **Flowers** with hypanthium obconical, distinctly 6-ribbed, ribs projecting into calyx lobes; calyx lobes 1.8–3.8 × 0.5–1 cm, margin entire, apex acute to acuminate, free, erect or spreading; corolla funnelform, 4.3–7.6 cm long, orange to orange–red with darker venation; stamens with broad shieldlike filament bases, anthers 0.5–1 cm long. **Fruits** baccate; seeds numerous, elliptic–oblong, 2–2.6 × 0.6–0.8 mm, dark brown, finely striate. **Distr:** E tropical Africa. **Ecol:** In upland or riverine forests, epiphytic or among rocks. **Zones:** 10–12.

Clermontia

Species 22. **Shrubs** branched, or **small trees**, terrestrial or epiphytic. **Leaves** callose-toothed. **Inflorescences** subumbellate racemes, 2–3(–10)-flowered, axillary. **Flowers** with hypanthium hemispherical, obovoid, obconical, or turbinate; calyx lobes either similar to corolla or similar to those found in other genera of the family, shorter than hypanthium to half as long as the corolla, triangular to oblong, firm, green, distinct or rarely connate at base, persistent; corolla zygomorphic, tube suberect to arcuate, dorsally cleft nearly to base, lobes erect, down-curved, spreading, or recurved; stamens connate, anthers with apical tufts of white hairs on the lower 2; ovary inferior, biloculed, stigma bilobed. **Fruits** berries, orange or yellow; seeds dark brown to black, smooth, shiny, 0.2–0.5 mm long. **Distr:** Hawai'i. **Cult:** Propagate from seeds or cuttings. Requires a fertile, moist, well-drained soil in full or partial sun.

C. micrantha (Hillebr.) Rock, *Fl. Hawai'ian Isl.* 242, 1888. (Syn: *C. multiflora* var. *micrantha, C. multiflora* f. *montana*). [Maui Clermontia]. **Shrubs**, terrestrial, 0.4–2 m high. **Leaves** elliptic to oblong, 4–10 × 1–2.5 cm, base cuneate or attenuate, margins callose-crenulate, apex acute or acuminate, glabrous; petiole 1.5–4 cm long. **Inflorescences** 2–3(–5)-flowered; peduncle 0.5–1 cm long; pedicels 0.7–1.4 cm long. **Flowers** with hypanthium obovoid to turbinate, 5–8 × 3–7 mm; perianth rose to dark pink, 1.8–2.8 × 0.3–0.5 cm, glabrous, tube suberect, lobes recurved, about as long as the tube; staminal column rose to dark pink, glabrous; anthers purplish, glabrous. **Fruits** berries orange, obovoid, 1.2–1.5 × 0.9–1 cm. **Distr:** Hawai'i (Lana'i, Maui). **Ecol:** In wetlands and bogs. Fl. Feb–Nov. **Zones:** 9–12.

Delissea

Species 9. **Shrubs** or **trees**, terrestrial, branched or unbranched, rarely palmlike or subherbaceous, unarmed. **Leaves** petiolate; lamina with margins toothed or irregularly laciniate. **Inflorescences** racemes, 5–20-flowered, axillary. **Flowers** with hypanthium ovoid, ellipsoid, cylindrical, or rarely campanulate; calyx lobes subulate to triangular, shorter than hypanthium; corolla zygomorphic, white or greenish-white, rarely suffused or striped with purple or blue, suberect to strongly curved, dorsally cleft halfway, with a small knob at the cleft, sometimes with 2 additional lateral knobs, usually glabrous; stamens connate, lower 2 anthers with apical tufts of white hairs; ovary inferior, biloculed, stigma bilobed. **Fruits** berries, purple; seeds white, dull, transversely rugose, 0.7–1.5 mm long. **Distr:** Hawai'i. **Cult:** Propagate from seeds or cuttings. Requires a fertile, moist, well-drained soil in full or partial sun.

D. rhytidosperma H. Mann, *Proc. Amer. Acad. Arts* 7: 180, 1867. (Syn: *D. kealiae*). **Shrubs**, 0.5–2.5 m high; stems moderately leafy toward apex; latex white. **Leaves** oblanceolate, elliptic, or narrowly elliptic, 8–19 × 2–5.5 cm, base cuneate or attenuate, margins serrulate or crenulate, apex acute or acuminate; petiole 2–6.5 cm long. **Inflorescences** racemes, 5–20-flowered; peduncle 0.3–2 cm long; rachis 0.5–3 cm long; bracts 0.8–2.7 mm long, rarely foliaceous, oblanceolate, 1.8–3 × 0.35–0.7 cm, glabrous; pedicels 0.8–1.3 cm long. **Flowers** with hypanthium broadly ellipsoid or obovoid, 4–6 mm long; calyx lobes 0.5–1 × 0.5–0.7 mm, glabrous; corolla 1.4–2.4 cm long, tube 1–1.4 cm long, dorsal lobes 0.4–1.1 cm × 0.8–1.5 mm, ventral lobes 0.7–1 cm × 1–1.5 mm; filament tube 1.2–2 cm long, shortly pubescent at base, dorsal anthers 4.5–5 mm long, ventral anthers 3.5–4 mm long, with tufts of hairs 1–1.5 mm long at apex. **Fruits** berries, globose or ellipsoid, 0.5–1.2 × 0.5–1.1 cm; seeds 0.7–1.2 × 0.5–0.8 mm. **Distr:** Hawai'i (Kaua'i). **Ecol:** In lowland dry forests; 300–1,000 m. Fl. Apr–May, Sep–Nov. **Zones:** 9–12.

Laurentia

Species 1. *Generic description same as specific.* **Cult:** Propagate from seeds. Requires a moist, well-drained soil in full or partial sun.

L. longiflora (L.) E. Wimm., *Publ. Field Mus. Nat. Hist.*, Bot. Ser. 13(6): 474, 1937. (Syn: *Hippobroma longiflora, Isotoma longiflora*). [Madam Fate, Star Flower, Star of Bethlehem]. **Herbs**, perennial, erect; stems branched, to 50 cm high. **Leaves** sessile, rosetted; lamina lanceolate to elliptic, 5–15 × 1–5 cm, margins unevenly toothed. **Inflorescences** with flowers solitary, axillary, flowers inverted by twisting of pedicels. **Flowers** with calyx lobes 5, narrowly triangular; corolla white, tube cylindrical, 6–9 × 0.2–0.4 cm, not longitudinally split, limb with lobes spreading; stamens 5, fused into a tube, anthers with a terminal tuft of bristles; ovary obconical, biloculed, stigma lobed, recurved at maturity. **Fruits** capsules, spherical, 1.2–1.7 cm long, dehiscing via apical slits; seeds light brown 6–8 mm long. **Distr:** Tropical America; naturalized in many parts of the Old World tropics. **Ecol:** In disturbed places, weedy in tropical situations. **Zones:** 9–12. **Info:** The plant is highly poisonous if ingested, and the latex causes severe skin irritation.

Lobelia

Species ~350. **Herbs** annual or perennial, or sometimes shrubs, pubescent or glabrous. **Leaves** alternate, cauline or sometimes all or mostly basal, petiolate or sessile; lamina with margins commonly serrate or dentate or if subentire, usually with minute callosities. **Inflorescences** few- or numerous-flowered, flowers commonly pedicellate in axils of upper, usually reduced leaves or bracts, but the area of inflorescence often appearing pedunculate and terminal. **Flowers** with calyx tube adherent to ovary, lobes 5; corolla irregular, tube cleft dorsally to base in most species, often fenestrate, limb ± bilabiate, 2 upper lobes usually erect, 3 lower ones spreading; stamens 5, usually free from corolla, anther tube partially closed at orifice by the incurved tips of longer anthers, the 2 shorter ones usually tufted at apex; corolla, with stamens and style, withering and persistent on fruit; ovary partially or wholly inferior. **Fruits** capsular, dehiscent by apical valves; seeds numerous, minute. **Distr:** Temperate to tropical regions worldwide. **Ecol:** Mostly in montane regions. **Cult:** Propagate annuals from seeds, and perennials from basal cuttings or by division. Requires a moist, well-drained soil in full sun.

L. laxiflora HBK., *Nov. Gen. and Sp.* 3: 311, 1819. [Mexican Cardinal Flower, Torch Lobelia]. **Herbs**, erect or ascending, usually ± woody or suffrutescent perennials, ± simple, usually >1(–2.5) m high; stems pubescent or glabrous. **Leaves** numerous, shortly petiolate or sessile; lamina usually lanceolate to oblong-lanceolate, mostly 3–15(–25) cm long, attenuate to base or base acute to subrounded, margins mostly serrate or serrulate (rarely subentire),

Below: *Canarina eminii, Clermontia micrantha, Delissea rhytidosperma, Laurentia longiflora.*

PL

F&KS

F&KS

Above: *Lobelia laxiflora, Nymphoides cristata, Nymphoides indica, Nymphoides peltata.*

apex acuminate. **Inflorescences** few- or numerous-flowered, axillary; pedicels erect or spreading, 1–8(–10) cm long, glabrous or villous. **Flowers** with hypanthium glabrous or pubescent, at anthesis campanulate to cupular; calyx lobes triangular, 1–6.5 mm, margin entire, apex acute; corolla 2–3 cm long, glabrous or pubescent, bright to pale red or orange-red, lobes often yellow, rarely the entire corolla yellow, tube fenestrate, lower lip 1–1.8 cm long; anther tube 5–9 mm long. **Fruits** 0.5–1 cm long; seeds ellipsoid or ovoid, ~0.6 mm long. **Distr:** Arizona and Mexico; Honduras and El Salvador to Panama; West Indies; Colombia and Peru. **Ecol:** In wet to dry thickets, brushy slopes and mixed forests, 600–2,900(–3,500) m. **Zones:** 9–11.

MENYANTHACEAE

Genera 5, **species** ~58. **Herbs**, annual or perennial, aquatic or paludal. **Leaves** alternate, exstipulate; lamina entire or trifoliolate, sometimes peltate; petiole sheathing. **Inflorescences** racemose, solitary or fasciculate or involucrate heads. **Flowers** bisexual, regular; calyx 5-parted; corolla 5-lobed, united, lobes valvate or induplicate valvate, fringed or barbed; stamens 5, alternating with corolla lobes, epipetalous, anthers dithecous, sagittate and versatile, glands 5, alternating with stamens; ovary unilocular, borne on a somewhat fleshy disk, placentation parietal, style 1, apex bifid. **Fruits** capsules, unilocular, 2- or 4-valved, few- to many-seeded; seeds winged or not, testa crustaceous. **Distr:** Worldwide.

Nymphoides

Species ~40. **Herbs**, perennials (or annuals), aquatic, usually carpeting surface of lakes and ponds, with short basal rhizomes producing many, slender, and petiolelike stolons (or lacking rhizomes); stems usually long, floating, sometimes producing rootlets from nodes. **Leaves** alternate, rarely apparently opposite; lamina floating, veins palmate. **Inflorescences** with flowers clustered at nodes, (4 or)5-merous, distylous or occasionally homostylous. **Flowers** with calyx lobed to near base; corolla rotate, lobed to near base, rarely less deeply lobed and campanulate, throat with 5 bundles of long fimbriae; stamens inserted on corolla tube; style linear; nectaries 5, attached at ovary base. **Fruits** capsules, indehiscent; seeds few, compressed or globose, smooth or ornamented. **Distr:** Temperate and tropical regions worldwide. **Cult:** Propagate from seeds or by division of rhizomes. Requires a rich soil mix (for planting the rhizomes), still water and full sun.

N. cristata (Roxburgh) Kuntze, *Revis. Gen. Pl.* 2: 429, 1891. (Syn: *Menyanthes cristata*). [Crested Floating-heart, White Water Snowflake]. **Herbs**, aquatic; rhizomes horizontal; stems cylindric, unbranched, with a terminal leaf. **Leaves** ovate-orbicular to suborbicular, 3–10 cm Ø, base cordate, margins entire, subcoriaceous, abaxially densely glandular, veins indistinct; petiole cylindric, 1–3 cm long. **Inflorescences** many-flowered, grouped on nodes; pedicels cylindric, 3–4.5 cm long. **Flowers** 5-merous; calyx 3–5.5 mm long, lobed to near base, lobes ovate-elliptic to ovate, obtuse; corolla white with a yellow center, 4–8 mm long, lobes ovate with a keeled fold from base to apex, obtuse; filaments linear, 1–2 mm long, anthers 0.8–1 mm long; style short, stigma lobes triangular. **Fruits** subglobose, 3–5 mm Ø; seeds few, light brown, globose, 1.3–1.5 mm Ø, testa smooth or scabrous. **Distr:** China, E India, Taiwan. **Ecol:** Grows in standing water. Fl. & Fr. Sep. **Zones:** 9–12.

N. indica (L.) Kuntze, *Revis. Gen. Pl.* 2: 429, 1891. (Syn: *N. humboldtiana*, etc.). [Floating Hearts, Water Snowflake]. **Herbs**, aquatic; rhizomes horizontal; stems cylindric, unbranched. **Leaves** broadly ovate to subcordate, 3–18 cm Ø, base cordate, margins entire, subcoriaceous, abaxially densely glandular, veins indistinct, palmate; petiole cylindric, 1–2 cm long. **Inflorescences** many-flowered, clustered at nodes; pedicels cylindric, 3–5 cm long. **Flowers** 5-merous, distylous; calyx 3–6 mm long, lobed to near base, lobes lanceolate to narrowly elliptic, obtuse; corolla white with a yellow center, 0.7–1.2 cm long, lobes ovate-elliptic, obtuse, outside densely fimbriate-pilose; filaments flattened, linear, 1.5–1.7 mm long, anthers sagittate, 2–2.2 mm long; style cylindric, stigma lobes triangular. **Fruits** ellipsoid, 3–5 mm long; seeds few, brown, globose, 1.2–1.5 mm Ø, testa smooth. **Distr:** Asia; Australia, Pacific Islands. **Ecol:** Grows in standing water; <100–1,600 m. Fl. & Fr. Aug–Oct. **Zones:** 9–12.

N. peltata (S.G. Gmelin) Kuntze, *Revis. Gen. Pl.* 2: 429, 1891. (Syn: *Limnanthemum peltatum, Menyanthes nymphoides*). [Yellow Floating Heart]. **Herbs**, aquatic; rhizomes horizontal; stem cylindric, unbranched. **Leaves** alternate at base, opposite at apex; lamina ovate-orbicular to orbicular, 1.5–8 cm Ø, base cordate, margins entire, subcoriaceous, adaxially glabrous, abaxially purple-brown, densely glandular; petiole cylindric, 5–10 cm long, base widened into an amplexicaul sheath. **Inflorescences** with flowers usually densely clustered at nodes; pedicel 3–7 cm long. **Flowers** 5-merous, distylous; calyx 7–9 mm long, lobed to near base, lobes elliptic-lanceolate to elliptic, obtuse; corolla golden-yellow, 2.5–3 cm long, rotate, lobed to near base, lobes obovate, margin broadly membranous and irregularly laciniate, apex rounded to emarginate; filaments sparsely pilose; *short-styled flowers* with ovary 5–7 mm long, style 1–2 mm long, stigma small; *long-styled flowers* with ovary 0.7–1.7 cm long, style to 1 cm long, stigma large, bilobed, suborbicular; glands golden-yellow. **Fruits** elliptic, 1.7–2.5 x 0.8–1.1 cm; seeds brown, compressed, elliptic, 4–5 mm long, densely ciliate. **Distr:** Europe to Asia. **Ecol:** In standing water; ~100–1,800 m. Fl. & Fr. Apr–Oct. **Zones:** 6–12.

GOODENIACEAE

Genera 12, **species** 440. **Herbs** perennial or **shrubs**. **Leaves** alternate, rarely opposite or basal, exstipulate; lamina simple. **Inflorescences** with flowers solitary, axillary, or in cymes, racemes or spikes. **Flowers** bisexual, zygomorphic; calyx usually 5-lobed; corolla 5-lobed, valvate in bud; stamens 5, epipetalous, free, alternating with corolla lobes, anthers free or connate in a ring around the style; ovary bicarpellate, syncarpous, inferior or superior, 1–biloculed, placentation axile or basal, ovules 1, 2 or many per locule, style simple or 2–3-fid with a pollen cup under the stigma. **Fruits** capsules, drupes or nuts. **Distr:** Pantropical.

Scaevola

Species 130. **Herbs** or **shrubs**. **Leaves** alternate, rarely opposite; lamina with margins entire or toothed. **Inflorescences** dichasial cymes; flowers sessile or pedicellate. **Flowers** with calyx lobes obsolete or conspicuous; corolla oblique, split to base on one side, lobes subequal, spreading; stamens with free anthers; ovary usually inferior, biloculed, locules 1–2-ovuled, erect, style simple with pollen cup enclosing a truncate or bilobed stigma. **Fruits** drupes, often fleshy. **Distr:** Tropical and subtropical regions worldwide; mostly in Australia and Polynesia. **Cult:** Propagate from seeds or cuttings. Requires any soil in full sun; salt-tolerant.

S. mollis Hook. and Arn., *Bot. Beechey Voy.* 89, 1832. (Syn: *Temminckia mollis*, etc.). **Shrubs**, large, 1–2.5 m high. **Leaves** lanceolate to oblanceolate or oblong, 5–15 x 1–5 cm, base acute-acuminate, margins minutely toothed, apex acute, coriaceous, dark green, adaxially sparsely pubescent or glabrate, abaxially densely gray-pubescent; petiole 0.5–2 cm long. **Inflorescences** cymes, (1–)3(–5)-flowered, 2 cm long, pubescent (hairs golden, gray or white); peduncle 2–7 mm long; terminal flower sessile, lateral ones on pedicels 1–2 mm long; flowers subtended by paired bracteoles. **Flowers** with calyx minute, or rarely with acute lobes 5, 0.5–1 cm long; corolla purple, rarely white, tube 1.2–2.2 cm long, densely pubescent, lobes 0.8–1.7 cm long, margins winged, pubescent outside; stamens ± as long as corolla tube; style 1.8–2.3 cm long, pubescent below. **Fruits** drupes, 2- or 4-celled, dark purplish-black, oblong, 0.6–1.2 x 0.4–1 cm, densely pubescent. **Distr:** Hawai'i (Kaua'i, O'ahu.). **Ecol:** In wet forests, slopes; to ~750 m. Fl. throughout most of the year. **Zones:** 10–12.

S. taccada (Gaertn.) Roxb., *Fl. Ind.* 1:527. 1824. (Syn: *S. sericea*, etc.). [Beach Berry, Half Flower, Hail Stones]. **Shrubs**, branched, 1–3.5 m high. **Leaves** alternate, sessile, axils with long white hairs; lamina obovate, 10–20 x 2–10 cm, base gradually tapered, margins entire, apex obtuse, surfaces glabrous. **Inflorescences** dichasial cymes, tending to become monochasial; peduncle 1–2.5 cm long; bracts

TR

Above: *Scaevola mollis, Scaevola taccada, Acmella oleracea, Ageratum conyzoides.*

3–4 mm long, with long white hairs inside; pedicels ~5 mm long, pubescent or glabrous. **Flowers** with calyx 5-lobed, linear-lanceolate or linear obovate, usually 4–6 mm long, enlarged in fruits; corolla white, pubescent or glabrous outside, tube 0.9–1.6 cm long, lobes elliptic-obovate, 0.9–1.2 cm long. **Fruits** drupes, subglobose, white, with persistent calyx teeth, ~1.8–2 cm Ø, 1-seeded. **Distr:** India, Maldives, Myanmar, Pakistan; along the African coast from Kenya to S. Africa; China, Hong Kong, Malaysia; Pacific Islands. **Ecol:** Mostly near the sea coast; salt tolerant. Fl. Jun–Dec. **Zones:** 10–12.

ASTERACEAE

Subfamilies 12, **genera** 1,620, **species** 23,600. **Herbs**, annual, biennial or perennial, sometimes shrubs, rarely trees, with or without milky latex, sometimes armed with spines. **Leaves** alternate, opposite or rosulate, exstipulate; lamina with margins entire, toothed, lobed or variously dissected. **Inflorescences** heads; individual flowers (florets) usually small and numerous, sessile and aggregated on a receptacle, surrounded by an involucre made up of 1 to several series of bracts (phyllaries or involucral bracts) and the whole structure forming a capitulum (head); phyllaries herbaceous; receptacle naked (epaleate) or bearing scales (paleate) or long hairs or bristles; heads with all florets of one sexual state, either ♂, ♀ or hermaphrodite (homogamous) and heterogamous; homogamous capitula discoid or ligulate; heterogamous capitula radiate, radiant or diskiform. **Florets** epigynous, bisexual, ♀ or functionally ♂ or sterile (neuter); calyx often represented by a pappus of 1 or more series of bristles or scales on apex of ovary, or sometimes pappus absent; corolla gamopetalous, tubular, filiform, ligulate or rarely bilabiate, usually 3 or 5-toothed; stamens 5, epipetalous, anthers connate forming a tube around style, filaments free, rarely anthers free; ovary inferior, 1-loculed, style usually 2-branched. **Fruits** cypselae, sometimes with distinct carpopodium (basal attachment area). **Distr:** Worldwide.

Acmella

Species 30. **Herbs**, annual or perennial, 10–20(–30+) cm high; stems prostrate to erect, usually branched ± throughout. **Leaves** cauline, opposite, petiolate (rarely ± sessile); lamina ovate to rhombic or lanceolate (rarely linear to filiform), base ± cuneate, margins entire or toothed, sparsely pilose to strigillose, glabrescent, usually trinerved. **Inflorescences** heads, radiate or discoid, borne singly at tips of branches (in corymbiform arrays); involucres ± hemispheric to ovoid, 3–6+ mm Ø; phyllaries persistent, 8–15+ in 1–3 series; receptacles conic, paleate. **Florets** *ray* absent or 5–20+, ♀, fertile, corollas yellow to orange (white or purplish), ovate to linear, rarely absent; *disk* florets 25–100(–200+) bisexual, fertile, corollas yellow or orange, tubes shorter than campanulate throats, lobes 4–5, deltate. **Fruits** cypselae, 2–3-angled or strongly compressed, ellipsoid to obovoid; pappi absent. **Distr:** S USA, Mexico, West Indies, C and S America. **Cult:** Propagate from seeds. Requires a moist, well-drained soil in full sun.

A. oleracea (L.) R.K. Jansen, *Syst. Bot. Mongr.* 8: 65, 1985. (Syn: *Spilanthes acmella* var. *oleracea*, etc.). [Toothache Plant]. **Herbs**, annual, prostrate to ascending, to 60–90 cm tall, or sometimes short-lived perennial herbs. **Leaves** opposite; lamina broadly ovate to deltoid, 5.3–10.6 x 4–7.9 cm, base truncate, attenuate in upper leaves, margins dentate, apex acute, surfaces sparsely hairy, mainly on upper midrib; petiole 2–6.4 cm long, flattened, adaxially grooved, narrowly winged, sparsely pilose. **Inflorescences** of solitary, terminal and axillary pedunculate capitula; capitula 1–2.4 cm high, 1.1–1.7 cm Ø, florets yellow with distinct purple to red paleae; phyllaries triseriate, imbricate, green, lanceolate, apices purple to red. **Florets** hermaphrodite, numerous (400–620), fertile; corolla 2.7–3.3 mm long, green, glabrous, constricted into a tube at base, corolla lobes (4–)5, 0.5–0.6 mm long, yellow, papillose inside; anthers 5, black, pollen bright orange; style glabrous. **Fruits** achenes 2–2.5 × 0.9–1.1 mm, black. **Distr:** Origin unknown (perhaps S America). **Ecol:** Fl. throughout most of the year. **Zones:** 10–12. **Info:** Mostly cultivated for medicinal purposes, as most parts of the plant causes a strong anaesthetic effect.

Ageratum

Species ~40. **Herbs**, erect. **Leaves** opposite or upper ones alternate, petiolate; lamina ovate, margins dentate or serrate, penninerved. **Inflorescences** heads, in corymbs or loose panicles, homogamous, discoid, many-flowered; involucre campanulate; bracts imbricate, 2–3 seriate, linear, acute to acuminate, subequal; receptacle flat or nearly so, naked or with caducous scales. **Flowers** with corollas all tubular, equal, regular, limb 5-fid; anthers with an apical appendage, base entire, obtuse; style arms long, slender, obtuse, pubescent at apex. **Fruits** achenes, oblong, 5-angular; pappus uniseriate, of 5 short free or connate scales, or of 10–20 narrow acuminate unequal scales. **Distr:** Tropical America, West Indies. **Cult:** Propagate from seeds. Requires a moist, well-drained soil in full sun.

A. conyzoides L., *Sp. Pl.* 839, 1753. (Syn: *A caeruleum, Carelia conyzoides, Eupatorium conyzoides*). [Billygoat's Weed, Mexican Ageratum]. **Herbs**, annual; stems erect, 30–60 cm high, hispid, branched, somewhat aromatic. **Leaves** petiolate; lamina ovate, 5–13 cm long, base truncate to rounded, rarely cordate, margins crenate, apex acute. **Inflorescences** capitula, to 6 mm Ø, in dense small terminal corymbs of 8–15 heads; involucres campanulate, bracts 2–3 seriate, linear, subequal, acute, sparsely pilose outside. **Flowers** many (60–75), pale blue or white; corollas tubular, 1–1.5 mm long, limb 5-cleft. **Fruits** achenes, linear-oblong, black, 5-angled; pappus of 5 short scales, scales often basally serrate, ending in a long awn. **Distr:** Pantropical. **Ecol:** In waste areas, often weedy. **Zones:** 10–12.

A. houstonianum Miller, *Gard. Dict.* ed. 8. Ageratum no. 2, 1768. (Syn: *A. conyzoides* var. *mexicanum*). [Bluemink, Floss Flower]. **Herbs**, annual, 30–80 cm high; roots fibrous; stems erect to decumbent, sparsely to densely pilose. **Leaves** deltate to ovate, mostly 3–8 × 2.5–4 cm, margins toothed, abaxially sparsely to densely pilose, not evidently gland-dotted. **Inflorescences** heads; peduncles viscid-puberulent, pilose, and stipitate-glandular; involucres ~4 x 5–6 mm; phyllaries narrowly lanceolate, 0.6–1 mm wide, stipitate-glandular, sparsely to densely pilose, eciliate or inconspicuously ciliate, tips gradually tapering, indurate-subulate, 0.8–2 mm long. **Florets** with corollas usually lavender, rarely white. **Fruits** cypselae, sparsely strigose-hispid, pappi of 5 distinct, oblong scales 2–3 mm long. **Distr:** Tropical N America, through Mexico to C America. **Ecol:** In disturbed sites, mostly coastal. Fl. Jun–Aug. **Zones:** 9–12.

Anaphalis

Species ~110. **Herbs**, perennial, erect, whitish woolly or tomentose, often stoloniferous, dioecious or polygamodioecious. **Leaves** alternate; lamina linear to lanceolate or oblong, base often decurrent, margins entire. **Inflorescences** heads, small, arranged in flat-topped corymbs; involucre oblong to campanulate; bracts scarious, imbricated in 5–8 series, often whitish, outer bracts shorter, outermost woolly; receptacles mostly convex. **Florets** *ray* with corolla filiform, 2–4-dentate, style undivided, anthers sagittate at base; *disk florets* with corolla tubular, 5-toothed, functionally ♂, style bifid. **Fruits** achenes, very small, oblong; pappus bristles uniseriate, slender, scabrid, free, caducous. **Distr:** Mainly in Asia, a few in N America. **Cult:** Propagate from seeds. Requires a moist, well-drained soil in full sun.

A. brevifolia DC., *Prod.* 6: 273, 1837. (Syn: *A. bournei, Gnaphalium brevifolium*). **Herbs**, annual or biennial, much-branched; stems 10–30 cm high, bearing numerous small leaves and covered with dense whitish tomentum. **Leaves** linear-lanceolate, 0.5–1.2 cm × 1.5–3 mm, semiamplexicaul, margins entire, apex acute, revolute, surfaces tomentose, 1-nerved. **Inflorescences** capitula, 7–8 mm Ø, in dense corymbs of 7–15; phyllaries 4–5 seriate, ovate-lanceolate, ~5 × 1.5 mm, lower half stiff, straw-colored, upper half white, scarious, spreading. **Flowers** yellow; ♀ corollas ~12, 2–2.5 mm long; hermaphrodite corollas ~60, 2.25–2.5 mm long, lobes bearing glandular hairs.

Above: *Ageratum houstonianum*, *Anaphalis brevifolia*, *Aster laevis*.

Fruits achenes 0.5 mm long, minutely puberulent; pappus hairs 2.25 mm long, deciduous. **Distr:** S India (Anamalai Hills), Sri Lanka. **Ecol:** In grassy banks and plains; 1,500–2,000 m. Fl. Sep–Apr. **Zones:** 10–12.

Aster

Species ~180. **Herbs** perennial, **subshrubs** or **shrubs**, rarely annual or biennial herbs. **Leaves** alternate, polymorphic; basal rosette leaves usually dissimilar to cauline leaves, peduncular leaves often bractlike, mostly dentate; lamina simple. **Inflorescences** heads, heterogamous, solitary and terminal or variously arranged in corymbs or panicles, radiate, rarely discoid; involucral bracts in 2–6 imbricate, usually graduated series, margins herbaceous; receptacle naked, often flat or convex, pitted or alveolate. **Florets** *ray* ♀ and fertile, 1(–2)-seriate, ligules white, pink, blue or purple, style filiform; *disk florets* mostly yellow, all bisexual and fertile, corolla regular, narrowly tubular or funnelform, limb sometimes abruptly dilated at throat, 5-lobed, anthers obtuse or subauriculate at base, with narrow scarious terminal appendages, style branched. **Fruits** achenes, oblong-obovoid, mostly somewhat compressed, usually pubescent; pappus usually in 1 row, similar in ray and disk florets, sometimes very short or absent, or only absent in ray florets, rarely in 2 rows with an outer row of shorter scalelike bristles. **Distr:** Cosmopolitan, especially N America and Eurasia. **Cult:** Propagate from seeds, cuttings or by division. Requires a moist, well-drained soil in full or partial sun.

A. laevis L., *Sp. Pl.* 876, 1753. [Michaelmas Daisy, Smooth Aster]. **Herbs**, stout, glabrous, erect perennials, usually 25–60 cm high, simple below, branched above. **Leaves** mostly sessile and amplexicaul; lamina lanceolate to oblong-lanceolate, 3–15 cm long, margins mostly entire, sometimes ± serrate, scabrous, apex acute, uppermost leaves usually much-reduced, basal and lowest cauline ones narrowed into winged petiole. **Inflorescences** varying from thyrsoid panicles to open panicles of capitula; involucres campanulate, 5–8 mm high; phyllaries several-seriate, rigid, acute, appressed, green-tipped. **Flowers** with *ray florets* 15–30, ligules lanceolate, 0.6–1 cm long, usually blue or violet, rarely white; *disk florets* yellow. **Fruits** achenes, glabrous or nearly so; pappus fulvous, 4–6 mm long. **Distr:** USA. **Ecol:** In dry open areas. Fl. Jul–Oct. **Zones:** 4–12.

Barnadesia

Species ~20. **Shrubs** or **small trees**, 0.6–20 m tall. **Leaves** alternate, often fasciculate in groups of 2–8, subsessile to petiolate; lamina lanceolate to oblong-elliptic, margins entire, penninerved, glabrous to sericeous. **Inflorescences** heads, solitary, terminal or axillary; heads heterogamous or homogamous, radiate; involucres campanulate to cylindric; receptacles plane, epaleate, fimbriate to villous; phyllaries 6–14-seriate, imbricate, linear or lanceolate to ovate, acute, acuminate, or long-attenuate. **Florets** *ray* florets ♀, 8–13, 1-seriate, corollas red to purple, occasionally pink or white, bilabiate with outer lip liguliform, 5-nerved, apically 4-toothed, the inner lip filiform, entire, stamens 5, filaments free, or connate and forming a tube, style branches short; *disk florets* hermaphroditic, 1–3, corollas tubular, limb 4–5-toothed, stamens 4–5, style shortly branched. **Fruits** achenes; *ray* cylindric to turbinate, villous, pappus of bristles, persistent, yellow to brownish; *disk* turbinate, densely villous, pappus of setae, yellowish. **Distr:** Colombia southwards to NW Argentina and SE Brazil. **Cult:** Propagate from seeds, cuttings or by air-layering. Requires a fertile, moist, well-drained soil in full or partial sun.

B. caryophylla (Vell.) S. F. Blake, *Proc. Biol. Soc. Wash.* 38: 86, 1925. (Syn: *Xenophontia caryophylla*, etc.). **Shrubs**, 0.6–4 m high, branched, hirsute, glabrescent, spiny; spines divaricate to 3 cm long, caducous on distal branchlets. **Leaves** fasciculate, 3–4 per node; lamina elliptic to obovate-lanceolate, 2.8–8 × 1.1–4.2 cm, apex acute, mucronate, abaxially sericeous; petiole 3–8 mm long. **Inflorescences** heads, usually solitary, terminal or axillary, sessile, or, occasionally cymose with 2–3 heads; involucres campanulate or turbinate, 3.3–4.6 × 0.8–3.2 cm; phyllaries ~95, 10–13-seriate, *outer* ovate-triangular, 5–5.5 × 2.2–2.5 mm, apex acute, mucronate, *inner* linear, ~3.5 cm × 1.2–2 mm, apex acute. **Florets** *ray* 13, corolla pink, 4.6–4.8 cm long, tube 3.3–3.5 cm long, *outer lip* 1.2–1.6 × 3.4 cm, *inner lip* filiform, ~1.2 cm long, pubescent, filaments free, anthers 0.95–1 cm long; *disk florets* 3, corollas ~1.8 cm long, tube ~7.5 mm long, pubescent, limb 5-toothed, anthers ~8.5 mm long. **Fruits** achenes; *ray* 3–5 × 1.4–1.6 mm; *disk* to 4.2 × 3.2 mm; pappus to 1.9 cm long. **Distr:** N Peru to Bolivia. **Ecol:** In rainforests, woodlands; 600–2,700 m. Fl. Feb–Nov. **Zones:** 10–12.

Bidens

Species ~240. **Herbs**, annual or perennial, rarely shrubs; stems erect. **Leaves** opposite or rarely whorled; lamina compound, lobed or simple. **Inflorescences** heads, solitary, in corymbs or panicles, radiate or discoid. **Florets** *ray* ligulate, neuter or ♀, usually yellow; *disk florets* perfect or rarely functionally ♂, (4)5-lobed, yellow or brownish-orange. **Fruits** achenes, linear-oblong to ellipsoid, 3- or 4-angled or compressed, sometimes winged; pappus of up to 4 usually barbed awns and/or bristles, or absent. **Distr:** Worldwide, especially N and S America. **Cult:** Propagate from seeds. Requires a well-drained soil in full sun.

B. pilosa L., *Sp. Pl.* 2: 832, 1753. (Syn: *B. alba*, *B. odorata*, etc.). [Spanish Needle]. **Herbs**, annuals (perennials), to 0.6(–2.5) cm high. **Leaves** compound, 5–15 cm long; *lateral leaflets* ovate, 4–9 × 1.5–3.5 cm, base ± abruptly attenuate into a short petiolule, 0.5–1 cm long, apex acute or shortly acuminate; *terminal leaflets* ovate-lanceolate,

Below: *Barnadesia caryophylla*, *Bidens pilosa*, *Cosmos bipinnatus*.

SA/SAJF

6.5–12 cm long, base attenuate into a petiolule, 1.5–2.5 cm long, apex gradually acuminate, surfaces glabrous; petioles 1.5–4 cm long. **Inflorescences** capitula, usually numerous in loose dichotomously branched panicles; involucre 4–5 mm Ø; outer phyllaries 8, spathulate, ~3 × 0.7–1 mm, slightly accrescent, margins sparsely ciliate, base puberulous; inner phyllaries oblong-lanceolate, ~5 × 1.5 mm, apex ciliate, margins membranous, yellowish. **Florets** *ray* 5–6, white, ~6 × 4 mm, sometimes absent; *disk florets* yellow, 3.5–5 mm long; receptacular scales ~4 mm long, accrescent. **Fruits** achenes, 7–8 × 0.7–1 mm, black, pappus bristles 2–4, straw-colored, 2–3 mm long. **Distr:** Tropical America, Africa, SE Asia, Australia, New Zealand. **Ecol:** Weedy throughout its range. Fl. Nov–Mar. **Zones:** 10–12.

Cosmos

Species ~26. **Herbs** annual or perennial, or **subshrubs**; stems usually 1, erect or ascending, branched distally or ± throughout. **Leaves** mostly cauline, opposite, petiolate or sessile; lamina usually pinnately 1–trilobed, or undivided, ultimate margins usually entire, usually glabrous or puberulent. **Inflorescences** heads, radiate, borne singly or in corymbiform arrays; calycles of (5–)8 basally connate, ± linear to subulate, herbaceous bractlets; involucres ± hemispheric, 0.3–1.5 cm Ø; phyllaries persistent, (5–)8, ± biseriate, distinct, lanceolate to oblong, ± equal, membranous or herbaceous, margins ± scarious; receptacles flat, paleate; paleae falling, linear, scarious or entire. **Florets** *ray* sterile, corollas white to pink or purple, or yellow to reddish-orange; *disk florets* bisexual, fertile, corollas yellow or orange, tubes shorter than throats, lobes 5, ± deltate. **Fruits** achenes, slender, somewhat fusiform, slightly arcuate, beaked, glabrous or hispid; pappi of 2–4(–8) retrorsely barbed awns, sometimes absent. **Distr:** Tropical and subtropical America. **Cult:** Propagate annuals from seeds and perennials from basal cuttings. Requires a moist, well-drained soil in full sun.

C. bipinnatus Cav., *Icon.* 1: 10, plate 14, 1791. [Garden Cosmos]. **Herbs**, 0.3–2 m high, glabrous or sparsely puberulent. **Leaves** pinnate, fernlike, 6–11 cm long, ultimate lobes very fine, to 1.5 mm wide, margins entire, apices acute; petiole 0–1 cm long. **Inflorescences** heads, solitary, 3–6 cm Ø; peduncles 10–20 cm long; calycles of spreading, linear to lanceolate bractlets 0.6–1.3 cm long, apices acuminate; involucres 0.7–1.5 cm Ø; phyllaries erect, lanceolate to ovate-lanceolate, 0.7–1.3 cm long, apices round or obtuse. **Florets** *ray* with corollas white, pink, or purplish, obovate to oblanceolate, 1.5–5 cm long, apices ± truncate, dentate; *disk florets* with corollas 5–7 mm long. **Fruits** achenes, 0.7–1.6 cm long, glabrous, papillose; pappus absent , or of 2–3 ascending to erect awns 1–3 mm long. **Distr:** Mexico and SW USA. **Ecol:** Fl. Jun–Aug. **Zones:** 7–12.

C. sulphureus Cav., *Icon.* 1: 56, plate 79, 1791. [Cosmos]. **Herbs**, 0.3–2 m high, glabrous or sparsely pilose to hispid. **Leaves** pinnate, 5–12(–35) cm long, ultimate lobes 2–5 mm wide, margins sparsely spinulose-ciliate, apex apiculate; petiole 1–7 cm long. **Inflorescences** heads; peduncles 10–20 cm long; calycles of spreading-ascending, linear-subulate bractlets 0.5–0.7(–1) cm long, apices acute; involucres 0.6–1 cm Ø; phyllaries erect, oblong-lanceolate, 0.9–1.3(–1.8) cm, apices acute to rounded-obtuse. **Florets** *ray* with corollas bright yellow to reddish-orange, obovate, 1.8–3 cm long, apices ± truncate, denticulate; *disk florets* with corollas 6–7 mm long. **Fruits** achenes 1.5–3 cm long, usually hispidulous, rarely glabrous; pappus absent, or of 2–3 widely divergent awns 1–7 mm long. **Distr:** Mexico. **Ecol:** Fl. Jun–Sep. **Zones:** 7–12.

Dahlia

Species ~24. **Herbs**, mostly perennial, sometimes shrubby, rarely epiphytes; stems branched or not, sometimes fistulose, mostly with thickened rootstocks. **Leaves** opposite or 3 per node, stipels and petiole present; lamina simple or pinnately 1–3-compound; leaflets ovate to linear, margins mostly serrate or dentate. **Inflorescences** terminal heads, solitary, sometimes aggregated into corymbs of panicles; capitula radiate, often showy; involucral bracts in 2 unlike series, outer series 4–7, green, porrect or reflexed, linear to ovate, fleshy, longitudinally striate-nerved, inner series 7–9, thinner, broader, sometimes margins scarious, imbricate; paleas scarious or slightly indurate. **Florets** *ray* numerous, showy, variously colored, limb often apically denticulate, ovary sterile; *disk florets* numerous, much shorter, corolla yellow, reddish or purplish, tubular, apically 5-lobed, anthers with deltoid apical appendages and obtuse bases, style branches narrow. **Fruits** achenes, mostly linear, apically contracted; pappus lacking or of 2 rudimentary awns or threads. **Distr:** Mostly in Mexico, C America, south to Colombia. **Cult:** Propagate from seeds, basal cuttings or by division of tubers. Requires a fertile, moist, well-drained soil in full or partial sun.

D. coccinea Cav., *Icon. Pl.* 3: 33, t. 266, 1794. (Syn: *D. gentryi*, etc.). [Red Dahlia]. **Herbs**, erect, 0.5–1.8 m high, ± simple, glabrous or pubescent. **Leaves** opposite, variable; lamina simple to tripinnate, mostly 12–35 cm long; primary leaflets 3–11, sessile or petiolulate, thin, ovate to oblong-lanceolate or elliptic, base acute, rounded, truncate, or rarely subcordate, margins serrate, apex acute to acuminate, sparsely villous, mostly on veins abaxially, or glabrous throughout. **Inflorescences** capitula, erect or cernuous, usually few (mostly 2–3); peduncle 2–30 cm long; outer phyllaries spreading or reflexed, herbaceous, linear to oblong or obovate to spathulate, 0.6–1.5 cm long, inner ones narrowly oblong, 1.1–2 cm long. **Florets** *ray* 1.6–4 cm long, yellow, orange, or scarlet, acute or denticulate at apex; *disk florets* 71–157, yellow or tipped with scarlet, 0.8–1 cm long. **Fruits** achenes, 0.8–1.3 cm long, pappus absent or reduced to 2 minute awns, these rarely elongated into delicate, filiform threads. **Distr:** Mexico, British Honduras. **Ecol:** Often rocky, open slopes, and in deciduous forests; 1,100–3,000 m. **Zones:** 9–11.

D. imperialis Roezl ex Ortgies, *Gartenflora* 12: 243, 1863. (Syn: *D. arborea*). [Giant Dahlia, Tree Dahlia]. **Herbs** or **shrubs**, perennial, rarely trees, mostly branched from base, to 4(–6) m high; stems stout. **Leaves** opposite; lamina to >35 cm long, 1–3-pinnate; leaflets ovate, base rounded or obtuse, margins evenly serrate, apex acuminate, adaxially puberulent on veins, abaxially pilose below; petiole stout, slightly shorter than lamina. **Inflorescences** capitula, terminal, several in panicles, branches subtended by foliaceous bracts and reduced leaves resembling abortive peduncles; peduncles glabrous, stout, of varying lengths, tardily articulating near apex; capitula large, radiate; involucral bracts in 2 unlike series, outermost oblong, thickened, reflexed, ~1 cm long, innermost twice as long, broadly ovate, scarious, drying dark brown with hyaline margins. **Florets** *ray* ~9, corollas white or lavender, ligule narrowly elliptic or oblanceolate, 2–4 mm long; *disk florets* very numerous, yellow. **Fruits** achenes, linear or slightly oblanceolate, 1.3–1.7 cm long. **Distr:** Mexico to Costa Rica, Panama, Colombia. **Ecol:** In damp thickets, steep slopes, wet meadows, oak–pine and coniferous forests; 1,200–3,800 m. Fl. Oct–Dec. **Zones:** 9–11.

Duhaldea

Species 12–14. **Shrubs**, **perennial herbs**, or small trees. **Leaves** alternate; lamina simple, hairy. **Inflorescences** heads, heterogamous, radiate or disklike, solitary or in terminal corymbs; receptacle epaleate, with scalelike ridges. **Florets** *ray* (marginal) ♀, corolla yellow to white, ± radiate; *disk florets* bisexual, corolla yellow or whitish, anthers ecalcarate, with branched tails, apical appendix truncate, almost emarginate, style branches with hairs. **Fruits** achenes, ellipsoid, pilose; pappus of barbellate, capillary bristles, 1-seriate. **Distr:** Mostly in Asia, 1 sp. in E Africa. **Cult:** Propagate from seeds or root cuttings. Requires a fertile, moist, well-drained soil in full or partial sun.

D. stuhlmannii (O. Hoffm.) Anderb., *Pl. Syst. Evol.* 176: 104, 1991. (Syn: *Inula stuhlmannii*). **Shrubs**, sometimes scandent, or woody herbs, 0.9–2 m high; stems scabridulous and glandular. **Leaves** ovate, 2.5–11 × 1–5.5 cm, base cuneate to rounded, margins crenate to irregularly dentate with callose tips, apex acute and mucronate, pilose on midrib and veins, glandular abaxially; petiole 1–3.5 cm long. **Inflorescences** capitula 6–9 mm long, heterogamous, many together in large leafy corymbs; capitula stalks 1–5 mm long; phyllaries ovate to lanceolate, 2–6 mm long, margins slightly erose, ciliate, acute, glandular. **Florets** yellow; *ray florets* with tube cylindric, 2.5–2.8 mm long, limb 4.5–6.5 × 1.5–2 mm,

Below: *Cosmos sulphureus*, *Dahlia coccinea*, *Dahlia imperialis*.

PL

Above: *Duhaldea stuhlmannii, Eclipta prostrata, Emilia sonchifolia, Erigeron karvinskianus.*

3-toothed, style 3.5–5 mm long; *disk florets* few (3–5), tube cylindric or slightly infundibuliform, 3–4.2 mm long, lobes 4, 1.3–1.4 mm long, glandular, anthers 2.2–2.8 mm long, style 4–6 mm long. **Fruits** achenes, 1.4–2 mm long, sparsely glandular; pappus 3–5 mm long. **Distr:** Tanzania. **Ecol:** In forests; 1,350–2,300 m. **Zones:** 10–12.

Eclipta

Species ~4. **Herbs**, annual, erect, branched, strigose. **Leaves** opposite; lamina with margins toothed. **Inflorescences** heads, terminal on stems and branches or axillary, peduncled, heterogamous; involucre campanulate, bracts imbricate, in ~2 series; receptacle flat or convex, paleaceous, paleae awnlike. **Florets** whitish; *ray florets* bisexual, mostly fertile, corolla tubular, white, 4(5)-lobed, anthers entire or very shortly bifid at base, style branches obtuse, mammillate at apex. **Fruits** achenes, thick, those of ray florets 3-angled, those of disk florets compressed, 4-angled, apex truncate and depressed, margins with 1–3 minute teeth, coarsely hairy, tuberculate; pappus none, or of a few short teeth. **Distr:** Tropical regions; often weedy. **Cult:** Propagate from seeds. Requires a well-drained soil in full sun.

E. prostrata (L.) L., *Mant. Pl.* 2: 286, 1771. (Syn: *E. alba*). [False Daisy, Thistles, Trailing Eclipta]. **Herbs**, annual, erect or prostrate, to 0.6(–1) m high, strigose-pilose, branched at base. **Leaves** sessile or shortly petiolate; lamina lanceolate, 3–10 × 0.5–2.5 cm, base narrowed, margins serrulate, apex gradually acuminate, chartaceous, surfaces densely strigose-pubescent. **Inflorescences** capitula, 6 mm Ø; peduncle slender, 2–4.5 cm long; involucre globose-campanulate, 5 × 6–7 mm, enlarging to 1.1 cm Ø in fruit; bracts 5 or 6, biseriate, oblong, acute, outer ones longer. **Florets** *ray* biseriate, corolla white, 2.5–3 × 0.4 mm, bifid or entire; *disk* florets many, corolla white, 1.5 mm long, 4-lobed. **Fruits** achenes, 2.8 × 1.5 mm, margins ribbed. **Distr:** Warm parts worldwide. **Ecol:** In disturbed places, usually weedy. **Zones:** 9–12.

Emilia

Species 50–100. **Herbs**, annual, sometimes persisting, usually monocarpic, rarely perennial, usually 0.2–1 m high; stems usually 1, erect or lax. **Leaves** mostly cauline, alternate, petiolate or sessile; lamina pinnately nerved, mostly ovate to obovate or oblanceolate, bases auriculate or clasping, ultimate margins entire or toothed, surfaces usually glabrous, sometimes villous or ± arachnose. **Inflorescences** heads discoid, in cymiform or corymbiform arrays; calyculi absent; involucres urceolate to campanulate or cylindric, mostly 2–8+ mm Ø; phyllaries persistent, usually 8 or 13 in 1–2 series, erect, mostly linear to oblong; receptacles flat to convex, smooth or obscurely foveolate, epaleate. **Florets** *ray* absent; *disk florets* 20–50(100+), all bisexual and fertile or inner functionally ♂, corollas usually pinkish, lavender, or purplish, rarely reddish, orange, white, yellow, tubes shorter than to equaling funnelform to cylindric throats, lobes 5, erect to spreading. **Fruits** achenes, fusiform-prismatic, 5-ribbed, glabrous; pappi fragile, of 80–100+, white, barbellulate bristles. **Distr:** Old World tropics. **Cult:** Propagate from seeds. Requires a well-drained soil in full sun.

E. sonchifolia (L.) de Candolle, *Contr. Bot. India.* 24, 1834. (Syn: *Cacalia sonchifolia*). [Flora's Paintbrush, Lilac Tasselflowers]. **Herbs**, glabrous or ± villous, 20–80 cm high, single-stemmed; stem erect or somewhat lax, simple or branched. **Leaves** mostly in proximal half, usually petiolate; lamina ovate to obovate or oblanceolate, mostly 5–12 × 1.5–4.5 cm (distal smaller, bractlike), margins often deeply lobed to lyrate-pinnatifid. **Inflorescences** capitula in loosely clustered corymbs; involucres urceolate to campanulate, 0.9–1.2 cm long, relatively slender, lengths mostly 3–4 times widths; phyllaries usually 8. **Florets** 15–30(–40), surpassing involucres by 0–1(–2) mm; corollas usually lavender, pinkish, or purplish, rarely reddish, 5-lobed, lobes 0.5–0.7(–1.5) mm long; stamens 5, style appendages 0–0.1 mm long, stigma bilobed. **Fruits** achenes, with a fluff of hairs. **Distr:** Tropical Africa and Asia. **Ecol:** In disturbed areas, often weedy. Fl. throughout most of the year. **Zones:** 9–12.

Erigeron

Species >200. **Herbs**, annual, biennial or usually perennial; stems erect or prostrate. **Leaves** alternate, frequently rosulate at base; lamina with margins entire, toothed or divided. **Inflorescences** heads, heterogamous, radiate, solitary, or few in loose racemes, sometimes numerous; involucre campanulate, bracts 2- or 3-seriate, narrow, varying from herbaceous to scarious; receptacles flat or convex, naked. **Florets** *ray* ♀, fertile, numerous, in 1 or more marginal rows, ligules long, narrow, rarely very short, white, pink or purplish, rarely yellow; *disk florets* bisexual, fertile, commonly several times as many as ray florets, corolla mostly yellow, 5-lobed, lobes sometimes purplish. **Fruits** achenes commonly narrowly ellipsoid or obovoid, sometimes compressed; pappus mostly double, outer row of scales or absent, inner row of fine, mostly fragile bristles; pappus rarely absent. **Distr:** Cosmopolitan. **Cult:** Propagate from seeds or by division. Requires a well-drained soil in full sun to shade.

E. karvinskianus de Candolle, *Prodr.* 5: 285, 1836. (Syn: *E. mucronatus*). [Karwinsky's Fleabane, Mexican Daisy]. **Herbs**, perennial, often short-lived, 0.1–1 m high; caudices woody, usually simple, stems sometimes rooting adventitiously; stems erect to sprawling or decumbent. **Leaves** basal and cauline; *cauline* lamina elliptic to obovate, mostly 1–4(–5) × 0.5–1.3 cm, usually relatively even-sized along stems, margins entire or with 1–2 distal pairs of acute, mucronulate teeth or lobes, surfaces sparsely and loosely strigose to hirsute-villous, eglandular. **Inflorescences** capitula, 1–5; involucres 2.5–3.5 × 0.7–1 cm; phyllaries 3–4-seriate, sparsely strigose to hirsute-villous or glabrate, sometimes minutely glandular. **Florets** *ray* 45–80, corollas 5–8 mm long, lamina not or slightly coiling, white, sometimes drying pinkish; *disk florets* with corollas 2–3 mm long. **Fruits** cypselae, 1–1.4 mm long, binerved, surfaces sparsely strigose, bristles 15–27. **Distr:** Mexico, C America; naturalized and weedy in the rest of the tropics and subtropics. **Ecol:** In moist disturbed sites, shaded rock walls; 300–1,100 m. Fl. Apr–Aug. **Zones:** 8–11.

Gaillardia

Species 15–17. **Herbs** annual or perennial, or **subshrubs**, 10–80+ cm high, sometimes rhizomatous; stems usually erect, usually branched. **Leaves** basal, basal and cauline, or mostly cauline, alternate, petiolate or sessile; lamina mostly elliptic, lanceolate, linear, oblong, obovate, ovate, or spathulate, often pinnately lobed, ultimate margins toothed or entire, often villous. **Inflorescences** heads, usually radiate or discoid, borne singly; involucres hemispheric to subrotate, mostly 0.9–2+ cm Ø; phyllaries persistent, 14–40+ in 2–3+ series; receptacles convex to hemispheric, smooth or pitted. **Florets** *ray* absent, or 5–15+, usually sterile, rarely ♀ and fertile, corollas yellow to white or orange to red, purplish, or brown, often bicolored; *disk florets* 20–100+, bisexual, fertile, corollas yellow or orange to red, purplish, or brown, tubes much shorter than ampliate, campanulate to cylindric or urceolate throats, lobes 5, broadly deltate or ovate (± equal, sometimes attenuate or attenuate-terete apically. **Fruits** achenes, obpyramidal to clavate, ± 4-angled, sparsely to densely hairy. **Distr:** Tropical America. **Cult:** Propagate from seeds, basal cuttings or by division. Requires a gritty, moist, well-drained soil in full sun.

G. pulchella Fougeroux, *Hist. Acad. Roy. Sci. Mém. Math. Phys.* (Paris). 1786: 5, 1788. (Syn: *G. drummondii*, etc.). [Firewheel, Indian Blanket]. **Herbs**, annuals, biennials, or sometimes longer-living, 5–35(–60+) cm high. **Leaves** cauline; lamina linear, oblong, or spathulate, 1–5(–12) × 0.4–1.2(–3.5) cm, margins usually entire, sometimes toothed or lobed, surfaces closely strigillose or hirtellous to ± villous; petiolar bases 0–3+ cm long. **Inflorescences** capitula; peduncle 3–10(–20) cm long; phyllaries 18–28+, narrowly triangular- to linear-attenuate, 0.6–1.4+ cm long, usually ciliate with jointed hairs; receptacular setae 1.5–3 mm long. **Florets** *ray* usually 8–14, rarely absent, corollas usually reddish to purplish proximally, yellow to orange distally, rarely yellow, reddish, or purplish throughout, 1.3–3+ cm long; *disk florets* 40–100+, corollas yellowish to purple or brown, often bicolored, tubes 0.8–1.2 mm long, throats campanulate to urceolate, 3–4 mm long. **Fruits** cypselae, obpyramidal, 2–2.5 mm long, hairs 1.5–2 mm long, inserted at

Above: *Gaillardia pulchella, Gynura aurantiaca, Helianthus annuus, Melampodium divaricatum.*

bases and on angles. **Distr:** Tropical N America to Mexico. **Ecol:** In disturbed places; 0–1,800 m. Fl. throughout the year. **Zones:** 8–12.

Gynura

Species ~40. **Herbs** succulent, rarely undershrubs, glabrous or hispid. **Leaves** alternate; lamina with margins entire, toothed or pinnatisect. **Inflorescences** heads, solitary or in corymbs, bracteolate at base, homogamous, discoid, yellow or purplish; involucre cylindric or subcampanulate, calyculate at base; bracts 10–12, 1-seriate, narrow, equal, with scarious margins; receptacle flat, pitted or shortly fimbrillate. **Florets** all bisexual, fertile, tubular, limb 5-toothed; anther base entire or subauriculate; style arms slender, tips long, subulate, hispid. **Fruits** achenes, narrow, many-ribbed; pappus copious, bristles slender, white. **Distr:** Tropical Africa and Asia, Madagascar and N Australia. **Cult:** Propagate from cuttings. Requires a moist, well-drained soil in full or partial sun.

G. aurantiaca (Blume) de Candolle, *Prodr.* 6: 300, 1838. (Syn: *Cacalia aurantiaca*). [Velvet Plant]. **Herbs**, succulent, base woody, to 90 cm high; stems somewhat lax or clambering, branching, fleshy, conspicuously fuzzy-hairy (± velutinous to villous, crosswalls of hairs purplish). **Leaves** lanceolate, 4–10(–15) × 2–5(–8) cm, margins raggedly toothed, surfaces ± velutinous to villous, hairs purplish; petiole 0–3+ cm long. **Inflorescences** capitula, 1–5; calyculi with bractlets 3–6+ mm long; phyllaries 1–1.2+ cm long, tips greenish or minutely darkened. **Florets** *disk* with corollas yellow or orange to brick-red, drying to purplish, 0.8–1.2+ cm long, exceeding the involucral bracts, style-branch appendages orange, 1–2 mm long. **Fruits** cypselae, glabrous, pappi 1–1.2 cm long. **Distr:** SE Asia. **Ecol:** In shaded, damp sites; 0–50 m. **Zones:** 10–12. **Info:** The flowers are malodorous, and in cultivation are often cut before they open, as the plants are mostly grown for their attractive foliage.

Helianthus

Species ~60. **Herbs**, annual or perennial; stems erect or ascending to decumbent or procumbent, usually branched distally. **Leaves** basal and/or cauline, opposite below, often alternate above, sessile or petiolate; lamina simple, mostly deltate, lanceolate, linear, or ovate, base cordate to narrowly cuneate, margins usually entire, serrate or rarely lobed, glabrous or hairy, often gland-dotted. **Inflorescences** heads, large, terminal and solitary or few in a cyme, pedunculate; involucral bracts herbaceous, triseriate; phyllaries persistent, in 2–3+ series, subequal to unequal; receptacle flat or convex, with non-persistent scales. **Florets** *ray* usually 5–30, rarely absent, sterile, corollas usually yellow; *disk florets* numerous, bisexual, fertile, corollas yellow or reddish, tubes short, lobes 5, triangular. **Fruits** achenes, usually purplish-black, sometimes mottled, ± obpyramidal; pappi absent or of 2 or more non-persistent scalelike awns. **Distr:** N America, Mexico. **Cult:** Propagate annuals from seeds, and perennials by division or from basal cuttings. Requires a fertile, moist, well-drained soil in full sun.

H. annuus L., *Sp. Pl.* 904, 1753. (Syn: *H. aridus*, etc.). [Sunflower]. **Herbs**, annual or biennial, with a taproot, 1–3 m high; stems erect, usually hispid, branching apical. **Leaves** mostly cauline, alternate; lamina ovate to ± cordate, 10–40 × 5–40 cm, base cuneate to ± cordate, margins serrate, abaxially usually ± hispid, sometimes gland-dotted; petioles 2–20 cm long. **Inflorescences** heads, 1–9; peduncles 2–20 cm long; involucres hemispheric or broader, 1.5–4(–20) cm Ø; phyllaries 20–30(–100), ovate to lance-ovate, 1.3–2.5 × (0.3–)0.5–0.8 cm. **Florets** *ray* (13–)17–30(–100+), laminas 2.5–5 cm long; *disk florets* 150–1000, corollas 5–8 mm, throats bulbous at bases, lobes usually reddish, sometimes yellow, anthers brownish to black, appendages yellow or dark. **Fruits** achenes, (0.3–)0.4–0.5(–1.5) cm long, glabrate; pappus of 2 lanceolate scales. **Distr:** N America. **Ecol:** Fl. Jul–Sep. Fr. Aug–Oct. **Zones:** 4–12.

Melampodium

Species ~36. **Annuals, perennials, or subshrubs,** to 0.5(–1.5+) m. **Stems** erect or prostrate. **Leaves** cauline, opposite, petiolate (petiole usually winged) or sessile; lamina deltate, lance-elliptic, lanceolate, lance-linear, linear, linear-oblong, ovate, or rhombic, sometimes pinnately lobed, ultimate margins entire or toothed, faces usually hairy, usually gland-dotted (at least abaxial). **Inflorescences** heads, radiate, borne singly (from forks of branches, peduncles often uncinate); involucres mostly hemispheric (sometimes ± cupulate), to 2(–3+) cm Ø; phyllaries persistent (outer) or falling with cypselae, 8–20+ in 2 series, outer (2–)5 often ± connate, herbaceous, inner each investing a ray ovary; receptacles flat or convex to conic; paleae lanceolate to linear, scarious, conduplicate, often with dilated, erose tips. **Florets** *ray* (3–)5–13+, ♀, fertile, corollas yellow, orange, or cream-white, sometimes purplish abaxially (tubes often wanting); *disk florets* (3–)45–70(–100+), functionally ♂, corollas ochroleucous, yellow, or orange, tubes shorter than funnelform throats, lobes 5, ± deltate. **Fruits** cypselae; pappi absent. **Distr:** USA, Mexico, West Indies, C and S America. **Cult:** Propagate from seeds. Requires a moist, well-drained soil in full sun.

M. divaricatum (Richard) de Candolle, *Prodr.* 5: 520, 1836. (Syn: *Dysodium divaricatum, M. paludosum*). [Butter Daisy]. **Herbs**, annual, to 0.2–1 m high; stems with 2 lines of pubescence on opposite sides, otherwise glabrous. **Leaves** broadly elliptic or ovate; lamina 3–15 × 1–9.5 cm, base attenuate, margins ± coarsely and irregularly sinuate-serrate, teeth rounded, apex acute, surfaces sparsely hispid, trinerved at base; petiole winged, 0.5–3 cm long, wing margins pubescent. **Inflorescences** capitula, campanulate, 2.5 cm Ø; outer phyllaries broadly ovate, ~5 × 4.5 mm, slightly accrescent in fruit, lower third connate, margins ciliate. **Florets** *ray* 8–13, yellow, basal tube minute, ligules elliptic, 6–8 × 3–4 mm; *disk florets* yellow, 4–70+, 3–3.5 mm long, turning brown with age, receptacular scales oblong, ~3 mm long, margin finely lacerate, apex rounded, flattened. **Fruits** achenes of ray florets (enclosed in inner phyllaries) trigonous, with a short point on posterior angle, top flattened, 3–3.5 × 2–2.5 mm, sides wrinkled. **Distr:** C America. **Ecol:** In disturbed areas. Fl. throughout the year. **Zones:** 10–12.

Montanoa

Species 50. **Shrubs, trees** or **vines**; stems minutely pubescent on herbaceous parts. **Leaves** opposite, petiolate; lamina with margins entire to deeply 3–5-lobed or coarsely pinnatifid, wings and auricles present or absent, main veins 3. **Inflorescences** heads, solitary or often in compound corymbs or panicles; involucral bracts ± biseriate, veins 3; receptacle convex or conic, scales persistent, often spine-tipped, enclosing and surpassing the achenes. **Florets** *ray* in 1 row or rarely absent, sterile, ligulate; *disk florets* few or numerous, fertile, ± tubular, 5-toothed; anthers auriculate at base, apical appendages acute; style branches linear, densely papillose-pilose. **Fruits** achenes, brownish-black, obovoid or 4-angled, usually compressed, smooth, glabrous; pappus a ring-shaped collar. **Distr:** C and S America. **Cult:** Propagate from seeds or root cuttings. Requires a fertile, moist, well-drained soil in full or partial sun.

M. bipinnatifida (Kunth) K. Koch, *Woch. Vereines Beförd. Gart. Königl.* 7: 407, 1864. (Syn: *Uhdea bipinnatifida*). [Mexican Tree Daisy]. **Shrubs**, large, often treelike, ~2–6 m high, evergreen; stems brittle. **Leaves** sparse; lamina pinnatifid or deeply dissected, mature leaves usually palmately lobed, 12–30 × 9–40 cm, lobes ovate to ± lanceolate, dark green, minutely pubescent; petiole 4–20 cm long, with 2 auricles and sometimes with a thin lamina margin. **Inflorescences** solitary heads, pendent, several to many in compound oppositely branched corymbs; peduncles 2–7 cm long, hairy; involucral bracts reflexed in fruit, biseriate, ± equal, ± ovate to lanceolate. **Florets** *ray* white, ligules 2.5–3.5 cm long, acute to 2-notched; *disk florets* green turning yellow with age. **Fruits** achenes, 3.5 × 2 mm. **Distr:** S Mexico. **Ecol:** In scrubs and forests by streams; to 1,750 m. Fl. Aug–Dec. **Zones:** 10–12.

Pseudogynoxys

Species 14. **Vines**, twining, herbaceous; stems striate, puberulous. **Leaves** alternate, petiolate; lamina

RF

Above: *Montanoa bipinnatifida, Pseudogynoxys chenopodioides, Senecio serpens, Sphagneticola trilobata.*

simple, margins entire or serrate. **Inflorescences** solitary or in terminal cymes; involucre campanulate or hemispherical, canaliculate; phyllaries uniseriate, oblong; capitula heterogamous, radiate, long-pedunculate, with numerous flowers. **Florets** *disk* bisexual, corollas actinomorphic, yellow, long-tubular, 5-lobed, stamens 5, anthers connate, sagittate, ovary inferior, style filiform, with 2 hirsute stigmatic branches on distal portion; *ray florets* ♀, corollas zygomorphic, ligulate, ligules orange. **Fruits** achenes, cylindrical, pappus of numerous bristles, without scales. **Distr:** Neotropics. **Cult:** Propagate from seeds, cuttings or runners. Requires a fertile, well-drained soil in full sun.

P. chenopodioides (Kunth) Cabrera, *Brittonia* 7: 54, 1950. (Syn: *Senecio chenopodioides, S. confusus*). [Mexican Flame Vine]. **Vines**, herbaceous, twining, 10–12 m long; stems striate, subcylindrical, glabrous or puberulous. **Leaves** alternate; lamina lanceolate, 5–8 × 2.5–5 cm, base truncate, obtuse, or slightly cordiform, frequently unequal, margins dentate-mucronate, apex acuminate or acute, membranous, adaxially dull, glabrous, abaxially dull, glabrous, with prominent venation; petiole 1.5–2.2 cm long. **Inflorescences** corymbiform cymes, terminal, capitula 2–6; peduncles 2–5 cm long, puberulous; involucre crateriform, ~6 mm long, phyllaries green, lanceolate, ~4 mm long. **Florets** *disk* with yellow tubular corollas, 0.9–1 cm long, stigmatic branches yellow; *ray florets* with corollas orange, ligulate, ligule elliptical, retuse at apex, 2–2.5 cm long. **Fruits** achenes, turbinate, hispidulous, ~4 mm long, pappus of numerous white bristles, 5–7 mm long, scabrous. **Distr:** C America. **Ecol:** In both moist and dry areas; 0–500 m. Fl. & Fr. Jan–Apr. **Zones:** 9–12.

Senecio

Species ~1,000. **Herbs** annual or perennial, **subshrubs**, **shrubs**, **climbers** or **small trees**, highly variable in habit and foliage; stems and leaves sometimes succulent. **Leaves** usually alternate or spiraled, without stipules, deciduous or evergreen. **Inflorescences** heads, radiate or discoid; involucre generally calyculate. **Florets** *ray* absent, or present and then ♀ only, yellow; *disk florets* bisexual, white, yellow, red or purple; filament collar dilated, anthers not caudate; style in disk florets with stigma truncate and mostly penicillate, or conical, with a crown of hairs, without appendage, with discrete stigmatic lines. **Fruits** achenes, terete to ellipsoid-obovoid, ribbed, glabrous or pubescent, sometimes with myxogenic hairs; pappus of many fine bristles, multiseriate, caducous. **Distr:** Worldwide, with the highest diversity in S America and Africa. **Cult:** Propagate from seeds or cuttings. Requires a moderately fertile, moist, well-drained soil in full sun.

S. serpens G.D. Rowley, *Nat. Cact. Succ. J.* 10(2): 31, 1955. (Syn: *Cacalia repens*, etc.). [Blue Chalksticks]. **Herbs**, evergreen, spreading and mat-forming, to 30 cm high, spreading to 60 cm Ø, bluish-white pruinose, succulent; stems creeping, suckering and rooting; branches fleshy, ascending at first, 5–7 mm Ø. **Leaves** oblong to oblanceolate, narrow, 3–5 × 0.7–0.8 cm, ± channeled above, apex obtuse with a brown mucro, with a brown window above and 10–19 obscure striations, thick, fleshy, succulent, glaucous with a powdery white coating. **Inflorescences** heads, 5–7, subcorymbose, terminal, 20–30 cm long, sparsely branched; phyllaries 10–14. **Florets** *ray* absent; *disk florets* 15–30, white, corolla lobes curved. **Distr:** S Africa (Cape Province). **Ecol:** On mountains, often on slopes and among rocks; <1,000 m. **Zones:** 9–12.

Sphagneticola

Species 4. **Herbs**, perennial, mostly 0.3–2 m high; stems prostrate, branched ± throughout, sometimes rooting at nodes, usually ± succulent. **Leaves** cauline, opposite, sessile or petiolate; lamina mostly trullate to lanceolate, often obscurely trilobed, bases ± cuneate, margins entire or toothed, usually trinerved, usually scabrous. **Inflorescences** heads, radiate, borne singly; involucres obconic, 0.6–1.2 cm Ø; phyllaries persistent, 12–15 in 2–3 series (outer larger, more foliaceous than inner); receptacles convex to conic; paleae conduplicate, scarious, acute. **Florets** *ray* 4–10+, ♀, fertile, corollas yellow to orange; *disk florets* 30–60+, bisexual, fertile, corollas yellow to orange, tubes shorter than campanulate throats, lobes 5, lance-triangular. **Fruits** cypselae, strongly biconvex to plumply 3-4-angled; pappi absent or persistent. **Distr:** Mexico, West Indies, C and S America; Asia, Indian Ocean islands, Pacific Islands, Australia. **Cult:** Propagate from cuttings or seeds. Requires a moist, well-drained soil in full or partial sun.

S. trilobata (L.) Pruski, *Mem. New York Bot. Gard.* 87: 114, 1996. (Syn: *Wedelia trilobata*, etc.). [Bay Biscayne Creeping-oxeye, Singapore Daisy]. **Herbs**, perennial, procumbent; stems stout, glabrous or pubescent, seldom scabrous. **Leaves** opposite; lamina elliptic or lanceolate, to 5–18 × ~1–5 cm, base cuneate, margins often with triangular lobes and conspicuous marginal teeth, apex acute, somewhat succulent, glabrous or sparsely pubescent, sometimes scabrous; petiole distinct <5 mm long. **Inflorescences** capitula, solitary on elongate peduncles, radiate; involucre green; bracts lanceolate, ciliate, obscurely nerved, 1–1.5 cm long, innermost bracts narrower. **Florets** *ray* 4–8, showy, yellow, 1.5–2 cm long, 3 or 4 denticulate, fertile; *disk florets* yellow, numerous, ~2 cm long, corolla 5–6 mm long. **Fruits** achenes, blackish, sometimes mottled, clavate, angled, 3–5 mm long, pappus of fused scales, coroniform. **Distr:** Tropical and subtropical America; widely cultivated and naturalized in the Old World tropics. **Ecol:** In moist places, forest margins, hillsides, weedy; 0–700 m. Fl. throughout the year. **Zones:** 10–12.

Stifftia

Species 6. **Trees** small, or **vines** to 30 m long, moderately branched. **Leaves** alternate; lamina elliptic or oblong to oblanceolate, 5–16 cm long, margins entire, apex obtuse to shortly apiculate, herbaceous to subcoriaceous, glabrous to sparsely arachnoid-pilose, venation pinnate; petiole short, slender, to ~1 cm long. **Inflorescences** 1 to many heads, terminal on leafy branches; flowers 1–40 per head, homogamous; involucre subimbricate with unequal persistent bracts; receptacle plane, glabrous. **Florets** with corollas whitish, subcarnose, actinomorphic, with narrow basal tubes, throats nearly lacking to elongate, lobes 5, linear, tightly rolling backward at maturity; anthers with long sterile tails, apical appendage oblong-lanceolate; style glabrous, 2-veined, style branches short with rounded to shortly acute tips, smoothly adaxially. **Fruits** achenes, cylindrical; carpopodium annuliform to shortly cylindrical; pappus showy, whitish, yellowish, or reddish, of ~100 capillary bristles in ~4–5 series, of unequal lengths, inner bristles longer. **Distr:** NE South America. **Cult:** Propagate from seeds, cuttings or by air-layering. Requires a fertile, moist, well-drained soil in full or partial sun.

S. chrysantha Mikan, *Del. Fl. Faun. Bras.* i. 1, t. 1., 1820. (Syn: *Aristomenia fruticosa*, etc.). [Golden-flowered Stifftia, Orange Powderpuff]. **Shrubs**, robust, 2–4 m high; stems woody, to ~12 cm Ø. **Leaves** alternate, exstipulate; lamina lanceolate, 6–15 × 2–4.5 cm, base obtuse to acute, margins entire, apex acuminate, adaxially smooth and glossy, primary veins anastomosing at margins; petiole 5 cm long, articulated with stem. **Inflorescences** capitula, solitary or rarely 2–3, terminal; florets usually 30–40, rarely 10–15; peduncles short, thickened apically, with small scales; involucre somewhat turbinate, coriaceous, scales 30–40, imbricate, arranged in several series, closely appressed before anthesis, spreading after corollas are shed; receptacle with milky juicy, nearly flat. **Florets** with corollas smooth, tubular, 4.5 cm long, pale orange below, becoming darker above, limb divided into 5 narrow, revolute circinnate segments, 5 mm long (when spread); filaments smooth, colored, inserted into upper part of corolla tube, alternating with lobes, anthers bilobed, much-exserted, tails not or scarcely papillose; ovary green, triangular, ~2 cm long, beak short, yellowish, style cylindrical, exserted ~2.5 cm beyond corolla, stigma bifid, pappus reaching near upper part of corolla tube, in several rows. **Distr:** Brazil. **Ecol:** In forests and forest margins. Fl. Mar–Aug. Fr. Sep–Nov. **Zones:** 10–12.

Tagetes

Species ~40. **Herbs** annual or perennial, **subshrubs**, or **shrubs**, mostly 0.1–0.8(–2) m high; stems erect,

Above: *Stifftia chrysantha*, *Tagetes erecta*, *Tithonia diversifolia*, *Tithonia rotundifolia*.

branched distally or ± throughout. **Leaves** cauline, mostly opposite (distal sometimes alternate), petiolate or sessile; lamina lanceolate to oblanceolate overall, usually 1–3-pinnately lobed or 1–3-pinnatisect, ultimate margins toothed or entire. **Inflorescences** heads, radiate or discoid, solitary or in ± corymbiform arrays; involucres narrowly cylindric or fusiform to turbinate or broadly campanulate, 0.1–1.2+ cm Ø; phyllaries persistent, 3–21+, 1–biseriate; receptacles convex to conic, smooth or finely pitted; paleae absent. **Florets** *ray* 0 or 1–8(–13+), ♀, fertile, corollas yellow or orange, red-brown, or white; *disk florets* 6–120+, bisexual, fertile, corollas greenish-yellow to orange, sometimes tipped with red or red-brown, tubes ± equal to the funnelform throats, lobes 5, deltate to lance-linear. **Fruits** cypselae, narrowly obpyramidal or fusiform-terete, glabrous or hairy; pappi persistent, of 2–5(–10) dissimilar, distinct or connate scales in ± 1-seriate. **Distr:** Tropical and warm-temperate America, especially Mexico. **Cult:** Propagate from seeds. Requires a fertile, well-drained soil in full sun.

T. erecta L., *Sp. Pl.* 2: 887, 1753. (Syn: *T. patula*, *T. tenuifolia*). [African Marigold, French Marigold]. **Herbs**, annuals, 0.1–1.2+ m high. **Leaves** pinnate, 3–12(–25+) cm long; principal lobes or leaflets 9–25, lanceolate to linear-lanceolate, 1.5–2.5(–4.5+) × 0.3–0.8(–1.2+) cm. **Inflorescences** capitula, borne ± singly; peduncles 3–10(–15+) cm long; involucres 1–2.2+ × (0.3–)0.5–1.2+ cm. **Florets** *ray* (3–)5–8+, yellow to orange, red-brown, white or sometimes multicolored, ± flabellate to oval-quadrate, (0.2–)1.2–1.8(–2.5+) cm; *disk florets* (10–)50–120+, corollas 0.7–1.2+ cm long. **Fruits** cypselae, 0.6–1.1 cm long, pappi of 0–2, ± subulate-aristate scales 0.6–1.2+ cm long, with 2–4 distinct or connate, linear-oblong, ± erose scales 2–6+ mm long. **Distr:** Tropical N America to Mexico; cultivated widely throughout the tropics and subtropics. **Ecol:** In disturbed places; 0–1,000+ m. Fl. throughout most of the year. **Zones:** 9–12.

Tithonia

Species 11. **Herbs** annual or perennial, **subshrubs**, or **shrubs**, rarely small trees, 0.7–5(–7) m high; stems erect, branched. **Leaves** alternate, opposite at stem base; lamina simple or lobed. **Inflorescences** heads, solitary or few together, radiate; peduncle apically broadened, hollow; involucral bracts in 2 to 5 rows; paleae rigid, acuminate to aristate, usually lobed. **Florets** *ray* 8–30, sterile, corollas yellow or orange; *disk florets* 40–120(–200+), bisexual, fertile, corollas yellow, tubes shorter than throats (bases of throats bulbous and hairy), lobes 5, ± triangular, anthers black or brownish, style branches relatively slender. **Fruits** achenes, oblong or obovate, ± 3- or 4-angled; pappus of connate scales, sometimes of 1 or 2 awns, or lacking. **Distr:** N Mexico to Panama. **Cult:** Propagate from seeds or cuttings. Requires a fertile, moist, well-drained soil in full sun.

T. diversifolia (Hemsl.) A. Gray, *Proc. Amer. Acad. Arts.* 19: 5, 1883. (Syn: *Mirasolia diversifolia*). [Mexican Sunflower, Tree Marigold]. **Herbs** perennial, **subshrubs**, or **shrubs**, to 2.5(–5) m high. **Leaves** alternate; lamina ± deltate to pentagonal, 3- or 5-lobed, 7–33 × 7–22 cm, base cuneate, lobes acute with broad, rounded sinuses, margins often crenate, apex acuminate or acute; petiole 2–6(–15) cm long. **Inflorescences** heads, solitary, radiate, ~10(–12.5) cm Ø; peduncles 7–24 cm long; involucre 2.4–4 cm Ø; bracts 3- or 4-seriate, imbricate, 1–2 cm long; phyllaries 16–28, (3–)4-seriate, oblong to ovate, *outer* 0.6–1 × 0.4–0.7 cm, rounded to acute, *inner* 1–2 × 0.3–1 cm, rounded to acute; paleae 1–1.3 × 0.2–0.3 cm. **Florets** *ray* 7–14, 1-seriate, ligules elliptic, 4–6 cm long, apically 2- or 3-dentate, abaxially puberulent, corollas yellow, laminas linear, 4.8–6.9 × 0.9–1.6 cm; *disk florets* 80–120+, ~1.5 cm long, corolla tubular, lobes yellow. **Fruits** achenes, dark brown, triquetrous, 4–6 mm long, pappus of 2 sturdy 4–5 mm long awns and several shorter. **Distr:** Mexico, C America. **Ecol:** In disturbed secondary forests; 0–1,000 m. **Zones:** 10–12.

T. rotundifolia (Mill.) Blake, *Contr. Gray Herb.* 52: 41, 1917. (Syn: *Tagetes rotundifolia*, etc.). [Mexican Sunflower, Red Sunflower]. **Herbs**, annual, erect, 1–2(–4) m high; stems initially densely hairy. **Leaves** alternate; lamina simple, sometimes trilobate, broadly ovate to triangular-ovate, mostly 7–38 × 7–30 cm, base cuneate, sometimes subtruncate, contracted and decurrent on petiole, margins serrate, apex acuminate, thin, hispid-pilose, abaxially glandular-punctate; petiole 2–7 cm long. **Inflorescences** heads; peduncles 11–27 cm long; involucres 2–3 cm wide; phyllaries 14–21, biseriate, 1.5–2.5 cm long, subequal or graduate, *outer* ones lance-oblong to ovate-oblong, acute or acuminate, finely pilosulous, apex herbaceous, often reflexed, *inner* ones similar but usually shorter; paleae 1.15–1.5 × 0.2–0.3 cm. **Florets** *ray* 8–13, ligules golden-yellow or orange-yellow, 2–3 cm long; *disk florets* 60–90, yellow, corollas ~9 mm long; paleae acuminate to cuspidate, 1.2–1.8 cm long. **Fruits** achenes, 6–7 mm long. **Distr:** Mexico, British Honduras to El Salvador and Panama. **Ecol:** On rocky or grassy slopes; 120–1,500 m. Fl. Dec–Jun. **Zones:** 9–12.

Tridax

Species ~30. **Herbs**, annual or perennial. **Leaves** opposite; lamina incised-toothed or pinnatisect, segments narrow. **Inflorescences** heads, long-peduncled, heterogamous, radiate; involucre ovoid to campanulate; bracts various, inner ones broader than outer; receptacle flat or concave, paleae membranous. **Florets** *ray* florets few, fertile, corolla with a slender glabrous tube and broad yellow or pale ligule, or 2-lipped with outer lip large, trilobed and inner lip small, bilobed or unlobed; *disk florets* bisexual, fertile, corolla tube narrowly funnelform, tube short, limb elongate, 5-lobed, anthers linear, base short auriculate, style branches of disk florets hairy apically, tips subulate. **Fruits** achenes, turbinate or oblong, much shorter than corolla; pappus of many short or long, aristate, feathery bristles. **Distr:** Tropical America. **Cult:** Propagate from seeds. Requires a well-drained soil in full sun.

T. procumbens L., *Sp. Pl.* 900, 1753. [Coat Buttons, Tridax Daisy]. **Herbs**, perennial, caulescent, decumbent; stems procumbent, branched at base; branches slender, spreading or ascending, 20–50 cm long, hirsute. **Leaves** few, shortly petiolate; lamina ovate to ovate-lanceolate, 2.5–5 cm long, base cuneate, margins deeply irregularly serrate, pinnatisect, segments few, narrow, apex acute or acuminate. **Inflorescences** capitula, solitary, 1–1.5 cm Ø; peduncle 10–30 cm long; involucre subcampanulate, 6–7 mm long; involucral bracts hispid, in few series; outer bracts densely grayish-white, elliptic, ~5 mm long, long-hirsute; inner bracts purplish-tinged, narrower, ~6 mm long. **Florets** *ray* 4, white, ligulate or 2-lipped; *disk florets* yellow, tube 5 mm long, limb 5-lobed, lobes reflexed, hairy. **Fruits** achenes, brown, oblong, ~2 mm long, densely silky-pubescent, pappus of 5–6 mm long shiny feathery bristles. **Distr:** Tropical America; naturalized widely. **Ecol:** In lowland regions; 0–500 m. **Zones:** 10–12.

Vernonia

Species ~1,000. **Subshrubs** or **herbs**, erect, branching, rarely scandent. **Leaves** alternate; lamina pubescent or glabrous. **Inflorescences** heads, discoid, in corymbs or sometimes panicles; involucre campanulate or subglobose; bracts imbricate, many-seriate, outer series smaller, sometimes spreading, acute, acuminate, awn-tipped, or obtuse; receptacle alveolate denticulate, rarely with crisped hairs. **Florets** pink, purple or blue, tubular, fertile; corolla regular, 5-cleft; anthers obtuse or sagittate at base, auricles connate; style arms linear, subacute, puberulent. **Fruits** achenes, 8–10-ribbed, truncate, callose at base, glabrous or pubescent; pappus bristles numerous, 1- or biseriate, inner row capillary, deciduous or persistent, outer row very short. **Distr:** Warmer regions of America, Asia and Africa. **Cult:** Propagate from seeds. Requires a well-drained soil in full sun.

V. zeylanica (L.) Less., *Linnaea* 4 (1): 344, 1829. [Pupula]. **Shrubs**, usually erect or scrambling, occasionally prostrate (0.5–)1–2.5 m high; young stems densely pubescent or tomentose, later glabrescent. **Leaves** pandurate or ovate, 5–8 × 2–3 cm, base attenuate, margins crenate-dentate, apex acute or obtuse, surfaces whitish-pubescent, sparsely so adaxially and later glabrescent, abaxially

TG

Above: *Tridax procumbens, Vernonia zeylanica, Xanthium indicum, Auranticarpa rhombifolia.*

tomentose; petiole 0.5–1 cm long, base auriculate. **Inflorescences** capitula, campanulate, numerous in terminal corymbs; involucres 3–4 × 2–3 mm, 4-seriate; phyllaries ovate, acute, pubescent and ciliate at apex. **Flowers** 6–8 per head; corollas 5–6 mm long, pale mauve or violet, glandular, otherwise glabrous. **Fruits** achenes, 2–2.5 mm long, brown, 4–5 ribbed, glandular and puberulous; pappus dirty-white, 4–5 mm long with a ring of short outer hairs ~0.5 mm long. **Distr:** Sri Lanka. **Ecol:** In montane grasslands and on dunes bordering forest scrubs; 0–900 m. Fl. Jun–Apr. **Zones:** 10–12.

Xanthium

Species 2–3. **Herbs**, hispid or pubescent, annual; stems generally erect, branched, elsewhere sometimes spiny. **Leaves** alternate, petiolate; lamina entire or variously lobed. **Inflorescences** capitula, discoid, unisexual, monoecious; ♂ involucres multiflorous, subsessile, borne at branches apices, phyllaries numerous, free, receptacle conical or cylindrical, paleaceous; ♀ involucres usually 2-flowered, sessile, axillary in lower parts of branches, outer phyllaries free, inner phyllaries connate, covered with uncinate spines and bearing 2 larger spines at apex. **Flowers** ♂ 5-lobed, anthers strongly exserted, rounded at base, styles none or rudimentary, pappus absent; ♀ with corollas absent or rudimentary, style branches reaching almost to apex of terminal spines. **Fruits** achenes, 2 per each head, included in the hardened spiny involucre at maturity, pappus absent. **Distr:** Neotropical. **Cult:** Propagate from seeds. Requires a moist, well-drained soil in full sun.

X. indicum Koenigin ex. Roxb., *Fl. Ind.* 3: 601, 1832. (Syn: *X. orientale, X. strumarium*). [Cocklebur]. **Herbs**; stems erect, 1–1.5 m high, hispidulous or glabrescent. **Leaves** broadly ovate, 3–15 × 2.5–15 cm, 3–5-lobed, base rounded or cordate, margins ± irregularly serrate, apex acute or acuminate, surfaces hispidulous, abaxially glandular; petioles 2–12 cm long. **Inflorescences** ♂ capitula globose, ~5 mm ∅, phyllaries 1-biseriate, lanceolate, 2–2.5 mm long, pubescent, corollas ~2.5 mm long, stamens exserted 1 mm beyond corolla mouth; paleae oblong, obtuse, 2.5 mm long, pubescent near apex; ♀ capitula ovoid, 5–6 mm long at flowering time, covered with spines and glands, flowers 2, each consisting of a narrowly ovoid ovary, 2–2.5 mm long, and style 2.5–4 mm long, divided almost to base. **Fruiting** heads 1.5–1.8 cm long including rostra, 5–6(–9) mm broad excluding spines, spines ~3 mm long, glandular and hispidulous, rostra thick, divergent, hooked at apex, about as long as spines. **Fruits** achenes, narrowly ovoid or elliptic, 1.3–1.5 × 0.3–0.4 cm, grayish-black, smooth, tapering to a fine point at apex. **Distr:** Neotropical; naturalized tropical Asia. **Ecol:** In moist places. Fl. Aug–Mar. **Zones:** 10–12.

APIALES

FAMILIES 7, GENERA 494, SPECIES 5,489.

Trees, **shrubs**, **woody lianas**, or **herbs**, dioecious, glabrous or with various types of trichomes. **Leaves** pinnatinerved (margins toothed or lobed). **Inflorescences** paniculate, terminal; pedicels articulated. **Flowers** small; calyx small; corolla apparently free; stamens free from corolla; ovary triloculed, abaxial carpel fertile, ovules 1–2 per carpel, apical, apotropous. **Fruits** drupes, single-seeded (Stevens 2001 onwards). **Distr:** Worldwide.

PITTOSPORACEAE

Genera 9, **species** ~250. **Trees** or **shrubs**, evergreen, glabrous or pubescent, occasionally spiny. **Leaves** alternate, occasionally opposite, estipulate; lamina with margins entire, rarely dentate or lobed, mostly leathery. **Inflorescences** umbellate, corymbose, paniculate, or flowers solitary, bracteate and bracteolate. **Flowers** usually bisexual, sometimes polygamous, actinomorphic, rarely zygomorphic, usually 5-merous (except ovary); sepals usually free or slightly connate; petals free or connate, white, yellow, blue, or red; stamens opposite sepals, filaments filiform, anthers basifixed or dorsifixed, biloculed, dehiscing longitudinally or by pores; ovary superior, of 2 or 3(–5) carpels, usually 1-loculed or incompletely 2–5-loculed, ovules numerous, anatropous, style short, simple or 2–5-lobed, persistent or deciduous. **Fruits** capsules, dehiscing by adaxial suture, or berries; seeds numerous. **Distr:** Tropical and subtropical regions of Africa, Asia, Australia, and Pacific Islands, Australia.

Auranticarpa

Species 6. **Trees**, small to medium, slender, 5–30 m tall; young shoots usually white, sometimes green-brown, glabrous. **Leaves** usually alternate or appearing whorled at ends of branchlets; lamina elliptic, obovate or angular obovate to rhombic, 6.5–14 × 1.5–3.5 cm, margins entire and undulate, or slightly crenulate to angular-toothed; petiole 1–3.5 cm long. **Inflorescences** terminal, condensed, corymbose, much-branched. **Flowers** bisexual; sepals usually 5, yellow-green, ovate, 1–3 mm long, margin sparsely hairy; petals linear-spathulate, 5–9 mm long, margin recurved, apex slightly recurved, cream to whitish, mostly glabrous; stamens with filaments widened at base, tapering to a very narrow connective, anthers versatile, sagittate, 2–3 mm long, yellow-orange; ovary globose, 2 or rarely more locules, style stout, stigma insignificant. **Fruits** pear-shaped to tetragonal, 0.6–1.6 × 0.6–2 cm, stipe 2–3 mm, orange-red, unequally 2(3)-loculed; seeds solitary or few. **Distr:** N Australia. **Cult:** Propagate from seeds. Requires a moderately fertile, moist, well-drained soil in full or partial sun.

A. rhombifolia (A.Cunn. ex Hook.) L.Cayzer, Crisp and I.Telford, *Austr. Syst. Bot.* 13(6): 907–909, 2000. (Syn: *Pittosporum rhombifolium*). [Diamond-leaved Pittosporum]. **Trees**, small to medium, 13–30 m high, almost glabrous; canopy narrow. **Leaves** alternate or clustered at end of branches; lamina angular-obovate to rhombic, 8.6–12 × 3.5–7 cm, margins flat or up-turned, sinuate, especially distally (teeth rounded), apex rounded, with drip tip 1 cm long, glabrous; petiole 1.5–2.5 cm long. **Inflorescences** branched corymbs, densely many-flowered, terminal. **Flowers** sepals 1–2 mm long, yellow, margins minutely ciliate; petals 5–6 mm long, cream to white; ovary longitudinally ridged, hairy. **Fruits** capsules, pear-shaped, 0.5–1 × 0.5–0.8 cm, orange-yellow, smooth, ± glabrous, yellowish, stipe 2 mm long; seeds usually 1 per locule, rarely more, black. **Distr:** E Australia. **Ecol:** In rainforests. **Zones:** 9–11.

Hymenosporum

Species 1. *Generic description same as specific.* **Cult:** Propagate from seeds, cuttings or by air-layering. Requires a humus-rich, moist, well-drained soil in full sun.

H. flavum (Hook.) F.Muell., *Frag. Phyt. Austral.*, 2, 1861. (Syn: *Pittosporum flavum*). [Australian Frangipani, Sweet Shade]. **Shrubs** or **small trees**, evergreen, slender, to 12 m tall. **Leaves** often crowded twig ends; lamina obovate-oblong to oblanceolate, 7.5–15 × 3–4.5 cm, base narrow, cuneate-decurrent, margins entire, apex acute-acuminate, glabrescent; petiole 1–1.5 cm long. **Inflorescences** pseudo-umbellate corymbs, few- to many-flowered, ~10(–20) × 10 cm, terminal; peduncle 3–6 cm long; bracts foliaceous, oblong, 1–1.5 × 0.3 cm, acute; bracteoles linear or lanceolate, ~1 cm long, acute; pedicels 3–4 cm long. **Flowers** fragrant, 5-merous, opening to creamish-white, aging through yellow to golden-orange; sepals free, caducous, oblong, 5–6 × 2.5–3 mm, margin thin, apex rounded or acute; petals suffused with reddish-violet inside at base, ± basally clawed, claw linear, 2.5 cm long, tomentose outside, lamina spreading, rhomboid-ovate, ~1–1.25 × 0.75 cm, veined; stamens 5, exserted, filaments linear, anthers ovate-oblong, ~2 mm long; ovary silky-tomentose, terete, ~1.75 cm long, biloculed, ovules numerous, style ~4–5 mm long, stigma ± bilobed. **Fruits** capsules, broadly elliptic to ± orbicular, compressed, ~2.5–3 cm ∅, 2-valved; seeds flat, reniform, ~4–6 mm long, surrounded by a membranous wing 2–3 mm wide. **Distr:** Australia (Queensland to NSW), New Guinea. **Ecol:** In or near evergreen rainforests, wet sclerophyll forests and tall open forests; 0–2,000 m. Fl. Sep–Nov. **Zones:** 9–11.

Above: *Hymenosporum flavum*, *Pittosporum eugenioides*, *Pittosporum tobira*, *Cussonia paniculata*.

Pittosporum

Species ~150. **Trees**, **shrubs**, or sometimes subshrubs, evergreen, pubescent or glabrous. **Leaves** alternate, appearing opposite or pseudo-verticillate, usually clustered at ends of branchlets; lamina with margins entire, undulate-dentate, rugose, leathery or sometimes membranous. **Inflorescences** umbellate, corymbose, paniculate, or flowers solitary, terminal or axillary. **Flowers** bisexual, rarely polygamous; sepals 5, free, usually short and small; petals 5, free or partly connate; stamens 5, filament glabrous, anther dorsifixed, ± sagittate, longitudinally dehiscent; ovary superior, usually stipitate, of 2 or 3(–5) carpels, 1-loculed or incompletely 2–5-loculed, pubescent or glabrous, ovules usually numerous, sometimes 1–4, style short, simple or 2–5-lobed, usually persistent. **Fruits** capsules, ellipsoid or globose, sometimes compressed, dehiscing by 2–5 valves, pericarp woody or leathery; seeds usually surrounded by glutinous or greasy material. **Distr:** Old World tropics. **Cult:** Propagate from seeds or cuttings. Requires a moist, well-drained soil in full or partial sun.

P. eugenioides A. Cunn., *Ann. Nat. Hist.* 4: 106, 1840. (Syn: *P. elegans*, *P. microcarpum*). [Lemonwood]. **Trees**, polygamodioecious, to 12 m tall; trunk to 60 cm dbh. **Leaves** alternate, often approximate; lamina elliptic to elliptic-oblong, 5–10(–15) × 2–4 cm, base narrowed to the petiole, margins usually undulate, apex acute to subacute, subcoriaceous, glossy; petiole 1–2 cm long, slender. **Inflorescences** subcorymbose umbels, terminal, large; peduncles 1–2 cm long, hairy; flowers ♀, functionally ♀ and bisexual; pedicels ~5 mm long. **Flowers** sweetly scented, strongly honeylike, 1–1.5 cm Ø; sepals ovate to narrow-ovate, pale, membranous, ~2 mm long; petals narrowly oblong, yellowish, 5–7 mm long. **Fruits** glabrous, finely granulate, ovoid to elliptic, 5–6 mm long, 2- or 3-valved; seeds immersed in viscid pulp, ± covered by the papery endocarp at dehiscence. **Distr:** New Zealand. **Ecol:** In forest clearings, forest margins and streambanks; 0–600 m. Fl. Oct–Dec. Fr. Oct–Jan. **Zones:** 9–11. **Info:** Leaves emit a lemonlike smell when crushed.

P. tobira (Thunberg) W.T. Aiton, *Hortus Kew.* 2: 27, 1811. [Japanese Pittosporum]. **Shrubs** or **small trees**, to 6 m high; young branchlets lenticellate. **Leaves** clustered at branchlet ends, biennial; lamina obovate or obovate-lanceolate, 4–9 × 1.5–4 cm, base narrowly cuneate, margins entire, revolute, apex rounded or obtuse, usually emarginate, adaxially dark green, shiny, leathery, lateral veins 6–8 pairs; petiole ~2 cm long. **Inflorescences** umbellate or corymbose, ± terminal; bracts lanceolate, 4–5 mm long; bracteoles 2–3 mm long; pedicels 1–2 cm long. **Flowers** fragrant; sepals lanceolate, 3–4 mm long; petals free, initially white, becoming yellow with age, oblanceolate, 1–1.2 cm long; stamens dimorphic; *reduced stamens* with filament 2–3 mm long, anther nearly sterile; *fertile stamens* with filaments 5–6 mm long, anthers yellow, oblong, ~2 mm long; ovary long-ovoid, densely pubescent, placentas 3, ovules numerous, in 2 rows. **Fruits** globose, angular, ~1.2 cm Ø, dehiscing via 3 valves, ± pubescent, yellow-brown, shiny, woody; seeds numerous, red, angular, ~4 mm long. **Distr:** S Japan and S Korea. **Ecol:** Fl. Mar–May. Fr. May–Oct. **Zones:** 9–11.

ARALIACEAE

Genera 43, **species** 1,450. **Trees** or **shrubs**, sometimes woody vines with aerial roots, rarely perennial herbs, hermaphroditic, andromonoecious or dioecious. **Stipules** absent or forming a ligule or membranous border of petiole. **Leaves** alternate, rarely opposite; lamina simple and often palmately lobed, palmately compound, or 1–3-pinnately compound, usually crowded toward apices of branches; petiole with base often broad and sheathing stem. **Inflorescence** terminal or pseudo-lateral, umbellate, compound-umbellate, racemose, racemose-umbellate, or racemose-paniculate, flowers rarely solitary. **Flowers** bisexual or unisexual, actinomorphic; calyx absent or forming a low rim; corolla of (3–)5(–20) petals, mostly valvate; stamens usually as many as and alternate with petals, sometimes numerous, distinct; disk epigynous; ovary inferior, (1 or)2–10(–many)-carpellate, carpels united, ovules pendulous, 2 per locule, 1 abortive, styles as many as carpels. **Fruits** drupes or berries, terete or sometimes laterally compressed; seeds 1 per pyrene. **Distr:** Mainly tropical, few in temperate regions.

Cussonia

Species 25. **Trees** or **shrubs**, generally glabrous, rarely tomentose. **Stipules** often intrapetiolar, connate with petiole for some distance. **Leaves** petiolate; lamina simple and palmately lobed or digitately compound, often crowded, borne toward ends of stem or main branches; leaflets with margins entire to crenate, or pinnatifid, rarely irregularly lobed. **Inflorescences** spikes, racemes, umbels or panicles of umbellules; bracts subtending flowers (floral bracts) scalelike or obsolete; pedicels not articulated beneath the flower. **Flowers** greenish, 4–8 mm Ø; calyx margin repand to (4–)5-dentate or subentire; petals (4–)5; stamens (4–)5, anthers ovate; disk flat, depressed or conical; ovary bilocular, very rarely with 1–2 carpels, styles 2, very rarely 3–4, basally connivent. **Fruits** drupes, laterally compressed or subglobose, often urceolate or subobconical to wedge-shaped, exocarp fleshy or submembranous; seeds ovoid-globose or laterally compressed. **Distr:** Tropical Africa, S Africa, Madagascar and the Mascarene Islands. **Cult:** Propagate from seeds or cuttings. Requires a moist, well-drained soil in full sun.

C. paniculata Eckl. & Zeyh., *Enum. Pl. Afr. Austr. Extratrop.*: 355, 1837. [Mountain Cabbage Tree]. **Trees**, small, to 6 m tall. **Leaves** digitately compound; leaflets sessile, 7–9(13), to 27 cm long, bases long, narrowly cuneate, sometimes becoming almost petiolate, deeply pinnatifid, very variable with the sinus extending sometimes almost to midrib, lobes equally variable in width, margin sometimes with a few small teeth, apex broadly acute with a short terminal mucro, glabrous, abaxially glaucous, shiny; petiole to 50 cm long. **Inflorescences** groups of panicles, each bearing racemosely arranged spikes in subopposite pairs; main axes to 35 cm long, glabrous, finely grooved; spikes subtended by inconspicuous basal bracts and with distinct peduncles to 5 cm long; fertile parts of spikes at first clearly shorter than peduncles, later elongating and often exceeding peduncle in fruit; floral bracts minute, ± lanceolate; flowers sessile, stylopodium depressed. **Fruits** subglobose, fleshy, pealike, sometimes borne irregularly on spikes, with many infertile flowers. **Distr:** Botswana, S Africa. **Ecol:** In open wooded grasslands; <2,100 m. **Zones:** 9–12.

Fatsia

Species 2 or 3. **Shrubs** or **small trees**, evergreen, andromonoecious, unarmed. **Stipules** united with petiole, sheathed at base. **Leaves** simple, palmately lobed, margins serrate. **Inflorescences** panicles of umbels, terminal; bracts large, membranous, caducous; pedicels not articulate below ovary. **Flowers** with calyx rim obsolete or 5-toothed; petals 5, valvate; stamens 5; ovary 5- or 10-carpellate, styles 5 or 10, free. **Fruits** drupes, subglobose; seed ± laterally compressed. **Distr:** Japan, Taiwan. **Cult:** Propagate from seeds, cuttings, suckers or by air-layering. Requires a moist, well-drained soil in full or partial sun. All species are salt and pollution tolerant.

F. japonica (Thunberg) Decaisne and Planchon, *Rev. Hort.* (Paris), sér. 4. 3: 105, 1854. (Syn: *Aralia japonica*). [Fatsia, Japanese Aralia]. **Shrubs**, to ~5 m high; young branches, leaves, and inflorescences densely woolly tomentose, later glabrescent. **Leaves** suborbicular, (5–)7–9(–11) cm wide, palmately lobed, lobes 7–9 deeply cleft, narrowly ovate-elliptic, base cordate to truncate-cordate, margins crenate to crenate-serrate, teeth rounded to blunt, apex acuminate, surfaces glabrous, dark green; petiole 10–30 cm long. **Inflorescences** panicles of umbels, rounded; primary axes 20–40 cm long; peduncles 10–15 cm long; umbels 3–4 cm Ø, with numerous flowers; pedicels 1–1.5 cm long. **Flowers** creamy-white; calyx rim indistinctly denticulate; petals ovate, 3–4 mm long; ovary 5-carpellate, styles 5, free, ~1.5 mm long. **Fruits** globose, ~5 mm Ø, green, ripening to black. **Distr:** Japan. **Ecol:** Fl. Oct–Nov. Fr. Feb–May. **Zones:** 8–11.

Above: *Fatsia japonica, Osmoxylon lineare, Polyscias fruticosa, Polyscias guilfoylei.*

Osmoxylon

Species ~50. **Trees** or **shrubs**, evergreen, hermaphroditic, unarmed, glabrous or pubescent. **Stipules** forming a ligule. **Leaves** simple, palmately lobed or compound, margins entire to crenate or serrate; petiole base expanded with 1 to several spiral or transverse crests or collars. **Inflorescences** compound umbels, terminal; secondary axes trifid, each with a central head or umbel of sterile, bacciform flowers ('pseudo-fruit') and 2 lateral heads or umbels of bisexual flowers; bracts deciduous; pedicels not articulate below ovary. **Flowers** with calyx obsolete or denticulate; petals few to many, valvate, united below into a short tube; stamens 4–30; ovary (4 or)5- to many-carpellate, styles united, forming a column, stigmas pustular. **Fruits** drupes; seeds triangular. **Distr:** Borneo and Philippines east to New Guinea; few species in Micronesia and Melanesia, one species extending north to Taiwan. **Cult:** Propagate from seeds or cuttings. Requires a fertile, moist, well-drained soil in full or partial sun.

O. lineare (Merr.) Philipson, *Blumea* 23(1): 106, 1976. (Syn: *Boerlagiodendron lineare*). [Miagos Bush]. **Shrubs**, erect, ~3 m high. **Leaves** crowded at apices of branches; lamina palmately cut into 4–6 linear segments which are free to base, segments 15–20 × 1–1.5 cm, base narrowly decurrent, appear petiolulate, this portion 1–1.5 cm long, margins irregularly and slightly repand, teeth minute, distant, narrowed above to an acuminate apex, nerves numerous, spreading, glabrous, submembranous; petiole 4–6 cm long, base with 2 or 3 prominent crests. **Inflorescences** compound umbels, ~10, with few, dark-brown, fimbriate scales; peduncles 2 cm long, bracteate at base and apex, each bearing at its apex a short-peduncled central head and two lateral branches, central head composed of numerous, dark-brown, fimbriate-lacerate bracteoles (of fallen sterile flowers), lateral branches ~3 cm long, with a pair of bracts at middle, each bearing a terminal head of perfect flowers 1 cm Ø or less, these heads also with numerous dark brown lacerate-fimbriate bracteoles; flowers sessile. **Flowers** with calyx ± funnelform, truncate, ~2 × 1 mm; ovary 5-loculed. **Fruits** ovoid, 5-ridged, 5-celled, ~3 mm long. **Distr:** Philippines (Luzon). **Ecol:** Natural habitat unknown. Fl. & Fr. throughout most of the year. **Zones:** 11–12.

Polyscias

Species ~150. **Shrubs** or **trees**, evergreen, hermaphroditic, andromonoecious or dioecious, unarmed, often glabrous, some strongly aromatic. **Stipules** sometimes intrapetiolar and adnate to inside of petiole or absent. **Leaves** 1–5-pinnately compound, base sometimes sheathing and clasping, margins entire to crenate or serrate. **Inflorescences** panicles of umbels, heads or spikes, terminal, sometimes with a terminal umbel of bisexual flowers and 1 to several lateral umbels of ♂ flowers; pedicels articulate below ovary. **Flowers** with calyx undulate or with 4 or 5(–8 or more) small lobes; petals 4 or 5(–8 or more), valvate; stamens as many as petals; ovary 4 or 5(–8 or more) carpellate, styles free or rarely basally connate. **Fruits** drupes, terete or laterally flattened; seeds compressed. **Distr:** Paleotropical. **Cult:** Propagate from seeds, cuttings or by division of suckers.

P. fruticosa (L.) Harms, *Nat. Pflanzenfam.* 3(8): 45, 1894. (Syn: *Panax fruticosus*, etc.). [Black/Ming Aralia]. **Shrubs** or **treelets**, to 3(–5) m high, andromonoecious. **Leaves** 3–5-pinnate, primary leaf divisions (7–)11–15, each further divided once or twice, sometimes variegated; leaflets usually lanceolate, (1–)2–18 × 0.2–5 cm, base narrowly cuneate to attenuate, margins laciniate to spinulose-serrate, teeth 0.5–1 cm long, apex long-acuminate, papery; petiole (2–)5–15 cm long, clasping at base, with membranous wings; petiolules 1–5 cm long. **Inflorescences** umbellate panicles, terminal, erect; primary axis 8–30(–60) cm long; secondary axes 5–15, scattered or subverticillate, 7–25(–30) cm long; tertiary axes 5–15 per secondary axis, mostly grouped in 2–4 verticils, with a terminal umbellule of bisexual flowers and 2–6 lateral umbellules of ♂ flowers; pedicels 1.5–5 mm long (shorter in ♂ flowers). **Flowers** with ovary 2- or 3(or 4)-carpellate, styles free nearly to base, 0.8–1.2 mm long at anthesis, recurving, expanding in fruit to 1.5 mm. **Fruits** laterally compressed or trigonous (rarely quadrangular), orbicular to ovate-orbicular, 4–5 × 4.5–6 mm, base rounded (sometimes shallowly subcordate). **Distr:** SW Pacific Islands. **Ecol:** Fl. Aug–Sep. **Zones:** 11–12.

P. guilfoylei (W. Bull) L. H. Bailey, *Rhodora.* 18: 153, 1916. (Syn: *Aralia guilfoylei*). [Geranium Aralia]. **Shrubs** or **treelets**, to 5 m high, andromonoecious. **Leaves** 1-pinnate (irregularly decompound or 2- or 3-pinnate in some forms); leaflets (5–)7–9, often variegated, elliptic, ovate, or obovate, often broadly so, 5–20 × 2.5–12 cm, base broadly cuneate to attenuate, margins spinulose-serrate, teeth (1–)1.5–5 mm long, obtuse to broadly acute or acuminate, papery; petiole 7–18 cm long, clasping at base, alate for 1.5–3 cm with membranous wings; petiolules 1–3.5 cm long. **Inflorescences** umbellate panicles, terminal, pendent; primary axis 3–6 cm long; secondary axes 5–10, scattered toward base, upper ones forming a terminal umbel, 20–60 cm long; tertiary axes 8–12 per secondary axis, grouped in 1–3 verticils, (1.5–)3–12 cm long; peduncles 1–2 cm long, with a terminal umbel of bisexual flowers and often 1(or 2) lateral umbels of ♂ flowers; pedicels 0.4–1 cm long. **Flowers** with ovary 3 or 4(or 5)-carpellate, styles free nearly to base, recurving in fruit. **Fruits** subglobose, 4–5 mm Ø. **Distr:** SW Pacific Islands. **Ecol:** In moist forest margins. Fl. anytime during the year. **Zones:** 11–12.

P. scutellaria (N.L. Burman) Fosberg, *Occas. Pap. Univ. Hawaii.* 46: 9, 1948. (Syn: *Aralia balfouriana*, etc.). [Plum/Shield Aralia]. **Treelets** or **small trees**, 2–6(–7) m high, andromonoecious. **Leaves** 1- or bipinnate; lamina broadly elliptic to oblate or reniform, occasionally ovate or obovate, 5–20(–24) × 5–20(–26) cm, base shallowly cordate or convex, margins subentire to coarsely crenulate or shallowly serrate, teeth short, rounded, papery to subleathery; petiole 30–35 cm long, enlarged and clasping at base, alate for 1–6 cm with membranous wings; petiolules 1–5 cm long; leaflets 1, 3, or 5 (rarely 2 or 4). **Inflorescences** panicles of umbels, terminal, erect; primary axis 0.3–1.2 m long; secondary axes 15–30, in 2–4 verticils, 15–50 cm long; tertiary axes (peduncles) 7–30 per secondary axis, in irregular verticils, 0.2–1.8 cm long, with a terminal umbel bisexual and lateral umbels of ♂ flowers; pedicels 1.5–7 mm long. **Flowers** with ovary (2 or)3–5-carpellate, styles free nearly to base, 0.4–0.6 mm long at anthesis, spreading, expanding in fruit to 0.8 mm long. **Fruits** subglobose to depressed-globose (triangular to quadrangular when 3- or 4-carpellate), 4–6 mm Ø. **Distr:** SW Pacific Islands. **Ecol:** In forests; to 300 m. Fl. anytime during the year. **Zones:** 11–12.

Pseudopanax

Species ~10. **Shrubs** or **trees**, dioecious, glabrous. **Leaves** sometimes polymorphic, seedling and juvenile forms usually simple, while adult forms are very variable, usually digitately compound (these sometimes simple by reduction). **Inflorescences** simple or compound umbels, often racemosely arranged on rays. **Flowers** with calyx truncate or minutely 5-toothed; petals 5, valvate; stamens 5; ovary 5- or rarely 3–4-loculed, style-branches often connate into short a column. **Fruits** somewhat fleshy, 5- or rarely fewer-loculed, seeds one per locule. **Distr:** S America, New Zealand. **Cult:** Propagate from seeds or half-hardened cuttings. Requires a fertile, moist, well-drained soil in full or partial sun.

P. arboreus (L. f.) Philipson, *New Zealand J. Bot.* 3: 338, 1965. (Syn: *Neopanax arboreus*, etc.). [Five Finger]. **Trees**, usually dioecious., to 8 m tall, glabrous; branches stout, spreading. **Leaves** 5–7-foliolate; leaflets on petiolules ~3–5 cm long, obovate-oblong to oblong-cuneate, margins coarsely serrate-dentate, apex acute or acuminate to obtuse, thinly coriaceous, midveins and main laterals evident above and below; terminal leaflet 10–20 × 4–7 cm; petioles ~15–20 cm long, stipulately sheathing at base. **Inflorescences** umbels, terminal, compound; flowers usually completely unisexual; primary rays 8–20, to 10 cm long; secondary rays 15–20; umbellules 10–15-flowered. **Flowers** with calyx truncate or obscurely 5-toothed;

Top row: *Polyscias scutellaria, Pseudopanax arboreus, Pseudopanax ferox, Schefflera actinophylla.*
Bottom row: *Schefflera setulosa, Trevesia palmata, Daucus carota, Eryngium foetidum.*

petals 5, ovate to triangular, acute; ovary biloculed, 2-ovuled, style-branches 2, spreading. **Fruits** 5–8 mm Ø, almost black, somewhat compressed. **Distr:** New Zealand. **Ecol:** In lowland forests. **Zones:** 8–11.

P. ***ferox*** Kirk, *Gard. Chron.* i. 432, 1890. (Syn: *Panax ferox*). [Toothed Lancewood]. **Shrubs** or **trees**, to 5 m tall; trunk slender. **Leaves** of *seedlings* linear-lanceolate, margins dentate; leaves of unbranched *juveniles* narrowly linear, deflexed, to ~50 cm long, very thick and coriaceous, with closely or distantly placed broad, rounded, usually hooked lobes, ~0.75–1.5 cm wide at base, midribs very stout, raised, ~2 mm wide, lobes often crowded at leaf apex; leaves at *branching stage* becoming shorter, ascending, often more sharply lobed, passing into those of adult stage; leaves of *adults* oblong to linear-obovate, ~5–15 × 1–2 cm, base gradually narrowed to a stout petiole ~1–2 cm long, margins bluntly serrate to entire, apex obtuse or mucronate-apiculate to retuse, veins evident above. **Inflorescences** umbels, compound, terminal; ♂ with 5–12 rays ~3–5 cm long, with flowers racemosely distributed; ♀ with rays 1–3 cm long, umbellules 2–5-flowered. **Flowers** with stamens 4–5; ovary 5-loculed, 5-ovuled, style branches 5, connate, tips sometimes free. **Fruits** broadly oblong, ~8–9 mm Ø. **Distr:** New Zealand. **Ecol:** In lowland forests and scrubs. Fr. Jan–Apr. **Zones:** 9–11.

Schefflera

Species ~1,100. **Shrubs** or **trees**, sometimes climbers or epiphytes, evergreen, hermaphroditic or andromonoecious, unarmed. **Stipules** united within petiole. **Leaves** palmately compound, rarely unifoliolate, margins entire to serrate. **Inflorescences** terminal or pseudo-lateral panicles or compound racemes; flowers arranged in umbels, heads, or racemes; bracts pubescent, deciduous or persistent; pedicels not articulate below ovary. **Flowers** with calyx rim entire or 5-toothed; petals 5–11, valvate; stamens 5–11; ovary (4 or)5–11-carpellate, styles partly or completely united into a column or stigmas sessile. **Fruits** drupes, globose or ovoid; seeds (4 or)5–11, laterally compressed. **Distr:** Tropics and subtropics of both hemispheres. **Cult:** Propagate from seeds, cuttings or by air-layering. Requires a moist, well-drained soil in full or partial sun.

S. ***actinophylla*** (Endl.) Harms, *Die Nat. Pflanz.* 3(8): 36, 1894. (Syn: *Brassaia actinophylla*). [Octopus Tree, Queensland Umbrella Tree, Schefflera]. **Shrubs** to **trees**, to 12 m tall; branches numerous with age. **Leaves** palmate, to 75 cm Ø; leaflets (5–)7–16 (fewer in juvenile plants), oblong, to 30 × 10 cm, margins entire (or apex few-toothed), adaxially glossy bright green, glabrous; petiole 3–80 cm long, stipular base elongate. **Inflorescences** compound panicles, very large, often to 1.5 m across, terminal, 11–14 flowers tightly clustered in shortly pedicellate capitula; branches 6–13 pink to red, radiating; flowers sessile, each subtended by an involucre of 4 bracts. **Flowers** red, small; calyx rim obscure; sepals absent; petals 11–15, subulate, 3–4 mm long, caducous, red; stamens 11–15, spreading, pink; disk ridged, red; ovary 10–11biloculed, stigmas sessile, 10–12, yellow. **Fruits** drupes, globose, ~7 mm Ø, red, ripening to black, partially inferior at maturity, stones (pyrenes) ~9–12, flattened. **Distr:** New Guinea, Indonesia, Australia. **Ecol:** In tropical and subtropical areas, often in forests and forest margins. Fl. & Fr. Dec–Mar. **Zones:** 10–12.

S. ***setulosa*** Harms, *Bot. Jahrb. Syst.* lvi. 392, 1920. **Shrubs** or **small trees**; branch apices densely setulose-hispid. **Leaves** digitate, 7–11-foliolate; leaflets lanceolate, oblanceolate, oblong-lanceolate or oblong-oblanceolate, 11–23 × 3–8 cm, base obtuse or rounded, apex caudate-acuminate, adaxially subglabrous; petiole 25–50 cm long, base setulose; petiolules 2–6 cm long, villose or glabrescent. **Inflorescences** panicles, racemose, 14–26+ cm long, flowers arranged in dense subglobose heads 0.7–1 cm Ø; rachis densely setulose-villose. **Flowers** very small; ovary obconic, ~4–6 mm long, 5–6-loculed, style simple, short, apex truncate or subtruncate. **Distr:** New Guinea. **Ecol:** ~2,000 m. Fl. May–Jun. **Zones:** 9–12.

Trevesia

Species ~10. **Shrubs** or **trees**, evergreen, hermaphroditic, with few to many spines, glabrous or stellate-pubescent. **Stipules** ligulate, partly or fully united. **Leaves** simple, palmately lobed or appearing almost palmately compound, with a fanlike base and lobes constricted to midvein, margins serrated. **Inflorescences** racemes or panicles of umbels, terminal or pseudo-lateral; bracts small to large, persistent or caducous; pedicels not articulate below ovary. **Flowers** with calyx margin entire or minutely lobed; petals 7–12, valvate, often cohering and calyptrate, falling as a single unit or abscising separately; stamens as many as petals; ovary 6–16-carpellate, styles united into a short column. **Fruits** drupes, globose to ovoid;

seeds depressed. **Distr:** SE Asia, Indo-China, India, Nepal. **Cult:** Propagate from seeds, softwood cuttings or by air-layering. Requires a moist, well-drained soil in full or partial sun.

T. palmata (Roxburgh ex Lindley) Visiani, *Mem. Reale Accad. Sci. Torino*, ser. 2. 4: 262, 1842. (Syn: *Fatsia cavaleriei*, etc.). [Snowflake Aralia].**Trees**, evergreen, to 8 m tall; trunk to ~15 cm dbh; branches prickly, farinose stellate-pubescent. **Stipules** united into a bilobed sheath. **Leaves** simple, large, 60–90 cm Ø, deeply 5–9-lobed, lobes narrowly ovate-lanceolate, margins serrate, apex acuminate, leathery, surfaces glabrous or hairy, lateral veins distinct on both surfaces; petiole 30–70(–90) cm long, often prickly. **Inflorescences** panicles of umbels, ~45 cm across, densely farinose stellate-pubescent when young, glabrescent; peduncles 4–17 cm long; umbels 4–5 cm Ø, 25–45-flowered; pedicels 1.5–2 cm long. **Flowers** off-white; calyx rim 1–2 mm Ø, farinose stellate-pubescent; stamens 7–12; ovary 7–1bicarpellate, styles united, conic, 2–4 mm long, stout, persistent. **Fruits** subglobose to compressed-globose, 1–1.8 cm Ø, smooth or ribbed. **Distr:** Bangladesh, Cambodia, China, India, Laos, Nepal, Thailand, Vietnam. **Ecol:** In mixed forests on mountain slopes; 600–2,000 m. Fl. Oct. Fr. May–Jul. **Zones:** 10–12.

APIACEAE

Subfamilies 4, **genera** 434, **species** 3,780. **Herbs**, annual or perennial, rarely woody at base; caulescent or acaulescent, stem hollow or solid. **Stipules** usually absent. **Leaves** alternate, rarely opposite or basal; lamina compound or sometimes simple, usually much-incised or -divided, pinnatifid to pinnatisect, or ternate-pinnately decompound; petiole usually sheathing at base. **Inflorescences** simple or compound umbels; umbellules few- to many-flowered, usually subtended by bracteoles forming an involucel; rays often subtended by bracts forming an involucre; pedicels long, short or obsolete. **Flowers** epigynous, small, bisexual or ♂, regular; calyx tube wholly adnate to the ovary, teeth small or obsolete, forming a ring around the top of the ovary; ovary inferior, biloculed, ovule 1 per locule, anatropous, styles 2, usually swollen at base forming a stylopodium which often secretes nectar. **Fruits** dry, of 2 united mericarps; mericarps variously flattened dorsally, laterally or terete, 1-seeded. **Distr:** Worldwide, especially N temperate regions.

Daucus

Species ~20. **Herbs**, biennial; stem solitary erect, branching, retrorsely hispid. **Leaves** *basal* petiolate, pinnately decompound, ultimate segments small and narrow; *upper leaves* reduced, becoming sessile, wholly sheathing, divisions narrow and elongate. **Inflorescences** umbels, terminal and axillary, loosely compound; bracts numerous, pinnate, rarely entire, usually reflexed; rays numerous, spreading or incurved after anthesis, tightly compact in fruit; bracteoles numerous, toothed or entire; umbellules many-flowered, central flowers usually sterile with enlarged purple petals; pedicels unequal. **Flowers** with calyx teeth obsolete to conspicuous; petals white or yellow, obcordate, with an inflexed apex, outer petals in outer flowers of an umbellule enlarged and radiant; stylopodium conic, styles short. **Fruits** ellipsoid, dorsally compressed; primary ribs filiform, bristly; secondary ribs winged, wings with glochidiate prickles; vittae 1 in furrows under the secondary ribs, 2 on commissure; seed face shallowly concave to nearly plane; carpophore entire or bifid at apex. **Distr:** N Africa, SW Asia, Europe. **Cult:** Propagate from seeds. Requires a rich, moist, compost in full or partial sun.

D. carota L., *Sp. Pl.* 1: 242, 1753. [Carrot, Wild Carrot]. **Herbs**, biennial, tuberous, 0.1–1.2 m high; tubers subterranean, cream, orange or reddish-orange, conical; stem glabrous to pilose; hairs white. **Leaves** oblong, 2–3-pinnate/pinnatisect; ultimate segments linear to lanceolate, 0.2–1.5 cm x 0.5–4 mm, apex acute, mucronate, glabrous to hispid especially on veins and margins. **Inflorescences** umbels; peduncles 10–55 cm long, retrorsely hispid; bracts foliaceous, pinnate, rarely entire, lobes linear, 0.3–3 cm long, margin scarious; rays 2–7.5 cm long, unequal; bracteoles 5–7, linear, entire or 2–trilobed, ± scarious and ciliate, equaling or exceeding flowers. **Flowers** with calyx teeth minute; petals white, sometimes yellow or pinkish, outer ones radiate; petals of the central flower of an umbel sometimes red; ovary hispid, styles 0.5–1 mm long. **Fruits** ovoid, 2–4 x ~2 mm, primary ridges not prominent, slightly bristly; secondary ridges winged, spiny; spines white; one vitta under each secondary ridge; commissure bivittate. **Distr:** N Africa, SW Asia, Europe. **Ecol:** On mountain slopes; 2,000–3,000 m. Fl. May–Jul. **Zones:** 3–10.

Eryngium

Species 220–250. **Herbs**, biennial or perennial, caulescent or acaulescent, usually glabrous; taproot fusiform or stout; stem solitary, creeping to erect, branched above. **Leaves** simple, pinnately or palmately parted or divided, margins entire, often ciliate to spinose, leathery or membranous, venation parallel or reticulate. petiole sheathing. **Inflorescences** umbels, simple, capitate, forming globose to cylindrical heads; heads solitary or in cymes, sometimes racemes; bracts 1 to several, entire or divided, subtending the head; bracteoles 1 to many, subtending individual flowers; flowers sessile. **Flowers** small, bisexual; calyx teeth prominent, persistent, ovate to lanceolate, acute to obtuse; petals white or purple, ovate to oblong, apex incurved; stylopodium absent; styles shorter than or exceeding calyx teeth. **Fruits** globose to obovoid, scarcely laterally flattened, variously covered with scales or tubercles, ribs obsolete, vittae mostly 5, inconspicuous; commissure broad; seed subterete, face plane or slightly concave; carpophore absent. **Distr:** Tropics and temperate regions worldwide, especially S America. **Cult:** Propagate from seeds or by division. Requires a moist, well-drained soil in shade to full sun.

E. foetidum L., *Sp. Pl.* 1: 232, 1753. [Mexican Cilantro]. **Herbs**, 8–40 cm high, from a basal rosette; taproot fusiform with fibrous roots; stem green. **Leaves** *basal* numerous, entire, lanceolate or oblanceolate, 5–25 x 1.2–4 cm, base cuneate to decurrent, margins callous, crenate to finely spinulose-serrate, apex obtuse, venation pinnately reticulate, petiole short or obsolete, sheath to 3 cm long; *cauline leaves* sessile, opposite, deeply spinulose-serrate to parted. **Inflorescences** divaricately trifurcate; lateral branches often continuing to form a monochasium, capitula numerous, shortly pedunculate; capitula cylindrical, 0.5–1.2 x 0.3–0.5 cm, bracts 4–7, foliaceous, lanceolate, 1.5–3.5 x 0.4–1 cm, spreading to reflexed, margin 1–3-spinulose-serrate; bracteoles lanceolate, 1.5–1.8 x ~0.6 mm, margins brightly scarious. **Flowers** with calyx teeth ovate-lanceolate, 0.5–1 mm long, acute, equaling petals; petals white or pale yellow; styles erect, ~1.1 mm long, exceeding calyx teeth. **Fruits** ovoid-globose, 1.1–1.3 mm Ø, covered with tubercles. **Distr:** C America. **Ecol:** In forests and moist places; 100–1,500 m. Fl. & Fr. Apr–Dec. **Zones:** 10–12.

Foeniculum

Species 1. *Generic description same as specific.* **Cult:** Propagate from seeds. Requires fertile, moist but well-drained soil in full sun.

F. vulgare (L.) Miller, *Gard. Dict.*, ed. 8. Foeniculum no. 1, 1768. (Syn: *Anethum foeniculum*, etc.). [Fennel, Indian Sweet Fennel, Sweet Cumin]. **Herbs**, perennial, aromatic, 0.4–2 m high; stem hollow. **Leaves** aromatic, smelling of aniseed when crushed; lamina fernlike, broadly triangular in outline, 4–30 x 5–40 cm, 4–5-pinnatisect, ultimate segments linear, 1–6 x ~0.1 mm; lower petioles 5–15 cm long. **Inflorescences** umbels, 5–9 cm Ø; peduncles 2–25 cm long; rays 6–29(–40), unequal, 1.5–10 cm long; umbellules 14–39-flowered; pedicels thin, 0.2–1 cm long, unequal. **Fruits** 4–6(–10) x 1.5–2.2(–2.5) mm. Fl. May–Jun. Fr. Jul–Sep. **Distr:** Mediterranean region; cultivated worldwide. **Ecol:** Mostly in dry open grasslands; 200–2,600 m. **Zones:** 5–12. **Info:** Cultivated widely and commercially for the spice 'fennel'.

DIPSACALES

FAMILIES 2 OR 7, GENERA 45, SPECIES 1,090.

Trees small, **shrubs**, ± woody **lianas**, or **herbs**; vessels with simple or sometimes scalariform perforations; buds with scales. **Leaves** opposite; lamina with base ± confluent, margins gland-toothed; petiole bundles arcuate. **Inflorescences** cymose. **Flowers** with calyx persistent in fruit; stamens with pollen grains trinucleate; gynoecium ± inferior. **Seeds** with testa vascularised (Stevens 2001 onwards). **Distr:** Worldwide, especially northern temperate regions.

ADOXACEAE

Genera 4, **species** ~220. **Shrubs**, less frequently perennial herbs or small trees. **Leaves** opposite; lamina simple or compound, ternate, biternate, ternate-pinnate, biternate-pinnate, or odd-pinnate. **Inflorescences** panicles or umbels of cymes, terminal. **Flowers** bisexual; calyx and corolla both connate, (3–)5-merous; stamens 5, alternate with corolla lobes, and inserted at corolla tube; fertile stamens 5, 4 or 3, cleft into 2 semistamens, anthers 1-celled, peltate, extrorse, longitudinally dehiscent; staminodes 5, 4, or 3, in inner whorl, opposite to corolla lobes; ovary semi-inferior to inferior, 1- or 3–5-loculed, styles 5, 4, or 3, connate or free, or absent, stigmas capitate or bi- or trifid. **Fruits** drupes, 1- or 3–5-seeded. **Distr:** Mainly in Northern Hemisphere.

Sambucus

Species ~20. **Shrubs** or **small trees**, gynodioecious, rarely herbaceous perennials; branches with stout pith. **Stipules** present or absent. **Leaves** imparipinnate or incompletely bipinnate, rarely laciniate; leaflets serrate or divided, opposite or alternate. **Inflorescences** flat or convex corymbs or panicles, terminal; flowers articulate with pedicel; bracts mostly absent; bracteoles 1 or absent. **Flowers** actinomorphic mostly white, sometimes dimorphic or some forming glandular nectaries; calyx tube short, limb 3–5-parted; corolla rotate, lobes 3–5, valvate; stamens 5, inserted on base of corolla, filaments erect, filiform, anthers oblong, cells free, attached at middle; ovary 3–5-loculed, locules 1-ovuled, stigmas 3 or 5, short, on a cushionlike style. **Fruits** berrylike drupes; seeds 3–5, trigonal or ellipsoid. **Distr:** Temperate and subtropical regions and tropical mountains. **Cult:** Propagate from seeds or cuttings. Requires a fertile, well-drained soil in full or partial sun.

S. nigra L., *Sp. Pl.* 269, 1753. [Black Elder, European Elder]. **Trees small** or **large shrubs**, to 10 m tall; stem lenticellate. **Stipules** absent or inconspicuous. **Leaves** pinnate; leaflets 5–7, oval to ovate, 3–7 x 1.5–4 cm, margins serrate, apex acute, surfaces glabrous to strigose, veins strigose. **Inflorescences** corymbose, to 15 cm Ø; pedicels jointed; bracteole minute, glandular below the joint; hypanthium turbinate, ~1 mm long. **Flowers** white, very small,

Top row: *Foeniculum vulgare, Sambucus nigra, Lonicera japonica.*
Bottom row: *Leycesteria formosa, Lonicera hildebrandtiana.*

mildly fragrant; calyx 5-toothed, minute; corolla rotate with fused petals, lobes 5, round, spreading, 1.5 mm long, trinerved; anthers oblong, 1 mm long, filament 2 mm long; stigmas 3, almost sessile. **Fruits** globose, 5–8 mm Ø, purplish-black; pyrenes oblong, 3.5 mm long, surface rugose. **Distr:** Europe and Asia. **Ecol:** Fl. May. **Zones:** 5–12.

CAPRIFOLIACEAE

Genera ~13, species ~500. **Shrubs** or **woody vines**, rarely small trees or herbs. **Stipules** absent or very small. **Leaves** opposite, rarely verticillate, simple or pinnately compound. **Inflorescences** basically cymose or with pairs of flowers sometimes joined toward bases of ovaries. **Flowers** bisexual, regular or irregular; calyx 4- or 5-toothed or -lobed, tube adnate to ovary; corolla epigynous, gamopetalous, lobes 5, spreading, sometimes 2-lipped, mostly imbricate in bud; stamens (4 or)5, arising from corolla tube, alternating with lobes, anthers free, 2-celled, opening by longitudinal slits, mostly introrse; carpels 3–5, united, ovary inferior, 1–5-loculed, placenta axile, ovules 1 to many per locule, pendulous, style solitary, stigma capitate or lobed. **Fruits** berries or achenelike drupes, rarely capsules; seeds 1 to many. **Distr:** Mostly in temperate regions of eastern N America and E Asia.

Leycesteria

Species 6. **Shrubs** erect or **undershrubs**; deciduous; branches hollow. **Stipules** absent or present. **Leaves** opposite; lamina simple, ovate-lanceolate, base rounded, margins entire or irregularly dentate, membranous, abaxially pale green. **Inflorescences** cymose whorls of 2–6 flowers in pendent spikes, subtended by bracts, axillary or terminal; bracts large foliaceous, colorful, sessile, basally connate; flowers sessile. **Flowers** with calyx tube glandular, pubescent, lobes 5, unequal, persistent; corolla funnelform, lobes 5, subequal; stamens 5, inserted on the throat of corolla tube, filaments short, equaling corolla lobes; ovary 5-locular, style filiform, stigma capitate. **Fruits** berries, globose to subglobose; seeds many, minute, compressed. **Distr:** NW Himalayas to SW China. **Cult:** Propagate from seeds or cuttings. Requires a moderately fertile, moist, well-drained soil in full or partial sun.

L. formosa Wall., *Fl. Ind.* ed. Carey. 2: 182, 1824. (Syn: *L. limprichtii*). [Himalayan Honeysuckle]. **Shrubs** or **undershrubs**, erect, to 2 m high, branching from base; branches erect, hollow, herbaceous, glabrous or slightly hairy. **Leaves** broadly ovate-lanceolate, 5–17 × 2.5–8 cm, margins entire or slightly serrate, apex acuminate; petiole 1–2 cm long. **Inflorescences** spikes, pendent, terminal, 3–10 cm long, flowers verticillate; bracts 1.5–4 cm long, purplish or purplish-red, exceeding the flowers. **Flowers** partially hidden within the bracts; corolla white, purple-tinged, or wholly reddish-white or violet, narrowly campanulate, ~2 cm long. **Fruits** berries, subglobose, ~6 mm long, black, glandular-hairy, with persistent calyx. **Distr:** NW Himalayas, eastward to India, Nepal, Myanmar, Bhutan, Assam, SW Tibet, W China. **Ecol:** In moist shady places as forest undergrowth; 500–3,000 m. Fl. Jun–Jul. **Zones:** 7–11. **Info:** This species is considered a noxious weed in some countries.

Lonicera

Species >200. **Shrubs**, deciduous or evergreen, erect or prostrate, sometimes twining, rarely creeping. **Leaves** petiolate or sessile; lamina entire or lobed or basally connate. **Inflorescences** with flowers in axillary pairs or terminal capitate heads, pedunculate or sessile, often connate by their ovaries, bracteate and bibracteolate; bracteoles distinct or connate or sometimes absent. **Flowers** 5-merous; calyx tube ovoid, limb 5-toothed or cup-shaped, deciduous or persistent; corolla tubular, gibbous or not at base; limb subequally 5-lobed or 2-lipped, upper lip erect, 4-toothed, lower lip reflexed, entire; stamens 5, inserted on corolla tube, anthers usually exserted; ovary 2–trilocular, style long-filiform, stigma capitate, usually exserted. **Fruits** berries, distinct or connate in pairs. **Distr:** Temperate and subalpine regions of Northern Hemisphere. **Cult:** Propagate from seeds, cuttings or by air-layering. Requires a fertile, humus-rich, moist, well-drained soil in full sun.

L. hildebrandtiana Collett & Hemsl., *J. Linn. Soc.*, Bot. 28: 64, t. 11, 1890. [Giant Burmese Honeysuckle]. **Shrubs**, scandent, lianalike, robust, to 4 m high. **Leaves** elliptic or ovate, 7–12 × 4.5–7 cm, base attenuate, margins entire, apex acute to cuspidate, surfaces glabrous, adaxially green, abaxially paler, secondary nerves 3–6 pairs; petiole 1–1.5 cm long, purple. **Inflorescences** monochasia, terminal. **Flowers** large, scented; calyx tube ~5 mm long, lobes 5, green; corolla to 18 cm long, bilabiate, lobes 5, upper lip 4-lobed, lower lip with 1-lobed, creamy-white, turning reddish-orange; stamens 5, inserted on corolla tube, filaments puberulous, 4–5 cm long, anthers versatile, linear, ~8 mm long; pistil solitary, filiform, tomentose, as long as corolla lobe, stigma capitate. **Fruits** berries, pyriform, ovoid, ~2.5 cm long; seeds few. **Distr:** S China, Upper Myanmar and N Thailand. **Ecol:** In evergreen forests; 1,000–1,900 m. **Zones:** 10–12.

L. japonica Thunb., *Fl. Jap.* 89, 1784. (Syn: *L. shintenensis*, etc.). [Japanese Honeysuckle]. **Vines**, twining to scandent, semievergreen, woody; branchlets pubescent. **Leaves** broadly lanceolate, ovate to elliptic, 3–8 × 2–3 cm, base rounded to subcordate, margins entire, apex acute to obtuse or rounded, chartaceous, adaxially slightly pubescent, abaxially densely so, midrib distinct; petiole 0.3–1 cm long, densely pubescent. **Inflorescences** cymes, 2-flowered, axillary; peduncles usually solitary, ~0.3–1.5 cm long, pubescent; bracts below each pair of flowers foliaceous, petiolate, densely pubescent; bracteoles suborbicular, pubescent. **Flowers** sweetly fragrant, with ample nectar; calyx tube conical, pubescent; corolla 3–4 cm long, pubescent and glandular outside, white, purple-tinged, turning yellow or golden-yellow with age, tube ~2 cm long, limb ~2 cm long; stamens 5, exserted; style 1. **Fruits** berries, globose, 3–4 mm Ø, black. **Distr:** China, Japan, Taiwan. **Ecol:** In thickets of low elevations; 0–1,000 m. **Zones:** 4–12.

GLOSSARY

abaxial Surface facing away from the axis of an axis (e.g. the abaxial surface of a leaf is the underside, or side facing away from the stem).
acanthophyll Leaf or leaflet modified into a spine.
acaulescent Lacking an aboveground stem or trunk, except the inflorescence axis.
accrescent Increasing in size after flowering (e.g. calyx).
accumbent Lying or leaning against something.
acerose Needlelike, having a sharp, stiff point.
acetabuliform Shaped like a shallow cup (saucer-shaped).
achene A small, dry, indehiscent one-seeded fruit with a thin wall (with the seed attached to the pericarp at one point only).
achlorophyllaceous Lacking chlorophyll.
acropetal Developing or maturing from the base toward the apex, so that the youngest are at the apex.
acrostichoid Densely covering the abaxial surface of the fertile frond (of fern sporangia), not in distinct groups.
acrotony Giving preference to the apical region; flowering (seasonal) shoots which produce leaves below the inflorescence.
actinomorphic Having radial symmetry, with three or more planes of symmetry (capable of being divided into equal halves along any diameter).
aculeate Having prickles, sharp appendages.
aculeolate Having small prickles or sharp points.
acuminate Long-pointed; apical margins abruptly incurved (concave).
acute Ending in a sharp but not prolonged point; apical margins approximately straight.
adaxial Surface facing toward or nearest the axis (e.g. the adaxial surface of a leaf is the upperside, or side facing toward the stem).
adnate United (or fused) to a part or organ of a different kind.
adpressed Pressed close to or lying flat against a surface.
adventitious Of plant parts, usually roots, in an atypical position.
adventive Species not native to and not fully established in a new habitat or environment.
aerenchyma Spongy tissue with large air spaces between the cells; of stems and leaves of aquatic plants, providing buoyancy (and allowing circulation of gases).
aerial roots Roots that originate and are primarily aboveground.
aggregate fruit A fruit of one flower with separate carpels developing in close contact on one receptacle, but not fused.
air-layering A propagation technique whereby a short section of young branch is stripped of its outer layer of bark, including the cambium layer, and wrapped with sphagnum moss (and sometimes with growth hormones) and covered tightly with plastic or sometimes in a porous material (and watered regularly).
alae Flattened border of some stems, fruits and seeds; either one of the two side petals of certain flowers in Fabaceae subfamily Faboideae.
alate Having winglike extensions or parts; winged.
albuminous Containing albumin (simple, water-soluble proteins).
alternipetalous Stamens with point of attachment between the sepals or calyx lobes.
alveolate Pitted with cell-like cavities (as a honeycomb).
amphitropous (hemitropous) Ovule type in which the curvature of the ovule development displaces the micropyle adjacent to the base of the funicle; of an ovule, having the raphe terminating about half way between the chalaza and the orifice.
amplexicaul Sessile and clasping (or enveloping) most of the stem circumference.
ampliate Widened or enlarged.
amylophilic Starch grains produced in seeds and vegetative organs.
anantherous Lacking anthers.
anastomose Come together or open into each other (e.g. leaf veins).
anatropous Completely inverted so that the micropyle is facing downward and situated near the base of the funicle.
anauxotelic Inflorescences, parts of inflorescences or axes that do not end in a flower, and in which growth does not continue beyond the flowering region.
ancipital (adj. ancipitous) Flattened and having two edges.
androdioecious Having male flowers on some individuals and perfect flowers on the others.
androecium Male organ(s) of a flower (collectively all the stamens).
androgynophore A stalklike structure bearing both the androecium and gynoecium of a flower, above the level of insertion of the perianth.
andromonoecious Having both male and perfect flowers on the same plant.
androphore A stalklike structure bearing the stamens of a flower.
anemophily (adj. anemophilous) Pollination by wind.
angulate Having angles or an angular shape.
angustate Narrowed.
anisophyllous Having unequal pairs of leaves.
annual A plant that completes its life cycle in one season and then dies.
annulate (adj. annuliform) Having or consisting of rings or ringlike segments.
annulus *See* corona.
anomocytic A stoma lacking morphologically differentiated subsidiary cells.
antepetalous (antipetalous) Opposite the petals.
anterior In front of and facing away from the axis or stem.
antesepalous (antisepalous) Opposite the sepals (applied to stamens).
anthelodium Anthelalike (cymose corymb with the lateral flowers higher than the central ones), indeterminate inflorescence (e.g. in Cyperaceae).
anther The pollen-bearing organs of seed plants, often borne on a filament.
antheridium The fertile organ of a male gametophyte (or the male organ) of a bisexual gametophyte, in which male gametes are formed.
antheriferous Producing anthers (supporting anthers, as the filaments).
antherode The non-functional anther of a staminode.
anthesis The time of full opening of a flower when pollen is presented and/or the stigma is receptive.
anthocarp A fruit in which one or more flower parts functions as accessory tissues.
anthocyanins (adj. anthocyanic) A class of water-soluble flavonoids (vacuolar pigments) forming red-, pink-, purple- and blue-colored pigments, according to pH.
anticous Facing toward the axis of the flower.
antipetalous Stamens which stand opposite to petals or corolla lobes.
antisepalous Stamens which stand opposite to sepals or calyx lobes.
antitropic Bilaterally symmetrical but in an opposite direction.
antrorse Bent or directed forward and upward (toward the apex), usually referring to small appendages.
apetalous Having no petals or corolla.
apex The distal or free end of a stem, root or leaf.
apical Of placentae, at the top of the ovary, thus the ovules pendulous.
apiculate Ending abruptly with a sharp, flexible tip.
apocarp (adj. apocarpous) A fruit with several separate carpels.
apogamy (adj. apogamous) The development of an embryo without the occurrence of fertilization.
apomixis (adj. apomictic) Plant reproduction without fertilization, meiosis, or the production of gametes.
apomorphy A derived condition or character state of a clade, representing an evolutionary novelty.
apostemonous With separate stamens.
apotropous Turned away (of the curvature of an ovule with respect to the ovary axis, abaxial).
appendiculate Having small appendages; forming an appendage.
appendix The sterile tip of a spadix in the family Araceae.
applanate Having a horizontally flattened form.
appressed Pressed closely against a surface, often that of another organ, but not united with it.
arachnoid (arachnose) Trichoms (hairs) resembling a spider's web.
arborescent Resembling a tree (applied to non-woody plants and shrubs attaining tree height); treelike.
arcuate Curved like a bow
areole (adj. areolar, areolate) A small, discretely bounded area (e.g. a cluster of hairs, spines or bristles borne at a node as in Cactaceae.
aril (adj. arillate) A fleshy outgrowth of the funicle, raphe or integuments, partly or wholly covering a seed.
arillode A fleshy structure partly or wholly covering a seed and developed from any part of the ovule.
arista (adj. aristate) Having a stiff, bristlelike awn or tip (at the apex of a structure).
aristulate Having a short beard or awn.
asperulous Having a surface with short, hard projections.
astylous Lacking a style.
asymmetrical Having a shape that cannot be divided into equal halves.
attenuate Having a long, gradually tapering point with concave sides.
atropous An erect, straight ovule with funicle, chalaza and micropyle in a straight line.
auricle (adj. auriculate, auricular) An ear-shaped appendage at the base of a leaf, leaflet or corolla lobe.
autotroph (adj. autotrophic) An organism independent of others in respect of organic nutrition (being able to fix CO_2 by photosynthesis to form carbohydrates).
auxotelic Inflorescences, parts of inflorescences or axes that do not end in a flower and in which growth continues beyond the flowering region (the ability of the apical bud to grow on into a leafy shoot).
awn A long bristlelike apical appendage; needlelike.
axil The region at the upper (adaxial) junction of leaf and stem.
axile Of the placentation of an ovary, usually (unless pseudomonomerous) with septae separating loculi, the ovules being borne on the central axis.
axillary On the side of a structure or at the nodes of an axis; infloresence positioned in the axil of the nearest vegetative leaf.
baccate (adj. bacciform) Fleshy or juicy (e.g. berrylike fruits).
baculate A pollen sculpturing with rod-shaped elements (baculi).
barbed/barbellate With minute terminal or lateral retrorse projections on a structure, each projection being a barb.
barbellulate Barbellate with diminutive hairs or barbs.
basifixed Attached at or by the base.
basipetal (adj. basipetalous) Produced or differentiated in succession towards the base of an organ.
basitonic Flowering seasonal shoots which produce

no leaves (except for bracts in some cases) below the inflorescence.
beaked Having the shape of a parrot's beak.
berry A fleshy fruit with a succulent pericarp.
bialate Having two wings.
biapiculate Ending abruptly with two sharp, flexible tips.
bibracteolate Having two bracteoles.
bicallose Having two callosities (calli or hard spots).
bicarinate Having two keel-like projections.
bicarpellate Having a two-celled fruit (formed of two carpels).
bicirrate Having or resembling two cirri.
bicornuate Having two horns or horn-shaped parts.
bicyclic Composed of or arranged in two distinct whorls (e.g. the petals of a flower).
bidentate Having two teeth or toothlike parts.
bidenticulate Having two minute teeth.
biennial A plant which lives for more than one but less than two years after germination.
bifarious Arranged in two rows.
bifid (bipartite) Forked or divided for about half the length of the structure, or into two parts.
bifoliate Having two leaves; having paired leaves.
bifoliolate Having two leaflets; having paired leaflets.
bifurcate Forked or divided into two parts or branches (e.g. the Y-shaped styles of certain flowers).
bigeminate In two pairs.
bigibbous Having two humps.
biglandular Having two glands.
biglobose Having double spheres.
bilabiate Having two lips.
bilamellate Formed of two plates; having two elevated ridges, as in the lip of some flowers.
bilaterally symmetric *See* zygomorphic.
bilobed Two-lobed; having two lobes.
bilobulate (bilobate) Two-lobuled; having two lobules.
bilocular (biloculate) Divided into or containing two locules (chambers).
binerved Having two nerves (veins).
bipartite *See* bifid.
bipinnate (bipinnately compound) Having pinnate leaflets on stems that grow opposite each other on a main stem; twice pinnate; a compound leaf with two orders of axes, the second axes (rachillae) bearing leaflets.
bipinnatifid (bipinnatipartite) Doubly pinnatifid; bipinnately-lobed to divided.
bipinnatisect Doubly pinnatisect.
biporate Having two pores.
bipulvinate Having two pulvini.
bisaccate Having two small bags, sacs or pouches.
biseriate Arranged in two rows or whorls (especially of the perianth and androecium).
bisexual Flowers having both carpel(s) and stamen(s); perfect flowers.
bisporangiate Having two different types of sporangia; anther with two microsporangia and typically one theca; a flower or cone producing both megaspores and microspores.
bistomal Micropyle of an ovule formed from both integuments; micropyle having two openings.
bisulcate Having two grooves (or furrows).
bitegmic Having two integuments, especially in reference to ovules, apomorphic for the angiosperms.
biternate (biternately compound) A compound leaf with three axes, each of which is ternately compound.
bithecous Having two thecae, cells or compartments.
bituberculate Having two warts (tubercles).
bivalvate (bivalved, bivalvular) Having two valves.
biventricose Having two swellings (one on each side).
boss A mound of aboveground roots at the base of the trunk.
brachyblast A shoot in which the internodes elongate slightly (with limited growth) or not at all, bearing reproductive structures and/or leaves, when well-developed.
bract (adj. bracteate) A modified, generally reduced leaf, usually different in form from the foliage leaves, generally subtending a flower (thus referred to as floral bract).
bracteole (bractlet) (adj. bracteolar) A small bract borne singly or two together and more or less opposite on the pedicel; technically any bract that arises on a pedicel instead of subtending it. *See also* prophyll.
bractiferous Having bracts.
bractlet *See* bracteole.
branch A secondary woody stem or limb growing from the trunk or main stem of a tree or shrub or from another secondary limb; a lateral division or subdivision of certain other plant parts, such as a root or flower cluster.
branchlet A small branch; the terminal or ultimate subdivision of a branch.
bulb (adj. bulbous) A storage organ, usually underground, made up of a relatively inconspicuous stem and large leaf bases, the food reserves being stored in the inner, fleshy leaf bases.
bulbil A small, deciduous bulb (of variable construction), formed in the axil of a leaf or replacing flowers in an inflorescence, enabling the plant to propagate vegetatively.
bulblet A small bulb.
bulla (pl. bullae) A rounded bony projection; a blister.
bullate Having a puckered or blistered appearance; having bullae.
bushveld A subtropical woodland ecoregion and vast savannas that cover much of Southern Africa.
buttress Bracing outgrowths at the base of the trunk of some large trees.
caducous Dropping off or shedding of an organ or part at an early stage of development (compared to those of other plants), but not withering.
calcar (adj. calcarate) A spur; a hollow projection or nectar spur (as at the base of a corolla).
calceolate Pouched like the toe of a slipper.
calciform In the form of chalk or lime.
callus (pl. calli, adj. callose) An isolated thickening of tissue (especially a stiff protuberance on the lip of an orchid).
callose (n.) A carbohydrate in plant cells that plugs the sieve pores when the sieve tubes stop functioning.
calycine Related to or resembling a calyx.
calycophyll When the calyx is more or less expanded and like a petal(s), as in some species of Rubiaceae.
calyculus (adj. calyculate) A structure varying in its morphological nature that looks like a small calyx (*see* epicalyx).
calyptra (adj. calyptriform) Any cap-shaped structure; in mosses and liverworts, a caplike structure covering or partly covering the capsule and derived from the venter of the archegonium.
calyx (pl. calyces) The sepals of one flower collectively.
cambium A layer of secondary growing tissue encircling woody plants between the bark and wood.
campanulate Bell-shaped, broadly tubular and then gradually widening towards the more or less spreading lobes (e.g. corolla of a flower).
campylotropous Having an ovule partially inverted and curved such that the micropyle nearly meets the funicle.
canaliculate Longitudinally grooved or channeled.
canescent More or less shortly gray-pubescent; hoary.
capitate Headlike, globose; enlarged at the head or tip; spherical with a short basal stalk.
capitellate Shaped like a small head; having a very small knoblike termination; collected into minute capitula.
capitulum (pl. capitula, adj. capituliform) An inflorescence with sessile flowers aggregated into a dense cluster (usually racemose or cymose, sometimes composed of disk and ray florets).
capsule A dry, dehiscent fruit.
carina (pl. carinae) A keel-shaped ridge or structure; a median ridge or fold on the abaxial surface of the structure.
carinate (carinal) Boat-shaped.
carnose Baccate, fleshy, succulent.
carpellary One of the structural units of a pistil, representing a modified, ovule-bearing leaf.
carpel The unit of the gynoecium of angiosperms; ancestrally, a modified, conduplicate megasporophyll (female) of a flower, at maturity enclosing one or more ovules.
carpellate With functional carpels only.
carpophore The central axis that persists between the carpels in a schizocarp where the ovary is inferior.
carpopodium The basal abscission zone of the cypsela or achene, consisting of one or two rows of cells that are distinct morphologically from other cells of the cypselal wall.
caruncle (adj. caruncled, carunculate) A more or less fleshy (often hard) outgrowth from the surface of a seed, usually near the micropyle.
caryotoid Palm species of the genera *Arenga*, *Caryota* and *Wallichia*.
castaneous Chestnut-colored.
cataphyll A modified leaf appearing before the production of new leaves or an inflorescence.
cauda (pl. caudae, adj. caudate) A tail-like structure.
caudex (pl. caudices) A more or less short, thick, erect trunk or stem (vertical or branched perennial), underground or at/near ground level.
caudiciform A plant with an enlarged, perennial, swollen, usually water-storing basal portion of the stem (from which arise annual or otherwise non-persistent photosynthetic shoots).
caudicle (caudicula) A slender and more or less elastic to stout structure derived from inside the anther to which the pollinium is attached (in orchids).
caulescent Having a well-developed aboveground stem.
cauliflory (adj. cauliflorous) The production of flowers and fruits on the older (well-developed) branches or trunks of woody plants.
cauline Borne on an aerial stem, usually separated by elongated internodes (mostly applied to leaves).
cespitose Growing in dense tufts or clumps, matlike clumps without creeping stems.
chalaza (adj. chalazal) The part of an ovule opposite the micropyle, to which the end of the funicle is attached and below the insertion of the integuments.
chartaceous Papery in texture; thin.
chasmogamy (chasmogamous) Flowers in which the perianth opens and exposes the sexual organs, with subsequent cross-pollination.
chitin (adj. chitinous) A tough, protective, semi-transparent substance.
chlorophyll (adj. chlorophyllous, chlorophyllose) The green pigment (found in almost all plants, algae and cyanobacteria) essential to the photosynthetic process.
choripetalous Having separate petals; polypetalous.
cilium (pl. cilia, ciliae, adj. ciliate) Microscopic, hairlike projection on the surfaces of some cells and of certain organisms.
cincinnus (pl. cincinni, adj. cincinnal) A scorpioid cyme; a monochasial cymose inflorescence branching alternately from a bracteole/prophyll on one side of a pedicel and then from one on the other side (the flowers being borne in two rows), the whole more or less zigzag but also coiled like the tail of a scorpion.
cincinnoid Resembling a cincinnus.
cinnamic aldehyde A yellow oil with a cinnamon odor and sweet taste, obtained from Ceylon Cinnamon (*Cinnamomum verum*).
circinate Spirally coiled downward from apex to base, with the tip innermost.
circumbasal Surrounding the base.
circumscissile Splitting or opening along a circumference, with the top coming off as a lid.
cirrus (pl. cirri, cirrhi, adj. cirrose) A slender, flexuous tendril-like appendage.
clade Any monophyletic group of organisms, consisting of a common ancestor and all lineages arising from that common ancestor.

cladistic(s) A method of classification that groups taxa hierarchically into discrete sets and subsets.
clambering Sprawling across objects, without specialized climbing structures; scandent.
clathrate Fern scales in which the cell walls of adjacent cells are thick; latticed or pierced with apertures like a trellis.
clavate (clavellate, claviform) Club-shaped, gradually widening toward the apex.
clavuncle Style head.
clawed Of a petal or sepal, that narrows abruptly into a stalk at the base.
cleft Divided almost to its axis.
cleistogamy (adj. cleistogamous) Flowers that never open but are self-fertilized and set fertile seed.
climber A plant that climbs by twining, tendrils or adhesive pads.
clinandrium A hollow containing the anther in the upper part of the column of an orchid flower.
coccus (pl. cocci) A division containing a single seed that splits apart from a many-lobed fruit; any of the carpels, containing one seed into which compound fruits split when ripe.
cochleate Shaped like a snail shell; spirally twisted.
colchicine A toxic natural product and secondary metabolite, originally extracted from species of the genus *Colchicum*.
colleter Usually a mucilaginous secretory hair, commonly found in groups or tufts near or at the adaxial base of the leaf and on the calyx.
colpus (pl. colpi, adj. colpate) A pollen grain aperture that is elongate with a length:width ratio greater than 2:1.
colporoidate (colporate) A pollen grain aperture type shaped like a colpus with a circular region in the center.
columella A small column-shaped anatomical structure.
column A tube consisting of fused stamen filaments that surround the style (as in Malvaceae); a knoblike fused pistil and stamens (as in Orchidaceae).
coma (adj. comose) A terminal tuft or cluster, especially a tuft of hairs on a seed.
commissure The surface or place along which two structures, such as carpels, are joined.
complanate Having a flattened or compressed surface.
complete Of a flower, possessing all the four floral whorls (calyx, corolla, stamens and carpels).
concolored (concolorous) Uniformly colored; both sides with the same color.
concrescent Growing together.
conduplicate Folded together (lengthwise) along the midrib, the sides flat and facing one another, as of leaf ptyxis or certain petals in a bud.
condyle An inward projection of the endocarp around which the seed is folded (as in Menispermaceae).
conflorescence A compound inflorescence consisting of two or more simple inflorescences.
confluent Merging or running together to form a mass.
coniform In the form of a cone.
connate Fused.
connective (adj. connectival) The part of the stamen that connects the pollen sacs/sporangia.
connivent Coming into contact, closely converging, but not fusing (not organically joined).
conoid Resembling a cone in shape; cone-shaped.
contortuplicate Plaited lengthwise, and twisted in addition; a variant of imbricate aestivation.
cordate (adj. cordiform) Broad and notched at the base; heart-shaped.
cordulate Slightly heart-shaped.
coriaceous Thick or leathery in texture.
corm (adj. cormous) A fleshy, swollen, bulblike base of a stem, usually underground and growing vertically in which food reserves are stored between growing seasons.
corniculus (adj. corniculate) A horn or hornlike projection.
corolla The petals collectively, usually the second whorl of the perianth inside the calyx.
corona A crownlike ring of modified floral tissue (often stamen filaments).
coroniform Having the form of a crown or coronet; resembling a crown.
corymb (adj. corymbiform, corymbose) A racemose inflorescence in which the pedicels of the lower flowers are longer than those of the flowers above, bringing all flowers to about the same level; a flat-topped or convex cluster, the outermost flowers being the first to open.
costa (adj. costate) A prominent, raised riblike structure (e.g. the midrib of a leaf).
costapalmate A palmate leaf in which the stem extends well into the frond, partially dividing it at the distal end of the petiole (leaf with a definte costa/midrib unlike the typical palmate leaf).
costule An axis that is a branch of a costa or another costule.
cowled Shaped like a cowl; cucullate; hooded or hoodlike.
crassinucellate (crassinucellar) Of an ovule with one or more layers of cells outside the embryo sac but distinct from the epidpermis of the ovule.
crassulescent Thick and fleshy; resembling the leaves of the genus *Crassula*.
crateriform Having the form of a shallow bowl; craterlike.
creeping Of plants, with horizontally spreading stems that produce roots and shoots from the nodes.
crenate (n. crenation) Having a margin with low, rounded or scalloped projections.
crenulate Having a margin with very small, low, rounded teeth; finely crenate.
crescentic Crescent-shaped; shaped like a crescent moon.
crista (pl. cristae, adj. cristate) A projection formed by infolding of a membrane or similar structure.
crown The branches and leafy canopy of a tree.
crownshaft A conspicuous cylinder formed by the tubular bases of leaf sheaths at the apex of the stem or trunk of a palm.
crozier The coiled portion of a young leaf resembling the top of a violin (as in the leaves of many ferns).
cruciate (cruciately) Having the form of a cross with equal arms.
crustose Having a thin thallus adhering closely to a substrate.
cucullate *See* cowled.
culm An aerial stem or stalk, especially the jointed and usually hollow stem of grasses, bamboos, etc.
cultigen A cultivated plant of unknown or obscure taxonomic origin.
cultivar A cultivated variety of a plant that has been selected for specific desirable characteristics.
cuneate Wedge-shaped; triangular at base and tapering to a point.
cuniculus (adj. cuniculate) A passage open at one end (e.g. the nectary of some orchids that can be seen as a hump or bulge on the surface of the ovary).
cupule (adj. cupular, cupulate, cupuliform) Resembling a small cup; a cup-shaped whorl of hardened, cohering bracts.
curvinerved Having the ribs or the veins of the leaves curved.
cusp (adj. cuspidate) A point, projection or elevation; a pointed end.
cuticle (adj. cuticular) A very thin hyaline film covering the surface of plants (derived from the outer surfaces of the epidermal cells).
cyathium (pl. cyathia, adj. cyathial) A congested cymose inflorescence consisting of a cup-shaped involucre enclosing an apetalous, female flower surrounded by several male flowers.
cyathophyll A bowl-shaped to tubular involucre subtended by a pair of bracts (as in the flowers of *Euphorbia* species).
cylindric Having the form of a cylinder.
cymbidiform Leaves resembling those of the genus *Cymbidium*.
cymbiform Boat-shaped; elongated and having the upper surface decidedly concave.
cyme (adj. cymiform, cymoid, cymose) An inflorescence in which the primary axis bears a single central or terminal flower that blooms first; a flat or convex inflorescence of this type.
cymule (adj. cymulose) A small cyme, generally with very few flowers.
cypsela An achene with an adherent calyx (as in the family Asteraceae); a dry, indehiscent, one-seeded fruit formed from an inferior ovary.
cystolith A mass of calcium carbonate occurring in the epidermal cells of certain plants, such as figs.
deciduous Of plants, shedding leaves or other parts naturally.
declinate Descending from the horizontal in a curve; angled downward or forward; drooping.
decompound A leaf that is more than once compound.
decrescent Diminishing; decreasing.
decumbent Of a reclining branch or stem, with an upturned tip.
decurrent (n. decurrence) Extending down the stem below the place of insertion (as certain leaves).
decurved Bent or curved downward.
decussate Having pairs of opposite leaves arranged perpendicularly to the pairs immediately above and below.
deflexed Bent downward and tilted forward.
defoliate An unnatural loss of leaves.
dehiscent Having a seed capsule that splits open, often explosively when ripe, releasing its seeds.
deltate (deltoid) Triangular, with the sides of about equal length.
dendritic Any structure with branches arising along the main axis.
dendroid Treelike; branching like a tree; arborescent.
dentate Of margins, with triangular teeth.
denticle A small tooth or toothlike structure.
dentiform Having the form of a tooth; tooth-shaped.
determinate Of an inflorescence, where flowers open from the top to the bottom and is not capable of further growth.
dextrorse (adj. dextrorsely) Clockwise; of the direction of twining, the stem taking an ascending clockwise course when viewed from above.
diadelphous Having the stamens connate and forming two groups, or all but one stamen connate and the other one free.
diad (dyad) A group of 2; a pair.
diaphragm (adj. diaphragmed) A membranous part that divides or separates.
diarch Having two strands of xylem.
dichasium (adj. dichasial, adv. dichasially) A cymose inflorescence with opposite branching below the flower which terminates each axis, each branch in turn terminating in a flower.
dichotomous Forking repeatedly by twos in a regular pattern.
dicot (pl. dicots) A dicotyledon; an angiosperm that has two cotyledons in the seed.
dictyosteles (adj. dictyostelic) A modified siphonostele in which the vascular tissue is dissected into a network of distinct strands.
didymous Borne in pairs.
didynamous An androecium consisting of four stamens, two being distinctly longer than the other two.
digitate Having radiating divisions or leaflets resembling the fingers of a hand.
digitiform Shaped like a finger.
dimerous Consisting of two parts; having flower parts, such as petals, sepals and stamens, in sets of two.
dimorphy (adj. dimorphic, dimorphous) Two different forms.
dioecious Having the staminate and carpellate reproductive structures on separate plants; having the male

and female organs in separate and distinct individuals; having separate sexes.
dipterocarp A tree of the family Dipterocarpaceae.
disk A flat structure or part; a plate or rim of tissue, derived from the receptacle of a flower, occurring between whorls of floral parts, commonly found between the androecium and gynoecium or on top of the gynoecium when the ovary is inferior, and often secreting nectar.
disk floret A specialized flower with a tubular corolla, with or without surrounding ray florets.
discolor (adj. discolorous) To change in color as by fading, streaking, or staining; of different colors; two surfaces different in color; of two colors.
disjunct Disjoined; separated.
dissected Having a cleft leaf or other organ.
distichous Arranged alternately in two vertical rows on opposite sides of an axis (e.g. as in leaves).
distinct Free, not fused or united.
distyly (adj. distylous) Of plants that have either long styles and short stamens, or vice versa.
dithecal (dithecous) Having two thecae, cells or compartments; anther consisting of two anther lobes.
divaricate Branching or diverging at a wide angle.
dolomite (adj. dolomitic) A mineral (calcium magnesium carbonate, $CaMg(CO_3)_2$) occurring in crystals and rocks.
domatium (pl. domatia) A small structure located in the axils of the primary veins on the lower surface of leaves in some woody dicotyledons, generally consisting of depressions and partly enclosed by leaf tissue or hairs.
dorsifixed Attached at or by the back.
dorsiventral Having tissues derived from the adaxial and abaxial surfaces of the leaf primordium; flattened and having distinct upper and lower surfaces.
double flower A flower with more than the normal complement of petals or other floral parts.
downcurved Curved downward, especially at the end.
druparium (drupe) A succulent fruit having the seed(s) enclosed by a stony endocarp.
ebracteate Lacking bracts.
ebracteolate Lacking bracteoles.
ecalcarate Lacking a spur or a spurlike process.
ecallose Lacking calli.
ecarinate Lacking a carina or keel.
echinate Having bristles or prickles.
echinulate Having a covering of small prickles or spines.
eciliate Lacking cilia.
ecorniculate Lacking horn(s) or hornlike projection(s).
ectophloic Of stems in which the phloem is external to the xylem.
eglandular Lacking glands.
elaiophore A specialized gland producing oil (as in Orchidaceae).
elastoviscin A highly viscous product of the degeneration of a limited number of tapetal cells keeping pollen together in orchid pollinia.
elliptic (adj. elliptical) Broadest in the middle, tapering uniformly toward the ends.
emarginate Notched at the apex (e.g. as in a petal or a leaf).
emergent Of a plant or plant part, rising above the water.
emersed Risen or standing out of water, surrounding leaves; (of the leaves or stems of aquatic plants) protruding above the surface of the water.
enantiomorph (adj. enantiomorphic, enantiomorphous) A mirror image of something (an opposite reflection); flowers with mirror image morphologies.
endarch Developing from the periphery (of a primary xylem or root); having the oldest cells closest to the core.
endemic Native to a limited geographic region.
endocarp (endocarpid) (adj. endocarpic) The inner, usually woody, layer of the pericarp of a fruit (e.g. the stone of a peach or cherry).
endophyte (adj. endophytic) A plant living within another plant, usually as a parasite.
endosperm (adj. endospermous) The tissue within the seed of an angiosperm that surrounds and nourishes the developing embryo; nutritive matter in angiosperm ovules derived from the embryo sac.
endospore (adj. endosporic, endosporous) The innermost wall of a spore or pollen grain.
endostomal The micropyle of an ovule formed from the inner integument alone.
endotestal Of seeds, with the inner epidermis of the testa becoming a mechanical layer.
endothecium The lining of the cavity of an anther.
ensiform Sword-shaped; xiphoid.
entire Having smooth margins, not lobed or toothed.
entomophilous Pollinated by or having spores distributed by insects.
eophyll The first fully-expanded leaf (or leaves).
epaleate Lacking a covering of chaffy scales.
epapillose Lacking a small nipplelike projection or a protuberance.
ephemerous (ephemeral) Lasting for a very short time; short-lived; lasting for a day (e.g. some flowers).
epicalyx A whorl of structures just below a flower, looking like a second calyx.
epicarp The outermost layer of a pericarp, as the rind or peel of certain fruits.
epichile The terminal part of the lip in some orchids, distinctly different from the form of the basal part.
epicormic Of buds, shoots or flowers, borne on the old wood of trees (especially those shoots arising from dormant buds after injury or fire).
epicortical On top of the bark.
epigeal (epigeous) Growing on or close to the ground; (of cotyledons) lifted above ground in germination.
epigynous Of flowers, having all floral parts conjoint and generally divergent from the ovary at or near its summit.
epilithic Plants growing on stones or rocks.
epiparasite (adj. epiparasitic) A parasite that feeds off another parasite; parasitic plant that forms an obligate haustorial attachment to another parasitic plant.
epipetalous On or attached to the petals, usually applied to stamens mounted at the bases of the petals.
epipetiolar With branches arising from buds on the petiole.
epipetric Growing on rocks; epilithic.
epiphyte (adj. epiphytic) A plant that grows upon another plant and uses it for support but does not derive nourishment from the host plant.
episepalous Having stamens growing on or adnate to the sepals.
episperm The often thick or hard outer coat of a seed.
epitepalous Having stamens attached to the tepals.
epitropous Of the curvature of an ovule with respect to the ovary axis, adaxial.
epunctate Lacking tiny spots, points or depressions.
equitant Fanned leaves whose folded bases enclose the base of the leaf above (commonly seen in Iridaceae).
equivalved Having both valves of equal size and shape.
erose Irregularly notched, toothed or indented.
estomate Lacking stomates.
eudicot One of the major divisions of angiosperms (flowering plants) with more than one cotyledon (seed leaf); tricolpate.
eustele (adj. eustelic) A stele that is medullated with highly dissected vascular tissue, the vascular bundles being either collateral or bicollateral.
evergreen A plant that normally retains most of its foliage throughout the year.
exalbuminous Without endosperm.
exappendiculate Lacking small appendages.
exarch Of a primary xylem or root, developing from the center; having the youngest cells closest to the core.
exareolate Lacking a small space or interstice in a tissue or part (e.g. the area bound by small veins in a leaf); a small, specialized, cushionlike area on a cactus from which hairs, glochids, spines, branches or flowers are borne.
exarillate Lacking aril.
excentric Not having the normally central portion in the true center; to one side, off center.
excurrent Having a single, undivided trunk with lateral branches; extending beyond the apex of a leaf (e.g. as in a midrib or vein).
exindusiate Lacking indusia.
exine The outer coat of a spore or pollen grain, highly resistant to strong acids and bases.
exocarp The 'skin', developing from the ovary/carpel wall. *See* epicarp.
exotesta Outer coat of a testa.
explanate Spreading or extending outwardly in a plane.
exsert (adj. exserted, n. exsertion) Protrude, stretch out (especially, stamens projecting beyond the corolla of a flower).
exstipellate (estipellate) Lacking stipels.
exstipular (exstipulate) Lacking stipules.
extrafloral A nectary situated on a leaf or stem rather than in a flower.
extramarginal Lying outside or beyond a margin.
extrapetiolar Lying outside or beyond the petiole.
extrastaminal Lying outside the stamen whorl; outside of the stamens.
extrorse Facing outward; turned away from the axis.
exudate A substance exuded; exudation.
eye A contrasting spot around the center or throat of a flower.
facultative Capable of functioning under varying environmental conditions.
falcate Curved and tapering to a point; sickle-shaped.
family A related collection of genera with defined characteristics.
fasciate Abnormally flattened or coalesced (e.g. certain stems).
fasciclode A sterile stamen fascicle.
fascicle A crest, an inflorescence or vegetative part in a compact, raised cluster.
fasciculate Arranged in or formed of fascicles; fascicular.
fastigiate Having erect and almost parallel branches tapering toward the top.
faucial Lying in/at the throat.
felty (felted) Having a feltlike texture.
ferruginous Rust-colored.
fetid Malodorous; having a foul odor.
filament (adj. filamentose, filiform) The stalklike portion of a stamen, supporting the anther.
filiferous Producing threads.
fimbria (pl. fimbriae, fimbriate, fimbricate) A fringe or fringed border.
fimbrillate Bordered by or having a small or fine fringe.
fistular (fistulose) Tubular; tubelike; hollow; reedlike.
flabellate Fan-shaped.
flaccid Limp; lax.
flagelliform Long, slender and flexible (similar to the lash of a whip).
flexuose (flexuous) Having a wavy or zigzag form; flexible.
flocculose Covered with tufts of small soft hair.
floral cup Hypanthium.
floriferous Flowering in profusion, or bearing flowers over an extended period.
flowerhead See capitulum.
foliaceous Leaflike.
foliolate Having or relating to leaflets.
follicetum A multiple fruit, dry, dehiscent formed by an aggregation of follicles.
follicle A dry, dehiscent fruit with one to many seeds, derived from one carpel and splitting open only along one side.
fossa A small longitudinal cavity or depression.
fovea (adj. foveate) A small pit or depression.
foveola (foveole) (adj. foveolate) A very small pit or depression.
fractiflex Broken or intermittent zigzag lines.
frond An often large, finely divided leaf, especially as

applied to ferns and certain palms; a leaflike expansion not differentiated into stem and foliage.
fruit The ripened ovary of a flower and any attached floral remnants.
fruitlet A small fruit, especially those forming an aggregate fruit; a seed-bearing structure derived from a single free carpel.
fruticose Shrublike.
fulvellous (fulvellus) Slightly tawny.
fulvous Tawny; dull brownish-yellow.
funicle (funiculus) (pl. funicles, funiculi) The stalk of an ovule or seed.
funnelform Shaped like a funnel; infundibuliform.
furcate Forked; branching.
furfuraceous Covered with scales.
fuscous Dark.
fused Connate; wholly or partially joined.
fusiform Spindle-shaped; rounded and tapering from the middle toward each end.
fynbos A natural shrubland or heathland vegetation occurring in a small belt of the Western Cape of South Africa with a Mediterranean climate; a collection of plants (mainly shrubs), from this region.
galea (adj. galeate) A helmetlike structure.
gall An abnormal swelling with an external cause, usually by insects.
gall flower An abortive female flower with short styles, as in *Ficus* species.
gamopetalous (sympetalous) Having the petals (perianth segments) united; having a corolla with partially or wholly fused petals; having petals united by their margins, at least at the base.
gamophyllous Having leaves united by their margins.
gamosepalous Having the sepals united.
gasteriform Swollen, inflated, markedly bulbous; round belly-shaped.
gemma (pl. gemmae, adj. gemmiferous) A bud; a cell or cluster of cells, or a leaflike or budlike structure, that separates from the parent plant to form a new organism.
geniculum (adj. geniculate) A small kneelike anatomical structure or abrupt (sharp) bend.
genus (pl. genera) The principal rank below family, and above species.
geophyte (adj. geophytic) A plant propagated by means of underground buds.
geoxylic Of woody plants, with most of their modified stems underground, often forming underground forests.
gibbose (gibbous) Having a bulge.
glabrate (glabrescent) Glabrous, but obviously having previously had an indumentum; becoming glabrous; somewhat glabrous.
glabrous Having a surface devoid of hair or pubescence; smooth; glossy.
gland A secretory organ.
glandular (glandulose, glandulous) Consisting of, containing or bearing glands; (of tapetum) the cells remain intact until the maturation of the microspores or pollen.
glanduliferous Bearing glandules.
glaucescent Becoming glaucous; somewhat glaucous.
glaucous Covered with a whitish bloom; bluish- or whitish-green in color.
globose (globular) Globelike; having the shape of a globe.
glochid (adj. glochidiate) Stiff bristles with barbed or anchorlike tips.
glomerate Compactly clustered.
glomerule (adj. glomerulate) Any small compact cluster; a cyme condensed into a headlike cluster.
glumaceous Glumelike; chaffy; consisting of or having glumes.
glume A chafflike bract in the inflorescence of a grass, sedge or similar plant; one of the pair of bracts at the base of a spikelet.
gradate (adv. gradately) To change or cause to change imperceptibly; to arrange in grades or ranks.
graft A scion, a bud or a twig spliced to a rootstock.
grevilleoid conflorescence A raceme or racemiform panicle of flower pairs.
gynobase An elevation of the receptacle of a flower, bearing the gynoecium.
gynodioecious Having female flowers on one plant and hermaphrodite flowers on another plant of the same species.
gynoecium (pl. gynoecia, adj. gynoecial) The pistil or pistils of a flower; the female parts.
gynomonoecious Having both female and hermaphrodite flowers on the same plant.
gynophore The elongated stalk of a pistil.
gynostegium A specialized caplike mass of tissue covering a gynoecium.
gynostemium The united stamens and pistil of an orchid.
habit The overall shape or appearance of a species.
hapaxanthic (monocarpic, semelparous) *See* monocarpic.
hastate Arrow-shaped; pointed at the apex with two spreading, pointed basal lobes.
hastula Of palms, a raised area or flap of tissue at the end of a petiole to which palmate leaflets are attached.
haustorium (pl. haustoria) The organ of a parasitic plant that penetrates the host tissues and absorbs food and water from them.
head *See* capitulum.
helicoid (adj. helicoidal) Coiled or curving like a spiral; a helicoid cyme is a monochasial cymose inflorescence.
hemiepiphyte (adj. hemiepiphytic) A plant which begins its life as an epiphyte but which later grows roots down into the ground (e.g. as in the genera *Ficus* and *Metrosideros*).
hemiparasite (adj. hemiparasitic) A plant (e.g. mistletoe) that obtains some nourishment from its host but also carries out photosynthesis; a heterotroph which lives on and derives part of its nourishment from a different organism, and is also partially self-supporting.
hemitropous *See* amphitropous.
herb (adj. herbaceous) A non-woody plant.
hermaphroditic Having pistils and stamens in the same flower; monoclinous.
heteranthous Flowering from special shoots that do not produce pseudobulbs or foliage leaves.
heterocarpy (adj. heterocarpic, heterocarpous) Of plants, characterized by the production of distinct types of fruits that usually differ in their ecological behavior; the production of more than one kind of fruit in one plant.
heterodistyly (adj. heterodistylous) Flowers having styles of two different lengths.
heteromorphic (heteromorphous) With two or more distinct forms.
heteromycotroph (adj. heteromycotrophic) A plant that is a mycotroph as part of its method of nutrition, usually with inadequate photosynthesis and hence often not green; a facultative mycotroph.
heterophyllous Of plants, having more than one type of leaf on the same plant.
heterosporous Of a plant, producing two kinds of spores, microspores and megaspores (as in all seed plants).
heterostyly (adj. heterostylous) Flowers of a species having different style lengths and different stamen lengths, the flowers of any one plant being identical (promoting cross-pollination).
heterotroph (autotroph) An organism which obtains part of its nourishment as complex compounds ultimately coming from other organisms.
hexapetaloideous Of flowers, having six petals.
hilum The mark or scar on a seed produced by separation from its funicle or placenta; the nucleus of a granule of starch.
hippocrepiform Shaped like a horseshoe.
hirsute Coarsely hairy; covered with long, stiff hairs.
hirsutulous (hirtellous) Minutely hirsute.
hispid Having bristly hairs.
hispidulous Covered with stiff, short hairs.
holomycotroph (adj. holomycotrophic) A plant that is a mycotroph as its sole method of nutrition, without chlorophyll and hence not green; an obligate mycotroph.
holotype The single specimen designated as the type of a species or lesser taxon.
homogamous Having flowers or florets that do not differ sexually.
homomorphy (adj. homomorphic) Possession of perfect flowers of only one kind.
homophyllous Having leaves or leaflets all alike.
homosporous Of plants, that produce spores of the same size and type; producing spores of one kind only.
homostylous Of a plant, having styles of the same form or length in all flowers.
homotropic All leaves rolled in the same direction in the bud; leaves on the plant all formed alike.
hybrid A natural or artificial cross between two individuals (plants) with different characteristics.
hydathode A specialized leaf structure through which water is exuded; a water pore or water gland.
hydrophyte (adj. hydrophytic) A plant that grows in water or very moist ground; an aquatic plant; a plant with resting buds in water, or with only the inflorescences being emergent.
hypanthium (pl. hypanthia, adj. hypanthial, hypanthoid) A cup-shaped or tubular structure bearing the perianth and stamens on its rim.
hypocarp Of accessory fruits, the swollen more or less fleshy tissue immediately below and more or less enveloping the fruit proper, usually derived from the pedicel.
hypochile The (often fleshy or otherwise modified) basal portion of the labellum (lip) in orchids.
hypocotyl (adj. hypocotylar) The part of the axis, or stem, below the cotyledons in the embryo of a plant.
hypocrateriform (salverform) Having a slender, tubular corolla with the lobes spreading at right angles to the tube; salver-shaped.
hypogynous Having the floral parts, such as sepals, petals and stamens, borne on the receptacle beneath the ovary.
hypotropous An ovule in which the micropyle points proximally.
hysterantherous When an apical inflorescence is produced after the pseudobulb and leaves on the same shoot. The inflorescence develops on the top of a fully developed pseudobulb with a fully grown leaf or leaves.
idioblast A unique cell, clearly distinct, specialized and/or differentiated; a plant cell (as a sclereid) that differs markedly from neighboring cells.
imbricate Leaves or other structures overlapping.
imparipinnate A pinnately compound leaf with a terminal leaflet, typically odd-pinnate; imparipinnately compound.
imperfect A unisexual flower that bears either stamens or carpels, but not both.
impresso-venose Having impressed veins.
inaperturate A pollen grain that lacks any recognizable aperture.
included Enclosed within, applied to stamens that do not extend beyond the edges of a tubular or funnel-shaped perianth.
indehiscent Of a fruit, does not split open to release its seeds.
indeterminate Of an inflorescence, with flowers opening from bottom to top, the stalk continuing to lengthen as the buds mature toward the tip, as in a panicle or raceme.
indumentum (indument) A covering of fine hairs or scales.
induplicate Folded upward; V-shaped in cross-section.
indurate Hardened and inflexible.
indusium (pl. indusia, adj. indusiate) A flap of tissue that covers a sorus, found in some leptosporangiate ferns.
inequilateral Having unequal sides; unsymmetrical.
inferior Of the ovary, located below the attachment of the perianth and stamens.

inflorescence A cluster of flowers and their arrangement on a specialized stalk (rachis).
infrafoliar Of palm inflorescences, that is positioned below the leaves of the crownshaft.
infragenera (adj. infrageneric) Taxa of lower taxonomic rank than genera (subgenera, species, etc.).
infraspecies (adj. infraspecific) Subspecies, varieties, or rarely forms of a species; taxa of lower taxonomic rank than species.
infrastaminal Below the stamens.
infructescence The complete inflorescence at the stage of fruiting.
infundibuliform (infundibular) See funnelform.
inrolled Incurved or rolled inwards.
inserted With stamens included within the perianth.
intercostal Of leaves, veinlets between lateral veins.
interfloral Occuring in between flowers (e.g. the bracts between the flowers in inflorescences of *Artocarpus*).
interfoliar Of a palm inflorescence, that is positioned among the leaves of the crownshaft.
intergeneric Occurring between different genera; of a hybrid, formed between plants of two different genera.
internode The region between two adjacent nodes of a shoot.
intersepalar Of a structure, borne between the sepals (as in the flowers of Melastomaceae).
interstaminal Of a structure, borne between the stamens; the interstaminal corona is the fleshy lobes, often connate into a tube, attached to the base of the staminal column in the interstaminal areas (e.g. in species of the family Asclepiadaceae).
intrafoliar Of palms with crownshafts, an inflorescence produced below the canopy of leaves.
intramarginal Situated within the margin.
intrapetiolar Enclosed by the base of the petiole; situated between the petiole and the stem.
intrastaminal A structure borne from the receptacle at the inside of the stamens and/or base of the ovary.
introrse Of anthers, dehiscing inward, toward the center of the flower.
invaginate (adj. invaginated) Of an organ or part, folding back upon itself; to form a pocket by turning in.
involucel (adj. involucellar) A group or cluster of bracts subtending a unit of an inflorescence.
involucre (involucrum) (adj. involucral) A group or cluster of bracts subtending an inflorescence.
involute Margins or outer portion of sides rolled inward or upward over the adaxial surface; valvate with each perianth part induplicate, folded longitudinally inward along a central axis.
iridiform Of leaves, resembling those of the family Iridaceae.
isobifacial Of flattened structures, especially leaves, having both surfaces structurally similar.
isodiametric Roughly spherical or polyhedral; having a similar diameter in all planes.
isomerous Having or composed of parts that are similar in number or position.
isophyllous Leaves equal in size and shape at any point on the branch.
isostylous Of flowers, with the styles and stamens similar in relation to each other by length or position.
isotype A duplicate specimen of the holotype, collected at the same time by the same collector from the same population.
iterauctant Of bamboos, an inflorescence with pseudo-spikelets with glumes, subtending axillary buds capable of partial or extensive spikelet ramification.
jaggery A sweet and often brown unrefined sugar.
keel Carinate; somewhat resembling the keel of a boat.
keiki In orchids, a distal vegetative branch of the main stem, which ultimately grows roots and separates.
labellum (lip) The median, modified petal of an orchid flower.
lacinia (lacina) (pl. laciniae, adj. laciniate) A jagged or irregular part of a leaf or petal; a narrow, pointed lobe; a narrow, ribbonlike lobe.
lactescent Secreting or producing a milky juice.
lageniform Flask-shaped; having an enlarged base tapering to a narrow neck.
lamella (pl. lamellae, adj. lamellate) A thin, platelike layer.
lamina (blade) (adj. laminar) The blade of a leaf or the expanded upper part of a petal, sepal or bract.
laminar Flattened.
lanate *See* villous.
lanceolate Of shape, narrow and tapering toward the apex or sometimes at the base; widest near the base, length:width ratio between 6:1 and 3:1.
lance-ovate Of shape, widest near the base, length:width ratio between 3:1 and 2:1.
lanceoloid The three-dimensional equivalent of lanceolate, widest below the middle and tapering at both ends; lance-shaped.
lanose Woolly; covered with something resembling wool; lanate.
lateral On either side of an axis.
latex A sticky clear, white or yellow sap characteristic of some plant groups.
laticifer (adj. laticiferous) A tubular structure through which latex circulates in a plant.
latrorse Of anthers, dehiscing laterally relative to the center of the flower.
laxiflorous Limp; not rigid.
laxiflorous Having loosely scattered flowers; loosely flowered.
leaf Vegetative organ of angiosperms (flowering plants).
leaflet A pinnule; a leaflike member of a compound leaf.
legume A pod; a fruit of the family Fabaceae.
lemma The lower of the two bracts enclosing the grass flower and together with the palea comprising a floret.
lenticel (adj. lenticellate) A pore in the bark, generally functioning in gas exchange; a cellular structure formed on the periderm of a stem, appearing as a lens-shaped spot and serving as a pore.
lenticular (lentiform) Resembling the seed of a lentil in form; lentil-shaped; lens-shaped.
lepidote Covered with scurfy scales or scaly spots.
leptocaulous Having a slender stem; slender-stemmed.
leptomorph Of bamboos, with rhizome monopodial, elongated, more slender than culms.
liana (liane) (adj. lianaceous, lianoid) Any of various usually woody vines that can climb as high as the tree canopy in a tropical forest.
lignescent Tending to be or become ligneous or woody; somewhat woody.
lignotuber (burl) (adj. lignotuberous) A starchy swelling of the root crown in some plants as a protection against destruction of the plant stem by fire.
lignous (ligneous) Of the nature of or resembling wood; woody.
ligule (adj. liguliform) A membrane or line of hairs on the inner (adaxial) side of the junction of the leaf sheath and leaf lamina. Bamboos sometimes have an external ligule on the abaxial side of the junction.
limen Tissue that is apparently derived from the stamens or staminal tissue between the nectary and the base of the androgynophore, or near the base of the androgynophore; threshold; any place or point of entering or beginning.
linear Slender with parallel margins.
lineate Marked with lines (especially parallel lengthwise lines); striped.
lineolate Marked with very fine parallel lines; finely lineate.
linguiform Tongue-shaped.
lingulate Formed like a tongue.
lip *See* labellum.
lithophyte (adj. lithophytic) A plant that grows on rock as its substrate.
littoral Of the seashore.
lobe (adj. lobate, lobed) The free, often flaring edges of a funnel- or tube-shaped calyx or corolla; a rounded projection of a leaf.
lobulate Consisting of, divided into, or having lobes.
locellus (pl. locelli, adj. locellate) One of the two cavities of a pollen sac; in some legumes, a secondary compartment of a unilocular ovary that is formed by a false partition; a secondary compartment (where one cavity is separated into several smaller ones).
locule (loculus) (pl. locules, loculi) A small chamber in plant tissue within which specialized structures may develop, such as the pollen-containing cavity within an anther.
loculicidal Of a capsule, splitting lengthwise so as to divide each locule into two parts.
lodicule A small scalelike or fleshy structure at the base of the stamens in a grass floret, usually two in each floret (often three or more in bamboos), they swell at anthesis, causing the floret to gape open.
loment A legume that splits crosswise, rather than lengthwise, into 1-seeded segments.
lorate Having the form of a thong or strap.
loriform Parallel-sided; narrow; filiform hair.
lunate Shaped like a crescent.
lyrate Of a pinnate leaf, divided transversely into several lobes, the smallest at the base.
malodorous *See* fetid.
malvoid A leaf tooth in which the medial vein forms a non-glandular persistent apex, lateral veins are not involved.
mammiform Having the form of a breast.
mammilla (pl. mammillae, adj. mammillate) A nipple-shaped structure.
mammilliform Having the form of a nipple.
marcescent Withering but not falling off, as a part of a plant.
marginate Having a margin.
mealy Granular.
medifixed Attached at the middle.
medullary bundle A vascular bundle in the pith.
medullary ray Of stems of woody plants, one of the vertical bands or plates of unspecialized tissue that radiate between the pith and the bark.
medulla (adj. medullated) The pith in the stems or roots of certain plants.
megagametophytes The female gametophyte in seed plants; the female gametophyte that arises from a megaspore of a heterosporous plant.
megasporangium A sporangium containing megaspores.
mentum A spurlike or chinlike extension of the flower composed of the variably united column foot, lip and lateral sepals.
mericarp One of the carpels of a schizocarp.
meristem Embryonic tissue in plants; undifferentiated, growing, actively dividing cells.
mesic Describing a moist habitat, or an organism adapted to such a habitat.
mesocarp The middle layer of pericarp; the fleshy part of certain fruits; the sarcocarp; the fleshy, often edible part of a fruit which lies between the epicarp and the endocarp.
mesochile The middle lobe of a lip that is differentiated into a hypochile, sometimes a mesochile, and an epichile.
mesophyll The parenchyma, usually containing chlorophyll, that forms the interior parts of a leaf.
metaxylem The part of the primary xylem that is the last to be formed, usually having weblike or pitted surfaces.
microgametophyte The male gametophyte that develops from the microspores of heterosporous plants; the pollen grains of gymnosperms and angiosperms.
micropyle The minute orifice or opening in the integuments of an ovule.
microsporangium (pl. microsporangia) A sporangium containing microspores.
microsporongenesis The formation of microspores inside the microsporangia (or pollen sacs) of seed plants.
microsporophyll A leaflike organ bearing microsporangia.

microverrucate Having minute protuberances.
midvein (midrib) The vein in the center of a leaf.
misapplied A valid scientific name applied to the wrong species; a homonym.
monadelphous Of stamens, united into one bundle or set by their filaments; of a plant or flower, having the stamens so united.
moniliform Consisting of or characterized by a series of beadlike swellings alternating with contractions (as in roots or stems of some plants).
monocarpic Producing flowers and fruit only once before dying.
monochasium A form of cymose inflorescence in which the main axis produces only a single branch.
monochlamydeous Of a flower, having a perianth of one whorl of members; not having a separate calyx and corolla.
monoclinous Of a plant, having both the stamens and pistils in the same flower.
monocot One of the primary divisions of angiosperms (flowering plants) characterized by a single cotyledon (seed leaf).
monoecious Of a plant, having the stamens and the pistils in separate flowers on the same plant.
monoembryony (adj. monoembryonic) The condition of an ovule having a single embryo.
monolete Describing a spore that has a simple linear scar marking the point at which it was joined in the tetrad.
monolocular Having a single cavity or chamber.
monomegaspory Selective megaspore abortion; the reduction of the contents of the megasporangium to a single viable megaspore.
monomerous Of flowers, having one member in each whorl.
monomorphic Having only one form; of a species, existing or having parts that exist in only one form.
monophyllous Having only a single leaf or leaflike part; having united sepals or petals; of a calyx, consisting of a single leaf.
monophyly (adj. monophyletic) A group or clade consisting of a common ancestor and all the descendents of that ancestor.
monopodium (pl. monopodia, adj. monopodial) Referring to a growth habit in which new leaves develop from the same meristem or growing point as all previous leaves.
monosymmetric Having bilateral symmetry; divisible lengthwise into similar or symmetrical halves; zygomorphic.
monothecal Having a single loculament.
monothecous Unilocular anthers.
montane Of, growing in or inhabiting mountainous areas.
mucro (pl. mucrones, adj. mucronate) A sharp, pointed part or organ, especially a sharp terminal point, as of a leaf.
mucronulate Having, or tipped with, a small point or points.
multiannual A plant that lives for three or more years before blooming and then dies.
multiseriate Arranged or occurring in multiple series.
multicellular Composed of several or numerous cells.
multifid Cleft into many parts, divisions or lobes.
multiflorous Bearing many flowers.
muricate Covered with short, sharp points.
muriculate Minutely muricate.
musoid Having erect leaves with long petioles like bananas (family Musaceae); having a bananalike habit.
muticous Lacking a point or awn; blunt.
mycoheterotrophy (n. mycoheterotroph) A symbiotic relationship between certain kinds of plants and fungi, in which the plant gets all or part of its food from parasitism upon fungi rather than from photosynthesis.
mycorrhiza (pl. mycorrhizae, adj. mycorrhizal) The association of fungi and roots of higher plants; symbiosis.
mycotroph (adj. mycotrophic) A plant that obtains part or all of its nutrition from organic substances provided by fungi.
myrcioid panicle An elongate axis bearing usually several decussate pairs of compound, several-flowered branches.
myrmecophilous Living symbiotically with ants.
myxogenic *See* papillose.
native An organism in association with its natural habitat.
navicular Boat-shaped.
nectar guides Colored veins in the throat of a flower, thought to direct pollinators toward the nectar source, thus enabling pollination.
nectary (adj. nectariferous, nectariform) An organ or part that secretes nectar.
neotropics (adj. neotropical) The New World extending from the Tropic of Cancer southward.
nerviform Characterized by a nerve.
nerve (adj. nervose, nerved) A vein, as in a leaf.
nervule A minute nerve.
nigrohirsute Having black hairs.
nitid Bright; lustrous; shiny.
node (adj. nodose) A joint in a stem; a part of a stem that normally bears a leaf.
nucellus The central portion of an ovule, containing the embryo sac; the megasporangium.
nut A hard, indehiscent, 1-seeded fruit.
nutant Drooping; nodding.
nutlet A small nut; a small nutlike fruit or seed; the stone of a drupe.
obclavate Inversely clavate.
obconic (obconical, obconiform, obconoid) Conical, with the attachment at the pointed end.
obcordate Heart-shaped, with the attachment at the pointed end, as a leaf.
obcordiform Inverted heart-shaped.
obcuneate Inversely cuneate.
obdeltate/obdeltoid Inversely deltate/deltoid.
obelliptic Inversely elliptic.
oblanceolate/oblanceoloid Inversely lanceolate/lanceoloid.
oblique Having unequal sides.
oblong (adj. oblongoid) Outlines and plane shapes with length:width ratio 2:1 to 3:2, the sides almost parallel.
obovate Outlines and plane shapes, with length:width ratio 2.5:1 to 3:2, the broadest part is above the middle.
obovoid Inversely ovoid; ovoid with the narrow end at the base, as certain fruits.
obpyramidal Inversely pyramidal.
obpyriform Shaped like an inverted pear, narrowed at the top and globose at the base.
obsagittate Inversely saggitate.
obtriangular Inversely triangular.
obtuse Having a broadly rounded, wide-angled or rather blunt end.
ocellated Of a spot or marking, eyelike; having ocelli, or eyelike spots.
ochrea (ocrea) (adj. ochreate, ochreous) A sheathing part, as a pair of stipules united about a stem; a cup-shaped structure that sheathes the stems of certain plants, formed from united stipules or leaf bases.
ochroleucous Yellowish white; having a faint tint of dingy yellow.
offset A shoot; a plantlet developing from the base of the parent plant.
open Of sepals or petals, that do not overlap; valvate.
opercule (operculum) (adj. opercular, operculate) The covering of the spore-bearing capsule of a moss.
oral setae Marginal setae inserted at junction of leaf sheath and lamina, on the auricles when these are present.
orbiculate (orbicular) Of a leaf or similar flat part, circular or nearly circular.
orthotropal (orthotropous) Having the axis of an ovule or seed straight from the hilum and chalaza to the orifice or the micropyle; atropous.
ostiole A small opening or orifice, as a pore (e.g. that of a fruiting body).
outcrossing To cross (to produce a hybrid) by breeding individuals of different strains but, usually, of the same breed.
ovary The seed-bearing organ of a flower at the base of the pistil.
ovate Having the shape of a longitudinal section of an egg; having such a shape with the broader end at the base.
ovoid Egg-shaped.
ovule A rudimentary seed; the plant part that contains the embryo sac and hence the female germ cell, which after fertilization develops into a seed; the megasporangium.
pachycaul (adj. pachycaulous) Trees, with particularly thick-stemmed trunks, often bottle-shaped, unbranched or sparingly branched and more or less succulent.
pachymorph Of bamboos, rhizome sympodial, thicker than culm.
palea (pl. paleae, adj. paleaceous, paleate) Flat or terete moveable appendages attached by a threadlike base, found on the sepals and petals of certain species of *Bulbophyllum*.
paleolate Having small paleae.
palmate Having four or more lobes or leaflets radiating from a single point.
palmatifid Palmate, with the divisions separated but slightly more than halfway to the common center; palmately lobed to divided.
palmatipartite Having lobes with incisions that extend over halfway toward the petiole.
palmatisect Of a leaf lamina, palmately divided almost to the petiole.
palminerved Having the venation arranged as the fingers of an open hand.
paludal (paludose) Pertaining to marshes, marshy, palustral.
pandurate (panduriform) Fiddle- or violin-shaped; obovate with the side margins concave.
panicle (adj. paniculate) An inflorescence with many branches, in which each successive branch becomes smaller; in grasses, an inflorescence in which the primary axis bears branched secondary axes with pedicellate spikelets.
pantropical Distributed worldwide in tropical areas.
papilionaceous (papilionoid) Of flowers, having a large posterior petal (banner or standard), two inner, lateral petals (wings), and two usually apically connate lower petals (keel); floral structure of the subfamily Faboideae (Papilionoideae).
papilla (pl. papillae, adj. papillose, papillous, papillate) A rounded protuberance; a small vascular process at the root of a hair.
pappus (pl. pappi) The calyx of the family Asteraceae.
papyraceous (papery) Like paper; thin or flimsy.
paracytic Situated between cells (but independent of them).
paraphyses Hairlike structures arising from the receptacle of a fern sorus.
paratype A specimen cited in a publication which is neither a holotype, isotype, or syntype.
parenchyma (adj. parenchymatous) Cells that, at maturity, are alive and potentially capable of cell division; the fundamental tissue of plants, composed of thin-walled cells able to divide.
paripinnate (paripinnately compound) A pinnately compound leaf without a terminal leaflet, typically even-pinnate.
parthenocarpic The production of fruit without fertilization of an egg in the ovary.
patent Horizontal; divaricate; divergent.
patelliform Shaped like a saucer or a patella (kneecap).
pectinate Comblike; pinnately divided with close, narrow lobes.
pedate A palmately divided leaf in which the lateral lobes are further divided (as in some ferns).
pedatisect Of leaves, pedate and cleft almost to the midrib.

pedicel (adj. pedicellate, pedicelled) The stalk of a flower; in grasses, the stalk of a single spikelet within an inflorescence.
pedoconnective Of connectives of Melastomaceae, the connectives of stamens are curved so that a knee is formed between the connective and filament; a basal-ventrally prolonged connective.
peduncle (adj. peduncular, pedunculate) The stalk of an inflorescence; in grasses the stalk of a raceme or cluster of spikelets.
pellucid Translucently clear.
peltate Having a stalk attached away from the margin of a flattened structure (e.g. a petiole attached on the underside of a leaf lamina); trichomes with stalk attached on the underside of a disk-shaped apical portion.
pendent (pendant, pendulous) Hanging downward loosely or freely.
penicillate (penicilliform) Pencil-shaped; having a tufted tip of fine hairs, like a painter's brush.
penninerved (penniveined) Pinnately veined or nerved.
pentacyclic Of a flower, which contains five whorls (one calyx, one corolla, two whorls of stamens, one whorl of carpels).
pentamerous Having a whorl with five members.
pentangular Having five angles.
pepo A non-septate, fleshy fruit with parietal placentation and a leathery exocarp derived from an inferior ovary.
percurrent Of or relating to a midrib that extends through the entire length of a leaf.
perennate (intr.v. perennating) Of plants, to live from one growing season to another, usually with a period of reduced activity between seasons.
perennial A plant that can live for more than two years.
perfect (hermaphroditic) Of flowers, bisexual.
perfoliate A leaf lamina that is sessile with the base completely surrounding the stem.
perforate With perforations or openings.
perforation plate The contact area of two adjacent vessel members, may be compound (if composed of several pores) or simple (if composed of a single opening).
pergameneous Having the texture of a parchment; thinly textured.
perianth The outermost, non-reproductive group of modified leaves of a flower, composed of the combined calyx and corolla of tepals.
pericarp The fruit wall, derived from the mature ovary wall, sometimes divisible into three layers (endocarp, mesocarp and exocarp); rind of a fruit.
pericarpel *See* hypanthium.
perichalazal Of ovules, being bitegmic, crassinucellate and anatropous.
perigone (adj. perigonal) Any organ enclosing the essential organs of a flower; a perianth.
perigonal nectary Nectaries on the perianth, usually at the base of sepals, petals or tepals.
perigonium The envelope of modified leaves surrounding the antheridia in mosses; perianth.
perigynium (adj. perigynous) An appendage about the pistil; hypanthium attached at base of ovary, with sepals, petals and stamens attached to hypanthium rim, the ovary superior.
perine A thick, outer layer in the spores of mosses and ferns.
perisperm The seed storage tissue that is derived from the nucleus and surrounds the embryo of the seed.
peristome The one or two circles of small, pointed, toothlike appendages around the orifice of an enclosure, appearing when the lid is removed (e.g. in pitcher plants).
perlite Amorphous volcanic glass that has a relatively high water content, used in horticulture to prevent water loss.
persistent Of leaves, bracts or flower parts that remain attached.
perula (pl. perulae, adj. perulate) One of the scales of a leaf bud; a pouchlike portion of the perianth in certain orchid species.
petal (adj. petaliferous, petaline) Any of the whorl of flower parts generally just inside the sepals, usually colorful and showy; a unit of the corolla, usually the second whorl from the outside of a flower.
petaloid Petal-like, of another organ that resembles a petal.
petiole (adj. petioled, petiolate) A leaf stalk.
petiolule (adj. petioluled, petiolulate) A leaflet stalk.
phenological The timing of flower emergence, sequence of bloom, fruiting and leaf drop.
phloem The part of a vascular bundle consisting of specialized sieve tubes, companion cells, parenchyma, and fibers, and forming the food-conducting tissue of a plant.
phyllanthoid Of branching, in which the vertical stems bear deciduous, floriferous (flower-bearing), plagiotropic (horizontal or oblique) stems.
phyllary (pl. phyllaries) A bract below the flowerhead in Asteraceae; one of bracts that forms the primary involucre in Asteraceae.
phylloclade (phylloclad) A flattened, photosynthetic branch or stem that resembles or performs the function of a leaf, as in certain cacti; cladode.
phyllode (phylodium) (pl. phyllodes, phyllodia, adj. phyllodic, phyllodinous) An expanded petiole resembling and having the function of a leaf, but without a true lamina.
phyllopodium (pl. phyllopodia) An outgrowth from the stem to which a stipe is articulated.
phyllotaxy (pl. phyllotaxis) The arrangement of leaves on a stem or axis.
phylogeny (adj. phylogenetic) A classification that is based on evolutionary history, or pattern of descent; relationships between groups of individuals at a given point in time; the evolutionary history of an organism or group of organisms.
phytochrome A plant pigment associated with the absorption of light that may regulate various types of growth and development.
phytomelan The black, carbonaceous, opaque material forming a crustlike covering of some seeds, mainly in the order Asparagales.
piliferous Bearing or producing hair.
pilose (pilous) Covered with soft hair.
pilosulose (pilosulous) Bearing minute, long, soft, straight hairs.
pinna (pl. pinnae, adj. pinnatiform, pinniferous) One of the primary divisions of a pinnate leaf.
pinnate (pinnately compound) Of a leaf, having leaflets or primary divisions arranged on each side of a common stalk (rachis).
pinnatifid Pinnately-lobed to divided.
pinnatilobate Of leaves, pinnately divided into lobes.
pinnatinerved Pinnately nerved (veined).
pinnatipartite Of leaves, pinnately divided into lobes reaching just over halfway to the midrib; having lobes with incisions that extend more than halfway toward the midrib.
pinnatisect Pinnately divided, almost into discrete leaflets but confluent at the midrib.
pinninerved *See* pinnatinerved.
pinnule (adj. pinnular) The ultimate divisions or leaflets of a leaf; a secondary pinna, one of the pinnately disposed divisions of a bipinnate leaf.
pistil The part of the gynoecium composed of an ovary, one or more styles (which may be absent), and one or more stigmas.
pistillate Of flowers, unisexual with carpel(s) only, lacking fertile stamens.
pistillode (pl. pistillodes, pistillodia) A sterile vestigial pistil remaining in a staminate flower.
placentation The position or arrangement of the placentae, and hence of ovules, within the ovary.
plagiotropic Growing more or less divergent from the vertical.
plasmodial (amoeboid) An anther tapetum type in which the tapetal cell walls break down, with release of the tapetal cells into the locule.
pleiothyrsoid (adj. pleiothyrsic) An inflorescence in upper leaf axils.
pleonanthic An indeterminate shoot that bears lateral flowers but that continues vegetative growth.
pleurogram The fine inconspicuous line on the surface of the seed coat that circumscribes the areole that occurs on the opposite faces of the seeds.
plicate (n. plication) With a series of longitudinal folds; pleated.
plinerved Of a leaf, with lateral nerves similar to the midrib arising near the insertion of the petiole.
plumose Featherlike.
plumule The bud of the ascending axis of a plant while still in the embryo; the young shoot above the cotyledon(s) of an embryo or seedling.
plurilocular Of an ovary, with two or more locules; divided into locules or chambers.
pluriovulate Having multiple ovules.
plurisulcate Having multiple grooves or channels.
pneumathode An aerating organ on some roots, groups of cells that allow gas exchange to occur.
pneumatophore Of some mangrove and swamp plants, roots that grow upwardly from the soil to air, functioning in obtaining additional oxygen for the plant.
pod A seed pod, usually of the family Fabaceae.
poikilohydry (adj. poikilohydric) The equilibration of cellular water content with that in the environment, which can be combined with tolerance to dessication, as in mosses and other resurrection plants.
pollen The grains containing spores, often brightly colored; male gametes.
pollinarium (pl. pollinaria) The functional unit of pollen transfer in orchid pollination, consisting of two or more pollinia (sometimes with caudicles), often a stipe and a viscidium.
pollinator An animate or inanimate agent that transmits pollen to a receptive stigma.
pollinium (pl. pollinia) A coherent mass of pollen grains.
polocytic A stomata type, with stomata attached to the wall of the guard cells at the anterior end.
polyad A group of pollen grains that are fused in precise units of more than four.
polyarch A root that has many xylem bundles.
polycarpic Producing fruit many times (as a perennial plant); having a gynoecium composed of two or more distinct ovaries.
polycylic Having two or more rings or whorls.
polyembryony (adj. polyembryonic) The production of more than one embryo from a single fertilized egg cell.
polygamodioecious Having bisexual and male flowers on some plants, and bisexual and female flowers on others, i.e. mostly dioecious, but with some perfect flowers.
polygamomonoecious Polygamous but chiefly monoecious, i.e. mostly monoecious, but with some perfect flowers.
polygamous Bearing both unisexual and hermaphrodite flowers on the same or different plants of the same species.
polymorphic Having more than one distinct form.
polyphyllous Many-leaved.
polystichous Arranged in two or more series or rows.
polysymmetric Of flowers, radially symmetric, that can be divided into two equal halves from any angle; actinomorphic.
pome (pl. pomes) A firm but fleshy fruit with the seeds enclosed by a thin, papery endocarp.
porate Of a pollen grain, with rounded apertures only.
poricidal Of anther, dehiscing by a pore at one end of the thecae; of fruits, dehiscing by a small hole (pore) at one end.
poriform Resembling a pore or a small puncture.
porrect Extended forward.
posterior Toward the back and near the main axis.
posterior lobes Of leaves, the lobes at the base of a leaf (e.g. in sagittate and hastate leaves).

postfloral The period after flowering (from corolla abscission to fruiting).
posticous Posterior.
praemorse Having the end abruptly and unevenly truncated, as if it has been bitten or broken off.
precocious Developing early, often applied to flowers of deciduous plants that open before the new leaves are fully developed.
premontane Mountainous regions of altitudes between 500 and 1,500 meters.
prickle A small, pointed process growing from the bark of a plant.
primordium (pl. primordia) An organ or part in the earliest stage of development.
probract A small, leaflike structure at the base of an inflorescence in members of the cucumber family (Curcurbitaceae), usually borne opposite a tendril.
procumbent Trailing along the ground without taking root.
prolate Of pollen shape, radially symmetrical, with the polar axis longer than the equatorial diameter, i.e. the grain is longer than it is broad.
prolepsis (adj. proleptic) Of the timing of axillary growth, development of an axillary bud only after a period of rest.
proliferous Having the ability to reproduce vegetatively from the shoot system, by plantlets developing on leaves or fronds or in the inflorescence, etc.
prominulous (prominulent) Slightly raised above the adjoining tissue.
propagule Any structure with the capacity to give rise to a new plant.
prophyll (adj. prophyllate) A leaflike structure, such as a bracteole, subtending a single flower or pedicel; lower bract on a peduncle.
protandrous Of or relating to a flower in which the anthers release their pollen before the stigma of the same flower is receptive; (of hermaphrodite or monoecious plants) maturing the anthers before the stigma.
protantherous When leaves appear before flowers.
proteranthous When an inflorescence is produced before the pseudobulb and leaves on the same vegetative shoot.
protogynous A kind of dichogamy in which the flower sheds pollen after the stigma has ceased to be receptive.
protostele (adj. protostelic) A variant of a stele in which the central vascular column is solid.
protoxylem The first-formed tissue of the primary xylem differentiating from the procambium (the tracheary cells having spiral or annular thickenings).
pruinose Covered with a frostlike bloom or powdery secretion.
pseudoberry A false berry.
pseudobulb (adj. pseudobulbous) The variously thickened portion of an aerial orchid stem.
pseudocalyx An epicalyx; an involucre of calyxlike bracts.
pseudocarp An accessory fruit; a fruit having enlarged accessory structures in addition to those formed from the ovary.
pseudocorolla A false corolla formed by the outer members of the androecium (as in the family Lecythidaceae).
pseudolateral Having the tendency to become lateral when it is normally terminal.
pseudopetiole The narrow basal portion of some leaf laminas, resembling a petiole.
pseudospikelet Of bamboos, a spikelet in which the outer glumes or bracts subtend axillary buds which can develop to form lateral spikelets or branches.
pseudostem A false stem composed of concentric rolled or folded laminae and sheaths that surround the growing point (as in the family Musaceae).
pseudostigma A false stigma formed from a modified style.
pseudostyle A false style formed from the stigmatic crest, or by lateral flattening.
pseudoterminal When an inflorescence is apparently terminal but is actually axillary on a very short terminal internode that is usually concealed by small bracts.
pseudowhorl An arrangement of leaves that arise so close together as to appear to arise at the same level (whorled); leaves attached at almost equal levels and separated by single fully elongated internodes.
pteridophyte Any plant of the division Pteridophyta.
ptyxis The posture of embryonic structures, such as cotyledons within a seed or immature leaves or leaf parts in a bud.
puberulent (puberulous) Minutely pubescent.
pubescent Covered with short, soft, erect hairs.
pulque A fermented milky drink made from the juice of certain species of *Agave* in Mexico.
pulverulent Powdery.
pulvinus (pl. pulvini, adj. pulvinar, pulvinate, pulviniform) A cushionlike swelling at the base of a leaf or leaflet at the point of junction with the axis.
punctate (punctiform) (n. punctation) Marked with dots or spots.
pungent Ending in a stiff, sharp point; having an acrid taste or smell.
pusticules (adj. puncticulate, punctulate) Minute dots.
pustulate (pustulose, pustular) Covered with small pustulelike or blisterlike elevations.
putamen A hard or stony endocarp, as a peach stone; a pyrene.
pyrene *See* putamen.
pyriform Pear-shaped.
pyxidium (pyxide, pyxis) (pl. pyxides) A seed vessel (capsule) that opens transversely, the top part acting as a lid.
quadrifarious Arranged in four rows or ranks.
quadrigonal Having four corners or edges.
quadripinnate (tetrapinnate) Pinnately compound leaves in which the leaflets are themselves tripinnate.
quaquaversal Directed outward from a common center.
quincuncial Of, relating to, or forming a quincunx (an arrangement of five-petaled flowers in which two petals are inferior, two are exterior, and one is partly interior and partly exterior).
quinquangular Having five angles or corners.
raceme (adj. racemose, racemiform) An indeterminate inflorescence in which the main axis produces a series of flowers on lateral stalks, the oldest at the base and the youngest at the top; a spike; in grasses, an unbranched axis bearing spikelets (the racemes may be solitary, digitate or scattered).
rachilla (pl. rachillae) Of grasses, the central axis of the spikelet which bears the floret.
rachis (pl: rachises, rachides) The axis of an inflorescence or a pinnate leaf.
radial (radially symmetric) *See* actinomorphic.
radiate Radiating symmetrically; of a capitulum, consisting of ray florets.
radicle The first root of a seed embryo.
rameal Of or pertaining to a branch; of inflorescences, borne on branches.
ramenta One of the thin, chafflike scales covering the shoots or leaves of certain ferns.
ramicaul A rhizome with an above-ground portion to which the leaves are attached.
ramiflorous Flowering on the branches.
rank One column of alternate or opposite leaves; one level of nomenclature.
raphe (adj. raphal) A ridge connecting the hilum with the chalaza; the portion of the funicle that is united to the ovule wall, commonly visible as a line or ridge on the seed coat.
raphide A needle-shaped crystal of an irritant substance such as oxalic acid formed in a plant.
ray floret The radiating florets around the periphery of the floral heads, having a tubular corolla with a flattened lobe (ligule).
receptacle (adj. receptacular) The expanded tip of a flower stalk or axis that bears the floral organs or the group of flowers in a head.
reclinate (reclining) Bent or turned downward toward the base.
recurved (v. recurving) Curved backward or inward.
reduplicate Of petals or sepals, having the margins curving outwards.
reflexed Bent backward or downward.
regular Of a radially symmetrical flower, actinomorphic.
reniform Kidney-shaped.
repand Having a wavy margin.
replum The framework of some fruits, which remains after the valves drop off.
resiniferous Yielding resin.
resupinate Of orchids, with the pedicel twisted so that the lip is always in the same position (usually at the bottom of the flower) regardless of the position of the inflorescence.
reticulate Having the veins or nerves disposed like the threads of a net.
retrorse Especially of plant parts, pointing backward or in a direction opposite to normal.
retuse Having an obtuse or rounded apex with a shallow notch.
revolute Rolled under or back.
rheophyte (adj. rheophytic) An aquatic plant that lives in fast-moving water.
rhipidium (pl. rhipidia, adj. rhipidiate) A fan-shaped inflorescence with cymose branching in which branches lie in the same plane and are suppressed.
rhizome The indeterminate stem or system of stems of many plants, which successively give rise to new shoots and flowers, often horizontal or underground but sometimes appressed to branches or rocks.
rhizophore A leafless, downward-growing dichotomous shoot developed in certain species of *Selaginella*, which bears the true roots. It has the external appearance of a root, but without a root-cap.
rhombic (rhombiform, rhomboid) Parallelogram-shaped, widest near the middle, with a length:width ratio of 2:1 to 3:2.
ricin A white, poisonous, protein powder obtained from the seeds of the castor-oil plant (*Ricinus communis*).
ring scar A ring around the trunk, typically of palms, where a leaf has separated.
ringent Having widely spread lips, as some corollas.
riparian *See* riverine.
riverine Of or pertaining to a river; riparian.
rodlet A minute rodlike structure.
rosette (adj. rosulate) A circular cluster of leaves or other organs.
rostellum (adj. rostellar, rostellate) Part of the median stigma lobe of orchid flowers. The rostellum remnant is the often cleft or bilobed part of the rostellum that remains after the viscidium has been removed by a pollinator.
rostrate (rostriform) Furnished with a rostrum (a beaklike process or extension of some part).
rotate Of a flower, wheel-shaped, with a very short corolla tube and spreading lobes.
rubiginose Rusty or rust-colored; brownish-red.
rufous Reddish or tinged with red; brownish red.
rugose (rugate) Wrinkled.
rugulose (rugulate) Finely rugose; having many small wrinkles.
saccate Having a sac or the form of a sac.
sagittate Shaped like an arrowhead.
salicoid Of a leaf tooth, in which the medial vein ends in dark but not opaque persistent spherical callosity, no laterals are involved.
salverform Of the corolla of flowers, consisting of a narrow tube with flat, spreading terminal petals.
samaracetum An aggregation of samaras.

samara (adj. samaroid) An indehiscent, usually one-seeded, winged fruit.
saprophyte (adj. saprophytic) A heterotroph obtaining complex nutrients from the decay of other organisms, usually lacking chlorophyll.
sarcocarp The fleshy mesocarp of certain fruits; any fruit of fleshy consistency.
sarcotesta (adj. sarcotestal) A layer of pulp in which one or more seeds are embedded.
sarmentose Of plants, having stems in the form of runners; stoloniferous.
savanna A grassland region with scattered trees grading into either open plain or woodland, usually in subtropical or tropical regions, especially where the rainfall is seasonal.
saxicolous Growing on, or living among, rocks.
scaberulous Minutely scabrous.
scabrate *See* spinule.
scabrescent *See* scaberulous.
scabrid Somewhat scabrous.
scabridulous *See* scaberulous.
scabrous Having a rough or scurfy surface or trichomes; resembling sandpaper.
scale A small, non-green leaf of a bud or underground rootstock; a bract of some inflorescences.
scalariform Ladderlike; resembling the rungs of a ladder.
scandent *See* clambering.
scape (adj. scapose, scapigerous) The stemlike 'naked' peduncle of a plant with radical or rosette leaves that lacks leaves along its length.
scariose (scarious) Of plant parts, membranous, dry, and brownish in color.
scarious Thin and appearing dry, usually whitish or brownish.
schizocarp (adj. schizocarpic) A dry, dehiscent fruit formed from two or more carpels and breaking apart septicidally into one-carpellate units (mericarps) when ripe, these containing one (or more) seeds.
schizogenous Of cavities in plants, formed by the separation of cells down their middle lamellae.
sclereid (adj. sclerified) A thick-walled lignified plant cell that is often branched.
sclerenchyma (adj. sclerenchymatous) Mechanical tissue made up of fibers and/or sclereids; supporting or protective tissue composed of thickened, dry and hardened cells.
sclerophyll A type of vegetation characterized by hard, leathery, evergreen foliage that is specially adapted to prevent moisture loss and short internodes.
scorpioid *See* cincinnus.
scrambling Of a plant, that climbs with the aid of tendrils or prickles.
scrobiculate Marked with many shallow depressions, grooves or pits.
scurfy Gritty.
scutiform Having the shape of a shield.
seasonally dormant Of herbs, that die back to a subterranean rhizome during cold or dry seasons; plants that remain evergreen, but stopping all growth seasonally.
secretory Pertaining to secretion; performing the process of secretion.
sectile Referring to pollinia comprising several 'packets' connected by elastoviscin.
secund Flowers, inflorescences or other structures arranged on one side of the axis, often by twisting of stalks; with all the parts grouped on one side or turned to one side.
semideciduous Of plants, that lose part of their foliage; of plants, that can lose their foliage for a very short period.
semiepiphytic A mostly epiphytic plant with aerial roots, but also able to grow on the ground.
semipetalous Having united petals; gamopetalous.
semiterrestrial A mostly terrestrial plant with the ability to thrive as an epiphyte in certain conditions.
sepal (adj. sepaline, sepaloid) A unit of the outer floral whorl collectively referred to as the calyx, often protecting the flower in bud; any of the outermost whorl of lower parts.
septicidal Of seed capsules, dehiscing by splitting along or through the septa.
septifragal Of seed capsules, dehiscing by the breaking away of the valves from its partitions.
septum (pl. septa, septae, adj. septal, septate) A partition (e.g. in the spur of the lip of some orchids).
seriate Arranged or occurring in a series or in rows.
sericeous Covered with soft silky hairs.
serotinous Late in developing or blooming.
serrate Saw-toothed; having leaf margins with teeth angled toward the apex.
serrature One of the teeth in a serrated edge; a serration.
serrulate Having a minutely serrate margin.
sessile Lacking a stalk (petiole, pedicel, etc.).
seta (pl. setae, adj. setiform) A stiff hair, bristle or bristlelike process or part of an organism; the stalk of a moss capsule.
setose Bristly; setaceous.
setulose Having small bristles or setae.
shrub (adj. shrubby) A woody plant lacking a main trunk or axis, branching near the base.
shrubland A plant community characterized by vegetation dominated by shrubs.
shrublet A dwarf shrub.
silicaceous (siliceous) Containing silica.
silique (adj. siliquiform) A dry, flattened, dehiscent, bicarpeled fruit that dehisces along two sutures, and has a persistent partition (septum).
simple Not divided.
sinuate (sinuous) Having the margin strongly or distinctly wavy.
sinuolate Having the margin minutely wavy.
sinus The space between two projecting lobes or teeth.
siphonostele A type of stem in which the central vascular column is medullated (i.e. pith surrounded by xylem and phloem).
sobole (adj. soboliferous) A shoot running along underground, forming new plants at short distances.
solenostele (adj. solenostelic) A siphonostele in which there is also a cylinder of phloem internal to the xylem.
sorus (pl. sori) A discrete cluster or aggregation of leptosporangia.
spadicles Club-shaped bracts borne on the spadix in the family Balanaphoraceae. They protect the flowers and sometimes secrete nectar.
spadix (pl. spadices) A spike with a thickened or fleshy central axis, typically with congested flowers.
spathe (adj. spathate, spathiform, spathoid, spatheate, spathaceous) A bract or modified leaf (without a lamina) subtending the inflorescence or part of it.
spatheole The uppermost spathe supporting the racemes within a compound inflorescence.
spathulate Spoon-shaped; broad at the tip and narrowed towards the base; like a spatula.
species The lowest basic division of plants (and other organisms).
spheroid (spheroidal) Shaped like but not exactly a sphere.
spicule (adj. spiculate) A small needlelike structure or part.
spike (adj. spicate, spiciform) An indeterminate inflorescence consisting of a single axis bearing sessile flowers.
spikelet The basic unit of a grass inflorescence, usually composed of two glumes and one or more florets on a rachill.
spindle-shaped *See* fusiform.
spine (adj. spinose, spinous, spinescent, spiniform) A sharp-pointed leaf or leaf part.
spinule (adj. spinulose, spinulate) A small spine.
spiralate Having the shape of a spiral.
sporangiophore A unit of the strobilus bearing sporangia.
sporocarp A multicellular spherical structure in which spores are formed; a fruiting body.
sporophore An organ that produces or carries spores.
sporophyll A modified leaf that bears sporangia.
sporophyte (adj. sporophytic) The diploid (2n) phase in the haplodiplontic life cycle of all land plants; the form of a plant in the alternation of generations that produces asexual spores.
spur A saccate or tubular extension of the lip (or other floral parts) in many orchids, often containing nectar.
squamate Covered with scales.
squamellae A diminutive scale or bractlet; a palea.
squamiform Having the shape of scales or a scale.
squamose Covered with or formed of scales; scaly.
squarrose Having tips projecting at right angles, or having spreading bracts.
stamen (adj. staminal) The pollen-bearing organs of seed plants.
staminate Having a stamen or stamens; having stamens but no pistils; a male flower.
staminodium (staminode) (pl. staminodia) A sterile or abortive stamen; a part resembling such a stamen.
standard An enlarged or otherwise distinctive erect petal.
stelidium (pl. stelidia) Of orchids, a discrete arm or projection borne on each side of the column, near the apex, middle, or base, often slender and elongated; usually interpreted as staminodia (sterile anthers).
stellate Star-shaped.
stigma (adj. stigmatose, stigmoid) The part of a pistil that receives the pollen.
stipe (pl. stipites, adj. stipitate) A pollinium stalk derived from the rostellum.
stipel (adj. stipellate) A small stipe.
stipule (adj. stipular, stipulate, stipuliform) One of the usually small paired appendages at the base of a petiole in certain plants.
stolon (adj. stoloniferous, stolonoid) A prostrate stem, at or just below the surface of the ground, that produces new plants from buds at its tips or nodes.
stomate (stoma, stomata) Any of various small apertures, especially one of the minute orifices or slits in the epidermis of leaves, stems, etc., through which gases are exchanged.
stomium The part of the sporangium of ferns that ruptures to release the spores.
stramineous Straw-colored; yellowish.
stria (pl. striae) A thin, narrow groove or channel.
striate Striped or streaked.
strigose Set with stiff bristles of hairs; hispid.
strigulose (strigillose) Minutely strigose; set with minute bristles of hairs.
striolate Minutely striate, or striped.
strobilus (pl. strobili, adj. strobilate, strobiliform, strobilaceous) A reproductive structure characterized by overlapping scalelike parts; a conelike structure composed of sporophylls (as in club mosses and horsetails).
strophiole (adj. strophiolate) A crestlike excrescence about the hilum of certain seeds; a caruncle.
struma A cushionlike swelling on an organ.
stylar The stalk of a carpel, bearing the stigma.
style A narrow, usually cylindrical and more or less filiform extension of the pistil, bearing the stigma at its apex.
stylid *See* stelidium.
stylode (pl. stylodia) A sterile style (often found in staminate flowers).
stylopodium A swelling at the base of the style in some plants.
stylule (pl. styluli) A small style.
suberose Relating to, resembling or consisting of cork; corky.
submontane Pertaining or belonging to the lower slopes of mountains.
subshrub A plant consisting of a woody, perennial base with annual, herbaceous shoots.

subspecies A taxon or division of a species, above variety, that has distinctive characteristics unique to plants from a specific geographic range.
subterete Cylindrical, with a groove down one side.
subtribe A category in the classification of organisms usually between a subfamily and a genus or sometimes between a suborder and a family.
subulate (subuliform) Slender, somewhat cylindrical, and tapering to a point; awl-shaped.
succulent Partially or wholly fleshy or juicy; moisture-storing stems, leaves or roots.
sucker An auxiliary shoot from the root.
suffrutice (adj. suffruticose, suffrutescent) A subshrub; suffrutex.
sulcus (pl. sulci, adj. sulcate) A furrow or groove.
superior With the ovary located above the attachment of the perianth and stamens.
superposed Placed on top of each other or at opposite ends.
supervolute Having a plaited and convolute arrangement in the bud.
suprafoliar Of inflorescenes, in which the peduncle is tall enough to support the umbel completely above the top of the leaves.
suprareticulate Of pollen grains, having a reticulate pattern due to suprategillar elements.
syconium A multiple fruit developed from a hollow fleshy receptacle containing numerous flowers, as in the fig.
symbiosis (adj. symbiotic) An ecological relationship between two different organisms in which both obtain mutual benefit.
sympetalous See gamopetalous.
sympodium (pl. sympodia, adj. sympodial) A discontinuous main axis where the stem is made up of a series of superposed branches, these imitating a single main axis with each new shoot developing from an axillary bud on the previous shoot unit; of stems, where growth is continued not by the main stem but by lateral branches.
synandrium An androecium with the anthers of the stamens cohering.
synandrodia Sterile flowers.
synangium A cluster of sporangia which have become fused in development.
synanthous When pseudobulb, leaf, and apical inflorescence are produced together.
syncarp (adj. syncarpous) An aggregate fruit; a fleshy compound fruit composed either of the fruits of several flowers (as in the pineapple) or of several carpels of a single flower (as in the raspberry).
syncolpate A pollen grain in which the colpi are joined.
synergid One of two small cells lying near the egg in the mature embryo sac of a flowering plant.
synflorescence A compound inflorescence composed of both terminal and lateral flowering branches.
syngenesia (adj. syngenesious) Plants with the anthers fused, forming a ring around the style.
synovarious Having free styles and stigmas.
synpetal A floral part formed by the partial or complete fusion of two or more petals.
synsepal (synsepalum) (adj. synsepalous) A floral part formed by the partial or complete fusion of two or more sepals.
synstylovarious Having partially fused styles and free stigmas.
syntepalous A floral part formed by the partial or complete fusion of two or more tepals.
syntype One of two or more specimens of a species or lesser taxon when no single specimen was selected as the holotype.
syrinx An inverted narrowed structure on the lower end of the calyx tube (especially in *Aristolochia*).
systematics The science that includes and encompasses traditional taxonomy, with its primary goal being the reconstruction of phylogeny.
tapetum (adj. tapetal) A layer of cells often investing the archespore in a developing sporangium and absorbed as the spores mature.
taxon (pl. taxa) An organism or group of organisms, ideally monophyletic and traditionally treated at a particular rank.
tegmic Pertaining to the inner coat of a seed.
tendril (adj. tendrilled) A leaf, stipule or stem modified into a slender spirally coiling sensitive organ, serving to attach a climbing plant to its support.
tenuinucellate An ovule in which the nucellus is composed of a single layer of cells, with no formation of a parietal cell.
tepal A component of the perianth in which the parts are not clearly differentiated into petals and sepals.
terete Cylindrical.
terminal At the distal end of a branch or stalk; at the apex.
terminalioid Of branching, in which the branches are arranged in tiers.
ternate Of leaves, consisting of three leaflets; of plants, having leaves arranged in whorls of three.
terrestrial Growing on the ground; growing with the roots anchored in the ground.
testa Seed coat.
tetrad A fused product of four pollen grains, developing from the four products of microsporogenesis.
tetradynamous With stamens in two groups, of four long and two short.
tetragonous Four-angled.
tetralocular Four-loculed.
tetraploid Having a chromosome number that is four times the basic or haploid number.
tetrasporangium (adj. tetrasporangiate) A sporangium containing four haploid spores.
thallus (adj. thalloid) A simple vegetative body undifferentiated into true leaves, stem and root, ranging from an aggregation of filaments to a complex plantlike form.
theca One half of a typical anther containing two microsporangia.
thermogenic Causing or pertaining to the production of heat.
thyrse (adj. thyrsic, thyrsiform, thyrsoid) A compact branching inflorescence in which the main axis is indeterminate and the lateral axes are determinate.
tomentulose (tomentellous, tomentellate) Slightly tomentose.
tomentum (adj. tomentose) Pubescence consisting of moderately long, soft, entangled hairs pressed close to the surface.
torulose Of a cylindrical or ellipsoid body, swollen and constricted at intervals.
tracheary Water-conducting.
tracheid An elongated cell in the xylem of vascular plants that serves in the transport of water and mineral salts.
trailing Of horizontal stems that do not take root, procumbent.
trapeziform Shaped like a trapezium.
tree A woody plant with one or several well-developed trunks.
treelet A small tree.
triad A group of three spikelets borne together.
tribracteolate Having three bracteoles.
tricarpellate Having three carpels.
trichasial Having three cymes.
trichome An external, hairlike plant structure.
trichomidia Minute hairlike plant structures.
trichosclereid Describes hard, long and slender (needlelike) cells found in some species of plants that protects the plant from being eaten.
trichotomous Division into three parts or elements.
tricoccal Having three cocci.
tricolpate Of pollen grains, with three, long, grooved apertures.
tricuspidate An organ or part, having three cusps.
tridenticulate Having three minute teeth.
trifid Divided or cleft into three narrow parts or lobes.
trifoliate Bearing three leaves.
trifoliolate A compound leaf bearing three leaflets.
trigonal Of a triangle; triangular.
trigonous Having three angles or corners.
trilamellate Having or composed of three scales.
trilete Spores with a three-branched laesura (the arm of a proximal fissure or scar of a spore).
trilobed Trilobed.
trilobulate (trilobate) Trilobuled.
trilocular Triloculed.
trimerous Having flower parts, such as petals, sepals and stamens, in sets of three.
trimorphism (adj. trimorphic) Having three distinct forms.
trindentate Having three teeth.
trinerved Having three nerves or veins.
trinucleate Having three nuclei.
triovulate Having three ovuli.
tripetaloideous Having three petaloids.
tripinnate Pinnately compound leaves in which the leaflets are themselves bipinnate.
tripinnatipartite 3 times pinnatipartite.
tripinnatisect 3 times pinnatisect
triplacentiferous Having three placentas.
triplinerved Leaves in which the midrib produces two slightly smaller side ribs near the base.
triquetrous Having three salient angles, the sides concave or channeled.
triseriate Arranged in three vertical or spiral rows.
trisulcate Having or characterized by three grooves.
trithecal Having three thecae.
trochlea A large swelling near or at the base of the androgynophore (especially in *Passiflora*).
trophophore (tropophore, trophophyll) Of fronds, the laminalike sterile photosynthetic portion.
trullate (trulliform) Parallelogram-shaped; shape with with the widest axis below the middle and with straight margins, with length:width 2:1 to 3:2; trowel-shaped.
truncate Apical or basal margin cut straight across.
tuber (tuberoid, tubiform) A thick, underground storage stem, usually not upright, typically bearing outer buds and lacking surrounding storage leaves or protective scales.
tubercule (tubercle) (adj. tuberculate) A small wartlike outgrowth.
tuberiferous Producing or bearing tubers.
tubulate Shaped like or having a tube.
tufted Plants with the leaves clustered at the apical end.
turbinate Top- or turban-shaped.
twining Climbing by encircling a support such as a tree limb.
type species The typical form of a plant, not a variant or cultivar; a species as originally described by the author.
ultramafic Of rocks, containing iron and magnesium, with little or no silica.
umbel (adj. umbellate, umbelliform) A determinate or indeterminate, flat-topped or convex inflorescence with pedicels attached at one point.
umbellule One of the secondary umbels in a compound umbel.
umbilicate Navel-like.
umbo (adj. umbonate) A blunt or rounded protuberance arising from a surface.
unciform (uncinate) Hook-formed.
underbark The inner layer of tree bark, beneath the overbark.
undershrub A low-growing, woody, bushy plant.
understory The area of a forest which grows at the lowest height level below the forest canopy.
undulate Having a wavelike or rippled form, surface, edge; wavy.
unguiculate Having a clawlike base.
unicarpellate With a single, free carpel.
unicespitose Of bamboos, culms in a single dense clump.

unifacial Of leaves, rounded or terete, i.e. with no adaxial and abaxial surfaces.
unifoliate Bearing one leaf.
unifoliolate A compound leaf bearing one leaflet (or reduced to a single leaflet).
unilabiate Having one lip only.
unilacunar Of nodes, with a single leaf gap.
unilocular Having or consisting of only one loculus, chamber, or cell.
uninucleate Of a cell, having one nucleus.
uniovulate Containing one ovule.
uniseriate Arranged in a single row or line.
unisexual Of flowers, that are either male (staminate) or female (pistillate).
united Of flower parts, fused or connate.
unitegmic An ovule having a single integument.
unsegmented A structure not divided into segments.
upcurved Curving upward.
urceolar (urceolate) Urn-shaped.
utriculate (utricular) Resembling a bladder or swollen like a bladder; inflated.
vaginate Forming or enclosed in a sheath; resembling a sheath.
vallecula (pl. valleculae, adj. vallecular) A small groove or furrow in a plant stem or fruit.
valvate Open; of petals or sepals, that do not overlap.
valve One of the sections into which the wall of a seedpod or other dehiscent fruit splits; a lidlike covering of an anther.
vandaceous Orchids having characteristics as species of *Vanda*.
vascular bundle A part of the transport system in vascular plants.
vegetative Having the power to produce or support growth in plants; of reproduction, asexual; noting the parts of a plant not specialized for reproduction; growing or developing as or like plants.
veinlet A small vein.
velamen The spongy outer layer of an orchid root, consisting of dead cells at maturity.
veld Characteristic parts of southern Africa where the open country contains grass, bushes or shrubs, or are thinly forested.
velutinous (velutinate) Covered with dense, soft, silky hairs.
venose (venous) Of a plant, covered with veins or similar ridges.
ventricidal Of seeds, dehiscence along the ventral suture of the carpel; marginicidal.
ventricose Having a swelling especially on one side, or unequally; protuberant.
venule (adj. venulate, venulose) Full of venules, or small veins.
vernation The arrangement of the young leaves within a bud.
verrucose Rough, warty.
verruculose Minutely verrucose.
verticel A circle of leaves or flowers about a stem at the same node.
verticillaster A typically elongate inflorescence in which paired, nearly sessile cymes surround the axis appearing as whorls (as seen in the family Lamiaceae).
verticil (adj. verticillate, adv. verticillately) A single (false) whorl of flowers in a verticillaster; a whorl of three or more parts at a node.
vestigial (adv. vestigially) An imperfectly developed remnant of what was once a fully developed, functional organ.
vexillum (adj. vexillar, vexillary) The large upper petal of a papilionaceous flower.
villosulose (villosulous) Covered with minute trichomes.
villous (villose, lanate) Covered with very long, soft, crooked trichomes; pubescent with long and soft hairs that are not interwoven.
vine A climbing plant.
violaceous Violet-colored.
virgate Straight, long and slender.
viscidium (viscid) (pl. viscidia) The sticky portion of the rostellum, which is often connected to pollinia.
viscin Elastic and/or somewhat sticky material often covering pollen grains.
vitiform Shaped like a grape leaf.
vivipary (adj. viviparous) Of plants, producing bulbils or young plants instead of flowers; of seeds, germinating before separating from the parent plant.
volva A cup-shaped structure that sheathes the base of a stalk.
whorl (adj. whorled) A cluster of leaves or other organs densely arranged around a point or short stalk.
wing A lateral organ, appendage or membrane, sometimes found on stems, petioles, capsules, seeds, etc.
xerophyte (adj. xerophytic, xeric) A plant adapted to life in dry (xeric) habitats, or have the ability to survive periods of drought.
zingiberoid Have horizontal leaves with short petioles resembling gingers (Zingiberaceae).
zonasulculus (adj. zonasulcate) A ringlike sulculus around the equator of a pollen grain.
zoniaperturate (zonoaperturate) Of pollen grains, with apertures occuring in the equatorial region.
zygomorphic Bilaterally symmetrical; divisible into two equal halves along only a single plane.

REFERENCES

Alverson, W.S., Whitlock, B.A., Nyffeler, R., Bayer, C., & Baum, D.A. (1999) Phylogeny of the core Malvales: evidence from ndhF sequence data. *American Journal of Botany, 86*(10), 1474–1486.

Argent, G. (2006) *Rhododendrons of subgenus* Vireya. Edinburgh: Royal Horticultural Society.

Asmussen, C.B., Dransfield, J., Deickmann, V., Barfod, A.S., Pintaud, J.-C., & Baker, W.J. (2006) A new subfamily classification of the palm family (Arecaceae): evidence from plastid DNA phylogeny. *Botanical Journal of the Linnean Society, 151*(1), 15–38.

Baker, R., & Simmonds, N. (1953) The genus *Ensete* in Africa. *Kew Bulletin, 8*(3), 405–416.

Baker, W.J. (2000) *Orania regalis* rediscovered. *Palms, 44*(4), 166–169.

Baker, W.J., Asmussen, C.B., Barrow, S.C., Dransfield, J., & Hedderson, T.A. (1999) A phylogenetic study of the palm family (Palmae) based on chloroplast DNA sequences from the trnL-trnF region. *Plant Systematics and Evolution, 219*(1), 111–126.

Barkman, T.J., Chenery, G., McNeal, J. R., Lyons-Weiler, J., Ellisens, W. J., Moore, G., et al. (2000) Independent and combined analyses of sequences from all three genomic compartments converge on the root of flowering plant phylogeny. *Proceedings of the National Academy of Sciences of the United States of America, 97*(24), 13166.

Barwick, M. (2004) *Tropical and subtropical trees: an encyclopedia*. Portland, Oregon: Timber Press.

Baum, D.A., Alverson, W.S., & Nyffeler, R. (1998) A durian by any other name: taxonomy and nomenclature of the core Malvales. *Harvard Papers in Botany, 3*(2), 315–330.

Baum, D.A., DeWitt Smith, S., Yen, A., Alverson, W.S., Nyffeler, R., Whitlock, B.A., et al. (2004) Phylogenetic relationships of Malvatheca (Bombacoideae and Malvoideae; Malvaceae sensu lato) as inferred from plastid DNA sequences. *American Journal of Botany, 91*(11), 1863.

Bayer, C., Fay, M.F., Bruijn, A.Y., Savolainen, V., Morton, C.M., Kubitzki, K., et al. (1999) Support for an expanded family concept of Malvaceae within a recircumscribed order Mai vales: a combined analysis of plastid atpB and rbcL DNA sequences. *Botanical Journal of the Linnean Society, 129*(4), 267–303.

Bechtel, H., Cribb, P., & Launert, E. (1992) *The manual of cultivated orchid species* (3 ed.). Cambridge, Massachusetts: The MIT Press.

Bennett, J. R., & Mathews, S. (2006) Phylogeny of the parasitic plant family Orobanchaceae inferred from phytochrome A. [Review]. *American Journal of Botany, 93*(7), 1039–1051.

Bennett, J.R., & Scotland, R.W. (2003) A Revision of *Strobilanthes* (Acanthaceae) in Java. *Kew Bulletin, 58*(1), 1–82.

Bharathan, G., & Zimmer, E. (1995) Early branching events in monocotyledons-partial 18S ribosomal DNA sequence analysis. In P.J. Rudall, P.J. Cribb & D. Cutler (Eds.), *Monocotyledons: systematics and evolution* (pp. 81–107). Richmond, Surrey: Royal Botanic Gardens, Kew.

Bouchenak-Khelladi, Y., Salamin, N., Savolainen, V., Forest, F., van der Bank, M., Chase, M.W., et al. (2008) Large multi-gene phylogenetic trees of the grasses (Poaceae): *Progress towards complete tribal and generic level sampling. Molecular Phylogenetics and Evolution, 47*(2), 488–505.

Bremer, B., Bremer, K., Chase, M.W., Fay, M.F., Reveal, J.L., Soltis, D.E., et al. (2009) An update of the Angiosperm Phylogeny Group classification for the orders and families of flowering plants: APG III. *Botanical Journal of the Linnean Society, 161*(2), 105–121.

Bremer, B., Bremer, K., Heidari, N., Erixon, P., Olmstead, R.G., Anderberg, A.A., et al. (2002) Phylogenetics of asterids based on 3 coding and 3 non-coding chloroplast DNA markers and the utility of non-coding DNA at higher taxonomic levels. *Molecular Phylogenetics and Evolution, 24*(2), 274–301.

Bremer, B., Olmstead, R.G., Struwe, L., & Sweere, J.A. (1994) rbcL sequences support exclusion of *Retzia*, *Desfontainea* and *Nicodemia* from the Gentianales. *Plant Systematics and Evolution, 190*(3–4), 213–230.

Bremer, B., & Struwe, L. (1992) Phylogeny of the Rubiaceae and the Loganiaceae: Congruence of Conflict between Morphological and Molecular Data? *American Journal of Botany, 79*(10), 1171–1184.

Bricker, J.S. (1991) A Revision of the Genus *Crinodendron* (Elaeocarpaceae). *Systematic Botany, 16*(1), 77–88.

Brummitt, R.K. (1968) The genus *Baphia* (Leguminosae) in East and North-East tropical Africa. *Kew Bulletin, 22*(3), 513–360.

Burger, W.C., & History., F.M.o.N. (1991) Flora Costaricensis/William Burger, editor (Vol. n.s. no.28(1991)). Chicago: Field Museum of Natural History.

Cameron, K.M. (2004) Utility of plastid psaB gene sequences for investigating intrafamilial relationships within Orchidaceae. *Molecular Phylogenetics and Evolution, 31*(3), 1157–1180.

Cameron, K.M., & Carmen Molina, M. (2006) Photosystem II gene sequences of psbB and psbC clarify the phylogenetic position of *Vanilla* (Vanilloideae, Orchidaceae). *Cladistics, 22*(3), 239–248.

Chase, M., Soltis, D., Soltis, P., Rudall, P., Fay, M., Hahn, W., et al. (2000) Higher-level systematics of the monocotyledons: an assessment of current knowledge and a new classification. In K.L. Wilson & D.A. Morrison (Eds.), *Monocots: Systematics and evolution* (pp. 3–16). Collingwood: CSIRO.

Chase, M.W. (2004) Monocot relationships: an overview. *American Journal of Botany, 91*(10), 1645.

Chase, M.W. (2005) Classification of Orchidaceae in the Age of DNA data. *Curtis's Botanical Magazine, 22*(1), 2–7.

Chase, M.W., Duvall, M.R., Hills, H.G., Conran, J.G., Cox, A.V., Eguiarte, L.E., et al. (1995a) Molecular phylogenetics of Lilianae. In P.J. Rudall, P.J. Cribb, D.F. Cutler & H.C.J. (Eds.), *Monocotyledons: systematics and evolution* (Vol. 2, pp. 109–137). Richmond, Surrey: Royal Botanic Gardens, Kew.

Chase, M.W., Fay, M.F., Devey, D., Maurin, O., Rønsted, N., Davies, J., et al. (2006) Multigene analyses of monocot relationships: a summary. In J.T. Columbus, E.A. Friar, J.M. Porter, L.M. Prince & M.G. Simpson (Eds.), *Monocots: Comparative Biology and Evolution. Excluding Poales.* (pp. 63–75). Claremont: Rancho Santa Ana Botanical Garden.

Chase, M.W., Morton, C.M., & Kallunki, J.A. (1999) Phylogenetic relationships of Rutaceae: a cladistic analysis of the subfamilies using evidence from RBC and ATP sequence variation. *American Journal of Botany, 86*(8), 1191.

Chase, M.W., Soltis, D.E., Olmstead, R.G., Morgan, D., Les, D.H., Mishler, B.D., et al. (1993) Phylogenetics of seed plants: an analysis of nucleotide sequences from the plastid gene rbcL. *Annals of the Missouri Botanical Garden, 80*(3), 528–580.

Chase, M.W., Stevenson, D., Wilkin, P., & Rudall, P. (1995b) Monocot systematics: a combined analysis. In P.J. Rudall, P.J. Cribb, D.F. Cutler & C.J. Humphries (Eds.), *Monocotyledons: systematics and evolution* (Vol. 2, pp. 685–730). Richmond, Surrey: Botanic Gardens, Kew.

Cheek, M., Jebb, M., & Malesiana, F.F. (2001) Flora Malesiana: Seed plants. *Nepenthaceae: Rijksherbarium/Hortus Botanicus.*

Christenson, E.A. (2001) Phalaenopsis: *a monograph*. Portland, Oregon: Timber Press.

Clausager, K., & Borchsenius, F. (2003) The Marantaceae of Sabah, northern Borneo. *Kew Bulletin, 58*(3), 647–678.

Clausing, G., & Renner, S. (2001) Molecular phylogenetics of Melastomataceae and Memecylaceae: implications for character evolution. *American Journal of Botany, 88*(3), 486.

Clayton, W.D., Harman, K.T., & Williamson, H. (2006 onwards) GrassBase – The Online World Grass Flora. Retrieved 05 Jan 2010; 13:30 GMT, from http://www.kew.org/data/grasses-db.html

Cootes, J., Banks, D.P., & Titmuss, D. (2001) *The orchids of the Philippines*. Portland, Oregon: Timber Press.

Couvreur, T.L. P., & Luke, W. (2010) A new species of Uvariopsis (Annonaceae), endemic to the Eastern Arc Mountains of Tanzania. *Blumea-Biodiversity, Evolution and Biogeography of Plants, 55*(1), 68-72.

Craven, L., Goetsch, L., Hall, B., & Brown, G. (2008) Classification of the Vireya group of Rhododendron (Ericaceae). *Blumea-Biodiversity, Evolution and Biogeography of Plants, 53*(2), 435-442.

Cribb, P. (1987) *The genus* Paphiopedilum. London and Portland, Oregon: Royal Botanic Gardens, Kew & Timber Press.

Cribb, P., Chan, C.L., Lamb, A., Shim, P.S., & Society, S. (2000) *Orchids of Borneo: Introduction and a selection of species*. Richmond, Surrey: Royal Botanic Gardens, Kew.

Croix, I. I. (2008) *The new encyclopedia of orchids: 1500 species in cultivation*. Portland, Oregon: Timber Press.

Dahlgren, R.M.T., Clifford, H.T., & Yeo, P.F. (1985) *The families of the monocotyledons: structure, evolution and taxonomy*. New York: Springer-Verlag.

Darwin, C. (1869) *On the origin of species by means of natural selection, or the preservation of favoured races in the struggle for life* (5 ed.). London: John Murray.

Dassanayake, M.D., & Clayton, W.D. (1980–2000) *A revised handbook to the flora of Ceylon* (Vols. 1–14). New Delhi: Oxford & IBH Publishing Co.

Dassanayake, M.D., & Shaffer-Fehre, M. (Eds.). (2006) *A revised handbook to the flora of Ceylon Vol. 15: Ferns and fern-allies (Part A)*. Enfield: Science Publishers.

Davis, C., Wurdack, K.J., Jaramillo, C.A., Webb, C.O., & Donoghue, M J. (2005) Explosive radiation of Malpighiales supports a mid-Cretaceous origin of modern tropical rain forests. *American Naturalist, 165*(3), E36–E65.

Davis, J.I., Stevenson, D.W., Petersen, G., Seberg, O., Campbell, L.M., Freudenstein, J.V., et al. (2004) A phylogeny of the monocots, as inferred from rbcL and atpA sequence variation, and a comparison of methods for calculating jackknife and bootstrap values. *Systematic Botany, 29*(3), 467–510.

De Block, P. (2007) Three new Madagascan *Ixora* species (Rubiaceae) with flowers up to 23 cm long. *Nordic Journal of Botany, 25*(1–2), 75–84.

De Groot, H., Wanke, S., & Neinhuis, C. (2006) Revision of the genus *Aristolochia* (Aristolochiaceae) in Africa, Madagascar and adjacent islands. *Botanical Journal of the Linnean Society, 151*(2), 219–238.

de Jussieu, A.-L. (1789) Genera Plantarum: *secundum ordines naturales disposita, juxta methodum in Horto Regio Parisiensi exaratam*. In F.A. Stafleu (Ed.), *Historiae naturalis classica*. Paris: Apud viduam Herissant et Theophilum Barrois.

Deginani, N.B. (2001) The species of *Passiflora* (Passifloraceae) of Argentina. *Darwiniana (San Isidro), 39*(1–2), 43–129.

Ding, H. (1996) *Flora Malesiana: Caesalpiniaceae, Geitonoplesiaceae, Hernandiaceae, Lowiaceae*. New York: Kluwer.

Dowe, J.L. (2009) A taxonomic account of *Livistona* R.Br. (Arecaceae). *Gardens' Bulletin (Singapore), 60*(2), 185–344.

Doyle, J.A., & Endress, P.K. (2000) Morphological phylogenetic analysis of basal angiosperms: Comparison and combination with molecular data. [Proceedings Paper]. *International Journal of Plant Sciences, 161*(6), S121–S153.

Dransfield, J., & Beentje, H. (1995) *The palms of Madagascar*. London: International Palm Society, Royal Botanic Gardens, Kew.

Dransfield, J., & Uhl, N.W. (1998) Palmae. In K. Kubitzki (Ed.), *The Families and Genera of Vascular Plants: Flowering Plants – Monocotyledons [Alismatanae and Commelinanae (except Gramineae)]* (Vol. IV, pp. 306–389). New York: Springer.

Dransfield, J., Uhl, N.W., Asmussen, C.B., Baker, W.J., Harley, M.M., & Lewis, C.E. (2005) A new phylogenetic classification of the palm family, Arecaceae. *Kew Bulletin, 60*(4), 559–569.

Dressler, R.L. (1993) *Phylogeny and classification of the orchid family*. Cambridge: Cambridge University Press.

Duvall, M.R., Clegg, M.T., Chase, M.W., Clark, W.D., Kress, W.J., Hills, H.G., et al. (1993) Phylogenetic hypotheses for the monocotyledons constructed from rbcL sequence data. *Annals of the Missouri Botanical Garden, 80*(3), 607–619.

eFloras (2008) Flora of China. *eFloras (published on the internet)*. Retrieved 22 February, 2011, from http://www.efloras.org

Eggli, U. (2005) Crassulaceae. In U. Eggli & H.E.K. Hartman (Eds.), *Illustrated Handbook of Succulent Plants*. New York: Springer-Verlag.
Endress, M.E., & Bruyns, P.V. (2000) A revised classification of the Apocynaceae sl. *The Botanical Review, 66*(1), 1–56.
Endress, P.K., & Matthews, M.L. (2006) First steps towards a floral structural characterization of the major rosid subclades. [Proceedings Paper]. *Plant Systematics and Evolution, 260*(2–4), 223–251.
Ezcurra, C. (2002) El Género *Justicia* (Acanthaceae) en Sudamérica Austral. *Annals of the Missouri Botanical Garden, 89*(2), 225–280.
Figlar, R.B., & Nooteboom, H.P. (2004) Notes on Magnoliaceae IV. *Blumea, 49*(1), 87–100.
Frame, D., & Durou, S. (2001) Morphology and biology of *Napoleonaea vogelii* (Lecythidaceae) flowers in relation to the natural history of insect visitors. *Biotropica, 33*(3), 458–471.
Garcês, H.M.P., Champagne, C.E.M., Townsley, B.T., Park, S., Malhó, R., Pedroso, M.C., et al. (2007) Evolution of asexual reproduction in leaves of the genus *Kalanchoe. Proceedings of the National Academy of Sciences, 104*(39), 15578–15583.
Goldblatt, P., & Mabberley, D.J. (2005) *Belamcanda* included in *Iris*, and the new combination *I. domestica* (Iridaceae : Irideae). *Novon, 15*(1), 128–132.
Graf, A.B. (1978) *Tropica: color cyclopedia of exotic plants and trees from the tropics and subtropics (for warm-region horticulture, in cool climate the sheltered indoors)* (1st ed.). New Jersey: Roehrs Co.
Graham, S.W., & Olmstead, R.G. (2000) Utility of 17 chloroplast genes for inferring the phylogeny of the basal angiosperms. *American Journal of Botany, 87*(11), 1712.
Graham, S.W., Zgurski, J.M., McPherson, M.A., Cherniawsky, D.M., Saarela, J.M., Horne, E.S.C., et al. (2006) Robust inference of monocot deep phylogeny using an expanded multigene plastid data set. *Aliso, 22*, 3–20.
Hahn, W.J. (2002) A molecular phylogenetic study of the palmae (Arecaceae) based on atpB, rbcL, and 18S nrDNA sequences. *Systematic Biology, 51*(1), 92–112.
Hasebe, M., Wolf, P.G., Pryer, K.M., Ueda, K., Ito, M., Sano, R., et al. (1995) Fern phylogeny based on rbcL nucleotide sequences. *American Fern Journal, 85*(4), 134–181.
Henderson, A., Fischer, B., Scariot, A., Whitaker Pacheco, M., & Pardini, R. (2000) Flowering phenology of a palm community in a central Amazon forest. *Brittonia, 52*(2), 149–159.
Hilu, K.W., Borsch, T., Muller, K., Soltis, D.E., Soltis, P. S., Savolainen, V., et al. (2003) Angiosperm phylogeny based on matK sequence information. *American Journal of Botany, 90*(12), 1758.
Hogan, S. (2003) *Flora: a gardener's encyclopedia* (Vol. 1–2). Portland, Oregon: Timber Press.
Hopkins, H.C.F. (1994) The Indo-Pacific species of *Parkia* (Leguminosae: Mimosoideae). *Kew Bulletin, 49*(2), 181–234.
Jansen, R.K., Cai, Z., Hansen, D., Dastidar, S., Peñaflor, C., Timme, R., et al. (2006b) Phylogeny of angiosperms based on whole chloroplast sequences. Paper presented at the Botany 2006 – Looking to the Future – Conserving the Past.
Jansen, R.K., Cai, Z., Raubeson, L.A., Daniell, H., DePamphilis, C.W., Leebens-Mack, J., et al. (2007) Analysis of 81 genes from 64 plastid genomes resolves relationships in angiosperms and identifies genome-scale evolutionary patterns. *Proceedings of the National Academy of Sciences, 104*(49), 19369.
Jansen, R.K., Kaittanis, C., Saski, C., Lee, S.B., Tomkins, J., Alverson, A.J., et al. (2006a) Phylogenetic analyses of *Vitis* (Vitaceae) based on complete chloroplast genome sequences: effects of taxon sampling and phylogenetic methods on resolving relationships among rosids. *BMC Evolutionary Biology, 6*(1), 32.
Jeeva, S., Kiruba, S., Lalhruaitluanga, H., Prasad, M.N.V., & Rao, R.R. (2009) Flowering of *Melocanna baccifera* (Bambusaceae) in northeastern India. [Letter]. *Current Science, 96*(9), 1165–1166.
Jian, S., Soltis, P.S., Gitzendanner, M.A., Moore, M.J., Li, R., Hendry, T.A., et al. (2008) Resolving an ancient, rapid radiation in Saxifragales. *Systematic Biology, 57*(1), 38.
Jones, D.L. (c1995) *Palms throughout the world*. Washington: Smithsonian Institution Press.
Judd, W.S., Campbell, C.S., Kellog, E.A., Stevens, P.F., & Donoghue, M.J. (2008) *Plant systematics: A phylogenetic approach* (3 ed.). Sunderland, Massachusetts: Sinauer Associates.
Judd, W.S., Campbell, C.S., Kellogg, E.A., Stevens, P.F., & Donoghue, M.J. (2002) *Plant Systematics: A Phylogenetic Approach* (2 ed.). Sunderland, Massachusetts: Sinauer Associates.
Käällersjö, M., Farris, J.S., Chase, M.W., Bremer, B., Fay, M.F., Humphries, C.J., et al. (1998) Simultaneous parsimony jackknife analysis of 2538rbcl dna sequences reveals support for major clades of green plants, land plants, seed plants and flowering plants. *Plant Systematics and Evolution, 213*(3), 259–287.
Kier, G., Kreft, H., Lee, T.M., Jetz, W., Ibisch, P.L., Nowicki, C., et al. (2009) A global assessment of endemism and species richness across island and mainland regions. *Proceedings of the National Academy of Sciences of the United States of America, 106*(23), 9322–9327.
Killip, E.P. (Ed.). (1938) *The American species of Passifloraceae* (Vol. 19). Chicago: Field Museum of Natural History.
Kostermans, A.J. G.H., & Bompard, J.-M. (c1993) *The mangoes: their botany, nomenclature, horticulture and utilization*. London: Academic Press (for the International Board for Plant Genetic Resources and the Linnean Society of London).
Kress, W.J. (1995) Phylogeny of the Zingiberanae: Morphology and molecules. In P.J. Rudall, P.J. Cribb, D.F. Cutler & C.J. Humphries (Eds.), *Monocotyledons: Systematics and Evolution* (Vol. 2, pp. 443–460). Richmond, Surrey: Royal Botanic Gardens, Kew.
Kress, W.J., Liu, A.Z., Newman, M., & Li, Q.J. (2005) The molecular phylogeny of *Alpinia* (Zingiberaceae): a complex and polyphyletic genus of gingers. *American Journal of Botany, 92*(1), 167.
Kress, W.J., Prince, L.M., & Williams, K.J. (2002) The phylogeny and a new classification of the gingers (Zingiberaceae): evidence from molecular data. *American Journal of Botany, 89*(10), 1682.
Kubitzki, K. (2007) Introduction to Gunnerales. In K. Kubitzki (Ed.), *The Families and Genera of Vascular Plants: Flowering Plants – Eudicots* (Vol. IX). New York: Springer.
Kuijt, J. (1988) Revision of *Tristerix* (Loranthaceae). *Systematic Botany Monographs, 19*, 1–61.
Lamb, A., & Cribb, P. (1991) *Orchids of Borneo*. The Sabah Society.
La-Ongsri, W., Trisonthi, C., & Balslev, H. (2009) A synopsis of Thai Nymphaeaceae. *Nordic Journal of Botany, 27*(2), 97–114.
Lewis, B., & Cribb, P. (1989) *Orchids of Vanuatu*. Richmond, Surrey: Royal Botanic Gardens, Kew.
Lewis, C.E., & Doyle, J.J. (2001) Phylogenetic utility of the nuclear gene malate synthase in the palm family (Arecaceae). *Molecular Phylogenetics and Evolution, 19*(3), 409–420.
Lewis, C.E., & Doyle, J.J. (2002) A phylogenetic analysis of tribe Areceae (Arecaceae) using two low-copy nuclear genes. *Plant Systematics and Evolution, 236*(1–2), 1–17.
Lewis, G., Schrire, B., Mackinder, B., & Lock, M. (Eds.). (2005) *Legumes of the world*. Richmond, Surrey: Royal Botanic Gardens, Kew.
Liu, N., Zhu, Y., Wei, Z.X., Chen, J., Wang, Q.B., Jian, S.G., et al. (2009) Phylogenetic relationships and divergence times of the family Araucariaceae based on the DNA sequences of eight genes. *Chinese Science Bulletin, 54*(15), 2648–2655.
Llamas, K.A. (2003) *Tropical Flowering Plants: a guide to identification and cultivation*. Portland, Oregon: Timber Press.
Lundberg, J., & Bremer, K. (2003) A phylogenetic study of the order Asterales using one morphological and three molecular data sets. *International Journal of Plant Sciences, 164*(4), 553–578.
Manning, J., Goldblatt, P., & Snijman, D. (2002) *The color encyclopedia of Cape bulbs*. Portland, Oregon: Timber Press.
Mathews, S., & Donoghue, M.J. (2000) Basal angiosperm phylogeny inferred from duplicate phytochromes A and C. *International Journal of Plant Sciences, 161*(6), 41–55.
Matthews, L.J. (2002) *The protea book: a guide to cultivated Proteaceae*. Portland, Oregon: Timber Press.
Matthews, M.L., & Endress, P.K. (2002) Comparative floral structure and systematics in Oxalidales (Oxalidaceae, Connaraceae, Brunelliaceae, Cephalotaceae, Cunoniaceae, Elaeocarpaceae, Tremandraceae). *Botanical Journal of the Linnean Society, 140*(4), 321–381.
Matthews, M.L., & Endress, P.K. (2004) Comparative floral structure and systematics in Cucurbitales (Corynocarpaceae, Coriariaceae, Tetramelaceae, Datiscaceae, Begoniaceae, Cucurbitaceae, Anisophylleaceae). *Botanical Journal of the Linnean Society, 145*(2), 129–185.
Matthews, M.L., & Endress, P.K. (2005) Comparative floral structure and systematics in Celastrales (Celastraceae, Parnassiaceae, Lepidobotryaceae). [Review]. *Botanical Journal of the Linnean Society, 149*(2), 129–194.
Middleton, D.J. (2004) A revision of Kopsia (Apocynaceae: Rauvolfioideae). *Harvard Papers in Botany, 9*, 89–142.
Morgan, R.J. (2007) *Impatiens: the vibrant world of Busy Lizzies, Balsams, and Touch-me-nots*. Portland, Oregon: Timber Press.
Nash, N., Croix, I.F.L., & Press, T. (2005) *Flora's orchids*. Portlandm Oregon: Timber Press.
Nicholas, A., & Baijnath, H. (1994) A consensus classification for the order Gentianales with additional details on the suborder Apocynineae. *The Botanical Review, 60*(4), 440–482.
Nickrent, D.L., Blarer, A., Qiu, Y.L., Soltis, D.E., Soltis, P.S., & Zanis, M. (2002) Molecular data place Hydnoraceae with Aristolochiaceae. *American Journal of Botany, 89*(11), 1809.
Nooteboom, H.P. (1987) Magnoliaceae. In M.D. Dassanayake (Ed.), *A Revised Handbook to the Flora of Ceylon* (Vol. VI, pp. 152–156). New Delhi: Oxford & IBH Publishing Co.
Oh, S.-H., & Manos, P.S. (2008) Molecular phylogenetics and cupule evolution in Fagaceae as inferred from nuclear CRABS CLAW sequences. [Review]. *Taxon, 57*(2), 434–451.
Ohrnberger, D. (1999) Subtribe Bambusinae in *The Bamboos of the World* (pp. 249–317). Amsterdam: Elsevier.
Plowman, T., Knapp, S., Press, J.R., & Alliance, O.C. (1998) *A revision of the South American species of* Brunfelsia *(Solanaceae)*. Chicago: Field Museum of Natural History.
Pridgeon, A., Cribb, P.J., Chase, M.W., & Rasmussen, F.N. (2003) *Genera Orchidacearum: Orchidoideae. Vanilloideae*. Oxford: Oxford University Press.
Pridgeon, A.M. (1992) *The Illustrated encyclopedia of orchids*. Portland, Oregon: Timber Press.
Pridgeon, A.M. (2001) *Genera Orchidacearum*. Oxford: Oxford University Press.
Pridgeon, A.M. (2009) *Genera Orchidacearum: Epidendroideae* (Part two). Oxford: Oxford University Press.
Pridgeon, A.M., Cribb, P., & Chase, M.W. (2005) *Genera Orchidacearum: Epidendroideae*. Oxford: Oxford University Press.
Pridgeon, A.M., Cribb, P.J., Chase, M.W., & Rasmussen, F.N. (2001) *Genera Orchidacearum* (Vol 2: Orchidoideae). New York: Oxford University Press.
Pridgeon, A.M., & Rasmussen, F. (1999) *General introduction, Apostasioideae, Cypripedioideae*. Oxford: Oxford University Press.
Pryer, K.M. (1999) Phylogeny of marsileaceous ferns and relationships of the fossil Hydropteris pinnata reconsidered. *International Journal of Plant Sciences, 160*(5), 931–954.
Pryer, K.M., Schneider, H., Smith, A.R., Cranfill, R., Wolf, P.G., Hunt, J.S., et al. (2001) Horsetails and ferns are a monophyletic group and the closest living relatives to seed plants. *Nature, 409*(6820), 618–622.
Pryer, K.M., Schuettpelz, E., Wolf, P.G., Schneider, H., Smith, A.R., & Cranfill, R. (2004) Phylogeny and evolution of ferns (monilophytes) with a focus on the early leptosporangiate divergences. *American Journal of Botany, 91*(10), 1582.
Pryer, K.M., Smith, A.R., & Skog, J.E. (1995) Phylogenetic relationships of extant ferns based on evidence from morphology and rbcL sequences. *American Fern Journal, 85*(4), 205–282.
Puff, C. (1988) Observations on *Carphalea* Juss. (Rubiaceae, Hedyotideae), with Particular Reference to the Madagascan Taxa and Its Taxonomic Position. *Bulletin du Jardin botanique national de Belgique/ Bulletin van de National Plantentuin van België,*

58(3/4), 271–323.
Punt, W., Hoen, P.P., Blackmore, S., Nilsson, S., & Le Thomas, A. (2007) Glossary of pollen and spore terminology. *Review of Palaeobotany and Palynology, 143*(1–2), 1–81.
Qiu, Y.L., Chase, M.W., Les, D.H., & Parks, C.R. (1993) Molecular phylogenetics of the Magnoliidae: cladistic analyses of nucleotide sequences of the plastid gene rbcL. *Annals of the Missouri Botanical Garden, 80*(3), 587–606.
Qiu, Y. L., Lee, J., Bernasconi-Quadroni, F., Soltis, D. E., Soltis, P. S., Zanis, M., et al. (1999) The earliest angiosperms: evidence from mitochondrial, plastid and nuclear genomes. *Nature, 402*(6760), 404-407.
Qiu, Y.L., Lee, J., Bernasconi-Quadroni, F., Soltis, D.E., Soltis, P.S., Zanis, M., et al. (2000) Phylogeny of basal angiosperms: Analyses of five genes from three genomes. [Proceedings Paper]. *International Journal of Plant Sciences, 161*(6), S3–S27.
Qiu, Y.-L., Dombrovska, O., Lee, J., Li, L., Whitlock, B.A., Bernasconi-Quadroni, F., et al. (2005) Phylogenetic analyses of basal angiosperms based on nine plastid mitochondrial and nuclear genes. *International Journal of Plant Sciences, 166*(5), 815–842.
Ravi, V., Khurana, J.P., Tyagi, A.K., & Khurana, P. (2007) Rosales sister to Fabales: Towards resolving the rosid puzzle. *Molecular Phylogenetics and Evolution, 44*(1), 488–493.
Renner, S.S. (1999) Circumscription and phylogeny of the Laurales: Evidence from molecular and morphological data. [Review]. *American Journal of Botany, 86*(9), 1301–1315.
Savolainen, V., Chase, M.W., Hoot, S.B., Morton, C.M., Soltis, D.E., Bayer, C., et al. (2000(a)) Phylogenetics of flowering plants based on combined analysis of plastid atpB and rbcL gene sequences. *Systematic Biology, 49*(2), 306.
Schneider, H., Smith, A.R., Cranfill, R., Hildebrand, T.J., Haufler, C.H., & Ranker, T.A. (2004) Unraveling the phylogeny of polygrammoid ferns (Polypodiaceae and Grammitidaceae): exploring aspects of the diversification of epiphytic plants. *Molecular Phylogenetics and Evolution, 31*(3), 1041–1063.
Schuiteman, A., and Vogel, E.d. (2002) *Flora Malesiana: Orchids of New Guinea Vol 1: Illustrated Checklist and Genera.* Amsterdam: Expert Center for Taxonomic Identification, University of Amsterdam.
Schuiteman, A., and Vogel, E.d. (2002) *Flora Malesiana: Orchids of New Guinea Vol 2: Dendrobium and allied genera.* Amsterdam: Expert Center for Taxonomic Identification, University of Amsterdam.
Shaffer-Fehre, M. (Ed.). (2006) *A revised handbook to the flora of Ceylon Volume XV: Ferns and fern-allies (Part B).* Enfield: Taylor & Francis.
Simmonds, N.W., & Shepherd, K. (1955) The taxonomy and origins of the cultivated bananas. *Journal of the Linnean Society, 55,* 302–312.
Simpson, M.G. (2006) *Plant Systematics.* Massachusetts: Academic Press.
Simpson, M.G. (2010) *Plant Systematics* (2 ed.). Massachusetts: Academic Press.
Skog, J., Zimmer, E., & Mickel, J. (2002) Additional support for two subgenera of Anemia (Schizaeaceae) from data for the chloroplast intergenic spacer region trnL-F and morphology. *American Fern Journal, 92*(2), 119–130.
Smith, A.R., Pryer, K.M., Schuettpelz, E., Korall, P., Schneider, H., & Wolf, P.G. (2006) A classification for extant ferns. *Taxon, 55*(3), 705–731.
Soladoye, M.O. (1985) A revision of *Baphia* (Leguminosae – Papilionoideae). *Kew Bulletin, 40*(2), 291–386.
Soltis, D.E., Albert, V.A., Leebens-Mack, J., Palmer, J.D., Wing, R.A., dePamphilis, C.W., et al. (2008) The *Amborella* genome: an evolutionary reference for plant biology. [Correspondence]. *Genome Biology, 9*(3), 402.
Soltis, D.E., Senters, A.E., Zanis, M.J., Kim, S., Thompson, J.D., Soltis, P.S., et al. (2003) Gunnerales are sister to other core eudicots: implications for the evolution of pentamery. *American Journal of Botany, 90*(3), 461.
Soltis, D.E., Soltis, P.S., Chase, M.W., Mort, M.E., Albach, D.C., Zanis, M., et al. (2000) Angiosperm phylogeny inferred from 18S rDNA, rbcL, and atpB sequences. [Review]. *Botanical Journal of the Linnean Society, 133*(4), 381–461.
Soltis, D.E., Soltis, P.S., Endress, P.K., & Chase, M.W. (2005) *Phylogeny and evolution of angiosperms.* Sunderland, Massachusetts Sinauer Associates.
Soltis, D.E., Soltis, P.S., Nickrent, D.L., Johnson, L.A., Hahn, W.J., Hoot, S.B., et al. (1997) Angiosperm phylogeny inferred from 18S ribosomal DNA sequences. *Annals of the Missouri Botanical Garden, 84*(1), 1–49.
Soltis, P.S., Brockington, S.F., Yoo, M.J., Piedrahita, A., Latvis, M., Moore, M.J., et al. (2009) Floral variation and floral genetics in basal angiosperms. *American Journal of Botany, 96*(1), 110–128.
Specht, C.D., Kress, W.J., Stevenson, D.W., & DeSalle, R. (2001) A molecular phylogeny of Costaceae (Zingiberales). *Molecular Phylogenetics and Evolution, 21*(3), 333–345.
Staples, G., & Herbst, D.R. (2005) *A tropical garden flora: plants cultivated in the Hawaiian Islands and other tropical places.* Honolulu: Bishop Museum Press.
Steenis, C.G.G.J. v. (1958) *Flora Malesiana Series I* (Vol. 5). Leyden: P. Noordhoff Ltd.
Steenis, C.G.G.J. v. (1976) *Flora Malesiana: being an illustrated systematic account of the Malaysian flora including keys for determination, diagnostic descriptions, references to the literature, synonymy, and distribution, and notes on the ecology of its wild and commonly cultivated plants.* Leyden: P. Noordhoff Ltd.
Stefanoviac, S., Jager, M., Deutsch, J., Broutin, J., & Masselot, M. (1998) Phylogenetic relationships of conifers inferred from partial 28S rRNA gene sequences. *American Journal of Botany, 85*(5), 688.
Stevens, P.F. (2001 onwards) Angiosperm Phylogeny Website. Retrieved 22 March, 2011, from http://www.mobot.org/MOBOT/research/APweb/
Stevenson, D., Davis, J., Freudenstein, J., Hardy, C.R., Simmonds, M., & Specht, C. (2000) A phylogenetic analysis of the monocotyledons based on morphological and molecular character sets, with comments on the placement of Acorus and Hydatellaceae. *Monocots: systematics and evolution* (pp 17–24). Collingwood: CSIRO.
Stevenson, D.W., & Loconte, H. (1995) Cladistic analysis of monocot families. In P.J. Rudall, Cribb, P.J., Cutler, D.F., & Humphries, C.J. (eds.), *Monocotyledons: Systematics and Evolution* (pp. 543–578). Richmond, Surrey: Royal Botanic Gardens, Kew.
Stone, R.D. (2006) Phylogeny of major lineages in Melastomataceae, subfamily Olisbeoideae: Utility of nuclear glyceraldehyde 3-phosphate dehydrogenase (GapC) gene sequences. *Systematic Botany, 31*(1), 107–121.
Struwe, L., Albert, V.A., & Bremer, B. (1994) Cladistics and family level classification of the Gentianales. *Cladistics, 10*(2), 175–206.
Sutter, D., Forster, P., & Endress, P. (2006) Female flowers and systematic position of Picrodendraceae (Euphorbiaceae s.l., Malpighiales). *Plant Systematics and Evolution, 261,* 187–215.
Takhtajan, A. (1986) *Floristic Regions of the world* (Translation of: Floristicheskie oblasti Zemli). Berkeley and Los Angeles: University of California Press.
Takhtajan, A. (2009) *Flowering Plants.* New York: Springer/Columbia University Press.
Tate, J.A., Fuertes Aguilar, J., Wagstaff, S.J., La Duke, J.C., Bodo Slotta, T.A., & Simpson, B.B. (2005) Phylogenetic relationships within the tribe Malveae (Malvaceae, subfamily Malvoideae) as inferred from ITS sequence data. *American Journal of Botany, 92*(4), 584.
The Board of Trustees of the Royal Botanic Gardens, Kew. (2010) World Checklist of Selected Plant Families. Retrieved 22 February 2010, 2010, from http://www.kew.org/wcsp/
Treutlein, J., & Wink, M. (2002) Molecular phylogeny of cycads inferred from rbcL sequences. *Naturwissenschaften, 89*(5), 221–225.
Tsai, C.C., Huang, S.C., & Chou, C.H. (2006) Molecular phylogeny of *Phalaenopsis* Blume (Orchidaceae) based on the internal transcribed spacer of the nuclear ribosomal DNA. *Plant Systematics and Evolution, 256*(1/4), 1–16.
Ulmer, T., & MacDougal, J.M. (2004) *Passiflora: passionflowers of the world.* Portland, Oregon: Timber Press.
University., H. (1955) *Botanical Museum leaflets, Harvard University.* Vol. v.17 (1955–1957). Cambridge, Massachusetts: Botanical Museum, Harvard University.
USDA, ARS, & National Genetic Resources Program (2010) Germplasm Resources Information Network (GRIN) [Online Database]. Retrieved 27 March 2010, 2010, from http://www.ars-grin.gov/cgi-bin/npgs/html/taxon.pl?407012
Vermeulen, J.J. (1987) *A taxonomic revision of the Continental African Bulbophyllinae* (Vol. 2). Leiden: E.J. Brill.
Wagenitz, G. (1959) Die systematische Stellung der Rubiaceae. [Journal]. *Bot. Jahrb. Syst., 79,* 17–35.
Wagenitz, G. (1992) The Asteridae: evolution of a concept and its present status. *Annals of the Missouri Botanical Garden, 79*(2), 209–217.
Wagner, W.L., Herbst, D.R., & Sohmer, S.H. (1990) *Manual of the flowering plants of Hawai'i.* Honolulu: University of Hawai'i Press.
Wang, H., Moore, M.J., Soltis, P.S., Bell, C.D., Brockington, S.F., Alexandre, R., et al. (2009) Rosid radiation and the rapid rise of angiosperm-dominated forests. *Proceedings of the National Academy of Sciences of the United States of America, 106*(10), 3853–3858.
Weber, A., & Vogel, S. (1986) The pollination syndrome of *Deplanchea tetraphylla* (Bignoniaceae). *Plant Systematics and Evolution, 154*(3–4), 237–250.
Weberling, F. (1988) The architecture of inflorescences in the Myrtales. *Annals of the Missouri Botanical Garden, 75*(1), 226–310.
Weston, P.H., & Barker, N.P. (2006) A new suprageneric classification of the Proteaceae, with an annotated checklist of genera. *Telopea, 11*(3), 314–344.
Whistler, W.A. (2004) *Rainforest trees of Samoa: a guide to the common native and naturalized lowland and foothill forest trees of the Samoan archipelago.* Honolulu: Isle Botanica.
Whitmore, T.C. (1977) *Palms of Malaya* (Rev. ed.). Kuala Lumpur: Oxford University Press.
Wikström, N., Kenrick, P., & Vogel, J.C. (2002) Schizaeaceae: a phylogenetic approach. *Review of Palaeobotany and Palynology, 119*(1–2), 35–50.
Wikström, N., Savolainen, V., & Chase, M.W. (2001) Evolution of the angiosperms: calibrating the family tree. *Proceedings of the Royal Society of London. Series B: Biological Sciences, 268*(1482), 2211.
Wood, J.J. (2003) *Orchids of Borneo* (Vol. 4). Kota Kinabalu: Sabah Society and Royal Botanic Gardens, Kew.
Worberg, A., Alford, M.H., Quandt, D., & Borsch, T. (2009) Huerteales sister to Brassicales plus Malvales, and newly circumscribed to include Dipentodon, Gerrardina, Huertea, Perrottetia, and Tapiscia. *Taxon, 58*(2), 468–478.
Wurdack, K.J., Hoffmann, P., Samuel, R., de Bruijn, A., van der Bank, M., & Chase, M.W. (2004) Molecular phylogenetic analysis of Phyllanthaceae (Phyllanthoideae pro parte, Euphorbiaceae sensu lato) using plastid rbcL DNA sequences. *American Journal of Botany, 91*(11), 1882.
Wyk, B.-E. V. (1996) *Guide to the aloes of South Africa* (1 ed.). Pretoria: Briza Publications.
Xia, N., & Gilbert, M.G. (2009) Santalaceae. *Flora of China Vol. 5 Page 208.* Retrieved 22/03/2011, 2011, from www.efloras.org
Yuan, Y.M., Song, Y., Geuten, K., Rahelivololona, E., Wohlhauser, S., Fischer, E., et al. (2004) Phylogeny and biogeography of Balsaminaceae inferred from ITS sequences. *Taxon, 53*(2), 391–392.
Zanis, M.J., Soltis, D.E., Soltis, P.S., Mathews, S., & Donoghue, M.J. (2002) The root of the angiosperms revisited. *Proceedings of the National Academy of Sciences of the United States of America, 99*(10), 6848.
Zanis, M.J., Soltis, P.S., Qiu, Y.L., Zimmer, E., & Soltis, D.E (2003) Phylogenetic analyses and perianth evolution in basal angiosperms. *Annals of the Missouri Botanical Garden, 90*(2), 129–150.
Zhang, L.B., Simmons, M.P., Kocyan, A., & Renner, S.S. (2006) Phylogeny of the Cucurbitales based on DNA sequences of nine loci from three genomes: Implications for morphological and sexual system evolution. [Review]. *Molecular Phylogenetics and Evolution, 39*(2), 305–322.
Zhu, X.-Y., Chase, M.W., Qiu, Y.-L., Kong, H.-Z., Dilcher, D.L., Li, J.-H., et al. (2007) Mitochondrial matR sequences help to resolve deep phylogenetic relationships in rosids. *BMC Evolutionary Biology, 7,* Article No.: 217.
Zona, S. (1996) *Flora Neotropica: Roystonea (Arecaceae: Arecoideae).* New York: New York Botanical Garden.
Zona, S. (1997) The genera of Palmae (Arecaceae) in the southeastern United States. *Harvard papers in botany.*

BOTANICAL NAME INDEX

COMMON NAME INDEX